TABLE OF CONTENTS

This edition covers specifications and service procedures on 1985-90 models available at time of publication. Data reported in this manual is subject to change. To report current additions or revisions between editions of the manual, a "Manual Update" page is published every month for your guidance in Motor magazine.

MOTOR FRONT END & BRAKE SERVICE

1985-90 DOMESTIC CAR EDITION

First Printing

Michael J. Kromida, SAE
Editor

John R. Lypen, SAE
Managing Editor

James A. Jablonski
Production Editor

Warren Schildknecht, SAE
Senior Editor

Christopher P. Jakubowski
Associate Editor

Lynda Slater
Production Assistant

Vita Green
Editorial Assistant

Mark L. Kaufman
Assistant Editor

Mark C. Ferrand
Assistant Editor

Timothy P. Kedzior
Assistant Editor

Eric P. Gordin
Assistant Editor

Rick Metcalf
Assistant Editor

Timothey L. Martin
Assistant Editor

Richard F. Cahoon
Assistant Editor

Richard G. Glover
Assistant Editor

Donald R. Cobb
Assistant Editor

Charles R. Burstall
Assistant Editor

James J. Whiteford
Assistant Editor

Daniel C. Rock
Assistant Editor

Andy Brian
Assistant Editor

Published by

MOTOR

**Hearst Books/Business Publishing Group,
A Division of The Hearst Corp.**

5600 Crooks Road, Troy, MI 48098

Printed in the U.S.A.

ISBN 0-87851-699-9

WHEEL ALIGNMENT SPECIFICATIONS

INDEX

Chrysler Motors/Eagle

TABLE OF CONTENTS

CHRYSLER MOTORS REAR WHEEL DRIVE MODELS EXCEPT CONQUEST

FRONT WHEEL ALIGNMENT

Year	Model	Caster Angle, Degrees		Camber Angle, Degrees				Toe-In Inch	Toe Out on Turns, Deg.	
		Limits	Desired	Limits Left	Limits Right	Desired Left	Desired Right		Outer Wheel	Inner Wheel
1985–89	All	+1.5 to +3.5	+2.5	0 to +1	0 to +1	+.5	+.5	1/8	18	20

CAR RIDING HEIGHT

Year	Model	Fig.	Riding Height Inches
CHRYSLER FIFTH AVENUE			
1985–89	All	A	12.5
DODGE DIPLOMAT			
1985–89	All	A	12.5
PLYMOUTH GRAND FURY			
1985–89	All	A	12.5

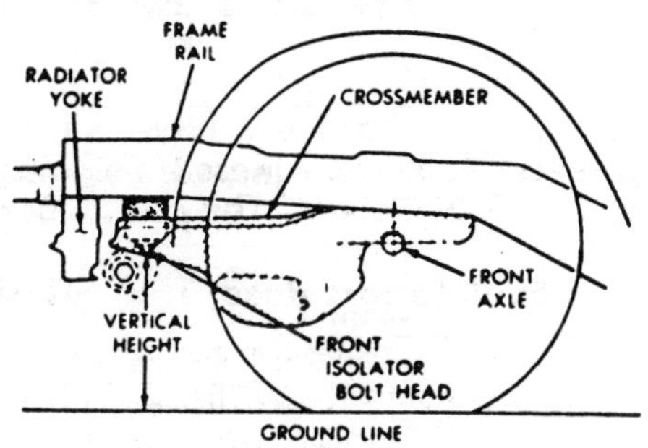

Fig. A

WHEEL ALIGNMENT SPECIFICATIONS

CHRYSLER MOTORS FRONT WHEEL DRIVE MODELS EXCEPT COLT, COLT VISTA & 1990 PLYMOUTH LASER

FRONT WHEEL ALIGNMENT

Year	Model	Caster Angle Desired, Degrees	Camber Angle, Degrees Limits Left	Camber Angle, Degrees Limits Right	Camber Angle, Degrees Desired Left	Camber Angle, Degrees Desired Right	Toe-In Inch
1985–87	Aries & Reliant	①	−.2 to +.8	−.2 to +.8	+.3	+.3	1/16
	Horizon & Omni	+1.4	−.2 to +.8	−.2 to +.8	+.3	+.3	1/16
	LeBaron & Town & Country	①	−.2 to +.8	−.2 to +.8	+.3	+.3	1/16
	Charger & Turismo	+1.9	−.2 to +.8	−.2 to +.8	+.3	+.3	1/16
	E Class & 600	+1.2	−.2 to +.8	−.2 to +.8	+.3	+.3	1/16
	Executive & New Yorker	+1.2	−.2 to +.8	−.2 to +.8	+.3	+.3	1/16
	Daytona & Laser	+1.2	−.2 to +.8	−.2 to +.8	+.3	+.3	1/16
	Caravelle, Lancer & LeBaron GTS	+1.2	−.2 to +.8	−.2 to +.8	+.3	+.3	1/16
1987	Shadow & Sundance	—	−.2 to +.8	−.2 to +.8	+.3	+.3	1/16
1988	Horizon & Omni	+1.4	−.2 to +.8	−.2 to +.8	+.3	+.3	1/16
	Station Wagons	+.9	−.2 to +.8	−.2 to +.8	+.3	+.3	1/16
	Except Horizon, Omni & Station Wagons	+1.2	−.2 to +.8	−.2 to +.8	+.3	+.3	1/16
1989	Horizon & Omni	+1.4	−.2 to +.8	−.2 to +.8	+.3	+.3	1/16
	Except Horizon & Omni	+1.2	−.2 to +.8	−.2 to +.8	+.3	+.3	1/16
1990	Horizon & Omni②	+1.4	−.2 to +.8	−.2 to +.8	+.3	+.3	1/16
	Except Horizon & Omni②	+1.2	−.2 to +.8	−.2 to +.8	+.3	+.3	1/16
	New Yorker 5th Ave & Imperial③	+1.3	−.3 to +.5	−.3 to +.5	+.1	+.1	1/16

①—Except wagon, +1.2 degrees; wagon, +.9 degree.
②—Less Air Suspension.
③—With Air Suspension.

REAR WHEEL ALIGNMENT

Year	Model	Camber Angle, Degrees Limits Left	Camber Angle, Degrees Limits Right	Camber Angle, Degrees Desired Left	Camber Angle, Degrees Desired Right	Toe-In Inch
1985–87	Aries & Reliant	−1 to 0	−1 to 0	−.5	−.5	0
	Horizon & Omni	−1.25 to −.25	−1.25 to −.25	−.75	−.75	3/32
	LeBaron & Town & Country	−1 to 0	−1 to 0	−.5	−.5	0
	Charger & Turismo	−1.25 to −.25	−1.25 to −.25	−.75	−.75	3/32
	E Class & 600	−1 to 0	−1 to 0	−.5	−.5	0
	Executive & New Yorker	−1 to 0	−1 to 0	−.5	−.5	0
	Daytona & Laser	−1 to 0	−1 to 0	−.5	−.5	0
	Caravelle, Lancer & LeBaron GTS	−1 to 0	−1 to 0	−.5	−.5	0
1987	Shadow & Sundance	−1 to 0	−1 to 0	−.5	−.5	0
1988	Horizon & Omni	−1.25 to −.25	1.25 to −.25	−.75	−.75	3/32
	Station Wagons	−1 to 0	−1 to 0	−.5	−.5	0
	Except Horizon, Omni & Station Wagons	−1 to 0	−1 to 0	−.5	−.5	0
1989	Horizon & Omni	−1.25 to −.25	1.25 to −.25	−.75	−.75	3/32
	Except Horizon & Omni	−1 to 0	−1 to 0	−.5	−.5	0
1990	Horizon & Omni	−1.3 to −.2	1.3 to −.2	−.75	−.75	3/32
	Except Horizon & Omni	−1.3 to +.3	−1.3 to +.3	−.5	−.5	0

CHRYSLER MOTORS COLT, COLT VISTA & CONQUEST

FRONT WHEEL ALIGNMENT

Year	Model	Caster Angle, Degrees		Camber Angle, Degrees				Toe In Inch
				Limits		Desired		
		Limits	Desired	Left	Right	Left	Right	
1985	Colt	+13/60 to +1 13/60	+43/60	−1/2 to +1/2	−1/2 to +1/2	0	0	①
	Colt Vista②	+3/10 to +1 3/10	+4/5	−1/12 to +1 1/12	−1/12 to +1 1/12	+5/12	+5/12	①
	Colt Vista③	+3/10 to +1 3/10	+4/5	+2/6 to +1 1/3	+2/6 to +1 1/3	+5/6	+5/6	①
	Conquest	—	+5 1/3	—	—	0	0	0
1986	Colt	+13/60 to +1 13/60	+43/60	−1/2 to +1/2	−1/2 to +1/2	0	0	①
	Colt Vista②	+3/10 to +1 3/10	+4/5	−1/12 to +1 1/12	−1/12 to +1 1/12	+5/12	+5/12	①
	Colt Vista③	+3/10 to +1 3/10	+4/5	+2/6 to +1 1/3	+2/6 to +1 1/3	+5/6	+5/6	①
1986–89	Conquest	+5 1/3 to +6 1/3	+5 5/6	−1 to 0	−1 to 0	−1/2	−1/2	④
1987	Colt	⑤	⑤	−1/2 to +1/2	−1/2 to +1/2	0	0	①
	Colt Vista②	+3/10 to +1 3/10	+4/5	−1/12 to +1 1/12	−1/12 to +1 1/12	+5/12	+5/12	⑥
	Colt Vista③	+1/3 to +1 1/3	+5/6	+11/30 to +1 1/30	+11/30 to +1 1/30	+13/15	+13/15	⑦
1988	Colt	⑤	⑤	−1/2 to +1/2	−1/2 to +1/2	0	0	①
	Colt Vista②	+3/10 to +1 3/10	+4/5	−1/12 to +1 1/12	−1/12 to +1 1/12	+5/12	+5/12	⑥
	Colt Vista③	+1/3 to +1 1/3	+5/6	+1/3 to +1 1/3	+1/3 to +1 1/3	+5/6	+5/6	①
1989–90	Colt Exc. Wagon	+1 5/6 to +2 5/6	+2 1/3	−1/2 to +1/2	−1/2 to +1/2	0	0	①
	Colt Wagon②	⑤	⑤	−1/2 to +1/2	−1/2 to +1/2	0	0	①
	Colt Wagon③	+1/2 to +1 1/2	+1	−1/2 to +1/2	−1/2 to +1/2	0	0	①
	Colt Vista②	+3/10 to +1 3/10	+4/5	−1/12 to +1 1/12	−1/12 to +1 1/12	+5/12	+5/12	①
	Colt Vista③	+1/3 to +1 1/3	+5/6	+1/3 to +1 1/3	+1/3 to +1 1/3	+5/6	+5/6	①

①—.12 inch toe in to .12 inch toe-out.
②—2 wheel drive.
③—4 wheel drive.
④—.2 inch toe in to .2 inch toe-out.
⑤—Limits: less power steering, +1/6 to +1 1/6; w/power steering, +5/6 to +1 5/6. Desired: less power steering, +1; w/power steering, +1 1/3.
⑥—At center of tread, .12 inch toe-out to .12 inch toe-in; At rim of disc wheel, .06 inch toe-out to .06 inch toe-in.
⑦—At center of tread, .12 inch toe-out to .12 inch toe-in; At rim of disc wheel, .07 inch toe-out to .07 inch toe-in.

REAR WHEEL ALIGNMENT

Year	Model	Camber Angle, Degrees				Toe-In
		Limits		Desired		
		Left	Right	Left	Right	
1985–87	Colt	—	—	−2/3	−2/3	0
	Colt Vista①	—	—	−2/3	−2/3	0
	Colt Vista②	—	—	0	0	0
	Conquest	—	—	0	0	③
1988	Colt	−1 1/4 to −1/12	−1 1/4 to −1/12	−2/3	−2/3	④
1988–89	Conquest	−3/4 to +1/4	−3/4 to +1/4	−1/4	−1/4	③
1988–90	Colt Vista①	−1 1/10 to 0	−1 1/10 to 0	−7/12	−7/12	.059
	Colt Vista②	−1/2 to +1/2	−1/2 to +1/2	0	0	⑤
1989–90	Colt Exc. Wagon	−1 1/10 to −1/10	−1 1/10 to −1/10	−2/3	−2/3	⑥
	Colt Wagon①	−7/12 to +7/12	−7/12 to +7/12	0	0	⑥
	Colt Wagon②	—	—	0	0	0

①—2 wheel drive.
②—4 wheel drive.
③—.08 inch toe in to .08 inch toe-out.
④—.30 inch toe in to .08 inch toe-out.
⑤—.12 inch toe in to .12 inch toe-out.
⑥—.18 inch toe in to .18 inch toe-out.

WHEEL ALIGNMENT SPECIFICATIONS

1990 PLYMOUTH LASER

FRONT WHEEL ALIGNMENT

Year	Model	Caster Angle, Degrees		Camber Angle, Degrees				Toe In Inch
				Limits		Desired		
		Limits	Desired	Left	Right	Left	Right	
1990	1.8L	+1 5/6 to +2 5/6	+2 1/3	−4/15 to +11/15	−4/15 to +11/15	+7/30	+7/30	0①
	2.0L	+1 27/30 to +2 27/30	+2 2/5	−5/12 to +7/12	−5/12 to +7/12	+1/12	+1/12	0①

①—Plus or minus .12 inch.

REAR WHEEL ALIGNMENT

Year	Model	Camber Angle, Degrees				Toe In Inch
		Limits		Desired		
		Left	Right	Left	Right	
1990	All	−1 1/4 to −1/4	−1 1/4 to −1/4	−3/4	−3/4	0①

①—Plus or minus .118 inch.

EAGLE 4 x 4

Year	Model	Caster Angle, Degrees		Camber Angle, Degrees				Toe-In Inch	Toe Out on Turns, Deg.	
				Limits		Desired				
		Limits	Desired	Left	Right	Left	Right		Outer Wheel	Inner Wheel
1985–88	All	+3 to +5	+4	−1/8 to +5/8	−1/8 to +5/8	+3/8	+3/8	1/16 to 3/16	①	38

①—Wheels at full turn.

EAGLE MEDALLION

FRONT WHEEL ALIGNMENT

The specifications listed below are for unloaded vehicles

Year	Model	Caster Angle, Degrees ①		Camber Angle, Degrees		Toe-Out, Inch		King Pin Inclination Degrees	
		Limits	Desired	Limits	Desired	Per Wheel	Total Over Two Wheels	Limits	Desired
1988–89	All	②	②	0 to +1	+1/2	.078 to .125	.04 to .12	12 1/6 to 13 1/6	12 2/3

①—Vehicle ride height must be checked and, if necessary adjusted before checking front wheel alignment.

②—3 1/2 degrees when vehicle ride height is 1.38 inch, 3 degrees vehicle ride height is 2.17 inch, 2 1/2 degrees when vehicle ride height is 2.95 inch, 2 degrees when vehicle ride height is 3.74 inch or 1 1/2 degrees when vehicle ride height is 4.53 inch.

REAR WHEEL ALIGNMENT

The specifications listed below are for unloaded vehicles

Year	Model	Camber Angle, Degrees		Toe-In Inch
		Limits	Desired	
1988–89	All	−1/2 to −5/6	−2/3	.078–.197

EAGLE PREMIER

FRONT WHEEL ALIGNMENT

Year	Model	Caster Angle, Degrees		Camber Angle, Degrees				Toe-In Inch
				Limits		Desired		
		Limits	Desired	Left	Right	Left	Right	
1988–90	All	+1.32 to +2.82	+2.07	−.08 to +.58	−.08 to +.58	+.25	+.25	①

①—Toe-out per wheel, 1/16 inch.

REAR WHEEL ALIGNMENT

Year	Model	Camber Angle, Degrees				Toe-In Inch
		Limits		Desired		
		Left	Right	Left	Right	
1988–90	All	−.4 to −.93	−.4 to −.93	−.67	−.67	①

①—1/32 to 1/4 inch toe-in.

EAGLE SUMMIT

FRONT WHEEL ALIGNMENT

Year	Model	Caster Angle, Degrees		Camber Angle, Degrees				Toe-In
				Limits		Desired		
		Limits	Desired	Left	Right	Left	Right	
1989–90	All	+1 5/6 to +2 5/6	+2 1/3	−1/2 to +1/2	−1/2 to +1/2	0	0	①

①—.12 inch toe in to .12 inch toe-out.

REAR WHEEL ALIGNMENT

Year	Model	Camber Angle, Degrees				Toe-In
		Limits		Desired		
		Left	Right	Left	Right	
1989–90	All	−1 1/10 to −1/10	−1 1/10 to −1/10	−2/3	−2/3	①

①—.18 inch toe in to .18 inch toe-out.

EAGLE TALON

FRONT WHEEL ALIGNMENT

Year	Model	Caster Angle, Degrees		Camber Angle, Degrees				Toe In Inch
				Limits		Desired		
		Limits	Desired	Left	Right	Left	Right	
1990	2WD①	+1 5/6 to +2 5/6	+2 1/3	−4/15 to +11/15	−4/15 to +11/15	+7/30	+7/30	0③
	2WD②	+1 27/30 to +2 27/30	+2 2/5	−5/12 to +7/12	−5/12 to +7/12	+1/12	+1/12	0③
	4WD	+1 4/5 to +2 4/5	+2 3/10	−1/3 to +2/3	−1/3 to +2/3	+1/6	+1/6	0③

2WD—Two wheel drive.
4WD—Four wheel drive.
①—Models w/1.8L engine.
②—Models w/2.0L engine.
③—Plus or minus .12 inch.

REAR WHEEL ALIGNMENT

Year	Model	Camber Angle, Degrees				Toe In Inch
		Limits		Desired		
		Left	Right	Left	Right	
1990	2WD	−1 1/4 to −1/4	−1 1/4 to −1/4	−3/4	−3/4	0①
	4WD	−2 1/20 to −1 1/20	−2 1/20 to −1 1/20	−1 11/20	−1 11/20	.14①

2WD—Two wheel drive. 4WD—Four wheel drive. ①—Plus or minus .118 inch.

Ford Motor Co.

INDEX

FORD CROWN VICTORIA & MERCURY GRAND MARQUIS

		Caster Angle, Degrees		Camber Angle, Degrees					Toe Out on Turns, Deg.	
				Limits		Desired				
Year	Model	Limits	Desired	Left	Right	Left	Right	Toe-In Inch	Outer Wheel	Inner Wheel
1985–86	All	+2.25 to +4	+3	−.25 to +1.25	−.25 to +1.25	+.5	+.5	1/16	18.51	20
1987	All①	+2.5 to +5	+3.75	.3 to +1.2	−.3 to +1.2	+.45	+.45	②	18.51	20
1987	All③	+2.5 to +5	+3.75	−1.2 to +.3	−1.2 to +.3	−.45	−.45	②	18.51	20
1988	All	+2.5 to +5	+3.75	−1.2 to +.3	−1.2 to +.3	−.45	−.45	②	18.51	20
1989–90	All	+2.5 to +4.5	+3.5	−.75 to +.25	−.75 to +.25	−.5	−.5	1/16	18.51	20

①—Models built prior to 1/87. ②—Total toe, −.12 to +.37 degrees. ③—Models built from 1/87.

FORD LTD & THUNDERBIRD & MERCURY COUGAR & MARQUIS

FRONT WHEEL ALIGNMENT

		Caster Angle, Degrees		Camber Angle, Degrees					Toe Out on Turns, Deg.	
				Limits		Desired				
Year	Model	Limits	Desired	Left	Right	Left	Right	Toe-In Inch	Outer Wheel	Inner Wheel
1985–86	Cougar	0 to +1 1/2	+3/4	−1/2 to +1	−1/2 to +1	+1/4	+1/4	3/16	19.73	20
	LTD Sedan	+1/8 to +2 1/8	+7/8	−3/8 to +1 1/8	−3/8 to +1 1/8	+3/8	+3/8	3/16	19.84	20
	LTD Sta. Wag.	0 to +2	+3/4	−5/16 to +1 3/16	−5/16 to +1 3/16	+7/16	+7/16	3/16	19.84	20
	Marquis Sedan	+1/8 to +2 1/8	+7/8	−3/8 to +1 1/8	−3/8 to +1 1/8	+3/8	+3/8	3/16	19.84	20
	Marquis Sta. Wag.	0 to +2	+3/4	−5/16 to +1 3/16	−5/16 to +1 3/16	+7/16	+7/16	3/16	19.84	20
	Thunderbird	0 to +1 1/2	+3/4	−1/2 to +1	−1/2 to +1	+1/4	+1/4	3/16	19.73	20
1987–88	Cougar & Thunderbird Less Turbo	+.5 to +2.1	+1.3	−1.4 to +.1	−1.4 to +.1	−.85	−.85	①	19.73	20
	Thunderbird Turbo	+.5 to +2.1	+1.3	−.7 to +.8	−.7 to +.8	+.05	+.05	①	19.73	20
1989–90	Cougar & Thunderbird	+4.75 to +6.25	+5.5	−1.25 to +.25	−1.25 to +.25	−.5	−.5	1/8	19.73	20

①—Total toe, +.12 to +.62 degrees.

REAR WHEEL ALIGNMENT

		Camber Angle, Degrees				
		Limits		Desired		
Year	Model	Left	Right	Left	Right	Toe-In Inch
1989–90	Cougar & Thunderbird	−1 to 0	−1 to 0	−.5	−.5	1/16

FORD MUSTANG & MERCURY CAPRI

Year	Model	Caster Angle, Degrees		Camber Angle, Degrees				Toe-In Inch	Toe Out on Turns, Deg.	
				Limits		Desired				
		Limits	Desired	Left	Right	Left	Right		Outer Wheel	Inner Wheel
1985–86	All	+1/4 to +1 3/4	+1	−3/4 to +3/4	−3/4 to +3/4	0	0	3/16	19.84	20
1987–88	Mustang Exc. 5.0L GT	+.5 to +2.1	+1.3	−.5 to +1	−.5 to +1	+.25	+.25	.19	19.84	20
	Mustang 5.0L GT	+.5 to +2.1	+1.3	−.6 to +.9	−.6 to +.9	+.15	+.15	.19	19.84	20
1989	Mustang Exc. 5.0L GT	+.4 to +1.9	+1.15	−.85 to +.65	−.85 to +.65	−.1	−.1	.19	19.84	20
	Mustang 5.0L GT	+.5 to +2	+1.27	−.6 to +.9	−.6 to +.9	+.14	+.14	.19	19.84	20
1990	Mustang 2.3L	+1.15 to +2.65	+1.9	−1.25 to +.25	−1.25 to +.25	−.5	−.5	①	19.84	20
	Mustang 5.0L	+1.15 to +2.65	+1.9	−1.35 to +.15	−1.35 to +.15	−.6	−.6	①	19.84	20

①—Total toe, −.12 inch.

FORD ESCORT, EXP & TEMPO & MERCURY LYNX & TOPAZ

FRONT WHEEL ALIGNMENT

Year	Model	Caster Angle, Degrees		Camber Angle, Degrees				Toe-In Inch	Toe Out on Turns, Deg.	
				Limits		Desired				
		Limits	Desired	Left	Right	Left	Right		Outer Wheel	Inner Wheel
1985	Escort, EXP, Lynx ②	+9/16 to +2 1/16	+1 5/15	+1 1/8 to +2 5/8	+11/16 to 2 3/16	+1 7/8	+1 7/16	①	④	20
	Escort, EXP, Lynx ③	+1/4 to +1 3/4	+1	+7/16 to +1 15/16	0 to +1 1/2	+1 3/16	+3/4	①	④	20
	Tempo, Topaz	+1/2 to +2	+1 1/4	+1 1/4 to +2 3/4	+3/4 to +2 1/4	+2	+1 1/2	①	④	20
1986	Escort, EXP, Lynx	+1.6 to +3.1	+2.35	+.39 to +1.79	−.01 to +1.49	+1.04	+.74	①	④	20
	Tempo, Topaz	+1.69 to +3.19	+2.44	+.66 to +2.16	+.22 to +1.72	+1.41	+.97	①	④	20
1987–88	Escort, EXP & Lynx	+1.3 to +2.8	+2.05	+.6 to +2.1	+.2 to +1.7	+1.35	+.95	①	④	20
	Tempo & Topaz	+1.7 to +3.2	+2.45	+.6 to +2.1	+.2 to +1.7	+1.35	+.95	①	④	20
1989	Escort	+1.57 to +3.17	+2.32	+.43 to +1.93	−.01 to +1.49	+1.18	+.74	①	④	20
	Tempo & Topaz	+1.69 to +3.19	+2.44	+.66 to +2.16	+.22 to +1.72	+1.41	+.97	①	④	20
1990	Escort Exc. GT	+1.57 to +3.17	+2.32	+.65 to +2.15	−.21 to +1.71	+1.4	+.96	①	④	20
	Escort GT	+1.57 to +3.17	+2.32	+.25 to +1.75	−.19 to +1.31	+1	+.56	①	④	20

①—Toe-out, .10 inch.
②—Models less turbocharger.
③—Models with turbocharger.
④—Left wheel, 20 degrees; right wheel, 18.2 degrees.

REAR WHEEL ALIGNMENT

Year	Model	Camber Angle, Degrees				Total Toe-In Inch
		Limits		Desired		
		Left	Right	Left	Right	
1985	Escort, EXP, Lynx	−1 3/32 to +19/32	−1 3/32 to +19/32	−1/4	−1/4	3/16
	Tempo, Topaz	−2 1/8 to −3/8	−1 1/2 to +1/4	−1 1/4	−5/8	3/16
1986	Escort, EXP, Lynx	−1.2 to +1.2	−1.2 to +1.2	0	0	3/16
	Tempo, Topaz	−.91 to +.59	−.91 to +.59	−.16	−.16	0
1987–88	Escort, EXP & Lynx	−1.2 to +.5	−1.2 to +.5	−.35	−.35	.18
	Tempo & Topaz ①	−.9 to +.6	−.9 to +.6	−.15	−.15	0
	Tempo & Topaz ②	−.04 to +1.1	−.4 to +1.1	+.35	+.35	0
1989–90	Escort	−1.2 to +.5	−1.2 to +.5	−.35	−.35	.18
	Tempo & Topaz ①	−.91 to +.59	−.91 to +.59	−.16	−.16	0
	Tempo & Topaz ②	−.04 to +1.09	−.4 to +1.09	+.34	+.34	0

①—Except all wheel drive models.
②—All wheel drive models.

FORD TAURUS & MERCURY SABLE

FRONT WHEEL ALIGNMENT

Year	Model	Caster Angle, Degrees		Camber Angle, Degrees				Toe-In Inch	Toe Out on Turns, Deg.	
				Limits		Desired				
		Limits	Desired	Left	Right	Left	Right		Outer Wheel	Inner Wheel
1986	Sable Sedan	+2.88 to +5.88	+3.88	−1.10 to +.10	−1.10 to +.10	−.50	−.50	①	—	—
	Taurus Sedan	+3 to +6	+4	−1.10 to +.10	−1.10 to +.10	−.50	−.50	①	—	—
	Sta. Wag.	+2.81 to +5.81	+3.81	−1.04 to +.16	−1.04 to +.16	−.44	−.44	①	—	—
1987–88	All	+2.9 to +5.9	+4.4	−1.1 to +.1	−1.1 to +.1	−.6	−.6	①	18.25	20
1989	Taurus Sedan	+2.8 to +5.8	+3.8	−1.1 to +.1	−1.1 to +.1	−.5	−.5	②	18.25	20
	Sable Sedan	+2.7 to +5.7	+3.7	−1.1 to +.1	−1.1 to +.1	−.5	−.5	①	18.25	20
	All Sta. Wag.	+2.8 to +5.8	+3.8	−1.04 to +.16	−1.04 to +.16	−.44	−.44	①	18.25	20
1990	Taurus Sedan	+2.8 to +5.8	+3.8	−1.1 to +.1	−1.1 to +.1	−.5	−.5	①	18.25	20
	Sable Sedan	+2.7 to +5.7	+3.7	−1.1 to +.1	−1.1 to +.1	−.5	−.5	①	18.25	20
	All Sta. Wag.	+2.6 to +5.6	+3.6	−1.05 to +.15	−1.05 to +.15	−.45	−.45	①	18.25	20

①—Total toe, −.10 inch.

REAR WHEEL ALIGNMENT

Year	Model	Camber Angle, Degrees				Toe-In Inch
		Limits		Desired		
		Left	Right	Left	Right	
1986	Sable Sedan	−1.58 to −.18	−1.58 to −.18	−.88	−.88	.06
	Taurus Sedan	−1.64 to −.24	−1.64 to −.24	−.94	−.94	.06
	Sta. Wag.	−1.32 to +.08	−1.32 to +.08	−.62	−.62	.06
1987–88	Sedan	−1.6 to −.2	−1.6 to −.2	−.9	−.9	.06
	Sta. Wag.	−1.9 to +.1	−1.9 to +.1	−.9	−.9	.06
1889	Sedan	−1.6 to −.2	−1.6 to −.2	−.9	−.9	.06
	Sta. Wag.	−1.32 to +.08	−1.32 to +.08	−.062	−.062	.06
1990	All	−1.6 to −.2	−1.6 to −.2	−.9	−.9	.06

FORD PROBE

FRONT WHEEL ALIGNMENT

Year	Model	Caster Angle, Degrees		Camber Angle, Degrees				Toe-In Inch	King Pin Inclination
				Limits		Desired			
		Limits	Desired	Left	Right	Left	Right		
1989–90	All	+.47 to +1.97	+1.22	−.47 to +1.03	−.47 to +1.03	+.28	+.28	0	12.78

REAR WHEEL ALIGNMENT

Year	Model	Camber Angle, Degrees				Toe-In Inch
		Limits		Desired		
		Left	Right	Left	Right	
1989–90	All	−.25 to +1.25	−.25 to +1.25	+.5	+.5	.12

FORD FESTIVA

FRONT WHEEL ALIGNMENT

Year	Model	Caster Angle, Degrees		Camber Angle, Degrees				Toe-In Inch	King Pin Inclination Degrees
				Limits		Desired			
		Limits	Desired	Left	Right	Left	Right		
1988–90	All	1/3 to 1 5/6	1 7/12	−1/4 to +1 7/12	−1/4 to +1 7/12	+2/3	+2/3	.14	14 11/60

REAR WHEEL ALIGNMENT

Year	Model	Camber		Toe-In Inch
		Limits	Desired	
1988–90	All	+1/15 to +13/30	+1/4	.12

LINCOLN

FRONT WHEEL ALIGNMENT

Year	Model	Caster Angle, Degrees		Camber Angle, Degrees				Toe-In Inch	Toe Out on Turns, Deg.	
				Limits		Desired				
		Limits	Desired	Left	Right	Left	Right		Outer Wheel	Inner Wheel
1985–86	Town Car	+2 1/4 to +4	+3	−1/4 to +1 1/4	−1/4 to +1 1/4	+1/2	+1/2	1/16	18.51	20
	Mark VII	+5/8 to +2 3/4	+1 1/2	−3/4 to +3/4	−3/4 to +3/4	0	0	1/8	17.14	20
	Continental	+5/8 to +2 3/4	+1 1/2	−3/4 to +3/4	−3/4 to +3/4	0	0	1/8	17.14	20
1987	Town Car ①	+2.5 to +5	3.75	−.3 to +1.2	−.3 to +1.2	.45	.45	②	18.51	20
	Town Car ③	+2.5 to +5	+3.75	−1.2 to +.3	−1.2 to +.3	+.45	+.45	②	18.51	20
	Mark VII	+.6 to +2.7	+1.65	−.8 to +.7	−.8 to +.7	−.05	−.05	④	17.14	20
	Continental	+.6 to +2.7	+1.65	−.8 to +.7	−.8 to +.7	−.05	−.05	④	17.14	20
1988	Town Car	+2.5 to +5	+3.75	−1.2 to +.3	−1.2 to +.3	+.45	+.45	②	18.51	20
	Mark VII	+.6 to +2.7	+1.65	−.8 to +.7	−.8 to +.7	−.05	−.05	④	17.14	20
	Continental	+4 to +5.5	+4.75	−1.5 to −.3	−1.5 to −.3	−.9	−.9	⑤	—	—
1989–90	Town Car	+2.5 to +4.5	+3.5	−.75 to +.25	−.75 to +.25	−.5	−.5	1/16	18.51	20
	Mark VII	+.6 to +2.7	+1.5	−.75 to +.75	−.75 to +.75	0	0	1/8	17.14	20
	Continental	+3.6 to +5.2	+4.4	−1.7 to −.5	−1.7 to −.5	−1.1	−1.1	⑤	18.21	20

①—Models built prior to 1/87.
②—Total toe, −.12 to +.37 degrees.
③—Models built from 1/87.
④—Total toe, 0 to +.5 degrees.
⑤—Total toe, −.20 degrees.

REAR WHEEL ALIGNMENT

Year	Model	Camber Angle, Degrees				Toe-In Inch
		Limits		Desired		
		Left	Right	Left	Right	
1988–90	Continental	−2 to −.6	−2 to −.6	−1.3	−1.3	①

①—Total toe, +.20 degrees.

WHEEL ALIGNMENT SPECIFICATIONS

MERKUR

FRONT WHEEL ALIGNMENT

Year	Model	Caster Angle, Degrees		Camber Angle, Degrees				Toe-In Inch	Wheel Turning Angle ①	
				Limits		Desired				
		Limits	Desired	Left	Right	Left	Right		Left	Right
1985–89	XR4Ti	+29/30 to +2 29/30	+1 29/30	−1 8/15 to +7/15	−1 8/15 to +7/15	−8/15	−8/15	5/64	—	—
1988–90	Scorpio	+1 to +3	+2	−1 7/16 to +9/16	−1 7/16 to +9/16	−7/16	−7/16	9/64	20	19.2

①—At outside wheel with inside wheel turned at 20 degrees.

REAR WHEEL ALIGNMENT

Camber angle varies with ride height. Measure ride height from center of rear wheel to bottom of wheel opening molding.

Model	Toe In Inch	Ride Height Inches	Camber Angle Degrees
XR4Ti	3/64	13.78 to 14.13	−2 5/6 to −1/2
	3/64	14.17 to 14.53	−2 7/15 to −2/15
	3/64	14.57 to 14.92	−2 1/10 to +1/4
	3/64	14.96 to 15.31	−1 43/60 to +37/60
	3/64	15.35 to 15.71	−1 7/20 to +1
	3/64	15.75 to 16.10	−29/30 to +1 11/30
	3/64	16.14 to 16.50	−3/5 to +1 3/4
	3/64	16.54 to 16.89	−13/60 to +2 7/60
Scorpio	1/20	14.17 to 14.53	−2 1/60 to −1/60
	1/20	14.56 to 14.92	−1 19/30 to +11/30
	1/20	14.96 to 15.31	−1 4/15 to +11/15
	1/20	15.35 to 15.70	−53/60 to +1 7/60
	1/20	15.75 to 16.10	−31/60 to +1 29/60
	1/20	16.14 to 16.50	−2/15 to +1 13/15
	1/20	16.54 to 16.89	+7/30 to +2 7/30
	1/20	16.92 to 17.28	+37/60 to +2 37/60

MERCURY TRACER

FRONT WHEEL ALIGNMENT

Year	Model	Caster Angle, Degrees		Camber Angle, Degrees		Toe-In Inch	King Pin Inclination	Steering Angle Degrees	
		Limits	Desired	Limits	Desired			Inner	Outer
1988–89	All	+5/6 to +2 1/3	+1 7/12	+1/20 to +1 23/60	+4/5	.08	+12 11/30	40	33

REAR WHEEL ALIGNMENT

Year	Model	Camber Angle, Degrees		Toe-In Inch
		Limits	Desired	
1988–89	All	−3/4 to +3/4	0	.08

General Motors Corp.

INDEX

BUICK REAR WHEEL DRIVE & 1985 RIVIERA

FRONT WHEEL ALIGNMENT

Year	Model	Caster Angle, Degrees		Camber Angle, Degrees				Toe-In Inch
				Limits		Desired		
		Limits	Desired	Left	Right	Left	Right	
1985	Estate Wagon	+2.5 to +3.5	+3	+.3 to +1.3	+.3 to +1.3	+.8	+.8	1/16 to 3/16
	LeSabre	+2.5 to +3.5	+3	+.3 to +1.3	+.3 to +1.3	+.8	+.8	1/16 to 3/16
	Regal	+2.5 to +3.5	+3	0 to +1	0 to +1	+.5	+.5	1/16 to 3/16
	Riviera	+2 to +3	+2.5	−.5 to +.5	−.5 to +.5	0	0	④
1986	Estate Wagon	+2 to +4	+3	0 to +1.6	0 to +1.6	+.8	+.8	③
	Regal	+2 to +4	+3	−.3 to +1.3	−.3 to +1.3	+.5	+.5	③
1987	Regal	+1.8 to +3.8	+2.8	−.3 to +1.3	−.3 to +1.3	+.5	+.5	②
1987–88	Estate Wagon	+1.8 to +3.8	+2.8	0 to +1.6	0 to +1.6	+.8	+.8	②
1989–90	Estate Wagon	+2.5 to +3.5	+3	+.3 to +1.3	+.3 to +1.3	+.8	+.8	①

①—Toe, +.05 degree.
②—Total Toe, +.1 degree.
③—Toe-in, .15 degree per wheel.
④—1/16 inch toe-in to 1/16 inch toe-out.

REAR WHEEL ALIGNMENT

Year	Model	Camber Angle, Degrees				Toe-In Inch
		Limits		Desired		
		Left	Right	Left	Right	
1985	Riviera	−.4 to +.4	−.4 to +.4	0	0	①

①—3/32 inch per wheel; both wheels, 5/32 inch.

WHEEL ALIGNMENT SPECIFICATIONS

1988–90 BUICK REGAL, OLDS CUTLASS SUPREME & PONT. GRAND PRIX & 1990 CHEVROLET LUMINA

FRONT WHEEL ALIGNMENT

		Caster Angle, Degrees		Camber Angle, Degrees				
				Limits		Desired		
Year	Model	Limits	Desired	Left	Right	Left	Right	Total Toe Degrees
1988	All	+1.5 to +2.5	+2	+.2 to +1.2	+.2 to +1.2	+.7	+.7	0
1989–90	All	+1.5 to +2.5	+2	+.2 to +1.2	+.2 to +1.2	+.7	+.7	0

REAR WHEEL ALIGNMENT

		Camber Angle, Degrees				
		Limits		Desired		
Year	Model	Left	Right	Left	Right	Total Toe Deg.
1988	All	−.44 to +.56	−.44 to +.56	+.06	+.06	−.1
1989	All①	−.44 to +.56	−.44 to +.56	+.06	+.06	0
	All②	+.1 to +1.1	+.1 to +1.1	+.6	+.6	0
1990	All①	−.4 to +.6	−.4 to +.6	+.1	+.1	0
	All②	−.65 to +.45	−.65 to +.45	−.15	−.15	0

①—Models with soft ride suspension.
②—Models with sport suspension.

BUICK ELECTRA, LESABRE & PARK AVENUE, CADILLAC DEVILLE & FLEETWOOD, OLDSMOBILE 88 & 98 & PONTIAC BONNEVILLE FRONT WHEEL DRIVE MODELS

FRONT WHEEL ALIGNMENT

		Caster Angle, Degrees		Camber Angle, Degrees				
				Limits		Desired		
Year	Model	Limits	Desired	Left	Right	Left	Right	Total Toe Degrees
1985	Electra & Park Ave.	+1.8 to +2.8	+2.3	−1 to 0	0 to +1	−.5	+.5	0
	DeVille & Fleetwood	—	—	−.1 to +1.1	−.1 to +1.1	+.6	+.6	+.125
	98	+1.8 to +2.8	+2.3	+.25 to +.75	+.25 to +.75	+.5	+.5	0
1986	Electra & LeSabre	+1.8 to +2.8	+2.3	−1 to 0	0 to +1	−.5	+.5	0
	DeVille & Fleetwood	+1.5 to +3.5	+2.5	−1 to 0	0 to +1	−.5	+.5	0
	88 & 98	+1.8 to +2.8	+2.3	−1 to 0	0 to +1	−.5	+.5	0
1987–88	Exc. Deville & Fleetwood	+2 to +3	+2.5	−.3 to +.7	−.3 to +.7	+.2	+.2	0
	Deville & Fleetwood	+2 to +3	+2.5	−1 to 0	0 to +1	−.5	+.5	0
1989–90	Exc. Deville & Fleetwood	+2.5 to +3.5	+3	−.3 to +.7	−.3 to +.7	+.2	+.2	0
	Deville & Fleetwood	+2.5 to +3.5	+3	−1 to 0	0 to +1	−.5	+.5	0

REAR WHEEL ALIGNMENT

		Camber Angle, Degrees				
		Limits		Desired		
Year	Model	Left	Right	Left	Right	Total Toe Degrees
1985	Electra & Park Ave.	−1 to +0	0 to +1	−.5	+.5	+.1
	DeVille & Fleetwood	−.3 to +.5	−.3 to +.5	+.1	+.1	+.2
	98	−.55 to −.05	−.55 to −.05	−.3	−.3	+.1
1986	Electra & LeSabre	−.8 to +.2	−.8 to +.2	−.3	−.3	+.1
	DeVille & Fleetwood	−.8 to +.2	−.8 to +.2	−.3	−.3	+.1
	88 & 98	−.8 to +.2	−.8 to +.2	−.3	−.3	+.1
1987–88	All	−.8 to +.2	−.8 to +.2	−.3	−.3	+.1
1989–90	All	−.8 to +.2	−.8 to +.2	−.3	−.3	+.1

1986–90 BUICK REATTA & RIVIERA, CAD. ELDORADO & SEVILLE, OLDS TORONADO & TROFEO

FRONT WHEEL ALIGNMENT

		Caster Angle, Degrees		Camber Angle, Degrees				
				Limits		Desired		Total Toe
Year	Model	Limits	Desired	Left	Right	Left	Right	Degrees
1986	Eldorado & Seville	+1.7 to +3.3	+2.5	−.5 to +.5	−.5 to + .5	0	0	0①
	Riviera	+1.5 to +3.5	+2.5	−.3 to +1.3	−.3 to +1.3	+.5	+.5	0
	Toronado	+1.3 to +3.3	+2.3	−.8 to +8	−.8 to +.8	0	0	0①
1987	Eldorado & Seville	+1.3 to +3.3	+2.3	−.8 to +.8	−.8 to +.8	0	0	0
	Riviera	+1.3 to +3.3	+2.3	−.8 to +.8	−.8 to +.8	0	0	0
	Toronado	+1.3 to +3.3	+2.3	−.8 to +.8	−.8 to +.8	0	0	0
1988	Eldorado & Seville	+1.3 to +3.3	+2.3	−.8 to +.8	−.8 to +.8	0	0	0
	Reatta	+1.8 to +3.8	+2.8	−.5 to +.5	−.5 to +.5	0	0	0
	Riviera	+1.5 to +3.5	+2.5	−.8 to +.8	−.8 to +.8	0	0	0
	Toronado	+1.5 to +3.5	+2.5	−.8 to +.8	−.8 to +.8	0	0	0
1989	Eldorado & Seville	+1.3 to +3.3	+2.3	−.8 to +.8	−.8 to +.8	0	0	+.1
	Reatta	+2.0 to +4.0	+3.0	−.8 to +.8	−.8 to +.8	0	0	0
	Riviera	+1.5 to +3.5	+2.5	−.8 to +.8	−.8 to +.8	0	0	0
	Toronado	+1.3 to +3.3	+2.3	−.8 to +.8	−.8 to +.8	0	0	+.1
1990	Eldorado & Seville	+1.3 to +3.3	+2.3	−.8 to +.8	−.8 to +.8	0	0	+.1
	Reatta	+2.0 to +4.0	+3.0	−.8 to +.8	−.8 to +.8	0	0	0
	Riviera	+1.5 to +3.5	+2.5	−.8 to +.8	−.8 to +.8	0	0	0
	Toronado & Trofeo	+1.3 to +3.3	+2.3	−.8 to +.8	−.8 to +.8	0	0	+.1

①—Per wheel.

REAR WHEEL ALIGNMENT

		Camber Angle, Degrees				
		Limits		Desired		
Year	Model	Left	Right	Left	Right	Total Toe Degrees
1986	Eldorado & Seville	−1 to −.4	−1 to −.4	−.7	−.7	+.1①
	Riviera	−.8 to −.6	−.8 to −.6	−.7	−.7	+.1①
	Toronado	−.7 to −.1	−.7 to −.1	−.4	−.4	+.1①
1987	Eldorado & Seville	−.4 to +.2	−.4 to +.2	−.1	−.1	+.2
	Riviera	−1 to −.4	−1 to −.4	−.7	−.7	+.2
	Toronado	−.7 to −.1	−.7 to −.1	−.4	−.4	+.2
1988	Eldorado & Seville	−.4 to +.2	−.4 to +.2	−.1	−.1	+.2
	Reatta	−.1 to +.5	−.1 to +.5	+.2	+.2	+.2
	Riviera	−1.3 to −.7	−1.3 to −.7	−1	−1	+.2
	Toronado & Trofeo	−1.3 to −.7	−1.3 to −.7	−1	−1	+.2
1989	Eldorado & Seville	−.4 to +.2	−.4 to +.2	−.1	−.1	+.2
	Reatta	−.1 to +.5	−.1 to +.5	+.2	+.2	+.2
	Riviera	−.4 to +.2	−.4 to +.2	−1	−1	+.2
	Toronado & Trofeo	−.4 to +.2	−.4 to +.2	−.1	−.1	+.2
1990	Reatta	−1.3 to −.7	−1.3 to −.7	−1	−1	+.2
	Riviera	−1.3 to −.7	−1.3 to −.7	−1	−1	+.2
	Toronado & Trofeo	−.4 to +.1	−.4 to +.1	−.1	−.1	+.2

①—Per wheel.

BUICK SOMERSET, OLDS CALAIS, PONT. GRAND AM & 1986–90 BUICK SKYLARK & 1988–90 OLDS CUTLASS CALAIS

FRONT WHEEL ALIGNMENT

Year	Model	Caster Angle, Degrees		Camber Angle, Degrees				Total Toe Degrees
				Limits		Desired		
		Limits	Desired	Left	Right	Left	Right	
1985–86	All	+.7 to +2.7	+1.7	+.2 to +1.5	+.2 to +1.5	+.85	+.85	①
1987	All	+.7 to +2.7	+1.7	+.2 to +1.4	+.2 to +1.4	+.8	+.8	0
1988	All	−.8 to +4.2	+1.7	−.2 to +1.8	−.2 to +1.8	+.8	+.8	0
1989	All	+.7 to +2.7	+1.7	+.1 to +1.5	+.1 to +1.5	+.8	+.8	0
1990	All②	+.7 to +2.7	+1.7	+.1 to +1.5	+.1 to +1.5	+.8	+.8	0
	All③	+.7 to +2.7	+1.7	−.7 to +.7	−.7 to +.7	0	0	0

①—Toe-out, .06 degree per wheel.
②—Models less 16 inch wheels.
③—Models with 16 inch wheels.

REAR WHEEL ALIGNMENT

Year	Model	Camber Angle, Degrees				Total Toe Degrees
		Limits		Desired		
		Left	Right	Left	Right	
1990	All	−.75 to +.25	−.75 to +.25	−.25	−.25	+.25

CADILLAC REAR WHEEL DRIVE & 1985 ELDORADO & SEVILLE

FRONT WHEEL ALIGNMENT

Year	Model	Caster Angle, Degrees		Camber Angle, Degrees				Toe-In Inch
				Limits		Desired		
		Limits	Desired	Left	Right	Left	Right	
1985	Eldorado & Seville	+2 to +3	+2.5	−.5 to +.5	−.5 to +.5	0	0	①
1985–86	Brougham	+2.5 to +3.5	+3	+.1 to +.9	+.1 to +.9	+.5	+.5	1/8
1987	Brougham	+2 to +4	+3	−.2 to +.8	−.2 to +.8	+.3	+.3	②
1988–90	Brougham	+2 to +4	+3	−.5 to +.5	−.5 to +.5	0	0	③

①—1/16 toe-in to 1/16 toe-out.
②—Total toe, .1 degree.
③—Total toe, 0 degrees.

REAR WHEEL ALIGNMENT

Year	Model	Camber Angle, Degrees				Toe-In Inch
		Limits		Desired		
		Left	Right	Left	Right	
1985	Eldorado & Seville	−.4 to +.4	−.4 to +.4	0	0	①

①—3/32 inch per wheel; both wheels, 5/32.

CHEVROLET CAMARO, CAPRICE, CORVETTE & MONTE CARLO

FRONT WHEEL ALIGNMENT

Year	Model	Caster Angle, Degrees		Camber Angle, Degrees				Toe-In Inch
				Limits		Desired		
		Limits	Desired	Left	Right	Left	Right	
1985	Camaro	+2.5 to +3.5	+3	+.5 to +1.5	+.5 to +1.5	+1	+1	③
	Caprice & Impala	+2.5 to +3.5	+3	+.3 to +1.3	+.3 to +1.3	+.8	+.8	1/8
	Corvette	+2.5 to +3.5	+3	+.3 to +1.3	+.3 to +1.3	+.8	+.8	④
	Malibu & Monte Carlo①	+.5 to +1.5	+1	0 to +1	0 to +1	+.5	+.5	1/8
	Malibu & Monte Carlo②	+2.5 to +3.5	+3	0 to +1	0to +1	+.5	+.5	1/8
1986	Camaro Exc. Z28	+2.5 to +3.5	+3	+.5 to +1.5	+.5 to +1.5	+1	+1	④
	Camaro Z28	+3 to +4	+3.5	+.5 to +1.5	+.5 to +1.5	+1	+1	④
	Caprice	+2.5 to +3.5	+3	+.3 to +1.3	+.3 to +1.3	+.8	+.8	1/8
	Corvette	+2.5 to +3.5	+3	+.3 to +1.3	+.3 to +1.3	+.8	+.8	④
	Monte Carlo①	+.5 to +1.5	+1	0 to +1	0 to +1	+.5	+.5	1/8
	Monte Carlo②	+2.5 to +3.5	+3	0 to +1	0 to +1	+.5	+.5	1/8
1987	Camaro	+4.2 to +5.2	+4.7	−.2 to +.8	−.2 to +.8	+.3	+.3	⑤
	Caprice	+1.8 to +3.8	+2.8	0 to +1.6	0 to +1.6	+.8	+.8	⑦
	Corvette	+4.7 to +6.3	+5.5	+.3 to +1.3	+.3 to +1.3	+.8	+.8	⑤
	Monte Carlo①	−.2 to +1.8	+.8	−.3 to +1.3	−.3 to +1.3	+.5	+.5	⑦
	Monte Carlo②	+1.8 to +3.8	+2.8	−.3 to +1.3	−.3 to +1.3	+.5	+.5	⑦
1988	Camaro	+4.2 to +5.2	+4.7	−.2 to +.8	−.2 to +.8	+.3	+.3	⑤
	Caprice	+1.8 to +3.8	+2.8	0 to +1.6	0 to +1.6	+.8	+.8	⑥
	Corvette	+5.5 to +6.5	+6	+.3 to +1.3	+.3 to +1.3	+.8	+.8	⑤
	Monte Carlo①	−.2 to +1.8	+.8	−.3 to +1.3	−.3 to +1.3	+.5	+.5	⑥
	Monte Carlo②	+1.8 to +3.8	+2.8	−.3 to +1.3	−.3 to +1.3	+.5	+.5	⑥
1989	Camaro	+4.2 to +5.2	+4.7	−.2 to +.8	−.2 to +.8	+.3	+.3	⑤
	Caprice	+2.5 to +3.5	+3	+.3 to +1.3	+.3 to +1.3	+.8	+.8	⑥
	Corvette	+5.3 to +6.3	+5.8	0 to +1	0 to +1	+.5	+.5	⑤
1990	Camaro	+4.5 to +5.5	+5	−.2 to +.8	−.2 to +.8	+.3	+.3	⑤
	Caprice	+2.5 to +3.5	+3	+.3 to +1.3	+.3 to +1.3	+.8	+.8	⑥
	Corvette	+5.5 to +6.5	+6	+.3 to +1.3	+.3 to +1.3	+.8	+.8	⑤

①—Man. steer.
②—Power steer.
③—Except Z28, .2 degree per wheel; Z28, .15 degree per wheel.
④—.15 degree per wheel.
⑤—Total toe, 0 degree.
⑥—Toe, +.05 degree.
⑦—Total toe, .1 degree.

REAR WHEEL ALIGNMENT

Year	Model	Camber Angle, Degrees				Total Toe Degrees
		Limits		Desired		
		Left	Right	Left	Right	
1985	Corvette	−.1 to +.9	−.1 to +.9	+.4	+.4	①
1986	Corvette	−.1 to +.9	−.1 to +.9	+.4	+.4	①
1987	Corvette	−.1 to +.9	−.1 to +.9	+.4	+.4	0
1988	Corvette	−.5 to +.5	−.5 to +.5	0	0	0
1989–90	Corvette	−.5 to +.5	−.5 to +.5	0	0	0

①—.15 degree per wheel.

CHEV. CAVALIER, BUICK SKYHAWK, CAD. CIMARRON, OLDS FIRENZA & PONT. SUNBIRD

FRONT WHEEL ALIGNMENT

Year	Model	Caster Angle, Degrees		Camber Angle, Degrees				Total Toe Degrees
				Limits		Desired		
		Limits	Desired	Left	Right	Left	Right	
1985	All	+.7 to +2.7	+1.7	+.2 to +1.5	+.2 to +1.5	+.85	+.85	①
1986	All	+.7 to +2.7	+1.7	+.2 to +1.5	+.2 to +1.5	+.85	+.85	①
1987	All	+.7 to +2.7	+1.7	+.2 to +1.4	+.2 to +1.4	+.8	+.8	②
1988	All	−.8 to +4.2	+1.7	−.2 to +1.8	−.2 to +1.8	+.8	+.8	②
1989–90	Exc. Cavalier Z24	+.7 to +2.7	+1.7	+.1 to +1.5	+.1 to +1.5	+.8	+.8	0
	Cavalier Z24	+.7 to +2.7	+1.7	−1 to +1	−1 to +1	0	0	0

①—.06 degree toe-out per wheel.
②—Except Cavalier Z24 & Sunbird W/60 series tires, total toe 0 degree; Cavalier Z24 & Sunbird W/60 series tires, −.12 degree.

REAR WHEEL ALIGNMENT

Year	Model	Camber Angle, Degrees				Total Toe Degrees
		Limits		Desired		
		Left	Right	Left	Right	
1989–90	All	−.75 to +.25	−.75 to +.25	−.25	−.25	+.25

CHEVROLET CHEVETTE & PONTIAC 1000

Year	Model	Caster Angle, Degrees		Camber Angle, Degrees				Total Toe Degrees
				Limits		Desired		
		Limits	Desired	Left	Right	Left	Right	
1985–86	All	+4 to +6	+5	−.2 to +.6	−.2 to +.6	+.2	+.2	+.12
1987	All	+2.7 to +6.7	+4.7	−.5 to +.9	−.5 to +.9	+.2	+.2	+.12

1985 CHEV. CITATION & BUICK SKYLARK

Year	Model	Caster Angle, Degrees		Camber Angle, Degrees				Toe-In Inch
				Limits		Desired		
		Limits	Desired	Left	Right	Left	Right	
1985	All	+.9 to +2.9	+1.9	−.5 to +.5	−.5 to +.5	0	0	0

CHEV. CELEBRITY, BUICK CENTURY, OLDS CUTLASS CIERA & CUTLASS CRUISER & PONT. 6000

FRONT WHEEL ALIGNMENT

Year	Model	Caster Angle, Degrees		Camber Angle, Degrees				Total Toe Degrees
				Limits		Desired		
		Limits	Desired	Left	Right	Left	Right	
1985–86	All	+.9 to +2.9	+1.9	−.5 to +.5	−.5 to +.5	0	0	0
1987–88	All	+.7 to +2.7	+1.7	−.5 to +.5	−.5 to +.5	0	0	0
1989	All	+.7 to +2.7	+1.7	−.5 to +.5	−.5 to +.5	0	0	0
1990	All	+1.2 to +2.6	+1.9	−.5 to +.5	−.5 to +.5	0	0	+.15

CHEV. CELEBRITY, BUICK CENTURY, OLDS CUTLASS CIERA & CUTLASS CRUISER & PONT. 6000—Continued

REAR WHEEL ALIGNMENT

Year	Model	Camber Angle, Degrees				Total Toe Degrees
		Limits		Desired		
		Left	Right	Left	Right	
1989	①	−.25 to +.25	−.25 to +.25	0	0	0
	②	—	—	−.5	−.5	+.15

①—Exc. Pontiac STE All Wheel Drive Models. ②—Pontiac STE All Wheel Drive Models.

CHEVROLET NOVA & GEO PRIZM

FRONT WHEEL ALIGNMENT

Year	Model	Caster Angle, Degrees		Camber Angle, Degrees				Toe-In Inch	Toe Out on Turns, Deg.	
		Limits	Desired	Limits		Desired			Outer Wheel	Inner Wheel
				Left	Right	Left	Right			
1985–86	All	+2/15 to +1 19/30	+53/60	−3/4 to +1/4	−3/4 to +1/4	−1/4	−1/4	0 to .08	①	②
1987	All	+1 19/30 to +8	+4 49/60	−3/4 to +1/4	−3/4 to +1/4	−1/4	−1/4	0 to .08	①	②
1988	All	+23/60 to +1 23/60	+53/60	−3/4 to +1/4	−3/4 to +1/4	−1/4	−1/4	0 to .08	—	—
1990	Exc. GSI	+5/6 to +1 5/6	+1 1/3	−2/3 to +1/3	−2/3 to +1/3	−1/6	−1/6	0 to .08	—	—
	GSI③	+11/12 to +1 11/12	+1 5/12	−2/3 to +1/3	−2/3 to +1/3	−1/6	−1/6	0 to .08	—	—
	GSI④	+5/6 to +1 5/6	+1 1/3	−3/4 to +1/4	−3/4 to +1/4	−1/4	−1/4	0 to .08	—	—

①—Less power steer., 33 degrees; with power steer., 32.5 degrees.
②—Less power steer., 40 degrees; with power steer., 39 degrees.
③—Models with auto. transaxle.
④—Models with man. transaxle.

REAR WHEEL ALIGNMENT

Year	Model	Camber Angle, Degrees				Toe-In Inch
		Limits		Desired		
		Left	Right	Left	Right	
1985–86	All	−1 4/15 to −1/10	−1 4/15 to −1/10	−31/60	−31/60	.07 to .23
1987	All	−1 4/15 to +7/30	−1 4/15 to +7/30	−51/60	−51/60	.07 to .23
1988	All	−1 1/60 to −1/60	−1 1/60 to −1/60	−31/60	−31/60	.07 to .23
1990	Exc. GSI	−1 1/12 to 1/12	−1 1/12 to 1/12	−7/12	−7/12	.16
	GSI	−1 1/10 to 1/10	−1 1/10 to 1/10	−2/3	−2/3	.16

CHEVROLET BERETTA & CORSICA

FRONT WHEEL ALIGNMENT

Year	Model	Caster Angle, Degrees		Camber Angle, Degrees				Total Toe Degrees
		Limits	Desired	Limits		Desired		
				Left	Right	Left	Right	
1987–88	All	+.7 to +2.7	+1.7	+.2 to +1.4	+.2 to +1.4	+.8	+.8	0
1989–90	All①	+.4 to +1.9	+1.15	0 to +1.2	0 to +1.2	+.6	+.6	0
	All②	+.4 to +1.9	+1.15	−.8 to +.4	−.8 to +.4	−.2	−.2	0

①—Less Sport Suspension. ②—With Sport Suspension.

REAR WHEEL ALIGNMENT

Year	Model	Camber Angle, Degrees				Total Toe Degrees
		Limits		Desired		
		Left	Right	Left	Right	
1989–90	Beretta	−.89 to +.39	−.89 to +.39	−.25	−.25	+.3
	Corsica	−.82 to +.32	−.82 to +.32	−.25	−.25	+.3

WHEEL ALIGNMENT SPECIFICATIONS

CHEVROLET & GEO SPECTRUM & STORM

FRONT WHEEL ALIGNMENT

		Caster Angle Degrees		Camber Angle Degrees					Steering Angle Degrees @ Full Lock	
				Limits		Desired				
Year	Model	Limits	Desired	Left	Right	Left	Right	Toe Degrees	Outer Wheel	Inner Wheel
1985	Spectrum	+1¾ to +2¾	+2¼	−5/12 to +1 1/12	−5/12 to +1 1/12	+⅓	+⅓	①	32½	37⅔
1986–87	Spectrum	+1¾ to +2¾	+2¼	−⅔ to +1⅓	−⅔ to +1⅓	+⅓	+⅓	①	32½	37⅔
1988–89	Spectrum	+1¾ to +2¾	+2¼	−.69 to +1.31	−.69 to +1.31	+.31	+.31	②	—	—
1990	Storm	+2.5 to +3.5	+3	−1.5 to +.5	−1.5 to +.5	−.5	−.5	0	—	—

①—Toe, −.08 to +.08.
②—Toe, −.34 to +.34.

REAR WHEEL ALIGNMENT

		Camber Degrees				
		Limits		Desired		
Year	Model	Left	Right	Left	Right	Toe
1985–86	Spectrum	—	—	0	0	①
1987	All	−¾ to +¾	−¾ to +¾	0	0	②
1988–89	Spectrum	−¾ to +¾	−¾ to +¾	0	0	③
1990	Storm	−1.5 to +.5	−1.5 to +.5	−.5	−.5	.16

①—Toe-in, .079 inch.
②—Toe, 0 to .20 inch.
③—Toe, −.43 to +.43 degrees.

CHEVROLET SPRINT & GEO METRO

FRONT WHEEL ALIGNMENT

		Caster Angle, Degrees		Camber Angles, Degrees			Turning Angle Degrees	
Year	Models	Limits	Desired	Limits	Desired	Toe-In Inch	Inner Wheel	Outer Wheel
1985–86	All	+1⅙ to +5⅙	+3⅙	0 to +2	+1	.08	38	32
1987–88	All	+1⅙ to +5⅙	+3⅙	-¾ to +1¼	+¼	0	38	32
1989–90	All	+1 to +5	+3	−1 to +1	0	0	38	32

REAR WHEEL ALIGNMENT

		Camber Degrees				
		Limits		Desired		
Year	Model	Left	Right	Left	Right	Toe In Inch
1985–88	All	—	—	0	0	0
1989–90	All	—	—	0	0	.079

OLDSMOBILE REAR WHEEL DRIVE & 1985 TORONADO

FRONT WHEEL ALIGNMENT

		Caster Angle, Degrees		Camber Angle, Degrees				
				Limits		Desired		
Year	Model	Limits	Desired	Left	Right	Left	Right	Toe-In Inch
1985	Cutlass	+2.5 to +3.5	+3	0 to +1	0 to +1	+.5	+.5	1/8
	88 & 98	+2.5 to +3.5	+3	+.25 to +1.25	+.25 to +1.25	+.75	+.75	1/8
	Toronado	+2 to +3	+2.5	−.5 to +.5	−.5 to +.5	0	0	③
1986–87	Cutlass	+1.8 to +3.8	+2.8	−.3 to +1.3	−.3 to +1.3	+.5	+.5	②
1986–88	Custom Cruiser	+1.8 to +3.8	+2.8	0 to +1.6	0 to +1.6	+.8	+.8	②
1989–90	Custom Cruiser	+2.5 to +3.5	+3	+.3 to +1.3	+.3 to +1.3	+.8	+.8	①

①—Toe, +.05 degrees.
②—Total toe, .1 degree.
③—1/16 inch toe-out to 1/16 inch toe-in.

REAR WHEEL ALIGNMENT

		Camber Angle, Degrees				
		Limits		Desired		
Year	Model	Left	Right	Left	Right	Toe-In Inch
1985	Toronado	−.4 to +.4	−.4 to +.4	0	0	①

①—3/32 inch per wheel; both wheels, 5/32 inch.

PONTIAC—EXCEPT FIERO, 1000 & FRONT WHEEL DRIVE

		Caster Angle, Degrees		Camber Angle, Degrees				
				Limits		Desired		
Year	Model	Limits	Desired	Left	Right	Left	Right	Toe-In Inch
1985	Firebird	+2.5 to +3.5	+3	+.5 to +1.5	+.5 to +1.5	+1	+1	3/16
	Grand Prix	+2.5 to +3.5	+3	0 to +1	0 to +1	+.5	+.5	1/8
	Bonneville	+2.5 to +3.5	+3	0 to +1	0 to +1	+.5	+.5	1/8
	Parisienne	+2.5 to +3.5	+3	+.3 to +1.3	+.3 to +1.3	+.8	+.8	1/8
1986	Firebird	+3 to +4	+3.5	+.5 to +1.5	+.5 to +1.5	+1	+1	①
	Bonneville	+1.8 to +3.8	+2.8	−.3 to +1.3	−.3 to +1.3	+.5	+.5	①
	Grand Prix	+1.8 to +3.8	+2.8	−.3 to +1.3	−.3 to +1.3	+.5	+.5	①
	Parisienne	+1.8 to +3.8	+2.8	0 to +1.6	0 to +1.6	+.8	+.8	①
1987	Firebird	+4.5 to +5.5	+5	+.5 to +1.5	+.5 to +1.5	+1	+1	②
	Grand Prix	+1.8 to +3.8	+2.8	−.3 to +1.3	−.3 to +1.3	+.5	+.5	②
	Safari	+1.8 to +3.8	+2.8	0 to +1.6	0 to +1.6	+.8	+.8	②
1988	Firebird	+4.5 to +5.5	+5	−.5 to +.5	−.5 to +.5	0	0	③
	Safari	+1.8 to +3.8	+2.8	0 to +1.6	0 to +1.6	+.8	+.8	④
1989	Firebird	+4.2 to +5.2	+4.7	−.2 to +.8	−.2 to +.8	+.3	+.3	③
	Safari	+2.5 to +3.5	+3	+.3 to +1.3	+.3 to +1.3	+.8	+.8	④
1990	Firebird	+4.5 to +5.5	+5	−.2 to +.8	−.2 to +.8	+.3	+.3	③

①—Toe-in, .15 degree per wheel.
②—Total toe-in,.1 degree.
③—Toe, 0 degree per wheel.
④—Toe, +.05 degree.

PONTIAC FIERO

FRONT WHEEL ALIGNMENT

Year	Model	Caster Angle, Degrees		Camber Angle, Degrees				Toe-In Inch
				Limits		Desired		
		Limits	Desired	Left	Right	Left	Right	
1985–86	All	+3 to +7	+5	−.3 to +1.3	−.3 to +1.3	+.5	+.5	1/16
1987	All	+3 to +7	+5	−.3 to +1.3	−.3 to +1.3	+.5	+.5	①
1988	Exc. GT & Formula	+2.5 to +3.5	+3	−.5 to +.5	−.5 to +.5	0	0	②
	GT & Formula	+4.5 to +5.5	+5	−.5 to +.5	−.5 to +.5	0	0	②

①—Total toe, 0 degree.
②—Toe +.3 degrees per wheel.

REAR WHEEL ALIGNMENT

Year	Model	Camber Angle, Degrees				Toe, Degree
		Limits		Desired		
		Left	Right	Left	Right	
1985–86	All	−1¼ to −¾	−1¼ to −¾	−1	−1	①
1987	All	−1.5 to −.5	−1.5 to −.5	−1	−1	①
1988	All	−1.5 to −.5	−1.5 to −.5	−1	−1	②

①—Total toe, 0 degree.
②—Toe +.15 degree per wheel.

PONTIAC LEMANS

FRONT WHEEL ALIGNMENT

Year	Model	Caster Angle, Degrees		Camber Angle, Degrees				Total Toe Degrees
				Limits		Desired		
		Limits	Desired	Left	Right	Left	Right	
1988–90	All	+.75 to +2.75	+1.75	+.45 to −1.05	+.45 to −1.05	−.3	−.3	0

REAR WHEEL ALIGNMENT

Year	Model	Camber Angle Degrees				Total Toe Degrees
		Limits		Desired		
		Left	Right	Left	Right	
1988	All	0 to −1	0 to −1	−.5	−.5	+.33
1989–90	All	—	—	−1	−1	+.33

WHEEL ALIGNMENT

INDEX

CHRYSLER MOTORS/EAGLE

TABLE OF CONTENTS

Chrysler Motors Colt, Colt Vista, Conquest & Eagle Summit

INDEX

PRELIMINARY CHECK

1. Road test vehicle noting any abnormal steering or handling characteristics.
2. Ensure tires are the proper type, correctly inflated and that tires on each axle are the same size.
3. Inspect ball joints, suspension arms, bushings and tie rods, and repair or replace any component that is damaged or excessively worn.
4. Ensure wheel run-out is not excessive, and that wheel bearings are properly adjusted.
5. Jounce vehicle several times to settle suspension.
6. Place vehicle on suitable alignment rack following manufacturer's instructions.
7. Check and correct alignment angles in the following sequence: Rear toe and camber, caster, front camber and front toe.
8. Correct any angle that is not within specifications. If no adjustment is possible, check for damaged or worn suspension components and/or damaged or distorted chassis and correct as needed.

FRONT WHEEL ALIGNMENT

CASTER

Colt & Summit

Caster is pre-set at the factory and is not adjustable. If caster is out of specifications, replace bent or otherwise damaged parts.

Colt Vista & Conquest

Caster normally requires no adjustment. However, slight adjustments can be made by moving the strut bar nut. After adjustment, ensure that variation between left and right sides is less than $1/2°$.

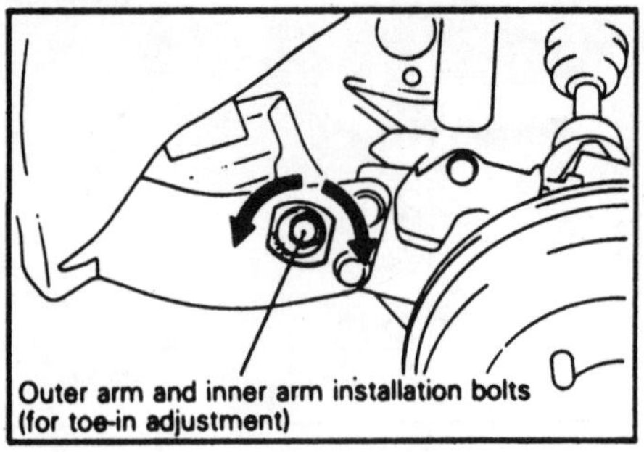

Fig. 1 Rear toe adjustment. Colt Vista

CAMBER

The specified camber is built into the steering knuckle, which is part of the strut assembly, and no adjustment is provided.

TOE-IN

Colt, Conquest & Summit

On Colt and Summit models, remove outer bellows clamp from tie rod before adjusting toe. After completing adjustment, reinstall clamp.

The amount of toe-in of the left front wheel is reduced by turning the tie rod turnbuckle toward the front of the car and the amount of toe-in on the right front wheel is reduced by turning it toward the rear of the the car. After adjustment, the difference in length between the two tie rods should not exceed .2 inch (5 mm).

Colt Vista

Adjust toe-in by removing the left and right tie rod turnbuckle retaining clips, then turning the left and right turnbuckles the same amount in opposite directions. To reduce toe-in, turn the left turnbuckle toward the front of the vehicle and the right turnbuckle toward the rear of the vehicle. For each half turn of the turnbuckle, toe-in will be adjusted by approximately .24 inch.

REAR WHEEL ALIGNMENT

CAMBER

Rear camber is pre-set during vehicle assembly and cannot be adjusted. If camber is not within specifications, check for worn or damaged suspension component, and damaged or deformed floor pan or body and repair as needed.

TOE-IN

Colt & Summit

Rear toe is pre-set during vehicle assembly and cannot be adjusted. If toe is not within specifications, check for worn or damaged suspension components, and damaged or deformed floor pan or body and repair as needed.

Colt Vista

Toe is adjustable by rotating the outer and inner arm mounting bolts. If toe-in is not within specifications, rotate left and right outer arm and inner arm bolts, each by the same amount, to perform adjustment. An adjustment of approximately .08 inch (2 mm) will be made when the outer arm and inner arm bolts are turned the equivalent of 1 alignment mark, **Fig. 1.**

Conquest

Toe-in is adjusted by rotating the lower control arm mounting bolts, **Fig. 2**, as follows:

1. Loosen locknut while holding mounting bolt.
2. Rotate mounting bolts until toe is within specifications. Movement of 1 graduation, **Fig. 2**, will change toe by approximately .08 inch (2 mm). **Rotate left and right mounting bolts equally.**
3. Tighten locknut and recheck setting.
4. Repeat adjustment until toe setting is equal for both wheels and total toe is within specifications.

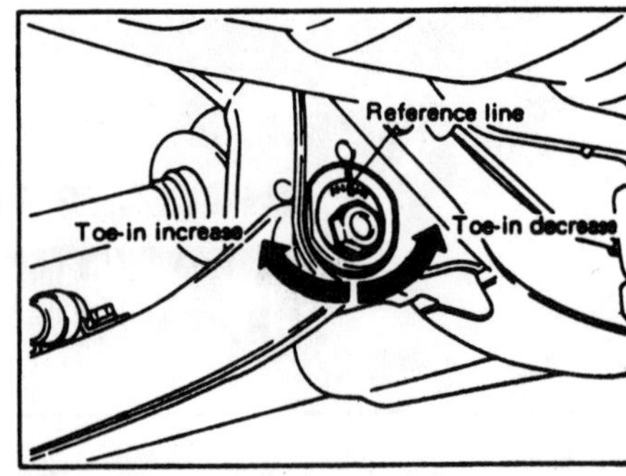

Fig. 2 Rear toe adjustment. Conquest

VEHICLE RIDING HEIGHT SPECIFICATIONS

Model & Year	Height In Inches ① Front Bumper To Ground	Height In Inches ① Rear Bumper To Ground
Colt H'back		
1985-86	10	13.2
1987-88	②	13.2
1989-90	—	—
Colt & Summit Sedan		
1985-88	9.8	13.6
1989-90	—	—
Colt Wagon		
1988-90	9.3	11
Colt Vista 2WD		
1985-90	15	15
Colt Vista 4WD		
1985-90	17.5	17.5
Conquest		
1985-89	14	14.6

①—If not within specifications, check suspension components for wear and damage.

②—Less air dam, 10 inches; with air dam, 9.1 inches.

Chrysler Motors Front Wheel Drive Models Except Colt, Colt Vista & Plymouth Laser

INDEX

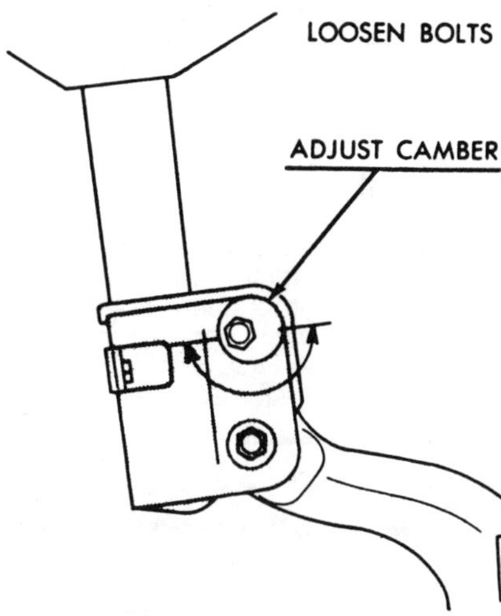

Fig. 1 Camber adjustment. 1985–87 Charger & Turismo & 1985–90 Horizon & Omni

FRONT WHEEL ALIGNMENT

Prior to wheel alignment ensure tires are at recommended pressure, are of equal size and have approximately the same wear pattern. Check front wheel and tire assembly for radial run-out and inspect lower ball joints and steering linkage for looseness. Check front and rear springs for sagging or damage. Front suspension inspections should be performed on a level floor or alignment rack with fuel tank at capacity and vehicle free of luggage and passenger compartment load. The vehicle should be bounced an equal number of times from the center of the bumper alternately, first from the rear, then the front, releasing at bottom of down cycle.

CASTER

The caster angle on these models cannot be adjusted.

CAMBER

To adjust camber, loosen the cam and through bolts, **Figs. 1 and 2.** On 1985 Lancer and LeBaron GTS, remove rear washer plate and one of the knuckle bolts and replace with a new cam bolt. On all models except 1986-89 Lancer and LeBaron, 1987-90 Shadow and Sundance and 1988-90 Dynasty, Imperial, New Yorker and New Yorker Fifth Avenue, rotate cam bolt to move top of wheel in or out to obtain specified camber angle. On 1986-89 Lancer and LeBaron, 1987-90 Shadow and Sundance and 1988-90 Dynasty, Imperial, New Yorker and New Yorker Fifth Avenue,, loosen knuckle bolts just enough to allow movement between strut and knuckle, then push or pull tire to obtain specified camber angle. On 1985-87 Charger and Turismo and 1985-90 Horizon and Omni, **torque** cam bolt nuts to 45 ft. lbs., then advance nuts an additional 1/4 turn. On all models except Charger, Horizon, Omni and Turismo, **torque** nuts to 75 ft. lbs., then advance an additional 1/4 turn.

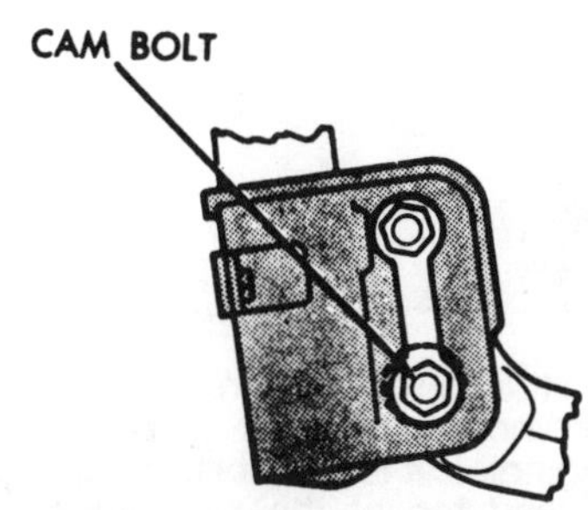

Fig. 2 Camber adjustment. Except Charger, Horizon, Omni & Turismo

TOE-IN

To adjust toe-in, center steering wheel and hold in position with a suitable tool. Loosen tie rod locknuts and rotate the rod, **Fig. 3**, to adjust toe-in to specifications. Use care not to twist steering gear rubber boots. **Torque** tie rod locknuts to 55 ft. lbs. Adjust position of steering gear rubber boots. Remove steering wheel holding tool.

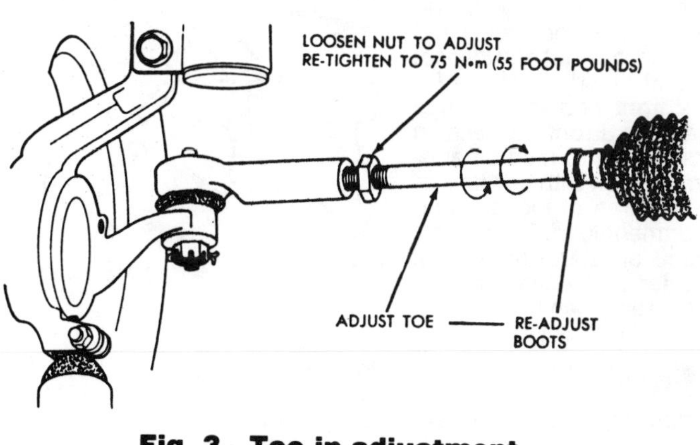

Fig. 3 Toe-in adjustment

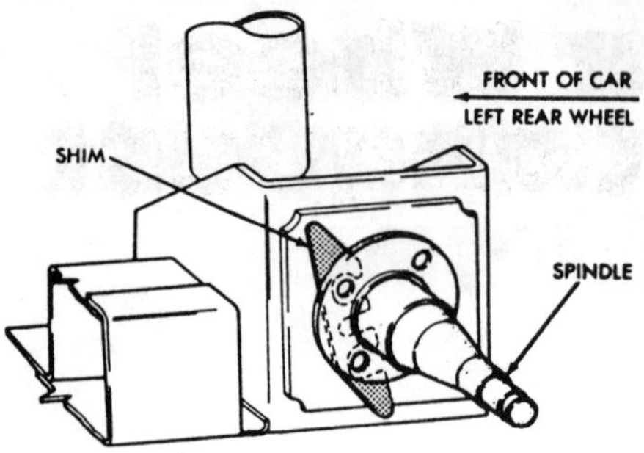

Fig. 4 Shim installation for toe-out

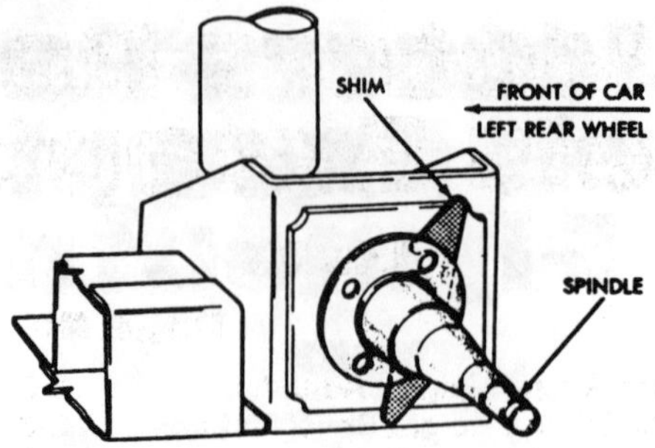

Fig. 5 Shim installation for toe-in

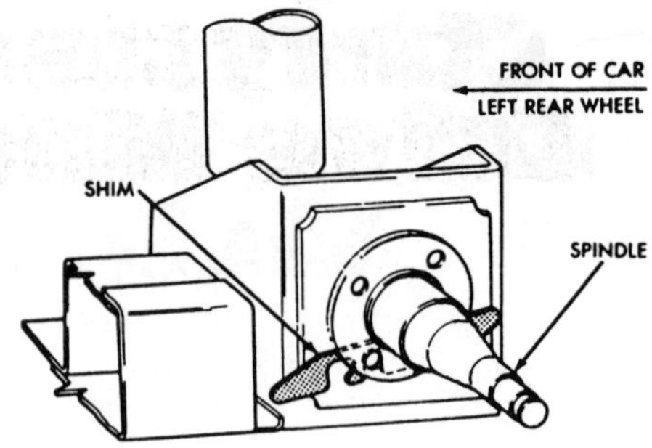

Fig. 6 Shim installation for positive camber

REAR WHEEL ALIGNMENT

Due to the design of the rear suspension and the incorporation of stub axles or wheel spindles, it is possible to adjust the camber and toe of the rear wheels on these models. Adjustment is controlled by adding shims approximately .010 inch thick between the spindle mounting surface and spindle mounting plate. The amount of adjustment is approximately 0° 18' per shim. Refer to **Figs. 4 through 7** for proper placement of shims.

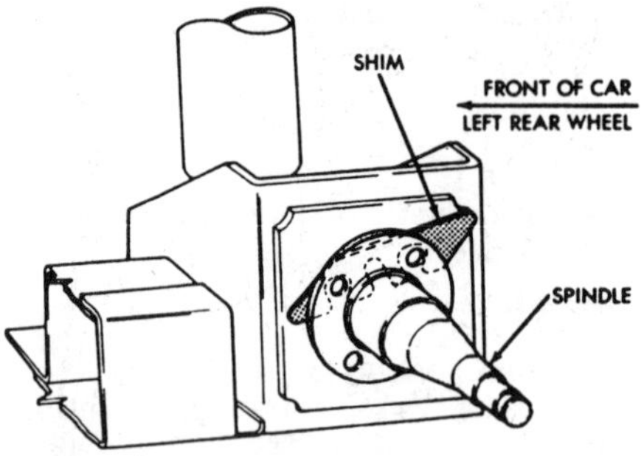

Fig. 7 Shim installation for negative camber

Chrysler Motors Plymouth Laser & Eagle Talon

INDEX

FRONT WHEEL ALIGNMENT

Prior to wheel alignment, ensure tires are at recommended pressure, are of equal size and have approximately the same wear pattern. Check front wheel and tire assembly for radial run-out and inspect lower ball joints and steering linkage for looseness. Check front and rear springs for sagging or damage. Front suspension inspections should be performed on a level floor or alignment rack with fuel tank at capacity and vehicle free of luggage and passenger compartment load. The vehicle should be bounced an equal number of times from the center of the bumper alternately, first from the rear, then the front, releasing at the bottom of down cycle.

CASTER & CAMBER

Caster and camber are preset at the factory and cannot be adjusted. If caster or camber are not within specifications, replace bent or damaged components.

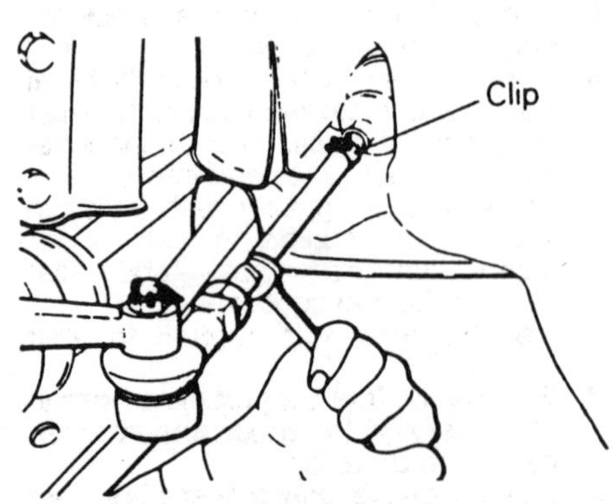

Fig. 1 Adjusting toe-in

TOE-IN

Adjust toe-in by undoing clips and turning each tie rod turnbuckle an equal amount in opposite directions, **Fig. 1.** Toe will move out as the left turnbuckle is turned toward front of vehicle and the right turnbuckle is turned toward rear of vehicle. For each half turn, toe-in will increase or decrease .24 inch.

REAR WHEEL ALIGNMENT

Camber and toe-in are preset at the factory and cannot be adjusted. If camber or toe-in are not within specifications, replace bent or damaged components.

Chrysler Motors Rear Wheel Drive Models Except Conquest

INDEX

Page No.

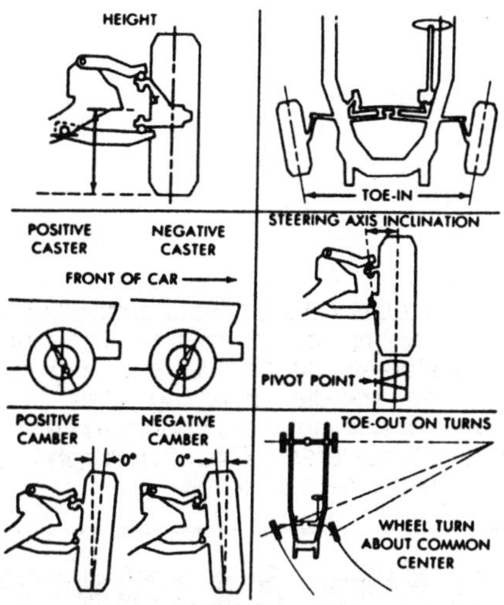

Fig. 1 Alignment factors

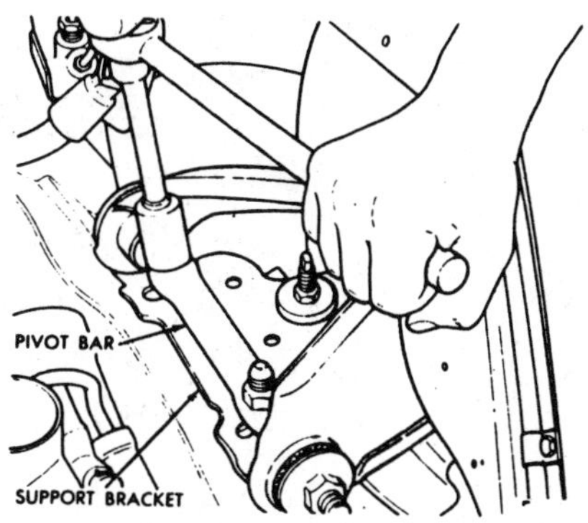

Fig. 2 Loosening pivot bolt nuts

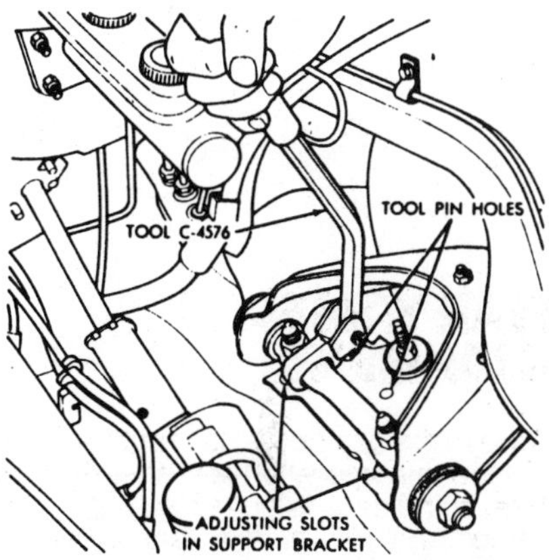

Fig. 3 Adjusting camber & caster using tool C-4576

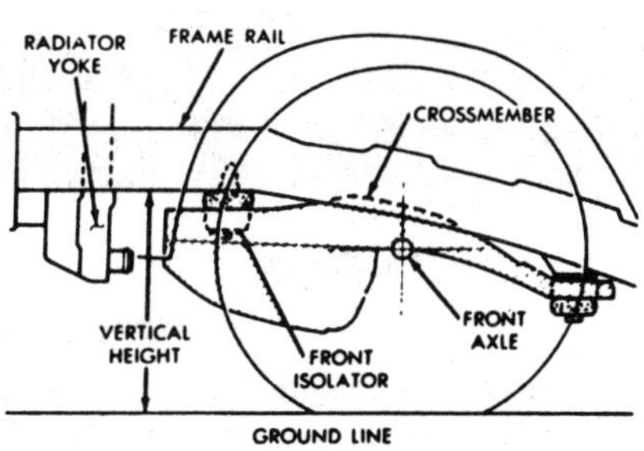

Fig. 4 Measuring front suspension height

CASTER & CAMBER

ADJUST

Front suspension height must be checked and corrected as necessary before performing wheel alignment.

1. Remove all foreign material from exposed threads of cam adjusting bolt nuts or pivot bar adjusting bolt nuts.
2. Record initial camber and caster readings before loosening cam bolt nuts or pivot bar bolt nuts, **Fig. 1.**
3. Adjusting tool C-4576 is required to adjust caster and camber. When performing adjustments, the camber settings should be held as close as possible to the "desired" setting, and the caster setting should be held as nearly equal as possible on both wheels, **Figs. 2 and 3.**

There may be cases when the vehicle may not have sufficient positive camber adjustment. Upper control arm plate spacers are available that will allow more positive camber adjustment, if required. If this condition is encountered, use spacer 1-4014352 for front suspension upper control arm front pivot support and spacer 1-4014353 for the rear pivot support. To install spacers, proceed as follows:

1. Loosen but do not remove caster/camber adjustment nut.
2. Raise vehicle and remove wheel and tire assembly.
3. Loosen but do not remove shock absorber upper mounting nut.
4. Remove two support plate bolts at front end of plate, then loosen two rear bolts enough to slide front spacer between support plate and frame.
5. Align holes in spacer with holes in support plate and frame.
6. Insert two front bolts and start threads. Do not tighten.
7. Repeat steps 4 through 6 for rear spacer.
8. **Torque** the four support plate bolts to 65 ft. lbs. and the shock absorber upper nut to 25 ft. lbs.
9. Lower vehicle and adjust alignment on side that spacers were installed.

TOE-IN

ADJUST

With the front wheels in straight ahead position, loosen the clamps at each end of both adjusting tubes. Adjust toe-in by turning the tie rod sleeve which will "center" the steering wheel spokes. If the steering wheel was centered, make the toe-in adjustment by turning both sleeves an equal amount. Position the clamps so they are on the bottom and tighten bolts to 15 ft. lbs.

CAR RIDING HEIGHT

ADJUST

Before taking measurements, grasp the bumpers at the center (rear bumper first) and jounce the car up and down several times. Jounce the car at the front bumper the same number of times and release the bumper at the same point in the cycle each time.

1. Ride height is measured from the head of the front suspension front crossmember insulator bolt to ground, **Fig. 4.**
2. If necessary, turn torsion bar adjusting bolt clockwise to increase height and counterclockwise to decrease height.
3. After completing adjustment, jounce vehicle and recheck riding height. Both sides must be measured even though only one side may have been adjusted. Front vehicle height should not vary more than 1/4 inch from the specified riding height. Riding height should also be within 1/4 inch side to side.

CAR RIDING HEIGHT SPECIFICATIONS

Year	Model	Height
CHRYSLER		
1985–89	Fifth Avenue	12½ inch
DODGE		
1985–89	Diplomat	12½ inch
PLYMOUTH		
1985–89	Gran Fury	12½ inch

Eagle 4 x 4

INDEX

Page No.

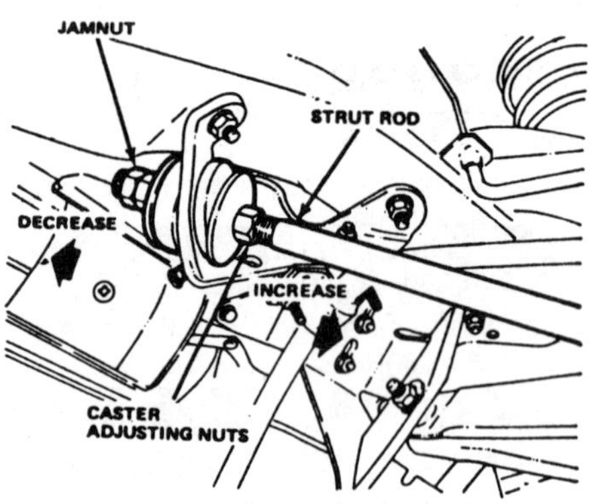

Fig. 1 Caster adjustment

WHEEL ALIGNMENT

CASTER

Caster is obtained by moving the two adjusting nuts on the threaded strut rod, **Fig. 1.** One nut is on each side of the mounting bracket. Therefore, moving the nuts on the rod will move the lower control arm to front or rear for desired caster angle. After adjustment, **torque** adjusting nuts to 65 ft. lbs. **Torque** locknuts to 75 ft. lbs.

CAMBER

Camber is obtained by turning on the eccentric lower control arm bolt, **Fig. 2.** After adjustment, **torque** locknut to 110 ft. lbs.

TOE-IN, ADJUST

To adjust toe-in, loosen the clamps at both ends of the adjustable tubes on each tie rod. Turn the tubes an equal amount until the toe-in is correct. Turning the right tube in the direction the wheels revolve when the car is going forward increases the toe-in and turning the left tube in the opposite direction increases toe-in. To decrease toe-in turn the right tube backward and the left tube forward. It is important that both tubes be turned an equal amount in order to maintain the correct position of the steering wheel. When adjustment is complete, tighten all clamp bolts.

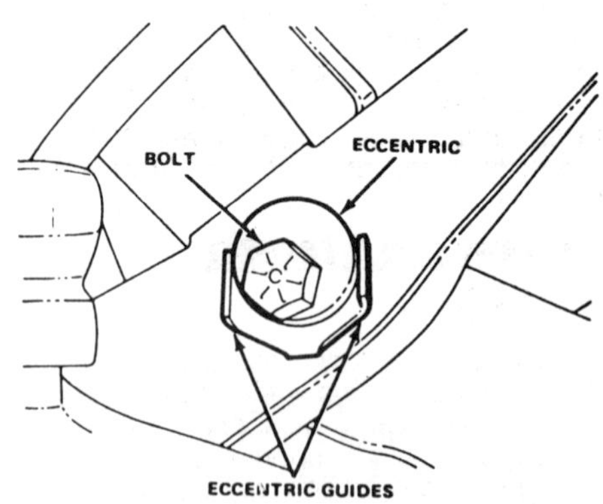

Fig. 2 Camber adjustment

In performing service operations on the steering linkage or when adjusting toe-in, be sure to square the tie rod ball sockets on the studs and align the tie rod stud in the center, or slightly above center, of the cross tube opening, before tightening the steering linkage adjusting tube. This will prevent the stud from contacting the side of the cross tube opening, which would otherwise result in noise problems or damage.

Eagle Medallion

INDEX

DESCRIPTION & PRELIMINARY INFORMATION

Before proceeding with any check/adjustment procedure, ensure vehicle ride height is checked and/or adjusted as described previously. Also the following information is for front wheel alignment only. Rear wheel alignment is factory set and cannot be adjusted.

CASTER & CAMBER ANGLES

Caster and camber are built-in angles and are not adjustable. If either angle is incorrect, it will be necessary to inspect and/or replace the suspension or steering component contributing to the incorrect angle. Refer to "Alignment Diagnosis."

The relationship between left and right caster/camber angle is critical. A side-to-side difference of more than 1 ° may result in a pull to one side.

If the vehicle pulls to one side, the condition usually is a result of excessive camber or insufficient caster on the pulling side. Incorrect caster/camber is not the only cause of vehicle pull. Brake drag, a damaged or under inflated tire, body damage or damage to the steering components can all cause a pulling condition.

TOE-OUT ANGLE

The front suspension is set/adjusted for front wheel toe-out. The toe setting ia adjusted only after the steering gear has been locked in a centered position using tool Dir.1067 or equivalent. Refer to "Toe-Out, Adjust" for service procedure.

To identify a incorrect toe condition, excessive toe-out will wear the inner edge of each front tire. Insufficient toe-out (or a toe-in condition), will wear the outer edge of each front tire.

ALIGNMENT DIAGNOSIS

The following information outlines the general causes of incorrect front alignment angles since caster/camber angles are not adjustable.

INCORRECT CASTER ANGLE

The possible cause(s) for incorrect caster angles are as follows:

1. Bent lower control arm.
2. Bent frame rails.
3. Bent or damaged body components.
4. Loose or worn lower control arm bushings.

INCORRECT CAMBER ANGLE

The possible cause(s) for incorrect camber angles are as follows:

1. Bent lower control arm.
2. Bent frame rails.
3. Bent or damaged body components.
4. Loose or worn lower ball joint.
5. Loose or worn lower control arm bushings.

TOE ANGLE INCORRECT BY MORE THAN 1/4 INCH (TOTAL TOE)

The possible cause(s) for incorrect toe angle exceeding 1/4 inch are as follows:

1. Bent or damaged steering knuckle.
2. Bent tie rod end.
3. Worn or loose tie rod end.
4. Worn or loose lower control arm bushings.

FRONT WHEEL ALIGNMENT CHECKS

Perform the front wheel alignment procedures in the following sequence: Preliminary Inspection, Steering Gear Centering, Alignment Angles Check, Toe Setting Adjust, and Steering Wheel Centering.

Perform the preliminary inspection before the vehicle is on an alignment rack. Center the steering gear and check alignment angles with the vehicle placed on an alignment rack.

PRELIMINARY INSPECTION

1. Ensure tires are of the same size and properly inflated.
2. Inspect tread surfaces and note type and degree of wear.
3. Measure and adjust vehicle ride height.
4. Check braking operation. Correct problems before alignment adjustment.
5. Check front wheel bearing end play.
6. Check wheel radial and lateral run-out.

STEERING GEAR CENTERING

1. Remove bolts and nuts attaching steering bracket to the rack assembly and move bracket (and tie rods) aside.
2. Place centering tool Dir.1067, **Fig. 1,** onto gear housing (A) and rack assembly (B) as shown. **Ensure centering studs on tool are firmly seated.**
3. With tool installed, the steering gear will .automatically be centered and locked in position.
4. Check position of the steering shaft coupling (C), **Fig. 1.** The coupling pinch bolt (D) must be located at the coupling underside and positioned 3 splines away from a horizontal position. Adjust coupling position, if necessary.

ALIGNMENT ANGLES CHECK

1. Jounce vehicle front end several

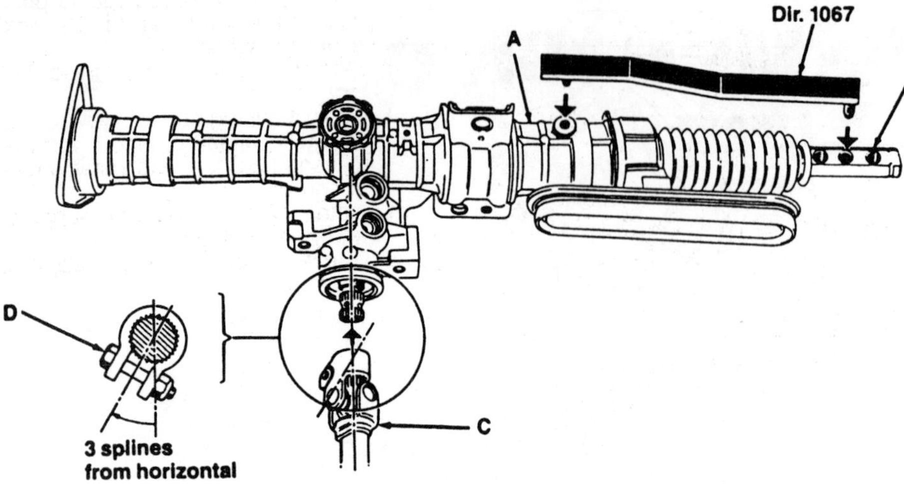

Fig. 1 Positioning centering tool Dir.1067 onto gear housing and rack assembly. Eagle Medallion

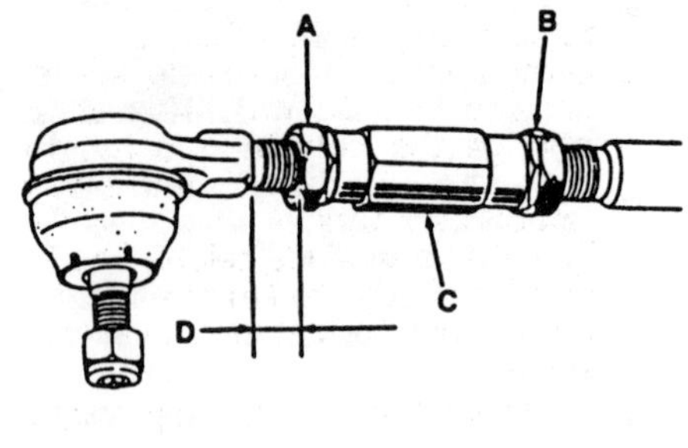

Fig. 2 Toe-out adjustment

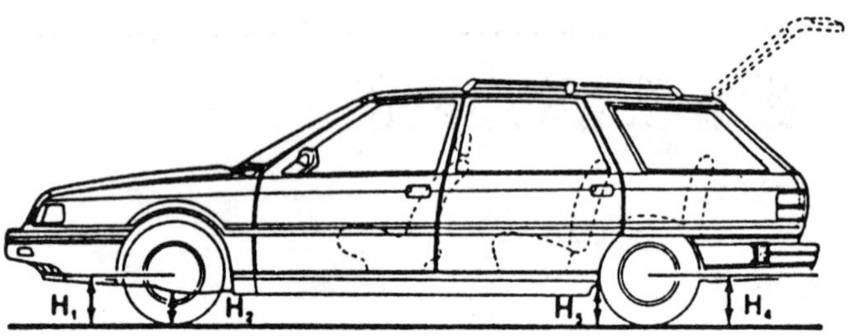

REAR

L48 (SEDAN) $H_4 - H_3$ = 30mm (1.18in.) ± 7.5mm (0.295in.)

K48A (WAGON) $H_4 - H_3$ = 10mm (0.393in.) ± 7.5mm (0.295in.)

Fig. 3 Vehicle ride height specifications

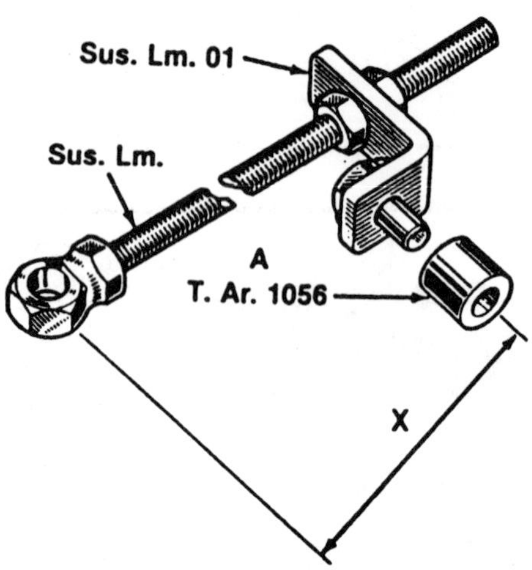

Fig. 4 Threaded rod tool assembly

times to settle the suspension components.

2. Check alignment angles in the following sequence: Caster, Camber and Toe-Out.
3. If the alignment angles are within specifications proceed to "Toe-Out Setting, Adjust."
4. If caster, camber or toe-out are not within specified limits, it will be necessary to repair or replace the steering, suspension or body component causing the incorrect alignment.
5. Remove tool Dir.1067 and install steering bracket onto the steering gear rack assembly.
6. Center the steering wheel, if necessary. Refer to "Steering Wheel Centering."
7. Remove vehicle from alignment rack.

TOE-OUT SETTING, ADJUST

1. Loosen tie rod locknut (A), **Fig. 2,** and steering arm locknut (B). **Locknut (A) has right hand threads. Locknut (B) has left hand threads.**
2. Turn each adjusting sleeve (C) inward or outward to obtain the required toe-out reading.
3. The number of threads (D) visible at each tie rod end should be approximately the same.
4. **Torque** locknuts (A) and (B) to 26 ft. lbs. **Do not allow the adjusting sleeve (C) to rotate while tightening the locknuts.**

STEERING WHEEL CENTERING

Do not use the tie rod adjusting sleeves to center the steering wheel. The correct method of centering the steering wheel is to remove the steering wheel and center it on the steering shaft as necessary. After centering torque steering wheel attaching nut to 30 ft. lbs.

VEHICLE RIDE HEIGHT ADJUST

Only the rear suspension height is adjustable. If front suspension ride height is not within specification, check suspension for worn and damaged components. The vehicle rear ride height is adjusted by rotating the suspension bars clockwise or counterclockwise from its original position.

1. Measure vehicle ride height and compare to specifications in **Fig. 3.**
2. Determine measurement necessary to raise or lower the vehicle, as follows:
   a. Raise and support vehicle.
   b. Remove both rear wheels.
   c. Remove shock absorbers as described previously.
   d. **Install, then adjust threaded rod tool Sus.Lm (A) with tool Sus.Lm.01 bracket until dimension X is 19.37 inches on sedan models or 18.97 inches on wagon models, Fig. 4.**
   e. Position tool assembly in place of the shock absorber and attach it at point C, **Fig. 5.** The spacer (A) from tool set T.Ar.1056 or equivalent must also be used.
   f. Slip spacer (A) into the lower shock absorber mounting hole.

g. Mark position of the suspension torsion bars before removal.
h. Using a suitable tool, remove suspension bar. **Do not attempt to change the vehicle ride height by use of the anti-sway (stabilizer) bars.**
i. To raise vehicle ride height, lengthen the threaded rod to position the torsion bar to the next available spline.
j. To lower the vehicle ride height, shorten the threaded rod to position the torsion bar to the next available spline.
k. One spline change is equal to .118 inch.
l. After obtaining correct vehicle ride height measurement, remove threaded rod tool and install shock absorber.
m. Install wheels, remove supports and lower vehicle.
n. Check vehicle ride height, adjust if necessary.

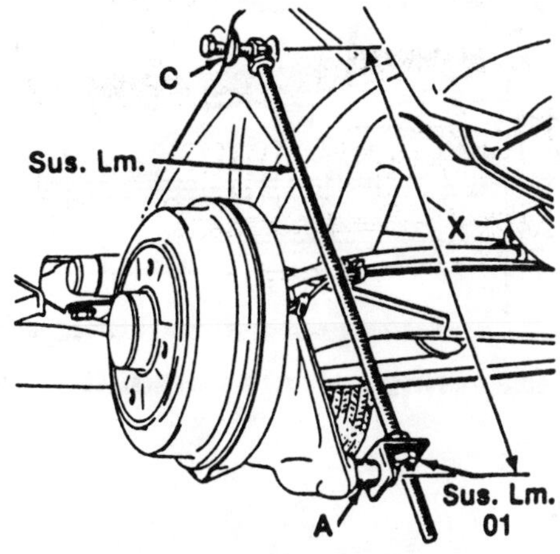

Fig. 5 Installing threaded rod tool assembly in place of shock absorber

Eagle Premier

INDEX

FRONT WHEEL ALIGNMENT

PRELIMINARY INSPECTION

Prior to front end alignment, check vehicle for proper tire size and inflation. Also check wheel and tire radial lateral run-out. Steering and suspension components should be inspected for damage and replaced if necessary. Wheel bearing end play and brake operation should be corrected if necessary. Center steering gear as follows:

1. Support vehicle on alignment rack.
2. Turn steering wheel completely to the left, then completely to the right and count number of turns to the stop.
3. Turn steering wheel to the left one-half the number of turns counted.
4. If steering wheel needs to be centered, adjust after toe adjustment. Refer to "Steering Wheel Alignment."

CASTER & CAMBER

Caster and camber built-in angles are not adjustable. A difference of more than one degree between left and right caster or camber angles may result in pulling to one side, or tire wear. Inspection and replacement of damaged suspension and steering parts should be performed to correct alignment angles. Brake drag or an under inflated tire, can also cause the vehicle to pull.

Bounce vehicle several times and check alignment angles of caster and camber then. Repair or replace vehicle suspension parts if alignment is out of specifications.

TOE-OUT

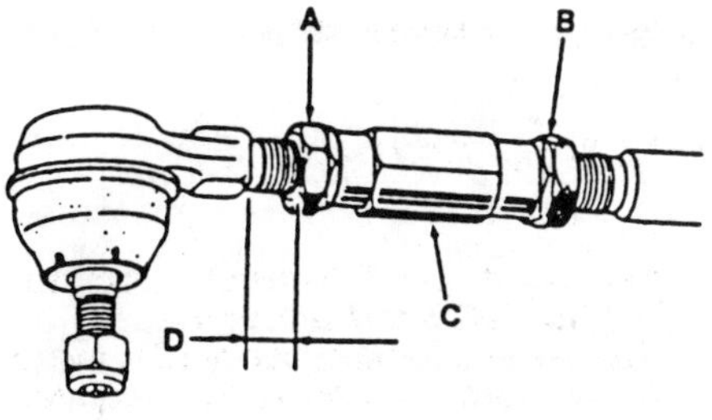

Fig. 1 Toe-out adjustment

Front wheel toe-out is adjusted only after the steering gear has been centered. Excessive toe-out will wear the inner edge of tires. Insufficient toe-out will wear outer edge of tires. To adjust toe-out, turn tie rods in or out.

1. Loosen tie rod locknut (A - right-hand threads), then (B - left had threads), **Fig. 1.**
2. Turn adjusting sleeve (C) to required length, ensuring both tie rod ends have an equal amount of threads showing.
3. **Torque** locknuts and to 26 ft. lbs.

STEERING WHEEL ALIGNMENT

Do not use the tie rod adjusting sleeves to center the steering wheel. The correct method of centering the steering wheel is to remove the steering wheel and center it on the steering shaft as necessary. After centering, torque steering wheel attaching nut to 52 ft. lbs.

RIDE HEIGHT

1. Measure height H-1 and H-4 from wheel hub horizontal center line to surface on right and left sides of vehicle, **Fig. 2.**
2. Measure height H-2 from engine cradle at wheel hub vertical center line to surface.
3. Measure height H-3 from front torsion bar horizontal center line to surface, **Fig. 2.**

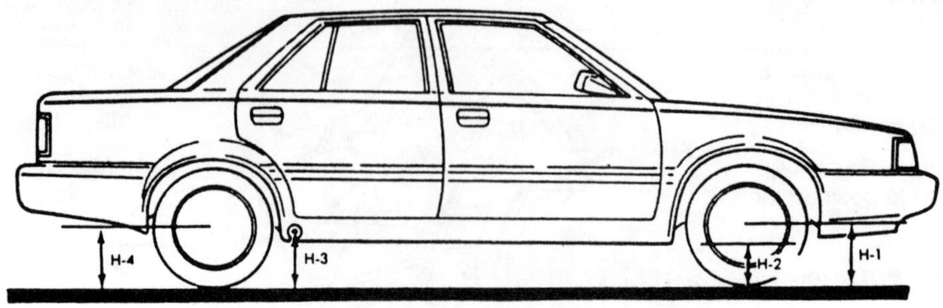

Fig. 2 Measuring vehicle height (Part 1 of 2)

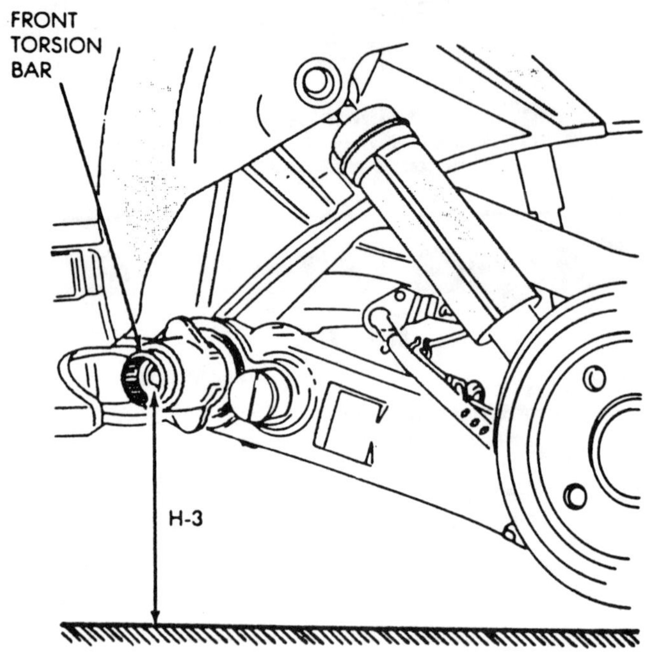

Fig. 2 Measuring vehicle height (Part 2 of 2)

4. Compare height measurements from each side of vehicle, as follows:
   a. Subtract H-2 from H-1, measurement should be 3.36 to 3.98 inches.
   b. Subtract H-3 from H-4, measurement should be 1.25 to 1.87 inches.

FORD MOTOR CO.

TABLE OF CONTENTS

Ford Crown Victoria, Lincoln Town Car & Mercury Grand Marquis

INDEX

WHEEL ALIGNMENT

Wheel balancing differs on vehicles with disc brakes, dynamic balancing of the wheel and tire assembly on the vehicle should not be attempted without first pulling back the shoe and lining assemblies from the rotor. If this is not done, brake drag may burn out the motor on the wheel spinner.

The drag can be eliminated by removing the wheel, taking out the two bolts holding the caliper splash shield, and detaching the shield. Then push the pistons into their cylinder bores by applying steady pressure on the shoes on each side of the rotor for at least a minute. If necessary, use water pump pliers to apply the pressure.

After the pistons have been retracted, reinstall the splash shield and wheel. The wheel and tire assembly can then be dynamically balanced in the usual way. After the balancing job has been completed, be sure to pump the brake pedal several times until the shoes are seated and a firm brake pedal is obtained.

Caster and camber can be adjusted by loosening the bolts that attach the upper suspension arm to the shaft at the frame side rail, and moving the arm assembly in or out in the elongated bolt holes, **Fig. 1.** Since any movement of the arm affects

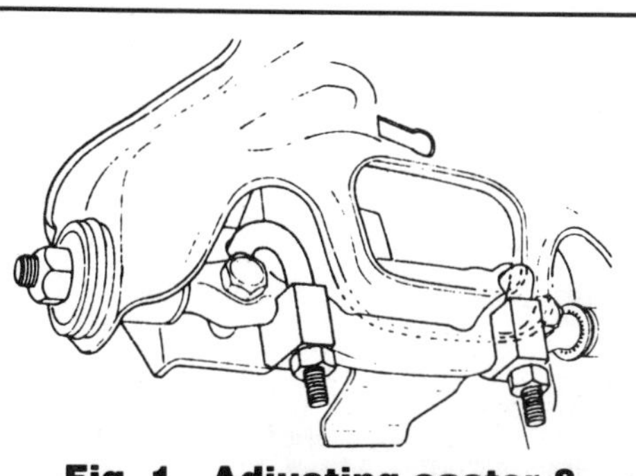

Fig. 1 Adjusting caster & camber

both caster and camber, both factors should be balanced against one another when making the adjustment.

Use alignment tool T79P-3000A or equivalent, **Fig. 1.** Install the tool with the pins in the frame holes and the hooks over the upper arm inner shaft. Tighten the hook nuts snug before loosening the upper arm inner shaft attaching bolts.

CASTER

1. Tighten the tool front hook nut or loosen the rear hook nut as required to increase caster to the desired angle.
2. To decrease caster, tighten the rear hook nut or loosen the front hook nut as required. **The caster angle can be checked without tightening the inner shaft retaining bolts.**
3. Check the camber angle to be sure it did not change during the caster adjustment and adjust if necessary.
4. **Torque** the upper arm inner shaft retaining bolts to 120-140 ft. lbs. and remove tool.

CAMBER

1. Loosen both inner shaft retaining bolts.
2. Tighten or loosen the hook nuts as necessary to increase or decrease camber.
3. Recheck caster and readjust if necessary.
4. **Torque** upper arm inner shaft retaining bolts to 120-140 ft. lbs.

TOE-IN

Position the front wheels in their straight-ahead position. Then turn both tie rod adjusting sleeves an equal amount until the desired toe-in setting is obtained. **Torque** tie rod sleeve clamp bolt to 20-22 ft. lbs.

Ford Escort, EXP & Tempo & Mercury Lynx & Topaz

INDEX

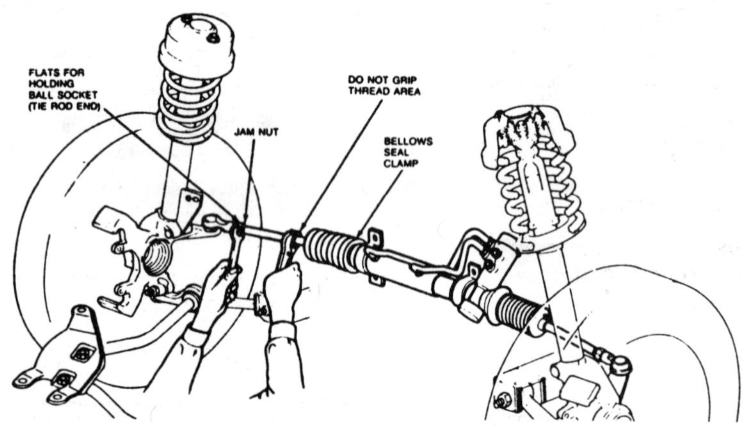

Fig. 1 Adjusting front wheel toe-in & toe-out

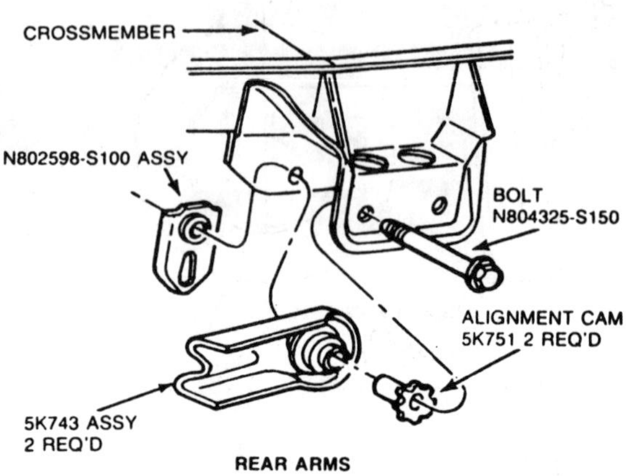

Fig. 2 Rear wheel toe-in & toe-out alignment cam

FRONT WHEEL ALIGNMENT

CASTER & CAMBER

Caster and camber angles are preset at the factory and cannot be adjusted.

TOE

To adjust toe-in, **Fig. 1,** lock steering wheel in the straight ahead position using suitable steering wheel holder. Remove small outer clamp from steering boot to prevent boot from twisting during adjustment procedure. Loosen tie rod adjusting nuts, then adjust left and right tie rods until each wheel has 1/2 the desired total toe specification. Tighten tie rod adjusting nuts, replace steering gear rubber boots and tighten clamp. Remove steering wheel holding tool.

REAR WHEEL ALIGNMENT

CAMBER

Camber cannot be adjusted and factory set.

TOE

To adjust toe of either wheel, loosen bolt attaching rear control arm to body, **Fig. 2,** and rotate alignment cam until the required alignment setting is obtained. **Torque** control arm attaching bolt to 40-55 ft. lbs. on 1985-87 models, or 52-74 ft. lbs. on 1988-90 models.

Ford Festiva

INDEX

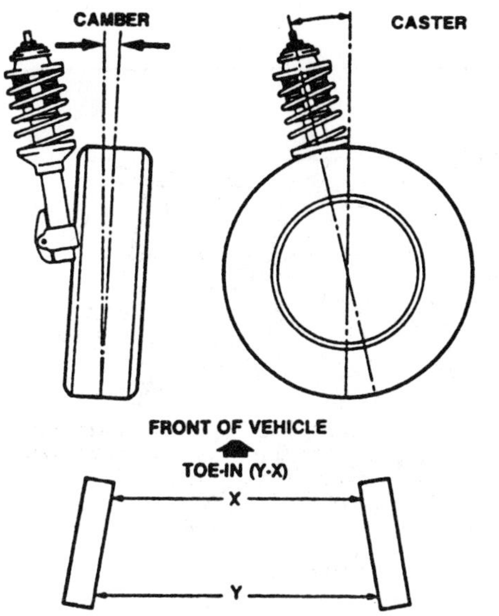

Fig. 1 Front wheel alignment settings

FRONT WHEEL ALIGNMENT

CASTER

Caster is not adjustable, **Fig. 1.** If caster angles are not within specification, check for damaged or bent suspension components, deteriorated bushings, or distorted body mounting points.

CAMBER

Camber angle, **Fig. 1,** is controlled by the position of the strut mounting block in the shock tower. The mounting block can be positioned in the tower in two ways, resulting in a camber variation of approximately $1/2^\circ$. If the camber setting has not been changed previously, the white alignment mark on the mounting block will be positioned on the outboard side of the strut assembly, **Fig. 2.** Changing the position of the alignment mark 180° will increase camber angle approximately $1/2^\circ$, **Fig. 3.** If the alignment mark is already positioned inboard of the strut, changing the mounting block position can only result in camber being reduced by approximately $1/2^\circ$. To adjust camber setting, proceed as follows:

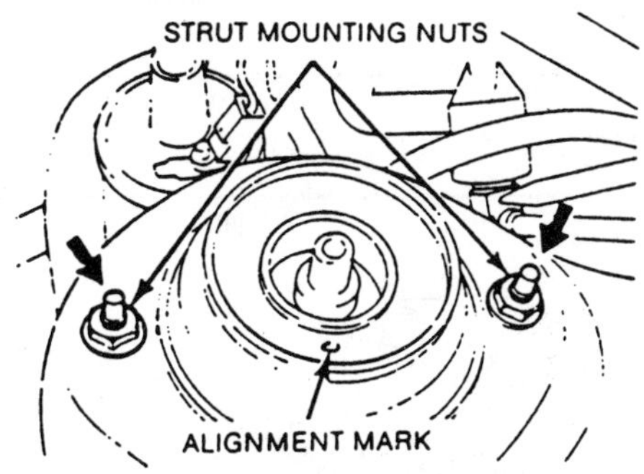

Fig. 2 Front strut camber alignment mark

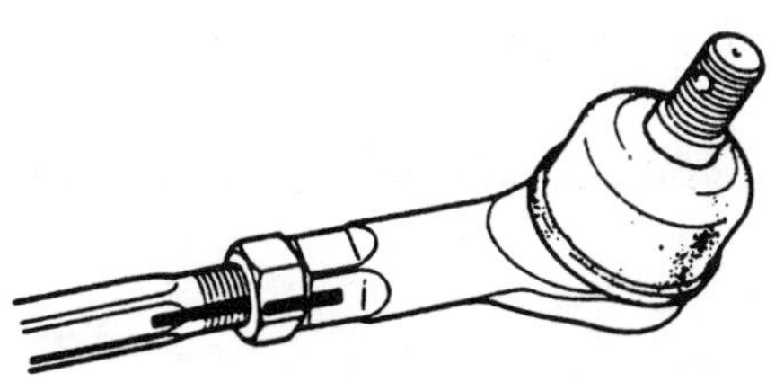

Fig. 4 Marking position of tie rod, tie rod end & jam nut

1. Raise and support vehicle and remove front wheels.
2. Remove upper strut attaching nuts.
3. Lower strut sufficiently so that mounting studs clear tower, then rotate mounting block 180° as necessary.
4. Reposition strut in tower, then install attaching nuts and **torque** to 32 to 45 ft. lbs.

TOE-IN

1. Ensure that tires are properly inflated, then position vehicle on suitable alignment rack.
2. Bounce vehicle several times to normalize ride height, then check toe setting. If toe setting is not within specification, proceed to next step.
3. Mark relationship between tie rods, tie rod ends and jam nuts, then loosen tie rod boot clamps to prevent twisting or damage to boots, **Fig. 4.**
4. Adjust toe-in by turning **both** tie rods in or out an equal amount until toe setting is within specification.
5. After adjustment is completed, tighten jam nuts and boot clamps.

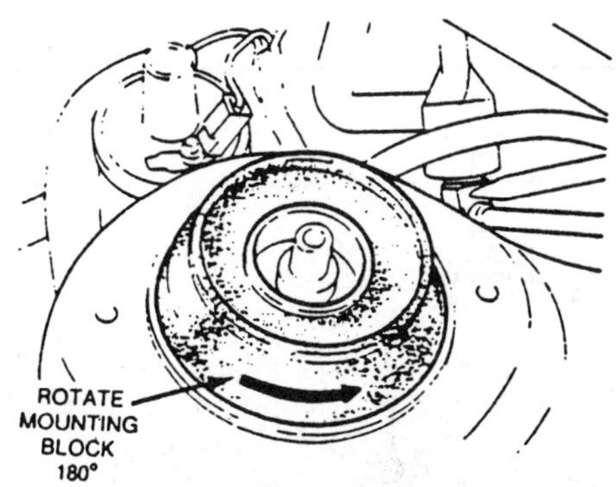

Fig. 3 Front camber adjustment

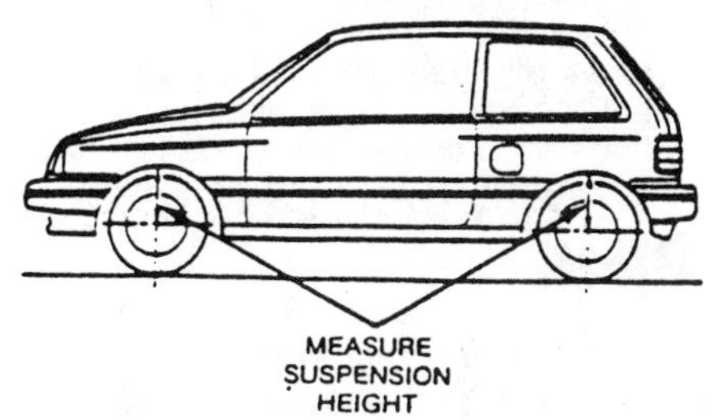

Fig. 5 Suspension height measurement

REAR WHEEL ALIGNMENT

Rear camber and toe-in are determined by the configuration of the trailing arms on the torsion beam rear axle. No provision for adjustment is provided.

SUSPENSION HEIGHT

Prior to checking suspension height, remove heavy items from vehicle, such as tool boxes etc. Suspension height is measured from ground to the fender cut out above the center of the wheel and tire assembly, **Fig. 5.** This measurement should be performed at all four wheels. Readings from side to side and front to rear should not vary by more than .4 inch. **It should be noted that uneven tire wear will increase the variance in the measurements and should be taken into consideration. If variance is more than .4 inch, check springs and suspension components for wear and damage.**

Ford LTD & Mustang, Lincoln Mark VII, Mercury Capri & Marquis, 1985–87 Lincoln Continental & 1985–88 Ford Thunderbird & Mercury Cougar

INDEX

Page No.

CASTER

The caster angle of this suspension is factory pre-set and cannot be adjusted.

CAMBER

1. Remove pop rivet from camber plate.
2. Loosen 3 camber plate-to-body apron nuts.
3. Move top of shock strut as needed to bring camber angle within specifications, then tighten nuts. **It is not necessary to replace the pop rivet.**

TOE-IN

1. Check to see that steering shaft and steering wheel marks are in alignment and in the top position.
2. Loosen clamp screw on the tie rod bellows and free the seal on the rod to prevent twisting of the bellows, **Fig. 1.**
3. Loosen tie rod jam nut.

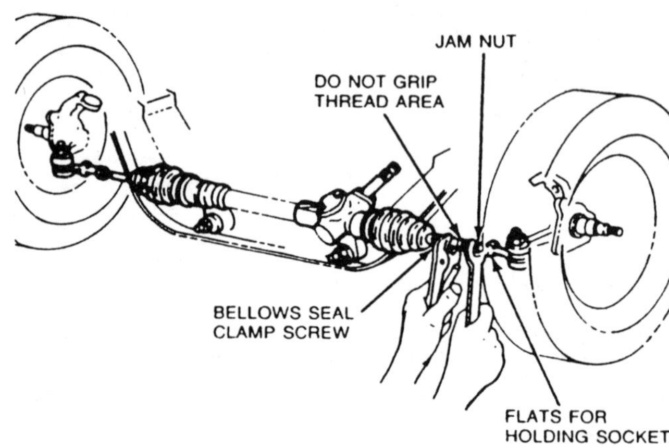

Fig. 1 Toe-in adjustment

4. Use suitable pliers to turn the tie rod inner end to correct the adjustment to specifications. Do not use pliers on tie rod threads. Turning to reduce number of threads showing will increase toe-in. Turning in the opposite direction will reduce toe-in. On all models, torque to 43-50 ft. lbs.

1989–90 Ford Thunderbird & Mercury Cougar

INDEX

CASTER

Caster is adjusted by moving the tension strut relative to the front sub-frame. Loosen nuts securing strut to sub-frame and adjust caster by running the nuts in the appropriate direction until the desired setting is achieved. The setting should be locked by holding the adjusting nut and **torquing** the locknut to 117 ft. lbs.

CAMBER

Front

Camber is adjusted by rotating a cam bolt at the lower control arm inner pivot. Loosen nut securing adjustment and adjust by rotating the head of the bolt. Lock the adjustment by holding the bolt and **torque** nut to 125 ft. lbs.

Rear

Camber is adjusted by rotating a cam bolt at the upper control arm inner pivot. Loosen nut securing adjustment and adjust by rotating the head of the bolt. Lock the adjustment by holding the bolt and **torque** nut to 98 ft. lbs.

TOE-IN

Front

1. Check to see that steering shaft and steering wheel marks are in alignment and in the top position.
2. Loosen clamp screw on the tie rod bellows and free the seal on the rod to prevent twisting of the bellows.
3. Place opened end wrench on flats of tie rod socket to prevent socket from turning, then loosen tie rod jam nuts.
4. Use suitable pliers to turn the tie rod inner end to correct the adjustment to specifications. Do not use pliers on tie rod threads. Turning to reduce number of threads showing will increase toe-in. Turning in the opposite direction will reduce toe-in.
5. **Torque** tie rod jam nuts to 43-50 ft. lbs.

Rear

The recommended adjustment sequence for rear alignment is to set toe then camber. Toe should be rechecked before final tightening.

Toe is adjusted by rotating a cam bolt at the lower arm inner pivot. Loosen nut securing adjustment and adjust by rotating the head of the bolt. Lock the adjustment by holding the bolt and **torque** nut to 170 ft. lbs.

Ford Probe

INDEX

PRELIMINARY ADJUSTMENT

Before measuring and setting front wheel alignment, rest front wheels on turn plates.

Before setting rear toe, rest rear wheels on slider plates or turn plates. Before setting any alignment angle, jounce the vehicle three times at each end to establish trim height.

Special adapters are available for using a magnetic hub gauge at rear wheels. Depending on type of equipment used, these may not be necessary. After removing hub cap and bearing cap, hub gauge will snap into place on brake drum. Magnetic mounting toe gauges may also be installed in the same manner.

Always perform wheel alignment on a level alignment rack. Do not attempt to check or adjust front wheel alignment without first inspecting front end components.

Check all factors of front wheel alignment except turning angle before making any adjustments. Check the turning angle only after camber and toe have been adjusted to specification. Check front wheel alignment under following curb load conditions:

1. Establish standing curb height.
2. Remove heavy weights from trunk.
3. Ensure all tires are inflated to specification (cold).

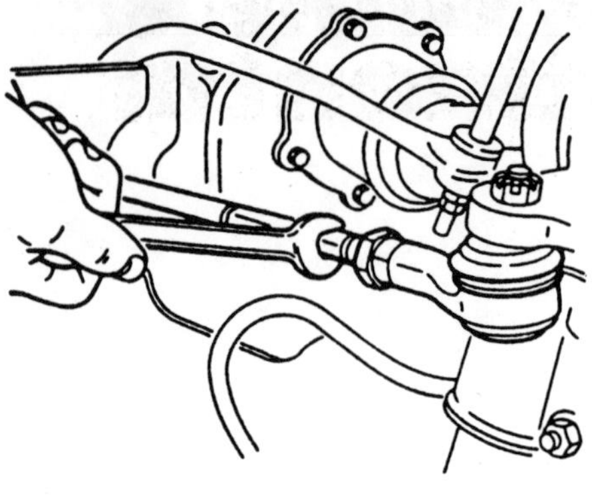

Fig. 1 Adjusting front toe-in

4. Ensure Fuel tank, oil reservoir and radiator are filled to specification. If necessary, add six pounds of weight to trunk for each gallon of gasoline missing from tank.
5. Place front seats in full rear position.
6. Check rear toe adjustment.
7. Always road test vehicle after adjusting alignment. If vehicle still pulls, switch front tires. If vehicle still pulls in same direction, recheck alignment and rear tracking. If vehicle pulls in opposite direction, rotate tires, then road test again.

FRONT WHEEL ALIGNMENT

Always adjust rear toe before setting front alignment angles.

CASTER

Front caster adjustment is not a separate procedure on this vehicle. Front caster should fall within specification when front camber is adjusted. If caster does not fall within specification, check control arms, stabilizers and bushings. If these components are satisfactory, check vehicle body for distortion at suspension mounting points.

CAMBER

1. Raise vehicle and support at body so that front suspension is unloaded.
2. Remove tire and wheel assembly.
3. Remove upper strut attaching nuts.
4. Lower strut, then rotate strut mounting block until camber is within specification.
5. Reinstall strut in strut tower.
6. **Torque** strut attaching nuts to 34-46 ft. lbs, then recheck camber adjustment.

TOE-IN

1. Loosen jam nuts at tie rod ends, then release clips at small ends of steering gear boots. **Ensure boots are free on tie rod ends so that they will not be twisted when tie rods are turned.**
2. Turn tie rods into or out of tie rod ends an equal amount on each side, **Fig. 1** . To increase toe-in, turn right tie rod toward front of vehicle, and turn left tie rod toward read of vehicle. To decrease toe-in, turn tie rods in opposite direction. One turn of each side tie rod changes toe-in by approximately .28 inch.
3. Check front tracking. **Always set tracking immediately after setting toe.** Set tracking by by using rear wheels as a reference point. Follow equipment manufacture's instructions to check tracking. The angle of each front wheel in relationship to rear wheels must be the same.
4. Ensure toe-in is still within specifications, then **torque** jam nuts to 51-72 ft. lbs. Ensure steering gear boot ends are correctly positioned on appropriate sections of tie rods, then install boot clips.

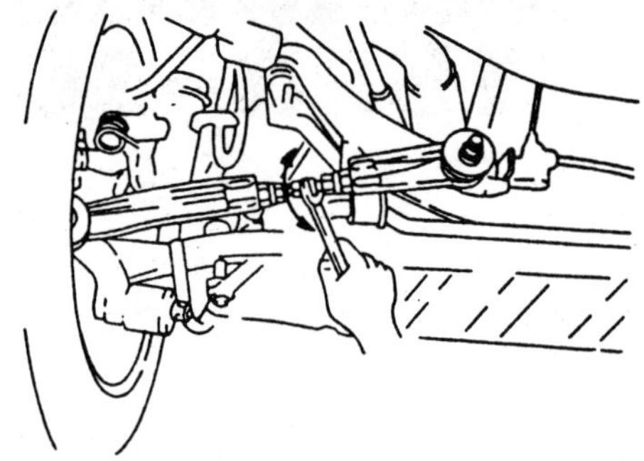

Fig. 2 Adjusting rear toe-in

5. Measure turning angle by placing front wheels on a turning-radius gauge and fully turning wheels to the left, then right. Inner turning angle should be 36.44° and outer turning angle should be 30.99°.

REAR WHEEL ALIGNMENT

TOE-IN

1. Loosen jam nuts clockwise on right control arm, then loosen jam nuts counterclockwise on left control arm.
2. To increase toe-in, turn right control arm rod counterclockwise and turn left control arm rod clockwise, **Fig. 2.**
3. To decrease toe-in, turn right control arm clockwise and turn left control arm counterclockwise. **Turn control arm rods into or out of control arm ends an equal amount on each side.** One turn of each control arm rod changes toe by .46 inch.
4. **Torque** control arm jam nuts to 41-59 ft. lbs.

Ford Taurus, Mercury Sable & 1988—90 Lincoln Continental

INDEX

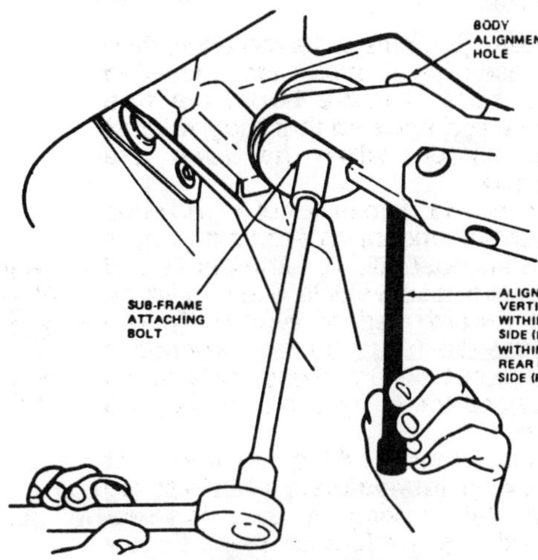

Fig. 1 Aligning front suspension

FRONT WHEEL ALIGNMENT

CASTER & CAMBER

On 1988-90 Continental, perform ride height setup and measurement procedure prior to setting caster and camber alignment.

1. Prior to aligning the front end, the sub-frame alignment must be checked using the following procedure.
   a. Loosen sub-frame to body attaching bolts.
   b. Install a 3/4 inch outside diameter pipe or similar tool into left front sub-frame and body alignment holes, **Fig. 1.**
   c. Align left front sub-frame and body alignment holes, then slightly tighten left front sub-frame attaching bolt.
   d. Repeat steps b and c on right front alignment holes, then recheck left front alignment.
   e. Torque sub-frame attaching bolts to specifications.
2. Center punch spot welds on both strut alignment plates, then loosen strut attaching nuts, **Fig. 2.**
3. Using Rotunda Spot-Eze or equivalent, remove spot welds. **Do not drill deeper than thickness of alignment plates.**
4. Remove strut attaching nuts, then the alignment plates.
5. Remove burrs from strut towers and alignment plates, then paint all exposed metal on strut towers and alignment plates.
6. Install alignment plates, then loosely install strut attaching nuts.
7. Align front end, then torque strut attaching nuts to specifications.
8. Drill three 1/8 inch holes as indicated in **Fig. 3,** through alignment plates and

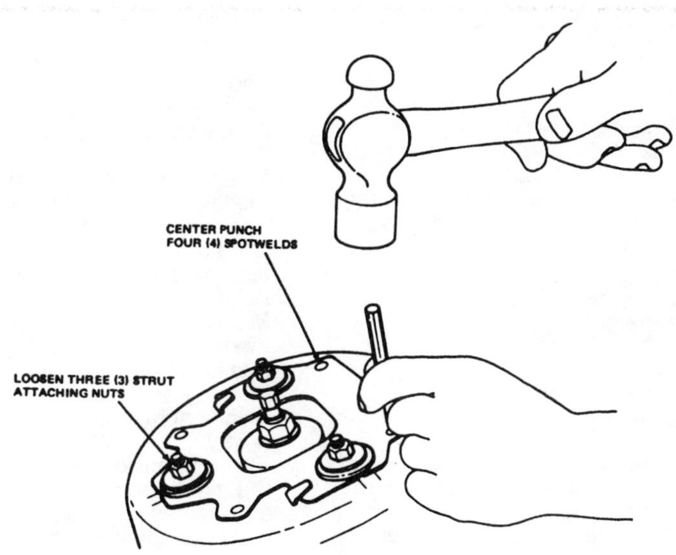

Fig. 2 Loosening alignment plate

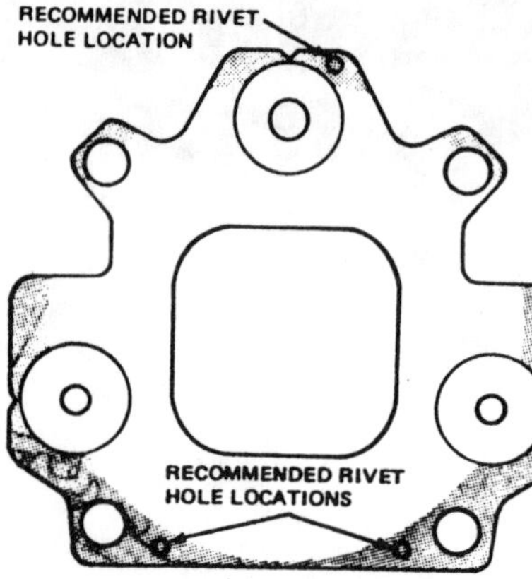

Fig. 3 Rivet hole location

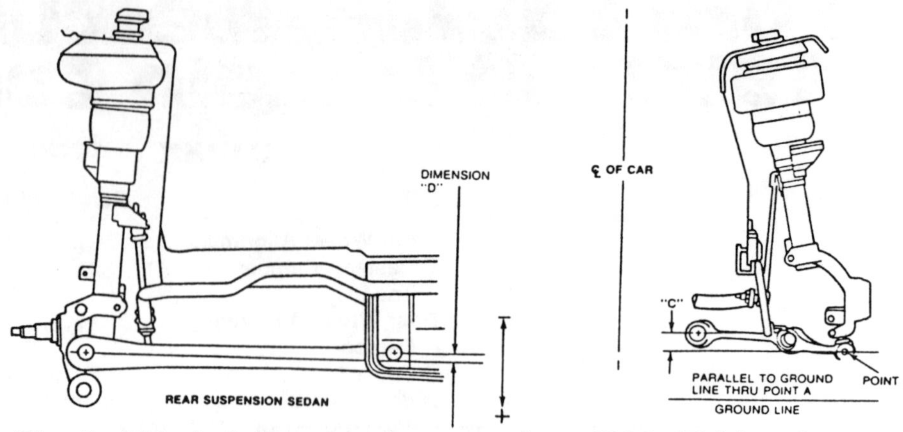

Fig. 4 Ride height setup & measuring. 1988–90 Lincoln Continental

Link Part Number		Height Sensor Link Change	Front Ride Height Affect With Respect To The Nominal Link
LH Front	RH Front		
E80F-3C11T-CA	E80F-3C111-GA	Plus One (Green)	+6mm (+0.24 IN)
E80F-3C111-BA	E80F-3C111-FA	Minus One (Red)	-6mm (-0.24 IN)
E80F-3C111-JA	E80F-3C111-KA	Nominal (Blue)	-0-
E80F-3C111-DA	E80F-3C111-HA	Plus Two (Yellow)	+12mm (+0.47 IN)
E80F-3C111-AA	E80F-3C111-EA	Minus Two (White)	-12mm (-0.47 IN)

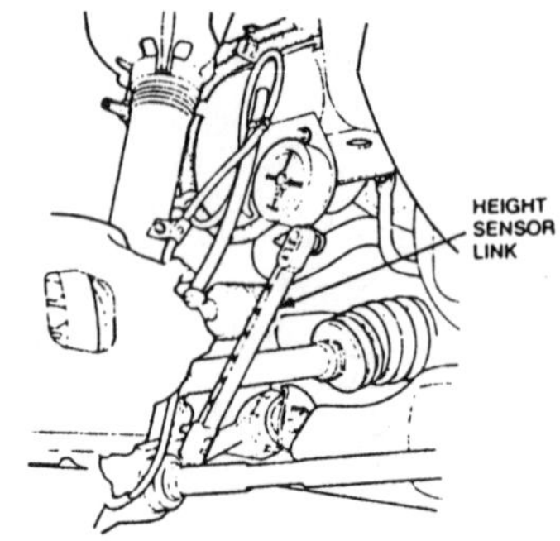

Fig. 5 Front ride height & sensor link chart. 1988–90 Lincoln Continental

strut towers, then paint exposed metal. **Do not drill deeper than 3/8 inch into strut tower.**

9. Install three 1/8 inch diameter pop rivets with a grip range of 1/4 inch into alignment plate/strut tower.

TOE-IN

To adjust toe-in, lock steering wheel in straight ahead position using suitable steering wheel holder. Loosen and slide off small outer clamps from steering boot to prevent boot from twisting during adjustment procedure. Loosen tie rod adjusting, then adjust left and right tie rods until each wheel has 1/2 the desired total toe specification. Tighten tie rod adjusting nuts and install clamps. Remove steering wheel holding tool.

REAR WHEEL ALIGNMENT

CAMBER

The camber angles are factory set and cannot be adjusted.

TOE-IN

On sedan models, toe-in is adjusted by rotating the cams located inside the rear inner lower control arm bushings.

On wagon models, toe-in is adjusted by rotating the cams located inside the outer lower control arm bushings.

RIDE HEIGHT SETUP & MEASUREMENT

1988–90 LINCOLN CONTINENTAL

This adjustment must be used prior to front caster and camber alignment or ride height checking.

1. Position vehicle on alignment rack.
2. To ensure ride heights are measured at a consistent point they should be measured only after the service bay diagnostic auto mode has successfully been completed and the Star tester displays a code 12. At this time, place the suspension power switch in the OFF position, remove the Star tester and leave the ignition switch in the RUN position.
3. At this point the vehicle is at the top of the trim band.
4. Measure the front suspension C dimension as shown in **Fig. 4.** The front ride height (C) is the vertical difference of the lower arm inner pivot attachment height minus the outer pivot height.
5. Measure the rear suspension D dimension as shown in **Fig. 4.** The rear ride height D is the vertical difference of the rear lower arm inner pivot attachment height minus the outer pivot height.
6. The suspension heights for the top of the trim band are:
   a. C dimension, 1.72 inch (43.6mm).
   b. D dimension, minus .36 inch (minus 9.3mm).
7. If the suspension heights are not within the listed specifications, ride height adjustment is required.
8. For reference purposes, the ride heights at the center of the trim band are approximately .35 inch (8.8mm) lower than the above suspension heights in the front and approximately .47 inch (12mm) lower in the rear.
9. If adjustment to front ride height is required, it should be performed by replacing the height sensor link according to chart, **Fig. 5.**
10. When removing a link, carefully separate it from the ball stud with a wide blade screwdriver. Do not damage the link or ball stud. When installing a link to the sensor, support the sensor lever from behind to avoid possible damage to the sensor.
11. Rear height sensor adjustment is performed by loosening and repositioning the height sensor lever adjustment screw. For adjustment purposes, each notch on the rear height sensor lever provides approximately .60 inch (15mm) of ride height adjustment.

Mercury Tracer

INDEX

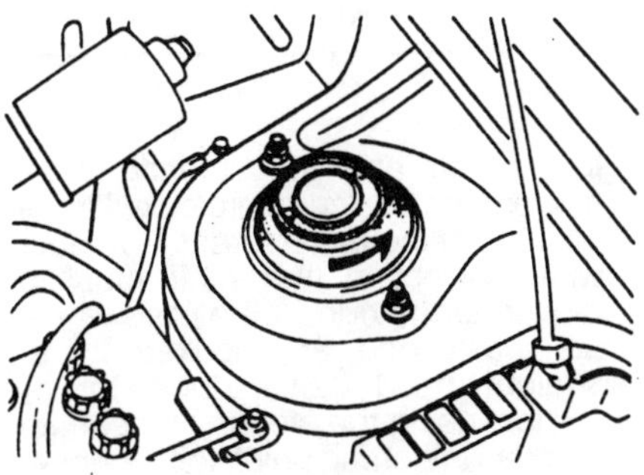

Fig. 1 Rotating strut bearing to adjust front camber

FRONT WHEEL ALIGNMENT

CASTER

When checking caster ensure tires are properly inflated. Caster angles are not adjustable. If caster is not within specifications, check for damaged or distorted control arm or stabilizer bar or worn bushings.

CAMBER

Camber angle can be changed by 28 minutes by the strut bearing one half turn.

1. Raise and support vehicle by body so that front suspension is unloaded.
2. Remove wheel and tire assembly.
3. Remove strut to strut tower attaching nuts, then lower strut from tower.
4. Rotate strut bearing 180 degrees, **Fig. 1.**
5. Position strut to strut tower, then install attaching nuts.
6. Install wheel and tire assembly, then recheck camber setting.

TOE-IN

1. Loosen jam nuts at tie rod ends, then release clips securing boots to tie rod.
2. Turn tie rods in or out of tie rod ends as necessary, **Fig. 2.** Rotate tie rods and equal amount on each side to maintain steering center.
3. Check vehicle front tracking.
4. After setting front track, recheck toe-in setting.

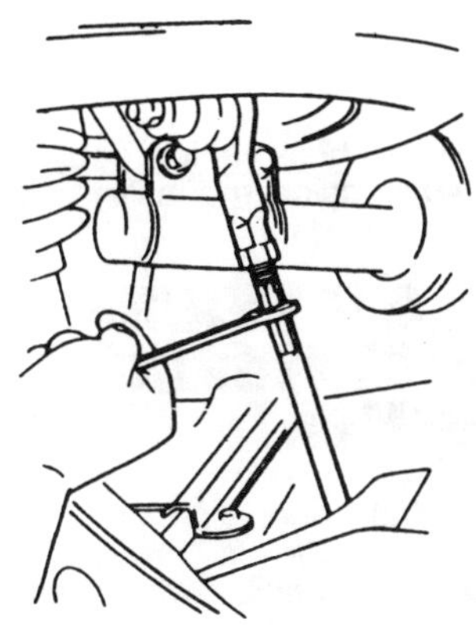

Fig. 2 Adjusting front toe-in

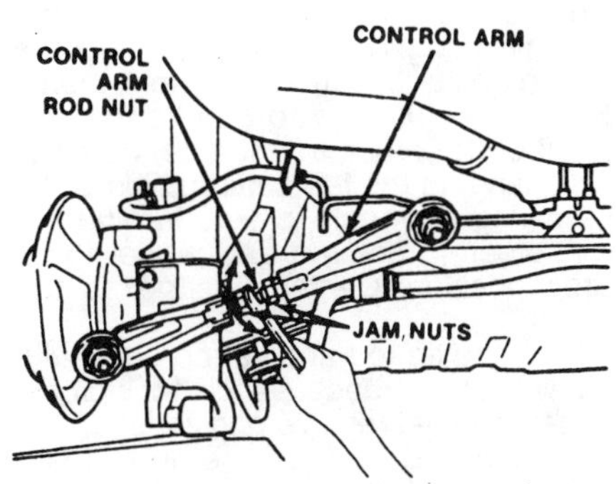

Fig. 4 Rear toe-in adjustment

5. Torque tie rod end jam nuts to 25.3 to 28.9 ft. lbs., then install clips retaining boots to tie rods.

CENTERING STEERING WHEEL

After completing front toe-in and front track adjustments, steering wheel should be centered, if not proceed as follows:

1. Using a grease pencil, place alignment marks on tie rod ends and tie rods.
2. Loosen tie rod end jam nuts, then remove clips securing boots to tie rods.
3. Rotate tie rods in equal amounts as indicated in **Fig. 3.** Refer to alignment marks made with grease pencil to ensure both tie rods are being rotated equal amounts.

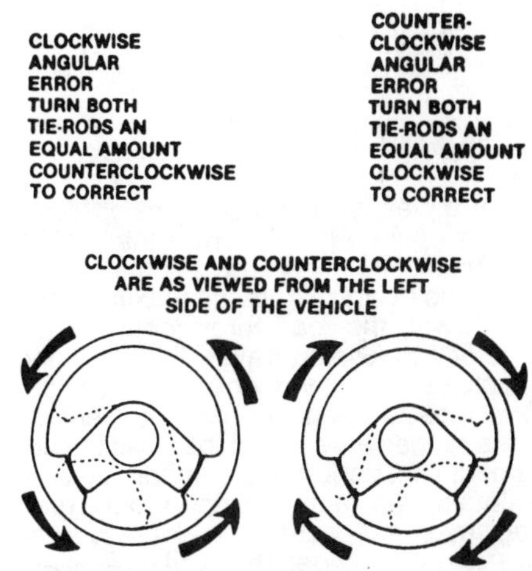

Fig. 3 Centering steering wheel

4. After completing adjustment, torque jam nuts to 25.3 to 28.9 ft. lbs., then install boot to tie rod retaining clips.

REAR WHEEL ALIGNMENT

CAMBER

Rear camber angle cannot be adjusted on these models. If rear camber is not within specifications, check chassis and body to chassis mountings for damage and distortion.

TOE-IN

Hatchback Models

Rear toe-in on these models is adjusted by rotating adjusting cams located on the control arms.

Station Wagon

1. Loosen jam nuts at control arms.
2. To increase toe-in, rotate right control arm clockwise and left control arm counterclockwise, **Fig. 4.** To decrease

toe-in, rotate right control arm counterclockwise and left control arm clockwise. Rotate both control arms by equal amounts.

3. After completing rear toe-in adjustment, check vehicle rear tracking as per alignment equipment manufacturers instructions.
4. Torque jam nuts to 41 to 47 ft. lbs.

Merkur

INDEX

DESCRIPTION

The angles at which suspension components operate in relation to the vehicle and wheel center lines, are calculated during vehicle design to provide maximum tire contact with the road throughout the full range of suspension travel. The checking of caster, camber and wheel toe-in or out is the measurement of these suspension angles at the wheels, and the possible adjustments are performed to ensure that the tires make maximum contact with the road at all times and cause the vehicle to track in a straight line when the front wheels are centered. Proper alignment of the front and rear wheels with the vehicle chassis and each other is essential for acceptable handling and to minimize tire wear.

Prior to checking wheel alignment, ensure that tires are properly matched (same size tires on each axle set), correctly inflated and uniformly worn. Ensure that wheel bearings are properly adjusted and that suspension components are not damaged or worn. Wheel alignment should be checked with the vehicle unloaded, as any abnormal or uneven loads affect ride height which, in turn, affects suspension operating angles (wheel alignment).

Only wheel toe-in or out is adjustable on these models. Caster and camber are not adjustable. However, they should be checked as possible causes of handling or tire wear complaints. If caster and camber are not within limits suspension components should be inspected for wear and damage, and replaced as needed. If control arms, stabilizers and bushings are in good condition, check vehicle body for distortion at suspension mounting points or for collision damage.

TOE-IN & STEERING WHEEL CENTERING ADJUSTMENT

If the steering wheel is not properly centered when vehicle is driven straight ahead, mark its position with a piece of tape across the gap between steering wheel hub and steering column shroud.

If toe-in is to be adjusted, the operation can be combined with steering wheel centering. But, to avoid complications, one should be completed before starting the other. To center the steering wheel, the tie rods must be turned into one tie rod end and out of the other, in equal amounts to avoid changing the toe setting. This shifts the steering rack right or left, turning the pinion, steering column and wheel to its desired position. Center steering as follows:

1. Mark the tie rod and tie rod ends with paint or a grease pencil to indicate their original relative positions.
2. Loosen and back off the tie rod end jam nuts, then release the steering gear boot clips. Ensure boots are free on the tie rods to avoid twisting.
3. Screw tie rods into one tie rod end and out of the other, depending on which way the steering wheel is to be moved and how much. Example: if left tie rod is screwed in and right tie rod is screwed out, the steering rack moves to the left and the steering wheel movement is counterclockwise as seen from the drivers seat. For clockwise correction, the rack must be moved to the right. If the road wheels are positioned and locked in the straight ahead position during this operation, using the steering wheel tape mark as a starting point, the wheel will turn during adjustment and the centered position can be judged visually. One revolution of the tie rods will result in a steering wheel correction of approximately 19°. Check marks on tie rods and tie rod ends, to ensure tie rods are turned equal amounts. Adjust toe-in as follows:
   a. Loosen jam nuts at tie rod ends and release clips at small ends of steering gear boots, ensuring boots are free on tie rods so they will not be twisted when tie rods are turned.
   b. Turn tie rods into or out of tie rod ends an equal amount on each side to keep steering wheel centered.
   c. When toe-in is within specifications, **torque** tie rod end jam nuts to 42-50 ft. lbs., then ensure steering gear boot ends are positioned in the reduced-diameter sections of the tie rods and install boot clips.

RIDE HEIGHT

Refer to "Wheel Alignment Specifications, Rear" for ride height specifications.

GENERAL MOTORS CORP.

TABLE OF CONTENTS

Buick, Cadillac, Chevrolet, Oldsmobile & Pontiac Rear Wheel Drive Models Except Chevette, 1000 & Fiero

INDEX

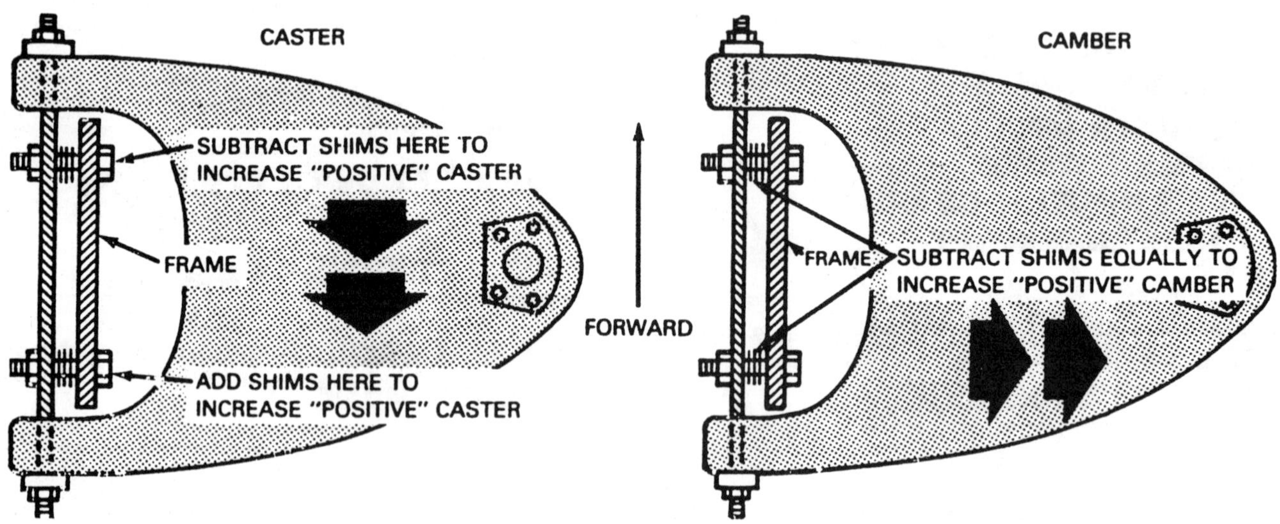

Fig. 1 Caster & camber adjustments. Except Camaro, Corvette & Firebird

FRONT WHEEL ALIGNMENT

Prior to checking or adjusting front suspension alignment, inspect suspension components for damage or excessive wear, and replace as needed. Ensure tire pressures and wheel bearings are properly adjusted, then raise and release front bumper several times to allow vehicle to assume normal ride height.

CASTER

Except Camaro & Firebird

Caster adjustments are made by means of shims between the upper control arm inner support shaft and the support bracket attached to the frame. Shims may be added, subtracted or transferred to change the readings.

Transfer shims from front to rear or rear to front. The transfer of one shim to the front bolt from the rear bolt will decrease positive caster. One shim transferred from the rear bolt to the front bolt will change caster about 1/2 degree. **Refer to Figs. 1 and 2,** and note the effect of shim placement.

Camaro & Firebird

Caster adjustments are made by moving the position of the upper strut mount assembly, **Fig. 3.** To make adjustment, remove dust cap and fender bolt and attach tool J-29724 to original fender bolt, **Fig. 4.** Tighten the turnbuckle and loosen the three strut mount attaching nuts.

Adjust caster by lightly tapping the mount assembly forward or rearward. Move mount forward to decrease caster, or rearward to increase caster, **Fig. 3.**

CAMBER

Except Camaro & Firebird

Camber adjustments are made by means of shims between the upper control arm inner support shaft and the support bracket attached to the frame. Shims may be added, subtracted or transferred to change the readings.

Change shims at both the front and rear of the shaft. Adding an equal number of shims at both front and rear of the support shaft will decrease positive camber. One shim at each location will move camber approximately 1/6 degree.

Camaro & Firebird

Camber adjustments are made by moving the position of the upper strut mount assembly, **Fig. 5.** To make adjustment, re-

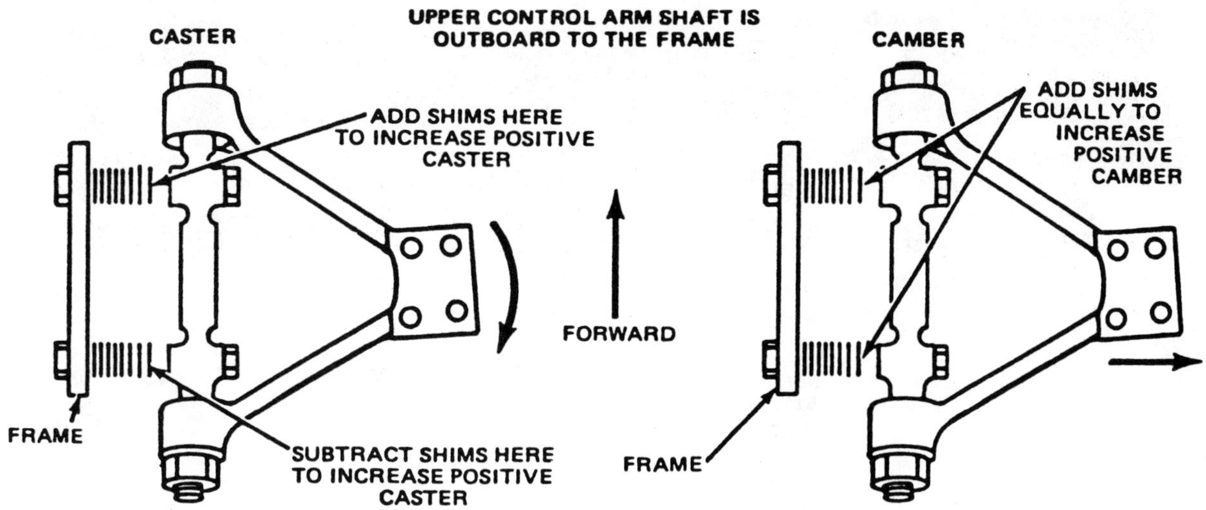

Fig. 2 Caster & camber adjustments. Corvette

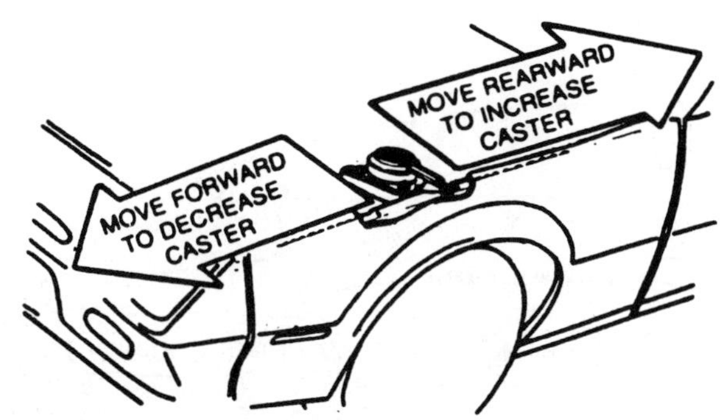

Fig. 3 Caster adjustment. Camaro & Firebird

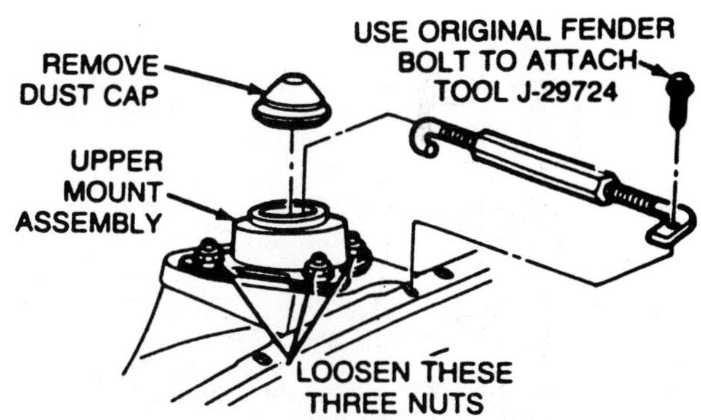

Fig. 4 Caster & camber adjustment tool installation. Camaro & Firebird

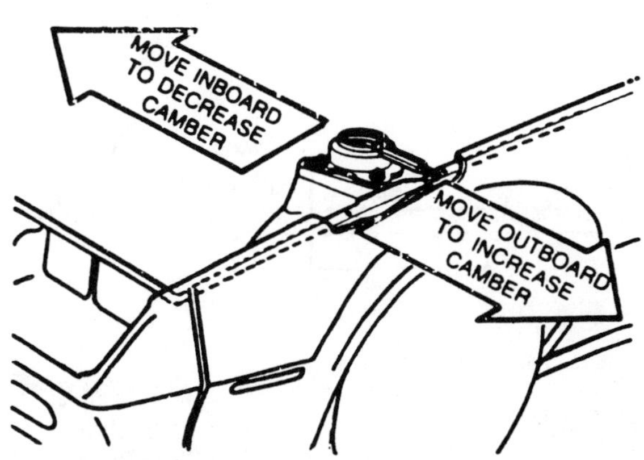

Fig. 5 Camber adjustment. Camaro & Firebird

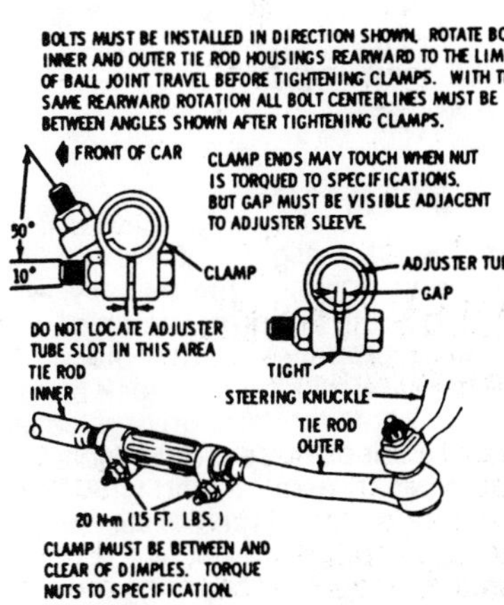

Fig. 6 Tie rod clamp & sleeve positioning. Except Camaro, Corvette & Firebird

move dust cap and fender bolt and attach tool J-29724 to original fender bolt, **Fig. 4.** Tighten the turnbuckle and loosen the three strut mount attaching nuts.

Adjust camber by rotating the turnbuckle to move mount assembly inward or outward. Move mount inboard to decrease camber, or outboard to increase camber, **Fig. 5.**

TOE-IN

Except Corvette

1. Loosen clamp bolts at each end of tie rod adjustable sleeves.
2. Set steering wheel in straight-ahead position, then turn tie rods as necessary until toe-in is within specifications.
3. After adjustment is complete, ensure threads exposed on each end of sleeve are equal and tie rod end housings are at right angles to steering arm.
4. Position tie rod clamps and sleeves as shown, **Figs. 6 and 7,** then tighten nuts.

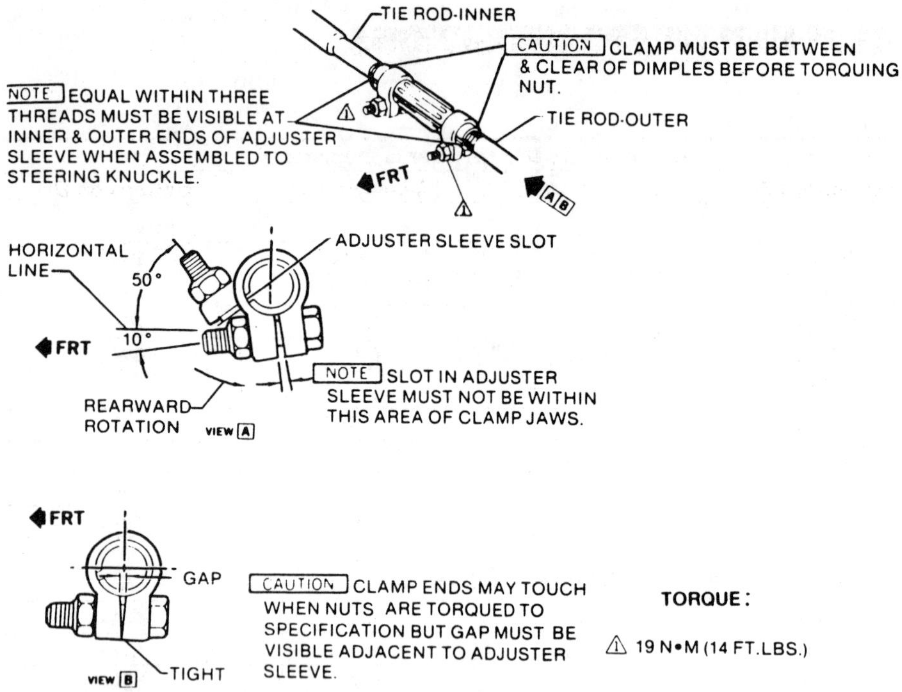

Fig. 7 Tie rod clamp & sleeve positioning. Camaro & Firebird

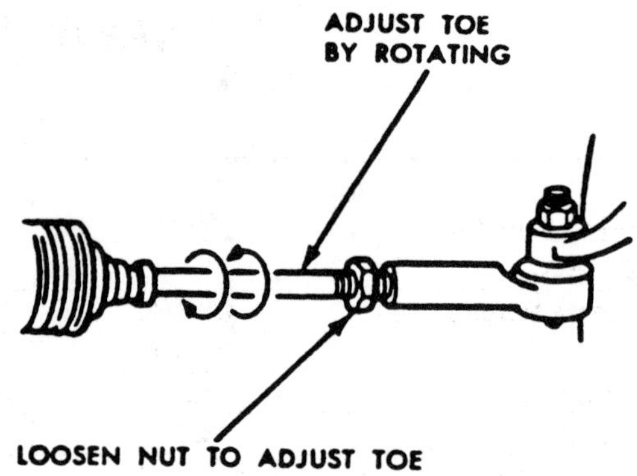

Fig. 8 Toe adjustment. Corvette

Toe-In

Toe-in is adjusted by loosening lock nuts on tie rod ends and rotating adjuster sleeves until desired setting is obtained.

Corvette

With steering gear in the straight ahead position, loosen the clamp bolts at each end of each tie rod. Rotate each tie rod to increase or decrease its length, as necessary. When proper toe-in is obtained, tighten tie rod clamp bolts, **Fig. 8.**

REAR WHEEL ALIGNMENT

CORVETTE

Rear wheel alignment should be checked and adjusted periodically, when rear tires indicate abnormal wear, or when suspension components are replaced. Prior to rear wheel alignment, check suspension components for damage or excessive wear and repair as necessary. Also ensure tires are properly inflated, and wheel bearing end play is within specifications.

Camber

Wheel camber is adjusted by rotating the eccentric cam and bolt located at the inboard end of the strut rod or spindle support rod. To check and adjust camber, proceed as follows:

1. Place rear wheels of vehicle on suitable alignment equipment following manufacturer's instructions, then check camber reading.
2. If wheel camber is not within specifications, loosen cam bolt retaining nut.
3. Rotate cam bolt until camber reading is within specifications.

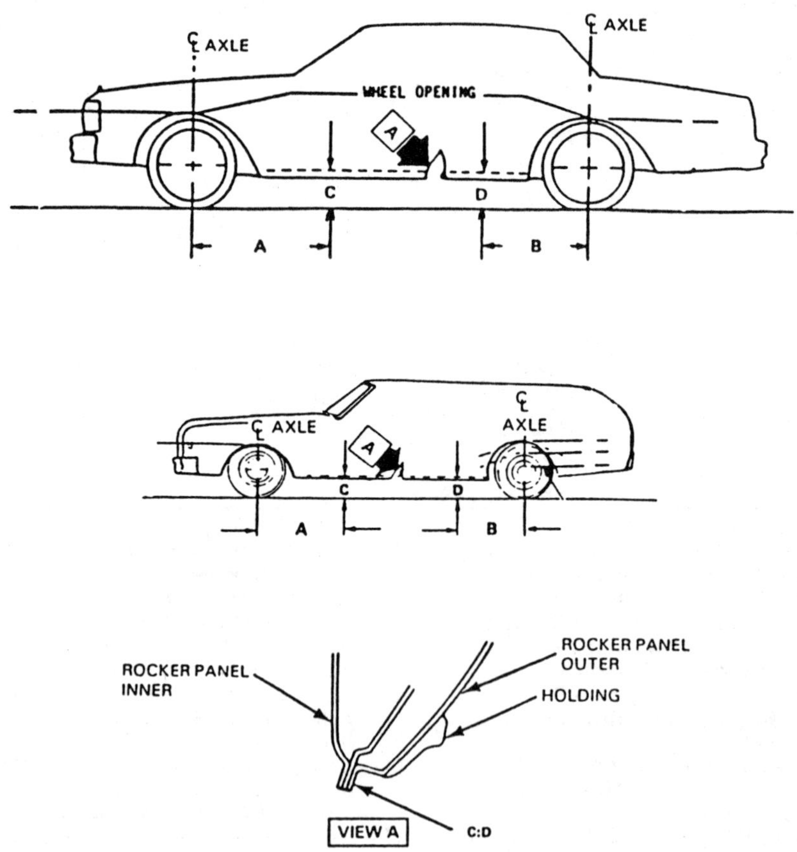

Fig. 9 Vehicle riding height measurement locations & specifications (Part 1 of 2). 1985–90 Buick, Chevrolet, Oldsmobile & Pontiac

VEHICLE RIDE HEIGHT SPECIFICATIONS

Make & Model	Year	Riding Height Measurements In Inches ①			
		Dimension A	Dimension B	Dimension C	Dimension D
BUICK					
Estate Wagon	1985–87	27 61/64	19 7/32	10 7/16	10 9/16
	1988–90	31 1/2	24 39/64	10 7/16	10 9/16
LeSabre	1985	27 61/64	19 7/32	10 5/16	10 5/16
Regal	1985–87	24 33/64	20 15/16	9 27/32	9 27/32
CHEVROLET					
Camaro	1985–87	32 19/32	17 15/64	⑧	⑨
	1988	32 19/32	17 15/64	⑩	⑪
	1989–90	32 19/32	17 15/64	⑩	⑫
Caprice Sedan	1985–88	31 1/2	24 39/64	10 1/2	10 5/8
	1989	31 1/2	24 39/64	②	③
	1990	31 1/2	24 39/64	④	⑤
Caprice Wagon	1985–88	31 1/2	24 39/64	10 9/16	10 5/8
	1989–90	31 1/2	24 39/64	10 33/64	10 9/16
Corvette	1985	25 21/32	19 1/8	⑬	⑭
	1986–87	25 21/32	19 1/8	⑮	⑭
	1988–90	25 21/32	19 1/8	⑯	⑰
Monte Carlo	1985–88	24 33/64	20 15/16	⑥	⑦
OLDSMOBILE					
Custom Cruiser	1985–88	29 21/64	18 11/16	10 1/2	10 5/8
	1989–90	31 1/2	24 39/64	10 1/2	10 5/8
Cutlass	1985–87	24 17/32	20 15/16	⑱	⑱
	1988	24 17/32	20 15/16	10	10
Delta 88	1985	29 21/64	18 11/16	10 3/8	10 9/16
PONTIAC					
Bonneville	1985	26 19/64	20 15/16	10 5/16	10 5/8
Firebird	1985–86	32 19/32	17 15/64	⑧	⑨
	1987–90	32 19/32	17 15/64	8 15/64	8 17/64
Grand Prix	1985	26 19/64	20 15/16	9 31/32	10 5/8
	1986–87	26 19/64	20 15/16	9 59/64	10 5/32
Parisienne Sedan	1985–86	31 1/2	24 39/64	10 1/2	10 5/8
Parisienne Wagon	1985–86	31 1/2	24 39/64	10 9/16	10 5/8
Safari	1987–89	31 1/2	24 39/64	10 9/16	10 5/8

①—Plus or minus 13/32 inch.
②—Models with P205-75 or P225-70 tires, 10 23/64 inches; models with P215-75 tires, 10 47/64 inches.
③—Models with P205-75 or P225-70 tires, 10 19/32 inches; models with P215-75 tires, 10 55/64 inches.
④—Models with V6 engine & P225-70 tires, 10 23/64 inches; other models, 10 7/16 inches.
⑤—Models with V6 engine & P225-70 tires, 10 19/32 inches; other models, 10 23/64 inches.
⑥—Models P195-75R14 or P205-70R14 tires, 9 23/32 inches; Models with P215-65R15 tires, 10 5/32 inches.
⑦—Models P195-75R14 or P205-70R14 tires, 9 11/16 inches; Models with P215-65R15 tires, 10 5/32 inches.
⑧—Except models with P245-50VR16 tires, 8 11/32 inches; models with P245-50VR16 tires, 7 33/64 inches.
⑨—Except models with P245-50VR16 tires, 8 19/64 inches; models with P245-50VR16 tires, 8 5/64 inches.
⑩—Except models with P245-50VR16 tires, 8 11/32 inches; models with P245-50VR16 tires, 8 1/32 inches.
⑪—Except models with P245-50VR16 tires, 8 19/64 inches; models with P245-50VR16 tires, 8 1/32 inches.
⑫—Except models with P245-50VR16 tires, 8 19/64 inches; models with P245-50VR16 tires, 8 7/64 inches.
⑬—Less heavy duty suspension, 7 13/32 inches; with heavy duty suspension, 7 17/32 inches.
⑭—Less heavy duty suspension, 7 43/64 inches; with heavy duty suspension, 7 5/8 inches.
⑮—Less heavy duty suspension, 7 1/2 inches; with heavy duty suspension, 7 5/8 inches.
⑯—Coupe, less heavy duty suspension, 7 51/64 inches; with heavy duty suspension, 7 19/32 inches. Convertible, 7 51/64 inches.
⑰—Coupe, less heavy duty suspension, 7 57/64 inches; with heavy duty suspension, 7 11/16 inches. Convertible, 8 1/8 inches.
⑱—2 Dr. 10 inches, 4 Dr., 10 3/16 inches.

Fig. 9 Vehicle riding height measurement locations & specifications (Part 2 of 2). 1985–90 Buick, Chevrolet, Oldsmobile & Pontiac

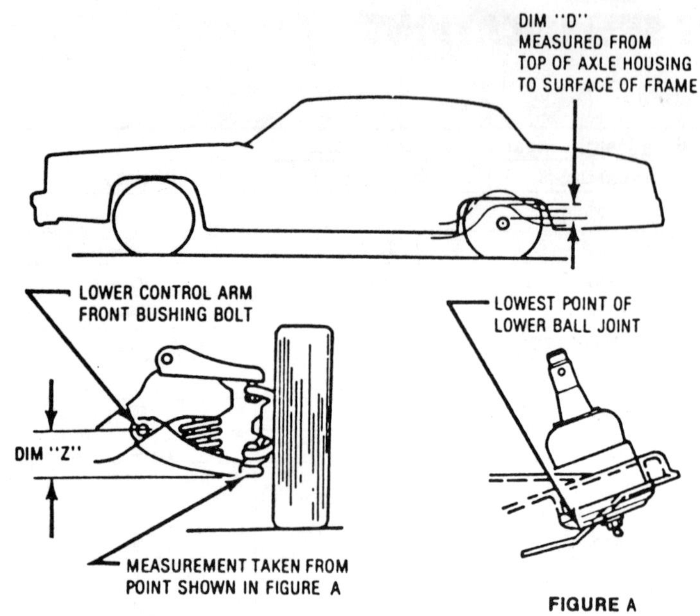

DIMENSIONS IN MILLIMETERS

SERIES	RIDE OPTION	ENGINE	FRONT-DIM "Z" MIN.	FRONT-DIM "Z" CURB TARGET	FRONT-DIM "Z" MAX.	REAR-DIM "D" MIN.	REAR-DIM "D" CURB TARGET	REAR-DIM "D" MAX.
BROUGHAM	ELC	GAS	45	55	65	122	132	142
		DIESEL	46	56	66	122	132	142

Fig. 10 Ride height measurement. 1985 Cadillac Brougham

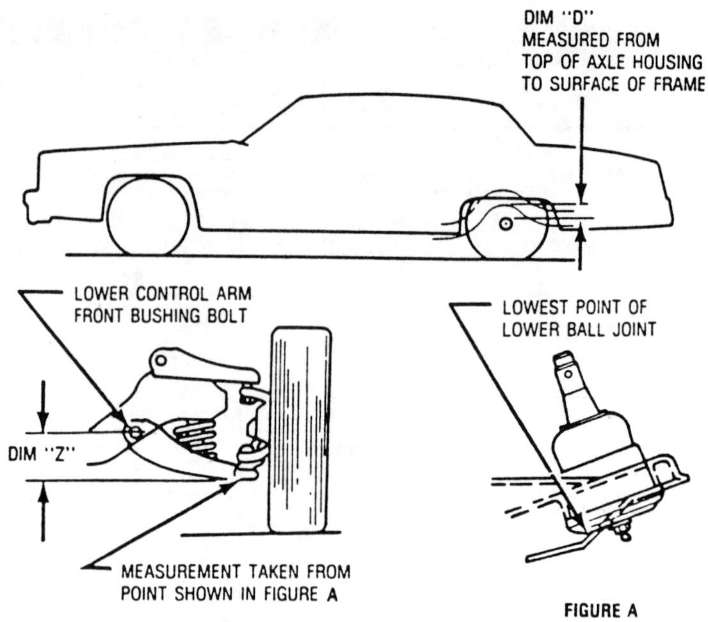

DIMENSIONS IN MILLIMETERS

MODEL	FRONT-DIM "Z" (MM) MIN.	FRONT-DIM "Z" (MM) CURB TARGET	FRONT-DIM "Z" (MM) MAX.	REAR-DIM "D" (MM) MIN.	REAR-DIM "D" (MM) CURB TARGET	REAR-DIM "D" (MM) MAX.	
W/O ELC	46 (1.81")	56 (2.21")	66 (2.60")	130 (5.11")	140 (5.51")	150 (5.91")	MAX. VARIATION IN TRIM HEIGHT FROM SIDE TO SIDE IS 9 MM (.35")
WITH ELC	44 (1.73")	54 (2.13")	64 (2.52")	121 (4.76")	131 (5.16")	141 (5.55")	
WITH HEAVY DUTY RIDE	44 (1.73")	54 (2.13")	64 (2.54")	121 (4.76")	131 (5.16")	141 (5.55")	
WITH TRAILER TOWING	35 (1.38")	45 (1.77")	55 (2.17")	112 (4.41")	122 (4.80")	132 (5.20")	

Fig. 11 Ride height measurement. 1986–90 Cadillac Brougham

RIDE HEIGHT

EXCEPT CADILLAC

Refer to **Fig. 9** for ride height measurements and specifications.

CADILLAC

Refer to **Figs. 10 and 11** for ride height measurements and specifications.

Buick Century & Skylark (1985), Chevrolet Celebrity & Citation, Oldsmobile Cutlass Ciera & Cutlass Cruiser & Pontiac 6000

INDEX

Camber and toe-in are the only adjustments that can be performed on these vehicles.

CAMBER

The camber angle is adjusted by loosening the strut cam bolt and the through bolt, then rotating the cam bolt to move the upper portion of the steering knuckle inboard or outboard. When correct camber angle is obtained, **torque** cam and through bolts to 140 ft. lbs.

When performing this adjustment, the top through bolt must be loosened to prevent damage to the outer cam guide.

TOE-IN

Toe-in is controlled by tie rod position. Adjustment is made by loosening the nuts at the steering knuckle end of the tie rods and rotating the rods to obtain proper toe-in setting. When adjusting toe-in, the tie rod boot clamps must be removed. After correct toe-in setting is obtained, **torque** tie rod nuts to 45 ft. lbs.

RIDE HEIGHT

Refer to **Fig. 1,** for ride height specifications and measurement location points.

Trim heights checked with correct tire pressures, fuel tank full or equivalent weight in the trunk. No passengers or added weight in car. Front seat in rear position. Trunk must be empty except for spare tire and jack or simulated fuel load. Measure from known level floor to rocker panel with steering wheel in the centered position.

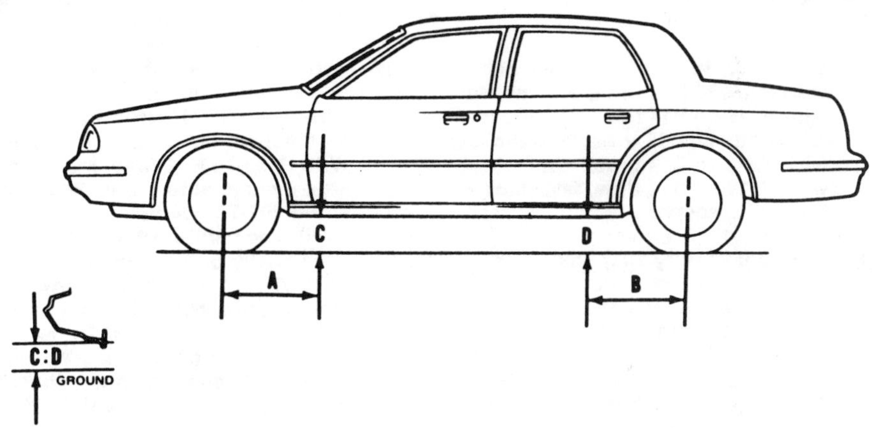

C & D DIMENSION

Lift center of front bumper up approximately 38 mm (1-1/2") and let vehicle settle gently. Repeat two more times, then measure "C" dimension. Push center of bumper down 38 mm (1-1/2") and let vehicle settle gently. Repeat two more times, then measure "C" dimension. The "C" dimension is an average of the high and low measurements. Repeat procedure on the rear bumper for the "D" dimension.

Fig. 1 Vehicle riding height measurement locations & specifications (Part 1 of 3)

VEHICLE RIDING HEIGHT SPECIFICATIONS

Make & Model	Year	Riding Height Measurements In Inches ①			
		Dimension A	Dimension B	Dimension C	Dimension D
BUICK					
Century ②	1985–90	19 7/8	18 1/16	9 3/32	9 11/16
Century ③	1985–90	19 7/8	18 1/16	9 11/64	9 3/8
Skylark	1985	19 7/8	18 5/64	9 1/64	9 3/32

Fig. 1 Vehicle riding height measurement locations & specifications (Part 2 of 3)

Continued

VEHICLE RIDING HEIGHT SPECIFICATIONS—Continued

Make & Model	Year	Riding Height Measurements In Inches ①			
		Dimension A	Dimension B	Dimension C	Dimension D
CHEVROLET					
Celebrity ② ⑥	1985–90	19 7/8	18 5/64	⑧	⑨
Celebrity ② ⑦	1985–90	19 7/8	18 5/64	⑩	⑪
Celebrity ③ ⑥	1985–90	19 7/8	18 5/64	⑧	⑬
Celebrity ③ ⑦	1985–90	19 7/8	18 5/64	⑬	⑭
Citation	1985–90	19 7/8	18 5/64	④	⑤
OLDSMOBILE					
Cutlass Ciera	1985–88	21 7/8	18 1/16	⑮	⑯
	1989–90	19 7/8	18 5/64	⑰	⑱
Cutlass Cruiser	1985–88	21 7/8	18 1/16	9 3/8	9 9/16
	1989–90	19 7/8	18 5/64	⑰	⑱
PONTIAC					
6000 ⑲	1985–90	—	—	⑳	㉑
6000 ㉒	1989–90	—	—	9 1/4	9 29/64

①—Plus or minus .4 inch.
②—Coupe & sedan.
③—Station wagon.
④—With P185-80 R13 tires, 9 15/16 inches; with P205-70 R13 tires, 9 1/16 inches; with P215-60 R14 tires, 9 7/32 inches.
⑤—With P185-80 R13 tires, 9 29/64 inches; with P205-70 R13 tires, 9 7/32 inches; with P215-60 R14 tires, 9 21/64 inches.
⑥—Except Eurosport.
⑦—Eurosport.
⑧—Less heavy duty suspension, with P185-75 R14 tires, man. trans. 9 1/16 inches, auto. trans. 9 9/32 inches; with P195-75 R14 tires, man. trans. 9 21/64 inches, auto. trans. 9 9/16 inches. Heavy duty suspension, with P185-75 R14 tires, 9 9/32 inches; with P195-75 R14 tires, 9 9/16 inches.
⑨—Less heavy duty suspension, with P185-75 R14 tires, man. trans. 9 1/2 inches, auto. trans. 9 33/64 inches; with P195-75 R14 tires, man. trans. 9 23/32 inches, auto. trans. 9 51/64 inches. Heavy duty suspension, with P185-75 R14 tires, 9 33/64 inches; with P195-75 R14 tires, 9 51/64 inches.
⑩—With P195-75 R14 tires, man. trans. 9 9/32 inches, auto. trans. 9 33/64 inches; with P195-70 R14 tires, man. trans. 8 15/16 inches, auto. trans. 9 1/8 inches.
⑪—With P195-75 R14 tires, man. trans. 9 23/32 inches, auto. trans. 9 63/64 inches; with P195-70 R14 tires, man. trans. 8 15/16 inches, auto. trans. 9 13/32 inches.
⑫—Less heavy duty suspension, with P185-75 R14 tires, man. trans. 9 29/64 inches, auto. trans. 9 1/2 inches; with P195-75 R14 tires, man. trans. 9 23/32 inches, auto. trans. 9 49/64 inches. Heavy duty suspension, with P185-75 R14 tires, 9 1/2 inches; with P195-75 R14 tires, 9 49/64 inches.
⑬—With P195-75 R14 tires, 9 1/2 inches; with P195-70 R14 tires, 9 1/8 inches.
⑭—With P195-75 R14 tires, 9 11/16 inches; with P195-70 R14 tires, 9 21/64 inches.
⑮—Except ES & GT, 9 9/32 inches; ES & GT, 2 door 8 7/8 inches, 4 door 9 3/16 inches.
⑯—Except ES & GT, 9 3/8 inches; ES & GT, 2 door 9 inches, 4 door 9 3/32 inches.
⑰—Less sport suspension, with P185-75R14 tires, 9 19/64 inches; with P195-74R14 tires, 9 37/64 inches; with P195-70R14 tires, 9 7/32 inches. With sport suspension, 8 15/16 inches.
⑱—Less sport suspension, with P185-75R14 tires, 9 29/64 inches; with P195-74R14 tires, 9 23/32 inches; with P195-70R14 tires, 9 13/32 inches. With sport suspension, 9 1/8 inches.
⑲—Except all wheel drive models.
⑳—Models with 14 inch wheels, 9 1/8 inches; models with 15 inch wheels, 9 13/32 inches.
㉑—Models with 14 inch wheels, 9 21/64 inches; models with 15 inch wheels, 9 41/64 inches.
㉒—All wheel drive models.

Fig. 1 Vehicle riding height measurement locations & specifications (Part 3 of 3)

Buick Electra, LeSabre & Park Avenue, Cadillac Deville & Fleetwood, Oldsmobile 88 & 98 & Pontiac Bonneville Front Wheel Drive Models

INDEX

Page No.

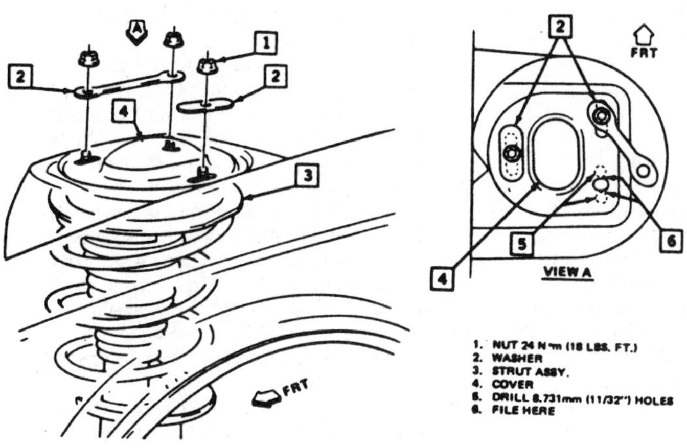

Fig. 1 Caster adjustment

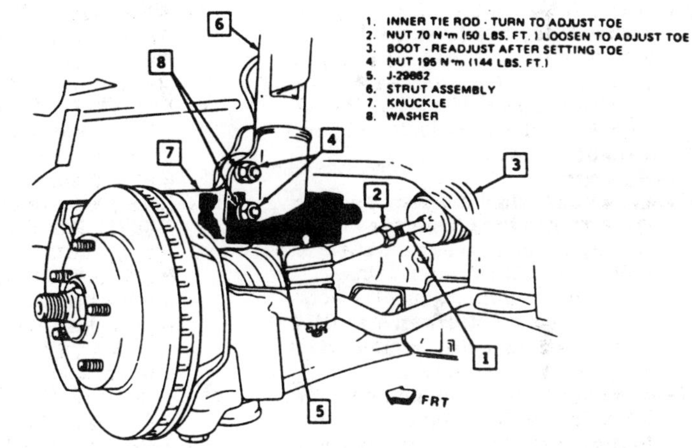

Fig. 2 Front camber adjustment

FRONT WHEEL ALIGNMENT

Camber and toe are the only adjustments normally required. After checking settings, perform adjustments in the following order: caster, camber and toe in.

CASTER

1. Loosen, but do not remove, two of the three top strut mounting nuts covering the slotted mounting holes, **Fig. 1.**
2. Remove the third nut over the oval strut mounting hole and move the washer away from the mounting hole.
3. Raise vehicle by the body to separate the strut from inner wheel house.
4. Drill 11/32 inch holes at front and rear of oval strut mounting hole and file excess metal, **Fig. 1.**
5. Lower body and insert strut into proper position.
6. Set caster to specifications by moving top of strut forward or backward as needed.
7. Install mounting nut and washer. **Torque** all mounting nuts to 18 ft. lbs.
8. Recheck caster.

CAMBER

1. Loosen both strut to knuckle attaching nuts, **Fig. 2.**
2. Install camber adjusting tool J-29862.
3. Set camber to specifications.
4. Remove adjusting tool and **torque** strut to knuckle attaching nuts to 144 ft. lbs.
5. Recheck camber.

TOE

1. Loosen locknuts on both inner tie rods, **Fig. 3.**
2. Adjust toe to specifications by rotating inner tie rod.
3. **Torque** locknuts to 50 ft. lbs.
4. Recheck toe setting.

REAR WHEEL ALIGNMENT

When checking rear wheel alignment, the electronic leveling system must have the Super Lift struts inflated with residual pressure only.

Place a weight in trunk and turn ignition on and move transmission selector from Park to Reverse position and back. This will activate the compressor. Turn ignition off and remove weight from trunk. Wait 30 seconds for the system to exhaust. Roll vehicle forward one complete wheel rotation. Jounce vehicle before checking alignment.

CAMBER

1. Loosen strut to knuckle attaching nuts.

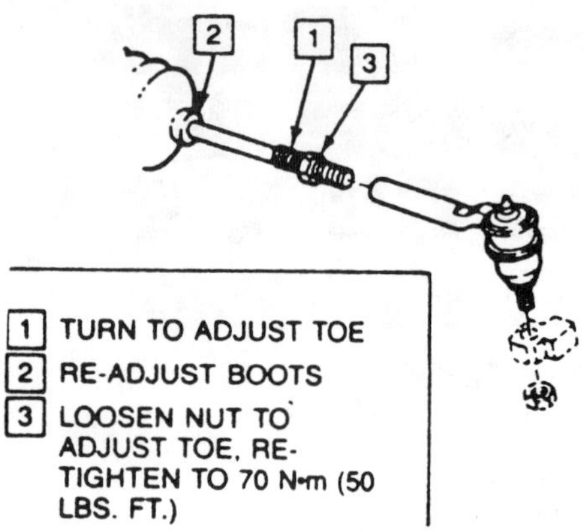

Fig. 3 Front toe adjustment

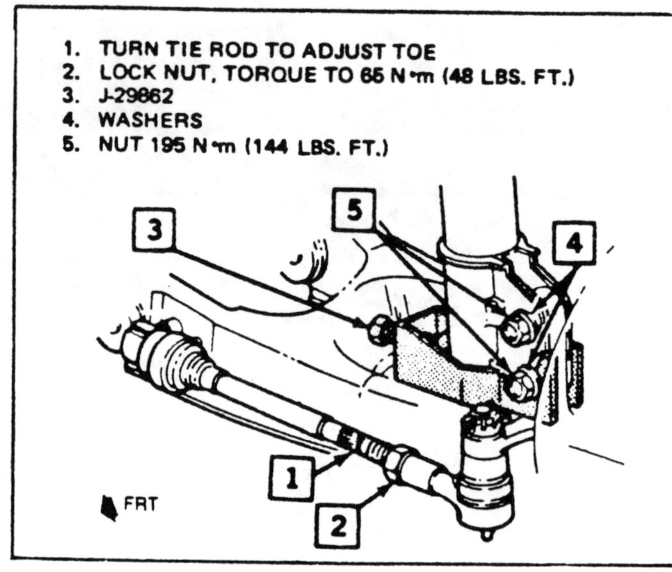

Fig. 4 Rear camber & toe adjustments

2. Install camber adjusting tool J-29862, **Fig. 4.**
3. Move strut to set camber to specifications.
4. Remove camber adjusting tool and **torque** strut to knuckle nuts to 144 ft. lbs.
5. Recheck camber setting.

TOE

Toe adjustment is made by loosening the locknut at tie rod end and turning inner tie rod to set toe to specifications, **Fig. 4.**

RIDE HEIGHT

Refer to **Figs. 5 through 7,** for ride height specifications and measurement locations. Check ride height as follows:

1. Ensure vehicle is on level ground.
2. Ensure tires are inflated to proper pressures.
3. Fuel tank should be full to obtain accurate readings.
4. Trunk should be empty except for spare tire and jack.
5. Place the front seat in the far rearward position, then turn ignition to On position to activate the electronic level control, if equipped.
6. Bounce the car three times at the front and rear to normalize suspension.
7. Measure from lowest point on the ball joint housing to control arm bolt center line, "D" and "Z" positions.
8. Measure from level floor to rocker panel at "J" and "K" positions on all models except SSE.
9. Measure from level floor to wheel opening at "P" and "R" positions.

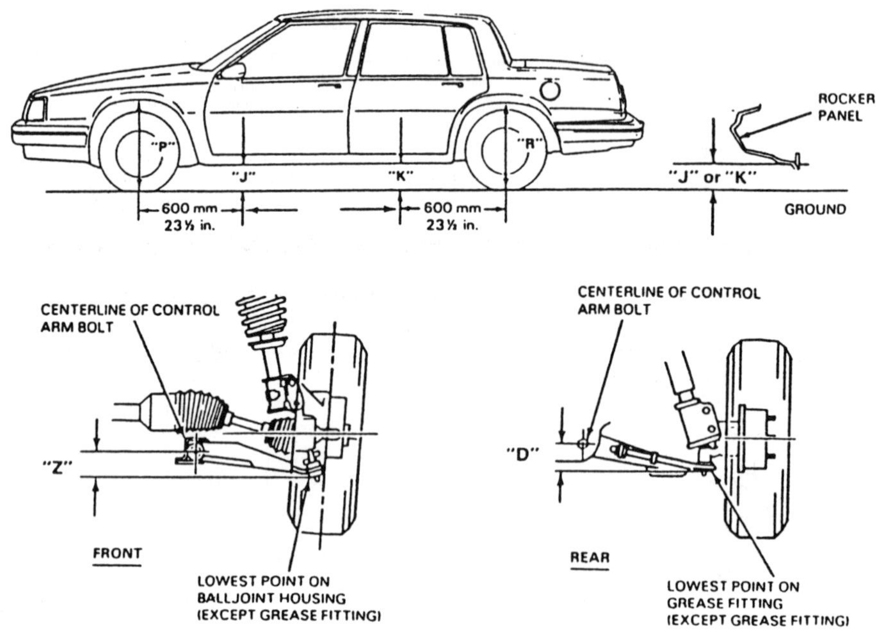

MAXIMUM VARIATION SIDE TO SIDE & FRONT TO REAR IS 19MM (3/4 IN)

LOAD CONDITIONS — COOLANT TO CAPACITY AND FULL TANK OF FUEL

	"Z"	"D"	"P"	"R"	"J"	"K"
C	56-76 mm 2 13/64-3 in.	70-90 mm 2 3/4-3 35/64 in.	699-719 mm 27 1/2-28 5/16 in.	691-711 mm 27 13/64-28 in.	228-248 mm 8 31/32-9 49/64 in.	230-250 mm 9 3/64-9 27/32 in.
H	50-70 mm 1 31/32-2 3/4 in.	77-97 mm 3 1/32-3 13/16 in.	699-719 mm 27 1/2-28 5/16 in.	699-719 mm 27 1/2-28 5/16 in.	229-249 mm 9 1/64-9 13/16 in.	236-256 mm 9 19/64-10 5/64 in.

Fig. 5 Ride height measurements. Electra (C), LeSabre (H), Park Ave. (C), 88 (H) & 98 (C) front wheel drive models

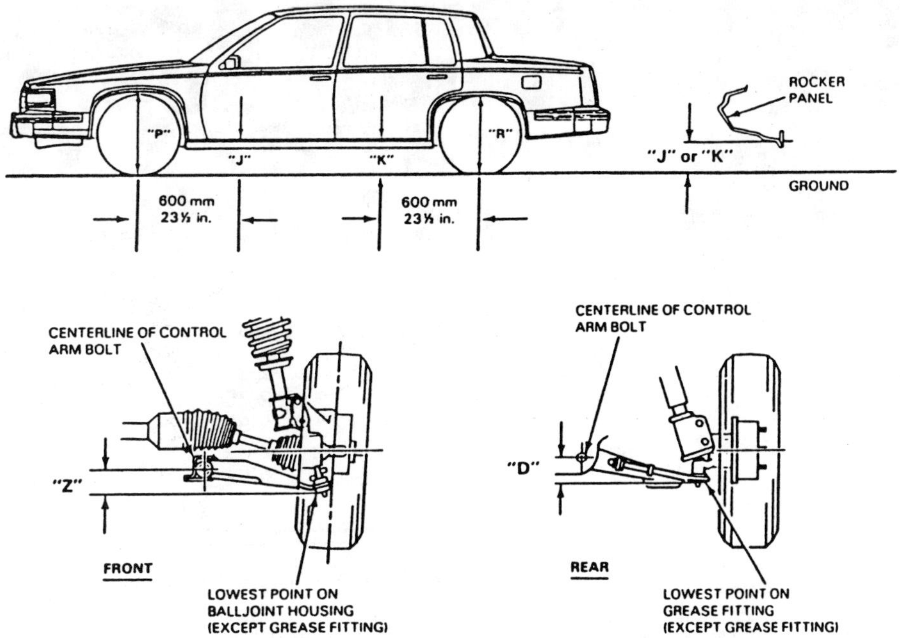

MAXIMUM VARIATION SIDE TO SIDE & FRONT TO REAR IS 19MM (¾ IN)

LOAD CONDITIONS – COOLANT TO CAPACITY AND FULL TANK OF FUEL

"Z"	"D"	"P"	"R"	"J"	"K"
64-84 mm 2 33/64-3 5/16 in.	71-91 mm 2 51/64-3 19/32 in.	706-726 mm 27 51/64-28 37/64 in.	656-676 mm 25 53/64-26 39/64 in.	230-250 mm 9 3/64-9 27/32 in.	230-250 mm 9 3/64-9 27/32 in.

Fig. 6 Ride height measurements. DeVille & Fleetwood front wheel drive models

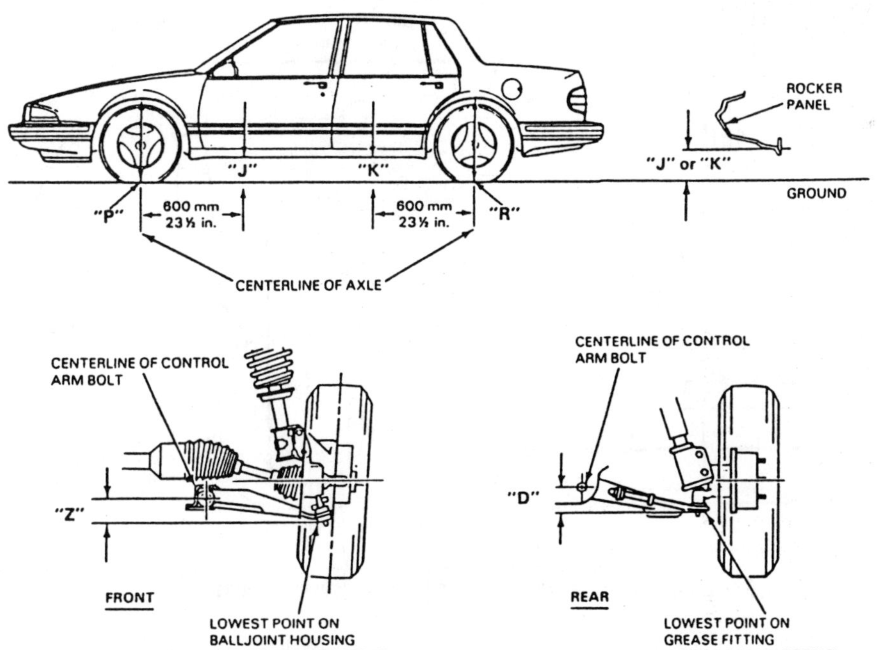

MAXIMUM VARIATION SIDE TO SIDE & FRONT TO REAR IS 19MM (¾ IN)

LOAD CONDITIONS – COOLANT TO CAPACITY AND FULL TANK OF FUEL

"Z"	"D"	"P"	"R"	"J"	"K"
55-75 mm 2 5/32-2 61/64 in.	70-90 mm 2 ¾-3 35/64 in.	708-728 mm 27 55/64-28 21/32 in.	710-730 mm 27 61/64-28 ¾ in.	231-251 mm 9 3/32-9 ⅞ in.	231-251 mm 9 3/32-9 ⅞ in.

Fig. 7 Ride height measurements. Bonneville front wheel drive models

1988–90 Buick Regal, Oldsmobile Cutlass Supreme & Pontiac Grand Prix & 1990 Chevrolet Lumina

INDEX

Page No.

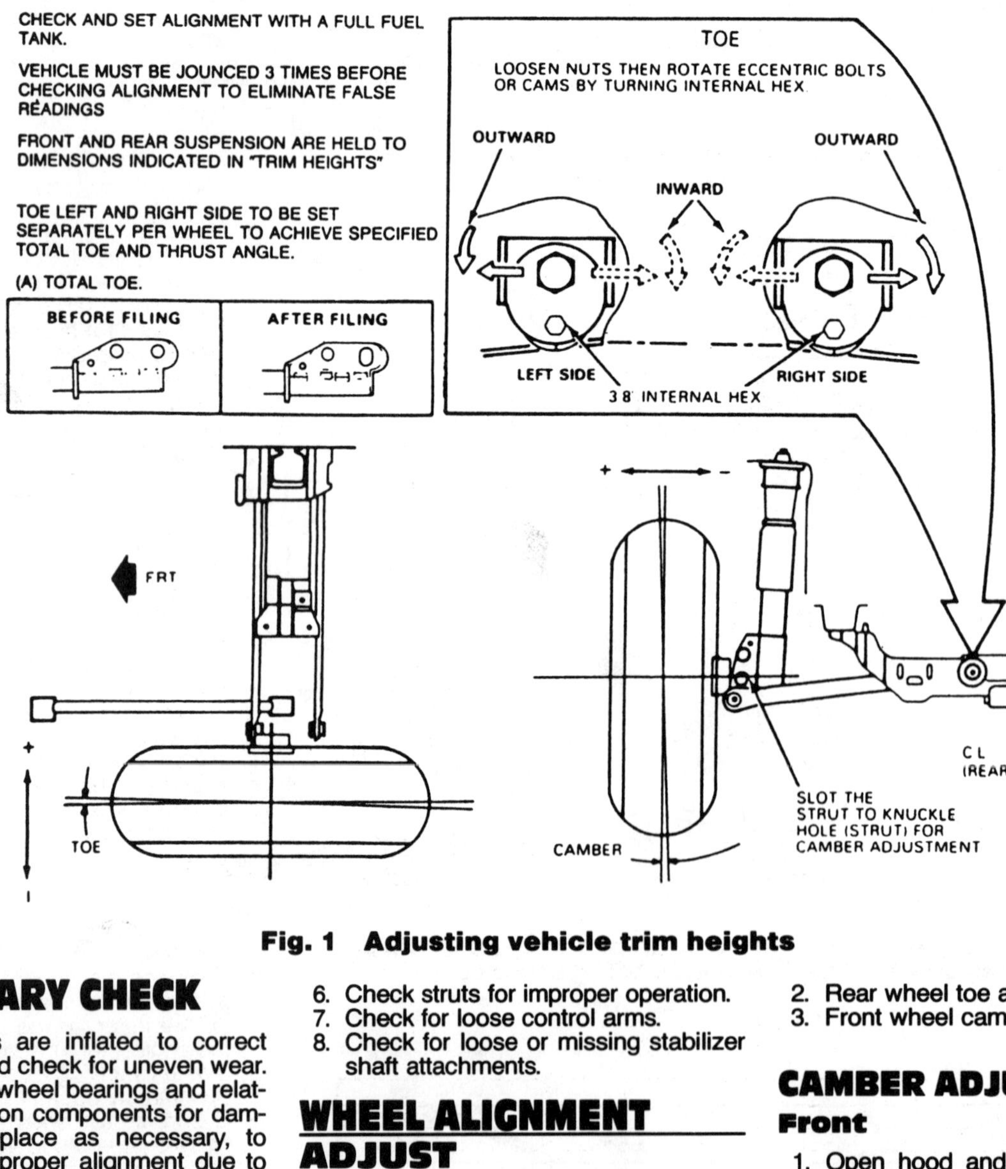

Fig. 1 Adjusting vehicle trim heights

PRELIMINARY CHECK

1. Ensure tires are inflated to correct pressure, and check for uneven wear.
2. Check front wheel bearings and related suspension components for damage and replace as necessary, to eliminate improper alignment due to faulty components.
3. Check ball joints and tie rods.
4. Check vehicle trim heights.
5. Check steering gear for looseness at frame.
6. Check struts for improper operation.
7. Check for loose control arms.
8. Check for loose or missing stabilizer shaft attachments.

WHEEL ALIGNMENT

ADJUST

When adjusting wheel alignment, always adjust both front and rear alignment, proceeding in the following order.

1. Rear wheel camber
2. Rear wheel toe and tracking
3. Front wheel camber and toe

CAMBER ADJUSTMENT

Front

1. Open hood and remove three strut cover plate nuts and the cover plate.
2. Lift front of vehicle just to the point that strut stud clears strut tower and cover top of strut. **Do not over extend drive axle. Do not lift by sus-**

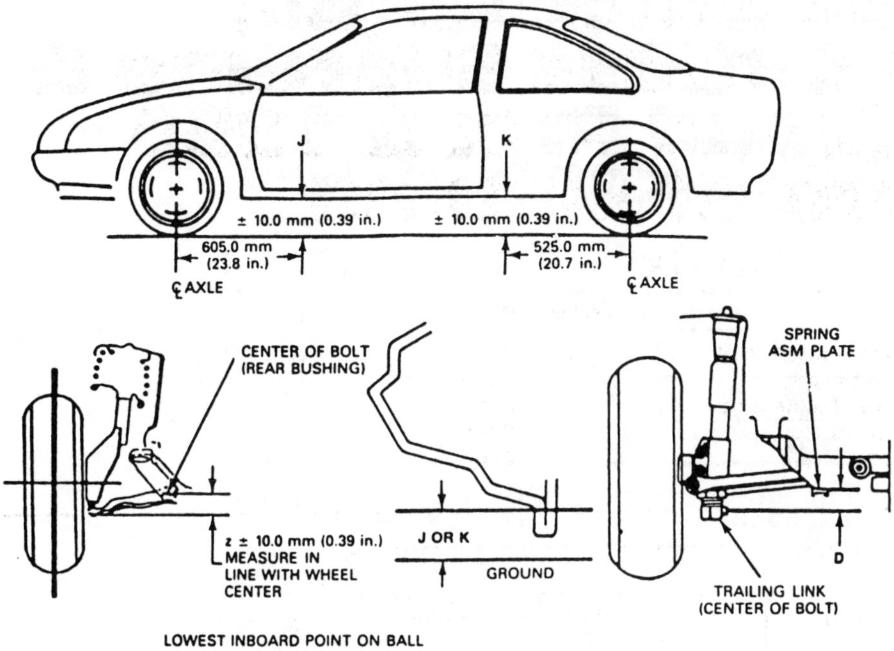

Fig. 2 Ride height measurement location. 1988 Cutlass Supreme, Grand Prix & Regal

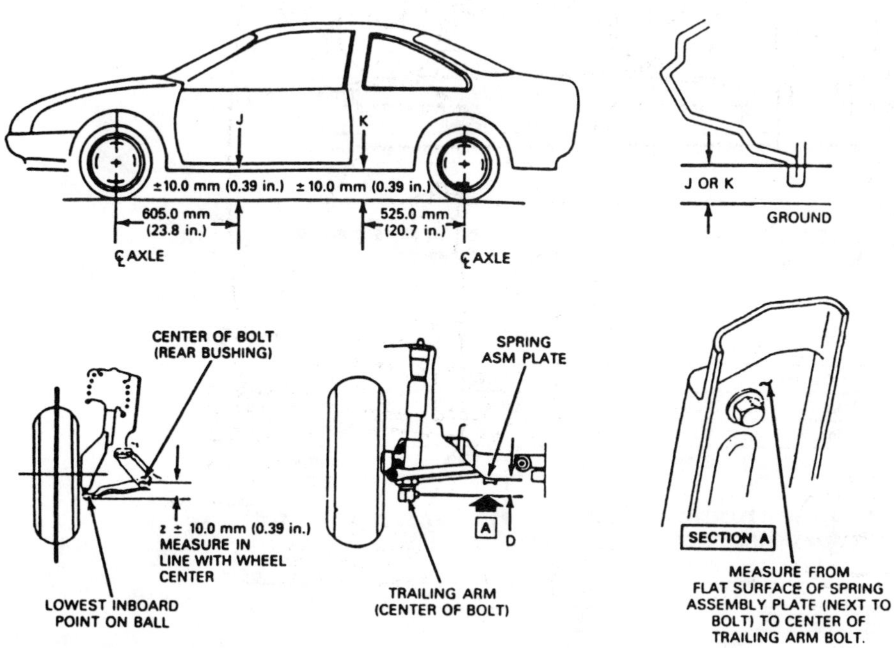

Fig. 3 Ride height measurement location. 1989–90 Cutlass Supreme, Grand Prix & Regal & 1990 Lumina

		Z	D	J	K
SUSP	TIRE SIZE	CURB	CURB	CURB	CURB
STD	P195/75R14	77	97	249	249
FE3	195/70 R15	82	103	249	249
F41	205/70 R14	82	103	249	249

Fig. 4 Ride height specification chart (FE1, std. suspension; FE3, sport suspension; F41, handling package). 1988 Regal

pension.

3. Use strut alignment templet tool No. J-36892 or equivalent to mark holes, then file three holes. File inboard or outboard of existing hole depending or camber requirement. Do not file more than .2 inch in either direction. **Paint exposed metal with red oxide primer and, after primer has dried, paint area with paint matching body color.**
4. Lower front of vehicle and guide strut studs into slotted holes.
5. Install three strut cover plate nuts.
6. Set camber to specifications by moving strut, then **torque** strut cover plate nuts to 17 ft. lbs.

Rear

1. Loosen lug nuts, then raise and support vehicle and remove lug nuts and the wheel and tire assembly.
2. Using suitable jack, support suspension under rear knuckle and hub assembly.
3. Thread tool J-37098 or equivalent into auxiliary spring assembly and tighten to hold assembly in compressed state.
4. Remove brake caliper, leaving hose attached, and support caliper out of way.
5. Remove brake rotor and the brake hose bracket bolt.
6. Scribe strut and knuckle to ensure installation in same position.
7. Remove strut bolts at body and let assembly drop down.
8. Remove stabilizer shaft bracket, then the strut to knuckle attaching bolts.
9. Remove rear auxiliary spring to rear lateral link attaching bolt and nut, then the rear auxiliary spring assembly.
10. Remove strut assembly.
11. Place strut in vise and file lower strut attaching hole oblong.
12. Place auxiliary spring assembly in vise and file lower strut attaching hole oblong.
13. Place stabilizer bracket in vise and file lower stabilizer bracket to strut attaching hole oblong.
14. Reverse procedure to install, **torquing** upper strut bolts to 34 ft. lbs. and the lug nuts to 100 ft. lbs.
15. adjust camber by moving strut and knuckle assembly, then **torque** strut to knuckle attaching nuts to 136 ft. lbs.

TOE

Front

1. Remove small seal clamps.
2. With steering wheel in straight ahead position, loosen jam nuts on tie rods.
3. Rotate inner tie rod to obtain proper toe angle, then ensure that number of threads showing on each tie rod is approximately equal.
4. Ensure tie rod ends are square, then **torque** jam nuts to 46 ft. lbs.
5. Ensure seals are not twisted and install seal clamps.

Rear

1. Loosen inboard cam nuts of rear support rod.

		Z	D	J	K
SUSP	TIRE SIZE	CURB	CURB	CURB	CURB
STD	P195/75R14	80	112	251	257
FE2	P205/70R14	82	114	251	257
FE3	P215/65R15	70	103	251	257

Fig. 5 Ride height specification chart (FE1, std. suspension; FE3, sport suspension; FE2, handling package). 1988 Cutlass Supreme

		Z		D		J		K	
SUSP	TIRE SIZE	SHIPPED	CURB	SHIPPED	CURB	SHIPPED	CURB	SHIPPED	CURB
STD	195-751R14	81	80	124	112	254	251	267	257
	195-70-R15	77	75	120	109	254	251	267	257
FE3	215-65-R15	71	70	115	103	254	251	267	257

Fig. 6 Ride height specification chart (FE1, std. suspension; FE3, sport suspension). 1988–89 Grand Prix

MODEL		Z	D	J	K
SUSP	TIRE SIZE	CURB	CURB	CURB	CURB
FE1	P205-70R14	83	105	249	249
FE3	P215-60R16	69	92	249	249

Fig. 7 Ride height specification chart (FE1, std. suspension; FE3, sport suspension). 1989 Regal

MODEL		Z	D	J	K
SUSP	TIRE SIZE	CURB	CURB	CURB	CURB
FEI	P195-75R14	81mm (3.2 in.)	115mm (4.5 in.)	251mm (9.9 in.)	257mm (10.1 in.)
FE3	P215-65R15	71mm (2.8 in.)	106mm (4.2 in.)	251mm (9.9 in.)	257mm (10.1 in.)
FE3	P215-60R16	69mm (2.7 in.)	103mm (4.0 in.)	251mm (9.9 in.)	257mm (10.1 in.)

Fig. 8 Ride height specification chart (FE1, std. suspension; FE3, sport suspension). 1989 Cutlass Supreme

MODEL		Z	D	J	K
SUSP	TIPE SIZE	CURB	CURB	. CURB	CURB
FE1	P205-70-R14	86mm (3.39 in.)	106mm (4.22 in.)	249mm (9.80 in.)	249mm (9.80 in.)
FE1	P205-70-R15	73mm (2.87 in.)	94mm (3.70 in.)	249mm (9.80 in.)	249mm (9.80 in.)
FE3	P215-60-R16	72mm (2.83 in.)	93mm (3.66 in.)	249mm (9.80 in.)	249mm (9.80 in.)

Fig. 9 Ride height specification chart (FE1, std. suspension; FE3, sport suspension). 1990 Regal

MODEL		Z	D	J	K
SUSP	TIRE SIZE	CURB	CURB	CURB	CURB
FE1 (Except Z7S)	P195-75R14	84mm (3.30 in.)	105mm (4.13 in.)	249mm (9.80 in.)	249mm (9.80 in.)
FE1 (Except Z7S)	P195-70R15	79mm (3.11 in.)	100mm (3.94 in.)	249mm (9.80 in.)	249mm (9.80 in.)
FE1 (Except Z7S)	P195-75R15	84mm (3.30 in.)	104mm (4.09 in.)	249mm (9.80 in.)	249mm (9.80 in.)
FE1 (Except Z7S)	P205-65R15	81mm (3.19 in.)	102mm (4.02 in.)	249mm (9.80 in.)	249mm (9.80 in.)
FE3 (Except Z7S)	P215-60R16	71mm (2.80 in.)	93mm (3.66 in.)	249mm (9.80 in.)	249mm (9.80 in.)
FE3 (Except Z7S)	P215-65R15	74mm (2.91 in.)	95mm (3.74 in.)	249mm (9.80 in.)	249mm (9.80 in.)
FE1 (Z7S)	P195-70R15	81mm (3.19 in.)	101mm (3.98 in.)	249mm (9.80 in.)	249mm (9.80 in.)
FE1 (Z7S)	P205-65R15	83mm (3.27 in.)	104mm (4.09 in.)	249mm (9.80 in.)	249mm (9.80 in.)
FE3 (Z7S)	P215-60R16	73mm (2.87 in.)	94mm (3.70 in.)	249mm (9.80 in.)	249mm (9.80 in.)

Fig. 10 Ride height specification chart (FE1, std. suspension; FE3, sport suspension; Z7S, convertible). 1990 Cutlass Supreme

MODEL		Z	D	J	K
SUSP	TIRE SIZE	CURB	CURB	CURB	CURB
FE1	P195-70R15	80mm (3.15 in.)	110mm (4.33 in.)	251mm (9.88 in.)	257mm (10.12 in.)
FE1	P195-75R14	84mm (3.31 in.)	114mm (4.49 in.)	251mm (9.88 in.)	257mm (10.12 in.)
FE1	P205-65R15	81mm (3.19 in.)	113mm (4.45 in.)	251mm (9.88 in.)	257mm (10.12 in.)
FE3	P215-60R16	72mm (2.83 in.)	103mm (4.06 in.)	251mm (9.88 in.)	257mm (10.12 in.)
FE3	P245-502R16	74mm (2.91 in.)	107mm (4.21 in.)	251mm (9.88 in.)	257mm (10.12 in.)

Fig. 11 Ride height specification chart (FE1, std. suspension; FE3, sport suspension). 1990 Grand Prix

MODEL		Z	D	J	K
Susp	Tire Size	Curb	Curb	Curb	Curb
FE1	P195-75R14	83	103	249	249
F41	P195-70R15	79	99	249	249
F41	P215-60R16	71	92	249	249
F41	P205-65R15	81	101	249	249

Fig. 12 Ride height specification chart (FE1, std. suspension; F41, handling package). 1990 Lumina

2. Rotate cam to obtain proper toe angle, **Fig. 1.**
3. **Torque** cam nuts to 140 ft. lbs.

RIDE HEIGHT

Check ride height as follows, referring to **Figs. 2 and 3** for ride height location and **Figs. 4 through 12** for specification charts:

1. Ensure vehicle is on level ground.
2. Ensure tires are inflated to proper pressures.
3. Fuel tank should be full to obtain accurate readings.
4. Trunk should be empty except for spare tire and jack.
5. Bounce the car three times at the front and rear to normalize suspension.
6. Measure from lowest point on the ball joint housing to control arm bolt center line, "D" and "K" positions.
7. Measure from level floor to rocker panel at "Z" and "J" positions.

1985 Buick Riviera, Cadillac Eldorado & Seville & Oldsmobile Toronado

INDEX

FRONT WHEEL ALIGNMENT

Prior to checking caster and camber settings, check front end standing height, refer to "Standing Height, Adjust."

CAMBER

While holding cam bolt in position, loosen cam bolt nut, **Fig. 1.** Rotate cam bolt to obtain a change in camber equal to 1/2 the needed correction, then hold cam bolt in position and tighten cam bolt nut to 70 ft. lbs. To obtain the remaining 1/2 of needed correction, apply the same procedure to other cam bolt.

CASTER

Record camber reading, then hold front cam bolt and loosen nut, **Fig. 1.** Turn front cam bolt to obtain 1/4 of the desired caster

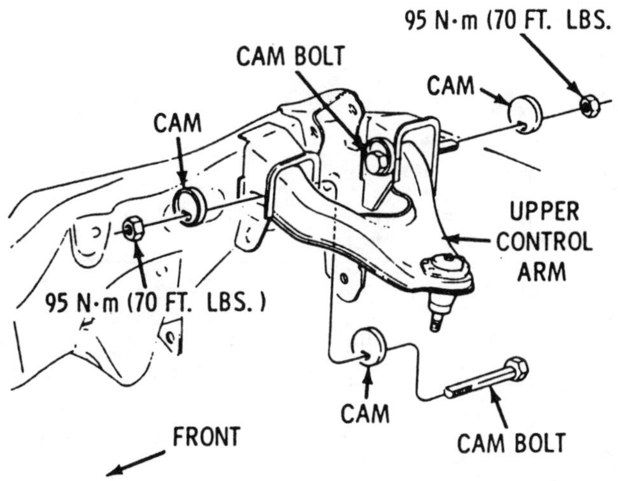

Fig. 1 Caster & camber adjustment

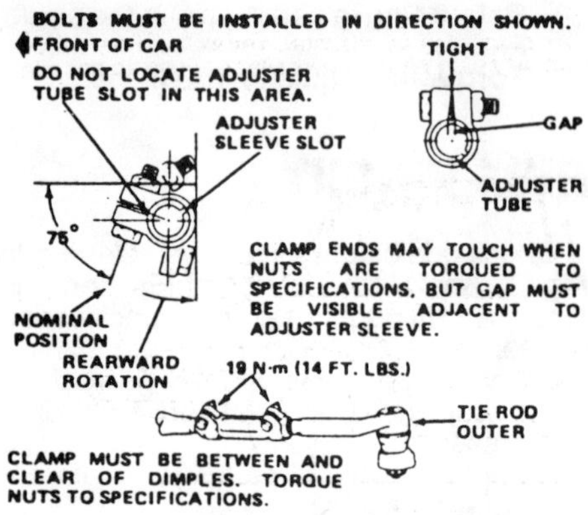

Fig. 2 Tie rod adjustment sleeve & clamp installation

change. At front cam bolt, a positive camber change produces a positive caster change, while a negative camber change produces a negative caster change. Hold cam bolt in position and tighten nut. Loosen rear cam bolt nut and rotate cam bolt to return camber to setting recorded earlier. Recheck caster and repeat procedure, if necessary. When adjustment is completed, hold rear cam bolt, then torque nut to 70 ft. lbs.

TOE-IN

Prior to checking or resetting toe-in, bounce the front end at least twice to allow vehicle to return to normal trim height.

1. Ensure wheel bearings are adjusted properly.
2. Using alignment equipment, with steering wheel in the straight ahead position, check toe-in.
3. If toe-in is not within specifications, loosen tie rod adjustment sleeve clamp bolts.
4. Turn tie rod adjustable sleeves an equal amount of turns but in opposite directions to bring toe-in within specifications.
5. Ensure tie rod end housings are at right angles to the steering arm, then position and torque tie rod adjustable sleeve clamps as shown in **Fig. 2.**

REAR WHEEL ALIGNMENT

CAMBER & TOE-IN

Camber and toe-in are checked at the same time following the procedure outlined below. Camber is factory set and is not adjustable. If camber deviates from specified settings, check for a bent control arm or frame, or for a improperly mounted hub and bearing assembly.

1. Check front and rear trim heights.
2. Check automatic level control system, if equipped, for proper operation.
3. Position vehicle on alignment rack as follows:
   a. Place masking tape on floor from wheel plate to rear of vehicle to use as a guide to square vehicle on alignment rack.
   b. Back rear of vehicle onto rack, positioning wheels on wheel plates.
   c. Place a ruler at same rib of tire at front and rear of vehicle, then measure distance from inside edge of ruler to edge of guide line, **Fig. 3.** When vehicle is square, measurement from guide line to ruler should be greater at rear tire by approximately 5/8 inch.
4. Attach alignment mirrors to rear wheels, then check camber and toe settings and compare to specifications. **When vehicle is backed onto alignment rack, toe-in and toe-out are reversed. That is, toe-in will be read as toe-out, while toe-out will be read as toe-in.**
5. As stated previously, camber is not adjustable. To adjust toe-in, loosen inner pivot bushing retaining nut and bolt. Move control arm rearward to increase toe-in, or toward front of vehicle to decrease toe-in. A pry bar can be used to facilitate adjustment. After adjustment is completed, torque bushing retaining nut to 70 ft. lbs.

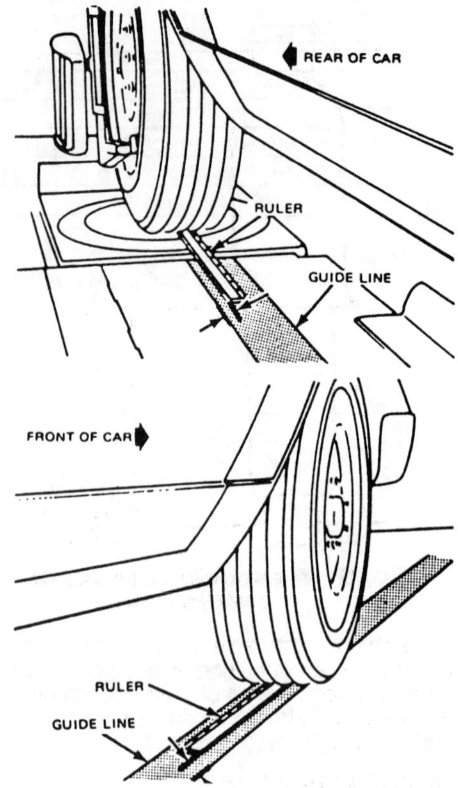

Fig. 3 Checking vehicle for squareness on alignment rack

STANDING HEIGHT, ADJUST

Before standing height adjustment is performed, vehicle should be on a flat, level surface, with front seat all the way back, fuel tank full and tires inflated to proper pressure. Air lines must be loosened at shock absorbers, if equipped with electronic level control, and system must be depressurized.

The standing height must be checked and adjusted if necessary before checking and adjusting front wheel alignment. The standing height is controlled by the adjustment setting of the torsion bar adjusting bolt, **Fig. 4.** Clockwise rotation of the bolt increases standing height: counterclockwise rotation decreases standing height.

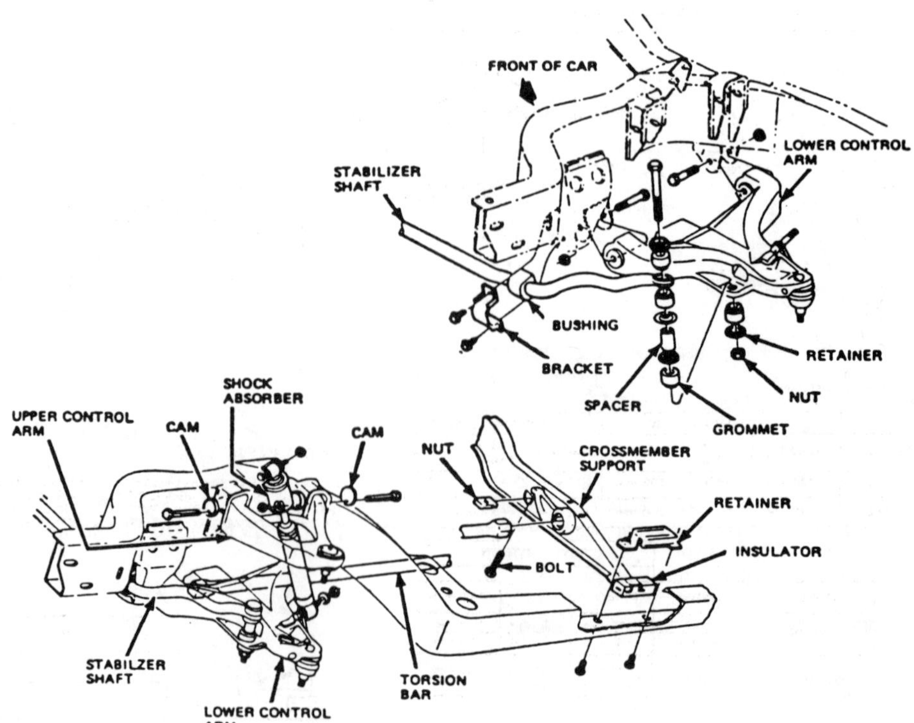

Fig. 4 Front suspension

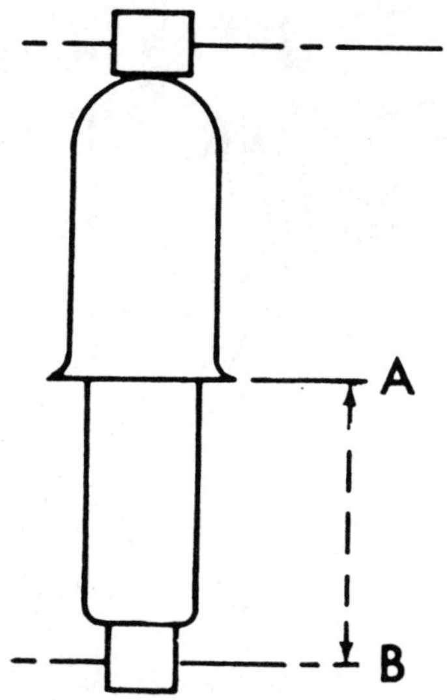

Fig. 5 Measuring standing height

To check vehicle height, measure from lower edge of front shock absorber dust tube (A) to center line of lower attachment (B), **Fig. 5.** This dimension between (A) and (B) should be 5½-6 inches.

RIDE HEIGHT

Refer to **Figs. 6 through 8** for ride height specifications and measurement locations.

TRIM HEIGHTS CHECKED WITH CORRECT TIRE PRESSURES, FUEL TANK FULL OR EQUIVALENT WEIGHT IN THE TRUNK. NO PASSENGERS OR ADDED WEIGHT IN CAR. FRONT SEAT IN REAR POSITION. TRUNK MUST BE EMPTY EXCEPT FOR SPARE TIRE AND JACK OR SIMULATED FUEL LOAD. MEASURE FROM KNOWN LEVEL FLOOR TO ROCKER PANEL WITH STEERING WHEEL IN THE CENTERED POSITION.

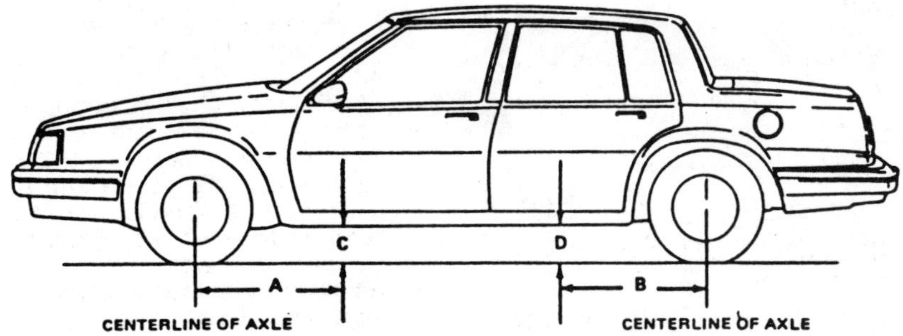

C & D DIMENSION

LIFT CENTER OF FRONT BUMPER UP APPROXIMATELY 38mm (1-½") AND LET VEHICLE SETTLE GENTLY. REPEAT TWO MORE TIMES, THEN MEASURE "C" DIMENSION. PUSH CENTER OF BUMPER DOWN 38mm (1-½") AND LET VEHICLE SETTLE GENTLY. REPEAT TWO MORE TIMES, THEN MEASURE "C" DIMENSION. THE "C" DIMENSION IS AN AVERAGE OF THE HIGH AND LOW MEASUREMENTS. REPEAT PROCEDURE ON THE REAR BUMPER FOR THE "D" DIMENSION.

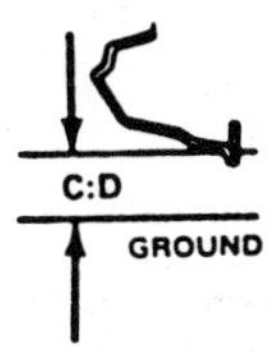

SERIES	DIMENSION A	DIMENSION B	DIMENSION C	DIMENSION D
E	620 (24-13/32")	475 (18-45/64")	246 (9-11/16")	243 (9-9/16")

Fig. 6 Ride height measurements. 1985 Riviera

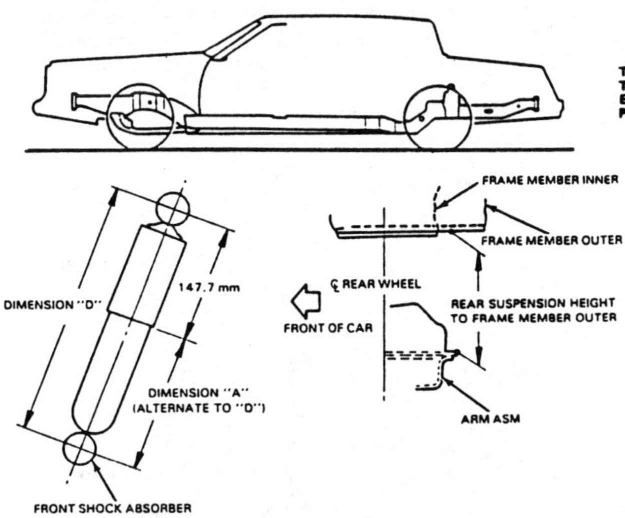

ENGINE	FRONT-DIMENSION "D" MINIMUM	CURB TARGET	MAXIMUM	FRONT DIMENSION "A" MINIMUM	CURB TARGET	MAXIMUM
GAS	279 mm	289 mm	299 mm	132 mm	142 mm	152 mm
DIESEL	280 mm	290 mm	300 mm	133 mm	143 mm	153 mm

SERIES	ENGINE	REAR DIMENSION MINIMUM	CURB TARGET	MAXIMUM
ELDORADO	GAS	123 mm	133 mm	143 mm
	DIESEL	123 mm	133 mm	143 mm
SEVILLE	GAS	123 mm	133 mm	143 mm
	DIESEL	124 mm	134 mm	144 mm

Fig. 7 Ride height measurements. 1985 Eldorado & Seville

TRIM HEIGHTS CHECKED WITH CORRECT TIRE PRESSURES, FUEL TANK FULL OR EQUIVALENT WEIGHT IN THE TRUNK. NO PASSENGERS OR ADDED WEIGHT IN CAR. FRONT SEAT IN REAR POSITION. TRUNK MUST BE EMPTY EXCEPT FOR SPARE TIRE AND JACK OR SIMULATED FUEL LOAD. MEASURE FROM KNOWN LEVEL FLOOR TO ROCKER PANEL WITH STEERING WHEEL IN THE CENTERED POSITION.

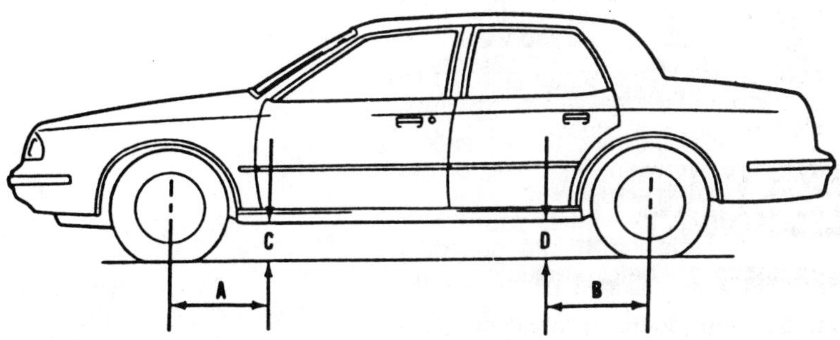

C & D DIMENSION

LIFT CENTER OF FRONT BUMPER UP APPROXIMATELY 38mm (1-½") AND LET VEHICLE SETTLE GENTLY. REPEAT TWO MORE TIMES, THEN MEASURE "C" DIMENSION. PUSH CENTER OF BUMPER DOWN 38mm (1-½") AND LET VEHICLE SETTLE GENTLY. REPEAT TWO MORE TIMES, THEN MEASURE "C" DIMENSION. THE "C" DIMENSION IS AN AVERAGE OF THE HIGH AND LOW MEASUREMENTS. REPEAT PROCEDURE ON THE REAR BUMPER FOR THE "D" DIMENSION.

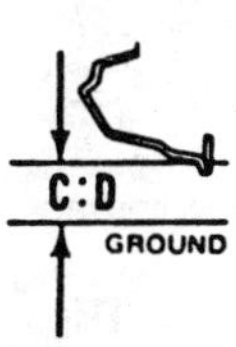

MAXIMUM VARIATION SIDE TO SIDE AND FRONT TO REAR IS 19mm (3/4")				
VIEW	A	B	C	D
TORONADO	580mm (22-27/32")	563mm (22-11/64")	245mm (9-5/8")	245mm (9-5/8")

Fig. 8 Ride height measurements. 1985 Toronado

1986–90 Buick Riviera, Cadillac Eldorado & Seville & Oldsmobile Toronado & Trofeo & 1988–90 Buick Reatta

INDEX

FRONT WHEEL ALIGNMENT

Steering and vibration complaints are not always the result of improper alignment. Another possibility is tire "lead" due to worn or improperly manufactured tires. "Lead" is the vehicles's deviation from a straight path on a level road without pressure on the steering wheel.

Before making any adjustment affecting wheel alignment, make the following checks to ensure correct alignment readings and alignment adjustments.

1. Check all tires for proper inflation pressures and ensure all tires have approximately equal tread wear.
2. Check hub and bearing assemblies for excessive wear, correcting as necessary.
3. Check ball joints and tie rod ends. If they are excessively loose, correct before making adjustment.
4. Check run out of wheels and tires.
5. Check vehicle trim height, correcting as necessary before adjusting alignment.
6. Check strut dampers for proper operation.
7. Check control arms for loose bushings.
8. Check stabilizer bar for loose or missing parts.

Consideration must also be given to excess loads, such as tool boxes or sample cases. If these items are normally carried in the vehicle, they should remain in the vehicle during alignment adjustments. Consideration should be given to condition of equipment being used to adjust alignment. Be sure to follow equipment manufacturer's instructions. Regardless of equipment being used, vehicle must be on level surface, both fore and aft and sideways.

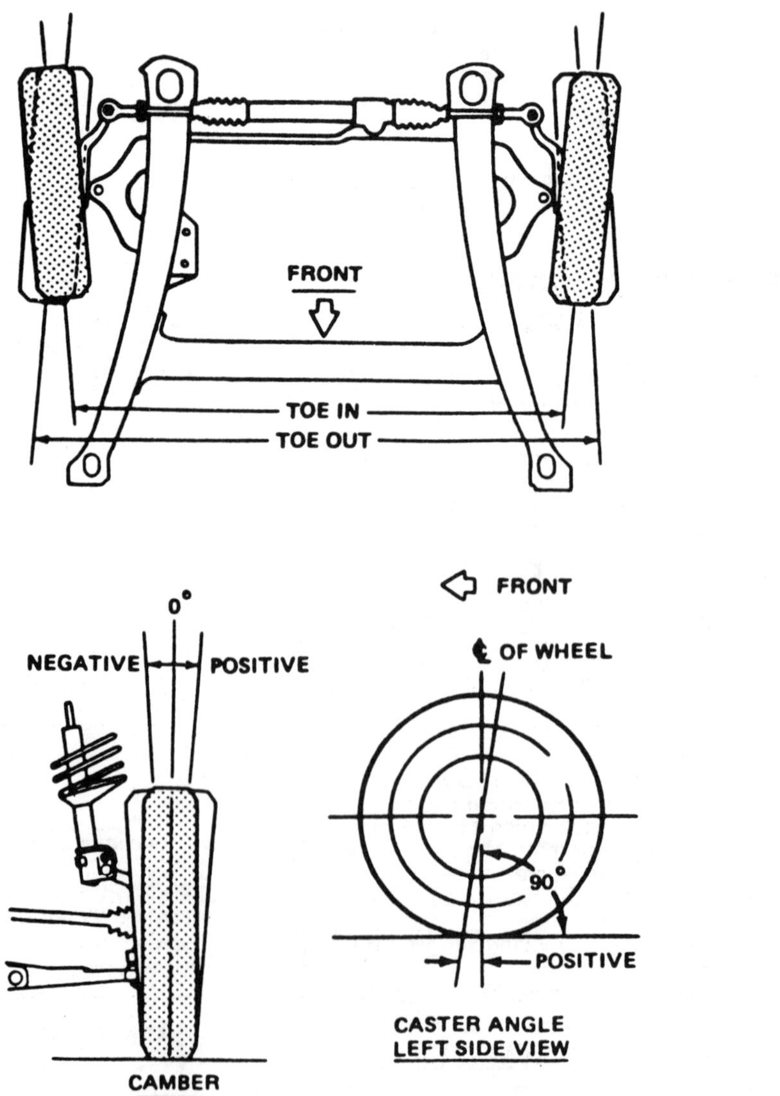

Fig. 1 Front end alignment angles

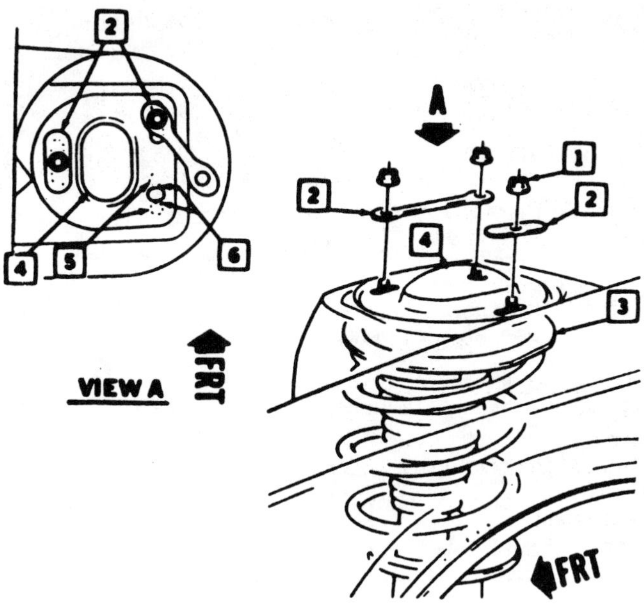

1. NUT 24 N·m (18 LBS. FT.)
2. WASHER
3. STRUT ASSY.
4. COVER
5. DRILL 8.731mm (11/32") HOLES
6. FILE HERE THEN PAINT WITH PRIMER

Fig. 2 Adjusting front wheel caster

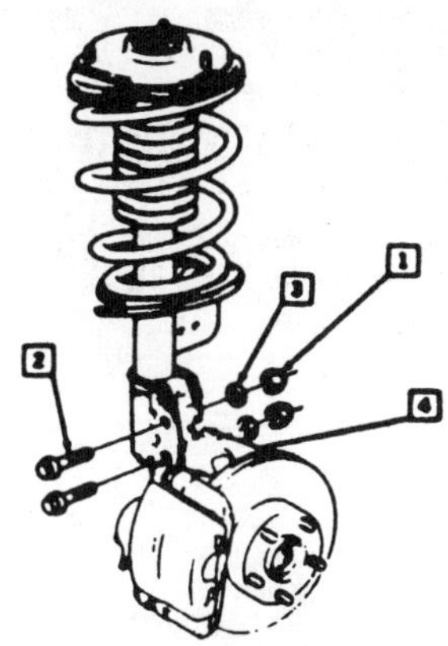

1. NUT 185 N·m (140 LBS. FT.)
2. BOLT
3. WASHER
4. CAMBER ADJUSTMENT BOLT

Fig. 3 Adjusting front wheel camber

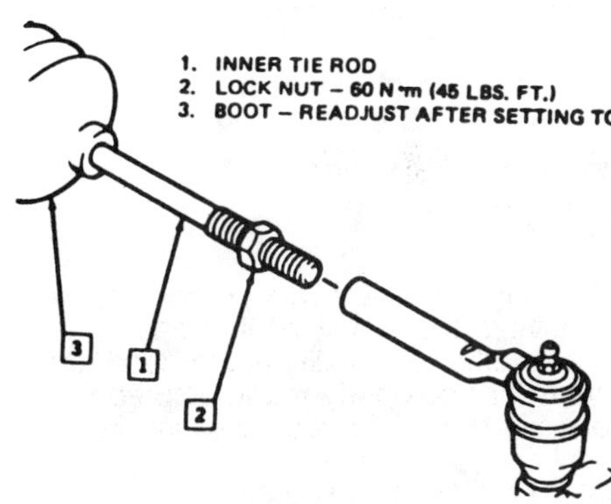

1. INNER TIE ROD
2. LOCK NUT – 60 N·m (45 LBS. FT.)
3. BOOT – READJUST AFTER SETTING TOE

Fig. 4 Adjusting front wheel toe

MEASURING FRONT ALIGNMENT ANGLES

Install alignment equipment following equipment manufacturer's instructions. Measure alignment angles, **Fig. 1**, and record the readings. If adjustments are necessary, they must be made in order; caster first, camber second and toe third. Jounce front and rear bumpers 3 times to normalize suspension prior to measuring angles.

CASTER ADJUSTMENT

1. Loosen but do not remove, 2 of 3 top strut attaching nuts covering slotted mounting holes, **Fig. 2**, then remove nut over remaining oval strut mounting hole.
2. Move washer away from oval strut mounting hole.
3. Lift front of vehicle by body to separate strut from inner wheel house.
4. Drill two 11/32 inch holes at front and rear of oval strut mounting hole, then file away excess metal.
5. Lower front of vehicle and install washer and nut.
6. Adjust caster by moving top of strut forward or rearward.
7. **Torque** top strut attaching nuts to 18 ft. lbs.

CAMBER ADJUSTMENT

1. Loosen both strut to knuckle bolts just enough to allow movement.
2. Adjust camber by turning camber adjustment bolt, **Fig. 3.**
3. **Torque** strut to knuckle nuts to 140 ft. lbs.
4. **Torque** camber adjustment bolt to 7 ft. lbs.

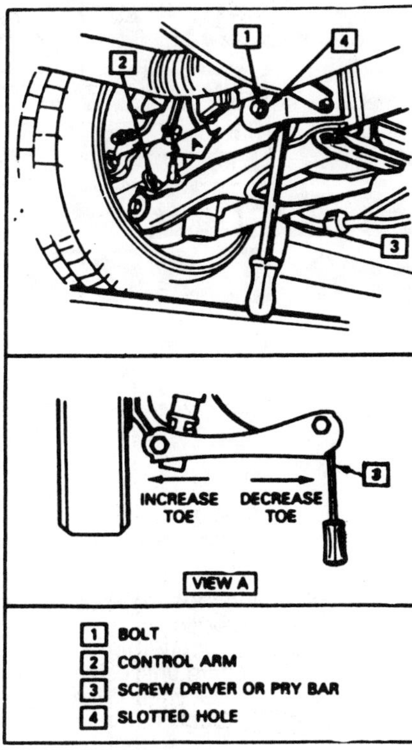

1 BOLT
2 CONTROL ARM
3 SCREW DRIVER OR PRY BAR
4 SLOTTED HOLE

Fig. 5 Adjusting rear toe

TOE ADJUSTMENT

1. Loosen locknut on inner tie rods, **Fig. 4.**
2. Adjust toe by turning inner tie rods.
3. Adjust boots so that they are not twisted.
4. **Torque** locknuts to 45 ft. lbs.

REAR WHEEL ALIGNMENT

Rear wheel toe is the only adjustable angle. If camber is not within specifications, inspect for worn or damaged rear suspension components.

Before checking rear trim height or measuring rear alignment angles, the following procedure should be performed to ensure the rear air-adjustable struts are filled with residual pressure only.

1. Place a weight of at least 300 lbs. in vehicle trunk.
2. Turn ignition on.
3. Turn ignition off and remove weight from trunk.
4. Wait 30 seconds for Electronic Level Control system to exhaust.
5. Roll car forward or backward at least 25 ft. to eliminate the effects of tire camber side thrust.

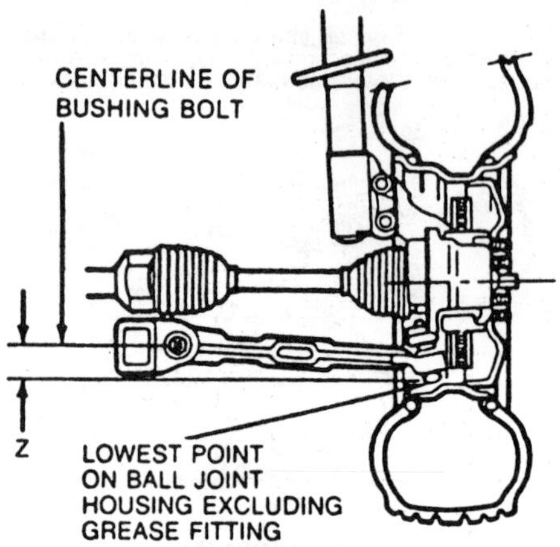

Fig. 6 Front ride height measurement

6. Jounce front and rear bumpers 3 times to normalize suspension prior to measuring angles.

TOE ADJUSTMENT

1. Loosen front and rear inside control arm mounting bolts, **Fig. 5.**
2. Insert a screwdriver or pry bar between the inside rear control arm mounting bolt and rear support assembly, then move control arm to set toe to specifications.
3. **Torque** inside front control arm mounting bolt to 66 ft. lbs.

RIDE HEIGHT

Check ride height as follows:

1. Using your arms, lift center of front bumper up approximately $1^1/_2$ inches, then gently release and let vehicle settle. Repeat twice.
2. Using your arms, lift center of rear bumper up approximately $1^1/_2$ inches, then gently release and let vehicle settle. Repeat twice.
3. Measure dimension Z as shown in **Fig. 6** and dimension D as shown in **Fig. 7.**
4. Dimension Z should be 62 mm ±6 mm on 1986-87 Eldorado and Seville models, 53 mm ±6 mm on 1988-90 Eldorado and Seville models, 51 mm ±6 mm om 1986-88 Riviera models, 46 mm ±6 mm on 1989-90 Riviera models, 42 mm ±6 mm on 1988 Reatta models, 43 mm ±6 mm on

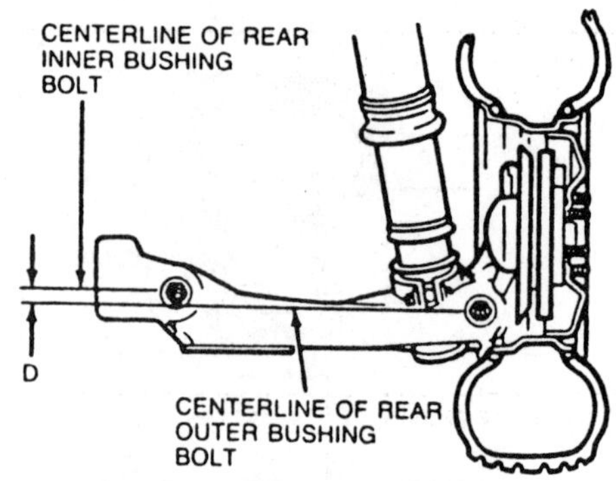

Fig. 7 Rear ride height measurement

1989-90 Reatta models and 49 mm ±6 mm on Toronado and Trofeo models.

5. Dimension D should be 42 mm ±7 mm on 1986-87 Eldorado and Seville models, 31 mm ±7 mm on 1988-90 Eldorado models, 35 mm ±7 mm on 1988-90 Seville models, 25 mm ±7 mm om 1986-88 Riviera models and all Reatta, Toronado and Trofeo models and 30 mm ±7 mm on 1989-90 Riviera models.

Buick Skyhawk, Skylark (1986–90), Somerset & Somerset Regal, Cadillac Cimarron, Chevrolet Cavalier, Oldsmobile Calais, Cutlass Calais & Firenza & Pontiac Grand Am & 2000

INDEX

Page No.

CAMBER

Toe setting is the only adjustment normally required. However, in special circumstances, such as damage due to road hazard or collision, camber may be adjusted by modifying the strut assembly.

1. Secure bottom of strut assembly in a suitable vise.
2. Enlarge bottom holes in outer flanges with a round file until holes in outer flanges match slots in inner flanges, **Fig. 1.**
3. Connect strut to steering knuckle and install bolts finger tight.
4. Grasp top of tire firmly, then move tire inboard or outboard until correct camber reading is obtained. Tighten retaining bolts enough to secure camber setting.
5. Remove wheel and tire and tighten strut to steering knuckle retaining bolts. On 1985 models, torque strut to steering knuckle retaining bolts to 140 ft. lbs. On 1986-90 models, torque strut to steering knuckle retaining bolts to 135 ft. lbs.

TOE-OUT

Toe-out is controlled by tie rod position. Adjustment is made by loosening the clamp bolts at the steering knuckle end of the tie rods and rotating the rods to obtain proper toe setting. After correct toe setting is obtained, tighten clamp bolts. **Torque** clamp bolts to 41 ft. lbs.

RIDE HEIGHT

Refer to **Fig. 2,** for ride height specifications and measurement locations.

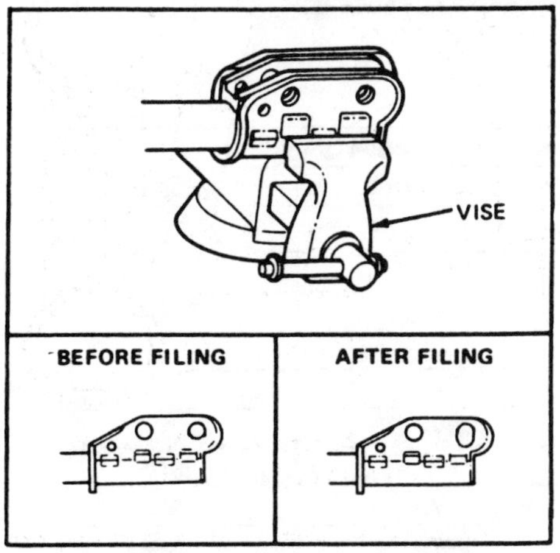

Fig. 1 Modifying strut bracket to adjust camber

TRIM HEIGHTS CHECKED WITH CORRECT TIRE PRESSURES, FUEL TANK FULL OR EQUIVALENT WEIGHT IN THE TRUNK. NO PASSENGERS OR ADDED WEIGHT IN CAR. FRONT SEAT IN REAR POSITION. TRUNK MUST BE EMPTY EXCEPT FOR SPARE TIRE AND JACK OR SIMULATED FUEL LOAD. MEASURE FROM KNOWN LEVEL FLOOR TO ROCKER PANEL WITH STEERING WHEEL IN THE CENTERED POSITION.

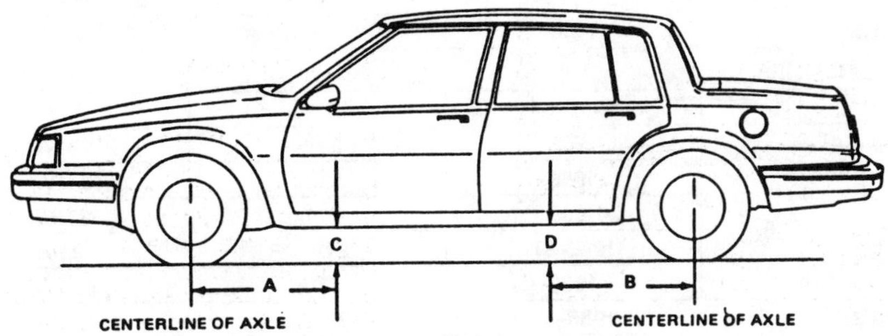

C & D DIMENSION

LIFT CENTER OF FRONT BUMPER UP APPROXIMATELY 38mm (1-½") AND LET VEHICLE SETTLE GENTLY. REPEAT TWO MORE TIMES, THEN MEASURE "C" DIMENSION. PUSH CENTER OF BUMPER DOWN 38mm (1-½") AND LET VEHICLE SETTLE GENTLY. REPEAT TWO MORE TIMES, THEN MEASURE "C" DIMENSION. THE "C" DIMENSION IS AN AVERAGE OF THE HIGH AND LOW MEASUREMENTS. REPEAT PROCEDURE ON THE REAR BUMPER FOR THE "D" DIMENSION.

Fig. 2 Ride height measurement locations & specifications (Part 1 of 3)

VEHICLE RIDE HEIGHT SPECIFICATION

Make & Model	Year	Riding Height Measurements In Inches ①			
		Dimension A	Dimension B	Dimension C	Dimension D
BUICK					
Skyhawk	1985–87	21 59/64	21 13/16	9 31/64	9 3/8
Skyhawk ②	1988–90	21 7/8	21 3/4	9 1/2	9 11/32
Skyhawk ③	1988–90	21 7/8	21 3/4	9 23/32	9 29/32
Skylark	1986–87	21 59/64	22 3/64	9 13/32	9 7/32
	1988–90	21 7/8	21 13/32	9 3/8	9 3/16
Somerset	1985–87	21 59/64	22 3/64	9 13/32	9 7/32
CADILLAC					
Cimarron	1985–87	21 7/8	22 3/4	9 27/64	9 19/64
Cimarron	1988	21 7/8	22 3/4	9 15/16	9 17/32
CHEVROLET					
Cavalier ②	1985	21 7/8	22 13/16	④	⑤
	1986	21 7/8	22 13/16	⑧	⑨
	1987	21 7/8	22 13/16	⑫	⑬
	1988–89	21 7/8	21 3/4	9 7/32	9 3/16
	1990	21 7/8	21 3/4	9 1/3	9 1/3
Cavalier ③	1985	21 7/8	22 13/16	⑥	⑦
	1986–87	21 7/8	22 13/16	⑩	⑪
	1988–89	21 7/8	21 3/4	9 17/32	9 5/8
	1990	21 7/8	21 3/4	9 19/32	9 3/4

Fig. 2 Ride height measurement locations & specifications (Part 2 of 3)

Continued

VEHICLE RIDE HEIGHT SPECIFICATIONS—Continued

Make & Model	Year	Riding Height Measurements In Inches ①			
		Dimension A	Dimension B	Dimension C	Dimension D
OLDSMOBILE					
Calais	1985–87	21 15/16	22 5/16	9 1/2	9 3/8
Cutlass Calais	1988	21 7/8	21 13/32	⑭	⑮
	1989	21 7/8	21 13/32	⑯	⑰
	1990	21 7/8	21 13/32	⑱	⑲
Firenza ②	1985–87	21 15/16	21 7/8	9 1/2	9 3/8
	1988	21 7/8	21 3/4	9 1/2	9 11/32
Firenza ③	1985–87	21 15/16	21 7/8	9 11/16	9 13/16
	1988	21 7/8	21 3/4	9 23/32	9 29/32
PONTIAC					
Grand Am	1985–87	21 15/16	22 5/16	9 9/16	9 39/64
	1988	21 7/8	21 13/32	⑱	⑲
	1989–90	21 7/8	21 13/32	⑳	㉑
Sunbird ②	1985–86	21 15/16	21 7/8	9 1/2	9 3/8
	1987	21 15/16	22 7/8	㉒	㉓
	1988–89	21 7/8	21 3/4	9 7/32	9 3/32
	1990	21 7/8	21 3/4	9 1/3	9 1/3
Sunbird ③	1985–86	21 15/16	21 7/8	9 11/16	9 13/16
	1987	21 15/16	22 7/8	⑩	⑪
	1988–89	21 7/8	21 3/4	9 17/32	9 5/8
	1990	21 7/8	21 3/4	9 19/32	9 3/4

①—Plus or minus 3/4 inch.
②—Except station wagon.
③—Station wagon.
④—With P175-80 R13 tires, 9 19/64 inches; with P195-80 R13 tires, 9 7/32 inches; with P215-60 R14 tires, 9 1/4 inches.
⑤—With P175-80 R13 tires, 9 11/64 inches; with P195-80 R13 tires, 9 1/16 inches; with P215-60 R14 tires, 9 1/8 inches.
⑥—With P175-80 R13 tires, 9 1/2 inches; with P195-80 R13 tires, 9 13/32 inches.
⑦—With P175-80 R13 tires, 9 41/64 inches; with P195-80 R13 tires, 9 9/16 inches.
⑧—With P175-80 R13 tires, 9 9/32 inches; with P195-80 R13 tires, 9 7/32 inches; with P215-60 R14 tires, 9 11/64 inches.
⑨—With P175-80 R13 tires, 9 1/8 inches; with P195-80 R13 tires, 9 1/16 inches; with P215-60 R14 tires, 9 3/32 inches.
⑩—With P175-80 R13 tires, 9 9/16 inches; with P195-80 R13 tires, 9 1/2 inches.
⑪—With P175-80 R13 tires, 9 11/16 inches; with P195-80 R13 tires, 9 39/64 inches.
⑫—With P175-80 R13 tires, 9 1/4 inches; with P195-80 R13 tires, 9 11/64 inches; with P215-60 R14 tires, 9 11/64 inches.
⑬—With P175-80 R13 tires, 9 1/8 inches; with P195-80 R13 tires, 9 1/16 inches; with P215-60 R14 tires, 9 1/16 inches.
⑭—Models less P215/60-R14 tires, 9 1/2 inches; models with P215/60-R14 tires, to VIN No. JM202625, 9 1/2 inches, after VIN No. JM202625, 9 7/32 inches.
⑮—Models less P215/60-R14 tires, 9 3/8 inches; models with P215/60-R14 tires, to VIN No. JM202625, 9 3/8 inches, after VIN No. JM202625, 9 3/32 inches.
⑯—Models less heavy duty suspension, 9 1/2 inches; models with heavy duty suspension, P215/60-R14 tires, 9 3/32 inches, P205/55-R16 tires, 9 3/32 inches.
⑰—Models less heavy duty suspension, 9 3/8 inches; models with heavy duty suspension, P215/60-R14 tires, 9 7/32 inches, P205/55-R16 tires, 9 3/32 inches.
⑱—Models less heavy duty suspension, 9 9/16 inches; models with heavy duty suspension, P215/60-R14 tires, 9 3/32 inches, P205/55-R16 tires, 9 9/16 inches.
⑲—Models less heavy duty suspension, 9 1/2 inches; models with heavy duty suspension, P215/60-R14 tires, 9 7/32 inches, P205/55-R16 tires, 9 1/2 inches.
⑳—Models less P215/60-R14 tires, 9 9/16 inches; models with P215/60-R14 tires, to VIN No. JC610485, 9 9/16 inches, after VIN No. JC610458, 9 21/64 inches.
㉑—Models less P215/60-R14 tires, 9 5/8 inches; models with P215/60-R14 tires, to VIN No. JC610485, 9 5/8 inches, after VIN No. JC610485, 9 11/32 inches.
㉒—Models less heavy duty suspension, 9 9/16 inches; models with heavy duty suspension, P215/60-R14 tires, 9 21/64 inches, P205/55-R16 tires, 9 9/16 inches.
㉓—Models less heavy duty suspension, 9 5/8 inches; models with heavy duty suspension, P215/60-R14 tires, 11 7/32 inches, P205/55-R16 tires, 9 5/8 inches.
㉔—With P175-80 R13 tires, 9 9/32 inches; with P195-80 R13 tires, 9 11/64 inches; with P215-60 R14 tires, 9 11/64 inches.
㉕—With P175-80 R13 tires, 9 1/8 inches; with P195-80 R13 tires, 9 1/16 inches; with P215-60 R14 tires, 9 3/32 inches.

Fig. 2 Ride height measurement locations & specifications (Part 3 of 3)

Chevrolet Beretta & Corsica

INDEX

FRONT WHEEL ALIGNMENT

Toe setting is the only adjustment normally required. However, in special circumstances, such as damage due to road hazard or collision, camber may be adjusted by modifying the strut assembly. Caster is not adjustable.

CAMBER

1. Secure bottom of strut assembly in a suitable vise.
2. Enlarge bottom holes in outer flanges with a round file until holes in outer flanges match slots in inner flanges, **Fig. 1.**
3. Connect strut to steering knuckle and install bolts finger tight.
4. Grasp top of tire firmly, then move tire inboard or outboard until correct camber reading is obtained. Tighten retaining bolts enough to secure camber setting.

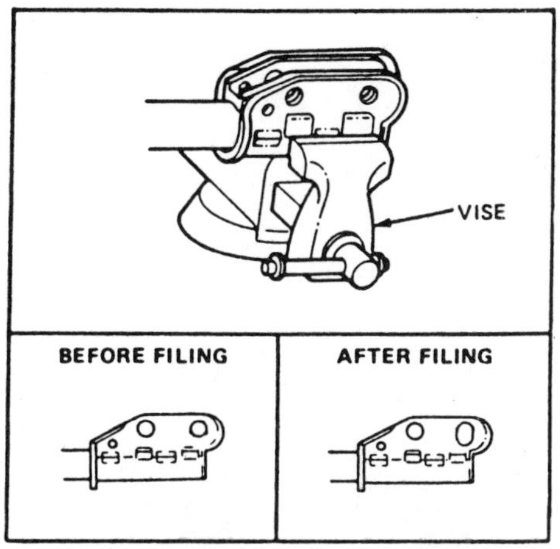

Fig. 1 Modifying strut bracket to adjust camber

5. Remove wheel and tire and torque strut to steering knuckle retaining bolts to 133 ft. lbs.

TOE

Toe is controlled by tie rod position. Adjustment is made by loosening the clamp bolts at the steering knuckle end of the tie rods and rotating the adjuster to obtain proper toe setting. After correct toe setting is obtained, torque clamp bolts to 35 ft. lbs.

REAR WHEEL ALIGNMENT

There are no adjustments to be made on this rear suspension. If alignment is outside specifications, check for broken or bent parts and replace as necessary.

RIDE HEIGHT

Refer to **Fig. 2,** for ride height specifications and measurement locations. Check ride height as follows:

1. Ensure vehicle is on level ground.
2. Ensure tires are inflated to proper pressures.

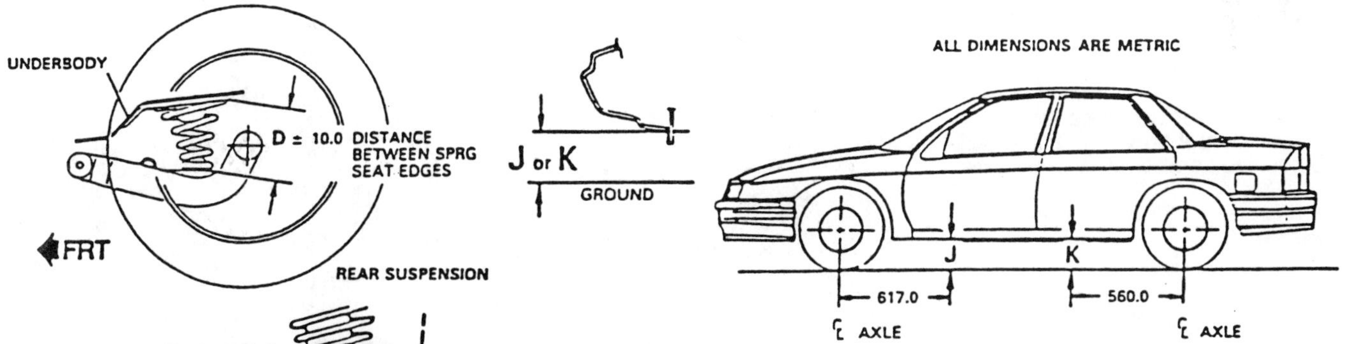

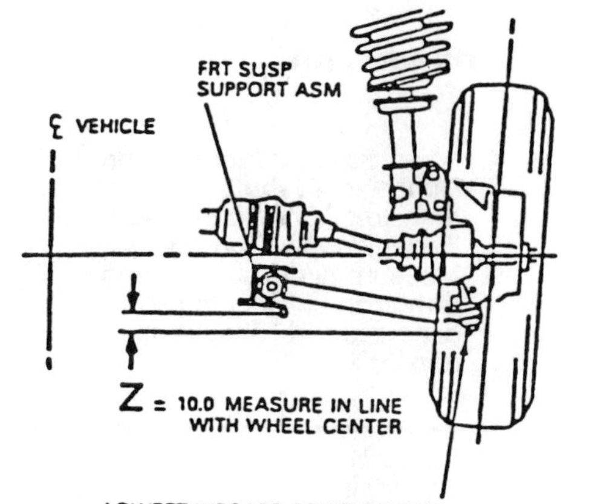

	Z		D		J		K	
TIRE SIZE	SHIPPED	CURB	SHIPPED	CURB	SHIPPED	CURB	SHIPPED	CURB
P195/70R14	21.6	19.5	189.6	176.4	248.9	244.0	264.3	253.0
P185/80R13	24.5	22.4	192.6	179.4	248.9	244.0	264.3	253.0
P195/70R14	20.6	19.0	186.7	176.4	247.9	244.0	261.9	253.0
P185/80R13	24.4	22.8	190.7	179.9	248.0	244.0	262.3	253.0
P195/70R14	16.2	14.6	182.0	171.7	242.9	239.0	256.9	248.0
P205/60R15	7.1	5.9	172.4	162.4	232.6	229.0	246.6	238.0
P205/60R15	6.5	5.2	172.3	162.3	232.6	229.0	246.6	238.0

Fig. 2 Ride height measurements & specifications

3. Fuel tank should be full to obtain accurate readings.
4. Trunk should be empty except for spare tire and jack.
5. Bounce the car three times at the front and rear to normalize suspension.
6. Measure from lowest point on the ball joint housing to control arm bolt center line, "D" and "Z" positions.
7. Measure from level floor to rocker panel at "J" and "K" positions.

Chevrolet Chevette & Pontiac 1000

INDEX

Page No.

CASTER

Caster angle is adjusted by rearranging washers located at both ends of the upper control arm, **Fig. 1.** A kit consisting of two washers, one of 3 mm thickness and one of 9 mm thickness, must be used when adjusting caster angle.

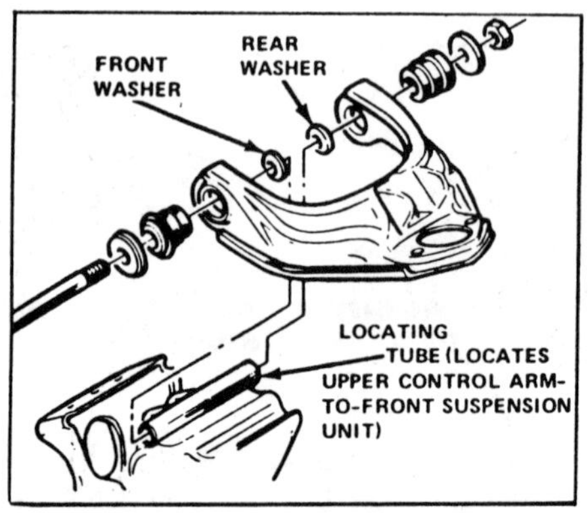

Fig. 1 Caster adjustment

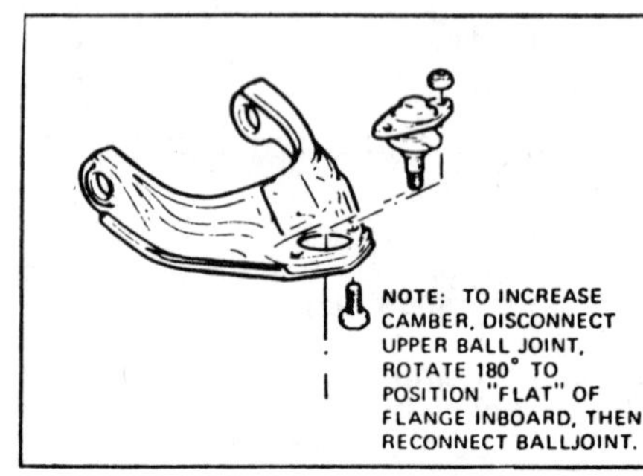

Fig. 2 Camber adjustment

CAMBER

Camber is adjusted by removing the upper ball joint, rotating it 1/2 turn and reinstalling it with flat of upper flange on inboard side of control arm, **Fig. 2.** This will increase positive camber by approximately 1°.

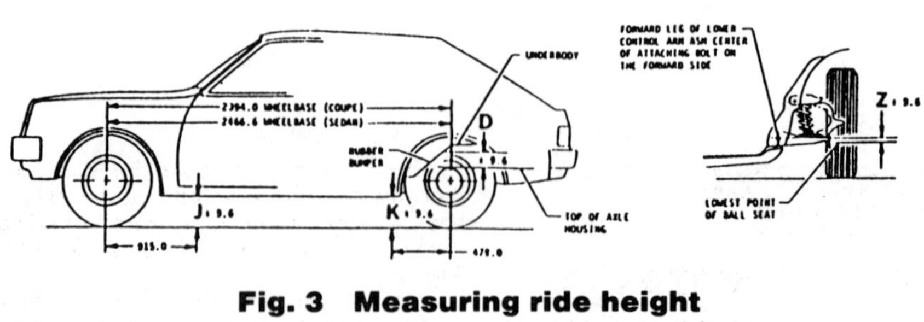

Fig. 3 Measuring ride height

TOE-IN

To adjust toe-in, loosen tie rod jam nuts at tie rod ends and loosen clamp at rubber bellows. Turn each tie rod to increase or decrease its length until proper toe-in is obtained. Torque jam nut to 50 ft. lbs.

RIDE HEIGHT

Check ride height as follows:

1. Using arms, lift front of vehicle approximately two inches.
2. Gently remove hands and let vehicle settle.
3. Repeat steps 1 and 2.
4. Measure dimensions Z and J, **Fig. 3.** Dimension Z should be 60 mm and dimension J should be 230 mm.
5. Repeat steps 1 through 3.
6. Measure dimensions D and K, **Fig. 3.** Dimension D should be 139 mm and dimension K should measure 226 mm.

Chevrolet Nova & Geo Prizm

INDEX

Page No.

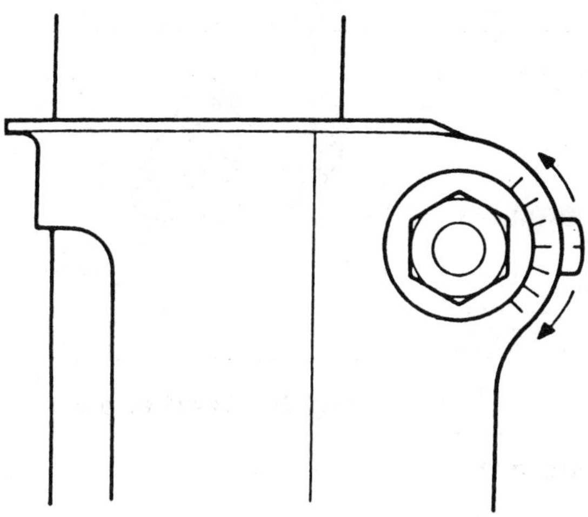

Fig. 1 Front camber

Fig. 2 Front camber adjustment

FRONT WHEEL ALIGNMENT

CASTER & CAMBER

Camber is adjusted by first loosening the upper and lower strut to knuckle nuts and bolts, then rotating cam to obtain correct specification.

Caster cannot be adjusted. Should caster be found out of specification, locate cause first. If components are damaged, bent, loose, dented or worn, they should be replaced. To prevent an incorrect reading of camber or caster, jounce the bumper three times before checking, **Fig. 1.**

After adjustment, tighten nuts to specification.

TOE-IN

Toe-in is adjusted by changing tie rod length. Loosen boot clamps and slide from the boot assembly. Loosen right and left tie rod end locknuts, then turn right and left tie rods to align toe-in to specification. Right and left tie rods must be equal in length, **Fig. 2.**

After adjustment, install boot clamps, tighten nuts to specification and ensure rack boots are not twisted.

Torque strut-to-knuckle nut to specifications, and tie rod locking nut to 35 ft. lbs.

REAR WHEEL ALIGNMENT

TOE

Toe-in can be adjusted by rotating the cam located on the rear lower control arm. Loosen bolt and rotate nut to give the correct specification. Measure distance between left and right wheel discs and center line of adjustment cam, **Fig. 3.**

CAMBER

Rear camber cannot be adjusted. Should camber be found out of specification, locate cause first. If components are damaged, bent loose, dented or worn, they should be replaced. To prevent an incorrect reading of camber, jounce the bumper three times before checking. If a tie rod or tie rod end is replaced, check toe and steering angle with turning radius gages. If steering angle is not correct, check right and left tie rods for equal length. If tie rod length is change to correct steering, reinspect toe.

RIDE HEIGHT

Refer to **Figs. 4 and 5** for vehicle height and **Fig. 6,** for specification chart.

REAR TOE ADJUSTMENT MECHANISM

Rear suspension arm No. 2 is equipped with toe adjustment cams.

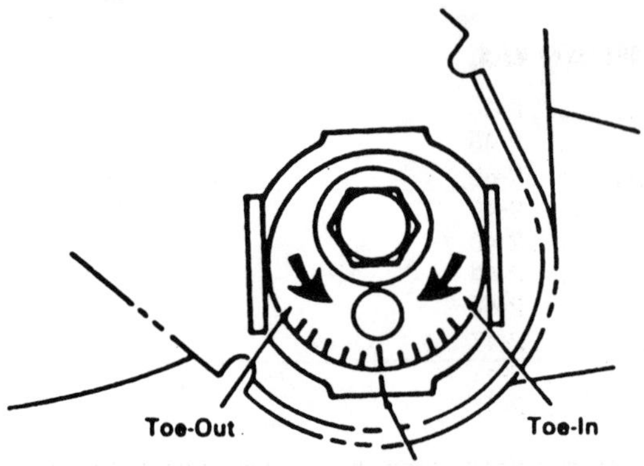

Toe Adjustment

(1) Measure the distance and find the difference between the left and right disc wheels and the centerline of the adjustment cam. (If the difference is not within standard specifications, check each part for deformation.)

(2) Turn each cam an equal amount in the opposite direction. (Toe changes about 2 mm for each turn of the cam on each side.)

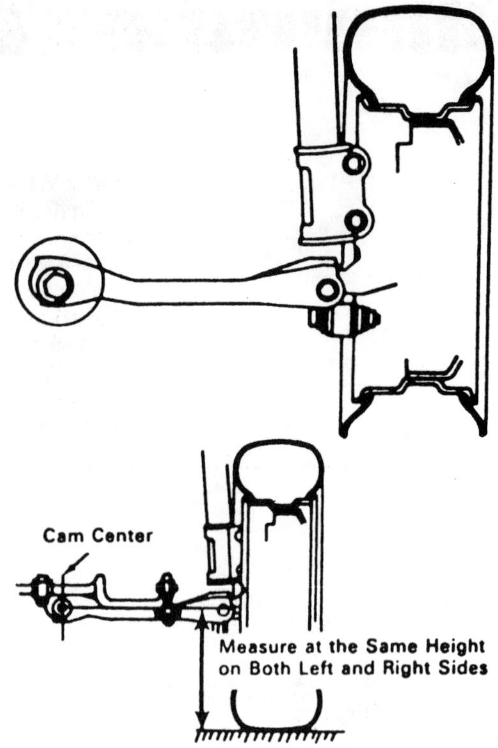

Fig. 3 Rear toe adjustment

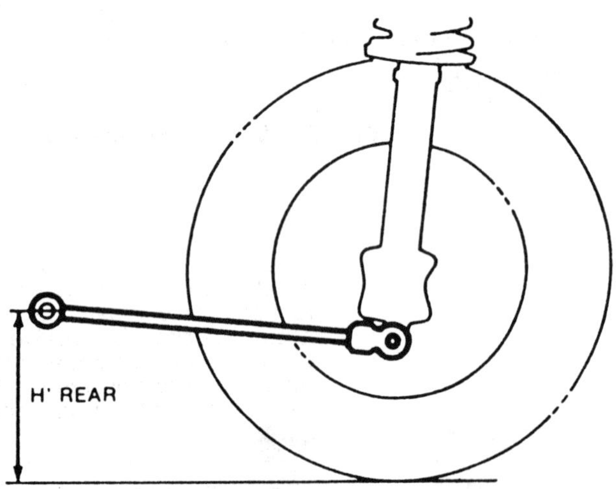

Fig. 4 Ride height. Front

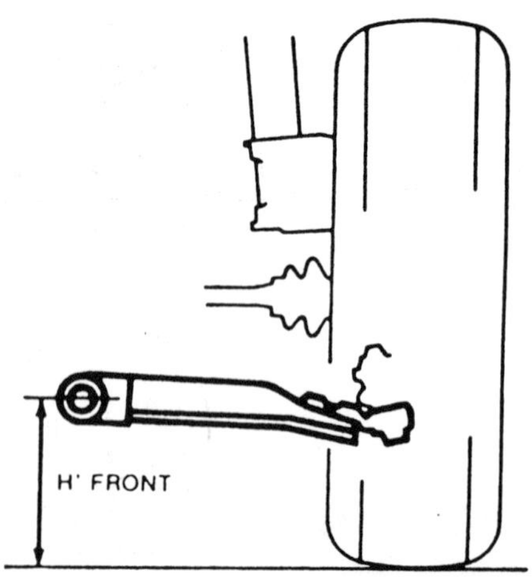

Fig. 5 Ride height. Rear

TIRE SIZE		FRONT	REAR
4A-LC	P175/70 R 13	192 mm	258 mm
	P155/80 R 13	191 mm	257 mm
4A-GE	P175/70 HR 13	183 mm	259 mm

Fig. 6 Ride height measurement chart

Chevrolet & Geo Spectrum

INDEX

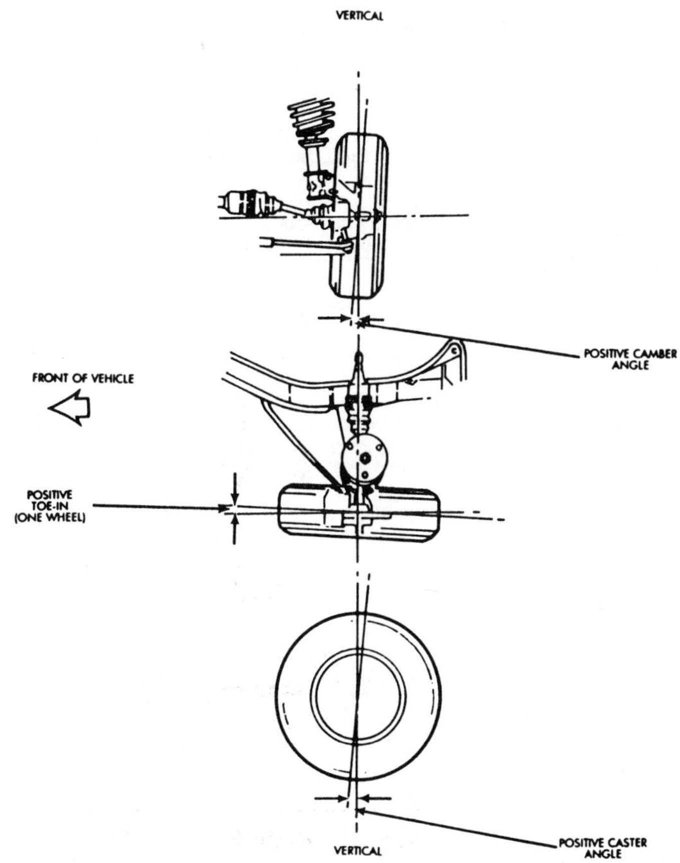

Fig. 1 Suspension geometry

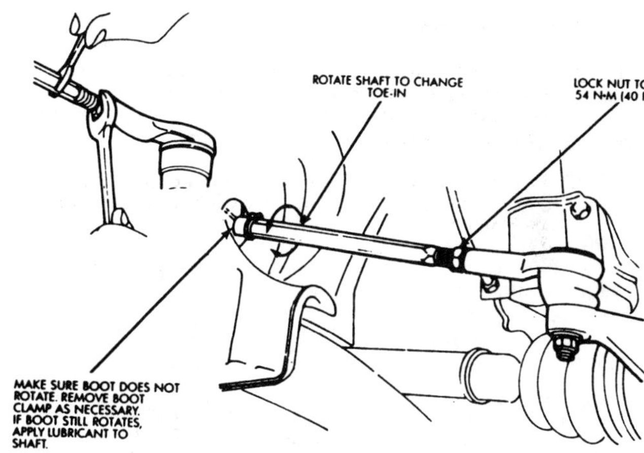

Fig. 2 Adjusting toe-in

DESCRIPTION

Wheel alignment, is the angular relationship between the wheels, suspension attaching parts and ground. The angle of the knuckle away from the vertical, pointing in or out of wheels, tilt of the wheels from vertical (when viewed from front of vehicle) and tilt of suspension members from vertical (when viewed from side of vehicle), all of these are involved in proper alignment, **Fig. 1.**

CASTER

Caster is tilting of the front steering axis either forward or backward from the vertical (when viewed from side of vehicle). A backward tilt is positive (+) and a forward tilt is negative (−). On short and long arm type suspensions you cannot see a caster angle without using a special instrument, but if you look straight down from the top of the upper control arm to the ground you would find that ball joints do not line up (fore and aft) when a caster angle other than 0° is present.

CAMBER

Camber is the tilting of front and rear wheels from the vertical when viewed from front of vehicle. When wheels tilt outward at top, camber is positive (+). When wheels tilt inward, camber is negative (MI). Amount of tilt is measured in degrees from the vertical and this is camber angle.

TOE

Toe is the turning in or out of wheels. The actual amount of toe is normally .078 inch (2 mm). The purpose of toe is to ensure parallel rolling of wheels. Excessive toe-in or toe-out may increase tire wear. Toe also serves to offset small deflections of the suspension which occurs when vehicle is moving.

PRELIMINARY CHECK

Steering and vibration problems are not always the result of alignment. An additional problem to be checked is tire lead due to worn or improperly manufactured tires. "Lead" is the deviation of the vehicle from a straight path on a level road without hand pressure on the steering wheel.

To insure correct alignment readings and alignment specifications, the following checks and inspections should be made.

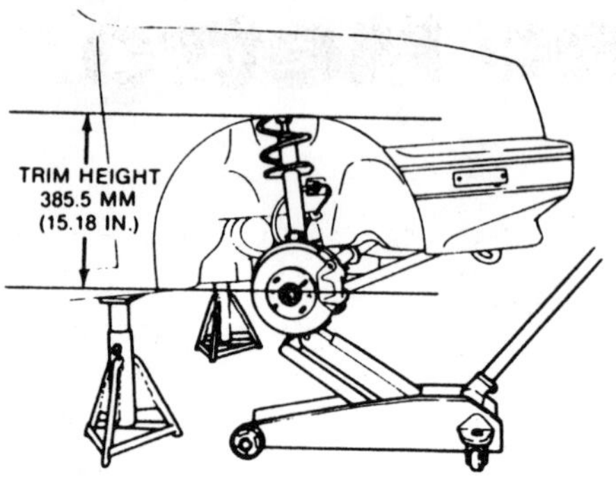

Fig. 3 Adjusting front trim height

1. Check tire for proper inflation and thread wear.
2. Check for loose ball joints and tie rod ends. If excessive looseness is noted, replace defective parts before adjusting toe.
3. Check for wheel and tire assembly run-out.
4. Check trim heights. If not within specifications, correct before adjusting toe.
5. Check for loose control arms.
6. Check for loose or missing stabilizer bar components.
7. Consideration must be given to excess loads, such as tool boxes, etc. If excess load is normally carried in vehicle, it should remain during alignment checks.

ADJUSTMENTS

CASTER & CAMBER

Caster and camber angles are pre-set and cannot be adjusted. If caster and/or camber are out of specifications, the cause must be found. If worn or damaged suspension components are the cause, they should be replaced. If problem is body related, the body should be repaired as needed.

TOE-IN

Toe-in is adjusted by altering tie rod length. To perform adjustment, loosen small end boot clamps and slide from the boot. Loosen right and left tie rod end locknuts, then turn both tie rods, **Fig. 2,** the same amount to bring toe-in within specifications. Ensure right and left tie rods are equal in length, then tighten locknuts. When adjustment is complete, ensure boots are not twisted.

STEERING ANGLE

When a tie rod or tie rod end is replaced, check toe and steering angle using suitable turning radius gauges. If steering angle is not $37^{2}/_{3}°$ inside and $32^{1}/_{2}°$ outside, check right and left tie rods for equal length. **If tie rod length is changed to correct steering angle, reinspect toe in.**

TRIM HEIGHT

Front

1. Loosen wheel lug nuts.
2. Raise and support vehicle using suitable jack stands.
3. Remove wheel and tire assemblies.
4. Loosen control arm to body, then tension rod attaching bolts.

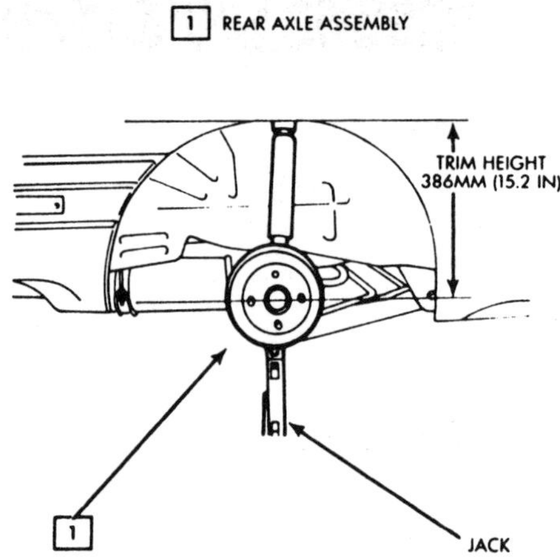

Fig. 4 Adjusting rear trim height

5. Using a suitable jack, jack up the lower part of control arm to 15.18 inches, **Fig. 3,** tighten control arm to body attaching bolts to 41 ft. lbs. and tension rod bolts to 80 ft. lbs.

Rear

1. Loosen wheel lug nuts.
2. Raise and support vehicle using suitable jack stands.
3. Remove wheel and tire assemblies.
4. Loosen rear axle attaching bolts.
5. Using a suitable jack, jack up lower part of rear axle and set rear trim height to 15.2 inches, **Fig. 4.,** then tighten rear axle bolts to 72 ft. lbs.

Chevrolet Sprint & Geo Metro

INDEX

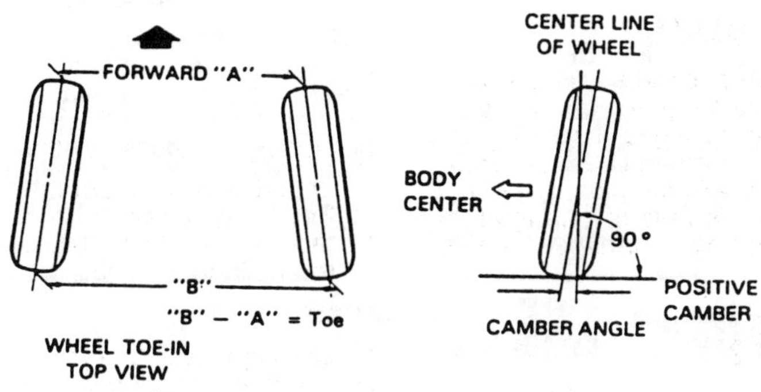

Fig. 1 Suspension geometry

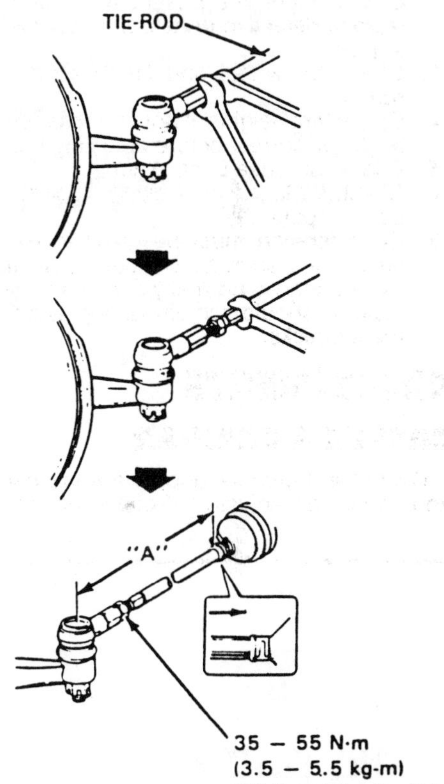

Fig. 2 Front wheel toe adjustment

DESCRIPTION

Wheel alignment, is the angular relationship between the wheels, suspension attaching parts and ground. The angle of the knuckle away from the vertical, pointing in or out of wheels, tilt of the wheels from vertical (when viewed from front of vehicle) and tilt of suspension members from vertical (when viewed from side of vehicle), all of these are involved in proper alignment, **Fig. 1.**

CASTER

Caster is tilting of the front steering axis either forward or backward from the vertical (when viewed from side of vehicle). A backward tilt is positive (+) and a forward tilt is negative (--). On short and long arm type suspensions you cannot see a caster angle without using a special instrument, but if you look straight down from the top of the upper control arm to the ground you would find that ball joints do not line up (fore and aft) when a caster angle other than 0° is present.

CAMBER

Camber is the tilting of front and rear wheels from the vertical when viewed from front of vehicle. When wheels tilt outward at top, camber is positive (+). When wheels tilt inward, camber is negative (MI). Amount of tilt is measured in degrees from the vertical and this is camber angle.

TOE

Toe is the turning in or out of wheels. The purpose of toe is to ensure parallel rolling of wheels. Excessive toe-in or toe-out may increase tire wear. Toe also serves to offset small deflections of the suspension which occurs when vehicle is moving.

PRELIMINARY CHECK

Steering and vibration problems are not always the result of alignment. An additional problem to be checked is tire lead due to worn or improperly manufactured tires. "Lead" is the deviation of the vehicle from a straight path on a level road without hand pressure on the steering wheel.

To insure correct alignment readings and alignment specifications, the following checks and inspections should be made.

1. Check tire for proper inflation and thread wear.
2. Check for loose ball joints and tie rod ends. If excessive looseness is noted, replace defective parts before adjusting toe.
3. Check for wheel and tire assembly run-out.
4. Check trim heights. If not within specifications, correct before adjusting toe.
5. Check for loose control arms.
6. Check for loose or missing stabilizer bar components.
7. Consideration must be given to excess loads, such as tool boxes, etc. If excess load is normally carried in vehicle, it should remain during alignment checks.

FRONT WHEEL ALIGNMENT

CAMBER & CASTER ADJUSTMENT

1. Position vehicle on an alignment fixture following manufacturer's instructions and check caster and camber angles. Bumper should be bounced three times before inspection, to prevent incorrect reading.
2. Camber and caster cannot be adjusted. Should either be found out of specification, locate the cause first.
3. If improper alignment is caused by damaged, worn or loose suspension parts, they should be replaced. If vehicle body or chassis is damaged, it should be repaired.

TOE ADJUSTMENT

1. Loosen right and left tie rod end lock nuts, **Fig. 2.**
2. Apply grease between tie rods and rack boots.
3. Turn right and left tie rods by the same amount to align toe to specification. **Right and left tie rods should become equal in length.**
4. After adjustment, tighten locknuts and ensure rack boots are not twisted.

STEERING ANGLE

When a tie rod or tie rod end is replaced, check toe and steering angle with turning radius gages. If steering angle is not correct, check right and left tie rods for equal length.

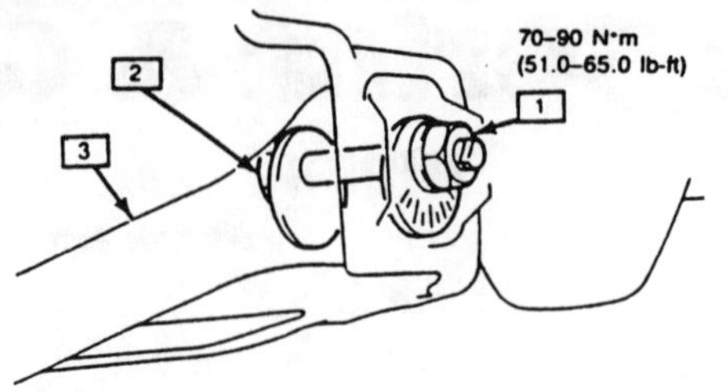

1 INSIDE NUT
2 INSIDE BOLT (CAM BOLT)
3 CONTROL ROD

Fig. 3 Rear wheel toe adjustment. 1989–90

REAR WHEEL ALIGNMENT

1989–90

Camber

Camber cannot be adjusted. Should camber be found out of specification, locate the cause. If improper alignment is caused by damaged, worn or loose suspension parts, they should be replaced. If vehicle body or chassis is damaged, it should be repaired.

Toe

The rear wheel toe is adjusted by the inner control arm cam bolt, **Fig. 3.** If toe is found to out of specification, loosen left and right inner control arm cam bolt nuts, then rotate cam bolts by equal amounts to correct toe setting. After completing adjustment, torque cam bolt nuts to 51 to 65 ft. lbs.

Pontiac Fiero

INDEX

Page No.

FRONT WHEEL ALIGNMENT

1985–87

Camber

1. Remove upper ball joint.
2. Camber may be increased approximately 1° by rotating 1/2 turn and reinstalling with flat of upper flange on inboard side of control arm, **Fig. 1.**

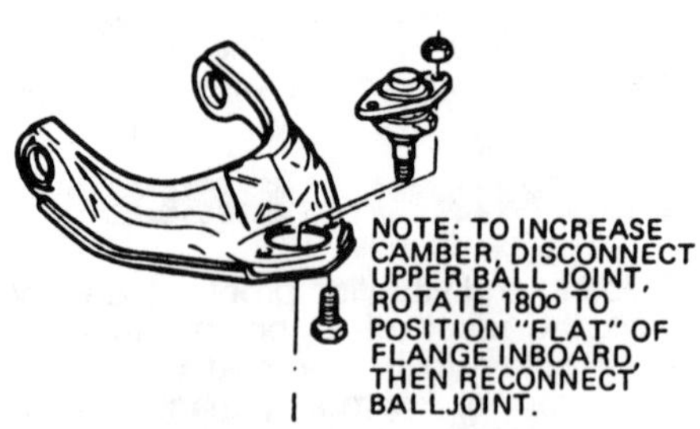

Fig. 1 Front camber adjustment. 1985–87

Caster Adjustment

1. Remove upper control arm.
2. Adjust caster by installing washers between legs of upper control arm, **Fig. 2. When caster is adjusted, two washers totalling .472 inch (12 mm) must be installed, with one at each end of locating tube.**

Toe-In Adjustment

1. Loosen jam nuts on toe link rod, then rotate rods until toe-in is within specifications, **Fig. 3.**
2. **Torque** jam nuts to 47 ft. lbs. **Use care not to twist or damage rubber boots.**

1988

Camber & Caster

1. Jounce front and rear bumpers three times.
2. Loosen upper control arm shaft bolts and move upper control arm as shown, **Fig. 4,** to adjust caster and camber. **If upper control arm shaft bolts are removed, paddle nut assembly must be replaced.**
3. **Torque** upper control arm shaft bolts to 52 ft. lbs. plus an additional 1/4 turn.

Toe-In

This adjustment is the same as on 1985-87 vehicles.

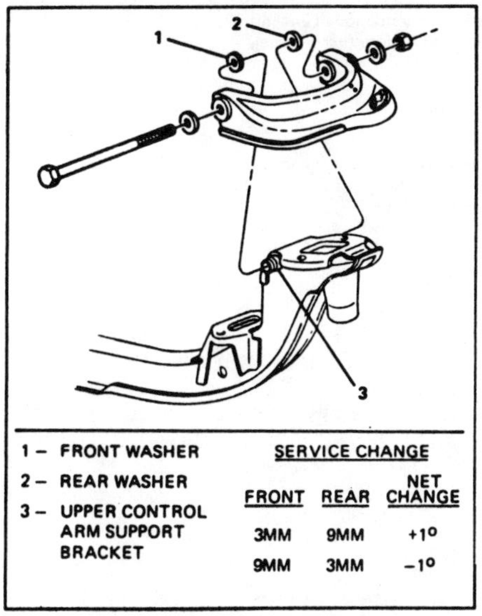

Fig. 2 Front caster adjustment. 1985–87

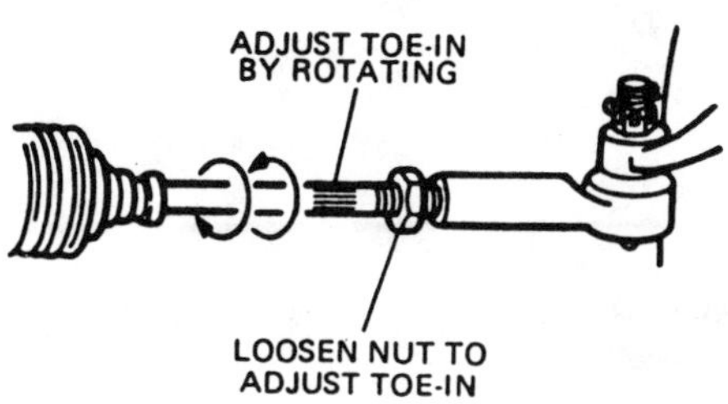

Fig. 3 Toe-in adjustment

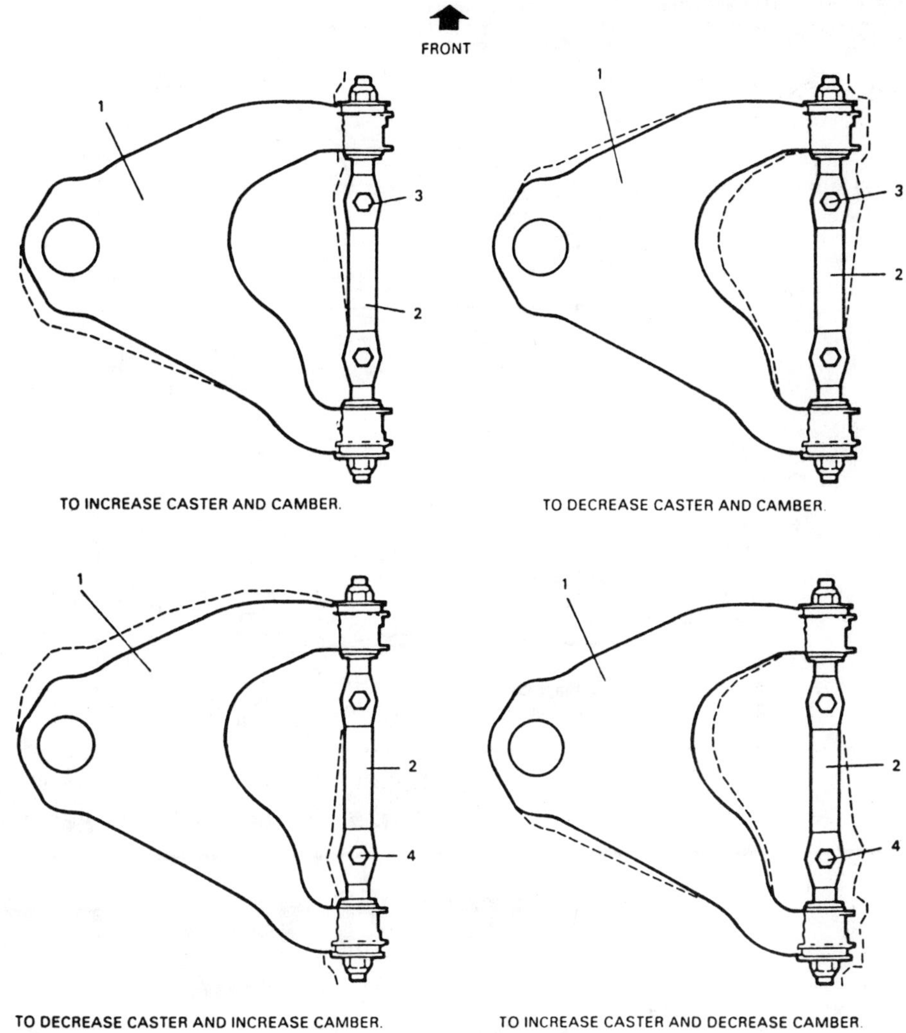

1 UPPER CONTROL ARM
2 U.C.A. SHAFT
3 FRONT BOLT
4 REAR BOLT

Fig. 4 Front camber & caster adjustment. 1988

REAR WHEEL ALIGNMENT

CAMBER

1. Loosen both strut-to-knuckle attaching bolts sufficiently to allow movement between strut and knuckle.
2. Move top of tire inboard or outboard until camber is within specifications, then torque both strut-to-knuckle bolts to 140 ft. lbs. **If complete torque cannot be applied to bolts due to inaccessibility, tighten bolts just enough to hold camber position, then remove wheel and tire and apply final torque.**

TOE-IN

1985–87

1. Loosen jam nuts on toe link rod, then rotate rods until toe-in is within specifications, **Fig. 3.**
2. **Torque** jam nuts to 47 ft. lbs. **Use care not to twist or damage rubber boots.**

1988

1. Loosen adjuster tube clamps.
2. Rotate adjuster tube to change toe-in angle.
3. **Torque** clamps to 47 ft. lbs.

RIDE HEIGHT

Check ride height as follows:

1. Using arms, lift front of vehicle approximately two inches.
2. Gently remove hands and let vehicle settle.
3. Repeat steps 1 and 2.
4. Measure dimensions Z and J, **Fig. 5.** Dimension Z should be .2716 inch and dimension J should be 7.0-7.7 inches.
5. Repeat steps 1 through 3.
6. Measure dimensions D and K, **Fig. 5.** Dimension D should be .984 inch on vehicles with P195/70R14 tires, .866 inch with P185/75R14 tires and .511 inch with 215/60HR15 tires. Dimension K should measure 6.9-7.7 inches.

Trim heights checked with correct tire pressures, fuel tank full or equivalent weight in the trunk. No passengers or added weight in car. Trunk must be empty except for spare tire and jack or simulated fuel load. Measure from known level floor to rocker panel with steering wheel in the center position.

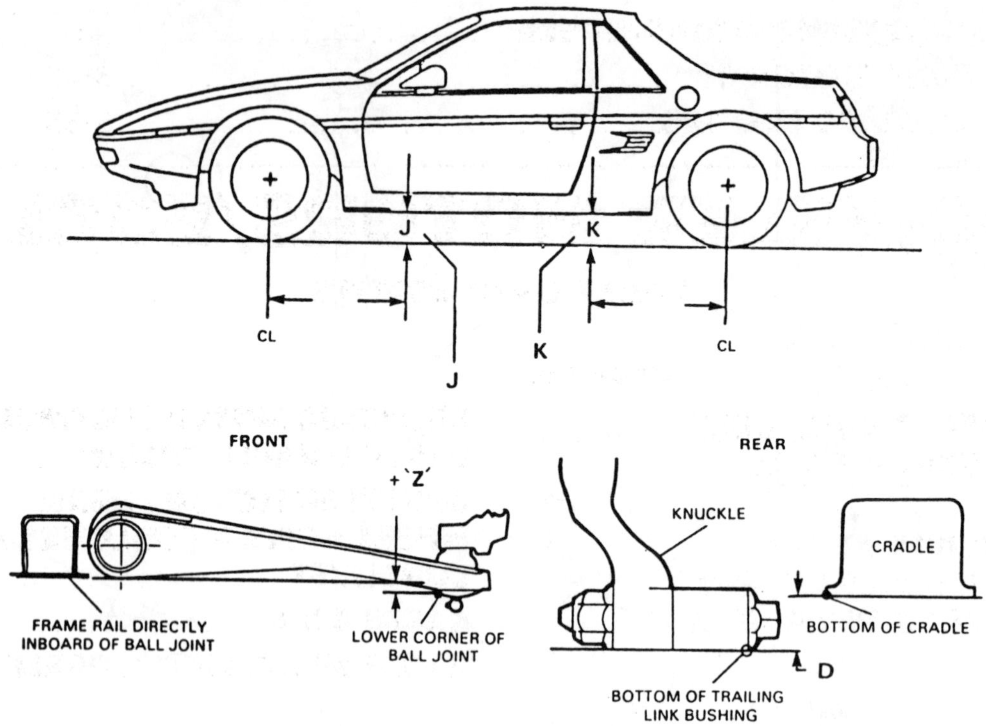

Fig. 5 Measuring ride height

Pontiac Lemans

INDEX

VEHICLE TRIM HEIGHT

Vehicle trim height is measured from the rocker panel to ground. From rocker panel front to ground should be 7.7 inches. From rocker panel rear to ground, trim height should be 7.5 inches. If trim height is not within limits, check control arms, struts and ball joints for wear and damage.

FRONT WHEEL ALIGNMENT

CASTER & CAMBER

Front caster and camber angles are not adjustable. If caster or camber is found to be out of specifications, inspect suspension and related body components for damage and repair or replace as necessary.

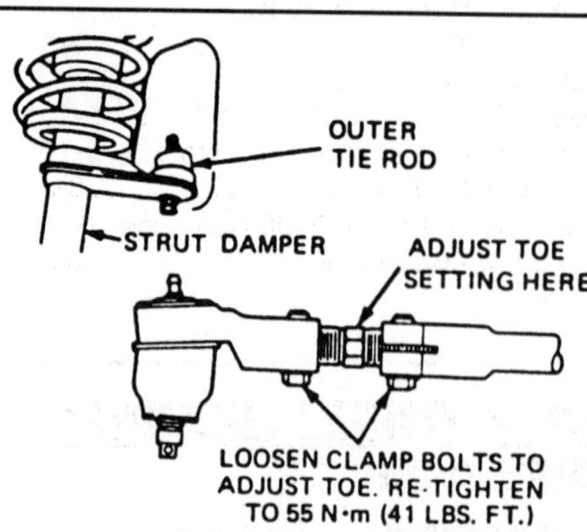

Fig. 1 Front toe adjustment

TOE

1. Loosen right and left tie rod end clamps, **Fig. 1.**
2. Turn right and left tie rod adjusters as necessary to bring toe within specifications.
3. Ensure right and left tie rods are equal in length, then **torque** clamp bolts to 41 ft. lbs.

REAR WHEEL ALIGNMENT

CAMBER

Rear camber angles are not adjustable. If camber is found to be out of specifications, inspect suspension and related body components for damage and repair or replace as necessary.

TOE

Rear toe is not adjustable. If toe is found to be out of specifications, inspect rear axle assembly and wheel spindle for damage and correct as necessary.

FRONT SUSPENSION & STEERING

INDEX

CHRYSLER MOTORS/EAGLE

TABLE OF CONTENTS

Chrysler Motors Colt, Colt Vista & Conquest & Eagle Summit

INDEX

WHEEL BEARINGS ADJUST

CONQUEST

After installing the bearings and wheel hub, tighten the nut to 14.5 ft. lbs. to seat all parts then loosen the nut. Tighten the nut to 4 ft. lbs. and after installing the cap, install the cotter pin. If the holes do not align after shifting the position of the installed cap, loosen the nut until a flute on the nut aligns with cotter pin hole in the spindle. Do not back off the nut more than 15 degrees.

HUB & WHEEL BEARING REPLACE

CONQUEST

1. Raise and support front of vehicle and remove wheel.
2. Remove bolts securing caliper support to steering knuckle, remove caliper and support as an assembly leaving hose connected and suspend caliper from front spring, taking care not to damage brake hose.
3. Remove hub cap, cotter pin, and pin retainer.
4. Remove hub nut, washer and outer bearing, then remove hub and rotor assembly from spindle.
5. Scribe matching marks between hub and rotor. On models less intercooler, remove bolts securing brake rotor to hub, then separate hub and rotor. On models with intercooler, remove retaining clips and bolts, then separate rotor from hub.
6. Remove oil seal and inner bearing, then drive bearing outer races from hub using suitable drift.
7. Clean hub and bearings with suitable solvent and blow dry with compressed air. Inspect components not-

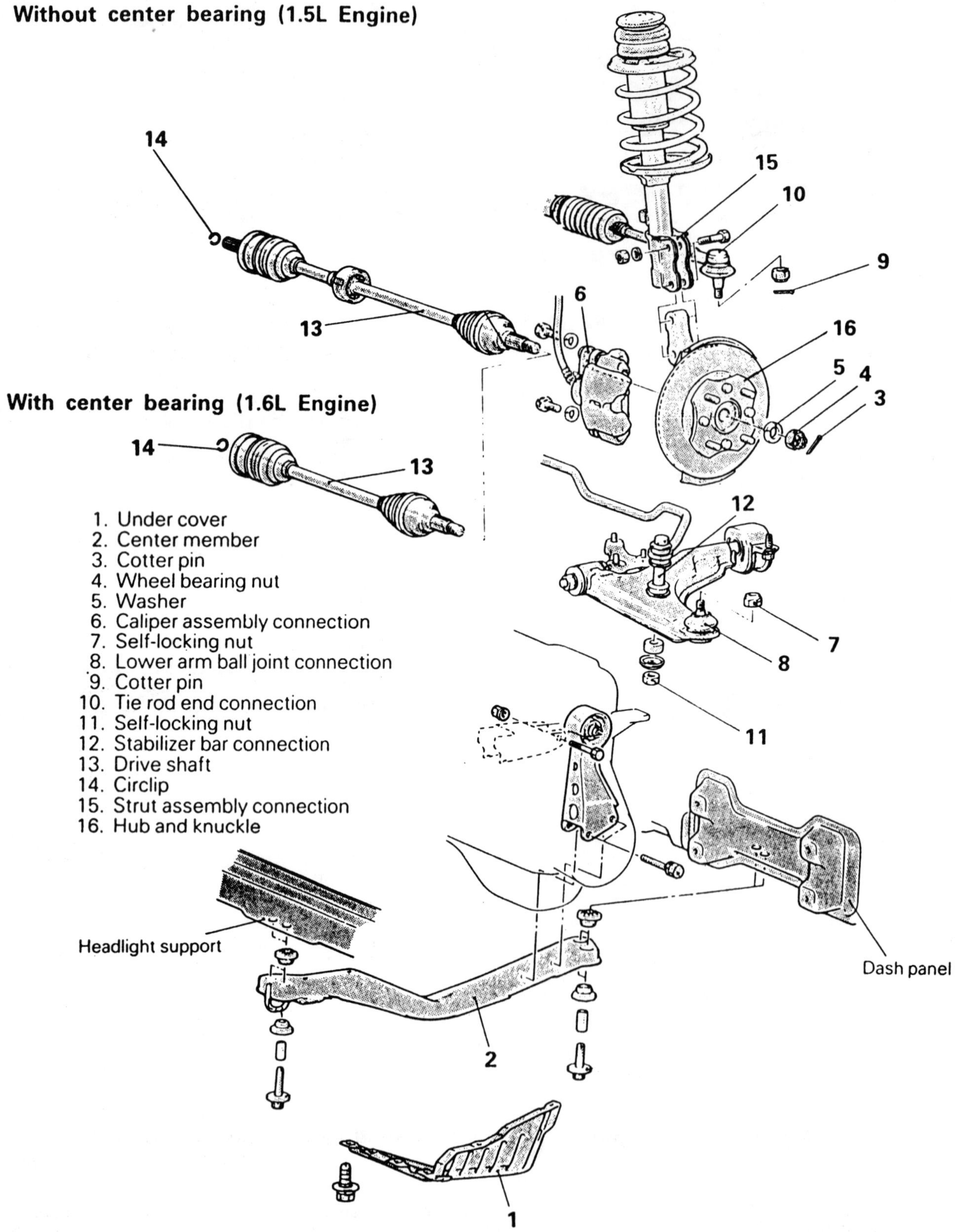

Fig. 1 Front suspension. 1985–88 Colt except 1988 Wagon

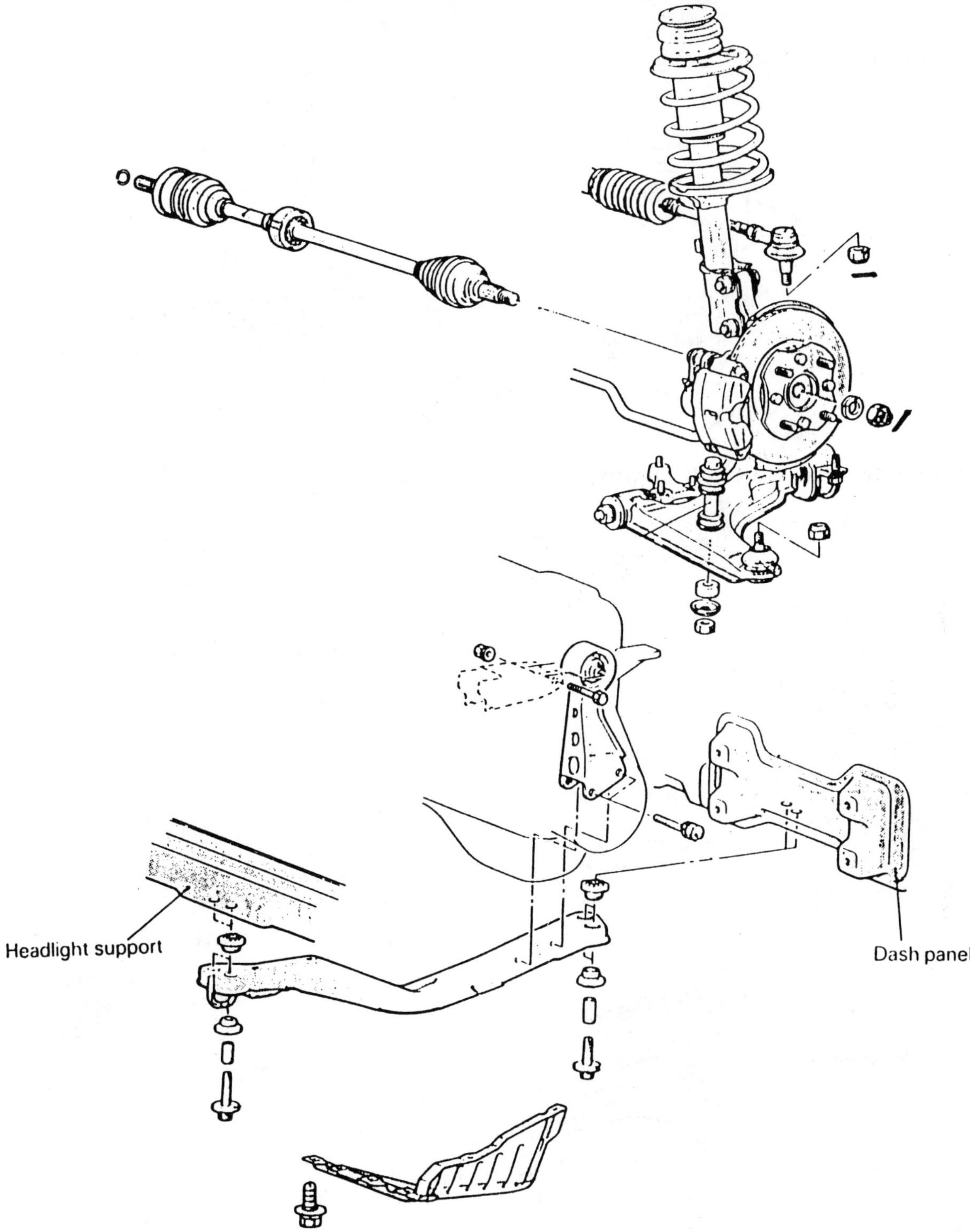

Fig. 2 Front suspension. 1988–90 Colt Wagon 2 wheel drive

ing the following:

a. Check fit of bearing outer races in hub. If bearings have "spun" and races are loose in hub, hub should be replaced.
b. Inspect bearings and replace if damaged, scored or excessively worn.

8. Install inner and outer bearing outer races in hub using suitable driver, then pack center of hub with grease.
9. Install brake rotor on hub and align matching marks. On models less intercooler tighten nuts. On models with intercooler, install hub bolts and secure rotor with retaining clips.
10. Pack wheel bearings, working grease through rollers from wide end of bearing with palm of hand. Keep bearings covered until installation.
11. Install inner wheel bearing in hub, then press new seal into hub using suitable driver.
12. Coat seal lips with grease, mount hub assembly on spindle.
13. Install outer bearing, washer and nut, then adjust wheel bearings as outlined.
14. Install caliper assembly.

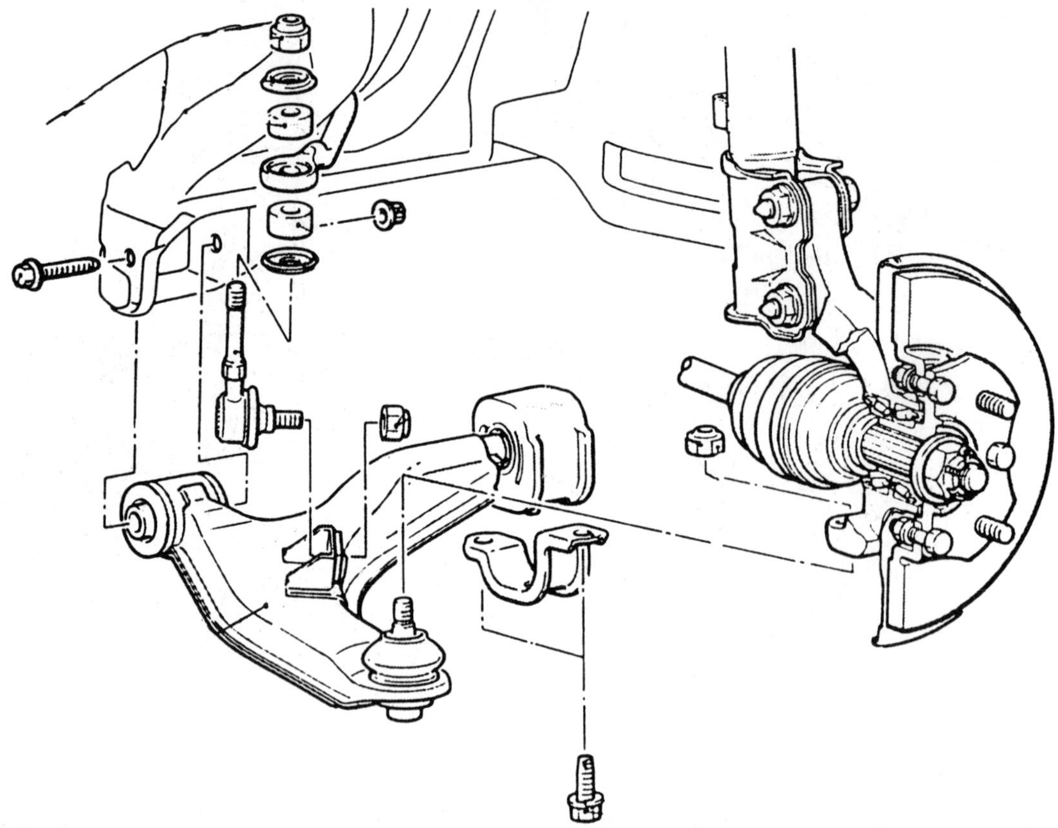

Fig. 3 Front suspension. 1989–90 Colt Hatchback & Sedan & Summit w/4-89.6/1.5L engine

HUB & KNUCKLE REPLACE

COLT & COLT VISTA

1. With vehicle on ground, remove cotter pin and loosen hub to drive shaft retaining nut.
2. Raise and support front of vehicle, then remove wheel and tire assembly, then remove hub to driveshaft nut and washer.
3. Remove front brake assembly from hub, then using suitable puller, disconnect tie rod end from steering knuckle.
4. Using tool No. CT-1003 or equivalent, detach driveshaft from hub.
5. Disconnect steering knuckle from strut, then remove knuckle and hub as an assembly.
6. Reverse procedure to install. After installing brake assembly, bleed brake system.

LOWER CONTROL ARM REPLACE

COLT

1. Raise and support vehicle.
2. On 1988-90 models, remove center member. On all models, remove chassis under cover, then disconnect stabilizer bar from lower control arm.
3. Disconnect ball joint from steering knuckle, **Figs. 1 through 5.**
4. Loosen lower control arm to chassis attaching bolts, then remove lower control arm.
5. Replace ball joint dust cover.
6. Reverse procedure to install.

COLT VISTA

1. Remove stabilizer bar to lower control arm attaching bolts, then the strut bar to lower control arm attaching bolts.
2. Loosen ball joint stud, then disconnect ball joint from knuckle, **Figs. 6 and 7.**
3. Remove lower arm from No. 2 crossmember. **Ball joint dust cover must be replaced during installation.**
4. Reverse procedure to install.

CONQUEST

1. Remove stabilizer bar and strut bar attaching bolts.
2. Disconnect tie rod from knuckle arm, **Fig. 8.**
3. Remove strut as outlined.
4. Remove lower control arm and knuckle arm from crossmember, then the ball joint from the knuckle arm using tool No. MB990635.
5. Reverse procedure to install.

LOWER BALL JOINT REPLACE

COLT

With load removed from lower ball joint, check starting torque with inch lb. wrench. If starting torque is not 21.7-86.8 inch lbs., replace lower control arm and ball joint as an assembly. Refer to "Lower Control Arm, Replace" for procedure.

COLT VISTA & CONQUEST

To replace lower ball joints, refer to "Lower Control Arm, Replace."

STRUT REPLACE

COLT

1. Raise and support front of vehicle.
2. Remove brake hose bracket to strut attaching screws, then the bracket.
3. Remove strut to knuckle attaching bolts.
4. Remove dust cover from top of strut.
5. Remove strut assembly attaching nuts, **Figs. 1 through 5,** then the strut.
6. Reverse procedure to install.

COLT VISTA

1. Raise and support front of vehicle,

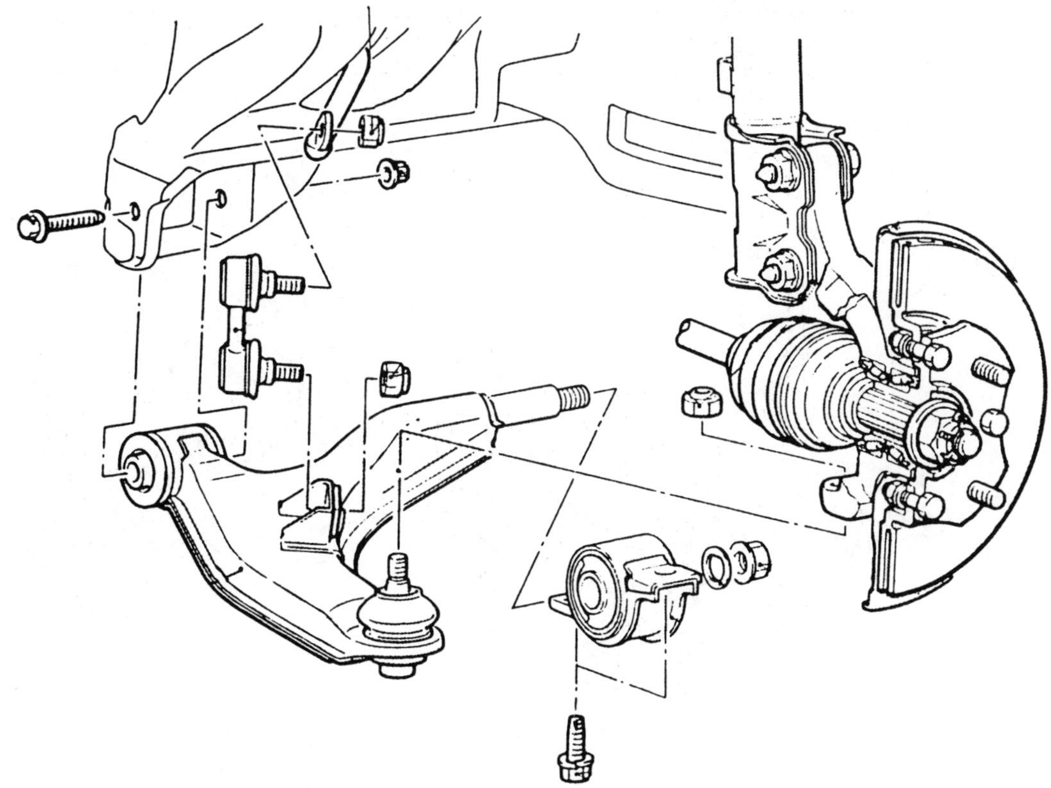

Fig. 4 Front suspension. 1989–90 Colt Hatchback & Sedan & Summit w/4-97.4/1.6L engine

then remove wheel and tire assembly.
2. Remove brake hose bracket from strut assembly, then the strut assembly from knuckle.
3. Remove strut insulator to strut housing panel attaching bolts, then remove strut from wheel housing, **Figs. 6 and 7.**
4. Reverse procedure to install.

CONQUEST

1. Raise and support front of vehicle.
2. Remove wheel, tire, brake caliper, rotor and dust cover from steering knuckle.
3. Remove knuckle arm to strut attaching bolts, then separate knuckle arm and strut, **Fig. 8.**
4. Remove strut to body attaching nuts, then the strut.
5. Reverse procedure to install.

STABILIZER & STRUT BAR

REPLACE

CONQUEST

1. Disconnect strut bars from lower control arms, then from strut bar bracket.
2. Disconnect stabilizer bar from lower control arm, then from frame.
3. Remove strut bar and/or stabilizer bar from vehicle.
4. Reverse procedure to install. Tighten stabilizer bar bushing attaching nuts until distance from top of threads to top of nuts is .59-.67 inch.
5. Tighten strut bar attaching nut until distance between the end of strut bar and the front surface of the locknut is 3.09 inches.

COLT VISTA

1. Disconnect stabilizer bar from lower control arm, then from frame.
2. Disconnect strut bars from lower control arms, then from frame.
3. Remove strut bar and/or stabilizer bar from vehicle.
4. Reverse procedure to install. Tighten stabilizer bar to frame bushing attaching nuts until distance from top of threads to top of nuts is .31-.39 inch. Tighten stabilizer bar to lower arm bushing attaching nuts until distance from top of threads to top of nuts is .31-.39 inch on two wheel drive models, or .51-.59 inch on four wheel drive models.
5. Tighten strut bar attaching nut until distance between the end of strut bar and the front surface of the locknut is 3.07 inches.

STABILIZER BAR

REPLACE

1985–88 COLT & 1989–90 COLT LESS 4-97.4/1.6L ENGINE

1. Raise and support vehicle, then remove under cover attaching bolt and the under cover. On 1988-90 models, remove the center member as well.
2. Remove stabilizer bar to lower control arm mounting bolt, nut and hardware.
3. Remove stabilizer bar bracket mounting bracket bolts, then the bracket and stabilizer bar.
4. Reverse procedure to install. Tighten stabilizer bar to lower control arm bushing attaching nuts until distance from top of threads to top of nuts is .83-.91 inch.

1989–90 COLT HATCHBACK & SEDAN WITH 4-97.4/1.6L ENGINE

1. Remove cotter pin and nut, then separate tie rod end from steering knuckle using tool No. MD990635 or equivalent. Suspend tie rod from chassis with cord.
2. Remove center member rear attaching bolts, then remove roll stopper mounting bolts.
3. Remove stabilizer link lock nuts. Use a suitable Allen wrench to hold link stud in position when removing lock nut.
4. Remove stabilizer bar link.
5. Remove stabilizer bar mounting bolts, then remove stabilizer bar through steering gear access provided in body.
6. Reverse procedure to install. When installing stabilizer link, tighten nut until .12 to .20 inch of stud thread protrudes above nut.

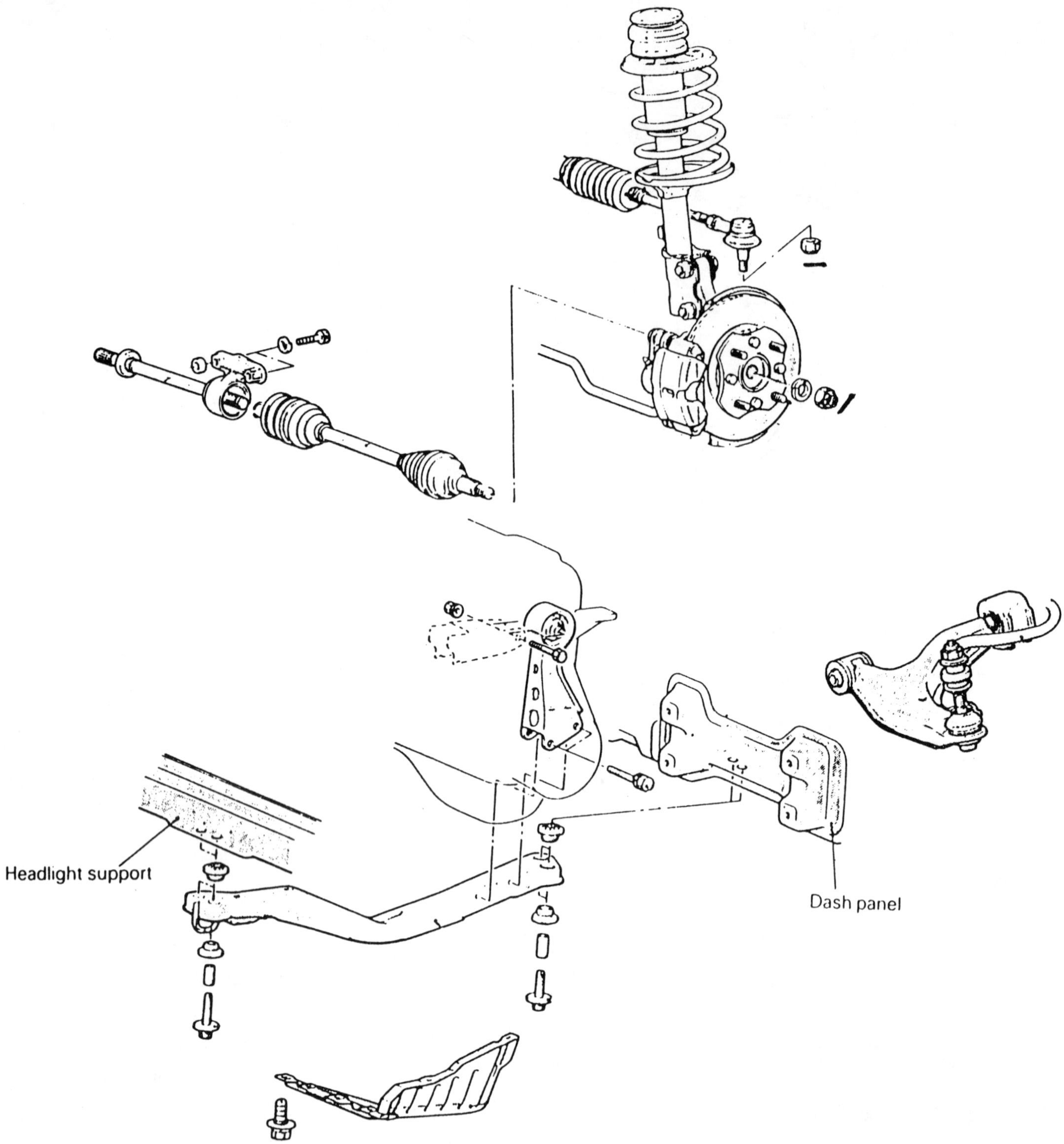

Fig. 5 Front suspension. 1989–90 Colt Wagon 4 wheel drive

MANUAL STEERING GEAR

REPLACE

1. Raise and support front of vehicle, then remove wheel and tire assembly.
2. Remove clamp bolt securing joint with pinion shaft.
3. Remove tie end stud nut, then using suitable puller disconnect tie rod from steering knuckle.
4. Remove steering gear clamp bolts from crossmember, then remove steering gear.
5. Reverse procedure to install.

POWER STEERING GEAR REPLACE

COLT

1. Raise and support front of vehicle.
2. Remove return hose from reservoir

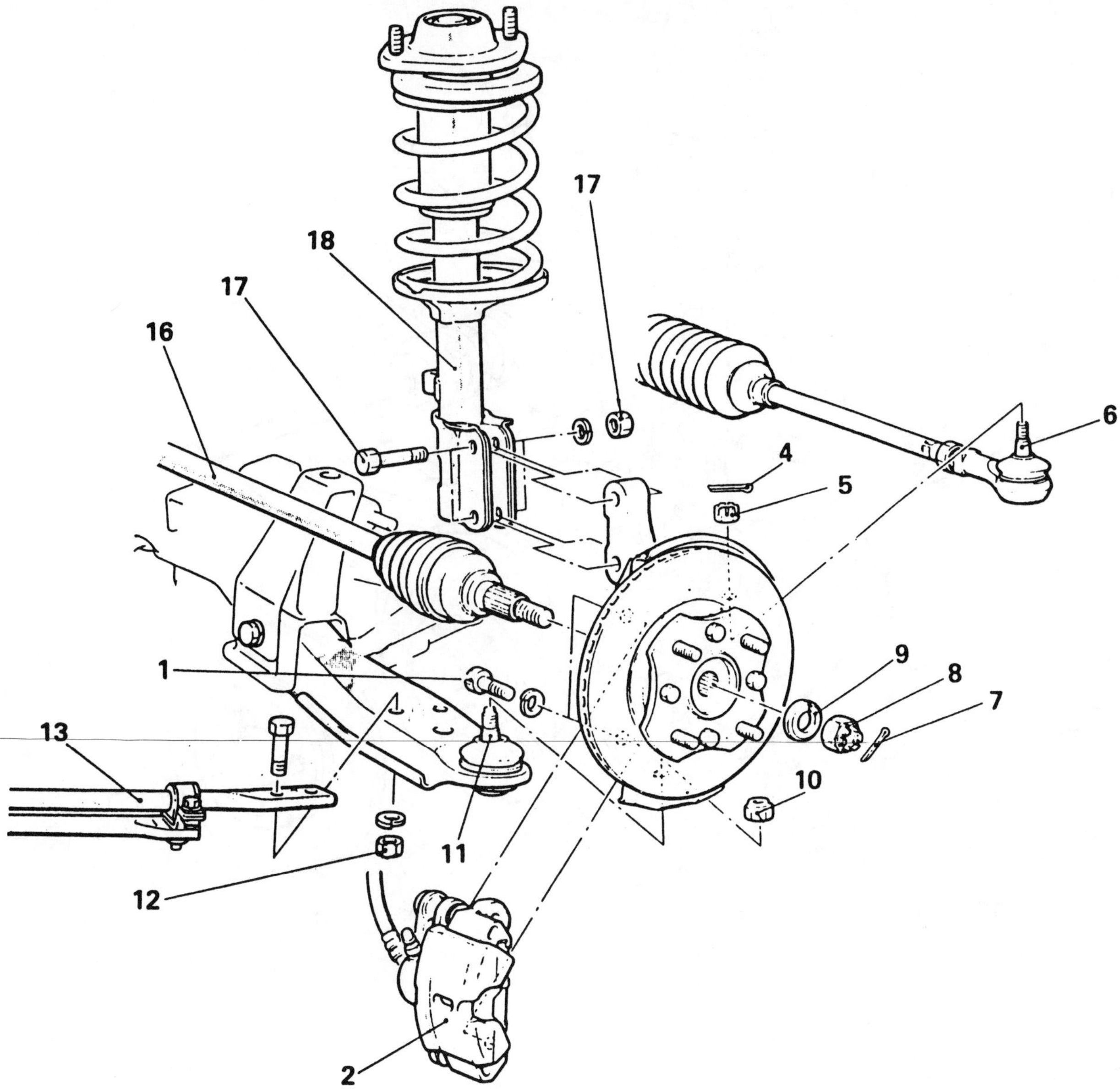

1. Front brake mounting bolts
2. Front brake assembly
4. Cotter pin
5. Tie-rod end mounting nut
6. Tie-rod end connection
7. Cotter pin
8. Drive shaft nut
9. Washer
10. Lower arm mounting nut
11. Lower arm ball joint connection
12. Strut bar mounting nuts
13. Strut bar
16. Drive shaft
17. Front strut mounting bolts and nuts
18. Front strut assembly

Fig. 6 Front suspension. Colt Vista 2 wheel drive

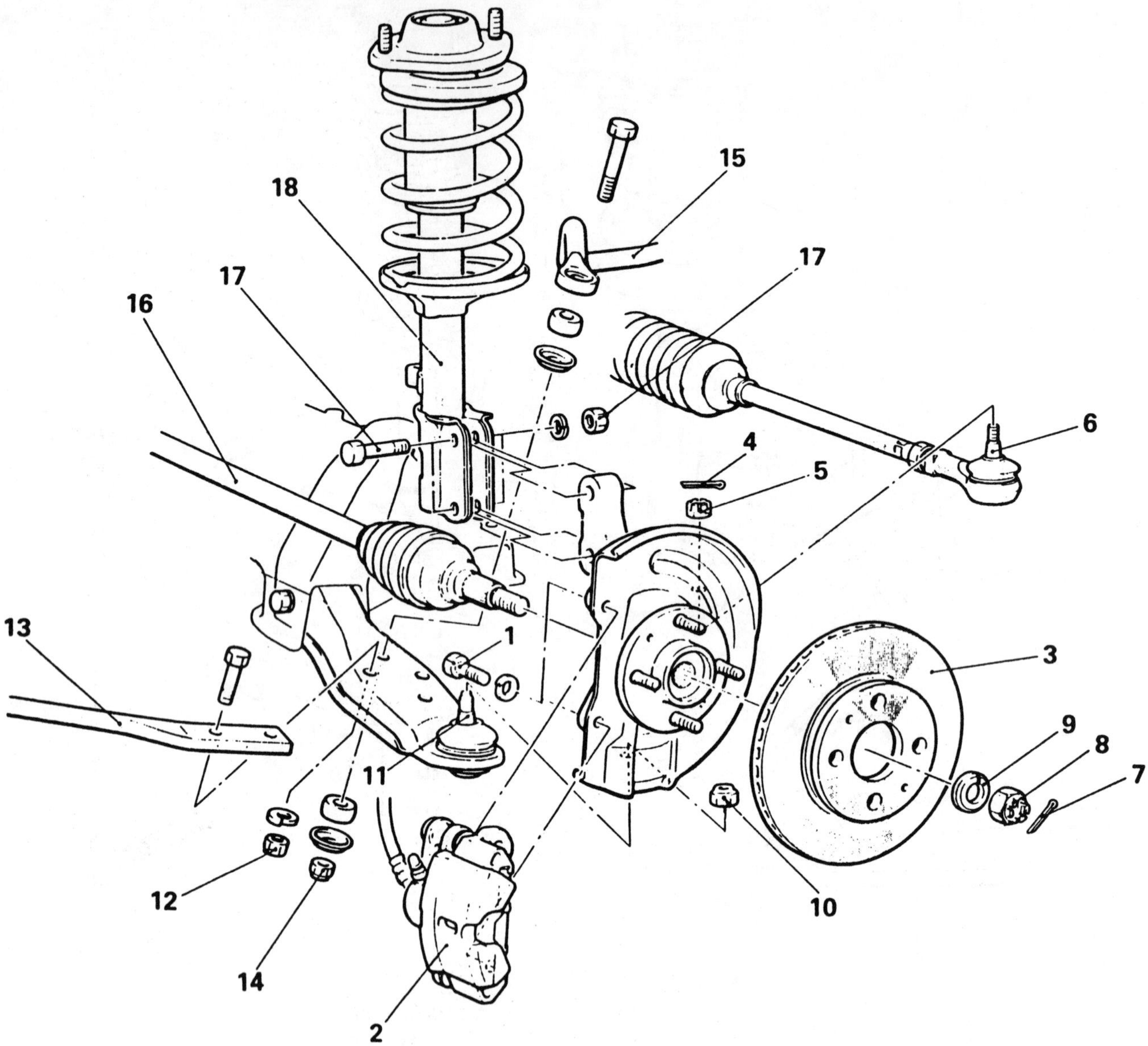

1. Front brake mounting bolts
2. Front brake assembly
3. Brake disc
4. Cotter pin
5. Tie-rod end mounting nut
6. Tie-rod end connection
7. Cotter pin
8. Drive shaft nut
9. Washer
10. Lower arm mounting nut
11. Lower arm ball joint connection
12. Strut bar mounting nuts
13. Strut bar
14. Stabilizer bar mounting nut
15. Stabilizer bar
16. Drive shaft
17. Front strut mounting bolts and nuts
18. Front strut assembly

Fig. 7 Front suspension. Colt Vista 4 wheel drive

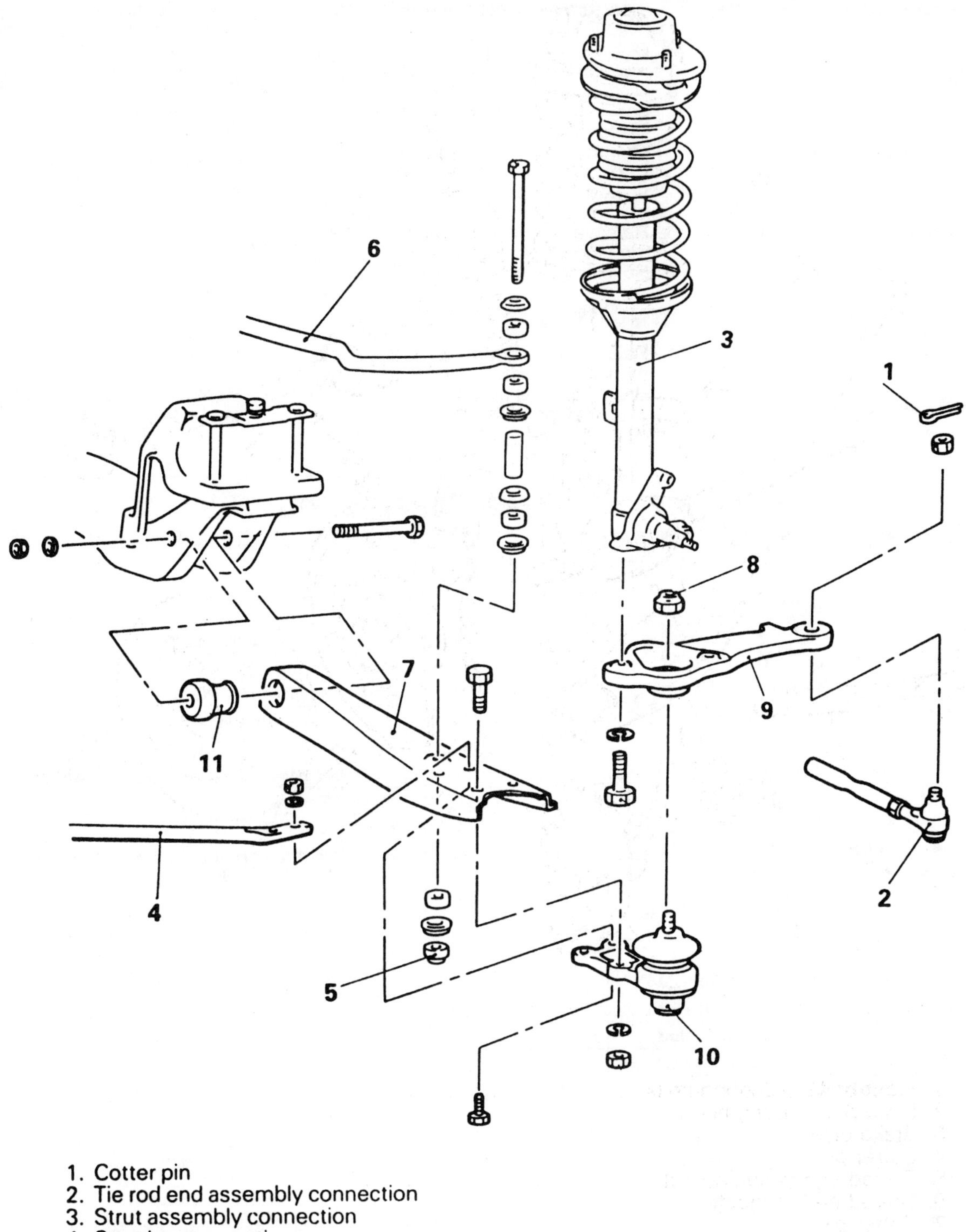

1. Cotter pin
2. Tie rod end assembly connection
3. Strut assembly connection
4. Strut bar connection
5. Stabilizer bar mounting nut (Self-locking nut)
6. Stabilizer bar connection
7. Lower arm
8. Self locking nut
9. Knuckle arm
10. Ball joint
11. Lower arm shaft bushing

Fig. 8 Front suspension. Conquest

and allow fluid to drain into suitable container.
3. Remove steering shaft to gear box pinion coupling bolt.
4. Remove tie rod ends from steering knuckles using tool No. MB991113 or equivalent.
5. Remove pressure and return lines from steering gear box.
6. Remove clamp from lower steering column dust cover.
7. Remove steering gear bracket bolts, then the brackets.
8. Remove mounting rubber, then the gear box assembly.
9. Reverse procedure to install. When installing mounting rubber, apply adhesive to joints to prevent opening.

COLT VISTA

1. Raise and support front of vehicle, then remove wheel and tire assemblies.
2. Disconnect tie rod from knuckle using tool No. C-3894-A.
3. Disconnect joint assembly from gear box, then drain power steering fluid.
4. Disconnect pressure and return hoses from power steering gear.
5. Remove crossmember support bracket attaching bolts, then the support bracket from No. 2 crossmember.
6. Remove steering gear attaching bolts, then the steering gear.
7. Reverse procedure to install.

CONQUEST

1. Raise and support front of vehicle, then remove wheel and tire assembly.
2. Remove steering shaft to steering gear mainshaft retaining clamp, then disconnect pressure and return hoses from steering gear.
3. Disconnect pitman arm from relay rod using tool No. MB990635, then remove steering gear assembly.
4. Remove pitman arm from steering gear using tool No. C-3894-A.
5. Reverse procedure to install.

POWER STEERING PUMP REPLACE

1. Disconnect power steering pressure switch electrical connector, if equipped.
2. Disconnect pressure and return hoses from power steering pump.
3. Loosen pump mounting bolts, then remove drive belt.
4. Remove power steering pump attaching bolts, then remove pump.
5. Reverse procedure to install.

FRONT SUSPENSION & STEERING TIGHTENING SPECIFICATIONS

COLT HATCHBACK & SEDAN; SUMMIT

Year	Component	Torque/ ft. lbs.
1985-90 ①	Caliper To Knuckle	58-72
	Center Bearing Bracket To Engine	26-33
	Center Member Rear Mounting Bolt	43-58
	Connector To Power Steering Pump Body	⑨
	Driveshaft Nut	144-188
	Hub To Brake Rotor	36-43
	Knuckle To Strut Assembly	②
	Lower Arm Ball Joint	43-52
	Lower Arm Bushing B Mounting Nut	90-112⑥
	Lower Arm Front Mounting Nut	69-87④
	Lower Arm Rear Mounting Bolt	43-58
	Lower Arm Shaft To Body	116-137⑤
	Lower Arm To Lower Arm Shaft	69-87⑤
	Power Steering Cooler Tube Clamp Bolt	2-3
	Power Steering Cooler Tube Clip Bolt	2.3-3.5⑤
	Power Steering Cooler Tube To Body	7-10④
	Power Steering Cooler Tube To Hood Stay	3-4④
	Power Steering Gear End Plug	36-51
	Power Steering Gear Feed Tubes	9-13
	Power Steering Gear Pinion & Valve Assembly To Locknut	14-22
	Power Steering Gear Valve Housing Bolts	12-19
	Power Steering Pressure Hose To Pressure Tube	22-29
	Power Steering Pressure Hose To Pump	10-15
	Power Steering Pressure Tube To Steering Gear	9-13
	Power Steering Pump Brace Bolt	18-24⑤
	Power Steering Pump Bracket To Engine	18-24④
	Power Steering Pump Cover To Pump Body	13-16

①—1985-90 Colt, 1989-90 Eagle Summit.
②—1985-88 models, 54-65 ft. lbs.; 1989-90 models, 80-94 ft. lbs.
③—1985-88 models, 33-43 ft. lbs.; 1989-90 models, 43-51 ft. lbs.
④—1989-90 models.
⑤—1985-88 models.

COLT HATCHBACK & SEDAN; SUMMIT—Cont'd

Year	Component	Torque/ft. lbs.
	Power Steering Pump Engine Mount Bracket	35-50⑤
	Power Steering Pump Heat Shield	7-10
	Power Steering Pump Reservoir Bolt	7-10
	Power Steering Pump Reservoir Bracket Bolt	7-10
	Power Steering Pump Support Bolt	18-24⑤
	Power Steering Pump To Pump Bracket	33-40④
	Power Steering Return Tube To Steering Gear	9-13
	Power Steering Suction Connector To Pump Body	4-7
	Rack Support Cover Locknut	36-51
	Rear Roll Stopper Mounting Nut	33-43
	Stabilizer Bar Mounting Bolt	12-19
	Stabilizer Link Mounting Nut	40-51
	Steering Column Lower Bracket	6-9④
	Steering Column Mounting Bolts & Nut	7-10⑤
	Steering Column Upper Bracket	6-9④
	Steering Gear To Body	43-58
	Steering Shaft Dust Cover	2.6-3.6
	Steering Shaft Joint To Steering Gear	⑦
	Steering Wheel Locknut	25-33
	Strut Top End Nut	③
	Strut Upper Mounting Nut	25-33
	Terminal Assembly To Power Steering Pump Body	18-22
	Tie Rod End Ball Joint	11-25
	Tie Rod End Locknut	25-36
	Tie Rod End To Rack	58-72
	Top Cover Locknut	36-51⑧

⑥—1989-90 models w/1.6L engine.
⑦—1985-88 models, 22-25 ft. lbs.; 1989-90 models, 11-14 ft. lbs.
⑧—Models w/manual steering.
⑨—1985-88 models, 29-43 ft. lbs.; 1989-90 models, 36-51 ft. lbs.

COLT VISTA

Year	Component	Torque/ ft. lbs.
1985-90	Center Bearing Bracket Mounting Bolt	29-40
	Center Member Front Mounting Bolt	22-29①
	Center Member Rear Mounting Nut	54-72①
	Connector To Power Steering Pump Body	29-43
	Driveshaft Mounting Nut	145-188
	Dust Cover Mounting Bolt	6-8②
	Flare Nuts Of Power Steering Gear Feed Tube	9-13
	Front Brake Assembly To Knuckle	58-72
	Front Exhaust Pipe To Catalytic Converter Bolt	29-43①
	Front Hub To Brake Rotor	36-43
	Front Roll Bracket To No. 1 Crossmember	29-36
	Hanger Bracket Mounting Bolt	7.2-14①
	Lower Arm Ball Joint To Knuckle	43-52
	Lower Arm Shaft	58-69①
	Lower Arm Shaft	80-101②
	No. 1 Crossmember To Body Side Member	69-87
	No. 2 Crossmember To Side Members	58-72
	Power Steering Cooler Tube To Body	5.0-7.9
	Power Steering Cooler Tube To Grill Bracket	6.5-10.1
	Power Steering Gear End Plug	36-51
	Power Steering Gear Pinion Housing Locknut	72-108①
	Power Steering Gear Self-Locking Nut	14-22
	Power Steering Hose Clamp Attaching Bolts	6-9
	Power Steering Pressure Hose To Connector	12-19②
	Power Steering Pressure Hose To Connector	29-36①
	Power Steering Pressure Hose To Steering Gear	9-13
	Power Steering Pressure Hose To Steering Gear	9-13
	Power Steering Pump Brace Bolt	18-24
	Power Steering Pump Bracket	14-20
	Power Steering Pump Cover	13-16
	Power Steering Pump Guide Bracket Locknut	22-29①
	Power Steering Pump Oil Reservoir	6-9

COLT VISTA—Cont'd

Year	Component	Torque/ ft. lbs.
	Power Steering Pump Reservoir Bracket	6-9
	Power Steering Pump Support Bolt	14-20
	Power Steering Return Hose To Steering Gear	9-13
	Power Steering Return Hose To Steering Gear	9-13
	Rear Insulator To Center Member	22-29①
	Rear Roll Insulator To Engine Bracket	22-29②
	Rear Roll Stopper Bracket To No. 2 Crossmember	43-58②
	Stabilizer Bar Hanger Bracket	7-9
	Stabilizer Bar To Stabilizer Bar Hanger	14-22②
	Steering Column Mounting Bolts	6-9
	Steering Gear End Housing Clamp	43-58
	Steering Gear Housing Clamp	43-58
	Steering Gear Mounting Bolts	43-58
	Steering Rack Support Cover Locknut	36-39
	Steering Shaft Joint To Steering Gear	③
	Steering Shaft To Joint Assembly	③
	Steering Wheel Mounting Nut	25-33
	Strut Assembly To Knuckle	54-64
	Strut Bar To Lower Arm	43-51
	Strut Bar To No. 1 Crossmember	98-116
	Strut Insulator To Piston Rod	43-51
	Strut Insulator To Wheel well	22-29
	Suction Connector To Power Steering Pump Body	4-7
	Terminal Assembly To Power Steering Pump Body	18-22
	Tie Rod End Ball Joint To Knuckle	17-25
	Tie Rod End Locknut	36-39
	Tie Rod End To Steering Rack	58-72
	Transfer Extension Stopper Bracket To No. 2 Crossmember	22-29①
	Transfer Extension Stopper Mounting Bolts	6-9①
	Transfer Extension Stopper Mounting Nuts	6-9①

①—Four wheel drive models.
②—Two Wheel drive models.
③—1985–88 models, 22–25 ft. lbs; 1989–90 models, 11–14 ft. lbs.

COLT WAGON

Year	Component	Torque/ft. lbs.
1989-90	Caliper To Knuckle	58-72
	Center Bearing Bracket To Engine	26-33
	Center Member To Body	43-58
	Hub To Brake Disc	36-43
	Knuckle To Strut	54-65
	Lower Arm Ball Joint	43—52
	Lower Arm Bushing Support Bracket To Body	43-58
	Lower Arm Shaft To Body	116-137
	Lower Arm To Shaft	69-87
	Roll Stopper Bracket To Center Member	33-43
	Stabilizer Bar Bracket To Body	12-19
	Strut Upper Nut	33-43
	Strut To Body	25-36
	Tie Rod End	11-25
	Wheel Bearing	144-188
	Wheel & Tire Assembly	65-80

CONQUEST

Year	Component	Torque/ft. lbs.
1985-89	Crossmember To Body	43-58
	Disc Cover To Strut Assembly	6-9
	Engine Mount Bracket To Crossmember	22-29
	Front Hub To Brake Disc	25-29
	Idler Arm Attaching Nut	29-43
	Idler Arm Support & Frame	25-29
	Knuckle Arm To Ball Joint	43-52
	Knuckle Arm To Strut Assembly	58-72
	Lower Arm Shaft Bolt	58-69
	Lower Arm To Ball Joint	43-51
	Pitman Arm To Steering Gear	94-108
	Power Steering Pressure Hose	22-29
	Power Steering Pump Brace Bolt	10-15
	Power Steering Pump Connector	36-51
	Power Steering Pump Cover	13-16
	Power Steering Pump Reservoir	6-9
	Power Steering Pump Suction Connector	4-7
	Power Steering Return Hose	29-36
	Relay Rod To Idler Arm	25-33
	Relay Rod To Pitman Arm	25-33
	Shock Absorber Ring Nut	101-108
	Stabilizer Bar Bracket	6-9
	Stabilizer Bar To Lower Control Arm	7-14

①—If special tool is used to measure torque, torque to 98–127 ft. lbs.

CONQUEST—Cont'd

Year	Component	Torque/ ft. lbs.
	Steering Column Support Plate	6.5-10
	Steering Column Tube Clamp	3.6-6.0
	Steering Gear Adjusting Bolt Locknut	22-33
	Steering Gear Attaching Bolts	25-29
	Steering Gear Breather Plug	2-3
	Steering Gear Locknut	130-166①
	Steering Gear Side Cover	33-40
	Steering Gear Valve Housing	33-40
	Steering Shaft To Steering Gear Bolt	14-18
	Steering Wheel Locknut	25-33
	Stopper Bolt Locknut For Adjustment Of Steering Angle	14
	Strut Assembly To Brake Assembly	58-72
	Strut Bar Bracket To Frame	25-33
	Strut Bar To Lower Control Arm	43-51
	Strut Bar To Strut Bar Bracket	54-61
	Strut Insulator To Body	18-25
	Strut Top End Nut	43-51
	Tie Rod End & Socket Relay Rod	25-33
	Tie Rod End Stud	36-40
	Tie Rod End To Knuckle Arm	25-33
	Tie Rod End	25-33
	Tilt Link Cover	6.5-10

Chrysler Motors Front Wheel Drive Models Except Colt, Colt Vista & Plymouth Laser

INDEX

DESCRIPTION

These models use a MacPherson type front suspension with vertical shock absorber struts attached to the upper fender reinforcement and steering knuckle, **Fig. 1.** The lower control arms are attached inboard to a crossmember and outboard to the steering knuckle through a ball joint to provide lower steering knuckle position. During steering maneuvers, the strut and steering knuckle rotate as an assembly.

The driveshafts are attached inboard to the transaxle output drive flanges and outboard to the driven wheel hub.

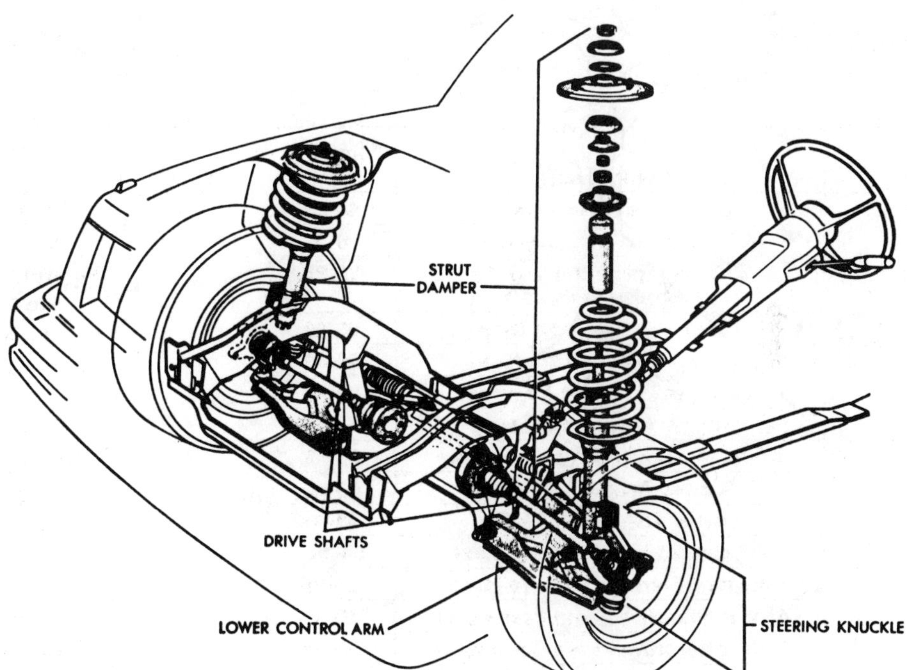

Fig. 1 Front suspension

STRUT DAMPER ASSEMBLY REPLACE

REMOVAL

1. Raise and support vehicle, then remove front wheels.
2. **On all models except 1988-90 Dynasty and New Yorker Landau, 1990 Imperial and New Yorker Fifth Avenue, 1985-89 Lancer and LeBaron GTS, 1987-90 Sundance and Shadow,** mark position of camber adjusting cam for proper alignment during installation.
3. **On 1988-90 Dynasty and New Yorker Landau, 1990 Imperial and New Yorker Fifth Avenue, 1985-89 Lancer and LeBaron GTS, 1987-90 Sundance and Shadow,** mark outline of strut on knuckle for proper alignment during installation.
4. **On all models,** remove cam bolt, knuckle bolt or bolts, washer plate or plates and brake hose to damper bracket attaching screw, **Figs. 2 and 3. On 1985 Lancer and LeBaron GTS, discard one knuckle bolt and washer plate.**
5. **On 1990 models with variable damping,** disconnect electrical connector from upper strut rod by pinching the two latching arms, then pulling connector straight off of rod end, **Fig. 4. Do not rotate connector.**
6. **On all models,** remove strut damper to fender shield attaching nut and washer assemblies.
7. Remove strut damper from vehicle.

INSTALLATION

1. Position strut assembly into fender reinforcement, then install retaining nuts and washers and **torque** to 20 ft. lbs.
2. Position steering knuckle and washer plate to strut, then install upper cam and lower through bolts. **On 1985 Lancer and LeBaron GTS, replace one knuckle bolt with a new cam bolt.**
3. Install brake hose retainer on damper,

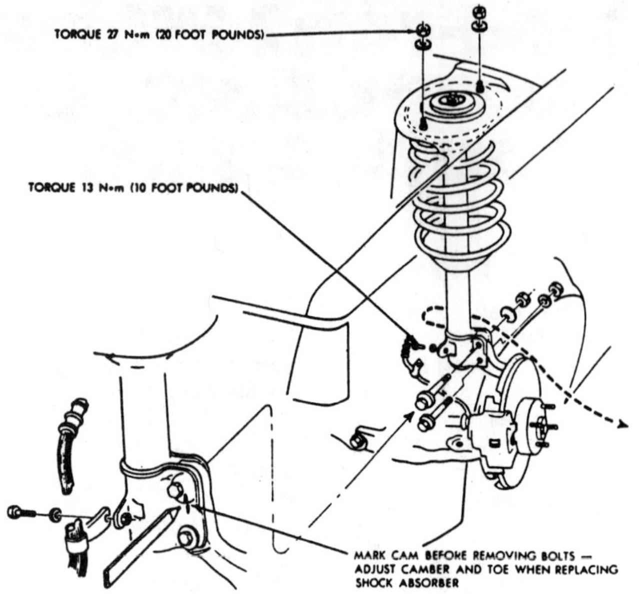

Fig. 2 Strut damper replacement. Charger, Horizon, Omni & Turismo

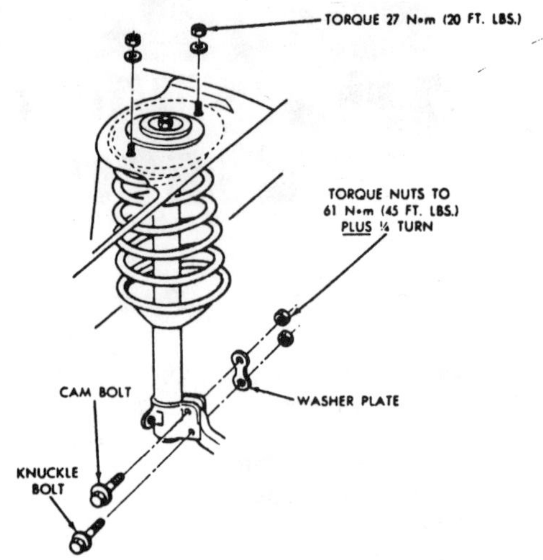

Fig. 3 Strut damper replacement. Except Charger, Horizon, Omni & Turismo

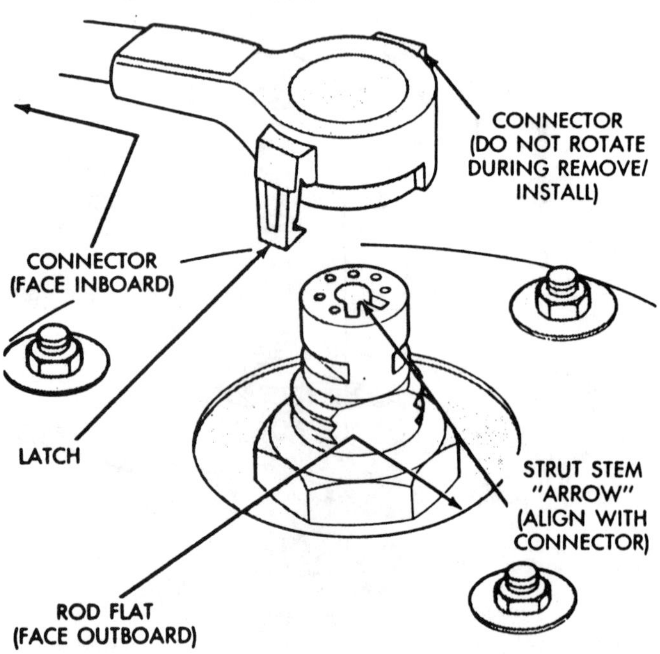

Fig. 4 Variable strut damper electrical connector

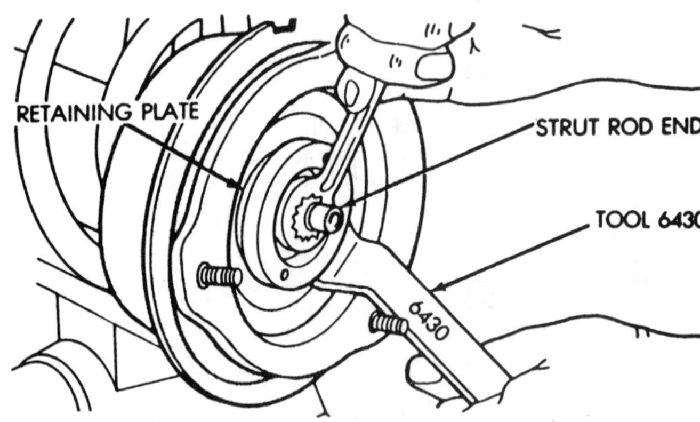

Fig. 5 Removing variable damping strut rod nut

then index alignment marks made during removal.

4. Position a four inch or larger C-clamp on steering knuckle and strut, then tighten clamp just enough to eliminate any looseness between strut and knuckle. Check alignment of marks made during removal. On 1985-87 Charger and Turismo and 1985-90 Horizon and Omni, torque cam bolt nuts to specifications, then advance nuts an additional 1/4 turn. On 1985-90 models except Charger, Turismo, Horizon and Omni, torque nuts to specifications., then advance an additional 1/4 turn.
5. **On all models,** remove C-clamp, then install wheel and tire assembly.
6. **On 1990 models with variable damping, attach electrical connector to top of strut rod, using caution to align key-way (wire should point toward vehicle center line), Fig. 4.** Connector is correctly installed when both latching fingers engage in strut stem.

COIL SPRING REPLACE

1. Remove strut damper assembly as outlined previously.
2. Using tool spring compressor tool No. 4838 or equivalent, compress coil spring.
3. **On all except 1990 models with variable damping,** remove strut rod nut while holding strut rod to prevent rotation.
4. **On 1990 models with variable damping,** hold strut rod retaining plate with tool No. 6430 or equivalent to prevent strut rod rotation, **Fig. 5.**
5. Remove mount assembly, **Figs. 6 and 7.**
6. Remove coil spring from strut damper.
7. Inspect mount assembly for deterioration of rubber isolator, retainers for cracks and distortion and bearings for binding.
8. Install bumper dust shield assembly.

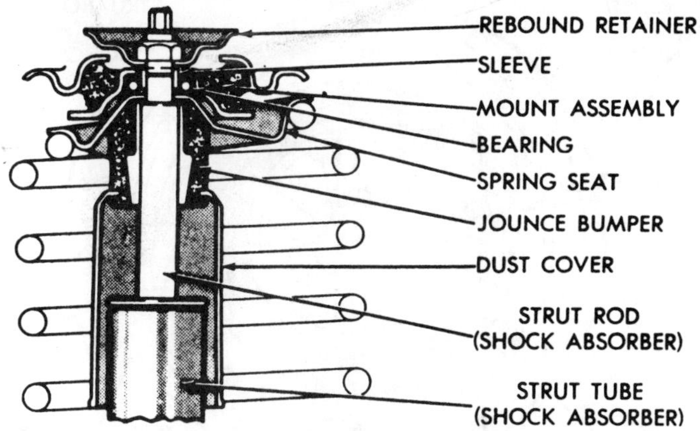

Fig. 6 Strut damper assembly. Charger, Horizon, Omni & Turismo

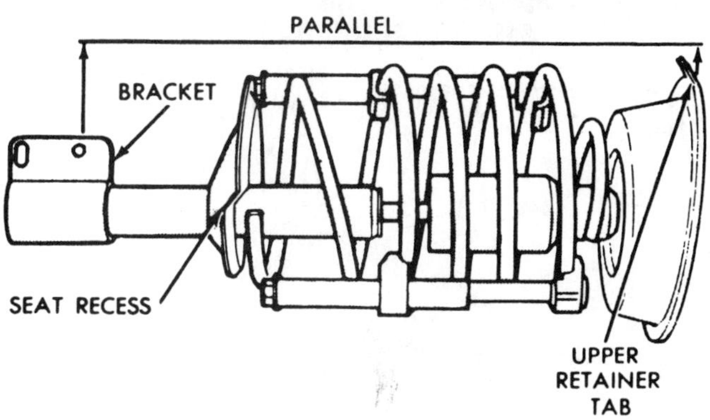

Fig. 8 Spring seat & retainer position. 1985–89 Except Charger, Horizon, Omni & Turismo

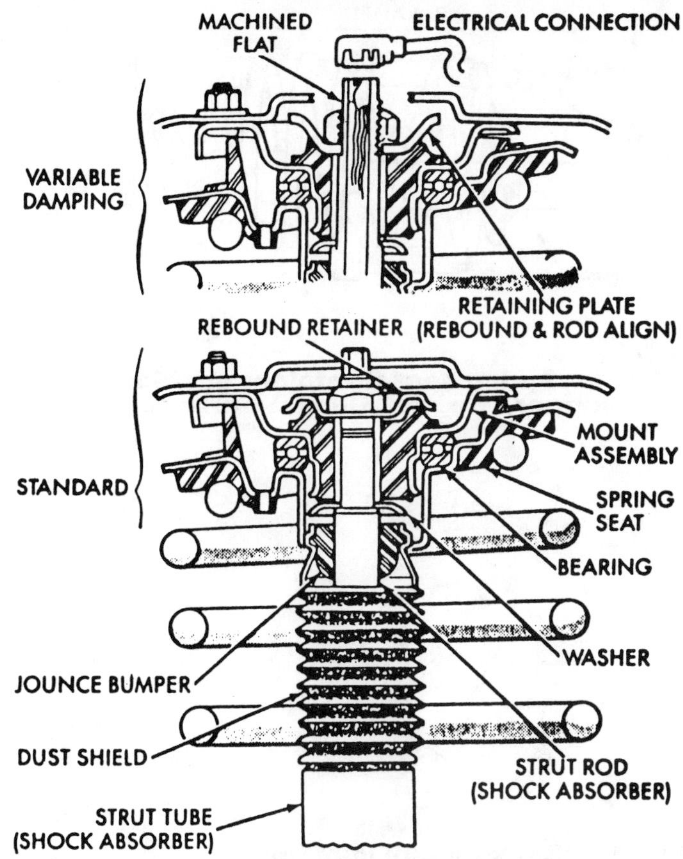

Fig. 7 Strut damper assembly. Except Charger, Horizon, Omni & Turismo

9. Install spring and seat, upper spring retainer, bearing and spacer, mount assembly, rebound bumper and retainer and rod nut. **On 1985-89 models except Charger, Horizon, Omni and Turismo, position spring upper retainer tab or notch parallel to damper lower attaching bracket, Fig. 8. On all 1990 models, 1985-87 Charger and Turismo and 1985-89 Horizon and Omni, position spring upper retainer tab 180° to damper lower attaching bracket, Fig. 9.**
10. **On 1990 models with variable damping,** align strut rod electrical connector as follows:
    a. Align flat on strut rod and one retaining stud on mount within 15° in opposite direction from spring seat alignment notch, **Fig. 10.**
11. **Torque** strut rod nut to 60 ft. lbs. on 1985-87 models and 55 ft. lbs. on 1988-90 models less variable damping. On 1990 models with variable damping, hold retaining plate and strut rod with tool No. 6340 and **torque** strut rod nut to 75 ft. lbs. **Do not release spring compressor before torquing nut.**
12. Remove spring compressor tool.

BALL JOINTS

The lower control arm ball joints operate with no freeplay. The ball joint is pressed into the lower control arm. Ball joints can be pressed from lower control arm using a 1 1/16 inch deep socket and tool No. C-4699-2. When pressing ball joint into lower control arm, use tool Nos. C-4699-1 and C-4699-2. Install ball joint seal using a 1 1/2 socket and tool No. C-4699-2. **On some models the ball joint is welded to the lower control arm. On these models the ball joint and lower control arm must be replaced as an assembly.**

CHECKING BALL JOINTS

With weight of vehicle resting on wheel and tire assembly, attempt to move grease fitting with fingers, **Fig. 11.** Do not use a tool or added force to attempt to move grease fitting. If grease fitting moves freely, then ball joint is worn and should be replaced.

LOWER CONTROL ARM REPLACE

1985–87 ALL; 1988 EXCEPT DYNASTY & NEW YORKER; 1989–90 HORIZON & OMNI

Removal

1. Raise and support vehicle.
2. Remove front inner pivot through bolt, rear stub strut nut, retainer and bushing and the ball joint to steering knuckle clamp bolt, **Fig. 12.**
3. Separate the ball joint from steering knuckle by prying between ball stud retainer and lower control arm. **Removing steering knuckle from vehicle after releasing from ball joint can separate inner C/V joint.**
4. Remove sway bar to control arm nut and reinforcement. Rotate control arm over sway bar. Remove rear stub strut bushing, sleeve and retainer.

Installation

1. Install retainer, bushing and sleeve on stub strut.
2. Position control arm over sway bar and install rear stub strut and front pivot into crossmember.

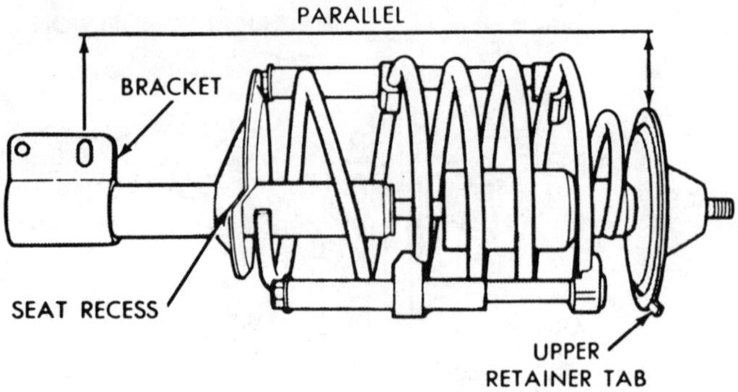

Fig. 9 Spring seat & retainer position. Charger, Horizon, Omni & Turismo

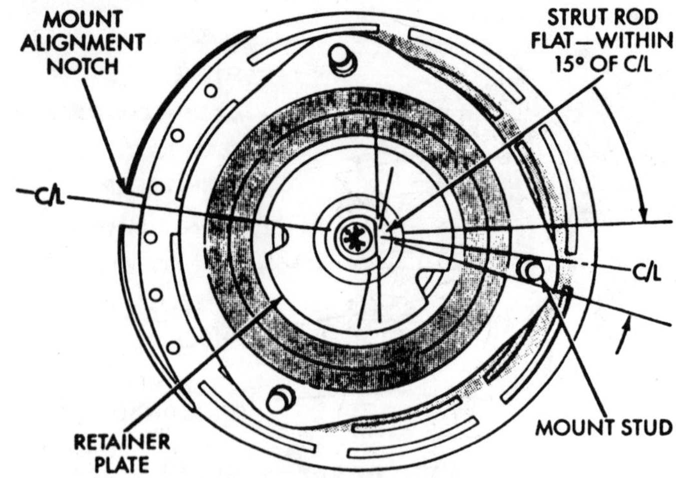

Fig. 10 Aligning variable damping strut rod electrical connector

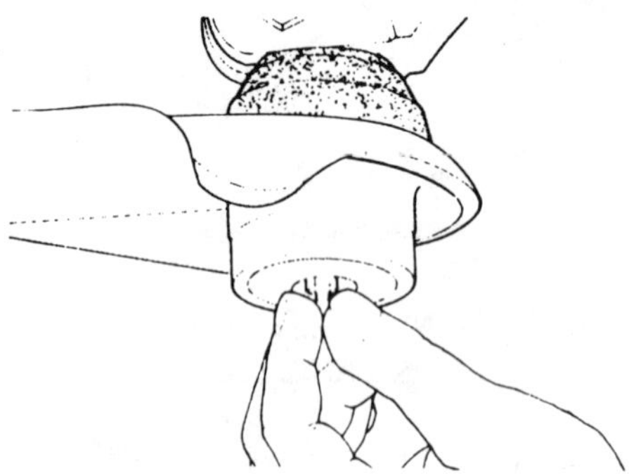

Fig. 11 Checking ball joint for wear

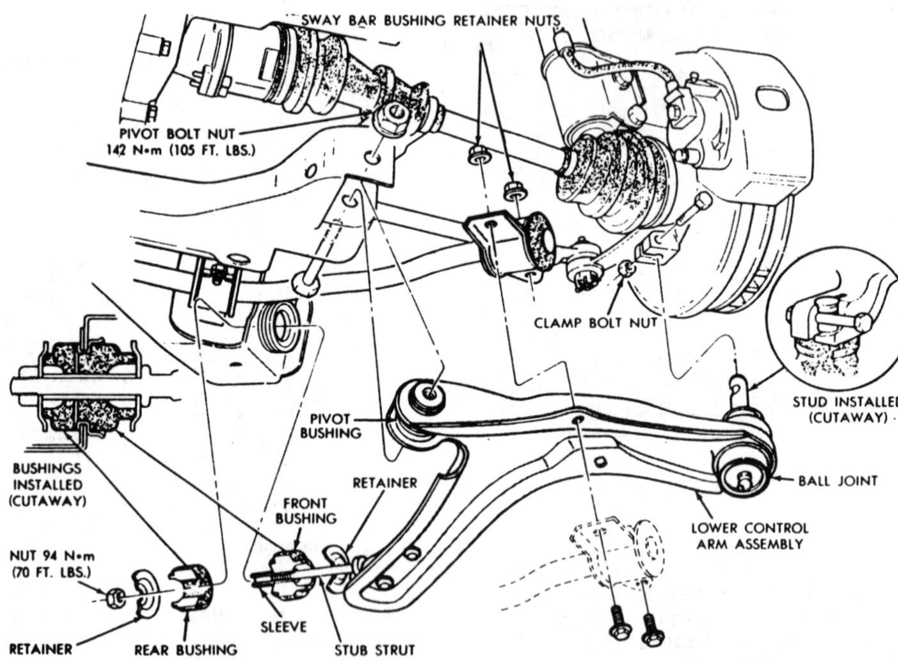

Fig. 12 Lower control assembly. 1985–87 All; 1988 except Dynasty & New Yorker; 1989–90 Horizon & Omni

3. Install front pivot bolt and loosely assemble nut, **Fig. 12.**
4. Install stub strut bushing and retainer and loosely assemble nut.
5. Install ball joint stud into steering knuckle, then the clamp bolt. **Torque** bolt to 70 ft. lbs.
6. Place sway bar bracket stud through control arm and install retainer and nut. Torque nut to specifications.
7. Lower vehicle and with suspension supporting vehicle, **torque** front pivot bolt to 105 ft. lbs. on 1985 models and 95 ft. lbs. on 1986-90 models. **Torque** stub strut nut to 70 ft. lbs.

1988 DYNASTY & NEW YORKER; 1989–90 EXCEPT HORIZON & OMNI

The lower control arm is serviced as a complete assembly with pivot bushings and ball joint.

Removal

1. Raise and support vehicle.
2. Remove front and rear inner pivot through bolt and ball joint to steering knuckle clamp bolt, **Fig. 13.**
3. Separate ball joint from steering knuckle. **Removing steering knuckle from vehicle after releasing from ball joint can separate inner C/V joint.**
4. Remove sway bar to control arm end bushing retainer nuts and rotate control arm over sway bar. Remove stub strut retainer. Inspect lower control arm for distortion and check bushings for deterioration.

Installation

1. Position control arm over sway bar, then install pivot bolts and loosely assemble nuts.
2. Install ball joint stud into steering knuckle, then install clamp bolt. **Torque** bolt to 70 ft. lbs.
3. Install sway bar end bushing to control arm, then install retainer bolts. Torque to specifications.
4. Lower vehicle and with suspension supporting vehicle, **torque** pivot bolts to 125 ft. lbs.

STEERING KNUCKLE REPLACE

REMOVAL

1. Remove cotter pin, locknut and spring washer.

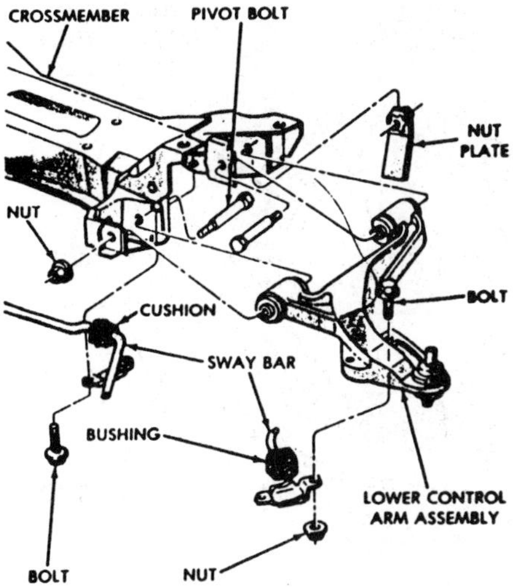

Fig. 13 Lower control assembly. 1988 Dynasty & New Yorker; 1989–90 except Horizon & Omni

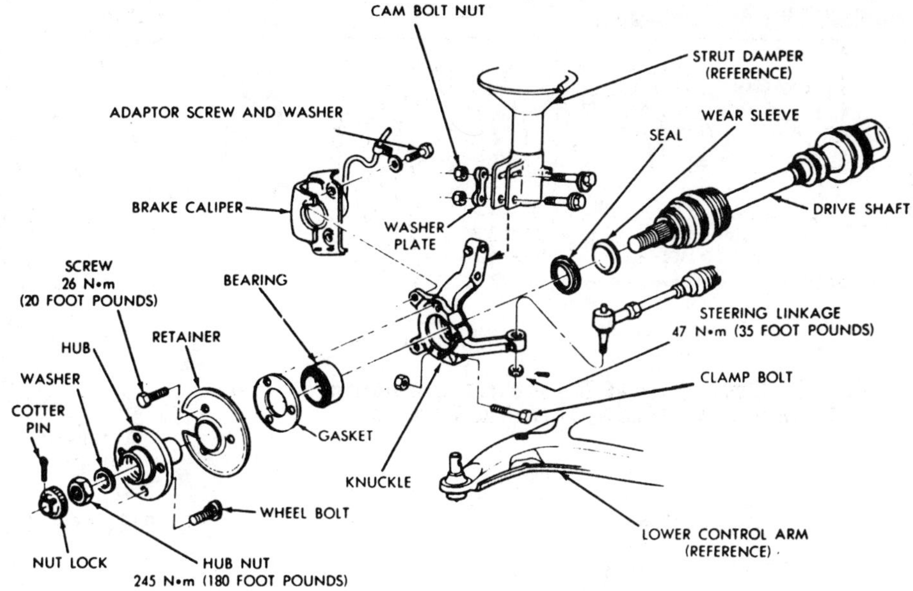

Fig. 14 Steering knuckle assembly

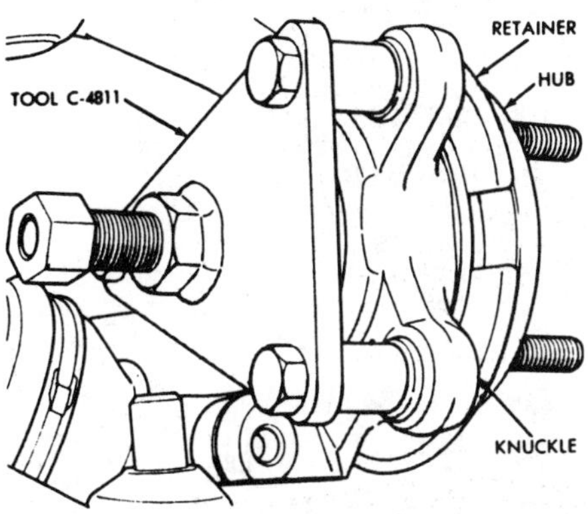

Fig. 15 Hub removal

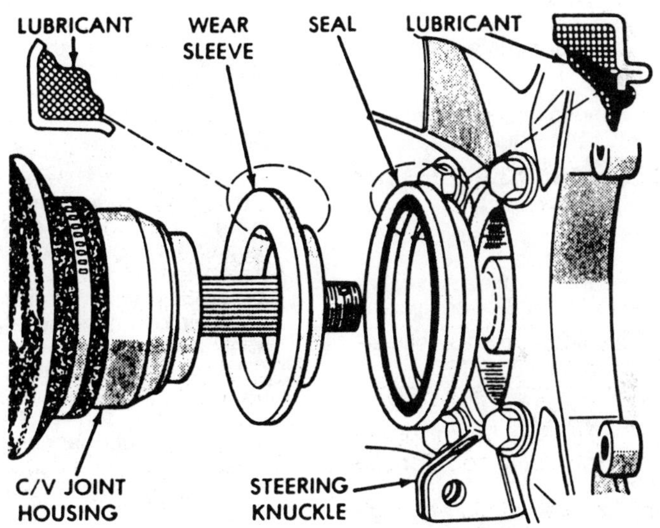

Fig. 16 Lubricating seal & wear sleeve

2. Loosen hub nut with brakes applied, Fig. 14. **The hub and driveshaft are splined together through the knuckle and retained by hub nut.**
3. Raise and support vehicle.
4. Remove front wheel and hub nut. Ensure that splined driveshaft is free to separate from spline in hub during knuckle removal. **A pulling force on the shaft can separate the inner C/V joint.** Tap lightly with a brass drift, if required.
5. Disconnect tie rod end from steering arm with puller tool No. C-3894-A or equivalent.
6. Disconnect brake hose retainer from strut damper.
7. Remove clamp bolt securing ball joint stud into steering knuckle and brake caliper adapter screw and washer assemblies.
8. Support caliper with a piece of wire. **Do not hang by brake hose.**
9. Remove rotor.
10. Separate ball joint stud from knuckle assembly, then pull knuckle out and away from driveshaft. **Do not permit driveshaft to hang after separating steering knuckle from vehicle.**

INSTALLATION

1. Place steering knuckle on lower ball joint stud and driveshaft through hub.
2. Install and **torque** ball joint to steering knuckle clamp bolt to 70 ft. lbs.
3. Install tie rod end into steering arm and torque nut to specifications. Install cotter pin.
4. Install rotor.
5. Install caliper over rotor and position adapter to steering knuckle. Install adapter to knuckle bolts and **torque** to 160 ft. lbs.
6. Attach brake hose retainer to strut damper and **torque** screw to 10 ft. lbs.
7. Install washer and hub nut, then with brakes applied, **torque** hub nut to 180 ft. lbs.
8. Install spring washer, locknut and new cotter pin.

HUB & BEARING REPLACE

If there is movement noted on one or both wheels but no bearing noise, measure movement at wheel outer rim diameter with a dial gauge. The maximum allowable movement on a 13 inch wheel at the rim lip is .020 inch, on a 14 inch wheel .023 inch and .025 inch on a 15 inch wheel. Do not replace bearings for looseness if movement is as specified. Also do not over torque the axle retaining nut beyond 180 ft. lbs. to minimize bearing freeplay.

Service procedures that involve hub removal require a new hub bearing be installed.

PRESSED-IN TYPE

Removal

1. Remove steering knuckle as outlined under "Steering Knuckle, Replace."
2. Using tool kit No. C-4811 or equivalent, remove hub from bearing as follows:
   a. **On 1985-87 models,** back out three bearing attaching screws from knuckle until hub is unseated, then install adapter screw into rear attaching screw threads, **Fig. 15.**
   b. **On 1988-89 models,** back out one of the bearing retainer screws, then install bracket between screw head and retainer, **Fig. 15.**
   c. **On all models,** place thrust button inside hub bore.
   d. Install two screws firmly into tapped brake adapter extensions, then the nut and washer on adapter screw.
   e. Tighten screw to remove hub from bearing.
3. Remove three screws and bearing retainer from knuckle.
4. Pry bearing seal from machined recess in knuckle.
5. Press bearing out of knuckle using tool No. C-4811 or equivalent.

Installation

1. Press new bearing into knuckle using tool No. C-4811 or equivalent.
2. **On all models,** install new seal and bearing retainer. **Torque** retainer screws to 20 ft. lbs.
3. Press hub into bearing using tool No. C-4811 or equivalent.
4. Install new bearing seal using seal installer tool No. C-4698 or equivalent.

BOLT-IN TYPE

Some models use a bolt-in type hub and bearing assembly. This unit is serviced as a complete assembly and is attached to the steering knuckle by four mounting screws that are removed from the rear of the steering knuckle.

1. Remove steering knuckle as outlined under "Steering Knuckle, Replace."
2. Remove four hub and bearing assembly attaching screws from rear of steering knuckle.
3. Remove hub and bearing assembly. **Replacement of grease seal is recommended whenever this service is performed.**
4. Install new hub and bearing assembly in steering knuckle and **torque** screws in a cross pattern to 45 ft. lbs. **Knuckle and bearing mounting surfaces must be smooth and free of foreign material and nicks.**
5. Install new seal using seal installer tool No. C-4698 or equivalent.
6. **During any service procedures where knuckle and driveshaft are separated, clean seal and wear sleeve, then lubricate both parts.** Lubricate full circumference of seal and wear sleeve with multi-purpose lubricant, part No. 4318063 or equivalent, as shown in **Fig. 16.**
7. Install steering knuckle as outlined under ""Steering Knuckle, Replace."

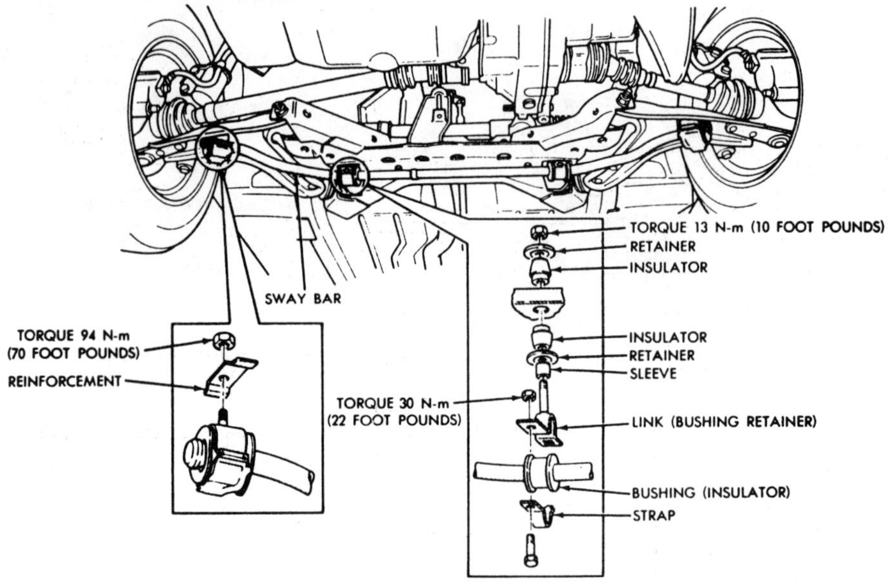

Fig. 17 Sway bar assembly. Except models with "D" shaped insulators

SWAY BAR REPLACE

REMOVAL

1. Raise and support front of vehicle.
2. **On except models with "D" shaped insulators,** remove end bushing to control arm nut and reinforcement plate, **Fig. 17.**
3. **On models with "D" shaped insulators,** remove nuts, bolts and retainer at control arm, **Fig. 18.**
4. **On except models with "D" shaped insulators,** remove nut, retainer and insulators at top of crossmember. Bushing are permanently installed on sway bars.
5. **On models with "D" shaped insulators,** remove bolts at crossmember clamps, then remove clamps.
6. **On all models,** remove sway bar.

INSTALLATION

Some models have a linkless sway bar in the front suspension. The new sway bar is nearly symmetrical looking and it is possible to install the sway bar improperly in the vehicle when it is removed for service. Always mark the sway bar prior to removal to assure proper installation. Sway bars used for production and service replacement are marked on the left side by a dab or stripe of paint.

1. Position crossmember bushings on bar with curved surface up and split to front of vehicle. Set upper clamps onto crossmember bushings, lift bar assembly into crossmember and install lower clamps and bolts.
2. Position retainers at control arms, then insert bolts and install nuts.
3. With lower control arms raised to design height, tighten bolts as follows:
   a. Refer to **Fig. 15** for bolt torques on except models with "D" shaped insulators.
   b. **On models with "D" shaped insulators,** torque bolts to specifications.

RACK & PINION STEERING GEAR REPLACE

1985–87

1. Raise and support vehicle, then remove front wheels.
2. Remove tie rod ends with a suitable puller.
3. **On all models except Omni, Horizon, Charger and Turismo,** remove steering column as follows:
   a. Disconnect battery ground cable.
   b. **On column shift models,** disconnect cable rod by prying rod out of grommet in shift lever.
   c. Disconnect all wiring connectors at steering column jacket and remove steering wheel center pad.
   d. Disconnect horn wires and horn switch, then pull steering wheel from column.
   e. Expose steering column bracket, remove instrument panel steering column cover and lower reinforcement. Remove bezel.
   f. Remove indicator set screw and shaft indicator pointer from shift housing.
   g. Remove nuts attaching steering column bracket to instrument panel support, then lower the bracket support to floor. **Do not remove roll pin to remove steering column assembly.**
   h. Pull steering column rearward, and

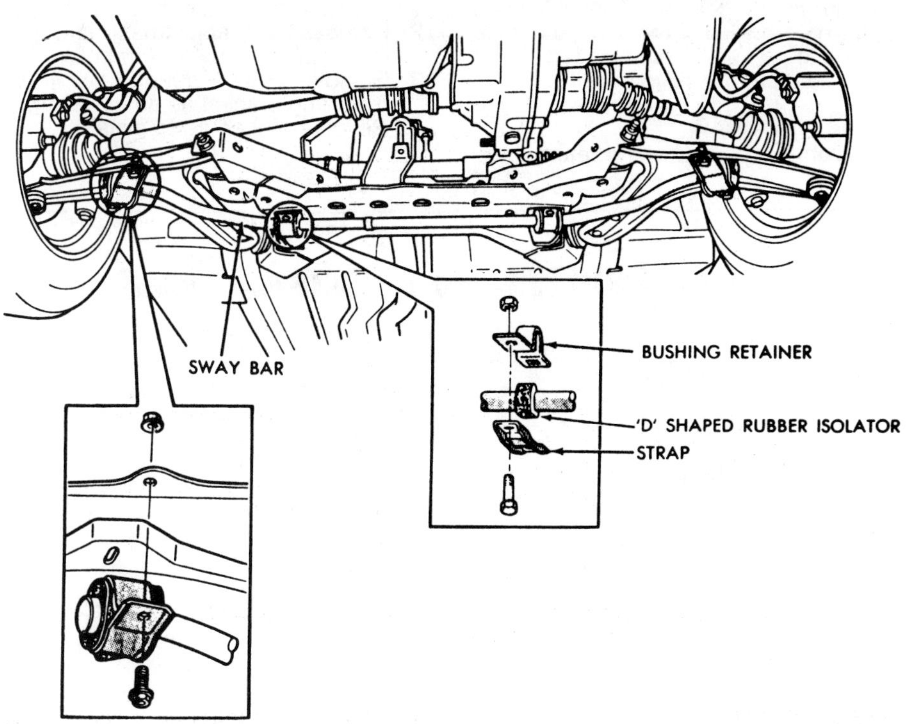

Fig. 18 Sway bar assembly. Models with "D" shaped insulators

disconnect lower shaft from coupling.

i. Reinstall anti-rattle clips into lower coupling tube slot.
j. Remove column assembly out through passenger compartment, being careful not to damage paint or trim.
k. Cut plastic grommets from shift levers and install new grommets from rod side of lever using pliers and a backup washer. Apply grease to grommets.

4. **On Omni, Horizon, Charger and Turismo,** drive out lower roll pin attaching pinion shaft to lower universal joint.
5. **On all models except Omni, Horizon, Charger and Turismo,** support the front suspension crossmember with a suitable jack, then remove bolts attaching steering gear to front crossmember. Lower crossmember from vehicle frame.
6. **On Omni, Horizon, Charger and Turismo,** remove 2 rear crossmember nuts and loosen two front bolts, then lower crossmember slightly to gain access to boot seal shields.
7. **On all models,** remove splash shields and boot seal shields.
8. **On models with power steering units,** disconnect hoses from steering gear.
9. **On all models,** disconnect tie rod ends from steering knuckles.
10. Remove gear to front suspension crossmember attaching bolts, then the gear from left side of engine.
11. Reverse procedure to install. On all models except Omni, Horizon, Charger and Turismo, reinstall steering column as follows:
    a. Align and insert lower stub shaft into coupling, raise column into position and loosely install bracket nuts. Pull column assembly rearward and torque nuts to specifications.
    b. With needle nose pliers, pull coupling spring upward until it touches the universal flange.
    c. Snap gearshift rods into grommets.
    d. Readjust gearshift linkage as necessary.
    e. Install steering wheel and torque nut to specifications.
    f. Install horn switch and horn switch wire.
    g. Connect all wiring connectors at steering column jacket and install steering wheel pad.
    h. Connect battery ground cable and test operation of lights and horn.
    i. **On column shift models,** connect gearshift indicator pointer to its approximate original location. Slowly move gear shift lever from 1 (low) to park, pausing briefly at each position. The indicator pointer must align with each selector position. If necessary, loosen and readjust pointer correctly.
    j. **On all models,** install instrument panel steering column cover.

1988–90

1. Raise and support vehicle, then remove front wheels.
2. Remove tie rod ends with a suitable puller.
3. Disconnect engine damper strut from crossmember, if equipped.
4. Remove four front suspension crossmember attaching bolts, then lower the crossmember using a transmission jack, so that steering gear can be disconnected from steering column.
5. Disconnect lower stub shaft from from steering gear coupling. **Do not remove roll pin.**
6. **On models with power steering,** disconnect hoses from steering gear.
7. **On all models,** remove bolts attaching steering gear to crossmember, then remove steering gear from crossmember.
8. Reverse procedure to install, noting the following:
    a. Using an assistant, from inside of vehicle, guide steering column coupling onto steering gear. **On models with manual steering, ensure master serrations are aligned.**
    b. **Right rear crossmember bolt is a pilot bolt that correctly locates crossmember. Tighten this bolt first. Torque** all crossmember attaching bolts to 90 ft. lbs.
    c. Torque bolts attaching steering gear to crossmember to specifications.
    d. Check for oil leaks.
    e. Adjust toe. Refer to procedure to outlined under "Wheel Alignment."

POWER STEERING PUMP REPLACE

EXCEPT 4-97 & 4-156 ENGINES

1. Disconnect vapor separator hose from carburetor or throttle body, then the two wires from A/C clutch cycling switch, if equipped.
2. Remove power steering pump drive belt adjusting bolt and nut, then the nut attaching pump end hose bracket, if equipped.
3. Raise and support vehicle, then remove nut attaching pump pressure hose bracket to crossmember.
4. Disconnect pressure hose from steering gear and allow fluid to drain into a suitable container.
5. Remove drive belt splash shield, then disconnect both pressure and return hoses at power steering pump. Cap hoses and fittings to prevent entry of dirt.
6. Remove lower stud nut and pivot bolt from power steering pump, then lower the vehicle.
7. Remove belt from pump, then move pump rearward to clear mounting bracket and remove adjusting bracket.
8. Rotate pump so pulley faces rear of vehicle, then lift pump assembly from vehicle.
9. Reverse procedure to install.

4-97 ENGINE

1. Loosen pressure hose connector at pump.
2. Remove drive belt adjustment nut from top rubber isolator stud, then

loosen three locknuts on rear studs.
3. Place a suitable container on top of radiator yoke to catch any spilled fluid.
4. Remove drive belt and three locknuts, then lift pump and bracket assembly from vehicle.
5. Remove pump reservoir cap and drain fluid, then disconnect hoses from pump. Plug pump ports and hose ends to prevent contamination.
6. Reverse procedure to install.

4-156 ENGINE

1. Disconnect pressure and return lines from pump. Cap all lines and fittings.
2. Remove belt adjustment and pivot bolts, and drive belt from pulley.
3. Remove power steering pump and mounting bracket as an assembly.
4. Reverse procedure to install.

FRONT SUSPENSION & STEERING TIGHTENING SPECIFICATIONS

Year	Component	Torque/ Ft. Lbs.
1985–90	Manual Steering Gear Clamp & Housing Pad Bolts	17-25
	Manual Steering Gear Tie Rod End Locknut	45-65
	Manual Steering Gear Tie Rod End Nut	25-50
	Power Steering Gear Inner Tie Rod	70
	Power Steering Gear To Crossmember Bolts	50
	Power Steering Gear Tie Rod End Locknut	55
	Power Steering Gear Tie Rod End Nut	38
	Power Steering Pressure Hose Locating Bracket	9
	Power Steering Pressure Hose Tube Nuts	25
	Power Steering Return Tube Locating Bracket	21
	Power Steering Return Tube Nut	25
	Power Steering Pump Bracket Mounting Fasteners	①
	Power Steering Pump Discharge Fitting	②
	Power Steering Pump Relief Valve Ball Seat	4③
	Steering Column Clamp Bolt	105④
	Steering Column Clamp Stud	105④
	Steering Column Clamp Stud Nut	20④
	Steering Wheel To Steering Shaft Nut	45
	Strut Damper To Knuckle Leg⑤	45⑥
	Strut Damper To Knuckle Leg⑦	75⑥
	Sway Bar Bushing & Cushion Bolt⑤	25
	Sway Bar Bushing & Cushion Bolt⑦	40
	Wheel Nuts	95

①—M8 bolts, 21 ft. lbs.; M10 bolts & nuts, 30 ft. lbs.; M10 stud, 35 ft. lbs.
②—Saginaw pump, 40 ft. lbs; ZF pump, 37 ft. lbs.
③—Saginaw pump.
④—Inch lbs.
⑤—Charger, Horizon, Omni & Turismo.
⑥—Turn an additional 1/4 turn after reaching specified torque.
⑦—All models except Charger, Horizon, Omni & Turismo.

Chrysler Motors Plymouth Laser & Eagle Talon

INDEX

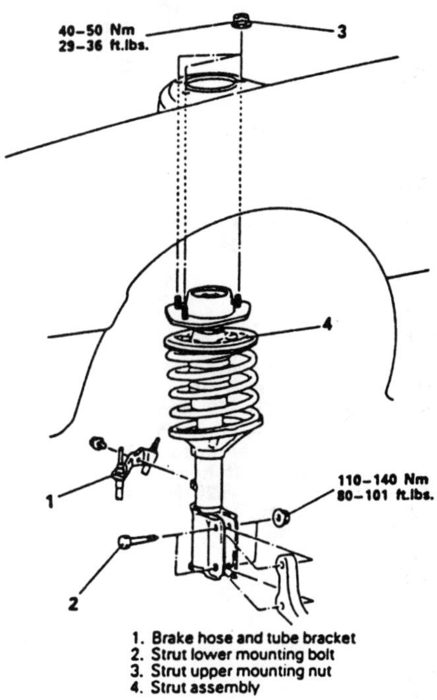

Fig. 1 Removing strut assembly

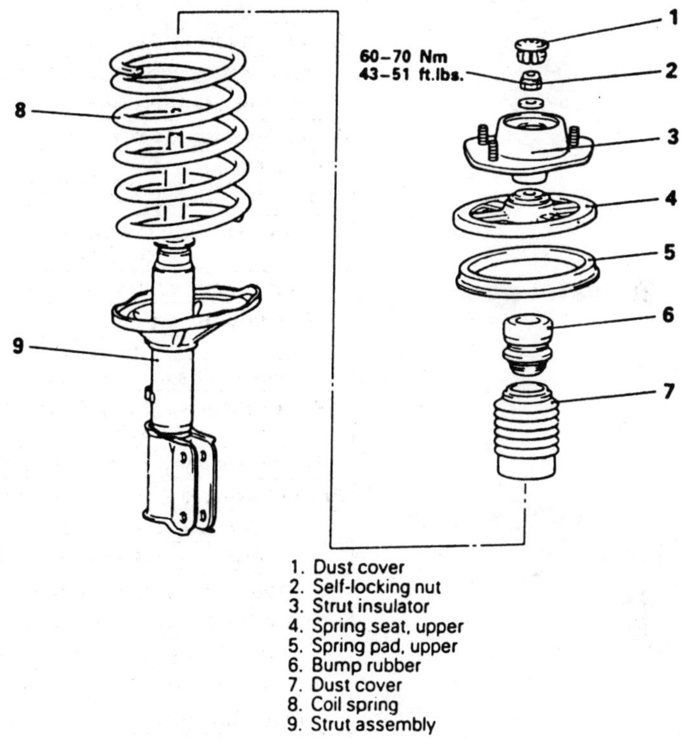

Fig. 2 Disassembling strut

STRUT ASSEMBLY REPLACE

REMOVAL

Remove strut assembly in numbered sequence, **Fig. 1,** noting the following:

1. Do not pry brake hose and line away from strut.

DISASSEMBLY & ASSEMBLY

Disassemble strut in numbered sequence, **Fig. 2,** noting the following:

1. Compress coil spring **Fig. 3,** then remove lock nut.
2. Join dust cover and bump rubber, **Fig. 4.**
3. Line up holes in spring upper and lower seats.

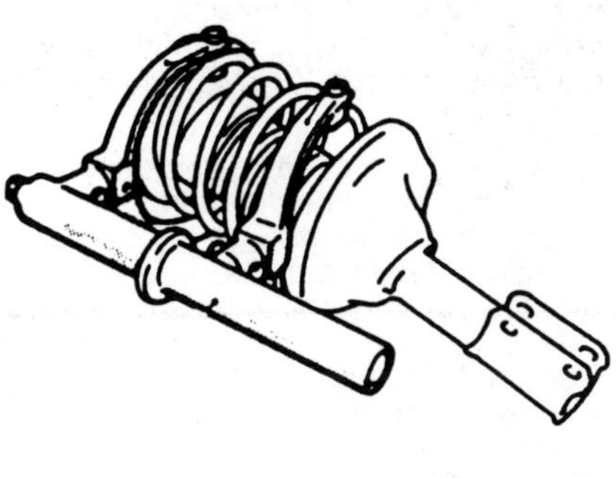

Fig. 3 Compressing spring

INSTALLATION

Reverse removal procedure to install, noting the following:

1. Install strut assembly insulator as shown, **Fig. 5.**

LOWER CONTROL ARM REPLACE

Remove lower control arm in numbered sequence, **Figs. 6 and 7,** noting the following:

1. Check ball joint starting torque; torque should be 26–87 inch lbs. If reading exceeds specified amount, replace lower arm assembly.
2. Reverse procedure to install.

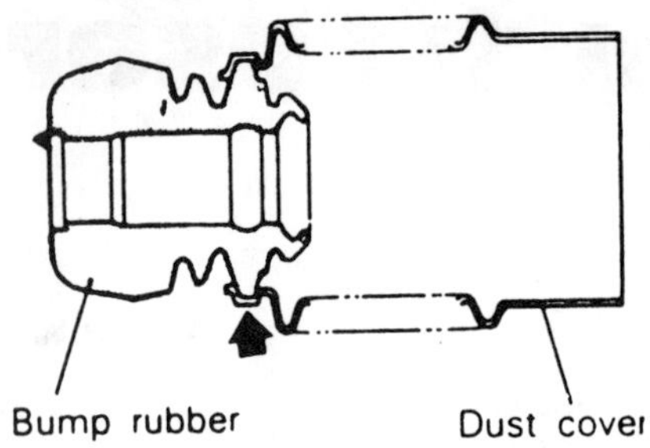

Fig. 4 Installing dust cover

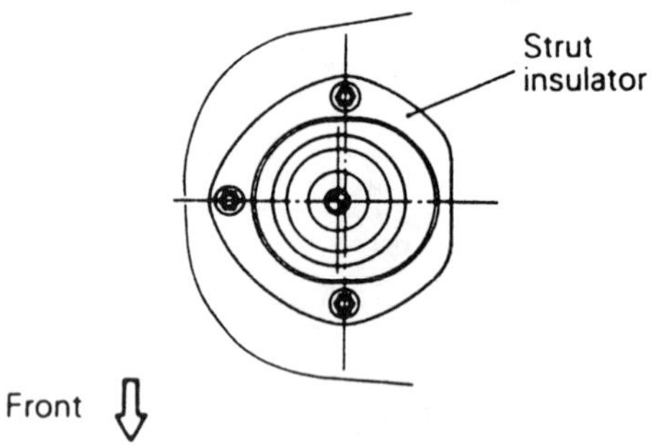

Fig. 5 Insulator position

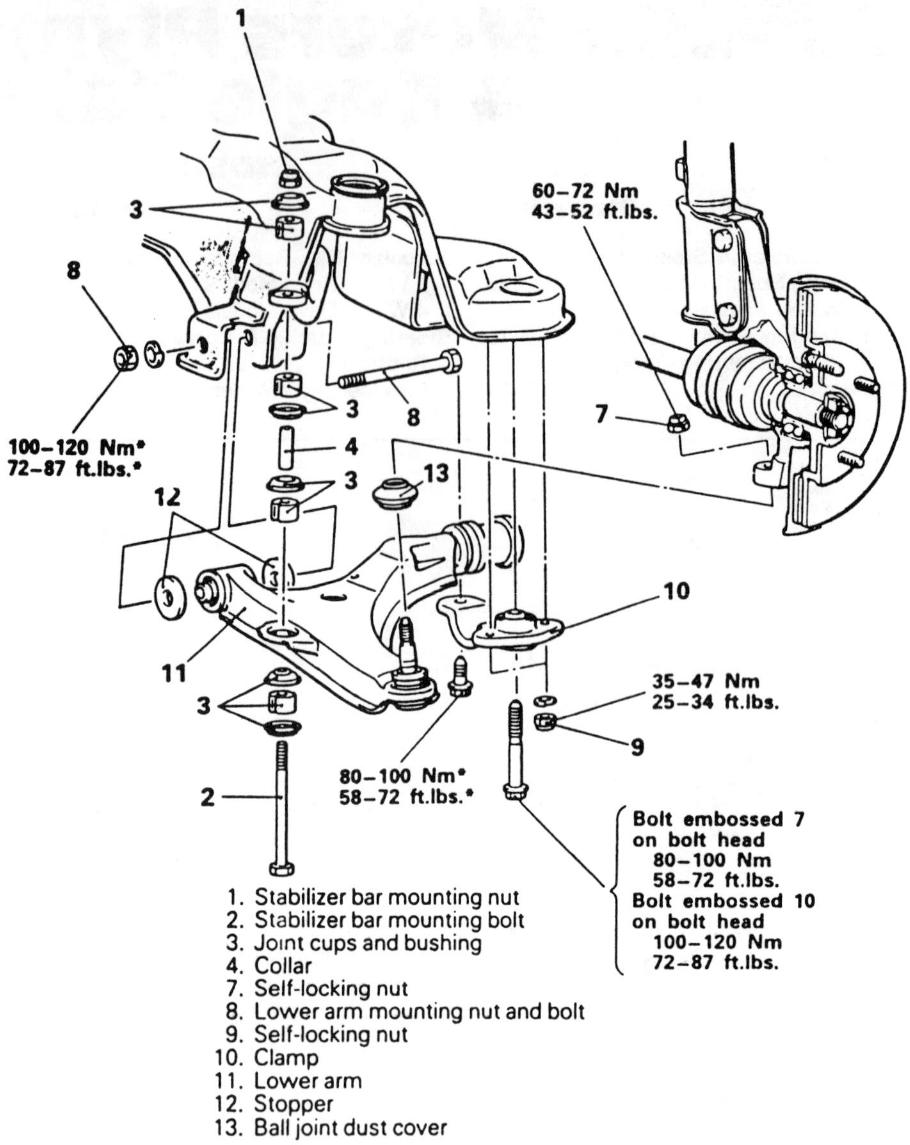

Fig. 6 Removing lower control arm. Models with rubber bushing stabilizer

STEERING KNUCKLE & HUB ASSEMBLY

REPLACE

REMOVAL

Remove knuckle and hub assembly in numbered sequence, **Fig. 8,** noting the following:

1. Loosen drive shaft nut with vehicle on floor and brakes applied.
2. Remove and suspend caliper assembly using a piece of wire.
3. Insert pry bar between transaxle case and drive shaft, **Fig. 9,** then pry drive shaft from transaxle. **Do not pull on drive shaft.**

DISASSEMBLY & ASSEMBLY

Disassemble knuckle and hub assembly in numbered sequence, **Fig. 10,** noting the following:

1. Use tool No. MB991001 or equivalent, to separate hub from knuckle. **Do not strike with hammer, as bearing damage could occur.**
2. Remove oil seal from knuckle.
3. Using puller, remove wheel bearing inner race from front hub. It may be necessary to crush oil seal so puller will catch inner race.
4. Remove snap ring from knuckle.
5. Remove bearings from knuckle.
6. Fill wheel bearings with grease, applying thin coating of grease to knuckle and bearing surfaces.
7. With wheel bearing inner race removed, press in bearing using tool No. MB990985 or equivalent.
8. Install wheel bearing inner race to wheel bearing.
9. Drive hub side of oil seal into knuckle until flush with knuckle end surface.
10. Mount hub assembly to knuckle and torque to specifications.
11. Rotate hub assembly to seat bearing.
12. Measure wheel bearing starting torque; torque should be 16 inch lbs.
13. Measure end play of hub; end play should be .008 inch.
14. If starting torque and hub end play are not within specification, it is possible that assembly has not been installed correctly. Repeat disassembly and assembly procedure.

INSTALLATION

Reverse removal procedure to install, noting the following:

1. Install washer and wheel bearing nut in direction shown, **Fig. 11.**
2. After installing wheel, lower vehicle to ground, then final tighten wheel bearing nut.
3. If cotter pin holes do not match, tighten nut up to 188 ft. lbs.

STABILIZER BAR

REPLACE

2 WHEEL DRIVE MODELS

Remove stabilizer bar in numbered sequence, **Figs. 12 and 13,** noting the following:

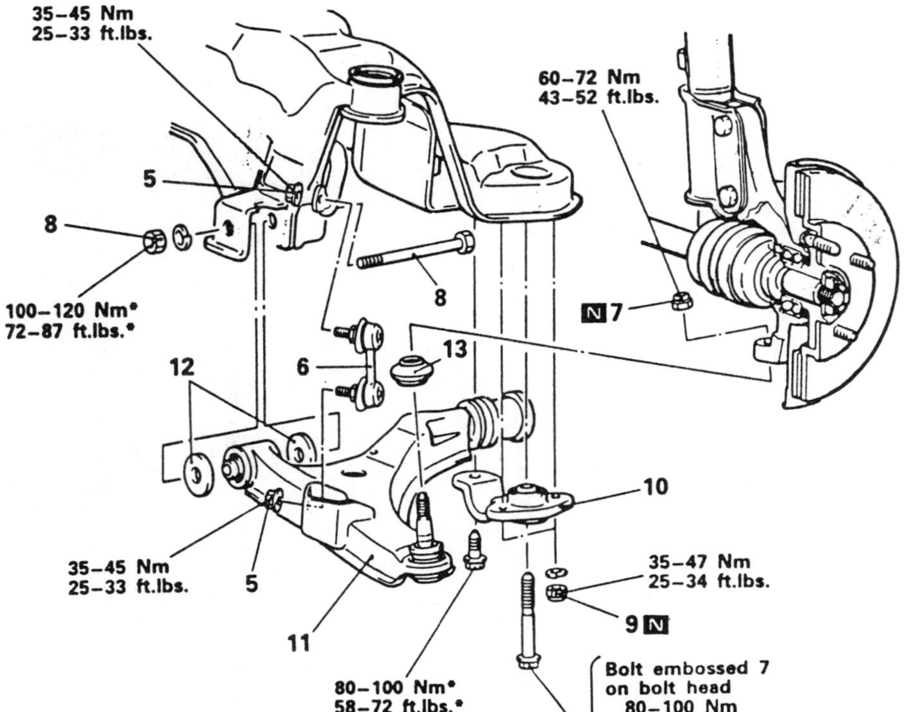

5. Stabilizer link mounting nut
6. Stabilizer link
7. Self-locking nut
8. Lower arm mounting nut and bolt
9. Self-locking nut
10. Clamp
11. Lower arm
12. Stopper
13. Ball joint dust cover

Fig. 7 Removing lower control arm. Models with pillow-ball stabilizer

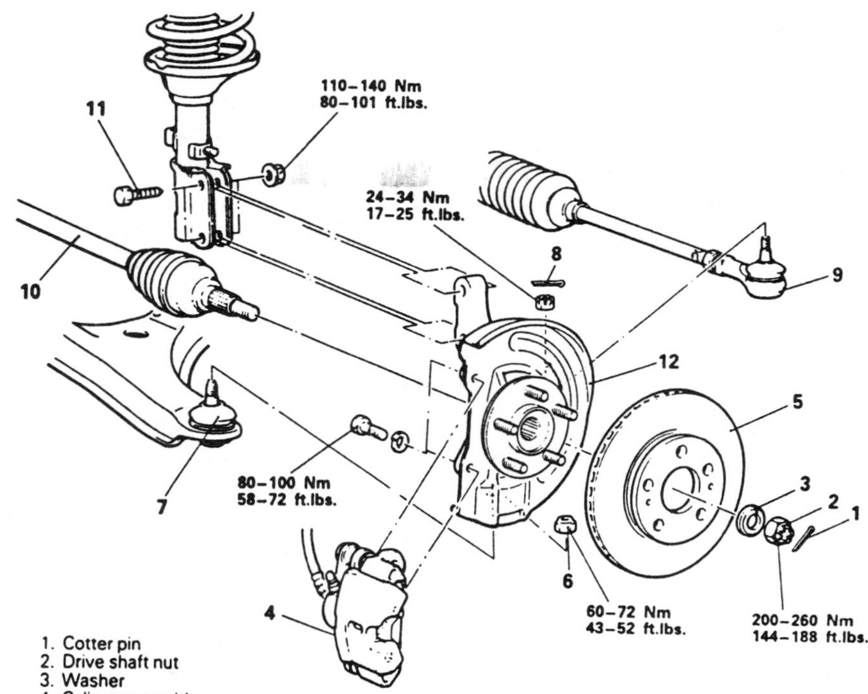

1. Cotter pin
2. Drive shaft nut
3. Washer
4. Caliper assembly
5. Brake disc
6. Self locking nut
7. Connection for lower arm ball joint
8. Cotter pin
9. Connection for tie rod end
10. Drive shaft
11. Front strut mounting bolt
12. Hub and knuckle

Fig. 8 Removing hub & knuckle assembly

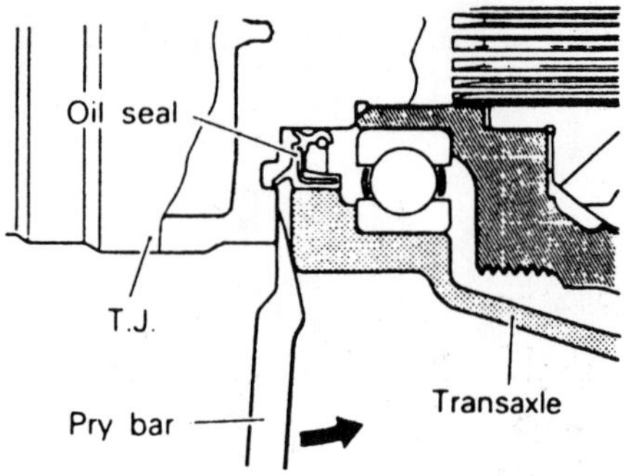

Fig. 9 Removing drive shaft

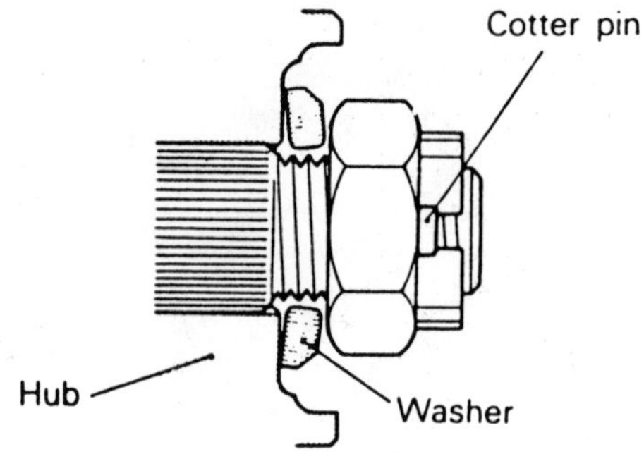

Fig. 11 Installing washer

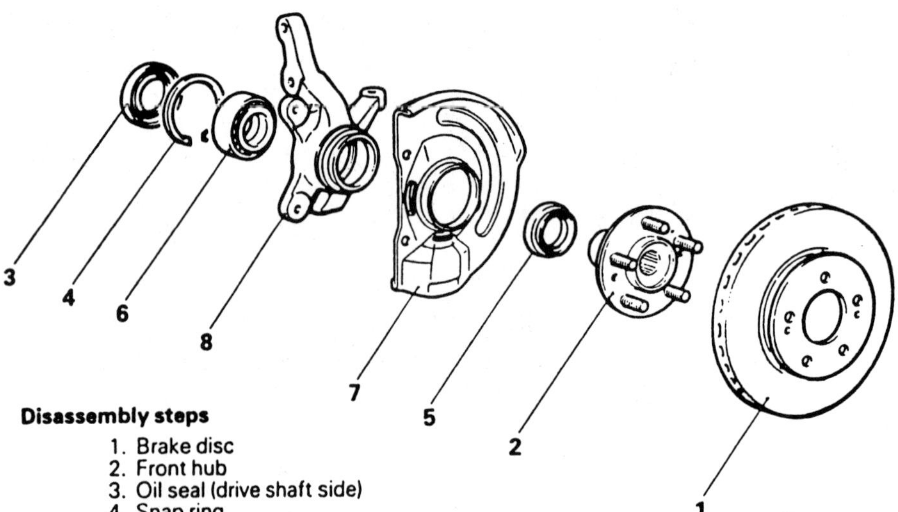

Disassembly steps

1. Brake disc
2. Front hub
3. Oil seal (drive shaft side)
4. Snap ring
5. Oil seal (hub side)
6. Wheel bearing
7. Dust shield
8. Knuckle

Reassembly steps

8. Knuckle
7. Dust shield
6. Wheel bearing
5. Oil seal (hub side)
4. Snap ring
2. Front hub
   Adjustment of wheel bearing starting torque
3. Oil seal (drive shaft side)
1. Brake disc

Fig. 10 Disassembling hub & knuckle

1. Front exhaust pipe must be removed.
2. Pull both ends of stabilizer bar to rear of drive shaft.
3. Move right side of bar until it clears lower arm.
4. Pull stabilizer bar out right side.
5. Check ball joint starting torque; torque should be 15-28 inch lbs. If starting torque exceeds specified torque, replace stabilizer link.
6. Reverse procedure to install.

4 WHEEL DRIVE MODELS

Remove stabilizer bar in numbered sequence, **Fig. 14,** noting the following:

1. Disconnect coupling of knuckle and lower arm on right side.
2. Pull sides of bar between drive shaft and lower arm.
3. Reverse procedure to install.

RACK & PINION STEERING GEAR

REPLACE

Remove rack & pinion in numbered sequence, **Figs. 15 and 16,** noting the following:

1. Turn rack completely to right, the disconnect gear box from crossmember.

1. Self-locking nut
2. Front exhaust pipe
3. Gasket
4. Stay
5. Center member rear installation bolt
6. Stabilizer bar mounting nut
7. Stabilizer bar mounting bolt
8. Joint cups and bushing
9. Collar
12. Stabilizer bar bracket mounting bolt
13. Stabilizer bar bracket
14. Bushing
15. Stabilizer bar

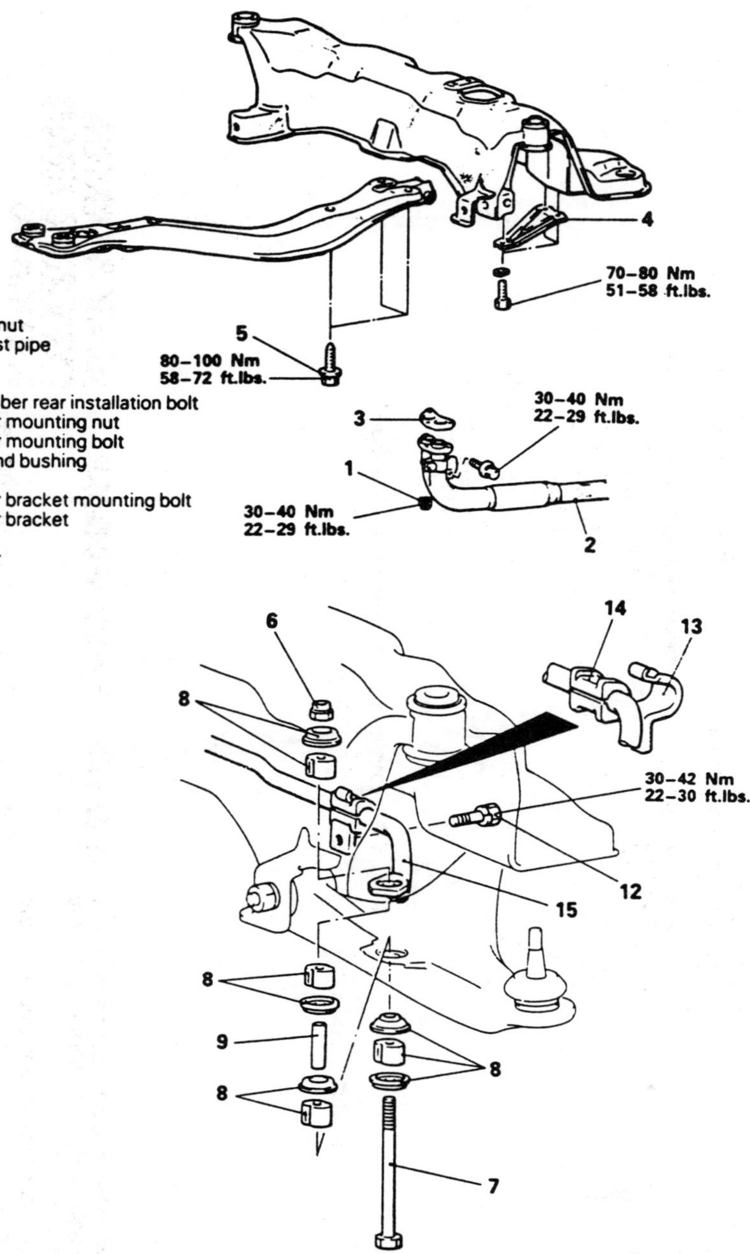

Fig. 12 Removing stabilizer bar (2WD models). Rubber bushing type

1. Self-locking nut
2. Front exhaust pipe
3. Gasket
4. Stay
5. Center member rear installation bolt
10. Stabilizer link mounting nut
11. Stabilizer link
12. Stabilizer bar bracket mounting bolt
13. Stabilizer bar bracket
14. Bushing
15. Stabilizer bar

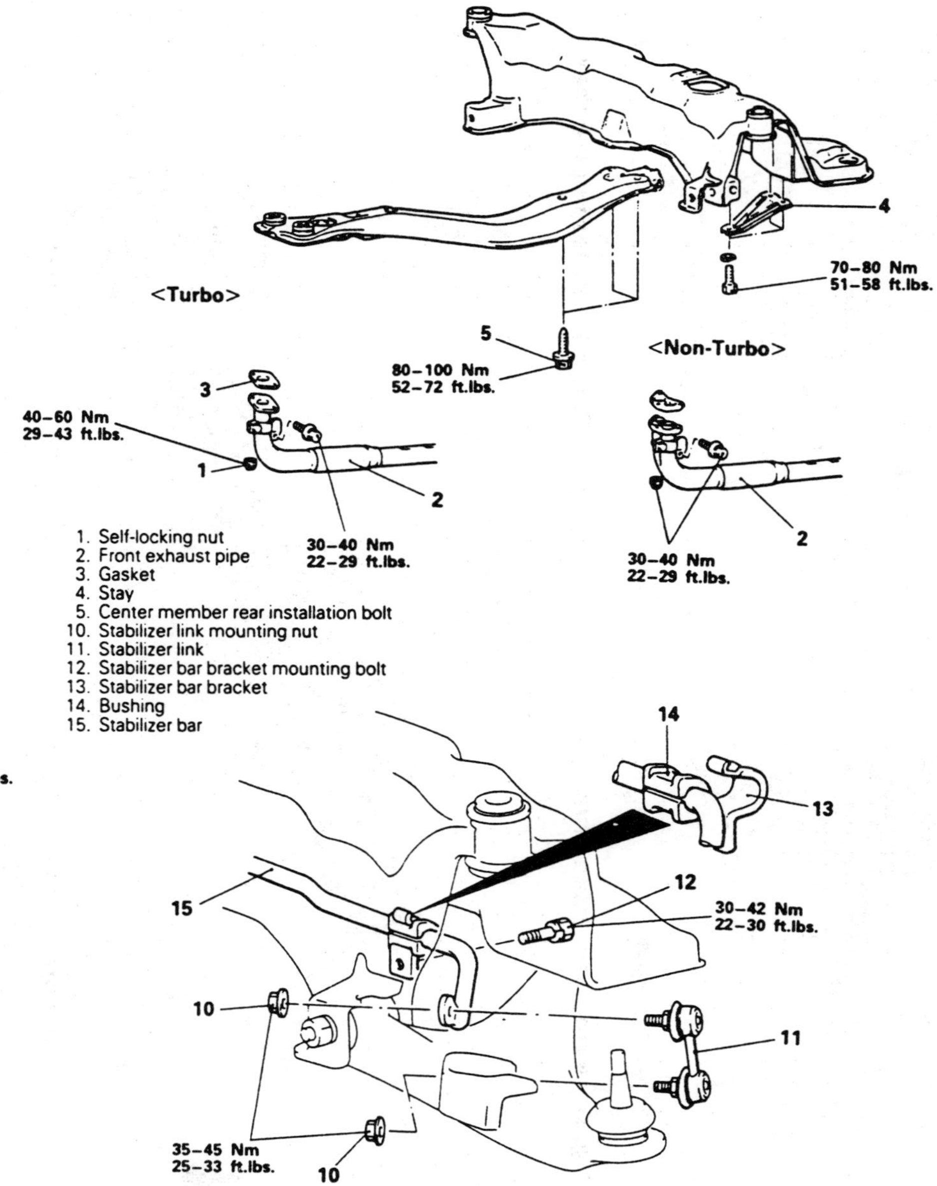

Fig. 13 Removing stabilizer bar (2WD models). Pillow-ball type

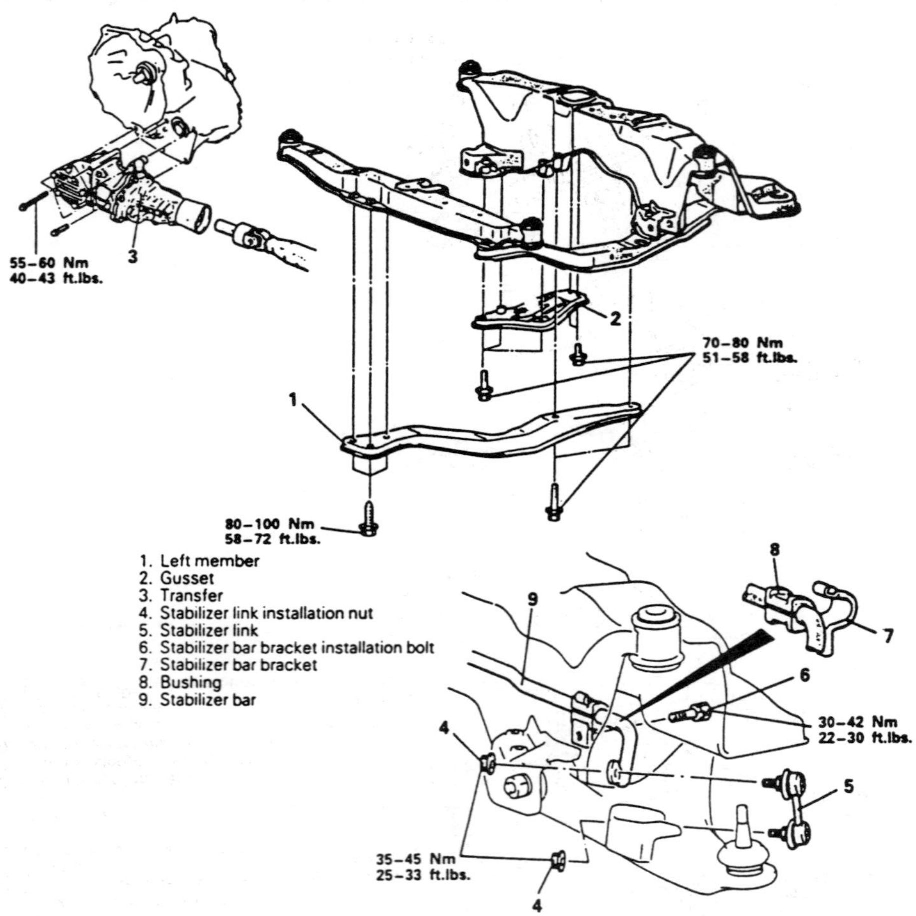

1. Left member
2. Gusset
3. Transfer
4. Stabilizer link installation nut
5. Stabilizer link
6. Stabilizer bar bracket installation bolt
7. Stabilizer bar bracket
8. Bushing
9. Stabilizer bar

Fig. 14 Removing stabilizer bar. Eagle Talon 4WD models

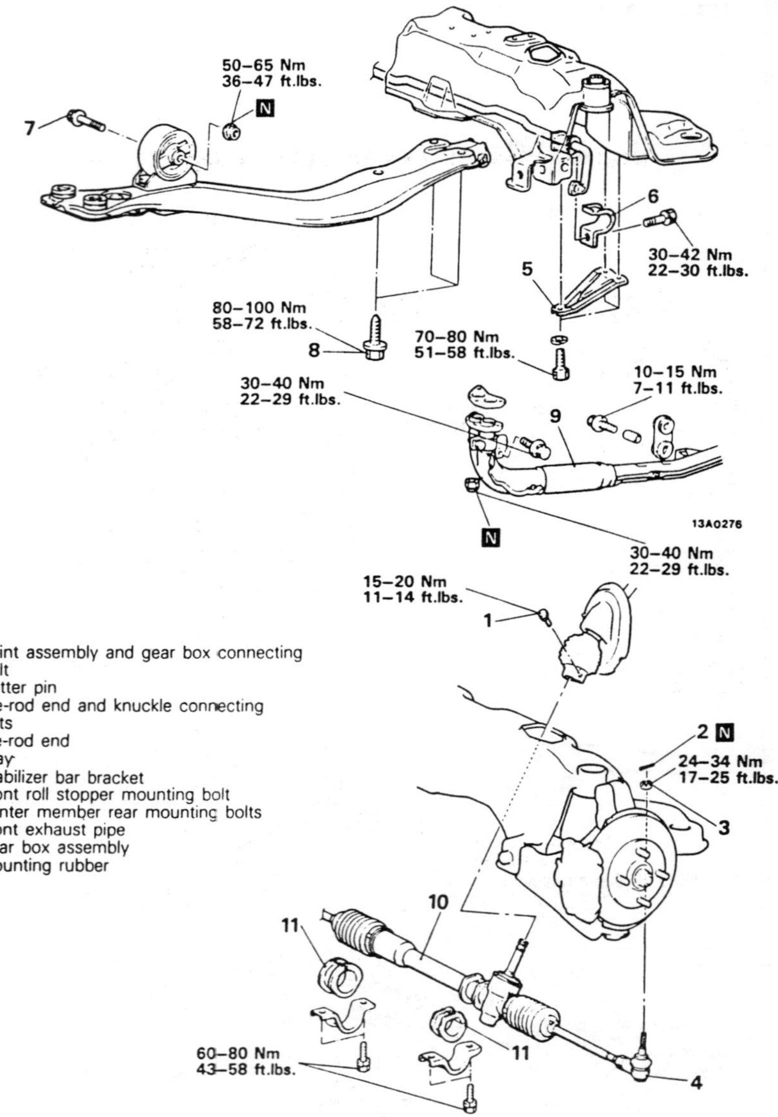

1. Joint assembly and gear box connecting bolt
2. Cotter pin
3. Tie-rod end and knuckle connecting nuts
4. Tie-rod end
5. Stay
6. Stabilizer bar bracket
7. Front roll stopper mounting bolt
8. Center member rear mounting bolts
9. Front exhaust pipe
10. Gear box assembly
11. Mounting rubber

Fig. 15 Removing rack & pinion. Models w/manual steering

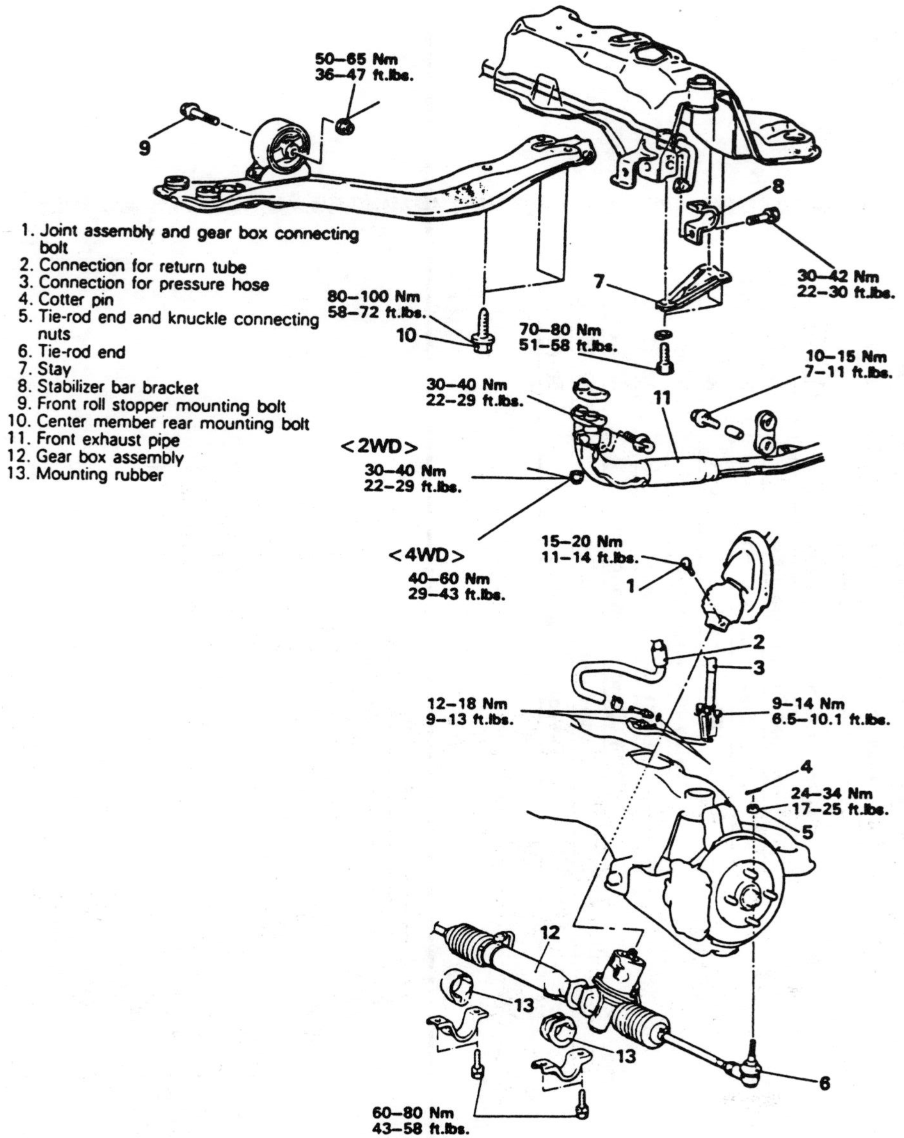

Fig. 16 Removing rack & pinion. Models w/power steering

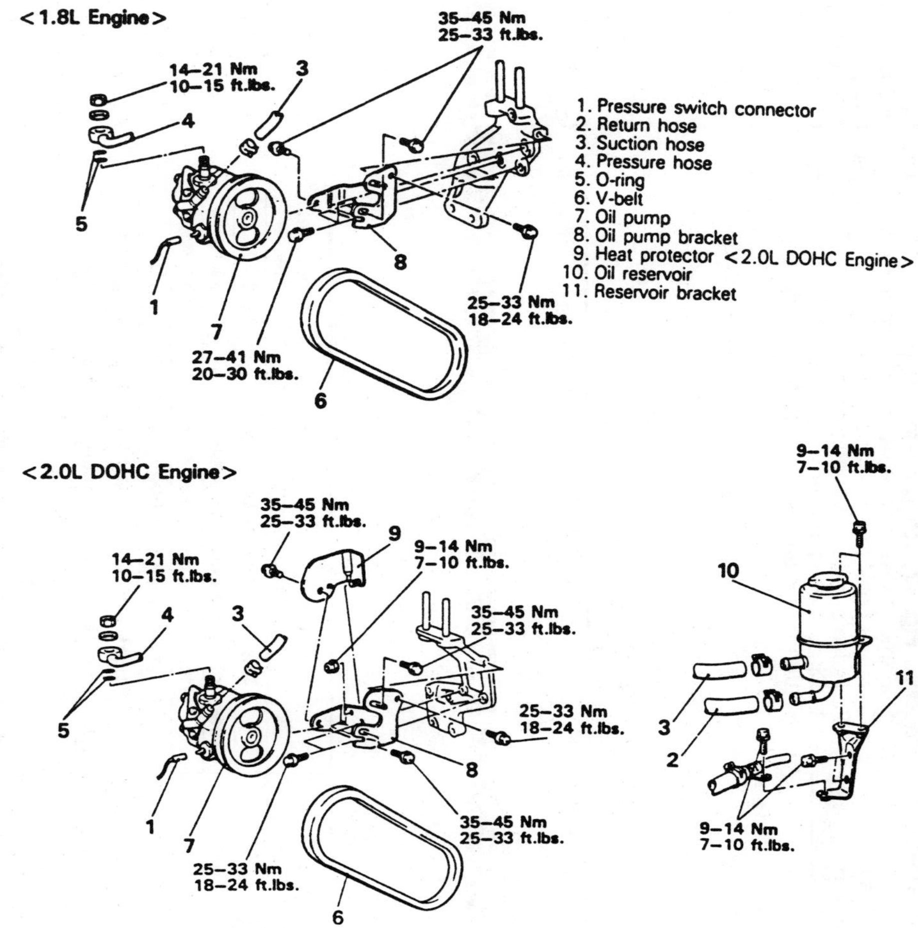

Fig. 17 Removing power steering pump

2. While tilting gear box downward, remove from left side.
3. Reverse procedure to install.

POWER STEERING PUMP REPLACE

Remove steering pump in numbered sequence, **Fig. 17**, noting the following:

1. Remove reservoir cap and disconnect return hose to drain fluid.
2. Raise and support vehicle.
3. Disconnect high tension cable, then turn crank engine to drain fluid from gear box.
4. Cover alternator (located under oil pump) if any hoses are removed
5. When connecting pressure hose, ensure slit contacts oil pump guide bracket.

FRONT SUSPENSION & STEERING TIGHTENING SPECIFICATIONS

Year	Component	Torque/ft. lbs.
1990	Center member to body	58-72
	Gear box to bracket to crossmember	43-58
	Hub assembly to knuckle	144-188
	Knuckle to ball joint	43-52
	Knuckle to strut assembly	80-101
	Lower arm clamp to crossmember (nut)	25-34
	Lower arm clamp to crossmember (bolt)	58-72
	Lower arm to crossmember	72-87
	Pressure hose to gear box	9-13
	Oil pump to bracket	25-33
	Pressure hose to oil pump	10-15
	Return tube to gear box	9-13
	Stabilizer bar bracket	22-30
	Stabilizer link	25-33
	Stay to crossmember	51-58
	Strut top end nut	43-51
	Strut upper mounting nut	29-39
	Tie rod end ball joint	17-25
	Tie rod end to rack	58-72

Chrysler Motors Rear Wheel Drive Models Except Conquest

INDEX

DESCRIPTION

This front suspension, **Figs. 1 and 2,** incorporates two transverse torsion bars which react on the outboard end of the lower control arms. The torsion bars are anchored in the front crossmember opposite the affected wheel. The torsion bars are mounted parallel to the front crossmember through a "Pivot Cushion Bushing," attached to the crossmember, and turns and extends rearward to the lower control arm. The torsion bar ends are provided with an isolated bushing, bolted to the lower control arm and sway bar, which acts as the lower control arm strut.

Riding height is controlled by the torsion bar adjusting bolts on the anchor end of the torsion bar. The right torsion bar is adjusted from the left side and the left torsion bar is adjusted from the right side.

The torsion bar assembly incorporates the "Pivot Cushion Bushing," **Fig. 3** and "Bushing to Lower Control Arm." The lower control arm inner ends are bolted to the crossmember and pivots through bushings.

Caster and camber settings are made by loosening the upper control arm pivot bar bolt nuts and adjusting as necessary.

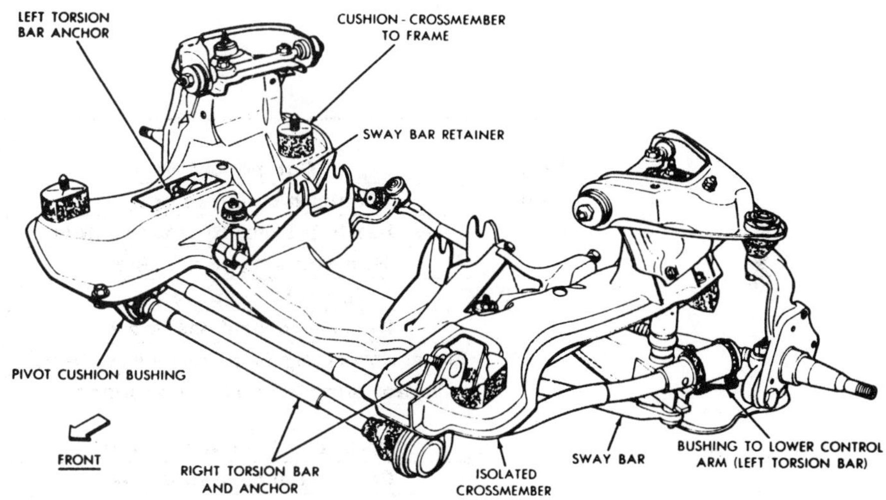

Fig. 1 Transverse torsion bar front suspension

WHEEL BEARINGS
ADJUST

1. Tighten adjusting nut to 20-25 ft. lbs. while rotating wheel.
2. Back off adjusting nut 1/4 turn.
3. Finger tighten adjusting nut while rotating wheel, then align nut lock with cotter pin slot and install cotter pin.
4. The resulting adjustment should be .0001-.003 end play.

WHEEL BEARINGS
REPLACE

1. Raise car and remove front wheels.
2. Remove grease cap, cotter pin, locknut and bearing adjusting nut.
3. Remove bolts that attach caliper to steering knuckle.
4. Slowly slide caliper up and away from disc and support caliper on steering knuckle arm. **Do not allow caliper to hang by brake hose.**
5. Remove thrust washer and outer bearing cone. Remove hub and disc assembly. Grease retainer and inner bearing can now be removed.
6. Reverse procedure to install.

CHECKING BALL JOINTS FOR WEAR

UPPER BALL JOINT

1. Position a suitable jack under lower control arm and raise wheel and tire assembly clear of floor, then remove wheel cover and wheel bearing dust cover and cotter pin.
2. Tighten wheel bearing adjusting nut just enough to remove all play between hub, bearings and spindle.
3. Lower jack positioned under lower control arm to allow tire to lightly contact floor.
4. Grasp top of tire and move wheel and tire assembly inward and outward. While moving tire inward and outward, check for movement at ball joints between steering knuckle and upper control arm.
5. If any lateral movement is present, the upper ball joint should be replaced.
6. After completing upper ball joint check, readjust wheel bearing as described under "Wheel Bearings, Adjust."

LOWER BALL JOINT

If loose ball joints are suspected, first make sure the front wheel bearings are properly adjusted and that the control arms are tight.

1. Raise front of vehicle and place jack stands underneath each lower control arm as far out as possible. **The upper control arms must not contact the rubber rebound bumpers.**
2. With weight of vehicle on lower control arms, attach dial indicator onto lower control arm, **Fig. 4.**
3. Place dial indicator plunger tip against ball joint housing and zero dial indicator.
4. Using a pry bar under the center of the tire, raise and lower the tire and measure the axial travel of the ball joint housing with respect to the ball joint.

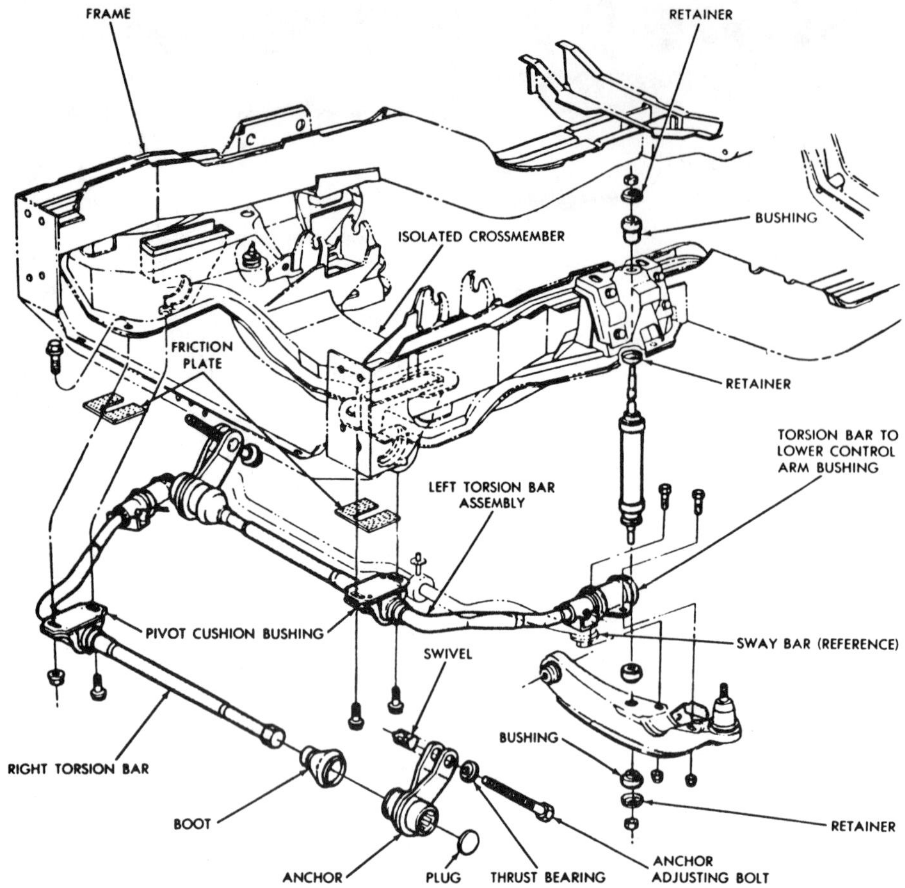

Fig. 2 Transverse torsion bar installation

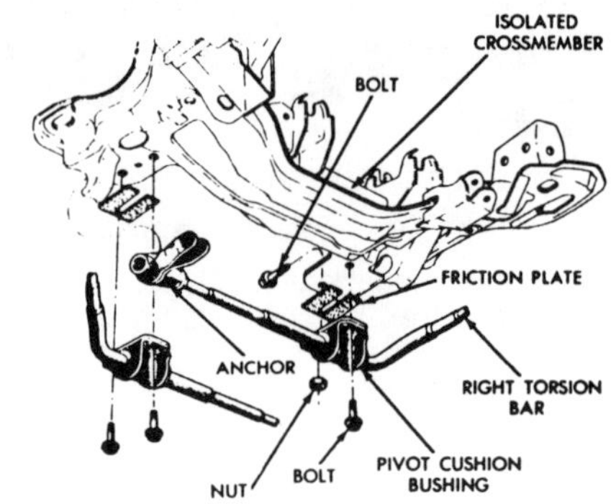

Fig. 3 Pivot cushion bushing & torsion bars

Fig. 4 Checking lower ball joint for wear

If the axial travel is .030 inch or more than specified, the ball joint should be replaced.

BALL JOINTS REPLACE

UPPER BALL JOINT

1. Place ignition switch in the "Off" position.
2. Using a suitable jack raise front of vehicle and position a jack stand under lower control arm as close to wheel and tire assembly as possible. Check to ensure that jack stand is not in contact with brake splash shield. Also check to ensure that rubber rebound bumper is not in contact with frame. **The torsion bar will remain in the loaded position.**
3. Remove wheel and tire assembly.
4. Remove cotter pin and nut from lower ball joint stud. Position tool No. C3564-A over lower ball joint stud, **Fig. 5,** allowing tool to rest on knuckle arm, then set tool securely against upper ball joint stud.
5. Tighten tool to apply pressure against upper ball joint stud, then strike knuckle with hammer to loosen stud.
6. Remove tool, then detach upper ball joint from knuckle. **Support knuckle and brake assembly to prevent damage to lower ball joint and brake hoses.**
7. Remove upper ball joint from upper control arm, using tool No. C3560.
8. Reverse procedure to install. Thread upper ball joint into control arm as far as possible by hand. Torque upper ball joint into control arm to specifications. After tightening lower ball joint stud nut, install cotter pin. **Ball joint seals should be replaced whenever they have been removed.**

LOWER BALL JOINT

1. Place ignition switch in the "Off" position.
2. Raise vehicle and support so front suspension is in the full rebound position. Position jack stands under front frame for additional support.
3. Remove wheel and tire assembly, then remove disc brake caliper and support with wire hook to prevent brake hose from becoming damaged.
4. Remove disc brake hub and rotor assembly and splash shield, then disconnect shock absorber at lower mounting.
5. Release load on torsion bar by rotating adjusting bolt counterclockwise.
6. Remove upper and lower ball joint stud nuts and cotter pin, then position tool No. C3564-A over upper ball joint stud so that tool is resting on steering knuckle, **Fig. 5.**
7. Rotate threaded portion of tool to lock it against lower ball joint stud. Tighten tool to place pressure on lower ball joint stud, then strike steering knuckle with a hammer to loosen stud. Remove tool and disconnect lower ball joint.
8. Use tool No. C4212 to press ball joint from lower control arm.
9. Position replacement ball joint on lower control arm, then press into control arm using tool No. C4212.
10. Install seal over lower ball joint. Use tool No. C4039 to press retainer portion of seal until it is locked in position.
11. Position lower ball joint to steering knuckle, then install upper and lower stud nuts and torque to specifications. After tightening stud nuts, install cotter pin.
12. Place tension on torsion bar by rotating adjusting bolt clockwise.
13. Install disc brake assembly and wheel and tire assembly, then adjust front wheel bearing as described under "Wheel Bearings, Adjust."
14. Lubricate ball joint, then lower vehicle and adjust vehicle riding height.

TORSION BAR
REPLACE
REMOVAL

1. Raise vehicle and support so front suspension is in full rebound position.
2. Rotate anchor adjusting bolts located in frame crossmember, counterclockwise to release load on both torsion bars. Then, remove anchor adjusting bolt from torsion bar to be removed.
3. Raise lower control arms until $2\frac{7}{8}$ inch clearance is obtained between crossmember ledge at jounce bumper and the torsion bar end bushing and support lower arms at this height. **This procedure will align the sway bar and the lower control arm attaching points for disassembly and component realignment and attachment during assembly.**
4. Remove sway bar to control arm attaching bolt and retainers, then the two bolts securing torsion bar end bushing to lower control arm.
5. Remove two bolts securing torsion bar pivot cushion bushing to crossmember, then the torsion bar and anchor assembly from crossmember.
6. Separate anchor from torsion bar.

INSPECTION

1. Inspect seal for damage and replace, if necessary.
2. Inspect bushing to lower control arm and pivot cushion bushing. Inspect seals on cushion bushing for cuts, tears or severe deterioration that may allow moisture to enter under cushion. If corrosion is evident, replace torsion bar assembly.
3. Inspect torsion bars for paint damage and touch up, if necessary.
4. Clean anchor hex openings and torsion bar hex ends.
5. Inspect torsion bar adjusting bolt and swivel for damage or corrosion and replace, if necessary.

INSTALLATION

1. Slide balloon seal over torsion bar end with cupped end facing toward hex.
2. Lubricate torsion bar hex end with lubricant, P/N 2525035, and install hex end into anchor bracket. With the torsion bar in horizontal position, the anchor bracket ears should be positioned nearly straight upward. Position swivel into anchor bracket ears.
3. Install torsion bar anchor bracket assembly into crossmember anchor retainer, then the anchor adjusting bolt and bearing.
4. Install two bolt and washer assemblies securing pivot cushion bushing to crossmember. Leave assemblies loose enough to install friction plates.
5. With lower control arms supported as outlined in step 3 under "Removal," install the two bolt and nut assemblies securing torsion bar bushing to lower control arm and torque nuts to specification.
6. Ensure that torsion bar anchor bracket is fully seated in crossmember. Then install friction plates between crossmember and pivot cushion bushing with open end of slot to rear and bottomed out on mounting bolt. Torque cushion bushing bolts to specification. Place balloon seal over anchor bracket.
7. Install new bolt through sway bar, retainer cushions and sleeve and attach to lower control arm end bushing, then torque bolt to specifications.
8. Rotate anchor adjusting bolt clockwise to load torsion bar.
9. Lower vehicle and adjust riding height.

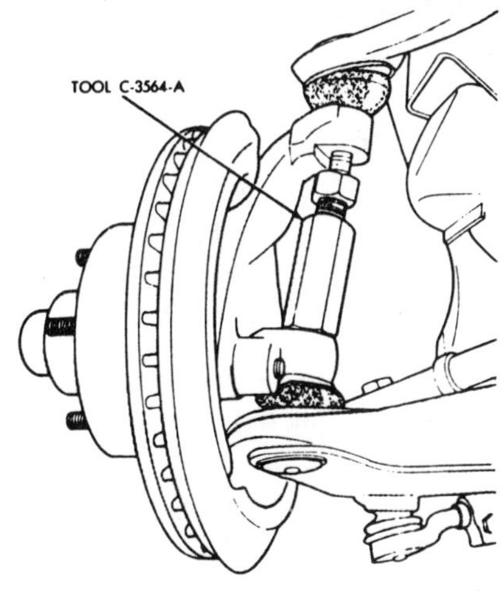

Fig. 5 Removing upper ball joint stud using tool C-3564-A

SWAY BAR
REPLACE

1. Raise and support front of vehicle. **Sway bar to lower control arm attaching points are aligned only when lower control arms are at design height. If frame contact or twin post hoist is used, release load on torsion bar by turning adjuster bolts counterclockwise, then raise lower control arms until clearance between crossmember ledge and torsion bar to lower control arm bushing is $2\frac{7}{8}$ inches. Support lower control arms with jack stand during sway bar removal and installation.**
2. With lower control arms properly supported, remove sway bar to torsion bar bushing attaching bolts, retainers, cushions and sleeves.
3. Remove retainer assembly strap bolts and retainer straps, then remove sway bar.
4. Reverse procedure to install. Inspect cushions and bushings for excessive wear or deterioration and replace as necessary.

POWER STEERING GEAR
REPLACE

1. Disconnect battery ground cable.
2. Remove steering column.
3. Disconnect fluid hoses from steering gear and support free ends above pump to avoid loss of fluid. Plug fittings on gear.
4. Disconnect steering arm from gear with suitable puller.
5. Remove gear to frame retaining bolts or nuts and remove gear.
6. Reverse procedure to install.

POWER STEERING PUMP
REPLACE

1. Loosen power steering pump mounting and locking bolts, then remove drive belt.
2. Disconnect pressure and return lines at power steering pump.
3. Remove pump mounting bolts, then remove pump and mounting bracket.
4. Reverse procedure to install.

FRONT SUSPENSION & STEERING TIGHTENING SPECIFICATIONS

Year	Component	Torque/ ft. lbs.
1985–89	Upper Control Adjusting Bolt Nut	150
	Ball Joint To Control Arm	125
	Ball Joint Stud Nut	100
	Rebound & Jounce Bumpers	200①
	Lower Control Arm Pivot Shaft Nut	75
	Steering Knuckle Bolts/Nuts	160
	Sway Bar Link Retainer Nut	100①
	Sway Bar Strap Nut	30
	Idler Arm Bolt Nut	70
	Tie Rod End	40
	Tie Rod Sleeve Clamps	150①
	Torsion Bar Bushing To Lower Control Arm Bolt/Nut	70
	Torsion Bar Pivot Cushion Retainer Nut/Bolt	85
	Shock Absorber Nut (Upper)	25
	Shock Absorber Nut (Lower)	35
	Sway Bar Cushion Bolt	65
	Upper Control Arm Pivot Bushing Nut	110

①—Inch lbs.

Eagle 4 x 4

INDEX

DESCRIPTION

These models utilize an independent front coil spring suspension system, **Fig. 1.** The wheels are suspended on upper and lower control arms with a coil spring located between the upper control arm and a spring seat beneath the wheel housing panels.

Direct acting telescoping shock absorbers are located inside the coil springs. A stabilizer bar attached to each control arm with stabilizer links provides stability, and a dampener dampens steering oscillations.

CHECKING BALL JOINTS FOR WEAR

Before checking ball joints for wear, make sure the front wheel bearings are properly adjusted and that the control arms are tight.

Raise wheel with a jack placed under the frame as shown. Then test by moving the wheel up and down to check axial play, and rocking it at the top and bottom to measure radial play.

The upper ball joint should be replaced if total travel when rocking and tire exceeds .160 inch.

The lower ball joint is spring loaded and should be replaced if there is any noticeable lateral shake.

If lower ball joint is worn excessively, the lower control arm and ball joint must be replaced as an assembly.

BALL JOINTS REPLACE

UPPER BALL JOINT

1. Position a 2x4x5 inch block of wood on side sill under upper control arm.
2. Raise vehicle and support at frame side sills.
3. Remove wheel, brake drum, caliper and rotor assembly. **When removing caliper do not damage brake tubing or hose. Secure caliper to frame with wire.**
4. Remove cotter pin and retaining nut from upper ball joint stud.
5. Install tool No. J-9656 on ball stud, then using a hammer strike tool to loosen ball stud in steering knuckle.

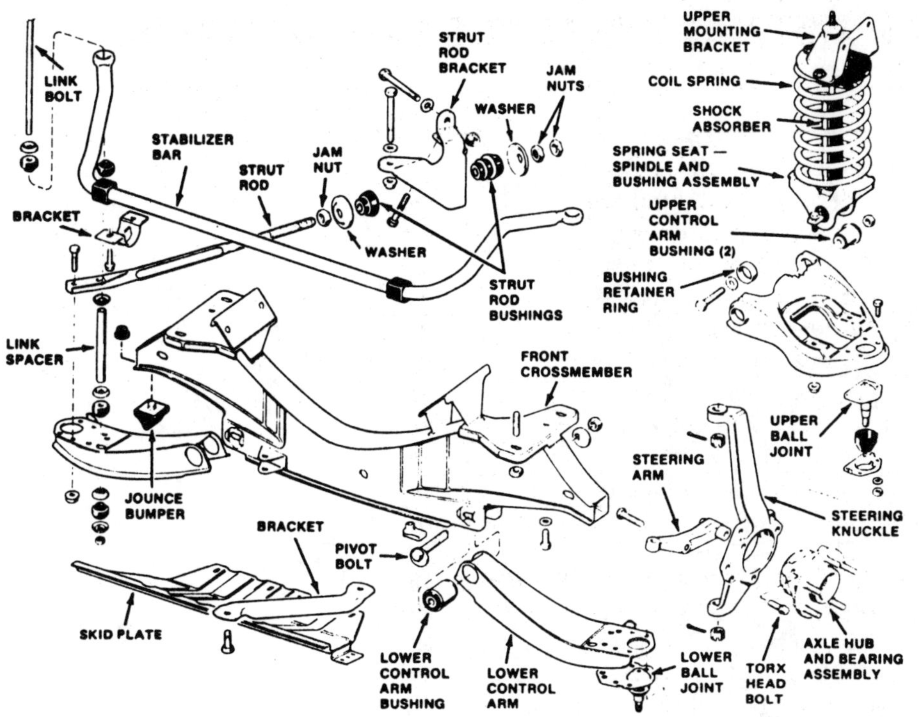

Fig. 1 Front suspension

6. Support lower control arm with a suitable jack.
7. Chisel heads from rivets attaching ball joint to upper control arm, then drive rivets out using a punch.
8. Remove tool from ball stud and ball joint from steering knuckle.
9. Position ball joint on control arm and install nuts and bolts. **Torque** bolts to 25 ft. lbs.
10. Reverse procedure to assemble. Torque ball joint stud nut to specifications.

LOWER BALL JOINT

The lower control arm and lower ball joint must be replaced as an assembly.

1. Remove cotter pin, nut lock and hub nut, then raise and support front of vehicle.
2. Remove wheel, caliper and rotor.
3. Remove lower ball joint cotter pin and retaining nut.
4. Using tool No. J-9656, disconnect lower ball joint from steering knuckle.
5. Remove half shaft flange bolts, then remove half shaft.
6. Remove bolts attaching strut rod to lower control arm, then disconnect stabilizer bar from lower control arm.
7. Remove inner pivot bolt, then remove lower control arm from crossmember.
8. Position control arm on crossmember, then install inner pivot bolt. Do not tighten inner pivot bolt at this time.
9. Position lower ball joint on steering knuckle, then install ball joint stud nut. **Torque** ball joint stud nut to specifications, then install cotter pin.
10. Connect stabilizer bar to lower control arm and torque locknut to specifications.
11. Connect strut rod to lower control arm and torque attaching bolts to specifications.
12. Install half shaft to axle flange and **torque** bolts to 45 ft. lbs.
13. Position a jack under control arm and raise control arm slightly to compress coil spring, then tighten inner pivot bolt to 110 ft. lbs.
14. Install rotor, caliper and hub nut. **Torque** hub nut to 180 ft. lbs., then install nut lock and cotter pin.
15. Install wheel and tire assembly, then lower vehicle and check wheel alignment.

SHOCK ABSORBER REPLACE

After disconnecting shock absorber from wheel well panel at top and lower spring seat at the bottom, withdraw shock absorber out of top of wheel well.

SPRING REPLACE

1. Remove shock absorber.
2. Install spring compressor tool J-23474 through upper spring seat opening, place tool lower attaching screws through shock absorber mounting holes in the lower spring seat. Install tool lower retainer.
3. Remove lower spring seat pivot retaining nuts.
4. Tighten compressor until spring is compressed approximately 1 inch.
5. Raise and support front of car under frame allowing control arms to fall free of lower spring seat. Remove wheel.
6. Pull lower spring seat away from car. Loosen compressor and allow lower spring seat to come out.
7. When all spring tension is released, remove tool lower retainer spring seat and spring.
8. Reverse procedure to install.

STABILIZER BAR REPLACE

1. Raise and support front of vehicle, then remove right side wheel.
2. Disconnect idler arm at frame side sill.
3. Remove stabilizer bar mounting clamps at the frame side sills.
4. Disconnect stabilizer bar link at lower control arms, then remove link bolts, rubber grommets, spacer and retainers.
5. Remove stabilizer bar from right side of vehicle. Move steering linkage and left side wheel, as necessary, to obtain clearance.
6. Reverse procedure to install.

STEERING GEAR REPLACE

MANUAL STEERING

1. Remove flexible coupling bolts.
2. Remove pitman arm, using a suitable puller.
3. Remove mounting screws and lower steering gear from vehicle.
4. Reverse procedure to install.

POWER STEERING

1. Disconnect pressure and return hoses from gear. Raise hoses above pump level to keep oil from draining out of pump.
2. Remove flexible coupling bolt nuts, noting the different nut sizes to insure correct assembly.
3. Remove skid plate, if equipped, then remove left hand crossmember to sill support brace and stabilizer bar bracket from frame.
4. Remove pitman arm with a suitable puller.
5. Remove gear attaching bolts.
6. Slide lower shaft free of coupling flange, then remove gear.
7. Reverse procedure to install.

POWER STEERING PUMP REPLACE

LESS A/C

1. Remove power steering pump adjusting stud nuts and pivot bolt, then remove pump drive belt.
2. Disconnect pressure and return lines, then remove bolts attaching power steering pump bracket to engine and remove pump.
3. Reverse procedure to install.

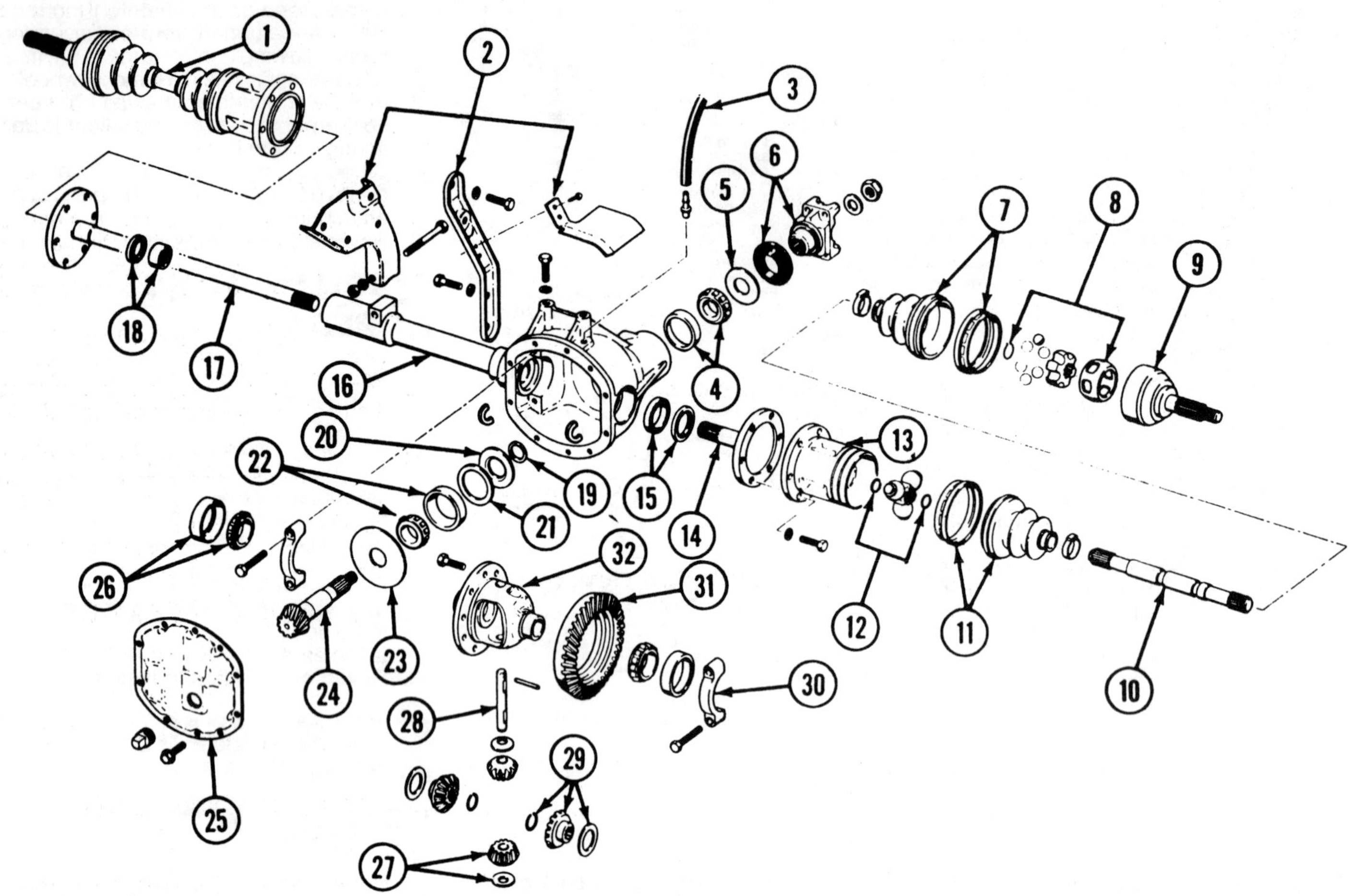

1. HALF-SHAFT ASSEMBLY
2. AXLE MOUNTING BRACKETS
3. VENT HOSE
4. PINION AND FRONT BEARING CUP
5. WASHER
6. YOKE AND SEAL
7. OUTER BOOT AND RETAINER
8. RZEPPA JOINT ASSEMBLY
9. SPINDLE
10. HALF-SHAFT
11. INNER BOOT AND RETAINER
12. TRI-POT JOINT ASSEMBLY
13. TRI-POT HOUSING
14. AXLE SHAFT (SHORT)
15. BALL BEARING AND SEAL
16. AXLE HOUSING
17. AXLE SHAFT (LONG)
18. NEEDLE BEARING AND SEAL
19. PRELOAD SHIM
20. WASHER
21. DEPTH SHIM
22. PINION REAR BEARING AND CUP
23. SLINGER
24. PINION GEAR
25. COVER
26. DIFFERENTIAL BEARING AND CUP
27. DIFFERENTIAL PINION AND THRUST WASHER
28. PINION MATE SHAFT
29. SIDE GEAR, THRUST WASHER AND LOCKRING
30. BEARING CAP
31. RING GEAR
32. DIFFERENTIAL CASE

Fig. 2 Front axle assembly

WITH A/C

1. Remove air induction hoses.
2. Remove air pump adjusting bolt and pivot stud, then remove air pump drive belt.
3. Remove A/C compressor drive belt, then remove compressor pulley and pulley bracket as an assembly.
4. Remove power steering pump pivot bolt attaching lower edge of pump to engine.
5. Remove power steering pump adjusting bracket stud nuts and bolt attaching pump to rear bracket, then slide drive belt from pulley.
6. Pull power steering pump and front mounting bracket forward until bracket clears air pump pivot stud.
7. Disconnect power steering pump pressure and return lines, then remove pump.
8. Reverse procedure to install.

FRONT AXLE SERVICE

Refer to **Fig. 2,** during front axle service procedures.

AXLE, REPLACE

Removal

1. Raise and support vehicle.
2. Install half shaft boot protectors J-28712 onto boots.
3. Remove half shaft to axle flange bolts and secure half shaft to vehicle underbody.
4. Insert wire through half shaft flange bolt holes, then compress half shaft and wrap wire between boots to prevent half shaft from separating.
5. Mark propeller shaft and axle yoke for reference during reassembly and remove propeller shaft.
6. Support axle assembly using a jack.
7. **On models equipped with Select Drive System,** remove brace rod and shift motor shield.
8. **On all models,** remove axle mounting bolts. On models less Select Drive System, the axle is secured to the engine block with five bolts. On models with Select Drive System, the axle is secured to the engine block by five bolts on the left hand side and one

bolt on the right hand side.

9. **On models equipped with Select Drive System,** disconnect vacuum hose from axle shift motor, then partially lower axle assembly and disconnect vent hose.
10. **On all models,** lower axle assembly to allow access to vent hose and remove hose.
11. Remove axle assembly from vehicle. **Do not apply any load or weight on the hub assembly whenever the half shafts are not securely attached to the axle shaft flanges.**

Installation

1. Support axle assembly on a jack, then position assembly under vehicle.
2. Raise axle assembly slightly, then install vent hose.
3. Continue raising axle assembly until properly positioned, then install axle mounting bolts. **Torque** bolts to 50 ft. lbs.
4. Connect propeller shaft to axle yoke. Ensure shaft and yoke are aligned.
5. Remove wire used to prevent half shafts from separating, then install half shaft to axle flange bolts and **torque** to 45 ft. lbs.
6. **On models equipped with Select Drive System,** connect vacuum harness on axle shift motor shield.
7. **On all models,** remove half shaft boot protectors.
8. **On models equipped with Select Drive,** connect brace rod to axle.

AXLE SHAFT SEAL, REPLACE

Removal of the right hand shaft seal is the same as that of the left hand shaft seal.

1. Remove axle assembly.
2. Remove axle housing cover and drain lubricant.
3. Remove axle shaft retaining clip, then the axle shaft.
4. Remove axle shaft seal using suitable screwdriver.
5. Install replacement axle shaft seal using Installer J-29152 for right hand side axle shaft and Installer J-29154 for left hand side axle shaft.
6. Install axle shaft, then the axle shaft retaining clip.
7. Apply sealant to axle housing cover and install cover. **Torque** cover bolts to 20 ft. lbs.
8. Fill axle with 2 1/2 pints of SAE 85W-90 lubricant, then install assembly onto vehicle. Refer to Axle, Installation for procedures.

AXLE HUB & BEARING

Removal

1. Raise and support vehicle, then remove wheel, caliper and rotor.
2. Remove bolts attaching axle shaft flange to half shaft. **Insert wire through half shaft flange bolt holes, then compress half shaft and wrap wire between boots to prevent half shaft from separating.**
3. Remove cotter pin, locknut and axle shaft nut, then the half shaft.
4. Remove steering arm from steering knuckle.
5. Remove caliper anchor plate from steering knuckle, then the three Torx head bolts retaining hub assembly using tool J-26359.
6. Remove hub assembly from steering knuckle and clean grease from steering knuckle cavity.

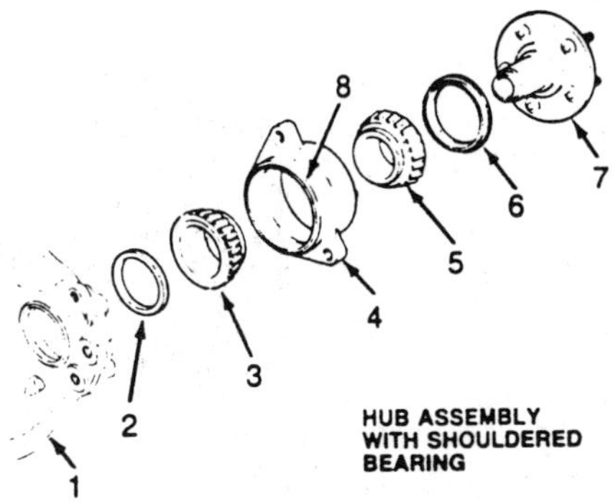

1. Steering Knuckle Pin
2. Hub Spacer
3. Shouldered Bearing
4. Hub and Bearing Carrier
5. Shouldered Bearing
6. Carrier to Hub Seal
7. Hub
8. Carrier O-Ring

Fig. 3 Exploded view of axle hub

Disassembly

1. Press hub from bearing carrier, then remove bearings, hub spacer, steering knuckle pin seal, carrier to hub seal, carrier O-ring and bearing spacer, if equipped, **Fig. 3.**
2. Clean all components and inspect for wear and damage and replace as necessary. **If hub incorporates ball bearings, the entire hub assembly must be replaced if damage to internal components is indicated. If hub incorporates tapered bearings, internal hub components may be replaced as necessary. Bearings and races must be replaced in matched sets.**
3. Press bearing races from hub carrier, if necessary. Install replacement bearing races using a brass drift or suitable press.

Assembly

1. Fill steering knuckle pin, hub, bearing carrier and lip type seal cavities with lithium base wheel bearing lubricant.
2. Pack bearings with lithium base wheel bearing lubricant, then install bearings into hub carrier, **Fig. 3.**
3. Install seal onto hub side of bearing carrier, then press hub into carrier and install hub spacer on end of hub shaft. **Do not install bearing carrier O-ring at this time.**
4. Install inner seal into steering knuckle pin.
5. Install splash shield onto hub and bearing carrier, then install bearing carrier O-ring.
6. Install splash shield and bearing carrier into steering knuckle pin, then install carrier attaching bolts and **torque** to 75 ft. lbs.

Installation

1. Partially fill hub cavity of steering knuckle with chassis lubricant and install hub assembly onto steering knuckle. **Torque** hub torx head bolts to 75 ft. lbs.
2. Install caliper anchor plate and torque plate retaining bolts to specifications.
3. Install steering arm onto steering knuckle and **torque** retaining bolts to specifications.
4. Install half shaft, then the axle flange to shaft bolts and hub nut. **Torque** half shaft to flange bolts to 45 ft. lbs. and hub nut 175 ft. lbs.
5. Install locknut, replacement cotter pin, rotor, caliper and wheel.

AXLE SHAFT, REPLACE

Removal of the right hand shaft is the same as that of the left hand shaft.

1. Refer to Axle Shaft Seal, Replace. Perform steps 1 through 3 for axle shaft removal and steps 6 through 8 for axle shaft installation.

AXLE SHAFT BEARING, REPLACE

Two different style axle shaft bearings are used on the Eagle front axle. The left side axle shaft uses a ball bearing. The right side axle shaft uses a needle bearing. Removal of the right hand side bearing is the same as that of the left hand bearing.

1. Remove axle assembly. Refer to Axle, Removal for procedure.
2. Remove axle housing cover and drain lubricant.
3. Remove axle shaft retaining clip, then the axle shaft.
4. Remove axle shaft seal using suitable screwdriver.
5. Remove needle bearing using Tools J-29173 and J-2619-1. Remove ball bearing using a brass punch and hammer. **If needle bearing tools are not available, remove the differential, then the bearing using a 15/16 inch socket and three foot extension bar.**
6. Install axle shaft bearings. Use tool J-29153 to install needle bearing and tool J-29154 to install ball bearing.
7. Install axle shaft seals using Installer J-29152 for the right hand side axle shaft and Installer J-29154 for left hand side axle shaft.
8. Install axle shafts, then the retaining clips.
9. Apply sealant to axle housing cover and install cover. **Torque** cover bolts to 20 ft. lbs.
10. Fill axle with 2 1/2 pints of SAE 85W-90 lubricant, then install assembly onto vehicle. Refer to Axle, Installation for procedure.

HALF SHAFT SERVICE

Care should be taken when removing or replacing drive shaft rubber boots. These boots are relatively fragile and precautions should be taken to avoid damage. Do not drop the drive shafts on the boots. Although no apparent damage is observed, internal damage could result in eventual rupture. Use care when working around the front axle drive shafts to avoid any contact with the boots. Cover the boots with shop towels.

Disassembly

1. Cut and remove both outer boot clamps, then roll outer boot from constant velocity joint.
2. Position a wooden block against constant velocity joint inner race and drive joint from shaft. If half shaft is to be positioned in a vise, use protective vise jaws.
3. Using a brass drift, tap constant velocity joint cage until cage is tilted outward far enough to remove first ball, then remove remaining balls in the same manner.
4. Align two outer constant velocity joint oblong holes with slots located on interior wall of spindle housing and remove cage and inner race.
5. Align shoulder between race grooves with inside of oblong cage holes, then rotate inner race out of cage using the larger of the two openings.
6. Remove two snap rings from shaft then remove outer boot.
7. Remove rubber retaining ring from small end of inner boot.
8. Cut inner boot retaining strap, then remove strap from tripod housing. Slide boot away from housing.
9. Remove tripod joint and shaft from housing.
10. Remove snap ring that retains tripod joint on shaft using suitable pliers, then remove tripod joint.
11. Remove inner boot from shaft if replacement is required.
12. Remove any remaining snap rings from shaft, if not removed previously.

Inspection

Clean all components in a suitable solvent and dry with compressed air. Check all components for wear and damage and joints for rough operation, replace components as necessary. **Outer constant velocity joint and inner tripod joint are replaced as an assembly.**

Assembly

1. Pack spindle hub with chassis grease.
2. Install outer constant velocity joint inner into cage, then install cage and race assembly into spindle hub. The smaller diameter of the cage and stopping groove of race must face outward.
3. Using a brass drift, tilt cage outward until first ball can be installed into cage, then install remaining balls in the same manner.
4. Pack constant velocity joint with chassis grease, then install outer boot on half shaft.
5. Install inner and outer snap rings, then constant velocity joint onto shaft until inner race contacts inner snap ring.
6. Position boot over constant velocity joint, then install both boot clamps using tool No. J-22716 or equivalent. **Check and adjust installed length of boot. Total installed length should be 4.134 inch. Move small end of boot in or out on shaft to obtain specified length.**
7. Pack tripod joint housing with chassis grease, then install inner small retainer ring, boot and large retainer ring onto half shaft.
8. Install tripod joint and snap ring onto shaft, then install tripod joint and shaft into housing.
9. Position boot and install rubber retainer ring over small end of boot.
10. Install large end boot clamp.

INTERMEDIATE SHAFT, OUTER AXLE SHAFT & SHIFT HOUSING, REPLACE

Models w/Select Drive System

1. Raise and support front of vehicle.
2. Remove front axle assembly, then drain lubricant.
3. Remove bolts securing shift housing and remove shift fork, shifting housing and shift motor as an assembly.
4. Working through access hole in the outer axle shaft flange, remove screws securing axle tube cover.
5. Remove outer axle shaft assembly by tapping shaft flange with a rubber or plastic mallet.
6. Remove intermediate shaft snap ring, then remove intermediate shaft and shift collar.
7. Using a puller and slide hammer, remove outer axle shaft bearing race.
8. Using adapter tool No. J-26225 and slide hammer J-6471-2, remove intermediate shaft needle bearing from axle tube end.
9. Using spreader tool No. J-29369-1 and slide hammer No. J-2619-01, remove intermediate shaft gear bearing. **The intermediate shaft gear is retained on the axle shaft by an internal type expandable snap ring. When removing the bearing from the gear be sure to support the gear face on vise jaws to avoid pulling the gear off the shaft. The gear and shaft are serviced only as an assembly.**
10. Remove outer axle shaft retaining E-ring.
11. Make a reference mark on outer axle shaft gear (for assembly purposes) and remove gear from shaft.
12. Remove outer axle shaft bearing using an arbor press.
13. Remove axle oil seal and tube cover.
14. Clean and carefully inspect all components for wear or damage. Replace all necessary components.
15. Install axle oil seal and tube cover on axle shaft.
16. Using an arbor press, install axle shaft bearing and race on shaft.
17. Install axle shaft gear on shaft using an arbor press. Be sure to align reference mark. Also ensure gear splines are facing outward.
18. Install axle shaft E-ring.
19. Using tool No. J-29153, install intermediate shaft needle bearing.
20. Install intermediate shaft gear needle bearing in gear bore using an arbor press.
21. Install intermediate shaft, then install shift collar, and seat shaft in differential.
22. Install intermediate shaft lock ring.
23. Install outer axle shaft assembly, torque axle tube cover bolts evenly and in a cross sequence. **Torque** bolts to 12 ft. lbs.
24. Install gasket on shift housing cover and install. **Check to ensure the shift fork and tabs are aligned in the shift collar.**
25. **Torque** shift housing cover bolts to 108 inch pounds.
26. Install axle cover using a new gasket. **Torque** bolts to 20 ft. lbs.
27. Fill shift housing with five ozs. of specified lubricant, then fill axle housing.

SELECT DRIVE SYSTEM

The Select Drive is a system which permits the full time four wheel drive power train to be operated in a two wheel drive mode. When the vehicle is stopped, the system can be changed from two wheel drive to four wheel or four wheel drive to two wheel drive, by pulling the instrument panel control switch pull release pin downward and moving the control switch lever to the desired position. The two major components of this system are a shifting device in the transfer case and a disconnecting mechanism in the front axle, which are controlled by the instrument panel switch through a series of vacuum actuators.

The vacuum actuated shifter mechanism located on the transfer case slides a splined clutch, located on the transfer case main shaft, rearward to engage or forward to disengage power to the front drive chain sprocket. In the two wheel drive position, the shifter mechanism also locks the transfer case differential. Also the main shaft and differential assembly will turn as a unit, transferring power to the rear driveshaft, while front output shaft, chain and sprocket remain stationary.

A vacuum actuated disconnect device is used to engage or disengage a splined clutch that connects the right hand driveshaft to the right hand half shaft. When four wheel drive is selected, the spline clutch is engaged and the front drive axle operates in a normal manner. When two wheel is selected, the vacuum actuator releases the spline clutch, which releases drive to the right hand half shaft and allows the right hand driveshaft to rotate freely. Due to the design of the differential, to equalize torque between left and right axle shafts, disconnecting the right axle shaft will allow the left axle shaft to rotate freely. The axle disconnect will not release all front drive components in the two wheel drive mode. Half shafts and different side gears will continue to rotate with the front wheels as the vehicle is driven.

Axle Shift Motor, Replace

1. Make a reference mark on fork and housing for assembly purposes.
2. Rotate shift motor and remove retaining snap rings, using a screwdriver.
3. Remove shift motor from housing and remove O-ring from motor.
4. Reverse procedure to install.

FRONT SUSPENSION & STEERING TIGHTENING SPECIFICATIONS

Year	Component	Torque/ft. lbs.
1985-89	Ball Joint Stud Nut	75
	Caliper Anchor Plate Bolt	80
	Crossmember To Sill Bolt/Nut	65①
	Jounce Bumper To Sill Bolt	25
	Lower Control Arm Pivot Bolt	110
	Shock Absorber Nut	8
	Shock Absorber Mounting Bracket Bolt	20
	Spindle To Anchor Plate Bolt	55
	Steering Arm Bolt	100
	Strut Rod To Bracket Jam Nut	75
	Strut Rod To Bracket Nut	65
	Strut Rod Bracket To Still Bolt	75
	Strut Rod Nut Bolt To Lower Control Arm	75
	Sway Bar Clamp To Still Bolt	25
	Sway Bar Link To Lower Control Arm Nut	7
	Tie Rod Adjusting Sleeve Clamp Bolt Nut	12
	Upper Control Arm Inner Pivot Bolt	80

①—Minimum.

Eagle Medallion & Premier

INDEX

COIL SPRING & STRUT SERVICE

MEDALLION

Removal

Do not support vehicle weight using the lower control arms or damage may result.

1. Disconnect battery ground cable.
2. Raise and support vehicle.
3. Remove tire and wheel assembly.
4. Using tool T.Av.476 or equivalent, remove outer tie rod, **Fig. 1.**
5. Remove two attaching bolts securing lower portion of strut. The attaching bolts are splined and require the use of a brass of lead mallet for removal. **Prior to removing the two lower strut attaching bolts, ensure front suspension is hanging free. There should be no pressure or tension on any front suspension component.**
6. Remove upper attaching bolts.
7. Press on lower control arm and guide strut/coil spring assembly past the drive axle to prevent the strut from coming into contact with the driveshaft boot.
8. To separate coil spring from strut, proceed as follows:
   a. Place the lower plate (C) from tool Sus.1052.99 into a suitable vise, **Fig. 2.**
   b. Place lower adapter plates (A, shoulder facing downward) into

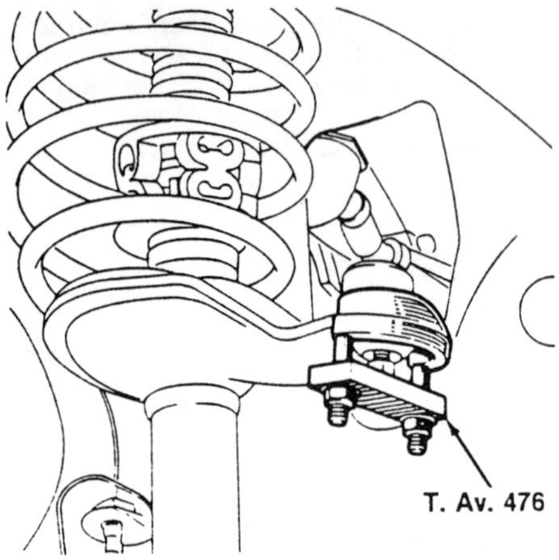

Fig. 1 Using tool T.Av.476 to remove outer tie rod end. Medallion

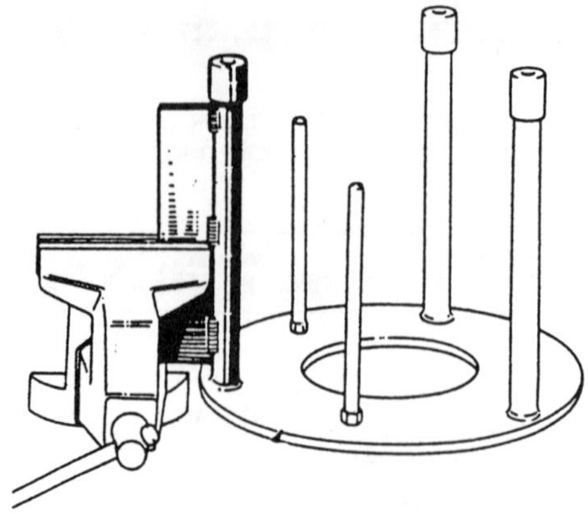

Fig. 2 Lower adapter plate assembly. Medallion & Premier

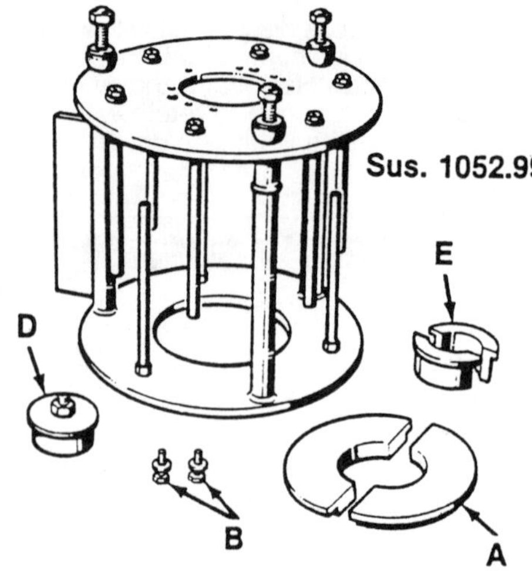

Fig. 3 Using tool Sus.1052.99. Medallion

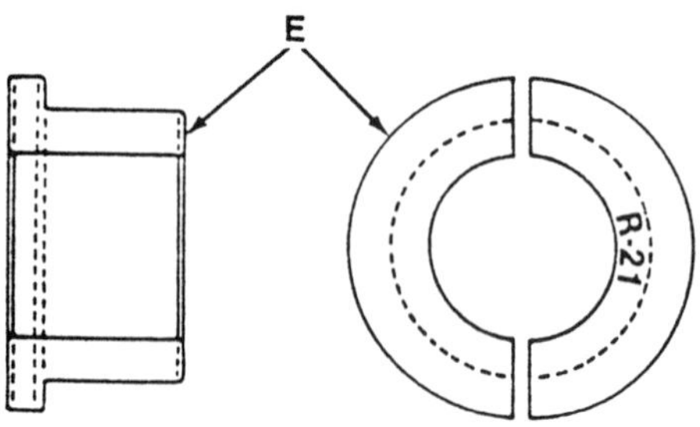

Fig. 4 Adapter plates R-21. Medallion

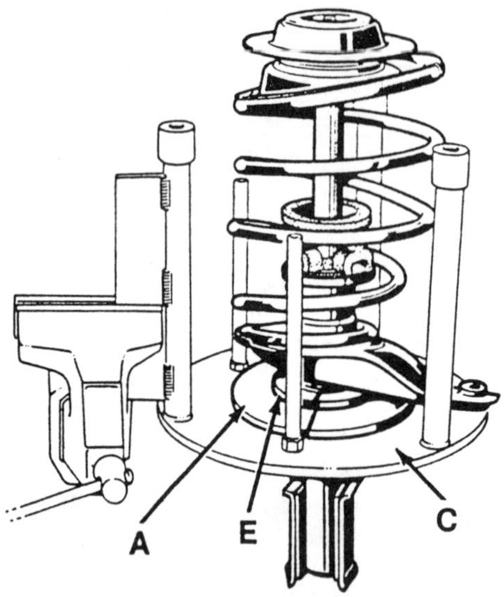

Fig. 5 Placing coil spring/strut assembly into tool. Medallion

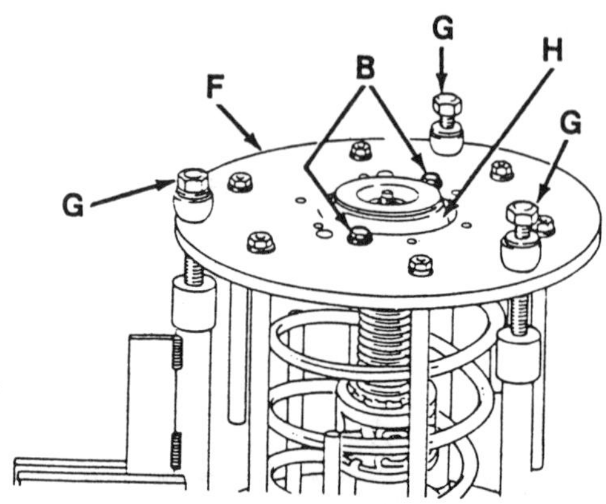

Fig. 6 Installing upper plate to lower plate & aligning plates using threaded rods. Medallion

the lower plate (C), **Fig. 3.** Tool set Sus.1052.99 is supplied with different adapter plates (E) to fit various Renault models. For use on the Medallion, locate and use adapter stamped R-21 and shaped as shown in **Fig. 4.**

c. Place adapter plates (E) around the lower portion of the strut assembly, then position lower strut assembly into plates (A), **Fig. 5.** Ensure strut assembly is properly seated before proceeding with next step.
d. Install upper plate (F) to lower plate (C) while aligning plates with the threaded rods (G), **Fig. 6.**
e. Install attaching bolts (B, from tool Sus.1052.99) through adapter plate (F) and into original strut mounting plate (H), **Fig. 6.** The upper plate (F) has multiple mounting holes used for various Renault models. Install bolts (B) through holes stamped R-21.
f. Liberally apply clean oil to points indicated (J) as shown in **Fig. 7.** The tool Sus.1052.99 is under extreme tension and oil application is necessary.
g. Using threaded rods (G), slowly compress the upper plate (F) approximately .393 inches.
h. Using a suitable Allen wrench, remove strut locknut.
i. Using the threaded rods (G), slowly remove all tension from coil spring.
j. Remove upper plate and strut components A through I, **Fig. 8.**

Installation

1. To attach coil spring to strut assembly, proceed as follows:
   a. Invert upper plate (F), **Fig. 9.**
   b. Insert bolt from adapter cup (D), through top of strut mounting plate, **Fig. 9.**
   c. Invert strut mounting plate on upper plate.
   d. Assemble (in order) the pivot bearing and spring retainer (E) using adapter cup (D). Install nut onto adapter cup finger tight.
   e. Position strut (H) into adapter plates (A) and (E). Position jounce bumper (G), **Fig. 10.**
   f. Place upper plate (F), with adapter cup (D) and strut components, onto lower plate (C). The upper portion of the coil spring must be rotated to fit the "stop" located on the spring retainer.
   g. Using the threaded rods (G), slowly compress upper plate (F) until dimension X is 15.70 inches, **Fig. 11.** Ensure upper portion of the coil spring is located in the stop on the spring retainer.
   h. Remove adapter cup (D) from the strut assembly.
   i. Continue to tighten threaded rods (G) until the threaded portion of the strut shaft has passed through the upper plate (F) enough to install the locknut, **Fig. 11.** If difficulty is experienced in bringing the strut shaft through the strut mounting, position a suitable screwdriver into the hex hole in the top end of the shaft for alignment.
   j. Install bearing and locknut.
   k. Tighten lock nut, then slowly release tension from the threaded rods until all tension is relieved from the coil spring.
   l. Remove upper plate (F) from strut assembly.
   m. Remove strut assembly from lower plate.
2. To install coil spring/strut assembly, proceed as follows:
   a. Install strut assembly in position

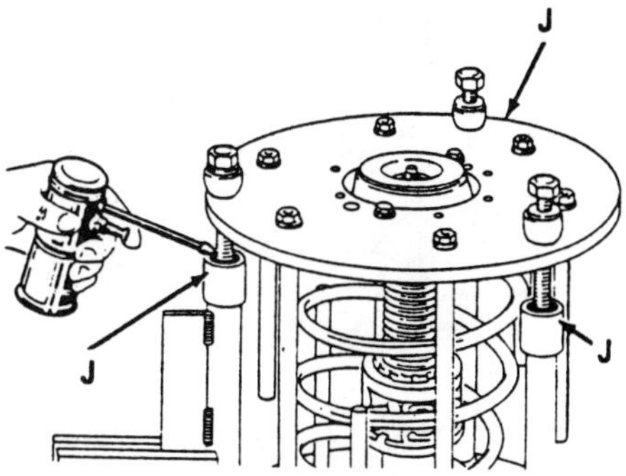

Fig. 7 Applying oil to points (J). Medallion

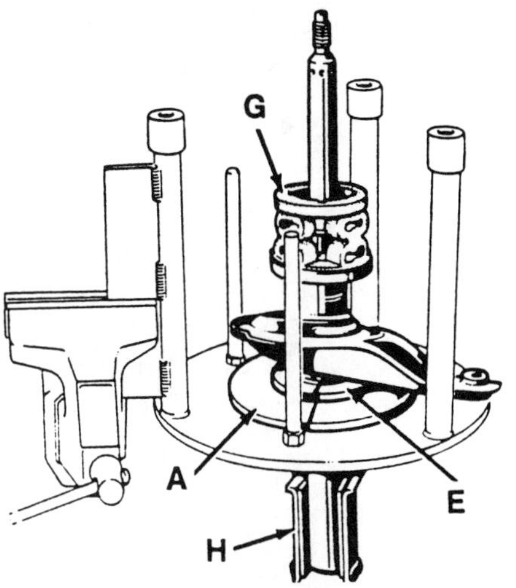

Fig. 10 Position jounce bumper, upper plate, adapter cup, strut components onto lower plate assembly. Medallion

using the three upper attaching bolts. Finger tighten bolts.

b. When installing do not damage driveshaft boot.
c. Install outer tie rod end.
d. Install lower attaching bolts.
e. Install upper attaching bolts.
f. Since the outer tie rod end is attaching to the strut body, check toe measurement if the strut has been replaced.

PREMIER

Removal

1. Raise and support vehicle. **Ensure vehicle weight is not supported under front lower control arms.**
2. Remove front tire and wheel assemblies.
3. Remove outer tie rod end securing nut, then the tie rod using suitable tool.
4. Remove three strut to body attaching nuts.
5. Remove securing nuts from bottom of shock absorber.
6. Remove lower shock absorber mounting bolts.
7. To prevent damage, wrap boot then

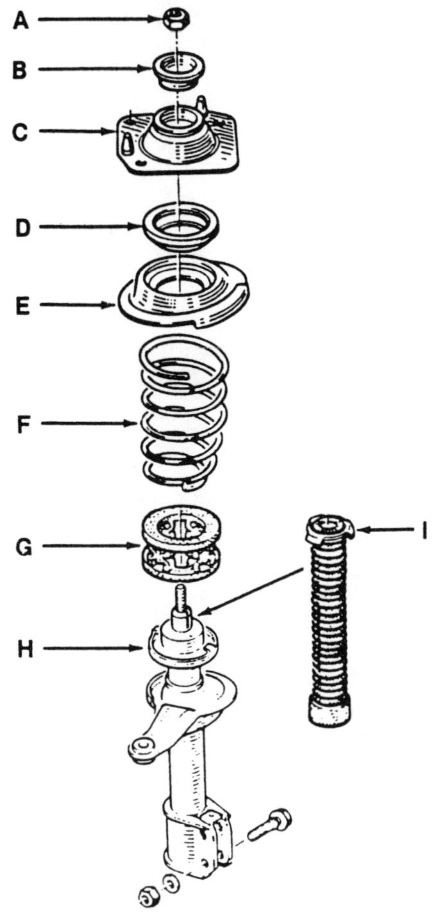

A-Locknut
B-Bearing
C-Shock Mounting
D-Pivot Bearing
E-Spring Retainer
F-Coil Spring
G-Jounce Bumper
H-Shock Absorber
I-Dust Cover

Fig. 8 Coil spring & strut components. Medallion

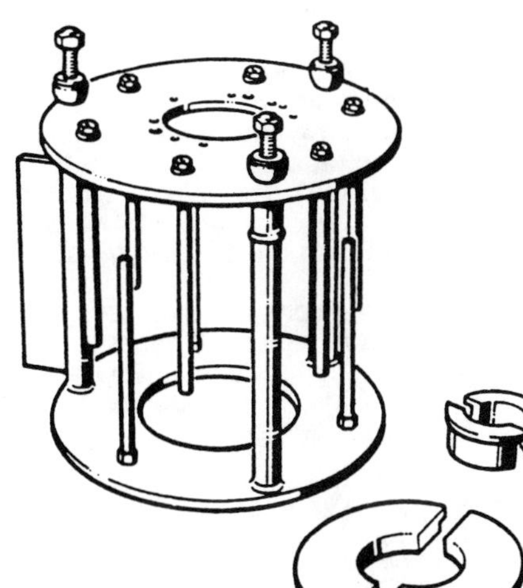

Fig. 12 Installing lower adapter plates. Premier

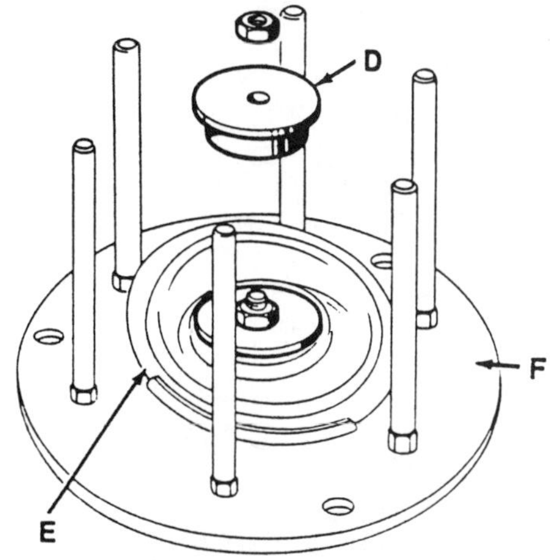

Fig. 9 Assembling pivot bearing, spring retainer & adapter cup. Medallion

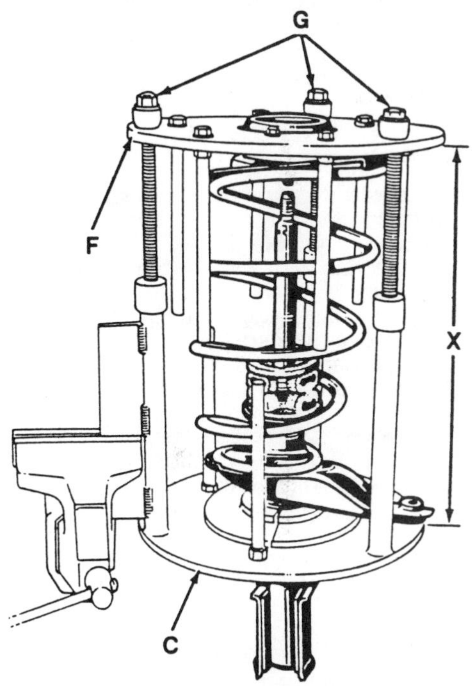

Fig. 11 Obtaining dimension X. Medallion

press down on lower control arm and guide assembly out.

Disassembly

Suspension Tool Kit 1052.99 is supplied with various adapter plates for different models. Premier models use adapter plates stamped R-21.

1. Place lower plate from spring removal tool into a vise, **Fig. 2.**
2. Place lower adapter plates (G) into lower plate, **Fig. 12.**
3. Place smaller adapter plates (F) around lower part of strut assembly and properly seat into lower plate that is secured in the vise.
4. Install top adapter plate (H), **Fig. 13,** onto strut assembly and rotate strut to align with holes on plate. Install three $\frac{5}{16}$ inch by 1.5 inch bolts (L & K) in

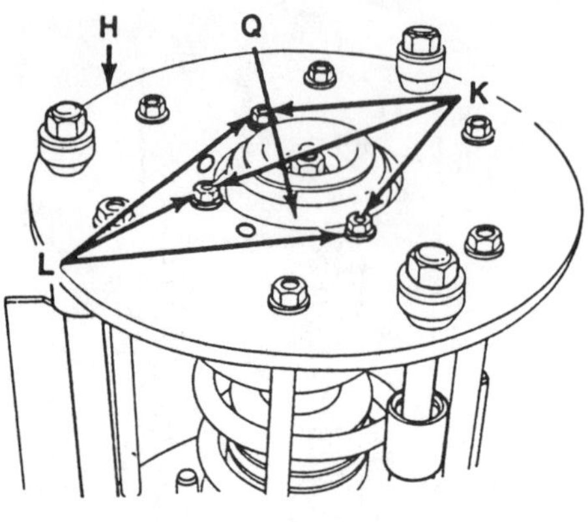

Fig. 13 Installing top adapter plate. Premier

length through top plate and tighten to cushion (Q).

5. Threading rods into mounts, slowly compress coil spring to approximately 13/32 inch.
6. Relieve spring tension evenly by loosening threaded bolts and remove spring.

Strut Cartridge Replacement

To avoid damage to strut assembly, mount strut in strut clamping vise No. YA-457 or equivalent.

1. Align tab (R) with opening (S) on jounce bumper, then remove jounce bumper and dust boot, **Fig. 14.**
2. Remove pressed-on cap. Use care to avoid damaging threads.
3. Remove 7/16 to 31/64 from top of strut body using a suitable cutting tool.
4. Remove and discard strut cartridge, then pour residual oil from strut body.
5. Remove excess metal from inside of strut tube caused by cutting operation.
6. Install new cartridge and threaded cap. Torque threaded cap to specifications.

Assembly

1. Install dust boot over strut body, **Fig. 15.**
2. Inspect rubber isolator tubes for any wear and replace if necessary.
3. Rotate spring on strut assembly so it bottoms into stop. Spring should be separated by isolator tubes.
4. Place spring retainer on top of spring so it bottoms on end of spring.
5. Install pivot bearing on top of spring retainer.
6. Place strut cushion on top of pivot bearing.
7. Place lower adapter plates (G) into lower plate, **Fig. 12.**
8. Place smaller adapter plates (F) around lower part of strut assembly and properly seat into lower plate that is secured in the vise.
9. Install top adapter plate (H), **Fig. 13,** onto strut assembly and rotate strut to align with holes on plate, install three 5/16 inch by 1.5 inch bolts (L) in length through top plate and tighten down.
10. Threading rods (M) into mounts (N) slowly compress adapter plate evenly until strut shaft appears through upper plate.

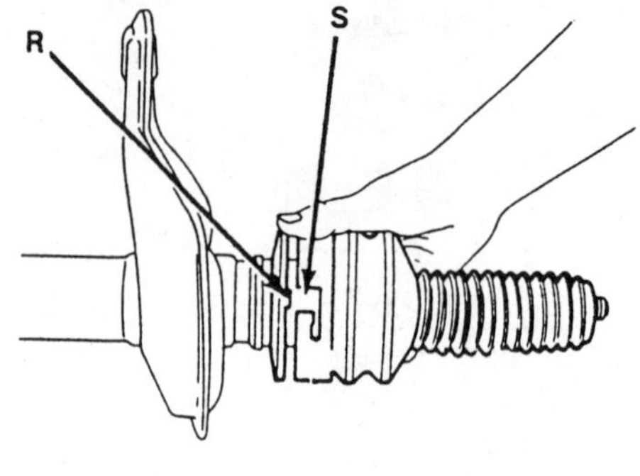

Fig. 14 Removing jounce bumper & dust boot. Premier

11. Install rebound cup and torque to specifications.
10. Remove from tool and install in vehicle.

Installation

1. Position spring/strut assembly on vehicle and finger tighten strut to body nuts.
2. Install lower mounting bolts and nuts, torquing nuts to specifications. **Bolts are splined and must not be tightened.**
3. Install upper mounting bolts and torque to specifications.
4. Install tie rod end to strut nut and torque to specifications.
5. Install wheel and tire assembly, lower vehicle.

STABILIZER BAR REPLACE

MEDALLION

Vehicle weight must rest on wheels for easier removal of bar.

1. Remove stabilizer bar bracket to body attaching bolts.
2. Remove stabilizer bar bracket to control arm attaching nuts. **The nuts and bolts securing the stabilizer bar lower bracket also retains the lower ball joint(s). Replace a nut on one of the two bolts to ensure ball joint does not move out of position.**
3. Check condition of stabilizer bar bushings and replace, if necessary.
4. Reverse procedure to install. Coat all new bushings using lubricant No. 8993630 or equivalent.

ANTI-SWAY BAR REPLACE

PREMIER

1. Remove anti-sway bar clamp bolts and nuts, then the anti-sway bar.
2. Temporarily install one nut to retain lower ball joint.
3. Check anti-sway bar bushings for damage or excessive wear and replace if necessary.
4. Reverse procedure to install. Torque inner and outer anti-sway bar clamp to specifications.

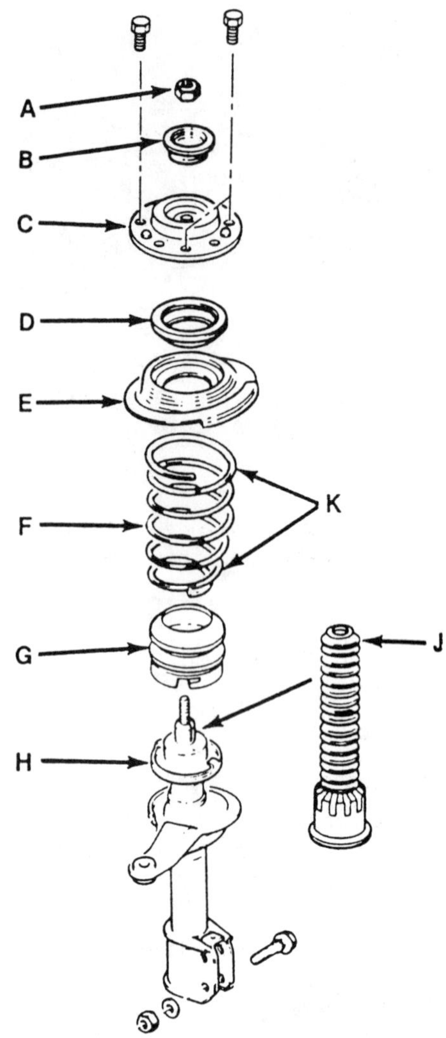

A-Locknut
B-Rebound Cup
C-Strut Cushion
D-Pivot Bearing
E-Spring Seat
F-Coil Spring
G-Jounce Bumper
H-Shock Absorber
J-Dust Boot
K-Rubber Spring Isolator Tubes

Fig. 15 Exploded view of spring strut assembly. Premier

LOWER BALL JOINT REPLACE

MEDALLION

1. Disconnect stabilizer bar from both lower control arms. The bracket bolts and nuts also secure the lower ball joints to the arms. Replace a nut on one of the two attaching bolts after re-

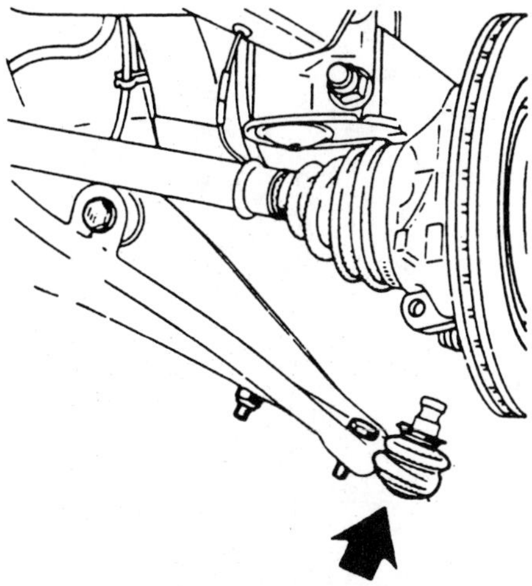

Fig. 16 Lower ball joint removal. Premier

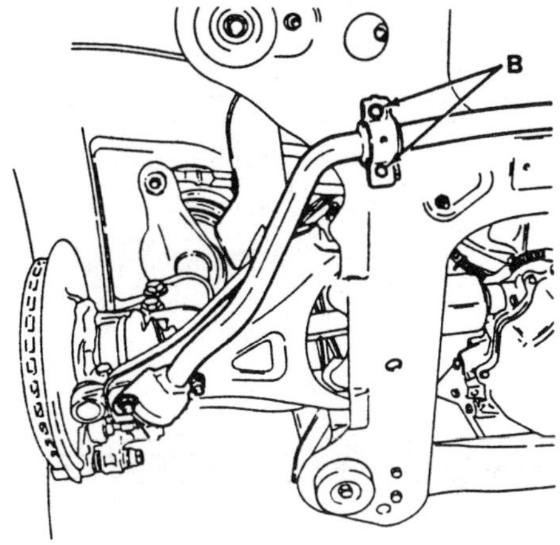

Fig. 17 Stabilizer bar inboard bracket bolts. Premier

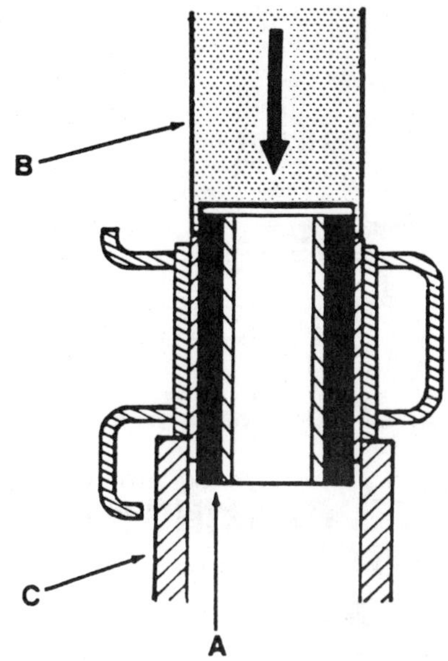

Fig. 18 Using driver/receiver for bushing removal. Medallion

moving the bracket(s).

2. Raise and support vehicle.
3. Remove wheel(s).
4. Remove lower ball joint key bolt and nut.
5. Loosen but do not remove bolts attaching lower control arm to the engine cradle.
6. Remove and retain plastic washer located on lower ball joint.
7. Remove ball joint attaching bolts, then the ball joint.
8. Reverse procedure to install.

PREMIER

Removal

1. Raise and support vehicle.
2. Remove wheel and tire assembly, then protect driveshaft boot with shop towel.
3. Loosen stabilizer bar inner bracket bolts and remove outer mounting brackets.
4. Remove lower ball joint key bolt, then loosen lower control arm bolts.
5. Lower stabilizer bar, then with brass or plastic mallet, tap out ball joint in direction of arrow, **Fig. 16.**

Installation

1. Install new ball joint in lower control arm and finger tighten bolts.
2. Insert ball stud in steering knuckle. Torque ball joint key bolt and nut to specifications.
3. Finger tighten stabilizer bar outer mounting brackets.
4. With vehicle weight supported by wheels, **torque** inner and outer stabilizer bar mounting brackets to specifications.

LOWER CONTROL ARM REPLACE

MEDALLION

Remove nuts from stabilizer bar mounting bracket bolts and slide bracket off the bolts. Disconnect the brackets from both lower control arms. The bracket bolts and nuts also secure the lower ball joints to the control arms. Replace a nut on one of the two bolts after removing the bracket.

1. Move stabilizer bar downward and away from the lower control arm.
2. Raise and support vehicle.
3. Remove wheel(s).
4. Remove lower ball joint key bolt and nut.
5. Remove bolts attaching lower control arm to the engine cradle, then the control arm.
6. Retain plastic washer located on the lower ball joint.
7. Reverse procedure to install.

PREMIER

1988 Premier vehicles built before fabrication sequence No. 8644 are equipped with stamped lower front suspension arms. All vehicles, sequence No. 8644 and beyond are equipped with cast lower front suspension arms.

If it should become necessary to replace a lower front suspension arm on a vehicle with the stamped style arms, both right and left suspension arms should be replaced with cast style suspension arms part No. 32003136 and 32003137.

1. Raise and support vehicle.
2. Remove wheel and tire assembly, then protect driveshaft boot with shop towel.
3. Loosen stabilizer bar inboard bracket bolts (B), **Fig. 17.**
4. Remove stabilizer bar outboard mounting brackets.
5. Remove key bolt from lower ball joint.
6. Remove control arm to engine cradle attaching bolts, then the control arm.
7. Inspect lower control arm bushings for damage or excessive wear and replace if necessary.
8. Reverse procedure to install. Torque ball joint key bolt nut and control arm to engine cradle bolt to specifications.

LOWER CONTROL ARM BUSHING SERVICE

MEDALLION

Removal

1. Using a suitable press, drive bushings (A), **Fig. 18,** out.
2. Use suitable size sockets of sections of pipe as a bushing driver (B) and a bushing receiver (C), **Fig. 18.**
3. The driver (B) outside diameter should be 1.34 inches.
4. The receiver (C) inside diameter should be 1.42 inches.

Installation

Install replacement bushings using a suitable press and the same bushing driver/receiver tools used during removal. The control arm bushing are not identical and should not be interchanged from one side to the next. **Press the bushings (D), Fig. 19, into the control arm equally and in small increments. Dimension (E) must be 7.48 inches plus or minus .019 inch. Dimension (F) must be the same at both bushing locations.**

PREMIER

1. Remove lower control arm as described under "Lower Control Arm, Replace."
2. Remove bushings using a suitable press and socket.
3. Install new bushings using same tools used for removal. **Bushings must be installed equally in small increments until dimension E, Fig. 19,** measures 7.461-7.499 and dimension F is the same at both bushings within .197 inch.

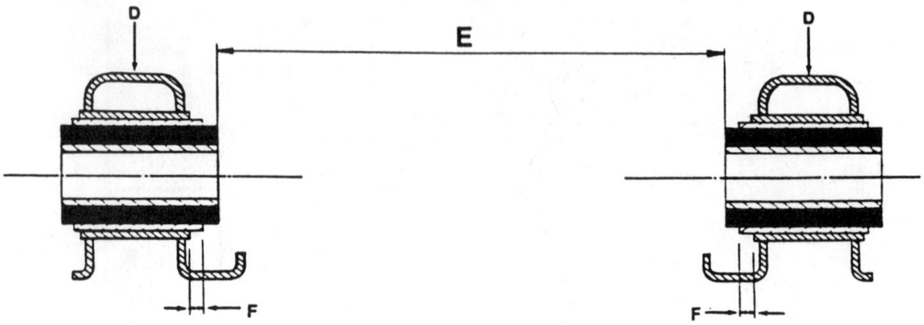

Fig. 19 Control arm bushing installation. Medallion & Premier

WHEEL BEARING & HUB ASSEMBLY

REPLACE

MEDALLION

1. Raise and support vehicle.
2. Remove wheel(s).
3. Remove brake rotor and caliper assembly.
4. Tie caliper to the spring or strut. Do not allow the brake line to support weight of the caliper.
5. Using toll Rou.604.01 or equivalent to secure wheel hub, remove driveshaft attaching nut.
6. Remove bolts attaching bearing to the steering knuckle.
7. Remove hub and bearing from knuckle and driveshaft assembly.
8. Reverse procedure to install.

PREMIER

Hub and bearing are serviced separately. The hub can be replaced without removing bearing from steering knuckle. In order to replace bearing from steering knuckle, the wheel hub must be removed as an assembly. If bearing components are worn or damaged, replace complete assembly.

WHEEL HUB, REPLACE

Removal

1. Raise and support vehicle.
2. Remove wheel and tire assembly, then unfasten brake caliper and position aside.
3. Remove hub nut and the hub using a suitable tool.
4. If replacing hub, remove bearing race using a suitable tool.

Installation

1. If necessary, install new bearing race in hub using a suitable press.
2. Lubricate bearing surface of hub with suitable chassis grease.
3. Insert hub on end of driveshaft and tap with a brass hammer so driveshaft extends past hub.
4. Install driveshaft nut and torque to specifications.
5. Install brake rotor and caliper, then the wheel and tire assembly.

WHEEL BEARING, REPLACE

Removal

1. Remove hub as described under "Front Wheel Hub, Replace."
2. Remove bearing assembly to steering knuckle attaching bolts, then the bearing assembly.
3. If necessary, remove bearing outer race using a suitable press and puller. **Bearing race and bearing should be replaced as a set.**

Installation

1. Pack new bearing with suitable grease.
2. Install bearing race in bearing.
3. Press other bearing race into front wheel hub.
4. Place bearing assembly in steering knuckle, then torque attaching bolt to specifications.
5. Install front wheel hub.

STEERING KNUCKLE

REPLACE

MEDALLION

1. Raise and support vehicle.
2. Remove wheel, brake caliper and rotor assemblies. Secure caliper to the strut or spring. Do not allow the brake hose to support the weight of the caliper.
3. Using tool Rou.604.01 or equivalent to secure wheel hub, remove driveshaft attaching nut.
4. Remove wheel bearing and hub as an assembly.
5. Remove steering knuckle to strut attaching bolts and the lower ball joint key bolt and nut.
6. Disconnect ball joint from steering knuckle and remove steering knuckle. Do not damage driveshaft boot when removing steering knuckle.
7. Reverse procedure to install.

PREMIER

When removing steering knuckle, the front wheel hub and wheel bearing assembly can be removed as an assembly.

Removal

1. Raise and support vehicle.

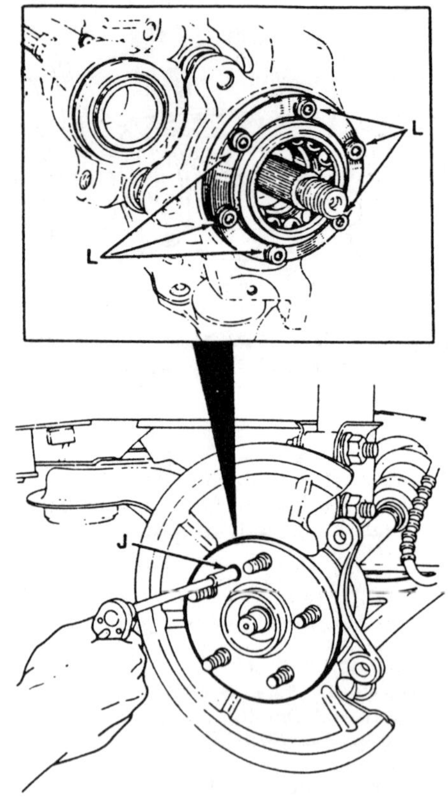

Fig. 20 Removing wheel bearing attaching bolts. Premier

2. Remove wheel and tire assembly, then protect driveshaft boot with shop towel.
3. Unfasten brake caliper and position aside, leaving hydraulic lines attached.
4. Remove driveshaft center nut and the rotor.
5. Remove wheel bearing to steering knuckle attaching bolts (L) through access hole (J) in front hub, **Fig. 20.**
6. Remove front hub and wheel bearing assembly using a suitable tool.
7. Remove ball joint key bolt and separate ball joint from knuckle.
8. Remove steering knuckle to strut attaching bolts, then the steering knuckle assembly.

Installation

1. Position ball joint in steering knuckle, then install ball joint key bolt and nut.
2. Install steering knuckle to strut attaching bolts.
3. Place front hub and wheel bearing assembly over driveshaft. Using access hole in wheel hub, attach wheel bearing to steering knuckle and torque to specifications.
4. **Torque** driveshaft nut to specifications.
5. Install brake rotor and caliper.

TIE ROD

REPLACE

MEDALLION

1. Raise and support vehicle.

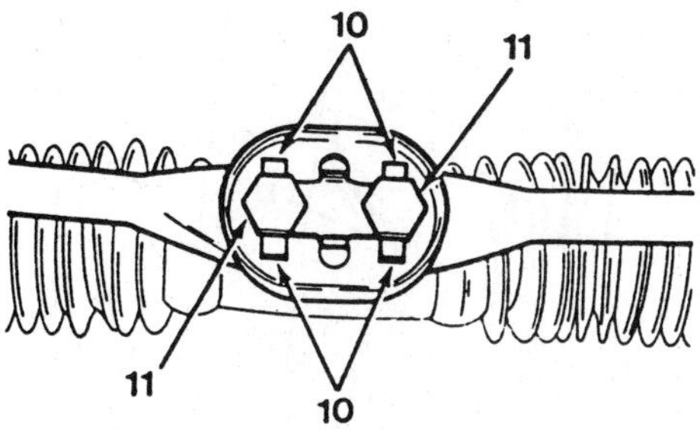

Fig. 21 Tie rod attaching bolt lock tabs. Premier

2. Remove tie rod end attaching nut.
3. Using tool T.Av.476 or equivalent, disconnect tie rod end from steering knuckle.
4. Remove nuts and bolts attaching steering bracket to the steering gear rack assembly. The attaching nuts are located on the underside of the rack assembly.
5. Remove nuts and bolts attaching tie rod to the steering bracket.
6. Remove tie rod assembly.
7. Reverse procedure to install.

STEERING GEAR

REPLACE

MEDALLION

1. Disconnect battery ground cable.
2. Remove instrument panel lower cover attaching screws, then the lower cover.
3. Remove steering shaft coupling pinch bolt. Mark position of coupling and steering shaft.
4. **On models equipped with A/C, carefully and slowly discharge refrigerant from system.**
5. On all models, raise and support vehicle, then remove both front wheel and tire assemblies.
6. Remove nuts securing tie rod ends, then using tool T.Av.476, disconnect rod ends.
7. Disconnect electrical connectors from ignition module.
8. Remove bolt and nut attaching diagnostic connectors to the dash panel and position connectors aside.
9. Remove steering bracket to steering gear rack attaching nuts and bolts.
10. **On models equipped with A/C, disconnect and cap lines from the expansion valve. Remove bolt attaching retainer to the side sill and position lines aside. A/C hoses and lines are under extreme pressure. Before disconnecting hoses and lines, ensure A/C system is completely discharged of refrigerant.**
11. On all models, remove crossmember attaching bolts, then the crossmember.
12. Install tool Mot.453.01 or equivalent, onto power steering pump reservoir hoses to prevent fluid leakage.

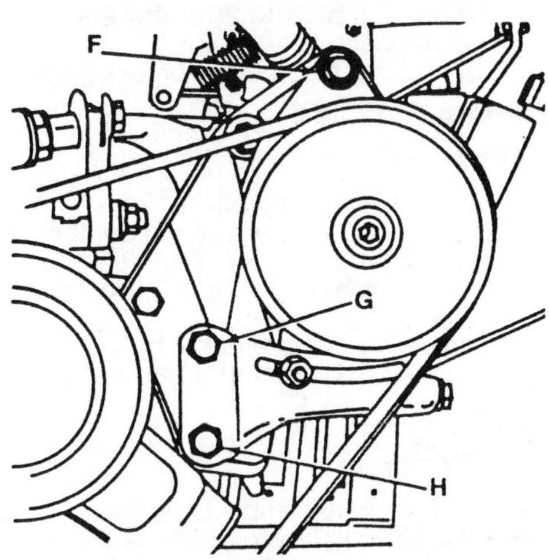

Fig. 22 Power steering attaching bolts. Premier

13. Remove power steering hose bracket and disconnect power steering hoses from gear assembly.
14. Remove nuts attaching steering gear support collar to body.
15. Remove steering gear mounting bracket nut, then the bracket.
16. Lower steering gear and move forward.
17. Mark position of the intermediate shaft coupling and the steering gear shaft for assembly reference, then remove coupling pinch bolt.
18. Tilt steering gear downward and remove from vehicle.
19. Reverse procedure to install.

PREMIER

Removal

1. Remove instrument panel lower cover.
2. Unfasten steering shaft boot flange and slide upward for access of U-joint.
3. Mark intermediate shaft and steering shaft for installation reference.
4. Remove shaft through bolt.
5. In engine compartment, pry out splash shield clips using a suitable screwdriver.
6. Fold back lock tab (10) on bolts (11) that attach tie rods to steering rack, then loosen bolts one or two turns, **Fig. 21.**
7. Remove power steering pump lines from rubber mounting block and disconnect lines from steering gear. **Replace O-rings on disconnected lines during installation.**
8. Remove mounting nut.
9. Raise and support vehicle.
10. Remove left front wheel and tire assembly, then disconnect tie rod ends from steering knuckles using a suitable tool.
11. Remove three bolts attaching steering gear to vehicle body and slide steering gear assembly through access opening in left fender panel.

Installation

1. Attach tie rods to steering gear. Do not tighten completely at this time.

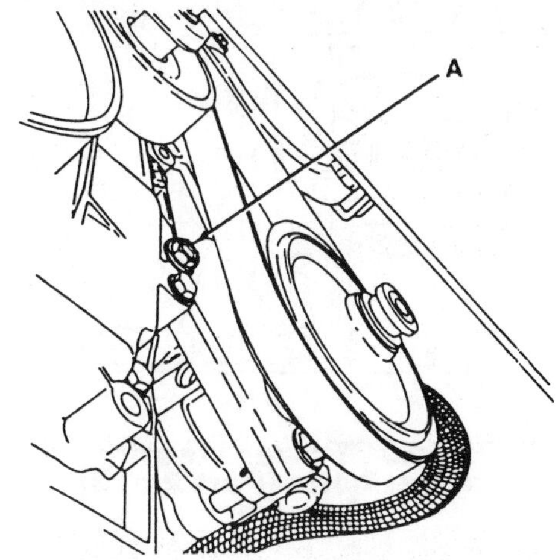

Fig. 23 Power steering bracket bolts (Part 1 of 3). Premier

2. Install gear assembly through access opening in left fender panel and install gear to body mounting bolts and nuts.
3. Connect tie rod ends to steering knuckle, torque retaining nut to specifications.
4. Connect fluid lines with new O-rings to steering gear.
5. Install fluid lines into rubber mounting block.
6. **Torque** tie rod to steering gear attaching bolts to specifications, then bend lock tabs over tie rod bolt heads.
7. Align and connect intermediate shaft and steering gear shaft with reference marks made during removal.
8. Install and **torque** intermediate shaft coupling bolt to specifications.
9. Install steering shaft boot.
10. Fill power steering system to proper level.

POWER STEERING PUMP

REPLACE

MEDALLION

The power steering pump must be removed from underneath vehicle.

1. Raise and support front of vehicle, then remove underbody splash shield.
2. Install suitable clamps on power steering pump return hose to prevent fluid leakage.
3. Remove power steering pump drive belt.
4. Disconnect power steering pump pressure and return lines at pump.
5. Remove power steering pump to bracket attaching bolts, then remove power steering pump.
6. Reverse procedure to install.

PREMIER

4-150 ENGINE

1. Disconnect battery ground cable.
2. Loosen pump drive belt.
3. Disconnect return hose and high pressure hose from pump.
4. Remove pump mounting bolts and

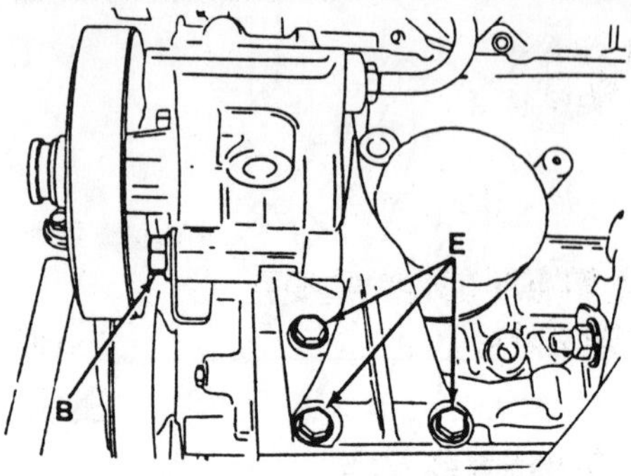

Fig. 23 Power steering bracket bolts (Part 2 of 3). Premier

the pump.

5. Reverse procedure to install, noting the following:
   a. Install and tighten attaching bolts (F, G & H) as indicated in **Fig. 22,** then remaining bolts. **Torque** bracket to intake manifold and water pump bolts to 30 ft. lbs. and all other bolts to 20 ft. lbs.
   b. Fill pump reservoir to proper level.

V6-180 ENGINE

Removal

1. Raise and support vehicle.
2. Remove underbody splash pan, then loosen drive belt.
3. Disconnect high pressure hose from pump.
4. Remove all pump bracket bolts, then the pump and bracket as an assembly.

Installation

1. Install pump assembly on engine.
2. Tighten pump bracket bolts and nuts A through E, **Fig. 23.** in alphabetical order. **Torque** rear bracket to reservoir bolt to 20 ft. lbs., front bracket to rear bracket stud to 15 ft. lbs. and all other nuts and bolts to 30 ft. lbs.
3. Connect return hose and pressure hose to pump.
4. Install and adjust drive belt.

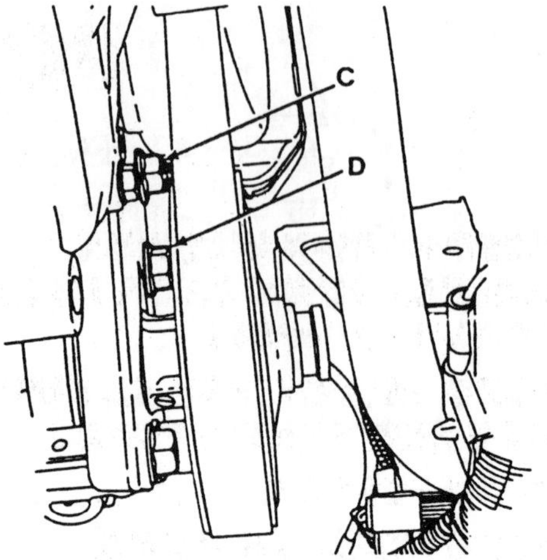

Fig. 23 Power steering bracket bolts (Part 3 of 3). Premier

5. Install underbody splash shield.
6. Lower vehicle and fill pump to proper level.

FRONT SUSPENSION & STEERING TIGHTENING SPECIFICATIONS

MEDALLION

Year	Component	Torque/ft. lbs.
1988-89	Disc Brake Caliper Attaching Bolts	74
	Driveshaft Nut	184
	Lower Ball Joint Pinch Bolt	44
	Lower Ball Joint Retaining Bolts & Nuts	59
	Lower Control Arm Top Cradle Bolts	59
	Stabilizer Bar Mounting Bracket	22
	Steering Gear Rack To Bracket Bolts & Nuts	30
	Steering Knuckle To Strut Bolts	148
	Steering Rear Mounting Bolts	37
	Steering Shaft Coupling Pinch Bolts	19
	Strut Upper Nut	44
	Strut Upper Mounting To Strut Tower Nuts	18
	Tie Rod End Adjusting Sleeve Lock Nuts	26
	Tie Rod To Bracket Bolts & Nuts	26
	Upper Ball Joint Nut	30
	Wheel Bearing To Knuckle Bolts	15
	Wheel & Tire Assembly Nuts	66

PREMIER

Year	Component	Torque/ft. lbs.
1988-90	Ball Joint Key Bolt & Nut	77
	Ball Joint Key Bolt Nut	77
	Bearing Assembly Attaching Bolts	11
	Control Arm To Engine Cradle Bolts	103
	Driveshaft Nut	181
	Inner Anti-Sway Bar Clamps	21
	Inner Stabilizer Bar Mounting Brackets	21
	Intermediate Shaft Coupling Bolt	25
	Lower Mounting Bolts	123
	Outer Anti-Sway Bar Clamps	60
	Outer Stabilizer Bar Mounting Brackets	60
	Rebound Cup Locknut	52
	Strut Rebound Cup	73
	Tie Rod End To Strut Nut	27
	Tie Rod Ends To Steering Knuckle Nuts	35
	Tie Rod To Steering Gear Attaching Bolts	55
	Upper Mounting Nuts	17
	Wheel Bearing To Steering Knuckle Bolts	11

FORD MOTOR CO.

TABLE OF CONTENTS

Ford Crown Victoria, Lincoln Town Car & Mercury Grand Marquis

INDEX

WHEEL BEARINGS
ADJUST

1. Raise and support vehicle so front wheels are free to turn.
2. Remove wheel cover, then the grease cap from hub.
3. Clean excess grease from end of spindle, then remove cotter pin and nut lock.
4. Back off adjusting nut three turns, then rock wheel assembly in and out several times to push shoe and linings away from rotor.
5. **Torque** adjusting to nut 17-25 ft. lbs. while rotating the wheel assembly.
6. Back off adjusting nut 1/2 turn, then retighten nut to 10-12 inch lbs.
7. Install nut lock on nut so castellations on lock are aligned with cotter pin hole in spindle, then install cotter pin.
8. Check wheel rotation. If wheel rotates roughly or makes noise, lubricate or replace bearings as necessary.

WHEEL BEARINGS
REPLACE

1. Raise and support front of vehicle, then remove tire and wheel assemblies.
2. Remove caliper mounting bolts. **It is not necessary to disconnect the brake lines for this operation.**
3. Slide caliper off of disc, inserting a spacer between the shoes to hold them in their bores after the caliper is removed. Position caliper assembly out of the way. **Do not allow caliper to hang by brake hose.**
4. Remove hub and disc assembly. Grease retainer and inner bearing can now be removed.
5. Reverse procedure to install.

CHECKING BALL JOINTS FOR WEAR

UPPER BALL JOINT

1. Raise car on floor jacks placed beneath lower control arms.
2. Grasp lower edge of tire and move wheel in and out.
3. As wheel is being moved in and out, observe upper end of spindle and upper arm.
4. Any movement between upper end of spindle and upper arm indicates ball joint wear and loss of preload. If any such movement is observed, replace upper ball joint.

During the foregoing check, the lower ball joint will be unloaded and may move. Disregard all such movement of the lower ball joint. Also, do not mistake loose wheel bearings for a worn ball joint.

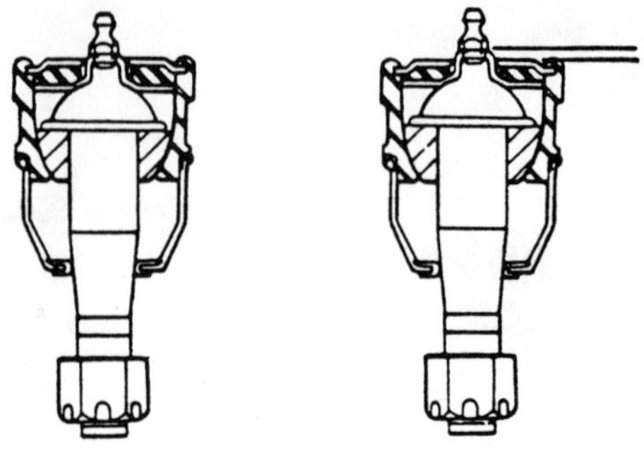

Fig. 1 Lower ball joint wear indicators

LOWER BALL JOINT

These models are equipped with lower ball joint wear indicators, **Fig. 1.** To check ball joint for wear, support vehicle in normal driving position with both ball joints loaded. Observe the checking surface of the ball joint. If the checking surface is inside the cover, **Fig. 1,** replace the ball joint.

BALL JOINTS
REPLACE

Ball joint repair kits which do not require control arm replacement, are available and can be installed using the following procedure.

The ball joints are riveted to the upper and lower control arms. The ball joints can be replaced on the car by removing the rivets and retaining the new ball joint to the control arm with the attaching bolts, nuts and washers furnished with the ball joint kit.

When removing a ball joint, use a suitable pressing tool to force the ball joint out of the spindle.

SHOCK ABSORBER
REPLACE

1. Remove nut, washer and bushing from shock absorber upper end.
2. Raise vehicle and install safety stands.
3. Remove two thread-cutting screws from lower end of shock absorber, then remove shock absorber.
4. Reverse procedure to install. **If threads in lower arm become damaged, reuse original thread-cutting screws along with 5/16-18 locknuts.**

COIL SPRING
REPLACE

1. Raise and support vehicle.
2. Remove wheel.
3. Disconnect stabilizer bar link from lower control arm.
4. Remove shock absorber.
5. Remove steering center link from pitman arm.
6. Compress coil spring with a suitable spring compressor, tool D-78P-5310-A or equivalent.
7. Remove two lower control arm pivot bolts and disengage arm from crossmember.
8. Remove spring from vehicle.
9. Reverse procedure to install. **Torque** stabilizer bar to lower control arm nuts and lower control arm to crossmember bolts to specifications.

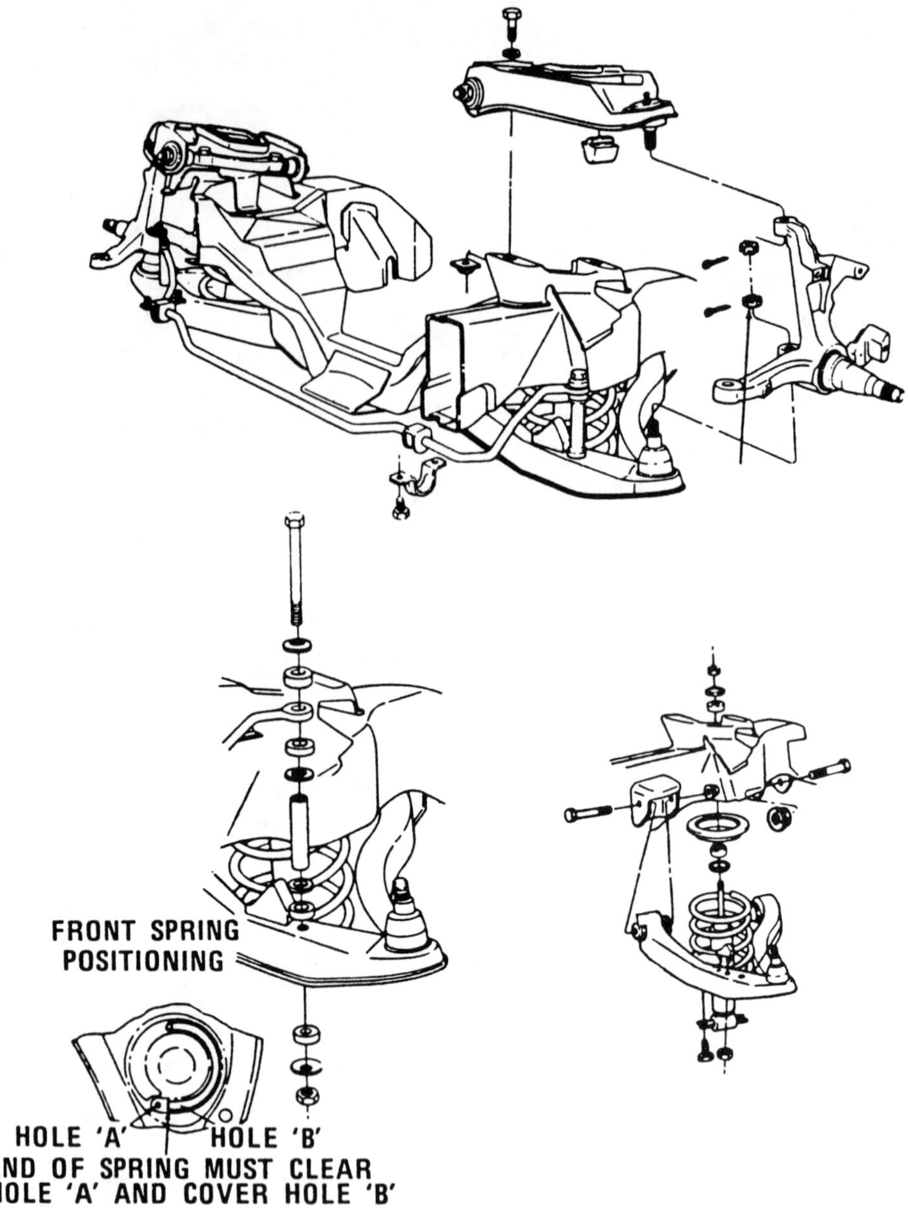

Fig. 2 Typical front suspension

STABILIZER BAR
REPLACE

1. Raise and support vehicle.
2. Remove stabilizer bar attaching clamps, then the stabilizer bar attaching bolts from each stabilizer link.
3. Remove stabilizer bar assembly.
4. Reverse procedure to install.

CONTROL ARM
REPLACE

LOWER CONTROL ARM

1. Raise and support front of vehicle, then remove front wheels.
2. Remove brake caliper, rotor and dust shield.
3. Remove jounce bumper, if equipped.
4. Remove shock absorber.
5. Disconnect stabilizer bar link from lower control arm.
6. Disconnect steering center link from pitman arm.
7. Remove lower ball joint attaching nut cotter pin, then loosen lower ball joint nut one or two turns. **Do not remove nut from stud at this time.**
8. Install ball joint press tool

T57P-3006-B or equivalent between upper and lower ball joint studs.

9. Compress ball joint with tool, then tap spindle, near lower stud, to loosen stud in spindle.
10. Remove ball joint press tool, then position a suitable jack under lower control arm.
11. Install suitable coil spring compression tool, then remove coil spring.
12. Remove ball joint nut, then the lower control arm assembly.
13. Reverse procedure to install, noting the following:
    a. **Torque** ball joint attaching nut to specification.
    b. Ensure coil spring is properly aligned, **Fig. 2.**
    c. **Torque** control arm-to-crossmember attaching bolts and nuts to specification.
    d. Check wheel alignment.

UPPER CONTROL ARM

1. Raise and support front of vehicle, then remove front wheels.
2. Remove upper ball joint attaching nut cotter pin, then loosen upper ball joint nut one or two turns. **Do not remove nut from stud at this time.**
3. Install ball joint press tool T57P-3006-B or equivalent between upper and lower ball joint studs.
4. Compress ball joint with tool, then tap spindle, near upper stud, to loosen stud in spindle.
5. Remove ball joint press tool, then position a suitable jack under lower control arm.
6. Remove upper control arm attaching bolts, then the upper control arm assembly.
7. Reverse procedure to install, noting the following:
    a. **Torque** ball joint attaching nut to specification.
    b. **Torque** upper control arm attaching bolts to specification.
    c. Check wheel alignment.

POWER STEERING GEAR REPLACE

1. Remove stone shield, if equipped.
2. Disconnect pressure and return lines from steering gear. Plug lines and ports in gear to prevent entry of dirt.
3. Remove two bolts that secure flex coupling to steering gear and to column.
4. Raise car and remove sector shaft nut.
5. Use a puller to remove pitman arm.
6. Support steering gear, then remove attaching bolts.
7. Work steering gear free of flex coupling and remove it from car.
8. Reverse procedure to install.

POWER STEERING PUMP REPLACE

1. Disconnect power steering pump return line and allow power steering pump fluid to drain into a suitable container.
2. Disconnect power steering pump pressure hose from pump fitting.
3. Disconnect drive belt from power steering pump pulley, remove pulley, then remove pump.
4. Reverse procedure to install. **Torque** pump to mounting bracket bolts to 30-45 ft. lbs. **On Ford model CII power steering pump, torque** pressure hose to pump fitting to 10-15 ft. lbs. End play on this fitting is normal and does not indicate a loose fitting.

FRONT SUSPENSION & STEERING TIGHTENING SPECIFICATIONS

Year	Component	Torque/ft. lbs
1985-90	Ball Joint To Spindle Lower	100-120
	Ball Joint To Spindle Upper	75-90
	Jounce Bumper (Nut)	30-35
	Lower Arm To Crossmember	120-140
	Shock Absorber To Lower Arm	12-18
	Shock Absorber Upper Attachment	14-25
	Stabilizer Bar To Frame	14-26
	Stabilizer Bar To Lower Arm	9-12
	Upper Arm To Frame	120-140

Ford Escort, EXP & Tempo & Mercury Lynx & Topaz

INDEX

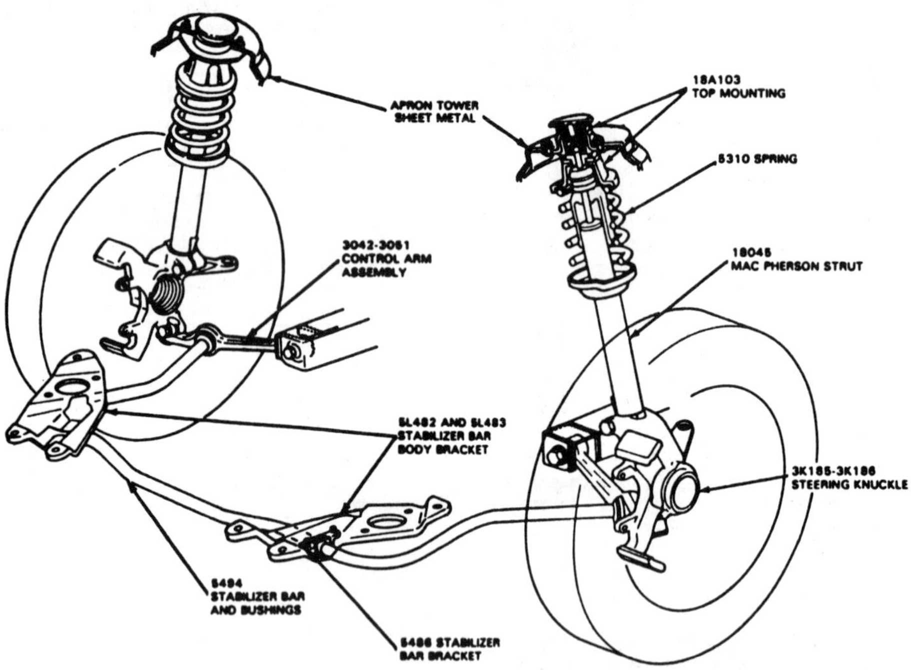

Fig. 1 Front suspension

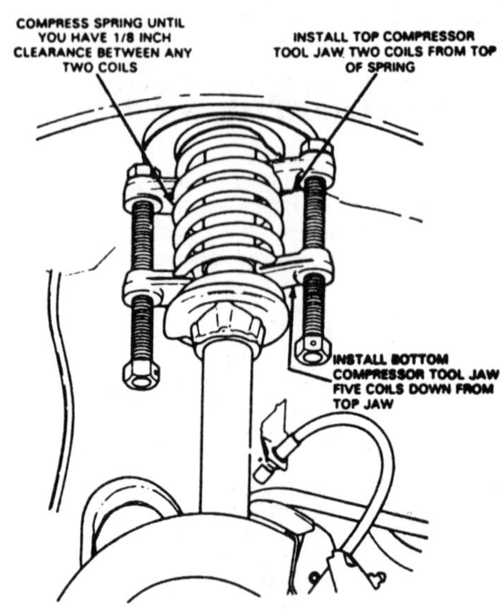

Fig. 2 Installing coil spring compressor. Escort, EXP & Lynx

DESCRIPTION

These vehicles use a MacPherson type front suspension with the vertical shock absorber struts attached to the upper fender reinforcements and the steering knuckle, **Fig. 1.** The lower control arms are attached inboard to a crossmember and outboard to the steering knuckle through a ball joint to provide lower steering knuckle position.

STRUT ASSEMBLY REPLACE

1. Raise and support vehicle, then remove tire and wheel assembly.
2. Remove brake hose retaining bracket from strut.
3. **On Tempo and Topaz models,** remove brake caliper, brake rotor, and tie rod end.
4. Place suitable jack under lower control arm and raise strut as far as possible without raising vehicle.
5. **On Escort, EXP and Lynx,** install spring compressor tool No. T81P-5310A or equivalent by placing top jaw on second coil from top and bottom jaw so as to grip a total of 5 coils, then compress spring until there is about 1/8 inch between any 2 coils, **Fig. 2.**
6. **On Tempo and Topaz models,** install spring compressor tool No. 14-0259 or 86-0016 by placing top jaw on 5th or 6th coil from bottom, then compress spring a minimum of 3 1/2 inches, **Fig. 3. Spring must be compressed before strut is removed to insure that no excessive force is applied to constant velocity universal joints.**
7. Remove steering knuckle-to-strut pinch bolt, then loosen two top mount to apron nuts and top strut shaft nut. **When loosening or tightening strut-to-mount nut, hold 8 mm hex with a deep socket 1/4-inch drive wrench, Figs. 4 and 5, and turn only the nut.**
8. Lower jack from control arm, then using a large screwdriver, spread knuckle-to-strut pinch joint.
9. Place block of wood 2 x 4 x 7 1/2 inches against knuckle shoulder. Using suitable pry bar between wood block and lower spring seat, separate strut from knuckle, **Fig. 6.**
10. Remove two top mounting nuts, then the strut, spring, and top mount assembly from vehicle.
11. **On 1985-86 Escort, EXP and Lynx,** place 18 mm deep socket tool No. D81P-18045-A1 or equivalent on strut shaft nut. Place an 8 mm deep socket, 1/4 inch drive onto shaft end and clamp mount into vise, **Fig. 4.**
12. **On Tempo and Topaz,** place 18 mm deep socket tool No. D81P-18045-A1 or equivalent on strut shaft nut. Insert an 8 mm hex deep socket with

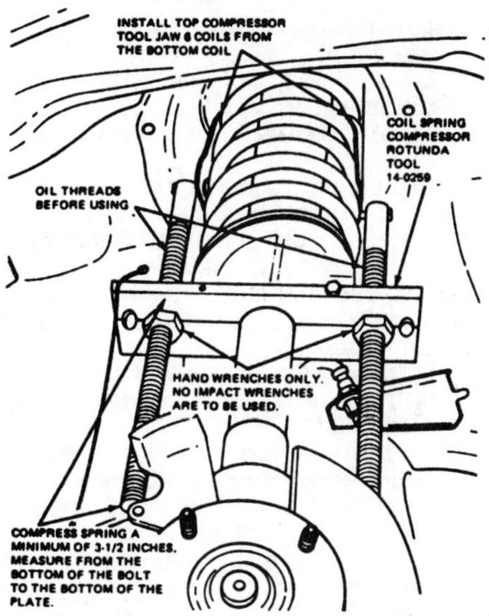

Fig. 3 Installing coil spring compressor. Tempo & Topaz

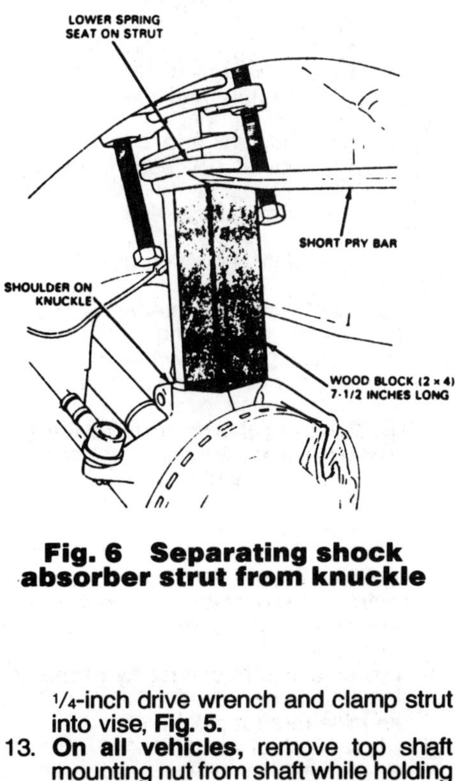

Fig. 6 Separating shock absorber strut from knuckle

1/4-inch drive wrench and clamp strut into vise, **Fig. 5.**

13. **On all vehicles,** remove top shaft mounting nut from shaft while holding Allen wrench or 1/4-inch drive socket. **Do not clamp directly on strut with vise as damage to strut may result.**
14. Remove strut top mount components and, if equipped, spring. **On Tempo and Topaz models,** check spring insulator for damage to splash shield.
15. **On Tempo and Topaz models,** remove spring compressor from strut and remove spring.
16. Reverse procedure to install. Refer to **Figs. 7 and 8** for proper installation sequence of top mount components. **On Escort, EXP and Lynx, torque** strut shaft nut to 48-62 ft. lbs., on Tempo & Topaz models, **torque** strut

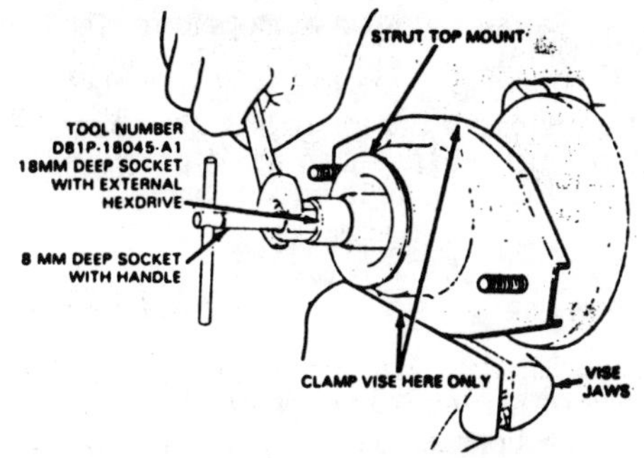

Fig. 4 Removing shock absorber strut top mounting nut. 1985–86 Escort, EXP & Lynx

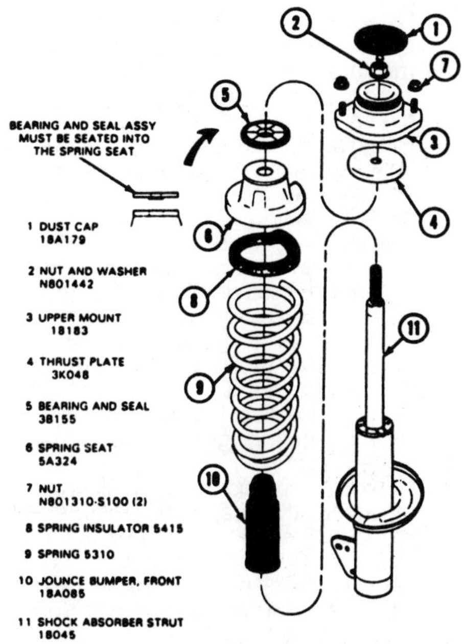

Fig. 7 Installation of shock absorber strut top mounting components. Escort, EXP & Lynx

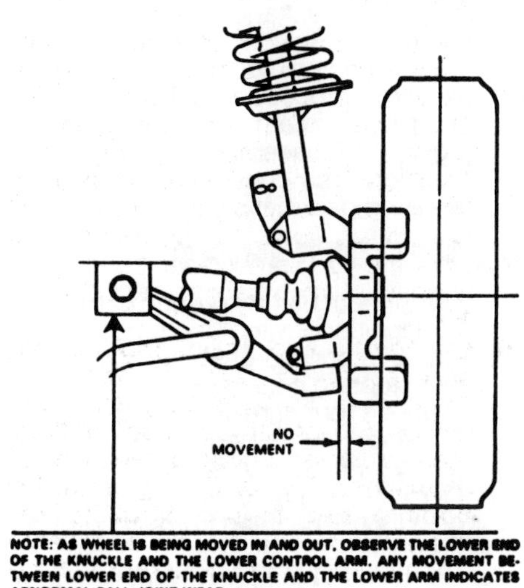

Fig. 9 Checking lower ball joint

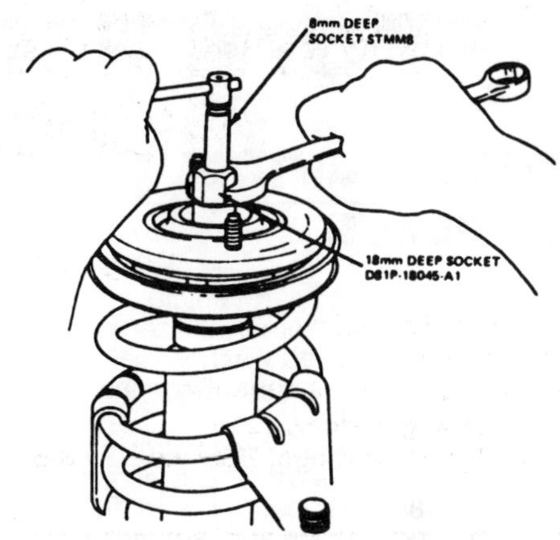

Fig. 5 Removing shock absorber strut top mounting nut. Tempo & Topaz

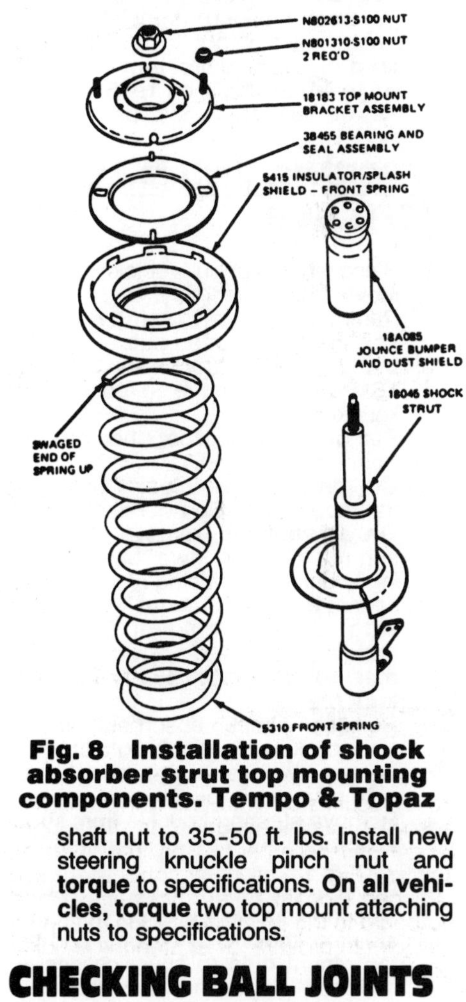

Fig. 8 Installation of shock absorber strut top mounting components. Tempo & Topaz

shaft nut to 35-50 ft. lbs. Install new steering knuckle pinch nut and **torque** to specifications. **On all vehicles, torque** two top mount attaching nuts to specifications.

CHECKING BALL JOINTS

1. Raise and support vehicle.
2. With suspension in full rebound position, grasp lower edge of tire and move wheel in and out, **Fig. 9.**
3. Observe lower end of knuckle and lower control arm as wheel is being moved in and out. Any movement between lower end of knuckle and lower arm indicates excessive ball joint wear.
4. If any movement is observed, install a new lower control arm assembly. The

lower ball joint and control arm are serviced as an assembly only. Refer to Lower Control Arm, Replace in this section.

LOWER CONTROL ARM

REPLACE

1. Raise and support vehicle.
2. Remove nut from stabilizer bar, then the large dished washer.
3. Remove lower control arm inner pivot bolt and nut.
4. Remove lower control arm ball joint pinch bolt, then using a screwdriver, separate the control arm from the steering knuckle and remove from vehicle. **Ensure steering column is in unlocked position. Do not use a hammer to separate ball joint from knuckle.**
5. Reverse procedure to install. **Torque** pinch bolt and nut to 38-45 ft. lbs. **Torque** lower control arm inner pivot nut to 44-55 ft. lbs. on 1985-87 models, or 98-115 ft. lbs. on 1988-90 models.

STEERING KNUCKLE

REPLACE

1. Raise and support vehicle, then remove wheel assembly.
2. Remove cotter pin from tie rod end stud, then the slotted nut.
3. Using tool 3290C and adapter T81P3504W, remove tie rod end from knuckle.
4. Remove brake caliper, then the hub from the driveshaft.
5. **On 1985-90 vehicles,** loosen two top mount nuts. Do not remove nuts.
6. **On all vehicles,** remove pinch bolt and nut securing lower arm to steering knuckle, then using a screwdriver, separate lower arm from knuckle. **Ensure steering column is in unlocked position. Do not use a hammer to separate ball joint from knuckle.**
7. Remove shock absorber strut to steering knuckle pinch bolt, then using a screwdriver, slightly open knuckle to strut pinch joint.
8. Remove steering knuckle from shock absorber strut, **Fig. 6,** then from the vehicle.
9. Reverse procedure to install. **Torque** steering knuckle, lower control arm to steering knuckle pinch bolt, and two top mount nuts to specifications. Install new slotted nut then **torque** to 23-35 ft. lbs. on 1985-87 models, or 28-32 ft. lbs. on 1988-90 models.

STABILIZER BAR

REPLACE

1. Raise and support vehicle.
2. Remove stabilizer insulator mounting bracket bolts.
3. Remove stabilizer bar to control arm attaching bolts, then the stabilizer bar assembly.
4. Remove worn insulators from stabilizer bar.
5. Reverse procedure to install. **Torque** stabilizer bar to control arm attaching bolt and stabilizer insulator mounting bracket bolts to specifications.

STEERING GEAR

REPLACE

MANUAL STEERING

1. Disconnect battery ground cable, then turn ignition switch to "On" position.
2. Remove access panel from dash below steering column.
3. Remove intermediate shaft bolts at gear input shaft and at steering column shaft.
4. Using a wide blade screwdriver, spread slots enough to loosen intermediate shaft at both ends.
5. Turn steering wheel fully left to allow clearance for tie rod removal.
6. Remove tie rod ends from steering knuckles using tool 3290D and adapter T81P3504W. Turn right wheel to full left position.
7. Remove left tie rod end from tie rod, then on vehicles equipped with automatic transmission disconnect speedometer cable at transmission.
8. Disconnect secondary air tube at check valve, then exhaust pipes from exhaust manifold.
9. Remove exhaust hanger bracket from below steering gear. Wire exhaust system aside.
10. Remove gear mounting brackets and insulators, then separate gear from intermediate shaft while simultaneously pulling upward on shaft from inside vehicle. **Right and left hand brackets and insulators are not interchangeable.**
11. Rotate gear forward and downward to clear input shaft.
12. Ensure input shaft is in full left turn position, then remove gear through right side apron opening until left tie rod clears shift linkage.
13. Lower left side of gear and remove gear from vehicle.
14. Reverse procedure to install. Ensure input shaft is at full left turn stop and right wheel assembly is in full left turn position. Use caution not to damage steering gear bellows.

POWER STEERING

1. Disconnect battery ground cable, then turn ignition switch to On position.
2. Remove access panel from dash below steering column.
3. Remove four screws from dash panel steering column boot, then slide boot along intermediate shaft.
4. Remove intermediate shaft bolts at gear input shaft and from steering column shaft.
5. Using wide blade screwdriver, spread slot wide enough to loosen intermediate shaft at both ends.
6. Turn steering wheel to full left stop to facilitate gear removal.
7. Remove pressure switch.
8. **On 1985-87 vehicles,** remove air cleaner.
9. **On vehicles equipped with air conditioning,** except Tempo and Topaz models, secure liquid line above dash opening.
10. Disconnect secondary air tube at check valve, then the exhaust pipes from exhaust manifold. Secure exhaust system to the side.
11. Remove exhaust hanger brackets from below steering gear and from side apron.
12. Disconnect pressure and return lines from intermediate connector and drain fluid.
13. Remove tie rod ends from steering knuckles using tool 3290D and adapter T81P3504W. Turn right wheel to full left turn position.
14. **On vehicles equipped with manual transmission,** remove left tie rod end from tie rod.
15. **On vehicles equipped with automatic transmission,** disconnect speedometer cable from transmission.
16. **On vehicles equipped with automatic transmission,** disconnect shift cable assembly from transmission.
17. **On Escort, EXP and Lynx,** remove screws securing heater water tube to brace below oil pan.
18. **On Escort, EXP and Lynx,** remove nut from lower bolt securing engine mounting bracket to transmission housing. Tap bolt out as far as possible.
19. Remove gear mounting brackets and insulators.
20. Remove gear from intermediate shaft by pushing upward on shaft with bar while pulling gear downward.
21. Rotate gear downward and forward to clear input shaft.
22. Ensure input shaft is in full left turn position, then move gear through right side apron opening until left tie rod clears opening. Use caution to avoid damaging bellows.
23. Lower left side of gear and remove gear from vehicle.
24. Reverse procedure to install.

POWER STEERING PUMP

REPLACE

DIESEL ENGINE

1. Remove drive belts.
2. **On vehicles with air conditioning,** remove alternator.
3. **On vehicles with air conditioning,** remove both braces from support bracket.
4. Disconnect power steering lines and drain fluid from pump.
5. Remove four bracket mounting bolts.
6. Install pulley removal tool No. T69L-10300-B or equivalent on pulley hub. Hold small hex head of tool and turn tool nut counterclockwise to remove pulley.
7. Remove pump from bracket.
8. Reverse procedure to install.

GASOLINE ENGINE

Escort, EXP & Lynx

1. Remove air cleaner, thermactor pump drive belt and thermactor pump.
2. Remove power steering reservoir filler extension. Cover opening to prevent entry of dirt.
3. From underneath vehicle, loosen one power steering pump adjusting bolt and remove one pump to mounting bracket bolt, then disconnect return hose.
4. Form engine compartment, loosen one power steering pump adjusting bolt, then loosen pivot and remove drive belts.
5. Remove the remaining two power steering pump to bracket retaining bolts, then remove pump from bracket by passing pulley through adjusting bracket opening.
6. Disconnect pressure hose from power steering pump, then remove pump.
7. Reverse procedure to install.

Tempo & Topaz

1. Remove alternator drive belt.
2. Place alternator in upper most position.
3. Remove radiator overflow bottle.
4. Remove power steering pump drive belt.
5. Disconnect return line from pump.
6. Completely back off power steering pump pressure line nut. The pressure line will separate when the pump bracket is removed.
7. Remove power steering pump mounting bolts and pump.
8. Reverse procedure to install.

FRONT WHEEL BEARINGS REPLACE

The front wheel bearings are cartridge design and are pre-greased, sealed and require no maintenance. The bearings are preset and cannot be adjusted.

1. Raise and support vehicle, then remove tire and wheel assembly.
2. Remove brake caliper and rotor.
3. Disconnect lower control arm and tie rod from knuckle (leave strut attached).
4. Loosen two strut top mount to apron attaching nuts.
5. Using suitable tools, remove hub bearing and knuckle assembly by pushing out constant velocity joint outer shaft until it is free of assembly.
6. Install tool D80L-1002-L and D80L-625-1 or equivalents, onto knuckle bosses and remove hub.
7. Remove snap ring retaining bearing in knuckle assembly. Discard snap ring.
8. Using a suitable press and tools T83P-1104-AH3 and T83P-1104-AH2, press bearing from knuckle assembly. Discard bearing.
9. Remove half shaft assembly. Place shaft into a suitable vise and remove bearing dust shield. Discard dust shield.
10. Reverse procedure to install.

FRONT SUSPENSION & STEERING TIGHTENING SPECIFICATIONS

Year	Component	Torque/ ft. lbs.
1985-90	Control Arm To Body	48-55
	Control Arm To Knuckle	38-45
	Stabilizer Bar Bracket Assembly To Body	48-55
	Stabilizer Bar Insulator U-Bracket Clamps To Bracket Assembly	①
	Stabilizer Bar To Control Arm	98-115
	Strut Top Mount To Body	25-30
	Strut To Knuckle	66-81
	Strut To Top Mount	35-50

①—Escort, 59-68; Tempo & Topaz, 85-100.

Ford Festiva

INDEX

DESCRIPTION

The front suspension, **Fig. 1**, is of the strut design, utilizing single pivot lower control arms with integral ball joints. The front wheels ride on forged steering knuckles, which are clamped to the ball joints on the lower end, and bolted to the lower bracket of the strut assembly on the upper end. A front stabilizer bar, connected to the lower control arms, is used to limit body sway during hard cornering and doubles as a trailing arm to maintain lower control arm alignment.

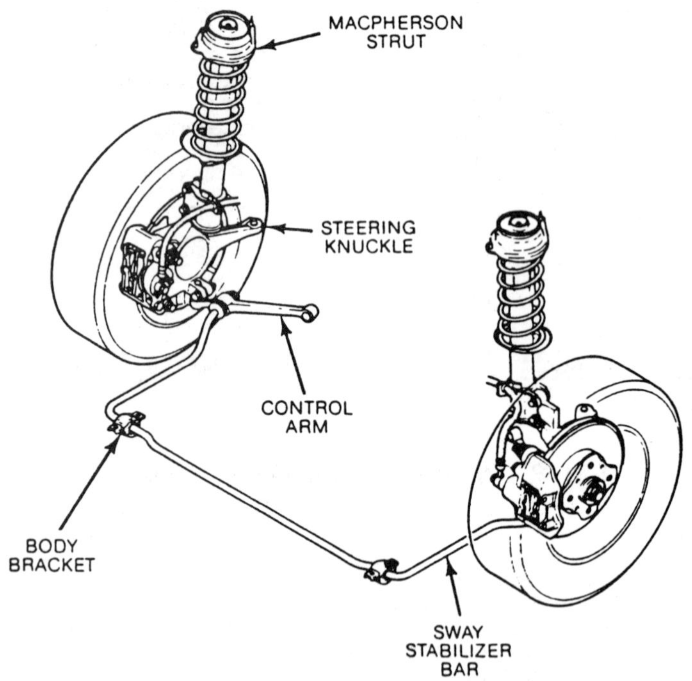

Fig. 1 Front suspension

STRUT ASSEMBLY REPLACE

REMOVAL

1. Raise and support vehicle, ensuring strut is fully extended, then remove front wheel.
2. Remove brake line clip, then disengage brake line from strut assembly.
3. Remove strut to steering knuckle attaching bolts and nuts.
4. Working from engine compartment, remove nuts securing strut mounting block to shock tower.
5. Pull lower end of strut outward to disengage strut from steering knuckle, then lower strut assembly and remove from vehicle.

INSTALLATION

1. Position strut, together with spacer plate, into shock tower, ensuring white alignment mark faces outward.
2. Install mounting block attaching nuts.
3. Connect steering knuckle to lower strut bracket, then install attaching bolts and nuts.
4. Position brake line in cutout in strut lower bracket, then install retaining clip.
5. Install front wheel, then lower vehicle.

STRUT ASSEMBLY SERVICE

DISASSEMBLY

1. Remove strut as outlined previously.
2. Compress spring using suitable spring compressor.
3. Pry out mounting block cap, then remove upper attaching nut and lockwasher, **Fig. 2.**
4. Remove spacer plate, mounting block, washer, seal and bearing from strut rod.
5. Remove upper spring seat, seat insulator and spring, then the jounce bumper and shield.
6. Inspect all components for excessive wear or damage. Replace components as necessary.

ASSEMBLY

1. Slide jounce bumper and shield over strut rod and onto main body of strut assembly.
2. Compress spring using suitable spring compressor, then install spring, seat insulator and upper seat. Ensure spring ends are positioned against steps in seats.
3. Install bearing, seal and washer on strut rod.
4. Install mounting block, ensuring that white alignment mark is positioned on same side of strut as steering knuckle mounting bracket.
5. Install spacer plate, lock washer and attaching nut.
6. Remove spring compressor, then install mounting block cap.
7. Reinstall strut assembly into vehicle.

CONTROL ARM REPLACE

REMOVAL

1. Raise and support vehicle, then remove pivot bolt at frame bracket.
2. Remove ball joint clamp bolt and nut at steering knuckle.
3. Remove stabilizer bar bushing retain-

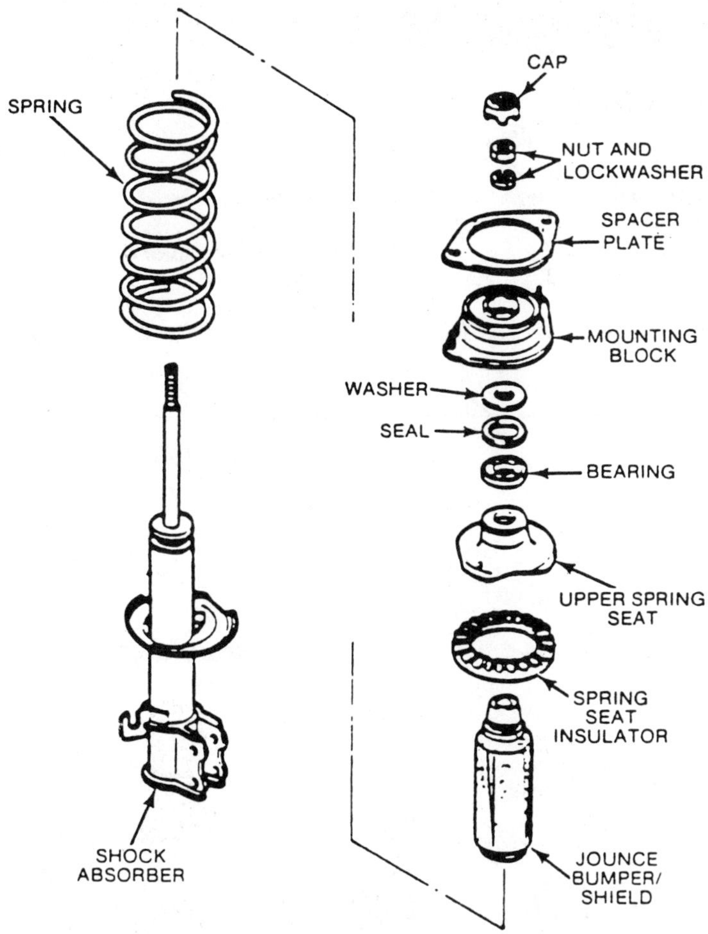

Fig. 2 Disassembled view of front strut assembly

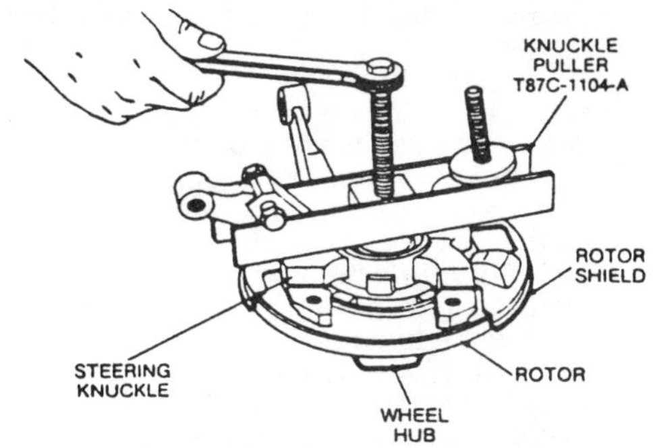

Fig. 3 Separating hub/rotor assembly from steering knuckle

ing nut from rear of control arm, then remove bushing washer and bushing.

4. Pry ball joint stud out of steering knuckle, then disengage control arm from stabilizer bar and remove from vehicle.

INSPECTION

1. Check control arm for cracks and pivot bushing for deterioration. If bushing requires replacement, remove and install bushing using suitable tools.
2. Check ball joint stud rotating torque with suitable torque wrench. Rotating torque should be 16-27 inch lbs. If rotating torque is not as specified, replace control arm.

INSTALLATION

1. Install stabilizer bar into control arm. If removed, ensure front bushing washer is positioned with dished side facing front of vehicle.
2. Raise inner part of control arm and install pivot bolt. Do not tighten bolt at this time.
3. Connect ball joint stud to steering knuckle, then install clamp bolt and nut.
4. Install rear stabilizer bar bushing and washer. Ensure rear washer is positioned with dished side facing rear of vehicle, then install attaching nut.
5. Tighten pivot bolt and ball joint clamp bolt attaching nuts.

STABILIZER BAR
REPLACE

1. Raise and support vehicle.
2. Remove stabilizer bar mounting bracket attaching nuts, then the brackets and split bushings.
3. Remove stabilizer bar to control arm attaching nuts, then the rear washers and bushings.
4. Pull stabilizer bar forward and remove from vehicle. If necessary, remove front bushings and washers from bar.
5. Reverse procedure to install, noting the following:
   a. When installing bushing washers, ensure that front washer is installed with dished side facing front of vehicle and rear washer is positioned with dished side facing rear of vehicle.
   b. When installing split bushings, ensure that split side faces forward and that bushings are positioned next to white locating marks on bar.

STEERING KNUCKLE, HUB & BEARING
REPLACE

1. Raise and support vehicle, then remove front wheel.
2. Unstake half shaft attaching nut using suitable chisel, apply brakes to prevent hub assembly from turning, then remove and discard attaching nut.
3. Remove brake hose to strut bracket retaining clip.
4. Remove cotter pin and tie rod attaching nut, then separate tie rod from steering knuckle using suitable tool.
5. Remove brake caliper attaching bolts, then lift caliper assembly off steering knuckle. **Do not allow caliper to hang unsupported from brake hose. Wire caliper in place as required.**
6. Remove ball joint to steering knuckle clamp bolt and nut, then pry downward on control arm and disconnect joint from knuckle.
7. Remove steering knuckle to strut bracket attaching bolts and nuts, then slide hub/rotor assembly together with steering knuckle off end of half shaft. If difficulty is encountered separating hub from half shaft, tap end of shaft with plastic mallet to facilitate removal.
8. Separate hub/rotor assembly from steering knuckle using tool shown in **Fig. 3.**
9. Remove bearing preload spacer from hub, **Fig. 4. The spacer is pre-selected to yield correct bearing preload. Save spacer to use during reassembly.**
10. Clamp hub/rotor assembly in soft jaw vise, scribe alignment marks on hub and rotor, then remove hub to rotor attaching bolts and separate rotor from hub.
11. Using suitable bearing remover, press hub shaft from outer bearing, then remove outer grease seal from hub.
12. Using suitable seal remover, pry inner grease seal from steering knuckle bore, then remove inner bearing.
13. Inspect bearings, hub, steering knuckle and dust shield for damage or ex-

cessive wear. Replace components as necessary.

14. If original bearings and steering knuckle are being used, proceed to step 15. If bearings or steering knuckle require replacement, proceed as follows to select proper bearing preload spacer:
    a. Drive new inner and outer bearing races into steering knuckle using suitable tools.
    b. Lubricate races and new bearings with engine oil, then install bearings into steering knuckle.
    c. Install spacer selection tool kit T87C-1104-B, original spacer and hardware shown in **Fig. 5** onto steering knuckle, then clamp steering knuckle in soft jaw vise.
    d. Tighten center bolt in increments. After bolt is tightened to specification, remove assembly from vise and rotate steering knuckle to seat bearings.
    e. Again clamp steering knuckle in vise, then measure the amount of torque necessary to start rotation of center bolt using an inch lb. torque wrench.
    f. If torque reading is 2.2-10.4 inch lbs., spacer is correct thickness. If torque reading is less than 2.2 inch lbs., a thinner spacer must be used. If torque reading is greater than 10.4 inch lbs., a thicker spacer must be used. Twenty one spacers of various thicknesses are available for service. Each spacer has a number stamped on it for identification purposes, **Fig. 6.** Changing the spacer by one number, either up or down, will result in a 1.7-3.5 inch lb. change in bearing preload.
15. Pack bearings and hub with suitable high temperature grease, then install inner bearing into steering knuckle.
16. Lubricate seal lip, then install new inner seal using suitable tool.
17. Place original bearing preload spacer, or spacer selected in step 14, into steering knuckle.
18. Lubricate seal lip, then install outer wheel bearing and new outer seal into knuckle.
19. Position rotor onto hub, aligning marks made during disassembly, then install attaching bolts.
20. Place hub/rotor assembly into steering knuckle bore, then press assembly fully into knuckle using suitable tools.
21. Follow steps 1 through 7 in reverse order to complete installation and note the following:
    a. Apply a thin coat of grease to half shaft splines before installing steering knuckle and hub/rotor assembly.
    b. When installing new half shaft attaching nut, apply brakes to prevent hub assembly from turning. Stake nut into groove on half shaft after tightening.
    d. Tighten tie rod end to steering knuckle attaching nut, then install new cotter pin. **If holes in attaching nut do not line up with hole in ball stud, tighten nut as required until cotter pin can be installed. Never loosen nut when installing cotter pin.**

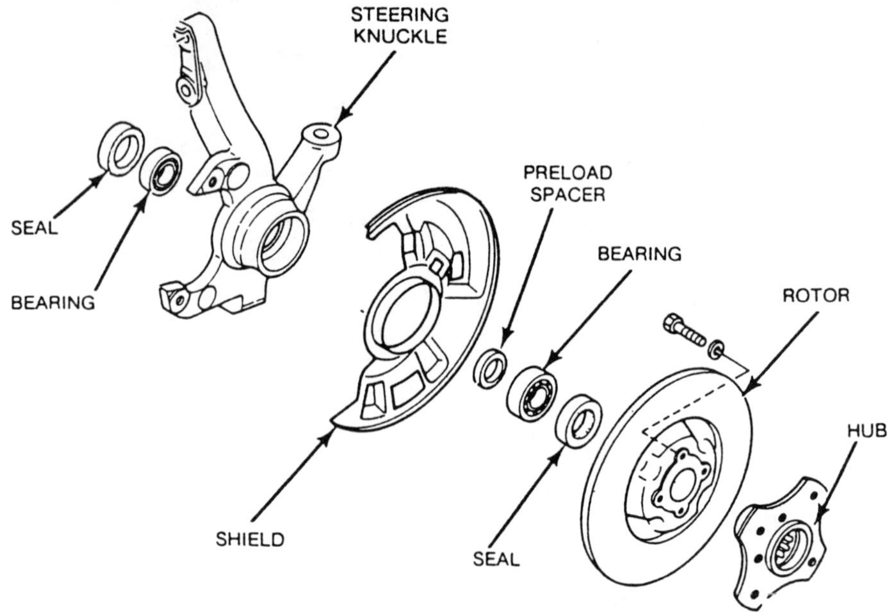

Fig. 4 Disassembled view of steering knuckle & hub/rotor assembly

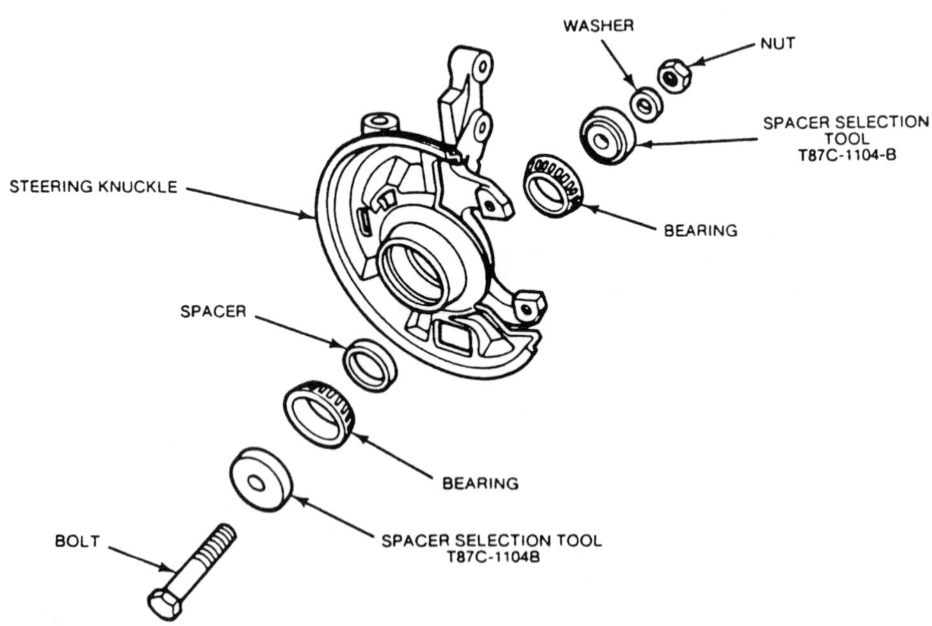

Fig. 5 Bearing preload spacer selection tool installation

MANUAL STEERING GEAR

REPLACE

1. Disconnect battery cables and remove battery.
2. Scribe alignment marks on steering column lower universal joint and steering gear pinion shaft, then remove steering column as follows:
    a. Working from underneath steering column, remove instrument panel brace and air duct.
    b. Remove upper and lower steering column covers, release harness clip, then disconnect harness connectors from back of combination switch.
    c. Remove ignition switch.
    d. Remove steering column upper mounting bracket to instrument panel attaching nuts, then lower column.
    e. Scribe alignment marks on column

Stamped mark	Thickness
1	6.285 mm (0.2474 in)
2	6.325 mm (0.2490 in)
3	6.365 mm (0.2506 in)
4	6.405 mm (0.2522 in)
5	6.445 mm (0.2538 in)
6	6.485 mm (0.2554 in)
7	6.525 mm (0.2570 in)
8	6.565 mm (0.2586 in)
9	6.605 mm (0.2602 in)
10	6.645 mm (0.2618 in)
11	6.685 mm (0.2634 in)
12	6.725 mm (0.2650 in)
13	6.765 mm (0.2666 in)
14	6.805 mm (0.2682 in)
15	6.845 mm (0.2698 in)
16	6.885 mm (0.2714 in)
17	6.925 mm (0.2730 in)
18	6.965 mm (0.2746 in)
19	7.005 mm (0.2762 in)
20	7.045 mm (0.2778 in)
21	7.085 mm (0.2794 in)

Fig. 6 Bearing preload spacer identification chart

shaft and intermediate shaft upper universal joint, then remove clamp screw from joint.

f. Remove steering column hinge bracket to pedal support attaching nuts, then pull steering column rearward and remove from vehicle.

3. Cut plastic tie strap securing steering column boot to steering gear.
4. Raise and support vehicle, then remove front wheels.
5. Remove tie rod to steering knuckle attaching nuts, then separate tie rods from knuckle.
6. Remove catalytic converter.
7. Remove tie rod splash shield from right inner fender.
8. Remove steering gear mounting bolts and lower gear until free from steering column boot.
9. Slide steering gear to the right until left tie rod is clear of inner fender, then lower gear and remove through left side of vehicle. **When sliding gear as required, guide gear carefully to prevent damage to boots.**
10. Reverse procedure to install, aligning all marks made during removal.

RACK YOKE PRELOAD ADJUST

1. Remove steering gear as previously described.
2. With rack centered, measure pinion rotating torque. Pinion torque should be 8-12 inch lbs. within 90 degrees of centered position, and should not exceed 13 inch lbs. when turned beyond 90° of center.

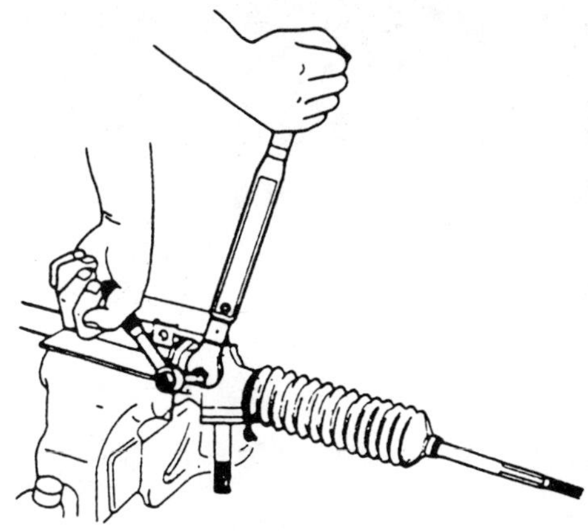

Fig. 7 Adjusting rack yoke preload

3. If pinion rotating torque is not as specified in previous step, loosen yoke adjusting bolt locknut and turn adjusting bolt as required until specified rotational torque is reached, **Fig. 7.**
4. After adjustment is completed, hold adjusting bolt in position and tighten locknut.
5. Reinstall steering gear.

FRONT SUSPENSION & STEERING TIGHTENING SPECIFICATIONS

Year	Component	Torque/ft. lbs.
1988-90	Ball Joint Pinch Bolt	32-40
	Control Arm Bushing Retaining Nuts	47-57
	Control Arm Frame Bracket Pivot Bolt	32-40
	Disc Brake Caliper Attaching Bolts	29-36
	Half Shaft Nut	116-174
	Rotor Attaching Bolts	33-40
	Stabilizer Mounting Bracket Nuts	40-50
	Stabilizer Rear Bushing Nut	43-52
	Steering Gear Mounting Bolts	23-34
	Steering Knuckle Bolts	69-86
	Steering Rack Support Cover	29-43
	Strut Rod Nut	40-50
	Tie Rod Ball Joints	43-58
	Tie Rod End Nut	22-33
	Upper Mounting Block Bolts	22-27
	Wheel & Tire Assembly	65-87

Ford LTD & Mustang, Mercury Capri & Marquis & 1985–88 Ford Thunderbird & Mercury Cougar

INDEX

FRONT SUSPENSION

The front suspension, **Fig. 1,** is of the modified MacPherson strut design, which uses shock struts and coil springs. The springs are mounted between the lower control and a spring pocket in the crossmember.

WHEEL BEARINGS

ADJUST

1. Raise vehicle until wheel and tire clear floor.
2. Remove wheel cover and dust cap from hub.
3. Remove cotter pin and locknut.
4. Loosen adjusting nut 3 turns, then rock wheel, hub and rotor assembly in and out several times to move shoe and linings away from rotor.
5. While rotating wheel assembly, **torque** the adjusting nut to 17-25 ft. lbs. to seat the bearings. **Fig. 2.**
6. Back off the adjusting nut one half turn. Retighten the nut to 10-15 inch lbs. with a torque wrench or finger tight.
7. Locate the nut lock on the adjusting nut so the castellations on the lock are aligned with the cotter pin hole in the spindle.
8. Install new cotter pin and replace dust cap and wheel cover.

WHEEL BEARINGS

REPLACE

1. Raise vehicle and remove front wheels.
2. Remove caliper mounting bolts. **It is not necessary to disconnect the brake lines for this operation.**
3. Slide caliper off of disc, inserting a clean spacer between the shoes to hold them in their bores after the caliper is removed. Position caliper out of the way. **Do not allow caliper to hang by brake hose.**
4. Remove hub and disc assembly. Grease retainer and inner bearing can now be removed. **Fig. 3.**
5. Reverse procedure to install.

CHECKING BALL JOINTS FOR WEAR

Support vehicle in normal driving position with both ball joints loaded. Clean area around grease fitting and checking surface. The checking surface is the round boss into which the grease fitting is installed. The checking surface should project outside the ball joint cover, **Fig. 4.** If checking surface is inside the cover replace the lower control arm assembly.

SHOCK STRUT

REPLACE

1. Place ignition switch in the unlocked position.
2. From engine compartment, remove upper shock absorber mounting nut.
3. Raise front of vehicle and support lower control arms. Position safety stands under frame jacking pads located rearward of wheels.
4. Remove wheel and tire assembly.
5. Remove caliper, rotor and dust shield.
6. Remove two bolts attaching shock absorber to spindle. **Gas pressurized struts must be held firmly during removal of the last spindle-to-strut bolt since gas pressure will cause strut to extend fully when bolt is removed.**
7. Lift strut upward from spindle to compress rod, then pull downward and remove shock absorber.
8. Remove jounce bumper, if equipped.
9. Reverse procedure to install. Remove suspension load from lower control arm by lowering front of vehicle, then torque lower mounting nuts to specification.

COIL SPRING

REPLACE

1. Raise front of vehicle and place safety stands under jack pads located rearward of wheels, then remove wheel and tire assembly.
2. Remove the brake caliper and wire it out of the way.
3. Disconnect stabilizer bar link from lower control arm.
4. Remove steering gear attaching bolts and position gear out of way.
5. Using tool 3290-C or equivalent, disconnect tie rod from spindle.
6. Install spring compressor No. T82P-5310A on models except 1989-90 with 5.0L engine, and No. D78P-5310-A on 1989-90 models with 5.0L engine, then compress coil spring until it is free of the spring seat. **Ensure spring compressor is properly installed before compressing spring. Also ensure spring is sufficiently compressed to permit removal of lower control arm pivot bolts.**
7. Remove two lower control arm pivot bolts, then disengage lower control arm and remove spring assembly, **Fig. 1. Measure compressed length of spring and amount of curvature to aid in compressing and installing spring.**
8. Reverse procedure to install. Ensure lower spring end is positioned between two holes in lower control arm spring pocket. Torque stabilizer bar and steering gear to No. 2 crossmember to specifications. **Torque** tie rod to steering spindle to 35 ft. lbs.

BALL JOINTS

REPLACE

The lower ball joint and lower control arm must be replaced as an assembly.

CONTROL ARM

REPLACE

1. Raise and support vehicle, allowing control arm to hang freely.
2. Remove brake caliper, rotor and dust shield, then disconnect tie rod from

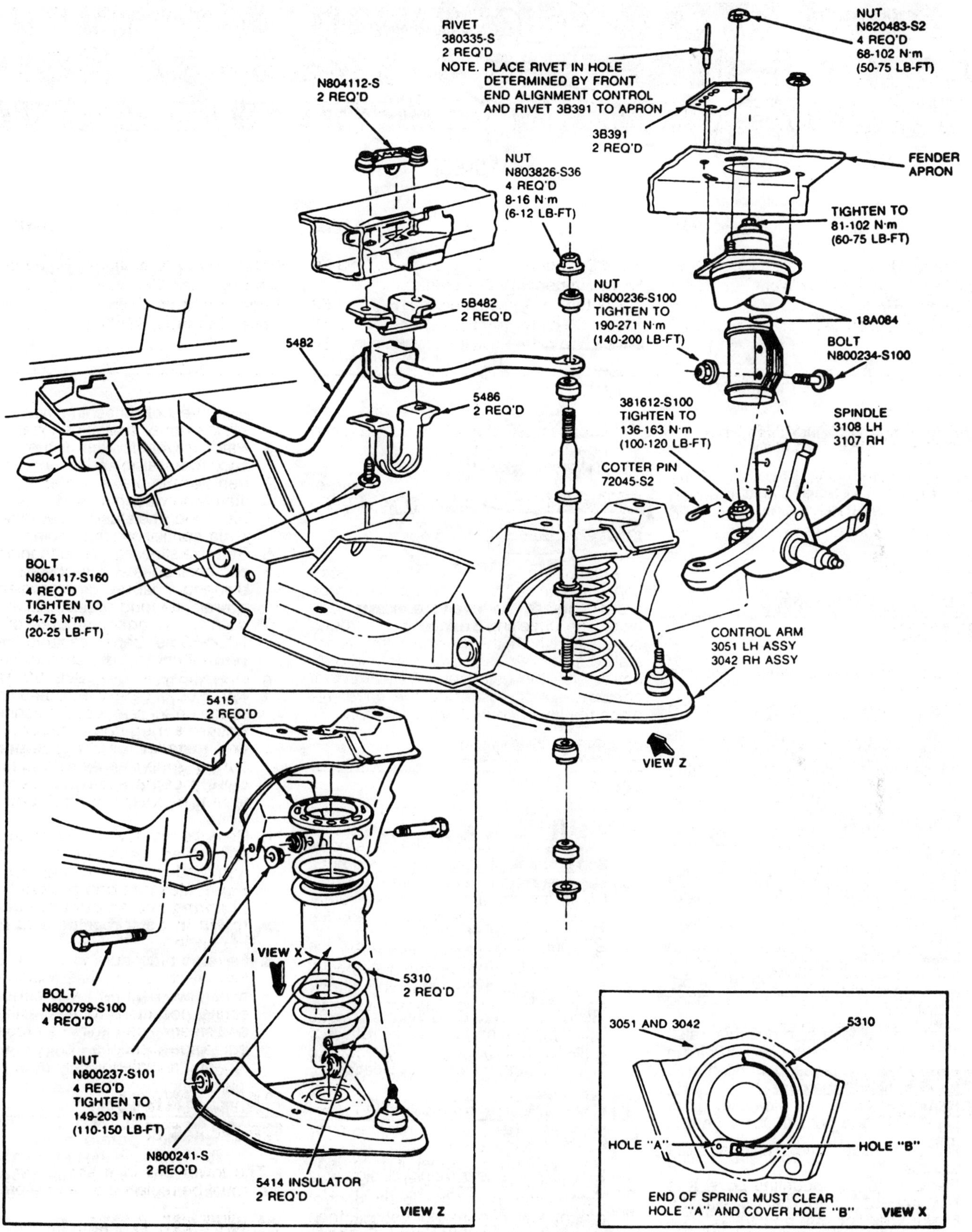

Fig. 1 Front suspension assembly

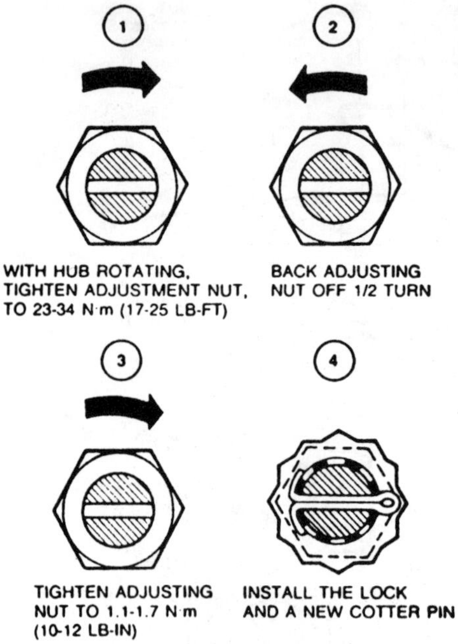

Fig. 2 Wheel bearing adjustment

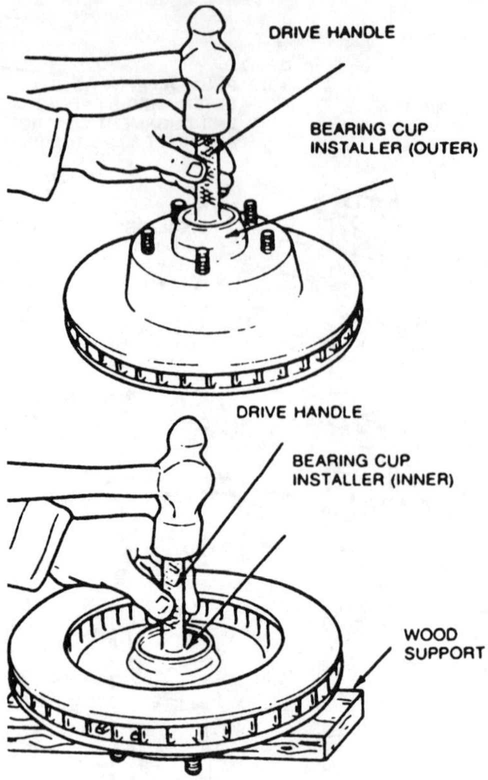

Fig. 3 Wheel bearing replacement

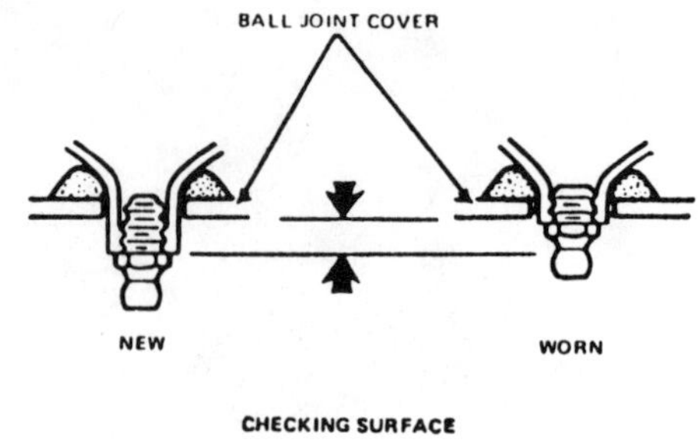

Fig. 4 Checking lower ball joint for wear

spindle using suitable tool.

3. If necessary, remove steering gear attaching bolts and position gear as necessary to gain access to control arm attaching bolts.
4. Remove cotter pin, then loosen ball joint stud nut approximately two turns. **Do not remove stud nut at this time.**
5. Tap spindle boss with suitable mallet to disengage ball joint stud from spindle.
6. Use spring compressor tool No. T82P-5310-A on all models except 1989-90 with 5.0L engine, or tool No. D78P-5310-A on 1989-90 5.0L models, then compress the coil spring slightly.
7. Remove ball joint stud nut, then raise strut and spindle assembly and wire in place to gain increased working area for control arm removal.
8. Remove control arm to crossmember pivot bolts and nuts, then the control arm.
9. Reverse procedure to install, noting the following:
   a. When installing coil spring, ensure lower spring end is positioned between the two holes in control arm spring pocket as shown in **Fig. 1.**
   b. Torque control arm attaching nuts to 130 ft. lbs., steering gear to crossmember attaching nuts to 95 ft. lbs., ball joint stud nut to 110 ft. lbs. and tie rod to spindle attaching nut to 35 ft. lbs.

STABILIZER BAR & INSULATOR

REPLACE

1. Raise and support vehicle.
2. Disconnect stabilizer bar from each link, then remove insulator attaching clamps, insulators and stabilizer bar from vehicle.
3. Reverse procedure to install. **Torque** attaching clamp to side rail retaining bolts to 48 ft. lbs. and stabilizer bar to link attaching nuts to specifications.

STEERING GEAR

REPLACE

A new steering gear mount design was introduced on the 1985 Mustang. The most important change in service procedure is the installation torque of 30-40 ft. lbs. Should the sleeve be damaged, it can be replaced. Remove the sleeve by collapsing it with a pair of locking pliers, then twisting and pulling. Install the new sleeve, drawing it into position with the mounting bolt during gear installation.

1. Disconnect battery ground cable.
2. Remove bolt attaching flexible coupling to input shaft.
3. Place ignition switch in the On position, then raise and support front of vehicle.
4. Remove cotter pins and nuts from tie rod ends, then using a suitable tool separate tie rods from spindle arms.
5. Support steering gear, then remove two nuts, bolts and washers attaching steering gear to crossmember. On power steering gears, lower gear slightly and disconnect pressure and return lines. Cap lines and fittings to prevent entry of dirt.
6. Remove steering gear from vehicle.
7. Reverse procedure to install. **Torque** flexible coupling to input shaft bolt to 20-30 ft. lbs., tie rod to spindle arm nuts to 35-47 ft. lbs. and steering gear to crossmember bolts to specifications. **Torque** pressure line fitting at gear housing to 10-15 ft. lbs.

POWER STEERING PUMP

REPLACE

1. Disconnect return hose from power steering pump reservoir and allow fluid to drain into a suitable container.
2. Disconnect pressure hose from power steering pump fitting, then remove pump mounting bracket and disconnect drive belt from pulley.
3. **On 2.3L, 3.8L and 5.0L engines which incorporate a fixed pump system,** remove the belt from the pulley, and remove the pulley, **Fig. 5.**
4. **On all models,** remove power steering pump.
5. Reverse procedure to install. **Torque** pump to mounting bracket bolts to 30-45 ft. lbs. and pressure hose to pump tube nut to 10-15 ft. lbs. **End play of pressure hose to pump fitting is normal and does not indicate a loose fitting. Do not over tighten.**

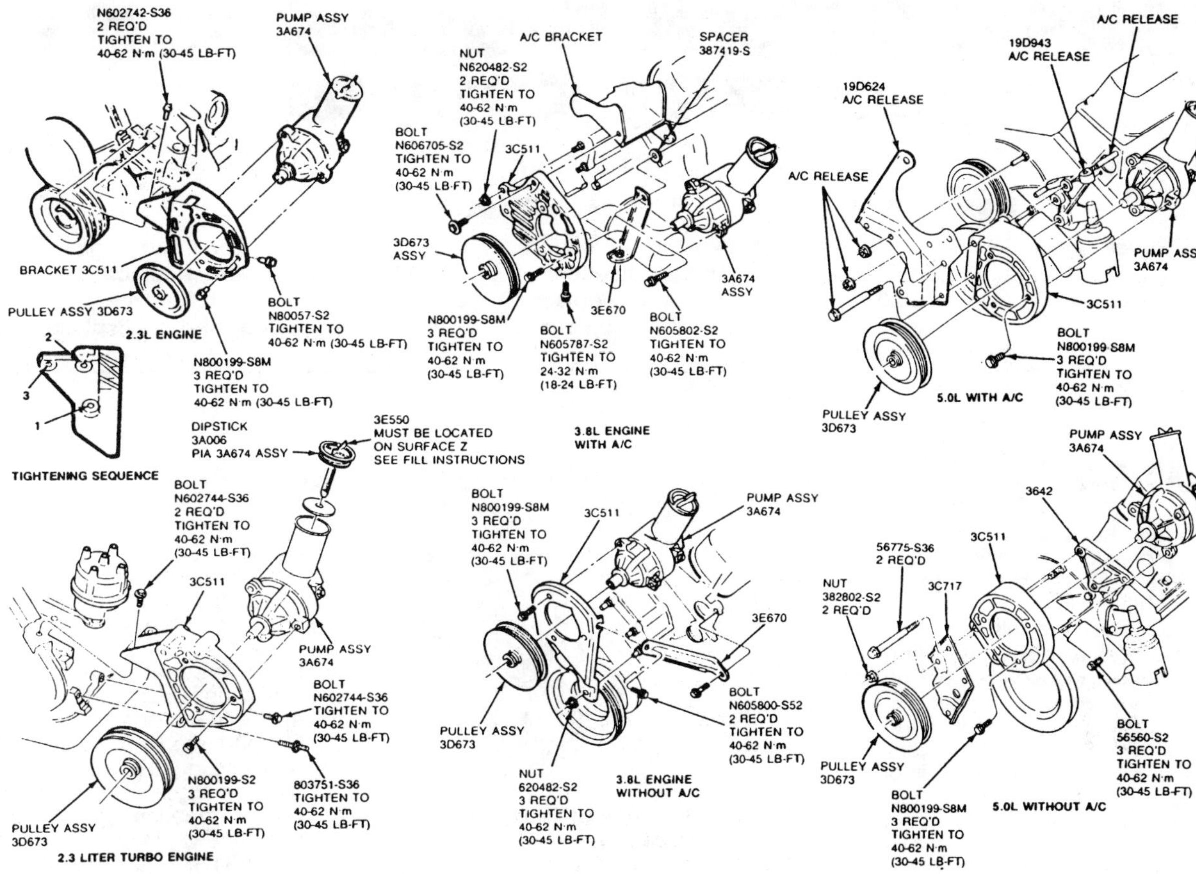

Fig. 5 Power steering pump installation

FRONT SUSPENSION & STEERING TIGHTENING SPECIFICATIONS

Year	Component	Torque/ Ft. Lbs.
CAPRI & MUSTANG		
1985-90	Ball Joint To Spindle	80-120
	Lower Arm To No. 2 Crossmember	110-150
	Stabilizer Bar Mounting Clamp To Bracket	37-50
	Stud & Washer Assembly To Stabilizer Bar & Lower Arm	6-12
	Spindle To Shock Strut	140-200
	Shock Strut To Upper Mount	50-75
	Shock Upper Mount To Body	50-75
	Steering Gear To No. 2 Crossmember	90-100
	Tie Rod End To Spindle	35-47
LTD & MARQUIS & 1985–88 COUGAR & THUNDERBIRD		
1985–88	Ball Joint To Spindle	80–120
	Lower Arm To No. 2 Crossmember	110–150
	Shock Strut To Upper Mount	55–92
	Shock Upper Mount To Body	50–75
	Spindle To Shock Strut	140–200
	Stabilizer Bar Mounting Clamp To Bracket	15–25
	Stabilizer Bar To Lower Arm	6–12
	Steering Gear To No. 2 Crossmember	90–100
	Tie Rod End To Spindle	35–47

1989–90 Ford Thunderbird & Mercury Cougar

INDEX

DESCRIPTION

These models have a long spindle short/long arm design suspension, **Fig. 1.** This configuration utilizes an upper A-arm, lower arm, double isolated tension strut and stabilizer bar. Base models have a 1.1 inch diameter stabilizer bar, while Thunderbird Super Coupe and Cougar XR7 utilize a 1.2 inch diameter stabilizer bar.

All models except the base model Thunderbird have a speed sensitive electronic variable assist power rack and pinion steering system. This system provides lower effort steering at low speeds and for parking, and higher effort steering at high speed to give a greater "feel" of the road. At 95 MPH, effort becomes constant.

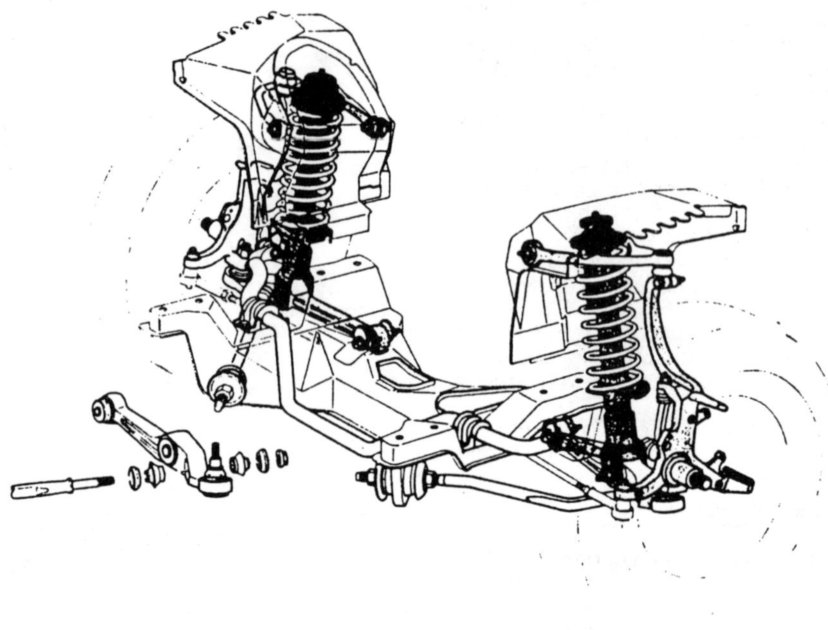

Fig. 1 Front suspension. 1989–90 Cougar & Thunderbird

STABILIZER BAR & BUSHINGS

1. Remove air inlet tube, then remove stabilizer bar retaining bracket bolts and brackets.
2. Remove serpentine drive belt, then raise and support vehicle.
3. Remove front tire and wheel assemblies, then remove crankshaft vibration damper.
4. Remove cotter pins and castellated nuts at the rod ends.
5. Separate tie rod ends from spindles using tie rod end removal tool No. 3290-D or equivalent.
6. Remove transmission oil cooler line bracket.
7. Remove stabilizer bar link from stabilizer bar using joint separator tool No. D88L-3006-A or equivalent.
8. Remove stabilizer bar from vehicles right side
9. Remove stabilizer bar bushings from stabilizer bar.
10. Reverse procedure to install.

COIL SPRING & SHOCK ASSEMBLY

REPLACE

1. Remove plastic cover at upper shock mount, **Fig. 2.**
2. **On models with auto level controls,** remove automatic ride actuator.

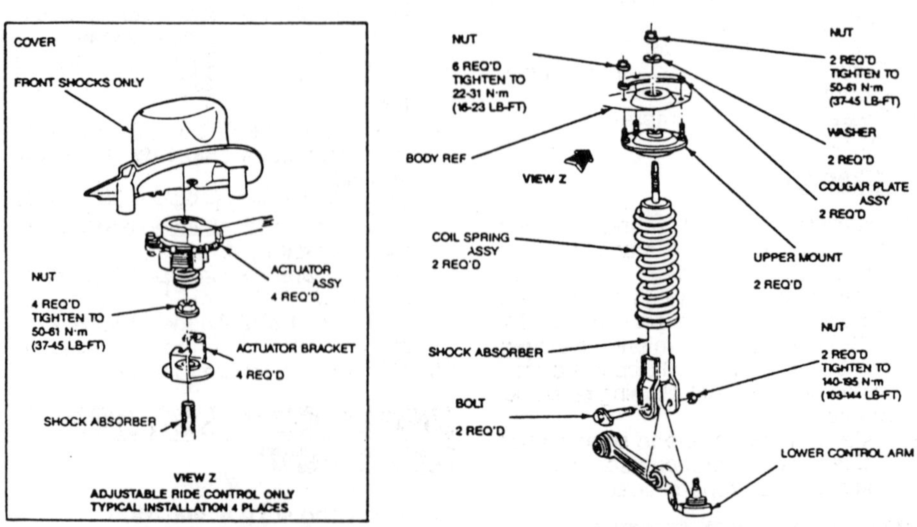

Fig. 2 Exploded view of coil spring & shock assembly. 1989–90 Cougar & Thunderbird

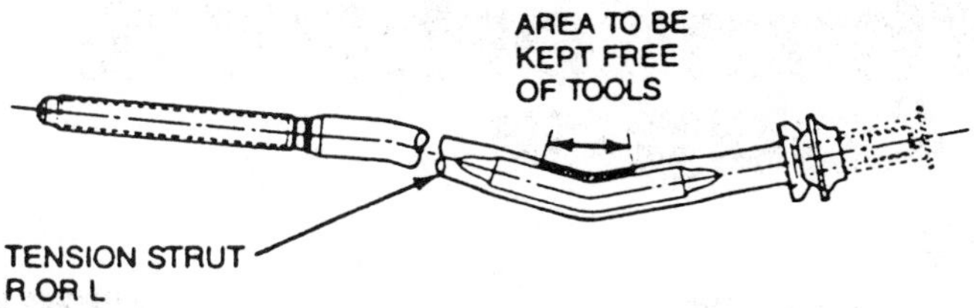

Fig. 3 Tension strut. 1989–90 Cougar & Thunderbird

3. **On all models,** raise and support vehicle.
4. Remove tire and wheel assemblies.
5. Remove lower shock mounting bolt and nut.
6. Remove nut at stabilizer link upper mounting stud.
7. Separate link from spindle using joint separator tool No. D88L-3006-A or equivalent.
8. Support lower control arm assembly with transmission jack.
9. Raise control arm and spindle with jack until stabilizer link can be completely separated from spindle. Position link out of the way.
10. Remove spindle to upper control arm attaching nut and bolt. Discard nut and bolt.
11. Lower jack to separate spindle from upper control arm. Do not allow spindle to hang free, support with wire or other means.
12. Remove support from lower control arm and remove shock and coil spring assembly.

DISASSEMBLY

The upper shock mount cannot be rotated when the shock and spring are assembled. Mark position of upper mount to coil spring with chalk, paint or grease pencil, prior to disassembly. If upper mount is not properly positioned during assembly it will not install in the vehicle.

1. Position shock assembly in Rotunda Spring Compressor tool No. 086-00029 or equivalent.
2. Compress spring and remove upper mount.
3. Release spring compressor to remove coil spring.

ASSEMBLY

If installing new coil spring or upper mount transfer reference marks from removed parts to new parts.

1. Position shock and coil assembly in Rotunda Spring Compressor tool No. 086-00029 or equivalent and compress spring to install upper mount.
2. Install upper mount aligning reference marks.
3. Install nut and torque to specification.
4. Release spring compressor, ensuring coil spring is properly seated.

SPINDLE ASSEMBLY REPLACE

1. Raise and support vehicle.
2. Remove tire and wheel assembly.
3. Remove brake caliper and rotor.
4. Remove hub and bearing assembly.
5. Remove brake anti-lock sensor and position out of the way.
6. Remove tie rod end cotter pin and loosen castellated nut. Separate tie rod end from spindle using tie rod end remover tool No. 3290-D or equivalent.
7. Remove stabilizer bar link at spindle using joint separator D88L-3006-A or equivalent.
8. Separate lower ball joint from spindle. Loosen nut and rap spindle with hammer. Remove nut.
9. Remove and discard upper spindle to upper control arm bolt and nut. Spread slot slightly and remove spindle from control arm and vehicle.
10. Reverse procedure to install.

UPPER CONTROL ARM REPLACE

1. Raise and support vehicle.
2. Remove tire and wheel assembly.
3. Remove and discard upper spindle to ball joint bolt and nut. Slightly spread spindle at slot and remove ball joint.
4. Lower vehicle, then brake off flags on upper control arm pivot bolt heads.

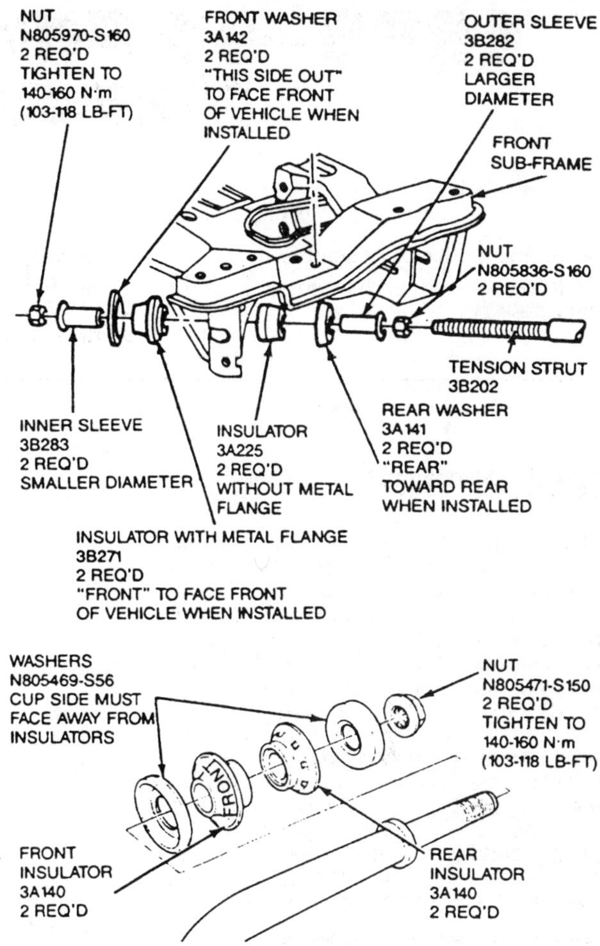

Fig. 4 Tension strut insulator replacement. 1989–90 Cougar & Thunderbird

5. Remove upper control arm bolts, then upper control arm.
6. Reverse procedure to install.

LOWER CONTROL ARM REPLACE

1. Raise and support vehicle.
2. Remove tire and wheel assembly.
3. Loosen lower ball joint nut three or four turns.
4. Rap spindle to separate ball joint. Leave nut attached.
5. Support spindle by wire to prevent excessive sagging of upper control arm.
6. Mark position of camber adjustment cam.
7. Remove nut attaching tension strut to control arm. hold strut by flats with wrench while turning nut. **Do not hold strut or damage surface in area shown in Fig. 3, damage to tension strut may result.**
8. Remove lower shock bolt and nut.
9. Remove pivot (camber) bolt and nut.
10. Remove nut at lower ball joint and remove control arm.
11. Reverse procedure to install, after installation check front end alignment and adjust as necessary.

STEERING GEAR
REPLACE

1. Raise and support vehicle.
2. Remove both front wheel and tire assemblies.
3. Remove cotter pins at outer tie rod ends and remove castellated nuts at each end. Discard cotter pins.
4. Separate tie rod ends from spindles, using tie rod end removal tool No. 3290-D or equivalent.
5. Place drain pan under vehicle, then disconnect and plug power steering return line hose.
6. Disconnect power steering pressure line at intermediate fitting and position out of the way.
7. Remove steering shaft retaining bolt.
8. Remove rack to sub-frame bolts and nuts, access nuts through hole in front crossmember.
9. Lower rack as necessary, to remove pressure line inlet tube. Remove and discard plastic seal on inlet tube.
10. Cut tie strap securing pressure line to each tube.
11. Remove steering rack from vehicle.
12. Reverse procedure to install.

TENSION STRUT
REPLACE
REMOVAL

1. Raise and support vehicle.
2. Remove tire and wheel assembly.
3. Hold tension strut on flats with wrench. Remove front sub-frame attaching nut and insulator. **Do not hold strut or damage surface in area shown in Fig. 3, damage to tension strut may result.**
4. Mark position, or note number of visible threads at tension strut rear sub-frame nut.
5. Back off tension strut rear sub-frame nut.
6. Remove tension strut to lower control arm attaching nut.
7. Remove lower shock bolt and nut.
8. Remove brake hose bracket to body attaching bolt.
9. Remove ABS sensor attaching bolt and position sensor out of the way.
10. Pry lower control arm rearward and remove tension strut.

INSTALLATION

1. If front tension strut insulators were removed, de-burr insulator sleeves before reassembly. If new sleeves are used, the inner sleeve must be shorter than the outer sleeve. Cut up to 1/4 inch off inner sleeve if necessary, **Fig. 4.**
2. Install tension strut retaining nut all the way down on threads. Install with nylon insert facing forward.
3. Install outer sleeve and rear washer. Word REAR must face outward.
4. Install rear insulator (without metal flange) with large end toward sub-frame.
5. Install tension strut in sub-frame.
6. Install front insulator, with the word FRONT toward front of vehicle and metal flange toward sub-frame.
7. Install front washer, with the words "THIS SIDE OUT" toward front of vehicle.
8. Install inner sleeve, loosely install front nut.
9. Install front washer and insulator on rear of tension strut, cup to face away from insulator.
10. Install tension strut into lower control arm.
11. Install rear insulator with small end toward control arm.
12. Install washer and cup facing rear of vehicle.
13. Return torsion strut-to-sub-frame nut to original position.
14. Install lower shock bolt and nut.
15. Holding tension strut with wrench, torque strut to sub-frame to specification.
16. Torque tension strut to lower control arm attaching nut to specification.
17. Install brake hose bracket, **torque** attaching bolt to 8-12 ft. lbs.
18. Install ABS sensor, **torque** attaching bolt to 40-60 inch lbs.
19. Install tire and wheel assembly, check front wheel alignment.

FRONT SUSPENSION & STEERING TIGHTENING SPECIFICATIONS

Year	Component	Torque/ft. lbs.
1989–90	Lower Control Arm To Shock	103–144
	Lower Control Arm To Spindle (Ball Joint)	80–120
	Lower Control Arm To Sub-frame	92–125
	Lower Control Arm To Tension Strut	90–120
	Upper Control Arm To Body	55–90
	Upper Control Arm To Spindle (Ball Joint	50–65
	Stabilizer Bar Link To Spindle	40–55
	Stabilizer Bar Link To Stabilizer Bar	40–55
	Tension Strut To Sub-frame	90–120
	Shock To Upper Mount	37–45
	Upper Shock Mount To Body	16–23
	Tie Rod End To Spindle	39–54

Ford Probe

INDEX

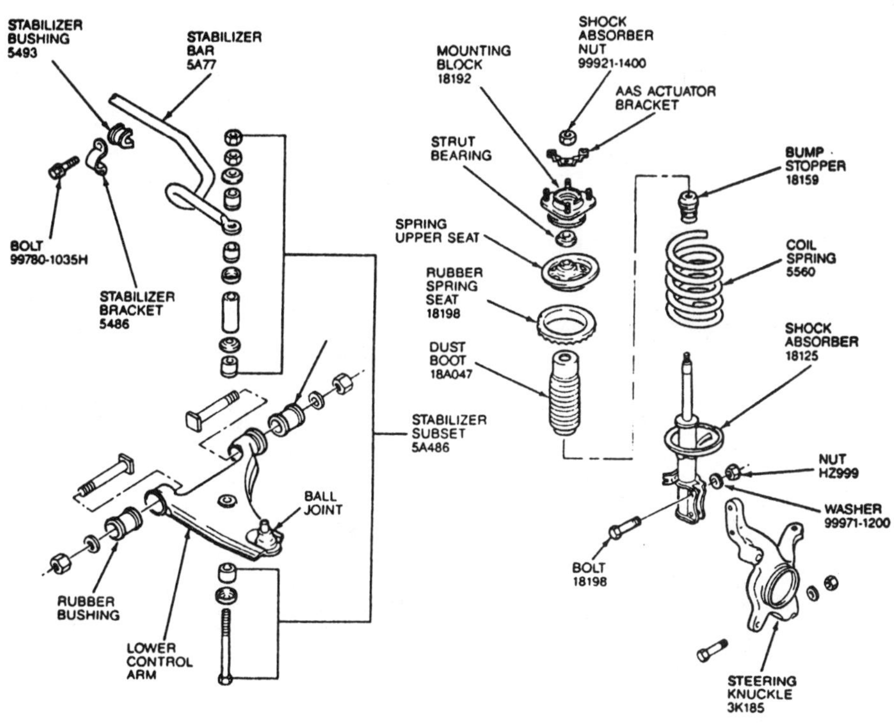

Fig. 1 Front suspension (Part 1 of 2)

DESCRIPTION

This suspension is of the MacPherson strut type, **Fig. 1.** The strut towers are located in the wheel wells and position the upper ends of the struts. If the vehicle is equipped with the optional Programmed Ride Control system, the ride control actuator bolts to the top of the strut mounting block that houses a rubber mounted strut bearing. The upper end of the coil spring rides in a rubber sprint seat. A forged steering knuckle is bolted to the shock absorber.

If the vehicle is not equipped with the Programmed Ride Control System, the struts used are the conventional non-adjustable type. These struts are not interchangeable.

The lower ball joints are pressed into the lower control arm that is attached to the steering knuckle. The control arms are supported by rubber bushings at each end. A hollow stabilizer bar is connected to the control arms.

STRUT ASSEMBLY REPLACE

1. Raise and properly support front of vehicle.
2. Remove rubber cap from strut mounting block. If equipped with Programmed Ride Control, disconnect control module connector.
3. Place an alignment mark between the inside of the strut mounting block and the chassis strut tower, **Fig. 2.**
4. If equipped with Programmed Ride Control, remove control module.
5. If equipped with anti-lock brake system, remove system harness and bracket.
6. **On all models,** remove brake caliper, the U-clip from brake line hose, then slide from strut bracket, **Fig. 3.**
7. Remove steering knuckle to strut attaching bolts.
8. Remove vane airflow meter assembly, then the ignition coil bracket from strut tower.
9. Remove strut mounting bolts from tower, then the strut assembly from vehicle.
10. Place strut assembly in a vise.
11. Loosen, but do not remove shock absorber nut, **Fig. 4.**
12. Remove strut assembly from vise, then compress coil spring, using a spring compressor or tool D85P-7178-A.
13. Remove shock absorber nut, gradually releasing spring tension.
14. If equipped with Programmed Ride Control, remove control module bracket.
15. **On all models,** remove strut mounting block, upper rubber spring seat, dust boot, bump stopper, then coil spring from strut, **Fig. 5.**
16. Reverse procedure to assemble, then install. When installing the strut mounting block, ensure that the notch on the mounting block is 180 degrees from the steering knuckle mounting bracket on the strut. **Torque** strut mounting bolts to specifications. **Torque** steering knuckle to strut attaching bolts to specifications.

STABILIZER BAR REPLACE

1. Raise and properly support vehicle.
2. Remove wheel and tire assembly.
3. Remove stabilizer bar link assembly from lower control arm, **Fig. 6.**
4. Remove mount bolt from stabilizer bar bushing.
5. Remove stabilizer bar from vehicle.
6. Reverse procedure to install. **Torque** stabilizer bar bushing bolt to specifications. Tighten stabilizer bar link nut until .79 inch of threads protrude above nut, **Fig. 7.**

LOWER CONTROL ARM REPLACE

1. Raise, then properly support vehicle.
2. Remove wheel and tire assembly.
3. Remove brake caliper, then secure to coil spring. **Do not allow to hang by the brake hose.**
4. Remove stabilizer bar as outlined elsewhere in this section.
5. Remove ball joint clamp bolt, then separate ball joint from steering knuckle, **Fig. 8.**

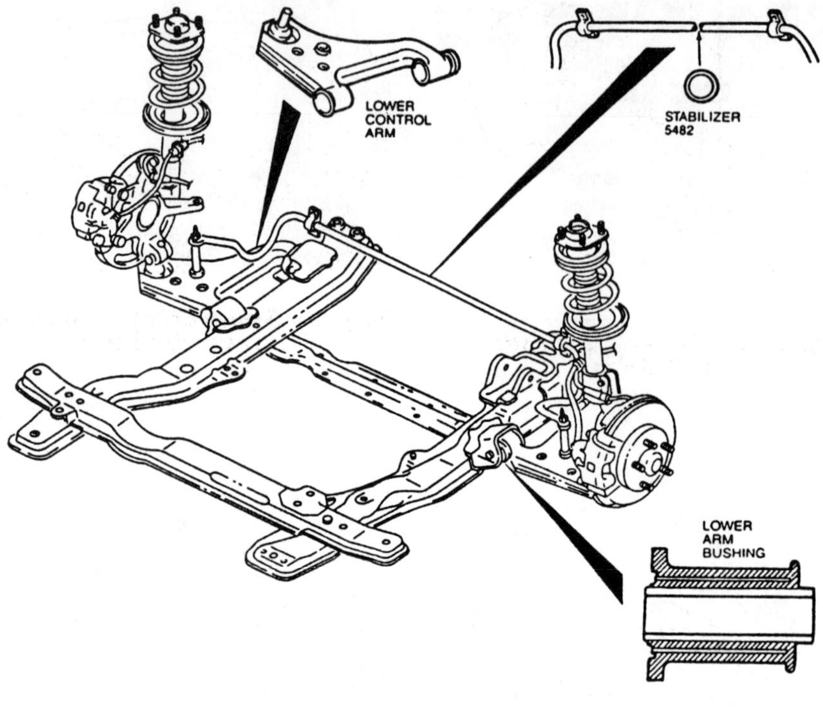

Fig. 1 Front suspension (Part 2 of 2)

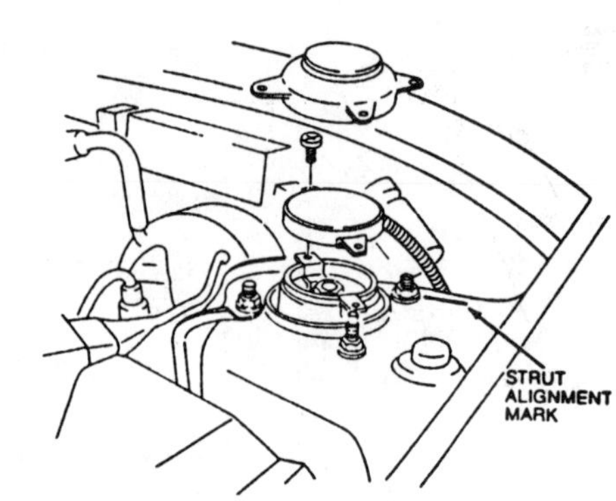

Fig. 2 Placing strut alignment mark

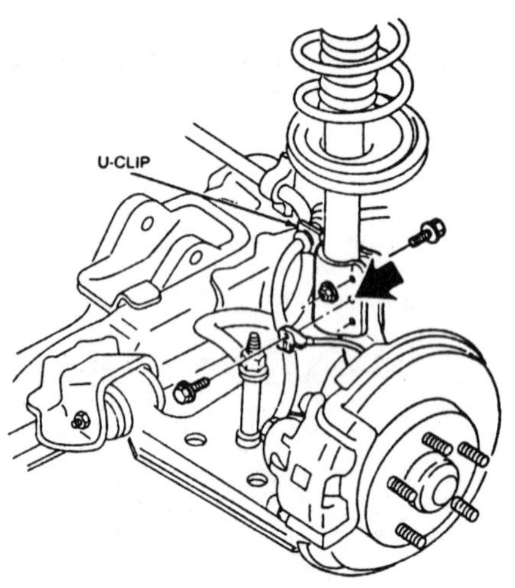

Fig. 3 Disconnecting brake line from strut

6. **On vehicles equipped with automatic transaxle,** remove harmonic damper from chassis sub-frame, left hand side, **Fig. 9.**
7. Remove lower control arm mounting bolts, then the control arm, **Fig. 10.**
8. Reverse procedure to install. **Torque** control arm mount bolts to specifications. **Torque** ball joint clamp bolt to specifications.

BALL JOINT SERVICE

The ball joints on this vehicle are not serviceable parts. Only the dust boots are replaceable.

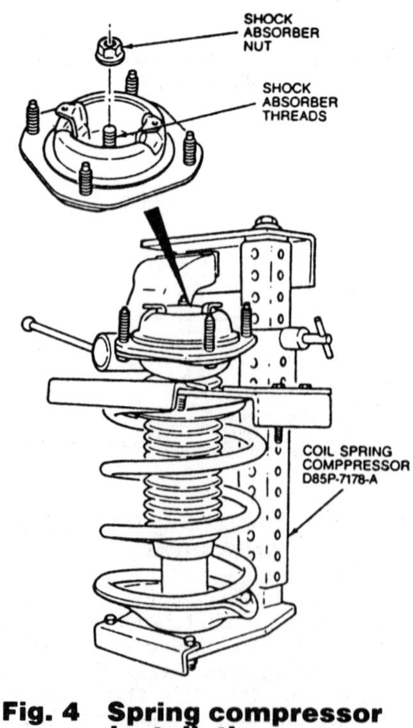

Fig. 4 Spring compressor installation

1. Remove lower control arm as outlined elsewhere in this section.
2. Place control arm in a vise.
3. Remove dust boot with a chisel, **Fig. 11. Use caution not to damage ball joint.**
4. Liberally coat inside of new dust boot with lubricant C1AZ-19590-B or equivalent.

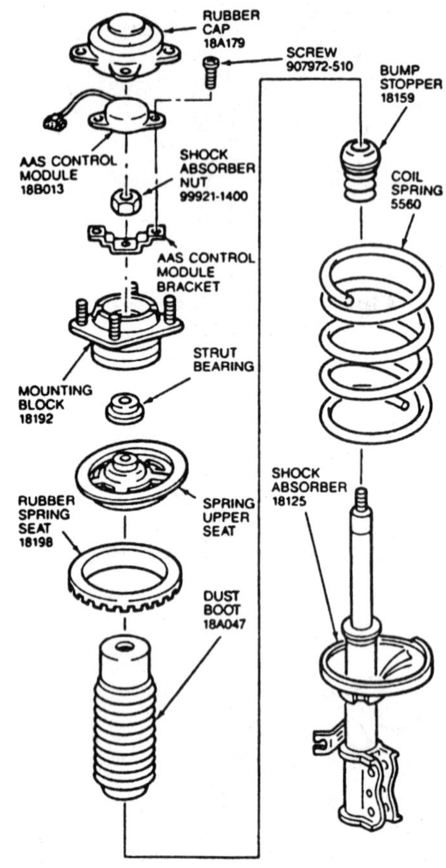

Fig. 5 Strut disassembled

5. Install dust boot with tool T88C-5493-AH or equivalent, **Fig. 12.**
6. Install lower control arm.

STEERING KNUCKLE REPLACE

1. Raise, then properly support vehicle.
2. Remove tire and wheel assembly.
3. Carefully raise the staked portion of the half shaft attaching nut using a small cape chisel, **Fig. 13.**
4. Remove half shaft attaching nut. When loosening the nut, lock the hub

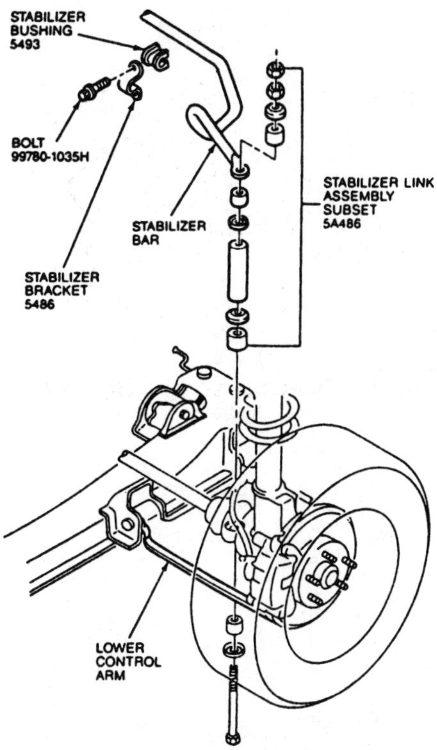

Fig. 6 Stabilizer bar installation

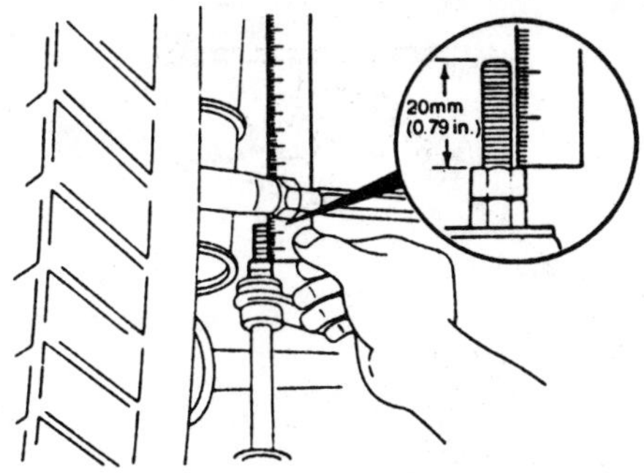

Fig. 7 Stabilizer bar link installation

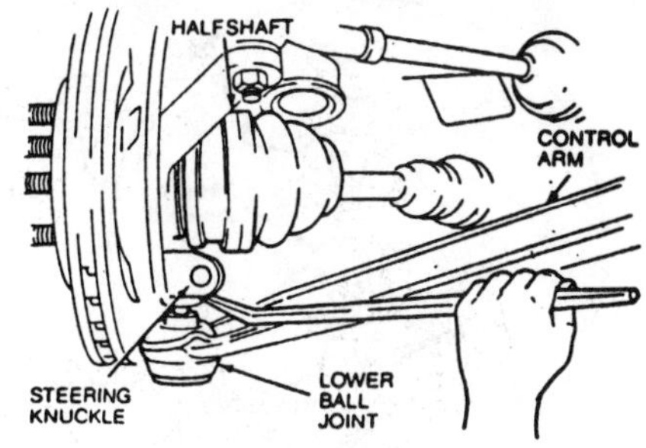

Fig. 8 Separating ball joint from control arm

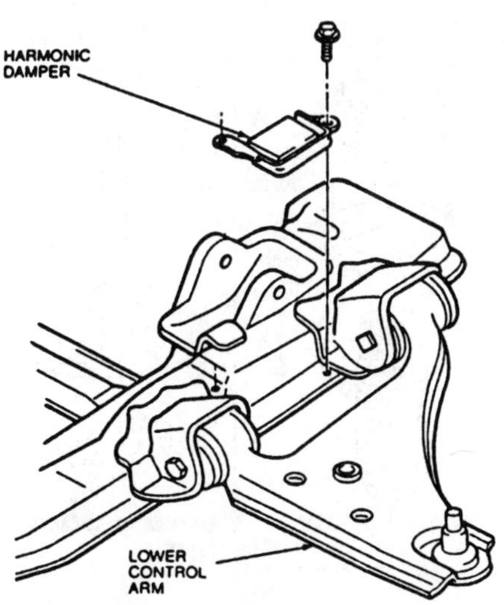

Fig. 9 Harmonic damper installation

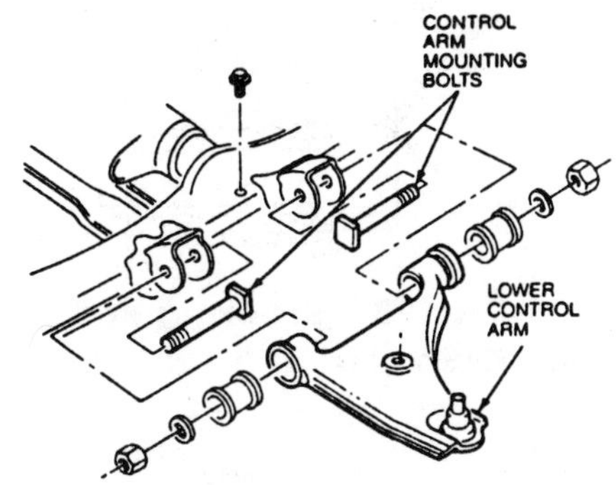

Fig. 10 Lower control arm installation

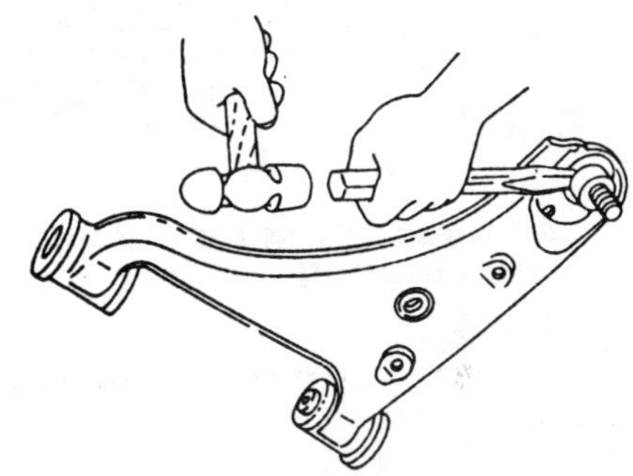

Fig. 11 Removing ball joint dust boot

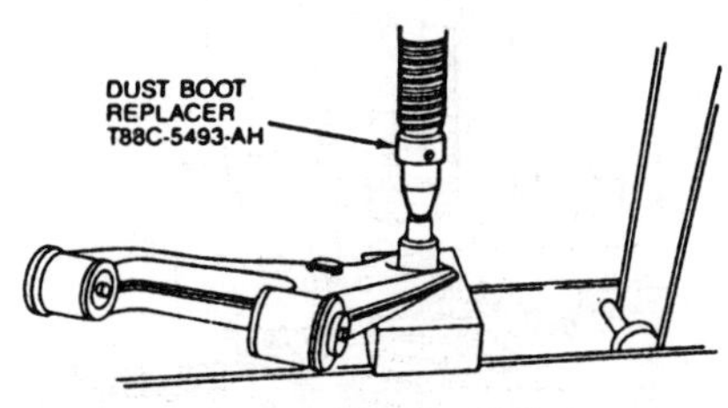

Fig. 12 Installing ball joint dust boot

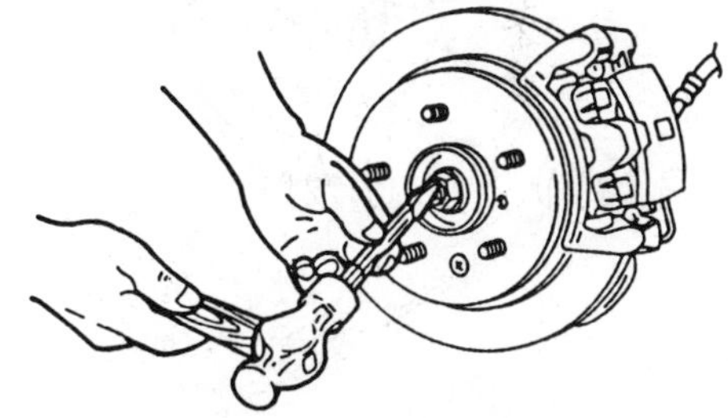

Fig. 13 Removing stake from half shaft attaching nut

in position by having a helper lock the brakes. **Discard the nut, it should not be reused.**

5. Remove stabilizer bar link bolts.
6. Separate tie rod end from steering knuckle, **Fig. 14.**
7. Remove caliper, then anchor bracket assembly. Suspend caliper from coil spring. **Do not allow to hang from brake line.**
8. Remove brake rotor, **Fig. 15.**
9. Remove lower ball joint clamp bolt, then separate ball joint from steering knuckle.
10. Remove steering knuckle to strut attaching bolts.
11. Slide front hub/steering knuckle assembly from strut bracket and half shaft, **Fig. 15. Use caution not to damage grease seals. If the hub binds on the half shaft splines, lightly tap end of half shaft with a plastic hammer. If the half shaft splines become rusted to the hub, use a two-jawed puller or a hub puller to separate.**
12. Reverse procedure to install. **Torque** steering knuckle to strut attaching bolts to specifications. **Torque** ball joint clamp bolt to specifications. **Torque** new half shaft attaching nut to specifications, then stake nut, **Fig. 16. Torque** tie rod ending attaching nut to 22-33 ft. lbs.

STEERING GEAR REPLACE

STANDARD POWER STEERING GEAR

1. Disconnect battery ground cable.
2. Raise, then properly support vehicle.
3. Remove wheel and tire assemblies.
4. Separate tie rods from steering knuckles.
5. Remove plastic dust shield from both sides of lower inner fender.
6. Pull back dust boot, have a helper rotate steering column shaft until clamp bolt becomes accessible, then lock steering column.
7. Place an alignment mark between

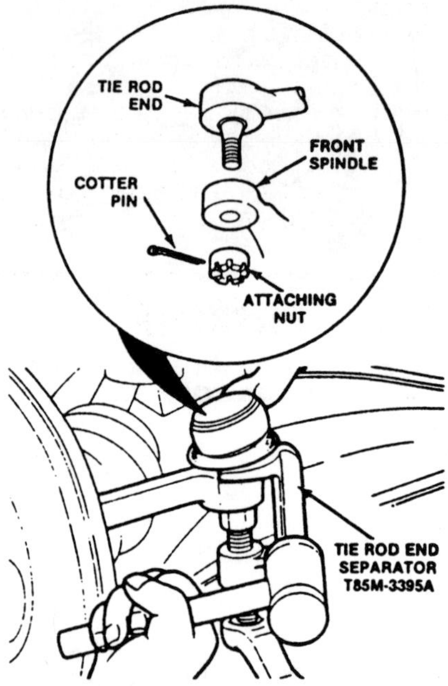

Fig. 14 Separating tie rod end from steering knuckle

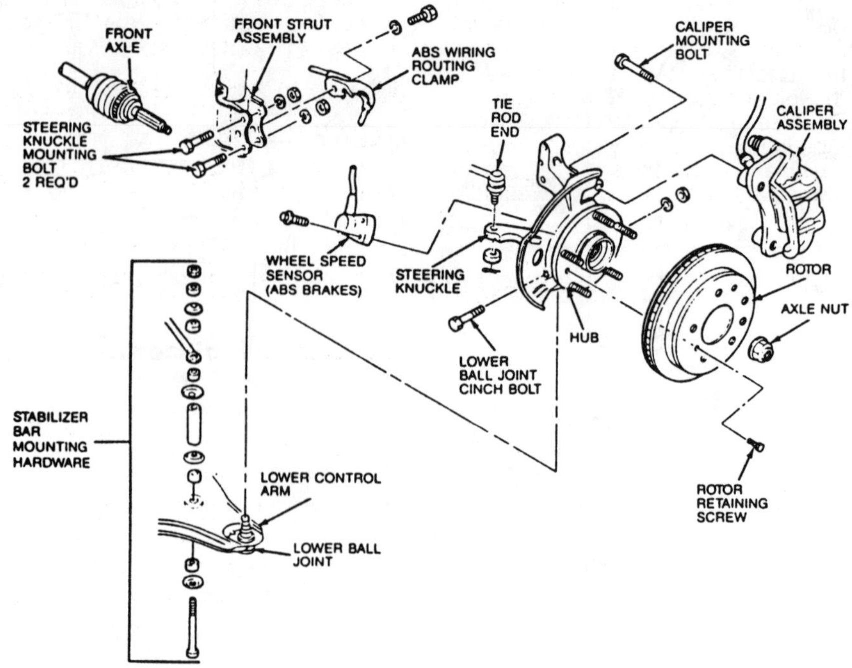

Fig. 15 Steering knuckle, hub & rotor installation

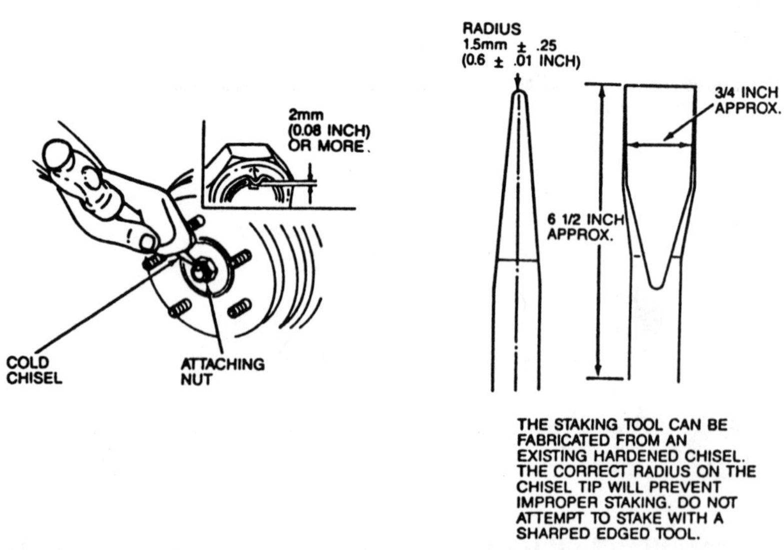

Fig. 16 Staking half shaft attaching nut

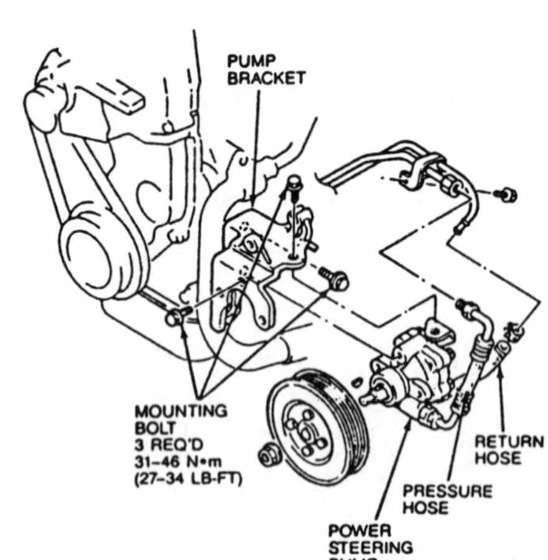

Fig. 17 Power steering pump installation

steering column pinion shaft and the intermediate shaft lower universal joint.

8. Remove clamp bolt from steering column intermediate shaft lower universal joint.
9. Disconnect two hydraulic lines from steering gear. Position hydraulic lines aside.
10. Remove steering gear mount bolts, then lower the steering gear until it clears bulkhead. Slide the steering gear toward the right until the left hand tie rod clears the left lower control arm. Slide steering gear to the left, then remove from vehicle.
11. Reverse procedure to install. **Torque** steering gear mount bolts to specifications. **Torque** steering column intermediate shaft lower universal joint clamp bolt to 13-20 ft. lbs.

ELECTRONIC POWER STEERING GEAR

1. Disconnect battery ground cable.
2. Raise, then properly support vehicle.
3. Remove wheel and tire assemblies.
4. Separate tie rods from steering knuckles.
5. Remove plastic dust shield from both sides of lower inner fender.
6. Pull back dust boot and have a helper rotate steering column shaft until clamp bolt becomes accessible, then lock steering column.
7. Place an alignment mark between steering column pinion shaft and the intermediate shaft lower universal joint.
8. Remove clamp bolt from steering column intermediate shaft lower universal joint.
9. Disconnect solenoid valve, then the power steering pressure switch electrical connectors.
10. Disconnect three hydraulic lines from steering gear. Discard the two copper washers from each fitting. Position hydraulic lines aside.
11. Remove steering gear mount bolts, then lower the steering gear until it

clears bulkhead. Slide the steering gear toward the right until the left hand tie rod clears the left lower control arm. Then, slide steering gear to the left and remove from vehicle.

12. Reverse procedure to install. **Torque** steering gear mount bolts to specifications. **Torque** steering column intermediate shaft lower universal joint clamp bolt to 13-20 ft. lbs. When connecting hydraulic lines to steering gear, install new copper washers.

POWER STEERING PUMP REPLACE

1. Disconnect battery ground cable.
2. Remove right hand inner fender splash shield.
3. Remove drive belt.
4. Disconnect power steering pressure and return hoses from pump.
5. Remove three power steering pump mounting bolts and the pump, **Fig. 17.**
6. Reverse procedure to install. **Torque** power steering pump mounting bolts to specifications.

FRONT SUSPENSION & STEERING TIGHTENING SPECIFICATIONS

Year	Component	Torque/ft. lbs.
1989-90	Ball Joint Clamp Bolt	32-40
	Control Arm Rod Jam Nuts	41-59
	Dynamic Damper Mounting Bolts	31-46
	Half shaft Attaching Nut	116-174
	Lower Control Arm Mounting Bolts	69-93
	Power Steering Pump Mounting Bolts	27-34
	Stabilizer Bar Bushing	27-40
	Steering Gear Mount Bolts	27-40
	Steering Knuckle To Strut	69-86
	Strut Attaching Nuts	34-46
	Strut Tower Nut	47-69
	Tie Rod End Jam Nuts	51-72
	Wheel Lug Nuts	65-87

Ford Taurus & Mercury Sable

INDEX

DESCRIPTION

This suspension is of the gas filled MacPherson strut type, **Fig. 1.** The strut top mount consists of a rubber insulated bearing and seat and coil spring insulator. The top mount is attached to the body side apron by three bolts. The lower part of the strut is mounted in the steering knuckle and is retained by a pinch bolt. A forged lower control arm is attached to the sub-frame and to the steering knuckle. A tension strut is connected to the lower control arm and to the forward part of the sub-frame.

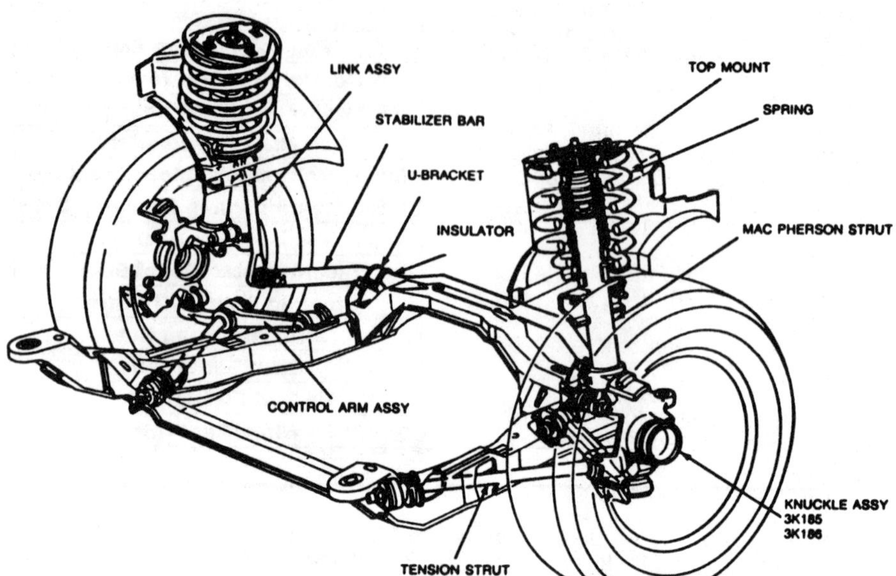

Fig. 1 Front suspension assembly

STRUT ASSEMBLY REPLACE

1. Place ignition switch in Off position and ensure steering wheel is not locked.
2. Remove hub nut and loosen three strut attaching nuts, then raise and support vehicle. Do nut raise vehicle with lower control arm.
3. Remove wheel and tire assembly.
4. Remove brake caliper and wire it aside.
5. Remove brake rotor and tie rod end.
6. Remove stabilizer bar link nut, then remove link from strut.
7. Remove lower control arm to steering knuckle pinch nut and bolt, then slightly spread joint and remove lower control arm.
8. Using suitable hub remover/installer, press axle from hub. Wire axle shaft to body to maintain level position. **Do not allow axle shaft to move outward. Over extension of the constant velocity (CV) joint could result in separation of internal parts, which could cause CV joint failure.**
9. Remove strut-to-steering knuckle pinch bolt, then spread joint slightly and remove steering knuckle and hub assembly.
10. Remove strut attaching nuts, then remove strut assembly from vehicle.
11. Compress strut spring with coil spring compressor D85P-7178-A or equivalent.
12. Using a 10 mm box wrench, restrain strut shaft, then remove strut mounting nut with suitable 21 mm crow foot socket. **Do not allow strut shaft to rotate.**
13. Loosen compressor tool, then remove strut top mount bracket assembly, bearing and seat assembly and spring, **Fig. 2.**
14. Reverse procedure to install, noting the following:
    a. **Torque** strut-to-steering knuckle pinch bolt to 70-95 ft. lbs.
    b. Torque lower control arm to steering knuckle pinch bolt to specifications.
    c. Torque stabilizer bar assembly to strut to specifications.
    d. **Torque** tie rod end attaching nut to 23-36 ft. lbs.
    e. Torque strut attaching nuts to specifications.
    f. With vehicle on ground, **torque** hub nut to 180-200 ft. lbs.

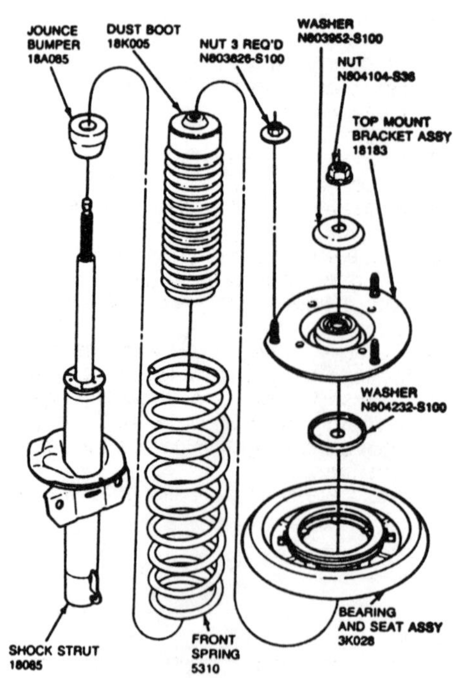

Fig. 2 MacPherson strut assembly

STEERING KNUCKLE REPLACE

1. Place ignition switch in Off position and ensure steering wheel is not locked.
2. Remove hub nut and loosen three strut attaching nuts, then raise and support vehicle. Do not raise vehicle

with lower control arm.
3. Remove wheel and tire assembly.
4. Remove brake caliper and wire it aside.
5. Remove brake rotor and tie rod end.
6. Remove stabilizer bar link nut, then remove link from strut.
7. Remove lower control arm to steering knuckle pinch nut and bolt, then slightly spread joint and remove lower control arm.
8. Using suitable hub remover/installer, press axle from hub. Wire axle shaft to body to maintain level position. **Do not allow axle shaft to move outward. Over extension of the constant velocity (CV) joint could result in separation of internal parts, which could cause CV joint failure.**
9. Remove rotor splash shield (if equipped).
10. Remove strut-to-steering knuckle pinch bolt, then spread joint slightly and remove steering knuckle and hub assembly.
11. Reverse procedure to install, noting the following:
    a. **Torque** strut-to-steering knuckle pinch bolt to 70-95 ft. lbs.
    b. **Torque** lower control arm to steering knuckle pinch bolt to 40-55 ft. lbs.
    c. Torque stabilizer bar assembly-to-strut to specifications.
    d. **Torque** tie rod end attaching nut to 23-36 ft. lbs.
    e. **Torque** strut attaching nuts to 20-30 ft. lbs.
    f. With vehicle on ground, **torque** hub nut to 180-200 ft. lbs.

LOWER CONTROL ARM REPLACE

1. Place ignition switch in Off position and ensure steering wheel is not locked.
2. Raise and support vehicle.
3. Remove wheel and tire assembly.
4. Remove tension strut nut, then the dished washer.
5. Remove lower control arm-to-steering knuckle pinch nut and bolt, then slightly spread joint and separate ball joint from steering knuckle. **Do not use a hammer to separate suspension pieces.**
6. Remove lower control arm inner pivot bolt and nut, then the control arm.
7. Reverse procedure to install, noting the following:
    a. **Torque** lower control arm pivot bolt to 70-95 ft. lbs.
    b. Torque lower control arm-to-steering knuckle attaching bolt to specifications.
    c. **Torque** tension strut nut to 70-95 ft. lbs.

STABILIZER BAR REPLACE

1. Raise and support vehicle.
2. Remove stabilizer bar link-to-strut attaching nuts.
3. Remove stabilizer bar link-to-stabilizer bar attaching nuts.
4. Remove steering gear-to-sub-frame attaching bolts, then move steering gear off sub-frame.
5. Support sub-frame with suitable safety stands, then remove rear sub-frame attaching bolts. Lower rear part of sub-frame to gain access to stabilizer bar mounting brackets.
6. Remove mounting brackets, then the stabilizer bar.
7. Reverse procedure to install, noting the following:
    a. Torque stabilizer bar mounting brackets to specifications.
    b. **Torque** sub-frame attaching bolts to 85-100 ft. lbs.
    c. **Torque** stabilizer bar link attaching nuts to 35-48 ft. lbs.

STEERING GEAR REPLACE

1. Disconnect battery ground cable.
2. Remove steering shaft weather boot-to-dash panel attaching bolts.
3. Remove intermediate shaft-to-steering column shaft attaching bolts.
4. Move weather boot aside, then remove steering gear input shaft pinch bolt and remove intermediate shaft.
5. Raise and support vehicle.
6. Remove left front wheel and heat shield.
7. Remove bundling strap retaining lines to gear.
8. Remove tie rod ends from steering knuckles.
9. Place suitable drain pan under steering gear, then remove pressure and return lines from gear and allow to drain into pan.
10. Remove nuts from gear mounting bolts. **The gear mounting bolts are pressed into the steering gear housing and should not be removed during normal service procedures.**
11. Push weather boot end into vehicle and lift steering gear out of mounting holes. Rotate gear as necessary so input shaft passes between brake booster and floor pan. Carefully start working steering gear out through left front fender apron opening.
12. Rotate input shaft as necessary so it clears left front fender apron opening and remove steering gear from vehicle.
13. Reverse procedure to install, noting the following:
    a. Prior to installing hydraulic hoses, install new plastic seals on fittings.
    b. **Torque** steering gear attaching nuts to 85-100 ft. lbs.
    c. **Torque** hydraulic pressure hose fitting to 20-25 ft. lbs. **Torque** hydraulic return hose fitting to 15-20 ft. lbs. **When hydraulic fittings are properly installed, the hoses are free to swivel.**
    d. Fill power steering system with automatic transmission fluid type F.
    e. Set toe-in to specifications.

POWER STEERING PUMP REPLACE

4-153 ENGINE

1. Disconnect battery ground cable.
2. Using suitable 1/2 inch drive socket wrench, rotate tensioner pulley clockwise and remove alternator and power steering belts.
3. Position suitable drain pan under pump, then disconnect pressure and return lines and allow fluid to drain.
4. Using hub puller T69L-10300-B or equivalent, remove pulley from shaft.
5. Remove three pump-to-bracket attaching bolts, then the pump.
6. Reverse procedure to install.

V6 ENGINE

1. Disconnect battery ground cable.
2. Loosen idler pulley, then remove power steering belt.
3. Using hub puller T69L-10300-B or equivalent, remove pulley from shaft.
4. Position suitable drain pan under pump, then disconnect return line and allow fluid to drain.
5. Completely back off pressure line fitting. Line will separate during pump removal.
6. Remove three pump-to-bracket attaching bolts, then the pump.
7. Reverse procedure to install.

FRONT SUSPENSION & STEERING TIGHTENING SPECIFICATIONS

Year	Component	Torque/ft. lbs.
1986-90	Control Arm to Knuckle	40-55
	Control Arm to Sub-Frame	70-95
	Stabilizer Bar Link Assembly to Stabilizer Bar	35-48
	Stabilizer Bar Link Assembly to Shock Strut	55-75
	Strut Top Mount to Body	22-32
	Strut to Top Mount	35-50
	Strut to Knuckle	70-95
	Tension Strut to Control Arm	70-95
	Tension Strut to Sub-Frame	70-95
	Tie Rod End to Steering Knuckle	23-25

Lincoln Continental & Mark VII

INDEX

Page No.

MARK VII & 1985–87 CONTINENTAL

DESCRIPTION

This suspension, **Fig. 1,** is a modified MacPherson strut design, which uses shock struts and coil springs. The springs are mounted between the lower control arm and a spring pocket in the crossmember.

WHEEL BEARINGS, ADJUST

1. With wheel rotating, tighten adjusting nut to 17-25 ft. lbs.
2. Back off adjusting nut 1/2 turn and retighten nut to 10-15 inch lbs.
3. Place nut lock on nut so that castellations on lock are aligned with cotter pin hole in spindle and install cotter pin.
4. Check front wheel rotation, if it rotates noisily or rough, clean, inspect or replace wheel bearings as necessary.

WHEEL BEARINGS, REPLACE

1. Raise car and remove front wheels.
2. Remove caliper mounting bolts. **It is not necessary to disconnect the brake line for this operation.**
3. Slide caliper off of the disc, inserting a spacer between the shoes to hold them in their bores after the caliper is removed. Position caliper assembly out of the way. **Do not allow caliper to hang by brake hose.**
4. Remove hub and disc. Grease retainer and inner bearing can now be removed.
5. Reverse procedure to install.

CHECKING BALL JOINTS FOR WEAR

LOWER BALL JOINT

1. Support vehicle in normal driving position with both ball joints loaded.
2. Clean area around grease fitting and checking surface. **The checking surface is the round boss into which the grease fitting is installed.**
3. The checking surface should project outside the cover, **Fig. 2.** If surface is inside cover replace lower arm assembly.

BALL JOINTS, REPLACE

These ball joints are not serviceable. If they require replacement, the control arm and ball joint must be replaced as an assembly. Torque ball joint stud to specifications

SHOCK STRUT, REPLACE

Turn air suspension switch off before removing shock strut.

1. Place ignition switch in the Unlocked position so that front wheels are free to move.
2. From engine compartment, remove one strut to upper mounting nut. Use

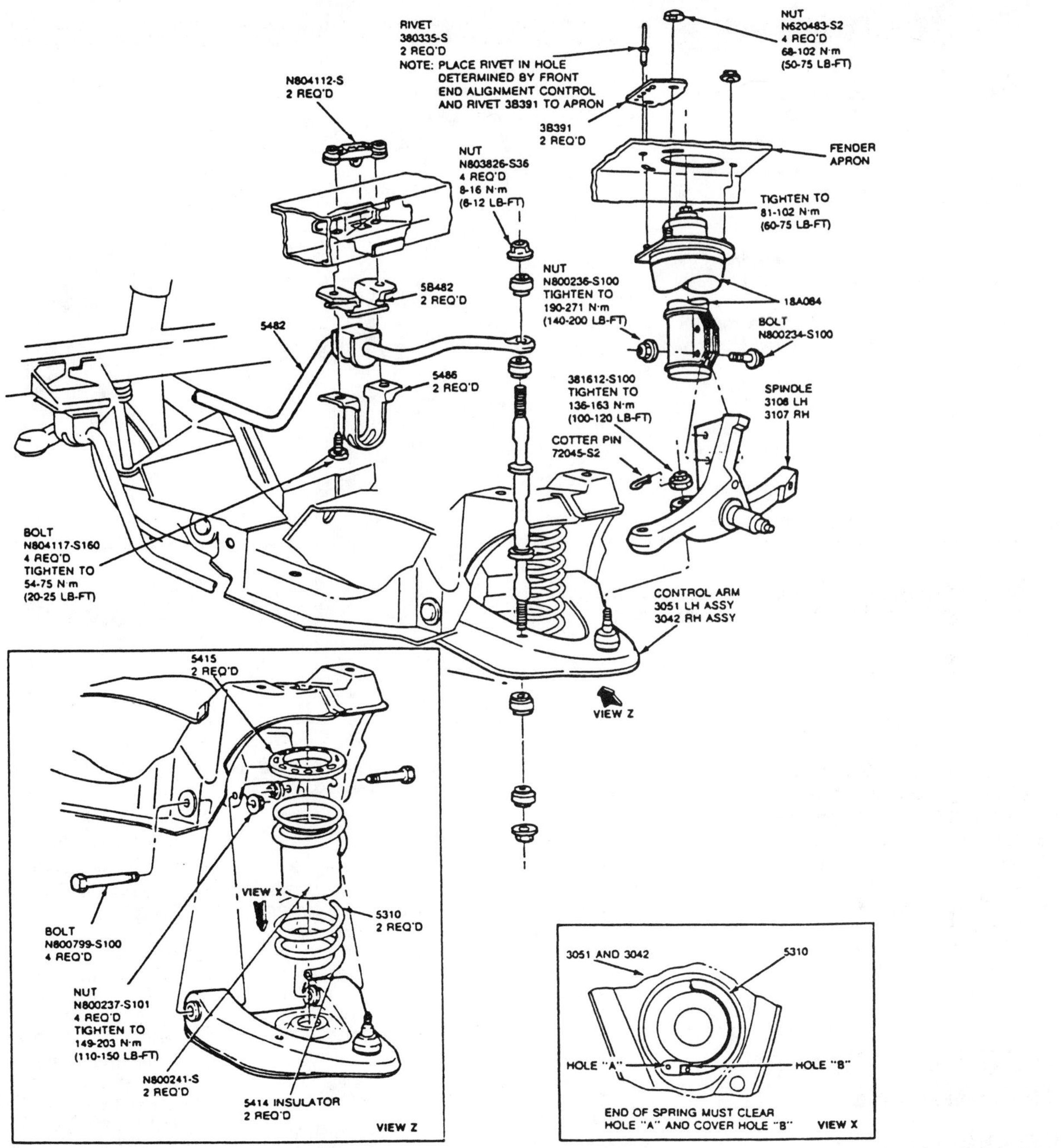

Fig. 1 Front suspension. Mark VII & 1985–87 Continental

a screwdriver in rod slot to hold rod stationary when removing nut.
3. Raise front of vehicle by lower control arms, then place safety stands under frame jack pads located rearward of wheels.
4. Remove wheel and tire assembly, then remove brake caliper, rotor assembly and dust shield.
5. Remove two nuts and bolts attaching lower strut to spindle. **When removing the second lower strut to spindle nut, hold strut firmly as gas pressure will cause strut to fully extend.**
6. Lift strut upward from spindle to compress rod, then pull downward and remove strut.

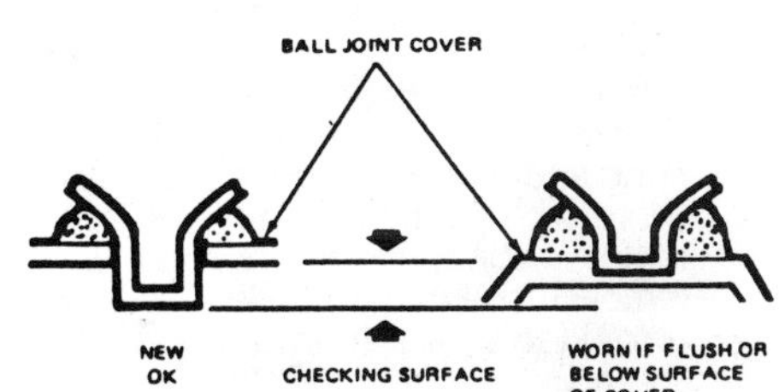

Fig. 2 Lower ball joint wear indicator. Mark VII & 1985–87 Continental

7. Reverse procedure to install.

POWER STEERING GEAR, REPLACE

INTEGRAL POWER RACK & PINION

1. Disconnect battery ground cable.
2. Remove bolt retaining flexible coupling to input shaft.
3. Turn ignition switch on and raise vehicle.
4. Remove tie rod end retaining nuts, then separate studs from spindle arms.
5. Support gear and remove attaching bolts, then lower gear enough to gain

access to pressure and return lines. Remove bolt attaching the hose bracket to the gear and bolts from the crossmember.
6. Disconnect and cap pressure and return lines, then remove steering gear.
7. Reverse procedure to install.

INTEGRAL POWER STEERING GEAR

1. Disconnect lines from steering gear and plugs lines and ports.
2. Remove the two bolts securing flex coupling to steering gear and to column.
3. Raise vehicle and remove sector shaft nut and Pitman arm. **Do not damage the seals.**
4. Support steering gear and remove three attaching bolts. Remove flex coupling clamp bolt and work steering gear free of coupling, then remove steering gear.
5. Reverse procedure to install.

POWER STEERING PUMP, REPLACE

1. Disconnect fluid return hose from reservoir and drain power steering fluid into a container.
2. Remove pressure hose from pump fitting. Do not remove fitting from pump.
3. Disconnect belt from pulley. If necessary, remove pulley from pump installing pulley removal tool No. T75L-3733-A or equivalent so small diameter threads engage in pump shaft. While holding small hex head, rotate tool nut to remove pulley. Do not apply in and out pressure on pump shaft as this will damage the internal thrust areas.
4. Remove pump.
5. If pulley was removed, install tool and while holding small hex head, turn tool nut clockwise to install pulley. Pulley must be flush within .010 inch of the end of the pump. Do not apply in and out pressure on shaft. Remove tool.

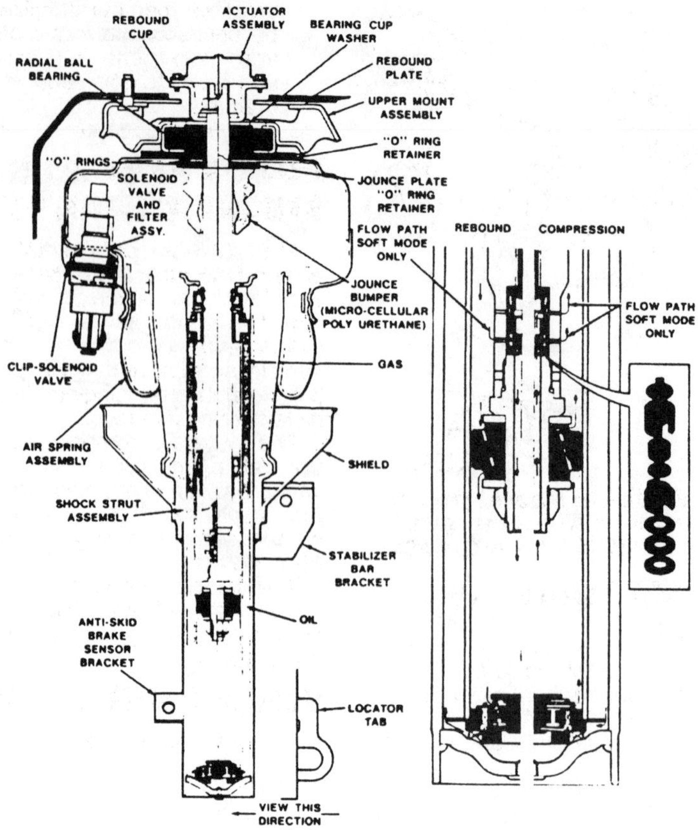

Fig. 3 Cross-sectional view of front strut assembly. 1988–90 Continental

1988–90 CONTINENTAL

DESCRIPTION

The 1988-90 Continental front suspension consists of MacPherson struts with integral air springs and dual-damping shock absorbers, and two height sensors, one at each control arm.

The air suspension and dual damping functions are controlled by a microcomputer based module which receives inputs for vehicle speed, door switch position, damping actuator feedback, steering wheel turning rate and angle, engine vacuum, throttle position, brake actuation, ignition switching, and vehicle ride height. The dual-damping function automatically switches from a soft to firm ride when the driving situation (hard cornering, acceleration or braking, etc.) dictates the need for increased damping effect.

The front struts, **Fig. 3,** are mounted to the body through a precision ball bearing and rubber mount system. The ball bearing provides a durable pivot for the strut/wheel assembly, while the rubber mount provides for a smooth ride on rough roads with minimal noise. The rotary design Hall effect type height sensor, identical to the rear sensor, permits multiple height positions to be defined, resulting in specialized leveling during all types of driving and load characteristics.

SERVICE PRECAUTION

Always place the air suspension switch in the Off position before performing any work, or whenever raising the front suspension.

STRUT ASSEMBLY, REPLACE

1. Remove hub nut and loosen the upper strut attaching nuts, then raise and support vehicle. **Do not lift vehicle from lower control arm.**
2. Disconnect air suspension electrical wiring and all related parts that will interfere with strut removal.
3. Remove wheel and tire assembly.
4. Remove brake caliper and support with suitable wire, then disconnect tie rod end.
5. Remove stabilizer bar link nut, then remove link from strut.
6. Remove lower control arm to steering knuckle pinch nut and bolt, then spread joint as necessary and disengage control arm from knuckle.
7. Using a suitable hub installation/removal tool, press axle from hub/rotor assembly. Wire axle shaft as necessary to maintain level position. **Do not permit axle shaft to move outward during disengagement from hub, since damage to CV joints could result.**
8. Remove strut to steering knuckle pinch bolt, then spread joint and remove steering knuckle and hub assembly.
9. Remove strut upper attaching nuts, then remove strut from vehicle.
10. Reverse procedure to install.

LOWER CONTROL ARM, REPLACE

1. Raise and support vehicle.
2. Disconnect air suspension electrical wiring and all related parts that will interfere with control arm removal.
3. Remove wheel and tire assembly, then the tension strut nut and washer.
4. Remove lower control arm to steering knuckle pinch nut and bolt, then spread joint and separate ball joint from knuckle.
5. Remove lower control arm inner pivot bolt and nut, then the lower control arm.
6. Reverse procedure to install.

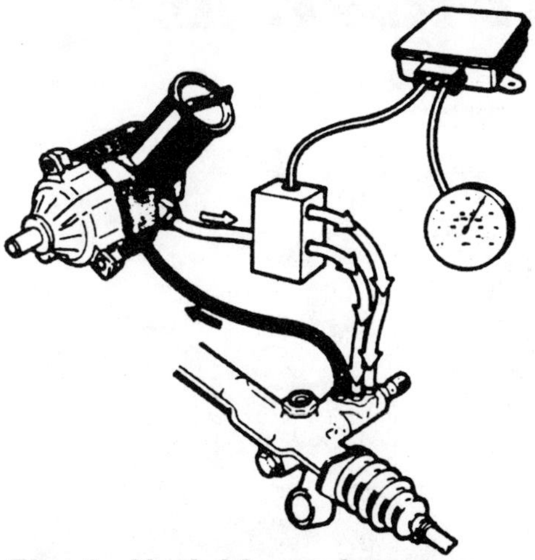

Fig. 4 Variable assist power rack & pinion steering gear system. 1988–90 Continental

STABILIZER BAR, REPLACE

1. Raise and support vehicle.
2. Remove stabilizer bar link to strut and link to bar attaching nuts.
3. Remove stabilizer bar mounting brackets, then the stabilizer bar. It may be necessary to move steering gear from sub-frame and lower rear of sub-frame to gain access to mounting brackets.
4. Reverse procedure to install.

POWER RACK & PINION STEERING GEAR SYSTEM

The 1988-90 Continental uses a new electro-hydraulic, speed sensitive variable assist rack and pinion power steering system. The system is designed to generate higher levels of power assist during low vehicle speeds and while parking, while progressively reducing power assist as vehicle speed increases.

The system is comprised of a new control valve, an electro-hydraulic actuator valve, and an electronic control module that is programmed to provide the right amount of assist during all vehicle speeds, **Figs. 4 and 5.**

The control module also has built-in diagnostic capabilities, which enables quicker troubleshooting in the event of a system malfunction.

SERVICE BULLETINS

On 1988 Continental a rattling or clunking noise from under the front of the vehicle may be present. If noise is present check for a loose tension strut bushing retainer.

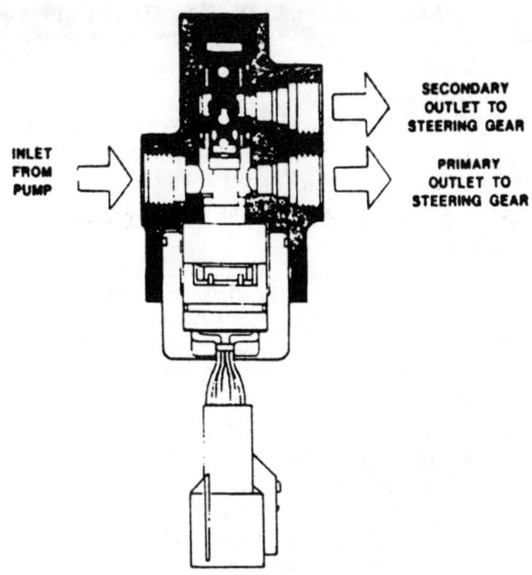

Fig. 5 Cross-sectional view of actuator valve. 1988–90 Continental

FRONT SUSPENSION & STEERING TIGHTENING SPECIFICATIONS

Year	Component	Torque/ft. lbs.
MARK VII & 1985–87 CONTINENTAL		
1985-89 ①	Lower Arm To No. 2 Crossmember	110-150
	Stabilizer Bar Mounting Clamp	40-55
	Stabilizer Bar To Lower Arm	9-12
	Spindle To Shock Strut	140-200
	Shock Strut Upper Mount	50-75
	Ball Joint To Spindle	80-120
	Shock Upper Mount To Body	50-75
	Steering Gear To No. 2 Crossmember	90-100
	Tie Rod End To Spindle Nut	35-47
1989-90 ②	Lower Arm To No. 2 Crossmember	110-150
	Stabilizer Bar Mounting Clamp	40-55
	Stabilizer Bar To Lower Arm	9-12
	Spindle To Shock Strut	140-200
	Shock Strut Upper Mount	55-92
	Ball Joint To Spindle	100-120
	Shock Upper Mount To Body	62-75
	Steering Gear To No. 2 Crossmember	90-100
	Tie Rod End To Spindle Nut	35-47
	Compressor Bracket To Frame	30-40 ③
	Air Compressor To Bracket	30-40 ③
	Sensor Upper Attachment To Frame	26-34
	Sensor Lower Attachment To Arm	8-12
1988–90 CONTINENTAL		
1988–90	Strut Top Mount To Body	22-32
	Strut To Top Mount	35-50
	Strut To Knuckle	70-95
	Control Arm To Knuckle	40-55
	Control Arm To Sub-Frame	70-95
	Tension Strut To Control Arm	70-95
	Tension Strut To Sub Frame	70-95
	Stabilizer Bar Bracket To Sub-Frame	21-32
	Stabilizer Bar Link Assembly To Bar	35-48
	Stabilizer Bar Link Assembly To Strut	55-75
	Tie Rod End To Steering Knuckle	23-25

①—Except air suspension.
②—Air suspension.
③—Inch lbs.

Mercury Tracer

INDEX

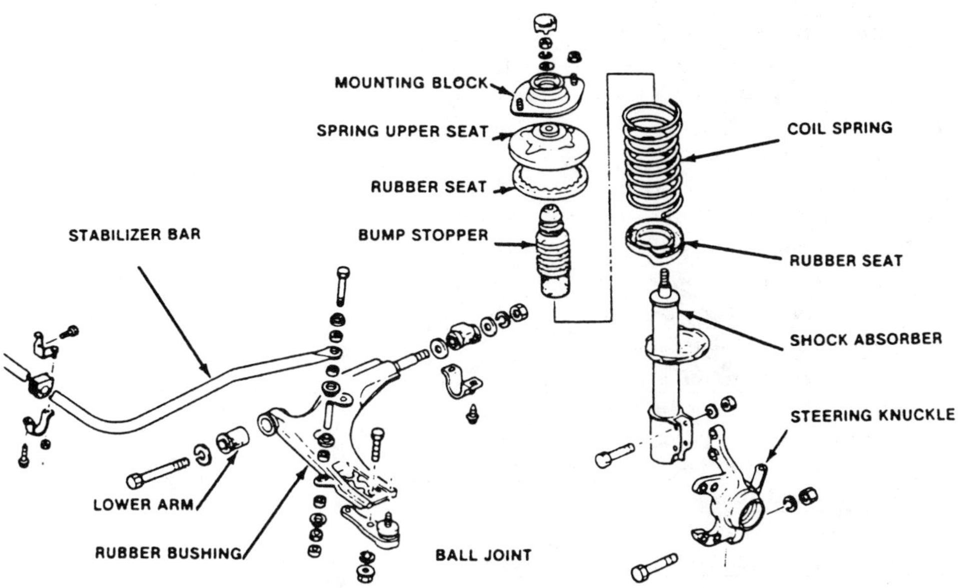

Fig. 1 Exploded view of front suspension

STRUT UNIT & COIL SPRING

REPLACE

1. Raise and support front vehicle, then remove wheel and tire assembly.
2. Remove rotor and hub as described in "Front Drive Axle."
3. Place alignment marks on inside of strut mounting block.
4. Remove steering knuckle to strut attaching bolts, **Fig. 1.**
5. Remove clip attaching brake hose to strut unit, then slide hose out of bracket.
6. Remove strut mounting bolts at strut tower, then remove strut and coil spring unit from vehicle.
7. Using a suitable spring compressor, compress coil spring.
8. Remove strut nut, then carefully release coil spring compressor.
9. Remove coil spring and upper mounting components.
10. Reverse procedure to install.

BALL JOINT

REPLACE

1. Raise and support front of vehicle, then remove wheel and tire assembly.
2. Remove caliper and suspend from coil spring with brake hose attached.
3. Remove stabilizer link from control arm, **Fig. 1.**
4. Remove tie rod end from steering knuckle using a suitable puller.
5. Remove ball joint to steering knuckle clamp bolt, then using a suitable pry bar, pull downward on lower control arm to separate ball joint stud from steering knuckle.
6. Remove two ball joint to control arm attaching nuts, then remove ball joint from lower control arm.
7. Reverse procedure to install. When installing stabilizer link, tighten until .43 inch of thread extends beyond nut.

CONTROL ARM

REPLACE

1. Raise and support front of vehicle, then remove wheel and tire assembly.
2. Remove brake caliper and suspend from coil spring with brake hose attached.
3. Remove stabilizer link, **Fig. 1.**
4. Place alignment marks on rear control arm bushing and mounting bracket, then place alignment marks on front control arm bushing and control arm.
5. Remove ball joint to steering knuckle clamp bolt.
6. Loosen lower control arm front bushing nut and rear bushing bolt.
7. Remove lower control front bushing bracket attaching bolts and remove bracket.
8. Remove lower control arm rear bushing bolt and remove lower control arm.
9. If necessary, remove lower control arm from bushing nut and bushing.
10. If necessary, replace control arm rear bushing as follows:
    a. Using a hacksaw remove rear flange from rear control arm bushing.
    b. Using tool No. T87C-5493-B, press rear bushing from control arm.
    c. Position replacement lower control arm rear bushing, then press in using tool.
11. Reverse procedure to install. When installing stabilizer link, tighten until .43 inch of thread extends beyond nut. Recheck stabilizer bar bushing bracket attaching bolts torque after vehicle has been lowered.

HUB & STEERING KNUCKLE

REPLACE

REMOVAL

1. Raise and support front of vehicle, then remove wheel and tire assembly.
2. Using a suitable chisel, carefully raise staked portion of axle nut.
3. With brakes applied, remove and discard axle nut, **Fig. 2.**
4. Remove stabilizer to control arm attaching bolt.
5. Remove cotter pin and retaining nut, then separate tie rod end from steering knuckle using a suitable tool.
6. Remove U-clip attaching brake hose to strut unit.
7. Remove caliper assembly with brake hose attached and suspend from coil spring using wire.
8. Remove bolt retaining ball joint to steering knuckle, then pry downward on lower control arm to separate ball joint from steering knuckle.
9. Remove steering knuckle to strut unit attaching bolts, then slide hub and steering knuckle out of strut unit bracket and off end of half shaft, **Fig. 3.** If hub and steering knuckle binds

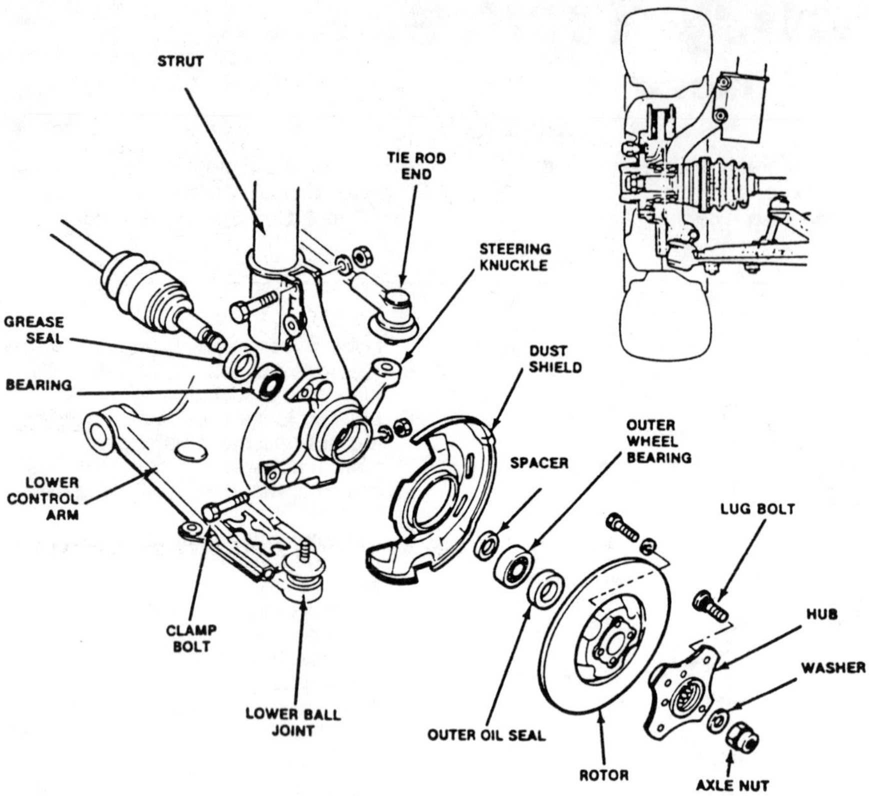

Fig. 2 Hub & steering knuckle assembly

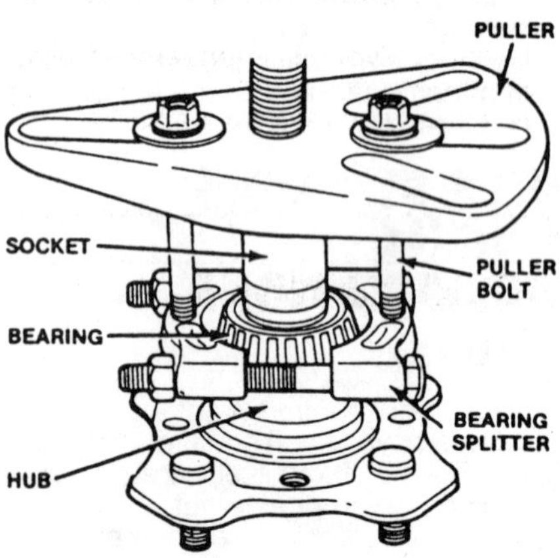

Fig. 4 Removing bearing from hub

on half shaft, tap end of half shaft with a soft faced mallet. If half shaft splines are rusted, a suitable two jawed or hub puller must be used to separate half shaft from hub. Use care when separating, as damage to grease seals may result.

10. If bearings are to be replaced, proceed as follows:
    a. Using tool No. T87C-1104-A, separate steering knuckle from hub and rotor.
    b. Remove bearing preload spacer from hub and rotor assembly and position aside.
    c. Place alignment marks on hub and rotor for use during installation, then remove rotor to hub attaching bolts and separate hub and rotor.
    d. Using a suitable puller and bearing splitter, remove bearing from hub, **Fig. 4.**
    e. After removing bearing, remove outer grease seal from hub.
    f. Using a suitable screwdriver, remove inner grease seal from steering knuckle, then remove bearing.
    g. If necessary, remove dust shield from steering knuckle.
    h. Using tool No. T87C-1175-B, install dust shield on steering knuckle, if removed.
    i. Pack bearings and hub cavity with lubricant C1AZ-1959D or equivalent. Also apply lubricant to seal lips prior to installation.
    j. Place inner bearing on steering knuckle, then install inner grease seal using a suitable seal installer.
    k. Install bearing preload spacer onto steering knuckle, then position bearing that was removed from hub onto steering knuckle.
    l. Install outer grease seal onto steering knuckle using a suitable seal installer.
    m. Position rotor to hub alignment marks, then install.
    n. Press hub and rotor assembly onto steering knuckle.
11. Reverse procedure to install hub and knuckle assembly. Tighten axle nut to

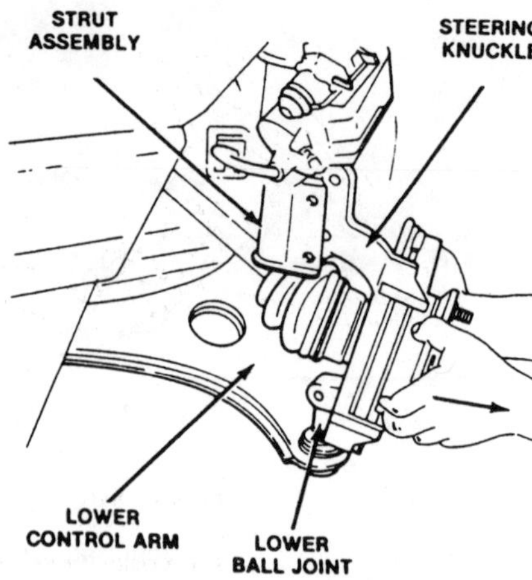

Fig. 3 Separating hub & steering knuckle assembly from half shaft

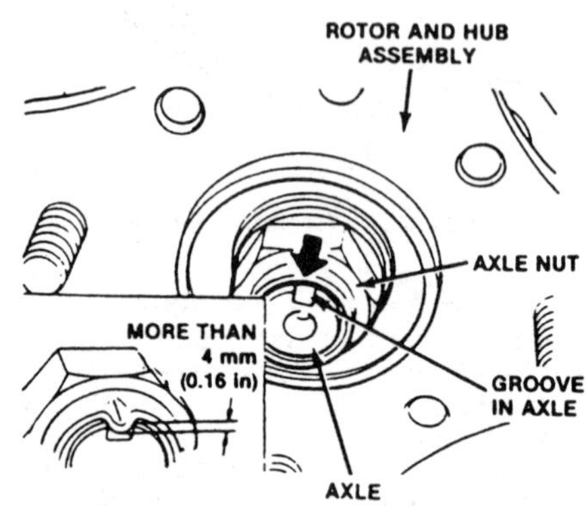

Fig. 5 Staking axle nut

torque listed at the end of this section. After tightening axle nut, stake nut in place, **Fig. 5.** Tighten stabilizer link until .43 inch of thread extends beyond nut.

STEERING GEAR REPLACE

1. Remove battery from vehicle.
2. Raise and support front of vehicle, then remove wheel and tire assemblies.
3. Separate tie rod ends from steering knuckle.
4. Remove dust shield from right hand lower inner fender.
5. Remove plastic tie securing steering column dust cover to steering gear, **Fig. 6.**
6. Place alignment marks on steering gear pinion flange and intermediate shaft lower lower joint.
7. Remove intermediate shaft lower joint bolt.
8. On models with power steering, dis-

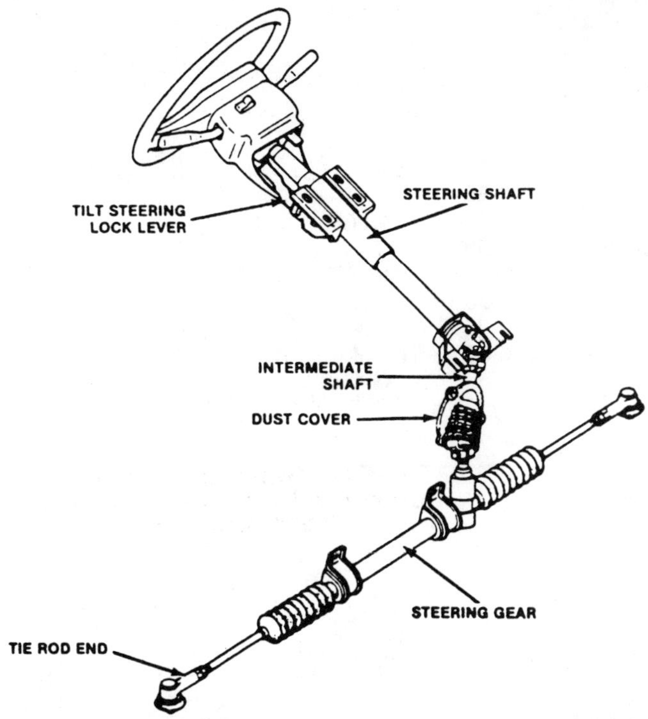

Fig. 6 Manual steering gear replacement

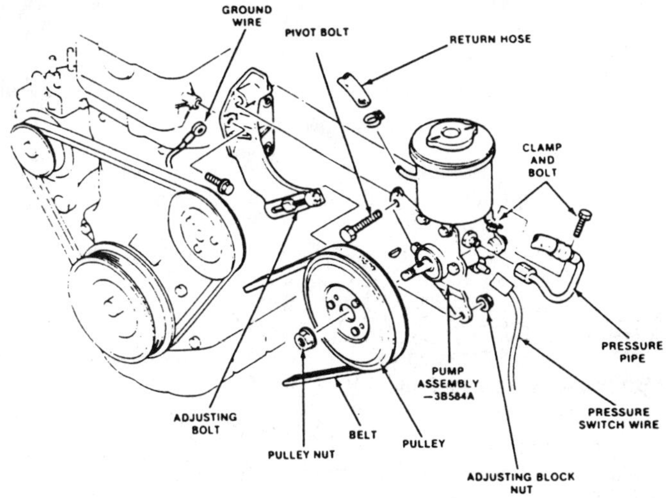

Fig. 7 Power steering pump replacement

connect pressure and return hoses from power steering gear.

9. On all models, lower steering until steering shaft clears intermediate shaft joint.
10. Carefully slide steering gear out through right hand side of fender well through tie rod opening.
11. Reverse procedure to install.

POWER STEERING PUMP REPLACE

1. Remove power steering pump drive belt, **Fig. 7.**
2. Disconnect ground wire from engine lifting eye.
3. Disconnect and plug power steering pump pressure and return hoses.
4. Disconnect electrical lead from power steering pump pressure switch.
5. Remove power steering pump from bracket.
6. Remove pivot bolt, then remove power steering pump.
7. Reverse procedure to install.

FRONT SUSPENSION & STEERING TIGHTENING SPECIFICATIONS

Year	Component	Torque/ft. lbs.
1988-89	Axle Nut	116-174①
	Ball Joint To Lower Control Arm	69-86
	Ball Joint To Steering Knuckle Clamp Bolt	32-40
	Brake Caliper Mounting Bolts	29-36
	Brake Disc To Hub	33-40
	Stabilizer Bar Bushing Bracket Attaching Bolts	44-54
	Strut Unit To Steering Knuckle.	69-72
	Tie Rod End To Steering Knuckle	22-33
	Tie Rod to Rack (Man. Steer.)	58-72
	Tie Rod to Rack (Power Steer.)	43.4-57.9

①—Stake nut using a suitable chisel. If axle nut is damaged during staking, replace nut.

Merkur

INDEX

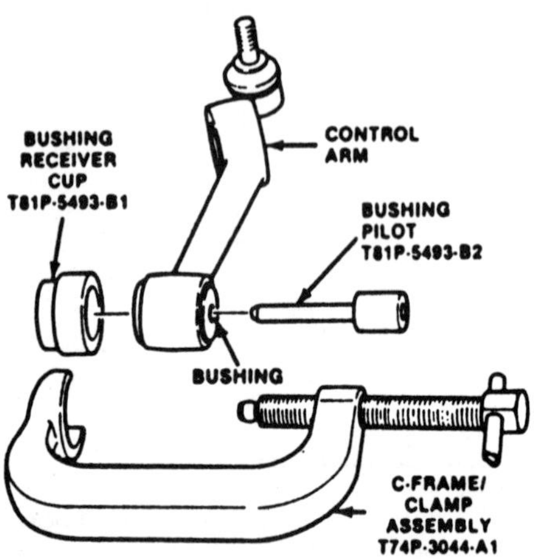

Fig. 1 Removing control arm bushings

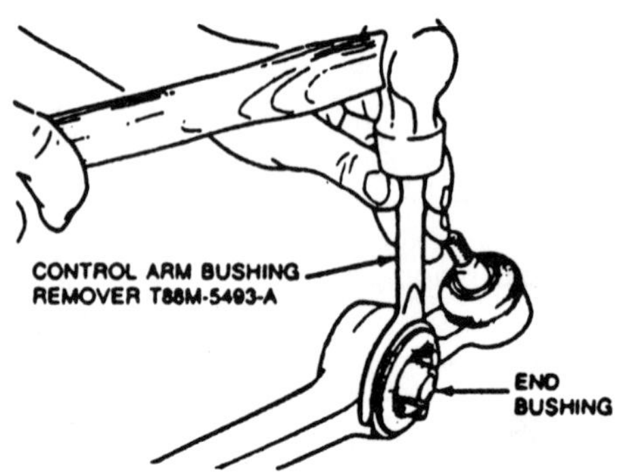

Fig. 2 Stabilizer bar bushing removal. Scorpio

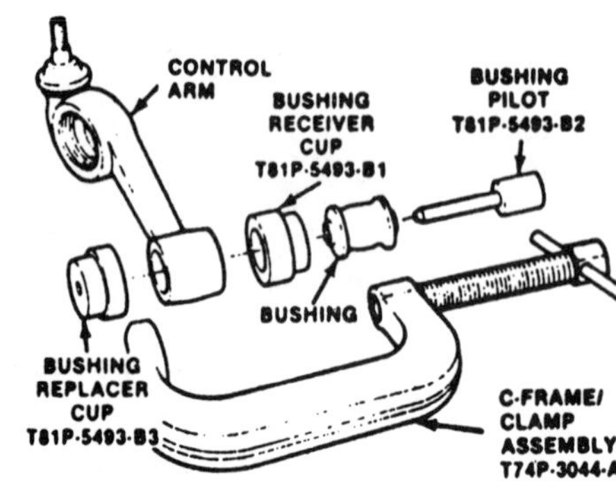

Fig. 3 Installing control arm bushings

CONTROL ARM &/OR STABILIZER BAR BUSHINGS REPLACE

SCORPIO

Removal

1. Remove pivot bolt, nut and washer securing control arm to crossmember.
2. Remove stabilizer bar-to-control arm attaching nut.
3. Remove front washer and cover from end of stabilizer bar.
4. Raise and support vehicle.
5. Remove and discard cotter pin and nut from control arm ball joint, then lower control arm from vehicle.
6. Remove control arm and stabilizer bar bushings, **Figs. 1 and 2.**

Installation

1. Install control arm bushing as follows:
   a. Apply rubber lubricant to control arm bore and outer surface of bushing.
   b. Install bushing using tools shown in **Fig. 3.** Install bushing with a quick, continuous motion to avoid damage.
2. Install stabilizer bar bushing as follows:
   a. Secure control arm in a vise.
   b. Install one bushing using tool T88M-5493-B or equivalent, then tighten screw to seat bushing.
   c. Invert control arm and install second bushing.
3. Slide control arm onto stabilizer bar, ensuring rear washer is properly positioned on stabilizer.
4. Install front washer cover and the stabilizer bar-to-control arm attaching nut. Do not tighten nut at this time.
5. Install ball joint into spindle carrier. **Torque** nut to 48-62 ft. lbs., then install new cotter pin.
6. Pull bottom of tire inward until control arm enters crossmember and aligns with bolt holes. Secure arm in this position using a drift punch.
7. Install control arm pivot bolt, washer and attaching nut. Snug nut but do not tighten at this time.
8. Lower vehicle, then **torque** control arm pivot bolt to 22 ft. lbs.
9. **Torque** stabilizer bar-to-control arm attaching nut to 52-81 ft. lbs.

XR4TI

Removal

1. Remove cotter pin and attaching nut and separate control arm from spindle carrier. **With spindle carrier and control arm disconnected, the spindle carrier can easily cause damage to control arm ball joint boot.**
2. Remove pivot bolt attaching control arm to crossmember.
3. Remove stabilizer bar to control arm attaching nut, **Fig. 4.**
4. Remove front washer/plastic cover from end of stabilizer bar.
5. Remove control arm and bushings as an assembly.
6. Remove rear washer/plastic cover from end of stabilizer bar.
7. Remove bushings from control arm as necessary, **Fig. 1.**

Installation

Stabilizer bar bushings are designed to allow control arm to move backward and forward somewhat.

1. Coat control arm bore and outer surface of bushing with non-petroleum lubricant.

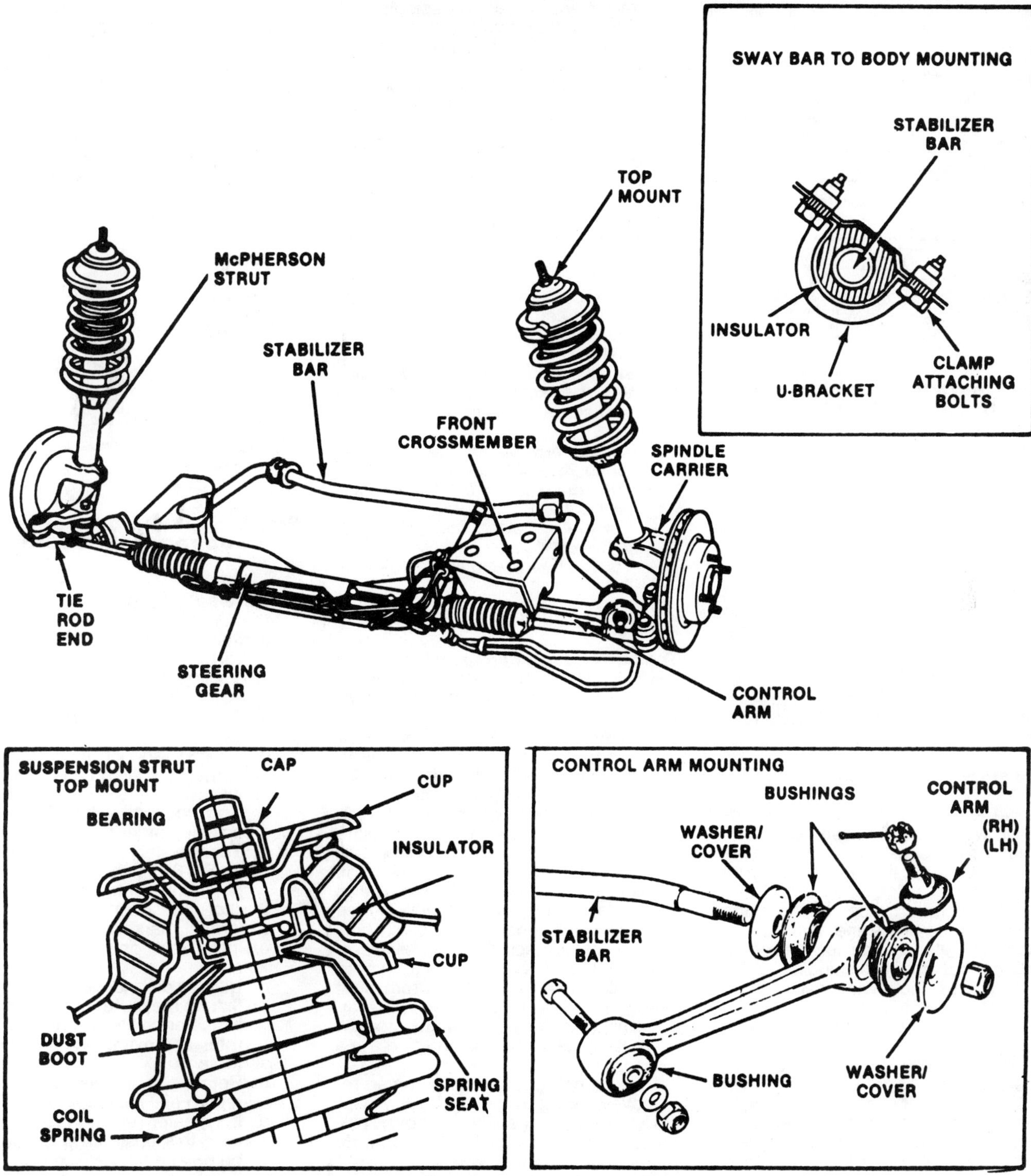

Fig. 4 Front suspension. XR4Ti

2. Install control arm bushing, **Fig. 3,** inserting bushing quickly and with a continuous motion so that bushing deforms only for a short time.
3. If removed, press stabilizer bar bushings into control arm.
4. Install rear washer/plastic cover on stabilizer bar. **Rear washer has a shallower dish than the front washer. When washer is installed, ensure that plastic cover is in place between dished steel washer and the bushing and that dished side of steel washer faces away from bushing, Fig. 5.**
5. Install control arm and bushing assembly on stabilizer bar.
6. Install front washer/plastic cover on stabilizer bar, ensuring dished side of washer faces away from bushing, **Fig. 5.**
7. Install stabilizer bar attaching nut, but do not tighten.
8. Position control arm ball joint stud in spindle carrier and install attaching nut, **torque** nut to 48-65 ft. lbs. and install cotter pin. **If slots in nut do not align with cotter pin hole, tighten nut to next slot. Never loosen nut for alignment.**
9. Grip bottom of tire and pull inward until control arm enters crossmember and aligns with bolt holes, then, using

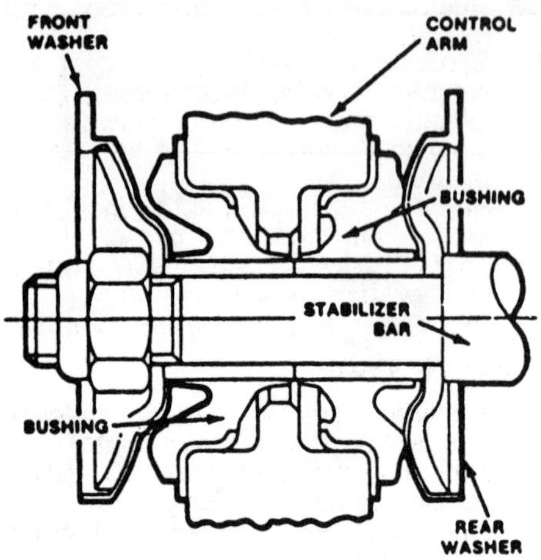

Fig. 5 Installing stabilizer bar bushings. XR4Ti

drift punch to hold arm in alignment, install attaching bolt.
10. Install control arm pivot bolt, plain washer and attaching nut, snugging but not tightening attaching nut.
11. Lower vehicle.
12. With vehicle weight on tires, **torque** control arm pivot bolt nut to 11 ft. lbs. then additional 90°, and the stabilizer bar attaching nut to 52-81 ft. lbs.

HUB, ROTOR & BEARING REPLACE

SCORPIO

1. Raise and support vehicle, then remove wheel and tire assembly.
2. Remove cotter pin and nut from tie rod end, then separate tie rod from spindle.
3. Unfasten brake caliper and position aside, leaving brake line attached.
4. Remove cotter pin and nut, then separate control arm from spindle.
5. Remove anti-lock wheel sensor from spindle carrier.
6. Remove brake rotor spring clip and the brake rotor.
7. Remove suspension strut pinch bolt, then using spindle carrier lever T85M-3206-A or equivalent, rotate 90° and remove spindle carrier.
8. Install nuts on each of the spindle wheel studs, then position spindle in a vise.
9. Using a drift punch and hammer, remove bearing plug.
10. Remove spindle bearing locknut. **The spindle bearing locknut will have right hand threads if spindle carrier is removed from left side of vehicle and if the spindle carrier is from the right side of vehicle, the locknut will have left hand threads.**
11. Lift spindle carrier and inner bearing off spindle shaft, then remove inner bearing and washer. If bearing is to be reused, tag it so it can be installed in its original position.
12. Install spindle carrier in vise, then using a screwdriver remove spindle seal and remaining bearing.
13. Remove spindle carrier from vise and position in a press with spacer.
14. Position bearing cup removal tools as shown in **Fig. 6,** then remove cup using a press.
15. Invert spindle carrier and remove opposite bearing cup.
16. Reverse procedure to install, noting the following:
    a. Install bearing cups using tool T88M-1225-C or equivalent with carrier supported by a block of wood.
    b. After installing bearing on outer side of carrier, install new grease retainer using tool No. T88M-1249-A or equivalent. **Add a small amount of grease to cavities between sealing lips of grease retainers prior to installation.**
    c. **Torque** new spindle bearing locknut to 288-331 ft. lbs.
    d. **Torque** lower control arm to spindle nut to 48-62 ft. lbs.
    e. **Torque** spindle carrier pinch bolt to 59-66 ft. lbs.
    f. **Torque** tie rod end nut to 15-23 ft. lbs.

XR4TI

The hub and rotor are a matched and balanced assembly. Before removing rotor locate the paint mark or etch mark that indicates proper hub to rotor alignment. If marks are not present, mark hub and rotor for assembly alignment.

1. Raise and support vehicle, then remove wheel and tire assembly.
2. Remove anchor plate plate attaching bolts, then lift caliper and anchor plate assembly off rotor and tie out of way.
3. Remove rotor retaining clip, then rotor.
4. Disconnect tie rod end, control arm and stabilizer bar from spindle carrier.
5. Remove suspension strut pinch bolt, then using spindle carrier lever T85M-3206-A or equivalent, spread spindle carrier to release from strut and remove spindle carrier.
6. Install nuts on each of the spindle wheel studs, then position spindle in a vise.
7. Using a drift punch and hammer, remove bearing plug.
8. Using a 41 mm socket, remove spindle bearing locknut. **The spindle bearing locknut will have right hand threads if spindle carrier is removed from left side of vehicle and if the spindle carrier is from the right side of vehicle, the locknut will have left hand threads.**
9. Lift spindle carrier and inner bearing off spindle shaft, then remove inner bearing and washer. If bearing is to be reused, tag it so it can be installed in its original position.
10. Install spindle carrier in vise, then using a screwdriver remove spindle seal.
11. Remove outer bearing from spindle carrier. If bearing is to be reused, tag it so it can be installed in its original position.

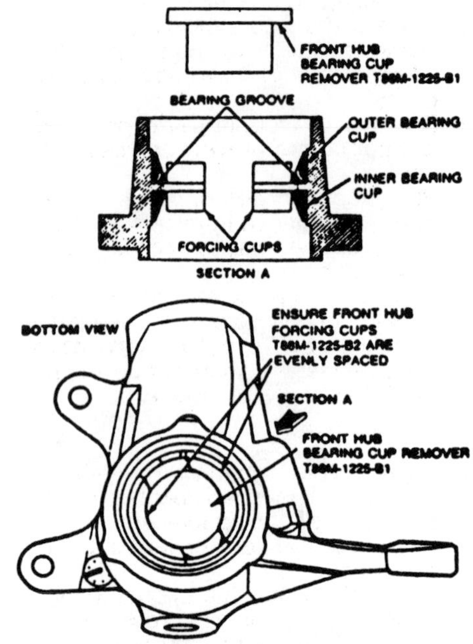

Fig. 6 Spindle carrier bearing cup removal. Scorpio

12. Reverse procedure to install, noting the following:
    a. Install spindle bearing locknut and **torque** to 229-250 ft. lbs.
    b. **Torque** tie rod end ball joints to 18-22 ft. lbs.
    c. **Torque** strut pinch bolt to 59-66 ft. lbs.
    d. **Torque** anchor plate to spindle carrier attaching bolts to 43-44 ft. lbs.
    e. **Torque** control ball joint stud in spindle carrier to 48-63 ft. lbs.
    f. **Torque** stabilizer bar attaching nut to 52-81 ft. lbs. and control arm pivot bolt nut to 11 ft. lbs.
13. If rotor has been removed without alignment marks, proceed as follows:
    a. Install rotor on hub, then wheel lug nuts.
    b. Using a dial indicator and a holding fixture, measure rotor run-out. The indicator stylus should contact the rotor approximately 7/16 inch from the end.
    c. Rotate hub and disc assembly and record dial indicator reading.
    d. If indicator reading is .003 inch or less, this positioning of rotor may be used during assembly.
    e. If indicator reading is greater than .003 inch, reposition rotor on hub in 90° increments until lowest reading is obtained.
    f. When a reading of no more than .003 inch is obtained, mark alignment of hub and rotor.

STABILIZER BAR REPLACE

SCORPIO

Removal

1. Remove four bolts securing two

U-brackets to chassis.
2. Raise and support vehicle.
3. Remove attaching nuts and front washers/covers from ends of stabilizer bar.
4. Remove one control arm pivot bolt and pull control arm out of crossmember.
5. Pull stabilizer out of lower control arms and remove from vehicle.
6. Remove rear washers/covers from stabilizer bar.
7. Remove insulators from stabilizer bar. It may be necessary to pry insulators slightly open. **Do not pry any more than necessary to avoid damaging the integral metal sleeve.**

Installation

1. Apply rubber lubricant to stabilizer bar, then install stabilizer bar insulators with split facing forward.
2. Install rear washers/covers onto stabilizer bar.
3. Position stabilizer bar through control arms, then install front washers/covers and attaching nuts. Do not tighten nuts at this time.
4. Install lower control arm to crossmember, then lower vehicle. Do not tighten pivot bolt at this time.
5. Attach stabilizer bar to chassis. Install U-brackets and **torque** attaching bolts to 52-66 ft. lbs.
6. With vehicle weight on tires, **torque** control arm pivot bolt to 22 ft. lbs. plus an additional 90°.
7. **Torque** stabilizer bar-to-control arm attaching nuts to 52-81 ft. lbs.

XR4Ti

Removal

1. Remove attaching nuts and front washers/covers from ends of stabilizer bar.
2. Remove four bolts securing two U-brackets and torque brace to body.
3. Remove one control arm pivot bolt and pull control arm out of crossmember.
4. Pull stabilizer out of lower control arms and remove from vehicle.
5. Remove rear washers/covers from stabilizer bar.
6. Remove insulators from stabilizer bar.

Installation

1. Coat inside of stabilizer bar bushings and bushing surfaces on stabilizer bar with non-petroleum lubricant, then install body insulators on stabilizer bar.
2. Install rear washers/plastic covers on stabilizer bar. **Rear washer has a shallower dish than front washer. When washer is installed, ensure plastic cover is in place between dished steel washer and the bushing and that dished side of steel washer faces away from bushing, Fig. 7.**
3. Install stabilizer bar into control arms bushings and install control arm into crossmember with pivot bolt, washer and nut, snugging but not tightening attaching nut.

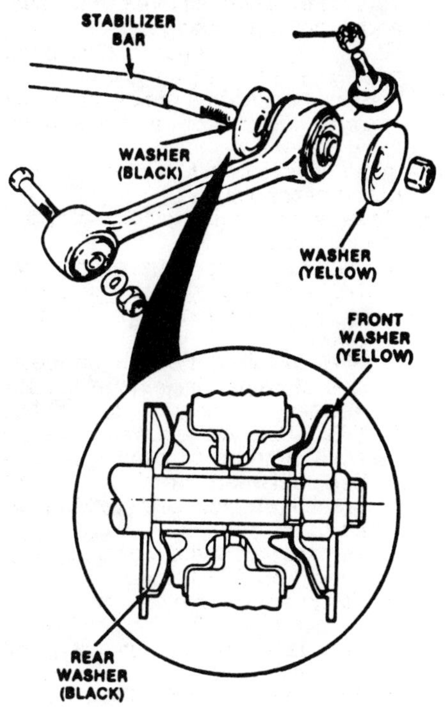

Fig. 7 Installing stabilizer bar. XR4Ti

4. Install U-brackets on insulators and install attaching bolts, **torquing** to 42-52 ft. lbs.
5. Install front washers/plastic covers on stabilizer bar. **Ensure dished side of steel washer faces away from bushing, with plastic cover in place between bushing and steel washer.**
6. Install stabilizer bar attaching nuts, snugging but not tightening.
7. Lower vehicle.
8. With vehicle weight on tires, **torque** stabilizer bar attaching nut to 52-81 ft. lbs. and the control arm pivot bolt nut to 11 ft. lbs. plus an additional 90°.

STRUT REPLACE

SCORPIO

Removal

1. Disconnect battery ground cable.
2. Raise and support vehicle.
3. Remove wheel and tire assembly, then unfasten brake caliper assembly and position aside, leaving brake line attached.
4. Remove wheel sensor from spindle carrier.
5. Remove cotter pin and nut, then separate tie rod end from spindle.
6. Remove control arm as previously described.
7. Using a pry bar, pry down on control arm and stabilizer bar and pull strut clear of control arm.
8. Remove brake pad sensor wiring from strut clip.
9. Remove dust cap from top strut mount.
10. Support strut and spindle carrier assembly, then remove three upper locknuts and washers and lower strut assembly and top mount from inner fender.
11. Remove suspension strut pinch bolt, then using spindle carrier lever T85M-3206-A or equivalent, rotate 90° and remove spindle carrier.
12. Remove strut assembly from vehicle.

Installation

1. Position spindle carrier lever T85M-3206-A or equivalent into spindle carrier slot and rotate 90°. Slide spindle carrier assembly over strut, then install new pinch bolt and **torque** to 59-66 ft. lbs.
2. Position strut assembly into inner fender, then install washers and locknuts. **Torque** locknuts to 15-17 ft. lbs.
3. Install control arm into spindle carrier, **torquing** nut to 48-62 ft. lbs.
4. Attach tie rod end to spindle carrier. **Torque** nut to 19-22 ft. lbs., then install cotter pin.
5. Secure brake pad sensor wiring into strut clip.
6. Install wheel sensor and caliper assembly.
7. Install wheel and tire assembly, then lower vehicle.
8. Install top mount dust cap, then connect battery ground cable.

XR4Ti

Removal

If twin post hoist is used to perform this procedure, front of vehicle must be supported on safety stands to allow lowering of front post. If front post is not lowered, the lower control may contact lift and prevent removal of strut.

1. Remove wheel and raise and support vehicle.
2. Remove strut pinch bolt from spindle carrier.
3. Insert spindle carrier lever T85M-3206-A or equivalent into slot in spindle carrier and rotate it through 90° to open slot.
4. Push downward on brake rotor to disengage spindle carrier from strut. **When releasing spindle carrier from strut, be careful not to damage brake hose. Place jack under control arm bushing boss to keep being pushed down too far.**
5. Remove cap from top mount attaching nut.
6. Hold piston shaft with 6 mm Allen wrench and loosen top mount attaching nut.
7. Support strut from below and remove top mount nut, retainer and strut, discarding nut.

Installation

1. Position strut through top mount insulation and install retainer and a new top mount attaching nut but do not tighten nut.
2. Position strut in spindle by pulling out-

ward on strut and lifting spindle using brake rotor. Use spindle carrier lever or equivalent to spread spindle carrier to allow strut to center correctly. When strut enters spindle, grip rotor and pivot spindle into position. **If strut is not pulled outward to match spindle angle, it will jam as the spindle is pivoted into position.**

3. Install strut pinch bolt, **torquing** to 59-66 ft. lbs.
4. Ensure top of strut is centered in suspension tower and **torque** top mount attaching nut to 29-38 ft. lbs., holding piston shaft as necessary with 6 mm Allen wrench.
5. Install cap on top mount attaching nut and install wheel.

TOP MOUNT INSULATOR REPLACE

XR4Ti

1. Remove cap from top mount attaching nut.
2. Hold piston shaft with 6 mm Allen wrench and remove and discard top mount attaching nut.
3. Remove retainer from piston shaft.
4. Using pry bar, force suspension assembly downward, being careful not to stretch brake hose.
5. Remove top mount insulator.
6. Reverse procedure to install, noting the following:
   a. Ensure top of strut is centered in suspension tower before installing retainer and attaching nut.
   b. **Torque** new top mount attaching nut to 29-38 ft. lbs., holding piston shaft with 6 mm Allen wrench as necessary.

CROSSMEMBER REPLACE

REMOVAL

Steering must be set straight ahead during following procedure to ensure that correct alignment is maintained.

1. Install engine support fixture, then raise and support vehicle.
2. Remove pivot bolts attaching control arms to crossmember.
3. Remove control arm end from crossmember by gripping bottom of tire and pulling outward.
4. Remove pinch bolt at steering column to steering gear coupling.
5. Remove steering gear to crossmember attaching bolts and washers.
6. Disengage steering gear coupling from steering column and pull steering gear forward away from crossmember, being careful not to stretch or bend power steering gear hoses or tubes.
7. Support steering gear by tying it out of way.
8. Remove engine mounts to crossmember attaching stud nuts.
9. **On Scorpio models,** remove brake line clips from crossmember and carefully slide brake lines out from locator slots.
10. **On all models,** position jack under crossmember and secure with safety chain.
11. Remove crossmember to side rail attaching bolts and washers.
12. Lower crossmember and remove from under vehicle.

INSTALLATION

1. Raise crossmember into position and install crossmember to side rail attaching bolts and washers.
2. **Torque** attaching bolts to 51-66 ft. lbs. and remove jack.
3. Position steering gear on crossmember and connect coupling to steering column, ensuring that block splines are properly mated.
4. Install steering gear attaching bolts and washers, snugging bolt to 11 ft. lbs. plus additional 90°.
5. Install steering coupling pinch bolt, **torquing** to 15-22 ft. lbs. on Scorpio models, or 18-22 ft. lbs. on XR4Ti models.
6. Install engine mount attaching bolts, **torquing** to 31-42 ft. lbs. on Scorpio models, or 38-47 ft. lbs. on XR4Ti models.
7. Grip bottom of tire and pull inward until control arm enters crossmember, aligning control arm with bolt holes and install pivot bolt. Use drift punch to hold control arm in alignment while pivot bolt is installed.
8. Install pivot bolt and washer, snugging but not tightening bolt.
9. Lower vehicle.
10. With weight of vehicle on front tires, **torque** control arm pivot bolt to 22 ft. lbs. plus an additional 90° on Scorpio models, or 11 ft. lbs. plus additional 90° on XR4Ti models.
11. Remove engine support fixture.

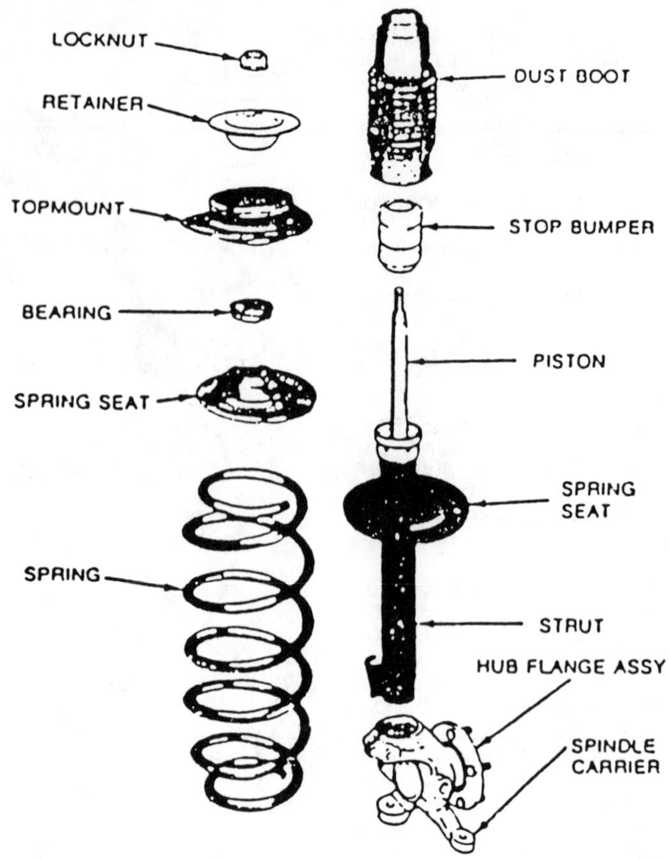

Fig. 8 Strut assembly exploded view. Scorpio

STRUT SERVICE

SCORPIO

Disassembly

1. Clamp strut spring compressor 086-00029 or equivalent in a vise and adjust tool to START position following manufacturer's instructions.
2. Position strut assembly in tool and carefully compress spring.
3. Hold piston rod with a 6 mm Allen wrench and remove coil spring locknut and retainer, **Fig. 8.**
4. Remove top mount, bearing and spring seat, then the spring and spring compressor.
5. Remove dust boot and stop bumper from piston.

Assembly

1. Position stop bumper and dust boot over piston.
2. Ensure both springs are free of dirt and grease, then position compressed spring onto strut lower spring seat.
3. Install upper spring seat, bearing, top

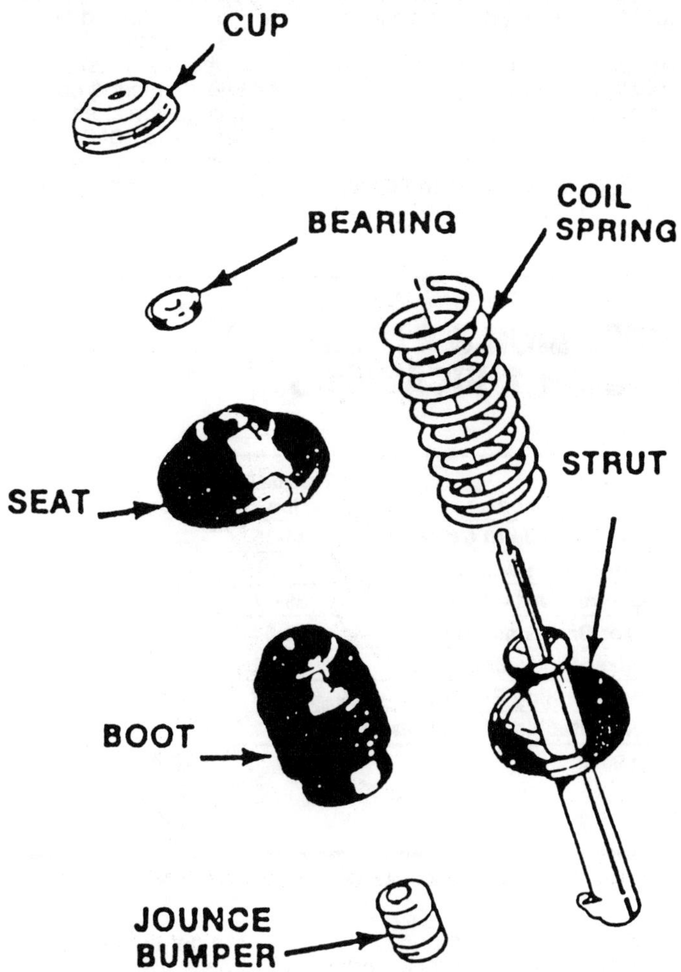

Fig. 9 Strut assembly exploded view. XR4Ti

mount and retainer. Ensure bearing is installed with small hole facing up.
4. Hold piston rod with a 6 mm Allen wrench, then install locknut and **torque** to 29-38 ft. lbs.
5. Carefully release strut spring compressor.
6. Ensure bearing is properly positioned in upper spring seat and spring is properly positioned in both seats, then remove assembly from spring compressor.

XR4TI

Disassembly

1. Clamp strut spring compressor 086-00016 or equivalent in a vise and adjust tool to START position following tool manufacturer's instructions.
2. Using tool, carefully compress coil spring.
3. Hold piston shaft with a 6 mm Allen wrench and remove coil spring retaining nut.
4. Release coil spring tension.
5. Remove top mount cup, bearing, upper spring seat and dust boot, **Fig. 9.**
6. Slide jounce bumper off piston shaft and remove coil spring.

Assembly

1. Install coil spring, then jounce bumper, dust boot, upper spring seat, bearing and top mount cup.
2. Compress spring using spring compressor.
3. Install coil spring retaining nut, **torquing** to 38-48 ft. lbs., holding piston shaft with 6 mm Allen wrench as necessary.
4. Release coil spring tension and remove strut from spring compressor.

POWER STEERING GEAR REPLACE

Scorpio models are equipped with a ZF power steering gear. XR4Ti models may be equipped with either a TRW or ZF steering gear.

TRW GEAR

Removal

1. Disconnect battery ground cable and turn ignition key to ON position.
2. Remove cotter pins and nuts attaching tie rod ends to spindle carriers.
3. Separate tie rod ends from spindle carriers.
4. Remove pinch bolt from flexible coupling, rotating coupling as necessary to gain access to pinch bolt.
5. Position drain pan and disconnect power steering hoses from steering gear by removing routing clamp and the washer-head screw securing pump line plate assembly to gear housing and pulling assembly free. Plug housing port to prevent entry of foreign matter.
6. Remove steering gear mounting bolts and the gear.

Installation

1. Center steering wheel and turn ignition key to OFF position.
2. Center steering gear as follows:
   a. Turn steering gear input shaft clockwise to full left turn stop.
   b. Turn input shaft counterclockwise and count number of turns required to move rack from full left turn stop to full right turn stop.
   c. From right turn stop, turn input shaft clockwise exactly 1/2 the turns previously counted.
3. Position steering gear on crossmember, engaging flexible coupling with input shaft, rocking coupling as necessary to align blind splines on shaft and coupling.
4. Turn ignition key to ON position, then turn steering column as necessary to align blind splines on coupling and the steering shaft.
5. Connect steering gear to steering column and position on crossmember.
6. Install steering coupling pin bolt, **torquing** to 18-22 ft. lbs.
7. Plug pump line assembly into steering gear supply and return ports and secure line plate with washer-head screw and routing clamp.
8. Install new steering gear mounting bolts and **torque** to approximately 11 ft. lbs. plus additional 90°.
9. Connect tie rods to spindle carriers and install castle nuts, **torque** nuts to 15-23 ft. lbs. and install new cotter pins. **If slots in nut do not align with cotter pin holes, tighten nut to next slot. Never loosen nut for alignment.**
10. Turn ignition key to OFF position.
11. Connect battery ground cable.
12. Fill power steering pump to correct level.

ZF GEAR

Removal

1. Turn steering wheel to straight-ahead position.
2. Raise and support vehicle.
3. Remove lower pinch bolt securing flex coupling to steering gear pinion shaft.
4. Remove nuts and cotter pins securing tie rod ends to spindle carriers.
5. Separate tie rod ends from spindle carriers.
6. Position drain pan and disconnect power steering hoses from steering gear by removing routing clamp and the bolt securing pump line plate as-

sembly to gear housing and pulling assembly free. Plug housing port to prevent contamination.

7. Remove steering gear attaching bolts and the steering gear.

Installation

1. Position steering gear on engine front crossmember. **Torque** two attaching bolts to 11 ft. lbs., then an additional 90°.
2. Remove plugs from fluid lines and gear housing ports.
3. Install new O-rings on pressure and return lines, then connect lines to gear.
4. Install routing clamp screw and clamp plate bolt. **Torque** bolt to .6-.9 ft. lbs.
5. Connect tie rod ends to spindle carriers. **Torque** attaching nuts to 19-22 ft. lbs., then install cotter pins.
6. Install steering column coupling shaft onto pinion shaft. Install pinch bolt and **torque** to 19-22 ft. lbs. **Ensure block splines on coupling and pinion shaft are properly aligned.**

FRONT SUSPENSION & STEERING TIGHTENING SPECIFICATIONS

Year	Component	Torque/ft. lbs.
1985-90	Control Arm Ball Joint Stud Nut	48-65
	Control Arm Pivot Bolt	11①
	Engine Mount Attaching Nut	38-47
	Stabilizer Bar Attaching Nut	52-81
	Stabilizer Bar U-Brackets	42-52
	Steering Coupling Pinch Bolt	18-22
	Strut Mount Attaching Nut (Upper)	29-38

①—Then an additional 90°.

GENERAL MOTORS CORP.

TABLE OF CONTENTS

Buick, Cadillac, Chevrolet, Oldsmobile & Pontiac Rear Wheel Drive Models Except Chevette, 1000 & Fiero

INDEX

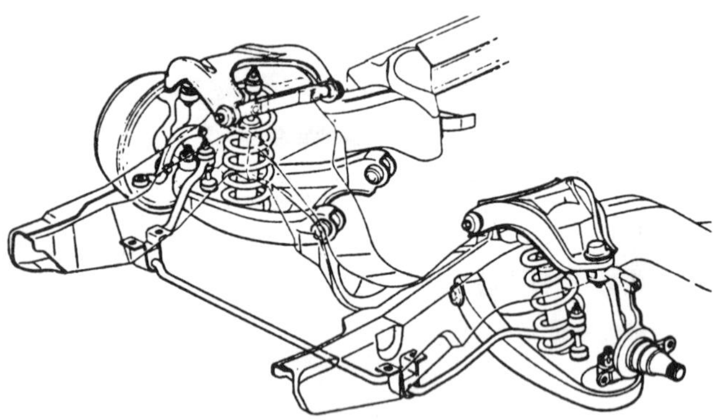

Fig. 1 Front suspension. Except Camaro, Corvette & Firebird

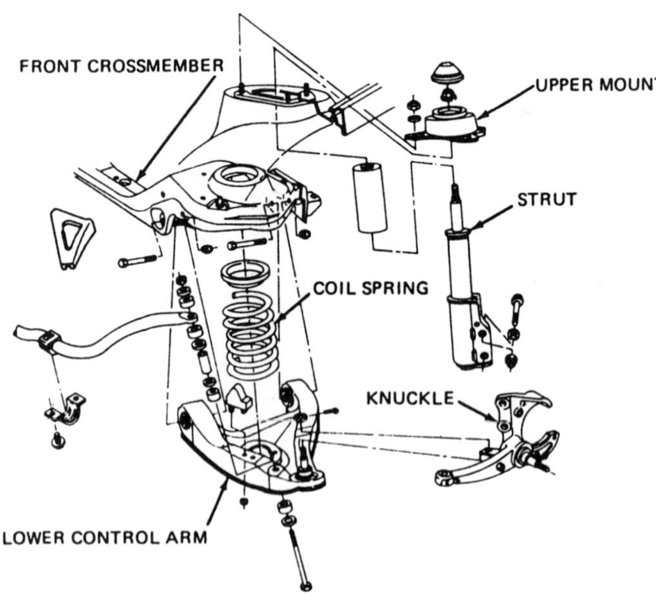

Fig. 2 Front suspension exploded view. Camaro & Firebird

DESCRIPTION

All models except Camaro, Corvette and Firebird use a Short-Long Arm (SLA) type front suspension with independent coil springs riding on lower control arms, **Fig. 1.** Ball joints link upper and lower control arms to a spindle assembly, and tubular shock absorbers are used to dampen spring action. On some models, a spring steel stabilizer shaft is connected between the chassis and lower control arms to control side roll.

A modified strut type suspension is used on Camaro and Firebird, **Fig. 2.** Each wheel is independently connected to the chassis by a lower control arm, spindle and a strut assembly which locates the spindle and controls ride by dampening spring action. Coil springs are mounted between the lower control arm and crossmember, and a stabilizer shaft is connected between the chassis and control arms to control side roll.

The front suspension used on Corvette consists of forged aluminum upper and lower control arms and steering knuckle, a fiberglass mono-leaf spring, shock absorbers and stabilizer bar, **Fig. 3.** The front spring is transverse mounted on the crossmember and bears against lower control arms. The stabilizer bar and shock absorbers are connected between the chassis and lower control arms, and control side roll and dampen spring action respectively. Upper and lower control arms are connected through the knuckle, which is specially designed to move the wheel center line rearward of the conventional ball joint center line.

WHEEL BEARINGS ADJUST

EXCEPT CORVETTE

1. While rotating wheel forward, torque spindle nut to specifications.
2. Back off nut until just loose then hand tighten nut and back it off again until either hole in spindle lines up with hole in nut. **Do not back off nut more than 1/2 flat.**
3. Install new cotter pin. With wheel bearing properly adjusted, there will be .001-.005 inch end play.

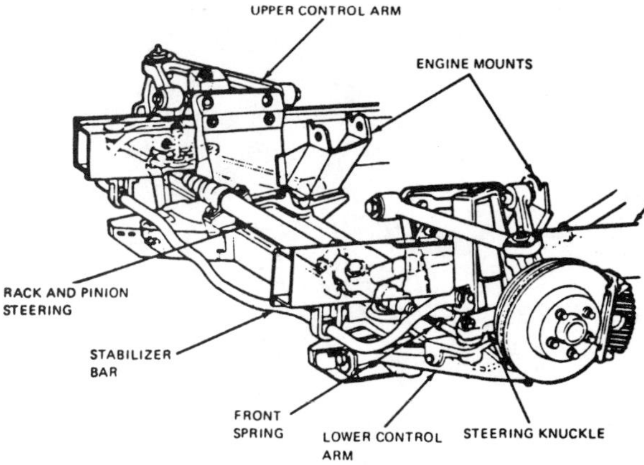

Fig. 3 Front suspension. Corvette

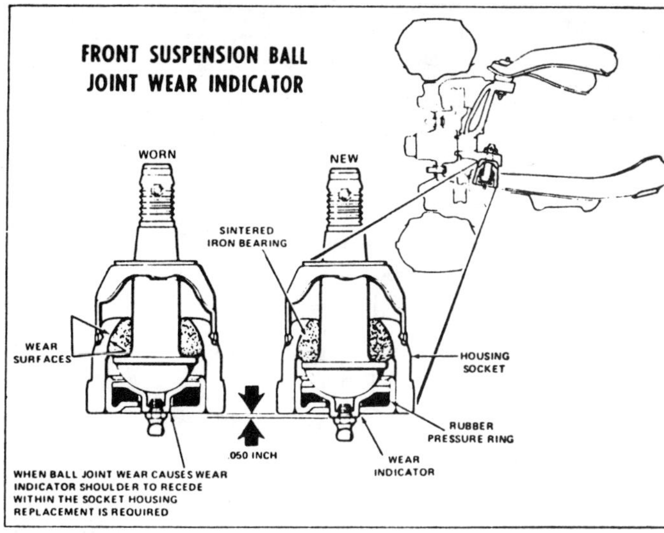

Fig. 4 Lower ball joint wear indicator

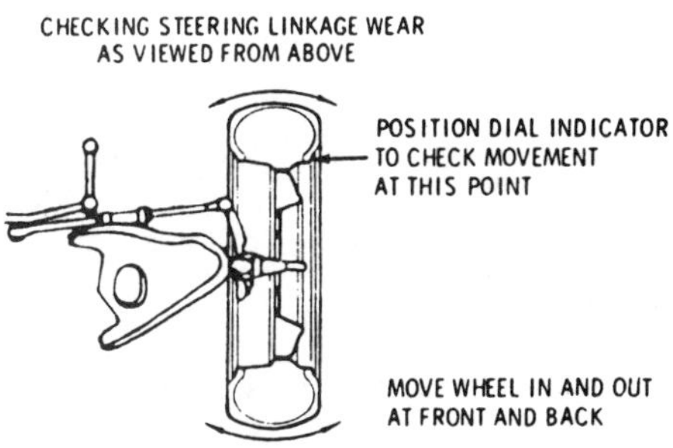

Fig. 5 Suspension & steering linkage check

WHEEL BEARINGS REPLACE

EXCEPT CORVETTE

1. Raise car and remove front wheels.
2. Remove bolts holding brake caliper to its mounting and insert a fabricated block ($1^1/_{16}$ x $1^1/_{16}$ x 2 inches in length) between the brake pads as the caliper is being removed. Once removed, the caliper can be wired or secured in some manner away from the disc.
3. Remove spindle nut and hub and disc assembly. Grease retainer and inner wheel bearing can now be removed.
4. Reverse procedure to install.

CORVETTE

The wheel bearing and hub assembly is a sealed unit. If end play exceeds .005 inch, or if noise or roughness is detected, unit must be replaced as an assembly.

1. Raise and support vehicle, and remove wheel and tire.
2. Remove 2 bolts securing brake caliper bracket to steering knuckle, and secure caliper assembly aside.
3. Remove brake rotor, bolts securing hub to knuckle and hub assembly.
4. Reverse procedure to install.

CHECKING BALL JOINTS FOR WEAR

UPPER BALL JOINT

1. Raise front of vehicle with jacks placed between the coil spring pocket and ball joint of lower control arm.
2. Shake top of wheel in and out. Observe steering knuckle for any movement relative to the control arm.
3. Replace upper ball joint if looseness is indicated.

LOWER BALL JOINT

Raise vehicle and support lower control arm so spring is compressed in the same manner as if the wheels were on the ground and check axial (up and down) play at ball joint. If play exceeds $^1/_8$ inch, replace the joint.

Another indication of lower ball joint excessive wear is when difficulty is experienced when lubricating the joint. If the liner has worn to the point where the lubrication grooves in the liner have been worn away, then abnormal pressure is required to force lubricant through the joint. Should this condition be evident, replace both lower ball joints.

All models except Camaro and Firebird have a wear indicator built into the lower ball joint, **Fig. 4.** When inspecting wear indicator, vehicle must be supported normally on wheels to properly load ball joint.

SUSPENSION & STEERING LINKAGE CHECK

1. Raise vehicle with jack placed under frame torque box behind front wheel.
2. Lock steering wheel with wheels in straight ahead position, then mount dial indicator on a suitable stand with pointer bearing against outer rim of

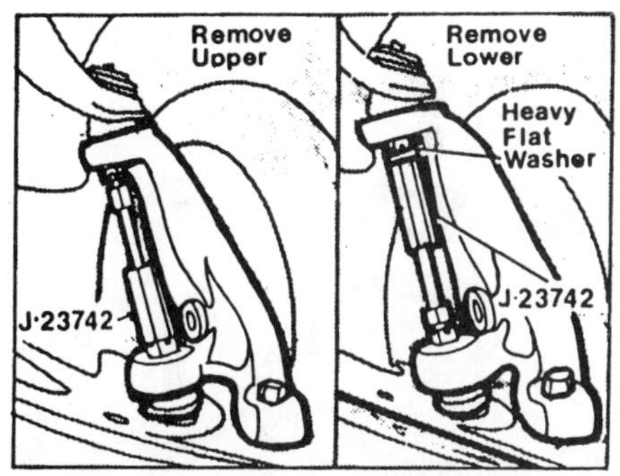

Fig. 6 Disconnecting ball joints from steering knuckle. Except Camaro & Firebird

wheel, **Fig. 5.**

3. Move wheel in and out at front and rear, without moving steering wheel, while observing gauge.
4. If gauge reading exceeds .108 inches, check steering linkage and suspension for excessive wear or damage.

UPPER BALL JOINT REPLACE

1. Raise vehicle and support with stands at outer ends of lower control arms.
2. Remove wheel and tire.
3. Remove cotter pin and retaining nut, then separate ball joint stud from knuckle using a suitable tool, **Fig. 6.**
4. Support upper control arm in a raised position.
5. Remove heads of rivets securing joint to arm, then drive out rivets to remove joint.
6. Position replacement joint on top of control arm, insert retaining bolts supplied with joint from under arm, install nuts and torque to specifications.
7. Remove upper control arm support, assemble ball joint to steering knuckle, install washer, if equipped, and retaining nut.
8. Torque retaining nut to specifications.

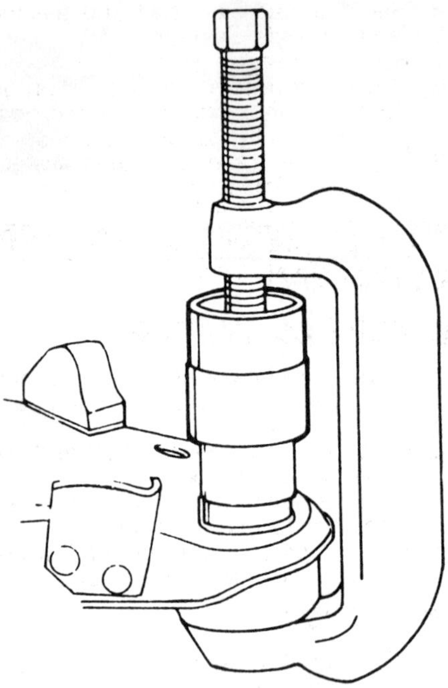

Fig. 7 Pressing ball joint from lower control arm

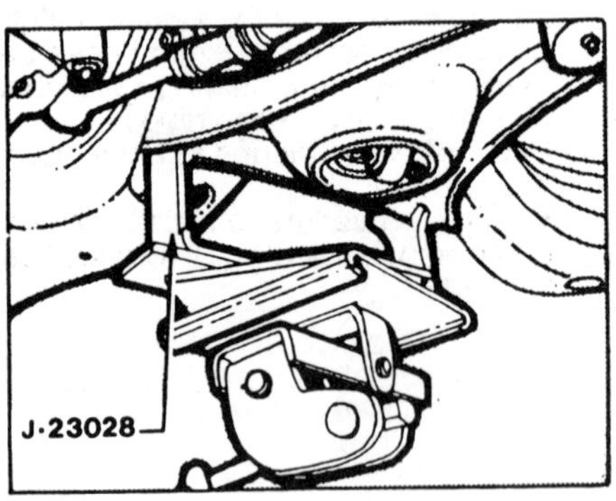

Fig. 8 Coil spring removal

9. Tighten retaining nut up to an additional 1/16 turn, if necessary, to align hole in ball stud with nut, then install cotter pin.

LOWER BALL JOINT
REPLACE

1. Raise vehicle and support at frame, and remove wheel and tire.
2. Position a suitable jack under lower control arm spring seat, and raise jack to compress coil spring. **Jack must remain in place during ball joint replacement to hold spring and lower control arm in position.**
3. Remove cotter pin and nut securing ball joint stud to steering knuckle, then disconnect joint from knuckle using a suitable tool, **Fig. 6.**
4. Lift knuckle assembly from ball stud, guiding control arm out of splash shield, then support knuckle aside to allow clearance for joint removal.
5. Remove grease fitting, then press ball joint assembly out of lower control arm using a suitable tool, **Fig. 7.**
6. Press replacement joint into arm by reversing removal tools, fit spindle over ball stud, install washer, if equipped, and retaining nut.
7. Torque retaining nut to specifications.
8. Tighten nut up to an additional 1/16 turn, if necessary, to align hole in ball stud with nut, then install cotter pin.

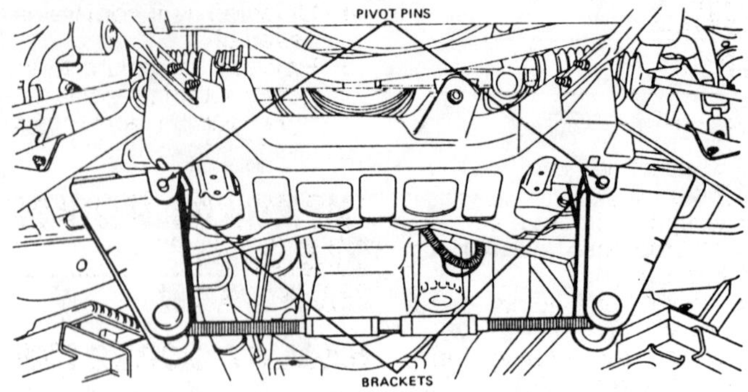

NOTE: PIVOT PINS ARE REMOVED SO THAT THE BRACKET MAY BE PLACED OVER THE TOP OF THE SPRING.

Fig. 9 Leaf spring removal. Corvette

SHOCK ABSORBER
REPLACE

On Camaro and Firebird, refer to "Strut, Replace" procedure.

1. Raise and support vehicle as needed, and remove wheel and tire.
2. Hold shock absorber shaft with a suitable wrench and remove upper retaining nut, washer and bushing.
3. Remove lower retaining bolts and shock absorber. On Corvette, remove shock mounting bracket, if necessary, to provide clearance for shock absorber removal.
4. Reverse procedure to install. Torque upper retaining nut and lower mounting bolts to specifications.

STRUT
REPLACE
CAMARO & FIREBIRD

1. Raise and support vehicle.
2. Remove wheel and support lower control arm with a suitable jack.
3. Remove brake hose bracket and two strut to knuckle bolts, **Fig. 2.**
4. Remove upper mounting assembly cover.
5. Remove nut from upper end of strut, then the strut and shield.
6. Reverse procedure to install. Tighten strut to knuckle bolts to 195 ft. lbs., and strut to upper mount nut to 50 ft. lbs.

COIL SPRING
REPLACE
EXCEPT CAMARO & FIREBIRD

1. Disconnect shock absorber from lower mounting, push shock absorber through hole in lower control arm and compress into spring.
2. Support vehicle by frame so control arms hang free.
3. Install a safety chain through spring and lower control arm.
4. Install tool J-23028 onto a suitable jack and position jack so control arm is supported by bushings seated in grooves of tool, **Fig. 8.**
5. Remove stabilizer to lower control arm attachment.
6. Raise jack to relieve tension on control arm bolts and remove bolts.
7. Lower jack until tension is removed from spring, remove chain and spring from vehicle.
8. Reverse procedure to install, installing front pivot bolt first. **To ensure adequate suspension clearance, install front pivot bolt from front, with nut toward rear of vehicle. Rear bolt can be installed from either direction.**
9. Torque pivot bolts to specifications.

CAMARO & FIREBIRD

1. Raise and support vehicle and remove wheel.
2. Remove stabilizer link and bushings from lower control arm.
3. Remove pivot bolt nuts, leaving the bolts installed.
4. Install tool J-23028 onto a suitable jack and position so that tool supports bushings, **Fig. 8.**
5. Raise jack to relieve tension on pivot bolts and remove bolts.
6. Carefully lower jack until tension is removed from spring and remove spring from vehicle.
7. Reverse procedure to install. Torque pivot bolt nuts and stabilizer link nut to specifications. **The spring force under compression is very great. Exercise every safety precaution when performing this operation to see that individuals and materials subject to damage are removed from the path of the spring when the control arm is being lowered. Also, the compressed spring should be relaxed immediately after lowering the control arm to reduce the time of exposure to the great compressive force.**

LEAF SPRING
REPLACE
CORVETTE

1. Raise and support vehicle, and remove front wheels.
2. Remove both spring protectors.
3. Install spring compressor J-33432 or equivalent, **Fig. 9.**
4. Disconnect lower ball joints from steering knuckles.
5. Compress spring by rotating turnbuckle on spring compressor.
6. Remove bolts securing shock brackets to lower control arms and spring mounting bolts.
7. Release tension on spring compressor and remove compressor.
8. Remove spring.
9. Reverse procedure to install. Torque spring mounting bolts to specifications, with vehicle on ground.

MANUAL STEERING GEAR
REPLACE

On models where shield is installed, remove shield from coupling.

1. Remove nuts, washers and bolts at steering coupling.
2. Remove pitman arm nut and washer from sector shaft and mark relation of arm position to shaft.
3. Use a suitable puller to remove pitman arm.
4. Unfasten gear from frame and remove assembly.
5. Reverse procedure to install. Torque gear mounting bolts, shaft coupling nuts, pitman arm retaining nut to specifications.

INTEGRAL POWER STEERING
REPLACE
EXCEPT CORVETTE

To remove gear assembly, disconnect pressure and return hoses from gear housing and cap both hoses and steering gear outlets to prevent foreign material from entering system, then follow procedure as outlined under "Manual Steering Gear, Replace".

RACK & PINION STEERING GEAR
REPLACE
CORVETTE

1. Raise and support vehicle, and remove left wheel and tire.
2. Disconnect hoses at steering gear, and plug lines and open ports.
3. Disconnect outer tie rods from steering knuckles.
4. Remove upper and lower mounting bolts on right side and single mounting bolt on left side.
5. Disconnect intermediate shaft universal joint from steering gear.
6. Remove stabilizer shaft and electric fan to provide clearance.
7. Remove steering gear.
8. Reverse procedure to install. Torque bolts securing right bracket and gear to crossmember to specifications.
9. Top off fluid reservoir, bleed system and check for leaks.

POWER STEERING PUMP
REPLACE
EXCEPT CORVETTE

1. Disconnect hoses at power steering pump, then plug pump parts and hoses.
2. Loosen pump adjusting bolt and remove pump drive belt.
3. Remove pump to support bracket attaching bolts and the pump.
4. Reverse procedure to install. Torque attaching bolts to specifications.

CORVETTE

1. Rotate belt tensioner counterclockwise, and remove serpentine drive belt.
2. Remove AIR pump pulley.
3. Remove bolts securing power steering reservoir bracket and bolts securing reservoir brace to intake manifold.
4. Disconnect power steering hoses between pump and steering gear, then plug hoses and open fittings.
5. Remove pump mounting bolts, pump and reservoir.
6. Reverse procedure to install, torquing pump and bracket bolts to specifications.

FRONT SUSPENSION & STEERING TIGHTENING SPECIFICATIONS

Year	Component	Torque/ft. lbs.
EXCEPT CAMARO, CORVETTE & FIREBIRD		
1985-90	Lower Ball Joint Stud	83
	Lower Control Arm	92
	Manual Steering Gear Mounting Bolts	70
	Pitman Arm Nut	185
	Power Steering Pump	35
	Shaft Coupling Nut	30
	Shock Absorber Lower Bolt	20
	Shock Absorber Upper Nut	8
	Stabilizer Link	13
	Tie Rod Stud	35
	Upper Ball Joint Stud	60
	Upper Control Arm Bolts	72
	Upper Control Arm Pivot Shaft	85
	Wheel Hub Spindle Nut	12
CAMARO & FIREBIRD		
1985-90	Ball Joint Stud	83
	Control Arm Pivot Bolt	66
	Crossmember Bolt	118
	Manual Steering Gear Mounting Bolts	70
	Pitman Arm Nut	185
	Power Steering Pump	35
	Shaft Coupling Nut	30
	Stabilizer Link	16
	Strut Bottom Lower Nut	203
	Strut Bottom Upper Nut	46
	Strut Upper Mounting Bolt	18
	Wheel Hub Spindle Nut	12
CORVETTE		
1985-90	Leaf Spring Mounting Bolt	46
	Lower Ball Stud To Knuckle	50
	Lower Control Arm Shaft To Crossmember	82
	Power Steering Pump Bolts	18
	Rack & Pinion Right Bracket	18
	Shock Absorber To Lower Control Arm	19
	Shock Absorber To Shock Towers	19
	Stabilizer Link To Lower Control Arm	35
	Steering Gear To Crossmember	25
	Tie Rod Stud To Knuckle	32
	Upper Ball Stud To Control Arm	19
	Upper Ball Stud To Knuckle	33
	Upper Control Arm Shaft To Crossmember	37

Buick Century & Skylark (1985), Chevrolet Celebrity & Citation, Oldsmobile Cutlass Ciera & Cutlass Cruiser & Pontiac 6000

INDEX

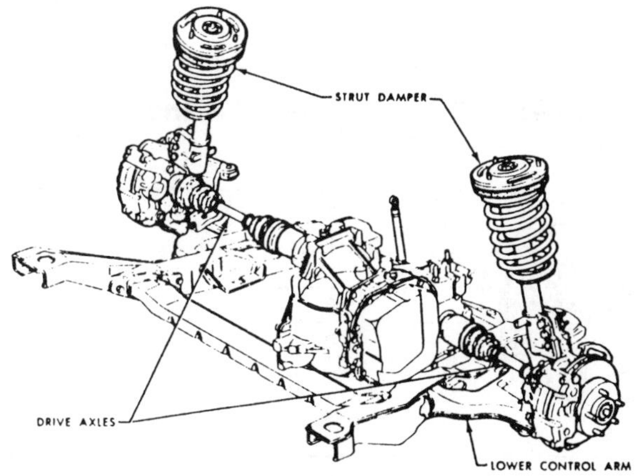

Fig. 1 Front suspension

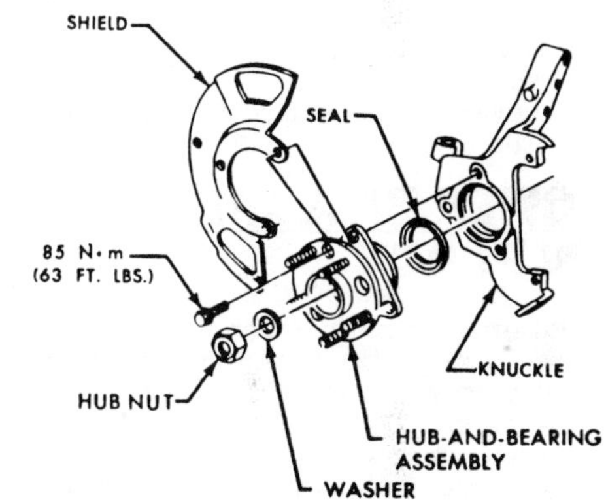

Fig. 2 Front wheel bearing assembly & seal

DESCRIPTION

The front suspension, **Fig. 1,** on these vehicles is a MacPherson strut design. The lower control arms pivot from the engine cradle. This engine cradle has isolation mounts to the body and conventional rubber bushings are used for the lower control arm pivots. The upper end of the strut is isolated by a rubber mount incorporating a bearing for wheel turning. The lower end of the steering knuckle pivots on a ball stud which is retained in the lower control arm with rivets and is clamped to the steering knuckle. Sealed wheel bearings are used and are bolted to the steering knuckle.

WHEEL BEARING REPLACE

REMOVAL

1. Loosen hub nut with vehicle on ground.
2. Raise and support vehicle, then remove front wheel.
3. Install drive axle seal protectors. **These protectors should be used any time service is performed on or near drive axle. Failure to use them may result in interior joint or seal damage.**
4. Remove and **discard** hub nut and washer.
5. Remove brake caliper from support and suspend caliper from frame with a length of wire. **Do not suspend caliper by the brake hose.** Remove rotor.
6. Remove three hub and bearing attaching bolts. If the old bearing is being reinstalled, mark attaching bolts and holes for reinstallation, **Fig. 2.**
7. Using tool J-28733, or equivalent, remove bearing, **Fig. 3. If excessive corrosion is present, ensure that the bearing is loose in the knuckle before using puller tool.**
8. If installing new bearing, replace steering knuckle seal.

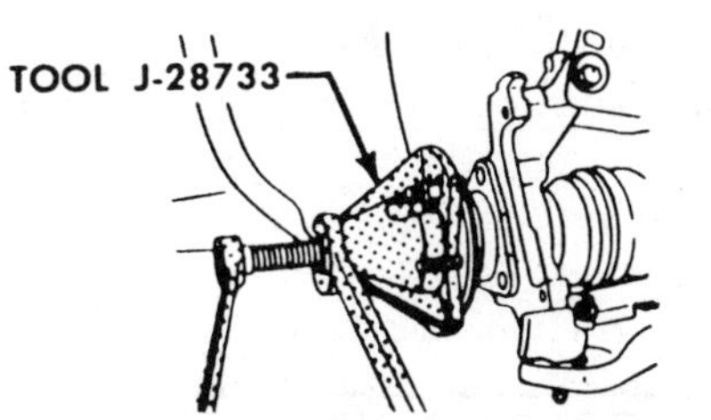

Fig. 3 Removing front wheel bearing & hub assembly

INSTALLATION

1. Clean and inspect bearing mating surfaces and steering knuckle bore for dirt, nicks and burrs.
2. If installing steering knuckle seal, use

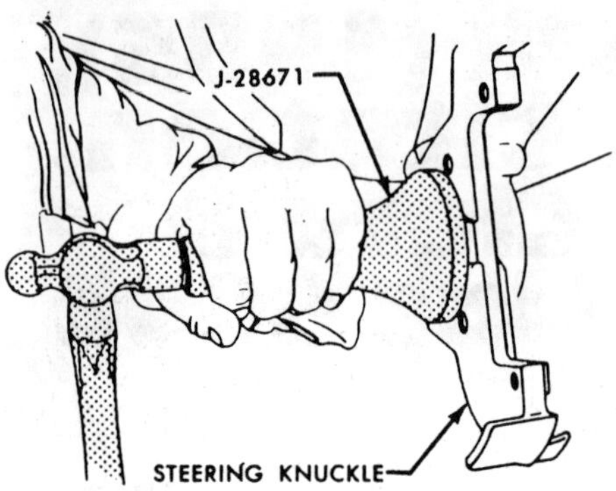

Fig. 4 Installing front wheel bearing seal

tool J-28671 or equivalent to install seal, **Fig. 4.** Apply grease to seal and knuckle bore.

3. Push bearing onto axle shaft.
4. Tighten hub nut to fully seat bearing.
5. Install bearing attaching bolts and **torque** to 63 ft. lbs., **Fig. 2.** If equipped with heavy duty brake option, **torque** attaching bolts to 70 ft. lbs.
6. Install rotor, caliper assembly and wheel.
7. Lower vehicle and **torque** hub nut to 185 ft. lbs.

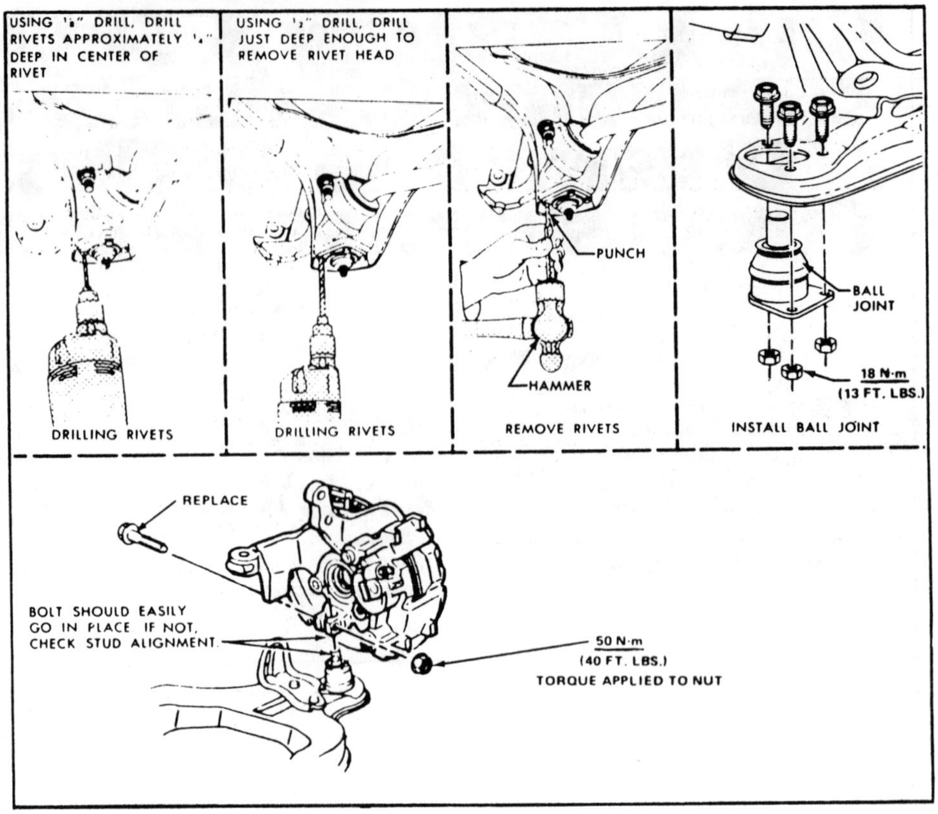

Fig. 5 Lower ball joint removal & installation

LOWER BALL JOINT

REPLACE

When servicing lower ball joint, be sure to align broached notch in ball joint stud with clamp bolt hole during reassembly. This prevents interference between stud and clamp bolt. To ensure proper bolt torque, replace clamp bolt and nut each time ball joint stud is separated from steering knuckle.

For removal and installation procedures, refer to **Fig. 5.**

LOWER CONTROL ARM & BUSHINGS

REPLACE

For removal and installation procedures, refer to **Fig. 6. Torque** ball joint clamping bolt to 40 ft. lbs. and control arm attaching bolts to 66 ft.lbs.

STEERING KNUCKLE

REPLACE

For removal and installation procedures, refer to **Fig. 7.**

STABILIZER BAR & BUSHINGS

REPLACE

For removal and installation procedures, refer to **Fig. 8.**

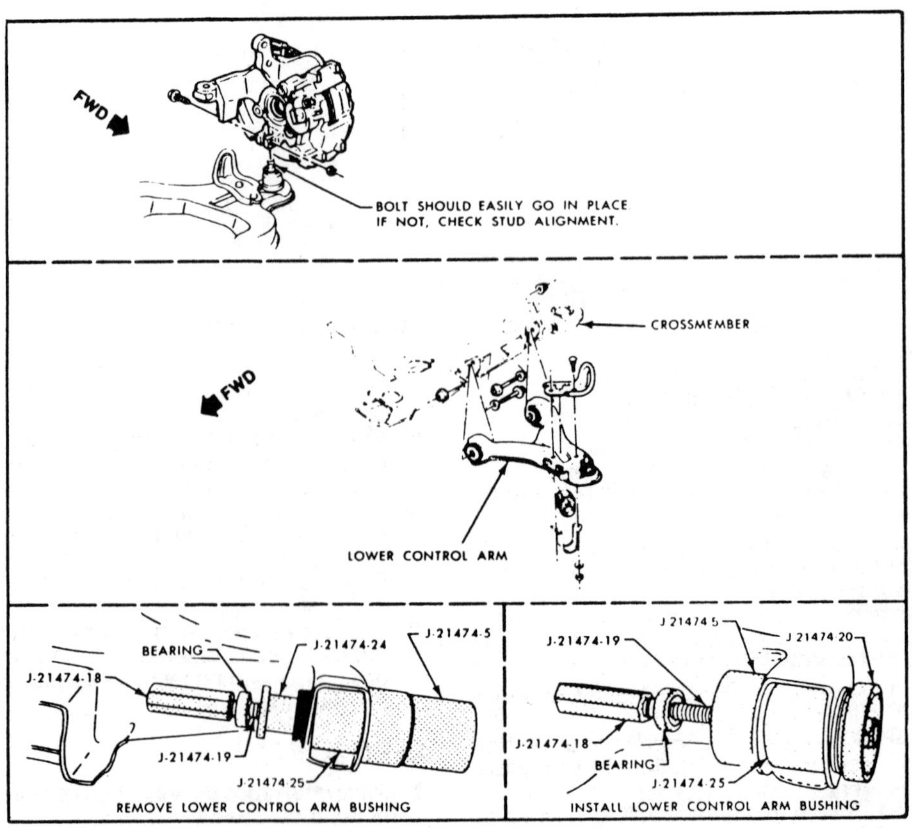

Fig. 6 Lower control arm & bushing removal & installation

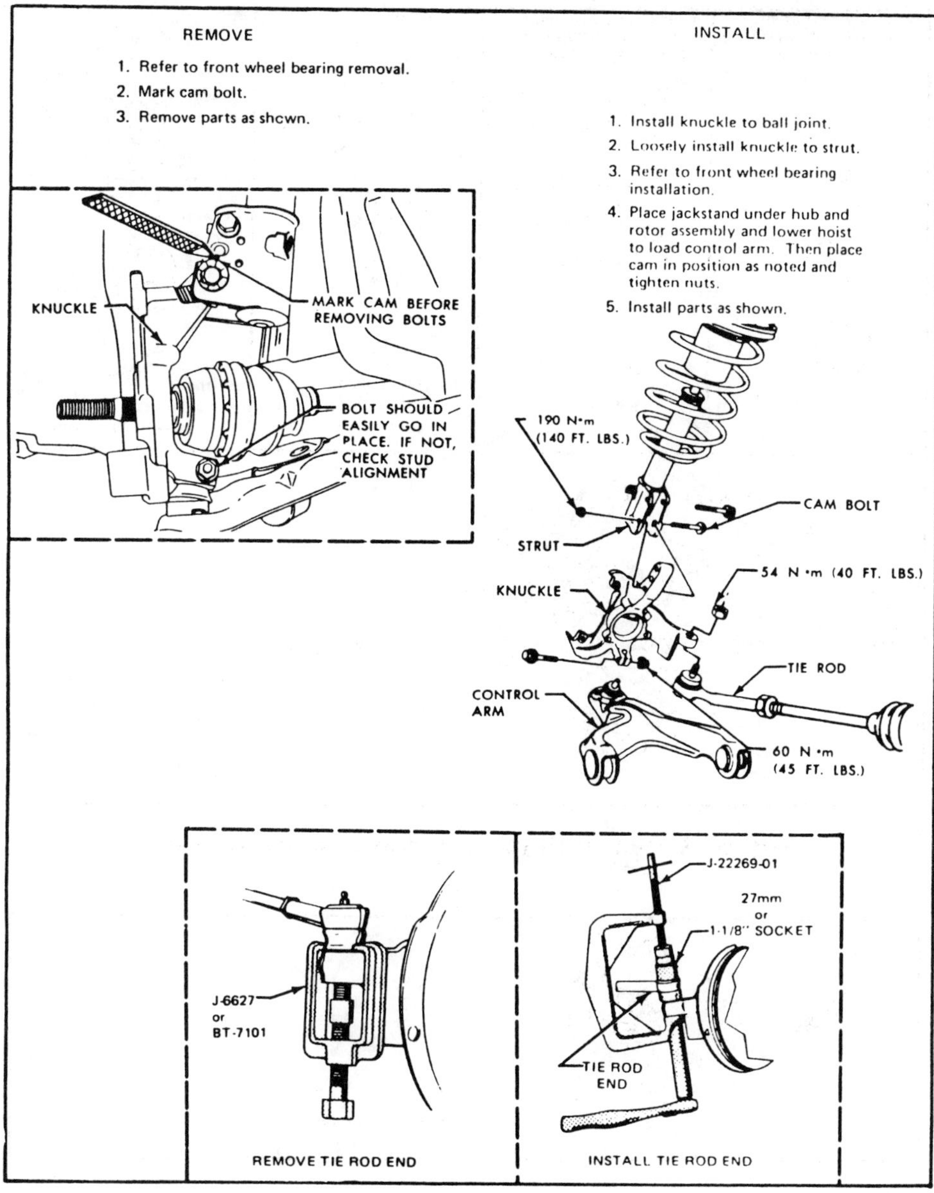

Fig. 7 Steering knuckle removal & installation

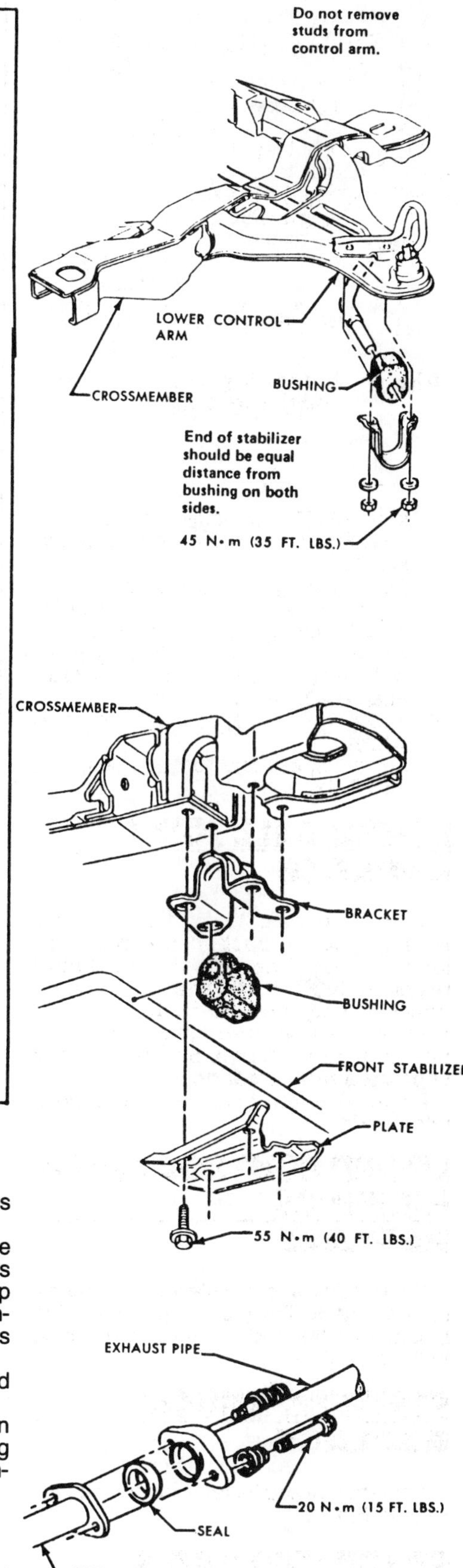

Fig. 8 Stabilizer bar & bushing removal & installation

STRUT ASSEMBLY REPLACE

For removal and installation procedures, refer to **Fig. 9.**

STRUT ASSEMBLY SERVICE

DISASSEMBLY

1. Mount compressor J-26584, or equivalent, in a vise.
2. Mount strut in compressor using bottom adapter J-26584-400 or equivalent.
3. Ensure that strut is fully contained by bottom adapter and that locating pins are fully engaged.
4. Rotate strut to align lip of top mount with notch in compressor support.
5. Install top adapter J-26584-430, aligning studs with properly marked holes on adapter.
6. Rotate compressor screw clockwise until top support contacts strut/adapter assembly, then keep turning screw until spring is compressed to approximately 1/2 of its height.
7. Remove nut from damper shaft and install alignment rod.
8. Gradually release spring tension, then remove and disassemble strut noting position of components for reassembly, **Fig. 10.**

ASSEMBLY

1. Perform steps 1 through 3 as outlined in the disassembly procedure.
2. Rotate strut assembly until mounting flange is facing outward opposite compressor forcing screw.
3. Install spring and related components on strut. Ensure spring is properly

seated on bottom spring plate.

4. Install strut spring seat assembly on top of spring. Ensure flat faces in same direction as lower flange.
5. Position top adapters over spring seat assembly.
6. Rotate compressor forcing screw until compressor top support makes contact with top adapters. Do not compress spring at this time.
7. Install strut alignment rod through top spring seat. Thread rod onto damper shaft hand tight, **Fig. 11. Install shims between lower spring seat and bottom adapter to keep alignment rod centered in upper spring seat opening.**
8. Compress spring by slowly rotating forcing screw. While rotating screw, observe position of damper shaft, as it must pass directly through center of opening in upper spring seat to provide proper operation. If spring is off-center, back off screw and proceed as follows:
   a. Reposition both top adapters in support fixture to provide proper spring seat-to-damper position.
   b. Install thin shim stock between lower spring seat and bottom adapter to tilt strut assembly into proper position.
   c. When damper shaft can be held in proper position, rotate screw until approximately 1 1/2 inch of shaft can be pulled through upper spring seat. Do not compress spring beyond this point.
9. Remove alignment rod, place strut mounting assembly over damper shaft and install nut. **Torque** nut to 65 ft. lbs.
10. Rotate forcing screw counterclockwise and remove strut assembly from compressor.

STEERING GEAR REPLACE

REMOVAL

1. **On models equipped with power steering,** remove air cleaner, then disconnect and cap hydraulic lines from steering gear.
2. Pull intermediate shaft seal upward, then remove intermediate shaft to steering gear stub shaft pinch bolt, **Fig. 12.**
3. Raise and support vehicle, then remove both front wheel and tire assemblies.
4. Using tool J6627, or equivalent, disconnect tie rod ends from steering knuckles.
5. Remove AIR pipe bracket to crossmember attaching bolt, if equipped.
6. Remove two rear frame cradle mounting bolts and lower rear of cradle 4 to 5 inches. **Do not lower cradle more than specified, as damage to engine components may result.**
7. Remove steering gear heat shield from crossmember.

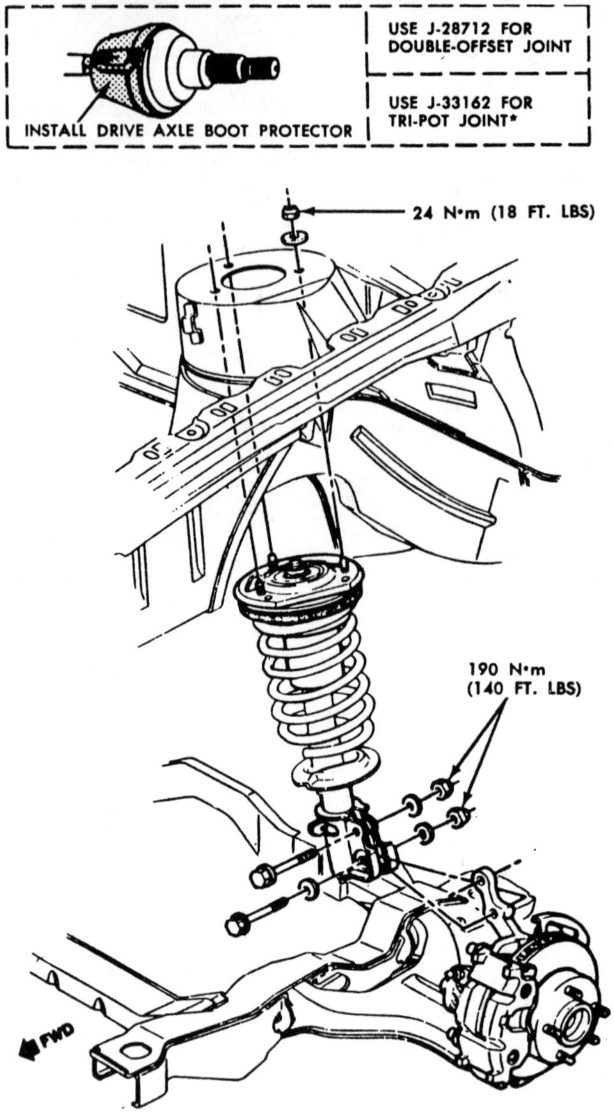

REMOVE

1. Remove (3) top attaching nuts.
2. Loosen wheel lug nuts.
3. Raise car and support on frame.
4. Remove wheel-and-tire; remove brake line clip.
5. Install boot protectors as shown in upper panel.
   *Whenever a tri-pot design joint is used on a drive axle shaft, it is necessary to disconnect the axle shaft from the trans-axle BEFORE separating the knuckle from the strut. Use J-33008.
6. Scribe the parts
7. Remove (2) bolts; separate the strut from the knuckle.
8. Remove the strut.

INSTALL

1. Install parts in reverse order of removal.
2. Place jack stand under hub and rotor assembly. Lower hoist and place car in position as noted in removal.
3. Raise hoist.
4. Install brake line clip.
5. Remove boot protectors.
6. Install wheel and lower car.

Fig. 9 Strut assembly removal & installation

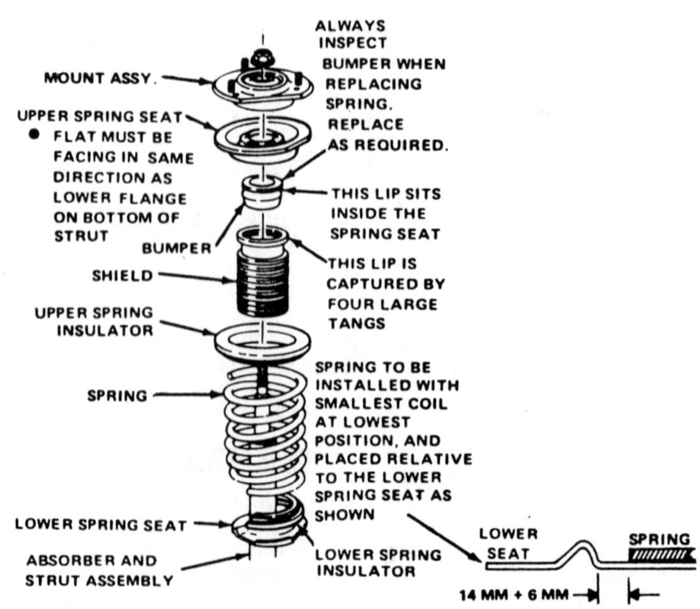

Fig. 10 Strut assembly exploded view

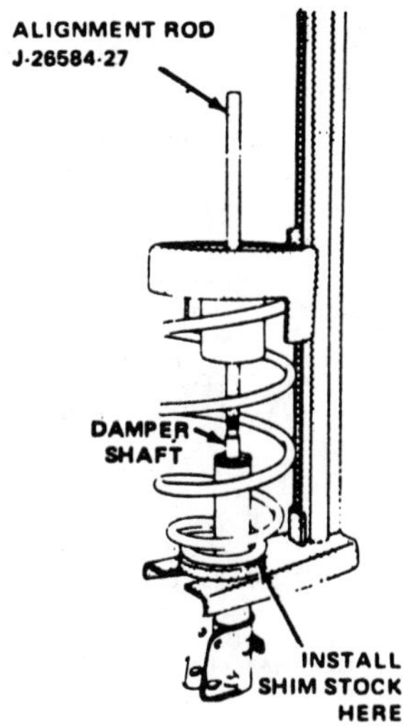

Fig. 11 Strut alignment rod installation (Typical)

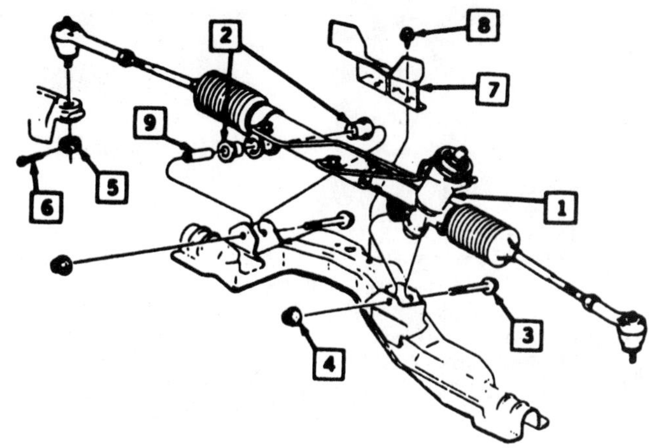

1. STEERING GEAR
2. BUSHING
3. BOLT
4. NUT – 95 N•m (70 LBS. FT.)
5. NUT – 40 N•m (30 LBS. FT.)
   70 N•m (52 LBS. FT.) MAX. TO INSTALL COTTER PIN
6. COTTER PIN
7. SHIELD – WITH LK9
8. BOLT – WITH LK9 – FULLY DRIVEN, SEATED AND NOT STRIPPED
9. SLEEVE

Fig. 12 Steering gear replacement

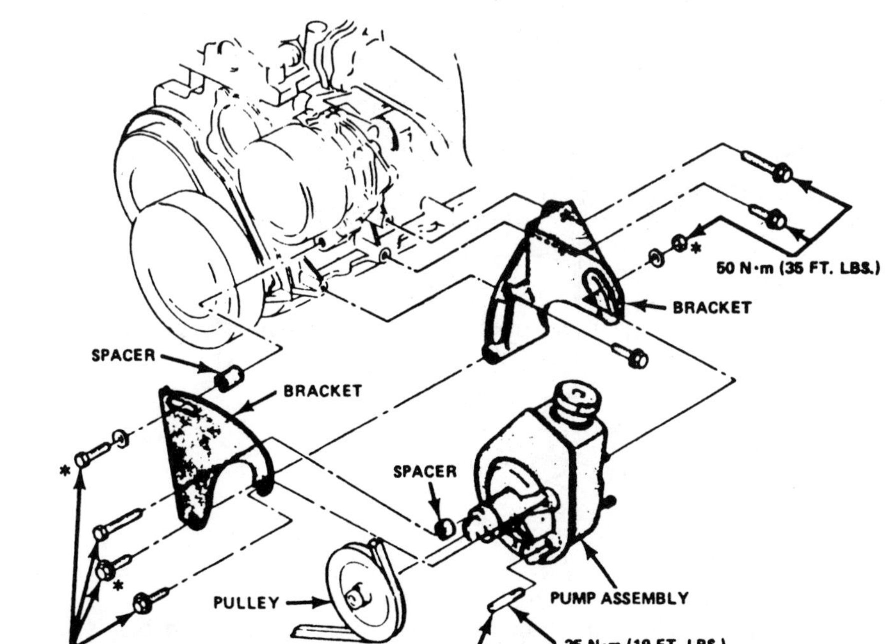

Fig. 13 Power steering pump replacement. 4-151

8. Remove steering gear assembly mounting bolts, then the steering gear assembly through the left wheel opening.

INSTALLATION

1. Place steering gear assembly into mounts on crossmember, then install and **torque** mounting bolts to 70 ft. lbs.
2. Install heat shield.
3. Raise frame cradle and install mounting bolts. **Torque** bolts to 75 ft. lbs.
4. Install AIR pipe bracket to crossmember, then connect tie rod ends to steering knuckles. **Torque** ball studs to 30 ft. lbs.
5. Install wheel and tire assemblies, then lower vehicle.
6. Connect intermediate shaft to steering gear stub shaft, install pinch bolt and **torque** to 45 ft. lbs.
7. Check and adjust toe-in as required.

POWER STEERING PUMP REPLACE

4-151

1. Raise and support vehicle, then remove radiator hose clamp bolt.
2. Disconnect and cap hydraulic lines from power steering pump.
3. Remove power steering pump to rear bracket attaching nut.
4. Remove front pump bracket to engine and front pump bracket to rear pump bracket attaching bolts, **Fig. 13.**
5. Remove pump and bracket assembly from vehicle.
6. Reverse procedure to install.

V6-173

1. Disconnect battery ground cable.
2. Disconnect electrical connector from blower motor, then remove blower motor from vehicle.
3. Disconnect heater hose from water pump and hydraulic lines from power steering pump. Cap lines to avoid fluid leakage.
4. Remove rear pump bracket to engine bracket attaching nut and front pump bracket to cylinder head attaching bolts, **Fig. 14.**
5. Remove pump and bracket assembly from vehicle.

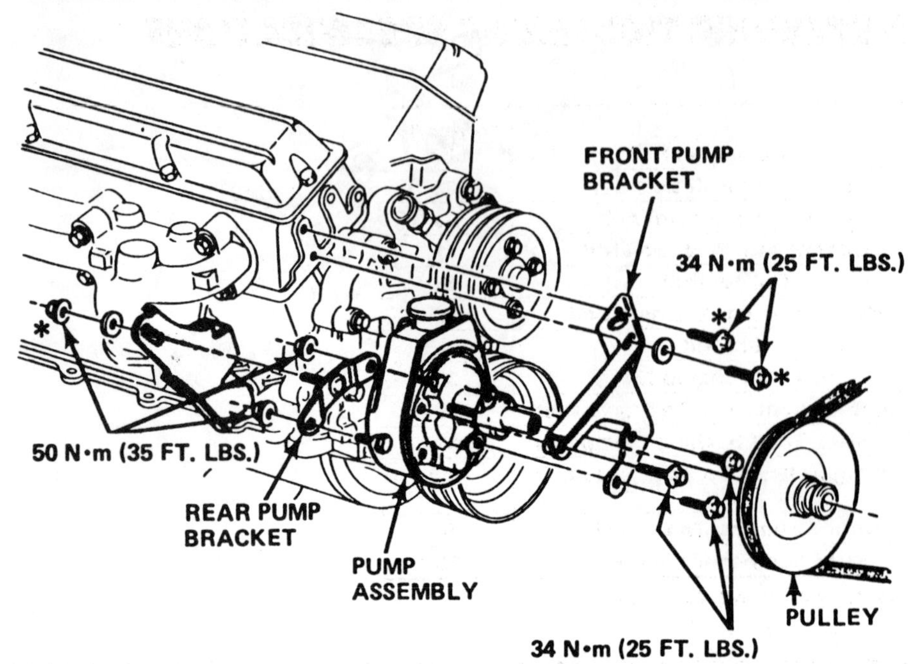

Fig. 14 Power steering pump replacement. V6-173

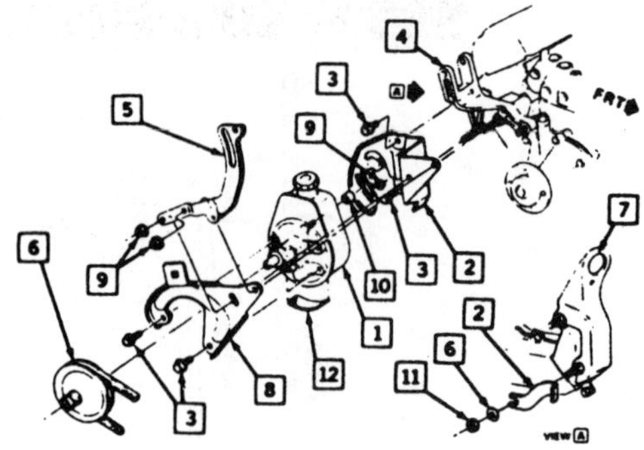

1. POWER STEERING PUMP
2. REAR ADJ. BRACKET
3. BOLT – 50 N•m (38 LBS. FT.)
4. GENERATOR MOUNTING BRACKET
5. GENERATOR ADJUSTING BRACKET
6. PULLEY
7. ENG. LIFT BRACKET AND SHIELD
8. FRONT ADJ. BRACKET
9. NUT – 50 N•m (38 LBS. FT.)
10. REAR BRACKET SPACER
11. NUT – 27 N•m (20 LBS. FT.)
12. PROTECTOR

Fig. 15 Power steering pump replacement. V6-181, 204 & 231

6. Reverse procedure to install.

V6-181, 204 & 231

1. Disconnect battery ground cable, then remove air cleaner.
2. Remove alternator drive belt and alternator, then raise and support vehicle.
3. Remove rear adjustment bracket to pump retaining nut, **Fig. 15.**
4. Remove alternator adjustment bracket and power steering pump drive belt.
5. Remove double nut and two studs, then the alternator mounting bracket.
6. Remove rear pump adjustment bracket, then disconnect hydraulic lines from power steering pump. Cap lines to prevent spillage of fluid.
7. Remove pump and bracket assembly from vehicle.
8. Reverse procedure to install.

V6-262 DIESEL

1985

1. Rotate drive belt tensioner counterclockwise, remove serpentine drive belt, then slowly release tensioner.
2. Raise and support vehicle, then remove right front wheel.
3. Place suitable drain pan under pump assembly, disconnect lines from pump, then plug lines and open ports.
4. Remove bolts and nut securing pump and the pump, **Fig. 16.**

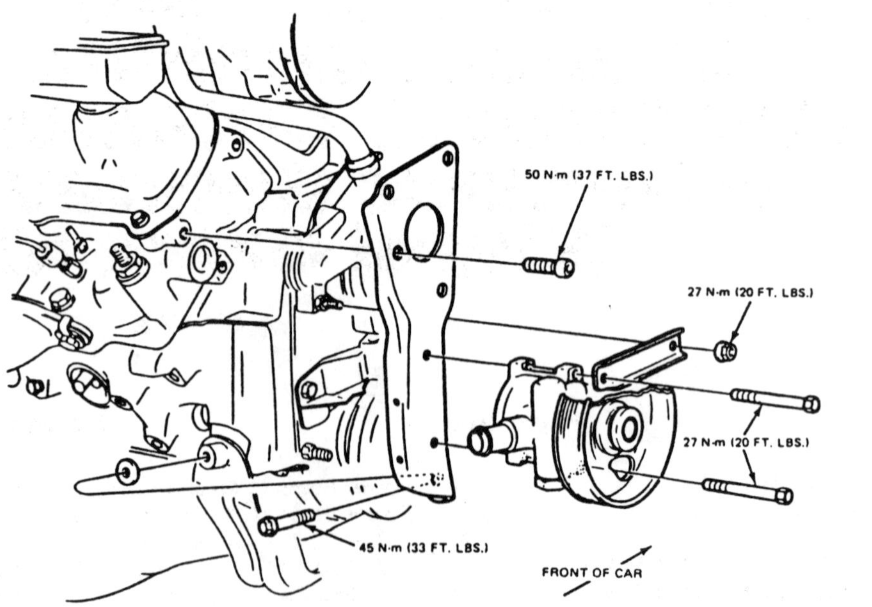

Fig. 16 Power steering pump removal & installation. 1985 V6-262 Diesel

5. Remove pulley using J-25034, or equivalent.
6. Install pulley on replacement pump using J-25033, or equivalent, then reverse remaining procedure to complete installation.

FRONT SUSPENSION & STEERING TIGHTENING SPECIFICATIONS

Year	Component	Torque/Ft. Lbs.
1985-90	Ball Joint Pinch Bolt Nut	33
	Brake Line Bracket Bolt	13
	Control Arm Pivot Bolt Nut	61
	Caliper Bolt	38
	Hub & Bearing Retaining Bolt	①
	Hub Nut	192
	Insulator Clamp Nut	33
	Reinforcement Plate To Frame Bolt	40
	Stabilizer Shaft Bushing Clamp Nut	32
	Steering Knuckle To Strut Assembly Bolt	140
	Strut Assembly To Body Nut	18
	Strut Dampener Shaft Nut	65

①—W/O JA2, 63 ft. lbs., W/JA2, 70 ft. lbs.

Buick Electra, LeSabre & Park Avenue, Cadillac Deville & Fleetwood, Oldsmobile 88 & 98 & Pontiac Bonneville Front Wheel Drive Models

INDEX

DESCRIPTION

The front suspension is of the MacPhereson design, **Fig. 1.** The control arm pivots from the cradle and is mounted in rubber bushings. The upper end of the strut is isolated by a rubber mount and contains a non-serviceable bearing to allow for rotation. The lower end of the steering knuckle pivots on a ball joint riveted to the control arm. The ball joint is fastened to the steering knuckle with a castle nut and cotter pin.

HUB & BEARING
REPLACE

1. Raise and support vehicle and remove wheel assembly.
2. Install drive axle boot protector tool J-28712 on outer joints and J-34754 on inner joints.
3. Insert a suitable drift through opening in caliper into rotor cooling fins to prevent assembly from rotating, then remove hub nut and washer.
4. Remove caliper bracket mounting bolts and the caliper and bracket assembly, then secure assembly aside taking care not to stretch brake hose. Use care not to damage brake hose.
5. Remove rotor, then press drive axle from hub using J-28733 or equivalent, **Fig. 2.**

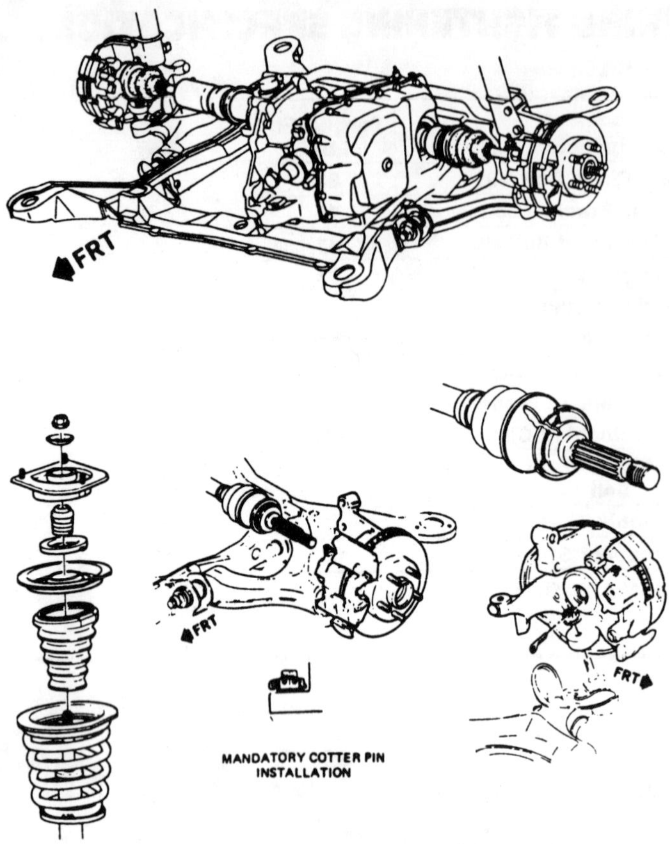

Fig. 1 Front suspension

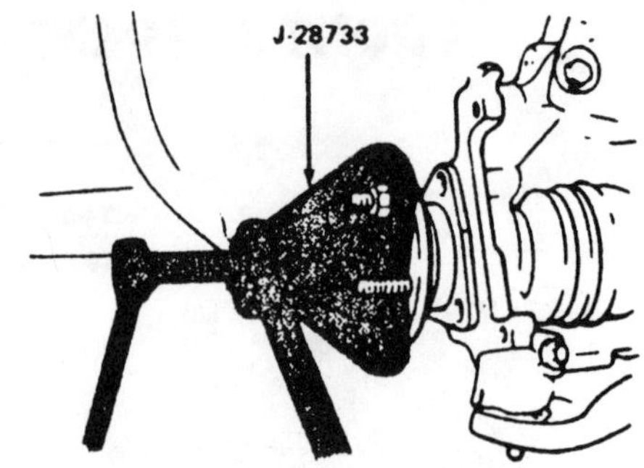

Fig. 2 Separating drive axle from hub

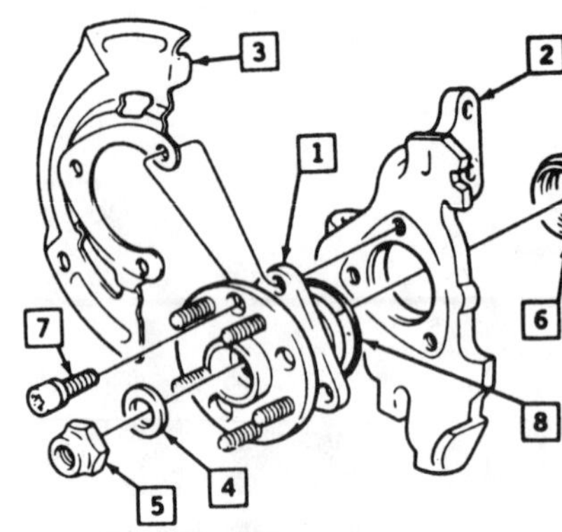

1. HUB AND BEARING ASSEMBLY
2. STEERING KNUCKLE
3. SHIELD
4. WASHER
5. HUB NUT 245 N·m (180 LBS. FT.)
6. SEAL
7. HUB AND BEARING RETAINING BOLT (55 TORX) 95 N·m (70 LBS. FT.)
8. "O" RING

Fig. 3 Hub & bearing assembly

6. Remove hub assembly retaining bolts, shield, hub and bearing assembly, and O-ring, **Fig. 3**
7. Reverse procedure to install. Torque hub attaching bolts to specifications, caliper bolts to 38 ft. lbs. and hub nut to specifications.

BALL JOINT
REPLACE

1. Raise and support vehicle and place jackstands under cradle. Vehicle weight should not be placed on the control arms.
2. Remove wheel assembly.
3. Install drive axle boot protectors.
4. Remove cotter key from ball joint nut and, using tool J-34505, separate ball joint from steering knuckle, **Fig. 4.**
5. Drill out ball joint retaining rivets.
6. Remove stabilizer bar bushing to control arm bolt.
7. Pull control arm downward and remove ball joint from steering knuckle and control arm.
8. Reverse procedure to install. **Torque** new ball joint attaching nuts to specifications, **Fig. 5,** and ball joint castle nut to 81 ft. lbs.

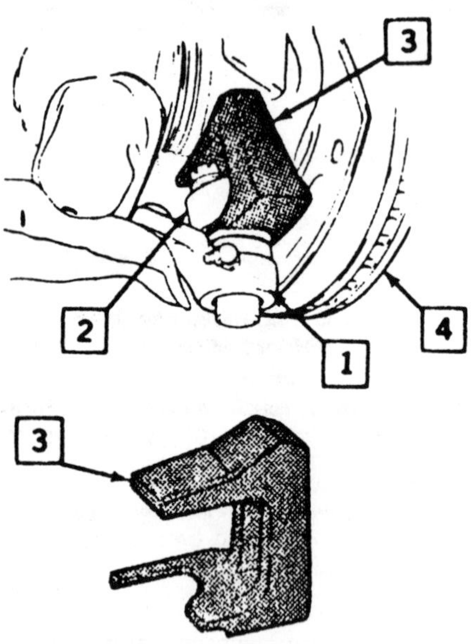

1. BALL JOINT
2. STEERING KNUCKLE
3. BALL JOINT SEPERATOR J-34505
4. ROTOR

Fig. 4 Separating ball joint from steering knuckle

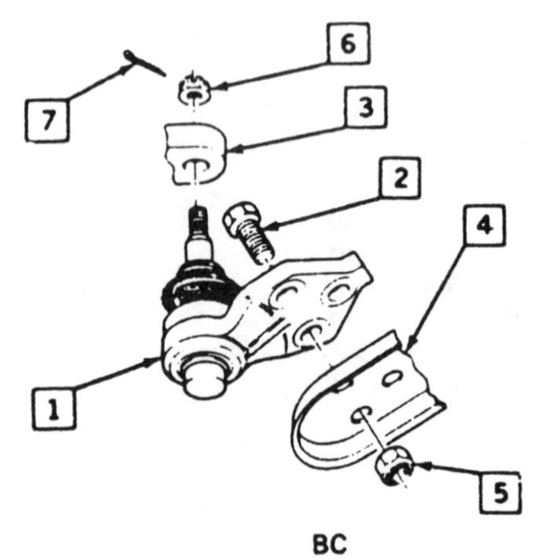

1. SERVICE BALL JOINT
2. BALL JOINT MOUNTING BOLTS MUST FACE DOWN
3. STEERING KNUCKLE
4. CONTROL ARM
5. BALL JOINT MOUNTING NUTS 68 N·m (50 LBS. FT.)
6. BALL JOINT TO STEERING KNUCKLE NUT 110 N·m (81 LBS. FT.) BEFORE COTTER PIN INSTALLATION
7. COTTER PIN

Fig. 5 Service ball joint attachment

CONTROL ARM
REPLACE

1. Raise and support vehicle and place jackstands under cradle. Vehicle

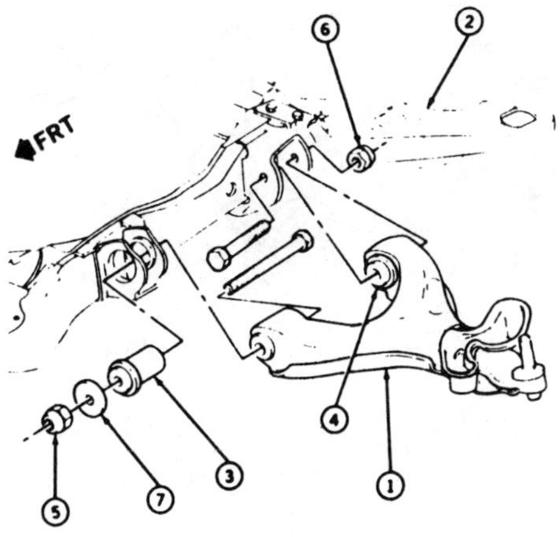

Fig. 6 Control arm assembly

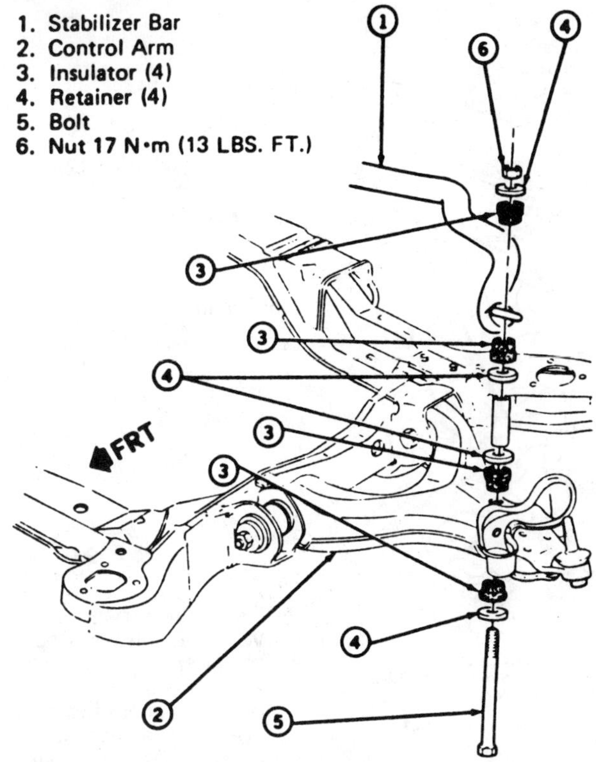

Fig. 7 Stabilizer bar bushing assembly

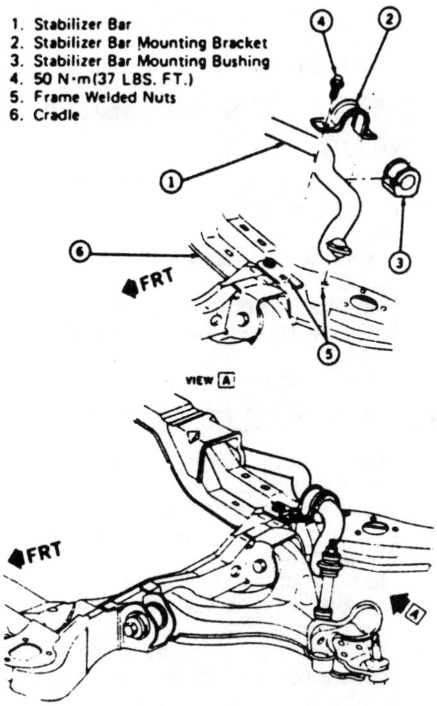

Fig. 8 Stabilizer bar mounting

weight should not be placed on the control arms.
2. Remove wheel assembly.
3. Install drive axle boot protectors.
4. Remove cotter key from ball joint nut and, using tool J-34505, separate ball joint from steering knuckle.
5. Remove stabilizer bar bushing to control arm bolt.
6. Remove control arm mounting bolts and the control arm, **Fig. 6.**
7. Reverse procedure to install. When installing control arms, install but do not torque control arm mounting bolts. Sufficiently tighten bolts to assure security but final torque is applied when vehicle weight is supported by control arms. Torque rear arm bolt and front bolt to specifications.

STABILIZER BAR REPLACE

1. Raise and support vehicle and place jackstands under cradle. Vehicle weight should not be placed on the control arms.
2. Remove wheel assembly.
3. Install drive axle boot protectors.
4. Remove nuts, washers, bushings and bolt securing stabilizer shaft to each control arm, **Fig. 7.**
5. Remove stabilizer bar mounting bolts, two bolts from each side, **Fig. 8.**
6. Disconnect tie rods from steering knuckles.
7. Remove exhaust pipe between exhaust manifold and catalytic converter. Remove intermediate pipe on diesel models.
8. **On vehicles with V6-181 and V8-250 engines,** remove air pipe to exhaust pipe bolt, if equipped.
9. Rotate right side strut assembly completely to the right.
10. Slide stabilizer bar to the right over the steering knuckle and pull downward on left side until stabilizer bar clears the cradle, **Fig. 9.**
11. Reverse procedure to install.

STRUT ASSEMBLY REPLACE

1. Remove three nuts attaching top of strut to body, **Fig. 10.**
2. Raise and support vehicle and place

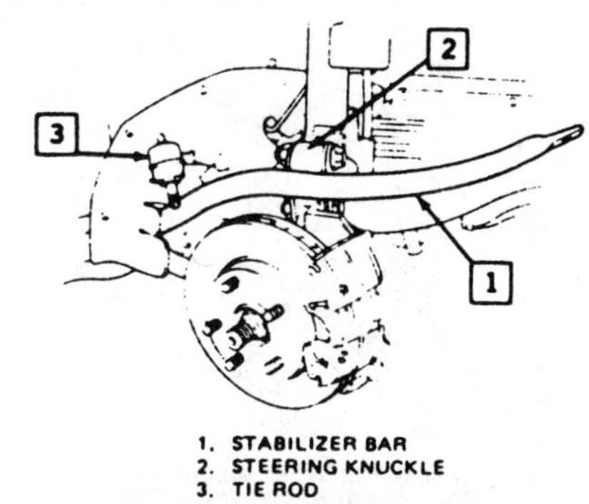

Fig. 9 Stabilizer bar replacement

jackstands under cradle. Vehicle weight should not be placed on the control arms.
3. Remove wheel assembly.
4. Install drive axle boot protectors.
5. Remove brake line bracket bolt from strut assembly.
6. Remove strut to steering knuckle attaching bolts.
7. Remove strut assembly from vehicle.
8. Reverse procedure to install. Torque strut mounting bolts to specifications.

STRUT SERVICE

1. Remove strut as outlined under "Strut Assembly, Replace" procedure.
2. Mount strut in compressor tool J-34013 and holding fixture J-3289-20.
3. Rotate compressor forcing screw until

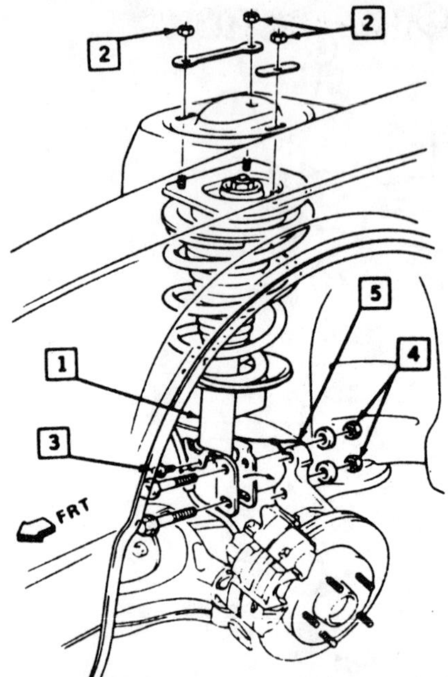

1. STRUT ASSEMBLY
2. STRUT TO BODY NUTS 24 N·m (18 LBS.FT.)
3. BRAKE LINE BRACKET BOLT 17 N·m (13 LBS. FT.)
4. STRUT TO STEERING KNUCKLE NUTS 196 N·m (144 LBS. FT.)
5. RETAIN STEERING KNUCKLE WITH WIRE ONCE STRUT ASSEMBLY IS REMOVED

Fig. 10 Strut assembly

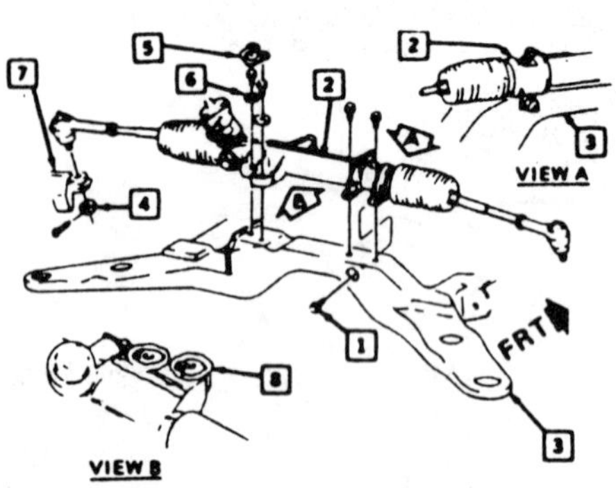

1. BOLT 68 N·m (50 LBS. FT.) AFTER SECOND REUSE OF BOLT, LOCKTITE THREAD LOCKING KIT NO. 1052624 MUST BE USED.
2. STEERING GEAR
3. FRAME
4. 40 N·m (30 LBS. FT.), 70 N·m (52 LBS. FT.) MAXIMUM PERMISSIBLE TORQUE TO ALIGN COTTER PIN SLOT. (1/6 TURN MAXIMUM) DO NOT BACK OFF FOR COTTER PIN INSERTION.
5. RETAINER
6. WASHER
7. STEERING KNUCKLE
8. RTV SEALER AROUND INSERTS

Fig. 11 Steering gear replacement

spring compresses slightly.

4. Hold damper shaft from rotating and remove nut from top of strut assembly.
5. Use tool J-34013-30 to guide damper shaft from assembly.
6. Loosen compressor forcing screw while guiding damper shaft from assembly. Continue to loosen nut until strut damper and spring can be removed.

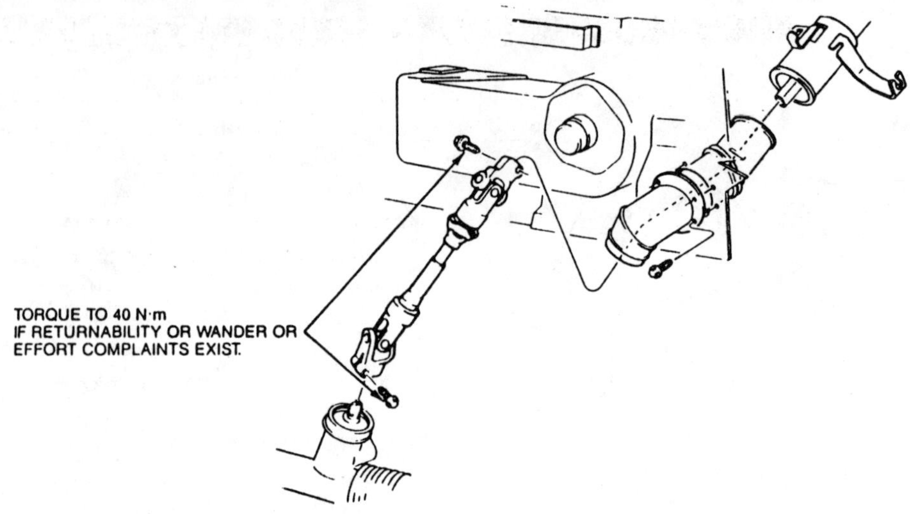

Fig. 12 Intermediate shaft pinch bolts

7. Reverse procedure to assemble. When assembling spring, the flat on upper spring seat must face outward 90 degrees from centerline of vehicle or when mounted in the strut compressor, the seat faces in the same direction as the steering knuckle mounting flange.

POWER STEERING GEAR REPLACE

1. Raise and support vehicle with weight resting on suspension.
2. Remove front wheel assemblies.
3. Disconnect intermediate shaft from steering gear stub shaft.
4. Disconnect both tie rod ends from steering knuckles.
5. Remove line retainers and disconnect hydraulic lines from steering gear.
6. Remove five steering gear attaching bolts, **Fig. 11.**
7. Remove steering gear from vehicle by sliding out to the side.
8. Reverse procedure to install. **Torque** steering gear attaching bolts to 50 ft. lbs.

POWER STEERING PUMP REPLACE

V6-181, 231 GASOLINE ENGINES

1. Disconnect battery ground cable.
2. Remove air cleaner on V6-181 engine, if necessary.
3. Remove alternator drive belt and alternator.
4. Raise and support vehicle.
5. Disconnect hydraulic lines from pump.
6. Remove rear pump adjustment bracket to pump nut.
7. Remove pump drive belt.
8. Lower vehicle.
9. Remove pump rear adjustment bracket.
10. Remove power steering pump from vehicle.
11. Reverse procedure to install.

V8-250 & 273 GASOLINE ENGINE

1. Disconnect battery ground cable.
2. Remove pump drive belt and the pulley.
3. Disconnect hydraulic lines from pump.
4. Remove two pump mounting bolts.
5. Remove power steering pump from vehicle.
6. Reverse procedure to install.

V6-262 DIESEL ENGINE

1. Disconnect battery ground cable.
2. Remove pump drive belt.
3. Raise and support vehicle.
4. Remove engine splash shield and crankshaft pulley.
5. Remove engine absorber.
6. Disconnect reservoir hose from pump and drain pump.
7. Remove high pressure support and disconnect high pressure hose from pump.
8. Remove two front bracket nuts.
9. Remove one rear bracket bolt.
10. Remove power steering pump with brackets attached.
11. Remove brackets from pump assembly.
12. Reverse procedure to install.

SERVICE BULLETINS

INCREASED STEERING EFFORT, SLIGHT WANDER OR POOR RETURNABILITY

On some 1989 Deville and Fleetwood models, a condition of increased steering effort, slight wander and returnability may be experienced.

To correct this condition, loosen pinch bolts at the intermediate shaft, **Fig. 12,** then **torque** pinch bolts to 29.5 ft. lbs.

FRONT SUSPENSION & STEERING TIGHTENING SPECIFICATIONS

Year	Component	Torque/ft. lbs.
1985-90	Ball Joint To Knuckle	88①
	Brake Caliper To Knuckle	38
	Brake Line Bracket To Underbody	13
	Control Arm Mounting Nut (Front)	140
	Control Arm Mounting Nut (Rear)	90
	Drive Axle Shaft Nut	180
	Hub & Bearing To Knuckle	70
	Service Ball Joint To Control Arm	50
	Stabilizer Shaft Bushing Nut	13
	Stabilizer Shaft Mounting Bracket	37
	Steering Knuckle To Strut Bolts	144
	Strut Assembly To Body Nuts	18
	Strut Mount Nut	55
	Tie Rod To Knuckle	35

①—Tighten an additional 120°.

1988–90 Buick Regal, Oldsmobile Cutlass Supreme & Pontiac Grand Prix & 1990 Chevrolet Lumina

INDEX

DESCRIPTION

The front suspension system on these vehicles is of the MacPherson strut design. This design incorporates MacPherson struts with coil springs and a one piece configuration with lower control arms. The use of tapered top coil springs on top of the struts provides a well controlled ride and allows a lower hood profile.

STRUT CARTRIDGE REPLACE

Do not service strut unless weight of vehicle of vehicle is on suspension.

REMOVAL

1. Scribe alignment marks on cover plate, remove cover plate retaining nuts, then the cover plate.
2. Using No. 50 Torx bit and strut rod remover/installer tool No. J-35669, or equivalents, remove strut shaft.
3. Remove strut mount bushing by prying with suitable tool.
4. **On 1988-89 models,** using strut mount plate wrench tool No. J-35670 or equivalent, remove jounce bumper retainer.
5. **On all models,** attach strut extension rod tool No. J-35668 or equivalent to strut shaft and compress shaft down into cartridge, then remove tool and pull out jounce bumper.
6. Attach strut extension rod tool No. J-35668 or equivalent to strut shaft and extend shaft, then remove tool and, using strut cap nut wrench tool No. J-35671 or equivalent, unscrew closure nut.
7. Remove strut cartridge.
8. Remove oil from strut tube using suction device.

INSTALLATION

1. Using strut cap nut wrench tool No. J-35671 or equivalent, install self contained replacement cartridge.
2. Install jounce bumper, then, on

1988-89 models, using strut mount plate wrench tool No. J-35670 or equivalent, the jounce bumper retainer.
3. **On all models,** install strut mount bushing. If necessary, use strut extension rod tool No. J-35668 or equivalent after bushing is partially installed and position strut shaft as required. Lubricate bushing with a soap solution to ease installation.
4. Using No. 50 Torx bit and strut rod remover/installer tool No. J-35669, or equivalents, install strut shaft nut, torquing to specifications.
5. Install strut mount cover, aligning scribe marks. Torque cover plate nuts to specifications.

HUB & BEARING ASSEMBLY
REPLACE

1. Remove wheel cover, then loosen drive axle shaft nut and washer one turn. **Do not remove nut at this time. Failure to follow proper removal sequence may cause bearing damage.**
2. Loosen lug nuts, then raise and support vehicle and remove wheel and tire assembly.
3. Remove brake caliper and bracket assembly, then the brake rotor, **Fig. 1.**
4. Remove drive axle shaft nut and washer.
5. Loosen hub and bearing assembly to knuckle attaching bolts.
6. Using front hub spindle remover tool No. J-28733-A or equivalent, push axle splines back out of hub and bearing assembly.
7. Remove hub and bearing assembly to knuckle attaching bolts.
8. **On models with anti-lock brakes (ABS),** remove ABS sensor mounting bolt and position sensor aside.
9. **On all models,** remove hub and bearing assembly. **Protect axle boots from damage during handling.**
10. Reverse procedure to install, torquing hub and bearing assembly to knuckle attaching bolts to specifications, lug nuts to 100 ft. lbs. and the drive axle nut to specifications.

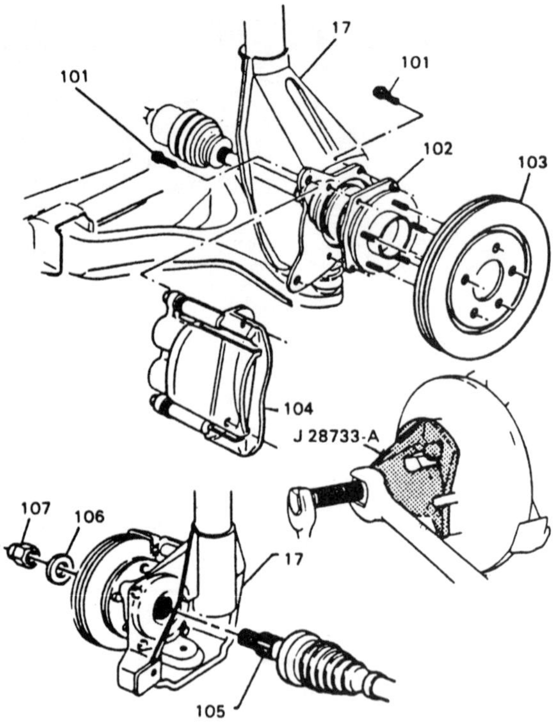

17 KNUCKLE/STRUT ASSEMBLY
101 BOLT 70 N·m (52 lbs. ft.)
102 HUB/BEARING ASSEMBLY
103 ROTOR
104 CALIPER AND BRACKET ASSEMBLY
105 DRIVE AXLE
106 WASHER
107 AXLE NUT 250 N•m (184 lbs. ft.)

Fig. 1 Removing hub & bearing assembly

STABILIZER SHAFT & INSULATORS
REPLACE

1. Loosen lug nuts, then raise and support vehicle and remove wheel and tire assembly.
2. Move steering shaft dust shield to gain access to pinch bolt.
3. Remove pinch bolt from lower intermediate steering shaft.
4. Loosen all insulator clamp attaching nuts and bolts, **Fig. 2.**
5. Place suitable jack under center of rear frame crossmember.
6. Loosen two front frame to body bolts four turns.
7. Remove two rear frame to body bolts.
8. Lower rear of frame just enough to gain access for stabilizer shaft removal.
9. Remove insulators and clamps from frame and control arms.
10. Pull stabilizer shaft rearward, swing down and remove from left side of vehicle.
11. Reverse procedure to install, noting the following:

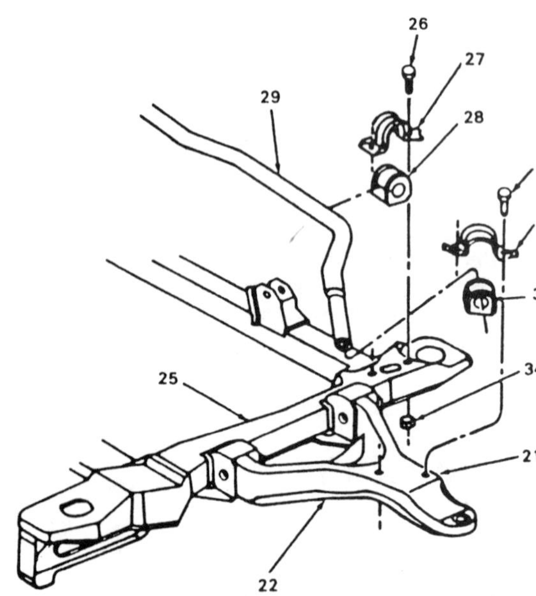

21 WELD NUTS
22 LOWER CONTROL ARM
25 FRAME
26 CLAMP BOLT
27 CLAMP
28 INSULATOR
29 STABILIZER SHAFT
30 BOLT 47 N•m (35 lbs. ft.)
31 CLAMP
32 INSULATOR
34 NUT 47 N•m (35 lbs. ft.)

Fig. 2 Replacing stabilizer shaft & insulators

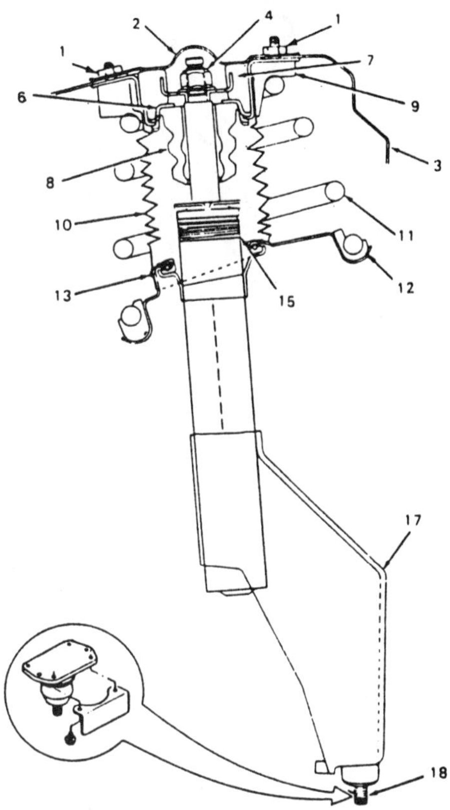

1 COVER PLATE NUT
2 COVER PLATE
3 SHOCK TOWER
4 STRUT SHAFT NUT
6 JOUNCE BUMPER RETAINER
7 STRUT MOUNT
8 JOUNCE BUMPER
9 SPRING INSULATOR
10 DUST SHIELD
11 SPRING
12 SPRING INSULATOR
13 SPRING SEAT AND BEARING
15 CLOSURE NUT
17 KNUCKLE STRUT ASSEMBLY
18 BALL JOINT

Fig. 3 Cross-sectional view of strut & knuckle assembly. 1988–89 models shown, 1990 models similar

a. Coat new insulators with rubber lubricant.
b. Torque clamp to frame nuts and clamp to lower control arm bolts to specifications.
c. **Torque** frame attaching bolts to 100 ft. lbs.
d. Torque pinch bolt to specifications.
e. **Torque** lug nuts to 100 ft. lbs.

STRUT & KNUCKLE ASSEMBLY SERVICE

REMOVAL

1. Disconnect battery ground cable.
2. Scribe alignment marks on cover plate, then loosen three cover plate retaining nuts.
3. Loosen lug nuts, then raise and support vehicle and remove wheel and tire assembly.
4. Remove brake caliper and bracket assembly, leaving hose attached, and support caliper out of way.
5. Remove brake rotor, then the hub and bearing to knuckle attaching bolts.
6. **On models with anti-lock brakes (ABS),** remove ABS sensor mounting bolt and position sensor aside.
7. **On all models,** separate axle from transaxle and carefully remove drive axle assembly.
8. Remove tie rod to knuckle attaching nut.
9. Using tie rod puller/ball joint remover tool No. J-35917 or equivalent, separate tie rod from knuckle.
10. Remove lower ball joint to knuckle attaching nut.
11. Using tool No. J-35917 or equivalent, separate ball joint from lower control arm.
12. Remove ball joint heat shield, then the cover plate retaining nuts.
13. Remove strut and knuckle assembly.

SERVICE

Springs are under high tension. Do not remove strut shaft nut without using a suitable spring compressing tool.

1. Mount strut and knuckle assembly into strut spring compressor tool No. J-34013-A and strut compressor adapter tool No. J-34013-88, or equivalents, then compress spring with compressor forcing screw just enough to release tension from upper spring insulator.
2. Using Torx bit and strut rod nut remover/installer tool No. J-35669, or equivalents, remove strut shaft nut.
3. Relieve all spring tension, then remove spring and strut components, **Fig. 3.**
4. Install spring seat and bearing.
5. Install lower spring insulator. Lower spring coil end must be visible between the step and the first retention tab of insulator.
6. Install front suspension spring.
7. Install dust shield to lower spring seat.
8. Install jounce bumper.
9. Install upper spring insulator. Upper spring coil end must be between step and location mark on insulator.
10. **On 1988-89 models,** using strut mount plate wrench tool No. J-35670 or equivalent, install jounce bumper retainer to strut mount.
11. **On all models,** install strut mount and upper strut mount bushing.
12. Using strut spring compressor and strut compressor adapter tools Nos. J-34013-A and J-34013-88, or equivalents, compress strut assembly.
13. Align strut cartridge shaft with strut extension rod tool No. J-35668 or equivalent.
14. Using strut rod nut remover/installer tool No. J-35669 or equivalent and Torx bit, install strut shaft nut, torquing to specifications.

INSTALLATION

1. Install strut mount cover plate and upper strut mount to body attaching nuts. Do not tighten nuts until vehicle is lowered to floor.
2. Install ball joint heat shield.
3. Install lower ball joint to control arm attaching nut, **torquing** to 15 ft. lbs., then tighten an addition 90°.
4. Further tighten to next slot in nut with cotter pin hole in stud and install new cotter pin. **Do not tighten more than an additional 60° to align with hole and do not loosen nut at any time during installation.**
5. Install tie rod to knuckle attaching nut, torquing to specifications, and install cotter pin.
6. Carefully install drive axle assembly through opening in knuckle and insert drive axle into transaxle.
7. Install hub and bearing to knuckle attaching bolts, torquing to specifications.
8. **On models with anti-lock brakes (ABS),** position ABS sensor and install mounting bolt.
9. **On all models,** install brake rotor, then the brake caliper and bracket assembly.
10. Install wheel and tire assembly and lower vehicle.
11. **Torque** lug nuts to 100 ft. lbs. and the cover plate nuts to specifications.
12. Connect battery ground cable.

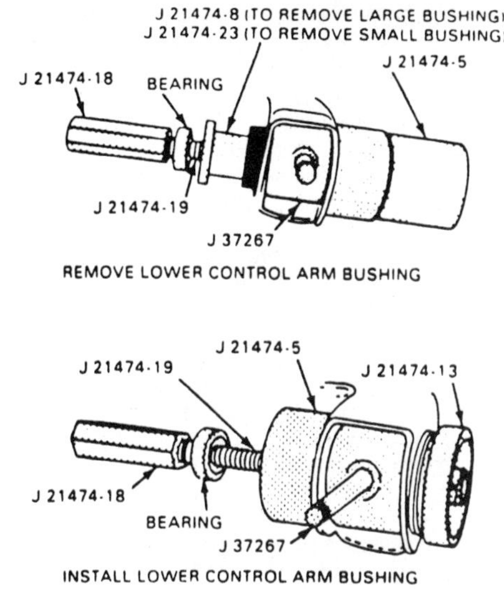

Fig. 4 Replacing lower control arm bushings

LOWER CONTROL ARM REPLACE

1. Loosen lug nuts, then raise and support vehicle and remove wheel and tire assembly.
2. Remove stabilizer shaft to lower control arm insulator bracket bolts.
3. Remove lower ball joint cotter pin and nut, then, using tie rod puller/ball joint remover tool No. J-35917 or equivalent, separate ball joint from lower control arm.
4. Remove lower control arm to frame attaching nuts and bolts, then the lower control arms.

5. If bushing replacement is necessary, refer to **Fig. 4.** Coat threads of control arm bushing service tool set No. J-21474-19 or equivalent with an extreme pressure lubricant. To facilitate installation, coat outer casing of bushing with suitable lubricant.
6. Reverse procedure to install, noting the following:
   a. Lower control arm to frame bolts must be installed as shown, **Fig. 5.**
   b. Torque lower control arm to frame bolts to specifications.
   c. **Torque** lower ball joint nut 15 ft. lbs. and tighten an additional 90°, then further tighten to align next slot in nut with cotter pin hole in stud. **Do not tighten more than 60° to align with hole and do not loosen nut at any time during installation.**
   d. Torque stabilizer shaft bolts to specifications.
   e. **Torque** lug nuts to 100 ft. lbs.

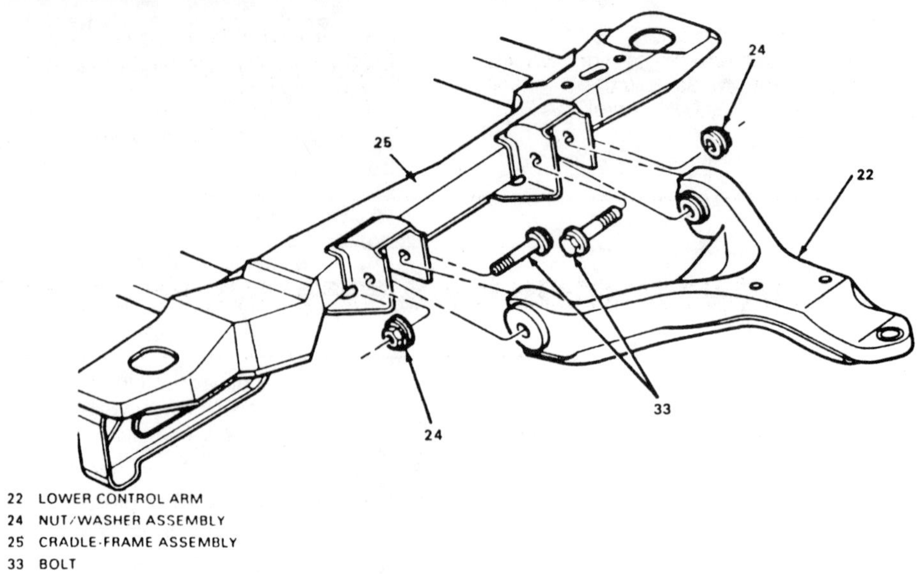

Fig. 5 Installing lower control arm

STEERING GEAR
REPLACE

1. Remove air cleaner, then raise and support vehicle. **Provide additional support at rear of vehicle.**
2. Remove front wheel and tire assemblies.
3. Remove intermediate shaft lower pinch bolt at steering gear.
4. Remove intermediate shaft from stub shaft. **Failure to disconnect intermediate shaft from rack and pinion shaft stub may result in damage to steering gear and/or intermediate shaft. This damage can cause loss of steering control.**
5. Disconnect electrical connector at switch.
6. Using tie rod puller/ball joint remover tool No. J-35917 or equivalent, disconnect tie rod ends from knuckle and strut assembly.
7. Support body with suitable stands, then remove rear frame mounting bolts and lower rear frame no more than five inches. **Do not lower rear of frame too far. Damage to engine components nearest to cowl may result.**
8. Remove heat shield, then the pipe retaining clip from steering gear.
9. Disconnect inlet pipes and outlet line from power steering gear.
10. Remove remaining brackets and clips.
11. Remove rack and pinion mounting nuts and bolts, then the rack and pinion assembly through the left wheel opening.
12. Reverse procedure to install, noting the following:
    a. Torque rack and pinion mounting bolts to specifications.
    b. Replace O-rings as necessary.
    c. Torque power steering line fitting to specifications.
    d. Torque heat shield attaching screws to specifications.
    e. **Torque** rear frame attaching bolts to 100 ft. lbs.
    f. Torque tie rod end castle nuts to specifications.
    g. Torque intermediate shaft lower pinch bolt to specifications.
13. After completing installation, bleed power steering system as follows:
    a. Turn wheels all the way to left.
    b. Add power steering to Cold mark on fluid level indicator.
    c. Start engine and run at fast idle, then add fluid, if necessary, to bring level to Cold mark.
    d. Bleed system by turning wheels from side to side without hitting stops. Keep fluid level at Cold mark.
    e. Return wheels to center position and continue running engine for 2-3 minutes.
    f. Road test vehicle to ensure steering functions normally and is free of noise.
    g. Check fluid level and ensure level is at Hot mark after system has stabilized at its normal operating temperature.

POWER STEERING PUMP
REPLACE

EXCEPT 4-138 ENGINES

1. Disconnect battery ground cable.
2. Disconnect inlet and outlet hoses from pump.
3. Remove serpentine belt guard as necessary.
4. Remove belt from pulley, using 1/2 inch breaker bar.
5. Remove pump assembly.
6. Transfer pulley as necessary.
7. Reverse procedure to install, them bleed power steering system as follows:
   a. Turn wheels all the way to left.
   b. Add power steering to Cold mark on fluid level indicator.
   c. Start engine and run at fast idle, then add fluid, if necessary, to bring level to Cold mark.
   d. Bleed system by turning wheels from side to side without hitting stops. Keep fluid level at Cold mark.
   e. Return wheels to center position and continue running engine for 2-3 minutes.
   f. Road test vehicle to ensure steering functions normally and is free of noise.
   g. Check fluid level and ensure level is at Hot mark after system has stabilized at its normal operating temperature.

4-138 ENGINES

1. Disconnect battery ground cable.
2. Disconnect left side torque strut from engine.
3. Separate throttle cable bracket from engine torque strut bracket and position aside. Do not remove cables.
4. Remove engine torque strut bracket.
5. Disconnect inlet and outlet hoses from pump.
6. Remove serpentine belt guard as necessary.
7. Remove belt from pulley, using 1/2 inch breaker bar.
8. Remove rear bracket to pump bolts.
9. **On models except high output,** remove rear bracket to transaxle bolts.
10. **On all models,** remove rear bracket.
11. Remove front bracket to engine bolt, then the pump assembly with bracket.
12. Transfer pulley and bracket as necessary.
13. Reverse procedure to install, them bleed power steering system as follows:
    a. Turn wheels all the way to left.
    b. Add power steering to Cold mark on fluid level indicator.
    c. Start engine and run at fast idle, then add fluid, if necessary, to bring

level to Cold mark.

d. Bleed system by turning wheels from side to side without hitting stops. Keep fluid level at Cold mark.

e. Return wheels to center position and continue running engine for 2-3 minutes.

f. Road test vehicle to ensure steering functions normally and is free of noise.

g. Check fluid level and ensure level is at Hot mark after system has stabilized at its normal operating temperature.

FRONT SUSPENSION & STEERING TIGHTENING SPECIFICATIONS

Year	Component	Torque/ ft. lbs.
1988–90	Adjuster Plug Locknut	50
	Ball Joint Heat Shield Retaining Nuts	54①
	Cylinder End Fittings	20
	Drive Axle Nut	184
	Engine Torque Strut Bracket Bolt	40④
	Front Bracket To Engine Bolt	67④
	Front Strut Closure Nut	82
	Front Strut Piston Shaft Nut	72
	Front Wheel Hub/Bearing To Knuckle Bolts	52
	Heat Shield Screws	54①
	Idle Speed Power Steering Pressure Switch	116①
	Inner Tie Rod To Rack	70
	Intermediate Shaft Pinch Bolt	35
	Lower Ball Joint Nut	②
	Lower Control Arm To Frame Bolts	56
	Pinion Locknut	26
	Pinion Preload	16① ⑤
	Power Steering Cooler Pipe Retaining Screw	71①
	Pump Mounting Bolt	21③
	Pump To Front Or Rear Bracket Bolt	19④
	Rack And Pinion Fittings	20
	Rack And Pinion Mounting Bolts	59
	Stabilizer Bar Insulator Clamp To Frame Nuts	35
	Stabilizer Bar Insulator Clamp To Lower Control Arm Bolt	39
	Strut Mount Cover Plate Nuts	18
	Tie Rod End Jam Nut	50
	Tie Rod End Nuts	40
	Union Fitting To Hydraulic Pump Housing	55
	Valve End Fittings	12

①—Inch lbs.
②—Refer to text.
③—Models with except 4-138 engines.
④—Models with 4-138 engines.
⑤—Maximum.

1985 Buick Riviera, Cadillac Eldorado & Seville & Oldsmobile Toronado

INDEX

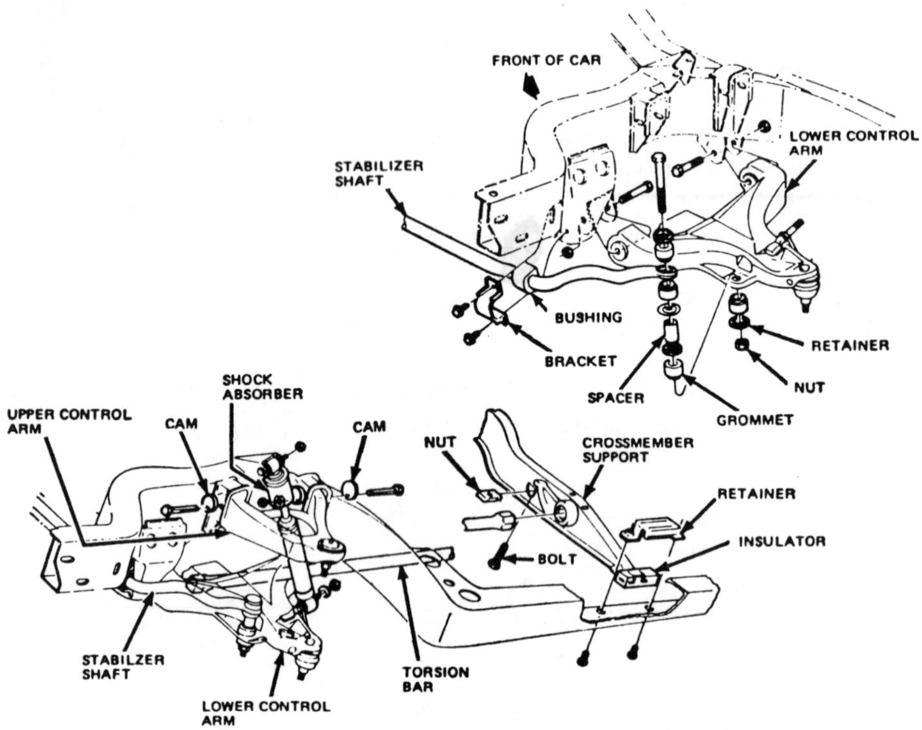

Fig. 1 Front suspension disassembled. Eldorado & Seville

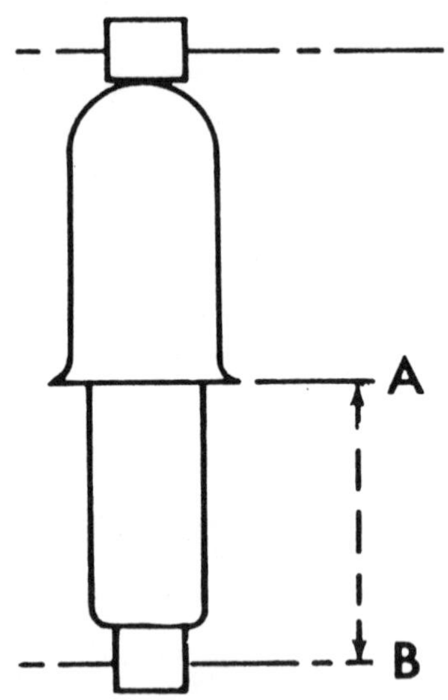

Fig. 2 Checking standing height. Eldorado & Seville

DESCRIPTION

The front suspension consists of two upper and two lower control arms, a stabilizer bar, shock absorbers and a right and left torsion bar, **Fig. 1**. Torsion bars are used instead of the conventional coil springs. The front end of the torsion bar is attached to the lower control arm. The rear of the torsion bar is mounted into an adjustable arm in the torsion bar crossmember. The standing height of the car is controlled by this adjustment.

STANDING HEIGHT

ADJUST

Before standing height adjustment is performed, vehicle should be on a flat, level surface, with front seat all the way back, fuel tank full and tires inflated to proper pressure. Air lines must be loosened at shock absorbers, if equipped with electronic level control, and system must be depressurized.

The standing height must be checked and adjusted if necessary before checking and adjusting front wheel alignment. The standing height is controlled by the adjustment setting of the torsion bar adjusting bolt, **Fig. 1.** Clockwise rotation of the bolt increases standing height: counterclockwise rotation decreases standing height.

To check vehicle height, measure from lower edge of front shock absorber dust tube (A) to center line of lower attachment (B), **Fig. 2.** This dimension between (A) and (B) should be 5$^{1}/_{2}$-6 inches.

WHEEL BEARING INSPECTION

The front wheel bearing is a sealed unit bearing. The bearing cannot be adjusted or repacked. There are darkened areas on the bearing assembly. These darkened areas are from a heat treatment process and do not indicate need for bearing replacement, **Fig. 3.**

To check wheel bearing assembly for looseness, free brake pads from disc or remove calipers. Install two lug nuts to secure disc to bearing. Mount dial indicator as shown in **Fig. 4,** then rock disc and note indicator reading. If looseness exceeds .005 inch, replace hub and bearing assembly.

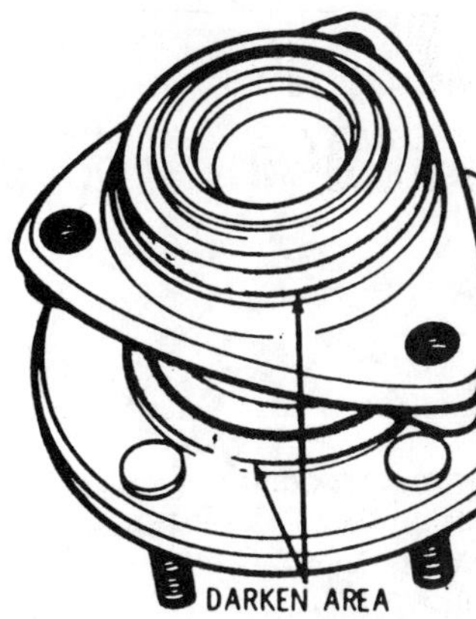

Fig. 3 Sealed wheel bearing assembly. Eldorado & Seville

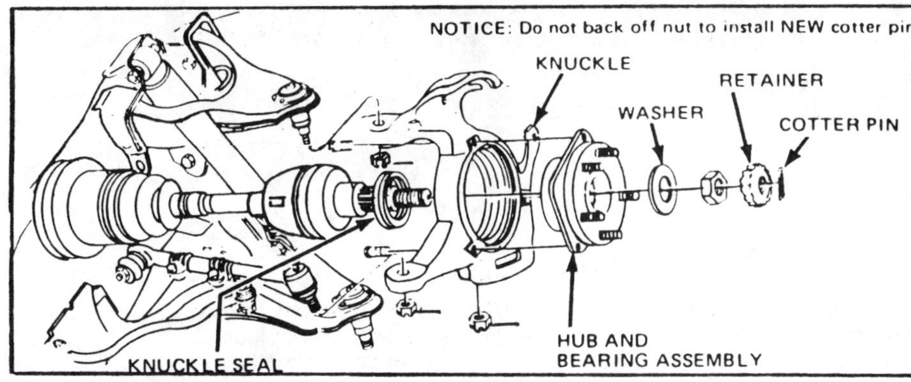

Fig. 5 Wheel bearing & steering knuckle assembly. Eldorado & Seville

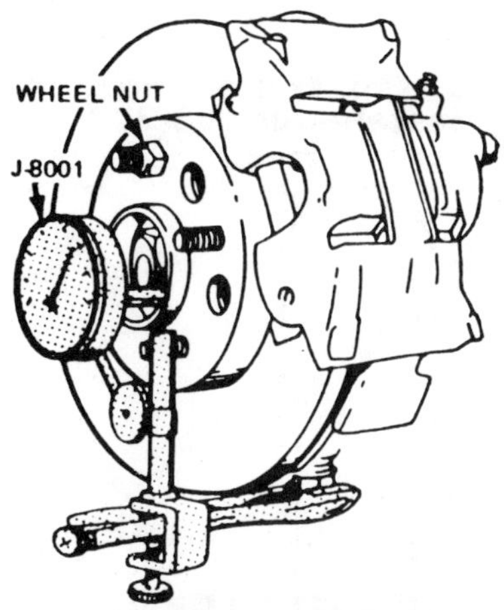

Fig. 4 Checking wheel bearing for looseness. Eldorado & Seville

WHEEL BEARING & STEERING KNUCKLE
REPLACE

1. Raise and support vehicle under lower control arms.
2. Remove drive axle nut and washer, then the wheel and tire assembly.
3. Remove brake hose clip from ball joint, replace nut, then remove the brake caliper from the disc, using a length of wire support caliper on suspension. **Do not allow caliper to hang from brake hose as this could cause damage and premature failure of hose.**
4. Mark hub and disc assembly for alignment during assembly. Remove disc, then strike steering knuckle in area of upper ball joint until upper ball joint is loose. **Use extreme care to prevent striking and damaging brake hose or ball joint seal.**
5. Place a short length of rubber hose over lower control arm torsion bar connector to avoid damage to inboard tri-pot joint seal when hub and knuckle are removed.
6. Using appropriate puller, disconnect tie rod end, upper and lower ball joints, then remove steering knuckle and hub assembly, **Fig. 5.**
7. Reverse procedure to install. Torque the following to specifications:
   a. drive axle nut to hub and bearing.
   b. lower ball joint nut.
   c. upper ball joint nut.
   d. tie rod to steering knuckle nut.
   e. hub and bearing to knuckle bolts.

CHECKING BALL JOINTS FOR WEAR

1. Raise car, then position jack stands under lower control arms as near as possible to each ball joint.
2. Clamp vise grips on end of drive axle, then position a dial indicator so that dial indicator ball rests on vise grip, **Fig. 6.**
3. Place a pry bar between lower control arm and outer race, then pry down on bar. Reading must not exceed $1/8$ inch.

UPPER BALL JOINT
REPLACE

1. Raise vehicle, then support under lower control arms.
2. Remove wheel and tire assembly.
3. Remove cotter pin and nut from upper ball joint stud, then disconnect brake hose clip from stud.
4. Using a hammer and brass drift, disengage ball joint stud from steering knuckle.
5. Place a block of wood between control arm and frame, then drill rivets with a $1/8$ inch drill bit, $1/4$ inch deep from top side of control arm.
6. Drill rivet heads off using a $1/2$ inch drill bit. Do not drill into control arm.
7. Using a hammer and punch, drive rivets out, then remove ball joint.
8. Install ball joint into control arm, then torque nuts and bolts to specifications.

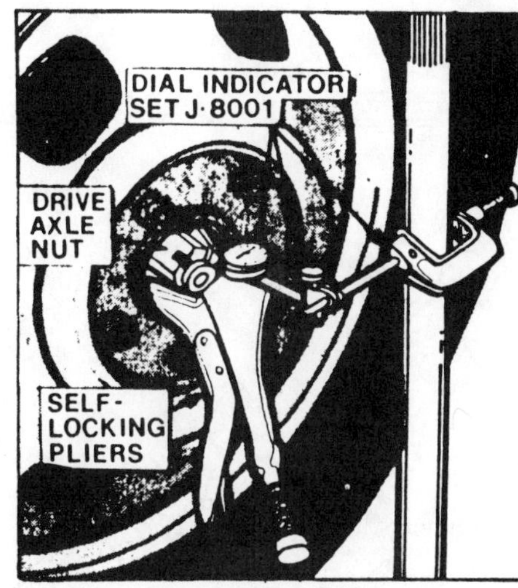

Fig. 6 Checking ball joints for wear. Eldorado & Seville

LOWER BALL JOINT
REPLACE

1. Remove knuckle.
2. Using $1/8$ inch drill bit, drill center of rivets $1/4$ inch deep, then using $1/2$ inch drill bit, drill deep enough to remove rivet heads.
3. Using a hammer and punch, drive rivets out, then remove ball joint.
4. Install ball joint into control arm, then torque nuts and bolts to specifications.

TORSION BAR
REPLACE

1. Raise and support vehicle, then remove parts as shown in **Fig. 7.**
2. Remove torsion bar adjusting screw as shown in **Fig. 7.** Count number of turns when removing screw. When installing, turn screw in the same number of turns to return vehicle to proper height.
3. Slide torsion bar forward in lower control arm until torsion bar clears sup-

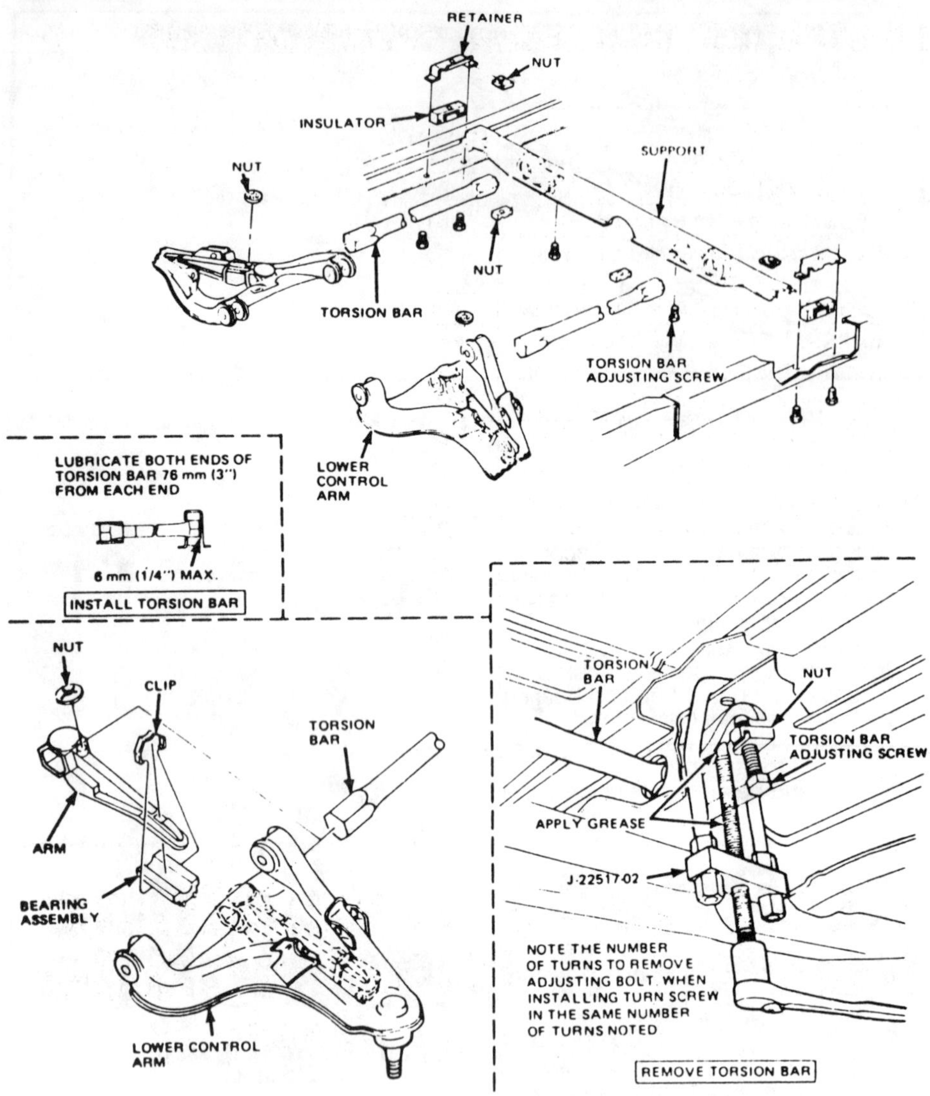

Fig. 7 Torsion bar removal. Eldorado & Seville

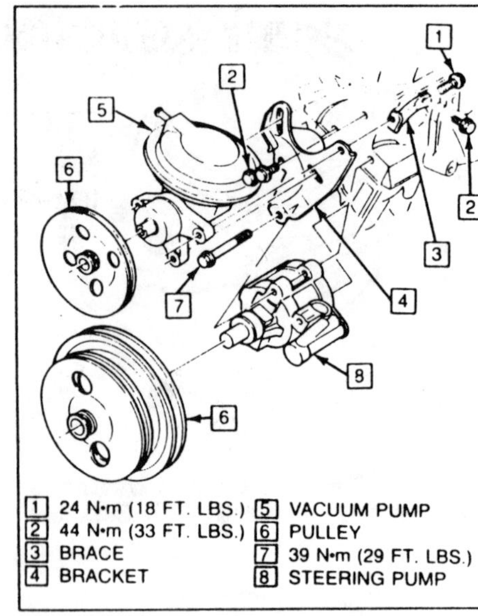

Fig. 8 Power steering pump installation. V8-250

port, pull down on bar, then remove from control arm.

4. Reverse procedure to install. Torque torsion bar support retainer bolts to specifications.

POWER STEERING GEAR REPLACE

1. Disconnect pressure and return lines from gear, then plug all lines and openings.
2. Remove stone shield, if equipped.
3. Remove pinch bolt, then flex coupling from gear.
4. Raise and support vehicle.
5. **On all models, except 1985 Eldorado and Seville,** remove pitman arm nut and lock washer, then using a puller remove pitman arm from steering gear.
6. Remove steering gear to side rail retaining bolts, then steering gear.
7. **On 1985 Eldorado and Seville,** remove pitman arm from steering shaft.
8. Reverse procedure to install. Torque the following items to specifications:
   a. Steering gear to side rail retaining bolts.
   b. Pitman arm to pitman shaft nut.
   c. flex coupling pinch bolt.
   d. steering gear pressure lines.

POWER STEERING PUMP REPLACE

1. Loosen A/C mounting bracket and vacuum pump bracket, then remove belts from pulley.
2. Disconnect pressure and return lines from pump, then plug all openings.
3. Remove power steering pump pulley using tool J-25034, or equivalent.
4. Remove two bolts holding pump to engine block, then the pump, **Fig. 8.**
5. Reverse procedure to install. Press pulley onto pump using tool J-25033, or equivalent, then bleed system.

SERVICE BULLETIN

TIRE SCUFFING ON FULL TURNS

Some vehicles may exhibit a scuffing condition during full turns. The tire may scuff the frame or stabilizer shaft. This condition may be heard and/or felt and leaves a shiny spot at the area of contact.

To repair this condition, the installation of steering stops is required. The installation of these stops increases the turning circle diameter by approximately three feet.

The parts required are as follows: one left hand stop, part No. 1612187; one right hand stop, part No. 1612188; two bolts, No. 25534181; two nuts, No. 9422295.

1. Raise and support vehicle.
2. Remove front tire and wheel assemblies.
3. Remove brake caliper, then position aside. **Do not let caliper hang from brake hose.**
4. Remove brake rotor.
5. Remove bottom dust shield to steering knuckle bolt, then replace with part number 25534181.
6. Install steering stop to knuckle, over new bolt, then secure with nut, part number 9422295. **Torque** nut to 18 ft. lbs. **Pointed end of steering stop should point toward rear of car.**
7. Repeat procedure to assemble.

FRONT SUSPENSION & STEERING TIGHTENING SPECIFICATIONS

Year	Component	Torque/ft. lbs.
1985-89	Ball Joint Nuts	20
	Ball Joint Nut (Upper)	61
	Ball Joint Stud (Lower) To Steering Knuckle	83
	Ball Joint To Control Arm Bolts/Nuts	8
	Ball Stud To Steering Knuckle	61①
	Control Arm To Frame Mounting Bolts	75
	Drive Axle Nut To Hub And Bearing	175
	Flex Coupling Pinch Bolt	30
	Hub And Bearing To Knuckle Bolts	75
	Lower Shock Absorber Mounting Bolts	22
	Pitman Arm To Pitman Shaft Nut	185
	Steering Gear Pressure Lines	20
	Steering Gear To Side Rail Retaining Bolts	70
	Tie Rod Nut	30
	Tie Rod To Steering Knuckle Nut	40
	Torsion Bar Support Retainer Bolts	20

①—May be torqued to a maximum of 100 ft. lbs. Do not back nut off to align cotter pin hole.

1986–90 Buick Riviera, Cadillac Eldorado & Seville & Oldsmobile Toronado & 1988–90 Buick Reatta

INDEX

STABILIZER SHAFT REPLACE

1. Raise vehicle and support with weight of vehicle resting on frame, not the lower control arms.
2. Remove right front wheel and tire.
3. Remove left and right insulators, retainers, spacers and bolts, **Fig. 1.**
4. Remove left and right bracket bolts, brackets and insulators.
5. Disconnect exhaust pipe from rear manifold, raise pipe to gain clearance, then withdraw stabilizer shaft.
6. Reverse procedure to install.

BALL JOINT REPLACE

Care must be taken not to overextend driveshaft Tri-Pot joints when replacing suspension components. Overextending joint could cause separation of internal

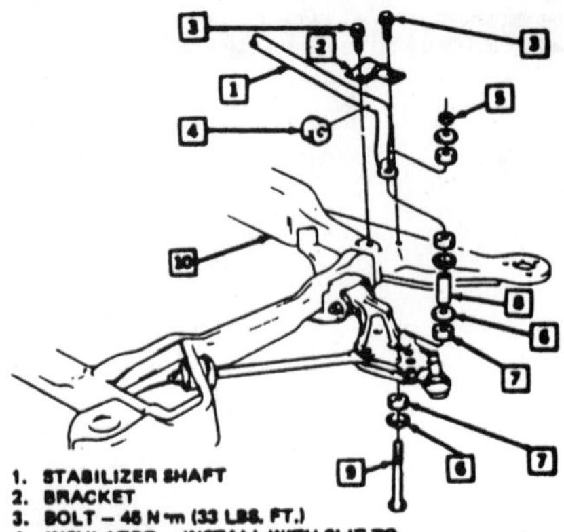

1. STABILIZER SHAFT
2. BRACKET
3. BOLT – 45 N·m (33 LBS. FT.)
4. INSULATOR – INSTALL WITH SLIT TO REAR OF CAR
5. NUT 17 N·m (13 LBS. FT.) OBTAIN TORQUE BY RUNNING NUT TO UNTHREADED PORTION OF BOLT
6. RETAINER
7. INSULATOR
8. SPACER
9. BOLT – INSTALL IN DIRECTION SHOWN
10. FRAME

Fig. 1 Stabilizer shaft installation

joint components, resulting in failure of the joint.

1. Raise vehicle and support frame with suitable stands.
2. Remove wheel and tire assembly.
3. Remove insulator, retainers, spacer and bolt securing stabilizer shaft to control arm, **Fig. 1**.
4. Cut inner tabs from boot protector J-34754 and install protector over drive axle outer joint.
5. Remove nut and cotter pin securing ball joint to steering knuckle.
6. Separate ball joint from knuckle using separator J-35315 or equivalent.
7. Center punch rivets securing ball joint to control arm, then drill out rivets starting with 1/4 inch drill bit and finishing with 1/2 inch bit.
8. Install service replacement joint in control arm, Insert bolts from underside of arm, and torque to specifications.
9. Insert ball stud into steering knuckle and install nut.
10. Torque ball joint nut to 84 inch lbs. using 90° adapter J-35551 or equivalent.
11. Tighten nut an additional 180° on Eldorado and Seville models, or 120° on Reatta, Riviera, Toronado and Trofeo models, noting torque reading, then tighten nut up to 60° additional to align holes and install cotter pin. **When tightening nut, minimum torque (as listed in "Tightening Specifications") must be obtained. If minimum torque cannot be obtained, check for stripped threads and repair as needed. If threads are satisfactory, replace ball joint and steering knuckle.**
12. Reverse remaining procedure to complete installation.

CONTROL ARM
REPLACE

1. Remove ball joint from steering knuckle as outlined in steps 1-6 of "Ball Joint, Replace."

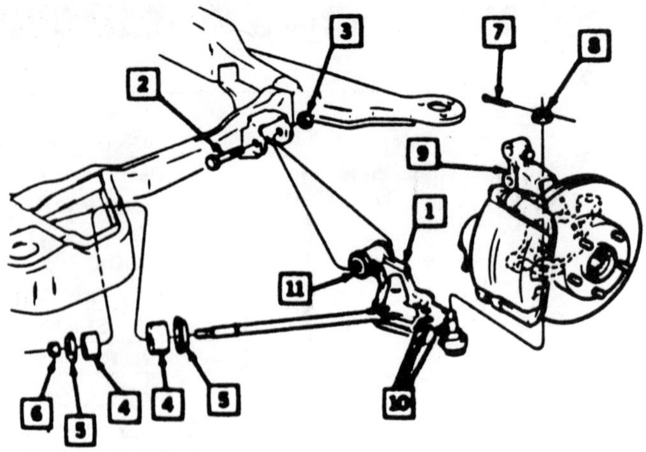

1. CONTROL ARM
2. BOLT – 140 N·m (100 LBS. FT.) TIGHTEN WITH CAR AT PROPER TRIM HEIGHT
3. NUT – 123 N·m (91 LBS. FT.) TIGHTEN WITH CAR AT PROPER TRIM HEIGHT
4. INSULATOR
5. RETAINER
6. NUT – 70 N·m (52 LBS. FT.)
7. PIN
8. NUT – TIGHTEN NUT TO 10 N·m (88 LBS. IN.). CONTINUE TIGHTENING BY ROTATING NUT AN ADDITIONAL 120°, DURING WHICH A MINIMUM TORQUE OF 50 N·m (37 LBS. FT.) MUST BE OBTAINED. INSTALL COTTER PIN
9. KNUCKLE
10. BALL JOINT ATTACHMENT RIVETS
11. BUSHING

Fig. 2 Lower control arm installation

2. Remove control arm bushing bolt and front nut, retainer and insulator, **Fig. 2**, then the control arm.
3. Install control arm on frame.
4. Install control arm bushing bolt and nut, retainer and insulator but do not tighten bolt.
5. Install lower ball joint in steering knuckle as outlined.
6. Reverse remaining procedure to complete installation. Torque bolts to specifications as shown in **Figs. 1 and 2.**

LOWER CONTROL ARM BUSHINGS
REPLACE

When replacing lower control arm bushings, refer to **Fig. 3.**

HUB & BEARING ASSEMBLY
REPLACE

1. Siphon 2/3 of the brake fluid from master cylinder, raise vehicle and support frame with suitable stands.
2. Remove wheel and tire assembly.
3. Insert suitable drift through rotor and remove hub nut and washer.
4. Remove caliper boots, lift caliper off rotor and bracket and secure caliper aside.
5. Remove bolts securing caliper bracket, bracket and the brake rotor. **Prevailing torque bolts used to secure mounting bracket cannot be reused. If bolts become loose or are removed, bolts must be replaced.**
6. Mount tool J-28733 or equivalent on hub, **Fig. 4,** then tighten screw just enough to separate drive axle from hub.
7. Remove hub retaining bolts and the hub and bearing assembly, **Fig. 5.**
8. If seal is being replaced, drive seal toward engine, then cut seal off drive axle.
9. Lubricate seal lip and install using suitable driver, install O-ring and fill cavity between seal and bearing with grease.
10. Install hub and bearing assembly over drive axle and torque bolts to specifications.
11. Install rotor and caliper mounting bracket and torque new bracket bolts to specifications.
12. Install caliper ensuring insulators are properly positioned, coat mounting

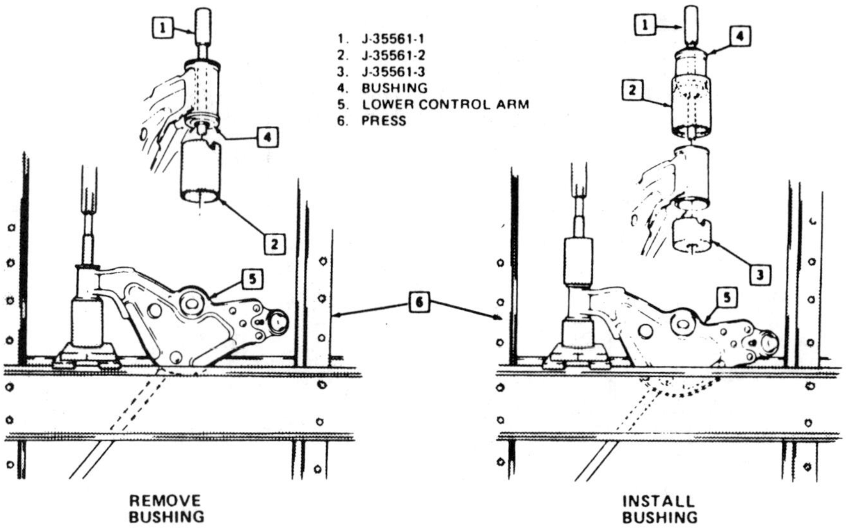

Fig. 3 Replacing lower control arm bushings

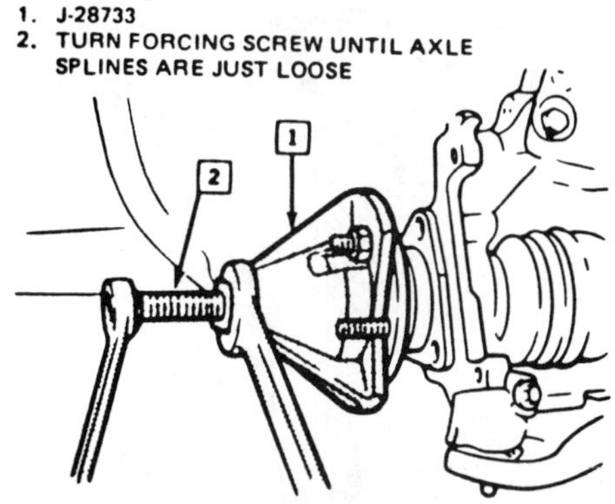

Fig. 4 Separating drive axle from hub

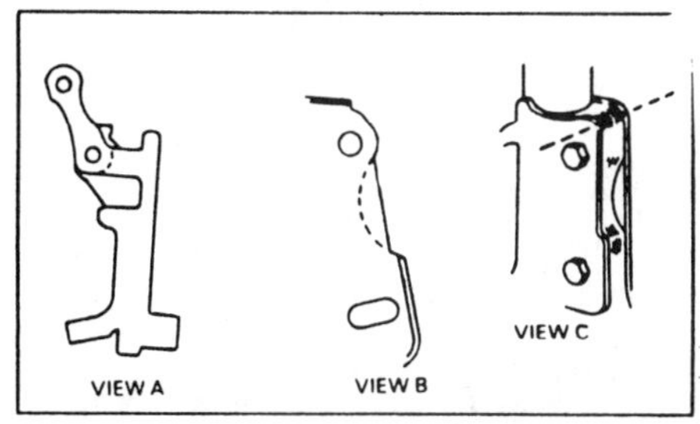

Fig. 7 Marking steering knuckle & strut alignment

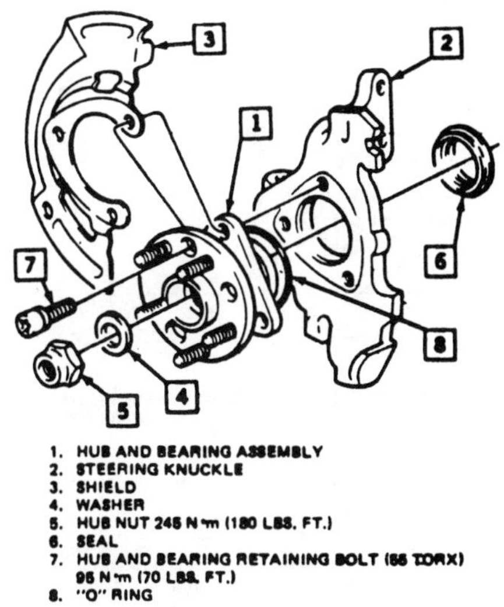

Fig. 5 Hub & bearing assembly installation

bolt shafts with silicone grease and torque to specifications.

13. Install hub nut and washer, insert drift through rotor and tighten nut securely.
14. Remove drift, install wheel and tire assembly and lower vehicle, then torque hub nut to specifications.

STEERING KNUCKLE REPLACE

1. Remove hub assembly as outlined.
2. Remove nut securing tie rod, then separate tie rod from knuckle.
3. Cut inner tabs from boot protector J-34754 and install protector over drive axle outer joint.
4. Remove cotter pin and nut securing lower ball joint and separate ball stud from steering knuckle.
5. Remove bolts securing strut to steering knuckle, **Fig. 6**, and the steering

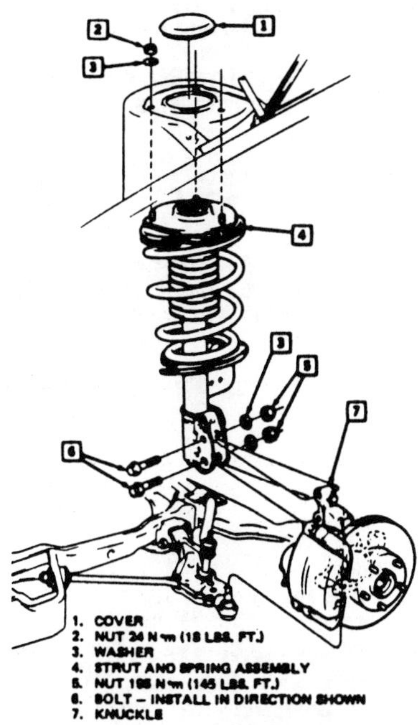

Fig. 6 Steering knuckle & strut installation

knuckle.

6. Reverse procedure to install.

STRUT ASSEMBLY REPLACE

Care must be taken not to overextend driveshaft Tri-Pot joints when replacing suspension components. Overextending joint could cause separation of internal joint components, resulting in failure of the joint.

1. Remove nuts securing top of strut assembly to body, **Fig. 6**.
2. Raise vehicle and support frame with suitable stands, then remove wheel and tire.
3. Scribe reference marks between strut and steering knuckle as follows:
   a. Scribe strut along lower outboard strut radius (A), **Fig. 7.**
   b. Scribe strut flange on inboard side along curve of knuckle (B).
   c. Scribe mark along strut/knuckle interface (C).
4. Disconnect brake line bracket from strut.
5. Remove bolts securing strut to steering knuckle, support knuckle with wire and remove strut.
6. Reverse procedure to install, aligning scribe marks made during disassembly. Torque nuts and bolts to specifications, then check alignment.

STRUT SERVICE

1. Place strut assembly in strut compressor J-34013, **Fig. 8.**
2. Turn compressor forcing screw until spring compresses slightly.
3. Keep dampener shaft from turning with socket and remove 24 mm nut on top of dampener shaft.
4. Using tool J-34013-38 or equivalent, guide dampener shaft out of assembly.
5. Loosen compressor screw while guiding dampener shaft out of assembly, continually loosening compressor screw till strut dampener and spring can be removed. **Be careful to avoid chipping or cracking the spring coating when handling the front suspension coil spring.**

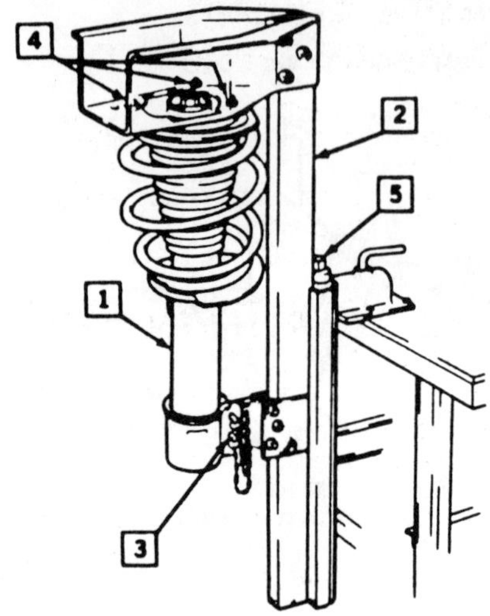

1. STRUT ASSEMBLY
2. STRUT COMPRESSOR J-34013
3. INSTALL LOCKING PINS THROUGH STRUT ASSEMBLY
4. TIGHTEN NUTS TILL FLUSH WITH STRUT COMPRESSOR
5. COMPRESSOR FORCING SCREW

Fig. 8 Installing strut assembly in strut compressor

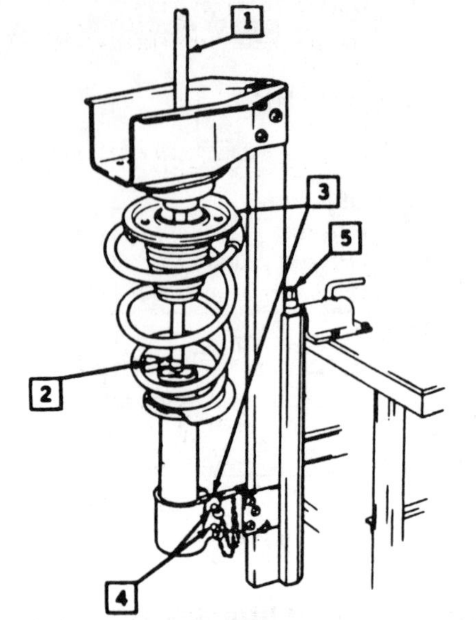

1. ROD J-34013-38 INSTALLED
2. CLAMP J-34013-20 INSTALLED
3. FLAT ON SPRING SEAT MUST FACE SAME DIRECTION AS STEERING KNUCKLE FLANGE
4. BOTH LOCKING PINS INSTALLED
5. COMPRESSOR FORCING SCREW

Fig. 9 Aligning flat on upper spring seat

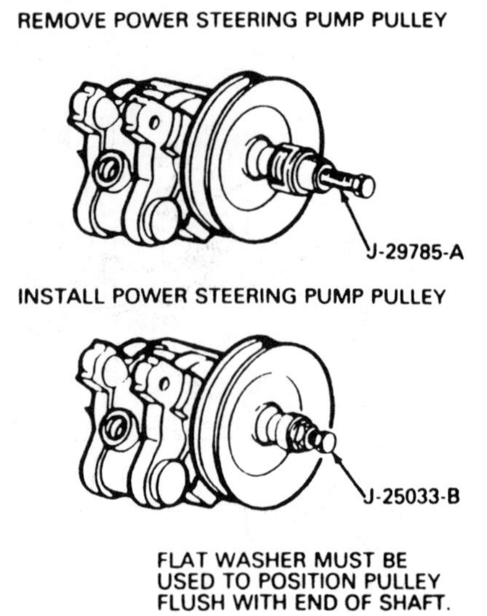

Fig. 10 Power steering pump pulley removal. Eldorado & Seville

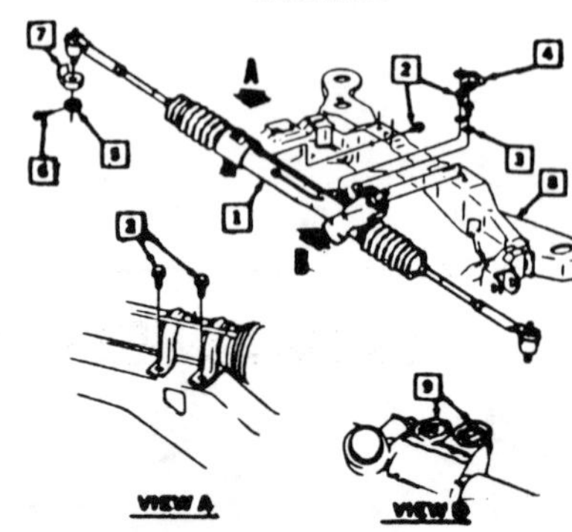

1. STEERING GEAR
2. BOLT – 68 N·m (50 LBS. FT.) AFTER SECOND REUSE OF BOLT, LOCKTITE THREAD LOCKING KIT NO. 1052624 MUST BE USED.
3. WASHER
4. RETAINER
5. NUT – TIGHTEN NUT TO 10 N·m (7 LBS. FT.) THEN TIGHTEN NUT AN ADDITIONAL 120 DEGREES (2 FLATS) DURING WHICH A MINIMUM TORQUE OF 45 N·m (33 LBS. FT.) IS TO BE OBTAINED. IF 45 N·m (33 LBS. FT.) IS NOT OBTAINED, INSPECT FOR STRIPPED THREADS. IF THREADS ARE SATISFACTORY, REPLACE KNUCKLE.
6. PIN
7. KNUCKLE
8. FRAME
9. RTV SEALER AROUND INSERTS

Fig. 13 Power steering gear installation. Reatta, Riviera, Toronado & Trofeo

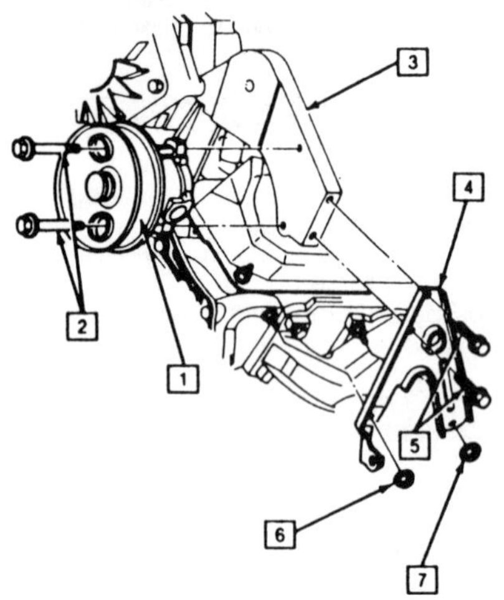

[1] PUMP
[2] PUMP MOUNTING BOLTS (25 N·m/18 FT. LBS.)
[3] PUMP AND TENSIONER BRACKET
[4] BRACE
[5] BOLTS (45 N·m/33 FT. LBS.)
[6] NUT (45 N·m/33 FT. LBS.)
[7] NUT (25 N·m/18 FT. LBS.)

Fig. 11 Power steering pump installation. Eldorado & Seville

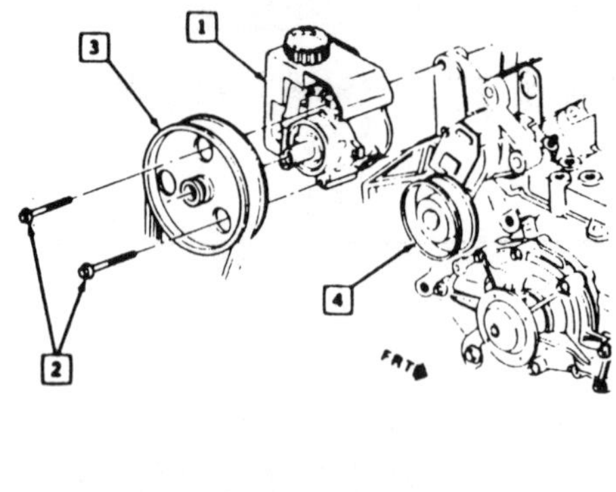

1. POWER STEERING PUMP
2. BOLT – 27 N·m (20 LBS. FT.)
3. PULLEY
4. BELT TENSIONER ASSEMBLY

Fig. 12 Power steering pump installation. Reatta, Riviera, Toronado & Trofeo

6. Install strut dampener in strut compressor J-34013 with clamp J-34013-20 clamped on dampener shaft.
7. Install spring over strut in correct position and move assembly upright in strut compressor and install upper locking pin. **Flat on upper spring seat must face out from centerline of vehicle, Fig. 9, or when mounted in strut compressor spring seat faces same direction as steering knuckle mounting flange.**
8. Install rod J-34013-38 or equivalent into strut assembly to guide dampener shaft on reassembly of strut.
9. Start turning compressor screw clockwise on J-34013 while guiding J-34013-38 which will center dampener shaft in the assembly.
10. Continue turning compressor screw until dampener shaft threads are visible through top of strut assembly.
11. Install washer and nut, then remove J-34013-20 from dampener shaft.
12. Hold dampener shaft with socket and, torque dampener shaft nut to specifications using a 24 mm flare nut crowsfoot.
13. Remove three nuts securing strut assembly to strut compressor, the two locking pins at bottom of strut compressor and the strut assembly.

POWER STEERING PUMP REPLACE

ELDORADO & SEVILLE

1. Remove serpentine drive belt.
2. Remove pump pulley using suitable puller, **Fig. 10.**
3. Disconnect pressure and return lines from pump, noting position for installation.
4. Remove 2 pump mounting bolts, **Fig. 11,** and the pump.
5. Reverse procedure to install.

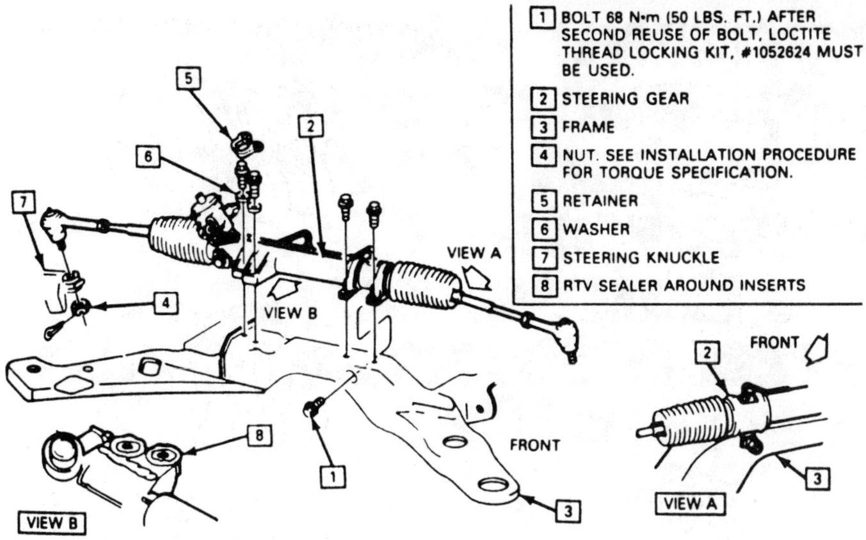

Fig. 14 Power steering gear installation. Eldorado & Seville

REATTA, RIVIERA, TORONADO & TROFEO

1. Remove drive belt.
2. Disconnect pressure and return lines from pump.
3. Remove pump mounting bolts, **Fig. 12,** and the pump.
4. Reverse procedure to install.

POWER STEERING GEAR REPLACE

1. Raise vehicle and support vehicle on frame contact hoist, then remove front wheels.
2. Remove pinch bolt and disconnect intermediate shaft from steering gear.
3. Disconnect tie rods from steering knuckles.
4. Remove power steering line retainer, then disconnect pressure and return lines from steering gear.
5. Remove steering gear mounting bolts, **Figs. 13 and 14,** then withdraw gear through wheel opening.
6. Reverse procedure to install. Tighten tie rod nuts to specifications. After tightening nut as specified, continue tightening nut to align cotter pin holes, then install new cotter pins.

FRONT SUSPENSION & STEERING TIGHTENING SPECIFICATIONS

Year	Component	Torque/ft. lbs.
1986–90	Ball Joint To Control Arm	50
	Ball Joint To Knuckle	①
	Brake Reaction Rod To Frame	58
	Caliper Attaching Bolts	63
	Caliper Mounting Bracket	83
	Control Arm Bushing Nuts	103
	Dampener Shaft	44
	Hub To Drive Axle	180
	Hub & Bearing To Knuckle	70
	Stabilizer Bracket To Frame	35
	Stabilizer Link Nut	13
	Strut Mount To Body	18
	Strut To Knuckle	136
	Strut To Strut Top Mount	②
	Tie Rod Nuts	③

①—Tighten to 37 ft. lbs. on Reatta, Riviera, Toronado and Trofeo models and to 48 ft. lbs. on Eldorado and Seville models.
②—Tighten old design (flanged) nut to 75 ft. lbs. and new design (non-flanged) nut to 55 ft. lbs.
③—Tighten to 7.5 ft. lbs., then an additional ⅓turn.

Buick Skyhawk, Skylark (1986–90), Somerset & Somerset Regal, Cadillac Cimarron, Chevrolet Cavalier, Oldsmobile Calais, Cutlass Calais & Firenza & Pontiac Grand Am & Sunbird

INDEX

DESCRIPTION

The front suspension, **Fig. 1,** on these vehicles is of the MacPherson strut design. The lower control arms pivot from the lower side rails through rubber bushings. The upper end of the strut is isolated by a rubber mount incorporating a non-serviceable bearing for wheel turning. The tie rods connect to the steering arm on the strut, below the spring seat. The lower end of the steering knuckle pivots on a ball stud which is retained to the lower control arm by rivets and is secured to the steering knuckle with a nut and cotter pin. The sealed wheel bearings are integral with the hub and are serviced as an assembly.

WHEEL BEARING REPLACE

REMOVAL

1985–88

1. Loosen hub nut with vehicle on ground.
2. Raise and support vehicle, then remove front wheel.
3. Install drive axle protective boot cover.
4. Remove hub nut.
5. Remove brake caliper from support and suspend caliper from flame with a length of wire. Do not suspend caliper by brake hose.
6. Remove three hub and bearing attaching bolts. If the old bearing is being reinstalled, mark attaching bolts and corresponding holes for reinstallation, **Fig. 2.**
7. Using tool J-28733 or equivalent, remove bearing from steering knuckle, **Fig. 3. If excessive corrosion is present, ensure that bearing is loose in the knuckle before using puller tool.**
8. If installing new bearing, replace steering knuckle seal. **Do not move drive axle until hub nut is installed and torqued to specifications.**

1989–90

1. Raise and support vehicle.
2. Remove tire and wheel assembly, then install modified outer seal protector No. J 34754 or equivalent, **Fig. 4.**
3. Insert a drift punch through the rotor, **Fig. 5,** then remove axle shaft nut and washer.
4. Using a plastic or rubber mallet, strike end of axle shaft to disengage axle from hub and bearing. Shaft nut can be partially installed to protect threads.
5. Move axle shaft inward, then remove caliper attaching bolts and support caliper.
6. Remove brake rotor, then hub and bearing assembly attaching bolts.
7. Remove hub and bearing assembly.

INSTALLATION

1985–88

1. Clean and inspect bearing mating surfaces and steering knuckle bore for dirt, nicks and burrs.
2. If installing new steering knuckle seal, apply grease to seal and knuckle bore, then press seal into steering knuckle.
3. Push bearing onto axle shaft and install two hub to steering knuckle attaching bolts. Install a longer bolt into third mounting hole and through hub cutout. Install hub to axle retaining nut and torque to specification, **Fig. 6.**
4. Remove long bolt and replace with original bolt. Torque hub to steering knuckle attaching bolts to specification.
5. Install brake caliper and wheel assembly.
6. Lower vehicle and torque hub to axle nut to specification.

1989–90

1. Install hub and bearing assembly, torque attaching bolts to specification.
2. Install hub and bearing seal, then the brake rotor.
3. Install caliper, torque attaching bolts to specification.
4. Move axle shaft outward, then insert drift punch through rotor, **Fig. 5.**
5. Install washer and new shaft nut, torque shaft nut to specification.
6. Remove drift punch and seal protector.
7. Install tire and wheel assembly, then lower vehicle.

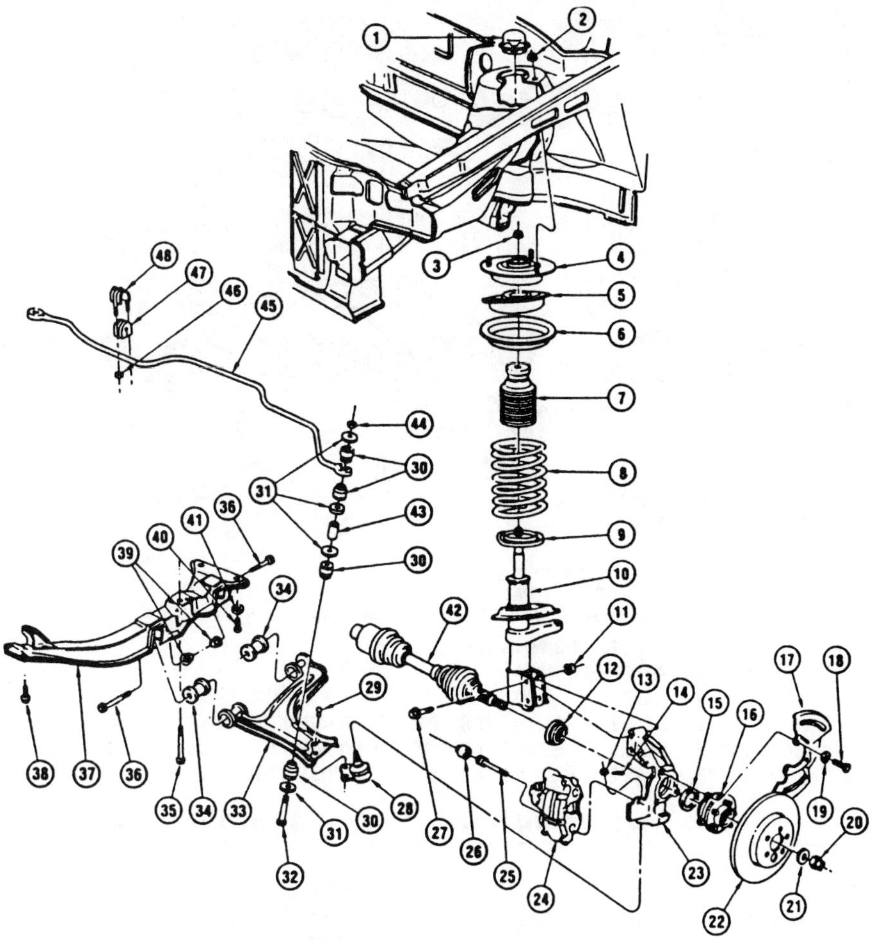

1. COVER
2. NUT
3. NUT
4. MOUNT
5. SEAT
6. INSULATOR
7. BUMPER
8. SPRING
9. INSULATOR
10. STRUT
11. NUT
12. SEAL
13. NUT
14. PIN
15. SEAL
16. HUB
17. SHIELD
18. BOLT
19. WASHER
20. NUT
21. WASHER
22. ROTOR
23. KNUCKLE
24. CALIPER
25. BOLT
26. COVER
27. BOLT
28. BALL JOINT
29. RIVET
30. GROMMET
31. WASHER
32. BOLT
33. ARM
34. BUSHING
35. BOLT
36. BOLT
37. SUPPORT
38. BOLT
39. NUT
40. BOLT
41. WASHER
42. AXLE
43. SPACER
44. NUT
45. SHAFT
46. NUT
47. INSULATOR
48. CLAMP

Fig. 1 Exploded view of front suspension

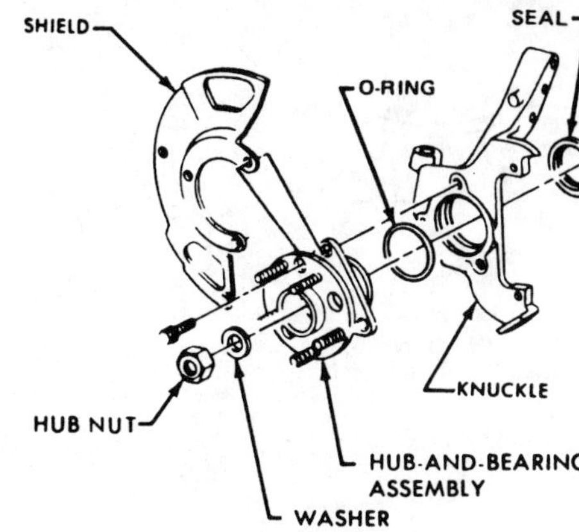

Fig. 2 Front hub & wheel bearing assembly

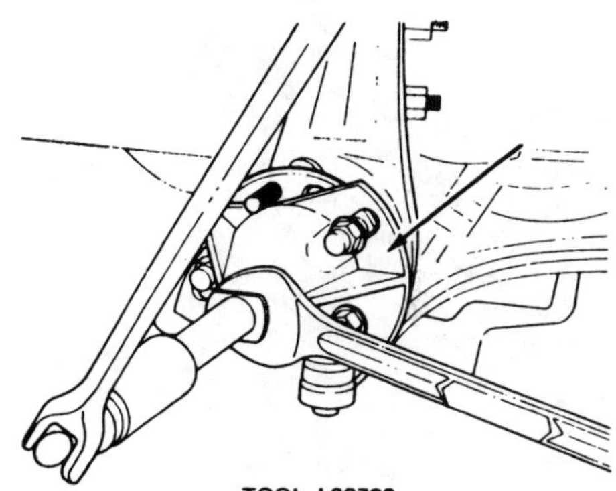

Fig. 3 Front hub & bearing assembly removal

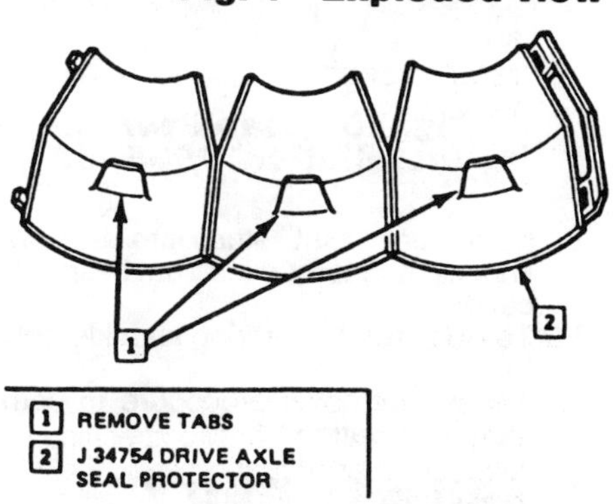

Fig. 4 Modified outer seal protector

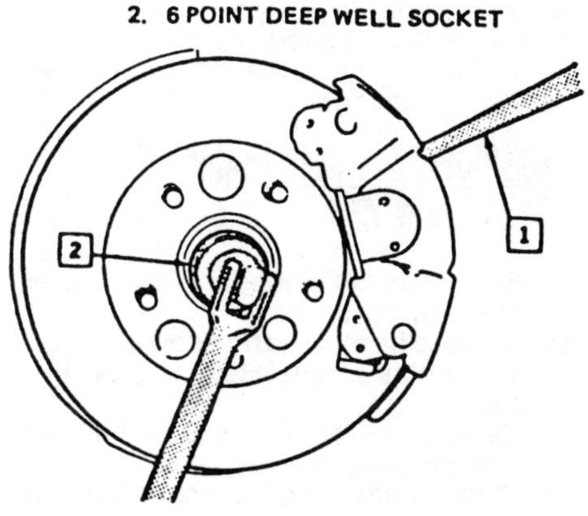

Fig. 5 Removing and installing shaft nut. 1989–90

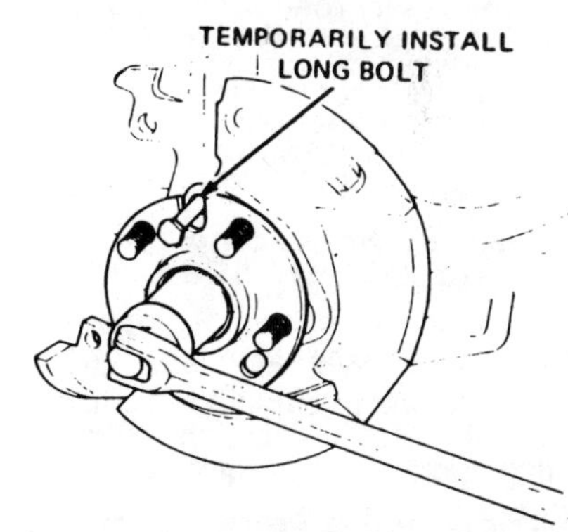

Fig. 6 Hub nut installation

LOWER BALL JOINT REPLACE

1. Raise and support vehicle, then remove wheel and tire.
2. Locate center of rivet body and mark with a center punch.
3. Using a 1/8 inch drill, drill pilot holes completely through the rivets. Using a 1/2 inch drill, drill final holes through rivets to ensure fitting of new ball joint.
4. Remove ball joint stud retaining nut, then, using tool J-29330, separate ball joint from steering knuckle. Remove ball joint from lower control arm.
5. Assemble new ball joint to lower control arm with bolts provided in service package, **Fig. 7.** Torque bolts to spec-

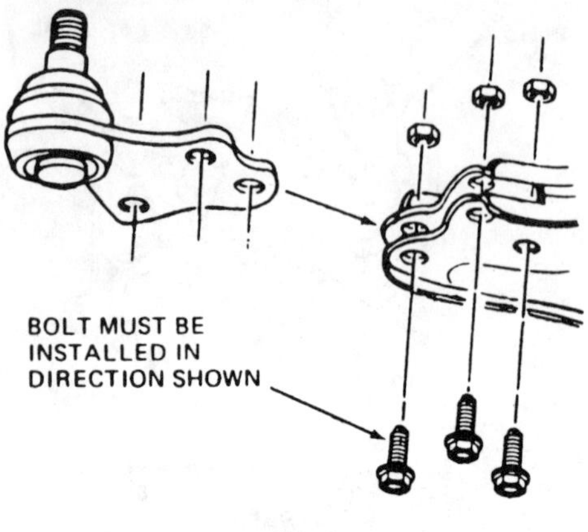

Fig. 7 Assembling lower ball joint to lower control arm

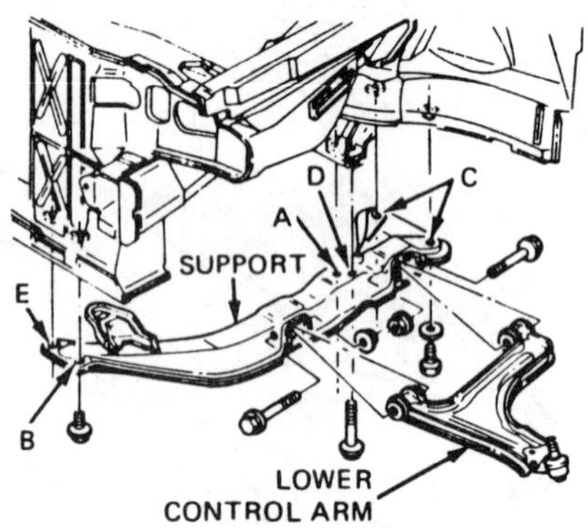

1. LOOSELY INSTALL CENTER BOLT INTO HOLE (A).
2. LOOSELY INSTALL TIE BAR BOLT INTO OUTBOARD HOLE (B).
3. INSTALL BOTH REAR BOLTS INTO HOLES (C) TORQUE REAR BOLTS.
4. INSTALL BOLT INTO CENTER HOLE(D), THEN TORQUE.
5. TORQUE BOLT IN HOLE (A).
6. INSTALL BOLT INTO FRONT HOLE (E), THEN TORQUE.
7. TORQUE BOLT IN HOLE (B).

Fig. 8 Replacing lower control arm. 1985–87

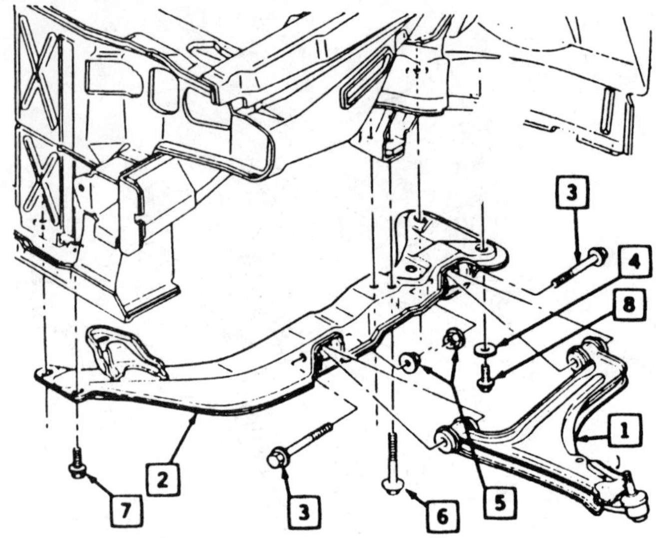

1. CONTROL ARM
2. SUSPENSION SUPPORT
3. BOLT – 83 N·m (61 LBS. FT.)
4. WASHER
5. NUT
6. BOLT – 90 N·m (66 LBS. FT.) TIGHTEN FIRST
7. BOLT – 88 N·m (65 LBS. FT.) TIGHTEN SECOND
8. BOLT – 88 N·m (65 LBS. FT.) TIGHTEN THIRD

Fig. 9 Replacing lower control arm. 1988–90

ification.

6. Insert ball joint stud into steering knuckle and torque nut to 55 ft. lbs.
7. Install wheel and tire, check toe setting and adjust as required.

LOWER CONTROL ARM & BUSHING
REPLACE

1. Raise and support vehicle, then remove wheel and tire.
2. Disconnect stabilizer bar at lower control arm and control arm support.
3. Using tool J-29330, separate ball joint from steering knuckle.
4. Remove control arm support to chassis retaining bolts and remove control arm support and control arm as an assembly.
5. Separate control arm from support, then using tool Nos. J29792-1 and J29792-2 or equivalent, remove bushings from control arm.
6. Lubricate new bushings and install into control arm using tool Nos. J29792-1, J29792-2 and J29792-3 or equivalent.
7. Attach lower control arm to control arm support. On 1985-86 models torque pivot bolts to 59 ft. lbs. On 1987-90 models, torque pivot bolts to 61 ft. lbs.
8. Install control arm support to chassis. On 1985-86 models, tighten attaching bolts in sequence shown in **Fig. 8.** On 1985-87 models, tighten attaching bolts to 59 ft. lbs. On 1988-90 models, refer to **Fig. 9,** for attaching bolt torques.
9. Reverse procedure to complete installation. Check toe setting and adjust as required.

STEERING KNUCKLE
REPLACE

1. Raise and support vehicle, then remove wheel and tire.
2. Remove front hub and bearing as outlined under "Wheel Bearing, Replace" procedure.
3. Using tool J-29330, separate ball joint from steering knuckle.
4. Remove strut to steering knuckle attaching bolts, then disconnect strut from steering knuckle.
5. Assemble strut to new steering knuckle and install attaching bolts finger tight.

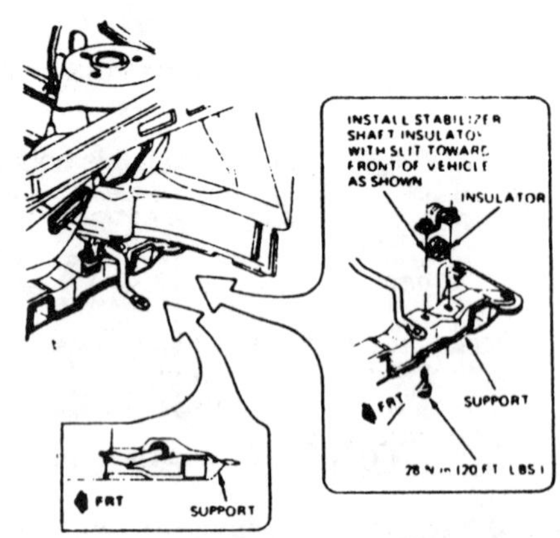

Fig. 10 Stabilizer bar installation. 1985–87

6. Insert ball joint stud into steering knuckle and torque stud nut to specification.
7. Torque strut to steering knuckle bolts to specification.
8. Reverse removal procedure to complete installation.

STABILIZER BAR & BUSHINGS
REPLACE

1986–87

On some 1985-86 Cimarron models, a squawking noise may occur when subjected to significant suspension motions. This condition may be caused by stabilizer

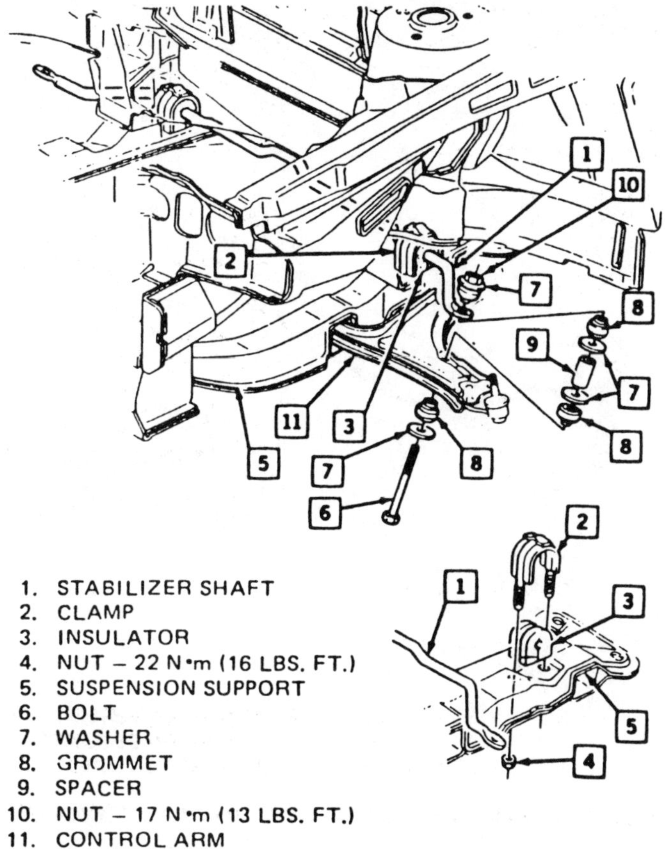

Fig. 11 Stabilizer bar installation. 1988–90

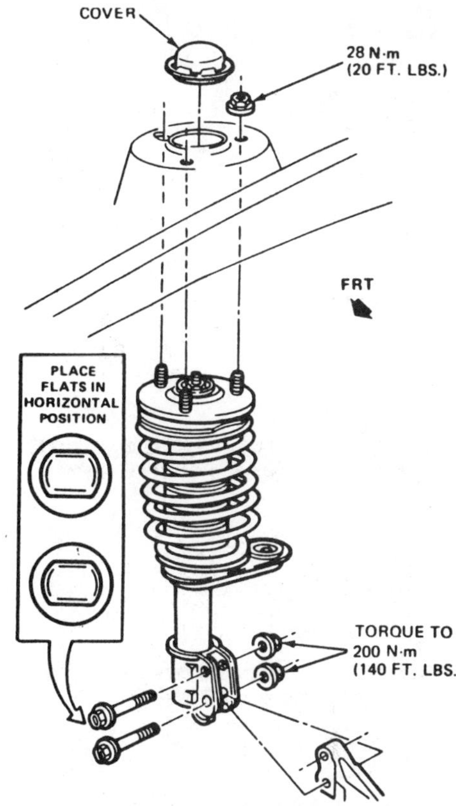

Fig. 12 Installing strut assembly

bushings. Vehicles experiencing this noise condition can be repaired by installing new insulators. Use insulators 22532212 for vehicles with 2.0L 4 cylinder engines, and 22532213 for vehicles with 2.8L V6 engines.

1. Raise and support vehicle, allowing control arms to hang freely.
2. Disconnect stabilizer bar from control arms, **Figs. 8 and 9.**
3. Remove stabilizer bar support assemblies.
4. Loosen front and remove rear and center bolts from suspension support assembly. Lower support enough to allow stabilizer bar removal.
5. Remove stabilizer bar with insulators.
6. Reverse procedure to install. Refer to "Front Suspension Tightening Specifications," for torque specifications.

1988–90

1. Raise and support vehicle, allowing control arms to hang freely.
2. Disconnect stabilizer bar at control arms and control arm supports.
3. Remove rear and center control arm support bolts, lower support assembly in a manner which will allow front suspension to hang free.
4. Remove left front wheel and tire assembly.
5. Disconnect stabilizer bar from control arms, **Figs. 10 and 11.**
6. Remove stabilizer bar support assemblies.
7. Loosen front and remove rear and center bolts from suspension support assembly. Lower support enough to allow stabilizer bar removal.
8. Remove stabilizer bar with insulators.
9. Reverse procedure to install. Torque suspension support assembly in a manner which will allow front suspension to hang free.

STRUT ASSEMBLY REPLACE

1. Raise hood and remove strut protective cap and three strut to body attaching nuts.
2. Raise and support vehicle, allowing suspension to hang freely.
3. Remove wheel and tire, then install drive axle protective cover, J-28712.
4. Using tool J-24319, disconnect tie rod from strut assembly.
5. Remove strut to steering knuckle attaching bolts, then remove strut from vehicle.
6. Reverse procedure to install. Position flats of strut mounting bolts as shown in **Fig. 12,** then torque to specification.

STRUT ASSEMBLY SERVICE

1985–87

Disassemble

1. Position strut compressor tool J-26584 in a suitable vise.
2. Place strut assembly in compressor tool and install bottom adapter J-26584-86, making sure that adapter captures strut and locating pins are fully engaged.
3. Rotate strut assembly to align top mounting assembly lip with strut compressor support notch.
4. Insert J-26584-88 top adapters between top mounting assembly and top spring seat. Position top adapters so that split line is perpendicular to spring compressor.
5. Rotate compressor forcing screw clockwise until top support flange contacts top adapters. Continue rotating forcing screw until strut spring is compressed approximately 1/2 inch. Use care not to bottom spring or strut damper rod.
6. Remove damper top nut, then remove strut mounting assembly from damper shaft.
7. Rotate forcing screw counterclockwise to relieve spring tension, then remove spring from strut assembly.

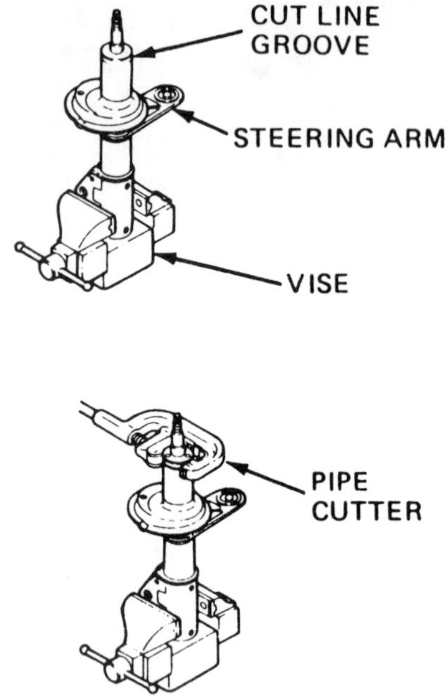

Fig. 13 Strut reservoir tube removal. 1985–87

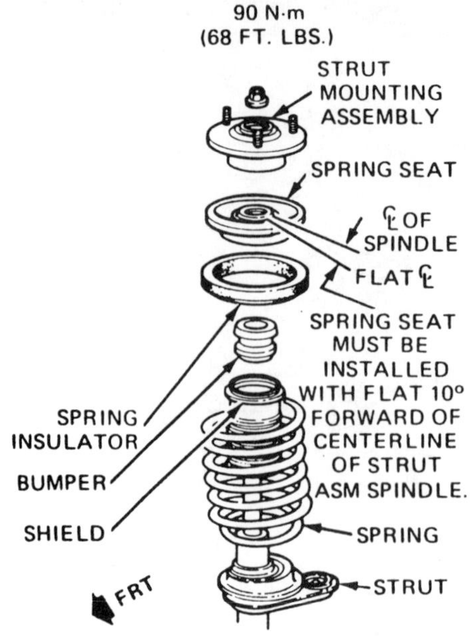

Fig. 14 Strut coil spring & upper mounting installation. 1985–87

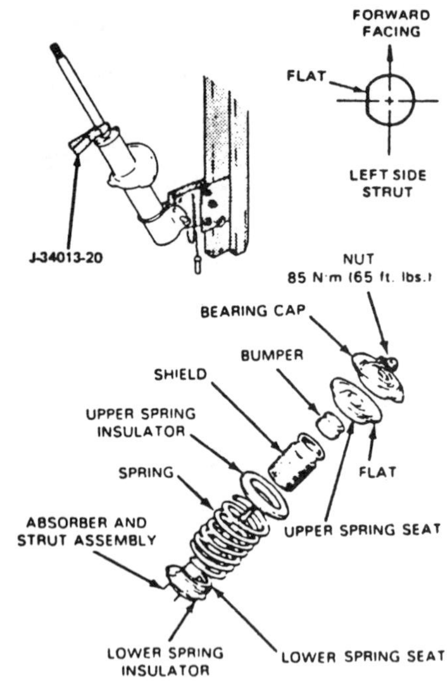

Fig. 15 Assembling strut unit. 1988–90

Strut Cartridge Replacement

1. Clamp strut assembly firmly in a suitable vise. Do not over tighten.
2. Install pipe cutter at cut line groove located at top of strut damper. Cut around groove until reservoir tube is cut completely through, **Fig. 13.**
3. Remove and discard end cap, cylinder and piston rod assembly.
4. Place flaring cup tool, provided in service package, onto open end of reservoir tube and strike with hammer until cup flat outer surface rests on reservoir tube.
5. Install strut carriage into reservoir tube and align grooves on cartridge base with pads at bottom of reservoir tube.
6. Using guiding tool J-34013 or equivalent, torque cartridge retaining hex nut to specification.

Assemble

1. Perform steps 1 and 2 as outlined in the "Disassemble" procedure.
2. Rotate strut assembly until mounting flange is facing outward, opposite compressor forcing screw.
3. Position spring on strut, making sure spring is properly seated on bottom spring plate, **Fig. 14.**
4. Install strut spring seat assembly and J26584-88 top adapters over strut spring.
5. Turn compressor forcing screw until compressor top support contacts top adapters.
6. Install a long extension and socket onto hex on damper shaft to align components during installation.
7. Compress spring until approximately 1 1/2 inch of damper shaft extends through spring plate.
8. Remove extension and socket, then position top mounting assembly over damper shaft and install retaining nut. Torque retaining nut to specification.
9. Turn forcing screw counterclockwise and remove strut assembly from compressor.

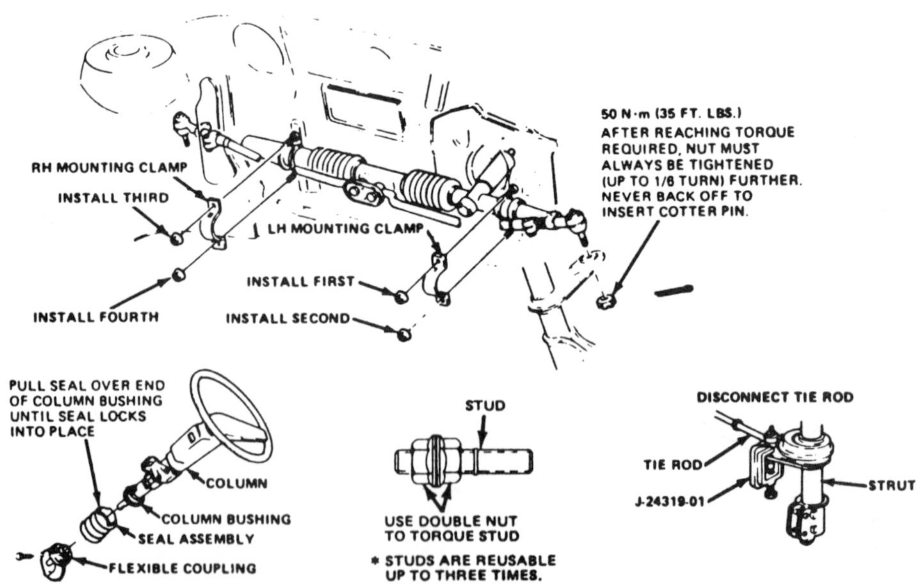

Fig. 16 Power rack & pinion steering gear removal

1988–90

Disassemble

1. Position strut compressor J-34013 in holding fixture J-3289-20.
2. Position strut in strut compressor, then compress strut approximately 1/2 of its height. Use care not to bottom spring or damper rod.
3. Remove nut from strut dampener shaft, then position guide rod J-34013-27 on dampener shaft. Use guide rod J-34013-27 to position dampener shaft down through bearing cap while compressing coil spring.
4. Remove components from coil strut unit.

Assemble

1. Position bearing cap on strut compressor.

2. Position strut to strut compressor and install compressor bottom locking pin.
3. Extend dampener shaft and install clamp J-34013-20 to hold shaft in position, **Fig. 15.**
4. Position spring over dampener, then position strut to strut compressor upper locking pin hole and install pin.
5. Install upper insulator, shield, bumper and upper spring seat. The flat on the upper spring seat should face in the same direction as the centerline of the strut assembly spindle, **Fig. 15.**
6. Install guide rod J-34013-27 onto dampener shaft, the compress strut unit until dampener shaft threads are visible. Remove guide rod and install retaining nut.
7. While holding dampener shaft in position with a suitable wrench, torque retaining nut to specification. After tightening nut, remove clamp from dampener shaft clamp J-34013-20.

MANUAL STEERING GEAR

REPLACE

1. Disconnect battery ground cable.
2. Remove left hand sound insulator.
3. From under instrument panel, pull downward on steering column seal, then remove upper pinch bolt from flexible coupling.
4. Remove air cleaner, then remove windshield washer reservoir attaching screws and position reservoir aside.
5. Raise and support front of vehicle, then remove both front wheel and tire assemblies.
6. Disconnect tie rods from struts using tool No. J24319-01, then lower vehicle.
7. Remove steering gear mounting clamps, **Fig. 16.**
8. Move steering gear assembly slightly forward, then remove lower pinch bolt from flexible coupling and detach coupling from steering gear stub shaft.
9. Remove dash panel seal from steering gear.
10. Raise and support front of vehicle, then remove splash shield from left inner fender.
11. Place left hand knuckle and hub assembly in the full left turn position, then remove steering gear through access hole in left hand inner fender.
12. Reverse procedure to install. If steering gear mounting clamp studs have backed out during removal, install double nuts on stud and **torque** stud to 15 ft. lbs. **Torque** coupling to stub shaft pinch bolt to 29 ft. lbs. and coupling to steering column shaft pinch bolt to 30 ft. lbs. Torque steering gear mounting clamp attaching nuts, tie rod-to-strut and strut attaching nuts to specifications. The tie rod to strut attaching nuts can be tightened up to an additional 1/6 turn to allow installation of cotter pin.

POWER STEERING GEAR

REPLACE

On some 1985-87 Firenza models, the power steering lines may come into contact with the HVAC case, causing a rattle or buzzing noise. To ensure that a rattle or buzzing noise will not occur, position power steering lines away from the case.

1. Disconnect battery ground cable.
2. Remove left hand sound insulator.
3. From under instrument panel, pull downward on steering column seal, then remove upper pinch bolt from flexible coupling.
4. Remove air cleaner, then remove windshield washer reservoir attaching screws and position reservoir aside.
5. Disconnect pressure line from steering gear and remove screw attaching line bracket to cowl.
6. Raise and support front of vehicle, then remove both front wheel and tire assemblies.
7. Disconnect tie rods from struts using tool No. J24319-01, then lower vehicle.
8. Remove steering gear mounting clamps, **Fig. 16.**
9. Move steering gear slightly forward, then disconnect return line from gear and drain power steering fluid.
10. Remove lower pinch bolt from flexible coupling, then detach coupling from steering gear stub shaft and remove dash seal.
11. Raise and support front of vehicle, then remove splash shield from left inner fender.
12. Place steering knuckle and hub assembly into the full left turn position, then remove steering gear through access hole in left hand inner fender.
13. Reverse procedure to install, noting the following:
    a. If steering gear mounting clamps have backed out during removal, install double nuts on stud and **torque** to 15 ft. lbs.
    b. **Torque** coupling to stub shaft pinch bolt to 37 ft. lbs.
    c. **Torque** coupling to steering column shaft pinch bolts to 30 ft. lbs. on 1985-86 models, to 34 ft. lbs. on 1987-90 models.
    d. **Torque** pressure and return line fittings to 20 ft. lbs.

POWER STEERING PUMP

REPLACE

1. Remove air cleaner assembly.
2. Disconnect reservoir to pump hose and pressure line from pump.
3. Remove clip attaching pressure line to pump.
4. Loosen pump pivot and adjusting bolts, then remove pump drive belt.
5. Remove three pump to bracket attaching bolts. On some models, the pump may be attached to bracket by special bolts. To remove these special bolts, use tool No. T-45 or equivalent.
6. Remove pump from mounting bracket.
7. Reverse procedure to install.

FRONT SUSPENSION & STEERING TIGHTENING SPECIFICATIONS

Year	Component	Torque/ft. lbs.
1985-90	Ball Joint To Knuckle	45
	Brake Line Bracket Bolt	13
	Caliper Bolts	38
	Control Arm Pivot Bolt	61
	Hub & Bearing Assembly	70
	Hub Nut	185
	Stabilizer Shaft To Control Arm	15
	Stabilizer To Support Assembly	18
	Steering Gear Mounting Clamp	28
	Steering Knuckle To Strut Assembly	133
	Strut Assembly To Body	18
	Suspension Support Assembly	①
	Strut Cartridge Retaining Nut	65
	Tie Rod To Strut	28

①—Torque suspension support assembly center bolts to 66 ft. lbs.; then front bolts to 65 ft. lbs; then rear bolts to 65 ft. lbs.

Chevrolet Beretta & Corsica

INDEX

DESCRIPTION

The front suspension, **Figs. 1 and 2,** on these vehicles is of the MacPherson strut design. The lower control arms pivot from the lower side rails through rubber bushings. The upper end of the strut is isolated by a rubber mount incorporating a non-serviceable bearing for wheel turning. The tie rods connect to the steering arm on the strut, below the spring seat. The lower end of the steering knuckle pivots on a ball stud which is retained to the lower control arm by rivets and is secured to the steering knuckle with a nut and cotter pin. The sealed wheel bearings are integral with the hub and are serviced as an assembly.

WHEEL BEARING REPLACE

REMOVAL

1. Loosen hub nut with vehicle on ground.
2. Raise and support vehicle, then remove front wheel.
3. Install drive axle protective boot cover J-28712.
4. Remove hub nut.
5. Remove brake caliper from support and suspend caliper from flame with a length of wire. Do not suspend caliper by brake hose.
6. Remove brake rotor.
7. Remove three hub and bearing attaching bolts, then the splash shield. If the old bearing is being reinstalled, mark attaching bolts and corresponding holes for reinstallation, **Fig. 3.**
8. Using tool J-28733 or equivalent, remove bearing from steering knuckle, **Fig. 4. If excessive corrosion is present, ensure that bearing is loose in the knuckle before using puller tool.**
9. If installing new bearing, replace steering knuckle seal. **Do not move drive axle until hub nut is installed and torqued to specifications.**

INSTALLATION

1. Clean and inspect bearing mating sur-

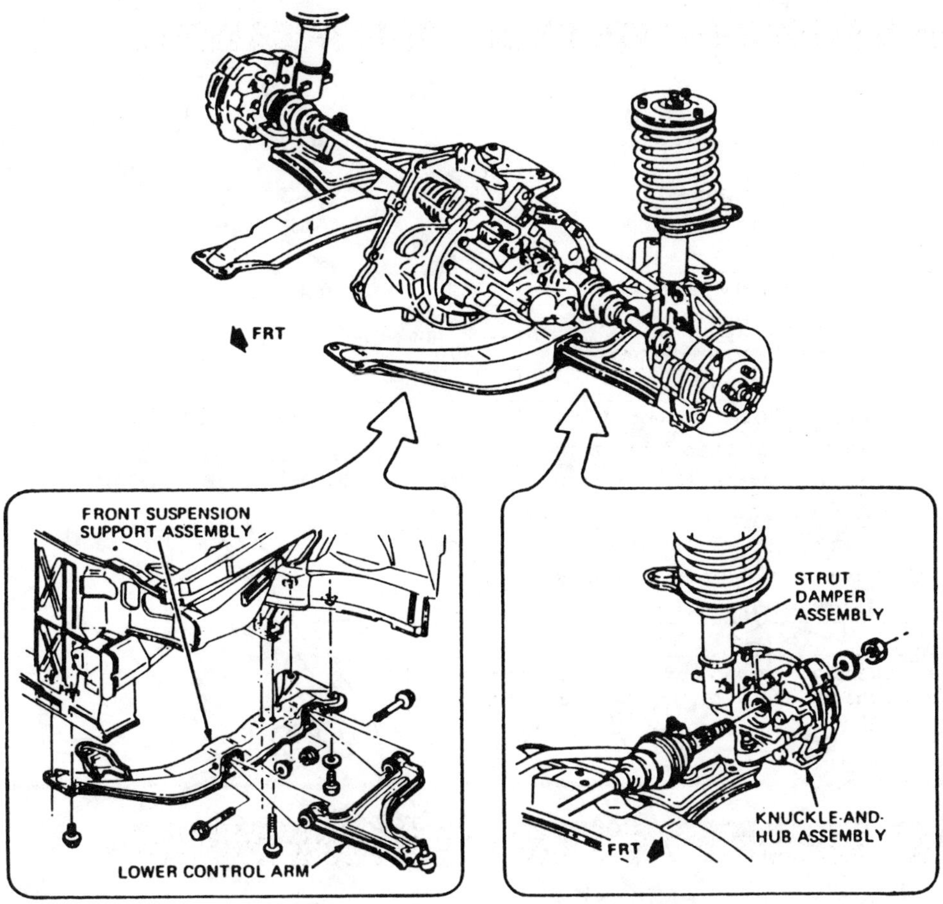

Fig. 1 Front suspension

faces and steering knuckle bore for dirt, nicks and burrs.
2. If installing new steering knuckle seal, apply grease to seal and knuckle bore, then press seal into steering knuckle.
3. Push bearing onto axle shaft, then install splash shield and hub to steering knuckle attaching bolts. Torque attaching bolts to specifications. Install hub to axle retaining nut and apply a partial **torque** of 74 ft. lbs., **Fig. 5.**
4. Install rotor, brake caliper and wheel assembly.
5. Lower vehicle and **torque** hub nut to 191 ft. lbs.

LOWER BALL JOINT
REPLACE

1. Raise and support vehicle, then remove wheel and tire.
2. Locate center of rivet body and mark with a center punch.
3. Using a 1/8 inch drill, drill pilot holes completely through the rivets. Using a 1/2 inch drill, drill final holes through rivets to ensure fitting of new ball joint.
4. Remove ball joint stud retaining nut, then, using tool J-29330, separate ball joint from steering knuckle. Remove ball joint from lower control arm.
5. Assemble new ball joint to lower control arm with bolts provided in service package, **Fig. 6,** and torque bolts to specifications.
6. Insert ball joint stud into steering knuckle and torque nut to specifications.
7. Install wheel and tire, check toe setting and adjust as required.

LOWER CONTROL ARM & BUSHING
REPLACE

1. Raise and support vehicle, then remove wheel and tire.
2. Disconnect stabilizer bar at lower control arm and control arm support.
3. Using tool J-29330, separate ball joint from steering knuckle.
4. Remove control arm support to chassis retaining bolts and remove control arm support and control arm as an assembly.
5. Separate control arm from support, then using tools shown in **Fig. 7,** remove bushings from control arm.
6. Lubricate new bushings and install into control arm using tools shown in **Fig. 7.**
7. Attach lower control arm to control arm support and torque pivot bolts to specifications.
8. Install control arm support to chassis, using attaching bolt tightening sequence shown in **Fig. 8.** Torque bolts to specifications.
9. Reverse procedure to complete installation. Check wheel alignment.

STEERING KNUCKLE
REPLACE

1. Raise and support vehicle, then remove wheel and tire.
2. Remove front hub and bearing as outlined under Wheel Bearing, Replace procedure.
3. Using tool J-29330, separate ball joint from steering knuckle.
4. Remove strut to steering knuckle attaching bolts, then disconnect strut from steering knuckle.
5. Assemble strut to new steering knuckle and install attaching bolts finger tight.
6. Insert ball joint stud into steering knuckle and torque stud nut to specifications.
7. Torque strut to steering knuckle attaching bolts to specifications.
8. Reverse removal procedure to complete installation.

STABILIZER BAR & BUSHINGS
REPLACE

1. Disconnect battery ground cable.
2. Install engine support tool J-28467 or equivalent.
3. Raise and support vehicle, allowing control arms to hang free.
4. Remove left front tire and wheel assembly.
5. Disconnect stabilizer bar from arms, then from support assemblies.
6. Loosen front support attaching bolts, then remove rear and center support bolts. Allow supports to drop down enough to remove stabilizer bar, **Fig. 9.**
7. Reverse procedure to install. Before tightening any attaching bolts, ensure stabilizer bar is centered in chassis. Refer to **Fig. 9,** for torque specifications.

STRUT ASSEMBLY
REPLACE

1. Raise hood and remove strut protective cap and three strut to body attaching nuts.
2. Raise and support vehicle, allowing suspension to hang freely.
3. Remove wheel and tire, then install drive axle protective cover, J-28712.
4. Using tool J-24319, disconnect tie rod from strut assembly.
5. Remove strut to steering knuckle attaching bolts, then remove strut from vehicle.
6. Reverse procedure to install. Position flats of strut mounting bolts as shown in **Fig. 10.** Torque all nuts and bolts to specifications.

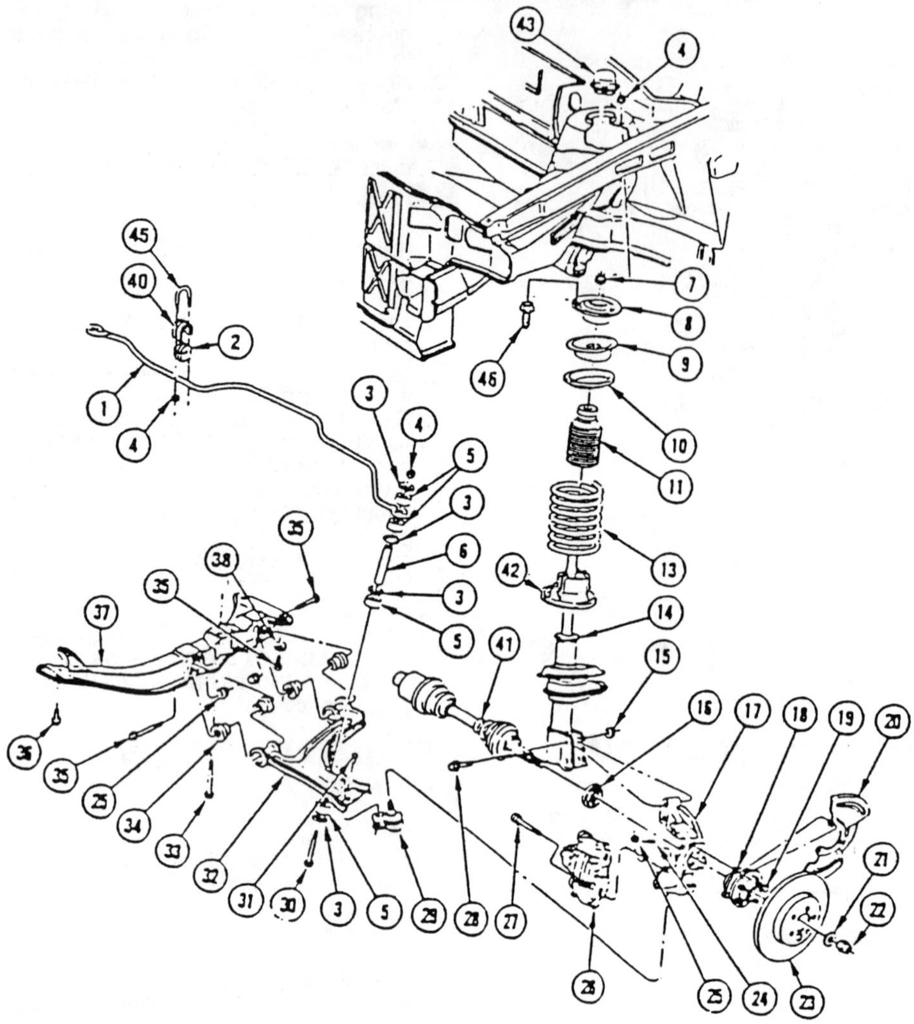

1. SHAFT, Frt Stab
2. INSULATOR, Frt Stab Shf
3. WASHER, Fl (.119 x .353)
4. NUT, Hex (M8 x 1.25)
5. GROMMET, Frt Stab Shf Link
6. SPACER, Unit Frt Stab Shf
7. NUT, Frt Susp Strut
8. MOUNT, Frt Susp Strut
9. SEAT, Frt Spr
10. INSULATOR, Frt Spr
11. BUMPER/SHIELD, Frt Susp Strut
13. SPRING, Frt Coil
14. ABSORBER, W/Strut, Frt Shk STRUT, Frt Shk-LH
15. NUT, Hex (M16 x 2)
16. SEAL, Frt Whl Inr Brg
17. KNUCKLE, Strg
18. BEARING, Frt Whl
19. BOLT, Frt Disc Brk Spl Shld
20. SHIELD, Frt Brk Disc Splash
21. WASHER, Frt Susp Reinf
22. NUT, Frt Whl Dr Shf
23. DISC, Frt Whl Brk
24. PIN, Cotter (3.2 x 25M)
25. NUT, Hex Lock (M12 x 1.75)
26. CALIPER, Frt Brk
27. BOLT, Frt Brk Caliper Hsg Ret
28. BOLT, Screw Frt Susp Strut
29. BALL JOINT, Strg Knu Lwr Cont Arm
30. BOLT, Hex (M8 x 1.25 x 170)
31. RIVET
32. ARM ASM, Frt Lwr Cont-RH
33. BOLT, Screw Frt Susp Supt
34. BUSHING, Strg Knu Lwr Cont Arm
35. BOLT, Hex Hd (M12 x 1.75 x 85)
36. BOLT, Screw Frt Susp Supt
37. SUPPORT, Supt Frt Susp
38. WASHER, Frt
40. CLAMP, Frt Stab Shf
41. SHAFT, Axle Frt Whl Dr
42. INSULATOR, Frt Spr Lwr
43. COVER, Frt Susp Strut Mt
45. BOLT, "U" Frt Stab Shaft Insl Clamp
46. SCREW, M8 1 1/4 x 16

Fig. 2 Exploded view of front suspension

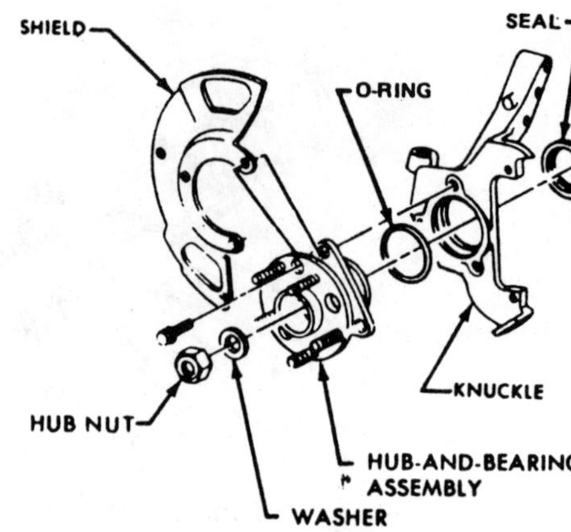

Fig. 3 Front hub & wheel bearing assembly

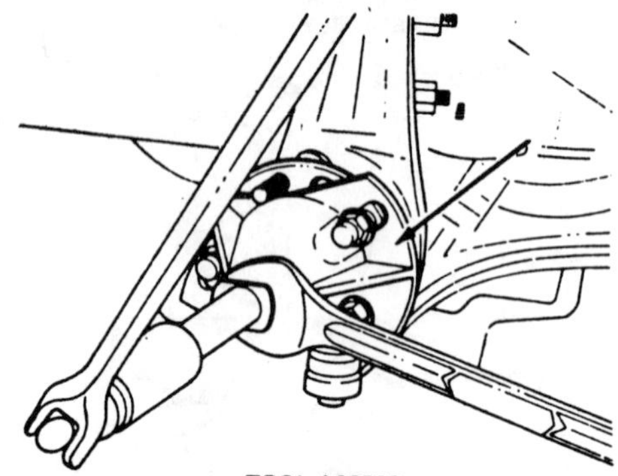

Fig. 4 Front hub & bearing assembly removal

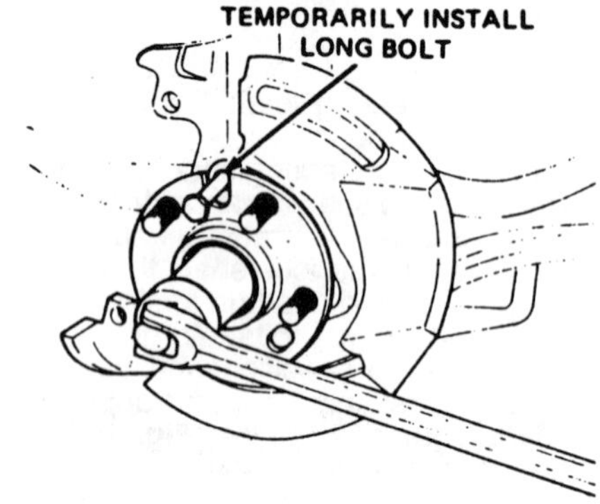

Fig. 5 Hub nut installation

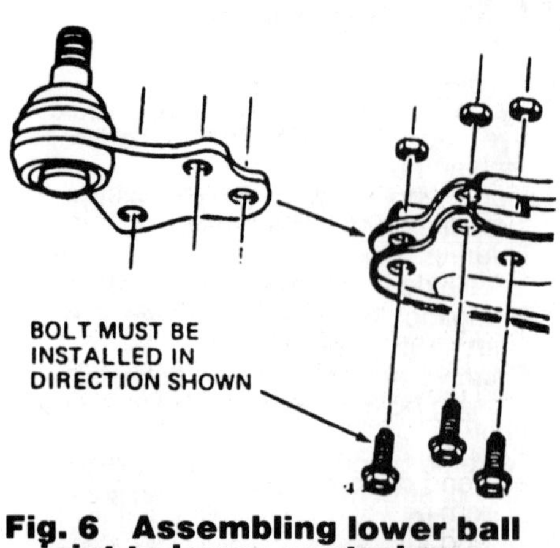

Fig. 6 Assembling lower ball joint to lower control arm

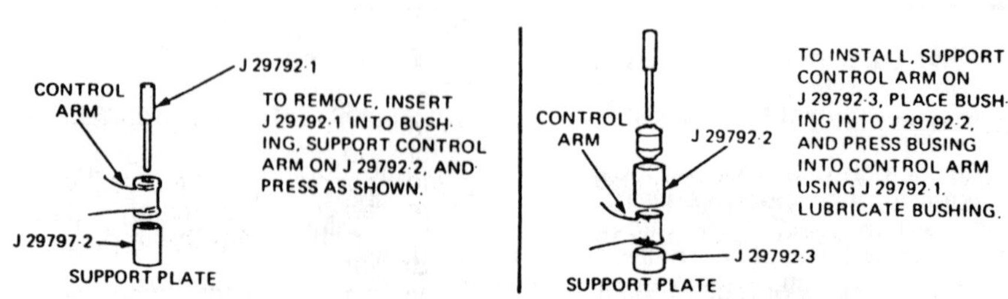

Fig. 7 Replacing lower control arm bushing

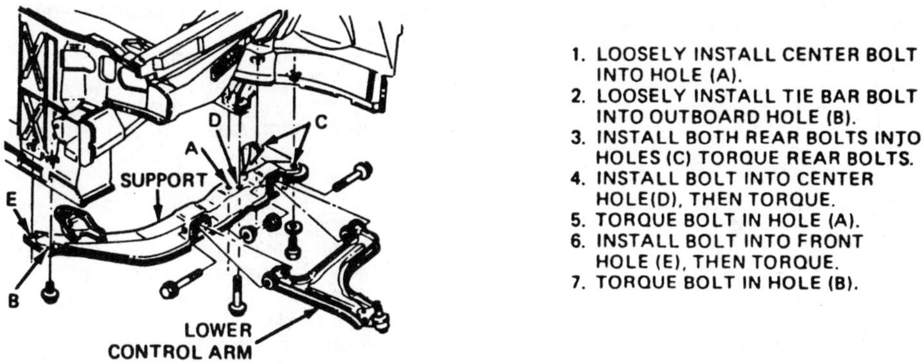

Fig. 8 Replacing lower control arm

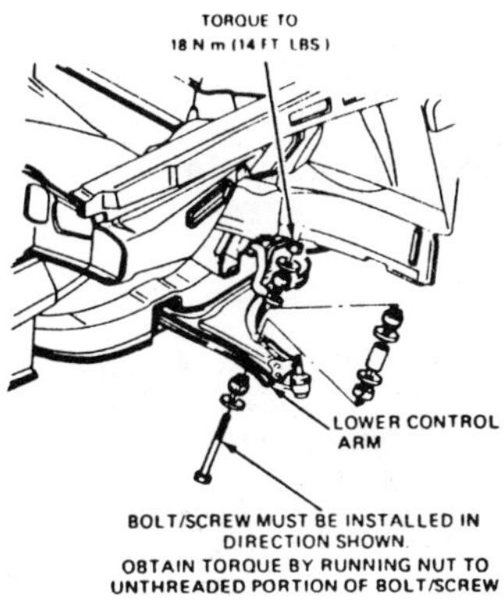

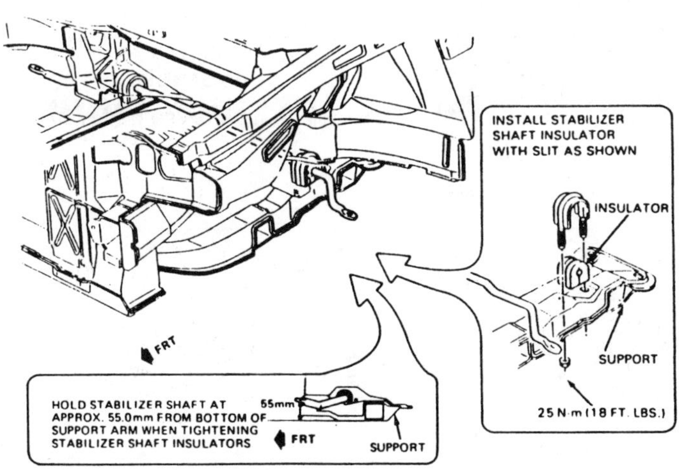

Fig. 9 Stabilizer bar installation

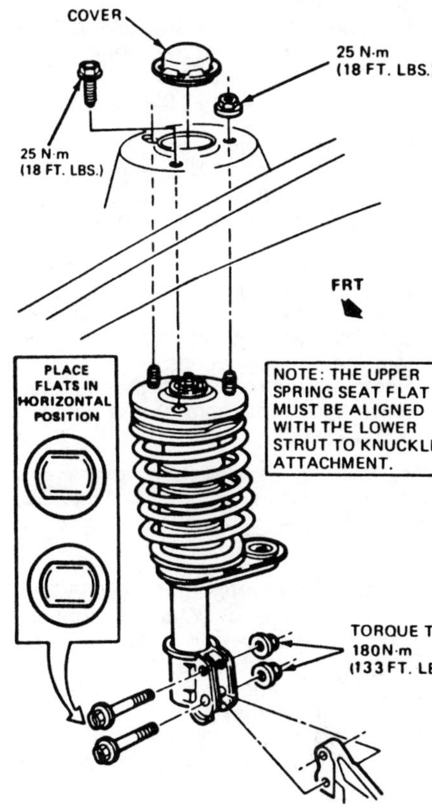

Fig. 10 Installing strut assembly

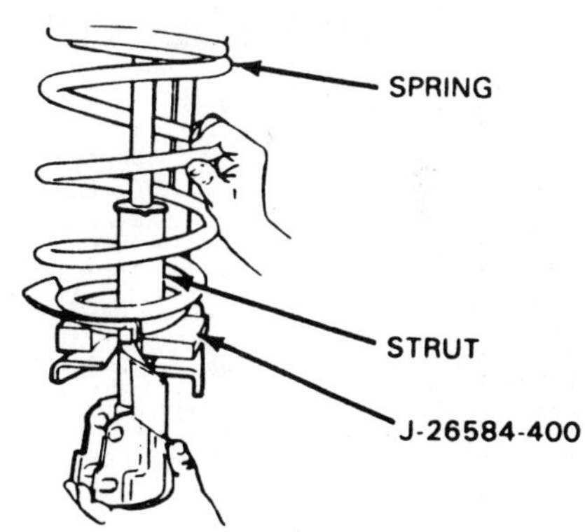

Fig. 11 Removing damper & coil spring from strut

STRUT ASSEMBLY SERVICE

DISASSEMBLY

Care must be taken not to damage special coating on coil springs. If special coating is damaged, coil spring damage could occur.

1. Clamp strut compressor tool J-26584 in a suitable vise.
2. Place strut assembly in compressor tool and install bottom adapter J-26584-400, making sure that adapter captures strut and locating pins are fully engaged, **Fig. 11**.
3. Position top adapter J-26584-430 on strut cap, aligning mounting holes as necessary.
4. Rotate compressor forcing screw clockwise until top support flange contacts top adapter. Continue rotating forcing screws until strut spring is compressed to approximately 1/2 its height. **Do not bottom the spring or strut damper rod.**
5. Remove damper top nut, then place alignment rod on strut shaft. Use alignment rod to guide shock shaft through spring cap during removal.
6. Remove strut components, then relieve compressor tension.

ASSEMBLY

1. Perform Steps 1 & 2 as outlined in the Disassembly procedure.
2. Position spring on strut, making sure spring is properly seated on bottom spring plate.
3. Install shields, bumpers and insulators on spring seat, then install coil spring seat on top of spring, **Fig. 12.**
4. Install bearing cap on spring seat, ensuring they are centered together and properly aligned.
5. Rotate spring as necessary to have upper end of spring located within 10 mm from end of groove in upper insulator and lower end of spring located 10-15 mm from end of groove in low-

er insulator.

6. Position top adapter J-26584-430 on strut cap, then engage compressor slightly. Pull up strut rod to its full extension and clamp in place with tool No. J-34013-20
7. Insert alignment rod J-34013-27 through bearing and spring cap and position on top of strut rod. Compress spring slowly while guiding strut rod through bearing cap with alignment rod.
8. Continue compressing spring until one inch of strut rod is above bearing cap. Install strut attaching nut and torque to specifications.
9. Remove strut rod clamp, then release tension on coil spring and remove compressor.

POWER STEERING GEAR REPLACE

1. Disconnect battery ground cable.
2. Remove left hand sound insulator.
3. Remove steering column coupling upper pinch bolt.
4. Remove hydraulic line retainer, then disconnect pump pressure and return lines from steering gear.
5. Raise and support vehicle.
6. Remove front wheel and tire assemblies.

Fig. 12 Strut coil spring & upper mounting installation

7. Disconnect tie rods from steering knuckles.
8. Lower vehicle, then remove steering gear mounting clamps.
9. Move steering gear forward, then remove steering column coupling lower pinch bolt.
10. Remove steering column coupling and dash seal from steering gear.
11. Raise and support vehicle, then remove steering gear through left front wheel opening.
12. Reverse procedure to install. Torque all bolts to specifications noting the following:
    a. When installing steering gear, attach clamps and finger tighten all attaching nuts prior to tightening. Ensure dash seal is indexed and flush with steering gear.

POWER STEERING PUMP REPLACE

1. Disconnect battery ground cable.
2. Place suitable container under pump to catch fluid.
3. Disconnect pressure and return lines at pump, then plug open ends at hoses and pump.
4. Remove auxiliary drive belt.
5. Working through holes in pump pulley, remove pump attaching bolts.
6. Remove power steering pump.
7. Reverse procedure to install.

FRONT SUSPENSION & STEERING TIGHTENING SPECIFICATIONS

Year	Component	Torque/ft. lbs.
1987-90	Ball Joint To Control Arm	50
	Ball Joint To Steering Knuckle	55
	Control Arm Pivot Bolts	63
	Control Arm Support To Chassis	66
	Hub to Axle Nut	①
	Hub To Steering Knuckle	70
	Steering Column Pinch Bolts	②
	Steering Gear Clamps	28
	Strut To Steering Knuckle	129
	Strut Rod To Bearing Cap	60
	Tie Rod To Steering Knuckle	44

①—See text.
②—Torque upper pinch bolt to 29 ft. lbs., torque lower pinch bolt to 37 ft. lbs.

Chevrolet Chevette & Pontiac 1000

INDEX

Fig. 1 Front suspension

DESCRIPTION

The front suspension system, **Fig. 1**, uses long and short control arms with coil springs mounted between the lower control arms and front suspension crossmember.

WHEEL BEARINGS
ADJUST

1. While rotating wheel in the forward direction, torque spindle nut to specifications.
2. Back spindle nut off to the just loose position.
3. Hand tighten spindle nut, then loosen until cotter holes are aligned and install cotter pin. **Do not loosen nut more than 1/2 flat to align cotter pin holes.**
4. With bearings properly adjusted, end play must be .001-.005 inch.

WHEEL BEARINGS
REPLACE

1. Raise vehicle, remove wheel and tire assembly.
2. Remove two caliper mounting bracket to steering knuckle attaching bolts, then the caliper.
3. Remove hub dust cap, cotter pin, spindle nut, washer, outer wheel bearing and hub.
4. Pry out grease seal, then remove inner bearing from hub.

SHOCK ABSORBER
REPLACE

1. Hold upper stem of shock absorber, and remove nut, retainer and grommet.
2. Raise vehicle and remove lower shock absorber retaining bolts.
3. Lower shock absorber from vehicle.
4. Reverse procedure to install. Torque shock absorber retaining bolts to specifications.

BALL JOINTS
REPLACE

UPPER BALL JOINT

1. Raise vehicle and remove wheel and tire assembly.
2. Support lower control arm using a suitable jack.
3. Loosen upper ball joint stud nut, however, do not remove nut.
4. Position tool J-26407 or equivalent with cupped end over lower ball joint stud and turn threaded end of tool until upper ball joint stud is free of steering knuckle, **Fig. 2.**
5. Remove tool, then the upper stud nut. **Discard the stud nut. The stud nut is of special design and must be replaced whenever removed or loosened.**
6. Remove two nuts securing ball joint to upper control arm and the ball joint. **Inspect tapered hole in steering knuckle. If out of round or damaged, the knuckle must be replaced.**
7. Install ball joint on upper control arm and torque bolts to specifications.
8. Position upper ball joint stud on steering knuckle, then install a standard nut to draw ball joint into position on knuckle. Torque standard nut to 35 ft. lbs., then remove standard nut. Install special nut for final assembly and torque to specifications.
9. Install wheel and tire assembly and lower vehicle.

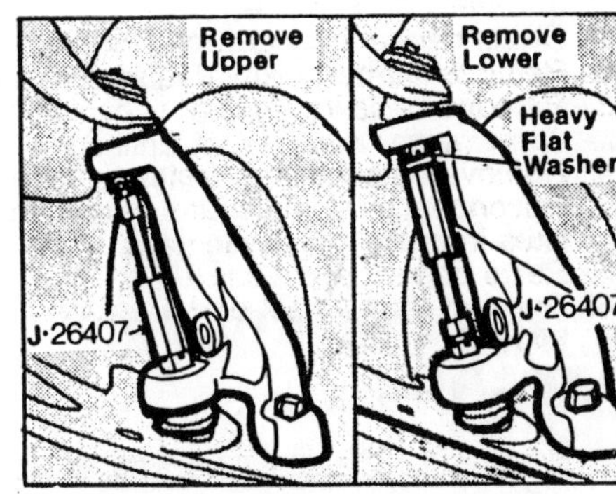

Fig. 2 Separating ball joint studs from steering knuckle

LOWER BALL JOINT

1. Raise vehicle and remove wheel and tire assembly.
2. Support lower control arm using a suitable jack.
3. Loosen lower ball joint stud nut, however, do not remove nut.
4. Position tool J-26407 or equivalent with cupped end of tool over upper ball joint stud and turn threaded end of tool until lower ball joint stud is free of steering knuckle, **Fig. 2.**
5. Remove tool, then the lower ball joint stud nut. **Discard the stud nut. The stud nut is of special design and must be replaced whenever removed or loosened.**
6. Remove ball joint from lower control arm. **Inspect tapered hole in steering knuckle. If out of round or damaged, the knuckle must be replaced.**
7. Insert ball joint through lower control arm and into steering knuckle.
8. Install a standard nut to draw ball joint into position on knuckle. Torque standard nut to 35 ft. lbs., then remove standard nut. Install special nut for final assembly and torque to specifica-

tions.
9. Install wheel and tire assembly and lower vehicle.

COIL SPRING

REPLACE

1. Support vehicle by frame, remove wheel and tire assembly.
2. Disconnect stabilizer bar from lower control arm and tie rod from steering knuckle.
3. Support lower control arm using a suitable jack.
4. Loosen lower ball joint stud nut, then use tool J-26407 or equivalent to free stud from steering knuckle. Position knuckle and hub out of way and remove stud nut. **Discard stud nut. The stud nut is of special design and must be replaced whenever removed or loosened.**
5. Loosen lower control arm pivot bolts.
6. Install a safety chain around spring and through lower control arm.
7. Slowly lower control arm until spring is extended as far as possible, then use pry bar to carefully lift spring over lower control arm spring seat.

LOWER CONTROL ARM

REPLACE

1. Remove coil spring as described in "Coil Spring, Replace."
2. Remove control arm pivot bolts, then the control arm.
3. Reverse procedure to install. Torque lower control arm pivot bolts to specifications.

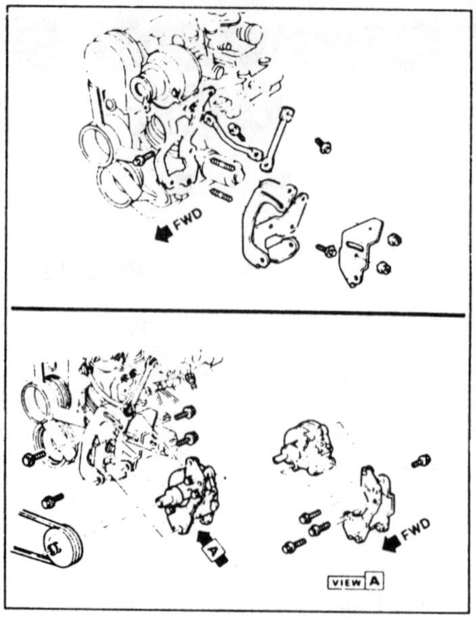

Fig. 3 Power steering pump installation (Typical)

UPPER CONTROL ARM

REPLACE

1. Support vehicle by frame, remove tire and wheel assembly.
2. Position jack to support lower control arm.
3. Disconnect upper ball joint as described in "Ball Joints, Replace."
4. Remove upper control arm pivot bolts, then remove control arm from vehicle.
5. Reverse procedure to install. Torque upper control arm pivot bolts to specifications.

MANUAL STEERING GEAR

REPLACE

1. Raise vehicle and remove bolts and shield.
2. Remove tie rod cotter pins and nuts from tie rod ends and separate tie rods from steering knuckles.
3. Remove flexible coupling pinch bolt.
4. Remove steering gear clamp bolts, then the steering gear assembly from vehicle.
5. Reverse procedure to install. Torque nuts and bolts to specifications.

POWER STEERING GEAR

REPLACE

The procedure for power steering gear replacement is identical to manual type gears. In addition, it is necessary to disconnect and connect the two hydraulic lines at the steering gear.

POWER STEERING PUMP

REPLACE

1. Remove pump adjusting bolt.
2. Remove lower brace to pump bracket attaching bolt, **Fig. 3.**
3. Remove crossmember to frame support brace, then disconnect hydraulic lines at pump.
4. Remove rear pump adjusting bracket, then the front pivot bolt.
5. Remove front pump bracket to engine attaching bolt, then the pump and bracket.
6. Reverse procedure to install.

FRONT SUSPENSION & STEERING TIGHTENING SPECIFICATIONS

Year	Component	Torque/ft. lbs.
1985-87	Ball Joint Bolts	29
	Flex Coupling Pinch Bolt	30
	Lower Ball Joint To Steering Knuckle	②
	Lower Control Arm Pivot Bolts	49
	Shock Bolt (Lower)	29
	Shock Bolt (Upper)	13
	Spindle Nut	12
	Steering Gear Clamp Bolt	14
	Steering Knuckle To Tie Rod	30
	Upper Ball Joint To Steering Knuckle	①
	Upper Control Arm Pivot Bolts	47

①—Torque standard nut to 35 ft. lbs., then final assembly nut to 36 ft. lbs. See text.

②—Torque standard nut to 35 ft. lbs., then final assembly nut to 41–54 ft. lbs. See text.

Chevrolet Nova & Geo Prizm

INDEX

DESCRIPTION

The front suspension, **Fig. 1,** on this vehicle is a MacPherson strut design. The upper end of the strut is anchored to the body by a strut support. The strut and strut support are isolated by a rubber mount. The lower end of the strut is connected to the upper end of the steering knuckle. The lower end of the knuckle is attached to the ball joint, which is attached to the suspension control arm assembly. Movement of the steering wheel is transmitted to the tie-rod end and then to the knuckle, turning the wheel and tire assembly.

STRUT

REPLACE

1. Open hood, then remove strut to body attaching nuts.
2. Loosen wheel lug nuts. Loosen axle shaft nut if knuckle will be removed.
3. Raise and support vehicle until wheels are clear from ground.
4. Remove tire and wheel assembly.
5. Disconnect brake flex hose clip at strut bracket.
6. Disconnect brake flex hose to brake pipe connection. Remove brake hose clip.
7. Pull brake hose back through strut bracket opening, then tape or cap brake hose and caliper.
8. Remove two brake caliper mounting bracket bolts, then the caliper.
9. Mark adjusting cam, then remove both strut to knuckle attaching nuts and bolts.
10. Remove strut assembly and camber adjusting cam from knuckle.
11. Reverse procedure to install. Torque strut to knuckle attaching nuts and bolts to specifications, upper mount bolts to 13 ft. lbs., and wheel nuts to 76 ft. lbs.

LOWER CONTROL ARM

REPLACE

1. Raise and support vehicle.
2. Remove lower control arm attaching nuts and bolts.
3. Remove control arm from vehicle.
4. Reverse procedure to install. **Torque** control arm to body front nuts and bolts to 83 ft. lbs., and rear nuts and bolts to 64 ft. lbs.

LOWER CONTROL ARM BUSHING

REPLACE

1. Remove control arm.
2. Using a suitable tool, remove lower control arm bushing from lower control arm.
3. Reverse procedure to install. **Torque** bushing nut to 76 ft. lbs.

LOWER BALL JOINT

REPLACE

1. Loosen wheel lug nuts.
2. Raise and support vehicle.
3. Remove tire and wheel assembly.
4. Using tool J-35413 or equivalent, separate ball joint from knuckle assembly.
5. Remove two nuts and bolts attaching ball joint and control arm.
6. Remove ball joint.
7. Reverse procedure to install. **Torque** knuckle to ball joint nut to 82 ft. lbs. and control arm attaching nuts to 47 ft. lbs. During installation, always replace a self-locking nut with a new one.

STEERING KNUCKLE

REPLACE

1. Loosen wheel and hub nuts.
2. Raise and support vehicle.
3. Remove tire and wheel assembly.
4. Remove brake hose retaining clip at strut.
5. Disconnect flex hose from brake line.
6. Remove caliper bracket to knuckle mounting bolts, then support caliper.
7. Remove disc assembly.
8. Remove drive axle unit. Use puller J-25287 or equivalent, to push out drive axle assembly.
9. Remove cotter pin and tie rod to knuckle attaching nut. Use tool J-24319-01 or equivalent to separate the tie rod assembly.
10. Remove ball joint to control arm attaching nuts and bolt.
11. 2Mark camber relationship for installation, then remove two strut to knuckle attaching nuts and bolts, then the knuckle.
12. Reverse procedure to install. **Torque** control arm nuts and bolts to 47 ft. lbs, ball joint to knuckle, 14 ft. lbs., strut to lower bracket bolts to 105 ft. lbs., tie rod to knuckle 38 ft. lbs. and brake caliper bracket to knuckle attaching bolts to 65 ft. lbs.

HUB

REPLACE

1. Remove knuckle from vehicle and mount into a suitable vise.
2. Using a suitable screwdriver, remove dust deflector.
3. Using seal remover J-26941 or equivalent, remove inner grease seal from knuckle, **Fig. 2.**
4. Remove inner bearing snap ring and disc brake dust shield.
5. Push out hub using hub remover J-25287 and J-35378 or equivalents.
6. Using tools mentioned in step 5, remove outer bearing race from hub.
7. Using tool J-26941 or equivalent, remove outer grease seal.
8. Using bearing remover J35399 and J-35379 or equivalents, remove bearing assembly.
9. Reverse procedure to install. Use tool J-8092 and J-35411 to install hub bearing assembly. Use tool J-35737 to install outer grease seal. Install hub using tools J-8092 and J-35399. Use tool J-35737 to install inner grease seal. Use tool J-35379 to install dust deflector ring. Ensure dust deflector ring faces down (open end down).

MANUAL STEERING GEAR

REPLACE

1. Disconnect battery ground cable.
2. Remove intermediate shaft cover.
3. Loosen upper pinch bolt. Remove lower pinch bolt from pinion shaft.
4. Loosen wheel lug nuts.
5. Raise and support vehicle.
6. Remove both front wheel and tire assemblies.

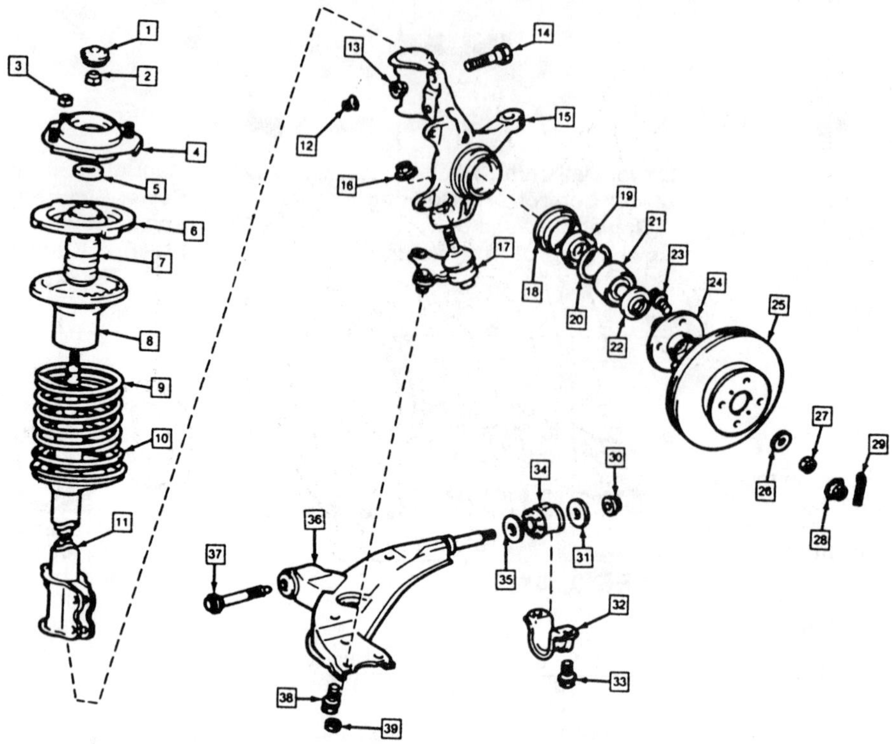

1. COVER, DUST
2. NUT, ABSORBER SHAFT
3. NUT, STRUT SUPPORT
4. SUPPORT, STRUT
5. SEAL, SUPPORT
6. SEAT, SPRING
7. BUMPER, SPRING
8. INSULATOR, UPPER
9. SPRING, COIL
10. INSULATOR, LOWER
11. ABSORBER, SHOCK/STRUT
12. NUT, STRUT TO KNUCKLE
13. CAM, CAMBER ADJUST
14. BOLT, STRUT TO KNUCKLE
15. KNUCKLE, STEERING
16. NUT, BALL JOINT TO KNUCKLE
17. JOINT, LOWER BALL
18. DEFLECTOR, DUST
19. SEAL, INNER HUB
20. RING, SNAP
21. BEARING, FRONT AXLE HUB
22. SEAL OUTER HUB
23. STUD, WHEEL
24. HUB, FRONT AXLE
25. DISC, FRONT BRAKE
26. WASHER, FRONT HUB
27. NUT, FRONT HUB
28. CAP, FRONT HUB
29. PIN, COTTER
30. NUT, CONTROL ARM
31. RETAINER, BUSHING
32. BRACKET, BUSHING
33. BOLT, BUSHING BRACKET
34. BUSHING, CONTROL ARM
35. RETAINER, BUSHING
36. ARM, LOWER CONTROL
37. BOLT, CONTROL ARM
38. BOLT, BALL JOINT
39. NUT, BALL JOINT

Fig. 1 Disassembled view of front suspension

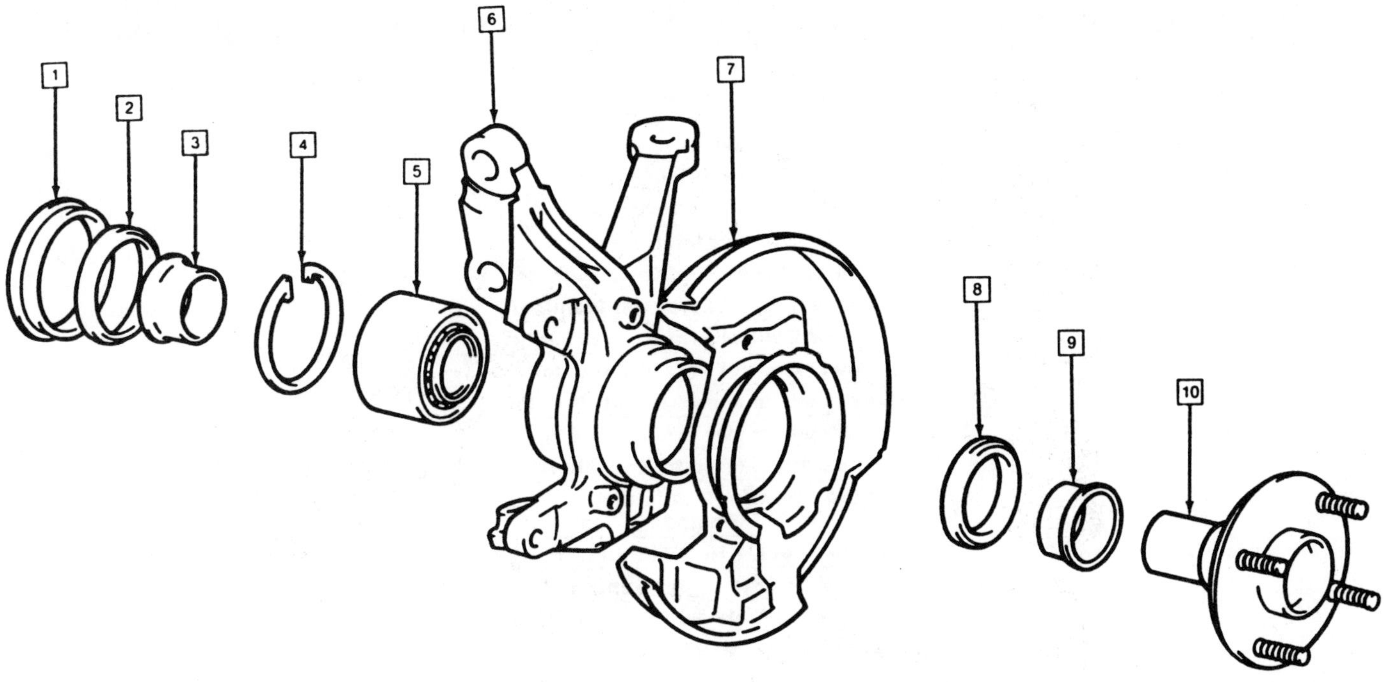

PARTS LIST

1 DEFLECTOR, DUST
2 SEAL, INNER GREASE
3 RACE, INNER BEARING
4 RING, SNAP
5 BEARING, AXLE HUB
6 KNUCKLE, STEERING FRONT
7 SHIELD, DUST
8 SEAL, OUTER GREASE
9 RACE, OUTER BEARING
10 HUB, DRIVE AXLE

Fig. 2 Disassembled view of hub & bearing assembly

7. Remove cotter pins from tie rod ends.
8. Disconnect both tie rod ends from steering knuckles. Use tool J-24319-01, or equivalent only, to separate tie rod joints.
9. Remove steering gear to body mounting nuts and bolts.
10. Remove steering assembly from vehicle.
11. Reverse procedure to install. **Torque** intermediate shaft pinch bolts to 26 ft. lbs., **torque** tie rod to knuckle nuts to 36 ft. lbs., gear housing to mounting nuts and bolts to 43 ft. lbs., tie rod end locking nut to 35 ft.lbs., inner tie rod to rack nuts to 61 ft. lbs., and wheel nuts to 76 ft. lbs.

POWER STEERING GEAR REPLACE

1. Remove intermediate steering shaft protector.
2. Loosen upper, and remove lower intermediate shaft pinch bolts.
3. Open hood and place drain pan under gear assembly.
4. Loosen wheel lug nuts.
5. Raise and support vehicle.
6. Remove both front wheel and tire assemblies.
7. Using tie rod remover J-24319-01 or equivalent, remove both tie rod ends from knuckles.
8. Support transaxle with a suitable jack.
9. Remove rear center engine mounting member to body mounting bolts.
10. Remove rear engine mount to mount bracket attaching nut and bolt. Raise and lower rear of transaxle to gain access to steering gear to body attaching nuts and bolts.
11. Disconnect pressure and return lines at steering gear.
12. Remove four gear to body mounting nuts and bolts.
13. Remove gear through access opening.
14. Reverse procedure to install.

POWER STEERING PUMP REPLACE

1. Remove air cleaner assembly.
2. Remove return hose clamp.
3. Disconnect and cap pressure and return lines from pump.
4. Loosen pump pulley.
5. Remove adjusting bolt, pivot bolt and drive belt.
6. Remove pump assembly.
7. Reverse procedure to install.

FRONT SUSPENSION & STEERING TIGHTENING SPECIFICATIONS

Year	Component	Torque/ ft. lbs.
1985-90	Axle Shaft Nut	137
	Ball Joint to Control Arm	59
	Ball Joint to Steering Knuckle	82
	Caliper Bracket to Knuckle Bolts	65
	Exhaust Pipe to Exhaust Manifold	46
	Lower Control Arm Bracket Bolts	72
	Lower Control Arm to Body Bolts	105
	Lower Control Arm Bushing (Rear) Nut	76
	Stabilizer Bar to Control Arm	13
	Steering Tie Rod to Knuckle Nut	36
	Strut to Knuckle Nuts (Except 16 Valve Engine)	105
	Strut to Knuckle Nuts (16 Valve Engine)	166
	Strut Shaft to Upper Support Nut	34
	Upper Strut to Body Nuts (16 Valve Engine)	29
	Upper Strut to Body Nuts (Except 16 Valve Engine)	23

Chevrolet & Geo Spectrum

INDEX

Page No.

DESCRIPTION

The front suspension, **Fig. 1,** is of the MacPherson strut, independent type. A strut support anchors the upper end of the strut to the body. The lower end of the strut is connected to the upper end of the steering knuckle. The lower end of the steering knuckle is attached to the ball joint, which is mounted to the suspension control arm.

STRUT SERVICE

REMOVAL

1. Remove both strut to body attaching nuts.
2. Raise and support front of vehicle.
3. Remove wheel and tire assembly.
4. Remove brake hose clip from strut, then disconnect hose from caliper.
5. Pull brake hose back through opening in strut bracket, then cap hose and open port in caliper to prevent contamination.
6. Remove both strut to knuckle attaching nuts and bolts, then the strut assembly, **Fig. 1.**

DISASSEMBLY

1. Secure strut spring compressor in a vise or suitable holding fixture.
2. Install strut into compressing tool, **Fig. 2,** and compress spring just enough to release spring tension.
3. Remove nut from strut shaft, then release spring and disassemble strut, **Fig. 3.**

INSPECTION

1. Ensure the following items meet specified measurements:
   a. **Strut:** piston diameter, 1.18 inch; stroke, 5.75 inches; compressed length, 14.65 inches; extended length, 20.39 inches.
   b. **Coil spring free length:** with auto. trans. and A/C, 13.5 inches; with auto. trans. less A/C 13.7; with manual trans. and A/C, 13.7 inch-

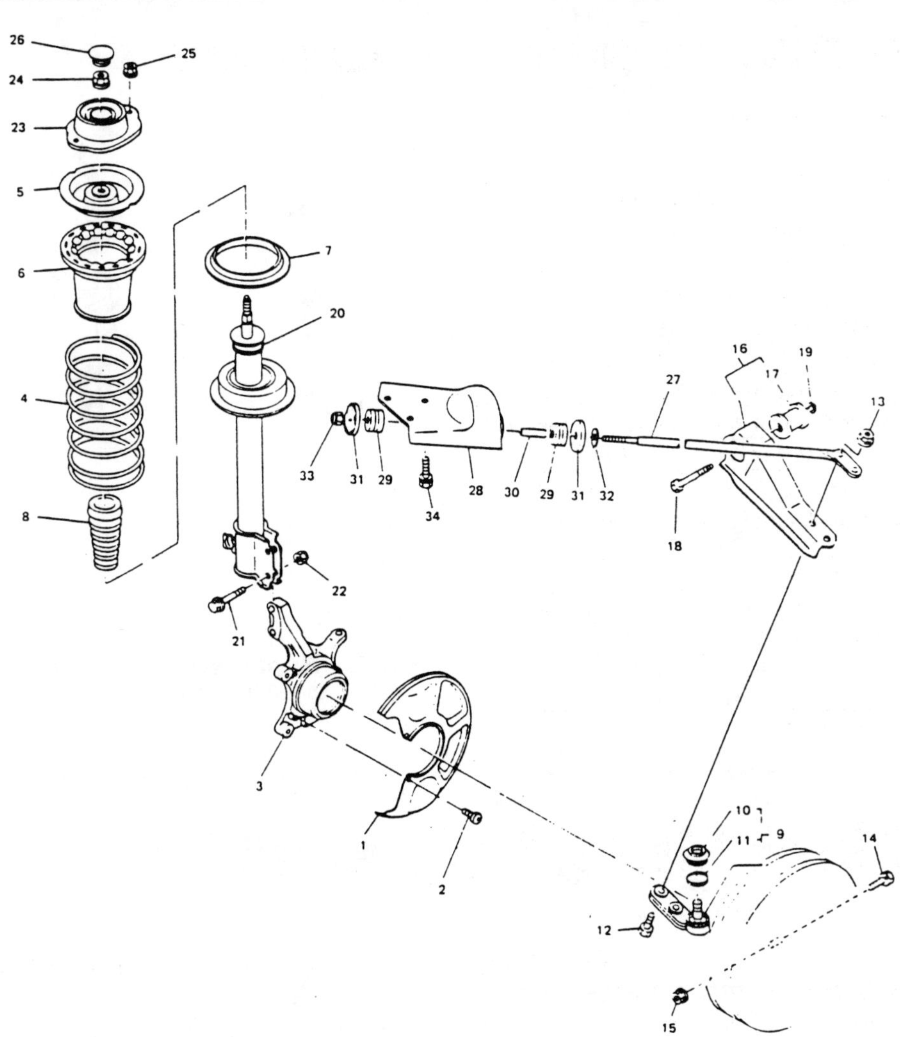

1. Cover, Dust
2. Screw, Dust
3. Knuckle, Steering
4. Spring, Coil
5. Seat, Upper
6. Seat, Rubber
7. Seat, Lower, Rubber
8. Rubber, Bumper
9. Ball Joint, Lower Control Arm
10. Boot, Ball Joint
11. Ring, Boot Clip
12. Bolt, Control Arm
13. Nut, Lock
14. Bolt, Arm End (Pinch)
15. Nut, Lock
16. Arm Asm., Lower
17. Bushing, Arm
18. Bolt, Arm
19. Washer, Lock
20. Strut Asm., Front
21. Bolt, Strut
22. Nut, Lock
23. Mount Asm., Strut Upper
24. Nut, Flange (Strut Shaft)
25. Nut, Strut
26. Cap, Strut Upper
27. Rod, Tension
28. Bracket, Support
29. Cushion, Rubber
30. Spacer, Tension Rod
31. Washer, Tension Rod
32. Washer, Tension Rod
33. Nut, Lock
34. Bolt, Support

Fig. 1 Exploded view of front suspension assembly

es; with man. trans. less A/C, 13.9 inches.

2. Check shock for oil leaks and defective operation.
3. Check coil spring for wear, cracks or distortion.
4. Check upper strut mounting for abnormal noise, binding or defective turning, or any other defects.

ASSEMBLY

1. Assemble strut in compressor tool, **Fig. 4,** ensuring "IN" marking on inside of spring upper seat faces inside of vehicle.
2. Compress spring just enough to allow installation of shaft nut.
3. Install shaft nut and tighten until shaft begins to rotate.
4. Release spring compressing tool, then torque shaft nut to specification listed at the end of this section.

INSTALLATION

1. Install strut to knuckle attaching bolts and torque to specification listed at the end of this section.
2. Route flex hose back through strut bracket opening. Connect fitting to caliper and torque to specification listed at the end of this section.
3. Install wheel and tire assembly.
4. Lower vehicle to a height where upper mount can be aligned with holes in body, then install and torque nuts to specification listed at the end of this section.
5. Lower vehicle to ground and tighten wheel nuts.

CONTROL ARM

REPLACE

1. Raise and support front of vehicle.
2. Remove lower control arm to tension rod attaching nuts and bolts.
3. Remove control arm to body attaching bolt and nut.
4. Remove control arm, **Fig. 1.**
5. Reverse procedure to install, noting the following:
   a. Replace self-locking nuts when installing arm.
   b. When attaching control arm to body, first jack up lower part of arm to 15.18 inches, **Fig. 5,** then torque attaching bolts to specifications listed at the end of this section, **Fig. 6.**

LOWER BALL JOINT

REPLACE

1. Raise and support front of vehicle.
2. Remove wheel and tire assembly.
3. Remove two nuts and bolts attaching ball joint, tension rod and control arm.
4. Remove ball joint to knuckle pinch bolt, then the ball joint, **Fig. 7.**
5. Reverse procedure to install.

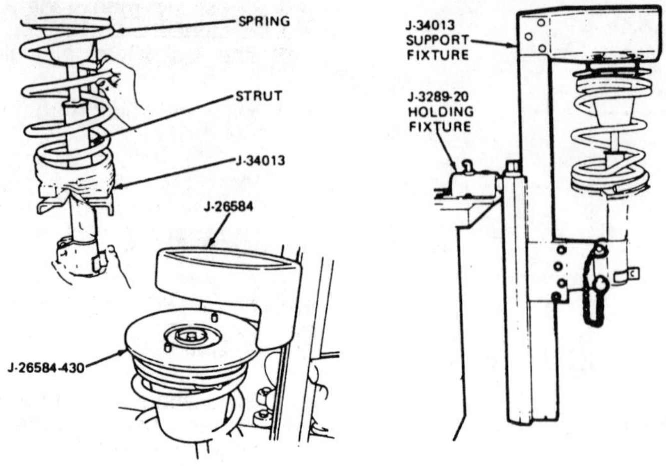

Fig. 2 Strut disassembly & assembly

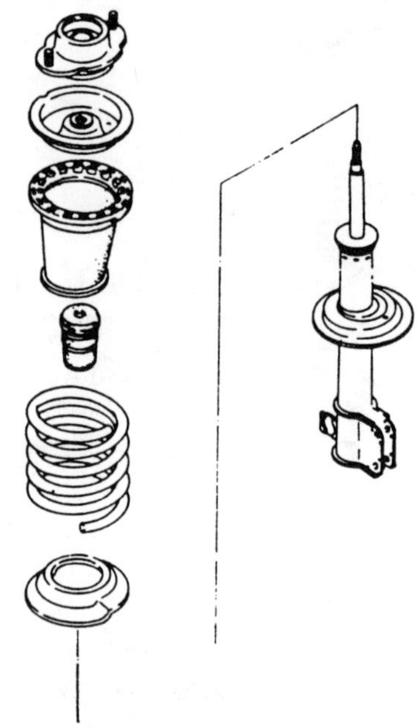

Fig. 3 Exploded view of strut assembly

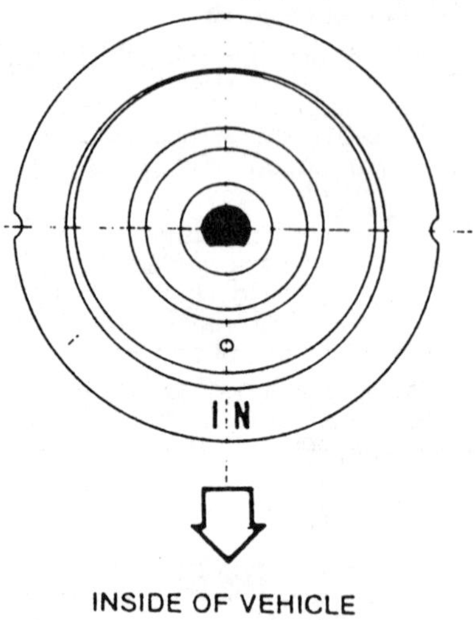

Fig. 4 Positioning spring upper seat

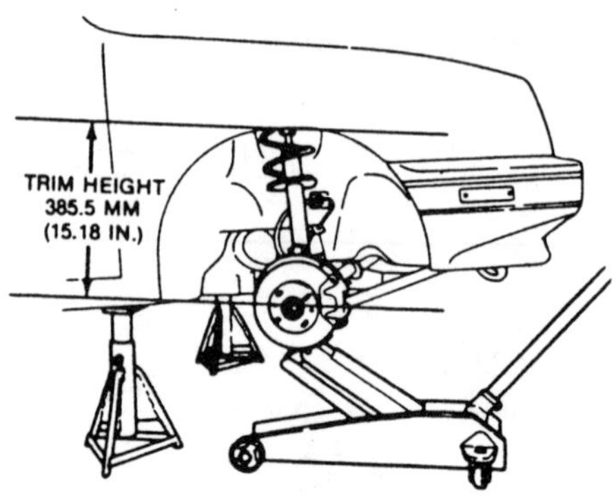

Fig. 5 Setting vehicle trim height

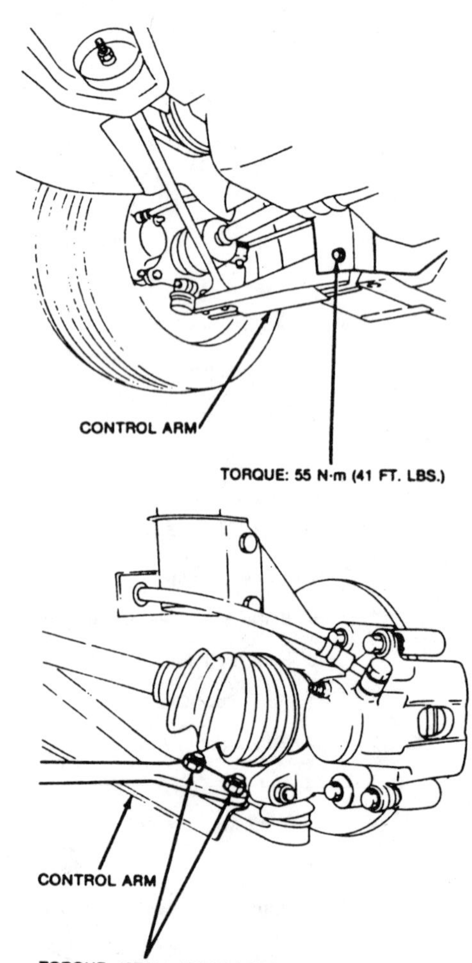

Fig. 6 Control arm installation

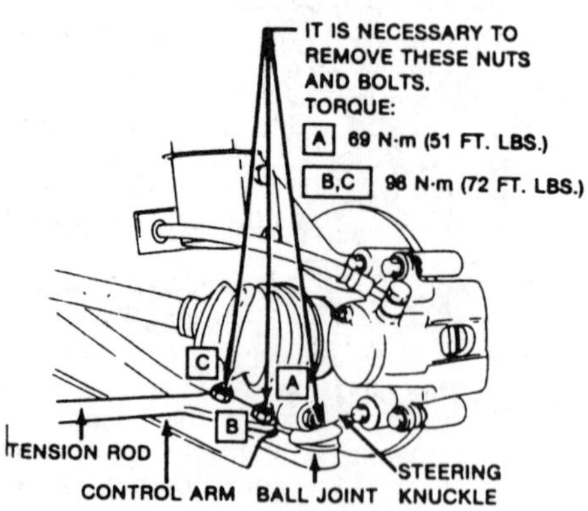

Fig. 7 Ball joint replacement

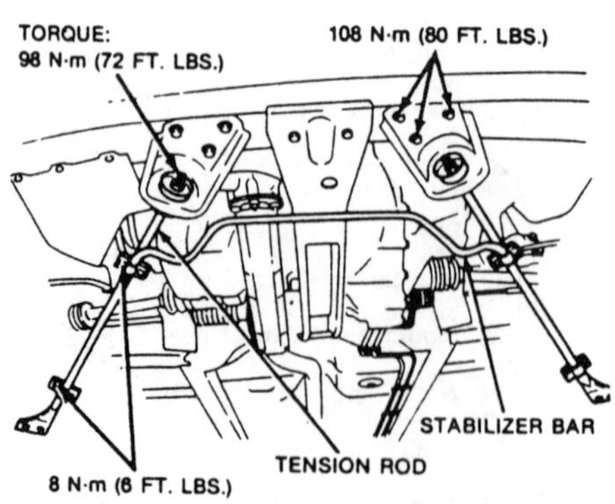

Fig. 8 Tension rod & stabilizer bar replacement

TENSION ROD REPLACE

1. Raise and support front of vehicle.
2. Remove stabilizer bar to tension rod bracket and insulator attaching nuts and bolts, if equipped.
3. Remove tension rod to body attaching nut and washer.
4. Remove tension rod to control arm attaching nuts and bolts and the tension rod, **Fig. 8.**
5. Reverse procedure to install, using new self-locking nut.

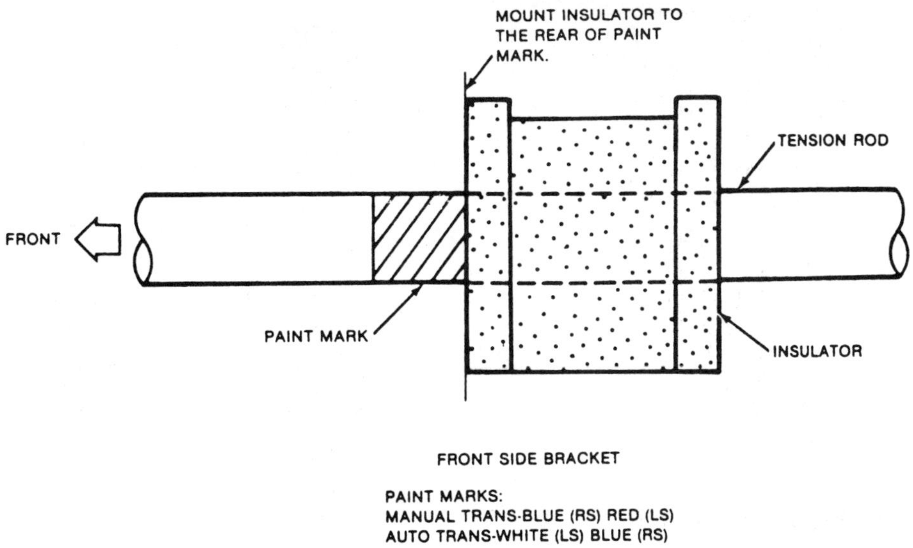

Fig. 9 Stabilizer bar insulator alignment

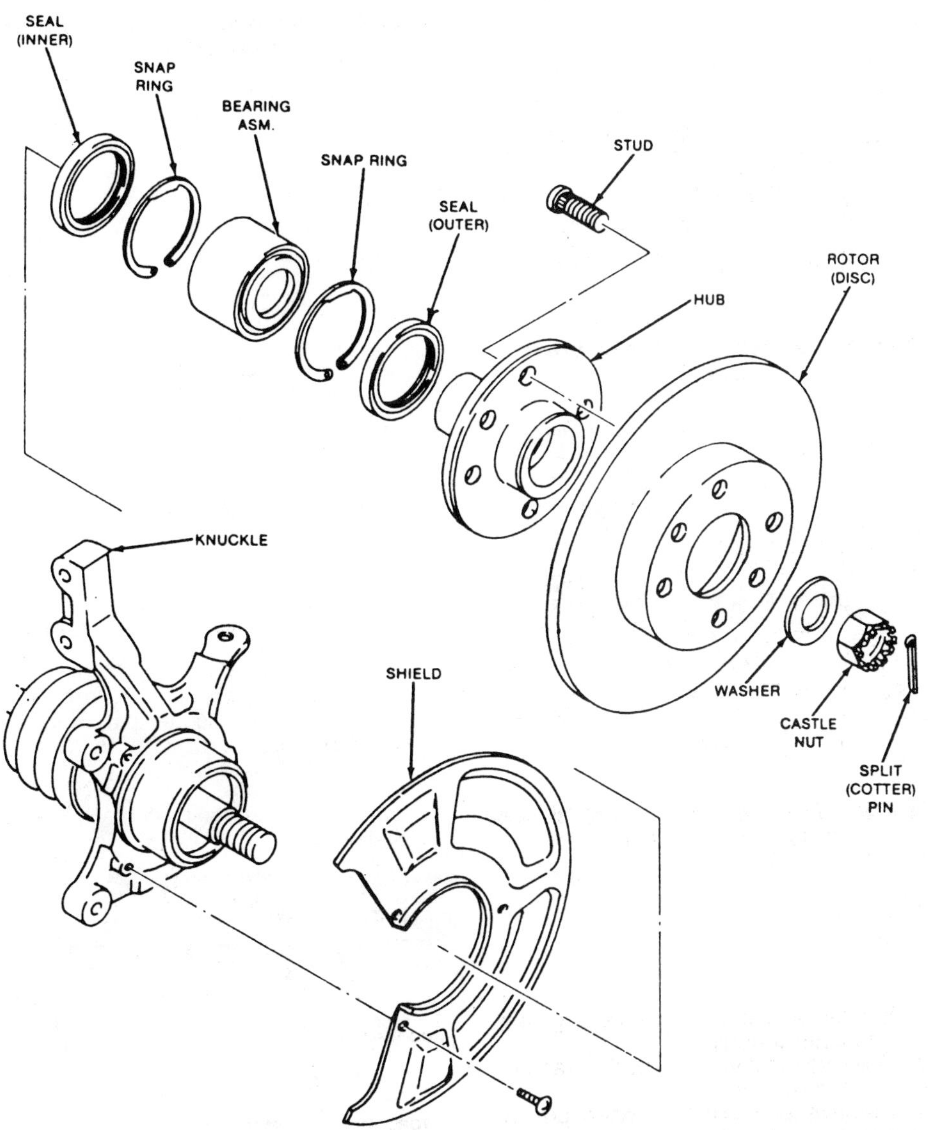

Fig. 10 Exploded view of steering knuckle & hub assembly. 1985–87

STABILIZER BAR
REPLACE

1. Raise and support front of vehicle.
2. Remove four stabilizer bar bracket and insulator attaching nuts and bolts, then the stabilizer bar, **Fig. 8.**
3. Reverse procedure to install. Align stabilizer bar front side insulator as shown, **Fig. 9.**

HUB & ROTOR
REPLACE

1. Raise and support front of vehicle.
2. Loosen wheel lug nuts, then remove wheel and tire assembly.
3. Remove caliper to knuckle attaching bolts, then position caliper aside.
4. **On 1985-87 models,** remove cotter pin, castle nut and washer, then remove rotor, **Fig. 10.**
5. **On 1988-89 models,** unstake nut, then remove nut and rotor, **Fig. 11.**
6. Using tool Nos. J-34866 and J-2619-01, remove hub, **Fig. 12.**
7. Reverse procedure to install. Replace all bearings, seals and self locking nuts. On 1985-87 models, after tightening spindle nut to specified torque, install cotter pin, **Fig. 13.** On 1988-89 models, after tightening nut to specified torque, stake nut, **Fig. 14.**

STEERING KNUCKLE
SERVICE

REMOVAL

1. Raise and support front of vehicle.
2. Remove wheel and tire assembly.
3. Remove brake hose retaining clip from strut, then disconnect hose from caliper.
4. Unfasten caliper from steering knuckle and support caliper.
5. Remove rotor, then the hub as mentioned previously.
6. Remove splash shield.
7. Remove tie rod using tool No. J-21687-02 or equivalent.
8. Remove two ball joint to control arm and tension rod attaching nuts and bolts.
9. Remove steering knuckle to strut attaching nuts and bolts and the steering knuckle, **Fig. 15.**

DISASSEMBLY

1. Remove inner seal, snap rings and inner bearing race from steering knuckle.
2. Press hub bearing from knuckle, **Fig. 16.**
3. Press outer bearing race from hub, **Fig. 17. The hub bearing, both inner and outer seals and races must not be reused.**

ASSEMBLY

1. Install outer snap ring to knuckle.
2. Install hub bearing assembly using a suitable press.

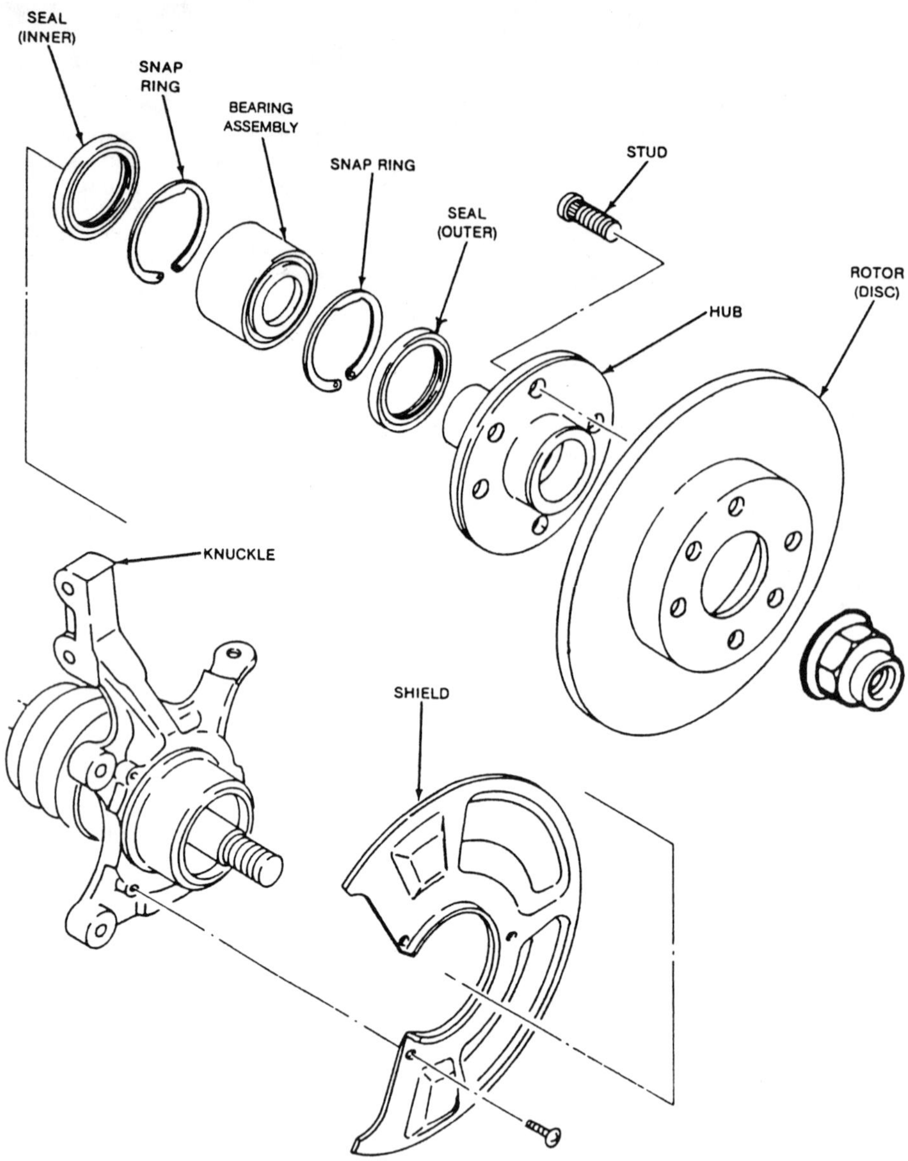

Fig. 11 Exploded view of steering knuckle & hub assembly. 1988–89

3. Install inner snap ring, then the inner and outer seals using tool J-35303 or equivalent.
4. Install outer race using tool No. J-3502 or equivalent.

INSTALLATION

1. Install ball joint to control arm attaching nuts and bolts and torque to specification listed at the end of this section.
2. Install strut to knuckle attaching nuts and bolts and torque to specification listed at the end of this section.
3. Install tie rod to knuckle attaching nut and torque to specification listed at the end of this section.
4. Install brake rotor.
5. Install brake caliper to knuckle attaching bolts and torque to specification listed at the end of this section.
6. Install brake hose to caliper and torque fitting to specification listed at the end of this section.

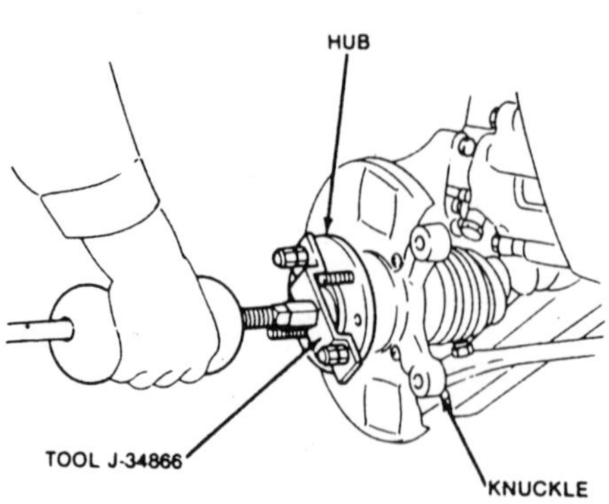

Fig. 12 Hub removal

7. Install wheel and tire assembly, then lower vehicle.
8. Apply suitable grease to axle shaft threads and nut. On 1985-87 models, after tightening spindle nut to specified torque, install cotter pin, **Fig. 13.** On 1988-89 models, after tightening nut to specified torque, stake nut, **Fig. 14.**
9. Tighten wheel nuts, then bleed brake system.

MANUAL STEERING GEAR

REPLACE

1. Remove intermediate shaft cover.
2. Loosen upper pinch bolt, then remove lower pinch bolt from pinion shaft.
3. Raise and support front of vehicle.
4. Remove both front wheel and tire assemblies.
5. Disconnect both tie rod ends from steering knuckles using tool No. J-21687-02 or equivalent.

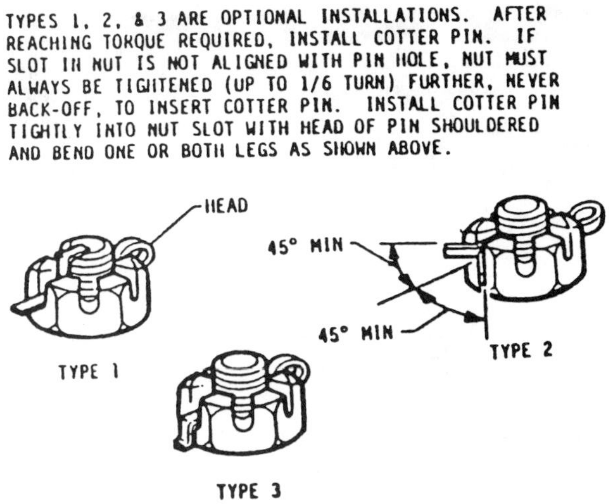

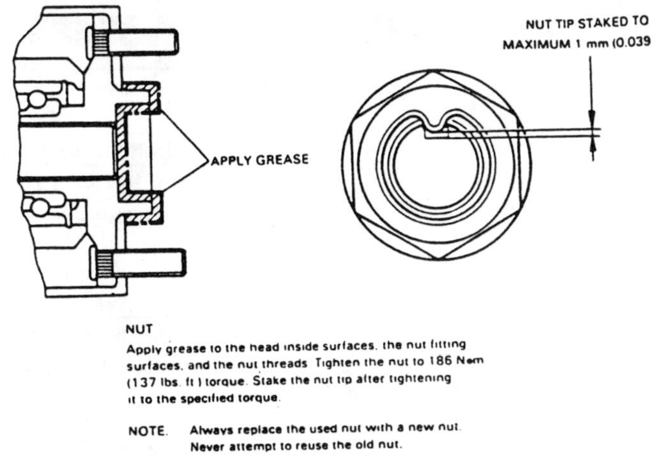

Fig. 14 Hub spindle nut installation. 1988–89

Apply grease to the axle shaft threads and nut. Install and torque axle to hub attaching nut washer and cotter pin. Torque to 186 N·m (137 ft. lbs.).

Fig. 13 Hub spindle nut installation. 1985–87

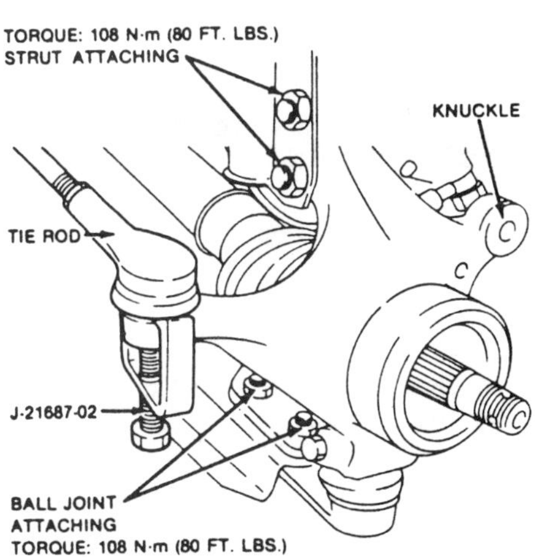

Fig. 15 Steering knuckle removal

6. Remove steering gear to body mounting nuts.
7. Remove band from left boot, then slide boot off steering gear.
8. Straighten lock washer between inner tie rod and rack, then separate tie rod from rack.
9. Reverse procedure to install.

POWER STEERING GEAR REPLACE

1. Remove intermediate steering shaft protector.
2. Loosen upper and remove lower intermediate shaft pinch bolts.
3. Position suitable container below gear.
4. Clean area around pressure and return hoses at gear valve.
5. Raise and support front of vehicle.
6. Remove both front wheel and tire assemblies.
7. Disconnect both tie rod ends from steering knuckles.
8. Remove steering gear to dash mounting nuts.
9. Cut plastic retaining straps from power steering pipes and hose.
10. Disconnect inner boot band from right rack boot.
11. Pull boot back to provide access to inner tie rod.
12. Straighten lock washer between inner tie rod and rack, then separate tie rod from rack.
13. Reverse procedure to install.

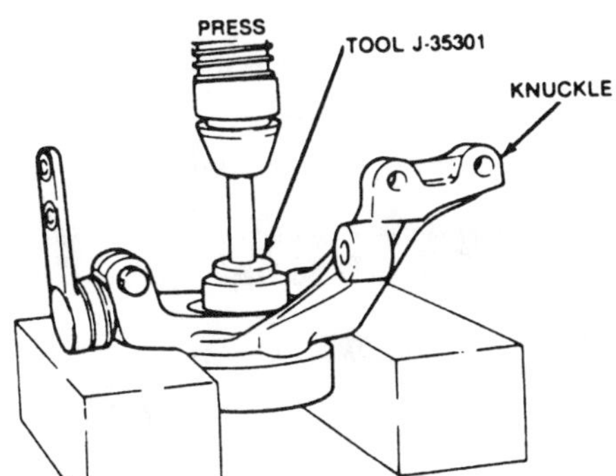

Fig. 16 Hub bearing removal

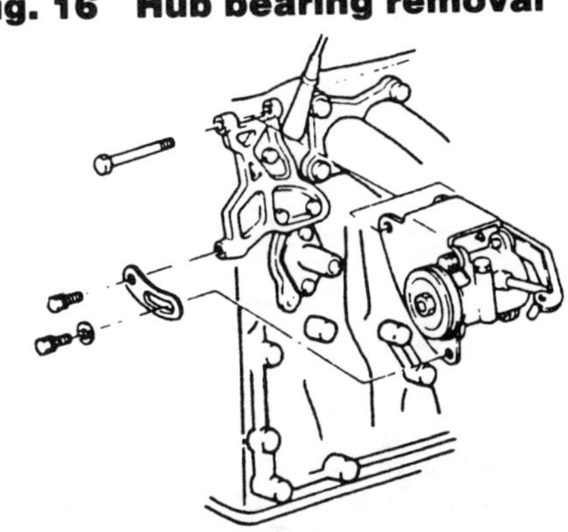

Fig. 18 Power steering pump & bracket removal

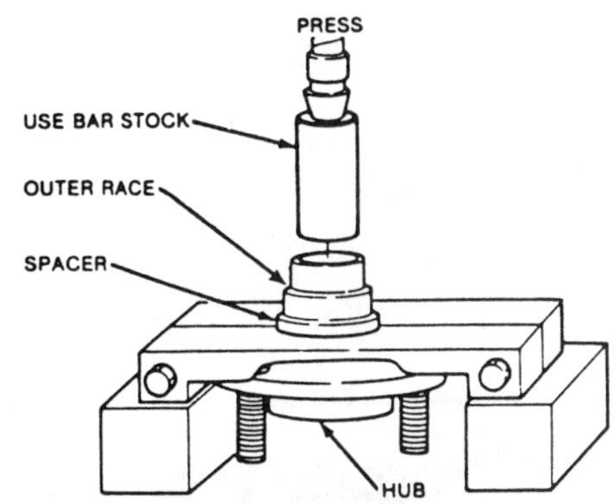

Fig. 17 Outer bearing race removal

POWER STEERING PUMP REPLACE

1. Position suitable container below pump.
2. Remove pressure hose clamp.
3. Disconnect pressure and return hoses from pump.
4. Remove adjusting bolt, pivot bolt and drive belt.
5. Remove pump and pump bracket, **Fig. 18.**
6. Remove pulley from pump.
7. Reverse procedure to install.

FRONT SUSPENSION & STEERING TIGHTENING SPECIFICATIONS

Year	Component	Torque/ft. lbs.
1985-89	Brake Hose To Disc Brake Caliper	13
	Control Arm To Crossmember Bolt	41
	Disc Brake Caliper To Steering Knuckle Bolts	41
	Front Hub Spindle Nut	137①
	Inner Tie Rod To Rack Nut	65
	Lower Ball Joint To Control Arm Nuts	80
	Lower Ball Joint To Steering Knuckle Pinch Bolt	51
	Power Steering Pump Mounting Bolts	15
	Steering Gear Mounting Bolts	30
	Strut Shaft Nut	43
	Strut To Steering Knuckle Attaching Bolts	87
	Strut Upper Attaching Nuts	41
	Tension Rod To Support Bracket Nut	72
	Tension Rod To Control Arm Nuts	80
	Tension Rod Support Bracket Attaching Bolts	40
	Tension Rod Clamp Bolts	6
	Tie Rod End To Steering Knuckle	29
	Wheel & Tire Assembly	65

①—On 1985–87 models, a cotter pin is used to hold nut in position. On 1988–89 models, nut must be staked after tightening.

Chevrolet Sprint & Geo Metro

INDEX

STRUT REPLACE

1. Raise and support vehicle allowing front suspension to hang free.
2. Remove front wheel and tire assembly.
3. Remove E-clip from brake hose and disengage hose from strut bracket, **Figs. 1 and 2.**
4. Support lower control arm and knuckle assembly, then remove strut to knuckle bolts.
5. Remove upper strut mount nuts, then remove strut from vehicle.
6. Reverse procedure to install.

STRUT SERVICE

1. Mount strut in a spring compressor.
2. **On 1985-88 models,** compress spring to about 1/2 its height. Do not bottom spring.
3. **On 1989-90 models,** compress spring approximately 1/2 inch.
4. **On all models,** remove nut from strut shaft and remove components, **Figs. 3 and 4.**
5. Reverse procedure to assemble. Compress spring so strut shaft protrudes through cap by about 1 inch. Torque nut to specification listed at the end of this section.

STABILIZER BAR REPLACE

1985–88

1. Raise and support vehicle.

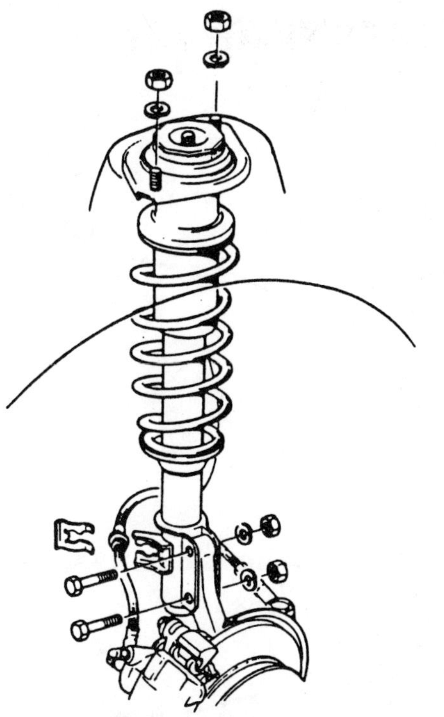

Fig. 1 Strut attaching nut bolt locations. 1985–88

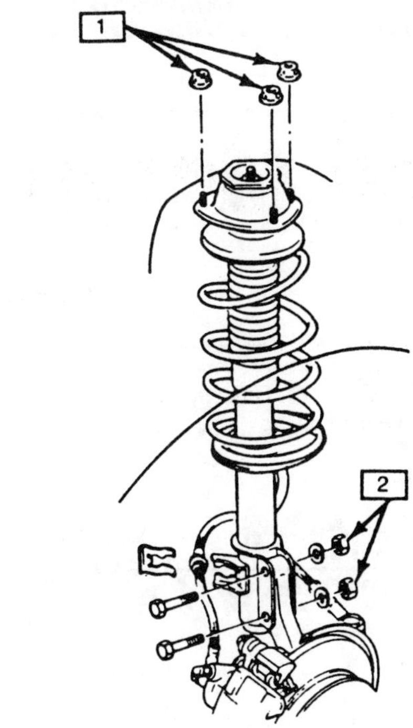

1 STRUT SUPPORT NUTS
2 STRUT BRACKET BOLTS

Fig. 2 Strut attaching nut bolt locations. 1989–90

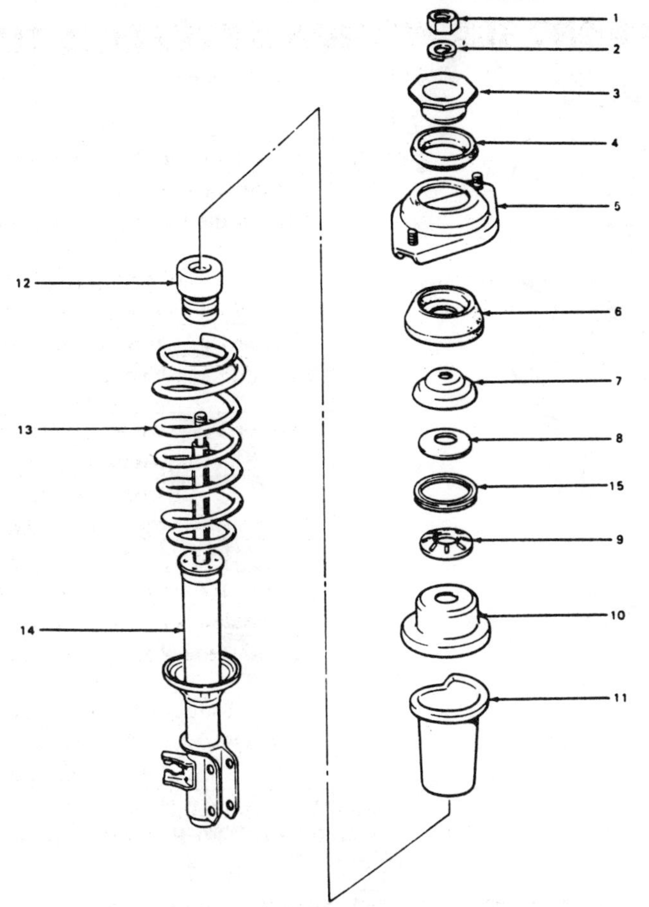

1. STRUT NUT
2. STRUT LOCK WASHER
3. STRUT INNER SUPPORT
4. STRUT REBOUND STOPPER
5. STRUT SUPPORT
6. STRUT MOUNT
7. STRUT MOUNT SEAT
8. STRUT BEARING SEAT
9. STRUT BEARING
10. SPRING UPPER SEAT
11. STRUT SPRING SEAT
12. BUMP STOPPER
13. COIL SPRING
14. STRUT ASSEMBLY
15. STRUT BEARING DUST SEAL

Fig. 3 Strut assembly exploded view. 1985–88

2. Remove both front wheel and tire assemblies.
3. Remove stabilizer bar bracket bolts, **Fig. 5.**
4. Remove nuts from stabilizer bar ends and disconnect stabilizer bar ends from suspension control arms.
5. Reverse procedure to install. Refer to **Fig. 6** for proper orientation of stabilizer bar.

WHEEL HUB, BEARING & KNUCKLE

REPLACE

When checking wheel bearings, raise and support front of vehicle, then rotate wheel and tire assembly to check bearing for smoothness of rotation and noise. Also check wheel bearing end play with a dial indicator. Wheel bearing end play should not exceed .016 inch.

1. Raise and support vehicle, then remove front wheel and tire assembly.
2. **On 1985-88 models,** remove cotter pin and driveshaft castle nut.
3. **On 1989-90 models,** unstake, then remove hub retaining nut.
4. **On all models,** remove caliper bolts, then caliper with brake line attached. Suspend caliper from chassis with wire.
5. Refer to **Fig. 7,** and measure dimension A for assembly reference. Pull hub out of knuckle.
6. Disconnect tie rod end from knuckle with tool J-21687-02 or equivalent.
7. Remove strut to knuckle bolts and then ball joint stud pinch bolt, **Fig. 8.**
8. Remove knuckle.
9. Using a drift, remove outer, then inner bearing from knuckle.
10. Apply suitable grease to balls and oil seal lips of wheel bearings. Fill area A, **Fig. 9,** to approximately 40 percent of capacity with suitable grease.
11. Install wheel bearings using bearing installer J-34856 or equivalent. **Install wheel bearings with sealed side facing outward. Also ensure spacer is spacer is snug and centered between inner and outer bearings.**
12. Install wheel bearing seal (5), **Fig. 9,** using tool J-34881 or equivalent.
13. Install spacer on hub with bevel side first, **Fig. 10.**
14. Check that wheel bearing spacer bore is aligned with bearing bores. If not, move spacer until aligned.
15. Using plastic hammer, tap hub lightly into knuckle, taking care that alignment is maintained.
16. Using installing tool J-34856 or

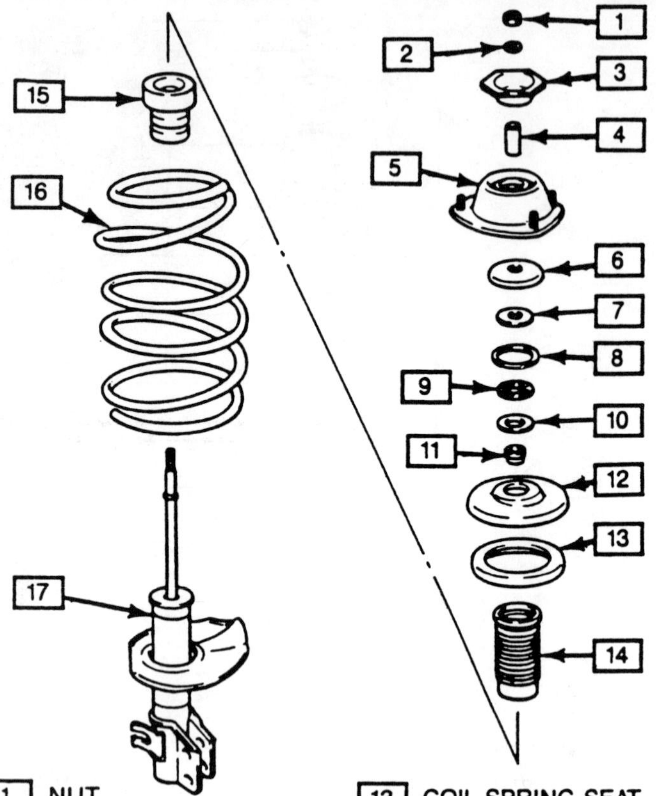

Fig. 4 Strut assembly exploded view. 1989–90

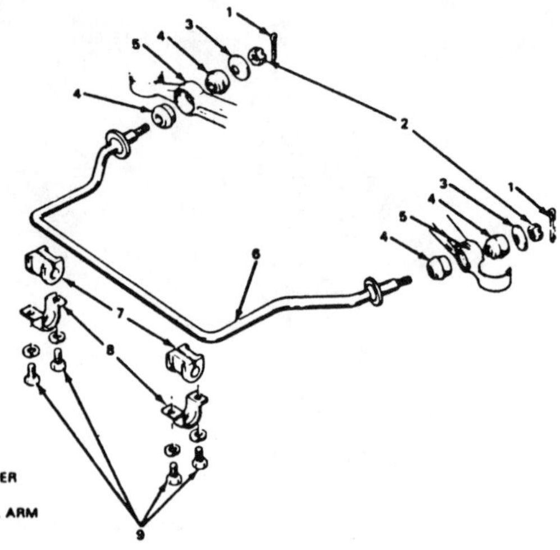

Fig. 5 Stabilizer bar & bushings. 1985–88

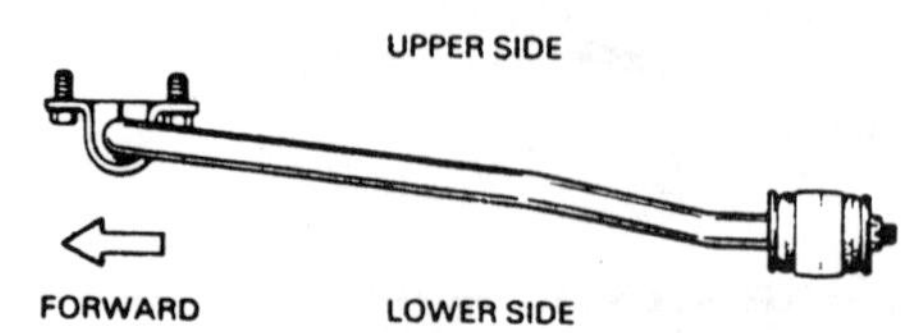

Fig. 6 Correct installation of stabilizer bar. 1985–88

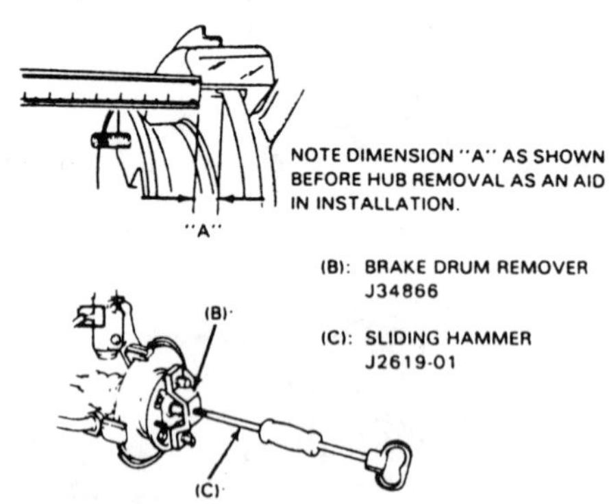

Fig. 7 Checking dimension "A" to assure proper installation of hub

J-7079-2 or equivalent, drive hub until dimension A, noted in step 4, is obtained.

17. Install brake caliper.
18. Torque caliper bolts to specification listed at the end of this section.
19. Torque drive shaft castle nut to specification listed at the end of this section.
20. **On 1985-88 models,** install cotter pin.
21. **On 1989-90 models,** stake nut in position.
22. Install wheel and tire assembly, then lower vehicle.

LOWER CONTROL ARM & BUSHING REPLACE

1985–88

1. Raise and support vehicle.
2. Remove stabilizer bar cotter pin, castle nut, washer and bushing, **Fig. 11.**
3. Remove stabilizer bar bracket bolts on same side as control arm to be removed.
4. Remove ball joint stud bolt (2), **Fig. 11.**
5. Remove control arm bolt (5), **Fig. 11,** and control arm.
6. If control arm bushing is to be replaced, proceed as follows:
   a. Remove lower control arm as described under "Lower Control Arm, Replace."
   b. Position control arm and bushing over tool No. J-34865 or equivalent, then press bushing from control arm, **Fig. 12.**
   c. Apply soapy water to bushing and

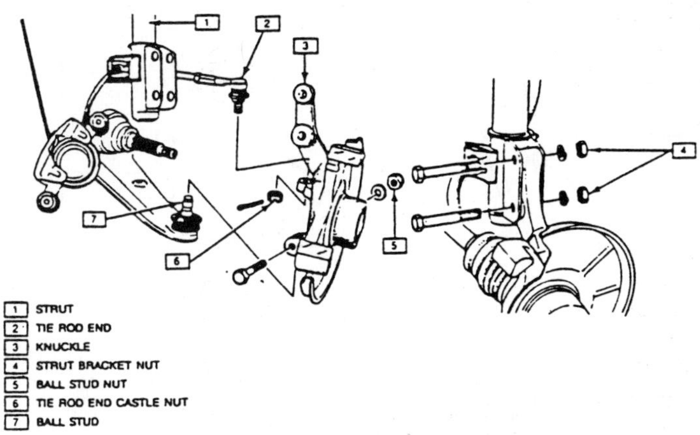

Fig. 8 Removing steering knuckle

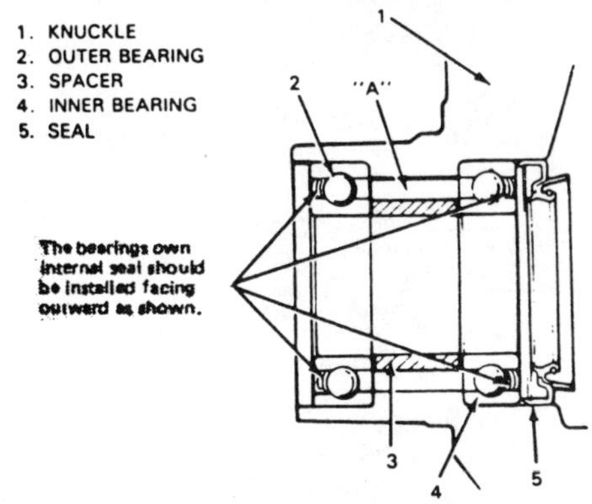

Fig. 9 Steering knuckle hub bearings & seal

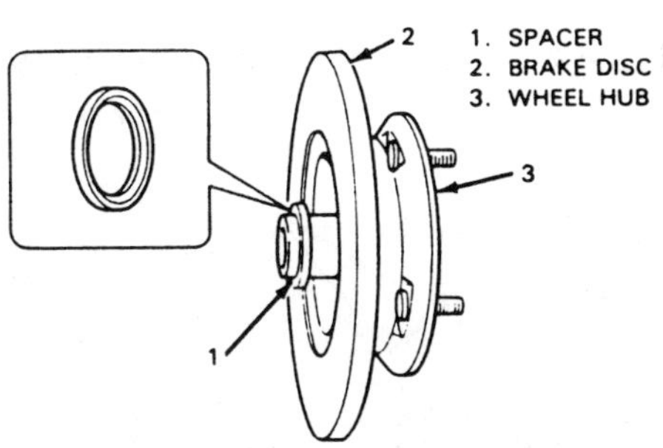

Fig. 10 Installing spacer on wheel hub

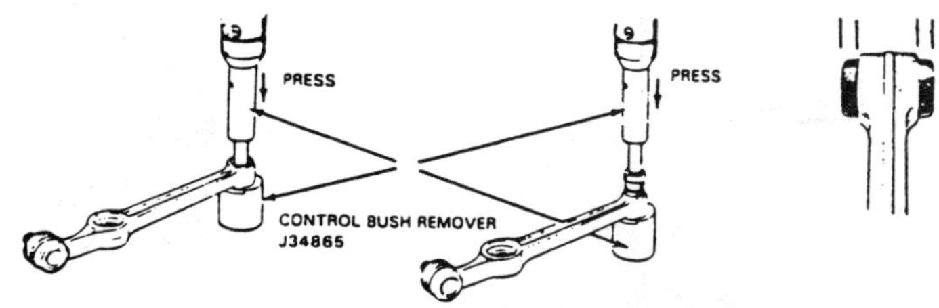

Fig. 12 Lower control arm bushing replacement. 1985–88

1. STABILIZER BAR
2. BALL STUD BOLT
3. KNUCKLE
4. BALL STUD
5. STABILIZER BAR MOUNT BRACKET BOLT
6. CONTROL ARM BOLT
7. SUSPENSION CONTROL ARM
8. STABILIZER BAR BUSHING
9. WASHER
10. CASTLE NUT
11. SPLIT PIN

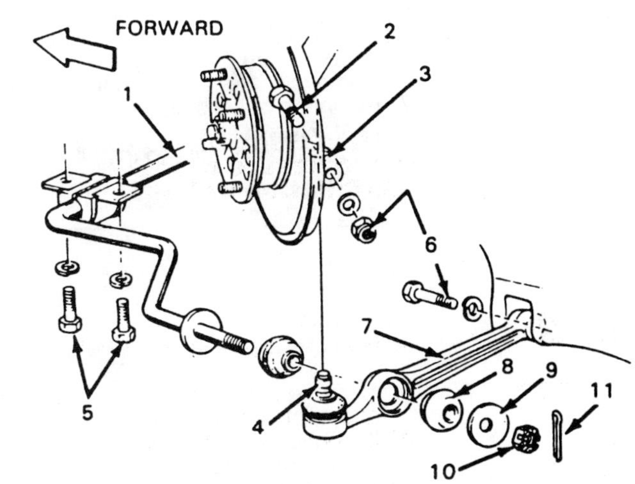

Fig. 11 Lower control arm replacement. 1985–88

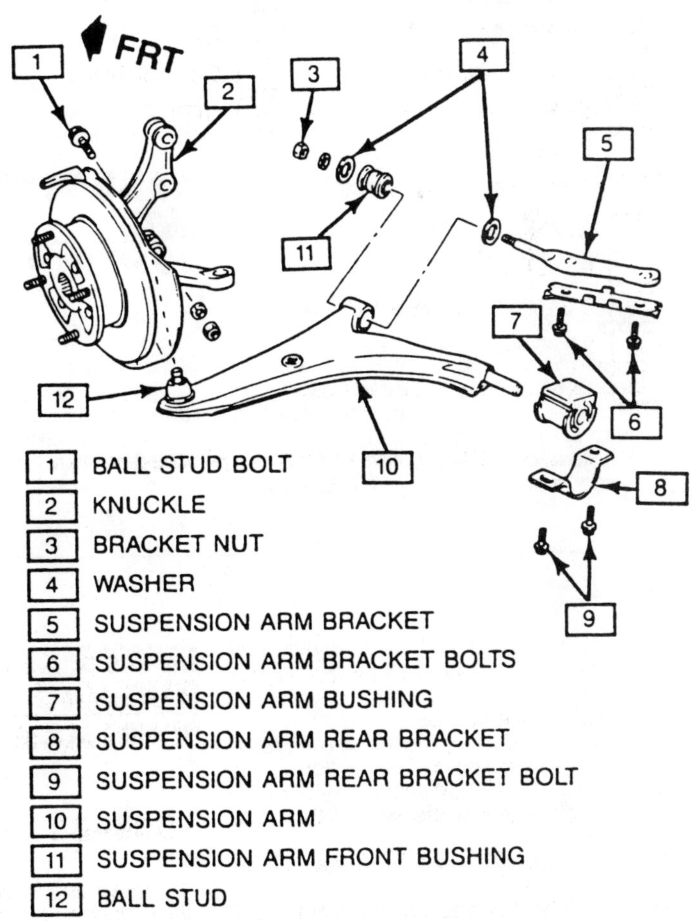

Fig. 13 Lower control arm replacement. 1989–90

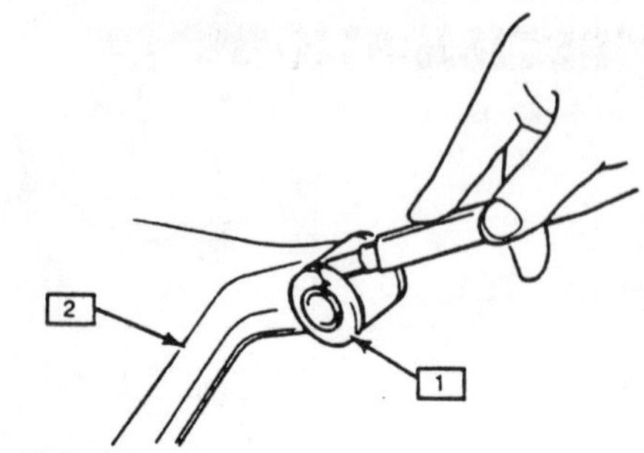

1 FRONT BUSHING
2 SUSPENSION ARM

Fig. 14 Cutting flange from control arm front bushing. 1989–90

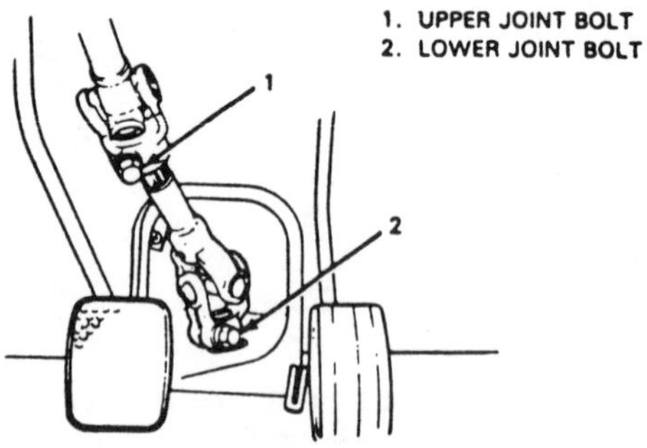

Fig. 17 Steering shaft upper & lower joint bolts

control arm bore, then position control arm and bushing to tool No. J-34865 or equivalent.
   d. Press bushing into control arm, until an equal amount of bushing is exposed from each side of control arm, **Fig. 12.**
7. Reverse procedure to install.

1989–90

1. Raise and support vehicle, then remove wheel and tire assembly.
2. Remove ball joint stud to steering knuckle pinch bolt, **Fig. 13.**
3. Remove control arm bracket nut, then remove bracket bolts.

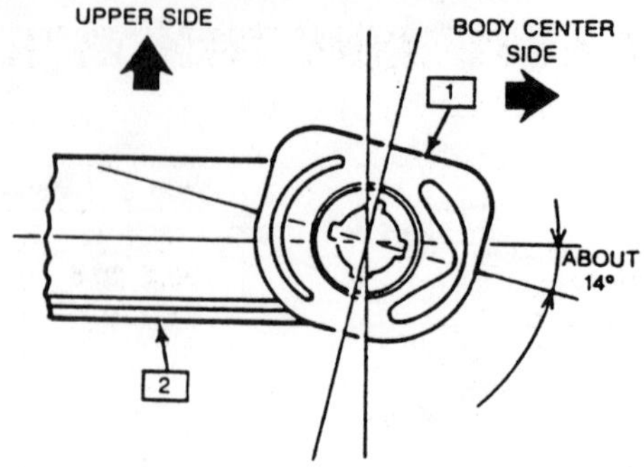

1 REAR BUSHING
2 SUSPENSION ARM

Fig. 15 Rear bushing to control arm installation. 1989–90

4. Remove control arm and bracket.
5. If control arm bushing are to be replaced, proceed as follows:
   a. Remove lower control arm as described under "Lower Control Arm, Replace."
   b. Press rear bushing from control arm, **Fig. 13.**
   c. Cut flange off front bushing, then press front bushing from control arm, **Fig. 14.**
   d. Apply soap and water to outer surface of front bushing, then press front bushing into control arm until it is centered.
   e. Position rear bushing to control arm, then drive bushing into control arm, **Figs. 15 and 16.**
6. Reverse procedure to install.

MANUAL STEERING GEAR

REPLACE

1. Slide driver's seat rearward.
2. Pull off front part of floor mat on driver's side, then remove steering shaft joint cover.
3. Loosen steering shaft upper joint bolt without removing, **Fig. 17.**
4. Remove steering shaft lower joint bolt and disconnect lower joint from pinion.

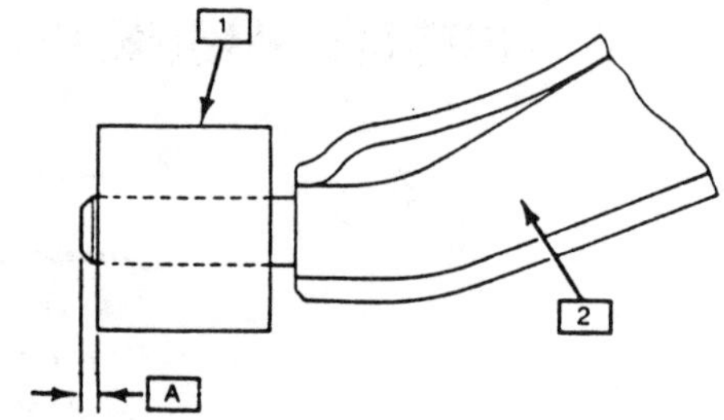

A 5mm ±1
1 REAR BUSHING
2 SUSPENSION ARM

Fig. 16 Rear bushing positioning. 1989–90

1. CAR BODY
2. STEERING GEAR CASE
3. CASE MOUNT BOLT
4. PINION SIDE BRACKET
5. RACK SIDE BRACKET

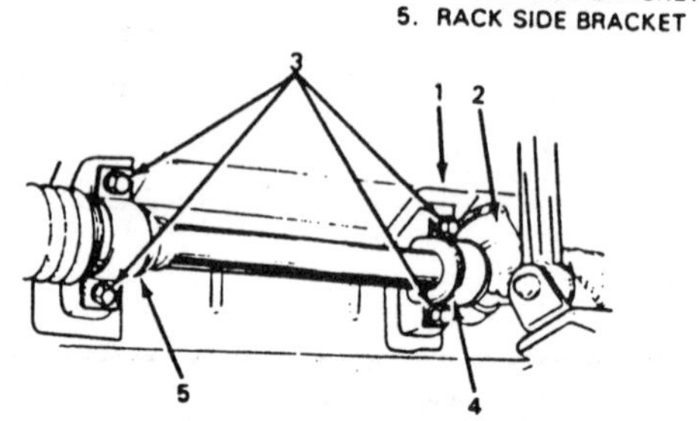

Fig. 18 Steering gear attaching bolts & brackets

5. Raise and support vehicle.
6. Remove front wheel and tire assemblies.
7. Remove cotter pins and castle nuts from tie rod ends.
8. Using tie rod end remover J-21687-02 or equivalent, disconnect tie rods from knuckles.
9. Remove steering gear housing mounting bolts, brackets and steering gear, **Fig. 18.**
10. Reverse procedure to install.

FRONT SUSPENSION & STEERING TIGHTENING SPECIFICATIONS

Component	Year & Torque/ft. lbs. 1985-88	1989-90
Ball Joint Stud Nut	36.5-50.5	44
Control Arm Bolt	36.5-50.5	—
Control Arm Bracket Bolt	—	69
Control Arm Rear Bracket Bolt	—	32
Control Arm Bracket Nut	—	92
Disc Brake Caliper Attaching Bolts	17.5-26	17.5-26
Hub Bearing Nut	108.5-195①	129②
Steering Shaft Coupling Bolts	14.5-21.5	18
Steering Gear Mounting Bracket Bolts	14.5-21.5	18
Strut Bracket Nut	50.5-65	59
Strut Nut	29-43	37
Strut Support Nut	13.5-20	20
Stabilizer Bar Attaching Bolt	22-39.5	—
Stabilizer Bar Castle Nut	29-65	—
Tie Rod End Castle Nut	22-39.5	32
Tie Rod End Lock Nut	25.5-39.5	33
Wheel & Tire Assembly	29-50.5	44

①—After tightening, install cotter pin.
②—After tightening, stake nut in position.

Pontiac Fiero

INDEX

Page No.

DESCRIPTION

1985–87

The front suspension, **Fig. 1** is a conventional short and long arm design with coil springs. The control arms are attached with bolts and bushings at the inner pivot points, and are attached to the steering knuckle/front wheel spindle assembly at the outer pivot points.

1988

The front suspension, **Fig. 2,** used on 1988 models features several improvements over the design used on earlier models. In conjunction with the elimination of the steering damper assembly, a larger stabilizer bar is used. Additionally, both the king pin angle and scrub radius have been reduced. Other notable changes include the use of a shorter spindle and an increase in both upper and lower control arm lengths. Besides producing lighter steering effort and a smoother ride, the redesigned suspension also results in a shorter turning radius.

WHEEL BEARINGS
ADJUST

1985–87

1. Raise and support vehicle.
2. Remove wheel and tire assembly.
3. Remove dust cap from hub, then the cotter pin from spindle and spindle nut.
4. Torque spindle nut to specifications while rotating wheel forward by hand.
5. Back off spindle nut until just loose, then hand tighten nut and back off again until either hole in spindle lines up with hole in nut. **Do not back off nut more than 1/2 flat.**
6. Install new cotter pin, then measure hub end play. With bearing properly adjusted, end play should measure .001-.005 inch.

WHEEL BEARINGS
REPLACE

REMOVAL

1. Raise and support vehicle.
2. Remove wheel and tire assembly, then disconnect brake caliper from steering knuckle and suspend from frame with a piece of wire. Do not let brake lines support weight of caliper.
3. Remove hub dust cap, cotter pin, spindle nut and washer.
4. Remove hub and bearing from spindle.
5. Remove outer bearing from hub, then pry out grease seal and remove inner bearing. Discard seal.

INSTALLATION

1. Clean all grease from hub, spindle and bearing.
2. Apply a thin film of suitable grease to spindle at outer bearing seat and the inner bearing seat, shoulder and seal seat.
3. Apply grease inboard of each bearing race in hub.
4. Completely fill bearing cone and roller assemblies with grease.
5. Install inner bearing into hub and apply additional grease outboard of bearing.
6. Install new grease seal flush with hub and lubricate seal lip with a thin coating of grease.
7. Install hub and rotor assembly onto spindle, then position outer bearing in race.
8. Install washer and nut, then adjust bearing as previously described.
9. Install brake caliper, then the wheel and tire assembly.

UPPER BALL JOINT
REPLACE

1985–87

1. Raise and support vehicle.
2. Remove wheel and tire assembly, then support lower control arm with a suitable jack.
3. Remove upper ball joint stud nut, then reinstall nut finger tight.
4. Install tool No. J-26407 with cup end over lower ball joint stud nut. Rotate threaded end of tool until upper stud is free of steering knuckle, then remove tool and nut from stud, **Fig. 3.**
5. Remove 2 ball joint-to-upper control arm attaching bolts and nuts.
6. Note position of flat on ball joint for proper installation, then remove ball joint from vehicle.
7. Reverse procedure to install. Torque ball joint-to-upper control arm attaching bolts to specifications. **Torque** stud nut to 35 ft. lbs., then tighten nut up to an additional 1/6 turn to align cotter pin hole.

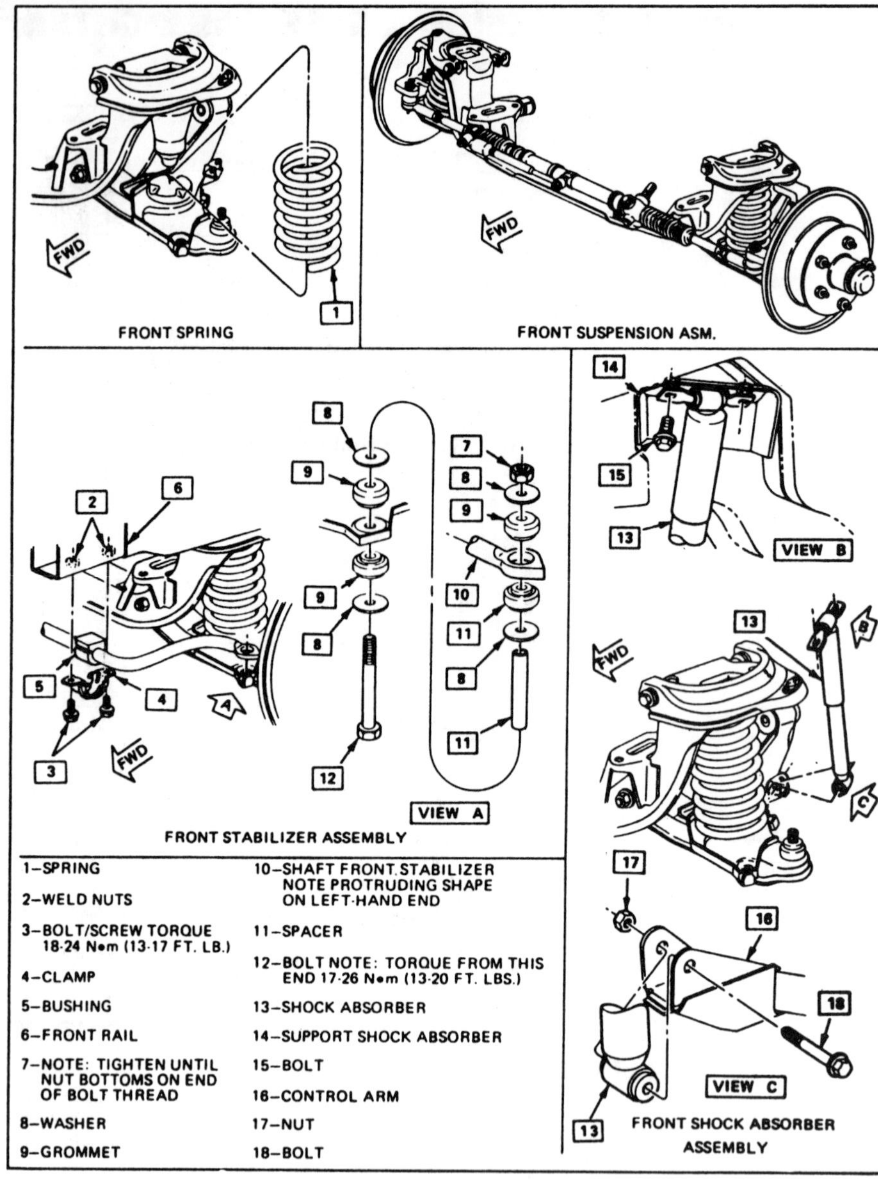

Fig. 1 Front suspension. 1985–87

1988

Removal

1. Raise and support vehicle, supporting lower control arm with suitable jack.
2. Remove wheel and tire assembly.
3. Disconnect brake line from upper control arm.
4. Disconnect tie rod end from steering knuckle and swing knuckle outboard.
5. Using tool J-26407 or equivalent, disconnect upper ball joint from steering knuckle.
6. Remove upper ball joint from control arm by drilling out three adjusting rivets, **Fig. 4.**

Installation

1. Install upper ball joint to control arm

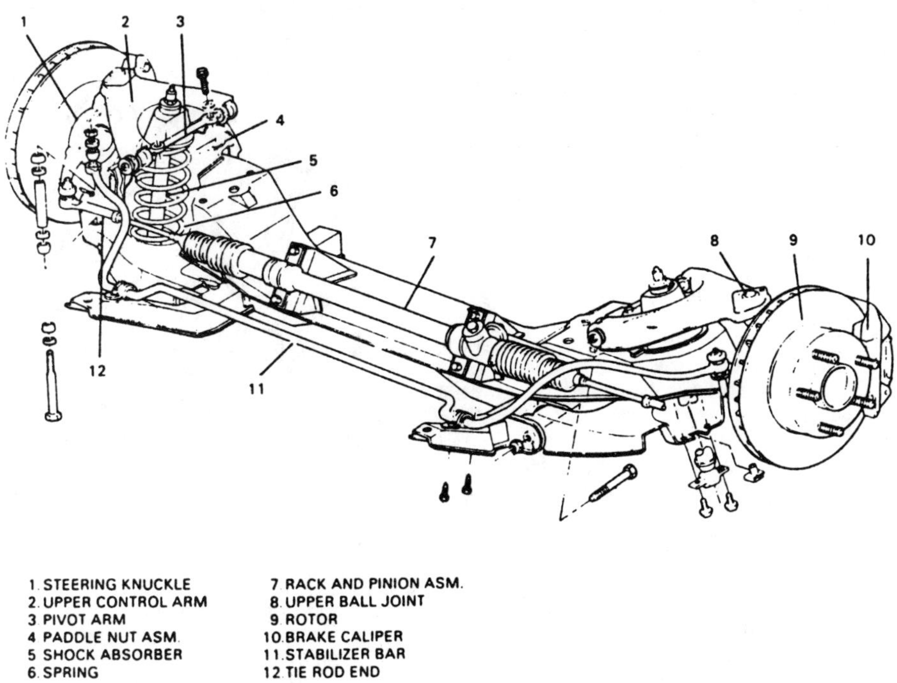

Fig. 2 Front suspension. 1988

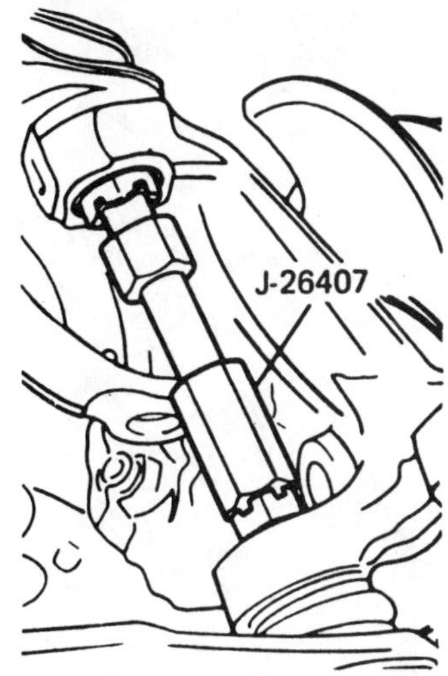

Fig. 3 Upper ball joint removal. 1985–87

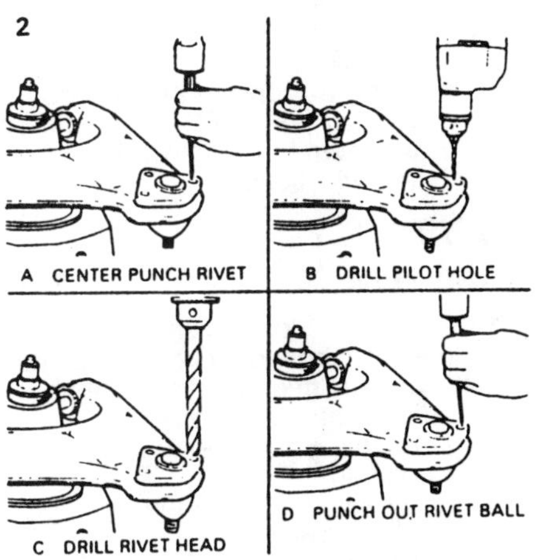

Fig. 4 Upper ball joint removal. 1988

with nuts and bolts, torquing to specifications provided in service repair package.

2. Inspect tapered hole in steering knuckle and remove any dirt. If any out of roundness, deformation or damage is noticed, replace knuckle.
3. Install upper ball joint to steering knuckle.
4. Torque nut to specifications. Nut may be turned up to an additional 1/6 turn to align cotter pin hole. **Do not exceed torque of 55 ft. lbs.**
5. Attach brake line to control arm and install tire and wheel assembly.

LOWER BALL JOINT REPLACE

1985–87

The lower ball joint is permanently attached to the lower control arm and cannot be serviced separately. If the lower ball joint requires replacement, the entire lower control arm must be replaced.

1988

Removal

1. Raise and support vehicle and support lower control arm with suitable jack.
2. Remove tire and wheel assembly and disconnect tie rod end from steering knuckle.
3. Using tool J-26407 or equivalent, disconnect lower ball joint from steering knuckle and swing knuckle with rotor, caliper and bearing out of way.
4. Inspect tapered hole in steering knuckle, removing any dirt.
5. Assemble clamp J-9519-10 and bolt J-9519-18 with removal adapters J-37161-1 and J-37161-3 on lower control arm, **Fig. 5.**
6. Press ball joint out of lower control arm.

Installation

1. Assemble clamp J-9519-10 and bolt J-9519-18 with installation adapters J-37161-2 and J-37161-4 on lower control arm, **Fig. 5.**
2. Press ball joint into lower control arm.
3. Position ball stud into steering knuckle boss and install ball nut.
4. Torque nut to specifications plus an additional 1/2 turn. Nut may be tightened up to an additional 1/6 turn to align cotter pin hole.
5. Connect tie rod to steering knuckle, torquing to 15 ft. lbs. plus an additional 1/2 turn. Nut may be tightened up to an additional 1/6 turn to align cotter pin hole.
6. Install tie and wheel assembly.

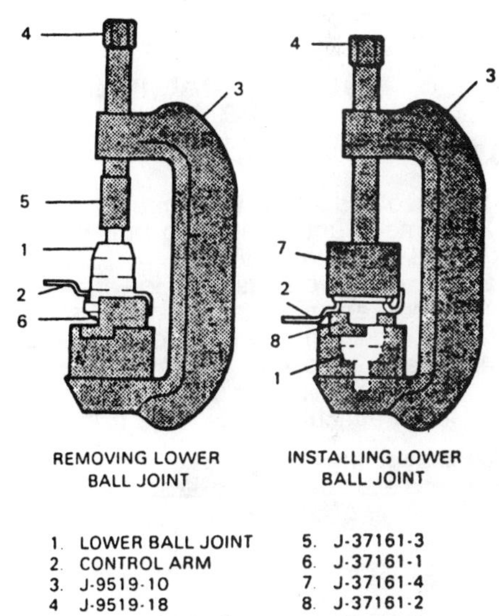

Fig. 5 Lower ball joint replacement. 1988

7. check alignment, adjusting as necessary.

SHOCK ABSORBER REPLACE

1985–87

1. Raise and support vehicle.
2. Remove wheel and tire assembly, then the 2 shock absorber upper attaching bolts.
3. Remove shock absorber lower attaching bolts and nut.
4. Remove shock absorber from vehicle.
5. Reverse procedure to install. Torque upper and lower attaching bolt to specifications.

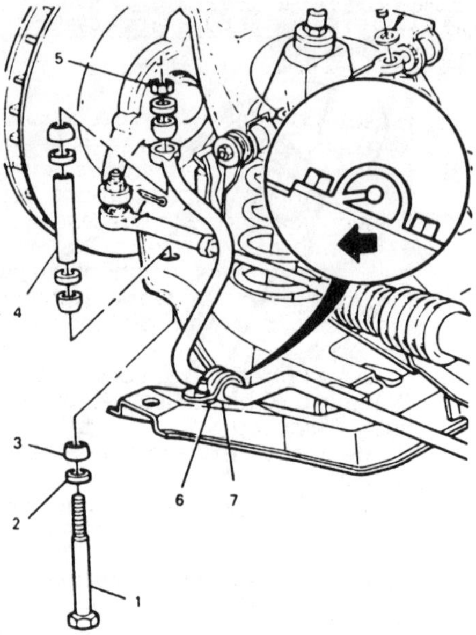

1 BOLT
2 WASHER (4)
3 INSULATOR (4)
4 SPACER
5 NUT
6 CLAMP
7 BUSHING *NOTE SLIT MUST BE TOWARD FRONT OF VEHICLE.

Fig. 6 Installing stabilizer bar. 1988

1988

1. Raise and support vehicle.
2. Remove upper retaining nut, washer and shock insulator.
3. Remove two bolts from lower end of shock absorber and remove shock absorber through lower control arm.
4. reverse procedure to install, torquing upper retaining nut to 8 ft. lbs. and the lower attaching bolts to 20 ft. lbs.

STABILIZER BAR
REPLACE

1985–87

1. Raise and support vehicle.
2. Remove stabilizer bar attaching bolt and nut and associated components from lower control arms, **Fig. 1**
3. Remove stabilizer bar clamp and the stabilizer bar from vehicle.
4. Reverse procedure to install. Torque clamp attaching bolts and stabilizer bar to control arm attaching bolt to specifications.

1988

Removal

1. Raise and support vehicle.
2. Mark stabilizer bar on right end to ensure proper installation.
3. Remove both front wheel and tire assemblies.
4. Disconnect tie rods at steering knuckles.
5. Disconnect brake line at upper control arm.
6. Remove left side caliper, rotor and splash shield, suspending caliper out of way.
7. Remove stabilizer bar to lower control arm links.
8. Remove two bolts at each stabilizer bar clamp and bushing assembly.
9. Remove stabilizer bar from underside of vehicle by moving bar toward left side of vehicle until right side clears frame rail.

Installation

1. Install stabilizer bar from underside of vehicle.
2. Install clamps and bushings and hand tighten bolts to hold stabilizer in place. **Slit in stabilizer bushing must be installed toward front of vehicle, Fig. 6.**
3. Install link assemblies from stabilizer bar to lower control arms, torquing bolts to 12 ft. lbs.
4. Torque bushing clamps to specifications.
5. Connect tie rod ends at steering knuckle. Torque nuts to specifications, plus an additional 1/2 turn. Nut may be tightened up to an additional 1/6 turn to align cotter pin hole.
6. Install left side rotor, caliper and splash shield.
7. Connect brake line to upper control arm and install tire and wheel assemblies.

LOWER CONTROL ARM & COIL SPRING
REPLACE

1985–87

1. Raise vehicle and support at frame crossmember.
2. Remove wheel and tire assembly, then disconnect stabilizer bar from lower control arm.
3. Disconnect tie rod from steering knuckle, then the shock absorber from lower control arm.
4. Support lower control arm with a suitable jack, then remove lower ball joint stud nut and press ball joint out of steering knuckle, using tool No. J-26407, **Fig. 7.** Position steering knuckle and hub aside.
5. Loosen lower control arm pivot bolts.
6. Install a safety chain through coil spring, then slowly lower jack and remove spring, **Fig. 1.**
7. Remove lower control arm attaching bolts and the lower control arm. **It may be necessary to loosen or remove steering gear attaching bolts to gain access to the control arm attaching bolt at the crossmember.**
8. Reverse procedure to install, noting the following:
   a. Install control arm attaching bolts finger tight. Do not torque bolts until all other components have been assembled and torqued to specifications.
   b. Torque ball joint stud nut to specifications.
   c. Torque stabilizer bar attaching bolt to specifications.
   d. Torque tie rod attaching nut to specifications.
   e. Torque shock absorber lower attaching bolt to specifications.
   f. If steering gear attaching bolts have been loosened or removed, install new bolts and torque to specifications.
   g. **Torque** control arm-to-body attaching bolt to 62 ft. lbs. and control arm-to-crossmember nut to specifications.

Fig. 7 Disconnecting lower ball joint

1988

Removal

1. Raise and support vehicle, supporting lower control arm with suitable jack.
2. Remove wheel and tire assembly.
3. Remove shock absorber.
4. Disconnect stabilizer bar from lower control arm.
5. Install suitable chain through spring.
6. Remove lower control arm pivot bolts and **slowly** lower jack, removing safety chain when spring is no longer under load.
7. Remove spring and spring insulator. **Any handling of springs must not cause any damage to the corrosion protection coating on springs. Hard steel contact must be avoided. Springs with a damaged corrosion protection coating should not be used until repaired.**
8. Remove cotter pin from lower ball joint stud and loosen ball stud nut one full turn.
9. Using tool J-26407 or equivalent, loosen lower ball joint from steering knuckle.
10. Remove lower ball stud nut and the lower control arm.
11. Inspect tapered hole in steering knuckle and remove any dirt. If any out of roundness, deformation or damaged is noticed, replace knuckle.

Installation

1. Place lower control arm with ball joint

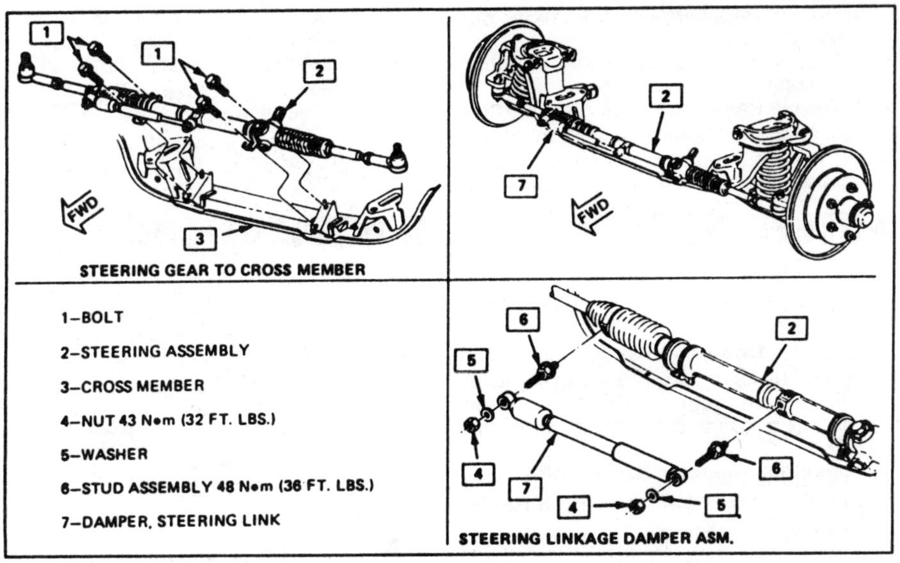

Fig. 8 Steering gear replacement. 1985–87

in position and insert lower ball stud into knuckle boss.

2. Install ball stud nut, torquing to 26 ft. lbs. plus an additional 1/2 turn. Nut may be tightened up to an additional 1/6 turn to align cotter pin hole.
3. Position spring and spring insulator into upper spring seat, then align bottom of spring with lower control arm seat and install.
4. Install a safety chain through spring.
5. Using suitable jack, slowly raise lower control arm and compress spring.
6. Align control arm bushings to crossmember and install pivot bolts, tightening slightly.
7. Remove safety chain.
8. Install shock absorber, then connect stabilizer bar, torquing bolt to 12 ft. lbs.
9. Install tire and wheel assembly, then remove jack supporting lower control arm and lower vehicle to ground.
10. With suspension system at its normal standing height, **torque** pivot bolts to 37 ft. lbs. plus an additional 3/4 turn.
11. Check alignment, adjusting as necessary.

UPPER CONTROL ARM

REPLACE

1985–87

1. Raise and support vehicle.
2. Remove wheel and tire assembly, then the rivet securing brake line clip to upper control arm.
3. Support lower control arm with a suitable jack, then remove upper ball joint as described under "Upper Ball Joint, Replace."
4. Remove control arm attaching bolt and the control arm from vehicle.
5. Reverse procedure to install. Torque control arm attaching bolt to specifications. **Washers and shims must be installed in their original positions unless a change in caster angle is desired.**

1988

1. Raise and support vehicle and support lower control arm with suitable jack.
2. Remove tire and wheel assembly.
3. Disconnect tie rod end from steering knuckle and swing knuckle outboard.
4. Using tool J-26407 or equivalent, disconnect upper ball joint from steering knuckle.
5. Remove two bolts and saddle nut assembly attaching upper control arm shaft to crossmember and remove control arm assembly.
6. Reverse procedure to install noting the following:
   a. Use new paddle nut assembly and do not torque upper control arm shaft bolts until after alignment is performed.
   b. Inspect tapered hole in steering knuckle, removing any dirt. If out of roundness, deformation or damage is noticed, replace knuckle.
   c. Torque upper ball joint to specifications. Nut may be tightened up to an additional 1/6 turn to align cotter pin hole as long as torque does not exceed 55 ft. lbs.
   d. Check alignment, adjusting as necessary.
   e. Torque upper control arm shaft to crossmember bolts to specifications, plus an additional 1/4 turn.

CROSSMEMBER BUMPER

REPLACE

1985–87

1. Remove coil spring as described under "Lower Control Arm & Coil Spring, Replace."
2. Remove crossmember bumper from vehicle.
3. Reverse procedure to install.

STEERING KNUCKLE

REPLACE

1. Raise and support vehicle. Support lower control arm with a suitable jack.
2. Remove wheel and tire assembly, then disconnect brake caliper from steering knuckle and suspend from frame with a piece of wire. Do not let brake lines support weight of caliper. **Install a block of wood between brake shoes to hold piston in caliper bore.**
3. Remove hub, disc and splash shield.
4. Remove both ball joint studs as previously described.
5. Disconnect tie rod end from steering knuckle.
6. Press both ball joint studs from steering knuckle using tool No. J-26407 or equivalent.
7. Remove ball joint stud nuts, then the steering knuckle.
8. **On 1985-87 models,** reverse procedure to install. Torque the following to specifications: splash shield bolts, tie rod nut, upper and lower ball joint stud nuts.
9. **On 1988 models,** reverse procedure to install. Torque the following to specifications: upper ball joint, plus an additional 1/6 turn to align cotter pin hole as long as torque does not exceed 55 ft. lbs.; lower ball joint, plus an additional 1/2 turn, turning nut up to an additional 1/6 turn to align cotter pin hole; tie rod end to steering knuckle, plus an additional 1/2 turn, turning nut up to an additional 1/6 turn to align cotter pin hole; hub and bearing assembly attaching bolts.

MANUAL STEERING GEAR

REPLACE

1985–87

1. Raise and support vehicle.
2. Remove both front crossmember braces.
3. Remove flex coupling pinch bolt from shaft.
4. Remove outer tie rod cotter pins and nuts on both sides, and disconnect tie rods from steering knuckles.
5. Remove 4 steering gear attaching bolts and the steering gear, **Fig. 8.**
6. Reverse procedure to install. Torque the following to specifications: flex coupling bolt, steering gear bolts, tie rod nuts, plus up to an additional 1/6 turn to align cotter pin hole and front crossmember brace bolts.

1988

1. Raise and support vehicle.
2. Remove flexible coupling to shaft pinch bolt.
3. Remove outer tie rod cotter pins and nuts on both sides.
4. Disconnect tie rods from steering knuckles.
5. Remove four bolts securing steering

assembly to crossmember and remove steering assembly.

6. reverse procedure to install, noting the following:
   a. Use new steering assembly to crossmember attaching nuts.
   b. Torque flexible coupling bolt to specifications.
   c. Torque steering assembly attaching bolts to specifications.
   d. Torque tie rod to knuckle attaching nuts to specifications, plus an additional 1/2 turn. Nut may be tightened up to an additional 1/6 turn (39 ft. lbs.) to align cotter pin hole.
   e. Check wheel alignment, adjusting as necessary.

FRONT SUSPENSION & STEERING TIGHTENING SPECIFICATIONS

Year	Component	Torque/ ft. lbs.
1985–87	Ball Joint To Upper Control Arm Attaching Bolt	28
	Flex Coupling Bolt	46
	Front Crossmember Brace Bolts	20
	Lower Control Arm Ball Joint Stud Nut	55
	Lower Control Arm Tie Rod Attaching Nut	29
	Lower Control Arm To Body Attaching Bolts	62
	Lower Control Arm To Crossmember Attaching Bolts	52
	Lower Shock Bolt	35
	Stabilizer Bar Clamp Attaching Bolts	15
	Stabilizer Bar To Lower Control Arm Attaching Bolts	16
	Spindle Nut	12
	Splash Shield	7
	Steering Gear Bolts	21
	Steering Knuckle To Upper Ball Joint Stud Nut	35
	Upper Control Arm Attaching Bolts	66
	Upper Shock Bolts	20
1988	Flex Coupling Bolt	46
	Lower Ball Joint To Steering Knuckle	26
	Stabilizer Bar Bushing Clamps	20
	Steering Assembly Attaching Bolts	20
	Tie Rod To Steering Knuckle	15
	Upper Ball Joint To Steering Knuckle	30-40
	Upper Control Arm Shaft To Steering Knuckle	52

Pontiac LeMans

INDEX

STEERING KNUCKLE & STRUT ASSEMBLY
REPLACE

1. Loosen two upper strut to body attaching nuts, **Fig. 1.**
2. Loosen wheel lug nuts, then remove drive axle shaft retaining pin, nut and washer.
3. Raise and support front of vehicle. Support vehicle on suitable jackstands.
4. Remove wheel and tire assembly.
5. Install modified outer seal protector J-28712 or equivalent.
6. Remove brake caliper and position aside, leaving brake line attached.
7. Remove brake disc attaching screws and the disc.
8. Remove outer tie rod to steering arm attaching nuts, then separate tie rod from arm using tool No. J-24319-01 or other suitable puller.
9. Remove lower ball joint retaining clip and stud nut, then separate ball joint from steering knuckle using tool J-36226 or equivalent. **Disregard "This Side Towards Wheel" marking on tool. Tool must be turned to opposite position when used on these vehicles.**
10. Support drive axle, then separate the axle from front wheel hub using tool No. J-37105 or equivalent and remove two strut to body attaching nuts and steering knuckle and strut assembly from vehicle.
11. Reverse procedure to install, noting the following:
    a. Torque strut to body attaching nuts to specification listed at the end of this section.
    b. Torque lower ball joint nut to specification listed at the end of this section.
    c. Torque tie rod attaching nut to specification listed at the end of this section.
    d. Torque disc to wheel hub bolt to specification listed at the end of this section.
    e. With weight of vehicle on wheels, **torque** new drive axle to hub nut to 74 ft. lbs., then back off nut and retorque to 15 ft. lbs. Tighten nut an additional 1/4 turn, then install cotter pin. If necessary, loosen nut slightly to align cotter pin holes.

STRUT ASSEMBLY SERVICE

1. Secure strut assembly in compressor tool No. J-34013 with adapter J-34013-87 or equivalents, **Fig. 2.**
2. Turn compressor forcing screw to compress spring slightly, then remove upper bearing dust cover.
3. Remove dampener shaft nut while using a suitable socket to prevent shaft from turning.
4. Guide dampener shaft out of strut assembly while loosening compressor screw.
5. Continue to loosen compressor screw, then remove strut dampener and spring.
6. Remove thrust washer, upper strut mount, raised edge washer and plastic mount from top of strut, **Fig. 1.**
7. Remove strut shield, upper spring insulator, spring and strut bumper.
8. Remove strut cartridge nut and the cartridge using tool No. J-36804 or equivalent. **Exercise extreme caution as strut cartridge nut is under high torque pressure.**
9. Reverse procedure to assemble, noting the following:
    a. Install new strut cartridge nut and tighten to torque listed at the end of this section. Do not remove wax lubricant from nut.
    b. Apply suitable lubricant to upper strut bearings prior to assembly.
    c. Position spring as shown, **Fig. 3,** and move assembly upright in strut compressor before installing upper locking pin.
    d. Tighten dampener shaft nut to torque listed at the end of this section, while preventing nut from turning.

CONTROL ARM
REPLACE

1. Raise and support front of vehicle. Support vehicle on suitable jackstands placed under frame.
2. Remove wheel and tire assembly.
3. Disconnect stabilizer bar from control arm.
4. Install modified outer seal protector J-28712 or equivalent.
5. Remove ball joint retaining clip and stud nut, then separate ball joint from steering knuckle using tool No. J-36226 or equivalent. **Disregard "This Side Towards Wheel" marking on tool. Tool must be turned to opposite position when used on these vehicles.**
6. Remove control arm attaching bolts and the control arm.
7. If necessary, replace control arm bushings as shown in **Figs. 4 through 7.**
8. Reverse procedure to install. Tighten attachments to torque listed at the end of this section.

BALL JOINT INSPECTION

1. Raise and support front of vehicle so that suspension is allowed to hang free.
2. Grasp wheel and tire assembly at 6 and 12 o'clock positions, then rock top of wheel and tire assembly inward and outward.
3. While rocking wheel and tire assembly, observe movement between steering knuckle and control arm. If any horizontal movement is present, replace ball joint.
4. If ball joint is disconnected from steering knuckle, use finger pressure to try to twist ball joint in its socket. If ball joint can be twisted in its socket, replace ball joint.

BALL JOINT
REPLACE

1. Remove control arm as previously described.
2. Drill heads off ball joint rivets, then punch out rivets using a suitable drift and remove ball joint.
3. Reverse procedure to install. Tighten ball joint retaining nuts from below the control arm to torque listed at the end of this section.

STABILIZER BAR
REPLACE

1. Raise vehicle and support so that

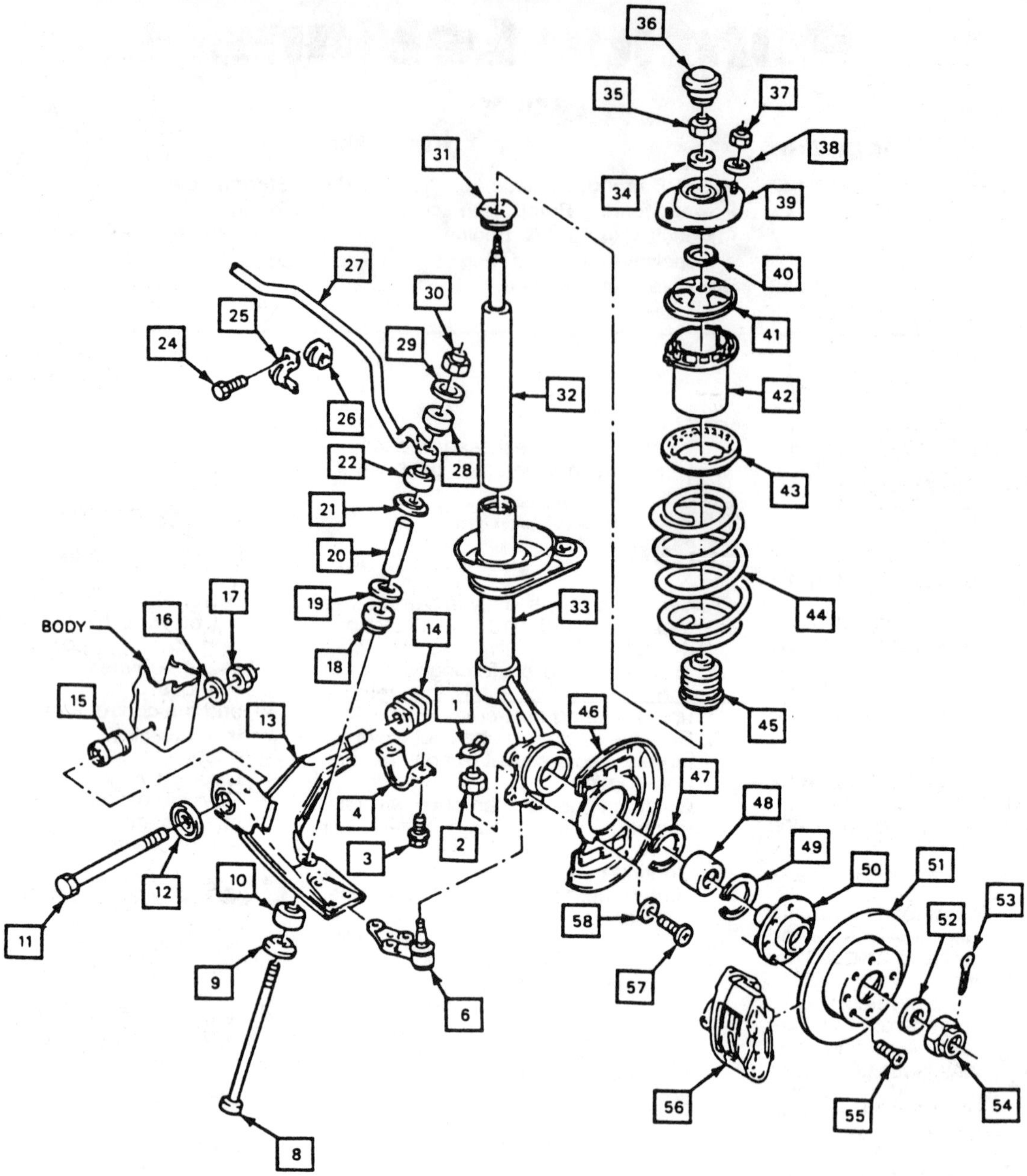

1. CLIP, RETAINING
2. NUT
3. BOLT
4. CLAMP, FRT STAB SHF
5.
6. BALL JOINT, LOWER CONT ARM
7.
8. BOLT
9. WASHER
10. GROMMET, STAB SHF LINK
11. BOLT
12. WASHER
13. ARM, LOWER CONTROL
14. BUSHING, LOWER CONTROL REAR
15. BUSHING, LOWER CONTROL FRONT
16. WASHER
17. NUT
18. GROMMET, STAB SHF LINK
19. WASHER
20. SPACER
21. WASHER
22. GROMMET, STAB SHF LINK
23.
24. BOLT
25. CLAMP
26. INSULATOR, STAB SHF
27. SHAFT, FRONT STAB
28. GROMMET, STAB SHF LINK
29. WASHER
30. NUT
31. NUT, STRUT CARTRIDGE CLOSURE
32. CARTRIDGE, STRUT
33. KNUCKLE/SUPT TUBE
34. WASHER, THRUST
35. NUT, PISTON ROD
36. COVER, UPPER BEARING DUST
37. NUT
38. WASHER
39. MOUNT (BEARING) UPPER STRUT
40. WASHER, RAISED EDGE
41. MOUNT (PLASTIC)
42. SHIELD, STRUT
43. INSULATOR, UPPER SPRING
44. SPRING
45. BUMPER, STRUT
46. SHIELD, FRONT BRAKE
47. RING, SNAP
48. BEARING, FRONT WHEEL
49. RING, SNAP
50. HUB, FRONT WHEEL
51. DISC, BRAKE
52. WASHER
53. PIN, COTTER
54. NUT, DRIVE SHAFT
55. SCREW, STOP
56. CALIPER, FRONT BRAKE
57. SCREW
58. WASHER

Fig. 1 Exploded view of front suspension

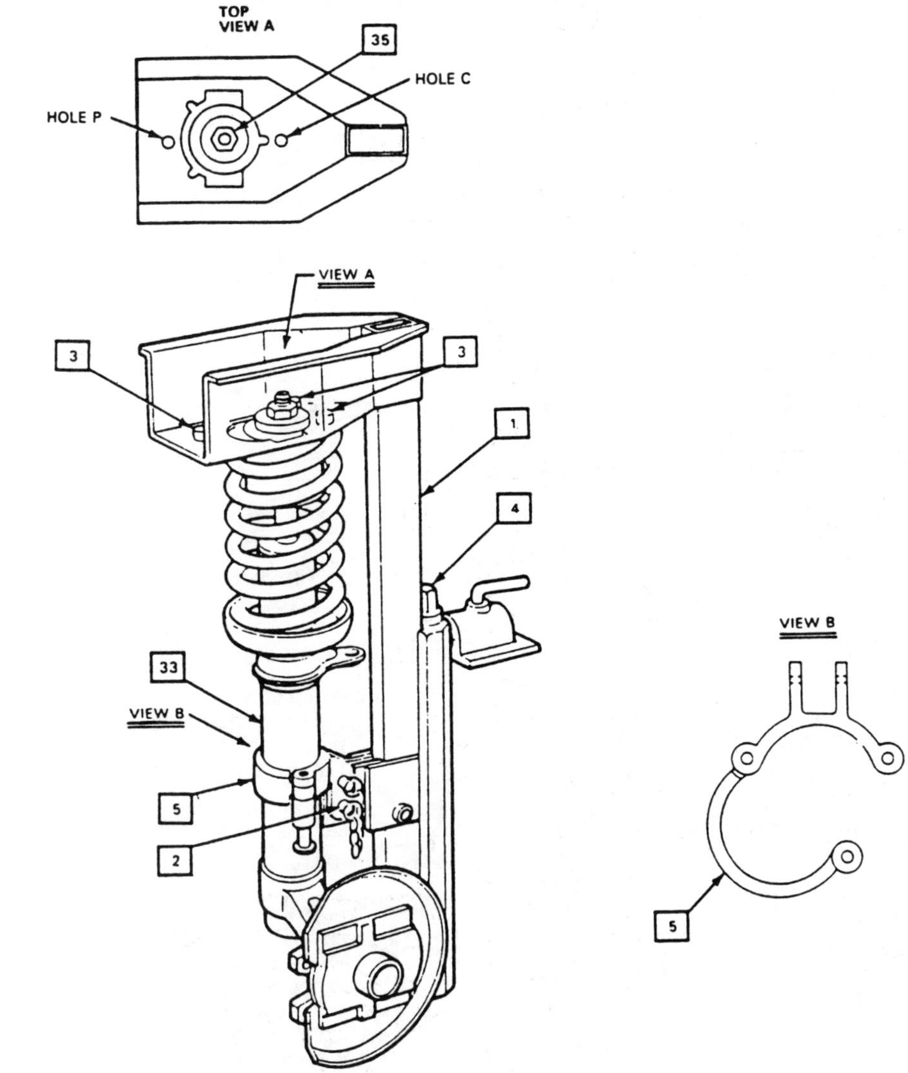

1. STRUT COMPRESSOR J 34013
2. INSTALL LOCKING PINS THROUGH STRUT ASSEMBLY
3. TIGHTEN NUTS TILL FLUSH WITH STRUT COMPRESSOR (SEE VIEW A)
4. COMPRESSOR FORCING SCREW
5. CLAMPING TOOL J 34013-87 (SEE VIEW B)
33. STRUT ASSEMBLY
35. NUT, PISTON ROD 30 N·m (22 LBS. FT.)

Fig. 2 Strut disassembly

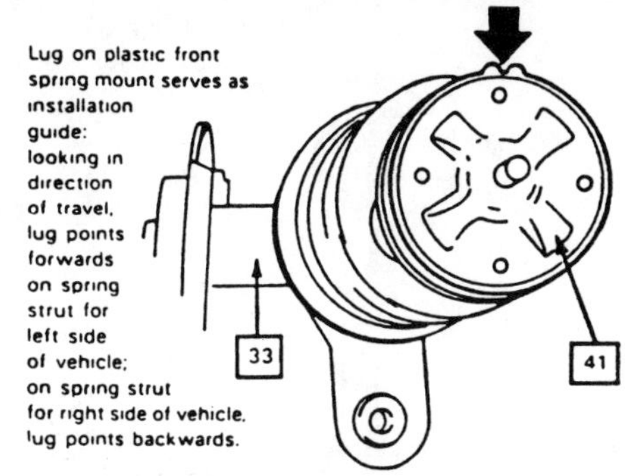

Fig. 3 Spring mount alignment

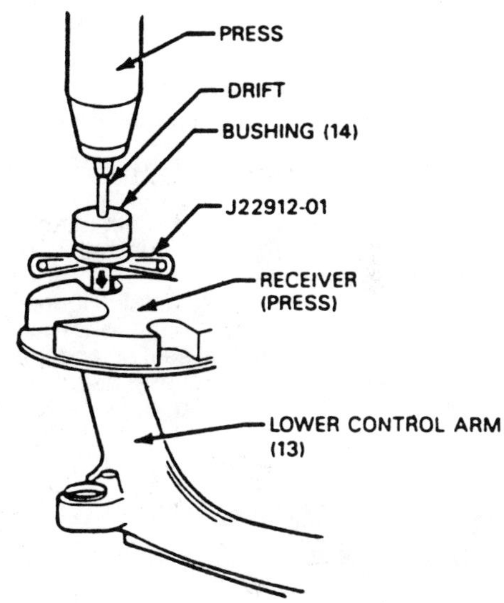

Fig. 4 Control arm rear bushing removal

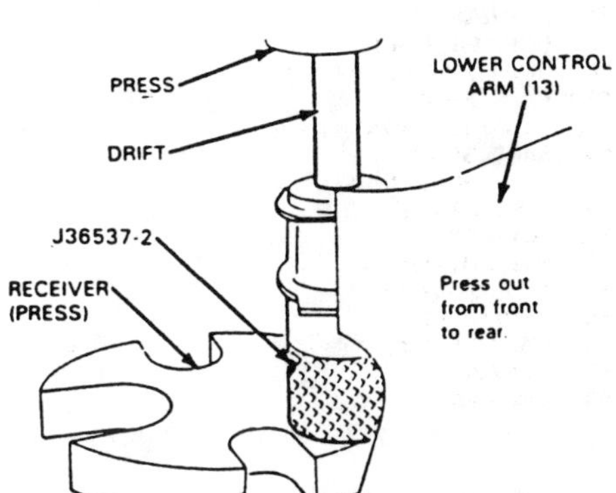

Fig. 5 Control arm front bushing removal

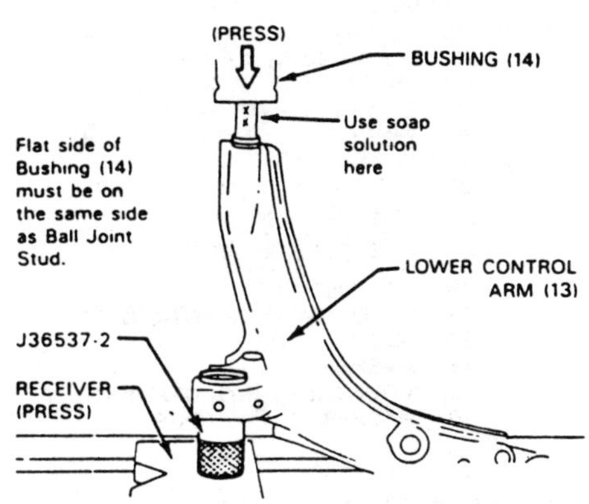

Fig. 6 Control arm rear bushing installation

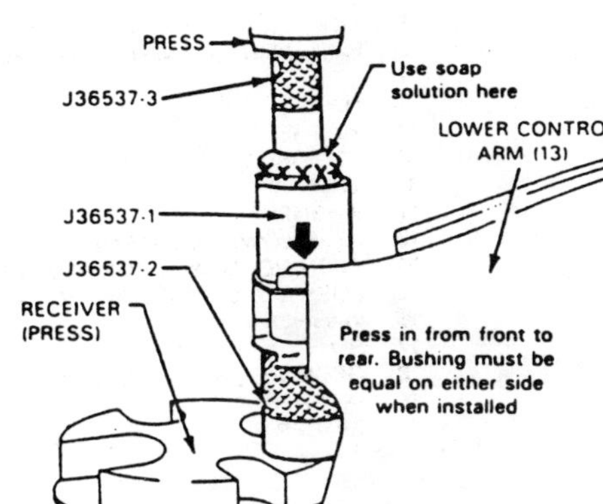

Fig. 7 Control arm front bushing installation

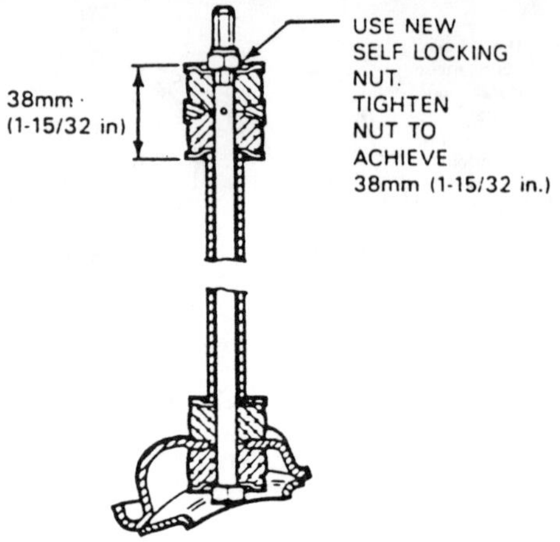

Fig. 8 Stabilizer bar attaching nut installation

front suspension hangs free.

2. Remove front wheel and tire assemblies.
3. Disconnect stabilizer bar link assemblies from control arms, **Fig. 1.**
4. Remove stabilizer bar bracket attaching bolts, then the bushings, brackets and stabilizer bar.
5. Reverse procedure to install. Tighten stabilizer bar to body bolts to torque listed at the end of this section and the stabilizer bar to control arm nuts to a length of 1 15/32 inch, **Fig. 8.**

HUB & BEARING REPLACE

1. Remove steering knuckle as previously described.
2. Remove hub from steering knuckle and wheel bearing using tools J-37105 or equivalent, **Fig. 9.**
3. Using tool Nos. J-22888 or equivalent, remove bearing race from hub, **Fig. 10.**
4. Remove snap rings retaining bearing to steering knuckle, then remove bearing using tool J-37105 or equivalent, **Fig. 11.**
5. Position snap ring into steering knuckle bore, then install bearing using tool No. J-37105 or equivalent.
6. Using tool No. J-37105 or equivalent, pull hub onto bearing assembly and steering knuckle, then install snap ring.

STEERING GEAR REPLACE

1. Disconnect battery ground cable, then remove air cleaner assembly.
2. Remove both pinch bolts from coupling assembly, then slide coupling up on steering spindle. Position steering wheel straight ahead as shown in **Fig. 12.**
3. **On models with power steering,** remove air cleaner assembly.
4. **On all models,** pry lock plate in half, then remove both tie rod bolts from center of steering gear housing.
5. **On models equipped with power steering gear,** disconnect fluid lines from gear.
6. **On all models,** remove two damper brackets and mounting clamps, as equipped.
7. Remove dash seal from gear, then remove steering gear through right hand wheel opening.
8. Reverse procedure to install, noting the following:
   a. After second reuse of mounting clamp studs, thread locking kit No. 1052624 or equivalent must be used.
   b. Tighten mounting clamp attaching nuts to torque listed at the end of this section.
   c. Tighten coupling to steering column attaching bolts to torque listed at the end of this section.

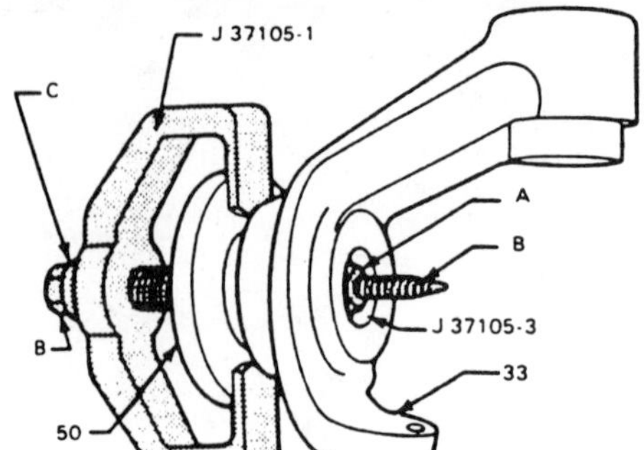

A NUT
B BOLT
C BEARING
33 KNUCKLE AND SUPPORT TUBE
50 HUB

Fig. 9 Removing hub from steering knuckle

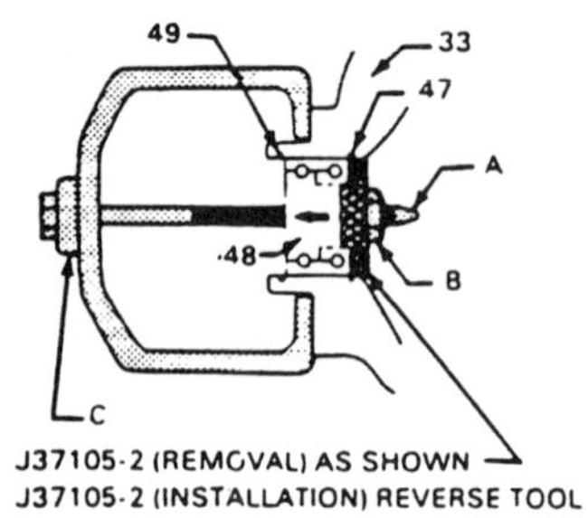

A BOLT
B NUT
C BEARING
33 KNUCKLE AND SUPPORT TUBE
47 RING-SNAP-MUST BE REMOVED FOR BEARING REMOVAL. INSTALL AFTER HUB (50) INSTALLATION
48 BEARING-FRONT WHEEL
49 RING-SNAP-MUST BE INSTALLED FOR BEARING INSTALLATION AND REMOVED FOR BEARING REMOVAL.

Fig. 11 Removing bearing from steering knuckle bore

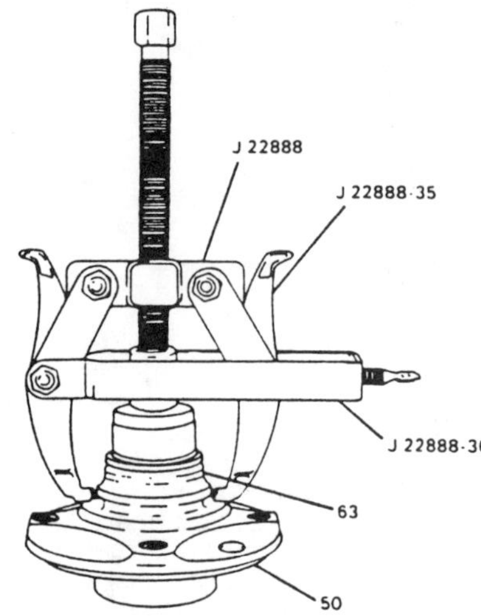

50 HUB-FRONT WHEEL
63 INNER RACE

Fig. 10 Remove bearing inner race from hub

POWER STEERING PUMP REPLACE

MODELS w/4-98/1.6L ENGINE

1. Disconnect battery ground cable.
2. Remove power steering pump drive belt, then three attaching bolts and power steering pulley.
3. Disconnect fluid lines from pump.
4. Remove pump to upper timing belt cover attaching clips, then remove clips.
5. Remove power steering pump mounting bolts, then remove pump from vehicle.
6. Reverse procedure to install. After completing installation, bleed system by cycling steering wheel from side to side without contacting stops. Keep fluid at proper level during bleeding procedure.

MODELS w/4-121/2.0L ENGINE

1. Disconnect battery ground cable.
2. Remove power steering pump drive belt, then disconnect fluid lines from pump.
3. Remove power steering pump to bracket attaching bolts through holes in pulley.
4. Remove power steering pump from vehicle.
5. Reverse procedure to install. After completing installation, bleed system by cycling steering wheel from side to side without contacting stops. Keep fluid at proper level during bleeding procedure.

STEERING, POSITION STRAIGHT AHEAD

After all necessary operations on the steering gear or the steering column assembly are completed (removing and reinstalling, disassembling and re-assembling), the exact straight ahead position of the steering is to be checked in each case

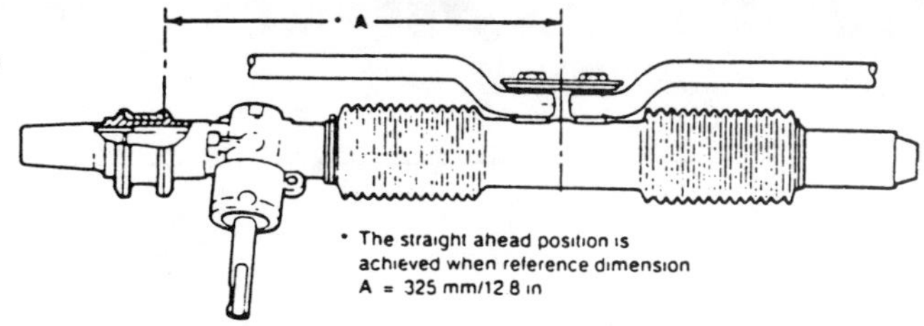

To make check, place Measuring Gage J36666 between end of tie rod and left hand mounting recess for steering gear (looking in direction of travel).

Hold steering wheel in this exact straight ahead position

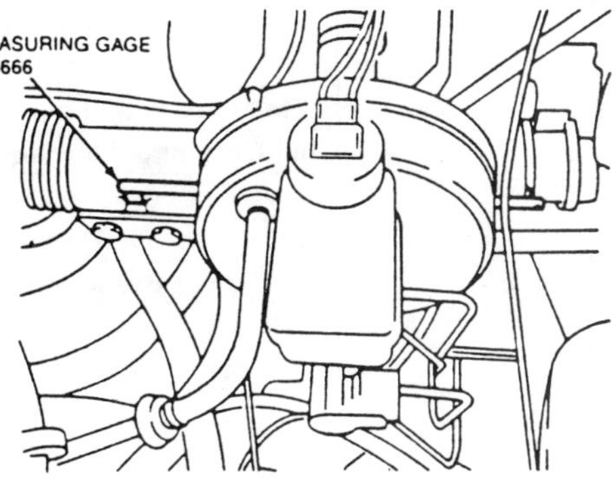

The following conditions must be fulfilled

1 The steering spindle clamp flange bolt must be lying horizontally on top

2 The steering wheel must be centered on the steering spindle

(Steering wheel spokes centered diagonally and pointing downwards)

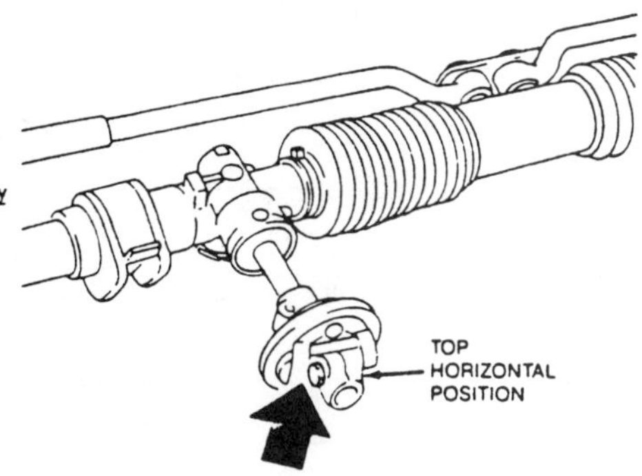

1a If clamp flange bolt is slanted pinion is displaced one tooth on rack (assembly error) — steering pinion position must be corrected

2a If steering wheel is off center by more than ± 5° (clamp flange bolt must lie horizontally on top), pull steering wheel off of steering spindle using J36541 and center on spindle splines.

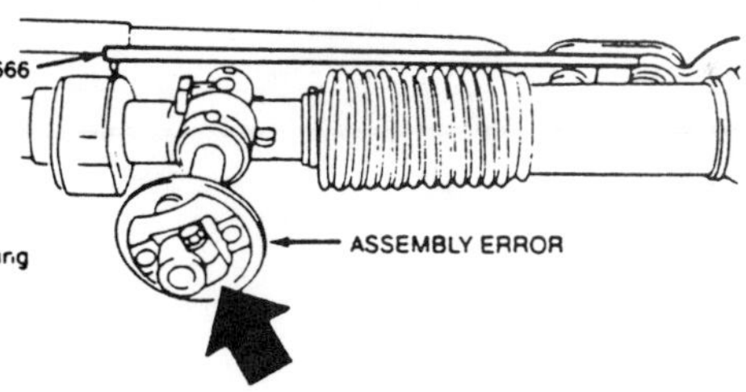

Fig. 12 Adjusting steering wheel position

FRONT SUSPENSION & STEERING TIGHTENING SPECIFICATIONS

Year	Component	Torque/ ft. lbs.
1988-90	Ball Joint To Strut/Knuckle Assembly Nut	50
	Control Arm Front Mounting Bolts	100
	Control Arm Rear Mounting Bolts	50
	Disc Brake Rotor To Hub Screw	3
	Flex Coupling To Stub Shaft Clamp Bolt	16
	Flex Coupling To Steering Shaft Clamp Bolt	16
	Hub & Bearing Nut	①
	Inner Tie Rod To Steering Gear	66
	Power Steering Fluid Line Fittings	20
	Power Steering Pump Attaching Bolts	18
	Power Steering Pump Pulley Attaching Bolts	20
	Stabilizer Bar To Body Clamp Bolts	29
	Steering Damper Bracket To Steering Gear Bolts	16
	Steering Gear Mounting Clamp Bolts & Nuts	28
	Steering Gear Retaining Bracket Nuts	16
	Strut To Body Nuts	18
	Strut Cartridge Retaining Nut	145
	Strut Push Rod Nut	40
	Tie Rod End To Steering Knuckle Assembly Nut	45
	Tie Rod Pinch Bolts	16
	Wheel & Tire Assembly	65

REAR SUSPENSION

INDEX

CHRYSLER MOTORS/EAGLE

TABLE OF CONTENTS

Chrysler Motors Colt, Colt Vista & Conquest & Eagle Summit

INDEX

SHOCK ABSORBER REPLACE

COLT, COLT VISTA 2 WHEEL DRIVE & SUMMIT

1. Raise rear of vehicle and position jack stands under frame side rails.
2. Remove wheel and tire assembly.
3. Using a suitable jack, support lower suspension arm, then disconnect shock absorber from upper and lower mounting and remove from vehicle, **Figs. 1 and 2.**
4. Reverse procedure to install.

CONQUEST

1. Raise and support rear of vehicle, then remove wheel and tire assembly.
2. Disconnect rear brake hose at shock assembly, then separate drive axle from companion flange.
3. Remove shock assembly to axle housing attaching bolts, then the shock assembly from axle housing, **Fig. 3.** Push axle housing down while opening coupling with a suitable tool.
4. Remove shock assembly attaching bolts from under side trim in rear hatch area, then the shock assembly from rear wheel housing.
5. Compress coil spring using tool L-4514 or equivalent, then remove top end nut.
6. Remove shock insulator, spring seat, dust cover, rubber helper, then the seat.
7. Reverse procedure to install.

COIL SPRING REPLACE

COLT HATCHBACK & SEDAN, COLT VISTA 2 WHEEL DRIVE & SUMMIT

1. Raise rear of vehicle and position

jackstands under frame side rails to support body.
2. Remove wheel and tire assembly, then remove rear drum brake assembly.
3. Disconnect muffler from exhaust pipe, then remove muffler assembly.
4. Position a suitable jack under suspension arm, then raise suspension arm slightly.
5. Disconnect shock absorber from upper and lower mounting, **Figs. 1 and 2.**
6. Carefully lower jack to relieve spring tension, then remove coil spring. When removing spring, note location of upper and lower spring seats.
7. Reverse procedure to install. After completing installation, bleed brake system and adjust rear wheel bearings.

1988–90 COLT WAGON

2 Wheel Drive

1. Raise and support vehicle, then remove wheel and tire assembly and rear drum brake assembly, **Fig. 4.**
2. Remove brake tube. Disconnect brake hose from torsion axle and arm assembly.
3. Position a piece of wood between a suitable jack and the torsion axle and arm assembly. Do not apply jack to lateral rod. After raising torsion axle and arm slightly, remove lateral rod from body.
4. Remove torsion axle and arm from body.
5. Disconnect shock absorber upper and lower mounts, then remove shock absorber from vehicle.
6. Remove coil spring and upper spring seat. Note location of spring seats.
7. Reverse procedure to install. Note the following:
   a. Coil springs have colored identification marks which indicate specific application. When replacing springs, ensure that replacement springs are appropriately marked.
   b. After completing installation, bleed brake system and adjust rear wheel bearings.

4 Wheel Drive

1. Raise and support rear of vehicle at body, then remove wheel and tire assembly. Support rear axle using a suitable jack.
2. Remove stabilizer bar link nut, **Fig. 5.**
3. Remove load sensing proportioning valve mounting bracket bolt.
4. Remove lateral rod to axle housing attaching bolt.
5. Remove shock absorber lower attaching bolt.
6. Carefully lower jack positioned under rear housing until coil spring can be removed.
7. Reverse procedure to install.

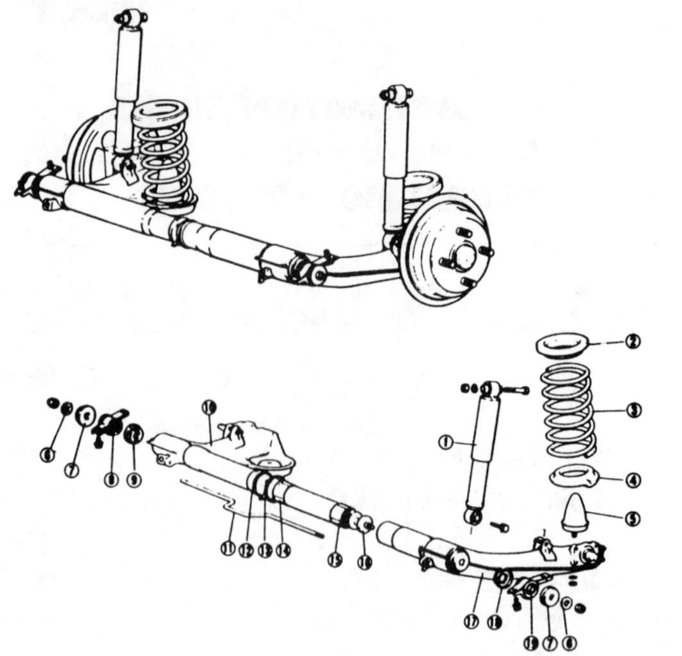

(1) Shock absorber
(2) Spring upper seat
(3) Coil spring
(4) Spring lower seat
(5) Bump stopper
(6) Plain washer
(7) Washer B
(8) Fixture assembly, R.H.
(9) Washer C
(10) Suspension arm, R.H.
(11) Stabilizer bar
(12) Dust cover
(13) Clamp
(14) Bushing A
(15) Bushing B
(16) Rubber stopper
(17) Suspension arm, L.H.
(18) Washer A
(19) Fixture assembly, L.H.

Fig. 1 Coil spring rear suspension. Colt Vista 2 wheel drive & 1985–88 Colt except 1988 Colt Wagon

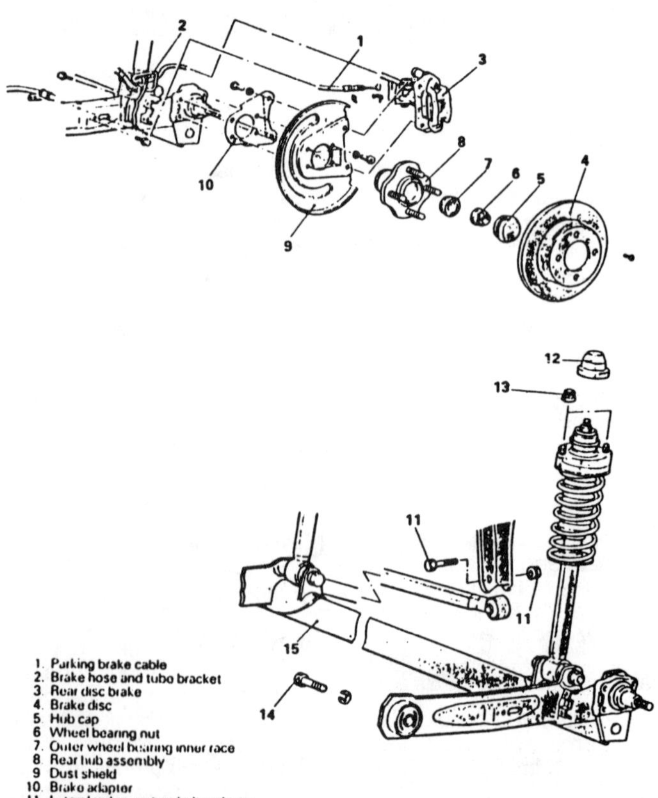

Fig. 2 Coil spring rear suspension. 1989–90 Colt Hatchback & Sedan & Summit

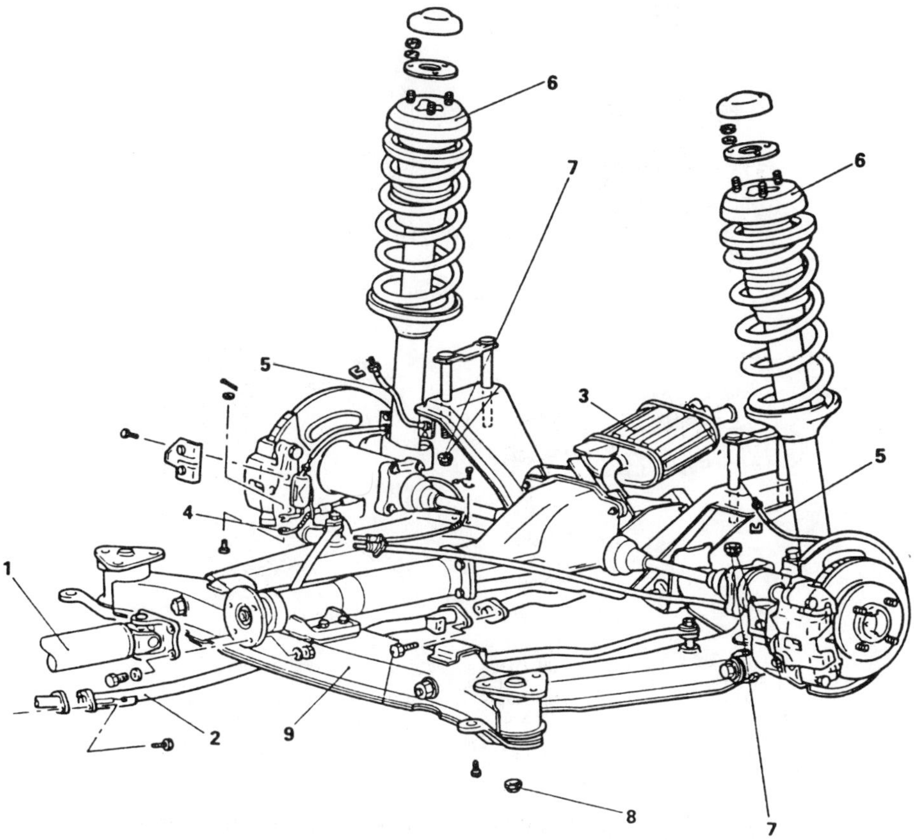

1 Propeller shaft
2 Center exhaust pipe
3 Main muffler
Parking brake cable adjustment
4. Parking brake cable connection
5 Rear brake hose connection
6. Strut assembly connection
7. Crossmember mounting nut (Self-locking nut)
8. Pin assembly to front support mounting nut (Self-locking nut)
9. Suspension assembly

Fig. 3 Coil spring rear suspension. Conquest

CONTROL ARMS REPLACE

CONQUEST

1. Raise and support rear of vehicle, then remove wheel and tire assembly.
2. Disconnect parking brake cable from lower control arm, then remove the stabilizer bar.
3. Remove lower control arm to axle housing attaching bolts, then the lower arm to front support attaching bolts, **Fig. 3.**
4. Remove lower control arm to crossmember attaching bolts, then the lower control arm from vehicle.
5. Reverse procedure to install. Apply a coat of suitable grease to shaft cutout between lower control arm and axle housing. Install shaft with identification mark facing downward.

1989–90 COLT WAGON 4 WHEEL DRIVE

Lower Arm

1. Remove coil spring as described under "Coil Spring, Replace."
2. Remove lower arm to axle and body bracket attaching bolts, then remove lower arm, **Fig. 5.**
3. When replacing lower arm bushings, refer to **Figs. 6 and 7,** for installation dimensions.
4. Reverse procedure to install.

Upper Arm

1. Raise and support rear of vehicle, then remove wheel and tire assembly.
2. Remove upper arm to axle housing and body bracket bolts, then remove upper arm, **Fig. 5.**
3. Reverse procedure to install.

SUSPENSION ARM REPLACE

COLT, COLT VISTA 2 WHEEL DRIVE & SUMMIT

Removal

1. Remove coil springs as outlined.
2. Disconnect brake hose at suspension arm assembly.
3. Remove suspension arm bracket attaching bolts at each side of arm, then remove right and left hand suspension arms as an assembly, **Figs. 1 and 2.**

Disassembly

Before disassembly, place alignment marks on suspension arms and components so they can be assembled in the same position.

1. Remove nut at each end of the suspension arm assembly, then remove mounting brackets and bushings, **Figs. 1 and 2.**
2. Remove dust cover retaining band, then on models equipped with a stabilizer bar, place an alignment mark at each end of the stabilizer bar in line with punch mark on stabilizer bracket.
3. Separate suspension arm assembly into right and left hand suspension arms, leaving the dust cover attached to the right hand suspension arm.
4. Remove rubber stopper from right hand suspension arm.
5. Using a screwdriver, remove bushing A from left hand suspension arm, **Fig. 8.**
6. Using a suitable drift, drive bushing B from left hand suspension arm, **Fig. 9.**
7. Inspect all components for wear and damage and replace as necessary.

Assembly

1. Apply grease to inner surface of left hand suspension arm.
2. Apply grease to outer surfaces of bushings, then using tools Nos. MB990779 and MB990780, drive bushing B in until tool contacts end of left suspension arm. Using tool No. MB990780, drive bushing A into left suspension arm, **Fig. 10.**
3. Position dust cover on right suspension arm, then coat suspension arm with grease from shoulder at mounting bracket end approximately 15.7 inches toward center of arm and install rubber stopper.
4. On models equipped, align stabilizer bar and stabilizer bar bracket marks.
5. Carefully push left and right suspension arms together, while wiping excess grease as necessary, then align suspension arm and component alignment marks.
6. Install washers and mounting brackets as shown in **Fig. 11.** Loosely install mounting bracket nuts. Do not tighten nuts until suspension assembly has been installed and vehicle lowered to floor.

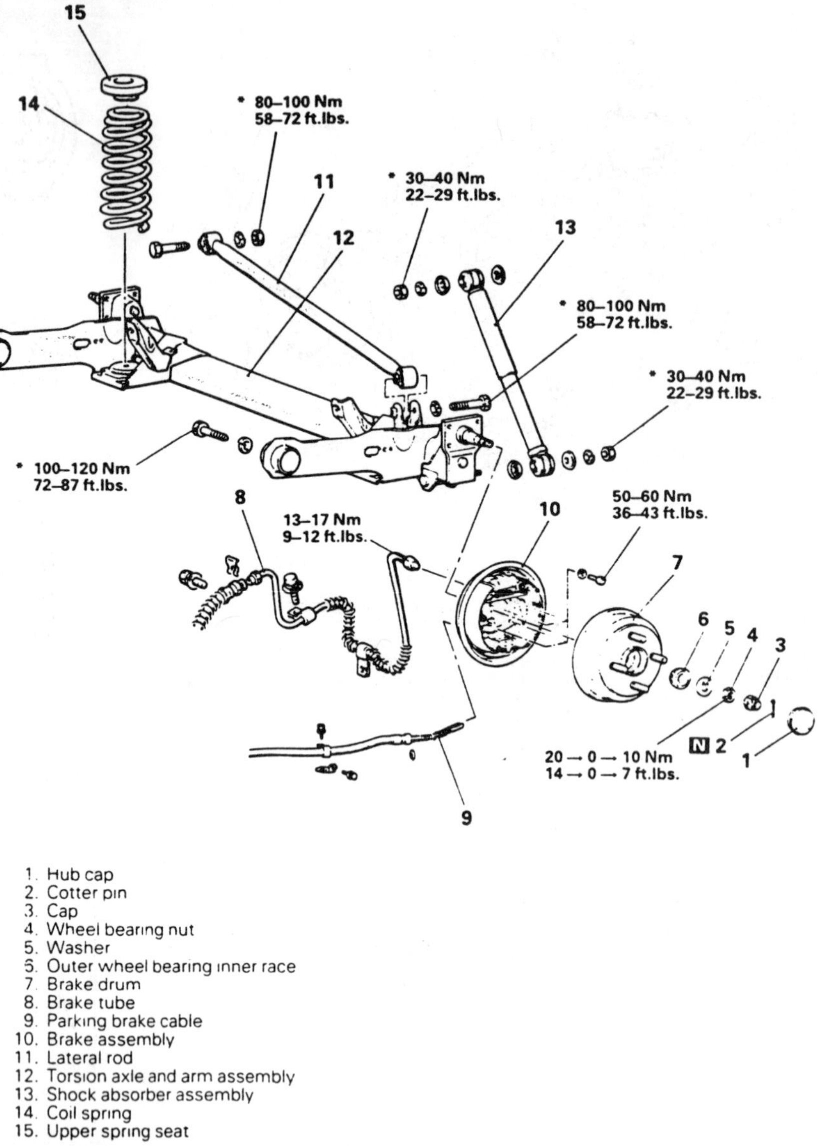

Fig. 4 Coil spring rear suspension. 1988–90 Colt Wagon 2 Wheel Drive

7. Pack dust cover with grease, then install cover retaining clamp.

Installation

1. Position suspension arm assembly to body, then install mounting bracket to body attaching bolts.
2. Install coil spring assembly.
3. Lower vehicle, then torque mounting bracket to suspension arm attaching nuts to specification listed at the end of this section.

CONQUEST

1. Raise and support rear of vehicle, then remove rear wheel and tire assemblies.
2. Secure shock assembly to crossmember with wire, then support rear suspension assembly with a suitable jack.
3. Disconnect propeller shaft from torque tube, then remove center exhaust pipe and main muffler.
4. Disconnect parking brake cable from rear disc brake and lower control arm.
5. Disconnect brake hose at rear floor, then remove the shock absorber attaching nuts.
6. Remove crossmember attaching nuts, then the front support attaching nuts and bolts.
7. Slowly lower rear suspension assembly from vehicle.
8. Reverse procedure to install.

STABILIZER BAR REPLACE

1989–90 COLT WAGON 4 WHEEL DRIVE

1. Raise and support rear of vehicle, then remove wheel and tire assembly.
2. Remove stabilizer bar link nuts.
3. Remove stabilizer bar bracket bolts, then remove stabilizer bar.
4. Reverse procedure to install. When installing stabilizer bar link, tighten nut until .74 to .82 inch of bolt thread is exposed from end of nut.

REAR SUSPENSION SERVICE

COLT VISTA 4 WHEEL DRIVE

Refer to **Figs. 12 and 13** when servicing rear suspension.

REAR WHEEL BEARING ADJUST

1985–90 COLT VISTA & 1988–90 COLT WAGON 2 WHEEL DRIVE MODELS

1. Raise and support rear of vehicle, then remove wheel and tire assembly.

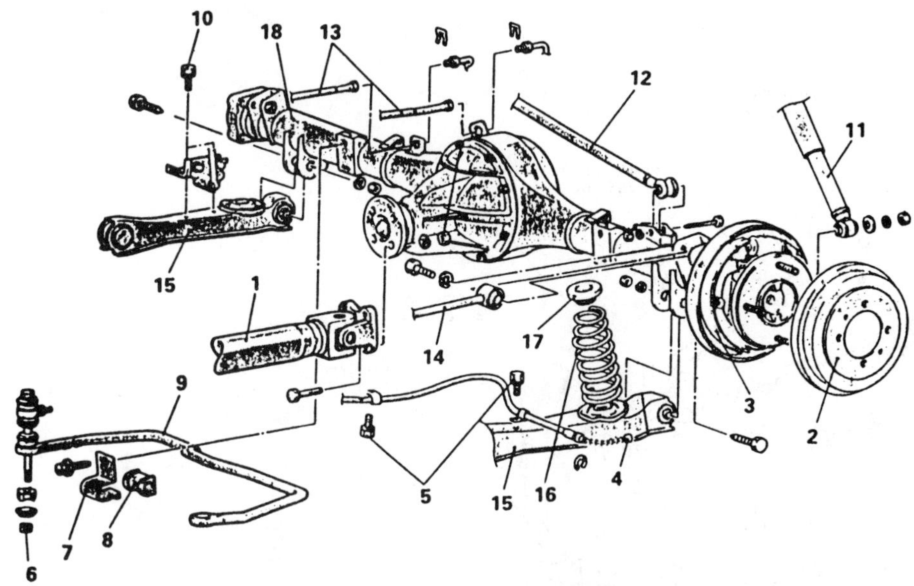

1. Propeller shaft
2. Brake drum
3. Shoe assembly
4. Parking brake cable connection
5. Parking brake cable attaching bolt
6. Self-locking nut
7. Fixture
8. Bushing
9. Stabilizer bar
10. Load sensing proportioning valve bracket mounting bolt
11. Shock absorber connection
12. Lateral rod connection
13. Rear brake tube connection
14. Upper arm connection
15. Lower arm connection
16. Coil spring
17. Spring seat
18. Axle assembly

Fig. 5 Rear axle & suspension. 1989–90 Colt Wagon 4 Wheel Drive

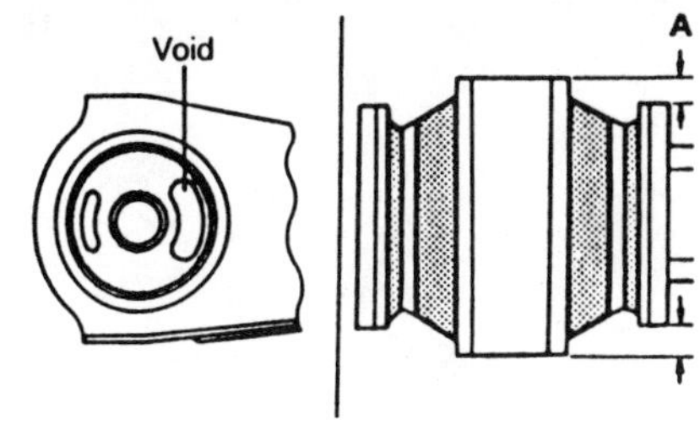

Fig. 6 Lower arm front bushing installation. 1989–90 Colt Wagon 4 wheel drive.

Fig. 7 Lower arm rear bushing installation. 1989–90 Colt Wagon 4 wheel drive.

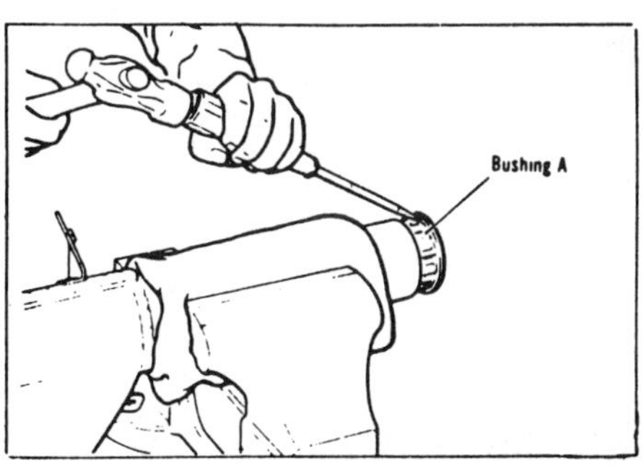

Fig. 8 Removing bushing "A" from left suspension arm. Colt Vista 2 wheel drive & 1985–88 Colt except 1988 Colt Wagon

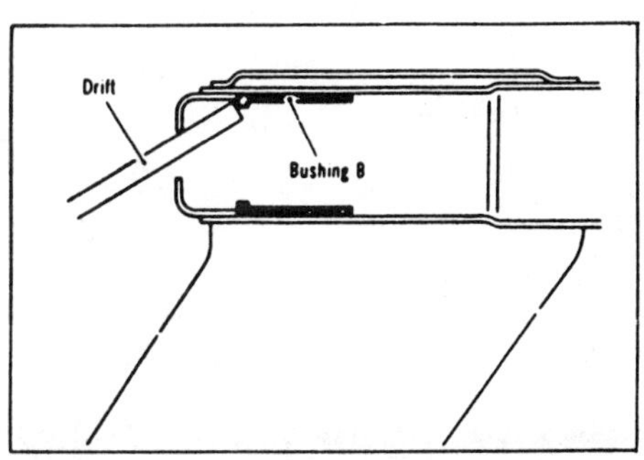

Fig. 9 Removing bushing "B" from left suspension arm. Colt Vista 2 wheel drive & 1985–88 Colt except 1988 Colt Wagon

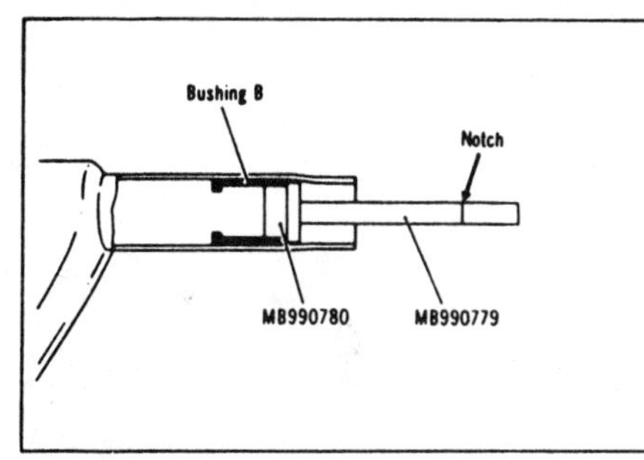

Fig. 10 Installing bushing "B" into left suspension arm. Colt Vista 2 wheel drive & 1985–88 Colt except 1988 Colt Wagon

2. Remove dust cap from brake drum, then remove cotter pin and adjusting nut retainer, **Fig. 14.**
3. Torque wheel bearing adjusting nut to 14 ft. lbs. and rotate drum several turns to seat bearings, then loosen adjusting nut.
4. Torque adjusting nut to 4 ft. lbs. for 1985-86 or 7 ft. lbs. on 1987-90, rotate drum several times, then retorque nut to 4 ft. lbs. for 1985-86 or 7 ft. lbs. for 1987-90.
5. Install nut retainer and cotter pin. If cotter pin holes are not aligned reposition retainer. If holes cannot be aligned, the adjusting nut may be backed off a maximum of 15° to align holes.
6. Install dust cap on brake drum, then install wheel and tire assembly.

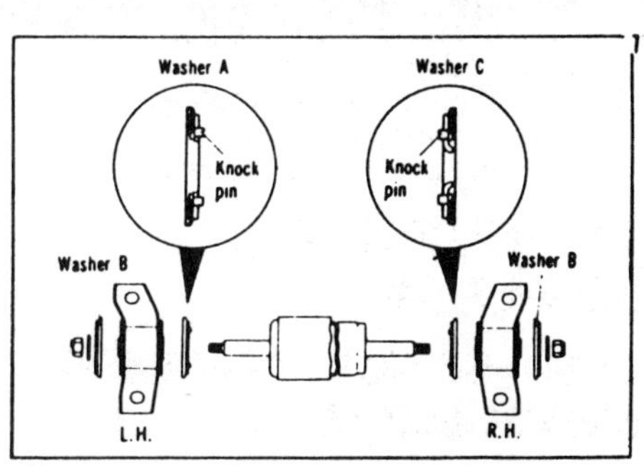

Fig. 11 Installing mounting bracket components. Colt Vista 2 wheel drive & 1985–88 Colt except 1988 Colt Wagon

1985–90 COLT HATCHBACK & SEDAN & SUMMIT

1. Raise and support rear of vehicle and remove wheel.
2. Ensure parking brake is fully released.
3. Check wheel bearing end play with suitable dial indicator. Maximum allowable end play is .008 inch.
4. Check hub and drum starting torque by rotating hub with suitable spring scale. Maximum allowable starting torque is 4.9 lbs.
5. If end play or starting torque exceed limits, loosen bearing retaining nut, then retorque nut to 72-108 ft. lbs for 1985-88 models, 108 to 145 ft. lbs. for 1989-90 models.
6. If end play or starting torque are still not within limits, replace wheel bearings as needed.

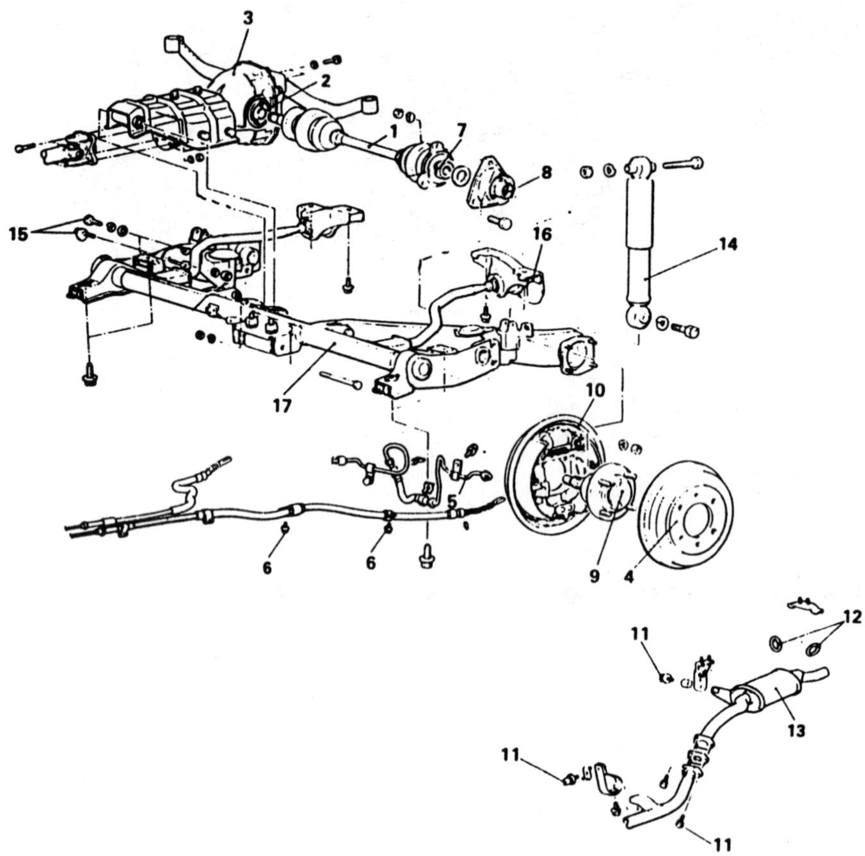

1. Drive shaft
2. Circlip
3. Rear differential carrier
4. Brake drum
5. Brake tube
6. Parking brake mounting bolts
7. Lock nut
8. Companion flange
9. Axle shaft
10. Rear brake assembly
11. Main muffler mounting bolts
12. O-rings
13. Main muffler
14. Shock absorber
15. Inner arm and outer arm mounting bolts and nuts
16. Protector
17. Rear suspension assembly

Fig. 12 Rear suspension removal & installation. Colt Vista 4 wheel drive

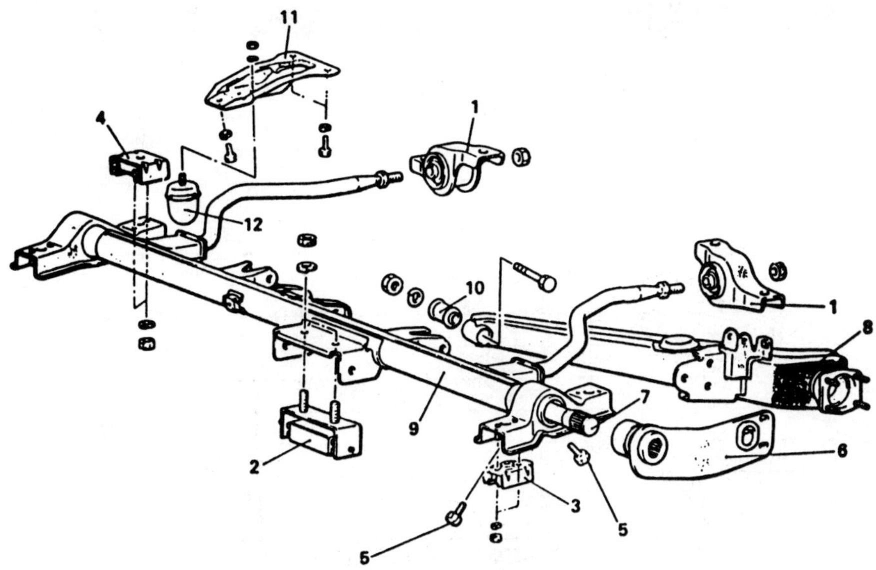

1. Fixture assembly
2. Dynamic damper assembly
3. Front insulator
4. Rear insulator
5. Lock bolts for securing outer arm bushing
6. Outer arm
7. Torsion bar
8. Inner arm
9. Crossmember assembly
10. Inner arm bushing
11. Stopper bracket
12. Bump stopper

Fig. 13 Rear suspension disassembly & assembly. Colt Vista 4 wheel drive

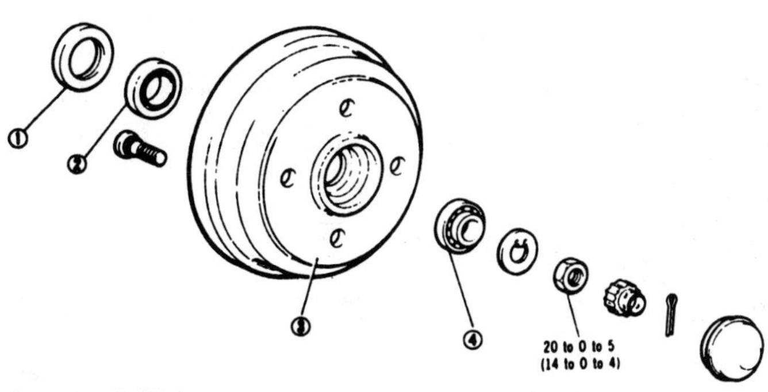

(1) Oil seal
(2) Inner bearing
(3) Brake drum
(4) Outer bearing

Fig. 14 Rear wheel bearing & hub. Colt Vista 2 wheel drive

REAR SUSPENSION TIGHTENING SPECIFICATIONS

Component	Year & Torque/ft. lbs.					
	1985	1986	1987	1988	1989	1990
COLT & SUMMIT						
Dust Shield To Trailing Arm	—	—	—	—	36-43	36-43
Brake Adapter, Disc Brakes	—	—	—	—	36-43	36-43
Lateral Rod To Axle	—	—	—	—	58-72	58-72
Lateral Rod To Body	—	—	—	—	58-72	58-72
Piston Rod Nut	—	—	—	—	14-22	14-22
Shock Unit Upper Attachment	47-58	47-58	47-58	47-48	18-25	18-25
Shock Unit Lower Attachment	47-48	47-58	47-58	47-58	58-72	58-72
Suspension Arm, Left To Right Hand	58-72	58-72	58-72	58-72	—	—
Suspension Arm To Body	36-51	36-51	36-51	36-51	—	—
Trailing Arm	—	—	—	—	98-108	98-109
Wheel Bearing Nut	72-108	72-108	72-108	72-108	108-145	108-145
Wheel & Tire Assembly	65-80	65-80	65-80	65-80	65-80	65-80
COLT WAGON 2 WHEEL DRIVE (1988–90)						
Lateral Rod To Axle	—	—	—	58-72	58-72	58-72
Lateral Rod To Body	—	—	—	58-72	58-72	58-72
Load Sensing Proportioning Valve	—	—	—	7-10	7-10	7-10
Shock Unit Upper Attachment	—	—	—	22-29	22-29	22-29
Shock Unit Lower Attachment	—	—	—	22-29	22-29	22-29
Trailing Arm	—	—	—	72-87	72-87	72-87
Wheel & Tire Assembly	—	—	—	65-80	65-80	65-80
COLT WAGON 4 WHEEL DRIVE (1989–90)						
Bracket To Lateral Rod	—	—	—	—	9-11	9-11
Lateral Rod To Axle	—	—	—	—	58-72	58-72
Lateral Rod To Body	—	—	—	—	58-72	58-72
Lower Arm	—	—	—	—	94-108	94-108
Propeller Shaft To Pinion Flange	—	—	—	—	22-25	22-25
Shock Unit Upper Attachment	—	—	—	—	22-29	22-29
Shock Unit Lower Attachment	—	—	—	—	22-29	22-29
Upper Arm	—	—	—	—	94-108	94-108
Upper Arm Bracket	—	—	—	—	36-51	36-51
Wheel & Tire Assembly	—	—	—	—	65-80	65-80
COLT VISTA 2 WHEEL DRIVE						
Shock Unit Lower Attachment	58-79	58-79	58-80	58-80	58-80	58-80
Shock Unit Upper Attachment	47-58	47-58	58-80	58-80	58-80	58-80
Suspension Arm To Body	51-65	51-65	87-108	87-108	87-108	87-108
Suspension Arm, Left To Right Hand	94-108	94-108	87-108	87-108	87-108	87-108
Wheel & Tire Assembly	①	①	①	①	①	①
COLT VISTA 4 WHEEL DRIVE						
Axle Shaft To Companion Flange	116-159	116-159	116-159	116-159	116-159	116-159
Crossmember Support To Body	58-87	58-87	58-87	58-87	58-87	58-87
Differential Front Support Bracket To Crossmember	87-101	87-101	87-101	87-101	87-101	87-101
Differential Rear Support To Differential Carrier	43-65	43-65	43-65	43-65	43-65	43-65
Drive Shaft To Axle Shaft Flange	36-43	36-43	36-43	36-43	40-47	40-47
Dynamic Damper Nuts	14-20	14-20	14-20	14-20	14-20	14-20
Extension Rod To Fixture	36-51	36-51	72-101	72-101	72-101	72-101
Fixture Assembly To Body	36-50	36-50	36-50	36-50	36-50	36-50
Front Insulator To Crossmember Bracket	7-11	7-11	7-11	7-11	7-11	7-11
Inner Arm Nut	51-65	51-65	51-65	51-65	51-65	51-65

Continued

REAR SUSPENSION TIGHTENING SPECIFICATIONS—Continued

Component	Year & Torque/ft. lbs.					
	1985	1986	1987	1988	1989	1990
Outer Arm Bushing Lock Bolts	16-22	16-22	16-22	16-22	16-22	16-22
Outer & Inner Arm Attaching Bolts	58-72	58-72	58-72	58-72	58-72	58-72
Outer & Inner Arm Toe Adjusting Nut	87-101	87-101	87-101	87-101	87-101	87-101
Shock Absorber To Body	47-58	47-48	58-80	58-80	58-80	58-80
Shock Absorber To Inner Arm	58-79	58-79	58-80	58-80	58-80	58-80
Stopper Bracket To Body	22-36	22-36	22-36	22-36	22-36	22-36
Stopper Bracket To Bumper Stop	14-22	14-22	14-22	14-22	14-22	14-22
Wheel & Tire Assembly	①	①	①	①	①	①
CONQUEST						
Front Support Lower Stopper Bolt To Body	29-36	29-36	29-36	29-36	29-36	—
Front Support Nut To Pin Assembly	51-61	51-61	51-61	51-61	51-61	—
Lower Control Arm Locking Pin	11-14	11-14	11-14	11-14	11-14	—
Lower Control Arm To Axle Housing	51-58	51-58	51-58	80-94	80-94	—
Lower Control Arm To Crossmember	94-108	94-108	94-108	94-108	94-108	—
Lower Control Arm To Front Support	94-108	94-108	94-108	94-108	94-108	—
Pin Assembly	51-61	51-61	51-61	51-61	51-61	—
Propeller Shaft To Pinion Flange	36-43	36-43	36-43	36-43	36-43	—
Rear Support To Differential Carrier	36-51	36-51	36-51	36-51	36-51	—
Stabilizer Bar To Front Support	22-29	22-29	22-29	22-29	22-29	—
Stabilizer Bar To Lower Control Arm	7-14	7-14	7-14	7-14	7-14	—
Strut Unit Upper Attachment	18-25	18-25	18-25	18-25	18-25	—
Strut Unit Lower Attachment	36-51	36-51	36-51	36-51	36-51	—
Support Insulator To Crossmember	18-22	18-22	18-22	18-22	18-22	—
Support Insulator To Rear Support	22-25	22-25	22-25	22-25	22-25	—
Wheel & Tire Assembly	65-80	65-80	65-80	65-80	65-80	—

①—Steel wheel, 50–57; aluminum wheel, 65–80.

Chrysler Motors Front Wheel Drive Models Except Colt, Colt Vista & Plymouth Laser

INDEX

SHOCK ABSORBER & COIL SPRING REPLACE

EXCEPT HORIZON, OMNI, CHARGER, TURISMO

Removal

1. Raise and support vehicle.
2. Support axle assembly and remove wheel and tire assemblies.
3. **On models with air shocks,** disconnect air lines. On all models, remove both upper and lower shock absorber attaching bolts, then the shock absorbers.
4. **On 1986-90 models equipped with automatic load leveling system,** remove link from track bar to sensor.
5. **On all models,** lower axle assembly until spring and spring upper isolator can be removed, **Figs. 1 through 3.**

Installation

1. Position jounce bumper to rail, install and **torque** attaching screws to 70 inch lbs.
2. Install isolator over jounce bumper and install spring.
3. Raise axle and install shock absorbers. Loosely assemble lower shock absorber attaching bolts and torque upper attaching bolts to specifications. On 1986-90 models with load leveling system, attach link to track bar.
4. With suspension supporting vehicle, torque shock absorber lower attaching screws to specifications.

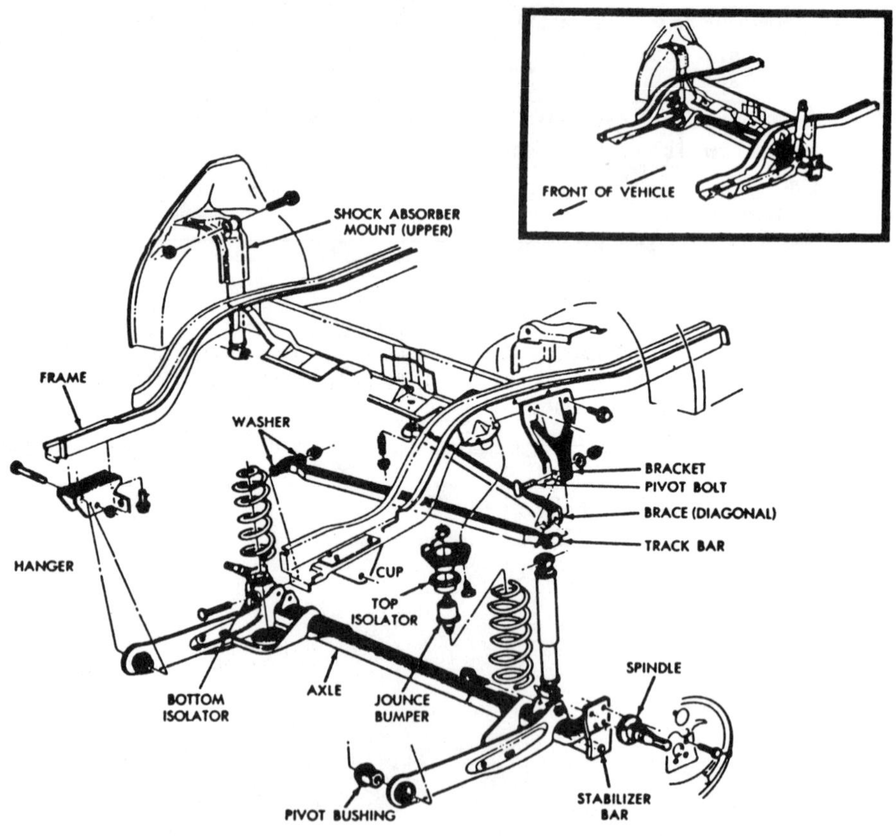

Fig. 1 Rear axle & suspension assembly. 1985 Except Charger, Horizon, Omni & Turismo.

HORIZON, OMNI, CHARGER & TURISMO

Replacement

1. Remove upper shock absorber mounting protective cap, located inside vehicle at upper rear wheel well area. On two door models, remove lower rear quarter trim panel for access.
2. **On all models,** remove upper shock absorber mounting nut, isolator retainer and upper isolator, **Fig. 4.**
3. Raise and support vehicle.
4. Remove shock absorber lower mounting bolt.
5. Remove shock absorber and coil spring assembly from vehicle.
6. Reverse procedure to install.

Service

1. Install coil spring retractor tool No. L-4838 or equivalent, on coil spring and support in a vise, **Fig. 5.** Grip four or five coils of the spring in the retractors. Also, do not extend retractors more than 9 1/4 inches.
2. Tighten retractors evenly until spring pressure is released from upper spring seat.
3. Hold flat end of push rod and loosen retaining nut. **Ensure that spring is properly compressed before loosening retaining nut since personal injury may result.**
4. Remove lower isolator, push rod sleeve and upper spring seat.
5. Remove shock absorber from coil spring.
6. Remove jounce bumper and dust shield from push rod, **Fig. 6.**
7. Remove lower spring seat. **Fig. 6.**

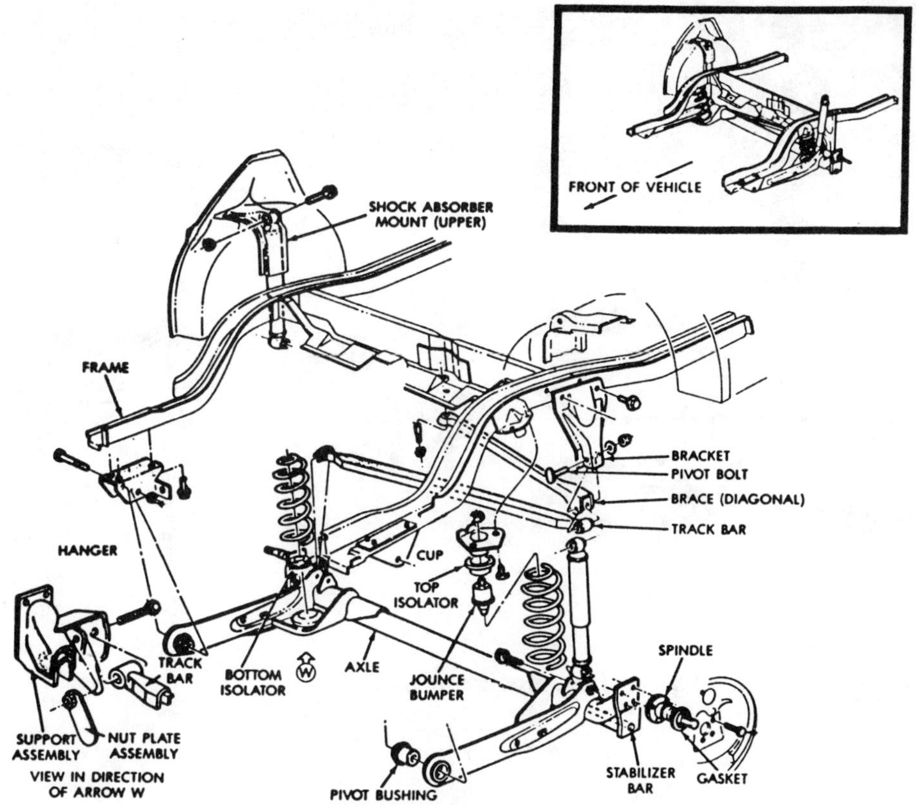

Fig. 2 Rear axle & suspension assembly. 1986–88 Except Charger, Horizon, Omni & Turismo.

TORQUE		
A	40 FT. LBS.	54 N•m
B	50 FT. LBS.	68 N•m
C	55 FT. LBS.	75 N•m
D	70 IN. LBS.	8 N•m
E	70 FT. LBS.	95 N•m
F	45 FT. LBS.	61 N•m

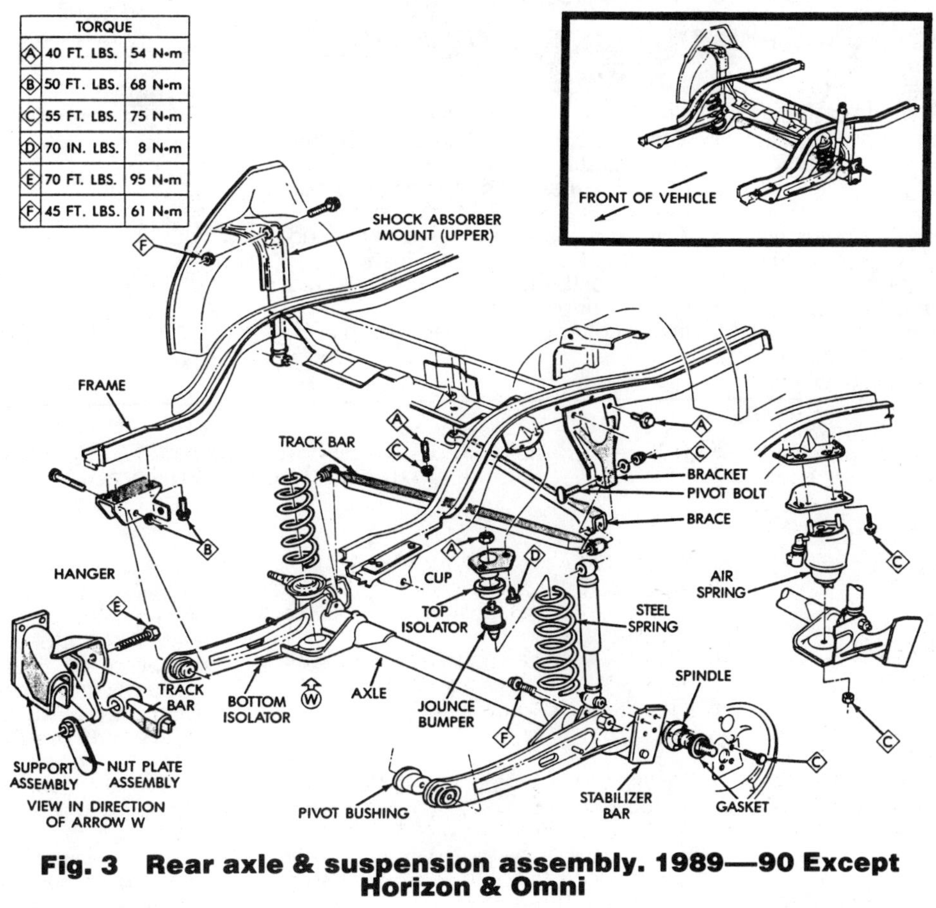

Fig. 3 Rear axle & suspension assembly. 1989—90 Except Horizon & Omni

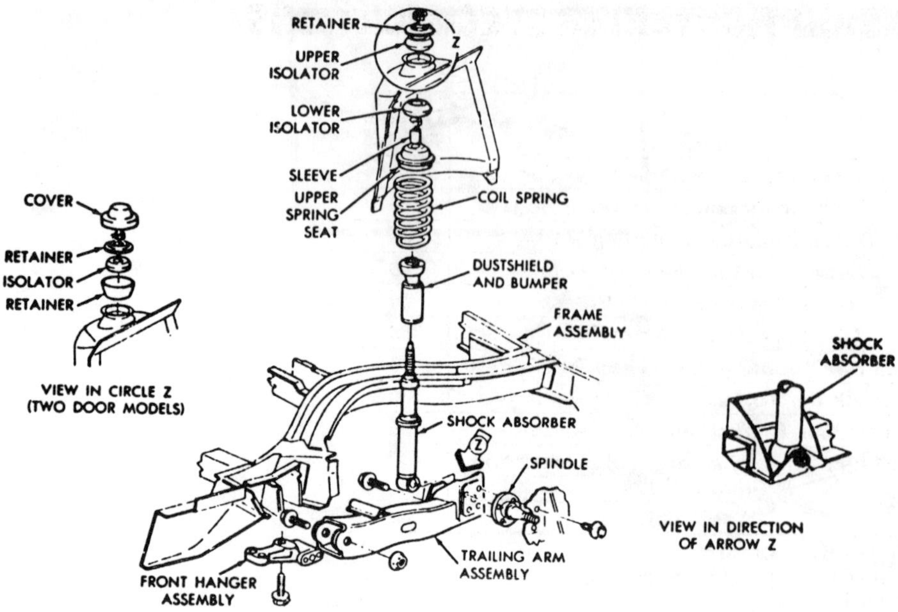

Fig. 4 Rear axle & suspension assembly. Horizon, Omni, Charger & Turismo

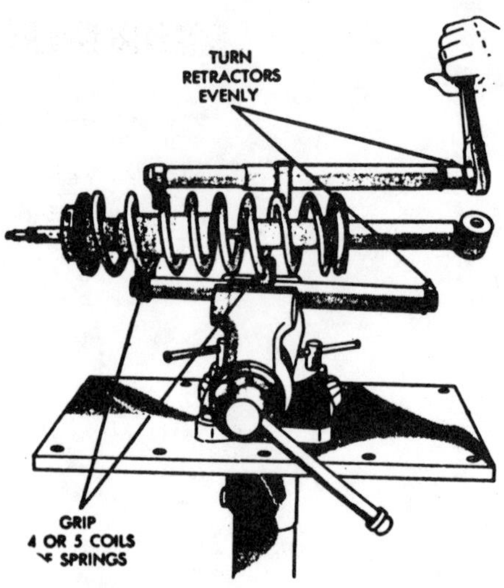

Fig. 5 Retracting coil spring

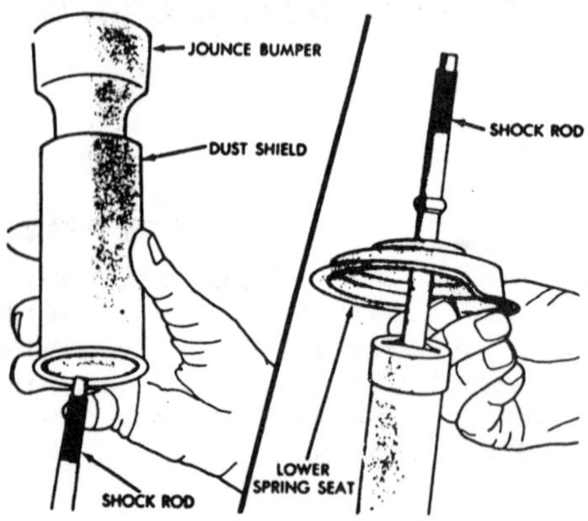

Fig. 6 Jounce bumper, dust shield & lower spring seat replacement

8. Reverse procedure to assemble.

REAR WHEEL BEARING ADJUST

1. **Torque** adjusting nut to 270 inch lbs. while rotating wheel.
2. Stop wheel and loosen adjusting nut, **Fig. 7.**
3. Tighten adjusting nut finger tight. end play should be .001-.003 inch.
4. Install castle lock with slots aligned with cotter pin hole.
5. Install cotter pin and grease cap.

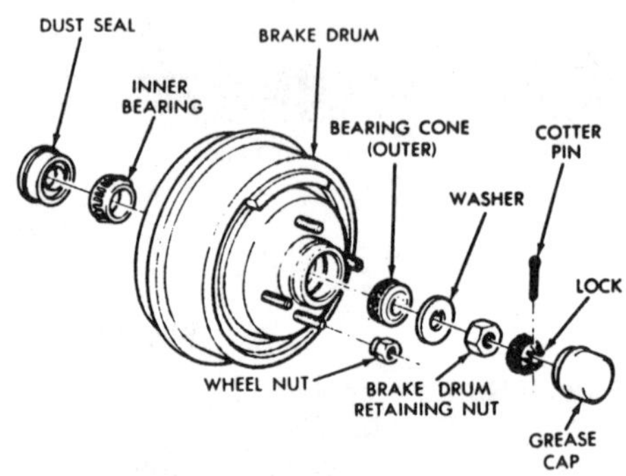

Fig. 7 Wheel bearing assembly

REAR SUSPENSION TIGHTENING SPECIFICATIONS

Year	Component	Torque/ Ft. Lbs.
1985–90	Brake Assembly And Spindle Mounting Bolts	55
	Shock Absorber Lower Mounting Bolts ①	40
	Shock Absorber Mounting Nut ②	45
	Shock Absorber Mounting Nuts ①	20
	Track Bar Brace/Body Stud ②	40
	Track Bar Brace To Stud Nut ②	55
	Track Bar Mounting Bracket To Rail Bolts ②	40
	Track Bar To Axle Attaching Bolts ②	70
	Track Bar To Mounting Bracket Bolt ②	55
	Trailing Arm To Hanger Bracket Mounting Bracket Bolts	40
	Wheel Nuts	95

①—Charger, Horizon, Omni & Turismo.
②—All models except Charger, Horizon, Omni & Turismo.

Chrysler Motors Plymouth Laser & Eagle Talon

INDEX

SHOCK ABSORBER & COIL SPRING REPLACE

2 WHEEL DRIVE MODELS

Remove shock absorber assembly in numbered sequence, **Fig. 1,** noting the following:

1. Jack up torsion axle and arm assembly in order to release tension.
2. Place jack at center of axle beam, ensuring jack does not contact lateral rod.
3. Reverse procedure to install.

4 WHEEL DRIVE MODELS

Remove shock absorber assembly in numbered sequence, **Fig. 2.** Reverse procedure to install.

LATERAL ROD REPLACE

4 WHEEL DRIVE MODELS

Remove lateral rod in numbered sequence, **Fig. 3,** noting the following:

1. Install lateral rod mounting bolt from top of axle beam with nut on bottom of lateral rod.
2. Reverse procedure to install.

TORSION AXLE & ARM REPLACE

2 WHEEL DRIVE MODELS

Remove torsion axle and arm in numbered sequence, **Fig. 4,** noting the following:

1. Suspend caliper assembly with a piece of wire. **Do not hang caliper by brake hose.**
2. Rear hub bearing cannot be disassembled.
3. Remove shock absorber as previously described.
4. Adjust parking brake.

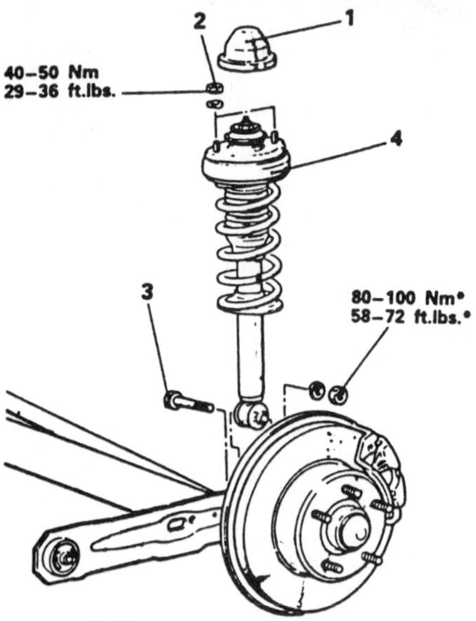

1. Cap
2. Shock absorber upper mounting nut
3. Shock absorber lower mounting bolt
4. Shock absorber

Fig. 1 Removing shock absorber assembly (2WD)

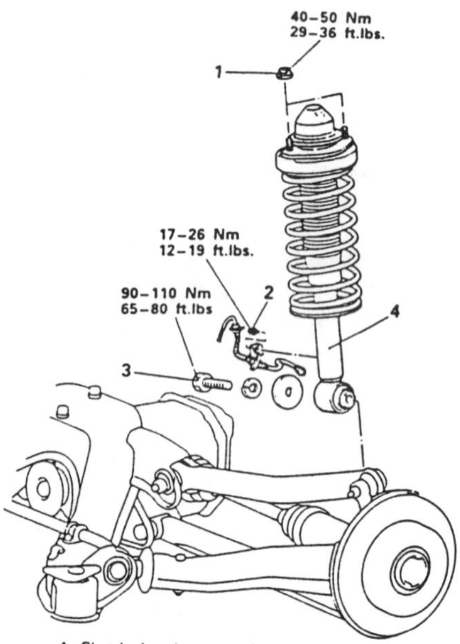

1. Shock absorber upper installation nut
2. Brake tube bracket installation bolt
3. Shock absorber lower installation bolt
4. Shock absorber assembly

Fig. 2 Removing shock absorber assembly (4WD)

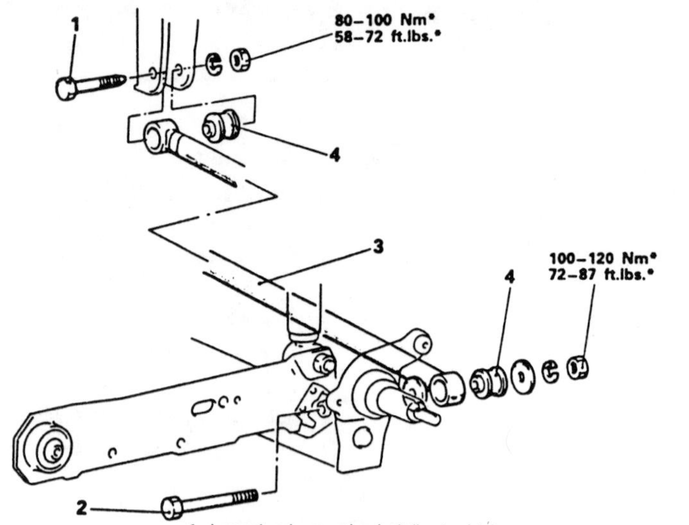

1. Lateral rod mounting bolt (body side)
2. Lateral rod mounting bolt (axle beam side)
3. Lateral rod
4. Lateral rod bushing

Fig. 3 Replacing lateral rod

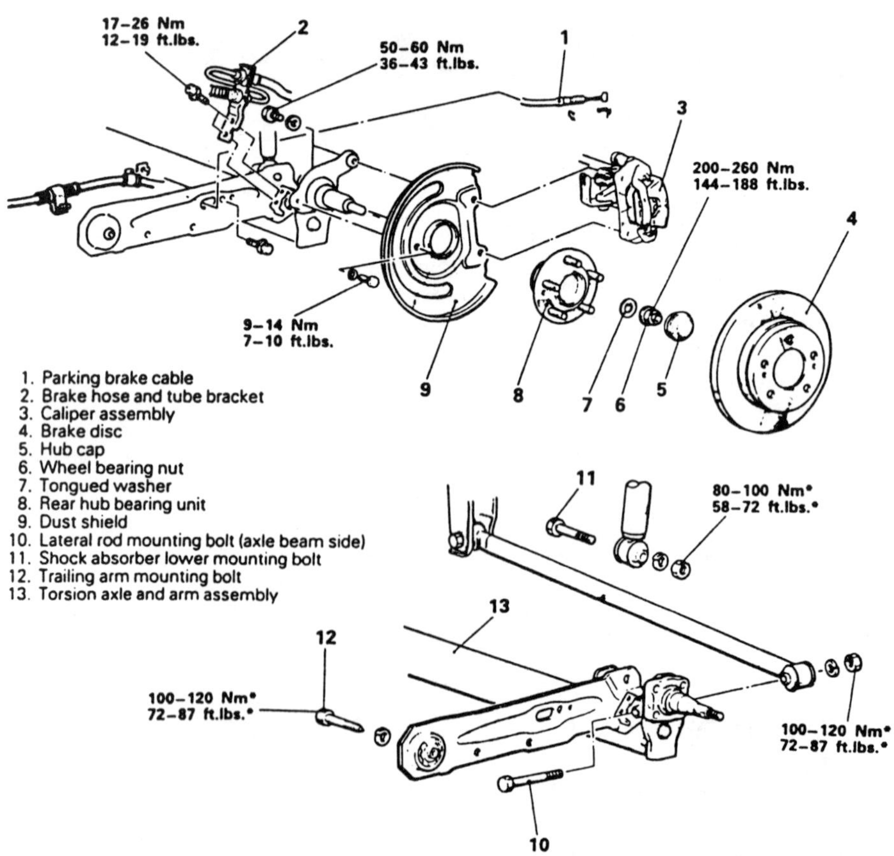

1. Parking brake cable
2. Brake hose and tube bracket
3. Caliper assembly
4. Brake disc
5. Hub cap
6. Wheel bearing nut
7. Tongued washer
8. Rear hub bearing unit
9. Dust shield
10. Lateral rod mounting bolt (axle beam side)
11. Shock absorber lower mounting bolt
12. Trailing arm mounting bolt
13. Torsion axle and arm assembly

Fig. 4 Replacing torsion axle and arm assembly

5. Install lateral rod mounting bolt as previously described.
6. Install hub unit until inner race contacts with spindle end.
7. Install wheel bearing and adjust as described under "Rear Wheel Bearing Adjust."
8. Reverse procedure to install.

REAR AXLE HUB REPLACE

2 WHEEL DRIVE MODELS

Remove rear axle hub in numbered sequence, **Fig. 5,** noting the following:

1. Suspend caliper assembly with a piece of wire. **Do not hang caliper by brake hose.**
2. Rear hub bearing cannot be disassembled.
3. Press inner race until it contacts with

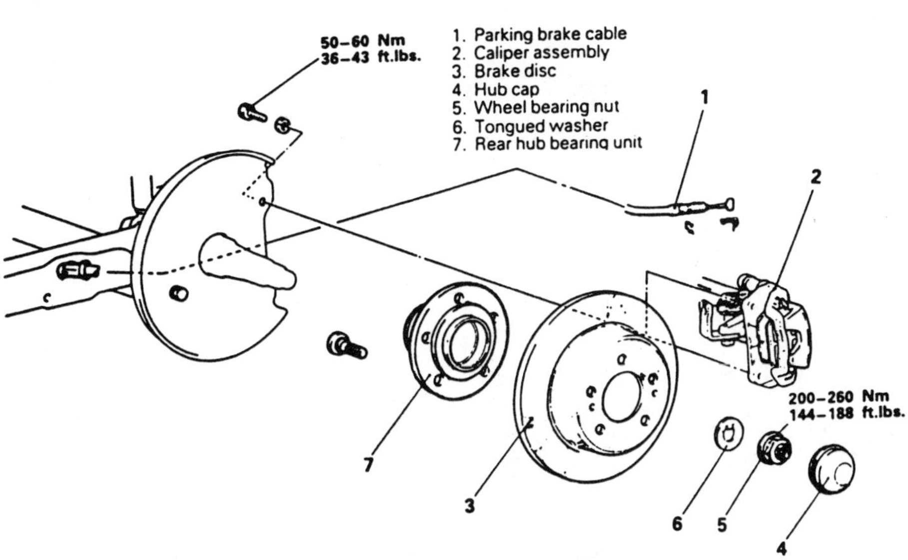

Fig. 5 Replacing rear axle hub

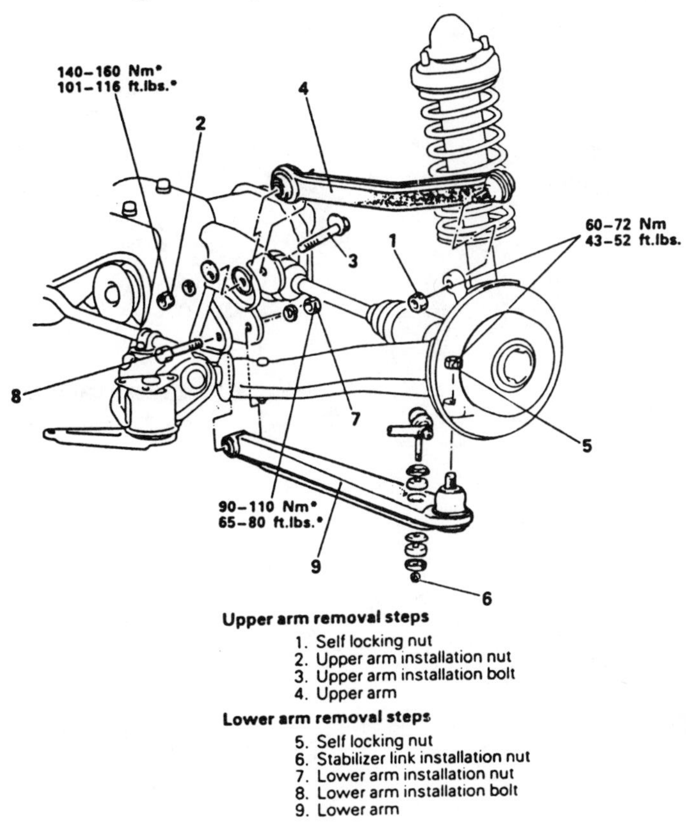

Fig. 6 Replacing upper and lower arm

spindle end.

4. Align wheel bearing nut with spindle indentation and crimp.
5. Reverse procedure to install.

UPPER & LOWER ARM REPLACE

4 WHEEL DRIVE MODELS

Remove upper and lower arm in numbered sequence, **Fig. 6,** noting the following:

1. Loosen, but do not remove self-locking nut
2. Press fit lower arm bushing until outer edge is flush with lower arm edge.
3. Reverse procedure to install.

TRAILING ARM REPLACE

4 WHEEL DRIVE MODELS

Remove trailing arm in numbered sequence, **Fig. 7,** noting the following:

1. Suspend caliper assembly with a piece of wire. **Do not hang caliper by brake hose.**
2. Secure rear axle shaft, then remove self locking nut.
3. Using puller, remove rear axle shaft.
4. On upper and lower arms, loosen, but do not remove self locking nut.
5. Replace connecting rod as follows:
   a. Remove trailing arm bushing, then bolt and nut, **Fig. 8.**
   b. Using tool No. MB991254, **Fig. 9,** set onto trailing arm as shown in **Fig. 10.**
   c. Apply lubricant to sliding portion of tool (A), then install bolt (B) in trailing arm, **Fig. 10,**
   d. Using suitable wrench, turn portion as marked (C), **Fig. 10,** to remove connection rod.
   e. Installation of body should be performed with screw shaft and guide shaft center lines aligned as shown in **Fig. 11.**
   f. Apply soapy water to rubber portion of connection rod.
   g. Reverse procedure to press fit, torquing bolt to specifications.
6. Temporarily assemble rear axle shaft to trailing arm.
7. Install companion flange to rear axle shaft, then install locking nut.
8. Ensure rear axle shaft does not turn when tightening lock nut.
9. Reverse procedure to install.

STABILIZER BAR REPLACE

4 WHEEL DRIVE MODELS

Remove stabilizer bar in numbered sequence, **Fig. 12,** noting the following:

1. Support rear suspension assembly with transaxle jack.
2. Remove crossmember brackets.
3. Lower transaxle jack slightly, maintaining gap between rear suspension and body.
4. Ensure stabilizer link ball joint starting torque is 15-28 inch lbs. If starting torque exceeds upper limit, replace link.
5. Reverse procedure to install.

REAR WHEEL BEARING ADJUST

1. To check bearing end play, proceed as follows:
   a. Place dial gauge against hub surface, then move hub in axial direction.
   b. If end play exceeds .004 inch, lock nut should be tightened.
   c. Torque lock nut to specifications.
2. Replace hub bearing unit if adjustment cannot be made to within limit.

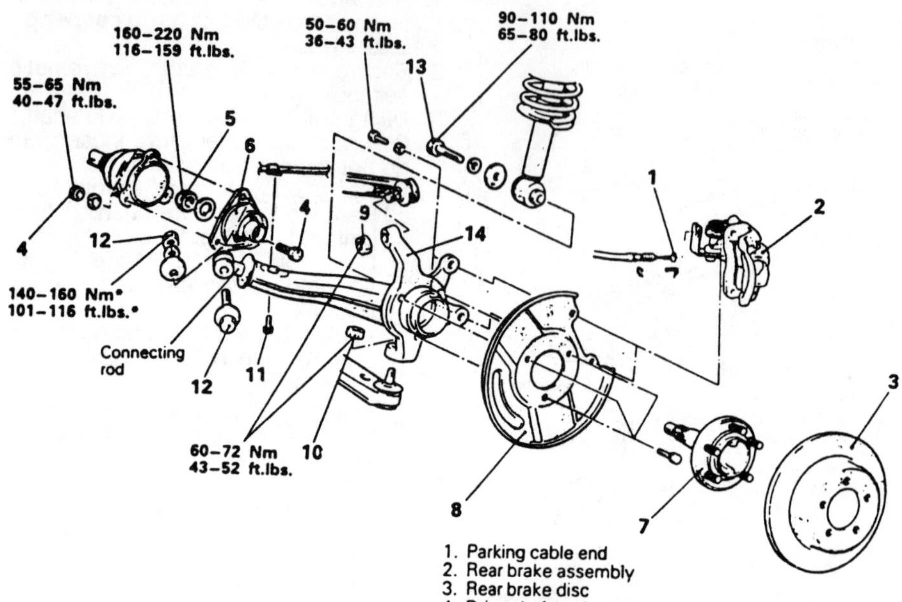

1. Parking cable end
2. Rear brake assembly
3. Rear brake disc
4. Drive shaft and companion flange installation bolt, nut
5. Self locking nut
6. Companion flange
7. Rear axle shaft
8. Dust shield
9. Self locking nut (upper arm)
10. Self locking nut (lower arm)
11. Parking brake cable installation bolt
12. Trailing arm installation bolt, nut
13. Rear shock absorber installation bolt
14. Trailing arm

Fig. 7 Replacing trailing arm

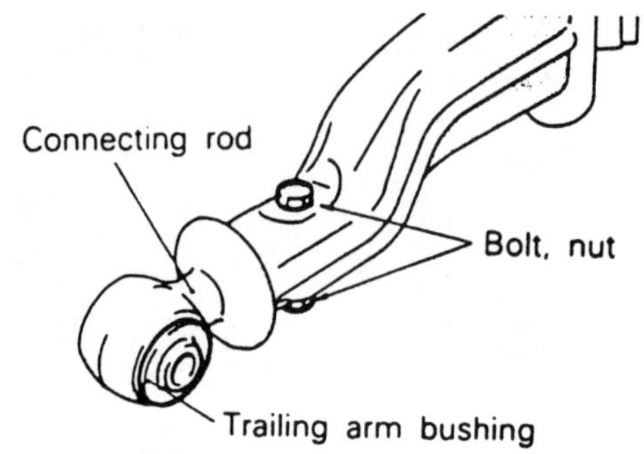

Fig. 8 Removing connecting rod

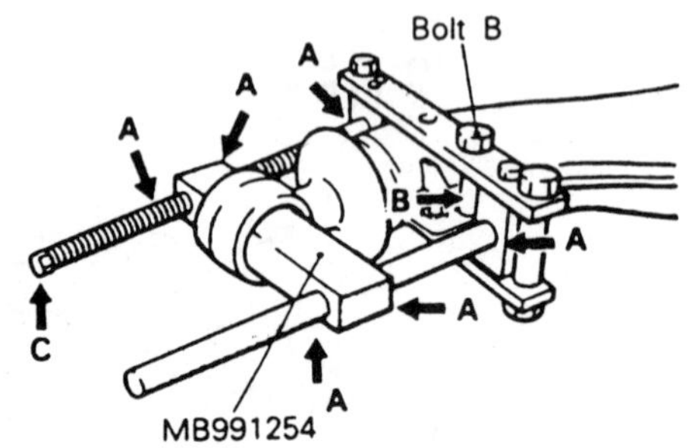

Fig. 10 Lubrication points

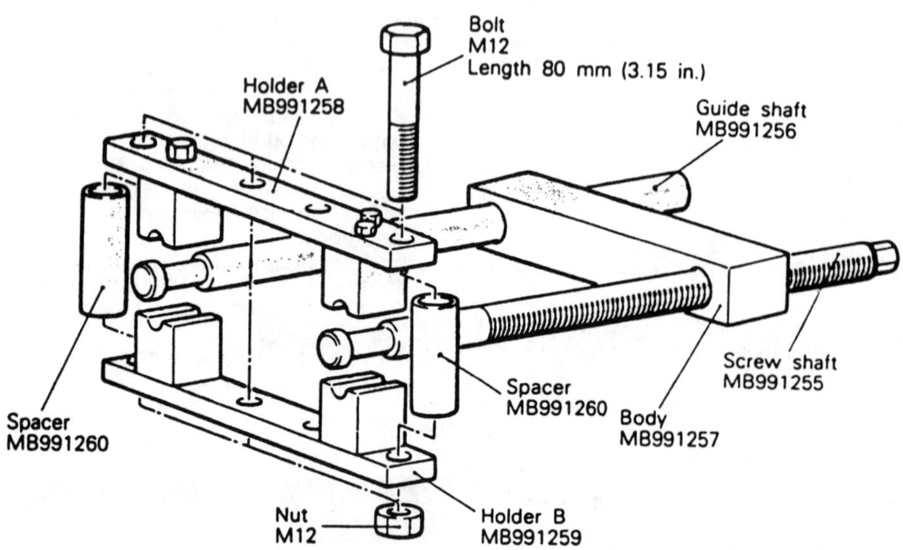

Fig. 9 Tool MB991254

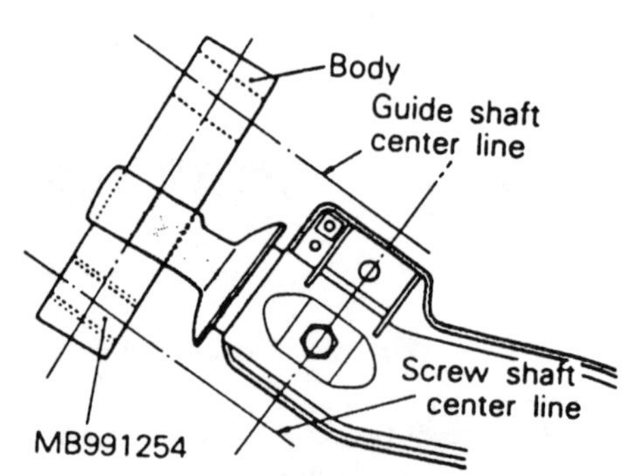

Fig. 11 Installing MB991254

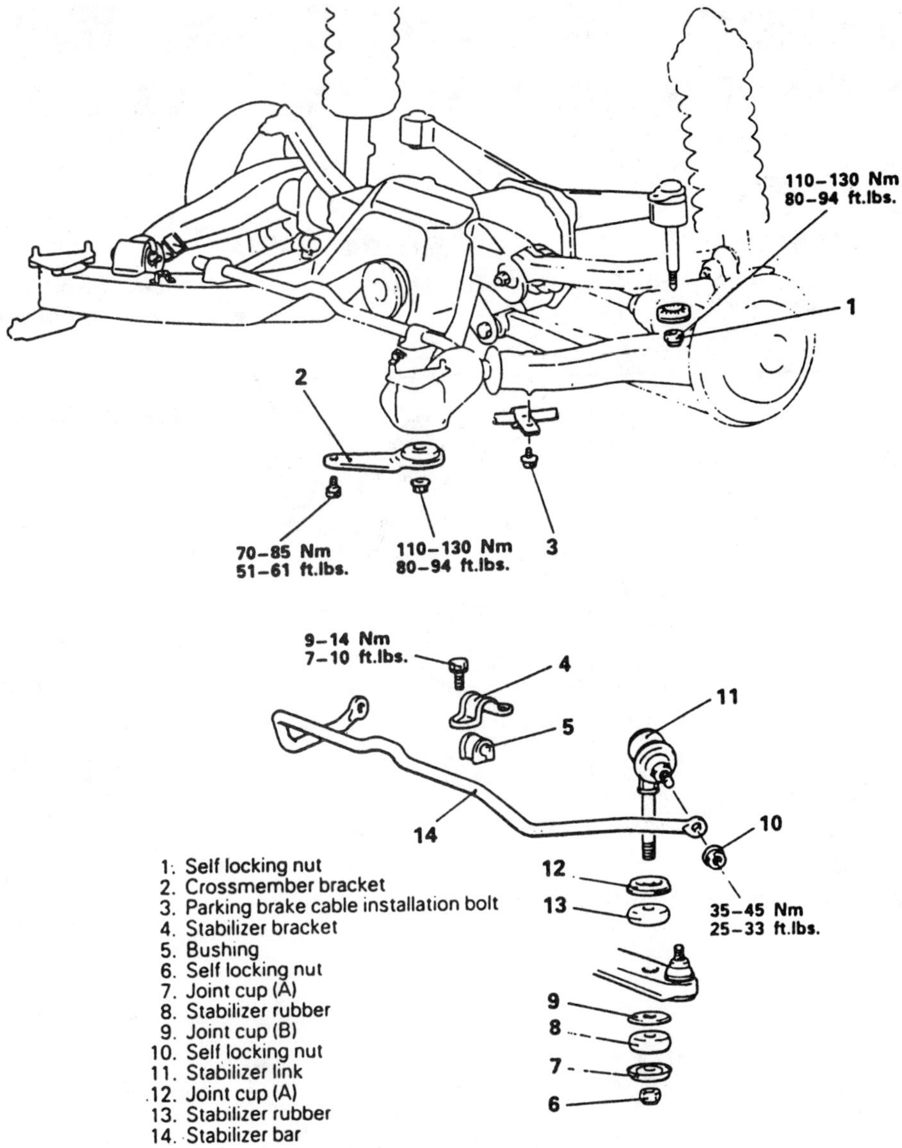

Fig. 12 Replacing stabilizer bar

REAR SUSPENSION TIGHTENING SPECIFICATIONS

Year	Component	Torque/ft. lbs.
1990	Wheel bearing nut	144-188
	Caliper assembly mounting bolt	36-43
	Lateral rod mounting nut (body side)	58-72
	Lateral rod mounting nut (axle beam side)	72-87
	Piston rod nut	14-18
	Shock absorber upper mounting nut	29-36
	Shock absorber lower mounting nut	58-72
	Trailing arm mounting bolt	72-87

Chrysler Motors Rear Wheel Drive Models Except Conquest

INDEX

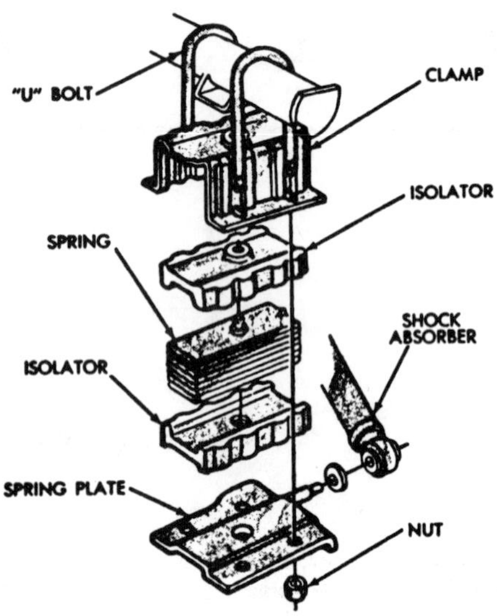

Fig. 1 Rear spring isolator

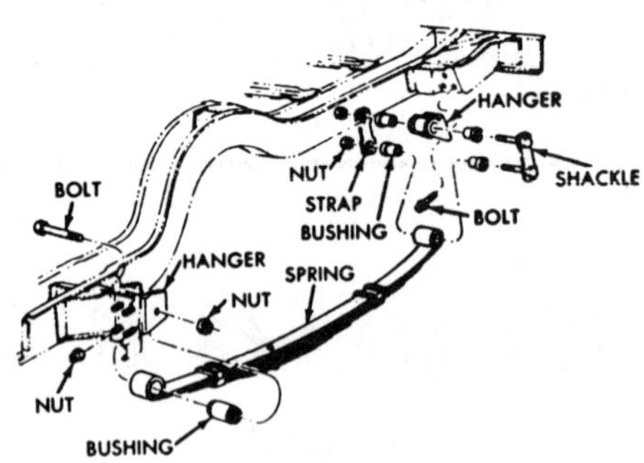

Fig. 2 Rear spring (typical)

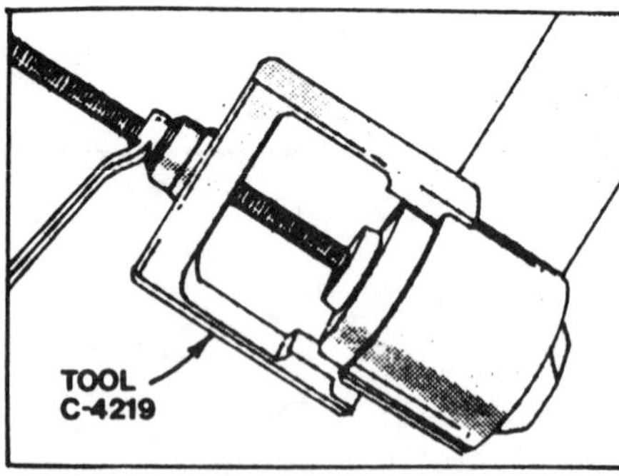

Fig. 3 Spring pivot bushing replacement

SHOCK ABSORBER REPLACE

To replace shock absorber, support rear axle properly and disconnect shock absorber at upper and lower mountings.

LEAF SPRINGS & BUSHINGS REPLACE

1. Support rear axle, relieving tension from spring.
2. Disconnect shock absorber from lower mounting.
3. Remove U-bolts and spring plate, or lower spring seat isolator retainer and isolator, **Fig. 1.**
4. Remove spring front hanger to body mount bracket nuts, **Fig. 2.**
5. Remove rear shackle bolts, lower spring, thus pulling spring front hanger bolts out of holes.
6. Remove front hanger and rear shackle from spring. To replace pivot bushings, refer to **Fig. 3.** Bushing replacement is accomplished in one operation.
7. Reverse procedure to install.

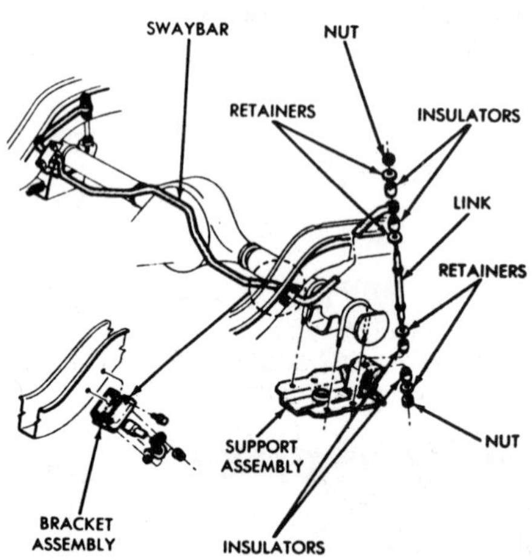

Fig. 4 Sway bar installation

LEAF SPRING SERVICE

To replace interliners, remove spring alignment clips and discard alignment clips. Separate spring leaves with a screwdriver or other suitable tool and remove interliners. Thoroughly clean spring surfaces before installation of new interliners.

To replace zinc interleaves, clamp spring in a vise and remove center bolt. Open vise carefully, allowing spring to expand. Interleaves can now be serviced. Install a drift through spring center bolt holes and clamp spring in a vise. Remove drift and install center bolt.

SWAY BAR REPLACE

1. Remove nuts, retainers and rubber insulators from sway bar upper links, **Fig. 4.**
2. Disconnect sway bar brackets from frame.
3. Remove link from support assembly and replace insulators. Reverse procedure to install.

REAR SUSPENSION TIGHTENING SPECIFICATIONS

Year	Component	Torque/ft. lbs.
1985–89	Spring Center Bolt Nut	40
	Spring Front Hanger Nut	35
	Spring Pivot Volt And Nut	105
	Spring Rear Hanger	35
	Spring Shackle Nut	35
	Spring U-Bolt Nut	45
	Shock Absorber Upper Nut	70
	Shock Absorber Lower Nut	35
	Sway Bar Shaft Bracket Screws	100①
	Sway Bar Link Nuts	200①

①—Inch lbs.

Eagle 4 x 4

INDEX

SHOCK ABSORBER REPLACE

1. With the rear axle supported properly, disconnect lower end of shock absorber from stud on mounting bracket.
2. Remove upper mounting bracket from underbody.
3. Reverse procedure to install.

LEAF SPRINGS REPLACE

1. Support rear axle, removing tension from springs.
2. Disconnect lower end of shock absorber from stud on mounting bracket, **Fig. 1.**
3. Remove U-bolts securing spring plate and spring to axle tube.
4. Disassemble rear shackle and remove eye bolt from spring forward mounting bracket.
5. Reverse procedure to install. Replace bushings as necessary.

CONTROL ARMS & BUSHINGS REPLACE

Replace control arms one at a time to prevent axle assembly misalignment. A misaligned axle assembly creates installation difficulties.

UPPER CONTROL ARMS

1. Support vehicle at frame.
2. Remove control arm bolts from frame crossmember and axle tube bracket.
3. To replace axle tube bracket bushings, refer to **Figs. 2 and 3.**
4. Reverse procedure to install.

LOWER CONTROL ARMS

1. Support vehicle at rear axle.
2. Remove stabilizer bar, if equipped.
3. Remove control arm mount bolts from frame and axle tube brackets.
4. Reverse procedure to install. **The lower control arm bushings are not serviceable.**

STABILIZER BAR REPLACE

1. Raise and support rear of vehicle, then remove nuts and grommets attaching stabilizer to connecting links, **Fig. 4.**
2. Remove bolts attaching stabilizer bar mounting clamps to spring clip plates and remove stabilizer bar.
3. Reverse procedure to install. **Torque** link locknuts to 7 ft. lbs. and mounting clamp bolts to 25 ft. lbs.

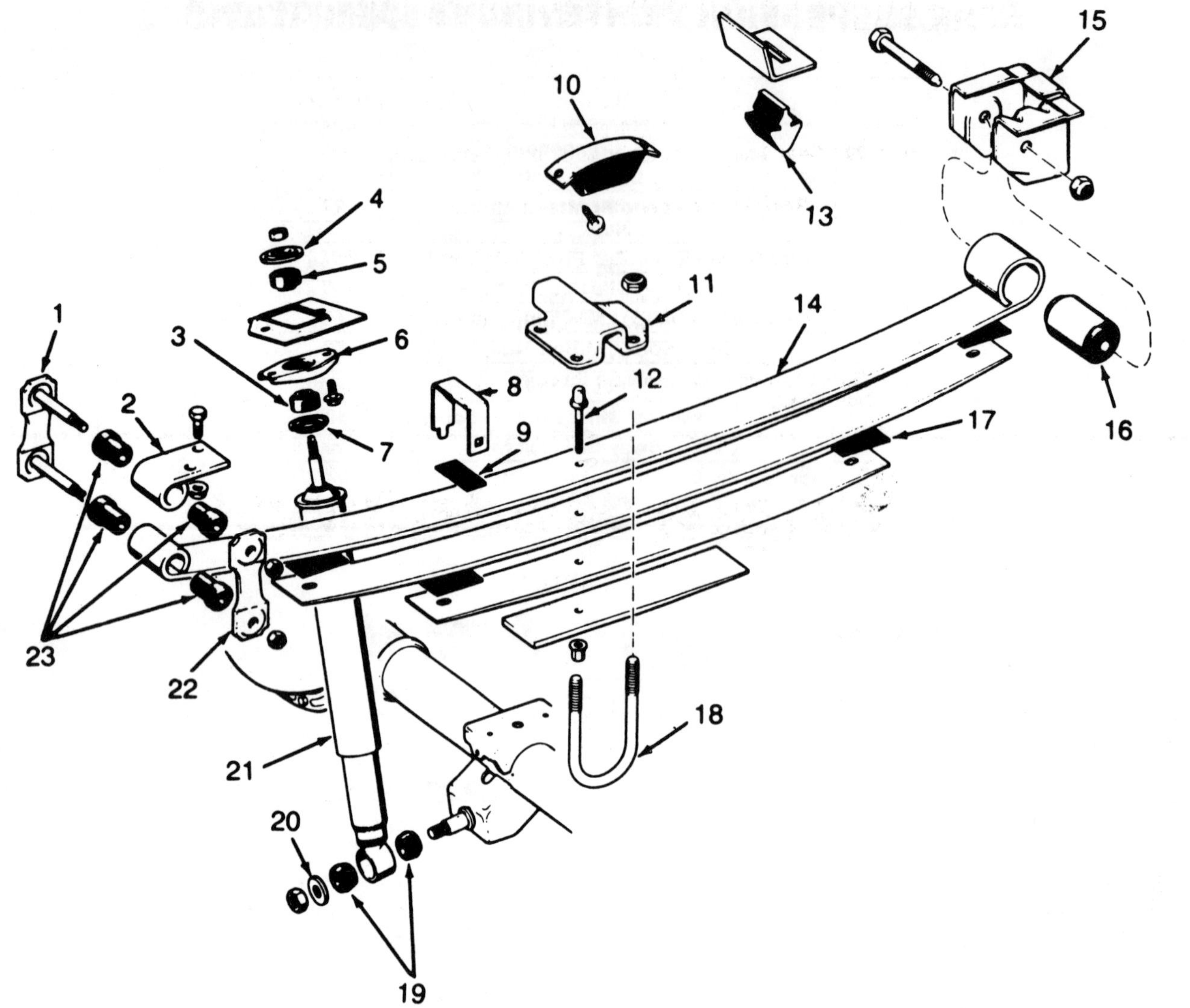

1 - Spring Shackle
2 - Hanger Bracket
3 - Grommet
4 - Retainer
5 - Grommet
6 - Shock Absorber Bracket
7 - Retainer
8 - Spring Clip
9 - Insert
10 - Bumper Cushion
11 - Plate
13 - Cushion
14 - Main Leaf Spring
15 - Mounting Bracket
16 - Front Bushing
17 - Insert
18 - U-Bolt
19 - Grommets
20 - Retainer
21 - Shock Absorber
22 - Shackle Plate
23 - Rear Bushing

Fig. 1 Rear suspension assembly

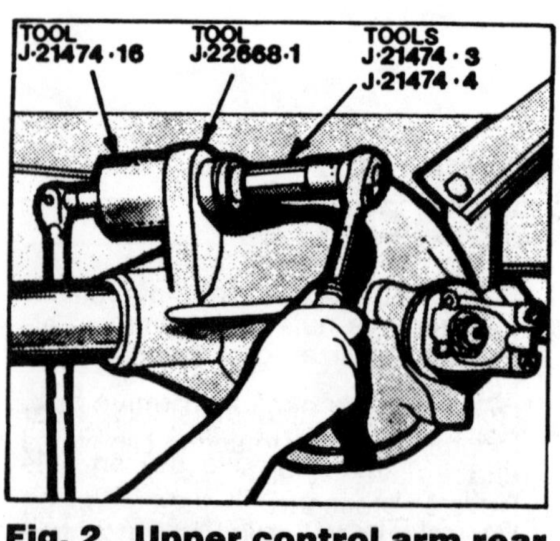

Fig. 2 Upper control arm rear bushing removal

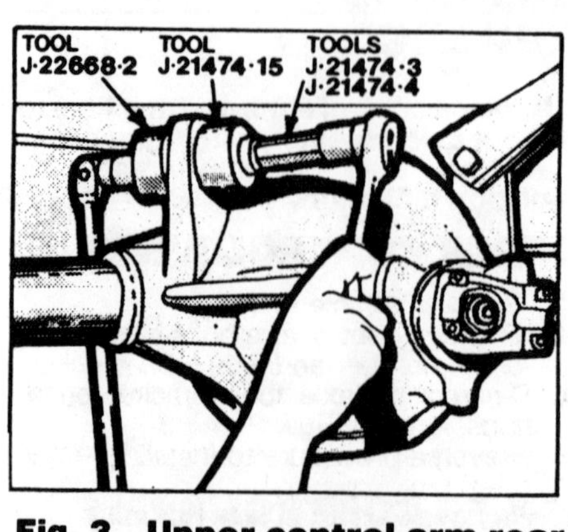

Fig. 3 Upper control arm rear bushing installation

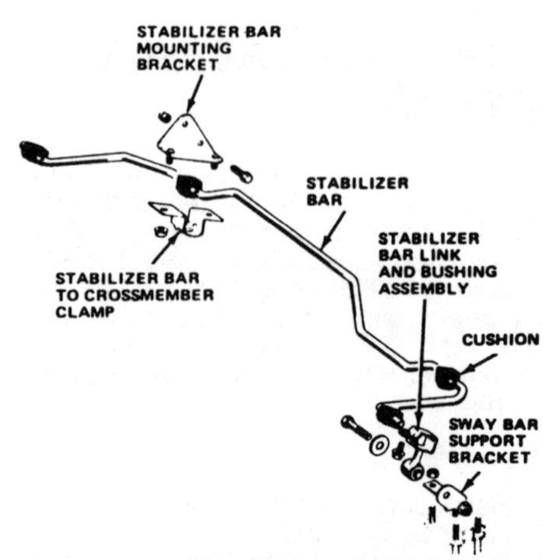

Fig. 4 Rear stabilizer bar

REAR SUSPENSION TIGHTENING SPECIFICATIONS

Year	Component	Torque/ ft. lbs.
1985-89	Rear Shock Absorber Locknut (Upper And Lower)	8
	Rear Shock Absorber Mounting Bracket Nuts	20
	Rear Spring Front Eye Pivot Bolt And Nut	110①
	Rear Spring Shackle Pin Nuts	30
	Rear Spring Hanger Mounting Stud Nuts	45
	Rear Spring U-Bolt Nuts	50

①—Tighten, then check torque through bolt head.

Eagle Medallion

INDEX

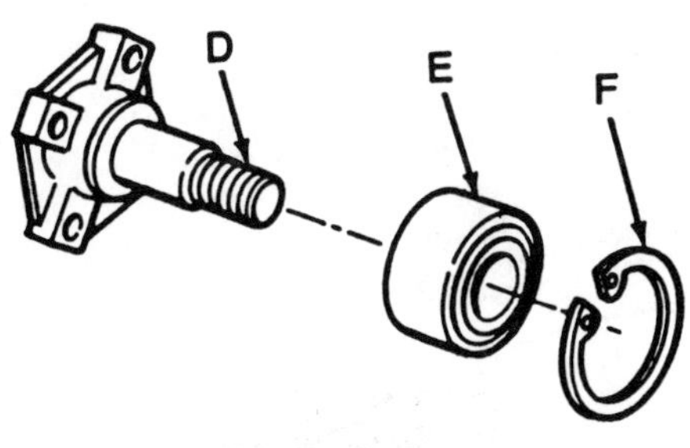

Fig. 1 Axle shaft hub assembly

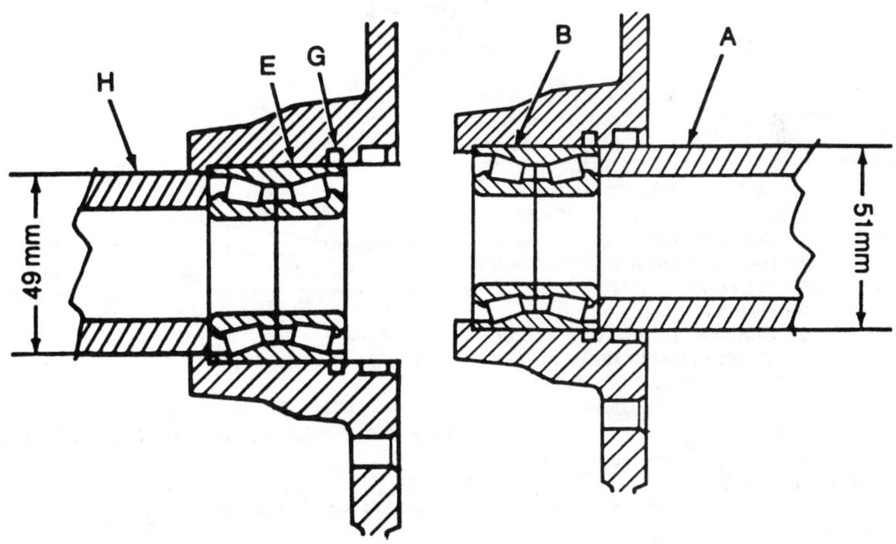

Fig. 2 Rear wheel bearing installation

REAR WHEEL BEARING SERVICE

Before removing hub bearing, check bearing axial play. Bolt a dial indicator onto the rear brake drum using dial indicator adapter Rou.541 or equivalent. Measure axial play at the end of the axle shaft. Axial play should not exceed .001 inch. If measurement obtained exceeds specification, replace hub bearing.

REMOVAL

1. Raise and support vehicle.
2. Remove wheel(s).
3. Remove screws attaching brake drum to axle shaft hub.
4. Remove brake drum from axle shaft hub.
5. Using a suitable screwdriver, remove plastic hub cover.
6. Remove axle shaft hub locknut.
7. Remove axle shaft hub from axle shaft. **If necessary use tool T.Av. 1050 or equivalent, to remove hub.**
8. Remove bearing retaining clip (F) from machined groove (G) on axle shaft hub, **Fig. 1.** Using a suitable section of pipe (H, approximately 1.92 inch outside diameter) and a press, remove wheel bearing (E) from hub.

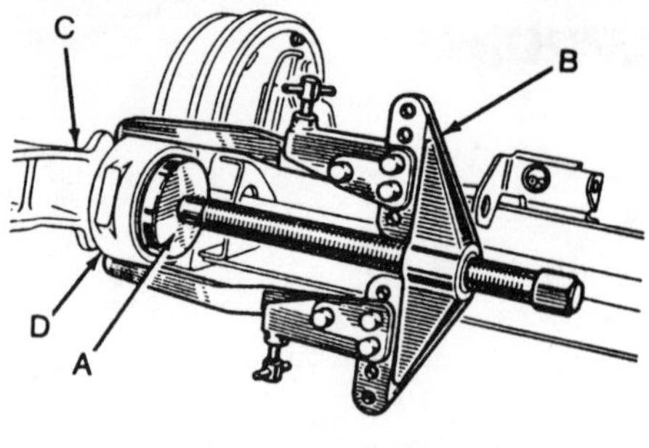

Fig. 3 Installing extractor tool onto axle pivot arm

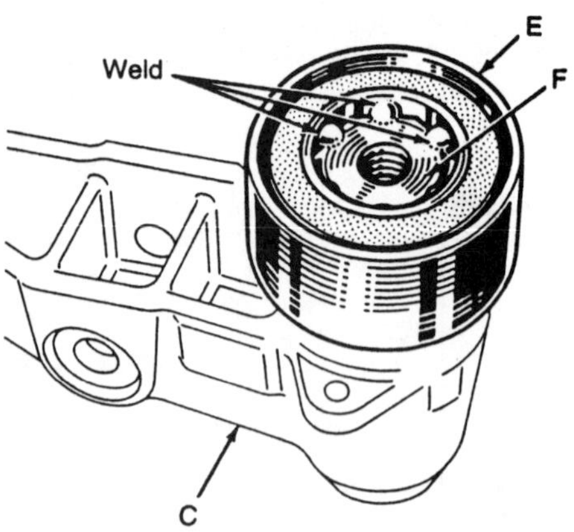

Fig. 4 Removing bushing from axle pivot arm

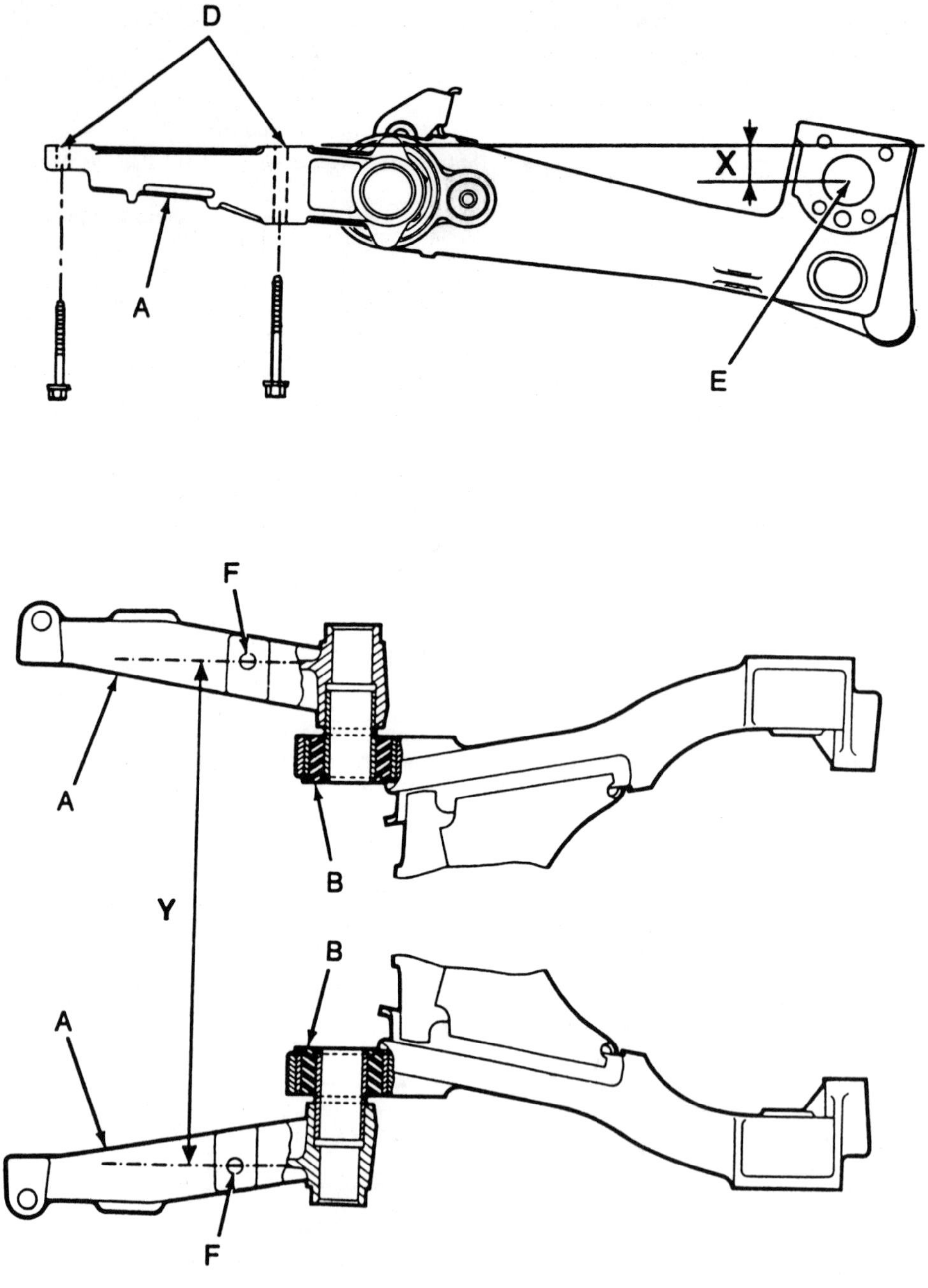

Fig. 5 Axle pivot arm bushing & pivot arm alignment

INSTALLATION

Do not install a used bearing. When installing bearing onto axle shaft hub, do not apply press load directly to the center bearing race. Apply load to the outer portion of the bearing only.

1. Thoroughly clean hub assembly.
2. Using a suitable press and a section of pipe (A, approximately 2 inches in diameter), press bearing (B) into hub assembly, **Fig. 2.**
3. Install a new bearing retaining clip.
4. Lightly lubricate axle shaft and install axle shaft hub onto axle shaft.
5. Install and tighten a new axle shaft hub locknut.
6. Install brake drum and attaching screws.
7. Install plastic hub cover.
8. Install wheel(s).
9. Check brakes from proper operation.
10. Lower vehicle.

AXLE PIVOT ARM SERVICE

REMOVAL

1. Remove rear axle assembly from vehicle as Follows:
   a. Raise and support vehicle.
   b. Remove rear wheel and tire assemblies.
   c. Remove shock absorbers.
   d. Disconnect parking brake cable from rear wheels.
   e. Disconnect and cap brake lines.
   f. Disconnect brake compensator linkage from axle assembly.
   g. Using a suitable jack support axle assembly, then remove axle assembly attaching bolts.
   h. Lower and remove axle assembly from vehicle.
2. Remove torsion bars from vehicle.
3. Install tool component (A) from tool T.Ar. 1056 or equivalent and a two jaw extractor (B) onto axle assembly (D), **Fig. 3.**
4. Drive axle pivot arm (C) and bushing from rear axle assembly, **Fig. 4.**
5. Remove bushing (E) from axle pivot arm (C). **A piece of metal (a 26 mm nut would be acceptable) and must be welded to the inside of the bushing, Fig. 4. The nut is used as a back-up to drive the bushing from the axle pivot arm.**
6. Using a suitable press and bearing splitter tool, J-22912-01 or equivalent, drive bushing from axle pivot arm. **When driving bushing, use a suitable section of pipe or socket to prevent damage. Do not reuse bushing after it has been removed.**

INSTALLATION

1. Position bushing into rear axle assembly.
2. Using tool T.Ar. 1056 or equivalent, draw bushing into axle until it is flush with the outer edge of the axle. Posi-

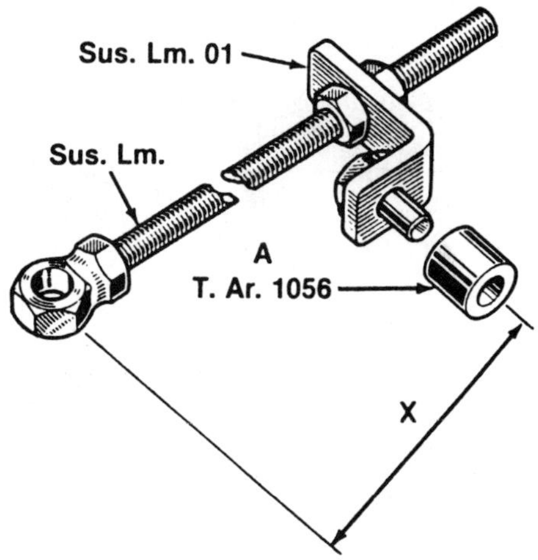

Fig. 6 Threaded rod tool assembly

tion pivot arm to the axle for proper bushing preload. Before pressing the axle pivot arm (A) onto the axle bushing (B), it must be properly positioned. Position a straight edge along the top of the axle pivot arm where it attaches to the body at points (D), **Fig. 5.** Move the pivot arm until dimension X is approximately .275 inch. Take measurement from the center line of the axle shaft (E) to the straightedge on the axle pivot arm mounting points.

3. With pivot arm correctly positioned to the axle assembly, press pivot arm to the rear axle using tool T.Ar. 1056 or equivalent.
4. Press axle pivot arm until dimension Y is 50.5 inches. Dimension Y is taken between the left and right pivot arms at bolt holes (F).
5. Install rear axle assembly. Check bearing play as follows:
   a. Position suitable dial indicator to brake drum.
   b. Measure axial play at end of axle shaft.
   c. Axial play should not exceed .001 inch.
   d. If axial play exceeds .001 inch, replace hub bearing.
6. Install torsion bars into axle.

AXLE SHAFT

REPLACE

1. Raise and support vehicle.
2. Remove tire and wheel assembly.
3. Remove brake drum to axle shaft hub attaching screws, then the brake drum.
4. Remove brake drum from axle shaft hub.
5. Using a suitable screwdriver, pry off plastic hub cover.
6. Remove axle shaft hub locknut.
7. Remove axle shaft hub from axle shaft.
8. Remove axle shaft to rear axle attaching bolts.
9. Remove bolt attaching brake backing

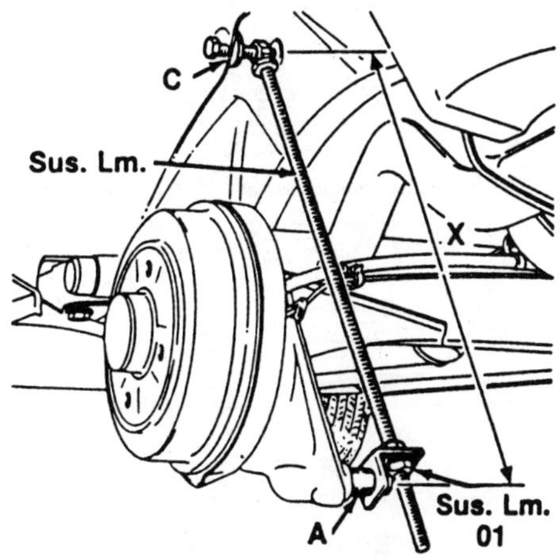

Fig. 7 Installing threaded rod tool assembly in place of shock absorber

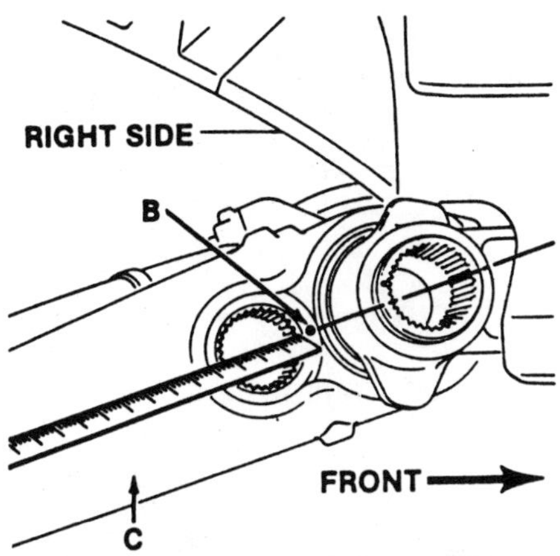

Fig. 9 Aligning suspension bar & lower control arm tooth marks

plate to axle shaft.

10. Remove axle shaft.
11. Reverse procedure to install.

SHOCK ABSORBER

REPLACE

Never raise vehicle from under the V shaped channel on the rear axle.

1. Raise and support vehicle.
2. Lift control arm to relieve tension from shock absorber. **The shock absorber mounting bolts are not interchangeable. Mark location of bolt before removing.**
3. Remove upper and lower shock absorber attaching bolts, then the shock absorber.
4. Reverse procedure to install.

TORSION BAR SERVICE

REMOVAL

Never raise vehicle from under the V shaped channel on the rear axle.

1. Raise and support vehicle.
2. Remove both rear tire and wheel assemblies.
3. Remove shock absorbers.
4. Remove dust caps covering ends of

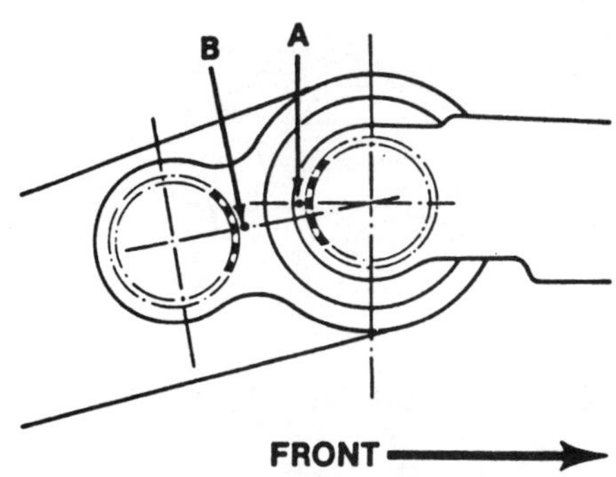

Fig. 8 Suspension bars mounting hole stamping marks

torsion bar by prying them off.

5. Mark and note splined positions of torsion bars before removal. Remove suspension bar or anti-sway bar (stabilizer bar).
6. Remove torsion bar connecting link only if all four torsion bars are being removed.

INSTALLATION

1 To 3 Bars

Before the torsion bars are installed, the rear trailing arms must be properly positioned. To accomplish this, proceed as follows:

1. Tools Sus.Lm, Sus.Lm. 01 and T.Ar. 1056 are required.
2. Adjust bracket tool Sus.Lm.01 until dimension X is obtained.
3. **On sedan models dimension X should be 19.37 inch and on wagon models dimension X should be 18.97 inch, Fig. 6.**
4. Position tool assembly in place of the shock absorber and bolt it to point (C), **Fig. 7.**
5. Slip spacer (A) into lower shock absorber mounting hole.
6. The suspension bar mounting holes have a stamped mark A which enables the suspension bars to be initially positioned. An additional mark B must be made on the lower control arm for the anti-sway bar, **Fig. 8.** Place a ruler on the center line of the two mounting holes and place a mark (B) at the bottom of a tooth on the lower control (C), **Fig. 9.** Ensure letter on the torsion bar matches the position it is being installed into. The left hand bar is marked G and the right hand bar is marked D, **Figs. 10 and 11.**
7. Apply lubricant No. 8993630 or equivalent, onto splines on the torsion bar and their mounting holes. Align torsion bar indexing mark to the mounting hole.
8. On sedan models, proceed as follows:
   a. The anti-sway (stabilizer) bar (D) is positioned 5 teeth up from the mark (B) made previously, **Fig. 12.**
   b. The suspension bar (E) is positioned 1 tooth down from the initial position marked (A).

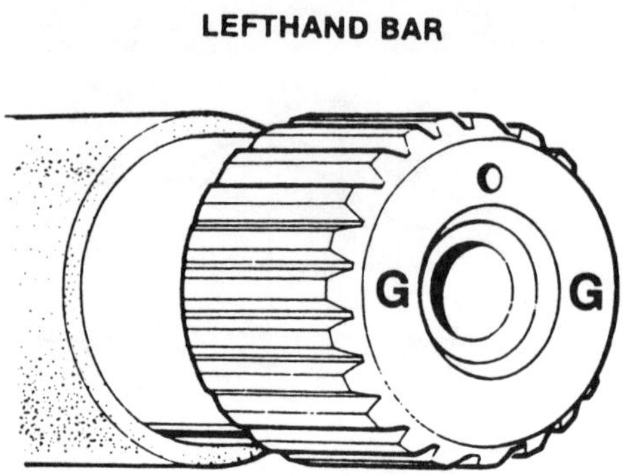

Fig. 10 Left hand bar identification letter G

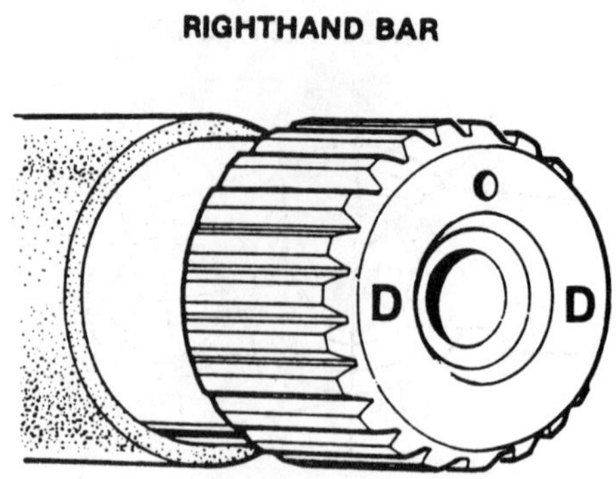

Fig. 11 Right hand bar identification letter D

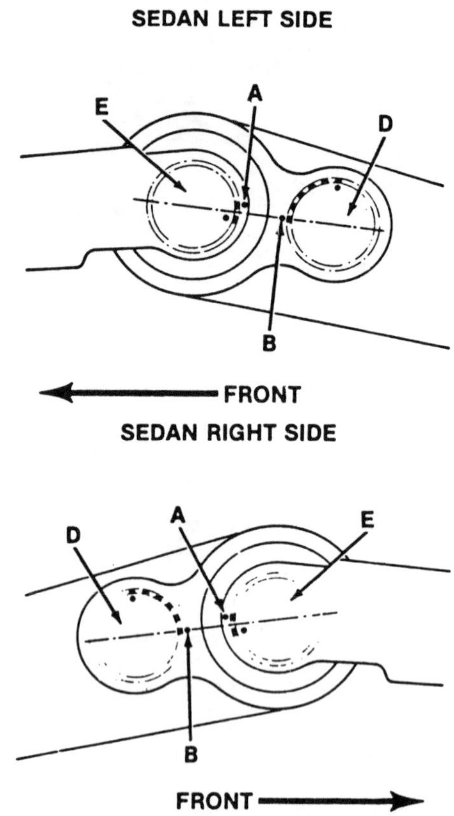

Fig. 12 Positioning anti-sway & suspension bars (torsion bar installation 1 to 3 bars). Sedan models

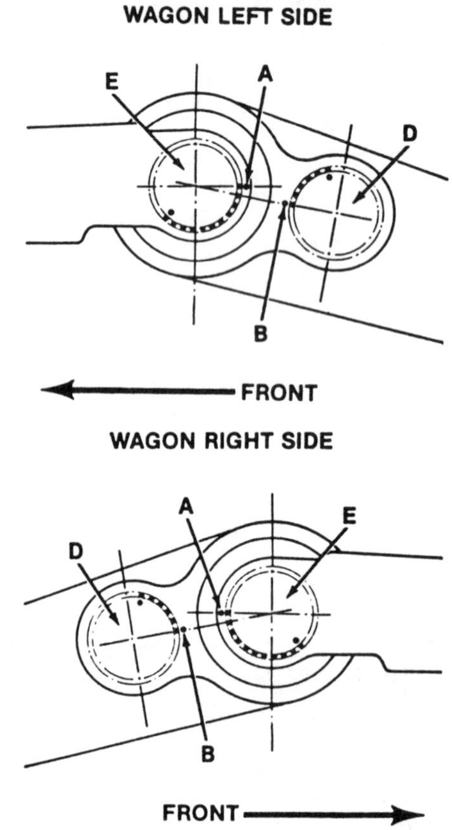

Fig. 13 Positioning anti-sway & suspension bars (torsion bar installation 1 to 3 bar). Station wagon models

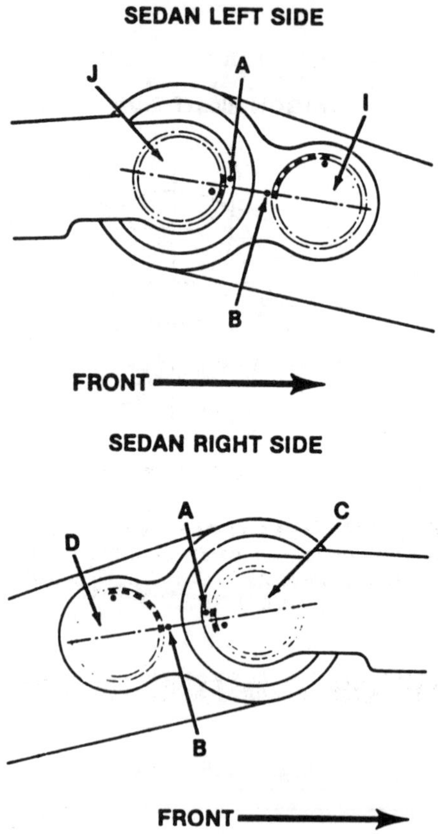

Fig. 14 Positioning anti-sway & suspension bars (torsion bar installation all 4 bars). Sedan models

9. **On wagon models, proceed as follows:**
   a. The anti-sway (stabilizer) bar (D) is positioned 5 teeth up from the mark (B) made previously.
   b. The suspension bar (E) is positioned 10 teeth down from the initial position mark (A), **Fig. 13.** Do not use a hammer to drive in torsion bars (other than brass) as damage to the splines will result.
10. When the bars are properly aligned, push into place. A brass hammer may be used for this purpose.
11. Remove threaded rod.
12. Install tire and wheel assemblies.
13. Lower vehicle.
14. Check and adjust vehicle height as required.

All 4 Bars

Before installing torsion bars, the rear trailing arms must be properly positioned. To accomplish this, proceed as follows:

1. Tools Sus.Lm, Sus.Lm. 01 and T.Ar. 1056 are required.
2. Refer to steps 1 through 7 under "1 To 3 Bars".
3. On sedan models, proceed as follows:
   a. The suspension bars (J and C) are positioned 1 tooth down from the initial position marked (A).
   b. The anti-sway (stabilizer) bars (D and I) are positioned 5 teeth up from the mark (B) made previously, **Fig. 14.**
4. **On wagon models, proceed as follows:**
   a. The anti-sway (stabilizer) bars (D and I) are positioned 5 teeth up from the mark (B) made previously.
   b. The suspension bars (J and C) are positioned 10 teeth down from the initial position mark (A), **Fig. 15.**
5. With anti-sway and suspension bars properly positioned, slide anti-sway (stabilizer) bar through the mounting hole.
6. Position bar into mounting hole as described previously.
7. Slip torsion bar connecting link over the two previously installed torsion bars. **The connecting link must fit into the V shaped section of the rear axle assembly.**
8. Slide suspension bar through mounting hole.
9. Position bar into mounting hole as described previously.
10. Slide anti-sway (stabilizer) bar through mounting hole, then position bar into hole as described previously and slide into connecting link. **If it is difficult to align the torsion bars, the connecting link can be slightly twisted to facilitate alignment of the torsion bars.**
11. Remove threaded rod tool assembly, install wheels, lower vehicle and check vehicle ride height.
12. Adjust vehicle ride height, if necessary.

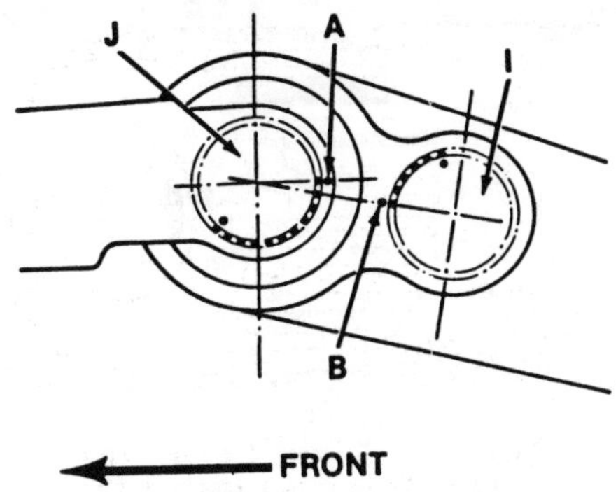

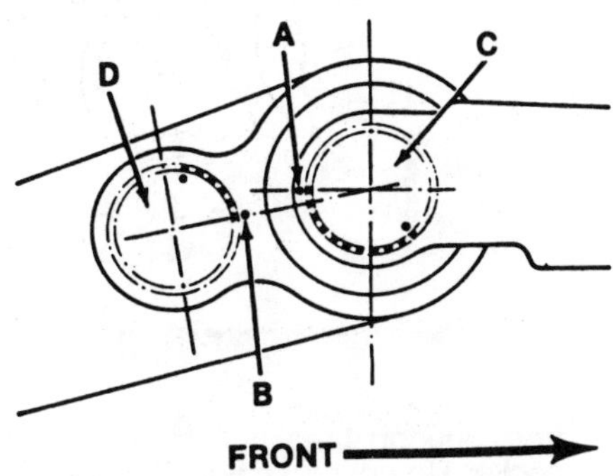

Fig. 15 Positioning anti-sway & suspension bars (torsion bar installation all 4 bars). Station wagon models

REAR SUSPENSION TIGHTENING SPECIFICATIONS

Year	Component	Torque/ft. lbs.
1988-89	Axle Spindle & Backing Plate To Axle	55
	Brake Bleeder Fittings	5.8
	Brake Drum To Hub Retaining Screws	14
	Brake Hose At Rear Axle	10
	Pivot Arm Bolts	63
	Rear Hub Attaching Spindle Nut	118
	Shock Absorber Lower Attachment	63
	Shock Absorber Upper Attachment	59
	Wheel & Tire Assembly Nuts	66

Eagle Premier

INDEX

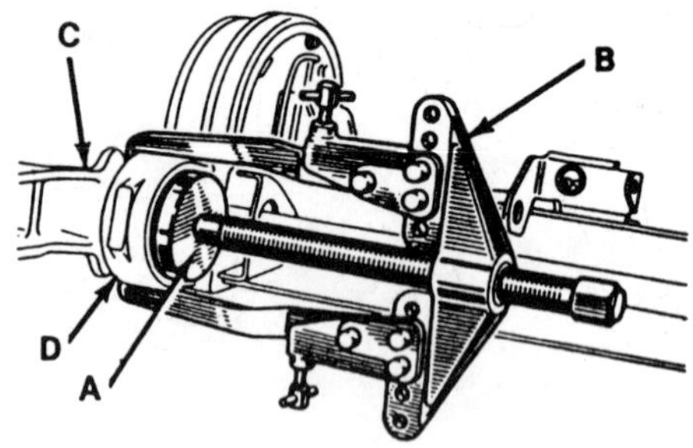

Fig. 1 Rear axle bushing removal

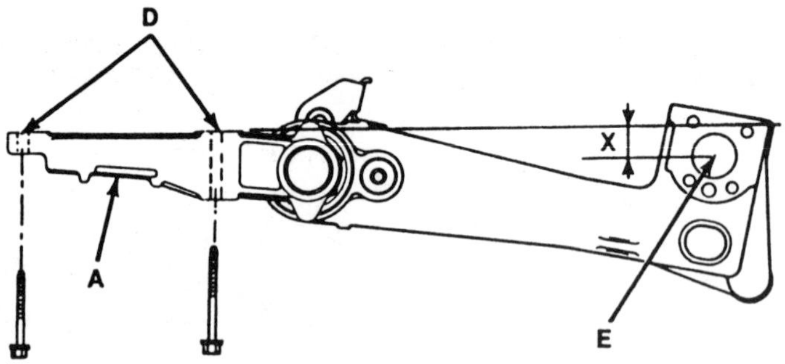

Fig. 2 Axle support bracket

REAR AXLE SUPPORT BUSHING REPLACE

REMOVAL

1. Remove rear axle assembly as follows:
   a. Raise and support vehicle.
   b. Remove rear wheel and tire assemblies.
   c. Disconnect shock absorbers from axle.
   d. Loosen adjustment nuts to allow balls on end of parking brake cables to pass through access holes. Remove cables from body support.
   e. Disconnect brake hoses from metal brake lines and axle.
   f. Position supporting jack under axle assembly and remove attaching bolts.
   g. Remove axle assembly by lowering support jack.
2. Remove torsion bars from axle as described under, "Torsion Bar, Replace."
3. Install component (A) from T.Ar.1056, bushing remover set, and a two jaw extractor (B) on axle, **Fig. 1.**
4. Drive support bracket and bushing from rear axle.
5. To remove bushing from support bracket, weld a 1.02 inch nut to inside of bushing. This nut is used as a back-up to drive bushing from support bracket.
6. Using a suitable hydraulic press and bearing splitter tool, drive bushing from support bracket with a suitable socket. Do not reuse bushing after removal.

INSTALLATION

1. Position new bushing into rear axle.
2. Using driver tool and bushing receiving cup from T.Ar.1056, draw bushing into axle until flush with outside edge of axle.
3. Position support bracket to axle for proper bushing preload, **Fig. 2.** Before pressing support bracket (A) to axle bushing, it must be properly positioned. Adjust support bracket so that dimension"X" measures $^{27}/_{32}$–$^{29}/_{32}$ inch.
4. Position straightedge along top of support bracket where it attaches to body at points (D), **Fig. 2.**
5. After support bracket is positioned to axle, press support bracket to rear axle using T.Ar.1056.
6. Press support bracket until left and right support brackets are at a distance of 51.87-51.98 inch. This measurement is taken between left and right support bracket bolt holes.
7. Install torsion bars to axle as described under, "Torsion Bar, Replace."
8. Install rear axle assembly.

REAR AXLE SHAFT REPLACE

1. Raise and support vehicle.
2. Remove rear wheel and tire assembly.
3. Remove and discard safety nuts securing brake drum to axle shaft hub.
4. Remove plastic hub cover using suitable screwdriver.
5. Remove brake drum from axle shaft hub.
6. Remove and discard axle shaft hub lock nut.
7. Remove axle shaft hub from axle shaft using a suitable puller if necessary.
8. Remove axle shaft-to-rear axle attaching bolts.
9. Remove brake backing plate-to-axle shaft attaching bolt.
10. Remove axle shaft from trailing arm.
11. Reverse procedure to install, noting the following:
    a. Tighten axle shaft-to-rear axle attaching bolts in sequence shown, **Fig. 3,** to specifications.
    b. **Torque** brake backing plate-to-axle shaft attaching bolt to specifications.
    c. Install and **torque** new axle shaft hub lock nut and torque to specifications.
    d. Install brake drum and new safety nuts to axle shaft hub.

HUB & BEARING SERVICE

BEARING INSPECTION

1. Raise and support vehicle.
2. Remove rear wheel and tire assembly.
3. Remove plastic cap from brake drum.
4. Bolt dial indicator to rear brake drum

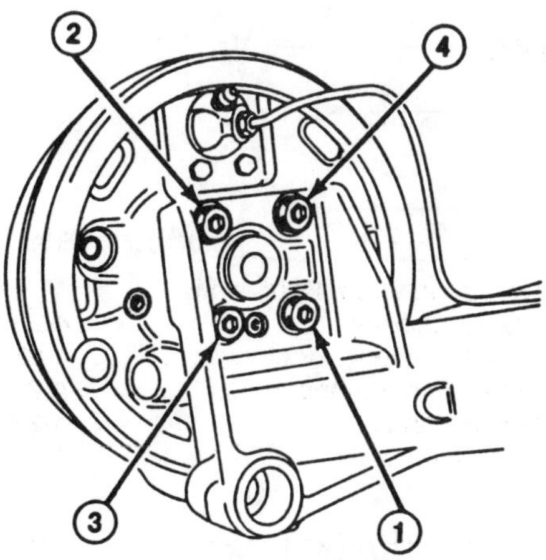

Fig. 3 Axle shaft bolt torque sequence

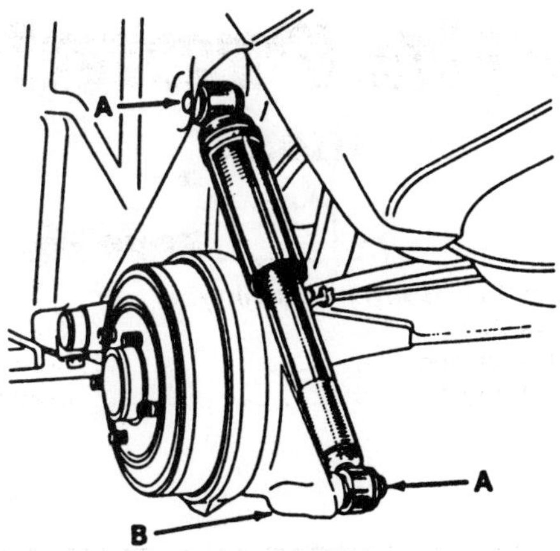

Fig. 4 Shock absorber bolt location

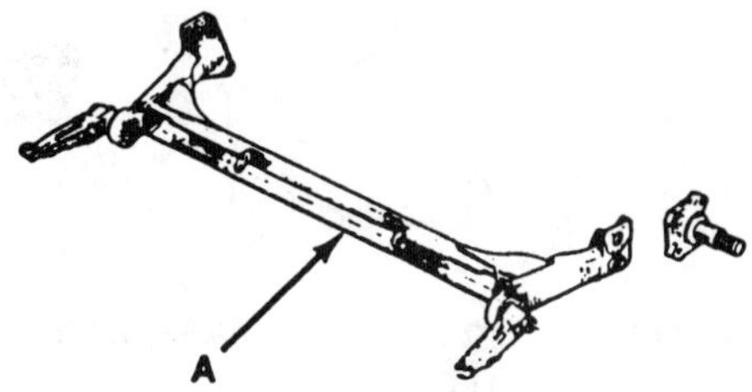

Fig. 5 Hydraulic jack positioning point

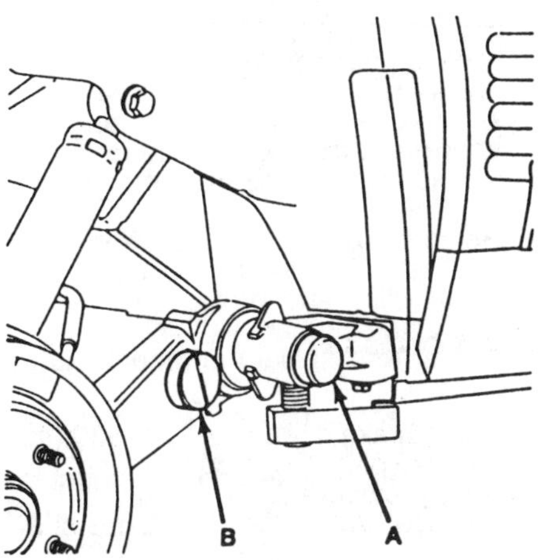

Fig. 6 Anti-sway bar protective end caps

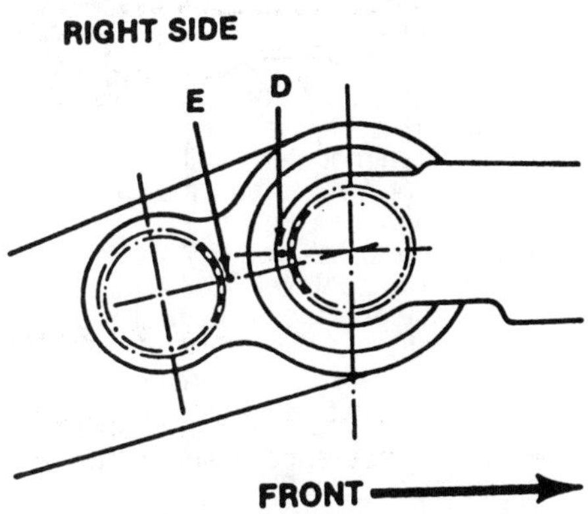

Fig. 7 Lower control arm reference marks

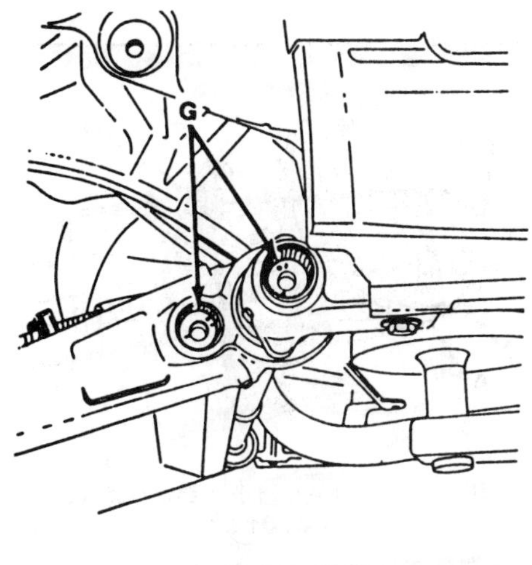

Fig. 8 Torsion bar reference & location marks

using dial indicator adapter Rou. 541.

5. Measure axial play at end of axle shaft. Axial play should not exceed .001 inch. If play exceeds this measurement, replace hub & bearing assembly.

HUB & BEARING, REPLACE

If wheel bearing must be replaced, the hub and bearing assembly must be replaced as a unit.

1. Raise and support vehicle.
2. Remove rear wheel and tire assembly.
3. Remove safety nuts securing brake drum to axle shaft hub.
4. Remove plastic hub cover using suitable screwdriver.
5. Remove brake drum from axle shaft hub.
6. Remove axle shaft hub lock nut.
7. Remove wheel hub and bearing assembly from vehicle.
8. Reverse procedure to install noting the following:
   a. **Torque** new hub nut to specifications.
   b. Install brake drum to hub using new safety nuts.

SHOCK ABSORBER

REPLACE

1. Raise and support vehicle. **Do not support vehicle weight under V-shaped channel on axle.**
2. Raise and support control arm at point B, **Fig. 4**, to relieve tension from shock absorber.
3. Remove shock absorber attaching bolts, then lower shock absorber from vehicle.
4. Reverse procedure to install. **Torque** upper and lower attaching bolt,

TORSION BAR

REPLACE

REMOVAL

1. Raise and support vehicle. **Do not support vehicle weight under V-shaped channel on axle. Axle weight must be supported at point A, Fig. 5,** using a suitable jack.
2. Remove both rear wheel and tire assemblies.
3. Remove rear shock absorbers as previously described.
4. Remove protective caps (A) and (B), **Fig. 6**, from both sides of vehicle.
5. Remove clips from ends of suspension and anti-sway bars using a suitable screwdriver.
6. Note alignment mark D, **Fig. 7,** on suspension bar mount and make a reference mark (position E) on lower control arm as shown.
7. Note and record positions of marks G, **Fig. 8,** on all four torsion bars relative to marks D and E. Location of marks on torsion bars (position G) will vary depending on application.
8. Lower rear axle, then pull suspension bars using a suitable slide hammer until splines disengage.
9. Pull anti-sway bars out using a suitable slide hammer. **The rear axle assembly will be damaged if a load is applied to jack supporting the axle.**
10. Support axle at point L, **Fig. 9,** then loosen bolts (M) approximately four turns and bolts (N) approximately ten turns.
11. Lower jack until axle has dropped approximately one inch.
12. Remove suspension bars, then the torsion bar connecting link from vehicle.

INSTALLATION

1. Install torsion bars into rear axle. **Do not engage splines at this time. En-**

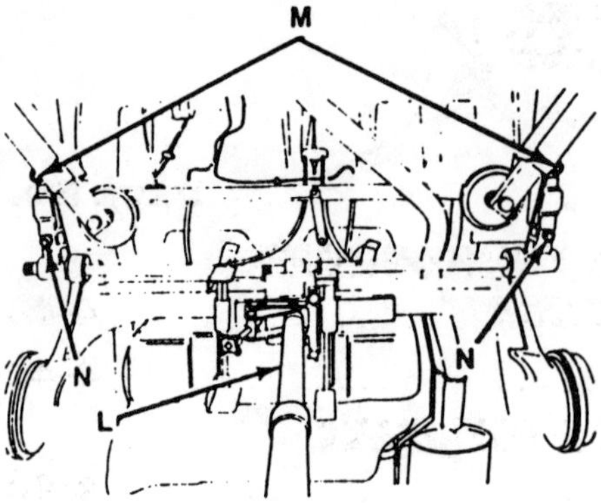

Fig. 9 Axle attaching bolt locations

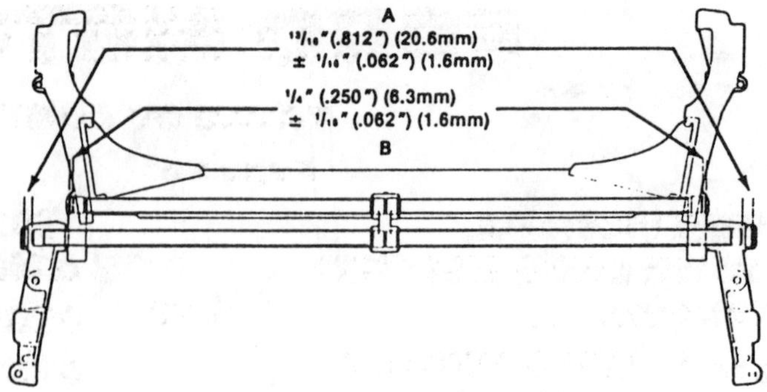

Fig. 11 Torsion bar spline dimensions

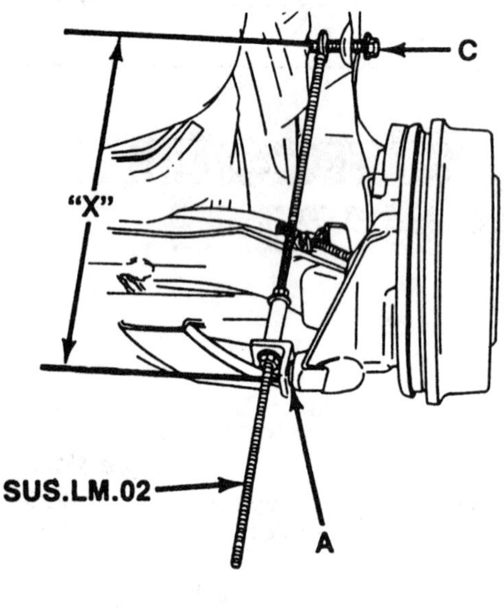

Fig. 10 Axle spacing tool position

sure torsion bars are installed in original position.

2. Raise axle and **torque** rear axle mounting bolts to specifications.
3. Position rear trailing arms as follows:
   a. Adjust dimension X, **Fig. 10,** on each of two tools Sus.Lm.02 to 17 15/16 inches.
   b. Loosely install bolt (C) and spacer (A) from tool set T.Ar. 1056 into rear lower shock absorber mounting hole.
4. Apply a suitable lubricant to torsion bar splines and install torsion bars in original position. Ensure recorded marks G, D, and E are properly aligned to avoid added stress to bars.
5. Use recorded marks to position anti-sway bar G into mounting.
6. Install connecting link to splines of anti-sway bar G. Ensure connecting link is centered in V-shaped channel in axle.
7. Install anti-sway bar D in outer mounting and secure in connecting link.
8. Install both suspension bars through outer mounting and secure in connecting link. **Ensure all torsion bars are installed in original position. If splines are difficult to engage, slightly twist link as necessary using suitable pliers on the connecting link.**
9. Remove threaded rod tools from vehicle, then install shock absorbers.
10. Install clips and protective caps on end of torsion bars.
11. Center torsion bars in mountings as follows:
    a. Tap ends of suspension bars until outer end of bar protrudes 3/4-7/8 inch from outer edge of mount on both sides, **Fig. 11, using a large brass drift. Do not hammer directly on torsion bar splines.**
    b. Tap ends of anti-sway bars until outer end of bar protrudes 3/16-5/16 inch from outer edge of mount on both sides, **Fig. 11. Do not hammer directly on torsion bar splines.**
12. Install wheel and tire assemblies.

REAR SUSPENSION TIGHTENING SPECIFICATIONS

Year	Component	Torque ft. lbs.
1988-90	Axle Shaft Hub Lock Nut	123
	Axle Shaft-To-Rear Axle Attaching Bolts	47
	Brake Backing Plate-To-Axle Shaft Attaching Bolt	63
	Hub Nut	123
	Lower Shock Absorber Bolt	85
	Rear Axle Mounting Bolts	68
	Upper Shock Absorber Attaching Bolt	60

FORD MOTOR CO.

TABLE OF CONTENTS

Ford Crown Victoria, Lincoln Town Car & Mercury Grand Marquis

INDEX

SHOCK ABSORBER REPLACE

1. With the rear axle supported properly disconnect shock absorber at upper mounting and compress it to clear hole in spring seat.
2. Disconnect shock absorber from stud on axle bracket.
3. Reverse procedure to install.

COIL SPRING REPLACE

1. Raise rear of vehicle and support at frame. Support rear axle with a suitable jack.
2. Disconnect shock absorbers at lower mountings.
3. Disconnect brake line from rear brake hose and remove hose to bracket clip.
4. Lower axle to remove springs. **On some models, it may be necessary to disconnect the right hand parking brake cable from right hand upper arm retainer before lowering axle.**
5. Reverse procedure to install. Install an insulator between upper seat and spring, if necessary.

CONTROL ARMS REPLACE

Control arms must be replaced in pairs.

1. Raise rear of vehicle and support at frame. Support axle with a suitable jack.
2. Remove stabilizer bar, if necessary.
3. Lower axle and install a second jack under differential pinion nose.
4. Disconnect parking brake cable from upper arm retainer.
5. Disconnect control arm from axle bracket. On upper arms, disconnect arm from crossmember and on lower arms, disconnect arm from frame attachment bracket.
6. Reverse procedure to install.

STABILIZER BAR REPLACE

1. Raise and support vehicle at frame side rails.
2. Support rear axle with a suitable jack and position axle so the shock absorbers are fully extended.
3. Remove bolts, nuts and spacers attaching stabilizer bar to lower arms, **Fig. 1.**
4. Remove stabilizer bar from vehicle.
5. Reverse procedure to install.

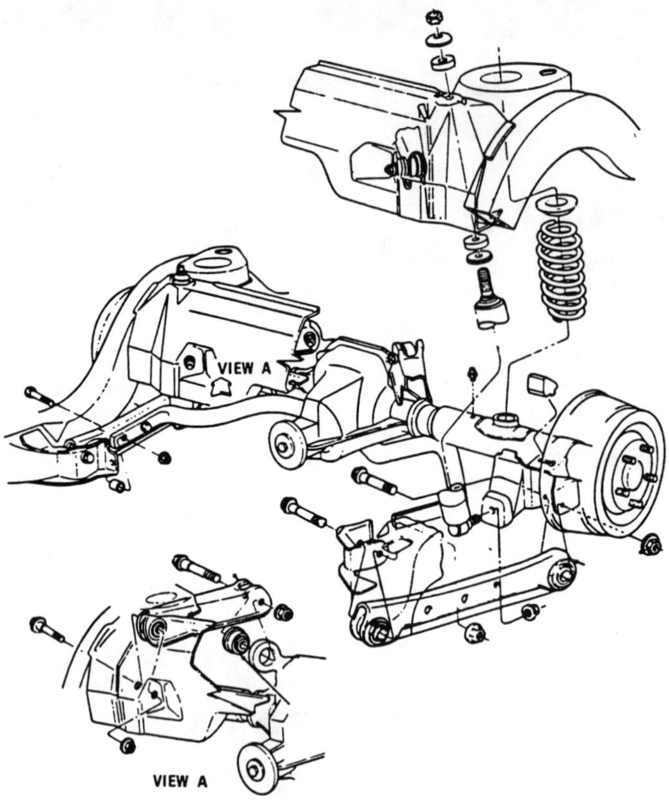

Fig. 1 Rear suspension (typical)

REAR SUSPENSION TIGHTENING SPECIFICATIONS

Year	Component	Torque/ft. lbs
1985-90	Lower Arm To Axle	103-133
	Lower Arm To Frame	120-150
	Shock Absorber To Axle Bracket	52-85
	Shock Absorber To Upper Mount Nut	14-26
	Stabilizer Bar	70-92
	Upper Arm To Axle	103-133
	Upper Arm To Frame	120-150

Ford Escort, EXP & Tempo & Mercury Lynx & Topaz

INDEX

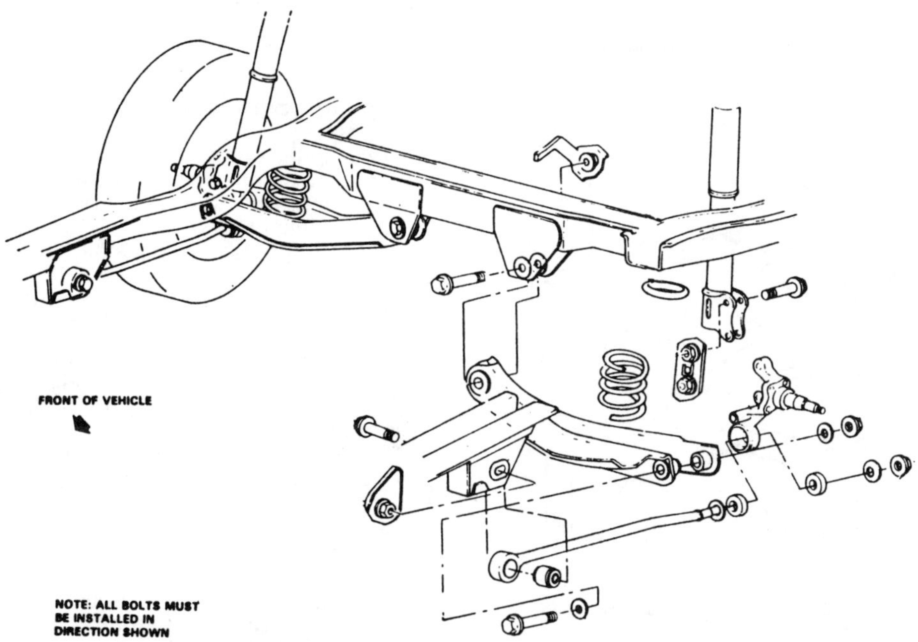

Fig. 1 Rear suspension, exploded view. Escort, EXP & Lynx

DESCRIPTION

ESCORT, EXP & LYNX

These vehicles use a modified MacPherson strut independent rear suspension, **Fig. 1.** Each side consists of a shock strut, lower control arm, tie rod, spindle and a coil spring mounted between the lower control arm and body crossmember side rail.

TEMPO & TOPAZ

These vehicles use a new MacPherson strut independent rear suspension, **Fig. 2.** Each side consists of a shock absorber strut assembly, two parallel control arms per side, tie rod, spindle and a jounce bumper and bracket.

The shock absorber strut assembly includes a rubber isolated top mount, upper spring seat, coil spring insulator, coil spring and a lower spring seat. the strut assembly is attached at the top by two studs, which retain the top mount of the strut to the inner body side panel. The lower end of the assembly is bolted to the spindle. The two control arms are attached to the underbody and spindle with nuts and bolts. The tie rod is attached to the underbody and the spindle. The jounce bumper bracket is bolted to the strut.

TEMPO & TOPAZ, W/ALL WHEEL DRIVE

HALF SHAFT, REPLACE

1. Remove rear suspension control arm attaching bolt.
2. Remove outboard U-joint attaching bolts and caps.
3. Remove inboard U-joint attaching bolts and caps.
4. Carefully slide shafts together. Do not allow splined shafts to contact with excessive force. Remove half shafts. **Do not drop the half shafts as the impact may damage U-joint bearing cups.**
5. Remove and retain bearing cups.
6. Inspect U-joint assemblies for wear and/or damage. Replace U-joints, if necessary.
7. Reverse procedure to install. Note the following:
   a. The inboard shaft has a larger diameter than the outboard shaft.
   b. **Torque** inboard U-joint attaching bolts to 15-17 ft. lbs.
   c. **Torque** outboard U-joint attaching bolts to 15-17 ft. lbs.
   d. **Torque** rear suspension control arm attaching bolt to 60-86 ft. lbs.

DRIVESHAFT, REPLACE

During removal and installation, support driveshaft using a suitable jack or hoist under the center bearing assembly.

1. To maintain driveshaft balance, mark U-joints so they may be installed in their original positions. **Do not use a sharp tool to place alignment marks on any component.**
2. Remove front U-joint attaching bolts and caps.
3. Slide driveshaft toward rear of vehicle and disengage driveshaft.
4. Remove rear U-joint attaching bolts and caps attaching driveshaft, from torque tube yoke flange.
5. Slide driveshaft toward front of vehicle and disengage driveshaft. **Do not allow splined shafts to contact with excessive force.**
6. Remove center bearing attaching bolts.
7. Remove driveshaft and retain bearing

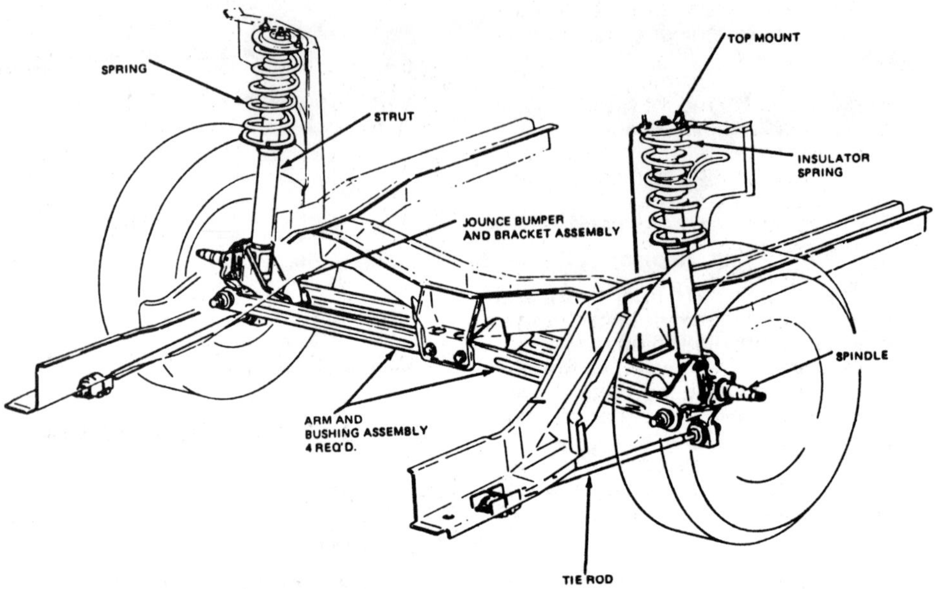

Fig. 2 Rear suspension components. Tempo & Topaz

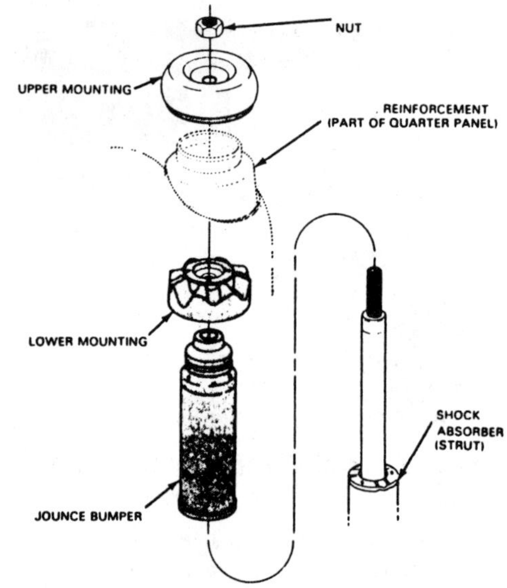

Fig. 3 Shock absorber strut upper mounting components. Escort, EXP & Lynx

cups with tape, if necessary.

8. Inspect U-joint assemblies for wear or damage. Replace, if necessary.
9. Reverse procedure to install. **Torque** U-joint retaining caps and bolts to 15-17 ft. lbs. **Torque** center bearing and attaching bolts to 23-30 ft. lbs.

SHOCK STRUT
REPLACE
ESCORT, EXP & LYNX

1. Raise and support vehicle.
2. Remove rear compartment access panels. **On four-door models,** remove quarter trim panel.
3. Loosen top shock strut attaching nut, then remove wheel assembly.
4. Remove clip retaining brake hose to rear shock and position hose aside.
5. Loosen but do not remove two nuts and bolts securing shock strut to spindle.
6. Remove top mounting nut, washer and rubber insulator, **Fig. 3.**
7. Remove two bottom mounting bolts, then the shock strut from the vehicle.
8. Reverse procedure to install.

SHOCK STRUT, UPPER MOUNT & SPRING
REPLACE
TEMPO & TOPAZ

1. Raise and support vehicle. Loosen upper strut mount to body nuts located in luggage compartment.
2. Remove wheel assembly.
3. Place a suitable jack under control arms.
4. Remove brake hose bracket to strut bolt and position brake hose bracket aside.
5. Remove jounce bumper bracket.

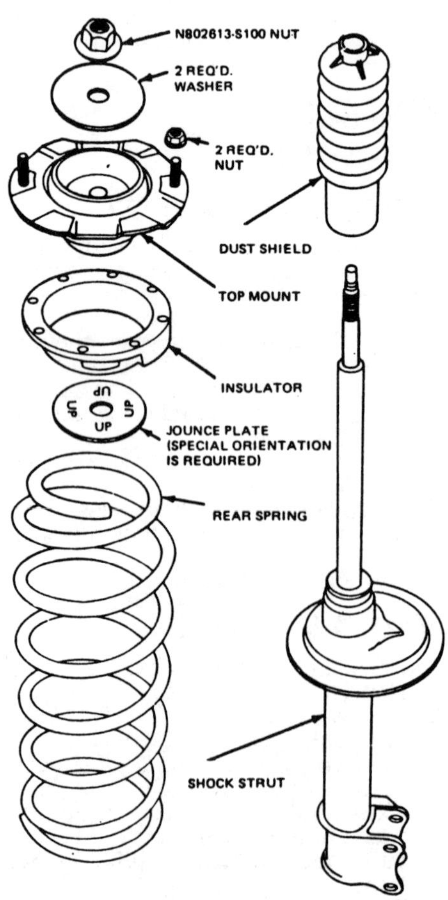

Fig. 4 Strut, spring & upper mount components. Tempo & Topaz

6. Remove two upper mount to body nuts, then the strut.
7. Place strut, spring and upper mount assembly into a suitable spring compressor tool. **Do not remove the spring from the strut without first compressing the spring.**
8. With spring compressed, remove strut shaft to mount nuts. Remove spring, strut and mount, **Fig. 4,** from spring compressor tool.
9. Reverse procedure to install. **Torque** shaft nut to 35-50 ft. lbs., torque jounce bumper bracket to strut mount bolts to 70-96 ft. lbs., **torque** top mount to body nuts to 25-30 ft. lbs.

LOWER CONTROL ARM
REPLACE
ESCORT, EXP & LYNX

1. Raise and support vehicle, then remove wheel assembly.
2. Place a suitable jack under lower control arm between spring and spindle mounting. **Rear suspension should be at full rebound and the shock strut fully extended.**
3. Remove control arm to body mounting nuts, then the control arm to spindle mounting nuts. Do not remove bolts.
4. Remove spindle end mounting bolt, then slowly lower jack until spring and spring insulator can be removed.
5. Remove bolts from body mountings, then the control arm from vehicle.
6. Reverse procedure to install.

TEMPO & TOPAZ

1. Raise and support vehicle.
2. Remove wheel assembly.
3. Remove control arm to spindle nut and bolt.
4. Remove center mounting nut and bolt.
5. Remove control arm from vehicle.
6. Reverse procedure to install. **Torque** control arm to body bolt and control arm to spindle nut to specifications. **When installing new control arms**

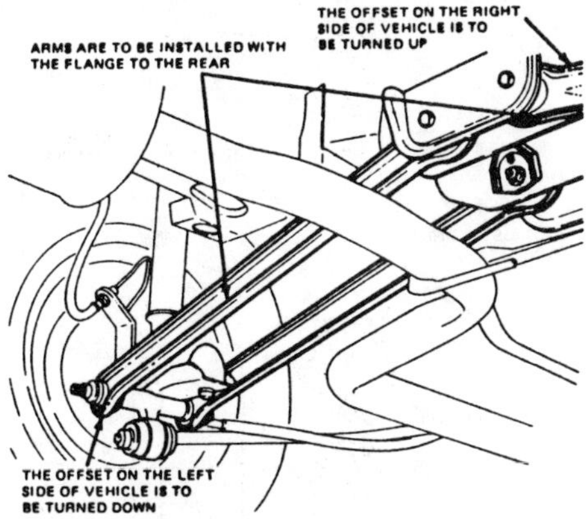

Fig. 5 Installing control arm. Tempo & Topaz

the bushing with the 10 mm hole is installed toward the center of the vehicle and the bushing with the 12 mm hole toward the spindle. The offset on the control arm must face up on the right side of the vehicle and down on the left side of the vehicle, Fig. 5. The flanged edge of the control arm stamping must face the rear of the vehicle.

TIE ROD REPLACE

ESCORT, EXP & LYNX

1. Raise and support vehicle.
2. Scribe a reference mark on tie rod front bracket at bolt head center line for use during reassembly.
3. Remove nut, washer and insulators attaching tie rod to spindle.
4. Remove nut and bolt attaching tie rod to body bracket, then the tie rod. **It may be necessary to pry front bracket sheet metal apart slightly to remove tie rod from body.**
5. Place a new dished washer over tie rod end with flange toward middle of rod, **Fig. 6.**
6. Install tie rod through spindle bushings and place eye of rod into body bracket. Secure rod to bracket using a new nut and bolt. Do not tighten.
7. Place another dished washer over end of rod with flange toward end of rod, then install a new nut and torque to specifications.
8. Using suitable jack, raise lower control arm to curb height.
9. Align center of bolt head with reference mark on body bracket, then **torque tie rod front bolt to specifications. This bolt must be installed with head inboard on vehicle.**
10. Lower vehicle to ground.

TEMPO & TOPAZ

1. Raise and support vehicle.
2. From inside of luggage compartment, loosen two strut top mount to body nuts.
3. Raise vehicle. Position a suitable jack under lower control arm with a piece of wood between jack and control arm.
4. Remove wheel assembly.
5. Remove two top mount studs, then tie rod to spindle retaining nut.
6. Remove tie rod to body retaining nut.
7. Lower jack until upper strut mount studs clear body mount holes.
8. Move spindle rearward until tie rod can be removed.
9. Place new washers and bushings on both ends of tie rod, **Fig. 6. Front and rear bushings are not interchangeable. The rear bushings have indentations incorporated in them.**
10. Insert tie rod into body bracket, then install new bushing, washer and nut. Do not tighten nut.
11. Pull back on spindle until tie rod can be installed into the spindle. Install new bushing, washer and nut. Do not tighten nut.
12. Raise jack enough to secure the two strut mounting studs in place.
13. Install two strut to body mount nuts. **Torque** nuts to specifications.
14. Using a suitable jack, raise lower control arm to curb height. Install tie rod nuts and torque to specifications.
15. Remove jack and install wheel assembly. Lower vehicle.

SPINDLE REPLACE

ESCORT, EXP & LYNX

If a frame contact hoist is used, a jack stand must be placed under lower control arm to raise it to curb height.

1. Raise and support vehicle.
2. Remove wheel assembly, then the brake drum and wheel bearings.
3. Remove brake backing plate assembly, then the tie rod retaining nut and dished washer.
4. Remove two nuts and bolts securing strut to spindle.
5. Remove nut and bolt securing lower control arm to spindle, then the spindle.
6. Reverse procedure to install.

TEMPO & TOPAZ

1. Raise and support vehicle.
2. Remove wheel assembly.
3. Remove brake drum. Remove brake flex hose bracket to strut bolt.
4. Remove brake backing plate to spindle bolts, then the brake backing plate. **Care should be taken to ensure that brake flex hose is not stretched and brake tube is not bent.**
5. Remove lower control arm to spindle bolt, washer and nut.
6. Remove tie rod nut, bushing and washer.
7. Remove spindle to strut bolts, then the spindle.
8. Reverse procedure to install. **Torque** spindle to strut bolts, tie rod nut and lower control arm to spindle nut to specifications.

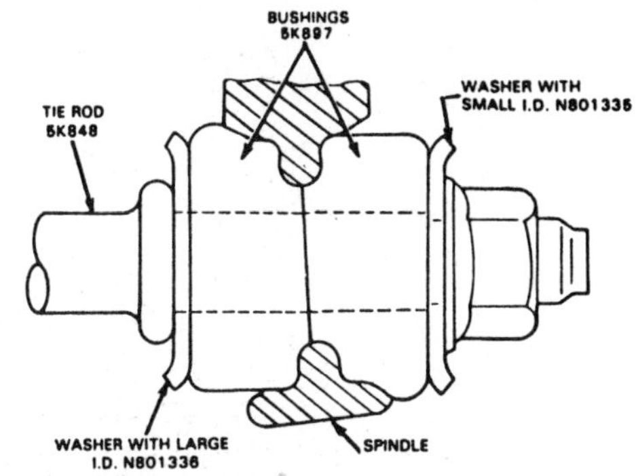

NOTE: WASHERS N801336 AND N801335 MUST BE INSTALLED IN THIS POSITION WITH DISH AWAY FROM BUSHINGS.

Fig. 6 Tie rod installation

COIL SPRING REPLACE

ESCORT, EXP & LYNX

1. Raise and support vehicle. Support lower control arm with suitable jack.
2. Remove tire and wheel assembly.
3. Remove nut, bolt and washer securing lower control arm to spindle.
4. Lower control arm until spring can be removed.
5. Reverse procedure to install. A new spring insulator must be used when replacing the spring.

REAR WHEEL BEARING ADJUST

EXCEPT ALL WHEEL DRIVE MODELS

1. Raise and support vehicle. Remove dust cover from hub. Remove wheel assembly, if necessary.
2. Remove cotter pin and nut retainer.
3. Back off adjusting nut 1 full turn.
4. **Torque** adjusting nut, **Fig. 7**, to 17-25 ft. lbs., while rotating drum assembly.
5. Back off adjusting nut 1/2 turn, then retighten adjusting nut to 10-15 inch lbs. Position adjusting nut retainer over nut so slots are aligned with cotter pin hole, then install cotter pin.
6. Install dust cover, wheel assembly, if necessary and lower vehicle to ground.

REAR WHEEL BEARING REPLACE

EXCEPT ALL WHEEL DRIVE MODELS

1. Raise and support vehicle.
2. Remove grease cap from hub. Remove cotter pin, nut retainer, adjusting nut and flat washer from spindle, **Fig. 7.** Discard cotter pin.
3. Pull hub and drum assembly off spindle being careful not to drop outer bearing assembly.
4. Remove outer bearing assembly.

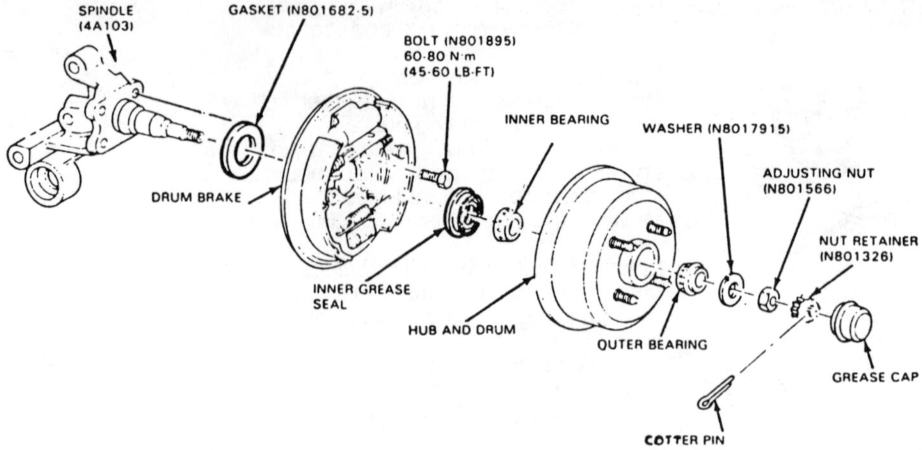

Fig. 7 Rear wheel bearing assembly. Except models with all wheel drive system

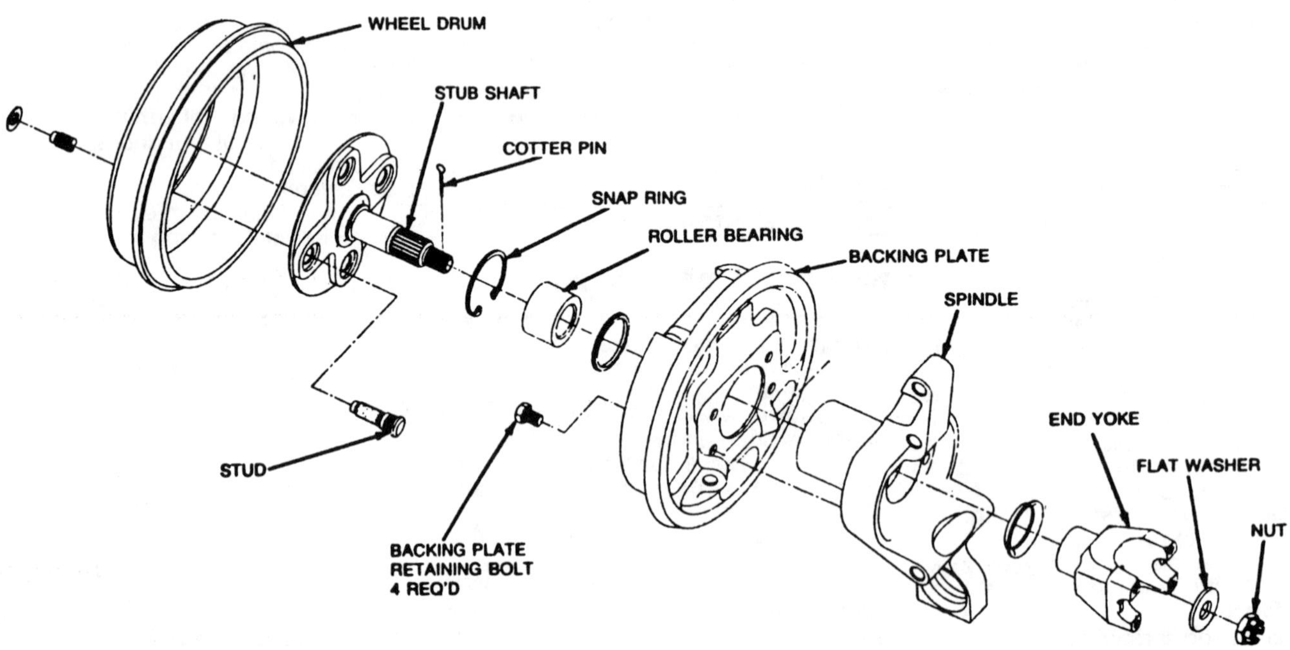

Fig. 8 Rear wheel bearing assembly. Models w/all wheel drive system

5. Using tool 1175-AC or equivalent, remove and discard grease seal. Remove inner bearing assembly from hub.
6. Reverse procedure to install.

MODELS w/ALL WHEEL DRIVE SYSTEM

1. Raise and support vehicle.
2. Remove wheel and tire assembly.
3. Remove brake drum, parking brake cable from brake backing plate.
4. Remove brake line from wheel cylinder.
5. Remove outboard U-joint attaching bolts and caps. Remove outboard end of half shaft from wheel stub shaft yoke and wire to control arm, **Fig. 8.**
6. Remove and discard control arm to spindle bolt, washers and nut.
7. Remove tie rod nut, bushing and washer. Discard nut.
8. Remove and discard two bolts attaching spindle to strut. Remove spindle from vehicle.
9. Mount spindle and backing plate assembly into a suitable vise.
10. Remove cotter pin and nut attaching stub shaft yoke to stub shaft. Discard cotter pin.
11. Remove spindle and backing plate assembly from vise. Remove stub shaft yoke using a suitable tool.
12. Mount spindle and backing plate assembly in a vise. Remove wheel stub shaft.
13. Remove snap ring retaining bearing.
14. Remove four bolts attaching spindle to backing plate. Remove backing plate.
15. Remove spindle from vise and mount into a suitable press. With spindle side facing upward, carefully press bearing out of spindle. Discard bearing after removal.
16. Reverse procedure to install.

REAR SUSPENSION TIGHTENING SPECIFICATIONS

Year	Component	Torque/Ft. Lbs.
1985-86①	Control Arm To Body	52-74
	Shock Absorber To Body	35-55
	Shock Absorber To Spindle	70-96
	Tie Rod To Body	6-12
	Tie Rod To Body Nut-Front	52-74
	Tie Rod To Spindle	52-72
1987-90①	Control Arm To Body	52-74
	Control Arm To Spindle	60-80
	Shock Absorber To Body	35-55
	Shock Absorber To Spindle	70-96
	Stabilizer Bar U-Bracket To Body	15-25
	Stabilizer Bar Link To Shock Bracket	41-55
	Stabilizer Bar To Link	6-12
	Tie Rod To Spindle	35-50
	Tie Rod To Body Nut-Front	52-74
1985-90②	Control Arm To Body	30-40
	Control Arm To Spindle	60-80
	Strut Top Mount To Body	20-30
	Strut To Spindle	70-96
	Strut To Top Mount	35-50

①—Escort.
②—Tempo & Topaz.

Ford Festiva

INDEX

DESCRIPTION

The rear suspension, **Fig. 1,** is of the MacPherson strut design, with the strut at each wheel seated in a tower in the cargo compartment. Suspension action is carried out through the use of semi-independent trailing arms, which are integral with the torsion beam type rear axle.

The torsion beam axle provides positive control of trailing arm alignment, while doubling as a stabilizer to limit sway during hard cornering. Bushings and rubber insulators, installed at key attachment points, keep vibration and road noise to a minimum.

The brake drum and bearing hub is an integral assembly which rides on a spindle attached to the trailing arms. The assembly is supported by opposed tapered roller bearings located within the bearing hub.

WHEEL BEARINGS & SEAL

REPLACE

1. Raise and support vehicle, then remove rear wheel.
2. Remove brake drum as follows:
   a. Remove grease cap, then pry upward on staked portion of locknut using suitable tool.
   b. Remove and discard locknut. **The locknut on the right side of the vehicle has left hand threads. Turn locknut clockwise to remove.**
   c. Remove washer, outer wheel bearing and brake drum from spindle.
3. Pry seal from drum using suitable tool, then remove inner wheel bearing from drum.
4. Clean bearings with suitable solvent and inspect for scoring, pitting or heat damage. Replace bearings as necessary.
5. Pack bearings and hub area with suitable grease, then install inner wheel bearing.

6. Lubricate new seal, then install into drum using suitable driver.
7. Install drum, outer wheel bearing, washer and locknut.
8. Adjust wheel bearing preload as outlined under "Wheel Bearings, Adjust" procedure.
9. Install grease cap and rear wheel, then lower vehicle.

WHEEL BEARINGS
ADJUST

1. Raise and support vehicle, then remove rear wheel and grease cap.
2. While rotating brake drum, torque locknut to 18-22 ft. lbs.
3. Loosen locknut until it can be turned by hand, then measure seal drag as follows:
   a. Install a lug nut onto brake drum and place nut at 12 o'clock position.
   b. Using an inch lb. torque wrench, measure amount of force needed to start rotation of drum. Record seal drag measurement.
4. To determine specified preload, add seal drag measurement recorded in previous step to required bearing preload of 1.3-4.3 inch lbs.
5. Tighten locknut slightly, then measure amount of force needed to start rotation of drum.
6. Continue to tighten locknut until preload determined in step 4 is reached.
7. After adjustment is completed, stake locknut to spindle with rounded chisel. **If splitting or cracking occurs after staking, replace locknut.**
8. Install grease cap and rear wheel, then lower vehicle.

SPINDLE
REPLACE

1. Raise and support vehicle, then remove brake drum.
2. Wire backing plate in position, then working from rear side of plate, remove spindle attaching nuts. **Do not confuse mounting stud nuts with spindle nuts. The spindle nuts are located underneath the vehicle on the inboard side of the trailing arm.**
3. Position backing plate, then install spindle. Tighten spindle attaching nuts.
4. Reinstall brake drum, then lower vehicle.

STRUT ASSEMBLY
REPLACE
REMOVAL

1. Raise and support vehicle, then remove rear wheel.
2. Compress spring using suitable tool.
3. Working from cargo compartment, remove rear quarter trim panel.
4. Remove jam and flanged nuts from strut rod, then the bushing washer and upper bushing, **Fig. 2.**

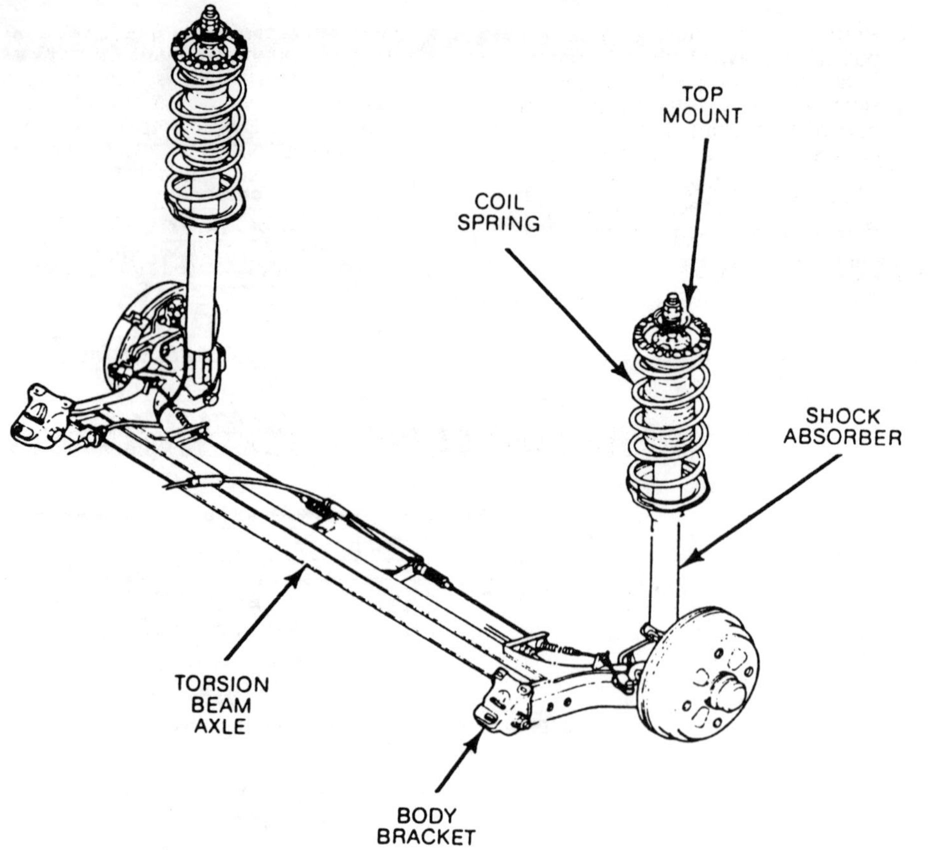

Fig. 1 Rear axle & suspension

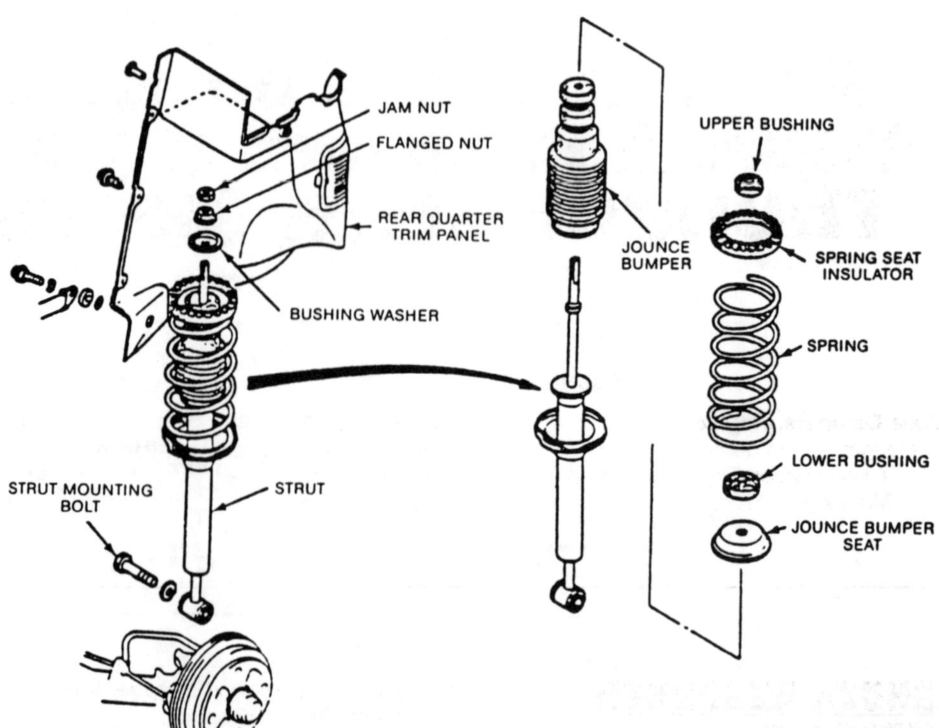

Fig. 2 Disassembled view of strut assembly

5. Remove lower mounting bolt, then pull strut downward and separate it from spring and seat insulator.
6. Remove lower bushing and jounce bumper seat from strut rod, then slide jounce bumper and shield off strut.
7. Inspect all bushings, bumpers and insulators for damage or deterioration. Replace components as necessary.

INSTALLATION

1. Slide jounce bumper, shield, bumper seat and lower bushing onto strut rod.
2. If upper spring seat insulator requires replacement, install insulator onto upper end of spring so that end of coil seats firmly against step in insulator.
3. Position spring on strut, ensuring end of coil seats firmly against step in strut spring seat.
4. Compress spring, then working from wheel well area, guide upper end of strut rod into tower mounting hole.
5. Line up lower end of strut with mounting hole in trailing arm, then install lower mounting bolt several turns to hold strut in position.
6. Install upper bushing, bushing washer and flanged nut onto upper end of strut rod. Tighten flanged nut, then lock in position with jam nut.
7. Tighten lower mounting bolt, then remove spring compressor.
8. Install wheel, then lower vehicle.

REAR SUSPENSION TIGHTENING SPECIFICATIONS

Year	Component	Torque/ft. lbs.
1988-90	Body Bracket Lower Mounting Bolt	69-87
	Body Bracket Upper Mounting Bolt	40-50
	Brake Backing Plate	32-45
	Master Cylinder To Power Brake Unit	7-12
	Power Brake Unit To Dash Panel Nuts	12-17
	Strut Lower Attaching Bolt	40-50
	Strut Rod Upper Nut	12-18
	Torsion Beam Pivot Bolt	69-87
	Wheel Spindle Support Nut	32-45
	Wheel & Tire Assembly	65-87

Ford LTD & Mustang, Mercury Capri & Marquis & 1985–88 Ford Thunderbird & Mercury Cougar

INDEX

SHOCK ABSORBER
REPLACE

1. **On hatchback models,** the upper shock absorber upper mounting is accessible from the luggage compartment. On sedans, open trunk to gain access to upper shock absorber attachment. On station wagons, remove side panel trim covers.
2. **On all models,** remove rubber cap from shock absorber stud.
3. **On all models,** remove shock absorber attaching nut, washer and insulator.
4. Raise vehicle and support rear axle.
5. Compress shock absorber to clear

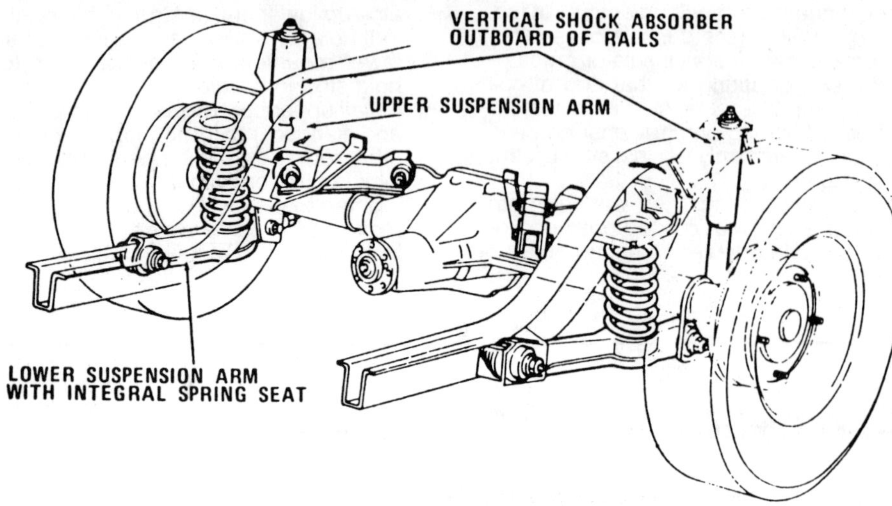

Fig. 1 Coil spring rear suspension (typical)

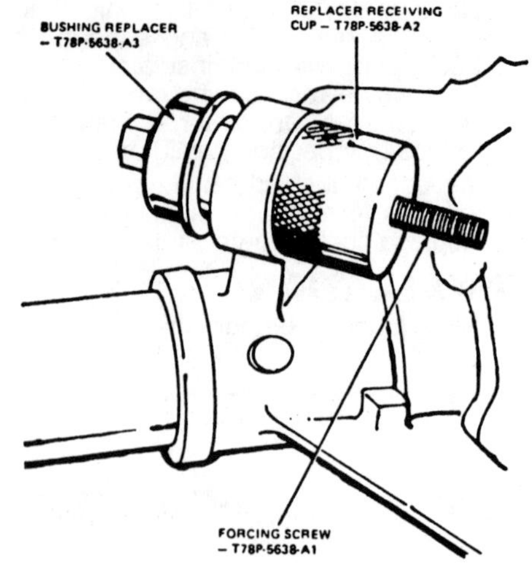

Fig. 2 Upper control arm axle bracket bushing installation

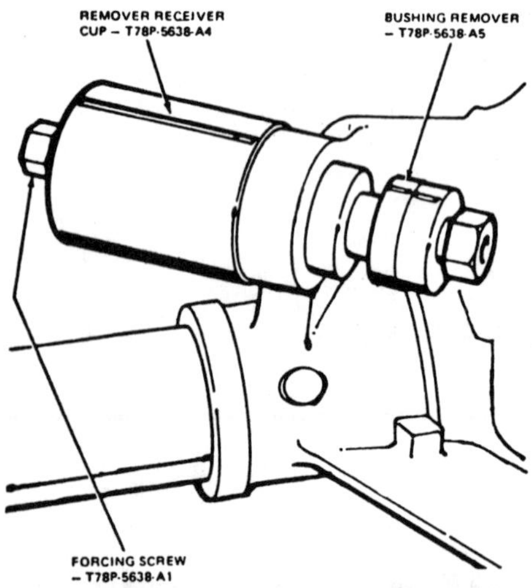

Fig. 3 Upper control arm axle bracket bushing removal

hole in upper shock absorber tower.

6. Disconnect shock absorber from lower mounting and remove shock absorber from vehicle.
7. Reverse procedure to install.

STABILIZER BAR
REPLACE

1. Raise and support rear of vehicle.
2. Remove four bolts attaching stabilizer bar to brackets on lower control arms.
3. Remove stabilizer bar from vehicle.
4. Reverse procedure to install. **Torque** all bolts to 45-50 ft. lbs.

COIL SPRING
REPLACE

1. Raise rear of vehicle and support at rear body crossmember.
2. Remove stabilizer bar, if equipped, **Fig. 1.**
3. Lower axle housing until shock absorbers are fully extended. **The axle**

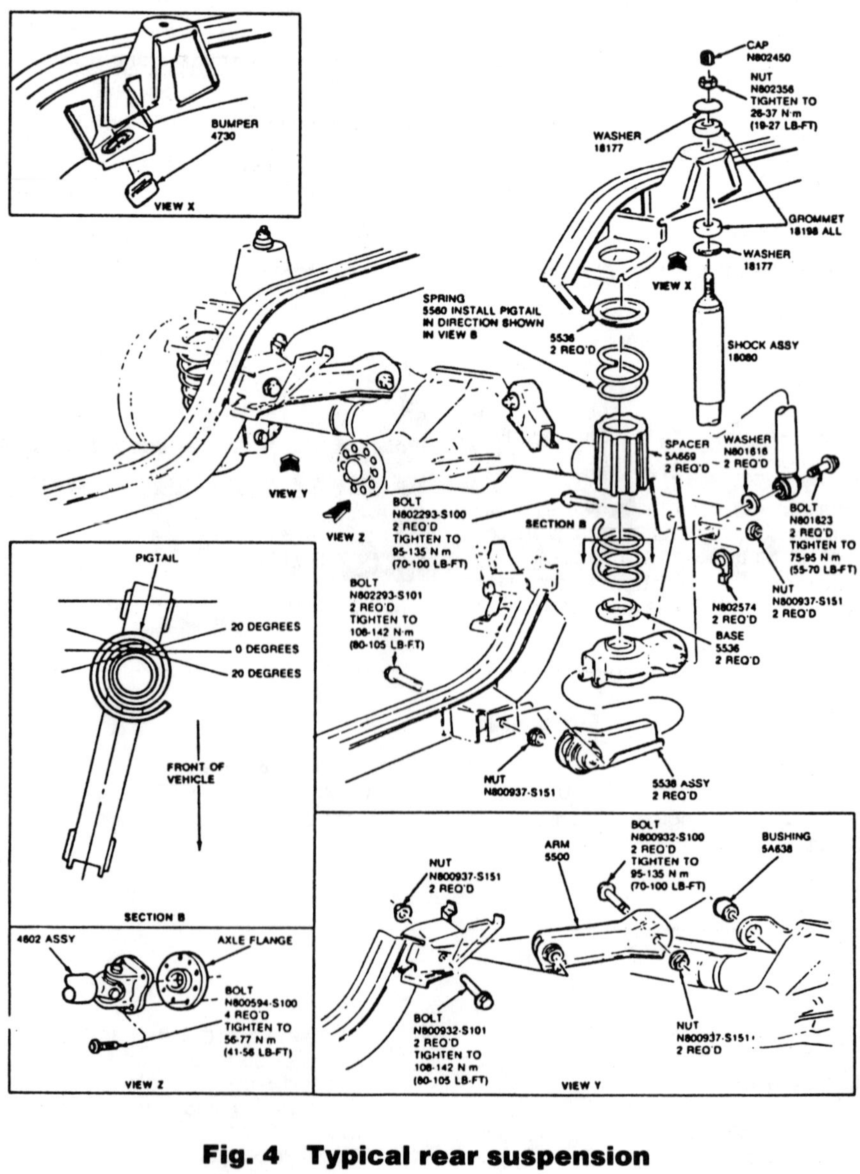

Fig. 4 Typical rear suspension

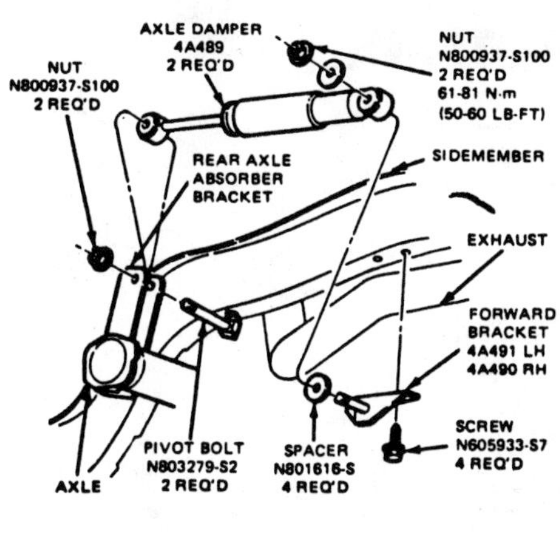

Fig. 5 Axle damper assembly

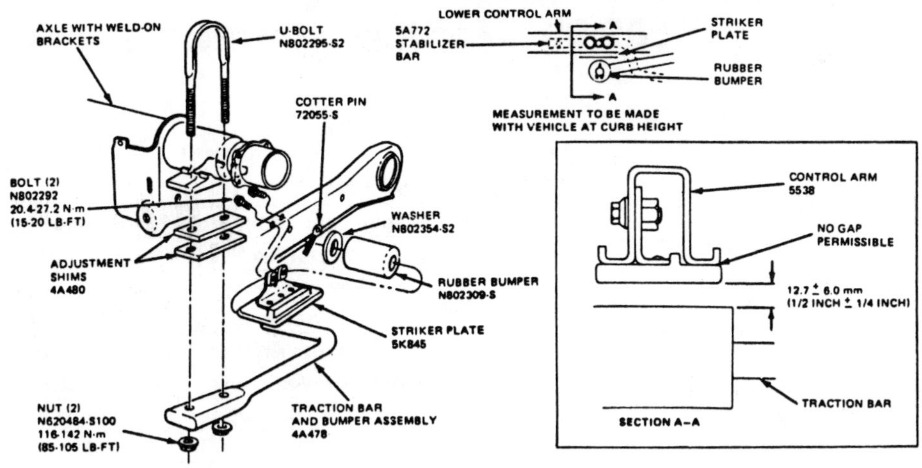

Fig. 6 Traction bar assembly. Capri & Mustang 4-140 turbocharged or V8-302 H.O. engines

housing must be supported with a suitable jack.

4. Position a suitable jack under lower control arm rear pivot bolt to support control arm, then remove pivot bolt.
5. Carefully lower the lower control arm until spring tension is relieved, then remove coil spring and insulator.
6. Reverse procedure to install. Torque lower control arm pivot bolt to specifications, with suspension at curb height.

CONTROL ARMS & BUSHINGS

REPLACE

UPPER CONTROL ARM

1. Raise rear of vehicle and support at rear body crossmember.
2. Remove upper control arm rear and front pivot bolts, then remove control arm.
3. If control arm axle bracket bushings are to be replaced, refer to **Figs. 2 and 3.**
4. Position upper control arm into side rail bracket, then install front pivot bolt. Do not tighten bolt at this time.
5. Raise rear axle until upper control arm rear pivot bolt hole is aligned with hole in axle housing, then install rear pivot bolt. Do not tighten bolt at this time.
6. Position suspension at curb height. Torque front pivot bolt to specifications.

LOWER CONTROL ARMS

1. Remove coil spring as described under "Coil Spring, Replace."
2. Remove lower control arm front pivot bolt and nut, then remove control arm, **Fig. 4.**
3. Reverse procedure to install. Torque front pivot bolt to specifications. Torque rear pivot bolt to specifications.

AXLE DAMPER

REPLACE

COUGAR & THUNDERBIRD W/4-140 TURBOCHARGED ENGINE & CAPRI & MUSTANG

1. Raise and support vehicle at rear axle.
2. Remove wheel.
3. Remove axle damper front attaching nut and pivot bolt, **Fig. 5.**
4. Remove rear attaching nut, then the axle damper and spacer.
5. Reverse procedure to install.

TRACTION BAR

REPLACE

1. Raise vehicle and support rear axle.
2. Remove traction bar to rear axle attaching bolts, then the traction bar, adjusting shims and "U" bolt, **Fig. 6.**
3. Remove cotter pin and washer, then the rubber bumper if necessary.
4. Reverse procedure to install, noting the following:
   a. If rubber bumper is replaced, use tire mounting solution ESAM-1B6B or equivalent to ease installation.
   b. **Torque** traction bar attaching nuts to 85-105 ft. lbs.
   c. Adjust traction bar, as needed, by adding or subtracting shims.

REAR SUSPENSION TIGHTENING SPECIFICATIONS

Year	Component	Torque/ft. lbs.
CAPRI & MUSTANG		
1985-90	Bracket Assembly Attaching Bolts	7-12
	Bumper To Bracket Nut	12-20
	Clevis Bracket To Axle	55-70
	Lower Arm To Axle	70-100
	Lower Arm To Frame	80-105
	Shock Absorber (lower attachment)	55-70
	Shock Absorber (upper attachment)	19-27
	Shock Absorber To Clevis Bracket	45-60
	Stabilizer Bar To Lower Arm	33-51
	Upper Arm To Axle	70-100
	Upper Arm To Frame	80-105
COUGAR, LTD, MARQUIS & THUNDERBIRD		
1985–88	Air Spring To Lower Arm	25–35
	Clevis Bracket To Axle	55–70
	Lower Arm To Axle	70–100
	Lower Arm To Frame	80–105
	Upper Arm to Axle	70–100
	Upper Arm To Frame	80–105
	Sensor Lower Bracket To Arm	7–10
	Sensor Upper Bracket To Frame	7–10
	Shock Absorber To Clevis Bracket	45–60
	Shock Absorber To Frame	17–27
	Stabilizer Bar To Axle	11–20
	Stabilizer Bar To Body	13–18

1989–90 Cougar & Thunderbird

INDEX

DESCRIPTION

The rear suspension is of the independent design with lower H-arm, upper arm and stabilizer bar, **Fig. 1.** Base stabilizer bar diameter is 1.0 inch and 1.1 inch for the Super Coupe and XR7 models.

Noise, vibration and harshness have been reduced by attaching the suspension arms to an isolated sub frame. A special toe link, attached from the control arm to the sub-frame steers the rear wheels under side load conditions, helping to maintain a neutral steer condition.

New suspension features make it possible to enlarge the usable trunk space, and increase suspension travel by .67 inch. Special teflon coated suspension bush-

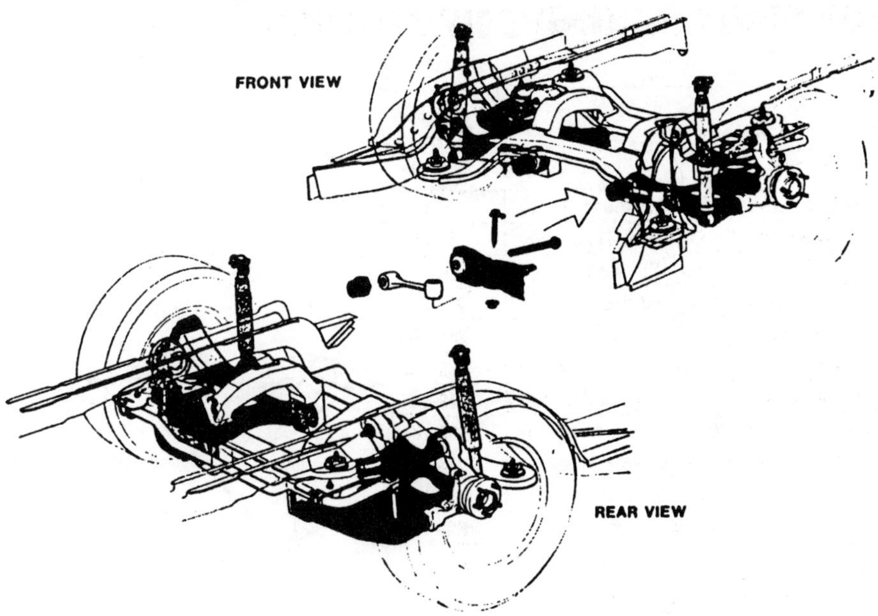

Fig. 1 Rear suspension. 1989–90 Cougar & Thunderbird

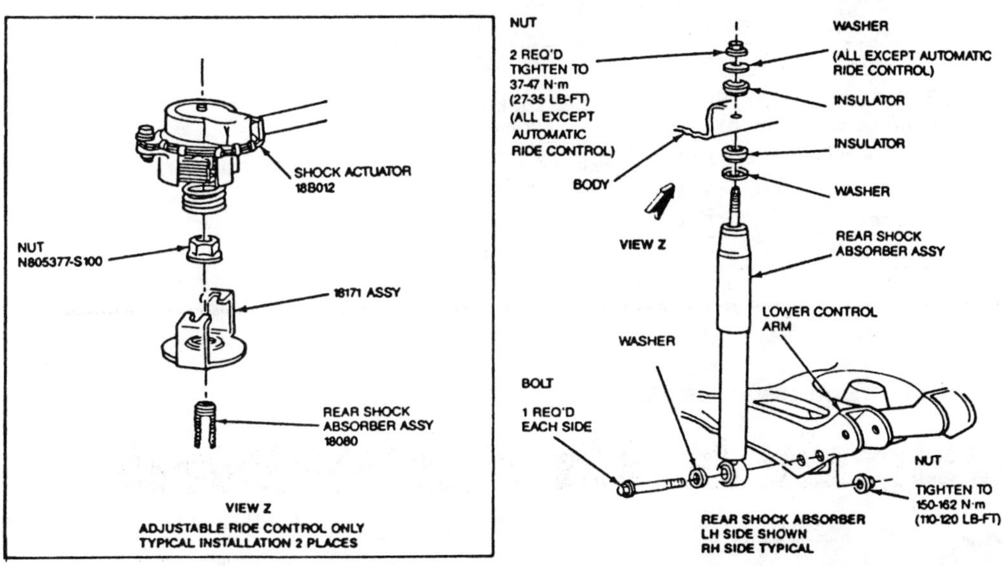

Fig. 2 Shock absorber replacement. 1989–90 Cougar & Thunderbird

ings reduce friction and allow easier movement.

SHOCK ABSORBERS
REPLACE

These vehicles are equipped with gas pressurized rear shock absorbers which will extend unassisted. Do not apply heat or flame to the shock tube during removal.

1. Position vehicle on drive-on hoist or alignment pit, so that rear suspension arms are supported during removal.
2. Remove all upper attaching parts from inside of luggage compartment, **Fig. 2.**
3. Remove attaching bolt and at lower arm, then remove shock absorber.
4. Reverse procedure to install.

COIL SPRING
REPLACE

1. Raise and support vehicle.
2. Remove rear wheel and tire assembly.
3. Remove rear stabilizer bar link nuts at both ends of stabilizer bar. Rotate bar up and out of the way.
4. Disconnect parking brake cable at brake caliper.
5. Install three Rotunda Spring Cages tool No. 086-00031 or equivalent, to rear springs as follows:
   a. Install one spring cage without an adjuster link to the inboard side or innermost bend of the spring, **Fig. 3.**
   b. Install two more spring cages, with adjusters, at 120° angles to the previously installed cage.
6. Place a transmission jack or suitable stand under the lower rear control arm as far outboard as possible.
7. Support rear knuckle and caliper assembly by wiring the upper control arm to the body.
8. Remove lower shock absorber mounting bolt and nut.
9. Mark position of toe adjustment cam to sub-frame for reference. Loosen both inboard pivot bolts on lower control arm. **Control arm must not be lowered until pivot bolts are loose. Do not attempt to remove plastic cap on front of pivot nut.**
10. Remove two bolts and nuts attaching lower control arm to knuckle, **Fig. 4.**
11. Lower control arm by lowering jack. Ensure spring cages properly seat on spring as control arm is dropped.
12. Remove jack, pull control arm down fully by hand and remove rear spring with cages in place.
13. Using coil spring compressor tool No. D78P-5310-A or equivalent, compress spring and remove spring cages.
14. Remove upper and lower spring insulators from spring.
15. Reverse procedure to install, compress spring to approximately 10.5 inches overall length prior to installation.

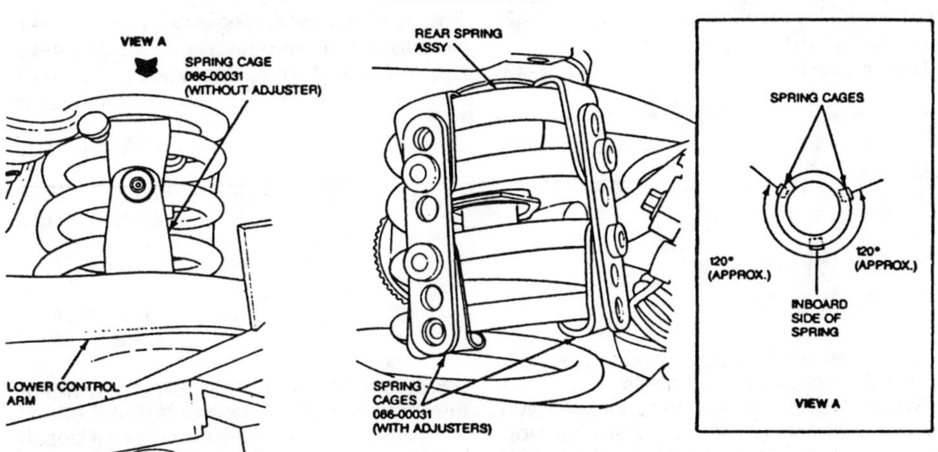

Fig. 3 Mounting spring cages. 1989–90 Cougar & Thunderbird

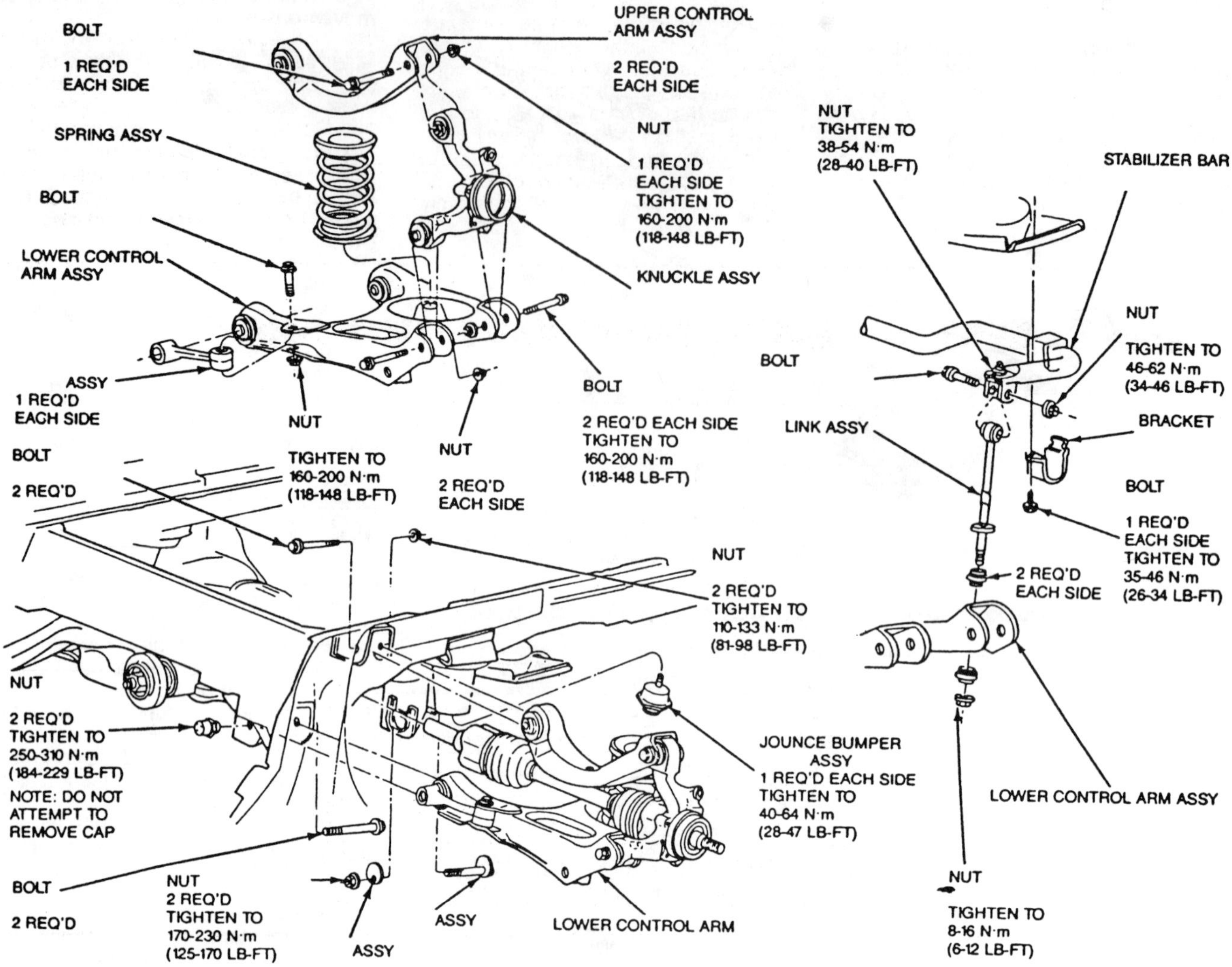

Fig. 4 Exploded view of rear suspension. 1989–90 Cougar & Thunderbird

LOWER CONTROL ARM REPLACE

1. Remove coil spring as described in "Coil Spring, Replace".
2. Remove inner control arm pivot bolts and nuts, then remove arm assembly.
3. Remove toe compensating link from control arm.
4. Reverse procedure to install.

UPPER CONTROL ARM REPLACE

1. Raise and support vehicle.
2. Remove rear wheel and tire assembly.
3. Support knuckle and hub assembly so that it cannot swing outward.
4. Remove inner and outer pivot bolts and nuts at upper control arm, then remove upper control arm.
5. Reverse procedure to install, noting that the inner pivot bolt used for camber adjustment has a specially-shaped washer under the bolt head. Ensure fasteners are used in correct locations. Set camber adjustment as described in "Wheel Alignment".

STABILIZER BAR REPLACE

1. Raise and support vehicle.
2. Remove both rear wheel and tire assemblies.
3. Remove both stabilizer bar link upper retaining bolts and nuts.
4. Remove stabilizer bar bracket bolts.
5. Remove rear muffler hanger retaining nuts.
6. Remove stabilizer bar from vehicle.
7. Reverse procedure to install.

REAR KNUCKLE REPLACE

Do not begin this procedure unless a new hub retainer nut is available. Once removed, this part cannot be reused during assembly. Its torque holding and/or retention capability is greatly diminished during removal.

1. Remove wheel cover/hub cover from wheel and tire assembly, then loosen wheel nuts.
2. Remove hub nut and washer and discard both parts.
3. Raise and support vehicle.
4. Remove wheel and tire assembly.
5. Pull back on parking brake cable release lever and at the same time pull on cable. This action will relax the cable so it can be removed from the brake caliper or backing plate.
6. **On models with disc brakes,** proceed as follows:
   a. Remove parking brake cable from the caliper.
   b. Remove upper and lower caliper attaching bolts and remove caliper assembly from rotor.
   c. Carefully wire caliper to brake junction bracket.
7. **On all models,** with push on nuts removed, remove brake rotor or drum assembly.
8. Remove three bolts retaining splash shield to knuckle and remove splash shield.
9. Disconnect parking brake cable and disconnect brake line from wheel cylinder.

10. Remove upper control arm nut and bolt.
11. Wire upper control arm to body to prevent damage to CV boots when knuckle and hub assembly is removed.
12. Install hub remover/replacer tool No. T81P-1104-C or equivalent to hub studs. Turn wrench counterclockwise until half shaft is free in hub.
13. Using a paint marker, mark the position of the control arm in relation to the knuckle with the bushings in a relaxed position. When upper control arm bolt is removed from knuckle, the lower arm bushings will return to a relaxed position. **Failure to mark position will cause bushing wind up upon assembly and incorrect ride height. These conditions can cause misalignment and premature tire wear.**
14. Note approximate angle of knuckle in relaxed position by measuring distance from upper bushing to any convenient point on the vehicle body.
15. Remove lower control arm to knuckle attaching bolts and nuts.
16. Remove knuckle assembly from half shaft.
17. Reverse procedure to install.

REAR SUSPENSION TIGHTENING SPECIFICATIONS

Year	Component	Torque/ ft. lbs.
1989–90	Hub Retainer Nut	250
	Jounce Bumper To Body	28–47
	Lower Control Arm To Toe Compensator Link Nut	118–148
	Lower Control Arm To Knuckle	118–148
	Lower Control Arm To Sub-frame (Front)	184–229
	Lower Control Arm To Sub-frame (Rear)	125–170
	Shock Absorber To Body	27–35
	Shock Absorber To Lower Control Arm	110–120
	Stabilizer Bar Link To Lower Control Arm	6–12
	Stabilizer Bar U-bracket To Sub-frame	26–34
	Stabilizer Clevis To Stabilizer Bar	28–40
	Stabilizer Link To Stabilizer Bar	34–46
	Upper Control Arm To Knuckle	118–148
	Upper Control Arm To Sub-frame	81–98

Ford Probe

INDEX

Page No.

DESCRIPTION

The rear suspension, **Fig. 1,** is fully independent utilizing rear MacPherson struts at each wheel. If the vehicle is equipped with the programmed ride control system, the rear strut towers locate the programmed ride control actuators and the strut assemblies. A forged rear spindle bolts to the shock absorber, double rear lateral links and a single trailing arm.

SHOCK STRUT REPLACE

1. Disconnect battery ground cable.
2. Remove tire and wheel assembly.
3. Remove upper trunk side garnish, then lower trunk side trim to gain access to the strut assembly.
4. Disconnect programmed ride control module electrical connector from top of strut assembly, if equipped.
5. Remove programmed ride control module, if equipped.
6. Remove ABS harness and bracket, if equipped.

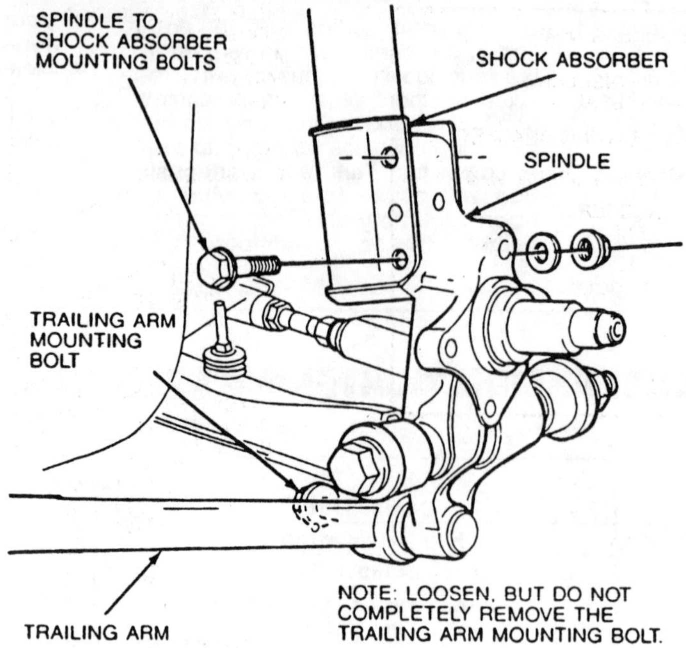

Fig. 1 Rear suspension components

7. Remove rear brake drum and backing plate assembly or rear disc brake caliper, then rotor assembly.
8. Remove brake line U-clip from the strut housing.
9. Loosen the trailing arm bolt, **Fig. 2.** Remove spindle to shock absorber attaching bolts.
10. Remove the strut attaching nuts from inside the vehicle. Remove strut assembly, **Fig. 3.**
11. Reverse procedure to install. **Torque** strut attaching nuts to 34-46 ft. lbs. **Torque** spindle to shock absorber mounting bolts to specifications. **Torque** trailing arm mounting bolt to specifications.

LATERAL LINK, TRAILING ARM & REAR CROSSMEMBER

REPLACE

1. Disconnect battery ground cable.
2. Remove spindle from vehicle.
3. Remove rear stabilizer.
4. Remove nut from lateral link mounting bolt at the rear crossmember. Remove lateral link.
5. Remove parking brake attaching bolts from trailing arm assembly.
6. Remove the trailing arm mounting bolt from the body mounting bracket.
7. Remove trailing arm from the vehicle.
8. Remove exhaust mounting bolts and brake line retaining bracket from rear crossmember.
9. Remove mounting bolts from end of the crossmember.
10. Remove rear crossmember and front lateral link as an assembly.
11. Remove common lateral link mounting bolt from the rear crossmember.
12. Remove front lateral link from rear crossmember, **Fig. 4.**
13. Reverse procedure to install. **Torque**

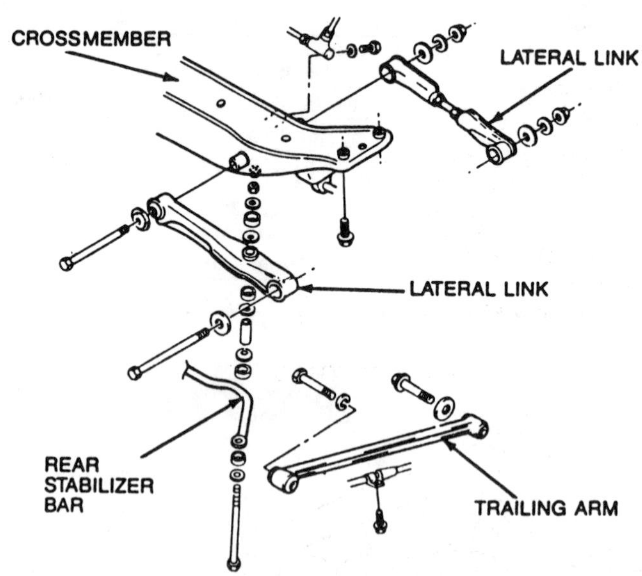

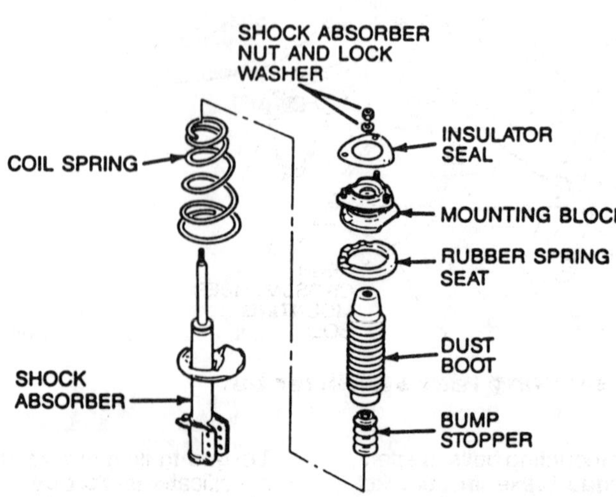

Fig. 2 Removing spindle

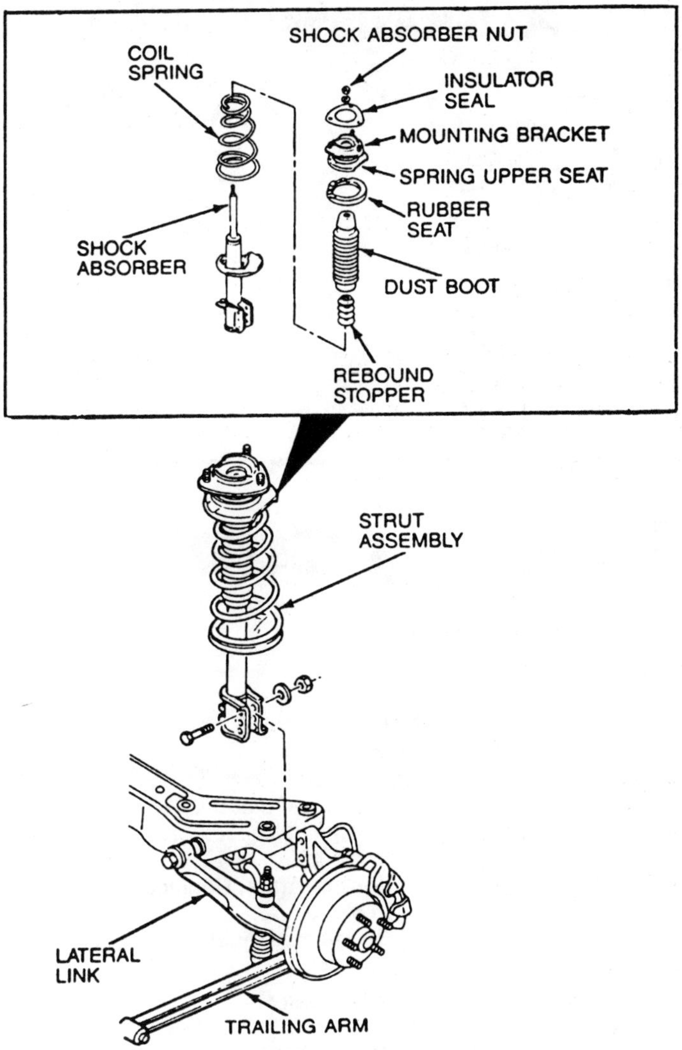

Fig. 3 Removing shock strut

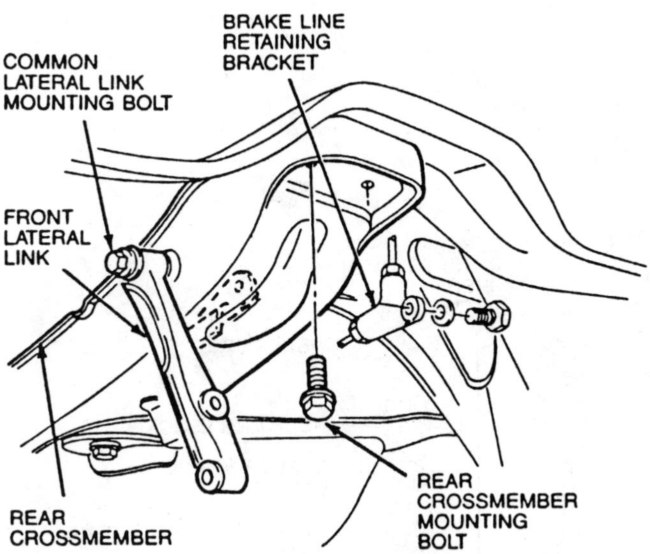

Fig. 5 Removing rear stabilizer bar

crossmember mounting bolts to specifications. **Torque** brake line bracket attaching bolt to specifications. **Torque** trailing arm mounting bolt to specifications. **Torque** common lateral link mounting bolt and nut at the

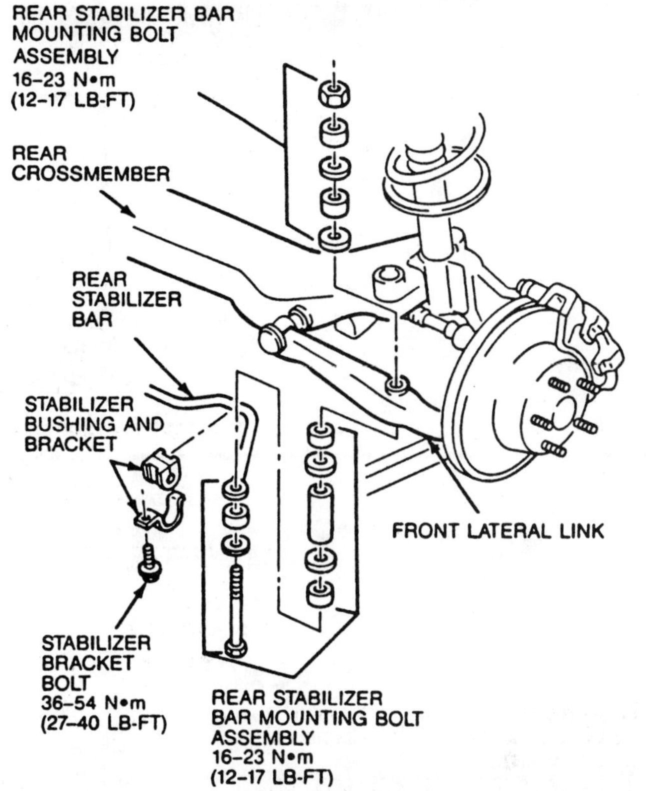

Fig. 4 Removing lateral link, trailing arm & rear crossmember

rear crossmember to specifications.

STABILIZER BAR
REPLACE

1. Remove mounting bolt assembly from front lateral link, **Fig. 5.**
2. Remove stabilizer bushing, then bracket from rear crossmember.
3. Reverse procedure to install. **Torque** bracket attaching bolt to specifications. **Torque** rear stabilizer bar mounting bolt assembly to front lateral link to specifications.

REAR WHEEL BEARING END-PLAY
ADJUST

1. Ensure parking brake is fully released.
2. Remove wheel and tire assembly.
3. Rotate brake drum to ensure no brake drag.
4. Using a dial indicator, check wheel bearing end play. End play should not exceed .008 inch.
5. If measurement exceeds specification, replace bearing.

REAR WHEEL BEARING
REPLACE

1. Remove rear brake drum or rotor from vehicle.
2. Remove dust seal and rear bearing.
3. Reverse procedure to install

REAR SUSPENSION TIGHTENING SPECIFICATIONS

Year	Component	Torque/ft. lbs.
1989-90	Brake Line Attaching Bolt	46-69
	Hub Spindle To Back Plate	33-43
	Hub Spindle To Shock Absorber	69-86
	Lateral Link Mounting Bolt	64-86
	Shock Absorber Tower Nut	47-67
	Stabilizer Bar Mounting Bolt	12-17
	Stabilizer Bracket Attaching Bolt	27-40
	Trailing Arm Mounting Bolt	46-69
	Wheel Lug Nuts	65-87

Ford Taurus & Mercury Sable

INDEX

DESCRIPTION

SEDAN MODELS

These models utilize an independent rear suspension. Each side consists of a MacPherson strut, an upper mount and washers, two parallel lower control arms, a tension strut, a spindle and a stabilizer bar mounted on the strut.

The top of the MacPherson strut is attached to the inner body side panel, while the lower end of the strut is attached to the spindle with a pinch clamp and bolt. The parallel lower control arms attach to the underbody with nuts and bolts. The tension strut attaches to the lower part of the spindle and to the underbody, **Fig. 1.**

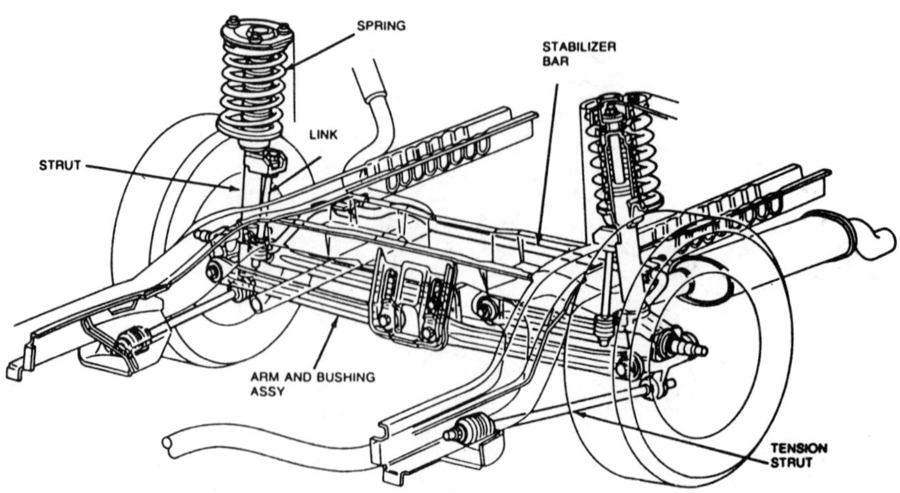

Fig. 1 Rear suspension. Sedan models

WAGON MODELS

These models also utilize an independent rear suspension. Each side consists of an upper and lower control arm, a shock absorber, a two piece spindle tension control strut and a coil spring.

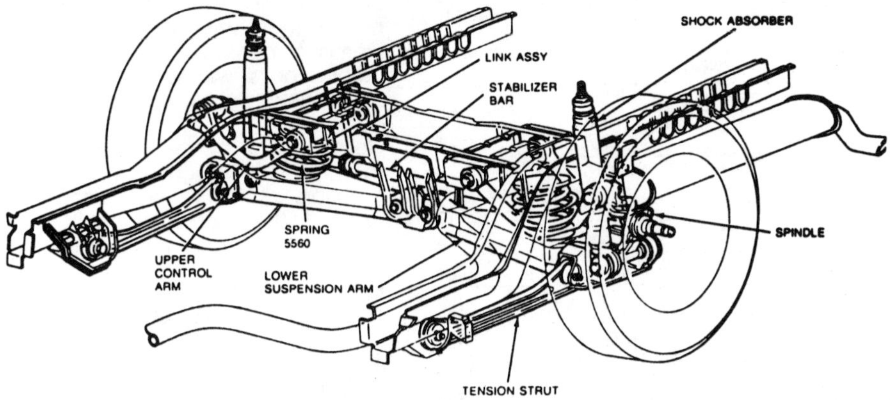

Fig. 2 Rear suspension. Wagon models

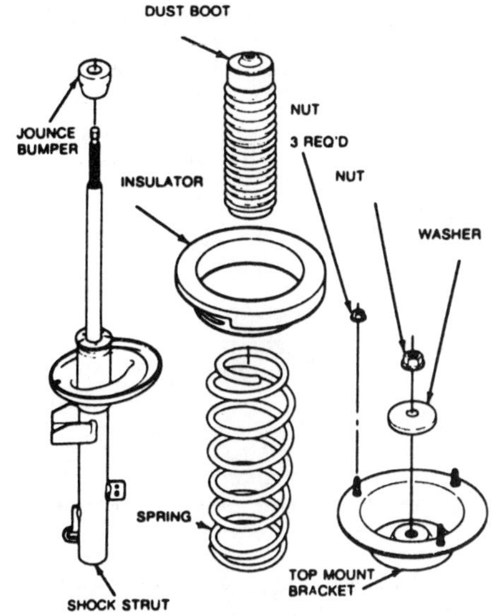

Fig. 3 MacPherson strut components. Sedan models

The top of the shock absorber is attached to the body side panel by a rubber insulated top mount assembly and to the lower control arms by two nuts. The upper control arm attaches to the crossmember and the upper part of the spindle. The lower control arm attaches to the underbody and lower part of the spindle. The coil spring operates against the lower control arm and is located inboard of the shock absorber, **Fig. 2.**

STRUT, UPPER MOUNT & SPRING REPLACE

SEDAN MODELS

1. Position suitable jack or hoist under vehicle, then raise just enough to contact body.
2. Working in trunk, loosen, but do not remove, the three strut-to-inner body attaching nuts.
3. Raise and support vehicle, then remove tire and wheel assembly from side being worked on.
4. Remove brake differential valve to control arm attaching bolt.
5. Using suitable wire suspend control arm to body to ensure proper support after strut removal.
6. Remove brake hose-to-shock strut bracket attaching clip and position hose aside.
7. If equipped with stabilizer bar, remove U-bracket from body, then the stabilizer bar attaching nut, washer and insulator. Separate stabilizer bar from link.
8. Remove tension strut-to-spindle attaching nut, washer and insulator, then move spindle rearward enough to separate it from tension strut.
9. Remove strut-to-spindle pinch bolt, then using a pry bar or other suitable tool, separate pinch joint as necessary to allow for strut removal.
10. Remove strut from pinch joint, then lower vehicle as necessary to allow removal of bolts loosened in step 2. Remove strut from vehicle. **During strut removal, use care not to stretch the rear brake hose or kink the steel brake line.**

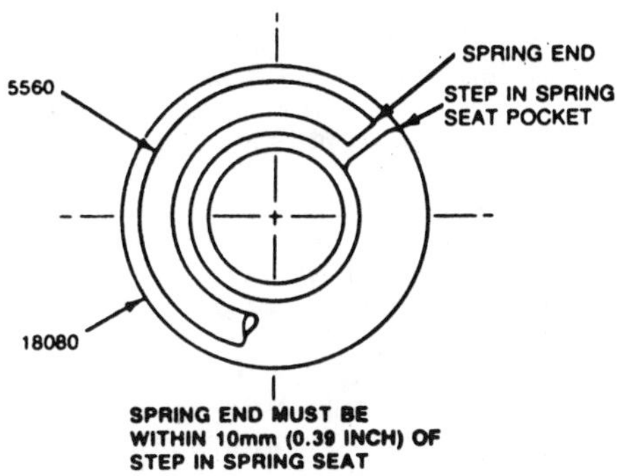

Fig. 4 Coil spring installation

11. Remove link attaching nut, washer and insulator, then the link from the strut.
12. Mark location of insulator to top mount, then place strut, spring and upper mount assembly in suitable spring compressor and compress spring.
13. While restraining strut shaft from turning, remove strut upper shaft mounting nut. **If strut is to be reused, do not use vise grips or pliers to hold strut shaft as damage will result.**
14. Carefully loosen spring compressor tool, then remove top mount bracket assembly, spring insulator and spring, **Fig. 3.**
15. Using new attaching parts, reverse procedure to install, noting the following:
    a. When installing spring on strut, ensure spring is properly located in upper and lower spring seats. Refer to **Fig. 4,** for spring end placement.
    b. When tightening strut nut, restrain strut shaft from turning. **Torque** strut nut to 35-50 ft. lbs.
    c. Torque stabilizer link-to-strut attaching nut to specifications.
    d. Torque strut pinch bolt to specifications.
    e. Torque tension strut-to-spindle attaching nut to specifications.
    f. Torque stabilizer link-to-stabilizer bar attaching nut to specifications.
    g. Torque stabilizer bar U-bracket attaching bolt to specifications.
    h. Tighten strut top mount-to-body attaching nuts to 19-26 ft. lbs.

SHOCK ABSORBER REPLACE

WAGON MODELS

These models use gas filled shock absorbers.

1. Remove rear compartment access panels.
2. While restraining shock absorber shaft, loosen, but do not remove top mounting nut. **If shock absorber is to be reused, do not use vise grips or pliers to hold shock absorber shaft as damage will result.**
3. Raise and support rear of vehicle, then remove tire and wheel assembly. **If a frame contact hoist is used, support lower control arm with floor jack. If a twin post lift is used, support body with floor jacks on lifting pads forward of tension strut body bracket.**
4. Loosen shock absorber-to-lower control arm attaching nuts. Do not remove nuts at this time.
5. Lower vehicle, then remove shock absorber top mounting nut, washer and insulator.
6. Raise and support rear of vehicle. **If a frame contact hoist is used, support lower control arm with floor jack. If a twin post lift is used, support body with floor jacks on lifting pads forward of tension strut body bracket.**
7. Remove shock absorber-to-lower control arm attaching nuts, then remove shock absorber from vehicle. **The shock absorbers are gas filled and will require an effort to collapse them for removal.**
8. Reverse procedure to install, noting the following:

a. Torque shock absorber top mounting nut to specifications.
b. Torque shock absorber-to-lower control arm mounting nuts to specifications.

STABILIZER BAR REPLACE

SEDAN MODELS

1. Raise and support vehicle.
2. Remove stabilizer bar-to-link attaching nuts, washers and insulators from both sides. Remove U-bracket attaching bolts, then the stabilizer bar.
3. Remove link-to-strut attaching nuts, washers and insulators.
4. Inspect attaching parts for damage and replace as necessary.
5. Using new attaching parts, reverse procedure to install, noting the following:
   a. Torque link-to-strut attaching nuts to specifications.
   b. Torque U-bracket attaching bolts to specifications.
   c. Torque stabilizer bar-to-link attaching nuts to specifications.

WAGON MODELS

1. Raise and support vehicle.
2. Remove U-bracket retaining nuts and bolts from either side, then slide U-brackets and insulators from stabilizer bar.
3. Remove link-to-body bracket attaching nuts and bolts, then the stabilizer and link assemblies.
4. Slide link assemblies from stabilizer bar.
5. Inspect attaching parts for damage and replace as necessary.
6. Reverse procedure to install, noting the following:
   a. Install new nuts and bolts, then torque link-to-body bracket attaching nuts and bolts to specifications.
   b. Install new nuts and bolts, then torque U-bracket attaching nuts and bolts to specifications.

TENSION STRUT REPLACE

SEDAN MODELS

1. Raise vehicle on frame contact hoist using lift pads located rearward of front wheels and forward of rear wheels. Raise hoist only enough to contact body.
2. Working inside trunk, loosen, but do not remove, three strut-to-inner body attaching nuts.
3. Raise vehicle, then remove tire and wheel assembly.
4. Remove tension strut-to-spindle attaching nut.
5. Remove tension strut-to-body attaching nut.
6. While moving spindle rearward, remove tension strut.
7. Install new inner washers and bushings on both ends of tension strut. Refer to **Fig. 5,** for bushing and washer identification.

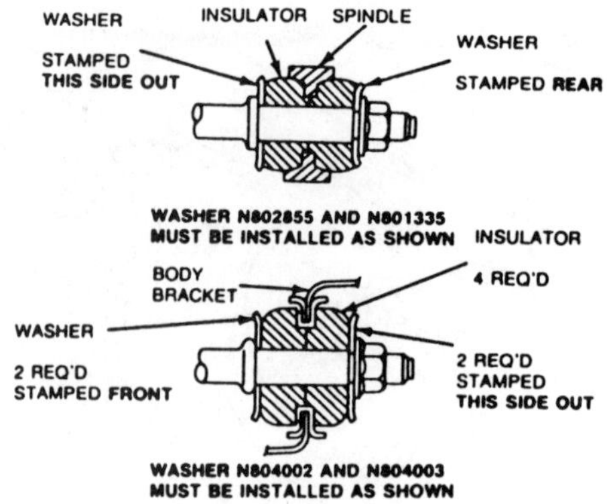

Fig. 5 Tension strut bushing installation. Sedan models

8. Install tension strut end into body bracket, then install outer bushing, washers and nut. Do not tighten nut at this time.
9. While moving spindle rearward, install tension strut in spindle, then install outer bushing, washer and nut.
10. Ensure bushings are correctly seated in mountings, **Fig. 5,** then **torque** attaching nuts 52-74 ft. lbs.
11. Support spindle with suitable jack stand, then working inside trunk, remove three strut-to-inner body attaching nuts. Install new nuts and **torque** to 19-26 ft. lbs.
12. Remove jack stand, then install tire and wheel assembly.
13. Lower vehicle.

WAGON MODELS

1. Raise vehicle on frame contact hoist, then position suitable jack under lower control arm and raise arm to normal curb height.
2. Remove wheel and tire assembly, then tension strut-to-lower control arm attaching nut and bolt.
3. Remove tension strut-to-body bracket attaching nut and bolt, then the tension strut.
4. Insert front end of replacement torsion strut in body bracket, then install new attaching nut and bolt. Do not tighten at this time.
5. Position rear end of torsion strut in lower control arm, then install new attaching nut and bolt. **Torque** to 40-55 ft. lbs.
6. **Torque** torsion strut-to-body bracket attaching nut and bolt to 40-55 ft. lbs.
7. Install wheel and tire assembly, then remove jack and lower vehicle.

SPINDLE REPLACE

SEDAN MODELS

1. Raise and support vehicle, then remove tire and wheel assembly.
2. Remove brake drum, then the brake hose-to-strut retaining clip.
3. Remove brake backing plate attaching bolts, then the plate. Wire backing plate out of the way.
4. Remove control arm-to-spindle attaching bolts, washers and nuts, then the tension strut nut, washer and bushing.
5. Remove spindle-to-strut pinch bolt, then the spindle.
6. Position replacement spindle onto tension strut, then onto shock strut.
7. Install new strut-to-spindle pinch bolt. Do not tighten at this time.
8. Install tension strut bushing, washer and new nut. Do not tighten at this time.
9. Install new control arm-to-spindle attaching bolts.
10. Using suitable jack, raise lower control arm to normal curb height, then tighten nuts and bolts as follows:
    a. **Torque** spindle-to-strut bolt to 55-81 ft. lbs.
    b. **Torque** tension strut nut to 52-74 ft. lbs.
    c. **Torque** control arm-to-spindle attaching nuts to 52-74 ft. lbs.
11. Install brake backing plate and brake hose-to-strut retaining clip.
12. Install brake drum and wheel and tire assembly.
13. Lower vehicle.

WAGON MODELS

1. Raise and support vehicle, then remove tire and wheel assembly. **If vehicle is raised on frame contact hoist, position suitable jack under lower control arm to raise arm to normal curb height.**
2. Remove brake drum and wheel bearings, then the brake backing plate.
3. Remove upper control arms to crossmember attaching nuts and bolts.
4. Remove bolt, one washer, adjusting cam and nut attaching spindle to lower control arm.
5. Remove spindle and upper control arm as an assembly, then remove upper control arm-to-spindle attaching nut and the spindle.
6. Install upper control arms to spindle using a new nut. Do not tighten at this time.
7. Position spindle and upper control arm assembly on lower control arm. Install new nut and washer, existing adjusting cam and new nut. Do not tighten at this time.
8. Position front and rear upper control arms to body bracket, then install new nuts and bolts. Do not tighten at this time.
9. Ensure lower control arm is at normal curb height, then proceed as follows:
   a. **Torque** upper control arms to body bracket attaching bolts to 70-95 ft. lbs.
   b. **Torque** upper control arms to spindle attaching nut to 150-190 ft. lbs.
   c. **Torque** spindle-to-lower control arm attaching nut to 60-86 ft. lbs.
   d. Install brake backing plate, brake drum and wheel bearings.
   e. Install tire and wheel assembly,

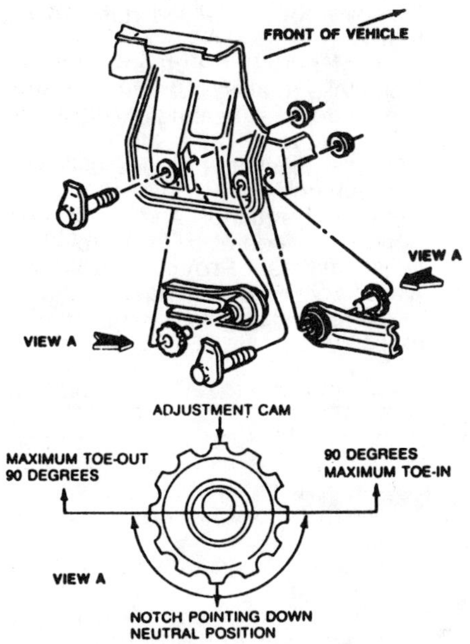

Fig. 6 Lower control arm bushing & cam installation. Sedan models

then remove jack assembly and lower vehicle.

LOWER CONTROL ARM REPLACE

SEDAN MODELS

1. Raise and support vehicle.
2. Disconnect brake proportioning valve from left side front control arm, then the parking brake cable from control arms.
3. Remove control arm-to-spindle attaching bolt, nut and washer.
4. Remove control arm-to-body bracket attaching bolt and nut, then the control arm.
5. Position control arm (and cam where required, **Fig. 6**), at body bracket, then install new nut and bolt. Do not tighten at this time. **When installing control arms, the offset must face up (the arms are stamped "bottom" on lower edge). The flange edge of the right side rear arm stamping must face the front of the vehicle. The other three must face the rear of the vehicle. During installation, note that the control arms have two adjustment cams that fit inside the bushings at the control arm-to-body attachment. The cam is installed from the rear on the left arm and from the front on the right arm, Fig. 6.**
6. Position outer end of arm at spindle, then install new bolt, washer and nut. **Torque** nut to 52-74 ft. lbs.
7. **Torque** control arm-to-body bracket attaching nut to 52-74 ft. lbs.
8. Attach parking brake cables and brake proportioning valve to control arms.
9. Lower vehicle, then check rear toe and reset as necessary.

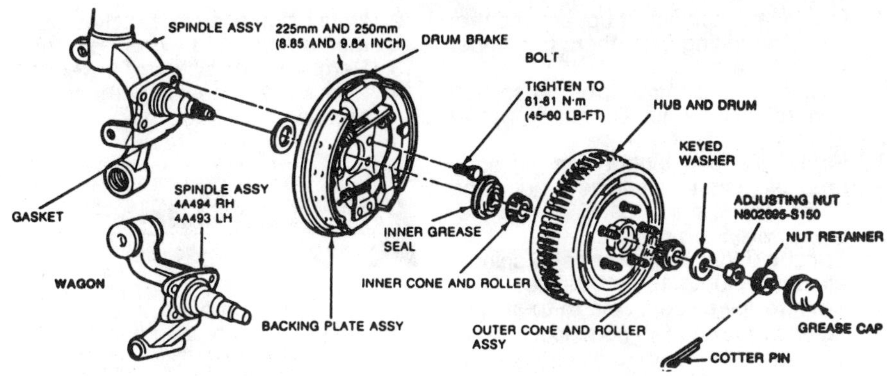

Fig. 7 Exploded view of rear hub & bearing assembly

WAGON MODELS

1. Raise and support rear of vehicle, then remove tire and wheel assembly.
2. Remove rear spring. Refer to" Spring, Replace" for procedure.
3. Remove lower control arm-to-body bracket attaching bolt, then the control arm.
4. Position lower control arm in body bracket, then install new nut and bolt with bolt head toward front of vehicle. Do not tighten at this time.
5. Install rear spring. Refer to" Spring, Replace" for procedure.
6. Using suitable jack, support lower control arm at normal curb height, then **torque** control arm-to-body bracket attaching bolt to 40-55 ft. lbs.
7. **Torque** lower control arm-to-spindle attaching bolt to 60-86 ft. lbs.
8. Install tire and wheel assembly, then lower vehicle.

SPRING REPLACE

WAGON MODELS

1. Raise vehicle on frame contact hoist, then using suitable floor jack, raise lower control arm to normal curb height.
2. Remove tire and wheel assembly.
3. Remove brake hose bracket from body, then the stabilizer bar U-bracket from lower control arm.
4. Remove shock absorber-to-lower control arm attaching nuts, then the parking brake cable and clip from lower control arm.
5. Remove tension strut-to-lower control arm attaching nut and bolt, then wire upper control arms and spindle to keep them from dropping down.
6. Remove lower control arm-to-spindle attaching nut, bolt and toe adjusting cam.
7. Carefully lower control arm with floor jack until spring can be removed.
8. Remove spring and insulators.
9. Install lower spring insulator on control arm. Ensure insulator is seated properly.
10. Position upper insulator on spring, then install spring on lower control arm. Ensure spring is properly seated.
11. Using floor jack, raise lower control arm and spring while guiding upper spring insulator onto upper spring seat.
12. Position spindle in lower control arm and install new bolt, nut and existing toe adjusting cam. Install bolt with head facing front of vehicle. Do not tighten at this time.
13. Remove wire from upper control arms and spindle assembly, then position tension strut in lower control arm and install new nut and bolt. Do not tighten at this time.
14. Install parking brake cable and clip to lower control arm, then install lower end of shock absorber. **Torque** nuts to 12-20 ft. lbs.
15. Install stabilizer bar and U-bracket. Install new bolt and **torque** to 20-30 ft. lbs.
16. Install brake hose bracket. **Torque** attaching bolt to 8-12 ft. lbs.
17. Using floor jack, raise lower control arm to normal curb height, then **torque** lower control arm-to-spindle nut to 60-86 ft. lbs. and tension strut-to-body bracket bolt to 40-55 ft. lbs.
18. Install tire and wheel assembly, then remove floor jack and lower vehicle.
19. Check rear end alignment.

UPPER CONTROL ARMS REPLACE

WAGON MODELS

1. Raise vehicle on frame contact hoist, then using suitable floor jack, raise lower control arm to normal curb height.
2. Remove wheel and tire assembly, then brake hose bracket from body.
3. Loosen spindle-to-upper control arms attaching nut.
4. Loosen spindle-to-lower control arm attaching nut.
5. Remove upper control arms-to-body brackets attaching nuts and bolts. Ensure spindle does not fall outward.
6. Carefully tilt upper part of spindle outward until upper control arms are clear of body brackets. Wire spindle in this position.

7. Remove spindle-to-upper control arms attaching nut, then the upper control arms.
8. Install upper control arms on spindle and install new nut. Do not tighten at this time.
9. Position upper control arms in body brackets, then install new nuts and bolts. **Torque** to 70-95 ft. lbs. Remove wire from spindle.
10. **Torque** upper control arms-to-spindle attaching nut to 150-190 ft. lbs.
11. **Torque** lower control arm-to-spindle attaching nut to 60-86 ft. lbs.
12. Install brake hose bracket on body, then the tire and wheel assembly.
13. Remove floor jack and lower vehicle.
14. Check rear wheel alignment and correct as necessary.

WHEEL BEARINGS

ADJUST

1. Raise and support rear of vehicle, then remove grease cap.
2. Remove cotter pin and nut retainer, **Fig. 7.**
3. Back of adjusting nut one full turn, then while rotating hub and drum assembly to seat bearings, **torque** adjusting nut to 17-25 ft. lbs.
4. Loosen adjusting nut one-half turn, then **retorque** to 10-15 inch lbs.
5. Install nut retainer and cotter pin. Bend over ends of cotter pin over retainer flange. **Properly adjusted wheel bearings may have a slightly loose feel which is considered normal.**

REAR SUSPENSION TIGHTENING SPECIFICATIONS

Year	Component	Torque/ ft. lbs.
1986-90 ①	Control Arm to Spindle	42-57
	Control Arm to Body	45-65
	Stabilizer Bar Link to Strut	5-7
	Stabilizer Bar Link to Stabilizer Bar	5-7
	Stabilizer U-Bracket to Body	25-37
	Strut Top Mount to Body	19-26
	Strut to Top Mount	35-50
	Strut to Spindle	50-70
	Tension Strut to Spindle	35-50
	Tension Strut to Body	35-50
1986-90 ②	Shock Absorber to Body	19-27
	Shock Absorber to Lower Suspension Arm	12.5-20
	Stabilizer Bar U-Bracket to Lower Suspension Arm	20-30
	Stabilizer Link Assembly to Body	40-55
	Tension Strut to Body	40-55
	Tension Strut to Lower Suspension Arm	40-55
	Upper Suspension Arms to Body	70-95
	Upper Suspension Arms to Spindle	150-190
	Upper Suspension Arm to Spindle	40-55

①—Sedan models.
②—Wagon models.

Lincoln Continental & Mark VII

INDEX

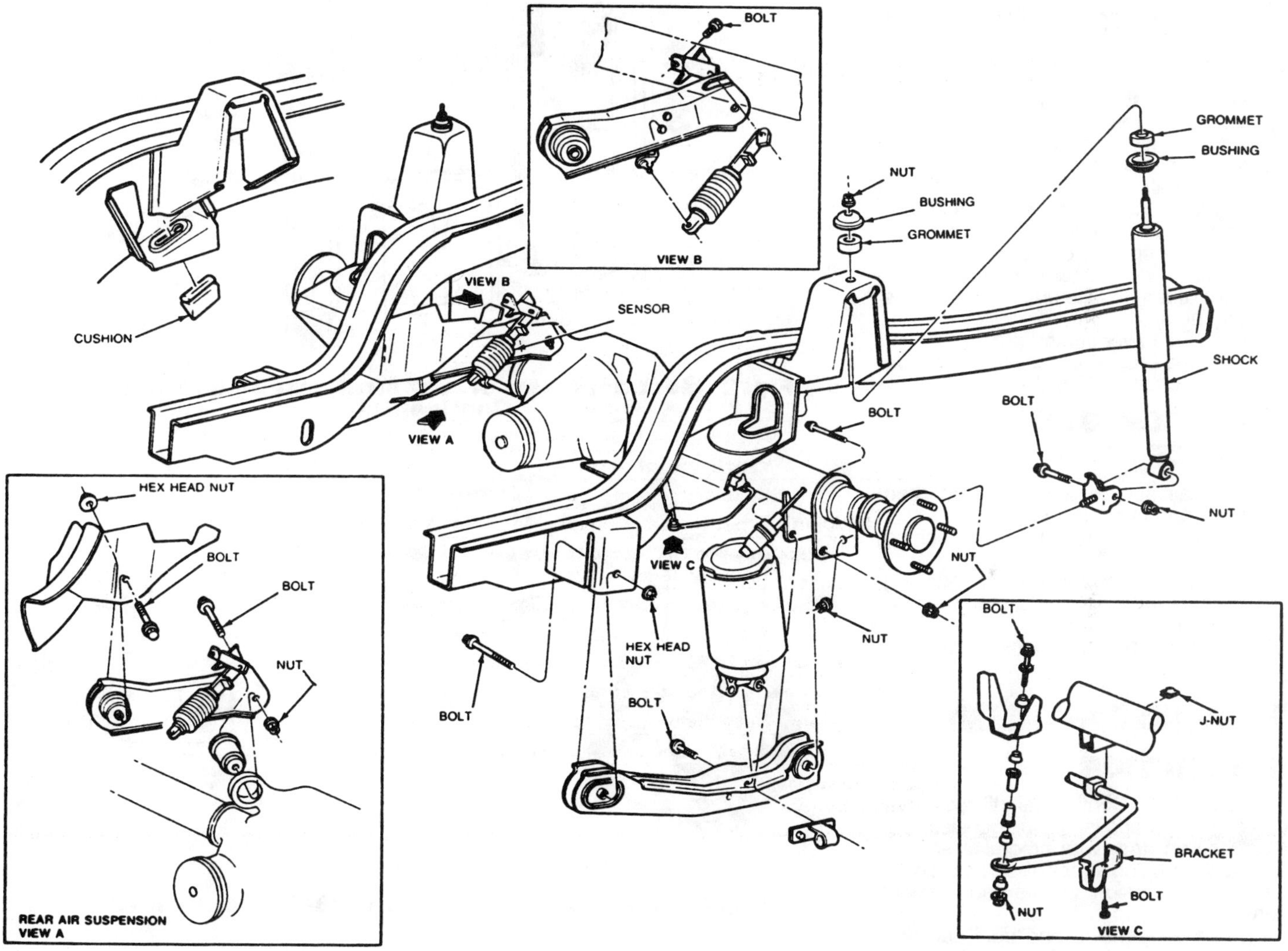

Fig. 1 Air spring rear suspension assembly. Mark VII & 1985–87 Continental

MARK VII & 1985–87 CONTINENTAL

SHOCK ABSORBER, REPLACE

Turn air suspension switch off before replacing shock absorber.

1. Open trunk to gain access to upper shock absorber attachment.
2. Remove rubber cap if equipped, from shock absorber stud, then remove nut, washer and insulator.
3. Raise vehicle and support rear axle.
4. Remove lower shock absorber protective cover, then remove cross bolt and nut from lower shock absorber mounting bracket.
5. From underneath vehicle, compress shock absorber to clear hole in upper shock tower, then remove shock absorber, **Fig. 1. These models are equipped with gas pressurized shock absorbers which extend unassisted during removal. Do not apply heat or flame to the shock absorber tube during removal.**
6. Reverse procedure to install. While holding shock absorber in position, **torque** lower cross bolt to 59 ft. lbs. Lower vehicle and install upper mounting nut, washer and insulator and **torque** nut to 24 to 26 ft. lbs.

CONTROL ARMS, REPLACE

Upper Arm

Always replace control arm in pairs. If one arm requires replacement, replace the same arm on the opposite side of the vehicle.

1. Turn air suspension switch off.
2. Raise and support vehicle, then disconnect rear height sensor from side arm. Note position of sensor adjustment bracket to aid in reassembly.
3. Remove upper arm to axle and upper arm to frame bracket pivot bolts and nuts.
4. Remove upper control arm.
5. Reverse procedure to install. **Torque pivot bolts to 100 ft. lbs.**

Lower Arm

1. Turn air suspension switch to off position, then raise and support vehicle and remove wheel assembly.
2. Vent air springs to atmosphere by removing air spring solenoid.
3. Remove the two air spring to lower control arm retaining bolts, and remove air spring from lower arm.
4. Remove control arm to frame and control arm to axle bracket pivot bolts and nuts.
5. Remove lower control arm.
6. Reverse procedure to install. **Torque pivot bolts to 100 ft. lbs.**

STABILIZER BAR, REPLACE

1. Turn air suspension switch off, then raise and support vehicle.
2. Remove stabilizer bar to link attaching nuts.
3. Remove stabilizer bar to bushing U-clamp attaching nuts, then the stabilizer bar.
4. Reverse procedure to install.

1988–90 CONTINENTAL

DESCRIPTION

The 1988-90 Continental utilizes a fully independent rear suspension consisting of MacPherson struts with integral air springs and dual-damping shock absorbers, counterbalancing torsion springs and a height sensor.

The air suspension and dual damping functions are controlled by a microcomputer based module which receives inputs for vehicle speed, door switch position, damping actuator feedback, steering wheel turning rate and angle, engine vacuum, throttle position, brake actuation, ignition switching, and vehicle ride height. The dual-damping function automatically switches from a soft to firm ride when the driving situation (hard cornering, acceleration or braking, etc.) dictates the need for increased damping effect.

The rear struts, **Fig. 2**, use a dual path mount which separates the strut and air spring mounting surfaces, to help provide for maximum isolation. The counterbalancing torsion springs, fitted between the strut and lower control arm, produce an outward force on the strut that tends to offset the binding forces induced by the rear wheels. The rotary design Hall effect type height sensor, **Fig. 3**, permits multiple height positions to be defined, resulting in specialized leveling during all types of driving and load characteristics.

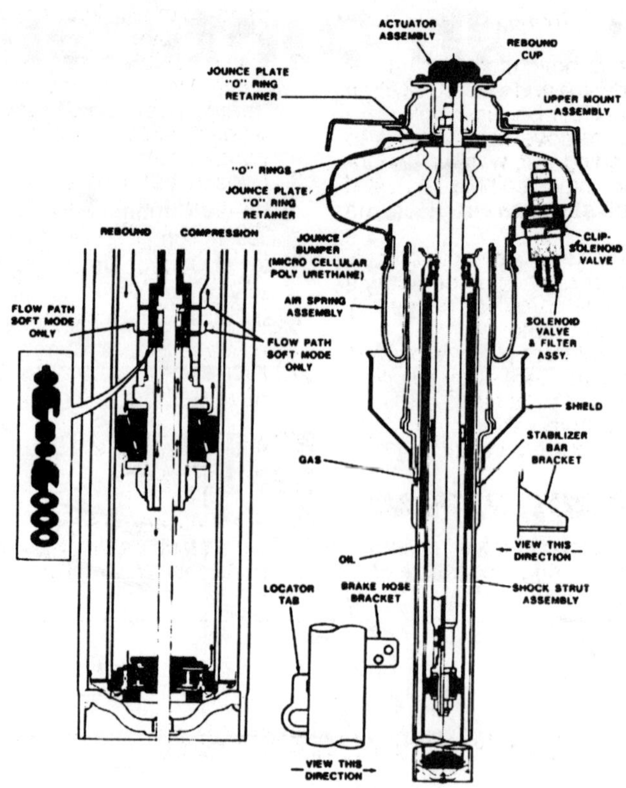

Fig. 2 Sectional view of rear strut assembly. 1988–90 Continental

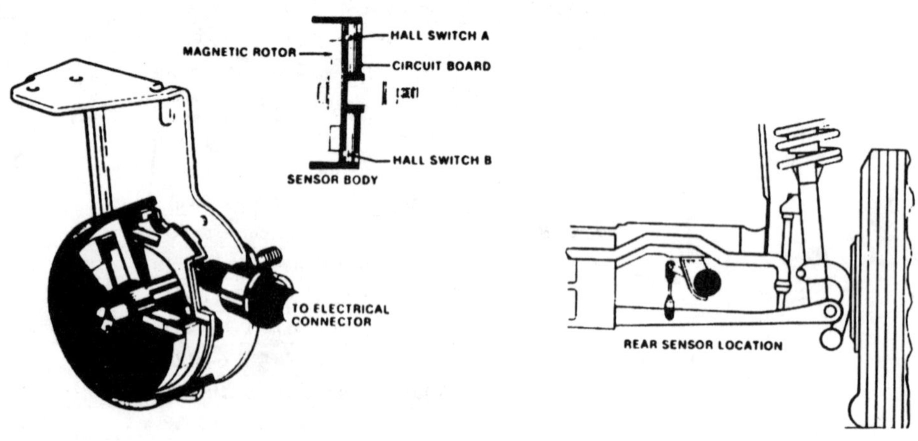

Fig. 3 Rotary height sensor. 1988–90 Continental

SERVICE PRECAUTION

Always place the air suspension switch in the Off position before performing any work, or whenever raising the rear suspension.

STRUT ASSEMBLY, REPLACE

1. Position suitable jack or hoist under vehicle, then raise just enough to contact body.
2. Disconnect air suspension electrical wiring and all related parts that will interfere with strut removal.
3. Loosen, but do not remove, the strut to inner body attaching nuts.
4. Raise and support vehicle, then remove wheel and tire assembly.
5. Remove brake hose to strut bracket

attaching clip, then position hose aside.

6. If applicable, remove stabilizer bar attaching hardware and insulators, then separate stabilizer bar from link.
7. If applicable, remove tension strut to spindle attaching nut, washer and insulator, then move spindle rearward until it can be separated from tension strut.
8. Remove strut to spindle pinch bolt, then using a pry bar or other suitable tool, separate pinch joint as necessary to allow for strut removal.
9. Disengage strut from pinch joint, then lower vehicle as necessary to allow removal of upper attaching nuts.
10. Remove strut from vehicle.
11. Reverse procedure to install.

TENSION STRUT, REPLACE

1. If necessary, disconnect air suspension electrical wiring and all related parts that will interfere with tension strut removal.
2. Raise vehicle on frame contact hoist using lift pads located rearward of front wheels and forward of rear wheels. Raise hoist only enough to contact body.
3. Loosen, but do not remove, the upper strut to inner body attaching nuts, then raise and support vehicle.
4. Remove wheel and tire assembly.
5. Remove tension strut to spindle and tension strut to body attaching nuts.
6. While moving spindle rearward, remove tension strut from vehicle.
7. Reverse procedure to install, using new washers and bushings.

STABILIZER BAR, REPLACE

1. If necessary, disconnect air suspension electrical wiring and all related parts that will interfere with stabilizer bar removal.
2. Raise and support vehicle.
3. Remove stabilizer bar to link attaching nuts, washers and insulators.
4. Remove U-bracket attaching bolts, then the stabilizer bar.
5. Reverse procedure to install, using new attaching parts.

LOWER CONTROL ARM, REPLACE

1. If necessary, disconnect air suspension electrical wiring and all related parts that will interfere with control arm removal.
2. Raise and support vehicle.
3. Remove control arm to spindle attaching bolt, nut, and washer.
4. Remove control arm to body attaching bolt and nut, then the control arm.
5. Reverse procedure to install, then check rear wheel alignment and adjust as necessary.

REAR SUSPENSION TIGHTENING SPECIFICATIONS

Year	Component	Torque/ft. lbs.
MARK VII & 1985–87 CONTINENTAL		
1985-90①	Shock Absorber (Upper)	19-27
	Upper Arm To Frame	80-105
	Upper Arm To Axle	70-100
	Lower Arm To Frame	80-105
	Lower Arm To Axle	70-100
	Stabilizer Bar To Lower Arm	32-52
1989-90②	Shock Absorber To Frame	17-27
	Upper Arm To Frame	80-105
	Upper Arm To Axle	70-100
	Lower Arm To Frame	80-105
	Lower Arm To Axle	90-100
	Shock Absorber To Clevis Bracket	45-60
	Clevis Bracket To Axle	55-70
	Stabilizer Bar To Axle	13-20
	Stabilizer Bar To Body	13-18
	Air Spring To Lower Arm	25-35
	Sensor Upper Bracket To Frame	110-150③
	Sensor Lower Bracket To Frame	7-10
1988–90 CONTINENTAL		
1988-90	Arm Inner Pivot Retainer	45-65
	Stud To Arm Nut	17-24
	Spring To Stud Nut	10-15
	Spring Clamp To Spindle Bolt	10-15
	Strut Top Mount To Body	19-26
	Strut To Top Mount	35-50
	Strut To Spindle	51-70
	Control Arm To Spindle	42-57

Continued

REAR SUSPENSION TIGHTENING SPECIFICATIONS—Continued

Year	Component	Torque/ft. lbs.
1988–90 CONTINENTAL—Cont'd		
	Control Arm To Body	45-65
	Tension Strut To Spindle	35-50
	Tension Strut To Body	35-50
	Stabilizer Bar Link	5-7
	Stabilizer U-Bracket	25-37

①—Except air suspension.
①—Air suspension.
①—Inch lbs.

Mercury Tracer

INDEX

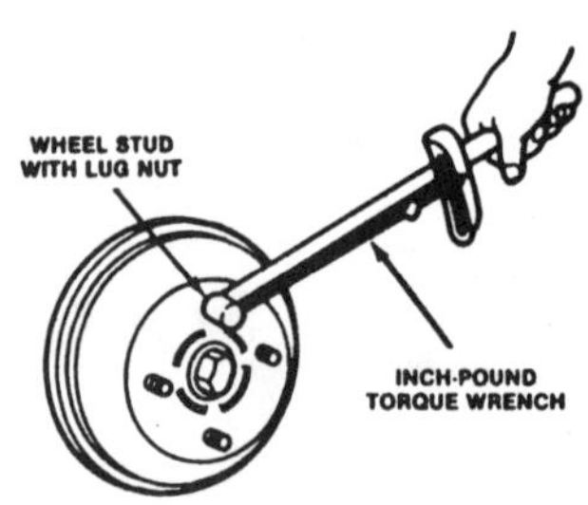

Fig. 1 Wheel bearing adjustment

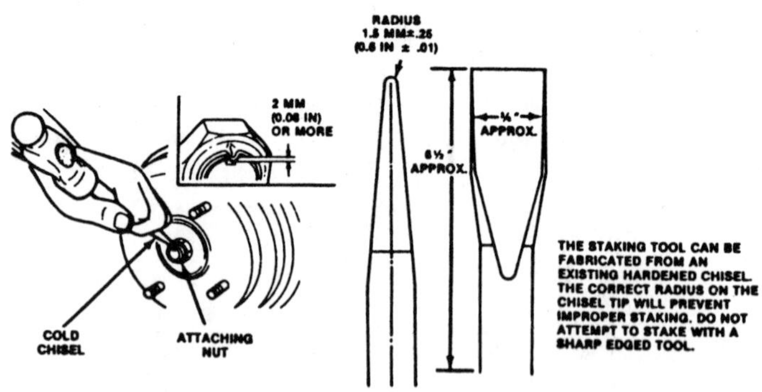

Fig. 2 Staking wheel bearing adjusting nut

WHEEL BEARING

ADJUST

1. Block front wheels, then raise and support rear of vehicle.
2. Remove wheel and tire assembly.
3. Ensure parking brake is fully released, then remove grease cap and raise staked portion of wheel bearing adjusting nut using a suitable chisel.
4. Remove wheel bearing adjusting nut and discard. To loosen adjusting nut on left hand side of vehicle, turn nut counterclockwise. To loosen adjusting nut on right hand side of vehicle, turn nut clockwise.
5. Position replacement adjusting nut on spindle, then torque nut to 18.1 to 21.7 ft. lbs., while rotating brake drum to seat bearings.
6. Loosen locknut slightly, so that nut can be rotated by hand.
7. Install lug nut on stud, then position stud at 12 o'clock position. Using an inch torque wrench on lug nut, measure and record amount of force required to start rotation of brake drum, **Fig. 1.** This measurement is seal drag.
8. To determine preload setting, add the required amount of preload, which is 1.3 to 4.3 inch lbs., to the seal drag reading obtained in step 7.
9. Tighten wheel bearing adjusting nut slightly, then using inch lb. torque wrench as described in step 7, measure wheel bearing preload. Continue to tighten wheel bearing adjusting nut a little at a time until wheel bearing preload determined in step 8 is obtained.
10. After proper preload setting has been obtained, stake wheel bearing adjusting nut, **Fig. 2.** If nut is damaged during staking, it must be replaced.
11. Install grease cap, then install wheel and tire assembly.

WHEEL BEARING

REPLACE

1. Raise and support rear of vehicle, then remove wheel and tire assembly.
2. Using a suitable chisel, carefully raise staked portion of wheel bearing adjusting nut.
3. Remove wheel bearing adjusting nut and discard, **Fig. 3.** To loosen adjusting nut on left hand side of vehicle, turn nut counterclockwise. To loosen adjusting nut on right hand side of vehicle, turn nut clockwise.
4. Remove outer wheel bearing from hub, then remove brake drum and hub assembly.
5. Remove grease seal, then remove inner bearing from hub.
6. If necessary, bearing races can be

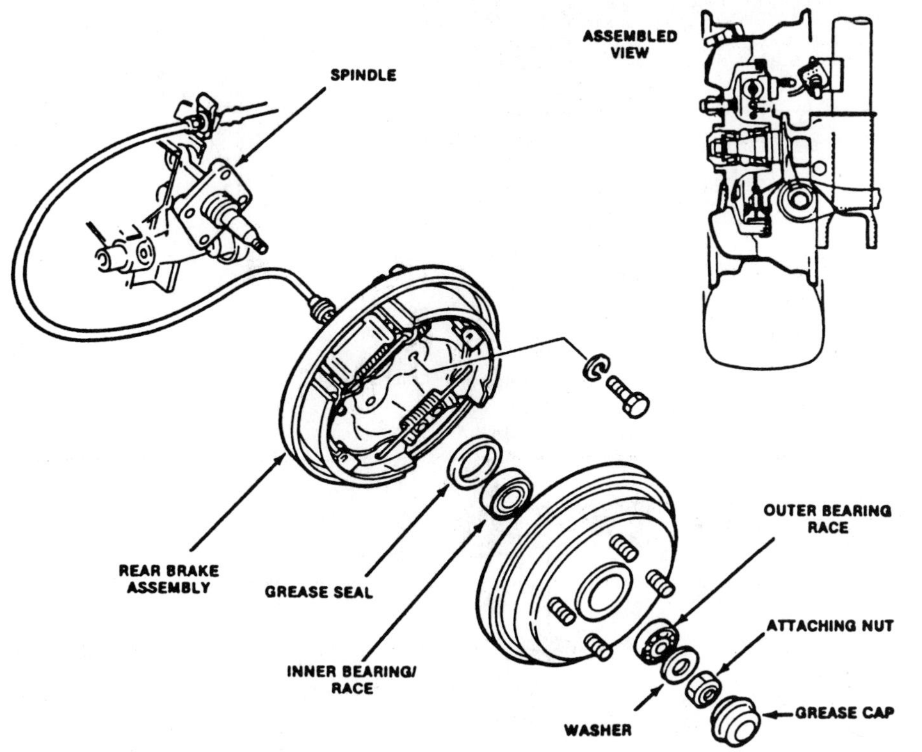

Fig. 3 Wheel bearings & hub

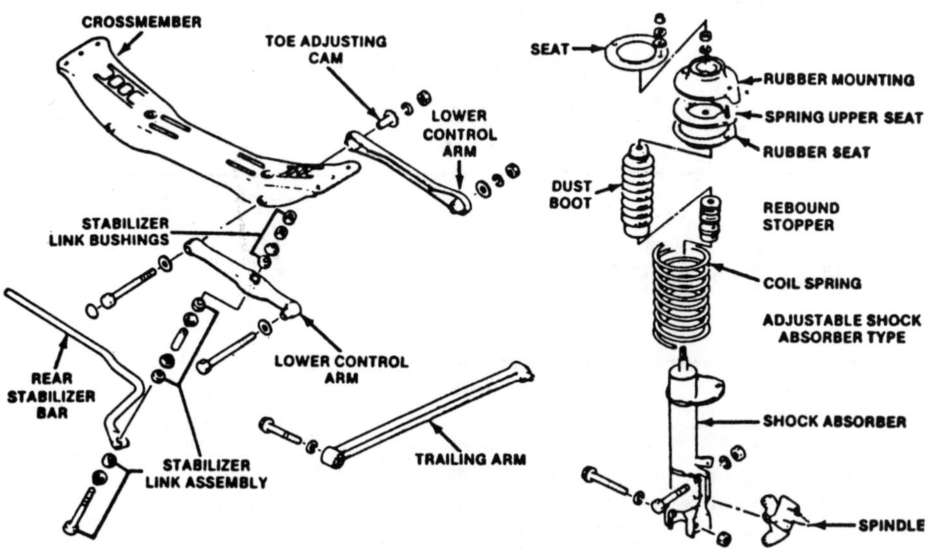

Fig. 4 Rear suspension. Hatchback models

driven from hub using a suitable brass drift.

7. Reverse procedure to install. Prior to installation, pack wheel bearings and hub cavity with suitable wheel bearing grease. Also apply grease to seal lip. After installing bearings, adjust wheel bearing preload as described under "Wheel Bearing, Adjust."

STRUT UNIT & COIL SPRING

REPLACE

1. Raise and support rear of vehicle, then remove wheel and tire assembly.
2. Remove rear brake drum and hub assembly as described under "Wheel Bearing, Replace."
3. Remove trailing arm and spindle to strut unit attaching bolt, **Figs. 4 and 5.**
4. Place alignment mark on strut rubber mounting bracket, then from inside vehicle, remove strut attaching nuts and remove strut from tower.
5. Using a suitable spring compressor, compress coil spring.
6. With coil spring compressed, remove remove strut nut, rubber mounting bracket, spring upper seal and spring seat.
7. Carefully release spring compressor and remove coil spring.
8. Reverse procedure to install. Tighten spindle to strut attaching bolts to torque listed at the end of this section.

CONTROL & TRAILING ARMS

REPLACE

1. Raise rear of vehicle and remove wheel and tire assembly.
2. Remove brake drum and hub as described under "Wheel Bearing, Replace."
3. On hatchback models, place alignment marks on rear toe adjusting cam and control arm, **Fig. 4.**
4. On all models, place alignment marks on control arm and control arm bushing.
5. On all models, place alignment marks on each side of trailing arm and crossmember.
6. Remove stabilizer bar link, then remove stabilizer bar and bushings, **Figs. 4 and 5.**
7. Loosen inner and outer trailing arm bolts, then loosen spindle to strut unit attaching bolts.
8. Remove parking brake attaching bolt from rear trailing arm assembly.
9. Loosen trailing arm to strut attaching bolts, then remove control and trailing arm.
10. Reverse procedure to install. When installing control and trailing arms, align all marks made during removal. When installing stabilizer link bolt, tighten until .31 inch of thread is exposed beyond nut. After vehicle has been lowered to ground, retighten stabilizer link bolt.

SPINDLE

REPLACE

1. Raise and support rear of vehicle, then remove wheel and tire assembly.
2. Remove brake drum and hub as described under "Wheel Bearing, Replace."
3. Remove spindle to strut unit attaching bolts, then remove control bolt, **Figs. 4 and 5.**
4. Remove spindle from strut unit.
5. Reverse procedure to install.

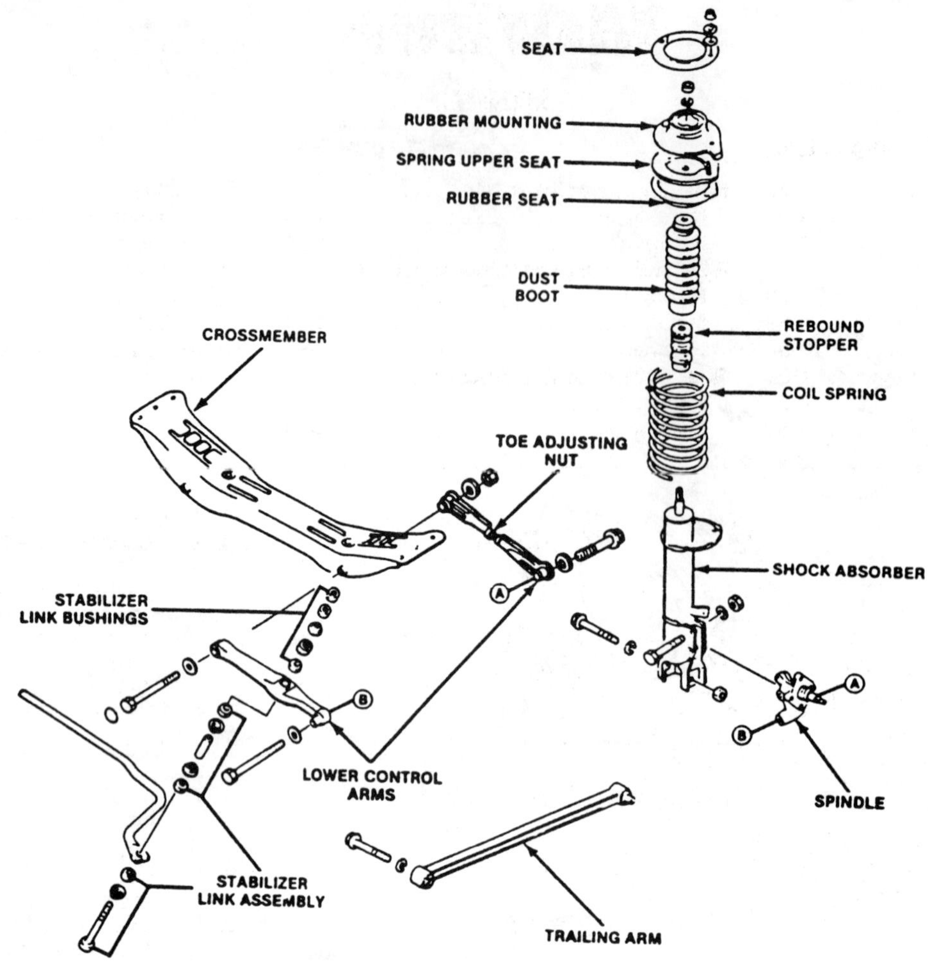

Fig. 5 Rear suspension. Station wagon

REAR SUSPENSION TIGHTENING SPECIFICATIONS

Year	Component	Torque/ft. lbs.
1988-89	Brake Line Fittings	9-16
	Control Arm Jam Nuts	41-47
	Control Arm Inner Bolt	69-86
	Control Arm To Spindle	69-86
	Lateral Link Through Bolt	69-86
	Master Cylinder Bracket	15-25
	Spindle To Backing Plate	33-49
	Spindle To Strut Unit Bolts	69-86
	Stabilizer Bar Bushing Bracket Bolts	32-40
	Wheel Cylinder To Backing Plate	7.2-9.4
	Wheel & Tire Assembly	65-87

Merkur

INDEX

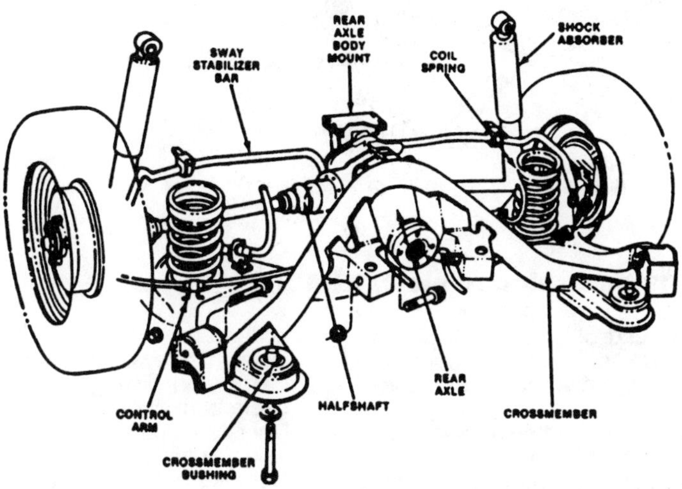

Fig. 1 Rear suspension exploded view. Scorpio

SHOCK ABSORBER REPLACE

SCORPIO

1. Remove luggage compartment floor covering.
2. Remove trim panels from wheel housing and side panel.
3. Remove shock absorber trim cover from rear wheel housing.
4. Raise and support rear of vehicle.
5. Position jack under control arm and raise it enough to relieve tension from shock absorber.
6. Remove nut and bolt attaching upper end of shock absorber.
7. Remove lower nut and bolt and the shock absorber.
8. Reverse procedure to install. **Torque** upper and lower attaching nuts to 30-36 ft. lbs. and lower bolt to 33-40 ft. lbs.

XR4TI

1. Remove rear parcel shelf.
2. Remove shock absorber trim cover from rear wheel housing.
3. Raise and support rear of vehicle.
4. Position jack under control arm and raise it enough to relieve coil spring tension from shock absorber.
5. Remove nut and bolt attaching upper end of shock absorber.
6. Remove cap from bottom of shock absorber.
7. Remove nut and bolt attaching lower end of shock absorber to control arm bracket.
8. Reverse procedure to install, **torquing** upper and lower attaching nuts and bolts to 30-37 ft. lbs.

COIL SPRING REPLACE

SCORPIO

Removal

1. Raise and support vehicle with rear wheels and suspension hanging free.
2. Remove bolts attaching half shaft to wheel stub shaft, **Fig. 1.** To prevent damage to constant velocity joints, tie half shaft to convenient underbody component.
3. Unfasten stabilizer bar from link rod.
4. Disconnect brake line from bracket on body and unclip line from floor assembly. **When removing left side spring, the brake line distribution piece must be unscrewed from floor panel.**
5. Remove clip attaching rear brake hose to routing bracket on control arm.
6. Using line wrenches, disconnect brake tube from brake hose.
7. Using jack, raise lower control arm enough to relieve coil spring tension from shock absorber.
8. Remove shock absorber-to-control arm bracket attaching nut and bolt.
9. Remove rear axle front crossmember guide plate-to-body attaching bolts and the bushing attaching bolts.
10. Slowly and carefully lower jack until spring and seat can be removed.

Installation

1. Install spring upper seat on spring end with color code and plastic sleeve, ensuring end of coil seats against step in spring seat and that seat tabs are positioned between first and second coil.
2. Position coil spring and seat assembly on control arm. **The spring and seat must be installed dry.**
3. Raise control arm to compress coil spring enough to allow installation of the shock absorber lower attaching bolt.
4. Install attaching nut on shock absorber bolt, **torquing** nut to 30-37 ft. lbs., or bolt to 33-40 ft. lbs.
5. Install bushing attaching bolt and **torque** to 59-73 ft. lbs.
6. Install rear axle front crossmember guide plate-to-body attaching bolts and **torque** to 31-37 ft. lbs.
7. Secure brake line to floor assembly and body bracket.
8. Position brake hose through routing bracket on control arm and connect it to brake tube.
9. Install brake hose retaining clip.
10. Position half shaft and install attaching bolts. **Torque** bolts to 28-31 ft. lbs.

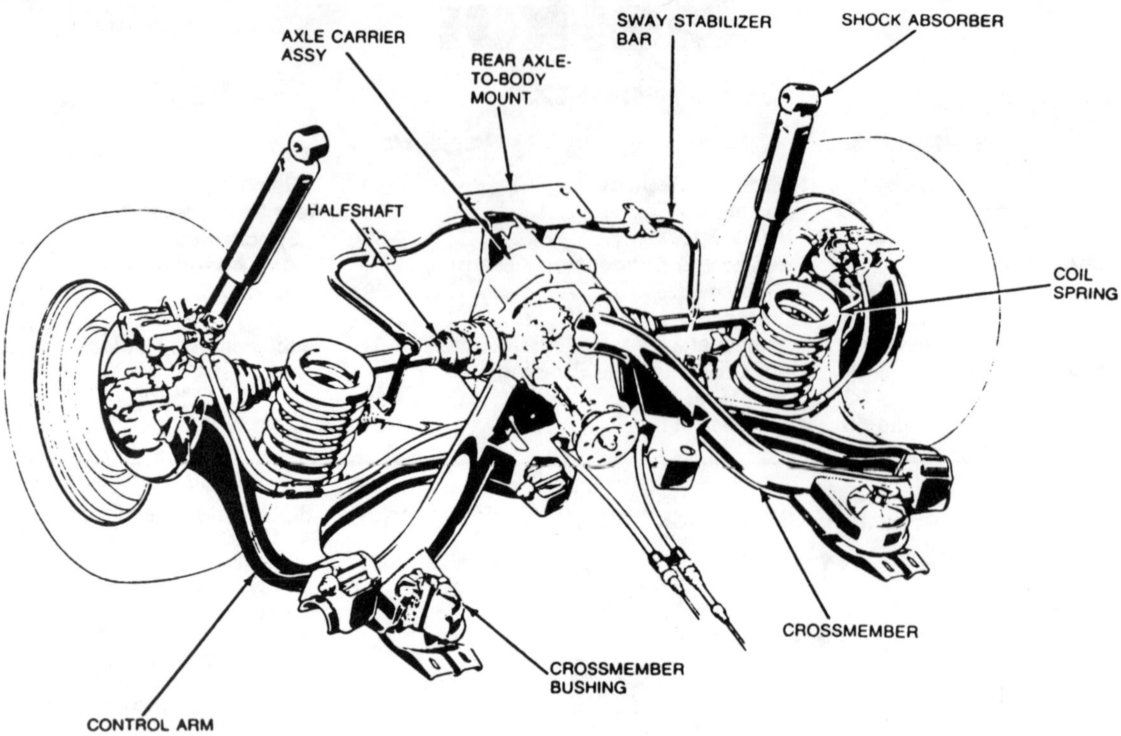

Fig. 2 Rear suspension exploded view. XR4Ti

11. Connect stabilizer bar link to control arm.
12. Bleed brake system.

XR4TI

Removal

1. Raise and support vehicle with rear wheels and suspension hanging free.
2. Remove bolts attaching half shaft to wheel stub shaft, **Fig. 2.** To prevent damage to constant velocity joints, tie half shaft to convenient underbody component.
3. Remove clip attaching rear brake hose to routing bracket on control arm.
4. Using line wrenches, disconnect brake tube from brake hose.
5. Remove cap from bottom of shock absorber.
6. Using jack, raise lower control arm enough to relieve coil spring tension from shock absorber.
7. Remove lower shock absorber attaching nut and bolt.
8. Slowly and carefully lower jack until it can be removed.
9. Support axle housing using jack.
10. Remove rear axle mount to body attaching bolts and disconnect axle vent tube.
11. Slowly and carefully lower jack until coil spring and its seat can be removed. Do not remove support from rear axle. Lower axle only enough to allow removal of coil spring.

Installation

1. Install spring upper seat on spring end with color code and plastic sleeve, ensuring that end of coil seats against step in spring seat and that seat tabs are positioned between first and second coil.
2. Install coil spring and seat assembly, then raise rear axle into position and install body mount attaching bolts. **Torque** two inner bolts to 37-41 ft. lbs. and four outer bolts to 37-50 ft. lbs. **The body mount bolts must be cleaned, and new Loctite applied.**
3. Remove axle support and position it under control arm, then raise jack until coil spring is compressed enough to allow installation of lower shock absorber attaching bolt.
4. Install attaching nut on shock absorber bolt, **torquing** nut to 30-37 ft. lbs., or bolt to 33-40 ft. lbs.
5. Remove control arm support and install shock absorber cap.
6. Position brake hose through routing bracket on control arm and connect it to brake tube.
7. Install brake hose retaining clip.
8. Position half shaft and install attaching bolts, **torquing** bolts to 28-31 ft. lbs.
9. Connect axle vent hose located at top right hand corner of axle housing.
10. Ensure that stabilizer bar link is connected to control arm.
11. Bleed brake system.

CONTROL ARM & BEARING HUB REPLACE

SCORPIO

1. Remove coil spring as previously described.
2. Using screwdriver, open routing clamp and disengage parking brake cable from control arm.
3. Unfasten brake caliper and position aside, leaving brake line attached.
4. Remove stub shaft nut, then pull off wheel hub.
5. Working from under vehicle, pull wheel stub shaft out of control arm.
6. Remove control arm inner and outer attaching bolts and the control arm.
7. Reverse procedure to install, noting the following:
   a. When installing control arm attaching bolts, ensure bolt heads face inboard. Use blue bolts and nuts for outboard side and gold bolts and nuts for inboard side.
   b. With vehicle weight on tires, **torque** control arm attaching bolts to 63-74 ft. lbs.

XR4TI

1. Remove coil spring as previously described.
2. Using screwdriver, open routing clamp and disengage parking brake cable from control arm.
3. Disconnect sway stabilizer link at control arm.
4. Remove wheel flange attaching nut and washer, then wheel flange, **Fig. 3.**
5. Remove rear bearing hub and tie brake backing plate out of way.
6. Working from under vehicle, pull wheel stub shaft out of control arm.
7. Remove control arm inner and outer attaching bolts and the control arm.
8. Reverse procedure to install, noting the following:

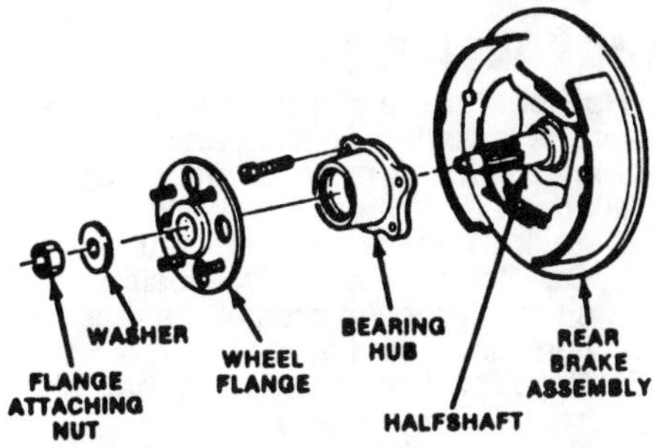

Fig. 3 Removing wheel flange. XR4Ti

a. When installing control arm attaching bolts, ensure bolt heads face inboard.
b. With vehicle weight on tires, **torque** control arm attaching bolts to 63-74 ft. lbs.

CONTROL ARM BUSHINGS

REPLACE

REMOVAL

1. Remove control arm as previously described.
2. Using draw bolt T78P-5638-A1, receiver cup T85M-5638-A2, a 7/16 inch 20 UNF hex nut and a 7/16 inch flat washer, or equivalents, remove larger control arm bushing.
3. Using draw bolt T78P-5638-A1, receiver cup T85M-5638-A1, a 7/16 inch 20 UNF nut and a 7/16 inch flat washer, or equivalents, remove smaller control arm bushing.

INSTALLATION

1. Lubricate bushings with non petroleum lubricant.
2. Using draw bolt T78P-5638-A1, replacer cup T85M-5638-A3, a 7/16 inch 20 UNF nut and a 7/16 inch flat washer, or equivalents install new larger control arm bushing.
3. Using draw bolt T78P-5638-A1, replacer cup T78P-5638-A3, a 7/16 inch 20 UNF nut and a 7/16 inch flat washer, or equivalents, install new smaller control arm bushing.

REAR CROSSMEMBER BUSHING

REPLACE

REMOVAL

1. Raise and support vehicle with suspension and tires hanging free.
2. Position jack under control arm and raise enough to support control arm against downward pressure of coil spring. **Ensure control arm is securely supported before removing bushing to body attaching bolt.**
3. Remove crossmember bushing to body attaching bolt, **Fig. 4.**
4. Remove bushing guide plate attaching bolts and the guide plate.

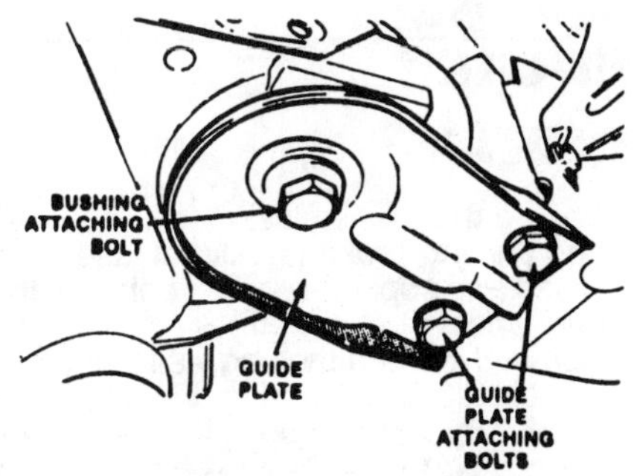

Fig. 4 Location of crossmember bushing to body attaching bolt

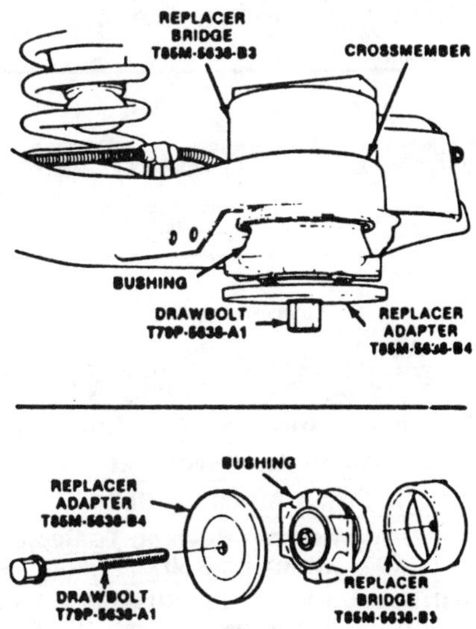

Fig. 6 Installing crossmember bushing

5. Carefully lower crossmember to provide clearance for bushing removal and installation tools.
6. Remove crossmember bushing, **Fig. 5.**

INSTALLATION

1. Lubricate replacement bushing with non-petroleum lubricant.
2. Install crossmember bushing, **Fig. 6.**
3. Carefully raise crossmember into position.
4. Position bushing guide plate and install attaching bolts, **torquing** to 30-37 ft. lbs.
5. Install bushing to body attaching bolt and washer, **torquing** bolt to 59-74 ft. lbs.

REAR AXLE & SUSPENSION ASSEMBLY

REPLACE

REMOVAL

1. Remove coil springs, driveshaft and muffler and silencer assembly.

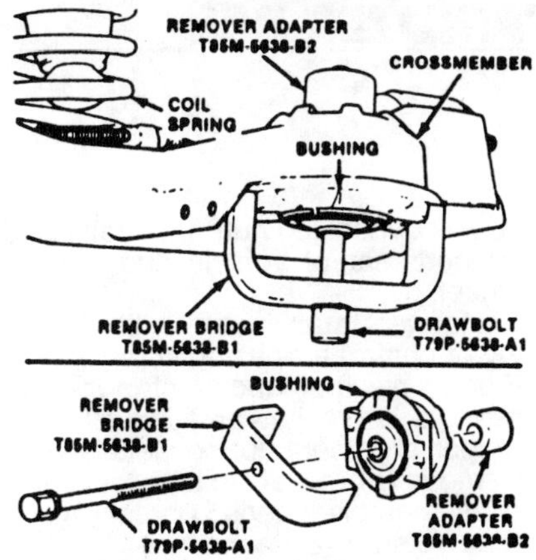

Fig. 5 Removing crossmember bushing

2. Remove nylon lock pin, if equipped, then loosen parking brake adjuster sleeve locknut, then thread adjuster sleeve away from body routing bracket.
3. Remove clip and clevis pin attaching parking brake cable equalizer to parking brake lever rod.
4. Disengage parking brake cable from body routing brackets.
5. Remove stabilizer bar U-brackets.
6. Position jack under rear axle and secure crossmember to jack.
7. Remove crossmember bushing attaching bolt on both sides of vehicle.
8. Remove crossmember bushing guide plate attaching bolts and the guide plate on both sides of vehicle.
9. Remove bolts attaching axle mount to body, then carefully lower crossmember out of vehicle.

INSTALLATION

1. Raise crossmember into position on vehicle.
2. Install but do not tighten axle mount attaching bolts.
3. Position crossmember insulator guide plates and install attaching bolts.
4. Install crossmember bushing attaching bolts and washers, **torquing** guide plate attaching bolts to 30-38 ft. lbs. and the crossmember bushing attaching bolts to 59-74 ft. lbs.
5. **On Scorpio models, torque** four axle mount attaching bolts to 37-49 ft. lbs. On XR4Ti models, **torque** two inner axle mount attaching bolts to 37-41 ft. lbs. and four outer bolts to 37-50 ft. lbs.
6. **On all models,** route parking brake cable through body brackets, then position parking brake equalizer and install clevis pin and retaining clip.
7. Install stabilizer bar U-brackets, **torquing** attaching bolts to 15-18 ft. lbs.
8. Install muffler and silencer assembly, driveshaft and coil springs.
9. Bleed rear brake system and adjust parking brake.

STABILIZER BAR REPLACE

1. Loosen wheel lug nuts, then raise and support rear of vehicle and remove wheel.
2. Disconnect stabilizer bar links from lower control arm using screwdriver.
3. Place a piece of tape on stabilizer bar next to U-bracket and insulator for proper alignment during installation.
4. Remove U-brackets to body attaching bolts, then disengage U-brackets and remove stabilizer bar.
5. Reverse procedure to install, **torquing** U-brackets to body attaching bolts to 15-18 ft. lbs.

HALF SHAFT REPLACE

REMOVAL

1. Ensure that transmission is in neutral and parking brake is fully released.
2. Raise and support rear of vehicle with rear wheels hanging free.
3. Remove bolts attaching half shaft to wheel stub shaft, turning driveshaft as necessary to bring bolts into accessible position.
4. **On Scorpio models,** position half shaft in grooved area of the rear suspension control arm.
5. **On XR4Ti models,** hang half shaft from floor of vehicle.
6. **On all models,** remove bolts attaching half shaft to rear axle stub shaft and remove half shaft.

INSTALLATION

Half shafts are different lengths and must be installed on the correct side of vehicle, with longer shaft being installed on right side of vehicle.

1. Pack CV joints with lubricant.
2. **On Scorpio models,** position half shaft in grooved area of the rear suspension control arm.
3. **On XR4Ti models,** hang half shaft from floor of vehicle.
4. **On all models,** position half shaft to axle stub shaft and install attaching bolts.
5. Release half shaft from vehicle floor and position it to axle stub shaft, installing attaching bolts.
6. **Torque** attaching bolts to 28-31 ft. lbs.

REAR SUSPENSION TIGHTENING SPECIFICATIONS

Year	Component	Torque/ ft. lbs.
1985-90	Axle Mount Attaching Bolts	22-26
	Coil Spring Seat Assembly Bolts	14-18
	Control Arm Attaching Nuts	61-72
	Control Arm Pivot Bolt	63-74
	Half shaft Attaching Bolts	38-43
	Rear Crossmember Bushing To Body Attaching Bolt	59-74
	Rear Crossmember Guide Plate Attaching Bolts	30-37
	Shock Absorber (Lower)	33-40
	Shock Absorber (Upper)	30-37
	Stabilizer Bar U-Bracket Bolts	14-19

GENERAL MOTORS CORP.

TABLE OF CONTENTS

Buick, Cadillac, Chevrolet, Oldsmobile & Pontiac Rear Wheel Drive Models Except Chevette, 1000 & Fiero

INDEX

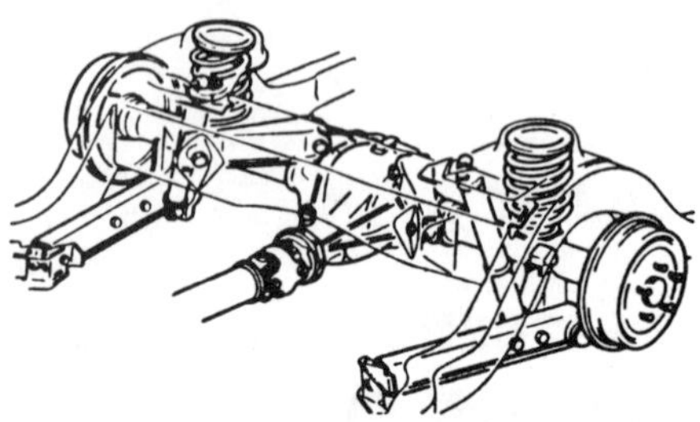

Fig. 1 Rear suspension. Except Camaro, Corvette & Firebird

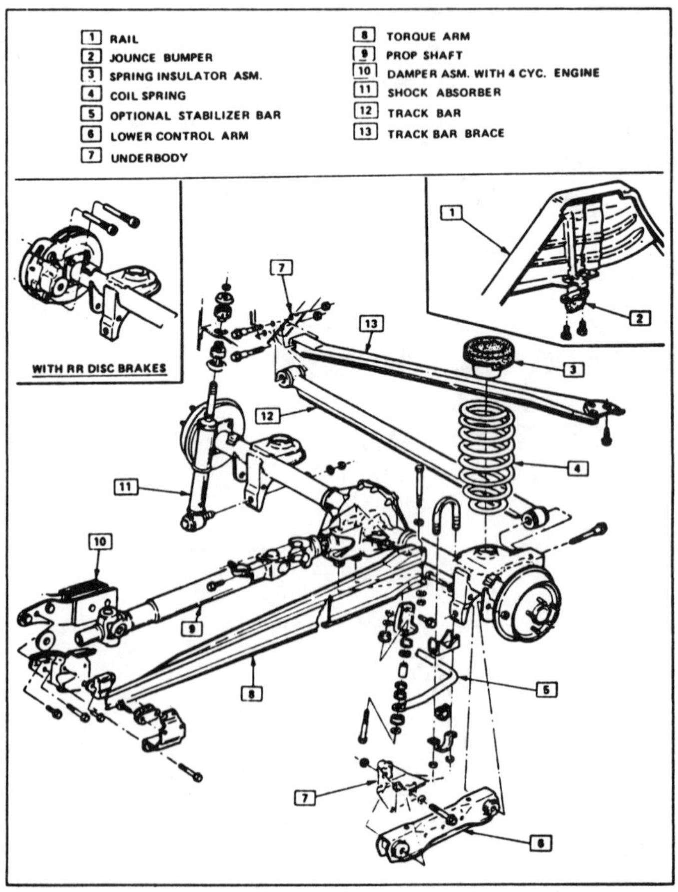

Fig. 2 Rear suspension. Camaro & Firebird

SHOCK ABSORBER
REPLACE

1. If equipped with Superlift shock absorbers, disconnect air lines from shock absorber fittings.
2. With rear axle properly supported, disconnect shock absorber from upper and lower mountings, **Figs. 1 through 3.**
3. Reverse procedure to install.

COIL SPRINGS
REPLACE

EXCEPT CAMARO & FIREBIRD

1. Support vehicle at frame and rear axle.
2. Disconnect shock absorbers at lower mountings.
3. Disconnect upper control arms from axle housing.
4. If equipped with a stabilizer bar, disconnect bar from either right or left hand side of control arm.
5. **On all models,** remove brake hose support bolt and support without disconnecting the brake lines.
6. Lower axle until it reaches end of its travel and using a suitable tool, pry lower pigtail over retainer on axle bracket. Remove spring and insulator.
7. Reverse procedure to install. Springs must be installed with an insulator between upper seat and spring and positioned properly, **Fig. 4.**

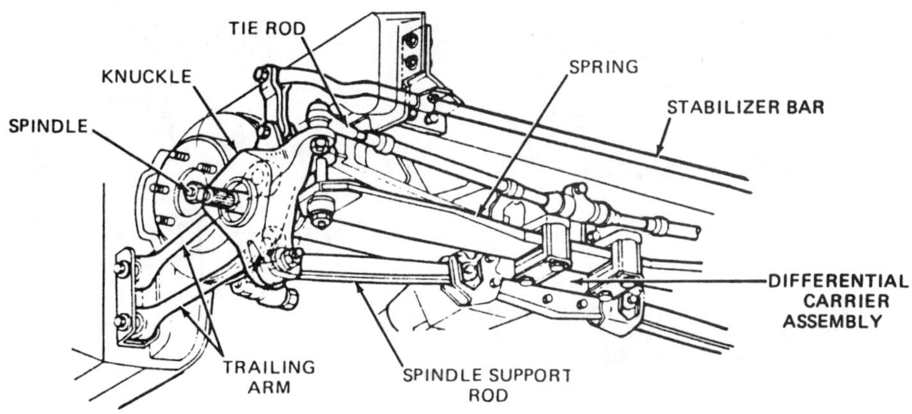

Fig. 3 Rear suspension. Corvette

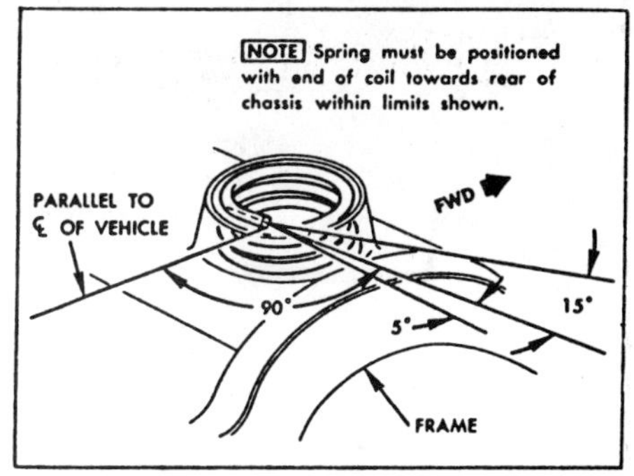

Fig. 4 Coil spring installation

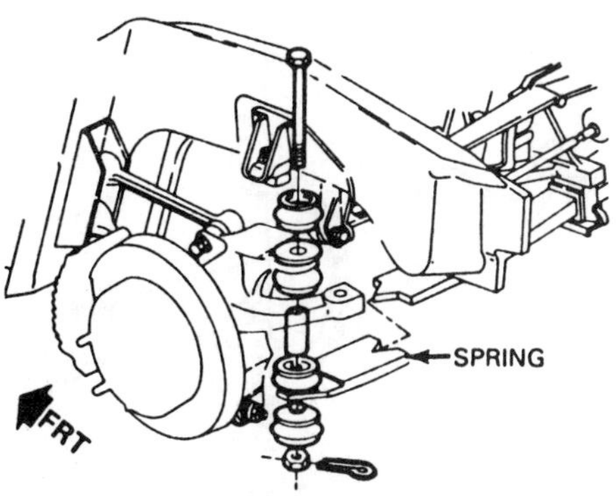

Fig. 5 Transverse leaf spring link assembly. Corvette

CAMARO & FIREBIRD

1. Raise and support vehicle and support rear axle with a suitable adjustable jack.
2. Remove track bar mounting bolt from axle and loosen track bar bolt at body brace.
3. Disconnect rear brake hose clip at underbody, then disconnect shock absorbers at lower mountings.
4. **On models equipped with 4-151 engine,** remove propeller shaft.
5. **On all models,** lower rear axle and remove springs and insulators.
6. Reverse procedure to install.

LEAF SPRINGS & BUSHINGS REPLACE

CORVETTE

1. Raise and support vehicle and remove one wheel and tire assembly.
2. Remove cotter pin, retaining nuts, bushings and link bolts securing spring to spindle support knuckles, **Fig. 5.**
3. Remove bolts securing spring to cover beam, spacers, insulators and spring, **Fig. 6.**
4. Reverse procedure to install. Torque bolts securing spring to cover beam to specifications.

LEAF SPRING SERVICE

The spring leaves are not serviced separately, however, the spring leaf inserts may be replaced.

1. Clamp spring in a vise and remove spring clips.
2. File peened end of center bolt to permit nut removal, remove nut and open vise slowly, allowing spring to expand.
3. Replace spring leaves or leaf inserts.
4. Use a drift to align center bolt holes, compress spring in vise and install new center bolt and nut. Peen end of bolt to retain nut.
5. Align springs and bend spring clips into position. **Over tightening of spring clips will cause spring binding.**

CONTROL ARMS & BUSHINGS REPLACE

EXCEPT CORVETTE

If more than one control arm is being replaced, remove and install one arm at a time to prevent axle assembly from slipping or twisting out of position.

CONTROL ARMS, REPLACE

1. Raise vehicle and support at frame pads. Support nose of axle housing to prevent assembly from twisting when control arm is removed.
2. If lower control arm is being replaced, remove bolts securing stabilizer bar to control arm, if equipped.
3. Remove bolts securing control arm to chassis and rear axle, and the control arm.
4. Reverse procedure to install, lower vehicle and torque control arm bolts to specifications, with vehicle at normal ride height. **All torque prevailing type fasteners must be torqued at the nut, not at the bolt, to ensure proper clamping force.**

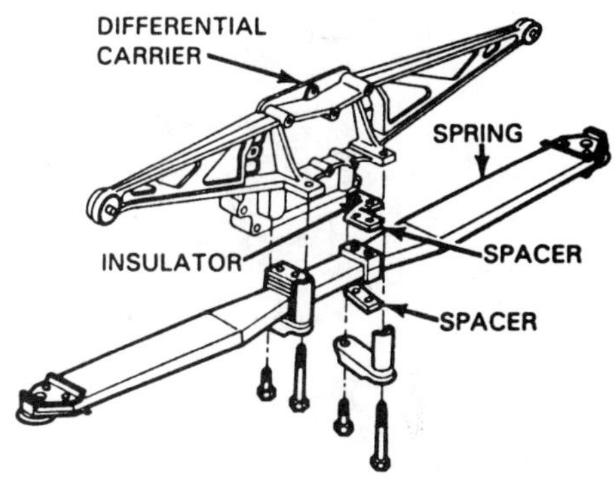

Fig. 6 Transverse leaf spring mounting. Corvette

BUSHING REPLACEMENT

Differential Carrier Bushings (Upper Control Arm Rear Bushing)

The upper control arm rear bushing, which is pressed into the differential carrier, can be replaced using the following procedure:

1. Raise vehicle and support at frame pads, and support nose of axle housing to prevent assembly from twisting.
2. Lower rear axle to obtain clearance, disconnect upper control arm from axle and position aside.
3. Install suitable bushing removal tool as shown in **Fig. 7,** tighten puller screw and press bushing out of housing.
4. To install replacement bushing, reverse position of removal tool and pull bushing into position by tightening screw, **Fig. 8.**

Control Arm Bushings

1. Raise and support vehicle and remove control arm as outlined previously.
2. Press bushings out of control arm using suitable tools as shown in **Fig. 9.**
3. Reverse procedure to install, ensuring bushing is properly seated in control arm, **Fig. 10. If replacement bushing fits loosely in control arm, or if mounting areas are damaged or deformed, control arm must be replaced.**

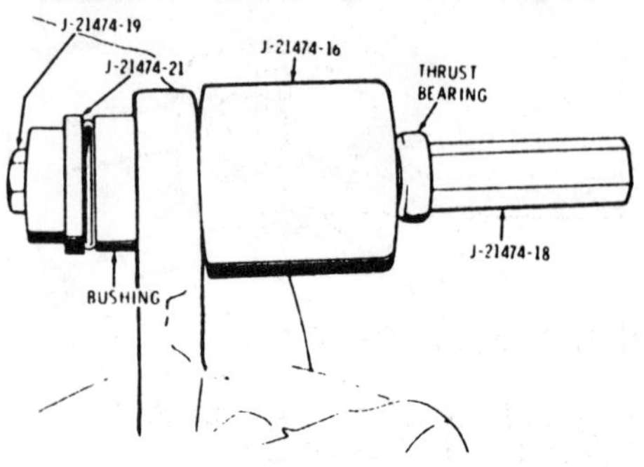

Fig. 7 Upper control arm rear bushing (differential carrier bushing) removal

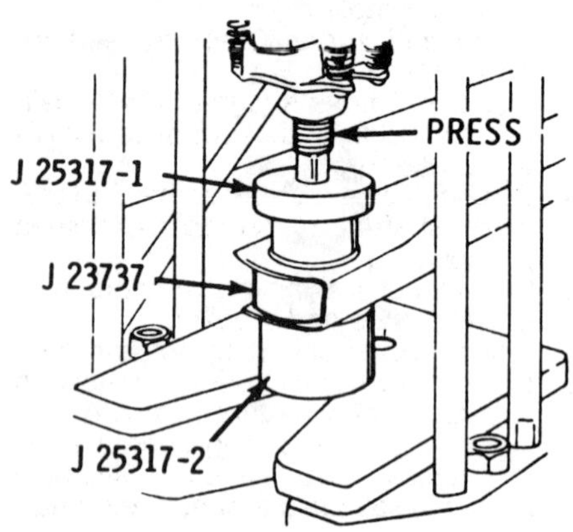

Fig. 10 Control arm bushing installation. Except Corvette

CORVETTE

Control Arms, Replace

1. Raise and support vehicle, and remove wheel and tire assembly.
2. Remove shock absorber. Use a back-up wrench on lower mounting stud when removing retaining nut.
3. Remove bolts securing control arm to spindle support knuckle.
4. Remove control arm bolt at mounting bracket and control arm.
5. Reverse procedure to install. Torque control arm to bracket bolt and control arm to knuckle bolt to specifications.

STABILIZER BAR REPLACE

EXCEPT CAMARO, CORVETTE & FIREBIRD

1. Support vehicle at rear axle.
2. Remove bolts securing stabilizer bar to lower control arms, **Fig. 11.**
3. Reverse procedure to install. Use spacer shims, if needed, placed equally on each side of stabilizer bar. Tighten attaching bolts with vehicle at curb height.

CAMARO & FIREBIRD

1. Raise and support vehicle.
2. Remove link bolt nuts, washers, bushings, spacers and link bolts securing

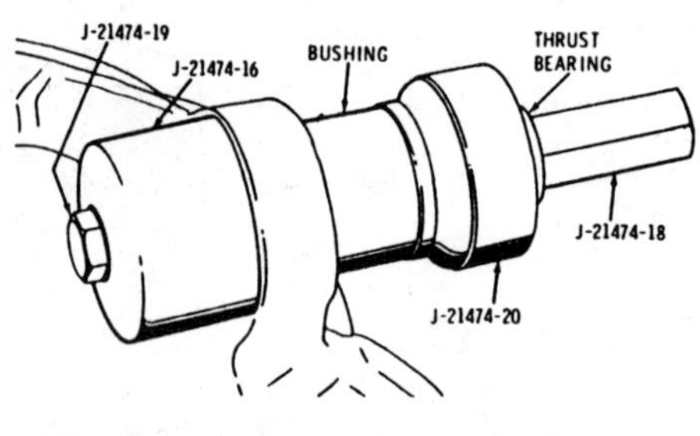

Fig. 8 Upper control arm rear bushing (differential carrier bushing) installation

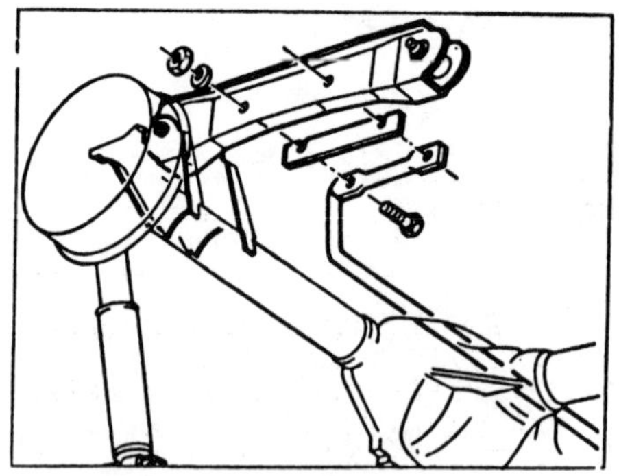

Fig. 11 Stabilizer bar installation. Except Camaro & Firebird

stabilizer to chassis, **Fig. 12.**
3. Remove clamps securing stabilizer shaft to rear axle and stabilizer shaft.
4. Reverse procedure to install. Torque link bolts and U-bolt nuts to specifications.

CORVETTE

1. Raise and support vehicle.
2. Remove spare tire and carrier.
3. Disconnect stabilizer links from spindle support knuckles.
4. Remove retainers securing shaft bushings to crossmember, bushings and stabilizer shaft.
5. Reverse procedure to install. Torque bushing retaining nuts and bolts securing stabilizer links to knuckles to specifications, and bolts securing links to stabilizer bar to specifications.

TRACK BAR & BRACE REPLACE

CAMARO & FIREBIRD

1. Raise vehicle and support rear axle at curb height.
2. Remove track bar mounting bolt and nut from rear axle and from body bracket, then remove track bar, **Fig. 13.**
3. Remove heat shield attaching screws from track bar brace.
4. Remove three track bar brace to body brace screws.
5. Remove nut and bolt from body

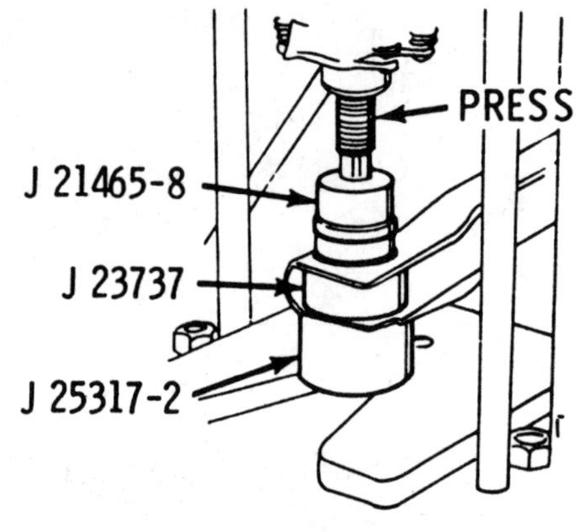

Fig. 9 Control arm bushing removal. Except Corvette

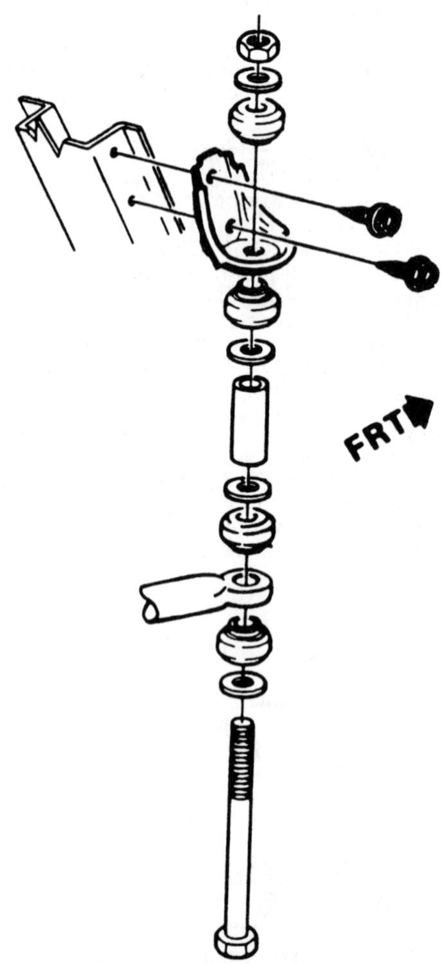

Fig. 12 Stabilizer link assembly. Camaro & Firebird

bracket, then remove track bar brace.
6. Reverse procedure to install.

STRUT ROD, SPINDLE SUPPORT ROD & TIE ROD REPLACE

CORVETTE

Spindle Support Rod

1. Raise and support vehicle, and re-

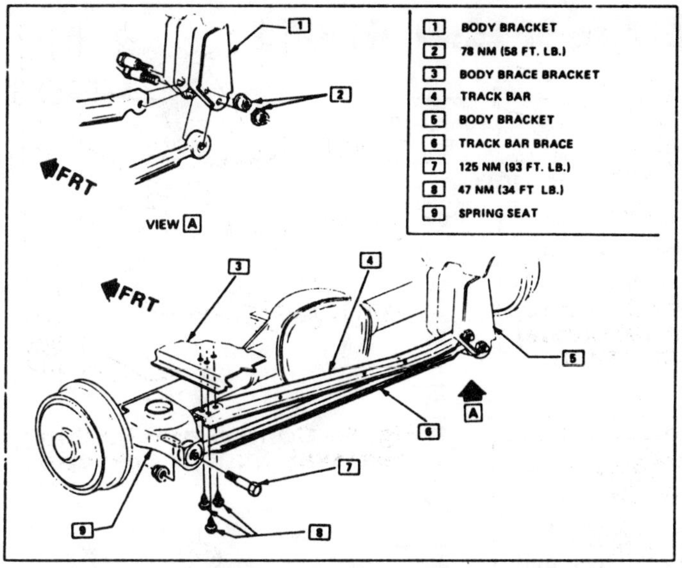

Fig. 13 Track bar & brace installation. Camaro & Firebird

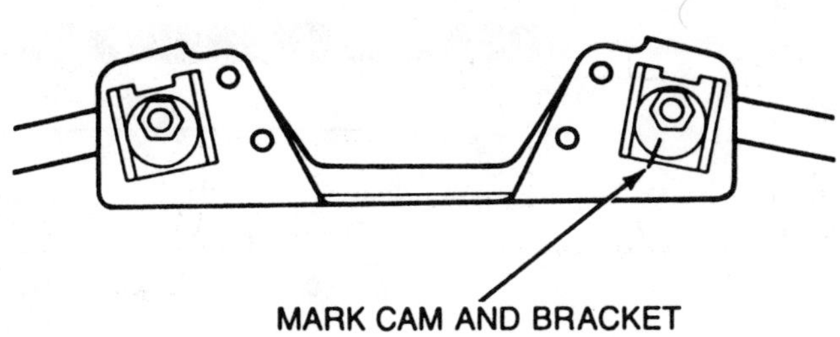

Fig. 14 Indexing adjustment cam bolt & bracket (Typical). Corvette

move wheel and tire.
2. Scribe mark between cam bolt and bracket for reassembly, **Fig. 14.**
3. Remove cam bolt and disconnect support rod from bracket.
4. Remove bolt securing spindle support rod to knuckle and rod.
5. Reverse procedure to install, then check rear suspension alignment. Torque retaining bolt at knuckle and cam bolt to specifications.

Tie Rod

1. Raise and support vehicle, and remove wheel and tire.
2. Remove cotter pin and nut securing tie rod to spindle support knuckle.
3. Press tie rod from knuckle using tool J-24319-01 or equivalent.
4. Remove tie rod from adjusting sleeve, counting number of turns necessary.
5. Reverse procedure to install, then check rear suspension alignment. Torque tie rod nut specifications.

SPINDLE SUPPORT KNUCKLE REPLACE

CORVETTE

1. Remove center cap from wheel, cotter pin and spindle nut.
2. Raise and support vehicle, and remove wheel and tire.
3. Remove 2 bolts securing brake caliper to knuckle, brake caliper assembly and brake rotor.
4. Disconnect tie rod, leaf spring and stabilizer shaft from knuckle as outlined previously.
5. Disconnect parking brake cable from backing plate and bracket.
6. Disconnect shock absorber and support rod from knuckle, using a back-up wrench on shock mounting stud.
7. Remove bolts securing control arms to knuckle, lower knuckle assembly and slide spindle out of hub and bearing.
8. Remove hub and bearing and parking brake assembly from knuckle, using a No. 45 Torx bit to remove hub retaining bolts, then remove splash shield.
9. Reverse procedure to install, then check rear suspension alignment. Torque all bolts to specifications given in individual component replacement procedures.

REAR SUSPENSION TIGHTENING SPECIFICATIONS

Year	Component	Torque/ft. lbs.
EXCEPT CAMARO, CORVETTE & FIREBIRD		
1985-90	Lower Control Arm Bushing Nut	92
	Lower Control Arm Bushing Bolt	122
	Shock Absorber Lower Nut (Exc. Cadillac)	48
	Shock Absorber Lower Nut (Cadillac)	65
	Shock Absorber Upper Nut	12
	Shock Absorber Upper Bolt	20
	Stabilizer Shaft To Bracket	20
	Upper Control Arm Front Bushing Nut	92
	Upper Control Arm Front Bushing Bolt	122
	Upper Control Arm Rear Bushing Nut	70
	Upper Control Arm Rear Bushing Bolt	80
CAMARO & FIREBIRD		
1985-90	Control Arm To Axle	85
	Control Arm To Underbody	85
	Shock Absorber To Axle Nut	70
	Shock Absorber Upper Mount Nut	13
	Stabilizer Shaft To Body Bracket	16
	Track Bar To Axle	59
	Track Bar To Body Bracket	80
	Torque Arm To Differential	98
CORVETTE		
1985-90	Cam Bolt To Spindle Rod	186
	Control Rods To Body	63
	Control Rods To Knuckle	140
	Differential Carrier To Body	100
	Shock Absorber To Body	22
	Shock Absorber To Knuckle	80
	Spindle Nut	164
	Spindle Rod To Knuckle	107
	Support Beam To Differential Carrier Bolt	60
	Support Beam To Transmission	37
	Stabilizer Link To Bracket	35
	Tie Rod Socket Nut	33
	Transverse Spring To Differential Carrier	37
	Trunnion Strap At Side Gear Yoke	26
	Trunnion Strap At Spindle Yoke	26
	Wheel Hub To Knuckle	66

Buick Century & Skylark (1985), Chevrolet Celebrity & Citation, Oldsmobile Cutlass Ciera & Cutlass Cruiser & Pontiac 6000

INDEX

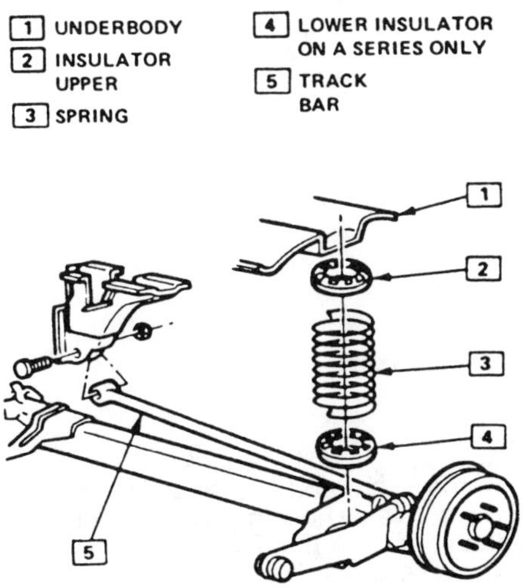

Fig. 1 Rear axle and suspension

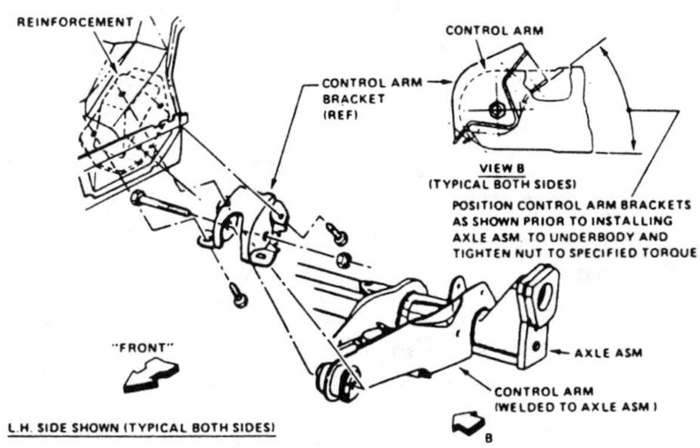

Fig. 2 Control arm bracket installation

DESCRIPTION

The rear suspension, **Fig. 1,** consists of a rear axle assembly, control arms, coil springs, shock absorbers, track bar and a track bar brace. The rear axle is of the trailing arm type design. A non-serviceable stabilizer bar is welded to the inside of the axle housing and is an integral part of the axle assembly. A single unit hub and bearing assembly is bolted to each end of the axle assembly. The hub and bearing assembly is a sealed unit and must be replaced as an assembly.

REAR AXLE

REPLACE

When removing rear axle assembly, do not use a twin-post type hoist. The swing arc tendency of this axle may cause it to slip from hoist. Perform axle removal with a floor jack, if necessary.

1. Raise rear of vehicle and support rear axle using a suitable jack.
2. Remove rear wheel assembly and brake drum. **Do not hammer on brake drum since damage to bearings may result.**
3. Disconnect parking brake cable from equalizer, then remove brake line brackets from frame.
4. Disconnect shock absorber from lower mountings on axle housing, then disconnect shock absorber from axle.
5. Remove track bar attaching nut and bolt, then disconnect track bar. **Do not suspend rear axle by brake hoses since damage to hoses may result.**
6. Carefully lower rear axle assembly, then remove coil spring and insulators.
7. Disconnect brake lines from control arm attachments.
8. Disconnect parking brake cable from rear axle attachments.
9. Remove hub attaching bolts, then the hub and bearing assembly. Position backing plate aside.
10. Remove control arm bracket to underbody attaching bolts (four each side), then lower axle assembly and remove from vehicle.
11. Reverse procedure to install, noting the following:
    a. If control arm brackets were removed from control arms, **torque** attaching nuts to 78 ft. lbs.
    b. Install control arm bracket at a 40-44° angle as shown in **Fig. 2.**
    c. **Torque** control arm-to-underbody attaching bolts to 28 ft. lbs.
    d. **Torque** track bar attaching nut to 35 ft. lbs.

HUB & BEARING ASSEMBLY

REPLACE

1. Raise and support rear of vehicle, then remove wheel and tire assembly and brake drum. **Do not hammer on brake drum since damage to bearing may result.**
2. Remove four hub and bearing assembly to rear axle attaching bolts, then

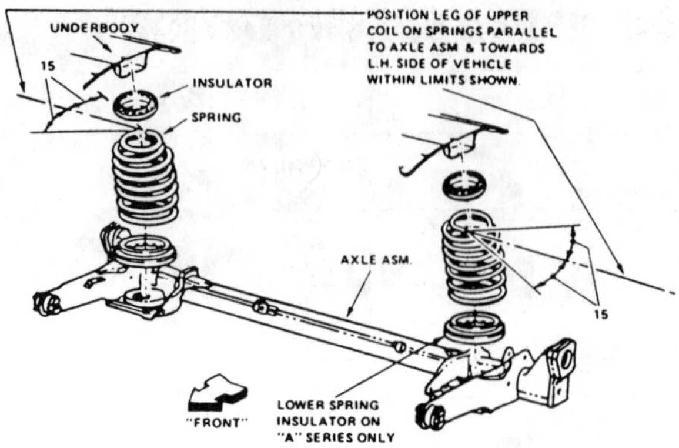

Fig. 3 Coil spring & insulator installation

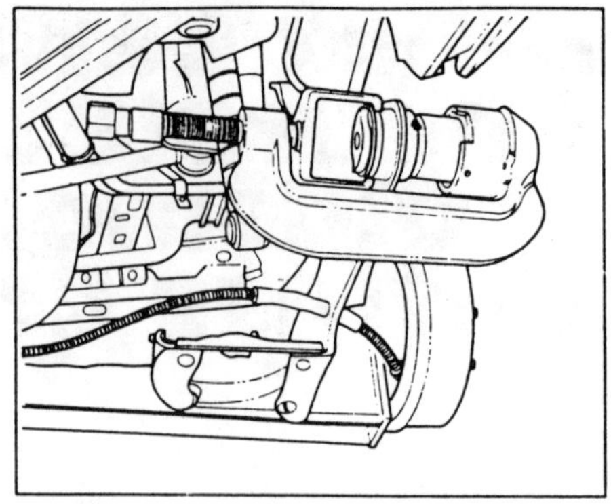

Fig. 4 Control arm bushing removal

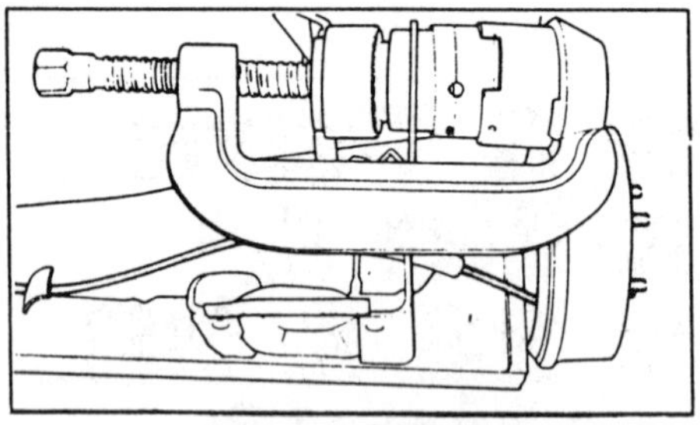

Fig. 5 Control arm bushing installation

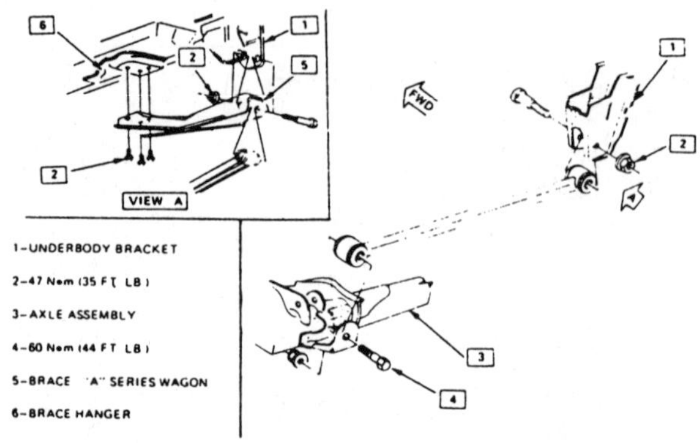

Fig. 6 Track bar installation

the hub and bearing assembly from axle.

3. Reverse procedure to install. **Torque hub and bearing assembly to rear axle attaching bolts to 45 ft. lbs. Use care not to drop hub and bearing assembly since damage to bearing may result.**

COIL SPRING
REPLACE

1. Raise rear of vehicle and support rear axle using a suitable jack.
2. Remove right and left brake line bracket attaching bolts from frame and allow brake lines to hang freely.
3. Remove track bar attaching nut and bolt, then disconnect track bar.
4. Disconnect shock absorbers at lower mountings. **Do not suspend rear axle by brake hoses since damage to hoses may result.**
5. Carefully lower rear axle assembly, then remove springs and insulators.
6. Reverse procedure to install, **Fig. 3.**

CONTROL ARM BUSHING
REPLACE

1. Raise rear of vehicle and support rear axle under front side of spring seat using a suitable jack.
2. If right hand side bushing is to be replaced, disconnect parking brake cable from equalizer.
3. Remove parking brake cables from bracket attachment, then position out of way.
4. Disconnect brake line bracket from frame.
5. Disconnect shock absorber from lower mounting, then pull spring out of way.
6. Remove four control arm to underbody attaching bolts, then allow control arm to rotate downward.
7. Remove nut and bolt from bracket attachment and the bracket.
8. The bushing can now be replaced using tools shown in **Figs. 4 and 5.** When installing bushing, press bushing inward until end of bushing is aligned with scribed line on tool J-28685-2, **Fig. 5.**
9. Reverse procedure to install control arm. Install bracket to control arm as shown in **Fig. 2.**

TRACK BAR
REPLACE

1. Raise rear of vehicle and support rear axle using a suitable jack.
2. Remove nut and bolt attaching track bar from axle housing and underbody, then the track bar, **Fig. 6.**
3. Reverse procedure to install, **Fig. 6. Open side of track bar must face rear of vehicle. Also, nut must be at rear of attachments at both axle and underbody attachments.**

SHOCK ABSORBER
REPLACE

1. Open deck lid and remove trim cover, then remove shock absorber upper attaching nut.
2. Raise rear of vehicle and support rear axle using a suitable jack.
3. Disconnect shock absorber from lower attachment and remove shock absorber from vehicle.
4. Reverse procedure to install. **Torque** shock absorber lower attaching nut to 44 ft. lbs. and shock absorber upper attaching nut to 16 ft. lbs.

REAR SUSPENSION TIGHTENING SPECIFICATIONS

Year	Component	Torque/Ft. Lbs.
1985–90	Control Arm Bracket To Underbody	28
	Control Arm To Bracket Nut	84
	Hub & Bearing Assembly Bolts	44
	Shock Absorber Lower Nut	44
	Shock Absorber To Mount	28
	Shock Absorber Upper Nut	16
	Track Bar Bolt At Axle	44
	Track Bar Nut At Brace Or Underbody Bracket	35

Buick Electra, LeSabre & Park Avenue, Cadillac Deville & Fleetwood, Oldsmobile 88 & 98 & Pontiac Bonneville Front Wheel Drive Models

INDEX

DESCRIPTION

These vehicles use an independent rear suspension consisting of lower control arms, coil springs, toe links, suspension knuckles, superlift struts and stabilizer bar, **Fig. 1.** The hub and wheel bearing is an assembly and does not require periodic lubrication.

HUB & BEARING ASSEMBLY

REPLACE

1. Raise and support rear of vehicle. Remove wheel assembly and brake drum. **Do not hammer on drum since damage to bearing may occur.**
2. Remove four hub and bearing assembly to axle attaching bolts and the assembly from axle, **Fig. 2.**
3. Reverse procedure to install. Torque attaching bolts to specifications.

COIL SPRING

REPLACE

1. Support vehicle so the rear wheel and control arm hang free. Remove rear wheel.
2. Disconnect rear stabilizer bar from knuckle bracket.
3. Disconnect electronic level control height sensor link if removing right control arm.
4. Disconnect parking brake cable clip from frame if removing left control arm.
5. Place tool J-23028-01 in position to cradle control arm bushings, **Fig. 3. Tool J-23028-01 should be secured to a suitable jack, otherwise, personal injury could result.**
6. Raise jack to relieve tension from control arm pivot bolts.
7. Place a chain around spring and control for safety.
8. Remove rear control arm pivot bolt and nut, **Fig. 4.**
9. Slowly lower jack until front bolt and

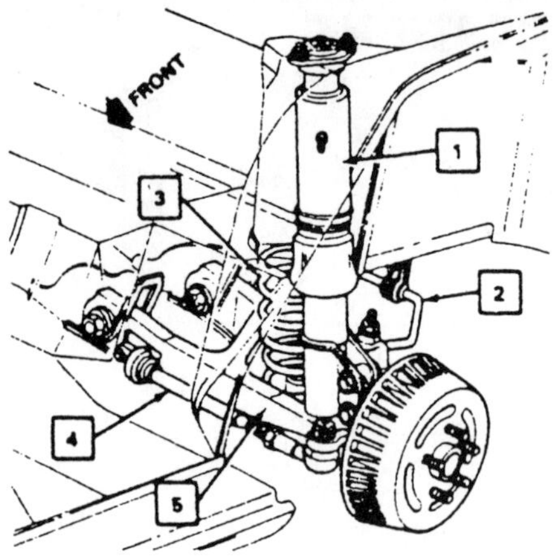

1 SUPERLIFT® STRUT
2 STABILIZER BAR
3 COIL SPRING
4 SUSPENSION ADJUSTMENT LINK
5 LOWER CONTROL ARM

Fig. 1 Rear suspension

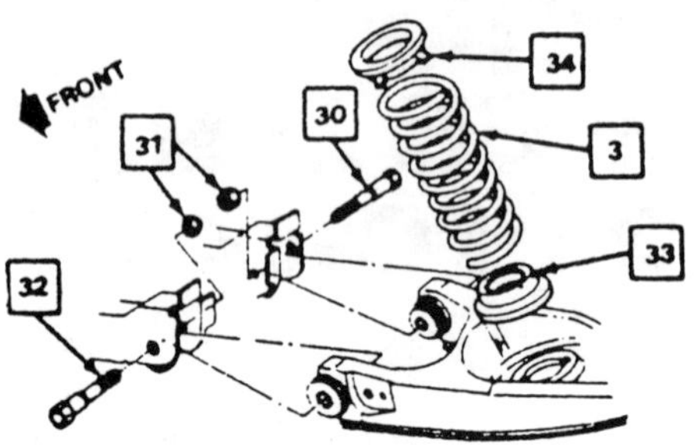

3 COIL SPRING
30 CONTROL ARM PIVOT BOLT-REAR (170 N·m/125 FT. LB.)
31 CONTROL ARM PIVOT NUTS (115 N·m/85 FT. LB.)
32 CONTROL ARM PIVOT BOLT-FRONT (170 N·m/125 FT. LB.)
33 LOWER COIL SPRING INSULATOR
34 UPPER COIL SPRING INSULATOR

Fig. 4 Coil spring replacement

nut can be removed.

10. Remove coil spring. **Do not apply force on lower arm and ball joint to remove spring. Maneuver spring to remove.**
11. Reverse procedure to install, noting the following:
    a. Replace spring insulators that are damaged or if vehicle has been in service for more than 50,000 miles
    b. Install springs so that upper pigtail end of left spring faces rear of vehicle, or upper pigtail end of right spring faces front of vehicle when vehicle is at normal ride height.
    c. Tighten suspension fasteners sufficiently to retain position of components, lower vehicle so that it rests on the wheels at normal ride height, then torque fasteners as follows: 1-Torque control arm pivot nuts to specifications 2-Torque control arm pivot bolts to specifications 3-Torque stabilizer support bolt to specifications. **Failure to torque fasteners to specifications in listed sequence may adversely affect ride and handling.**

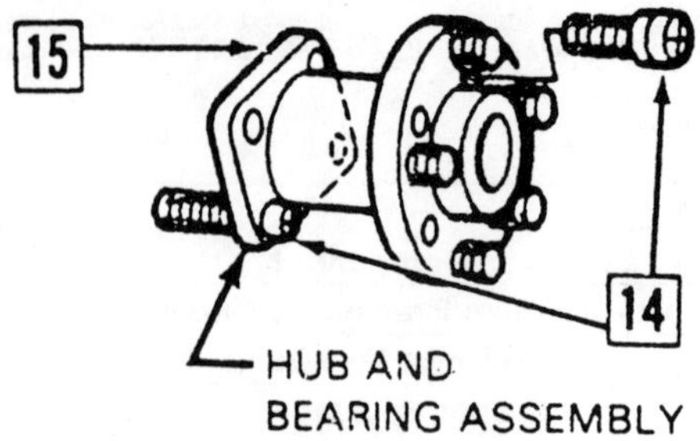

Fig. 2 Hub & bearing assembly

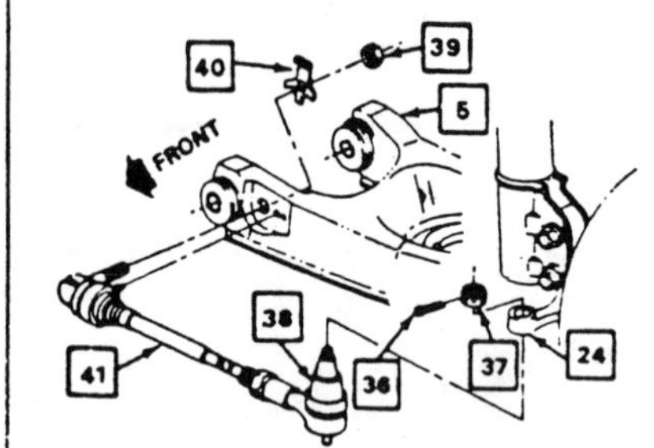

5 REAR CONTROL ARM
24 KNUCKLE
36 COTTER PIN
37 CASTELLATED NUT (50 N·m/37 FT. LB.)
38 OUTER SUSPENSION ADJUSTMENT LINK
39 LINK RETAINING NUT (85 N·m/63 FT. LB.)
40 LINK RETAINER
41 SUSPENSION ADJUSTMENT LINK ASSEMBLY

Fig. 5 Tie rod/adjustment link installation

REAR TIE ROD/ADJUSTMENT LINK REPLACE

1. Raise and support vehicle.
2. Remove wheel assembly, cotter key and castle nut, **Fig. 5.**
3. Using tool J-24319-01, disconnect outer tie rod/adjustment from knuckle. **When disconnecting the tie rod/adjustment from knuckle, do not use a wedge since seal damage will occur.**
4. Remove rod/link assembly retaining nut and retainer.
5. Remove rod/link assembly from lower control arm.
6. Reverse procedure to install. **Torque** link retaining nut to specifications and castellated nut securing ball stud to 37 ft. lbs. Install cotter pin retaining castellated nut, tightening nut as needed to insert pin through hole in stud. Do not loosen nut to align slots with hole.

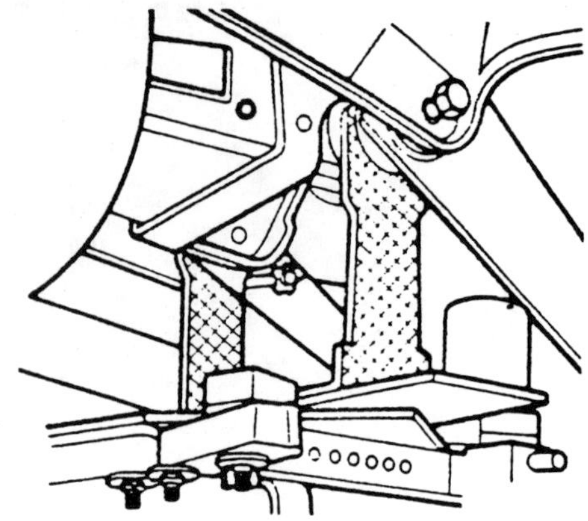

Fig. 3 Installing tool J-23028-01

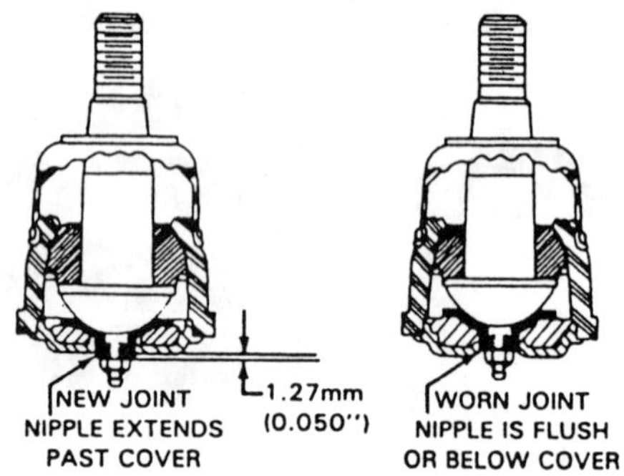

Fig. 6 Ball joint inspection

BALL JOINT INSPECTION

The ball joint has a visual wear indicator. Checking the condition of the ball joint is a simple procedure but must be followed accurately to prevent unnecessary ball joint replacement.

The vehicle must be supported by the wheels during inspection to ensure that the vehicle weight is properly loading the ball joints.

The ball joint is inspected for wear by visual observation alone. Wear is indicated by retraction of the 1/2 inch diameter nipple into the ball joint cover (the ball joint grease fitting is threaded into this nipple). Raise rear of vehicle and support rear axle using a suitable jack.2.Remove nut and bolt attaching track bar from axle housing and underbody, then the track bar, **Fig. 6.3.** Reverse procedure to install, **Fig. 6. Open side of track bar must face rear of vehicle. Also, nut must be at rear of attachments at both axle and underbody attachments.**

SHOCK ABSORBER REPLACE

1. Open deck lid and remove trim cover,

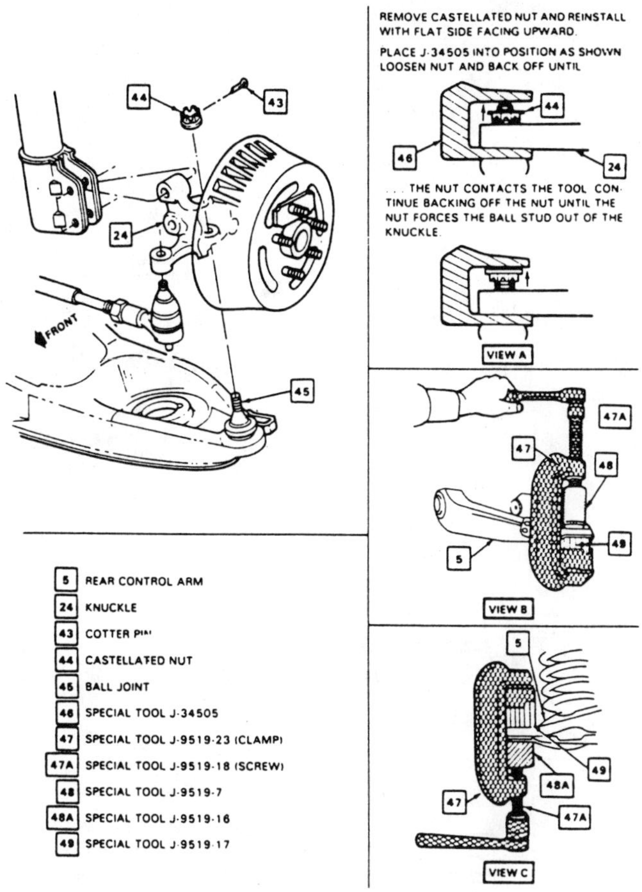

Fig. 7 Ball joint replacement

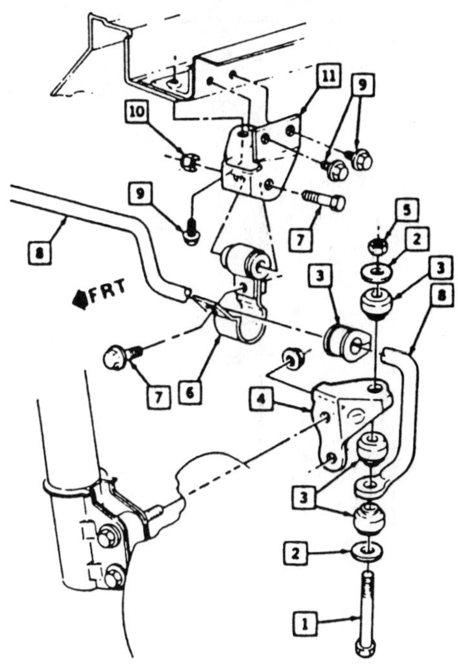

1. BOLT
2. RETAINER
3. INSULATOR
4. STABILIZER SHAFT BRACKET
5. NUT – 17 N·m (13 LBS. FT.)
6. LINK
7. BOLT 50 N·m (37 LBS. FT.)
8. STABILIZER SHAFT
9. BOLT – 17 N·m (13 LBS. FT.)
10. NUT – 47 N·m (35 LBS. FT.)
11. BRACKET

Fig. 8 Stabilizer bar & bushing assembly

then remove shock absorber upper attaching nut.
2. Raise rear of vehicle and support rear axle using a suitable jack.
3. Disconnect shock absorber from lower attachment and remove shock absorber from vehicle.
4. Reverse procedure to install. **Torque** shock absorber lower attaching nut to 44 ft. lbs. and shock absorber upper attaching nut to 16 ft. lbs.

The nipple protrudes .050 inch beyond the surface of the ball joint cover on a new unworn joint. Normal wear will result in the surface of this nipple retracting very slowly inward. The ball joint should be replaced if the nipple is flush or below the cover surface, **Fig. 6.**

Ball stud tightness in the knuckle boss should also be checked when inspecting the ball joint. This may be done by shaking the wheel and feeling for movement of the stud end or castellated nut at the knuckle boss.

Checking the fastener tightness at the castellated nut is an alternative method of inspecting (a loose nut can indicate a bent stud or an "opened up" hole in the knuckle boss). If worn, the ball joint and knuckle must be replaced.

If the ball joint is separated from the knuckle for suspension service, the ball joint seal should be inspected for damage. A damaged seal will cause joint failure. If seal damage is found the ball joint should be replaced.

LOWER CONTROL ARM BALL JOINT
REPLACE

1. Raise and support vehicle.
2. Remove wheel assembly, cotter key and castle nut.
3. Disconnect outer tie rod/adjustment link from knuckle as outlined previously.
4. Disconnect electronic level control height sensor link from right side control arm.
5. Support control arm with a suitable jack.
6. Remove ball joint cotter pin and nut.
7. Using tool J-29330, disconnect ball joint from knuckle, **Fig. 7.**
8. Using tool J-9519-7, remove ball joint from control arm.
9. Reverse procedure to install.

STABILIZER BAR
REPLACE

1. Remove rear wheels and tires.
2. Remove stabilizer shaft support bolt, nut, retainer and insulators from knuckle bracket, **Fig. 8.**
3. Remove bushing clip bolt.
4. Bend open end of support assembly downward.
5. Remove stabilizer shaft and bushings.

LOWER CONTROL ARM
REPLACE

1. Remove coil spring as outlined under "Coil Spring, Replace" procedure.
2. Remove control arm front pivot bolt.
3. Remove control arm from vehicle.
4. Reverse procedure to install.

REAR STRUT
REPLACE

1. Raise and support vehicle.
2. Remove trunk side cover.
3. Remove wheel assembly.
4. Disconnect air tube from superlift strut, **Fig. 9.**
5. Remove two strut tower mount nuts.
6. Remove two strut anchor bolts, washers and nuts from knuckle, then the knuckle bracket.
7. Remove strut from vehicle.
8. Reverse procedure to install.

CONTROL ARM BUSHINGS
REPLACE

These vehicles incorporate control arm configuration with two different sized

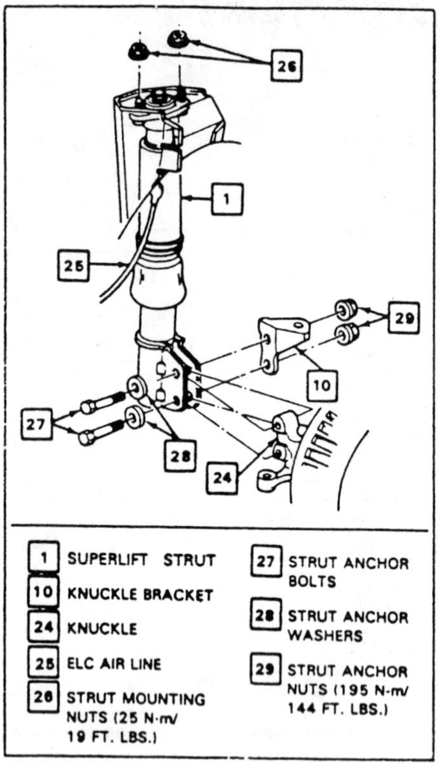

Fig. 9 Strut assembly

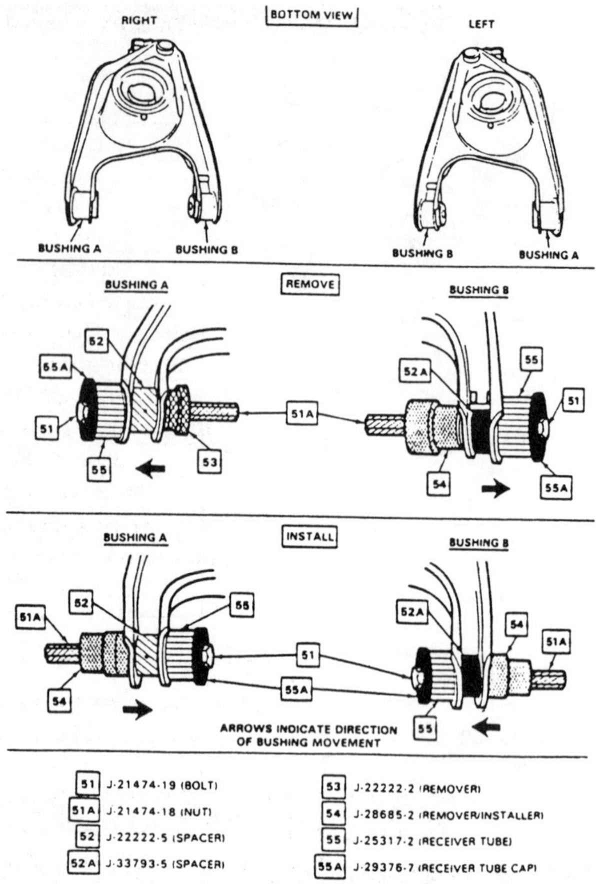

Fig. 10 Control arm bushing removal/installation

bushings, **Fig. 10.** These bushings are not interchangeable, and require different combinations of special tools for removal and installation. The following service procedure refer to these bushings as A and B.

REMOVAL

1. Remove control arm from vehicle.
2. Place control arm onto a suitable work bench, up side down.
3. For bushing A, install spacer tool J-22222-5, for bushing B, install spacer tool J-33793-5 onto control arm.
4. Position receiver tube J-25317-2 and cap tool J-29376-7 on outer side of control arm. Ensure receiver tube does not contact the bushing flange.
5. Coat threaded area of long bolt J-21474-19 with extreme pressure lubricant J-23444-A, then install bolt through receiver, cap and bushing.
6. Install remover tool J-22222-2 for bushing A or tool J-28685-2 for bushing B onto bolt at inner side of control arm. Position remover so small diameter portion contacts bushing.
7. Place bearing on bolt and install long nut tool J21474-18. Bearing must be installed between nut and remover.
8. Draw bushing out of control arm by tightening the nut.

INSTALLATION

1. With control arm up side down on work bench, position new bushing into control arm. **When installing new bushings, the flanged end of the bushing must face outward when installed in the control arm.**
2. Install spacer tool J-22222-5 for bushing A or tool J-33793-5 for bushing B.
3. Position receiver tube J-25317-2 and cap tool J-29376-7 onto inner side of control arm. Receiver tube should be centered over the hole.
4. Coat threaded area of long bolt with extreme pressure lubricant J-23444-A and install through receiver, cap and bushing.
5. Position installer J-28685-2 onto bolt at outer side of control arm. Installer should be positioned with large diameter end contacting bushing flange.
6. Place bearing on bolt and install long nut J-21474-18. Bearing must be positioned between nut and installer.
7. Draw bushing into control arm by tightening nut. Tighten nut until bushing flange seats firmly against control arm.
8. Install control arm into vehicle, adjust rear wheel alignment, if necessary.

REAR SUSPENSION TIGHTENING SPECIFICATIONS

Year	Component	Torque/ft. lbs.
1985-90	Ball Joint Nut	88①②
	Control Arm Pivot Bolts	135
	Control Arm Pivot Nuts	85
	Hub & Bearing Assembly Bolts	52
	Stabilizer Shaft Bushing Assembly Nut	37
	Stabilizer Shaft Bushing Clip Bolt	37
	Stabilizer Shaft Link Nut	160
	Stabilizer Shaft Mounting Bracket Bolt	160
	Strut To Knuckle Bolts	147
	Strut Mount To Underbody	18
	Suspension Adjustment Link At Knuckle	33
	Suspension Adjustment Link At Control Arm	63
	Suspension Adjustment Link Locknut	48
	Top Strut Mount Nut	220

①—Inch lbs.
②—Tighten nut an additional 2/3 turn after torquing.

1988–90 Buick Regal, Oldsmobile Cutlass Supreme & Pontiac Grand Prix & 1990 Chevrolet Lumina

INDEX

DESCRIPTION

These vehicles use a tri-link independent rear suspension system with a transverse leaf spring and tubular struts with large lateral links attached to the body crossmember. The three mounting points are the crossmember, strut tower and trailing arm. The cross member is stamped steel and the composite fiberglass mono leaf spring is transversely mounted to the under side of the crossmember, with its padded ends free riding on the cast knuckle assembly.

REAR HUB & BEARING ASSEMBLY

REPLACE

The rear hub and bearing assembly is not serviceable. If the hub and/or bearing is damaged, the complete assembly must be replaced.

1. Raise and support vehicle, then remove tire and wheel assembly.
2. Remove brake caliper, leaving hose attached, and suspend out of way.
3. Remove brake rotor.
4. Disconnect anti-lock brake system electrical harness connector, if equipped.
5. Remove hub and bearing assembly mounting bolts, then the assembly.
6. Reverse procedure to install, torquing hub and bearing assembly mounting bolts to specifications.

AUXILIARY SPRING ASSEMBLY

REPLACE

1. Raise and support vehicle, then re-

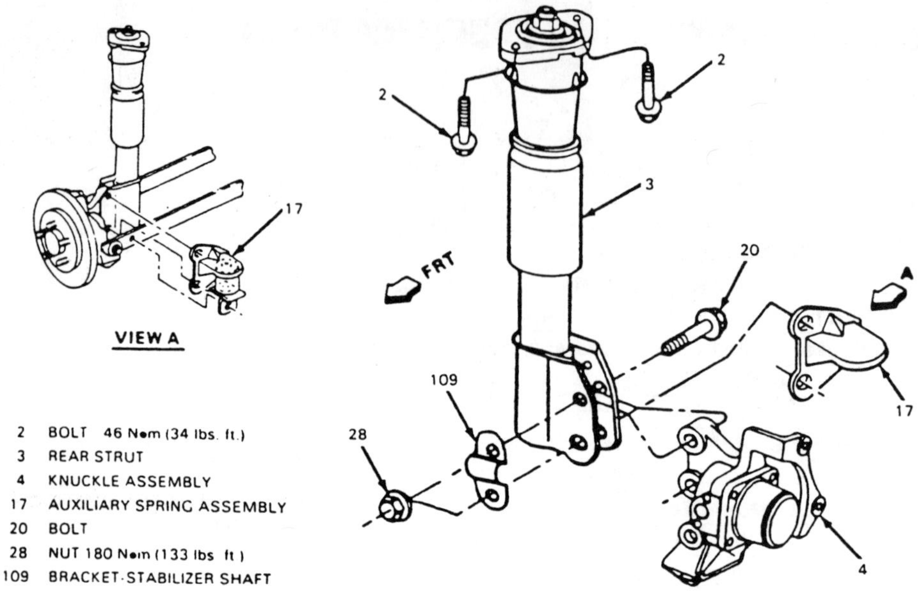

Fig. 1 Auxiliary spring replacement

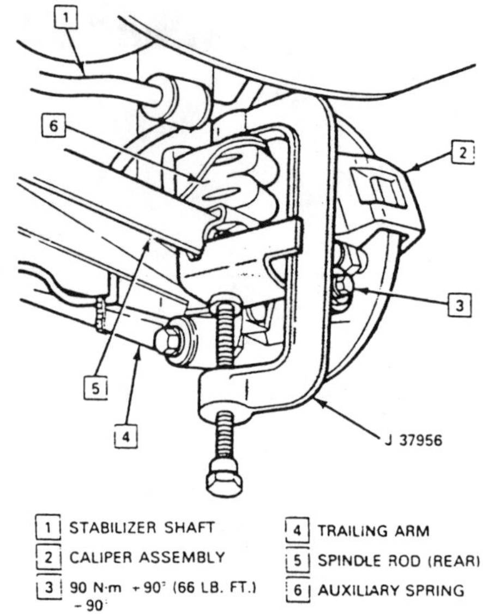

Fig. 2 Compressing auxiliary spring

move wheel and tire assembly.

2. Remove leaf spring rear retention plate bolt.
3. Remove leaf spring retention bolt only enough to rotate plate clear of rod.
4. Install rear auxiliary spring compressor tool No. J-37956 or equivalent ensuring pin in stationary end of clamp is inserted in hole of upper auxiliary spring bracket, **Figs 1 and 2.**
5. Remove plug from upper bracket, then seat rod in tool channel and hand tighten.
6. Remove rod-to-knuckle bolt, then loosen tool forcing screw to allow spring to expand.
7. Remove auxiliary spring attaching bolts, then the spring and tool. **When removing auxiliary spring, ensure rod/bushing clears transverse spring and boss on knuckle.**
8. Reverse procedure to install, noting the following:
   a. Compress auxiliary spring with spring compressor tool enough to install rod-to-knuckle bolt.
   b. Install rod-to-knuckle bolt, using Loctite or equivalent. Torque to specifications.
   c. Properly position leaf spring retention plate, install bolt and torque to specifications.

TRANSVERSE SPRING ASSEMBLY

REPLACE

Do not use corrosive cleaning agents, engine degreaser or solvents on or near the fiberglass leaf spring.

1. Raise and support vehicle.
2. Remove three jack pad attaching bolts and the pad.
3. Remove left and right spring retention plate attaching bolts.
4. Remove right side trailing link to knuckle attaching nut and bolt.

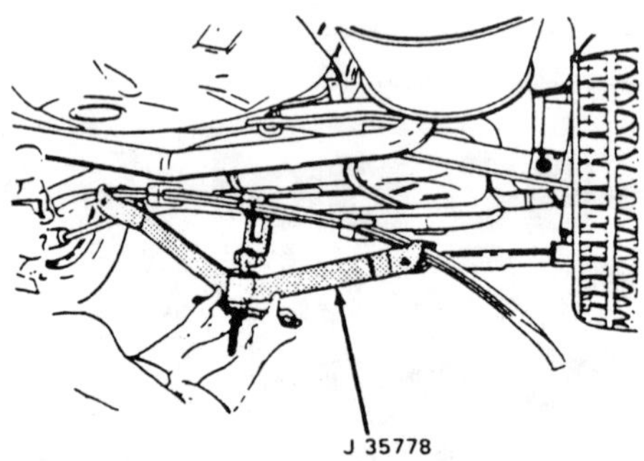

Fig. 3 Removing transverse spring assembly

5. Disconnect anti-lock brake system electrical harness connector, if equipped.
6. Assemble rear spring leaf spring compressor tool No. J-35778 or equivalent to transverse spring.
7. Loosen tool handle to base of tool shank.
8. Hang center shank of tool at spring center. It is not necessary to separate tool body and shank. **Attach from front side of vehicle only.**
9. Attach tool body to spring. **Always center spring on roller of tool.**
10. Fully compress spring, then slide spring to left side. It may be necessary to pry the spring, using a screwdriver/pry bar, to the opposite side, using the wheel for leverage.
11. Relax the spring until removal clearance is achieved, then remove spring, **Fig. 3.**
12. Reverse procedure to install, noting the following:
    a. Torque trailing link to knuckle attaching bolt to specifications.
    b. Torque spring retention plate mounting bolts to specifications. **Rear spring retention plates are designed with tabs on one end. Tabs must be aligned with support assembly.**
    c. Torque jack pad mounting bolts to specifications.

STABILIZER SHAFT ASSEMBLY

REPLACE

1. Raise and support vehicle, then remove rear wheel and tire assemblies.
2. Remove right and left hand stabilizer shaft link bolts, then open brackets to remove insulator.
3. Remove right and left hand strut to knuckle to stabilizer shaft nuts, **Fig. 4. Do not remove strut to knuckle bolts.**
4. Remove insulator brackets from bolts and from stabilizer shaft, then the stabilizer shaft.
5. Reverse procedure to install, noting the following:
   a. It may be necessary to pry stabilizer shaft to one side for installation clearance at strut. Use caution when prying.
   b. Torquing stabilizer shaft link bolts and strut to knuckle bolts to specifications.

TRI-LINK SUSPENSION ASSEMBLY

REPLACE

TRAILING LINK

1. Raise and support vehicle.
2. Disconnect anti-lock brake system electrical harness connector, if equipped.
3. Remove trailing link to knuckle attaching nut and bolt, then the trailing link to body attaching nut and bolt, **Figs. 5 and 6.**

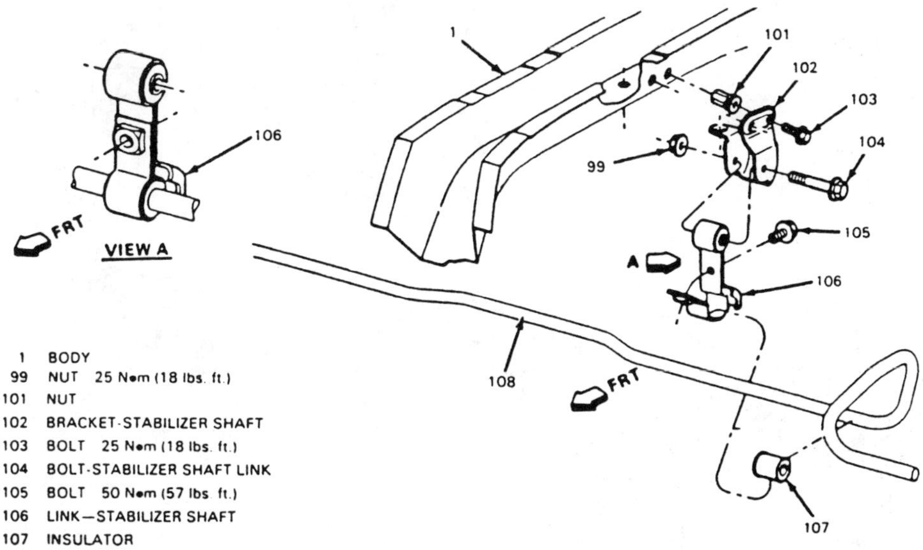

1 BODY
99 NUT 25 N•m (18 lbs. ft.)
101 NUT
102 BRACKET-STABILIZER SHAFT
103 BOLT 25 N•m (18 lbs. ft.)
104 BOLT-STABILIZER SHAFT LINK
105 BOLT 50 N•m (57 lbs. ft.)
106 LINK—STABILIZER SHAFT
107 INSULATOR
108 STABILIZER SHAFT

Fig. 4 Stabilizer shaft assembly replacement

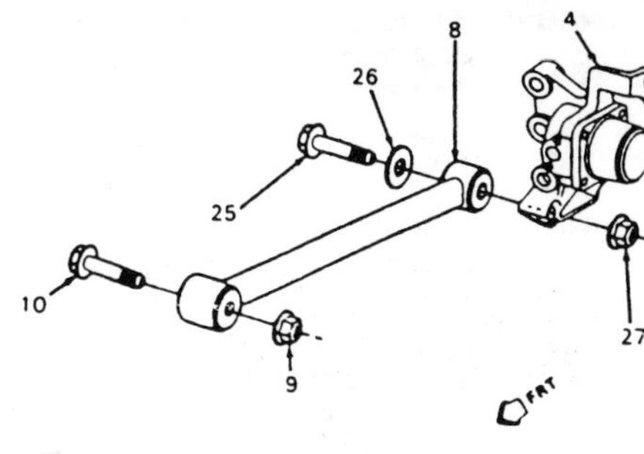

4 KNUCKLE ASSEMBLY
8 TRAILING LINK
9 65 N•m (48 lbs. ft.)
10 BOLT
25 BOLT
26 WASHER
27 NUT 260 N•m (192 lbs. ft.)

Fig. 6 Exploded view of trailing link

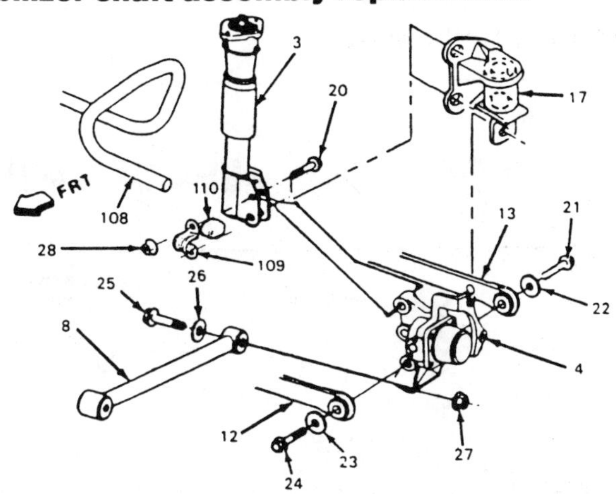

3 REAR STRUT
4 KNUCKLE ASSEMBLY
8 TRAILING ARM
12 FRONT LATERAL LINK
13 REAR LATERAL LINK
17 AUXILIARY SPRING
20 BOLT
21 BOLT 213 N•m (157 lbs. ft.)
22 WASHER
23 WASHER
24 BOLT 213 N•m (157 lbs. ft.)
25 BOLT
26 WASHER
27 NUT 260 N•m (192 lbs. ft.)
28 NUT 180 N•m (133 lbs. ft.)
108 STABILIZER SHAFT
109 BRACKET-STABILIZER SHAFT
110 INSULATOR

Fig. 5 Exploded view of tri-link suspension assembly

4. Remove trailing link.
5. Reverse procedure to install, torquing trailing link to body nut and trailing link to knuckle nut to specifications.

FRONT LATERAL LINK

1. Raise and support vehicle, then remove tire and wheel assembly.
2. Remove rod to knuckle bolt, then the exhaust pipe heat shield.
3. Lower to fuel tank to gain access.
4. Remove lateral link to knuckle attaching bolt.
5. Remove lateral link to suspension crossmember attaching nut and bolt, then the lateral link.
6. Reverse procedure to install, **torquing** lateral link to suspension crossmember nut to 140 ft. lbs. and the lateral link to knuckle attaching bolt to 157 ft. lbs.

REAR LATERAL LINK

1. Raise and support vehicle, then remove tire and wheel assembly.
2. Remove transverse spring as previously outlined.
3. Remove lower auxiliary spring bracket from rod.
4. Remove rear lateral link attaching nut from crossmember.
5. Push bolt forward enough to provide clearance for link removal, then remove rear lateral link.
6. Reverse procedure to install, **torquing** lateral link to crossmember attaching nut to 157 ft. lbs. and the rear lateral link to knuckle attaching bolt to 140 ft. lbs.

KNUCKLE ASSEMBLY REPLACE

1. Raise and support vehicle, then remove tire and wheel assembly.
2. Place scribe marks on strut and knuckle to ensure installation in same position.
3. Remove jack pad.
4. Install rear leaf spring compressor tool No. J-35778 or equivalent onto transverse spring assembly and tighten to hold spring pressure. **Do not remove spring or retention plates.**
5. Remove auxiliary spring as previously outlined.
6. Disconnect front link from knuckle.
7. Remove brake caliper, leaving hose attached, and suspend caliper out of way.
8. Remove brake rotor, then the hub and bearing assembly as previously described.
9. Disconnect anti-lock brake system electrical harness connector, if equipped.
10. Disconnect trailing link from knuckle, **Fig.7.**
11. Remove strut to knuckle attaching nuts, then the stabilizer shaft bracket.
12. Remove strut to knuckle attaching bolts, then the knuckle.
13. Reverse procedure to install, torquing strut to knuckle attaching bolts and trailing link retaining nut to specifications.

STRUT ASSEMBLY REPLACE

1. Remove upper strut mount cover from inside rear compartment, then raise and support vehicle and remove tire and wheel assembly.
2. Install rear leaf spring compressor tool No. J-35778 or equivalent onto transverse spring assembly and tighten to hold spring pressure. **Do not re-**

move spring or retention plates.

3. Scribe strut and knuckle to ensure installation in same position.
4. Remove jack pad.
5. Remove auxiliary spring as previously outlined.
6. Remove brake caliper, leaving hose attached, and suspend caliper out of way.
7. Remove brake rotor, then the brake hose bracket attaching bolt.
8. Remove upper strut attaching bolts at body and allow assembly to drop down.
9. Remove stabilizer shaft bracket by removing strut to knuckle attaching nuts.
10. Remove strut to knuckle attaching bolts, then the strut assembly.
11. Reverse procedure to install, torquing upper strut attaching bolts to specifications.

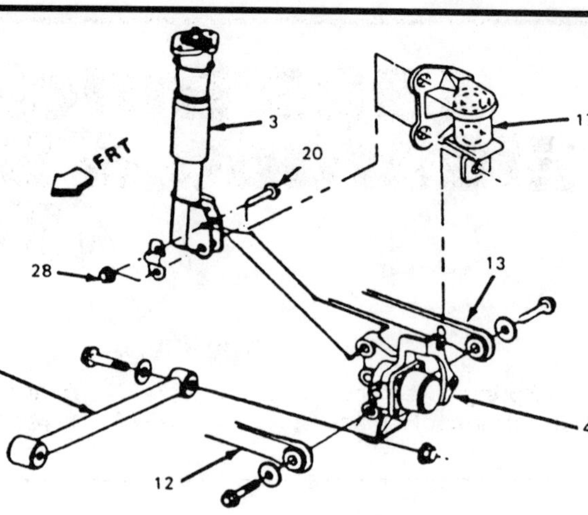

3 REAR STRUT
4 KNUCKLE ASSEMBLY
8 TRAILING LINK
12 REAR WHEEL SPINDLE ROD (FRONT)
13 REAR WHEEL SPINDLE ROD (REAR)
17 AUXILIARY SPRING ASSEMBLY
20 BOLT
28 NUT

Fig. 7 Knuckle assembly replacement

REAR SUSPENSION TIGHTENING SPECIFICATIONS

Year	Component	Torque/ft. lbs.
1988–90	Jack Pad Bolt	18
	Rod To Knuckle	66②
	Rod To Support Crossmember	81①
	Spring Retention Plate Bolt	15
	Stabilizer Shaft Link Bolt	40
	Stabilizer Shaft Link To Body Bracket Nut	18
	Strut To Knuckle Nut	133
	Suspension Crossmember To Body Bolt	85
	Trailing Arm To Knuckle	192
	Trailing Link To Body Nut	48
	Upper Strut Bolt	34
	Wheel/Hub/Bearing To Knuckle Bolt	52

①—Turn an additional 60° after reaching specified torque.
②—Turn an additional 90° after reaching specified torque.

1985 Buick Riviera, Cadillac Eldorado & Seville & Oldsmobile Toronado

INDEX

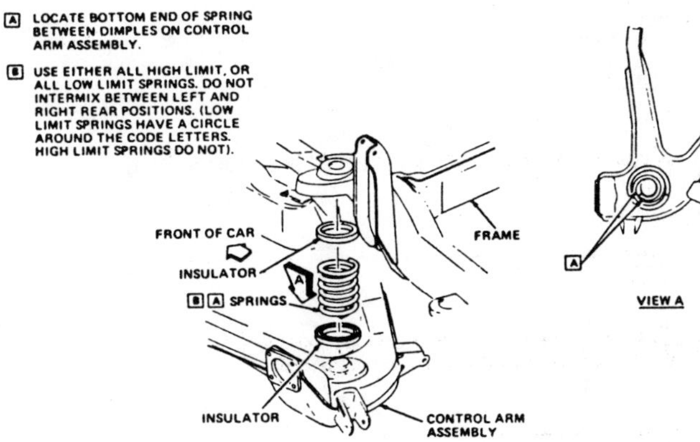

Fig. 1 Coil spring installation

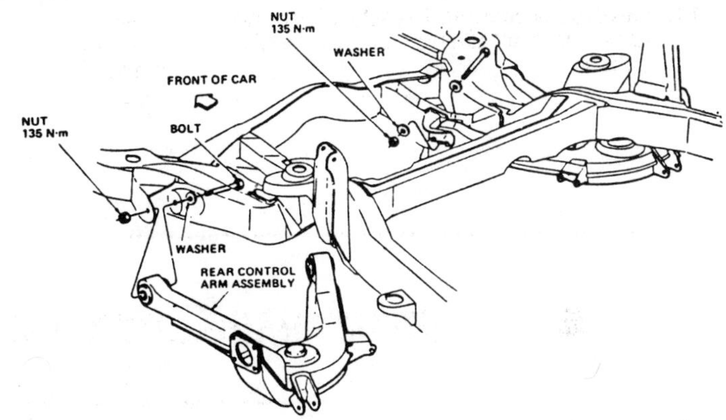

Fig. 2 Control arm assembly

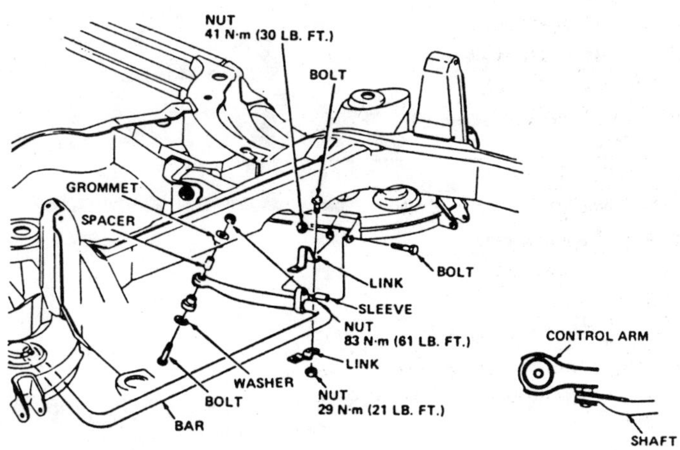

Fig. 3 Rear stabilizer installation

SHOCK ABSORBER
REPLACE

1. With the rear axle supported properly, disconnect shock absorber from lower mounting bracket. On models equipped with automatic level control, disconnect air hose from shock absorber.
2. Disconnect shock absorber upper end from underbody attachment.
3. Reverse procedure to install.

COIL SPRINGS
REPLACE

1. Raise and support rear of vehicle.
2. Using a jack support lower control arm.
3. Disconnect automatic level control air hose from shock absorber. If removing spring from left hand side of vehicle, disconnect level control link from control arm.
4. Disconnect shock absorber from upper mounting.
5. Lower the control arm until all spring tension is relieved, then remove spring.
6. Reverse procedure to install. Locate spring as shown in **Fig. 1.**

CONTROL ARMS
REPLACE

Remove and replace one control arm at a time as axle assembly may slip sideways, making installation difficult.

1. Raise and support rear of vehicle, then remove wheel and tire assembly.
2. Disconnect brake line from bracket on control arm.
3. Remove stabilizer bar. Refer to "Stabilizer Bar, Replace."
4. Remove rear wheel bearing and spindle.
5. If removing left hand control arm, disconnect automatic level control link from control arm.
6. Using a jack, support control arm.
7. Disconnect air hose from shock absorber, then detach shock absorber from upper and lower mountings.
8. Lower control arm until all spring tension is relieved, then remove spring and insulators.
9. Remove two bolts attaching control arm to frame, then remove control arm, **Fig. 2.**

STABILIZER BAR REPLACE

1. Raise and support rear of vehicle.
2. Remove bolts attaching front of stabilizer bar to control arms, **Fig. 3.**
3. Remove inside nut and bolt from each side of stabilizer bar link, then loosen nut and bolt on outside of link.
4. Rotate bottom parts of link to one side and slide stabilizer bar out of bushings.

REAR SUSPENSION TIGHTENING SPECIFICATIONS

Year	Component	Torque/ft. lbs.
1985	Ball Joint Castellated Nut	①
	ELC Compressor Bracket Bolts	3②
	ELC Head Bolts	35
	Hub/Bearing Assembly Torx Bolts	52
	Rear Control Arm Pivot Nuts	85
	Stabilizer Bar Bushing Assembly Nut	37
	Stabilizer Bar Bushing Clip Bolt	37
	Stabilizer Bar Link Nut	13
	Stabilizer Bar Mounting Bracket Bolt	13
	Superlift Strut Anchor Nuts (at knuckle)	144
	Superlift Strut Mount Nuts (top)	19
	Suspension Adjustment Link Castellated Nut	37
	Suspension Adjustment Link Retaining Nut	63

①—7.5 ft. lbs., then an additional 2/3 turn.
②—Torque opposing bolts, from center outward to edges.

1986–90 Buick Riviera, Cadillac Eldorado & Seville & Oldsmobile Toronado & 1988–90 Buick Reatta

INDEX

DESCRIPTION

All rear suspension components are mounted on a suspension support assembly, **Fig. 1,** which is attached to the body at four points. The transverse leaf spring is fully isolated from the crossmember and is held in position by left and right retainers and insulators. On their inboard end, the control arms pivot on the suspension support, allowing each end of the leaf spring to act directly on the control arm surface. A suspension knuckle pivots on the outboard end of each control arm, and each knuckle contains a sealed hub and wheel bearing assembly. Each knuckle is located at the top by an air adjustable strut assembly which is connected to the suspension support. On some models, a stabilizer shaft is used to further locate the knuckle

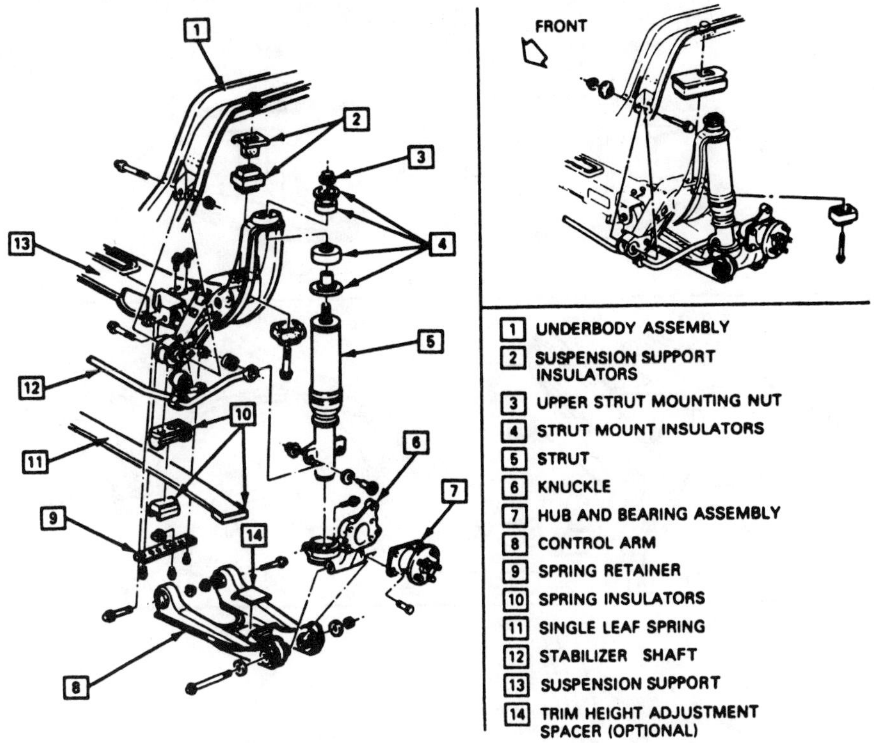

Fig. 1 Rear suspension exploded view. Left side shown

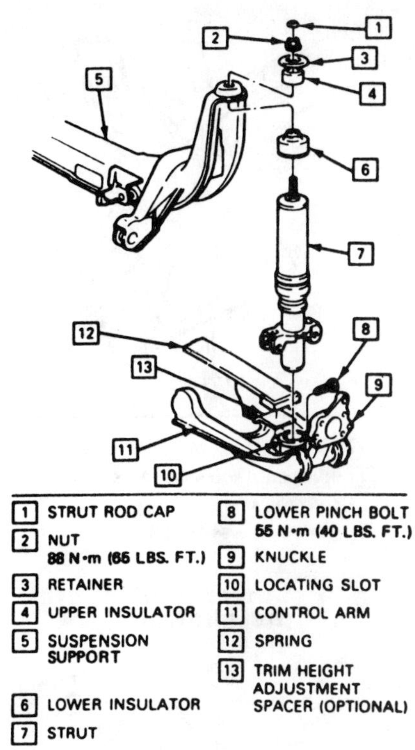

Fig. 2 Rear strut installation

assemblies an provide increased roll stiffness. Strut charging is controlled by an Electronic Level Control (ELC) system, which maintains a standard rear trim height under a variety of load conditions.

HUB & BEARING
REPLACE

1. Disconnect battery ground cable.
2. Using suitable lifting equipment, raise and support vehicle.
3. Remove tire and wheel assembly.
4. Loosen brake caliper attaching bolts and position caliper aside. **Do not suspend caliper by brake line.**
5. Remove rotor retainers, if equipped.
6. Remove rotor.
7. Remove four hub mounting bolts, then the hub and bearing assembly.
8. Reverse procedure to install, torquing to specifications.

STRUT
REPLACE

1. Raise and support vehicle on frame contact type hoist, remove wheel and tire assembly and reinstall 2 wheel nuts to retain brake rotor.
2. If left strut is to be replaced, disconnect ELC height sensor link.
3. Remove stabilizer shaft mounting bolt at strut, if equipped.
4. Remove brake caliper and secure aside leaving hoses connected.
5. Loosen but do not remove knuckle pivot bolt at outboard end of control arm.
6. Remove clip and disconnect air line from strut.
7. Remove upper strut rod cap, mounting nut and insulator, **Fig. 2.**
8. Compress strut by hand and remove lower insulator.
9. Rotate strut and knuckle assembly outward, remove pinch bolt, then separate strut from knuckle.
10. Reverse procedure to install, ensuring strut is fully seated with tang on strut bottomed in steering knuckle slot. When suspension is fully assembled, torque bolts to specifications.

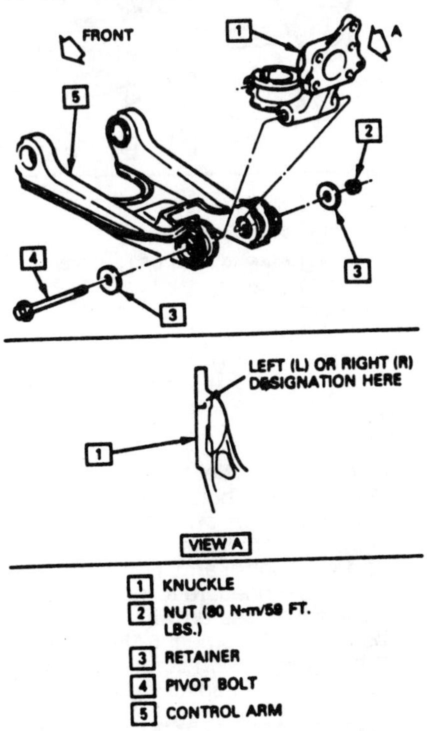

Fig. 3 Knuckle installation

KNUCKLE
REPLACE

1. Raise and support vehicle on frame contact type hoist, remove wheel and tire assembly.
2. If left knuckle is to be replaced, disconnect ELC height sensor link.
3. Remove stabilizer shaft mounting bolt at strut, if equipped.
4. Remove brake caliper and secure aside leaving hoses connected.
5. Remove brake rotor, hub retaining bolts and the hub and bearing assembly.
6. Loosen but do not remove knuckle pivot bolt at outboard end of control arm.
7. Remove clip and disconnect air line from strut.
8. Remove upper strut rod cap, mounting nut and insulator, **Fig. 2.**
9. Compress strut by hand and remove lower insulator.
10. Rotate strut and knuckle assembly outward, remove pinch bolt, then separate strut from knuckle.
11. Remove knuckle pivot bolt and the knuckle, **Fig. 3.**
12. Reverse procedure to install, ensuring strut is fully seated with tang on strut bottomed in steering knuckle slot. When suspension is fully assembled, torque bolts to specifications.

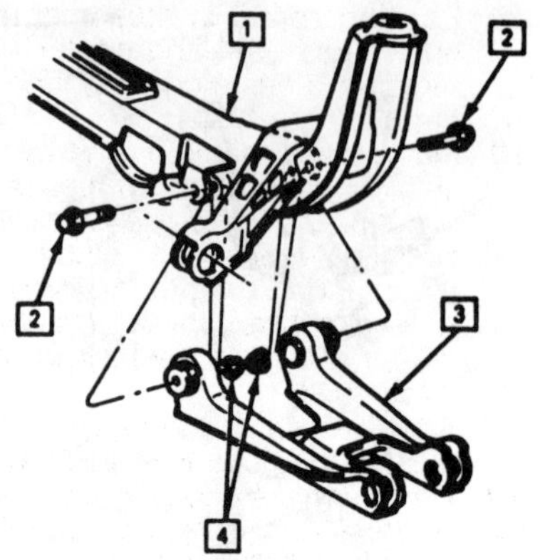

1 SUSPENSION CROSSMEMBER ASSEMBLY
2 INNER CONTROL ARM BOLTS
3 CONTROL ARM
4 NUTS (90 N·m/66 FT. LBS.)

Fig. 4 Lower control arm installation

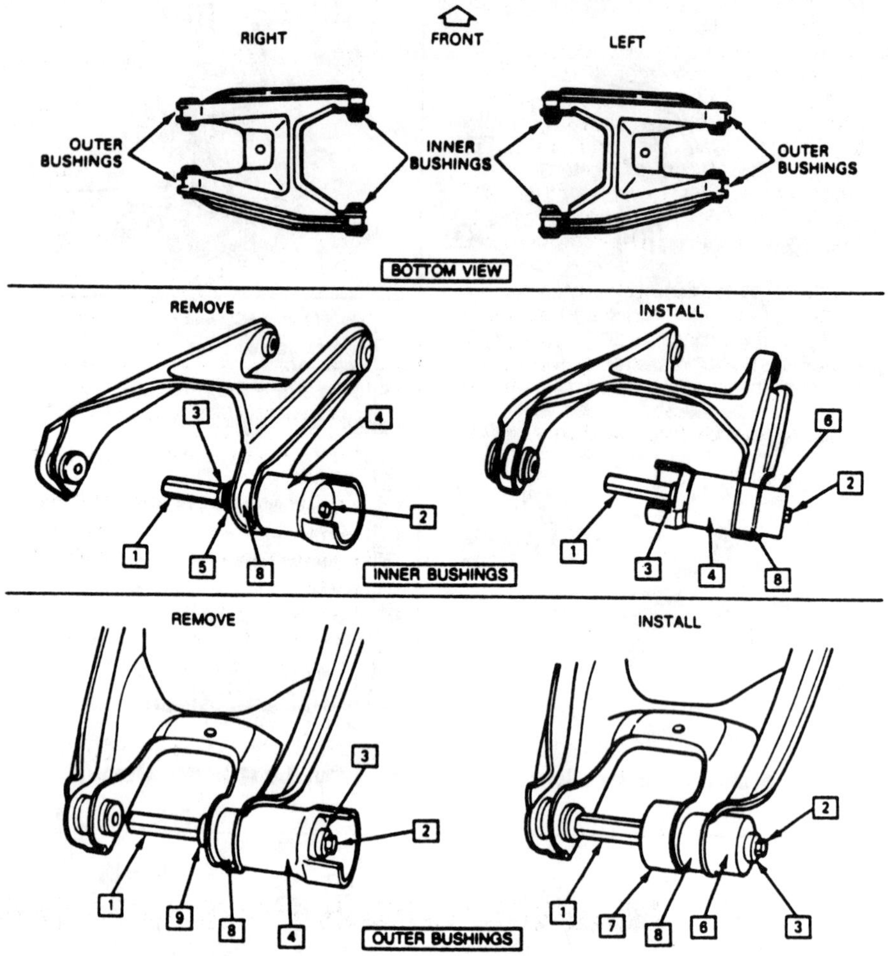

1 J-21474-18 NUT
2 J-21474-19 BOLT
3 BEARING (PART OF J-21474-19)
4 J-21474-5
5 J-21474-23
6 J-28576-1
7 J-35739-3
8 J-35739-1 SPACER — USE WIDER SIDE ON OUTER BUSHINGS, NARROW SIDE ON INNER BUSHINGS
9 J-35739-2

Fig. 5 Replacing lower control arm bushings

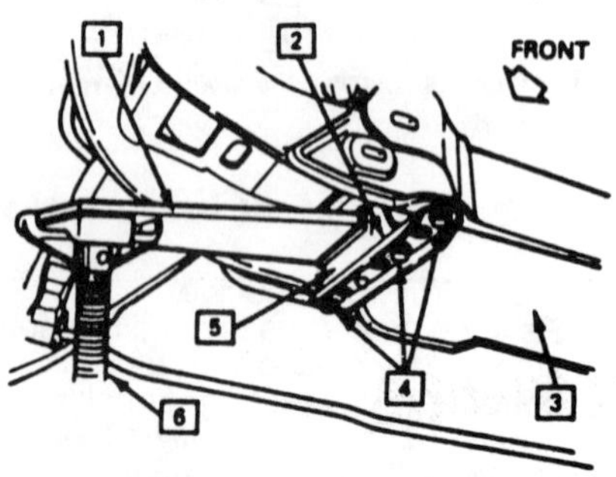

1 SPRING
2 SPRING RETAINER
3 SUSPENSION CROSSMEMBER ASSEMBLY
4 RETAINER BOLTS (28 N·m/21 FT. LBS.)
5 LOWER OUTBOARD INSULATOR
6 JACKSTAND

Fig. 6 Removing spring

LOWER CONTROL ARM
REPLACE

1. Raise and support vehicle on frame contact type hoist, remove wheel and tire assembly and reinstall 2 wheel nuts to retain brake rotor.
2. If left control arm is to be replaced, disconnect ELC height sensor link.
3. Remove stabilizer shaft mounting bolt at strut, if equipped.
4. Remove brake caliper and secure aside leaving hoses connected.
5. Loosen but do not remove knuckle pivot bolt at outboard end of control arm.
6. Remove clip and disconnect air line from strut.
7. Remove upper strut rod cap, mounting nut and insulator, **Fig. 2.**
8. Compress strut by hand and remove lower insulator.
9. Support knuckle assembly, remove knuckle pivot bolt, then remove strut and knuckle as an assembly.
10. Remove both inner control arm bolts and the control arm, **Fig. 4.**
11. Reverse procedure to install, ensuring strut is fully seated with tang on strut bottomed in steering knuckle slot. When suspension is fully assembled, torque bolts to specifications.

LOWER CONTROL ARM BUSHINGS
REPLACE

Outer control arm bushings can be replaced without removing control arm from vehicle after performing steps 1-9 of "Control Arm, Replace" procedure. If inner control arm bolts are not loosened, or removed, alignment will not be disturbed and it will not be necessary to perform wheel alignment after outer bushing replacement.

Control arm must be removed from vehicle to replace inner control arm bushings. After replacing inner bushings, check wheel alignment.

When replacing lower control arm bushings, refer to **Fig. 5.**

SPRING
REPLACE

Removal of rear spring requires disassembly of either left or right side of suspension while leaving opposite side intact. Spring may be removed from either side of vehicle.

1. Remove lower control arm as outlined.
2. Place suitable support under end of spring as shown in **Fig. 6,** then lower vehicle so weight of vehicle compresses spring. **Ensure jack stand can support weight of vehicle and**

is properly positioned under spring.

3. Remove 3 spring retainer bolts, retainer and lower insulator, **Fig. 7**, from end of spring nearest support stand.
4. Slowly raise vehicle, allowing spring to deflect downward until spring no longer exerts force on stand.
5. Remove retainer bolts, retainer and insulator from opposite end of spring, then withdraw spring from crossmember.
6. Remove upper spring insulators.
7. Inspect spring insulators and retainers and replace as needed.
8. Install center and upper outboard insulators in crossmember and torque bolts to specifications. **Ensure arrows on upper outboard insulators point toward centerline of vehicle.**
9. Position spring in crossmember, ensuring outboard and center insulator bands are centered on spring insulators.
10. Install lower insulator and retainer on side opposite disassembled portion of suspension, torque retaining bolts to specifications, then place suitable stand under free end of spring.
11. Lower vehicle, allowing weight to load spring and deflect free end of spring into position.
12. Install lower insulator and retainer, and torque bolts to specifications.
13. Raise vehicle and remove stand.
14. Reverse remaining procedure to complete installation.

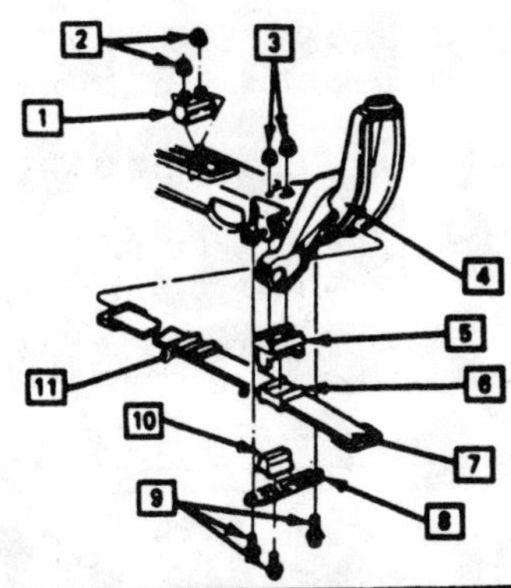

1 CENTER INSULATOR
2 CENTER INSULATOR NUTS (28 N·m/21 FT. LBS.)
3 OUTBOARD INSULATOR NUTS (28 N·m/21 FT. LBS.)
4 SUSPENSION CROSSMEMBER ASSEMBLY
5 UPPER OUTBOARD INSULATOR
6 OUTBOARD INSULATOR LOCATING BAND
7 SPRING
8 SPRING RETAINER
9 RETAINER BOLTS (28 N·m/21 FT. LBS.)
10 LOWER OUTBOARD INSULATOR
11 CENTER INSULATOR LOCATING BAND

Fig. 7 Exploded view of spring installation

REAR SUSPENSION TIGHTENING SPECIFICATIONS

Year	Component	Torque/ft. lbs.
1986–90	Control Arm Bolts	66
	Hub Mounting Bolt	52
	Knuckle Pivot Bolt	59
	Spring Insulator Retaining Bolts	21
	Stabilizer Bolt	43
	Strut Pinch Bolt	40
	Strut Upper Nut	65
	Wheel Nuts	100

Buick Skyhawk, Skylark (1986–90), Somerset & Somerset Regal, Cadillac Cimarron, Chevrolet Cavalier, Oldsmobile Calais, Cutlass Calais & Firenza & Pontiac Grand Am & Sunbird

INDEX

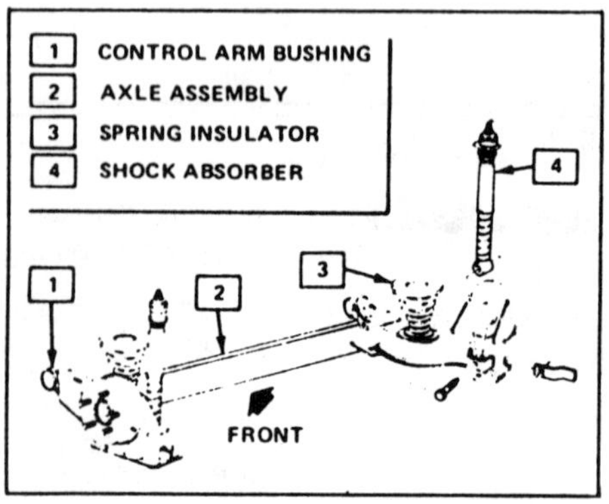

Fig. 1 Rear suspension

DESCRIPTION

The rear suspension, **Fig. 1**, is a semi-independent type suspension consisting of an axle assembly with trailing arms and twisting cross beam, coil springs and double action shock absorbers. A stabilizer bar is available and is attached to the inside of the axle beam and to the lower end of the control arms. A single unit hub and bearing assembly is bolted to each end of the axle assembly. The hub and bearing assembly is a sealed, non-serviceable unit and must be replaced as an assembly.

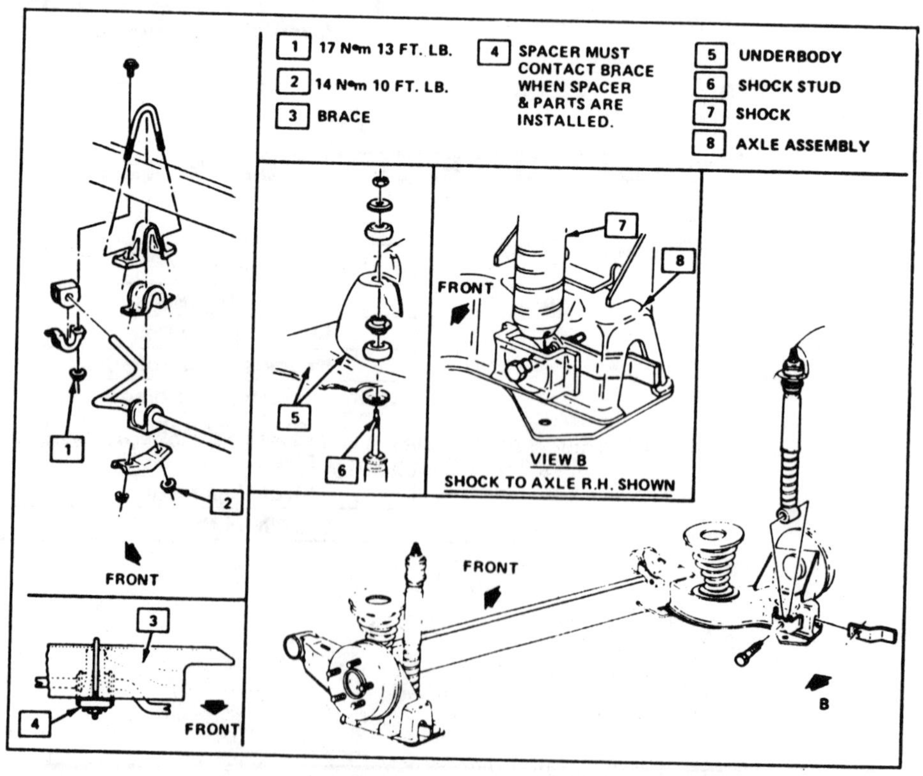

Fig. 2 Rear shock absorber upper attachment components. 1985–88 models

REAR AXLE
REPLACE

1. Raise vehicle and support vehicle. Support rear suspension with suitable jack.
2. Disconnect stabilizer bar at axle assembly, if equipped, **Figs. 2 and 3.**
3. Remove rear wheel assembly and brake drum. Do not hammer on brake drum since damage to bearings may result.
4. Remove shock absorber to lower mounting bracket attaching bolts, then disconnect shock absorbers from axle assembly, **Figs. 2, 3 and 4.**
5. Disconnect parking brake cable and brake lines at axle brackets.
6. Carefully lower rear axle assembly and remove coil springs and insulators.
7. Remove control arm to under body

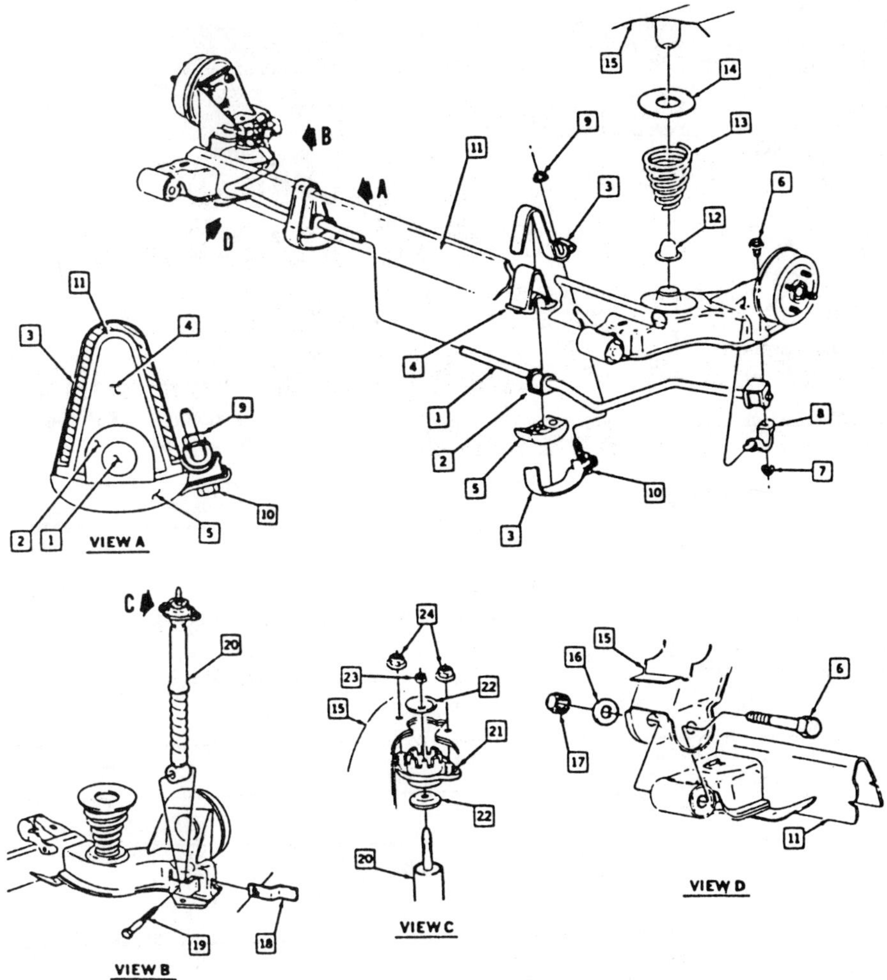

1. STABILIZER SHAFT
2. INSULATOR
3. CLAMP
4. UPPER SPACER
5. LOWER SPACER
6. BOLT
7. NUT – 22 N·m (16 LBS. FT.)
8. CLAMP
9. NUT
10. BOLT – 18 N·m (13 LBS. FT.)
11. AXLE
12. BUMPER
13. SPRING
14. INSULATOR
15. UNDERBODY
16. WASHER
17. NUT – 83 N·m (61 LBS. FT.)
18. TAB NUT
19. BOLT – 47 N·m (35 LBS. FT.)
20. SHOCK ABSORBER
21. MOUNT
22. RETAINER
23. NUT – 29 N·m (21 LBS. FT.)
24. NUT – 17 N·m (13 LBS. FT.)

Fig. 3 Rear shock absorber upper attachment components. 1989–90 models

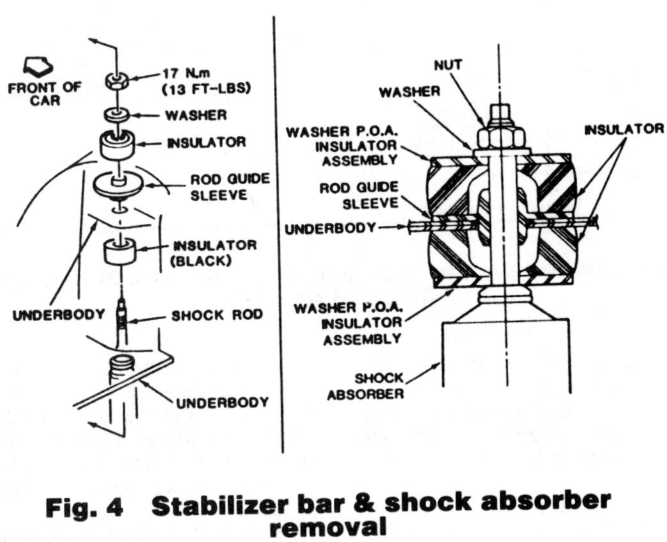

Fig. 4 Stabilizer bar & shock absorber removal

bracket bolts, then lower the axle assembly and remove from vehicle.

8. Remove hub to rear axle attaching bolts, then the hubs, bearings and backing plates from rear axle assembly.
9. Reverse procedure to install and bleed brake system.

HUB & BEARING ASSEMBLY

REPLACE

1. Raise and support vehicle, then remove wheel and tire assembly and brake drum. **Do not hammer brake drum since damage to bearing may result.**
2. Remove four hub/bearing assembly to rear axle attaching bolts, then the hub/bearing assembly from axle. **The upper rear hub attaching bolt may not clear brake shoe when removing hub and bearing assembly. Partially remove hub and bearing assembly prior to removing this bolt.**
3. Reverse procedure to install. Torque hub attaching bolts to specification. **Use care not to drop hub/bearing assembly since damage to bearing may result.**

SHOCK ABSORBER

REPLACE

1. Open deck lid, then remove trim cover and shock absorber upper retaining nut.
2. Raise rear of vehicle and support rear axle using a suitable jack.
3. Remove shock absorber lower attaching bolt, then disconnect shock absorber from mounting bracket, **Figs. 2 and 3.** Remove shock absorber from vehicle.
4. Reverse procedure to install. Torque attaching bolt to specification. **When installing upper shock absorber attachment components, refer to Fig. 4,** for proper installation order.

COIL SPRING

REPLACE

1. Raise and support rear of vehicle. Support rear axle using a suitable jack.
2. Remove wheel and tire assemblies.
3. Remove brake line bracket attaching bolts from frame, **Figs. 2 and 3,** and allow brake lines to hang freely.
4. Remove shock absorber to lower mounting bracket bolts, then disconnect shock absorbers from axle assembly. **Do not suspend rear axle by brake hoses since damage to hoses may result.**
5. Carefully lower rear axle assembly and remove springs and insulators.
6. Reverse procedure to install. Position ends of upper coil in seat of body and within limits shown in **Fig. 5.**

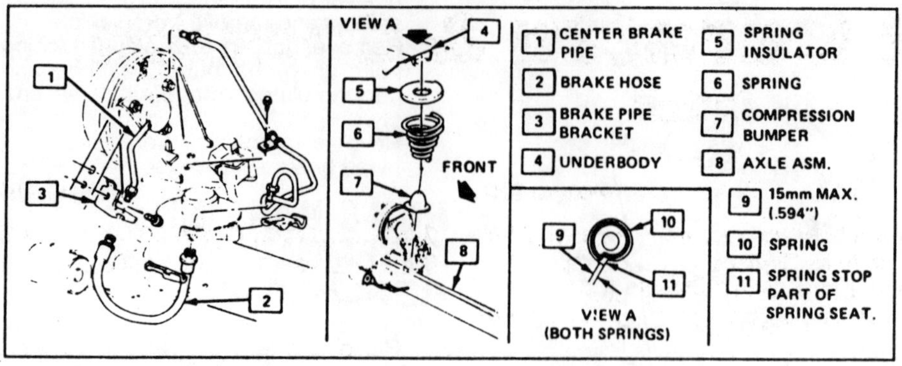

Fig. 5 Coil spring installation

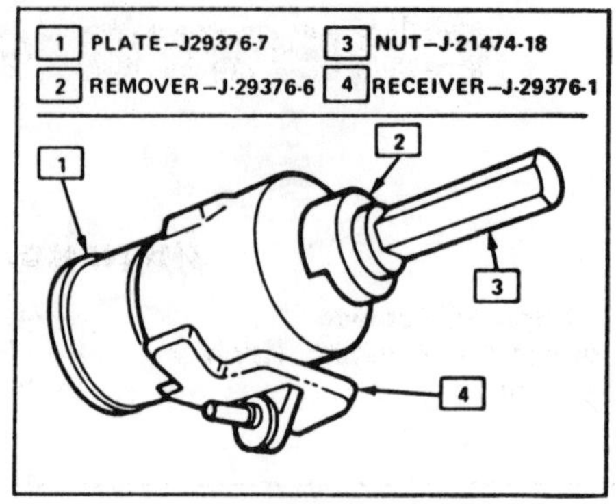

Fig. 6 Control bushing removal

CONTROL ARM BUSHING REPLACE

1. Raise rear of vehicle and support rear axle under front side of spring seat using a suitable jack.
2. Remove wheel and tire assembly.
3. If right hand side bushing is to be replaced, disconnect brake line bracket from body. If left hand side bushing is to be replaced, disconnect brake line bracket from frame and parking brake cable at hook guide.
4. Remove control arm to mounting bracket attaching nut, bolt and washer, then allow control arm to rotate downward.
5. The bushing can now be replaced using tools shown in **Figs. 6 and 7.** When installing bushing, the arrow on the installer must align with arrow on the receiver, **Fig. 6.**
6. Reverse procedure to complete installation. **The control arm attaching bolt must be torqued after vehicle is lowered to floor and is in its standing height position. Torque attaching bolt to specifications.**

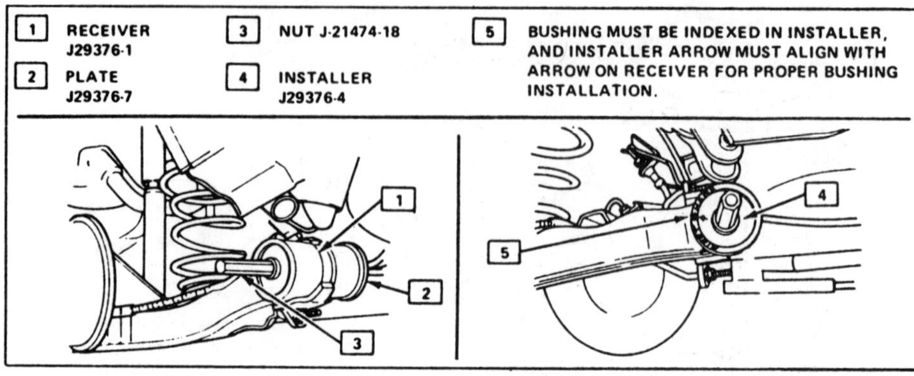

Fig. 7 Control arm bushing installation

REAR SUSPENSION TIGHTENING SPECIFICATIONS

Year	Component	Torque/ft. lbs.
1985–88	Control Arm To Body Bracket	67
	Hub & Bearing To Rear Axle	39
	Shock Absorbers (Lower)	41
	Shock Absorbers (Upper)	13
	Stabilizer Bar To Axle	10
	Stabilizer Bar To Control Arm	13
	Upper Shock Mount To Shock Tower	13
1989-90	Axle To Body Bracket	61
	Shock Absorber Mount To Body	21
	Shock Absorber To Mount	21
	Shock Absorber Bolt At Axle	35
	Stabilizer Shaft Clamp Bolts At Axle	13
	Stabilizer Shaft Clamp Nuts	16

Chevrolet Beretta & Corsica

INDEX

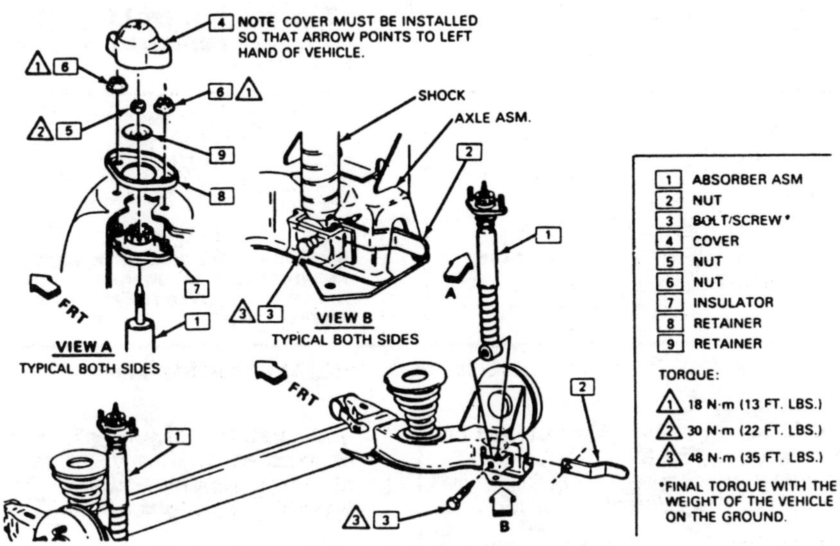

Fig. 1 Rear suspension. Corsica

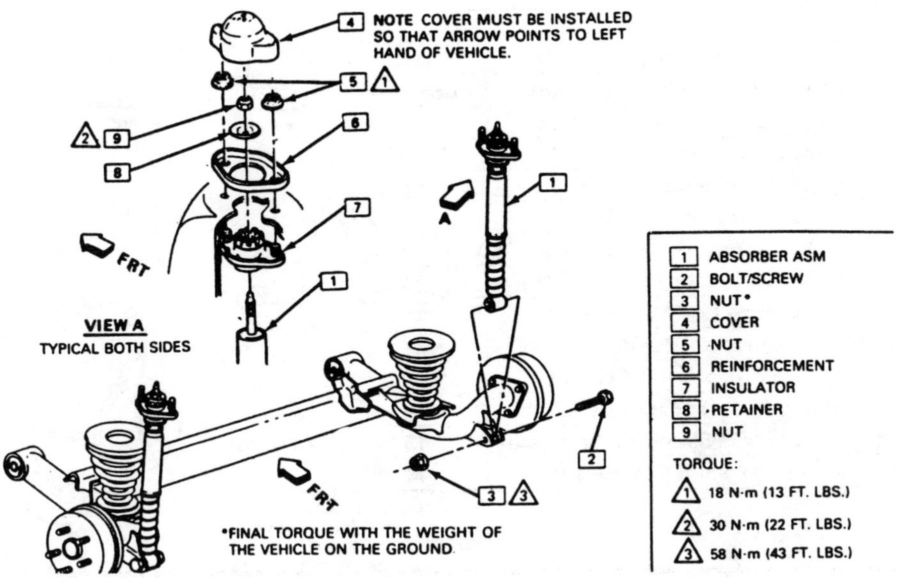

Fig. 2 Rear suspension. Beretta

DESCRIPTION

This rear suspension is a semi-independent type suspension consisting of an axle assembly with trailing arms and twisting cross beam, coil springs and double action shock absorbers. A stabilizer bar is available and is attached to the inside of the axle beam and to the lower end of the control arms. A single unit hub and bearing assembly is bolted to each end of the axle assembly. The hub and bearing assembly is a sealed, non-serviceable unit and must be replaced as an assembly.

REAR AXLE REPLACE

1. Raise vehicle and support vehicle. Support rear suspension with suitable jack.
2. Disconnect stabilizer bar at axle assembly, if equipped.
3. Remove rear wheel assembly and brake drum. **Do not hammer on brake drum since damage to bearings may result.**
4. Remove shock absorber to lower mounting bracket attaching bolts, then disconnect shock absorbers from axle assembly, **Figs. 1 and 2.**
5. Disconnect parking brake cable and brake lines at axle brackets.
6. Carefully lower rear axle assembly and remove coil springs and insulators.
7. Remove control arm to underbody bracket bolts, then lower the axle assembly and remove from vehicle.
8. Remove hub to rear axle attaching bolts, then the hubs, bearings and backing plates from rear axle assembly.
9. Reverse procedure to install and bleed brake system.

HUB & BEARING ASSEMBLY REPLACE

1. Raise and support vehicle, then remove wheel and tire assembly and brake drum. **Do not hammer brake drum since damage to bearing may result.**

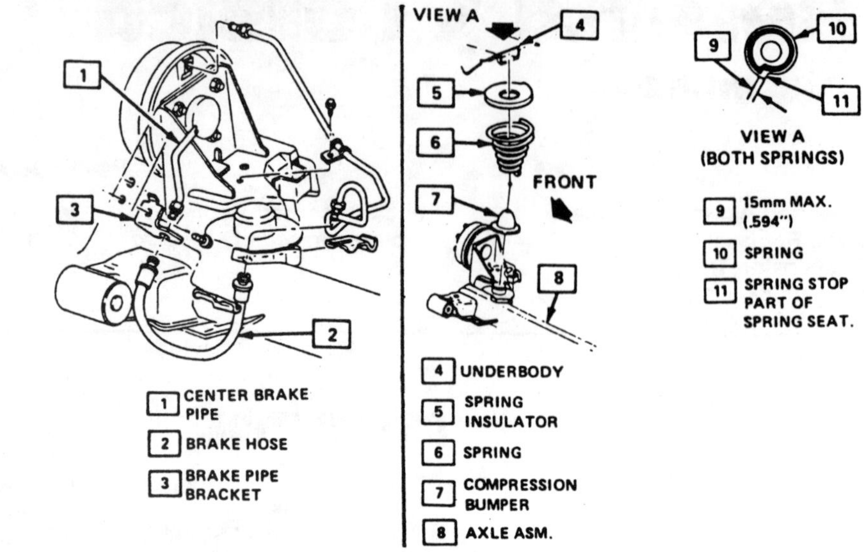

Fig. 3 Coil spring and brake line installation

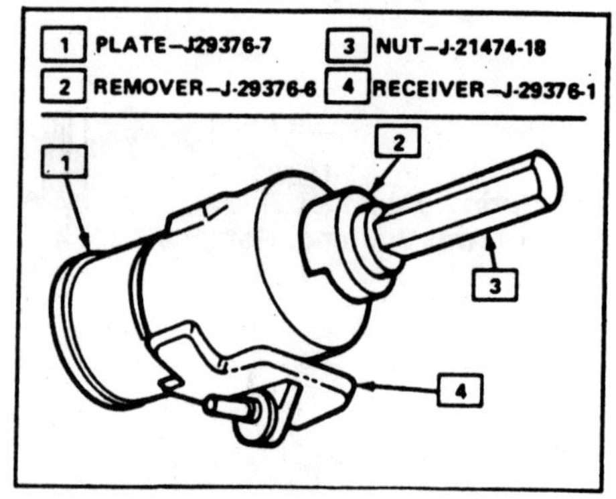

Fig. 4 Control bushing removal

2. Remove four hub/bearing assembly to rear axle attaching bolts, then the hub/bearing assembly from axle. **The upper rear hub attaching bolt may not clear brake shoe when removing hub and bearing assembly. Partially remove hub and bearing assembly prior to removing this bolt.**
3. Reverse procedure to install. Torque hub to axle attaching bolts to specifications. **Use care not to drop hub/bearing assembly since damage to bearing may result.**

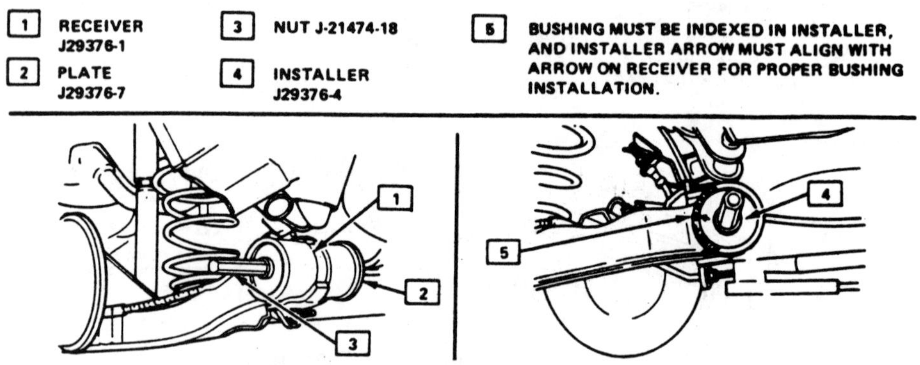

Fig. 5 Control arm bushing installation

SHOCK ABSORBER
REPLACE

1. Open deck lid, then remove trim cover (if equipped) and shock absorber upper retaining nut.
2. Raise rear of vehicle and support rear axle using a suitable jack.
3. Remove shock absorber lower attaching bolt, then disconnect shock absorber from mounting bracket, **Figs. 1 and 2.** Remove shock absorber from vehicle.
4. Reverse procedure to install. Torque attaching bolts to specifications.

COIL SPRING
REPLACE

1. Raise and support rear of vehicle. Support rear axle using a suitable jack.
2. Remove wheel and tire assemblies.
3. Remove brake line bracket attaching bolts, **Fig. 3,** from frame and allow brake lines to hang freely.
4. Remove shock absorber to lower mounting bracket bolts, then disconnect shock absorbers from axle assembly. **Do not suspend rear axle by brake hoses since damage to hoses may result.**

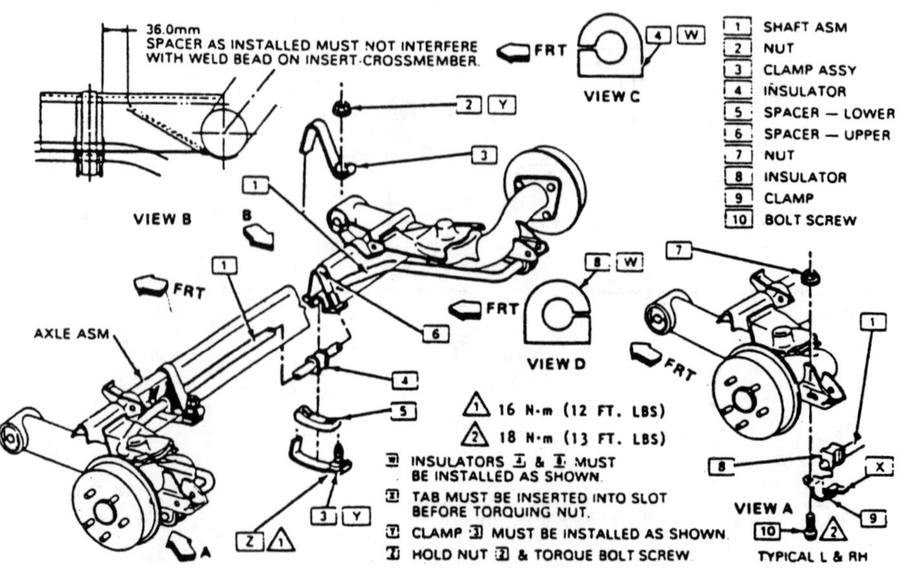

Fig. 6 Stabilizer assembly. Beretta

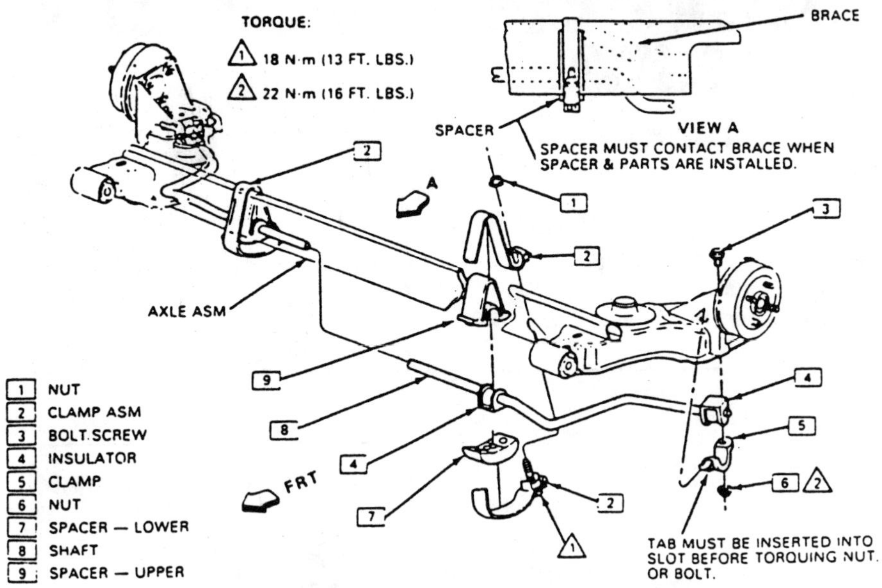

Fig. 7 Stabilizer assembly. Corsica

5. Carefully lower rear axle assembly and remove springs and insulators.
6. Reverse procedure to install. Position ends of upper coil in seat of body and within limits shown in **Fig. 3.**

CONTROL ARM BUSHING REPLACE

1. Raise and support rear of vehicle.
2. Remove wheel and tire assembly, then support body with suitable jack stands
3. Remove brake line bracket to body attaching screws from both sides.
4. Remove control arm to mounting bracket attaching nut, bolt and washer, then allow control arm to rotate downward.
5. The bushing can now be replaced using tools shown in **Figs. 4 and 5.** When installing bushing, the arrow on the installer must align with arrow on the receiver, **Fig. 4.**
6. Reverse procedure to complete installation. **The control arm attaching bolt must be torqued to specifications after vehicle is lowered to floor and is in its standing height position.**

STABILIZER ASSEMBLY REPLACE

1. Raise vehicle on hoist and support body with jack stands.
2. Remove nuts and bolts at both the axle and control arm attachments and remove bracket, insulator and stabilizer shaft, **Figs. 6 and 7.**
3. Reverse procedure to install.

REAR SUSPENSION TIGHTENING SPECIFICATIONS

Year	Component	Torque/ft. lbs.
1987–90	Control Arm Pivot Bolts	66①
	Hub-to-Axle	38
	Shock Absorber Lower②	12
	Shock Absorber Lower③	12
	Shock Absorber Upper Shaft Nut	22
	Stabilizer Bar To Axle	12
	Stabilizer Bar To Control Arm	13
	Upper Shock Mount	13

①—Torque control arm attaching bolt after vehicle is lowered to floor and is in its' standing height position.
②—Beretta
③—Corsica

Chevrolet Chevette & Pontiac 1000

INDEX

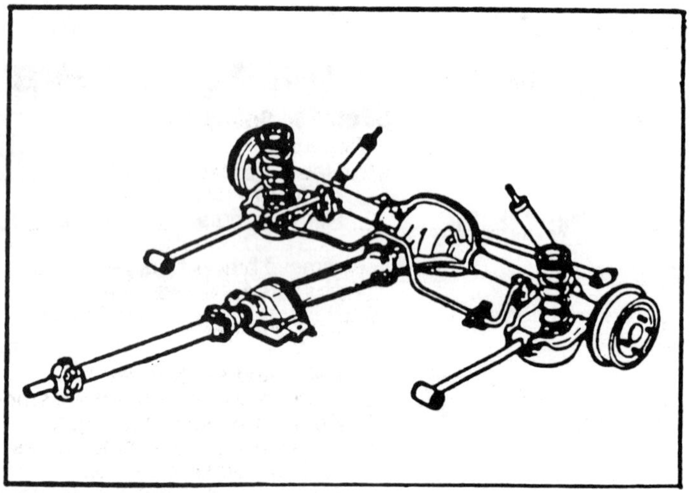

Fig. 1 Rear suspension

DESCRIPTION

This suspension system, **Fig. 1**, incorporates two tubular lower control arms, a straight track rod, two shock absorbers, two coil springs and a stabilizer bar.

SHOCK ABSORBER REPLACE

1. Raise vehicle and support rear axle with a suitable jack.
2. Disconnect shock absorber from upper and lower mounting.
3. Reverse procedure to install.

COIL SPRING REPLACE

1. Raise vehicle and support rear axle with a suitable jack.
2. Disconnect both shock absorbers from lower brackets.
3. Disconnect rear axle extension bracket. **Ensure axle is properly supported before disconnecting extension bracket.**
4. Slowly lower axle until springs and insulators can be removed. **When lowering axle ensure brake hoses do not stretch or become damaged.**
5. Reverse procedure to install.

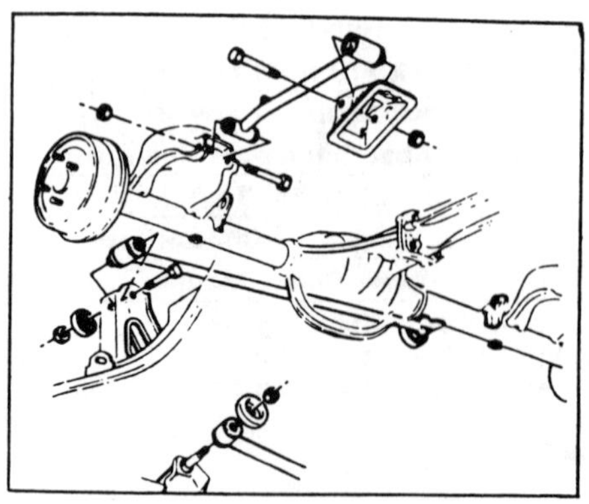

Fig. 3 Track rod and lower control arm

TRACK ROD REPLACE

1. Raise vehicle and support rear axle with a suitable jack.
2. Disconnect stabilizer bar, if equipped, **Fig. 2.**
3. Remove control arm front and rear attaching bolts and control arm, **Fig. 3.**
4. Remove attaching bolts and track rod.
5. Reverse procedure to install. **Vehicle must be at curb height when tightening pivot bolts.**

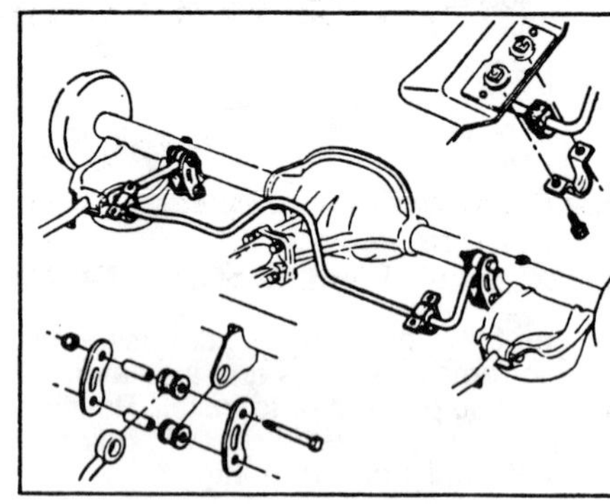

Fig. 2 Stabilizer bar installation

LOWER CONTROL ARMS REPLACE

1. Raise vehicle and support rear axle.
2. Disconnect stabilizer bar, if equipped, **Fig. 2.**
3. Disconnect the front and rear control arm attaching bolts and remove control arm from vehicle, **Fig. 3. Replace one control arm at a time to prevent axle from rolling or slipping sideways.**
4. Reverse procedure to install.

STABILIZER BAR REPLACE

1. Raise vehicle and support rear axle.
2. Disconnect stabilizer bar from underbody and axle tube connections, **Fig. 2,** then remove stabilizer bar from vehicle.
3. Reverse procedure to install.

REAR SUSPENSION TIGHTENING SPECIFICATIONS

Year	Component	Torque/ft. lbs.
1985–87	Lower Control Arm Pivot Bolts	33
	Shock Bolt (Lower)	21
	Shock Bolt (Upper)	7
	Stabilizer Bar	15
	Track Rod To Axle	34
	Track Rod To Body	67

Chevrolet Nova & Geo Prizm

INDEX

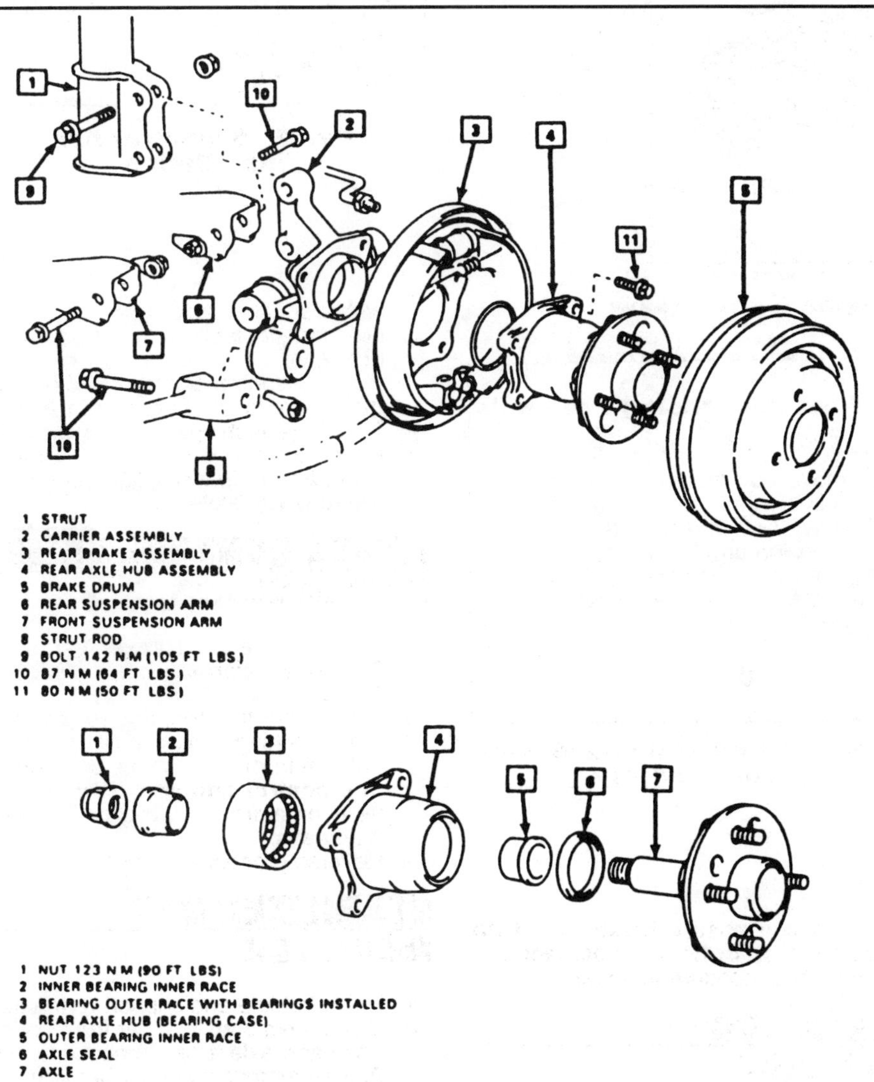

Fig. 1 Rear hub assembly

REAR AXLE HUB
REPLACE

1. Raise and support vehicle.
2. Remove wheel and tire assembly.
3. Remove brake drum, **Fig. 1.**
4. Remove four bolts attaching axle hub and rear brake assembly.
5. Remove hub, brake assembly and O-ring.
6. Reverse procedure to install.

STRUT/COIL SPRING ASSEMBLY
REPLACE

Do not attempt to disassemble strut, since cylinder is filled with high pressure gas. When discarding strut, drill a .090 inch hole in bottom of strut to relieve all gas pressure.

1. Remove quarter window garnish molding and back window panel.
2. Raise and support vehicle, then remove tire and wheel assembly.
3. Remove brake line at flex hose on strut, **Fig. 2.**
4. Disconnect flexible hose from shock absorber.
5. Remove bolts attaching strut to axle carrier.
6. Reverse procedure to install.

REAR SUSPENSION ARM
REPLACE

1. Raise and support vehicle.
2. Remove bolt and nut attaching rear suspension arm to axle carrier.
3. Remove cam and bolt attaching rear suspension arm to body, then the rear suspension arm.

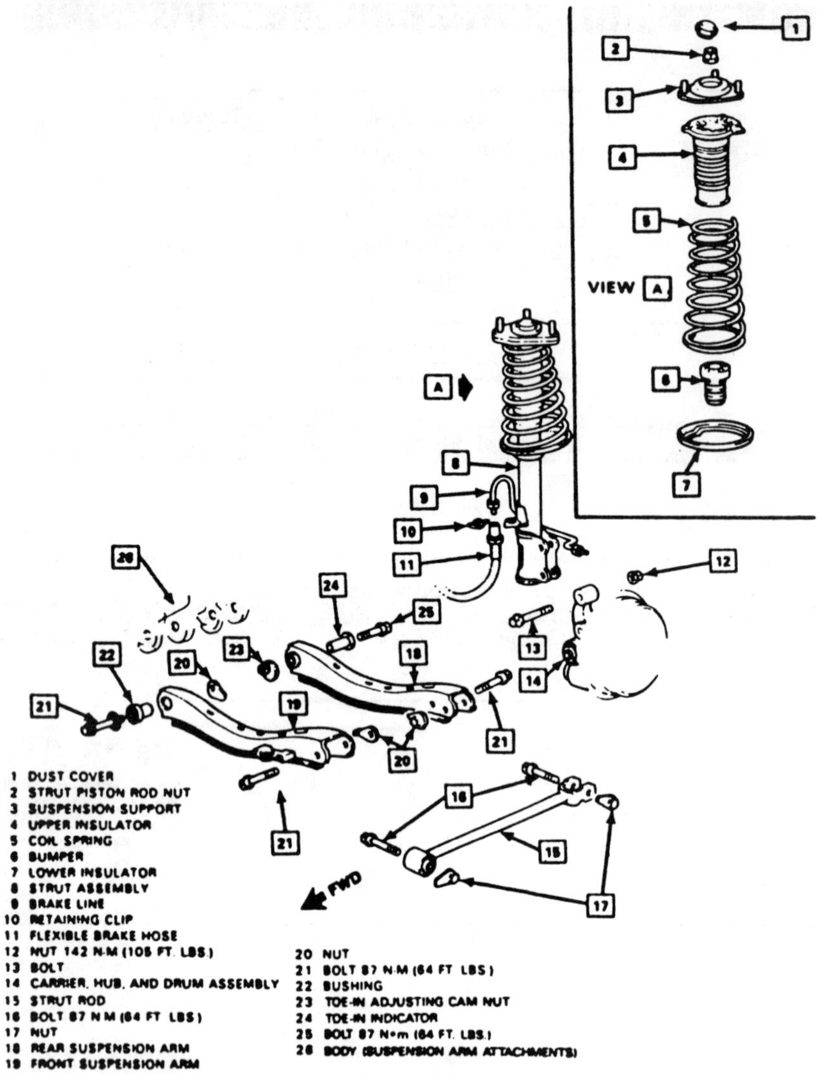

Fig. 2 Rear suspension assembly

4. Note location of cam plate mark before removing bolt and cam.
5. Reverse procedure to install.

FRONT SUSPENSION ARM

REPLACE

1. Raise and support vehicle.
2. Remove bolt and nut attaching front suspension arm to axle carrier.
3. Remove bolt and nut attaching front suspension arm to the body, then the front suspension arm.
4. Reverse procedure to install.

STRUT ROD

REPLACE

1. Raise and support vehicle.
2. Remove bolt and nut attaching strut rod to axle carrier.
3. Remove bolt and nut attaching strut rod to body, then the strut rod assembly from the vehicle.
4. Reverse procedure to install.

REAR SUSPENSION TIGHTENING SPECIFICATIONS

Year	Component	Torque/ft. lbs.
1985-90	Mounting Bracket Bolt	34
	Rear Axle Carrier Bolts	105
	Rear Suspension Arm Bolts	64
	Stabilizer Bar to Stabilizer Links	22
	Stabilizer Bracket Bolts	9
	Strut to Axle Carrier Nut	105
	Strut Piston Rod Nut	36
	Suspension Arm Bolts	64
	Stabilizer Link to Suspension Arm	11
	Suspension Support Nuts	17

Chevrolet & Geo Spectrum

INDEX

DESCRIPTION

The semi-independent suspension, **Fig. 1,** consists of an axle with trailing arms and twisting cross beam, two coil springs, shock absorbers, upper spring insulators and spring compression bumpers. The axle assembly is attached to the underbody through rubber bushings located at the front end of rear axle. The rear axle brackets are integral with the underbody side rails. The two coil springs support the weight of the vehicle in the rear. Each spring is retained between a seat in the underbody and a seat welded to the top of the rear axle. A rubber cushion isolates the coil spring upper end from the underbody seat.

REAR AXLE & COIL SPRINGS

REPLACE

1. Raise and support rear of vehicle.
2. Remove both rear wheel and tire assemblies.
3. Remove brake line, retaining clip and flexible hose from center of axle.
4. Remove tension spring, then disconnect parking brake cable and cable joint.
5. Support axle with a suitable jack.
6. Remove lower shock absorber bolt and disconnect shock from axle, **Fig. 1.**
7. Lower jack supporting axle and remove coil spring, **Fig. 2.**
8. Remove axle to body attaching bolts, then lower rear axle assembly from vehicle.
9. Reverse procedure to install, noting the following:
   a. When installing coil springs, the thinner end of spring must be attached at upper end and the bottom of spring must be positioned as shown, **Fig. 3.**
   b. Rear trim height, **Fig. 4,** must be set to 15.2 inches prior to tightening rear axle bolts.
   c. The lower shock absorber attaching bolt is coated with an adhesive agent and should be replaced whenever removed.

SHOCK ABSORBER

REPLACE

1. **On hatchback models,** remove trim cover from upper shock nut.
2. **On all models,** remove upper shock nut.
3. Remove lower shock absorber bolt, **Fig. 5,** then the shock absorber.
4. Reverse procedure to install. **The lower shock absorber attaching bolt is coated with an adhesive agent and should be replaced whenever removed.**

REAR WHEEL BEARING

ADJUST

1. Raise and support rear of vehicle.
2. Remove wheel and tire assembly.
3. Remove bearing dust cap, then remove cotter pin and loosen spindle nut.
4. While rotating hub and drum assembly, **torque** spindle nut to 22 ft. lbs.
5. Loosen spindle nut, then tighten finger tight.
6. Install spindle nut retainer and cotter pin. **If cotter pin holes are not aligned, tighten nut slightly to align.**
7. Apply suitable wheel bearing grease to inner area of bearing dust cap, then install cap.
8. Install wheel and tire assembly and lower vehicle and torque lug nuts to

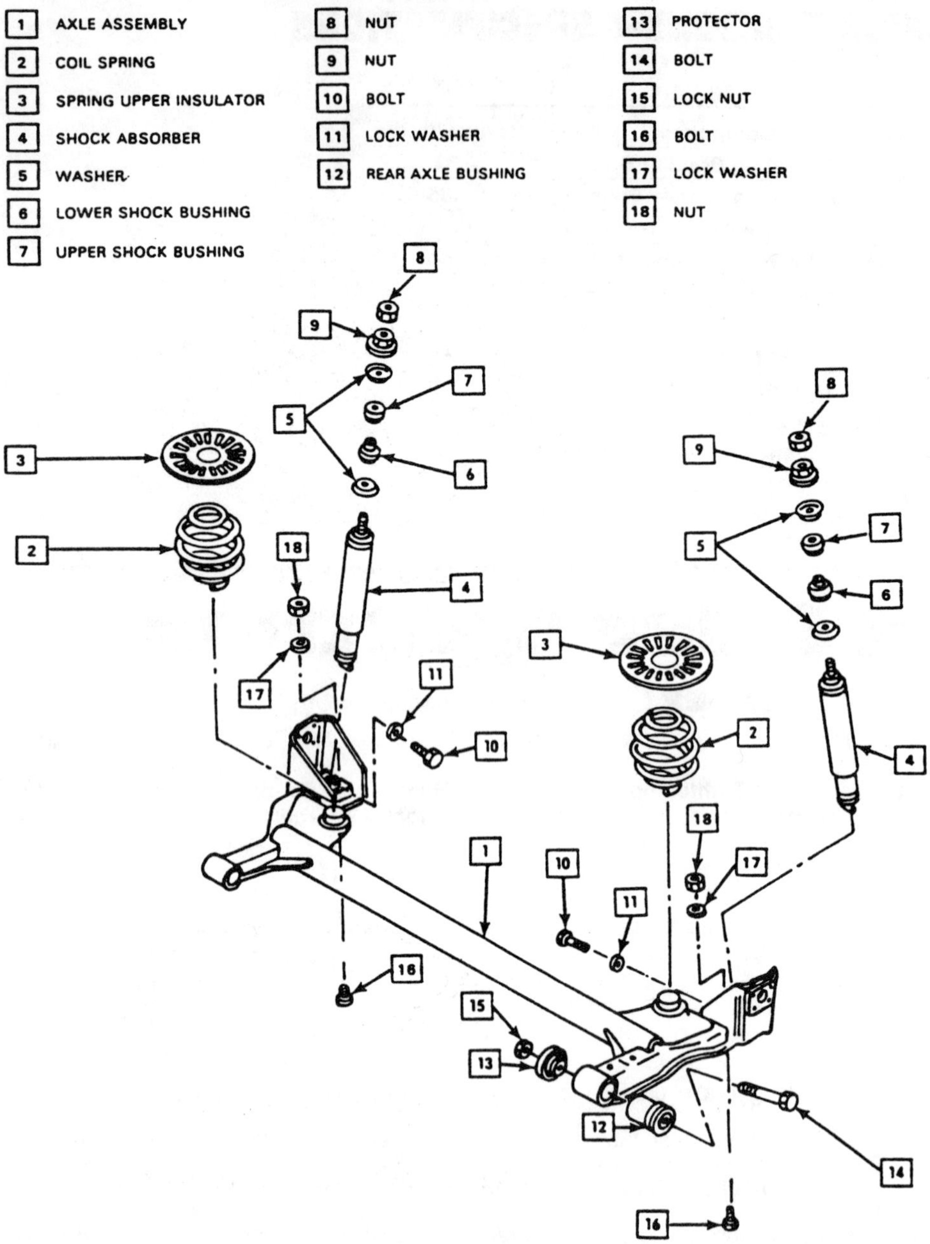

Fig. 1 Exploded view of rear axle & suspension assembly

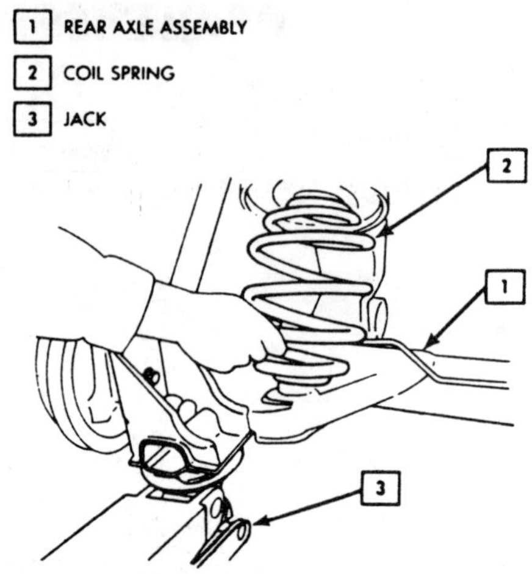

Fig. 2 Coil spring removal

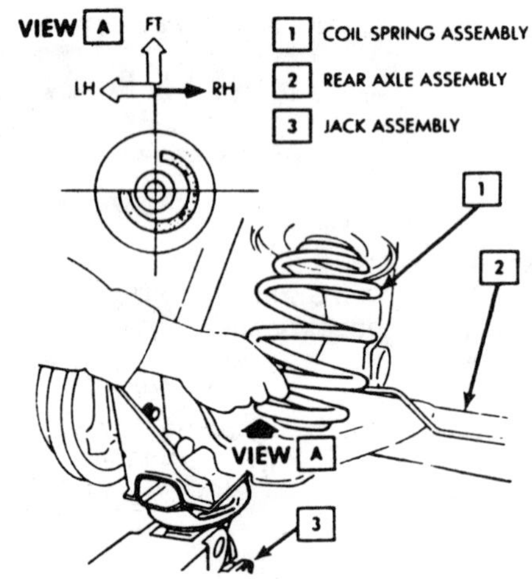

Fig. 3 Coil spring installation

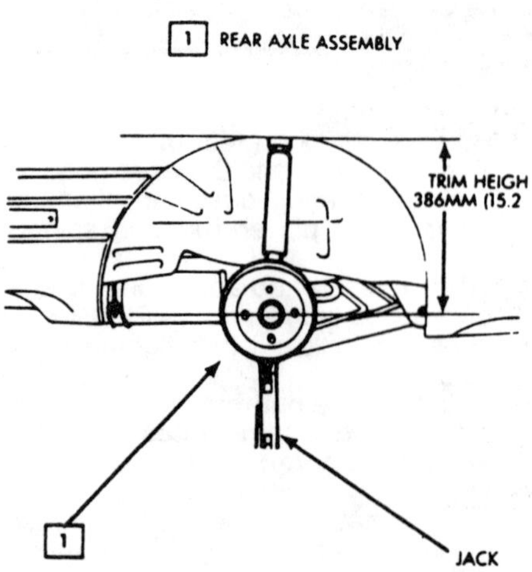

Fig. 4 Setting rear trim height

specification listed at the end of this section.

REAR HUB & WHEEL BEARINGS
REPLACE

1. Raise and support rear of vehicle.
2. Remove wheel and tire assembly.
3. Remove hub cap, cotter pin and nut and washer, then the hub and brake drum assembly.
4. Remove outer bearing from hub.
5. Remove oil seal from hub using tool No. J-26941 or equivalent, **Fig. 6**, then remove inner bearing from hub.
6. Tap inner bearing outer race and outer bearing outer race out of hub, **Fig. 7.**
7. Install inner and outer bearing races

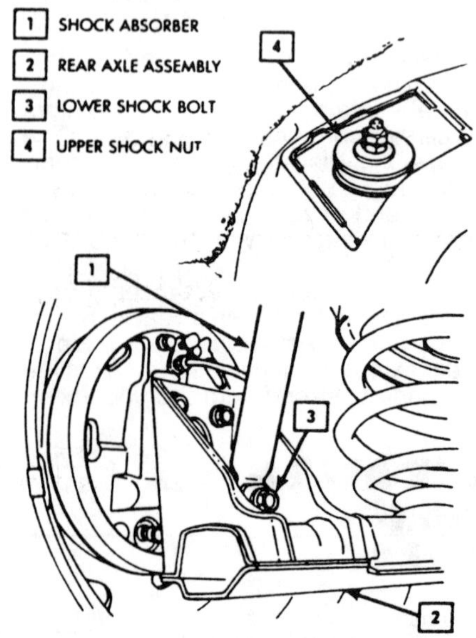

Fig. 5 Shock absorber replacement

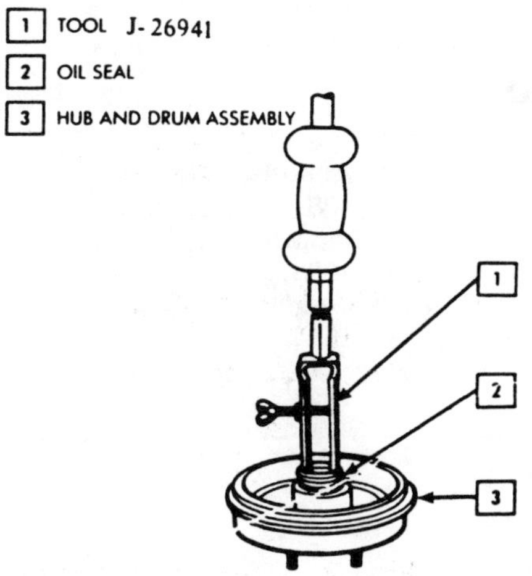

Fig. 6 Axle seal removal

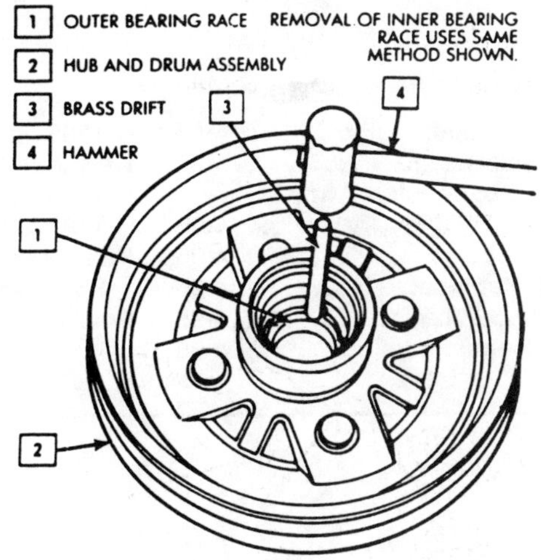

Fig. 7 Bearing race removal

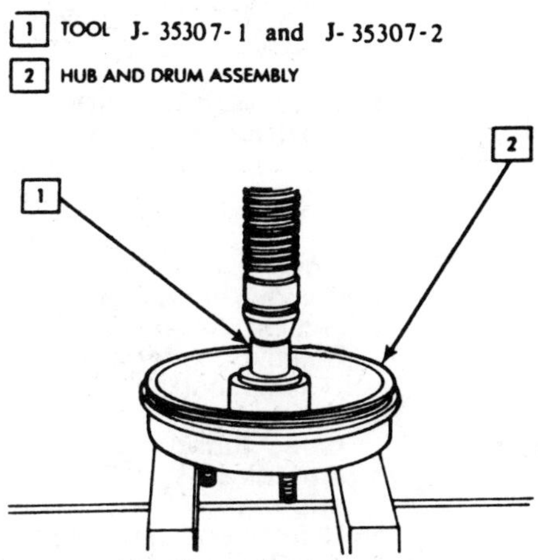

Fig. 8 Bearing race installation

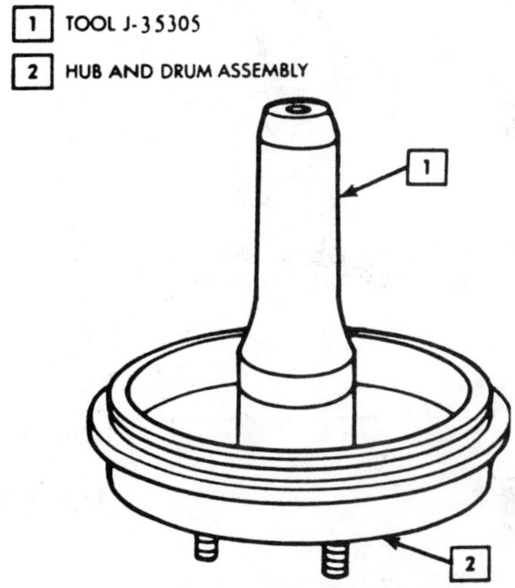

Fig. 9 Axle seal installation

using tool J-35307-1 and J-35307-2 respectively, **Fig. 8.**

8. Install replacement oil seal in inner side of hub using tool No. J-35305 or equivalent, **Fig. 9.**
9. Pack bearings using suitable grease.
10. Reverse procedure to install. Adjust wheel bearings as described under "Rear Wheel Bearing, Adjust."

REAR SUSPENSION TIGHTENING SPECIFICATIONS

Year	Component	Torque/ft. lbs.
1985–89	Control Arm To Body Bracket Attaching Bolt	72
	Rear Hub & Drum Spindle Nut	①
	Shock Absorber Lower Attaching Bolt	30
	Wheel & Tire Assembly	65

①—Refer to "Rear Wheel Bearing, Adjust" procedure.

Chevrolet Sprint & Geo Metro

INDEX

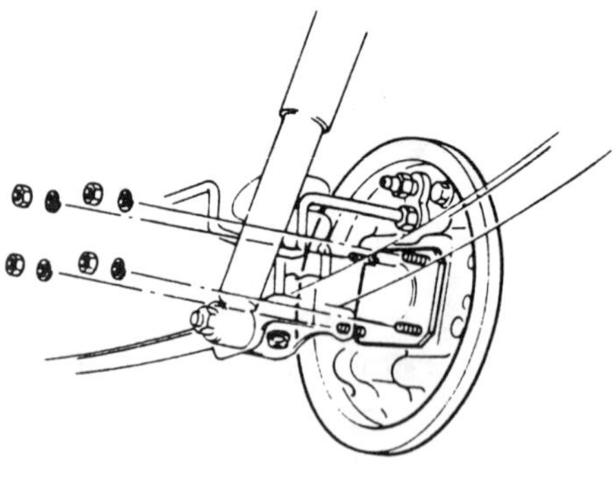

Fig. 1 Removing backing plate to axle attaching nuts

REAR AXLE REPLACE

1985–86

1. Raise and support vehicle.
2. Remove rear brake drums.
3. Disconnect brake hoses from lines.
4. Disconnect brake hoses and retainers from axle brackets.
5. Plug brake hoses to prevent loss of brake fluid.
6. Remove backing plate to axle nuts, **Fig. 1.**
7. Remove U-bolt nuts, U-bolts and jounce stop, **Fig. 2.**
8. Remove rear axle.
9. Reverse procedure to install. Note that alignment pin in leaf spring must engage in hole in axle bracket, **Fig. 2.** Tighten U-bolt nuts so equal amounts of threaded area are exposed.

1987–88 MODELS

1. Raise and support vehicle.
2. Remove rear brake drums.
3. Disconnect brake hoses from lines.
4. Disconnect brake hoses and retainers from trailing arm.
5. Plug brake hoses to prevent loss of brake fluid.
6. Remove brake shoe hold down spring by turning shoe hold down pin as shown in **Fig. 3.**
7. Disconnect parking brake cable from

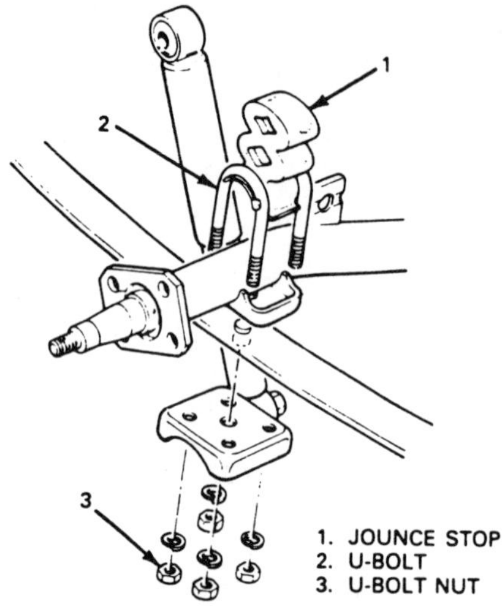

Fig. 2 Axle attachment parts removal. 1985–86

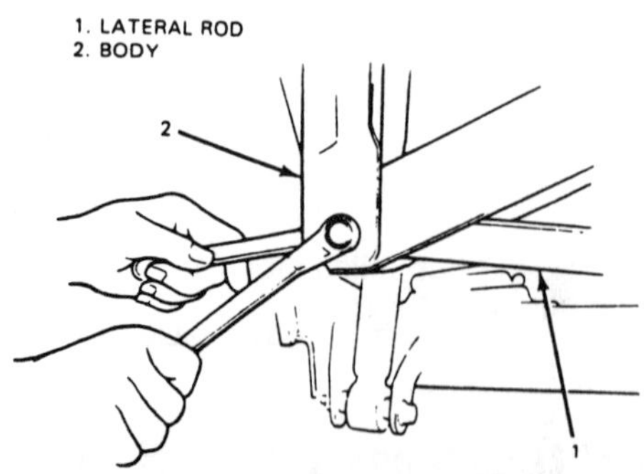

Fig. 4 Lateral rod body side bolt location. 1987–88

parking brake shoe lever and remove brake shoes. Remove cable from back plate.

8. Remove brake back plate from rear axle.
9. Support center of axle with a jack.
10. Remove lateral rod body side bolt, **Fig. 4.**
11. Remove shock absorber lower mounting bolt.
12. Lower rear axle slowly, until tension of suspension spring is relieved. Remove spring.

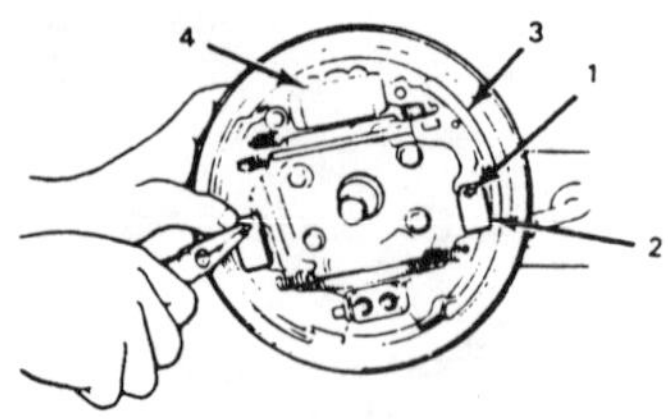

Fig. 3 Brake shoe hold down spring removal. 1987–88

13. Remove trailing arm front side bolt and lower rear axle.
14. Remove trailing arm bushing with suitable tool.
15. Reverse procedure to install. **When press fitting bushing into trailing arm, direct its slit side toward front and rear.**
16. Refer to end of this section for torque specifications.

SHOCK ABSORBER REPLACE

1985–88

1. Raise and support vehicle.
2. Remove shock absorber lower mounting bolt and lock washer.
3. Remove upper mounting locknut and absorber nut.
4. Remove shock absorber.
5. Reverse procedure to install.

STRUT REPLACE

1989–90

1. Raise and support rear of vehicle, then remove wheel and tire assembly.
2. Support suspension using a jack.
3. Remove strut upper support nuts, then push downward on strut, **Fig. 5.**
4. Remove strut lower mounting bolt, **Fig. 6.**
5. Separate strut from rear suspension knuckle, by compressing strut. **If strut is difficult to remove, open slit on**

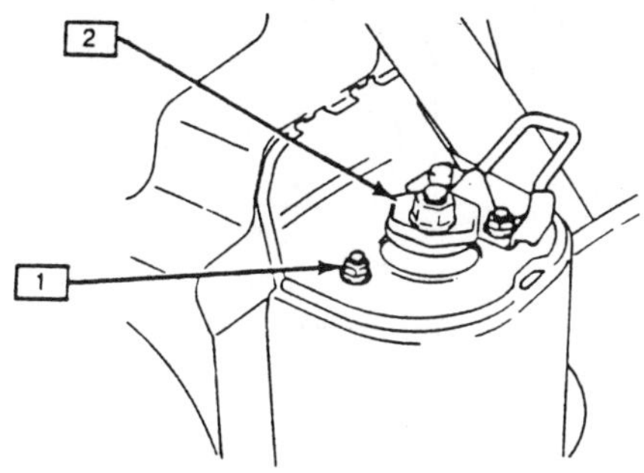

Fig. 5 Strut upper attaching nuts. 1989–90

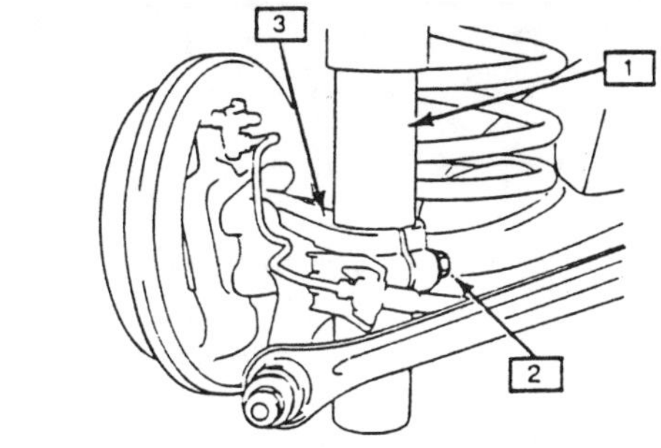

Fig. 6 Strut lower mounting bolt. 1989–90

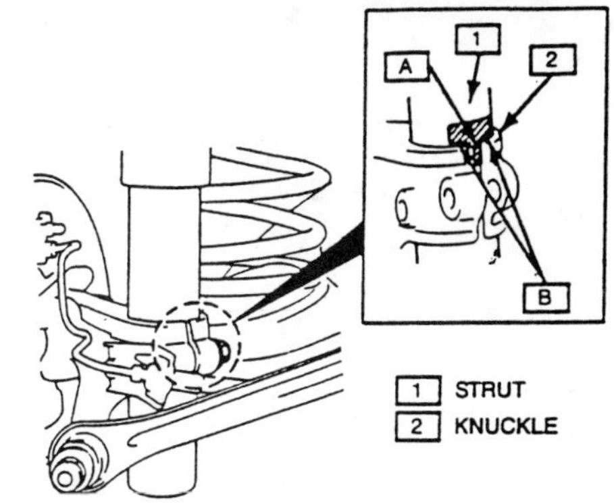

Fig. 7 Position strut to rear suspension knuckle. 1989–90

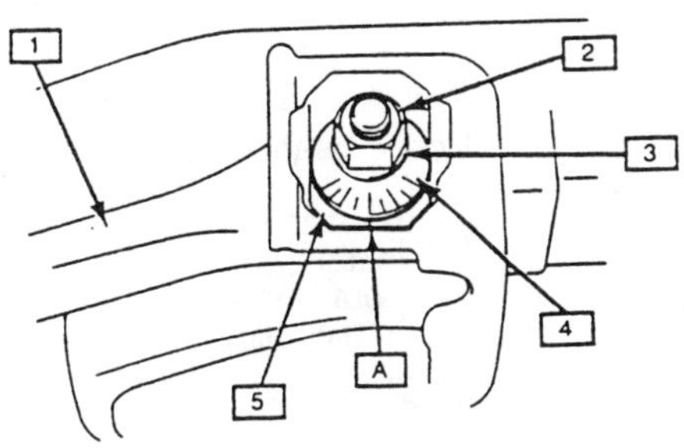

Fig. 8 Control rod inner bolt. 1989–90

knuckle just enough to allow strut removal.

6. Reverse procedure to install. When installing strut, align projection on strut with slit on rear suspension knuckle, **Fig. 7.**

LEAF SPRING
REPLACE
1985–86

1. Raise and support vehicle.
2. Support axle and remove U-bolt nuts, **Fig. 2.**
3. Remove shackle nuts and front pivot nut.
4. Remove front leaf spring pivot bolt and disengage leaf spring from shackle pin.
5. Reverse procedure to install. Lubricate leaf spring bushings with silicone grease. Tighten U-bolt nuts so equal amounts of threaded area are exposed.

COIL SPRING
REPLACE
1987–88 MODELS

Refer to "Rear Axle, Replace 1987-88 Models," and repeat steps 1 through 12 for coil spring removal.

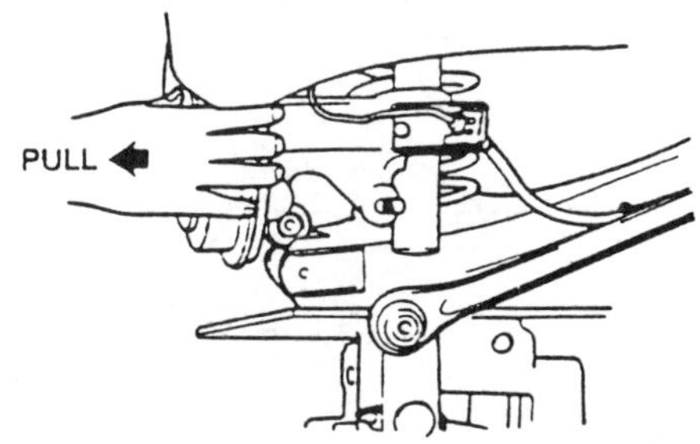

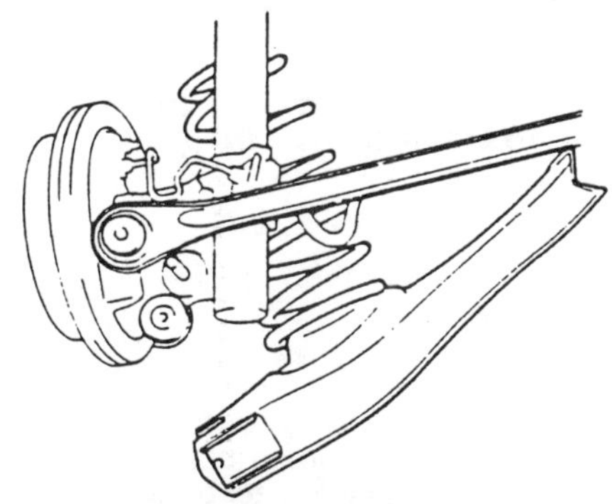

Fig. 9 Removing coil spring. 1989–90

COIL SPRING & SUSPENSION ARM
REPLACE
1989–90

1. Raise and support rear of vehicle, then remove wheel and tire assembly.
2. Place alignment marks on control rod and control rod washer (A), **Fig. 8,** for setting toe during installation.
3. Remove control rod to body attaching bolt, then separate control rod from bracket.
4. From wheel side of control rod, remove nut from rear suspension knuckle stud, then disconnect control rod.
5. Loosen, but do not remove suspension arm rear attaching nut.
6. Loosen rear suspension knuckle lower mounting nut. Position a jack under

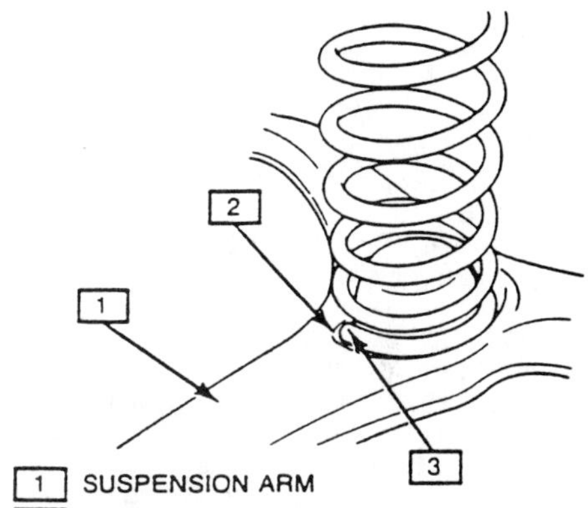

Fig. 10 Positioning coil spring to suspension arm. 1989–90

suspension arm, then remove knuckle lower mounting nut.

7. Raise lower arm slightly to allow removal of rear suspension knuckle lower mounting bolt.
8. Disengage rear suspension knuckle from suspension arm, then carefully lower suspension arm and remove coil spring, **Fig. 9.**
9. Remove remaining suspension arm attaching bolts and nuts and remove suspension arm, if necessary.
10. Reverse procedure to install. When installing spring, position spring end to stepped portion of suspension arm, **Fig. 10.** When installing control rod to body bracket, align marks made on washer and control rod during removal. **Do not torque control rod or suspension arm attaching nuts and bolts until after vehicle has been lowered.**

CONTROL ROD
REPLACE
1989–90

1. Raise and support rear of vehicle, then remove wheel and tire assembly.
2. Remove E-ring, then detach brake hose from control rod.
3. Place alignment marks on control rod

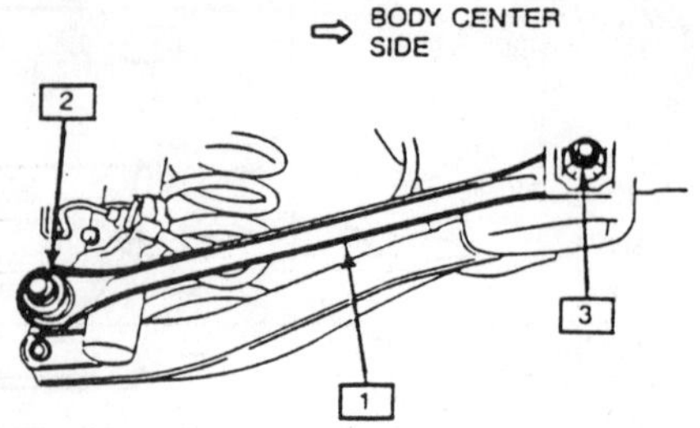

Fig. 11 Control arm installation. 1989–90

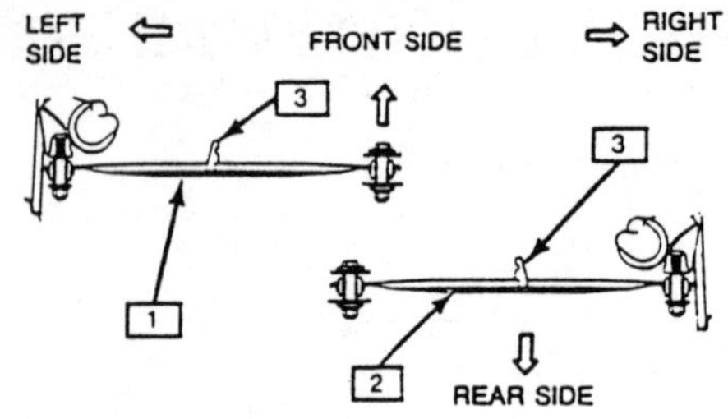

Fig. 12 Positioning control arm to vehicle. 1989–90

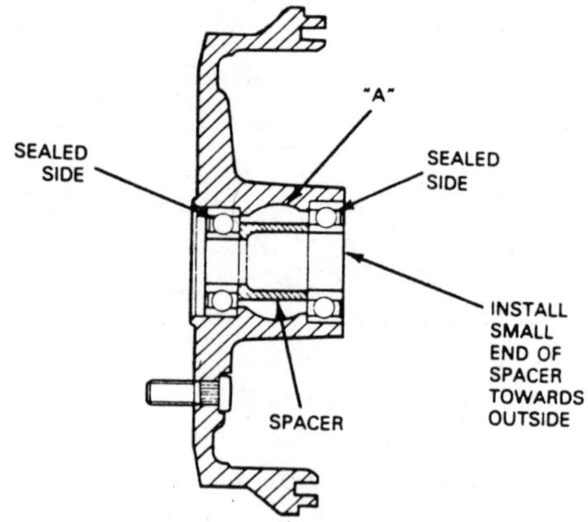

Fig. 13 Wheel bearing installation. 1985–88

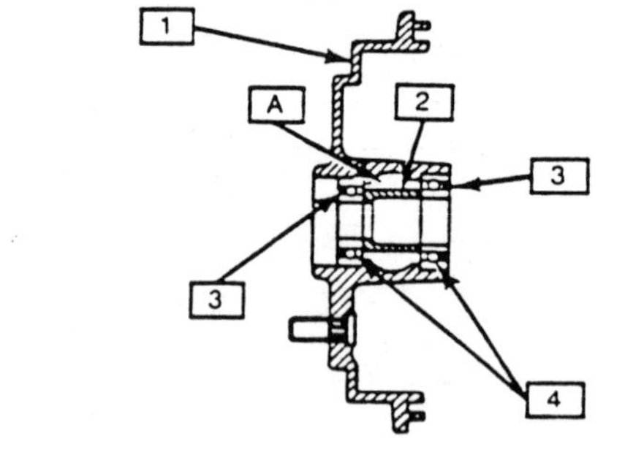

Fig. 14 Wheel bearing installation. 1989–90

and control rod washer at body bracket for setting toe during installation, **Fig. 8.**

4. Remove control rod to rear suspension knuckle attaching nut, **Fig. 11.**
5. Remove control rod to body bracket bolt, then remove control rod.
6. Reverse procedure to install. Position control rod to vehicle as shown in **Fig. 12.** When installing control rod to body bracket, align marks made on washer and control rod during removal. **Do not torque control rod attaching nuts and bolts until after vehicle has been lowered.**

REAR SUSPENSION KNUCKLE

REPLACE

1989–90

1. Raise and support rear of vehicle, then remove wheel and tire assembly.
2. Remove brake drum, then disconnect brake hose from rear suspension knuckle bracket.
3. Disconnect brake line from wheel cylinder. Cap brake line and wheel cylinder fitting bore.
4. Remove brake backing plate attaching bolts, then remove brake backing plate.
5. Position a jack under suspension arm.
6. Place alignment marks on control rod and control rod washer at body bracket for setting toe during installation.
7. Remove control rod to body bracket attaching nut and washer.
8. Remove control rod to rear suspension knuckle stud nut, then remove control rod.
9. Remove strut to rear suspension knuckle attaching bolt.
10. Remove strut to suspension arm attaching bolt, then separate rear suspension knuckle from suspension arm and strut. **If strut is difficult to remove, open slit on knuckle just enough to allow strut removal.**
11. Reverse procedure to install. When installing control rod to body bracket, align marks made on washer and control rod during removal. **Do not torque control rod or suspension arm attaching nuts and bolts until after vehicle has been lowered.** Prior to installation, apply sealer to mating surface of brake backing plate and rear suspension knuckle. When installing brake drum, torque wheel bearing nut to specification listed at the end of this section, then stake nut in place using a chisel. After completing installation, adjust and bleed brake system, then check for proper brake operation before moving vehicle.

WHEEL BEARING

REPLACE

1985–88

1. Raise and support vehicle.
2. Remove brake drum.
3. Using drift, drive bearings out of brake drum.
4. Using bearing installer J-34482 or equivalent, install bearings with stamped side outward. Refer to **Fig. 13** for spacer positioning.
5. Install brake drum.
6. Lower vehicle.

1989–90

Removal

1. Raise and support rear of vehicle, then remove wheel and tire assembly.
2. Remove wheel bearing dust cap from brake drum.
3. Using a chisel, unstake wheel bearing nut, then remove wheel bearing nut and washer.
4. Loosen parking brake cable adjusting nuts, then remove plug from backing plate and back off drum brake adjustment.
5. Using a suitable puller, remove brake. drum.
6. Using a brass drift, remove wheel bearings from brake drum.
7. Reverse procedure to install, noting the following:
   a. Fill Hub cavity (A), **Fig. 14,** with lithium wheel bearing grease.
   b. Install wheel bearings and spacer using tool Nos. J-7079-2 and J-34842 or equivalent. **Wheel bearing should be installed with sealed side facing outward, Fig. 14.**
   c. Install brake drum on spindle. Torque wheel bearing nut to specification listed at the end of this section, then stake nut in position.
   d. After completing installation, adjust drum brake and parking brake.
   e. Bleed and check brake system for proper operation prior to moving vehicle.

REAR SUSPENSION TIGHTENING SPECIFICATIONS

Component	Year & Torque/ft. lbs.		
	1985-86	1987-88	1989-90
Control Rod Nuts	—	—	59
Lateral Rod Bolt & Nut	—	32.5-50.5	—
Leaf Spring Front Mounting Nut	33-50.5	—	—
Leaf Spring Shackle Nut	22-39.5	—	—
Leaf Spring To Axle U-Bolts	22-32.5	—	—
Rear Suspension Knuckle Arm Lower Mounting Nut	—	—	37
Shock Absorber Upper Attachment	33-50.5	13-20	—
Shock Absorber Lower Attachment	8-12	32.5-50.5	—
Suspension Arm Mounting Bracket Bolt	—	—	33
Suspension Arm Front Nut	—	—	44
Suspension Arm Rear Nut	—	—	37
Strut Lower Mounting Bolt	—	—	44
Strut Support Nut	—	—	24
Strut Upper Nut	—	—	37
Trailing Arm Nut	—	50.5-65	—
Wheel Bearing Retaining Nut	58-86.5①	58-86.5②	74②
Wheel & Tire Assembly	29-50.5	29-50.5	44

①—After tightening nut, install cotter pin.
②—After tightening nut, stake nut in position.

Pontiac Fiero

INDEX

DESCRIPTION

1985–87

The drive axles are completely flexible assemblies which consist of an inner and outer constant velocity joint connected by an axle shaft. The inner constant velocity joint has the capability of moving in and out, whereas the outer joint does not.

The rear suspension, **Fig. 1,** is a MacPherson strut design. The lower control arms pivot from the engine cradle. The cradle uses isolation mounts to the body and conventional rubber bushings at the lower control arm pivots. A rubber mount isolates the upper end of the strut.

1988

The drive axles are completely flexible assemblies consisting of an inner Tri-Pot joint and an outer constant velocity joint connected by an axle shaft. The inner joint is completely flexible and has the capability of in and out movement. The outer joint is also flexible, but can not move in and out.

All drive axles except the right hand inboard joint of automatic transaxles incorporate a male spline and interlocks with the transaxle gears through the use of barrel type snap rings. The left hand inboard shaft attachment on automatic transaxles utilizes a female spline which installs over a stub shaft protruding from the transaxle.

The drive axle spline end mating with the steering knuckle and hub assembly incorporates a slight helix to ensure a tight press fit. This ensures that there is no radial play between hub and drive axle assembly.

The rear suspension, **Fig. 2,** features several improvements over the design used on earlier models. Major changes include the use of a "tri-link" design which allows for specific tuning of each component. In addition, a stabilizer bar is available on some models and the chassis cradle has been revised for suspension attachments.

DRIVE AXLE REPLACE

It is important that the axle not be over-extended. When one or both ends of the shaft are disconnected, over-extending the joint may cause separation of internal components, which could lead to failure of the joint.

1985–87

1. Shift transaxle to Neutral position.
2. Raise and support vehicle.
3. Remove wheel and tire assembly.
4. Remove lower ball joint bolt, then loosen tie rod end nut.
5. Position suitable punch through rotor and remove hub nut and washer. **A new nut must be used during installation.**
6. Separate tie rod shaft from tie rod end, then remove brake caliper and rotor assembly.
7. Depress lower control arm and separate strut from drive axle.
8. Remove drive axle from hub.
9. Remove drive axle from transaxle using tool No. J-28468 or J-33008 with J-29794 and J-2619-01 or equivalents.
10. Reverse procedure to install, noting the following:
    a. Install boot protectors (tool No. J-33162 or equivalent) prior to axle installation, **Fig. 3.**
    b. Torque lower ball joint bolt to specifications.
    c. Torque hub nut to specification.

1988

1. Shift transaxle into Neutral position and disconnect battery ground cable, then raise and support vehicle.
2. Remove wheel and tire assemblies.
3. Install drift punch through rotor and remove hub nut and washer, discarding nut.
4. Remove caliper and rotor, then disconnect trailing arm at knuckle.
5. Referring to **Fig. 4,** scribe strut and knuckle assembly as follows:
    a. Using a sharp tool, scribe knuckle along lower outboard strut radius (view A).
    b. Scribe strut flange on inboard side, along curve of knuckle (view B).
    c. Make a chisel mark across the strut/knuckle interface (view C).
6. Remove strut mounting bolts, then the hub from the drive axle.
7. Install tool J-28468 or J-33008 with J-29794 and J-2619-01 slide and remove drive axle from transaxle, **Fig. 3.**
8. Install drive axle seal boot protectors J-33162 or equivalent on all Tri-Pot inner joints with silicone (gray) boots.
9. Start splines of drive axle into transaxle and push drive axle until it snaps into place.
10. Align hub assembly with drive axle and install.
11. Loosely install strut mounting bolts with bolt head to rear of vehicle.
12. Install fixed adjusting link, lateral control arm through bolt with bolt head to front of vehicle and **torque** to 37 ft. lbs. plus an additional 90°.
13. Connect trailing arm at knuckle with bolt head inboard and **torque** to 44 ft. lbs. plus an additional 90°.
14. Install strut mounting bolts at knuckle and torque to specification while observing scribe marks.
15. Install caliper and rotor.
16. Insert drift punch through rotor and install washer and new hub nut, torquing to 183-208 ft. lbs.
17. Remove drive axle seal booster protector, if used.
18. Install tire and wheel assemblies.

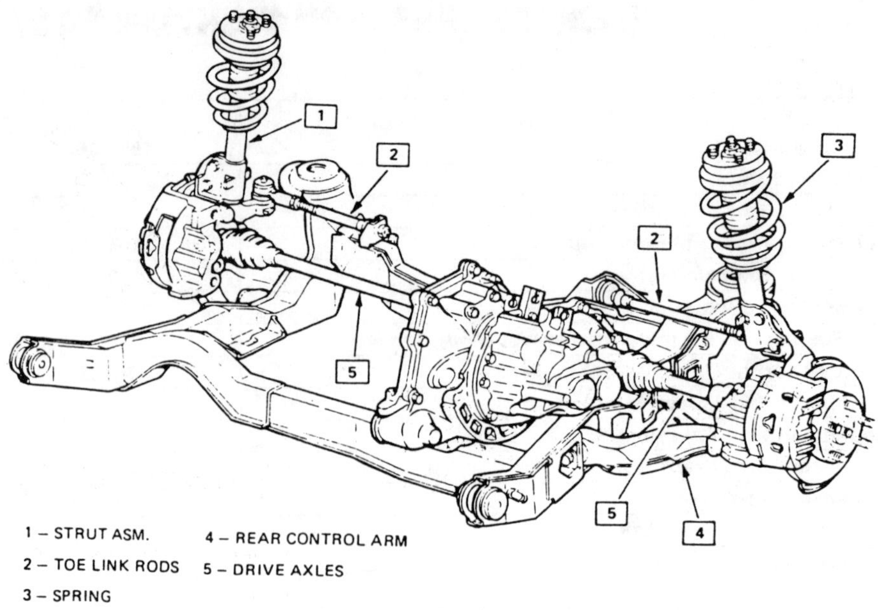

Fig. 1 Rear suspension. 1985–87

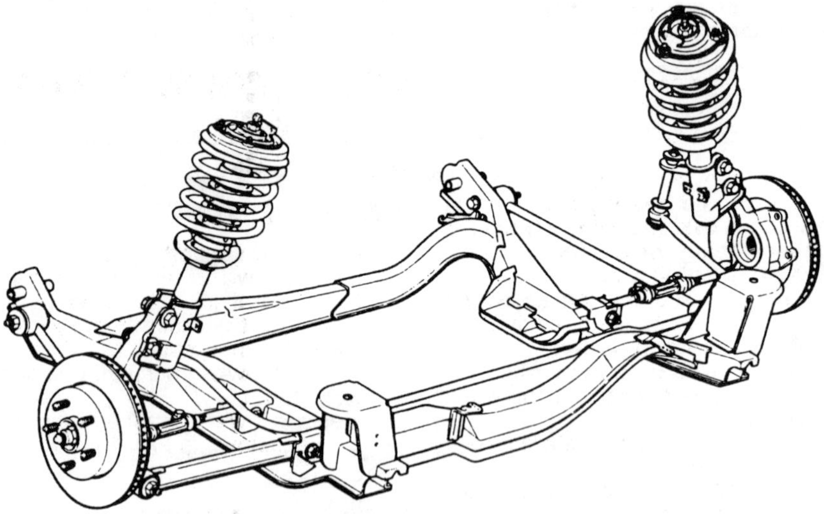

Fig. 2 Rear suspension. 1988

DRIVE AXLE SERVICE

Refer to **Figs. 5 and 6** for service procedures on drive axle assembly.

WHEEL BEARING REPLACE

REMOVAL

1. **On vehicles equipped with steel wheels,** remove hub cap and loosen hub nut, then raise and support vehicle and remove wheel and tire assembly.

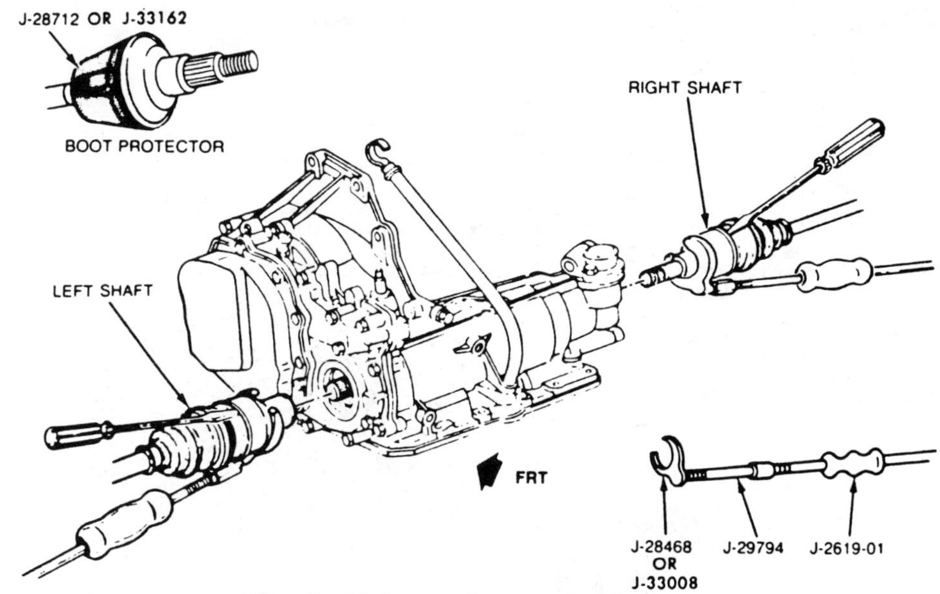

Fig. 3 Drive axle removal

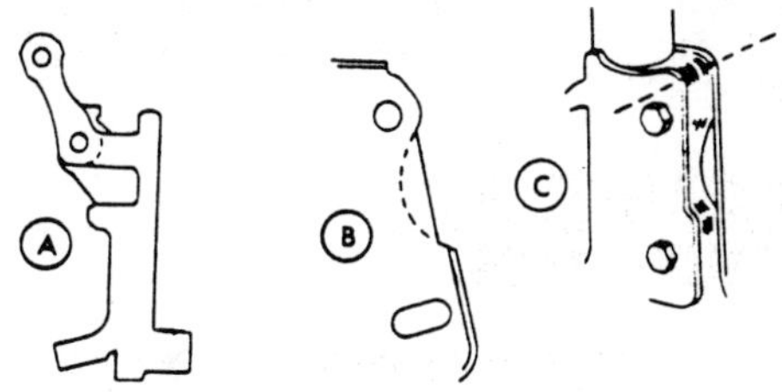

Fig. 4 Scribing strut and knuckle assembly

2. **On vehicles equipped with 14 inch aluminum wheels,** set parking brake, then raise and support vehicle and remove wheel and tire assembly.
3. **On all models,** install drive axle boot protector J-33162.
4. Remove and discard hub nut.
5. Remove brake caliper and rotor. Suspend caliper from frame with a piece of wire.
6. Remove hub and bearing attaching bolts. **If the old bearing is being reinstalled, mark attaching bolts and corresponding holes for installation reference, Fig. 7.**
7. Remove hub and bearing assembly using tool No. J-28733 or equivalent, **Fig. 8. If assembly is heavily corroded, ensure hub and bearing are loose in knuckle before using puller tool.**
8. Replace knuckle seal if installing new bearing. **Do not move drive axle until hub nut is installed and torqued to specifications.**

INSTALLATION

1. Clean and inspect knuckle bore and bearing mating surfaces for dirt, nicks and burrs.
2. If installing new knuckle seal, apply suitable grease to seal and knuckle bore, then press seal into knuckle using tool No. J-28671 or equivalent.
3. Install hub and bearing assembly onto axle shaft. Torque attaching bolts to specification.
4. Install hub nut and **torque** to 74 ft. lbs.
5. Install brake rotor and caliper.
6. Install wheel and tire assembly, then lower vehicle and **torque** hub nut to 200 ft. lbs.

LOWER BALL JOINT REPLACE

1985–87

1. Raise and support vehicle.
2. Remove wheel and tire assembly, then the ball stud clamp bolt.
3. Disconnect ball joint from knuckle by tapping with a mallet.
4. Replace ball joint as shown in **Fig. 9.**

LOWER CONTROL ARM & BUSHINGS REPLACE

1985–87

1. Raise and support vehicle.
2. Remove wheel and tire assembly, then the ball joint clamp bolt.
3. Disconnect ball joint from knuckle by tapping with a mallet.
4. Remove lower control arm pivot bolts from frame.
5. Replace lower control arm and bushings as shown in **Fig. 10.**

REAR KNUCKLE REPLACE

1985–87

Refer to **Fig. 11** for removal and installation procedures.

1988

Removal

1. Raise and support vehicle.
2. Install drive axle boot protector.
3. Remove wheel and tire assembly, then remove and discard hub nut.
4. Remove caliper and rotor, then disconnect trailing arm at knuckle.
5. Remove fixed adjusting link/lateral control arm through bolt, then the strut mounting bolts.
6. Remove hub/knuckle assembly from drive axle.

Installation

1. Install hub/knuckle assembly to drive axle.
2. Loosely install strut mounting bolts with bolt head to rear.
3. Install fixed adjusting link/lateral control arm through bolt.
4. Connect trailing arm at knuckle with bolt head inboard.
5. Install strut mounting bolts at knuckle.
6. Install rotor and caliper.
7. **Torque** new hub nut and washer to 74 ft. lbs., then remove drive axle seal boot protector.
8. Install tire and wheel assembly and lower vehicle.
9. **Torque** hub nut and washer to 200 ft. lbs.
10. Adjust camber and toe in as necessary.

STRUT ASSEMBLY REPLACE

1. Remove engine compartment cover.
2. Remove three upper strut nuts and washers, **Fig. 12.**
3. Loosen wheel lug nuts, then raise vehicle and support at rear control arm.
4. Remove wheel and tire assembly, then the brake line retaining clip.
5. Scribe strut and knuckle, **Fig. 4,** for assembly reference. When servicing the jounce bumper, strut mount, strut shield, spring seal or spring insulator, the strut and knuckle must be scribed as shown in **Fig. 4** to maintain the original camber setting. However, it will be necessary to check toe-in setting and adjust as necessary. **When servicing the strut damper, knuckle or rear ride spring, the scribe marks should not be made. However, it will be necessary to check both toe-in and camber settings and correct as necessary.**
6. Remove strut attaching nuts and bolts, then the strut assembly and spacer plate.
7. Reverse procedure to install. Torque knuckle attaching nuts and upper strut attaching nuts to specifications.

STRUT ASSEMBLY SERVICE

DISASSEMBLY

1. Clamp strut compressor, tool No. J-26854 (1985-87) or J-34013 (1988), in a suitable vise.
2. Install strut assembly into bottom

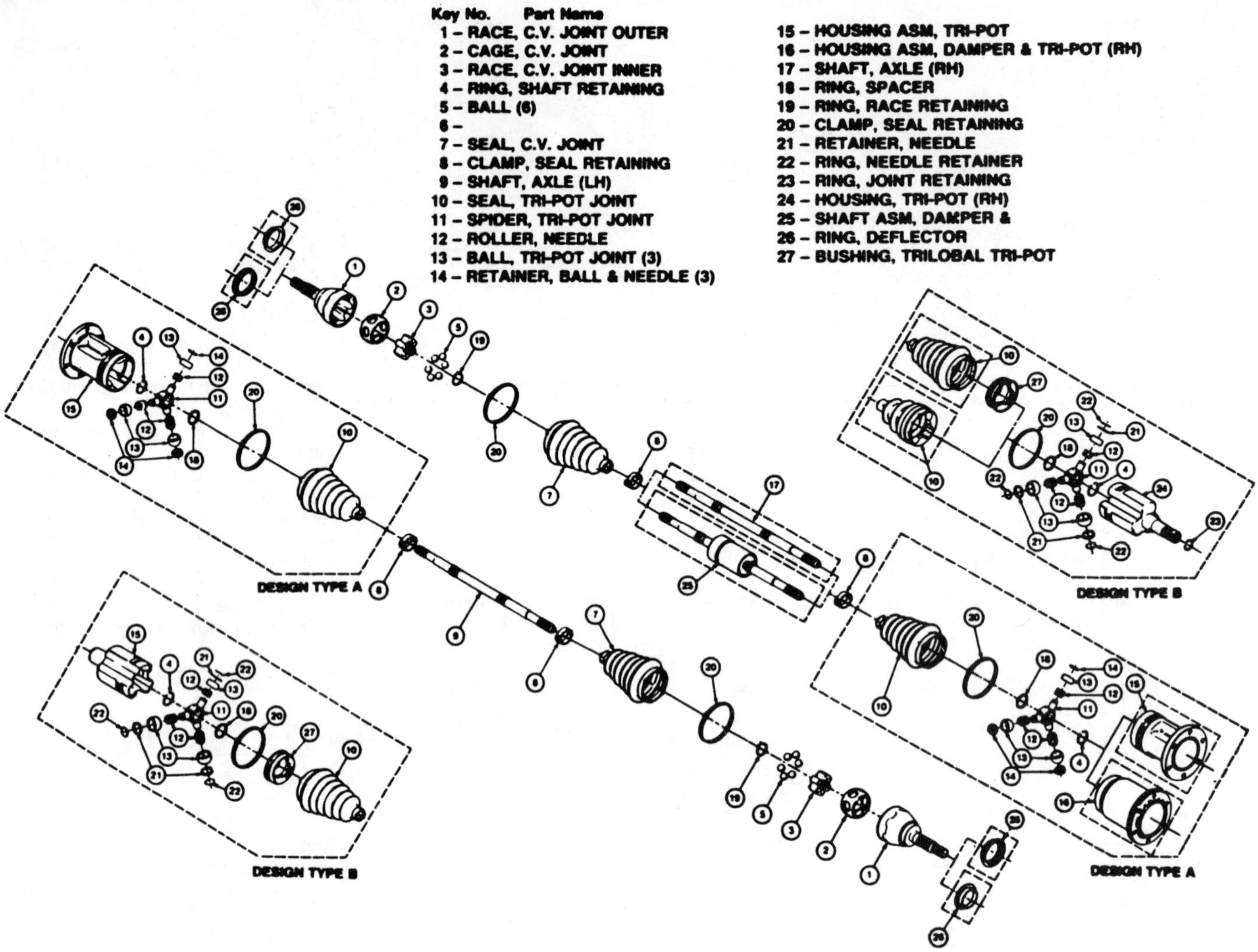

Fig. 5 Exploded view of drive axle

adapter of compressor and install bottom adapter, **Fig. 13.** Ensure strut and locating pins are fully engaged.

3. Rotate strut assembly until top mounting assembly lip is aligned with compressor support notch.
4. Install upper adapter onto top spring seal, **Fig. 13,** so long stud is at high location to strut flange.
5. Rotate compressor forcing screw clockwise until top support flange contacts top adapter, and continue turning screw to compress strut spring.
6. Install second upper adapter over spring seat assembly, then rotate forcing screw counterclockwise until strut spring tension is relieved.
7. Remove top adapters, bottom adapter and strut.

ASSEMBLY

1. Perform steps 1 and 2 as outlined in the "Disassembly" procedure.
2. Rotate strut assembly until mounting flange is facing outward, opposite compressor forcing screw.
3. Install strut components, **Fig. 14.** Ensure spring is properly seated on bottom spring plate.
4. Install strut spring seat assembly on top of spring with long stud positioned 180° from strut mounting flange.
5. Install top adapter over spring seat assembly.
6. Rotate compressor forcing screw until compressor top support just contacts top adapter. Do not compress spring.
7. Install strut alignment rod through top spring seat and thread onto damper shaft hand tight, **Fig. 14.**
8. Rotate compressor forcing screw clockwise to compress spring until damper shaft is exposed enough so nut can be threaded securely, then install the nut. Ensure damper shaft comes through center of spring seat opening to prevent damage. **Do not compress spring until bottomed.**
9. Remove alignment rod, then install mount and **torque** nut to 65 ft. lbs.
10. Rotate compressor forcing screw counterclockwise and remove strut assembly from compressor.

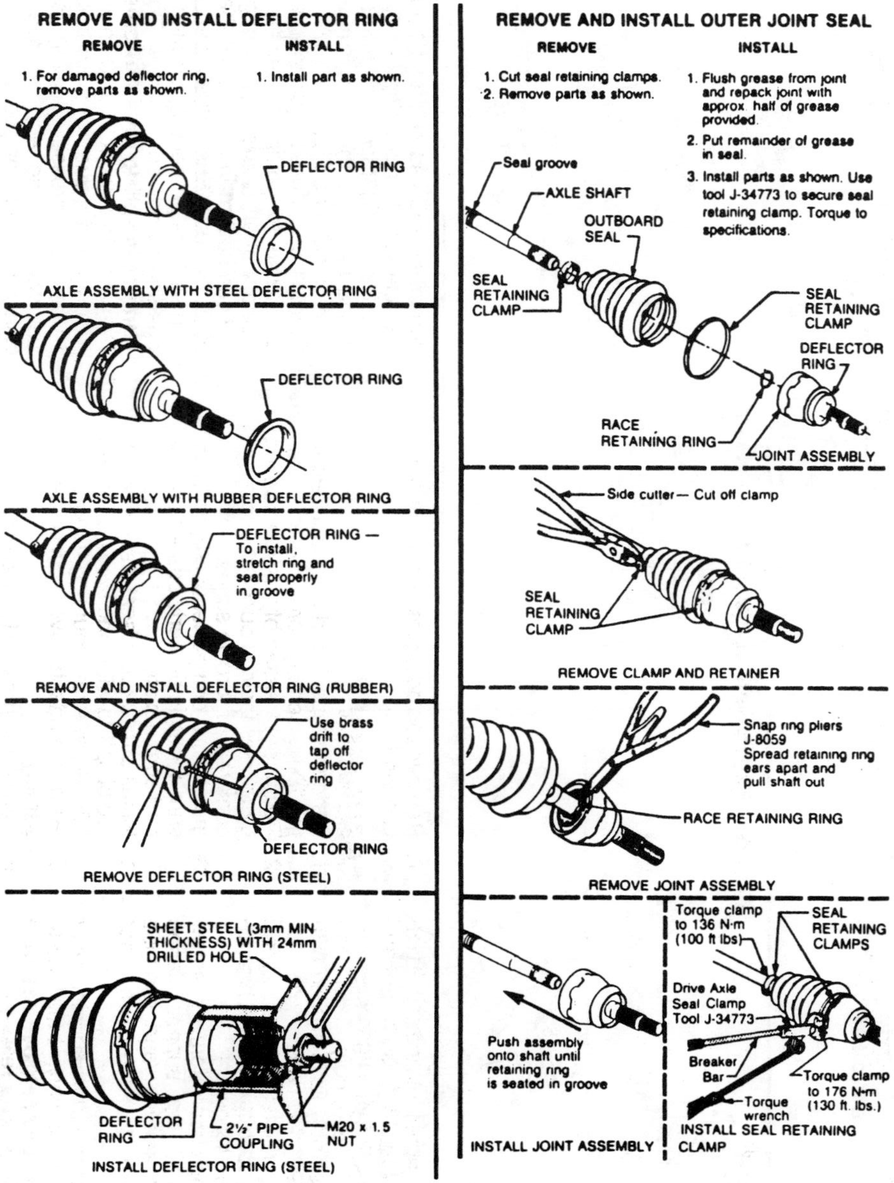

Fig. 6 Drive axle service (Part 1 of 2)

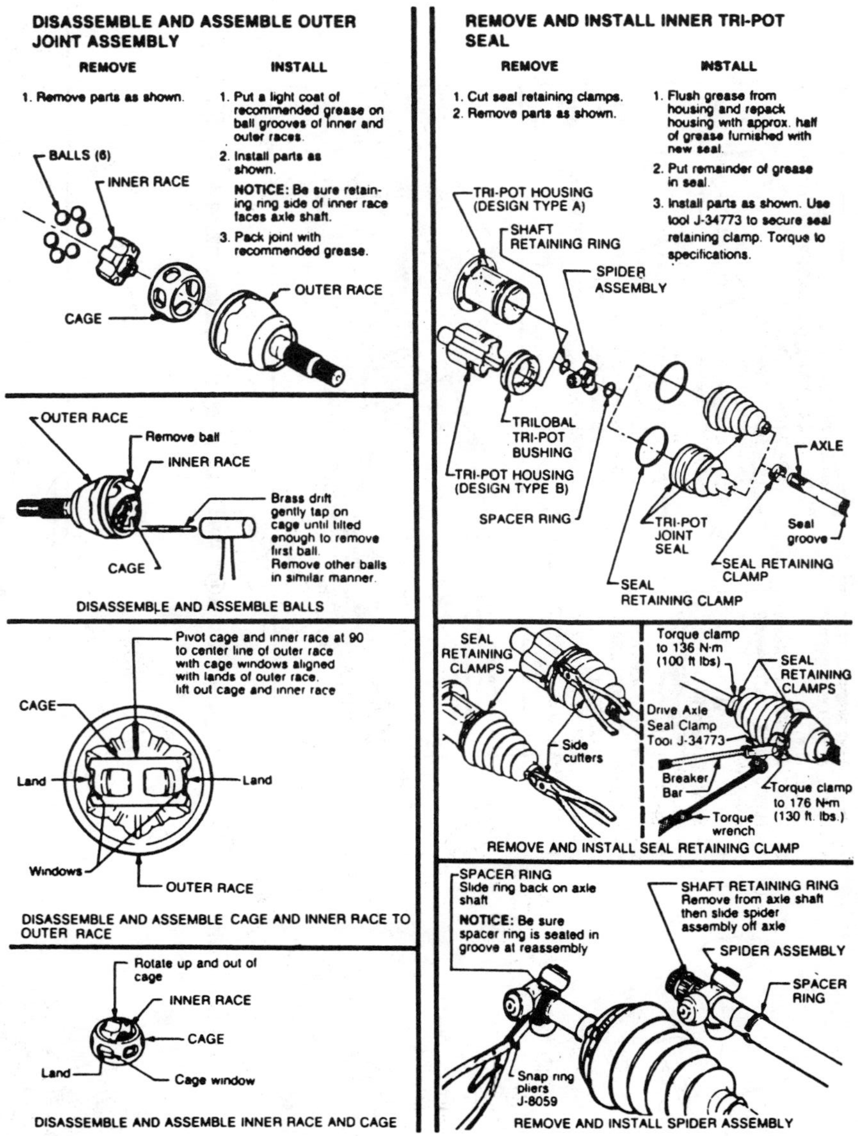

Fig. 6 Drive axle service (Part 2 of 2)

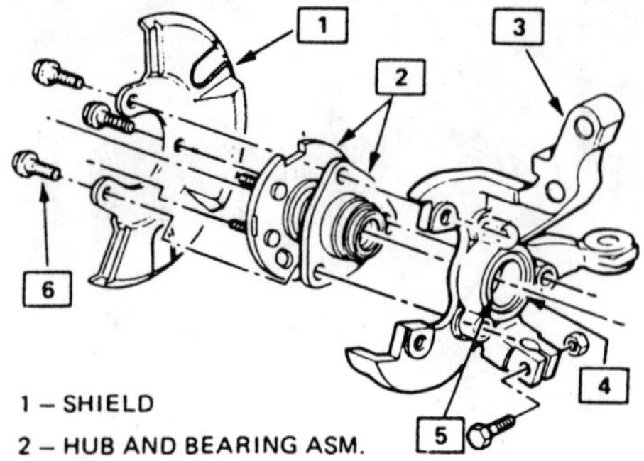

1 – SHIELD
2 – HUB AND BEARING ASM.
3 – KNUCKLE
4 – KNUCKLE SEAL ASM.
5 – FILL HUB BEARING CAVITY BETWEEN SEALING LIPS WITH .8 GRAMS OF CHASSIS LUBRICANT.
6 – BOLT 75-95 N·m (55-70 FT. LB.)

Fig. 7 Rear wheel hub & bearing assembly

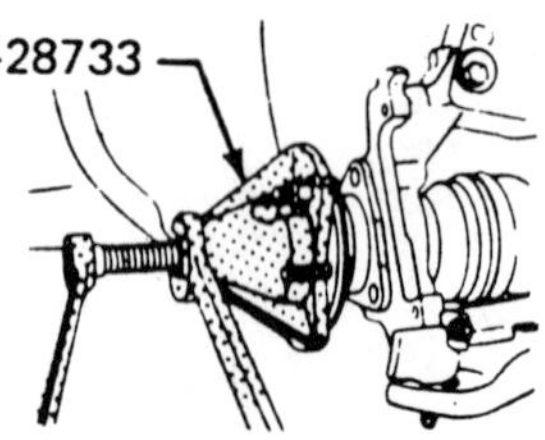

Fig. 8 Hub & bearing assembly removal

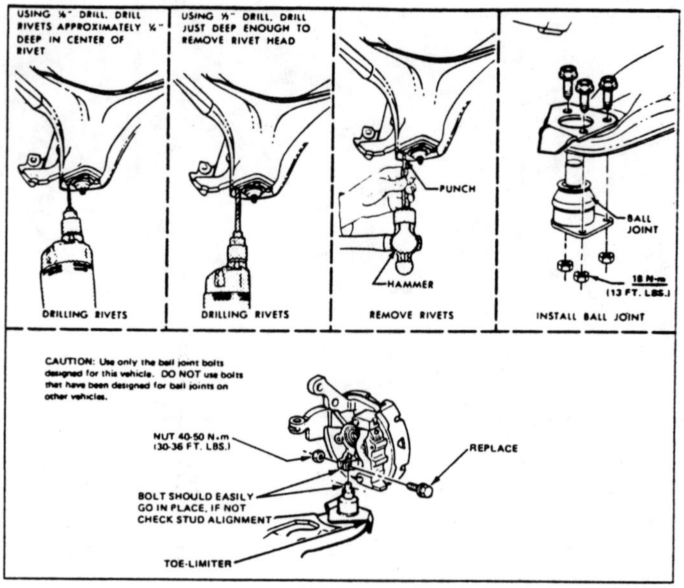

Fig. 9 Lower ball joint replacement. 1985–87

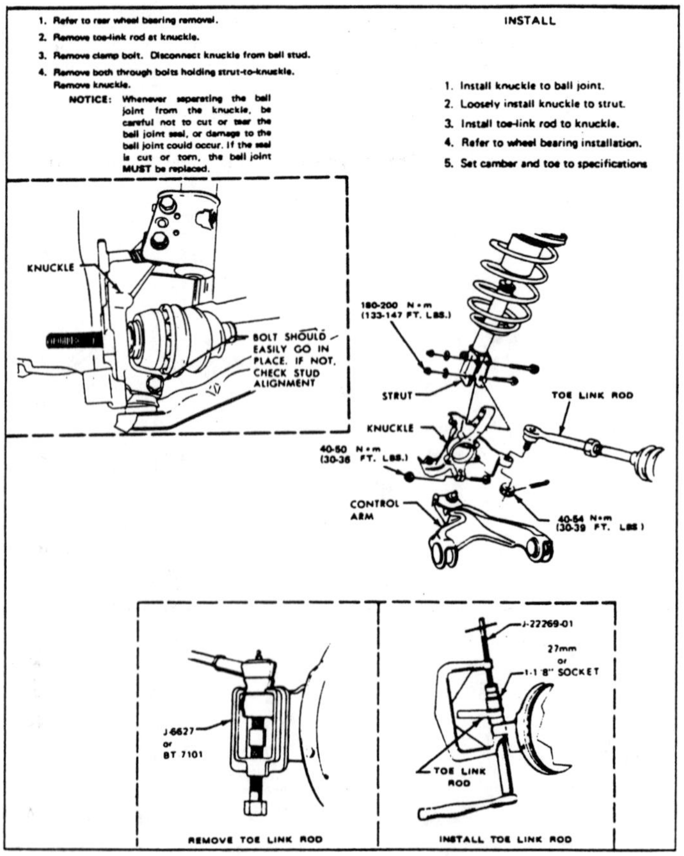

Fig. 11 Rear knuckle replacement. 1985–87

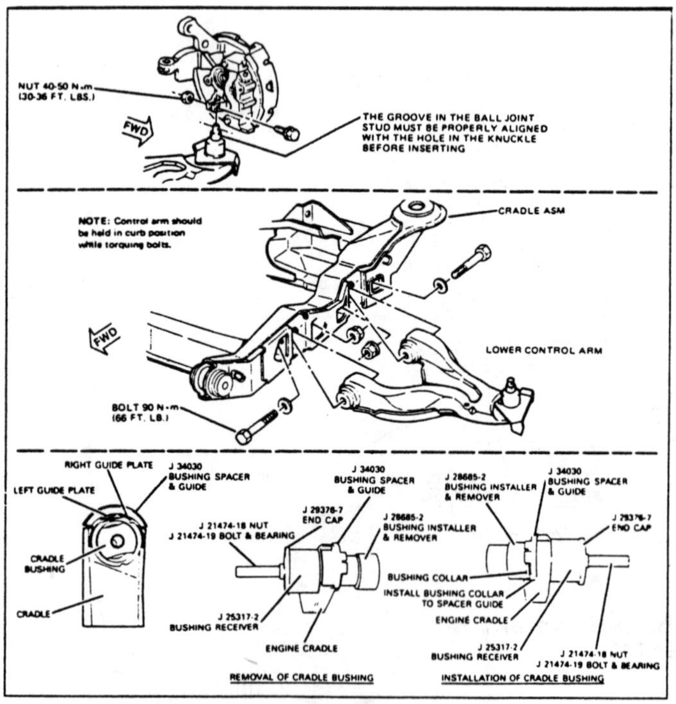

Fig. 10 Lower control arm & bushing replacement. 1985–87

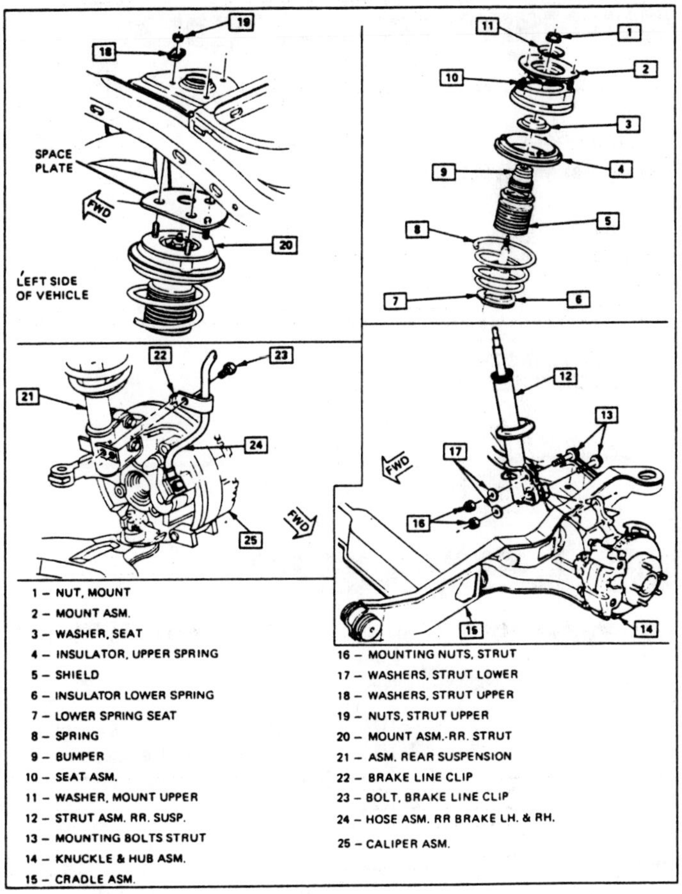

Fig. 12 Strut assembly replacement. 1985–87

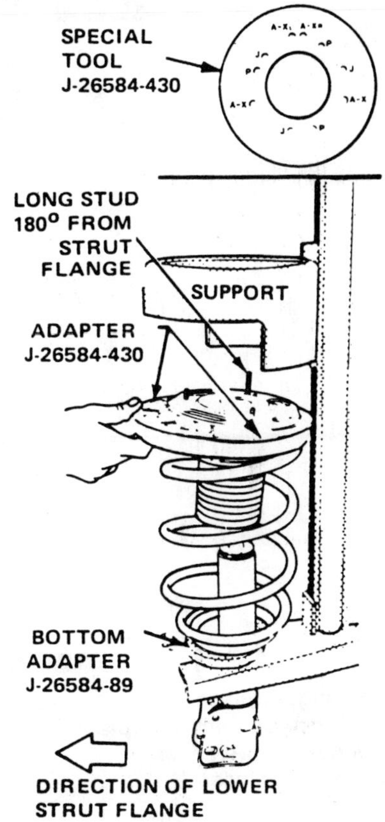

Fig. 13 Removing damper & coil spring from strut

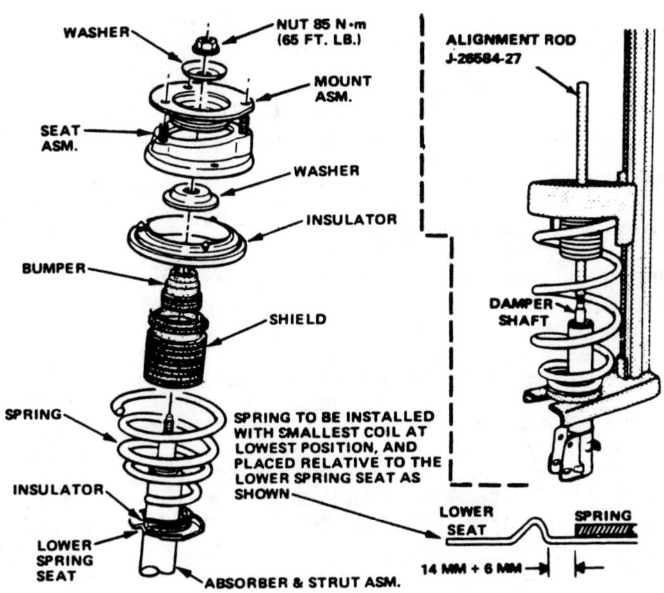

Fig. 14 Strut assembly alignment & components

REAR SUSPENSION TIGHTENING SPECIFICATIONS

Year	Component	Torque/ft. lbs.
1985–88	Hub And Bearing Assembly Attaching Nuts	55-70
	Hub Nut	183-208
	Lower Ball Joint	30-36
	Strut To Knuckle	140
	Upper Strut Attaching Nuts	18

①—1985–87, 17, ft. lbs; 1988, 10-15 ft. lbs.
②—1985–87, 22-30 ft. lbs; 1988, 20 ft. lbs.

Pontiac LeMans

INDEX

REAR AXLE REPLACE

REMOVAL

1. Raise and support vehicle.
2. Remove rear wheel and tire assemblies.
3. Remove exhaust heat shield from vehicle underbody.
4. Measure and record length of thread on parking brake lever push rod.
5. Remove self-locking nut from parking brake lever push rod and the lock plate from brake cable compensation yoke.
6. Disconnect parking brake cable from brake cable compensation yoke.
7. Disconnect parking brake cable from guides on transaxle tunnel and plastic sleeves on fuel tank and position toward muffler.
8. Raise one rear axle arm using a suitable jack.
9. Disconnect brake lines from brake pressure hoses on rear axle bracket and remove lock clips. **Cap brake line fittings.**
10. Disconnect brake pressure hoses from bracket.
11. Remove shock absorbers as described under "Shock Absorber, Replace."
12. Remove rear axle arm attaching bolt, then lower the arm and remove coil spring. If necessary, depress rear axle arm using a pry bar positioned in shock absorber bracket.
13. Repeat step 12 on opposite side.
14. Support center of rear axle assembly using a suitable jack.
15. Remove rear axle attaching bolts and carefully lower axle assembly from vehicle, **Fig. 1.**
16. If rear axle assembly is to be replaced, proceed as follows:
    a. Remove brake drum set screw and the drum.
    b. Remove dust cap from wheel hub.
    c. Remove wheel hub retaining nut and the wheel hub.
    d. Bend open brake line retaining clamps.
    e. Remove brake anchor plates and paper seal from rear axle arm.
    f. Remove stabilizer bar attaching bolts and the stabilizer bar.
    g. Disconnect parking brake cable from mounting.

INSTALLATION

1. If rear axle assembly is being replaced, proceed as follows:
    a. Attach stabilizer bar to rear axle and tighten attaching bolts to torque listed at the end of this section.
    b. Apply suitable sealant to brake anchor plate, then position new paper seals on plate.
    c. Apply suitable locking compound to anchor plate attaching bolts, then install the bolts and tighten to torque listed at the end of this section.
    d. Position brake lines in rear axle clamps and bend clamps closed.
    e. Apply suitable grease to wheel spindle, then install wheel hub and **torque** attaching nut to 12 ft. lbs.
    f. Adjust wheel bearings as described under "Wheel Bearings, Adjust."
    g. Install brake drums and set screw.
2. Raise rear axle assembly and secure to vehicle underbody. Raise one rear axle arm slightly using a suitable jack.
3. Install coil spring, ensuring lower rubber damper is properly positioned in spring seat.
4. Raise axle arm and install shock absorber. Tighten attaching bolts to torque listed at the end of this section.
5. Repeat steps 3 and 4 on opposite side.
6. Fasten brake pressure hoses to rear axle bracket and retaining clips.
7. Attach brake pressure lines to pressure hoses.
8. Install parking brake cable and adjust to dimension recorded at removal.
9. Install rear wheel and tire assemblies, then adjust wheel bearings as de-

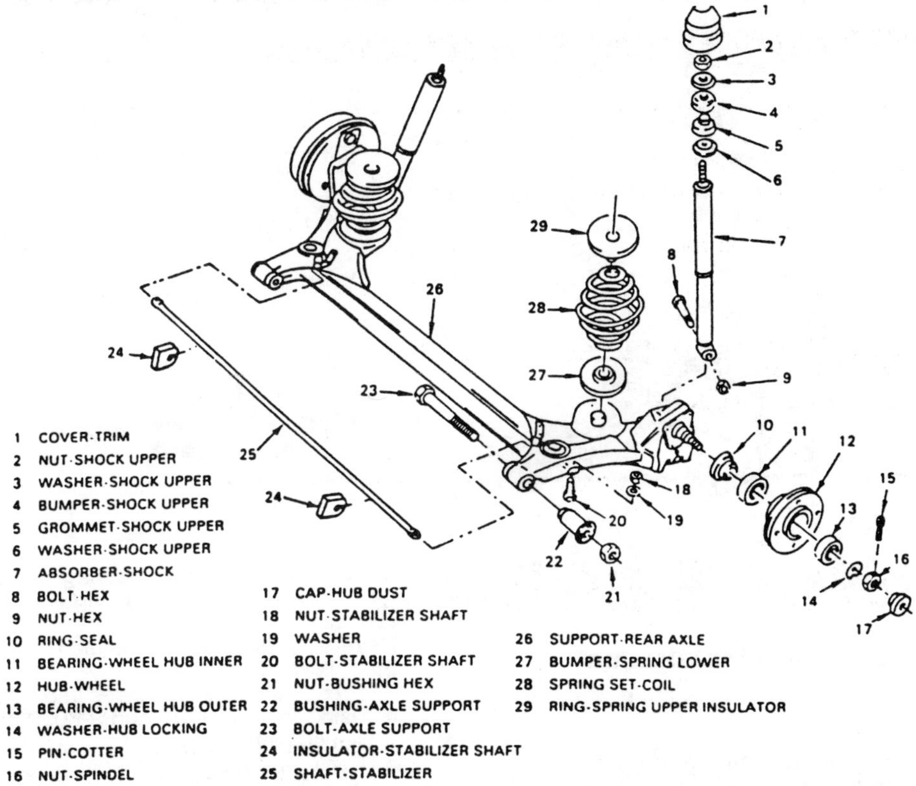

Fig. 1 Exploded view of rear axle & suspension assembly

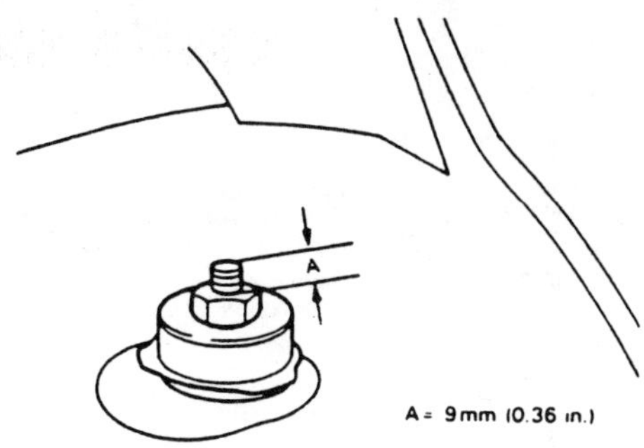

Fig. 2 Shock absorber upper nut installation

scribed under "Wheel Bearings, Adjust."

10. Tighten rear axle attaching bolts to torque listed at the end of this section, with weight of vehicle on wheels.
11. Bleed brake system, then adjust parking brake as necessary.

WHEEL BEARINGS
ADJUST

1. Raise and support rear of vehicle.
2. Remove hub dust cap, then the spindle nut cotter pin and nut.
3. **Torque** spindle nut to 18 ft. lbs. while turning wheel assembly forward by hand.
4. Back off spindle nut until just loose, then hand-tighten the nut.
5. Loosen spindle nut until either hole in spindle aligns with slot in nut, then install new cotter pin. **Do not loosen nut more than 1/2 turn.**
6. Ensure bearing end play measures .001 to .005 inch, then install dust cap on hub.

SHOCK ABSORBER
REPLACE

1. Open deck lid and remove trim cover, if equipped.
2. Remove upper shock absorber attaching nut.
3. Raise rear of vehicle and support rear axle using a suitable jack.
4. Remove lower attaching bolt and nut and the shock absorber. **Do not remove both shock absorbers at the same time.**
5. Reverse procedure to install, noting the following:
   a. Tighten lower attaching bolt to torque listed at the end of this section.
   b. Tighten upper attaching nut until thread extends .36 inch above the nut, **Fig. 2.**

STABILIZER BAR
REPLACE

1. Raise and support rear of vehicle.
2. Remove one rear wheel and tire assembly.
3. Remove stabilizer bar attaching bolts and nuts from both sides of axle, then remove stabilizer bar toward side of vehicle less wheel.
4. Reverse procedure to install. Tighten stabilizer bar attaching bolts to torque listed at the end of this section.

COIL SPRING
REPLACE

1. Raise and support rear of vehicle. Support control arms with suitable jackstands.
2. Remove brake line clips from body bracket and allow brake line to hang free.
3. Disconnect parking brake cable at equalizer.
4. Disconnect brake hoses and cap fittings.
5. Remove right and left shock absorber lower attaching bolts.
6. Carefully lower rear axle and remove springs and insulators.
7. Reverse procedure to install, noting the following:
   a. Prior to installing springs, secure upper insulators to body with suitable adhesive.
   b. Tighten shock absorber lower attaching bolts to torque listed at the end of this section.

CONTROL ARM BUSHINGS
REPLACE

1. Raise and support rear of vehicle.
2. Raise one rear axle arm using a suitable jack and remove shock absorber lower attaching bolt.
3. Raise opposite rear axle arm and remove shock absorber lower attaching bolt.
4. Lower axle arm slightly and remove coil spring with rubber damper, then raise axle and install shock absorber lower attaching bolt.
5. Repeat step 4 on opposite side of axle.
6. Support center of rear axle assembly using a suitable jack.
7. Remove brake line retaining clips from vehicle underbody and position brake lines in underbody recess.
8. Remove rear axle attaching bolts, then lower axle slightly and support with tool hangar No. J-29376-13 or equivalent.
9. Heat control arm bushings to 122-158°F, then remove the bushings using tools shown in **Fig. 3.**
10. Install new bushings as shown in **Fig. 3.**
11. Reverse remainder of removal procedure to complete installation. Ensure bushings are properly aligned, **Fig. 4.**

HUB & BEARING
REPLACE

1. Raise and support rear of vehicle.

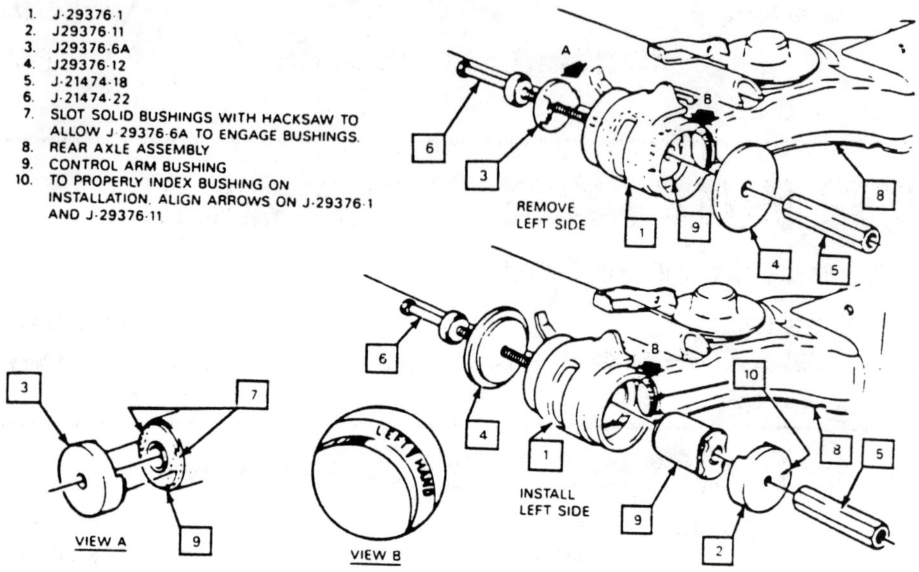

Fig. 3 Control arm bushing removal & installation

2. Remove wheel and tire assembly.
3. Remove brake drum set screw and the drum. If necessary, loosen parking brake cable and press parking brake lever in with a screwdriver.
4. Remove dust cap from hub, then the spindle nut cotter pin and nut.
5. Remove wheel hub and lockwasher.
6. Remove seal from wheel hub using a suitable screwdriver.
7. Remove inner tapered roller bearing from hub.

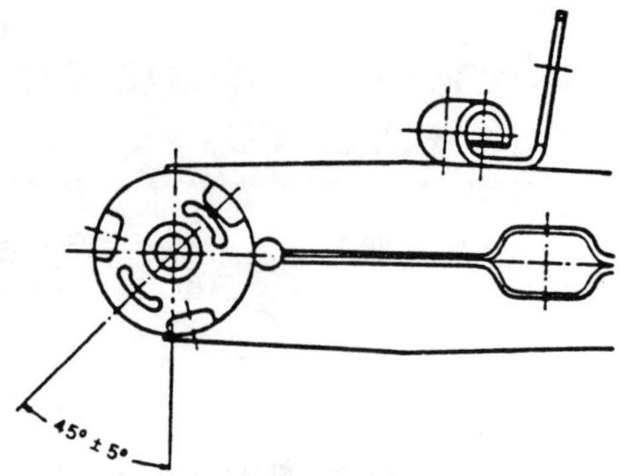

Fig. 4 Control arm bushing alignment

8. Drive inner and outer bearing outer races out of wheel hub using a suitable drift.
9. Reverse procedure to install, noting the following:
   a. Press bearing outer races into hub using tools J-6791 and J-8092 or equivalents.
   b. Pack wheel bearings, seal lip and wheel hub with suitable wheel bearing grease.
   c. Following completion of installation, adjust wheel bearings and parking brake as needed.

REAR SUSPENSION TIGHTENING SPECIFICATIONS

Year	Component	Torque/ft. lbs.
1988–90	Control Arm To Body Bracket Nut	70
	Shock Absorber To Body Mount	①
	Shock Absorber To Axle Bolt	51
	Stabilizer Shaft To Axle Nuts	59
	Wheel Bearing Spindle Nut	②
	Wheel & Tire Assembly	66

①—Tighten until .36 inch of thread is exposed above top of nut.
②—Refer to "Wheel Bearing, Adjust" procedure.

FRONT WHEEL DRIVE AXLES

INDEX

CHRYSLER MOTORS/EAGLE

TABLE OF CONTENTS

Chrysler Except Imports

INDEX

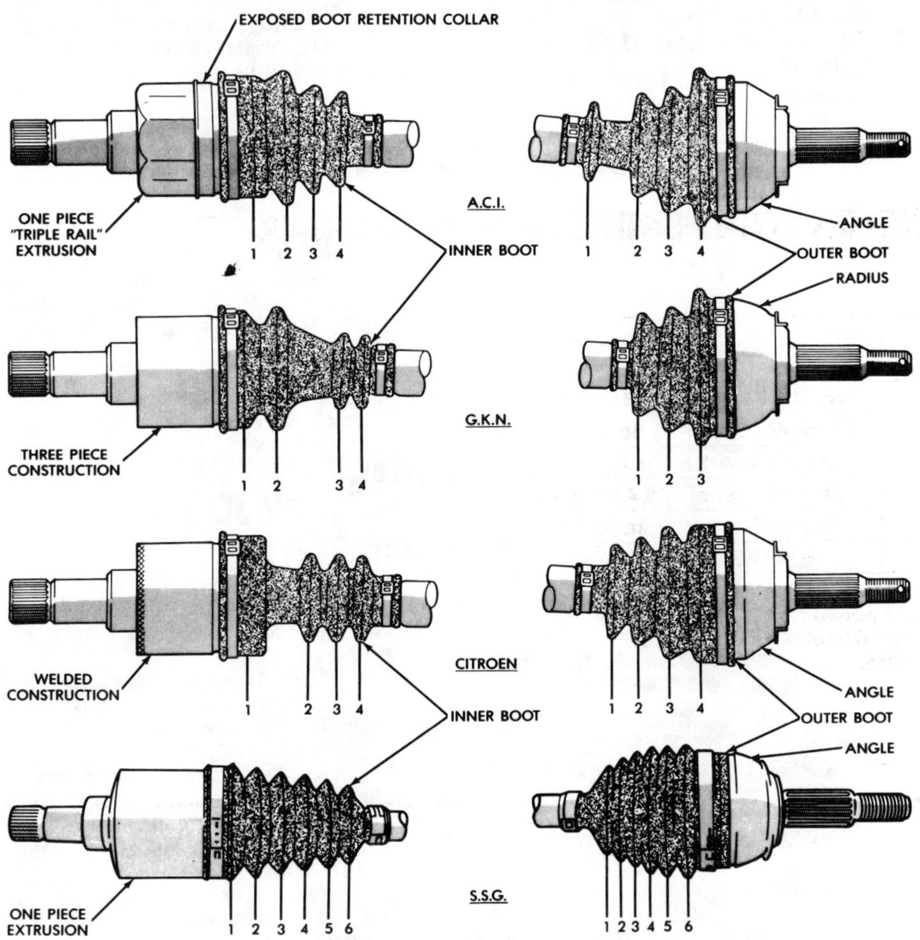

Fig. 1 Driveshaft identification. 1985–87

DRIVESHAFT IDENTIFICATION

Driveshafts are identified as "A.C.I.," "CITROEN," "G.K.N." or "S.S.G." assemblies, **Figs. 1 and 2.** Vehicles can be equipped with any of these assemblies, however they should not be intermixed.

Two different driveshaft systems are used. All 4-135/2.2L and some turbocharged models use an "equal length" system while all others use an "unequal length" system. The "equal length" system has short solid interconnecting shafts of equal length on the left and right sides. The "unequal length" system has a short solid interconnecting shaft on left side with a longer tubular interconnecting shaft on right.

Procedures for installation and removal of driveshafts are essentially the same for all types of assemblies used.

DRIVESHAFTS REPLACE

REMOVAL

Inboard C/V joints have stub shafts splined into differential side gears, or splined into the intermediate shaft on right side of an equal length system. Driveshafts are spring loaded and are retained to side gears by constant spring pressure provided by spring contained in C/V joints.

1. If removing the righthand driveshaft, speedometer pinion must be removed prior to driveshaft removal, **Fig. 3.**
2. Remove clamp bolt securing ball joint clamp bolt to steering knuckle, then separate ball joint stud from steering knuckle. **Do not damage ball joint or C/V joint boots.**

3. Separate outer C/V joint splined shaft from hub by holding C/V housing while moving knuckle/hub assembly away from C/V joint. **Do not damage slinger on outer C/V joint. Do not attempt to remove, repair or replace.**
4. Support assembly at C/V joint housings and remove by pulling outward on the inner C/V joint housing. Do not pull on shaft. **If removing lefthand driveshaft assembly, removal may be aided by inserting a screwdriver blade between differential pinion shaft and carefully prying against end face of stub.**
5. Remove driveshaft assembly from vehicle.

INSTALLATION

1. Hold inner joint assembly at housing while aligning and guiding inner joint spline into transaxle. **On turbocharged models, ensure rubber washer seal is in place on right inner C/V joint.**
2. **On all models,** push knuckle/hub assembly out and install splined outer C/V joint shaft into hub.
3. Install knuckle assembly on ball joint stud.
4. Install clamp bolt. **Torque** to 70 ft. lbs.
5. Install speedometer pinion, **Fig. 3.**
6. Fill differential to bottom of filler plug hole with Dexron automatic transaxle fluid.
7. Install washer and hub nut. **Torque** hub nut to 180 ft. lbs. Install nut lock and cotter pin.
8. If, after attaching driveshaft assembly in vehicle, the inboard boot appears collapsed or deformed, vent the inner boot by inserting a round tipped, small diameter rod between boot and shaft. As venting occurs, the boot will return to normal shape. **After installation of driveshaft, check driveshaft length as outlined in "Driveshaft Length, Adjust."**

DRIVESHAFT LENGTH
ADJUST

If the vertical bolts on both engine mounts have been loosened, or vehicle has experienced front structural damage, driveshaft length must be checked.

The engine mounts incorporate slotted bolt holes to permit side-to-side positioning of the engine, thereby affecting length of driveshaft. To check driveshaft length proceed as follows.

1. Position vehicle with wheels straight ahead and body weight distributed on all 4 tires.
2. Measure direct distance between inner edge of outboard boot to inner edge of inboard boot on both driveshafts, **Fig. 4.**
3. Driveshaft length must be within specifications in chart, **Figs. 5 through 10.** If measurement is not within specifications, engine position must be corrected as follows:
   a. Remove load from engine mounts by carefully supporting engine and transaxle assembly with a suitable jack.
   b. Loosen right and left engine mount vertical bolts, then loosen only the right and front engine mount bracket-to-crossmember attaching bolts.
   c. Pry engine to right or left as necessary to bring driveshaft length within specifications. **Ensure left engine mount is sleeved over long support bolt and shaft, Fig. 11, to provide lateral adjustment whether or not engine weight is removed.**
   d. **Torque** engine mount vertical bolts to 27 ft. lbs. and front and center left engine mount bolts to 40 ft. lbs.
   e. **On 1989 New Yorker Landau and Dynasty models, torque** vertical bolts to 60 ft.lbs.

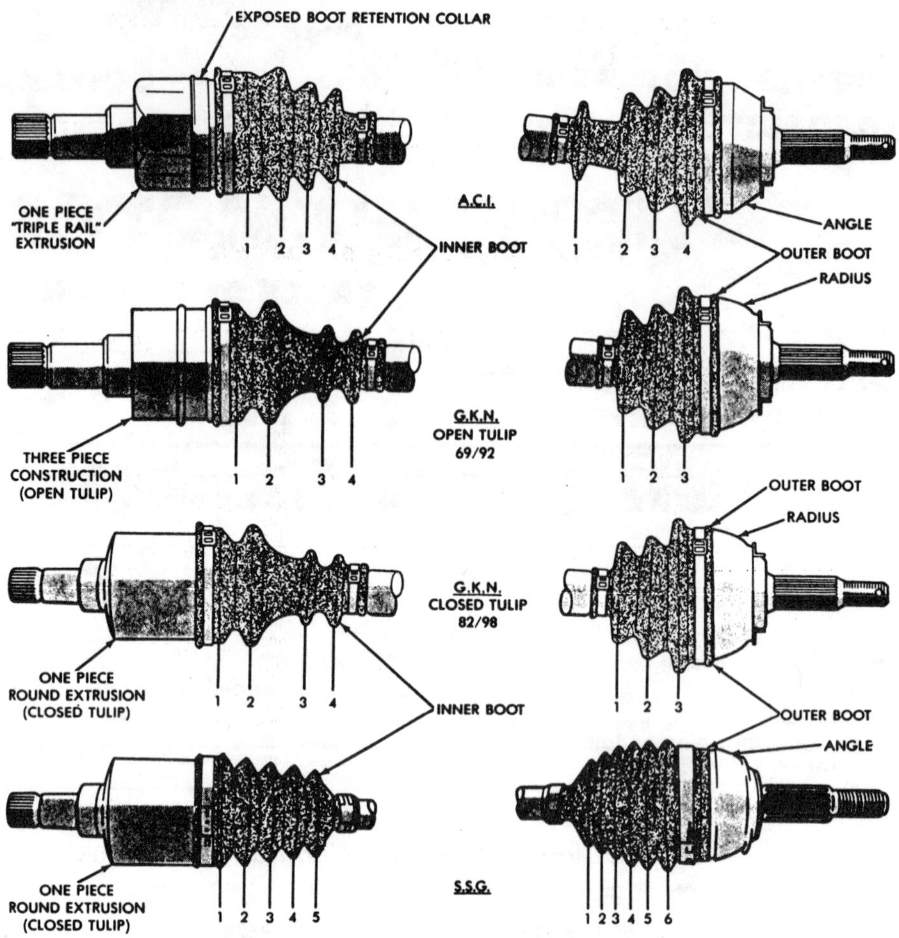

Fig. 2 Driveshaft identification. 1988–89

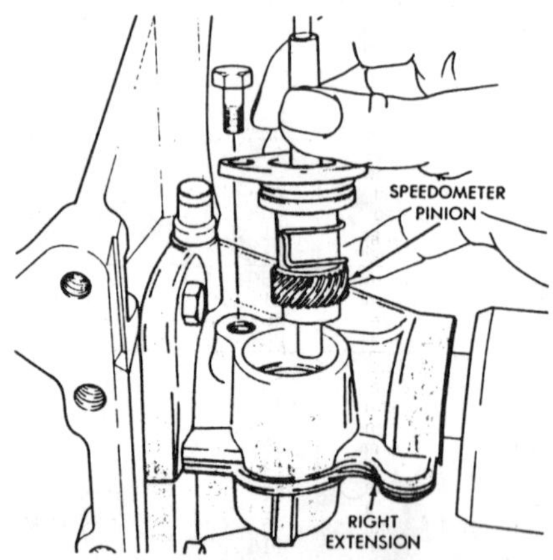

Fig. 3 Speedometer pinion replacement

INNER CONSTANT VELOCITY JOINT SERVICE
DISASSEMBLY

1. Remove clamp and boot from joint and discard, **Fig. 12.**
2. **On Citroen units,** separate tripod from housing by slightly deforming retaining ring at three locations, **Fig. 13.** If necessary, cut retaining ring from housing and install replacement retaining ring with inner flange rolled, then stake into machined groove with a suitable punch. **When removing tripod from housing, secure rollers. After tripod has been removed, secure assembly with tape.**
3. **On G.K.N. units,** bend tabs on joint cover using needle nose pliers, then remove tripod from housing, **Fig. 14.**
4. **On A.C.I. units,** hold housing and

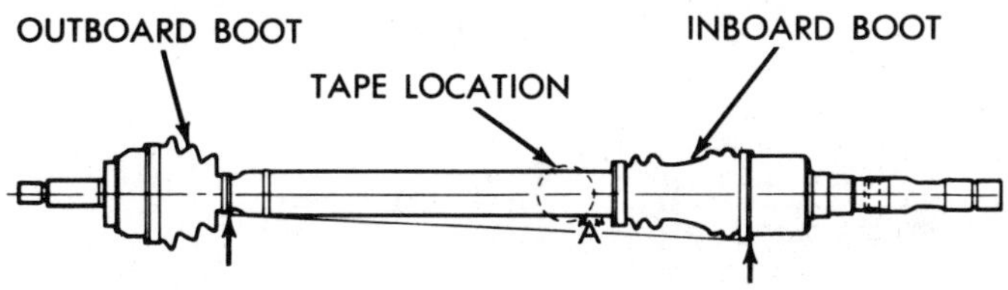

Fig. 4 Measuring driveshaft

Body	Engine	Driveshaft Identification Type	Side	Tape Color	"A" Dimension mm	Inch
CHARGER, HORIZON, OMNI & TURISMO	1.6L/2.2L	G.K.N.	Right	Yellow	498-509	19.6-20.0
			Left	Yellow	240-253	9.5-10.0
		A.C.I.	Right	Red	469-478	18.5-19.0
			Left	Red	208-218	8.2- 8.6
	2.2L Turbo	Citroen	Right	Orange	211-220	8.3-8.7
			Left	Orange	211-220	8.3-8.7
CHARGER, DAYTONA, HORIZON, LANCER, LASER, LEBARON GTS, OMNI & TURISMO	2.2L	Citroen	Right	Green	465-477	18.3-18.8
			Left	Green	211-220	8.3-8.7
		A.C.I.	Right	Blue	463-472	18.2-18.6
			Left	Blue	204-213	8.0-8.4
EXC. CHARGER, HORIZON, OMNI & TURISMO	2.2L	G.K.N.	Right	Blue	505-515	19.9-20.3
			Left	Blue	259-277	10.2-10.9
		A.C.I.	Right	Green	477-485	18.8-19.1
			Left	Green	229-244	9.0- 9.6
		G.K.N./A.C.I.	Right	Orange	492-500	19.4-19.7
			Left	Orange	243-258	9.6-10.2
		Citroen	Right	White	480-492	18.9-19.4
			Left	White	238-255	9.4-10.0
	2.6L	G.K.N.	Right	Silver	501-510	19.7-20.1
			Left	Silver	254-269	10.0-10.6
		Citroen	Right	Yellow	480-492	18.9-19.4
			Left	Yellow	238-255	9.4-10.0
EXC. CHARGER, HORIZON, OMNI & TURISMO	2.2L Turbo	G.K.N.	Right	Tan	257-265	10.1-10.4
			Left	Silver	254-269	10.0-10.6
		Citroen	Right	Red	241-251	9.5- 9.9
			Left	Yellow	238-255	9.4-10.0

Fig. 5 Driveshaft length specifications. 1985

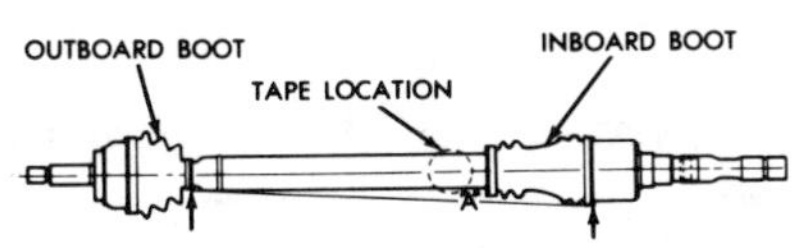

Body	Engine	Driveshaft Identification Type	Side	Tape Color	"A" Dimension mm	Inch
CHARGER, HORIZON OMNI & TURISMO	1.6L/2.2L	G.K.N.	Right	Yellow	498-509	19.6-20.0
			Left	Yellow	240-253	9.5-10.0
		A.C.I.	Right	Red	469-478	18.5-19.0
			Left	Red	208-218	8.2- 8.6
	2.2L Turbo	Citroen	Right	Orange	211-220	8.3-8.7
			Left	Orange	211-220	8.3-8.7
CHARGER, DAYTONA, HORIZON, LANCER, LASER, LEBARON GTS, OMNI & TURISMO	2.2L	Citroen	Right	Green	465-477	18.3-18.8
			Left	Green	211-220	8.3-8.7
		A.C.I.	Right	Blue	463-472	18.2-18.6
			Left	Blue	204-213	8.0-8.4
EXC. CHARGER, HORIZON, OMNI, & TURISMO	2.2L/2.5L	G.K.N.	Right	Blue	505-515	19.9-20.3
			Left	Blue	259-277	10.2-10.9
		A.C.I.	Right	Green	477-485	18.8-19.1
			Left	Green	229-244	9.0- 9.6
		G.K.N./A.C.I.	Right	Orange	492-500	19.4-19.7
			Left	Orange	243-258	9.6-10.2
		Citroen	Right	White	480-492	18.9-19.4
			Left	White	238-255	9.4-10.0
		S.S.G	Right	Gold	457-469	17.9-18.5
			Left	Gold	216-232	8.5- 9.1
	2.6L	G.K.N.	Right	Silver	501-510	19.7-20.1
			Left	Silver	254-269	10.0-10.6
		Citroen	Right	Yellow	480-492	18.9-19.4
			Left	Yellow	238-255	9.4-10.0
EXC. CHARGER, HORIZON, OMNI, & TURISMO	2.2L Turbo	G.K.N.	Right	Tan	257-265	10.1-10.4
			Left	Silver	254-269	10.0-10.6
		Citroen	Right	Red	241-251	9.5- 9.9
			Left	Yellow	238-255	9.4-10.0

Fig. 6 Driveshaft length specifications. 1986

lightly compress C/V joint retention spring while bending tabs back with a pair of pliers, **Fig. 15.** Support housing, then separate tripod from housing.

5. **On S.S.G. units,** use a flathead screwdriver to pry wire ring out of groove and slide tripod from housing,

Fig. 16. When removing housing from tripod, hold rollers in place on the trunnion studs to prevent rollers and needle bearings from falling out. After tripod is out of housing, secure rollers in place with tape.

6. Remove snap ring from end of shaft, then the tripod using a brass drift.

INSPECTION

Remove grease from assembly and inspect bearing race and tripod components for wear and damage and replace as necessary. On spring loaded joints inspect spring, spring cup and spherical end of connecting shaft for wear and damage and replace as necessary. **Components of spring loaded and non-spring loaded inner C/V joints cannot be interchanged.**

ASSEMBLY

Do not use the A.C.I. or G.K.N. boot clamps on the Saginaw hard plastic C/V boots, as these clamps do not have the load capacity to withhold the much greater force needed to clamp plastic boots.

1. **On right side of turbocharged models,** slide rubber seal over stub shaft and into groove.
2. **On all models,** slide small end of boot over shaft. On Tubular type shafts, align boot lip with mark on shaft outer diameter. On solid type shafts, position small end of boot in groove on shaft.
3. Place rubber clamp over groove on boot, if equipped.
4. **On A.C.I and G.K.N. units,** install tripod on shaft with non-chamfered face of tripod body facing shaft retainer groove. On S.S.G. units, place wire ring retainer over interconnecting shaft, then on all models slide tripod on shaft.
5. **On all models,** lock tripod assembly on shaft by installing snap ring in shaft groove.
6. **On G.K.N. units,** distribute two of three provided grease packets into boot and remaining packet into housing, on A.C.I. units, one of two packets into boot and remaining packet into housing, on Citroen units, 2/3 of packet into boot and remaining amount into housing and on S.S.G. units, 1/2 of packet into housing and remaining amount into boot.
7. **On all models,** position spring, with spring cup attached to exposed end, into spring pocket, **Fig. 17.** Place a small amount of grease on spring cup, then position housing over tripod. Slip tripod into housing. On Citroen units, reform or replace retainer ring. On G.K.N. units, bend retaining tabs down to original position. On A.C.I. units, reattach boot to hold housing onto shaft without bending back retaining tabs. On S.S.G. units, install tripod wire retaining ring into position. **On all spring units, ensure tripod is securely retained in housing and spring remains centered in housing spring pocket when tripod is installed and seated in spring cup.**
8. **On all units,** position boot over boot retaining groove in housing, then install clamp.

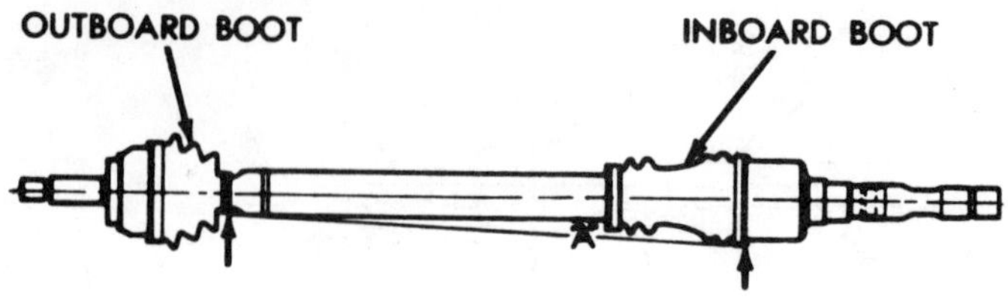

Body	Engine	Driveshaft Identification			"A" Dimension	
		Type	Side	Transaxle	mm	Inch
Charger, Horizon, Omni & Turismo	2.2L	G.K.N.	Right	Auto.	498-504	19.6-20.0
			Left	Auto.	208-221	8.2- 8.7
		G.K.N.	Right	Manual	498-504	19.6-20.0
			Left	Manual	240-253	9.4-10.0
	2.2L Turbo I	G.K.N.	Right	All	211-220	8.3-8.7
			Left	All	211-220	8.3-8.7
		Citroen	Right	All	227-232	8.9- 9.1
			Left	All	227-232	8.9- 9.1
Exc. Charger, Horizon, Omni & Turismo	2.2L/2.5L	G.K.N.	Right	All	505-515	19.9-20.3
			Left	All	227-245	8.9- 9.6
		A.C.I.	Right	All	477-485	18.8-19.1
			Left	All	197-212	7.8- 8.3
		G.K.N./A.C.I.	Right	All	492-500	19.4-19.7
			Left	All	211-226	8.3- 8.9
		Citroen	Right	All	480-492	18.9-19.4
			Left	All	206-223	8.1- 8.8
		S.S.G	Right	All	457-469	17.9-18.5
			Left	All	184-200	7.2- 7.9
	2.6L	G.K.N.	Right	All	501-510	19.7-20.1
			Left	All	222-237	8.7- 9.3
		Citroen	Right	All	480-492	18.9-19.4
			Left	All	206-223	8.1- 8.8
Exc. Charger, Horizon, Omni & Turismo	2.2L Turbo I	G.K.N.	Right	All	257-265	10.1-10.4
			Left	All	257-265	10.1-10.4
		Citroen	Right	All	241-251	9.5- 9.9
			Left	All	241-251	9.5- 9.9
		S.S.G.	Right	All	219-228	8.6- 9.0
			Left	All	219-228	8.6- 9.0
Daytona	2.2L Turbo II	G.K.N.	Right	Manual	227-234	8.9- 9.2
			Left	Manual	227-234	8.9- 9.2

Fig. 7 Driveshaft length specifications. 1987 except Shadow & Sundance

OUTER CONSTANT VELOCITY JOINT SERVICE

DISASSEMBLY

1. Cut boot clamps from boot and discard boot and clamps, **Fig. 18.**
2. Clean grease from joint.
3. **On except S.S.G. units,** support shaft in a soft jawed vise, support outer joint and tap with a mallet to dislodge joint from internal circlip installed in a groove at outer end of shaft, **Fig. 19.** Do not remove slinger from housing.
4. **On S.S.G. units,** loosen damper weight bolts and slide weight and boot toward inner joint, then expand circlip with suitable pliers, slide joint from shaft and reinstall damper weight.
5. **On all units,** remove circlip from shaft groove and discard, **Fig. 20.**
6. **On A.C.I. and G.K.N. units,** do not remove heavy lock ring from shaft, **Fig. 20,** unless shaft requires replacement.
7. **On all units,** if constant velocity joint was operating satisfactorily and grease does not appear contaminated, proceed to "Assembly" procedure, Step 8.
8. If constant velocity joint is noisy or badly worn, replace entire unit. The repair kit will include boot, clamps, circlip and lubricant. Clean and inspect joint as outlined in the following steps.
9. Clean surplus grease and mark relative position of inner cross, cage and housing with a dab of paint.
10. Hold joint vertically in a soft jawed vise.
11. Press downward on one side of inner race to tilt cage and remove ball from opposite side, **Fig. 21.** If joint is tight, use a hammer and a brass drift to tap inner race. Do not strike cage. Repeat this step until all six balls are removed. A screwdriver may be used to pry balls loose.
12. Tilt cage assembly vertically and position two opposing, elongated cage windows in area between ball grooves. Remove cage and inner race assembly by pulling upward from housing, **Fig. 22.**
13. Rotate inner cross 90 degrees to cage and align one of race spherical lands with an elongated cage window. Raise land into cage window and remove inner race by swinging outward, **Fig. 23.**

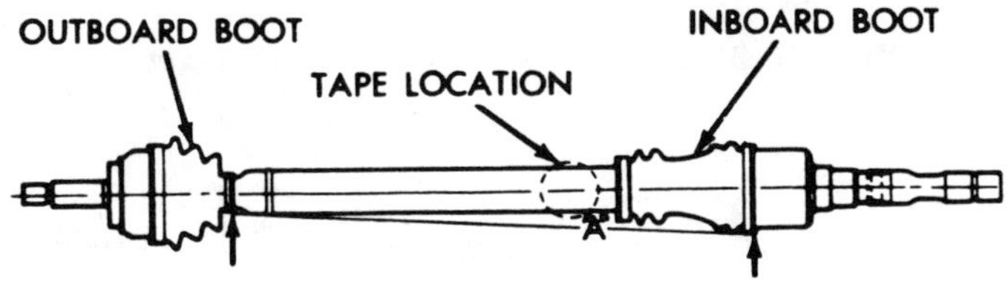

Engine	Driveshaft Identification			"A" Dimension	
	Type	Side	Tape Color	mm	Inch
2.2L	G.K.N.	Right	Blue	505-515	19.9-20.3
		Left	Blue	259-277	10.2-10.9
	A.C.I.	Right	Green	477-485	18.8-19.1
		Left	Green	229-244	9.0-9.6
	G.K.N./A.C.I.	Right	Orange	492-500	19.4-19.7
		Left	Orange	243-258	9.6-10.2
	Citroen	Right	White	480-492	18.9-19.4
		Left	White	238-255	9.4-10.0
	S.S.G.	Right	Gold	457-469	17.9-18.5
		Left	Gold	216-232	8.5-9.1
2.2L Turbo	G.K.N.	Right	Tan	257-265	10.1-10.4
		Left	Silver	254-269	10.1-10.6
	Citroen	Right	Red	241-251	9.5-9.9
		Left	Yellow	238-255	9.4-10.0
	S.S.G.	Right	Turquoise	219-228	8.6-9.0
		Left	Gold	216-232	8.5-9.1

Fig. 8 Driveshaft length specifications. 1987 Shadow & Sundance

INSPECTION

1. Check housing ball races for excessive wear.
2. Check splined shaft and nut threads for damage.
3. Inspect balls for pitting, cracks, scouring and wear. Dulling of the surface is normal.
4. Inspect cage for excessive wear on inner and outer spherical surfaces, heavy brinnelling of cage, window cracks and chipping.
5. Inspect inner race (cross) for excessive wear or scouring of ball races.
6. If any of the defects listed in Steps 1 through 5, are found, replace C/V assembly as a unit. **Polished areas in races (cross and housing) and on cage spheres are normal and do not indicate a need for joint replacement unless they are suspected of causing noise and vibration.**

ASSEMBLY

Do not use the A.C.I. or G.K.N. boot clamps on the Saginaw hard plastic C/V boots, as these clamps do not have the load capacity to withhold the much greater force needed to clamp plastic boots.

1. If removed, position wear sleeve on joint housing, then tap sleeve onto housing using seal wear sleeve No. C-4698 or equivalent.
2. Lightly oil components, then align marks made during disassembly.
3. Align one of the inner race lands with elongated window of cage, then insert race into cage and pivot 90°.
4. Align elongated cage windows with housing land, then pivot cage 90°.
5. Lubricate ball races with one packet of grease from kit.
6. Tilt cage and inner race assembly and insert balls.
7. With shaft supported in a soft jawed vise, install boot.
8. Install snap ring on shaft. Do not over-expand snap ring during installation.
9. Position joint housing on shaft, then engage by tapping sharply with a soft faced mallet.
10. Ensure snap ring is properly seated by attempting to pull joint from shaft.
11. Locate large end of boot over housing and secures boot clamps.

INTERMEDIATE SHAFT ASSEMBLY

The intermediate shaft assembly, Fig. 24, used on front wheel drive turbocharged models is the same for manual and automatic transaxles.

REMOVAL

1. Remove right driveshaft, referring to "Driveshaft, Replace" procedure.
2. Remove speedometer pinion from extension housing.
3. Remove two screws from bearing assembly bracket to engine block, **Fig. 24.**
4. Remove intermediate shaft assembly from transaxle extension by pulling yoke outward.

SUBASSEMBLY SERVICE

UNIVERSAL JOINT & ROLLER

Disassembly

1. Mark relationship of shafts to insure proper alignment during assembly. Apply penetrating oil to bushing, then remove snap rings.
2. Support yoke in vise, then position 1 1/8 inch socket over bushing on top of yoke.
3. Strike socket with a suitable hammer until bushing moves up out of yoke into socket.
4. Turn assembly in vise and remove remaining bushings in same manner.

Assembly

1. Hold cross in position between yoke ears with one hand and start one bushing assembly into yoke with other hand.
2. Hammer bushing assembly into yoke, then install snap ring.
3. Install remaining bushing assemblies in same manner.

BRACKET, BEARING & SLINGER ASSEMBLY

Disassembly

1. Remove two bearing assembly to support bracket attaching screws, then separate bearing from bracket.

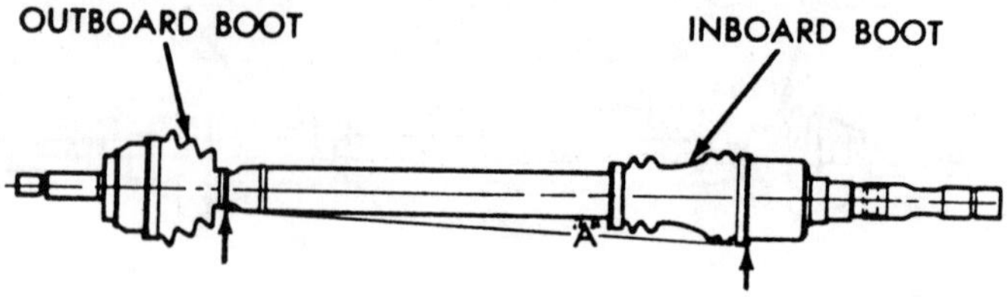

Body	Engine	Driveshaft Identification			"A" Dimension	
		Type	Side	Transaxle	mm	Inch
Horizon, Omni	2.2L	G.K.N.	Right	Auto.	498-504	19.6-19.8
		69/92*	Left	Auto.	208-221	8.2- 8.7
		69/92*	Right	Manual	498-504	19.6-19.8
			Left	Manual	240-253	9.4-10.0
Exc. Dynasty, New Yorker, New Yorker Landau, Horizon & Omni	2.2L	G.K.N.	Right	All	505-515	19.9-20.3
		69-92*	Left	All	227-245	8.9- 9.6
		G.K.N.	Right	All	490-498	19.3-19.6
		82-98*	Left	All	227-234	8.9- 9.2
		A.C.I.	Right	All	477-485	18.8-19.1
			Left	All	197-212	7.8- 8.3
		S.S.G.	Right	All	457-469	18.0-18.5
			Left	All	184-200	7.2- 7.9
	2.5L	G.K.N.	Right	All	490-498	19.3-19.6
		82-98*	Left	All	227-234	8.9- 9.2
		S.S.G.	Right	All	457-469	18.0-18.5
			Left	All	184-200	7.2- 7.9
	2.2L Turbo I	S.S.G.	Right	All	187-196	7.4- 7.7
			Left	All	187-196	7.4- 7.7
		G.K.N.	Right	All	227-234	8.9- 9.2
		82-98*	Left	All	227-234	8.9- 9.2
Daytona, Lancer, Lebaron GTS	2.2L Turbo II	G.K.N.	Right	All	227-234	8.9- 9.2
		82-98*	Left	All	227-234	8.9- 9.2
Dynasty, New Yorker & New Yorker Landau	3.0L	G.K.N.	Right	All	481-489	18.9-19.2
		82-98*	Left	All	216-224	8.5- 8.8

*OPEN TULIP DESIGN—THREE PIECE CONSTRUCTION.

Fig. 9 Driveshaft length specifications. 1988

2. Press intermediate shaft out of bearing assembly and outer slinger. **Do not dent or damage inner slinger or end of stub shaft.**
3. If either slinger is damaged, it should be replaced by carefully pressing shaft through slinger. **The bearing assembly is not serviceable and must be replaced as an assembly.**

Assembly

1. Place new slinger on stub shaft, then using a suitable tool, drive slinger down until it bottoms out on shoulder of shaft. Ensure slinger is properly seated.
2. Press bearing assembly onto shaft, leaving a minimum of 1/32 inch clearance between slinger and bearing assembly. **Apply pressure only to inner race of bearing assembly during installation.**
3. Press outer slinger into position using a suitable tool. The slinger must bottom out on shoulder of shaft.

INSTALLATION

1985–87

1. Attach bracket loosely to bearing assembly.
2. Hold stub yoke and install spline into transaxle.
3. Attach bracket to engine. **Torque** attaching screws to 40 ft. lbs.
4. Push intermediate shaft assembly into transaxle as far as it can travel, then **torque** bracket to bearing attaching screws to 21 ft. lbs.
5. Apply suitable grease inside spline and pilot bore on bearing end of intermediate shaft.
6. Install speedometer pinion, then the right driveshaft, referring to "Driveshaft, Replace" procedure.

1988–89

1. Attach bracket to bearing assembly and **torque** screws to 21 ft. lbs.
2. Hold stub yoke and install spline into transaxle.
3. Attach bracket to engine, then loosely install attaching screws.
4. Push intermediate shaft assembly into transaxle as far as possible, then **torque** bracket-to-engine attaching screws to 40 ft. lbs.
5. Apply suitable grease inside spline and pilot bore on bearing end of intermediate shaft.
6. Install speedometer pinion, then the right driveshaft. Refer to "Driveshaft, Replace" procedure.

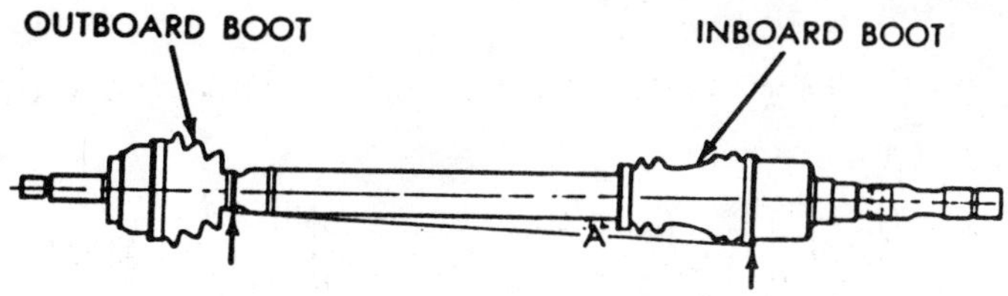

Body	Engine	Driveshaft Identification Type	Side	Transaxle	"A" Dimension mm	Inch
Omni & Horizon	2.2L	A.C.I.	Right	Auto	469-478	18.5-19.0
			Left	N/A	—	—
		69-92	Right	Auto	498-504	19.6-19.8
			Left	Auto	208-221	8.2- 8.7
		69-92	Right	Manual	498-504	19.6-19.8
			Left	Manual	240-253	9.4-10.0
Except Daytona, Dynasty, Horizon, New Yorker & Omni	2.2L	A.C.I.	Right	All	477-485	18.8-19.1
			Left	All	197-212	7.8- 8.3
		S.S.G.	Right	All	457-469	18.0-18.5
			Left	All	184-200	7.2- 7.9
		82-98	Right	All	481-489	18.9-19.2
			Left	All	216-224	8.5- 8.8
	2.5L	S.S.G.	Right	All	457-469	18.0-18.5
			Left	All	184-200	7.7- 7.9
		82.98	Right	All	481-489	18.9-19.2
			Left	All	216-224	8.5- 8.8
Except Daytona, Dynasty, Horizon, New Yorker & Omni	2.5L Turbo 1	S.S.G.	Right	Auto	457-469	18.0-18.5
			Left	Auto	184-200	7.2- 7.9
		82-98	Right	Auto	481-489	18.9-19.2
			Left	Auto	216-224	8.5- 8.8
		S.S.G.	Right	Manual	187-196	7.4- 7.7
			Left	Manual	187-196	7.4- 7.7
		82-98	Right	Manual	216-224	8.5- 8.8
			Left	Manual	216-224	8.5- 8.8
Dynasty & New Yorker	3.0L	82-98	Right	Auto	481-489	18.9-19.2
			Left	Auto	216-224	8.5- 8.8
Daytona	2.2L Turbo II	82-98	Right	Manual	216-224	8.5- 8.8
			Left	Manual	216-224	8.5- 8.8

Fig. 10 Driveshaft length specifications. 1989

Fig. 11 Left engine mount adjustment

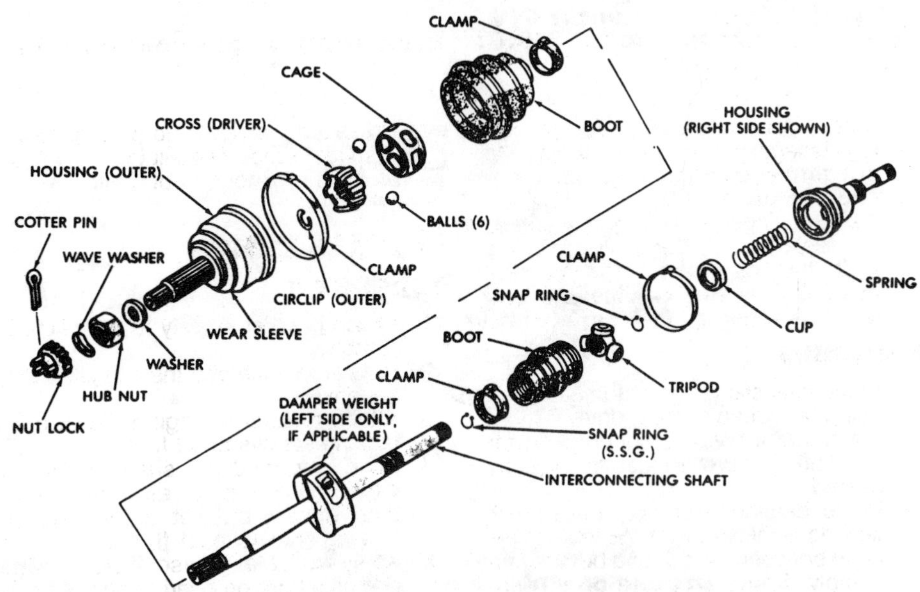

Fig. 12 Driveshaft components

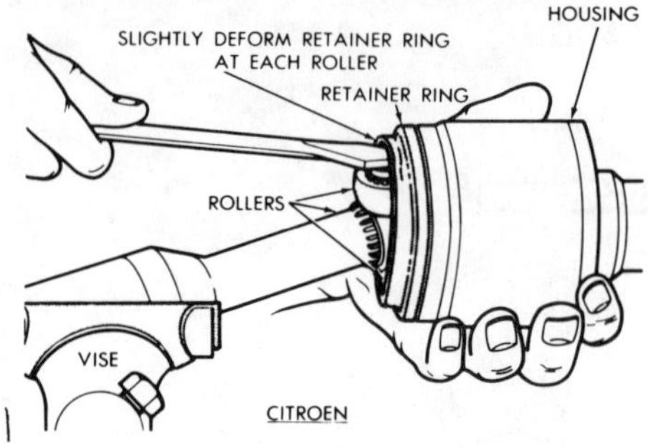

Fig. 13 Removing tripod from housing. Citroen inner C/V joint

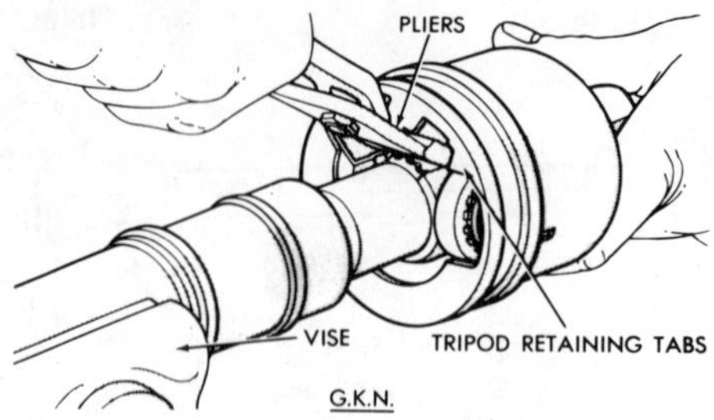

Fig. 14 Removing tripod from housing. G.K.N. inner C/V joint

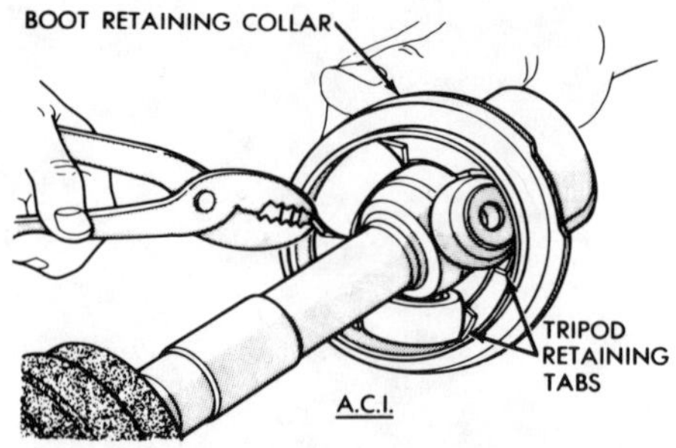

Fig. 15 Removing tripod from housing. A.C.I. inner C/V joint

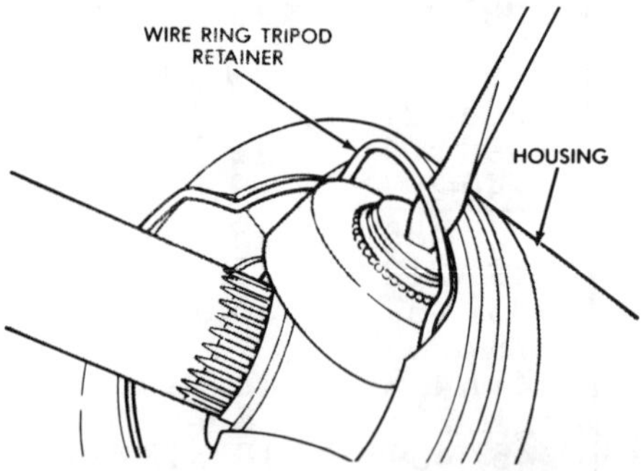

Fig. 16 Removing tripod from housing. S.S.G. inner C/V joint

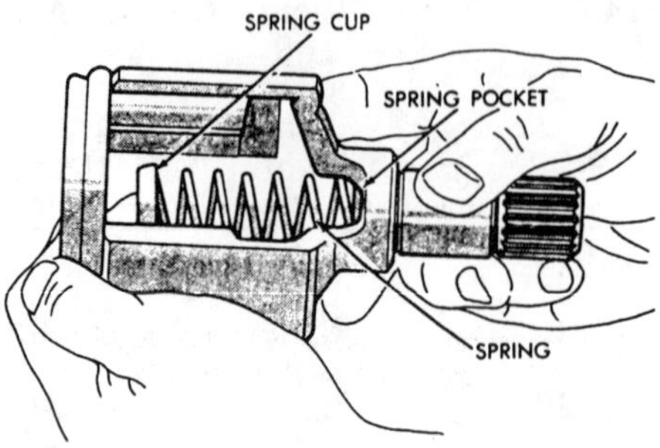

Fig. 17 C/V joint retention spring

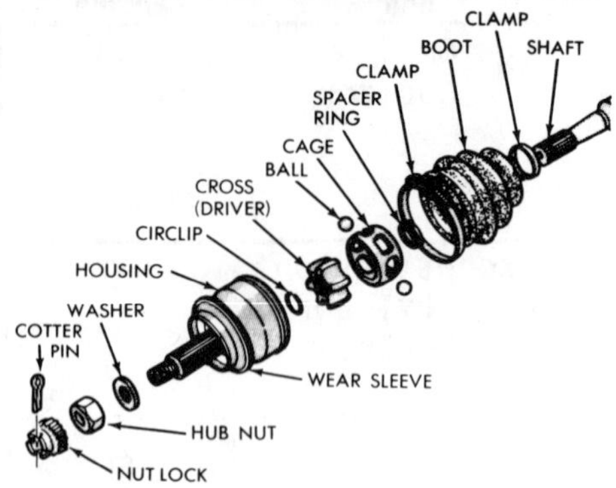

Fig. 18 Disassembled view of outer C/V joint

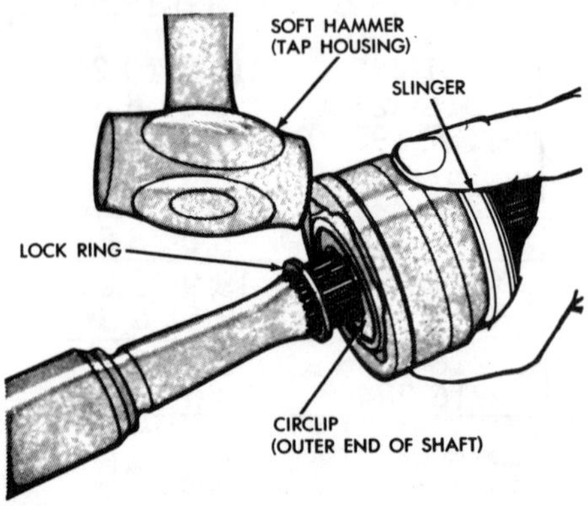

Fig. 19 Removing outer C/V joint from shaft. Except S.S.G. units

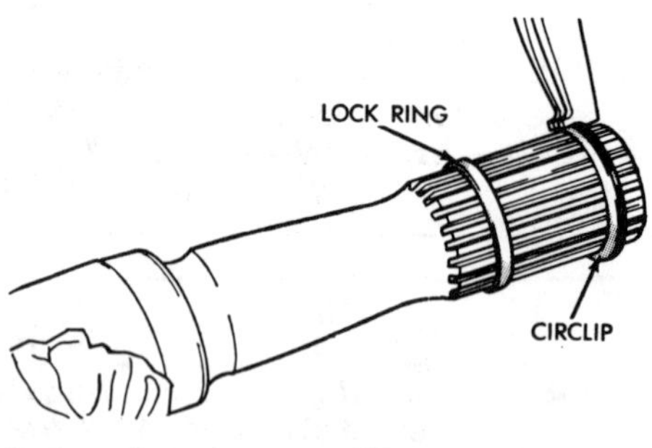

Fig. 20 Outer C/V joint circlip removal

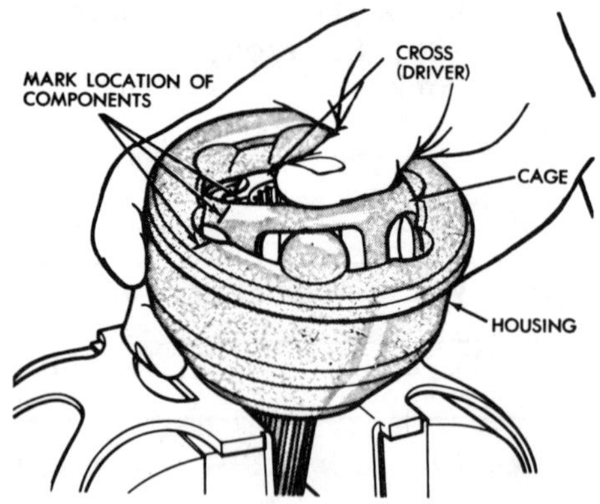

Fig. 21 Outer C/V joint ball removal

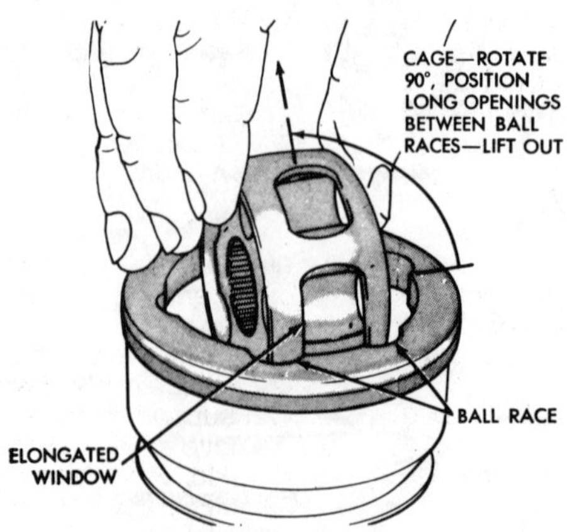

Fig. 22 Outer C/V joint cage & cross assembly removal

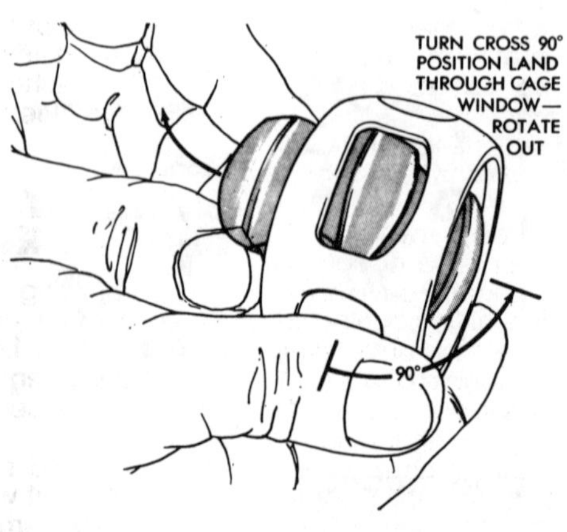

Fig. 23 Removing cross from cage. Outer C/V joints

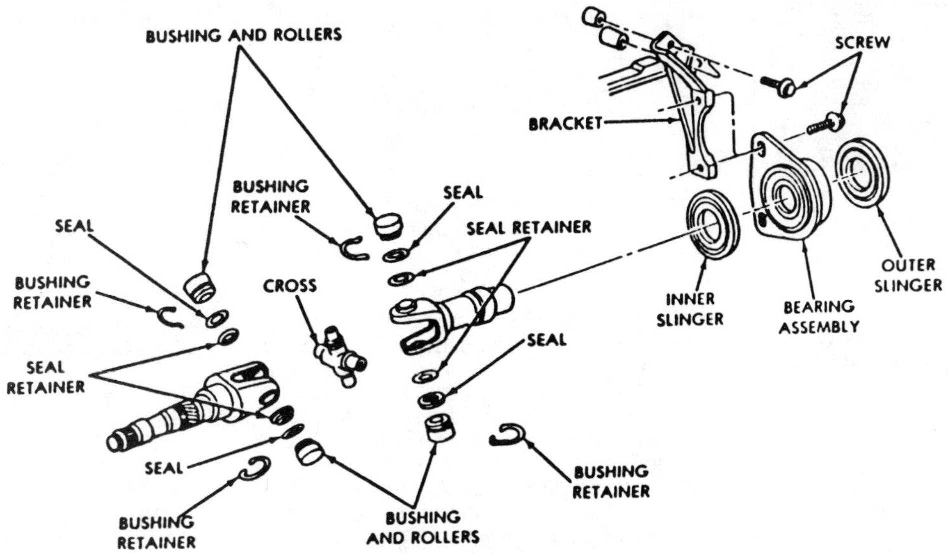

Fig. 24 Intermediate shaft assembly. Turbocharged models

Chrysler Imports

INDEX

FRONT DRIVE AXLE REPLACE

COLT VISTA w/TWO WHEEL DRIVE

1. Remove hub dust cap and loosen drive axle nut, then raise and support front of vehicle and remove wheel and tire assembly.
2. Remove under cover, then remove strut bar and ball joint from lower control arm. **Use care not to damage ball joint dust boot.**
3. Drain transaxle fluid, then on models with turbocharger, disconnect center bearing snap ring.
4. Insert a suitable pry bar between transaxle case and outer case of double offset joint, then withdraw drive axle, **Fig. 1.** Cover drive axle opening in transaxle. When removing drive axle, support at tripod or double offset joint and pull shaft straight out to prevent damage to boot or joint. After disconnecting drive axle from transaxle, support shaft in proper position. **Pry bar should not be inserted more than .28 inch between transaxle case and outer case of offset joint, as damage to oil seal may result. The double offset joint retainer ring should be replaced whenever the drive axle is removed from transaxle case.**
5. Press drive axle, from hub using axle shaft puller No. CT-1003, **Fig. 2,** then on models with turbochargers, lightly tap on double offset joint outer race with a suitable mallet to remove. **When pressing drive axle from hub, use care to prevent the spacer from moving out of position.**
6. Reverse procedure to install. Position drive axle so that raised inner diameter of washer is facing nut, then install and **torque** drive axle nut to 145-188 ft. lbs.

COLT VISTA w/FOUR WHEEL DRIVE

The following procedure is only for the left side drive axle. Refer to Colt Vista w/Two Wheel Drive for right side drive axle replacement procedure.

1. Remove center cap and drive axle nut.
2. Raise and support vehicle, then remove front wheel.
3. Drain transmission fluid.
4. Disconnect lower arm ball joint and knuckle coupling, then remove strut bar and stabilizer bar from lower arm.
5. Remove center bearing snap ring from bearing bracket.
6. Lightly tap double offset joint outer race with a wooden hammer and disconnect drive axle from cardan joint assembly.
7. Disconnect drive axle from bearing bracket.
8. Remove drive axle from hub.
9. Remove bearing bracket attaching bolts, then the bearing bracket.
10. Lightly tap yoke of cardan joint with a wooden hammer and remove it from transaxle assembly.
11. Reverse procedure to install. **Torque** bearing bracket bolts to 29-39 ft. lbs. **Torque** drive axle nut to 145-188 ft. lbs. After installation, ensure boot length is 3.23-3.47 inch.

COLT

1. **On 1988-89 models,** remove under cover and center member from vehicle.
2. **On all models,** remove hub nut cotter pin, hub nut and washer, then raise and support vehicle.
3. Remove lower ball joint from steering knuckle, using steering linkage puller No. MB991113 or equivalent.
4. Remove tie rod from steering knuckle using tool No. MB991113 or equivalent.
5. **On drive axles equipped with center bearing,** remove center bearing snap ring.

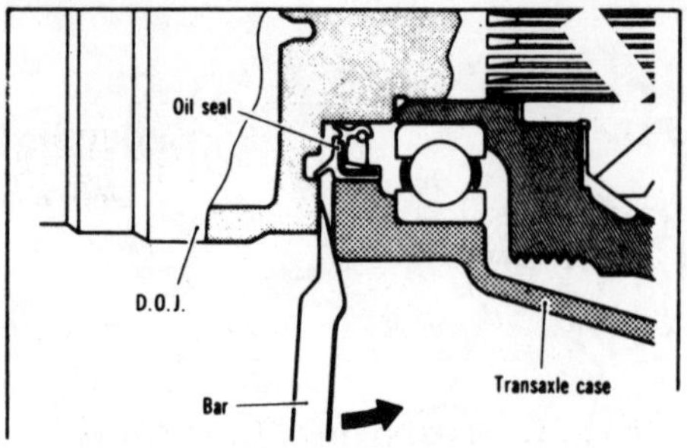

Fig. 1 Removing front drive axle from transaxle case

6. **On drive axles not equipped with center bearing,** insert pry bar between transaxle case and drive shaft, then pry drive axle from transaxle case, **Fig. 1.** Remove drive axle from hub using axle shaft puller No. CT-1003 or equivalent, **Fig. 2. Do not pull on drive axle or joints will be damaged. Do not insert pry bar further than necessary or damage to oil seal will result.**
7. **On drive axle equipped with center bearing,** remove shaft from transaxle by lightly tapping Double Offset Joint (DOJ) with plastic hammer, then drive the drive axle from hub by lightly tapping with plastic hammer. **Do not pull on drive axle or joints will be damaged.**

FRONT AXLE SERVICE

COLT & COLT VISTA W/TWO WHEEL DRIVE

DOUBLE OFFSET JOINT TYPE DRIVESHAFT

When servicing the drive axle, do not disassemble the Birfield bell type constant velocity joint, as components on this type joint are precision fitted.

Disassembly

1. Remove double offset joint boot band, then remove snap ring using a suitable screwdriver, **Figs. 3 and 4.**
2. Remove drive axle from double offset joint, then clean grease from shaft and joint.
3. Remove snap ring, then remove double offset joint inner race, cage and bearings as an assembly. Clean bearing assembly and disassemble bearing if necessary.
4. Wind tape around drive axle splines, then remove Birfield joint boot band and slide boot from drive axle. When inspecting Birfield joint, note amount of grease removed for reassembly.

Assembly

A special grease containing Molybdenum is used to lubricate the drive axle double offset joint and the Birfield joint. This special grease is included in the drive axle repair kit and must be used.

1. Apply gear oil to drive axle, then install

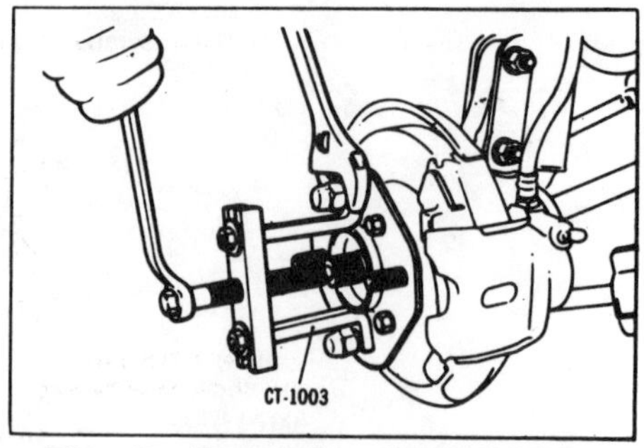

Fig. 2 Pressing front drive axle from hub

double offset joint and Birfield joint boots. Wrap tape around drive axle splines to prevent damage to boots.
2. Apply special Molybdenum grease to double offset joint inner race and cage, then assemble race and cage, aligning raised portions on race with bearing holes on cage.
3. Apply special Molybdenum grease to cage, then position ball bearings into cage.
4. Install inner race cage and bearings on drive axle with chamfered side of cage facing splined end of shaft, then install snap ring.
5. Apply 1.4-2.1 ounces of special Molybdenum grease to double offset joint outer race, then position outer race on drive axle.
6. Apply an additional .7-1.4 ounce of special Molybdenum grease to outer race, then install clip.

1. Cardan joint assembly
2. Dust seal
3. Bearing bracket
4. Bearing retainer
5. Center bearing
6. Oil seal
7. Oil seal retainer
8. O-ring
9. Snap ring
10. D.O.J. outer race
11. Center bearing assembly
12. Circlip
13. Snap ring
14. D.O.J. inner race
15. D.O.J. cage
16. Ball
17. D.O.J. boot
18. D.O.J. boot band
19. Boot band(small)
20. B.J. boot band
21. B.J. boot
22. B.J. Assembly
23. Dust cover

	Nm	ft.lbs.
A	40–55	29–39
B	200–260	145–188

B.J.—Birfield joint
D.O.J.—Double offset joint

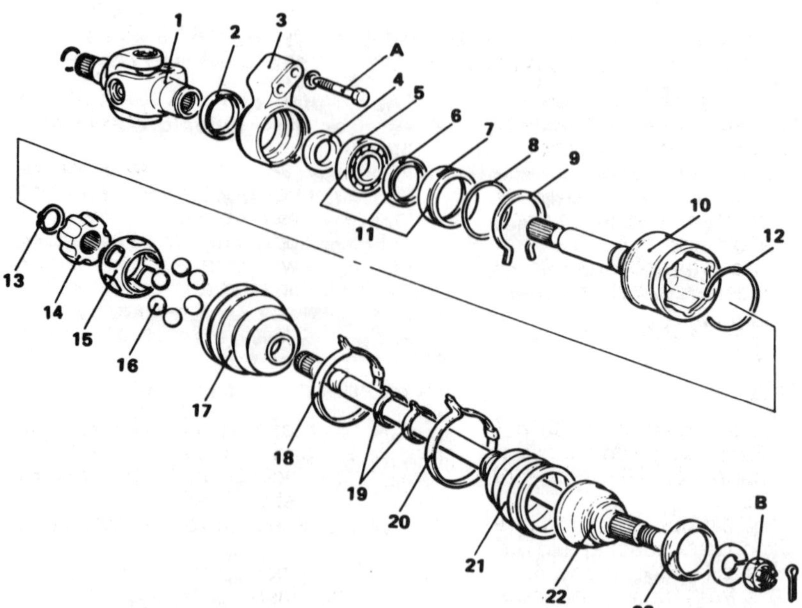

Fig. 3 Double offset joint drive axle exploded view. Colt Vista w/four wheel drive

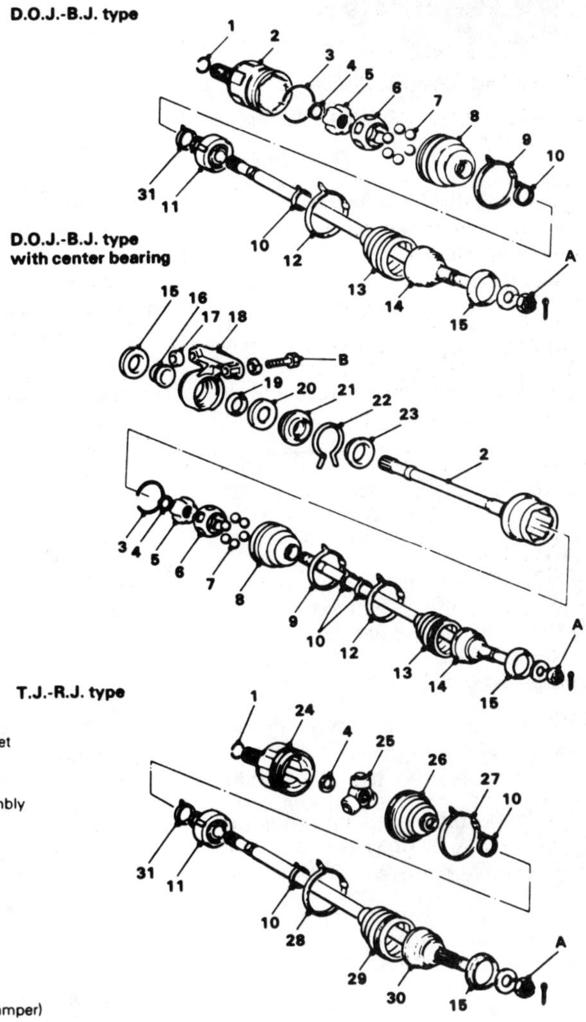

1. Circlip (small)
2. D.O.J. outer race
3. Circlip
4. Snap ring
5. D.O.J. inner race
6. D.O.J. cage
7. Ball
8. D.O.J. boot
9. D.O.J. boot band
10. Boot band (small)
11. Dynamic damper
12. B.J. boot band
13. B.J. boot
14. B.J. assembly
15. Dust cover
16. Sleeve
17. Spacer
18. Center bearing bracket
19. Bearing retainer
20. Dust cover
21. Center bearing assembly
22. Snap ring
23. Dust cover
24. T.J. case
25. Spider assembly
26. T.J. boot
27. T.J. boot band
28. R.J. boot band
29. R.J. boot
30. R.J. assembly
31. Band (for dynamic damper)

	Nm	ft.lbs.
A	200–260	145–188
B	35–40	25–29

Fig. 4 Double offset joint & tripod type front drive axle assemblies exploded view. Colt & Colt Vista 2 wheel drive

7. Add amount of special Molybdenum grease to Birfield joint as removed during disassembly and inspection.
8. Install boots and boot bands for each joint. When installing boot bands for double offset joint, position bands, **Fig. 5,** so that dimension A is 3.2 inches on models less turbocharger or 3.3 inches on models with turbocharger.

TRIPOD JOINT TYPE DRIVESHAFT

Disassembly

When servicing the drive axle, do not disassemble the Rzeppa type constant velocity joint, as components of this type joint are precision fitted.

1. Remove boot clamps, then remove boot from tripod joint housing and position on drive axle, **Figs. 3 and 4.**
2. Pull drive axle from tripod joint housing, then remove snap ring and lift tripod joint spider from housing. Clean tripod joint spider and check for wear and damage. Also check joint needle roller bearings for smooth operation. **Do not disassemble tripod joint spider.**
3. Wind tape around drive axle splines, then remove bands for Rzeppa type joint and remove boots from drive axle. If Rzeppa joint is to be reused, do not wipe away grease. Check grease for contamination and clean and replace grease only if necessary. Note amount of grease removed for use during reassembly.

Assembly

A special grease is used to lubricate the drive axle tripod joint and Rzeppa joint. This grease is included in the drive axle repair kit and must be used.

1. Apply grease to drive axle, then install boots.
2. Position tripod joint spider on drive axle, then install snap ring.
3. Apply 2.8-3.2 ounces of special grease to tripod joint housing, then insert drive axle and spider into housing.

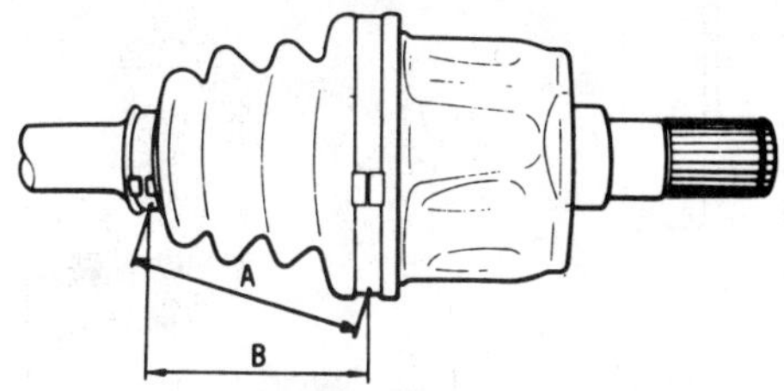

Fig. 5 Double offset boot installation

4. Apply another 2.8-3.2 ounces of special grease to tripod joint housing.
5. Apply as much special grease as removed to Rzeppa joint, if necessary.
6. Install bands and boots for each joint. When installing boot bands for tripod joint, bands must be positioned 3.1 inches apart (dimension A), **Fig. 5.**

COLT VISTA w/FOUR WHEEL DRIVE

The following procedure is only for the left side drive axle. Refer to Colt Vista w/Two Wheel Drive for service procedure on right side drive axle.

CENTER BEARING, REPLACE

1. Remove double offset joint outer race from drive axle.
2. Using a suitable grinder, partially grind the certain point of the circumference of the bearing retainer to a thickness of .04-.06 inch.
3. Using a suitable hammer and chisel, break then remove bearing retainer. When the break is made in the bearing retainer, tap the chisel in between the bearing and bearing retainer and pry to remove the bearing retainer. **Do not damage the drive axle in any way.**
4. Remove center bearing assembly using bearing remover MB990560 or equivalent and a press.
5. To install proceed as follows:
   a. After passing the snap ring through the double offset joint outer race, use tool MB990560 or equivalent and a press to install center bearing assembly onto the shaft of the double offset joint outer race.
   b. Face polished surface of the bearing retainer toward the bearing side and, after placing it onto the shaft of the double offset joint outer face, use the tool mentioned previously and the press to install bearing retainer.
   c. Install drive axle onto double offset joint outer race.

CENTER BEARING DUST SEAL, REPLACE

1. Using a suitable screwdriver, remove dust seal from bearing bracket assembly.
2. Using a handle and adapter installer tools No. MB990938 and 990930 or equivalents and a mallet, press dust seal in until seal is flush with bearing bracket end.
3. Coat inside circumference of seal with suitable grease.

Eagle

INDEX

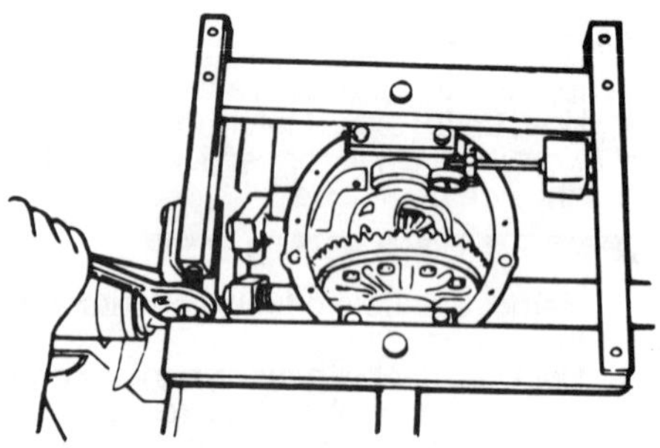

Fig. 1 Spreader tool installation. Eagle

EAGLE 4 X 4

DIFFERENTIAL SERVICE

Removal

1. Raise and support front of vehicle.
2. Remove axle assembly as described under "Axle, Replace."
3. Remove axle housing assembly and allow lubricant to drain, then using suitable punch, mark differential bearing caps for reference during reassembly.
4. Loosen differential bearing cap bolts leaving two to three bolt threads engaged in housing.
5. Install spreader tool No. J-24385-01, adapter No. J-24161 and dial indicator set No. J-8001, **Fig. 1,** or equivalent. Zero indicator and ensure indicator stylus contacts one side of housing opening.
6. While observing micrometer, spread housing with tool. Do not spread housing more than .020 inch as distortion could result.
7. Remove dial indicator, bearing cap bolts and bearing caps. Tag bearing caps for assembly reference.
8. Position one suitable pry bar under a ring gear bolt head while positioning another pry bar under differential case. Pry differential assembly upward to remove.
9. Remove spreader tool to prevent housing from taking set.
10. Mount differential assembly in suitable vise.

Disassembly

1. Using suitable puller, remove differential case side bearings.
2. Remove and discard ring gear attaching bolts.
3. Using a brass drift and a hammer, tap ring gear from differential case. **Do not attempt to wedge or chisel gear from case.**
4. Using a small pin punch, remove pinion mate shaft lockpin.
5. Remove pinion mate shaft and thrust block.
6. Rotate pinion gears until aligned with case opening, then remove pinion gears and thrust washers.
7. Remove side gears and thrust washers.

Cleaning & Inspection

Clean all parts in suitable solvent. Dry all parts except bearings with compressed or and shop towels. **Do not use compressed air to clean bearings as damage may result.**

Inspect all bearings and cups for wear, pitting, galling, flat spots or cracks. Any bearing or cup showing any signs of wear or damage must be replaced. Bearings and respective cups must be replaced as an assembly only. **Do not attempt to interchange bearings and cups as bearing life will be affected.**

Inspect non-machined differential case surfaces for nicks and burrs which can be removed with an oil stone or fine tooth file. Inspect pinion shaft bore to ensure it is not elongated or worn. If damage is evident, differential must be replaced. Inspect machined differential surfaces and counterbores. They must be smooth and free of nicks, gouges, cracks and other visible damage. If damage is evident, differential case must be replaced.

Inspect pinion shaft for excessive wear, scoring or galling. Ensure shaft is smooth and concentric. If any wear or damage is evident, replace ring gear and pinion as a matched set.

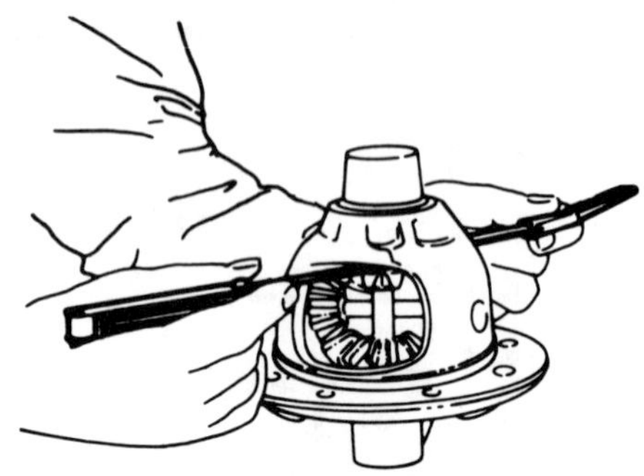

Fig. 2 Measuring side gear-to-case clearance. Eagle

Inspect pinion and ring gears for worn or chipped teeth, cracks, damaged bearing journals or attaching bolt threads. If any of the above are evident, replace ring gear and pinion as a matched set.

Inspect pinion and side gears. Gears must exhibit a uniform contact pattern without any signs of cracks, wear, scoring or galling. If any of the above are evident, replace all gears. Inspect thrust washers for wear and replace as necessary.

Inspect thrust block for excessive wear, distortion and cracks. Replace as necessary.

Installation & Adjustment

1. Install thrust washers on side gears, then install side gears in case.
2. Install thrust washers on pinion gears, then install pinion gears in case.
3. Install thrust block, then the pinion mate shaft lockpin in case. Position case on end.
4. Tap case lightly on flat surface to settle gears in position, then measure side gear-to-case clearance using two equal thickness feeler gauges inserted simultaneously between side gear and case, **Fig. 2.** Clearance should be .006 inch. If clearance exceeds specifications, replace differential case.
5. Install ring gear on case, then start two bolts 180° apart and tighten bolts

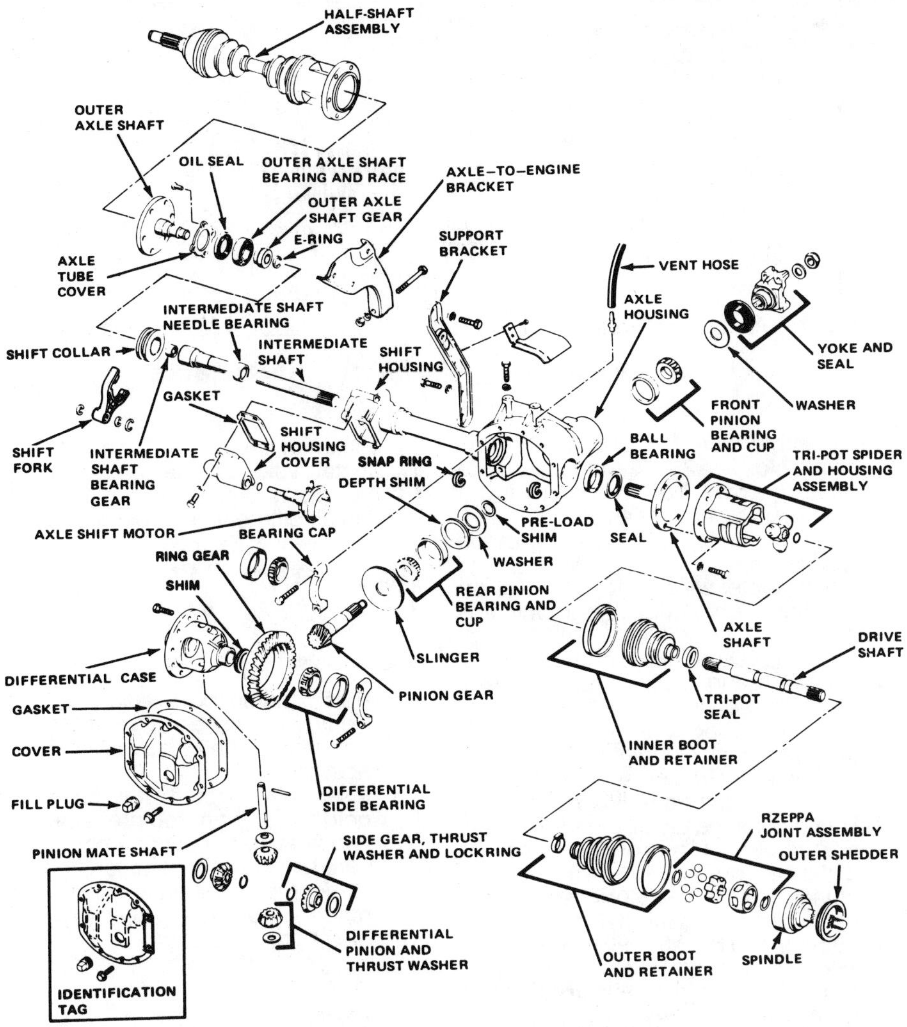

Fig. 3 Front axle assembly. Eagle

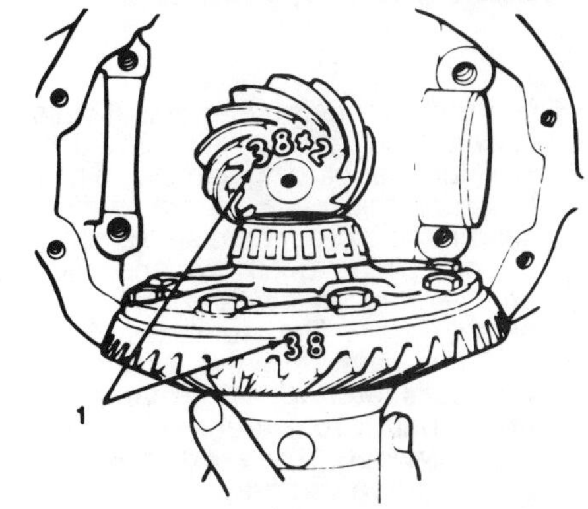

Fig. 4 Pinion depth variance numbers. Eagle

evenly to seat gear. Install remaining bolts and **torque** to 55 ft. lbs.

6. Using driver handle No. J-7079-02 and installer No. J-22175, install differential bearings.
7. Install differential bearing cups onto bearings, then install differential in axle housing. Install bearing caps and bolts and tighten securely, but not to specifications.
8. While holding ring gear in contact with pinion gear, pry differential bearing caps towards center of case, then insert feeler gauges of various thickness between each bearing cup and axle housing until proper ring gear backlash of 0.001-0.002 is established. **Feeler gauges must be installed at both ends of differential at same time to obtain an accurate measurement.**
9. Assemble shim pack to provide ring gear backlash as established in step 8, then insert shim packs between bearing cups and housing. Check ring gear backlash with feeler gauges. If satisfactory, remove, tag and set shims aside. If not satisfactory, adjust shims as necessary to bring backlash within specifications.
10. Remove differential case, then add an additional 0.015 inch shim to shim pack to be installed on drive tooth side of ring gear.
11. Remove differential bearings, then install shim packs on appropriate sides of differential case and install bearings.
12. Install spreader tool No. J-24385-01, adapter No. J-24161 and dial indicator set No. J-8001, **Fig. 1.** Zero indicator and ensure indicator stylus contacts one side of housing opening.
13. While observing micrometer, spread housing with tool. Do not spread housing more than 0.020 inch as distortion could result.
14. Lubricate axle bearing with gear lube, then install bearing cups on bearings.
15. Install differential assembly in housing. Use suitable rawhide mallet to seat bearings. Ensure ring and pinion gear teeth mesh properly.
16. Remove spreader tool, then install bearing caps. Ensure reference marks made during disassembly are aligned. Apply sealing compound to bearing cap bolts and install. **Torque** bolts to 40 ft. lbs.
17. Measure ring gear backlash at two points with dial indicator. Backlash should be 0.005-0.010. If backlash is not within specifications, add or subtract shims from differential bearing shim packs as necessary until correct backlash is obtained. **Changing position of a .005 inch shim from one side to the other will change amount of backlash 0.003 inch.**
18. Measure ring gear run-out. If run-out exceeds 0.006 inch, case may be distorted or dirt may be trapped between ring gear to case flange. Check and correct as necessary.
19. Install front axle assembly. Refer to "Front End & Steering" section for procedure.
20. Install gear housing cover using new gasket or RTV sealer. **Torque** bolts to 20 ft. lbs.
21. Fill axle with 2.5 pints of SAE 75-90 gear lubricant.
22. Road test vehicle.

DRIVE PINION SERVICE

Removal

1. Remove differential as described under "Differential Service."
2. Remove drive pinion nut, yoke and yoke dust cap, then using a suitable rawhide mallet, tap yoke end of pinion to drive pinion out of housing, **Fig. 3.** Collect and tag pinion bearing preload adjusting shims for use during reassembly.
3. Using a wooden 2 x 2 or suitable piece of pipe, drive front bearing, slinger and seal from axle housing.
4. Remove pinion rear bearing with suitable puller.
5. Remove rear bearing cup from housing bore, then the pinion depth adjustment shims. Tag shims for reference during reassembly.
6. Remove front bearing cup from axle housing.

Cleaning & Inspection

Refer to cleaning and inspection procedure as described under "Differential Service."

Assembly & Adjustment

To correctly measure and adjust pinion depth, ensure oil slinger is installed between pinion rear bearing and pinion head.

1. Measure thickness of pinion depth shim removed during disassembly.
2. Record pinion depth variance numbers etched on old and new pinion gears, **Fig. 4. The numbers etched on ring gear and the first set of numbers etched on drive pinion must match to ensure gears are a matched set. The second number on the drive pinion preceded by a plus or minus sign is the pinion depth variance, Fig. 4. The plus or minus signs and number denotes, in thousands of an inch, the amount the gear varied from the standard setting. The standard pinion depth setting is 2.250 inch.**
3. Refer to **Fig. 5**, to determine the amount, in thousands, to be added or subtracted from original depth shim to arrive at pinion height. **The pinion height arrived at is not the final shim thickness. An actual pinion depth measurement must be performed to determine final shim thickness.**
4. Using driver handle No. J-7079-02 and installer No. J-25101, install pinion front bearing cup in axle housing bore.
5. Install selected shim in rear bearing bore of housing. Ensure shim is centered in cup bore. If shim is chamfered, chamfer must face housing bore, not toward drive pinion head.
6. Using tools called for in step 4, install rear bearing cup in housing bore.
7. Install oil slinger on pinion gear. Ensure slinger is seated against pinion head before proceeding to step 8.
8. Using bearing installer sleeve No. J-5590, install rear bearing on drive pinion, then install pinion gear in axle housing.
9. Install pinion front bearing, yoke, washer and old pinion nut. Tighten pinion nut enough to remove endplay and provide 10-15 inch lbs. of drag torque when pinion is rotated. **Do not install pinion seal, dust cap or new pinion nut at this time.**
10. Assemble gauge arbor tool No. J-5223-4 and discs No. J-5223-26, then install in differential bearing cup bores. Ensure discs are firmly seated in cup bores, then install differential bearing caps and bolts and tighten securely.
11. Remove standard plunger from gauge block tool No. J-5223-20 and install plunger tool No. J-5223-27.
12. Compress plunger, then tighten gauge block thumbscrews securely. Install gauge block tool in such a way that plunger is directly under arbor tool installed in step 10. Ensure flat surface on anvil side of block is seated on end face of pinion. **To avoid false readings, do not allow anvil to contact pinion gear at any point.**

Old Pinion Marking	New Pinion Marking								
	−4	−3	−2	−1	0	+1	+2	+3	+4
+4	+0.008	+0.007	+0.0006	+0.005	+0.004	+0.003	+0.002	+0.001	0
+3	+0.007	+0.006	+0.005	+0.004	+0.003	+0.002	+0.001	0	−0.001
+2	+0.006	+0.005	+0.004	+0.003	+0.002	+0.001	0	−0.001	−0.002
+1	+0.005	+0.004	+0.003	+0.002	+0.001	0	−0.001	−0.002	−0.003
0	+0.004	+0.003	+0.002	+0.001	0	−0.001	−0.002	−0.003	−0.004
−1	+0.003	+0.002	+0.001	0	−0.001	−0.002	−0.003	−0.004	−0.005
−2	+0.002	+0.001	0	−0.001	−0.002	−0.003	−0.004	−0.005	−0.006
−3	+0.001	0	−0.001	−0.002	−0.003	−0.004	−0.005	−0.006	−0.007
−4	0	−0.001	−0.002	−0.003	−0.004	−0.005	−0.006	−0.007	−0.008

Fig. 5 Pinion depth variance chart. Eagle

13. Assemble bolt No. J-5223-29 and clamp No. J-5223-24, then using axle housing cover bolt, mount tools on axle housing cover flange.
14. Extend clamp bolt until pressed against gauge block. Align gauge block pointer with center of gauge arbor, then tighten clamp bolt until it is pressed against block with enough force to prevent block from moving.
15. Loosen gauge block thumbscrews and release plunger. When plunger contacts arbor tool, tighten thumbscrews to lock plunger in position. Do not disturb plunger setting. Remove bolt and clamp from axle housing.
16. Remove gauge block, then using suitable three inch micrometer, measure distance from end of anvil to end of plunger. This measurement represents pinion measured depth and is to be recorded for later use. **If measurement recorded above equ als standard pinion depth setting of 2.250 inch, the installed shim thickness is correct and no further adjustment is required.**
17. Remove bearing cap bolts and caps, then the arbor tool assembly.
18. Remove pinion nut, pinion gear, rear bearing cap and depth shim from axle housing.
19. Measure thickness of depth shim removed in previous step, then add this measurement to measurement obtained in step 16. From this total, subtract desired pinion depth. The result represents shim thickness required to adjust pinion depth. **Desired pinion depth is standard depth setting of 2.250 inch, plus or minus depth variance.**
20. Install proper thickness shim in housing bearing cup bore, then the rear bearing cup.
21. Install pinion bearing preload shims on pinion shaft, then install pinion gear, front bearing, oil slinger, yoke, washers and old pinion nut. **Torque** pinion nut to 260 ft. lbs.
22. Using suitable inch lbs. torque wrench, measure pinion rotational torque. The rotating **torque** should be 20-40 inch lbs. with new bearings or 10-20 inch lbs. with used bearings. Add preload as necessary to bring preload within specifications.
23. Remove pinion nut, washer and yoke, then using tool No. J-25104, install pinion oil seal.
24. Install yoke, washer and new pinion nut. **Torque** pinion nut ot 210 ft. lbs.
25. Install differential as described under "Differential Service."

AXLE, REPLACE

Removal

1. Raise and support vehicle.
2. Remove half shaft to axle flange bolts and secure half shaft to vehicle underbody.
3. Insert wire through half shaft flange bolt holes, then compress half shaft and wrap wire between boots to prevent half shaft from separating.
4. Mark propeller shaft and axle yoke for reference during reassembly and remove propeller shaft.
5. Support axle assembly using suitable jack.
6. Remove brace rod and shift motor shield.
7. Remove axle mounting bolts. The axle is secured to the engine block by five bolts on the lefthand side and one bolt on the righthand side.
8. Disconnect vacuum hose from axle shift motor, then partially lower axle assembly and disconnect vent hose.
9. Remove axle assembly from vehicle. **Do not apply any load or weight on the hub assembly whenever the half shafts are not securely attached to the axle shaft flanges.**

Installation

1. Support axle assembly on suitable jack and position assembly under vehicle.
2. Raise axle assembly slightly and install vent hose.
3. Install axle mounting bolts and **torque** to 50 ft. lbs.
4. Connect propeller shaft to axle yoke. Ensure shaft and yoke are aligned.
5. Remove wire used to prevent half shafts from separating, then install half shaft to axle flange bolts and **torque** to 45 ft. lbs.
6. Connect vacuum harness on axle shift motor shield.
7. Connect brace rod to axle and lower vehicle.

AXLE HUB & BEARING SERVICE

Removal

1. Raise and support vehicle, then remove wheel, caliper and rotor.
2. Remove bolts attaching axle shaft flange to half shaft. **Insert wire through half shaft flange bolt holes, then compress half shaft and wrap wire between boots to prevent half shaft from separating.**
3. Remove cotter pin, locknut and axle shaft nut, then the half shaft.
4. Remove steering arm from steering knuckle.
5. Remove caliper anchor plate from steering knuckle, then the three torx head bolts retaining hub assembly.
6. Remove hub assembly from steering knuckle.

Disassembly

1. Press hub from bearing carrier, then remove bearings, hub spacer, steering knuckle pin seal, carrier to hub seal, carrier O-ring and bearing spacer, if equipped, **Fig. 6.**
2. Clean all components and inspect for wear and damage and replace as necessary. **Replace bearing and bearing cups in matched sets only. If hub incorporates ball bearings, the entire hub assembly must be replaced if damage to internal components is indicated. If hub incorporates tapered bearings, internal hub components may be replaced as necessary. Bearings and races must be replaced in matched sets.**
3. If hub carrier bearing cups require replacement, remove and install cups in carrier using brass drift or suitable press.

Assembly

1. Fill steering knuckle pin, hub, bearing carrier and lip type seal cavities with lithium base wheel bearing lubricant.
2. Pack bearings with lithium base wheel bearing lubricant, then install bearings into hub carrier, **Fig. 6.**
3. Install seal onto hub side of bearing carrier, then press hub into carrier and install hub spacer on end of hub shaft. **Do not install bearing carrier O-ring at this time.**
4. Install inner seal into steering knuckle pin.
5. Install splash shield onto hub and bearing carrier, then install bearing carrier O-ring.
6. Install splash shield and bearing carrier into steering knuckle pin, then install carrier attaching bolts and **torque** to 75 ft. lbs.

Installation

1. Partially fill hub cavity of steering knuckle with chassis lubricant and install hub assembly onto steering knuckle. **Torque** hub Torx head bolts to 75 ft. lbs.
2. Install caliper anchor plate and **torque** plate retaining bolts to 100 ft. lbs.
3. Install steering arm onto steering knuckle and **torque** retaining bolts to 100 ft. lbs.
4. Install half shaft, then the axle flange to shaft bolts and hub nut. **Torque** half shaft to flange bolts to 45 ft. lbs and hub nut to 175 ft. lbs.
5. Install locknut, replacement cotter pin, rotor, caliper and wheel.

AXLE SHAFT, REPLACE

On 1985-88 models, removal of the righthand shaft is the same as that of the lefthand shaft. Refer to "Axle Shaft Seal, Replace." Perform steps 1 through 3 for axle shaft removal and steps 6 through 8 for axle shaft installation.

AXLE SHAFT BEARING, REPLACE

Two different style axle shaft bearings are used on the Eagle front axle. The left side axle shaft uses a ball bearing. The right side axle shaft uses a needle bearing. **On 1985-88 models, removal of the righthand shaft is the same as that of the lefthand shaft.**

1. Remove axle assembly as described under "Axle, Replace."
2. Remove axle housing cover and drain lubricant.
3. Remove axle shaft retaining clip, then the axle shaft.
4. Remove axle shaft seal using suitable screwdriver.
5. Remove needle bearing using bearing remover tools No. J-29173 and No. J-2619-1. Remove ball bearing using suitable brass punch and hammer. **If needle bearing tools are not available, remove the differential, then the bearing using a 15/16 inch socket and three foot extension bar.**
6. Install axle shaft bearings. Use bearing installer No. J-29153 to install needle bearing and tool No. J-29154 to install ball bearing.
7. Install axle shaft seals using Installer No. J-29152 for the righthand side axle shaft and Installer No. J-29154 for lefthand side axle shaft.
8. Install axle shafts, then the retaining clips.
9. Apply suitable sealant to axle housing cover and install cover. **Torque** cover bolts to 20 ft. lbs.
10. Fill axle with 2 1/2 pints of SAE 85W-90 lubricant. Install axle assembly as described under "Axle, Replace."

AXLE SHAFT SEAL, REPLACE

On 1985-88 models, removal of the righthand shaft is the same as that of the lefthand shaft.

1. Remove axle assembly as described under "Axle, Replace."
2. Remove axle housing cover and drain lubricant.
3. Remove axle shaft retaining clip, then the axle shaft.
4. Remove axle shaft seal using suitable screwdriver.
5. Install replacement axle shaft seal using Installer No. J-29152 for righthand side axle shaft and Installer No. J-29154 for lefthand side axle shaft.
6. Install axle shaft, then the axle shaft retaining clip.
7. Apply suitable sealant to axle housing cover and install cover. **Torque** cover bolts to 20 ft. lbs.
8. Fill axle with 2 1/2 pints of SAE 85W-90 lubricant. Install axle assembly as described under "Axle, Replace."

HALF SHAFT SERVICE

4 X 4

Disassembly

1. Using a suitable tool cut and remove both outer boot clamps, then roll outer boot from constant velocity joint.
2. Position a wooden block against constant velocity joint inner race and drive joint from shaft. If half shaft is to be positioned in a vise, use protective vise jaws.
3. Using a brass drift, tap constant velocity joint cage until cage is tilted outward far enough to remove first ball, then remove remaining balls in the same manner.
4. Align two outer constant velocity joint oblong holes with slots located on interior wall of spindle housing and remove cage and inner race.
5. Align shoulder between race grooves with inside of oblong cage holes, then rotate inner race out of cage using the larger of the two openings.
6. Remove two snap rings from shaft, then remove outer boot.
7. Remove rubber retaining ring from small end of inner boot.
8. Cut inner boot retaining strap, then remove strap from tripod housing. Slide boot away from housing.
9. Remove tripod joint and shaft from housing.
10. Remove snap ring that retains tripod joint on shaft using suitable pliers, then remove tripod joint.
11. Remove inner boot from shaft if replacement is required.
12. Remove any remaining snap rings from shaft, if not removed previously.

Inspection

Clean all components in a suitable solvent and dry with compressed air. Check all components for wear and damage and joints for rough operation, replace components as necessary. **Outer constant velocity joint and inner tripod joint are replaced as an assembly.**

Assembly

1. Pack spindle hub with chassis grease.
2. Install outer constant velocity joint inner race into cage, then install cage and race assembly into spindle hub. The smaller diameter of the cage and stopping groove of race must face outward.
3. Using a brass drift, tilt cage outward until first ball can be installed into

cage, then install remaining balls in the same manner.

4. Pack constant velocity joint with chassis grease, then install outer boot on half shaft.
5. Install inner and outer snap rings, then constant velocity joint onto shaft until inner race contacts inner snap ring.
6. Position boot over constant velocity joint, then install both boot clamps using boot clamp installer No. J-22716 or equivalent. **Check and adjust installed length of boot. Total installed length should be 4.134 inches. Move small end of boot in or out on shaft to obtain specified length.**
7. Pack tripod joint housing with chassis grease, then install inner small retainer ring, boot and large retainer ring onto half shaft.
8. Install tripod joint and snap ring onto shaft, then install tripod joint and shaft into housing.
9. Position boot and install rubber retainer ring over small end of boot.
10. Install large end boot clamp using a suitable tool.

INTERMEDIATE SHAFT, OUTER AXLE SHAFT & SHIFT HOUSING, REPLACE

Models w/Select Drive System

1. Raise and support front of vehicle.
2. Remove front axle assembly, then drain lubricant.
3. Remove bolts securing shift housing and remove shift fork, shifting housing and shift motor as an assembly.
4. Working through access hole in the outer axle shaft flange, remove screws securing axle tube cover.
5. Remove outer axle shaft assembly by tapping shaft flange with a rubber or plastic mallet.
6. Remove intermediate shaft snap ring, then remove intermediate shaft and shift collar.
7. Using a suitable puller and slide hammer, remove outer axle shaft bearing race.
8. Using adapter tool No. J-26225 and slide hammer No. J-6471-2, remove intermediate shaft needle bearing from axle tube end.
9. Using spreader tool No. J-29369-1 and slide hammer No. J-2619-01, remove intermediate shaft gear bearing. **The intermediate shaft gear is retained on the axle shaft by an internal type expandable snap ring. When removing the bearing from the gear be sure to support the gear face on vise jaws to avoid pulling the gear off the shaft. The gear and shaft are serviced only as an assembly.**
10. Remove outer axle shaft retaining E-ring.
11. Make a reference mark on outer axle shaft gear (for assembly purposes) and remove gear from shaft.
12. Remove outer axle shaft bearing using an arbor press.
13. Remove axle oil seal and tube cover.

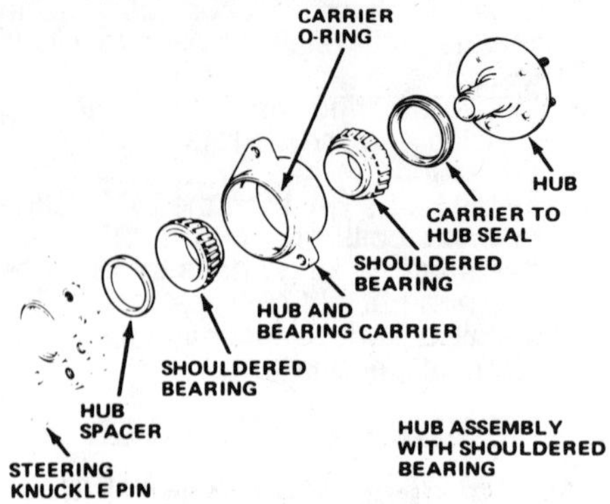

Fig. 6 Disassembled view of axle hub. Eagle

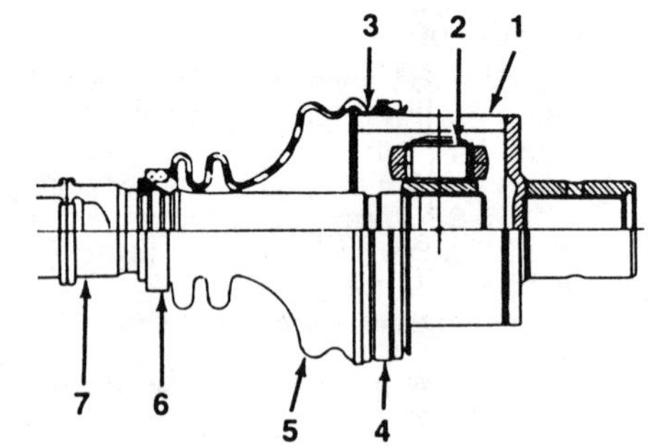

1 — Yoke
2 — Spider
3 — Metal Cover
4 — Retaining Clamp
5 — Boot
6 — Retaining Ring
7 — Driveshaft

Fig. 7 Inner C/V joint assembly. Eagle Premier

14. Clean and carefully inspect all components for wear or damage. Replace all necessary components.
15. Install axle oil seal and tube cover on axle shaft.
16. Using an arbor press, install axle shaft bearing and race on shaft.
17. Install axle shaft gear on shaft using an arbor press. Be sure to align reference mark. Also ensure gear splines are facing outward.
18. Install axle shaft E-ring.
19. Using tool No. J-29153, install intermediate shaft needle bearing.
20. Install intermediate shaft gear needle bearing in gear bore using an arbor press.
21. Install intermediate shaft, then install shift collar, and seat shaft in differential.
22. Install intermediate shaft lock ring.
23. Install outer axle shaft assembly, **torque** axle tube cover bolts evenly and in a cross sequence. **Torque** bolts 144 inch pounds.
24. Install gasket on shift housing cover and install. **Check to ensure the shift fork and tabs are aligned in the shift collar.**
25. **Torque** shift housing cover bolts to 108 inch pounds.
26. Install axle cover using a new gasket. **Torque** bolts to 20 ft. lbs.
27. Fill shift housing with 5 oz. of specified lubricant, then fill axle housing.

SELECT DRIVE SYSTEM

The Select Drive is a system which permits the full time four wheel drive power train to be operated in a two wheel drive mode. When the vehicle is stopped, the system can be changed from two wheel drive to four wheel or four wheel drive to two wheel drive, by pulling the instrument panel control switch pull release pin downward and moving the control switch lever to the desired position. The two major components of this system are a shifting device in the transfer case and a disconnecting mechanism in the front axle, which are controlled by the instrument panel switch through a series of vacuum actuators.

The vacuum actuated shifter mechanism located on the transfer case slides a splined clutch, located on the transfer case main shaft, rearward to engage or forward to disengage power to the front drive chain sprocket. In the two wheel drive position, the shifter mechanism also locks the transfer case differential. Also the main shaft and differential assembly will turn as a unit, transferring power to the rear driveshaft, while front output shaft, chain and sprocket remain stationary.

A vacuum actuated disconnect device is used to engage or disengage a splined clutch that connects the righthand driveshaft to the righthand half shaft. When four wheel drive is selected, the spline clutch is engaged and the front drive axle operates in a normal manner. When two wheel is selected, the vacuum actuator releases the spline clutch, which releases drive to the righthand half shaft and allows the righthand driveshaft to rotate freely. Due to the design of the differential, to equalize torque between left and right axle shafts, disconnecting the right axle shaft will allow the left axle shaft to rotate freely. The axle disconnect will not release all front drive components in the two wheel drive mode. Half shafts and different side gears will continue to rotate with the front wheels as the vehicle is driven.

Axle Shift Motor, Replace

1. Make a reference mark on fork and housing for assembly reference.
2. Rotate shift motor and remove retaining snap rings using a suitable screwdriver.
3. Remove shift motor from housing and remove O-ring from motor.
4. Reverse procedure to install.

EAGLE PREMIER

DRIVESHAFT, REPLACE

Use caution when servicing driveshafts. Do not strike the end of the shaft or drop it. Inspect C/V boots and replace if damaged

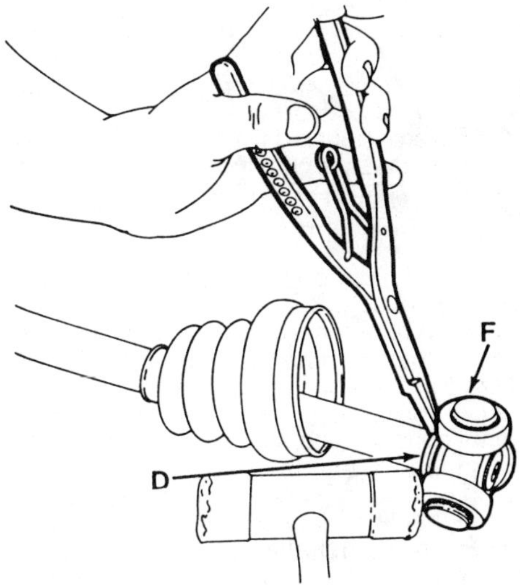

Fig. 8 Removing trunnion from driveshaft. Eagle Premier

or worn.

1. Raise and support vehicle.
2. Remove tire and wheel assembly, then the brake caliper. Do not disconnect brake caliper hose or suspend caliper from hose.
3. Using suitable tool to hold wheel hub in place, remove driveshaft nut.
4. Loosen driveshaft in hub. If driveshaft does not come loose by hand, use hub puller No. T.Av. 1050 or equivalent. **Do not attempt to loosen driveshaft by striking it with a hammer.**
5. Using a punch, remove driveshaft roll pin.
6. Remove tie rod end attaching nut, then disconnect tie rod end using ball joint extractor No. T.Av 476 or equivalent.
7. Remove ball joint keybolt nut, leaving keybolt in place.
8. Remove steering knuckle-to-strut attaching bolts. **Do not turn bolts, as they are splined just under bolt head.** Hold bolt head with a suitable wrench, then unscrew nut to end of bolt. Tap nut with brass hammer to loosen bolt, then remove nut and slide bolt from knuckle.
9. Tilt steering knuckle out and away from strut, then remove driveshaft.
10. Reverse procedure to install. **Torque** steering knuckle-to-strut attaching nuts to 123 ft. lbs. and driveshaft nut to 181 ft. lbs. Check and adjust transaxle fluid level if necessary.

INNER CONSTANT VELOCITY JOINT SERVICE

Removal & Disassembly

1. Remove driveshaft as described under "Driveshaft, Replace."
2. Cut and remove boot clamps, then

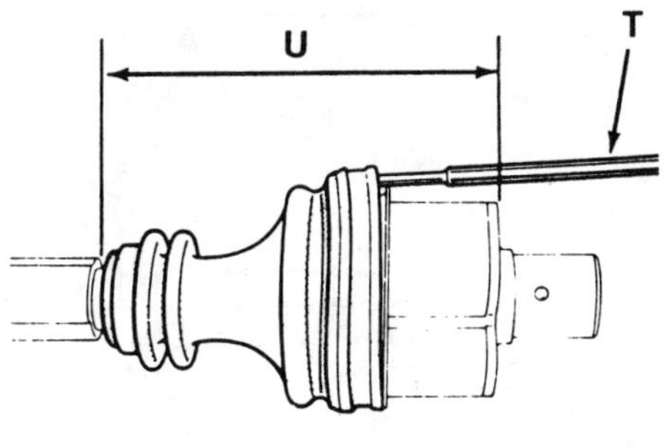

Fig. 9 Installing inner C/V boot. Eagle Premier

slide boot from C/V joint yoke, **Fig. 7.** If boot is to be reused, avoid damaging boot when cutting clamps.
3. Remove C/V joint yoke by pulling it straight off the trunnion.
4. Using snap ring pliers, spread plastic retaining ring (D), then tap trunnion off shaft with plastic mallet, **Fig. 8.**
5. Remove boot from shaft. Inspect C/V joint yoke, boot, plastic retaining ring, trunnion and trunnion bearings. Replace worn or damaged components.

Assembly & Installation

1. Install new retaining ring by pushing non-tapered end into trunnion until it snaps into groove.
2. Install boot on driveshaft.
3. Tap trunnion onto driveshaft until retaining ring seats in shaft groove.
4. Lubricate C/V joint yoke, trunnion bearings and the interior of boot with grease supplied in service kit, then slide yoke onto trunnion and bearings.
5. Seat boot on yoke and driveshaft, then bleed air from boot by inserting a smooth rod (T) between boot and yoke, **Fig. 9. Do not damage boot while purging air.**
6. Extend or retract C/V joint yoke until distance (U) from driveshaft end of yoke to small end of boot is 6.101-6.179 inches, as shown in **Fig. 9.**
7. Using boot clamp pliers No. T.Av 1034 or equivalent, install and tighten boot clamps.
8. Install driveshaft as described under "Driveshaft, Replace."

OUTER CONSTANT VELOCITY JOINT SERVICE

Removal & Disassembly

1. Remove driveshaft as described under "Driveshaft, Replace."
2. Cut and remove boot clamps, then slide boot rearward for access to plastic retaining ring. If boot is to be reused, avoid damaging boot when cutting clamps.
3. Using snap ring pliers, spread plastic retaining ring, then tap C/V joint with plastic mallet to free shaft from retaining ring.
4. Remove C/V joint, then slide boot off shaft. Inspect retaining ring, and replace if necessary.

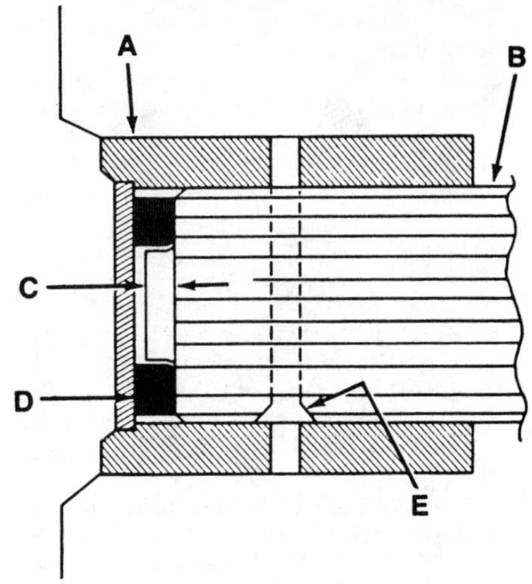

Fig. 10 Driveshaft identification. Eagle Medallion

Installation

1. Install new retaining ring with tapered end into C/V joint and segmented end onto driveshaft.
2. Slide boot onto driveshaft.
3. Lubricate C/V joint with grease supplied in service kit. Align driveshaft with plastic retaining ring and C/V joint, then tap joint onto shaft until retaining ring snaps into place.
4. Position boot in clamp groves of C/V joint and driveshaft, then using boot clamps pliers No. T.Av 1034 or equivalent, install and tighten boot clamps.
5. Install driveshaft as described under "Driveshaft, Replace."

EAGLE MEDALLION

FRONT DRIVESHAFT SERVICE

Removal

Three different types of side gear (output) shafts may be used. Different outer and inner constant velocity joints are used on the driveshafts. After removing the driveshaft boot, identify the type of constant velocity joint used and refer to the appropriate procedures for each joint. Separate service procedures are provided for type 1 and 2 constant velocity outer joints and the type 3 and 4 constant velocity inner joints.

1. Raise and support vehicle.
2. Remove front wheel and brake caliper assembly. **Secure caliper to the body or spring. Do not allow the brake line to support the weight of the caliper assembly.**
3. Using hub locking bar Rou.604.01 or equivalent to secure hub in position, remove driveshaft attaching nut.
4. Using pin drifts, B.Vi.31.01 or equivalent, remove double roller pins attaching driveshaft to the side gear shaft.
5. Using ball joint extractor T.Av.476 or equivalent, disconnect tie rod end.

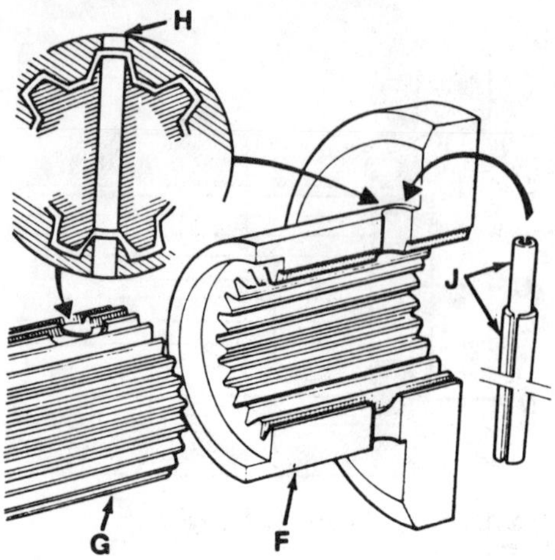

Fig. 11 Aligning roll pin holes in driveshaft side gear. Eagle Medallion

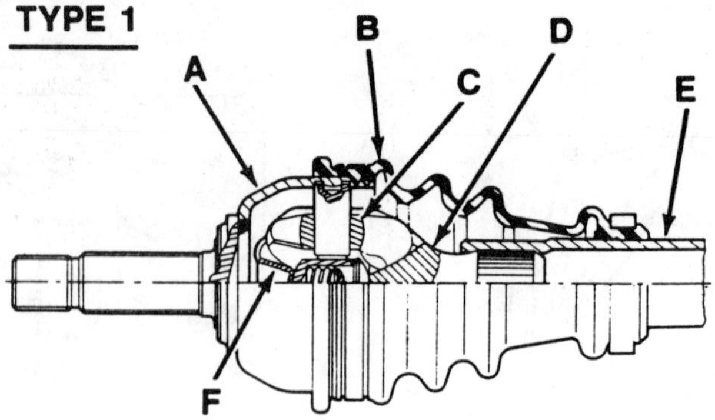

A — Housing
B — Protective Boot
C — Spider
D — Yoke
E — Shaft
F — Starplate

Fig. 12 Cross-sectional view of type 1 C/V joint. Eagle Medallion

6. Remove upper bolt attaching steering knuckle to shock absorber. Loosen but do not remove lower attaching bolt.
7. Tilt brake rotor/steering knuckle assembly outward. **The driveshaft splines are secured to the hub splines with Loctite. Use hub puller T.Av.1050 only to break the shaft loose from the hub for removal. Do not attempt to loosen the shaft with a hammer.**
8. Remove driveshaft as follows:
   a. Position tool T.Av.1050 (only) onto the rotor and hub assembly.
   b. Tighten tool attaching bolt to break the driveshaft splines loose from the hub.
   c. Disconnect the shaft at the transaxle and carefully remove the shaft from the vehicle. Do not damage the driveshaft boot during removal.

IDENTIFICATION & INSTALLATION

Clean driveshaft and hub splines thoroughly using a wire brush. Lubricate seals and splines on the transaxle side gear shaft and driveshaft with clean chassis grease only. Inspect the driveshaft boot. Replace boot if damaged. Before connecting the driveshaft to the transaxle side gear shaft, identify the type of side gear shaft in the transaxle. Three different types are used as described previously.

Type 1 shafts have a shoulder (C), **Fig. 10,** that is .039 inch (1mm) long. A .118 inch (3mm) thick rubber washer (D) is used between the end of the side gear shaft and driveshaft.

Type 2 shafts have a shoulder (C), **Fig. 10,** that is .118 inch (3mm) long. A .197 inch thick rubber washer (D) is used between the ends of the side gear shaft and driveshaft.

Type 3 shafts do not have a machined shoulder and do not require a rubber washer.

Verify the type of side gear shaft and install the appropriate thickness rubber washer (or no washer in type 3 shafts), before installing.

All of the side gear shaft types have a chamfer (E), **Fig. 10,** machined into one side of the roll pin hole. The chamfer is included to make installation of the double roller pins easier.

To install, proceed as follows:

1. Install driveshaft (F), **Fig. 11,** onto side gear shaft (G).
2. Align roll pin holes (H) in the driveshaft (F) side gear (G).
3. Install driveshaft onto side gear shaft and insert double roll pins (one pin fits inside the other). Position slits into pins 180° apart. Tap pins (J) into place with pin drifts B.Vi.31.01 or equivalent.
4. Seal open ends of the roll pins with silicone sealer after installation.
5. Apply a few drops of Loctite 271 to the driveshaft splines and insert shaft into hub assembly. Use installer tool T.Av.602 of the shafts are difficult to install into hub. **Do not damage the driveshaft boot while installing the shaft.**
6. Apply Loctite 271 onto driveshaft nut and install nut. Using hub locking bar Rou.604.01 or equivalent, tighten nut to 185 ft. lbs.
7. Connect tie rod end to shock arm.
8. Install brake caliper, wheel and lower vehicle.
9. Apply brakes several times to seat pads before moving vehicle.

CONSTANT VELOCITY JOINT & BOOT SERVICE

TYPE 1 OUTER C/V JOINT

Disassembly

Refer to Fig. 12 for cross-sectional view of type 1 constant velocity joint assembly.

1. Remove driveshaft.
2. Cut, then remove boot clamps.
3. Slide boot away from joint and remove as much grease as possible from constant velocity joint. **Do not use cleaning solvents.**
4. Using a suitable screwdriver, lift star plate tabs upward to release the yoke from the spider. Do not bend the tabs.
5. Remove yoke and shaft from constant velocity joint.
6. Remove and retain the thrust ball (C), spring (D) and shim (E) from the spider, **Fig. 13.**
7. Cut the boot in half and remove it from the shaft.
8. Remove all grease from constant velocity joint. **Do not use cleaning solvents.**

Assembly

1. Install shaft into a suitable vise. Position shaft so yoke is pointing upward. **Do not attempt to install the boot without using the proper tool for that purpose, as the boot can be damaged.**
2. Install expander tool T.Av.586.01 or equivalent onto shaft yoke.
3. Lubricate entire length of expander tool with clean oil.
4. Lubricate inside of the replacement boot with clean oil. Cover the small opening in the replacement boot with your thumb and pour oil into the boot. Spread the oil evenly around the boot interior.
5. Start the boot onto the expander tool.
6. Grip the boot so the first fold (J), **Fig. 14,** will stretch as the boot is worked onto the expander tool.
7. Pull the boot onto the tool as far as possible. Allow the boot to relax and return part way. Ensure the first fold (J) does not slip back. Repeat this procedure three or four times to stretch

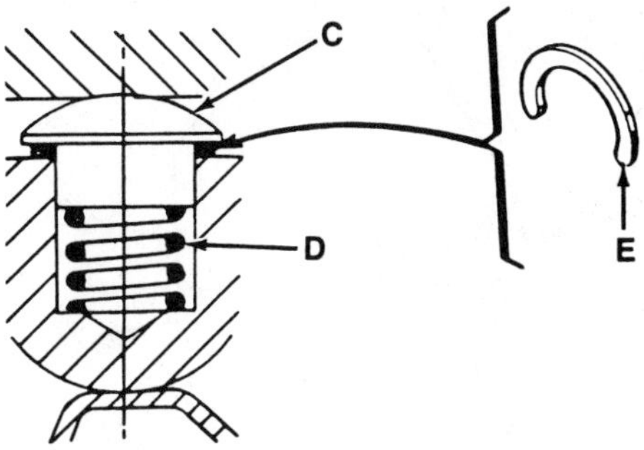

Fig. 13 Removing thrust ball (C), spring (D) & shim (E) Type 1. Eagle Medallion

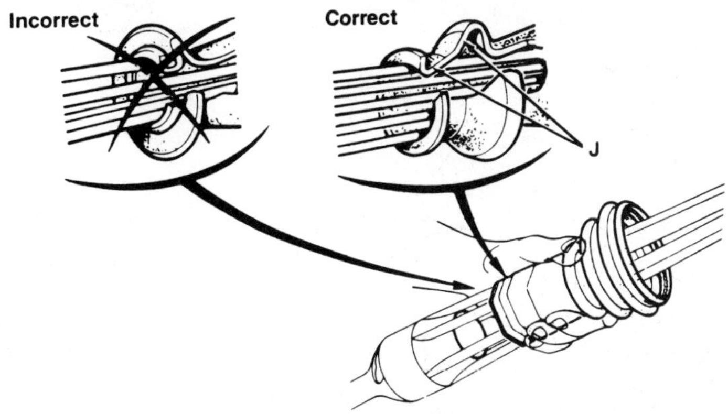

Fig. 14 Correct boot installation Type 1. Eagle Medallion

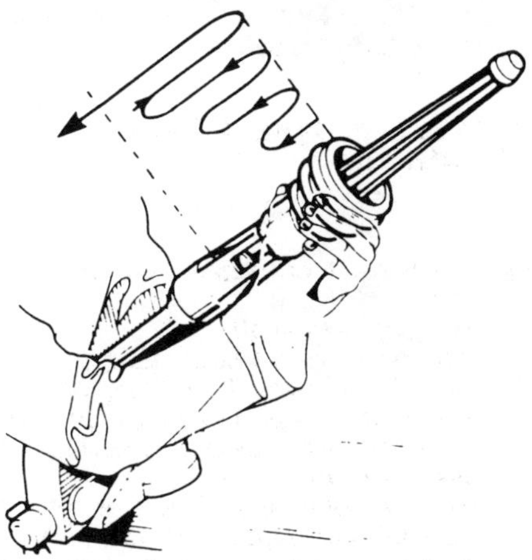

Fig. 15 Turning boot to aid installation Type 1. Eagle Medallion

and expand the boot. **To ease installation, wrap a dry cloth around the boot and use a two handed grip to gradually work the boot onto the tool. Turn the boot back and forth to aid installation, Fig. 15.**

8. When the boot feels most pliable (flexible), use one continuous motion to draw it up to the end of the tool.
9. Slide boot onto shaft and remove expander tool.
10. Lubricate the constant velocity joint components with suitable grease supplied in the service kit. Distribute the grease evenly between boot inner surface, bearing rollers, yoke and housing.
11. Install spring, shim and ball into the spider. **Lubricate these components before installation.**
12. Fabricate and install a spacer (N) under the spider and star plate to hold them upward during assembly. A Spacer can be made from an old shock absorber grommet. Loop a length of wire through it as shown in **Fig. 16,** then insert the spacer under the star plate (P). Move the roller (Q) toward the center and position the star plate arms (P) so that each arm is centered between each spider roller, **Fig. 17.**

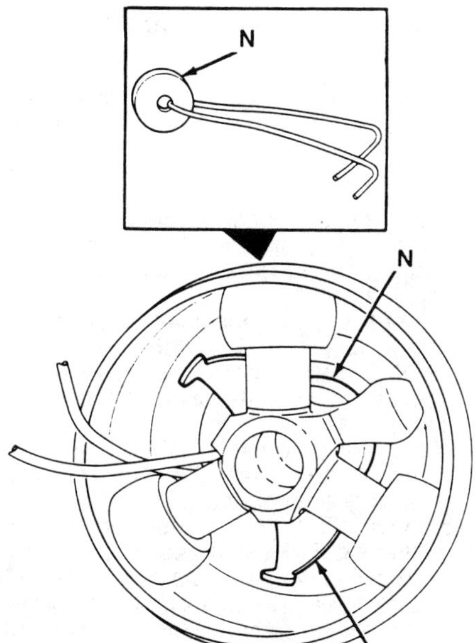

Fig. 16 Tool fabrication to secure spider and star plate Type 1. Eagle Medallion

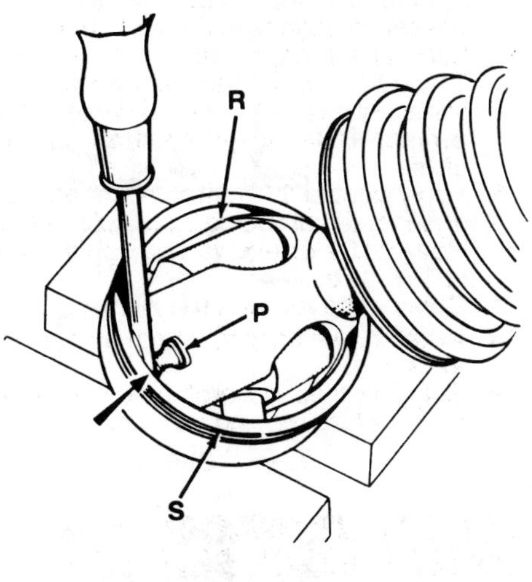

Fig. 18 Inserting shaft yoke into housing Type 1. Eagle Medallion

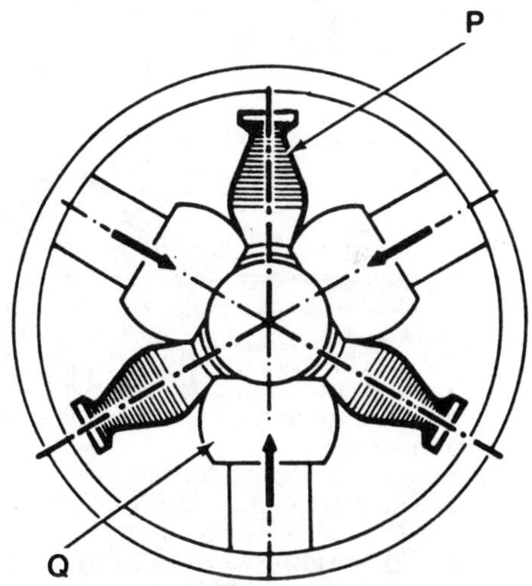

Fig. 17 Roller & star plate installation Type 1. Eagle Medallion

13. Insert the shaft yoke (R) into the housing (S), **Fig. 18.**
14. Tilt the shaft and insert one star plate arm (P) into the yoke. Then press the arm into the yoke slot using a screwdriver.
15. Remove spacer from under the star plate by pulling it out with the attached wire.
16. Press the remaining star plate arms into the yoke with a modified screwdriver (T), **Fig. 19.** Grind approximately a .20 by .12 inch notch in the screwdriver tip. Then seat the remaining star plate arms.
17. Ensure each star plate arm is properly seated.
18. Check spider movement. It should move freely in all directions.
19. Lubricate all constant velocity joint components.
20. Position boot into housing and shaft retaining grooves.
21. Purge air from boot. **Do not damage boot when purging air.**
22. Using tool T.Av.1034 or equivalent, install and tighten boot clamps. If a

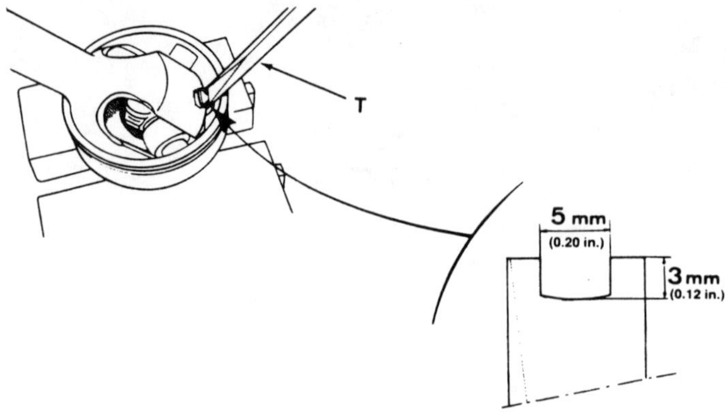

Fig. 19 Installing remaining star plates into yoke Type 1. Eagle Medallion

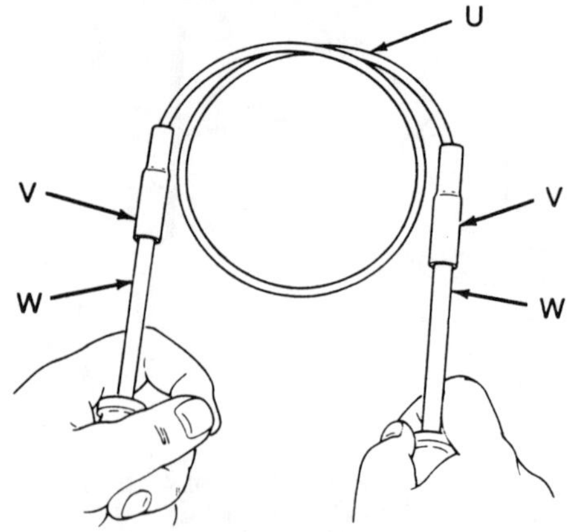

Fig. 20 Spring type clamp installation Types 1 & 2. Eagle Medallion

TYPE 2

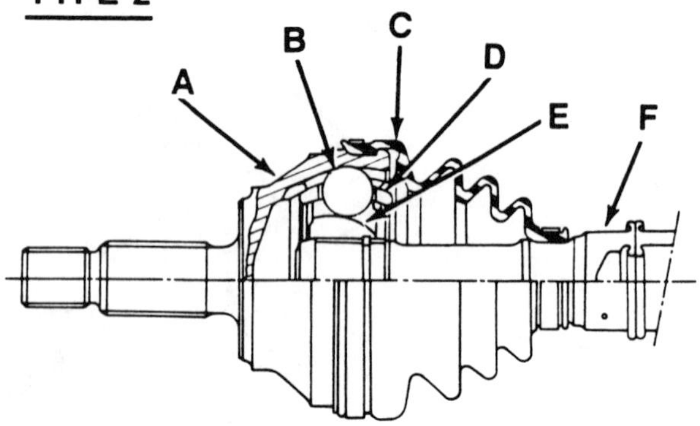

A — Housing
B — Bearing Balls (6)
C — Boot
D — Ball Cage
E — Ball Hub
F — Shaft

Fig. 21 Cross-sectional view of type 2 outer C/V joint. Eagle Medallion

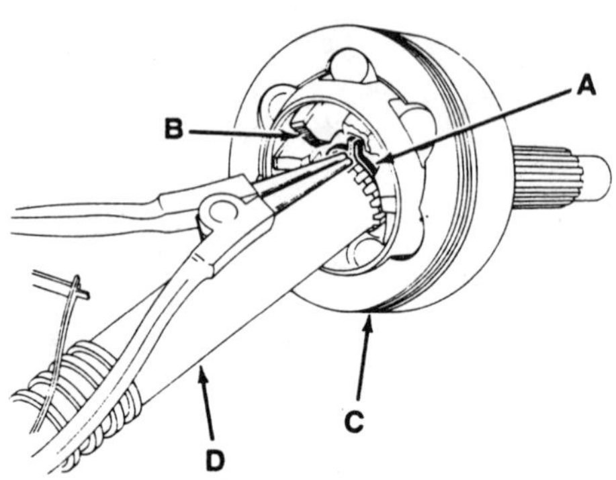

Fig. 22 Removing retaining ring, ball hub & joint from shaft Type 2. Eagle Medallion

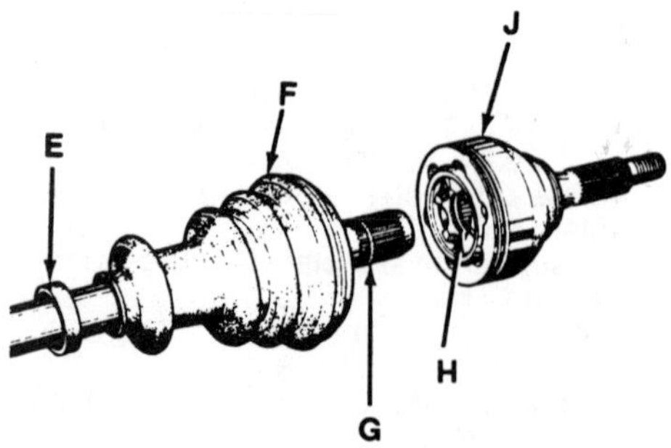

Fig. 23 Outer joint assembly Type 2. Eagle Medallion

TYPE 3

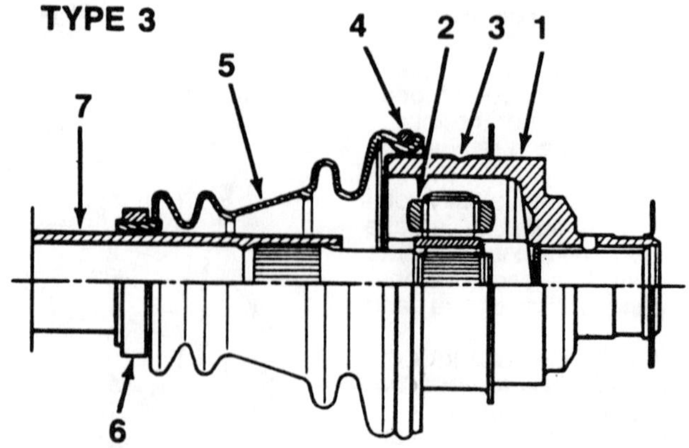

1 — Yoke
2 — Spider
3 — Metal Cover
4 — Retaining Spring
5 — Boot
6 — Retaining Ring
7 — Driveshaft

Fig. 24 Corss-sectional view of type 3 inner C/V joint. Eagle Medallion

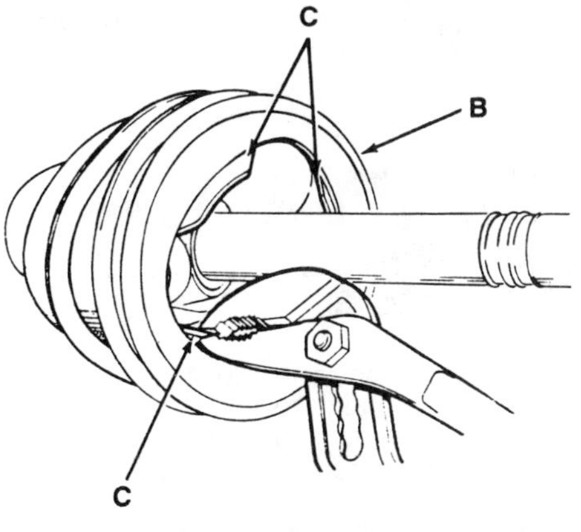

Fig. 25 Removing anti-separation plates from joint Type 3. Eagle Medallion

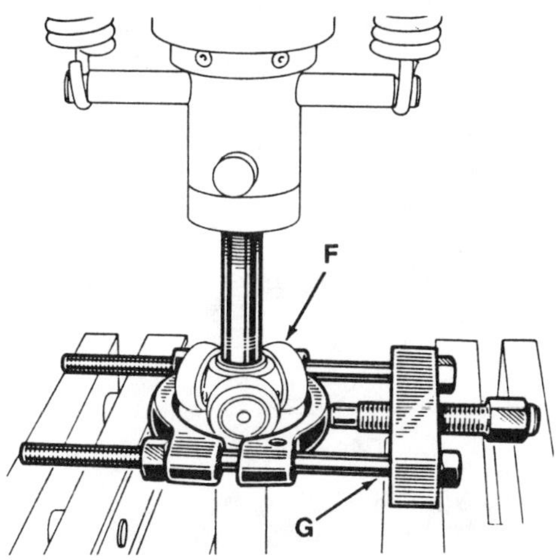

Fig. 26 Pressing bearing rollers from shaft Type 3. Eagle Medallion

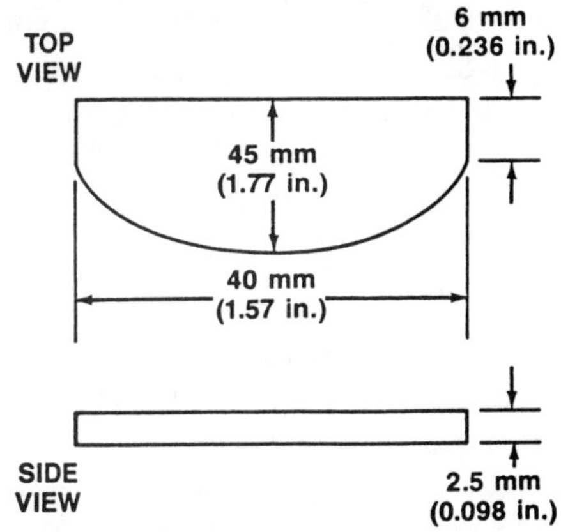

Fig. 27 Wedge fabrication dimensions Type 3. Eagle Medallion

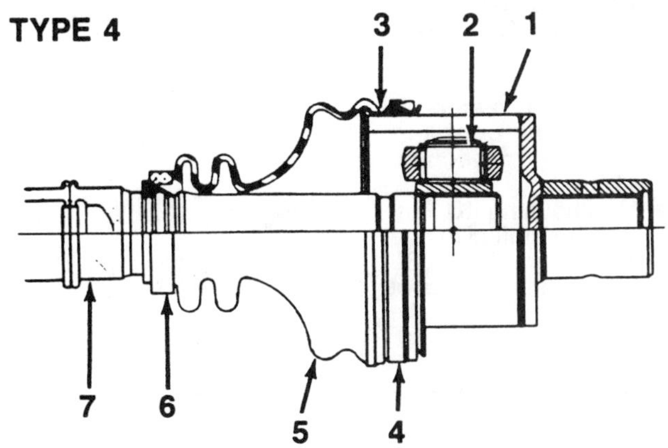

1 — Yoke
2 — Spider
3 — Metal Cover
4 — Retaining Clamp
5 — Boot
6 — Retaining Ring
7 — Driveshaft

Fig. 28 Cross-sectional view of type 4 inner C/V joint. Eagle Medallion

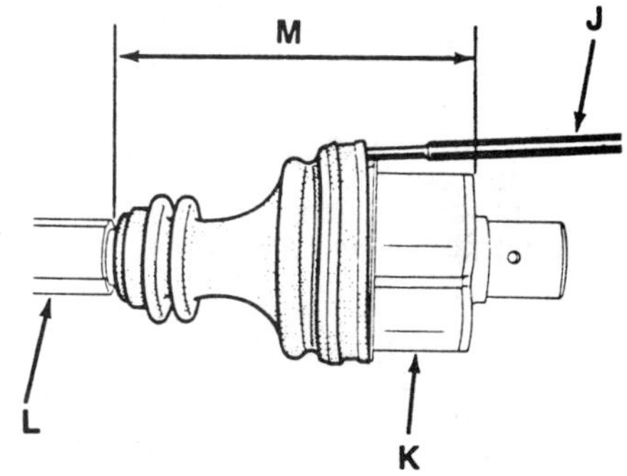

Fig. 29 Obtaining coupling housing distance (M) Type 4. Eagle Medallion

spring type clamp (U), **Fig. 20,** is used at the large end of the boot, use two 1/4 inch drive deep sockets (V) and extensions (W) to spread and install clamp.

TYPE 2 OUTER C/V JOINT

Disassembly

Refer to Fig. 21 for cross-sectional view of type 2 outer constant velocity joint assembly.

1. Remove clamp at large end of boot.
2. Remove retaining ring (A), **Fig. 22.** Spread ring, tap the ball hub (B) and slide the joint (C) off shaft (D).
3. Remove boot from shaft.
4. Remove as much grease as possible from assembly. **Do not use any cleaning solvents.**
5. Inspect constant velocity joint. Replace entire joint as an assembly, if required.

Assembly

1. Install replacement boot retainer (E) and boot (F) onto shaft, **Fig. 23.**
2. Install a replacement retaining ring (A) into constant velocity joint ring groove (H).
3. Lubricate constant velocity joint (J) and interior of boot (F) with suitable grease supplied in the service kit.
4. Install the constant velocity joint (J) onto shaft (G). Ensure retaining ring is completely seated in its groove.
5. Seat boot in constant velocity joint housing and shaft grooves, then install boot clamps.
6. If a spring type clamp (U), **Fig. 20,** is used at the large end of the boot, use two 1/4 inch drive deep sockets (V) and extensions (W) to spread and install clamp.

TYPE 3 INNER C/V JOINT

Disassembly

Refer to Fig. 24 for cross-sectional view of type 3 inner constant velocity joint assembly.

1. Remove boot retaining spring.
2. Cut boot lengthwise and remove boot from shaft.
3. Using a suitable tool, lift and remove anti-separation plates (C) from joint (B), **Fig. 25.**
4. Remove snap ring securing bearing roller onto shaft.
5. Using a suitable press and press tool U-53P (G), drive bearing rollers (F) off shaft assembly, **Fig. 26.**
6. Clean constant velocity joint assembly. **Do not use cleaning solvents.**
7. If the boot bearing requires replacement, refer to "Inner Constant Velocity Joint Boot Bearing, Replace."

Assembly

1. Lubricate driveshaft and install replacement boot and boot clamp onto shaft assembly.
2. Using a suitable press, drive replacement bearing roller assembly onto shaft.

3. Install a replacement snap ring onto shaft.
4. Fabricate a wedge to the dimensions shown in **Fig. 27.** The wedge will be used to install the anti-separation plates.
5. Insert wedge between anti-separation plate and yoke.
6. Tap each plate into position and remove wedge.
7. Purge air from boot. **Do not damage boot while purging air.**

TYPE 4 INNER C/V JOINT

Disassembly

Refer to Fig. 28 for cross-sectional view of type 4 inner constant velocity joint assembly.

1. Remove boot clamps.
2. Remove yoke by pulling it straight off the bearing rollers.
3. Remove bearing roller snap ring.
4. Using a suitable press and press tool U-53P or equivalent, remove bearing rollers from shaft.
5. Remove boot from shaft.
6. Clean grease from assembly. **Do not use cleaning solvents.**
7. If the boot bearing requires replacement, refer to "Inner Constant Velocity Joint Boot Bearing, Replace."

Assembly

1. Lubricate driveshaft with suitable grease and install boot and clamp onto shaft.
2. Press bearing rollers onto shaft.

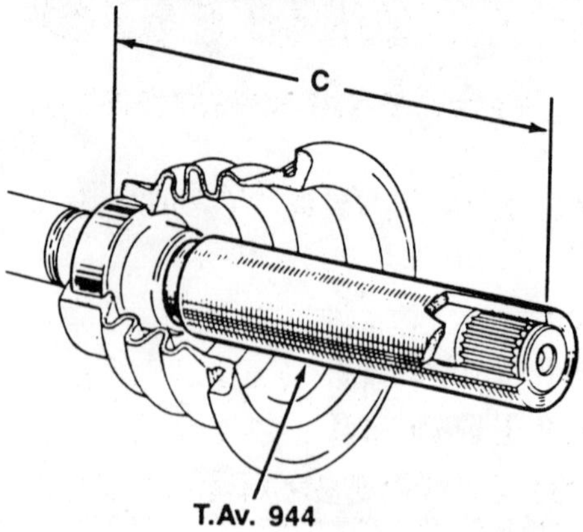

Fig. 30 Pressing boot bearing onto shaft. Eagle Medallion

3. Secure rollers using a replacement snap ring or stake rollers in position by center punching the ends of the shaft at three equally spaced points.
4. Lubricate all constant velocity joint components including the boot interior.
5. Install boot onto housing.
6. Purge air from boot. **Insert a smooth rod (J), Fig. 29, between boot and housing and allow air to escape. Do not damage boot.**
7. Extend or retract the housing (K), **Fig. 29,** on the coupling (L) until distance (M) is 6.14 inches plus or minus .039 inch. Remove rod (J) and install boot retainers.
8. Using tool T.Av.1034 or J-22610, tighten retainers.

INNER CONSTANT VELOCITY JOINT BOOT BEARING REPLACE

1. Remove constant velocity joint housing and bearing rollers.
2. Remove bearing snap ring.
3. Press bearing and boot off shaft assembly.
4. Using a suitable press and tool T.Av.944 or equivalent, install bearing and boot.
5. Install snap ring.
6. Press boot and bearing onto shaft until tool T.Av.944 bottoms against shaft. At this point, distance (C), **Fig. 30,** should be approximately 5.85 inches. **Do not attempt to install the bearing with a hammer or damage may result.**
7. Assemble constant velocity joint.

EAGLE SUMMIT

Refer to "Colt" in the "Chrysler Imports" section for service procedures on front drive axle.

FORD MOTOR CO.

TABLE OF CONTENTS

Escort, EXP, Lynx, Tempo & Topaz

INDEX

TROUBLESHOOTING

NOISE & VIBRATION IN TURNS

Clicking, popping or grinding noises while turning may be caused by the following:

1. Cut or damaged CV joint boots, resulting in contaminated lube in outboard or inboard CV joints.
2. Loose CV joint clamps.
3. Worn, damaged or improperly installed wheel bearings.
4. Foreign object contacting halfshaft assembly.

VIBRATION AT HIGHWAY SPEEDS

1. Out of balance front wheels or tires.
2. Improperly seated outboard CV joint in front wheel hub.
3. Bent interconnecting shaft.
4. Front tires out of round.

SHUDDER OR VIBRATION DURING ACCELERATION

1. Excessively worn or damaged inboard or outboard CV joint.

HALFSHAFT OR CV JOINT PULL-OUT

Engine Or Transaxle Misaligned

1. Check engine mounts for damage.

Front Suspension Components Worn Or Damaged

1. Check for worn bushings or bent front suspension components.

Improperly Installed Or Missing Retainers

1. Check for CV joint circlip missing or not properly seated in transaxle side gear.

DRIVESHAFTS REPLACE

If removing both right and left side halfshafts, plugs T81P-1177B or equivalent must be installed. Failure to do so may result in dislocation of differential side gears, necessitating transaxle disassembly to re-align the gears. Also, halfshaft removal and installation procedures are the same for manual and automatic transaxles except for the following: due to automatic transaxle case configuration the right side halfshaft assembly must be removed first. Tool T81P-4026A or equivalent is then inserted into transaxle to remove left side inner constant velocity joint assembly from transaxle. If only the left side halfshaft is to be removed from the vehicle, remove right side halfshaft assembly from the transaxle case only and secure to underside of vehicle, then remove left side halfshaft assembly. The hub nut and lower control arm to steering knuckle attaching bolt and nut must be discarded after removal and new nuts and bolts installed.

Driveshaft assembly removal and installation procedures are the same for ATX/FLC (automatic transaxle) applications as for MTX (manual transaxle) applications except the ATX/FLC case configuration requires that the righthand halfshaft assembly be removed first. Differential rotator T81P-4026-A or equivalent, should then be inserted into the transaxle to drive the lefthand inboard constant velocity joint assembly from the transaxle. If only the lefthand halfshaft assembly is to be removed for service, remove the righthand halfshaft assembly from the transaxle only. After removal, support it with a length of wire, then drive the lefthand halfshaft assembly from the transaxle case. **Do not begin this removal procedure unless the following parts are known to be available, a new hub retainer nut, a new longer lower control arm to steering knuckle attaching bolt and nut and a new inboard constant velocity joint stub shaft snap ring. Once removed, these components must not be reused during the assembly procedure. Their torque holding ability or retention capability is greatly diminished during removal.**

1. Loosen hub nut without unstaking. Use of a chisel or similar tool to unstake nut may damage spindle threads.
2. Raise and support vehicle and remove wheel assemblies.
3. Remove bolt attaching brake hose routing clip to suspension strut.
4. Remove nut from ball joint to steering knuckle attaching bolt, then drive bolt from knuckle using suitable punch and hammer.

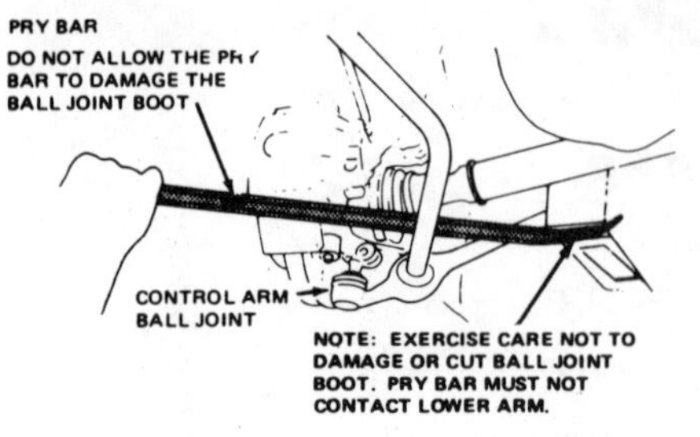

Fig. 1 Separating ball joint from steering knuckle

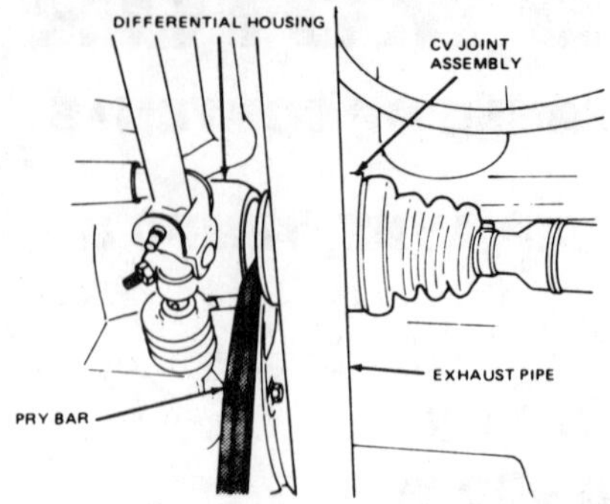

Fig. 2 Removing halfshaft from differential housing

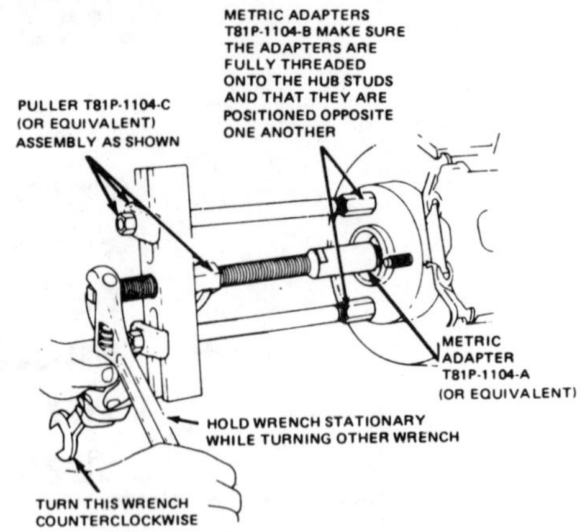

Fig. 3 Separating outer constant velocity joint from hub

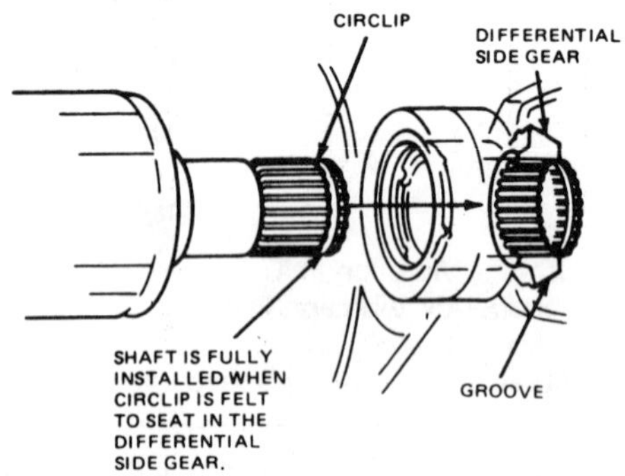

Fig. 4 Installing inner constant velocity joint into differential side gear

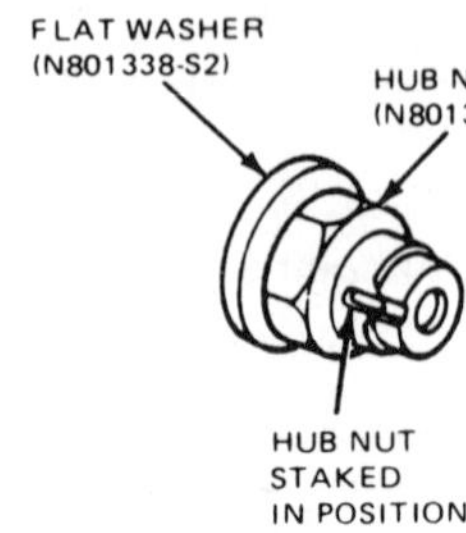

Fig. 5 Hub nut staking tool fabrication

5. Separate ball joint from steering knuckle using pry bar, Fig. 1. **Lower ball joints fit into a pocket formed in the plastic disc brake shield. The shield must be positioned away from the ball joint while removing ball joint from steering knuckle.**
6. Remove halfshaft from differential housing using suitable pry bar. Use caution not to damage dust deflector located between shaft and case, **Fig. 2.** If an automatic transaxle halfshaft assembly cannot be removed from differential by using a pry bar, insert a large bladed screwdriver between differential pinion shaft and inboard constant velocity joint stub shaft. Sharply tap on screwdriver handle, to free halfshaft from differential. **Use caution not to damage differential oil seal, constant velocity joint boot or constant velocity joint dust deflector.**
7. Separate outer constant velocity joint from hub using puller T81P-1104C or equivalent, **Fig. 3,** and adapters T81P-1104B and T81P-1104A or equivalent. **Do not use a hammer to separate outboard constant velocity joint stub shaft from hub as damage to internal components may result.**
8. Reverse procedure to install. Install new circlip on inboard constant velocity joint stub shaft. Align splines of inboard constant velocity joint stub shaft with splines in differential. Push joint into differential until circlip seats in side gear, **Fig. 4. Torque** new control arm to steering knuckle nut to 40 ft. lbs. **Torque** new hub nut to 180-200 ft. lbs., and stake nut, **Fig. 5. If the hub nut cracks or splits during staking, remove and replace hub nut.**

CONSTANT VELOCITY JOINT SERVICE

REMOVAL

Joints Except 5 Speed Manual Transaxle Inboard Constant Velocity Joint & Boot

1. Place halfshaft in suitable vise. Use caution not to damage boot or clamp.
2. Cut large boot clamp and remove from boot, **Figs. 6 through 8,** then position boot upward on shaft. If boot only is being replaced due to damage, check joint grease for contamination. If joints were operating satisfactorily and grease is not contaminated, add grease and install new boot. If grease is contaminated, joint must be completely disassembled.
3. Place interconnecting shaft in a suitable vise and angle constant velocity joint so that inner bearing race is exposed, **Fig. 9.**
4. Using suitable drift and hammer, tap inner bearing race to dislodge internal circlip and separate constant velocity joint from interconnecting shaft, being careful not to drop joint.
5. Remove boot from shaft, cutting remaining clamp as necessary.
6. Remove circlip from end of shaft and discard. Inspect stop ring for damage and replace as necessary.

Inboard Constant Velocity Joint & Boot, 5 Speed Manual Transaxle

1. Remove large boot clamp, roll boot back, and wipe away excess grease.
2. Remove wire ring bearing retainer from outer race, then remove outer race.
3. Pull inner race and bearing assembly out until it rests on circlip, then, using suitable pliers, spread stop ring and move it back on shaft.
4. Slide inner race and bearing assembly down shaft to expose circlip, then remove circlip.
5. Remove inner race and bearing assembly and, if necessary, remove boot.

INSTALLATION

Joints Except 5 Speed Manual Transmission Inboard Constant Velocity Joint & Boot

1. Install new stop ring, if removed. Ensure that stop ring is properly seated in groove.
2. Install new circlip in groove nearest end of shaft. To avoid over-expansion or twisting of circlip, start one end in groove and work circlip over stub shaft end and into groove. **Interconnecting shafts are different depending on application. These shafts are non-symmetrical. Outboard end is approximately 1/4 inch longer, from end of shaft to end of boot groove, than inboard end. Be careful to install inboard and outboard constant velocity joints to proper ends of shaft.**

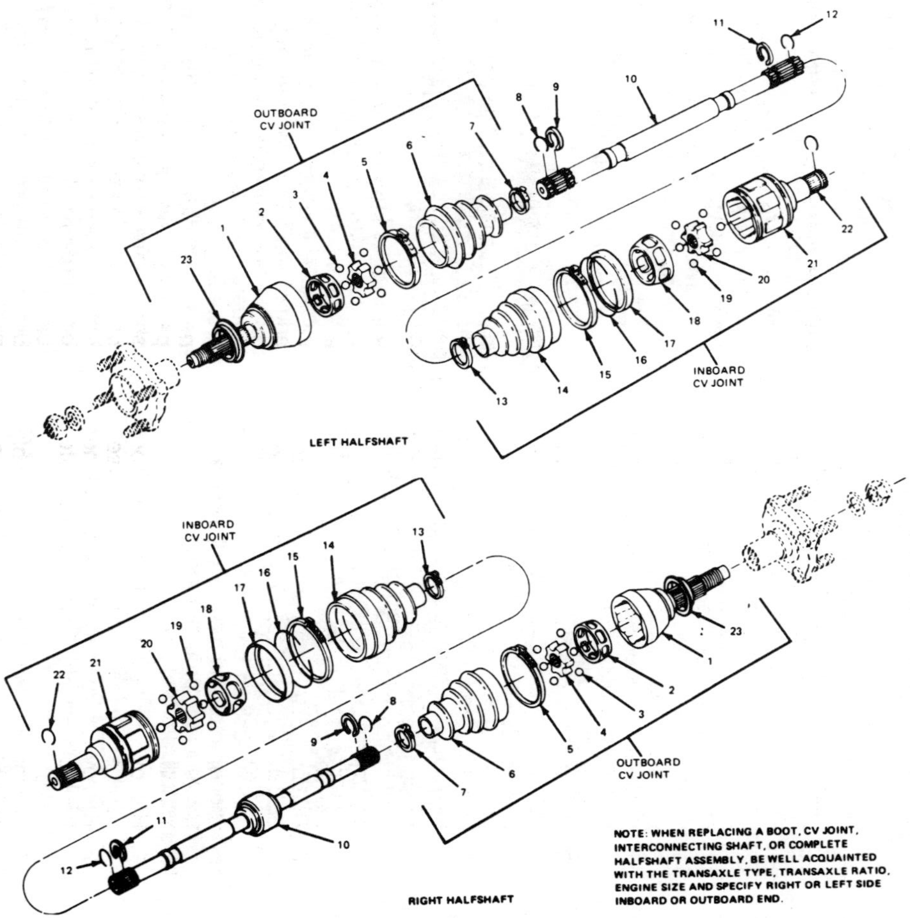

LEGEND:

1. OUTER BEARING RACE AND STUB SHAFT ASSEMBLY
2. BEARING CAGE
3. BALL BEARINGS (6)
4. INNER BEARING RACE
5. BOOT CLAMP (LARGE)
6. BOOT
7. BOOT CLAMP (SMALL)
8. CIRCLIP
9. STOP RING
10. INTERCONNECTING SHAFT
11. STOP RING
12. CIRCLIP
13. BOOT CLAMP (SMALL)
14. BOOT
15. BOOT CLAMP (LARGE)
16. BEARING RETAINER (MTX 5-SPEED ONLY)
17. BEARING RETAINER (MTX 4-SPEED AND ATX)
18. BEARING CAGE
19. BALL BEARINGS (6)
20. INNER BEARING RACE
21. OUTER BEARING RACE AND STUB SHAFT ASSEMBLY
22. CIRCLIP
23. DUST SEAL

Fig. 6 Halfshaft assemblies. Late 1985 models

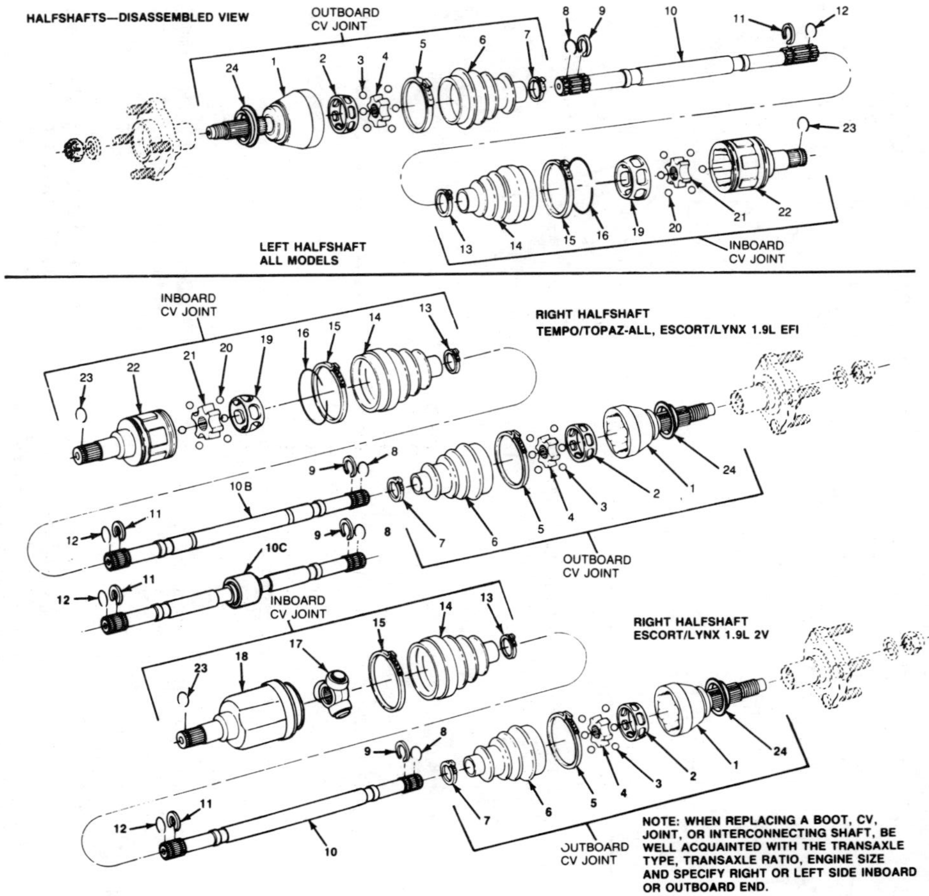

LEGEND:

1. OUTBOARD JOINT OUTER RACE AND STUB SHAFT
2. BALL CAGE
3. BALLS (SIX)
4. OUTBOARD JOINT INNER RACE
5. BOOT CLAMP (LARGE)
6. BOOT
7. BOOT CLAMP (SMALL)
8. CIRCLIP
9. STOP RING
10A. INTERCONNECTING SHAFT
10B. INTERCONNECTING SHAFT TEMPO/TOPAZ (MTX) ESCORT/LYNX (ALL)
10C. INTERCONNECTING SHAFT TEMPO/TOPAZ (ATX)
11. STOP RING
12. CIRCLIP
13. BOOT CLAMP (SMALL)
14. BOOT
15. BOOT CLAMP (LARGE)
16. WIRE RING BALL RETAINER
17. TRIPOD ASSY
18. TRIPOD OUTER RACE
19. BALL CAGE
20. BALLS (SIX)
21. INBOARD JOINT INNER RACE
22. INBOARD JOINT OUTER RACE AND STUB SHAFT
23. CIRCLIP
24. DUST SEAL

Fig. 7 Halfshaft assemblies. 1986 models

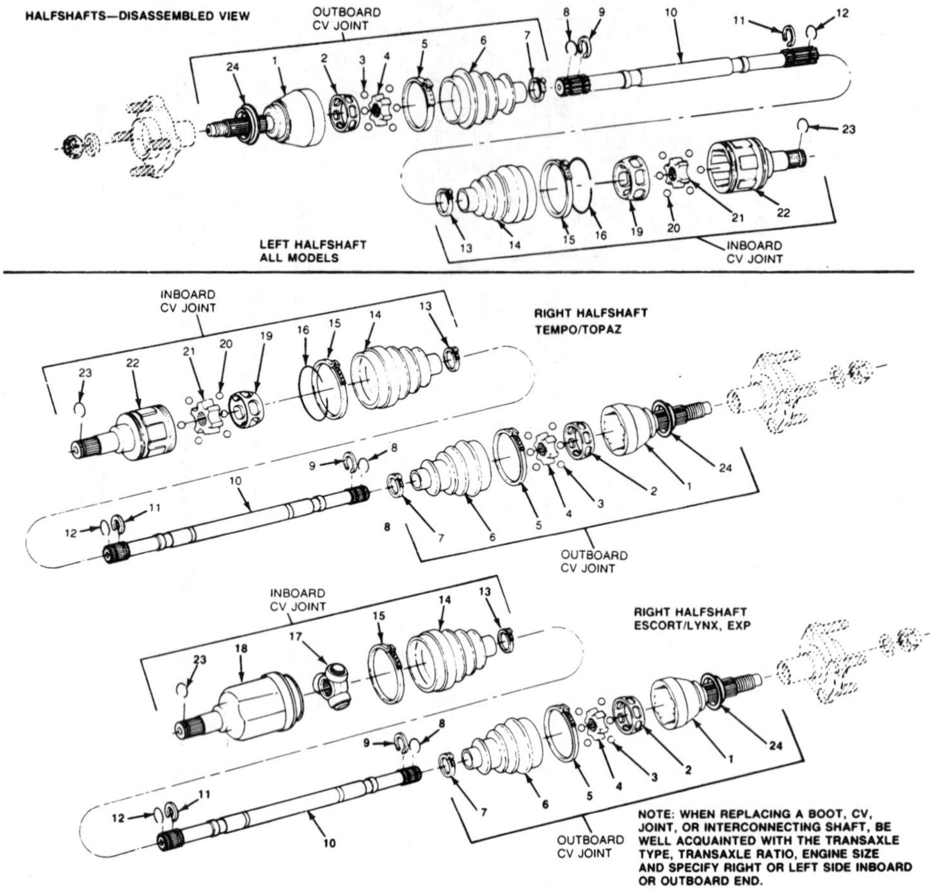

LEGEND:

1.	OUTBOARD JOINT OUTER RACE AND STUB SHAFT	13.	BOOT CLAMP (SMALL)
2.	BALL CAGE	14.	BOOT
3.	BALLS (SIX)	15.	BOOT CLAMP (LARGE)
4.	OUTBOARD JOINT INNER RACE	16.	WIRE RING BALL RETAINER
5.	BOOT CLAMP (LARGE)	17.	TRIPOD ASSY
6.	BOOT	18.	TRIPOD OUTER RACE
7.	BOOT CLAMP (SMALL)	19.	BALL CAGE
8.	CIRCLIP	20.	BALLS (SIX)
9.	STOP RING	21.	INBOARD JOINT INNER RACE
10.	INTERCONNECTING SHAFT	22.	INBOARD JOINT OUTER RACE AND STUB SHAFT
11.	STOP RING	23.	CIRCLIP
12.	CIRCLIP	24.	DUST SEAL

Fig. 8 Halfshaft assemblies. 1987–89 models

3. Install constant velocity joint boot, if removed, ensuring that boot is seated in groove. Tighten clamp securely, but not too tight.
4. Before positioning boot over constant velocity joint, pack joint and boot as follows:
   a. **On inboard constant velocity joint,** fill boot with 45 grams of grease and pack joint with 90 grams of grease.
   b. **On outboard constant velocity joint,** fill boot with 45 grams of grease and pack joint with 45 grams of grease. **Use only lubricant E2FZ-19590-A or equivalent.**
5. Position boot upward toward end of shaft, then position constant velocity joint onto shaft and tap into position using plastic mallet. Joint is fully seated when circlip locks in groove cut into joint bearing inner race. Check for proper seating by trying to pull joint from shaft.
6. Remove all excess grease from external surfaces of constant velocity joint, then position boot over constant velocity joint and move joint in or out to adjust to proper length, **Figs. 10 through 12.**
7. Before installing boot clamp, insert dulled screwdriver blade between boot and outer bearing race to allow trapped air to escape.
8. Ensure that boot is seated in groove, then install clamp securely but not too tight.

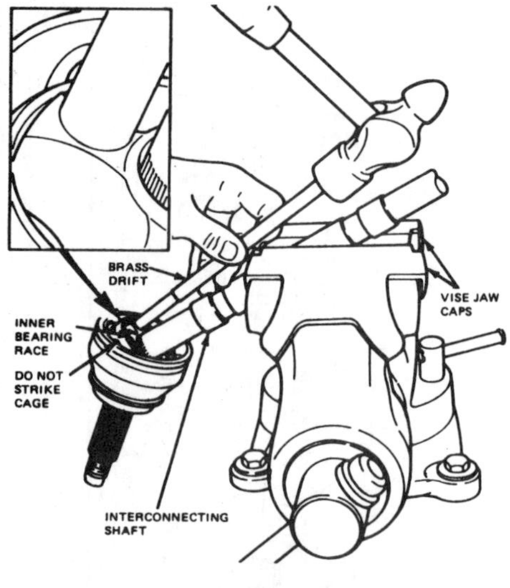

Fig. 9 Separating constant velocity joint from shaft. Late 1985–89 models

Inboard Constant Velocity Joint & Boot, 5 Speed Manual Transmission

1. Move circlip and stop ring back into their respective grooves on shaft. **Lefthand interconnecting shaft is symmetrical and inboard and outboard constant velocity joints may be installed on either end. Righthand interconnecting shaft is non-symmetrical and care must be taken so that inboard and outboard constant velocity joints are correctly installed, Figs. 13 and 14.**
2. Install constant velocity joint boot, if removed. Ensure that boot is seated in groove, then install clamp securely but not too tight.
3. Install new circlip in groove nearest end of shaft. To avoid over-expansion or twisting of circlip, start one end in groove and work circlip over stub shaft end and into groove.
4. Fill boot with 45 grams of grease and fill outer race with 90 grams of grease. Use only lubricant E2FZ-19590-A or equivalent.
5. Push inner race and bearing assembly into outer race by hand.
6. Install ball retainer into groove inside outer race.
7. With boot positioned upward toward end of shaft, install constant velocity joint using suitable hammer. Ensure that splines are aligned before hammering constant velocity joint onto shaft.
8. Remove all excess grease from external surfaces of constant velocity joint, then position boot over constant velocity joint and move joint in or out to adjust to proper length, **Figs. 10 through 12.**
9. Before installing boot clamp, insert dulled screwdriver blade between boot and outer bearing race to allow

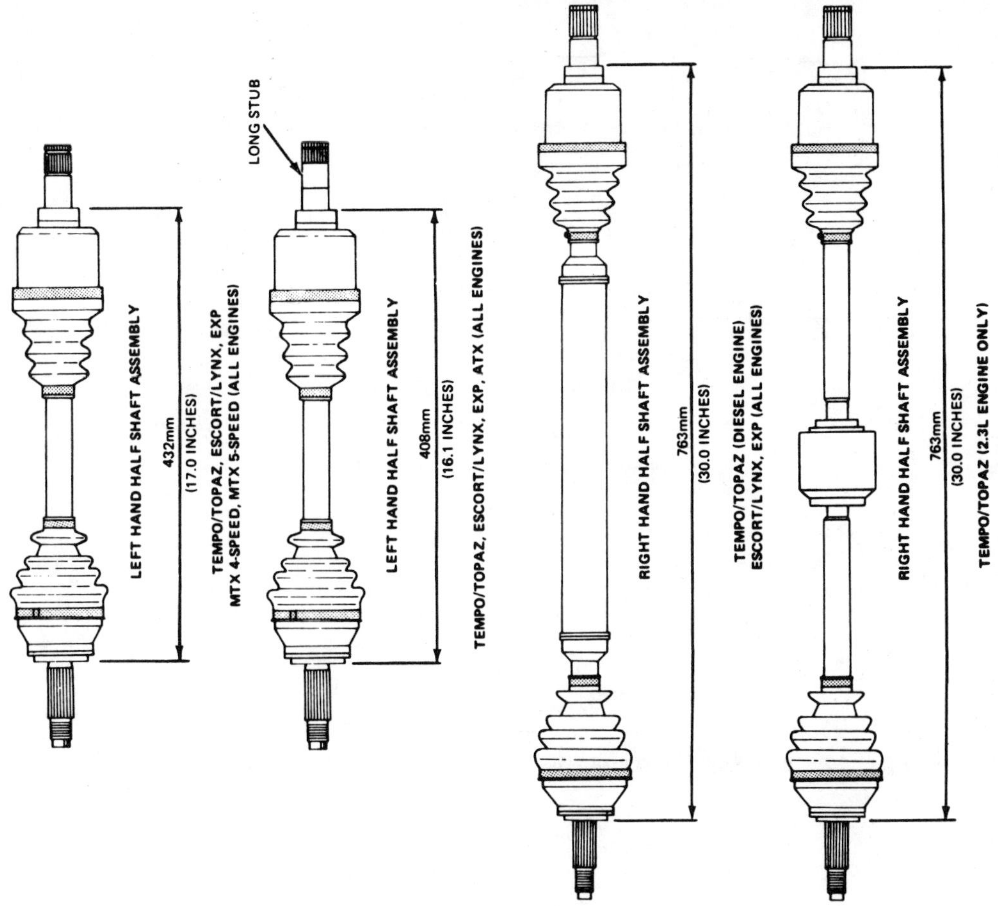

Fig. 11 Halfshaft assembled lengths. 1986 models

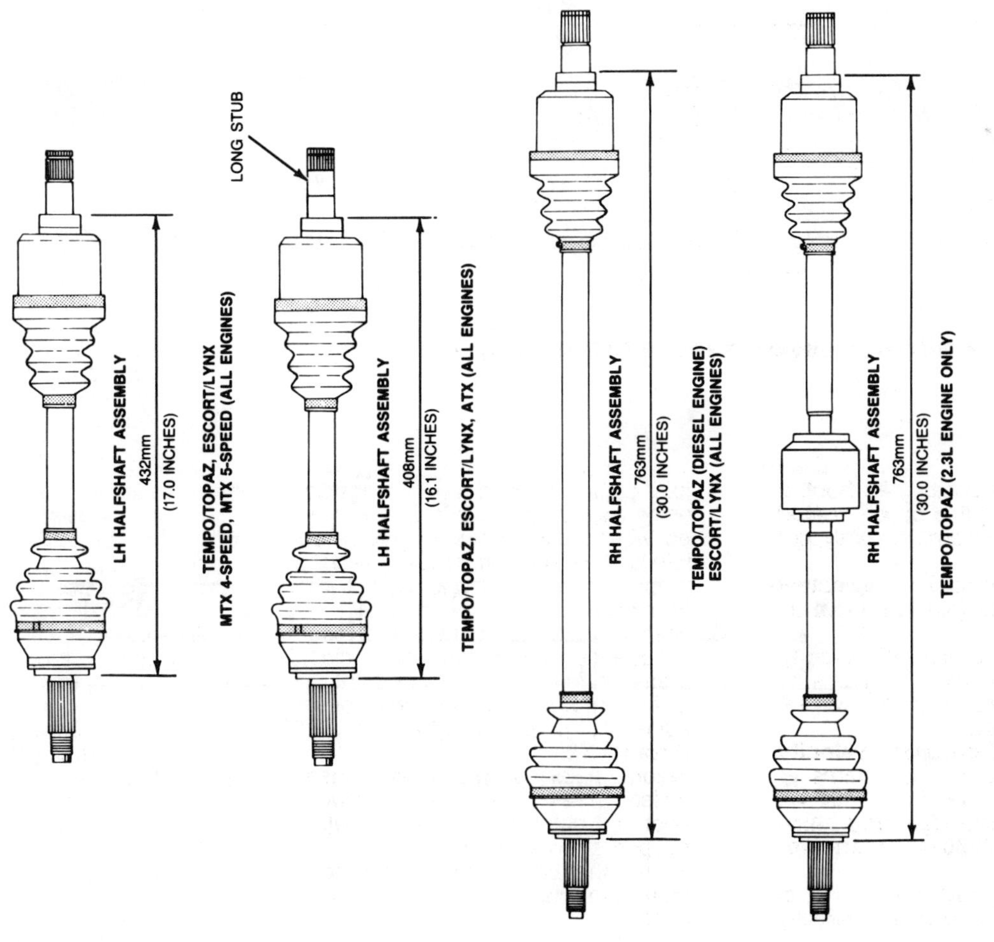

Fig. 10 Halfshaft assembled lengths. Late 1985 models

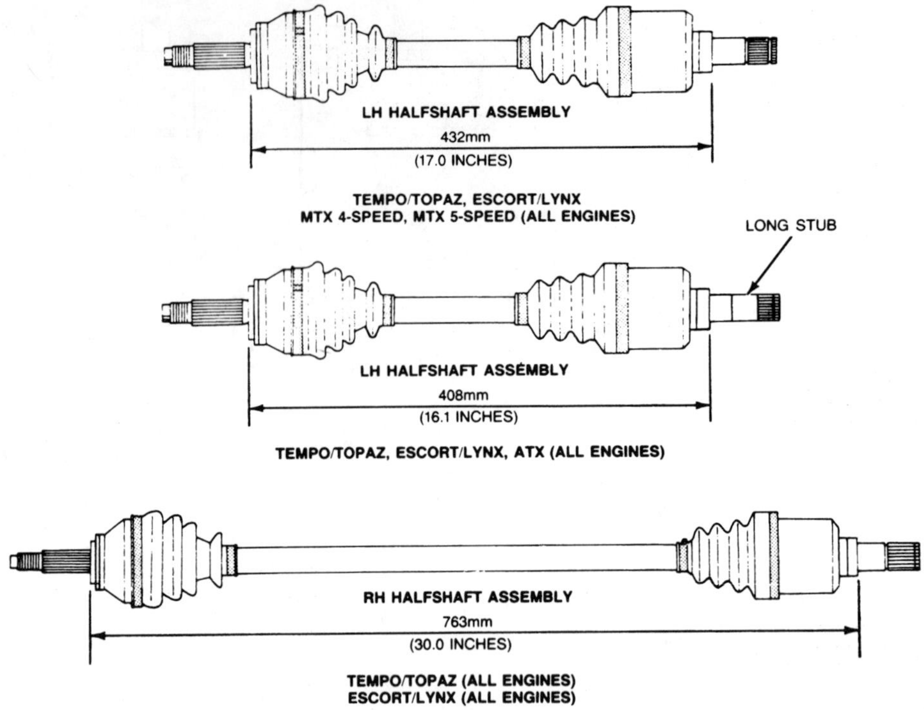

Fig. 12 Halfshaft assembled lengths. 1987–89 models

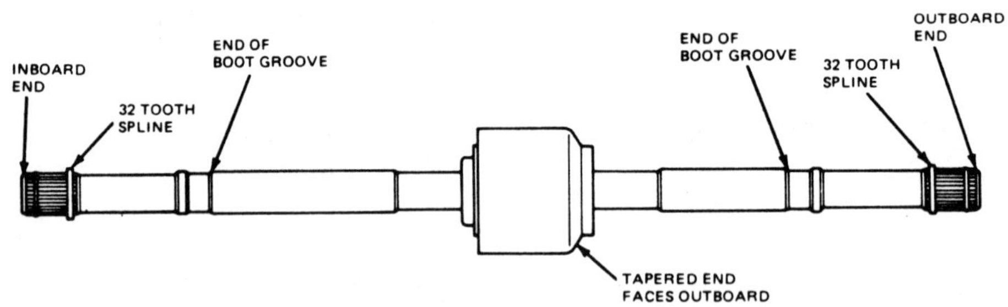

Fig. 13 Righthand interconnecting shaft. Late 1985–86 models

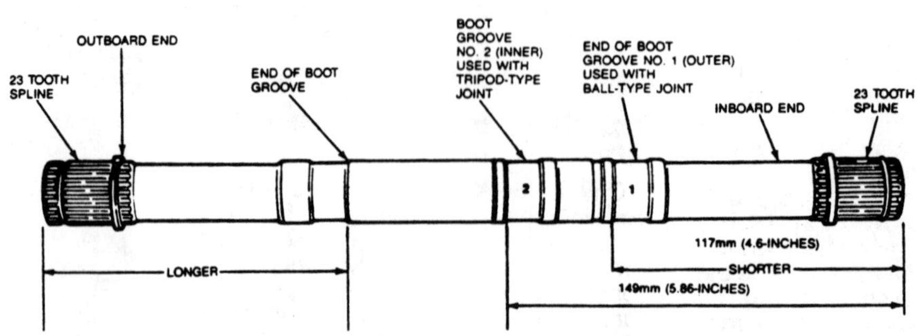

Fig. 14 Interconnecting shaft. 1987–89 models

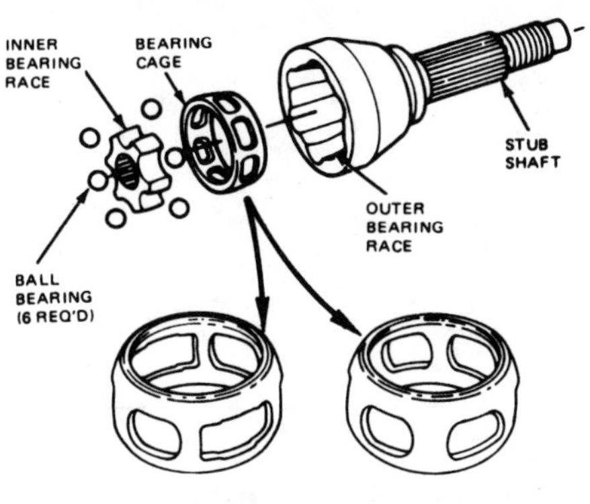

Fig. 15 Outer constant velocity joint bearing cage configuration

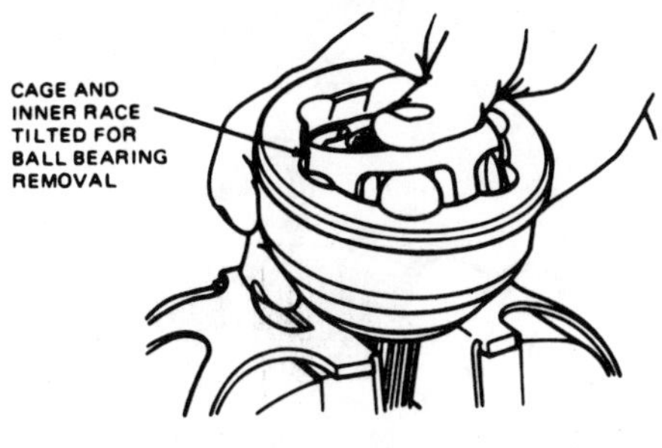

Fig. 16 Removing outer constant velocity joint bearings

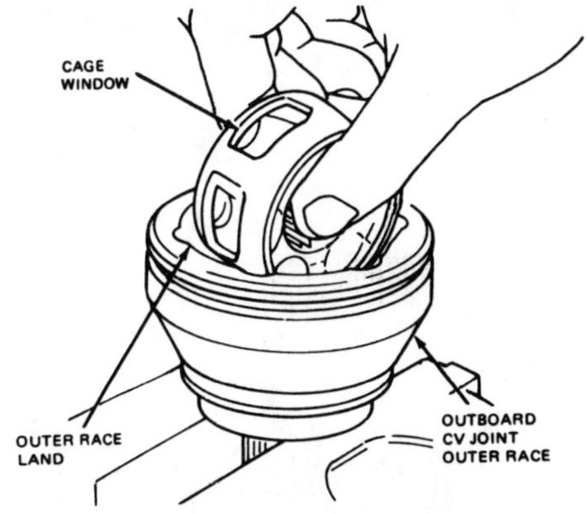

Fig. 17 Removing bearing cage & inner race assembly from outer constant velocity joint

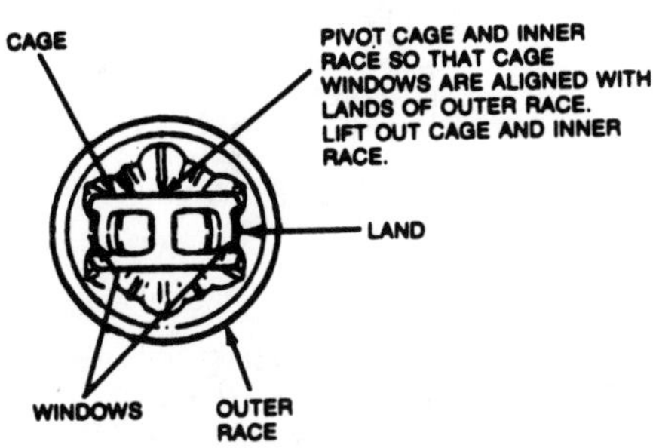

Fig. 18 Aligning inner cage & bearing race

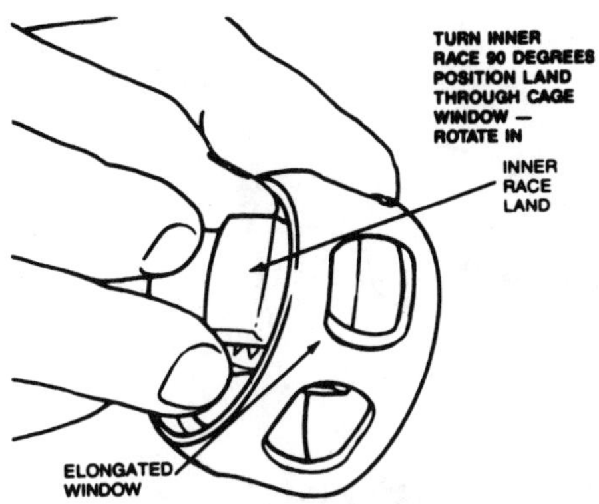

Fig. 19 Removing inner race from bearing cage

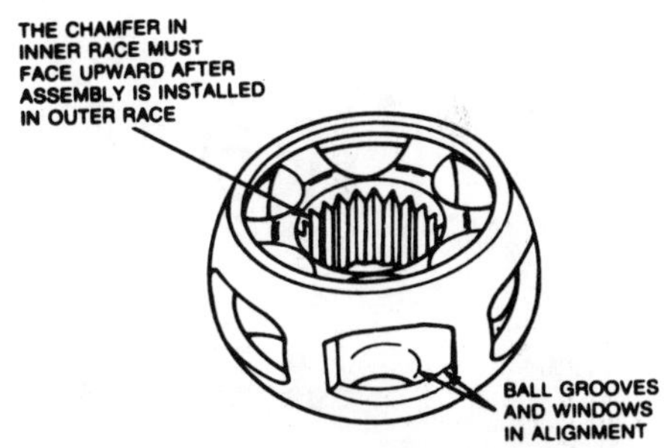

Fig. 20 Counterbore positioning & ball groove & window alignment

trapped air to escape.

10. Ensure that boot is seated in groove, then install clamp securely but not too tight.

OUTER JOINT SERVICE

Disassembly & Assembly

Different bearing cages are used on the outer joints. One type contains four equal sized windows and two elongated windows while the other type contains six windows of equal size, **Fig. 15.**

1. Position stub shaft in suitable vise with bearing facing upward.
2. Press downward on inner race until bearing can be removed, **Fig. 16.** Remove all six bearings in this manner.
3. Pivot bearing cage and inner race assembly into position shown in **Fig. 17.** Align cage windows with outer race lands while pivoting cage, **Fig. 18,** then remove from outer race.
4. To separate inner race from cage, determine cage design and proceed as follows: on cages with six equal windows rotate inner race upward and remove from cage. **On cages with two elongated windows,** pivot inner race until it is in position shown in **Fig. 17,** then align one inner race band with one elongated window and position race through the window. Rotate inner race upward and remove from cage, **Fig. 19.**
5. Reverse procedure to assemble. Refer to **Fig. 20** for ball groove and window alignment and proper counterbore positioning.

INNER JOINT SERVICE

Roll Crimp Ball Retainer Type, Fig. 21

On models with 5 speed manual transmission, disassembly of inboard constant velocity joint, **Fig. 22,** is performed during removal. Assembly of inboard constant velocity joint is performed during installation as previously described.

1. Remove circlip from end of joint stub shaft, then using suitable cutters, cut and remove ball retainer, **Fig. 23.** Discard retainer since a new retainer is not required for assembly.
2. Gently tap joint on work surface until assembly can be removed by hand, **Fig. 24.**
3. Remove bearings from cage by prying with a dulled screwdriver. Use caution not to damage or scratch any components, **Fig. 25.**
4. Rotate inner race to align lands with cage windows, then remove race from bearing cage through wider end of cage, **Fig. 26.**
5. Reverse procedure to assemble.

Wire Ring Ball Retainer Type

1. Remove large clamp, then slide boot back and wipe excess grease.
2. Using a suitable tool, remove wire ring ball retainer from race.
3. Remove outer race.
4. Pull inner race and bearing assembly out until race contacts snap ring.
5. Using suitable pliers, spread then slide snap ring back onto shaft.
6. Slide inner race and bearing assembly down the shaft to allow access to the snap ring.
7. Using a suitable screwdriver, remove snap ring.
8. Remove inner race and bearing assembly.
9. Reverse procedure to assemble.

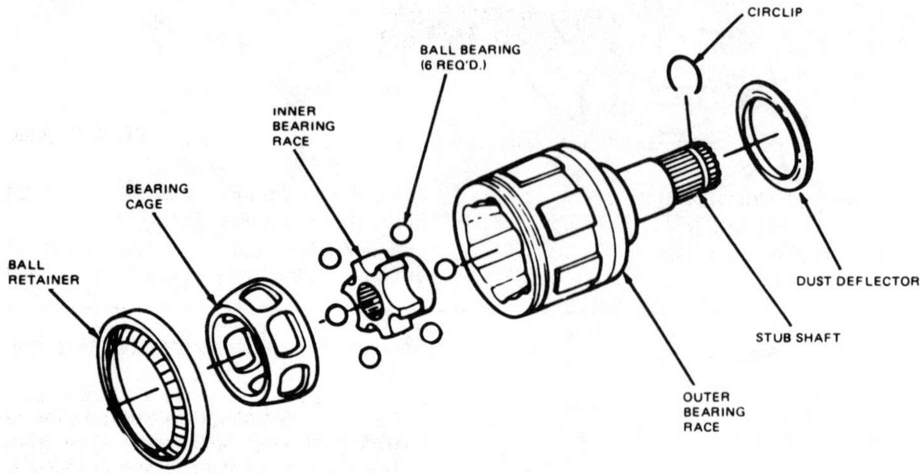

Fig. 21 Inner constant velocity joint assembly. Roll crimp ball retainer type

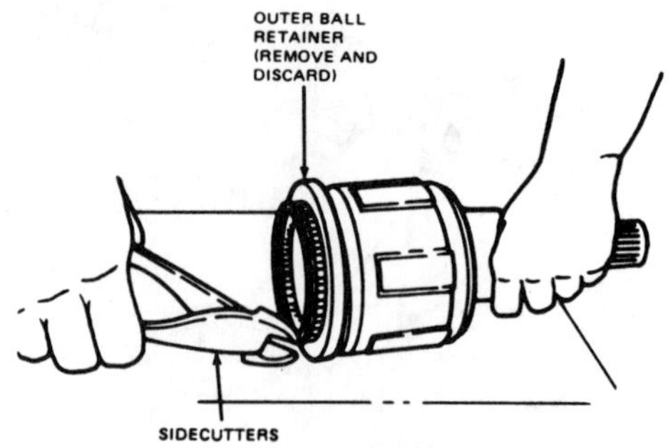

Fig. 23 Removing inner constant velocity joint outer ball retainer

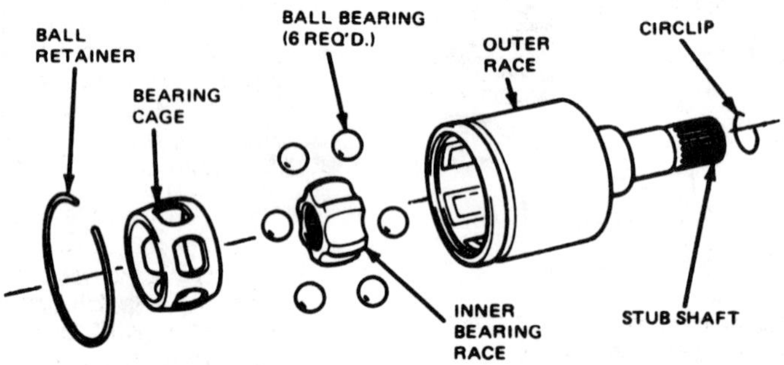

Fig. 22 Inner constant velocity joint assembly. Wire ring ball retainer type

Fig. 24 Removing inner constant velocity joint inner race & bearing assembly

Tripot CV Joint, RH Side, 1986 Escort & Lynx & 1987–89 Escort, Lynx & EXP

1. Remove large clamp, then slide boot back and wipe off excess grease. **Inspect CV joint grease for contamination by rubbing a small amount between two fingers. Any gritty feeling indicates contamination. If grease is contaminated, proceed with disassembly. If grease is not contaminated and joint was operating satisfactorily, add grease and replace boot.**
2. Using suitable pliers, bend retaining tabs back and separate outer race from tripot assembly.
3. Using suitable snap ring pliers, slide snap ring back on shaft, **Fig. 27.**
4. Push tripot assembly back on shaft, to gain access to circlip, then remove circlip from shaft.
5. Remove tripot assembly from shaft, and boot if necessary.
6. Reverse procedure to install noting the following:
   a. Fill CV joint outer race with 3.5 ounces of grease and CV boot with 2.1 ounces of grease. Use Ford Constant Velocity Joint Grease/High Temperature E43Z-19590-A or equivalent.

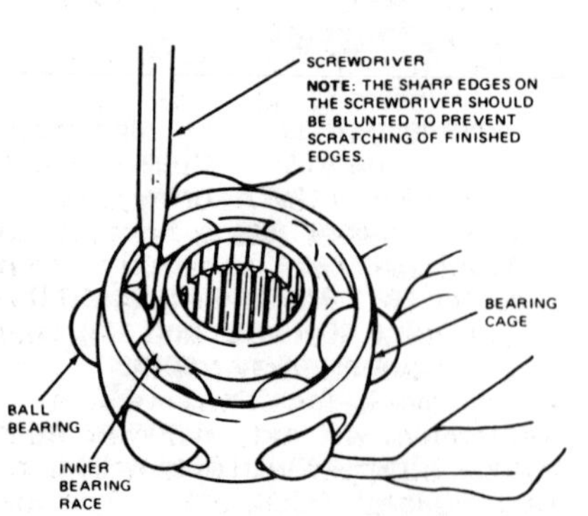

Fig. 25 Removing bearings from cage

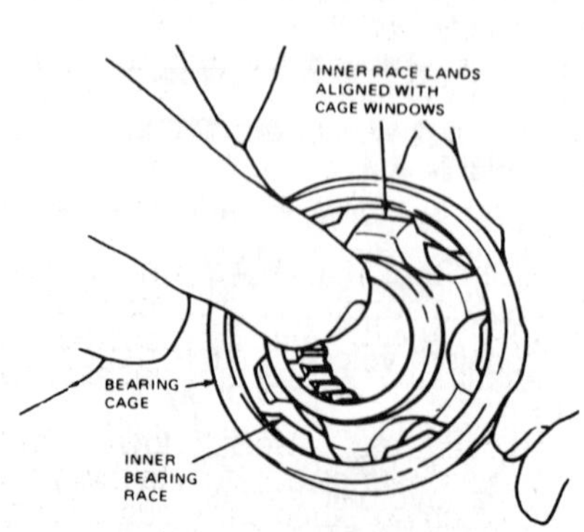

Fig. 26 Removing inner race from bearing cage

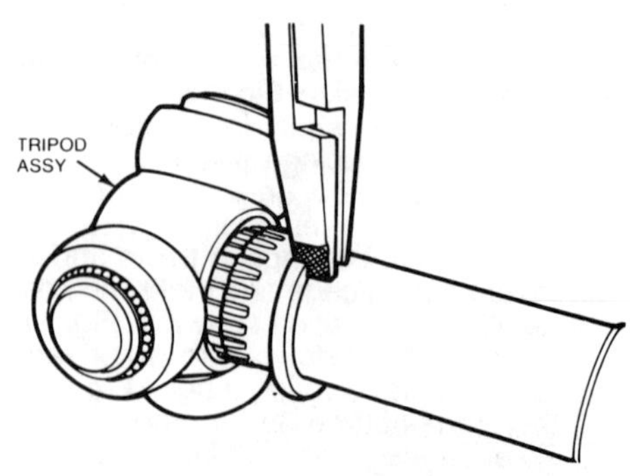

Fig. 27 Removing snap ring from shaft

Sable & Taurus

INDEX

TROUBLESHOOTING

NOISE & VIBRATION IN TURNS

Clicking, popping or grinding noises while turning may be caused by the following:

1. Cut or damaged CV joint boots, resulting in contaminated lube in outboard or inboard CV joints.
2. Loose CV joint clamps.
3. Worn, damaged or improperly installed wheel bearings.
4. Foreign object contacting halfshaft assembly.

VIBRATION AT HIGHWAY SPEEDS

1. Out of balance front wheels or tires.
2. Improperly seated outboard CV joint in front wheel hub.
3. Bent interconnecting shaft.
4. Front tires out of round.

SHUDDER OR VIBRATION DURING ACCELERATION

1. Excessively worn or damaged inboard or outboard CV joint.

HALFSHAFT OR CV JOINT PULL-OUT

Engine Or Transaxle Misaligned

1. Check engine mounts for damage.

Front Suspension Components Worn Or Damaged

1. Check for worn bushings or bent front suspension components.

Improperly Installed Or Missing Retainers

1. Check for CV joint circlip missing or not properly seated in transaxle side gear.

DRIVESHAFTS REPLACE

If removing both right and left side halfshafts, plugs T81P-1177-B or equivalent must be installed. Failure to do so may result in dislocation of differential side gears, necessitating transaxle disassembly to realign the gears. Also, halfshaft removal and installation procedures are the same for automatic and manual transaxles except for the following: due to the automatic transaxle case configuration, the right side halfshaft and linkshaft must be removed first. Tool T81P-4026-A or equivalent is then inserted into transaxle to remove left side inner Constant Velocity (CV) joint assembly from transaxle. If only the left side halfshaft is being removed from the vehicle, remove right side halfshaft assembly from the transaxle case and secure it in a horizontal position to the underside of vehicle, then remove left side halfshaft assembly.

Whenever removed, the hub nut, lower control arm-to-steering knuckle attaching nut and bolt and inboard CV joint stub shaft circlip must be replaced as their torque holding ability is destroyed during removal.

1. Loosen hub nut and lug nuts, then raise and support front of vehicle.
2. Remove wheel and tire assemblies, then remove hub nut and washer and discard nut.
3. Remove and discard lower ball joint-to-steering knuckle attaching nut and pinch bolt, then using suitable pry bar, separate ball joint from steering knuckle. **When separating ball joint from steering knuckle, use caution to avoid cutting or damaging ball joint boot.**

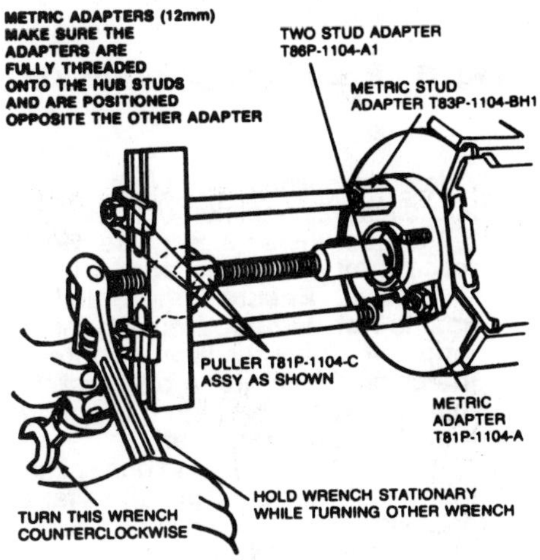

Fig. 1 Separating front hub from outer CV joint

4. Remove stabilizer bar link at stabilizer bar.
5. To remove right side halfshaft and linkshaft from all models except those equipped with Automatic Overdrive Transaxle (AOD), proceed as follows:
   a. Remove bearing support-to-bracket attaching bolts, then slide link shaft out of transaxle. Support end of shaft in horizontal position with suitable wire. **Do not allow shaft to hang unsupported as damage to the outboard CV joint can result.**
   b. Separate hub assembly from outer CV joint using hub remover T81P-1104-C, adapters T83P-1104-BH, T86P-1104-A1 and T81P-1104-A or equivalent, **Fig. 1. Never use a hammer to separate hub assembly from outer CV joint as damage to the CV joint threads and internal components may result.**
   c. Remove right side halfshaft and linkshaft from vehicle as an assembly.
6. To remove both halfshafts on models equipped with Automatic Overdrive Transaxle (AOD) or left side halfshaft on models equipped with manual transaxle, proceed as follows:
   a. Turn steering hub to one side or wire and/or wire strut assembly aside.
   b. Using puller tools shown in **Fig. 2**, attached to the inboard side of the inboard CV joint, remove CV joint from transaxle.
   c. Support end of shaft in horizontal position with suitable wire. **Do not allow shaft to hang unsupported as damage to the outboard CV joint can result.**
   d. Separate hub assembly from outer CV joint using hub remover T81P-1104-C, adapters T83P-1104-BH, T86P-1104-A1 and T81P-1104-A or equivalent, **Fig. 1. Never use a hammer to separate hub assembly from outer CV joint as damage to the CV joint threads and internal components may result.**
   e. Remove halfshaft from vehicle.
7. To remove left side halfshaft from models equipped with automatic transaxle (except AOD), proceed as follows. **If removing both right and left side halfshafts, plugs T81P-1177-B or equivalent must be installed. Failure to do so may result in dislocation of differential side gears, ne-**

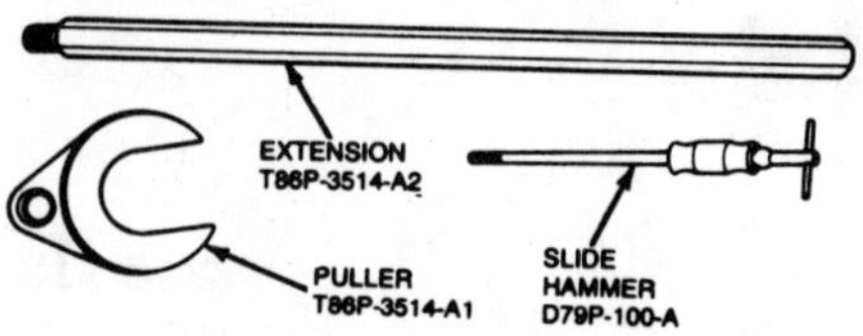

Fig. 2 Inboard CV joint removal tools. Except models equipped w/AOD

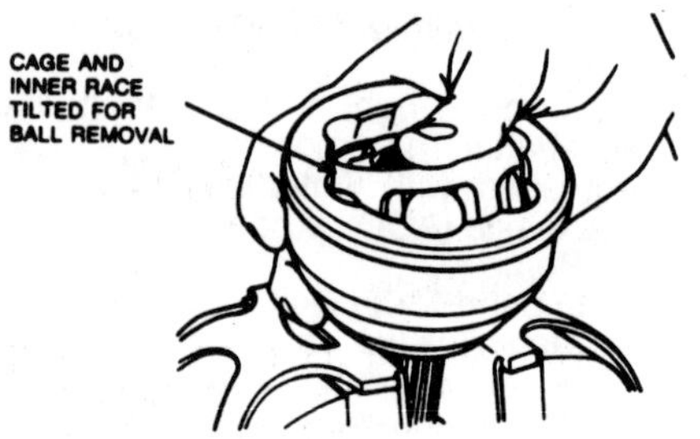

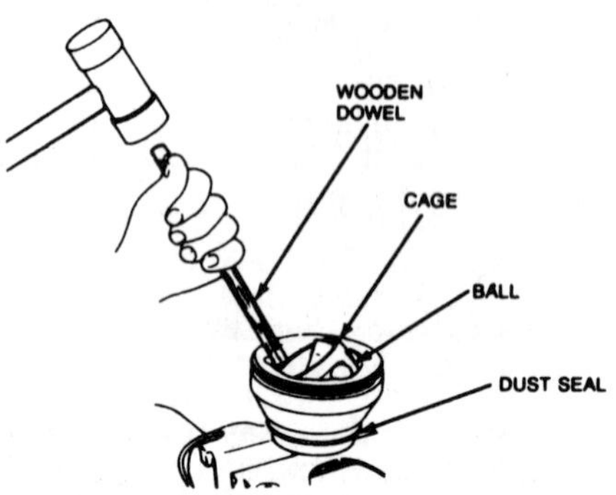

Fig. 5 CV joint ball removal

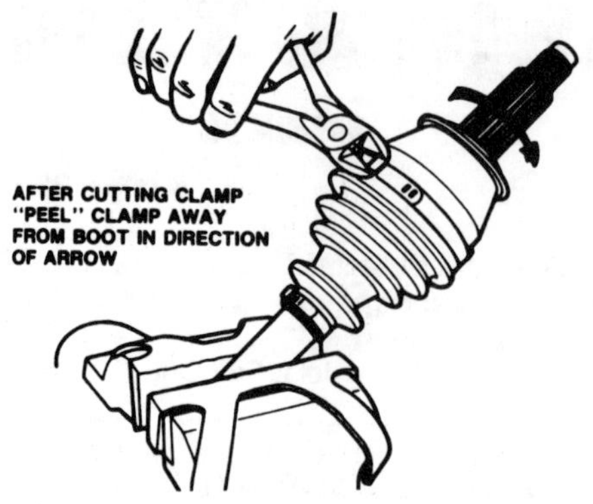

Fig. 3 Boot clamp removal

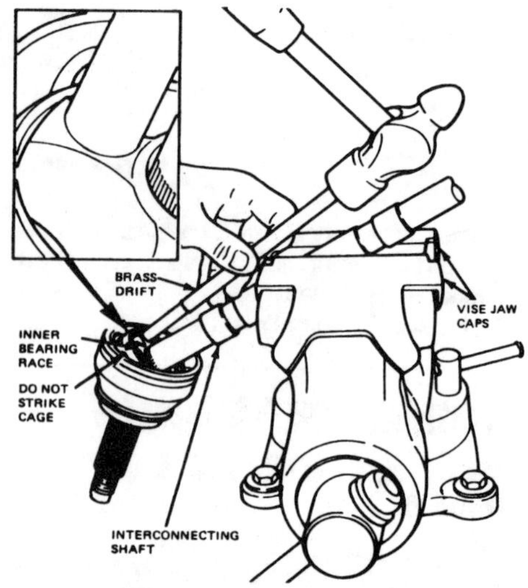

Fig. 4 Removing internal snap ring

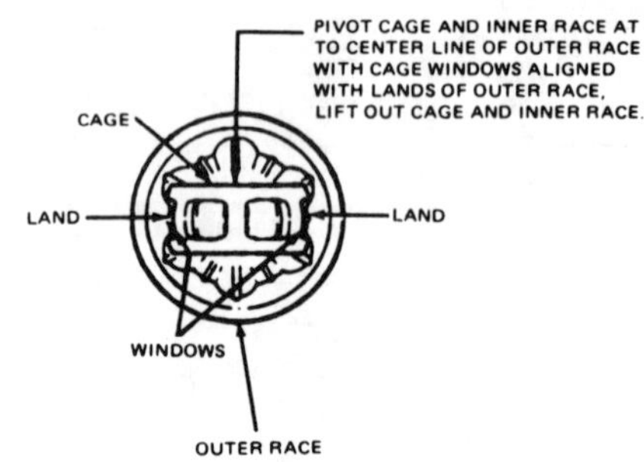

Fig. 6 Cage & inner race removal

cessitating transaxle disassembly to re-align the gears.

a. Remove right side halfshaft assembly from the transaxle case and secure it in a horizontal position to the underside of vehicle, then remove left side halfshaft by inserting driver T81P-4026-A or equivalent into right side halfshaft opening and driving left side halfshaft and CV joint from transaxle.
b. Support end of shaft in horizontal position with suitable wire. **Do not allow shaft to hang unsupported as damage to the outboard CV joint can result.**
c. Separate hub assembly from outer CV joint using hub remover T81P-1104-C, adapters T83P-1104-BH, T86P-1104-A1 and T81P-1104-A or equivalent, **Fig. 1. Never use a hammer to separate hub assembly from outer CV joint as damage to the CV joint threads and internal components may result.**
d. Remove halfshaft from vehicle.

8. **On all models,** prior to installation install new circlip on inboard CV joint stub shaft and/or linkshaft. **The original circlip cannot be reused.**
9. Align CV joint splines with transaxle differential splines, then push CV joint into differential splines until circlip is felt to seat inside side gears. **Some force may be necessary to insert CV joints. Ensure differential oil seal is not damaged during installation. If difficulty is encountered installing CV joints, a non-metallic mallet may be used on the outside joint CV joint stub shaft.**
10. Align CV joint splines with hub splines, then install stub shaft in hub as far as possible.
11. Temporarily fasten rotor to hub with two lug nuts and suitable washers. Install steel rod between lug nuts and use to prevent rotor from turning.
12. Install hub washer and new hub nut, then manually thread nut onto CV joint stub shaft as far as possible.
13. Connect steering knuckle to lower ball joint stud, then install new nut and bolt and **torque** to 37-44 ft. lbs.
14. Connect stabilizer bar to stabilizer bar link.
15. Install wheel and tire assembly, then lower vehicle to ground. **Torque** hub nut to 180-200 ft. lbs.
16. Top off transaxle with lubricants listed below:
    a. **On automatic overdrive transaxle (AOD),** use Motorcraft Type H XT-4-H (except 1988-89 models)
    b. **On automatic (except AOD) and manual transaxle,** use Dexron II (except 1988-89 models).
    c. **On automatic overdrive and FLC transaxle (1988-89 models),** use ESP-M2C185-A Mercon.
    d. **On manual transaxle (1988-89 models),** use ESP-M2C185-A Mercon.

CONSTANT VELOCITY JOINT SERVICE

OUTBOARD CV JOINT & BOOT

Disassembly

During manufacture, CV joints components are matched and cannot be interchanged with components of other CV joint. If a CV joint component is defective, the entire CV joint should be replaced.

1. Install soft vise jaw caps in vise to prevent damage to halfshaft, then position halfshaft in vise. Do not allow the vise to contact the CV joint boot or clamps.
2. Using suitable side cutting pliers, cut large boot clamp and peel away from boot. Roll boot back over halfshaft, **Fig. 3.**
3. Turn halfshaft over in vise, then angle CV joint so that inner bearing race is exposed, **Fig. 4.** Using suitable brass drift and hammer, give a sharp rap to inner bearing race to dislodge internal snap ring. Separate CV joint from halfshaft. Take care not to drop the CV joint. Remove CV boot from shaft.
4. Inspect CV joint grease for contamination. If grease is contaminated, proceed with disassembly. If grease is not contaminated and joint was operating satisfactorily, add grease and replace boot.
5. Remove and discard circlip from end of shaft. Inspect stop ring located below circlip, if it is worn or damaged, replace it.
6. Clamp CV joint stub axle in vise with soft vise jaw caps. Be careful not to damage dust seal.
7. Push down on CV joint inner race until it tilts enough to allow ball removal, **Fig. 5. If inner race is tight, it can be**

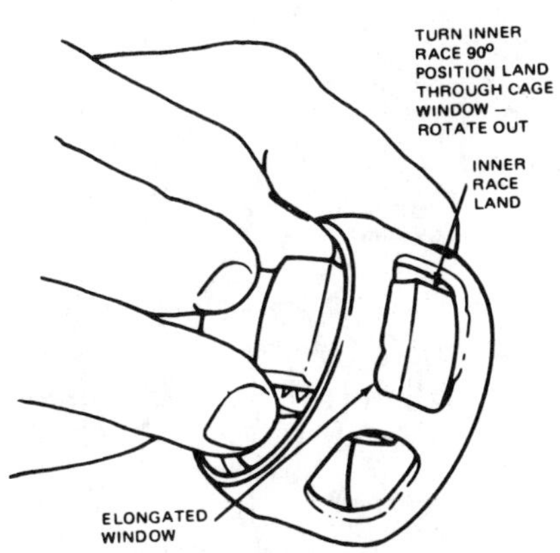

Fig. 7 Inner race removal & installation

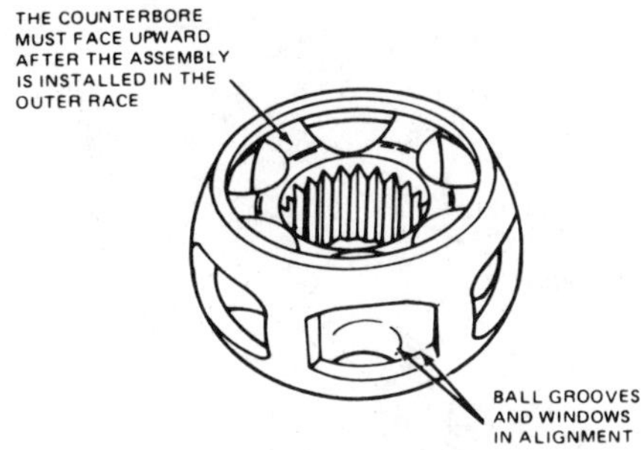

Fig. 8 Inner race & cage assembly

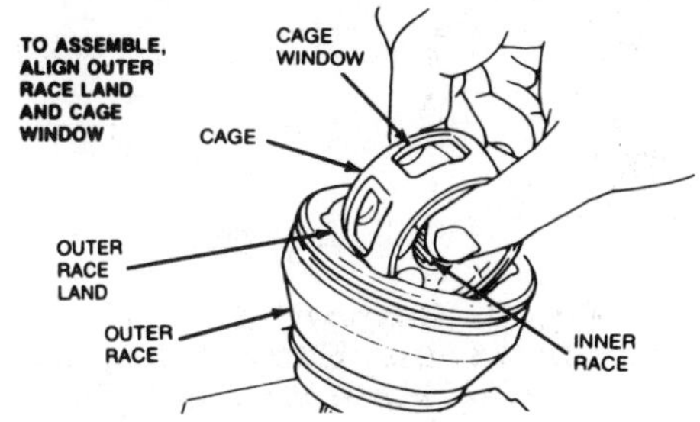

Fig. 9 Inner race & cage assembly to outer race installation

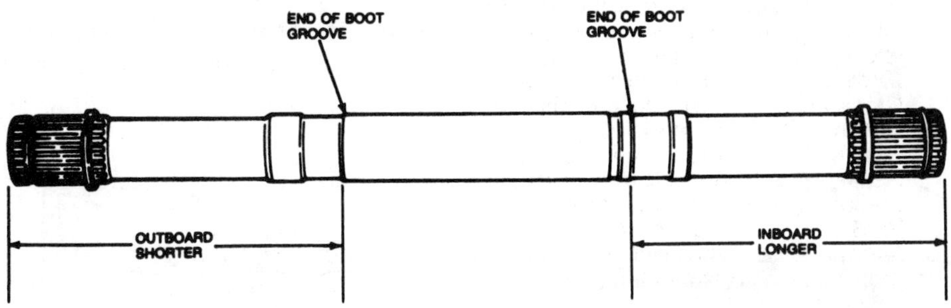

Fig. 10 Halfshaft end identification

tilted by tapping inner race with wooden dowel and hammer. Do not hit cage.

8. Remove balls from cage. If balls are tight, use blunt screwdriver to pry balls from cage.
9. Pivot cage and inner race assembly until its straight up, **Fig. 6.** Align cage windows with outer race lands while pivoting bearing cage, then lift out cage and inner race.
10. Rotate inner race up and out of cage, **Fig. 7.**

Inspection

During manufacture, CV joints components are matched and cannot be interchanged with components of other CV joint. If a CV joint component is defective, the entire CV joint should be replaced.

Inspect all parts. If any parts are cracked, broken, severely pitted, worn or otherwise unserviceable, replace CV joint. If any parts appear polished, do not replace CV joint as this is a normal condition.

Assembly

1. Apply light coating of Ford constant velocity joint grease No. E2FZ-19590-A or equivalent on inner and outer races, then install inner race in bearing cage, **Fig. 7.**
2. Install inner race and cage assembly in outer race land, **Fig. 8.**
3. Install CV joint assembly into outer race and pivot 90° into position, **Fig. 9.**
4. Align bearing cage and inner race with outer race, then tilt inner race and install a ball, followed by remaining five balls.
5. Determine which end of halfshaft is for outboard CV joint. The outboard joint side has a shorter end of boot groove to end of shaft dimension, **Fig. 10.**
6. Install CV joint boot and small boot clamp. If stop ring was removed, install at this time. If stop ring was not removed, ensure it is seated properly in groove.
7. Install new circlip, **Fig. 11. Do not over expand or twist the circlip during installation. To install properly, start one end in the groove and work the circlip over the stub shaft and into the circlip groove.**
8. Pack CV joint with Ford CV joint grease No. E2FZ-19590-A or equivalent. Correct quantity is 3.17 ounces on 1986-87 models or 3.52 ounces on 1988-89 models. Any remaining grease is to be spread evenly inside CV boot.
9. With boot peeled back, position CV joint on halfshaft and tap into position with suitable plastic hammer. The CV joint is properly seated when the circlip locks into position. Check for proper retention by attempting to pull off CV joint.
10. Remove all excess grease from CV external surfaces, then position boot over CV joint. Ensure boot is seated in its groove and install clamp.

OUTBOARD CV JOINT DUST SEAL

Disassembly

1. With halfshaft removed, use a light duty hammer and screwdriver to tap evenly around seal until unseated and remove seal, **Fig. 12.**

Assembly

Using spindle/axle seal tool T83T-3132-A1 and dust seal installer tool T83P-3425-AH or equivalent, install dust seal, **Fig. 13.** The dust seal flange must face outboard.

INBOARD CV JOINT

Disassembly

These models use two different types of inboard CV joints and boots, **Fig. 14.** The first one is of a conventional boot design that uses a crimped can on the large end. The other is a tri-lobe design CV joint that does not require a crimped can on the large end. Although both designs are similar, they are not interchangeable.

1. Cut and remove both large and small boot clamps from CV joint, then slide boot back on shaft. **All right side inboard CV joints use a re-usable low profile large boot clamp. Do not cut this clamp as it will be re-used.**
2. Slide outer race off tripot assembly. Inspect CV joint grease for contamination by rubbing a small amount between two fingers. Any gritty feeling indicates contamination. If grease is contaminated, proceed with disassembly. If grease is not contaminated and joint was operating satisfactorily, add grease and replace boot.
3. Using suitable snap ring pliers, slide stop ring back on shaft.
4. Slide tripot assembly back on shaft to provide clearance to circlip, then remove circlip, tripot assembly and boot.

Assembly

1. Install CV boot on shaft. Ensure boot small end is seated properly in halfshaft groove, then tighten small end clamp.
2. Install tripot assembly on shaft with chamfered side toward stop ring, then install new circlip. **Circlips cannot be re-used. They must be replaced with new ones.**

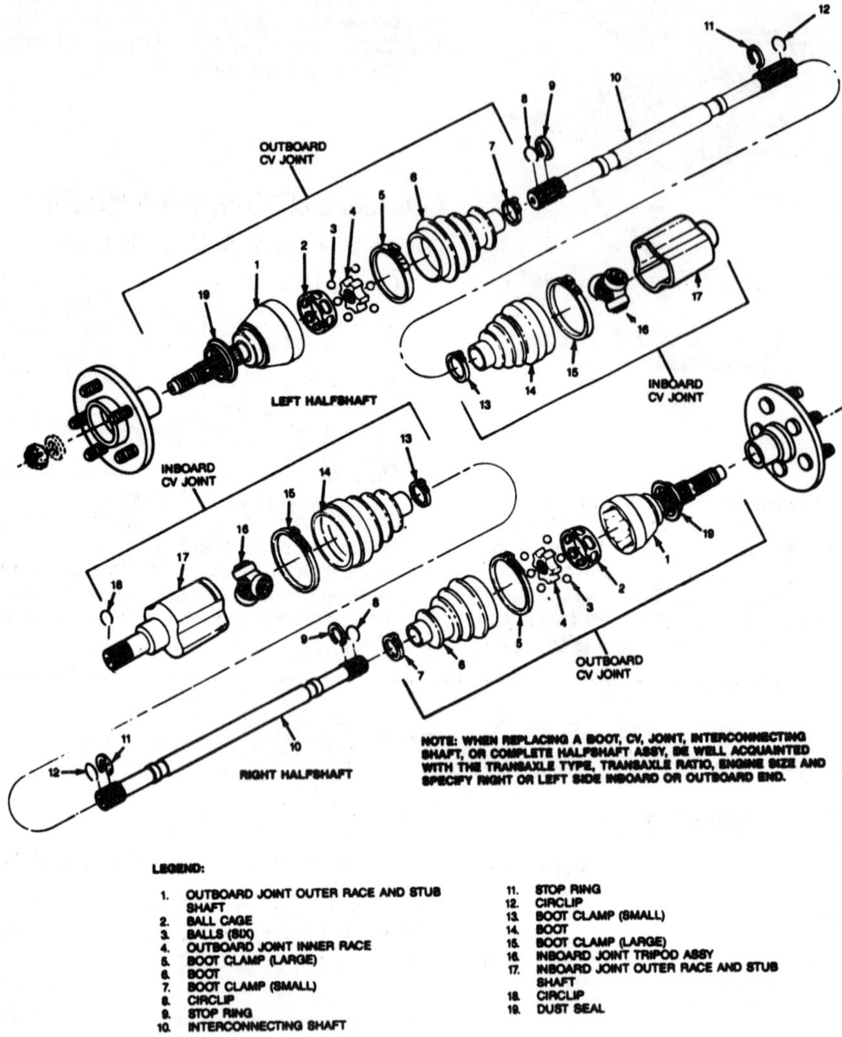

Fig. 11 Exploded view of halfshaft assemblies

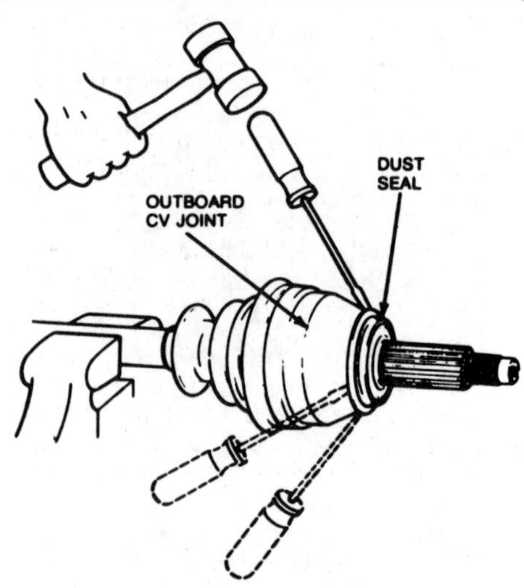

Fig. 12 Dust seal removal

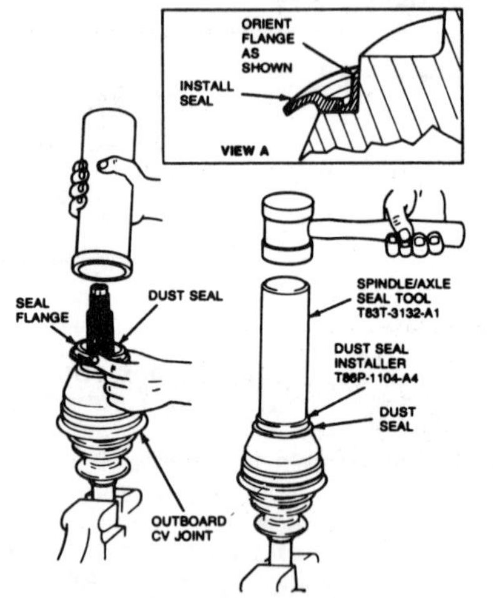

Fig. 13 Dust seal installation

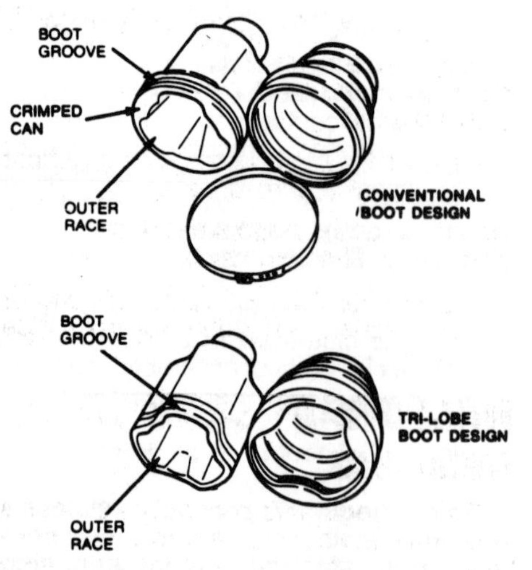

Fig. 14 CV joint identification

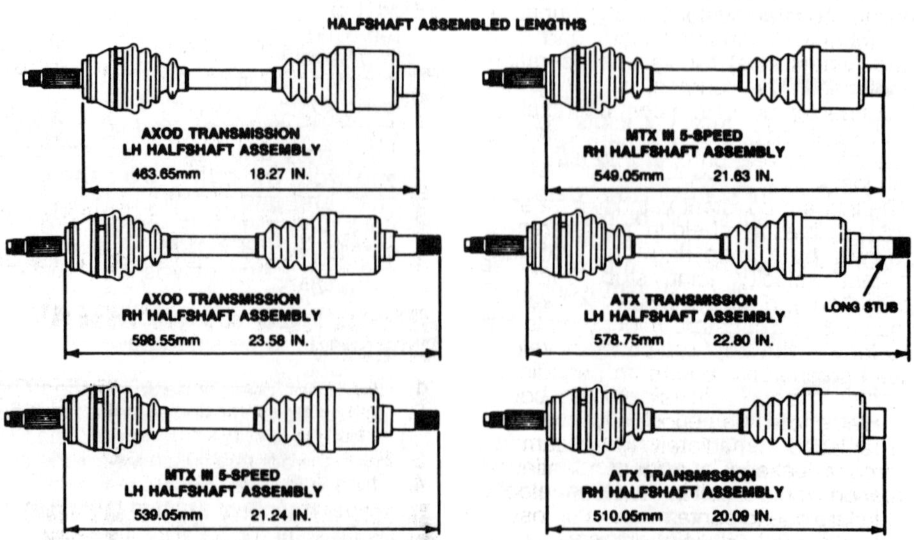

Fig. 15 Halfshaft assembled length

3. Slide tripot assembly forward to expose stop ring groove, then slide stop ring into position. Ensure stop ring is fully seated in groove.
4. **On models with conventional design boots, Fig. 14,** fill CV joint outer race with 3.5 ounces and CV boot with 2.5 ounces of Ford CV joint grease-high temperature No. E43Z-19590-A or equivalent. **On models with tri-lobe design boots, Fig. 14,** fill CV joint outer race with 4.4 ounces and CV boot with 4.4 ounces of Ford CV joint grease-high temperature No. E43Z-19590-A or equivalent.
5. Install outer race over tripot assembly, then position boot over outer race. Ensure boot is properly seated in its groove.
6. Remove excess grease from boot exterior. Move CV joint in or out as necessary to adjust halfshaft to length as shown in **Fig. 15.** After halfshaft length has been determined, expel any built up air pressure from boot by inserting a dull screwdriver between boot and outer bearing and allowing air to escape.

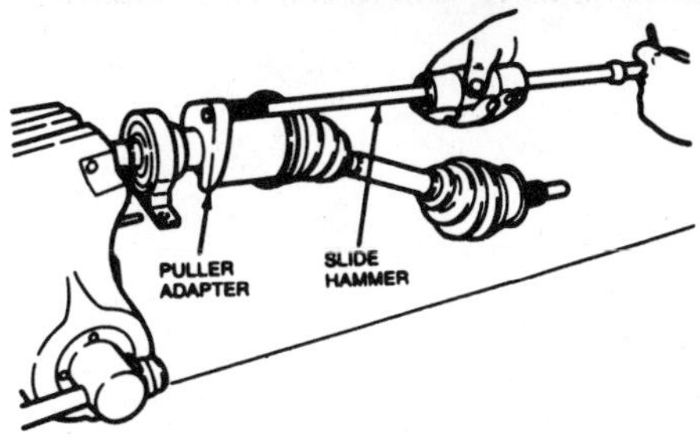

Fig. 16 Linkshaft removal

7. Install large boot clamp with suitable crimping pliers. **All right side inboard CV joints use a re-usable low profile large boot clamp. Do not install with crimping pliers.**
8. Install new circlip. **Do not over expand or twist the circlip during installation. To install properly, start one end in the groove and work the circlip over the stub shaft and into the circlip groove.**

LINKSHAFT/HALFSHAFT

Disassembly & Assembly

1. Clamp linkshaft in vise with halfshaft supported on workbench. Using puller adapter T86P-3514-A or equivalent and slide hammer D79P-100-A or equivalent, separate linkshaft from halfshaft, **Fig. 16.**
2. Pry seal from linkshaft with screwdriver.
3. Position linkshaft in suitable arbor press, then press off bearing.
4. Press on new bearing and bearing seal with arbor press.
5. Coat shaft splines with Ford CV joint grease No. E2FZ-19590-A or equivalent, then assemble linkshaft to halfshaft.

1988–89 Continental

INDEX

DESCRIPTION

Each front wheel driveshaft (halfshaft) employs constant velocity (CV) joints at both inboard (differential side) and outboard (wheel side) for vehicle operating smoothness. The constant velocity joints are connected by an interconnecting shaft which is splined at both ends and retained in the inboard and outboard constant velocity joints by snap rings, **Fig. 1.**

The inboard constant velocity joint stub shaft is splined and held in the differential side gear by a snap ring. The outboard constant velocity joint stub shaft is pressed on and secured with a prevailing torque nut. The constant velocity joints are lube-for-life with a special constant velocity joint grease and require no periodic lubrication. The constant velocity joint boots, however, should be periodically inspected and replaced immediately when damage or grease leakage is evident. Continued operation would result in constant velocity joint failure due to contamination or loss of the constant velocity joint grease.

Halfshaft removal is accomplished by applying a load to the back face of the inboard constant velocity joint assembly.

TROUBLESHOOTING

NOISE & VIBRATION IN TURNS

Clicking, popping or grinding noises while turning may be caused by the following:

1. Cut or damaged CV joint boots, resulting in contaminated lube in outboard or inboard CV joints.
2. Loose CV joint clamps.
3. Worn, damaged or improperly installed wheel bearings.
4. Foreign object contacting halfshaft assembly.

VIBRATION AT HIGHWAY SPEEDS

1. Out of balance front wheels or tires.
2. Improperly seated outboard CV joint in front wheel hub.
3. Bent interconnecting shaft.
4. Front tires out of round.

SHUDDER OR VIBRATION DURING ACCELERATION

1. Excessively worn or damaged inboard or outboard CV joint.

HALFSHAFT OR CV JOINT PULL-OUT

Engine Or Transaxle Misaligned

1. Check engine mounts for damage.

Front Suspension Components Worn Or Damaged

1. Check for worn bushings or bent front suspension components.

Improperly Installed Or Missing Retainers

1. Check for CV joint circlip missing or not properly seated in transaxle side gear.

HALFSHAFT SERVICE

REMOVAL

Do not begin this procedure unless a new hub retainer nut, a new lower control arm to steering knuckle attaching bolt and nut and a new inboard constant velocity joint stub shaft snap ring are available.

Once removed, these components must not be reused during assembly/installation. Their torque holding ability or retention capability is diminished during removal.

1. Remove hub retainer nut and washer. Discard nut after removal. **The nut is torqued to 180-200 ft. lbs.**
2. Raise and support vehicle. Remove tire and wheel assemblies.
3. Remove attaching nut from ball joint to steering knuckle attaching bolt.
4. Using a suitable punch and hammer, drive bolt out of steering knuckle. **Discard bolt and nut.**
5. Remove anti-lock brake sensor from steering knuckle. Remove height sensor link at lower arm ball stud attachment.
6. Remove stabilizer bar link at stabilizer bar.
7. Separate ball joint from steering knuckle using a suitable pry bar. **Position end of pry bar outside of bushing pocket to avoid damage to the bushing.**

Tools T86P-3514-A2, T86P-3514-A1 and D79P-100-A or equivalents, must be used to perform the following steps.

8. Install tool T86P-3514-A1 or equivalent, between constant velocity joint and transaxle case. Turn steering hub and/or wire strut assembly out of the way.
9. Screw extension T86P-3514-A2 or equivalent, into constant velocity joint puller and hand tighten. Screw impact slide hammer D79P-100-A or equivalent onto extension.
10. Remove constant velocity joint from transaxle.
11. Support end of shaft by suspending it from a conventional underbody component with a length or wire. **Do not allow shaft to hang unsupported, damage to the outboard constant velocity joint may result.**
12. Separate outboard constant velocity joint from hub using front hub remover tools T81P-1104-C, T83P-1104-BH, T86P-1104-A1 and T81P-1104-A or equivalents. **Never use a hammer to separate the outboard constant velocity joint stub shaft from the hub. Damage to the constant velocity joint threads and internal components may result.**
13. Remove halfshaft assembly from vehicle.

INSTALLATION

1. Install a new snap ring onto inboard constant velocity joint stub shaft and/or link shaft. **The outboard constant velocity joint stub shaft does not have a snap ring. To install the snap ring correctly, start one end in the groove and work the snap ring over the stub shaft end and into the groove. This will avoid over expanding the snap ring. The old snap ring must not be reused. A new snap ring must be installed each time the inboard constant velocity joint is installed into the transaxle differential.**
2. Carefully align splines of inboard constant velocity joint stub shaft or link shaft with splines in differential. Exerting force, push constant velocity joint into differential until snap ring seats in differential side gear. **Use care to prevent damage to the differential oil seal. A plastic hammer or equivalent, may be used to aid in seating the snap ring into the differential side gear groove.**
3. Carefully align splines of outboard constant velocity joint stub shaft with splines in hub and push the shaft into the hub as far as possible.
4. Temporarily attach rotor to hub with washers and two wheel lug nuts. Insert a steel rod into rotor and rotate clockwise to contact the steering knuckle to prevent rotor from turning during constant velocity joint installation.
5. Install hub nut washer and a new hub retainer nut. Manually thread retainer onto constant velocity joint shaft as far as possible. **A new hub retainer nut must be installed.**
6. Connect control arm to steering knuckle. Install a new nut and bolt. **Torque** nut to 40-55 ft. lbs. **A new nut and bolt must be installed.**
7. Connect stabilizer bar link to stabilizer bar. **Torque** attaching nut to 35-48 ft. lbs.
8. Connect ride height sensor link.
9. Install anti-lock sensor link into control arm and tighten retaining bolt.
10. **Torque** hub retainer nut to 180-200 ft. lbs.
11. Install tire and wheel assembly, then lower vehicle.
12. **Torque** lug nuts to 80-105 ft. lbs.
13. Fill transaxle to proper level with correct fluid as required.

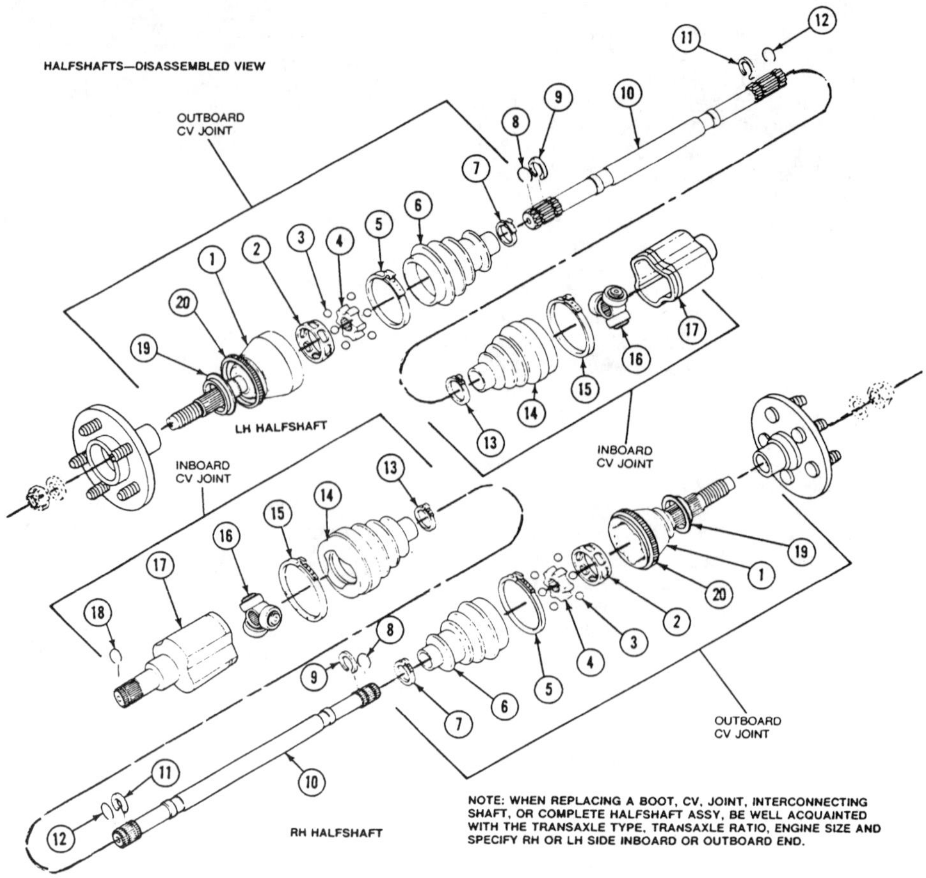

ITEM	DESCRIPTION
1.	OUTBOARD JOINT OUTER RACE AND STUB SHAFT
2.	BALL CAGE
3.	BALLS (SIX)
4.	OUTBOARD JOINT INNER RACE
5.	BOOT CLAMP (LARGE)
6.	BOOT
7.	BOOT CLAMP (SMALL)
8.	CIRCLIP
9.	STOP RING
10.	INTERCONNECTING SHAFT
11.	STOP RING
12.	CIRCLIP
13.	BOOT CLAMP (SMALL)
14.	BOOT
15.	BOOT CLAMP (LARGE)
16.	INBOARD JOINT TRIPOD ASSY
17.	INBOARD JOINT OUTER RACE AND STUB SHAFT
18.	CIRCLIP
19.	DUST SEAL
20.	SPEED INDICATOR RING (ANTI-LOCK BRAKES)

Fig. 1 Disassembled view of halfshaft assemblies

OUTBOARD CONSTANT VELOCITY JOINT & BOOT

Disassembly

The constant velocity joint components are matched during manufacture and cannot be interchanged with components from another constant velocity

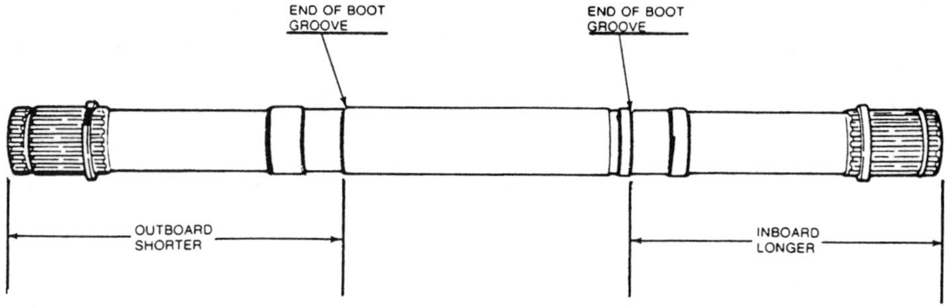

Fig. 2 Interconnecting shaft identification

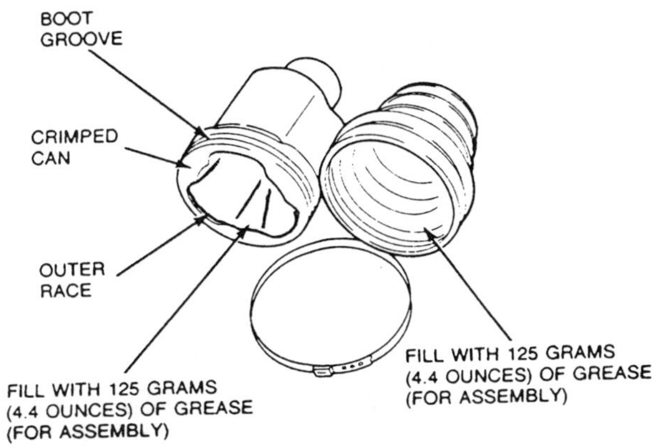

Fig. 3 Constant velocity joint & boot

joint. Extreme care should be taken not to intermix or substitute like components between constant velocity joints.

1. Clamp halfshaft into a suitable vise. Do not allow vise jaws to contact boot or clamp.
2. Cut large boot clamp and peel away from boot.
3. Support interconnecting shaft in a soft jaw vise and angle constant velocity joint to expose inner bearing race.
4. Using a brass drift and hammer, give a sharp tap to the inner bearing race to dislodge the internal snap ring and separate the constant velocity joint from the interconnecting shaft. The boot can now be removed from the shaft.
5. inspect constant velocity joint grease for contamination. If constant velocity joints are operating satisfactorily, and grease does not appear to be contaminated, add grease and replace boot. If lubricant appears contaminated, proceed with a complete constant velocity joint disassembly and inspection.
6. Remove snap ring located near end of shaft. Discard snap ring. A new snap ring is supplied with both boot replacement kit and constant velocity joint. **The stop ring, located just below the snap ring, should only be removed if it is damaged.**
7. Clamp the constant velocity joint stub shaft in a vise with outer facing pointing upward. Care should be taken not to damage the dust seal.
8. Press down on inner race until it tilts enough to allow removal of the ball.
9. With cage tilted, remove ball from cage. Repeat until all six balls are removed.
10. Pivot cage and inner race assembly until it is facing straight up and down in outer race. Align cage windows with the outer race lands while pivoting bearing race. With cage pivoted and aligned, lift assembly from outer race.
11. Rotate inner race up and out of cage.

Assembly

Because the constant velocity joint components are matched as a set during assembly, individual components are not available for service. If inspection determines a part to be worn or damaged, the constant velocity joint should be replaced as an assembly. Do not replace a joint because the components appear polished. Shiny areas in ball races and the cage spheres are normal. A constant velocity joint should be replaced only if inspection determines a component to be cracked, broken, severely pitted, worn or otherwise unserviceable.

1. Apply a light coat of grease onto inner and outer ball races. **Use only Ford CV grease E2FZ-19590-A or equivalent.**
2. Install inner race into bearing cage.
3. Install inner race and cage assembly into outer race.
4. Install assembly vertically and pivot 90° into position.
5. Align bearing cage and inner race with outer race.
6. Tilt inner race and cage, then install a ball. Repeat this step until all six balls are installed.
7. The lefthand and righthand interconnecting shafts are not the same end-for-end. The outboard end is shorter from end of shaft to end of boot groove than the inboard end. Take a measurement to ensure correct inboard and outboard constant velocity joint to shaft installation, **Fig. 2.**
8. If removed, install constant velocity joint boot after removing stop ring.
9. Ensure boot is properly seated in its groove and clamp into position.
10. If removed, install stop ring. If not removed, ensure stop ring is properly seated in groove.
11. Install a new snap ring, supplied with service kit, in groove nearest end of shaft.
12. Do not over expand or twist snap ring during installation.
13. Before positioning boot over constant velocity joint, pack constant velocity joint and boot with grease supplied in service kit. Add 3.52 ounces.
14. With boot peeled back, position constant velocity joint on shaft and tap into position. **The constant velocity joint is completely seated when the snap ring locks in groove cut into constant velocity joint inner race. Check for snap ring seating by trying to pull joint from shaft.**
15. Remove all excess grease from the constant velocity joint external surfaces.
16. Position boot over constant velocity joint.
17. Ensure boot is seated in its groove and clamp in position.

OUTBOARD CONSTANT VELOCITY JOINT DUST SEAL

Disassembly

Using a suitable hammer, gently and uniformly tap around dust seal until unseated.

Assembly

Using tools T83T-3132-A1 and T86P-1104-A4 or equivalents, install dust seal. The dust seal flange must face outboard.

SPEED INDICATOR RING

Disassembly

1. Remove outboard constant velocity joint as described previously.
2. Position tool T88P-2020-A or equivalent, onto a suitable press.
3. Position constant velocity joint onto tool.
4. With constant velocity joint position on tool, use press ram to apply pressure to the constant velocity joint and remove speed indicator ring.

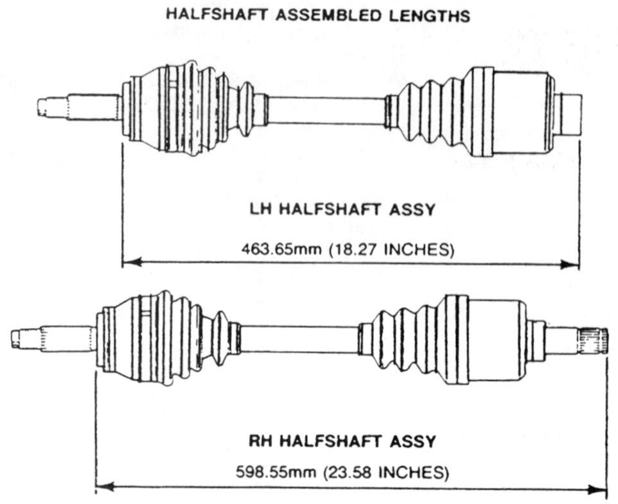

Fig. 4 Halfshaft assembled lengths

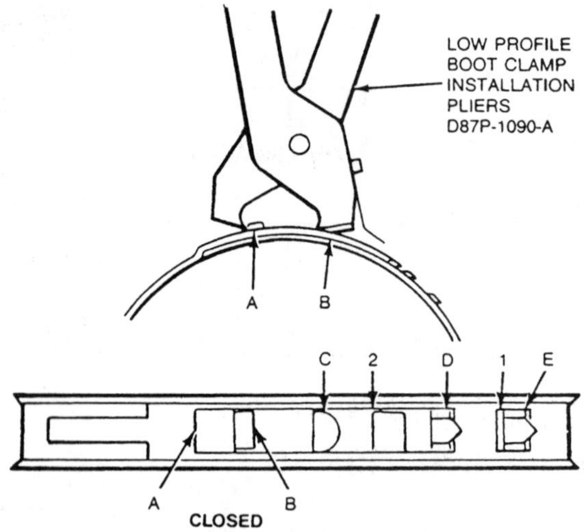

Fig. 5 Boot clamp installation

Assembly

1. With tool T88P-2020-A or equivalent, positioned on press, place sensor ring on tool.
2. Position constant velocity joint into speed indicator ring tool T88P-2020-A or equivalent. Allow constant velocity joint to rest on ring.
3. With constant velocity joint installed on tool, place a steel plate across the constant velocity joint back face. Press constant velocity joint until constant velocity joint bottoms out in tool. The ring will be properly installed when bottomed out in tool.

INBOARD CONSTANT VELOCITY JOINT

Disassembly

Two different types of inboard constant velocity joints and boots are used. The conventional style uses a crimped can on the large end. The tri-lobe style constant velocity joint does not require a crimped can on the large end. Although the designs are similar, there are no interchangeable components between the two designs, **Fig. 3.** The constant velocity joint tripot, outer race, boot and interconnecting shaft are unique for each style.

1. Cut and remove both boot clamps and slide boot back onto shaft. **The righthand inboard constant velocity joint requires a reuseable low profile large boot clamp. A special tool is required to remove and install the clamp.**
2. Remove clamp by using tool D87P-1090-A or equivalent. Engage jaws of tool in closing hooks and draw hooks on clamp together. Disengage windows and locking hooks, then remove the clamp.
3. Slide outer race off of the tripot.
4. When replacing a damaged constant velocity joint boot, the grease should be checked for contamination. If the constant velocity joints are operating satisfactory and grease does not appear to be contaminated, add grease and replace the boot. If grease appears contaminated, proceed with a complete constant velocity joint disassembly and inspection.
5. Using suitable pliers, move stop ring back on shaft.
6. Move tripot assembly back on shaft to allow access to snap ring.
7. Remove snap ring from shaft.
8. Remove tripot assembly from shaft. Remove boot, if necessary.

Assembly

1. Install constant velocity joint boot on shaft, if removed during disassembly. Ensure boot is seated in boot groove on shaft. Tighten clamp using crimping pliers.
2. Install tripot assembly onto shaft with chamfered side toward stop ring.
3. Install a new snap ring.
4. Compress snap ring and slide tripot assembly forward over the snap ring to expose stop ring groove.
5. Move stop ring into groove. Ensure stop ring is completely seated in groove.
6. Fill constant velocity joint outer race and constant velocity boot with suitable grease as shown in **Fig. 3.**
7. Install outer race over tripot assembly and position boot over outer race. Ensure boot is properly seated in groove.
8. Remove all excess grease from constant velocity joint external surfaces. Position boot over constant velocity joint. Move constant velocity joint inward and outward as necessary to length shown in **Fig. 4. Before installing boot clamp, ensure that any air pressure which may have built up in boot is relieved. Insert a dulled tip screwdriver blade between boot and outer bearing race to allow trapped air to escape from boot. The air should be released from the boot only after adjusting to specified dimension.**
9. Seat boot into groove and clamp in position. Install clamp as follows:
   a. With boot seated in groove, place clamp over boot.
   b. Engage hook C in window, **Fig. 5.**
   c. Using tool D87P-1090-A or equivalent, place tool pincer jaws in closing hooks A and B.
   d. Secure clamp by drawing closing hooks together. When windows 1 and 2 are above locking hooks D and E spring tab will press windows over locking hooks and engage clamp.
10. Install a new snap ring, supplied with service kit, in groove nearest shaft end. Start one end in groove and work snap ring over stub shaft end and into groove.

Festiva

INDEX

TROUBLESHOOTING

NOISE & VIBRATION IN TURNS

Clicking, popping or grinding noises while turning may be caused by the following:

1. Cut or damaged CV joint boots, resulting in contaminated lube in outboard or inboard CV joints.
2. Loose CV joint clamps.
3. Worn, damaged or improperly installed wheel bearings.
4. Foreign object contacting halfshaft assembly.

VIBRATION AT HIGHWAY SPEEDS

1. Out of balance front wheels or tires.
2. Improperly seated outboard CV joint in front wheel hub.
3. Bent interconnecting shaft.
4. Front tires out of round.

SHUDDER OR VIBRATION DURING ACCELERATION

1. Excessively worn or damaged inboard or outboard CV joint.
2. Ensure axle shaft is not twisted or cracked.
3. If equipped with vibration damper, ensure damper is not damaged or loose on shaft.

HALFSHAFT OR CV JOINT PULL-OUT

Engine Or Transaxle Misaligned

1. Check engine mounts for damage.

Front Suspension Components Worn Or Damaged

1. Check for worn bushings or bent front suspension components.

Improperly Installed Or Missing Retainers

1. Check for CV joint circlip missing or not properly seated in transaxle side gear.

FAULTY OPERATION OF DRIVESHAFT

1. Check for broken ball joint, tripot joint or worn or seized joint.

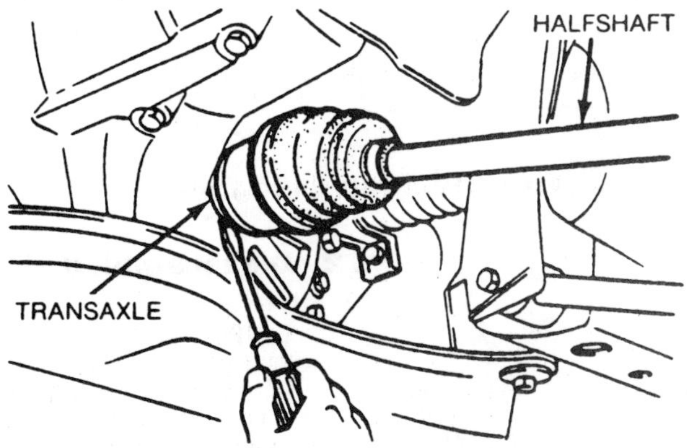

Fig. 1 Disengaging halfshaft from transaxle

HALFSHAFT REPLACE

REMOVAL

1. Raise and support front of vehicle, then drain transaxle.
2. Remove front wheel, then the splash shield, if applicable.
3. Unstake halfshaft attaching nut, apply brakes, then loosen nut. Do not remove attaching nut at this time.
4. Disconnect stabilizer bar from lower control arm.
5. Remove ball joint to steering knuckle clamp bolt and nut, then pry downward on control arm and disengage ball joint from steering knuckle.
6. Using a large screwdriver, disengage halfshaft from transaxle, Fig. 1. **Disengage halfshaft gradually to prevent damage to oil seal.**
7. Remove halfshaft attaching nut and lockwasher, then pull outward on steering knuckle and disengage halfshaft from hub. If difficulty is encountered separating hub from halfshaft, tap end of shaft with plastic mallet to facilitate removal.
8. Withdraw halfshaft from vehicle, then install differential plug T87C-7025-C or equivalent into transaxle to prevent movement of side gears.

INSTALLATION

1. Install new circlip on inboard end of halfshaft.
2. Lubricate inboard splines of halfshaft with grease, remove differential plug, then carefully install halfshaft into transaxle to avoid damaging oil seal.
3. Lubricate outboard splines of halfshaft with grease, then carefully install halfshaft into hub to avoid damaging oil seal.
4. Install new lockwasher and attaching nut. Do not tighten attaching nut at this time.
5. Push up on control arm, reconnect ball joint to steering knuckle, then install clamp bolt and nut. **Torque** nut to 36 ft. lbs.
6. Reconnect stabilizer bar to control arm and **torque** attaching nut to 48 ft. lbs.
7. Apply brakes to prevent hub assembly from turning, then **torque** halfshaft attaching nut to 145 ft. lbs. Stake nut into groove on halfshaft after tightening.
8. Install splash shield and front wheel, then lower vehicle.
9. Fill transaxle with lubricant as required.

HALFSHAFT SERVICE

The halfshafts on these vehicles use two different types of CV joints. The inboard joints are repairable Rzeppa type joints, while the outboard joints are non-repairable Birfield joints.

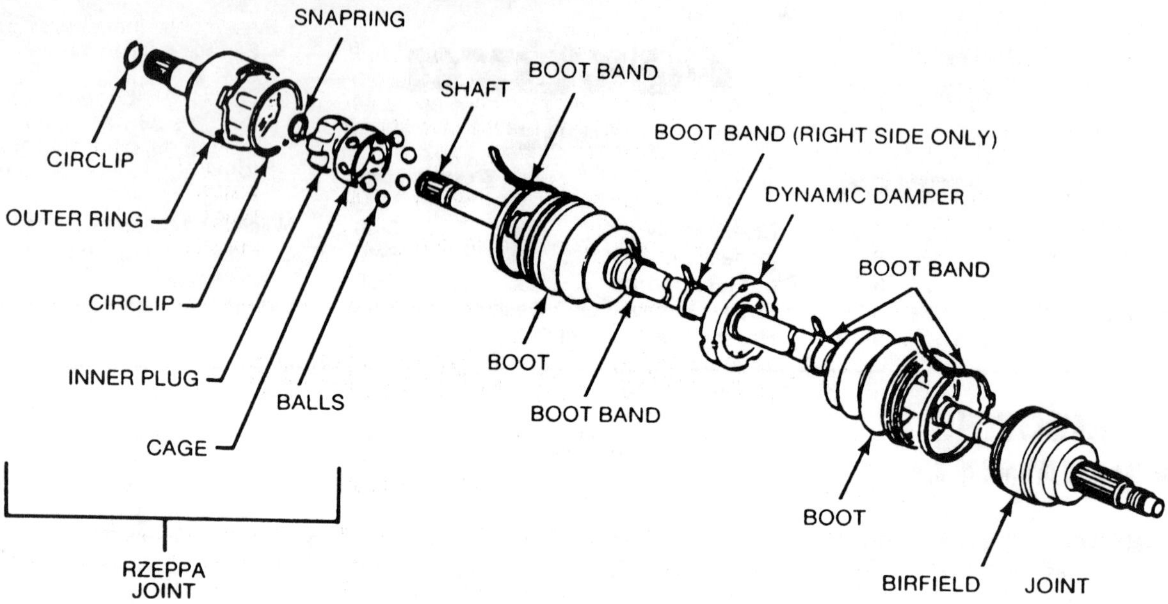

Fig. 2 Disassembled view of halfshaft

DISASSEMBLY

1. Clamp shaft assembly in soft jaw vise.
2. Using suitable screwdriver, pry up locking clips of boot bands, then cut bands off boots and discard.
3. Slide boot off Rzeppa joint, **Fig. 2**, then paint alignment marks on outer ring and shaft to aid assembly.
4. Remove large circlip retaining Rzeppa joint to outer ring. If no damage is evident, leave circlip installed.
5. Withdraw outer ring from Rzeppa joint.
6. Place punch marks on shaft and joint inner plug, then remove snap ring retaining shaft to plug and withdraw Rzeppa joint assembly from shaft.
7. Paint alignment marks on inner plug and cage to aid assembly, then position blade of screwdriver between inner plug and cage and remove balls.
8. Turn cage approximately 30°, then separate cage from inner plug.
9. Remove Rzeppa joint inner boot and dynamic damper (right shaft only) from shaft.
10. Remove Birfield joint inner boot from shaft. **Do not remove Birfield joint from shaft unless damage or excessive wear is evident. If necessary, replace complete Birfield joint assembly as required.**

INSPECTION

1. Clean all metal parts in suitable solvent and dry with compressed air.
2. Check shaft for twisting and/or worn or scored splines, joint components for rust, damage or excessive wear, and boots for cracks, damage or deterioration.
3. Replace all components as necessary.

ASSEMBLY

1. Install new Birfield joint, if necessary.

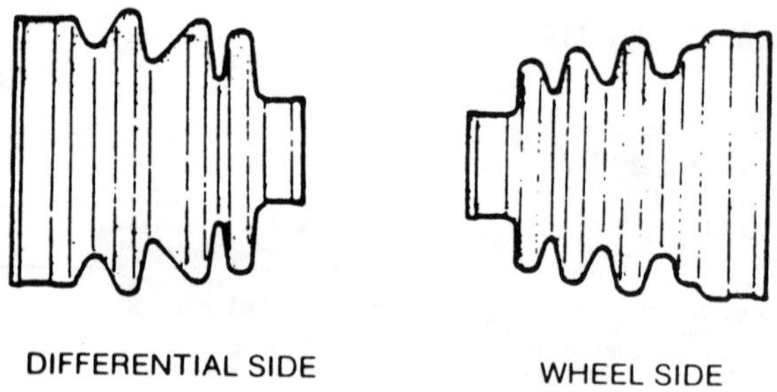

Fig. 3 Boot identification

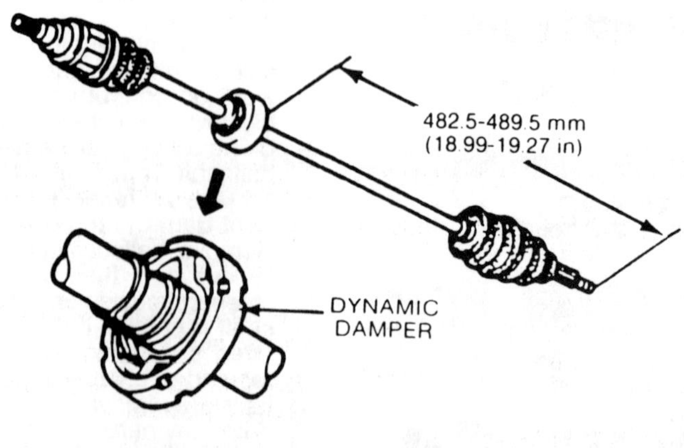

Fig. 4 Installing dynamic damper (right shaft only)

2. Place tape on shaft splines, then install wheel side boot. **Wheel side and differential side boots are not interchangeable. Refer to Fig. 3 for proper identification.**
3. If working on right halfshaft, install dynamic damper at a distance of 18.99-19.27 inches from outboard end of shaft, **Fig. 4.** Ensure outboard joint is fully engaged onto shaft before taking measurement.
4. Install differential side boot onto shaft.
5. Noting marks made during disassembly, reassemble cage, balls and inner plug. Coat all components of joint with molybdenum disulfide grease.
6. Noting marks made during disassembly, position joint onto halfshaft and retain with snap ring.
7. Again noting marks made during disassembly, install joint into outer ring. Secure joint to outer ring using new circlip.
8. Carefully fit boots in grooves on joints, then install new boot bands around boot in opposite direction of halfshaft forward rotation, using pliers to apply proper tension.
9. Bend locking tabs down to secure bands, then remove tape from shaft splines.

Tracer & Probe

INDEX

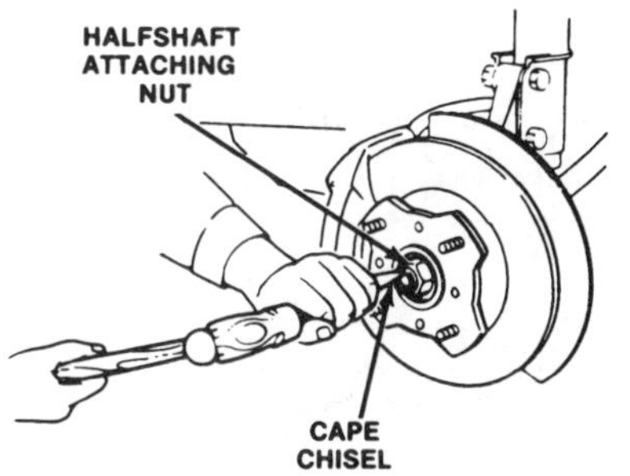

Fig. 1 Unstaking axle nut

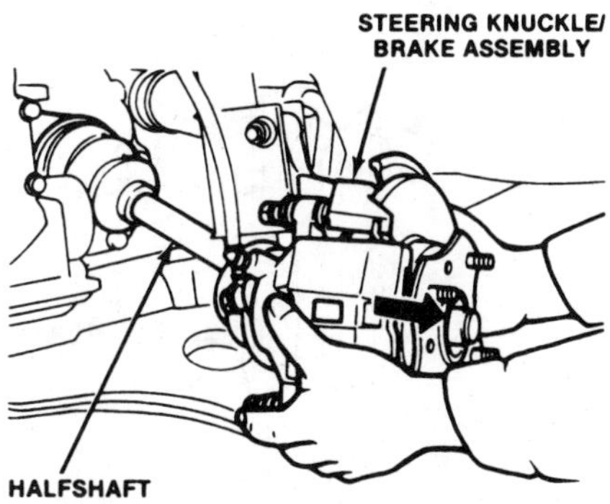

Fig. 2 Separating halfshaft from differential side gear. Manual transaxle models

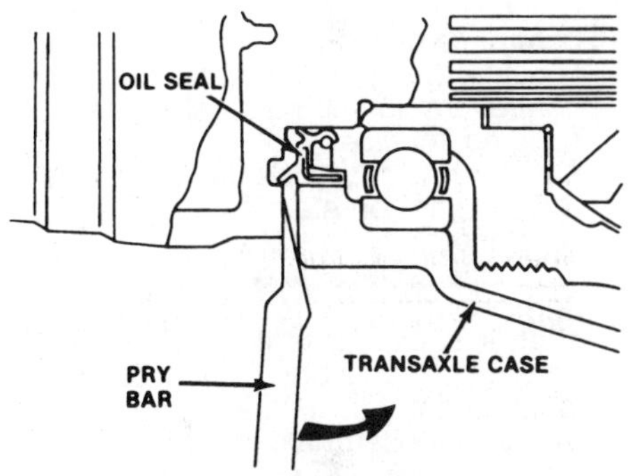

Fig. 3 Inserting pry bar to separate halfshaft from differential side gear

TROUBLESHOOTING

NOISE & VIBRATION IN TURNS

Clicking, popping or grinding noises while turning may be caused by the following:

1. Cut or damaged CV joint boots, resulting in contaminated lube in outboard or inboard CV joints.
2. Loose CV joint clamps.
3. Worn, damaged or improperly installed wheel bearings.
4. Foreign object contacting halfshaft assembly.

VIBRATION AT HIGHWAY SPEEDS

1. Out of balance front wheels or tires.
2. Improperly seated outboard CV joint in front wheel hub.
3. Bent interconnecting shaft.
4. Front tires out of round.

SHUDDER OR VIBRATION DURING ACCELERATION

1. Excessively worn or damaged inboard or outboard CV joint.

HALFSHAFT OR CV JOINT PULL-OUT

Engine Or Transaxle Misaligned

1. Check engine mounts for damage.

Front Suspension Components Worn Or Damaged

1. Check for worn bushings or bent front suspension components.

Improperly Installed Or Missing Retainers

1. Check for CV joint circlip missing or not properly seated in transaxle side gear.

FAULTY OPERATION OF DRIVESHAFT

1. Check for broken ball joint, tripot joint or worn or seized joint.

HALFSHAFT REPLACE

REMOVAL

1. Raise and support front of vehicle, then remove wheel and tire assembly.

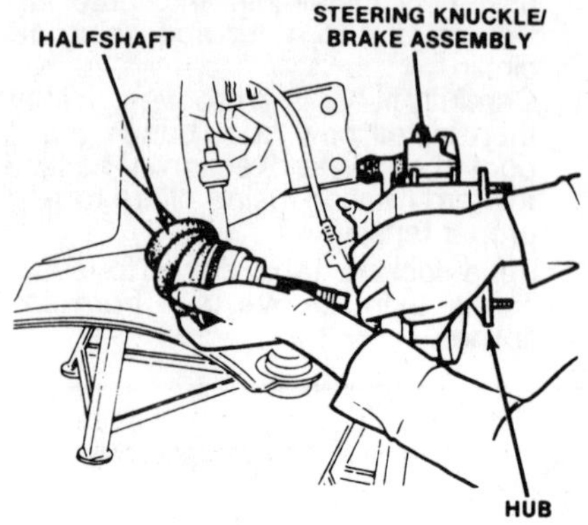

Fig. 4 Separating halfshaft from hub & steering knuckle

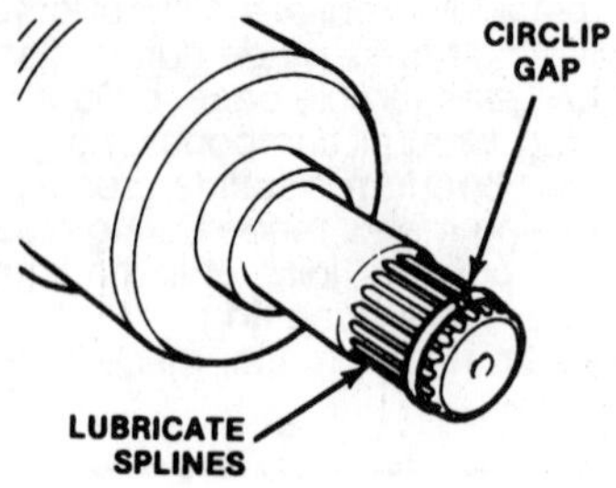

Fig. 5 Position circlip on halfshaft spline

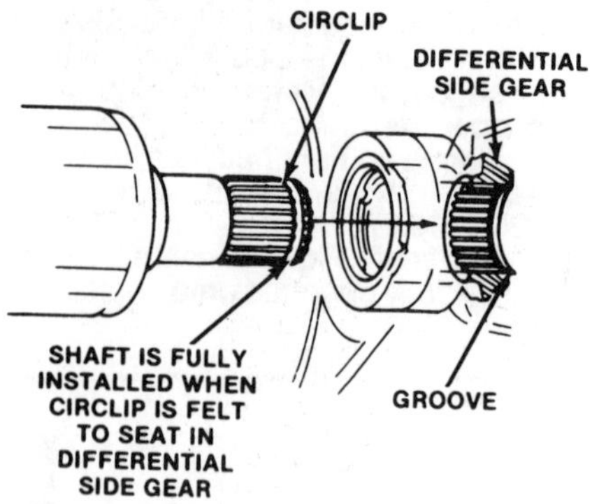

Fig. 6 Installing halfshaft to differential side gear

2. Remove underbody covers, then remove stabilizer bar link from control arm.
3. Using a suitable chisel, raise staked portion of axle nut, then with brakes applied, loosen but do not remove nut, **Fig. 1.**
4. Remove clamp bolt attaching lower control arm ball joint to steering knuckle, then pry downward on lower control arm to separate ball joint from steering knuckle.
5. **On models with manual transaxle,** pull outward on steering knuckle and hub assembly to separate halfshaft from differential side gear, **Fig. 2.** Use only enough force to loosen the halfshaft. In some cases, it may be necessary to position a pry bar between halfshaft and transaxle and light tap end of bar to loosen halfshaft from differential side gear, **Fig. 3.** When using a pry bar, use care not to damage transaxle case, oil seal or CV joint or boot. Do not pull halfshaft completely out of transaxle, as damage to oil seal may result.
6. **On models with automatic transaxle,** position pry bar between transaxle case and halfshaft, then lightly tap end of bar to loosen halfshaft from differential side gear, **Fig. 3.** Use care not damage transaxle case, oil seal or CV joint or boot.
7. **On all models,** remove axle nut and washer, then pull halfshaft from hub and steering knuckle assembly, **Fig. 4.** If binding is encountered, use puller D80L-1002-L to separate halfshaft from hub and knuckle assembly.
8. Support halfshaft and carefully slide from transaxle. Use care not to damage transaxle case oil seal when removing halfshaft.
9. Cap transaxle halfshaft openings.

INSTALLATION

1. Position replacement circlip halfshaft groove at transaxle end, **Fig. 5.** To prevent over expanding, position one end of circlip into groove then work other end over shaft and into groove. When installed, gap in circlip should be at top of halfshaft splines.

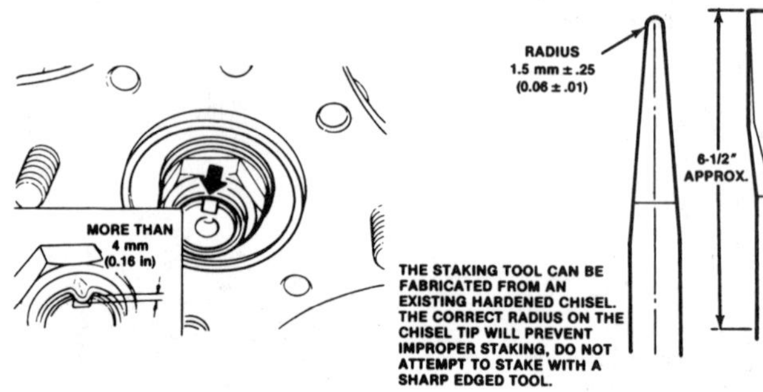

Fig. 7 Staking axle nut

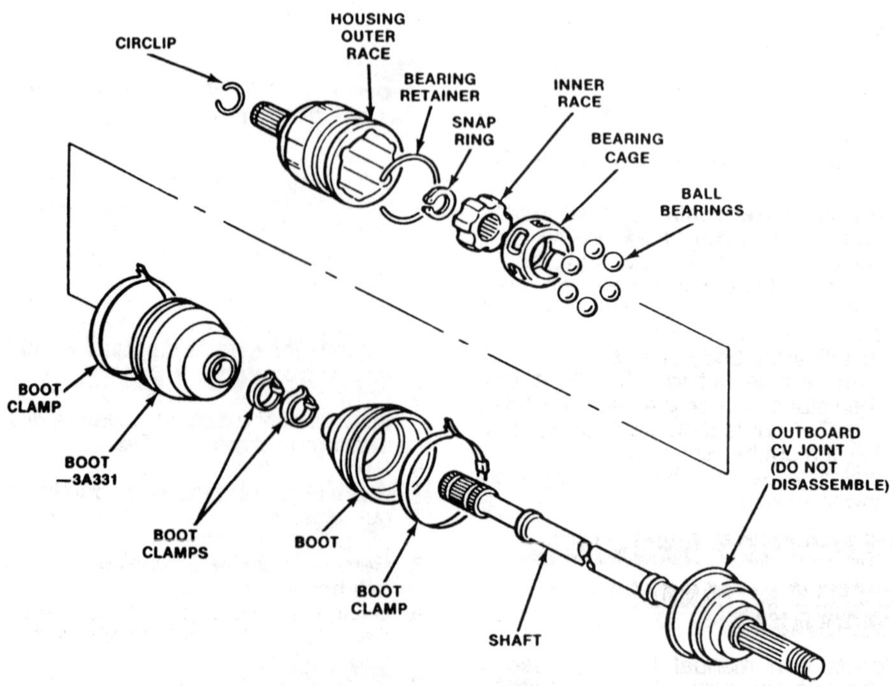

Fig. 8 Halfshaft with Rzeppa type constant velocity joint. Models with manual transaxle

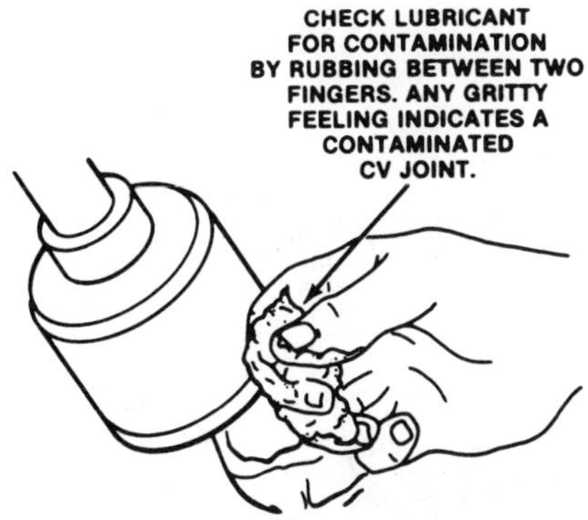

Fig. 9 Checking constant velocity joint lubricant for contamination

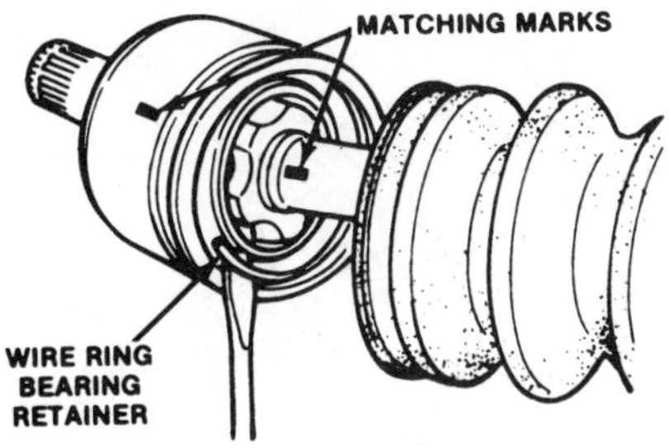

Fig. 10 Removing bearing retaining ring

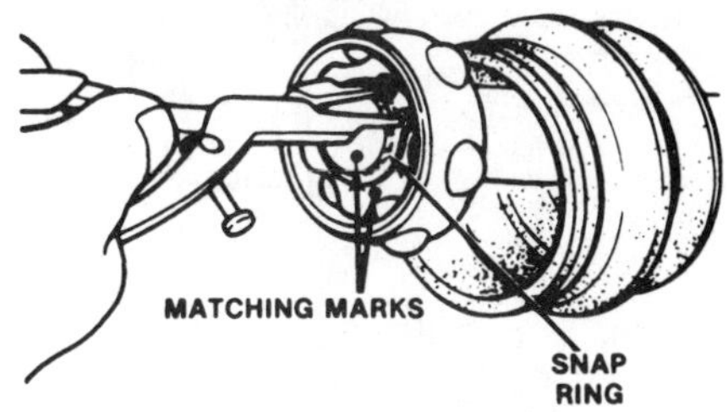

Fig. 11 Removing inner bearing race snap ring

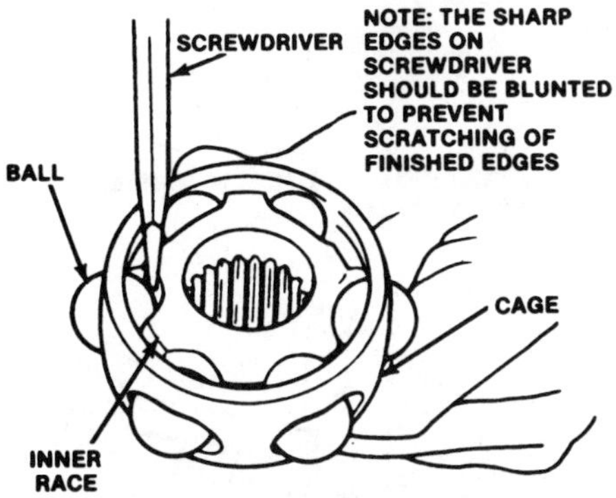

Fig. 12 Removing ball bearings from cage

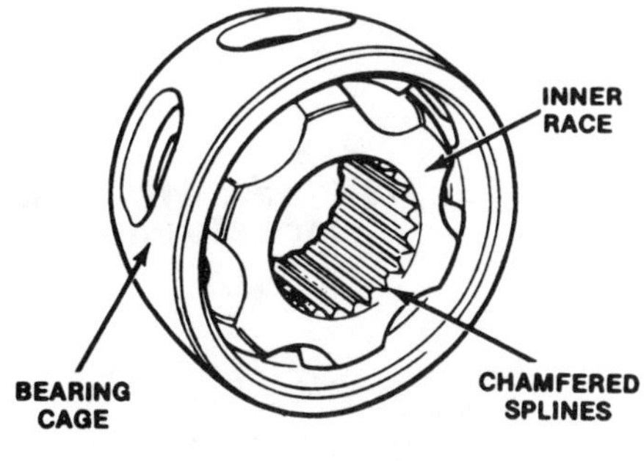

Fig. 13 Positioning inner bearing race to cage

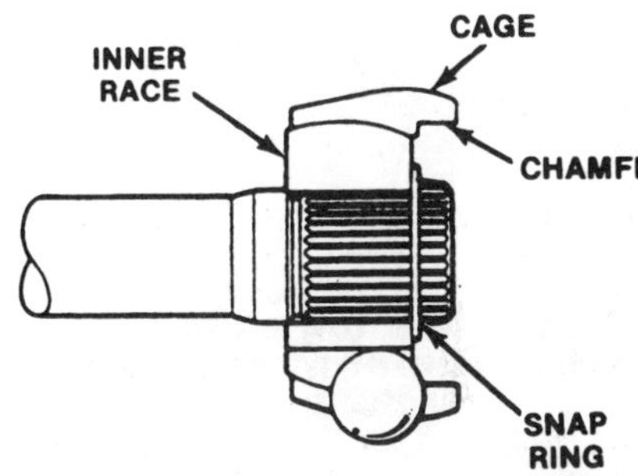

Fig. 14 Positioning bearing cage to halfshaft

2. Lightly lubricate halfshaft splines with lubricant C1AZ-1959D or equivalent, then position halfshaft splines to differential side splines, **Fig. 6.**
3. Push halfshaft splines into differential until circlip snaps into differential side gear groove, **Fig. 6.**
4. Position halfshaft through hub and steering knuckle and loosely install replacement axle nut and washer, **Fig. 4.**
5. Position lower control arm ball joint to steering knuckle, then install and **torque** clamp bolt to 32-40 ft. lbs.
6. Install stabilizer bar link to lower control arm. Tighten nut until .43 inch of thread extends beyond nut.
7. Install under body covers.
8. **Torque** axle nut to 116-174 ft. lbs., then stake nut using a suitable chisel, **Fig. 7.** If nut is damaged during staking, it must be replaced.
9. Install wheel and tire assembly, then lower vehicle.

HALFSHAFT SERVICE

MODELS w/MANUAL TRANSAXLE

Models with manual transaxle use a Rzeppa type constant velocity joint at the transaxle end of the halfshaft and a Birfield type constant velocity joint at the hub side of the halfshaft, **Fig. 8.** The Rzeppa type constant velocity type joint can be disassembled and serviced. The Birfield type constant velocity joint should not be disassembled and is serviced only as an assembly with the halfshaft. To service the Rzeppa type constant velocity joint, proceed as follows:

1. Remove halfshaft as described under "Halfshaft, Replace."
2. Position halfshaft in a soft jawed vise.
3. Remove large boot clamp, the roll boot back over shaft, **Fig. 8.**
4. Check joint grease for contamination, **Fig. 9.** If grease is not contaminated and joint has been operating satisfactorily, add required amount of lubricant and install a replacement bolt. If grease is contaminated, then constant velocity joint must be disassembled and inspected.
5. Place alignment marks on bearing outer race and halfshaft, then remove bearing retaining ring, **Fig. 10.**
6. Remove outer race.
7. Place alignment marks on bearing inner race and halfshaft, then remove snap ring, **Fig. 11.**
8. Remove inner race, bearing cage and ball bearing assembly.
9. Using a screwdriver with sharp edges filed down, pry ball bearings from cage **Fig. 12.**
10. Place alignment marks on bearing inner race and bearing cage.
11. Rotate bearing cage inner race to align bearing lands, then remove inner race through large end of cage.
12. If necessary, remove small clamp and boot from halfshaft. If boot is to be reused, wrap halfshaft splines with tape before removing.
13. Check all bearing components for wear and damage. Bearing components are matched during the manufacturing process, therefore components from another constant velocity should not be interchanged. If bearing is found to unsatisfactory, replace the constant velocity joint assembly.
14. Cover halfshaft spline with tape and install boot, if removed.
15. Lubricate inner race, bearing cage and ball bearings with lubricant E43Z-19590-A or equivalent.
16. Aligning mark made during disassembly, position inner race into bearing cage. Inner race chamfered splines should face large end of cage, **Fig. 13.**
17. Using hand pressure, press ball bearings into cage windows.
18. Aligning marks on inner race and halfshaft made during disassembly, position inner race, cage and ball bearing assembly onto halfshaft, then insert snap ring into halfshaft groove. When install bearing assembly on halfshaft, chamfered end of cage should face snap ring, **Fig. 14.**
19. Apply 1.4 to 2.1 ounces of lubricant E43Z-19590-A or equivalent to outer race, then install outer race over inner race, cage and ball bearing assembly.
20. Add approximately .7 to 1 ounce of lu-

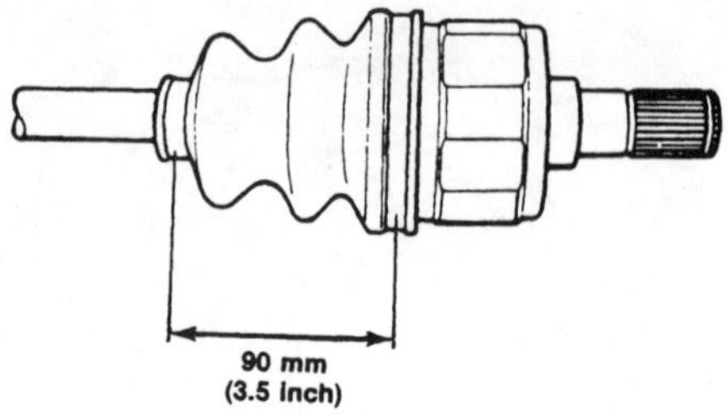

Fig. 15 Position constant velocity joint on halfshaft

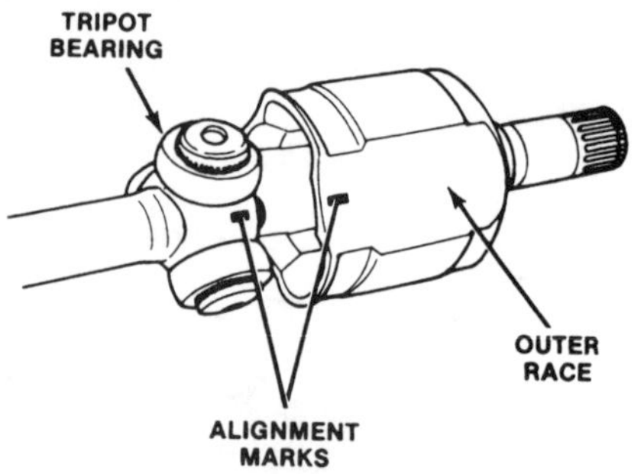

Fig. 17 Placing alignment marks in outer race & tripot joint

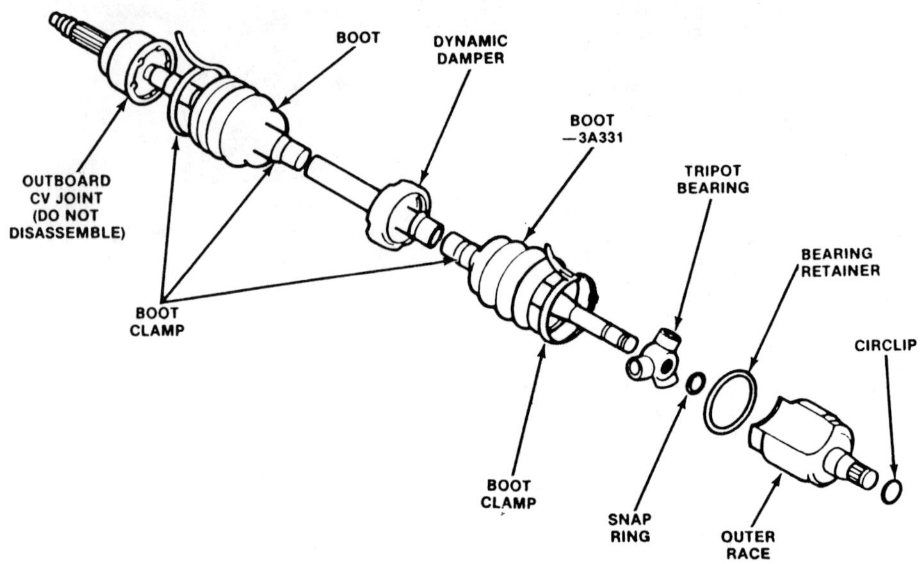

Fig. 16 Halfshaft with tripot type constant velocity joint. Models with automatic transaxle

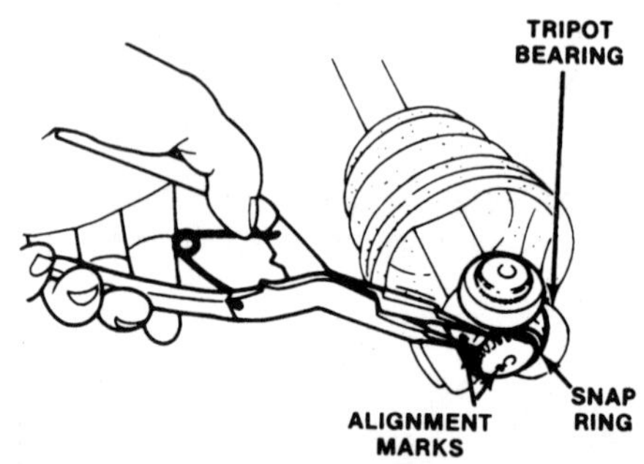

Fig. 18 Removing tripot joint snap ring

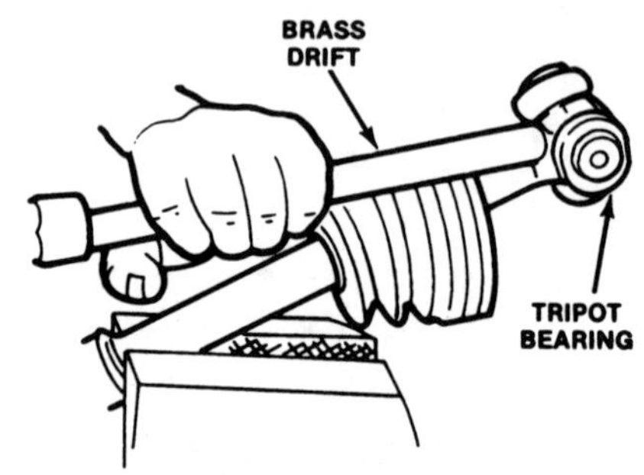

Fig. 19 Removing tripot joint from halfshaft

bricant specified in step 19 to outer race, then install outer race bearing retaining ring.

21. Seat boot in grooves on halfshaft and outer race, then extend constant velocity joint until distance between boot grooves is 3½ inches, **Fig. 15.** Keep joint at this distance until after boot clamps have been installed.
22. Using a screwdriver with sharp edges filed off, pry up on boot at outer race to allow trapped air to escape.
23. Install and lock boot clamps.
24. Check constant velocity joint for smoothness of operation through its full range of travel.
25. Install halfshaft on vehicle as described under "Halfshaft, Replace."

MODELS w/AUTOMATIC TRANSAXLE

Models with automatic transaxle use a tripot type constant velocity joint at the transaxle end of the halfshaft and a Birfield type constant velocity joint at the hub side of the halfshaft, **Fig. 16.** The tripot type constant velocity type joint can be disassembled and serviced. The Birfield type constant velocity joint should not be disassembled and is serviced only as an assembly with the halfshaft. To service the tripot type constant velocity joint, proceed as follows:

1. Remove halfshaft as described under "Halfshaft, Replace."
2. Position halfshaft in a soft jawed vise.
3. Remove large boot clamp, the roll boot back over shaft, **Fig. 16.**
4. Check joint grease for contamination, **Fig. 9.** If grease is not contaminated and joint has been operating satisfactorily, add required amount of lubricant and install a replacement bolt. If grease is contaminated, then constant velocity joint must be disassembled and inspected.
5. Remove bearing retaining ring.
6. Place alignment marks on outer race and tripot joint, then remove outer race, **Fig. 17.**
7. Place alignment mark on tripot bearing and halfshaft, then remove snap ring, **Fig. 18.**
8. Using a brass drift, carefully drive tripot joint from half shaft, **Fig. 19.**
9. If necessary, remove boot small clamp, then wrap tape around halfshaft splines and remove boot.
10. Check tripot joint for wear and damage and replace as necessary.
11. If removed, wrap halfshaft splines with tape and install boot to halfshaft groove.
12. Align marks on halfshaft and tripot joint made during removal.
13. Using a brass drift, drive tripot joint into position on halfshaft, then install snap ring.
14. Fill outer race with 3.5 ounces of lubricant E43Z-19590-A or equivalent, then position outer race over tripot joint, aligning marks made during removal.
15. Install bearing retaining ring, then position boot in groove on outer race.
16. Position joint so that distance between halfshaft boot groove and outer race groove is 3½inches, **Fig. 15.**
17. Using a screwdriver with sharp edges filled off, pry up on boot at outer race end to allow trapped air to escape.
18. Install and lock boot clamps, then check joint for smoothness of operation through its full range of travel.
19. Install halfshaft on vehicle as described under "Halfshaft, Replace."

FRONT WHEEL DRIVE AXLES

INDEX

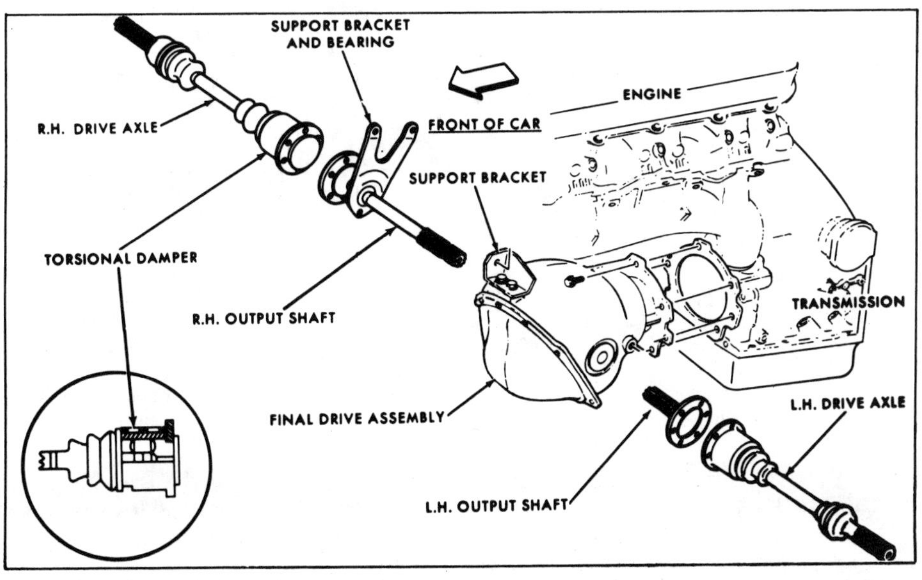

Fig. 1 Front wheel drive components. 1985 Eldorado, Riviera, Seville & Toronado

TROUBLESHOOTING

CLICKING NOISE IN TURNS

Worn or damaged outboard joint.

CLUNK WHEN ACCELERATING FROM COAST TO DRIVE

Worn or damaged constant velocity joint.

SHUDDER OR VIBRATION DURING ACCELERATION

1. Excessive C.V. joint angle.
2. Incorrect toe in or out.
3. Incorrect trim height.
4. Worn or damaged inboard or outboard C.V. joints.
5. Sticking tri-pot joint spider assembly.

VIBRATION AT HIGHWAY SPEEDS

1. Out-of-balance front tires or wheels.
2. Out-of-round front tires.
3. Worn C.V.joint.
4. Binding or tight C.V. joint.

DESCRIPTION

The front wheel drive system, **Figs. 1 , 2, and 3,** consists of a final drive unit, left and righthand output shafts and drive axles. **On 1985 Eldorado, Riviera, Seville and Toronado models,** the output shafts are splined to the side gears and are retained by a retaining ring. Each drive axle, **Fig. 4,** consists of an axle shaft, with a ball type constant velocity joint at the outboard end and a tri-pot joint at the inboard end.

On all models except Geo, Lemans, Nova, Spectrum, Sprint and 1985 Eldorado, Riviera, Seville and Toronado, a ball type constant velocity joint is used at the outboard end, while a tri-pot type joint is used at the inboard end, **Fig. 5.** Snap rings are used to lock the male splines of the axle shafts into the transaxle gears, except for the left side inboard joint used with the automatic transaxles. The left side inboard joint used on automatic transaxle models utilizes a female spline which installs over a transaxle stub shaft.

On some manual transmissions, a constant velocity joint is used on the inboard end of the drive axle, **Fig. 6.**

On Nova and Lemans models, a ball type constant velocity joint is used at the outboard end, while a tri-pot type joint is used at the inboard end. Snap rings are used to lock the male splines of the axle shafts into the transaxle gears, **Fig. 5.**

On Geo, the drive axles are flexible assemblies consisting of inner and outer constant velocity ball joints connected by an axle shaft, **Fig. 7.**

On Spectrum models, the drive axles are completely flexible assemblies containing an inner and outer constant velocity joint joined by an axle shaft, **Fig. 8.** Three types of constant velocity joints are used: the manual transaxle utilizes a Birfield outer and double offset inner, and the automatic uses a tripod inner and a Birfield outer.

On Sprint models, a constant velocity ball joint is used on the outboard side of the drive axle and a constant velocity tripod joint on the inboard side, **Fig. 9.**

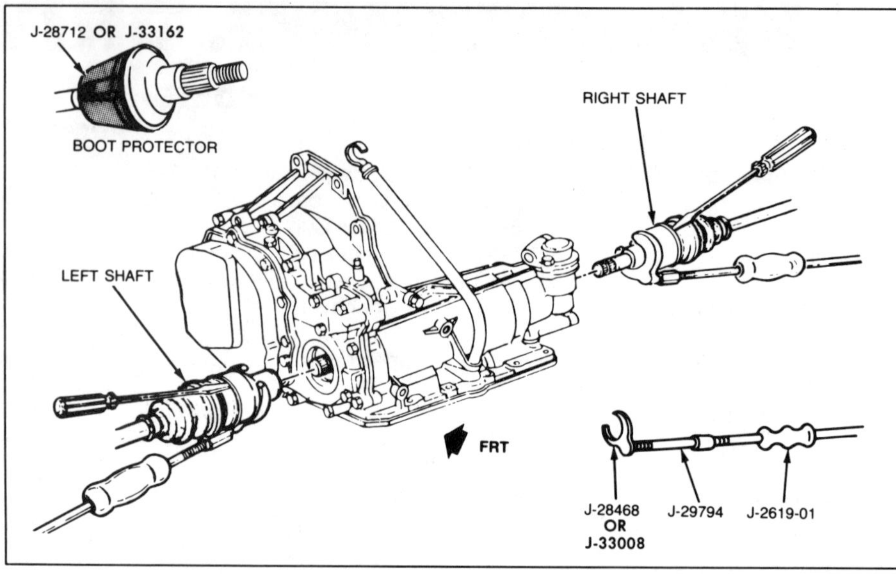

Fig. 2 Front wheel drive components and axle shaft removal. Except Geo, Lemans, Sprint and Spectrum (Typical)

DRIVE AXLE
REPLACE

On models equipped with tri-pot joints on inboard axles, care must be taken not to overextend joints. When either or both ends are disconnected, overextending the tri-pot joint could result in internal joint separation.

On vehicles equipped with ball type constant velocity inboard joints, install inner drive joint seal protector J-34754, on inboard seal and install axle boot protector J-28712 on outboard seal if necessary. On vehicles equipped with tri-pot inboard joints,install axle boot seal protector J-28712 on outboard seal and J-33162 on inboard seal, if necessary.

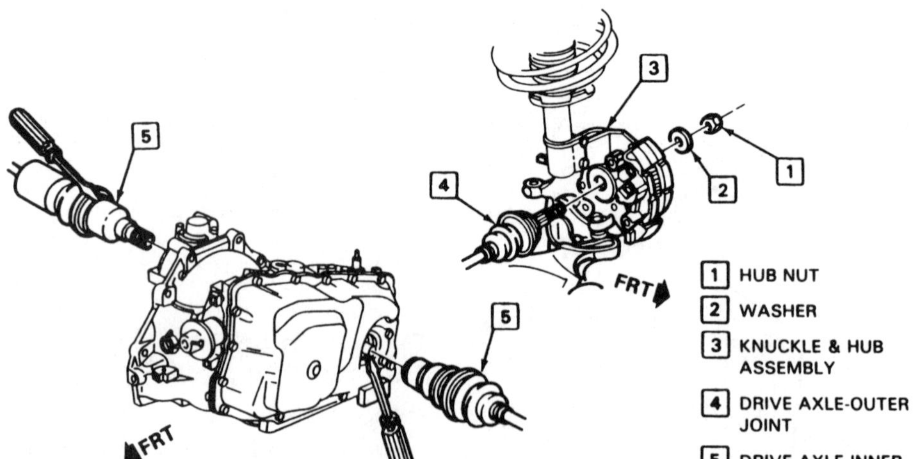

Fig. 3 Front wheel drive components and axle shaft installation. Except Geo, Lemans, Sprint and Spectrum (Typical)

ELDORADO, RIVIERA, REATTA, SEVILLE, TORONADO, 1986–88 DEVILLE & FLEETWOOD

Removal

1. Remove hub nut, then raise and support vehicle and remove wheel and tire assembly.
2. Remove brake caliper and rotor assembly.
3. Remove stabilizer link or stabilizer bar from control arm.
4. Remove tie rod end at steering knuckle, then remove lower ball joint stud at steering knuckle.
5. Remove axle from transaxle using a prybar or screwdriver and a wood block as a fulcrum to protect case cover.
6. Drive axle from hub and bearing assembly using drive axle spindle remover tool No. J-28733.
7. Remove axle from vehicle.

Installation

1. Install drive axle at steering knuckle.
2. Install lower ball joint stud at steering knuckle.
3. Install stabilizer link or stabilizer bar to lower control arm.
4. Install tie rod end at steering knuckle.
5. Install brake caliper, rotor and new caliper mounting bolts.
6. Install new torque prevailing nut and washer.
7. Insert a screwdriver or drift into caliper and rotor to prevent rotor from turning, then install hub nut and **torque** to 74 ft .lbs., **Fig. 10.**
8. Seat the drive axle into the transaxle by placing a screwdriver into the groove on the joint housing and tapping until seated, then grasp the inner housing and pull outward. If the snap ring is properly seated, the axle will remain in place, **Fig. 3.**
9. Install the front wheel and tire assembly and lower vehicle.

BONNEVILLE, CAVALIER, CELEBRITY, CENTURY, 1988 CIMARRON, CUTLASS CALAIS, DELTA 88 & 98, ELECTRA, FIRENZA, GRAND AM, LESABRE, PARK AVENUE, SKYHAWK, SKYLARK & SUNBIRD

Removal

1. Raise and support vehicle, then remove tire and wheel assembly.

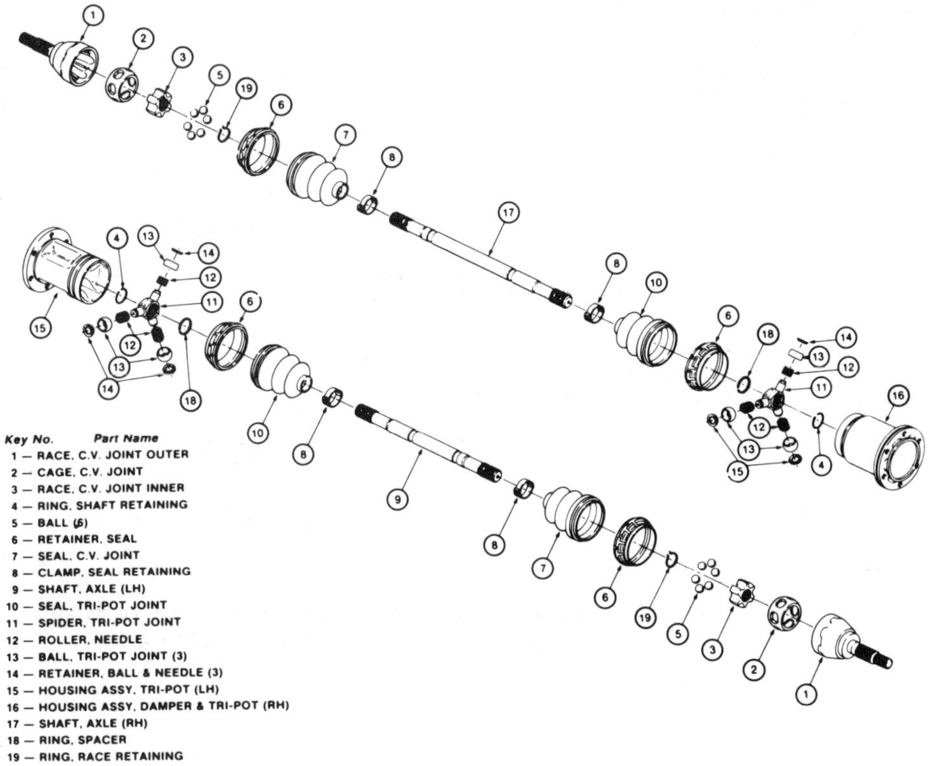

Fig. 4 Exploded view of drive axle. 1985 Eldorado, Riviera, Seville & Toronado

2. Install a brass drift or a screwdriver to prevent the rotor from turning, **Fig. 10.**
3. Remove shaft nut and washer using shaft nut socket tool No. J-34826.
4. Remove brake caliper from steering knuckle and suspend caliper assembly.
5. Remove rotor from hub and bearing assembly, then stabilizer shaft or link from control arm.
6. Remove ball joint from steering knuckle.
7. Remove drive axle from transaxle using axle shaft remover tool No. J-33008, axle shaft remover tool No. J-29794 and slide hammer tool No. J-2619-01 if necessary, **Fig. 2.**
8. Remove drive axle from hub and bearing assembly using front hub spindle remover tool No. J-28733-A, **Fig. 11.**

Installation

1. Install drive axle into hub and bearing assembly and transaxle.
2. Install lower ball joint to steering knuckle.
3. Install stabilizer shaft or link to control arm.
4. Install rotor to hub and bearing assembly.
5. Insert a screwdriver or drift into caliper and rotor to prevent rotor from turning, **Fig. 10.**
6. Install a new hub nut and washer, **torque** to 185 ft. lbs. **On Bonneville, Electra, LeSabre, Park Avenue and Sunbird models,torque** to 191 ft. lbs.
7. Seat the drive axle into the transaxle by placing a screwdriver into the groove on the joint housing and tapping until seated.
8. Ensure drive axle is seated by pulling outward on the housing. Do not pull on the drive axle.
9. Install tire and wheel assembly and lower vehicle.

CUTLASS SUPREME, GRAND PRIX & REGAL

Removal

1. Raise and support vehicle, then remove tire and wheel assembly.
2. Remove brake caliper and bracket assembly, then the rotor.
3. Remove the four bolts holding the hub/bearing assembly to the knuckle.
4. **On vehicles equipped with The Anti-Lock Brake System (ABS),** remove the ABS sensor mounting bolt, then position the sensor aside to avoid damage.
5. **On vehicles equipped with 125C and HM-282 transaxles and right shaft of vehicles with 440-T4 transaxles,** separate the drive axle from the transaxle using axle shaft remover tool No. J-3308, extension tool No. J-29794 and puller tool No. J-2619-01.**Fig. 2. On the left shaft of vehicles equipped with 440-T4 transmissions,** using the frame for leverage, separate the drive axle from the transaxle with a screwdriver or prybar in the groove provided on the inner joint, **Fig. 3.**
6. **On all transmissions,** remove the axle/bearing assembly through the knuckle.

Installation

1. **On all transmissions,** Install the axle/bearing assembly through the knuckle and into the transaxle. **If equipped with ABS,** install the ABS assembly.
2. Loosely secure the bearing to knuckle bolts.
3. Seat the drive axle into the transaxle by placing a screwdriver into the groove on the joint housing and tapping until seated, then grasp the inner housing and pull outward. If the snap ring is properly seated, the axle will remain in place, **Fig 3.**
4. Using a new nut and washer, install the hub and bearing assembly to axle, **torque** hub and bearing assembly to knuckle attaching bolts to 52 ft. lbs.
5. Install rotor and brake assembly.
6. Install tire and wheel assembly.
7. **Torque** drive axle nut to 184 ft. lbs, then lower vehicle.

1989 DEVILLE & FLEETWOOD

Removal

1. Raise and support vehicle, then remove tire and wheel assembly.
2. Install a brass drift or a screwdriver to prevent the rotor from turning, **Fig. 10.**
3. Remove shaft nut and washer.
4. Remove lower ball joint cotter and nut, then loosen ball joint using ball joint separator tool No. J-29330. **If removing right axle,** turn wheel to left, **if removing left axle,** turn wheel to the right.
5. Separate the ball joint by using a pry bar between the suspension support and lower control arm.
6. Pull on lower knuckle area and with a plastic or rubber mallet strike the end of the axle shaft to disengage axle from hub and bearing.
7. Separate hub and bearing assembly from drive axle and move strut and knuckle assembly rearward.
8. Remove inner joint from transaxle using axle shaft remover tool No. J-28468 or J-33008 if equipped with intermediate shaft, **Fig. 2.**

Installation

1. Seat the drive axle into the transaxle by placing a screwdriver into the groove on the joint housing and tapping until seated, then grasp the inner housing and pull outward. If the snap ring is properly seated, the axle will remain in place, **Fig. 3.**
2. Drive axle into hub and bearing assembly.
3. Install lower ball joint to knuckle, **torque** to 45 ft. lbs., then install cotter pin.
4. Install washer and new shaft nut.
5. Insert a screwdriver or drift into caliper and rotor to prevent rotor from turning. **Fig. 10.**
6. Install hub nut, **torque** to 185 ft. lbs.

Key No.	Part Name
1	RACE, C.V. JOINT OUTER
2	CAGE, C.V. JOINT
3	RACE, C.V. JOINT INNER
4	RING, SHAFT RETAINING
5	BALL (6)
6	PROTECTOR, CLAMP (2)
7	SEAL, C.V. JOINT
8	CLAMP, SEAL RETAINING
9	SHAFT, AXLE (LH)
10	SEAL, TRI-POT JOINT
11	SPIDER, TRI-POT JOINT
12	ROLLER, NEEDLE
13	BALL, TRI-POT JOINT (3)
14	RETAINER, BALL & NEEDLE (3)
15	HOUSING ASM, TRI-POT
16	HOUSING ASM, DAMPER & TRI-POT (RH)
17	SHAFT, AXLE (RH)
18	RING, SPACER
19	RING, RACE RETAINING
20	CLAMP, SEAL RETAINING
21	RETAINER, NEEDLE
22	RING, NEEDLE RETAINER
23	RING, JOINT RETAINING
24	HOUSING, TRI-POT (RH)
25	SHAFT ASM, DAMPER &
26	RING, DEFLECTOR
27	BUSHING, TRILOBAL TRI-POT

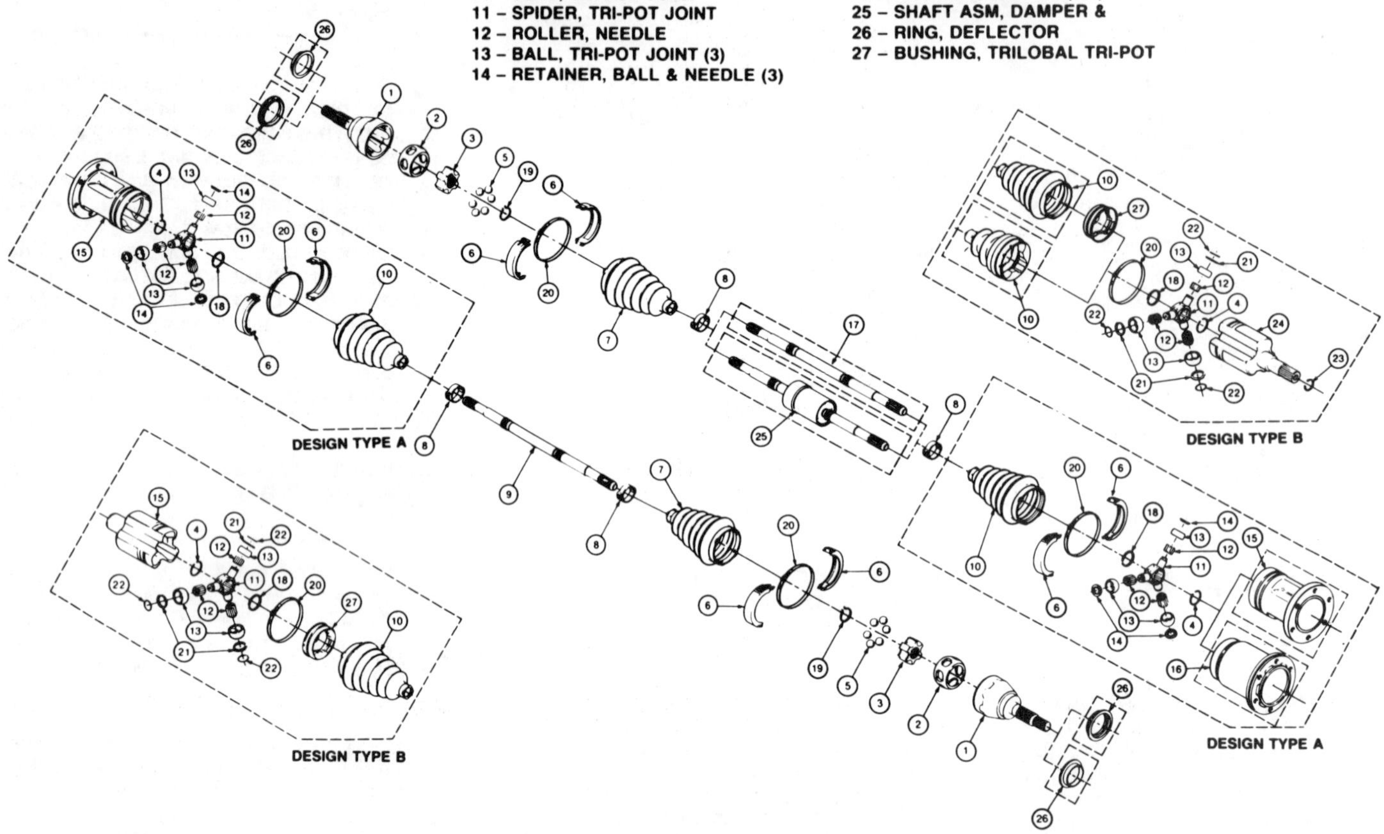

Fig. 5 Exploded view of drive axle with tri-pot inboard joint. Except Geo, Lemans, Spectrum and Sprint.

7. Install tire and wheel assembly and lower vehicle.

GEO

Removal

1. Raise and support vehicle, then remove wheel and tire assembly.
2. Drain transmission.
3. Remove locking hub, then drive shaft circlip, **Fig. 7.**
4. Remove stabilizer ball joint, then tie-rod end castle nut.
5. Remove brake caliper assembly and suspend caliper assembly.
6. Remove knuckle ball joint stud nut and support lower arm with jack.
7. Remove strut bracket bolts, then knuckle and wheel hub assembly by lowering jack.
8. **Right side,** to remove the differential side joint snap ring, pry the differential side joint away from the left differential assembly using shaft removal fork tool No. J-37780, **Fig. 12. Left side,** remove the driveshaft bolts.
9. Remove the driveshaft.

Installation

1. Using driveshaft oil seal installer tool Nos. J-6133-01 and J-36183, push in oil seal until its end contacts the surface of the drive axle.
2. Install **right side,** differential side joint with snap ring, **left side** side joint with drive shaft bolts.
3. Reverse removal procedure to install.

LEMANS

Removal

1. Loosen both upper strut mount attaching bolts.
2. Remove wheel cover, then loosen wheel lug nuts and remove cotter pin and drive axle shaft-to-hub nut and washer.
3. Raise and support vehicle.
4. Remove wheel and tire assembly.
5. Remove ball joint retaining clip and stud nut, then separate ball joint from steering knuckle using ball joint separator tool No. J-36226 or equivalent.
   - **Disregard "This Side Towards Wheel" marking on tool. Tool must be turned to opposite position when used on these vehicles.**
6. Remove tie rod nut, then separate tie rod from steering knuckle using tie rod separator tool No. J-24319-01 or equivalent.
7. Separate drive axle shaft from wheel hub using axle shaft to hub separator tool No. J-3666-1 or equivalent.
8. Position a drain pan under transaxle, then remove drive axle from transaxle using axle shaft to transaxle separator tool No. J-36639 and slide hammer tool No. J-23907 or equivalents.

Installation

1. Install drive axle into transaxle, using care to avoid damaging seal. Pull on outer joint race to ensure it will not come out of transaxle.
2. Install drive axle into knuckle hub bearing.
3. Install steering knuckle onto lower ball joint, then connect tie rod to the knuckle. **Torque** tie rod nut to 45 ft. lbs. and ball joint stud nut to 50 ft. lbs.,

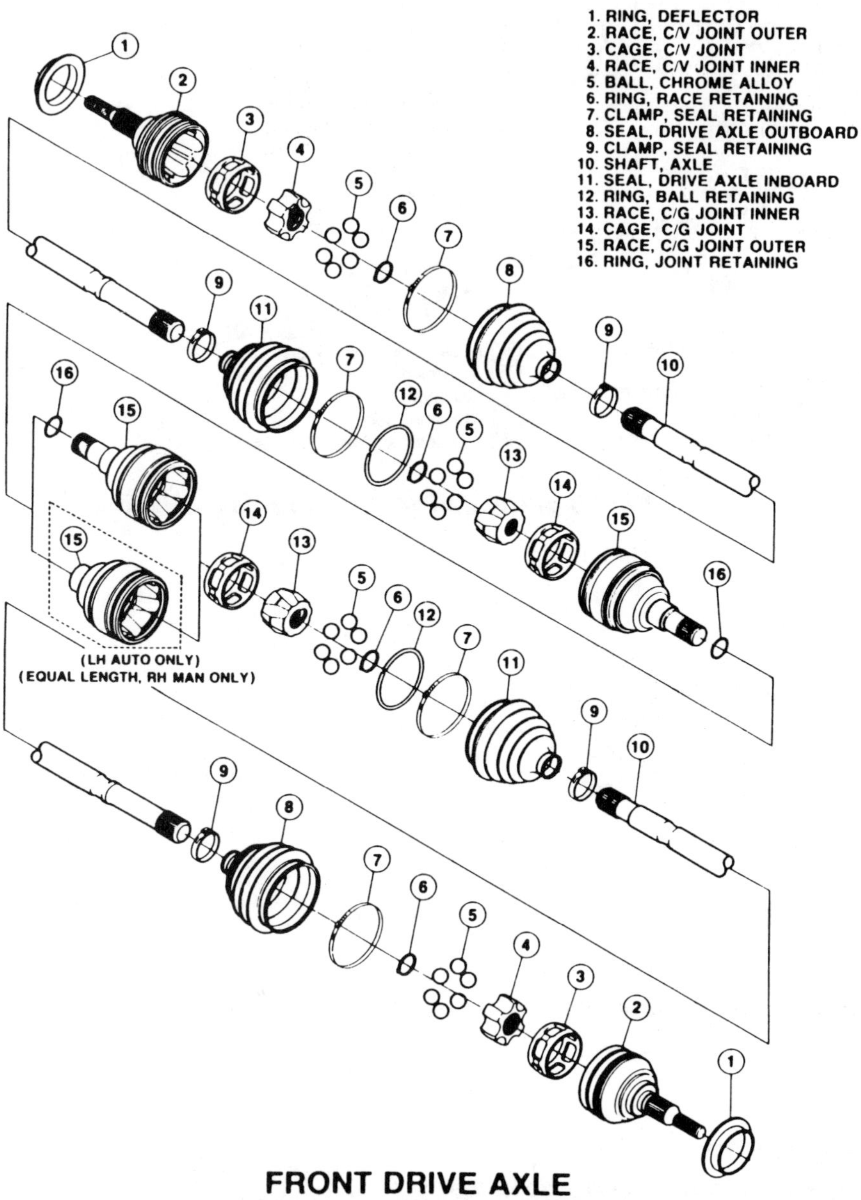

Fig. 6 Exploded view of cross-groove design drive axle. Models w/manual transmission except Geo, Lemans, Spectrum and Sprint

then tie rod nut cotter pin and stud nut retaining clip.
4. Install drive axle shaft-to-wheel hub washer and nut.
5. Install wheel and tire assembly, then lower vehicle.
6. With weight of vehicle on wheels, **torque** new drive axle-to-hub nut to 74 ft. lbs., then back off nut and **retorque** to 15 ft. lbs. Tighten nut an additional 1/4 turn, then install cotter pin. If necessary, loosen nut slightly to align cotter pin holes.
7. **Torque** upper strut mount attaching nuts to 22 ft. lbs.
8. Check transaxle fluid and replenish as needed.

6000

Removal

1. Raise and support vehicle, then remove tire and wheel assembly.
2. Insert drift or screwdriver into caliper to prevent rotor from turning, **Fig. 10.**
3. Remove hub nut and washer using hub nut socket tool No. J-34826 and discard the shaft nut,**Fig. 10.**
4. Remove brake caliper from steering knuckle and suspend caliper assembly.
5. Remove rotor from hub and bearing assembly.
6. Remove stabilizer shaft from control arm.
7. Remove ball joint from steering knuckle.
8. Using axle shaft remover tool No. J-33008 with extension J-29794 on slide hammer, remove drive axle from transaxle, then vehicle, **Fig. 2.**
9. Remove drive axle from hub and bearing assembly using puller No. J-28733, **Fig. 11.**

Installation

1. Install drive axle into hub and bearing assembly, then the transmission.
2. Install the lower ball joint to the steering knuckle.
3. Install the stabilizer shaft to the control arm.
4. Install caliper to steering knuckle, then install washer and a new torque prevailing shaft nut.
5. Insert a screwdriver or drift into caliper and rotor to prevent rotor from turning, **Fig. 10.**
6. Install hub nut, **torque** to 185 ft. lbs. **Fig. 10.**
7. Seat the drive axle into the transaxle by placing a screwdriver into the groove on the joint housing and tapping until seated, then grasp the inner housing and pull outward. If the snap ring is properly seated, the axle will remain in place, **Fig. 3.**
8. Install tire and wheel assembly and lower vehicle.

SPECTRUM

Replace

1. Remove wheel cover, cotter pin and driveshaft nut, then loosen wheel nuts.
2. Raise and support vehicle.
3. Remove front wheel and tire assembly.
4. Drain transaxle oil.
5. Detach snap ring installed on spline of inboard joint using a large screwdriver.
6. Remove knuckle and hub assembly.
7. Unfasten inboard joint from differential side gear and outboard joint from steering knuckle, then remove drive axle assembly.
8. Reverse procedure to install.

SPRINT

Replace

1. Remove wheel center cap.
2. Remove cotter pin, then drive shaft nut.
3. Loosen wheel lug nuts.
4. Raise and support vehicle.
5. Remove front wheel.
6. Drain transaxle fluid.
7. Using a suitable lever, pry inboard drive shaft joint from differential side gear to disengage snap ring on shaft spline, **Fig. 9.**
8. Remove 2 stabilizer bar mount bracket bolts and ball joint stud bolt. Pull down on stabilizer bar to disconnect ball joint stud from steering knuckle, **Fig. 13.**
9. Disconnect inboard joint from differ-

ential side gear, then outboard joint from steering knuckle. Remove drive shaft.
10. Reverse procedure to install. **Torque** drive shaft nut to 108.5-195 ft. lbs.

CORSICA, BERETTA & 1985–87 CIMARRON

Removal

1. **On all year Corsica and Beretta models and 1985-87 Cimarron models,** remove hub nut, then raise and support vehicle and remove wheel and tire assembly.
2. **On all Corsica and Beretta models and 1985–87 Cimarron models,**remove disc brake caliper and caliper support.
3. **On 1989 Corsica, models,** remove lower control arm ball joint nut and cotter pin.**On 1985-88 Corsica and 1985-87 Cimarron models,** remove strut to steering knuckle attaching bolts then pull steering knuckle assembly out of strut bracket.
4. **On 1989 models,** use front hub spindle remover tool No. J-28733 and remove drive axle from hub and bearing assembly, **Fig. 11.**
5. **On 1989 models,** drain transaxle.
6. **On 1985-88 Corsica and Cimarron models** using axle shaft remover tool No. J-33008 with extension No. J-29794 remove drive axle from transaxle, then temporarily support with a wire or rope, **Fig. 2. On 1989 models,** using axle shaft remover No. J-33008 with extension No. J-29794 on slide hammer, remove drive axle from transaxle, then vehicle,**Fig. 2.**
7. **On 1985-88 Corsica and Cimarron models,** use front hub spindle remover tool No. J-28733 and remove drive axle from hub and bearing assembly, **Fig. 11.**

Installation

1. **On all models,** loosely install drive axle to strut and transaxle.
2. **On 1985-88 Corsica and Cimarron models,** position steering knuckle to strut bracket and install bolts.
3. **On all models,**install rotor and brake caliper.
4. Install a brass drift or a screwdriver to prevent the rotor from turning, **Fig. 10,** then install hub nut and washer and **torque** to 71 ft. lbs. Remove drift or screwdriver.
5. Using screwdriver in groove provided on inner retainer, install axle shaft on transaxle, **Fig. 3.** Tap on screwdriver until axle shaft is seated in transaxle.
6. Remove boot seal protector, then install wheel and tire.
7. Lower vehicle, then **torque** hub nut to 191 ft. lbs. **on 1989 models,** fill axle with proper fluid.

NOVA

Removal

1. Remove hub nut cotter pin, the the hub nut and washer.
2. Loosen wheel lug nuts, then raise and

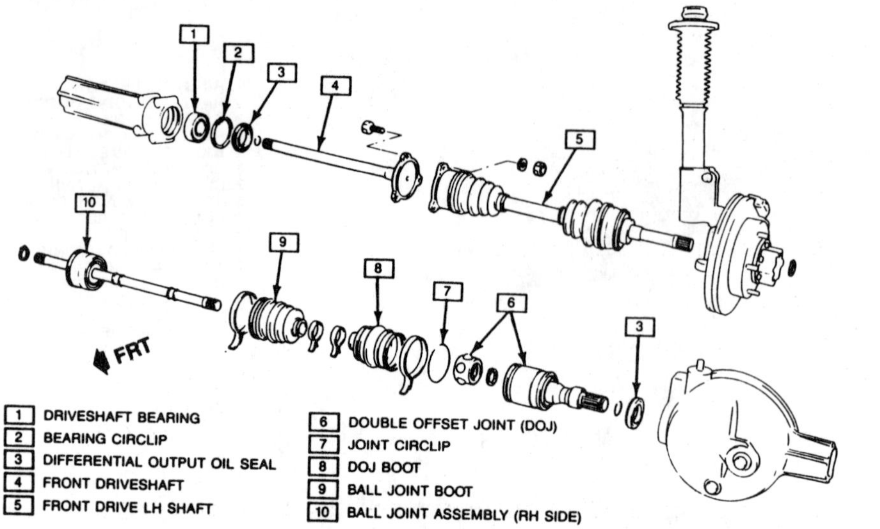

Fig. 7 Exploded view of drive axle. Geo

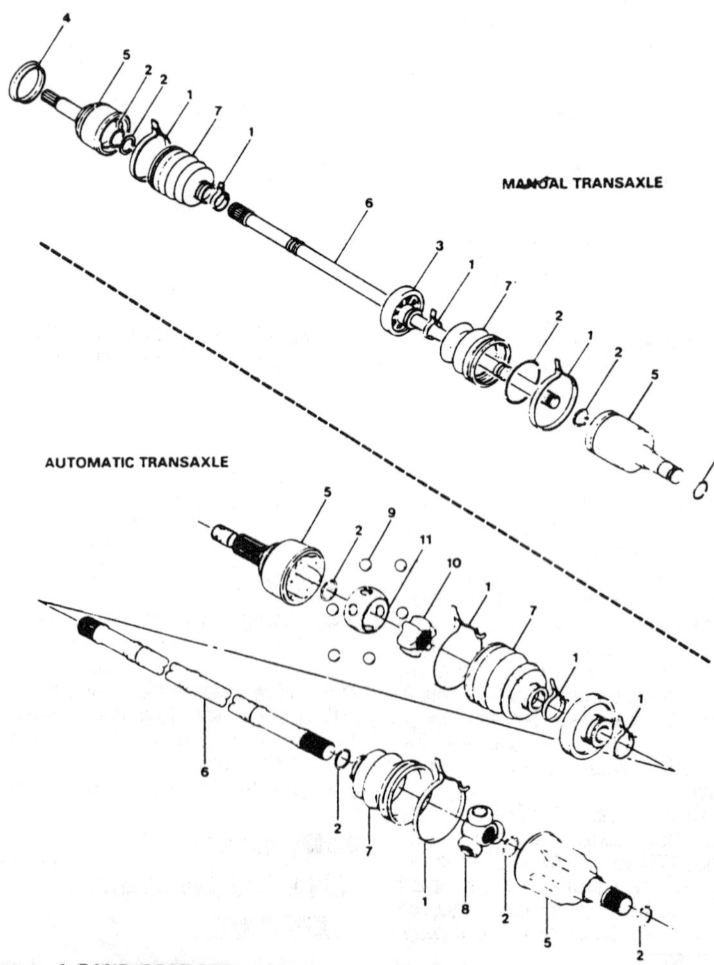

Fig. 8 Exploded view of drive axle. Spectrum

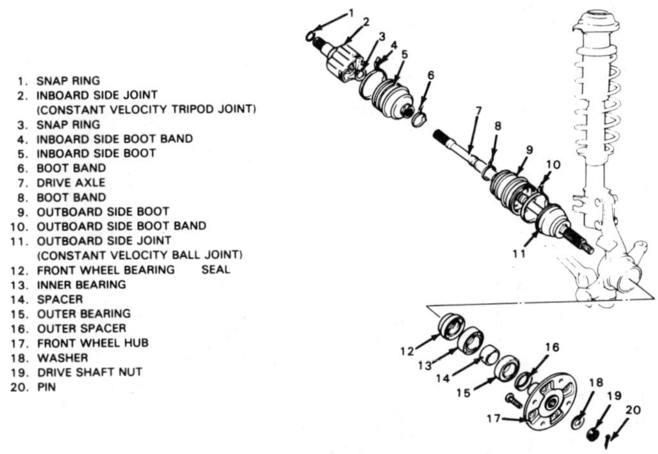

Fig. 9 Exploded view of drive axle. Sprint

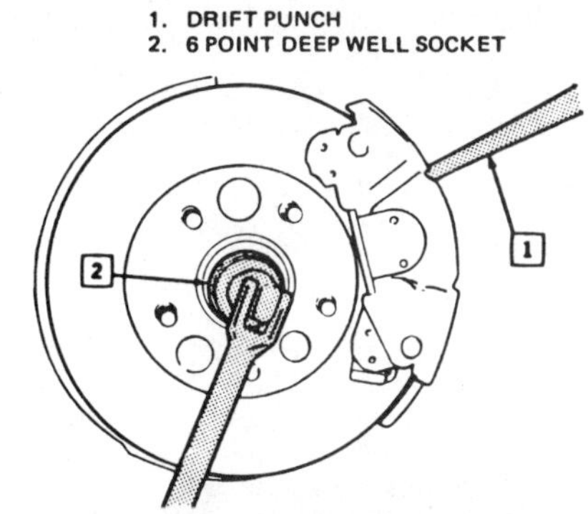

Fig. 10 Removing and installing axle shaft nut. Except Geo, Lemans, Sprint and Spectrum

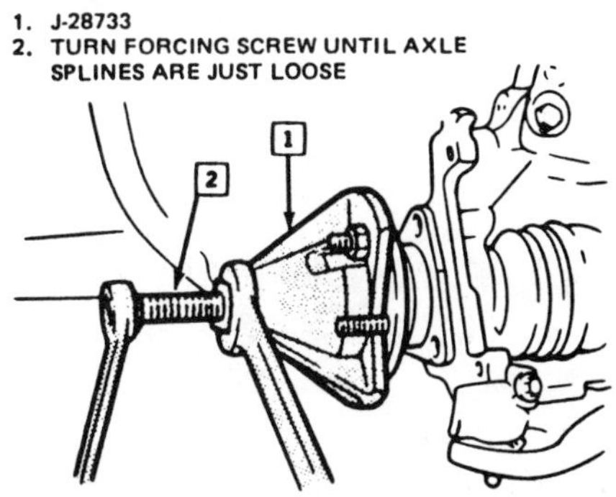

Fig. 11 Removing drive axle from hub and bearing assembly

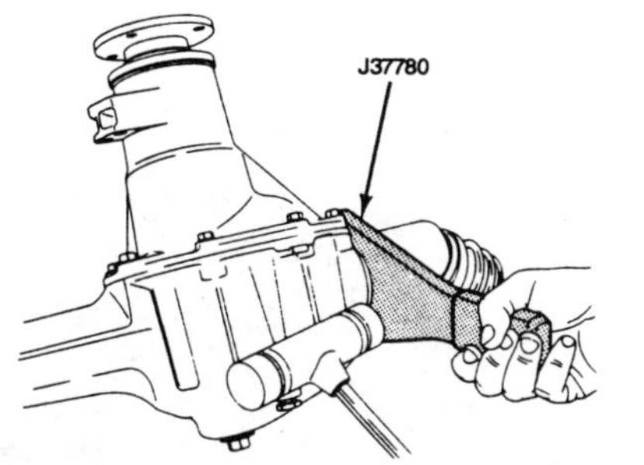

Fig. 12 Removing differential side joint snap ring. Geo

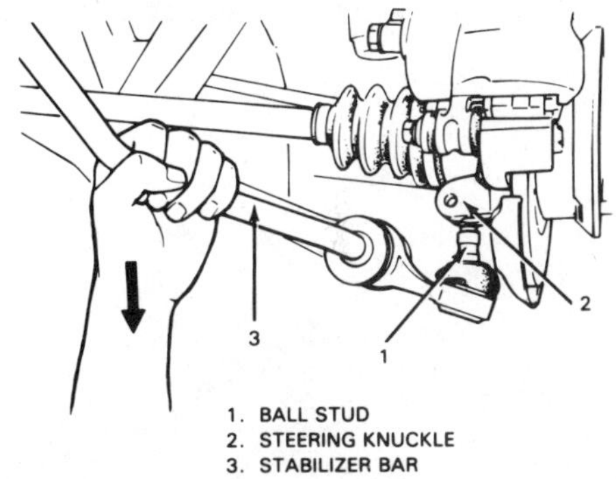

Fig. 13 Removing ball joint stud from steering knuckle. Sprint

support vehicle.

3. Remove wheel and tire assembly, then lower control arm to ball joint attaching nuts and bolts.
4. Using ball remover tool No. J-24319-01, remove tie rod end from steering knuckle.
5. Remove brake caliper support-to-steering knuckle attaching bolts, then the caliper. Support caliper to prevent brake hose damage.
6. Remove rotor, then using puller tool No. J-25287, push axle assembly from hub.
7. Using tool Nos. J-2619-01 and J-35762, pull axle assembly from transaxle. **When removing axle assembly, do not pull on axle shaft. To prevent axle boot damage, do not allow them to contact any parts during removal.**

Installation

Prior to installing axle assembly, inspect knuckle inner grease seal. Replace if distorted or worn. Also check for foreign material at hub bearing area.

1. Install axle assembly into transaxle. Use a long brass drift and hammer positioned on the tri-pot housing ribs to drive in.
2. Install axle into wheel hub, then lower control arm to ball joint. **Torque** ball joint nuts and bolts to 47 ft. lbs.
3. Install tie rod end. **Torque** nut to 38 ft. lbs.
4. Install rotor, then the brake caliper. **Torque** attaching bolts to 65 ft. lbs.
5. Install tire and wheel assembly, then drive axle nut and washer.
6. Lower vehicle to floor, then **torque** lug nuts to 76 ft. lbs. and drive axle nut to 137 ft. lbs.
7. Install nut cap and cotter pin.

DRIVE AXLE DISASSEMBLY

LEMANS

1. Secure axle shaft in a soft-jawed vise.
2. Cut off seal clamps and remove seals, then spread bearing retaining rings, **Fig. 14.**
3. Mark relationship between shaft and joints for assembly reference, then tap joints off shaft using a plastic mallet.
4. Inspect all components for excessive wear or damage and replace as necessary.

SPRINT

1. Remove boot band from inboard joint, **Fig. 9.**
2. Remove housing from inboard joint.
3. Remove snap ring, then spider from shaft.
4. Remove inside and outside boots from shaft. **Do not disassemble outboard joint. If joint is defective, it must be replaced as an assembly. Do not disassemble inboard joint spider as it is serviced as an assembly.**

SPECTRUM

Double Offset Joint

1. Secure axle assembly, **Fig. 8,** in a soft-jawed vise.
2. Remove boot band from differential side joint, **Fig. 15.**
3. Move boot slightly and remove circular clip, **Fig. 16.**
4. Separate case from shaft and remove 6 balls with a screwdriver, **Fig. 17.**
5. Turn ball guide on an angle and move it to center shaft side.
6. Remove snap ring from shaft, **Fig. 18,** then slide out ball guide and retainer.
7. Remove boot from shaft.

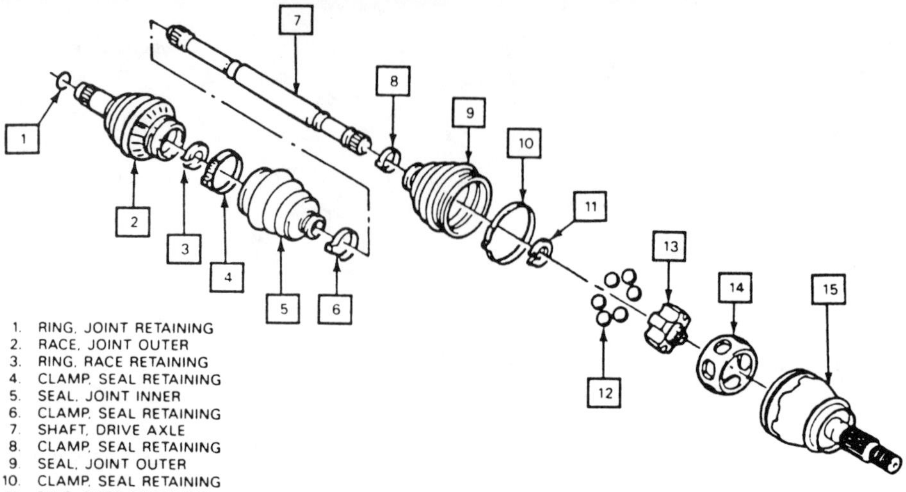

1. RING, JOINT RETAINING
2. RACE, JOINT OUTER
3. RING, RACE RETAINING
4. CLAMP, SEAL RETAINING
5. SEAL, JOINT INNER
6. CLAMP, SEAL RETAINING
7. SHAFT, DRIVE AXLE
8. CLAMP, SEAL RETAINING
9. SEAL, JOINT OUTER
10. CLAMP, SEAL RETAINING
11. RING, RACE RETAINING
12. BALLS, BEARING
13. RACE, INNER
14. CAGE, BEARING
15. RACE, JOINT OUTER

Fig. 14 Exploded view of front drive axle assembly. Lemans

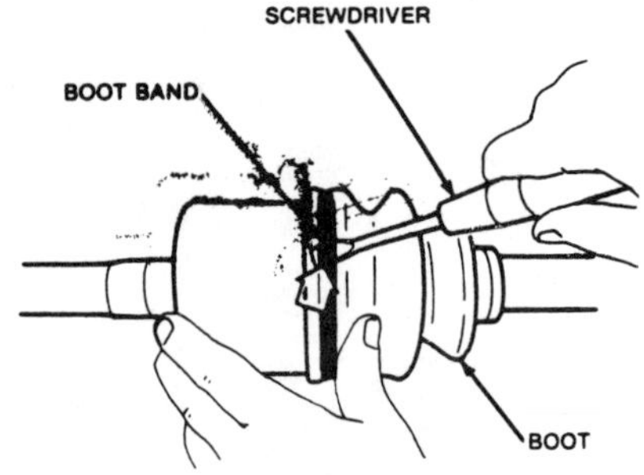

Fig. 15 Boot band removal. Spectrum

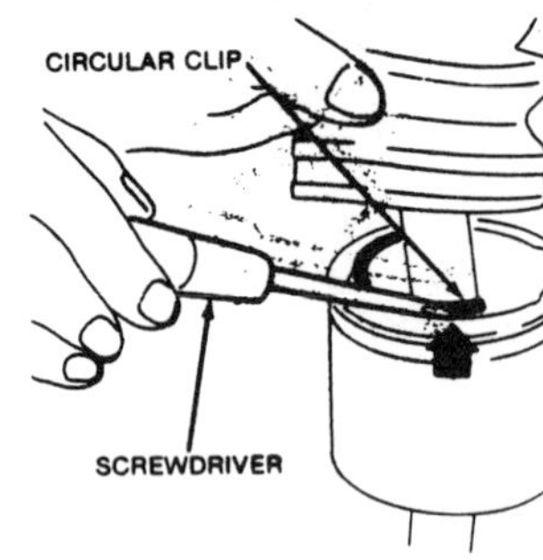

Fig. 16 Circular clip removal. Spectrum

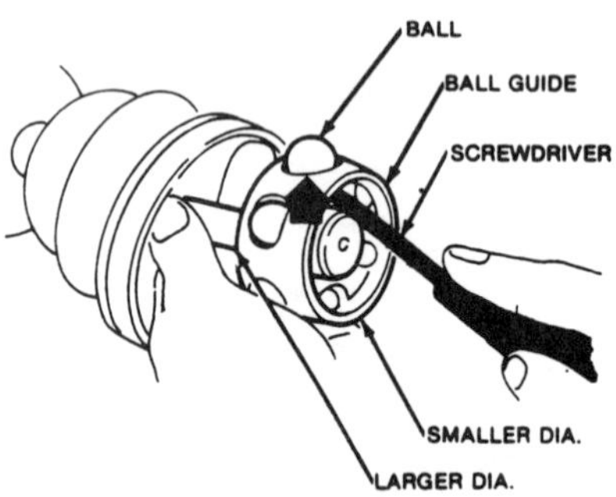

Fig. 17 Bearing removal. Spectrum

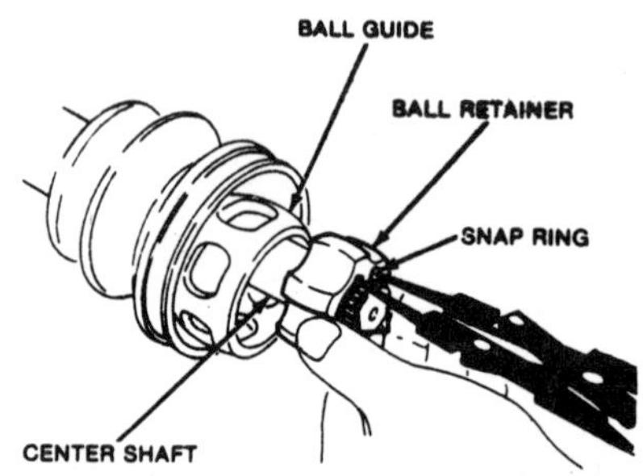

Fig. 18 Snap ring removal. Spectrum

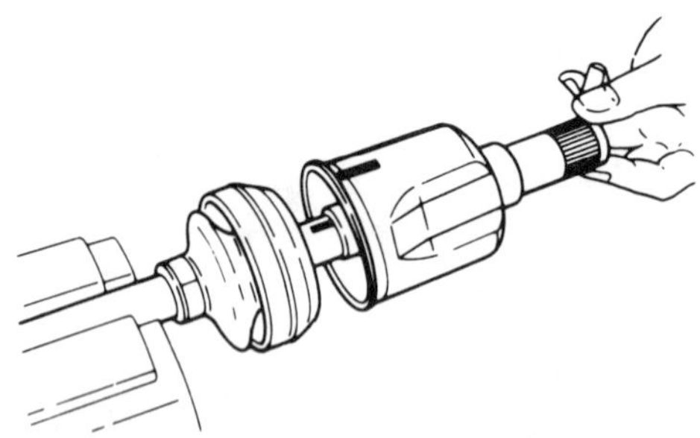

Fig. 19 Removing housing from shaft. Spectrum

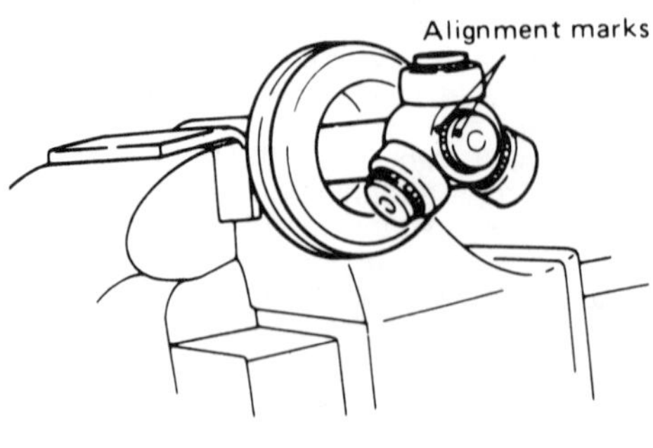

Fig. 20 Tripod removal. Spectrum

Birfield Joint

The Birfield joint must be serviced as an assembly only, however, then boot may be removed in the same manner as described for the Double Offset type joint.

Tripod Joint

1. Secure axle assembly, **Fig. 8**, in a soft-jawed vise.
2. Using a screwdriver, remove boot band from differential side joint.
3. Make alignment marks on both housing and shaft, then remove housing from shaft, **Fig. 19.**
4. Remove snap ring from shaft, then make alignment marks on both shaft and tripod, **Fig. 20.**
5. Using a hammer and brass drift, remove tripod from shaft. **Take care not to hit roller.**
6. Remove boot from shaft, and if necessary, dynamic damper.

GEO

Do not disassemble outboard joint or differential side ball joint. If any malfunction is detected in these components, replace as an assembly.

1. Remove boot band from differential side joint.
2. Remove circlip, then housing of differential side joint. **Fig. 21.**
3. Remove the snap ring, then the ball joint.
4. Remove the inside and outside boots from the shaft.

AXLE SHAFT ASSEMBLY

GEO

1. Pack outboard joint with grease using GM grease kit 7805942 or equivalent.
2. Install wheel side boot on shaft, then pack inside of boot with 90 grams of grease and fit boot bands.
3. Install differential side on shaft then pack with grease.
4. Install ball joint differential side ball joint on shaft, ensuring the joint's flush side faces the outboard joint.
5. Install snap ring in groove of shaft.
6. Pack differential side boot with grease, then install housing. Secure boot to housing with a boot band. Ensure boot band end is in reverse direction of rotation of driveshaft.

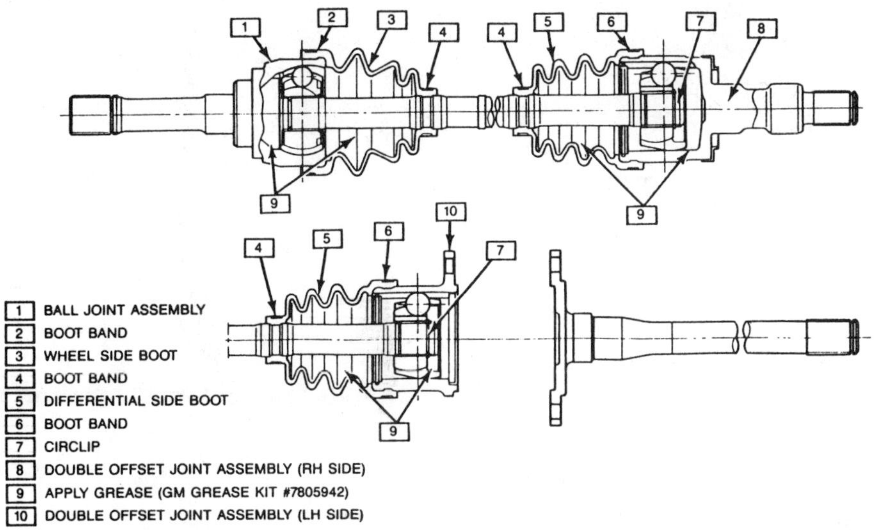

Fig. 21 Drive axle assembly. Geo

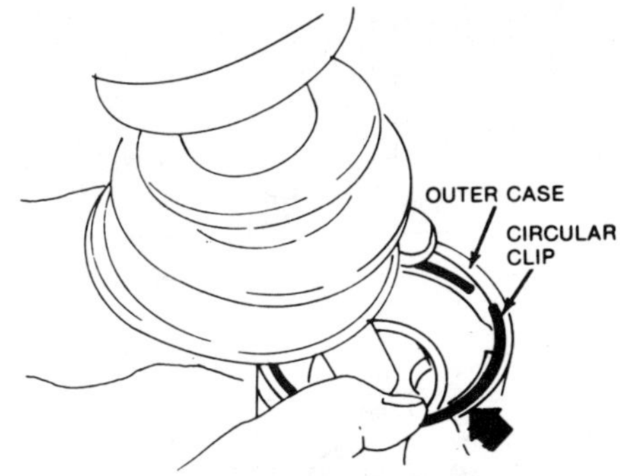

Fig. 22 Circular clip installation. Spectrum

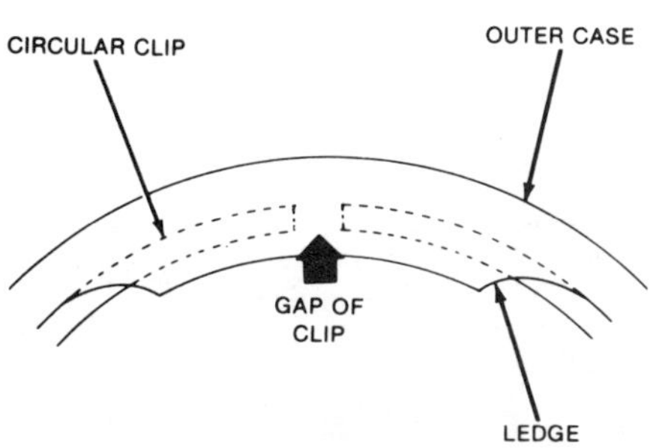

Fig. 23 Circular clip alignment. Spectrum

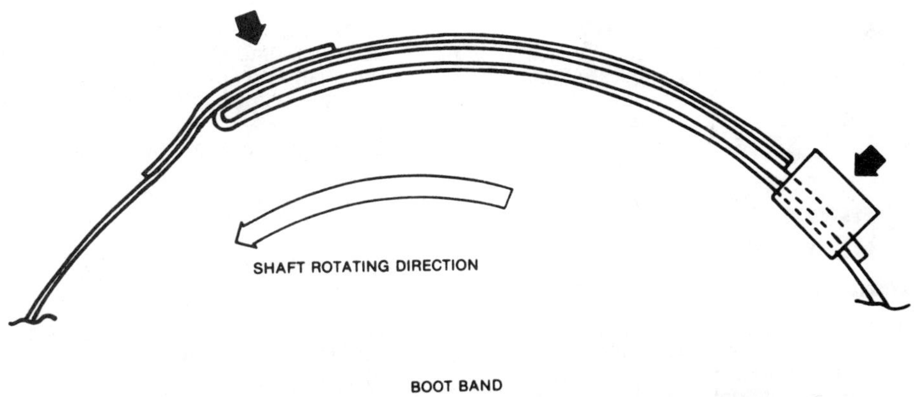

Fig. 24 Boot band installation (double offset joint). Spectrum

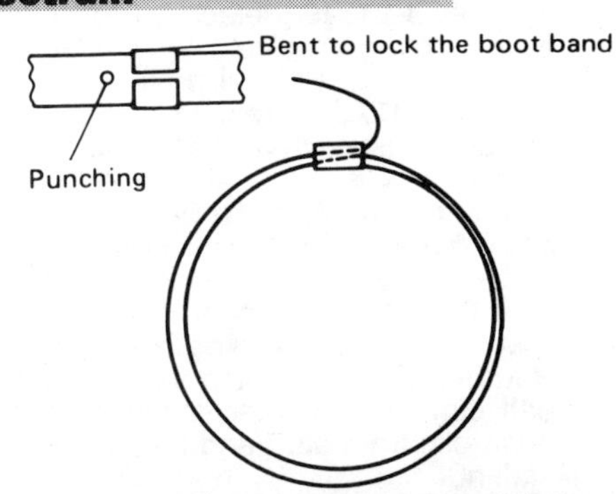

Fig. 25 Boot band installation (tripod joint). Spectrum

LEMANS

1. Pack joints with suitable grease, then tap joint assemblies into shaft splines until retaining rings are properly seated.
2. Install inner joint on long side of shaft and outer joint on short side of shaft.
3. Install joint seals and with new clamps. Secure clamps using seal clamp installer tool No. J-22610 or equivalent.

SPECTRUM

Double Offset Joint

1. Assemble in the reverse order of disassembly, noting the following:
   a. Use care to avoid damaging boot when installing boot on shaft.
   b. Install and align circular clip as shown, **Figs. 22 and 23.**
   c. To install boot band, first insert a screwdriver into end of boot, then install band as shown, **Fig. 24.** Ensure band end clip is securely tightened around the band and boot is not twisted.

Tripod Joint

Assemble in the reverse order of disassembly, noting the following:

1. Tape spline section of shaft with tape to prevent damage to boot when installing.
2. Install tripod onto shaft with shorter spline positioned on the outer side.
3. Ensure all alignment marks align before assembling.
4. Wind boot band twice, **Fig. 25,** then tighten band using pliers. Finally, lock band using a punch.

SPRINT

1. Liberally apply black grease included in outboard boot kit to outboard joint.
2. Install outboard boot on shaft. Fill inside of boot with about 2.8 oz. of joint grease, **Fig. 26.**
3. Install inboard boot on shaft.
4. Rally apply yellow joint grease included in inboard boot kit to inboard joint.
5. Install inboard spider on drive shaft with chamfered side toward outboard joint as shown in **Fig. 26.**
6. Install snap ring in shaft groove, **Fig. 27.**
7. Fill inside of inboard boot with about 4.5 oz. of joint grease and install housing, **Fig. 26.** Attach boot to housing with boot band.
8. Check boots for distortion and correct as necessary.

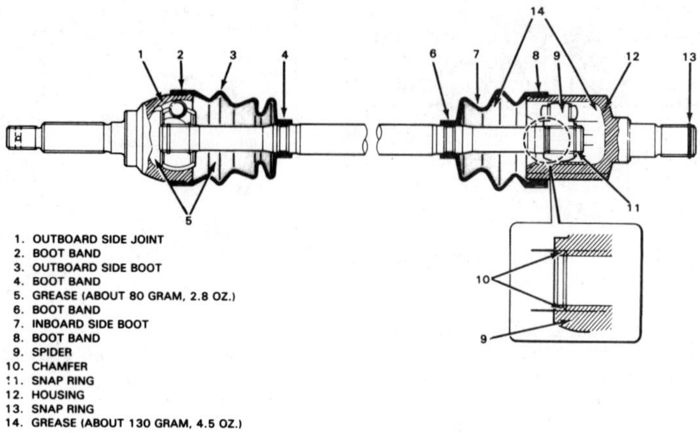

Fig. 26 Drive axle assembly. Sprint

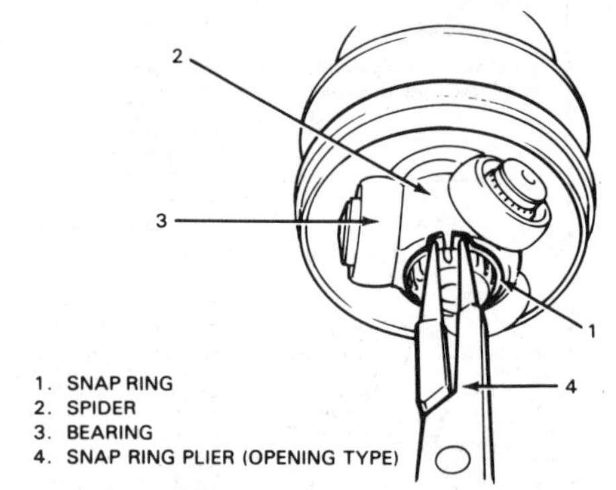

Fig. 27 Installing axle snap ring. Sprint

RIGHTHAND DRIVE AXLE, OUTPUT SHAFT & SEAL

REPLACE

1985 ELDORADO, RIVIERA, SEVILLÉ & TORONADO

Removal

1. Disconnect battery ground cable.
2. Raise front of vehicle and place jack stands under front frame horns.
3. Remove wheel and tire assembly.
4. Remove cotter pin, nut and shield from tie rod pivot, then using puller tool No. J-24319, detach tie rod end from steering knuckle.
5. Install drive shaft seal protector tool No. J-28712, then remove cotter pin, nut and washer from drive axle, then remove six screws attaching drive axle to output shaft. **To prevent axle shaft from rotating when removing nut or attaching screws, insert a drift through opening on top of caliper into corresponding rotor vane.**
6. Remove cotter pin and nut from upper ball joint stud nut, then remove brake hose clip from stud and loosely reinstall nut.
7. Using a hammer and brass drift, rap on steering knuckle to free upper ball joint stud. Use care not to damage brake hose or steering knuckle.
8. Remove nut and separate upper ball joint from steering knuckle.
9. Guide drive axle out of steering knuckle and remove from vehicle.
10. Remove two screws attaching battery cable retainer to support and remove two screws attaching output support to engine, then rotate support downward.
11. Remove front nut and bolt from righthand frame brace, then pivot brace outward to provide clearance.
12. Using a plastic mallet, drive on flange end of output shaft until shaft releases from retaining ring, then remove output shaft and support. Use care not to damage output shaft seal surfaces or splines.
13. Using a pry bar, pry output shaft seal out of housing. Pry at two or three different places to avoid cocking seal. Use care not to damage housing.

Installation

1. Using output shaft seal installer tool No. J-28518, install output shaft seal. Rotate tool to maintain proper alignment when installing seal.
2. Apply wheel bearing grease between lips of seal.
3. Index splines of output shaft with splines of side gear in final drive assembly, then install shaft by tapping on flange end with a soft faced mallet until retaining ring snaps into shaft groove. Ensure shaft is securely locked into position. **When installing output shaft, use care not to damage seal.**
4. Align output shaft support and engine block attaching screw holes, then install two support attaching screws. **Torque** screws to 50 ft. lbs.
5. Install two screws attaching battery cable retainer to support.
6. Guide drive axle into position and install splined end axle into steering knuckle.
7. Position upper ball joint stud into steering knuckle, then place brake hose clip on stud and install stud nut. **Torque** stud nut to 60 ft. lbs., then install cotter pin. The nut may be tightened an additional 1/6 turn to align cotter pin slots.
8. Install six screws attaching output shaft to drive axle. **Torque** screws to 60 ft. lbs.
9. Install drive axle washer, nut, retainer and cotter pin. **Torque** nut to 175 ft. lbs. Align cotter pin slot by rotating retainer and bend cotter pin so that retainer is held snugly.
10. Install tie rod pivot on steering knuckle. **Torque** nut to 44 ft. lbs. The nut may be tightened an additional 1/6 turn to align cotter pin slots.
11. Install righthand frame brace bolt and nut. **Torque** belt to 50 ft. lbs.
12. Install front wheel and tire assembly, then lower vehicle and connect battery ground cable.
13. Check output shaft seal for leakage.

LEFTHAND DRIVE AXLE, OUTPUT SHAFT & SEAL

REPLACE

1985 ELDORADO, RIVIERA, SEVILLÉ & TORONADO

Removal

1. Raise front of vehicle and place jack stands under front frame horns.
2. Remove wheel and tire assembly.
3. Remove cotter pin, nut and shield from tie rod pivot, then using puller tool No. J-24319, detach tie rod end from steering knuckle.
4. Remove cotter pin, nut and washer from drive axle, then remove six screws attaching drive axle to output shaft. **To prevent drive axle from rotating when removing nut or attaching screws, insert a drift through opening on top of caliper into corresponding rotor vane.**
5. Remove cotter pin and nut from upper ball joint stud nut, then remove brake hose clip from stud and loosely reinstall nut.
6. Using a hammer and brass drift, rap on steering knuckle to free upper ball joint stud. Use care not to damage brake hose or steering knuckle.
7. Remove nut and separate upper ball joint from steering knuckle.
8. Guide drive axle out of steering knuckle and remove from vehicle.
9. Remove front nut and bolt from lefthand frame brace, then pivot brace outward to provide clearance.
10. Using a hammer and brass drift, drive

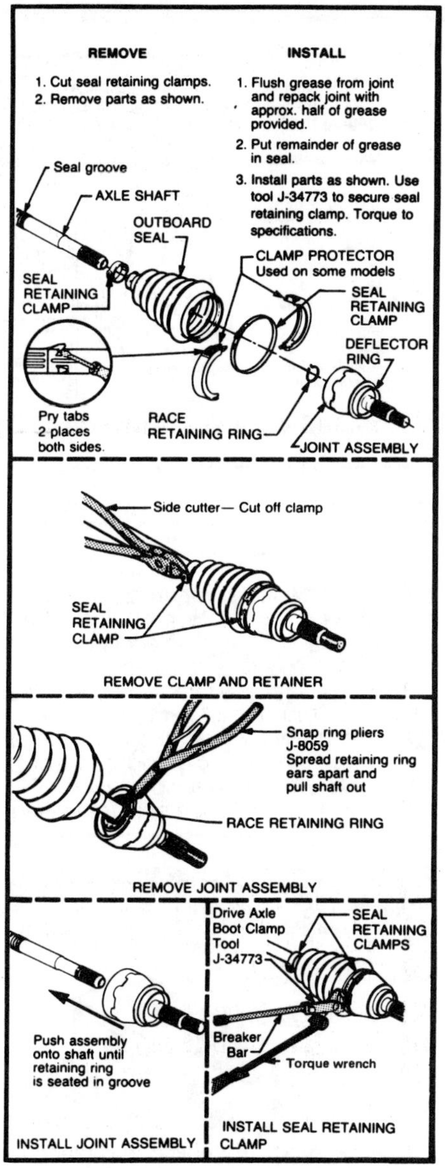

Fig. 28 Outer constant velocity joint seal removal & installation. Except Geo, Lemans, Spectrum and Sprint

on flange end of output shaft until shaft releases from retaining ring, then remove output shaft.

11. Using a pry bar, pry output shaft seal from housing. Pry at two or three locations to avoid cocking seal. Use care not to damage housing.

Installation

1. Using tool output shaft seal installer tool No. J-28518, install output shaft seal. Rotate tool to maintain proper alignment when installing seal.
2. Apply wheel bearing grease between lips of seal.
3. Index splines of output shaft with splines of side gear in final drive assembly then install shaft by tapping center of flange end with a soft faced mallet until retaining ring snaps into shaft groove. Ensure shaft is securely locked into position. **When installing output shaft, use care not to damage seal.**
4. Guide drive axle into position and install splined end into steering knuckle.
5. Position upper ball joint stud into steering knuckle, then place brake hose clip on stud and install nut. **Torque** nut to 60 ft. lbs., then install cotter pin. The nut may be tightened an additional 1/6 turn to align cotter pin slots.
6. Install six screws attaching output shaft to drive axle. **Torque** screws to 60 ft. lbs.
7. Install drive axle water nut, retainer and cotter pin. **Torque** nut to 175 ft. lbs. and bend cotter pin so that retainer is held snugly.
8. Install tie rod pivot on steering knuckle. **Torque** nut to 44 ft. lbs. The nut may be tightened an additional 1/6 turn to align cotter pin slots.
9. Install lefthand frame brace bolt and nut. **Torque** bolt to 50 ft. lbs.
10. Install wheel and tire assembly, then lower vehicle and check output shaft seal for leakage.

RIGHTHAND OUTPUT SHAFT SUPPORT BEARING

REPLACE

1985 ELDORADO, RIVIERA, SEVILLE & TORONADO

1. Remove righthand output shaft as described under Righthand Drive Axle, Output Shaft and Seal, Replace.
2. Remove three screws securing bearing retainer to support.
3. Install bearing remover tool No. J-22912 between flange end of output shaft and flat area of shaft support, with flat surface of tool against flat area of shaft support. Position assembly on a press and press shaft support, bearing, retainer and slinger from output shaft.
4. Remove bearing from output shaft support.
5. Lubricate output shaft support and bearing, then position bearing into support.
6. Pack bearing with wheel bearing grease, then install retainer and three attaching screws.
7. Place assembled components and slinger on output shaft, then position components and output shaft on a press. Using a standard 1 1/4 inch inside diameter pipe, press bearing and assembled components onto shaft.
8. Check to ensure bearing and support rotate smoothly, then install righthand output shaft.

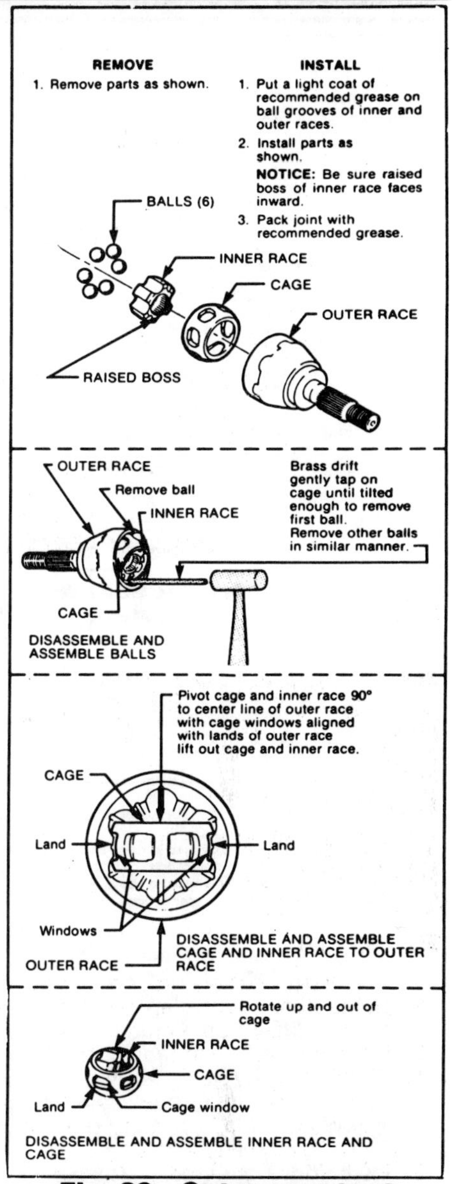

Fig. 29 Outer constant velocity joint disassembly & assembly. Except Geo, Lemans, Spectrum and Sprint.

OUTER CONSTANT VELOCITY JOINT & SEAL

REPLACE

EXCEPT GEO, LEMANS, SPECTRUM AND SPRINT

For removal and installation procedures refer to **Figs. 28, and 29.**

GEO, LEMANS, SPECTRUM AND SPRINT

For replacement procedure, refer to "Drive Axle Disassembly."

INNER TRI-POT SEAL

REPLACE

1985 ELDORADO, RIVIERA, SEVILLE & TORONADO

For removal and installation procedures

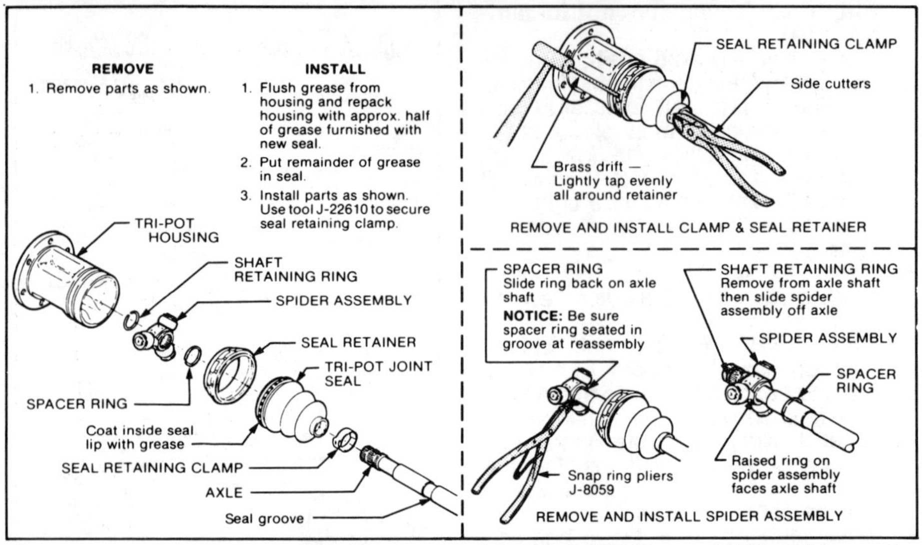

Fig. 30 Inner tri-pot seal removal & installation. 1985 Eldorado, Riviera, Seville & Toronado

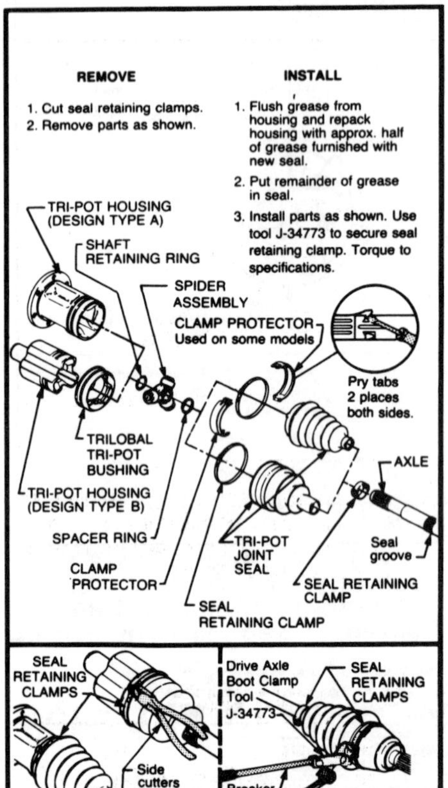

Fig. 31 Inner tri-pot seal removal & installation. Except Geo, Lemans, Spectrum and Sprint

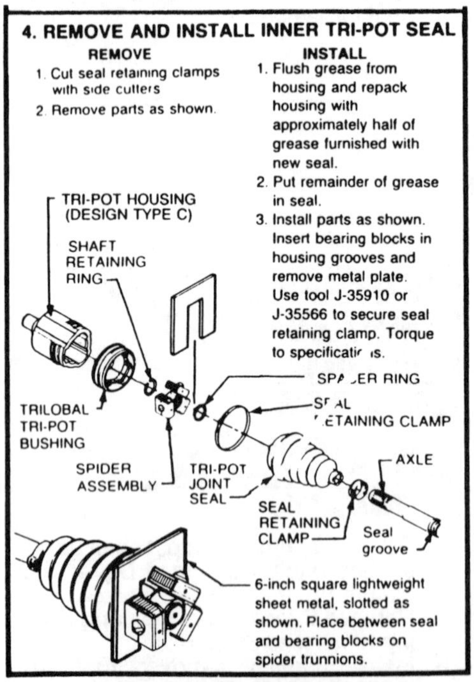

Fig. 32 Inner tri-pot seal removal & installation. Models w/bearing blocks except Geo, Lemans, Spectrum and Sprint (Part 1 of 2)

refer to Fig. 30.

EXCEPT GEO, LEMANS, SPECTRUM, SPRINT & 1985 ELDORADO, RIVIERA, SEVILLE & TORONADO

For removal and installation procedures, refer to **Figs. 31** and **32.** On 1986-88 models, when installing the thermo-plastic seal, the drive axle must be collapsed as shown in **Figs. 33 and 34, and 35** to prevent distortion of the seal.

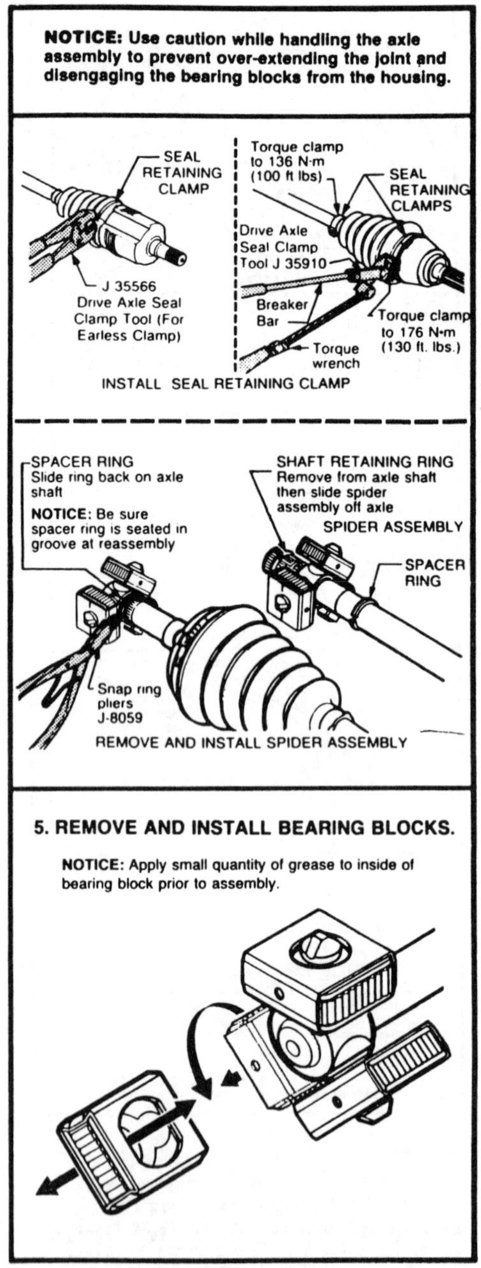

Fig. 32 Inner tri-pot seal removal & installation. Models w/bearing blocks except Geo, Lemans, Spectrum and Sprint (Part 2 of 2)

GEO, LEMANS, SPECTRUM AND SPRINT

For replacement procedure, refer to "Drive Axle Disassembly."

INNER CONSTANT VELOCITY JOINT & SEAL REPLACE

CROSS GROOVE DESIGN FRONT AXLE

For removal and installation procedures refer to **Fig. 36.**

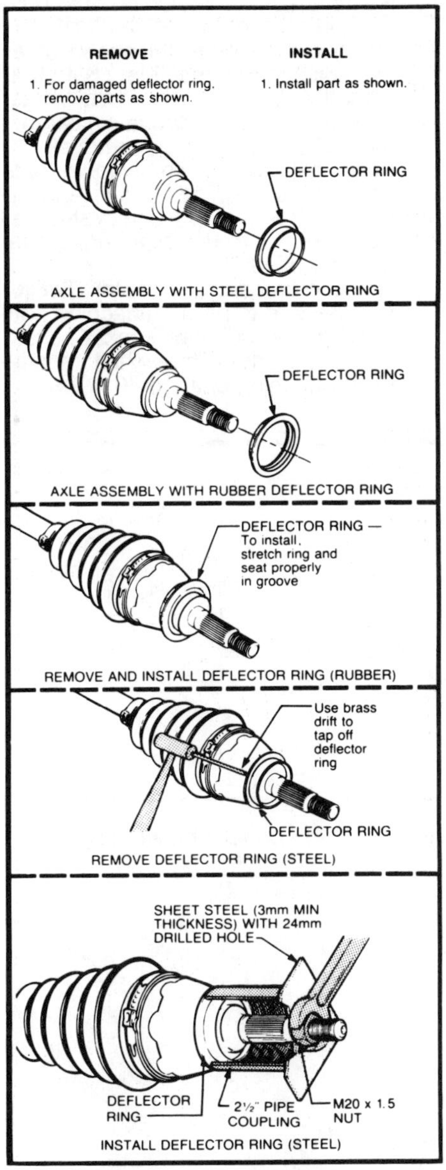

Fig. 33 Inboard thermo-plastic seal installation. Models except Geo, Lemans, Spectrum and Sprint

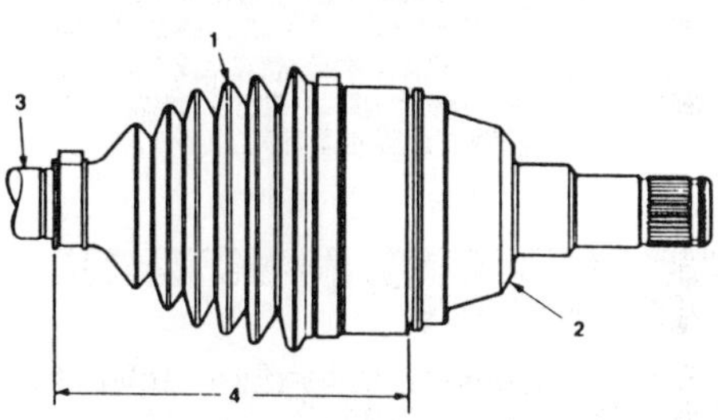

1. INBOARD THERMOPLASTIC SEAL
2. CROSS-GROOVE JOINT
3. AXLE SHAFT
4. 133 mm (5¼") – JOINT AND SEAL ARE TO BE COMPRESSED TO THIS DIMENSION BEFORE CRIMPING CLAMPS.

Fig. 34 Inboard thermo-plastic seal collapsed dimension. Models w/cross-groove joint except Geo, Lemans, Spectrum and Sprint

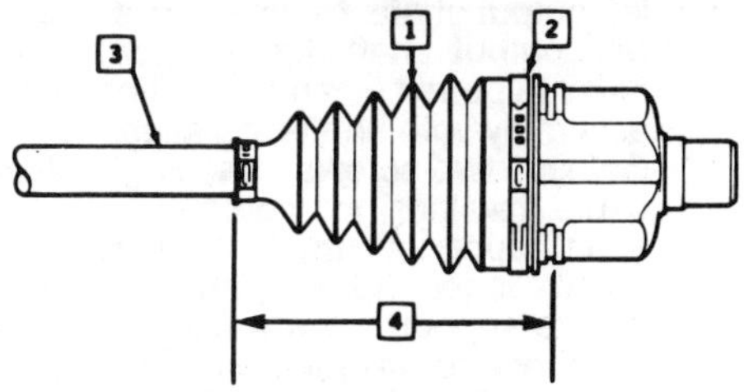

1. INBOARD THERMOPLASTIC SEAL
2. TRILOBAL TRIPOT JOINT
3. AXLE SHAFT
4. 130 MM (5 1/16") – JOINT AND SEAL ARE TO BE COMPRESSED TO THIS DIMENSION BEFORE CRIMPING CLAMPS.

Fig. 35 Inboard thermo-plastic seal collapsed dimension. Models w/trilobal trishaft joint except Geo, Lemans, Spectrum and Sprint.

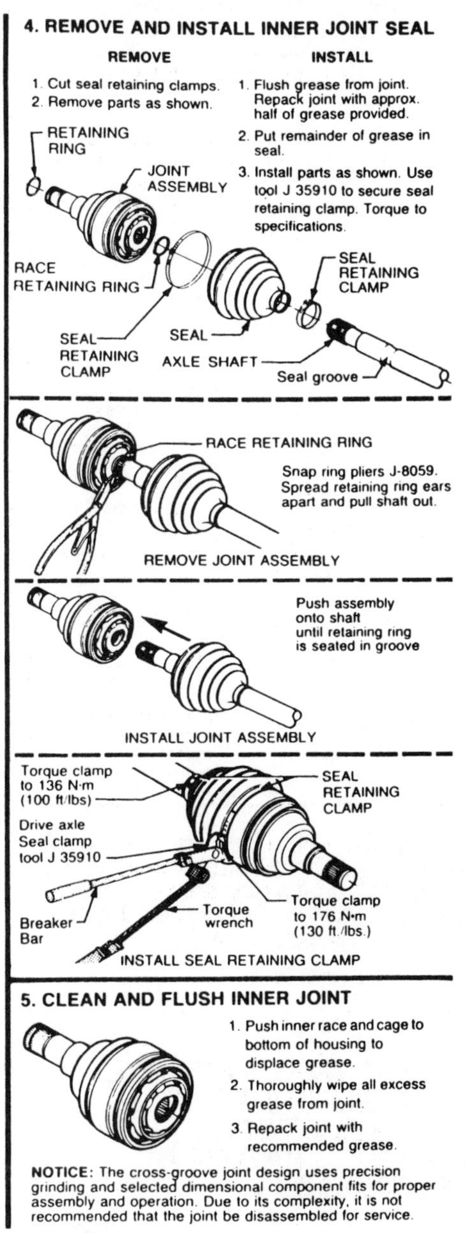

Fig. 36 Cross groove inner joint seal removal and installation. Models w/cross groove inner joint except Geo, Lemans, Spectrum and Sprint.

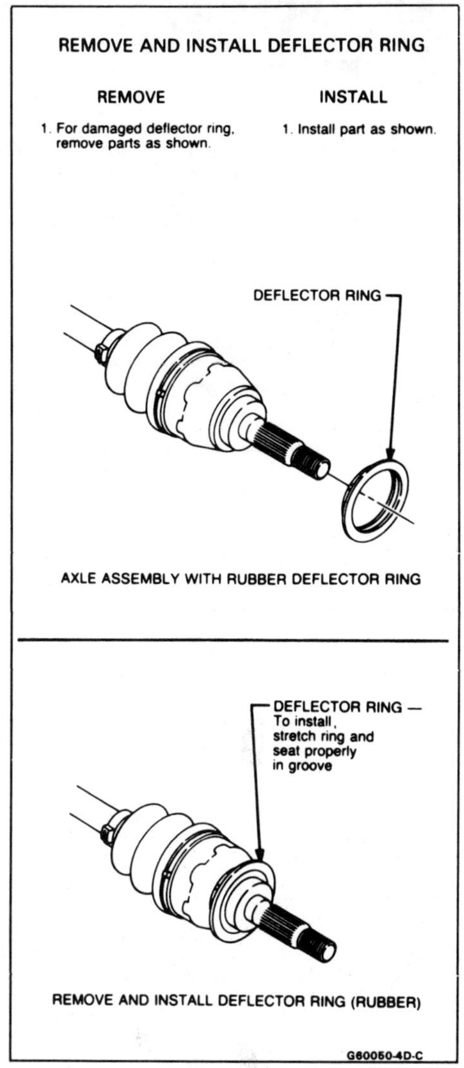

Fig. 37 Deflector ring removal & installation. Models except Geo, Lemans, Nova, Spectrum and Sprint

DEFLECTOR RING REPLACE

MODELS EXCEPT GEO, LEMANS, NOVA, SPECTRUM & SPRINT

For removal and installation procedures, refer to **Fig. 37.**

FINAL DRIVE REPLACE

EXCEPT 1985 ELDORADO, RIVIERA, SEVILLE & TORONADO

On these models, the final drive unit is an integral component of the transaxle assembly.

1985 ELDORADO, RIVIERA, SEVILLE & TORONADO

1. Disconnect battery ground cable.
2. Raise front of vehicle and position jack stands under front frame horns.
3. Remove righthand and lefthand frame brace front attaching bolts, then position braces to provide clearance.
4. Position drain pan under final drive cover, then loosen cover and allow lubricant to drain.
5. Remove final drive cover attaching screws, cover and gasket.
6. Remove screws attaching right and left output shafts to drive axles. Separate output shaft flanges from drive axles to permit clearance of final drive assembly with shafts installed.
7. Remove two screws attaching battery cable retainer to righthand output shaft support, then remove two screws securing support to engine. Rotate support downward to provide clearance for removal.
8. Remove final drive to transmission attaching screws that secures rear of final drive shield, then loosen support bracket screws that secure front of shield and remove shield.
9. Remove five remaining final drive to transmission attaching screws.
10. Remove screws attaching final drive support bracket to engine block.
11. Using tool J-24319, disconnect steering linkage intermediate shaft from pitman and idler arms. Push linkage forward to ensure adequate clearance for final drive assembly removal.
12. Slide final drive assembly forward, off transmission splined shaft and remove unit with output shafts attached. Use care not to damage output shaft seal or splines.
13. Reverse procedure to install. **Torque** final drive to transaxle attaching bolts to 30 ft. lbs., support bracket bolts to 50 ft. lbs. and output shaft to drive shaft attaching bolts to 60 ft. lbs.

DISC BRAKES

INDEX

CHRYSLER MOTORS/EAGLE

TABLE OF CONTENTS

Application

CHRYSLER

	Type No.
Conquest:	
Front:	
1987-89	10
Rear:	
1987-89	11
Executive:	
1985	2
1985	5
Fifth Avenue/Newport:	
1985-89	1
Laser:	
1985-86	2
1985-86	5
LeBaron:	
Front:	
1985-88	2
1985-88	5
1989 Except Turbo	5
1989 Turbo	6
Rear:	
1989	14
LeBaron GTS:	
1985-89	5
1987-88	2
New Yorker:	
Front:	
1985-88	2
1987-88	5
1988-89	6
Rear:	
1988-89	4
Town & Country:	
1985-87	2
1987-88	5

DODGE

	Type No.
Aries:	
1985-88	2
1987-89	5
1989	6
Charger:	
1985-87	5
Colt:	
Front:	
1985-89 w/1.6L Engine	9
1985-89 w/1.5L Engine	10
Rear:	
1989	13
Colt Vista:	
1985-89	9
Conquest:	
Front:	
1985-86	10
Rear:	
1985-86	11
Daytona:	
Front:	
1985-88	2
1985-88	5
1989 Except Turbo	5
1989 Turbo	6
Rear:	
1987-88	3
1989	14
Diplomat:	
1985-89	1
Dynasty:	
Front:	
1988	2
1988	5
1988-89	6
Rear:	
1988-89	4
Lancer:	
1985-89	5
1987	2
Omni:	
1985-86:	
Except GLH	5
GLH	2

APPLICATIONS—Continued

DODGE—Continued

Type No.

1987-89............................5

Shadow:

Front:

1987-88..........................2

1987-89..........................5

1989..............................6

Rear w/14 inch wheels:

1989..............................14

Shelby Charger:

1985-87............................2

600:

1985-88............................2

1985-88............................5

Spirit:

1989..............................5

EAGLE

Eagle 4 X 4:

1985-88............................7

Eagle Medallion:

1987-89............................12

Type No.

Eagle Premier:

1988-89............................8

Eagle Summit:

Front:

1989 w/1.6L Engine...............9

1989 w/1.5L Engine...............10

Rear:

1989..............................13

PLYMOUTH

Acclaim:

1989..............................5

Caravelle:

1985-87............................2

1987-88............................5

Colt:

1985-89 w/1.5L Engine.............10

1985-89 w/1.6L Engine.............9

Conquest:

Front:

1985-86..........................10

Type No.

Rear:

1985-86..........................11

Grand Fury:

1985-89............................1

Horizon:

1985-89............................5

Reliant:

1985-88............................2

1985-89............................5

1988-89............................6

Sundance:

Front:

1987-88..........................2

1987-89..........................5

1988-89..........................6

Rear w/14 inch wheels:

1989..............................14

Turismo:

1985-86............................5

Troubleshooting

This section covers general troubleshooting procedures for both disc and drum brake systems. Refer to **Figs. 1 through 6** for these procedures.

Symptom	Probable cause	Remedy
Noise and vibration when applying brakes	Caliper improperly mounted	Correct
	Loose caliper mounting bolts	Retighten
	Unevenly worn or cracked brake disc	Replace
	Worn pad friction material	Replace
	Excessive caliper to pad assembly clearance	Correct
	Uneven pad contact	Correct
	Lack of lubrication in sliding parts	Lubricate
	Loose suspension parts	Retighten
Vehicle pulls to one side when applying brakes	Difference in left and right tire inflation pressure	Adjust
	Inadequate contact of pad	Correct
	Grease or oil on pad surface	Replace
	Auto adjuster malfunction	Correct
Insufficient braking force	Low or deteriorated brake fluid	Replenish or change
	Air in brake system	Bleed the system
	Brake booster malfunction	Correct
	Inadequate contact of pads	Correct
	Grease or oil on pad surface	Replace
	Auto adjuster malfunction	Correct
	Overheated brake rotor due to dragging of pad	Correct
	Clogged brake line	Correct
	Proportioning valve or combination valve malfunction	Replace
Increased pedal stroke (Reduced pedal to floor clearance)	Air in brake system	Bleed the system
	Brake fluid leaks	Correct
	Auto adjuster malfunction	Correct
	Excessive push rod to master cylinder clearance	Adjust

Fig. 1 Brake system troubleshooting (Part 1 of 2). Conquest

Symptom	Probable cause	Remedy
Brake drag	Incomplete release of parking brake	Correct
	Incorrect parking brake adjustment	Adjust
	Worn brake pedal return spring	Replace
	Clogged master cylinder return port	Correct
	Lack of lubrication in sliding parts	Lubricate
	Defective master cylinder check valve or piston return spring	Replace
	Excessive push rod to master cylinder clearance	Adjust
Insufficient parking brake function	Worn brake pad Grease or oil on pad surface Parking brake cable sticking	Replace
	Auto adjuster malfunction	Correct
	Excessive parking brake lever stroke	Adjust the parking brake lever stroke or check the parking brake cable arrangement

Fig. 1 Brake system troubleshooting (Part 2 of 2). Conquest

Symptom	Probable cause	Remedy
Vehicle pulls to one side when brakes are applied	Grease or oil on pad or lining surface	Replace
	Inadequate contact of pad or lining	Correct
	Auto adjuster malfunction	Adjust
	Drum eccentricity or uneven wear	Repair or replace as necessary
Improper braking power	Low or deteriorated brake fluid	Replenish or change
	Air in brake system	Bleed air
	Overheated brake rotor due to dragging of pad or lining	Correct
	Grease or oil on pad surface	Replace
	Inadequate contact of pad or lining	Correct
	Brake booster malfunction	Replace
	Auto adjuster malfunction	Adjust
	Clogged brake line	Correct
	Proportioning valve malfunction	Replace
Increased pedal stroke (Reduced pedal to firewall clearance)	Air in brake system	Bleed air
	Worn lining or pad	Replace
	Broken vacuum hose	Replace
	Brake fluid leaks	Correct
	Auto adjuster malfunction	Adjust
	Excessive push rod to master cylinder clearance	Adjust
	Faulty master cylinder	Replace
Brake drag	Incomplete release of parking brake	Correct
	Incorrect parking brake adjustment	Adjust
	Worn brake pedal return spring	Replace
	Broken rear drum brake shoe return spring	Replace
	Lack of lubrication in sliding parts	Lubricate
	Improper push rod to master cylinder clearance	Adjust
	Faulty master cylinder piston return spring	Replace
	Clogged master cylinder return port	Correct

Fig. 2 Brake system troubleshooting (Part 1 of 3). Colt series vehicles including 1989 Colt & Eagle Summit

Symptom	Probable cause	Remedy
Improper parking brake function	Worn brake lining	Replace
	Excessive parking brake lever stroke	Adjust the parking brake lever stroke or check the parking brake cable routing
	Grease or oil on lining surface	Replace
	Auto adjuster malfunction	Adjust
	Parking brake cable sticking	Replace
	Sticked wheel cylinder or caliper piston	Replace
Scraping or grinding noise when brakes are applied	Worn brake linings	Replace
	Caliper to wheel interference	Correct or replace
	Dust cover to drum interference	Correct or replace
	Bent brake backing plate	Correct or replace
	Cracked drums or brake disc	Correct or replace
Squealing, groaning or chattering noise when brakes are applied	Disc brakes-missing or damaged disc pad outer shim	Replace
	Brake drums and lining, discs and pads worn or scored	Correct or replace
	Improper lining parts	Correct or replace
	Disc brakes-burred or rusted calipers	Clean or deburr
	Dirty, greased, contaminated or glazed linings	Clean or replace
	Drums brakes-weak damaged or incorrect shoe hold-down springs, loose or damaged shoe hold-down pins and springs	Correct or replace
	Incorrect adjustment of brake pedal or booster push rod	Adjust
Squealing noise when brakes are not applied	Bent or warped backing plate causing interference with drum	Replace
	Improper machining of drum causing interference with backing plate or shoe	Replace drum
	Disc brakes-rusted, stuck	Lubricate or replace
	Worn, damaged or insufficiently lubricated wheel bearings Drum brakes-weak, damaged or incorrect shoe return spring	Lubricate or replace
	Loose or extra parts in brakes	Retighten
	Improper positioning of pads in caliper	Correct
	Improper installation of support mounting and caliper body	Correct
	Poor return of brake booster or master cylinder or wheel cylinder	Replace
	Incorrect adjustment of brake pedal or booster push rod	Adjust

Fig. 2 Brake system troubleshooting (Part 2 of 3). Colt series vehicles including 1989 Colt & Eagle Summit

Symptom	Probable cause	Remedy
Groaning, clicking or rattling noise when brakes are not applied	Stones or foreign material trapped inside wheel covers	Remove stones, etc.
	Loose wheel nuts	Retighten
	Disc brakes-failure of shim	Replace
	Disc brakes-loose installation bolt	Retighten
	Worn, damaged or dry when bearings	Lubricate or replace
	Incorrect adjustment of brake pedal or booster push rod	Adjust
Poor parking brake function	Worn brake lining Poor condition of brake lining surface Parking brake cable sticking	Replace
	Auto adjuster malfuncion	–
	Excessive parking brake lever stroke	Adjust the parking brake lever stroke or check the parking brake cable arrangement

Fig. 2 Brake system troubleshooting (Part 3 of 3). Colt series vehicles including 1989 Colt & Eagle Summit

Use Guides A and B to identify a problem(s).

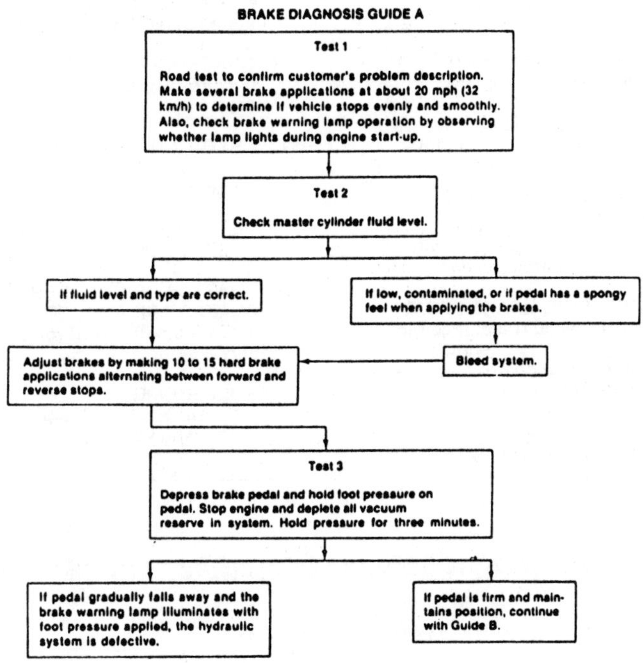

Fig. 3 Brake system troubleshooting (Part 1 of 16). Eagle 4 X 4

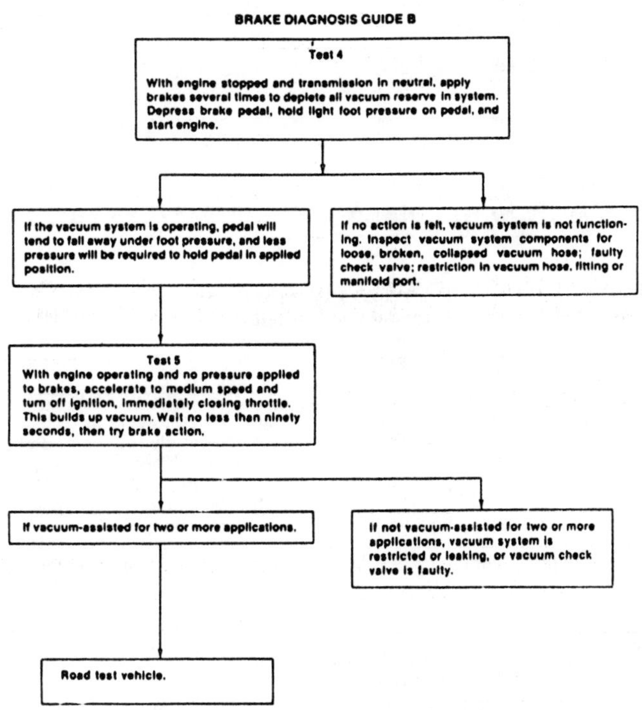

Fig. 3 Brake system troubleshooting (Part 2 of 16). Eagle 4 X 4

The diagnosis involves the following malfunctions:

A. LOW BRAKE PEDAL (Excessive pedal travel required for braking action.)

B. LOW BRAKE PEDAL (Pedal may go to floor with steady pressure applied.)

C. LOW BRAKE PEDAL (Pedal goes to floor on first application — o.k. on subsequent applications.)

D. FADING BRAKE PEDAL (Pedal height decreases with steady pressure applied.)

E. DECREASING BRAKE PEDAL TRAVEL (Pedal travel required for braking action decreases and may be accompanied by a hard pedal.)

F. SPONGY BRAKE PEDAL (Pedal has abnormally soft, springy, spongy feel when depressed.)

G. HARD BRAKE PEDAL (Excessive pedal pressure required to stop vehicle. May be accompanied by brake fade.)

H. GRABBING BRAKES (Severe reaction to brake pedal pressure.)

I. DRAGGING BRAKES (Slow or incomplete release of brakes.)

J. VEHICLE MOVES TO ONE SIDE WHEN BRAKES ARE APPLIED

K CHATTER OR SHUDDER WHEN BRAKES ARE APPLIED (Pedal pulsation and roughness may also occur.)

L. NOISY BRAKES (Squealing, clicking, scraping sound when brakes are applied.)

M. PULSATING BRAKE PEDAL

Fig. 3 Brake system troubleshooting (Part 3 of 16). Eagle 4 X 4

A. LOW BRAKE PEDAL (Excessive pedal travel required for braking action.)

POSSIBLE CAUSES	CORRECTIONS
(1) Excessive clearance between rear linings and drums caused by inoperative automatic adjusters.	(1) Make 10 to 15 alternate forward and reverse brake stops to adjust brakes. If brake pedal does not come up, repair or replace adjuster parts as necessary.
(2) Worn rear brakelining.	(2) Inspect and replace lining if worn beyond minimum thickness specification.
(3) Bent, distorted brakeshoes, front or rear.	(3) Replace brakeshoes in axle sets.
(4) Air in hydraulic system.	(4) Remove air from system

Fig. 3 Brake system troubleshooting (Part 4 of 16). Eagle 4 X 4

B. LOW BRAKE PEDAL (Pedal may go to floor with steady pressure applied.)

POSSIBLE CAUSES	CORRECTIONS
(1) Fluid leak in hydraulic system.	(1) Fill master cylinder to fill line; have helper apply brakes and check calipers, wheel cylinders, differential valve tubes, hoses and fittings for leaks. Repair or replace as necessary.
(2) Air in hydraulic system.	(2) Remove air from system.
(3) Incorrect or non-recommended brake fluid (fluid evaporates at below normal temp).	(3) Flush hydraulic system with clean brake fluid. Refill with correct-type fluid.
(4) Master cylinder piston seals worn, or master cylinder bore is scored, worn or corroded.	(4) Repair or replace master cylinder.

Fig. 3 Brake system troubleshooting (Part 5 of 16). Eagle 4 X 4

C. LOW BRAKE PEDAL (Pedal goes to floor on first application — o.k. on subsequent applications.)

POSSIBLE CAUSES	CORRECTIONS
(1) Disc brake pads sticking on abutment surfaces of anchor plate. Caused by a build-up of dirt, rust, or corrosion on abutment surfaces.	(1) Clean abutment surfaces.

Fig. 3 Brake system troubleshooting (Part 6 of 16). Eagle 4 X 4

D. FADING BRAKE PEDAL (Pedal height decreases with steady pressure applied.)

POSSIBLE CAUSES	CORRECTIONS
(1) Fluid leak in hydraulic system.	(1) Fill master cylinder reservoirs to fill mark, have helper apply brakes, check calipers, wheel cylinders, differential valve, tubes, hoses, and fittings for fluid leaks. Repair or replace parts as necessary.
(2) Master cylinder piston seals worn, or master cylinder bore is scored, worn or corroded.	(2) Repair or replace master cylinder.

Fig. 3 Brake system troubleshooting (Part 7 of 16). Eagle 4 X 4

E. DECREASING BRAKE PEDAL TRAVEL (Pedal travel required for braking action decreases and may be accompanied by a hard pedal.)

POSSIBLE CAUSES	CORRECTIONS
(1) Caliper or wheel cylinder pistons sticking or seized.	(1) Repair or replace the calipers, or wheel cylinders.
(2) Master cylinder compensator ports blocked (preventing fluid return to reservoirs) or pistons sticking or seized in master cylinder bore.	(2) Repair or replace the master cylinder.
(3) Power brake unit binding internally.	(3) Test unit according to the following procedure. (a) Shift transmission into neutral and start engine. (b) Increase engine speed to 1500 rpm, close throttle and fully depress brake pedal. (c) Slowly release brake pedal and stop engine. (d) Have helper remove vacuum check valve and hose from power unit. Observe for backward movement of brake pedal. (e) If the pedal moves backward, the power unit has an internal bind — replace power unit.

Fig. 3 Brake system troubleshooting (Part 8 of 16). Eagle 4 X 4

F. SPONGY BRAKE PEDAL (Pedal has abnormally soft, springy, spongy feel when depressed.)

POSSIBLE CAUSES	CORRECTIONS
(1) Air in hydraulic system.	(1) Remove air from system.
(2) Brakeshoes bent or distorted.	(2) Replace brakeshoes.
(3) Brakelining not yet seated with drums and rotors.	(3) Burnish brakes.
(4) Rear drum brakes not properly adjusted.	(4) Adjust brakes

Fig. 3 Brake system troubleshooting (Part 9 of 16). Eagle 4 X 4

G. HARD BRAKE PEDAL (Excessive pedal pressure required to stop vehicle. May be accompanied by brake fade.)

POSSIBLE CAUSES	CORRECTIONS
(1) Loose or leaking power brake unit vacuum hose.	(1) Tighten connections or replace leaking hose.
(2) Incorrect or poor quality brakelining.	(2) Replace with lining in axle sets.
(3) Bent, broken, distorted brakeshoes.	(3) Replace brakeshoes.
(4) Calipers binding or dragging on mounting pins. Rear brakeshoes dragging on support plate.	(4) Replace mounting pins and bushings. Clean rust or burrs from rear brake support plate ledges and lubricate ledges with molydisulfide grease. **NOTE:** If ledges are deeply grooved or scored, do not attempt to sand or grind them smooth — replace support plate.
(5) Caliper, wheel cylinder, or master cylinder pistons sticking or seized.	(5) Repair or replace parts as necessary.
(6) Power brake unit vacuum check valve malfunction.	(6) Test valve according to the following procedure. (a) Start engine, increase engine speed to 1500 rpm, close throttle and immediately stop engine. (b) Wait at least 90 seconds then depress brake pedal. (c) If brakes are not vacuum assisted for 2 or more applications, check valve is faulty.
(7) Power brake unit has internal bind.	(7) Test unit according to the following procedure. (a) With engine stopped, apply brakes several times to exhaust all vacuum in system. (b) Shift transmission into neutral, depress brake pedal and start engine. (c) If pedal height decreases with foot pressure and less pressure is required to hold pedal in applied position, power unit vacuum system is operating normally. Test power unit as outlined in item (3) within E, Decreasing Brake Pedal Travel. If power unit exhibits a bind condition, replace the power unit.
(8) Master cylinder compensator ports (at bottom of reservoirs) blocked by dirt, scale, rust, or have small burrs (blocked ports prevent fluid return to reservoirs).	(8) Repair or replace master cylinder. **CAUTION:** Do not attempt to clean blocked ports with wire, pencils, or similar implements. Use compressed air only
(9) Brake hoses, tubes, fittings clogged or restricted.	(9) Use compressed air to check or unclog parts. Replace any damaged parts.
(10) Brake fluid contaminated with improper fluids (motor oil, transmission fluid, causing rubber components to swell and stick in bores.	(10) Replace all rubber components, combination valve and hoses. Flush entire brake system with DOT 3 brake fluid or equivalent.
(11) Low engine vacuum.	(11) Adjust or repair engine.

Fig. 3 Brake system troubleshooting (Part 10 of 16). Eagle 4 X 4

H. GRABBING BRAKES (Severe reaction to brake pedal pressure.)

POSSIBLE CAUSES	CORRECTIONS
(1) Brakelining(s) contaminated by grease or brake fluid.	(1) Determine and correct cause of contamination and replace brakeshoes in axle sets.
(2) Parking brake cables incorrectly adjusted or seized.	(2) Adjust cables. Replace seized cables.
(3) Incorrect brakelining or lining loose on brakeshoes.	(3) Replace brakeshoes in axle sets.
(4) Caliper anchor plate bolts loose.	(4) Tighten bolts.
(5) Rear brakeshoes binding on support plate ledges.	(5) Clean and lubricate ledges. Replace support plate(s) if ledges are deeply grooved. Do not attempt to smooth ledges by grinding.
(6) Incorrect or missing power brake reaction disc.	(6) Install correct disc.
(7) Rear brake support plates loose.	(7) Tighten mounting bolts.

Fig. 3 Brake system troubleshooting (Part 11 of 16). Eagle 4 X 4

I. DRAGGING BRAKES (Slow or incomplete release of brakes)

POSSIBLE CAUSES	CORRECTIONS
(1) Brake pedal binding at pivot.	(1) Loosen and lubricate.
(2) Power brake unit has internal bind.	(2) Inspect for internal bind as outlined in item (3) within E, Decreasing Brake Pedal Travel. Replace unit if internal bind exists.
(3) Parking brake cables incorrectly adjusted or seized.	(3) Adjust cables. Replace seized cables.
(4) Rear brakeshoe return springs weak or broken.	(4) Replace return springs. Replace brakeshoe if necessary in axle sets.
(5) Automatic adjusters malfunctioning.	(5) Repair or replace adjuster parts as required.
(6) Caliper, wheel cylinder or master cylinder pistons sticking or seized.	(6) Repair or replace parts as necessary.
(7) Master cylinder compensating ports blocked (fluid does not return to reservoirs).	(7) Use compressed air to clear ports. Do not use wire, pencils, or similar objects to open blocked ports.

Fig. 3 Brake system troubleshooting (Part 12 of 16). Eagle 4 X 4

J. VEHICLE MOVES TO ONE SIDE WHEN BRAKES ARE APPLIED

POSSIBLE CAUSES	CORRECTIONS
(1) Incorrect front tire pressure.	(1) Inflate to recommended cold (reduced load) inflation pressure.
(2) Worn or damaged wheel bearings.	(2) Replace worn or damaged bearings.
(3) Brakelining on one side contaminated.	(3) Determine and correct cause of contamination and replace brakelining in axle sets.
(4) Brakeshoes on one side bent, distorted, or lining loose on shoe.	(4) Replace brakeshoes in axle sets.
(5) Support plate bent or loose on one side.	(5) Tighten or replace support plate.
(6) Brakelining not yet seated with drums or rotors.	(6) Burnish brakelining.
(7) Caliper anchor plate loose on one side.	(7) Tighten anchor plate bolts.
(8) Caliper piston sticking or seized.	(8) Repair or replace caliper.
(9) Brakelinings water soaked.	(9) Drive vehicle with brakes lightly applied to dry linings.
(10) Loose suspension component attaching or mounting bolts.	(10) Tighten suspension bolts. Replace worn suspension components.
(11) Brake combination valve failure.	(11) Replace combination valve.

Fig. 3 Brake system troubleshooting (Part 13 of 16). Eagle 4 X 4

K. CHATTER OR SHUDDER WHEN BRAKES ARE APPLIED (Pedal pulsation and roughness may also occur.)

POSSIBLE CAUSES	CORRECTIONS
(1) Brakeshoes distorted, bent, contaminated, or worn.	(1) Replace brakeshoes in axle sets.
(2) Caliper anchor plate or support plate loose.	(2) Tighten mounting bolts.
(3) Excessive thickness variation of rotor(s).	(3) Refinish or replace rotors in axle sets.

Fig. 3 Brake system troubleshooting (Part 14 of 16). Eagle 4 X 4

L. NOISY BRAKES (Squealing, clicking, scraping sound when brakes are applied.)

POSSIBLE CAUSES	CORRECTIONS
(1) Bent, broken, distorted brakeshoes.	(1) Replace brakeshoes in axle sets.
(2) Excessive rust on outer edge of rotor braking surface.	(2) Remove rust.
(3) Brakelining worn out — shoes contacting drum or rotor.	(3) Replace brakeshoes and lining in axle sets. Refinish or replace drums or rotors.
(4) Broken or loose holdown or return springs.	(4) Replace parts as necessary.
(5) Rough or dry drum brake support plate ledges.	(5) Lubricate support plate ledges.
(6) Cracked, grooved, or scored rotor(s) or drum(s).	(6) Replace rotor(s) or drum(s). Replace brakeshoes and lining in axle sets if necessary.
(7) Incorrect brakelining and/or shoes (front or rear).	(7) Install specified shoe and lining assemblies.

Fig. 3 Brake system troubleshooting (Part 15 of 16). Eagle 4 X 4

M. PULSATING BRAKE PEDAL

POSSIBLE CAUSE	CORRECTION
(1) Out of round drums or excessive lateral runout in disc brake rotor(s).	(1) Refinish or replace drums, re-index rotors or replace.

Fig. 3 Brake system troubleshooting (Part 16 of 16). Eagle 4 X 4

Incident	Possible Causes
"Hard pedal" Great effort needed and only a slight deceleration noticed in vehicle.	- Power servo defective or insufficient vacuum - Linings/pads - oily - glazed, not to specification - overheating (due to excessive braking) or not to specification - seized piston - pinched brake line - pads/linings worn
"Soft pedal" Note: Since the power servo system on current vehicles is very effective, the impression may be given that the pedal is "soft." To find out whether an incident has occurred or the braking system is operating normally, two checks must be performed. 1. Vehicle moving — check relation between pedal travel and start of vehicle deceleration. 2. Vehicle stationary with ignition off — depress the brake pedal 5 times to empty the brake servo, before assessing the result of the test.	- Air in system: poor bleeding - Internal leakage in braking system - External leak in braking system (lack of fluid in reservoir).
"Long" pedal travel Test to be performed with vehicle stationary and ignition off. Note: The brake pedal must be depressed 5 times in order to empty the brake servo before taking account of the test results.	- Incorrect shoe adjustment Drum brakes - Manual adjustments: shoes too far from drum surface. Disc and drum brakes - Handbrake cable too tight. Note: Automatic wear take-up is performed by the brake pedal, provided the handbrake is not abnormally tight in the off position. - Excessively worn pads/linings of pads/linings not symmetrical (or crossed) - Excessive master cylinder operating clearance. - Brake fluid boiling or has heated up.
Pedal "travels to floor" Test to be performed on stationary vehicle with ignition off. Note: The brake pedal must be depressed 5 times in order to empty the brake servo before taking account of the test results.	- Fluid loss (check for leaks) - Faulty sealing cup between the two master cylinder circuits. - Brake fluid boiling.

Fig. 4 Brake system troubleshooting (Part 1 of 2). Eagle Medallion & Premier

Incident	Possible Causes
Brakes binding or grabbing	- Linings not chamfered - Oil or grease on linings - Caliper seized - Weak return springs
Brakes pulsating	- Oval drums - Excessive disc run-out - Discs not of even thickness - Abnormal deposit on discs (corrosion between lining and disc). - Linings cracked or broken
Front brakes pulling to one side	- Check front end alignment - Check front axle, suspension and steering - Piston seized* - Tires (worn - incorrect inflation pressure or different tire size) - Pinched brake-line - Proportioning valve defective *ATTENTION: Pulling to one side indicates a seized piston on the opposite side.
Rear brakes pulling to one side	- Proportioning valve defective - Piston seized - Incorrect shoe adjustment: shoe too far from drum - Automatic adjustment: handbrake cable too tight. Note: Automatic wear take-up is performed by the brake pedal provided the handbrake cable is not abnormally tight when in off position. - Return spring defective
Overheating brakes	- Master cylinder operating clearance insufficient to allow master cylinder to return to neutral position. - Piston seized or not returning properly - Pinched brake line - Handbrake mechanism seized - Incorrect handbrake adjustment - Incorrect brake fluid

Fig. 4 Brake system troubleshooting (Part 2 of 2). Eagle Medallion & Premier

BASIC DIAGNOSIS GUIDE

SYMPTOM	CHART 1 MISC. COND.	CHART 2 ACTUATION	CHART 3 POWER BRAKES	CHART 4 BRAKE NOISE	CHART 5 WHEEL BRAKES
BRAKE WARNING LIGHT "ON"		X	NO		
EXCESSIVE PEDAL TRAVEL		X	NO		O
PEDAL GOES TO FLOOR	6O	X			
STOP LIGHT "ON" WITHOUT BRAKES	3X				
ALL BRAKES DRAG	5X	O			
REAR BRAKES DRAG	2X	NO	NO		
GRABBY BRAKES			O		X
SPONGY BRAKE PEDAL		X	NO		
PREMATURE REAR LOCKUP	4O	NO	NO		O
EXCESSIVE PEDAL EFFORT	1X		O		
ROUGH ENGINE IDLE			O		
BRAKE CHATTER (ROUGH)		NO	NO		X
SURGE DURING BRAKING					X
NOISE DURING BRAKING				X	
RATTLE OR CLUNKING NOISE		NO		X	
PEDAL PULSATES DURING BRAKING		NO	NO		X
PULL TO RIGHT OR LEFT		NO	NO		X

NO: NOT POSSIBLE CAUSE X: MOST LIKELY CAUSE O: POSSIBLE CAUSE

Fig. 5 Brake system troubleshooting (Part 1 of 6). F.W.D Chrysler Corp. vehicles

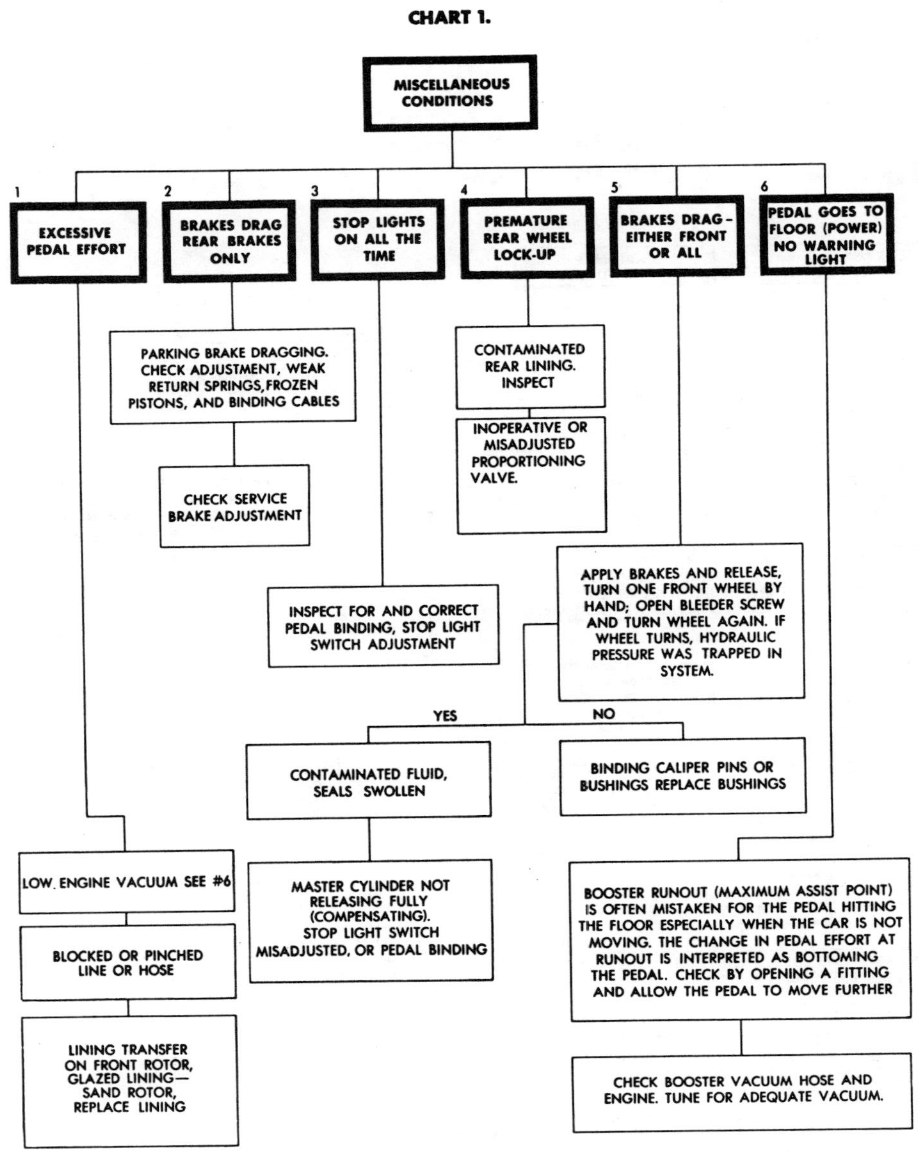

Fig. 5 Brake system troubleshooting (Part 2 of 6). F.W.D Chrysler Corp. vehicles

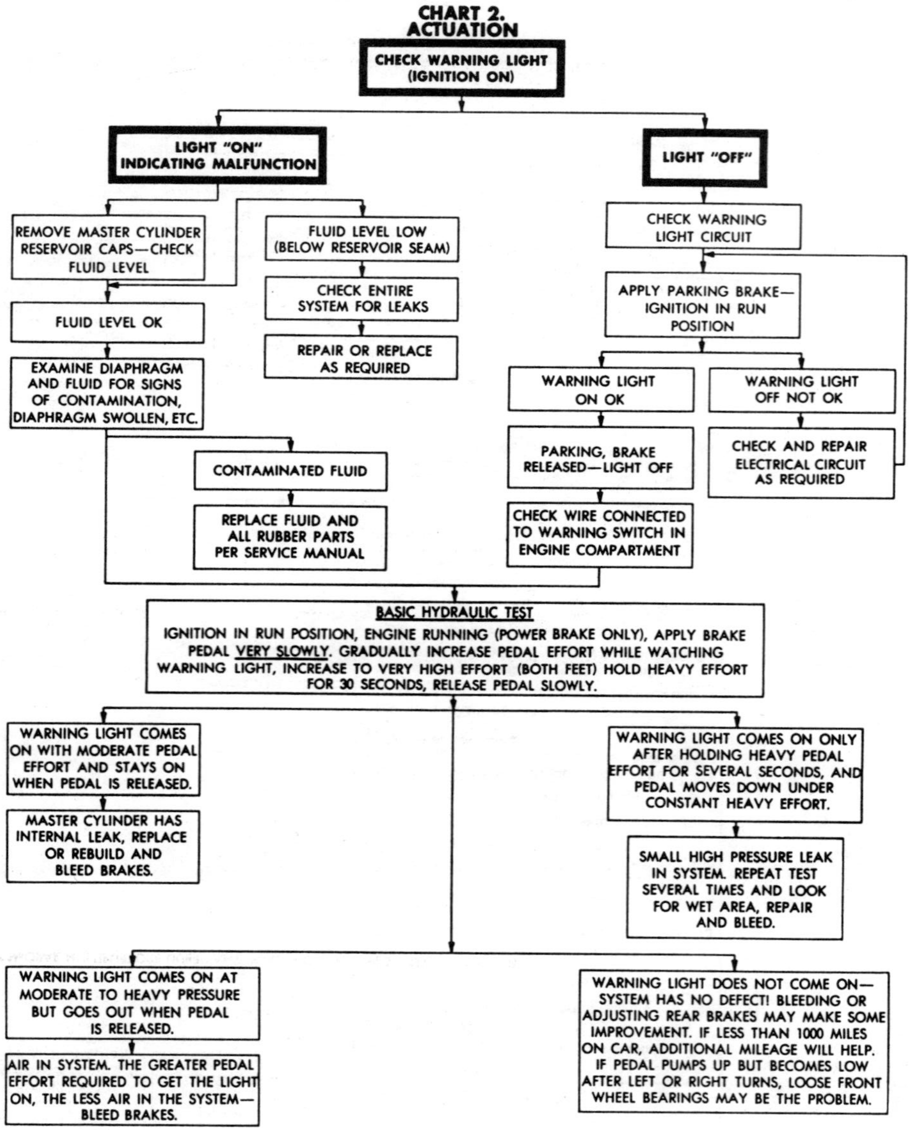

Fig. 5 Brake system troubleshooting (Part 3 of 6). F.W.D Chrysler Corp. vehicles

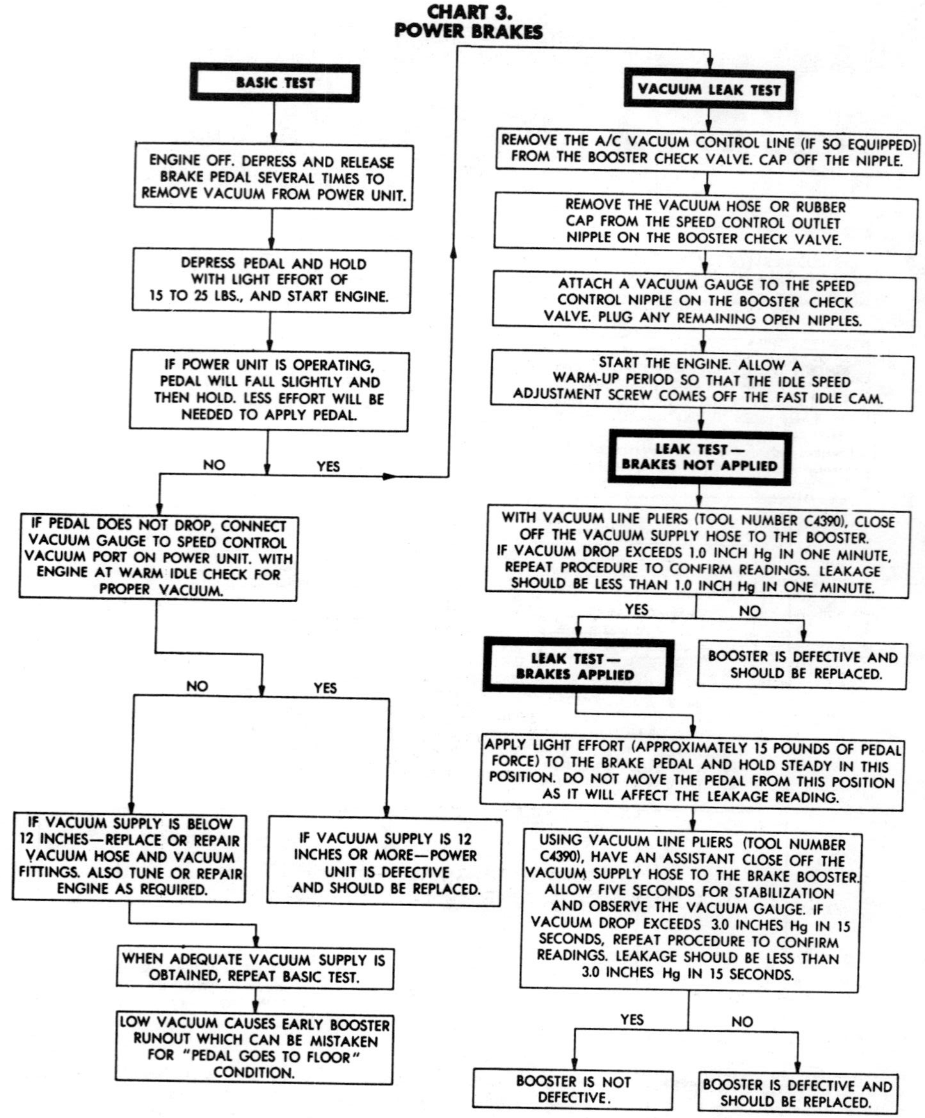

Fig. 5 Brake system troubleshooting (Part 4 of 6). F.W.D Chrysler Corp. vehicles

CHART 4. BRAKE NOISE

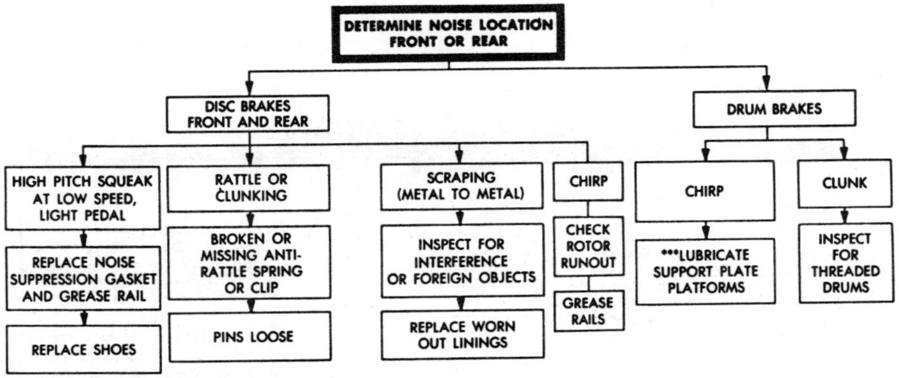

***MOPAR MULTIPURPOSE LIUBRICANT, PART NUMBER 4318062, OR EQUIVALENT.

Fig. 5 Brake system troubleshooting (Part 5 of 6). F.W.D Chrysler Corp. vehicles

CHART 5. WHEEL BRAKES

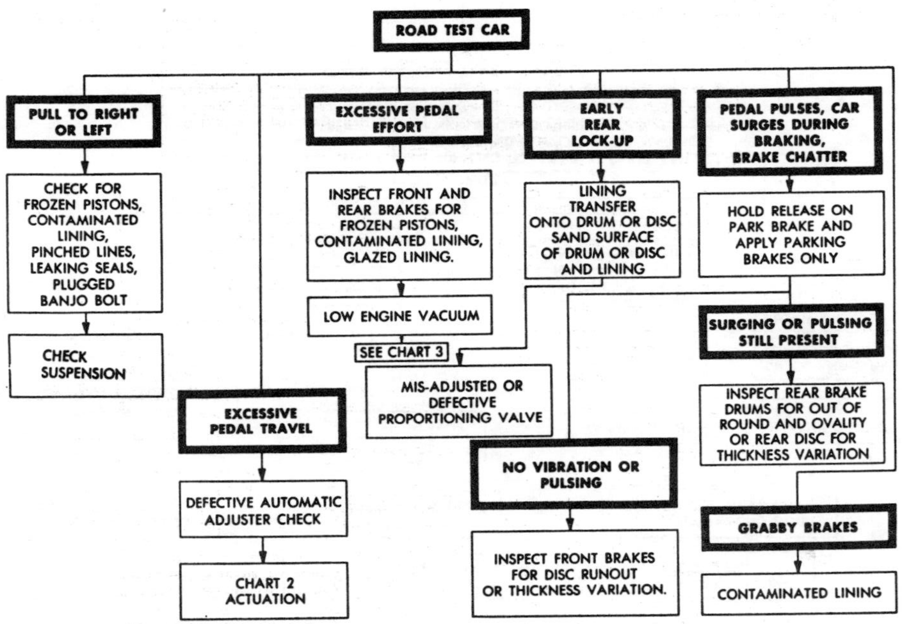

Fig. 5 Brake system troubleshooting (Part 6 of 6). F.W.D Chrysler Corp. vehicles

BASIC DIAGNOSIS GUIDE

SYMPTOM	CHART 1 MISC. COND.	CHART 2 ACTUATION	CHART 3 POWER	CHART 4 NOISE	CHART 5 WHEEL BRAKES	CHART 6 PULL
BRAKE WARNING LIGHT "ON"		X	NO		O	
EXCESSIVE PEDAL TRAVEL-SPONGY PEDAL		X	NO		O	
PEDAL GOES TO FLOOR	6	X	O			
STOP LIGHT "ON" ALL THE TIME	3		O			
BRAKES DRAG-FRONT OR ALL	5	X	O			
REAR BRAKES DRAG	2		NO			
GRABBY BRAKES					X	
EXCESSIVE PEDAL EFFORT	1		X		O	
ROUGH ENGINE IDLE			O			
BRAKE CHATTER (ROUGH)		NO	NO		X	
SURGE DURING BRAKING		NO	NO		X	
NOISE DURING BRAKING		NO		X		
PEDAL PULSES DURING BRAKING		NO	NO		X	
RATTLE OR CLUNKING NOISE		NO		X		
PREMATURE REAR WHEEL LOCKUP	4	NO	NO			
PULL TO RIGHT OR LEFT		NO	NO		X	X

X-MOST LIKELY CAUSE O-POSSIBLE CAUSE NO-NOT POSSIBLE CAUSE

Fig. 6 Brake system troubleshooting (Part 1 of 7). R.W.D Chrysler Corp. vehicles

BRAKE CHART 1 DIAGNOSIS

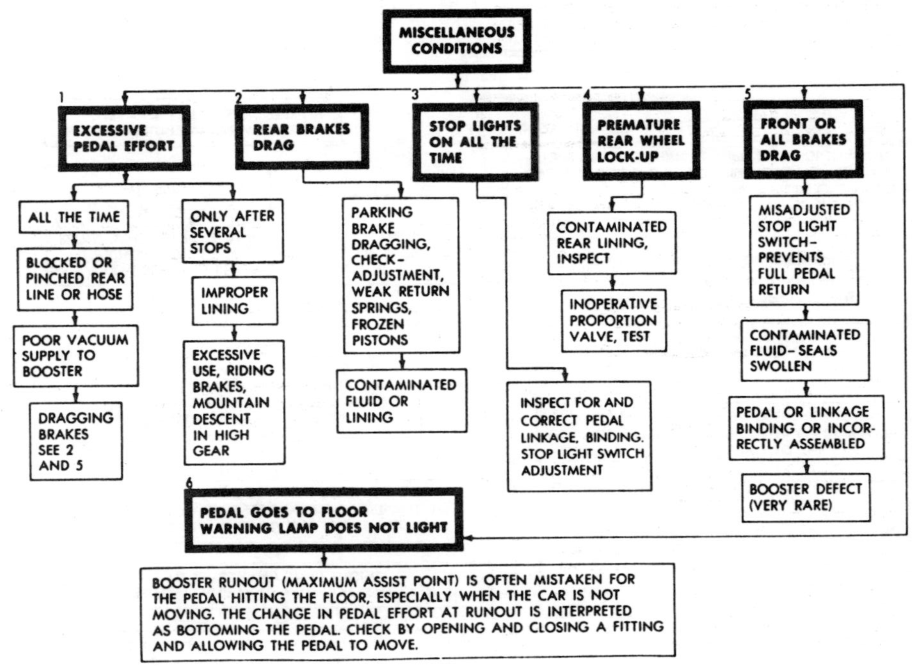

Fig. 6 Brake system troubleshooting (Part 2 of 7). R.W.D Chrysler Corp. vehicles

BRAKE CHART 2 ACTUATION DIAGNOSIS

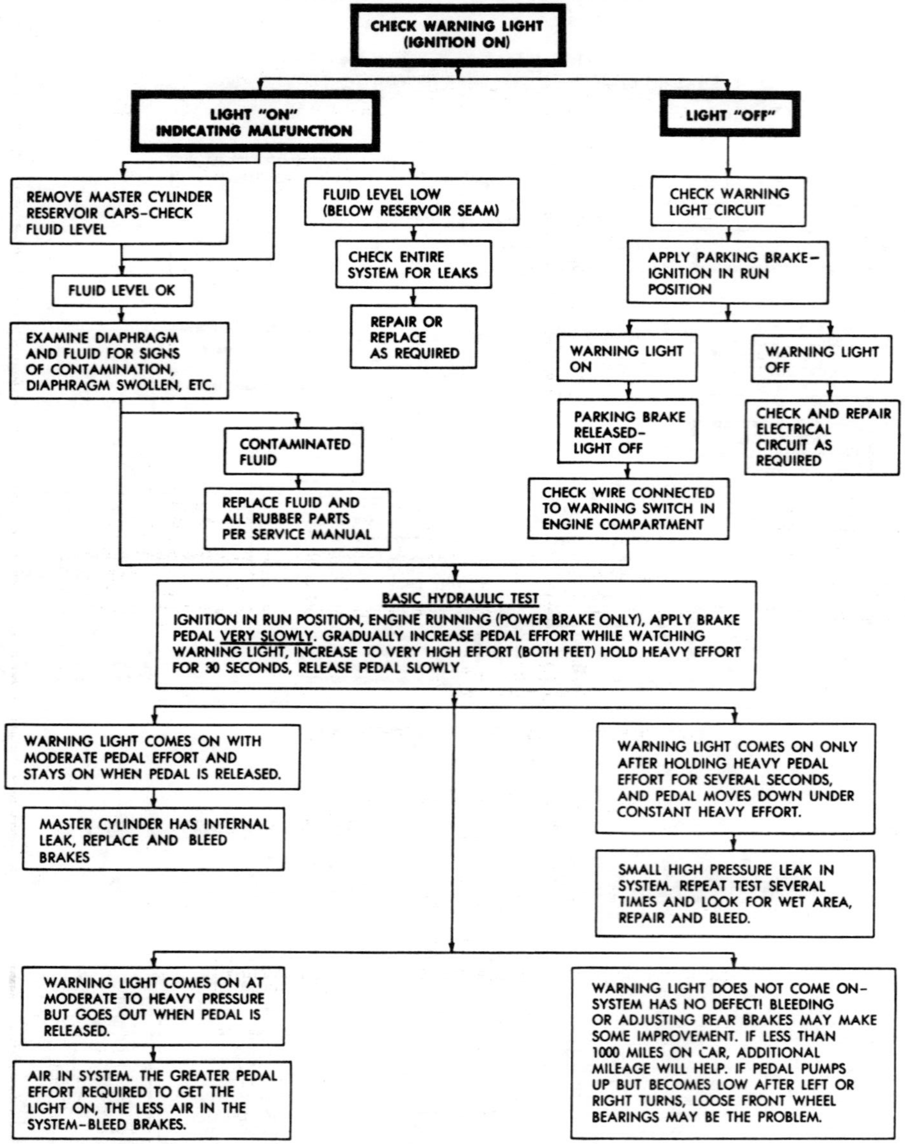

Fig. 6 Brake system troubleshooting (Part 3 of 7). R.W.D Chrysler Corp. vehicles

BRAKE CHART 3 TROUBLESHOOTING POWER BRAKE DIAGNOSIS

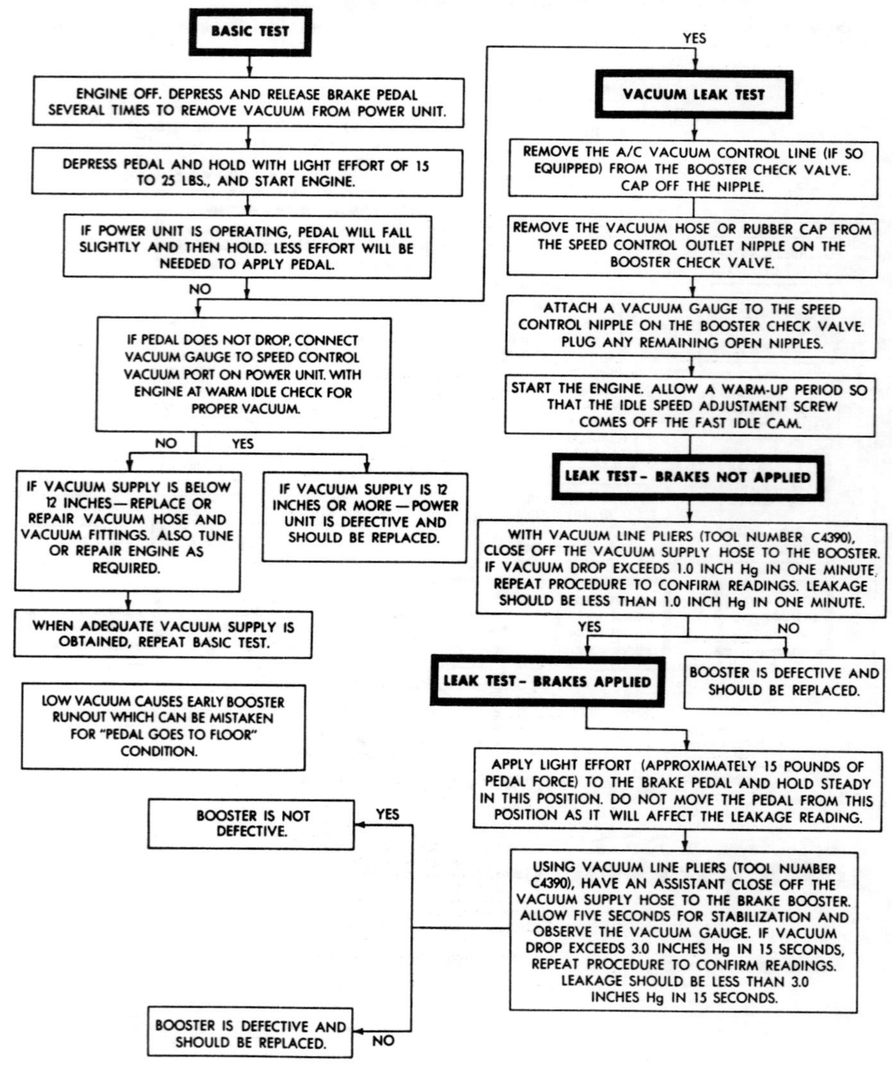

Fig. 6 Brake system troubleshooting (Part 4 of 7). R.W.D Chrysler Corp. vehicles

BRAKE CHART 4 BRAKE NOISE DIAGNOSIS

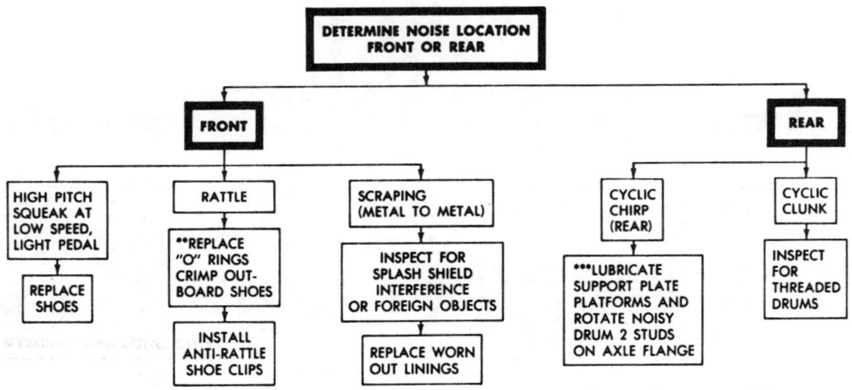

**INSTALL "O" RINGS (AROUND ADAPTER) BETWEEN MACHINED "WAYS" OF CALIPER AND ADAPTER.
***MOPAR MULTIPURPOSE LUBRICANT, PART NUMBER 4318063, OR EQUIVALENT.

Fig. 6 Brake system troubleshooting (Part 5 of 7). R.W.D Chrysler Corp. vehicles

BRAKE CHART 5 WHEEL BRAKES DIAGNOSIS

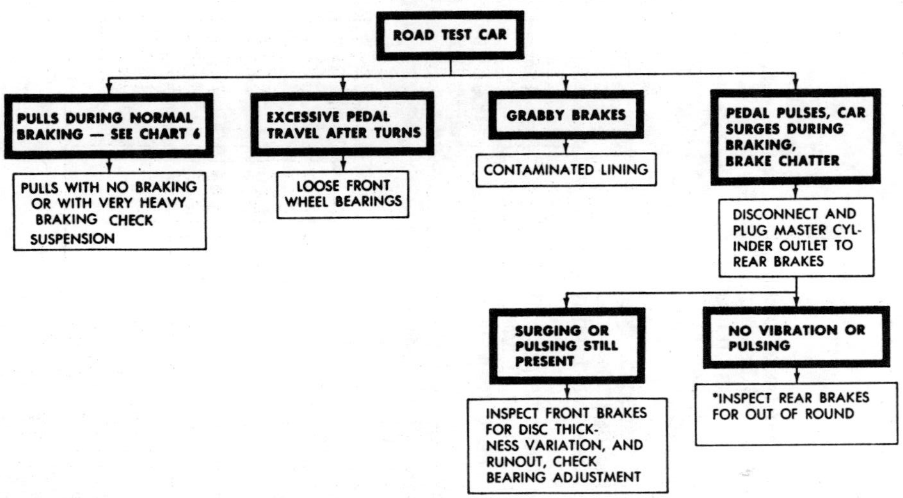

*PULSATION CAUSED BY REAR BRAKES MAY BE ELIMINATED OR REDUCED BY INDEXING DRUM (REAR) TWO STUDS ON AXLE FLANGE.

Fig. 6 Brake system troubleshooting (Part 6 of 7). R.W.D Chrysler Corp. vehicles

BRAKE CHART 6 BRAKE PULL DIAGNOSIS

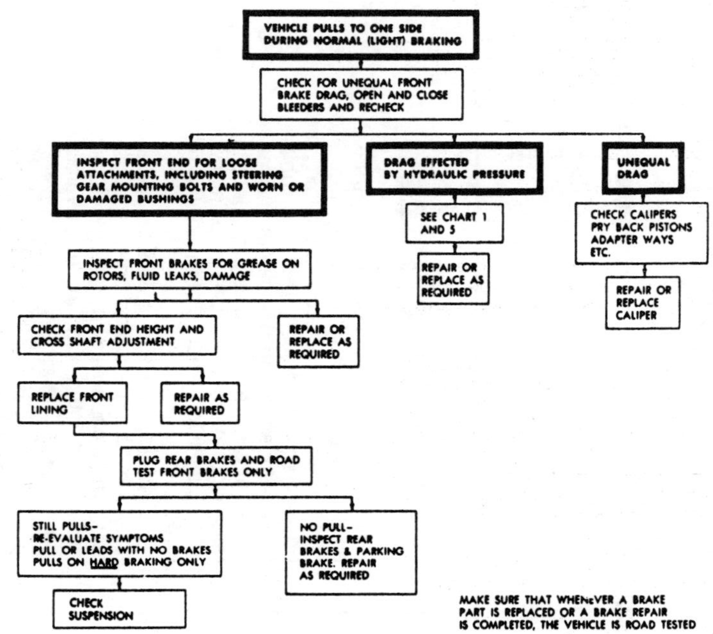

MAKE SURE THAT WHENEVER A BRAKE PART IS REPLACED OR A BRAKE REPAIR IS COMPLETED, THE VEHICLE IS ROAD TESTED

Fig. 6 Brake system troubleshooting (Part 7 of 7). R.W.D Chrysler Corp. vehicles

General Information

INDEX

Fig. 1 Checking rotor for lateral runout

BRAKE PADS & CALIPERS

Remove wheels and inspect brake rotor, caliper and linings. The wheel bearings should be inspected at this time and repacked if necessary. Use caution not to contaminate linings with grease or other foreign substances.

If an on-vehicle visual inspection does not adequately determine the condition of the linings, they should be removed and inspected. Brake linings should be replaced if cracked, damaged, or worn as per individual manufacturer limitations. If linings do not require replacement, reinstall in original positions.

Calipers assemblies should be inspected for damage, cracks, fluid leakage through the casting (porosity), or improper operation. Repair or replace as needed.

BRAKE ROUGHNESS

The most common cause of brake chatter on disc brakes is a variation in disc thickness. If roughness, vibration, or pedal pulsation is experienced the disc may have excessive thickness variation. Check rotor parallelism (thickness variation) as described under "Disc Brake Service."

Excessive lateral runout of the braking disc may cause a "knocking back" of the caliper pistons, possibly creating increased pedal travel and vibration when brakes are applied. Before checking runout, if applicable ensure wheel bearings are adjusted properly.

DISC BRAKE SERVICE

Maintenance of disc brakes is extremely critical for proper brake operation due to close tolerances required in machining the brake disc. In addition, the disc rubbing surface must be non-directional and maintained at a micro inch finish to control erratic performance, promote long lining life, and equal lining wear of both left and right brakes.

Refinishing of the rubbing surfaces should not be attempted unless precision equipment, capable of measuring in micro inches (millionths of an inch) is available.

LATERAL RUNOUT CHECK

Refer to "Rotor Specifications" for runout specifications.

1. If applicable, ensure wheel bearings are properly adjusted.
2. Mount a dial indicator to a convenient part of the vehicle (steering knuckle, tie rod, disc brake caliper housing).
3. Position dial indicator plunger so it contacts the disc at a point one inch from the outer edge, **Fig. 1.**
4. Rotate the rotor and note the dial indicator readings. Perform this check on both inboard and outboard rotor faces.
5. **On hub type rotors,** if runout exceeds specifications proceed as follows:
   a. If the rotor cannot be machined, replace it.
   b. If the rotor can be machined, resurface it, then proceed to step c.
   c. If runout still exceeds specifications, replace the rotor.
6. **On hubless type rotors,** if runout exceeds specifications proceed as follows:
   a. Reposition rotor on the hub, then recheck runout.
   b. If runout still exceeds specifications, replace the rotor.

PARALLELISM CHECK

Refer to "Rotor Specifications" for parallelism specifications.

1. Using a suitable micrometer, measure the rotor at 12 equally spaced points at a radius approximately one inch from edge of disc.
2. If measurements exceed specifications and the rotor cannot be machined, replace it.
3. If measurements exceed specifications and the rotor can be machined, resurface it.

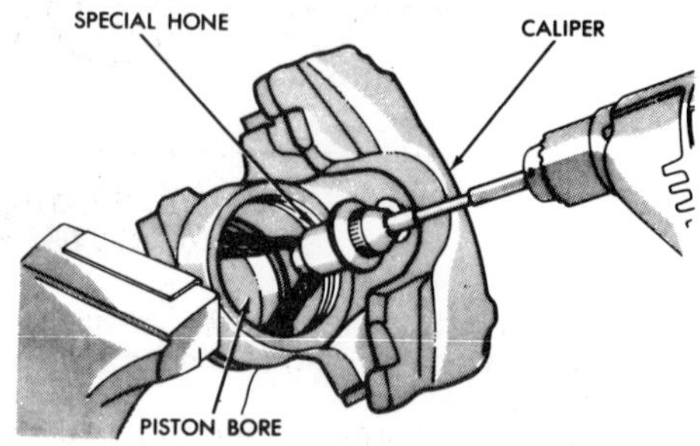

Fig. 2 Honing caliper piston bore

4. Recheck parallelism. If measurements still exceed specifications, replace the rotor.

GENERAL PRECAUTIONS

1. Grease or any other foreign material must be kept off the caliper, linings, surfaces of the disc and external surfaces of the hub, during service procedures. Handling the brake disc and caliper should be done in a way to avoid deformation of the disc and nicking or scratching brake linings.
2. If inspection reveals that the rubber piston seals are worn or damaged, they should be replaced.
3. During removal and installation of a wheel assembly, exercise care so as not to interfere with or damage the caliper splash shield or bleeder screw.
4. Wheel bearings should be adjusted to specifications.
5. Be sure vehicle is centered on hoist before servicing any of the front end components to avoid bending or damaging the disc splash shield on full right or left wheel turns.
6. Before the vehicle is moved after any brake service work, be sure to obtain a firm brake pedal.
7. The assembly bolts of the two caliper housings (if a two piece caliper) should not be disturbed unless the caliper requires service.

INSPECTION OF CALIPER

Should it become necessary to remove the caliper for installation of new parts, clean all parts in denatured alcohol, wipe dry using lint-free cloths. Using an air hose, blow out drilled passages and bores.

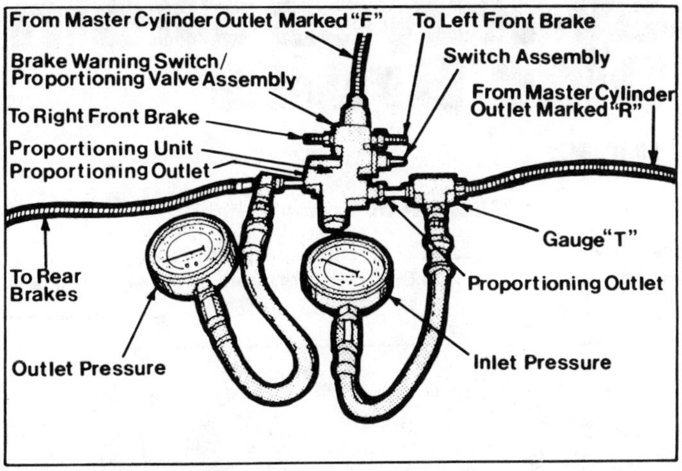

Fig. 3 Common gauge hook-up for testing proportioning valve

Check dust boots for punctures or tears. Generally, new boots should be installed upon reassembly.

Inspect piston bores for scoring or pitting. Bores that show light scratches or corrosion can usually be cleaned with crocus cloth. However, bores that have deep scratches or scoring may be honed, provided the diameter of the bore is not increased more than .002 inch. If the bore does not clean up to within this specification, a new caliper housing should be installed. Black stains on the bore walls are caused by piston seals and will do no harm.

When using a hone, **Fig. 2,** be sure to install the hone baffle before honing bore. The baffle is used to protect the hone stones from damage. Use extreme care in cleaning the caliper after honing. Remove all dust and grit by flushing the caliper with denatured alcohol. Wipe dry with clean lint-free cloth and then clean a second time in the same manner.

BLEEDING DISC BRAKES

Pressure bleeding is recommended for all hydraulic disc brake systems.

The disc brake hydraulic system can be bled manually or with pressure bleeding equipment. On vehicles with disc brakes, the brake pedal will require more pumping and frequent checking of fluid level in master cylinder during bleeding operation.

Never use brake fluid that has been drained from hydraulic system when bleeding the brakes. Be sure the disc brake pistons are returned to their normal positions and that the disc pads are properly seated. Before driving the vehicle, check brake operation to be sure that a firm pedal has been obtained.

PROPORTIONING VALVE

The proportioning valve (when used), **Fig. 3,** provides balanced braking action between front and rear brakes under a wide range of braking conditions. The valve regulates the hydraulic pressure applied to the rear wheel cylinders, thus limiting rear braking action when high pressures are required at the front brakes. In this manner, premature rear wheel skid is prevented.

TESTING PROPORTIONING VALVE

When a premature rear wheel slide is obtained on a brake application, it usually is an indication that the fluid pressure to the rear wheels is above the 50% reduction ratio for the rear line pressure and that malfunction has occurred within the proportioning valve.

To test the valve, install gauge set, **Fig. 3** in brake line between master cylinder and proportioning valve and at output end of proportioning valve and brake line as shown. Be sure all joints are fluid tight.

Have a helper exert pressure on brake pedal (holding pressure). Obtain a reading on master cylinder output of approximately 700 psi. While pressure is being held as above, reading on valve outlet should be 550-610 psi. If the pressure readings do not meet these specifications, the valve should be removed and a new valve installed.

Type 1—Kelsey-Hayes Sliding Caliper Disc Brake

INDEX

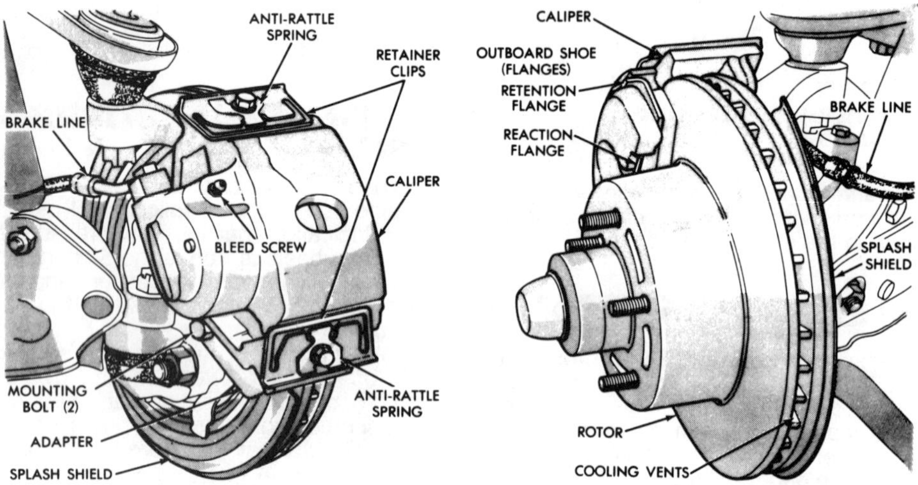

Fig. 1 Kelsey-Hayes sliding caliper disc brake assembly

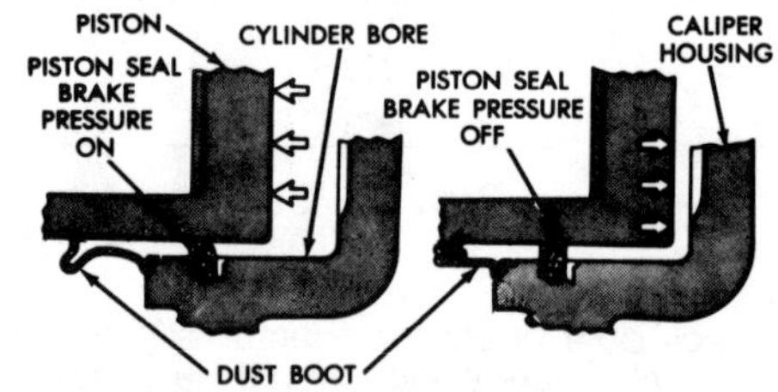

Fig. 2 Piston seal function

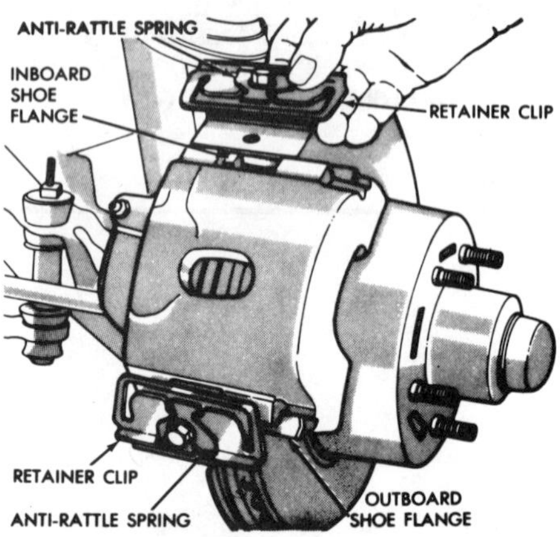

Fig. 3 Retainer clips & anti-rattle spring replacement

OPERATION

This sliding caliper single piston system uses a one piece hub and is actuated by the hydraulic system and disc assembly, **Fig. 1.** Alignment and positioning of the caliper is achieved by two machined guides or ways on the adapter, while caliper retaining clips allow lateral movement of the caliper. Outboard pad flanges are used to position and locate the pad on the caliper fingers while the inboard pad is retained by the adapter. Braking force applied onto the outboard pad is transferred to the caliper, while braking force applied onto the inboard pad is transferred directly to the adapter.

A square cut piston seal provides a hydraulic seal between the piston and the cylinder bore, **Fig. 2.** A dust boot with a wiping lip installed in a groove in the cylinder bore and piston prevents contamination in the piston and cylinder bore area. Adjustment between the disc and the pad is obtained automatically by the outward relocation of the piston as the inboard lining wears and inward movement of the caliper as the outboard lining wears.

CALIPER REMOVAL

1. Raise and support vehicle.
2. Remove front wheel and tire assembly.
3. Remove caliper retaining clips and anti-rattle springs, **Fig. 3.**
4. Remove caliper from disc by slowly sliding caliper assembly out and away from disc. **Support caliper to avoid damage to brake hose.**

BRAKE PAD REMOVAL

1. Remove caliper assembly as outlined above.
2. Remove outboard pad by prying between the pad and caliper fingers, **Fig. 4. Support caliper to avoid damage to brake hose.**
3. Remove inboard brake pad from adapter, **Fig. 5.**

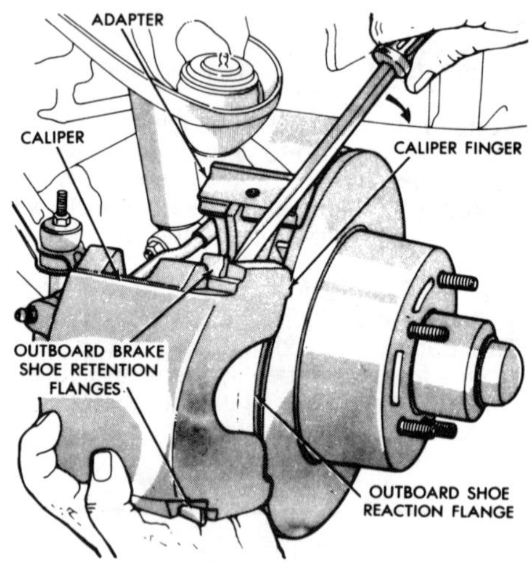

Fig. 4 Removing outboard pad

BRAKE PAD INSTALLATION

Remove approximately 1/3 of brake fluid from reservoir.

1. Push piston back into bore until bottomed.
2. Install new outboard pad in recess of caliper. **No freeplay should exist between brake pad flanges and caliper fingers, Fig. 6. If up and down (vertical) movement of the pad shows freeplay, pad must be removed and flanges bent to provide a slight interference fit, Fig. 7. Reinstall pad after modification. If pad cannot be finger snapped into place, use light C-clamp pressure, Fig. 8.**
3. Position inboard pad with flanges inserted in adapter ways, **Fig. 5.**
4. Carefully slide caliper assembly into

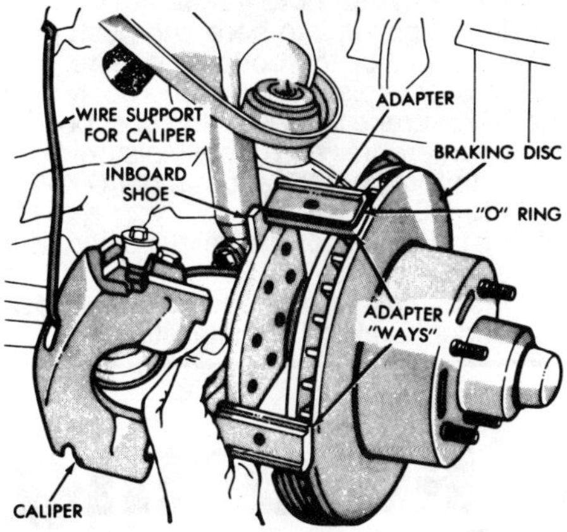

Fig. 5 Replacing inboard pad

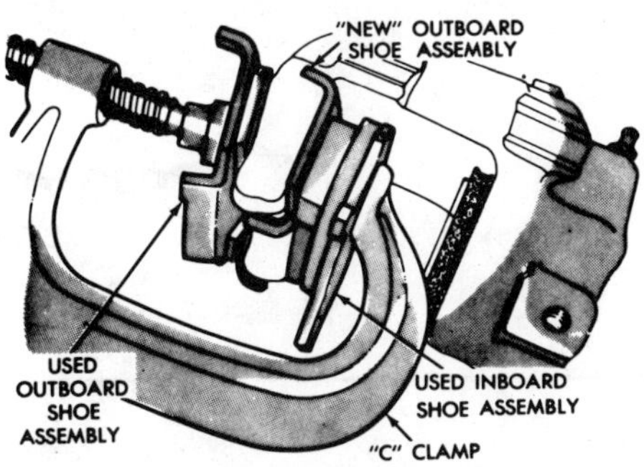

Fig. 8 Installing outboard pad using C-clamp

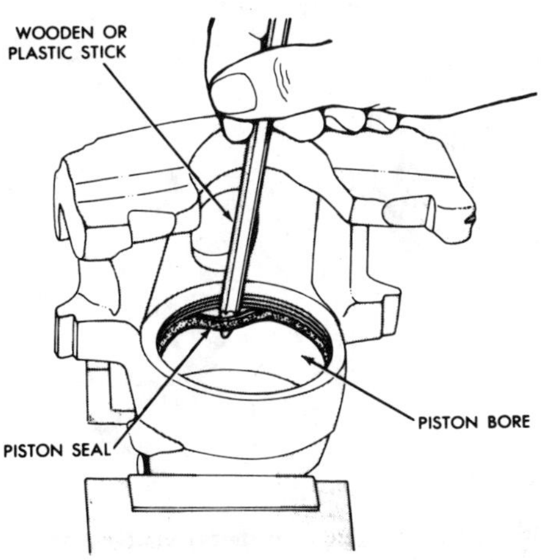

Fig. 11 Piston seal removal

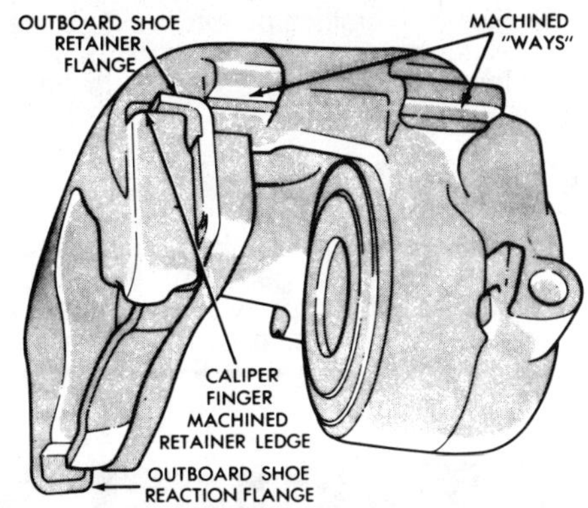

Fig. 6 Caliper finger & outboard pad retainer flange

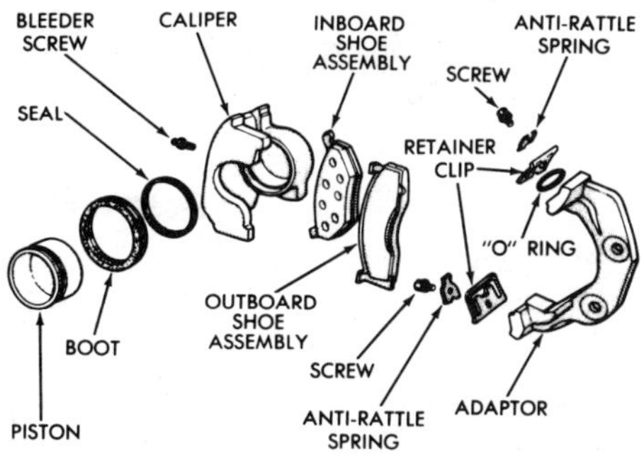

Fig. 9 Disassembled view of Kelsey-Hayes disc brake

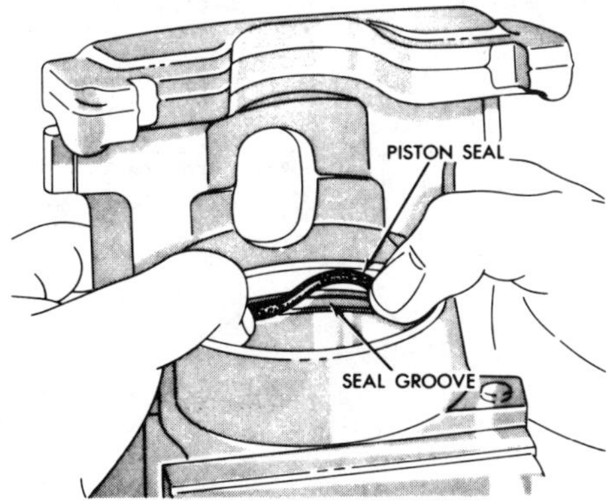

Fig. 12 Piston seal installation

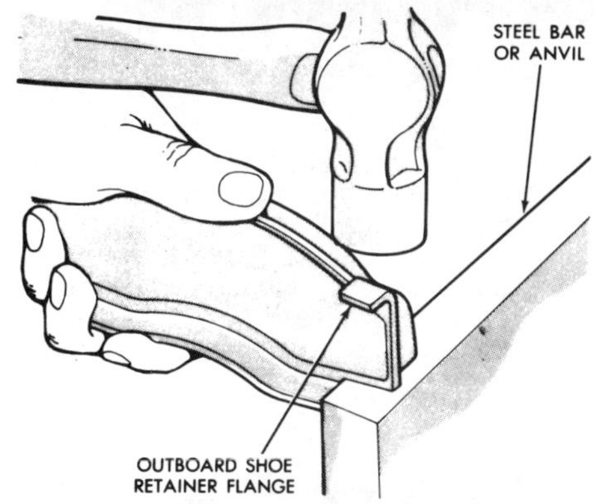

Fig. 7 Bending outboard pad retaining flange

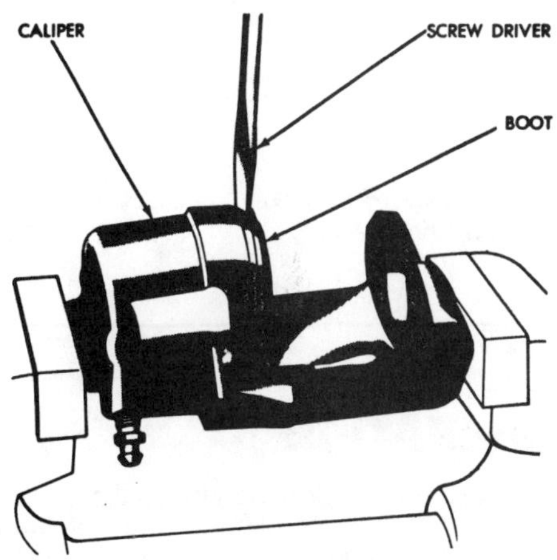

Fig. 10 Dust boot removal

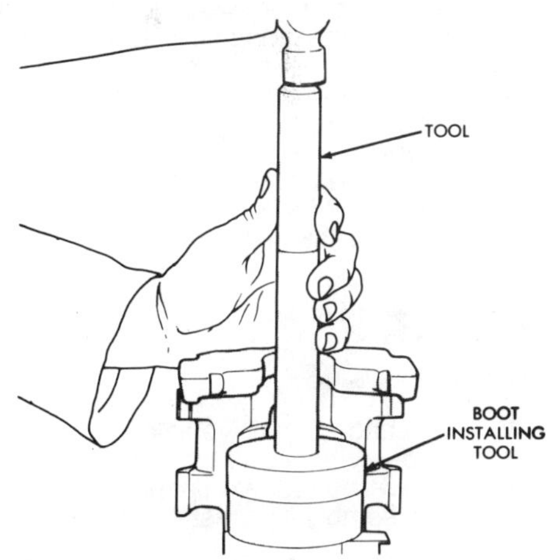

Fig. 13 Dust boot installation

adapter and over the disc while aligning caliper on machined ways of adapter. **Ensure dust boot is not pulled out from groove when piston and boot slide over the inboard pad.**

5. Install anti-rattle springs and retaining clips and **torque** retaining screws to 180 inch pounds. **The inboard pad anti-rattle spring is to be installed on top of the retainer spring plate, Fig 3.**

CALIPER DISASSEMBLY

1. With caliper and pads removed as described previously, place caliper onto the upper control arm and slowly depress brake pedal to hydraulically push piston from bore, **Fig. 9.** The pedal will fall when piston passes bore opening.
2. Support pedal below first inch of pedal travel to prevent excessive fluid loss.
3. To remove piston from the opposite caliper, disconnect flexible brake line at frame bracket at vehicle side where piston has been removed previously and plug tube to prevent pressure loss. By depressing brake pedal, this piston can also be hydraulically pushed out.
4. Mount caliper in a soft-jawed vise. **Excessive vise pressure will distort caliper bore.**
5. Using a suitable screwdriver, remove the dust boot, **Fig. 10.**
6. Insert a tool such as a small, pointed wooden or plastic object between the cylinder bore and the seal and work seal out of the groove in the piston bore, **Fig. 11. A metal tool such as a screwdriver should not be used, as it can cause damage to the piston bore or burr the edges of the seal groove.**

CALIPER INSPECTION

1. Clean all parts with a suitable brake cleaner and blow dry using compressed air.
2. Inspect piston bore for scoring or pitting. If scratches or corrosion cannot be cleaned with crocus cloth, hone bore with suitable brake hone using brake fluid as a lubricant. Do not increase bore more than .002 inch while honing. Replace caliper if corrosion or pitting is excessive.
3. Replace caliper piston if it is excessively pitted, scored or worn, or if honing the caliper bore was necessary.

CALIPER ASSEMBLY

1. Dip new piston seals in clean brake fluid. Work seal gently into the groove until seal is properly seated, ensuring seal is not twisted or rolled, **Fig. 12. Do not reuse old seals.**
2. Install piston into bore and push past seal until piston bottoms in bore.
3. Lubricate piston boot generously with clean brake fluid. Position dust boot in counterbore and install piston boot with tool C-4689 and handle C-4171, **Fig. 13.**
4. Install brake hose to caliper using new seal washers.
5. Install caliper and pads as described under "Brake Pad Installation."

Type 2—A.T.E. Dual Pin Floating Caliper (Front) Disc Brake

INDEX

OPERATION

The single piston floating caliper disc brake assembly consists of the hub and disc brake rotor assembly, caliper, pads, splash shield and adapter, **Fig. 1.**

The caliper assembly floats on two rubber bushings riding on two steel guide pins threaded into the adapter. The bushings are inserted on the inboard portion of the caliper. Two machined abutments on the adapter position and align the caliper fore and aft. Guide pins and bushings control caliper and piston seal movement to assist in maintaining proper pad clearance. All braking force is taken directly by the adapter.

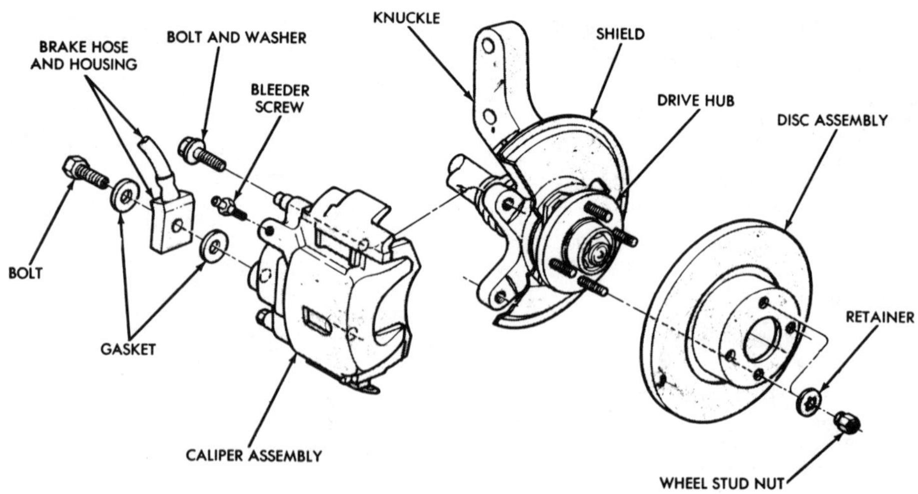

Fig. 1 A.T.E. dual pin floating caliper front disc brake assembly

BRAKE PADS
REPLACE

1. Raise and support vehicle.
2. Remove wheel and tire assembly.
3. Remove hold-down spring from caliper assembly by pressing spring outward.
4. Loosen, but do not remove caliper guide pins, then remove caliper from disc. Inboard pad will remain inside caliper. **Support caliper assembly to prevent damage to hydraulic brake hose. Remove caliper guide pins only if bushings or sleeves are to be replaced.**
5. Remove inboard pad from caliper and outboard pad from adapter.
6. Push caliper piston into bore. Remove some brake fluid from master cylinder reservoir to prevent overflowing when piston is pushed into bore.
7. Install new inboard pad into caliper with retainer positioned in piston bore.
8. Install outboard pad onto adapter.
9. Position caliper over brake disc and adapter, then **torque** guide pins to 18-26 ft. lbs.
10. Install hold-down spring, then install tire and wheel assembly and lower vehicle.
11. Check master cylinder reservoir for proper brake fluid lever.

CALIPER OVERHAUL
DISASSEMBLY

1. Remove caliper assembly as described under "Brake Pads, Replace."
2. Place a wood block between caliper piston and caliper fingers. With brake hose attached to caliper, **Fig. 2,** carefully depress brake pedal to push piston out of caliper bore. Prop brake pedal to any position below first inch of brake pedal travel to prevent brake fluid loss.
3. If pistons are to be removed from both calipers, disconnect brake hose at frame bracket after removing piston, then cap brake line and repeat procedure to remove piston from other caliper.
4. Disconnect brake hose from caliper.
5. Mount caliper in a soft jawed vise.
6. Support caliper, then remove and discard dust boot.
7. Using a small wooden or plastic stick, remove seal from groove in piston bore and discard. **Do not use a**

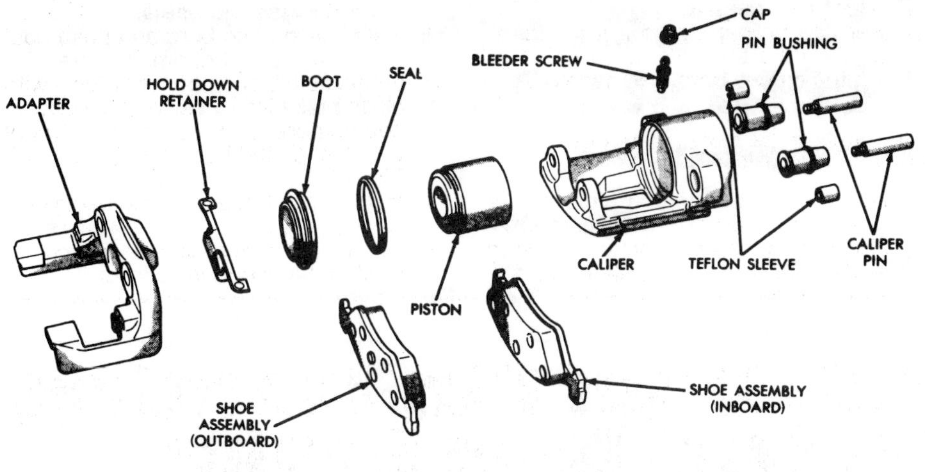

Fig. 2 Disassembled view of A.T.E. disc brake

screwdriver or other metal tool, as this may scratch caliper bore.

8. **On all models except 1988 Dynasty, New Yorker and New Yorker Landau,** remove caliper bushings. On 1988 Dynasty, New Yorker and New Yorker Landau, caliper bushing and sleeve assemblies are permanently sealed.

ASSEMBLY

1. Mount caliper in a soft jawed vise.
2. Lubricate piston seal with clean brake fluid and install seal in caliper bore groove, **Fig. 3.** Ensure seal is properly seated.
3. Lubricate piston boot with clean brake fluid and position over piston.
4. Install piston and boot assembly, pushing it past piston seal until it bottoms into caliper bore.
5. Using a hammer and tool No. C-4689 with C-4171 handle, drive dust boot into counterbore until properly seated, **Fig. 10.**
6. **On models where bushings require replacement,** compress flanges of bushings and install on caliper housing. Ensure bushing flanges extend evenly over caliper housing on both sides. Remove Teflon sleeves from guide pin bushings prior to installing bushings into caliper. After bushings are installed into caliper, reinstall Teflon sleeves into bushings.
7. **On all models,** connect brake hose to brake line at frame bracket.
8. Install caliper on vehicle as described under "Brake Pads, Replace."
9. Check brake fluid level of master cylinder reservoir, then open caliper bleed screw and bleed brake system. Continue bleeding procedure until firm pedal is obtained.

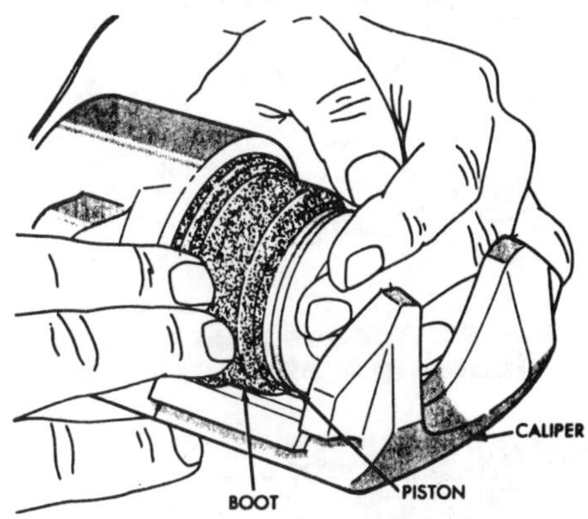

Fig. 4 Installing caliper piston & dust boot

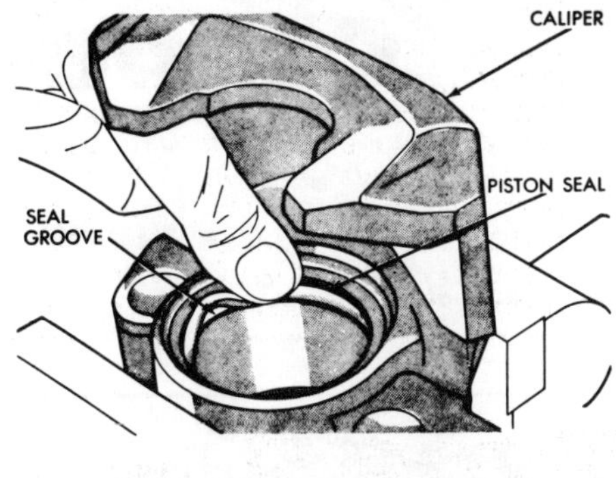

Fig. 3 Installing piston seal

Type 3—A.T.E. Dual Pin Floating Caliper (Rear) Disc Brake

INDEX

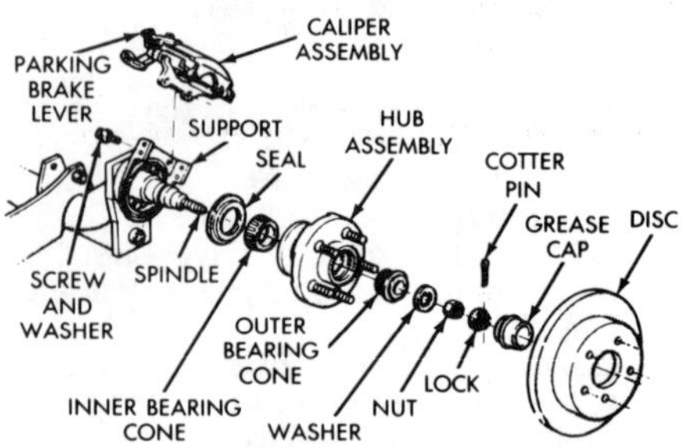

Fig. 1 A.T.E. floating caliper rear disc brake assembly

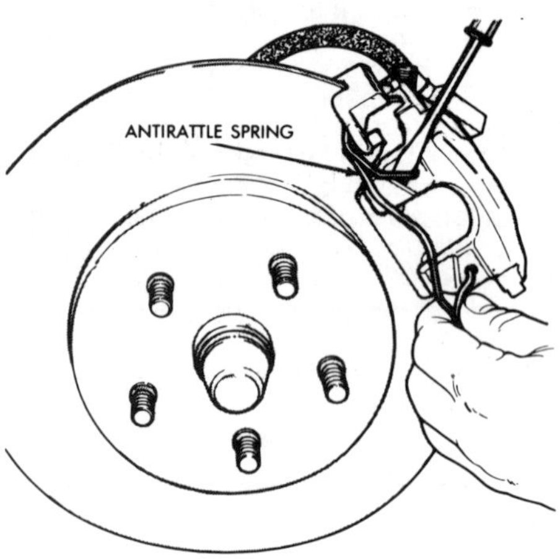

Fig. 4 Anti-rattle spring removal

DESCRIPTION

This disc brake, **Fig. 1,** is a single piston, floating caliper type that incorporates a mechanically actuated parking brake mechanism. The caliper is a one piece casting with the single piston located on the inboard side. An automatic adjuster within the piston maintains a constant running clearance between the brakes and disc. The parking brake mechanism is self-adjusting.

The caliper floats on rubber bushings, with Teflon inner sleeves, on the two guide pins threaded into the adapter. Two machined abutments on the adapter position and align the caliper and brake pads for fore and aft movement.

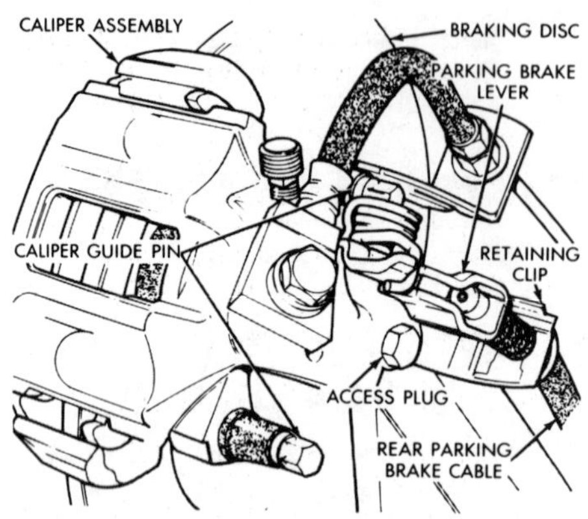

Fig. 2 Access plug location (parking brake retractor)

BRAKE PADS
REPLACE

1. Raise and properly support vehicle.
2. Remove rear wheels, then clean caliper in area of access plug, **Fig. 2.**
3. Remove access plug and insert a 4mm Allen wrench through access hole, and engage retraction shaft, **Fig. 3.** Rotate retraction shaft with Allen wrench counterclockwise to increase clearance between brake pads and disc.
4. Remove outboard anti-rattle spring, **Fig. 4. Use care not to deform spring.**
5. Back out the caliper guide pins to free caliper from adapter. **Remove guide pins only if bushings or sleeve require replacement.**
6. Lift caliper upward and away from disc, **Fig. 5.** The inboard pad will remain in caliper, attached to piston. Suspend caliper from a suspension member with suitable wire. **Do not allow caliper to hang by brake hose.**
7. Remove inboard pad from piston (caliper), **Fig. 6,** and outboard pad from adapter.
8. When installing pads and the caliper, note the following:
   a. It will be necessary to retract the piston further with the Allen wrench due to the thickness of the new pads. When retracting the piston, stop rotation of Allen wrench when slight resistance is encountered. This indicates that the piston has bottomed. **Do not use excessive force, as damage to retraction and actuation shafts may occur.**
   b. The inboard pads are interchangeable and may be used on either side of the vehicle. The outboard pads are not interchangeable and are marked "L" for left side or "R" for right side of vehicle, **Fig. 7.**
   c. Fit the retainer on inboard pad into piston. Install proper outboard pad on adapter. Lower caliper over disc

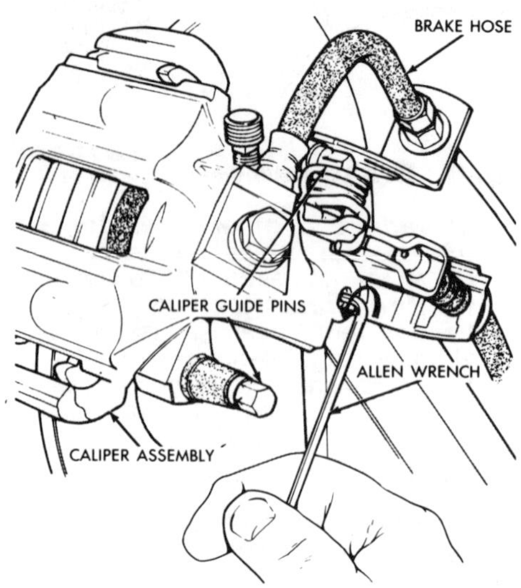

Fig. 3 Retracting brake pads from disc

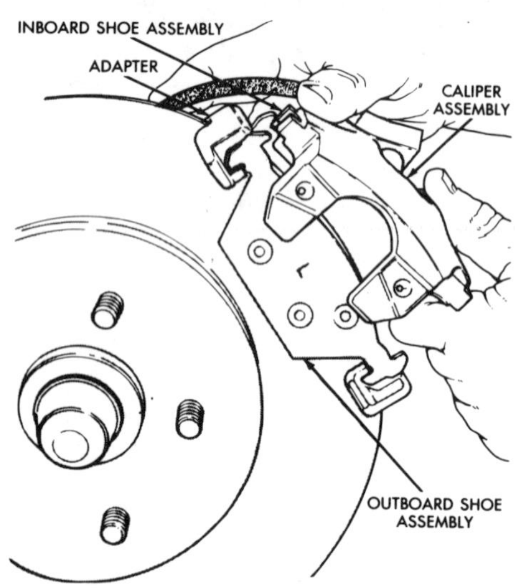

Fig. 5 Caliper removal

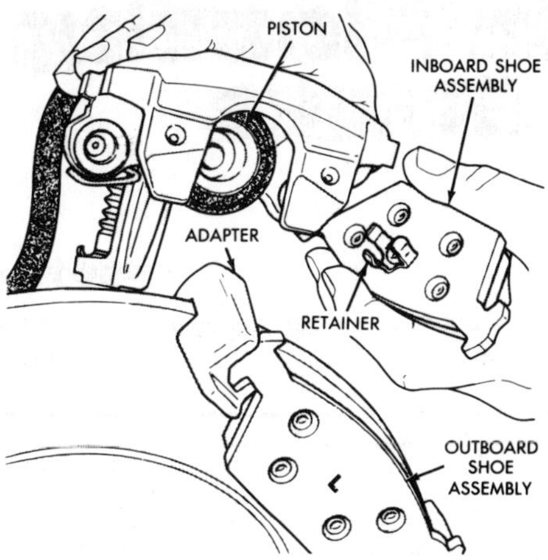

Fig. 6 Brake pad replacement

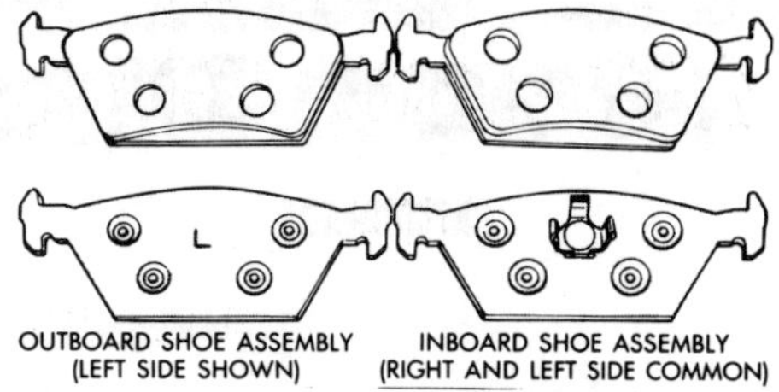

Fig. 7 Brake pad identification

and outboard pad. Install and **torque** guide pins to 18-26 ft. lbs. **Do not crossthread the pins.** Install anti-rattle spring.

d. To set the initial clearance between the brake pads and disc, rotate the retraction shaft with the Allen wrench clockwise until snug, then back off 1/3 turn.

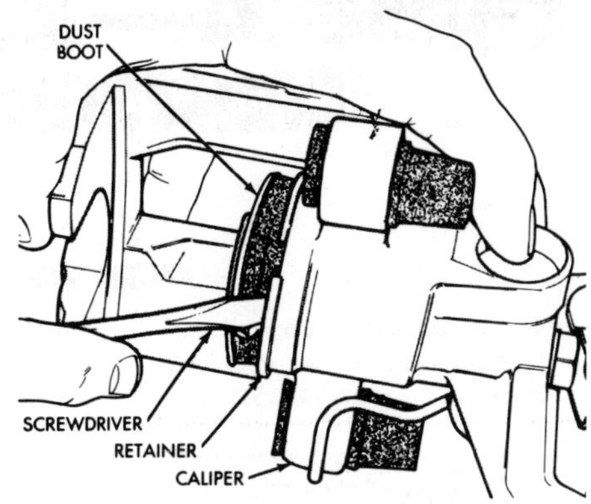

Fig. 8 Removing dust boot retainer

CALIPER SERVICE

DUST BOOT, REPLACE

1. Remove dust boot retainer with a suitable screwdriver, **Fig. 8.**
2. Remove dust boot from caliper and piston grooves.
3. Clean piston and caliper grooves, then coat new boot with clean brake fluid.
4. Place boot over piston and into the piston and caliper grooves, **Fig. 9.**
5. Install dust boot retainer into caliper groove.

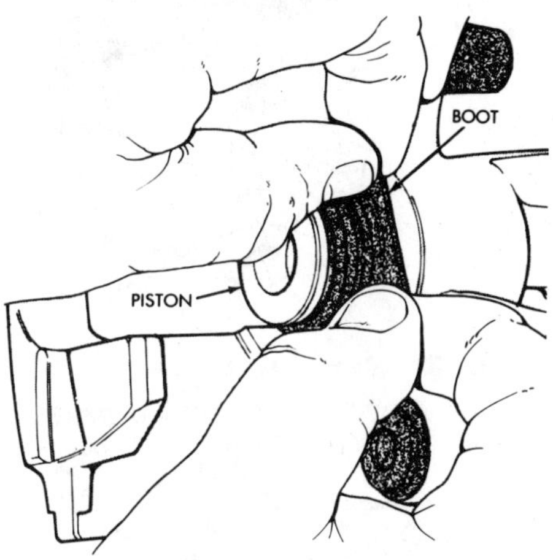

Fig. 9 Dust boot installation

GUIDE PIN BUSHING, REPLACE

1. Remove bushings from caliper with a suitable tool. Discard bushings and Teflon sleeves.

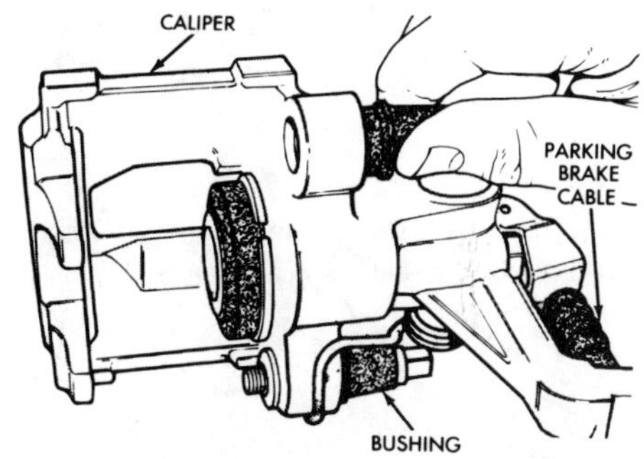

Fig. 10 Guide pin bushing installation

2. Remove Teflon sleeves from new bushings prior to bushing installation.
3. Compress bushing flange with fingers and push bushing into seated position, **Fig. 10.**
4. Install Teflon sleeves into bushings.

Type 4—A.T.E. Dual Pin Floating Caliper (Rear) Disc Brake

INDEX

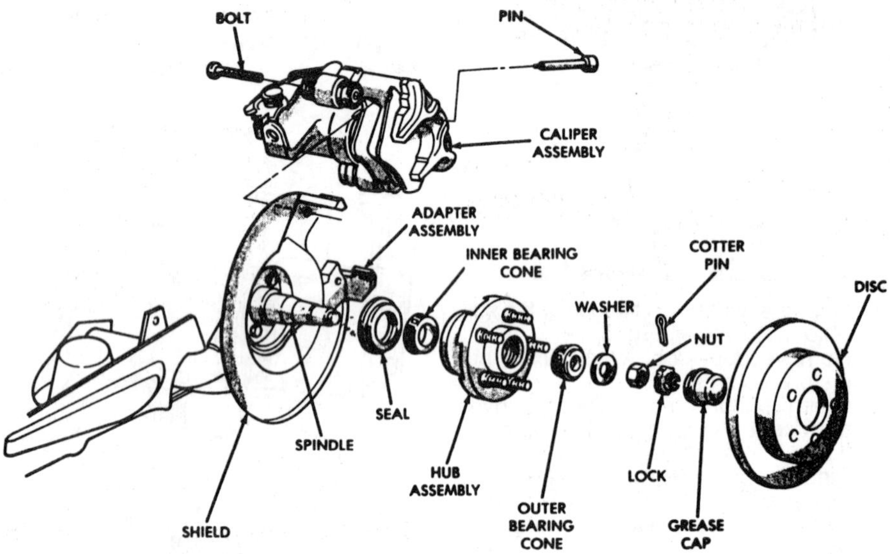

Fig. 1 A.T.E. floating caliper rear disc brake assembly

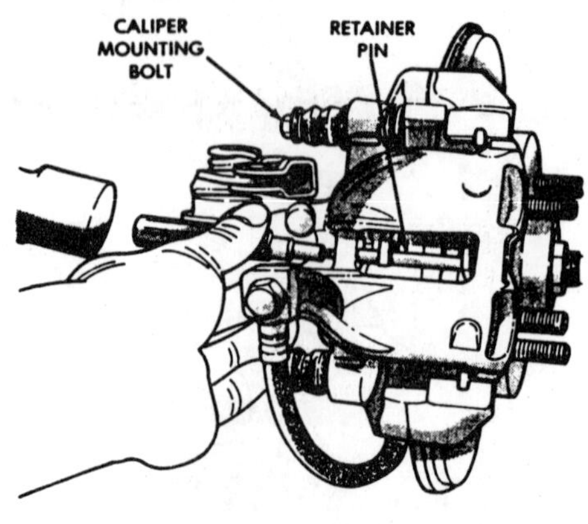

Fig. 2 Removing brake pad retainer pin

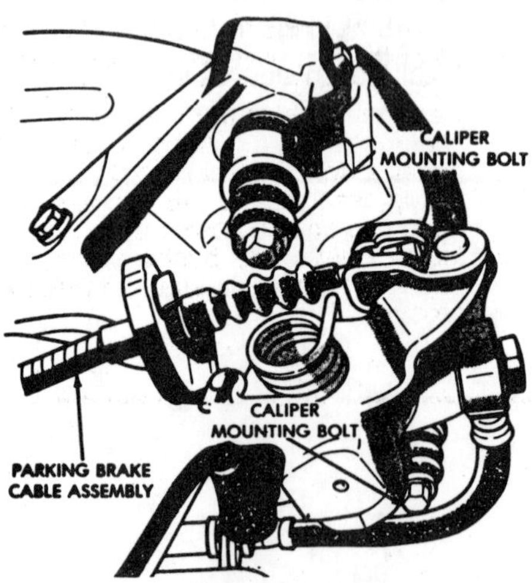

Fig. 3 Caliper mounting bolt locations

DESCRIPTION

This single piston, floating caliper rear disc brake assembly, **Fig. 1,** includes a hub assembly, rotor, caliper, pads, an adapter and a mechanically operated parking brake. The caliper has a 36mm piston located on the inboard side.

The caliper floats on rubber bushings with metal sleeves on two bolts that are threaded into the adapter. Two machined abutments on the adapter position and align the caliper and brake pads for movement fore and aft.

BRAKE PADS REPLACE

REMOVAL

1. Raise and properly support vehicle.
2. Remove rear wheel and tire assembly.
3. Drive brake pad retainer pin out of caliper, **Fig. 2.**
4. Remove caliper mounting bolts, **Fig. 3,** then the caliper by lifting upward and away from disc. Support caliper with suitable wire. **Do not allow caliper to hang by brake hose.**
5. Remove brake pads and anti-rattle clip.

INSTALLATION

1. Lubricate adapter ways with suitable grease.
2. Install inboard pad first, then the outboard pad on adapter assembly, **Fig. 4.**
3. Place caliper with lower end inserted first over disc and pads onto adapter. **Do not install mounting bolts at this time.**
4. Insert anti-rattle clip through opening in caliper as shown in **Fig. 5.** Place lower portion of clip in position first, then release. **Ensure clip falls into correct position, Fig. 6.**
5. Install brake pad retainer pin through caliper and pads. Drive into position using hammer and suitable drift, **Fig. 6.**
6. Push down on upper portion of caliper to load anti-rattle clip, **Fig. 7.**

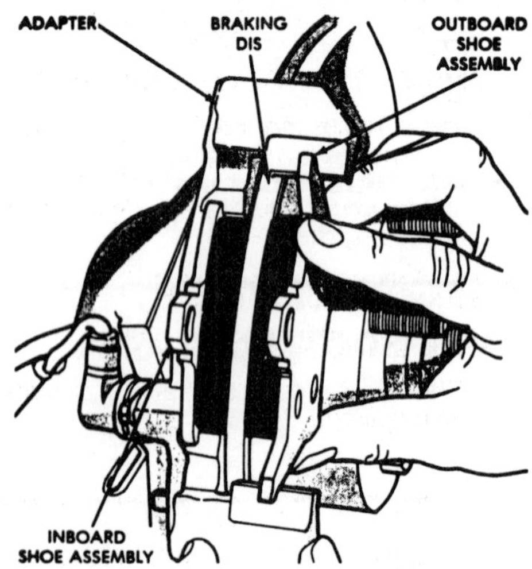

Fig. 4 Brake pad replacement

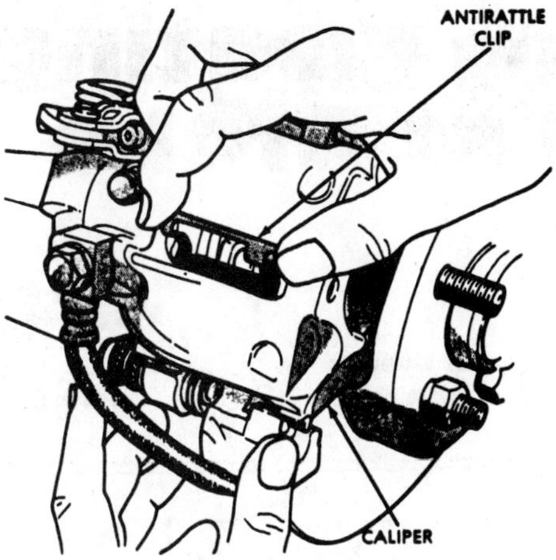

Fig. 5 Anti-rattle clip installation

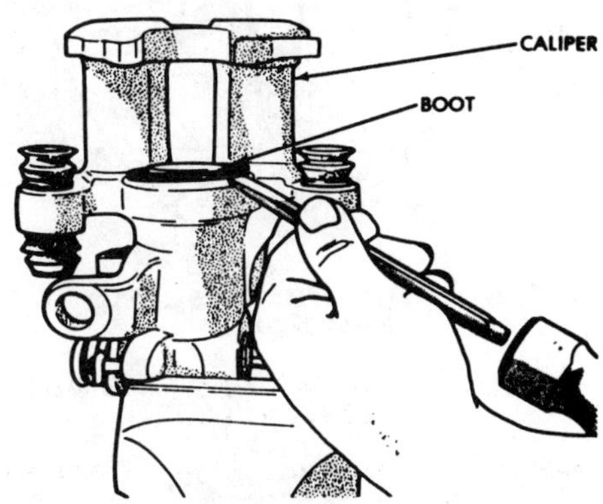

Fig. 8 Removing dust boot

7. Install upper mounting bolt first, then the lower bolt. **Torque** bolts to 250 inch lbs.
8. Install wheel and tire assembly, then lower vehicle.

CALIPER SERVICE

DUST BOOT, REPLACE

1. Clean area around dust boot with suitable solvent.

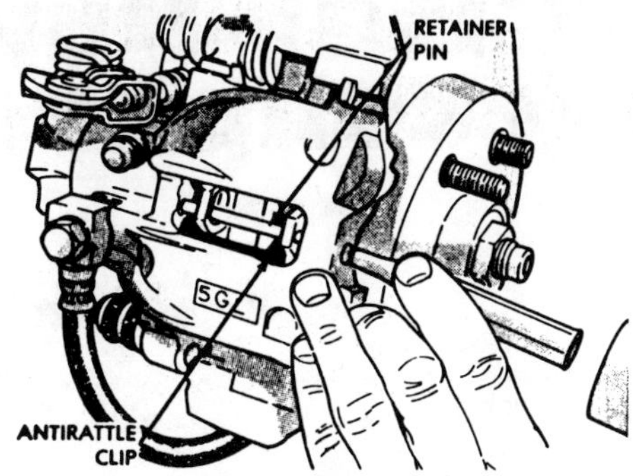

Fig. 6 Anti-rattle clip & retainer pin positioning

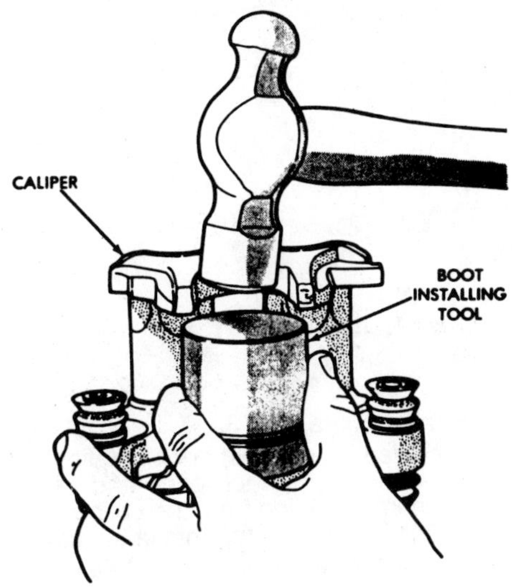

Fig. 9 Dust boot installation

2. Remove dust boot from caliper and piston grooves, **Fig. 8.**
3. Clean piston and caliper grooves, then coat new boot with clean brake fluid.
4. Place boot over piston and into piston groove, then position in counterbore of caliper.
5. Using suitable tools, drive dust boot into counterbore, **Fig, 9.**

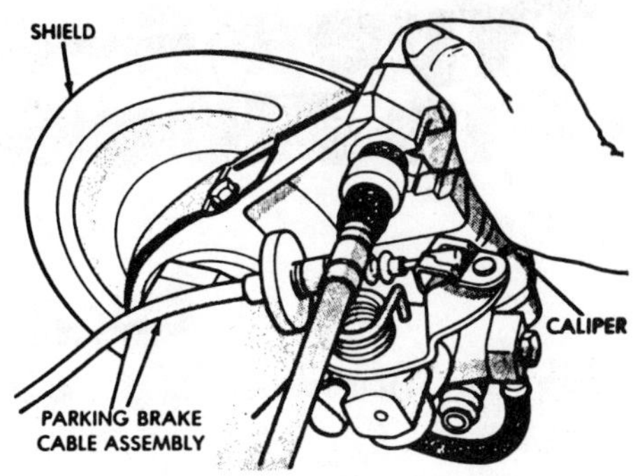

Fig. 7 Applying downward pressure to caliper

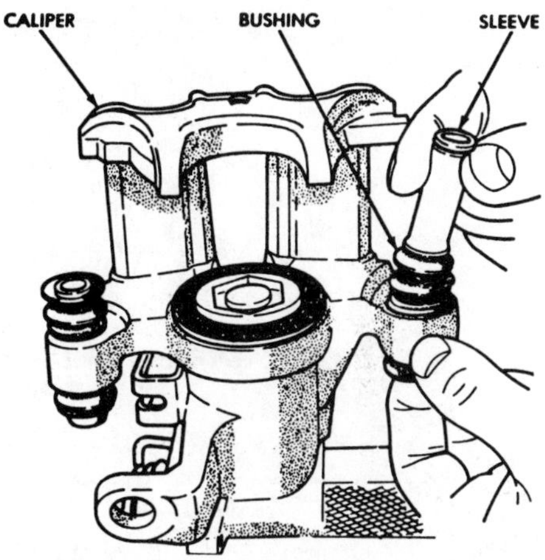

Fig. 10 Removing sleeve from bushing

BUSHING & SLEEVE, REPLACE

1. Unseat bushing from groove in metal sleeve, then remove metal sleeve from bushing.
2. Remove bushing from caliper.
3. Clean bushing mounting surface with suitable solvent and wipe dry.
4. Compress bushing with fingers and push into seated position, **Fig. 10.**
5. Install sleeve into bushing, ensuring bushing is properly seated in sleeve grooves.

Type 5—Kelsey-Hayes Single Pin Floating Caliper Disc Brake

INDEX

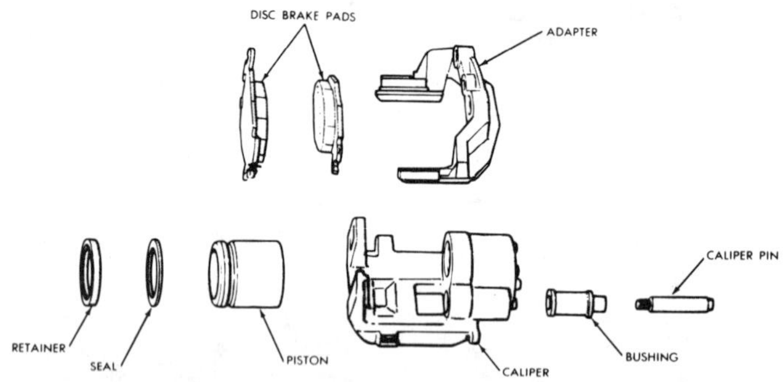

Fig. 1 Disassembled view of Kelsey-Hayes disc brake

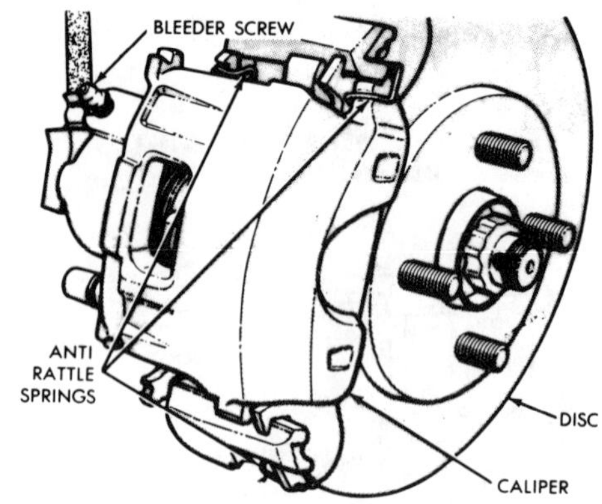

Fig. 2 Anti-rattle spring location

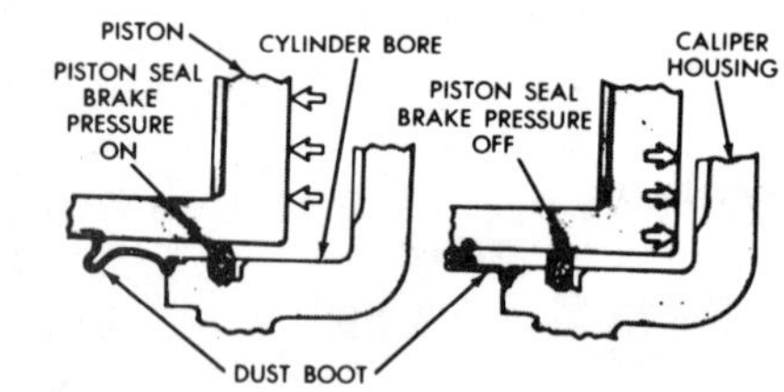

Fig. 3 Cross-sectional view of piston seal & dust boot

OPERATION

The caliper assembly consists of a rotor, caliper, pad, and adapter, **Fig. 1.** The single piston caliper assembly floats through a rubber bushing on a single pin, threaded into the adapter. The bushing is inserted into the inboard portion of the caliper. Two machined abutments on the adapter, position and align the caliper fore and aft. The guide pin and bushing controls the movement of the caliper and the piston seal to assist in maintaining proper pad clearance.

This assembly has three anti-rattle clips. One is on top of the inboard pad, one clip is on the bottom of the outboard pad, and one clip is on top of the caliper, **Fig. 2.**

All of the braking force is taken directly by the adapter. The caliper is a one piece casting with the inboard side containing a single piston cylinder bore.

A square cut rubber piston seal is located in a machined groove in the caliper bore and provides a seal between piston and caliper bore, **Fig. 3.**

A molded rubber dust boot installed in a groove in the cylinder bore and piston keeps contamination from the caliper bore and piston. The boot mounts in the caliper bore and in a groove in the piston, **Fig. 3.**

BRAKE PADS REPLACE

REMOVAL

1. Remove brake fluid until reservoir is half full.
2. Raise and support front of vehicle.
3. Remove wheel and tire assembly.
4. Remove caliper guide pin and anti-rattle clips.
5. Remove caliper from disc by sliding caliper assembly out and away from braking disc, **Fig. 4. Suspend caliper with wire so as not to damage flexible brake hose.**
6. Remove outboard brake lining, then lift off rotor and remove inboard brake lining.

INSTALLATION

1. Lubricate adapter ways with suitable grease.
2. Push piston back into cylinder bore with uniform pressure until it is bottomed.
3. Position inboard pad on adapter, then install rotor.
4. While holding outboard pad in position on adapter, carefully position caliper over disc brake rotor.
5. Lower caliper over rotor and adapter.
6. Install guide pin through bushing, caliper and adapter.
7. Press in on guide pin and thread pin into adapter. **Torque** pin to 25 to 35 ft. lbs. on 1985-88 models.
8. Install wheel and tire assembly, then lower vehicle.

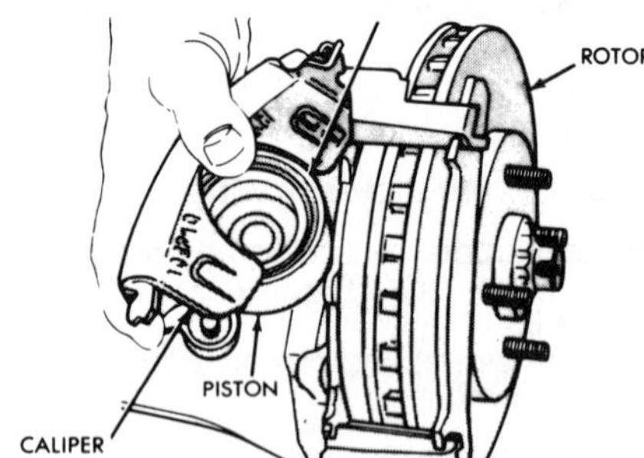

Fig. 4 Removing caliper

CALIPER OVERHAUL

Refer to "Caliper Overhaul" under "Type 2—A.T.E. Dual Pin Floating Caliper (Front) Disc Brake" for procedure.

Type 6—Kelsey-Hayes Dual Pin Floating Caliper Disc Brake

INDEX

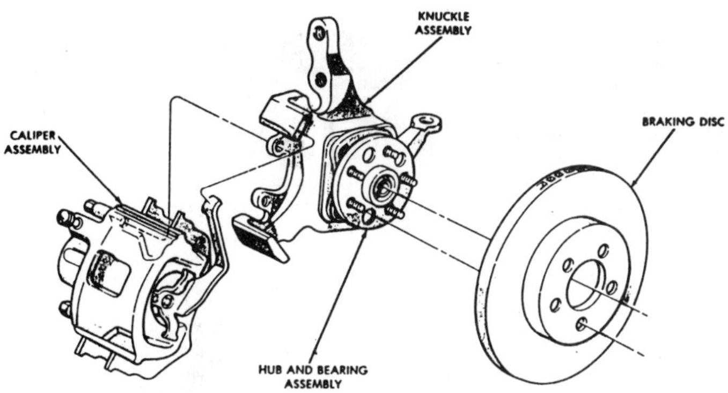

Fig. 1 Kelsey-Hayes dual pin floating caliper disc brake assembly

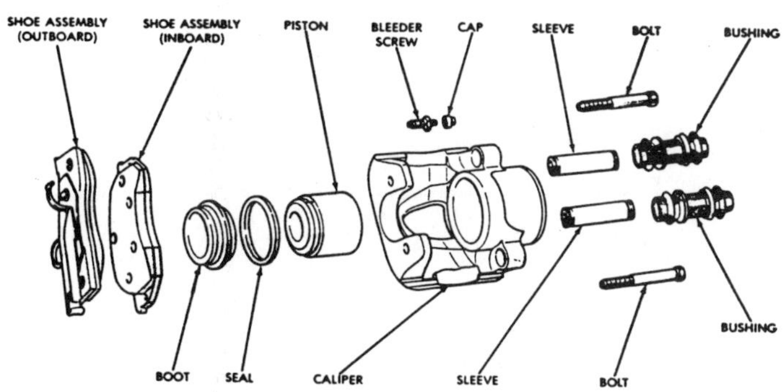

Fig. 2 Disassembled view of Kelsey-Hayes disc brake

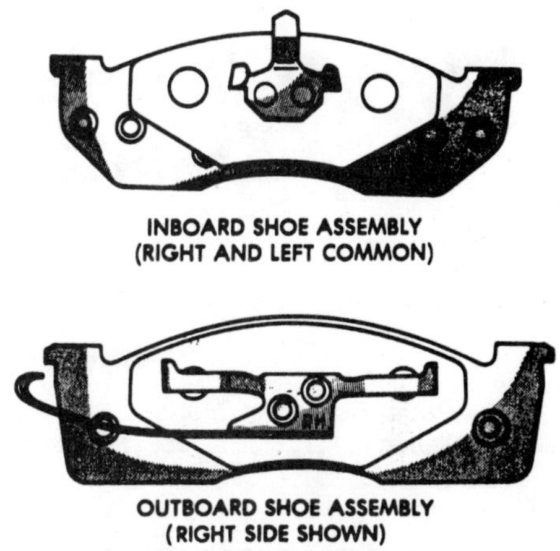

Fig. 3 Brake pad identification

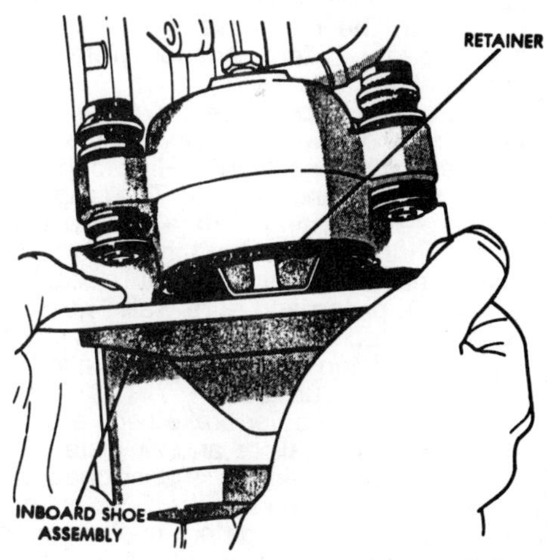

Fig. 4 Installing inboard pad

OPERATION

The single piston, floating caliper disc brake assembly, **Fig. 1,** consists of the rotor, caliper, pads and the driving hub. The caliper, **Fig. 2,** is mounted through bushings and sleeves by two through bolts threaded into the steering knuckle. Two machined abutments on the steering knuckle position and align the caliper fore and aft. The mounting bolts and bushings control the movement of the caliper and the piston seal to assist in maintaining proper pad clearance.

This assembly has an anti-rattle clip attached to the outer pad and an inner pad-piston retainer clip.

All of the braking force is taken directly by the steering knuckle. The caliper is a one piece casting with the inboard side containing a single piston cylinder bore.

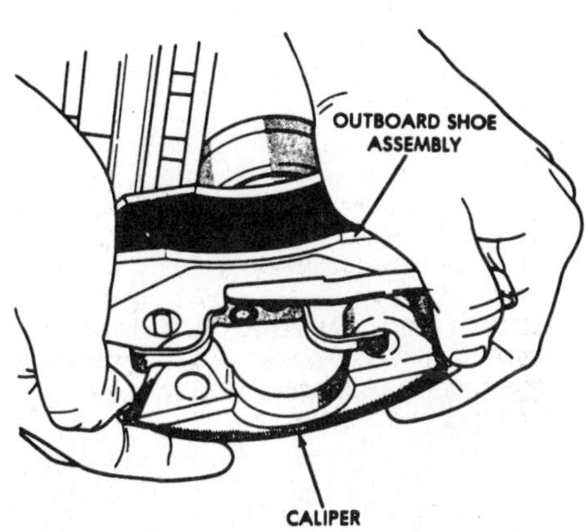

Fig. 5 Installing outboard pad

A square cut rubber piston seal is located in a machined groove in the caliper bore and provides a seal between piston and caliper bore.

A molded rubber dust boot installed in a groove in the cylinder bore and piston keeps contamination from the caliper bore and piston. The boot mounts in the caliper bore and in a groove in the piston.

BRAKE PADS REPLACE

REMOVAL

1. Remove $1/2$ of brake fluid from reservoir.
2. Raise and support front of vehicle.
3. Remove wheel and tire assembly.
4. Remove caliper-to-steering knuckle attaching bolts.
5. Remove caliper and brake pads as an assembly by pulling lower end of caliper outward from steering knuckle, then sliding assembly away from braking disc. **Suspend caliper with wire to avoid damaging brake hose.**
6. Remove brake pads from caliper.

INSTALLATION

The inboard pads are interchangeable and may be used on either side of vehicle. The outboard pads are not interchangeable and are marked "L" for left side or "R" for right side of vehicle, **Fig. 3.**

1. Install inboard pad in caliper, entering retainer into bore of piston, **Fig. 4.**
2. Position properly marked outboard pad hold-down spring into caliper as shown in **Fig. 5.**
3. Lower caliper over braking disc, then install attaching bolts. **Torque** bolts to 18-25 ft. lbs.
4. Install tire and wheel assembly and lower vehicle.

CALIPER OVERHAUL

Refer to "Caliper Overhaul" as outlined under "Type 2—A.T.E. Dual Pin Floating Caliper (Front) Disc Brake" for overhaul procedure.

Type 7—Dual Pin Sliding Caliper Disc Brake, Eagle 4 X 4

INDEX

OPERATION

The dual pin sliding caliper disc brake assembly incorporates a rotor which mounts to the hub assembly, a caliper, brake pads, caliper anchor plate and splash shield, **Fig. 1.**

The caliper is positioned over the rotor and slides on two mounting pins which maintain caliper position relative to the rotor and caliper anchor plate. The caliper is a one-piece casting with a groove machined in the bore to hold the square cut piston seal to maintain a hydraulic seal between the piston and bore wall. The dust boot is seated in a machined recess in the top of the piston bore and a groove in the piston exterior surface.

The inner and outer brake pads are positioned by the caliper anchor plate. The brake pad anti-rattle clip is positioned between the brake pad and caliper anchor plate. The brake linings are riveted to the pads and the inner and outer brake pads are not interchangeable.

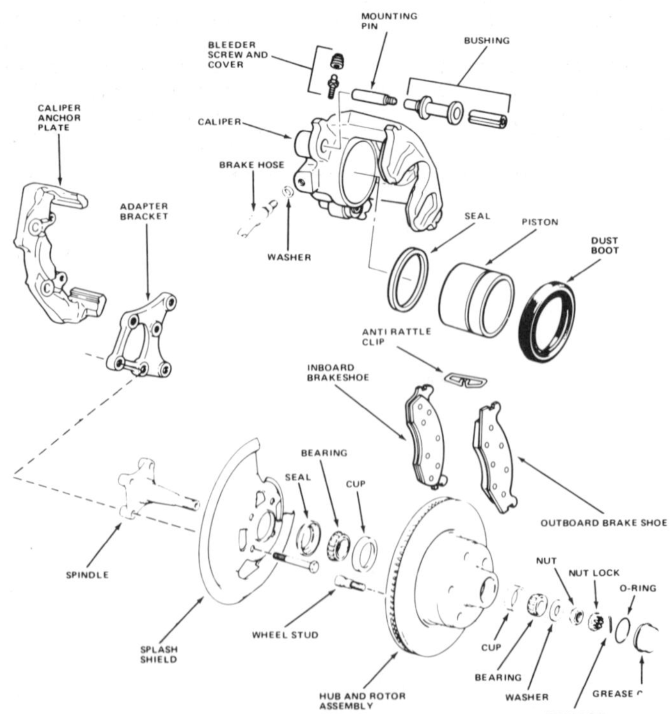

Fig. 1 Dual pin sliding caliper disc brake assembly. Eagle 4 X 4

BRAKE PADS, REPLACE

REMOVAL

1. Drain and discard approximately $2/3$ of the brake fluid from the larger brake fluid reservoir.
2. Raise and support vehicle.
3. Remove tire and wheel assembly.
4. Using a suitable screwdriver, pry piston fully into caliper bore. If piston cannot be bottomed in cylinder bore using a screwdriver, use a C-clamp.
5. Using a 7mm Allen wrench, remove caliper mounting pins.
6. Lift caliper from anchor plate and off rotor.
7. **Suspend caliper with suitable wire to prevent damage to brake hose.**
8. Hold anti-rattle clip against caliper anchor plate and remove outer brake pad. **Note position of anti-rattle clip for assembly reference.**
9. Remove inner brake pad from caliper anchor plate, then anti-rattle clip.
10. Wipe inside of caliper with dry, clean cloth and inspect piston bore for leakage. If leakage is present, refer to "Caliper Overhaul." **Do not clean caliper with compressed air as damage to the dust boot may result.**

INSTALLATION

1. Inspect caliper and anchor plate abutment surfaces for rust and corrosion. If rust and corrosion are evident, clean surfaces with wire brush.
2. Lightly lubricate caliper and anchor plate abutment surfaces with Molydisulfide grease.
3. Install anti-rattle clip on trailing end of anchor plate. Ensure split end of clip faces away from rotor.
4. While holding anti-rattle clip in position, install inner and outer brake pads.
5. Install caliper over rotor and into position on adapter.
6. Install caliper mounting pins. **Torque** to 30 ft. lbs.
7. Fill master cylinder with clean brake fluid, then pump brake pedal several times to position caliper piston and brake pads.
8. Install tire and wheel assembly, then lower vehicle.
9. Check brake fluid and refill as necessary.

CALIPER OVERHAUL

REMOVAL

1. Follow steps 1 through 3 of Removal procedure under "Brake Pads, Replace."
2. Clean dirt from brake hose fittings.
3. Disconnect brake hose from caliper, then cap open lines to prevent entry of dirt.
4. Discard hose fitting washer.
5. Remove caliper following steps 4, 5, 7 & 8 of Removal procedure under "Brake Pads, Replace."

DISASSEMBLY

1. Clean caliper exterior with suitable brake cleaning solvent.
2. Drain caliper, then place on clean work surface.
3. Pad caliper interior with clean shop cloths, then using compressed air, gently apply just enough air pressure into caliper fluid inlet hole to ease piston out of bore. **Do not place fingers in front of piston in an attempt to catch or protect it when applying compressed air. This could result in serious injury.**
4. Using a suitable screwdriver, pry dust boot from bore using care not to damage piston bore, **Fig. 1.**
5. Using a pencil or other suitable wooden object, remove piston seal.
6. Remove bushings from caliper ear and discard.

INSPECTION

1. Clean all parts in brake cleaning solvent or brake fluid, then blow out caliper fluid passages with filtered compressed air.
2. Inspect caliper mounting pins for corrosion, replace as necessary. **Do not attempt to clean or polish caliper mounting pins with abrasives as protective plating will be removed.**
3. Inspect caliper piston. If nicked, scratched, corroded or protective plating is worn off, replace. **Do not attempt to refinish piston in any way as protective plating will be removed, leading to corrosion and eventual failure.**
4. Inspect caliper bore. If nicked, scratched, worn, cracked, or badly corroded, replace caliper. **Minor corrosion and stains can be removed from caliper bore with crocus cloth. Do not attempt to clean caliper bore with any other abrasives.**

ASSEMBLY

1. Lubricate piston bore and seal with brake fluid, then work seal into piston groove using fingers only.
2. Lubricate piston with brake fluid, then slide metal portion of dust seal over open end of piston and pull rearward until seal boot seats in piston groove, then push metal portion of seal forward until retainer is flush with rim and seal fold snaps into position.
3. Insert piston and seal assembly into bore. Do not unseat piston seal.
4. Using a hammer handle, press piston in bore.
5. Using tool No. J-33028, install dust seal in caliper housing.
6. Install bleeder screw and tighten securely.
7. Install replacement plastic sleeves and rubber bushings in caliper ears.

INSTALLATION

1. Follow steps 1 through 6 of Installation procedures under "Brake Pads, Replace."
2. Install replacement hose fitting washer on brake hose, then install hose in caliper and **torque** to 25 ft. lbs.
3. Fill master cylinder with clean brake fluid and bleed brakes.
4. Install tire and wheel assembly, then lower vehicle.
5. Check brake fluid and refill as necessary.

Type 8—Dual Pin Sliding Caliper Disc Brake, Eagle Premier

INDEX

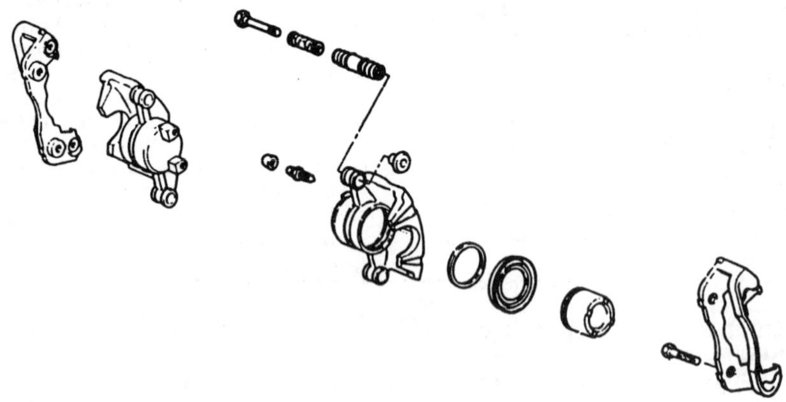

Fig. 1 Disassembled view of disc brake assembly

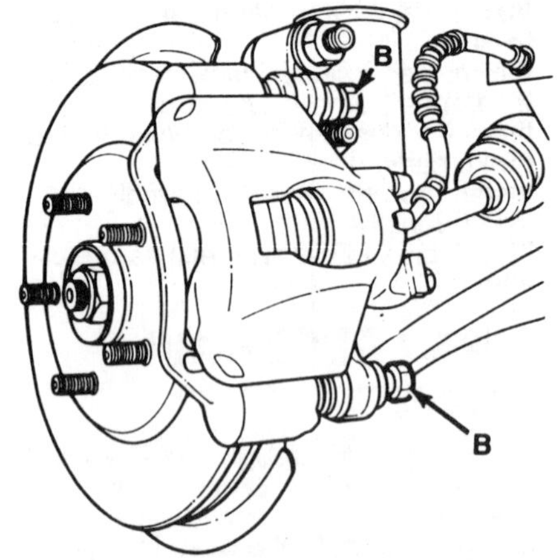

Fig. 2 Caliper mounting pin locations

OPERATION

This dual pin sliding caliper disc brake system uses a rotor mounted on a drive hub, brake pads, a mounting bracket and a single piston caliper, **Fig. 1.**

Two pins which are mounted through sleeves and bushings in the caliper onto a mounting bracket control the movement of the caliper and the piston seal, to assist in maintaining proper pad clearance.

This assembly uses two anti-rattle clips, one mounted on the bottom end of each brake pad.

A molded rubber and metal dust seal protects the piston and caliper bore from corrosion, and is held in place by an inner lip which is seated into a piston groove.

BRAKE PADS REPLACE

REMOVAL

1. Drain and discard approximately 1/2 of the brake fluid from the master cylinder reservoir.
2. Raise and support vehicle.
3. Remove wheel and tire assembly.
4. Using a suitable screwdriver, slightly pry piston into caliper bore.
5. Remove caliper mounting pins, **Fig. 2,** then lift caliper from mounting bracket and off rotor. **Suspend caliper with suitable wire to prevent damaging brake hose.**
6. Remove outer brake pad first, then the inner pad, noting position of anti-rattle clips for assembly reference.
7. Wipe inside of caliper with dry, clean cloth and inspect piston bore for leakage. If leakage is present, refer to "Caliper Overhaul." **Do not clean caliper with compressed air as damage to the dust boot may result.**

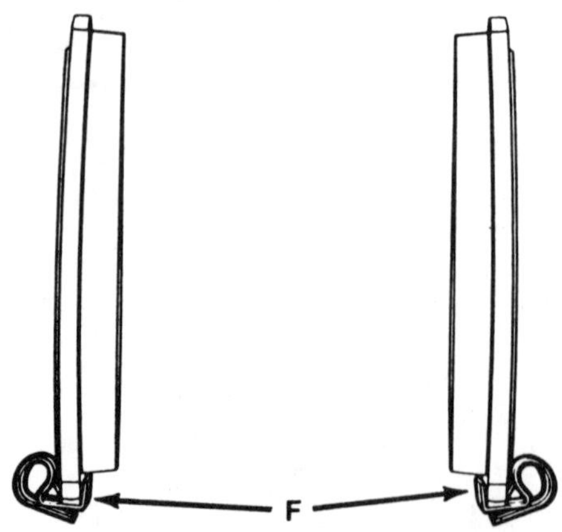

Fig. 3 Anti-rattle clip locations

INSTALLATION

1. Using a C-clamp, push caliper piston back until it is bottomed into cylinder bore.
2. Using a wire brush, thoroughly clean brake pad contact points on caliper mount.
3. Assemble anti-rattle clips on brake pads as shown in **Fig. 3.**
4. Position inner pad on caliper mount first, then the outer pad.
5. Lower caliper over brake pads onto caliper mounting bracket, ensuring brake hose is not twisted.
6. Lubricate mounting pins with suitable grease, then install through caliper sleeves and bushings. Thread pins into mounting bracket and **torque** to 18 1/2 ft. lbs.
7. Install wheel and tire assembly, then lower vehicle.
8. Slowly pump brake pedal to fully seat pads, then top off brake fluid level in reservoir as necessary.

CALIPER OVERHAUL

REMOVAL & DISASSEMBLY

1. Follow steps 1 through 5 of removal procedures under "Brake Pads, Replace."
2. Disconnect brake hose from caliper, capping open line to prevent dirt from entering system. Discard brake hose copper washer.
3. Clean caliper exterior with suitable brake cleaning solvent.
4. Drain caliper, then place on clean work surface.
5. Using screwdriver, loosen metal lip of caliper piston dust seal.
6. Place a block of wood between caliper and caliper piston, then using

compressed air, gently apply air pressure into caliper fluid inlet hole to ease piston out of bore. **Do not place fingers in front of piston in an attempt to catch or protect it when applying compressed air. This could result in serious injury.**

7. Using a feeler gauge or other suitable tool, remove piston inner seal, **Fig. 4.**

INSPECTION

1. Inspect caliper bore for scoring or corrosion. Light corrosion and scoring can be removed with brake hone, using brake fluid for lubricant. **Do not attempt to clean caliper bore with any other abrasives.**
2. Clean caliper with suitable brake cleaning solvent, then dry with compressed air.
3. Inspect caliper hardware for corrosion and replace as necessary.
4. Inspect caliper piston for corrosion, scoring or wear, replacing as necessary. Light corrosion on piston may be cleaned with mineral spirits. **Never sand or apply abrasive cleaners to piston.**

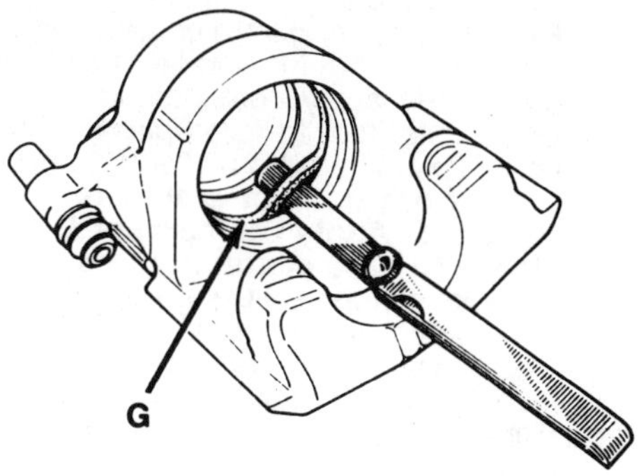

Fig. 4 Removing caliper piston inner seal

ASSEMBLY

1. Lubricate piston bore and new inner seal with brake fluid, then work seal into caliper bore groove by hand.
2. Lubricate piston with brake fluid, then slide metal portion of new dust seal over open end of piston and pull rearward until seal inner lip seats in piston groove, then push outer portion of seal forward until it is flush with end of piston.
3. Insert piston and seal assembly into bore. Do not unseat inner seal. With a slight rocking motion, bottom piston into bore.
4. Center seal lip into counterbore on caliper, then seat seal into caliper housing using tool No. Fre. KM.02 and adapter No. Fre. ML.02 or equivalents.

INSTALLATION

1. Position caliper over brake pads onto caliper mounting bracket.
2. Lubricate mounting pins with suitable grease, then install through caliper sleeves and bushings. Thread pins into mounting bracket and **torque** to 18 1/2 ft. lbs.
3. Using a new copper washer, install brake hose in caliper and **torque** to 13 ft. lbs. Ensure hose is not twisted.
4. Fill master cylinder with clean brake fluid and bleed brakes.
5. Install wheel and tire assembly, then lower vehicle.

Type 9—AD54 One Piece Pin Type Caliper (Front) Disc Brake

INDEX

1. Lid
2. Lock pin
3. Sleeve
4. Caliper body
5. Guide pin boot
6. Piston
7. Lock pin boot
8. Piston seal
9. Piston boot
10. Boot ring
11. Inner shim
12. Pad assembly
13. Pad clip B
14. Pad clip C
15. Anti-squeak shim
16. Guide pin
17. Sleeve
18. Support mounting
19. Brake disc
20. Hub

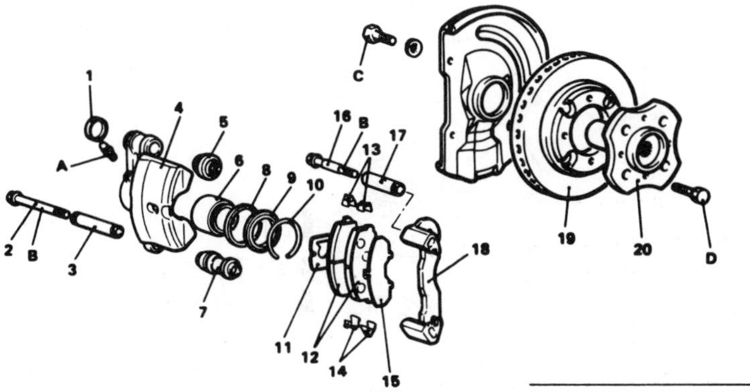

	Nm	ft.lbs.
A	7–9	5–7
B	22–32	16–23
C	80–100	58–72
D	50–60	36–43

Fig. 1 AD54 one piece pin type caliper front disc brake assembly

DISC BRAKE SERVICE

BRAKE PADS, REPLACE

1. Remove approximately 1/3 of brake fluid from master cylinder.
2. Raise and support front of vehicle, then remove tire and wheel assembly.
3. Remove lock pin, **Fig. 1,** then lift caliper assembly upwards. Support caliper assembly with suitable piece of wire to prevent damage to brake hose.
4. Remove inner shim, anti-squeak shim and brake pad assemblies from support mounting.
5. Press caliper piston into caliper bore with suitable tool.
6. Install pad clips "B" and "C," **Fig. 1,** pad assemblies, inner shims and anti-squeak shims into support mounting.
7. Lower caliper body, then install lock pin. Refer to **Fig. 1,** for torque specifications.
8. Depress brake pedal several times to seat pads, then check and replenish brake fluid as necessary.
9. Install wheel and tire assembly and lower vehicle.

CALIPER, REPLACE

1. Raise and support vehicle, then remove wheel and tire assembly.
2. Disconnect brake hose at caliper strut, then at the brake caliper.
3. Remove support mounting to steering knuckle attaching bolts, then the front brake assembly.
4. Remove brake pads, then separate caliper body from support mounting.
5. Reverse procedure to install. Refer to **Fig. 1**, for torque specifications.

CALIPER OVERHAUL

1. Remove caliper as described under "Caliper, Replace."
2. Remove boot ring from caliper body.
3. Place shop towel in saddle of caliper body, then apply compressed air to brake fluid inlet to force piston from bore. Remove piston. **Keep hands away from front of piston during removal. Apply compressed air slowly to ease piston from bore.**
4. Remove piston seal using caution not to scratch cylinder walls.
5. Clean piston and cylinder wall surfaces with brake fluid or alcohol.
6. Remove sleeve and boot from caliper body.
7. Inspect cylinder and piston for wear, damage and/or corrosion. Inspect caliper body and sleeve for wear.
8. Apply brake fluid to caliper body cylinder walls.
9. Apply an even coat of brake grease to piston seal, then install seal in caliper bore.
10. Carefully install caliper piston. Ensure seal does not twist during installation.
11. Apply brake grease to piston boot, then install boot and boot ring.
12. Apply brake grease to lock pin and guide pin sleeves.
13. Install caliper assembly. Refer to **Fig. 1**, for torque specifications.

Type 10—FS17 Conquest & PFS15 Colt, One Piece Pin Type Caliper (Front) Disc Brake

INDEX

DISC BRAKE SERVICE

BRAKE PADS, REPLACE

1. Raise and support vehicle, then remove wheel assembly.
2. Remove lower caliper slide pin, **Figs. 1 and 2, then support caliper with wire to prevent damage to brake hose.**
3. Remove brake pads along with inner and outer shims.
4. Reverse procedure to install. Refer to **Figs. 1 and 2,** for torque specifications.

CALIPER, REPLACE

1. Raise and support vehicle, then remove wheel assembly.
2. Disconnect brake hose at caliper strut, then at brake caliper.
3. Remove caliper support to steering knuckle attaching bolts, then front brake assembly.
4. Remove brake pads and shims, then separate caliper body from caliper support.
5. Reverse procedure to install. Refer to **Figs. 1 and 2,** for torque specifications.

CALIPER OVERHAUL

1. Remove caliper assembly as outlined under "Caliper, Replace."
2. Remove bushing from caliper support using slide pin, then cap and pin boot.
3. Position a shop towel in caliper body, then apply compressed air through the brake hose fitting hole to remove piston and dust boot.

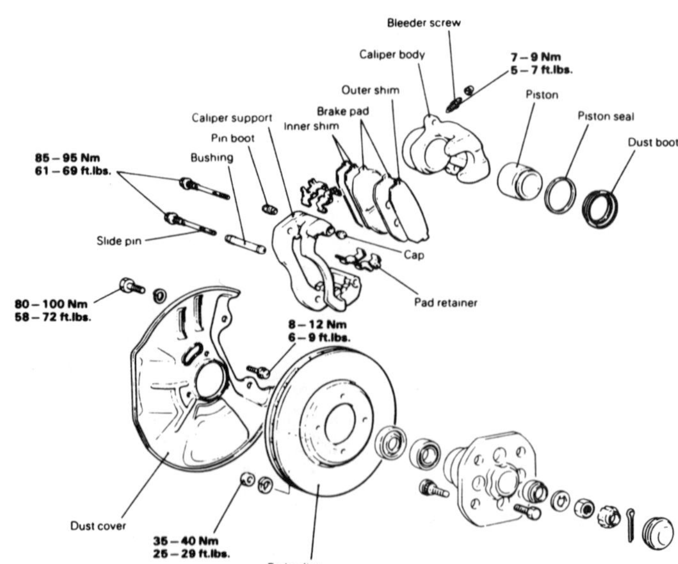

Fig. 1 FS17 Conquest one piece pin type caliper front disc brake assembly

1. Sleeve bolt B
2. Sleeve
3. Caliper
4. Bushing
5. Sleeve bolt A
6. Pin boot
7. Piston
8. Piston seal
9. Dust boot
10. Inner shim
11. Pad assembly
12. Pad liner
13. Outer shim
14. Torque member
15. Brake disc
16. Hub

	Nm	ft lbs.
A	80 – 100	58 – 72
B	22 – 32	16 – 23
C	7 – 9	5 – 7
D	50 – 60	36 – 43

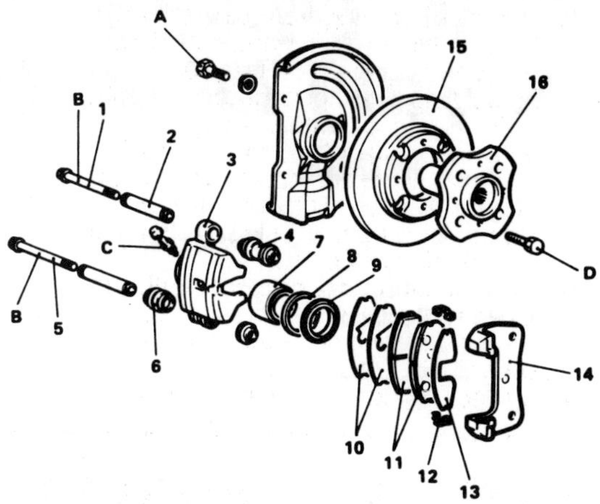

Fig. 2 PFS15 Colt one piece pin type caliper front disc brake assembly

4. Using a suitable screwdriver, remove piston seal.
5. Inspect cylinder and piston for wear or damage and/or corrosion. Inspect caliper body and sleeve for wear.
6. Install new piston seal into cylinder. The piston seal in repair kit is coated with a special grease, do not wipe it off.
7. Apply suitable grease to lip of cylinder, then brake fluid to external surface of piston.
8. Install new dust boot onto piston.
9. Position end of dust boot into caliper body groove and install piston into caliper by hand. Take care not to damage piston. Ensure end of dust boot is fitted into piston groove.
10. Apply a suitable grease to contact surface of slide pin, seat surface of cap for caliper support, and inside surface of pin boot.
11. Install cap and pin boot to caliper support, then apply suitable grease to inside surface of bushing.
12. Apply a suitable adhesive to lip of slide pin bushing, then install bushing into caliper support using slide pin.
13. Apply brake fluid to threads of slide pin and a suitable grease to shank of slide pin, then install caliper body to caliper support.
14. Install caliper assembly. Refer to **Figs. 1 and 2,** for torque specifications.

Type 11—One Piece Pin Type Caliper (Rear) Disc Brake

INDEX

DISC BRAKE SERVICE

BRAKE PADS, REPLACE

1. Raise and support vehicle, then remove wheel assembly.
2. Disconnect parking brake cable.
3. Remove lower caliper lock pin, then support caliper with wire to prevent damage to brake hose.
4. Remove brake pads and shims, **Fig. 1.**
5. Reverse procedure to install. Refer to **Fig. 1** for torque specifications.

CALIPER, REPLACE

1. Raise and support vehicle, then remove wheel assembly.
2. Disconnect brake hose, then parking brake cable from brake assembly.
3. Remove brake assembly attaching bolts, then brake assembly.
4. Remove brake pads and shims, then separate caliper body from caliper support, **Fig. 1.**
5. Reverse procedure to install. Refer to **Fig. 1** for torque specifications.

CALIPER OVERHAUL

1. Remove caliper assembly as outlined.
2. Using a suitable screwdriver, remove cap ring and garter spring from lever cap.
3. Remove lever cap from caliper groove, then retaining ring and pull out parking lever assembly.
4. Unscrew spindle from caliper body, then remove boot ring.
5. Push out piston from caliper body using a suitable screwdriver.
6. Using a suitable screwdriver, remove piston seal. Take care not to damage caliper bore.
7. Using tool No. MB990665 or equivalent, and a suitable vise, remove bearings from caliper.
8. Remove guide pin boot and lock pin boot from caliper body.
9. Inspect cylinder and piston for wear or damage and/or corrosion. Inspect caliper body and sleeve for wear.
10. Apply a suitable grease to caliper bearings.
11. Using a suitable vise and tool No. MB990665 or equivalent, press bearings in until it is flush with caliper body.
12. Install new piston seal into cylinder, then coat piston seal and inside surface of cylinder.

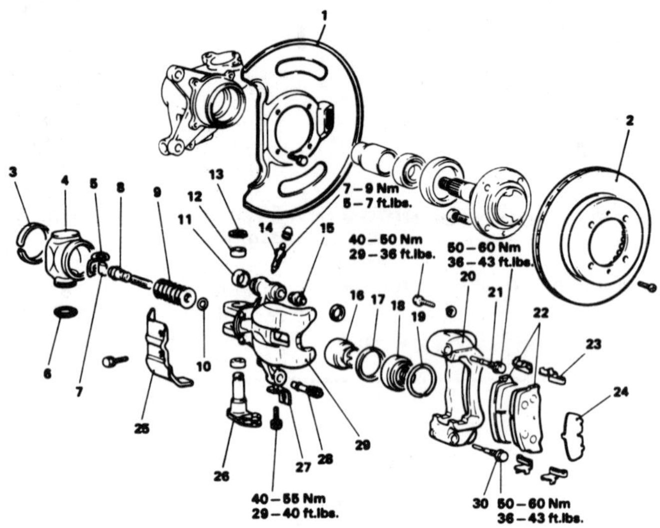

1. Dust cover
2. Brake disc
3. Cap ring
4. Lever cap
5. Return spring
6. Garter spring
7. Connecting link
8. Spindle
9. Spring washer
10. Spindle seal
11. Lid
12. Bearing
13. Retaining ring
14. Bleeder screw
15. Guide pin boot
16. Piston assembly
17. Piston seal
18. Piston boot
19. Boot ring
20. Caliper support
21. Guide pin
22. Brake pad
23. Pad clip
24. Outer shim
25. Dust cover
26. Parking lever assembly
27. Parking cable bracket
28. Lock pin boot
29. Caliper body
30. Lock pin

Fig. 1 One piece pin type caliper rear disc brake assembly

13. Install piston assembly into cylinder.
14. Apply suitable grease to dust boot fitting groove in caliper body, then install new dust boot and boot ring.
15. Coat spindle seal with brake fluid, then install washers on spindle.
16. Coat contact surface of caliper body and spring washers with suitable grease, then screw spindle into caliper body until it rotates freely.
17. Using tool No. MB990666 or equivalent and a suitable vise, compress spring washers and screw spindle into caliper.
18. Position connecting link and return spring onto spindle and install lever cap to parking lever assembly, then install assembly into caliper body. Hold parking lever assembly with retaining ring.
19. Apply suitable grease to lever cap and lip, then install lever cap to caliper body.
20. Coat contact point of caliper support guide pin and inside of lock pin boot with suitable grease, then install caliper body to caliper support.
21. Install caliper assembly. Refer to **Fig. 1** for torque specifications.

ADJUSTMENTS

PARKING BRAKE, ADJUST

1. Apply parking brake while counting number of clicks.
2. Lever should click 4-5 times. If not within specifications, remove console and release parking brake lever, then turn adjusting nut located behind parking brake lever as necessary to bring parking brake within specifications.

Type 12—One Piece Pin Type Caliper (Front) Disc Brake

INDEX

DISC BRAKE SERVICE

BRAKE PADS, REPLACE

1. Remove some fluid from master cylinder reservoir.
2. Raise and support vehicle.
3. Remove front wheels.
4. Disconnect brake pad wear warning lamp electrical connector.
5. Return piston into its bore by sliding caliper outward.
6. Remove lower caliper guide attaching bolt. Remove caliper by rotating it up from guide pivot and sliding it from upper guide pivot.
7. Remove brake pads.
8. Reverse procedure to install noting the following:
   a. **After removal of brake pads, inspect disc brake rotor. Do not machine rotor. Replace rotor if excessive wear or scoring is evident.**
   b. **Torque** caliper guide attaching bolt to 18 ft. lbs.

CALIPER, REPLACE

1. Raise and support vehicle.
2. Remove front wheels.
3. Disconnect and cap brake line from caliper.
4. Disconnect brake pad wear warning lamp electrical connector.
5. Return piston into its bore by sliding caliper outward.
6. Remove lower caliper guide attaching bolt. Remove caliper by rotating it up from guide pivot and sliding it from upper guide pivot.
7. Slide caliper onto upper guide pivot, then rotate caliper downward until seated over brake pads.
8. Install lower caliper guide attaching bolt. **Torque** bolt to 18 ft. lbs.
9. Reconnect brake pad wear warning lamp electrical connector.
10. Install caliper brake line. Properly bleed brake hydraulic system.
11. Install front wheels. Ensure brake operation is proper before test driving the vehicle.

CALIPER OVERHAUL

1. Remove caliper assembly as outlined under "Caliper, Replace."
2. Position a suitable block of wood between caliper and caliper piston. **Keep fingers away from piston.**
3. Using low pressure compressed air, extract piston from caliper piston bore.
4. Using a suitable tool, remove piston to caliper seal.
5. Install a new seal into caliper. Coat seal using clean brake fluid.
6. Install piston and dust cover onto caliper. Do not damage seal.
7. Check condition of anti-rattle spring. Replace spring if necessary.
8. Install caliper assembly, **torque** guide bolt to 18 ft. lbs. Ensure hydraulic brake system is properly bled.

Type 13—AD30P One Piece Pin Type Caliper (Rear) Disc Brake

INDEX

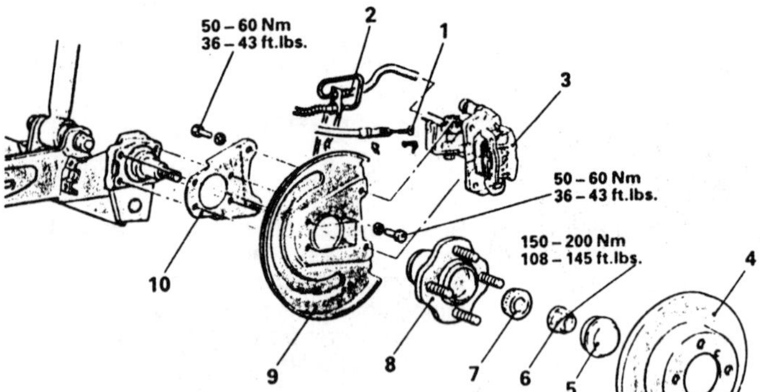

1. Connection of parking brake cable
2. Connection of brake hose
3. Rear brake assembly
4. Rear brake disc
5. Hub cap
6. Wheel bearing nut
7. Outer bearing inner race
8. Rear hub assembly
9. Dust shield
10. Disc brake adapter

Fig. 1 AD30P one piece pin type caliper rear disc brake assembly

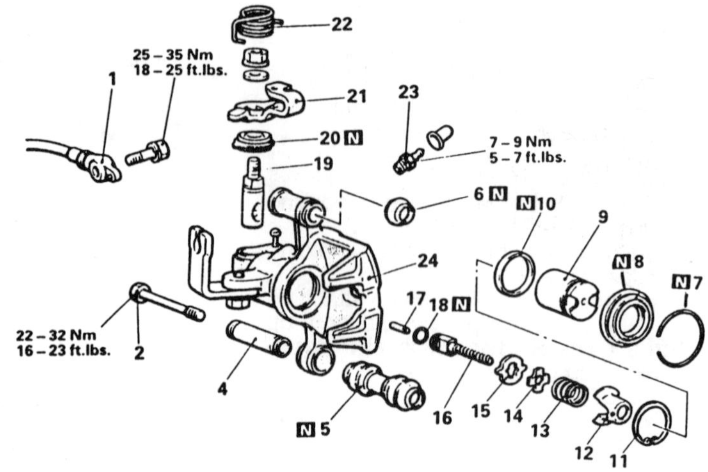

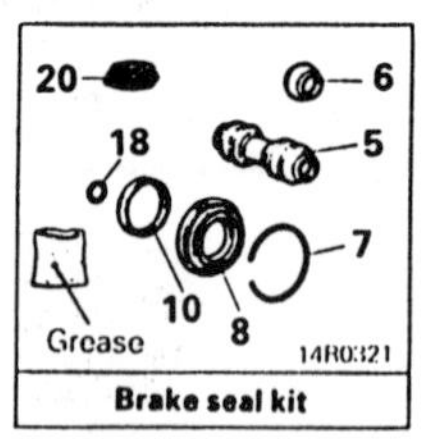

1. Connection for brake hose
2. Lock pin
4. Lock pin sleeve
5. Lock pin boot
6. Guide pin boot
7. Boot ring
8. Piston boot
9. Piston assembly
10. Piston seal
11. Snap ring
12. Spring case
13. Return spring
14. Stopper plate
15. Stopper
16. Auto-adjuster spindle
17. Connecting link
18. O-ring
19. Spindle lever
20. Lever boot
21. Parking brake lever
22. Return spring
23. Bleeder screw
24. Caliper body

Fig. 2 Disassembled view of AD30P disc brake

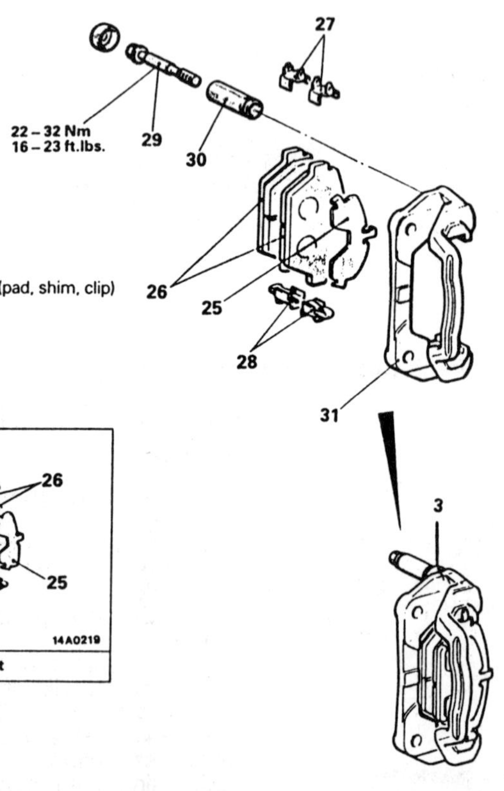

1. Lock pin
3. Support mounting (pad, shim, clip)
25. Outer shim
26. Pad assembly
27. Pad clip C
28. Pad clip B
29. Guide pin
30. Guide pin sleeve
31. Support mounting

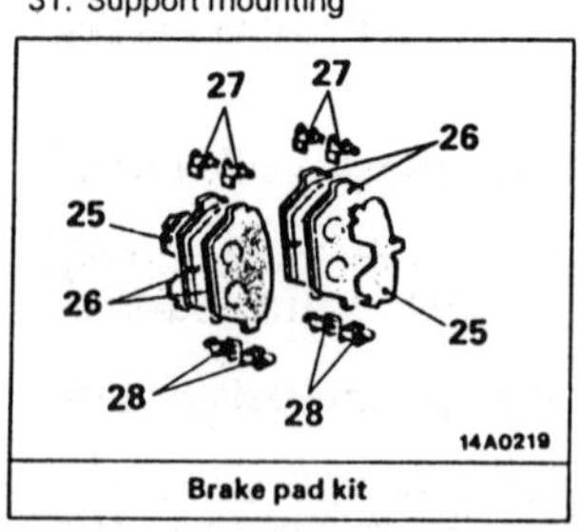

Fig. 3 Disc brake pad support mounting

DISC BRAKE SERVICE

CALIPER & BRAKE PAD ASSEMBLY, REPLACE

Refer to **Figs. 1 through 3**, when performing the following procedure.

1. Raise and support vehicle.
2. Remove rear wheel and tire assembly.
3. Disconnect parking brake cable, and brake hose from the caliper assembly.
4. Remove lower lock pin, then the caliper assembly from the vehicle.
5. Remove upper lock pin, then separate brake pad support mounting from the caliper assembly.
6. Remove brake pads from support mounting.
7. Remove outer shim, and pad clips B and C from brake pads.
8. Reverse procedure to install noting the following:
   a. **Torque** bolts to specifications as shown in **Figs. 1 through 3.**
   b. Ensure hydraulic brake system is

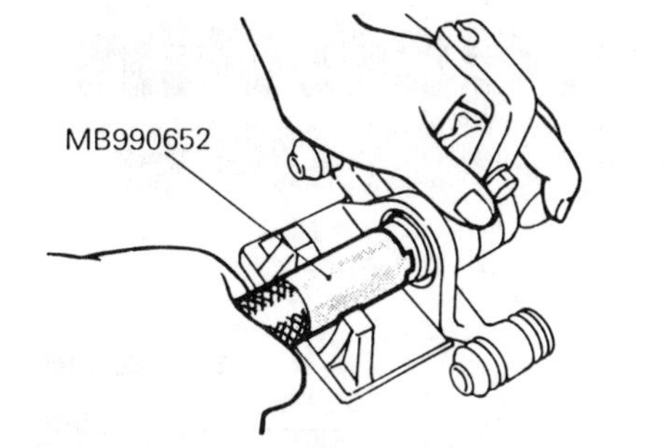

Fig. 4 Caliper piston removal

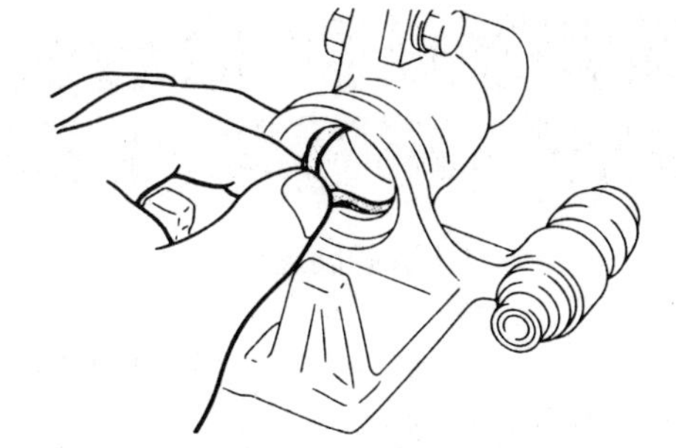

Fig. 5 Piston seal removal

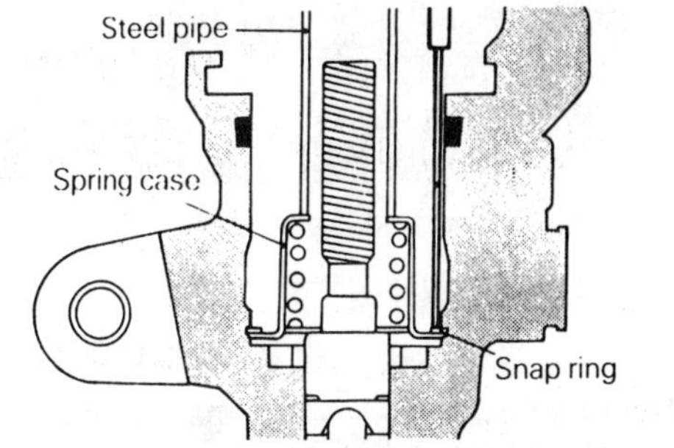

Fig. 6 Compressing spring case

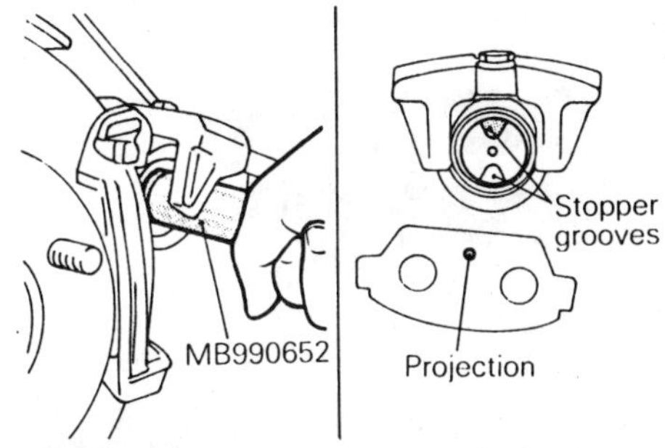

Fig. 7 Installing and aligning caliper piston

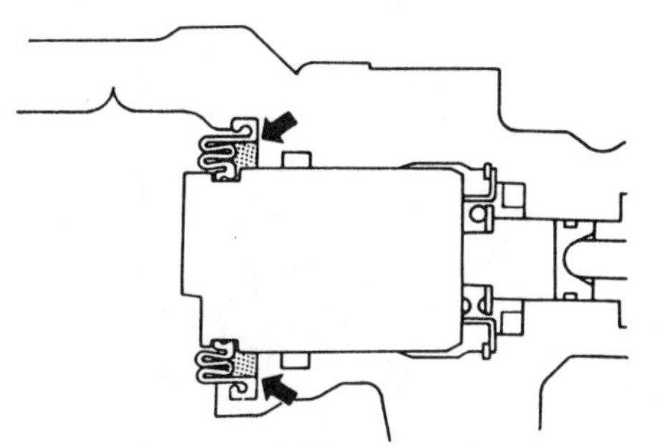

Fig. 8 Dust boot installation

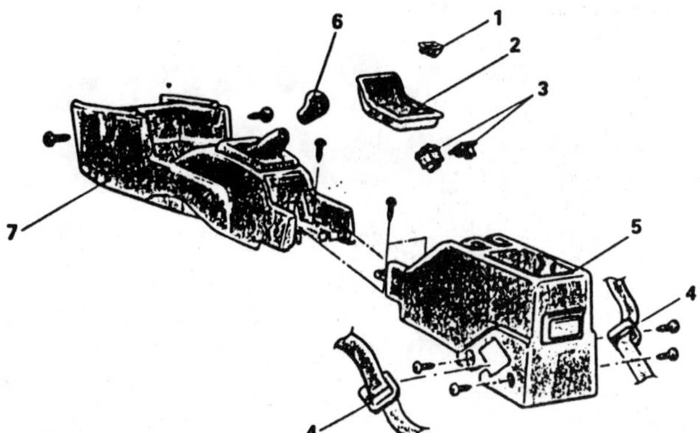

Disassembly steps

1. Coin holder
2. Console box tray
3. Remote control mirror switch or cover
4. Bezzel <Vehicles for U.S.>
5. Rear console box assembly
6. Shift lever knob <M/T>
7. Front console box assembly
8. Shift lever panel <M/T>
9. Shift lever boot <M/T>
10. Ashtray
11. Cup holder
12. Rear console box
13. Front console box

NOTE
Reverse the disassembly procedures to reassemble.

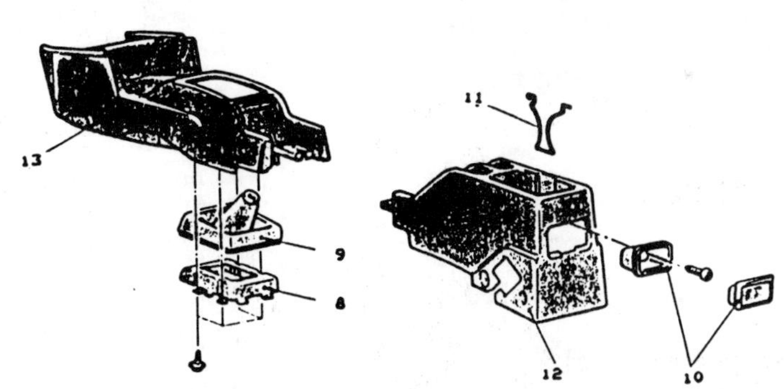

Fig. 9 Floor console assembly

properly bled.

c. Adjust parking brake lever stroke as described under "Adjustments."

CALIPER OVERHAUL

Refer to **Figs. 2 through 8** when performing the following procedure.

Disassembly

1. Remove caliper assembly as described under "Caliper & Brake Pad Assembly, Replace."
2. Remove piston with a twisting motion, using special tool as shown **Fig. 4.**
3. Remove piston seal as shown **Fig. 5.**
4. Compress spring case into caliper body using a 3/4 inch steel pipe as shown in **Fig. 6.** Remove snap ring using suitable snap ring pliers.
5. Remove guide pin boot, lock pin sleeve and boot from caliper assembly.

Inspection

1. Check for rust on piston and spindle lever shaft.
2. Check piston boot and seal for deterioration and wear.
3. Check caliper body for rust or cracks.

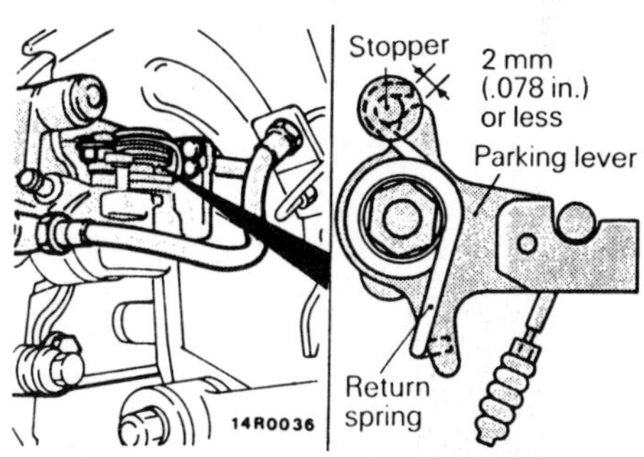

Fig. 10 Checking parking brake lever to stopper clearance

Assembly

1. Using a suitable grease, lubricate the following:
   a. Lever boot, spindle lever, O-ring, connecting link and auto-adjuster spindle.
   b. Cylinder walls, piston seal and piston.
   c. Guide pin boot lip inner surface.
   d. Lock pin boot lip inner surface, and lock pin sleeve. **Do not apply grease to sleeve threads.**
2. Compress spring case into caliper body using a 3/4 inch steel pipe as shown in **Fig. 6.** Install snap ring so its opening faces bleeder.

3. Install piston seal into caliper body.
4. Using special tool, press piston into caliper body with stopper grooves aligned, **Fig. 7.**
5. Install piston boot, **Fig. 8.**
6. Install guide pin boot, lock pin boot and sleeve to caliper body.

ADJUSTMENTS

PARKING BRAKE, ADJUST

1. Pull the parking brake lever and ensure it moves five to seven notches.
2. If lever movement is not as specified, make adjustments as follows:
   a. Remove rear console box as shown **Fig. 9.**
   b. Using two suitable wrenches, loosen the parking brake cable locknut while holding the adjusting nut.
   c. Back adjusting nut out to the end of the cable rod.
   d. Start the engine and allow it to idle.
   e. Forcefully depress brake pedal five to six times, until pedal stroke stops changing.
   f. Check stopper to parking brake lever clearance at the caliper side is as specified in **Fig. 10.** If clearance exceeds the specified amount, check for a sticking parking brake cable, or caliper automatic adjuster malfunction.
   g. Tighten the cable adjusting nut until parking brake lever movement is five to seven notches. **Use caution not to over adjust the cable, as excessive pull will cause caliper automatic adjuster mechanism failure.**
   h. Ensure no play exist between parking brake lever and the adjusting nut, then tighten the lock nut.
   i. Raise and support rear of vehicle. Ensure wheels spin freely with parking brake released.

Type 14—Dual Pin Floating Caliper (Rear) Disc Brake

INDEX

DISC BRAKE SERVICE

Refer to **Fig. 1,** when performing the following procedures.

CALIPER & BRAKE PADS, REPLACE

1. Raise and support vehicle.
2. Remove rear wheel and tire assembly.
3. Disconnect brake hose from the caliper.
4. Remove disc brake caliper retaining bolts.
5. Pull caliper upward and away from rotor.
6. Remove outer brake pad by sliding it off the caliper.
7. Remove inner brake pad by pulling it away from the piston.
8. Reverse procedure to install noting the following: **Torque** caliper retaining bolts to 155–230 inch lbs. **Torque** brake hose-to-caliper retaining bolts to 19–29 ft. lbs.

CALIPER OVERHAUL

1. Remove caliper as described under "Caliper & Brake Pads, Replace."
2. Separate bushing from groove in the metal sleeve, then pull sleeve out of the bushing.
3. Compress the bushing, then pull it out of the caliper housing.
4. Using a suitable tool, remove dust boot retainer.
5. Remove dust boot from piston and caliper grooves.

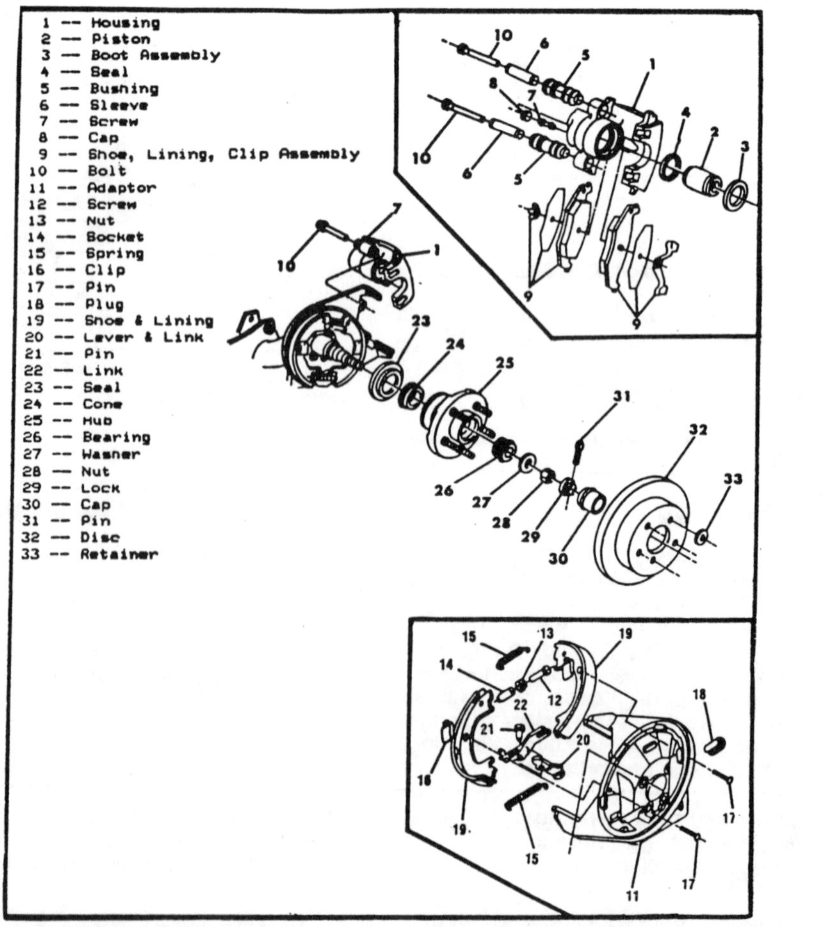

Fig. 1 Dual pin floating caliper rear disc brake assembly

6. Using denatured alcohol or equivalent, thoroughly clean piston and caliper grooves, caliper housing and bushing mounting surfaces.
7. Liberally apply new brake fluid to the dust boot.
8. Install dust boot to piston and caliper grooves.
9. Install dust boot retainer.
10. Compress the bushing and position it in the caliper mounting. Using a suitable tool, ensure bushing is properly seated.
11. Press sleeve into the bushing. Ensure bushing seats in the metal sleeve grooves.

PARKING BRAKE SERVICE

Refer to **Fig. 1**, when performing the following procedure.

PARKING BRAKE SHOES, REPLACE

1. Raise and support vehicle.
2. Remove rear wheel and tire assembly.
3. Remove disc brake caliper retaining bolts.
4. Pull caliper upward and away from rotor. **Do not allow caliper to hang by brake hose.**
5. Remove rotor from hub.
6. Remove grease cap, cotter pin, lock nut, retaining nut, washer, hub and bearing.
7. Using a suitable tool, remove primary shoe hold-down clip from hold down pin.
8. Loosen and remove the adjuster.
9. Remove upper shoe to shoe spring.
10. Pull primary shoe away from anchor. Separate primary shoe from the lower shoe to shoe spring.
11. Using a suitable tool, remove secondary shoe hold-down clip from hold down pin.
12. Remove secondary shoe.
13. Reverse procedure to install noting the following:
    a. Adjust shoe diameter to 6.75 inches.
    b. Tighten wheel bearing adjusting nut as follows:
        1. Rotate hub while **torquing** adjusting nut to 240-300 inch lbs.
        2. Back off adjusting nut 1/4 turn.
        3. Hand tighten adjusting nut. Align nut-to-spindle holes for cotter pin insertion.

ADJUSTMENTS

PARKING BRAKE, ADJUST

1. Release parking brake.
2. Raise and support vehicle.
3. Adjust parking brake cable until there is slack in the cable.
4. Tighten adjusting nut until a slight drag is felt when rotating the rear wheels.
5. Back off adjusting nut two full turns past the point when both rear wheels rotate freely.
6. Check parking brake operation.

Caliper Specifications

Year	Model	Caliper Bore Dia. In.
1985–88	Eagle 4 X 4	2.600
1985–89	Colt①	2.01
	Colt②	2.12
	Colt Vista	2.12
	Conquest③	2.25
	Conquest④	1.63
	Chrysler Front Wheel Drive③	2.130
	Chrysler Rear Wheel Drive	2.755
1987–88	Daytona④	1.300
1987–89	Eagle Medallion	2.125
1988	Dynasty, New Yorker & New Yorker Landau ④	1.420
1988–89	Eagle Premier	2.244
1989	New Yorker Landau④⑤	1.30
	New Yorker Landau④⑥	1.42
	Daytona, Lebaron④⑤	1.34
	Daytona, Lebaron④⑥	1.42
	Eagle Summit③	2.01⑦
	Eagle Summit③	2.12⑧
	Eagle Summit④	1.86
	Colt③	2.01⑦
	Colt③	2.12⑧
	Colt④	1.86

①—Models with 1.5L engine.
②—Models with 1.6L engine
③—Front wheel disc brake.
④—Rear wheel disc brake.
⑤—Models with 14 inch solid rotor.
⑥—Models with 15 inch vented rotor.
⑦—Models with 9.6 inch rotor.
⑧—Models with 10.5 inch rotor.

Rotor Specifications

Year	Model	Nominal Thickness (Inches)	Minimum Refinish Thickness (Inches)	Thickness Variation Parallelism (Inches)	Lateral Runout (T.I.R.)	Finish (Micro-Inch)
1985–88	Eagle 4 X 4	.880	.815	.0005	.004	15-80
	Colt	.51	.45	—	.006	—
1985–89	Colt Vista, 2WD	.71	.65	—	.006	—
	Colt Vista, 4WD	.94	.88	—	.004	—
	Conquest③	.94	.88	—	.006	—
	Conquest④	.71	.65	—	.006	—
	Chrysler Rear Wheel Drive	1.005	.940	.0005	.004	15–80
	Chrysler Front Wheel Drive③	.935	.882	.0005	.005	15–80
1987–88	Daytona④	.344	.291	.0005	.005	15–80
1987–89	Eagle Medallion	.775	.697⑤	—	.002	—
1988–89	Eagle Premier	.866	.807	.001	.003	—
	Dynasty, New Yorker & New Yorker Landau④	.354	.339	.0005	.003	15–80
1989	Daytona, Lebaron④ ⑥	.472	.409	.0005	.003	15-80
	Daytona, Lebaron④ ⑦	.866	.797	.0005	.003	15-80
	Eagle Summit① ③	.449	⑥	—	.006	—
	Eagle Summit② ③	.882	⑥	—	.006	—
	Eagle Summit④	.331	⑤	—	.006	—
	Colt① ③	.449	⑥	—	.006	—
	Colt② ③	.882	⑥	—	.006	—
	Colt④	.331	⑤	—	.006	—

①—Models with 1.5L engine.
②—Models with 1.6L engine.
③—Front wheel disc brake.
④—Rear wheel disc brake.
⑤—Do not machine brake disc rotor. Replace brake rotor if excessive wear and/or scoring is evident.
⑥—Solid rotor.
⑦—Vented rotor.

FORD MOTOR CO.

TABLE OF CONTENTS

General Information

INDEX

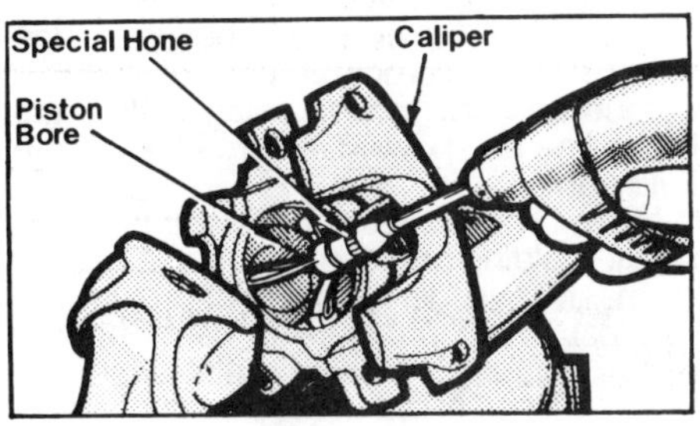

Fig. 1 Honing caliper piston bore

GENERAL PRECAUTIONS

1. Grease or any other foreign material must be kept off the caliper, surfaces of the disc and external surfaces of the hub, during service procedures. Handling the brake disc and caliper should be done in a way to avoid deformation of the disc and nicking or scratching brake linings.
2. If inspection reveals rubber piston seals are worn or damaged, they should be replaced immediately.
3. During removal and installation of a wheel assembly, exercise care so as not to interfere with or damage the caliper splash shield, or bleeder screw.
4. Front wheel bearings should be adjusted to specifications.
5. Be sure vehicle is centered on hoist before servicing any of the front end components to avoid bending or damaging the disc splash shield on full right or left wheel turns.
6. Before the vehicle is moved after any brake service work, be sure to obtain a firm brake pedal.
7. The assembly bolts of the two caliper housings should not be disturbed unless the caliper requires service.

INSPECTION

Remove wheels and inspect brake disc, caliper and linings. The wheel bearings should be inspected at this time and repacked if necessary. Do not get any grease on the linings.

If the caliper is cracked or fluid leakage through the casting is evident, it must be replaced as a unit.

Should it become necessary to remove the caliper for installation of new parts, clean all parts in alcohol, wipe dry using lint free cloths. Using an air hose, blow out drilled passages and bores. Check dust boots for punctures or tears. If punctures or tears are evident, new boots should be installed upon reassembly.

Inspect piston bores in both housings for scoring or pitting. Bores that show light scratches or corrosion can usually be cleaned with crocus cloth. However, bores that have deep scratches or scoring may be honed, provided the diameter of the bore is not increased more than .002 inch. If the bore does not clean up within this specification, a new caliper housing should be installed (black stains on the bore walls are caused by piston seals and will do no harm).

When using a hone, **Fig. 1**, be sure to install the hone baffle before honing bore. The baffle is used to protect the hone stones from damage. Use extreme care in cleaning the caliper after honing. Remove all dust and grit by flushing the caliper with alcohol. Wipe dry with clean lint free cloth and then clean a second time in the same manner.

BRAKE DISC SERVICE

Servicing of disc brakes is extremely critical due to the close tolerances required in machining the brake disc to insure proper brake operation.

Fig. 2 Checking rotor for lateral run-out

The maintenance of these close controls of the shape of the rubbing surfaces is necessary to prevent brake roughness. In addition, the surface finish must be non-directional and maintained at a micro inch finish. This close control of the rubbing surface finish is necessary to avoid pulls and erratic performance and promote long lining life and equal lining wear of both left and right brakes.

In light of the foregoing remarks, refinishing of the rubbing surfaces sh ould not be attempted unless precision equipment, capable of measuring in micro inches (millionths of an inch) is available.

To check lateral run-out of a disc, mount a dial indicator on a convenient part (steering knuckle, tie rod, disc brake caliper housing) so that the plunger of the dial indicator contacts the disc at a point one inch from the outer edge, **Fig. 2.** If the total

Fig. 3 Checking rotor parallelism (thickness variation)

indicated run-out exceeds specifications, install a new disc.

To check parallelism (thickness variation), mount dial indicators, **Fig. 3,** so the plunger contacts rotor approximately 1 inch from outer edge. If parallelism exceeds specifications, replace rotor.

BRAKE ROUGHNESS

The most common cause of brake chatter on disc brakes is a variation in thickness of the disc. If roughness or vibration is encountered during highway operation or if pedal pumping is experienced at low speeds, the disc may have excessive thickness variation. To check for this condition, measure the disc at 12 points with a micrometer at a radius approximately one inch from edge of disc. If thickness measurements vary by more than .0005 inch, the disc should be replaced with a new one.

Excessive lateral run-out of braking disc may cause a "knocking back" of the pistons, possibly creating increased pedal travel and vibration when brakes are applied.

Before checking the run-out, wheel bearings should be adjusted. The readjustment is very important and will be required at the completion of the test to prevent bearing failure.

BLEEDING DISC BRAKES

Pressure bleeding is recommended for all hydraulic disc brake systems.

The disc brake hydraulic system can be bled manually or with pressure bleeding equipment. **On vehicles with disc brakes,** the brake pedal will require more pumping and frequent checking of fluid level in master cylinder during bleeding operation.

Never use brake fluid that has been drained from hydraulic system when bleeding the brakes. Be sure the disc brake pistons are returned to their normal positions and that the shoe and lining assemblies are properly seated. Before driving the vehicle, check brake operation to be sure that a firm pedal has been obtained.

PROPORTIONING VALVE

DESCRIPTION

The proportioning valve (when used), **Fig. 4,** provides balanced braking action between front and rear brakes under a wide range of braking conditions. The valve regulates the hydraulic pressure applied to the rear wheel cylinders, thus limiting rear braking action when high pressures are required at the front brakes. In this manner, premature rear wheel skid is prevented.

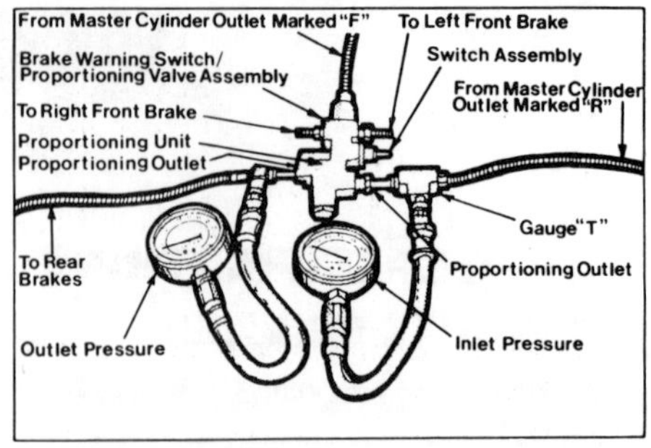

Fig. 4 Gauge hook-up for testing proportioning valve (typical)

TESTING

When a premature rear wheel slide is obtained on a brake application, it usually is an indication that the fluid pressure to the rear wheels is above the 50% reduction ratio for the rear line pressure and that malfunction has occurred within the proportioning valve.

To test the valve, install gauge set shown in **Fig. 4** in brake line between master cylinder and proportioning valve, and at output end of proportioning valve and brake line as shown. Be sure all joints are fluid tight.

Have a helper exert pressure on brake pedal (holding pressure). Obtain a reading on master cylinder output of approximately 700 psi. While pressure is being held as above, reading on valve outlet should be 550-610 psi. If the pressure readings do not meet these specifications, the valve should be removed and a new valve installed.

Front Disc Brakes

INDEX

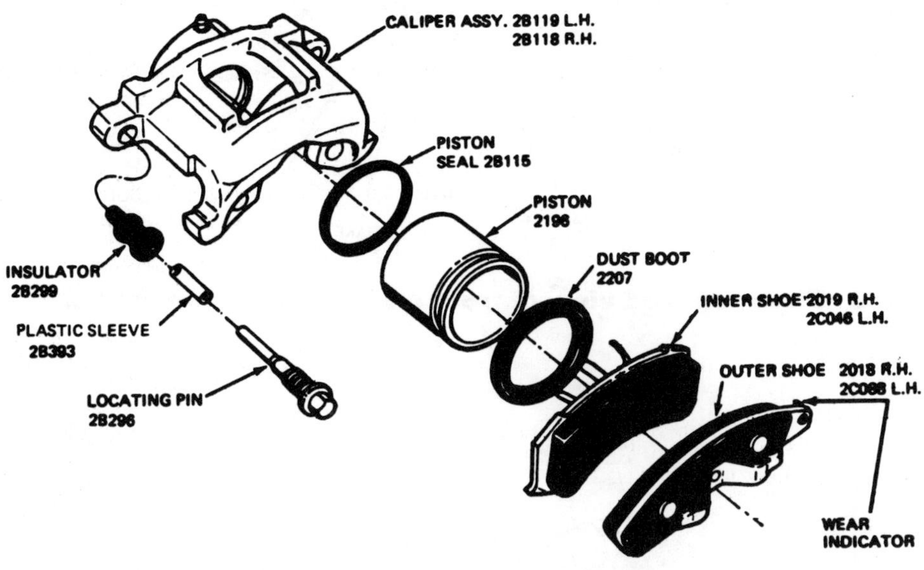

Fig. 1 Front pin sliding disc brake caliper. Capri, Cougar, Crown Victoria, Grand Marquis, LTD, Mark VI & VII, Marquis, Mustang, Thunderbird & 1985–87 Continental

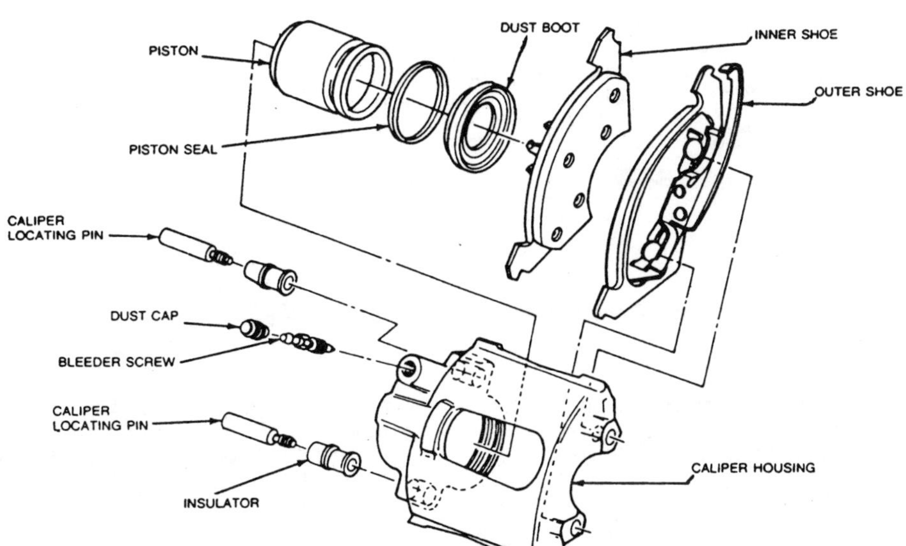

Fig. 2 Front pin sliding disc brake caliper. Escort, EXP, Lynx, Sable, Taurus, Tempo, Topaz & 1988–89 Continental

OPERATION

The caliper assembly consists of a pin slider caliper housing, inner and outer shoe and lining assemblies and a single piston, **Figs. 1 and 2.** The caliper slides on two pins which also act as attaching bolts between caliper and the combination anchor plate and spindle. The outer brake shoe and lining assembly is longer than the inner brake shoe and lining assembly. Inner and outer shoe and lining assemblies are attached to the caliper by spring clips riveted to the shoe surfaces. The inner shoe is attached to the caliper by installing the spring clip to the inside of the caliper piston. The outer shoe clips directly to the caliper housing. A wear indicator is incorporated which emits a noise when the lining is worn to a point when replacement is necessary. Inner and outer shoes are of left and right hand and are not interchangeable.

The inner shoe and lining on Capri, Fairmont, Mustang, and Zephyr with V8-302 engine has a replaceable single finger anti-rattle clip and an insulator held in position by the clip. The shoe is slotted to accept the snap on clip which loads the assembly against the caliper bridge. The inner shoe on the Capri and Mustang with 2300cc engine has a single finger anti-rattle clip similar to the Fairmont and Zephyr inner shoe, holding the shoe down against the spindle ledge. The clip does not lock into the piston. The insulator is also riveted to the shoe and is not replaceable.

BRAKE SHOE & LINING REPLACE

REMOVAL

Except Festiva, Tracer, Scorpio & XR4Ti

1. Remove brake fluid until reservoir is half full.
2. Raise and support front of vehicle, then remove wheel and tire assembly.
3. Remove caliper locating pins.
4. Lift caliper assembly from adapter plate, then remove outer shoe from caliper assembly. **On some models, slip shoe down caliper leg until clip is disengaged.**
5. Remove inner shoe and lining assembly. **On some models, pull shoe straight out of piston. This should**

require a force as high as 20-30 lbs.

6. Suspend caliper from inner fender housing with wire to avoid damaging brake hose.
7. Remove and discard locating pin insulators and plastic sleeves.

Festiva

1. Drain approximately 1/3 of brake fluid from master cylinder.
2. Raise and support front of vehicle, then remove wheel.
3. Remove brake pad pin retainer, then disengage anti-rattle spring from brake pads, **Fig. 3.**
4. Remove brake pad pins and anti-rattle spring, then pull brake pads and shims out of caliper. **Do not discard shims found behind inner brake pad.**

Tracer

1. Remove approximately 2/3 of brake fluid from master cylinder.
2. Raise and support front of vehicle.
3. Remove wheel and tire assembly, then, using suitable needle nose pliers, remove retainer locking brake pads to retainer pins, **Fig. 4.**
4. Using a suitable punch, remove brake pad retaining pins.
5. Using a screwdriver, pry caliper outward, then remove outboard pad and shim. **Do not damage caliper piston boots.** Note location of shim for assembly reference.
6. Push caliper inward, then remove inboard pad and shim.
7. Remove anchor plate clips and tag for assembly reference.

Scorpio & XR4TI

1. Raise and support vehicle.
2. Remove wheel and tire assembly.
3. Disconnect brake pad wear sensor electrical connector from main harness. Remove connector from clip, leaving clip in position under bleed screw.
4. Remove spring retaining clip from caliper housing.
5. Unfasten caliper assembly and position aside. **Do not disconnect brake line.**
6. Remove inboard brake pad from piston.
7. Force piston into caliper housing using tool No. D79L-2196-A or equivalent, **Fig. 5,** then remove outboard pad.

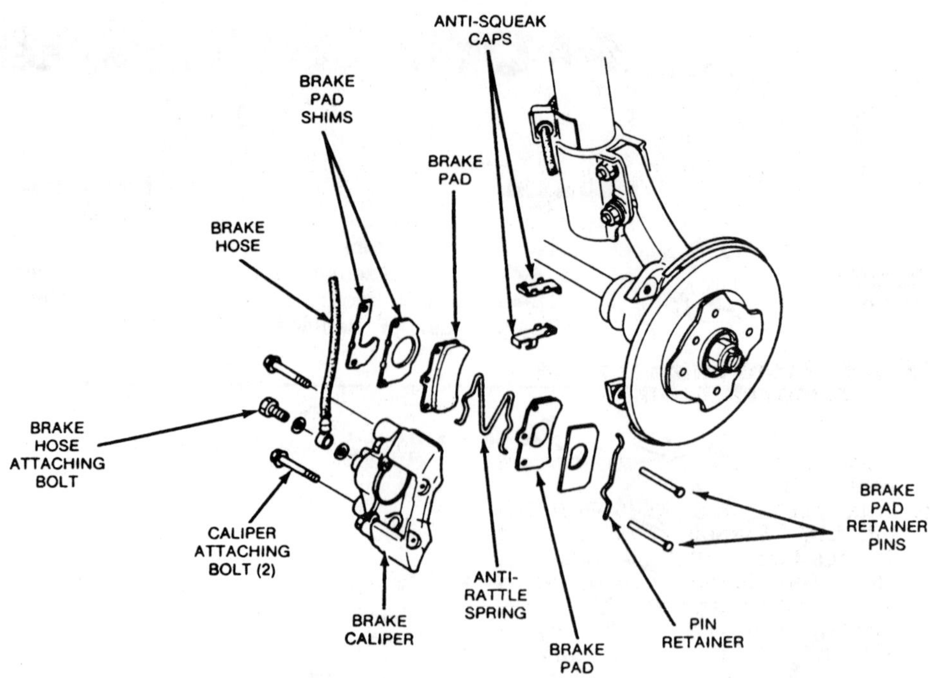

Fig. 3 Disassembled view of front disc brake assembly. Festiva

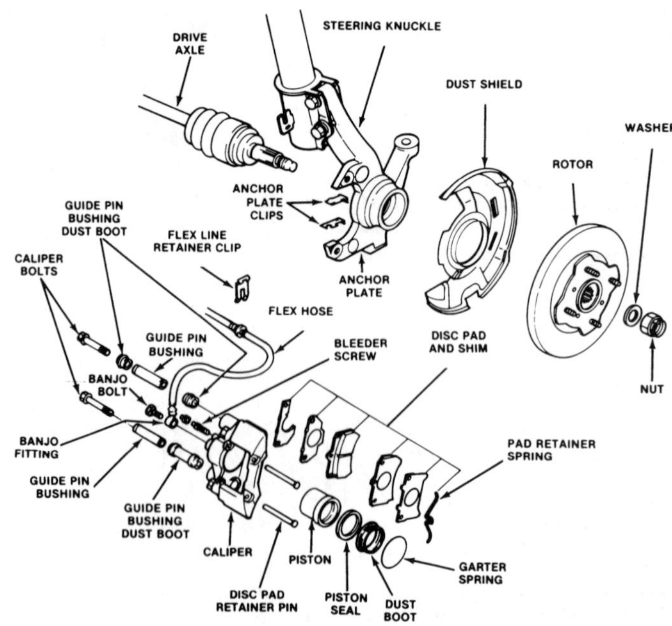

Fig. 4 Front disc brake assembly. Tracer

INSTALLATION

Except Festiva, Tracer, Scorpio & XR4TI

1. Using a 4 inch C-clamp and a block of wood 2 3/4 x 1 inch and approximately 3/4 inch thick, seat caliper piston in bore, then remove C-clamp and wooden block. **On some models,** the piston is made of phenolic material. Do not seat piston in bore by applying C-clamp directly to piston. Extra care must be taken during this procedure to prevent damage to the piston. Metal or sharp objects cannot come into direct contact with the piston or damage may result.
2. Install locating pin insulators and plastic sleeves on caliper housing. Ensure insulators and sleeves are properly positioned.
3. Install inner shoe and lining assembly on caliper piston, **Fig. 6. Some inner brake shoes are marked LH (left hand) and RH (right hand) and must be installed on the proper caliper. Use care to not bend spring clips too far during installation in piston, otherwise distortion and rattles may result.**
4. Install outer brake shoe and lining assembly, **Fig. 7.** Ensure that shoes are installed on proper caliper. Make sure that clip and buttons on shoe are properly seated. **The outer shoe can be identified as left hand and right hand by the wear indicator which**

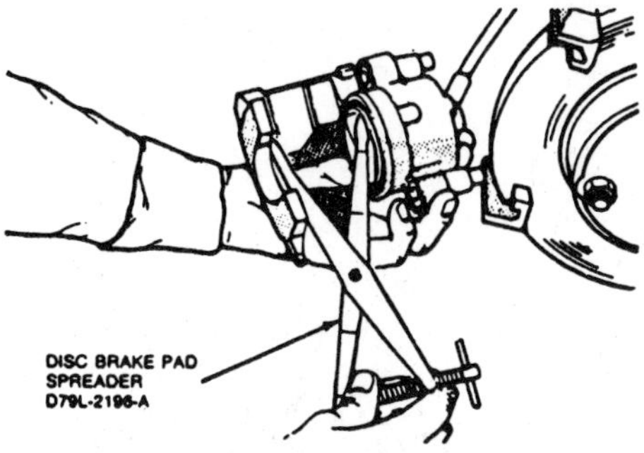

Fig. 5 Front outboard pad removal. Scorpio

must be installed toward front of vehicle.

5. Install locating pins. **On 1985-86 Capri, LTD, Marquis, Mustang and Thunderbird, 1985-87 Cougar, 1987 Thunderbird except turbo 1987 Mustang less V8-302, torque locating pins to 30-40 ft. lbs. On 1985-87 Lincoln Continental, 1985-87 Mark VI and VII, 1987 Mustang V8-302, 1987 Thunderbird turbo and 1985-89 Ford Crown Victoria, Mercury Grand Marquis and Lincoln Town Car models, torque locating pins to 40-60 ft. lbs. On 1988-89 Cougar, Mark VII, Mustang and Thunderbird, torque locating pins to 45 to 65 ft. lbs. On 1985-89 Escort, EXP and Lynx, 1985-89 Tempo and Topaz, 1986-89 Sable and Taurus and 1988-89 Lincoln Continental, torque to 18-25 ft. lbs.**
6. Refill master cylinder, then install wheel and tire assembly and lower vehicle.
7. Pump brake pedal several times to position brake linings before moving vehicle.

Festiva

1. Push piston back fully into caliper bore.
2. Apply suitable brake grease to both sides of inner shim and to back of inner brake pad.
3. Install brake pads and shims, ensuring shims are positioned correctly.
4. Install brake pad pins, anti-rattle spring and pin retainer.
5. Install wheel and lower vehicle, then check and adjust master cylinder fluid level as required.

Tracer

1. Install anchor plate clips in same order as removed.
2. Push caliper inboard, then install inboard brake pad and shims.
3. Pry caliper outboard, then install outboard brake pad and shim.
4. Install brake pad retaining pins and springs.
5. Install tire and wheel assembly, then pump brake pedal several times to seat brake pads.

Scorpio & XR4Ti

1. Install inboard pad into caliper piston, then the outboard pad into anchor bracket.
2. Install pad sensor wiring clip, then the caliper assembly. Torque caliper bolts to 15-18 ft. lbs.
3. Install wheel and tire assembly, then lower vehicle.
4. Pump brake pedal several times to properly position brake pads before moving vehicle.
5. Verify proper operation of pad wear warning lamp, then road test vehicle.

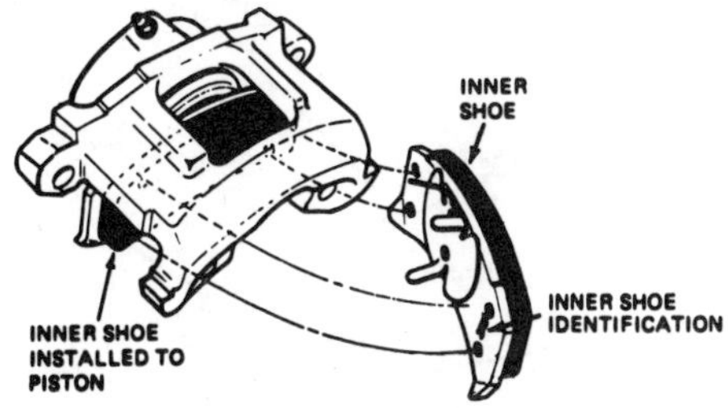

Fig. 6 Installing inner brake shoe on caliper

CALIPER REPLACE

REMOVAL

Except Festiva, Tracer, Scorpio & XR4Ti

Before removing calipers, mark left and right hand calipers so they can be installed in the same position.

1. Raise and support front of vehicle, then remove wheel and tire assembly.
2. Loosen brake tube fitting which connects brake tube to fitting on frame and plug brake tube. Remove retaining clip from brake hose and bracket, then disconnect brake hose from caliper.
3. Remove caliper locating pins.
4. Lift caliper from rotor and spindle anchor plate assembly. **On models equipped with phenolic caliper piston,** do not pry directly against the piston or damage may result.

Festiva

1. Raise and support vehicle, then remove wheel.
2. Remove brake pads as previously outlined.
3. Remove brake hose attaching bolt, then disconnect hose from caliper. Discard seal washers located between brake hose connection and caliper.
4. Remove caliper attaching bolts and the caliper.
5. Reverse procedure to install using new seal washers. **Torque** caliper attaching bolts to 33 ft. lbs., then bleed brake system.

Tracer

1. Raise and support front of vehicle.
2. Remove wheel and tire assembly, then remove brake pads as described previously.
3. Remove brake hose-to-caliper fitting, Fig. 4.
4. Remove two caliper attaching bolts, then lift caliper from rotor.

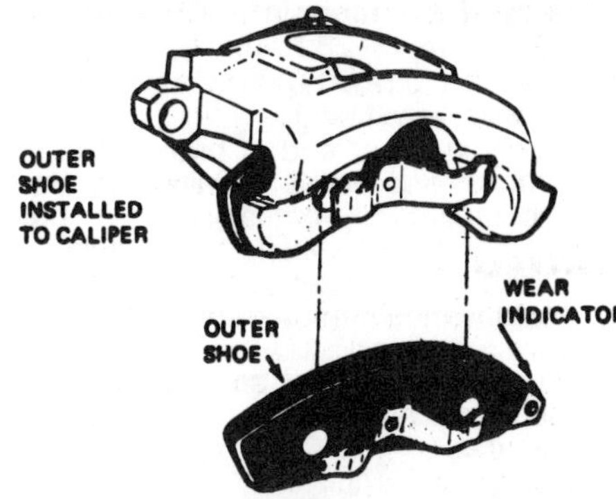

Fig. 7 Installing outer brake shoe on caliper

Scorpio & XR4Ti

1. Raise and support vehicle.
2. Remove wheel and tire assembly.
3. Disconnect electrical connector from brake pad wear sensor.
4. Remove spring retaining clip from caliper housing.
5. Remove piston housing bolts and the housing from anchor bracket.
6. Disconnect brake hose from caliper. Plug line to prevent leakage and contamination.
7. Remove caliper anchor bracket attaching bolts and the bracket, if necessary.
8. Reverse procedure to install, noting the following:
   a. **Torque** caliper attaching bolts to 37-48 ft. lbs., brake hose to 12-13 ft. lbs. and anchor bracket attaching bolts to 15-18 ft. lbs.
   b. Following completion of installation, bleed front brake system.

INSTALLATION

Except Festiva, Tracer, Scorpio & XR4Ti

1. Install caliper assembly over rotor with outer shoe against rotor braking surface during installation on spindle and anchor plate to prevent pinching of piston boot between inner brake shoe and piston. **Ensure calipers are installed in the correct position.**
2. Install locating pins. **On 1985-86 Capri, LTD, Marquis, Mustang and Thunderbird, 1985-87 Cougar, 1987 Thunderbird except turbo 1987 Mustang less V8-302, torque locating pins to 30-40 ft. lbs. On 1985-87 Lincoln Continental, 1985-87 Mark VI and VII, 1987 Mustang V8-302, 1987 Thunderbird turbo and 1985-89 Ford Crown Victoria, Mercury Grand Marquis and Lincoln Town Car models, torque locating pins to 40-60 ft. lbs. On 1988-89 Cougar, Mark VII, Mustang and Thunderbird, torque locating pins to 45 to 65 ft. lbs. On 1985-89 Escort, EXP and Lynx, 1985-89 Tempo and Topaz, 1986-89 Sable and Taurus and**

1988-89 Lincoln Continental, torque to 18-25 ft. lbs.

3. Connect brake hose to caliper and tighten hose fitting.
4. Position upper end of brake hose in bracket and install retaining clip. Remove plug from brake line, then connect brake hose fitting to brake line.
5. Bleed brake system.
6. Install wheel and tire assembly, then lower vehicle.
7. Pump brake pedal several times to position brake shoes before moving vehicle.

Tracer

Before installing caliper, remove guide pin bushing dust boots and push out caliper guide pin bushings. Lubricate guide pin bushings with high temperature grease D7AZ-19590-A or equivalent.

1. Install brake pads and shims as described previously.
2. Position caliper over rotor, then install caliper attaching bolts. **Torque** caliper bolts to 29-36 ft. lbs.
3. Position flex hose on caliper, then install two new copper washers and banjo bolt on flex hose. **Torque** bolt to 16-22 ft. lbs.
4. Bleed front brakes, then install wheel and tire assembly.

Scorpio & XR4Ti

1. Install caliper anchor bracket attaching bolts and the bracket.
2. Connect caliper flex hose to caliper.
3. Install piston housing to anchor bracket retaining bolts.
4. Install spring retaining clip to caliper housing.
5. Connect wear sensor electrical connector.
6. **Torque** caliper attaching bolts to 37-48 ft. lbs., brake hose to 12-13 ft. lbs. and anchor bracket attaching bolts to 15-18 ft. lbs.

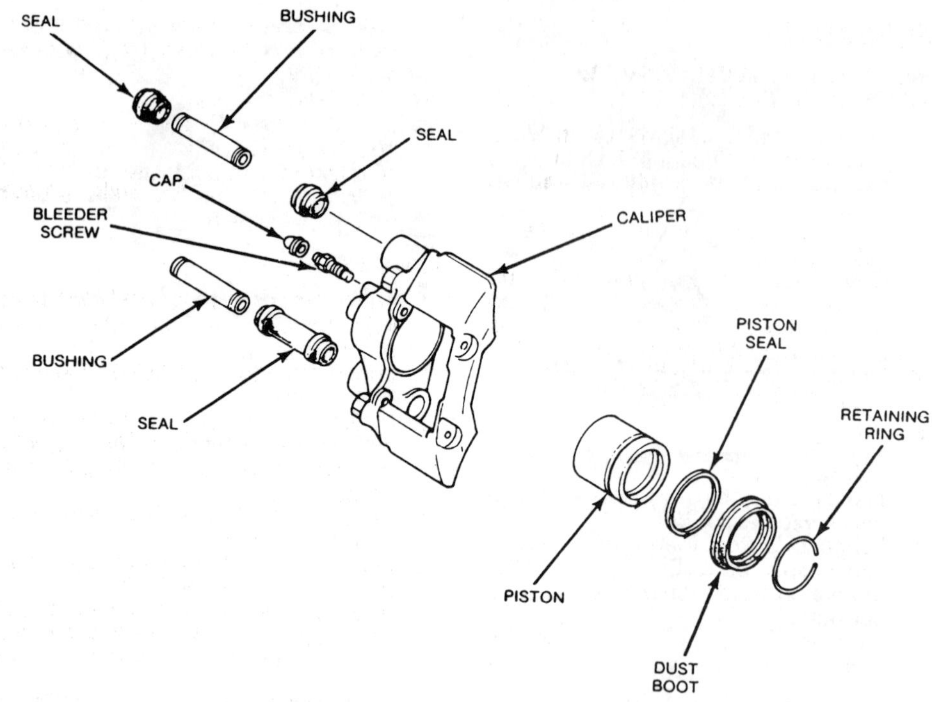

Fig. 8 Disassembled view of brake caliper. Festiva

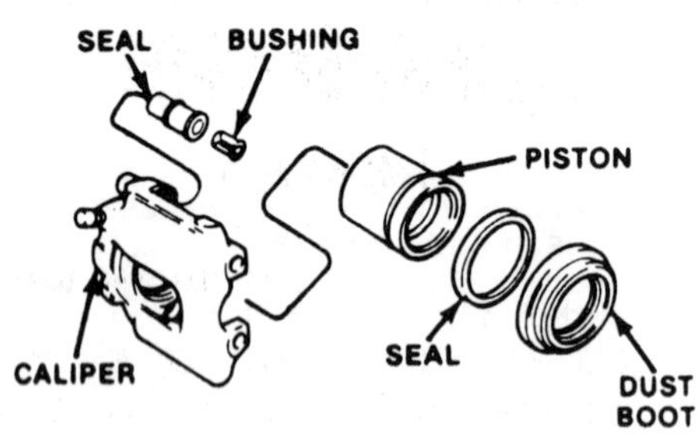

Fig. 9 Front brake caliper exploded view. Scorpio

CALIPER OVERHAUL

DISASSEMBLY

Except Festiva, Tracer, Scorpio & XR4Ti

1. Remove caliper assembly from vehicle as described under "Caliper, Replace."
2. Position fiber block and shop towels between caliper piston and caliper housing, then apply compressed air to caliper brake line fitting bore to force piston from caliper.
3. Remove dust boot from caliper assembly, **Figs. 1 and 2.**
4. Remove piston seal from cylinder and discard.

Festiva

1. Remove brake pads and caliper as outlined previously.
2. Remove bleeder screw, drain remaining fluid from caliper, then reinstall screw.
3. Remove dust boot retaining ring, **Fig. 8.**
4. Position a block of wood between piston and caliper, then apply compressed air to fluid inlet port and blow piston from caliper bore. **Apply only enough pressure to ease piston out of bore.**
5. Remove dust boot from piston, then using a suitable wooden pick, carefully pry piston seal from caliper bore.
6. Remove bleeder screw, then the caliper bushings and bushing seals.
7. Wash all parts in denatured alcohol and dry with compressed air.

Tracer

1. Open caliper bleed screw and drain caliper, then close bleed screw.
2. Remove caliper as described under as previously outlined.
3. Remove caliper guide bushing and dust boots, **Fig. 4.**
4. Remove snap ring from caliper piston dust boot.
5. Position a wooden block or shop towels between caliper and piston, then apply air pressure to brake hose fitting to remove piston from caliper. Use only enough air pressure to ease piston from caliper bore. Keep hands and fingers away from piston, as personal injury may result.
6. Remove dust boot from caliper, then using a wooden or plastic pick, remove piston seal from caliper bore.

Scorpio & XR4Ti

1. Remove caliper assembly and brake pads as previously described.
2. Open bleeder screw and drain fluid from caliper, then close bleeder screw.
3. Position block of wood between piston and caliper, then apply air pressure through fluid inlet port to remove piston. **Apply only enough air pressure to ease piston out of caliper.**
4. Remove dust boot from piston and discard, **Fig. 9.**
5. Remove piston seal from caliper and discard.
6. Remove caliper anchor pin bushing, then the anchor pin seal.

INSPECTION

1. Check piston for scratches, scoring or damage. Replace, if necessary.
2. Check caliper bore for scratches, scoring or corrosion. Light scratches or slight corrosion can be polished out using crocus cloth.
3. Check that bleeder screw and bleeder screw bore hole in caliper are fully open.
4. Check caliper bushings for corrosion and dust boot retaining ring for damage or tension loss. Replace parts as necessary.

ASSEMBLY

Except Festiva, Tracer, Scorpio & XR4Ti

1. Lubricate piston seal with clean brake fluid, then install seal in caliper bore. **Ensure seal is firmly seated in groove.**
2. Install dust boot in outer groove of caliper bore, **Figs. 1 and 2.**
3. Coat piston with clean brake fluid and install piston in caliper bore. Spread dust boot over piston as it is installed. Seat dust boot in piston groove.
4. Install caliper assembly as described under "Caliper, Replace."

Festiva

1. Lubricate piston seal with suitable brake grease or clean brake fluid, then install into caliper bore.
2. Lubricate piston, then partially install into caliper bore. Ensure dust boot groove on piston remains above caliper bore.
3. Lubricate dust boot with suitable brake grease, then install onto piston.
4. Press piston fully inward until it bottoms in bore, then install dust boot retaining ring.
5. Install bleeder screw.
6. Install caliper bushings and new bushing seals. Lubricate inner surface of bushing seal and outer surface of bushings with suitable brake grease before installation.

Tracer

1. Lubricate piston seal with brake fluid, then position seal in caliper bore groove, **Fig. 4.**
2. Lubricate piston and caliper bore with brake fluid.
3. Install dust boot on piston, then install piston into caliper bore. Use a gentle rocking motion to bottom piston in caliper bore.
4. Slide dust boot over caliper bore boss and install retaining ring.
5. Install caliper assembly as described under "Caliper, Replace."
6. After completing installation, bleed brake system, then cycle brake pedal several times to seat brake pads and to ensure system is function properly.

Scorpio & XR4Ti

1. Install caliper anchor pin seal, spraying seal with silicone lubricant to aid installation.
2. Install caliper anchor pin bushing, overlapping edges of bushing slightly to fit bushing into seal. After installation, press bushing against seal to remove overlapping. When properly installed, the ends of the bushing must abut against each other.
3. Lubricate new piston seal with brake fluid and install in seal groove, ensuring that seal does not become twisted but is firmly seated in groove.
4. Position dust boot at bottom of piston and, holding dust boot on piston, pull on seal lip until seal unfolds, allowing lip seal to extend beyond bottom of piston.
5. While holding dust boot on piston, fit seal lip in caliper bore and push piston into caliper. As piston enters bore, the dust boot will refold to its original shape.

Rear Disc Brakes & Parking Brakes

INDEX

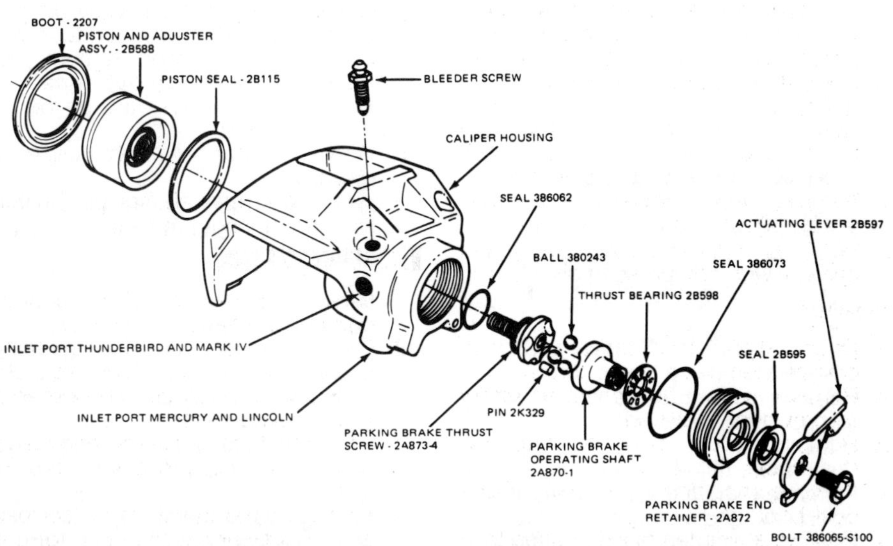

Fig. 1 Rear disc brake caliper assembly. 1985–87 Continental, 1985–86 Mustang SVO & 1985–89 Mark VII

After performing any service work, obtain a firm brake pedal before moving vehicle.

OPERATION

Sliding caliper rear disc brakes are used on 1985-89 Continental, 1985-89 Mark VII, 1987-89 Thunderbird Turbo & 1985-86 Mustang SVO, **Figs. 1 and 2.** The caliper is basically the same as the larger front wheel caliper, however, a parking brake mechanism and a larger inner brake shoe anti-rattle spring have been added. **On all models,** except Mustang SVO, a hydraulically powered brake booster (Hydroboost) provides the power assist for this four wheel disc brake system. **On Mustang SVO,** a vacuum brake booster is used.

The parking brake lever, located at the rear of the caliper, is actuated by a cable system similar to rear drum brake applications. When the parking brake is applied, the cable rotates the lever and operating

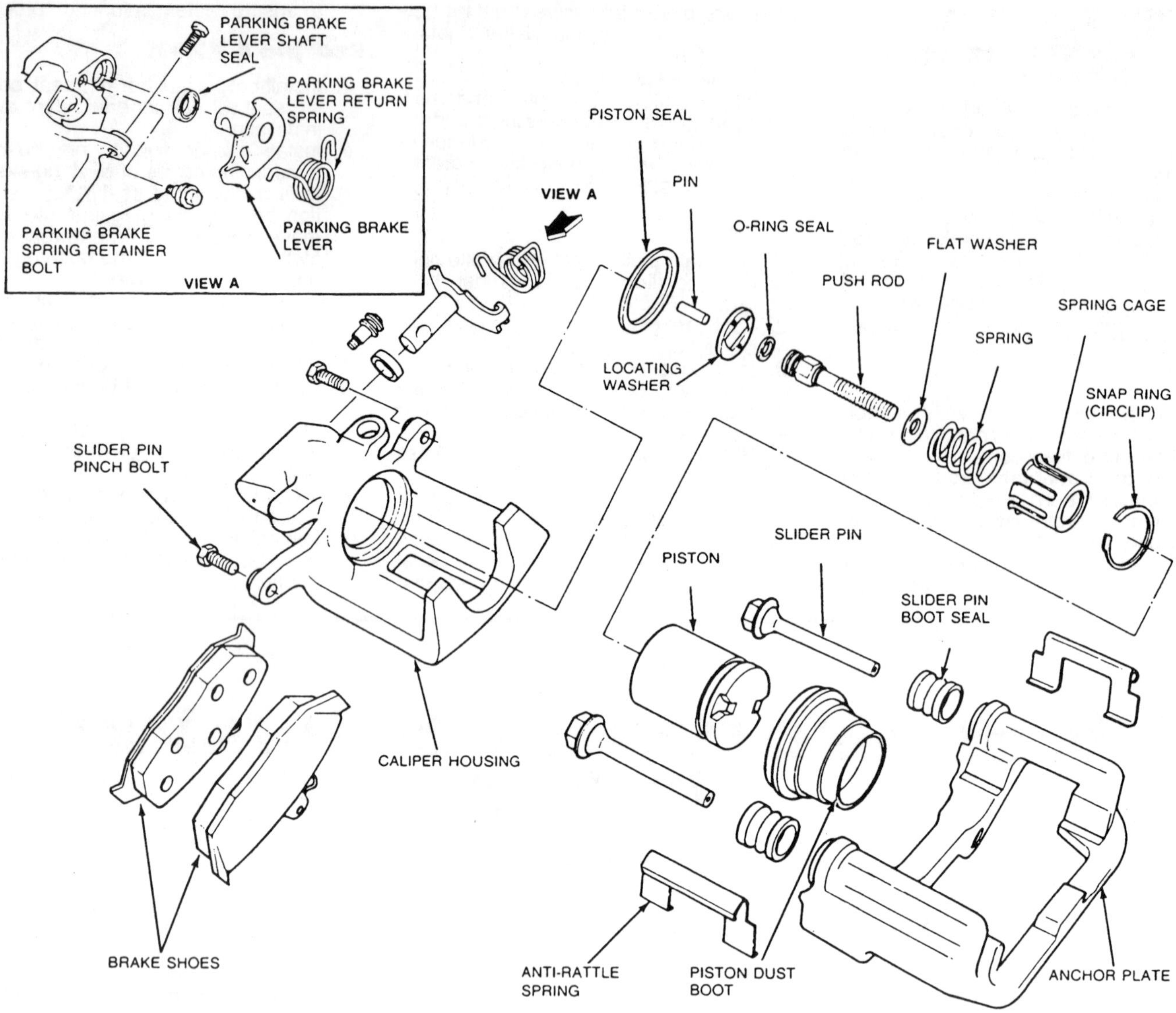

Fig. 2 Rear disc brake caliper assembly. 1987–89 Thunderbird Turbo & 1988–89 Continental

shaft, driving the caliper piston and brake shoe assembly against the rotor. An automatic adjuster in the assembly compensates for lining wear and maintains proper clearance in the parking brake mechanism.

The cast iron rotors are ventilated by curved fins located between the braking surfaces and are designed to cause the rotor to act as an air pump when the vehicle is traveling forward. The rotors are not interchangeable and are identified by a Right or Left marking cast inside the hat section of the rotor. The rotor is secured to the axle flange in the same manner as a rear brake drum. A splash shield is bolted to a forged axle adapter to protect the inboard rotor surface.

REAR DISC BRAKE SERVICE

MARK VII, MUSTANG SVO & 1985–87 CONTINENTAL

CALIPER, REPLACE

Removal

1. Raise vehicle and support on safety stands, then remove tire and wheel assemblies.
2. Disconnect fitting on rear brake tube from hose end fitting at frame mounted bracket and plug end of brake tube to prevent loss of fluid and entry of dirt. Remove horse shoe retaining clip from hose fitting and disengage hose from bracket.
3. Disconnect parking cable from lever, **Fig. 3,** using care to avoid kinking or cutting cable or return spring, then remove retaining screw from caliper retaining key, **Fig. 4,** then remove caliper locating pins.
4. Slide caliper retaining key and support spring from anchor plate, **Fig. 4.** If necessary, use a hammer and brass drift, being careful to avoid damaging key on sliding ways or hitting parking brake lever. **If caliper cannot be removed due to rust build-up on outer edge of rotor, scrape off loose scale, being careful not to damage braking surfaces. If rotor wear or scoring prevents removal of cali-**

Fig. 3 Parking lever & cable installation

per, it will be necessary to loosen caliper end retainer 1/2 turn maximum, to allow piston to be forced back into its bore. To loosen end retainer, remove parking brake lever and mark or scribe end retainer and caliper housing to be sure that end retainer is not loosened more than 1/2 turn, then force piston back in its bore, Fig. 5, and move caliper back and forth to center rotor and remove caliper. If retainer must be loosened more than 1/2 turn, use caution, as the seal between the thrust screw and housing may be broken and brake fluid will enter parking brake mechanism chamber. In this case, the end retainer must be removed and the internal parts cleaned and lubricated.

5. Remove outer shoe and lining assembly from anchor plate, then remove rotor retainer nuts and rotor from axle shaft.
6. Remove inner brake shoe and lining assembly from anchor plate and mark each shoe for identification if they are to be reused.
7. Remove anti-rattle clip from anchor plate, then remove flexible hose from caliper by removing hollow retaining bolt.

Cleaning & Inspection

Clean caliper, anchor plate and rotor assembly and inspect for signs of brake fluid leakage, excessive wear or damage. The caliper must be inspected for leakage both in piston boot area and operating shaft seal area. Lightly sand or wire brush any rust or corrosion from caliper and anchor plate sliding surfaces and inner brake shoe abutment surfaces in anchor plate. Inspect brake shoes for wear. Linings must not be worn to within less than 1/8 inch of shoe surface.

Installation

1. If end retainer has been loosened only 1/2 turn, reinstall caliper in anchor plate using key. Do not install shoe and lining assembly. **Torque** end retainer to 75-95 ft. lbs., then install parking brake actuating lever on its keyed spline. Lever arm must point down and rearward so that parking brake cable will pass freely under axle. **Torque** retainer screw to 16-22 ft. lbs.

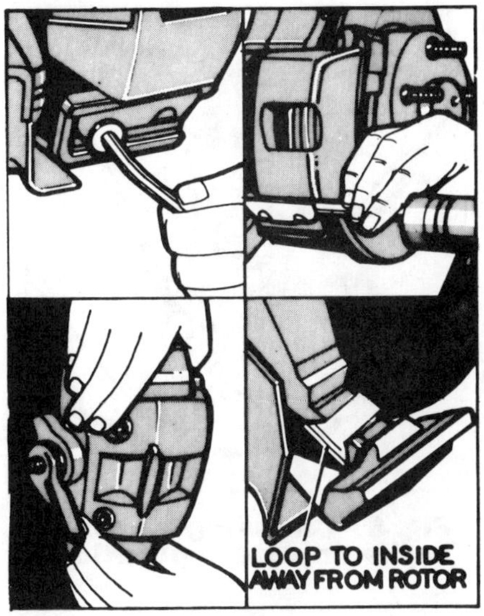

Fig. 4 Removing rear caliper assembly

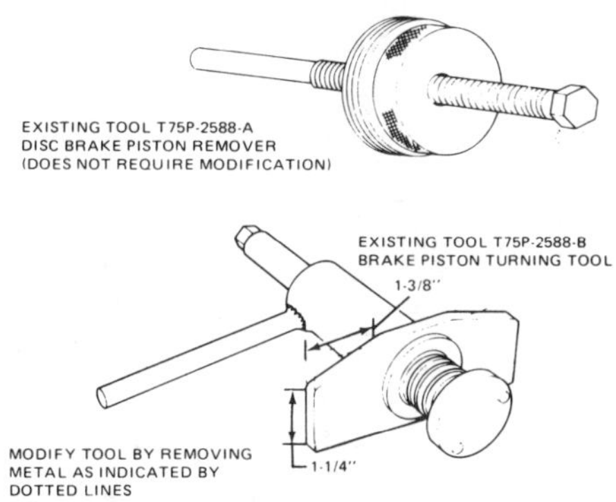

Fig. 6 Disc brake tool modification

Parking brake lever must rotate freely after torquing retainer screw.

2. Remove caliper from anchor plate. If new shoe and lining assemblies are to be installed, the piston must be bottomed in caliper bore using tool T75P-2588-B to provide clearance. Remove rotor and install caliper without lining and shoe assemblies in anchor plate using key only. Install tool and while holding shaft, rotate tool handle counterclockwise until the tool seats firmly against piston, **Fig. 5.** Loosen handle about 1/4 turn, and while holding handle rotate tool shaft clockwise until piston is fully bottomed in bore (piston will continue to turn even after it is bottomed). Turn tool handle until there is no further inward movement of piston and there is a firm seating force, then remove caliper from mounting plate and reinstall rotor. **For use on some vehicles, tool T75P-2588-B must be slightly modified, Fig. 6.**
3. Making certain that brake shoe anti-rattle clip is in place in lower inner brake shoe support on anchor plate with loop of clip toward inside of anchor plate, **Fig. 4,** position inner brake

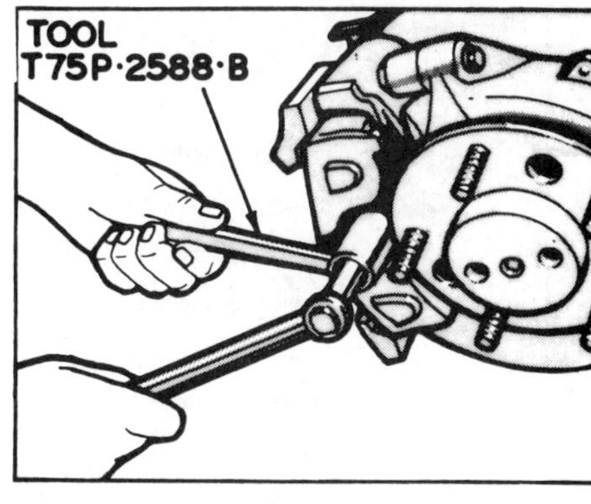

Fig. 5 Adjusting piston depth for lining installation

shoe and lining assembly on anchor plate, then install rotor and two retaining nuts.

4. Install outer brake shoe with lower flange ends against caliper abutments and brake shoe upper flanges over shoulders on caliper legs. The shoe upper flanges fit tightly against machined shoulder surfaces. **If old brake shoes and lining assemblies are reused, be certain the shoes are installed in their original positions as marked for identification during removal.**
5. Lubricate caliper and anchor sliding ways with D7AE-019590 grease, using care to prevent lubricant from getting on braking surfaces, then position caliper housing lower V-groove on anchor plate lower abutment surfaces.
6. Rotate caliper until it is completely over rotor, being careful not to damage piston dust boot, then pull caliper outboard until inner shoe and lining is firmly seated against rotor. Measure clearance between outer lining and rotor, clearance must be between 1/32 and 3/32 inch. If it is greater, remove caliper and move piston outward to narrow gap. Follow procedure in step 2 and note that 1/4 turn of the shaft counterclockwise, moves piston about 1/16 inch. **A clearance greater than specified limit may allow adjuster to be pulled out of piston when service brake is applied, causing parking brake to fail to adjust. It will then be necessary to replace piston/adjuster assembly.**
7. While holding caliper against anchor plate upper abutment surfaces, center caliper over lower anchor plate abutment, then position caliper support spring and key in slot and slide them into opening between lower end of caliper and lower anchor plate abutment until key semi-circular slot is centered over retaining screw threaded hole in anchor plate.
8. Install key retaining screw and **torque** to 12-16 ft. lbs., then reinstall brake hose on caliper. Place a new gasket on each side of the fitting outlet, then install the attaching bolt through the washers and fitting and **torque** to 20-30 ft. lbs.

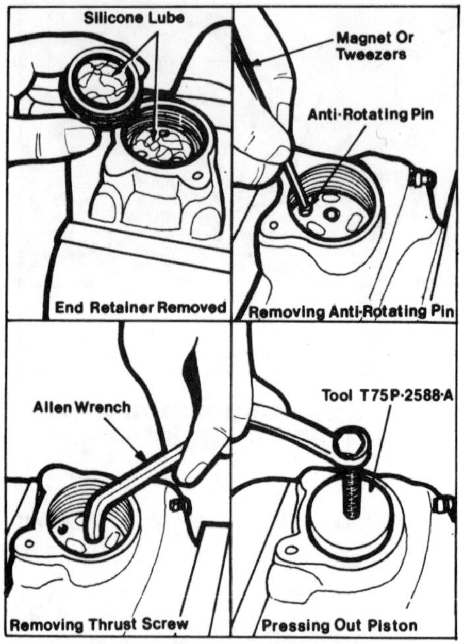

Fig. 7 Disassembling rear disc brake caliper

9. Lubricate pins and inside of insulator with D7AZ-19A331-A or equivalent silicone grease and add one drop of Loctite EOAC-19554-A, or equivalent, to locating pin threads. Install locating pins through caliper insulators and into anchor plate and **torque** to 29-37 ft. lbs.
10. Connect parking brake lever to lever on caliper.
11. Bleed brake system, then with engine running pump brake pedal lightly about 40 times allowing 1 second between pedal applications. An alternate with engine off is to pump brake pedal lightly about 10 times to discharge accumulator, then pump brake pedal firmly about 30 times. Check parking brake for excessive travel or very light effort, if so, repeat pumping brake pedal, and if necessary check parking brake cable tension.
12. Install wheel and **torque** nuts to 70-115 ft. lbs. **Before moving vehicle, make certain that a firm brake pedal has been obtained.**

DISC BRAKE PADS, REPLACE

To remove shoe and lining assemblies, follow "Caliper Removal" procedure and omit step 2 as it is not necessary to disconnect brake hose. After removing caliper, support it with a length of wire to avoid damaging brake hose. To install shoe and lining assemblies, follow "Caliper Installation" procedure, making certain that proper parking brake adjustment is obtained.

CALIPER OVERHAUL

Disassembly

1. Remove caliper assembly as described previously.
2. Remove caliper end retainer, operating shaft, thrust bearing and balls, **Fig. 1.**

Fig. 8 Checking parking brake adjuster operation

3. Remove thrust screw anti-rotation pin with a magnet or tweezers. If pin cannot be removed with a magnet or tweezers, proceed with the following procedure:
   a. With tool T75P-2588B, force piston approximately one inch from caliper bore.
   b. Push piston back into caliper housing with tool, then with tool in position, hold tool shaft in place and rotate handle counterclockwise until thrust screw clears anti-rotation pin. Remove thrust screw and anti-rotation pin.
4. Remove thrust screw by rotating with 1/4 inch Allen wrench.
5. Install Tool No. T75P-2588-A through back of caliper housing and remove piston assembly, **Fig. 7. Use care not to damage polished surface in thrust screw bore and do not attempt to remove or press adjuster can, as it is a press fit in piston.**
6. Remove and discard piston seal, boot, thrust O-ring seal, end retainer, O-ring and end retainer lip seal.

Cleaning & Inspection

1. Clean all metal parts with alcohol, then using clean, dry compressed air, blow out and dry all grooves and passages making sure the caliper bore and component parts are free of any foreign material.
2. Inspect caliper bore for damage or excessive wear. The thrust screw must be smooth and free of pits. If piston is pitted, scored or chrome plating is worn, replace piston and adjuster assembly.
3. Adjuster can must be bottomed in piston to be properly seated and provide consistent brake operation. If adjuster can is loose, appears high in piston, is damaged, or if brake adjustment is usually too tight, too loose or not functioning, replace piston/adjuster assembly. Check adjuster operation by assembling thrust screw into piston/adjuster assembly, then pull the two parts apart about 1/4 inch and release them, **Fig. 8.** When pulling on the two parts, the brass drive ring must remain stationary causing the nut to rotate. When releasing the two parts, the nut must remain stationary and drive ring must rotate. If action does not follow this pattern, replace piston/adjuster assembly.

Fig. 9 Filling piston/adjuster assembly

4. Inspect ball pockets, threads, grooves, bearing surfaces of thrust screw, operating shaft, balls and anti-rotation pin for wear, brinelling or pitting. Replace operating shaft, balls, thrust screw and anti-rotation pin if any of these parts are worn or damaged. A polished appearance on the ball paths is acceptable if there is no sign of wear into the surface.
5. Inspect thrust bearing for corrosion, pitting or wear and replace as necessary.
6. Inspect end plug bearing surface for wear or Brinelling and replace as necessary. A polished appearance on bearing surface is acceptable if there is no sign of wear into surface.
7. Inspect operating lever for damage and replace as necessary.

Assembly

1. Coat new caliper piston seal with clean brake fluid and install it in caliper making certain that seal is not twisted and is fully seated in groove.
2. Install new dust boot by seating flange squarely in outer groove of caliper bore, then coat piston/adjuster assembly with clean brake fluid and install it in caliper bore. Spread dust boot over piston as it is installed and seat dust boot in piston groove.
3. Install caliper in vise, **Fig. 9,** and fill piston/adjuster assembly with clean brake fluid.
4. Coat new thrust screw O-ring with clean brake fluid and install it in thrust screw groove, then install thrust screw into piston adjuster assembly until top surface of thrust screw is flush with bottom of threaded bore,

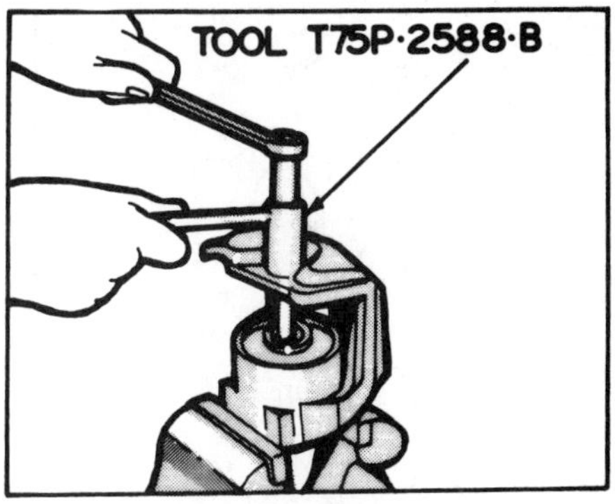

Fig. 10 Bottoming piston in caliper

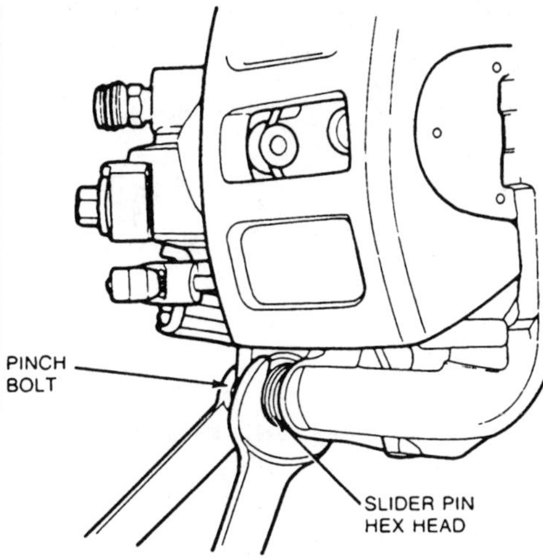

Fig. 11 Removing slider pin

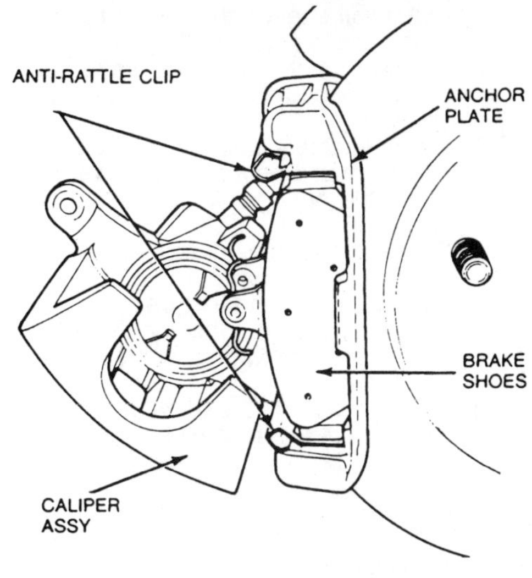

Fig. 12 Positioning anti-rattle clips

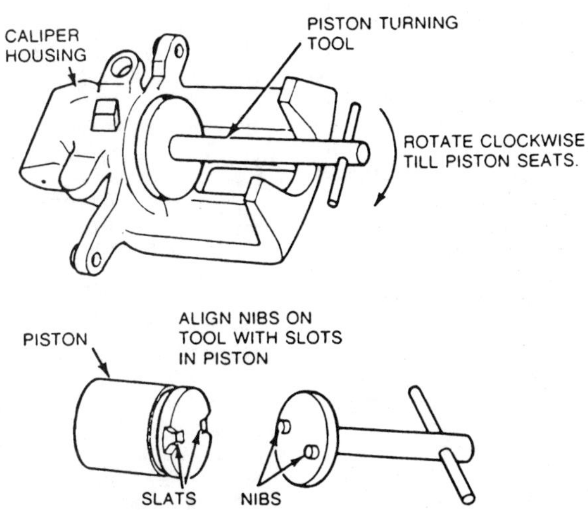

Fig. 13 Seating caliper piston

being careful to avoid cutting O-ring seal. Index notches on thrust screw and caliper housing and install anti-rotation pin. **The thrust screw and operating shafts are not interchangeable from side to side since the ramp direction in the ball pockets are different. The pocket surfaces of the operating shaft and thrust screws are stamped "R" (Right) and "L" (Left).**

5. Place a ball in each of three pockets of thrust screw and apply a liberal amount of silicone grease M1C-169-A on parking brake components, then install operating shaft on balls.
6. Coat thrust bearing with silicone grease and install it on operating shaft, then install a new lip seal and O-ring on end retainer.
7. Lightly coat O-ring seal and lip seal with silicone grease and install end retainer in caliper. Firmly hold operating shaft against internal mechanism while installing end retainer to prevent misalignment of balls. If lip seal moves out of position, reseat seal. **Torque** end retainer to 75-95 ft. lbs. **Parking brake lever must rotate freely after torquing.**
8. Install parking brake lever on keyed spline facing down and rearward. **Torque** retaining screw to 16-22 ft. lbs.
9. Bottom piston using tool T75P-2588-B, **Fig. 10,** and install caliper as described previously.

1987–88 THUNDERBIRD TURBO & 1988 CONTINENTAL

CALIPER, REPLACE

Removal

1. Raise and support rear of vehicle, then remove wheel and tire assembly.
2. Disconnect brake hose from caliper assembly.
3. Remove retaining clip, then disconnect parking brake cable from lever arm.
4. Using an open end wrench to hold slider pin in position, remove pinch bolts, **Fig. 11.**
5. Lift caliper assembly from anchor plate, then remove slider pins and boots.

Installation

1. Apply a suitable silicone dielectric compound to slider pins and inside of boots.
2. Place slider pins and boots on anchor plate, then position caliper assembly on anchor plate. Check to be sure that brake pads and anti-rattle springs are properly positioned, **Fig. 12.**
3. Apply suitable sealer and thread locking compound to pinch bolt threads, then install pinch bolts. While using an open end wrench to hold slider pin in position, **torque** pinch bolts to 30 to 35 ft. lbs.
4. Attach parking brake cable to lever arm, then install retaining clip.
5. Using replacement washers, connect brake hose to caliper. **Torque** retaining bolt to 30 to 44 ft. lbs.
6. Bleed brake system, then install wheel and tire assembly.
7. Cycle brake pedal several times to position brake pads and caliper piston.

DISC BRAKE PADS, REPLACE

Removal

1. Raise and support rear of vehicle, then remove wheel and tire assembly.
2. Remove screw attaching brake hose bracket to shock unit bracket.
3. Remove retaining clip, then disconnect parking brake cable from lever.
4. Using an open end wrench to hold slider pin in position, remove upper pinch bolt, **Fig. 11.** Loosen, but do not lower slider pin pinch bolt.
5. Carefully rotate caliper away from rotor, then remover inner and outer brake pads and anti-rattle springs from anchor plate.

Installation

1. Using tool No. T87P-2588-A or outer suitable tool, rotate caliper piston clockwise until fully seated, **Fig. 13.** Position one of the two piston slots so

that it will engage the nib on the rear of the brake pad, **Fig. 14.**
2. Position inner and outer brake pads on anchor plate, then install anti-rattle springs.
3. Carefully rotate caliper assembly over brake rotor. Ensure brake pads and anti-rattle springs are properly positioned, **Fig. 12.**
4. Apply a suitable thread sealer and locking compound to pinch bolt threads. Install and **torque** pinch bolts to 30 to 35 ft. lbs., while holding slider pin in position with a suitable open end wrench.
5. Position parking brake cable to lever, then install retaining clip.
6. Position brake hose and bracket to shock unit bracket and install attaching bolt.
7. Install wheel and tire assembly, then lower vehicle.
8. Cycle brake pedal several times to position brake pads and caliper piston.

CALIPER OVERHAUL

Disassembly

1. Remove caliper assembly from vehicle as described under "Caliper, Replace."
2. Position caliper assembly in a soft jawed vise.
3. Using tool No. T75P-2588-B or equivalent, rotate caliper piston counterclockwise to remove from caliper bore, **Fig. 13.**
4. Remove piston dust boot and seal from caliper piston bore, **Fig. 2.**.
5. Remove snap ring retaining pushrod to caliper. Use care when removing, as the snap ring and spring cover are under spring load.
6. Remove spring cover, spring, washer, key plate and pushrod and strut pin from caliper.
7. Remove O-ring from pushrod.
8. Remove parking brake lever return spring, then remove brake lever stop bolt and pull lever from caliper.

Cleaning & Inspection

1. Clean all metal components with isopropyl alcohol.
2. Use compressed air to clean out passages and grooves.
3. Inspect caliper bore for damage and excessive wear.
4. Inspect caliper piston for pitting, scoring or worn plating and replace as necessary.

Assembly

1. Apply a light coating of silicone dielectric compound to parking brake lever bore and parking brake lever seal, then position seal into caliper bore, **Fig. 2.**
2. Apply silicone dielectric compound to parking brake lever shaft, then insert shaft into caliper housing bore.
3. Install O-ring into groove on pushrod, then apply silicone dielectric compound to recesses in pushrod.
4. Place strut pin into caliper housing and into recess of parking brake lever shaft.
5. Position pushrod into caliper housing bore, ensuring strut pin is properly located between shaft recesses and recess at end of pushrod.
6. Position key plate over pushrod, so that washer nib is located in hole in caliper housing.
7. Install flat washer, spring and spring cage into caliper bore.
8. Install snap ring using tool No. T87P-2588-P or equivalent. Ensure snap ring is properly seated in recess.
9. Lubricate replacement piston seal with clean brake fluid, then install seal into caliper bore groove.
10. Lubricate piston and dust boot with clean brake fluid, then install dust boot into caliper bore.
11. Position piston into dust boot, seating dust boot in piston groove.
12. Using tool No. T75P-2588-B or equivalent, turn piston in clockwise direction until piston is fully seated in caliper bore, **Fig. 13.**
13. Position one of the two slots on the piston so that it will engage the nib on the rear of the disc pad when the caliper is installed, **Fig. 14.**
14. Install caliper assembly as described under "Caliper, Replace."

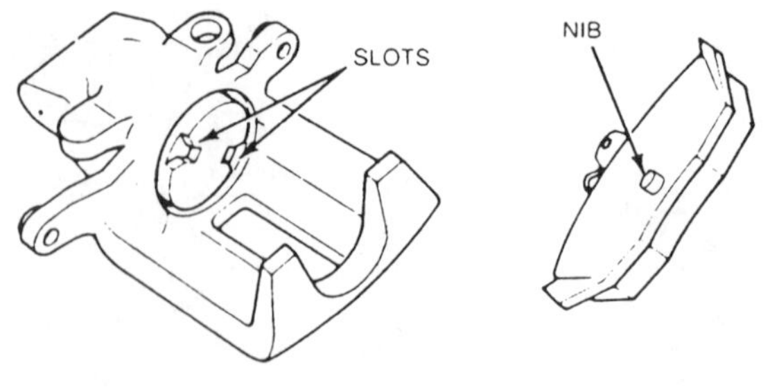

Fig. 14 Positioning caliper piston to brake pad nib

PROBE & TRACER

DISC BRAKE PADS, REPLACE

1. Raise and support vehicle.
2. Remove wheel and tire assembly, then, using a pair of needle nose pliers, remove parking brake return springs at back of caliper.
3. Loosen parking brake cable housing adjusting nut, then remove cable housing from bracket on rear lower control arm.
4. Loosen parking brake cable bracket to rear caliper retaining bolt.
5. Remove parking brake cable from rear caliper, them loosen lower caliper retaining bolt, **Fig. 15.**
6. Pivot caliper upward on upper caliper guide pin to gain access to brake pads.
7. Remove disc pad retaining spring, then the disc pads and shims.
8. Remove anchor plate clips from caliper anchor plate.
9. Remove and resurface rotor if necessary.

CALIPER, REPLACE

Removal

1. Raise and support vehicle.
2. Remove wheel and tire assembly, then the brake pads as described previously.
3. Remove brake hose retaining clip.
4. Remove brake hose banjo retaining bolt.
5. Remove lower caliper attaching bolt.
6. Remove upper caliper guide pin dust cap, then using a suitable Allen wrench, loosen and remove upper caliper guide pin.
7. Lift caliper from rotor, **Fig. 15.**

Installation

Before installation, remove upper and lower guide pin bushings and lubricate with high temperature grease D7AZ-19590-A or equivalent.

1. Install brake pads and shims as previously outlined.
2. Position caliper over rotor, then install caliper retaining bolts and **torque** to 29-36 ft. lbs.
3. Install two new copper washers and banjo bolt on flex hose banjo fitting.
4. Position flex hose on caliper and install banjo bolt. **Torque** bolt to 16-22 ft. lbs.
5. Bleed brake system, then install wheel and tire assembly.

CALIPER OVERHAUL

Disassembly

1. Remove disc pads as previously outlined, then disconnect brake hose from caliper.
2. Open bleeder screw and drain brake fluid from caliper.
3. Remove caliper guide bushing and dust boots, **Fig. 15.**
4. Pry retaining spring off dust boot, then remove piston.
5. Remove dust boot and discard, then remove piston seal from caliper and discard.
6. Remove parking brake mechanism from caliper housing.

Assembly

Clean caliper, anchor plate and rotor assembly and inspect for signs of brake fluid leakage, excessive wear or damage. The caliper must be inspected for leakage both

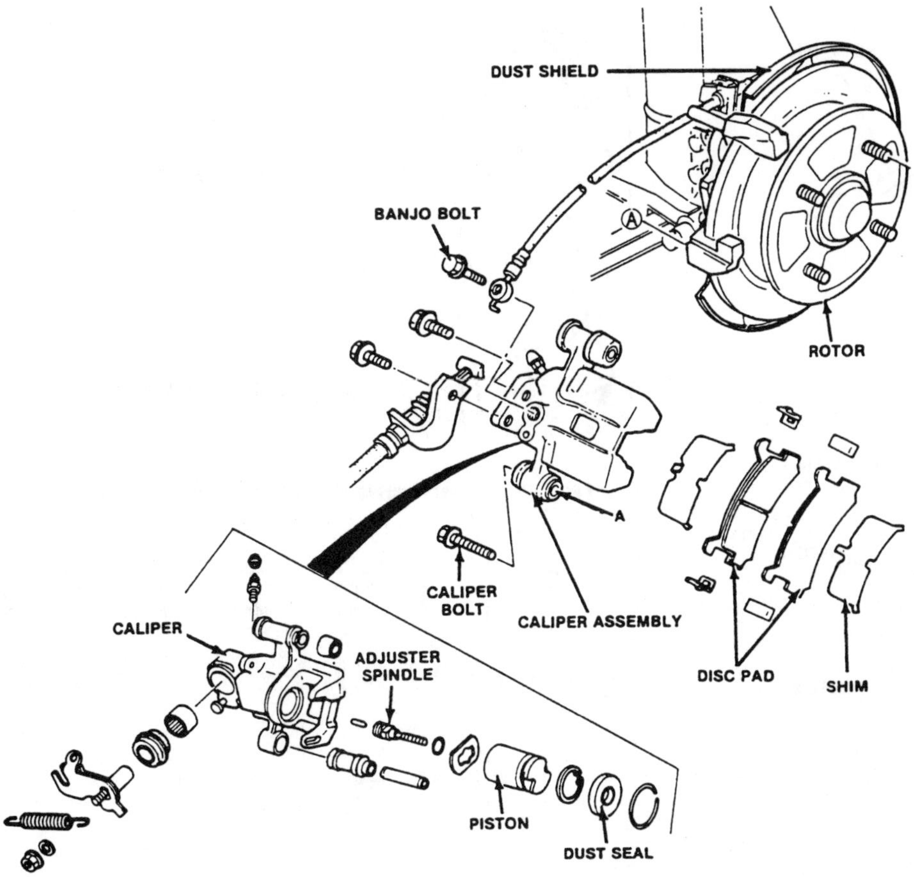

Fig. 15 Exploded view of caliper assembly. Tracer & Probe

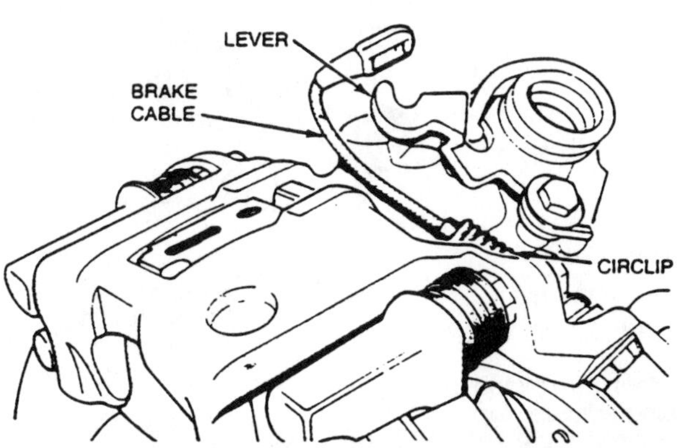

Fig. 16 Removing parking brake cable

SCORPIO

DISC BRAKE PAD, REPLACE

Removal

1. Raise and support vehicle.
2. Remove wheel and tire assembly.
3. Disconnect parking brake cable from bracket, **Fig. 16.**
4. Remove front caliper slide pin bolt securing piston housing to anchor bracket, **Fig. 17.**
5. Separate piston assembly from brake rotor, then remove brake pads from carrier bracket.
6. Turn piston clockwise into housing using piston adjuster tool No. T87P-2588-A or equivalent. **Ensure one slot in piston will engage with tab on back of brake pad during installation.**

Installation

1. Position brake pads in carrier bracket, then install caliper assembly over brake rotor. Ensure anti-rattle springs are properly positioned in housing and tab in brake pad aligns with slot in piston. **Adjuster tool No. T87P-2588-A or equivalent may be used to align piston if necessary.**
2. Install front caliper bolt, **Fig. 17,** and **torque** to 37-48 ft. lbs.
3. Connect parking brake cable, then install wheel and tire assembly.
4. Lower vehicle, then pump brake pedal several times to properly position brake pads before operation.

CALIPER REPLACE

1. Apply brake pedal several times to relieve system pressure.
2. Raise and support vehicle.
3. Remove wheel and tire assembly.
4. Disconnect electrical connector from brake pad wear sensor, **Fig. 18.**
5. Disconnect brake line from flex hose and flex hose from housing. Plug lines to prevent leakage and contamination.
6. Disconnect parking brake cable from retaining bracket.
7. Remove caliper assembly attaching bolts and the caliper.
8. Reverse procedure to install, noting the following:
   a. **Torque** caliper attaching bolts to 37-48 ft. lbs.
   b. Following completion of installation, bleed rear brake system.

CALIPER OVERHAUL

Disassembly

1. Remove caliper assembly and brake pads as previously described.
2. Remove slide pin bolts and the slide pins from anchor bracket.
3. Secure caliper assembly in a suitable vise.
4. Turn caliper piston counterclockwise using piston adjuster tool No. T87P-2588-A or equivalent, until it protrudes approximately .8 inch, **Fig. 19.**
5. Disengage piston boot cover from pis-

in piston boot area and operating shaft seal area. Lightly sand or wire brush any rust or corrosion from caliper and anchor plate sliding surfaces and inner brake shoe abutment surfaces in anchor plate. Inspect brake shoes for wear. Linings must not be worn to less than 1/8 inch of shoe surface.

Lubricate all new seals with brake fluid before installation.

1. Install needle bearings, dust boot and parking brake lever, **Fig. 15.**
2. Install adapter spindle in caliper.
3. Install dust boot into caliper bore, then install retaining spring.
4. Install caliper piston.
5. Install upper and lower guide pin dust boots.
6. Install anchor plate clips, brake pads, shims and spring retainer.
7. Install caliper on anchor plate, then tighten upper guide pin and install dust cap.
8. Install lower guide pin and **torque** to 29-36 ft. lbs.
9. Bleed brake system, them pump brake pedal several times to seat brake pads.
10. Check brake fluid level, adding fluid as necessary, then install wheel and tire assembly.

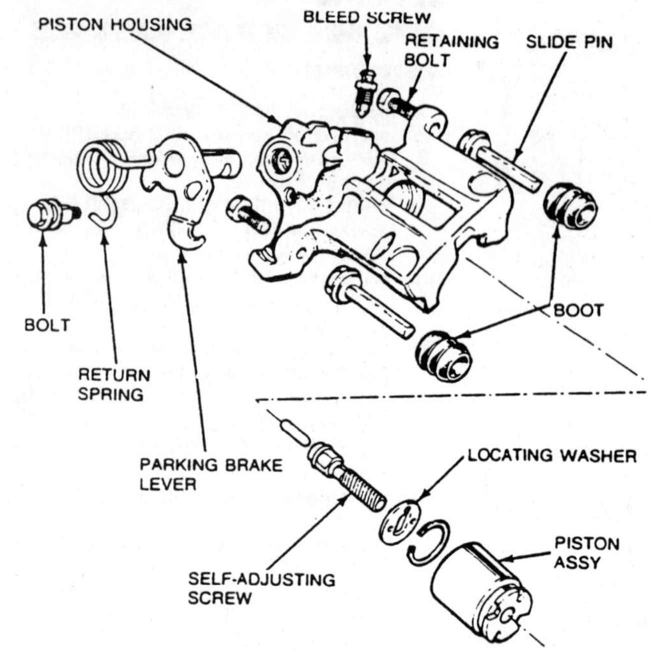

Fig. 17 Exploded view of caliper assembly. Scorpio

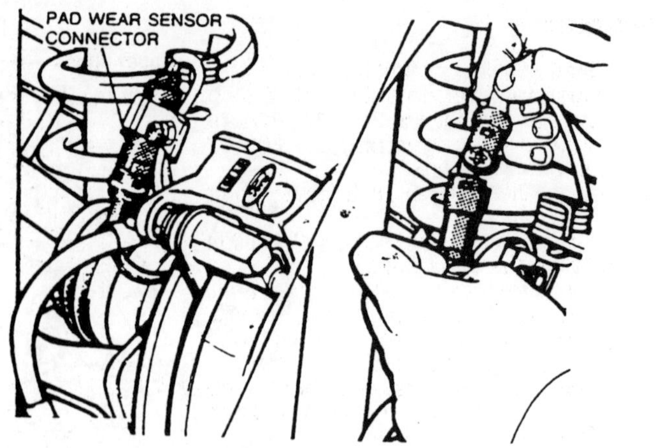

Fig. 18 Disconnecting brake pad wear sensor

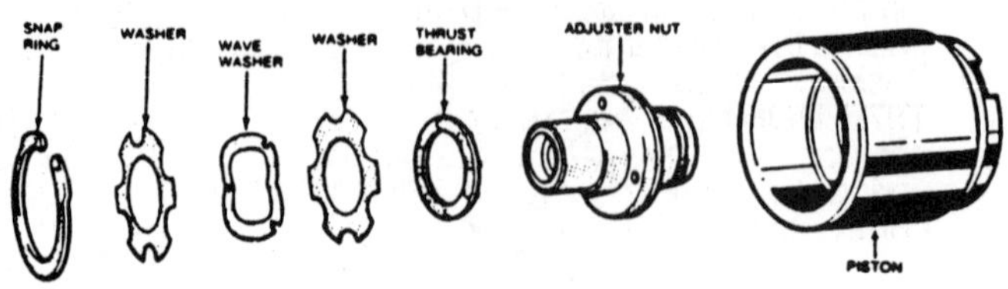

Fig. 20 Exploded view of caliper piston

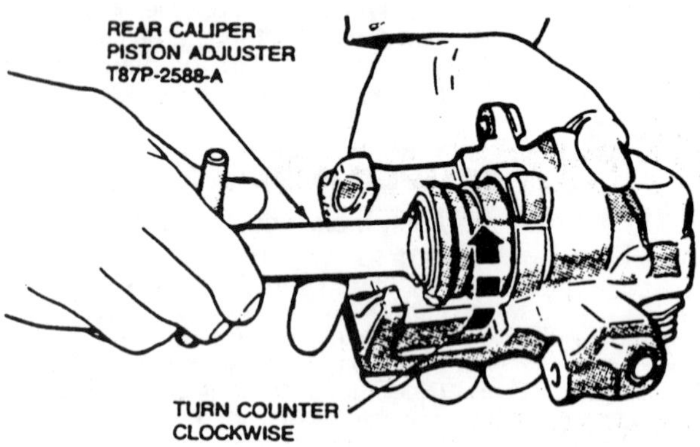

Fig. 19 Removing caliper piston

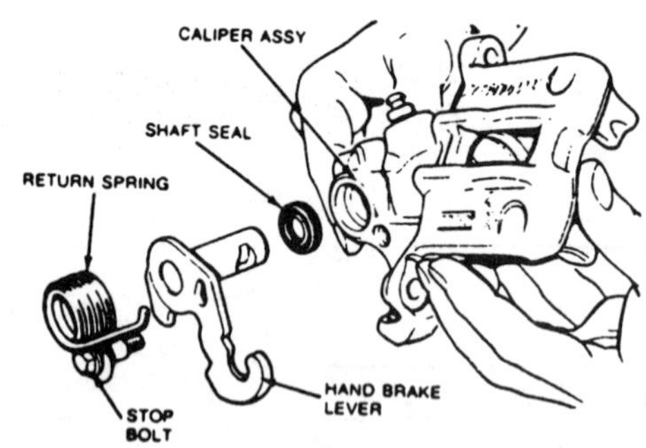

Fig. 21 Parking brake lever return spring

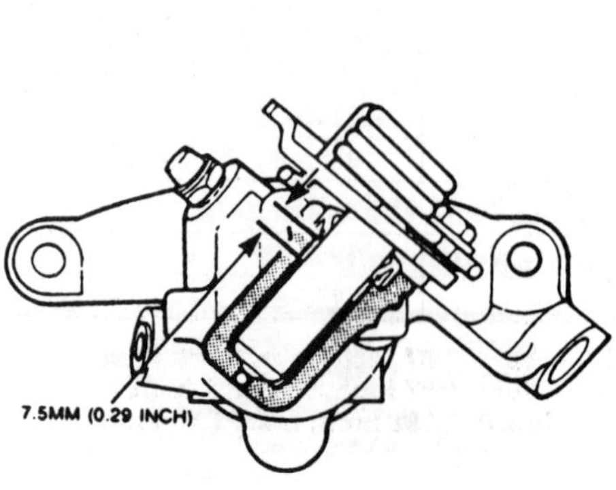

Fig. 22 Rear caliper lever shaft bushing installation

ton groove, then turn piston fully out of housing using adjuster tool.

6. Remove piston boot, then the piston body snap ring, **Fig. 20.**
7. Remove adjuster nut with thrust washers, wave washers and thrust bearing.
8. Remove and discard adjuster nut seal. **Note direction of seal lip for installation reference.**
9. Clean adjuster nut clutch surface and ensure piston breather port in piston groove is not restricted.
10. Remove piston seal from caliper body.
11. Remove circlip, then the washer, spring and spring cover housing from caliper.
12. Remove fast thread adjuster and snap ring retaining key plate from caliper. Remove and discard O-ring from adjuster.
13. Remove parking brake strut from bore.

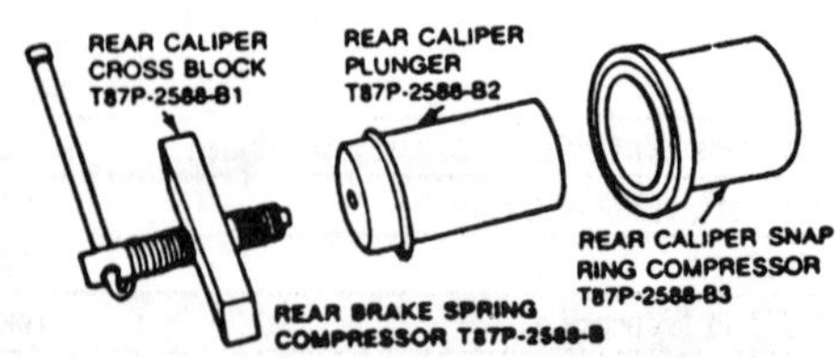

Fig. 23 Rear caliper assembly tools

14. Remove parking brake lever return spring from lever, then remove stop bolt, **Fig. 21.**
15. Remove parking brake lever return spring from caliper bore. Remove and discard shaft seal.
16. Clean and inspect all caliper components, replacing as necessary.

Assembly

1. Install new lever shaft bushing, if previously removed, to a depth of .29 inch from lower seal lip, **Fig. 22.** Ensure cutout in bushing is properly aligned with pushrod bore.
2. Apply suitable grease to lever shaft and bushing.
3. Press shaft seal into housing until it is fully seated on shoulder.
4. Slide lever into housing through seal, using care to avoid damaging seal lip.
5. Install parking brake stop bolt and spring, torquing bolt to 8-9 ft. lbs.
6. Apply suitable grease to bore area around strut, then position parking brake strut into caliper bore.
7. Install new O-ring onto fast thread adjuster, then position adjuster in caliper bore.
8. Install key plate and snap ring. Ensure dimple on plate is properly positioned prior to installing snap ring.
9. Position snap ring compressor T87P-2588-B3 or equivalent into caliper bore, then insert washer, spring, spring retainer and snap ring into compressor.

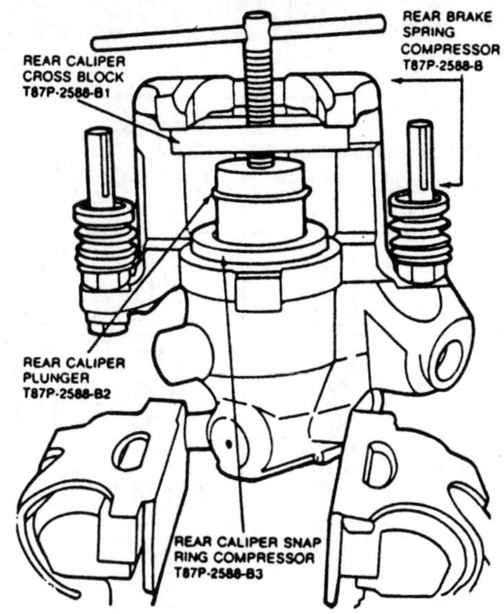

Fig. 24 Rear caliper brake spring compressor tool installation

10. Position rear brake spring compressor T87P-2588-B or equivalent, **Fig. 23,** as shown in **Fig. 24,** and turn screw clockwise until snap ring is firmly seated.
11. Lubricate caliper body bore with clean hydraulic fluid, then install piston seal into groove.
12. Assemble piston assembly, then install new seal on adjuster nut.
13. Apply brake fluid to piston adjuster nut contact face, then install thrust bearing, thrust washers, wave washer and snap ring.
14. Install piston protective boot into caliper body groove. Stretch boot over piston, then start thread of adjuster nut on fast thread.
15. Screw piston onto fast thread adjuster using piston adjuster T87P-2588-A or equivalent, then compress into caliper base.
16. Secure piston protective cover into piston groove.
17. Apply suitable grease to slide pin bores and inside of boot, then install slide pins and slide pin bolts into anchor bracket. Secure protective boot in grooves.

Caliper Specifications

Model	Year	Caliper Bore Dia. Inch
FORD & MERCURY		
Capri & Mustang	1985-86①	2.36
	1985-86②	③
	1987-89④	2.36
	1987-89⑤	2.87
Cougar & Thunderbird	1985-86	2.36
	1987-89⑥	2.36
	1987-89⑦	⑧
Crown Victoria & Grand Marquis	1985-89	2.88
Escort, EXP & Lynx	1985-89	2.362
Festiva	1988-89	2.12
LTD & Marquis	1985-86	2.36
Sable & Taurus	1986-89	2.598

Model	Year	Caliper Bore Dia. Inch
FORD & MERCURY		
Tempo & Topaz	1985-89	2.362
Tracer	1988-89	—
LINCOLN		
Continental	1985-87	⑧
	1988-89	⑩
	Continental	2.38②
Continental & Mark VII	1985-87	2.87②
Mark VII	1985-89	⑨
Town Car	1985-89	2.88
MERKUR		
Scorpio	1986-89	—
XR4Ti	1988-89	—

①—Except Mustang SVO.
②—Mustang SVO.
③—Front, 2.87 inch; rear, 2.1 inch.
④—Except Mustang w/V8-302 engine.
⑤—Mustang w/V8-302 engine.
⑥—Except Thunderbird Turbo.
⑦—Thunderbird Turbo.
⑧—Front, 2.87 inch; rear, 1.789 inch.
⑨—Front, 2.87 inch; rear, 2.1 inch.
⑩—Front, 2.598 inch; rear, 1.79 inch

Rotor Specifications

Model	Year	Nominal Thickness	Minimum Refinish Thickness	Thickness Variation (Parallelism)	Lateral Run-out (T.I.R.)	Finish (Micro-Inch)
FORD & MERCURY						
Capri & Mustang	1985-86 ①	.870	.810	.0005	.003	10-81
	1985-86 ②	1.030	.972	.0005	.003	15-125
	1987-89 ③	.870	.810	④	.003	10-81
	1987-89 ⑤	1.030	.972	.0005	.003	15-125
Cougar & Thunderbird	1985-86	.870	.810	④	.003	15-125
	1987-89 ⑥	.870	.810	④	.003	10-81
	1987-89 ⑦	1.030	.972	.0005	.003	15-125
Crown Victoria & Grand Marquis	1985-89	1.030	.972	.0005	.003	10-80
Escort, EXP & Lynx	1985-87	.945	.882	.0004	.002	20-140
	1988-89	.945	.882	.0005	.003	10-80
Festiva	1988-89	—	.460	—	—	—
LTD & Marquis	1985-86	.870	.810	.0003	.003	10-81
Sable & Taurus	1986-87	.945	.896	.0005	.003	10-80
	1988-89	1.024	.974	.0005	.003	10-80
Tempo & Topaz	1985-89	.945	.882	.0005	.003	10-80
Tracer	1988-89	0.71	0.63	—	.004	—
LINCOLN						
Continental	1985-87 ⑧	1.030	.972	.0005	.003	15-125
	1985-87 ⑨	.945	.895	.0005	.004	15-80
	1988-89 ⑧	1.020	.970	.0004	.002	10-80
	1988-89 ⑨	1.020	.974	.0005	.002	10-80
Mark VII	1985-89 ⑧	1.030	.972	.0005	.003	15-125
	1985-87 ⑨	.945	.895	.0005	.004	15-80
	1988-89 ⑨	.945	.895	.0005	.004	16-125
Town Car	1985-89	1.030	.972	.0005	.003	10-80
MERKUR						
Scorpio	1988-89 ⑧	—	0.90	.0039	—	—
	1988-89 ⑨	—	0.35	.0039	—	—
XR4Ti	1986-89	0.95	0.898	—	—	—

①—Except Mustang SVO.
②—Mustang SVO.
③—Except Mustang w/V8-302 engine.
④—Models w/steel wheels, .0005 inch; models w/aluminum wheels, .0003 inch.
⑤—Mustang w/V8-302 engine.
⑥—Except Thunderbird Turbo.
⑦—Thunderbird Turbo.
⑧—Front.
⑨—Rear.

GENERAL MOTORS CORP.

NOTE: Refer to "Applications" to determine which type brakes are used on vehicle being serviced.

TABLE OF CONTENTS

Applications

Page No.

AISIN SEIKI Single Piston Caliper:

TYPE 1 6-61
- Chevrolet Spectrum (Front), 1985-88
- GEO Spectrum (Front), 1989

TYPE 2 6-63
- Chevrolet Sprint (Front), 1985-88
- GEO Metro (Front), 1989

DAEWOO Single Piston w/Dual Mounting Bolts .. 6-64

Pontiac Lemans (Front), 1988-89

DELCO-MORAINE Dual Piston w/Dual Mounting Bolts 6-66

- Buick Regal (Front), 1988-89
- Oldsmobile Cutlass Supreme (Front), 1988-89
- Pontiac Grand Prix (Front), 1988-89

DELCO-MORAINE Single Piston w/Dual Mounting Bolts:

TYPE 1 6-67
- Buick Estate Wagon, 1985-89
- Buick LeSabre, 1985
- Buick LeSabre Wagon, 1989
- Buick Regal, 1985-87
- Buick Riviera, 1985
- Cadillac Brougham, 1985-89
- Cadillac Eldorado, 1985
- Cadillac Seville, 1985
- Chevrolet Camaro (Front), 1985-89
- Chevrolet Caprice, 1985-89
- Chevrolet Impala, 1985
- Chevrolet Monte Carlo, 1985-88
- Oldsmobile Custom Cruiser, 1985-89
- Oldsmobile Cutlass Except Ciera, 1985-87
- Oldsmobile Cutlass Supreme Classic, 1988
- Oldsmobile Toronado, 1985
- Oldsmobile 88, 1985
- Pontiac Bonneville, 1985-86
- Pontiac Firebird (Front), 1985-89
- Pontiac Grand Prix, 1985-87
- Pontiac Parisienne, 1985-86
- Pontiac Safari, 1987-89

TYPE 2 6-70
- Buick Century, 1985-89
- Buick Electra, 1985-89
- Buick LeSabre, 1986-89
- Buick Park Avenue, 1988-89
- Buick Skyhawk, 1985-89
- Buick Skylark, 1985-89
- Buick Somerset, Somerset Regal, 1985-88
- Cadillac Cimarron, 1985-88
- Cadillac Deville, 1985-89
- Cadillac Fleetwood, 1985-89
- Chevrolet Beretta, 1987-89
- Chevrolet Cavalier, 1985-89
- Chevrolet Celebrity, 1985-89
- Chevrolet Chevette, 1985-87
- Chevrolet Citation, 1985
- Chevrolet Corsica, 1987-89
- Oldsmobile Calais, 1985-89
- Oldsmobile Cutlass Ciera, 1985-89
- Oldsmobile Firenza, 1985-88
- Oldsmobile 88, 1986-89

General Information

INDEX

TROUBLESHOOTING

When troubleshooting brake system, refer to **Fig. 1**.

BRAKE SHOES, LININGS & CALIPERS

Remove wheels and inspect brake disc, caliper and linings. The wheel bearings should be inspected at this time and repacked if necessary. Do not get any grease on the linings.

Do not replace individual pads or replace pads on one wheel of an axle set alone, as braking performance will be adversely affected.

If a visual inspection does not adequately determine the condition of the linings, the brake shoe and lining assemblies should be removed and inspected. If shoes do not require replacement, reinstall them in their original positions. Brake shoes and linings should also be replaced if cracked or damaged.

If the caliper is cracked or fluid leakage through the casting is evident, it must be replaced as a unit.

BRAKE ROUGHNESS

The most common cause of brake chatter on disc brakes is a variation in thickness of the disc. If roughness or vibration is encountered during highway operation or if pedal pulsation is experienced at low speeds, the disc may have excessive thickness variation. To check for this condition, measure the disc at 12 points with a micrometer at a radius approximately one inch from edge of disc. If thickness measurements vary by more than .0005 inch, the disc should be replaced with a new one.

Excessive lateral runout of braking disc may cause a "knocking back" of the pistons, possibly creating increased pedal travel and vibration when brakes are applied.

Before checking the runout, wheel bearings should be adjusted. The readjustment is very important and will be required at the completion of the test to prevent bearing failure.

BRAKE DISC SERVICE

Servicing of disc brakes is extremely critical due to the close tolerances required in machining the brake disc to insure proper brake operation.

The maintenance of these close controls of the shape of the rubbing surfaces is necessary to prevent brake roughness. In addition, the surface finish must be non-directional and maintained at a micro inch finish. This close control of the rubbing surface finish is necessary to avoid pulls and erratic performance, and to promote long lining life and equal lining wear of both left and right brakes.

In light of the foregoing remarks, refinishing of the rubbing surfaces should not be attempted unless precision equipment,

CAUSE / SYMPTOM	Excessive Brake Pedal Travel	Brake Pedal Travel Gradually Increases	Excessive Brake Pedal Effort	Excessive Braking Action	Brakes Slow To Respond	Brakes Slow To Release	Brakes Drag	Uneven Braking Action (Side To Side)	Uneven Braking Action (Front To Rear)	Scraping Noise From Brakes	*Brakes Squeak During Application	*Brakes Squeak During Stop	**Brakes Pulsate (Roughness)	*Brakes Groan At End of Stop	"BRAKE" Warning Lamp Glows
Leaking Brake Line or Connection	X	XX						X							XX
Leaking Wheel Cylinder or Piston Seal	X	XX		X				X							XX
Leaking Master Cylinder	X	XX													XX
Air in Brake System	XX							X							X
Contaminated or Improper Brake Fluid					X	X	X								X
Leaking Vacuum System			XX		X										
Restricted Air Passage in Power Head			X		XX	X									
Damaged Power Head			X	X	X	X	X								
Improperly Assembled Power Head Valving			X	X	X	X	XX								
Worn Out Brake Lining - Replace			X	X				X	X	X	X	X		X	
Uneven Brake Lining Wear - Replace and Correct	X			X				X	X	X	X	X		X	X
Incorrect Lining Material-Replace			X	X				X	X			X		X	
Contaminated Brake Lining - Replace				XX				XX	X	X	X	X		X	
Linings Damaged by Abusive Use-Replace			X	XX				X	X	X	X	X		X	
Heat Spotted or Scored Brake Drums or Rotors				X				X	X		X	X	X	X	
Out-of-Round Brake Drums													X		
Improper Rotor Thickness Variation													XX		
Excessive Lateral Runout	X										X		X		
Faulty Automatic Adjusters	X						X	X	X						X
Incorrect Wheel Cylinder Sizes	X		X	X				X	X						
Weak or Incorrect Brake Shoe Retention Springs				X		X	XX	X	X	XX	X	XX			
Brake Assembly Attachments - Missing or Loose	X						X	X	X	X		X	X	X	
Insufficient Brake Shoe Guide Lubricant						X	X	X	X	XX	XX				
Restricted Brake Fluid Passage or Sticking Wheel Cylinder Piston		X	X		X	X	X	X	X						
Incorrect Stoplamp Switch Adjustment		X		X			X								
Brake Pedal Linkage Interference or Binding			X		X	XX	XX								X
Improperly Adjusted Parking Brake							X								X
Parking Brake Cable Siezed						X	X		X						X
Drums Tapered or Threaded										XX					
Incorrect Front End Alignment								XX							
Incorrect Wheel Bearing Runout	X									X					
Loose Front Suspension Attachments								X		XX				X	
Operator Riding Brake Pedal	X	X	X				X		X			X		X	
Improper Booster Pushrod Length	X					X	XX								
Sticking Wheel Cylinder or Caliper Pistons			X			X	X	X	X						
Caliper Not Sliding Freely	X						X			X			XX		
Faulty Proportioning Valve			X	X	X	X	X								

XX — Indicates more probable cause(s)
X — Indicates other causes

*May be a normal condition.
**Can usually be corrected by servicing front brakes.

Fig. 1 Troubleshooting brake system chart

Fig. 2 Checking rotor for lateral runout

Fig. 3 Checking rotor parallelism (thickness variation)

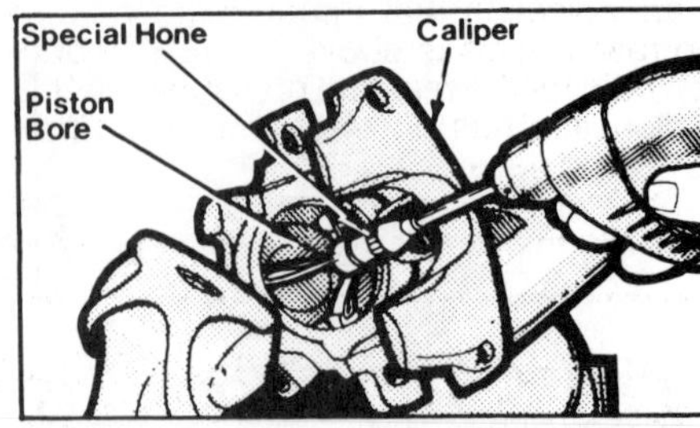

Fig. 4 Honing caliper piston bore

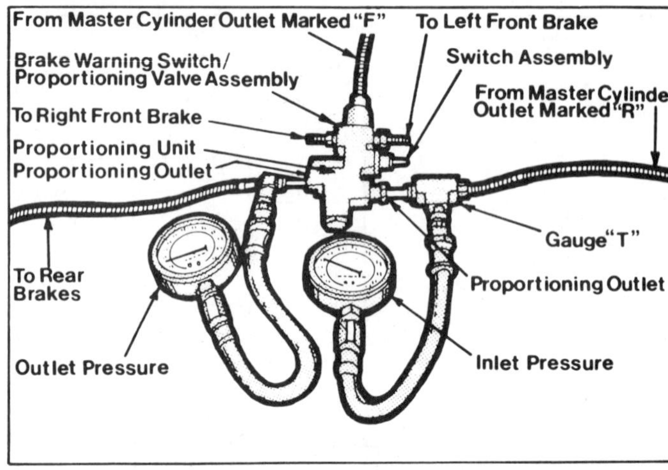

Fig. 5 Gauge hook-up for testing proportioning valve (typical)

capable of measuring in micro inches (millionths of an inch) is available.

To check lateral runout of a disc, mount a dial indicator on a convenient part (steering knuckle, tie rod, disc brake caliper housing) so that the plunger of the dial indicator contacts the disc at a point one inch from the outer edge, **Fig. 2.** If the total indicated runout exceeds specifications, install a new disc.

To check parallelism (thickness variation), mount dial indicators, **Fig. 3,** so the plunger contacts rotor approximately 1 inch from outer edge. If parallelism exceeds specifications, replace rotor.

GENERAL PRECAUTIONS

1. Grease or any other foreign material must be kept off the brake linings, caliper, surfaces of the disc and external surfaces of the hub, during service procedures. Handling the brake disc and caliper should be done in a way to avoid deformation of the disc and nicking or scratching brake linings.
2. If inspection reveals rubber piston seals are worn or damaged, they should be replaced immediately.
3. During removal and installation of a wheel assembly, exercise care so as not to interfere with or damage the caliper splash shield, or bleeder screw.
4. Front wheel bearings preload should be adjusted to specifications .
5. Be sure vehicle is centered on hoist before servicing any of the front end components to avoid bending or damaging the disc splash shield on full right or left wheel turns.
6. Before the vehicle is moved after any brake service work, be sure to obtain a firm brake pedal.
7. The assembly bolts of two piece caliper housings should not be disturbed unless the caliper requires service.

INSPECTION OF CALIPER

Should it become necessary to remove the caliper for installation of new parts, clean all parts in alcohol, wipe dry using lint free cloths, using an air hose to blow out drilled passages and bores. Check dust boots for punctures or tears. If punctures or tears are evident, new boots should be installed upon reassembly.

Inspect piston bores for scoring or pitting. Bores that show light scratches or corrosion can usually be cleaned with crocus cloth. However, bores that have deep scratches or scoring may be honed, provided the diameter of the bore is not increased more than .002 inches. If the bore does not clean up within this specification, a new caliper housing should be installed (black stains on the bore walls are caused by piston seals and will do no harm).

When using a hone, **Fig. 4,** be sure to install the hone baffle before honing bore. The baffle is used to protect the hone stones from damage. Use extreme care in cleaning the caliper after honing. Remove all dust and grit by flushing the caliper with alcohol. Wipe dry with clean lint free cloth and then clean a second time in the same manner.

BLEEDING DISC BRAKES

Pressure bleeding is recommended for all hydraulic disc brake systems.

The disc brake hydraulic system can be bled manually or with pressure bleeding equipment. **On vehicles with disc brakes** the brake pedal will require more pumping and frequent checking of fluid level in master cylinder during bleeding operation.

Never use brake fluid that has been drained from hydraulic system when bleeding the brakes. Be sure the disc brake pistons are returned to their normal positions and that the shoe and lining assemblies are properly seated. Before driving the vehicle, check brake operation to be sure that a firm pedal has been obtained.

PROPORTIONING VALVE

The proportioning valve (when used), **Fig. 5,** provides balanced braking action between front and rear brakes under a wide range of braking conditions. The valve regulates the hydraulic pressure applied to the rear wheel cylinders, thus limiting rear braking action when high pressures are required at the front brakes. In this manner, premature rear wheel skid is prevented.

TESTING PROPORTIONING VALVE

When a premature rear wheel slide is obtained on a brake application, it usually is an indication that the fluid pressure to the rear wheels is above the 50% reduction ratio for the rear line pressure and that malfunction has occurred within the proportioning valve.

To test the valve, install gauge set shown in **Fig. 5** in brake line between master cylinder and proportioning valve, and at output end of proportioning valve and brake line as shown. Be sure all joints are fluid tight.

Have a helper exert pressure on brake pedal (holding pressure). Obtain a reading on master cylinder output of approximately 700 psi. While pressure is being held as above, reading on valve outlet should be 550-610 psi. If the pressure readings do not meet these specifications, the valve should be removed and a new valve installed.

Aisin-Seiki Single Piston w/Dual Mounting Bolts (Type 1)

INDEX

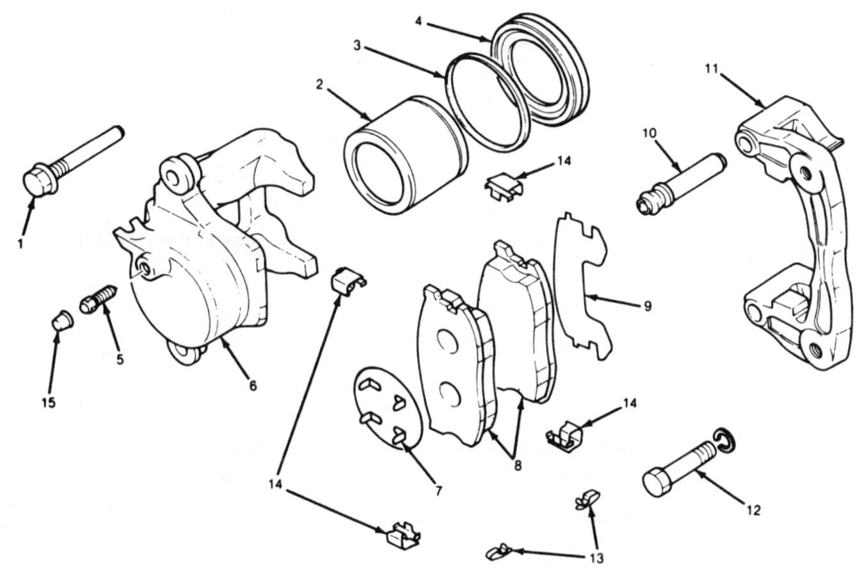

1. SLIDE PIN BOLT
2. PISTON
3. PISTON SEAL
4. PISTON BOOT
5. BLEEDER SCREW
6. CALIPER BODY
7. INNER SHIM
8. PADS
9. OUTER SHIM
10. SLIDE PIN BOOT
11. BRACKET
12. BOLT
13. WEAR INDICATOR
14. RETAINER
15. CAP

Fig. 1 Exploded view of disc brake caliper assembly. Except 1987–88 turbocharged models

Description

This single piston sliding caliper assembly, **Figs. 1 and 2**, is mounted to a support bracket by two slide pins. The caliper assembly slides on the two mounting pins. Upon brake application, fluid pressure against the piston forces the inboard shoe and lining assembly against the inboard side of the disc. This action causes the caliper assembly to slide until the outboard lining comes into contact with the disc.

SERVICE

CALIPER REMOVAL

1. Remove approximately 2/3 of brake fluid from master cylinder.
2. Raise and support vehicle.
3. Mark relationship between front wheel and axle, then remove wheel and tire assembly.
4. If caliper assembly is to be serviced, remove inlet fitting attaching bolt, copper washer, and inlet fitting from caliper housing.Plug opening in inlet fitting to prevent fluid loss and contamination. **Do not crimp brake hose, as this may damage internal structure of hose. If only shoe and lining assemblies are to be replaced, do not disconnect brake line fitting from caliper.**
5. Remove caliper slide pins and the caliper. If only shoe and lining assemblies are to be replaced, suspend caliper from chassis using suitable hanger. **Do not allow caliper to hang by brake hose.**
6. Remove shoe and lining assembly.
7. Remove bracket attaching bolts and the bracket.
8. Remove slide pin boot from bracket.

BRAKE SHOE & LINING, REPLACE

1. Remove approximately 2/3 of brake fluid from master cylinder.
2. Raise and support vehicle.
3. Mark relationship between front wheel and axle, then remove wheel and tire assembly.
4. Remove 2 caliper slide pins from bracket.
5. Unfasten caliper, and support with a length of wire, leaving hydraulic lines connected.
6. Remove brake pads, shims, wear indicators and retainers, **Fig. 1 and 2.**
7. Reverse procedure to install.

CALIPER DISASSEMBLY

1. Remove caliper assembly as outlined previously, then drain brake fluid from caliper.
2. Use clean shop towels to pad interior of caliper assembly, then remove piston by directing compressed air into caliper brake hose inlet hole, **Fig. 3. Use just enough air pressure to ease piston out of bore. Do not place fingers in front of piston for any reason when applying com-**

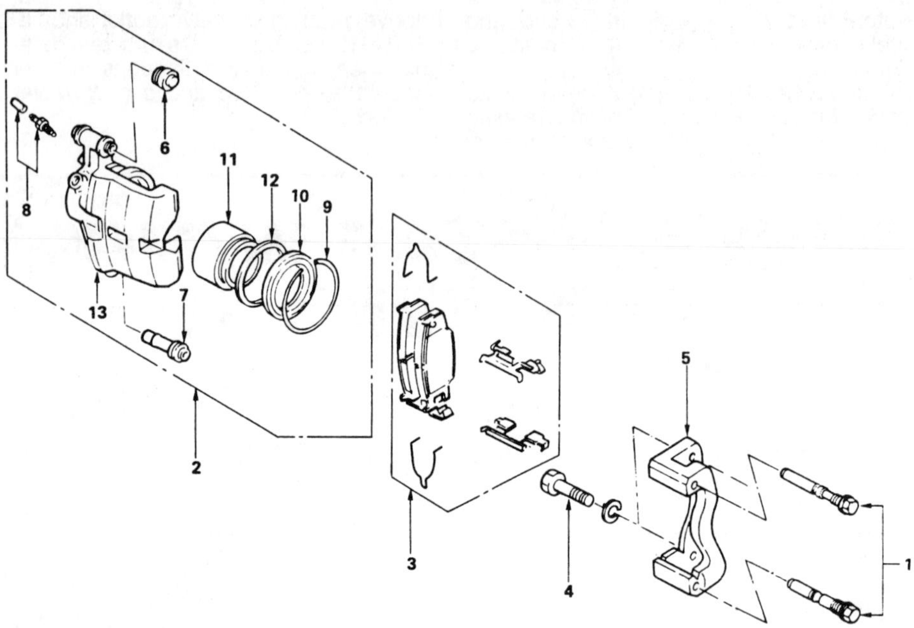

1. Slide pins
2. Caliper assembly
3. Pad assembly
4. Bolt
5. Bracket
6. Slide pin boot
7. Slide pin boot
8. Bleeder with cap
9. Boot ring
10. Piston boot
11. Piston
12. Piston seal
13. Caliper body

Fig. 2 Exploded view of disc brake caliper assembly. 1987–88 turbocharged models

pressed air. This could result in serious personal injury.

3. Remove dust boot from piston.
4. Using a small piece of wood or plastic, remove piston seal from bore. **Do not use a metal tool of any kind to remove seal as it may damage bore.**
5. Remove bleeder valve.
6. Inspect piston for scoring, nicks, corrosion, and wear and replace as needed.
7. Inspect caliper housing and seal groove for corrosion, nicks, scoring and excessive wear, and use crocus cloth to polish away corrosion from housing bore. Replace caliper housing if corrosion in and around seal groove will not clean up with crocus cloth.
8. Clean all parts with denatured alcohol. Dry with unlubricated compressed air. Blow out all passages in housing and bleeder valve.

CALIPER ASSEMBLY

1. Apply suitable grease to piston seal and cylinder wall, then install the seal. Check to ensure piston seal is not twisted.
2. Apply suitable grease to sliding portion of piston and install dust boot.
3. Insert edge of dust boot into boot groove, then slowly force piston fully into cylinder.
4. Install bleeder valve.

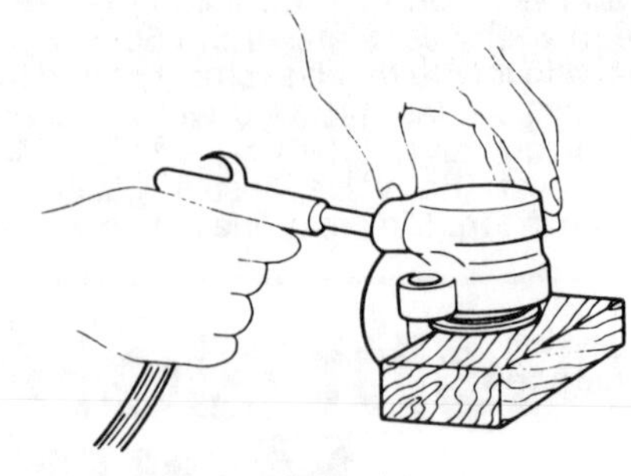

CAUTION: Do not place fingers in front of the piston in an attempt to catch or protect it when applying compressed air. This could result in serious injury.

NOTICE: Use just enough air to ease the piston out of the bore. If the piston is blown out, even with padding provided, it may be damaged.

Fig. 3 Removing caliper piston

CALIPER INSTALLATION

1. Apply suitable grease to inner face of slide pin boot.
2. Install slide pin boot to bracket.
3. Install bracket and **torque** attaching bolts to 40.5 ft. lbs.
4. Install shoe and lining assembly, ensure wear indicators are located on trailing edge of shoe assemblies during forward wheel rotation.
5. Install caliper assembly to bracket. **Torque** attaching bolts to 36 ft. lbs.
6. Attach flexible hose to caliper and **torque** fitting to 13.4 ft. lbs.
7. Install wheel and tire assembly, then lower vehicle.
8. Fill master cylinder to proper level and bleed brakes. **Before moving vehicle, pump brake pedal several times to be sure it is firm. Do not move vehicle until a firm pedal is obtained.**

Aisin-Seiki Single Piston w/Dual Mounting Bolts (Type 2)

INDEX

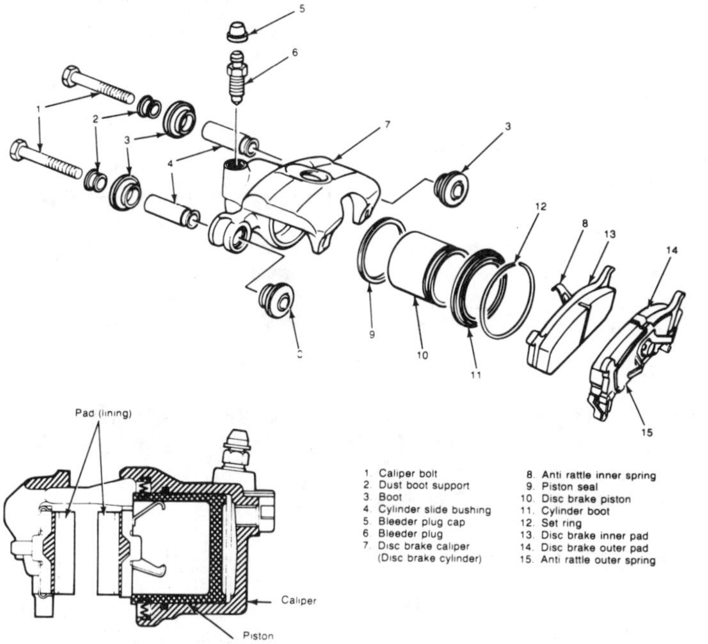

Fig. 1 Exploded view of brake caliper

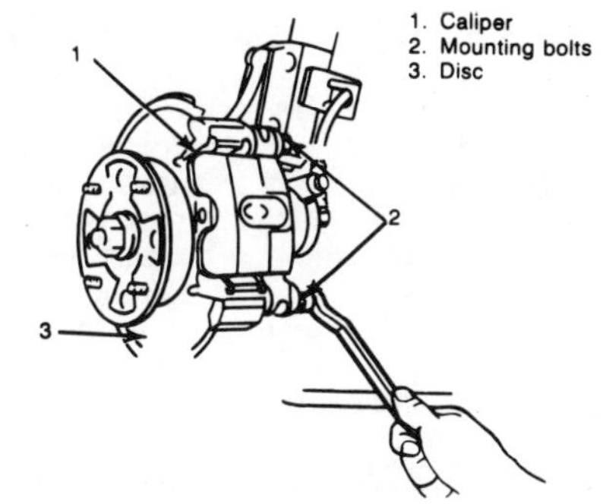

Fig. 2 Removing caliper mounting bolts

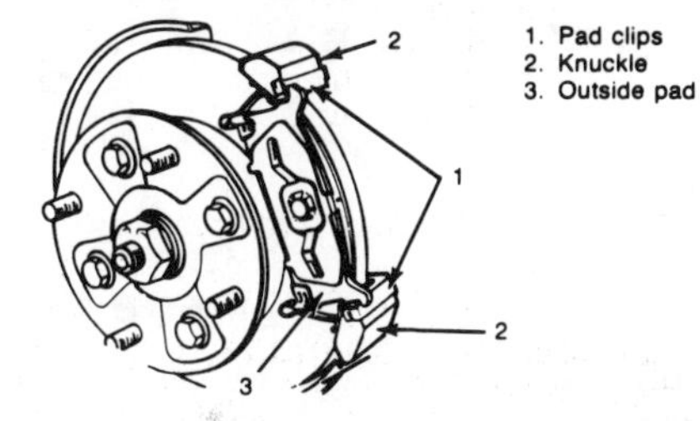

Fig. 3 Installing outside pad

DESCRIPTION

This single piston sliding caliper assembly, **Fig. 1,** is mounted to a support bracket by two slide pins. The caliper assembly slides on the two mounting pins. Upon brake application, fluid pressure against the piston forces the inboard shoe and lining assembly against the inboard side of the disc. This action causes the caliper assembly to slide until the outboard lining comes into contact with the disc.

SERVICE

CALIPER REMOVAL

1. Remove approximately 2/3 of brake fluid from master cylinder.
2. Raise and support vehicle.
3. Mark relationship between front wheel and axle, then remove wheel and tire assembly.
4. If caliper assembly is to be serviced, remove inlet fitting attaching bolt, copper washer, and inlet fitting from caliper housing.Plug opening in inlet fitting to prevent fluid loss and contamination. **Do not crimp brake hose, as this may damage internal structure of hose. If only shoe and lining assemblies are to be replaced, do not disconnect brake line fitting from caliper.**
5. Remove caliper mounting bolts.
6. Remove caliper from knuckle. **Be careful not to drop inside pad which will come off with caliper.**

BRAKE SHOE & LINING, REPLACE

1. Raise and support vehicle and remove wheel.
2. Remove caliper mounting bolts, **Fig. 2.**
3. Remove caliper from knuckle, taking care not to drop inside pad. If only shoe and lining assemblies are to be replaced, suspend caliper from chassis using suitable hanger. **Do not allow caliper to hang by brake hose.**
4. Remove outside pad.
5. Install pad clips, No. 1 & 2, then outside pad to knuckle, **Fig. 3.**
6. Install caliper and inside pad to knuckle, **Fig. 4.** Position outer springs over caliper as shown.
7. **Torque** caliper bolts to 17.5-26 ft. lbs.
8. Install front wheel.
9. Lower vehicle and perform brake test.

CALIPER DISASSEMBLY

1. Remove caliper assembly as outlined previously, then drain brake fluid from caliper.
2. Remove cylinder boot set ring and boot from cylinder groove, **Fig. 1.**
3. Use clean shop towels to pad interior of caliper assembly, then remove piston by directing compressed air into caliper brake hose inlet hole. **Use just enough air pressure to ease piston out of bore. Do not place fingers in front of piston for any reason when applying compressed air. This could result in serious personal injury.**
4. Using a small piece of wood or plastic, remove piston seal from bore. **Do not use a metal tool of any kind to remove seal as it may damage bore.**
5. Remove bleeder valve.
6. Inspect piston for scoring, nicks, cor-

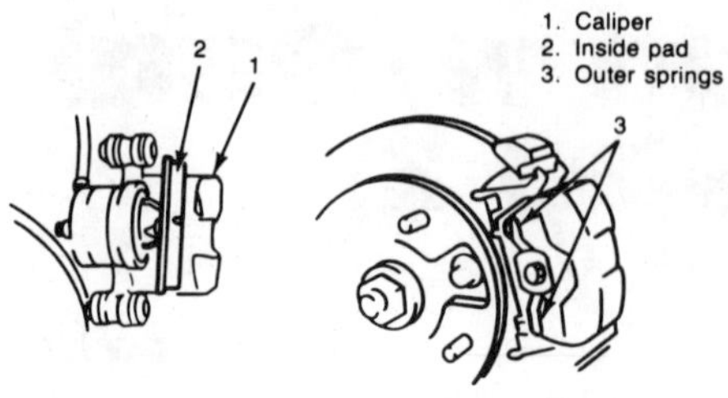

Fig. 4 Installing inside pad & caliper

rosion, and wear and replace as needed.

7. Inspect caliper housing and seal groove for corrosion, nicks, scoring and excessive wear, and use crocus cloth to polish away corrosion from housing bore. Replace caliper housing if corrosion in and around seal groove will not clean up with crocus cloth.
8. Clean all parts with denatured alcohol. Dry with unlubricated compressed air. Blow out all passages in housing and bleeder valve.

CALIPER ASSEMBLY & INSTALLATION

1. Apply suitable grease to piston seal and cylinder wall, then install the seal. Check to ensure piston seal is not twisted.
2. Install new boot on piston so side with 2 grooves faces in toward cylinder and side with 3 grooves faces out, **Fig. 5.**
3. Install piston in caliper so it projects about .4 inch to ease installation of boot on caliper housing. Install boot on caliper housing, then install set ring.
4. Install bleeder valve.
5. Check that caliper slide bushings are lubricated with rubber grease. Check that slide bushing in each carrier hole can be moved smoothly in each direction.
6. Install caliper and inside pad on knuckle.
7. Tighten caliper bolts.
8. Connect brake hose, then bleed hydraulic system.
9. Install wheel.
10. Lower vehicle and test brakes. **Before moving vehicle, pump brake pedal several times to be sure it is firm. Do not move vehicle until a firm pedal is obtained.**

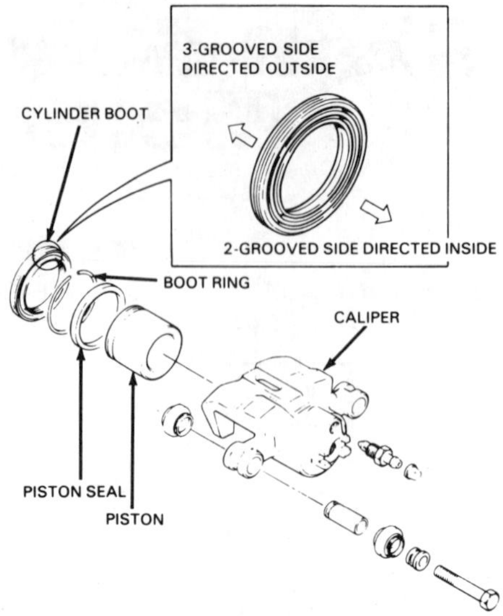

Fig. 5 Installing caliper boot

Daewoo Single Piston w/Dual Mounting Bolts

INDEX

DESCRIPTION

This caliper has a single bore, **Fig. 1,** and is mounted to the steering knuckle with two mounting bolts. Upon brake application, hydraulic pressure is converted by the caliper to stopping force. This force acts equally against the piston and the bottom of the caliper bore to move the piston outward and to slide the caliper inward resulting in a clamping action on the rotor. This clamping action forces the lining against the rotor, creating friction to stop the vehicle.

SERVICE

BRAKE SHOE & LINING, REPLACE

Caliper removal is not required to replace shoe and linings.

1. Remove approximately 2/3 of brake fluid from master cylinder.
2. Raise and support vehicle.
3. Mark relationship between front wheel and hub, then remove wheel and tire assembly.
4. Carefully drive pad retaining pins from inside out, **Fig. 2.**
5. Bottom piston into caliper bore using suitable pliers, then remove brake pads.
6. Reverse procedure to install, noting the following:
   a. Install brake pads with wear sensor to the piston side.
   b. Ensure pads can be moved back and forth slightly when installed.
   c. Install retaining pins from outside in, ensuring openings align.
   d. Upon completion of installation, lower vehicle and pump brake pedal several times to properly position brake pads.

CALIPER, REPLACE

1. Remove approximately 2/3 of brake fluid from master cylinder.
2. Raise and support vehicle.
3. Mark relationship between front wheel and hub, then remove wheel and hub assembly.
4. Bottom piston into caliper bore using suitable pliers to provide clearance between brake pads and rotor.
5. Disconnect brake hose from caliper. Plug openings in brake hose and caliper to prevent leakage and contamination. **Do not crimp brake hose, as this may damage internal structure of hose.**
6. Remove outer protective caps from caliper housing seats.
7. Remove caliper attaching bolts, and the caliper.
8. Reverse procedure to install, noting the following:
   a. Original equipment caliper attaching bolts are micro-encapsulated. After removal of these bolts, threads in steering knuckle must be re-cut using a M12 X 1.5 thread tap.
   b. Install new caliper attaching bolts not micro-encapsulated and **torque** to 70 ft. lbs.
   c. **Torque** brake hose inlet fitting attaching bolt to 18 ft. lbs.
   d. Install new outer protective caps. Ensure caps seat evenly on cali-

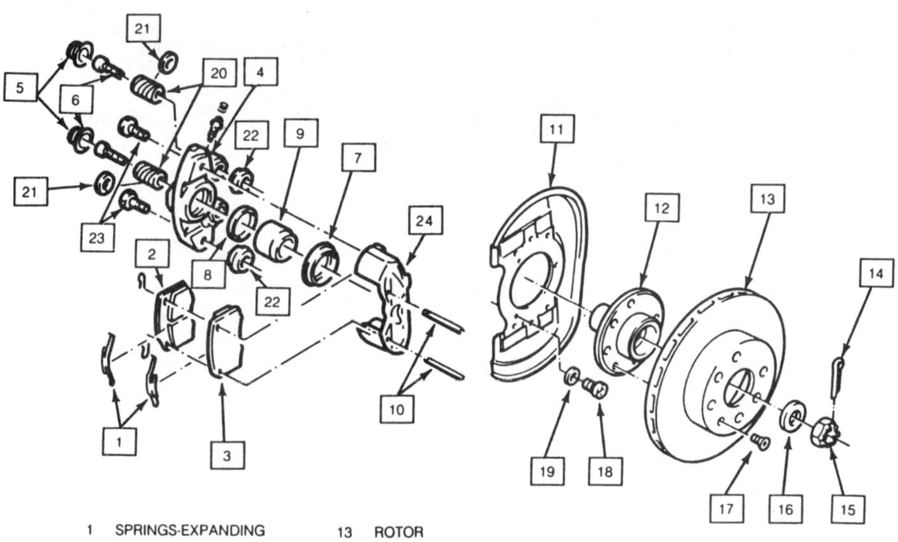

1	SPRINGS-EXPANDING	13	ROTOR
2	SHOE-INBOARD	14	PIN-COTTER
3	SHOE-OUTBOARD	15	NUT-WHEEL HUB
4	HOUSING-CALIPER	16	WASHER-WHEEL HUB
5	CAP-BLEEDER VALVE	17	SCREW-DETENT
6	VALVE-BLEEDER	18	SCREW-SHIELD
7	BOOT-PISTON	19	WASHER-SHIELD
8	SEAL-PISTON	20	PLUG-BLEEDER VALVE
9	PISTON-CALIPER	21	WASHER-BLEEDER VALVE
10	PINS-RETAINING	22	SPACER-CALIPER HOUSING
11	SHIELD	23	BOLT-CALIPER HOUSING
12	HUB-WHEEL	24	RETAINING FRAME

Fig. 1 Exploded view of brake caliper assembly

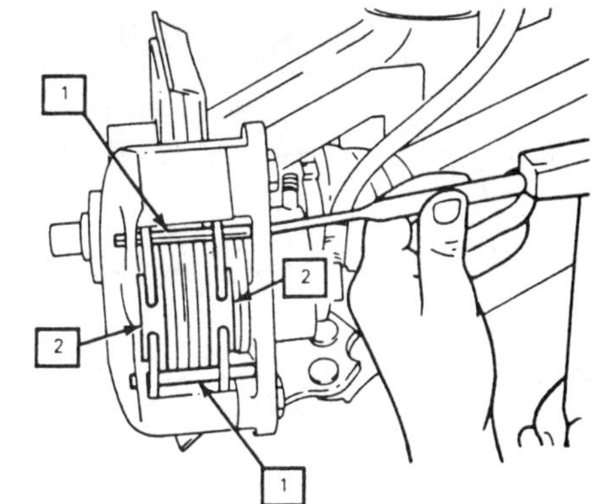

1 RETAINING PINS
2 EXPANDING SPRINGS

Fig. 2 Brake pad retaining pin removal

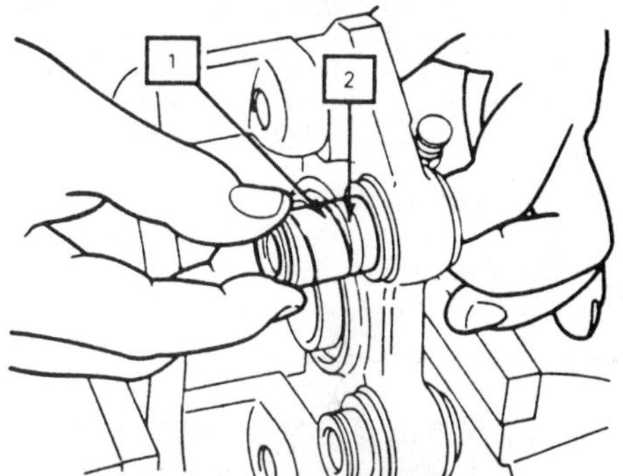

1. SLIDING SLEEVE
2. SEAL

Fig. 3 Caliper sliding sleeve removal

per.

e. Upon completion of installation, bleed hydraulic system, then lower vehicle. **Before moving vehicle, pump brake pedal several times to be sure it is firm. Do not move vehicle until a firm pedal is obtained.**

CALIPER DISASSEMBLY

1. Remove caliper and brake pads as previously described.
2. Remove inner protective caps for sliding sleeves.
3. Remove piston boot from housing.
4. Move sliding sleeve slightly toward protective cap and remove cap from groove in sleeve.
5. Remove boot from piston and housing.
6. Press sliding sleeves out of housing, **Fig. 3**, then remove seals using a suitable wood or plastic tool.
7. Clean outside of caliper, then drain brake fluid from caliper.
8. Use clean shop towels to pad interior of caliper assembly, then remove piston by directing compressed air into caliper brake hose inlet hole, **Use just enough air pressure to ease piston out of bore. Do not place fingers in front of piston for any reason when applying compressed air. This could result in serious personal injury.**
9. Secure caliper assembly in a suitable vise, then remove caliper retaining frame attaching bolts and the frame.
10. Remove piston seal from groove in caliper bore using a suitable wood or plastic tool. **Do not use a metal tool of any kind to remove seal as it may damage bore.**
11. Remove bleeder valve from caliper.
12. Inspect piston for scoring, nicks, corrosion, and wear and replace as needed.
13. Inspect caliper housing and seal groove for corrosion, nicks, scoring and excessive wear, and use crocus cloth to polish away corrosion from housing bore. Replace caliper housing if corrosion in and around seal groove will not clean up with crocus cloth.
14. Clean all parts with denatured alcohol. Dry with unlubricated compressed air. Blow out all passages in housing and bleeder valve.

CALIPER ASSEMBLY

1. Lubricate, then install new piston seal into caliper bore. Ensure seal is not twisted.
2. Install lubricated piston into caliper bore, ensuring piston does not jam and piston seal is properly seated. **Leave piston exposed enough to permit installation of boot in groove.**
3. Position boot in piston groove and press onto caliper housing. Seat boot using dust boot seal installer No. J36538 or equivalent.
4. Install caliper retaining frame, torquing attaching bolts to 70 ft. lbs.
5. Apply suitable grease to sliding sleeves, then install seals in center groove of sliding sleeves.
6. Position sliding sleeves in caliper with inner protective cap groove facing piston. **Leave sleeves exposed enough to permit installation of protective caps.**
7. Install new protective caps into sliding sleeve grooves and press onto caliper housing collar. Seat caps using sliding sleeve dust boot installer No. J36647 or equivalent.
8. Push sliding sleeves into housing, then install bleeder valve.
9. Install brake pads and caliper as previously described. Bleed hydraulic system, **Before moving vehicle, pump brake pedal several times to be sure it is firm. Do not move vehicle until a firm pedal is obtained.**

Delco-Moraine Dual Piston w/Dual Mounting Bolts

INDEX

DESCRIPTION

This dual piston sliding caliper, **Fig. 1,** is comprised of two interconnected bores and is attached to a mounting bracket with two mounting bolts. Hydraulic pressure acting on the bottom of the caliper bores forces the pistons outward, enabling the caliper to slide inward, thereby clamping the brake shoes against the rotor.

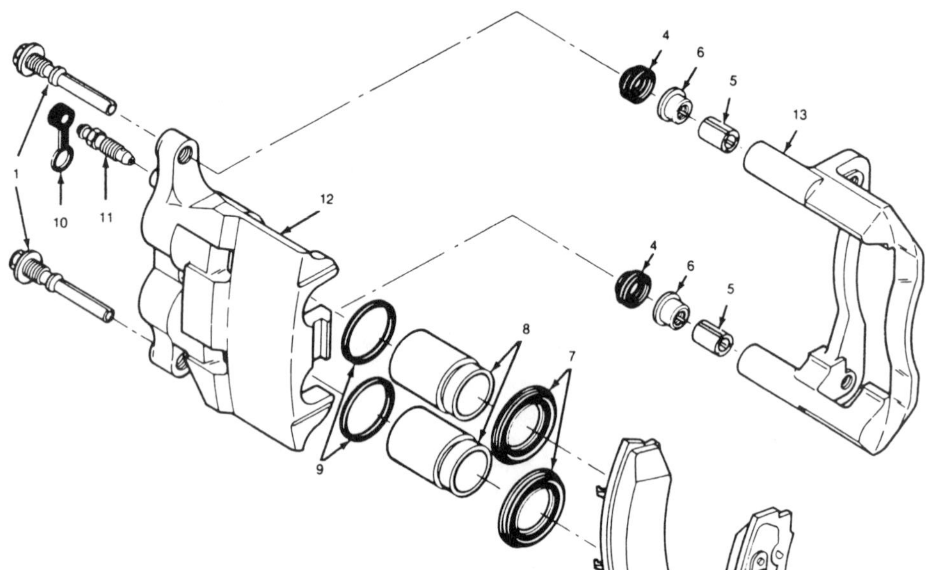

Fig. 1 Exploded view of Delco-Moraine 3242 series dual piston caliper

SERVICE

CALIPER REMOVAL

1. Siphon enough brake fluid out of the master cylinder to bring fluid level to 1/3 full to avoid fluid overflow when the caliper pistons are pushed back into bores.
2. Raise vehicle and remove front wheels.
3. Mark relationship between wheel to hub and bearing assembly. Remove tire and wheel assembly.
4. Using a suitable C-clamp and block of wood, push pistons back together into caliper bores.
5. If caliper assembly is to be serviced, remove inlet fitting attaching bolt, copper washer, and inlet fitting from caliper housing. Plug opening in inlet fitting to prevent fluid loss and contamination. **Do not crimp brake hose, as this may damage internal structure of hose. If only shoe and lining assemblies are to be replaced, do not disconnect brake line fitting from caliper.**
6. Remove two mounting bolts, then remove caliper. If only shoe and lining assemblies are to be replaced, suspend caliper from chassis using suitable hanger. **Do not allow caliper to hang by brake hose.**

BRAKE SHOE & LINING REMOVAL

1. Remove caliper as outlined previously.
2. Lift upward on outward shoe retaining spring of outboard shoe until it clears center lug, then remove shoe from caliper.
3. Pull inboard shoe outward to disengage retainer springs from pistons, then remove inboard shoe from caliper.

CALIPER DISASSEMBLY

1. Remove caliper assembly as outlined previously, then drain brake fluid from caliper.
2. Position shop towel in interior part of caliper, then slowly apply compressed air to inlet port and remove pistons. **It is imperative that one piston be partially installed to facilitate removal of second piston from bore. A pad or wooden spacer may be used to prevent complete removal of first piston.**
3. Remove piston boots from caliper bores, then using suitable wooden or plastic tool, pry piston seals from caliper bore grooves. **Do not use a metal tool of any kind to remove seal as it may damage bore.**
4. Remove bleeder valve from caliper.
5. Inspect piston for scoring, nicks, corrosion, and wear and replace as needed.
6. Inspect caliper housing and seal grooves for corrosion, nicks, scoring and excessive wear, and use crocus cloth to polish away corrosion from housing bore. Replace caliper housing if corrosion in and around seal grooves will not clean up with crocus cloth.
7. Clean all parts with denatured alcohol. Dry with unlubricated compressed air. Blow out all passages in housing and bleeder valve.

MOUNTING BRACKET SERVICE

Removal

1. Remove caliper as outlined previously.
2. Using suitable torx bit, remove mounting bracket attaching bolts and the mounting bracket.

Bushing & Boot Replacement

1. Remove mounting bracket from vehicle.
2. Remove bolt boots from support bushings, **Fig. 1.**
3. Clamp bracket in a suitable vise, then pry support bushings from inner bushings in bracket ears with small screwdriver.
4. Using a paper clip, pull inner bushings from mounting bracket ears.
5. Lubricate inner bushings with silicone based grease, then install bushings flush with bracket ears.
6. Position .080-.090 inch thick shim stock on bracket ear face, then drive support bushings into inner bushings, **Fig. 2.** When properly installed, bushing should protrude .080-.090 inch above bracket ear face.
7. Snap new bolt boots over support bushing lip.

Installation

1. Coat mounting bracket attaching bolt threads with Loctite sealant or equivalent.
2. Align mounting bracket holes, then install mounting bracket and **torque** attaching bolts to 148 ft. lbs.

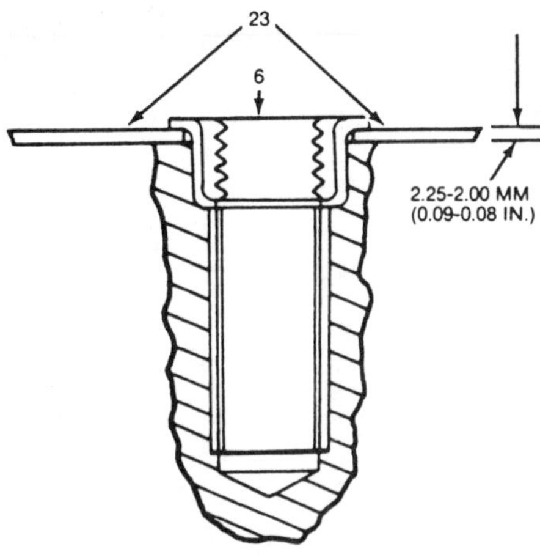

6. SUPPORT BUSHING
23. SHIM STOCK

Fig. 2 Installing support bushing

CALIPER ASSEMBLY

1. Install bleeder valve and **torque** to 10 ft. lbs.
2. Lubricate piston seals with clean brake fluid, then carefully install seals into caliper bore grooves. **Ensure seals are not twisted.**
3. Lubricate boots and install onto pistons, then push pistons fully into caliper bores.
4. Seat boots into caliper bores using boot seal installer No. J-36349 or equivalent.

BRAKE SHOE & LINING INSTALLATION

1. Install inboard shoe into caliper, ensuring retainer spring tangs are fully positioned into pistons.
2. Snap outboard shoe retaining spring over housing center lug, then install outboard shoe into caliper.
3. Install brake caliper as outlined further on.

CALIPER INSTALLATION

1. Position caliper over rotor and onto mounting bracket.
2. Lubricate entire length of mounting bolts with silicone based grease, then install bolts and **torque** to 79 ft. lbs.
3. If inlet fitting was removed, install fitting using new copper washer and **torque** bolt to 32 ft. lbs.
4. Install front wheels, fill master cylinder to proper level, then bleed brake system as required. **Before moving vehicle, pump brake pedal several times to be sure it is firm. Do not move vehicle until a firm pedal is obtained.**

Delco-Moraine Single Piston w/Dual Mounting Bolts (Type 1)

INDEX

DESCRIPTION

This single piston sliding caliper assembly, **Fig. 1,** incorporates a one piece housing with the inboard side of the housing bored for the piston. A seal within the housing bore provides a hydraulic seal between the piston and housing wall.

A spring steel scraper (wear sensor) is incorporated on each inboard shoe. When the shoe lining has worn to within .030 inch of the shoe, the sensor scrapes the rotor and emits an audible high frequency sound indicating that the linings should be replaced.

The caliper assembly slides on the mounting bolts. Upon brake application, fluid pressure against the piston forces the inboard shoe and lining assembly against the inboard side of the disc. This action causes the caliper assembly to slide until the outboard lining comes into contact with the disc. As pressure builds up, the linings are pressed against the disc with increased force.

SERVICE

CALIPER REMOVAL

1. Siphon enough brake fluid out of the master cylinder to bring fluid level to 1/3 full to avoid fluid overflow when the caliper piston is pushed back into its bore.
2. Raise vehicle and remove front wheels.
3. Using a C-clamp, as illustrated in **Fig. 2,** push piston back into its bore.
4. **If caliper assembly is to be serviced,** remove inlet fitting attaching bolt, copper washer, and inlet fitting from caliper housing.Plug opening in inlet fitting to prevent fluid loss and contamination. **Do not crimp brake hose, as this may damage internal structure of hose. If only shoe and lining assemblies are to be replaced, do not disconnect brake line fitting from caliper.**
5. Remove two mounting bolts, **Fig. 3,** and lift caliper away from disc. If only shoe and lining assemblies are to be replaced, suspend caliper from chassis using suitable hanger. **Do not allow caliper to hang by brake hose.**

BRAKE SHOE & LINING REMOVAL

1. Remove caliper assembly as outlined above.
2. Remove inboard shoe. Dislodge outboard shoe and position caliper on the front suspension so the brake

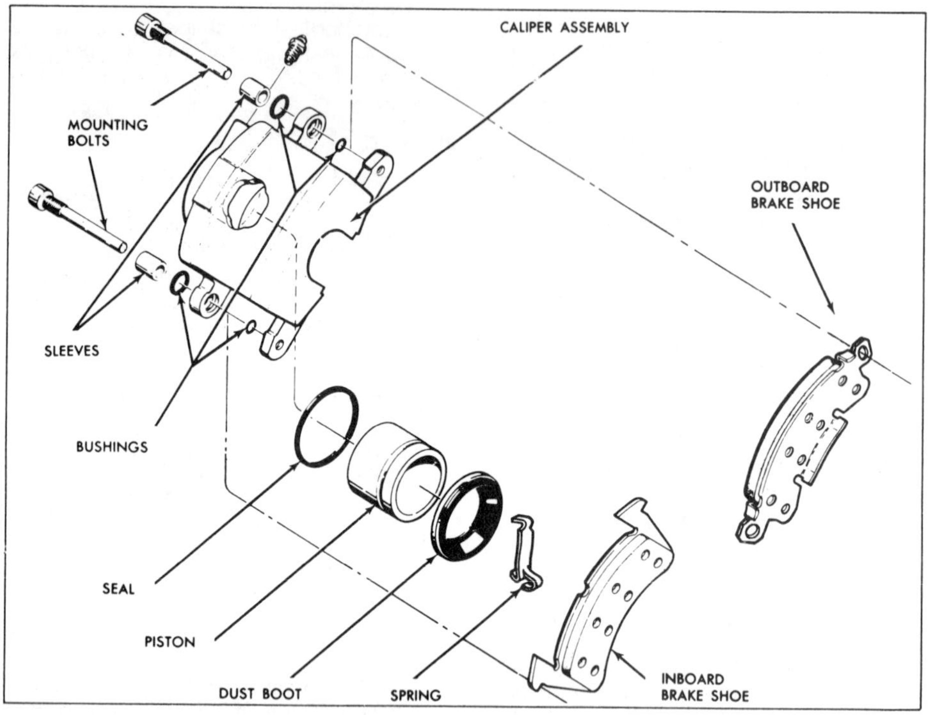

Fig. 1 Typical Delco-Moraine 3000/3075/3100 series caliper exploded view

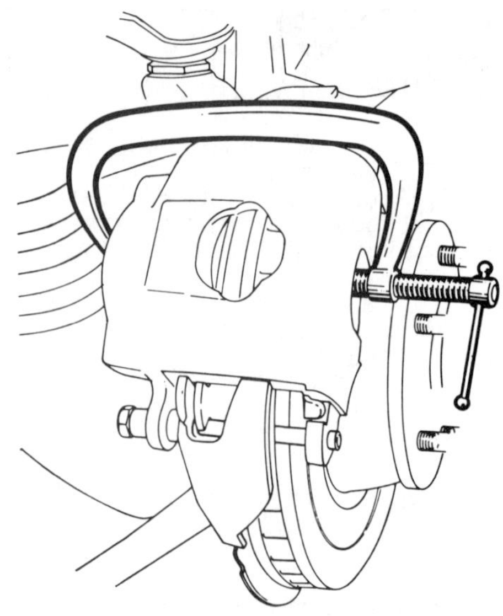

Fig. 2 Compressing piston and shoes with C-clamp

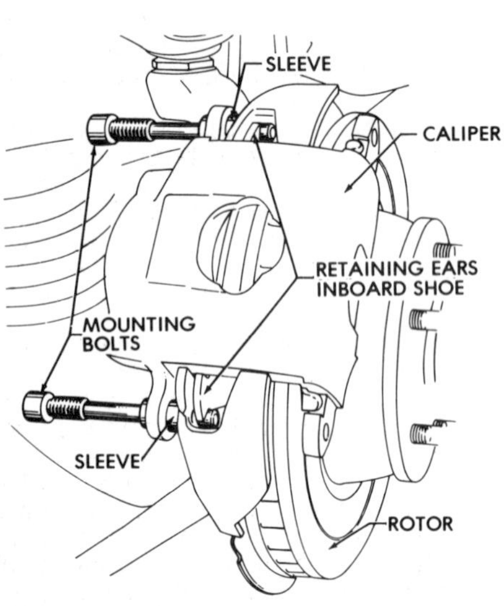

Fig. 3 Caliper & mounting bolts

hose will not support the weight of the caliper.

3. Remove shoe support spring from piston.
4. Remove two sleeves from inboard ears of the caliper.
5. Remove four rubber bushings from the grooves in each of the caliper ears.

CALIPER DISASSEMBLY

1. Remove caliper as outlined above.
2. Clean outside of caliper, then drain brake fluid from caliper.
3. Use clean shop towels to pad interior of caliper assembly, then remove piston by directing compressed air into caliper brake hose inlet hole, **Fig. 4. Use just enough air pressure to ease piston out of bore. Do not place fingers in front of piston for any reason when applying compressed air. This could result in serious personal injury.**

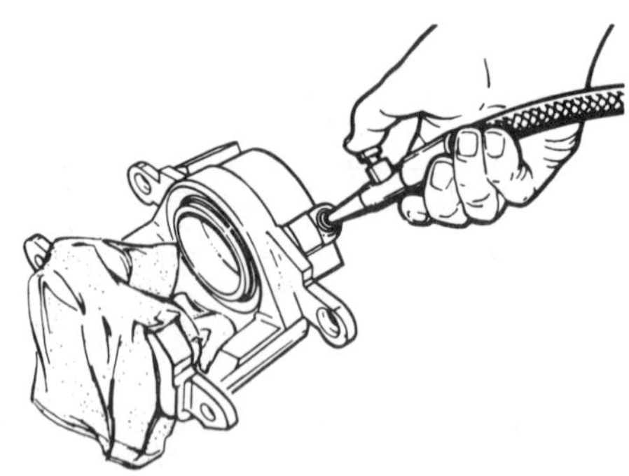

Fig. 4 Removing piston from caliper

4. Carefully pry dust boot out of bore.
5. Using a small piece of wood or plastic, remove piston seal from bore. **Do not use a metal tool of any kind to remove seal as it may damage bore.**
6. Remove bleeder valve.
7. Inspect piston for scoring, nicks, corrosion, and wear and replace as needed.
8. Inspect caliper housing and seal groove for corrosion, nicks, scoring and excessive wear, and use crocus cloth to polish away corrosion from housing bore. Replace caliper housing if corrosion in and around seal groove will not clean up with crocus cloth.

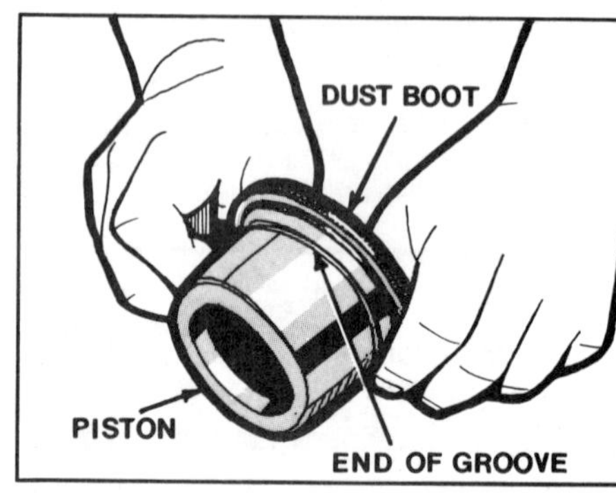

Fig. 5 Installing boot to piston

9. Clean all parts with denatured alcohol. Dry with unlubricated compressed air. Blow out all passages in housing and bleeder valve.

CALIPER ASSEMBLY

1. Lubricate caliper piston bore and new piston seal with clean brake fluid. Position seal in bore groove. **Ensure seal is not twisted.**
2. Lubricate piston with clean brake fluid and assemble a new boot into the groove in the piston so the fold faces the open end of the piston, **Fig. 5.**
3. Using care not to unseat the seal, insert piston into bore and force the piston to the bottom of the bore.
4. Position dust boot in caliper counterbore and install, using suitable seal installer, **Fig. 6. Check the boot installation to be sure the retaining ring molded into the boot is not bent and that the boot is installed below the caliper face and evenly all around. If the boot is not fully installed, dirt and moisture may enter the bore and cause corrosion.**
5. Install the brake hose in the caliper using a new copper gasket.
6. Install shoes and reinstall caliper assembly.

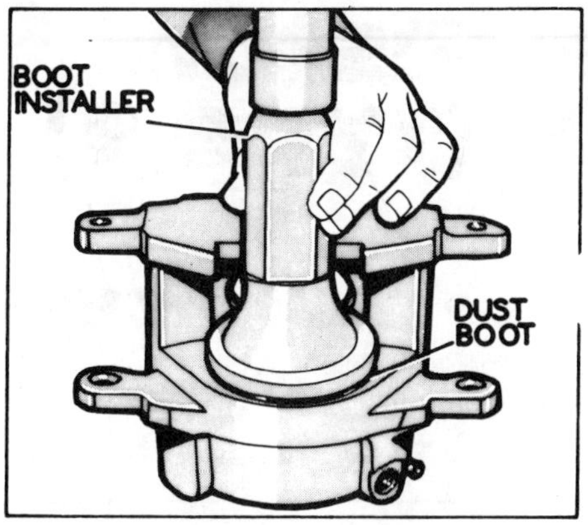

Fig. 6 Installing boot to caliper

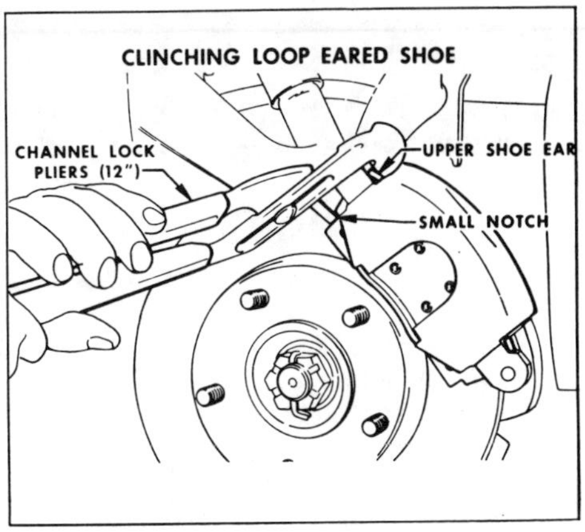

Fig. 7 Clinching loop eared brake shoe

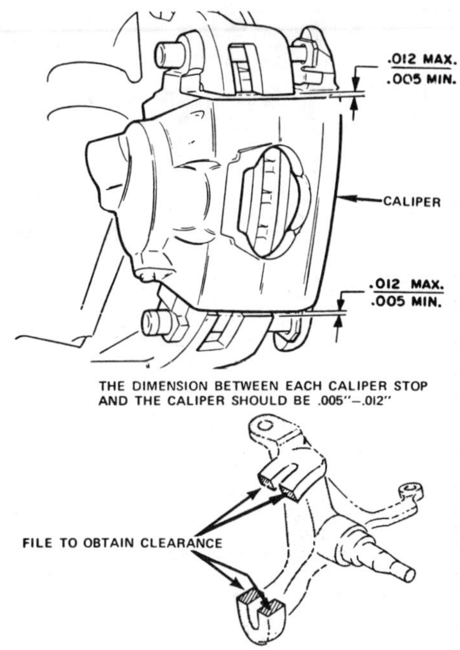

Fig. 8 Checking clearance between caliper & stops

BRAKE SHOE & LINING INSTALLATION

1. Lubricate new sleeves, rubber bushings, bushing grooves and mounting bolt ends with Delco Silicone Lube or its equivalent.
2. Install new bushings and sleeves in caliper ears. **Position the sleeve so that the end toward the shoe is flush with the machined surface of the ear.**
3. Install shoe support spring by positioning single tang end of spring into notch cut at top of inboard shoe. Press remaining end of spring over bottom edge of shoe until shoe is engaged securely.
4. Position inboard shoe with spring attached into caliper with ear end facing downward and bottom end facing upward with spring resting on inside diameter of piston. Press downward on both ends of shoe until shoe contacts piston and support spring contacts piston inside diameter. **Some inboard replacement brake pads incorporate wear sensors and have a specific left and righthand assembly. Properly installed, the wear sensor will face toward the rear of caliper.**
5. Position outboard shoe in caliper with shoe ears over caliper ears and tab at bottom of shoe engaged in caliper cutout.
6. With shoes installed, lift caliper and rest bottom edge of outboard lining on outer edge of brake disc to be sure there is no clearance between outboard shoe tab and caliper abutment.
7. Install caliper and **torque** mounting bolts to 38 ft. lbs.
8. Clinch upper ears of outboard shoe by positioning pliers with one jaw on top of upper ear and one jaw in notch on bottom shoe opposite ear, **Fig. 7.** Ears are to be flat against caliper housing with no radial clearance. If clearance exists, repeat clinching procedure. **Before moving vehicle, pump brake pedal several times to be sure it is firm. Do not move vehicle until a firm pedal is obtained. On some models with low drag calipers, apply approximately 175 pounds of pressure to the brake pedal three times to properly seat the caliper and related components.**

CALIPER INSTALLATION

1. Position caliper over disc, lining up holes in caliper with holes in mounting bracket. If brake hose was not disconnected during removal, be sure not to kink it during installation.
2. Start mounting bolts through sleeves in inboard caliper ears and the mounting bracket, making sure ends of bolts pass under ears on inboard shoe. **Right and left calipers must not be interchanged.**
3. Push mounting bolts through to engage holes in the outboard ears. Then thread mounting bolts into bracket.
4. **Torque** mounting bolts to 38 ft. lbs.
5. Check the dimensions between each caliper stop and caliper, **Fig. 8.**
6. If brake hose was removed, reconnect it and bleed the calipers.
7. Replace front wheels, lower vehicle and add brake fluid to master cylinder to bring level to 1/4 inch from top. **Before moving vehicle, pump brake pedal several times to be sure it is firm. Do not move vehicle until a firm pedal is obtained. On some models with low drag calipers,** apply approximately 175 pounds of pressure to the brake pedal three times to properly seat the caliper and related components.

Delco-Moraine Single Piston w/Dual Mounting Bolts (Type 2)

INDEX

DESCRIPTION

The caliper has a single piston and is mounted to the support bracket by two mounting bolts, **Figs. 1 through 4.** The caliper assembly slides on the two mounting bolts. Upon brake application, fluid pressure against the piston forces the inboard shoe and lining assembly against the inboard side of the disc. This action causes the caliper assembly to slide until the outboard lining comes into contact with the disc. As pressure builds up the linings are pressed against the disc with increased force.

SERVICE

CALIPER REMOVAL

1. Remove approximately 2/3 of brake fluid from master cylinder.
2. Raise and support front of vehicle, then remove wheel and tire assembly.
3. Position suitable pliers over inboard pad and housing as shown in **Fig. 5,** and squeeze pliers to compress caliper piston.
4. If caliper assembly is being removed for service, remove inlet fitting attaching bolt, copper washer, and inlet fitting from caliper housing.Plug opening in inlet fitting to prevent fluid loss and contamination. **Do not crimp brake hose, as this may damage internal structure of hose. If only shoe and lining assemblies are to be replaced, do not disconnect brake line fitting from caliper.**
5. Remove Allen head caliper mounting bolts, **Fig. 6.** If bolts show signs of corrosion, use new bolts when installing caliper assembly.
6. Remove caliper assembly from disc. If only shoe and lining assemblies are to be replaced, using a length of wire suspend caliper from spring coil. Never allow caliper to hang from brake hose.

BRAKE SHOE & LINING REMOVAL

1. Remove caliper assembly as outline above.
2. Remove shoe and lining assemblies from caliper, **Figs. 1 through 4,** noting the following:

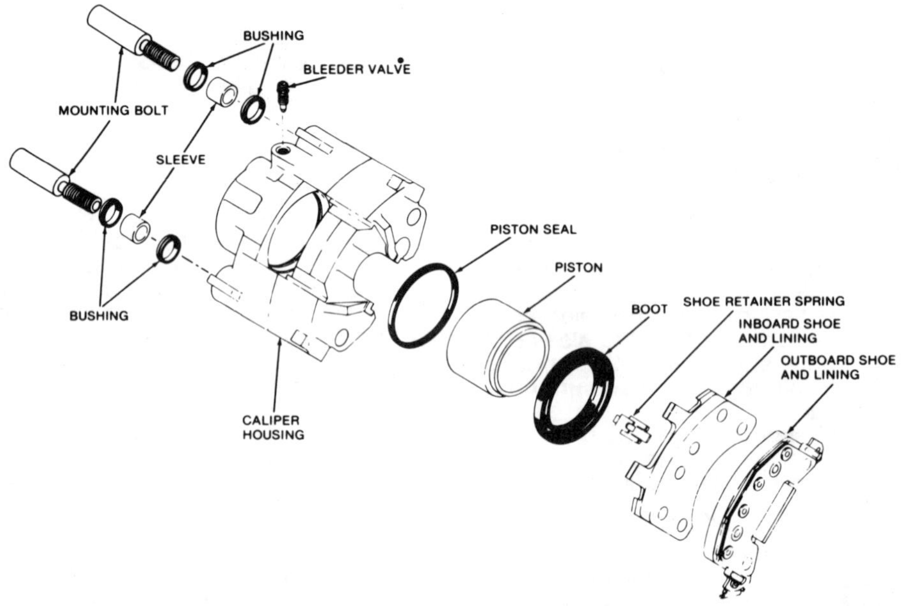

Fig. 1 Delco-Moraine 3200 series caliper exploded view.

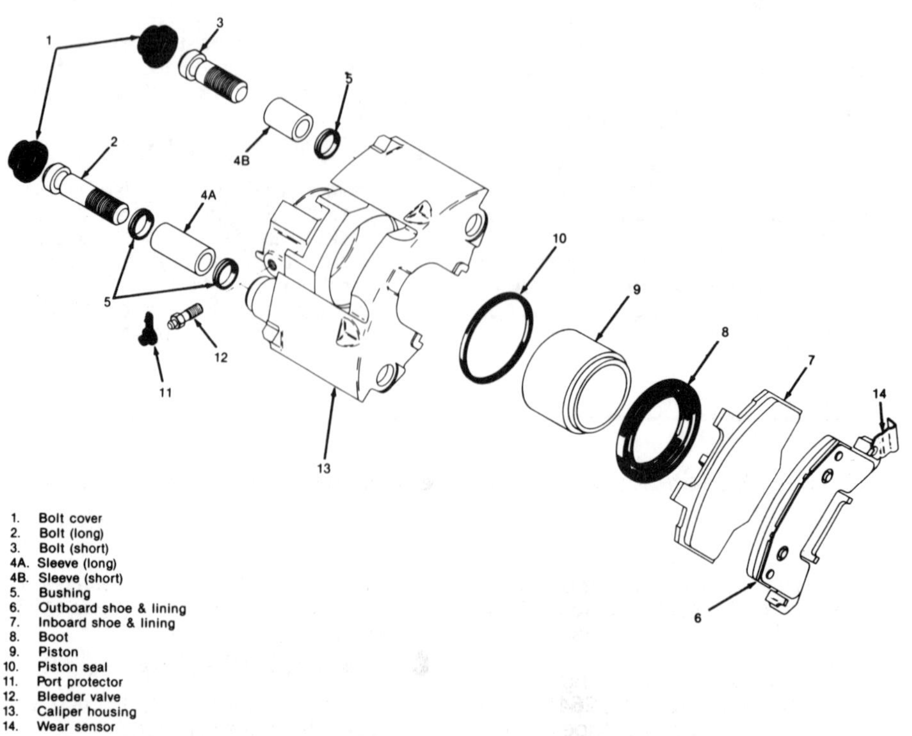

Fig. 2 Delco-Moraine 3352 series caliper exploded view. 1985–87 Chevette & 1000

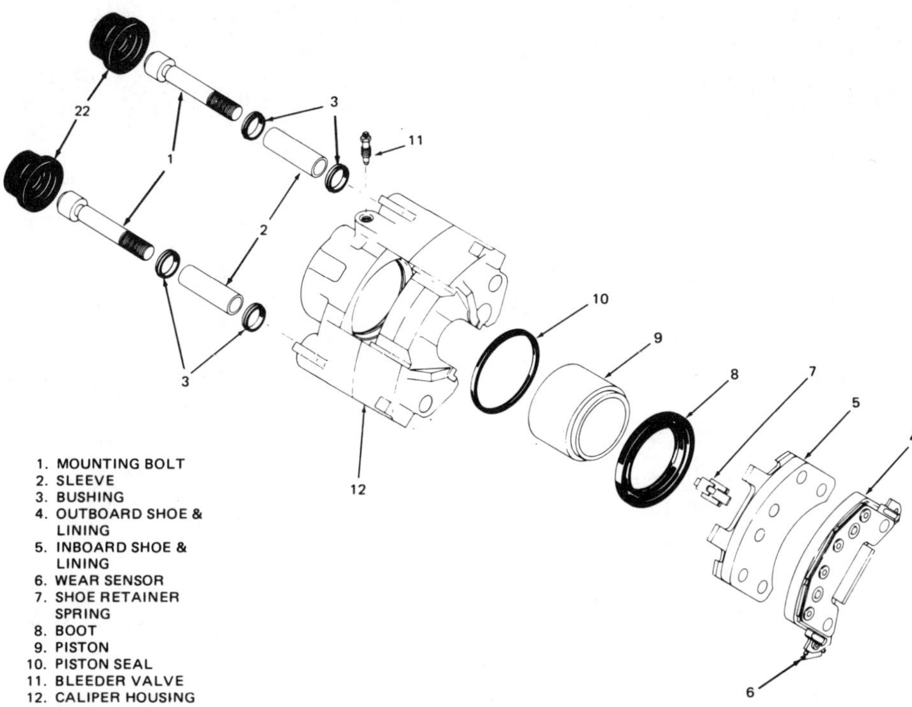

Fig. 3 Delco-Moraine 3257 series caliper exploded view.

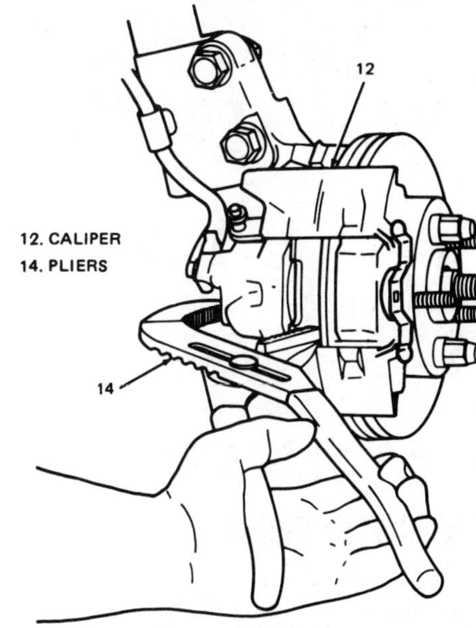

Fig. 5 Compressing caliper piston with pliers

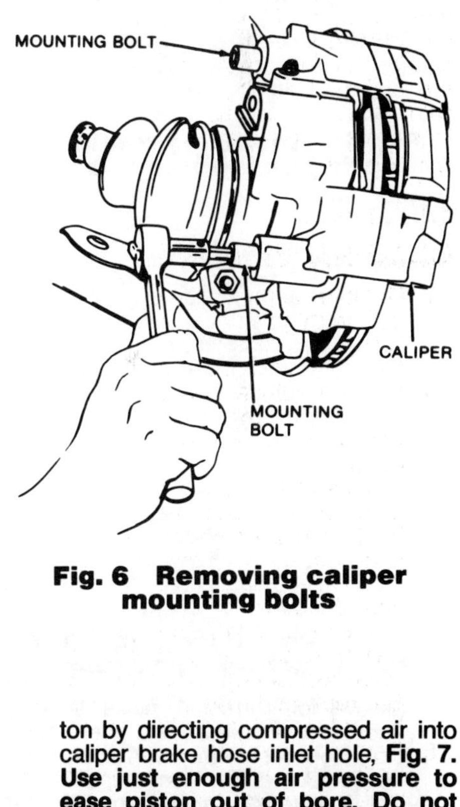

Fig. 6 Removing caliper mounting bolts

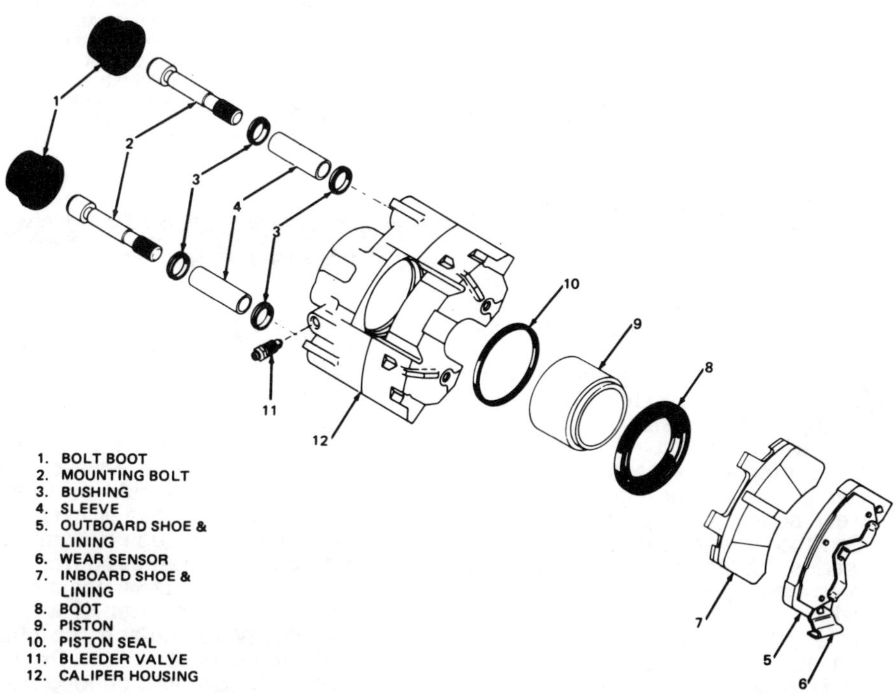

Fig. 4 Delco-Moraine 3264 series caliper exploded view.

a. **On models with tab mount outer shoe (all except series 3264),** straighten bent over shoe tabs with suitable pliers.
b. **On models with spring mount outer shoe (series 3264),** insert screwdriver between shoe and caliper and disengage buttons on shoe from holes in caliper.

3. Remove sleeves and bushings from grooves in caliper mounting bolt holes, **Figs. 1 through 4.**

CALIPER DISASSEMBLY

1. Clean outside of caliper, then drain brake fluid from caliper.
2. Use clean shop towels to pad interior of caliper assembly, then remove piston by directing compressed air into caliper brake hose inlet hole, **Fig. 7. Use just enough air pressure to ease piston out of bore. Do not place fingers in front of piston for any reason when applying compressed air. This could result in serious personal injury.**
3. Using a screwdriver, remove dust boot from caliper bore, **Fig. 8.**
4. Using a small piece of wood or plastic, remove piston seal from bore. **Do not use a metal tool of any kind to remove seal as it may damage bore.**
5. Remove bleeder valve.
6. Inspect piston for scoring, nicks, corrosion, and wear and replace as needed.
7. Inspect caliper housing and seal

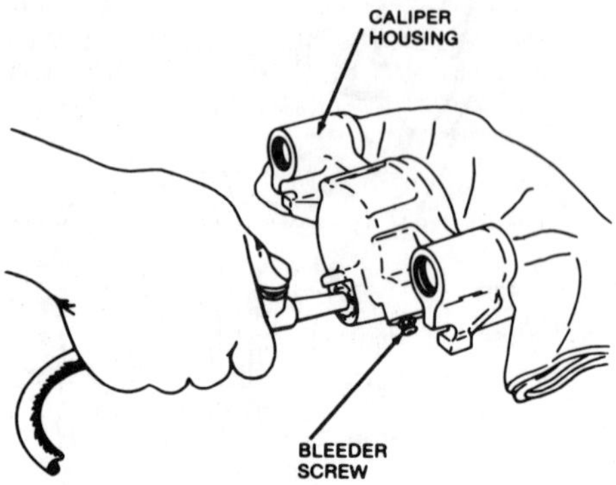

Fig. 7 Applying compressed air to caliper line port

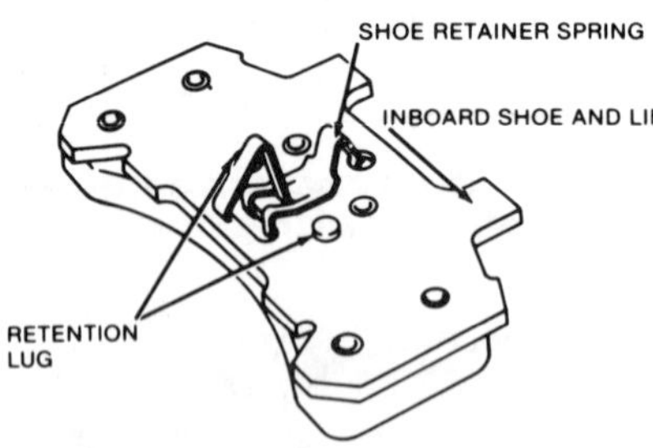

Fig. 10 Installing retainer spring on inboard shoe

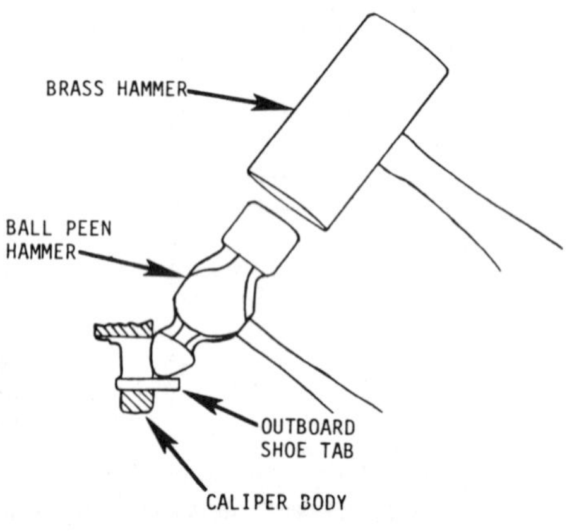

Fig. 12 Positioning hammer to clinch brake pad tabs

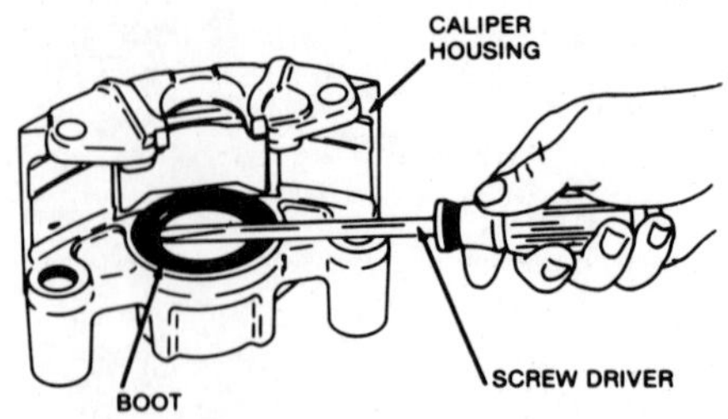

Fig. 8 Removing dust boot

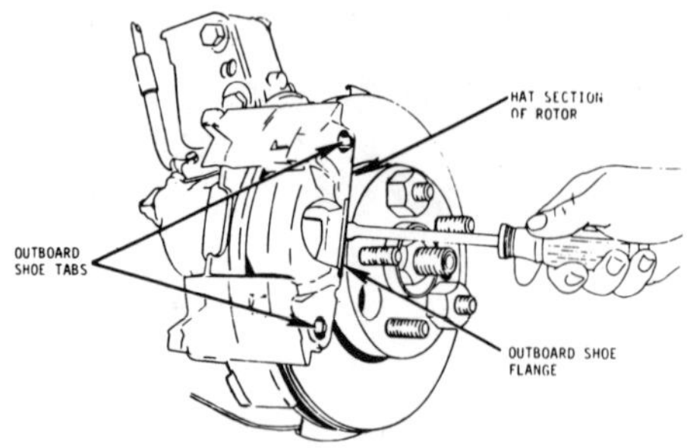

Fig. 11 Positioning screwdriver between outboard shoe flange & hat section of rotor

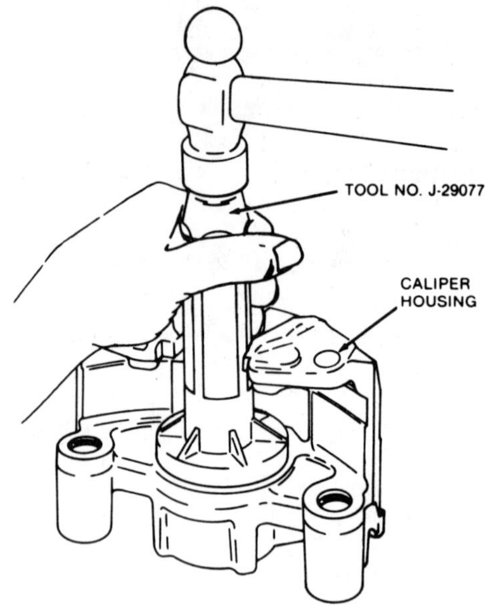

Fig. 9 Seating dust boot in caliper

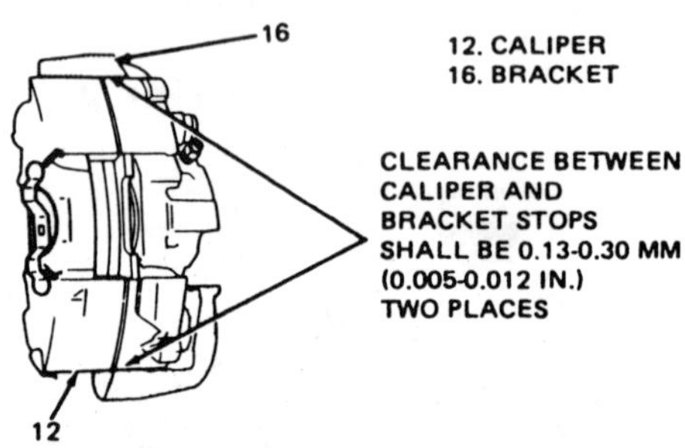

Fig. 13 Checking clearance between caliper & stops

groove for corrosion, nicks, scoring and excessive wear, and use crocus cloth to polish away corrosion from housing bore. Replace caliper housing if corrosion in and around seal groove will not clean up with crocus cloth.

8. Clean all parts with denatured alcohol. Dry with unlubricated compressed air. Blow out all passages in housing and bleeder valve.

CALIPER ASSEMBLY

1. Lubricate piston seal with clean brake fluid, then install piston seal into caliper bore groove. Check to ensure piston seal is not twisted.
2. Lubricate caliper bore with clean brake fluid.
3. Insert piston into caliper bore, then force piston down until piston bottoms in bore.
4. Position outer diameter of dust boot in caliper housing counterbore, then seat boot as shown in **Fig. 9.**
5. Install bleeder screw on caliper housing.

BRAKE SHOE & LINING INSTALLATION

1. Lubricate new bushings with suitable grease, install bushings in caliper grooves, then insert sleeves through bushings.
2. Install retaining spring onto inboard shoe and lining assembly, **Fig. 10,** if equipped, then install inboard shoe into caliper housing.
3. Install outboard shoe and lining assembly into caliper housing ensuring wear sensor is properly positioned. **On models with spring mount shoe (series 3264),** ensure buttons on shoe are properly engaged in holes in caliper.
4. Install caliper assembly as outlined.
5. **On models with tab mount outer shoe (except series 3264),** clinch outboard shoe to brake caliper as follows:
   a. Apply brakes several times to ensure caliper piston is extended and shoes are fully seated.
   b. Wedge a large flat blade screwdriver between outboard shoe flange and hat section of rotor, **Fig. 11.**
   c. Hold outer pad against caliper with suitable clamp or by applying moderate pressure on brake pedal.
   d. Position a ball peen hammer on outboard shoe tab, **Fig. 12,** then using a larger brass hammer, lightly tap the ball peen hammer to bend the outboard shoe tab. Tabs must be bent around casting to approximately 45°.
   e. After both tabs have been bent pressure should be released and outboard shoe should be locked into position. If shoe is loose, repeat steps a. through d. **If an outboard shoe is removed from the caliper, or the tabs unclinched for any reason, then it will be necessary to replace the shoe and lining assemblies. Do not re-clinch outboard shoe locking tabs after having removed shoe from caliper.**

CALIPER INSTALLATION

1. Position caliper assembly over disc and align mounting bolt holes. If brake hoses were not disconnected during removal, use care not to kink hoses during installation.
2. Install mounting bolts and **torque** to 38 ft. lbs., **Fig. 6.**
3. Check the dimensions between each caliper stop and caliper, **Fig. 13. If** necessary remove caliper and file

ends of bracket to provide proper clearance.
4. If brake hose fitting was disconnected during removal, install brake hose fitting and **torque** retaining bolt to 18 to 33 ft. lbs.
5. Fill master cylinder. Bleed brake system if brake line was disconnected and recheck master cylinder fluid level.
6. Install wheel and tire assembly on vehicle, then lower vehicle and check brake system operation. **Before moving vehicle, pump brake pedal several times to be sure it is firm. Do not move vehicle until a firm pedal is obtained.**

Delco-Moraine Single Piston w/Dual Mounting Bolts (Type 3)

INDEX

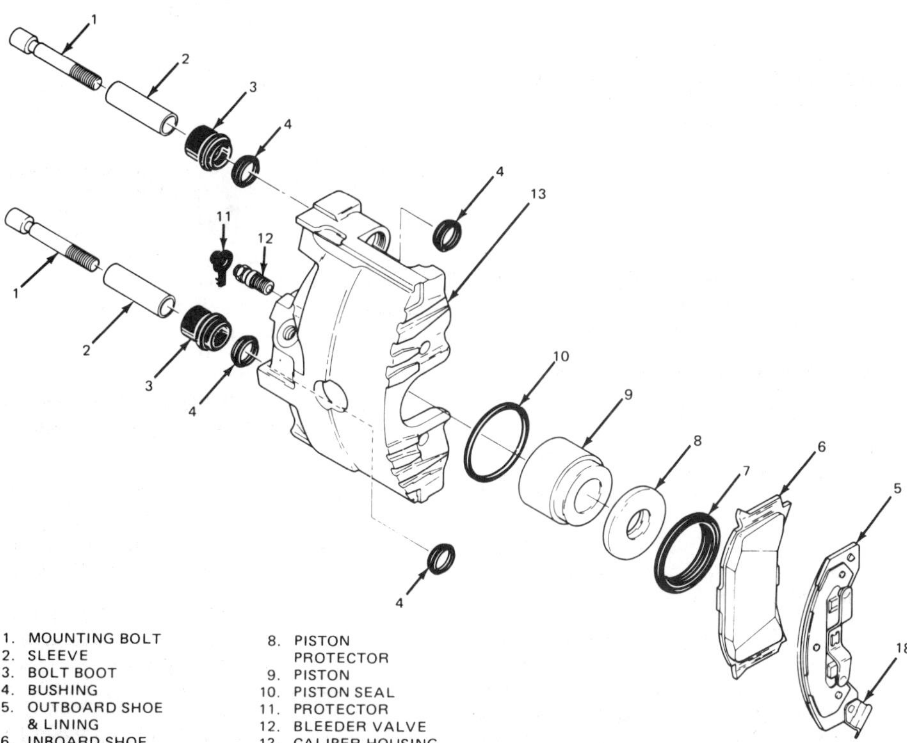

1. MOUNTING BOLT
2. SLEEVE
3. BOLT BOOT
4. BUSHING
5. OUTBOARD SHOE & LINING
6. INBOARD SHOE & LINING
7. CALIPER BOOT
8. PISTON PROTECTOR
9. PISTON
10. PISTON SEAL
11. PROTECTOR
12. BLEEDER VALVE
13. CALIPER HOUSING
18. WEAR SENSOR

Fig. 1 Delco-Moraine series 5349 front disc brake caliper exploded view

DESCRIPTION

This caliper, **Figs. 1 and 2,** has a single bore and is mounted to the support bracket with two mounting bolts. Hydraulic pressure, created by applying force to the brake pedal, is converted by the caliper into friction. The hydraulic pressure is applied equally against the piston and the bottom of the caliper bore moving the piston outward, resulting in a clamping action on the brake rotor.

Rear disc brake calipers include an integral parking brake mechanism. When the parking brake is applied, the lever turns an actuator screw which is threaded into a nut in the caliper piston. As the actuator screw turns, the piston is forced out, applying force on the rotor. The piston contains a self adjusting mechanism to keep the parking brake in proper adjustment and to ensure proper clearance when the parking brake is released.

FRONT BRAKE ASSEMBLY

CALIPER REMOVAL

1. Drain approximately 2/3 of brake fluid from master cylinder assembly.
2. Raise and support vehicle, then remove tire and wheel assembly.
3. If caliper is to be serviced or replaced, remove bolt securing inlet fitting, then disconnect brake hose, and plug end to prevent loss of fluid and contamination. **Do not crimp brake hose, as this may damage internal structure of hose. If only shoe and lining assemblies are being replaced, do not disconnect brake hose from caliper.**
4. Remove caliper retaining bolts and boots, **Fig. 1.**
5. Compress piston, placing suitable adjustable pliers over inboard surface of caliper housing and support bracket as shown in **Fig. 3.**
6. Remove caliper from disc. If shoe and lining assemblies are to be replaced, use a length of wire to suspend caliper. Do not allow caliper to hang from brake line.

BRAKE SHOE & LINING REMOVAL

1. Remove caliper as described under "Caliper Removal."
2. To remove outboard shoe and lining assembly, use a suitable screwdriver to disengage shoe buttons from caliper holes.
3. Remove inboard shoe and lining assembly from caliper.
4. Remove sleeves from mounting bolt holes.

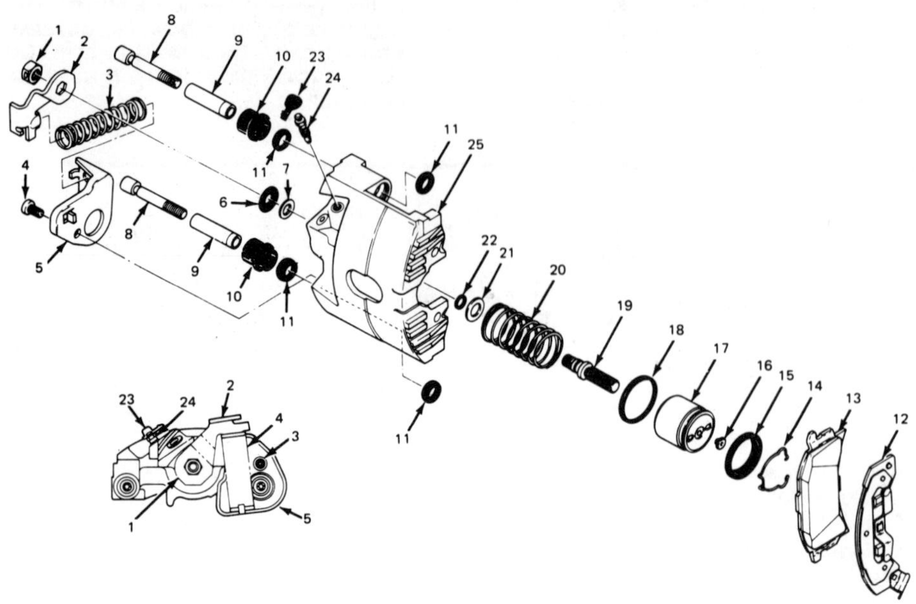

1. NUT
2. LEVER
3. RETURN SPRING
4. BOLT
5. BRACKET
6. LEVER SEAL
7. ANTI-FRICTION WASHER
8. MOUNTING BOLT
9. SLEEVE
10. BOLT BOOT
11. BUSHING
12. OUTBOARD SHOE & LINING
13. INBOARD SHOE & LINING
14. SHOE DAMPENING SPRING
15. CALIPER BOOT
16. TWO WAY CHECK VALVE
17. PISTON ASSEMBLY
18. PISTON SEAL
19. ACTUATOR SCREW
20. BALANCE SPRING
21. THRUST WASHER
22. SHAFT SEAL
23. PROTECTOR
24. BLEEDER VALVE
25. CALIPER HOUSING
26. WEAR SENSOR

Fig. 2 Delco-Moraine series 5748 rear disc brake caliper exploded view

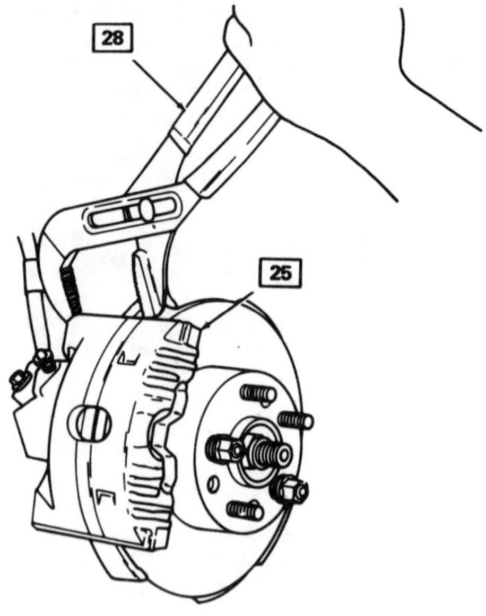

25. CALIPER HOUSING
28. PLIERS

Fig. 3 Compressing front caliper piston for caliper removal

5. Remove bushings from mounting bolt hole grooves.

CALIPER DISASSEMBLY

1. Clean outside of caliper, then drain brake fluid from caliper.
2. Position a clean shop towel on inside of caliper, then using compressed air, remove piston from caliper. **Use just enough air pressure to ease piston out of bore. Do not place fingers in front of piston for any reason when applying compressed air. This could result in serious personal injury.**
3. Remove dust boot from piston taking care not to scratch lousing bore.
4. Using wood or plastic tool, remove piston seal from caliper bore. **Do not use metal tool since this may damage caliper bore or seal groove.**
5. Remove bleeder valve and protector.
6. Inspect piston for wear, corrosion or damage. Replace piston, if necessary.
7. Inspect caliper housing for corrosion. If corrosion exists, use a crocus cloth to remove light surface corrosion. If corrosion cannot be removed with crocus cloth, replace caliper housing.
8. Clean all parts with denatured alcohol. Dry with unlubricated compressed air. Blow out all passages in housing and bleeder valve.

CALIPER ASSEMBLY

1. Install bleeder valve onto caliper. **Torque** valve to 80-140 inch lbs.
2. Using clean brake fluid, lubricate caliper housing bore and bore seals.
3. Install piston seal into caliper bore groove. Ensure seal is not twisted in groove.
4. Install boot and piston protector onto piston.
5. Install piston into caliper bore and push piston completely to the bottom.
6. Using dust boot installer No. J-29381 or equivalent, install boot into caliper housing counterbore.

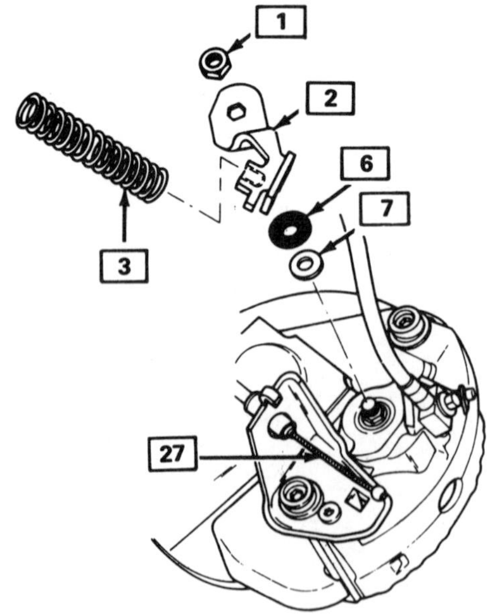

1. NUT
2. LEVER
3. RETURN SPRING
6. LEVER SEAL
7. ANTI-FRICTION WASHER
27. CABLE

Fig. 4 Parking brake lever assembly

BRAKE SHOE & LINING INSTALLATION

1. Lubricate, then install bushings into mounting bolt hole grooves.
2. Lubricate, then install sleeves into mounting bolt holes.
3. Install inboard shoe and lining onto caliper. **Ensure tabs on inboard shoe are positioned so that they fit into corresponding holes in face of piston.**
4. Install outboard shoe and lining onto caliper. Ensure wear sensor is positioned at leading edge of brake shoe.

CALIPER INSTALLATION

1. Install caliper onto brake rotor mounting bracket.
2. Using silicone grease or equivalent, liberally coat shoulder and mounting bolt threads so caliper cavity will be filled.
3. Install caliper mounting bolts. **Torque** bolts to 21-35 ft. lbs.

REAR BRAKE ASSEMBLY

CALIPER REMOVAL

1. Drain approximately 2/3 of brake fluid from master cylinder assembly.
2. Raise and support vehicle, then remove tire and wheel assembly.
3. Loosen tension on parking brake cable at equalizer, remove cable and spring from parking brake lever, **Fig. 4**
4. While holding parking brake lever, remove locknut, parking brake lever, lever seal and anti-friction washer, **Fig. 4. Parking brake lever assembly must be removed to provide clearance for compressing caliper pis-**

ton. Failure to remove lever will cause piston to be damaged when it is bottomed in bore.

5. Remove brake line fittings from caliper if caliper is to be removed for service or replacement. **If only the shoes and linings are to be replaced, do not disconnect brake hose from caliper.**
6. Remove caliper mounting bolts, then position C-clamp onto caliper as shown in **Fig. 5,** and tighten clamp until piston bottoms in piston bore. **Do not come in contact with actuator screw.**
7. Lift caliper off rotor and mounting bracket and support caliper, as needed, to prevent stretching brake hose.

BRAKE SHOE & LINING REMOVAL

1. Remove caliper as described under "Caliper Removal."
2. Remove inboard shoe and lining assembly from caliper.
3. To remove outboard shoe and lining assembly, use a suitable screwdriver to disengage shoe buttons from caliper holes.
4. Remove sleeves from mounting bolt holes.
5. Remove bolt boots and bushings from caliper.
6. Using a suitable screwdriver, remove two-way check valve from piston end.

CALIPER DISASSEMBLY

1. Remove dampening spring from piston end by turning retainer until inside tabs line up with notches in piston, **Fig. 2.**
2. Remove nut and lever assembly.
3. Remove lever seal and anti-friction washer.
4. Place caliper into a soft jaw vise. Place a clean shop towel on inside of caliper. Remove piston using suitable wrench to rotate actuator screw to work piston out of caliper. **Rotate in parking brake apply direction**
5. Press on threaded end of actuator screw, then remove screw.
6. Remove shaft seal and thrust washer from actuator screw.
7. Remove boot from caliper housing bore. Do not use a sharp tool to remove boot from caliper.
8. Using a piece of wool or plastic, remove piston seal from groove in caliper bore. **Do not use any type of metal tool to remove piston seal, since damage to caliper bore may result.**
9. Inspect piston for wear and damage. Replace piston, if necessary.
10. Inspect piston bore for scoring, nicks, corrosion, or wear. Use a crocus cloth to polish out light corrosion. Replace caliper housing if corrosion will not clean up with crocus cloth.
11. Remove bleeder valve and protector.
12. Inspect caliper mounting bracket. If mounting bracket is damaged, replace as required.

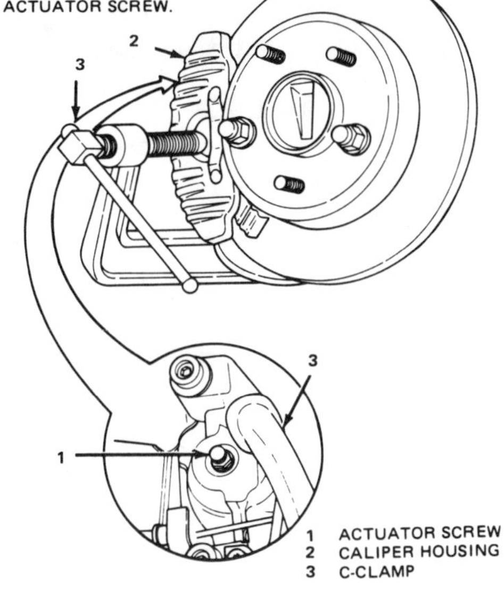

Fig. 5 Positioning C-clamp to compress rear caliper piston

13. Clean all parts with denatured alcohol. Dry with unlubricated compressed air. Blow out all passages in housing and bleeder valve.

CALIPER ASSEMBLY

1. Install bleeder valve. **Torque** valve to 80-140 inch lbs.
2. Install protector onto bleeder valve.
3. Install mounting bracket, if removed. **Torque** bracket bolt to 24-38 ft. lbs.
4. Lubricate new piston seal with clean brake fluid and install is caliper bore groove. **Make sure seal is not twisted.**
5. Lubricate and install boot onto piston with inside lip of boot in piston groove and boot fold toward end of piston that contacts inboard brake shoe.
6. Install thrust washer onto actuator screw with bearing, grayish, surface toward caliper housing.
7. Lubricate shaft seal with clean brake fluid, then install seal onto actuator screw.
8. Lubricate actuator screw, then install screw into piston.
9. Install balance spring into piston recess.
10. Lubricate piston assembly with clean brake fluid, then install assembly into caliper bore. Use piston installer No. J-23072 or equivalent to push piston completely down into bore.
11. Lubricate and install anti-friction washer and lever seal onto actuator screw. **Ensure sealing bead on lever is against housing.** Install lever onto actuator screw. Rotate lever away from stop slightly and hold while install nut. **Torque** nut to 30-40 ft. lbs. After nut is torqued, rotate lever back to contact stop.
12. Seat boot into caliper housing counterbore. Using dust boot installer No. J-28678 or equivalent.
13. Install dampening spring into groove end of piston. It may be necessary to move parking brake lever off the stop to extend piston and make spring groove accessible.

BRAKE SHOE & LINING INSTALLATION

1. Lubricate, then install bushings into mounting bolt hole grooves.
2. Lubricate, then install bolt boots into mounting bolt holes.
3. Lubricate, then install sleeves into mounting bolt holes.
4. Install two-way check valve into piston end.
5. Install inboard shoe and lining into caliper. Noting the following:
   a. Ensure D-shaped tab on shoe engages D shaped notch in piston.
   b. If tab and notch do not align, use spanner wrench or equivalent to turn piston.
   c. Slide edge of metal shoe under ends of dampening spring and snap shoe into position, flat against the piston.
6. Install outboard shoe and lining assembly into piston. Ensure spring ends on outboard shoe snap into piston recess. **Brake wear sensor must be at leading edge of shoe during forward wheel rotation.**
7. After installation of caliper assembly, apply brakes slowly and firmly three times to seat linings.

CALIPER INSTALLATION

1. Liberally fill cavities in housing between bushings with silicone grease.
2. Install sleeves in caliper, then caliper over brake rotor into mounting bracket.
3. **Torque** caliper mounting bolts to 30-45 ft. lbs.
4. Install brake line fitting onto caliper, if removed. **Torque** fitting to 30 ft. lbs.
5. Install anti-friction washer.
6. Lubricate seal, then install into caliper housing. Ensure sealing bead on seal contacts housing.
7. Install lever onto actuator screw. **Lever must be pointing down.**
8. While holding lever, install nut. **Torque** nut to 30-40 ft. lbs.
9. Rotate lever back against caliper stop, then install spring.
10. Install parking brake cable. Tighten cable at equalizer until lever starts to move off caliper stop. Loosen adjustment until lever moves back against caliper stop.
11. Adjust parking brake, then install tire and wheel assembly, then lower vehicle.
12. Fill master cylinder with brake fluid.
13. Bleed brake system, if necessary. **Before moving vehicle, pump brake pedal several times to be sure it is firm. Do not move vehicle until a firm pedal is obtained.**

Delco-Moraine Single Piston w/Dual Mounting Bolts (Type 4)

INDEX

DESCRIPTION

This caliper, **Figs. 1 and 2,** has a single bore and is mounted to the support bracket with two mounting bolts. Hydraulic pressure, created by applying force to the brake pedal, is converted by the caliper into friction. The hydraulic pressure is applied equally against the piston and the bottom of the caliper bore moving the piston outward and the housing inward, resulting in a clamping action on the brake rotor.

Rear disc brake calipers include an integral parking brake mechanism. When the parking brake is applied, the lever turns an actuator screw which is threaded into a nut in the caliper piston. As the actuator screw turns, the piston is forced out, applying force on the rotor. The piston contains a self adjusting mechanism to keep the parking brake in proper adjustment and to ensure proper clearance when the parking brake is released.

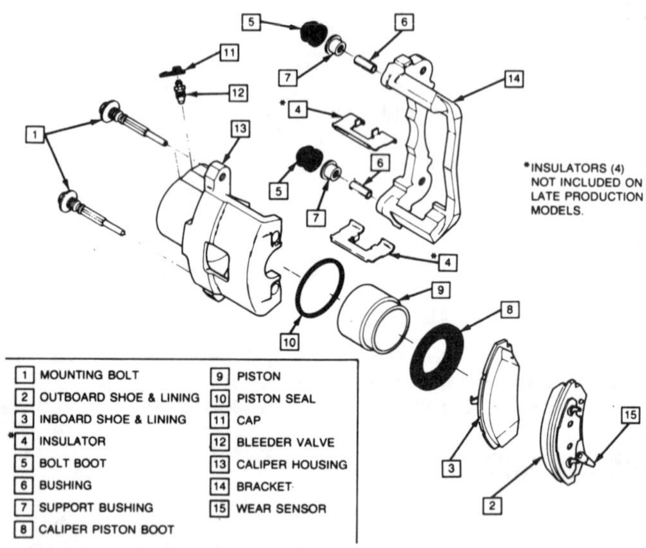

Fig. 1 Delco-Moraine series 3264M front disc brake caliper exploded view.

FRONT BRAKE ASSEMBLY

CALIPER REMOVAL

1. Drain approximately 2/3 of brake fluid from master cylinder assembly.
2. Raise and support vehicle, then mark position of wheel and remove tire and wheel assembly.
3. If caliper is to be serviced or replaced, remove bolt securing inlet fitting, then disconnect brake hose, and plug end to prevent loss of fluid and contamination. **Do not crimp brake hose, as this may damage internal structure of hose. If only shoe and lining assemblies are being replaced, do not disconnect brake hose from caliper.**
4. Remove caliper to support mounting bolts, then lift caliper off support. If only shoe and lining assemblies are to be replaced, suspend caliper from chassis using suitable hanger. **Do not allow caliper to hang by brake hose.**
5. Remove insulators (4) from caliper support if equipped, **Fig. 1.**

BRAKE SHOE & LINING REMOVAL

1. Remove caliper assembly as outlined.
2. Remove outboard shoe and lining assembly, using screwdriver to disengage shoe springs from holes in caliper housing.
3. Remove inboard shoe and lining assembly, unsnapping shoe retainer spring from caliper piston bore.
4. If new shoe and lining assemblies are to be installed, compress piston into housing with suitable clamp or pliers, taking care not to damage piston.

CALIPER DISASSEMBLY

1. Drain brake fluid from caliper and clean thoroughly.
2. Use clean shop towels to pad inside of caliper, then remove piston by applying compressed air to fluid inlet port. Use only enough air pressure to ease piston from bore. **Keep fingers away from piston to avoid injury as piston is forced out.**
3. Pry dust boot from caliper, taking care not to damage housing bore.
4. With a small wood or plastic tool remove piston seal from caliper bore. **Do not use metal tool since this may damage caliper bore or seal groove.**
5. Inspect piston for scoring nicks, corrosion, wear and damage and replace as needed.
6. Inspect caliper housing and seal groove for corrosion, nicks, scoring and excessive wear. Use crocus cloth to polish away corrosion from housing bore. Replace caliper housing if corrosion in and around seal groove will not clean up with crocus cloth.
7. Remove bleeder screw from caliper housing, **Fig. 1.**
8. Clean components with denatured alcohol and dry with unlubricated compressed air. Blow out all passages in housing and bleeder valve.

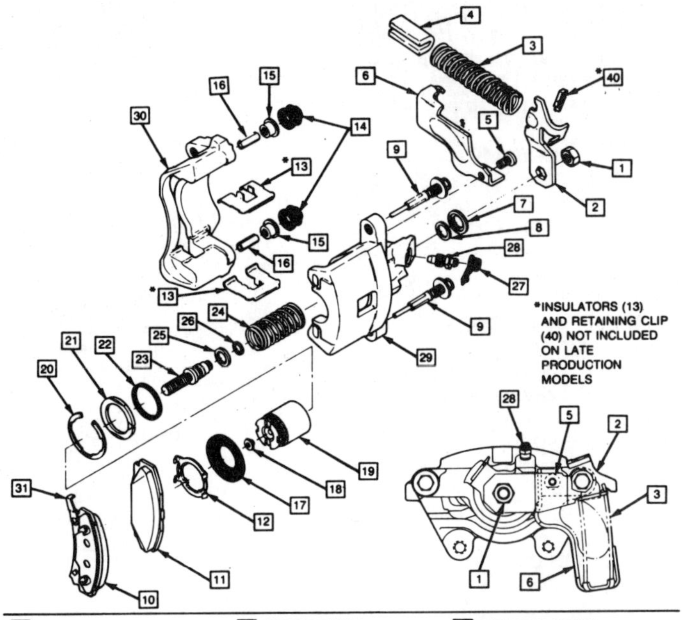

1 NUT	12 SHOE RETAINER	23 ACTUATOR SCREW
2 PARK BRAKE LEVER	*13 INSULATOR	24 BALANCE SPRING & RETAINER
3 RETURN SPRING	14 BOLT BOOT	25 THRUST WASHER
4 DAMPER	15 SUPPORT BUSHING	26 SHAFT SEAL
5 BOLT	16 BUSHING	27 CAP
6 BRACKET	17 CALIPER PISTON BOOT	28 BLEEDER VALVE
7 LEVER SEAL	18 TWO-WAY CHECK VALVE	29 CALIPER HOUSING
8 ANTI-FRICTION WASHER	19 PISTON ASSEMBLY	30 BRACKET
9 MOUNTING BOLT	20 RETAINER	31 WEAR SENSOR
10 OUTBOARD SHOE & LINING	21 PISTON LOCATOR	*40 RETAINING CLIP
11 INBOARD SHOE & LINING	22 PISTON SEAL	

Fig. 2 Delco-Moraine series 3738M rear disc brake caliper exploded view

MOUNTING BRACKET SERVICE

Removal

1. Remove caliper as outlined previously.
2. Remove mounting bracket attaching bolts, then mounting bracket.

Bushing & Boot Replacement

1. Remove mounting bracket from vehicle.
2. Remove bolt boots from support bushings, **Fig. 1.**
3. Clamp bracket in a suitable vise, then pry support bushings from inner bushings in bracket ears with small screwdriver.
4. Using a paper clip, pull inner bushings from mounting bracket ears.
5. Lubricate inner bushings with silicone based grease, then install bushings flush with bracket ears.
6. Position .080-.090 inch thick shim stock on bracket ear face, then tap support bushings, with plastic mallet, into inner bushings, **Fig. 3.** When properly installed, bushing should protrude .080-.090 inch above bracket ear face.
7. Snap new bolt boots over support bushing lip.

Installation

1. Coat mounting bracket attaching bolt threads with Loctite sealant or equivalent. **Failure to use sealant on mounting bolts may result in loosening of caliper to knuckle attachment.**
2. Align mounting bracket holes, then install mounting bracket and **torque** attaching bolts to 83 ft. lbs.

CALIPER ASSEMBLY

1. Install bleeder screw and **torque** to 110 inch lbs.
2. Lubricate piston seal, piston, caliper bore and dust boot with new brake fluid.
3. Roll piston seal into seal groove in caliper bore, ensuring seal is fully seated and not twisted.
4. Install dust boot over piston, ensuring lip of boot engages piston groove.
5. Install piston in caliper bore, press piston to bottom of bore, then seat boot in caliper counterbore with suitable installer.

BRAKE SHOE & LINING INSTALLATION

A revised shoe and lining assembly is used **on late production 1987 and all 1988-89 models.** The revised assembly includes a new lining material and extended shoes, eliminating the need for the

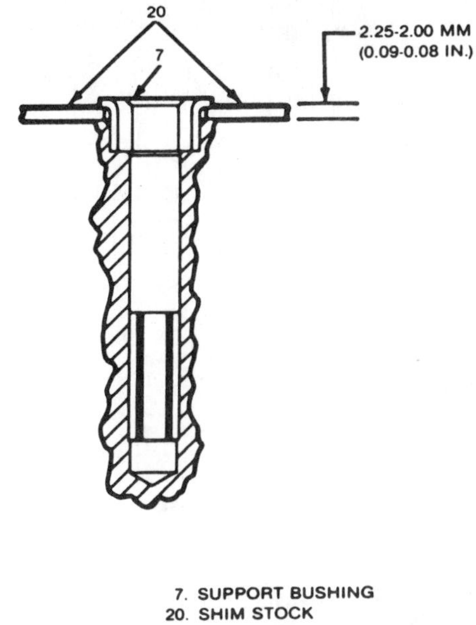

Fig. 3 Installing support bushing

stainless steel shoe abutment insulators (4), **Fig. 1.** This shoe and lining assembly can also be identified by the lining edge code: The original lining code is DM 121 FE, while the revised lining code is DM 128 FE. The revised shoe and lining assembly can be used on earlier models, however, the insulators must be removed and discarded.

1. If new shoe and lining assemblies are being installed, compress piston into caliper bore, taking care not to damage piston.
2. Install inboard shoe, snapping shoe retainer spring into piston bore, and ensure shoe is flat against piston.
3. Install outboard shoe assembly, snapping shoe springs into holes in caliper, and ensure shoe is flat against caliper housing.

CALIPER INSTALLATION

1. If first design shoe and lining assemblies are being installed, mount insulators (4), **Fig. 1**, on caliper bracket.
2. Position caliper on mounting bracket, ensuring insulators, first design, and mounting bolt boots remain in place.
3. Apply thin coating of silicone grease to caliper mounting bolts, install bolts and **torque** to 63 ft. lbs.
4. Install brake hose, if removed, and **torque** fitting bolt to 24 ft. lbs.
5. Install wheels and lower vehicle.
6. Fill master cylinder and bleed brakes as needed. **Do not attempt to move vehicle until brake pedal has been pumped several times and a firm brake pedal has been obtained.**

REAR BRAKE ASSEMBLY

CALIPER REMOVAL

1. Drain approximately 2/3 of brake fluid from master cylinder assembly.

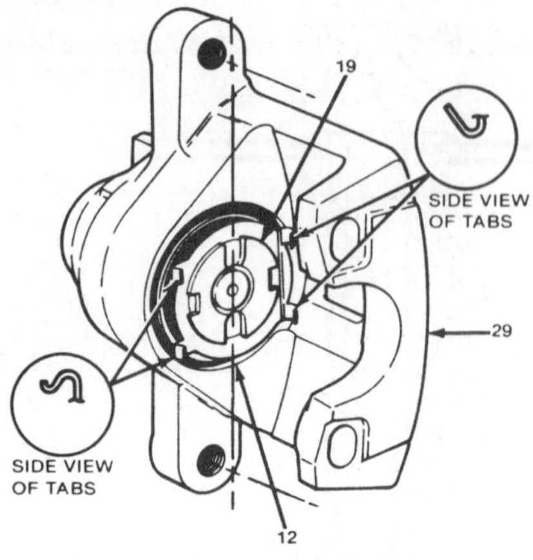

Fig. 4 Positioning caliper piston and pad retainer. Rear disc brake

2. Raise and support vehicle, then remove tire and wheel assembly.
3. Loosen tension on parking brake cable at equalizer.
4. Remove retaining clip from parking brake lever, if equipped, disconnect cable, then remove spring and damper, **Fig. 2.**
5. Hold parking brake lever, remove lock nut.
6. Remove lever, lever seal, and anti-friction washer.
7. Bottom piston into caliper bore to provide clearance between linings and rotor.
8. Reinstall anti-friction washer, lever seal (sealing bead against housing), lever, and lock nut.
9. If caliper is to be serviced, remove inlet fitting attaching bolt, then disconnect and plug inlet fitting. **Do not crimp brake hose, as this may damage internal structure of hose. If only shoe and lining assemblies are to be replaced, do not disconnect brake line.**
10. Remove caliper mounting bolts, then lift caliper from bracket. If only shoe and lining assemblies are being replaced, suspend caliper from chassis with suitable hanger. **Do not allow caliper to hang from brake hose.**
11. Remove insulators from caliper bracket, if equipped.

BRAKE SHOE & LINING REMOVAL

1. Remove caliper assembly as outlined.
2. Remove outboard shoe assembly, unsnapping springs from holes in caliper.
3. Remove inboard shoe assembly, pressing in on edge of shoe from open side of caliper and tilting shoe outward to release it from retainer.
4. Remove bushings from caliper mounting bracket holes, **Fig. 2,** and remove two-way check valve from end of piston with screwdriver. **If leakage is noted from piston hole after removing check valve, replace actuator and overhaul caliper.**

CALIPER DISASSEMBLY

1. Remove caliper assembly as outlined.
2. Rotate shoe retainer until tabs are aligned with recesses in piston, then remove shoe retainer, **Fig. 2.**
3. If installed remove lock nut, lever, lever seal, and anti-friction washer.
4. Secure caliper assembly in vise and insulate interior of caliper with shop towels.
5. Remove piston by rotating actuator screw in parking brake apply direction to work piston from caliper bore.
6. Remove balance spring (24), then remove actuator screw by pressing on threaded end, **Fig. 2.**
7. Remove shaft seal and thrust washer from actuator screw.
8. Pry dust boot from caliper taking care not to damage caliper.
9. Remove retainer snap ring, then the locator.
10. Using wooden or plastic tool remove piston seal from caliper bore. **Do not use metal tool since this may damage caliper bore or seal groove.**
11. Inspect piston for scoring, nicks, corrosion, and wear and replace as needed.
12. Inspect caliper housing and seal groove for corrosion, nicks, scoring and excessive wear, and use crocus cloth to polish away corrosion from housing bore. Replace caliper housing if corrosion in and around seal groove will not clean up with crocus cloth.
13. Remove bleeder screw. Remove bracket only is damaged.
14. Clean components with denatured alcohol and dry with unlubricated compressed air. Blow out all passages in housing and bleeder valve.

MOUNTING BRACKET SERVICE

Removal

1. Remove caliper as outlined previously.
2. Remove mounting bracket attaching bolts, then mounting bracket.

Bushing & Boot Replacement

1. Remove mounting bracket from vehicle.
2. Remove bolt boots from support bushings, **Fig. 2.**
3. Clamp bracket in a suitable vise, then pry support bushings from inner bushings in bracket ears with small screwdriver.
4. Using a paper clip, pull inner bushings from mounting bracket ears.
5. Lubricate inner bushings with silicone based grease, then install bushings flush with bracket ears.
6. Position .080-.090 inch thick shim stock on bracket ear face, then tap support bushings, with plastic mallet, into inner bushings, **Fig. 3.** When properly installed, bushing should protrude .080-.090 inch above bracket ear face.
7. Snap new bolt boots over support bushing lip.

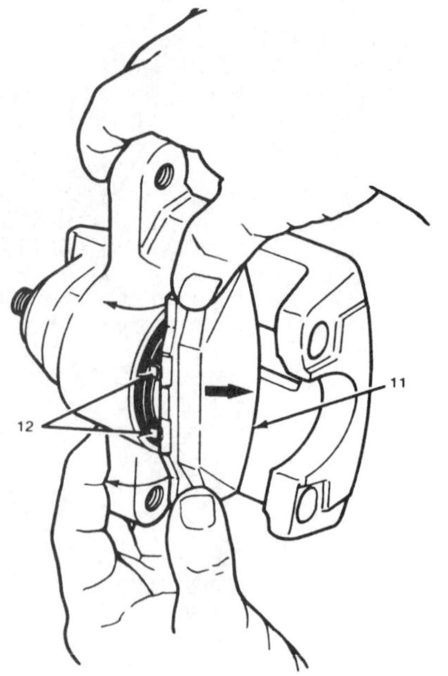

Fig. 5 Inboard shoe installation. Rear disc brake

Installation

1. Coat mounting bracket attaching bolt threads with Loctite sealant or equivalent. **Failure to use sealant on mounting bolts may result in loosening of caliper to knuckle attachment.**
2. Align mounting bracket holes, then install mounting bracket and **torque** attaching bolts to 83 ft. lbs.

CALIPER ASSEMBLY

1. Install bleeder valve, and **torque** to 110 inch lbs.
2. Install bracket, if removed, and **torque** to 31 ft. lbs.
3. Lubricate piston seal, piston, caliper bore and dust boot with new brake fluid.
4. Roll piston seal into seal groove in caliper bore, ensuring seal is fully seated and not twisted.
5. Lubricate piston locator with new brake fluid, then position locator on piston using piston locating tools No. J135588-1 and No. J135588-2 or equivalent.
6. Install thrust washer, grayish side towards caliper housing, on actuator screw.
7. Install lubricated shaft seal on actuator screw, then install actuator screw assembly into caliper housing.
8. Install balance spring in piston recess.
9. Install piston assembly, pressing pis-

ton in until locator is past snap ring groove, turn actuator screw as necessary to allow piston assembly to move toward bottom of caliper bore, then install snap ring.

10. Install lubricated dust boot over piston with inside lip in groove in piston and boot fold toward end of piston that contacts brake shoe, then bottom piston in caliper bore.
11. Install anti-friction washer and seal over end of actuator screw, ensuring seal bead is against caliper housing.
12. Install parking brake lever over actuator screw, rotate lever away from stop, hold lever and **torque** retaining nut to 35 ft. lbs., then rotate lever back to stop.
13. Seat dust boot in caliper counterbore using suitable driver.
14. Install shoe retainer on end of piston, ensuring piston and retainer are positioned as shown in **Fig. 4.**

BRAKE SHOE & LINING INSTALLATION

A revised shoe and lining assembly is used **on late production 1987 and all 1988-89 models.** The revised assembly includes a new lining material and extended shoes, eliminating the need for the stainless steel shoe abutment insulators (13), **Fig. 2.** This shoe and lining assembly can also be identified by the lining edge code: The original lining code is DM 121 FE, while the revised lining code is DM 128 FE. The revised shoe and lining assembly can be used on earlier models, however, the insulators must be removed and discarded.

1. Install lubricated bushings into bolt holes in caliper bracket.
2. Bottom piston in caliper bore using suitable pliers, protecting piston to prevent damage. **Do not allow pliers to contact actuator screw.**
3. Lubricate new two-way check valve, then insert valve into piston.
4. Install inboard shoe assembly as follows:
   a. Engage inboard shoe edge in straight tabs of retainer, then press down and snap shoe under S-shaped tabs, **Fig. 5.**
   b. Ensure shoe lays flat against piston and that retainer and piston are properly aligned. Note that tab on retainer are different, **Fig. 4,** and rotate retainer as needed.
   c. Ensure buttons on back of shoe engage D-shaped notches in piston. Rotate piston as needed with spanner wrench No. J-7624 or equivalent. Piston is properly aligned when D-shaped notches are aligned with caliper mounting bolt holes, **Fig. 4.**
5. Install outboard shoe assembly, snap shoe springs into holes in caliper and ensure pad lays flat against caliper. **Wear sensor should be at trailing edge of shoe during forward wheel rotation.**

CALIPER INSTALLATION

1. Lubricate sliding surfaces and mount insulators, if equipped, on caliper bracket.
2. Position caliper on mounting bracket ensuring insulators and retaining bolt boots remain in position.
3. Coat mounting bolts with thin film of silicone grease, install bolts and **torque** bolts to 63 ft. lbs.
4. Install brake hose using new sealing washers and **torque** fitting bolt to 15 ft. lbs.
5. Install parking brake lever as follows:
   a. Lubricate lever seal and anti-friction washer.
   b. Install washer, seal and parking brake lever. **Seal should be against caliper housing. Parking brake lever should be pointing downward.**
   c. Move lever off stop, hold lever and **torque** retaining nut to 35 ft. lbs., then rotate lever against stop.
6. Install spring and damper, connect parking brake cable, then secure assembly with retaining clip, if used.
7. Tighten brake cable at equalizer until parking brake lever moves off stop, then loosen cable until lever just contacts stop.
8. Install wheels and lower vehicle, fill master cylinder, and bleed brakes as needed. **Do not attempt to move vehicle until brake pedal has been moved several times a firm brake pedal is obtained.**

Delco-Moraine Single Piston w/Dual Mounting Bolts (Type 5)

INDEX

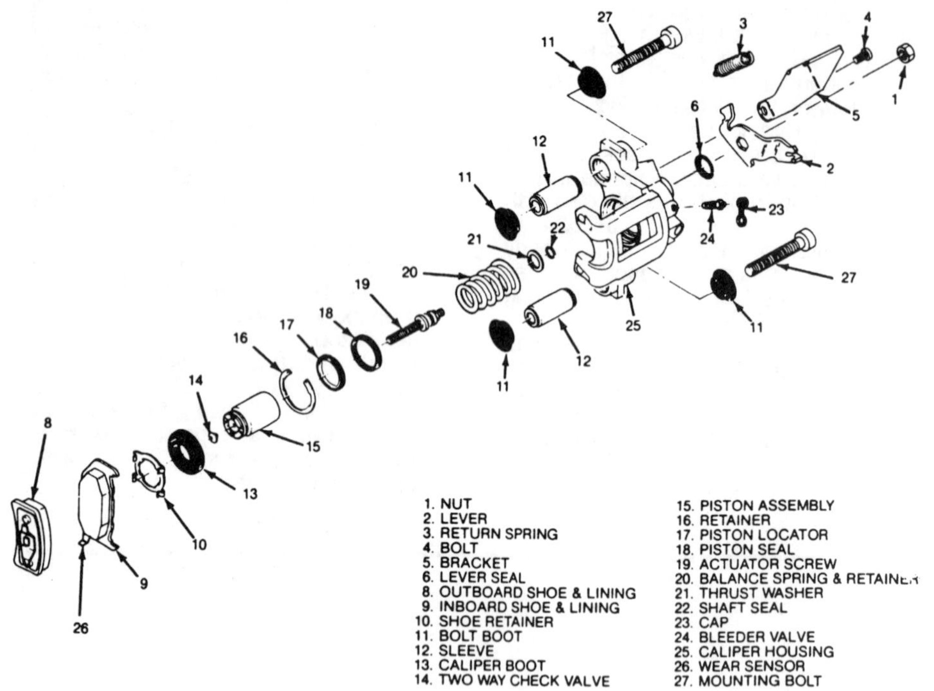

Fig. 1 Exploded view of Delco-Moraine 7735 series rear caliper

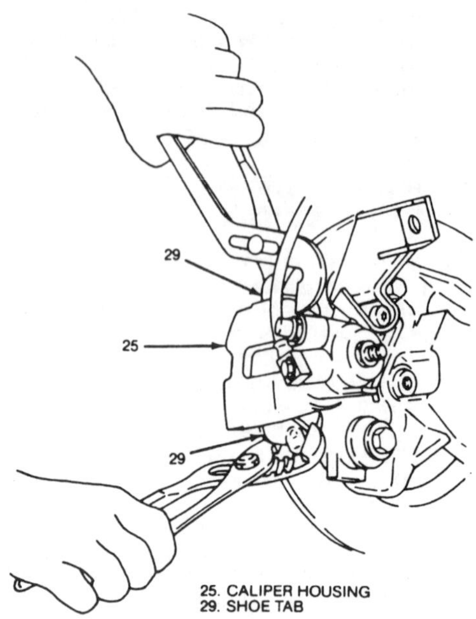

Fig. 2 Bottoming piston in caliper bore

DESCRIPTION

This rear caliper, **Fig. 1,** has a single bore and is mounted to the support bracket with two mounting bolts. Hydraulic pressure, created by applying force to the brake pedal, is converted by the caliper into friction. The hydraulic pressure is applied equally against the piston and the bottom of the caliper bore moving the piston outward and the caliper housing inward, resulting in a clamping action on the brake rotor.

The caliper includes an integral parking brake mechanism. When the parking brake is applied, the lever turns an actuator screw which is threaded into a nut in the caliper piston. As the actuator screw turns, the piston is forced out, applying force on the rotor. The piston contains a self-adjusting mechanism to keep the parking brake in proper adjustment and to ensure proper clearance when the parking brake is released.

SERVICE

CALIPER REMOVAL

1. Drain approximately 2/3 of brake fluid from master cylinder assembly.
2. Raise and support vehicle, then remove tire and wheel assembly.
3. Loosen tension on parking brake cable at equalizer, then disconnect cable and remove return spring from parking brake lever, **Fig. 1.**
4. Hold parking brake lever in position, loosen retaining nut, then remove nut, lever and seal.
5. Position two adjustable pliers over inboard shoe tabs and flanges on caliper housing as shown, **Fig. 2,** then bottom piston in caliper bore.
6. Reinstall lever seal, lever and retaining nut.
7. If caliper assembly is to be serviced, remove inlet fitting attaching bolt, copper washer, and inlet fitting from caliper housing.Plug opening in inlet fitting to prevent fluid loss and contamination. **Do not crimp brake hose, as this may damage internal structure of hose. If only shoe and lining assemblies are to be replaced, do not disconnect brake line fitting from caliper.**
8. Remove caliper mounting bolts, then lift caliper from mounting bracket. If only shoe and lining assemblies are being replaced, suspend caliper from chassis with suitable hanger. **Do not allow caliper to hang from brake hose.**

BRAKE SHOE & LINING REMOVAL

1. Remove caliper as outlined previously.
2. Remove outboard shoe assembly, disengaging buttons on shoe from holes in caliper.
3. Remove inboard shoe assembly, pressing in on edge of shoe from open side of caliper and tilting shoe outward to release it from retainer.
4. Remove two-way check valve from end of piston with screwdriver. **If leakage is noted from piston hole after**

removing check valve, caliper must be overhauled.

CALIPER DISASSEMBLY

1. Remove caliper assembly as outlined previously.
2. Secure caliper assembly in vise and drive out sleeves and bolt boots. Insulate interior of caliper with shop towels.
3. Rotate shoe retainer until tabs are aligned with notches in piston, then remove shoe retainer, **Fig. 1.**
4. Remove lock nut, lever, and lever seal, if installed.
5. Remove piston by rotating actuator screw in parking brake apply direction to work piston from caliper bore.
6. Remove balance spring (20), then remove actuator screw by pressing on threaded end, **Fig. 1.**
7. Remove seal and thrust washer from actuator screw.
8. Pry dust boot from caliper taking care not to damage caliper bore.
9. Remove piston locator retainer, then the locator.
10. Using a small piece of wood or plastic, remove piston seal from bore. **Do not use a metal tool of any kind to remove seal as it may damage bore.**
11. Remove bleeder screw. Remove bracket (5) and bolt (4), **Fig. 1,** if damaged.
12. Inspect piston for scoring, nicks, corrosion, and wear and replace as needed.
13. Inspect caliper housing and seal groove for corrosion, nicks, scoring and excessive wear, and use crocus cloth to polish away corrosion from housing bore. Replace caliper housing if corrosion in and around seal groove will not clean up with crocus cloth.
14. Clean all parts with denatured alcohol. Dry with unlubricated compressed air. Blow out all passages in housing and bleeder valve.

CALIPER ASSEMBLY

1. Install bleeder screw and **torque** to 116 inch lbs.
2. If removed, install bracket (5) and bolt (4), **Fig. 1,** then **torque** bolt to 32 ft. lbs.
3. Lubricate piston seals and bore grooves with new brake fluid, then roll piston seal into seal groove in bore, ensuring seal is fully seated and not twisted.
4. Lubricate piston locator with new brake fluid, then position locator on piston using piston locator installer No. J-36627 or equivalent.
5. Install thrust washer on actuator screw, with grayish side towards caliper housing, then install lubricated shaft seal onto actuator screw.
6. Install actuator screw into piston assembly.
7. Install balance spring in piston recess in housing.
8. Install piston assembly, pressing piston in until locator is past retainer groove, turn actuator screw as necessary to allow piston assembly to move toward bottom of caliper bore, then install retainer ring.
9. Install lubricated dust boot over piston with inside lip in groove in piston and boot fold toward end of piston that contacts brake shoe, then bottom piston in caliper bore.
10. Install lubricated lever seal over end of actuator screw, ensuring seal bead is against caliper housing.
11. Install parking brake lever over actuator screw, rotate lever away from stop, then hold lever and **torque** retaining nut to 32 ft. lbs. After tightening retaining nut, rotate lever back to stop.
12. Seat dust boot in caliper counterbore using suitable driver.
13. Install shoe retainer in groove on end of piston, ensuring retainer tabs are properly aligned with piston notches.
14. Install sleeves and bolt boots into caliper housing as follows:
    a. Lubricate sleeves, bolt boots and caliper mounting bolt holes with silicone based grease.
    b. Install one bolt boot into groove in mounting bolt hole of caliper housing.
    c. Push sleeve into mounting bolt hole, past inner diameter of bolt boot, and position as shown in view A of **Fig. 3.**
    d. Install second bolt boot in groove at other end of mounting bolt hole as shown in view B of **Fig. 3.**
    e. Push sleeve in opposite direction until grooves of sleeve seat in inner diameter of both bolt boots as shown in view C of **Fig. 3.**
    f. Repeat steps b through e to install remaining bolt boots and sleeve.
15. Install caliper as outlined further on.

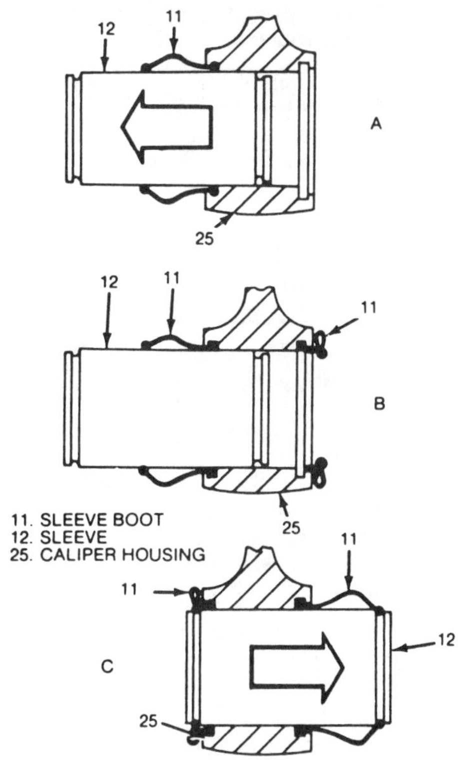

Fig. 3 Bolt boot & sleeve installation

BRAKE SHOE & LINING INSTALLATION

1. Lubricate, then install two-way check valve into piston.
2. Install inboard shoe assembly as follows:
    a. Engage inboard shoe edge in retainer tabs closest to caliper bridge, then press down and snap shoe tabs at open side of caliper. Ensure shoe lays flat against piston.
    b. Ensure buttons on back of shoe engage D-shaped notches in piston. Rotate piston as needed with spanner wrench No. J-7624 or equivalent. Piston is properly aligned when one of the D-shaped notches is nearest caliper bridge.
    c. Wear sensor should be at the leading edge of shoe during forward wheel rotation.
3. Install outboard shoe assembly, snapping shoe springs into holes in caliper. Ensure pad lays flat against caliper.

CALIPER INSTALLATION

1. Push sleeves inward, away from inboard brake shoe.
2. Reconnect brake hose bracket to suspension, then position caliper over rotor and onto mounting bracket.
3. Install mounting bolts and **torque** to 79 ft. lbs. (92 ft. lbs. on 1989 models).
4. Reinstall bracket (5) and bolt (4), **Fig. 1, torque** bolt to 32 ft. lbs., then reconnect brake line to caliper using new copper washers.
5. Remove retaining nut, parking brake lever and lever seal, then clean area around seal and actuator screw.
6. Lubricate, then install lever seal. Ensure seal is fully seated against caliper.
7. Install parking brake lever onto actuator screw hex, ensuring lever faces downward, then install retaining nut.
8. While holding lever against stop on caliper, **torque** retaining nut to 35 ft. lbs.
9. Install lever return spring, then reconnect parking brake cable.
10. Install wheels and lower vehicle, then fill master cylinder and bleed brakes as needed. **Before moving vehicle, pump brake pedal several times to be sure it is firm. Do not move vehicle until a firm pedal is obtained.**
11. After installation is completed, adjust parking brake as follows:
    a. Depress brake pedal several times to seat linings.
    b. Fully apply and release parking brake approximately three times. Each full application may require two full pedal strokes.
    c. Raise and support vehicle, then remove rear wheels. Reinstall two wheel nuts to retain rotor.
    d. Turn ignition switch on, and observe brake warning light. Light

should be Off. If light remains on, operate manual brake release and pull downward on front parking brake cable to remove slack from pedal assembly.

e. Check that parking brake levers on both calipers are against caliper stops. If not, check for binding at rear cables. If cables are not binding proceed to next step.

f. Loosen adjustment at equalizer until both parking brake levers are against caliper stops.

g. Tighten cable at equalizer until either parking brake lever moves off caliper stop, then loosen cable until lever just contacts stop. With cable properly adjusted, a firm pedal should be obtained by pumping parking brake two full strokes. Rear wheels should be fully locked.

h. After adjustment is completed, install wheels and lower vehicle.

Delco-Moraine Single Piston w/Dual Mounting Bolts (Type 6)

INDEX

DESCRIPTION

These calipers, **Figs. 1 and 2,** have a single bore and are mounted to a support bracket with two mounting bolts. Hydraulic force, created by applying force to the brake pedal, is converted by the caliper into friction. The hydraulic force is applied equally against the piston and the bottom of the caliper bore moving the piston outward, resulting in a clamping action on the brake rotor.

The rear caliper includes an integral parking brake mechanism. When the parking brake is applied, the lever turns an actuator screw which is threaded into a nut in the caliper piston. As the actuator screw turns, the piston is forced out, applying force on the rotor. The piston contains a self-adjusting mechanism to keep the parking brake in proper adjustment and to ensure proper clearance when the parking brake is released.

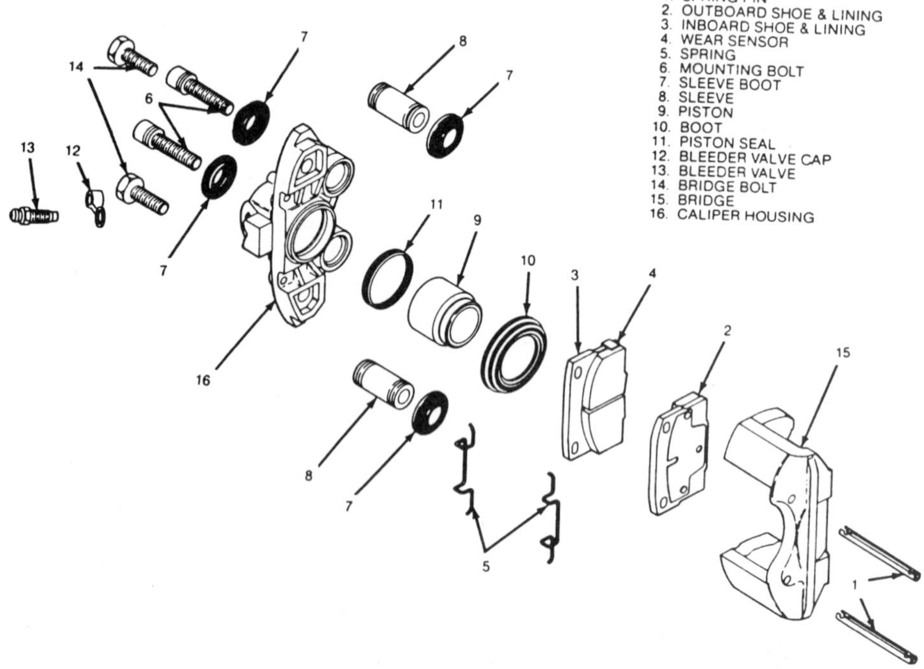

Fig. 1 Exploded view of Delco-Moraine 48H series front caliper

FRONT BRAKE ASSEMBLY

BRAKE SHOE & LINING REMOVAL

1. Drain approximately 2/3 of brake fluid from master cylinder assembly.
2. Raise and support vehicle, then mark position of wheel and remove tire and wheel assembly. Reinstall two wheel nuts to hold rotor in position.
3. Position adjustable pliers over caliper housing and flange of inboard brake shoe, **Fig. 3,** then squeeze pliers and bottom piston in caliper bore.
4. Using suitable tools, drive spring pins (1) from caliper housing, **Fig. 1. Use caution when performing this procedure, as retainer springs (5) may fly out and cause personal injury.**
5. Remove retainer springs from inboard and outboard shoe flanges.
6. Remove outboard, then the inboard brake shoe and lining through opening in caliper. If inboard brake shoe is difficult to remove, move caliper housing inboard to provide additional clearance for removal.

CALIPER REMOVAL

1. Remove brake shoes and linings as outlined previously.
2. If caliper requires overhaul, disconnect brake line from caliper and plug openings to prevent fluid loss or contamination.
3. Remove caliper mounting bolts (6), **Fig. 1,** using suitable torx bit. Do not confuse mounting bolts with bridge bolts (14) during removal.
4. Lift up on caliper and remove from brake rotor. If overhaul is not required, suspend caliper from suspension to avoid damaging brake hose.

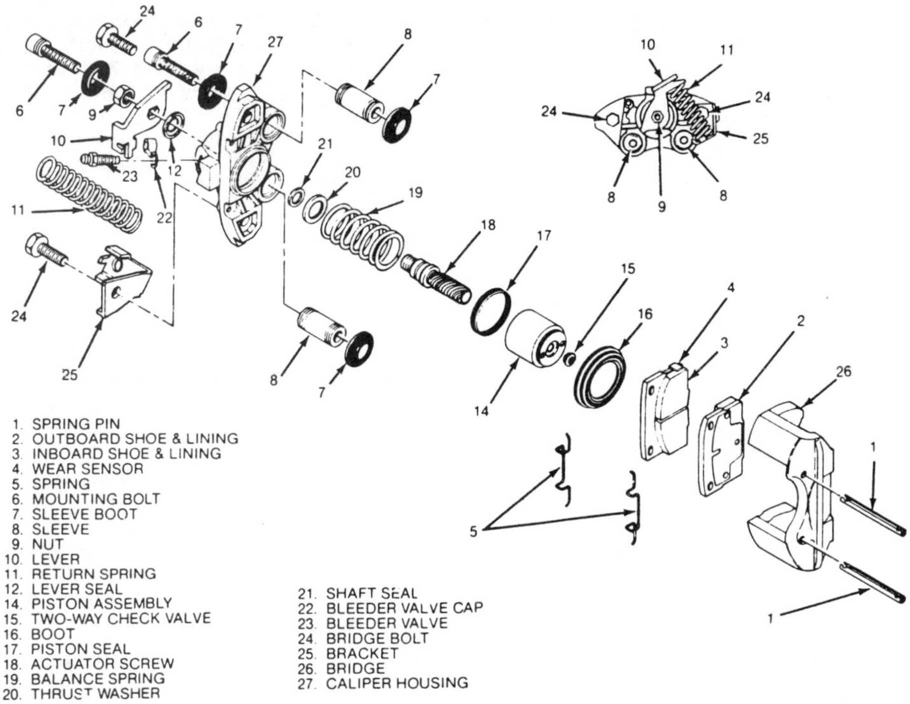

Fig. 2 Exploded view of Delco-Moraine 48H series rear caliper

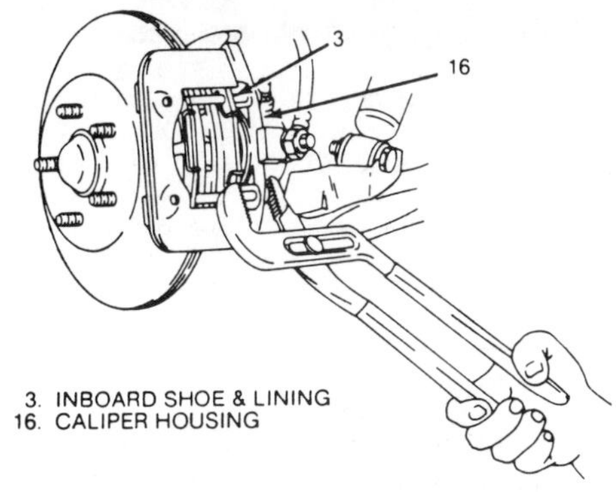

Fig. 3 Bottoming piston in front caliper bore

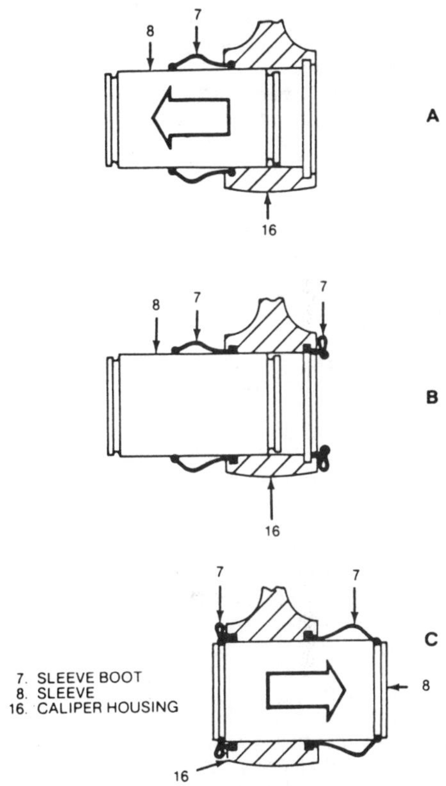

Fig. 4 Sleeve boot & sleeve installation. Front caliper (rear caliper similar)

5. Remove sleeves and sleeve boots from caliper.

CALIPER OVERHAUL

Caliper can be overhauled without removing caliper bridge. If bridge is damaged, remove bridge bolts and bridge as required.

Disassembly & Inspection

1. Remove caliper as outlined previously.
2. Use clean shop towels to pad inside of caliper, then remove piston by applying compressed air to fluid inlet port. Use only enough air pressure to ease piston from bore. **Keep fingers away from piston to avoid injury as piston is forced out.**
3. Pry dust boot from caliper, then remove piston seal, taking care not to damage housing.
4. Clean metal components with suitable solvent and blow dry with compressed air.
5. Inspect piston for scoring nicks, corrosion, wear and damage, and replace as needed.
6. Inspect caliper housing and seal groove for damage, corrosion, nicks, scoring and excessive wear, and use crocus cloth to polish away corrosion from housing bore. Replace caliper housing if damaged, or if corrosion is excessive.
7. Remove bleeder screw from caliper housing, **Fig. 1.**

Assembly

1. Reinstall caliper bridge, if removed during disassembly. **Torque** bridge bolts to 74 ft. lbs.
2. Lubricate piston seal, piston, caliper bore and dust boot with new brake fluid.
3. Roll piston seal into seal groove in caliper bore, ensuring seal is fully seated and not twisted.
4. Install dust boot over piston, ensuring lip of boot engages piston groove.
5. Install piston in caliper bore, press piston to bottom of bore, then seat boot in caliper counterbore with suitable driver.
6. Install bleeder screw and **torque** to 116 inch lbs.

CALIPER INSTALLATION

1. Install sleeves and sleeve boots into caliper as follows:
   a. Lubricate sleeves, boots and caliper mounting bolt holes with silicone based grease.
   b. Install one sleeve boot into groove in mounting bolt hole of caliper housing.
   c. Push sleeve into mounting bolt hole, past inner diameter of sleeve boot, and position as shown in view A of **Fig. 4. Ensure sleeves are installed with large hole facing inboard, or damage to mounting bracket threads may result.**
   d. Install second sleeve boot in groove at other end of mounting bolt hole as shown in view B of **Fig. 4.**
   e. Push sleeve in opposite direction until grooves of sleeve seat in inner diameter of both sleeve boots as shown in view C of **Fig. 4.**
   f. Repeat steps b through e to install remaining boots and sleeve.
2. Position caliper over rotor and onto mounting bracket, then install mounting bolts and **torque** to 74 ft.lbs.
3. If caliper was overhauled, reconnect brake line using new copper washers.
4. Install brake shoes and linings as outlined further on.
5. If caliper was overhauled, fill master cylinder to proper level, then bleed brake system.

BRAKE SHOE & LINING INSTALLATION

1. Install inboard shoe and lining, ensuring wear sensor is positioned at leading edge of shoe during forward rotation of wheel.
2. Install outboard shoe and lining.
3. Using a brass drift and hammer, drive in one spring pin until pin just begins

to emerge from inboard face of caliper.

4. Tap in other spring pin as noted above, but stop when pin is just through outboard section of caliper housing.
5. Position pins so slots face each other, then install retainer springs as follows:
   a. Hook one end of spring under pin installed in step 3, ensuring mid-section of spring is positioned over shoe flange. Repeat procedure for remaining spring.
   b. Using a screwdriver, press down on other end of springs while sliding remaining spring pin inboard as outlined in step 3.
   c. After installation, ensure retainer springs are equally centered on shoe flanges, and that spring ends project under pins an equal amount.
6. Install wheels, lower vehicle, then fill master cylinder to proper level.
7. Apply brake pedal several times to correctly seat linings.

REAR BRAKE ASSEMBLY

BRAKE SHOE & LINING REMOVAL

1. Drain approximately 2/3 of brake fluid from master cylinder assembly.
2. Raise and support vehicle, release parking brake, then mark position of wheel and remove tire and wheel assembly. Reinstall two wheel nuts to hold rotor in position.
3. Using suitable tools, drive spring pins (1) from caliper housing, **Fig. 2. Use caution when performing this procedure, as retainer springs (5) may fly out and cause personal injury.**
4. Remove retainer springs from inboard and outboard shoe flanges.
5. Remove outboard, then the inboard brake shoe and lining through opening in caliper. If inboard brake shoe is difficult to remove, move caliper housing inboard to provide additional clearance for removal.
6. Pry two-way check valve from piston using small screwdriver. If leakage is observed from hole in piston, overhaul caliper as outlined further on.
7. Bottom piston into caliper bore using piston adjusting tool No. J-36621, **Fig. 5. Turn lefthand piston counterclockwise and righthand piston clockwise to move piston assembly into bore.**

CALIPER REMOVAL

1. Remove brake shoes and linings as outlined previously.
2. Loosen parking brake cable adjustment, then disconnect parking brake cable and remove return spring from parking brake lever.
3. If caliper assembly is to be serviced, remove inlet fitting attaching bolt, copper washer, and inlet fitting from caliper housing.Plug opening in inlet fitting to prevent fluid loss and contamination. **Do not crimp brake hose, as this may damage internal structure of hose. If only shoe and lining assemblies are to be replaced, do not disconnect brake line fitting from caliper.**

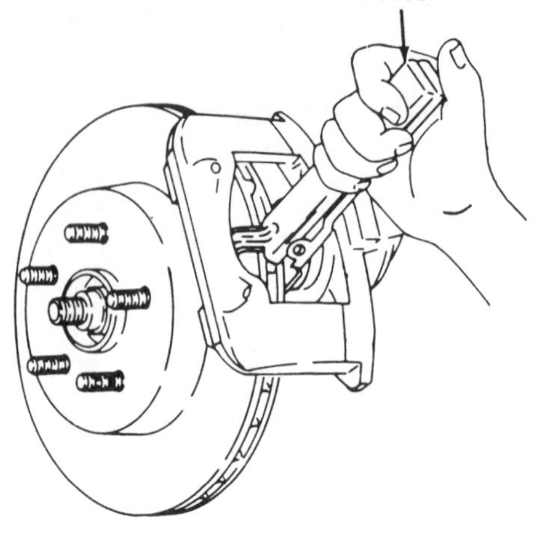

Fig. 5 Bottoming piston in rear caliper bore

4. Remove caliper mounting bolts (6), **Fig. 2,** using suitable torx bit. Do not confuse mounting bolts with bridge bolts (24) during removal.
5. Lift up on caliper and remove from brake rotor. If only shoe and lining assemblies are to be replaced, suspend caliper from chassis using suitable hanger. **Do not allow caliper to hang by brake hose.**
6. Remove sleeves and sleeve boots from caliper.

CALIPER OVERHAUL

Caliper can be overhauled without removing caliper bridge. If bridge is damaged, remove bridge bolts and bridge as required.

Disassembly & Inspection

1. Remove caliper as outlined previously.
2. Remove parking brake lever retaining nut, lever and lever seal.
3. Insulate interior of caliper with shop towels, then remove piston by rotating actuator screw in parking brake apply direction to work piston from caliper bore.
4. Remove actuator screw (18) by pressing on threaded end, then the balance spring (19), **Fig. 2.**
5. Remove seal and thrust washer from actuator screw.
6. Pry dust boot from caliper taking care not to damage caliper bore.
7. Using a small piece of wood or plastic, remove piston seal from bore. **Do not use a metal tool of any kind to remove seal as it may damage bore.**
8. Remove bleeder valve.
9. Inspect piston for scoring, nicks, corrosion, and wear and replace as needed.
10. Inspect caliper housing and seal groove for corrosion, nicks, scoring and excessive wear, and use crocus cloth to polish away corrosion from housing bore. Replace caliper housing if corrosion in and around seal groove will not clean up with crocus cloth.
11. Clean all parts with denatured alcohol. Dry with unlubricated compressed air. Blow out all passages in housing and bleeder valve.

Assembly

1. Reinstall caliper bridge, if removed during disassembly. **Torque** bridge bolts to 74 ft. lbs., then install bleeder valve.
2. Lubricate piston seal, piston, caliper bore and dust boot with new brake fluid.
3. Roll piston seal into seal groove in caliper bore, ensuring seal is fully seated and not twisted.
4. Install thrust washer on actuator screw, with copper side of washer facing toward piston assembly, then install shaft seal onto actuator screw.
5. Install lubricated shaft seal on actuator screw.
6. Install boot onto piston assembly.
7. Install actuator screw and balance spring into caliper bore. Ensure end of spring fully bottoms in recess at bottom of bore.
8. Install piston and boot assembly, push piston into bore, then turn actuator screw as necessary until piston fully bottoms in caliper.
9. Install lever seal over end of actuator screw, ensuring rubber sealing bead is against lever and copper colored side is against caliper housing.
10. Install parking brake lever over actuator screw, rotate lever away from stop, then hold lever and **torque** retaining nut to 35 ft. lbs. After tightening retaining nut, rotate lever back to stop.
11. Seat dust boot in caliper counterbore using suitable driver.

CALIPER INSTALLATION

1. Install sleeves and sleeve boots into caliper as follows:
   a. Lubricate sleeves, boots and caliper mounting bolt holes with silicone based grease.
   b. Install one sleeve boot into groove in mounting bolt hole of caliper housing.
   c. Push sleeve into mounting bolt hole, passed inner diameter of sleeve boot, and position as shown in view A of **Fig. 4. Ensure sleeves are installed with large hole facing inboard, or damage to mounting bracket threads may result.**
   d. Install second sleeve boot in groove at other end of mounting bolt hole as shown in view B of **Fig. 4.**
   e. Push sleeve in opposite direction until grooves of sleeve seat in inner diameter of both sleeve boots as shown in view C of **Fig. 4.**
   f. Repeat steps b through e to install remaining boots and sleeve.
2. Position caliper over rotor and onto

mounting bracket, then install mounting bolts and **torque** to 74 ft.lbs.

3. If caliper was overhauled, reconnect brake line using new copper washers.
4. Install brake shoes and linings as outlined further on.
5. If caliper was overhauled, fill master cylinder to proper level, then bleed brake system.
6. Install lever return spring, then reconnect parking brake cable.
7. Install wheels and lower vehicle. **Do not attempt to move vehicle until brake pedal has been moved several times and full braking action is restored.**
8. After installation is completed, adjust parking brake as follows:
   a. Depress brake pedal several times to seat linings.
   b. Fully apply and release parking brake approximately three times.
   c. Turn ignition switch on, and observe brake warning light. Light should be Off. If light remains on, operate manual brake release and pull downward on front parking brake cable to remove slack from pedal assembly.
   d. Raise and support vehicle.
   e. Check that parking brake levers on both calipers are against caliper stops. If not, check for binding at rear cables. If cables are not binding proceed to next step.
   f. Tighten cable at equalizer until either parking brake lever moves off caliper stop, then loosen cable until lever just contacts stop. With cable properly adjusted, parking brake lever should not travel more than six ratchet clicks and rear wheels should be fully locked at three to six ratchet clicks.
   g. After adjustment is completed, install wheels and lower vehicle.

BRAKE SHOE & LINING INSTALLATION

1. Lubricate, then install new two-way check valve into piston assembly.
2. Install inboard shoe and lining, ensuring wear sensor is positioned at leading edge of shoe during forward rotation of wheel.
3. Install outboard shoe and lining.
4. Using a brass drift and hammer, drive in one spring pin until pin just begins to emerge from inboard face of caliper.
5. Tap in other spring pin as noted above, but stop when pin is just through outboard section of caliper housing.
6. Position pins so slots face each other, then install retainer springs as follows:
   a. Hook one end of spring under pin installed in step 3, ensuring mid-section of spring is positioned over shoe flange. Repeat procedure for remaining spring.
   b. Using a screwdriver, press down on other end of springs while sliding remaining spring pin inboard as outlined in step 3.
   c. After installation, ensure retainer springs are equally centered on shoe flanges, and that spring ends project under pins an equal amount.
7. Install wheels, lower vehicle, then fill master cylinder to proper level.
8. Apply brake pedal several times to correctly seat linings.

Delco-Moraine Single Piston Rear Disc Brake

INDEX

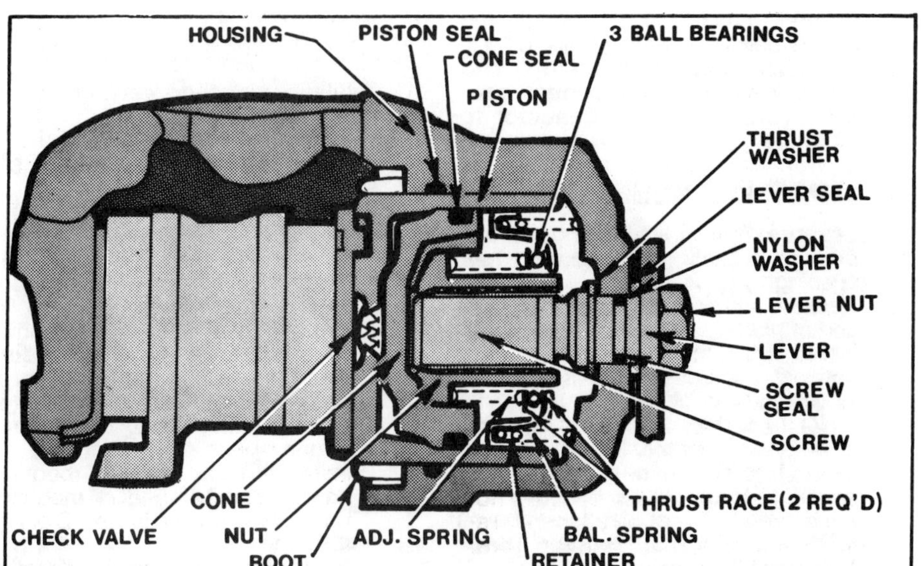

Fig. 1 Typical Delco-Moraine rear disc brake cross-sectional view

DESCRIPTION

Upon application of the brakes, **Figs. 1 & 2,** the cone and piston move out as one part. The nut remains stationary on the high lead screw and a gap develops between the cone and nut. When lining wear occurs, the cone and piston do not return to their original position, thereby leaving a small gap equal to the lining wear between the nut and cone. The adjusting spring causes the nut to rotate on the high lead screw to close the gap and adjust the caliper.

Upon application of parking brake, the lever rotation causes the high lead screw to turn and the nut to move down the screw, thereby loading through the cone and the cone-clutch interface of the piston, resulting in a clamp load on the linings. When the parking brake is released, the cone rotates on the clutch interface to adjust the caliper. The clutch interface prevents the cone from turning when the parking brake is applied.

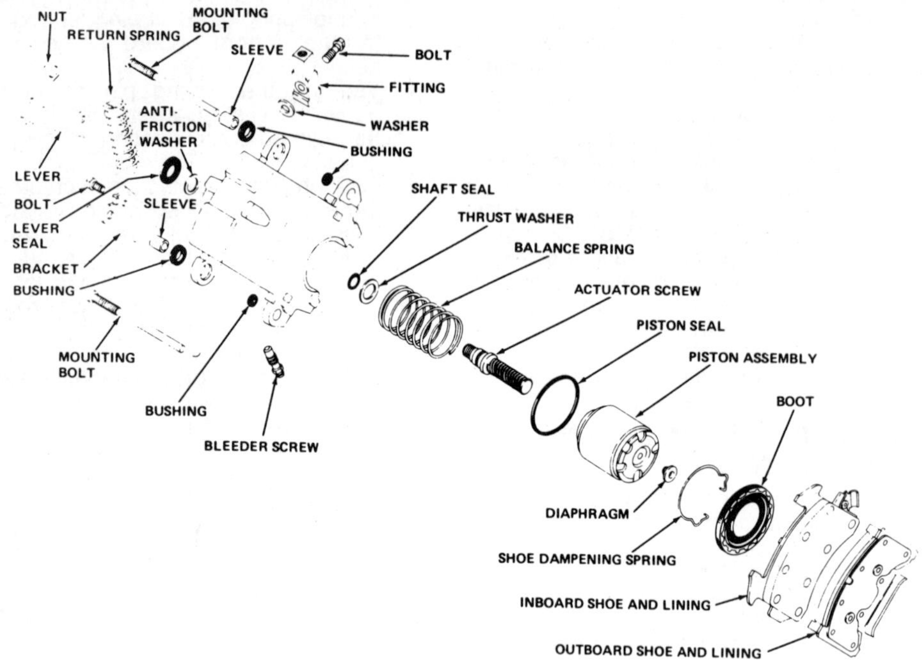

Fig. 2 Rear disc brake caliper exploded view (Typical)

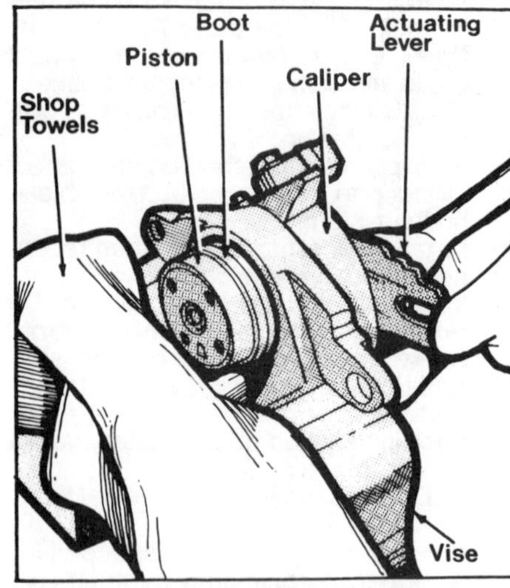

Fig. 3 Removing piston from bore

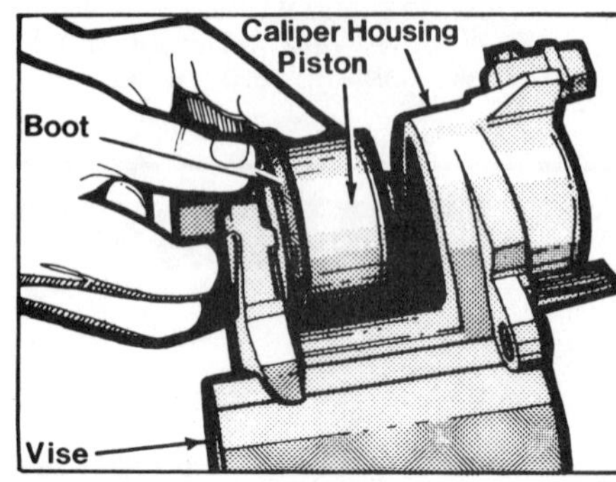

Fig. 4 Positioning piston in caliper

SERVICE

CALIPER REMOVAL

Do not mix power steering fluid with brake fluid. If brake seals contact steering fluid or steering seals contact brake fluid, damage will result.

1. Remove two thirds of the total brake fluid capacity from the master cylinder front reservoir, to prevent overflow of brake fluid.
2. Support vehicle on a hoist and remove tire and wheel assembly.
3. Install one nut with flat side facing rotor to prevent rotor from falling out when caliper is removed.
4. Loosen parking brake cable tension at equalizer, then remove cable from parking brake lever and remove return spring, locknut, lever, lever seal and anti-friction washer. **Lever must be held in place while removing nut.**
5. Clean surface in area of lever seal, then using a 7 inch (or larger) C-clamp, with the solid end on lever stop and screw end on back of outboard lining, turn clamp until piston is bottomed in caliper. **Do not position C-clamp on actuator screw.**
6. Before removing clamp, lubricate housing surface under lever seal with silicone lubricant.
7. Install anti-friction washer, a new lever seal and lever. **Install lever on hex with arm pointing downward.**
8. Rotate lever toward front of vehicle and while holding in this position, install nut and **torque** to 25 ft. lbs. and rotate lever back to stop.
9. Disconnect brake line from caliper and plug openings to prevent loss of fluid and contamination. **Do not crimp brake hose, as this may damage internal structure of hose.**
10. Remove the brass bolt from block, if applicable. **If brake line nut is seized, brass bolt and block can be removed with brake line attached by removing bolt. Plug openings to prevent loss of fluid and entry of dirt.**
11. Remove caliper mounting bolts and remove caliper.
12. Reverse procedure to install and **torque** caliper mounting bolts to 30 ft. lbs. **When installing brass bolt and block, use two new copper gaskets. Torque bolt or connector to 30 ft. lbs.**

CALIPER DISASSEMBLY

1. Clamp caliper in a vise and remove the two mounting sleeves and four bushings, **Fig. 2.**
2. Remove shoe dampening spring from end of piston.
3. If installed, remove locknut, lever, lever seal, and anti-friction washer.
4. Rotate parking brake lever in parking brake apply direction to remove piston from housing, **Fig. 3.** If piston will not move, Use a 9/16 inch wrench, rotate actuator screw clockwise on right hand caliper or counterclockwise on lefthand caliper until the piston moves from housing. **Pad caliper with shop cloths when removing piston.**
5. Remove balance spring, then remove actuator screw by pressing on thread end.
6. Remove shaft seal, and thrust washer from actuator screw.
7. Remove dust boot. **Take care not to scratch caliper housing or bore.**
8. Using a small piece of wood or plastic, remove piston seal from bore. **Do not use a metal tool of any kind to remove seal as it may damage bore.**
9. Remove bleeder screw, fitting, copper washers, and bolt. Remove bracket only if damaged.
10. Inspect piston for scoring, nicks, corrosion, and wear and replace as needed.
11. Inspect caliper housing and seal groove for corrosion, nicks, scoring and excessive wear, and use crocus cloth to polish away corrosion from housing bore. Replace caliper housing if corrosion in and around seal groove will not clean up with crocus cloth.
12. Clean all parts with denatured alcohol. Dry with unlubricated compressed air. Blow out all passages in housing and bleeder valve.

CALIPER ASSEMBLY

1. Install bleeder screw and **torque** to 116 inch lbs., then install bracket, if removed, and **torque** to 31 ft. lbs.
2. Install fitting and bolt, if equipped, using new copper washer and **torque** to 24 ft. lbs.
3. Lubricate and install new piston seal. **Make sure is not twisted.**

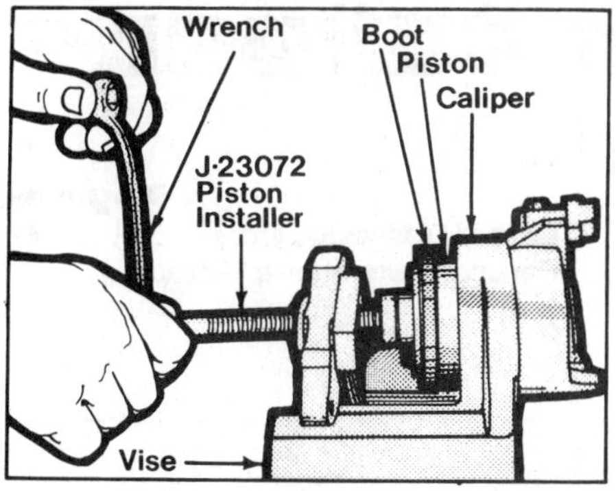

Fig. 5 Installing piston into caliper

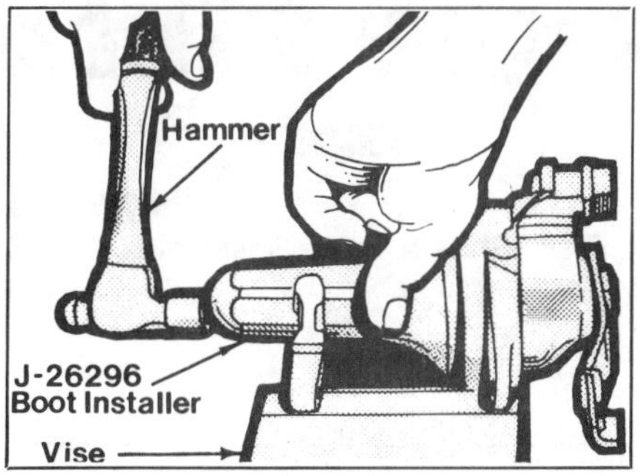

Fig. 6 Driving boot into caliper

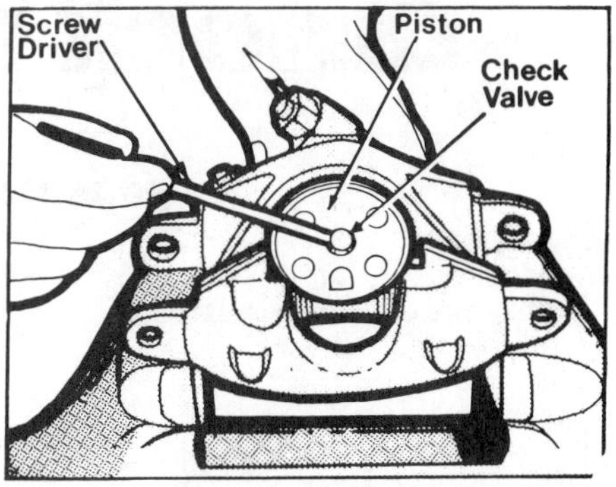

Fig. 7 Removing piston check valve

4. Install new boot onto piston assembly with lip of boot located in piston groove and boot fold toward end of piston that contact inboard brake shoe.
5. Install new thrust washer on actuator screw, with copper side of washer towards the piston assembly.
6. Install lubricated shaft seal on actuator screw, then install actuator screw in piston.
7. Coat piston seal with clean brake fluid. Install balance spring into piston and install assembly into caliper housing, **Fig. 4.**
8. With piston installer No. J-23072 or equivalent push piston fully into caliper housing, **Fig. 5. The piston must be pushed straight into caliper to prevent damage to the actuator screw seal as it passes through hole in rear of piston bore.**
9. Before removing piston installer No. J-23072, install lubricated anti-friction washer, new lever seal, lever and locknut. Position lever away from stop, rotate forward and hold lever in position, then **torque** nut to 35 ft. lbs.
10. Remove piston installer No. J-23072, rotate lever back to stop and install return spring. **On Cadillac the return springs are color coded red for righthand and black for lefthand.**
11. With dust boot installer No. J-26296 or equivalent, drive boot until seal bottoms in caliper housing, **Fig. 6.**
12. Install dampening spring in groove in end of piston.

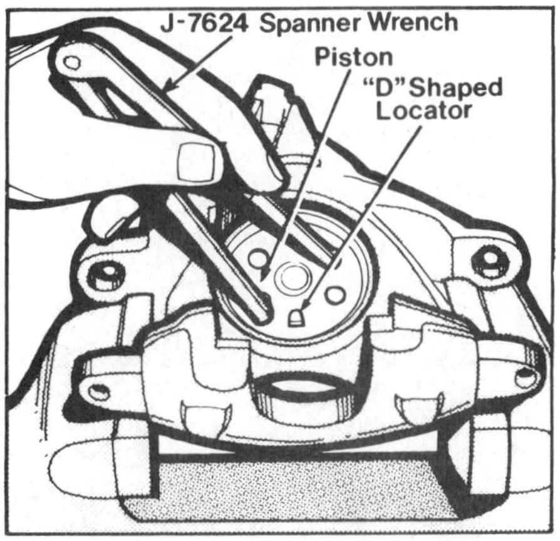

Fig. 8 Rotating piston in bore

BRAKE SHOE & LINING REPLACEMENT

1. Remove caliper as described previously and remove shoe and lining.
2. Remove and discard the two caliper mounting sleeves and the four bushings. Using silicone lubricant, install new bushings and seals. **Sleeves are installed in inner bushings.**
3. Remove and discard piston two-way check valve and install a new one, **Fig. 7. If leakage is noted from end of piston after check valve is removed caliper must be overhauled.**
4. Position new inboard shoe assembly on piston. The D-shaped tab must fit into indentation in piston. If piston requires rotation, use spanner wrench to rotate piston, **Fig. 8. Install new spring retainers as applicable.**
5. Install new outboard shoe assembly onto caliper. Install caliper and **torque** mounting bolts to 30 ft. lbs.
6. Position channel lock pliers over brake shoe ears, if equipped, and bottom edge of caliper. While holding moderate force on brake pedal, clinch outboard shoe.

Girlock Single Piston Caliper

INDEX

DESCRIPTION

This caliper, **Fig. 1,** has a single piston in an aluminum housing suspended in the mounting bracket on two slide pins. Hydraulic pressure, created by applying force to the brake pedal, acts equally against the piston and the bottom of the caliper bore to move the piston outward and to slide the caliper inward resulting in a clamping action on the brake rotor.

SERVICE

BRAKE SHOE & LINING REMOVAL

1. Drain approximately 2/3 of brake fluid from master cylinder.
2. Raise and support vehicle.
3. Mark relationship of wheel to axle assembly, then remove wheel(s).
4. Position a suitable C-clamp into caliper, with one end of clamp against inlet fitting bolt head and other end of clamp against outboard shoe.
5. Tighten clamp screw until caliper piston bottoms in piston bore.
6. Remove and discard upper caliper self locking bolt.
7. Rotate caliper until shoe and lining assemblies are exposed.
8. Remove shoe and lining assemblies.

BRAKE SHOE & LINING INSTALLATION

1. Install new shoe and lining assemblies onto mounting bracket.
2. Rotate caliper into position and install a new self locking bolt. **Torque** bolt to 22-25 ft. lbs.
3. Install wheel(s) and lower vehicle. Bleed brake system, if necessary.

PARKING BRAKE SHOES, REPLACE

1. Remove caliper as described under "Brake Shoe & Lining Removal."
2. Remove rotor, **Fig. 2.**
3. Spread shoes, then remove adjuster assembly and adjuster return spring.
4. Remove hold-down springs and pins, then shoe return spring using suitable pliers.
5. Remove shoe and lining assemblies.
6. Reverse procedure to install.

CALIPER DISASSEMBLY

1. Remove caliper as described under "Brake Shoe & Lining Removal."

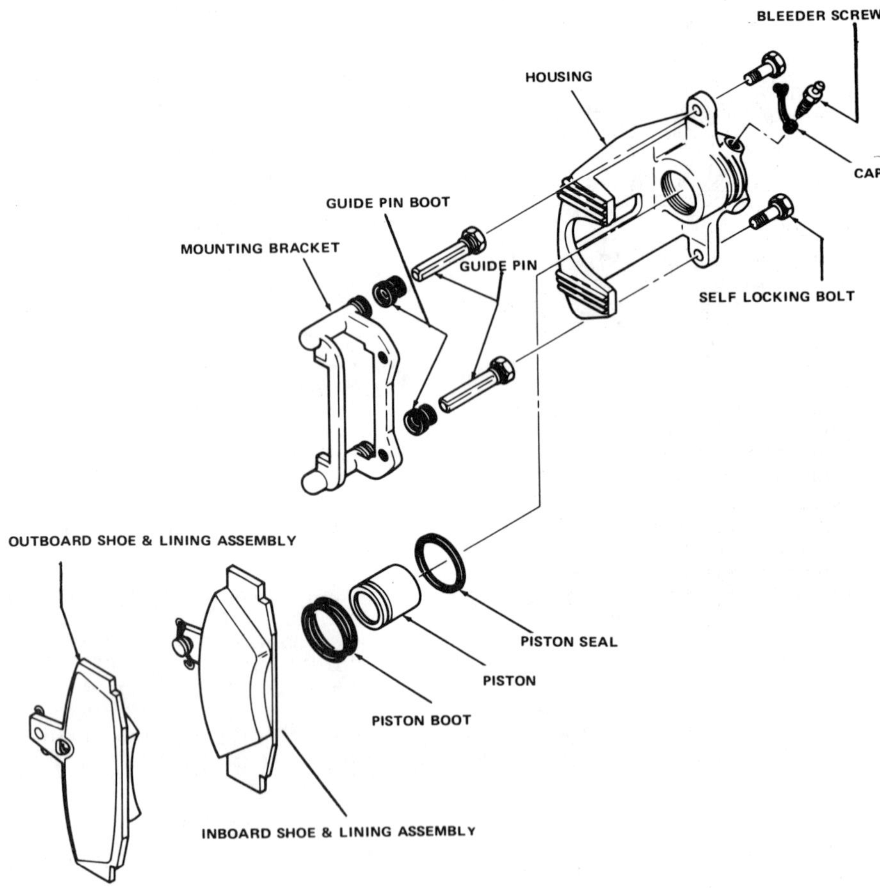

Fig. 1 Girlock single piston caliper exploded view

2. If caliper requires machining, remove inlet fitting.
3. Remove and discard self locking bolts.
4. Remove bracket bolts. **Remove brake shoe and lining assemblies only if they are to be replaced or if the mounting bracket is being removed for rotor machining.**
5. Position a clean shop towel on the inside of the caliper housing, then using compressed air, apply air to brake hose port and remove piston.
6. Inspect piston for wear and/or damage. Replace, if necessary.
7. Remove piston boot from housing.
8. Remove piston seal from caliper bore groove. **Do not use a sharp tool to remove piston seal.**
9. Inspect caliper bore for scoring or corrosion. If scoring or corrosion exists, replace caliper as required.
10. Remove bleeder screw and rubber cap from bleeder screw.
11. Inspect guide pins for corrosion. If corrosion exists, replace guide pins as required.

CALIPER ASSEMBLY

1. Clean all brake components in clean denatured alcohol. Use dry, filtered compressed air to dry components and blow out caliper housing passages.
2. Check piston to bore fit, by sliding piston in and out of piston bore. Piston should slide in and out smoothly.
3. Install rubber cap onto bleeder screw, then insert bleeder screw into caliper.
4. Using clean brake fluid, lubricate pis-

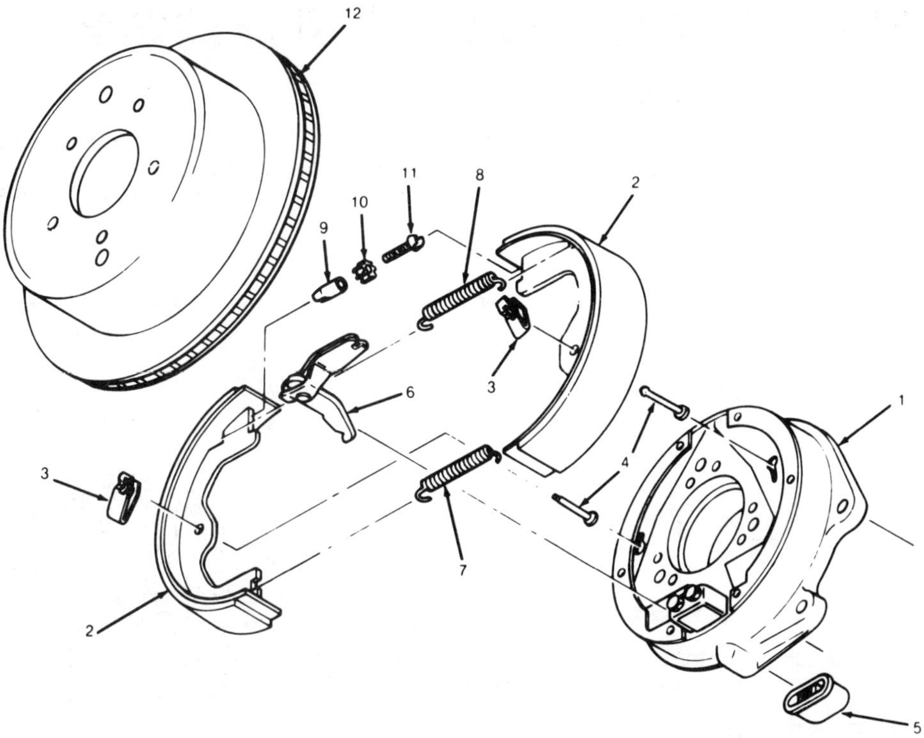

1. MOUNTING PLATE ASSEMBLY
2. SHOE & LINING
3. HOLD DOWN SPRING
4. HOLD DOWN PIN
5. BOOT
6. LEVER & STRUT ASSEMBLY
7. SHOE RETURN SPRING
8. ADJUSTER RETURN SPRING
9. ADJUSTER SOCKET
10. ADJUSTER NUT
11. ADJUSTER SCREW
12. ROTOR

Fig. 2 Parking brake assembly exploded view

ton seal, then install seal into caliper bore groove. Ensure seal is not twisted in piston bore groove.

5. Using clean brake fluid, lubricate caliper bore and piston assembly.
6. Install boot over end of piston. Place piston into caliper bore. Push piston completely down into caliper bore. Ensure boot is properly seated into groove around piston and into groove in caliper bore.
7. Using silicone grease or equivalent, lubricate caliper guide pins.
8. Install new guide pin boots over guide pins.
9. Install guide pins into mounting bracket.
10. Install caliper over rotor and into mounting bracket. Ensure brake shoe springs are positioned properly.
11. Install new self locking bolts. **Torque** bolts to 26-40 ft. lbs.
12. Pump brake pedal firmly and slowly three times to bring pads into contact with brake rotor.
13. Bleed brake system, if necessary.

PBR Dual Piston Caliper

INDEX

DESCRIPTION

This front caliper, **Fig. 1**, consists of dual pistons and an aluminum housing which is suspended on the shoe and lining assemblies. Hydraulic pressure, created by applying force to the brake pedal, acts equally against the pistons and the bottom of the caliper bores to move the pistons outward. This action slides the caliper inward, resulting in a clamping action on the brake rotor. This clamping action forces the linings against the rotor, creating the friction necessary to stop the vehicle.

SERVICE

CALIPER REMOVAL

1. Remove 2/3 of the total brake fluid capacity from master cylinder reservoir.
2. Raise and support vehicle, then remove tire and wheel assembly.
3. Install two wheel retaining nuts to retain rotor in position.
4. If caliper requires overhaul, remove inlet fitting attaching bolt, then disconnect inlet fitting from caliper housing. Discard the two gaskets, then plug openings in inlet fitting and caliper to prevent loss or contamination of fluid. **Do not crimp brake hose, as this may damage internal structure of hose.**
5. Remove circlip (11) and retaining pin (7), **Fig. 1,** then pull caliper off rotor and mounting bracket. If only shoe

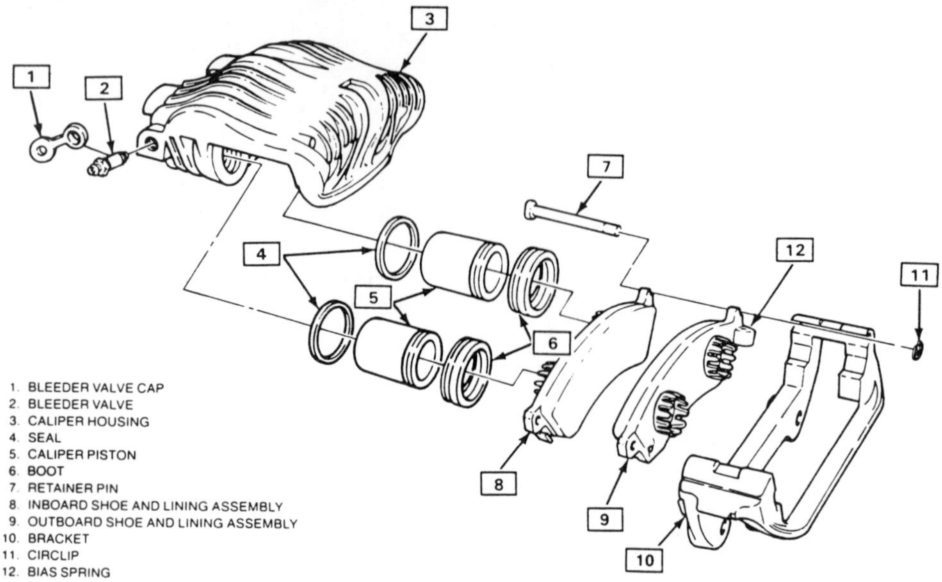

Fig. 1 Exploded view of PBR dual piston front caliper

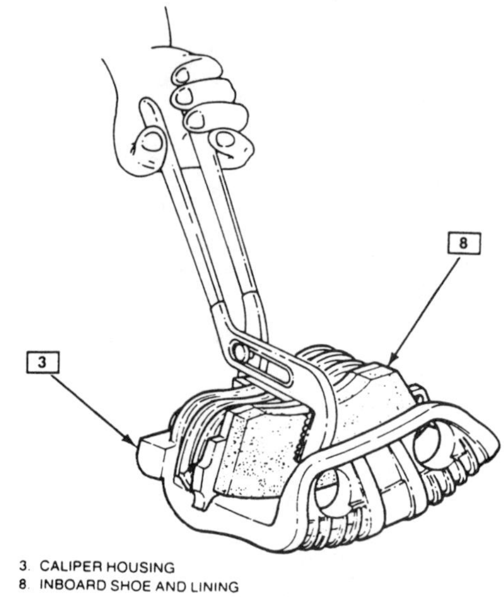

Fig. 2 Bottoming pistons in caliper bore

and linings require replacement, suspend caliper from upper control arm to prevent damage to brake hose.

BRAKE SHOE & LINING REMOVAL

1. Remove brake caliper as outlined previously.
2. Position suitable pliers over caliper and center of inboard shoe and lining as shown, **Fig. 2,** then squeeze pliers to bottom pistons in caliper bores.
3. Remove shoe and lining assemblies.

CALIPER DISASSEMBLY

1. Remove shoe and lining assemblies as previously described.
2. Pad interior of caliper housing with shop towels to prevent damage to pistons during removal.
3. Position shop towel in interior part of caliper, then slowly apply compressed air to inlet port and remove pistons. **It is imperative that one piston be partially installed to facilitate removal of second piston from bore. A pad or wooden spacer may be used to prevent complete removal of first piston.**
4. Remove dust boots from pistons.
5. Using a small piece of wood or plastic, remove piston seal from bore. **Do not use a metal tool of any kind to remove seal as it may damage bore.**
6. Remove bleeder valve cap and bleeder valve.
7. Inspect piston for scoring, nicks, corrosion, and wear and replace as needed.
8. Inspect caliper housing and seal groove for corrosion, nicks, scoring and excessive wear, and use crocus cloth to polish away corrosion from housing bore. Replace caliper housing if corrosion in and around seal groove will not clean up with crocus cloth.
9. Clean all parts with denatured alcohol. Dry with unlubricated compressed air. Blow out all passages in housing and bleeder valve.

CALIPER ASSEMBLY

1. Install cap onto bleeder valve, then install bleeder valve into caliper.
2. Using clean brake fluid, lubricate piston seals, then install seals into caliper bore grooves. Ensure seals are not twisted during installation.
3. Using clean brake fluid, lubricate caliper bores and piston assemblies.
4. Install boot over end of piston. Place piston into caliper bore, then push downward until fully bottomed in bore. Ensure boot is properly seated into groove around piston and into groove in caliper bore.
5. Repeat step 4 for remaining piston.
6. Install shoes and linings as outlined further on, then bleed brake system.

BRAKE SHOE & LINING INSTALLATION

1. Install inboard shoe and lining, ensuring tangs on shoe fully engage pistons. When properly installed, shoe should be flush with piston.
2. Install outboard shoe and lining into caliper housing, ensuring insulators are fully seated into holes in outboard side of housing.
3. Install caliper as outlined further on.

CALIPER INSTALLATION

1. Ensure guiding surfaces on shoe and lining assemblies and mounting bracket are seated correctly, then position caliper over rotor and onto mounting bracket.
2. Press caliper housing downward to compress bias springs (12), **Fig. 1,** then install new retainer pin and circlip. **Two sets of retainer pins are available for service, one for base calipers and one for heavy duty. Since the circlip grooves are cut in different positions, ensure the correct retainer pin is used.**
3. If caliper was overhauled, reconnect inlet fitting using new gaskets, then bleed brake system as required.
4. Reinstall wheel and tire assembly, then lower vehicle.
5. Fill master cylinder to proper level, then pump brake pedal firmly and slowly three times to bring pads into contact with brake rotor.

PBR Single Piston Caliper

INDEX

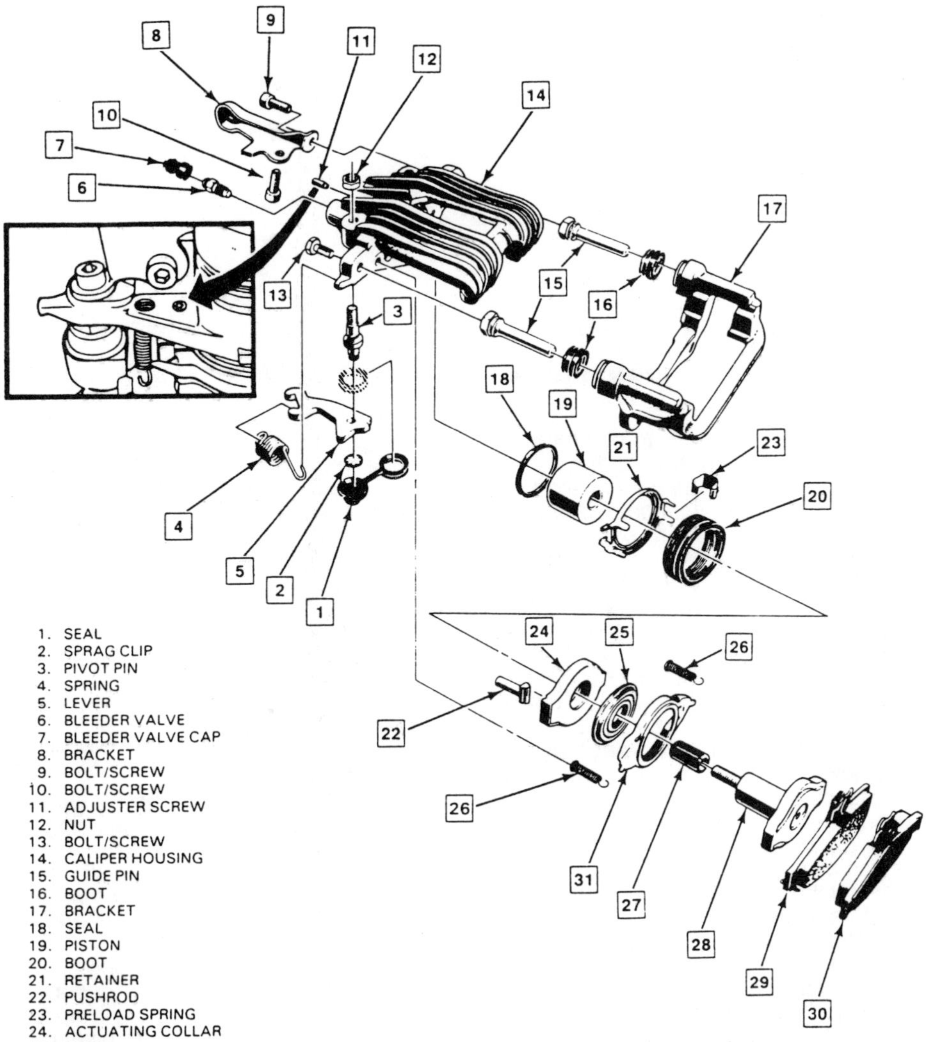

Fig. 1 Exploded view of PBR single piston rear caliper

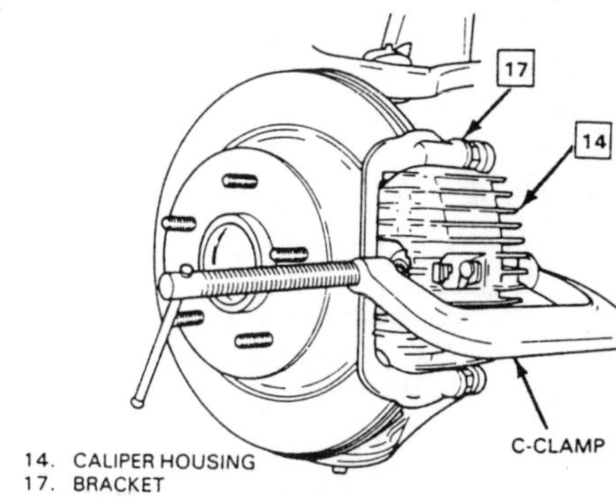

Fig. 2 Bottoming piston in caliper bore

DESCRIPTION

This rear caliper, **Fig. 1,** consists of a single piston and an aluminum housing which is suspended in a mounting bracket through two slide pins. Hydraulic pressure, created by applying force to the brake pedal, acts equally against the piston and the bottom of the caliper bore to move the piston outward. This action slides the caliper inward, resulting in a clamping action on the brake rotor. This clamping action forces the linings against the rotor, creating the friction necessary to stop the vehicle.

The parking brake mechanism on this caliper is completely independent of the hydraulic brake system. When the parking brake is applied, the lever on the caliper causes the pushrod, actuating collar and clamp rod assembly to move outward. This causes the caliper to move inward, mechanically forcing the linings against the rotor.

SERVICE

BRAKE SHOE & LINING REMOVAL

1. Remove ⅔ of the total brake fluid capacity from master cylinder reservoir.
2. Raise and support vehicle, then remove tire and wheel assembly.
3. Install two wheel retaining nuts to retain rotor in position.
4. Position one end of a suitable C-clamp against inlet fitting bolt, and the other end against outboard shoe and lining, then tighten clamp as shown, **Fig. 2,** until piston fully bottoms in caliper bore.
5. Remove upper guide pin bolt and discard.
6. Loosen lower guide pin bolt, then pivot caliper downward on the lower guide pin bolt to expose the shoe & lining assemblies. Use care to avoid damaging brake hose.
7. Remove shoes and linings from mounting bracket.

BRAKE SHOE & LINING INSTALLATION

1. Install outboard shoe and lining onto mounting bracket, ensuring insulator on shoe is positioned toward caliper housing.
2. Install inboard shoe and lining. Ensure wear sensor is positioned nearest caliper piston. When properly installed, sensor should be in trailing position

when wheel is rotated in forward direction.
3. Pivot caliper into position over shoes and linings. Ensure springs on outboard shoe do not protrude through inspection hole in housing. If protrusion is evident, lift caliper housing and readjust position of outboard shoe and lining.
4. Install new upper guide pin bolt and **torque** to 26 ft. lbs., then **torque** lower bolt to 16 ft. lbs.
5. Fill master cylinder to proper level, then pump brake pedal firmly and slowly three times to bring pads into contact with brake rotor.

CALIPER REMOVAL

1. Disable parking brake automatic adjuster as follows:
   a. Working from inside vehicle, remove driver seat cushion, then the parking brake lever cover screws and cover.
   b. Using .080 inch gauge wire, fabricate a tool to disengage drive pawl from sector, **Fig. 3.**
   c. Using tool mentioned above, disengage drive pawl from sector, then insert a nail through anchor plate to keep drive pawl in disengaged position.
   d. Pull up on lever until it aligns with pawl, then depress button until lever is fully downward.
   e. Visually inspect that anchor plate is fully against stud. If not, repeat procedure as needed.
   f. Pull front parking brake cable rearward to slacken cable at caliper assembly.
2. Raise and support vehicle, then remove tire and wheel assembly.
3. Install two wheel retaining nuts to retain rotor in position.
4. If caliper requires overhaul, remove inlet fitting attaching bolt, then disconnect inlet fitting from caliper housing. Discard the two gaskets, then plug openings in inlet fitting and caliper to prevent loss or contamination of fluid.
5. Remove caliper lever return spring. Discard spring if coils are open.
6. Disconnect parking brake cable from lever (5) and caliper bracket (8), **Fig. 1.**
7. Remove upper and lower guide pin bolts, then remove caliper from rotor and mounting bracket. If caliper does not require overhaul, suspend it from suspension to prevent damage to brake line.

CALIPER DISASSEMBLY

1. Remove caliper as previously outlined.
2. Remove the two return springs from actuating collar, then pull collar out of caliper housing and remove clamp rod (28) and bushing (27), **Fig. 1.** Discard bushing.
3. Bend back boot retainer tabs, then remove retainers (21, 31), boots (20, 25) and pushrod (22) from actuating collar. Remove preload spring (23) from retainer (31), then discard retainers and boots.
4. Use clean shop towels to pad interior of caliper assembly, then remove piston by directing compressed air into caliper brake hose inlet hole, **Use just enough air pressure to ease piston out of bore. Do not place fingers in front of piston for any reason when applying compressed air. This could result in serious personal injury.**
5. Using a small piece of wood or plastic, remove piston seal from bore. **Do not use a metal tool of any kind to remove seal as it may damage bore.**
6. Remove bleeder valve cap and bleeder valve.
7. Remove seal (1), sprag clip (2) and lever (5) from pivot pin (3). Discard sprag clip.
8. Clean all metal components with suitable solvent, then dry with compressed air.
9. Inspect parking brake lever components, piston, caliper bore and mounting bracket for scoring, excessive wear or corrosion. Replace parts as necessary.

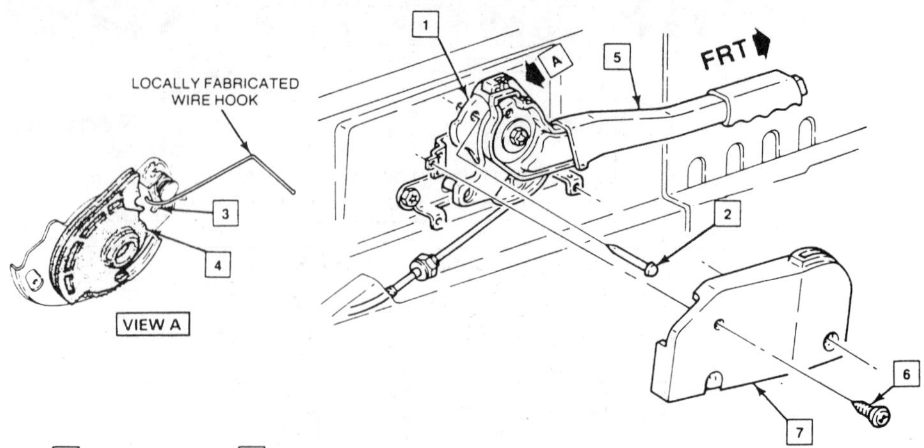

Fig. 3 Disabling parking brake automatic adjuster mechanism

CALIPER ASSEMBLY

1. Using clean brake fluid, lubricate piston seal, then install seal into caliper bore groove. Ensure seal is not twisted during installation.
2. Using clean brake fluid, lubricate caliper bore and piston.
3. Place piston into caliper bore, then push downward until fully bottomed in bore.
4. Apply lubricant provided in repair kit to actuating collar (24), then install pushrod (22), new boots (20, 25) and new retainers (21, 31) onto collar, **Fig. 1.** Clamp retainers firmly against collar, then bend tabs on retainer (21) to hold assembly together.
5. Reconnect preload spring (23) onto retainer (31).
6. Apply lubricant provided in repair kit to clamp rod (28), then slide rod through holes in boot (25) and actuating collar (24). Ensure boot is firmly positioned against reaction plate on clamp rod.
7. Lubricate new compliance bushing (27), then install bushing onto clamp rod (28).
8. Lubricate grooved bead of inner boot (20), boot groove in caliper housing and actuating collar with lubricant provided in repair kit.
9. Push clamp rod to bottom of piston mating hole, then pull actuating collar (24) and seat inner boot (20) into boot groove in caliper housing.
10. Ensure pushrod (22) is positioned in hole in caliper housing, then install bleeder cap and valve.
11. If removed, install pivot pin (3) and new nut (12) onto caliper, **torque** nut to 16 ft. lbs., then lubricate parking brake lever (5) and pivot pin.
12. Install pivot pin seal (1), parking brake lever and new sprag clip (2), ensuring teeth of sprag clip face away from lever, then snap seal cap over pivot pin.
13. Install the two collar return springs (26) onto retainer (31). Ensure retainer enters springs at end of second coil.
14. Install adjustment screw (11) into caliper housing until actuating collar is parallel to piston bore face of housing.
15. Lubricate guide pins with suitable grease, then slide boots onto pins.
16. Fill boots with grease, then install into mounting bracket. Ensure boots are properly positioned in grooves in pins and mounting bracket.
17. Install caliper and bleed brake system, then adjust parking brake free travel as outlined under "Caliper Installation."

CALIPER INSTALLATION

1. Install shoe and lining assemblies, if removed, as outlined previously.
2. Position caliper over rotor and onto

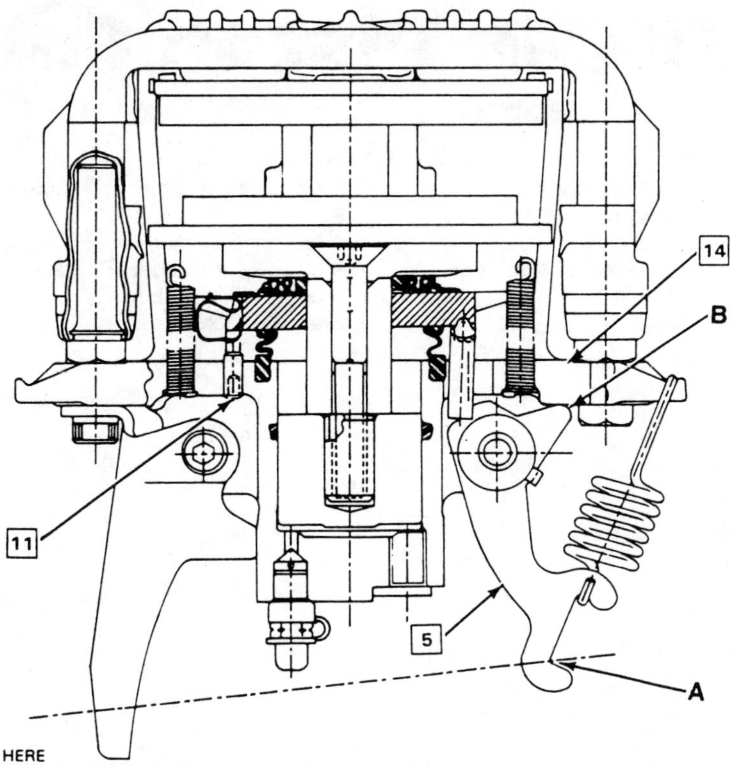

A. APPLY PRESSURE HERE
B. MEASURE HERE
5. LEVER
11. ADJUSTER SCREW
14. CALIPER HOUSING

Fig. 4 Parking brake free travel adjustment

mounting bracket, then install upper and lower guide pin bolts. **Torque** upper bolt to 26 ft. lbs. and lower bolt to 16 ft. lbs.

3. Attach parking brake cable to caliper bracket and lever, then install lever return spring.
4. If removed, install inlet fitting using new gaskets, then bleed brake system.
5. If caliper was overhauled, adjust parking brake free travel as follows:
   a. Have an assistant apply a light load to brake pedal until rotor can no longer be turned by hand.
   b. Apply pressure to caliper lever in direction shown in **Fig. 4.**
   c. Measure free travel between caliper lever and housing. Free travel should be .024-.028 inch. If free travel is not as specified, proceed to next step.
   d. Remove adjustment screw, then clean thread adhesive from threads.
   e. Coat threads with new adhesive, reinstall adjustment screw, then turn as required until specified free travel is obtained. **Turning screw clockwise increases free travel, while counterclockwise rotation decreases travel.**
   f. Release brake pedal, then firmly apply brake pedal three times and recheck free travel. Repeat adjustment procedure as necessary.
6. Remove nail from anchor plate installed during caliper removal procedure.
7. Apply and release parking brake three times, then lift lever upward and ensure parking brake fully engages at 7-9 clicks.
8. Release parking brake. No brake drag should exist and no gap between caliper housings and parking levers should be evident.
9. Install parking brake lever cover, screws and driver's cushion.
10. Reinstall tire and wheel assembly, then check and refill master cylinder as required.
11. Start engine and pump brake pedal several times to seat brake linings.

Toyota/GM Single Piston Caliper

INDEX

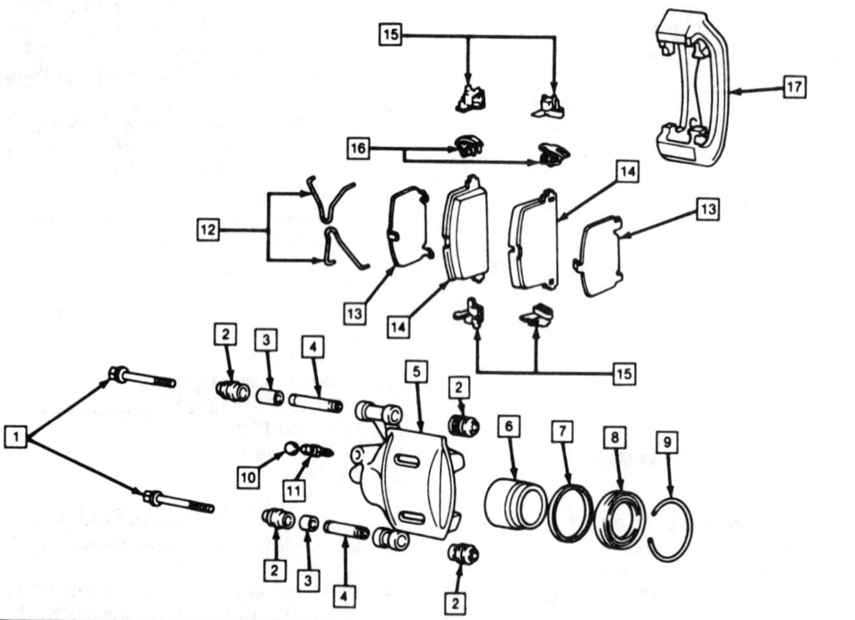

1. MOUNTING BOLT
2. DUST BOOT
3. COLLAR
4. SLIDE BUSHING
5. CALIPER HOUSING
6. PISTON
7. PISTON SEAL
8. BOOT
9. SET RING
10. CAP
11. BLEEDER SCREW
12. ANTI-RATTLE SPRING
13. ANTI-SQUEAL SHIM
14. PAD
15. PAD SUPPORT PLATE
16. PAD WEAR INDICATOR PLATE
17. MOUNTING BRACKET

Fig. 1 Toyota/GM front brake caliper exploded view

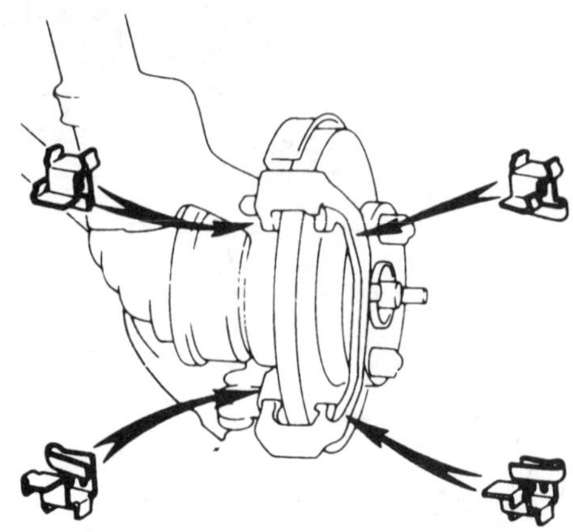

Fig. 2 Front caliper support installation

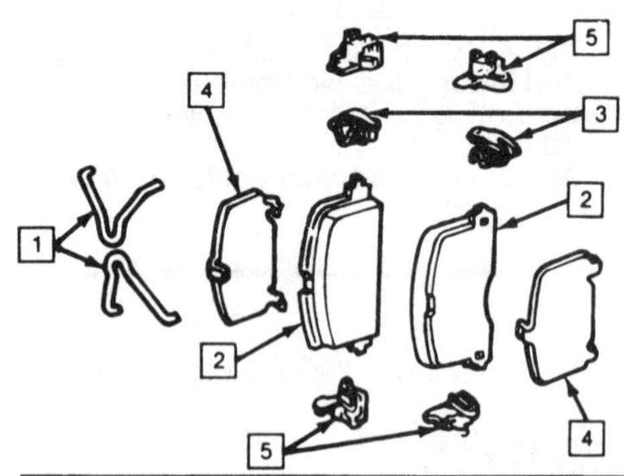

1. ANTI-RATTLE SPRINGS
2. BRAKE PADS
3. WEAR INDICATOR
4. ANTI-SQUEAL SHIMS
5. SUPPORT PLATES

Fig. 3 Front brake pad assembly

FRONT CALIPER REPLACE

REMOVAL

1. Siphon 2/3 of the brake fluid from master cylinder, raise and support vehicle, and remove wheels.
2. Reinstall 2 wheel nuts to retain rotor, then remove caliper mounting bolts, **Fig. 1.**
3. Remove union nut securing brake hose to caliper and drain fluid into suitable container. **If caliper is only being removed for brake pad replacement, do not disconnect brake hose.**
4. Compress piston as needed, then remove caliper. If brake hose remains connected, secure caliper aside to prevent hose from being stretched.

INSTALLATION

1. Seat piston in caliper bore, taking care not to damage piston.
2. Ensure support plates, **Fig. 2,** and anti-rattle springs are properly positioned, then mount caliper over rotor onto mounting bracket.
3. Install caliper mounting bolts and **torque** bolts to 18 ft. lbs.
4. Install brake hose and mounting bolts, using new copper gaskets, then **torque** bolt to 17 ft. lbs.
5. Refill master cylinder and bleed brake system.

FRONT BRAKE PADS REPLACE

Replace brake pads on one wheel at a time to prevent opposite side caliper piston from being forced out of bore.

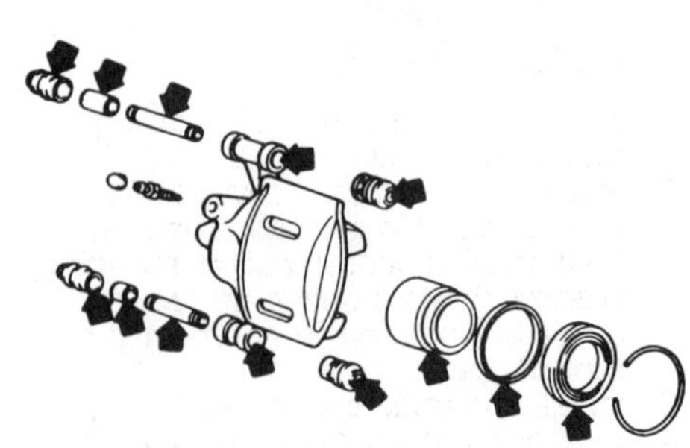

Fig. 4 Lubrication points for front caliper assembly

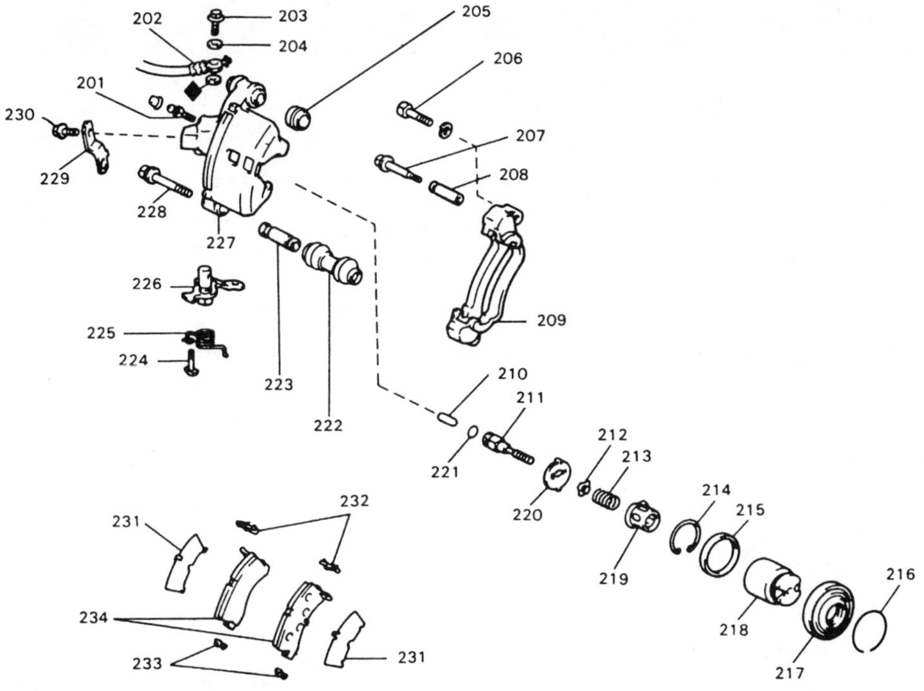

kg•cm (FT LB. N•m)

ILL NO.	DESCRIPTION	ILL NO.	DESCRIPTION
201.	BLEEDER SCREW	218.	PISTON
202.	BRAKE HOSE	219.	ADJUSTING BOLT SPRING RETAINER
203.	UNION BOLT	220.	ADJUSTING BOLT STOPPER
204.	GASKET	221.	O-RING
205.	MAIN PIN BOOT	222.	DUST BOOT
206.	MOUNTING BRACKET BOLT	223.	SLIDE BUSHING
207.	MAIN PIN	224.	STOPPER PIN
208.	SLIDING BUSHING	225.	SPRING
209.	MOUNTING BRACKET	226.	PARKING BRAKE CRANK
210.	PARKING BRAKE STRUT	227.	CALIPER HOUSING
211.	ADJUSTING BOLT	228.	MOUNTING BOLT
212.	ADJUSTING BOLT SPRING PLATE	229.	CABLE SUPPORT BRACKET
213.	SPRING	230.	CABLE SUPPORT BRACKET BOLT
214.	SNAP RING	231.	ANTI-SQUEAL SHIM
215.	PISTON SEAL	232.	ANTI-RATTLE SPRING
216.	SET RING	233.	PAD SUPPORT PLATE
217.	BOOT	234.	PAD

Fig. 5 Toyota/GM rear brake caliper exploded view

REMOVAL

1. Remove caliper as outlined, leaving brake hose connected, and secure caliper aside.
2. Remove 2 anti-rattle clips, then the brake brake pads, **Fig. 3.**
3. Remove pad wear indicator plates and anti-squeal shims.
4. Remove support plates.

INSTALLATION

1. Install new support plates on caliper mounting bracket, **Fig. 2.**
2. Install new wear indicators and anti-squeal shims on each pad, **Fig. 3,** then position pads in caliper mounting bracket. **Ensure arrow on wear indicator is pointing in rotating direction of rotor.**
3. Install anti-rattle springs.
4. Seat piston in caliper bore, then install caliper and mounting bolts, and **torque** bolts to 18. ft lbs.
5. Refill master cylinder and bleed brakes as needed.

FRONT CALIPER OVERHAUL

DISASSEMBLY

1. Remove 2 caliper slide bushings, 4 dust boots and spacer collars, **Fig. 1.**
2. Pry out caliper dust boot retaining ring and remove dust boot.
3. Place clean shop towels in caliper web to protect piston, then apply compressed air to caliper fluid inlet to force piston from bore. **Keep fingers clear of caliper web when removing piston. Use only enough air pressure to ease piston out of bore, or piston may be damaged.**
4. Remove piston seal from caliper bore, taking care not to mar machined surface of caliper.
5. Remove bleeder valve.
6. Clean components with alcohol and wipe dry with clean, lint free shop towels. Blow out caliper body and fluid passages with clean, filtered compressed air.
7. Inspect caliper and piston for damage, distortion, excessive wear and pitting, and replace as needed.
8. Replace mounting bolts, collars and caliper slides if they are damaged or worn.

ASSEMBLY

1. Apply lithium soap base glycol grease to components shown in **Fig. 4.**
2. Install piston seal in caliper, ensuring seal is squarely seated in groove.
3. Press piston into bore, ensuring piston enters bore straight.
4. Seat piston dust boot in caliper groove, then install retaining ring.
5. Install 2 collars and 4 slide bushing dust boots, **Fig. 1,** rotating boots as they are pressed in to ensure they are fully seated.
6. Install slide bushings through dust boots, ensuring boots remain seated in caliper grooves.

REAR CALIPER REPLACE

REMOVAL

1. Siphon 2/3 of the brake fluid from master cylinder, raise and support vehicle, and remove wheels.
2. Reinstall 2 wheel nuts to retain rotor.
3. Remove union nut securing brake hose to caliper and drain fluid into suitable container.
4. Remove clip from parking brake and disconnect parking brake cable.
5. Remove caliper mounting bolt, lift caliper, then remove parking brake pin clip.
6. While pushing parking brake crank, **Fig. 5,** remove pin.
7. Lift caliper and remove from vehicle.

INSTALLATION

1. Position caliper over mounting bracket, then install mounting bolt and **torque** to 14 ft. lbs.
2. Install parking brake clip.
3. Install brake hose and union bolt, using new copper gaskets, then **torque** bolt to 22 ft. lbs.
4. Refill master cylinder and bleed brake system.
5. Adjust parking brake as follows:
   a. Pull upward on parking brake lever several times, then fully release lever.
   b. Depress brake pedal several times to automatically adjust brake.
   c. When properly adjusted, parking brake crank should just contact stopper pin.

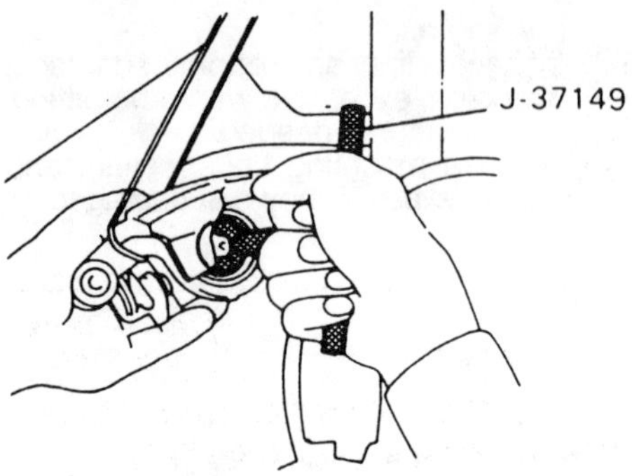

Fig. 6 Pressing piston into caliper bore

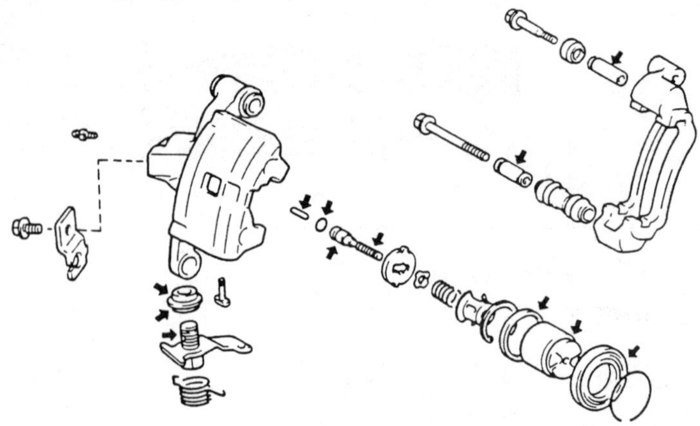

Fig. 7 Lubrication points for rear caliper assembly

REAR BRAKE PADS

REPLACE

Replace brake pads on one wheel at a time to prevent opposite side caliper piston from being forced out of bore.

REMOVAL

1. Siphon 2/3 of the brake fluid from master cylinder, raise and support vehicle, and remove wheels.
2. Reinstall 2 wheel nuts to retain rotor.
3. Remove caliper mounting bolt, then swing caliper upward to expose brake compone nts. **Do not remove main pin from caliper.**
4. Remove brake pads, anti-squeal shims, anti-rattle springs and the two support plates, **Fig. 5.**

INSTALLATION

1. Install new pad support plates to lower part of mounting bracket.
2. Install new anti-rattle springs to upper side of mounting bracket.
3. Position new anti-squeal shims to backside of each brake pad, then install pads onto mounting bracket. Ensure pads are positioned so wear indicators are on top side.
4. Using tool J-37149 or equivalent, **Fig. 6,** slowly turn piston clockwise and press into caliper bore until it locks in position at bottom of bore.
5. With pad protrusion fitted into piston stopper groove, swing caliper downward and install mounting bolt. **Torque** bolt to 14 ft. lbs.
6. Fill master cylinder to proper level, then depress brake pedal several times to seat linings and to actuate parking brake automatic adjuster.

REAR CALIPER OVERHAUL

DISASSEMBLY

1. Remove sliding bushing and the four dust boots, **Fig. 5.**
2. Using a suitable screwdriver, remove caliper set ring and boot.
3. Using tool J-37149 or equivalent, rotate piston out of caliper bore.
4. Pry piston seal from caliper bore.
5. Position tool J-37150 or equivalent over adjusting bolt, tighten tool, then remove snap ring from caliper bore and pull out adjusting bolt and components.
6. Remove parking brake strut and cable support bracket, then the parking brake crank torsion spring.
7. Remove parking brake crank, then the crank boot by tapping on metal portion of boot with screwdriver or similar tool. **Do not remove crank boot unless boot is damaged or excessively worn.**
8. Using a suitable punch, drive stopper pin from caliper.
9. Inspect all components for damage, wear or corrosion, and replace as necessary.

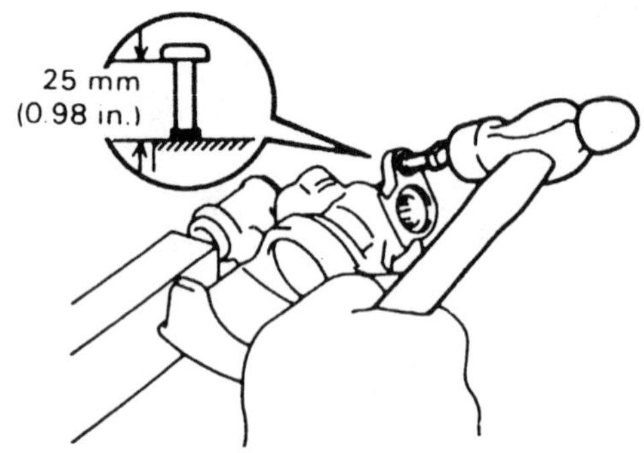

Fig. 8 Installing stopper pin

ASSEMBLY

1. Clean all parts not included in repair kit and dry with compressed air.
2. Lubricate all components indicated in **Fig. 7** with lithium based grease.
3. Install stopper pin into caliper until pin protrudes .98 inch outward as shown in **Fig. 8.**
4. If removed during disassembly, tap parking brake crank boot into position in caliper using a 24mm socket.
5. Install parking brake crank. Ensure crank boot aligns with groove of crank seal.
6. Install cable support bracket, then press surface of bracket flush against caliper and **torque** bolt to 34 ft. lbs. When properly installed, clearance between parking brake crank and support bracket should be .236 inch.
7. Install torsion spring, then check that parking brake crank subassembly touches stopper pin.
8. Install parking brake strut. **Before adjusting strut, ensure needle bearing rollers do not catch on caliper bore.**
9. Position new O-ring onto adjusting bolt, then assemble stopper, plate, spring and spring retainer onto bolt. Hand tighten assembly using tool J-27150 or equivalent. When properly assembled, inscribed surface of stopper should face upward and notches of spring retainer should align with notches of stopper.
10. Position adjusting bolt subassembly into caliper, then install snap ring, ensuring opening faces toward bleeder side of caliper. Pull upward on adjusting bolt and ensure bolt does not move.
11. Check operation of parking brake by moving crank by hand. Adjusting bolt should move smoothly, with no evidence of binding.
12. Install piston seal into caliper bore, then using tool J-37149, install piston and bottom in bore by turning tool clockwise.
13. Align center of piston stopper groove with protrusion of caliper bore.
14. Install boot and set ring into caliper.
15. Using a 21mm socket and suitable vise, press new main pin boot into caliper housing.
16. Install sliding bushing and bolt, then install caliper as outlined previously.

Disc Brake Caliper Specifications

Year	Model	Caliper Bore Dia. Inch
BUICK: REAR WHEEL DRIVE & 1985 RIVIERA		
1985	Riviera	2 1/2
	Except Century, Regal & Riviera	2 15/16
1985-87	Regal	2 1/2
1986-88	Estate Wagon	2 15/16
1989	Electra & Lasabre Wagons	2 15/16
BUICK SOMERSET REGAL, OLDS. CALAIS, PONT. GRAND AM & 1986–89 BUICK SKYLARK		
1985-89	All	2.24
CADILLAC REAR WHEEL DRIVE & 1985 ELDORADO & SEVILLE		
1985	Eldorado & Seville	④
1985-89	Except Eldorado & Seville	2 15/16
1986–89 RIVIERA, CAD. ELDORADO & SEVILLE, OLDS. TORONADO & 1988–89 BUICK REATTA		
1986-89	All	⑥
1985-89 BUICK ELECTRA, CAD. DEVILLE, FLEETWOOD & OLDS. 98; 1986-89 BUICK LESABRE & OLDS. 98; 1987-89 PONT. BONNEVILLE		
1985-89	All	2.52
CHEVROLET (CAMARO, CHEVROLET, CORVETTE & MONTE CARLO)		
1985-88	Camaro	①
1989	Camaro	⑤
1985-89	Chevrolet	2 15/16
1985-88	Monte Carlo	2.5
1985-87	Corvette	②

Year	Model	Caliper Bore Dia. Inch
CHEV. BERETTA, CAVALIER & CORSICA. BUICK SKYHAWK, CAD. 1985–88 CIMARRON, OLDS. 1985–88 FIRENZA, PONTIAC SUNBIRD		
1985-89	All	2.24
CHEVROLET CHEVETTE • PONTIAC 1000		
1985-87	All	2.05
CHEV. CITATION & 1985 BUICK SKYLARK		
1985	All	2.24
CHEV. CELEBRITY, BUICK CENTURY, OLDS. CUTLASS CIERA, PONT. 6000 & 1985-89 CUTLASS CRUISER		
1985-89	All	③
OLDSMOBILE: REAR WHEEL DRIVE & 1985 TORONADO		
1985	Except Cutlass & Toronado	2 15/16
	Cutlass & Toronado	2 1/2
1986-89	Cutlass RWD	2 1/2
	Custom Cruiser	2 15/16
PONTIAC (EXCEPT 1000 & FRONT WHEEL DRIVE)		
1985-89	Except Firebird & Fiero	2.9375
1985-88	Firebird	①
1989	Firebird	⑥
1985-87	Fiero	2.52

①—Front, 2.5 inches; rear, 1.89 inches.
②—Front, 2.1 inches; rear, 1.6 inches.
③—Except heavy duty, 2.24 inches; heavy duty, 2.50 inches.
④—Piston diameter: front, 2.5 inches; rear, 2.125 inches.
⑤—Front standard, 2.5 inches; front heavy duty, 1.5 inches; rear, 1.6 inches.
⑥—Front, 2.5 inches; rear, 1.5 inches.

Disc Brake Rotor Specifications

Car	Year	Nominal Thickness	Minimum Refinish Thickness ①	Thickness Variation (Parallelism)	Lateral Runout (T.I.R.)	Finish (Microinch)
BUICK: REAR WHEEL DRIVE & 1985 RIVIERA						
All	1985-89	1.043	.980	.0005	.005	—
1986–89 BUICK RIVIERA, CAD. ELDORADO & SEVILLE, OLDS. TORONADO & 1988–89 BUICK REATTA						
All②	1986-89	1.035	.971	.0005	.004	10-50
All③	1986-89	.494	.444	.0005	.003	10-50
BUICK SOMERSET, OLDS. CALAIS, PONT. GRAND AM & 1986-89 BUICK SKYLARK						
All	1985-89	.885	.830	.0005	.004	—
1985-89 BUICK ELECTRA & PARK AVE., CAD. DEVILLE & FLEETWOOD & OLDS. 98; 1986-89 BUICK LESABRE & OLDS. 88; 1987-89 PONT. BONNEVILLE						
All	1985-88	1.043	.980	.0005	.004	10-50
	1989	1.043	.972	.0005	.004	—
1988–89 BUICK REGAL, OLDS CUTLASS SUPREME & PONT. GRAND PRIX						
All②	1988	1.040	1.019	.0005	.003	—
	1989	1.040	.984	.0005	.004	—
All③	1988	.492	.426	.0005	.003	—
	1989	.492	.441	.0005	.004	
CADILLAC: REAR WHEEL DRIVE & 1985 ELDORADO & SEVILLE						
Eldorado & Seville	1985	1.035	.965	.0005	.004	—
Brougham	1985-89	1.031	.972	.0005	.004	—
CHEVROLET (CAMARO, CHEVROLET, CORVETTE & MONTE CARLO)						
Camaro②	1985-89	1.043	.980	.0005	.004	30-80
Camaro③	1985-88	1.042	.986	.0005	.005	40-80
	1989	.795	.744	—	.005	—
Corvette	1985-87	.780	.724	.0005	.006	—
	1988–89④	.795	.744	.0005	.006	—
	1988–89⑤	1.110	1.059	.0005	.006	—
Except Camaro & Corvette	1985-89	1.043	.980	.0005	.004	30-80
CHEV. BERETTA, CAVALIER & CORSICA. BUICK SKYHAWK, CAD. 1985–88 CIMARRON, OLDS. 1985–88 FIRENZA, PONT. SUNBIRD						
All	1985-89	.885	.830	.0005	.002	—
CHEVROLET CHEVETTE, PONTIAC 1000						
All	1985-87	.4334	.390	.0005	.005	20-60
CHEV. CELEBRITY & 1985 CITATION; BUICK CENTURY & 1985 SKYLARK; OLDS. CUTLASS CRUISER, CUTLASS CIERA; PONT. 6000						
All	1985-88	.885	.830	.0005	.004	12-50
Except 6000 STE	1989⑥	.885	.830	.0005	.004	—
	1989⑦	1.043	.972	.0005	.004	—
6000STE	1989③	.756	.744	.0005	.005	—
1985–88 CHEVROLET SPECTRUM & SPRINT; 1989 GEO SPECTRUM & METRO						
Spectrum	1985–89	.433	.378	—	.006	—
Sprint & Metro	1985–89	.394	.315	—	.003	—
CHEVROLET NOVA						
All②	1985-88	.531	.492	—	.006	—
All③	1988	.354	.315	—	.006	—
PONTIAC LEMANS						
All	1988–89	.500	.420	.0004	.004	—

Continued

DISC BRAKE ROTOR SPECIFICATIONS—Continued

OLDSMOBILE: REAR WHEEL DRIVE & 1985 TORONADO						
Except Cutlass & Toronado	1985-88	1.035	1.020	.0005	.004	—
	1989	1.043	.980	.0005	.004	—
Cutlass	1985-88	1.020	.984	.0005	.004	—
Toronado	1985	1.020	.984	.0005	.004	19-80
PONTIAC (EXCEPT 1000 & FRONT WHEEL DRIVE)						
Fiero ②	1985-87	.445	.386	.0005	.004	—
Fiero ③	1985-87	.500	.440	.0005	.004	—
Fiero ② ③	1988	.756	.702	.0005	.003	—
Firebird ②	1985-89	1.043	.980	.0005	.004	30-80
Firebird ③	1985-88	1.042	.986	.0005	.005	40-80
	1989	.795	.744	.0005	.005	—
Grand Prix	1985-87	1.043	.980	.0005	.005	30-80
Parisienne	1985-86	1.043	.980	.0005	.004	—
Safari	1987-89	1.043	.980	.0005	.004	40-80

①—All brake rotors have a discard dimension cast into them. This is a wear dimension, not a refinish dimension. Any rotor that does not meet specifications should be discarded.

②—Front.

③—Rear.

④—Except heavy duty front rotors.

⑤—Heavy duty front rotors.

⑥—Light duty.

⑦—Medium and heavy duty.

DRUM BRAKES

INDEX

CHRYSLER MOTORS/EAGLE

TABLE OF CONTENTS

Applications

	Type No.
Chrysler Corp. Front Wheel Drive Models	2
Chrysler Corp. Rear Wheel Drive Models	1
Eagle, Except Medallion & Premier	3
Eagle Premier	4
Colt & Eagle Summit	5
Colt Vista	6
Eagle Medallion	7

Troubleshooting

Refer to "Troubleshooting" in the "Disc Brake" section for general disc and drum brake troubleshooting procedures.

General Information

INDEX

SERVICE PRECAUTIONS

When working on or around brake assemblies, care must be taken to prevent breathing asbestos dust, as many manufacturers incorporate asbestos fibers in the production of brake linings. During routine service operations, the amount of asbestos dust from brake lining wear is at a low level due to a chemical breakdown during use. A few precautions will minimize exposure.

Do not sand or grind brake linings unless suitable local exhaust ventilation equipment is used to prevent excessive asbestos exposure.

1. Wear a suitable respirator approved for asbestos dust use during all repair procedures.
2. When cleaning brake dust from brake parts, use a vacuum cleaner with a highly efficient filter system. If a suitable vacuum cleaner is not available, use a water soaked rag. **Do not use compressed air or dry brush to clean brake parts.**
3. Keep work area clean using same equipment as for cleaning brake parts.
4. Properly dispose of rags and vacuum cleaner bags by placing them in plastic bags.
5. Do not smoke or eat while working on brake systems.
6. Never use any fluid containing mineral oil to clean brake system components. This will damage the rubber caps and seals. If system contamination is suspected, check brake fluid in the reservoir for dirt, discoloration, or separation (breakdown) of the brake fluid into distinct layers. Drain and flush the hydraulic system with clean brake fluid if contamination is suspected.

GENERAL INSPECTION

BRAKE DRUMS

Any time the brake drums are removed for brake service, the braking surface diameter should be checked with a suitable brake drum micrometer at several points to determine if they are within the safe oversize limit stamped on the brake drum outer surface. If the braking surface diameter exceeds specifications, the drum must be replaced. If the braking surface diameter is within specifications, drums should be cleaned and inspected for cracks, scores, deep grooves, taper, out of round and heat spotting. If drums are cracked or heat spotted, they must be replaced. Scoring and grooves in the braking surface can only be removed by machining with special equipment, as long as the braking surface is within specifications. Any brake drum showing taper or sufficiently out of round to cause vehicle vibration or noise while braking should also be machined, removing only enough stock to true up the drum.

After a brake drum is machined, wipe the braking surface diameter with a denatured alcohol soaked cloth. If one brake drum is machined, the other should also be machined to the same diameter to maintain equal braking forces.

BRAKE LININGS & SPRINGS

Inspect brake linings for excessive wear, damage, oil, grease or brake fluid contamination. If any of the above conditions exist, brake linings should be replaced as an axle set to maintain equal braking forces. Examine brake shoe webbing, hold-down and return springs for signs of overheating indicated by a slight blue color. Any component which exhibits overheating signs should be replaced. Overheated springs lose their pull and could cause brake linings to wear out prematurely. Inspect all springs for sags, bends and external damage and replace as necessary.

Inspect hold-down retainers and pins for bends, rust and corrosion. If any of the above is found, replace as required.

BACKING PLATE

Inspect backing plate shoe contact surface for grooves that may restrict shoe movement and cannot be removed by lightly sanding with emery cloth or other suitable abrasive. If backing plate exhibits above condition, it should be replaced. Also inspect for signs of cracks, warpage and excessive rust, indicating need for replacement.

ADJUSTER MECHANISM

Inspect all components for rust, corrosion, bends and fatigue. Replace as necessary. On adjuster mechanism equipped with adjuster cable, inspect cable for kinks, fraying or elongation of eyelet and replace as necessary.

PARKING BRAKE CABLE

Inspect parking brake cable end for kinks, fraying and elongation and replace as necessary. Use a small hose clamp to compress clamp where it enters backing plate to remove.

Type 1

INDEX

REMOVAL

1. Raise and support rear of vehicle.
2. Remove tire and wheel assembly, then remove brake drum. If brake lining is dragging on brake drum, back off brake adjustment by rotating adjustment screw. **If brake drum is rusted or corroded to axle flange and cannot be removed readily, lightly tap axle flange to drum mounting surface with a suitable hammer.**
3. Using brake spring pliers or equivalent, remove primary and secondary shoe return springs, **Fig. 1.**
4. Remove automatic adjuster cable from anchor plate and unhook from adjuster lever.
5. Remove adjuster cable, overload spring, cable guide and anchor plate.
6. Unhook adjuster lever spring from lever and remove spring and lever.
7. Remove shoe to shoe spring from secondary shoe web, then the primary shoe.
8. Spread shoes apart and remove parking brake strut and spring.
9. Using suitable tool, remove shoe retainers, then the springs and nails.
10. Disconnect parking brake cable from lever and remove brake shoes.
11. Remove parking brake lever from secondary shoe.
12. Clean dirt from brake drum, backing plate and all other components. **Do not use compressed air or dry brush to clean brake parts. Many brake parts contain asbestos fibers which, if inhaled, can cause serious injury. To clean brake parts, use a water soaked rag or a suitable vacuum cleaner to minimize airborne dust.**

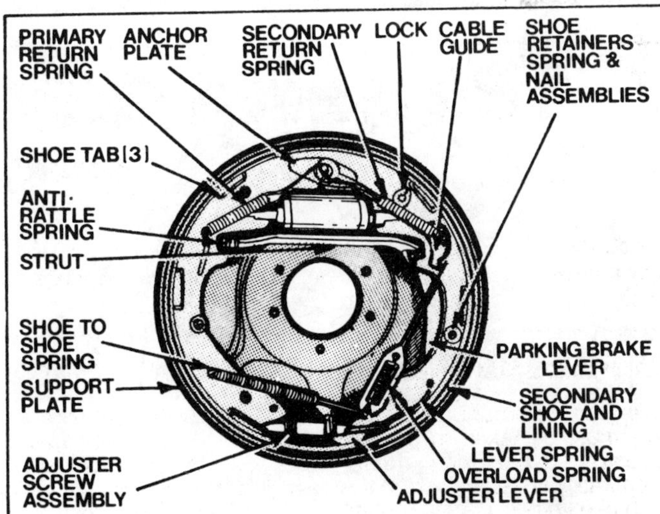

Fig. 1 Rear drum brake assembly. Type 1

INSPECTION

Refer to "General Information" at the front of this chapter.

INSTALLATION

1. Lubricate parking brake lever fulcrum with suitable brake lube, then attach lever to secondary brake shoe. Ensure the lever operates smoothly.
2. Lightly lubricate backing plate shoe contact surfaces with suitable brake lube.
3. Connect parking brake lever to cable and slide secondary brake shoe into position.
4. Connect wheel cylinder link to brake shoe.
5. Slide parking brake lever strut behind axle flange and into parking brake lever slot, then place parking brake anti-rattle spring over strut.
6. Position primary brake shoe on backing plate, then connect wheel cylinder link and parking brake strut.
7. Install anchor plate and position adjuster cable eye over anchor pin.
8. Install primary shoe return spring using brake spring pliers or equivalent.
9. Place protruding hole rim of cable guide in secondary shoe web hole, then holding guide in position, install secondary shoe return spring through cable guide and secondary shoe. Install spring on anchor pin using brake spring pliers or equivalent. **Ensure cable guide remains flat against secondary shoe web during and after return spring installation. Also ensure secondary spring end overlaps primary spring end on anchor pin.**
10. Using suitable pliers, squeeze spring ends around anchor pin until parallel.
11. Install adjuster screw assembly between primary and secondary brake shoes with star wheel on secondary shoe side. **The left side adjuster assembly stud is stamped "L" and is cadmium-plated. The right side adjuster assembly is not stamped and is colored black.**
12. Install shoe to shoe spring, then position adjusting lever spring over pivot pin on shoe web.
13. Install adjusting lever under spring and over pivot pin, then slide lever slightly rearward.
14. Install nails, springs and retainers.
15. Thread adjuster cable over guide and hook end of overload spring in lever. Ensure eye of cable is pulled tight against anchor and in a straight line with guide.
16. Install brake drum, tire and wheel assembly.
17. Adjust brakes.
18. If any hydraulic connections have been opened, bleed brake system.
19. Check master cylinder fluid level, and replenish as necessary.
20. Check brake pedal for proper feel and return.
21. Lower vehicle and road test. **Do not severely apply brakes immediately after installation of new brake linings or permanent damage may occur to linings, and/or brake drums may become scored. Brakes must be used moderately during first few hundred miles of operation to ensure proper burnishing of linings.**

Type 2

INDEX

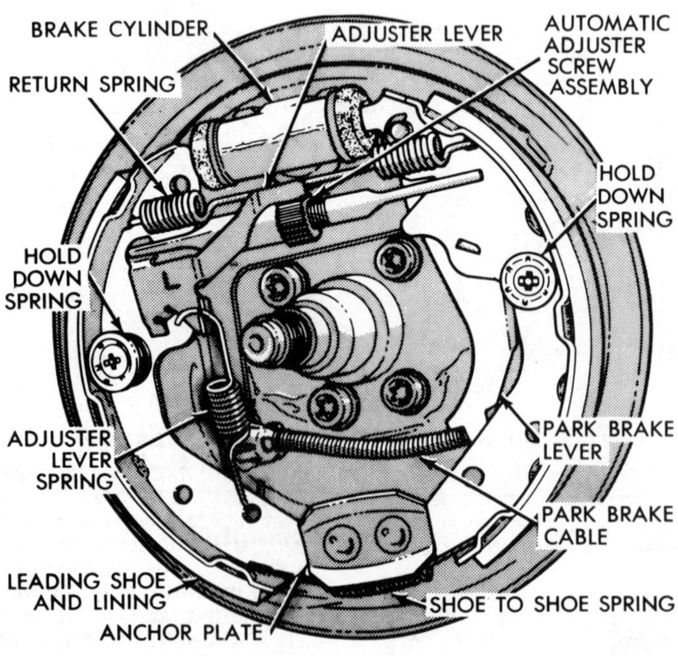

Fig. 1 Rear drum brake assembly. Type 2

REMOVAL

1. Raise and support rear of vehicle.
2. Remove tire and wheel assembly, then remove brake drum. If brake lining is dragging on brake drum, back off brake adjustment by rotating adjustment screw.
3. Using suitable pliers, remove adjuster lever spring, **Fig. 1.**
4. Remove adjuster lever.
5. Turn automatic adjuster screw out to expand shoes past wheel cylinder boot.
6. Using suitable tool, remove hold-down springs.
7. Pull brake shoe assembly down and away from anchor plate.
8. Remove C-clip retaining parking brake lever to trailing brake shoe webbing.
9. Disassemble shoe assembly.
10. Clean dirt from brake drum, anchor plate and all other components. **Do not use compressed air or dry brush to clean brake parts. Many brake parts contain asbestos fibers, which, if inhaled, can cause serious injury. To clean brake parts, use a water soaked rag or a suitable vacuum cleaner to minimize airborne dust.**

INSPECTION

Refer to "General Information" at the front of this chapter.

INSTALLATION

1. Lightly lubricate anchor plate shoe contact surfaces with suitable brake lube.
2. Remove brake drum hub grease seal and bearings, then clean and repack bearings and reinstall. Install new grease seal.
3. Assemble automatic adjuster screw assembly, return spring and shoe-to-shoe spring to brake shoe assembly.
4. Position lining assembly near anchor plate, then assemble parking brake lever to trailing shoe webbing. Secure with C-clip.
5. Install lining assembly onto anchor plate. When positioned, back off adjuster nut to seat brake shoe ends in wheel cylinder.
6. Install hold-down springs.
7. Position adjuster lever, then using suitable pliers, install adjuster lever spring.
8. Install brake drum and bearings.
9. Adjust brakes.
10. Install tire and wheel assembly.
11. If any hydraulic connections have been opened, bleed brake system.
12. Check master cylinder level, and replenish as necessary.
13. Check brake pedal for proper feel and return.
14. Lower vehicle and road test. **Do not severely apply brakes immediately after installation of new brake linings or permanent damage may occur to linings and/or brake drums may become scored. Brakes must be used moderately during first several hundred miles of operation to ensure proper burnishing.**

Type 3

INDEX

REMOVAL

1. Raise and support rear of vehicle.
2. Remove tire and wheel assembly, then the brake drum. If brake lining is dragging on brake drum, back off brake adjustment by rotating adjustment screw. Refer to individual car chapter for procedure. If brake drum is rusted or corroded to axle flange and cannot be removed, lightly tap axle flange to drum mounting surface with a suitable hammer.
3. Install suitable wheel cylinder clamp over ends of wheel cylinder to retain pistons in bore.
4. Remove parking brake lever retaining clip, **Fig. 1.**
5. Remove adjuster lever spring, primary and secondary shoe return springs using a suitable pair of brake spring pliers.
6. Remove shoe guide plate (if equipped), adjuster cable and guide plate.
7. Using suitable tool, compress hold-down springs, then remove spring retainers, hold down springs and pins.
8. Separate springs and remove from backing plate.
9. Remove parking brake lever from cable.
10. Separate all components from brake shoes.
11. Clean dirt from brake drum, backing plate and all other components. **Do not use compressed air or dry brush to clean brake parts. Many brake parts contain asbestos fibers which, if inhaled, can cause serious injury. Clean brake parts with a water soaked rag or a suitable vacuum cleaner to minimize airborne dust.**

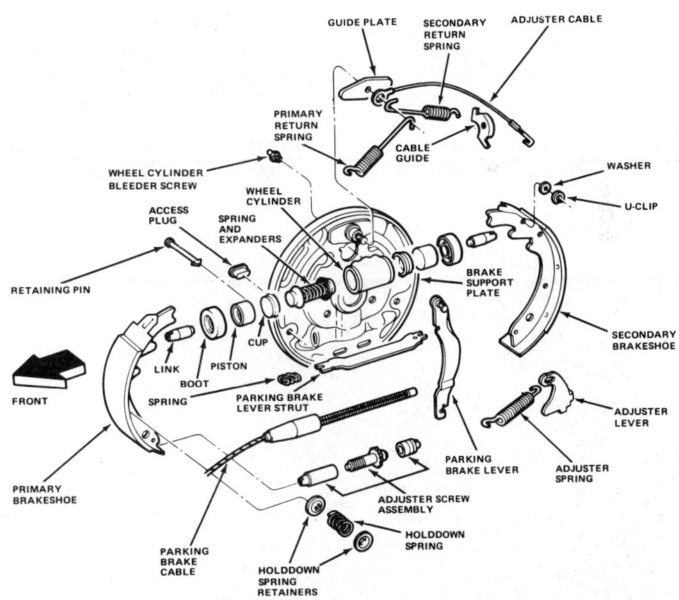

Fig. 1 Rear drum brake assembly. Type 3

INSPECTION

1. Inspect components for damage and unusual wear. Replace as necessary.
2. Inspect wheel cylinders. Boots which are torn, cut, or heat damaged indicate need for wheel cylinder replacement. Fluid spilling from boot center hole, or wetness around wheel cylinder ends indicates cup leakage and need for wheel cylinder replacement. **A small amount of fluid is always present and is considered normal, acting as a lubricant for the cylinder pistons.**
3. Inspect backing plate for evidence of seal leakage. If leakage exists, refer to individual car chapters for axle seal replacement procedure.

4. Inspect backing plate attaching bolts and ensure they are tight.
5. Check adjuster screw operation. If satisfactory, lightly lubricate adjusting screw and washer with suitable brake lube. If operation is unsatisfactory, replace.
6. Using fine emery cloth or other suitable abrasive, clean rust and dirt from shoe contact surfaces on backing plate.

INSTALLATION

1. Lightly lubricate backing plate shoe contact surfaces with suitable brake lube.
2. Assemble parking brake lever to secondary shoe and secure with spring washer and retaining clip. Crimp ends of clip with suitable pliers.
3. Position brake shoes on backing plate, primary (short lining) shoe facing front of vehicle and secondary (long lining) facing rear. Secure brake shoes with hold down springs, pins and retainers.
4. Install parking brake link and spring between shoes.
5. Loosen parking brake adjustment nut, then install parking brake cable on parking brake lever.
6. Install shoe guide plate and adjuster cable eyelet on anchor. Ensure adjuster cable crimp faces out.
7. Ensure parking brake link is properly positioned between brake shoes and wheel cylinder links are engaged in shoe web.
8. Using suitable brake spring pliers, install primary return spring from brake shoe to anchor, then secondary return spring from brake shoe to anchor.
9. Remove wheel cylinder clamp installed during removal of brake shoes.

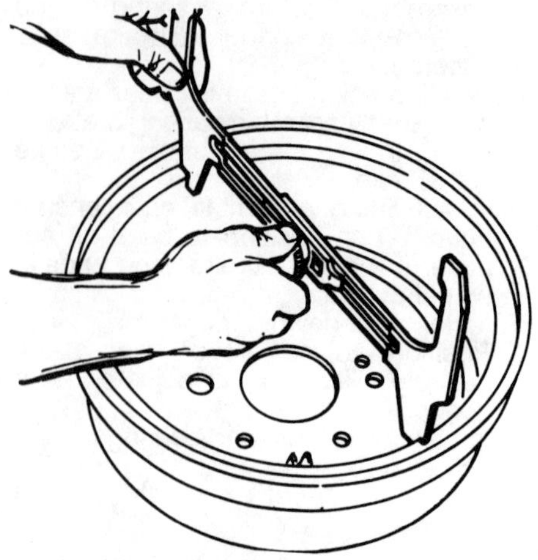

Fig. 2 Measuring brake drum inside diameter

10. Tighten adjuster screw assembly to thread limit and back off one-half turn.
11. Install adjuster screw assembly between shoes. Ensure toothed wheel is on secondary shoe side. **Adjuster screw assemblies are stamped R (right) and L (left). To ensure proper adjuster operation, they must be installed on their respective sides.**
12. Hook adjuster cable hook into adjuster lever hole, then position adjuster spring hook in large hole in primary shoe web. Using suitable brake spring pliers, install adjuster spring in adjuster lever hole.
13. Ensure adjuster cable is properly seated in cable guide, then pull adjuster lever, cable and adjuster spring down and towards the rear, engaging lever pivot hook in the large hole of secondary shoe web.
14. After installation, check adjuster operation by pulling adjuster cable between cable guide and adjuster lever towards secondary shoe sufficiently to lift adjuster lever past one tooth on adjuster screw assembly. The adjuster lever should snap into position behind the next tooth, then upon release of adjuster cable, rotate toothed wheel one notch. If operation is not satisfactory, recheck installation.
15. Ensure brake shoe upper ends are seated against anchor pin and shoe assemblies are centered on backing plate. If not, back off parking brake adjustment.
16. Using suitable brake drum to shoe gauge, **Fig. 2,** measure brake drum inside diameter. Adjust brake shoes to dimension obtained on outside portion of gauge using adjuster screw.
17. Install brake drum, wheel and tire assembly.
18. If any hydraulic brake connections have been opened, bleed brake system.
19. Adjust parking brake.
20. Inspect all hydraulic lines and connections for leakage and repair as necessary.
21. Check master cylinder fluid level and replenish as necessary.
22. Check brake pedal for proper feel and return.
23. Lower vehicle and road test. **Do not severely apply brakes immediately after installation of new brake linings or permanent damage may occur to linings, and/or brake drums may become scored. Brakes must be used moderately during first several hundred miles of operation to ensure proper burnishing of linings.**

Type 4

INDEX

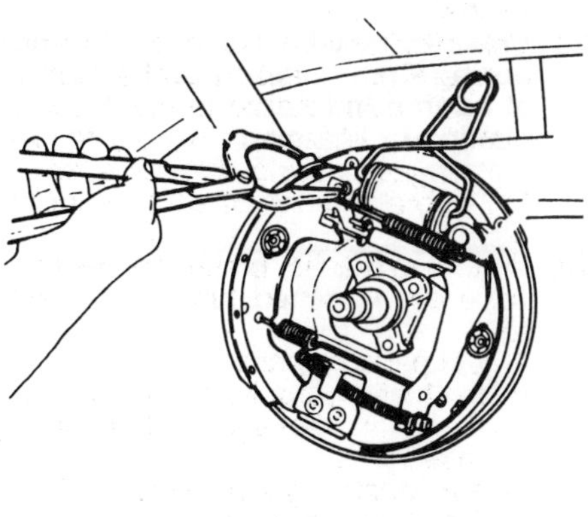

Fig. 1 Removing upper & lower return springs

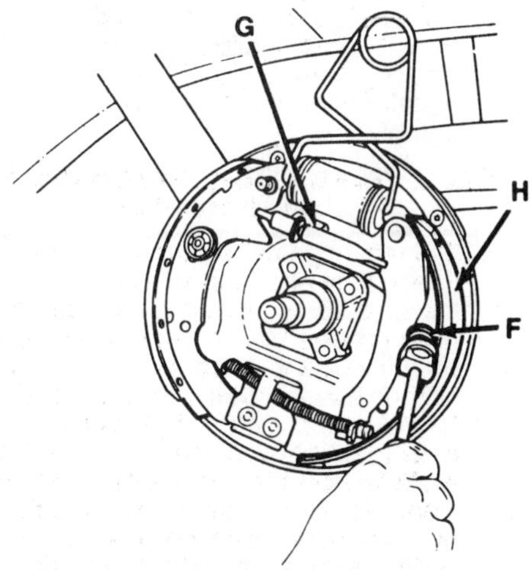

Fig. 2 Removing primary shoe, hold-down spring & self-adjusting lever

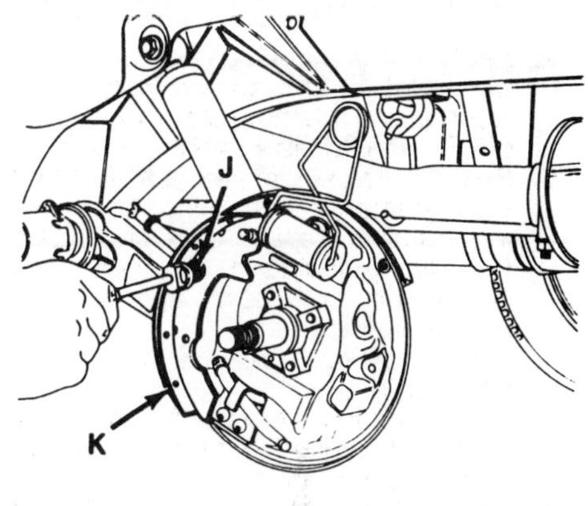

Fig. 3 Removing secondary shoe & hold-down spring

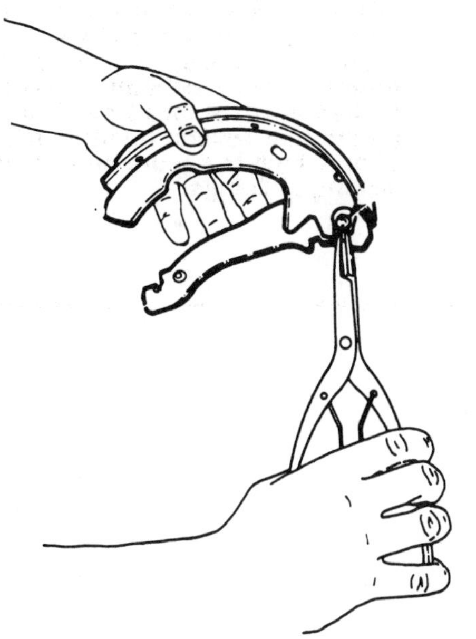

Fig. 4 Separating park brake lever from primary shoe

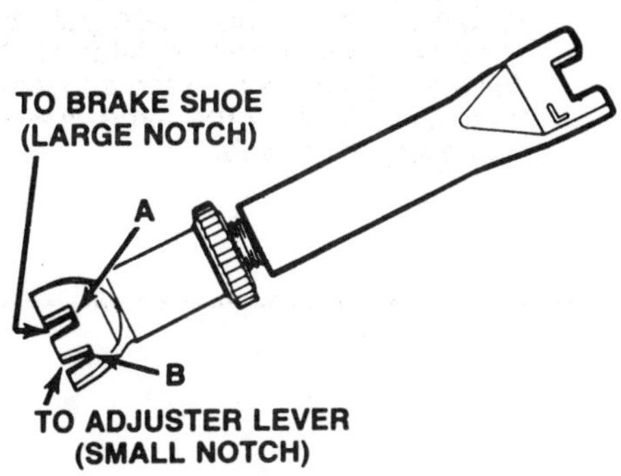

Fig. 5 Self-adjusting screw notch locations

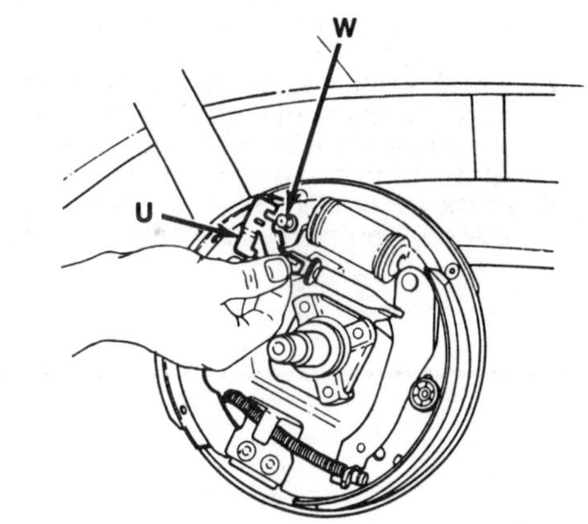

Fig. 6 Installing adjuster lever

REMOVAL

1. Raise and support vehicle.
2. Remove rear wheel and tire assembly.
3. Remove brake drum. If brake lining is dragging on brake drum, remove access plug from backing plate and back off brake adjustment by rotating adjustment screw.
4. Install suitable wheel cylinder clamp over ends of wheel cylinder to retain pistons in bore.
5. Remove upper and lower return springs, **Fig. 1**, using suitable brake spring pliers.
6. Remove adjuster lever, then the parking brake cable from parking brake lever.
7. Using suitable tool, compress and remove rear hold-down spring (F), then remove self-adjusting screw (G) and primary brake shoe (H), **Fig. 2**.
8. Remove front hold-down spring (J), then the secondary brake shoe (K), **Fig 3**.
9. Using snap ring pliers, remove clip from pivot pin on primary shoe, **Fig 4**, then separate park brake lever from shoe.
10. Clean dirt from backing plate, brake drum and hardware.

INSPECTION

1. Inspect components for damage and unusual wear. Replace as necessary.
2. Inspect wheel cylinders. Boots which are torn, cut, or heat damaged indicate need for wheel cylinder replacement. Fluid spilling from boot center hole, or wetness around wheel cylinder ends indicates cup leakage and need for wheel cylinder replacement. **A small amount of fluid is always present and is considered normal, acting as a lubricant for the cylinder pistons.**

3. Inspect backing plate for evidence of seal leakage. If leakage exists, refer to individual car chapters for axle seal replacement procedure.
4. Inspect backing plate attaching bolts and ensure they are tight.
5. Check adjuster screw operation. If satisfactory, lightly lubricate adjusting screw and washer with suitable brake lube. If operation is unsatisfactory, replace.
6. Using fine emery cloth or other suitable abrasive, clean rust and dirt from shoe contact surfaces on backing plate. Apply brake lube to contact surfaces and to park brake lever pivot pin.

INSTALLATION

1. Install park brake lever on primary brake shoe, compressing clip around pivot pin.
2. Using brake spring pliers, install park brake cable to park brake lever.
3. Install primary brake shoe, hold-down spring and self-adjusting screw.
4. Install secondary brake shoe and hold-down spring. **Ensure shoe is installed into larger (A) of two notches in adjuster lever, Fig. 5.**
5. Remove wheel cylinder clamp, then install adjuster lever (U) to pin (W), **Fig. 6**, and into self-adjusting screw. **Ensure lever is installed into smaller (B) of two notches in adjuster lever, Fig. 5.**
6. Install upper and lower return springs.
7. Ensure brake shoes are centered on backing plate.
8. Install brake drum, wheel and tire assembly.
9. If any hydraulic brake connections have been opened, bleed brake system.
10. Adjust parking brake.
11. Inspect all hydraulic lines and connections for leakage and repair as necessary.
12. Check master cylinder fluid level and replenish as necessary.
13. Check brake pedal for proper feel and return.
14. Lower vehicle and road test. **Do not severely apply brakes immediately after installation of new brake linings or permanent damage may occur to linings, and/or brake drums may become scored. Brakes must be used moderately during first several hundred miles of operation to ensure proper burnishing of linings.**

Types 5 & 6

INDEX

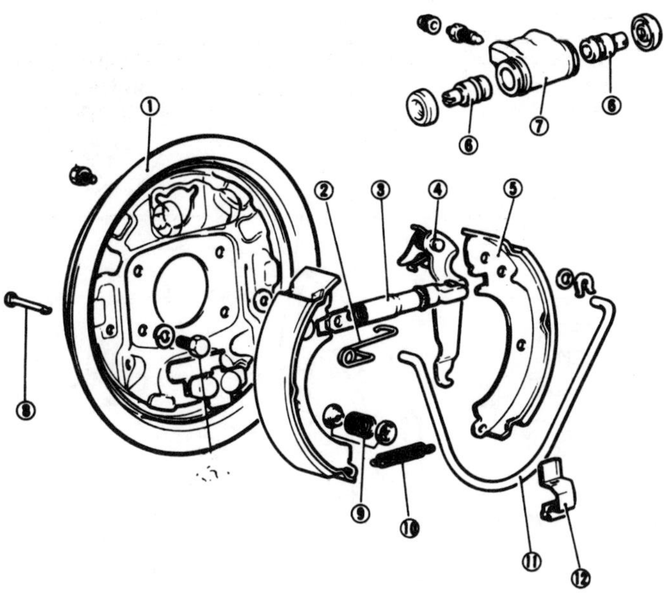

(1) Backing plate
(2) Spring
(3) Adjuster
(4) Parking lever
(5) Shoe and lining assembly
(6) Piston
(7) Wheel cylinder body
(8) Shoe hold spring pin
(9) Shoe hold-down spring
(10) Shoe to shoe spring
(11) Shoe return spring
(12) Clip spring

Fig. 1 Disassembled view of rear drum brake assembly. Type 5, Colt & Eagle Summit

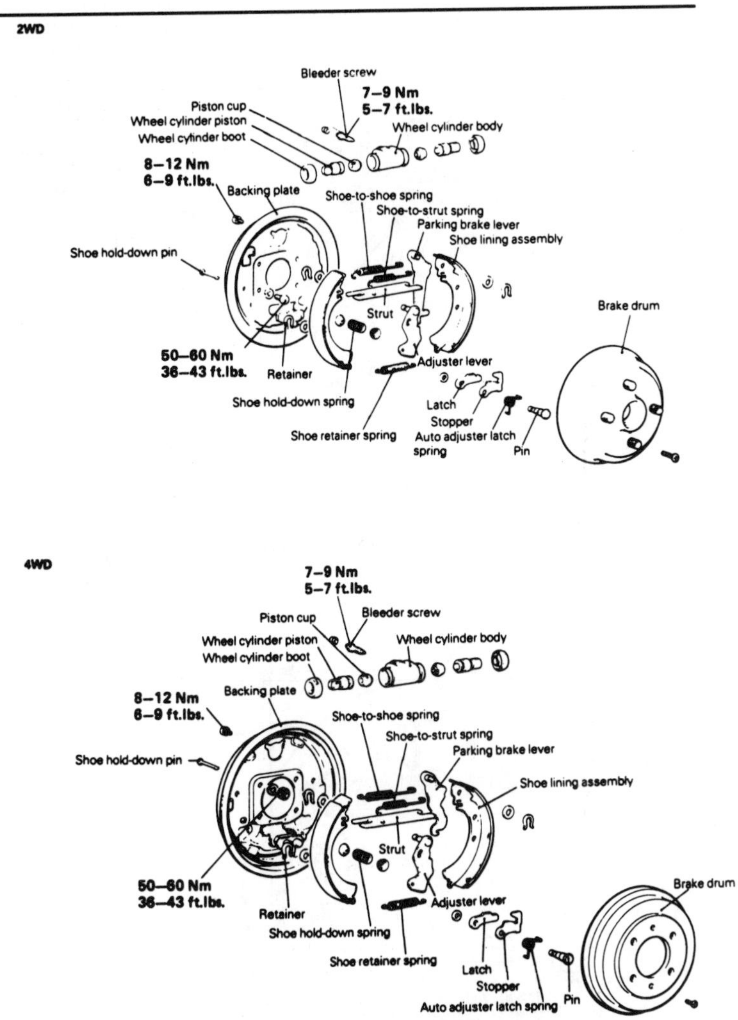

Fig. 2 Disassembled view of rear drum brake assembly. Type 6, Colt Vista

BRAKE SERVICE

For drum brake service procedures, refer to **Figs. 1 and 2.**

ADJUSTMENTS

SERVICE BRAKES, ADJUST

These brakes are equipped with self adjusting mechanisms, therefore periodic adjustments are not necessary. If stopping power is insufficient, or if brake pedal travel is excessive, brakes should be cleaned and inspected, and the self-adjusting mechanisms should be checked.

After performing brake service, adjust brake shoes as follows:

1. Ensure shoes are centered on backing plate and measure width of brake shoes using suitable caliper. On Colt Vista models, also measure inside diameter of brake drum.
2. Adjust width of brake shoes to approximately 7.06 inches for Colt. On Colt Vista, adjust brake shoes to a width approximately .060 inch less than inside diameter of brake drum.
3. Install brake drums and adjust parking brake as outlined.
4. After adjusting parking brake, release parking brake lever and ensure shoe actuating lever is not being pulled by parking brake cable. **If shoe actuating lever is pulled by parking brake cable, self-adjusters will not operate.**

PARKING BRAKE, ADJUST

1. Apply parking brake lever with a force of approximately 45 lbs. while counting number of clicks. Lever should click 5-7 times.
2. If not within specifications, release parking brake lever and remove center console, if equipped.
3. Loosen adjusting nut on parking brake lever to free parking brake cables, then depress brake pedal several times to ensure shoe to drum clearance is properly maintained by self-adjusters.
4. Tighten adjusting nut until brake lever can be raised 5-7 notches with a force of approximately 45 lbs. **If adjusting nut is tightened excessively, self-adjuster mechanism will be inoperative.**
5. After adjustment, raise rear of vehicle, ensuring brakes do not drag with parking brake lever released.

Type 7

INDEX

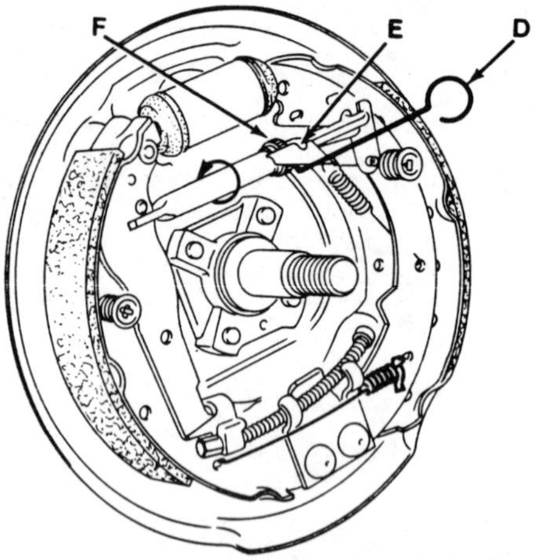

Fig. 1 Removing brake drum

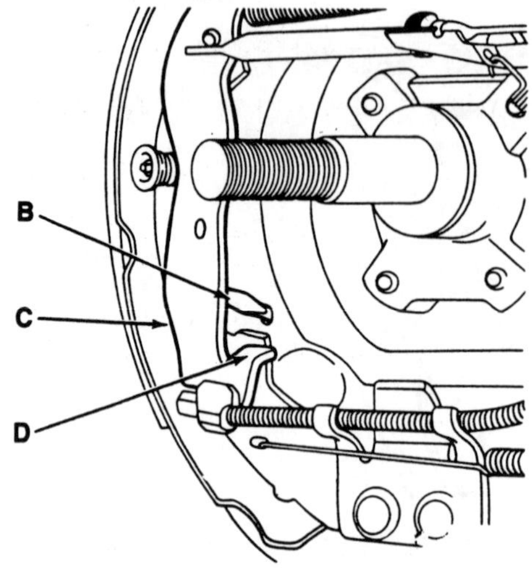

Fig. 2 Removing brake drum with screwdriver

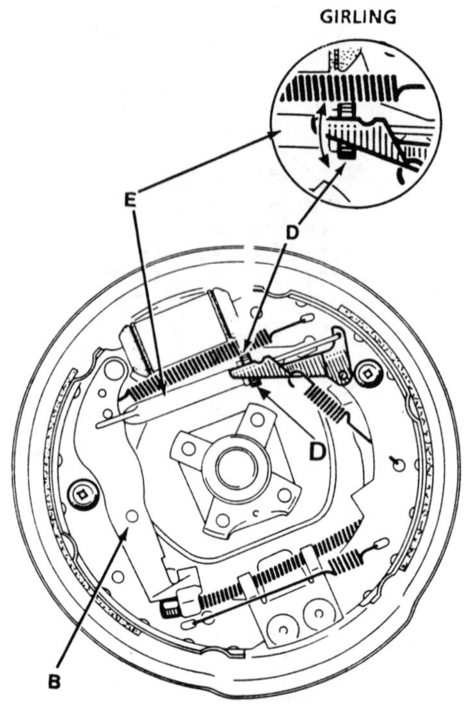

Fig. 3 Self-adjuster mechanism

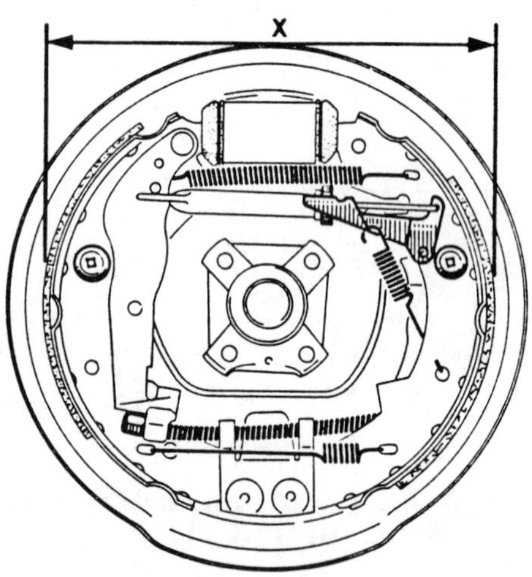

Fig. 4 Obtaining dimension X

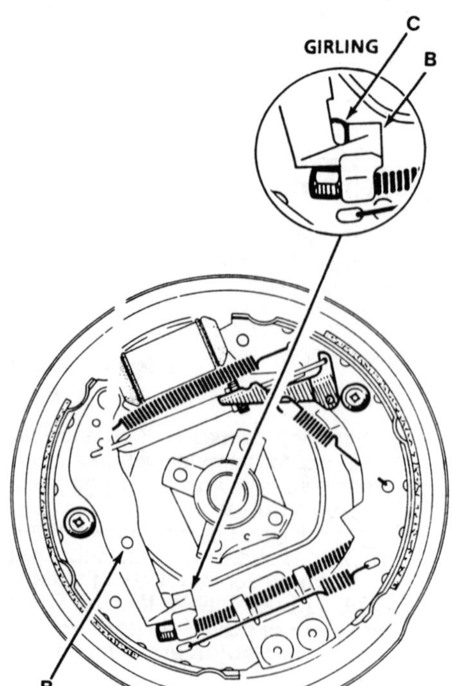

Fig. 5 Positioning brake shoes in normal position

BRAKE SERVICE

BRAKE DRUMS, REPLACE

Both rear brake drum assemblies must have the same inside diameter. If one brake drum is resurfaced, the other brake drum must also be resurfaced. Maximum inside diameter after machining (resurfacing) should be 9.035 inch. Also, do not allow brake dust to fly. Do not use compressed air or dry brush to clean components. Use approved breathing equipment and a water dampened cloth to clean or remove dirt and dust from brake components.

1. Raise and support vehicle.
2. Remove tire and wheel assemblies.
3. Remove plastic dust cover.
4. Remove the two machine screws and rear brake drum. If brake drum cannot be removed, back off the brake self adjuster. Fabricate a small hook from a piece of coat hanger or heavy wire. Insert the hook tool through one of the bolt hole openings and pull slightly on self adjuster lever while rotating star wheel adjuster using a suitable screwdriver in direction of arrow shown in **Fig. 1.** If brake drum still cannot be removed, loosen the parking brake cables and remove plastic access plug on the rear of the brake backing plate. Insert a suitable screwdriver (B) through access hole and push outward and inward on the parking brake lever (C) so tang (D) will clear the rear brake shoe, **Fig. 2.** This will separate the self adjuster star

wheel mechanism from the parking brake lever (C).

5. Reverse procedure to install.

BRAKE SHOES, REPLACE

Never replace brake linings (shoes) on only one wheel. Replace linings on both sides of axle assembly. Always replace linings with linings of the same make and grade.

1. Raise and support vehicle.
2. Remove brake drums as described previously.
3. Using a suitable tool, remove upper and lower spring.
4. Install attaching clip Fre. 05 or equivalent, onto wheel cylinder pistons.
5. Remove spring and adjusting lever.
6. Remove hand brake cable from lever.
7. Remove hold down spring and one brake shoe.
8. Remove self adjuster.
9. Remove other shoe hold down spring, then the brake shoe.
10. Using a suitable vacuum cleaner, dust out drums and backing plates.
11. Reverse procedure to install, noting the following:
    a. The internal components of the brakes are different for the lefthand and righthand sides. Do not interchange from side to side.
    b. Lightly grease self adjuster threads and mark it righthand or lefthand.
    c. The end the lefthand adjuster is metallic silver colored and has righthand threads.
    d. The end of the righthand adjuster is gold colored and has lefthand threads.
    e. Adjust brake shoes to 8.97-9.00 inch.

ADJUSTMENTS

PARKING BRAKE, ADJUST

1. Raise and support vehicle.
2. Remove rear tire and wheel assemblies.
3. Remove rear brake drums.
4. Loosen attaching nut at equalizer bracket.
5. Loosen attaching nut to the end of the threaded rod.
6. Verify operation of the self adjuster mechanism (E) by rotating the star wheel adjusters (D) in both directions, **Fig. 3.**
7. Using star wheel adjuster (D), adjust rear brake shoes to obtain dimension X, **Fig. 4.** Dimension X should be 8.97-9.0 inch.
8. With rear brake shoes adjusted to 8.97-9.0 inch, back off star wheel adjuster (D) 5-6 teeth.
9. Verify stop on lever (B), **Fig. 5,** is against brake shoe at point (C).
10. This will verify that the brake cables are returning to their normal position.
11. Install rear brake drums.
12. Install tire and wheel assemblies.
13. Pull up on parking brake lever to the second notch or tooth.
14. Adjust nut on equalizer bracket until both rear wheels are locked.
15. With parking brake lever still on the second notch, back off nut until both rear wheels turn freely.
16. Tighten attaching locknut.
17. Lower vehicle and verify parking brake operation.

Drum Brake Specifications

Year	Model	Brake Drum Inside Dia. In.	Rear Wheel Cylinder Bore Dia. In.
1985–86	Aries, Lebaron, Reliant, Town & Country①	7.87	.626
	Aries, Lebaron, Reliant, Town & Country②	8.66	.563
	Caravelle, Daytona, Laser, New Yorker, & 600	8.66①	.563②
1985–88	Colt	7.00	.750
	Colt Vista, 2WD	8.00	.810
	Colt Vista, 4WD	9.00	.750
	Eagle 4 X 4	10.0	.945
1985–89	Chrysler rear wheel drive①	10.0	.9385
	Chrysler rear wheel drive②	11.0	.9385
1987–88	Aries, Lebaron, Reliant, Town & Country①	7.87	.626
	Aries, Lebaron, Reliant, & Town & Country②	8.66	.563
	Caravelle, Daytona, New Yorker & 600	8.66	.563
1987–89	Eagle Medallion	9.00	.866
1988–89	Eagle Premier	8.858	.945
1989	Chrysler FWD①	7.87	.626
	Chrysler FWD②	8.66	.626
	Acclaim & Spirit	8.66	.626
	Colt Vista, 2WD	8.00	.810
	Colt Vista, 4WD	9.00	.750
	Colt	7.10	.750
	Eagle Summit	7.10	.750

①—Standard.
②—Heavy duty.

FORD MOTOR CO.

Refer to "Applications" to determine which type brakes are used on vehicle being serviced.

TABLE OF CONTENTS

Applications

Type No.	
1985-86 Capri	2
1985-88 Cougar Less Heavy Duty Brakes	2
1985-89 Cougar w/Heavy Duty Brakes	1
1985-89 Crown Victoria	1
1985 Escort 3 Door Less Styled Steel Wheels	3
1985 Escort Except 3 Door Less Styled Steel Wheels	4
1986-89 Escort 2 Door Less Styled Steel Wheels	3
1986-89 Escort Except 2 Door Less Styled Steel Wheels	4
1985-87 EXP	4
1988-89 Festiva	6
1985-89 Grand Marquis	1
1985-86 LTD Except Station Wagon	2
1985-86 LTD Station Wagon	1
1985 Lynx 3 Door Less Styled Steel Wheels	3
1985 Lynx Except 3 Door Less Styled Steel Wheels	4
1986-87 Lynx 2 Door Less Styled Steel Wheels	3
1986-87 Lynx Except 2 Door Less Styled Steel Wheels	4
1985-86 Marquis Except Station Wagon	2
1985-86 Marquis Station Wagon	1
1985-89 Mustang	2
1989 Probe	7
1986-89 Sable	5
1988-89 Scorpio	8
1986-89 Taurus	5
1985-89 Tempo	4
1985-89 Thunderbird Less Heavy Duty Brakes	2
1985-89 Thunderbird w/Heavy Duty Brakes	1
1985-89 Topaz	4
1985-89 Town Car	1
1988-89 Tracer	7

General Information

INDEX

	Page No.
General Inspection:	6-111
Adjuster Mechanism	6-112
Backing Plate	6-112
Brake Drums	6-111
Brake Linings & Springs	6-111
Parking Brake Cable	6-112
Service Precautions	6-111

SERVICE PRECAUTIONS

When working on or around brake assemblies, care must be taken to prevent breathing asbestos dust, as many manufacturers incorporate asbestos fibers in the production of brake linings. During routine service operations the amount of asbestos dust from brake lining wear is at a low level, due to a chemical break down during use and a few precautions will minimize exposure. **Do not sand or grind brake linings unless suitable local exhaust ventilation equipment is used to prevent excessive asbestos exposure.**

The brake shoe and lining assemblies should be replaced if the lining is worn to within 1/32 inch of rivet heads (riveted linings) or brake shoe (bonded linings). It is recommended that both front and/or rear wheel sets be replaced whenever a respective shoe and lining assembly is replaced.

If a visual inspection does not adequately determine the condition of the linings, the brake shoe and lining assemblies should be removed and inspected. If shoes do not require replacement, reinstall them in their original positions. Brake shoes and linings should also be replaced if cracked or damaged.

1. Wear a suitable respirator approved for asbestos dust use during all repair procedures.
2. When cleaning brake dust from brake parts, use a vacuum cleaner with a highly efficient filter system. If a suitable vacuum cleaner is not available, use a water soaked rag. **Do not use compressed air or dry brush to clean brake parts.**
3. Keep work area clean using same equipment as for cleaning brake parts.
4. Properly dispose of rags and vacuum cleaner bags by placing them in plastic bags.
5. Do not smoke or eat while working on brake systems. **Never use gasoline, kerosene, alcohol, motor oil, transmission fluid, or any fluid containing mineral oil to clean brake system components. These fluids will damage the rubber caps and seals. If system contamination is suspected, check brake fluid in the reservoir for dirt, discoloration, or separation (break down) of the brake fluid into distinct layers. Drain and flush the hydraulic system with clean brake fluid if contamination is suspected.**

GENERAL INSPECTION

BRAKE DRUMS

Any time the brake drums are removed for brake service, the braking surface diameter should be checked with a suitable brake drum micrometer at several points to determine if they are within the safe oversize limit stamped on the brake drum outer surface. If the braking surface diameter exceeds specifications, the drum must be replaced. If the braking surface diameter is within specifications, drums should be cleaned and inspected for cracks, scores, deep grooves, taper, out of round and heat spotting. If drums are cracked or heat spotted, they must be replaced. Minor scores should be removed with sandpaper. Grooves and large scores can only be removed by machining with special equipment, as long as the braking surface is within specifications stamped on brake drum outer surface. Any brake drum sufficiently out of round to cause vehicle vibration or noise while braking or showing taper should also be machined, removing only enough stock to true up the brake drum.

After a brake drum is machined, wipe the braking surface diameter with a denatured alcohol soaked cloth. If one brake drum is machined, the other should also be machined to the same diameter to maintain equal braking forces.

BRAKE LININGS & SPRINGS

Inspect brake linings for excessive wear, damage, oil, grease or brake fluid contamination. If any of the above conditions exists, brake linings should be replaced. Do not attempt to replace only one set of brake shoes. They should be replaced as an axle set only to maintain equal braking forces. Examine brake shoe webbing, hold-down and return springs for signs of overheating indicated by a slight blue color. If any component exhibits overheating signs, replace hold-down and return springs with new ones. Overheated springs lose their pull and could cause brake linings to wear out prematurely. Inspect all springs for sags, bends and external damage and replace as necessary.

Inspect hold-down retainers and pins for bends, rust and corrosion. If any of the above is found, replace as required.

BACKING PLATE

Inspect backing plate shoe contact surface for grooves that may restrict shoe movement and cannot be removed by lightly sanding with emery cloth or other suitable abrasive. If backing plate exhibits above condition, it should be replaced. Also inspect for signs of cracks, warpage and excessive rust, indicating need for replacement.

ADJUSTER MECHANISM

Inspect all components for rust, corrosion, bends and fatigue. Replace as necessary. **On adjuster mechanism equipped with adjuster cable,** inspect cable for kinks, fraying or elongation of eyelet and replace as necessary.

PARKING BRAKE CABLE

Inspect parking brake cable end for kinks, fraying and elongation and replace as necessary. Use a small hose clamp to compress clamp where it enters backing plate to remove.

Types 1 & 2

INDEX

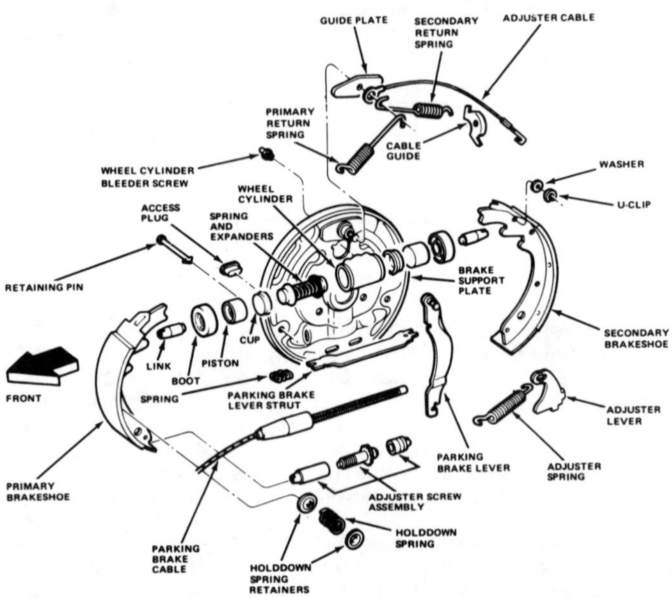

Fig. 1 Drum brake assembly. Type 1

REMOVAL

1. Raise and support rear of vehicle, then remove tire and wheel assembly.
2. Remove brake drum. If brake lining is dragging on brake drum, back off brake adjustment by rotating adjustment screw. **If brake drum is rusted or corroded to axle flange and cannot be removed, lightly tap axle flange to drum mounting surface with a suitable hammer.**
3. Install suitable wheel cylinder clamp over ends of wheel cylinder to retain pistons in bore.
4. **On type 1 brakes, Fig 1,** remove parking brake lever retaining clip.
5. **On both types, Figs. 1 and 2,** remove adjuster lever spring, primary and secondary shoe return springs using a suitable pair of brake spring pliers.
6. Remove shoe guide plate, if equipped and adjuster cable and guide plate.
7. Using suitable tool, compress hold-down springs, then remove spring retainers, hold-down springs and pins.
8. Separate springs and remove from backing plate.
9. **On type 2 brake,** disengage parking brake lever from secondary shoe.
10. **On all types,** remove parking brake lever from cable.
11. Separate all components from brake shoes.
12. Clean dirt from brake drum, backing plate and all other components. **Do not use compressed air or dry brush to clean brake parts. Many brake parts contain asbestos fibers which, if inhaled, can cause serious injury. Clean brake parts with a water soaked rag or a suitable vacuum cleaner to minimize airborne dust.**

INSPECTION

1. Inspect components for damage and unusual wear. Replace as necessary.
2. Inspect wheel cylinders. Boots which are torn, cut, or heat damaged indicate need for wheel cylinder replacement. Fluid spilling from boot center hole, or wetness around wheel cylinder ends indicates cup leakage and need for wheel cylinder replacement. **A small amount of fluid is always present and is considered normal, acting as a lubricant for the cylinder pistons.**
3. Inspect backing plate for evidence of seal leakage. If leakage exists, refer to individual car chapters for axle seal replacement procedure.
4. Inspect backing plate attaching bolts and ensure they are tight.
5. Check adjuster screw operation. If satisfactory, lightly lubricate adjusting screw and washer with suitable brake lube. If operation is unsatisfactory, replace.
6. Using fine emery cloth or other suitable abrasive, clean rust and dirt from shoe contact surfaces on backing plate.

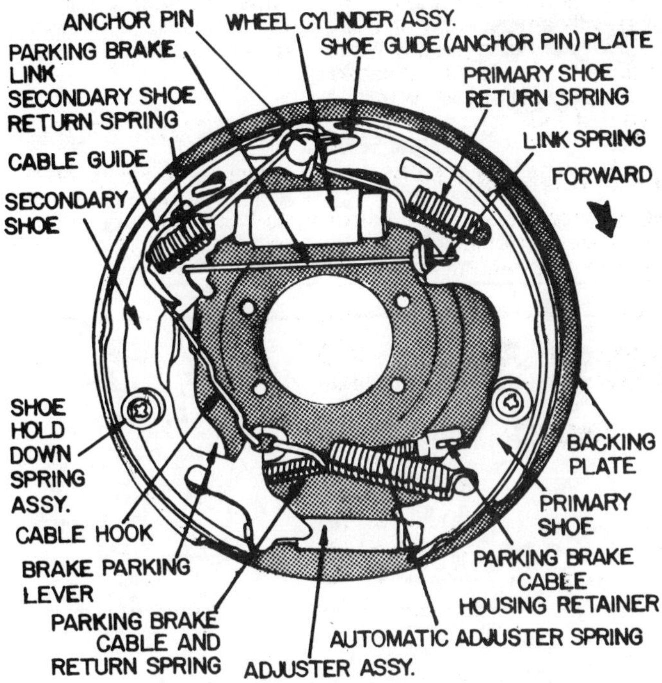

Fig. 2 Drum brake assembly. Type 2

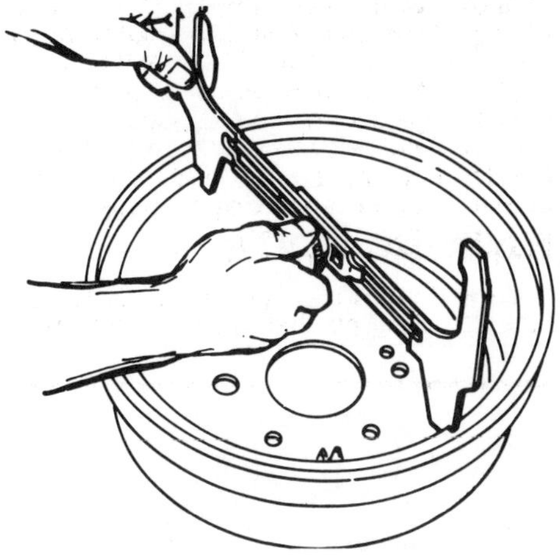

Fig. 3 Measuring brake drum inside diameter

INSTALLATION

1. Lightly lubricate backing plate shoe contact surfaces with suitable brake lube.
2. **On type 1 brakes,** assemble parking brake lever to secondary shoe and secure with spring washer and retaining clip. Crimp ends of clip with suitable pliers. **On type 2 brakes,** engage parking brake lever tang with secondary shoe.
3. Position brake shoes on backing plate, primary (short lining) shoe facing front of vehicle and secondary (long lining) facing rear. Secure brake shoes with hold-down springs, pins and retainers.
4. Install parking brake link and spring between shoes.
5. Loosen parking brake adjustment nut, then install parking brake cable on parking brake lever.
6. Install shoe guide plate and adjuster cable eyelet on anchor. Ensure adjuster cable crimp faces out.
7. Ensure parking brake link is properly positioned between brake shoes and wheel cylinder links are engaged in shoe web.
8. Using suitable brake spring pliers, install primary return spring from brake shoe to anchor, then secondary return spring from brake shoe to anchor.
9. **On all types,** remove wheel cylinder clamp installed during removal of brake shoes.
10. Tighten adjuster screw assembly to thread limit and back off 1/2 turn.
11. Install adjuster screw assembly between shoes. Ensure toothed wheel is on secondary shoe side. **Adjuster screw assemblies are stamped R (right) and L (left). To ensure proper adjuster operation, they must be installed on their respective sides.**
12. Hook adjuster cable hook into adjuster lever hole, then position adjuster spring hook in large hole in primary shoe web. Using suitable brake spring pliers, install adjuster spring in adjuster lever hole.
13. Ensure adjuster cable is properly seated in cable guide, then pull adjuster lever, cable and adjuster spring down and towards the rear, engaging lever pivot hook in the large hole of secondary shoe web.
14. After installation, check adjuster operation by pulling adjuster cable between cable guide and adjuster lever towards secondary shoe sufficiently to lift adjuster lever past one tooth on adjuster screw assembly. The adjuster lever should snap into position behind the next tooth, then upon release of adjuster cable, rotate toothed wheel one notch. If operation is not satisfactory, recheck installation.
15. Ensure brake shoe upper ends are seated against anchor pin and shoe assemblies are centered on backing plate. If not, back off parking brake adjustment.
16. Using suitable brake drum to shoe gauge, **Fig. 3,** measure brake drum inside diameter. Adjust brake shoes to dimension obtained on outside portion of gauge using adjuster screw.
17. Install brake drum, wheel and tire assembly.
18. If any hydraulic brake connections have been opened, bleed brake system.
19. Adjust parking brake.
20. Inspect all hydraulic lines and connections for leakage and repair as necessary.
21. Check master cylinder fluid level and replenish as necessary.
22. Check brake pedal for proper feel and return.
23. Lower vehicle and road test. **Do not severely apply brakes immediately after installation of new brake linings or permanent damage may occur to linings, and/or brake drums may become scored. Brakes must be used moderately during first several hundred miles of operation to ensure proper burnishing of linings.**

Type 3

INDEX

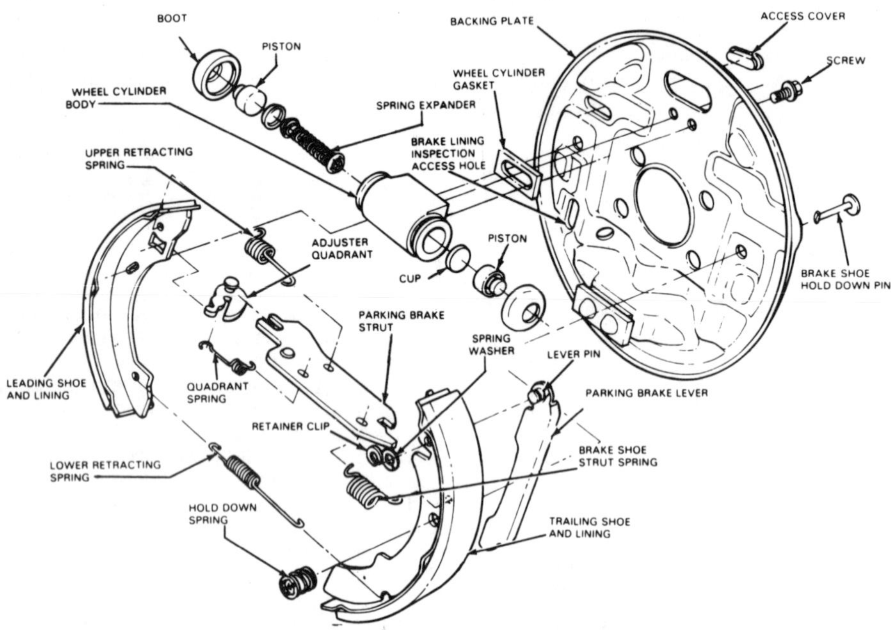

Fig. 1 Drum brake assembly. Type 3

REMOVAL

1. Raise and support rear of vehicle, then remove tire and wheel assembly.
2. Remove brake drum. If brake lining is dragging on brake drum, back off brake adjustment.
3. Using suitable tool, remove hold-down retainers, springs and pins, **Fig. 1.**
4. Remove brake shoe and adjuster assemblies from backing plate by lifting up and away from anchor block and shoe guide, **Fig. 2. When removing brake shoe and adjuster assemblies, use care not to damage wheel cylinder boots.**
5. Remove parking brake cable from parking brake lever.
6. Remove lower retracting spring from leading and trailing shoes.
7. While holding brake shoe and adjuster assemblies, remove leading shoe upper retracting spring by rotating leading shoe over adjuster quadrant until spring is slack, then remove spring. Remove leading shoe from adjuster assembly.
8. Remove parking brake strut from trailing shoe by pulling strut outward from shoe assembly, then twisting strut downward until spring tension is released. Unhook brake shoe strut spring, then remove parking brake strut and adjuster assembly from trailing shoe.
9. If adjuster disassembly is required, pull adjuster quadrant away from knurled pin in parking brake strut and rotate quadrant in either direction until quadrant teeth are disengaged from strut pin. Remove spring and slide quadrant out of strut. **Do not over stress quadrant spring during removal.**
10. Remove parking brake lever retaining clip and spring washer, then the parking brake lever.
11. Clean dirt from brake drum, backing plate and all other components. **Do not use compressed air or dry brush to clean brake parts. Many brake parts contain asbestos fibers which, if inhaled, can cause serious injury. Clean brake parts with a water soaked rag or a suitable vacuum cleaner to minimize airborne dust.**

INSPECTION

1. Inspect components for damage and unusual wear. Replace as necessary.

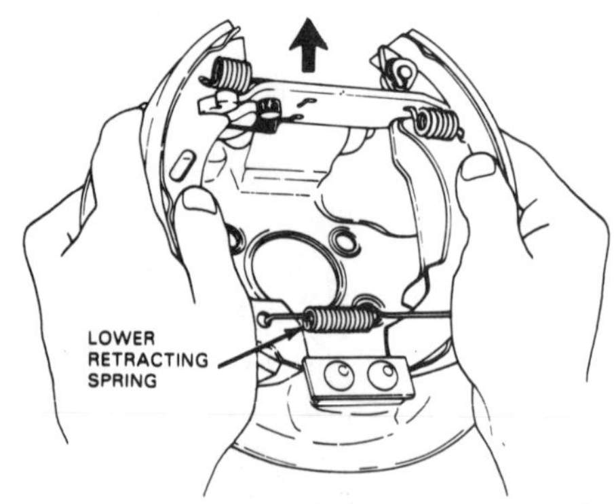

Fig. 2 Removing brake shoe and adjuster assemblies. Type 3

2. Inspect wheel cylinders. Boots which are torn, cut or heat damaged indicate need for wheel cylinder replacement. Peel back lower edge of boot. If fluid spills out, cup leakage is indicated and wheel cylinder should be replaced. **A slight amount of fluid is always present and is considered normal, acting as a lubricant for the cylinder pistons.**
3. Inspect backing plate attaching bolts and ensure they are tight.
4. Using fine emery cloth or other suitable abrasive, clean rust and dirt from shoe contact surfaces on backing plate.

INSTALLATION

1. Lightly lubricate backing plate shoe contact surfaces with suitable brake lube.
2. Remove brake drum hub grease seal and bearings, then clean and repack bearings and reinstall. Install new grease seal.
3. Lightly lubricate strut to adjuster quadrant contact surfaces with suitable brake lube.
4. Position adjuster quadrant pin in strut slot and install quadrant spring, then pivot quadrant until it engages with strut knurled pin in third or fourth notch of outboard end of quadrant.
5. Assemble parking brake lever to trailing shoe, then install spring washer and retaining clip. Using suitable pliers, crimp retaining clip until securely fastened.

6. Assemble parking brake strut to trailing shoe by attaching brake shoe strut spring to slots in shoe web and strut, and pivoting strut into position, tension spring and holding assembly in place. **Ensure end of spring with hook parallel to the center line of spring coils is installed in shoe web hole. Installed spring should be flat against shoe web and parallel to parking brake strut.**
7. Install lower retracting spring between shoes. Ensure spring hook with longest straight piece fits into trailing shoe hole, **Fig. 1.**
8. Install upper retracting spring by installing hooks in leading shoe web and other end in parking brake strut, then pivot leading shoe over adjuster quadrant and into position.
9. Spread shoe and strut assemblies sufficiently to fit over anchor plate and wheel cylinder piston inserts, and install onto backing plate. **When installing brake shoe and adjuster assemblies, use care not to damage wheel cylinder boots.**

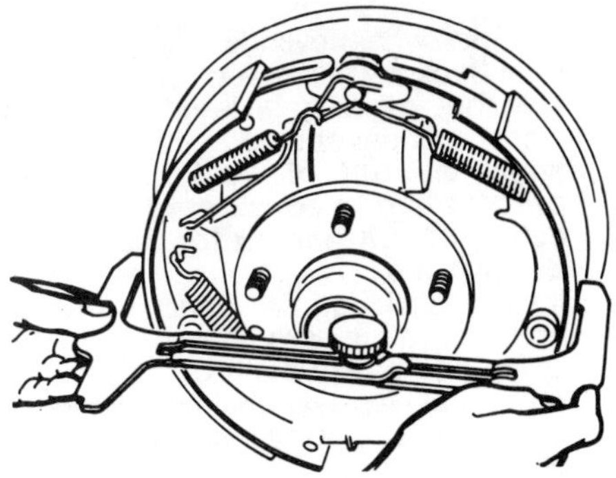

Fig. 3 Adjusting brake shoes to brake drum inside diameter. Type 3

10. Connect parking brake cable to parking brake lever.
11. Using suitable tool, install hold-down springs, retainers and pins.
12. Using suitable brake drum to shoe gauge, measure brake drum inside diameter. Adjust brake shoes to dimension obtained on outside portion of gauge, **Fig. 3.**
13. Install brake drum.
14. Install tire and wheel assembly.
15. If any hydraulic connections have been opened, bleed brake system.
16. Adjust parking brake.
17. Inspect all hydraulic lines and connections for leakage and repair as necessary.
18. Check master cylinder fluid level and replenish as necessary.
19. Check brake pedal for proper feel and return.
20. Lower vehicle and road test. **Do not severely apply brakes immediately after installation of new brake linings or permanent damage may occur to lining, and/or brake drums may become scored. Brakes must be used moderately during first several hundred miles of operation to ensure proper burnishing of linings.**

Type 4

INDEX

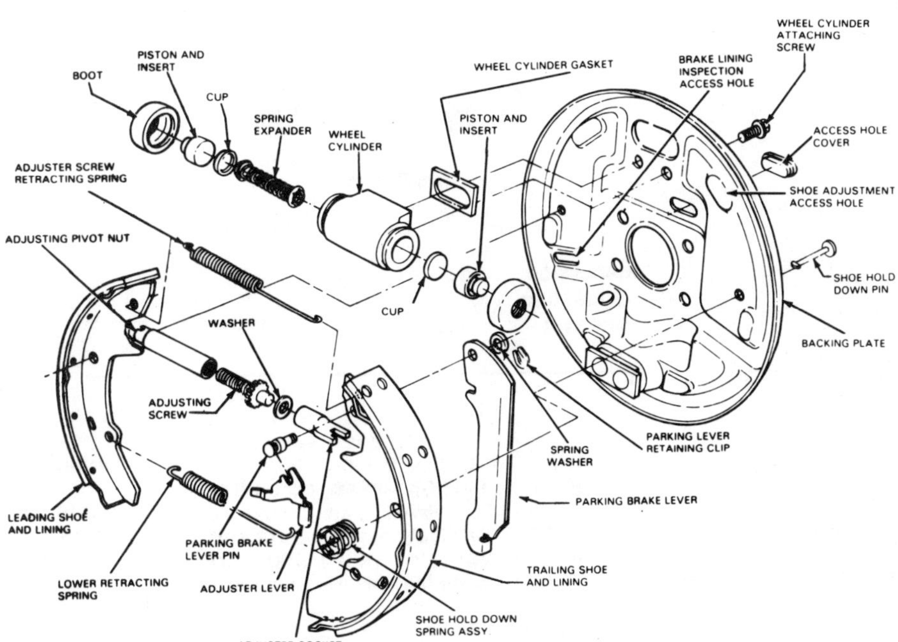

Fig. 1 Drum brake assembly. Type 4

REMOVAL

1. Raise and support rear of vehicle, then remove tire and wheel assembly.
2. Remove brake drum. If brake lining is dragging on brake drum, back off brake adjustment.
3. Using suitable tool, remove hold-down retainers, springs and pins, **Fig. 1.**
4. Remove brake shoes and adjuster assemblies from backing plate by lifting up and away from wheel cylinder assembly. **When removing brake shoe and adjuster assemblies, use care not to bend adjusting lever.**
5. Remove parking brake cable from parking brake lever.
6. Remove lower retracting spring, adjuster screw retracting spring and adjuster lever.
7. Separate brake shoes, then remove parking brake lever retaining clip and spring washer and slide lever off parking brake lever pin on the trailing shoe.
8. Clean dirt from brake drum, backing plate and all other components. **Do not use compressed air or dry brush to clean brake parts. Many brake parts contain asbestos fibers which, if inhaled, can cause serious injury. Clean brake parts with a water soaked rag or a suitable vacuum cleaner to minimize airborne dust.**

INSPECTION

1. Inspect components for damage and unusual wear. Replace as necessary.
2. Inspect wheel cylinders. Boots which are torn, cut or heat damaged indicate need for wheel cylinder replacement. Peel back lower edge of boot. If fluid spills out, cup leakage is indicated and wheel cylinder should be replaced. **A small amount of fluid is always present and is considered normal, acting as a lubricant for the cylinder pistons.**
3. Inspect backing plate attaching bolts and ensure they are tight.
4. Using fine emery cloth or other suitable abrasive, clean rust and dirt from shoe contact surfaces on backing plate.
5. Check adjuster screw operation. If satisfactory, lightly lubricate adjusting screw and washer with suitable brake lube. If operation is unsatisfactory, replace.

INSTALLATION

1. Lightly lubricate backing plate shoe contact surfaces with suitable brake lubrication.
2. Remove brake drum hub grease seal and bearings, then clean and repack bearings and reinstall. Install new grease seal.
3. Assemble parking brake lever to trailing shoe, then install spring washer and retaining clip. Using suitable pliers, crimp retaining clip until securely fastened.
4. Attach parking brake cable to parking brake lever.
5. Assemble lower retracting spring to leading and trailing shoe assemblies, then spread lower part of shoes and install on backing plate.
6. Using suitable tool, install hold-down springs.
7. Tighten adjuster assembly, then back off 1/2 turn. Install adjuster assembly between leading shoe slot and trailing shoe/parking brake lever slot. The adjuster socket end slot must fit into trailing shoe/parking brake lever. **Adjuster assemblies are stamped R (right) and L (left). To ensure proper adjuster operation, they must be installed on their respective sides. The letter must be installed in the upright position, facing wheel cylinder to ensure the deeper of the two slots in the adjuster socket fits in the parking brake lever.**
8. Install adjuster lever in the parking brake lever groove and into the adjuster socket slot.
9. Using suitable brake spring pliers, install adjusting screw retracting spring from leading shoe slot to adjuster lever notch.
10. Install brake drum.
11. Install tire and wheel assembly.
12. If any hydraulic connections have been opened, bleed brake system.
13. Adjust parking brake.
14. Inspect all hydraulic lines and connection for leakage, and repair as necessary.
15. Check master cylinder fluid level and replenish as necessary.
16. Check brake pedal for proper feel and return.
17. Lower vehicle and road test. **Do not severely apply brakes immediately after installation of new brake linings or permanent damage may occur to linings, and/or brake drums may become scored. Brakes must be used moderately during first several hundred miles of operation to ensure proper burnishing of linings.**

Type 5

INDEX

Page No.

REMOVAL

1. Raise and support rear of vehicle, then remove tire and wheel assembly.
2. Remove brake drum. If brake lining is dragging on brake drum, back off brake adjustment.
3. Using suitable tool, remove shoe hold-down springs and pins, **Fig. 1.**
4. Lift brake shoes, springs and adjuster assembly off backing plate and wheel cylinder assembly, being careful not to bend adjusting lever.
5. Remove parking brake cable from parking brake lever.
6. Remove retracting springs from lower brake shoe attachments and upper shoe to adjusting lever attachment points, then separate shoes and disengage adjuster mechanism.
7. Clean dirt from brake drum, backing plate and all other components. **Do not use compressed air or dry brush to clean brake parts. Many brake parts contain asbestos fibers which, if inhaled, can cause serious industry. Clean brake parts with a water soaked rag or a suitable vacuum cleaner to minimize airborne dust.**

INSPECTION

1. Inspect components for damage and unusual wear. Replace as necessary.
2. Inspect wheel cylinders. Boots which are torn, cut or heat damaged indicate need for wheel cylinder replacement.

Peel back lower edge of boot. If fluid spills out, cup leakage is indicated and wheel cylinder should be replaced. **A small amount of fluid is always present and is considered normal, acting as a lubricant for the cylinder pistons.**
3. Inspect backing plate attaching bolts and ensure they are tight.
4. Using fine emery cloth or other suitable abrasive, clean rust and dirt from shoe contact surfaces on backing plate.
5. Check adjuster screw operation. If satisfactory, lightly lubricate adjusting screw and washer with suitable brake lube. If operation is unsatisfactory, replace.

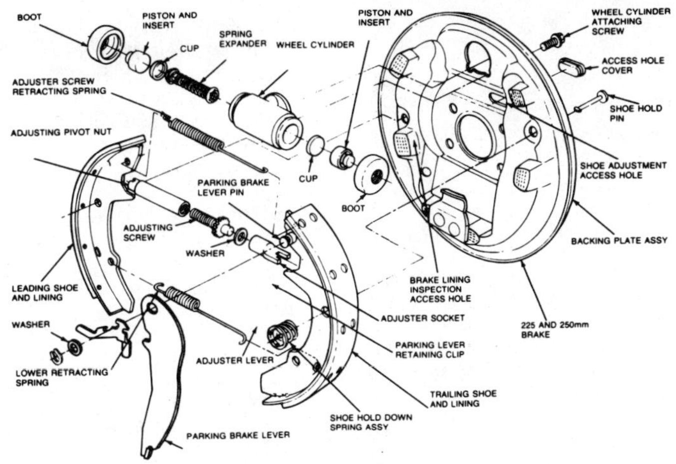

Fig. 1 Drum brake assembly. Type 5

INSTALLATION

1. Lightly lubricate backing plate shoe contact surfaces with suitable brake lubrication.
2. Apply a thin uniform coat of suitable brake lube to adjuster screw threads and socket end of adjusting screw.
3. Install stainless steel washer over socket end of adjusting screw and install socket, then turn adjusting screw fully into adjusting pivot nut and back off 1/2 turn.
4. Assemble parking brake lever to trailing shoe and lining assembly by installing spring washer and a new horseshoe retaining clip. Crimp the clip until it retains lever to shoe securely.
5. Attach parking brake cable to parking brake lever.
6. Attach lower shoe retracting spring to leading and trailing shoe assemblies and install on backing plate. It will be necessary to stretch retracting spring as shoes are installed downward over anchor plate to inside of shoe retaining plate.
7. Install adjuster screw assembly between leading shoe slot and the slot in the trailing shoe and parking brake lever. The adjuster socket end slot must fit into the trailing shoe and parking brake lever. **The adjuster socket blade is marked "R" or "L" for right and left side brake assemblies. The R or L adjuster blade must be installed with the letter R or L in the upright position (facing wheel cylinder) on the correct side to ensure that the deeper of two slots in adjuster sockets fits into the parking brake lever.**
8. Assemble adjuster lever in groove located in parking brake lever pin and into slot of adjuster socket that fits into trailing shoe web.
9. Attach upper retracting spring to leading shoe slot and, using suitable tool, stretch other end of spring into notch on adjuster lever. **If adjuster lever does not contact star wheel after installing spring, adjuster socket may be improperly installed.**
10. Install brake drum.
11. Install tire and wheel assembly.
12. If any hydraulic connections have been opened, bleed brake system.
13. Adjust parking brake.
14. Inspect all hydraulic lines and connections for leakage, repairing as necessary.
15. Check master cylinder fluid level and replenish as necessary.
16. Check brake pedal for proper feel and return.
17. Lower vehicle and road test. **Do not severely apply brakes immediately after installation of new brake linings or permanent damage may occur to linings., and/or brake drums may become scored. Brakes must be used moderately during first several hundred miles of operation to ensure proper burnishing of linings.**

Type 6

INDEX

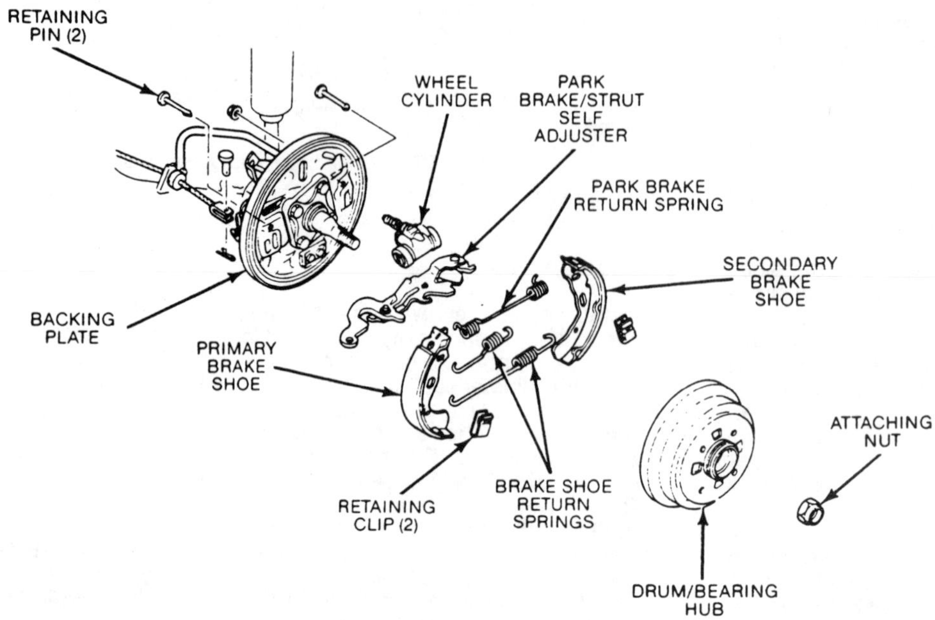

Fig. 1 Disassembled view of drum brake assembly

REMOVAL

1. Raise and support vehicle, then remove rear wheel
2. Remove locknut and brake drum. **The locknut on the right side of the vehicle has lefthand threads. Turn locknut clockwise to remove.**
3. Remove hold down clips and retaining pins, **Fig. 1.**
4. Remove all return springs, then the primary and secondary brake shoes.
5. Remove cotter pin, then the clevis pin from parking brake strut/self adjuster. Disengage parking cable from self adjuster.

INSTALLATION

1. Clean backing plate with a factory approved vacuum cleaner.
2. Lubricate backing plate shoe pads with suitable high temperature brake grease.
3. Install and position parking brake strut/self adjuster onto backing plate.
4. Install upper return spring to primary brake shoe, then position shoe on backing plate and install retaining pin and hold down clip, **Fig. 1.**
5. Connect upper return spring to secondary brake shoe, then position secondary shoe on backing plate and install retaining pin and hold down clip.
6. Install parking brake return spring.
7. Install lower return spring onto primary and secondary shoes.
8. Connect parking cable onto self adjuster, then install clevis and cotter pins.
9. Push on adjuster cam with screwdriver to position self adjuster in fully released position.
10. Install brake drum and locknut, push brake pedal down several times to set self adjuster, then install rear wheel.
11. Adjust parking brake as required.

Type 7

INDEX

REMOVAL

1. Raise and support rear of vehicle.
2. Remove wheel and tire assembly, then the grease cap.
3. Using a suitable chisel, raise staked portion of wheel bearing locknut.
4. Remove rear bearing locknut. To do so, turn the locknut at the lefthand rear wheel bearing counterclockwise and the locknut at the righthand rear wheel bearing clockwise.
5. Remove outer wheel bearing, then remove brake drum and bearing hub assembly.
6. Remove brake shoe hold-down springs, then pull front shoe away from backing plate, disconnect return springs and remove front shoe, **Fig. 1.**
8. Remove return springs from rear shoe, then disconnect anti-rattle spring attaching rear shoe to parking brake strut and remove rear shoe.
9. If necessary, remove adjuster assembly.

INSPECTION

Inspect lining material for wear beyond service limits, uneven wear, cracks, scoring, gouges, and contamination from brake fluid or grease. Inspect wheel cylinders for signs of leakage. Inspect the anti-rattle and return springs for signs of heat damage, bends, or damage to coils, and loss of tension. Inspect brake drum for scratches, scoring and out of round conditions.

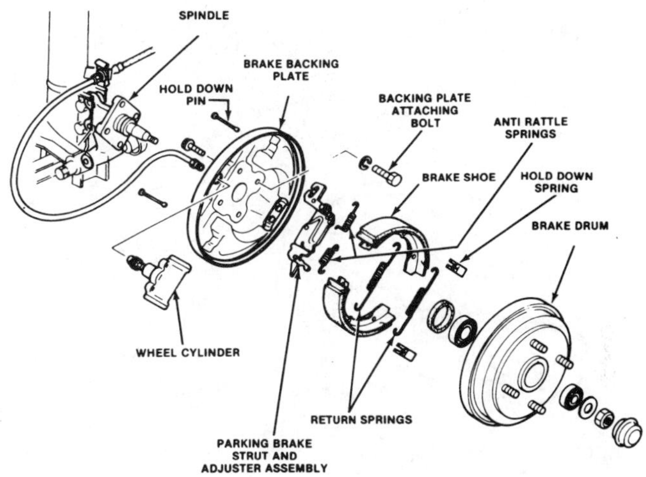

Fig. 1 Exploded view of rear drum brake

INSTALLATION

1. Lubricate the six shoe contact pads with high temperature grease D7AZ-19590-A or equivalent, then the adjuster mechanism toothed quadrant and the anchor plate.
2. Position rear brake shoe in the parking brake strut and install rear hold down pin and spring.
3. Hook brake shoe return springs in position on rear brake shoe.
4. Connect brake shoe return springs to front shoe and push into place against backing plate.
5. Install front brake shoe hold down pin and spring.
6. Insert screwdriver between knurled quadrant and parking brake strut, then twist screwdriver until quadrant touches backing plate.
7. Install drum, then firmly apply brakes two or three times to adjust rear brakes.

Type 8

INDEX

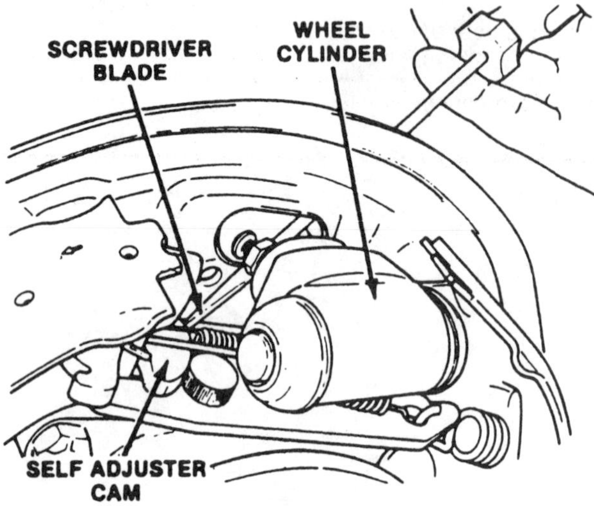

Fig. 1 Releasing self-adjuster mechanism for brake drum removal

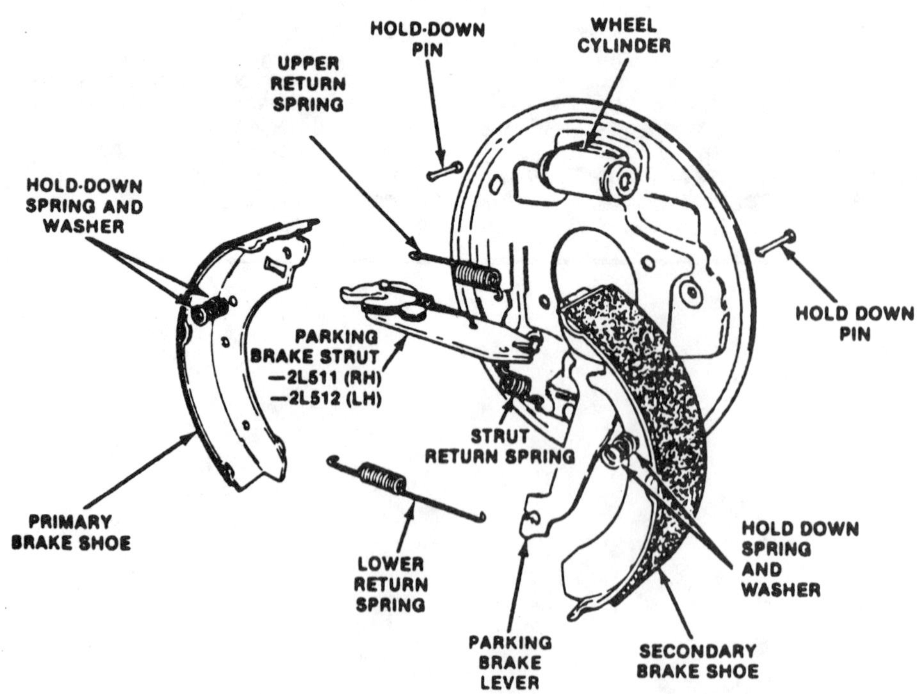

Fig. 2 Exploded view of brake drum assembly

REMOVAL

1. Disconnect battery ground cable.
2. Raise and support vehicle.
3. Remove rear tire and wheel assembly.
4. Remove brake drum. Discard brake drum retaining clip.
5. Should difficulty be experienced in removal of the brake drum due to the action of the automatic self-adjuster mechanism, the self-adjuster may be released and brake drum removed, as follows:
   a. Remove wheel cylinder attaching bolts.
   b. Push wheel cylinder away from the brake backing plate to provide an access opening in the backing plate.
   c. Insert a thin-bladed screwdriver through the backing plate and rotate self-adjuster cam to the released position, **Fig. 1.**
   d. Remove brake drum.
   e. Correctly position wheel cylinder and install attaching bolts. **Torque** bolts to 5-7 ft. lbs.
6. Remove both brake shoe hold-down springs, **Fig. 2.** To prevent rotation during removal, hold each hold-down pin head with a suitable tool.
7. Pry lower end of the primary shoe from its position against the anchor.
8. Remove lower return spring.
9. Remove shoes and strut by passing the strut between the wheel cylinder and hub assembly.
10. Pull top of primary shoe away from the secondary shoe to disconnect strut from the secondary shoe.
11. Disconnect the parking brake cable from the secondary shoe lever.
12. Remove strut return spring from the secondary shoe.
13. Remove adjuster cam spring.
14. Pull primary shoe away from the strut while rotating the cam to the fully released position.
15. Remove the primary shoe spring.
16. Remove the primary shoe from the strut.

INSPECTION

Inspect lining material for wear beyond service limits, uneven wear, cracks, scoring, gouges, and contamination from brake fluid or grease. Inspect wheel cylinders for signs of leakage. Inspect the anti-rattle and return springs for signs of heat damage, bends, or damage to coils, and loss of tension. Inspect brake drum for scratches, scoring and out of round conditions.

INSTALLATION

Before installing the brake shoes, ensure the wheel cylinder and wheel hub are tightened to specification.

1. Apply a light coating of grease supplied with the new brake shoes or high temperature grease D7AZ-19590-A, or equivalent, to support ledges where the brake shoes contact the backing plate.
2. Connect parking brake cable to the secondary shoe lever. To connect the cable, position the cable end through the lever and grip the cable end with locking type pliers, then push lever against the spring until it can be rotated over the cable. When properly installed, the plastic washer will be between the spring and lever.
3. Position the secondary shoe and install the hold-down spring.
4. Install the strut and cam assembly on the primary shoe.
5. Rotate the cam to the fully released position.
6. Install the adjuster cam spring.
7. Install the primary shoe spring.
8. Install the strut spring in the secondary shoe and then into the strut.
9. Place the strut on the parking brake lever and move the primary shoe toward the backing plate. The strut will then "click" into place over the parking brake lever and secondary shoe web.
10. Install the lower shoe spring with the longer leg on the secondary shoe.
11. Pry the lower end of the primary shoe into position against the anchor while holding the top of the shoe against the backing plate, in contact with the cylinder piston. Ensure not to damage the cylinder boot.
12. Install the primary shoe hold-down pin, spring and washer.
13. Ensure heel of each brake shoe is located behind the anchor plate. Check each brake component for proper installation.
14. If necessary, push adjuster cam to the released position.
15. Install brake drum(s).
16. Push brake pedal hard twice to set the self-adjuster cam position. The cam will make a ratcheting sound as it resets.
17. Adjust parking brake cable as necessary.

Drum Brake Specifications

Year	Model	Brake Drum Inside Dia. Inch	Brake Drum Bore Limit (Max.) Inch	Brake Drum Max. Run-out Inch
FORD CROWN VICTORIA & MERCURY GRAND MARQUIS				
1985–89	Except Sta. Wag.	10.00	10.03	.007
FORD ESCORT & EXP; MERCURY LYNX				
1985–87	EXP	8.00	8.06	.005
1985	Escort & Lynx ①	7.08	7.14	.005
	Escort & Lynx ②	8.00	8.06	.005
1986–89	Escort & Lynx ③	7.08	7.14	.005
	Escort & Lynx ④	8.00	8.06	.005
FORD FESTIVA				
1988–89	All	9.00	—	—
FORD LTD & MERCURY MARQUIS				
1985–86	Except Sta. Wag.	9.00	9.06	.007
	Sta. Wag.	10.00	10.06	.007
FORD MUSTANG & MERCURY CAPRI				
1985–89	All	9.00	9.06	.007
FORD TAURUS & MERCURY SABLE				
1986–89	Except Sta. Wag.	8.85	8.91	.005
	Sta. Wag.	9.84	9.90	.005
FORD TEMPO & MERCURY TOPAZ				
1985–89	All	8.00	8.06	.005
FORD THUNDERBIRD & MERCURY COUGAR				
1985–89	Except Heavy Duty	9.00	9.06	.007
	Heavy Duty	10.00	10.06	.007
LINCOLN				
1985–89	Town Car	10.00	10.06	.007
MERCURY TRACER				
1988–89	All	7.87	7.91	—
MERKUR XR4Ti				
1986–89	All	10.00	10.06	—

①—Escort & Lynx 3 door less styled steel wheels

②—Escort & Lynx except 3 door less styled steel wheels.

③—Escort & Lynx 2 door less styled steel wheels.

④—Escort & Lynx except 2 door less styled steel wheels.

Service Bulletins

INDEX

REAR BRAKE SQUEAL

1985–86 CAPRI, COUGAR, LTD, MARQUIS, MUSTANG & THUNDERBIRD

On these models, rear brake squeal can be corrected by replacing the existing riveted brake shoes with bonded brake shoes.

CHIRPING NOISE

1986 SABLE & TAURUS

A chirping noise from the rear of the vehicle may be caused by the rear brake cable spring rubbing the hub at the inside surface of the brake drum. To verify this condition, drive the vehicle slowly with the parking brake slightly applied. If the noise stops, replace both rear brake cables with newly designed parking brake cables, part No. E6DZ-2A635-D (Sedan right side), E6DZ-2A635-F (Sedan left side), E6DZ-2A635-E (Station Wagon right side) and E6DZ-2A635-G (Station Wagon left side).

RATTLE NOISE

1986 SABLE & TAURUS

A rattle noise heard in the rear of the vehicle may be caused by a broken or stripped rear brake pressure control valve to lower suspension arm self-tapping screw which. This condition can be corrected by drilling a hole in the suspension arm and installing a new self-tapping screw.

CREAKING NOISE

TRACER

On 1987-88 Tracer models, a creaking or groaning noise from the front disc brakes while the vehicle is stopped or slightly moving forward may be caused by the metallic front disc brake pads. This noise will not affect brake performance. Installing a new design, semi-metalic front disc brake pads will reduce the noise.

BRAKE SHUDDER

XR4Ti

On 1985-87 XR4Ti, a shudder during light brake application from highway speeds may be caused by the front wheel and tire assembly, brake rotor, or front hub assembly. Check the following to determine the cause of shudder:

1. Check tire air pressure and balance.
2. Check wheel and tire assembly runout. Runout should not exceed .063 inch.
3. Check wheel runout. Wheel runout should not exceed .0197 inch.
4. Check brake rotor thickness variation. If thickness varies more than .0004 inch, replace rotor.
5. Check hub face runout. If runout exceeds .002 inch, replace hub.
6. Check rotor runout, if reading exceeds .003 inch, replace rotor.

GENERAL MOTORS CORP.

NOTE: Refer to "Applications" to determine which type brakes are used on vehicle being serviced.

TABLE OF CONTENTS

Applications

	Type No.
BUICK	
Century	5
Electra:	
1985-89 Estate Wagon	1
1985-86 Except Estate Wagon	5
1987-89 Except Estate Wagon	6
LeSabre:	
1985	1
1986-89	6
Regal (RWD)	3
Riviera:	
1985	1
Skyhawk	5
Skylark	5
Somerset	5
CADILLAC	
Cimarron:	
1985-88	5
Deville (FWD):	
1985-86	5
1987-89	6
Fleetwood (FWD):	
1985-86 & 1988	5
1987 & 1989	6
Rear Wheel Drive	1
CHEVROLET	
Beretta	5
Camaro	3
Cavalier	5
Celebrity	5
Chevette:	
1985-87	5
Citation:	
1985	5
Corsica	6
Full Size	1
Malibu	3
Monte Carlo	3
Nova	7
Spectrum	8
Sprint	9
GEO	
Spectrum	8
Metro	9
OLDSMOBILE	
Calais	5
Custom Cruiser	1
Cutlass Except Ciera (RWD)	3
Cutlass Ciera	5
Cutlass Cruiser	5
Firenza:	
1985-88	5
Toronado:	
1985	1
88:	
1985	1
1986-89	6
98:	
1985-86	5
1987-89	6
PONTIAC	
Bonneville:	
1985-86	3
1987-89	5
Firebird	3
Grand Am	5
Grand Prix (RWD)	3
Lemans	4
Parisienne	1
Safari	1
Sunbird	5
1000:	
1985-87	5
6000	5

General Information

INDEX

SERVICE PRECAUTIONS

When working on or around brake assemblies, care must be taken to prevent breathing asbestos dust, as many manufacturers incorporate asbestos fibers in the production of brake linings. During routine service operations the amount of asbestos dust from brake lining wear is at a low level due to a chemical breakdown during use, and a few precautions will minimize exposure.

1. Do not sand or grind brake linings unless suitable local exhaust ventilation equipment is used to prevent excessive asbestos exposure.
2. Wear a suitable respirator approved for asbestos dust use during all repair procedures.
3. When cleaning brake dust from brake parts, use a vacuum cleaner with a highly efficient filter system. If a suitable vacuum cleaner is not available, use a water soaked rag. **Do not use compressed air or dry brush to clean brake parts.**
4. Keep work area clean using same equipment as for cleaning brake parts.
5. Properly dispose of rags and vacuum cleaner bags by placing them in plastic bags.
6. Do not smoke or eat while working on brake systems. **Never use gasoline, kerosene, alcohol, motor oil, transmission fluid, or any fluid containing mineral oil to clean brake system components. These fluids will damage the rubber caps and seals. If system contamination is suspected, check brake fluid in the reservoir for dirt, discoloration, or separation (breakdown) of the brake fluid into distinct layers. Drain and flush the hydraulic system with clean brake fluid if contamination is suspected.**

GENERAL INSPECTION

BRAKE DRUMS

Any time the brake drums are removed for brake service, the braking surface diameter should be checked with a suitable brake drum micrometer at several points to determine if they are within the safe oversize limit stamped on the brake drum outer surface. If the braking surface diameter exceeds specifications, the drum must be replaced. If the braking surface diameter is within specifications, drums should be cleaned and inspected for cracks, scores, deep grooves, taper, out of round and heat spotting. If drums are cracked or heat spotted, they must be replaced. Minor scores should be removed with sandpaper. Grooves and large scores can only be removed by machining with special equipment, as long as the braking surface is within specifications stamped on brake drum outer surface. Any brake drum sufficiently out of round to cause vehicle vibration or noise while braking or showing taper should also be machined, removing only enough stock to true up the brake drum.

After a brake drum is machined, wipe the braking surface diameter with a denatured alcohol soaked cloth. If one brake drum is machined, the other should also be machined to the same diameter to maintain equal braking forces.

BRAKE LININGS & SPRINGS

Inspect brake linings for excessive wear, damage, oil, grease or brake fluid contamination. If any of the above conditions exists, brake linings should be replaced. Do not attempt to replace only one set of brake shoes; they should be replaced as an axle set only to maintain equal braking forces. Examine brake shoe webbing, hold-down and return springs for signs of overheating indicated by a slight blue color. If any component exhibits overheating signs, replace hold-down and return springs with new ones. Overheated springs lose their pull and could cause brake linings to wear out prematurely. Inspect all springs for sags, bends and external damage and replace as necessary.

Inspect hold-down retainers and pins for bends, rust and corrosion. If any of the above is found, replace as required.

BACKING PLATE

Inspect backing plate shoe contact surface for grooves that may restrict shoe movement and cannot be removed by lightly sanding with emery cloth or other suitable abrasive. If backing plate exhibits above condition, it should be replaced. Also inspect for signs of cracks, warpage and excessive rust, indicating need for replacement.

ADJUSTER MECHANISM

Inspect all components for rust, corrosion, bends and fatigue. Replace as necessary. On adjuster mechanism equipped with adjuster cable, inspect cable for kinks, fraying or elongation of eyelet and replace as necessary.

PARKING BRAKE CABLE

Inspect parking brake cable end for kinks, fraying and elongation and replace as necessary. Use a small hose clamp to compress clamp where it enters backing plate to remove.

Types 1, 2 & 3

INDEX

REMOVAL

1. Raise and support rear of vehicle, then remove tire and wheel assembly.
2. Remove brake drum. If brake lining is dragging on brake drum, back off brake adjustment by rotating adjustment screw. **If brake drum is rusted or corroded to axle flange and cannot be removed, lightly tap axle flange to drum mounting surface with a suitable hammer.**
3. Using brake spring pliers or equivalent, unhook primary and secondary return springs, **Figs. 1, 2 and 3. Observe location of brake parts being removed to aid during installation.**
4. Remove brake hold-down springs with suitable tool.
5. Lift actuating lever, then unhook actuating link from anchor pin and remove.
6. Remove actuating lever(s) and return spring.
7. Spread shoes apart and remove parking brake strut and spring.
8. Disconnect parking brake cable from lever, then remove brake shoes from backing plate.
9. Separate brake shoes by removing adjusting screw and spring, then unhook parking brake lever from shoe assembly.
10. Clean dirt from brake drum, backing plate and all other components. **Do not use compressed air or dry brush to clean brake parts. Many brake parts contain asbestos fibers which, if inhaled, can cause serious injury. Clean brake parts with a water soaked rag or a suitable vacuum cleaner to minimize airborne dust.**

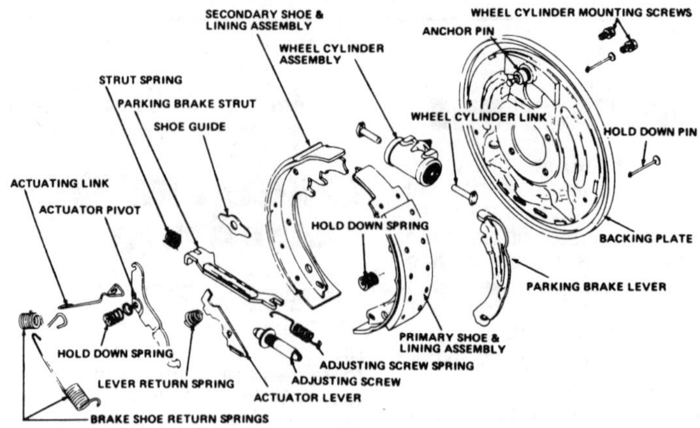

Fig. 1 Drum brake assembly. Type 1 typical

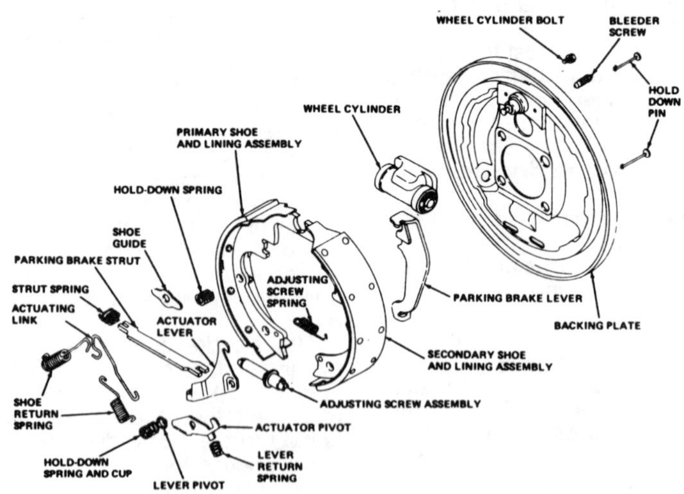

Fig. 2 Drum brake assembly. Type 2 typical

INSPECTION

1. Inspect components for damage and unusual wear. Replace as necessary.
2. Inspect wheel cylinders. Boots which are torn, cut or heat damaged indicate need for wheel cylinder replacement. **On type 1 brakes,** remove wheel cylinder links. Fluid spill from boot center hole indicates cup leakage and need for wheel cylinder replacement. **On type 2 and 3 brakes,** use a small screwdriver to pry center hole of boot away from piston. If fluid spills from center hole, cup leakage is indicated and wheel cylinder should be replaced. **On all types,** light fluid coatings on piston within cylinder is considered normal.
3. Inspect backing plate for evidence of axle seal leakage. If leakage exists, refer to individual car chapters for axle seal replacement procedures.
4. Inspect backing plate attaching bolts, and ensure they are tight.
5. Using fine emery cloth or other suitable abrasive, clean rust and dirt from shoe contact surface on backing plate.

INSTALLATION

1. Lubricate parking brake lever fulcrum with suitable brake lube, then attach lever to brake shoe. Ensure lever operates smoothly.
2. Connect brake shoes with adjusting screw spring, then position adjusting screw. **Ensure adjusting screw star wheel does not contact adjusting screw spring after installation and also ensure righthand thread adjusting screw is installed on left side of vehicle and lefthand thread adjusting screw is installed on right side of vehicle. When brake shoe installation is completed, ensure starwheel lines up with adjusting hole in backing plate.**
3. Lightly lubricate backing plate shoe contact surfaces with suitable brake lube, then the area where parking brake cable contacts backing plate.
4. Install brake shoes on backing plate while engaging wheel cylinder links (if

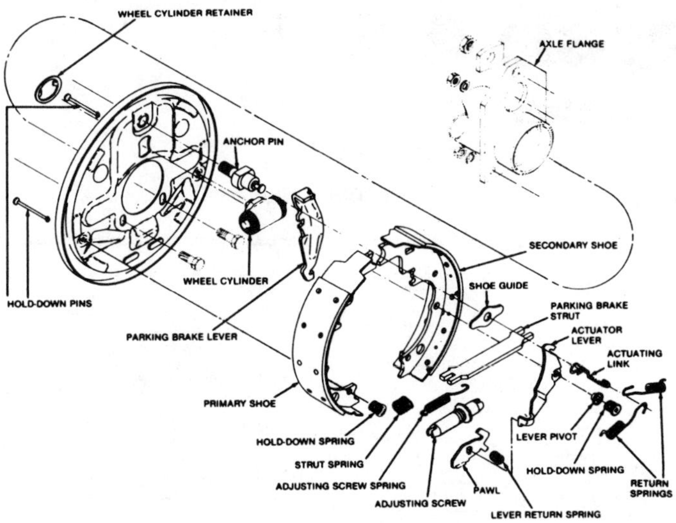

Fig. 3 Drum brake assembly. Type 3

equipped) with shoe webbing. Connect parking brake cable to parking brake lever. **The primary shoe (short lining) faces towards front of vehicle.**

5. Install actuating levers, actuating link and return spring, **Figs. 1, 2 and 3.**
6. Install hold-down springs with suitable tool.
7. Install primary and secondary shoe return springs using brake spring pliers or equivalent.
8. Using suitable brake drum to shoe gauge, measure brake drum inside diameter. Adjust brake shoes to dimension obtained on outside portion of gauge.
9. Install brake drum, wheel and tire assembly.
10. If any hydraulic connections have been opened, bleed brake system.
11. Adjust parking brake.
12. Inspect all hydraulic lines and connections for leakage and repair as necessary.
13. Check master cylinder fluid level and replenish as necessary.
14. Check brake pedal for proper feel and return.
15. Lower vehicle and road test. **Do not severely apply brakes immediately after installation of new brake linings or permanent damage may occur to linings, and/or brake drums may become scored. Brakes must be used moderately during first several hundred miles of operation to ensure proper burnishing of linings.**

Type 4

INDEX

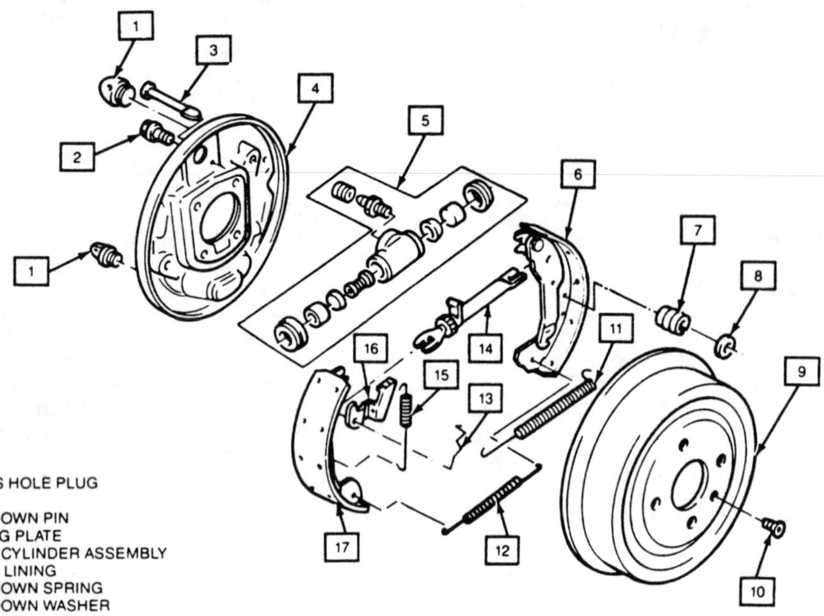

1. ACCESS HOLE PLUG
2. BOLT
3. HOLD-DOWN PIN
4. BACKING PLATE
5. WHEEL CYLINDER ASSEMBLY
6. SHOE & LINING
7. HOLD-DOWN SPRING
8. HOLD-DOWN WASHER
9. BRAKE DRUM
10. DETENT SCREW
11. UPPER RETURN SPRING
12. LOWER RETURN SPRING
13. SPRING CONNECTING LINK
14. ADJUSTER ASSEMBLY
15. ACTUATOR SPRING
16. ADJUSTER ACTUATOR
17. ADJUSTER SHOE & LINING

Fig. 1 Drum brake assembly. Type 4

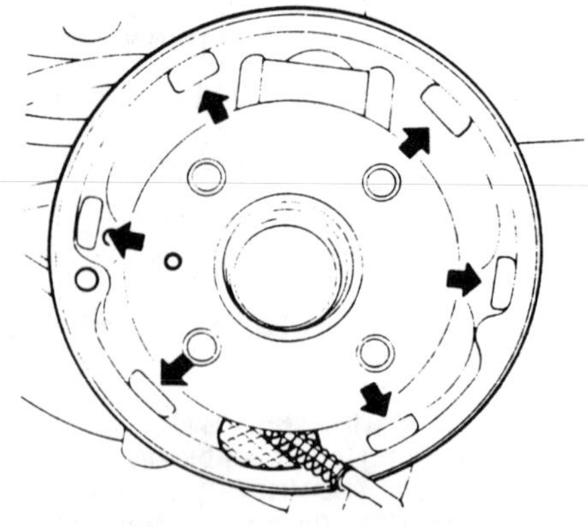

Fig. 2 Backing plate lubrication points

REMOVAL

1. Raise and support vehicle.
2. Mark relationship between rear wheel and hub, then remove wheel and tire assembly.
3. Remove brake drum detent screw, then the brake drum. If drum is difficult to remove, proceed as follows:
   a. Ensure parking brake cable is released.
   b. Loosen parking brake cable.
   c. Remove access hole plug from backing plate, then move parking brake lever until lever stop rests on brake shoe.
4. Remove upper and lower return springs, then the adjuster actuator and actuator spring, **Fig. 1.**
5. Press brake shoes outward slightly and remove adjuster assembly.
6. Remove hold-down springs, washers and pins using a suitable tool.
7. Remove brake shoes from backing plate and parking brake cable.

INSPECTION

1. If any parts are of doubtful strength or quality due to heat discoloration, or are worn, replace them.
2. Inspect wheel cylinder dust boots for signs of excessive wear or damage.
3. Inspect adjusting screw for smooth operation over full length.
4. Clean adjusting screw components in denatured alcohol.
5. Apply suitable lubricant to adjuster screw threads, inside diameter of socket and socket face. Adequate lubrication is achieved when a continuous bead of lubricant is at open end of adjuster nut and socket when threads are fully engaged.
6. Clean dirt and/or rust from brake drum, backing plate and all other components. **Do not use compressed air or dry brush to clean brake parts. Many brake parts contain asbestos fibers which, if inhaled, can cause serious injury. Clean brake parts with a water soaked rag or a suitable vacuum cleaner to minimize airborne dust.**
7. Ensure adjuster nut turns freely on adjuster screw.

INSTALLATION

1. Lubricate backing plate at points shown, **Fig. 2.**
2. Connect parking brake lever to parking brake cable.
3. Ensure parking brake cable is properly routed, then install shoes and lining to backing plate with hold-down spring, washer and pin.
4. Position adjuster shoe and lining and lower return spring against backing plate. Ensure spring is properly positioned under anchor plate.
5. Install hold-down spring, washer and pin.
6. Turn adjuster in to stop, then install adjuster assembly, **Fig. 3.** Ensure spring clip is positioned toward backing plate.
7. Install adjuster actuator onto pin and actuator spring.
8. Install spring connecting link to pin and press into brake hole.
9. Connect upper spring from connecting link to brake shoe.
10. Install brake drum and detent screw, then the wheel and tire assembly.
11. Adjust service and parking brakes as necessary.

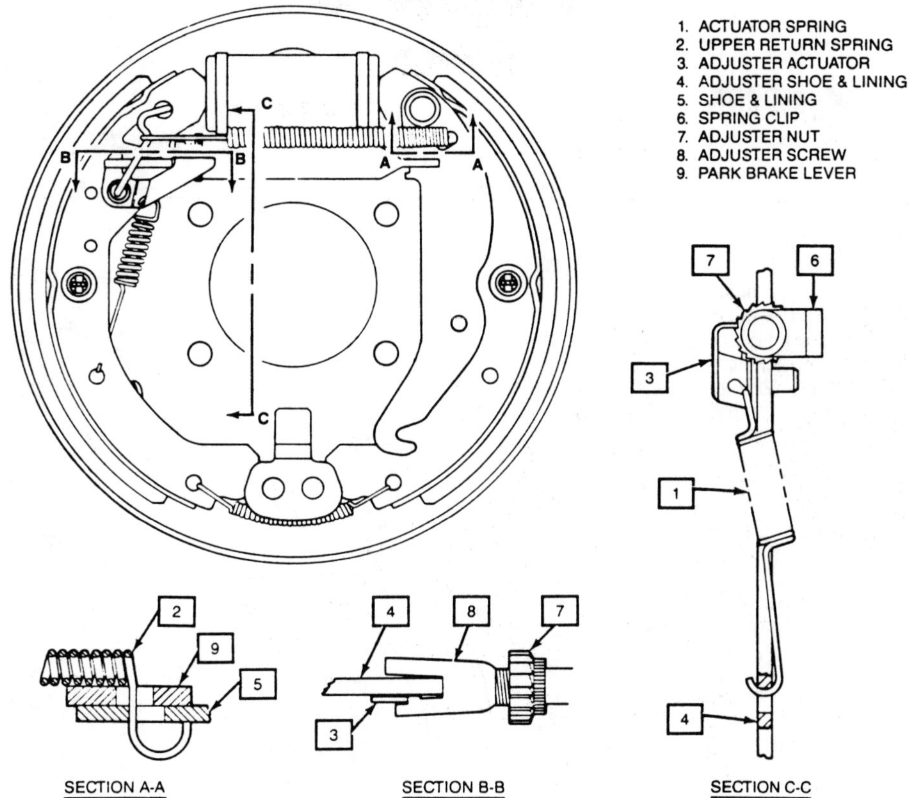

Fig. 3 Cross-sectional view of drum brake assembly

Type 5

INDEX

REMOVAL

1. Raise and support rear of vehicle, then remove tire and wheel assembly.
2. Remove brake drum. If brake lining is dragging on brake drum, back off brake adjustment by rotating adjusting screw. **If brake drum is rusted or corroded to axle flange and cannot be removed, lightly tap axle flange to drum mounting surface with a suitable hammer.**
3. Using brake spring pliers or equivalent, remove brake return springs, **Fig. 1.**
4. Using suitable pliers, remove hold-down springs and lever pivot.
5. While lifting up on actuator lever, disconnect actuating link.
6. Remove actuator lever and return spring.
7. Remove parking brake strut and spring.
8. Disconnect parking brake cable, then remove brake shoe and lining assembly.
9. Remove adjusting screw assembly and spring as follows:
   a. Note position of adjusting spring.
   b. Remove retaining ring, pin, and parking brake lever from secondary shoe.
   c. Remove adjusting screw and spring. **Do not interchange adjusting screws from right and left brake assemblies.**

INSPECTION

1. Inspect brake components for damage and/or wear. Replace as necessary.
2. Inspect wheel cylinders. Excessive fluid indicates cup leakage and need for wheel cylinder replacement.
3. Inspect backing plate for evidence of axle seal leakage.
4. Inspect backing plate attaching bolts. Ensure bolts are tight.
5. Check adjuster operation. If adjusters are worn, frozen or loose, replace adjuster and backing plate assembly as required.
6. Using fine emery cloth or other suitable abrasive, clean rust and/or dirt from shoe contact surface on backing plate.

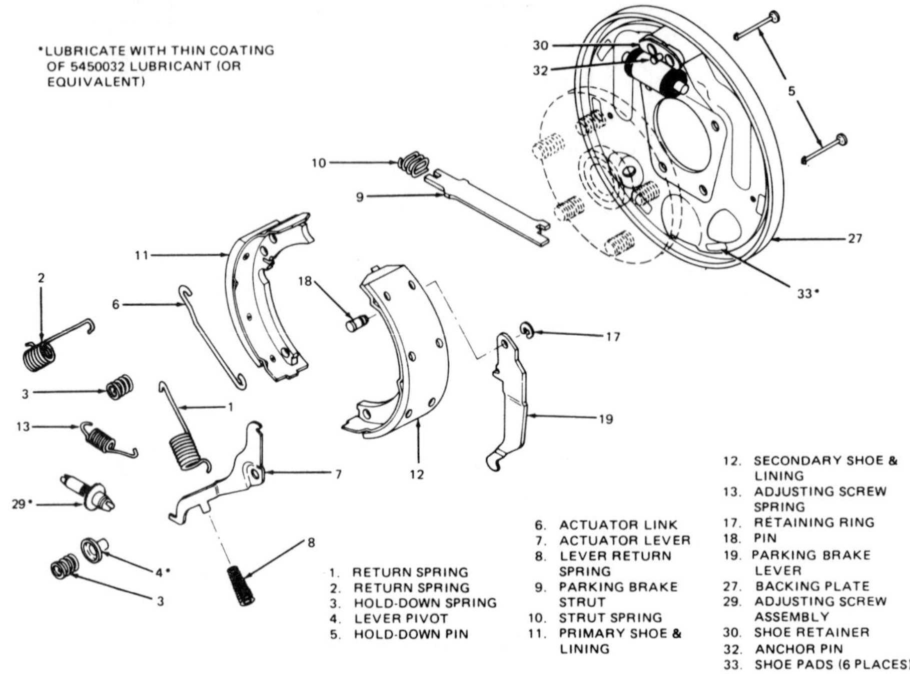

Fig. 1 Drum brake assembly. Type 5

INSTALLATION

1. Install parking brake lever, pin and retaining ring onto secondary shoe.
2. Apply silicone brake lubricant onto adjuster screw threads, inside diameter of socket and socket face. Adequate lubrication is achieved when a continuous bead of lubricant is at open end of pivot nut and socket when adjuster threads are completely engaged.
3. Install adjusting screw assembly and spring. **Spring coils must not overlap star wheel. Do not interchange right and lefthand springs.**
4. Connect parking brake cable, then install shoe and lining assembly.
5. Using a suitable tool, spread brake shoes apart, then install parking brake strut and spring. **Ensure parking brake strut is properly positioned. Ensure strut end without spring engages parking brake lever. Ensure strut end with spring engages primary brake shoe.**
6. Install actuator lever and return spring.
7. Install actuator link onto anchor pin.
8. While holding up on actuator lever, install link onto lever.
9. Install hold-down pins, lever pivot and hold-down springs.
10. Install shoe return springs.
11. Install brake drum, tire and wheel assembly.
12. If any hydraulic connections have been opened, bleed brake system.
13. Adjust brakes.

Type 6

INDEX

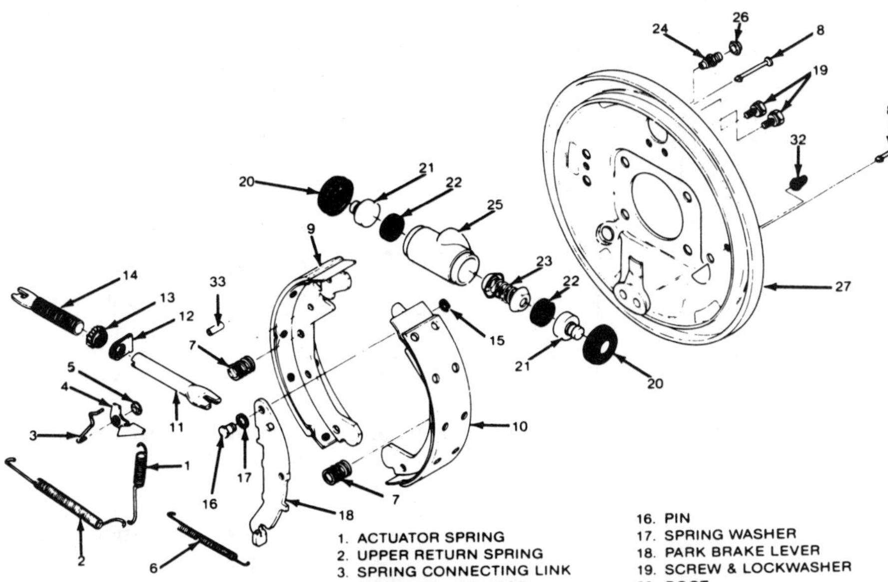

1. ACTUATOR SPRING
2. UPPER RETURN SPRING
3. SPRING CONNECTING LINK
4. ADJUSTER ACTUATOR
5. SPRING WASHER
6. LOWER RETURN SPRING
7. HOLD-DOWN SPRING ASSEMBLY
8. HOLD-DOWN PIN
9. ADJUSTER SHOE & LINING
10. SHOE & LINING
11. ADJUSTER SOCKET
12. SPRING CLIP
13. ADJUSTER NUT
14. ADJUSTER SCREW
15. RETAINING RING
16. PIN
17. SPRING WASHER
18. PARK BRAKE LEVER
19. SCREW & LOCKWASHER
20. BOOT
21. PISTON
22. SEAL
23. SPRING ASSEMBLY
24. BLEEDER VALVE
25. WHEEL CYLINDER
26. BLEEDER VALVE CAP
27. BACKING PLATE ASSEMBLY
32. ACCESS HOLE PLUG
33. ADJUSTER PIN

Fig. 1 Drum brake assembly. Type 6

REMOVAL

1. Raise and support vehicle.
2. Mark relationship of wheel to axle flange, then remove wheel and tire assembly.
3. Mark relationship of brake drum to axle flange, then remove drum. **If brake drum is difficult to remove, back off parking brake cable adjustment, remove inspection hole plug from backing plate, insert screwdriver through hole and press in to push parking brake lever off stop.**
4. Using suitable tool, remove actuator spring, then upper return spring, **Fig. 1.**
5. Remove spring connecting link, adjuster actuator, and spring washer.
6. Remove hold-down springs and pins.
7. Disconnect parking brake cable, then remove shoe and lining assembly together with lower return spring.
8. Remove adjusting screw assembly, and lower return spring.
9. Remove retaining ring, pin, spring washer and parking brake lever from shoe and lining assembly.

INSPECTION

1. If any parts are of doubtful strength or quality due to heat discoloration, or are worn, replace them.
2. Inspect wheel cylinder dust boots for signs of excessive wear or damage.
3. Inspect adjusting screw for smooth operation over full length.
4. Clean adjusting screw components in denatured alcohol. **Ensure spring clip is installed in correct position, Fig. 1.**
5. Apply suitable lubricant to adjuster screw threads, inside diameter of socket and socket face. Adequate lubrication is achieved when a continuous bead of lubricant is at open end of adjuster nut and socket when threads are fully engaged.
6. Clean dirt and/or rust from brake drum, backing plate and all other components. **Do not use compressed air or dry brush to clean brake parts. Many brake parts contain asbestos fibers which, if inhaled, can cause serious injury. Clean brake parts with a water soaked rag or a suitable vacuum cleaner to minimize airborne dust.**
7. Ensure adjuster nut turns freely on adjuster screw.

INSTALLATION

1. Press adjuster pin and spring washer into shoe and lining so that pin projects .275-.283 inch from the side on which the adjuster actuator is installed.
2. Install adjuster shoe and lining, to front of vehicle on left side or to rear of vehicle on right side.
3. Install parking brake lever on shoe and lining with spring washer, pin and retaining ring. Install spring washer with concave side against parking brake lever.
4. Attach adjuster shoe and lining to remaining shoe with lower return spring.
5. Install both shoes on backing plate, securing with hold-down pins and springs.
6. Install adjuster actuator, spring connecting link and actuator spring onto adjuster shoe and lining.
7. Install adjusting screw assembly, ensuring adjuster screw engages notch in adjuster shoe. Ensure adjuster screw spring clip is properly positioned and the adjuster actuator is properly engaged in notch in adjuster screw. **Do not over extend upper return spring. The spring will be damaged if extended length exceeds 5.49 inches.**
8. Install upper return spring by inserting angled hook end of spring through parking brake lever and shoe lining, using suitable tool to pull spring straight across and then down to hook into crook in spring connecting link.
9. Back off parking brake adjustment until parking brake lever rests on brake shoe.
10. Install brake drum and tire and wheel assembly, then lower vehicle.
11. Apply and release service brake at least 25 times, then continue to pump brake pedal until clicking noise at drum brake self adjusters stops on both sides.
12. Adjust parking brake.

Type 7

INDEX

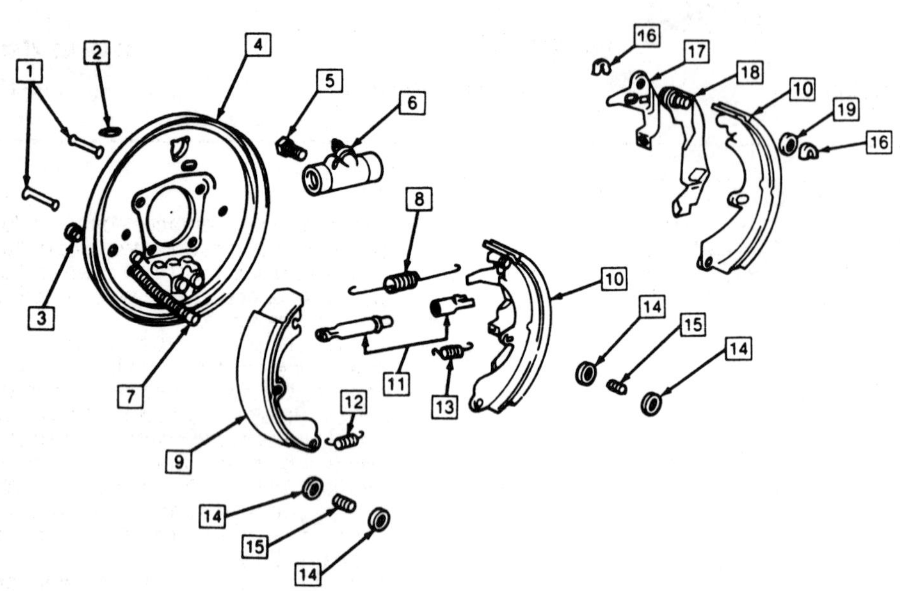

1. HOLD-DOWN PIN
2. PLUG
3. INSPECTION HOLE PLUG
4. BACKING PLATE
5. BOLT
6. WHEEL CYLINDER
7. PARKING BRAKE CABLE
8. RETURN SPRING
9. FRONT SHOE
10. REAR SHOE
11. STRUT
12. ANCHOR SPRING
13. ADJUSTING LEVER SPRING
14. RETAINER
15. HOLD-DOWN SPRING
16. C-WASHER
17. AUTOMATIC ADJUSTING LEVER
18. PARKING BRAKE LEVER
19. SHIM

Fig. 1 Drum brake assembly. Type 7

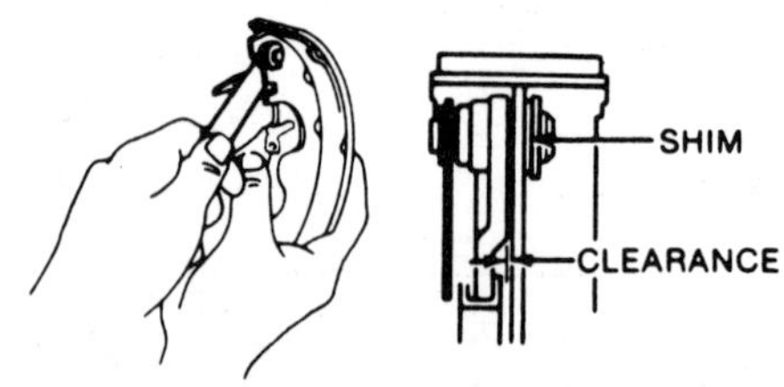

SHIM THICKNESS	
THICKNESS	THICKNESS
0.2 MM (0.008 IN.)	0.5 MM (0.020 IN.)
0.3 MM (0.012 IN.)	0.6 MM (0.024 IN.)
0.4 MM (0.016 IN.)	0.9 MM (0.035 IN.)

Fig. 2 Checking lever to shoe clearance

REMOVAL

1. Raise and support vehicle.
2. Mark relationship of wheel to axle, then remove wheel and tire assembly.
3. Remove brake drum. If drum is difficult to remove, insert screwdriver through hole in backing plate and hold automatic adjusting lever away from adjusting bolt, then, using a second screwdriver, reduce brake shoe adjustment.
4. Remove return spring, **Fig.1**
5. Remove hold-down spring, retainers and pin retaining front shoe.
6. Disconnect anchor spring from front shoe, then remove front shoe.
7. Remove anchor spring.
8. Remove hold-down spring, retainers and pin retaining rear shoe.
9. Using screwdriver, disconnect parking brake cable from anchor plate.
10. Using pliers, disconnect parking brake cable from lever and remove rear shoe together with strut.
11. Remove adjusting lever spring, then the strut together with return spring.
12. Remove parking brake lever and automatic adjusting lever from rear shoe by prying out "C" washer and removing shims and levers.
13. Clean dirt from brake drum, backing plate and all other components. **Do not use compressed air or dry brush to clean brake parts. Many brake parts contain asbestos fibers which, if inhaled, can cause serious injury. Clean brake parts with a water soaked rag or a suitable vacuum cleaner to minimize airborne dust.**

INSPECTION

1. Check brake drum, shoes, strut, auto adjuster lever, springs and backing plate for wear, distortion, cracks or other abnormal conditions.
2. If any parts are of doubtful strength or quality due to damage, heat discoloration, stress or wear, replace them.
3. Measure brake drum inside diameter and the brake shoe lining thickness.
4. Inspect lining and drum for proper contact.

INSTALLATION

1. Apply suitable lubricant to backing plate brake shoe contact points, anchor plate brake shoe contact points, strut and adjusting bolt contact points and the strut and brake shoe contact points.
2. Install parking brake lever and automatic adjusting lever to rear shoe as follows:
   a. Temporarily install levers and shim with a new "C" washer.
   b. Using feeler gauge, measure clearance between shoe and lever, **Fig. 2.**
   c. If clearance is not 0-.0138 inch, adjust by installing replacement shim.
   d. Using pliers, stake "C" washer.
3. Set strut and return spring in place on rear shoe and install adjusting lever spring.
4. Install rear shoe as follows:
   a. Using pliers, connect parking brake cable to lever.
   b. Pass parking brake cable through notch in anchor plate.
   c. Set rear shoe in place with end of shoe inserted in wheel cylinder and other end in anchor plate.
   d. Install hold-down spring, retainers and pin.
5. Install front shoe as follows:
   a. Install anchor spring between front and rear shoes.
   b. Set front shoe in place with end of shoe inserted in wheel cylinder and the strut in place.
   c. Install hold-down spring, retainers and pin.
   d. Connect return spring.
6. Check operation of automatic adjuster mechanism as follows:
   a. Move parking brake lever of rear shoe back and forth and check whether adjusting bolt turns. If bolt does not turn, check brakes for incorrect installation.
   b. Adjust strut length to shortest possible distance.
   c. Install brake drum.

d. Pull parking brake lever all the way up until a clicking sound can no longer be heard.
7. Check clearance between brake shoes and drum. Remove drum and measure brake drum inside diameter and diameter of brake shoes.
8. If clearance is not .024 inch, check parking brake system.
9. Install brake drum and the wheel and tire assembly.
10. Fill master cylinder as necessary and bleed brake system.
11. Check for fluid leakage.

Type 8

INDEX

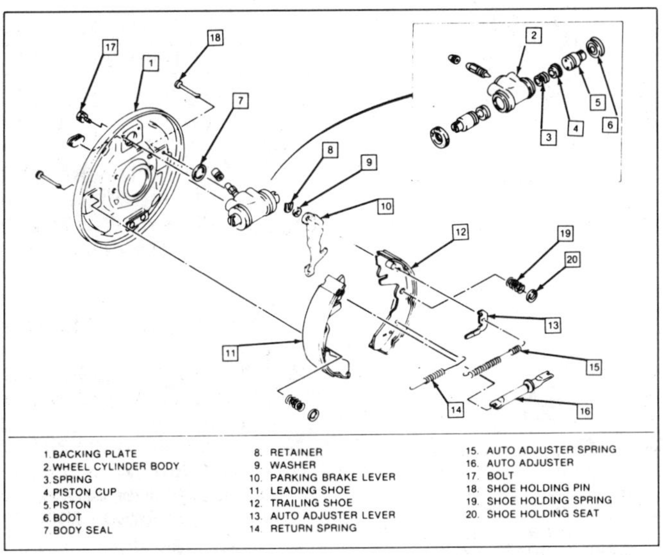

Fig. 1 Rear drum brake assembly. Type 8

REMOVAL

1. Raise and support rear of vehicle.
2. Remove wheels, then brake hubs and drums.
3. Remove return spring and auto adjuster spring.
4. Remove leading shoe hold-down spring and pin, **Fig. 1.**
5. Remove leading shoe and auto adjuster.
6. Remove trailing shoe hold-down spring and pin, then shoe.
7. Disconnect parking brake cable from lever, then remove trailing shoe.
8. Remove parking brake lever from trailing shoe.

INSPECTION

1. If any parts are of doubtful strength or quality due to heat discoloration, or are worn, replace them.
2. Inspect wheel cylinder dust boots for signs of excessive wear or damage.
3. Inspect auto adjuster gear for smooth operation over full length.
4. Clean adjusting gear components in denatured alcohol.
5. Apply suitable lubricant to adjuster screw threads, inside diameter of socket and socket face. Adequate lubrication is achieved when a continuous bead of lubricant is at open end of adjuster nut and socket when threads are fully engaged.
6. Clean dirt and/or rust from brake drum, backing plate and all other components. **Do not use compressed air or dry brush to clean brake parts. Many brake parts contain asbestos fibers which, if inhaled, can cause serious injury. Clean brake parts with a water soaked rag or a suitable vacuum cleaner to minimize airborne dust.**
7. Ensure adjuster gear turns freely on adjuster screw.

INSTALLATION

1. Apply suitable lubricant to backing plate and cylinder piston brake shoe contact points, adjuster and parking lever pin contact points.
2. Install parking brake lever to trailing shoe, then connect parking brake cable to lever.
3. Install trailing shoe to backing plate with hold-down pin and spring, then install brake adjuster.
4. Install leading shoe to backing plate with hold-down pin and spring.
5. Install auto adjuster lever and auto adjuster spring, then install return spring.
6. Install brake drum, then adjust brakes.
7. Adjust bearing preload.
8. If hydraulic connections have been opened, bleed system

Type 9

INDEX

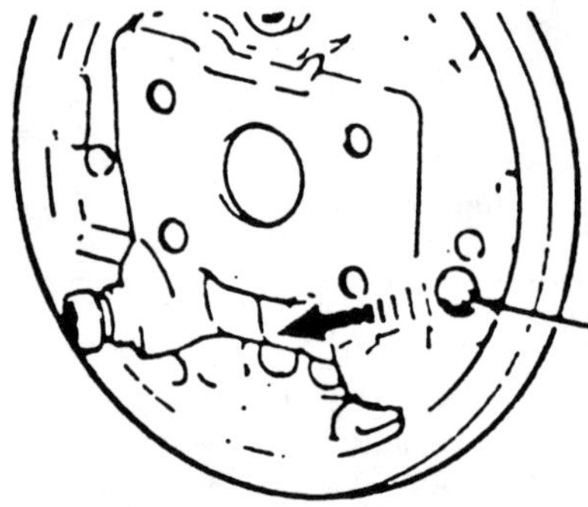

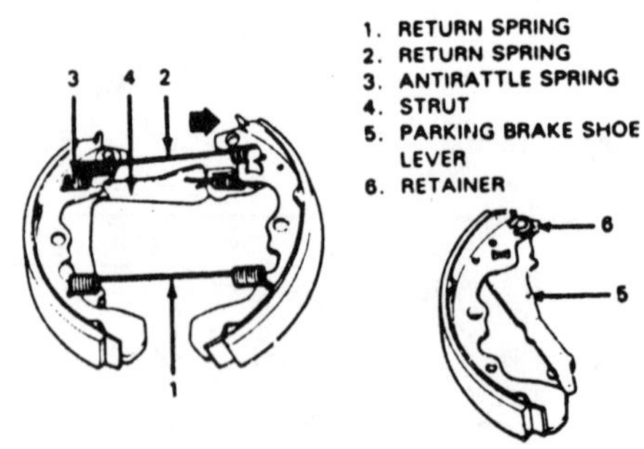

Fig. 2 Brake shoe return spring identification

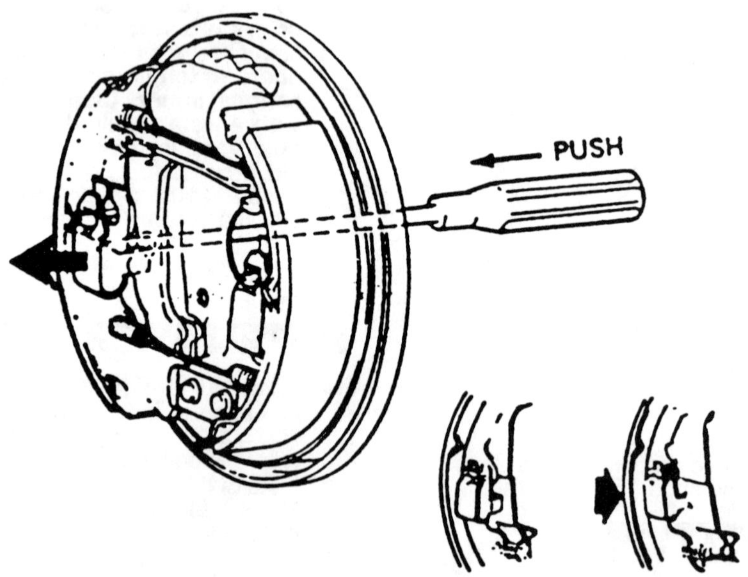

Fig. 1 Removing backing plate plug

REMOVAL

1. Raise and support vehicle.
2. Remove spindle cap by hammering lightly at 3 points around cap.
3. Remove cotter pin or unfasten staked portion of nut, then remove castle nut and washer.
4. Loosen parking brake cable adjusting nuts.
5. Remove backing plate plug, **Fig. 1.**
6. Insert screwdriver into plug hole until it contacts shoe hold down spring and push in direction shown.**Fig. 1.** This pushes hold down spring up and releases parking brake shoe lever from hold down spring, resulting in added clearance between shoe and drum.
7. Using slide hammer No. J-2619-01 or equivalent and brake drum remover No. J-34866 or equivalent, pull off brake drum.
8. Remove brake shoe hold down springs by turning hold down pins.
9. Disconnect parking brake cable from parking brake shoe lever and remove brake shoes.
10. Remove spring 1 in **Fig. 2,** pull primary shoe in direction of arrow and disengage strut 4 and return spring 2.
11. Disconnect return spring 3 from shoe.
12. Disconnect parking brake shoe lever from shoe.

INSPECTION

1. If any parts are of doubtful strength or quality due to heat discoloration, or are worn, replace them.
2. Inspect wheel cylinder dust boots for signs of excessive wear or damage. If any leakage is apparent replace or rebuild wheel cylinder.
3. Clean dirt and/or rust from brake drum, backing plate and all other components. **Do not use compressed air or dry brush to clean brake parts. Many brake parts contain asbestos fibers which, if inhaled, can cause serious injury. Clean brake parts with a water soaked rag or a suitable vacuum cleaner to minimize airborne dust.**

INSTALLATION

1. Assemble brake shoes, levers and springs as shown in **Fig. 3.**
2. Push shoe hold down springs down into place. Turn hold down pins to engage springs.
3. To minimize dimension 'A-A,' **Fig. 4,** push strut towards backplate while pushing out on shoe as shown.
4. Position tab of spring clip behind parking lever.
5. Install brake drum and torque castle nut to 58-86.5 ft. lbs. Install cotter pin.
6. Install spindle cap.
7. Install wheel.
8. Depress brake pedal several times to obtain proper drum to shoe clearance and adjust parking brake.
9. Check that brake drum does not drag.
10. Lower vehicle and test brake operation.

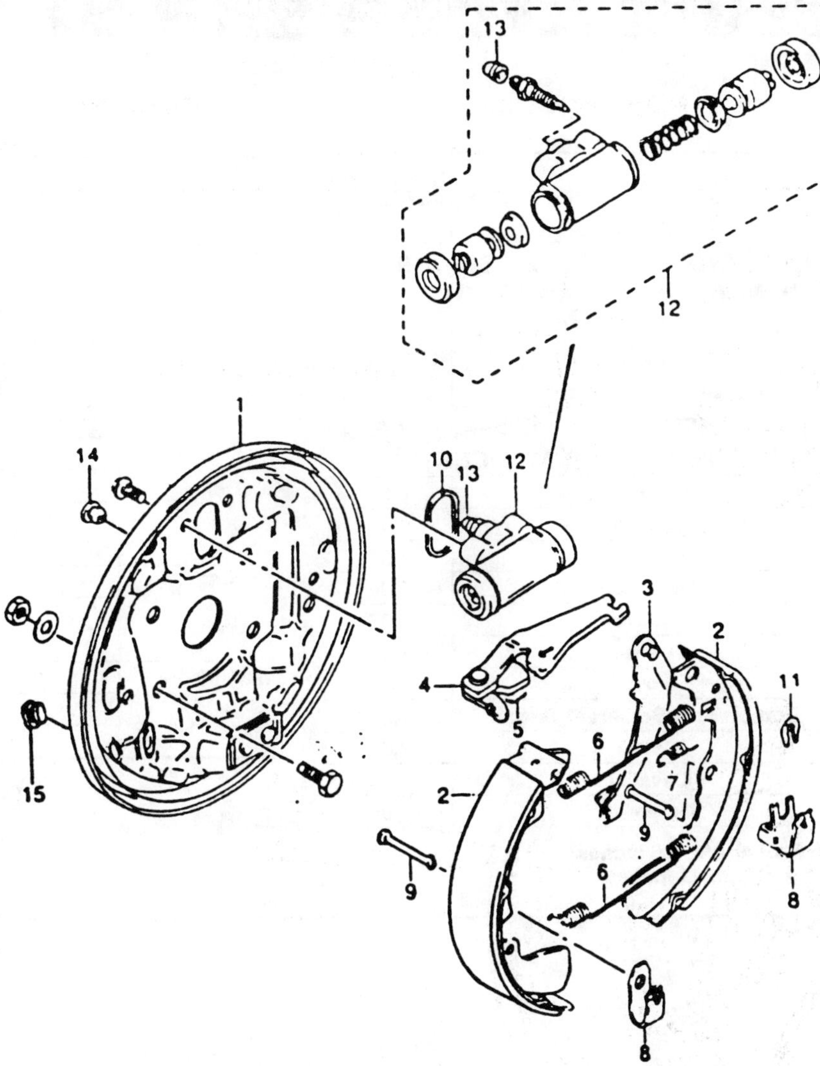

1. Brake back plate
2. Brake shoe
3. Parking brake shoe lever
4. Brake strut
5. Quadrant spring
6. Shoe return spring
7. Antirattle spring
8. Shoe hold down spring
9. Shoe hold down pin
10. Packing
11. Parking level retainer
12. Wheel cylinder
13. Bleeder plug cap
14. Rubber plug
15. Rubber plug

Fig. 3 Assembling rear brake unit. Type 9

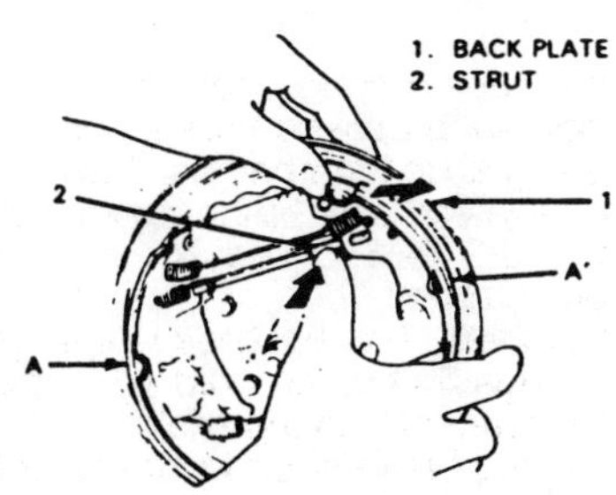

Fig. 4 Brake lining installation

Drum Brake Specifications

Year	Model	Brake Drum Inside Dia. Inch
BUICK, CADILLAC, CHEVROLET, OLDSMOBILE & PONTIAC (REAR WHEEL DRIVE MODELS)		
1985-89	Except Chevette, 1000 & Cadillac	①
	Chevette & 1000	7.88
1985-89	Cadillac Brougham	11
BUICK, CADILLAC, CHEVROLET, OLDSMOBILE & PONTIAC (FRONT WHEEL DRIVE MODELS)		
1985	Citation	7.87
	Riviera	9.50
	Skylark	7.87
	Toronado	9.50
1985–88	Cimarron	7.87
	Firenza	7.87
	Nova	7.87
	Somerset & Somerset Regal	7.87
1985-89	Calais	7.87
	Cavalier	7.87
	Celebrity	8.86
	Century	8.86
	Cutlass Ciera	8.86
	Cutlass Cruiser	8.86
	DeVille & Fleetwood	8.86
	Electra & Park Ave.	8.86
	Grand Am	7.87
	Skyhawk	7.87
	Sunbird	7.87
	Ninety-Eight	8.86
	6000	8.86
1986-89	LeSabre	8.86
	Eighty-Eight	8.86
	Skylark	7.87
1987-89	Beretta	8.86
	Bonneville	7.87
	Corsica	7.87

①—With 4¾ inch bolt circle, 9.5 inches; with 5 inch bolt circle, 11 inches.

ANTI-LOCK BRAKES

INDEX

CHRYSLER MOTORS/EAGLE

TABLE OF CONTENTS

Anti-Lock Brake System (Bosch ABS-3)

INDEX

DESCRIPTION

The Anti-Lock Braking System (ABS) prevents the wheels from locking up when braking, regardless of the surface conditions. This allows the car to stop in a shorter distance, and allows the driver to maintain directional control of the vehicle during heavy braking.

During normal braking conditions, the ABS operates like a conventional diagonally split, hydraulic power assist system. During heavy braking, however, each wheel's braking pressure is modulated according to its speed. To maintain vehicle stability, both rear wheels receive the same signal.

There are four major components, **Fig. 1,** in the ABS which act in unison to control brake operation.

HYDRAULIC ASSEMBLY

The hydraulic assembly acts as an integral master cylinder and hydraulic booster and contains the wheel circuit valves for brake pressure modulation.

WHEEL SPEED SENSOR (WSS)

Each wheel is equipped with a wheel speed sensor to transmit speed information to the control module during ABS operation.

ANTI-LOCK BRAKE CONTROL MODULE (ABCM)

The ABCM receives wheel speed information, controls anti-lock operation and monitors system performance.

PUMP/MOTOR ASSEMBLY

The pump/motor assembly is an electric pump that takes low pressure fluid from the hydraulic assembly and stores it under pressure in an accumulator for power assist and ABS operation.

DIAGNOSIS

The ABS system has a self-diagnostic mode which helps isolate fault codes which are displayed by the flashing red brake warning lamp on the instrument panel. One fault code is stored at a time, and is erased when ignition is turned off.

VISUAL INSPECTION

A visual inspection is the first step in the diagnostic procedure. All components must be checked for proper operation or condition as listed in **Fig. 2.**

SELF-DIAGNOSIS

Before using the self-diagnostic procedure, or symptom diagnosis charts, a visual inspection of system components should be made to verify that they are in good working order **Fig. 2.**

1. With ignition on, ensure the parking brake is fully released so that it is not causing brake warning lamp to light.
2. Check that brake warning or anti-lock warning lamp is illuminated, indicating a fault has been detected.
3. When a fault is detected, with ignition on, depress brake pedal firmly. After about five seconds if a fault code is present the brake warning light will begin to flash. The number of times the lamp flashes is the number of the code. **If ignition is turned off, fault code will be erased.** If brake pedal is held while lamp is flashing trouble code, the code will repeat after a ten second delay. Releasing the brake pedal stops code from repeating.
4. If lamp does not come on at all, crank the engine and perform a brake warning lamp bulb check and repair as necessary. If the lamp does light and no fault codes are present, the complaint should be verified and diagnosed with appropriate symptom diagnosis charts, **Figs. 3 through 14.**
5. To exit from diagnostic mode, release the brake pedal.

FUNCTIONAL CHECK

The functional check, **Fig. 15,** is done to determine the nature of system troubles. The process will isolate the proper system diagnosis procedure, **Figs. 3 through 14,** or trouble code diagnostic procedure, **Figs. 16 through 21,** to be followed.

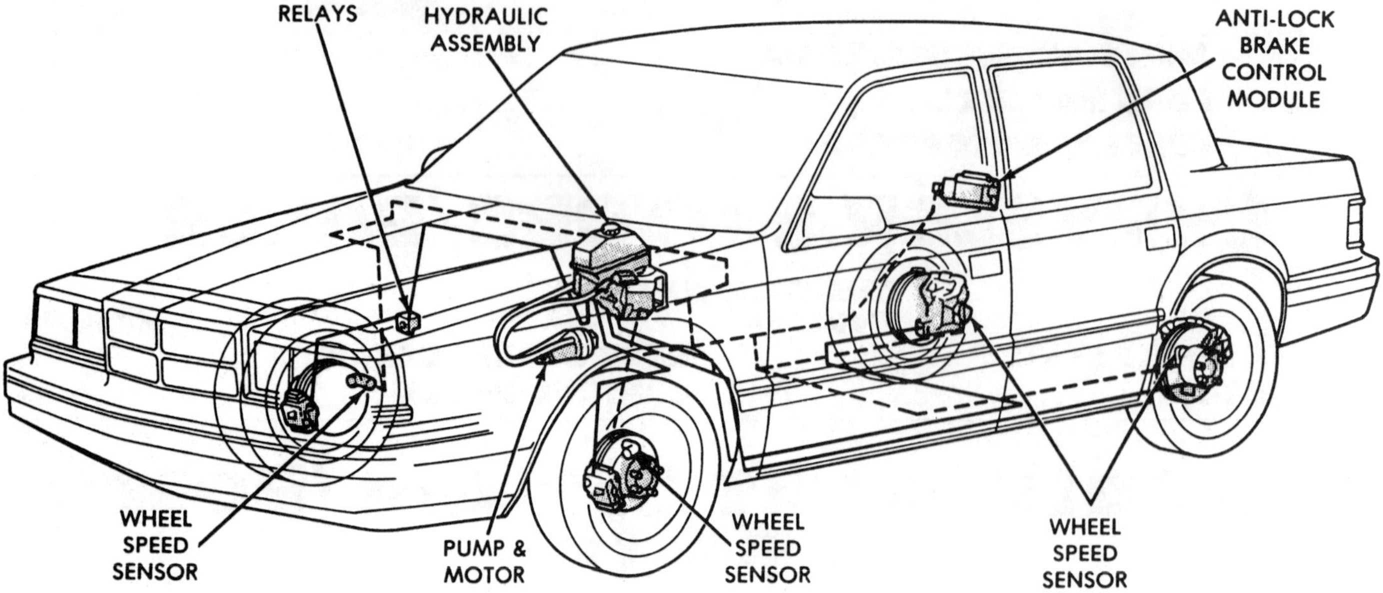

Fig. 1 Anti-Lock Brake System (ABS) major components

ITEM	INSPECT FOR	CORRECTIVE ACTION
Brake Fluid Reservoir	Low fluid Level	Add fluid as required. Determine cause of fluid loss and repair.
ABCM	Proper connector engagement. External Damage	Repair as required Verify defect and repair as required.
Hydraulic Assembly	Proper Connector Engagement External Leaks Damaged wiring/connectors Control Pressure Switch Fluid Level Sensor Defect	Repair as required Repair Leaks as required Repair as required Repair as required Repair as required
Pump/Motor Assembly and Hoses	Proper assembly Damaged/Leaking Hoses Damaged wiring/connector	Install components properly Repair hoses as required Repair as required
Parking Brake	Full release Park Brake Switch	Operate manual release lever to verify operation. Adjust cable or repair release mechanism as required Verify correct operation. Repair as required
Front & Rear Wheel Speed Sensors	Proper connector engagement Broken or damaged wires	Repair as required Go to appropriate ABS Code chart and verify fault. Correct as required
OVPR & Pump/Motor Relays	Proper connector engagement Loose wires or terminals	Repair as required Repair as required
Pump/Motor Ground	Corroded, broken or loose eyelets	Repair as required

Fig. 2 Visual inspection chart

SYMPTOM	REFER TO
No power assist (with or without BRAKE warning lamp on) High pedal effort (with or without BRAKE warning lamp on) Pump does not run Pumps runs for extended periods of time Pump runs every brake apply BRAKE warning lamp stays on for more than 20 seconds after key-on BRAKE warning lamp comes on while braking BRAKE warning lamp on continuously Brake boost system will not hold pressure (excessive pressure leakdown)	Chart C
BRAKE warning lamp inoperative	Chart D
LOW BRAKE FLUID message displayed and BRAKE warning lamp on continuously (EVID equipped cars only)	Chart E
BRAKE warning lamp on and power assist available	Chart F
No system power	Chart G
ANTI-LOCK warning lamp on (BRAKE warning lamp off) and no Fault Codes set	Chart H
Low or spongy pedal	Chart J

Fig. 3 Symptom diagnosis table

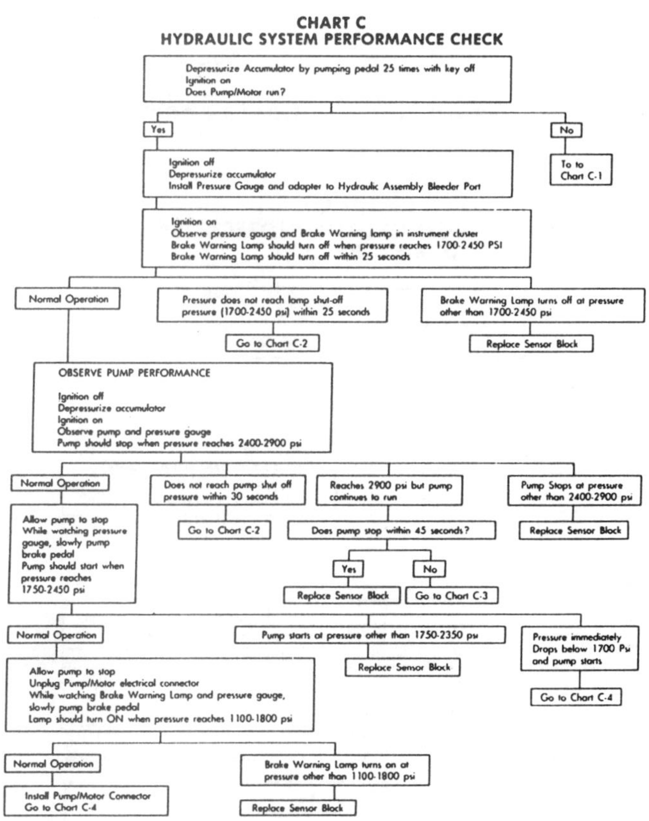

Fig. 4 Symptom diagnosis chart C

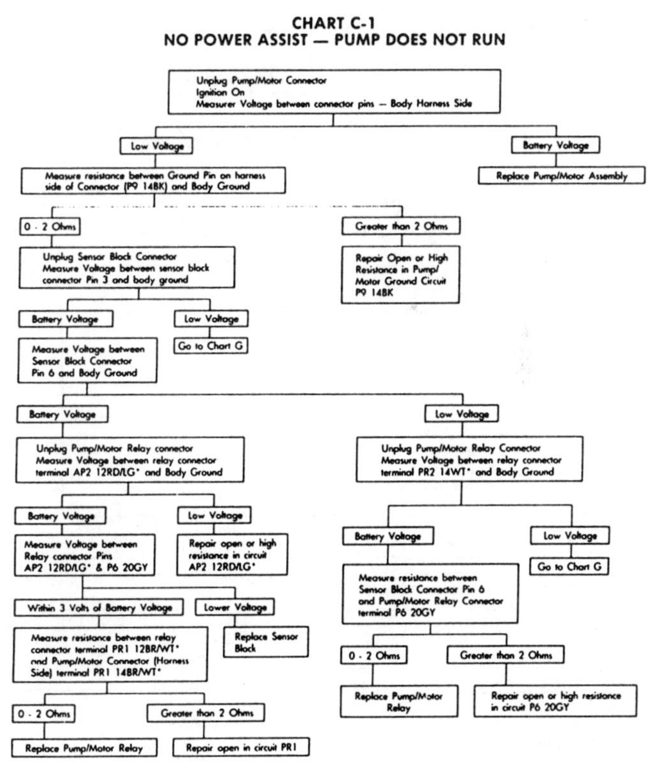

Fig. 5 Symptom diagnosis chart C-1

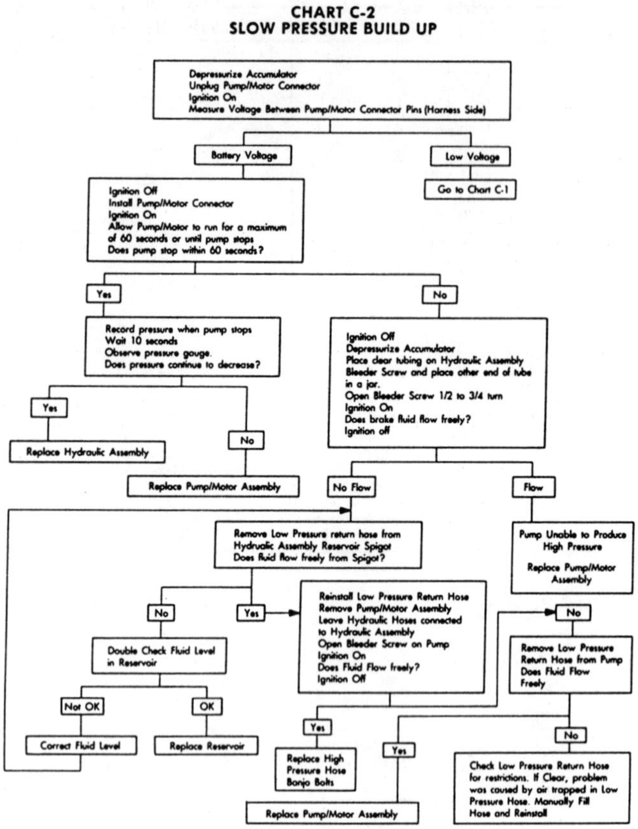

Fig. 6 Symptom diagnosis chart C-2

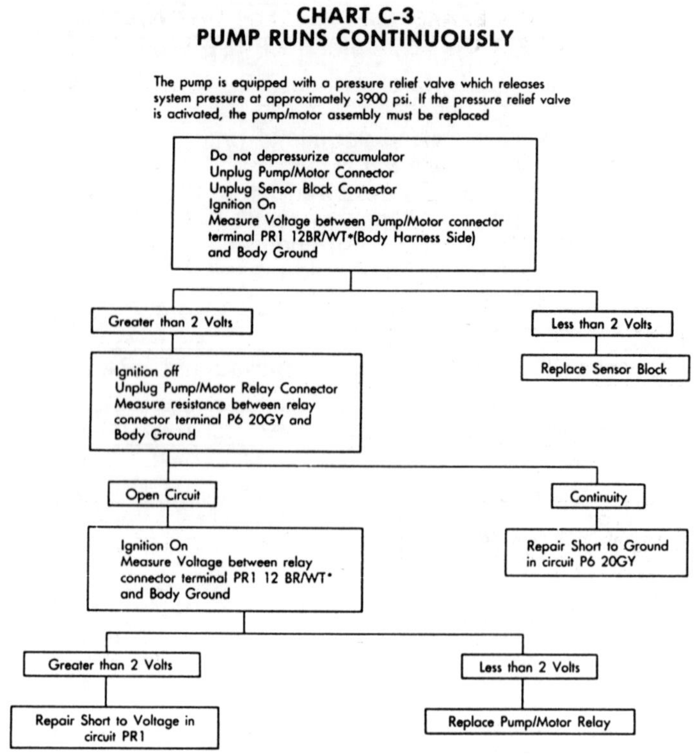

Fig. 7 Symptom diagnosis chart C-3

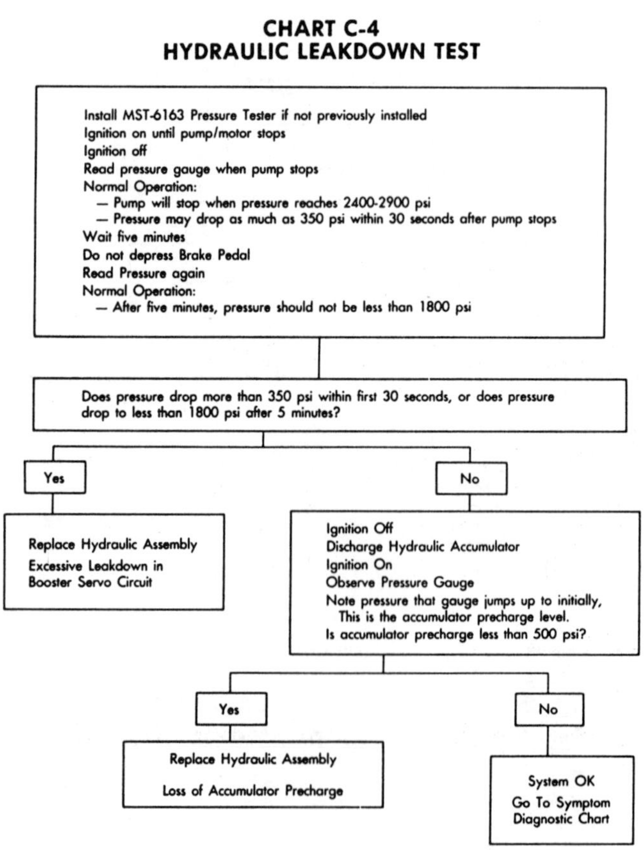

Fig. 8 Symptom diagnosis chart C-4

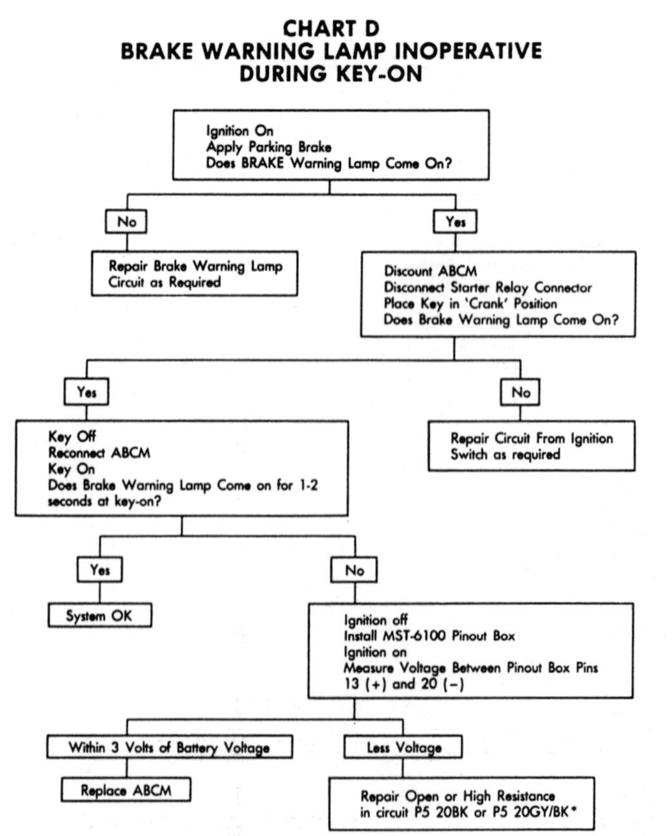

Fig. 9 Symptom diagnosis chart D

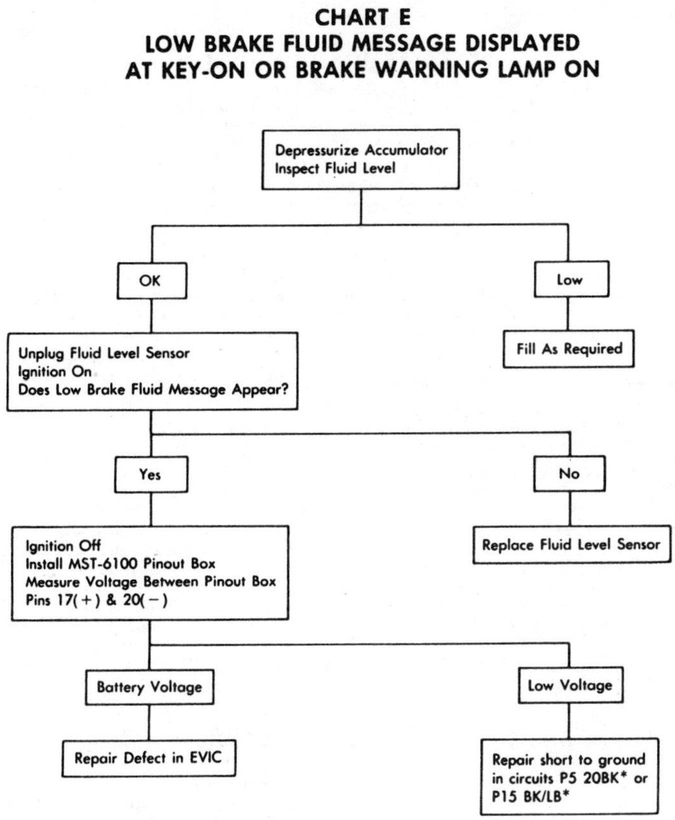

Fig. 10 Symptom diagnosis chart E

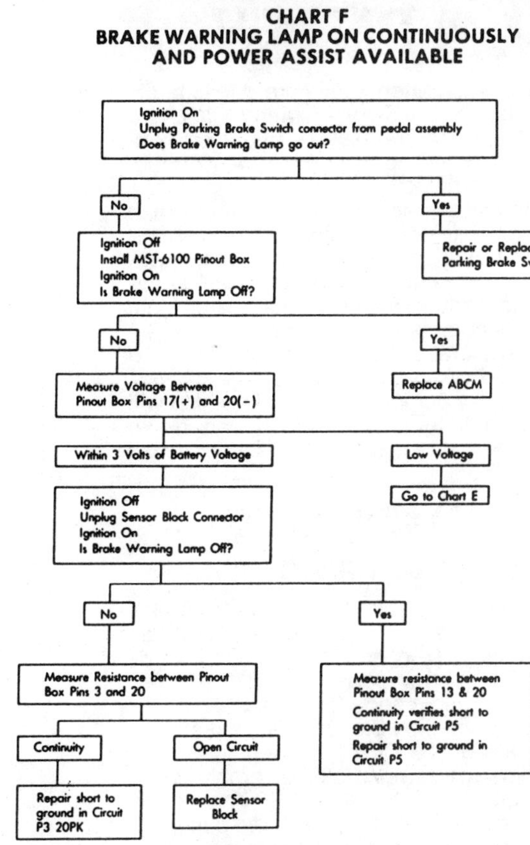

Fig. 11 Symptom diagnosis chart F

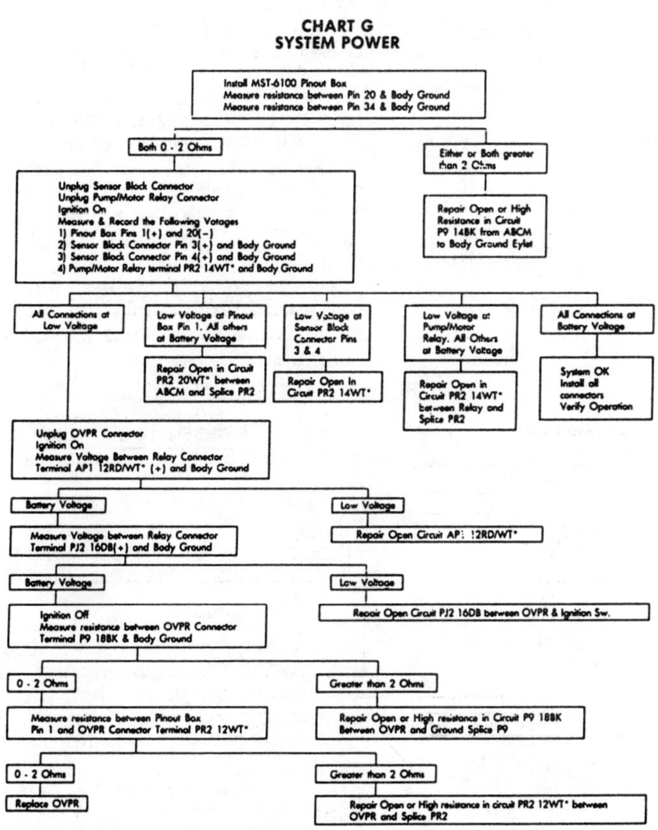

Fig. 12 Symptom diagnosis chart G

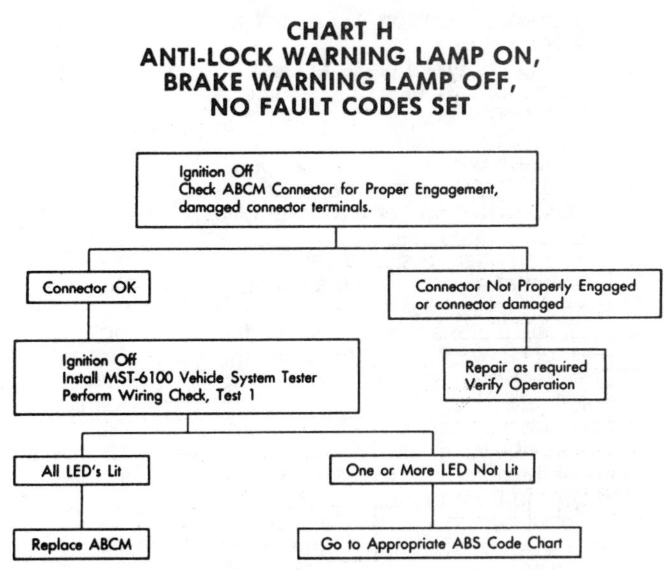

Fig. 13 Symptom diagnosis chart H

INTERMITTENT FAULT CODES

Most intermittent faults are due to faulty electrical connections or wiring. Check for poorly seated connectors, damaged connectors or wiring, or any other connector/wiring defects.

Intermittent illumination of the brake and/or anti-lock warning lamps can be caused by the following conditions:

1. Low system voltage, which will illuminate the anti-lock warning lamp until voltage returns to normal.
2. Low brake fluid, which will illuminate the brake warning lamp.
3. Low accumulator pressure, which will illuminate both anti-lock and brake warning lamps.

SYSTEM SERVICING

BRAKE BLEEDING

Any time air is allowed to enter the brake system, due to damage or service, the system must be bled. Air in the system will not cause a spongy pedal, but instead may cause a trouble code to appear.

When bleeding brakes, maintain an appropriate level of DOT 3 brake fluid, and do not allow the pump/motor assembly to run continuously for more than 60 seconds. If it becomes necessary to run the pump for more than 60 seconds, allow it to cool for several minutes before running it again.

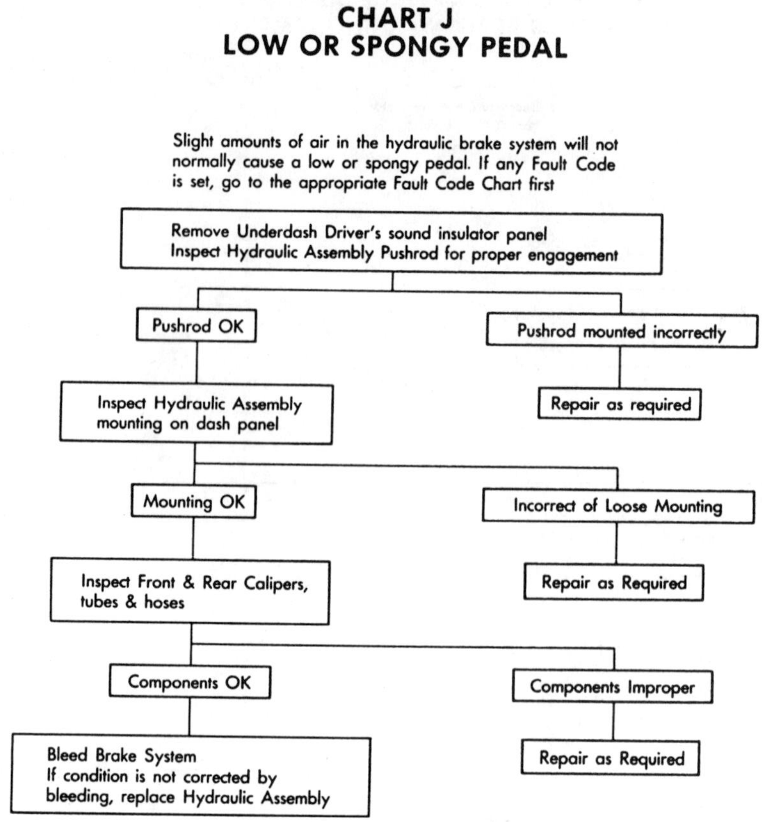

Fig. 14 Symptom diagnosis chart J

Booster Bleeding

1. Depressurize hydraulic accumulator as follows:
   a. Disconnect battery ground cable, and ensure ignition is in the Off position.
   b. Pump the brake pedal at least 25 times with about 50 lbs. of force. A noticeable change in pedal feel should be evident.
   c. When a definite increase in pedal effort is felt, pump pedal a few more times to remove hydraulic pressure from lines.
2. Connect battery negative cable.
3. Check that all brake lines and hoses are tight, and fill fluid reservoir.
4. Connect a transparent hose to the bleeder screw, **Fig. 22,** and place the other end into a clear container to hold brake fluid.
5. Open bleeder screw $1/2$ -$3/4$ of a turn.
6. Turn on the ignition switch to force fluid out the open bleeder. When the fluid flowing from the tube is free of air bubbles, turn off ignition. **If fluid does not flow when ignition is turned on, the pump may need to be primed. First try shaking the pump return hose, with the ignition still on to break up air bubbles. If necessary, turn off ignition, remove pump return hose at one end, fill with brake fluid and reconnect. Attempt bleeding operation again.**
7. Remove bleeder hose, and **torque** bleeder screw to 7.5 ft. lbs. **Do not over torque.**
8. Top off reservoir to correct level, then turn on ignition to allow pump to charge the accumulator. The pump should stop after about 30 seconds.

Pressure Bleeding

The brake lines can be bled using a standard diaphragm type pressure bleeding system. Use only a diaphragm type system to prevent contaminants from entering the system.

1. Ensure ignition remains off for the entire procedure.
2. Depressurize the hydraulic accumulator as outlined under "Booster Bleeding."
3. Disconnect electrical connector from fluid level sensor, then remove reservoir cap.
4. Install pressure bleeder adapter, **Fig. 23,** then attach bleeding equipment to adapter. Charge pressure bleeder to 20 psi.
5. Connect a transparent hose to caliper bleed screw, and submerge the other end in a clear container, partially filled with clean brake fluid **Fig. 24.**
6. Turn pressure bleeder on, then open caliper bleed screw $1/2$-$3/4$ turn allowing fluid to escape until no more air bubbles are present. If reservoir has been drained, or hydraulic assembly removed prior to bleeding, pump the brake pedal slowly one or two times with bleed screw open to purge air from hydraulic assembly.
7. Repeat step 6 for each caliper in the following order: LR-RR-LF-RF.
8. After bleeding all calipers, close the pressure bleeder valve and **torque** to 7.5 ft. lbs., then slowly unscrew the bleeder adapter from the hydraulic assembly reservoir.
9. Use a syringe or other method to remove excess fluid from reservoir and adjust to the full mark.
10. Install reservoir cap, connect the fluid level sensor connector, and turn on ignition to allow pump to charge the accumulator.

Manual Bleeding

1. Depressurize the hydraulic accumulator as outlined under "Booster Bleeding."
2. Connect a transparent host to caliper bleed screw, and place the other end of hose in a transparent container partially filled with clean brake fluid **Fig. 24.**
3. Slowly pump brake pedal several times with full strokes of the brake pedal. Allow about five seconds between pedal strokes. After two or three strokes, continue holding pedal under pressure at the bottom of its travel.
4. Open the bleed screw $1/2$-$3/4$ turn. Keep screw open until fluid no longer comes out from bleeder, then tighten bleeder screw and release brake pedal.

FUNCTIONAL CHECK

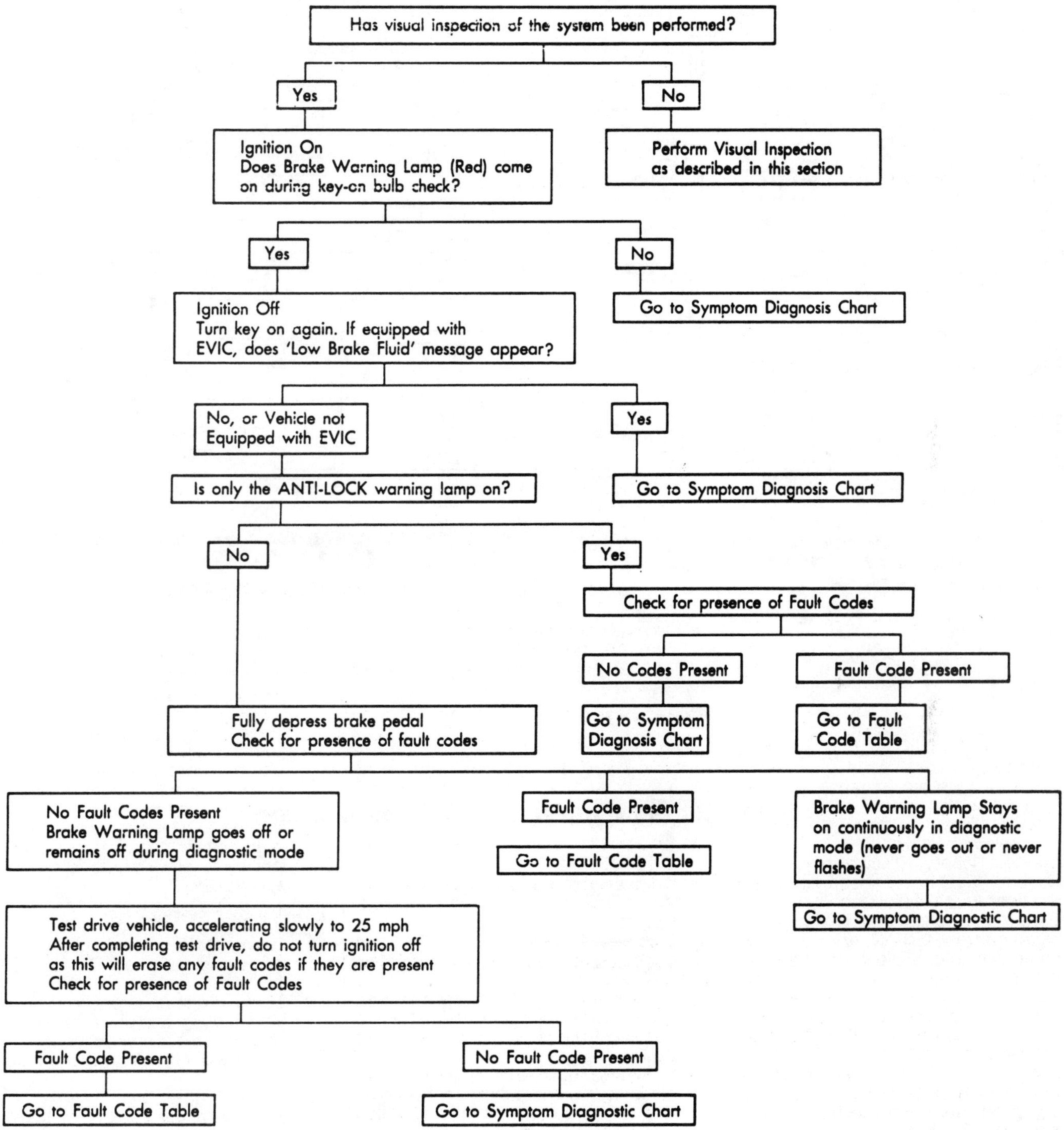

Fig. 15 Functional check chart

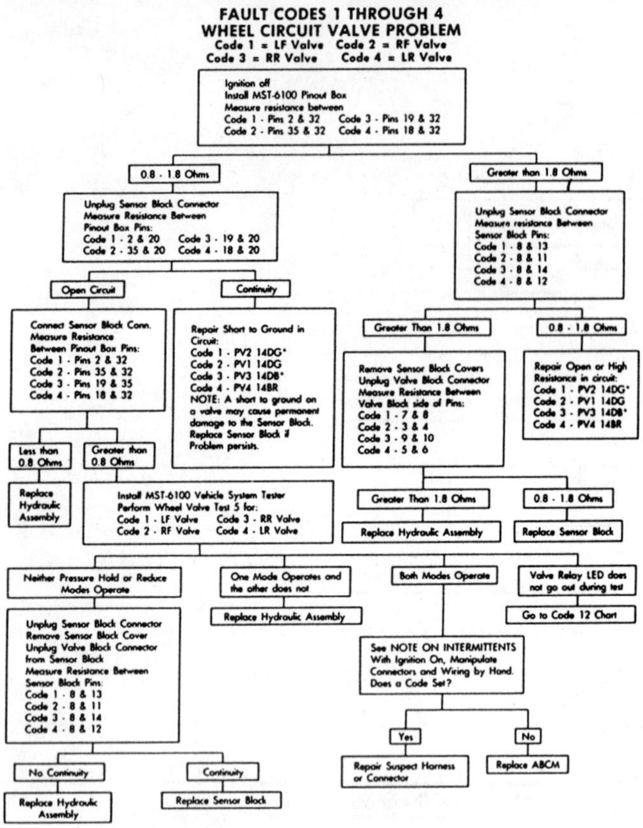

Fig. 16 Diagnostics for fault codes 1 through 4 (Part 1 of 3). 1988 models

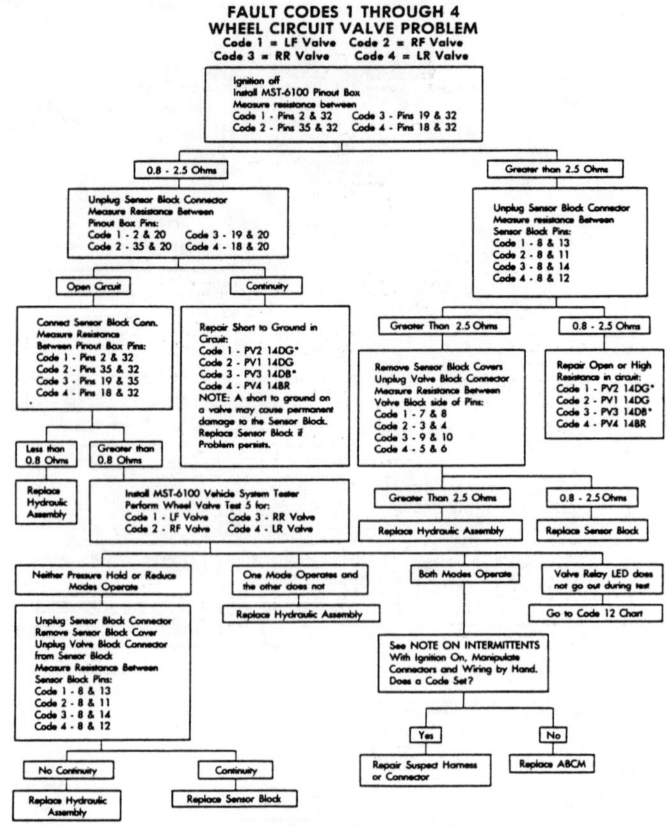

Fig. 16 Diagnostics for fault codes 1 through 4 (Part 2 of 3). 1989 models

FAULT CODES 1-4: WHEEL CIRCUIT VALVE PROBLEM

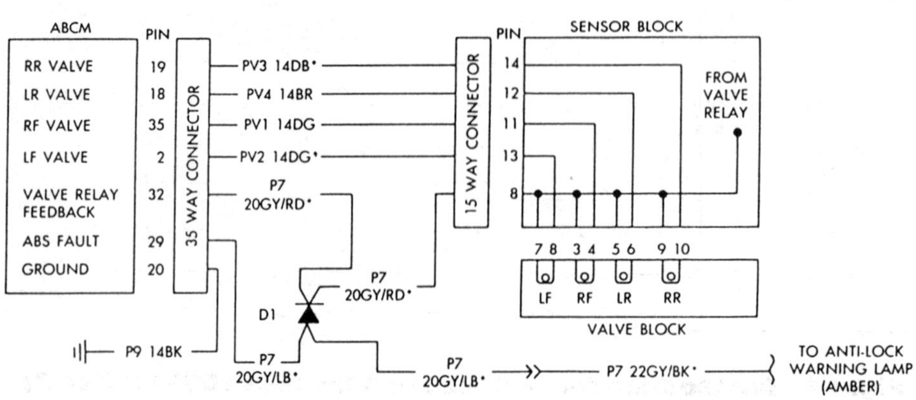

Fig. 16 Diagnostics for fault codes 1 through 4 (Part 3 of 3)

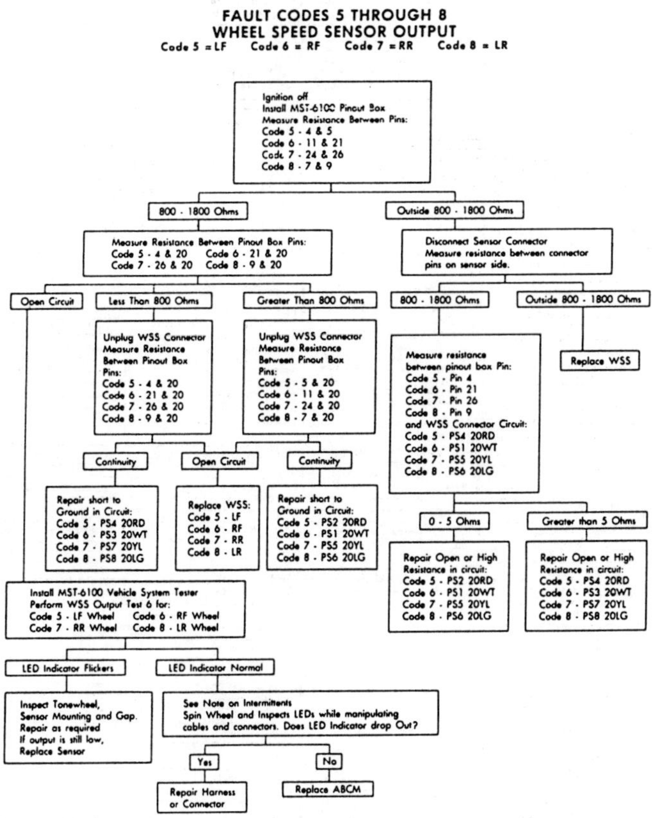

Fig. 17 Diagnostics for fault codes 5 through 10 (Part 1 of 3)

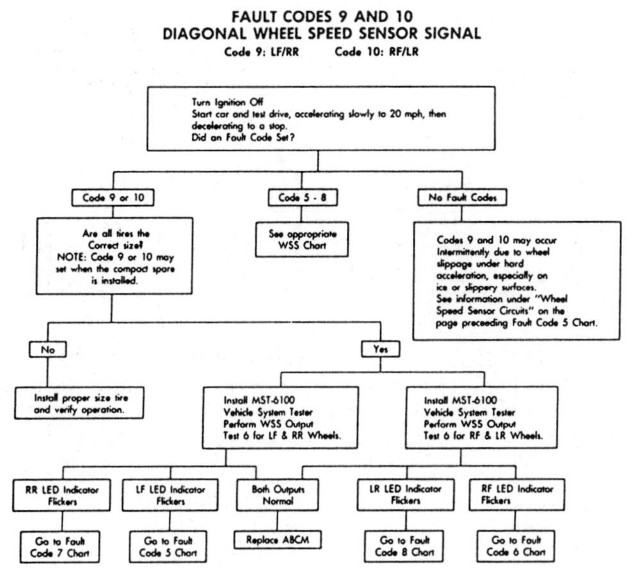

Fig. 17 Diagnostics for fault codes 5 through 10 (Part 2 of 3)

FAULT CODES 5-10: WHEEL SPEED SENSOR OUTPUT

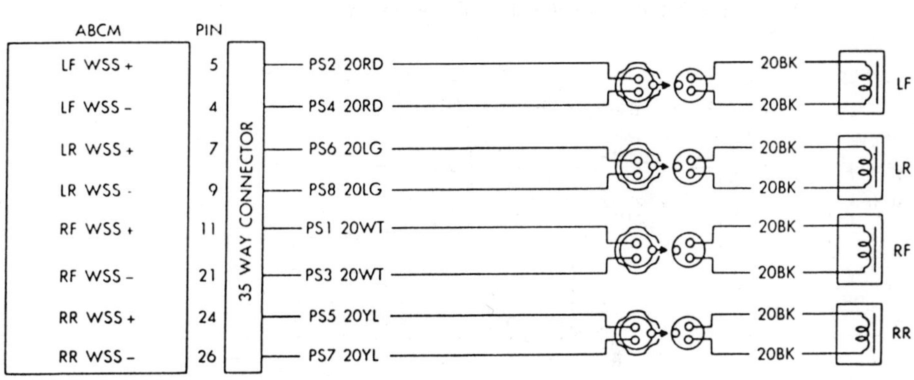

Fig. 17 Diagnostics for fault codes 5 through 10 (Part 3 of 3)

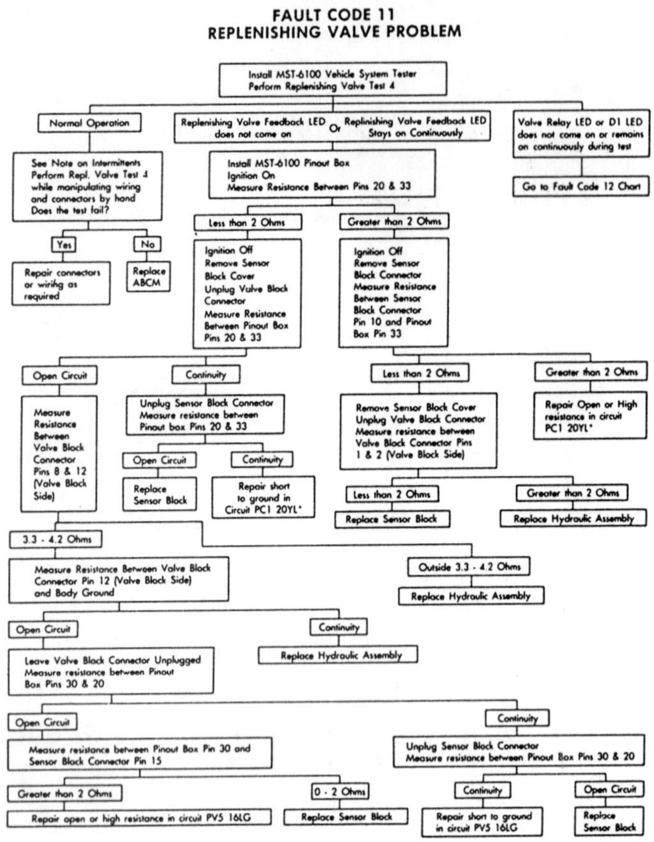

Fig. 18 Diagnostics for fault code 11 (Part 1 of 2)

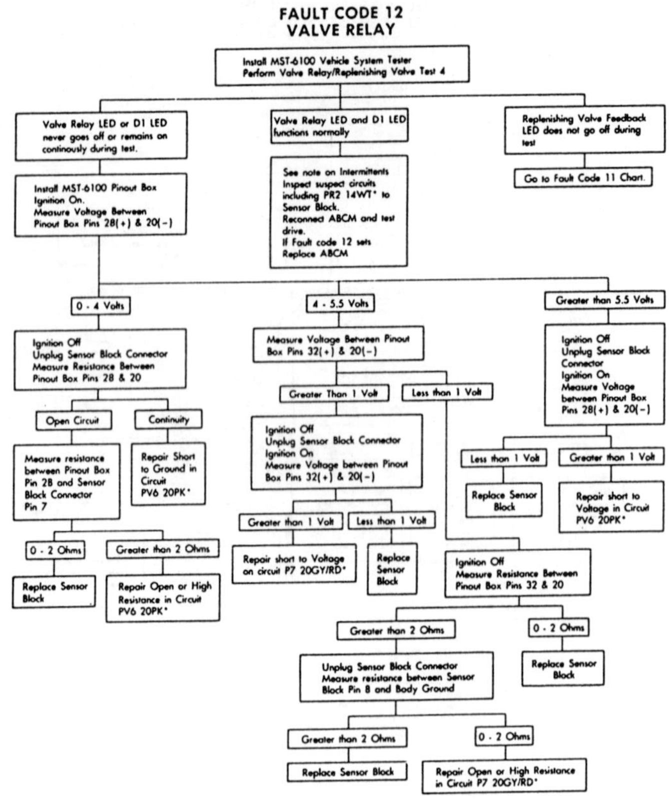

Fig. 19 Diagnostics for fault code 12 (Part 1 of 2)

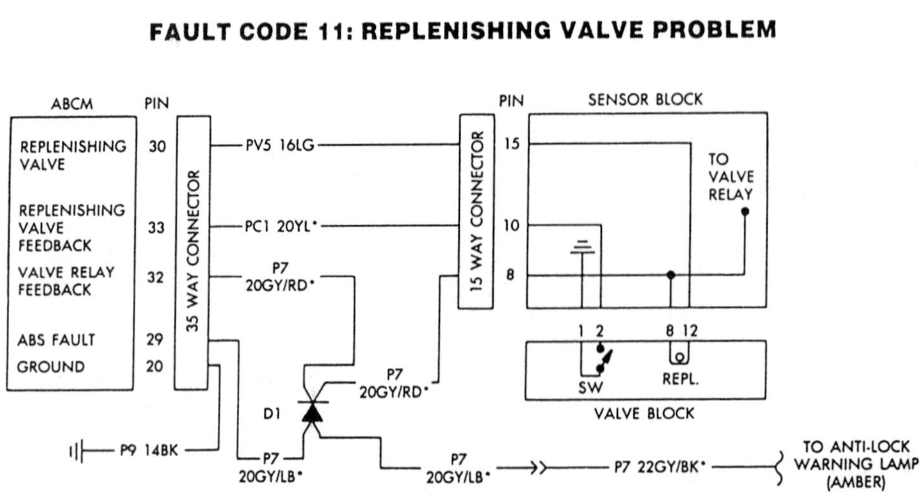

Fig. 18 Diagnostics for fault code 11 (Part 2 of 2)

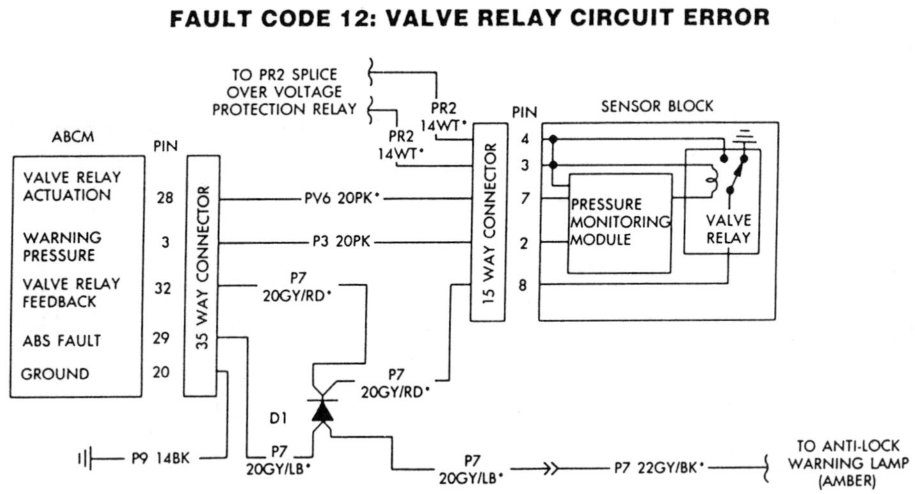

Fig. 19 Diagnostics for fault code 12 (Part 2 of 2)

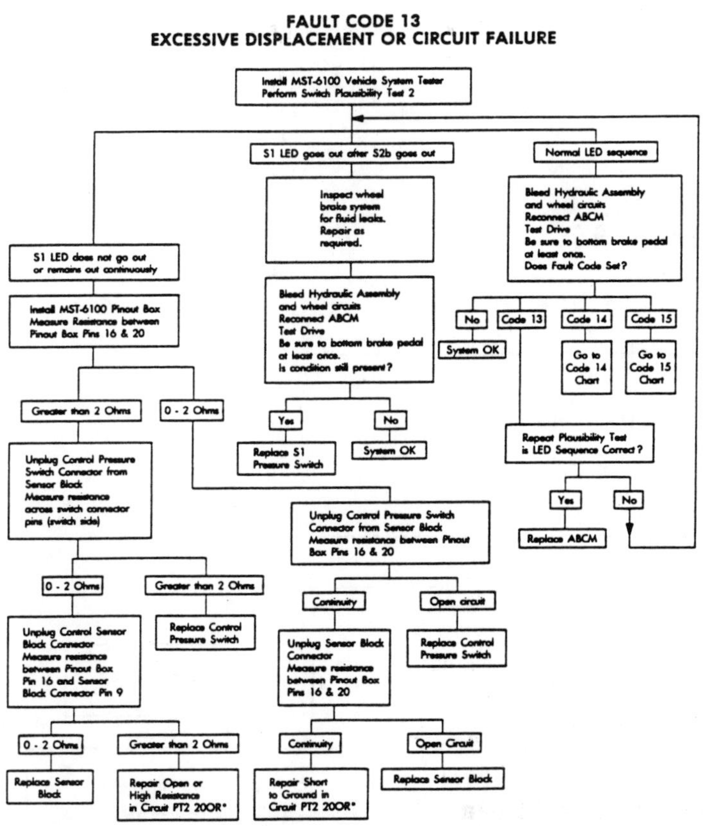

Fig. 20 Diagnostics for fault codes 13 through 15 (Part 1 of 4)

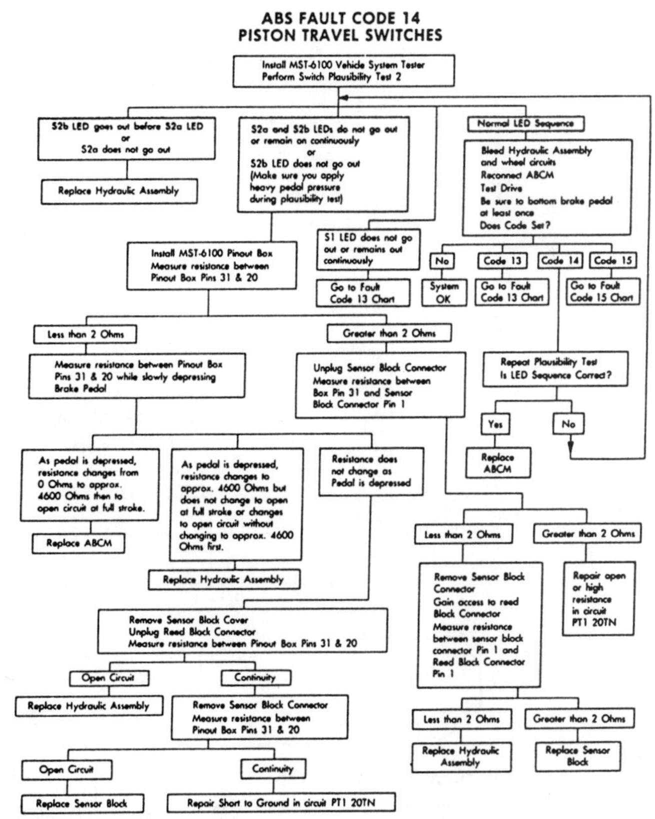

Fig. 20 Diagnostics for fault codes 13 through 15 (Part 2 of 4)

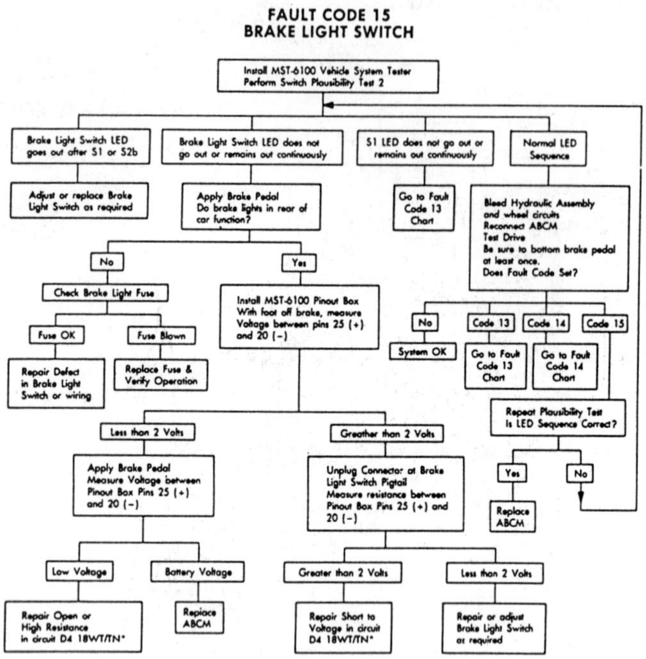

Fig. 20 Diagnostics for fault codes 13 through 15 (Part 3 of 4)

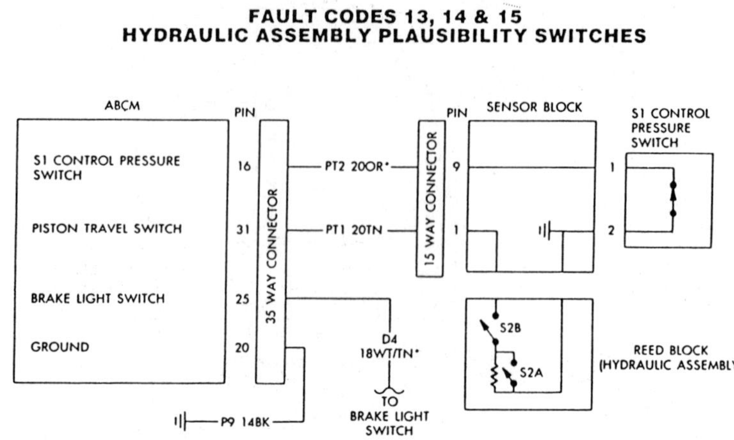

Fig. 20 Diagnostics for fault codes 13 through 15 (Part 4 of 4)

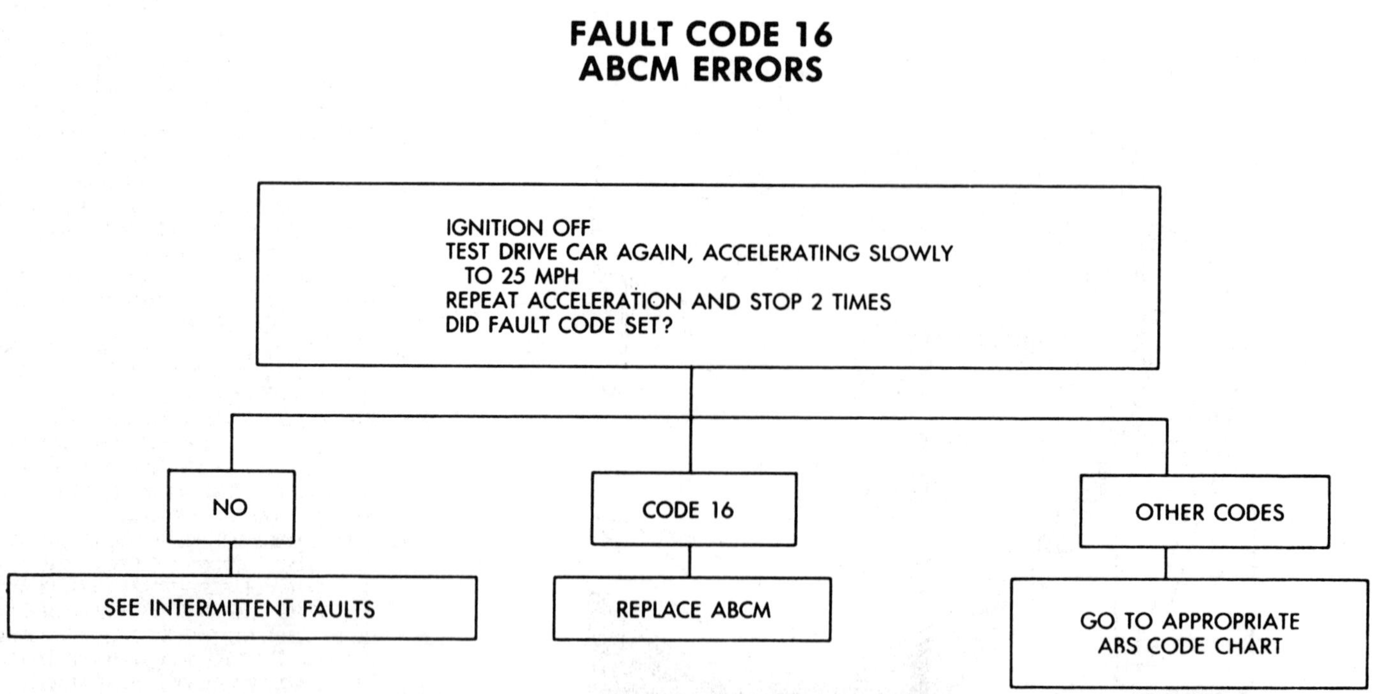

Fig. 21 Diagnostics for fault code 16

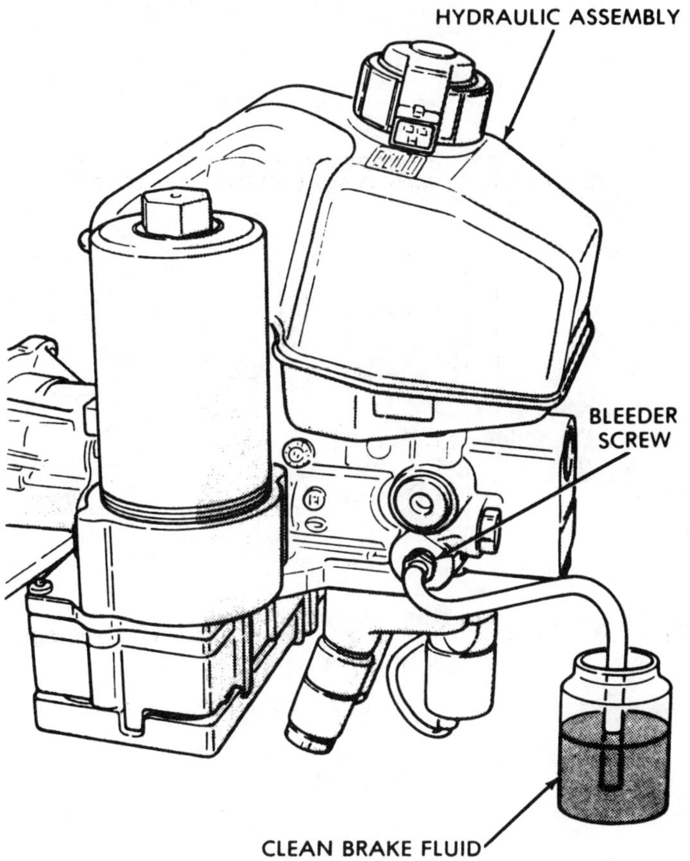

Fig. 22 Booster bleeding

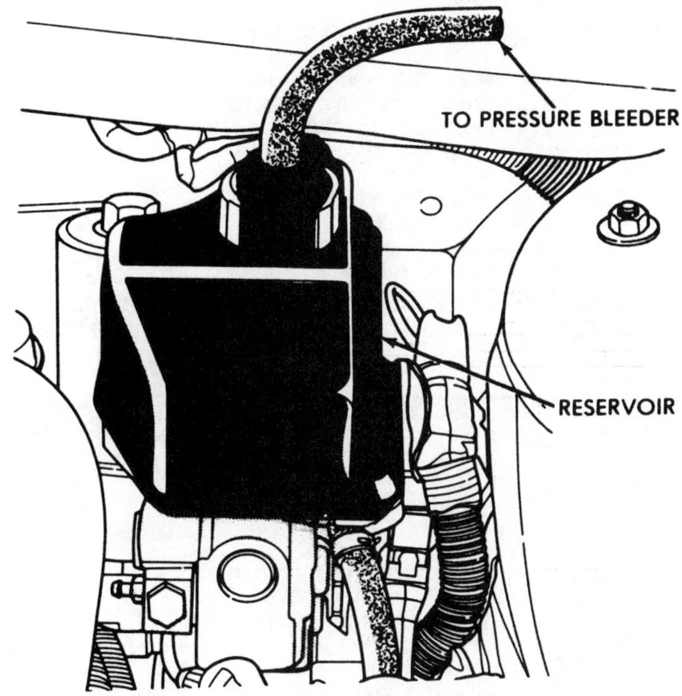

Fig. 23 Pressure bleeding

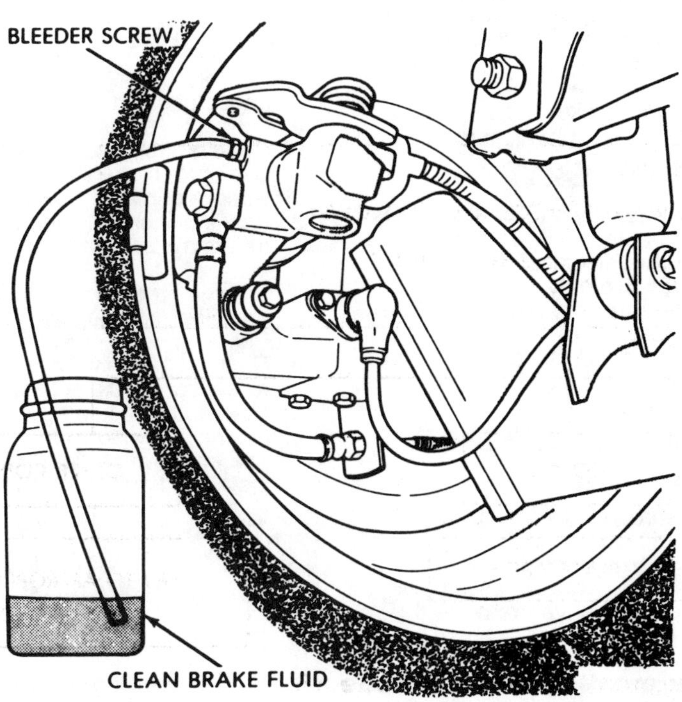

Fig. 24 Bleeding brake at caliper

5. Repeat this procedure until air bubbles are no longer present in fluid coming from bleeder hose.
6. Repeat for all wheels in the following order: LR-RR-LF-RF.

PUMP/MOTOR, REPLACE

1. Depressurize hydraulic accumulator as outlined under "Booster Bleeding," then remove the two fresh air intake ducts from engine induction system.
2. Disconnect the five electrical connectors in the vicinity of the pump/motor high and low pressure hoses.
3. Disconnect the high and low pressure hoses from hydraulic assembly. Cap the spigot on reservoir.
4. **On 1988 models,** disconnect transmission shift cable from transmission and position aside.
5. **On all models,** loosen two retaining nuts holding pump/motor to transmission differential cover.
6. **On 1988 models,** remove two hose bracket to transmission retaining bolts. **Do not disturb the inlet water extension pipe as an engine coolant leak may develop.**
7. Disconnect wiring harness retaining clip from hose bracket, then lift pump/motor assembly out of vehicle.
8. **On 1989 models,** remove retaining bolt attaching pump/motor to transmission, then lift pump/motor assembly off studs and out of vehicle.
9. **On all models,** remove heat shield from pump/motor and discard.
10. Reverse procedure to install, installing a new heat shield and adjusting gearshift linkage as necessary.

PRESSURE & RETURN HOSES, REPLACE

1. Remove pump/motor as previously outlined.
2. Cut the tie straps that secure hoses and wiring harness to pump mounting bracket, then open bracket tabs enough to remove wiring harness **Fig. 25.**
3. Remove banjo bolt from pump/motor and remove hose assembly.
4. Reverse procedure to install taking care to lubricate all rubber and steel O-rings with clean brake fluid and to replace tie straps.

HYDRAULIC ASSEMBLY, REPLACE

1. Depressurize hydraulic accumulator as outlined under "Booster Bleeding," then remove fresh air ducts from vehicle.
2. Disconnect electrical connectors from hydraulic assembly, then remove as much fluid as possible from reservoir on hydraulic assembly.
3. Remove pressure hose banjo bolt from hydraulic assembly taking care not to drop the two washers.
4. Disconnect return hose from reservoir nipple and cap spigot on reservoir.
5. Disconnect all brake lines from hydraulic assembly **Fig. 26.**
6. Use a screwdriver to remove the retaining clip from the brake pedal pin under the instrument panel **Fig. 27.**
7. Remove driver's side insulator panel, then four hydraulic assembly mounting nuts and hydraulic assembly.
8. Reverse procedure to install, noting the following:
   a. **Torque hydraulic assembly nuts to 210 inch. lbs.**
   b. **Use Lubriplate or equivalent lubricant to coat pedal pin bearing surface, then install a new retaining clip.**
   c. **Ensure brake tubes are correctly installed. The longer tube goes on the inboard proportioning valve, Fig. 26.**
   d. **Torque** banjo bolt to 160 inch lbs.
   e. Bleed entire brake system.

RESERVOIR, REPLACE

1. Depressurize hydraulic accumulator as outlined under "Booster Bleeding," then disconnect fluid level sensor.
2. Remove as much brake fluid as possible from reservoir, then disconnect return hose from reservoir nipple.
3. Remove reservoir by gently prying between it and hydraulic assembly, then remove grommets from hydraulic assembly, **Fig. 28.**
4. Reverse procedure to install, noting the following:
   a. Lubricate new grommets with clean brake fluid.
   b. Install reservoir by hand using a slight rocking motion.

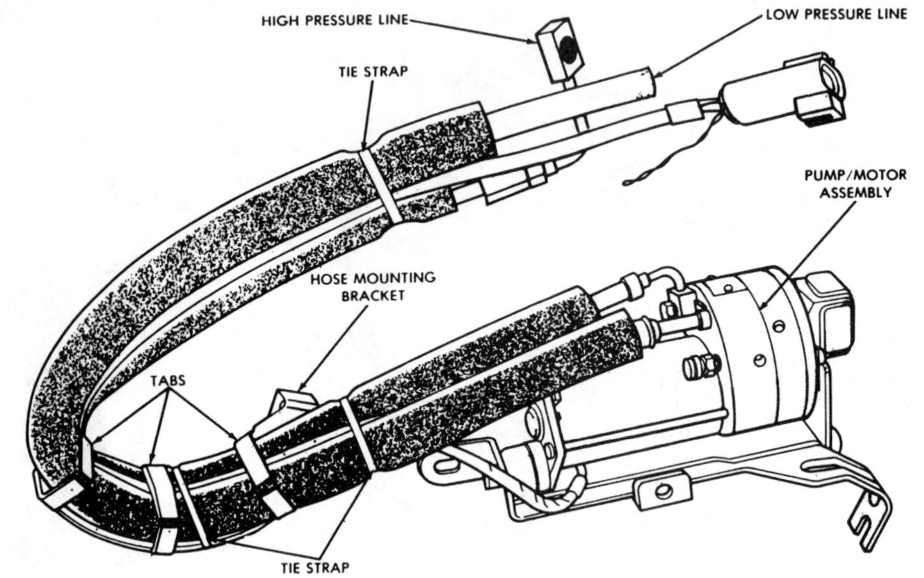

Fig. 25 Pressure hose assembly

FLUID LEVEL SENSOR, REPLACE

The fluid level sensor, **Fig. 28,** is part of the reservoir cap which must be replaced as an assembly.

PROPORTIONING VALVES, REPLACE

Proportioning valves should never be disassembled.

1. Remove brake tube and fitting from proportioning valve, then remove valve from hydraulic assembly, Fig. 28.
2. Install new proportioning valve. **Torque** valve to 20 ft. lbs. and brake tube to 11 ft. lbs.
3. Bleed the affected brake line as outlined under "Brake Bleeding."

CONTROL PRESSURE SWITCH, REPLACE

1. Depressurize hydraulic accumulator as outlined under "Booster Bleeding," then remove electrical connector from hydraulic assembly, **Fig. 28.**
2. Unscrew pressure switch from hydraulic assembly.
3. Reverse procedure to install. Lubricate O-ring with clean brake fluid before installation.

SENSOR BLOCK, REPLACE

Removal

1. Depressurize the hydraulic accumulator as outlined under "Booster Bleeding," then disconnect electrical connectors from hydraulic assembly.
2. Using a screwdriver, remove the retaining clip at the brake pedal pin under the instrument panel **Fig. 27.**
3. Remove driver's side sound insulator panel, then the four hydraulic assembly mounting nuts.
4. From under the hood, remove hydraulic assembly enough to gain access to sensor block, **Fig. 28. Do not remove or deform brake lines.**
5. Remove sensor block cover taking care not to damage gasket.
6. Disconnect the 12 pin valve block connector from sensor block.
7. Pull outward on the orange reed block connector (marked Push) to disengage it. The connector retaining clip will only move approximately 1/2 inch.
8. Remove three sensor block retaining bolts, then disengage sensor block pressure port from hydraulic and sensor block from car.
9. Inspect O-rings and sensor block for any damage.

Installation

1. Pull reed block connector to disengaged position.
2. Lubricate pressure port O-ring with clean brake fluid, then insert pressure port into hydraulic assembly orifice without damaging O-ring.
3. Position sensor block, then install mounting bolts and **torque** to 11 ft. lbs.
4. Engage reed block connector by pressing orange connector marked Push.
5. Connect 12 pin valve block connector to sensor block, sensor block cover, gasket and mounting bolt.
6. Connect sensor block and control pressure switch connectors, then reinstall hydraulic assembly.

ANTI-LOCK BRAKE CONTROL MODULE (ABCM), REPLACE

1. Remove rear seat bulkhead trim panel from trunk.
2. Pull back the lock plate and rotate connector away from ABCM, then remove retaining screws and ABCM.
3. Reverse procedure to install. **Install ground terminal securely.**

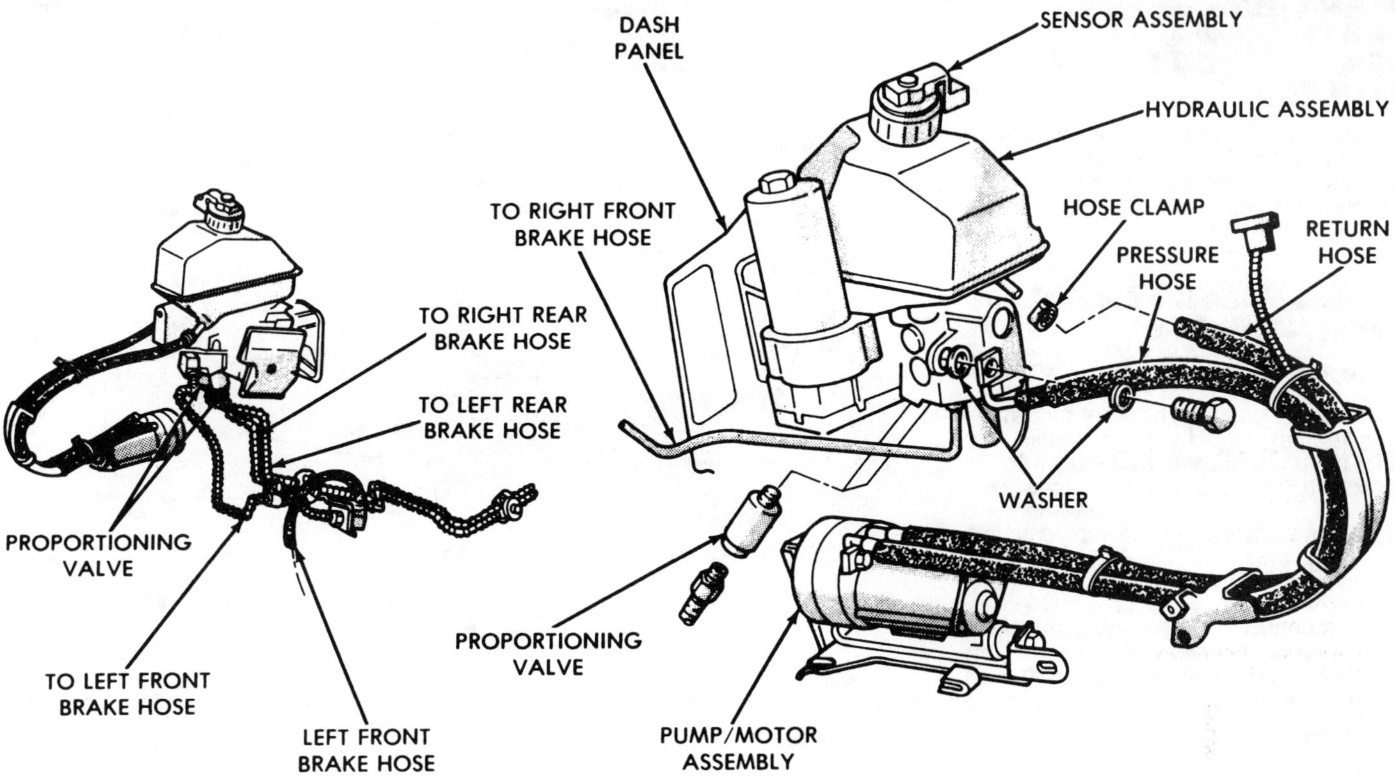

Fig. 26 Hydraulic assembly

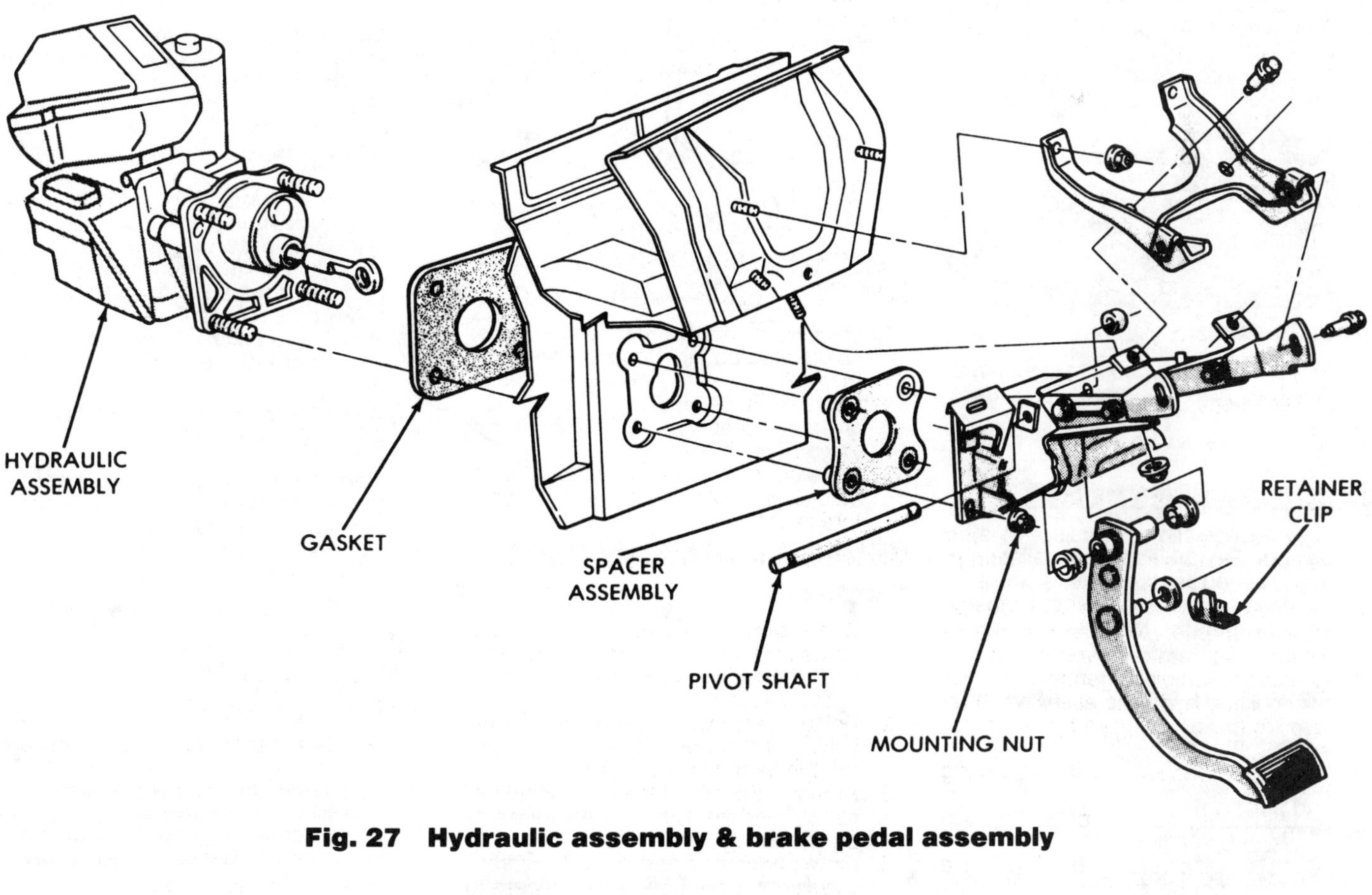

Fig. 27 Hydraulic assembly & brake pedal assembly

OVER VOLTAGE PROTECTION & PUMP/MOTOR RELAYS, REPLACE

1. Remove radiator overflow bottle and relay bracket retaining screw.
2. Disconnect relay from connector.
3. Reverse procedure to install.

WHEEL SPEED SENSOR (WSS), REPLACE

The wheel speed sensor generates a small AC voltage by magnetic induction as a tonewheel rotates. The front tonewheel is an integral part of the outer CV joint and must be replaced as an assembly with the outer CV joint **Fig. 29.** The rear tonewheel is an integral part of the hub assembly and must be replaced as an assembly with hub assembly.

Inspect tonewheel for missing or damaged teeth which can cause erratic operation. There should not be any contact with wheel speed sensor. Runout of the tonewheel should not exceed .010 inch. Replace assembly if runout exceeds specifications.

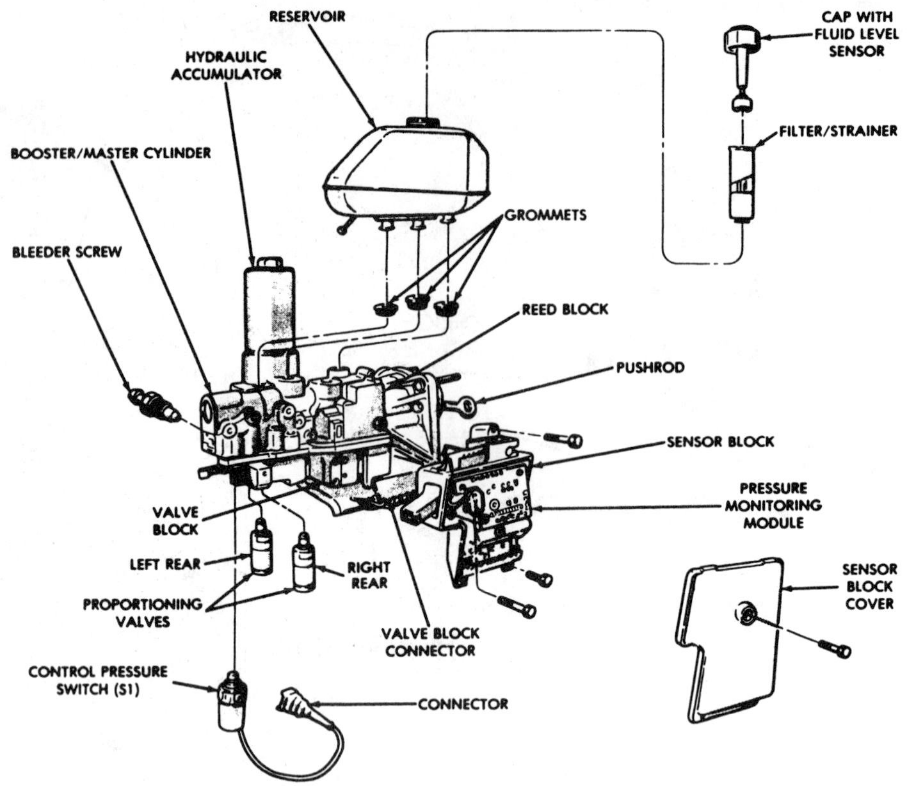

Fig. 28 Hydraulic assembly & reservoir

FRONT WHEEL SPEED SENSOR

Removal

1. Raise and support vehicle.
2. Remove wheel and tire assembly, then the retaining screw and clip from fender shield **Fig. 30.**
3. Carefully remove sensor assembly grommet from fender shield, then remove sensor connector lock and disconnect sensor.
4. Remove three assembly grommets, then the sensor head screw.
5. Remove sensor head from steering knuckle. If difficult to remove, use a hammer and punch to tap the sensor gently back and forth until free. **Do not use pliers on sensor head.**

Installation

1. Before installing, coat sensor with suitable high temperature grease. **Torque** to 60 inch lbs.
2. Connect sensor connector to harness, then install connector lock, assembly grommet, clip and screw.
3. Install sensor grommets in brackets, and retainer clip on retainer at strut damper, **Fig. 30.** Proper installation of wheel speed sensor cable is critical to continued system operation. Reinstall as shown in **Fig. 30.**

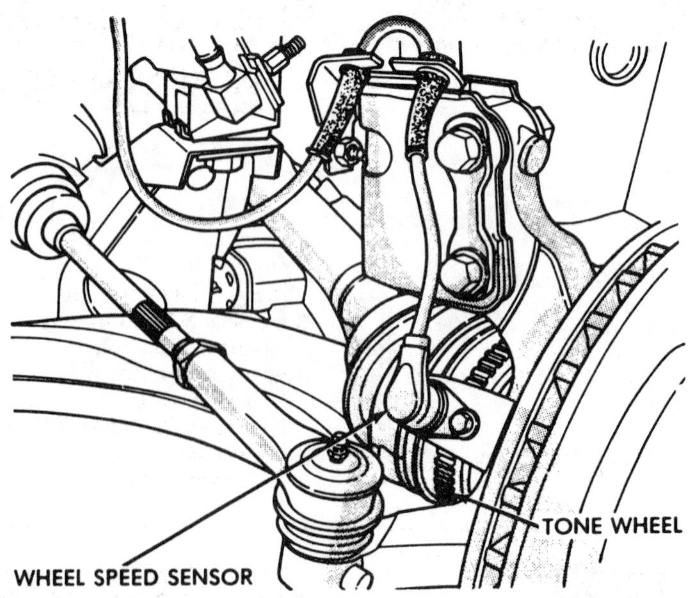

Fig. 29 Front wheel speed sensor & tonewheel assembly

REAR WHEEL SPEED SENSOR

1. Raise and support vehicle.
2. Remove wheel and tire assembly, then the sensor assembly grommet and pull harness through hole.
3. Remove connector lock and unplug from harness.
4. Remove spool grommet clip retaining screw from hose bracket just in front of trailing arm bushing
5. Remove assembly clip located at the inboard side of trailing arm, then remove outboard sensor assembly retainer nut.
6. Remove sensor head nut and carefully remove sensor head from adapter. If sensor is difficult to remove, use a hammer and punch to gently tap sensor back and forth until free. **Do not use pliers on sensor head.**
7. Reverse procedure to install being sure to coat sensor with suitable high temperature grease. **Torque** screw to 60 inch lbs.

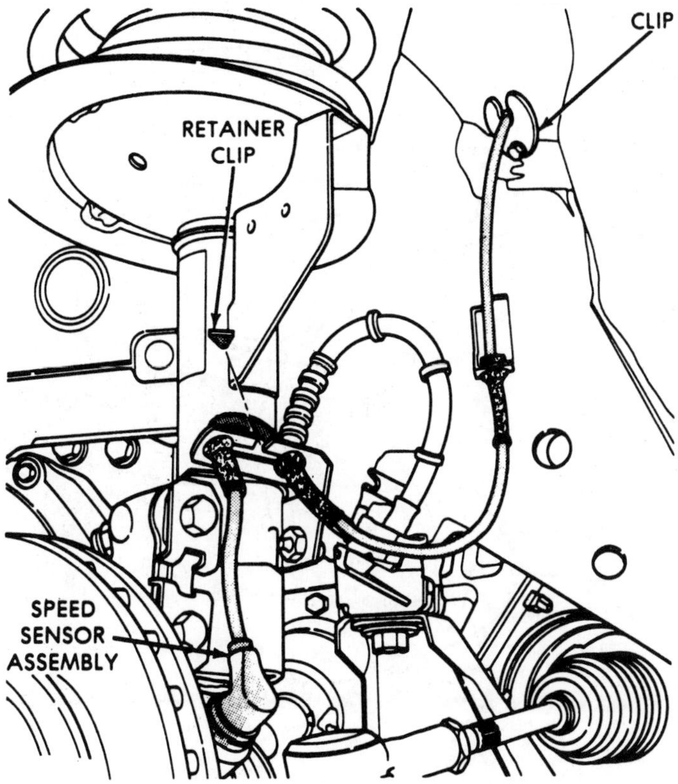

Fig. 30 Front wheel speed sensor assembly

Rear Brake Lock-Up Control System

INDEX

DESCRIPTION

The rear brake lock-up control system is an automatic brake control system used on Conquest models, designed for maximum braking efficiency on wet or icy road surfaces, and to reduce the possibility of the vehicle skidding. The rear brake lock-up control system automatically maintains optimum control of braking according to road conditions. However, this system is designed for rear wheel control only. If the front wheels become locked, the front brakes will not be automatically controlled.

The rear brake lock-up control system is designed to control the relationship between the tires and the road surface (friction factor between tires, road surface and slip rate of tires). By maintaining the maximum friction factor, braking distance can be reduced and lateral stability can be maintained.

There are four major components, **Fig. 1**, in the rear brake lock-up control system which act together to control rear brake operation.

The electronic control unit (ECU) determines the ideal vehicle speed reduction rate using input signals from the G-sensor and pulse generator. The ECU compares the ideal vehicle speed reduction rate and the actual speed reduction rate of the rear wheels. If the actual speed reduction rate of the rear wheels is greater (rotation speed of the wheels is decreasing too rapidly), brake fluid pressure for the rear brakes is decreased, and the ideal rate of vehicle speed reduction is restored. If the actual speed reduction rate of the rear wheels is smaller (rotation speed of the wheels is decreasing too slowly), brake fluid pressure for the rear brakes is increased, and the ideal rate of vehicle speed reduction is restored. In this man-

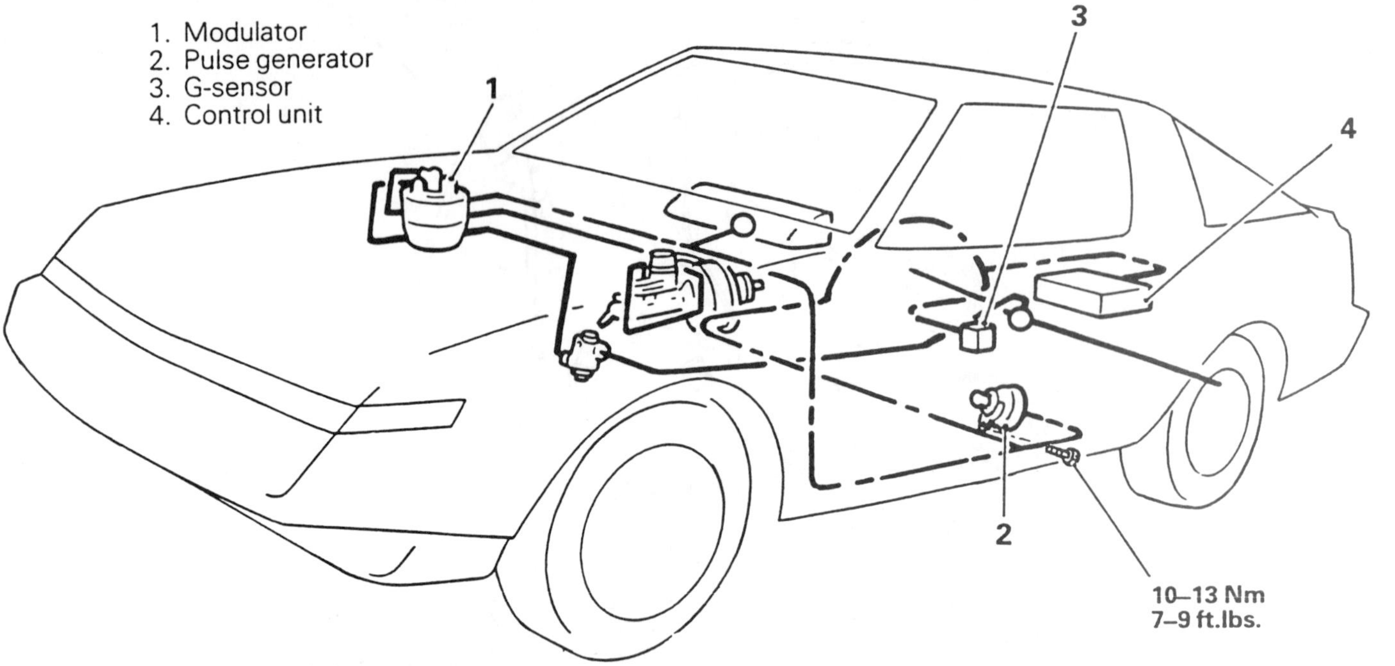

Fig. 1 Rear Brake Lock-Up Control System major components

ner, the rear wheels are controlled to maintain the ideal rate of speed reduction.

In the event of a malfunction in the rear brake lock-up control system, a signal from the ECU causes the warning lamp on the dashboard to illuminate and system operation is interrupted. Normal brake system operation (no rear lock-up control) continues. In addition, the warning light will illuminate for approximately three seconds when the ignition is turned to the On position. If the light does not illuminate, there is a malfunction of the light or the light circuit. If the light remains on, there is a malfunction in the rear brake lock-up control system. Refer to procedures outlined under "Troubleshooting" and "Component Testing."

PULSE GENERATOR

The pulse generator is located on the speedometer exit port of the transmission and is used for detection of rear wheel speed. The pulse generator consists of a permanent magnet, coil and rotor. The rotor is turned by the speedometer drive gear. The magnetic flux generated from the permanent magnet varies according to rotation of the rotor, and AC voltage is generated in the coil.

The voltage generated is proportional to the rotating speed of the rotor, and the frequency varies. The speed of the rear wheels is detected by using the frequency variations of the voltage generated by the pulse generator. The frequency of the generated voltage is the average value of the speeds of the left rear and right rear wheels.

G-SENSOR

The G-sensor is located on the floor of the luggage compartment and is used to detect reduction of vehicle speed. The G-sensor consists of a differential transformer and a control circuit (printed circuit board). The core within the differential transformer is normally stationary at the center of the coil. When a reduction of speed occurs, the core moves and a voltage proportional to the amount of core displacement is generated. This voltage is used to detect the extent of reduction in vehicle speed.

ELECTRONIC CONTROL UNIT (ECU)

The electronic control unit (ECU) is located inside the luggage compartment and receives signals from the pulse generator, G-sensor and the stop lamp switch. The ECU transmits a brake fluid pressure control signal to the modulator.

The signal to lower the brake fluid pressure of the rear brakes causes the modulator release valve to operate when the amount of slippage of the tires on the road surface is greater than the specified value. The specified value is calculated according to to the relationship between the speed reduction of the wheels and the vehicle. **There is no lock-up control of the rear wheels when the vehicle speed is approximately five mph. or less.**

MODULATOR

The modulator is a vacuum servo type brake fluid pressure control device located on the righthand side of the engine compartment. Using the control signal (voltage) from the ECU, the modulator controls brake fluid pressure to the rear brakes.

The modulator consists of a brake fluid pressure control section (to control brake fluid pressure for the rear brakes), a vacuum pressure drive section (to drive the brake fluid control section), and a solenoid valve (to control the vacuum pressure of the vacuum pressure drive section).

TROUBLESHOOTING

If a malfunction occurs in the rear brake lock-up control system, follow the troubleshooting procedures to check system operation. Check all circuits and components (including wiring) except the ECU. If no malfunctions are found in circuits or components, replace ECU.

HYDRAULIC SYSTEM

1. Brakes lock and vehicle skids: Restriction in modulator hydraulic circuit (clogged orifice). Check fluid pressure as outlined under "Component Testing, Modulator."
2. Brakes lock and vehicle skids: Faulty release solenoid valve in modulator. Check modulator as outlined under "Component Testing, Modulator."
3. Low braking force and efficiency: Restriction in modulator hydraulic circuit (clogged orifice). Check fluid pressure as outlined under "Component Testing, Modulator."
4. Low braking force and efficiency: Faulty build-up solenoid valve in modulator. Check modulator as outlined under "Component Testing, Modulator."

POWER SUPPLY CIRCUIT

Power is supplied to the ECU through the rear brake lock-up control relay, **Fig. 2.** The rear brake lock-up relay is switched on with the ignition switch in the Accessory or On positions. Supply voltage to the ECU is interrupted with the ignition switch

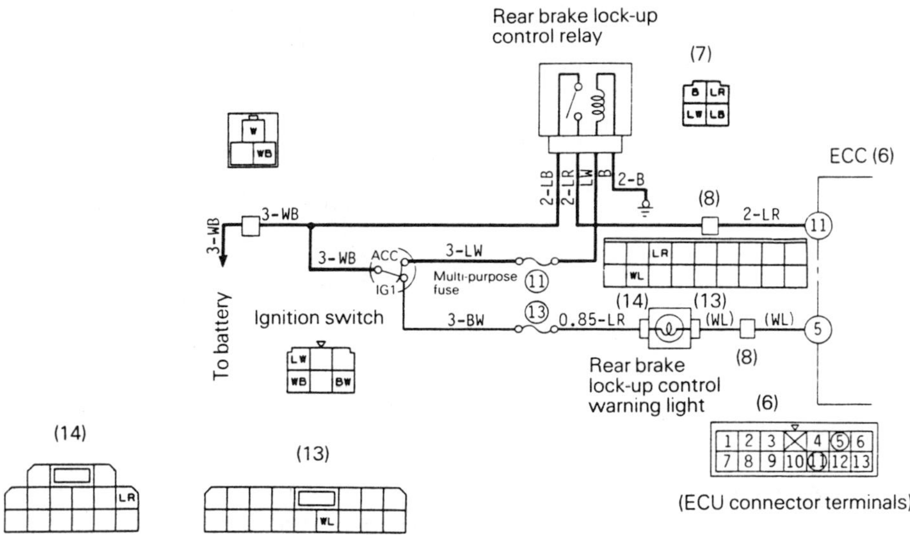

Fig. 2 Power supply circuit diagram

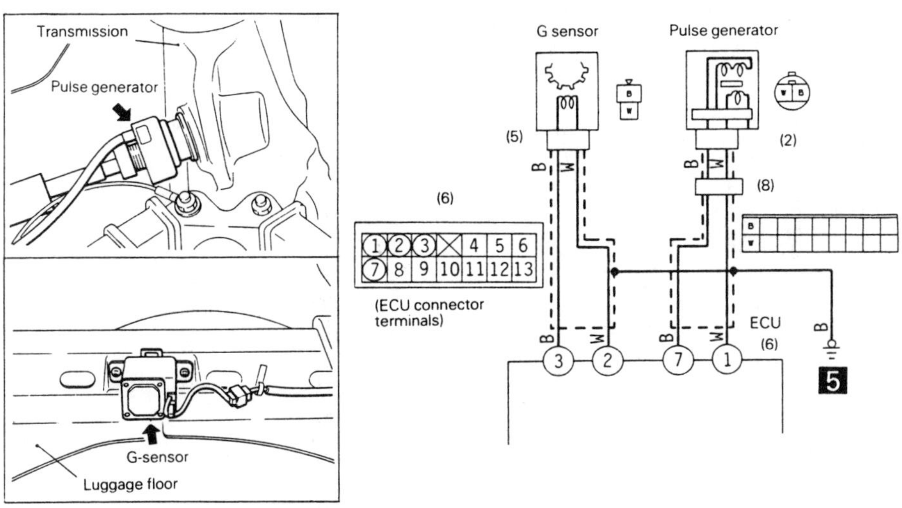

Fig. 3 Sensor circuit diagram

in the Start position.

If there is an open circuit in the power supply circuit (indicated by solid lines on the rear brake lock-up control relay), the warning lamp illuminates when the ignition switch is turned on.

If the supply voltage drops below 10.5 volts, the rear brake lock-up control warning lamp illuminates and the system operation is interrupted. When the supply voltage returns to 10.5 volts, the system operation resumes.

1. Check supply voltage to terminal No. 11 (ECU power supply) of ECU connector as follows:
   a. Disconnect ECU electrical connector.
   b. Connect a suitable multimeter between terminal No. 11 of disconnected ECU connector and ground, **Fig. 2.**
   c. With ignition switch in Accessory or On position, voltage indicated should be approximately battery voltage (12 volts).
   d. With ignition switch in Off position, voltage indicated should be approximately 0 volts.
   e. With ignition switch in Start position, voltage indicated should be approximately 0 volts.

SENSOR CIRCUIT

Refer to **Fig. 3,** when troubleshooting the sensor circuit.

1. Rear brake lock-up control warning light illuminates, and system operation is interrupted: Sensor signals are not input to the ECU, open circuit in G-sensor, pulse sensor or wiring. Check G-sensor and/or pulse generator as outlined under "Component Testing."
2. Rear brake lock-up control warning light illuminates, and system operation is interrupted: Abnormal sensor signal output (rotor deformation). Check G-sensor and/or pulse generator as outlined under "Component Testing."
3. Rear brake lock-up control warning light illuminates, and system operation is interrupted: Noise (interference) is easily picked up, and speed sensor may generate abnormal signals, damaged or disconnected ground of the shielded line. Check ground connection of the shielded line.

STOP LIGHT CIRCUIT

When the brake pedal is depressed (stop light switch on), input voltage (12 volts) is applied through the diode to ECU connector terminal No. 8 **Fig. 4,** through circuit indicated by solid lines, and the system operates.

If there is an open in the circuit from the stop light switch to the ECU, or if all the stop light bulbs are burned out, the current flows to the circuit with a resistor (indicated by broken lines). This resistor causes input voltage to ECU connector terminal No. 8 to decrease to approximately 7.5-8.5 volts. This interrupts system operation and the warning lamp will illuminate. Normal brake system operation (no rear lock-up control) continues.

1. Check supply voltage to terminal No. 8 (stop light switch) of ECU connector as follows:
   a. Disconnect ECU electrical connector.
   b. Connect a suitable multimeter between terminal No. 8 of disconnected ECU connector and ground, **Fig. 4.**
   c. With brake pedal depressed, voltage indicated should be approximately 0 volts.
   d. With brake pedal depressed, voltage indicated should be approximately battery voltage (12 volts).

SOLENOID VALVE CIRCUIT

If an open circuit is present in the release solenoid valve, build-up solenoid valve or wiring harness **Fig. 5,** the warning lamp will illuminate and system operation will be interrupted.

If the output transistor for ECU connector terminal Nos. 5 and/or 6 is shorted, the warning lamp will illuminate and system operation will be interrupted.

If the release solenoid valve and/or build-up solenoid valve has continued to operate for more than five seconds, the warning lamp will illuminate and system operation will be interrupted.

1. Check supply voltage to terminal No. 5 (build-up solenoid valve drive) of ECU connector as follows:
   a. Disconnect ECU electrical connector.
   b. Connect a suitable multimeter between terminal No. 5 of disconnected ECU connector and ground, **Fig. 5.**
   c. With ignition switch in Accessory or On position, voltage indicated should be approximately battery voltage (12 volts).

d. With ignition switch in Off position, voltage indicated should be approximately 0 volts.

2. Check supply voltage to terminal No. 6 (release solenoid valve drive) of ECU connector as follows:
   a. Disconnect ECU electrical connector.
   b. Connect a suitable multimeter between terminal No. 6 of disconnected ECU connector and ground, **Fig. 5.**
   c. With ignition switch in Accessory or On position, voltage indicated should be approximately battery voltage (12 volts).
   d. With ignition switch in Off position, voltage indicated should be approximately 0 volts.

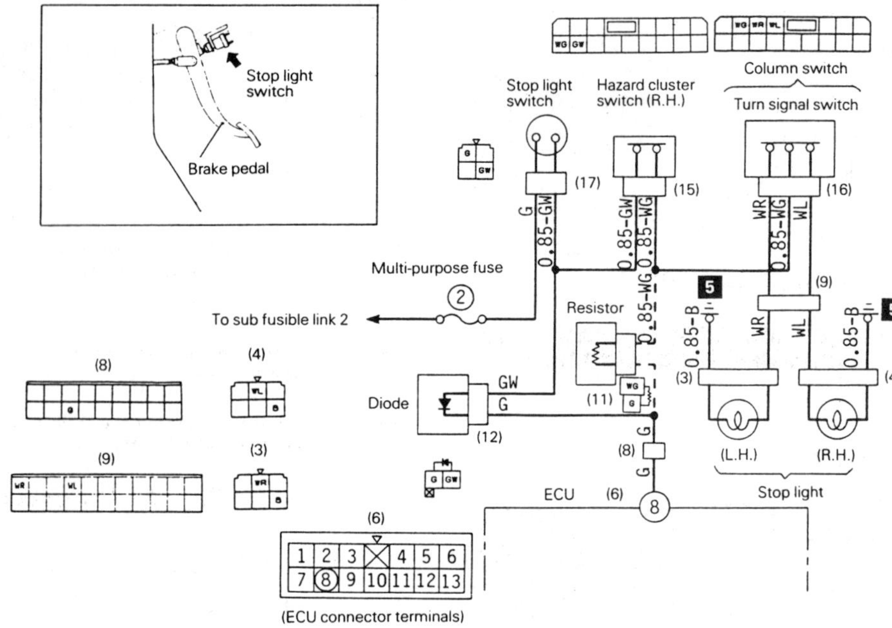

Fig. 4 Stop lamp circuit diagram

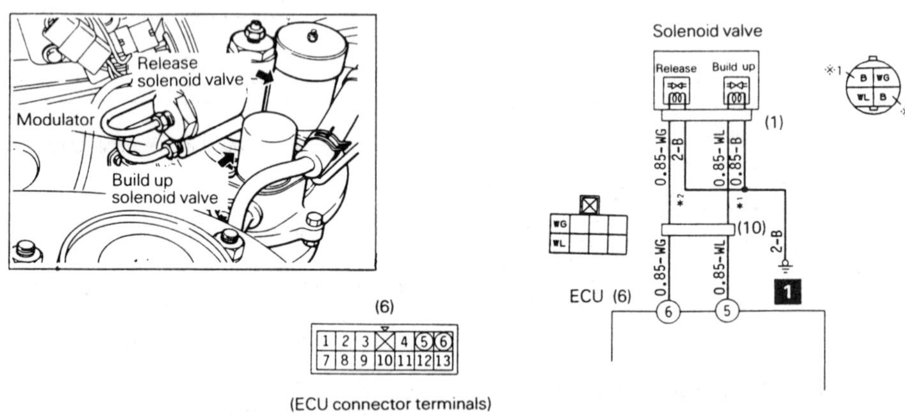

Fig. 5 Solenoid valve circuit diagram

DIAGNOSIS & TESTING

SYSTEM TESTS

SELF-DIAGNOSIS CHECK

Self-diagnosis system is checked by causing the release solenoid valve to operate.

1. Start and run engine for at least five seconds with vehicle not moving, turn ignition to Lock position, then depress brake pedal.
2. With brake pedal still depressed, turn ignition from Lock to On position, and ensure the modulator audibly operates.
3. A dull clicking sound indicates the solenoid valve is operating correctly.

Function Check

1. Raise and support vehicle, ensure rear wheels are completely off the ground, then block front wheels.
2. With engine at normal operating temperature, place shift lever in 2nd gear and accelerate until speedometer indicates approximately 19 mph.
3. While maintaining accelerator pedal position (approximately 19 mph.), suddenly depress brake pedal.
4. The brakes will attempt to stop the rotation of the rear wheels. As the rear brake lock-up control system begins to operate, brake fluid pressure will be reduced, and rotation of the rear wheels will continue. This process should repeat as long as brake pedal is depressed.

COMPONENT TESTING

MODULATOR

ELECTRICAL TEST

1. Disconnect modulator electrical connector.
2. Connect a suitable multimeter between terminals Nos. 1 and 3 (release solenoid valve) of modulator connector **Fig. 6,** and measure resistance.
3. Resistance indicated should be 3.8-4.8 ohms.
4. If resistance indicated is not as specified, replace release solenoid valve.
5. Measure resistance between modulator connector terminals Nos. 2 and 4 (build-up solenoid valve).
6. Resistance indicated should be 4.5-5.5 ohms.
7. If resistance indicated is not as specified, replace build-up solenoid valve.

Pressure Test

1. Connect suitable pressure gauges (A) and (B) with ranges of 0-2100 psi. or more, as shown in **Fig. 7.** Connect one pressure gauge to the rear brake exit point of modulator (B), and the other gauge between master cylinder and modulator (A).
2. Start engine and allow to run at idle.
3. Depress brake pedal far enough to obtain a pressure reading of approximately 711 psi. on pressure gauge (B), connected to rear brake exit port of modulator.
4. While observing pressure gauge (B), operate release solenoid valve as follows:
   a. Disconnect modulator electrical connector.
   b. Connect 12 volts to terminal No. 1 of modulator connector **Figs. 6 and 7,** then connect terminal No. 3 to ground. **Do not apply 12 volts for more than one minute.**
5. Pressure indicated on pressure gauge (B) should drop to approximately 0 psi. when release solenoid valve is operated.
6. With release solenoid valve still connected, operate build-up solenoid as follows:
   a. Connect 12 volts to terminal No. 2 of modulator connector **Figs. 6 and 7,** then connect terminal No. 4 to ground. **Do not apply to 12 volts for more than one minute.**
7. Disconnect release solenoid valve while observing pressure gauge (B).
8. Pressure indicated should increase to approximately 711 psi. when release solenoid is disconnected.

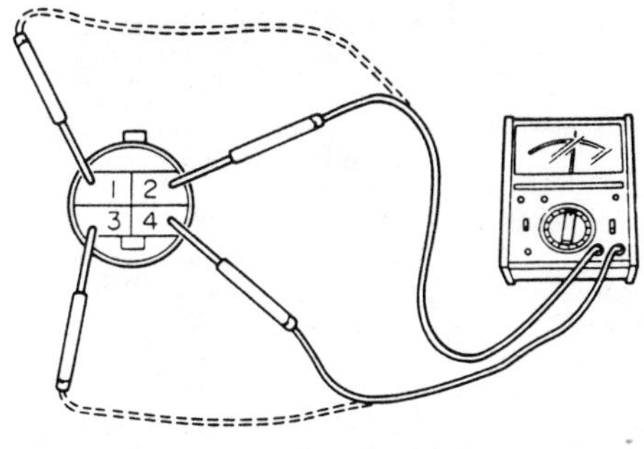

Fig. 6 Modulator connector terminal identification

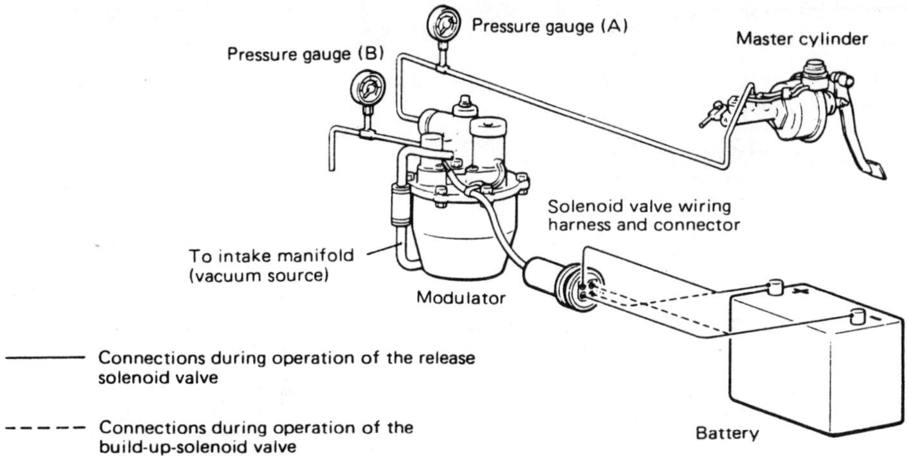

Fig. 7 Connecting pressure gauges

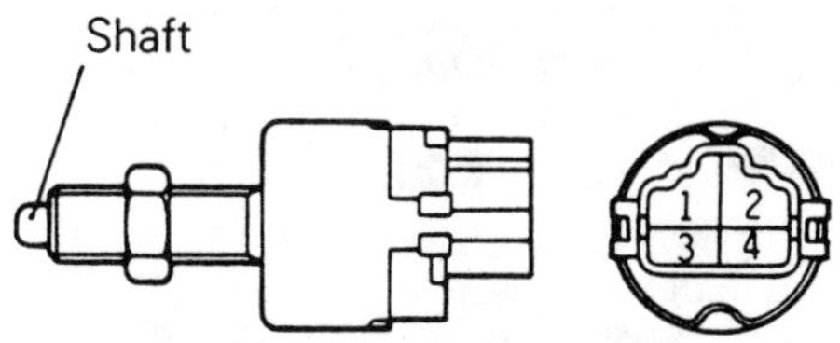

Fig. 8 Stop light switch connector terminal identification

Circuit \ Terminal — Shaft dimension mm (in.)		Free 5.8–6.8 (.23–.27) Full stroke		
			3.4–5.0 (.13–.20)	
				3–5 (.12–.20)
Circuit	Terminal			
Stop light	2 – 3	Continuity		No continuity
Speed control	1 – 4	No continuity	Continuity	

Fig. 9 Stop light switch resistance test

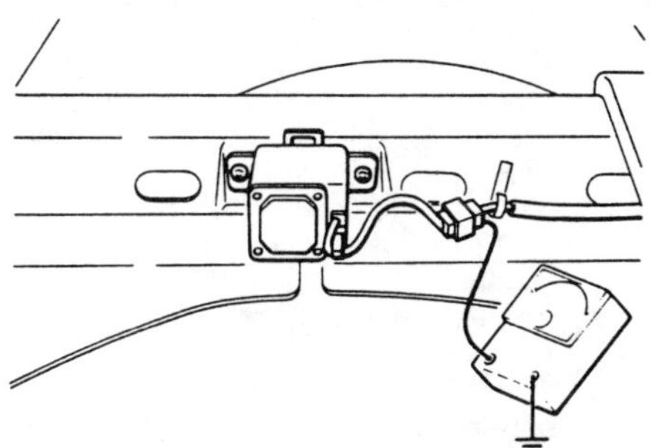

Fig. 10 Testing G-sensor

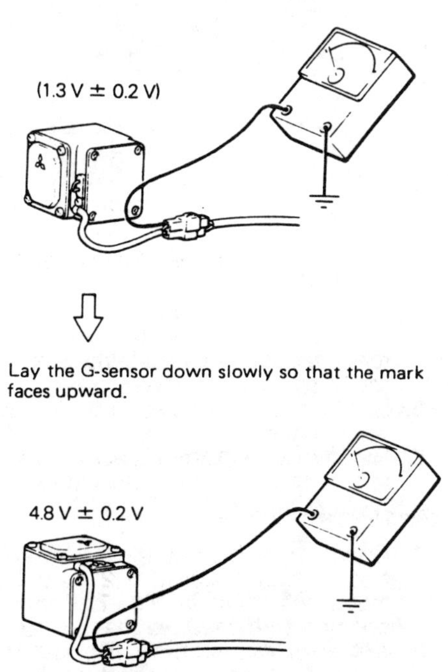

Fig. 11 Voltage characteristics of G-sensor

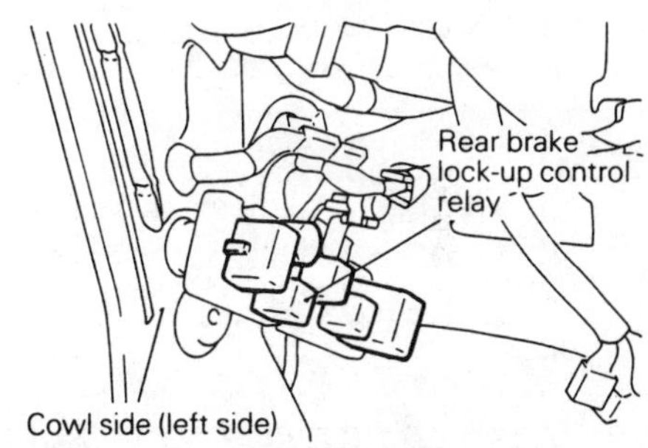

Fig. 12 Rear brake lock-up control relay location

9. Connect release solenoid valve, then disconnect build-up solenoid valve.
10. While observing pressure gauge (B), disconnect release solenoid valve.
11. Pressure indicated should increase to approximately 710 psi. within one second.
12. With both solenoid valves disconnected, increase fluid pressure of the master cylinder (A), and check the relationship with rear brake fluid pressure (B).
13. Ensure pressure gauge (B) indicates 1,422 psi. with brake pedal depressed until pressure gauge (A) indicates 1,707 psi.
14. If specified values cannot be obtained, replace modulator.

STOP LIGHT SWITCH

Disconnect stop light switch electrical connector, then operate stop lamp switch and check for continuity as shown in **Figs. 8 and 9.**

PULSE GENERATOR

1. Disconnect pulse generator electrical connector.
2. Using a suitable multimeter, measure resistance between pulse generator terminals.
3. Resistance indicated should be 600-800 ohms.
4. If resistance indicated is not as specified, replace pulse generator.
5. Measure resistance between pulse generator terminals and pulse generator case.
6. Infinite ohms should be indicated.
7. If continuity is indicated, replace pulse generator.

G-SENSOR

1. Ensure G-sensor is correctly mounted as follows:
   a. Position the vehicle, unloaded, on a level surface.
   b. Using a suitable level, ensure G-sensor is within 1° of level in both lateral and longitudinal directions.
   c. Adjust if necessary by installing a suitable shim to bring within 1° of level

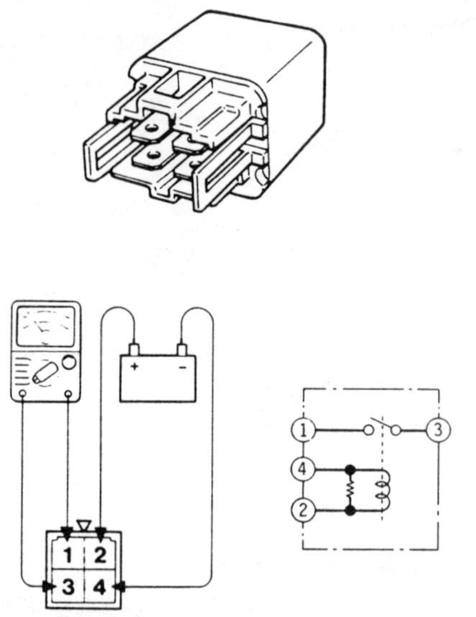

Fig. 13 Testing rear brake lock-up control relay

2. Ensure no oil leakage is present. If any leaks exist, replace G-sensor.
3. Using a suitable multimeter, measure voltage across red wire of G-sensor and ground, **Fig. 10. Ensure multimeter is set to voltage measurement range before connecting to protect the ECU from possible damage.**
4. Voltage indicated should be 7.0-7.5 volts.
5. If voltage indicated is not as specified, ECU is faulty. Replace ECU.
6. Remove G-sensor, leaving harness connected, then connect G-sensor to ground with a suitable wire.
7. With G-sensor mark positioned as shown in **Fig. 11,** measure voltage across green wire of G-sensor and ground.
8. Voltage indicated should be 1.1-1.5 volts.
9. Turn G-sensor with mark upward, **Fig. 11,** then measure voltage across green wire of G-sensor and ground.
10. Voltage indicated should be 4.6-5.0 volts.
11. If specified voltages are not obtained, replace G-sensor.
12. Install G-sensor, ensuring it is level.

REAR BRAKE LOCK-UP CONTROL RELAY

1. Remove rear brake lock-up control relay from relay panel, **Fig. 12.**
2. Apply battery voltage (12 volts) to relay terminal No. 2 **Fig. 13,** then connect terminal No. 4 to ground.

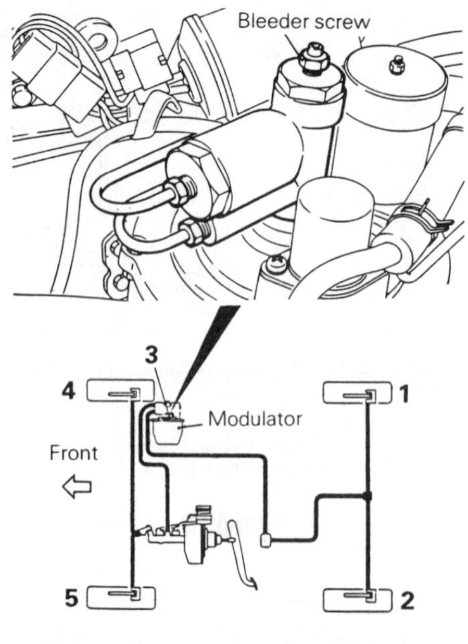

Fig. 14 Brake system bleeding sequence

3. Using a suitable multimeter, measure resistance between relay terminals Nos. 1 and 3.
4. Continuity should be indicated.
5. Disconnect voltage supply from relay terminal No. 2, then measure resistance between relay terminals Nos. 1 and 3.
6. Infinite ohms should be indicated.
7. With voltage supply disconnected, measure resistance between relay terminals Nos. 2 and 4.
8. Continuity should be indicated.
9. If specified resistance values are not obtained, replace rear brake lock-up control relay.

SYSTEM SERVICING

BRAKE BLEEDING

The brake hydraulic system should be bled any time air is allowed to enter the brake system due to damage or service, or if the brake pedal feels spongy when depressed. Bleed the brake system in the sequence shown in **Fig. 14. Bleed modulator prior to bleeding front brakes.**

MODULATOR, REPLACE

1. Remove two attaching bolts, then the heat protector.
2. Disconnect vacuum hose, brake tube, and solenoid valve connector from modulator.

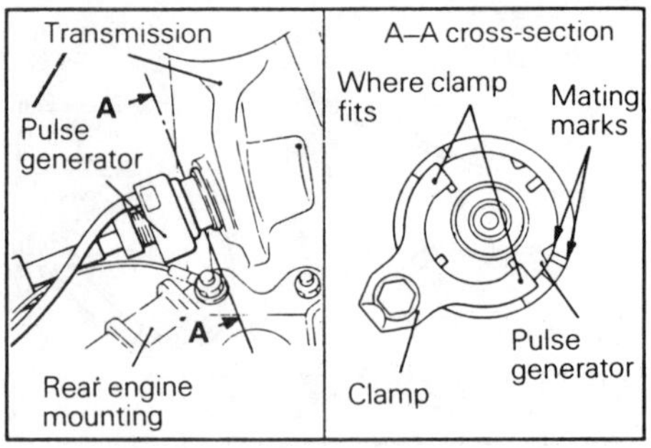

Fig. 15 Installing pulse generator

3. Remove modulator bracket from toeboard, then remove modulator.
4. Reverse procedure to install. Bleed brake system as necessary.

G-SENSOR, REPLACE

1. Disconnect G-sensor electrical connector.
2. Remove two G-sensor attaching screws, then the G-sensor. **Do not subject G-sensor to any impact or violent shaking.**
3. Reverse procedure to install, noting the following:
   a. Position the vehicle, unloaded, on a level surface.
   b. Using a suitable level, ensure G-sensor is within 1° of level in both lateral and longitudinal directions.
   c. Adjust if necessary by installing a suitable shim to bring within 1° of level.

PULSE GENERATOR, REPLACE

1. Disconnect speedometer cable at pulse generator side.
2. Remove clamp attaching bolt, then the pulse generator.
3. Reverse procedure to install, noting the following:
   a. Align mating marks of pulse generator and transmission as shown in **Fig. 15.**
   b. Securely fit clamp in grooves in pulse generator body, **Fig. 15.**
   c. **Torque** clamp attaching bolt to 7-9 ft. lbs.

ELECTRONIC CONTROL UNIT (ECU), REPLACE

1. Disconnect ECU electrical connector.
2. Remove four attaching screws (located under the high floor side panel on the right side of luggage compartment), then remove control unit.
3. Reverse procedure to install.

FORD MOTOR CO.

TABLE OF CONTENTS

1985–88, Except 1988 Continental

INDEX

DESCRIPTION

The anti-lock brake system, **Fig. 1,** is used on 1985-87 Continental and 1985-89 Cougar, Mark VII and Thunderbird Turbo Coupe models. The system uses brake fluid for both the braking function and the hydraulic boost to the brake system. The braking system is divided into three circuits, a combined rear circuit and individual front circuits.

OPERATION

Accumulator pressure is maintained between 2030 and 2610 psi, by the hydraulic pump, which is connected by a high pressure line to the booster chamber and a control valve. When the brakes are applied, the control valve is activated and a pressure, proportional to pedal travel, enters the booster chamber. This pressure travels through the normally open solenoid valve through the proportioning valve to the rear brakes. The same pressure moves the booster piston against the master cylinder piston, shutting off the central valves in the master cylinder and applying pressure to the front wheels through the two normally open solenoid valves.

The electronic controller monitors the electromechanical components of the system. A system malfunction will cause the controller to shut off or inhibit anti-lock brake control, while maintaining normal power assisted braking. Malfunctions are indicated by one or two indicator lamps inside the vehicle. A loss of hydraulic fluid, or other cause of low boost pressure, will disable the anti-lock system.

This anti-lock brake system is self-monitoring. When the ignition switch is placed in the the Run position, the electronic controller will perform a preliminary self check on the electrical system, indicated by a 3-4 second illumination of the amber "Check Anti-Lock Brakes" lamp. The indicator lamp is in the overhead console or instrument cluster. The electronic controller continues to monitor all electrical and many hydraulic anti-lock functions under all operating conditions.

SYSTEM COMPONENTS

MASTER CYLINDER & HYDRAULIC BOOSTER

The master cylinder and hydraulic brake booster are arranged with the booster behind the master cylinder, **Fig. 2.** The booster control valve is installed in a bore above the master cylinder centerline and operated by a lever connected to the brake pedal pushrod.

ELECTRIC PUMP & ACCUMULATOR

The electric high-pressure pump, **Fig. 3**, runs for short periods to charge the hydraulic accumulator supplying the service brake system pressure.

The accumulator, **Fig. 4**, is a gas filled pressure chamber which is part of the pump and motor assembly. The entire assembly is shock-mounted to the master cylinder and booster.

VALVE BODY ASSEMBLY

The valve body, **Fig. 5,** contains three pairs of solenoid valve, one pair for each front wheel and the third pair for both rear wheels combined. The solenoid valves are inlet and outlet valves, with the inlet valve normally open and the outlet valve normally closed. The valve assembly is mounted on the inboard side of the master cylinder/booster assembly.

RESERVOIR

The reservoir assembly, **Fig. 6,** is a clear plastic container, having two main chambers and is mounted on the hydraulic actuation unit. Integral with the reservoir cap assembly, are fluid level switches with electrical connectors at either end. Two low pressure hoses lead from the reservoir, one to the hydraulic pump assembly

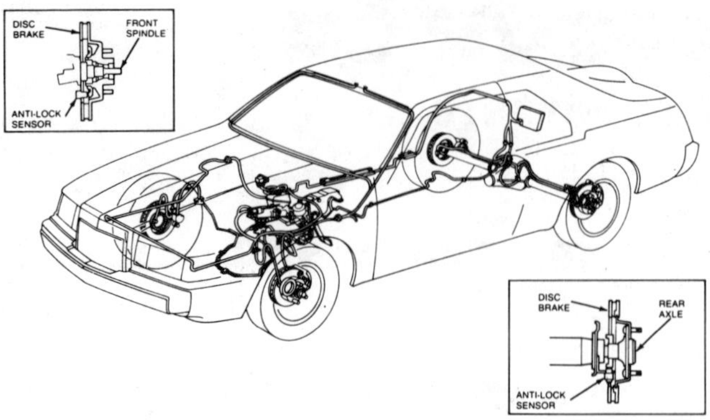

Fig. 1 Anti-lock brake system

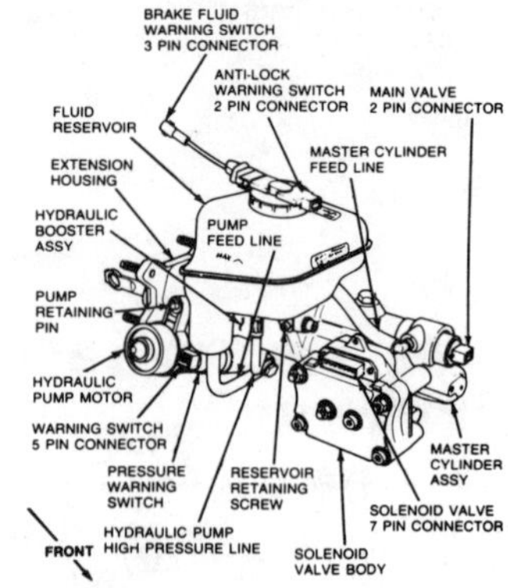

Fig. 2 Master cylinder & hydraulic booster assembly

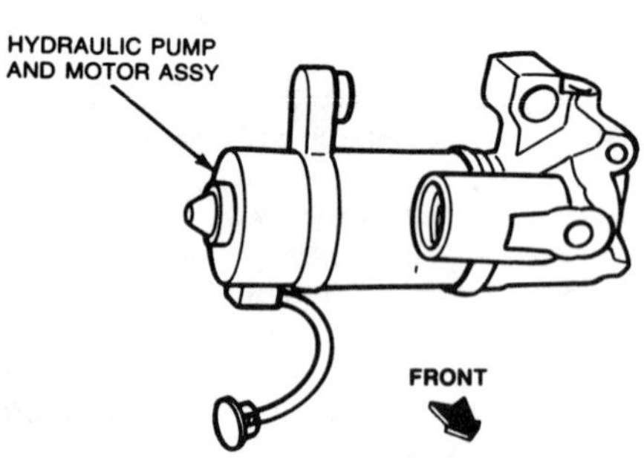

Fig. 3 Hydraulic pump assembly

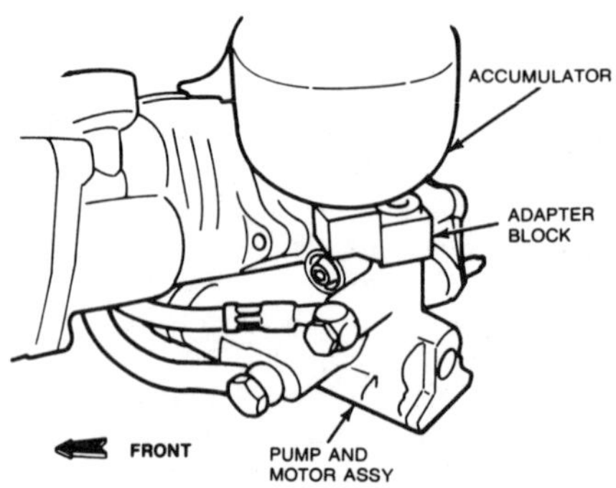

Fig. 4 Accumulator assembly

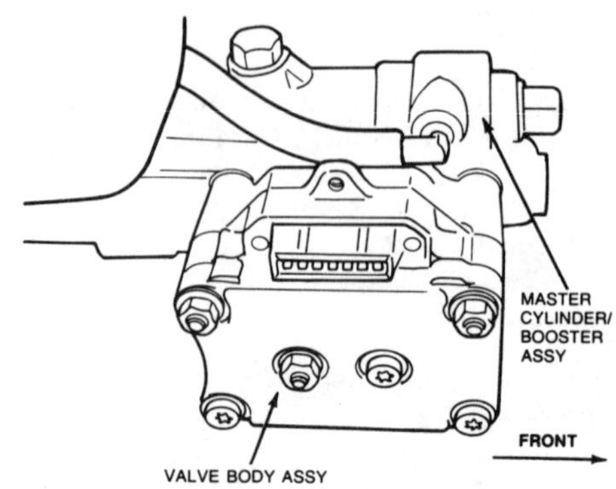

Fig. 5 Valve body assembly

and the other to the master cylinder housing. The reservoir is mounted to the hydraulic unit with a screw, bracket and a push-in tube outlet seated in a grommet in the brake booster housing. **The unit shown in Fig. 6 is used on 1985 Continental and Mark VII models. The 1986-87 Continental, Mark VII and 1987 Thunderbird Turbo Coupe use the identical unit with the exception of the connector located on top of the reservoir cap. The 1985 Continental and Mark VII models use a two pin connector, 1986-87 Continental, Mark VII and 1987 Thunderbird Turbo Coupe models use a 5 pin connector.**

WHEEL SENSORS

Four electronic wheel sensor assemblies, **Fig. 7**, are used in the brake system. Each sensor is connected to the electronic controller by electrical wiring and has an adjustable air gap between the sensor head and tooth surface.

The rear sensors are mounted on brackets bolted to the rear disc brake axle adapters. The toothed rear sensor rings are pressed on the axle shafts, inboard of the axle shaft flange.

The front sensors are mounted on brackets bolted to the front spindles. The front toothed sensor rings are pressed onto the inside of the front rotors. **Each sensor has an adjustable air gap between the sensor head and toothed sensor ring. The front sensor air gap is .043 inch on Continental and Mark VII models and .035 inch on Thunderbird Turbo Coupe models and is set using a paper spacer. The rear sensors are bolted to brackets which are bolted to the rear disc brake axle adapters. The toothed rear sensor rings are pressed on the axle shafts, inboard of the axle shaft flange. On Continental and Mark VII models the rear sensor air gap is .026 inch and can be set using either a paper spacer or a feeler gauge. On Thunderbird Turbo Coupe models, the rear sensors have a non-adjustable air gap.**

ELECTRONIC CONTROLLER

The electronic controller consists of two microprocessors and the necessary circuitry for their operation. The controller is a self-contained, non-repairable unit, which monitors system operation during normal driving as well as anti-lock braking. Under normal driving conditions, the microprocessors transmit short test pulses to the solenoid valves to check the electrical system. Under wheel lock conditions, the controller produces signals to open and close the appropriate solenoid valves and open the main valve which supplies pressure to a reaction sleeve. During anti-lock braking, moderate pulsation in the brake pedal are accompanied by a change in pedal height.

SYSTEM SERVICING

BRAKE SYSTEM BLEEDING

FRONT BRAKES

The front brakes can be bled in the same manner as conventional brakes, with or without the accumulator being charged.

REAR BRAKES

If the pump motor is allowed to operate continuously for approximately 20 minutes, a thermal safety switch (inside the motor) will automatically shut off the motor. A 2 to 10 minute cool down period is typical before normal operation can resume. Exercise care when opening the rear caliper bleeder screws, due to the high pressure available from a completely charged accumulator at the bleeder screws. Serious injury may result if care is not taken.

Bleeding w/Accumulator Charged

1. With accumulator pressure applied to the system, bleed rear brakes by opening rear brake caliper bleed screws for 10 seconds at a time while applying brake pedal with ignition switch in Run position.
2. Repeat step 1 until air-free flow of brake fluid is observed at each caliper.
3. Close bleed screws, then pump brake pedal several times.

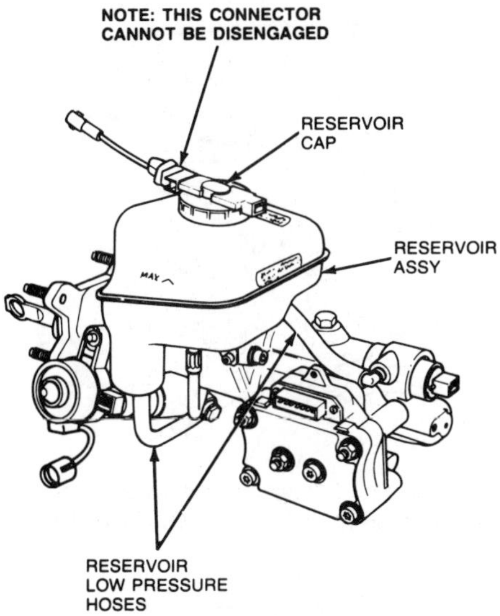

Fig. 6 Reservoir assembly

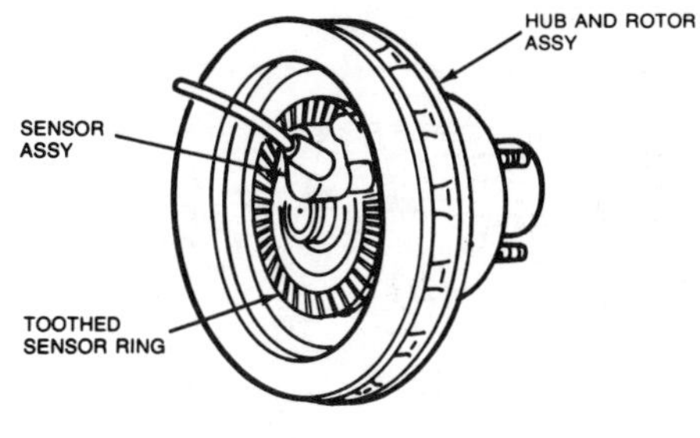

Fig. 7 Wheel sensor

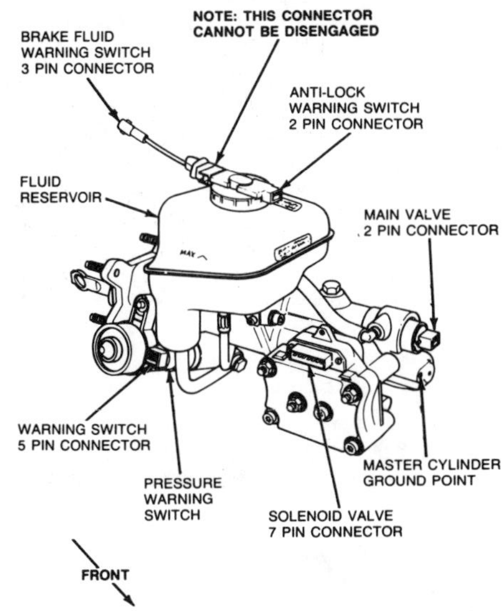

Fig. 8 Hydraulic unit replacement

4. Adjust brake fluid quantity in reservoir to MAX level with accumulator fully charged.

Bleeding w/Pressure Bleeder

1. Attach Rotunda bleeder No. 104-00064, or equivalent, to reservoir cap opening and apply a minimum pressure of 35 psi to system.
2. With brake pedal at rest and ignition switch in Off position, open rear caliper bleed screws for 10 seconds at a time.
3. When an air-free flow of brake fluid is observed at each caliper, close bleed screws and turn ignition switch to Run position.
4. Pump brake pedal several times, then adjust brake fluid quantity in reservoir to MAX level with accumulator fully charged.

HYDRAULIC UNIT, REPLACE

1. Relieve system hydraulic pressure by turning ignition off and pumping brake pedal at least 20 times until an increase in pedal force is felt.
2. Disconnect battery ground cable.
3. Disconnect electrical connectors from reservoir cap, main valve, solenoid valve body, pressure warning switch and hydraulic pump motor, **Fig. 8.**
4. Disconnect ground connector from master cylinder.
5. Disconnect and plug brake tubes from solenoid valve body.
6. Disconnect stop lamp switch electrical connector from brake pedal.
7. Remove hairpin connector from stop lamp switch on brake pedal, then slide switch far enough for switch outer hole to clear the pin. Carefully remove switch using a twisting motion.
8. Remove 4 retaining nuts from inside passenger compartment.
9. Remove hydraulic unit from dash panel through engine compartment.
10. Reverse procedure to install. **Torque** hydraulic unit locknuts to 13-25 ft. lbs.

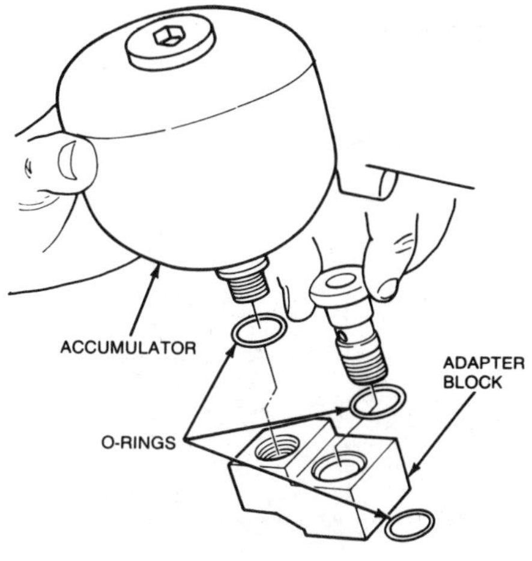

Fig. 9 Hydraulic accumulator replacement

HYDRAULIC ACCUMULATOR, REPLACE

1. Disconnect electrical connector from hydraulic pump motor.
2. Remove accumulator, **Fig. 9**, ensuring no debris falls into open port.
3. Remove accumulator adapter block, if necessary.
4. Reverse procedure to install. **Torque** adapter block bolt to 25-34 ft. lbs. and accumulator to 30-34 ft. lbs.

HYDRAULIC PUMP MOTOR, REPLACE

1. Disconnect battery ground cable.
2. Disconnect electrical connectors from hydraulic pump motor and warning switch.
3. Remove accumulator, ensuring no debris falls into open port.
4. Position a large vacuum nipple over reservoir opening to prevent fluid loss, then disconnect suction line between reservoir and pump at pump end by twisting and pulling.
5. Remove banjo bolt securing pump high pressure line, **Fig. 10.**
6. Remove Allen head bolt securing pump and motor assembly to extension housing, located directly under the accumulator. **Retain thick spacer between extension housing and shock mount for use during installation.**
7. move pump toward engine, the remove assembly from retainer pin on inboard side of extension housing.
8. Reverse procedure to install. **Torque** Allen head bolt to 5-7 ft. lbs., banjo bolt to 12-15 ft. lbs. and the accumulator to 30-34 ft. lbs.

HYDRAULIC RESERVOIR, REPLACE

1. Disconnect battery ground cable.
2. Disconnect electrical connectors from reservoir cap, then remove the cap.
3. Remove as much brake fluid as possible from reservoir.
4. Disconnect suction line between reservoir and pump at pump end by twisting and pulling.
5. Disconnect return line between reservoir and master cylinder housing at reservoir end by twisting and pulling.
6. Remove Allen head reservoir attaching screw, then pry reservoir from booster housing. **Ensure short sleeve and O-ring are removed from booster grommet hole following reservoir removal.**
7. Reverse procedure to install, noting the following:
   a. **Torque** reservoir attaching screw to 35-53 inch lbs.
   b. Fill reservoir to MAX line, then bleed brake system as previously described.
   c. After bleeding, top off reservoir to MAX line with a fully charged accumulator.

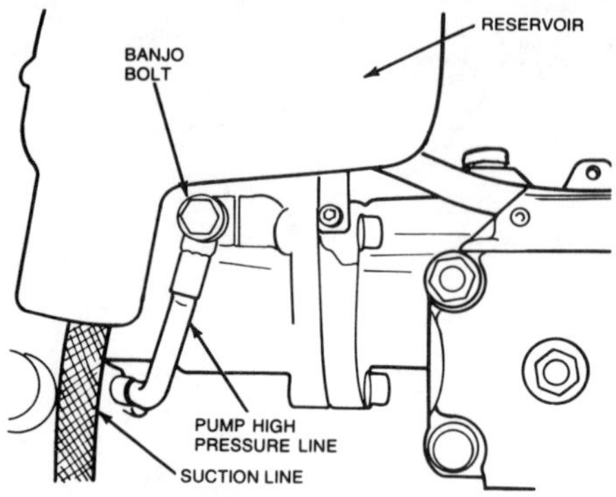

Fig. 10 Hydraulic pump motor replacement

ELECTRONIC CONTROLLER, REPLACE

1. Disconnect battery ground cable.
2. Disconnect electrical connector from electronic controller, **Fig. 11.**
3. Remove electronic controller attaching screws and the controller.
4. Reverse procedure to install.

PRESSURE SWITCH, REPLACE

1. Relieve system hydraulic pressure by turning ignition off and pumping brake pedal at least 20 times until an increase in pedal force is felt.
2. Disconnect battery ground cable.
3. Disconnect electrical connector from solenoid valve body, **Fig. 12.**
4. Remove pressure switch using a suitable tool.
5. Reverse procedure to install. **Torque** pressure switch to 15-25 ft. lbs.

FRONT WHEEL SENSOR, REPLACE

1. Disconnect battery ground cable.
2. Disconnect electrical connector from right or left front sensor, **Fig. 13,** in glove compartment.
3. Raise and support vehicle.
4. Remove wire grommet from right or lift shock tower and pull sensor cable electrical connector through hole.
5. Remove sensor wire from bracket on shock strut and side rail.
6. Loosen screw securing sensor to bracket post, then remove sensor through hole in disc brake splash shield.
7. Remove sensor bracket or sensor bracket post, if necessary, by first removing caliper and hub and rotor assembly.
8. Reverse procedure to install, noting the following:
   a. **Torque** post retaining bolt to 40-60 ft. lbs.
   b. **Torque** splash shield attaching bolts to 10-15 ft. lbs.
   c. **Torque** sensor setscrew to 21-26 inch lbs.
   d. If sensor is being reused, remove all foreign material from pole face, then glue a new front paper spacer on pole face.

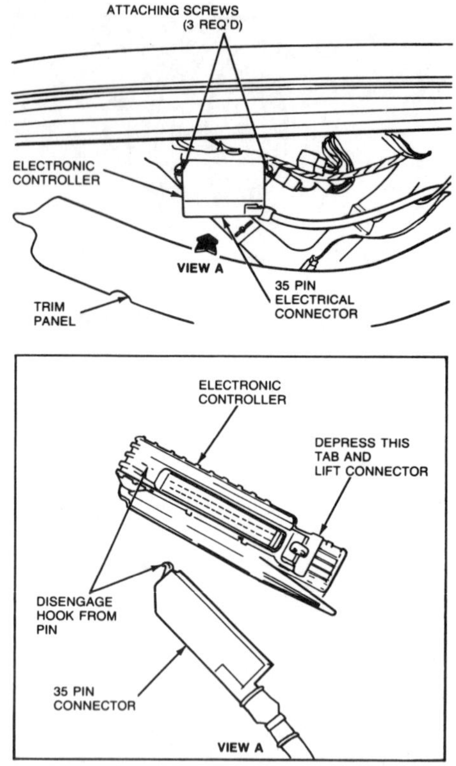

Fig. 11 Electronic controller replacement

REAR WHEEL SENSOR, REPLACE

1. Disconnect battery ground cable.
2. Disconnect wheel sensor electrical connector, **Fig. 14,** located behind forward luggage compartment trim panel.
3. Raise and support vehicle.
4. Remove wheel and tire assembly from side being worked on.
5. Carefully disconnect wheel sensor electrical connector from axle shaft housing. **The wiring harness uses three different types of retainers. The inboard retainer is a clip located on top of the differential housing which must be bent to one side to release the connector. The second retainer is a C-clip in the center of the axle housing which must be pulled rearward to disengage from housing. The third retainer is located at the union of the rear wheel brake tube and flexible hose. To remove connector from this retainer, remove hold-down bolt and open the clip.**
6. Remove rear wheel caliper and rotor assemblies.
7. Remove wheel sensor retaining bolt and the sensor, **Fig. 15.**
8. Remove grommet from rear brake splash shield and pull sensor wire out through hole.
9. Remove sensor bracket, if necessary.
10. Reverse procedure to install, noting the following:

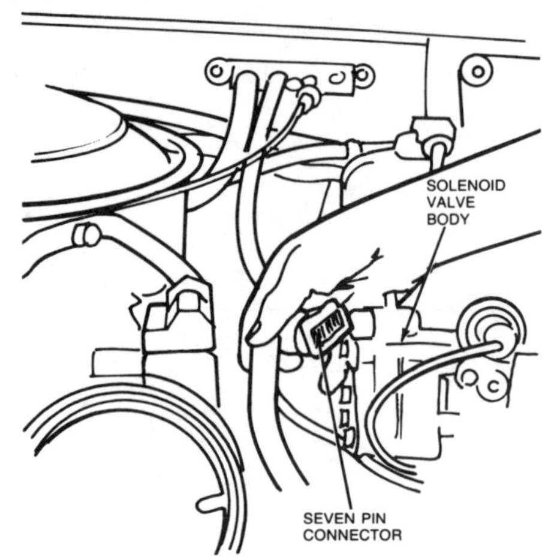

Fig. 12 Pressure switch replacement

   a. **Torque** sensor bracket attaching screws to 11-15 ft. lbs.
   b. Prior to installing sensor, loosen setscrew on sensor and ensure sensor slides freely on bracket post.
   c. If sensor is being reused, remove all foreign material from pole face, then glue a new front paper spacer on pole face.
   d. **Torque** sensor retaining bolt to 40-60 inch lbs.
   e. Push sensor toward toothed ring until paper spacer contacts sensor ring and **torque** setscrew to 21-26 inch lbs.

FRONT TOOTHED SENSOR RING, REPLACE

1. Remove wheel and tire assembly.
2. Remove brake caliper and rotor assemblies.
3. Position rotor assembly on an arbor press with studs facing up, then carefully press each stud individually until they contact sensor ring surface.
4. Press studs and sensor ring out of rotor simultaneously, **Fig. 16.**
5. Install studs into rotor one at a time.
6. Position sensor ring onto rotor, then press ring until fully seated, **Fig. 17.**
7. Install rotor and caliper assemblies, then the wheel and tire.

REAR TOOTHED SENSOR RING, REPLACE

1. Remove rear axle shaft as described elsewhere in this manual.
2. Install pinion bearing cone remover tool, **Fig. 18,** between axle shaft flange and sensor ring. Ensure recessed portion of inner surface of tool is toward axle flange.
3. Position axle in an arbor press and press axle out of sensor ring.
4. To install, position sensor on axle shaft with recessed side facing inboard.
5. Install pinion bearing cone remover tool, **Fig. 19,** on axle shaft.

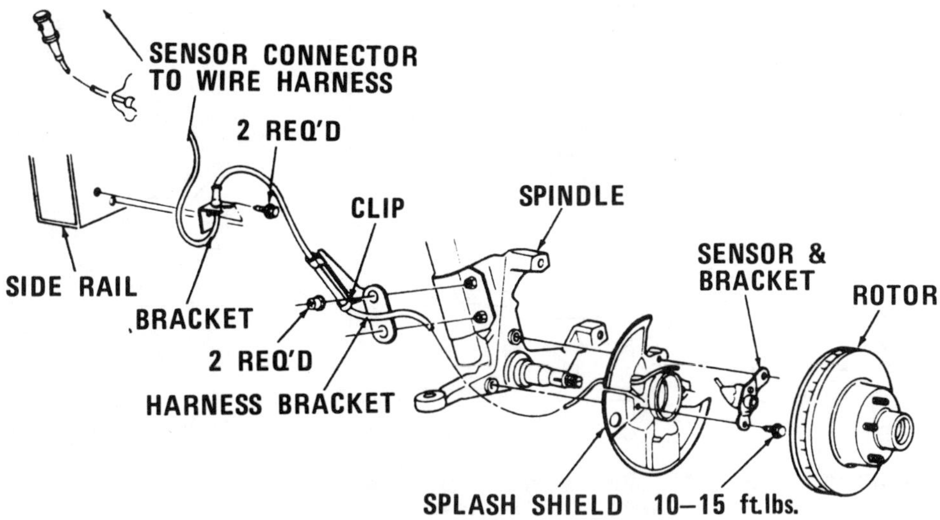

Fig. 13 Front wheel sensor replacement

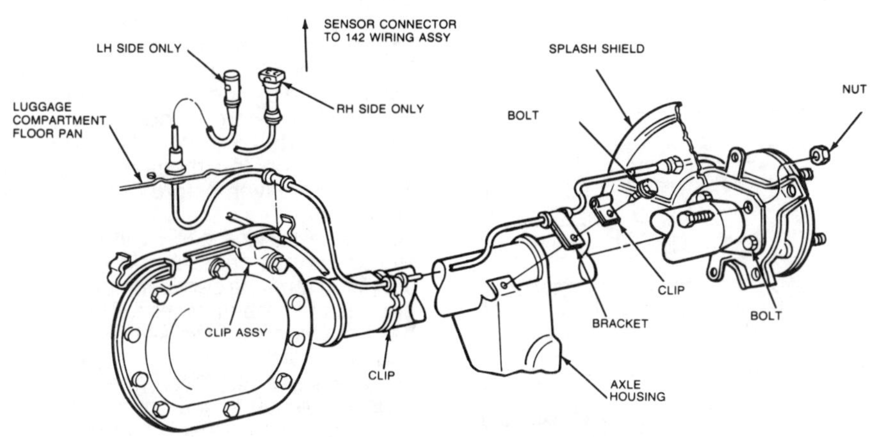

Fig. 14 Rear wheel sensor wiring

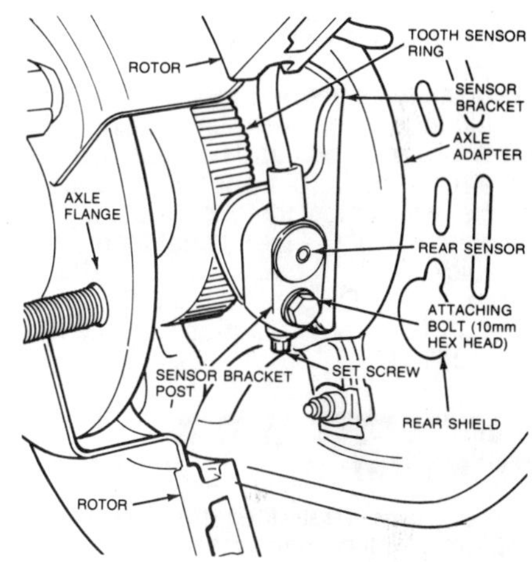

Fig. 15 Rear wheel sensor replacement

6. Position a piece of bar stock on top of axle flange, then press sensor ring onto axle shaft until a 1.8 inch gap between face of sensor ring and flange is obtained.
7. Install axle shaft.

CONTINENTAL & MARK VII SYSTEM DIAGNOSIS & TROUBLESHOOTING

The following procedures applies to 1985-87 Continental & Mark VII models only. Ensure the following information and diagnostic procedures are used in sequence and step by step order as indicated. Following the wrong sequence or bypassing steps will only lead to unnecessary replacement of components and/or incorrect resolution of the symptom. The diagnostic procedure consists of four sections, Pre-Test Checks, Quick Tests, Lamp Symptom Chart and the Diagnostic Tests (including electrical schematic(s).

PRE-TEST CHECKS

Verify that the parking brake is completely released, brake warning lamp switch is not grounded and the battery has been checked (completely charged). Ensure that the contacts of the 35 pin electronic controller harness electrical connector are properly inserted and are not damaged. Check the 35 contacts of the electronic controller assembly. If there are any damaged contacts replace the electronic controller assembly. Verify that all of the following electrical connectors or wires are connected/attached:

1. 1985-87 models:
   a. 7 pin plug of the valve unit.
   b. 2 pin plug of the main valve.
   c. 5 pin plug of the combined pressure warning switch.
   d. 2 and 3 pin plug of the reservoir warning switch on 1985 models, 5 pin plug of the fluid reservoir cap on 1986-87 models.
   e. 2 pin plugs on all 4 sensors.
   f. 2 and 4 pin plugs on each end of the motor jumper wire, if used.
   g. 2 relay plugs located near the brake fluid reservoir.
2. 1986-87 models only:
   a. Ground attachment on front of hydraulic booster assembly.
   b. Ground attachment in luggage compartment near the 35 pin connector.
   c. Ground attachment wire from negative battery terminal to fender sheet metal, to wiring harness through the 2 pin connector.

Ensure all relays, diodes and fuse links are intact and/or properly inserted. Ensure battery cable connections are clean and tight. Check ground terminals of the electronic controller and the hydraulic unit.

ANTI-LOCK QUICK TEST CHECKS

Required Equipment

1. EEC-IV breakout box, tool T83L-50-EEC-IV, Rotunda model 014-00322 or equivalent.

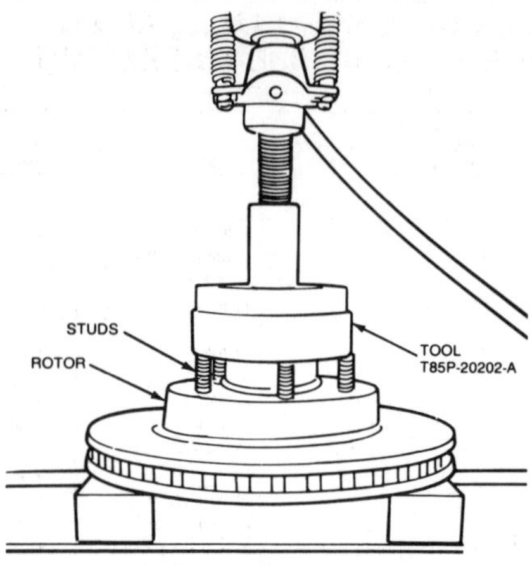

Fig. 16 Front toothed sensor ring removal

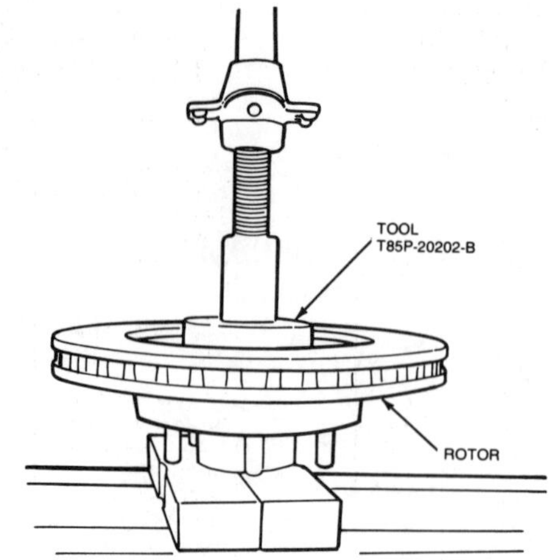

Fig. 17 Front toothed sensor ring installation

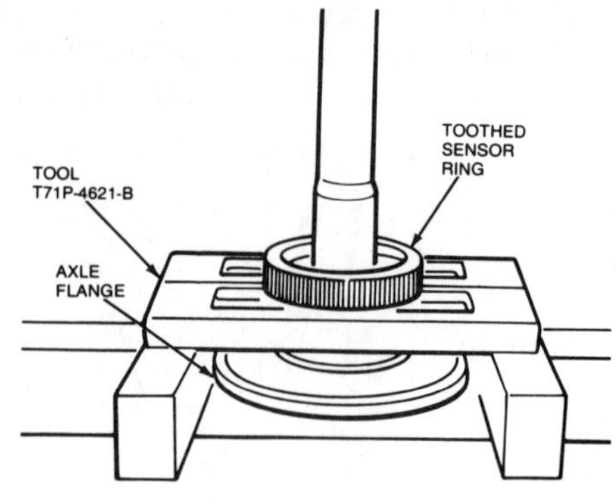

Fig. 18 Rear toothed sensor ring removal

2. Anti-lock harness adapter, T85P-50-ASA or equivalent.
3. Multimeter, tool 007-00001 or equivalent.

All quick tests are performed in the vehicle luggage compartment using the EEC-IV breakout box and harness adapter. These tests leads to s specific diagnostic Test that will in most instances, identify the fault condition. If the fault is not located by the Quick Test procedure, **Figs. 20 and 21,** proceed to Diagnostic Light Symptom to identify the proper diagnostic procedure to be applied.

DIAGNOSTIC LIGHT SYMPTOM

If the Quick Test does not isolate the symptom, it will be necessary to check the operation of the brake and anti-lock warning lamps. Observe the lamps and compare their ON/OFF operation to the conditions listed in **Figs. 22 and 23.** Once the actual warning lamp pattern has been matched to one of the conditions listed, perform the specific test indicated.

DIAGNOSTIC TESTS

Required Equipment

1. Anti-lock harness adapter tool T85P-50-ASA or equivalent.
2. EEC-IV breakout box T83L-50-EEC-IV, Rotunda model 014-00322 or equivalent.
3. Anti-lock pressure gauge T85P-20215-A or equivalent.
4. Multimeter Rotunda model 007-00001 or equivalent.

Refer to wiring diagrams, **Figs. 24 and 25,** when performing the Diagnostic Tests to locate wire circuits indicated in the test procedures. Each diagnostic test is completely independent of the other tests and within the tests are sequences that can locate the problem without having to complete the entire test procedure.

On 1985 models, refer to **Figs. 26 through 35** for diagnostic test procedures. On 1986-87 models, refer to **Figs. 36 through 44** for diagnostic test procedures.

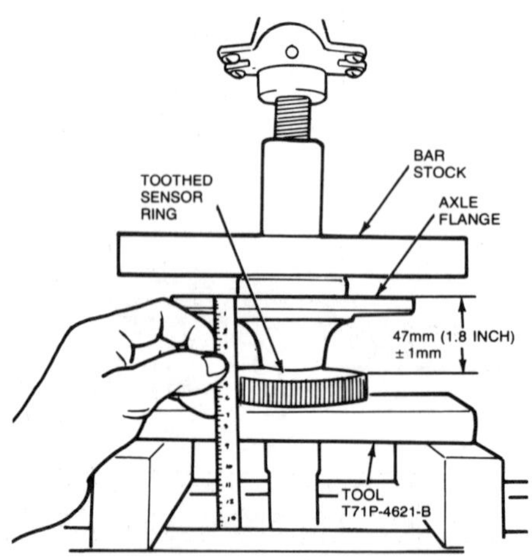

Fig. 19 Rear toothed sensor ring installation

THUNDERBIRD TURBO SYSTEM DIAGNOSIS & TROUBLESHOOTING

The following information and diagnostic procedures applies to the 1987 Thunderbird Turbo Coupe. Ensure the diagnosis procedures are used in sequence and step by step order as indicated. Following the wrong sequence or bypassing steps will only lead to unnecessary replacement of components and/or incorrect resolution of the symptom. The diagnostic procedure consists of four sections, on-board self-tests, manual quick tests, warning lamp symptom and diagnostic tests.

ON-BOARD SELF-TESTS

The anti-lock brake electronic control module is capable of performing a self-test using Rotunda Star Tester model 007-00017, or Super Star Tester 007-00019, or equivalents. The procedures for using the Star Tester are covered in the anti-lock self-test diagnosis chart, **Fig. 45.** If the star tester is not available the anti-lock Quick Check chart can be used as described further on.

The anti-lock control module monitors system operation and stores up to seven different service codes in its memory. However, it cannot store two service codes with the first (multiples of ten) digit. For example: The module cannot store a code 25 and a code 26 at the same time (it can however, store a code 25 and a code 35.) The code 25 is stored first. It must be serviced, then the memory cleared, and the vehicle driven before the code 26 can be stored.

A valve service code will override any stored service code during self-test. The condition must be serviced before any other codes can be displayed. After the valve has been serviced, the self-test can be run and any stored codes will be displayed. If no codes are displayed, the system has passed the self-test. If any codes are displayed during self-test, refer to the self-test code index, **Fig. 46.** The index will direct you to a specific pinpoint test. After the pinpoint tests have been completed, the components should be serviced. After servicing these codes the self-test should be repeated to see if any other codes have been stored in memory. These should then be serviced and the self-test repeated until no codes are displayed. After all codes have been displayed and serviced, the memory will be erased by driving the vehicle above 25 mph. If the self-test has not been completed and all codes outputted, the memory cannot be erased.

STAR TESTER CONNECTION & BATTERY CHECK

1. Turn ignition switch to OFF position.
2. Lower right hand module panel located in the luggage compartment.
3. Connect Star tester connectors to vehicle connectors located near electronic controller assembly.
4. Turn the power switch located on the right hand side of the Star tester to the ON position. A steady 00 will appear to signify that the Star tester is ready to start the self-test and receive service codes. **If the message LO BAT**

Anti-Lock Quick Check Sheet Using 60-Pin EEC-IV Breakout Box, Tool T83L-50-EEC-IV

Item to be Tested		Ignition Mode	Measure Between Pin Numbers	Tester Scale/ Range	Specification	Test Step
Battery Check		On	40 + 2	Volts	10 minimum	A-1
Main Relay		Off	40 + 8	Ohms	45 Ohms — 105 Ohms	A-6
			Place a jumper between pins 2 & 8			
		On	40 + 3	Volts	10 minimum	A-7
Power from Main Relay		On	40 + 20	Volts	10 minimum	A-3
			Remove jumper from pins 2 & 8			
Main Relay Circuit		Off	40 + 3	Continuity	Continuity	A-2
Main Relay Circuit		Off	20 + 40	Continuity	Continuity	A-3a
Sensor Resistance	(RR)	Off	4 + 22	K Ohms	0.8K to 1.4K Ohms	A-8
Sensor Resistance	(LF)	Off	5 + 23	K Ohms	0.8K to 1.4K Ohms	A-9
Sensor Resistance	(LR)	Off	6 + 24	K Ohms	0.8K to 1.4K Ohms	A-10
Sensor Resistance	(RF)	Off	7 + 25	K Ohms	0.8K to 1.4K Ohms	A-11
Main Valve Resistance		Off	11 + 18	Ohms	2 Ohms to 5 Ohms	A-12
Inlet & Outlet Valves		Off	11 + 40	Continuity	Continuity	A-19
		Off	11 + 15	Ohms	5 Ohms to 8 Ohms	A-13
		Off	11 + 17	Ohms	5 Ohms to 8 Ohms	A-14
		Off	11 + 35	Ohms	5 Ohms to 8 Ohms	A-15
		Off	11 + 33	Ohms	3 Ohms to 6 Ohms	A-16
		Off	11 + 16	Ohms	3 Ohms to 6 Ohms	A-17
		Off	11 + 34	Ohms	3 Ohms to 6 Ohms	A-18
Reservoir Warning		On	9 + 10	Ohms	Less than 5 Ohms	A-4a
Lift Fluid Level Indicator from Reservoir (Float at bottom position)		Off	9 + 10	Ohms	Infinite (Open Circuit)	A-5a
Sensor Cable Continuity Shielding to Ground	(RR)	Off	40 + 4	Continuity	No Continuity	B-1a
	(LF)	Off	40 + 5	Continuity	No Continuity	B-2a
	(LR)	Off	40 + 6	Continuity	No Continuity	B-3a
	(RF)	Off	40 + 7	Continuity	No Continuity	B-4a
Sensor Voltage (Rotate wheels at 1 revolution per second minimum)	(RR)	Off	4 + 22	AC Millivolts	50-700 Millivolts	C-5
	(LF)	Off	5 + 23	AC Millivolts	50-700 Millivolts	C-6
	(LR)	Off	6 + 24	AC Millivolts	50-700 Millivolts	C-7
	(RF)	Off	7 + 25	AC Millivolts	50-700 Millivolts	C-8

Fig. 20 Anti-lock quick check chart. 1985–86 Continental & Mark VII models

Anti-Lock Quick Check Sheet Using 60-Pin EEC-IV Breakout Box, Tool T83L-50-EEC-IV①

Item to be Tested		Ignition Mode	Measure Between Pin Numbers	Tester Scale/ Range	Specification	Test Step
No Boost/No Power Brakes (Hard Brake Pedal)		—	—	—	—	D-1
Battery Check		On	40 + 2	Volts	10 minimum	A-1
Main Power Relay		Off	40 + 8	Ohms	45 Ohms — 105 Ohms	A-6
			Place a jumper between pins 2 & 8			
		On	40 + 3	Volts	10 minimum	A-7
Power from Main Power Relay		On	40 + 20	Volts	10 minimum	A-3
			Remove jumper from pins 2 & 8			
Main Power Relay Circuit		Off	40 + 3	Continuity	Continuity	A-2
Main Power Relay Circuit		Off	20 + 40	Continuity	Continuity	A-3a
Sensor Resistance	(RR)	Off	4 + 22	K Ohms	800 to 1400 Ohms	A-8
Sensor Resistance	(LF)	Off	5 + 23	K Ohms	800 to 1400 Ohms	A-9
Sensor Resistance	(LR)	Off	6 + 24	K Ohms	800 to 1400 Ohms	A-10
Sensor Resistance	(RF)	Off	7 + 25	K Ohms	800 to 1400 Ohms	A-11
Main Valve Resistance		Off	11 + 18	Ohms	2 Ohms to 5.5 Ohms	A-12
Inlet & Outlet Valves		Off	11 + 40	Continuity	Continuity	A-13
		Off	11 + 15	Ohms	5 Ohms to 8 Ohms	A-14
		Off	11 + 17	Ohms	5 Ohms to 8 Ohms	A-15
		Off	11 + 35	Ohms	5 Ohms to 8 Ohms	A-16
		Off	11 + 33	Ohms	3 Ohms to 6 Ohms	A-17
		Off	11 + 16	Ohms	3 Ohms to 6 Ohms	A-18
		Off	11 + 34	Ohms	3 Ohms to 6 Ohms	A-19
Reservoir Warning		On	9 + 10	Ohms	Less than 5 Ohms	A-4a
Lift Fluid Level Indicator from Reservoir (Float at bottom position)		Off	9 + 10	Ohms	Infinite (Open Circuit)	A-5a
Sensor Cable Continuity Shielding to Ground	(RR)	Off	40 + 4	Continuity	No Continuity	B-1a
	(LF)	Off	40 + 5	Continuity	No Continuity	B-2a
	(LR)	Off	40 + 6	Continuity	No Continuity	B-3a
	(RF)	Off	40 + 7	Continuity	No Continuity	B-4a
Sensor Voltage (Rotate wheels at 1 revolution per second minimum) (Shut off air suspension switch in luggage compartment with vehicle on hoist if so equipped).	(RR)	Off	4 + 22	AC Millivolts	50-700 Millivolts	C-5
	(LF)	Off	5 + 23	AC Millivolts	50-700 Millivolts	C-6
	(LR)	Off	6 + 24	AC Millivolts	50-700 Millivolts	C-7
	(RF)	Off	7 + 25	AC Millivolts	50-700 Millivolts	C-8

Fig. 21 Anti-lock quick check chart. 1987 Continental & Mark VII models

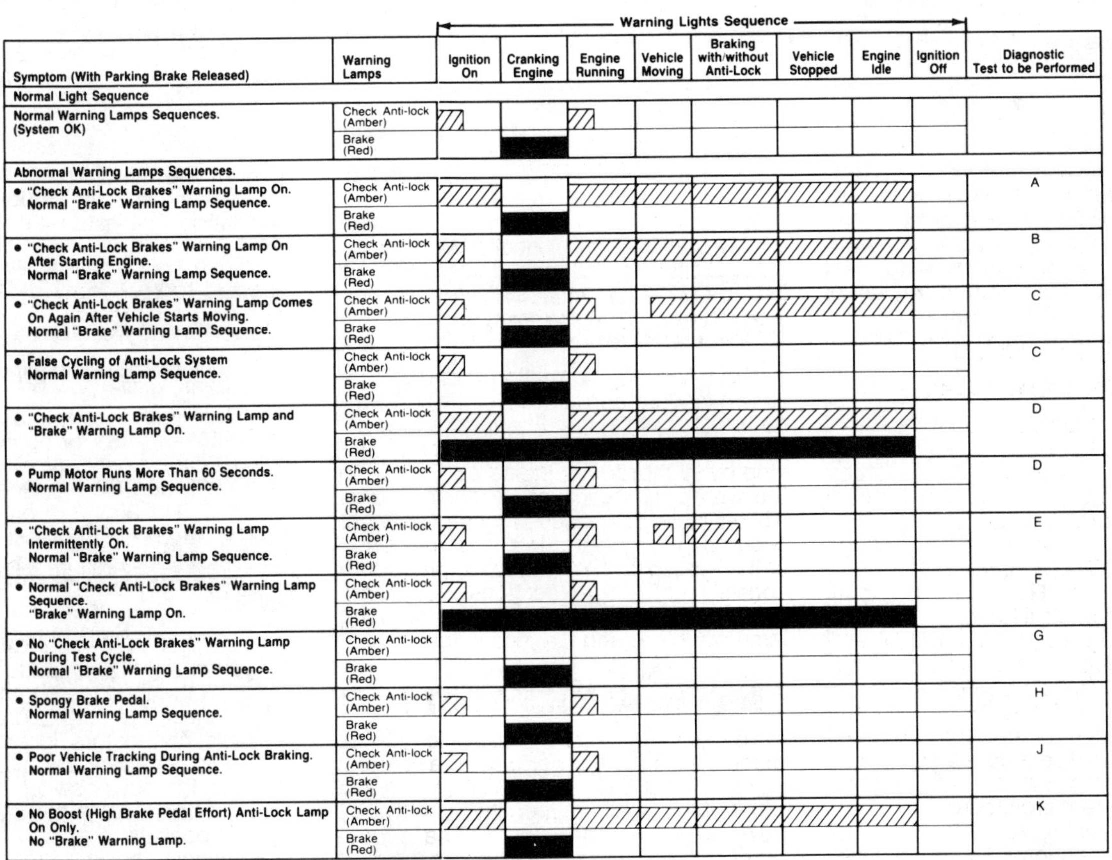

Fig. 22 Light symptom chart. 1985–86 Continental & Mark VII models

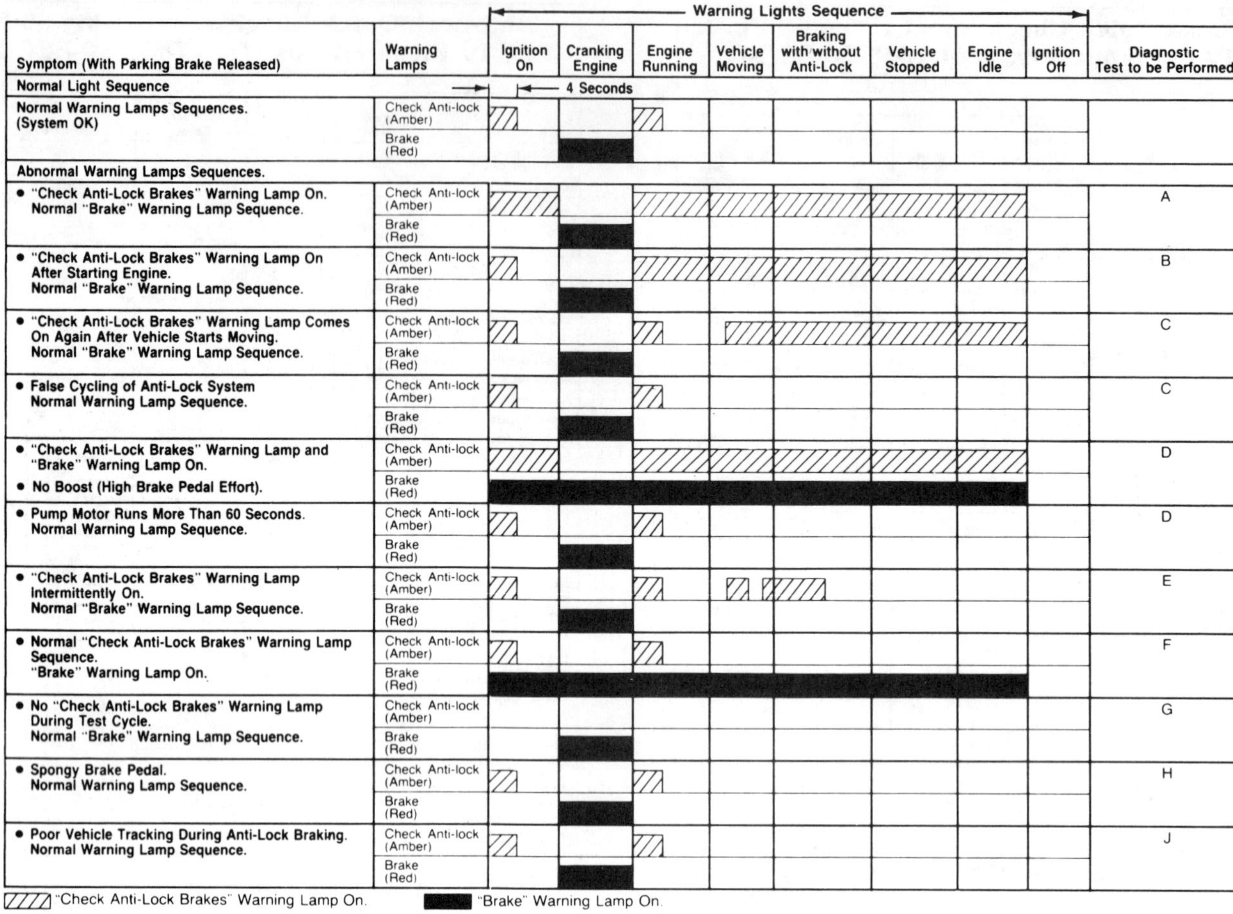

Fig. 23 Light symptom chart. 1987 Continental & Mark VII models

appears in the upper left hand corner of the read out display and stays on, replace the Star tester's 9 volt battery before continuing with the self-test. If the message LO BAT will appear momentarily when the power switch is turned OFF.

5. With the ignition switch in the OFF position, push the self-test button in the center of the Star tester. A colon should appear in the left hand side of the read out display.
6. Push Self-test button again. This deactivates the self-test sequence. The colon should disappear.
7. If the Star tester passes steps 1 through 6, proceed with the on-board self-test portion of the diagnostic procedure. If any codes appear during the self-test, refer to the on-board self-test service code index, **Fig. 46.**
8. Refer to **Figs. 47 through 52,** for on-board self-test diagnostic procedures.

QUICK TEST CHECKS

Required Equipment

The following equipment is necessary to perform the following quick tests:

1. EEC-IV breakout box T83L-50-EEC-IV, Rotunda breakout box 014-00322 or equivalent.
2. Anti-lock harness adapter T85P-50-ASA or equivalent.
3. Digital volt-ohmmeter 007-00001 or equivalent. All quick tests are performed in the vehicle luggage compartment using the EEC-IV breakout box and harness adapter. Use anti-lock quick check chart, **Fig. 53,** to locate particular problem areas. These tests lead to a specific diagnostic pinpoint test (which will follow these particular series of procedures) that will in most instances, identify the fault. If the fault is not identified by the quick test procedures use the Diagnostic Lamp Symptom chart, **Fig. 54,** to identify the proper diagnostic procedure.

DIAGNOSTIC LAMP SYMPTOM

If the quick test does not isolate the symptom, it will be necessary to check the operation of the brake and anti-lock warning lamps. Observe the lamps and compare their ON/OFF operation to the conditions listed on the chart. Once the actual warning lamp pattern has been matched to one of the conditions listed in the chart, perform the specific pinpoint test indicated in Diagnostic Tests.

DIAGNOSTIC TESTS

Required Equipment

The following test equipment is required to conduct the diagnostic tests:

1. Anti-lock harness adapter T85P-50-ASA or equivalent.
2. EEC-IV breakout box T83L-50-EEC-IV, Rotunda breakout box 014-00322 or equivalent.
3. Anti-lock pressure gauge T85P-20215-A or equivalent.
4. Rotunda Digital volt-ohmmeter 007-00001 or equivalent.
5. Rotunda Star tester 007-00019 or equivalent.

Refer to wiring diagram, **Fig. 55,** when performing the diagnostic test procedures to locate wire circuits indicated in the tests. Each test is completely independent of the other tests and within the tests are sequences that can find the problem without having to complete then entire test procedure.

Refer to **Figs. 56 through 64,** for diagnostic procedures.

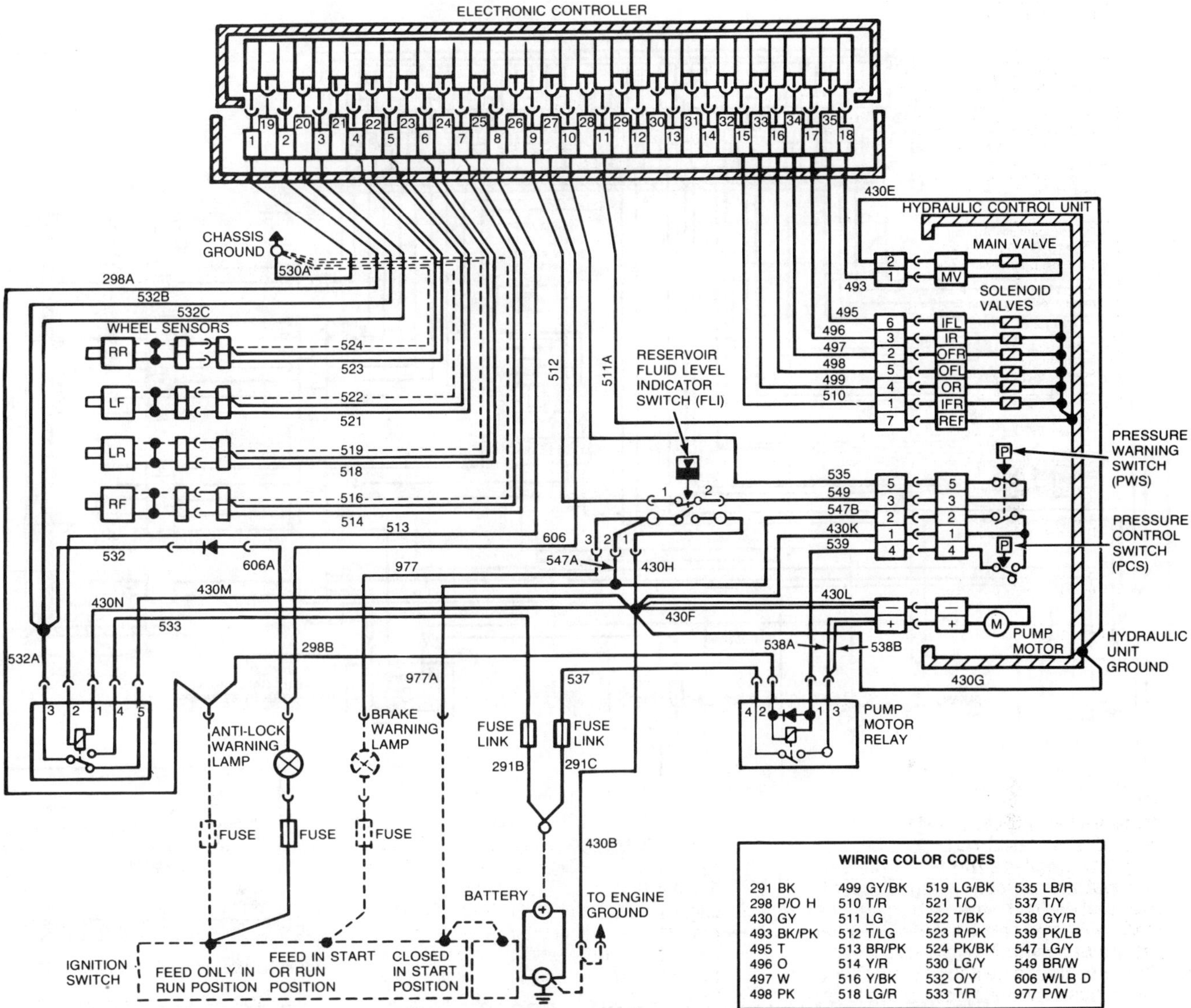

WIRING COLOR CODES

291 BK	499 GY/BK	519 LG/BK	535 LB/R
298 P/O H	510 T/R	521 T/O	537 T/Y
430 GY	511 LG	522 T/BK	538 GY/R
493 BK/PK	512 T/LG	523 R/PK	539 PK/LB
495 T	513 BR/PK	524 PK/BK	547 LG/Y
496 O	514 Y/R	530 LG/Y	549 BR/W
497 W	516 Y/BK	532 O/Y	606 W/LB D
498 PK	518 LG/R	533 T/R	977 P/W

Fig. 24 Anti-lock wiring circuit. 1985 Continental & Mark VII

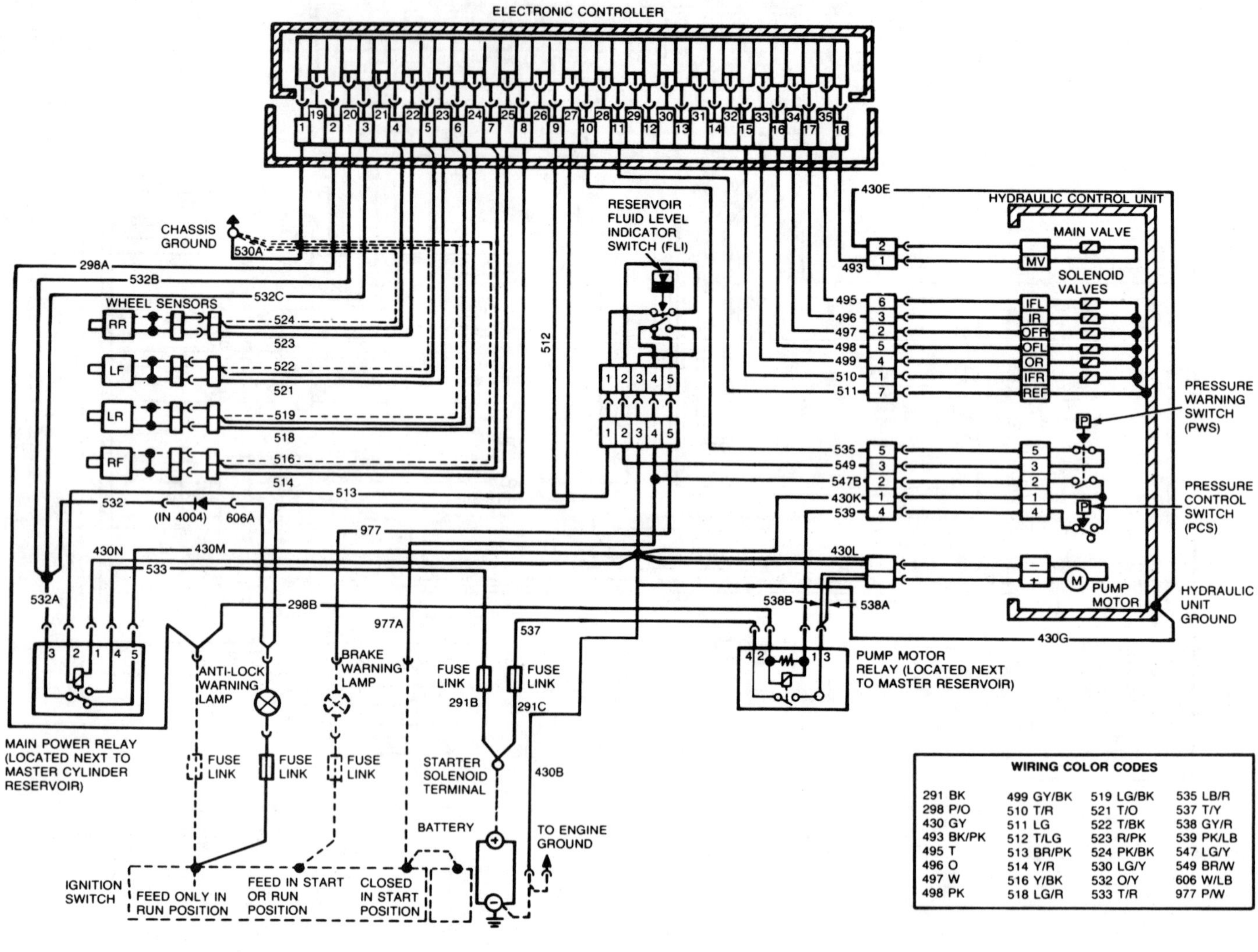

WIRING COLOR CODES

291 BK	499 GY/BK	519 LG/BK	535 LB/R
298 P/O	510 T/R	521 T/O	537 T/Y
430 GY	511 LG	522 T/BK	538 GY/R
493 BK/PK	512 T/LG	523 R/PK	539 PK/LB
495 T	513 BR/PK	524 PK/BK	547 LG/Y
496 O	514 Y/R	530 LG/Y	549 BR/W
497 W	516 Y/BK	532 O/Y	606 W/LB
498 PK	518 LG/R	533 T/R	977 P/W

Fig. 25 Anti-lock wiring circuit. 1986–87 Continental & Mark VII

Anti-Lock Warning Lamp On (With Brake Warning Lamp Off) — Test A

WARNING LIGHTS SEQUENCE

Warning Lamps	Ignition On	Cranking Engine	Engine Running	Vehicle Moving	Braking with/without Anti-Lock	Vehicle Stopped	Engine Idle	Ignition Off
Check Anti-Lock (Amber)	▨	▨	▨	▨	▨	▨	▨	
Brake (Red)		■						

TEST STEP	RESULT ▶	ACTION TO TAKE
A1 35-PIN PLUG TESTING		
• Turn ignition switch Off. • Disconnect 35-Pin plug from electronic controller. 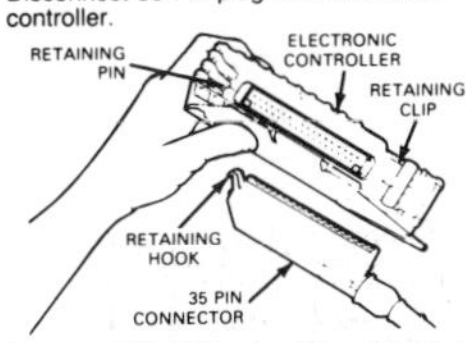 • Connect EEC-IV Breakout box with Tool T83L-50-EEC-IV or equivalent, anti-lock test adapter Tool T85P-50-ASA or equivalent, to the anti-lock 35-Pin plug wiring harness. 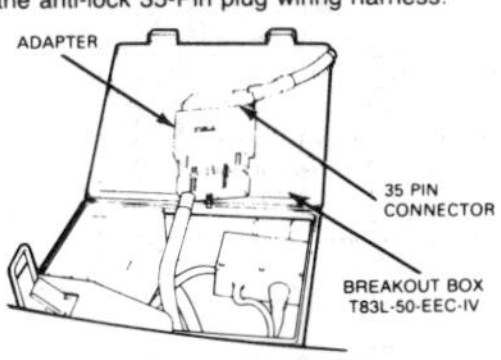 • Set multi-meter to read volts DC. • Turn ignition switch On. • Measure voltage between Breakout box Pins 40 and 2.	Over 10V (OK) ▶ Under 10V (NOT OK) ▶	GO to Step **A2**. GO to Step **A1a**.

Fig. 26 Diagnosis, warning lamp ON w/brake warning lamp OFF (Part 1 of 24). 1985

Anti-Lock Warning Lamp On (With Brake Warning Lamp Off) — Test A

TEST STEP	RESULT ▶	ACTION TO TAKE
A1a CHECK ELECTRONIC CONTROLLER TO GROUND WIRE		
• Check: — fuse (F5). — battery. • Remove positive battery cable. • Check continuity between Breakout box pin 40 and body ground.	Continuity (OK) ▶ No continuity (NOT OK) ▶	GO to Step **A1b**. REPAIR or REPLACE cable harness (Circuit 530 or 530A).
A1B CHECK IGNITION TO ELECTRONIC CONTROLLER WIRE		
• Check continuity between Breakout box Pin 2 and ignition switch wire plug G 298.	Continuity (OK) ▶ No continuity (NOT OK) ▶	RECONNECT positive battery cable. CHECK for power at ignition switch pin with switch On. If okay, connect electronic controller and reverify symptom. REPAIR or REPLACE cable harness (Circuit 298A).
A2 CHECK MAIN RELAY SECONDARY CIRCUIT (NORMAL)		
• Turn ignition switch Off. • Check for continuity between Breakout box Pins 40 and 3.	Continuity (OK) ▶ No continuity (NOT OK) ▶	GO to **A3**. GO to **A2a**.

Fig. 26 Diagnosis, warning lamp ON w/brake warning lamp OFF (Part 2 of 24). 1985

Anti-Lock Warning Lamp On (With Brake Warning Lamp Off) — Test A

TEST STEP	RESULT ▶	ACTION TO TAKE
A2a CHECK MAIN RELAY SECONDARY CIRCUIT (NORMAL)		
• Disconnect main relay from socket. • Check for continuity between main relay socket Pins 3 and 5. 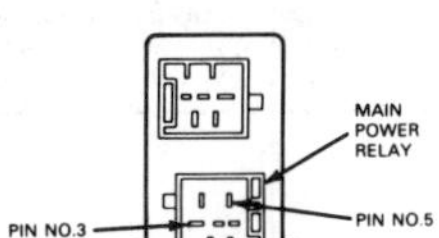	Continuity (OK) ▶ No continuity (NOT OK) ▶	GO to Step **A2b**. REPLACE main relay.
A2b CHECK MAIN RELAY SECONDARY CIRCUIT WIRING HARNESS		
• Disconnect positive battery cable. • Check for continuity between main relay socket Pin 3 and Breakout box Pin 3. 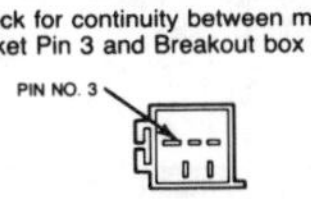	Continuity (OK) ▶ No Continuity (NOT OK) ▶	GO to Step **A2C**. REPAIR or REPLACE cable harness (Circuit 532C or 532A).
A2c CHECK MAIN RELAY SECONDARY CIRCUIT WIRING HARNESS		
• Check for continuity between main relay socket Pin 5 and body ground. 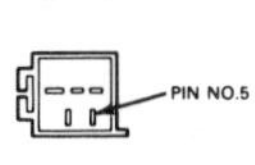	Continuity (OK) ▶ No continuity (NOT OK) ▶	RECONNECT main relay, electronic controller and battery cable and reverify symptom. REPAIR or REPLACE cable harness (Circuit 522).

Fig. 26 Diagnosis, warning lamp ON w/brake warning lamp OFF (Part 3 of 24). 1985

With Anti-Lock Warning Lamp On (With Brake Warning Lamp Off) — Test A

TEST STEP	RESULT ▶	ACTION TO TAKE
A3 CHECK MAIN RELAY SECONDARY CIRCUIT (NORMAL)		
• Check for continuity between Breakout box Pins 40 and 20.	Continuity (OK) ▶ No continuity (NOT OK) ▶	GO to Step **A4**. GO to Step **A3a**.
A3a CHECK MAIN RELAY SECONDARY CIRCUIT WIRING HARNESS		
• Remove main relay. • Check for continuity between main relay socket Pin 3 and Breakout box Pin 20. 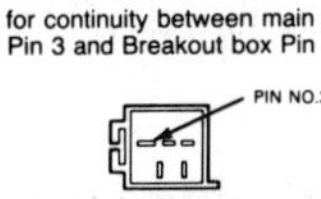	Continuity (OK) ▶ No continuity (NOT OK) ▶	CONNECT main relay and electronic controller and reverify symptom. REPAIR or REPLACE cable harness (Circuit 532B).
A4 CHECK FLI AND PWS CIRCUIT		
• Turn ignition switch On. • Set multi-meter to read resistance. • Measure the resistance between Breakout box Pins 9 and 10.	Less than 5 ohms (OK) ▶ Greater than 5 ohms (NOT OK) ▶	GO to Step **A5**. GO to Step **A4a**.
A4a CHECK FLI ANTI-LOCK WARNING CIRCUIT		
• Disconnect 2-Pin plug on reservoir Fluid Level Indicator (FLI). • Measure resistance between fluid level indicator electrical socket Pins 1 and 2 (with brake fluid level at maximum mark on reservoir). 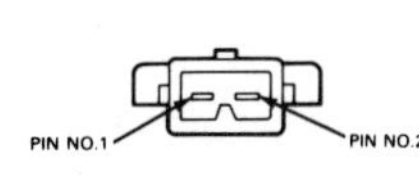	Less than 2 ohms (OK) ▶ Greater than 2 ohms (NOT OK) ▶	GO to Step **A4b**. REPLACE fluid level indicator.

Fig. 26 Diagnosis, warning lamp ON w/brake warning lamp OFF (Part 4 of 24). 1985

Anti-Lock Warning Lamp On (With Brake Warning Lamp Off) — **Test A**

TEST STEP	RESULT ▶	ACTION TO TAKE
A4b CHECK PWS ANTI-LOCK WARNING CIRCUIT		
• Disconnect 5-Pin plug at pressure warning switch (PWS). • Check for continuity between pressure warning switch socket Pins 3 and 5. 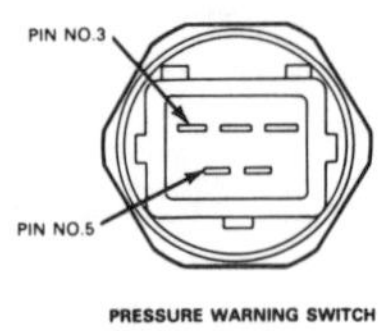	Continuity (OK) ▶	GO to Step **A4c**.
	No continuity (NOT OK) ▶	REPLACE pressure warning switch.
A4c CHECK ELECTRONIC MODULE TO FLI WIRE		
• Check for continuity between Breakout box Pin 9 and FLI 2-Pin plug (harness side) Pin 1. 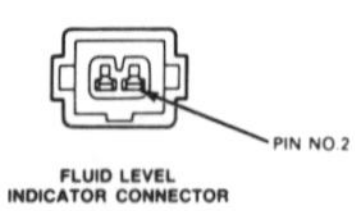	Continuity (OK) ▶	GO to Step **A4d**.
	No continuity (NOT OK) ▶	REPAIR or REPLACE cable harness (Circuit 512).

Fig. 26 Diagnosis, warning lamp ON w/brake warning lamp OFF (Part 5 of 24). 1985

Anti-Lock Warning Lamp On (With Brake Warning Lamp Off) — **Test A**

TEST STEP	RESULT ▶	ACTION TO TAKE
A4d CHECK FLI TO PWS WIRE		
• Check for continuity between Pin 2 of 2-Pin fluid level plug (harness side) 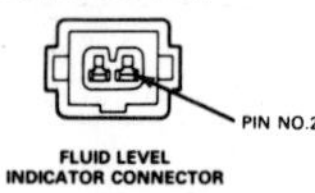 and 5-Pin pressure warning switch plug (harness side) Pin 3. 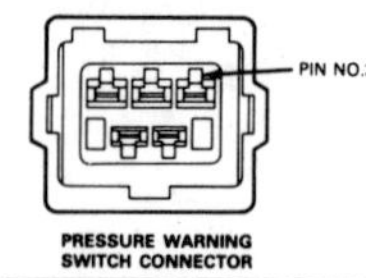	Continuity (OK) ▶	GO to Step **A4e**.
	No continuity (NOT OK) ▶	REPAIR or REPLACE cable harness (Circuit 549).
A4e PWS TO ELECTRONIC CONTROLLER WIRE		
• Check for continuity between 5-Pin pressure warning switch plug (harness side) Pin 5 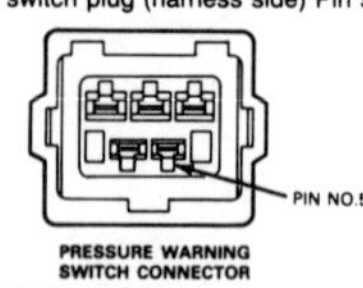 and Breakout box Pin 10.	Continuity (OK) ▶	TURN ignition Off. CONNECT all electrical connections. Reverify symptom.
	No continuity (NOT OK) ▶	REPAIR or REPLACE cable harness (Circuit 535).
A5 ISOLATION TEST FLI AND PWS		
• Check for continuity between Breakout box Pin 9 and body ground.	Continuity (NOT OK) ▶	GO to Step **A5a**.
	No continuity (OK) ▶	GO to Step **A6**.

Fig. 26 Diagnosis, warning lamp ON w/brake warning lamp OFF (Part 6 of 24). 1985

Anti-Lock Warning Lamp On (With Brake Warning Lamp Off) — **Test A**

TEST STEP	RESULT ▶	ACTION TO TAKE
A5d CHECK CONTINUITY BETWEEN PWS PINS 3 AND 5 AND BODY GROUND		
• Check for continuity from PWS 5-Pin socket Pins (pressure warning switch side) 3 and body ground and 5 and body ground. 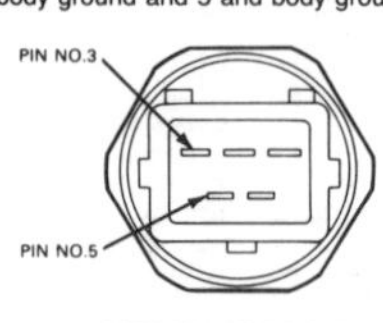	Continuity (NOT OK) ▶	REPLACE PWS.
	No continuity (OK) ▶	GO to Step **A5e**.
A5e CHECK PWS CONNECTOR PIN 5 AND GROUND		
• Check for continuity between 5-Pin PWS plug, Pin 5 (harness side) and body ground. 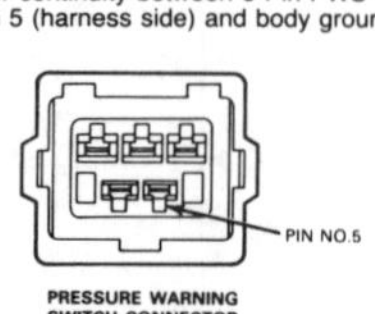	Continuity (NOT OK) ▶	REPAIR or REPLACE Circuit 535.
	No continuity (OK) ▶	CONNECT all plugs and reverify symptom.
A6 CHECK MAIN RELAY PRIMARY CIRCUIT RESISTANCE		
• Turn ignition switch Off. • Set multi-meter to read resistance. • Measure resistance between Breakout box Pins 40 and 8.	Resistance between 45 and 105 ohms (OK) ▶	GO to Step **A7**.
	Any other reading (NOT OK) ▶	GO to Step **A6a**.

Fig. 26 Diagnosis, warning lamp ON w/brake warning lamp OFF (Part 7 of 24). 1985

Anti-Lock Warning Lamp On (With Brake Warning Lamp Off) — **Test A**

TEST STEP	RESULT ▶	ACTION TO TAKE
A5a CHECK FLUID LEVEL INDICATOR PIN NO. 2		
• Disconnect FLI 2-Pin plug and check for continuity between FLI socket Pin 2 and body ground and Pin 1 and body ground. 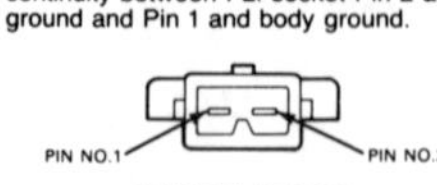	Continuity (NOT OK) ▶	REPLACE FLI.
	No continuity (OK) ▶	GO to Step **A5b**.
A5b CHECK FLUID LEVEL INDICATOR PIN NO. 1		
• Check for continuity between FLI plug Pin 1 (harness side) and body ground. 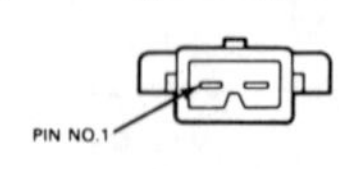	Continuity (NOT OK) ▶	REPAIR or REPLACE Circuit 512.
	No continuity (OK) ▶	GO to Step **A5c**.
A5c DISCONNECT POWER WARNING SWITCH PLUG		
• Disconnect 5-Pin plug from PWS. Check for continuity between FLI plug Pin 2 (harness side) and body ground. 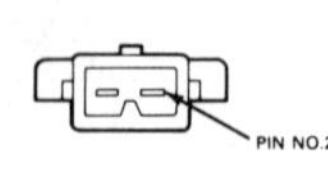	Continuity (NOT OK) ▶	REPAIR or REPLACE Circuit 549.
	No continuity (OK) ▶	GO to Step **A5d**.

Fig. 26 Diagnosis, warning lamp ON w/brake warning lamp OFF (Part 8 of 24). 1985

Anti-Lock Warning Lamp On (With Brake Warning Lamp Off) — Test A

TEST STEP	RESULT ▶	ACTION TO TAKE
A6a CHECK MAIN RELAY PRIMARY COIL RESISTANCE		
• Disconnect main relay from socket. • Set multi-meter on resistance. • Measure resistance between main relay Pins 1 and 2. 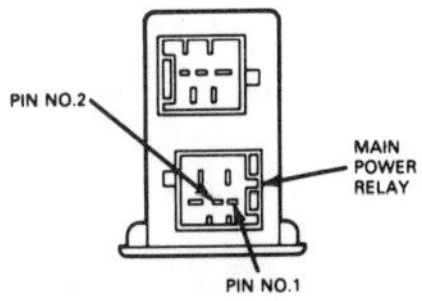	Resistance between 45 and 105 ohms (OK) ▶ Any other reading (NOT OK) ▶	GO to Step **A6b**. REPLACE main relay.
A6b CHECK MAIN RELAY PRIMARY TO ELECTRONIC CONTROLLER WIRE		
• Check continuity between main relay socket Pin 2. 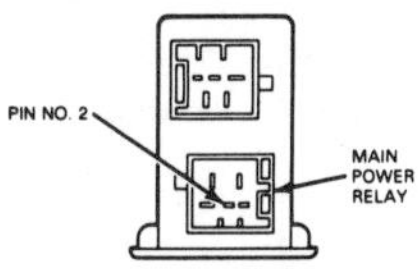 and Breakout box Pin 8.	Continuity (OK) ▶ No continuity (NOT OK) ▶	GO to Step **A6c**. REPAIR or REPLACE cable harness (Circuit 513).

Fig. 26 Diagnosis, warning lamp ON w/brake warning lamp OFF (Part 9 of 24). 1985

Anti-Lock Warning Lamp On (With Brake Warning Lamp Off) — Test A

TEST STEP	RESULT ▶	ACTION TO TAKE
A6c CHECK MAIN RELAY PRIMARY TO GROUND WIRE		
• Check continuity between main relay socket Pin 1 and ground. 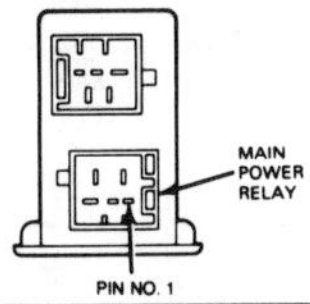	Continuity (OK) ▶ No continuity (NOT OK) ▶	RECONNECT all electrical connections and reverify symptom. REPAIR or REPLACE cable harness (Circuit 430N).
A7 CHECK MAIN RELAY SECONDARY (ACTIVATED)		
• Place a jumper wire between Breakout box Pins 2 and 8. • Set multi-meter to read volts DC. • Turn ignition switch On. • Measure voltage between Breakout box Pins 40 and 3.	Over 10 V DC (OK) ▶ Under 10 V DC (NOT OK) ▶	GO to Step **A8**. CHECK fuse F5. REMOVE jumper wire and GO to Step **A7a**.
A7a CHECK MAIN RELAY SECONDARY CIRCUIT (ACTIVE)		
• Disconnect main relay from socket. • Apply power (battery positive and ground) to main relay Pins 1 and 2. • Check for continuity between main relay Pins 3 and 4. 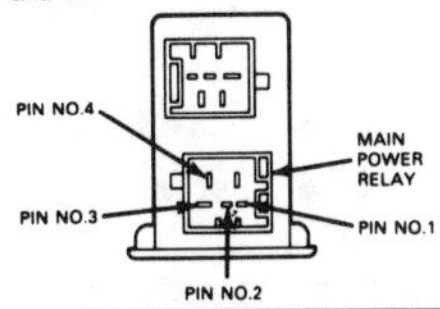	Continuity (OK) ▶ No continuity (NOT OK) ▶	GO to Step **A7b**. REPLACE main relay.

Fig. 26 Diagnosis, warning lamp ON w/brake warning lamp OFF (Part 10 of 24). 1985

Anti-Lock Warning Lamp On (With Brake Warning Lamp Off) — Test A

TEST STEP	RESULT ▶	ACTION TO TAKE
A7b CHECK MAIN RELAY SECONDARY CIRCUIT POWER WIRE		
• Check continuity between main relay socket Pin 4 and positive battery terminal. 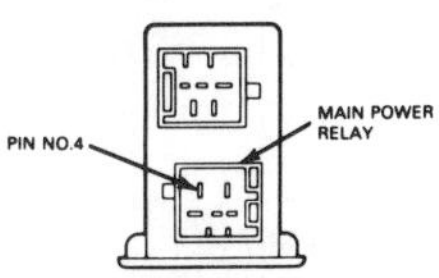	Continuity (OK) ▶ No continuity (NOT OK) ▶	RECONNECT main relay and reverify symptom. REPAIR or REPLACE cable harness (Circuit 533) or fuse link (Circuit 291B).
A8 MEASURE RIGHT REAR SENSOR CIRCUIT RESISTANCE		
• Turn ignition switch Off. • Set multi-meter to read resistance. • Measure resistance between Breakout box Pins 4 and 22.	800 to 1400 ohms (0.8 to 1.4K ohms) (OK) ▶ Any other reading (NOT OK) ▶	GO to Step **A9**. Go to Step **A8a**.
A8a MEASURE RIGHT REAR SENSOR RESISTANCE		
• Disconnect sensor plug (right rear). • Set multi-meter to read resistance. • Measure resistance of sensor at sensor plug. 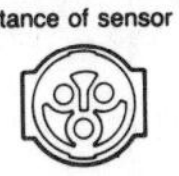 RH REAR SENSOR	800 to 1400 ohms (0.8 to 1.4K ohms) (OK) ▶ Any other reading (NOT OK) ▶	REPAIR or REPLACE cable harness (Circuit 523 or 524). REPLACE right rear wheel sensor.
A9 MEASURE FRONT LEFT SENSOR CIRCUIT RESISTANCE		
• Measure resistance between Breakout box Pins 5 and 23.	800 to 1400 ohms (0.8 to 1.4K ohms) (OK) ▶ Any other reading (NOT OK) ▶	GO to Step **A10**. GO to Step **A9a**.

Fig. 26 Diagnosis, warning lamp ON w/brake warning lamp OFF (Part 11 of 24). 1985

Anti-Lock Warning Lamp On (With Brake Warning Lamp Off) — Test A

TEST STEP	RESULT ▶	ACTION TO TAKE
A9a MEASURE FRONT LEFT SENSOR RESISTANCE		
• Disconnect sensor plug (left front). • Measure resistance of sensor at sensor plug. 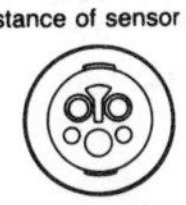 LH FRONT SENSOR	800 to 1400 ohms (0.8 to 1.4K ohms) (OK) ▶ Any other reading (NOT OK) ▶	REPAIR or REPLACE cable harness (Circuit 521 or 522). REPLACE left front wheel sensor.
A10 MEASURE LEFT REAR SENSOR CIRCUIT RESISTANCE		
• Measure resistance between Breakout box Pins 6 and 24.	800 to 1400 ohms (0.8 to 1.4K ohms) (OK) ▶ Any other reading (NOT OK) ▶	GO to Step **A11**. GO to Step **A10a**.
A10a MEASURE LEFT REAR SENSOR RESISTANCE		
• Disconnect sensor plug (left rear). • Measure resistance of sensor at left rear sensor plug.  LH REAR SENSOR	800 to 1400 ohms (0.8 to 1.4K ohms) (OK) ▶ Any other reading (NOT OK) ▶	REPAIR or REPLACE cable harness (Circuit 518 or 519). REPLACE left rear wheel sensor.
A11 MEASURE RIGHT FRONT SENSOR CIRCUIT RESISTANCE		
• Measure resistance between Breakout box Pins 7 and 25.	800 to 1400 ohms (0.8 to 1.4K ohms) (OK) ▶ Any other reading (NOT OK) ▶	GO to Step **A12**. GO to Step **A11a**.

Fig. 26 Diagnosis, warning lamp ON w/brake warning lamp OFF (Part 12 of 24). 1985

Anti-Lock Warning Lamp On (With Brake Warning Lamp Off) — Test A

TEST STEP	RESULT ▶	ACTION TO TAKE
A11a MEASURE RIGHT FRONT SENSOR RESISTANCE		
• Disconnect sensor plug (right front) • Measure resistance of sensor at sensor plug.  RH FRONT SENSOR	800 to 1400 ohms (0.8 to 1.4K ohms) (OK) ▶	REPAIR or REPLACE cable harness (Circuit 514 or 516).
	Any other reading (NOT OK) ▶	REPLACE right front wheel sensor.
A12 MEASURE MAIN VALVE CIRCUIT RESISTANCE		
• Turn ignition switch Off. • Set multi-meter to read resistance. • Measure the resistance between Breakout box Pins 11 and 18.	2 to 5 ohms (OK) ▶	GO to Step **A13**.
	Any other reading (NOT OK) ▶	GO to Step **A12a**.
A12a MEASURE MAIN VALVE RESISTANCE		
• Disconnect main valve 2-Pin plug. • Measure resistance between the main valve electrical Pins 1 and 2. 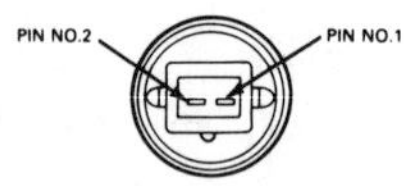	2 to 5 ohms (OK) ▶	REPLACE or REPAIR cable harness (Circuit 430E or 493).
	Any other reading (NOT OK) ▶	REPLACE hydraulic unit.
A13 MEASURE RIGHT FRONT INLET VALVE CIRCUIT RESISTANCE		
• Measure resistance between Breakout box Pins 11 and 15.	5 to 8 ohms (OK) ▶	GO to Step **A14**.
	Any other reading (NOT OK) ▶	GO to Step **A13a**.

Fig. 26 Diagnosis, warning lamp ON w/brake warning lamp OFF (Part 13 of 24). 1985

Anti-Lock Warning Lamp On (With Brake Warning Lamp Off) — Test A

TEST STEP	RESULT ▶	ACTION TO TAKE
A13a MEASURE RIGHT FRONT INLET VALVE RESISTANCE		
• Disconnect valve block 7-Pin plug. • Measure resistance between valve block socket electrical Pins 7 and 1. 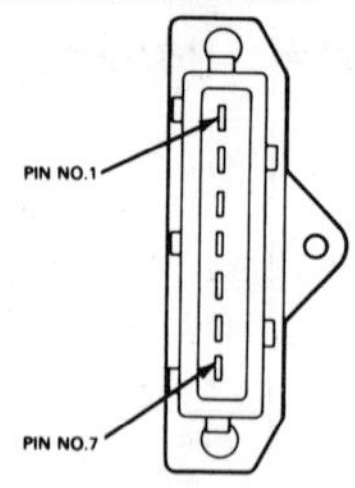 VALVE BLOCK	5 to 8 ohms (OK) ▶	REPLACE or REPAIR cable harness (Circuit 510).
	Any other reading (NOT OK) ▶	REPLACE hydraulic unit. RECONNECT 7-Pin plug.
A14 MEASURE REAR INLET VALVE CIRCUIT RESISTANCE		
• Measure resistance between Breakout box Pins 11 and 17.	5 to 8 ohms (OK) ▶	GO to Step **A15**.
	Any other reading (NOT OK) ▶	GO to Step **A14a**.

Fig. 26 Diagnosis, warning lamp ON w/brake warning lamp OFF (Part 14 of 24). 1985

Anti-Lock Warning Lamp On (With Brake Warning Lamp Off) — Test A

TEST STEP	RESULT ▶	ACTION TO TAKE
A14a MEASURE REAR INLET VALVE RESISTANCE		
• Disconnect valve block 7-Pin plug. • Measure resistance between valve block electrical Pins 7 and 3. 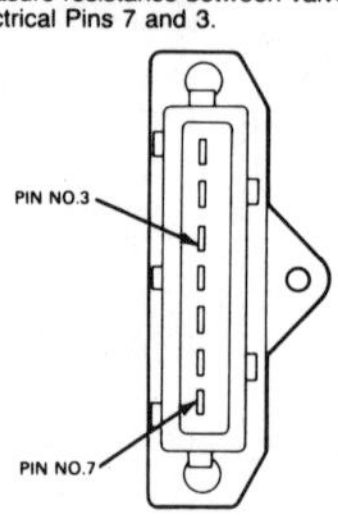 VALVE BLOCK	5 to 8 ohms (OK) ▶	REPLACE or REPAIR cable harness (Circuit 496).
	Any other reading (NOT OK) ▶	REPLACE hydraulic unit. RECONNECT 7-Pin plug.
A15 MEASURE RIGHT FRONT INLET VALVE CIRCUIT RESISTANCE		
• Measure resistance between Breakout box Pins 11 and 35.	5 to 8 ohms (OK) ▶	GO to Step **A16**.
	Any other reading (NOT OK) ▶	GO to Step **A15a**.

Fig. 26 Diagnosis, warning lamp ON w/brake warning lamp OFF (Part 15 of 24). 1985

Anti-Lock Warning Lamp On (With Brake Warning Lamp Off) — Test A

TEST STEP	RESULT ▶	ACTION TO TAKE
A15a MEASURE RIGHT FRONT INLET VALVE RESISTANCE		
• Disconnect valve block 7-Pin plug. • Measure resistance between valve block electrical Pins 7 and 6. 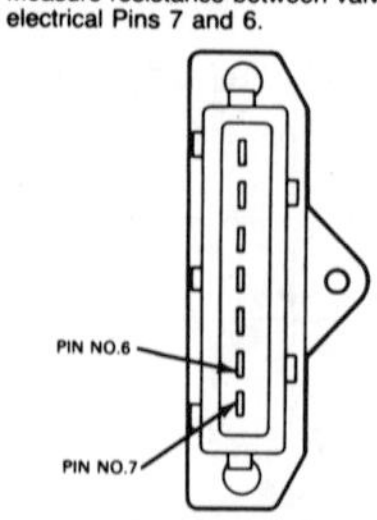 VALVE BLOCK	5 to 8 ohms (OK) ▶	REPLACE or REPAIR cable harness (Circuit 495).
	Any other reading (NOT OK) ▶	REPLACE hydraulic unit. RECONNECT 7-Pin plug.
A16 MEASURE REAR OUTLET VALVE CIRCUIT RESISTANCE		
• Measure resistance between Breakout box Pins 11 and 33.	3 to 6 ohms (OK) ▶	GO to Step **A17**.
	Any other reading (NOT OK) ▶	GO to Step **A16a**.

Fig. 26 Diagnosis, warning lamp ON w/brake warning lamp OFF (Part 16 of 24). 1985

Anti-Lock Warning Lamp On (With Brake Warning Lamp Off) — **Test A**

TEST STEP	RESULT ▶	ACTION TO TAKE
A16a MEASURE REAR OUTLET VALVE RESISTANCE		
• Disconnect valve block 7-Pin plug. • Measure resistance between valve block electrical Pins 7 and 4. 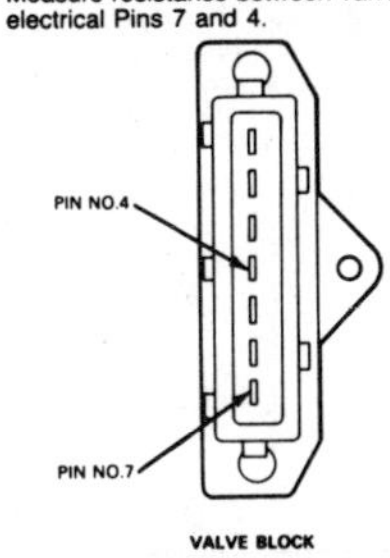 	3 to 6 ohms (OK) ▶	REPLACE or REPAIR cable harness (Circuit 499).
	Any other reading (NOT OK) ▶	REPLACE hydraulic unit. RECONNECT 7-Pin plug.
A17 MEASURE LEFT FRONT OUTLET VALVE CIRCUIT RESISTANCE		
• Measure resistance between Breakout box Pins 11 and 16.	3 to 6 ohms (OK) ▶	GO to Step **A18**.
	Any other reading (NOT OK) ▶	GO to Step **A17a**.

Fig. 26 Diagnosis, warning lamp ON w/brake warning lamp OFF (Part 17 of 24). 1985

Anti-Lock Warning Lamp On (With Brake Warning Lamp Off) — **Test A**

TEST STEP	RESULT ▶	ACTION TO TAKE
A17a MEASURE LEFT FRONT OUTLET VALVE RESISTANCE		
• Disconnect valve block 7-Pin plug. • Measure resistance between valve block electrical Pins 7 and 5. 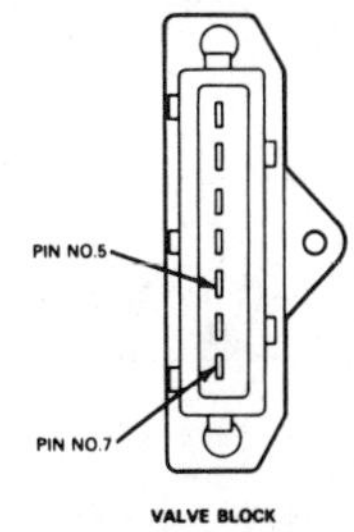 	3 to 6 ohms (OK) ▶	REPLACE or REPAIR cable harness (Circuit 498).
	Any other reading (NOT OK) ▶	REPLACE hydraulic unit. RECONNECT 7-Pin plug.
A18 MEASURE FRONT RIGHT OUTLET VALVE CIRCUIT RESISTANCE		
• Measure resistance between Breakout box Pins 11 and 34.	3 to 6 ohms (OK) ▶	GO to Step **A19**.
	Any other reading (NOT OK) ▶	GO TO Step **A18a**.

Fig. 26 Diagnosis, warning lamp ON w/brake warning lamp OFF (Part 18 of 24). 1985

Anti-Lock Warning Lamp On (With Brake Warning Lamp Off) — **Test A**

TEST STEP	RESULT ▶	ACTION TO TAKE
A18a MEASURE FRONT RIGHT OUTLET VALVE RESISTANCE		
• Disconnect valve block 7-Pin plug. • Measure resistance between valve block electrical Pins 7 and 2. 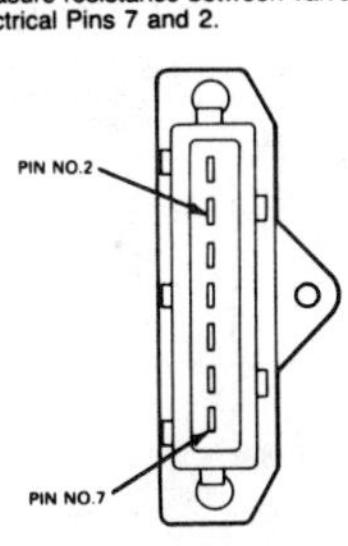 	3 to 6 ohms (OK) ▶	REPLACE or REPAIR cable harness (Circuit 497).
	Any other reading (NOT OK) ▶	REPLACE hydraulic unit. RECONNECT 7-Pin plug.
A19 CHECK VALVE BODY GROUND CIRCUIT		
• Measure resistance between Breakout box Pins 11 and 40.	Less than 2 ohms (OK) ▶	GO to Step **A20**.
	Greater than 2 ohms (NOT OK) ▶	GO TO Step **A19a**.

Fig. 26 Diagnosis, warning lamp ON w/brake warning lamp OFF (Part 19 of 24). 1985

Anti-Lock Warning Lamp On (With Brake Warning Lamp Off) — **Test A**

TEST STEP	RESULT ▶	ACTION TO TAKE
A19a TEST VALVE PLUG PIN NO. 7		
• Disconnect valve block 7-Pin plug. • Measure resistance between valve block electrical Pin 7 and body ground. 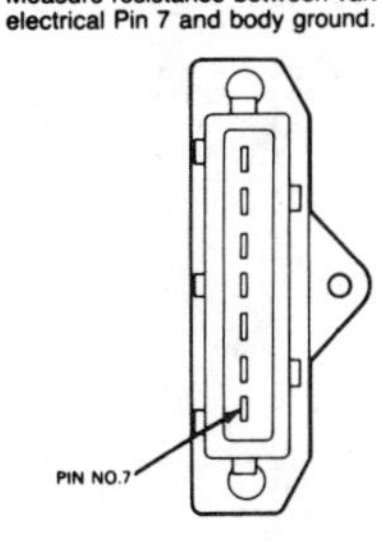 	Less than 2 ohms (OK) ▶	REPLACE or REPAIR cable harness (Circuit 511).
	Greater than 2 ohms (NOT OK) ▶	GO to Step **A19b**.
A19b CHECK VALVE BODY GROUND WIRE		
• Check for continuity between valve body and body ground.	Continuity (OK) ▶	REPLACE hydraulic unit (internal ground problem).
	No or Poor continuity (NOT OK) ▶	REPAIR or REPLACE hydraulic unit ground strap (Circuit 430E).

Fig. 26 Diagnosis, warning lamp ON w/brake warning lamp OFF (Part 20 of 24). 1985

Anti-Lock Warning Lamp On (With Brake Warning Lamp Off) — **Test A**

TEST STEP	RESULT	ACTION TO TAKE
A20 CHECK PWS BRAKE LAMP CIRCUIT (NO PRESSURE)		
• Vehicle must be cooled to room temperature. • Discharge the brake system as follows: a. Turn ignition switch Off. b. Pump brake pedal at least 20 times until you feel the pedal become hard. • Disconnect pressure warning switch 5-Pin plug. • Check continuity between pressure warning switch electrical Pins 1 and 2.	Continuity (OK) ▶ No continuity (Not OK) ▶	GO to Step **A21**. REPLACE pressure warning switch.

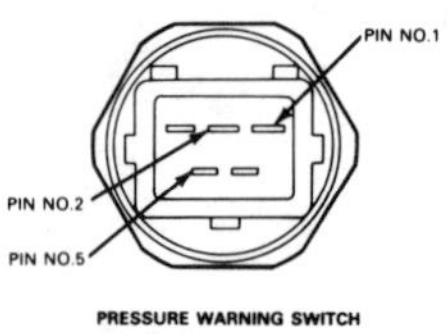

Fig. 26 Diagnosis, warning lamp ON w/brake warning lamp OFF (Part 21 of 24). 1985

Anti-Lock Warning Lamp On (With Brake Warning Lamp Off) — **Test A**

TEST STEP	RESULT	ACTION TO TAKE
A21 CHECK PWS BRAKE LAMP CIRCUIT (WITH PRESSURE)		
• Reconnect pressure warning switch 5-Pin plug. • Turn ignition switch On. • When pump motor stops, turn ignition switch Off. • Disconnect pressure warning switch 5-Pin plug again. • Check for continuity between pressure warning switch electrical Pins 1 and 2.	Continuity (Not OK) ▶ No continuity (OK) ▶	REPLACE pressure warning switch. GO to Step **A22**.

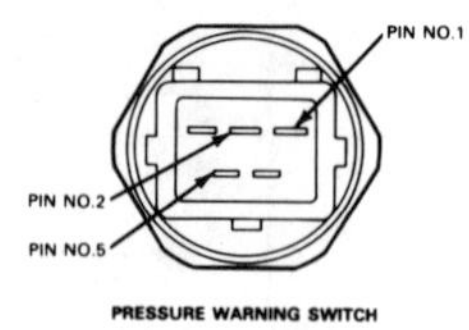

Fig. 26 Diagnosis, warning lamp ON w/brake warning lamp OFF (Part 22 of 24). 1985

Anti-Lock Warning Lamp On (With Brake Warning Lamp Off) — **Test A**

TEST STEP	RESULT	ACTION TO TAKE
A22 CHECK PWS BRAKE LAMP CIRCUIT THRESHOLD		
WARNING: Before disconnecting any hydraulic lines, you must ensure that the pressure system is discharged. • Discharge the brake system as follows: a. Turn ignition switch Off. b. Pump break pedal at least 20 times until you feel it become hard. • Remove high pressure banjo bolt below brake fluid reservoir on brake booster. • Disconnect quick-connect nipple from Anti-Lock High Pressure Gauge (Tool T85P-20215-A, or equivalent) and install nipple in place of the high pressure banjo bolt. NOTE: Be sure O-rings are installed correctly. • Connect Anti-Lock High Pressure Gauge to gauge nipple. **WARNING: DO NOT disconnect Anti-Lock High Pressure Gauge while system is under pressure.** • Reconnect pressure warning switch (PWS) 5-Pin connector. • Turn ignition switch On. • When pump motor stops, disconnect pressure warning switch (PWS) 5-Pin plug. • Lower hydraulic accumulator pressure by slowly pumping the brake pedal until you have continuity between pressure warning switch electrical Pins 1 and 2. • Observe the hydraulic pressure gauge when continuity is reached. • If you missed the reading or want to reverify the reading: — Reconnect the 5-Pin plug. — Turn the ignition switch On until the pump stops. — Disconnect the 5-Pin plug and reverify the readings.	100-110 Bar (1,450-1,595 psi) (OK) ▶ Any other reading (Not OK) ▶	GO to Step **A23**. REPLACE pressure warning switch.

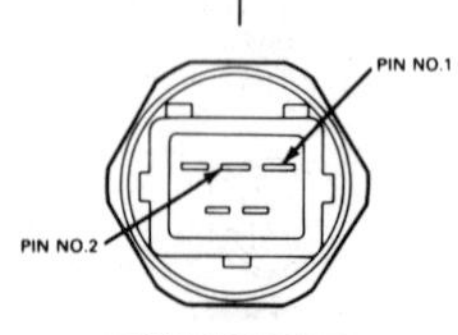

Fig. 26 Diagnosis, warning lamp ON w/brake warning lamp OFF (Part 23 of 24). 1985

Anti-Lock Warning Lamp On (With Brake Warning Lamp Off) — **Test A**

TEST STEP	RESULT	ACTION TO TAKE
A23 CHECK PWS HARNESS GROUND		
• Check for continuity between pressure warning switch 5-Pin plug Pin 1 and body ground.	Continuity (OK) ▶ No continuity (Not OK) ▶	GO to Step **A24**. REPLACE or REPAIR cable harness (Circuit 430K).
A24 REVERIFY SYSTEM SYMPTOM		
• Reconnect all electrical connections. • Reinstall high pressure banjo bolt. **WARNING: Before disconnecting any hydraulic lines, you must ensure that the brake hydraulic pressure system is discharged.** • Discharge the brake system as follows: a. Turn ignition switch Off. b. Pump brake pedal at least 20 times until you feel the pedal become hard. • Reverify symptom.	Symptom not present ▶ Symptom still present ▶	FAULT may have been a loose electrical connection. REPLACE electronic control module.

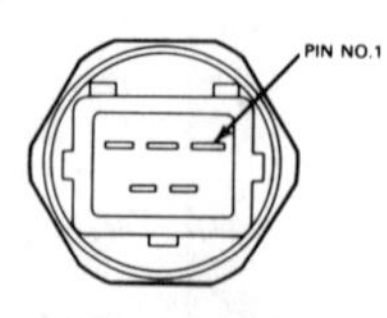

Fig. 26 Diagnosis, warning lamp ON w/brake warning lamp OFF (Part 24 of 24). 1985

Anti-Lock Lamp On After Engine Starts (Brake Warning Lamp Off) — Test B

WARNING LIGHTS SEQUENCE

Warning Lamps	Ignition On	Cranking Engine	Engine Running	Vehicle Moving	Braking with/without Anti-Lock	Vehicle Stopped	Engine Idle	Ignition Off
Check Anti-Lock (Amber)	▨		▨	▨	▨	▨	▨	
Brake (Red)		■						

TEST STEP	RESULT ▶	ACTION TO TAKE
B1 CHECK CONTINUITY OF CIRCUIT 524		
• Ignition switch Off. • Disconnect 35-Pin plug from controller. • Connect EEC-IV Breakout box with Anti-Lock Test Adapter to the Anti-Lock 35-Pin plug wiring harness. • Check for continuity between Breakout box Pins 40 and 4.	Continuity ~~OK~~ ▶	GO to Step **B1a**.
	No continuity OK ▶	GO to Step **B2**.
B1a CHECK CONTINUITY OF RIGHT REAR SENSOR		
• Disconnect wheel sensor plug (right rear). • Check for continuity between each sensor plug pin (sensor side) and vehicle ground.	Continuity ~~OK~~ ▶	REPLACE sensor (right rear).
	No continuity OK ▶	REPLACE or REPAIR cable harness. RECONNECT sensor plug.
 RH REAR SENSOR		
B2 CHECK CONTINUITY OF CIRCUIT 522		
• Check for continuity between Breakout box Pins 40 and 5.	Continuity ~~OK~~ ▶	GO to Step **B2a**.
	No continuity OK ▶	GO to Step **B3**.

Fig. 27 Diagnosis, warning lamp ON after engine starts w/brake warning lamp OFF (Part 1 of 3). 1985

Anti-Lock Lamp On After Engine Starts (Brake Warning Lamp Off) — Test B

TEST STEP	RESULT ▶	ACTION TO TAKE
B2a CHECK CONTINUITY OF LEFT FRONT SENSOR		
• Disconnect wheel sensor plug (left front). • Check for continuity between each sensor plug pin (sensor side) and vehicle ground.	Continuity ~~OK~~ ▶	REPLACE sensor (left front).
	No continuity OK ▶	REPLACE or REPAIR cable harness. RECONNECT sensor plug.
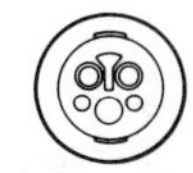 LH FRONT SENSOR		
B3 CHECK CONTINUITY OF CIRCUIT 519		
• Check for continuity between Breakout box Pins 40 and 6.	Continuity ~~OK~~ ▶	GO to Step **B3a**.
	No continuity OK ▶	GO to Step **B4**.
B3a CHECK CONTINUITY OF LEFT REAR SENSOR		
• Disconnect sensor plug (left rear). • Check for continuity between each sensor plug pin (sensor side) and vehicle ground.	Continuity ~~OK~~ ▶	REPLACE sensor (left rear).
	No continuity OK ▶	REPLACE or REPAIR cable. RECONNECT sensor plug.
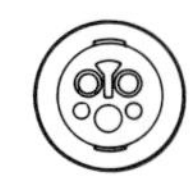 LH REAR SENSOR		

Fig. 27 Diagnosis, warning lamp ON after engine starts w/brake warning lamp OFF (Part 2 of 3). 1985

Anti-Lock Lamp On After Engine Starts (Brake Warning Lamp Off) — Test B

TEST STEP	RESULT ▶	ACTION TO TAKE
B4 CHECK CONTINUITY OF CIRCUIT 516		
• Check for continuity between Breakout box Pins 40 and 7.	Continuity ~~OK~~ ▶	GO to Step **B4a**.
	No continuity OK ▶	Test complete. If Anti-Lock lamp pattern remains, REPEAT Test B.
B4a CHECK CONTINUITY OF RIGHT FRONT SENSOR		
• Disconnect wheel sensor plug (right front). • Check for continuity between each sensor plug pin (sensor side) and vehicle ground.	Continuity ~~OK~~ ▶	REPLACE right front sensor.
	No continuity OK ▶	REPLACE or REPAIR cable harness. RECONNECT sensor plug.
 RH FRONT SENSOR		

Fig. 27 Diagnosis, warning lamp ON after engine starts w/brake warning lamp OFF (Part 3 of 3). 1985

Anti-Lock Warning Lamp On After Vehicle Starts To Move Or False Cycling Of Anti-Lock System — Test C

WARNING LIGHTS SEQUENCE

Warning Lamps	Ignition On	Cranking Engine	Engine Running	Vehicle Moving	Braking with/without Anti-Lock	Vehicle Stopped	Engine Idle	Ignition Off
Check Anti-Lock (Amber)	▨		▨	▨	▨	▨	▨	
Brake (Red)		■						

WARNING LIGHTS SEQUENCE

Warning Lamps	Ignition On	Cranking Engine	Engine Running	Vehicle Moving	Braking with/without Anti-Lock	Vehicle Stopped	Engine Idle	Ignition Off
Check Anti-Lock (Amber)	▨		▨					
Brake (Red)		■						

TEST STEP	RESULT ▶	ACTION TO TAKE
C1 MEASURE RIGHT REAR SENSOR CIRCUIT RESISTANCE		
• Turn ignition switch Off. • Set multi-meter to read resistance. • Measure resistance between Breakout box Pins 4 and 22.	800 to 1400 ohms (0.8 to 1.4K ohms) OK ▶	GO to Step **C2**.
	Any other reading ~~OK~~ ▶	Go to Step **C1a**.
C1a MEASURE RIGHT REAR SENSOR RESISTANCE		
• Disconnect sensor plug (right rear). • Set multi-meter to read resistance. • Measure resistance of sensor at sensor plug.	800 to 1400 ohms (0.8 to 1.4K ohms) OK ▶	REPAIR or REPLACE cable harness (Circuit 523 or 524).
	Any other reading ~~OK~~ ▶	REPLACE right rear wheel sensor.
 LH REAR SENSOR		

Fig. 28 Diagnosis, warning lamp ON after vehicle starts to move or false cycling of system (Part 1 of 5). 1985

Anti-Lock Warning Lamp On After Vehicle Starts To Move Or False Cycling Of Anti-Lock System	Test C

TEST STEP	RESULT ▶	ACTION TO TAKE
C2 MEASURE FRONT LEFT SENSOR CIRCUIT RESISTANCE		
• Measure resistance between Breakout box Pins 5 and 23.	800 to 1400 ohms (0.8 to 1.4K ohms) OK ▶	GO to Step **C3**.
	Any other reading ~~OK~~ ▶	GO to Step **C2a**.
C2a MEASURE FRONT LEFT SENSOR RESISTANCE		
• Disconnect sensor plug (left front). • Measure resistance of sensor at sensor plug.  LH FRONT SENSOR	800 to 1400 ohms (0.8 to 1.4K ohms) OK ▶	REPAIR or REPLACE cable harness (Circuit 521 or 522).
	Any other reading ~~OK~~ ▶	REPLACE left front wheel sensor.
C3 MEASURE LEFT REAR SENSOR CIRCUIT RESISTANCE		
• Measure resistance between Breakout box Pins 6 and 24.	800 to 1400 ohms (0.8 to 1.4K ohms) OK ▶	GO to Step **C4**.
	Any other reading ~~OK~~ ▶	GO to Step **C3a**.

Fig. 28 Diagnosis, warning lamp ON after vehicle starts to move or false cycling of anti-lock system (Part 2 of 5). 1985

Anti-Lock Warning Lamp On After Vehicle Starts To Move Or False Cycling Of Anti-Lock System	Test C

TEST STEP	RESULT ▶	ACTION TO TAKE
C3a MEASURE LEFT REAR SENSOR RESISTANCE		
• Disconnect sensor plug (left rear). • Measure resistance of sensor at left rear sensor plug. 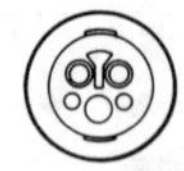 RH REAR SENSOR	800 to 1400 ohms (0.8 to 1.4K ohms) OK ▶	REPAIR or REPLACE cable harness (Circuit 518 or 519).
	Any other reading ~~OK~~ ▶	REPLACE left rear wheel sensor.
C4 MEASURE RIGHT FRONT SENSOR CIRCUIT RESISTANCE		
• Measure resistance between Breakout box Pins 7 and 25.	800 to 1400 ohms (0.8 to 1.4K ohms) OK ▶	GO to Step **C5**.
	Any other reading ~~OK~~ ▶	GO to Step **C4a**.
C4a MEASURE RIGHT FRONT SENSOR RESISTANCE		
• Disconnect sensor plug (right front). • Measure resistance of sensor at sensor plug.  RH FRONT SENSOR	800 to 1400 ohms (0.8 to 1.4K ohms) OK ▶	REPAIR or REPLACE cable harness (Circuit 514 or 516).
	Any other reading ~~OK~~ ▶	REPLACE right front wheel sensor.

Fig. 28 Diagnosis, warning lamp ON after vehicle starts to move or false cycling of anti-lock system (Part 3 of 5). 1985

Anti-Lock Warning Lamp On After Vehicle Starts To Move Or False Cycling Of Anti-Lock System	Test C

TEST STEP	RESULT ▶	ACTION TO TAKE
C5 CHECK RIGHT REAR SENSOR		
• Turn ignition switch Off. • Place vehicle on hoist and raise wheels clear of ground. • Disconnect 35-Pin plug from electronic controller and connect EEC-IV Breakout box with Anti-Lock Test Adapter to the Anti-Lock 35-Pin wiring harness connector. • Set multi-meter on voltage range (2V-AC). • Measure voltage between Breakout box Pins 4 and 22 while spinning the right rear wheel at approximately 1 revolution per second.	Between 0.05 and 0.70 Vac OK ▶	GO to Step **C6**.
	Less than 0.05 or more than 0.70 Vac ~~OK~~ ▶	CHECK sensor mounting, air gap, or toothed wheel mounting. CORRECT as required.
C6 CHECK LEFT FRONT SENSOR		
• Measure voltage between Breakout box Pins 5 and 23 while spinning left front wheel at approximately 1 revolution per second.	Between 0.05 and 0.70 Vac ▶	GO to Step **C7**.
	Less than 0.05 or more than 0.70 Vac ▶	CHECK sensor mounting, air gap, or toothed wheel mounting. CORRECT as required.
C7 CHECK LEFT REAR SENSOR		
• Measure voltage between Breakout box Pins 6 and 24 while spinning the left rear wheel at approximately 1 revolution per second.	Between 0.05 and 0.70 Vac ▶	GO to Step **C8**.
	Less than 0.05 or more than 0.07 Vac ▶	CHECK wheel sensor mounting, air gap, or toothed wheel mounting. CORRECT as required.

Fig. 28 Diagnosis, warning lamp ON after vehicle starts to move or false cycling of system (Part 4 of 5). 1985

Anti-Lock Warning Lamp on After Vehicle Starts to Move or False Cycling of Anti-Lock System	Test C

TEST STEP	RESULT ▶	ACTION TO TAKE
C8 CHECK RIGHT FRONT SENSOR		
• Measure voltage between Breakout box Pins 7 and 25 while spinning the right front wheel at approximately 1 revolution per second.	Between 0.05 and 0.70 Vac ▶	GO to Step **C9**.
	Less than 0.05 or more than 0.70 Vac ▶	CHECK wheel sensor mounting, air gap, or toothed wheel mounting. CORRECT as required.
C9 CHECK FRONT WHEEL BEARINGS		
• Check front wheel bearing end play. • Inspect each toothed sensor ring visually for damaged teeth.	Loose or damaged parts ▶	ADJUST bearings or REPLACE faulty parts.
	Not loose or damaged ▶	REVERIFY symptom.

Fig. 28 Diagnosis, warning lamp ON after vehicle starts to move or false cycling of system (Part 5 of 5). 1985

Anti-Lock Warning Lamp And Brake Warning Lamp On And/Or Pump Motor Runs More Than 60 Seconds	Test D

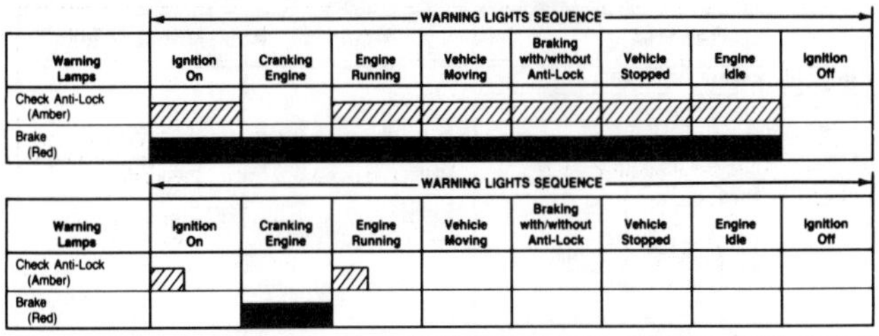

TEST STEP	RESULT ▶	ACTION TO TAKE
D1 ACCUMULATOR: PRE-CHARGE		
• Vehicle must be cooled to room temperature. **WARNING: Before disconnecting any hydraulic lines, you must ensure that the pressure system is discharged.** • To discharge hydraulic accumulator pressure, turn ignition Off, pump brake pedal at least 20 times until you feel it becomes hard. • Remove high pressure banjo bolt below brake fluid reservoir on brake booster. • Install Anti-Lock Pressure Gauge Tool T85P-20215A or equivalent. • Turn ignition switch On and read accumulator precharge pressure. (Gauge needle will spring to this point). Note: Gauge needle reading should spring to 40-90 bar (600-1325 psi) and climb to 160-190 bar (2350-2800 psi). • If reading was missed, discharge accumulator as described and repeat ignition On sequence.	40-90 bar (600-1325 psi) (OK) ▶	GO to Step **D2**.
	Under 40 bar (600 psi) (NOT OK) ▶	REPLACE accumulator.
	Over 90 bar (1325 psi) (NOT OK) ▶	REPLACE accumulator.
	No pressure increase (NOT OK) ▶	GO to Step **D1a**.

Fig. 29 Diagnosis, warning lamp & brake warning lamp ON and/or pump motor runs more than 60 seconds (Part 1 of 10). 1985

Anti-Lock Warning Lamp And Brake Warning Lamp On And/Or Pump Motor Runs More Than 60 Seconds	Test D

TEST STEP	RESULT ▶	ACTION TO TAKE
D1a PUMP MOTOR UNIT		
• Disconnect 2-Pin plug on pump motor unit. • Turn ignition switch On. • Set multi-meter on 20 Volt DC. • Connect meter to the 2-Pin plug on the harness side. • Observe voltmeter.	More than 10V DC (OK) ▶	CHECK brake reservoir filter and/or suction hose. REPLACE if necessary. GO to **D2**.
	Less than 10V DC (NOT OK) ▶	GO to **D1b**.

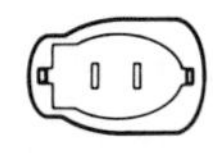

2-PIN MOTOR CONNECTOR

TEST STEP	RESULT ▶	ACTION TO TAKE
D1b CHECK FUSE LINK		
• Check for voltage at Circuits 537, 538a and 538b.	(OK) ▶	GO to **D1c**.
	(NOT OK) ▶	SERVICE short to ground on Circuits 537, 538a and 538b. REPLACE fuse link as necessary.

Fig. 29 Diagnosis, warning lamp & brake warning lamp ON and/or pump motor runs more than 60 seconds (Part 2 of 10). 1985

Anti-Lock Warning Lamp And Brake Warning Lamp On And/Or Pump Motor Runs More Than 60 Seconds	Test D

TEST STEP	RESULT ▶	ACTION TO TAKE
D1c PRESSURE WARNING SWITCH		
• Turn ignition switch Off. • Pump brake pedal at least 20 times until pedal becomes hard (pressure discharged). • Disconnect pressure warning switch 5-Pin plug. • Check for continuity between pressure warning switch Pins 1 and 4.	Continuity (OK) ▶	Go to Step **D1d**.
	No continuity (NOT OK) ▶	REPLACE pressure warning switch and pump motor relay.

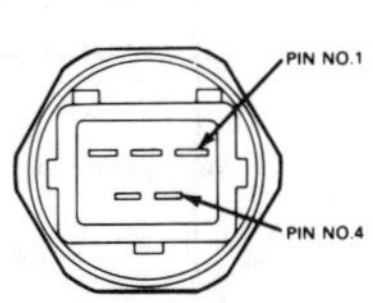

PRESSURE WARNING SWITCH

Fig. 29 Diagnosis, warning lamp & brake warning lamp ON and/or pump motor runs more than 60 seconds (Part 3 of 10). 1985

Anti-Lock Warning Lamp And Brake Warning Lamp On And/Or Pump Motor Runs More Than 60 Seconds	Test D

TEST STEP	RESULT ▶	ACTION TO TAKE
D1d CHECK PUMP MOTOR LINE CONTINUITY		
• Turn ignition switch Off. • Remove positive battery cable from battery. • Disconnect pump motor relay.	Continuity (OK) ▶	RECONNECT battery positive cable and GO to Step **D1e**.
	No continuity (NOT OK) ▶	REPAIR or REPLACE cable harness to pump motor relay.

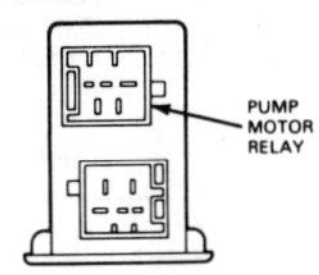

• Check continuity between battery positive cable and motor relay socket Pin 4 (harness side).

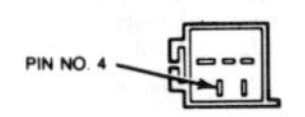

PUMP MOTOR RELAY CONNECTOR

Fig. 29 Diagnosis, warning lamp & brake warning lamp ON and/or pump motor runs more than 60 seconds (Part 4 of 10). 1985

Anti-Lock Warning Lamp And Brake Warning Lamp On And/Or Pump Motor Runs More Than 60 Seconds — **Test D**

TEST STEP	RESULT ▶	ACTION TO TAKE
D1e CABLE HARNESS PRESSURE WARNING SWITCH		
• Check for continuity between pressure warning switch 5-Pin plug Pin 1 (harness side) and ground. 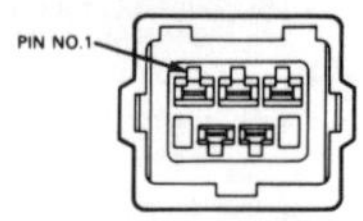	Continuity (OK) ▶ No continuity (NOT OK) ▶	GO to Step **D1f**. REPLACE or REPAIR cable harness (Circuit 430K).
D1f CHECK CONTINUITY CIRCUIT 539		
• Connect meter to 5-Pin plug Pin 4 (harness side) and motor relay socket Pin 1 (harness side). • Check for continuity. 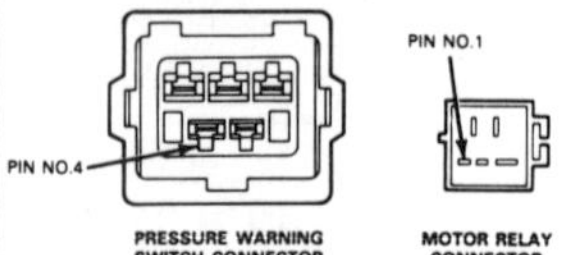	Continuity (OK) ▶ No continuity (NOT OK) ▶	GO to Step **D1g**. REPLACE or REPAIR cable harness (Circuit 539).
D1g CHECK CONTINUITY CIRCUIT 298		
• Connect multi-meter to motor relay socket Pin 2 (harness side) and ignition switch lock pin (On). • Check continuity. 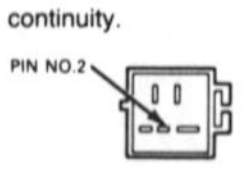	Continuity (OK) ▶ No continuity (NOT OK) ▶	GO to Step **D1h**. REPLACE or REPAIR cable harness (Circuit 298B).

Fig. 29 Diagnosis, warning lamp & brake warning lamp ON and/or pump motor runs more than 60 seconds (Part 5 of 10). 1985

Anti-Lock Warning Lamp And Brake Warning Lamp On And/Or Pump Motor Runs More Than 60 Seconds — **Test D**

TEST STEP	RESULT ▶	ACTION TO TAKE
D1h CHECK PUMP MOTOR RELAY		
• Set multi-meter on 200 ohm scale. • Connect meter to motor relay Pins 1 and 2. 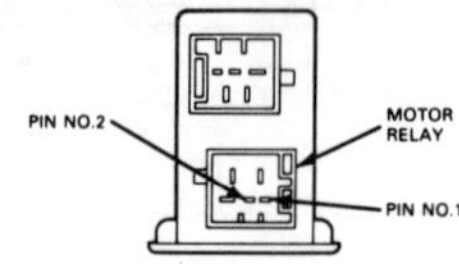	45 to 105 Ohms (OK) ▶ Other (NOT OK) ▶	GO to Step **D1i**. REPLACE motor relay.
D1i CHECK PUMP MOTOR RELAY (CONTINUED)		
• Connect meter and check continuity between motor relay Pins 3 and 4. 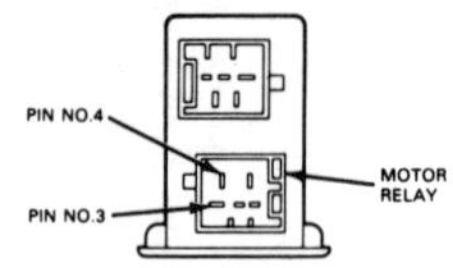	No continuity (OK) ▶ Continuity (NOT OK) ▶	GO to Step **D1j**. REPLACE motor relay.

Fig. 29 Diagnosis, warning lamp & brake warning lamp ON and/or pump motor runs more than 60 seconds (Part 6 of 10). 1985

Anti-Lock Warning Lamp And Brake Warning Lamp On And/Or Pump Motor Runs More Than 60 Seconds — **Test D**

TEST STEP	RESULT ▶	ACTION TO TAKE
D1j CHECK PUMP MOTOR RELAY (CONTINUED)		
• Connect battery to motor relay terminals 1 and 2. • Check continuity between relay Pins 3 and 4 with multi-meter. 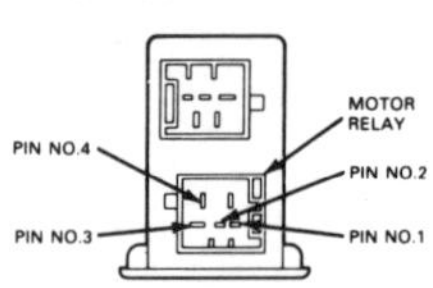	Continuity (OK) ▶ No continuity (NOT OK) ▶	END of Test. RECONNECT all electrical plugs and relay. REVERIFY symptom. REPLACE motor relay.
D2 CHECK PUMP MOTOR		
• Turn ignition switch Off. • Pump brake pedal at least 20 times, until brake pedal becomes hard. • Connect 2-Pin motor plug. • Turn ignition switch On. • Measure time pump takes to shut Off.	Under 60 seconds (OK) ▶ Over 60 seconds (or motor never turns on) (NOT OK) ▶	GO to Step **D3**. CHECK for corroded connections at: • Motor 2-Pin plug. • Body ground (Circuit 430F). • Motor relay Pin 4 socket (Circuit 537). • Battery to Pin 4. GO to Step **D2a**.
D2a CHECK PUMP MOTOR UNIT		
• Turn ignition switch Off. • Pump brake pedal 20 times to discharge system. • Connect ammeter between battery positive cable and battery positive terminal. • Turn Off any electrical components. • Turn ignition switch On. • Measure pump motor current.	Current more than 25 amps (NOT OK) ▶ Current less than 25 amps (OK) ▶	REPLACE pump motor unit. GO to Step **D2b**.

Fig. 29 Diagnosis, warning lamp & brake warning lamp ON and/or pump motor runs more than 60 seconds (Part 7 of 10). 1985

Anti-Lock Warning Lamp And Brake Warning Lamp On And/Or Pump Motor Runs More Than 60 Seconds — **Test D**

TEST STEP	RESULT ▶	ACTION TO TAKE
D2b CHECK VOLTAGE TO PUMP MOTOR		
• Turn ignition switch Off. • Pump brake pedal to discharge pressure system. • Connect voltmeter in parallel at 2-Pin motor plug. • Set meter to 20 Volts DC range. • Turn ignition switch On.	With pump running: Voltage over 8 Volts DC (OK) ▶ Voltage under 8 Volts DC (NOT OK) ▶	GO to Step **D3**. CHECK Circuits 430F, 538A, 538B, Pin 4 to battery and relay Pins 3 and 4 for voltage drop. REPAIR or REPLACE as necessary.
D3 CHECK HYDRAULIC UNIT		
• Turn ignition switch On; wait until pump motor stops. Wait 3 more minutes to stabilize gauge pressure. • Read pressure gauge. • Wait 5 minutes and read pressure gauge again to determine the pressure loss over those 5 minutes.	Pressure loss less than 10 bar (140 psi) on gauge (OK) ▶ More than 10 bar (140 psi) (NOT OK) ▶	GO to Step **D4**. CHECK for external leakage at hydraulic unit and REPAIR. If no external leakage is found, GO to Step **D3a**.

Fig. 29 Diagnosis, warning lamp & brake warning lamp ON and/or pump motor runs more than 60 seconds (Part 8 of 10). 1985

Anti-Lock Warning Lamp And Brake Warning Lamp On And/Or Pump Motor Runs More Than 60 Seconds — Test D

TEST STEP	RESULT ▶	ACTION TO TAKE
D3a CHECK FOR LEAKAGE		
• Turn ignition switch Off. • Pump brake pedal to discharge pressure system. • Turn ignition switch On. • When pump motor stops, turn ignition switch Off. • Quickly remove low pressure pump suction line and plug reservoir fitting with large rubber vacuum nipple. • Observe if brake fluid seeps from the low pressure line.	If line seeps fluid ▶	REPLACE pump/ motor assembly.
	If line does not seep fluid ▶	REPLACE hydraulic unit.

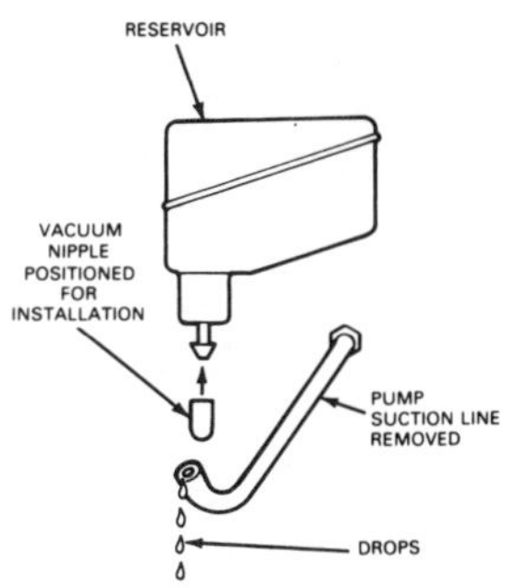

Fig. 29 Diagnosis, warning lamp & brake warning lamp ON and/or pump motor runs more than 60 seconds (Part 9 of 10). 1985

Anti-Lock Warning Lamp And Brake Warning Lamp On And/Or Pump Motor Runs More Than 60 Seconds — Test D

TEST STEP	RESULT ▶	ACTION TO TAKE
D4 CHECK PUMP PRESSURE		
• With the pressure gauge still attached and ignition switch still On, pump brake pedal to decrease pressure until pump motor restarts.	130-150 bar (1900-2200 psi) when pump starts (OK) ▶	GO to Step **D5**.
	Less than or more than 130-150 bar (1900-2200 psi) (NOT OK) ▶	REPLACE pressure warning switch.
D5 CHECK PRESSURE WARNING SWITCH		
• With the ignition switch still On and the pressure gauge connected, observe the pressure when the pump motor stops running.	160-190 bar (2350-2800 psi) when pump motor stops (OK) ▶	If pump motor takes longer than 60 seconds to reach 160-190 bar (2350-2800 psi), REPLACE pump/ motor assembly. If pump motor takes less than 60 seconds, REVERIFY symptom.
	Less than 160 bar (2350 psi) or over 190 bar (2800 psi) (NOT OK) ▶	REPLACE pressure warning switch.

Fig. 29 Diagnosis, warning lamp & brake warning lamp ON and/or pump motor runs more than 60 seconds (Part 10 of 10). 1985

Anti-Lock Warning Lamp Intermittently On — Test E

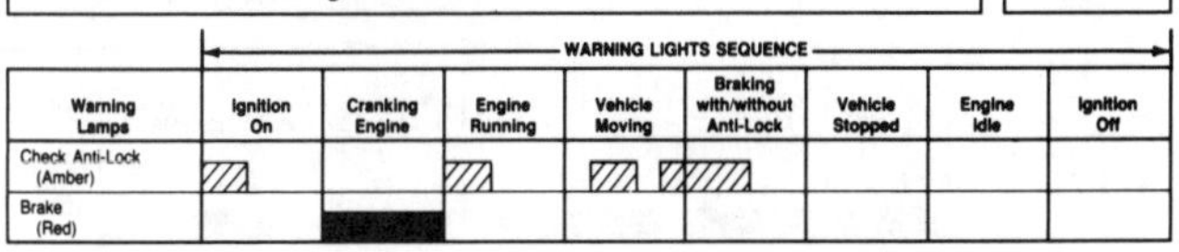

TEST STEP	RESULT ▶	ACTION TO TAKE
E1 CHECK FLI AND PWS CIRCUIT		
• Disconnect 35-Pin plug from electronic controller. • Connect EEC-IV Breakout box, T83L-50-EEC-IV with Anti-Lock Test Adapter T85P-50-ASA or equivalent to the Anti-Lock 35-Pin plug wiring harness. • Turn ignition switch On. • Set multi-meter to read resistance. • Measure the resistance between Breakout box Pins 9 and 10.	Less than 5 ohms (OK) ▶	GO to Step **E2**.
	Greater than 5 ohms (NOT OK) ▶	GO to Step **E1a**.
E1a CHECK FLI ANTI-LOCK WARNING CIRCUIT		
• Disconnect 2-Pin plug on reservoir fluid level indicator (FLI). • Measure resistance between fluid level indicator electrical Pins 1 and 2 (with brake fluid at maximum fluid level).	Less than 2 ohms (OK) ▶	GO to Step **E1b**.
	Greater than 2 ohms (NOT OK) ▶	REPLACE fluid level indicator.

Fig. 30 Diagnosis, warning lamp intermittently ON (Part 1 of 5). 1985

Anti-Lock Warning Lamp Intermittently On — Test E

TEST STEP	RESULT ▶	ACTION TO TAKE
E1b CHECK PWS ANTI-LOCK WARNING CIRCUIT (NO PRESSURE)		
• Turn ignition switch Off. • Pump brake pedal to discharge pressure system. • Disconnect 5-Pin plug at pressure warning switch (PWS). • Check for continuity between pressure warning switch Pins 3 and 5.	No Continuity (OK) ▶	GO to Step **E1c**.
	Continuity (NOT OK) ▶	REPAIR pressure warning switch.

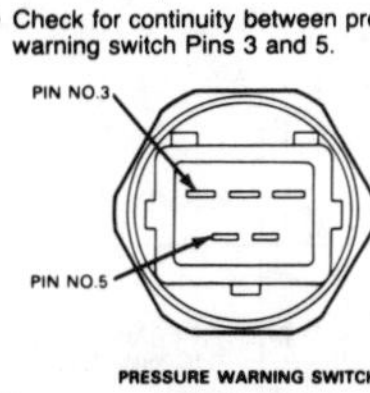

TEST STEP	RESULT ▶	ACTION TO TAKE
E1c CHECK ELECTRONIC MODULE TO FLI WIRE		
• Check for continuity between Breakout box Pin 9 and FLI 2-Pin plug Pin 1 (harness side).	Continuity (OK) ▶	GO to Step **E1d**.
	No continuity (NOT OK) ▶	REPLACE or REPAIR cable harness (Circuit 512).

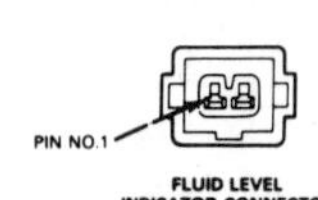

Fig. 30 Diagnosis, warning lamp intermittently ON (Part 2 of 5). 1985

Anti-Lock Warning Lamp Intermittently On — Test E

TEST STEP	RESULT ▶	ACTION TO TAKE
E1d CHECK FLI TO PWS WIRE		
• Check for continuity between 2-Pin fluid level plug Pin 2 (harness side) 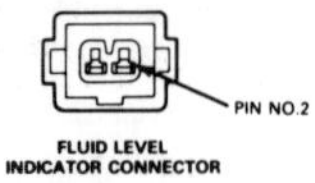 and 5-Pin pressure warning switch plug Pin 3 (harness side). 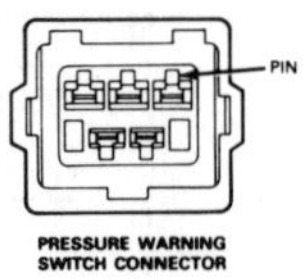	Continuity (OK) ▶	GO to Step **E1e**.
	No continuity (NOT OK) ▶	REPAIR or REPLACE cable harness (Circuit 549).
E1e CHECK PWS TO ELECTRONIC CONTROLLER WIRE		
• Check for continuity between 5-Pin pressure warning switch plug Pin 5 (harness side) 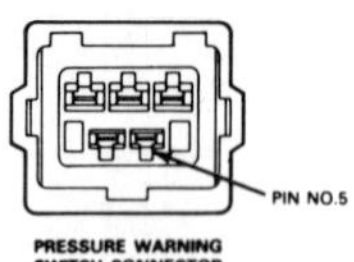 and Breakout box Pin 10.	Continuity (OK) ▶	TURN Ignition Off. CONNECT all electrical connections. REVERIFY symptom.
	No continuity (NOT OK) ▶	REPAIR or REPLACE cable harness (Circuit 535).

Fig. 30 Diagnosis, warning lamp intermittently ON (Part 3 of 5). 1985

Anti-Lock Warning Lamp Intermittently On — Test E

TEST STEP	RESULT ▶	ACTION TO TAKE
E2 CHECK ISOLATION TEST FLI AND PWS		
• Check for continuity between Breakout box Pin 9 and body ground.	Continuity (NOT OK) ▶	GO to Step **E2a**.
	No continuity (OK) ▶	REVERIFY symptom.
E2a CHECK CONTINUITY OF FLI SWITCH		
• Disconnect FLI plug and check for continuity between socket Pin 1 and body ground and Pin 2 and body ground. 	Continuity (NOT OK) ▶	REPLACE FLI.
	No continuity (OK) ▶	GO to Step **E2b**.
E2b CHECK CONTINUITY CIRCUIT 512		
• Check for continuity between FLI plug Pin 1 (harness side) and body ground. 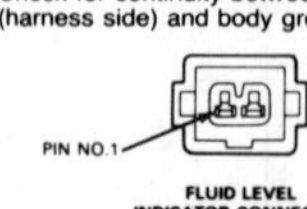	Continuity (NOT OK) ▶	REPAIR or REPLACE (Circuit 512).
	No continuity (OK) ▶	GO to Step **E2c**.
E2c CHECK CONTINUITY CIRCUIT 549		
• Disconnect 5-Pin plug from PWS. Check for continuity between FLI plug Pin 2 (harness side) and body ground. 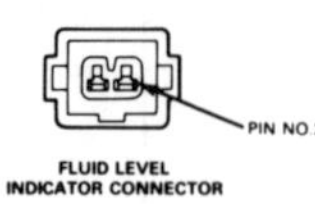	Continuity (NOT OK) ▶	REPAIR or REPLACE (Circuit 549).
	No continuity (OK) ▶	GO to Step **E2d**.

Fig. 30 Diagnosis, warning lamp intermittently ON (Part 4 of 5). 1985

Anti-Lock Warning Lamp Intermittently On — Test E

TEST STEP	RESULT ▶	ACTION TO TAKE
E2d CHECK CONTINUITY OF PWS		
• Check for continuity from PWS 5-Pin socket Pin 3 to body ground, and from Pin 5 to body ground. 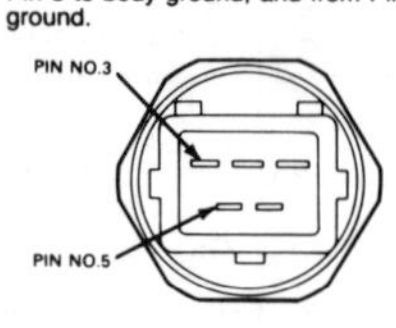	Continuity (NOT OK) ▶	REPLACE PWS.
	No continuity (OK) ▶	GO to Step **E2e**.
E2e CHECK CONTINUITY CIRCUIT 535		
• Check for continuity between 5-Pin PWS plug (harness side) Pin 5 and body ground. 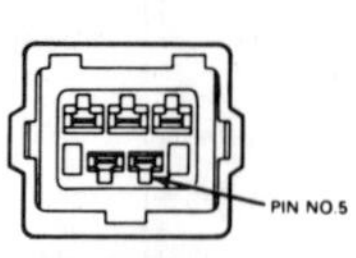	Continuity (NOT OK) ▶	REPAIR or REPLACE (Circuit 535).
	No continuity (OK) ▶	CONNECT all plugs and REVERIFY symptom.

Fig. 30 Diagnosis, warning lamp intermittently ON (Part 5 of 5). 1985

Brake Warning Lamp On (With Anti-Lock Lamp Off, Parking Brake Released And Brake Lining Wear Checked) — Test F

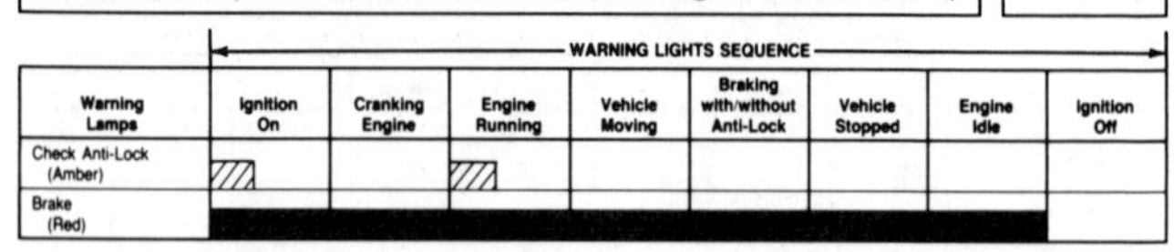

TEST STEP	RESULT ▶	ACTION TO TAKE
F1 CHECK BRAKE FLUID LEVEL		
• Turn ignition switch On. • Pump brake pedal until pump motor starts. • When pump motor stops check brake fluid level.	Low (NOT OK) ▶	CHECK system for external leaks. REPAIR as required.
	Normal (OK) ▶	GO to Step **F2**.
F2 CHECK CONTINUITY FLI SWITCH		
• Disconnect 3-Pin plug on brake reservoir switch. • Set multi-meter on 200 ohm range. Connect to reservoir warning switch Pins 1 and 2. 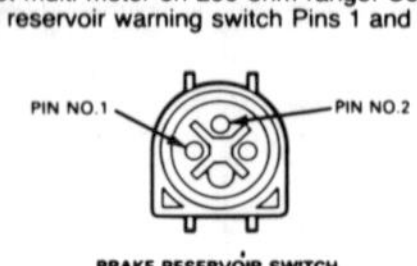	Above 10 ohms (OK) ▶	GO to Step **F3**.
	Below 10 ohms (NOT OK) ▶	REPLACE reservoir warning switch.

Fig. 31 Diagnosis, brake warning lamp ON w/anti-lock OFF, parking brake released & brake lining wear checked (Part 1 of 2). 1985

Brake Warning Lamp On (With Anti-Lock Lamp Off, Parking Brake Released And Brake Lining Wear Checked) — **Test F**

TEST STEP	RESULT ▶	ACTION TO TAKE
F3 CHECK CONTINUITY PWS SWITCH (CONTINUED)		
• Turn ignition switch On, wait until motor stops running. • Disconnect Pressure Warning Switch 5-Pin plug and connect multi-meter to Pins 1 and 2 (switch side). • Check for continuity.	Continuity — Not OK ▶	REPLACE pressure warning switch.
	No continuity — OK ▶	GO to Step **F4**.
F4 CHECK WIRING CONTINUITY		
• Check cable harness 430H, 512, 547B, 547A, 549, 977 Circuits. (Refer to vehicle Wiring Diagram.)	OK ▶	CHECK parking brake electrical circuit workshop manual. REVERIFY symptom.
	Not OK ▶	REPAIR or REPLACE as required.

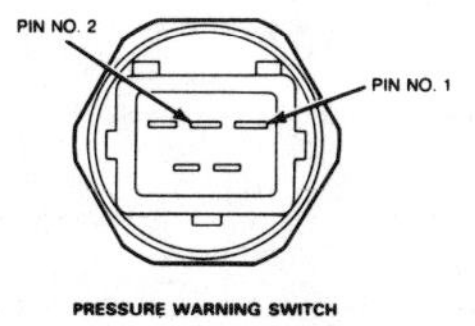

Fig. 31 Diagnosis, brake warning lamp ON w/anti-lock OFF, parking brake released & brake lining wear checked (Part 2 of 2). 1985

No Anti-Lock Warning Lamp On When Ignition Switch Turned On — **Test G**

WARNING LIGHTS SEQUENCE

Warning Lamps	Ignition On	Cranking Engine	Engine Running	Vehicle Moving	Braking with/without Anti-Lock	Vehicle Stopped	Engine Idle	Ignition Off
Check Anti-Lock (Amber)								
Brake (Red)								

TEST STEP	RESULT ▶	ACTION TO TAKE
G1 CHECK FUSE AND FUSE LINKS		
• Check fuse Number 5, two in-line fuse links and power to fuse with ignition turned On.	Fuse — OK ▶	GO to Step **G2**.
	Fuse — Not OK ▶	REPAIR or REPLACE as required.
G2 CHECK WARNING LAMP BULB		
• Check warning lamp bulb.	OK ▶	GO to Step **G3**.
	Not OK ▶	REPLACE.
G3 CHECK WARNING LAMP OPERATION		
• Turn ignition switch On. • Disconnect 35 Pin connector from control module.	If Anti-Lock lamp goes On ▶	REPAIR or REPLACE Circuit 606.
	If Anti-Lock lamp is not On ▶	REPAIR or REPLACE connector to 14401 wire harness.

Fig. 32 Diagnosis, no anti-lock warning lamp On when ignition switch is turned ON. 1985

Spongy Brake Pedal With/Without Anti-Lock Function (No Warning Lamp) — **Test H**

WARNING LIGHTS SEQUENCE

Warning Lamps	Ignition On	Cranking Engine	Engine Running	Vehicle Moving	Braking with/without Anti-Lock	Vehicle Stopped	Engine Idle	Ignition Off
Check Anti-Lock (Amber)								
Brake (Red)								

TEST STEP	RESULT ▶	ACTION TO TAKE
H1 CHECK COMPONENT MOUNTING		
• Check for proper brake pedal and hydraulic unit attachment. • Bleed brakes	Pedal still spongy — Not OK ▶	GO to Step **H2**.
	Pedal feels normal — OK ▶	Condition corrected.
H2 BLEED BRAKE SYSTEM		
• Rebleed brake system.	Pedal still spongy — Not OK ▶	REPLACE hydraulic unit.
	Pedal feels normal — OK ▶	Condition corrected.

Fig. 33 Diagnosis, spongy brake pedal with/without anti-lock function, no warning lamp. 1985

Poor Vehicle Tracking During Anti-Lock Function (Warning Lamp Off) — **Test J**

WARNING LIGHTS SEQUENCE

Warning Lamps	Ignition On	Cranking Engine	Engine Running	Vehicle Moving	Braking with/without Anti-Lock	Vehicle Stopped	Engine Idle	Ignition Off
Check Anti-Lock (Amber)								
Brake (Red)								

TEST STEP	RESULT ▶	ACTION TO TAKE
J1 VERIFY CONDITION		
• Verify condition exists as reported. • Bleed brake system per shop manual for Anti-Lock brake system.	Vehicle tracks properly ▶	Condition corrected.
	Vehicle still tracks poorly ▶	GO to **J2**.
J2 CHECK ANTI-LOCK OPERATION — LEFT FRONT WHEEL		
• Lift vehicle and rotate wheels to assure they turn freely. • Turn ignition switch Off. • Disconnect 35-Pin plug from electronic controller. • Connect EEC-IV Breakout box, Tool T83L-50-EEC-IV with Anti-Lock test adapter, Tool T85P-50-ASA to the Anti-Lock 35-Pin plug wiring harness. • Short Pins 2, 16 and 35 to each other at Breakout box. • Apply moderate brake pedal effort and check that left front wheel will not turn. • Check to see that left front wheel turns freely with ignition switch On. **CAUTION: DO NOT LEAVE IGNITION ON FOR MORE THAN 1 MINUTE MAXIMUM, OR SOLENOID VALVE DAMAGE MAY RESULT.**	If wheel turns freely ▶	TURN ignition switch Off. DISCONNECT wire leads. Go to Step **J3**.
	If wheel does not turn freely or pedal drops ▶	REPLACE hydraulic unit.

Fig. 34 Diagnosis, poor vehicle tracking during anti-lock function, warning lamp OFF (Part 1 of 2). 1985

Poor Vehicle Tracking During Anti-Lock Function (Warning Lamp Off) — Test J

TEST STEP	RESULT ▶	ACTION TO TAKE
J3 CHECK ANTI-LOCK OPERATION — RIGHT FRONT WHEEL		
• Short Pins 2, 15 and 34 to each other at Breakout box. • Apply moderate brake effort. Check that right front wheel does not turn. • Check that right front wheel turns freely with ignition switch On. **CAUTION: DO NOT LEAVE IGNITION ON MORE THAN 1 MINUTE MAXIMUM OR SOLENOID VALVE DAMAGE MAY RESULT.**	Wheel turns freely ▶	TURN ignition switch Off. DISCONNECT wire leads. GO to Step **J4**.
	Wheel does not turn freely or brake pedal drops ▶	REPLACE hydraulic unit.
J4 CHECK ANTI-LOCK OPERATION — REAR WHEELS		
• Short Pins 2, 17 and 33 to each other at Breakout box. • Apply moderate brake pedal pressure. Check that rear wheels will not turn. • Check that rear wheels turn freely with ignition switch On. **CAUTION: DO NOT LEAVE IGNITION ON MORE THAN 1 MINUTE MAXIMUM OR SOLENOID VALVE DAMAGE MAY RESULT.**	Wheels turn freely ▶	TURN ignition switch Off. DISCONNECT wire lead and Breakout box. LOWER vehicle. REVERIFY symptom.
	Wheels do not turn freely or brake pedal drops ▶	REPLACE hydraulic unit.

Fig. 34 Diagnosis, poor vehicle tracking during anti-lock function, warning lamp OFF (Part 2 of 2). 1985

No Booster; Anti-Lock Warning Lamp On — Test K

WARNING LIGHTS SEQUENCE

Warning Lamps	Ignition On	Cranking Engine	Engine Running	Vehicle Moving	Braking with/without Anti-Lock	Vehicle Stopped	Engine Idle	Ignition Off
Check Anti-Lock (Amber)								
Brake (Red)								

TEST STEP	RESULT ▶	ACTION TO TAKE
K1 CHECK CONTINUITY PWS SWITCH		
• Turn ignition switch On, wait until motor stops running. • Disconnect 5-Pin plug from Pressure Warning Switch (PWS). • Check continuity between Pins 1 and 4 (switch side).	Continuity (OK) ▶	REPAIR wire from Pin 1 to ground.
	No continuity (NOT OK) ▶	REPLACE PWS.

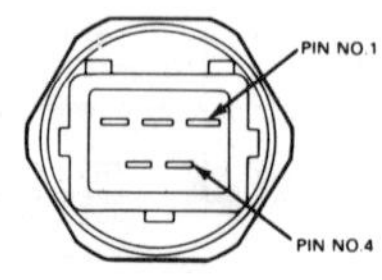

PRESSURE WARNING SWITCH

Fig. 35 Diagnosis, no booster, anti-lock warning lamp ON. 1985

Anti-Lock Warning Lamp On (With Brake Warning Lamp Off) — Test A

WARNING LIGHTS SEQUENCE

Warning Lamps	Ignition On	Cranking Engine	Engine Running	Vehicle Moving	Braking with without Anti-Lock	Vehicle Stopped	Engine Idle	Ignition Off
Check Anti-Lock (Amber)								
Brake (Red)								

TEST STEP	RESULT ▶	ACTION TO TAKE
A1 35-PIN PLUG TESTING		
• Turn ignition switch OFF. • Disconnect 35-Pin plug from electronic controller.	Over 10V (OK) ▶	GO to Step **A2**.
	Under 10V (NOT OK) ▶	GO to Step **A1a**.

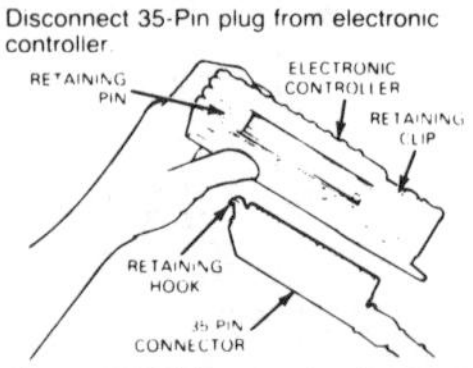

• Connect EEC-IV Breakout box Tool T83L-50-EEC-IV and anti-lock test adapter Tool T85P-50-ASA or equivalents, to the anti-lock 35-Pin plug wiring harness.

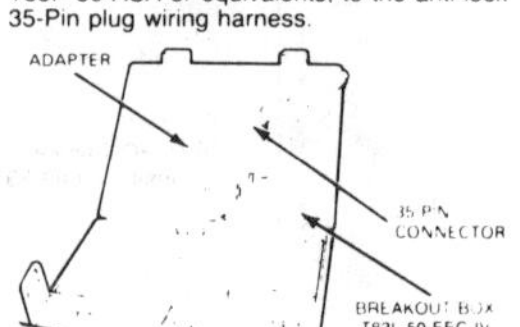

• Set multi-meter to read volts DC.
• Turn ignition switch ON.
• Measure voltage between Breakout box Pins 40 and 2.

Fig. 36 Diagnosis, warning lamp ON w/brake warning lamp OFF (Part 1 of 24). 1986–87 except Thunderbird

Anti-Lock Warning Lamp On (With Brake Warning Lamp Off) — Test A

TEST STEP	RESULT ▶	ACTION TO TAKE
A1a CHECK ELECTRONIC CONTROLLER TO GROUND WIRE		
• Check: — fuse link to anti-lock warning lamp. — battery. • Remove positive battery cable. • Check continuity between Breakout box pin 40 and body ground.	Continuity (OK) ▶	GO to Step **A1b**.
	No continuity (NOT OK) ▶	SERVICE or REPLACE cable harness (Circuit 530 or 530A).
A1b CHECK IGNITION TO ELECTRONIC CONTROLLER WIRE		
• Check continuity between Breakout box Pin 2 and ignition switch wire 298.	Continuity (OK) ▶	RECONNECT positive battery cable. CHECK for power at ignition switch pin with switch ON. If okay, connect electronic controller and reverify symptom.
	No continuity (NOT OK) ▶	SERVICE or REPLACE cable harness (Circuit 298A).
A2 CHECK MAIN POWER RELAY SECONDARY CIRCUIT (NORMAL)		
• Turn ignition switch OFF. • Check for continuity between Breakout box Pins 40 and 3.	Continuity (OK) ▶	GO to **A3**.
	No continuity (NOT OK) ▶	GO to **A2a**.

Fig. 36 Diagnosis, warning lamp ON w/brake warning lamp OFF (Part 2 of 24). 1986–87 except Thunderbird

Anti-Lock Warning Lamp On (With Brake Warning Lamp Off)	Test A

TEST STEP	RESULT ▶	ACTION TO TAKE
A2a CHECK MAIN POWER RELAY SECONDARY CIRCUIT (NORMAL)		
• Disconnect main relay from socket. • Check for continuity between main power relay socket Pins 3 and 5. 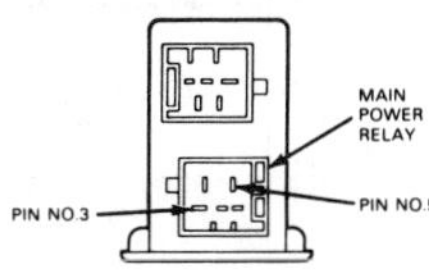	Continuity (OK) ▶ No continuity (NOT OK) ▶	GO to Step **A2b**. REPLACE main power relay.
A2b CHECK MAIN POWER RELAY SECONDARY CIRCUIT WIRING HARNESS		
• Disconnect positive battery cable. • Check for continuity between main power relay socket Pin 3 and Breakout box Pin 3.  MAIN POWER RELAY CONNECTOR	Continuity (OK) ▶ No Continuity (NOT OK) ▶	GO to Step **A2C**. SERVICE or REPLACE cable harness (Circuit 532C or 532A).
A2c CHECK MAIN POWER RELAY SECONDARY CIRCUIT WIRING HARNESS		
• Check for continuity between main power relay socket Pin 5 and body ground. 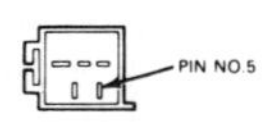 MAIN POWER RELAY CONNECTOR	Continuity (OK) ▶ No continuity (NOT OK) ▶	RECONNECT main power relay, electronic controller and battery cable and reverify symptom. SERVICE or REPLACE cable harness (Circuit 430M).

Fig. 36 Diagnosis, warning lamp ON w/brake warning lamp OFF (Part 3 of 24). 1986–87 except Thunderbird

With Anti-Lock Warning Lamp On (With Brake Warning Lamp Off)	Test A

TEST STEP	RESULT ▶	ACTION TO TAKE
A3 CHECK MAIN POWER RELAY SECONDARY CIRCUIT (NORMAL)		
• Check for continuity between Breakout box Pins 40 and 20.	Continuity (OK) ▶ No continuity (NOT OK) ▶	GO to Step **A4**. GO to Step **A3a**.
A3a CHECK MAIN POWER RELAY SECONDARY CIRCUIT WIRING HARNESS		
• Remove main power relay. • Check for continuity between main power relay socket Pin 3 and Breakout box Pin 20. 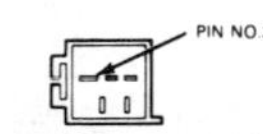 MAIN POWER RELAY CONNECTOR	Continuity (OK) ▶ No continuity (NOT OK) ▶	CONNECT main power relay and electronic controller and reverify symptom. SERVICE or REPLACE cable harness (Circuit 532B).
A4 CHECK FLI AND PWS CIRCUIT		
• Turn ignition switch ON. • Set multi-meter to read resistance. • Measure the resistance between Breakout box Pins 9 and 10.	Less than 5 ohms (OK) ▶ Greater than 5 ohms (NOT OK) ▶	GO to Step **A5**. GO to Step **A4a**.
A4a CHECK FLI ANTI-LOCK WARNING CIRCUIT		
• Disconnect 5-Pin plug on reservoir Fluid Level Indicator (FLI). • Measure resistance between fluid level indicator electrical socket Pins 1 and 2 (with brake fluid level at maximum mark on reservoir). 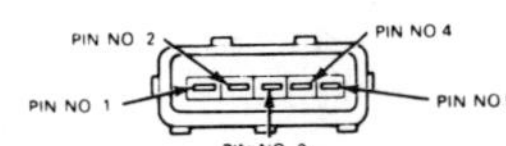 5-PIN FLUID LEVEL INDICATOR	Less than 2 ohms (OK) ▶ Greater than 2 ohms (NOT OK) ▶	GO to Step **A4b**. REPLACE fluid level indicator.

Fig. 36 Diagnosis, warning lamp ON w/brake warning lamp OFF (Part 4 of 24). 1986–87 except Thunderbird

Anti-Lock Warning Lamp On (With Brake Warning Lamp Off)	Test A

TEST STEP	RESULT ▶	ACTION TO TAKE
A4b CHECK PWS ANTI-LOCK WARNING CIRCUIT		
• Disconnect 5-Pin plug at pressure warning switch (PWS). • Check for continuity between pressure warning switch socket Pins 3 and 5. **NOTE: System must be pressurized during this test.** 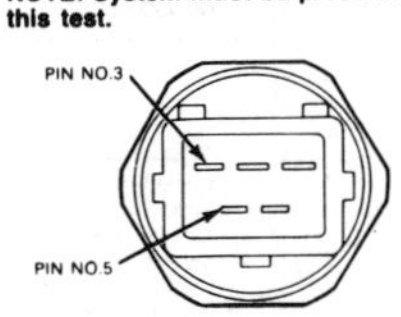 PRESSURE WARNING SWITCH	Continuity (OK) ▶ No continuity (NOT OK) ▶	GO to Step **A4c**. REPLACE pressure warning switch and pump motor relay.
A4c CHECK ELECTRONIC MODULE TO FLI WIRE		
• Check for continuity between Breakout box Pin 9 and FLI 5-Pin plug (harness side) Pin 1. 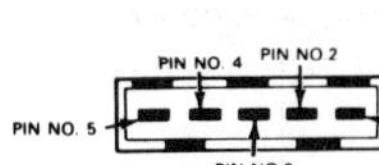 FLUID LEVEL INDICATOR CONNECTOR	Continuity (OK) ▶ No continuity (NOT OK) ▶	GO to Step **A4d**. SERVICE or REPLACE cable harness (Circuit 512).

Fig. 36 Diagnosis, warning lamp ON w/brake warning lamp OFF (Part 5 of 24). 1986–87 except Thunderbird

Anti-Lock Warning Lamp On (With Brake Warning Lamp Off)	Test A

TEST STEP	RESULT ▶	ACTION TO TAKE
A4d CHECK FLI TO PWS WIRE		
• Check for continuity between Pin 2 of 5-Pin fluid level plug (harness side) 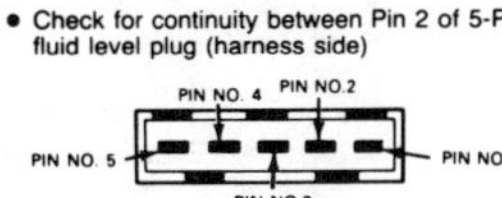 FLUID LEVEL INDICATOR CONNECTOR and 5-Pin pressure warning switch plug (harness side) Pin 3. 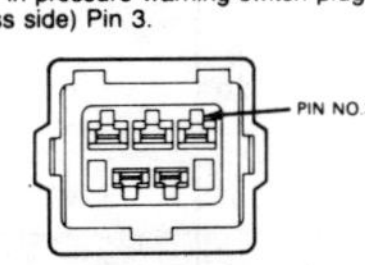 PRESSURE WARNING SWITCH CONNECTOR	Continuity (OK) ▶ No continuity (NOT OK) ▶	GO to Step **A4e**. SERVICE or REPLACE cable harness (Circuit 549).
A4e PWS TO ELECTRONIC CONTROLLER WIRE		
• Check for continuity between 5-Pin pressure warning switch plug (harness side) Pin 5 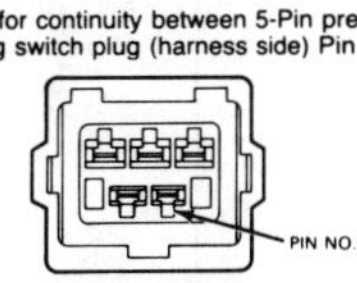 PRESSURE WARNING SWITCH CONNECTOR and Breakout box Pin 10.	Continuity (OK) ▶ No continuity (NOT OK) ▶	TURN ignition OFF. CONNECT all electrical connections. Reverify symptom. SERVICE or REPLACE cable harness (Circuit 535).
A5 ISOLATION TEST FLI AND PWS		
• Check for continuity between Breakout box Pin 9 and body ground.	Continuity (NOT OK) ▶ No continuity (OK) ▶	GO to Step **A5a**. GO to Step **A6**.

Fig. 36 Diagnosis, warning lamp ON w/brake warning lamp OFF (Part 6 of 24). 1986–87 except Thunderbird

Anti-Lock Warning Lamp On (With Brake Warning Lamp Off)	Test A

TEST STEP	RESULT ▶	ACTION TO TAKE
A5a CHECK FLUID LEVEL INDICATOR PIN NO. 2		
• Disconnect FLI 5-Pin plug and check for continuity between FLI socket Pin 2 and body ground and Pin 1 and body ground. **NOTE: Ensure that brake fluid is at max level.** 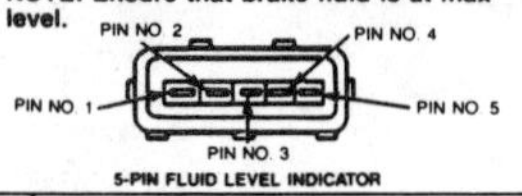	Continuity ~~OK~~ ▶ No continuity OK ▶	REPLACE FLI. GO to Step **A5b**.
A5b CHECK FLUID LEVEL INDICATOR PIN NO. 1		
• Check for continuity between FLI plug Pin 1 (harness side) and body ground. 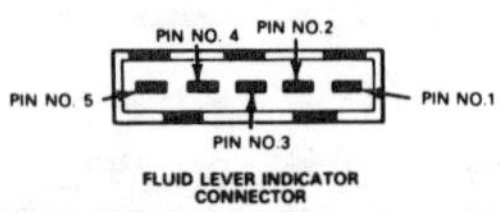	Continuity ~~OK~~ ▶ No continuity OK ▶	SERVICE or REPLACE Circuit 512. GO to Step **A5c**.
A5c CHECK FLUID LEVEL INDICATOR PIN NO. 2		
• Disconnect 5-Pin plug from PWS. Check for continuity between FLI plug Pin 2 (harness side) and body ground. 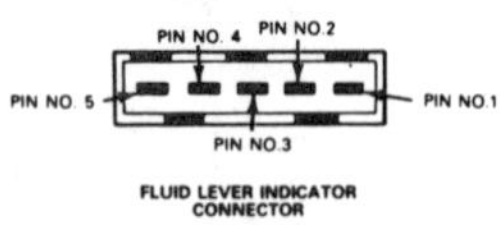	Continuity ~~OK~~ ▶ No continuity OK ▶	SERVICE or REPLACE Circuit 549. GO to Step **A5d**.

Fig. 36 Diagnosis, warning lamp ON w/brake warning lamp OFF (Part 7 of 24). 1986–87 except Thunderbird

Anti-Lock Warning Lamp On (With Brake Warning Lamp Off)	Test A

TEST STEP	RESULT ▶	ACTION TO TAKE
A5d CHECK CONTINUITY BETWEEN PWS PINS 3 AND 5 AND BODY GROUND		
• Check for continuity from PWS 5-Pin socket Pins (pressure warning switch side) 3 and body ground and 5 and body ground. **NOTE: System should be depressurized for this test.** 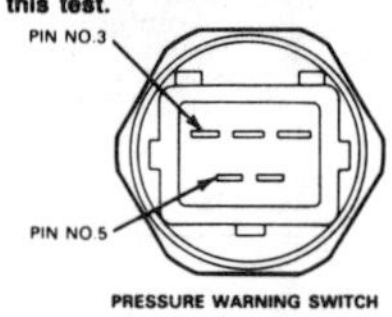	Continuity ~~OK~~ ▶ No continuity OK ▶	REPLACE Pressure Warning Switch and Pump Motor Relay. GO to Step **A5e**.
A5e CHECK PWS CONNECTOR PIN 5 AND GROUND		
• Check for continuity between 5-Pin PWS plug, Pin 5 (harness side) and body ground. 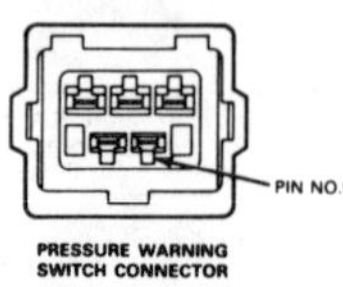	Continuity ~~OK~~ ▶ No continuity OK ▶	SERVICE or REPLACE Circuit 535. CONNECT all plugs and reverify symptom.
A6 CHECK MAIN RELAY PRIMARY CIRCUIT RESISTANCE		
• Turn ignition switch OFF. • Set multi-meter to read resistance. • Measure resistance between Breakout box Pins 40 and 8.	Resistance between 45 and 105 ohms OK ▶ Any other reading ~~OK~~ ▶	GO to Step **A7**. GO to Step **A6a**.

Fig. 36 Diagnosis, warning lamp ON w/brake warning lamp OFF (Part 8 of 24). 1986–87 except Thunderbird

Anti-Lock Warning Lamp On (With Brake Warning Lamp Off)	Test A

TEST STEP	RESULT ▶	ACTION TO TAKE
A6a CHECK MAIN POWER RELAY PRIMARY COIL RESISTANCE		
• Disconnect main power relay from socket. • Set multi-meter on resistance. • Measure resistance between main power relay Pins 1 and 2. 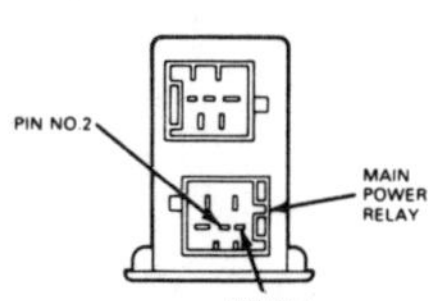	Resistance between 45 and 105 ohms OK ▶ Any other reading ~~OK~~ ▶	GO to Step **A6b**. REPLACE main power relay.
A6b CHECK MAIN POWER RELAY PRIMARY TO ELECTRONIC CONTROLLER WIRE		
• Check continuity between main power relay socket Pin 2 (harness side) and Breakout box Pin 8. 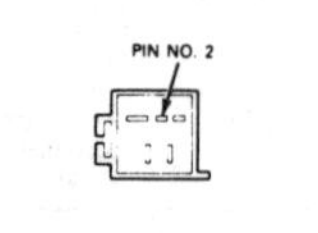	Continuity OK ▶ No continuity ~~OK~~ ▶	GO to Step **A6c**. SERVICE or REPLACE cable harness (Circuit 513).

Fig. 36 Diagnosis, warning lamp ON w/brake warning lamp OFF (Part 9 of 24). 1986–87 except Thunderbird

Anti-Lock Warning Lamp On (With Brake Warning Lamp Off)	Test A

TEST STEP	RESULT ▶	ACTION TO TAKE
A6c CHECK MAIN POWER RELAY PRIMARY TO GROUND WIRE		
• Check continuity between main power relay socket Pin 1 and ground. 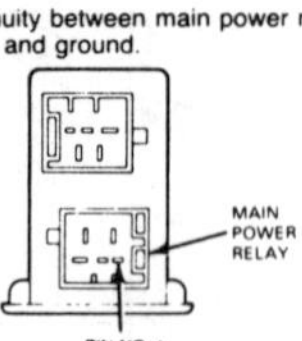	Continuity OK ▶ No continuity ~~OK~~ ▶	RECONNECT all electrical connections and reverify symptom. SERVICE or REPLACE cable harness (Circuit 430N).
A7 CHECK MAIN POWER RELAY SECONDARY (ACTIVATED)		
• Place a jumper wire between Breakout box Pins 2 and 8. • Set multi-meter to read volts DC. • Turn ignition switch ON. • Measure voltage between Breakout box Pins 40 and 3.	Over 10 V DC OK ▶ Under 10 V DC ~~OK~~ ▶	GO to Step **A8**. CHECK fuse F5. REMOVE jumper wire and GO to Step **A7a**.
A7a CHECK MAIN POWER RELAY SECONDARY CIRCUIT (ACTIVE)		
• Turn ignition switch OFF. • Disconnect main power relay from socket. • Apply power (battery positive and ground) to main relay Pins 1 and 2. • Check for continuity between main power relay Pins 3 and 4. 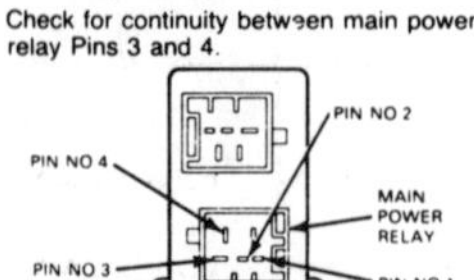	Continuity OK ▶ No continuity ~~OK~~ ▶	GO to Step **A7b**. REPLACE main power relay.

Fig. 36 Diagnosis, warning lamp ON w/brake warning lamp OFF (Part 10 of 24). 1986–87 except Thunderbird

Anti-Lock Warning Lamp On (With Brake Warning Lamp Off) — Test A

TEST STEP	RESULT ▶	ACTION TO TAKE
A7b CHECK MAIN POWER RELAY SECONDARY CIRCUIT POWER WIRE		
• Check continuity between main power relay socket Pin 4 (harness side) and positive battery terminal. 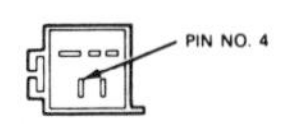 MAIN POWER RELAY CONNECTOR	Continuity (OK) ▶	RECONNECT main power relay and reverify symptom.
	No continuity (NOT OK) ▶	SERVICE or REPLACE cable harness (Circuit 533) or fuse link (Circuit 291B).
A8 MEASURE RH REAR SENSOR CIRCUIT RESISTANCE		
• Turn ignition switch OFF. • Set multi-meter to read resistance. • Measure resistance between Breakout box Pins 4 and 22.	800 to 1400 ohms (0.8 to 1.4K ohms) (OK) ▶	GO to Step **A9**.
	Any other reading (NOT OK) ▶	Go to Step **A8a**.
A8a MEASURE RH REAR SENSOR RESISTANCE		
• Disconnect sensor plug (right rear). • Set multi-meter to read resistance. • Measure resistance of sensor at sensor plug.  RH REAR SENSOR	800 to 1400 ohms (0.8 to 1.4K ohms) (OK) ▶	SERVICE or REPLACE cable harness (Circuit 523 or 524).
	Any other reading (NOT OK) ▶	REPLACE right rear wheel sensor.
A9 MEASURE LH FRONT SENSOR CIRCUIT RESISTANCE		
• Measure resistance between Breakout box Pins 5 and 23.	800 to 1400 ohms (0.8 to 1.4K ohms) (OK) ▶	GO to Step **A10**.
	Any other reading (NOT OK) ▶	GO to Step **A9a**.

Fig. 36 Diagnosis, warning lamp ON w/brake warning lamp OFF (Part 11 of 24). 1986–87 except Thunderbird

Anti-Lock Warning Lamp On (With Brake Warning Lamp Off) — Test A

TEST STEP	RESULT ▶	ACTION TO TAKE
A9a MEASURE LH FRONT SENSOR RESISTANCE		
• Disconnect sensor plug (left front). • Measure resistance of sensor at sensor plug. 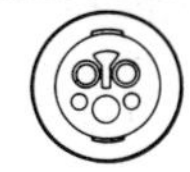 LH FRONT SENSOR	800 to 1400 ohms (0.8 to 1.4K ohms) (OK) ▶	SERVICE or REPLACE cable harness (Circuit 521 or 522).
	Any other reading (NOT OK) ▶	REPLACE left front wheel sensor.
A10 MEASURE LH REAR SENSOR CIRCUIT RESISTANCE		
• Measure resistance between Breakout box Pins 6 and 24.	800 to 1400 ohms (0.8 to 1.4K ohms) (OK) ▶	GO to Step **A11**.
	Any other reading (NOT OK) ▶	GO to Step **A10a**.
A10a MEASURE LH REAR SENSOR RESISTANCE		
• Disconnect sensor plug (left rear). • Measure resistance of sensor at left rear sensor plug. 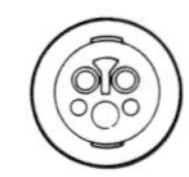 LH REAR SENSOR	800 to 1400 ohms (0.8 to 1.4K ohms) (OK) ▶	SERVICE or REPLACE cable harness (Circuit 518 or 519).
	Any other reading (NOT OK) ▶	REPLACE left rear wheel sensor.
A11 MEASURE RH FRONT SENSOR CIRCUIT RESISTANCE		
• Measure resistance between Breakout box Pins 7 and 25.	800 to 1400 ohms (0.8 to 1.4K ohms) (OK) ▶	GO to Step **A12**.
	Any other reading (NOT OK) ▶	GO to Step **A11a**.

Fig. 36 Diagnosis, warning lamp ON w/brake warning lamp OFF (Part 12 of 24). 1986–87 except Thunderbird

Anti-Lock Warning Lamp On (With Brake Warning Lamp Off) — Test A

TEST STEP	RESULT ▶	ACTION TO TAKE
A11a MEASURE RH FRONT SENSOR RESISTANCE		
• Disconnect sensor plug (right front). • Measure resistance of sensor at sensor plug. 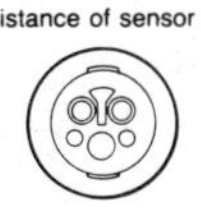 RH FRONT SENSOR	800 to 1400 ohms (0.8 to 1.4K ohms) (OK) ▶	SERVICE or REPLACE cable harness (Circuit 514 or 516).
	Any other reading (NOT OK) ▶	REPLACE right front wheel sensor.
A12 MEASURE MAIN VALVE CIRCUIT RESISTANCE		
• Turn ignition switch OFF. • Set multi-meter to read resistance. • Measure the resistance between Breakout box Pins 11 and 18.	2 to 5 ohms (OK) ▶	GO to Step **A13.**
	Any other reading (NOT OK) ▶	GO to Step **A12a.**
A12a MEASURE MAIN VALVE RESISTANCE		
• Disconnect main valve 2-Pin plug. • Measure resistance between the main valve electrical Pins 1 and 2. 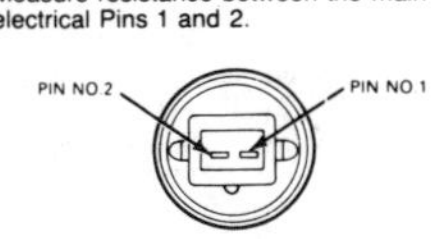	2 to 5.5 ohms (OK) ▶	REPLACE or SERVICE cable harness (Circuit 430E or 493).
	Any other reading (NOT OK) ▶	REPLACE actuation assembly.
A13 CHECK VALVE BLOCK GROUND CIRCUIT		
• Measure resistance between Breakout box Pins 11 and 40.	Less than 2 ohms (OK) ▶	GO to Step **A14**.
	Greater than 2 ohms (NOT OK) ▶	GO to Step **A13a**.

Fig. 36 Diagnosis, warning lamp ON w/brake warning lamp OFF (Part 13 of 24). 1986–87 except Thunderbird

Anti-Lock Warning Lamp On (With Brake Warning Lamp Off) — Test A

TEST STEP	RESULT ▶	ACTION TO TAKE
A13a TEST VALVE PLUG PIN NO. 7		
• Disconnect valve block 7-Pin plug. • Measure resistance between valve block electrical Pin 7 and body ground. 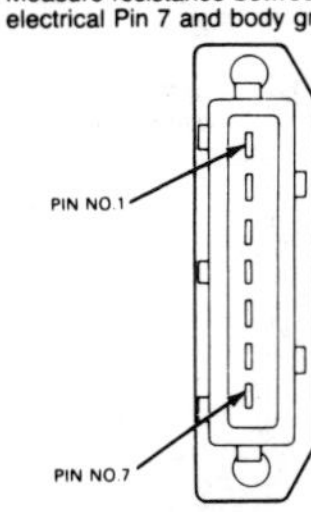 VALVE BLOCK 7-PIN CONNECTOR	Less than 2 ohms (OK) ▶	REPLACE or SERVICE cable harness (Circuit 511).
	Greater than 2 ohms (NOT OK) ▶	GO to Step **A13b**.
A13b CHECK VALVE BLOCK GROUND WIRE		
• Remove negative (−) ground strap from battery. • Check for continuity between valve block and body ground.	Continuity (OK) ▶	REPLACE valve block unit (internal ground problem).
	No or Poor continuity (NOT OK) ▶	SERVICE or REPLACE actuation assembly ground strap (Circuit 430G).
A14 MEASURE RH FRONT INLET VALVE CIRCUIT RESISTANCE		
• Measure resistance between Breakout box Pins 11 and 15.	5 to 8 ohms (OK) ▶	GO to Step **A15**.
	Any other reading (NOT OK) ▶	GO to Step **A14a**.

Fig. 36 Diagnosis, warning lamp ON w/brake warning lamp OFF (Part 14 of 24). 1986–87 except Thunderbird

Anti-Lock Warning Lamp On (With Brake Warning Lamp Off) — Test A

TEST STEP	RESULT ▶	ACTION TO TAKE
A14a MEASURE RH FRONT INLET VALVE RESISTANCE		
• Disconnect valve block 7-Pin plug. • Measure resistance between valve block electrical Pin 7 and 1.	5 to 8 ohms (OK) ▶	REPLACE or SERVICE cable harness (Circuit 510).
	Any other reading (NOT OK) ▶	REPLACE solenoid valve block unit. RECONNECT 7-Pin plug.
A15 MEASURE REAR INLET VALVE CIRCUIT RESISTANCE		
• Measure resistance between Breakout box Pins 11 and 17.	5 to 8 ohms (OK) ▶	GO to Step **A16**.
	Any other reading (NOT OK) ▶	GO to Step **A15a**.

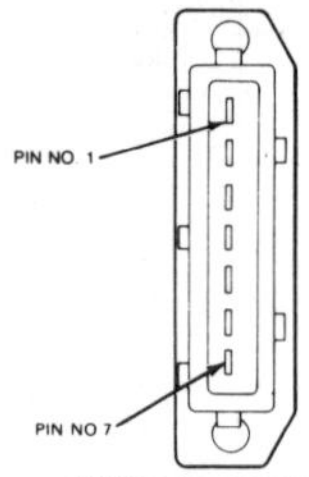

VALVE BLOCK 7-PIN CONNECTOR

Fig. 36 Diagnosis, warning lamp ON w/brake warning lamp OFF (Part 15 of 24). 1986–87 except Thunderbird

Anti-Lock Warning Lamp On (With Brake Warning Lamp Off) — Test A

TEST STEP	RESULT ▶	ACTION TO TAKE
A15a MEASURE REAR INLET VALVE RESISTANCE		
• Disconnect valve block 7-Pin plug. • Measure resistance between valve block electrical Pins 7 and 3.	5 to 8 ohms (OK) ▶	REPLACE or SERVICE cable harness (Circuit 496).
	Any other reading (NOT OK) ▶	REPLACE valve block unit. RECONNECT 7-Pin plug.
A16 MEASURE LH FRONT INLET VALVE CIRCUIT RESISTANCE		
• Measure resistance between Breakout box Pins 11 and 35.	5 to 8 ohms (OK) ▶	GO to Step **A17**.
	Any other reading (NOT OK) ▶	GO to Step **A16a**.

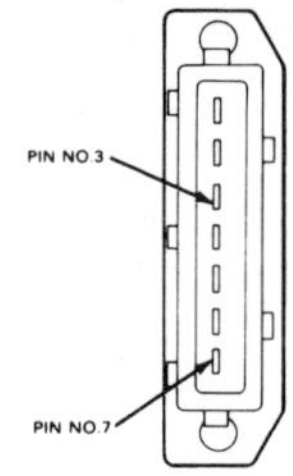

VALVE BLOCK 7-PIN CONNECTOR

Fig. 36 Diagnosis, warning lamp ON w/brake warning lamp OFF (Part 16 of 24). 1986–87 except Thunderbird

Anti-Lock Warning Lamp On (With Brake Warning Lamp Off) — Test A

TEST STEP	RESULT ▶	ACTION TO TAKE
A16a MEASURE LH FRONT INLET VALVE RESISTANCE		
• Disconnect valve block 7-Pin plug. • Measure resistance between valve block electrical Pins 7 and 6.	5 to 8 ohms (OK) ▶	REPLACE or SERVICE cable harness (Circuit 495).
	Any other reading (NOT OK) ▶	REPLACE valve block unit. RECONNECT 7-Pin plug.
A17 MEASURE REAR OUTLET VALVE CIRCUIT RESISTANCE		
• Measure resistance between Breakout box Pins 11 and 33.	3 to 6 ohms (OK) ▶	GO to Step **A18**.
	Any other reading (NOT OK) ▶	GO to Step **A17a**.

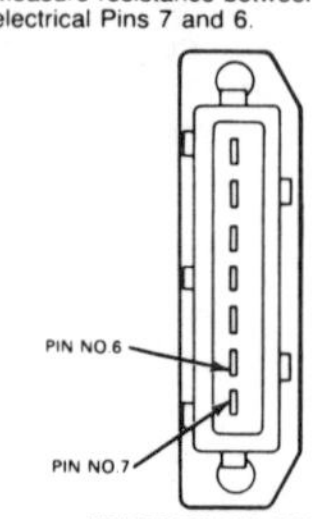

VALVE BLOCK 7-PIN CONNECTOR

Fig. 36 Diagnosis, warning lamp ON w/brake warning lamp OFF (Part 17 of 24). 1986–87 except Thunderbird

Anti-Lock Warning Lamp On (With Brake Warning Lamp Off) — Test A

TEST STEP	RESULT ▶	ACTION TO TAKE
A17a MEASURE REAR OUTLET VALVE RESISTANCE		
• Disconnect valve block 7-Pin plug. • Measure resistance between valve block electrical Pins 7 and 4.	3 to 6 ohms (OK) ▶	REPLACE or SERVICE cable harness (Circuit 499).
	Any other reading (NOT OK) ▶	REPLACE valve block unit. RECONNECT 7-Pin plug.
A18 MEASURE LH FRONT OUTLET VALVE CIRCUIT RESISTANCE		
• Measure resistance between Breakout box Pins 11 and 16.	3 to 6 ohms (OK) ▶	GO to Step **A19**.
	Any other reading (NOT OK) ▶	GO to Step **A18a**.

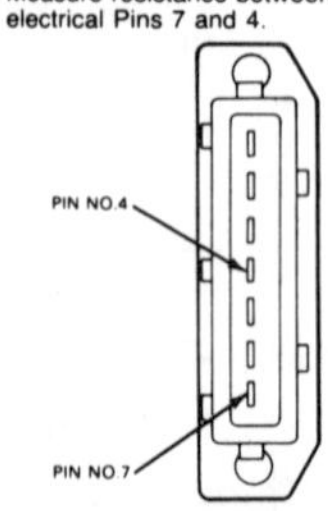

VALVE BLOCK 7-PIN CONNECTOR

Fig. 36 Diagnosis, warning lamp ON w/brake warning lamp OFF (Part 18 of 24). 1986–87 except Thunderbird

Anti-Lock Warning Lamp On (With Brake Warning Lamp Off)	Test A

TEST STEP	RESULT ▶	ACTION TO TAKE
A18a MEASURE LH FRONT OUTLET VALVE RESISTANCE		
• Disconnect valve block 7-Pin plug. • Measure resistance between valve block electrical Pins 7 and 5. 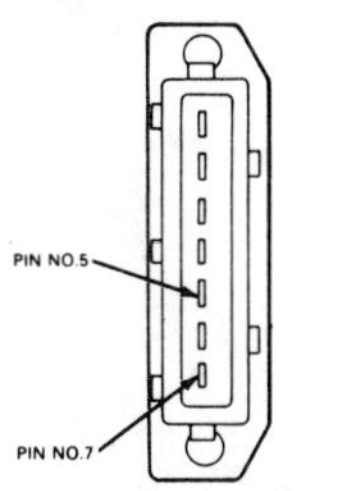 VALVE BLOCK 7-PIN CONNECTOR	3 to 6 ohms (OK) ▶	REPLACE or SERVICE cable harness (Circuit 498).
	Any other reading (NOT OK) ▶	REPLACE valve block unit. RECONNECT 7-Pin plug.
A19 MEASURE RH FRONT OUTLET VALVE CIRCUIT RESISTANCE		
• Measure resistance between Breakout box Pins 11 and 34.	3 to 6 ohms (OK) ▶	GO to Step **A20.**
	Any other reading (NOT OK) ▶	GO to Step **A19a.**

Fig. 36 Diagnosis, warning lamp ON w/brake warning lamp OFF (Part 19 of 24). 1986–87 except Thunderbird

Anti-Lock Warning Lamp On (With Brake Warning Lamp Off)	Test A

TEST STEP	RESULT ▶	ACTION TO TAKE
A19a MEASURE RH FRONT OUTLET VALVE RESISTANCE		
• Disconnect valve block 7-Pin plug. • Measure resistance between valve block electrical Pins 7 and 2. 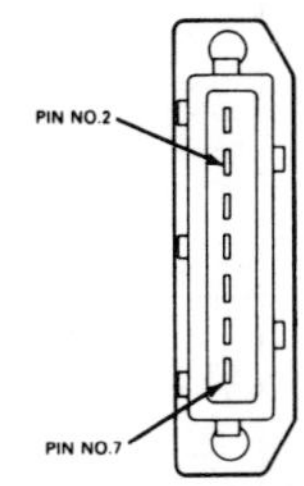 VALVE BLOCK 7-PIN CONNECTOR	3 to 6 ohms (OK) ▶	REPLACE or SERVICE cable harness (Circuit 497).
	Any other reading (NOT OK) ▶	REPLACE valve block unit. RECONNECT 7-Pin plug.

Fig. 36 Diagnosis, warning lamp ON w/brake warning lamp OFF (Part 20 of 24). 1986–87 except Thunderbird

Anti-Lock Warning Lamp On (With Brake Warning Lamp Off)	Test A

TEST STEP	RESULT ▶	ACTION TO TAKE
A20 CHECK PWS BRAKELAMP CIRCUIT (NO SYSTEM PRESSURE)		
• Vehicle must be cooled to room temperature. • Discharge the brake system as follows: a. Turn ignition switch OFF. b. Pump brake pedal at least 20 times until you feel the pedal become hard. • Disconnect pressure warning switch 5-Pin plug. • Check continuity between pressure warning switch electrical Pins 1 and 2. 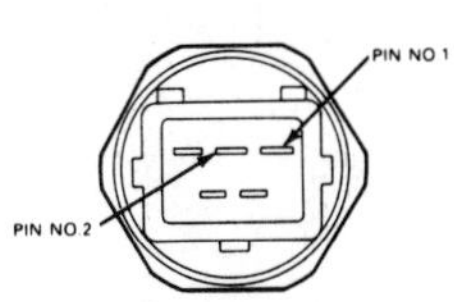 PRESSURE WARNING SWITCH	Continuity (OK) ▶	GO to Step **A21**.
	No continuity (NOT OK) ▶	REPLACE pressure warning switch and pump motor relay.

Fig. 36 Diagnosis, warning lamp ON w/brake warning lamp OFF (Part 21 of 24). 1986–87 except Thunderbird

Anti-Lock Warning Lamp On (With Brake Warning Lamp Off)	Test A

TEST STEP	RESULT ▶	ACTION TO TAKE
A21 CHECK PWS BRAKELAMP CIRCUIT (WITH SYSTEM PRESSURE)		
• Reconnect pressure warning switch 5-Pin plug. • Turn ignition switch ON. • When pump motor stops, turn ignition switch OFF. • Disconnect pressure warning switch 5-Pin plug again. • Check for continuity between pressure warning switch electrical Pins 1 and 2. 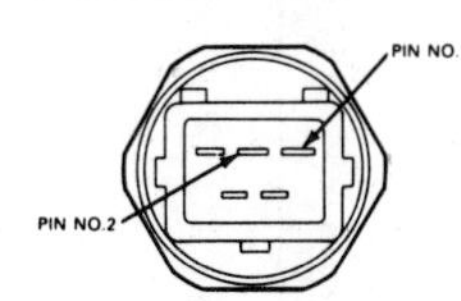 PRESSURE WARNING SWITCH	Continuity (NOT OK) ▶	REPLACE pressure warning switch.
	No continuity (OK) ▶	GO to Step **A22** and pump motor relay.

Fig. 36 Diagnosis, warning lamp ON w/brake warning lamp OFF (Part 22 of 24). 1986–87 except Thunderbird

Anti-Lock Warning Lamp On (With Brake Warning Lamp Off) — Test A

TEST STEP	RESULT ▶	ACTION TO TAKE
A22 CHECK PWS BRAKE LAMP CIRCUIT THRESHOLD		
WARNING: Before disconnecting any hydraulic lines, you must ensure that the pressure system is discharged. • Discharge the brake system as follows: a. Turn ignition switch OFF. b. Pump break pedal at least 20 times until you feel it become hard. • Remove high pressure banjo bolt below brake fluid reservoir on brake booster. • Disconnect quick-connect nipple from Anti-Lock High Pressure Gauge (Tool T85P-20215-A, or equivalent) and install nipple in place of the high pressure banjo bolt. NOTE: Be sure O-rings are installed correctly. • Connect Anti-Lock High Pressure Gauge to gauge nipple. **WARNING: DO NOT disconnect Anti-Lock High Pressure Gauge while system is under pressure.** • Reconnect pressure warning switch (PWS) 5-Pin connector. • Turn ignition switch ON. • When pump motor stops, disconnect pressure warning switch (PWS) 5-Pin plug. • Lower hydraulic accumulator pressure by slowly pumping the brake pedal until you have continuity between pressure warning switch electrical Pins 1 and 2. • Observe the hydraulic pressure gauge when continuity is reached. • If you missed the reading or want to reverify the reading: — Reconnect the 5-Pin plug. — Turn the ignition switch ON until the pump stops. — Disconnect the 5-Pin plug and reverify the readings.	100-110 Bar (1,450-1,595 psi) (OK) ▶ Any other reading (NOT OK) ▶	GO to Step **A23**. REPLACE pressure warning switch and pump motor relay.

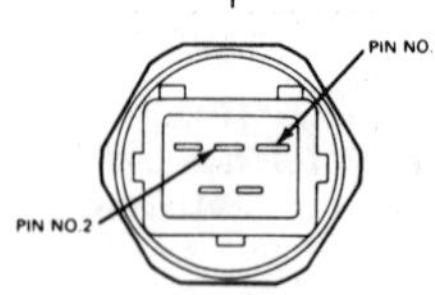

PRESSURE WARNING SWITCH

Fig. 36 Diagnosis, warning lamp ON w/brake warning lamp OFF (Part 23 of 24). 1986–87 except Thunderbird

Anti-Lock Warning Lamp On (With Brake Warning Lamp Off) — Test A

TEST STEP	RESULT ▶	ACTION TO TAKE
A23 CHECK PWS HARNESS GROUND		
• Check for continuity between pressure warning switch 5-Pin plug Pin 1 and body ground.	Continuity (OK) ▶ No continuity (NOT OK) ▶	GO to Step **A24**. REPLACE or SERVICE cable harness (Circuit 430K).
A24 REVERIFY SYSTEM SYMPTOM		
• Reconnect all electrical connections. • Reinstall high pressure banjo bolt. **WARNING: Before disconnecting any hydraulic lines, you must ensure that the brake hydraulic pressure system is discharged.** • Discharge the brake system as follows: a. Turn ignition switch OFF. b. Pump brake pedal at least 20 times until you feel the pedal become hard. • Reverify symptom.	Symptom not present ▶ Symptom still present ▶	FAULT may have been a loose electrical connection. REPLACE electronic control module.

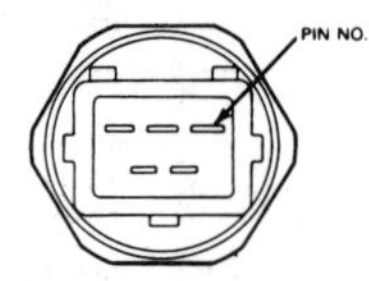

PRESSURE WARNING SWITCH

Fig. 36 Diagnosis, warning lamp ON w/brake warning lamp OFF (Part 24 of 24). 1986–87 except Thunderbird

Anti-Lock Lamp On After Engine Starts (Brake Warning Lamp Off) — Test B

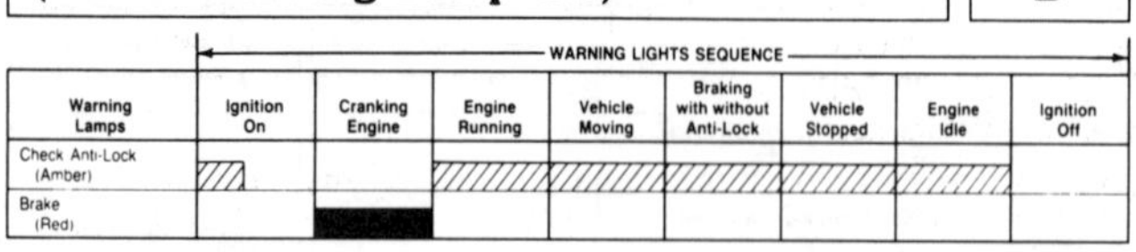

TEST STEP	RESULT ▶	ACTION TO TAKE
B1 CHECK CONTINUITY OF CIRCUIT 524		
• Ignition switch Off. • Disconnect 35-Pin plug from controller. • Connect EEC-IV Breakout box with Anti-Lock Test Adapter to the Anti-Lock 35-Pin plug wiring harness. • Check for continuity between Breakout box Pins 40 and 4.	Continuity (NOT OK) ▶ No continuity (OK) ▶	GO to Step **B1a**. GO to Step **B2**.
B1a CHECK CONTINUITY OF RH REAR SENSOR		
• Disconnect wheel sensor plug (RH rear). • Check for continuity between each sensor plug pin (sensor side) and vehicle ground.	Continuity (NOT OK) ▶ No continuity (OK) ▶	REPLACE sensor (RH rear). REPLACE or REPAIR cable harness. RECONNECT sensor plug.
B2 CHECK CONTINUITY OF CIRCUIT 522		
• Check for continuity between Breakout box Pins 40 and 5.	Continuity (NOT OK) ▶ No continuity (OK) ▶	GO to Step **B2a**. GO to Step **B3**.

RH REAR SENSOR

Fig. 37 Diagnosis, warning lamp ON after engine starts/brake warning lamp OFF (Part 1 of 3). 1986–87 except Thunderbird

Anti-Lock Lamp On After Engine Starts (Brake Warning Lamp Off) — Test B

TEST STEP	RESULT ▶	ACTION TO TAKE
B2a CHECK CONTINUITY OF LH FRONT SENSOR		
• Disconnect wheel sensor plug (LH front). • Check for continuity between each sensor plug pin (sensor side) and vehicle ground.	Continuity (NOT OK) ▶ No continuity (OK) ▶	REPLACE sensor (LH front). REPLACE or SERVICE cable harness. RECONNECT sensor plug.
B3 CHECK CONTINUITY OF CIRCUIT 519		
• Check for continuity between Breakout box Pins 40 and 6.	Continuity (NOT OK) ▶ No continuity (OK) ▶	GO to Step **B3a**. GO to Step **B4**.
B3a CHECK CONTINUITY OF LH REAR SENSOR		
• Disconnect sensor plug (LH rear). • Check for continuity between each sensor plug pin (sensor side) and vehicle ground.	Continuity (NOT OK) ▶ No continuity (OK) ▶	REPLACE sensor (LH rear). REPLACE or SERVICE cable. RECONNECT sensor plug.

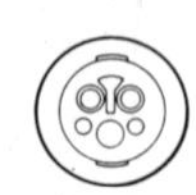
LH FRONT SENSOR

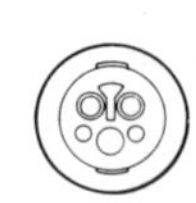
LH REAR SENSOR

Fig. 37 Diagnosis, warning lamp ON after engine starts/brake warning lamp OFF (Part 2 of 3). 1986–87 except Thunderbird

Anti-Lock Lamp On After Engine Starts (Brake Warning Lamp Off) — Test B

TEST STEP	RESULT ▶	ACTION TO TAKE
B4 CHECK CONTINUITY OF CIRCUIT 516		
• Check for continuity between Breakout box Pins 40 and 7.	Continuity (NOT OK) ▶	GO to Step **B4a**.
	No continuity (OK) ▶	Test complete. If Anti-Lock lamp pattern remains, REPEAT Test B.
B4a CHECK CONTINUITY OF RH FRONT SENSOR		
• Disconnect wheel sensor plug (RH front). • Check for continuity between each sensor plug pin (sensor side) and vehicle ground.  RH FRONT SENSOR	Continuity (NOT OK) ▶	REPLACE RH front sensor.
	No continuity (OK) ▶	REPLACE or SERVICE cable harness. RECONNECT sensor plug.

Fig. 37 Diagnosis, warning lamp ON after engine starts/brake warning lamp OFF (Part 3 of 3). 1986–87 except Thunderbird

Anti-Lock Warning Lamp On After Vehicle Starts To Move Or False Cycling Of Anti-Lock System — Test C

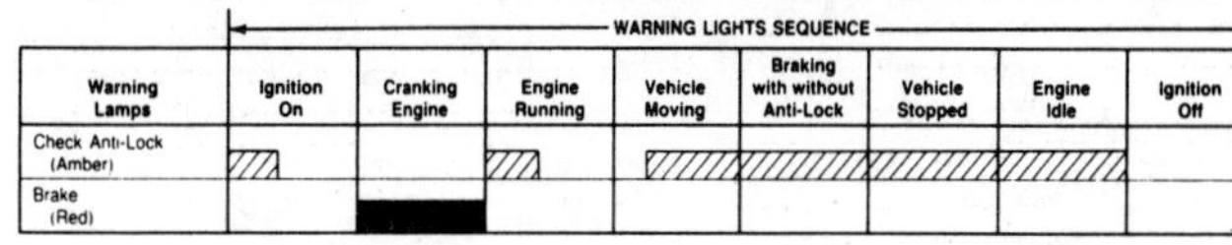

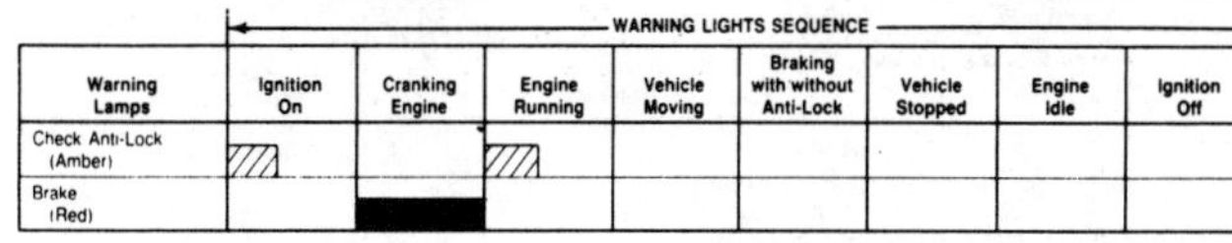

TEST STEP	RESULT ▶	ACTION TO TAKE
C1 MEASURE RH REAR SENSOR CIRCUIT RESISTANCE		
• Turn ignition switch OFF. • Disconnect 35-Pin plug from electronic controller. • Connect EEC-IV Breakout box tool T83-50-EEC-IV and anti-lock test adapter tool T85P-50-ASA or equivalents to anti-lock 35-Pin connector. • Set multi-meter to read resistance. • Measure resistance between Breakout box Pins 4 and 22.	800 to 1400 ohms (0.8 to 1.4K ohms) (OK) ▶	GO to Step **C2**.
	Any other reading (NOT OK) ▶	Go to Step **C1a**.
C1a MEASURE RH REAR SENSOR RESISTANCE		
• Disconnect sensor plug (RH rear). • Set multi-meter to read resistance. • Measure resistance of sensor at sensor plug. 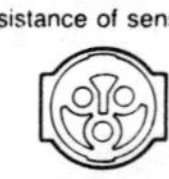 RH REAR SENSOR	800 to 1400 ohms (0.8 to 1.4K ohms) (OK) ▶	SERVICE or REPLACE cable harness (Circuit 523 or 524).
	Any other reading (NOT OK) ▶	REPLACE RH rear wheel sensor.

Fig. 38 Diagnosis, warning lamp ON after vehicle starts to move or false cycling of system (Part 1 of 5). 1986–87 except Thunderbird

Anti-Lock Warning Lamp On After Vehicle Starts To Move Or False Cycling Of Anti-Lock System — Test C

TEST STEP	RESULT ▶	ACTION TO TAKE
C2 MEASURE LH FRONT SENSOR CIRCUIT RESISTANCE		
• Measure resistance between Breakout box Pins 5 and 23.	800 to 1400 ohms (0.8 to 1.4K ohms) (OK) ▶	GO to Step **C3**.
	Any other reading (NOT OK) ▶	GO to Step **C2a**.
C2a MEASURE LH FRONT SENSOR RESISTANCE		
• Disconnect wheel sensor plug (LH rear). • Measure resistance of sensor at sensor plug. 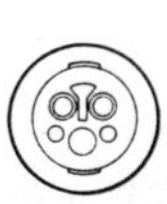 LH FRONT SENSOR	800 to 1400 ohms (0.8 to 1.4K ohms) (OK) ▶	SERVICE or REPLACE cable harness (Circuit 521 or 522).
	Any other reading (NOT OK) ▶	REPLACE LH front wheel sensor.
C3 MEASURE LH REAR SENSOR CIRCUIT RESISTANCE		
• Measure resistance between Breakout box Pins 6 and 24.	800 to 1400 ohms (0.8 to 1.4K ohms) (OK) ▶	GO to Step **C4**.
	Any other reading (NOT OK) ▶	GO to Step **C3a**.

Fig. 38 Diagnosis, warning lamp ON after vehicle starts to move or false cycling of system (Part 2 of 5). 1986–87 except Thunderbird

Anti-Lock Warning Lamp On After Vehicle Starts To Move Or False Cycling Of Anti-Lock System — Test C

TEST STEP	RESULT ▶	ACTION TO TAKE
C3a MEASURE LH REAR SENSOR RESISTANCE		
• Disconnect sensor plug (LH rear). • Measure resistance of sensor at left rear sensor plug. 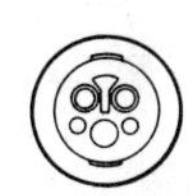 LH REAR SENSOR	800 to 1400 ohms (0.8 to 1.4K ohms) (OK) ▶	SERVICE or REPLACE cable harness (Circuit 518 or 519).
	Any other reading (NOT OK) ▶	REPLACE LH rear wheel sensor.
C4 MEASURE RH FRONT SENSOR CIRCUIT RESISTANCE		
• Measure resistance between Breakout box Pins 7 and 25.	800 to 1400 ohms (0.8 to 1.4K ohms) (OK) ▶	GO to Step **C5**.
	Any other reading (NOT OK) ▶	GO to Step **C4a**.
C4a MEASURE RH FRONT SENSOR RESISTANCE		
• Disconnect sensor plug (RH front). • Measure resistance of sensor at sensor plug. 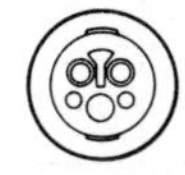 RH FRONT SENSOR	800 to 1400 ohms (0.8 to 1.4K ohms) (OK) ▶	SERVICE or REPLACE cable harness (Circuit 514 or 516).
	Any other reading (NOT OK) ▶	REPLACE RH front wheel sensor.

Fig. 38 Diagnosis, warning lamp ON after vehicle starts to move or false cycling of system (Part 3 of 5). 1986–87 except Thunderbird

Anti-Lock Warning Lamp On After Vehicle Starts To Move Or False Cycling Of Anti-Lock System	Test C

TEST STEP	RESULT ▶	ACTION TO TAKE
C5 CHECK RH REAR SENSOR		
• Turn ignition switch OFF. • Turn air suspension switch in luggage compartment OFF. • Place vehicle on hoist and raise wheels clear of ground. Refer to Pre-Delivery Manual Section 50-04. • Set multi-meter on voltage range (2V-AC). • Measure voltage between Breakout box Pins 4 and 22 while spinning RH rear wheel at approximately 1 revolution per second.	Between 0.05 and 0.70 Vac (OK) ▶ Less than 0.05 or more than 0.70 Vac (NOT OK) ▶	GO to Step **C6**. CHECK sensor mounting, air gap, or toothed wheel mounting. CORRECT as required.
C6 CHECK LH FRONT SENSOR		
• Measure voltage between Breakout box Pins 5 and 23 while spinning LH front wheel at approximately 1 revolution per second.	Between 0.05 and 0.70 Vac ▶ Less than 0.05 or more than 0.70 Vac ▶	GO to Step **C7**. CHECK sensor mounting, air gap, or toothed wheel mounting. CORRECT as required.
C7 CHECK LH REAR SENSOR		
• Measure voltage between Breakout box Pins 6 and 24 while spinning LH rear wheel at approximately 1 revolution per second.	Between 0.05 and 0.70 Vac ▶ Less than 0.05 or more than 0.07 Vac ▶	GO to Step **C8**. CHECK wheel sensor mounting, air gap, or toothed wheel mounting. CORRECT as required.

Fig. 38 Diagnosis, warning lamp ON after vehicle starts to move or false cycling of system (Part 4 of 5). 1986–87 except Thunderbird

Anti-Lock Warning Lamp on After Vehicle Starts to Move or False Cycling of Anti-Lock System	Test C

TEST STEP	RESULT ▶	ACTION TO TAKE
C8 CHECK RH FRONT SENSOR		
• Measure voltage between Breakout box Pins 7 and 25 while spinning RH front wheel at approximately 1 revolution per second.	Between 0.05 and 0.70 Vac ▶ Less than 0.05 or more than 0.70 Vac ▶	GO to Step **C9**. CHECK wheel sensor mounting, air gap, or toothed wheel mounting. CORRECT as required.
C9 CHECK FRONT WHEEL BEARINGS		
• Check front wheel bearing end play. • Inspect each toothed sensor ring visually for damaged teeth. NOTE: Turn air suspension switch ON when vehicle is off hoist.	Loose or damaged parts ▶ Not loose or damaged ▶	ADJUST bearings or REPLACE faulty parts. REVERIFY symptom.

Fig. 38 Diagnosis, warning lamp ON after vehicle starts to move or false cycling of system (Part 5 of 5). 1986–87 except Thunderbird

Anti-Lock Warning Lamp And Brake Warning Lamp On And/Or Pump Motor Runs More Than 60 Seconds	Test D

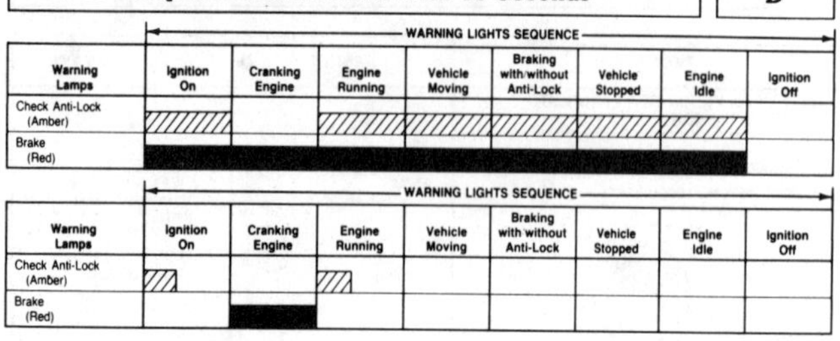

TEST STEP	RESULT ▶	ACTION TO TAKE
D1 CHECK PUMP MOTOR OPERATION		
• Turn ignition OFF. • Pump brake pedal at least 20 times until it becomes hard. • Turn ignition switch ON. • Pump motor should run. **NOTE: If pump motor is allowed to run continuously for approximately 20 minutes, a thermal safety switch (inside motor) will shut off motor. A 2-to-10 minute cool-down period is typical before normal operation can resume.**	(OK) ▶ (NOT OK) ▶	GO to **D2**. GO to **D1a**.

Fig. 39 Diagnosis, warning lamp & brake warning lamp ON and/or pump motor runs more than 60 seconds (Part 1 of 10). 1986–87 except Thunderbird

Anti-Lock Warning Lamp And Brake Warning Lamp On And/Or Pump Motor Runs More Than 60 Seconds	Test D

TEST STEP	RESULT ▶	ACTION TO TAKE
D1a PUMP MOTOR UNIT		
• Disconnect 4-Pin plug on pump motor unit. • Turn ignition switch ON. • Set multi-meter on 20 Volt DC. • Connect meter to the 4-Pin plug on the harness side. (Use one negative and one positive pin). • Observe voltmeter. 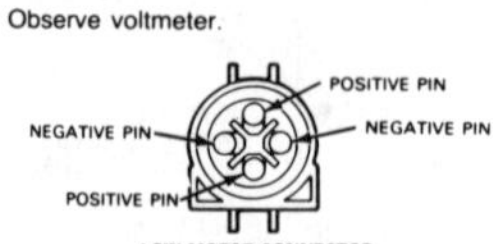	More than 10V DC (OK) ▶ Less than 10V DC (NOT OK) ▶	GO to **D2**. GO to **D1b**.
D1b CHECK CONTINUITY CIRCUITS 538a AND 538b		
• Disconnect pump motor relay. • Check for continuity between motor relay socket Pin 3 (harness side) and each positive pin of the 4-Pin motor connector (harness side). 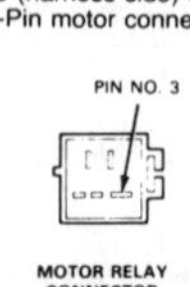 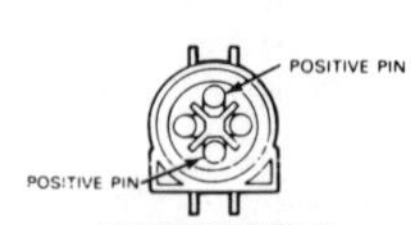	(OK) ▶ (NOT OK) ▶	GO to **D1c**. SERVICE Circuits 537, 538a and 538b.

Fig. 39 Diagnosis, warning lamp & brake warning lamp ON and/or pump motor runs more than 60 seconds (Part 2 of 10). 1986–87 except Thunderbird

Anti-Lock Warning Lamp And Brake Warning Lamp On And/Or Pump Motor Runs More Than 60 Seconds	Test D

TEST STEP	RESULT ▶	ACTION TO TAKE
D1c CHECK CONTINUITY CIRCUITS 430D AND 430F		
• Check for continuity between ground and each negative pin of 4-Pin motor connector (harness side). 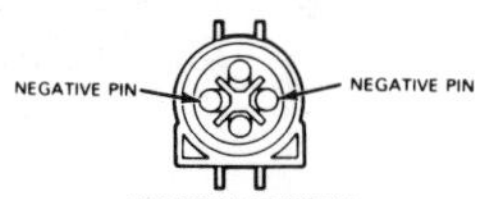 4-PIN MOTOR CONNECTOR	OK ▶ Not OK ▶	GO to **D1d**. SERVICE or REPLACE Circuit 4301.
D1d CHECK CONTINUITY CIRCUIT 537		
• Turn ignition switch OFF. • Remove positive battery cable from battery. • Disconnect pump motor relay. 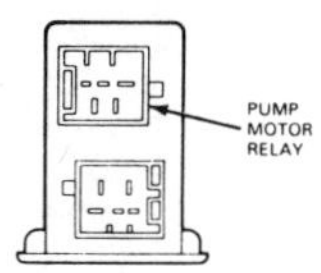 • Check continuity between battery positive cable and motor relay socket Pin 4 (harness side). 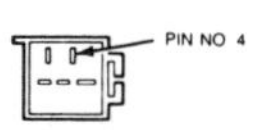 PUMP MOTOR RELAY CONNECTOR	Continuity OK ▶ No continuity Not OK ▶	RECONNECT battery positive cable and GO to Step **D1e**. SERVICE or REPLACE fuse link and/or Circuit 537.

Fig. 39 Diagnosis, warning lamp & brake warning lamp ON and/or pump motor runs more than 60 seconds (Part 3 of 10). 1986–87 except Thunderbird

Anti-Lock Warning Lamp And Brake Warning Lamp On And/Or Pump Motor Runs More Than 60 Seconds	Test D

TEST STEP	RESULT ▶	ACTION TO TAKE
D1e CHECK CONTINUITY CIRCUIT 298		
• Connect multi-meter to motor relay socket Pin 2 (harness side) and ignition switch lock pin (On). • Check continuity. 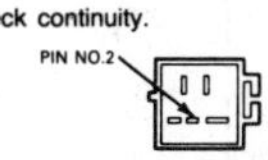 MOTOR RELAY CONNECTOR	Continuity OK ▶ No continuity Not OK ▶	GO to Step **D1f**. REPLACE or SERVICE Circuit 298A and/or fuse link.
D1f CHECK CONTINUITY CIRCUIT 430K		
• Disconnect PWS 5-pin plug. • Check for continuity between pressure warning switch 5-Pin plug Pin 1 (harness side) and ground. 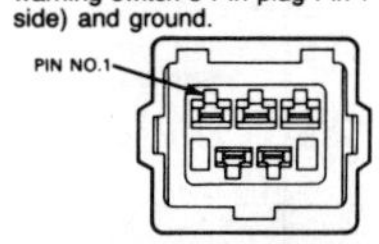 PRESSURE WARNING SWITCH CONNECTOR	Continuity OK ▶ No continuity Not OK ▶	GO to Step **D1g**. REPLACE or SERVICE Circuit 430K.
D1g CHECK CONTINUITY CIRCUIT 539		
• Connect meter to 5-Pin plug Pin 4 (harness side) and motor relay socket Pin 1 (harness side). • Check for continuity. 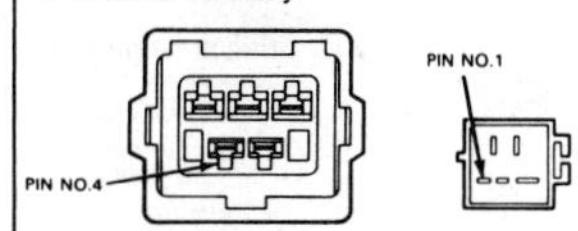 PRESSURE WARNING SWITCH CONNECTOR MOTOR RELAY CONNECTOR	Continuity OK ▶ No continuity Not OK ▶	GO to Step **D1h**. REPLACE or SERVICE Circuit 539.

Fig. 39 Diagnosis, warning lamp & brake warning lamp ON and/or pump motor runs more than 60 seconds (Part 4 of 10). 1986–87 except Thunderbird

Anti-Lock Warning Lamp And Brake Warning Lamp On And/Or Pump Motor Runs More Than 60 Seconds	Test D

TEST STEP	RESULT ▶	ACTION TO TAKE
D1h PRESSURE WARNING SWITCH		
• Check for continuity between pressure warning switch Pins 1 and 4. **NOTE: System must be depressurized.** 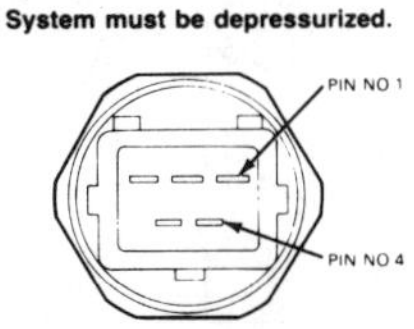 PRESSURE WARNING SWITCH	Continuity OK ▶ No continuity Not OK ▶	GO to Step **D1j**. REPLACE pressure warning switch and pump motor relay.
D1j CHECK PUMP MOTOR RELAY		
• Set multi-meter on 200 ohm scale. • Connect meter to motor relay Pins 1 and 2. 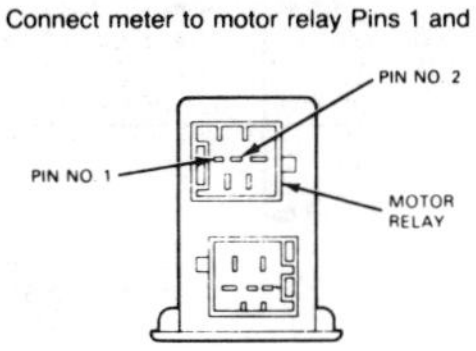	45 to 105 Ohms OK ▶ Other Not OK ▶	GO to Step **D1k**. REPLACE motor relay and pressure warning switch.

Fig. 39 Diagnosis, warning lamp & brake warning lamp ON and/or pump motor runs more than 60 seconds (Part 5 of 10). 1986–87 except Thunderbird

Anti-Lock Warning Lamp And Brake Warning Lamp On And/Or Pump Motor Runs More Than 60 Seconds	Test D

TEST STEP	RESULT ▶	ACTION TO TAKE
D1k CHECK PUMP MOTOR RELAY (CONTINUED)		
• Connect meter and check continuity between motor relay Pins 3 and 4. 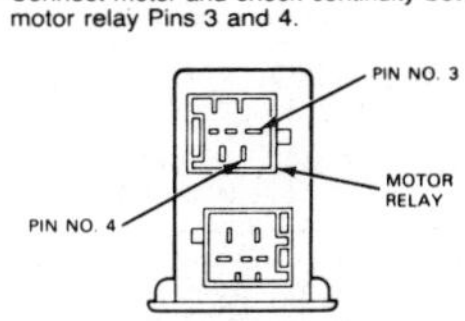	No continuity OK ▶ Continuity Not OK ▶	GO to Step **D1m**. REPLACE motor relay and pressure warning switch.
D1m CHECK PUMP MOTOR RELAY (CONTINUED)		
• Connect battery to motor relay terminals 1 and 2. • Check continuity between relay Pins 3 and 4 with multi-meter. 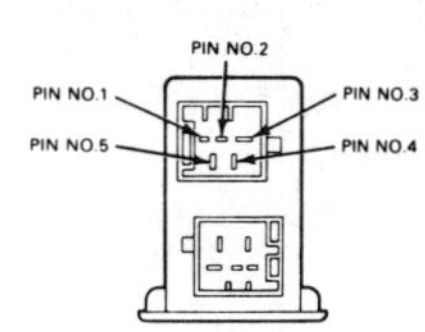	Continuity OK ▶ No continuity Not OK ▶	GO to Step **D1n**. REPLACE motor relay and pressure warning switch.

Fig. 39 Diagnosis, warning lamp & brake warning lamp ON and/or pump motor runs more than 60 seconds (Part 6 of 10). 1986–87 except Thunderbird

Anti-Lock Warning Lamp And Brake Warning Lamp On And/Or Pump Motor Runs More Than 60 Seconds	Test D

TEST STEP	RESULT ▶	ACTION TO TAKE
D1n CHECK PUMP MOTOR		
• Apply 12 volts to 4-Pin plug of motor (use one negative and one positive pin). • Pump motor should run.	(OK) ▶	End of test. CONNECT all electrical plugs and relay. REVERIFY symptom.
	(~~OK~~) ▶	REPLACE pump/motor assembly and pump motor relay.

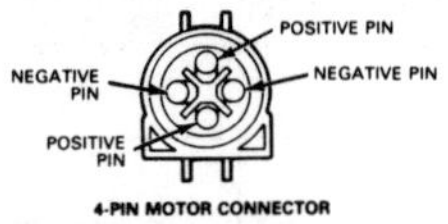

TEST STEP	RESULT ▶	ACTION TO TAKE
D2 CHECK PUMP MOTOR UNIT		
• Turn ignition switch OFF. • Pump brake pedal 20 times to discharge system. • Connect ammeter between battery positive cable and battery positive terminal. • Turn Off any electrical components. • Connect 4-Pin motor plug. • Turn ignition switch ON. • Measure pump motor current.	Current more than 25 amps (~~OK~~) ▶	REPLACE pump motor unit.
	Current less than 25 amps (OK) ▶	GO to Step **D2a**.
D2a CHECK PUMP MOTOR		
• Turn ignition switch OFF. • Pump brake pedal at least 20 times, until brake pedal becomes hard. • Turn ignition switch ON. • Measure time pump takes to shut OFF.	Under 60 seconds (OK) ▶	GO to Step **D2b**.
	Over 60 seconds (or motor never turns on) (~~OK~~) ▶	CHECK for corroded connections at: • Motor 4-Pin plug. • Body ground (Circuit 430L). • Motor relay Pin 4 socket (Circuit 537). • Battery to Pin 4. GO to Step **D2b**.

Fig. 39 Diagnosis, warning lamp & brake warning lamp ON and/or pump motor runs more than 60 seconds (Part 7 of 10). 1986–87 except Thunderbird

Anti-Lock Warning Lamp And Brake Warning Lamp On And/Or Pump Motor Runs More Than 60 Seconds	Test D

TEST STEP	RESULT ▶	ACTION TO TAKE
D2b CHECK LOW PRESSURE FLOW		
• Turn ignition switch OFF. • Disconnect low pressure hose from pump and allow fluid to flow into a suitable container. **NOTE: Discard fluid after test.**	Free fluid flow (OK) ▶	CONNECT low pressure hose. FILL reservoir to MAX. GO to **D2c**.
	Restricted flow (~~OK~~) ▶	SERVICE or REPLACE reservoir and/or low pressure hose as required.
D2c CHECK VOLTAGE TO PUMP MOTOR		
• Turn ignition switch OFF. • Pump brake pedal to discharge pressure system. • Set meter to 20 Volts DC range. • Connect voltmeter in parallel at 4-Pin motor plug. • Turn ignition switch ON.	With pump running: Voltage over 8 Volts DC (OK) ▶	GO to Step **D3**.
	Voltage under 8 Volts DC (~~OK~~) ▶	CHECK Circuits 430L, 538A, 538B, Pin 4 to battery and relay Pins 3 and 4 for voltage drop. SERVICE or REPLACE as necessary.

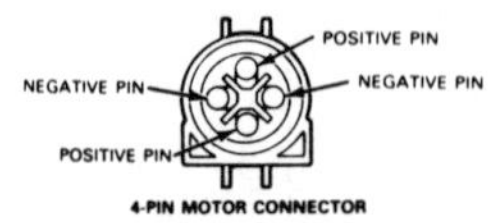

Fig. 39 Diagnosis, warning lamp & brake warning lamp ON and/or pump motor runs more than 60 seconds (Part 8 of 10). 1986–87 except Thunderbird

Anti-Lock Warning Lamp And Brake Warning Lamp On And/Or Pump Motor Runs More Than 60 Seconds	Test D

TEST STEP	RESULT ▶	ACTION TO TAKE
D3 ACCUMULATOR: PRE-CHARGE		
• Vehicle must be cooled to room temperature. **WARNING: Before disconnecting any hydraulic lines, you must ensure that the pressure system is discharged.** • To discharge hydraulic accumulator pressure, turn ignition OFF, pump brake pedal at least 20 times until you feel it become hard. • Remove high pressure banjo bolt below brake fluid reservoir on brake booster. • Install Anti-Lock Pressure Gauge Tool T85P-20215A or equivalent. • Turn ignition switch ON and read accumulator precharge pressure. (Gauge needle will spring to this point). **Note: Gauge needle reading should spring to 40-90 bar (600-1325 psi) and climb to 16,203-19,306 kPa bar (2350-2800 psi).** • If reading was missed, discharge accumulator as described and repeat ignition ON sequence.	4137-9135 kPa (600-1325 psi) (OK) ▶	GO to Step **D4**.
	Under 4137 kPa (600 psi) (~~OK~~) ▶	REPLACE accumulator.
	Over 9135 kPa (1325 psi) (~~OK~~) ▶	REPLACE accumulator.
D4 CHECK HYDRAULIC UNIT		
• Turn ignition switch ON; wait until pump motor stops. Wait 3 more minutes to stabilize gauge pressure. • Read pressure gauge. • Wait 5 minutes and read pressure gauge again to determine the pressure loss over those 5 minutes.	Pressure loss less than 10 bar (140 psi) on gauge (OK) ▶	GO to Step **D5**.
	More than 10 bar (140 psi) (~~OK~~) ▶	CHECK for external leakage at hydraulic unit and SERVICE. If no external leakage is found, GO to Step **D4a**.

Fig. 39 Diagnosis, warning lamp & brake warning lamp ON and/or pump motor runs more than 60 seconds (Part 9 of 10). 1986–87 except Thunderbird

Anti-Lock Warning Lamp And Brake Warning Lamp On And/Or Pump Motor Runs More Than 60 Seconds	Test D

TEST STEP	RESULT ▶	ACTION TO TAKE
D4a CHECK FOR HYDRAULIC LEAKS		
• Check for brake fluid leaks in the following areas: a. Brake lines and calipers. b. High and low pressure hoses on actuation assembly. c. Reservoir seals and seams. d. Accumulator. **NOTE: A small amount of leakage from pressure warning switch is allowable. If any leakage is found below switch, wipe off excess fluid and check for excessive leakage.**	No leakage found (OK) ▶	GO to **D5**.
	Leakage found (~~OK~~) ▶	SERVICE or REPLACE components. CHECK system operation.
D5 CHECK PUMP PRESSURE		
• With the pressure gauge still attached and ignition switch still ON, pump brake pedal to decrease pressure until pump motor restarts.	13100-15169 kPa (1900-2200 psi) when pump starts (OK) ▶	GO to Step **D6**.
	Less than or more than 13100-15169 kPa (1900-2200 psi) (~~OK~~) ▶	REPLACE pressure warning switch and pump motor relay.
D6 CHECK PRESSURE WARNING SWITCH		
• With the ignition switch still ON and the pressure gauge connected, observe the pressure when the pump motor stops running.	16203 kPa (2350-2800 psi) when pump motor stops (OK) ▶	If pump motor takes longer than 60 seconds to reach 16203 kPa (2350-2800 psi), REPLACE pump/motor assembly. REVERIFY symptom.
	Less than 16203 kPa (2350 psi) or over 19306 kPa (2800 psi) (~~OK~~) ▶	REPLACE pressure warning switch and pump motor relay.

Fig. 39 Diagnosis, warning lamp & brake warning lamp ON and/or pump motor runs more than 60 seconds (Part 10 of 10). 1986–87 except Thunderbird

Anti-Lock Warning Lamp Intermittently On — Test E

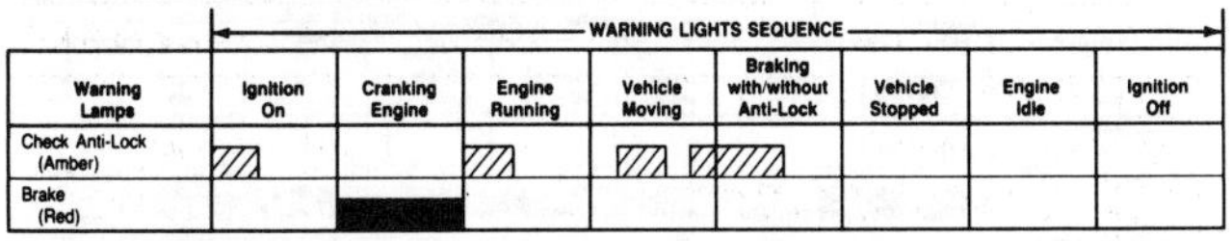

TEST STEP	RESULT ▶	ACTION TO TAKE
E1 CHECK FLI AND PWS CIRCUIT		
• Disconnect 35-Pin plug from electronic controller. • Connect EEC-IV Breakout box, T83L-50-EEC-IV with Anti-Lock Test Adapter T85P-50-ASA or equivalent to the Anti-Lock 35-Pin plug wiring harness. • Turn ignition switch ON. • Set multi-meter to read resistance. • Measure the resistance between Breakout box Pins 9 and 10.	Less than 5 ohms (OK) ▶ Greater than 5 ohms (NOT OK) ▶	GO to Step **E2**. GO to Step **E1a**.
E1a CHECK FLI ANTI-LOCK WARNING CIRCUIT		
• Disconnect 5-Pin plug on reservoir fluid level indicator (FLI). • Measure resistance between fluid level indicator electrical Pins 1 and 2 (with brake fluid at maximum fluid level). 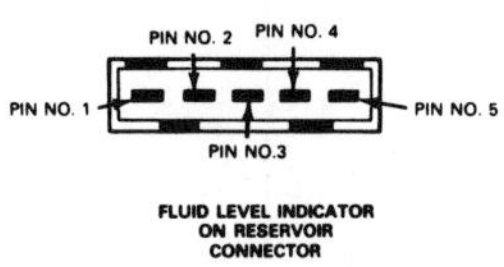	Less than 2 ohms (OK) ▶ Greater than 2 ohms (NOT OK) ▶	GO to Step **E1b**. REPLACE fluid level indicator.

Fig. 40 Diagnosis, warning lamp intermittently ON (Part 1 of 5). 1986–87 except Thunderbird

Anti-Lock Warning Lamp Intermittently On — Test E

TEST STEP	RESULT ▶	ACTION TO TAKE
E1b CHECK PWS ANTI-LOCK WARNING CIRCUIT (NO SYSTEM PRESSURE)		
• Turn ignition switch OFF. • Pump brake pedal to discharge pressure system. • Disconnect 5-Pin plug at pressure warning switch (PWS). • Check for continuity between pressure warning switch Pins 3 and 5. 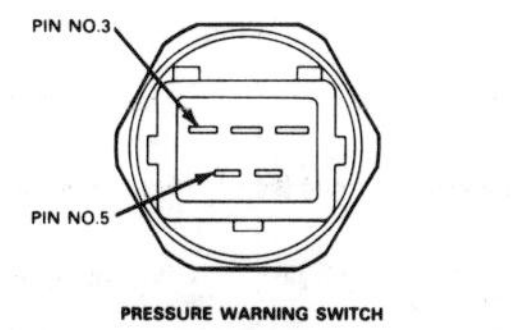	No Continuity (OK) ▶ Continuity (NOT OK) ▶	GO to Step **E1c**. REPLACE pressure warning switch and pump motor relay.
E1c CHECK ELECTRONIC MODULE TO FLI WIRE		
• Check for continuity between Breakout box Pin 9 and FLI 5-Pin plug Pin 1 (harness side). 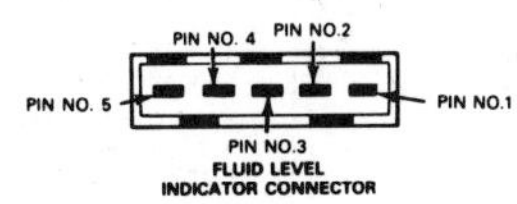	Continuity (OK) ▶ No continuity (NOT OK) ▶	GO to Step **E1d**. REPLACE or SERVICE cable harness (Circuit 512).

Fig. 40 Diagnosis, warning lamp intermittently ON (Part 2 of 5). 1986–87 except Thunderbird

Anti-Lock Warning Lamp Intermittently On — Test E

TEST STEP	RESULT ▶	ACTION TO TAKE
E1d CHECK FLI TO PWS WIRE		
• Check for continuity between 5-Pin fluid level plug Pin 2 (harness side) 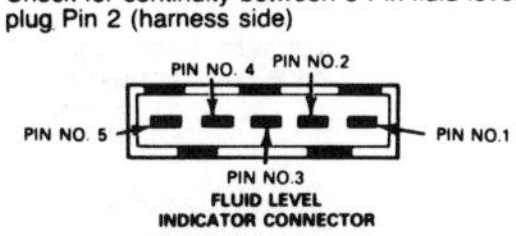 and 5-Pin pressure warning switch plug Pin 3 (harness side). 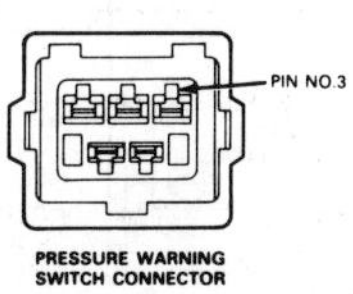	Continuity (OK) ▶ No continuity (NOT OK) ▶	GO to Step **E1e**. SERVICE or REPLACE cable harness (Circuit 549).
E1e CHECK PWS TO ELECTRONIC CONTROLLER WIRE		
• Check for continuity between 5-Pin pressure warning switch plug Pin 5 (harness side) 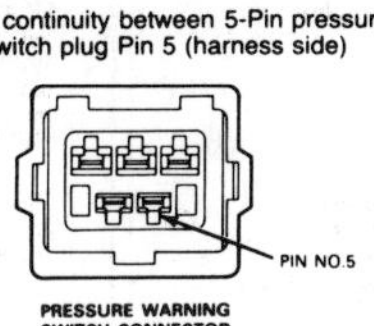 and Breakout box Pin 10.	Continuity (OK) ▶ No continuity (NOT OK) ▶	TURN Ignition OFF. CONNECT all electrical connections. REVERIFY symptom. SERVICE or REPLACE cable harness (Circuit 535).

Fig. 40 Diagnosis, warning lamp intermittently ON (Part 3 of 5). 1986–87 except Thunderbird

Anti-Lock Warning Lamp Intermittently On — Test E

TEST STEP	RESULT ▶	ACTION TO TAKE
E2 CHECK ISOLATION TEST FLI AND PWS		
• Check for continuity between Breakout box Pin 9 and body ground.	Continuity (NOT OK) ▶ No continuity (OK) ▶	GO to Step **E2a**. REVERIFY symptom.
E2a CHECK CONTINUITY OF FLI SWITCH		
• Disconnect FLI plug and check for continuity between socket Pin 1 and Pins 3, 4, and 5 and Pin 2 and Pins 3, 4, and 5. 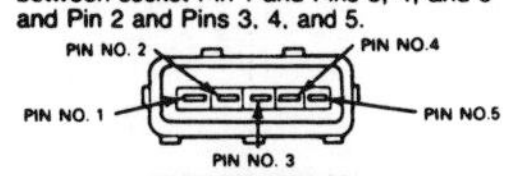	Continuity (NOT OK) ▶ No continuity (OK) ▶	REPLACE FLI. GO to Step **E2b**.
E2b CHECK CONTINUITY CIRCUIT 512		
• Check for continuity between FLI plug Pin 1 (harness side) and body ground. 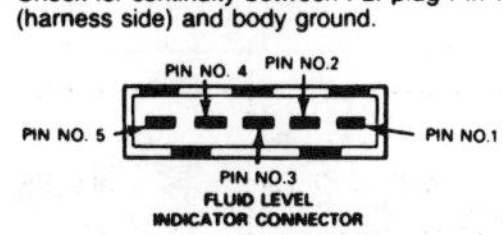	Continuity (NOT OK) ▶ No continuity (OK) ▶	SERVICE or REPLACE (Circuit 512). GO to Step **E2c**.
E2c CHECK CONTINUITY CIRCUIT 549		
• Disconnect 5-Pin plug from PWS. Check for continuity between FLI plug Pin 2 (harness side) and body ground. 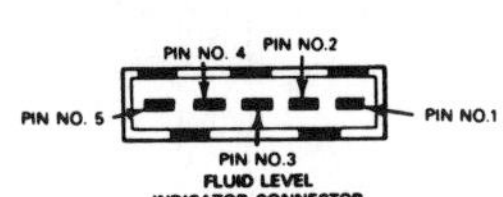	Continuity (NOT OK) ▶ No continuity (OK) ▶	SERVICE or REPLACE (Circuit 549). GO to Step **E2d**.

Fig. 40 Diagnosis, warning lamp intermittently ON (Part 4 of 5). 1986–87 except Thunderbird

Anti-Lock Warning Lamp Intermittently On	Test E

TEST STEP	RESULT ▶	ACTION TO TAKE
E2d CHECK CONTINUITY OF PWS		
• Check for continuity from PWS 5-Pin socket Pin 3 to body ground, and from Pin 5 to body ground.	Continuity ~~OK~~ ▶	REPLACE pressure warning switch and pump motor relay.
	No continuity OK ▶	GO to Step **E2e**.
E2e CHECK CONTINUITY CIRCUIT 535		
• Check for continuity between 5-Pin PWS plug (harness side) Pin 5 and body ground.	Continuity ~~OK~~ ▶	SERVICE or REPLACE Circuit 535.
	No continuity OK ▶	CONNECT all plugs and REVERIFY symptom.

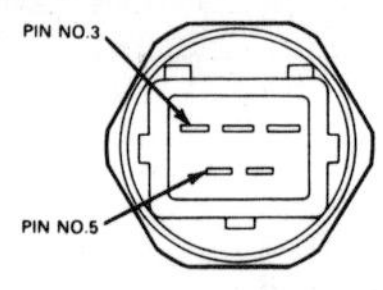

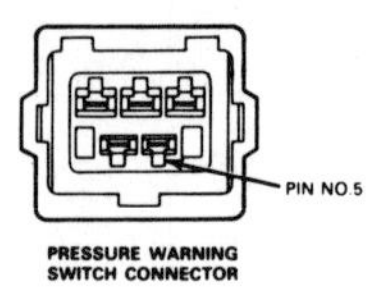

Fig. 40 Diagnosis, warning lamp intermittently ON (Part 5 of 5). 1986–87 except Thunderbird

Brake Warning Lamp On (With Anti-Lock Lamp Off, Parking Brake Released And Brake Lining Wear Checked)	Test F

WARNING LIGHTS SEQUENCE

Warning Lamps	Ignition On	Cranking Engine	Engine Running	Vehicle Moving	Braking with/without Anti-Lock	Vehicle Stopped	Engine Idle	Ignition Off
Check Anti-Lock (Amber)								
Brake (Red)								

TEST STEP	RESULT ▶	ACTION TO TAKE
F1 CHECK BRAKE FLUID LEVEL		
• Turn ignition switch ON. • Pump brake pedal until pump motor starts. • When pump motor stops check brake fluid level.	Low ~~OK~~ ▶	CHECK system for external leaks. SERVICE as required.
	Normal OK ▶	GO to Step **F2**.
F2 CHECK CONTINUITY FLI SWITCH		
• Disconnect 5-Pin plug on fluid reservoir cap. • Set multi-meter on 200 ohm range. Connect to reservoir warning cap Pins 3 and 4.	Above 10 ohms OK ▶	GO to Step **F3**.
	Below 10 ohms ~~OK~~ ▶	REPLACE 5-Pin fluid level indicator cap.

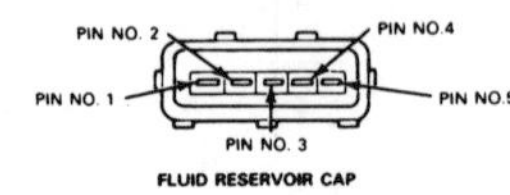

Fig. 41 Diagnosis, brake warning lamp ON w/anti-lock lamp off, parking brake released & brake lining wear checked (Part 1 of 2). 1986–87 except Thunderbird

Brake Warning Lamp On (With Anti-Lock Lamp Off, Parking Brake Released And Brake Lining Wear Checked)	Test F

TEST STEP	RESULT ▶	ACTION TO TAKE
F3 CHECK CONTINUITY PWS SWITCH (CONTINUED)		
• Turn ignition switch ON, wait until motor stops running. • Disconnect Pressure Warning Switch 5-Pin plug and connect multi-meter to Pins 1 and 2 (switch side). • Check for continuity.	Continuity ~~OK~~ ▶	REPLACE pressure warning switch and pump motor relay.
	No continuity OK ▶	GO to Step **F4**.
F4 CHECK WIRING CONTINUITY		
• Check cable harness 430H, 512, 547B, 547A, 549, 977 Circuits. (Refer to vehicle Wiring Diagram.)	OK ▶	CHECK parking brake electrical circuit including parking brake switch. REVERIFY symptom.
	~~OK~~ ▶	SERVICE or REPLACE as required.

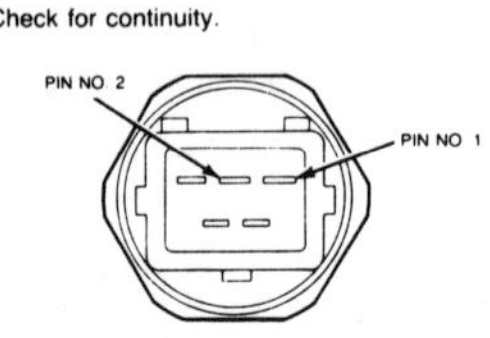

Fig. 41 Diagnosis, brake warning lamp ON w/anti-lock lamp off, parking brake released & brake lining wear checked (Part 2 of 2). 1986–87 except Thunderbird

No Anti-Lock Warning Lamp On When Ignition Switch Turned On	Test G

WARNING LIGHTS SEQUENCE

Warning Lamps	Ignition On	Cranking Engine	Engine Running	Vehicle Moving	Braking with/without Anti-Lock	Vehicle Stopped	Engine Idle	Ignition Off
Check Anti-Lock (Amber)								
Brake (Red)								

TEST STEP	RESULT ▶	ACTION TO TAKE
G1 CHECK FUSE AND FUSE LINKS		
• Check fuse Three in-line fuse links with ignition turned ON.	Fuse Links OK ▶	GO to Step **G2**.
	Fuse Links ~~OK~~ ▶	SERVICE or REPLACE as required.
G2 CHECK WARNING LAMP BULB		
• Check warning lamp bulb.	OK ▶	GO to Step **G3**.
	~~OK~~ ▶	REPLACE.
G3 CHECK WARNING LAMP OPERATION		
• Turn ignition switch ON. • Disconnect 35 Pin connector from control module.	If Anti-Lock lamp goes ON ▶	SERVICE or REPLACE Circuit 606.
	If Anti-Lock lamp is not ON ▶	SERVICE or REPLACE connector to 14401 wire harness.

Fig. 42 Diagnosis, no anti-lock warning lamp ON when ignition switch turned ON. 1986–87 except Thunderbird

Spongy Brake Pedal With/Without Anti-Lock Function (No Warning Lamp) — Test H

WARNING LIGHTS SEQUENCE

Warning Lamps	Ignition On	Cranking Engine	Engine Running	Vehicle Moving	Braking with/without Anti-Lock	Vehicle Stopped	Engine Idle	Ignition Off
Check Anti-Lock (Amber)								
Brake (Red)								

TEST STEP	RESULT ▶	ACTION TO TAKE
H1 CHECK COMPONENT MOUNTING		
• Check for proper brake pedal and hydraulic unit attachment. • Bleed brakes as outlined.	Pedal still spongy (~~OK~~) ▶	GO to Step **H2**.
	Pedal feels normal (OK) ▶	Condition corrected.
H2 BLEED BRAKE SYSTEM		
• Turn off air suspension switch in luggage compartment. • Rebleed brake system. • Turn on air suspension switch when vehicle is off hoist.	Pedal still spongy (~~OK~~) ▶	REPLACE actuation assembly.
	Pedal feels normal (OK) ▶	Condition corrected.

Fig. 43 Diagnosis, spongy brake pedal with/without anti-lock function, no warning lamp. 1986–87 except Thunderbird

Poor Vehicle Tracking During Anti-Lock Function (Warning Lamp Off) — Test J

WARNING LIGHTS SEQUENCE

Warning Lamps	Ignition On	Cranking Engine	Engine Running	Vehicle Moving	Braking with/without Anti-Lock	Vehicle Stopped	Engine Idle	Ignition Off
Check Anti-Lock (Amber)								
Brake (Red)								

TEST STEP	RESULT ▶	ACTION TO TAKE
J1 VERIFY CONDITION		
• Verify condition exists as reported. • Turn air suspension off. • Bleed brake system per shop manual for Anti-Lock brake system. • Turn air suspension on when vehicle is off hoist.	Vehicle tracks properly ▶	Condition corrected.
	Vehicle still tracks poorly ▶	GO to **J2**.
J2 CHECK ANTI-LOCK OPERATION — LEFT FRONT WHEEL		
• Turn air suspension off. • Lift vehicle and rotate wheels to assure they turn freely. • Turn ignition switch OFF. • Disconnect 35-Pin plug from electronic controller. • Connect EEC-IV Breakout box, Tool T83L-50-EEC-IV with Anti-Lock test adapter, Tool T85P-50-ASA to the Anti-Lock 35-Pin plug wiring harness. • Short Pins 2, 16 and 35 to each other at Breakout box. • Apply moderate brake pedal effort and check that left front wheel will not turn. • Check to see that left front wheel turns freely with ignition switch ON. • Turn air suspension on when vehicle is off hoist. **CAUTION: DO NOT LEAVE IGNITION ON FOR MORE THAN 1 MINUTE MAXIMUM, OR SOLENOID VALVE DAMAGE MAY RESULT.**	If wheel turns freely ▶	TURN ignition switch Off. DISCONNECT wire leads. Go to Step **J3**.
	If wheel does not turn freely or pedal drops ▶	REPLACE solenoid valve block.

Fig. 44 Diagnosis, poor vehicle tracking during anti-lock function warning lamp OFF (Part 1 of 2). 1986–87 except Thunderbird

Poor Vehicle Tracking During Anti-Lock Function (Warning Lamp Off) — Test J

TEST STEP	RESULT ▶	ACTION TO TAKE
J3 CHECK ANTI-LOCK OPERATION — RIGHT FRONT WHEEL		
• Turn air suspension off. • Short Pins 2, 15 and 34 to each other at Breakout box. • Apply moderate brake effort. Check that right front wheel does not turn. • Check that right front wheel turns freely with ignition switch ON. **CAUTION: DO NOT LEAVE IGNITION ON MORE THAN 1 MINUTE MAXIMUM OR SOLENOID VALVE DAMAGE MAY RESULT.**	Wheel turns freely ▶	TURN ignition switch OFF. DISCONNECT wire leads. GO to Step **J4**.
	Wheel does not turn freely or brake pedal drops ▶	REPLACE solenoid valve block.
J4 CHECK ANTI-LOCK OPERATION — REAR WHEELS		
• Turn air suspension off. • Short Pins 2, 17 and 33 to each other at Breakout box. • Apply moderate brake pedal pressure. Check that rear wheels will not turn. • Check that rear wheels turn freely with ignition switch ON. • Turn air suspension on when vehicle is off hoist. **CAUTION: DO NOT LEAVE IGNITION ON MORE THAN 1 MINUTE MAXIMUM OR SOLENOID VALVE DAMAGE MAY RESULT.**	Wheels turn freely ▶	TURN ignition switch OFF. DISCONNECT wire lead and Breakout box. LOWER vehicle. REVERIFY symptom.
	Wheels do not turn freely or brake pedal drops ▶	REPLACE solenoid valve block.

Fig. 44 Diagnosis, poor vehicle tracking during anti-lock function warning lamp OFF (Part 2 of 2). 1986–87 except Thunderbird

On-Board Self-Test

TEST STEP	RESULT ▶	ACTION TO TAKE
1.0 CONNECT STAR TESTER		
• Turn ignition switch OFF. • Connect Rotunda Star Tester 007-00017 or equivalent to Anti-Lock self-test connectors. • Turn tester On. **NOTE: Do not move vehicle or turn steering during test.** • Depress Star Tester push-button to test position. Display should show: COLON DISPLAY : 00 COLON MUST BE DISPLAYED TO RECEIVE SERVICE CODES • Turn ignition switch to RUN position. Wait 45 seconds. • Observe tester display.	Pass code (: 00) displayed and warning lamps off (OK) ▶	Vehicle OK. RETURN vehicle to customer.
	Pass code (: 00) displayed and warning lamp turns on (~~OK~~) ▶	PERFORM anti-lock Quick Test.
	Service code displayed (~~OK~~) ▶	GO to **2.0**.
2.0 CHECK SERVICE CODES		
• Record service code. **NOTE: If first displayed code indicates a valve failure (first digit is 2), service that valve. Then, repeat Step 1.0.** • Release, then depress Star Tester push-button. Wait 45 seconds. • Observe tester display.	Service code displayed ▶	REPEAT Step 2.0 each time a service code is displayed. RECORD each service code as it is displayed.
	Pass code (: 00) displayed ▶	SERVICE indicated fault(s). GO to **3.0** **NOTE: Cross off each service code from list as it is serviced.**
3.0 CHECK FOR OTHER FAULT CODES		
• Repeat Step 1.0 to see if any other new service codes are stored in memory.	If no new service codes are displayed ▶	GO to **4.0**.
	Service code displayed ▶	GO to **2.0**

Fig. 45 On-board self-test (Part 1 of 2). 1987 Thunderbird

On-Board Self-Test

TEST STEP	RESULT ▶	ACTION TO TAKE
4.0 CHECK SYSTEM OPERATION		
• Turn ignition to OFF position. • Disconnect Star Tester. • Turn ignition to RUN position and check warning lamp sequence. • Test drive vehicle.	"CHECK ANTI-LOCK" and "BRAKE" warning lamps Off. OK ▶	RETURN vehicle to customer.
	"CHECK ANTI-LOCK" and or "BRAKE" lamp On. NOT OK ▶	REPEAT Step **1.0**.

Fig. 45 On-board self-test (Part 2 of 2). 1987 Thunderbird

On-Board Self-Test Service Code Index

SERVICE CODE (COMPONENT)	PINPOINT TEST STEP
11 (Electronic Controller)	AA1
12 (Electronic Controller-Replacer)	AA2
21 (Main Valve)	BB1
22 (LH Front Inlet Valve)	CC1
23 (LH Front Outlet Valve)	CC2
24 (RH Front Inlet Valve)	CC3
25 (RH Front Outlet Valve)	CC4
26 (Rear Inlet Valve)	CC5
27 (Rear Outlet Valve and Ground)	CC6
31 (LH Front Sensor)	DD1
32 (RH Front Sensor)	DD5
33 (RH Rear Sensor)	DD9
34 (LH Rear Sensor)	DD13
35 (LH Front Sensor)	DD1
36 (RH Front Sensor)	DD5
37 (RH Rear Sensor)	DD9
38 (LH Rear Sensor)	DD13
41 (LH Front Sensor)	DD1
42 (RH Front Sensor)	DD5
43 (RH Rear Sensor)	DD9
44 (LH Rear Sensor)	DD13
51 (LH Front Outlet Valve)	EE1
52 (RH Front Outlet Valve)	EE3
53 (Rear Outlet Valve)	EE5
54 (Rear Outlet Valve)	EE7
55 (LH Front Sensor)	DD1
56 (RH Front Sensor)	DD5
57 (RH Rear Sensor)	DD9
58 (LH Rear Sensor)	DD13
61 (FLI and PWS Circuit)	FF1
71 (LH Front Sensor)	EE1
72 (RH Front Sensor)	EE3
73 (RH Rear Sensor)	EE5
74 (LH Rear Sensor)	EE7
75 (LH Front Sensor)	DD1
76 (RH Front Sensor)	DD5
77 (RH Rear Sensor)	DD9
78 (LH Rear Sensor)	DD13
99 (Electronic Controller)	AA1

Fig. 46 On-board self-test service code index. 1987 Thunderbird

Electronic Controller Diagnosis — Test AA

TEST STEP	RESULT ▶	ACTION TO TAKE
AA1 SERVICE CODE 11 AND OR 99: ELECTRICAL DISTURBANCE		
• Read all service codes by releasing and pressing the Star Tester push-button. • Allow a minimum of 45 seconds to pass after each time the push-button is pressed. • Write down on paper all service codes. • After all service codes are read and written down, drive vehicle above 40 km h (25 mph) to clear memory. • Read again all service codes.	Service code 11 and or 99 repeated. NOT OK ▶	REPLACE electronic controller.
	Memory erased or other service codes present except code 11 and or 99. OK ▶	PERFORM test step associated with service code or codes. REFER to On-Board Self-Test Service Code Index.
AA2 SERVICE CODE 12: REPLACE ELECTRONIC CONTROLLER		
• No test step to be performed.	NOT OK ▶	REPLACE electronic controller.

Fig. 47 Electronic controller diagnosis. 1987 Thunderbird

Main Valve Diagnosis — Test BB

TEST STEP	RESULT ▶	ACTION TO TAKE
BB1 SERVICE CODE 21: CHECK MAIN VALVE		
• Disconnect main valve 2-Pin plug. • Measure resistance between the main valve electrical Pins 1 and 2.	2 to 5.5 ohms OK ▶	REPLACE or SERVICE cable harness (Circuit 430E or 493).
	Any other reading NOT OK ▶	REPLACE actuation assembly.

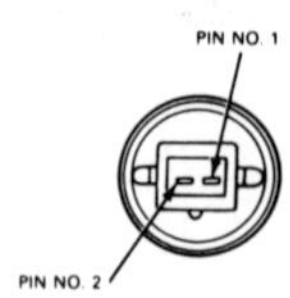

Fig. 48 Main valve diagnosis. 1987 Thunderbird

Solenoid Valve Diagnosis — Test CC

TEST STEP	RESULT ▶	ACTION TO TAKE
CC1 SERVICE CODE 22: CHECK LH FRONT INLET VALVE		
• Disconnect valve block 7-Pin plug. • Measure resistance between valve block electrical Pins 7 and 6. 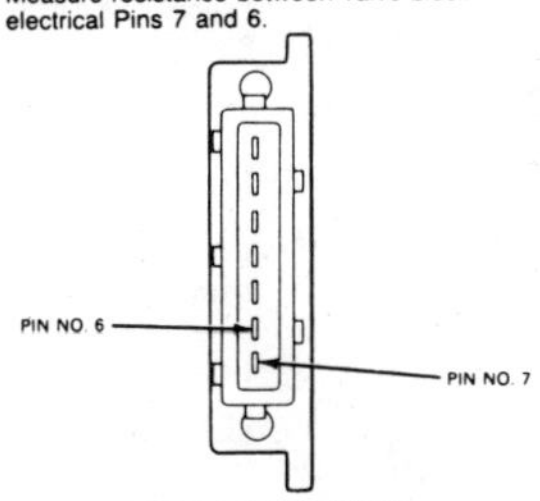 VALVE BLOCK 7-PIN CONNECTOR	5 to 8 ohms (OK) ▶ Any other reading (~~OK~~) ▶	REPLACE or SERVICE cable harness (Circuit 495). REPLACE valve block unit. RECONNECT 7-Pin plug.
CC2 SERVICE CODE 23: CHECK LH FRONT OUTLET VALVE		
• Disconnect valve block 7-Pin plug. • Measure resistance between valve block electrical Pins 7 and 5. 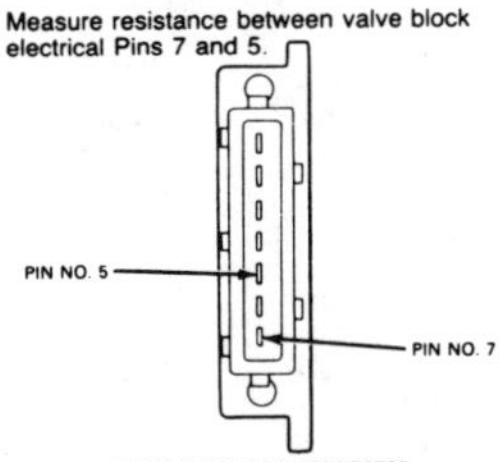 VALVE BLOCK 7-PIN CONNECTOR	3 to 6 ohms (OK) ▶ Any other reading (~~OK~~) ▶	REPLACE or SERVICE cable harness (Circuit 498). REPLACE valve block unit. RECONNECT 7-Pin plug.

Fig. 49 Solenoid valve diagnosis (Part 1 of 6). 1987 Thunderbird

Solenoid Valve Diagnosis — Test CC

TEST STEP	RESULT ▶	ACTION TO TAKE
CC3 SERVICE CODE 24: CHECK RH FRONT INLET VALVE		
• Disconnect valve block 7-Pin plug. • Measure resistance between valve block socket electrical Pins 7 and 1. 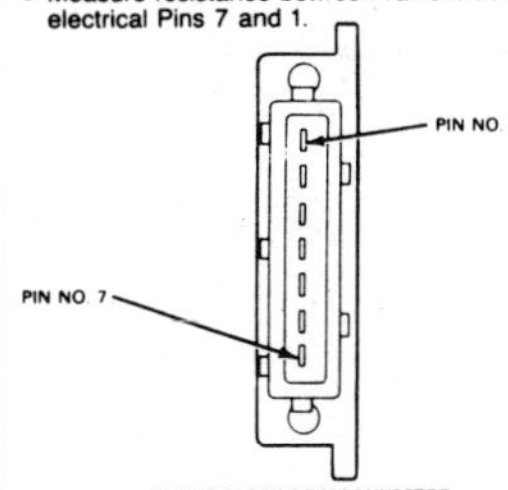 VALVE BLOCK 7-PIN CONNECTOR	5 to 8 ohms (OK) ▶ Any other reading (~~OK~~) ▶	REPLACE or REPAIR cable harness (Circuit 510). REPLACE solenoid valve block unit. RECONNECT 7-Pin plug.
CC4 SERVICE CODE 25: CHECK RH FRONT OUTLET VALVE		
• Disconnect valve block 7-Pin plug. • Measure resistance between valve block electrical Pins 7 and 2. 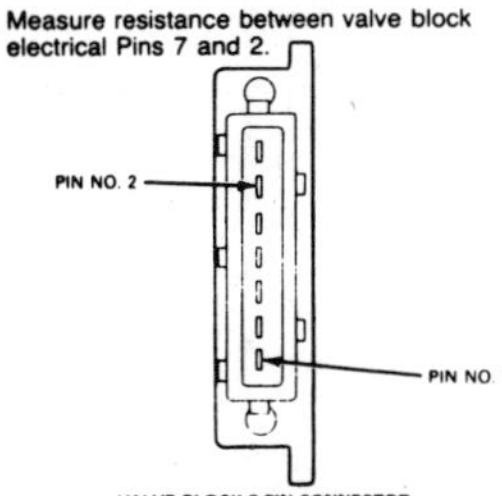 VALVE BLOCK 7-PIN CONNECTOR	3 to 6 ohms (OK) ▶ Any other reading (~~OK~~) ▶	REPLACE or SERVICE cable harness (Circuit 497). REPLACE valve block unit. RECONNECT 7-Pin plug.

Fig. 49 Solenoid valve diagnosis (Part 2 of 6). 1987 Thunderbird

Solenoid Valve Diagnosis — Test CC

TEST STEP	RESULT ▶	ACTION TO TAKE
CC5 SERVICE CODE 26: CHECK REAR INLET VALVE		
• Disconnect valve block 7-Pin plug. • Measure resistance between valve block electrical Pins 7 and 3. 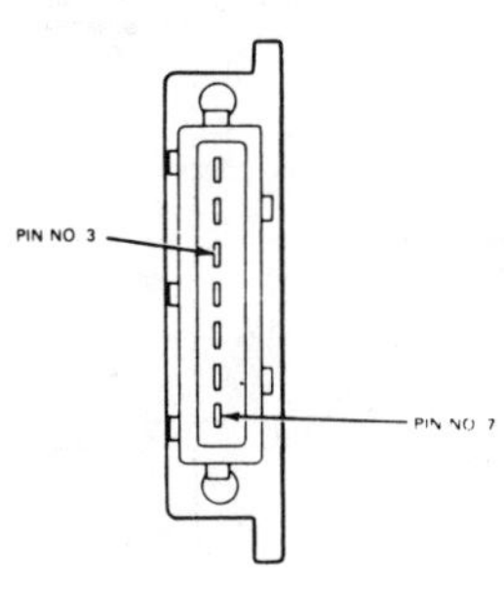 VALVE BLOCK 7-PIN CONNECTOR	5 to 8 ohms (OK) ▶ Any other reading (~~OK~~) ▶	REPLACE or SERVICE cable harness (Circuit 496). REPLACE valve block unit. RECONNECT 7-Pin plug.

Fig. 49 Solenoid valve diagnosis (Part 3 of 6). 1987 Thunderbird

Solenoid Valve Diagnosis — Test CC

TEST STEP	RESULT ▶	ACTION TO TAKE
CC6 SERVICE CODE 27: CHECK REAR OUTLET VALVE CIRCUIT		
• Turn ignition switch OFF. • Disconnect 32-Pin plug from electronic controller. 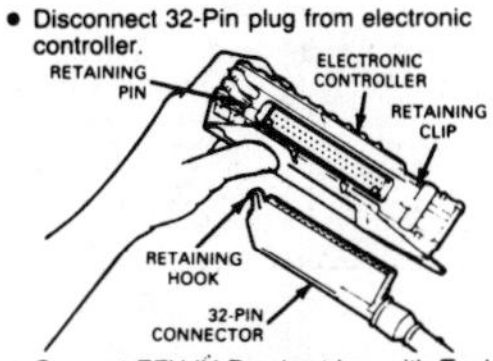 • Connect EEV-IV Breakout box with Tool T83L-50-EEC-IV or equivalent, Anti-Lock Test Adapter or equivalent, to the anti-lock 32-Pin plug wiring harness. 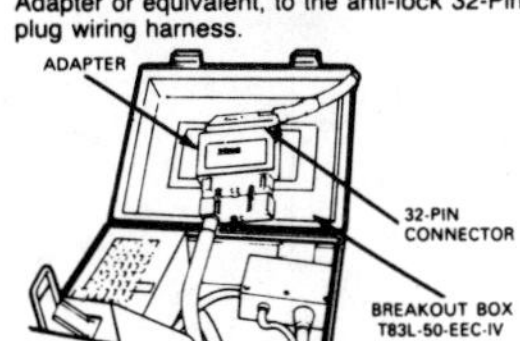 • Measure resistance between Breakout box Pins 11 and 12.	3 to 6 ohms (OK) ▶ Any other reading (~~OK~~) ▶	GO to Step **CC7**. GO to Step **CC6a**.

Fig. 49 Solenoid valve diagnosis (Part 4 of 6). 1987 Thunderbird

Solenoid Valve Diagnosis — Test CC

TEST STEP	RESULT ▶	ACTION TO TAKE
CC6a CHECK REAR OUTLET VALVE RESISTANCE		
• Disconnect valve block 7-Pin plug. • Measure resistance between valve block electrical Pins 7 and 4.	3 to 6 ohms (OK) ▶	REPLACE or SERVICE cable harness (Circuit 499).
	Any other reading (NOT OK) ▶	REPLACE valve block unit. RECONNECT 7-Pin plug.
CC7 CHECK VALVE BODY GROUND CIRCUIT		
• Measure resistance between Breakout box Pins 11 and 40.	Less than 2 ohms (OK) ▶	REVERIFY symptom.
	Greater than 2 ohms (NOT OK) ▶	GO to Step **CC7a**.

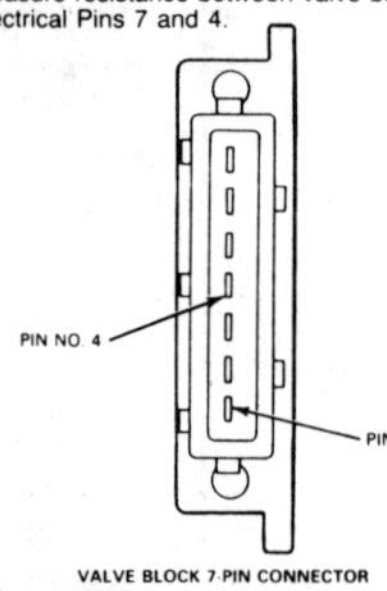

Fig. 49 Solenoid valve diagnosis (Part 5 of 6). 1987 Thunderbird

Solenoid Valve Diagnosis — Test CC

TEST STEP	RESULT ▶	ACTION TO TAKE
CC7a CHECK VALVE BODY GROUND		
• Disconnect valve block 7-Pin plug. • Measure resistance between valve block electrical Pin 7 and valve block body.	Less than 2 ohms (OK) ▶	REPLACE or SERVICE cable harness (Circuit 685).
	Greater than 2 ohms (NOT OK) ▶	GO to Step **CC7b**.
CC7b CHECK VALVE BODY GROUND WIRE		
• Remove negative (−) ground strap from battery. • Check for continuity between valve body and body ground.	Continuity (OK) ▶	REPLACE valve block unit (internal ground problem).
	No or poor continuity (NOT OK) ▶	SERVICE or REPLACE hydraulic unit ground strap (Circuit 430G).

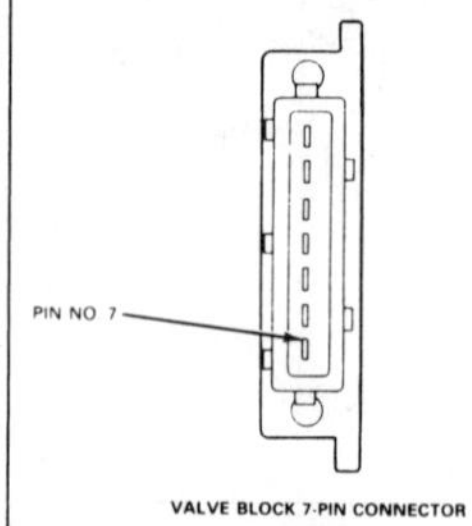

Fig. 49 Solenoid valve diagnosis (Part 6 of 6). 1987 Thunderbird

Wheel Sensor Diagnosis — Test DD

TEST STEP	RESULT ▶	ACTION TO TAKE
DD1 SERVICE CODES 31 35 41 55 OR 75 CHECK LH FRONT SENSOR		
• Turn ignition switch Off. • Disconnect 32-Pin plug from electronic controller. • Connect EEC-IV Breakout box with Tool T83L-50-EEC IV or equivalent, Anti-Lock Test Adapter T85P-50-ASA or equivalent, to the anti-lock 32-Pin plug wiring harness. • Measure resistance between Breakout box Pins 5 and 22.	800 to 1400 ohms (0.8 to 1.4K ohms) (OK) ▶	GO to Step **DD2**.
	Any other reading (NOT OK) ▶	GO to Step **DD1a**.
DD1a CHECK LH FRONT SENSOR RESISTANCE		
• Disconnect sensor plug (LH front). • Measure resistance of sensor at sensor plug.	800 to 1400 ohms (0.8 to 1.4K ohms) (OK) ▶	SERVICE or REPLACE cable harness (Circuit 521 or 522).
	Any other reading (NOT OK) ▶	REPLACE LH front wheel sensor.

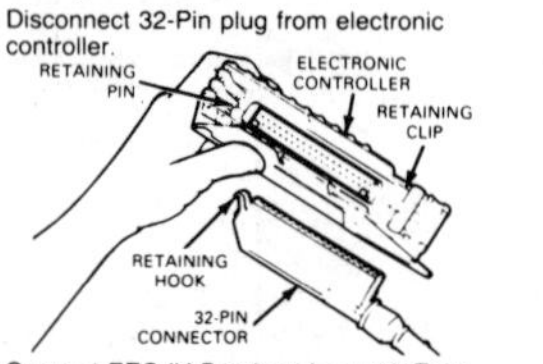

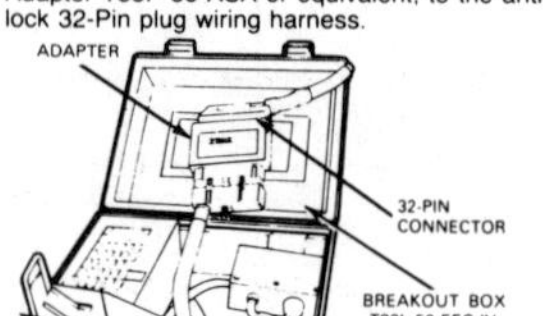

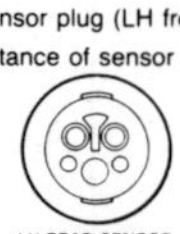

Fig. 50 Wheel sensor diagnosis (Part 1 of 8). 1987 Thunderbird

Wheel Sensor Diagnosis — Test DD

TEST STEP	RESULT ▶	ACTION TO TAKE
DD2 CHECK LH FRONT SENSOR VOLTAGE		
• Turn ignition switch Off. • Turn air suspension switch in luggage compartment Off, if so equipped. • Place vehicle on hoist and raise wheels clear of ground. Refer to Pre-Delivery Manual, Section 50-04. • Set multi-meter on voltage range (2V-AC). • Measure voltage between Breakout box Pins 5 and 22 while spinning LH front wheel at approximately 1 revolution per second.	Between 0.05 and 0.70 Vac ▶	GO to Step **DD3**.
	Less than 0.05 or more than 0.70 Vac ▶	CHECK sensor mounting, air gap, or toothed wheel mounting. CORRECT as required.
DD3 CHECK ELECTRONIC CONTROLLER TO GROUND WIRE		
• Check continuity between Breakout box Pin 40 and body ground.	Continuity (OK) ▶	GO to Step **DD4**.
	No continuity (NOT OK) ▶	SERVICE or REPLACE cable harness (Circuit 530A).
DD4 CHECK LH FRONT WHEEL BEARING		
• Check front wheel bearing end play. • Inspect toothed sensor ring visually for damaged teeth. **NOTE: Turn air suspension switch On when vehicle is off hoist, if so equipped.**	Loose or damaged parts ▶	ADJUST bearings or REPLACE faulty parts.
	Not loose or damaged ▶	REVERIFY symptom.

Fig. 50 Wheel sensor diagnosis (Part 2 of 8). 1987 Thunderbird

Wheel Sensor Diagnosis — Test DD

TEST STEP	RESULT ▶	ACTION TO TAKE
DD5 SERVICE CODES 32/36/42/56 OR 76 CHECK RH FRONT SENSOR		
• Turn ignition switch Off. • Disconnect 32-Pin plug from electronic controller.	800 to 1400 ohms (OK) ▶ (0.8 to 1.4K ohms)	GO to Step **DD6**.
	Any other reading (NOT OK) ▶	GO to Step **DD5a**.

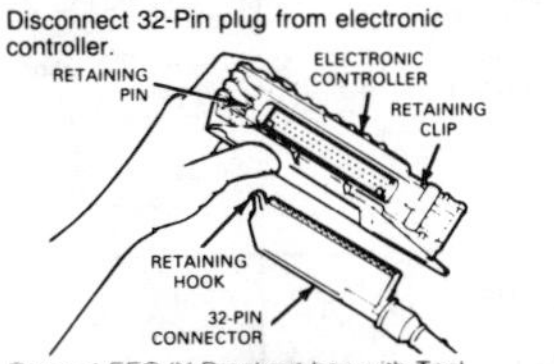

• Connect EEC-IV Breakout box with Tool T83L-50-EEC-IV or equivalent, Anti-Lock Test Adapter T85P-50-ASA or equivalent, to the anti-lock 32-Pin plug wiring harness.

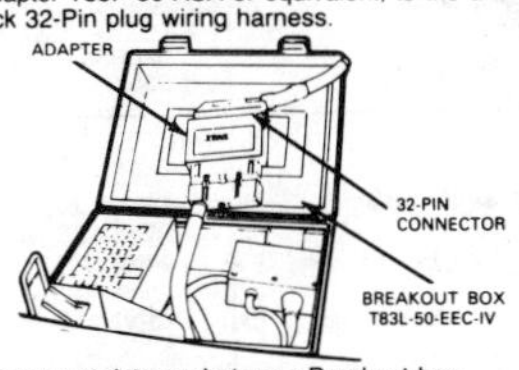

• Measure resistance between Breakout box Pins 3 and 20.

TEST STEP	RESULT ▶	ACTION TO TAKE
DD5a CHECK RH FRONT SENSOR RESISTANCE		
• Disconnect sensor plug (RH front). • Measure resistance of sensor at sensor plug.	800 to 1400 ohms (OK) ▶ (0.8 to 1.4K ohms)	SERVICE or REPLACE cable harness (Circuit 514 or 516).
	Any other reading (NOT OK) ▶	REPLACE RH front wheel sensor.

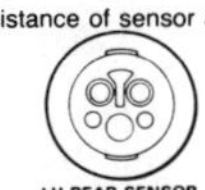

LH REAR SENSOR

Fig. 50 Wheel sensor diagnosis (Part 3 of 8). 1987 Thunderbird

Wheel Sensor Diagnosis — Test DD

TEST STEP	RESULT ▶	ACTION TO TAKE
DD6 CHECK RH FRONT SENSOR VOLTAGE		
• Turn air suspension switch in luggage compartment Off, if so equipped. • Place vehicle on hoist and raise wheels clear of ground. Refer to Pre-Delivery Manual, Section 50-04. • Set multi-meter on voltage range (2V-AC). • Measure voltage between Breakout box Pins 3 and 20 while spinning the RH front wheel at approximately 1 revolution per second.	Between 0.05 and 0.70 Vac ▶	GO to Step **DD7**.
	Less than 0.05 or more than 0.70 Vac ▶	CHECK wheel sensor mounting, air gap, or toothed wheel mounting. CORRECT as required.
DD7 CHECK ELECTRONIC CONTROLLER TO GROUND WIRE		
• Check continuity between Breakout box Pin 40 and body ground.	Continuity (OK) ▶	GO to Step **DD8**.
	No continuity (NOT OK) ▶	SERVICE or REPLACE cable harness (Circuit 530A).
DD8 CHECK RH FRONT WHEELING BEARING		
• Check front wheel bearing end play. • Inspect toothed sensor ring visually for damaged teeth. **NOTE: Turn air suspension switch On when vehicle is off hoist, if so equipped.**	Loose or damaged parts ▶	ADJUST bearings or REPLACE faulty parts.
	Not loose or damaged ▶	REVERIFY symptom.

Fig. 50 Wheel sensor diagnosis (Part 4 of 8). 1987 Thunderbird

Wheel Sensor Diagnosis — Test DD

TEST STEP	RESULT ▶	ACTION TO TAKE
DD9 SERVICE CODES 33/37/43/57 OR 77 CHECK RH REAR SENSOR		
• Turn ignition switch Off. • Disconnect 32-Pin plug from electronic controller.	800 to 1400 ohms (OK) ▶ (0.8 to 1.4K ohms)	GO to Step **DD10**.
	Any other reading (NOT OK) ▶	GO to Step **DD9a**.

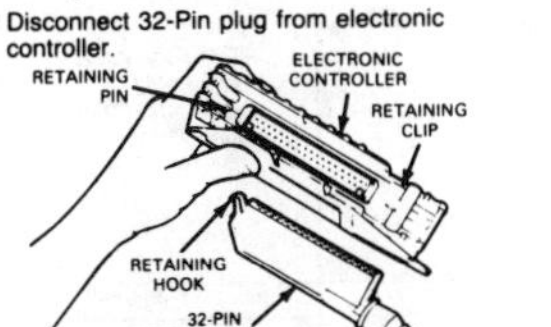

• Connect EEC-IV Breakout box with Tool T83L-50-EEC-IV or equivalent, Anti-Lock Test Adapter T85P-50-ASA or equivalent, to the anti-lock 32-Pin plug wiring harness.

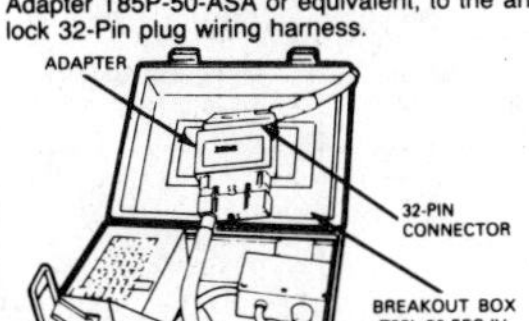

• Measure resistance between Breakout box Pins 6 and 23.

TEST STEP	RESULT ▶	ACTION TO TAKE
DD9a CHECK RH REAR SENSOR RESISTANCE		
• Disconnect sensor plug (RH rear). • Set multi-meter to read resistance. • Measure resistance of sensor at sensor plug.	800 to 1400 ohms (OK) ▶ (0.8 to 1.4K ohms)	SERVICE or REPLACE cable harness (Circuit 523 or 524).
	Any other reading (NOT OK) ▶	REPLACE RH rear wheel sensor.

RH REAR SENSOR

Fig. 50 Wheel sensor diagnosis (Part 5 of 8). 1987 Thunderbird

Wheel Sensor Diagnosis — Test DD

TEST STEP	RESULT ▶	ACTION TO TAKE
DD10 CHECK RH REAR SENSOR VOLTAGE		
• Turn air suspension switch in luggage compartment Off, if so equipped. • Place vehicle on hoist and raise wheels clear of ground. Refer to Pre-Delivery Manual, Section 50-04. • Set multi-meter on voltage range (2V-AC). • Measure voltage between Breakout box Pins 6 and 23 while spinning the RH rear wheel at approximately 1 revolution per second.	Between 0.05 and 0.70 Vac ▶	GO to Step **DD11**.
	Less than 0.05 or more than 0.70 Vac ▶	CHECK sensor mounting, air gap, or toothed wheel mounting. CORRECT as required.
DD11 CHECK ELECTRONIC CONTROLLER TO GROUND WIRE		
• Check continuity between Breakout box Pin 40 and body ground.	Continuity (OK) ▶	GO to Step **DD12**.
	No continuity (NOT OK) ▶	SERVICE or REPLACE cable harness (Circuit 530A).
DD12 CHECK FOR EXCESSIVE AXLE VIBRATION		
• Check differential housing for excessive play. • Check rear axle bearings for excessive play. • Inspect toothed sensor ring visually for damaged teeth. **NOTE: Turn air suspension switch On when vehicle is off hoist, if so equipped.**	Loose or damaged parts ▶	SERVICE or REPLACE faulty parts.
	Not loose or damaged ▶	REVERIFY symptom.

Fig. 50 Wheel sensor diagnosis (Part 6 of 8). 1987 Thunderbird

Wheel Sensor Diagnosis — Test DD

TEST STEP	RESULT ▶	ACTION TO TAKE
DD13 SERVICE CODES 34/38/44/58 OR 78 CHECK LH REAR SENSOR		
• Turn ignition switch Off. • Disconnect 32-Pin plug from electronic controller. 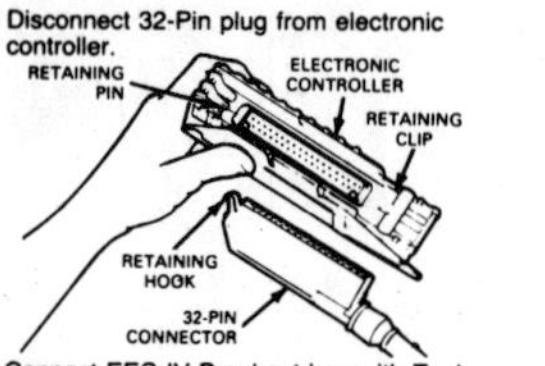 • Connect EEC-IV Breakout box with Tool T83L-50-EEC-IV or equivalent, Anti-Lock Test Adapter T85P-50-ASA or equivalent, to the anti-lock 32-Pin plug wiring harness. 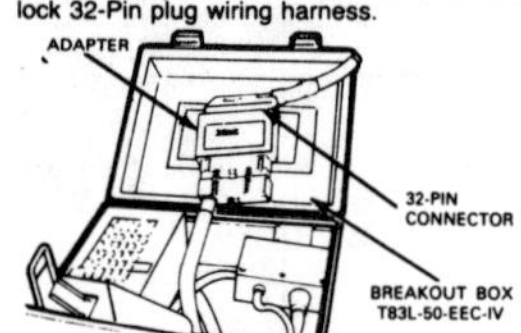 • Measure resistance between Breakout box Pins 4 and 21.	800 to 1400 ohms (0.8 to 1.4K ohms) (OK) ▶ Any other reading (NOT OK) ▶	GO to Step **DD14**. GO to Step **DD13a**.
DD13a CHECK LH REAR SENSOR RESISTANCE		
• Disconnect sensor plug (LH rear). • Measure resistance of sensor at sensor plug. 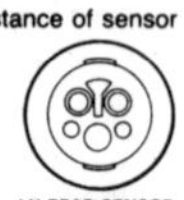	800 to 1400 ohms (0.8 to 1.4K ohms) (OK) ▶ Any other reading (NOT OK) ▶	SERVICE or REPLACE cable harness (Circuit 518 or 519). REPLACE LH rear wheel sensor.

Fig. 50 Wheel sensor diagnosis (Part 7 of 8). 1987 Thunderbird

Wheel Sensor Diagnosis — Test DD

TEST STEP	RESULT ▶	ACTION TO TAKE
DD14 CHECK LH REAR SENSOR VOLTAGE		
• Turn air suspension switch in luggage compartment Off, if so equipped. • Place vehicle on hoist and raise wheels clear of ground. Refer to Pre-Delivery Manual, Section 50-04. • Set multi-meter on voltage range (2V-AC). • Measure voltage between Breakout box Pins 4 and 21 while spinning the LH rear wheel at approximately 1 revolution per second.	Between 0.05 and 0.70 Vac ▶ Less than 0.05 or more than 0.70 Vac ▶	GO to Step **DD15**. CHECK sensor mounting, air gap, or toothed wheel mounting. CORRECT as required.
DD15 CHECK ELECTRONIC CONTROLLER TO GROUND WIRE		
• Check continuity between Breakout box Pin 40 and body ground.	Continuity (OK) ▶ No continuity (NOT OK) ▶	GO to Step **DD16**. SERVICE or REPLACE cable harness (Circuit 530A).
DD16 CHECK FOR EXCESSIVE AXLE VIBRATION		
• Check differential housing for excessive play. • Check rear axle bearings for excessive play. • Inspect toothed sensor ring visually for damaged teeth. **NOTE: Turn air suspension switch On when vehicle is off hoist, if so equipped.**	Loose or damaged parts ▶ Not loose or damaged ▶	SERVICE or REPLACE faulty parts. REVERIFY symptom.

Fig. 50 Wheel sensor diagnosis (Part 8 of 8). 1987 Thunderbird

Outlet Valve Diagnosis — Test EE

TEST STEP	RESULT ▶	ACTION TO TAKE
EE1 SERVICE CODE 51 AND/OR 71 CHECK LH FRONT SENSOR CIRCUIT CONTINUITY		
• Turn ignition switch Off. • Disconnect 32-Pin plug from electronic controller. 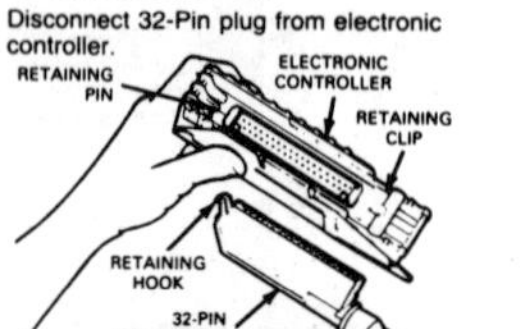 • Connect EEC-IV Breakout box with Tool T83L-50-EEC-IV or equivalent, Anti-Lock Test Adapter T85P-50-ASA or equivalent, to the anti-lock 32-Pin plug wiring harness. 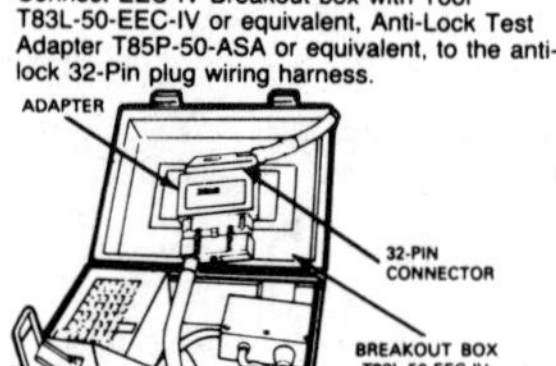 • Check for continuity between Breakout box Pins 40 and 5.	Continuity (NOT OK) ▶ No continuity (OK) ▶	GO to Step **EE1a**. GO to Step **EE2**.

Fig. 51 Outlet valve diagnosis (Part 1 of 6). 1987 Thunderbird

Outlet Valve Diagnosis — Test EE

TEST STEP	RESULT ▶	ACTION TO TAKE
EE1a CHECK LH FRONT SENSOR CONTINUITY		
• Disconnect wheel sensor plug (LH front). • Check for continuity between each sensor plug pin (sensor side) and vehicle ground. 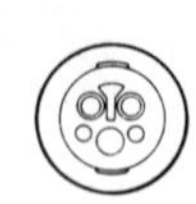	Continuity (NOT OK) ▶ No continuity (OK) ▶	REPLACE sensor (LH front). SERVICE or REPLACE cable harness. RECONNECT sensor plug.
EE2 CHECK ANTI-LOCK OPERATION LH FRONT WHEEL		
• Turn air suspension Off, if so equipped. • Lift vehicle and rotate wheels to ensure they turn freely. • Short Pins 18, 14 and 31 to each other at Breakout box. • Apply moderate brake pedal effort and check that LH front wheel will not turn. • Check to see that LH front wheel turns freely with ignition switch On. • Turn air suspension on when vehicle is off hoist, if so equipped. **CAUTION: DO NOT LEAVE IGNITION ON FOR MORE THAN 1 MINUTE MAXIMUM, OR SOLENOID VALVE DAMAGE MAY RESULT.**	If wheel turns freely ▶ If wheel does not turn freely or pedal drops ▶	REVERIFY symptom. REPLACE solenoid valve block.

Fig. 51 Outlet valve diagnosis (Part 2 of 6). 1987 Thunderbird

Outlet Valve Diagnosis — Test EE

TEST STEP | RESULT ▶ | ACTION TO TAKE

EE3 SERVICE CODE 52 AND/OR 72 CHECK RH FRONT SENSOR CIRCUIT CONTINUITY

- Turn ignition switch Off.
- Disconnect 32-Pin plug from electronic controller.

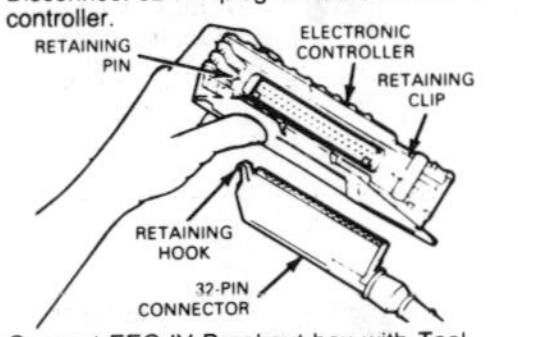

- Connect EEC-IV Breakout box with Tool T83L-50-EEC-IV or equivalent, Anti-Lock Test Adapter T85P-50-ASA or equivalent, to the anti-lock 32-Pin plug wiring harness.

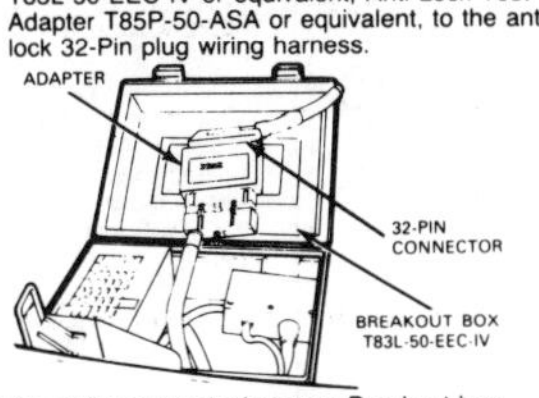

- Check for continuity between Breakout box Pins 40 and 3.

RESULT		ACTION TO TAKE
Continuity	(OK crossed out) ▶	GO to Step **EE3a**.
No continuity	(OK) ▶	GO to Step **EE4**.

Fig. 51 Outlet valve diagnosis (Part 3 of 6). 1987 Thunderbird

Outlet Valve Diagnosis — Test EE

TEST STEP | RESULT ▶ | ACTION TO TAKE

EE3a CHECK RH FRONT SENSOR CONTINUITY

- Disconnect wheel sensor plug (RH front).
- Check for continuity between each sensor plug pin (sensor side) and vehicle ground.

RH REAR SENSOR

RESULT		ACTION TO TAKE
Continuity	(OK crossed out) ▶	REPLACE right front sensor.
No continuity	(OK) ▶	SERVICE or REPLACE cable harness. RECONNECT sensor plug.

EE4 CHECK ANTI-LOCK OPERATION RH FRONT WHEEL

- Turn air suspension Off, if so equipped.
- Lift vehicle and rotate wheels to ensure they turn freely.
- Short Pins 18, 32 and 13 to each other at Breakout box.
- Apply moderate brake effort. Check that RH front wheel does not turn.
- Check to see that RH front wheel turns freely with ignition switch On.
- Turn air suspension on when vehicle is off hoist, if so equipped.

CAUTION: DO NOT LEAVE IGNITION ON FOR MORE THAN 1 MINUTE MAXIMUM, OR SOLENOID VALVE DAMAGE MAY RESULT.

RESULT		ACTION TO TAKE
If wheel turns freely	▶	REVERIFY symptom.
If wheel does not turn freely or pedal drops	▶	REPLACE solenoid valve block.

Fig. 51 Outlet valve diagnosis (Part 4 of 6). 1987 Thunderbird

Outlet Valve Diagnosis — Test EE

TEST STEP | RESULT ▶ | ACTION TO TAKE

EE5 SERVICE CODES 53 AND/OR 73 CHECK RH REAR SENSOR CIRCUIT CONTINUITY

- Turn ignition switch Off.
- Disconnect 32-Pin plug from electronic controller.

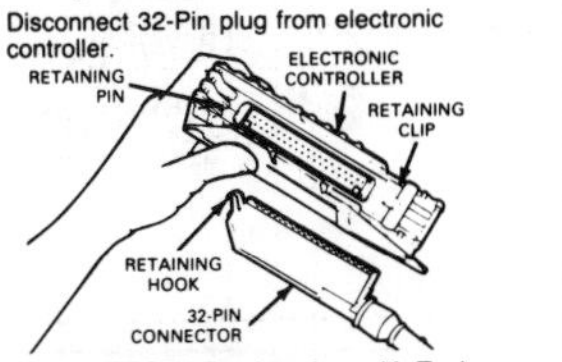

- Connect EEC-IV Breakout box with Tool T83L-50-EEC-IV or equivalent, Anti-Lock Test Adapter T85P-50-ASA or equivalent, to the anti-lock 32-Pin plug wiring harness.

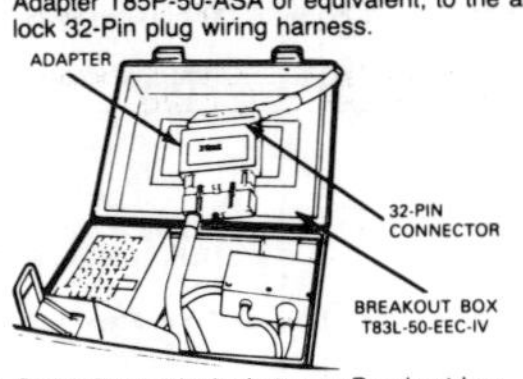

- Check for continuity between Breakout box Pins 40 and 6.

RESULT		ACTION TO TAKE
Continuity	(OK crossed out) ▶	GO to Step **EE5a**.
No continuity	(OK) ▶	GO to Step **EE6**.

Fig. 51 Outlet valve diagnosis (Part 5 of 6). 1987 Thunderbird

Outlet Valve Diagnosis — Test EE

TEST STEP | RESULT ▶ | ACTION TO TAKE

EE5a CHECK RH REAR SENSOR CONTINUITY

- Disconnect wheel sensor plug (RH rear).
- Check for continuity between each sensor plug pin (sensor side) and vehicle ground.

RH REAR SENSOR

RESULT		ACTION TO TAKE
Continuity	(OK crossed out) ▶	REPLACE sensor (RH rear).
No continuity	(OK) ▶	SERVICE or REPLACE cable harness. RECONNECT sensor plug.

EE6 CHECK ANTI-LOCK OPERATION: REAR WHEELS

- Turn air suspension Off, if so equipped.
- Lift vehicle and rotate wheels to ensure they turn freely.
- Short Pins 18, 30 and 12 to each other at Breakout box.
- Apply moderate brake pedal pressure. Check that rear wheels will not turn.
- Check that rear wheels turn freely with ignition switch On.
- Turn air suspension On when vehicle is off hoist, if so equipped.

CAUTION: DO NOT LEAVE IGNITION ON FOR MORE THAN 1 MINUTE MAXIMUM OR SOLENOID VALVE DAMAGE MAY RESULT.

RESULT		ACTION TO TAKE
If wheel turns freely	▶	REVERIFY symptom.
If wheel does not turn freely or pedal drops	▶	REPLACE solenoid valve block.

Fig. 51 Outlet valve diagnosis (Part 6 of 6). 1987 Thunderbird

Warning Circuit Diagnosis — Test FF

TEST STEP	RESULT ▶	ACTION TO TAKE
FF1 SERVICE CODE 61: CHECK FLI AND PWS CIRCUIT		
• Turn ignition switch Off. • Disconnect 32-Pin plug from electronic controller. 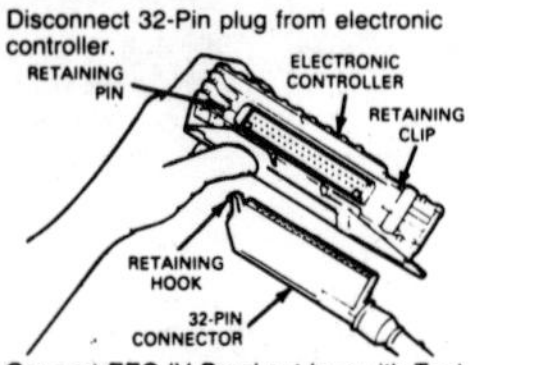 • Connect EEC-IV Breakout box with Tool T83L-50-EEC-IV or equivalent. Anti-Lock Test Adapter T85P-50-ASA or equivalent. to the anti-lock 32-Pin plug wiring harness. 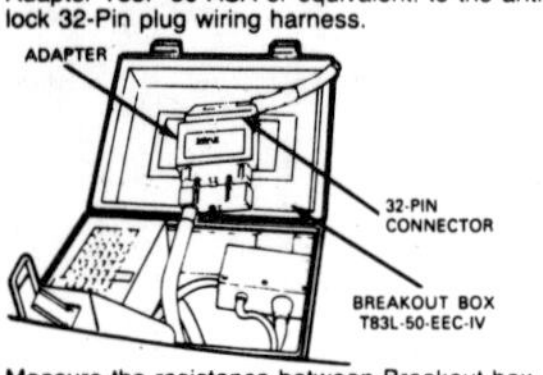 • Measure the resistance between Breakout box Pins 25 and 27.	Less than 5 ohms (OK) ▶ Greater than 5 ohms (NOT OK) ▶	GO to Step **FF2**. GO to Step **FF1a**.

Fig. 52 Warning circuit diagnosis (Part 1 of 6). 1987 Thunderbird

Warning Circuit Diagnosis — Test FF

TEST STEP	RESULT ▶	ACTION TO TAKE
FF1a CHECK FLI ANTI-LOCK WARNING CIRCUIT		
• Disconnect 5-Pin plug on reservoir Fluid Level Indicator (FLI). • Measure resistance between fluid level indicator electrical socket Pins 1 and 2 (with brake fluid level at maximum mark on reservoir). 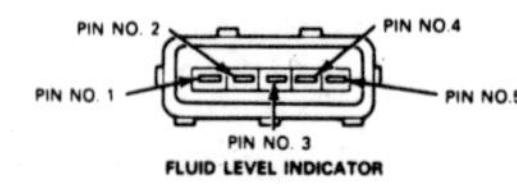	Less than 2 ohms (OK) ▶ Greater than 2 ohms (NOT OK) ▶	GO to Step **FF1b**. REPLACE fluid level indicator.
FF1b CHECK PWS ANTI-LOCK WARNING CIRCUIT		
• Disconnect 5-Pin plug at pressure warning switch (PWS). • Check for continuity between pressure warning switch socket Pins 3 and 5. **NOTE: System must be pressurized for this test.** 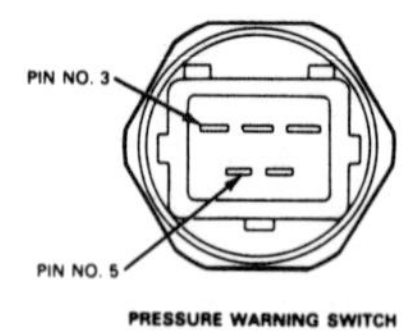	Continuity (OK) ▶ No continuity (NOT OK) ▶	GO to Step **FF1c**. REPLACE pressure warning switch and pump motor relay.

Fig. 52 Warning circuit diagnosis (Part 2 of 6). 1987 Thunderbird

Warning Circuit Diagnosis — Test FF

TEST STEP	RESULT ▶	ACTION TO TAKE
FF1c CHECK ELECTRONIC MODULE TO FLI WIRE		
• Check for continuity between Breakout box Pin 25 and FLI 5-Pin plug (harness side) Pin 1. 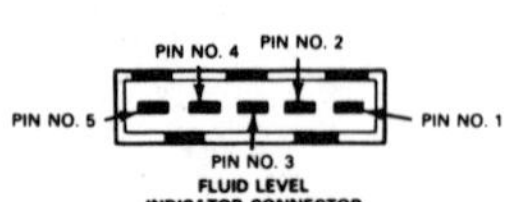	Continuity (OK) ▶ No continuity (NOT OK) ▶	GO to Step **FF1d**. SERVICE or REPLACE cable harness (Circuit 512).
FF1d CHECK FLI TO PWS WIRE		
• Check for continuity between Pin 2 of 5-Pin fluid level plug (harness side) 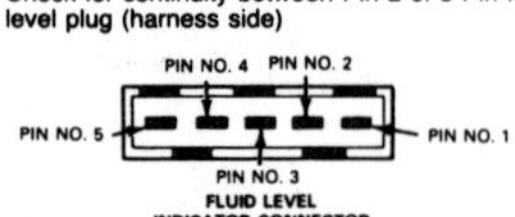 and 5-Pin pressure warning switch plug (harness side) Pin 3. 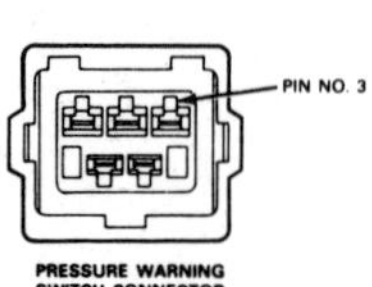	Continuity (OK) ▶ No continuity (NOT OK) ▶	GO to Step **FF1e**. SERVICE or REPLACE cable harness (Circuit 549).

Fig. 52 Warning circuit diagnosis (Part 3 of 6). 1987 Thunderbird

Warning Circuit Diagnosis — Test FF

TEST STEP	RESULT ▶	ACTION TO TAKE
FF1e PWS TO ELECTRONIC CONTROLLER WIRE		
• Check for continuity between 5-Pin pressure warning switch plug (harness side) Pin 5 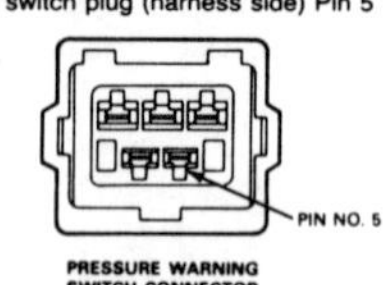 and Breakout box Pin 27.	Continuity (OK) ▶ No continuity (NOT OK) ▶	TURN ignition OFF. CONNECT all electrical connections. REVERIFY symptom. SERVICE or REPLACE cable harness (Circuit 535).
FF2 ISOLATION TEST FLI AND PWS		
• Check for continuity between Breakout box Pin 25 and body ground and Pin 27 and body ground.	Continuity (NOT OK) ▶ No continuity (OK) ▶	GO to Step **FF2a**. REVERIFY Symptom.
FF2a CHECK FLUID LEVEL INDICATOR PIN NO. 1		
• Check for continuity between FLI plug Pin 1 (harness side) and body ground. 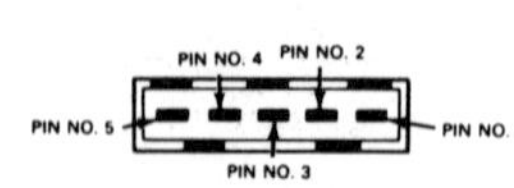	Continuity (NOT OK) ▶ No continuity (OK) ▶	SERVICE or REPLACE Circuit 512. GO to Step **FF2b**.

Fig. 52 Warning circuit diagnosis (Part 4 of 6). 1987 Thunderbird

Warning Circuit Diagnosis — Test FF

TEST STEP	RESULT ▶	ACTION TO TAKE
FF2b CHECK FLUID LEVEL INDICATOR PIN NO. 2		
• Disconnect 5-Pin plug from PWS. Check for continuity between FLI plug Pin 2 (harness side) and body ground.	Continuity (OK crossed out) ▶	SERVICE or REPLACE Circuit 549.
	No continuity (OK) ▶	GO to Step **FF2c**.

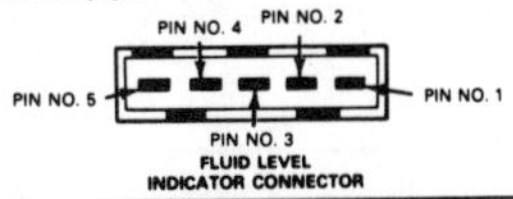

TEST STEP	RESULT ▶	ACTION TO TAKE
FF2c CHECK CONTINUITY BETWEEN PWS PINS 3 AND 5 AND BODY GROUND		
• Check for continuity from PWS 5-Pin socket Pins (pressure warning switch side) 3 and body ground and 5 and body ground.	Continuity (OK crossed out) ▶	REPLACE Pressure Warning Switch and Pump Motor Relay.
	No continuity (OK) ▶	GO to Step **FF2d**.

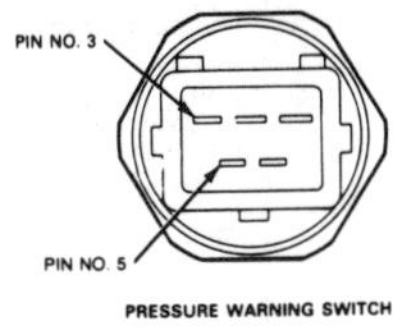

Fig. 52 Warning circuit diagnosis (Part 5 of 6). 1987 Thunderbird

Warning Circuit Diagnosis — Test FF

TEST STEP	RESULT ▶	ACTION TO TAKE
FF2d CHECK PWS CONNECTOR PIN 5 AND GROUND		
• Check for continuity between 5-Pin PWS plug, Pin 5 (harness side) and body ground.	Continuity (OK crossed out) ▶	SERVICE or REPLACE Circuit 535.
	No continuity (OK) ▶	CONNECT all plugs and reverify symptom.

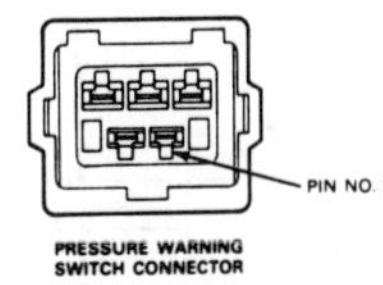

Fig. 52 Warning circuit diagnosis (Part 6 of 6). 1987 Thunderbird

Anti-Lock Quick Check Sheet Using 60-Pin EEC-IV Breakout Box, Tool T83L-50-EEC-IV①

Item to be Tested		Ignition Mode	Measure Between Pin Numbers	Tester Scale Range	Specification	Test Step
No Boost/No Power Brakes (Hard Brake Pedal)		—	—	—	—	D-1
Battery Check		On	40 - 18	Volts	10 minimum	A-1
Main Power Relay		Off	40 - 9	Ohms	45 Ohms — 105 Ohms	A-6
			Place a jumper between pins 9 & 18			
		On	40 - 16	Volts	10 minimum	A-7
Power from Main Power Relay		On	40 - 15	Volts	10 minimum	A-3
			Remove jumper from pins 9 & 18			
Main Power Relay Circuit		Off	40 - 16	Continuity	Continuity	A-2
Main Power Relay Circuit		Off	15 - 40	Continuity	Continuity	A-3a
Sensor Resistance	(RR)	Off	6 - 23	K Ohms	800 to 1400 Ohms	A-8
Sensor Resistance	(LF)	Off	5 - 22	K Ohms	800 to 1400 Ohms	A-9
Sensor Resistance	(LR)	Off	4 - 21	K Ohms	800 to 1400 Ohms	A-10
Sensor Resistance	(RF)	Off	3 - 20	K Ohms	800 to 1400 Ohms	A-11
Main Valve Resistance		Off	11 - 29	Ohms	2 Ohms to 5.5 Ohms	A-12
Inlet & Outlet Valves		Off	11 - 40	Continuity	Continuity	A-13
		Off	11 - 32	Ohms	5 Ohms to 8 Ohms	A-14
		Off	11 - 30	Ohms	5 Ohms to 8 Ohms	A-15
		Off	11 - 31	Ohms	5 Ohms to 8 Ohms	A-16
		Off	11 - 12	Ohms	3 Ohms to 6 Ohms	A-17
		Off	11 - 14	Ohms	3 Ohms to 6 Ohms	A-18
		Off	11 - 13	Ohms	3 Ohms to 6 Ohms	A-19
Reservoir Warning		On	25 - 27	Ohms	Less than 5 Ohms	A-4a
Lift Fluid Level Indicator from Reservoir (Float at bottom position)		Off	25 - 27	Ohms	Infinite (Open Circuit)	A-5a
Sensor Cable Continuity Shielding to Ground	(RR)	Off	40 - 6	Continuity	No Continuity	B-1a
	(LF)	Off	40 - 5	Continuity	No Continuity	B-2a
	(LR)	Off	40 - 4	Continuity	No Continuity	B-3a
	(RF)	Off	40 - 3	Continuity	No Continuity	B-4a
Sensor Voltage (Rotate wheels at 1 revolution per second minimum) (Shut off air suspension switch in luggage compartment with vehicle on hoist if so equipped).	(RR)	Off	6 - 23	AC Millivolts	50-700 Millivolts	C-5
	(LF)	Off	5 - 22	AC Millivolts	50-700 Millivolts	C-6
	(LR)	Off	4 - 21	AC Millivolts	50-700 Millivolts	C-7
	(RF)	Off	3 - 20	AC Millivolts	50-700 Millivolts	C-8

① If Quick Test does not isolate symptom, refer to Diagnostic Lamp Symptom Chart.

Fig. 53 Anti-lock quick check chart. 1987 Thunderbird

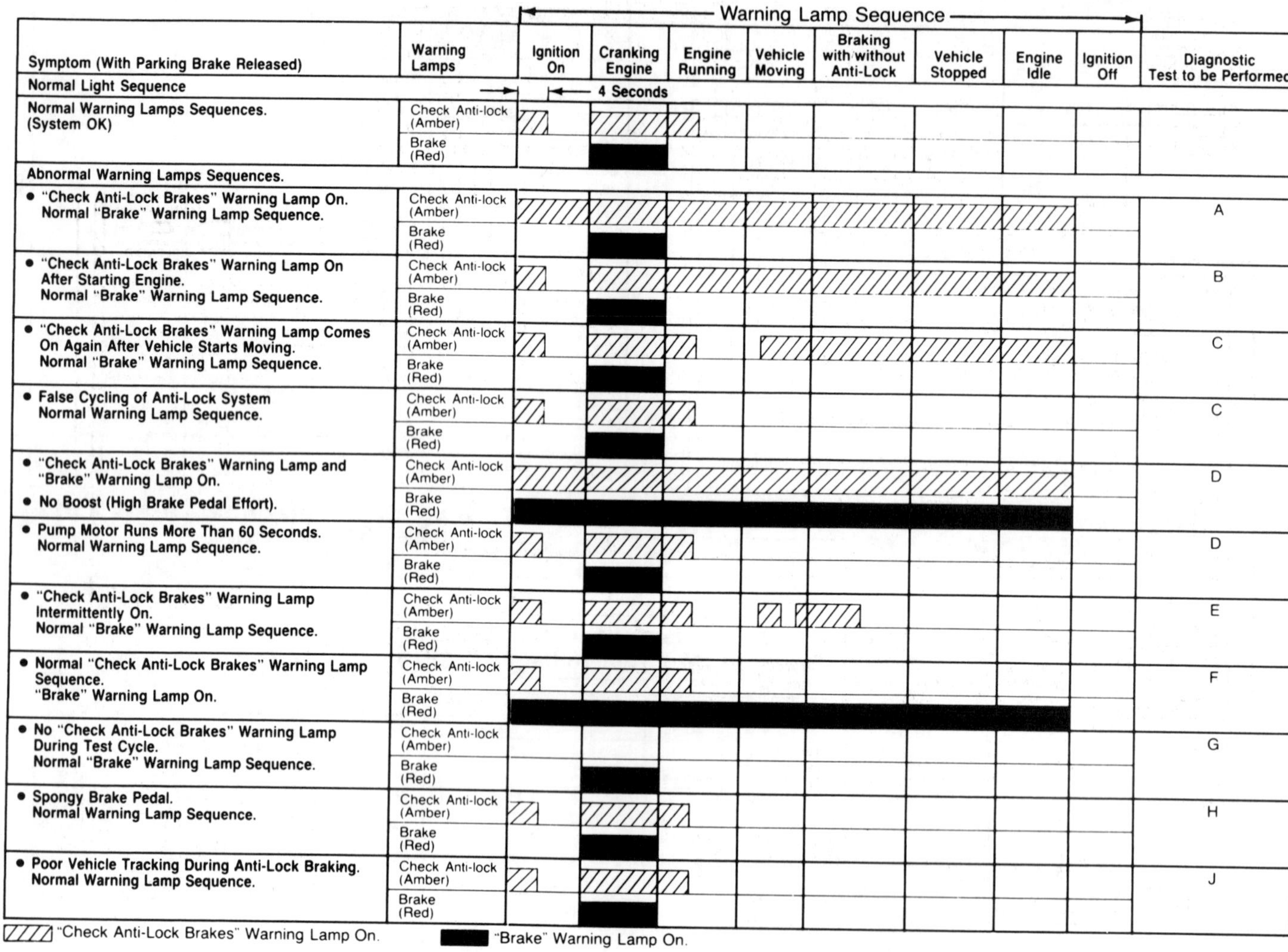

Fig. 54 Light symptom chart. 1987 Thunderbird

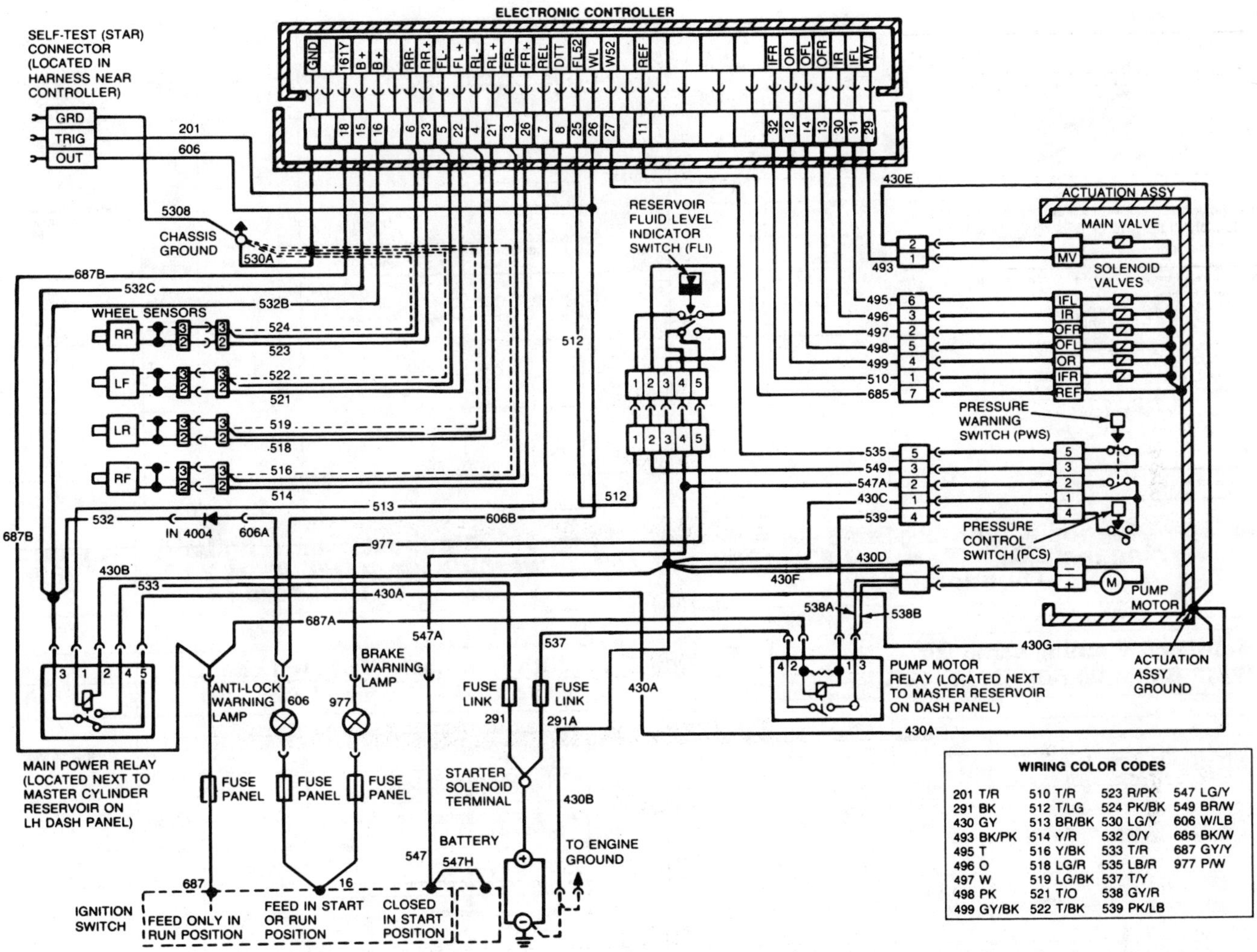

Fig. 55 Anti-lock system wiring diagram. 1987 Thunderbird

WIRING COLOR CODES

201 T/R	510 T/R	523 R/PK	547 LG/Y
291 BK	512 T/LG	524 PK/BK	549 BR/W
430 GY	513 BR/BK	530 LG/Y	606 W/LB
493 BK/PK	514 Y/R	532 O/Y	685 BK/W
495 T	516 Y/BK	533 T/R	687 GY/Y
496 O	518 LG/R	535 LB/R	977 P/W
497 W	519 LG/BK	537 T/Y	
498 PK	521 T/O	538 GY/R	
499 GY/BK	522 T/BK	539 PK/LB	

Anti-Lock Warning Lamp On (With Brake Warning Lamp Off) — Test A

WARNING LIGHTS SEQUENCE

Warning Lamps	Ignition On	Cranking Engine	Engine Running	Vehicle Moving	Braking with/without Anti-Lock	Vehicle Stopped	Engine Idle	Ignition Off
Check Anti-Lock (Amber)								
Brake (Red)								

TEST STEP	RESULT ▶	ACTION TO TAKE
A1 32 PIN PLUG TESTING		
• Turn ignition switch OFF. • Disconnect 32-Pin Plug from electronic controller. 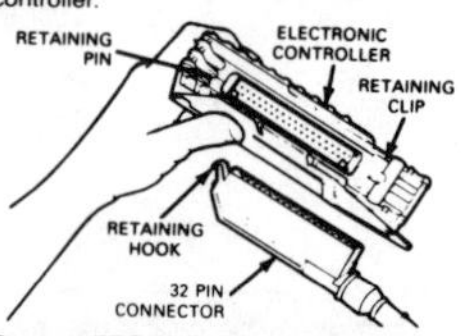 • Connect EEC-IV Breakout box Tool T83L-50-EEC-IV and anti-lock test adapter Tool T85P-50-ASA or equivalents, to the anti-lock 32-Pin plug wiring harness. 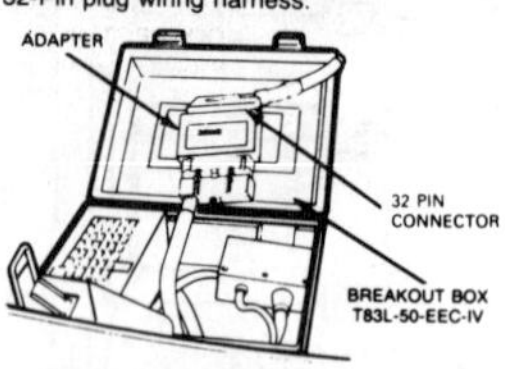 • Set multi-meter to read volts DC. • Turn ignition switch ON. • Measure voltage between Breakout box Pins 40 and 18.	Over 10V (OK) ▶ Under 10V (NOT OK) ▶	GO to Step **A2**. GO to Step **A1a**.

Fig. 56 Diagnosis, warning lamp ON w/brake warning lamp OFF (Part 1 of 23). 1987 Thunderbird

Anti-Lock Warning Lamp On (With Brake Warning Lamp Off) — Test A

TEST STEP	RESULT ▶	ACTION TO TAKE
A1a CHECK ELECTRONIC CONTROLLER TO GROUND WIRE		
• Check: — fuse link to anti-lock warning lamp. — battery. • Remove positive battery cable. • Check continuity between Breakout box pin 40 and body ground.	Continuity (OK) ▶ No continuity (NOT OK) ▶	GO to Step **A1b**. SERVICE or REPLACE cable harness (Circuit 530A).
A1b CHECK IGNITION TO ELECTRONIC CONTROLLER WIRE		
• Check continuity between Breakout box Pin 18 and ignition switch wire 687B.	Continuity (OK) ▶ No continuity (NOT OK) ▶	RECONNECT positive battery cable. CHECK for power at ignition switch pin with switch ON. If okay, connect electronic controller and reverify symptom. SERVICE or REPLACE cable harness (Circuit 687B).
A2 CHECK MAIN POWER RELAY SECONDARY CIRCUIT (NORMAL)		
• Turn ignition switch OFF. • Check for continuity between Breakout box Pins 40 and 16.	Continuity (OK) ▶ No continuity (NOT OK) ▶	GO to **A3**. GO to **A2a**.

Fig. 56 Diagnosis, warning lamp ON w/brake warning lamp OFF (Part 2 of 23). 1987 Thunderbird

Anti-Lock Warning Lamp On (With Brake Warning Lamp Off) — Test A

TEST STEP	RESULT ▶	ACTION TO TAKE
A2a CHECK MAIN POWER RELAY SECONDARY CIRCUIT (NORMAL)		
• Disconnect main relay from socket. • Check for continuity between main power relay socket Pins 3 and 5. 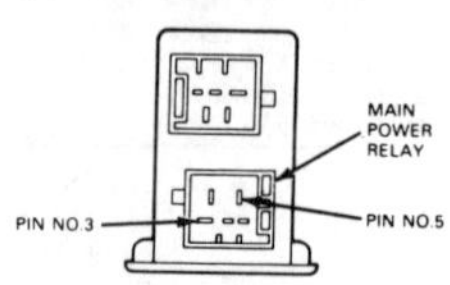 	Continuity (OK) ▶ No continuity (NOT OK) ▶	GO to Step **A2b**. REPLACE main power relay.
A2b CHECK MAIN POWER RELAY SECONDARY CIRCUIT WIRING HARNESS		
• Disconnect positive battery cable. • Check for continuity between main power relay socket Pin 3 and Breakout box Pin 16. 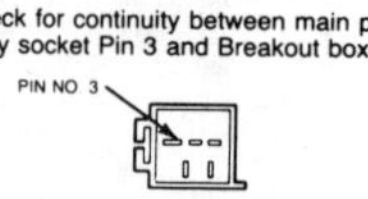 MAIN POWER RELAY CONNECTOR	Continuity (OK) ▶ No Continuity (NOT OK) ▶	GO to Step **A2C**. SERVICE or REPLACE cable harness (Circuit 532A or 532B).
A2c CHECK MAIN POWER RELAY SECONDARY CIRCUIT WIRING HARNESS		
• Check for continuity between main power relay socket Pin 5 and body ground. 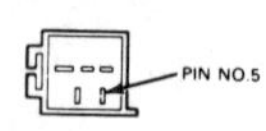 MAIN POWER RELAY CONNECTOR	Continuity (OK) ▶ No continuity (NOT OK) ▶	RECONNECT main power relay, electronic controller and battery cable and reverify symptom. SERVICE or REPLACE cable harness (Circuit 430A).

Fig. 56 Diagnosis, warning lamp ON w/brake warning lamp OFF (Part 3 of 23). 1987 Thunderbird

With Anti-Lock Warning Lamp On (With Brake Warning Lamp Off) — Test A

TEST STEP	RESULT ▶	ACTION TO TAKE
A3 CHECK MAIN POWER RELAY SECONDARY CIRCUIT (NORMAL)		
• Check for continuity between Breakout box Pins 40 and 15.	Continuity (OK) ▶ No continuity (NOT OK) ▶	GO to Step **A4**. GO to Step **A3a**.
A3a CHECK MAIN POWER RELAY SECONDARY CIRCUIT WIRING HARNESS		
• Remove main power relay. • Check for continuity between main power relay socket Pin 3 and Breakout box Pin 20. 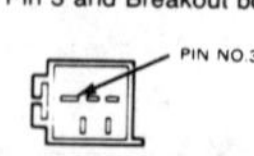 MAIN POWER RELAY CONNECTOR	Continuity (OK) ▶ No continuity (NOT OK) ▶	CONNECT main power relay and electronic controller and reverify symptom. SERVICE or REPLACE cable harness (Circuit 532A or 532C).
A4 CHECK FLI AND PWS CIRCUIT		
• Turn ignition switch ON. • Set multi-meter to read resistance. • Measure the resistance between Breakout box Pins 25 and 27.	Less than 5 ohms (OK) ▶ Greater than 5 ohms (NOT OK) ▶	GO to Step **A5**. GO to Step **A4a**.
A4a CHECK FLI ANTI-LOCK WARNING CIRCUIT		
• Disconnect 5-Pin plug on reservoir Fluid Level Indicator (FLI). • Measure resistance between fluid level indicator electrical socket Pins 1 and 2 (with brake fluid level at maximum mark on reservoir). 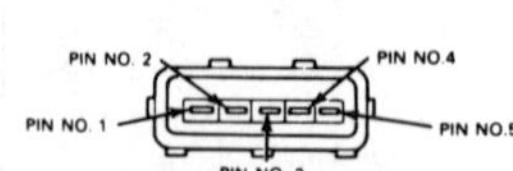 5-PIN FLUID LEVEL INDICATOR	Less than 2 ohms (OK) ▶ Greater than 2 ohms (NOT OK) ▶	GO to Step **A4b**. REPLACE fluid level indicator.

Fig. 56 Diagnosis, warning lamp ON w/brake warning lamp OFF (Part 4 of 23). 1987 Thunderbird

Anti-Lock Warning Lamp On (With Brake Warning Lamp Off) — Test A

TEST STEP	RESULT ▶	ACTION TO TAKE
A4b CHECK PWS ANTI-LOCK WARNING CIRCUIT		
• Disconnect 5-Pin plug at pressure warning switch (PWS). • Check for continuity between pressure warning switch socket Pins 3 and 5. **NOTE: System must be pressurized during this test.** 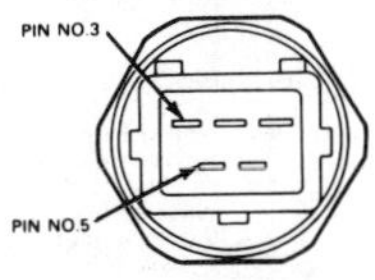	Continuity (OK) ▶	GO to Step **A4c**.
	No continuity (NOT OK) ▶	REPLACE pressure warning switch and pump motor relay.
A4c CHECK ELECTRONIC MODULE TO FLI WIRE		
• Check for continuity between Breakout box Pin 25 and FLI 5-Pin plug (harness side) Pin 1. 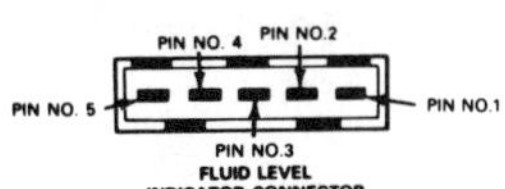	Continuity (OK) ▶	GO to Step **A4d**.
	No continuity (NOT OK) ▶	SERVICE or REPLACE cable harness (Circuit 512).

Fig. 56 Diagnosis, warning lamp ON w/brake warning lamp OFF (Part 5 of 23). 1987 Thunderbird

Anti-Lock Warning Lamp On (With Brake Warning Lamp Off) — Test A

TEST STEP	RESULT ▶	ACTION TO TAKE
A4d CHECK FLI TO PWS WIRE		
• Check for continuity between Pin 2 of 5-Pin fluid level plug (harness side) 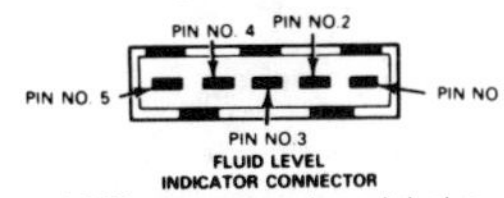 and 5-Pin pressure warning switch plug (harness side) Pin 3. 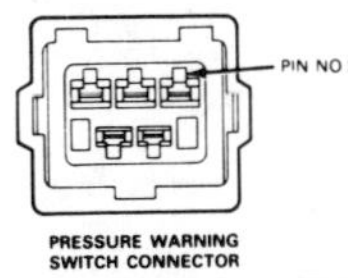	Continuity (OK) ▶	GO to Step **A4e**.
	No continuity (NOT OK) ▶	SERVICE or REPLACE cable harness (Circuit 549).
A4e PWS TO ELECTRONIC CONTROLLER WIRE		
• Check for continuity between 5-Pin pressure warning switch plug (harness side) Pin 5 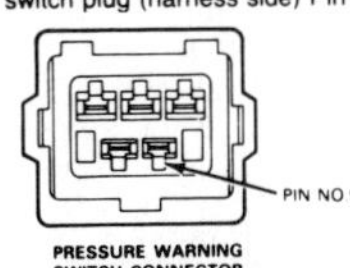 and Breakout box Pin 27.	Continuity (OK) ▶	TURN ignition OFF. CONNECT all electrical connections. Reverify symptom.
	No continuity (NOT OK) ▶	SERVICE or REPLACE cable harness (Circuit 535).
A5 ISOLATION TEST FLI AND PWS		
• Check for continuity between Breakout box Pin 9 and body ground.	Continuity (NOT OK) ▶	GO to Step **A5a**.
	No continuity (OK) ▶	GO to Step **A6**.

Fig. 56 Diagnosis, warning lamp ON w/brake warning lamp OFF (Part 6 of 23). 1987 Thunderbird

Anti-Lock Warning Lamp On (With Brake Warning Lamp Off) — Test A

TEST STEP	RESULT ▶	ACTION TO TAKE
A5a CHECK FLUID LEVEL INDICATOR PIN NO. 2		
• Disconnect FLI 5-Pin plug and check for continuity between FLI socket Pin 2 and body ground and Pin 1 and body ground. **NOTE: Ensure that brake fluid is at max level.** 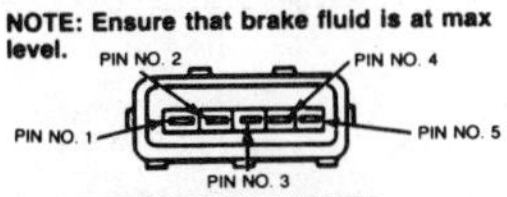	Continuity (NOT OK) ▶	REPLACE FLI.
	No continuity (OK) ▶	GO to Step **A5b**.
A5b CHECK FLUID LEVEL INDICATOR PIN NO. 1		
• Check for continuity between FLI plug Pin 1 (harness side) and body ground. 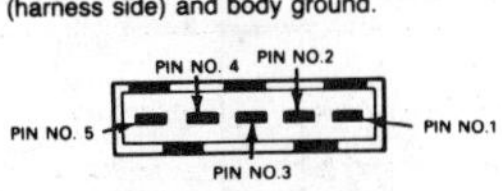	Continuity (NOT OK) ▶	SERVICE or REPLACE Circuit 512.
	No continuity (OK) ▶	GO to Step **A5c**.
A5c CHECK FLUID LEVEL INDICATOR PIN NO. 2		
• Disconnect 5-Pin plug from PWS. Check for continuity between FLI plug Pin 2 (harness side) and body ground. 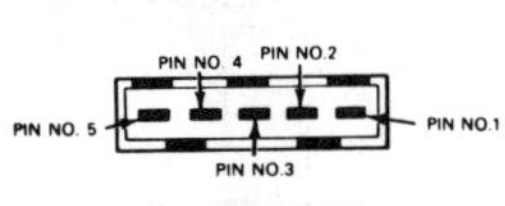	Continuity (NOT OK) ▶	SERVICE or REPLACE Circuit 549.
	No continuity (OK) ▶	GO to Step **A5d**.

Fig. 56 Diagnosis, warning lamp ON w/brake warning lamp OFF (Part 7 of 23). 1987 Thunderbird

Anti-Lock Warning Lamp On (With Brake Warning Lamp Off) — Test A

TEST STEP	RESULT ▶	ACTION TO TAKE
A5d CHECK CONTINUITY BETWEEN PWS PINS 3 AND 5 AND BODY GROUND		
• Check for continuity from PWS 5-Pin socket Pins (pressure warning switch side) 3 and body ground and 5 and body ground. **NOTE: System should be depressurized for this test.** 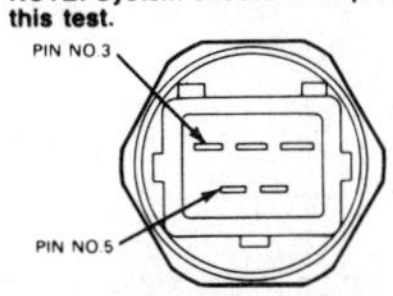	Continuity (NOT OK) ▶	REPLACE Pressure Warning Switch and Pump Motor Relay.
	No continuity (OK) ▶	GO to Step **A5e**.
A5e CHECK PWS CONNECTOR PIN 5 AND GROUND		
• Check for continuity between 5-Pin PWS plug. Pin 5 (harness side) and body ground. 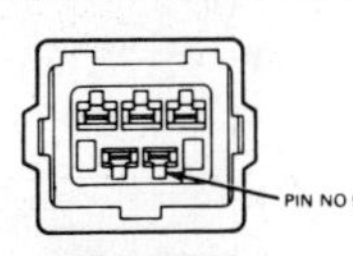	Continuity (NOT OK) ▶	SERVICE or REPLACE Circuit 535.
	No continuity (OK) ▶	CONNECT all plugs and reverify symptom.
A6 CHECK MAIN RELAY PRIMARY CIRCUIT RESISTANCE		
• Turn ignition switch OFF. • Set multi-meter to read resistance. • Measure resistance between Breakout box Pins 40 and 9.	Resistance between 45 and 105 ohms (OK) ▶	GO to Step **A7**.
	Any other reading (NOT OK) ▶	GO to Step **A6a**.

Fig. 56 Diagnosis, warning lamp ON w/brake warning lamp OFF (Part 8 of 23). 1987 Thunderbird

Anti-Lock Warning Lamp On (With Brake Warning Lamp Off) — Test A

TEST STEP	RESULT ▶	ACTION TO TAKE
A6a CHECK MAIN POWER RELAY PRIMARY COIL RESISTANCE		
• Disconnect main power relay from socket. • Set multi-meter on resistance. • Measure resistance between main power relay Pins 1 and 2. 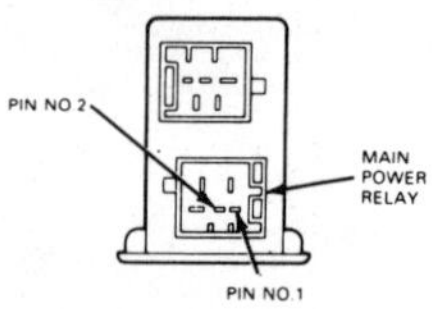	Resistance between 45 and 105 ohms (OK) ▶ Any other reading (NOT OK) ▶	GO to Step **A6b**. REPLACE main power relay.
A6b CHECK MAIN POWER RELAY PRIMARY TO ELECTRONIC CONTROLLER WIRE		
• Check continuity between main power relay socket Pin 2 (harness side) and Breakout Box Pin 9. 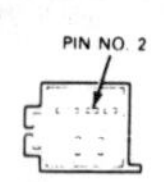 MAIN POWER RELAY CONNECTOR	Continuity (OK) ▶ No continuity (NOT OK) ▶	GO to Step **A6c**. SERVICE or REPLACE cable harness (Circuit 513).

Fig. 56 Diagnosis, warning lamp ON w/brake warning lamp OFF (Part 9 of 23). 1987 Thunderbird

Anti-Lock Warning Lamp On (With Brake Warning Lamp Off) — Test A

TEST STEP	RESULT ▶	ACTION TO TAKE
A6c CHECK MAIN POWER RELAY PRIMARY TO GROUND WIRE		
• Check continuity between main power relay socket Pin 1 and ground. 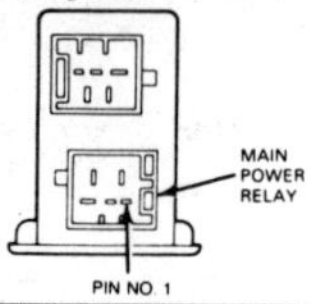	Continuity (OK) ▶ No continuity (NOT OK) ▶	RECONNECT all electrical connections and reverify symptom. SERVICE or REPLACE cable harness (Circuit 430B).
A7 CHECK MAIN POWER RELAY SECONDARY (ACTIVATED)		
• Place a jumper wire between Breakout box Pins 18 and 9. • Set multi-meter to read volts DC. • Turn ignition switch ON. • Measure voltage between Breakout box Pins 40 and 3.	Over 10 V DC (OK) ▶ Under 10 V DC (NOT OK) ▶	GO to Step **A8**. CHECK fuse F5. REMOVE jumper wire and GO to Step **A7a**.
A7a CHECK MAIN POWER RELAY SECONDARY CIRCUIT (ACTIVE)		
• Turn ignition switch OFF. • Disconnect main power relay from socket. • Apply power (battery positive and ground) to main relay Pins 1 and 2. • Check for continuity between main power relay Pins 3 and 4. 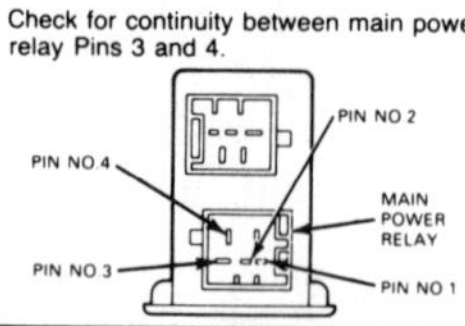	Continuity (OK) ▶ No continuity (NOT OK) ▶	GO to Step **A7b**. REPLACE main power relay.

Fig. 56 Diagnosis, warning lamp ON w/brake warning lamp OFF (Part 10 of 23). 1987 Thunderbird

Anti-Lock Warning Lamp On (With Brake Warning Lamp Off) — Test A

TEST STEP	RESULT ▶	ACTION TO TAKE
A7b CHECK MAIN POWER RELAY SECONDARY CIRCUIT POWER WIRE		
• Check continuity between main power relay socket Pin 4 (harness side) and positive battery terminal. 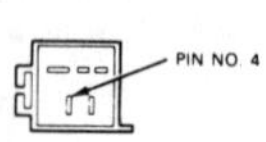 MAIN POWER RELAY CONNECTOR	Continuity (OK) ▶ No continuity (NOT OK) ▶	RECONNECT main power relay and reverify symptom. SERVICE or REPLACE cable harness (Circuit 533) or fuse link (Circuit 291).
A8 MEASURE RH REAR SENSOR CIRCUIT RESISTANCE		
• Turn ignition switch OFF. • Set multi-meter to read resistance. • Measure resistance between Breakout box Pin 6 and 23.	800 to 1400 ohms (0.8 to 1.4K ohms) (OK) ▶ Any other reading (NOT OK) ▶	GO to Step **A9**. Go to Step **A8a**.
A8a MEASURE RH REAR SENSOR RESISTANCE		
• Disconnect sensor plug (right rear). • Set multi-meter to read resistance. • Measure resistance of sensor at sensor plug. 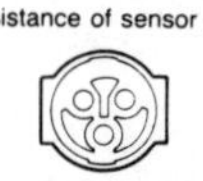 RH REAR SENSOR	800 to 1400 ohms (0.8 to 1.4K ohms) (OK) ▶ Any other reading (NOT OK) ▶	SERVICE or REPLACE cable harness (Circuit 523 or 524). REPLACE right rear wheel sensor.
A9 MEASURE LH FRONT SENSOR CIRCUIT RESISTANCE		
• Measure resistance between Breakout box Pins 5 and 22.	800 to 1400 ohms (0.8 to 1.4K ohms) (OK) ▶ Any other reading (NOT OK) ▶	GO to Step **A10**. GO to Step **A9a**.

Fig. 56 Diagnosis, warning lamp ON w/brake warning lamp OFF (Part 11 of 23). 1987 Thunderbird

Anti-Lock Warning Lamp On (With Brake Warning Lamp Off) — Test A

TEST STEP	RESULT ▶	ACTION TO TAKE
A9a MEASURE LH FRONT SENSOR RESISTANCE		
• Disconnect sensor plug (left front). • Measure resistance of sensor at sensor plug. 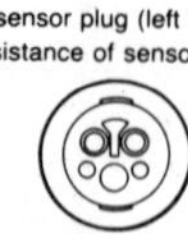 LH FRONT SENSOR	800 to 1400 ohms (0.8 to 1.4K ohms) (OK) ▶ Any other reading (NOT OK) ▶	SERVICE or REPLACE cable harness (Circuit 521 or 522). REPLACE left front wheel sensor.
A10 MEASURE LH REAR SENSOR CIRCUIT RESISTANCE		
• Measure resistance between Breakout box Pins 4 and 21.	800 to 1400 ohms (0.8 to 1.4K ohms) (OK) ▶ Any other reading (NOT OK) ▶	GO to Step **A11**. GO to Step **A10a**.
A10a MEASURE LH REAR SENSOR RESISTANCE		
• Disconnect sensor plug (left rear). • Measure resistance of sensor at left rear sensor plug. 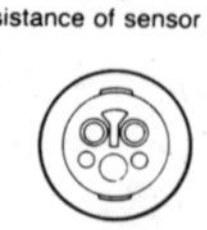 LH REAR SENSOR	800 to 1400 ohms (0.8 to 1.4K ohms) (OK) ▶ Any other reading (NOT OK) ▶	SERVICE or REPLACE cable harness (Circuit 518 or 519). REPLACE left rear wheel sensor.
A11 MEASURE RH FRONT SENSOR CIRCUIT RESISTANCE		
• Measure resistance between Breakout box Pins 3 and 20.	800 to 1400 ohms (0.8 to 1.4K ohms) (OK) ▶ Any other reading (NOT OK) ▶	GO to Step **A12**. GO to Step **A11a**.

Fig. 56 Diagnosis, warning lamp ON w/brake warning lamp OFF (Part 12 of 23). 1987 Thunderbird

Anti-Lock Warning Lamp On (With Brake Warning Lamp Off) — **Test A**

TEST STEP	RESULT ▶	ACTION TO TAKE
A11a MEASURE RH FRONT SENSOR RESISTANCE		
• Disconnect sensor plug (RH front). • Measure resistance of sensor at sensor plug. 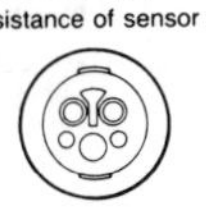 RH FRONT SENSOR	800 to 1400 ohms (0.8 to 1.4K ohms) OK ▶	SERVICE or REPLACE cable harness (Circuit 514 or 516).
	Any other reading ~~OK~~ ▶	REPLACE RH front wheel sensor.
A12 MEASURE MAIN VALVE CIRCUIT RESISTANCE		
• Turn ignition switch OFF. • Set multi-meter to read resistance. • Measure the resistance between Breakout box Pins 11 and 29.	2 to 5 ohms OK ▶	GO to Step **A13.**
	Any other reading ~~OK~~ ▶	GO to Step **A12a.**.
A12a MEASURE MAIN VALVE RESISTANCE		
• Disconnect main valve 2-Pin plug. • Measure resistance between the main valve electrical Pins 1 and 2. 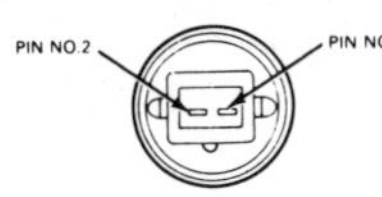 	2 to 5.5 ohms OK ▶	REPLACE or SERVICE cable harness (Circuit 430E or 493).
	Any other reading ~~OK~~ ▶	REPLACE actuation assembly.
A13 CHECK VALVE BLOCK GROUND CIRCUIT		
• Measure resistance between Breakout box Pins 11 and 40.	Less than 2 ohms OK ▶	GO to Step **A14**.
	Greater than 2 ohms ~~OK~~ ▶	GO to Step **A13a**.

Fig. 56 Diagnosis, warning lamp ON w/brake warning lamp OFF (Part 13 of 23). 1987 Thunderbird

Anti-Lock Warning Lamp On (With Brake Warning Lamp Off) — **Test A**

TEST STEP	RESULT ▶	ACTION TO TAKE
A13a TEST VALVE PLUG PIN NO. 7		
• Disconnect valve block 7-Pin plug. • Measure resistance between valve block electrical Pin 7 and valve block body. 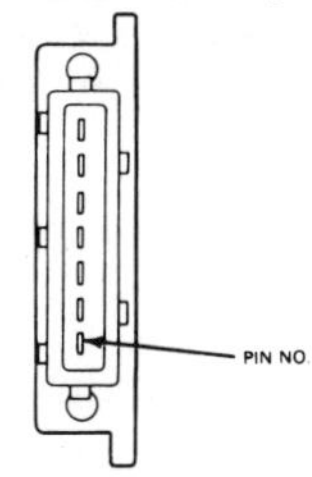 VALVE BLOCK 7-PIN CONNECTOR	Less than 2 ohms OK ▶	SERVICE or REPLACE cable harness (Circuit 511).
	Greater than 2 ohms ~~OK~~ ▶	GO to Step **A13b**.
A13b CHECK VALVE BLOCK GROUND WIRE		
• Remove negative (−) ground strap from battery. • Check for continuity between valve block and body ground.	Continuity OK ▶	REPLACE valve block unit (internal ground problem).
	No or Poor continuity ~~OK~~ ▶	SERVICE or REPLACE actuation assembly ground strap (Circuit 430G).
A14 MEASURE RH FRONT INLET VALVE CIRCUIT RESISTANCE		
• Measure resistance between Breakout box Pins 11 and 32.	5 to 8 ohms OK ▶	GO to Step **A15**.
	Any other reading ~~OK~~ ▶	GO to Step **A14a**.

Fig. 56 Diagnosis, warning lamp ON w/brake warning lamp OFF (Part 14 of 23). 1987 Thunderbird

Anti-Lock Warning Lamp On (With Brake Warning Lamp Off) — **Test A**

TEST STEP	RESULT ▶	ACTION TO TAKE
A14a MEASURE RH FRONT INLET VALVE RESISTANCE		
• Disconnect valve block 7-Pin plug. • Measure resistance between valve block electrical Pin 7 and Pin 1. 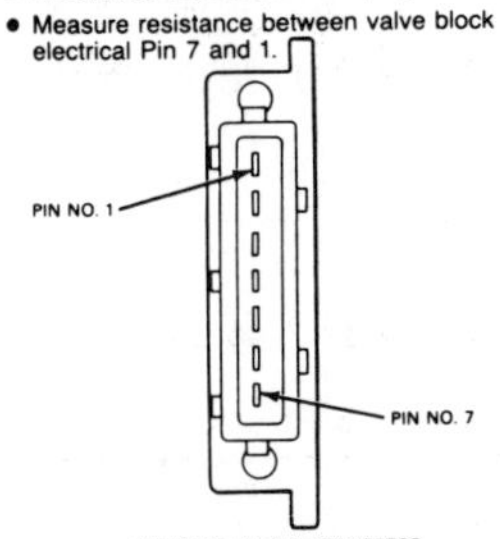 VALVE BLOCK 7-PIN CONNECTOR	5 to 8 ohms OK ▶	SERVICE or REPLACE cable harness (Circuit 510).
	Any other reading ~~OK~~ ▶	REPLACE solenoid valve block unit. CONNECT 7-Pin plug.
A15 MEASURE REAR INLET VALVE CIRCUIT RESISTANCE		
• Measure resistance between Breakout box Pins 11 and 30.	5 to 8 ohms OK ▶	GO to Step **A16**.
	Any other reading ~~OK~~ ▶	GO to Step **A15a**.

Fig. 56 Diagnosis, warning lamp ON w/brake warning lamp OFF (Part 15 of 23). 1987 Thunderbird

Anti-Lock Warning Lamp On (With Brake Warning Lamp Off) — **Test A**

TEST STEP	RESULT ▶	ACTION TO TAKE
A15a MEASURE REAR INLET VALVE RESISTANCE		
• Disconnect valve block 7-Pin plug. • Measure resistance between valve block electrical Pins 7 and 3. 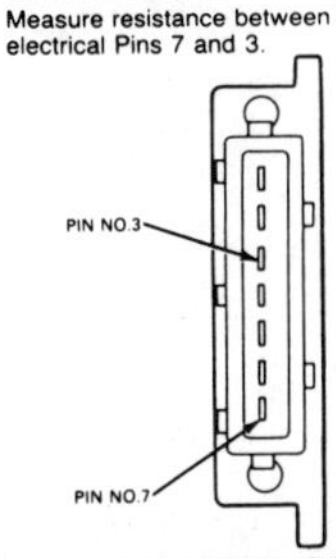 VALVE BLOCK 7-PIN CONNECTOR	5 to 8 ohms OK ▶	SERVICE or REPLACE cable harness (Circuit 496).
	Any other reading ~~OK~~ ▶	REPLACE valve block unit. CONNECT 7-Pin plug.
A16 MEASURE LH FRONT INLET VALVE CIRCUIT RESISTANCE		
• Measure resistance between Breakout box Pins 11 and 31.	5 to 8 ohms OK ▶	GO to Step **A17.**
	Any other reading ~~OK~~ ▶	GO to Step **A16a.**

Fig. 56 Diagnosis, warning lamp ON w/brake warning lamp OFF (Part 16 of 23). 1987 Thunderbird

Anti-Lock Warning Lamp On (With Brake Warning Lamp Off) — **Test A**

TEST STEP	RESULT ▶	ACTION TO TAKE
A16a MEASURE LH FRONT INLET VALVE RESISTANCE		
• Disconnect valve block 7-Pin plug. • Measure resistance between valve block electrical Pins 7 and 6.	5 to 8 ohms (OK) ▶	SERVICE or REPLACE cable harness (Circuit 495).
	Any other reading (NOT OK) ▶	REPLACE valve block unit. CONNECT 7-Pin plug.

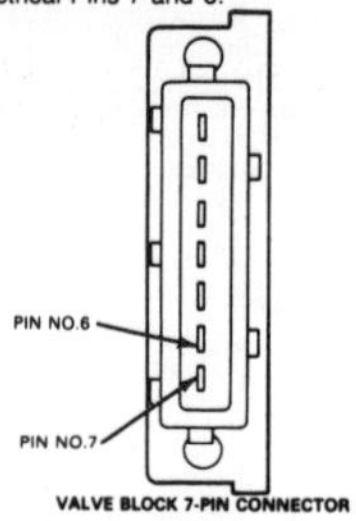

TEST STEP	RESULT ▶	ACTION TO TAKE
A17 MEASURE REAR OUTLET VALVE CIRCUIT RESISTANCE		
• Measure resistance between Breakout box Pins 11 and 12.	3 to 6 ohms (OK) ▶	GO to Step **A18.**
	Any other reading (NOT OK) ▶	GO to Step **A17a.**

Fig. 56 Diagnosis, warning lamp ON w/brake warning lamp OFF (Part 17 of 23). 1987 Thunderbird

Anti-Lock Warning Lamp On (With Brake Warning Lamp Off) — **Test A**

TEST STEP	RESULT ▶	ACTION TO TAKE
A17a MEASURE REAR OUTLET VALVE RESISTANCE		
• Disconnect valve block 7-Pin plug. • Measure resistance between valve block electrical Pins 7 and 4.	3 to 6 ohms (OK) ▶	SERVICE or REPLACE cable harness (Circuit 499).
	Any other reading (NOT OK) ▶	REPLACE valve block unit. CONNECT 7-Pin plug.

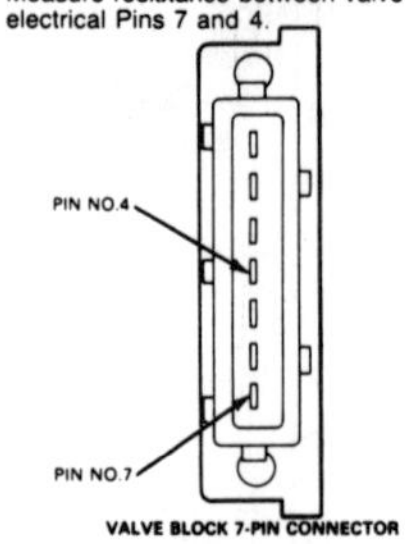

TEST STEP	RESULT ▶	ACTION TO TAKE
A18 MEASURE LH FRONT OUTLET VALVE CIRCUIT RESISTANCE		
• Measure resistance between Breakout box Pins 11 and 14.	3 to 6 ohms (OK) ▶	GO to Step **A19.**
	Any other reading (NOT OK) ▶	GO to Step **A18a.**

Fig. 56 Diagnosis, warning lamp ON w/brake warning lamp OFF (Part 18 of 23). 1987 Thunderbird

Anti-Lock Warning Lamp On (With Brake Warning Lamp Off) — **Test A**

TEST STEP	RESULT ▶	ACTION TO TAKE
A18a MEASURE LH FRONT OUTLET VALVE RESISTANCE		
• Disconnect valve block 7-Pin plug. • Measure resistance between valve block electrical Pins 7 and 5.	3 to 6 ohms (OK) ▶	SERVICE or REPLACE cable harness (Circuit 498).
	Any other reading (NOT OK) ▶	REPLACE valve block unit. CONNECT 7-Pin plug.

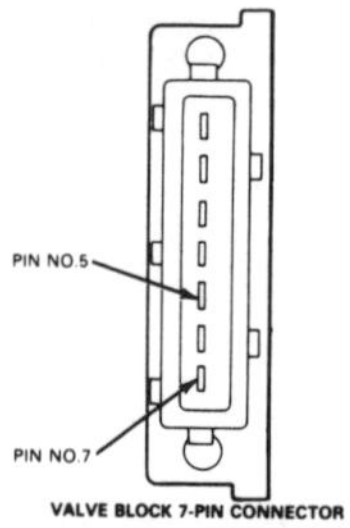

TEST STEP	RESULT ▶	ACTION TO TAKE
A19 MEASURE RH FRONT OUTLET VALVE CIRCUIT RESISTANCE		
• Measure resistance between Breakout box Pins 11 and 13.	3 to 6 ohms (OK) ▶	GO to Step **A20.**
	Any other reading (NOT OK) ▶	GO to Step **A19a.**

Fig. 56 Diagnosis, warning lamp ON w/brake warning lamp OFF (Part 19 of 23). 1987 Thunderbird

Anti-Lock Warning Lamp On (With Brake Warning Lamp Off) — **Test A**

TEST STEP	RESULT ▶	ACTION TO TAKE
A19a MEASURE RH FRONT OUTLET VALVE RESISTANCE		
• Disconnect valve block 7-Pin plug. • Measure resistance between valve block electrical Pins 7 and 2.	3 to 6 ohms (OK) ▶	SERVICE or REPLACE cable harness (Circuit 497).
	Any other reading (NOT OK) ▶	REPLACE valve block unit. CONNECT 7-Pin plug.

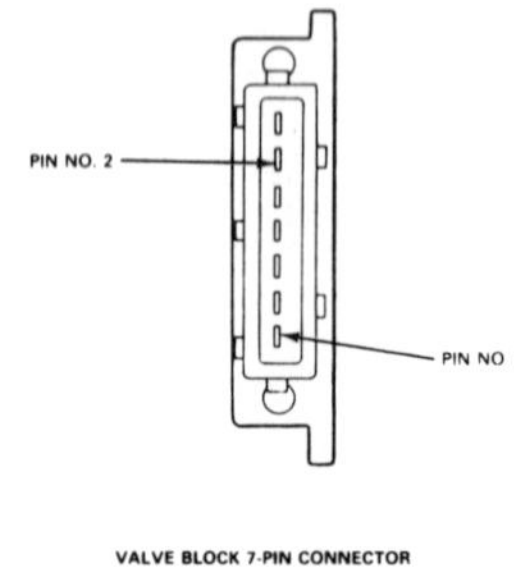

Fig. 56 Diagnosis, warning lamp ON w/brake warning lamp OFF (Part 20 of 23). 1987 Thunderbird

Anti-Lock Warning Lamp On (With Brake Warning Lamp Off) — **Test A**

TEST STEP	RESULT ▶	ACTION TO TAKE
A20 CHECK PWS BRAKELAMP CIRCUIT (NO SYSTEM PRESSURE)		
• Vehicle must be cooled to room temperature. • Discharge the brake system as follows: a. Turn ignition switch OFF. b. Pump brake pedal at least 20 times until you feel the pedal become hard. • Disconnect pressure warning switch 5-Pin plug. • Check continuity between pressure warning switch electrical Pins 1 and 2. 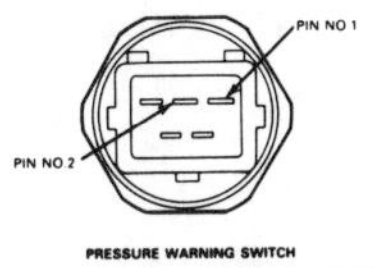	Continuity (OK) ▶ No continuity (~~OK~~) ▶	GO to Step **A21**. REPLACE pressure warning switch and pump motor relay.
A21 CHECK PWS BRAKELAMP CIRCUIT (WITH SYSTEM PRESSURE)		
• Reconnect pressure warning switch 5-Pin plug. • Turn ignition switch ON. • When pump motor stops, turn ignition switch OFF. • Disconnect pressure warning switch 5-Pin plug again. • Check for continuity between pressure warning switch electrical Pins 1 and 2. 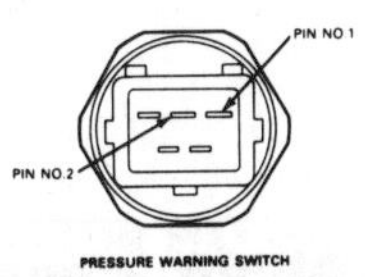	Continuity (~~OK~~) ▶ No continuity (OK) ▶	REPLACE pressure warning switch. GO to Step **A22** and pump motor relay.

Fig. 56 Diagnosis, warning lamp ON w/brake warning lamp OFF (Part 21 of 23). 1987 Thunderbird

Anti-Lock Warning Lamp On (With Brake Warning Lamp Off) — **Test A**

TEST STEP	RESULT ▶	ACTION TO TAKE
A22 CHECK PWS BRAKE LAMP CIRCUIT THRESHOLD		
WARNING: Before disconnecting any hydraulic lines, you must ensure that the pressure system is discharged. • Discharge the brake system as follows: a. Turn ignition switch OFF. b. Pump break pedal at least 20 times until you feel it become hard. • Remove high pressure banjo bolt below brake fluid reservoir on brake booster. • Disconnect quick-connect nipple from Anti-Lock High Pressure Gauge (Tool T85P-20215-A, or equivalent) and install nipple in place of the high pressure banjo bolt. NOTE: Be sure O-rings are installed correctly. • Connect Anti-Lock High Pressure Gauge to gauge nipple. **WARNING: DO NOT disconnect Anti-Lock High Pressure Gauge while system is under pressure.** • Reconnect pressure warning switch (PWS) 5-Pin connector. • Turn ignition switch ON. • When pump motor stops, disconnect pressure warning switch (PWS) 5-Pin plug. • Lower hydraulic accumulator pressure by slowly pumping the brake pedal until you have continuity between pressure warning switch electrical Pins 1 and 2. • Observe the hydraulic pressure gauge when continuity is reached. • If you missed the reading or want to reverify the reading: — Reconnect the 5-Pin plug. — Turn the ignition switch ON until the pump stops. — Disconnect the 5-Pin plug and reverify the readings.	100-110 Bar (1,450-1,595 psi) (OK) ▶ Any other reading (~~OK~~) ▶ 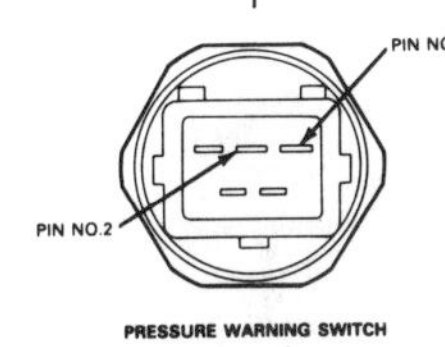	GO to Step **A23**. REPLACE pressure warning switch and pump motor relay.

Fig. 56 Diagnosis, warning lamp ON w/brake warning lamp OFF (Part 22 of 23). 1987 Thunderbird

Anti-Lock Warning Lamp On (With Brake Warning Lamp Off) — **Test A**

TEST STEP	RESULT ▶	ACTION TO TAKE
A23 CHECK PWS HARNESS GROUND		
• Check for continuity between pressure warning switch 5-Pin plug Pin 1 and body ground. 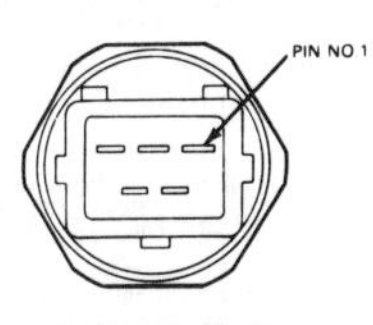	Continuity (OK) ▶ No continuity (~~OK~~) ▶	GO to Step **A24**. SERVICE or REPLACE cable harness (Circuit 430C).
A24 REVERIFY SYSTEM SYMPTOM		
• Reconnect all electrical connections. • Reinstall high pressure banjo bolt. **WARNING: Before disconnecting any hydraulic lines, you must ensure that the brake hydraulic pressure system is discharged.** • Discharge the brake system as follows: a. Turn ignition switch OFF. b. Pump brake pedal at least 20 times until you feel the pedal become hard. • Reverify symptom.	Symptom not present ▶ Symptom still present ▶	FAULT may have been a loose electrical connection. REPLACE electronic control module.

Fig. 56 Diagnosis, warning lamp ON w/brake warning lamp OFF (Part 23 of 23). 1987 Thunderbird

Anti-Lock Lamp On After Engine Starts (Brake Warning Lamp Off) — **Test B**

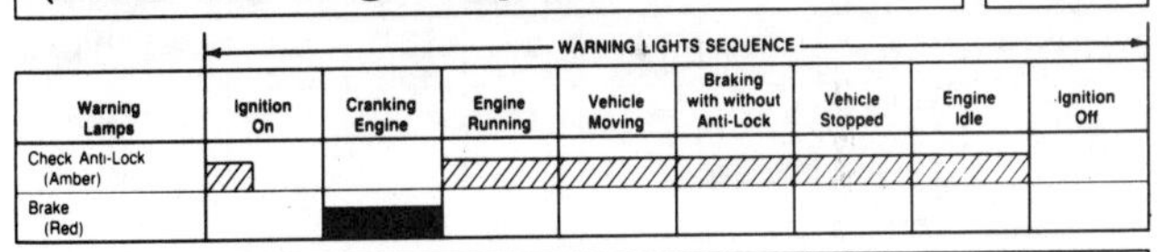

TEST STEP	RESULT ▶	ACTION TO TAKE
B1 CHECK CONTINUITY OF CIRCUIT 524		
• Ignition switch Off. • Disconnect 32-Pin plug from controller. • Connect ECV-IV Breakout box with Anti-Lock Test Adapter to the Anti-Lock 32-Pin plug wiring harness. • Check continuity between Breakout box Pins 40 and 6.	Continuity (~~OK~~) ▶ No continuity (OK) ▶	GO to Step **B1a**. GO to Step **B2**.
B1a CHECK CONTINUITY OF RH REAR SENSOR		
• Disconnect wheel sensor plug (RH rear). • Check for continuity between each sensor plug pin (sensor side) and vehicle ground. 	Continuity (~~OK~~) ▶ No continuity (OK) ▶	REPLACE sensor (RH rear). REPLACE or SERVICE cable harness. RECONNECT sensor plug.
B2 CHECK CONTINUITY OF CIRCUIT 522		
• Check for continuity between Breakout box Pins 40 and 5.	Continuity (~~OK~~) ▶ No continuity (OK) ▶	GO to Step **B2a**. GO to Step **B3**.

Fig. 57 Diagnosis, anti-lock ON after engine starts/brake warning lamp OFF (Part 1 of 3). 1987 Thunderbird

Anti-Lock Lamp On After Engine Starts (Brake Warning Lamp Off) — Test B

TEST STEP	RESULT ▶	ACTION TO TAKE
B2a CHECK CONTINUITY OF LH FRONT SENSOR		
• Disconnect wheel sensor plug (LH front). • Check for continuity between each sensor plug pin (sensor side) and vehicle ground.  LH FRONT SENSOR	Continuity (OK crossed) ▶	REPLACE sensor (LH front).
	No continuity (OK) ▶	REPLACE or SERVICE cable harness. CONNECT sensor plug.
B3 CHECK CONTINUITY OF CIRCUIT 519		
• Check for continuity between Breakout box Pins 40 and 4.	Continuity (OK crossed) ▶	GO to Step **B3a**.
	No continuity (OK) ▶	GO to Step **B4**.
B3a CHECK CONTINUITY OF LH REAR SENSOR		
• Disconnect sensor plug (LH rear). • Check for continuity between each sensor plug pin (sensor side) and vehicle ground.  LH REAR SENSOR	Continuity (OK crossed) ▶	REPLACE sensor (LH rear).
	No continuity (OK) ▶	REPLACE or SERVICE cable. CONNECT sensor plug.

Fig. 57 Diagnosis, anti-lock ON after engine starts/brake warning lamp OFF (Part 2 of 3). 1987 Thunderbird

Anti-Lock Lamp On After Engine Starts (Brake Warning Lamp Off) — Test B

TEST STEP	RESULT ▶	ACTION TO TAKE
B4 CHECK CONTINUITY OF CIRCUIT 516		
• Check for continuity between Breakout box Pins 40 and 3.	Continuity (OK crossed) ▶	GO to Step **B4a**.
	No continuity (OK) ▶	Test complete. If Anti-Lock lamp pattern remains, REPEAT Test B.
B4a CHECK CONTINUITY OF RH FRONT SENSOR		
• Disconnect wheel sensor plug (RH front). • Check for continuity between each sensor plug pin (sensor side) and vehicle ground. 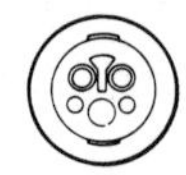 RH FRONT SENSOR	Continuity (OK crossed) ▶	REPLACE RH front sensor.
	No continuity (OK) ▶	REPLACE or SERVICE cable harness. CONNECT sensor plug.

Fig. 57 Diagnosis, anti-lock ON after engine starts/brake warning lamp OFF (Part 3 of 3). 1987 Thunderbird

Anti-Lock Warning Lamp On After Vehicle Starts To Move Or False Cycling Of Anti-Lock System — Test C

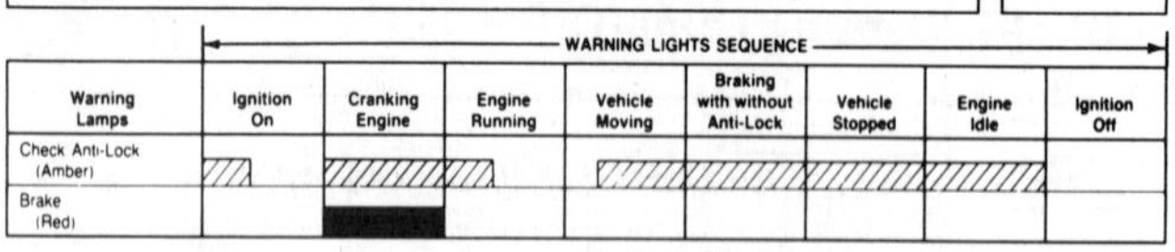

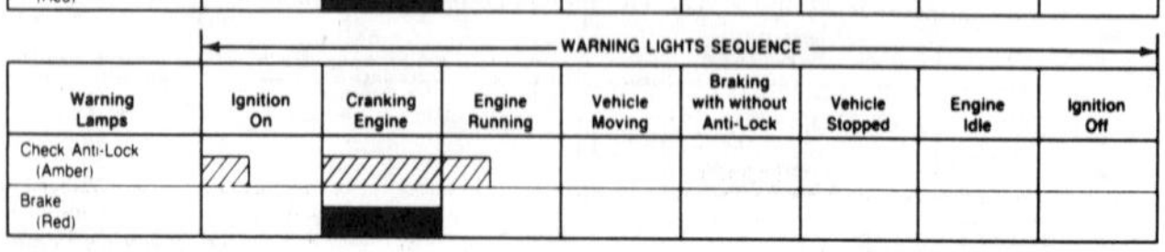

TEST STEP	RESULT ▶	ACTION TO TAKE
C1 MEASURE RH REAR SENSOR CIRCUIT RESISTANCE		
• Turn ignition switch OFF. • Disconnect 35-Pin plug from electronic controller. • Connect EEC-IV Breakout box tool T83-50-EEC-IV and anti-lock test adapter tool T85P-50-ASA or equivalents to anti-lock 35-Pin connector. • Set multi-meter to read resistance. • Measure resistance between Breakout box Pins 4 and 23.	800 to 1400 ohms (0.8 to 1.4K ohms) (OK) ▶	GO to Step **C2**.
	Any other reading (OK crossed) ▶	Go to Step **C1a**.
C1a MEASURE RH REAR SENSOR RESISTANCE		
• Disconnect sensor plug (RH rear). • Set multi-meter to read resistance. • Measure resistance of sensor at sensor plug. 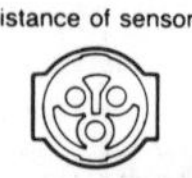 RH REAR SENSOR	800 to 1400 ohms (0.8 to 1.4K ohms) (OK) ▶	SERVICE or REPLACE cable harness (Circuit 523 or 524).
	Any other reading (OK crossed) ▶	REPLACE RH rear wheel sensor.

Fig. 58 Diagnosis, warning lamp ON after vehicle starts to move or false cycling of system (Part 1 of 5). 1987 Thunderbird

Anti-Lock Warning Lamp On After Vehicle Starts To Move Or False Cycling Of Anti-Lock System — Test C

TEST STEP	RESULT ▶	ACTION TO TAKE
C2 MEASURE LH FRONT SENSOR CIRCUIT RESISTANCE		
• Measure resistance between Breakout box Pins 5 and 22.	800 to 1400 ohms (0.8 to 1.4K ohms) (OK) ▶	GO to Step **C3**.
	Any other reading (OK crossed) ▶	GO to Step **C2a**.
C2a MEASURE LH FRONT SENSOR RESISTANCE		
• Disconnect wheel sensor plug (LH rear). • Measure resistance of sensor at sensor plug.  LH FRONT SENSOR	800 to 1400 ohms (0.8 to 1.4K ohms) (OK) ▶	SERVICE or REPLACE cable harness (Circuit 521 or 522).
	Any other reading (OK crossed) ▶	REPLACE LH front wheel sensor.
C3 MEASURE LH REAR SENSOR CIRCUIT RESISTANCE		
• Measure resistance between Breakout box Pins 4 and 21.	800 to 1400 ohms (0.8 to 1.4K ohms) (OK) ▶	GO to Step **C4**.
	Any other reading (OK crossed) ▶	GO to Step **C3a**.

Fig. 58 Diagnosis, warning lamp ON after vehicle starts to move or false cycling of system (Part 2 of 5). 1987 Thunderbird

Anti-Lock Warning Lamp On After Vehicle Starts To Move Or False Cycling Of Anti-Lock System	Test C

TEST STEP	RESULT ▶	ACTION TO TAKE
C3a MEASURE LH REAR SENSOR RESISTANCE		
• Disconnect sensor plug (LH rear). • Measure resistance of sensor at left rear sensor plug.  LH REAR SENSOR	800 to 1400 ohms (0.8 to 1.4K ohms) OK ▶	SERVICE or REPLACE cable harness (Circuit 518 or 519).
	Any other reading ~~OK~~ ▶	REPLACE LH rear wheel sensor.
C4 MEASURE RH FRONT SENSOR CIRCUIT RESISTANCE		
• Measure resistance between Breakout box Pins 3 and 20.	800 to 1400 ohms (0.8 to 1.4K ohms) OK ▶	GO to Step **C5**.
	Any other reading ~~OK~~ ▶	GO to Step **C4a**.
C4a MEASURE RH FRONT SENSOR RESISTANCE		
• Disconnect sensor plug (RH front). • Measure resistance of sensor at sensor plug.  RH FRONT SENSOR	800 to 1400 ohms (0.8 to 1.4K ohms) OK ▶	SERVICE or REPLACE cable harness (Circuit 514 or 516).
	Any other reading ~~OK~~ ▶	REPLACE RH front wheel sensor.

Fig. 58 Diagnosis, warning lamp ON after vehicle starts to move or false cycling of system (Part 3 of 5). 1987 Thunderbird

Anti-Lock Warning Lamp On After Vehicle Starts To Move Or False Cycling Of Anti-Lock System	Test C

TEST STEP	RESULT ▶	ACTION TO TAKE
C5 CHECK RH REAR SENSOR		
• Turn ignition switch OFF. • Turn air suspension switch in luggage compartment OFF, if so equipped. • Place vehicle on hoist and raise wheels clear of ground. Refer to Pre-Delivery Manual Section 50-04. • Set multi-meter on voltage range (2V-AC). • Measure voltage between Breakout box Pins 6 and 23 while spinning RH rear wheel at approximately 1 revolution per second.	Between 0.05 and 0.70 Vac OK ▶	GO to Step **C6**.
	Less than 0.05 or more than 0.70 Vac ~~OK~~ ▶	CHECK sensor mounting, air gap, or toothed wheel mounting. CORRECT as required.
C6 CHECK LH FRONT SENSOR		
• Measure voltage between Breakout box Pins 5 and 22 while spinning LH front wheel at approximately 1 revolution per second.	Between 0.05 and 0.70 Vac ▶	GO to Step **C7**.
	Less than 0.05 or more than 0.70 Vac ▶	CHECK sensor mounting, air gap, or toothed wheel mounting. CORRECT as required.
C7 CHECK LH REAR SENSOR		
• Measure voltage between Breakout box Pins 4 and 21 while spinning LH rear wheel at approximately 1 revolution per second.	Between 0.05 and 0.70 Vac ▶	GO to Step **C8**.
	Less than 0.05 or more than 0.07 Vac ▶	CHECK wheel sensor mounting, air gap, or toothed wheel mounting. CORRECT as required.

Fig. 58 Diagnosis, warning lamp ON after vehicle starts to move or false cycling of system (Part 4 of 5). 1987 Thunderbird

Anti-Lock Warning Lamp on After Vehicle Starts to Move or False Cycling of Anti-Lock System	Test C

TEST STEP	RESULT ▶	ACTION TO TAKE
C8 CHECK RH FRONT SENSOR		
• Measure voltage between Breakout box Pins 3 and 20 while spinning RH front wheel at approximately 1 revolution per second.	Between 0.05 and 0.70 Vac ▶	GO to Step **C9**.
	Less than 0.05 or more than 0.70 Vac ▶	CHECK wheel sensor mounting, air gap, or toothed wheel mounting. CORRECT as required.
C9 CHECK FRONT WHEEL BEARINGS		
• Check front wheel bearing end play. • Inspect each toothed sensor ring visually for damaged teeth. NOTE: Turn air suspension switch ON when vehicle is off hoist.	Loose or damaged parts ▶	ADJUST bearings or REPLACE faulty parts.
	Not loose or damaged ▶	REVERIFY symptom.

Fig. 58 Diagnosis, warning lamp ON after vehicle starts to move or false cycling of system (Part 5 of 5). 1987 Thunderbird

Anti-Lock Warning Lamp And Brake Warning Lamp On And/Or Pump Motor Runs More Than 60 Seconds	Test D

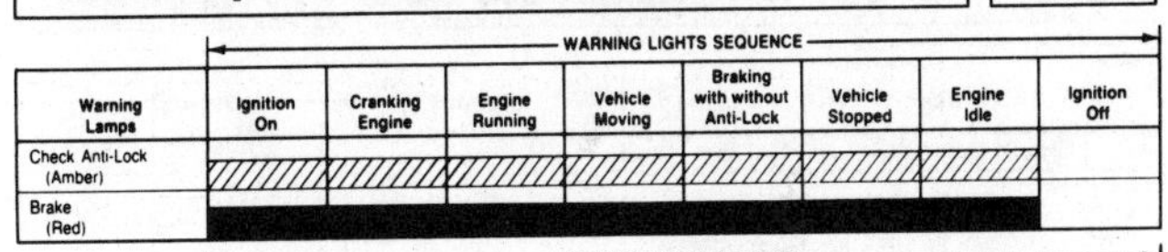

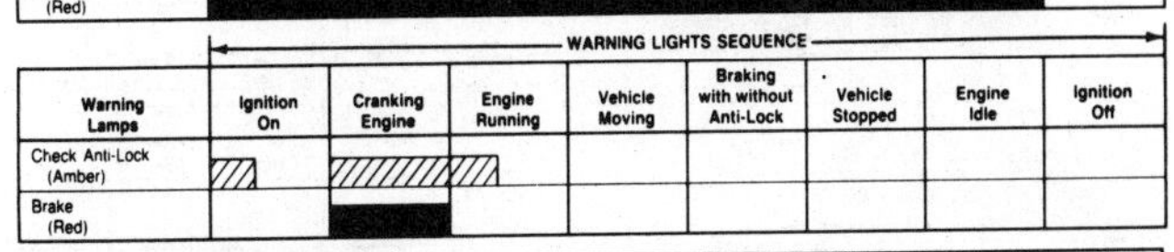

TEST STEP	RESULT ▶	ACTION TO TAKE
D1 CHECK PUMP MOTOR OPERATION		
• Turn ignition OFF. • Pump brake pedal at least 20 times until it becomes hard. • Turn ignition switch ON. • Pump motor should run. **NOTE: If pump motor is allowed to run continuously for approximately 20 minutes, a thermal safety switch (inside motor) will shut off motor. A 2-to-10 minute cool-down period is typical before normal operation can resume.**	OK ▶	GO to **D2**.
	~~OK~~ ▶	GO to **D1a**.

Fig. 59 Diagnosis, warning lamp & brake warning lamp ON and/or pump motor runs more than 60 seconds (Part 1 of 10). 1987 Thunderbird

Anti-Lock Warning Lamp And Brake Warning Lamp On And/Or Pump Motor Runs More Than 60 Seconds — **Test D**

TEST STEP	RESULT ▶	ACTION TO TAKE
D1a PUMP MOTOR UNIT		
• Disconnect 4-Pin plug on pump motor unit. • Turn ignition switch ON. • Set multi-meter on 20 Volt DC. • Connect meter to the 4-Pin plug on the harness side. (Use one negative and one positive pin). • Observe voltmeter. 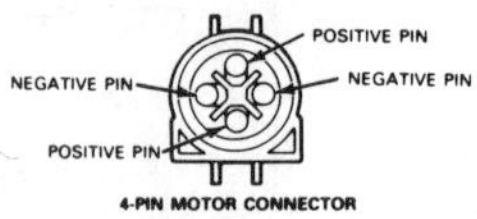 	More than 10V DC (OK) ▶ Less than 10V DC (~~OK~~) ▶	GO to **D2**. GO to **D1b**.
D1b CHECK CONTINUITY CIRCUITS 538a AND 538b		
• Disconnect pump motor relay. • Check for continuity between motor relay socket Pin 3 (harness side) and each positive pin of the 4-Pin motor connector (harness side). 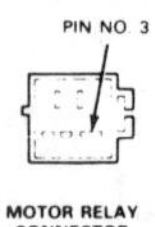 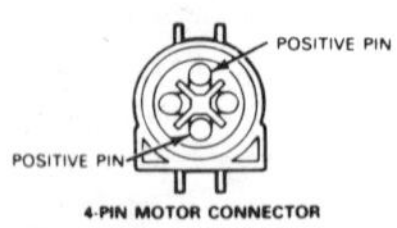 	(OK) ▶ (~~OK~~) ▶	GO to **D1c**. SERVICE Circuits 537, 538a and 538b.

Fig. 59 Diagnosis, warning lamp & brake warning lamp ON and/or pump motor runs more than 60 seconds (Part 2 of 10). 1987 Thunderbird

Anti-Lock Warning Lamp And Brake Warning Lamp On And/Or Pump Motor Runs More Than 60 Seconds — **Test D**

TEST STEP	RESULT ▶	ACTION TO TAKE
D1c CHECK CONTINUITY CIRCUITS 430D AND 430F		
• Check for continuity between ground and each negative pin of 4-Pin motor connector (harness side). 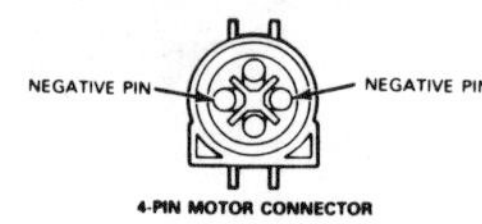 	(OK) ▶ (~~OK~~) ▶	GO to **D1d**. SERVICE or REPLACE Circuit 4301.
D1d CHECK CONTINUITY CIRCUIT 537		
• Turn ignition switch OFF. • Remove positive battery cable from battery. • Disconnect pump motor relay. 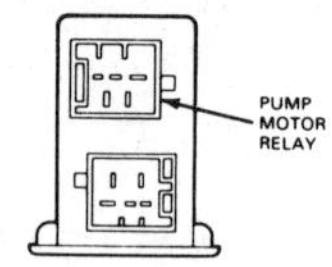 • Check continuity between battery positive cable and motor relay socket Pin 4 (harness side). 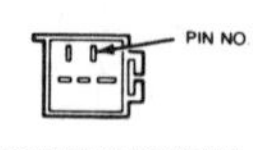 	Continuity (OK) ▶ No continuity (~~OK~~) ▶	RECONNECT battery positive cable and GO to Step **D1e**. SERVICE or REPLACE fuse link and/or Circuit 537.

Fig. 59 Diagnosis, warning lamp & brake warning lamp ON and/or pump motor runs more than 60 seconds (Part 3 of 10). 1987 Thunderbird

Anti-Lock Warning Lamp And Brake Warning Lamp On And/Or Pump Motor Runs More Than 60 Seconds — **Test D**

TEST STEP	RESULT ▶	ACTION TO TAKE
D1e CHECK CONTINUITY CIRCUIT 298		
• Connect multi-meter to motor relay socket Pin 2 (harness side) and ignition switch lock pin (On). • Check continuity. 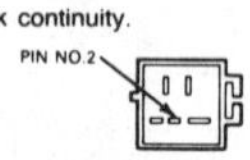 	Continuity (OK) ▶ No continuity (~~OK~~) ▶	GO to Step **D1f**. REPLACE or SERVICE Circuit 298A and/or fuse link.
D1f CHECK CONTINUITY CIRCUIT 430K		
• Disconnect PWS 5-pin plug. • Check for continuity between pressure warning switch 5-Pin plug Pin 1 (harness side) and ground. 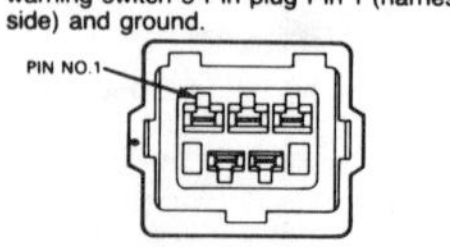 	Continuity (OK) ▶ No continuity (~~OK~~) ▶	GO to Step **D1g**. REPLACE or SERVICE Circuit 430K.
D1g CHECK CONTINUITY CIRCUIT 539		
• Connect meter to 5-Pin plug Pin 4 (harness side) and motor relay socket Pin 1 (harness side). • Check for continuity. 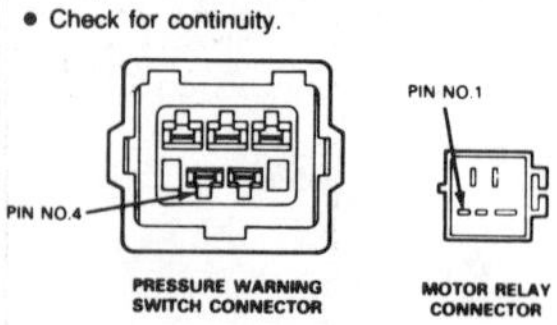 	Continuity (OK) ▶ No continuity (~~OK~~) ▶	GO to Step **D1h**. REPLACE or SERVICE Circuit 539.

Fig. 59 Diagnosis, warning lamp & brake warning lamp ON and/or pump motor runs more than 60 seconds (Part 4 of 10). 1987 Thunderbird

Anti-Lock Warning Lamp And Brake Warning Lamp On And/Or Pump Motor Runs More Than 60 Seconds — **Test D**

TEST STEP	RESULT ▶	ACTION TO TAKE
D1e CHECK CONTINUITY CIRCUIT 298		
• Connect multi-meter to motor relay socket Pin 2 (harness side) and ignition switch lock pin (On). • Check continuity. 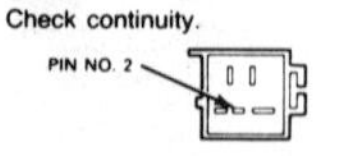 	Continuity (OK) ▶ No continuity (~~OK~~) ▶	GO to Step **D1f**. REPLACE or SERVICE Circuit 298B and/or fuse link.
D1f CABLE HARNESS PRESSURE WARNING SWITCH		
• Disconnect 5-Pin plug of PWS. • Check for continuity between pressure warning switch 5-Pin plug Pin 1 (harness side) and ground. 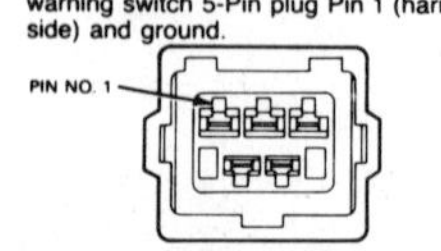 	Continuity (OK) ▶ No continuity (~~OK~~) ▶	GO to Step **D1g**. REPLACE or SERVICE Circuit 430K.
D1g CHECK CONTINUITY CIRCUIT 539		
• Connect meter to 5-Pin plug Pin 4 (harness side) and motor relay socket Pin 1 (harness side). • Check for continuity. 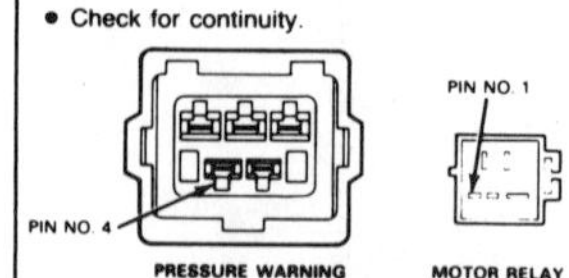 	Continuity (OK) ▶ No continuity (~~OK~~) ▶	GO to Step **D1h**. REPLACE or SERVICE Circuit 539.

Fig. 59 Diagnosis, warning lamp & brake warning lamp ON and/or pump motor runs more than 60 seconds (Part 5 of 10). 1987 Thunderbird

Anti-Lock Warning Lamp And Brake Warning Lamp On And/Or Pump Motor Runs More Than 60 Seconds — **Test D**

TEST STEP	RESULT ▶	ACTION TO TAKE
D1k CHECK PUMP MOTOR RELAY (CONTINUED)		
• Connect meter and check continuity between motor relay Pins 3 and 4.	No continuity OK ▶	GO to Step **D1m**.
	Continuity ~~OK~~ ▶	REPLACE motor relay and pressure warning switch.
D1m CHECK PUMP MOTOR RELAY (CONTINUED)		
• Connect battery to motor relay terminals 1 and 2. • Check continuity between relay Pins 3 and 4 with multi-meter.	Continuity OK ▶	GO to Step **D1n**.
	No continuity ~~OK~~ ▶	REPLACE motor relay and pressure warning switch.

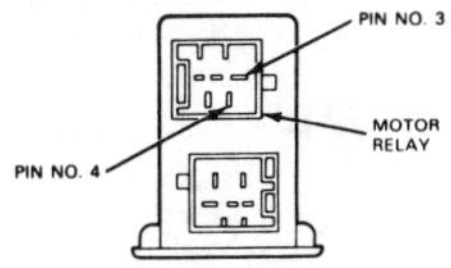

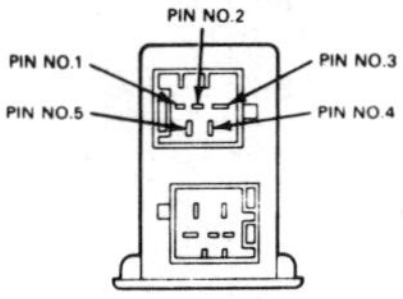

Fig. 59 Diagnosis, warning lamp & brake warning lamp ON and/or pump motor runs more than 60 seconds (Part 6 of 10). 1987 Thunderbird

Anti-Lock Warning Lamp And Brake Warning Lamp On And/Or Pump Motor Runs More Than 60 Seconds — **Test D**

TEST STEP	RESULT ▶	ACTION TO TAKE
D1n CHECK PUMP MOTOR		
• Apply 12 volts to 4-Pin plug of motor (use one negative and one positive pin). • Pump motor should run.	OK ▶	End of test. CONNECT all electrical plugs and relay. REVERIFY symptom.
	~~OK~~ ▶	REPLACE pump/ motor assembly and pump motor relay.
D2 CHECK PUMP MOTOR UNIT		
• Turn ignition switch OFF. • Pump brake pedal 20 times to discharge system. • Connect ammeter between battery positive cable and battery positive terminal. • Turn Off any electrical components. • Connect 4-Pin motor plug. • Turn ignition switch ON. • Measure pump motor current.	Current more than 25 amps ~~OK~~ ▶	REPLACE pump motor unit.
	Current less than 25 amps OK ▶	GO to Step **D2a**.
D2a CHECK PUMP MOTOR		
• Turn ignition switch OFF. • Pump brake pedal at least 20 times, until brake pedal becomes hard. • Turn ignition switch ON. • Measure time pump takes to shut OFF.	Under 60 seconds OK ▶	GO to Step **D2b**.
	Over 60 seconds (or motor never turns on) ~~OK~~ ▶	CHECK for corroded connections at: • Motor 4-Pin plug. • Body ground (Circuit 430L). • Motor relay Pin 4 socket (Circuit 537). • Battery to Pin 4. GO to Step **D2b**.

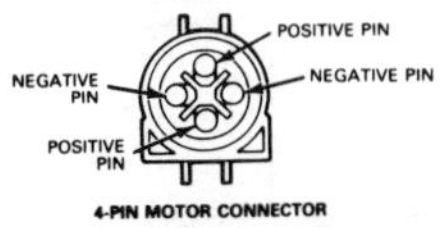

Fig. 59 Diagnosis, warning lamp & brake warning lamp ON and/or pump motor runs more than 60 seconds (Part 7 of 10). 1987 Thunderbird

Anti-Lock Warning Lamp And Brake Warning Lamp On And/Or Pump Motor Runs More Than 60 Seconds — **Test D**

TEST STEP	RESULT ▶	ACTION TO TAKE
D2b CHECK LOW PRESSURE FLOW		
• Turn ignition switch OFF. • Disconnect low pressure hose from pump and allow fluid to flow into a suitable container. **NOTE: Discard fluid after test.**	Free fluid flow OK ▶	CONNECT low pressure hose. FILL reservoir to MAX. GO to **D2c**.
	Restricted flow ~~OK~~ ▶	SERVICE or REPLACE reservoir and/or low pressure hose as required.
D2c CHECK VOLTAGE TO PUMP MOTOR		
• Turn ignition switch OFF. • Pump brake pedal to discharge pressure system. • Set meter to 20 Volts DC range. • Connect voltmeter in parallel at 4-Pin motor plug. • Turn ignition switch ON.	With pump running: Voltage over 8 Volts DC OK ▶	GO to Step **D3**.
	Voltage under 8 Volts DC ~~OK~~ ▶	CHECK Circuits 430D, 430F, 538A, 538B, Pin 4 to battery and relay Pins 3 and 4 for voltage drop. SERVICE or REPLACE as necessary.

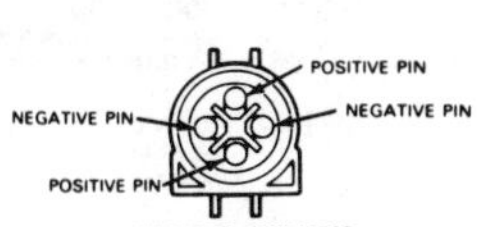

Fig. 59 Diagnosis, warning lamp & brake warning lamp ON and/or pump motor runs more than 60 seconds (Part 8 of 10). 1987 Thunderbird

Anti-Lock Warning Lamp And Brake Warning Lamp On And/Or Pump Motor Runs More Than 60 Seconds — **Test D**

TEST STEP	RESULT ▶	ACTION TO TAKE
D3 ACCUMULATOR: PRE-CHARGE		
• Vehicle must be cooled to room temperature. **WARNING: Before disconnecting any hydraulic lines, you must ensure that the pressure system is discharged.** • To discharge hydraulic accumulator pressure, turn ignition OFF, pump brake pedal at least 20 times until you feel it become hard. • Remove high pressure banjo bolt below brake fluid reservoir on brake booster. • Install Anti-Lock Pressure Gauge Tool T85P-20215A or equivalent. • Turn ignition switch ON and read accumulator precharge pressure. (Gauge needle will spring to this point). **Note: Gauge needle reading should spring to 40-90 bar (600-1325 psi) and climb to 16,203-19,306 kPa bar (2350-2800 psi).** • If reading was missed, discharge accumulator as described and repeat ignition ON sequence.	4137-9135 kPa (600-1325 psi) OK ▶	GO to Step **D4**.
	Under 4137 kPa (600 psi) ~~OK~~ ▶	REPLACE accumulator.
	Over 9135 kPa (1325 psi) ~~OK~~ ▶	REPLACE accumulator.
D4 CHECK HYDRAULIC ACTUATION UNIT		
• Turn ignition switch ON; wait until pump motor stops. Wait 3 more minutes to stabilize gauge pressure. • Read pressure gauge. • Wait 5 minutes and read pressure gauge again to determine the pressure loss over those 5 minutes.	Pressure loss less than 10 bar (140 psi) on gauge OK ▶	GO to Step **D5**.
	More than 10 bar (140 psi) ~~OK~~ ▶	CHECK for external leakage at actuation assembly and SERVICE. If no external leakage is found, GO to Step **D4a**.

Fig. 59 Diagnosis, warning lamp & brake warning lamp ON and/or pump motor runs more than 60 seconds (Part 9 of 10). 1987 Thunderbird

Anti-Lock Warning Lamp And Brake Warning Lamp On And/Or Pump Motor Runs More Than 60 Seconds — Test D

TEST STEP	RESULT ▶	ACTION TO TAKE
D4a CHECK FOR HYDRAULIC LEAKS		
• Check for brake fluid leaks in the following areas: a. Brake lines and calipers. b. High and low pressure hoses on actuation assembly. c. Reservoir seals and seams. d. Accumulator. **NOTE: A small amount of leakage from pressure warning switch is allowable. If any leakage is found below switch, wipe off excess fluid and check for excessive leakage.**	No leakage found (OK) ▶	GO to **D5**.
	Leakage found (NOT OK) ▶	SERVICE or REPLACE components. CHECK system operation.
D5 CHECK PUMP PRESSURE		
• With the pressure gauge still attached and ignition switch still ON, pump brake pedal to decrease pressure until pump motor restarts.	13100-15169 kPa (1900-2200 psi) when pump starts (OK) ▶	GO to Step **D6**.
	Less than or more than 13100-15169 kPa (1900-2200 psi) (NOT OK) ▶	REPLACE pressure warning switch and pump motor relay.
D6 CHECK PRESSURE WARNING SWITCH		
• With the ignition switch still ON and the pressure gauge connected, observe the pressure when the pump motor stops running.	16203 kPa (2350-2800 psi) when pump motor stops (OK) ▶	If pump motor takes longer than 60 seconds to reach 16203 kPa (2350-2800 psi), REPLACE pump/motor assembly. REVERIFY symptom.
	Less than 16203 kPa (2350 psi) or over 19306 kPa (2800 psi) (NOT OK) ▶	REPLACE pressure warning switch and pump motor relay.

Fig. 59 Diagnosis, warning lamp & brake warning lamp ON and/or pump motor runs more than 60 seconds (Part 10 of 10). 1987 Thunderbird

Anti-Lock Warning Lamp Intermittently On — Test E

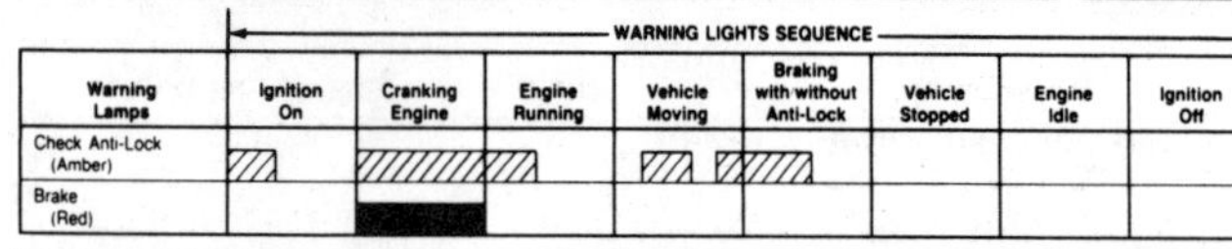

TEST STEP	RESULT ▶	ACTION TO TAKE
E1 CHECK FLI AND PWS CIRCUIT		
• Disconnect 32-Pin plug from electronic controller. • Connect EEC-IV Breakout box, T83L-50-EEC-IV with Anti-Lock Test Adapter T85P-50-ASA or equivalent to the Anti-Lock 32-Pin plug wiring harness. • Turn ignition switch ON. • Set multi-meter to read resistance. • Measure the resistance between Breakout box Pins 25 and 27.	Less than 5 ohms (OK) ▶	GO to Step **E2**.
	Greater than 5 ohms (NOT OK) ▶	GO to Step **E1a**.
E1a CHECK FLI ANTI-LOCK WARNING CIRCUIT		
• Disconnect 5-Pin plug on reservoir fluid level indicator (FLI). • Measure resistance between fluid level indicator electrical Pins 1 and 2 (with brake fluid at maximum fluid level).	Less than 2 ohms (OK) ▶	GO to Step **E1b**.
	Greater than 2 ohms (NOT OK) ▶	REPLACE fluid level indicator.

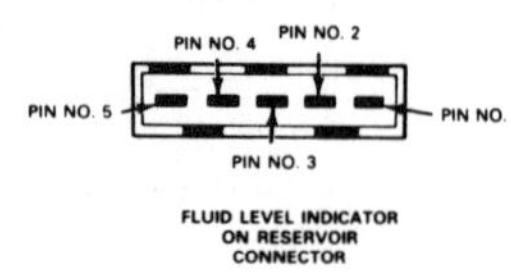

Fig. 60 Diagnosis, warning lamp intermittently ON (Part 1 of 5). 1987 Thunderbird

Anti-Lock Warning Lamp Intermittently On — Test E

TEST STEP	RESULT ▶	ACTION TO TAKE
E1b CHECK PWS ANTI-LOCK WARNING CIRCUIT (NO SYSTEM PRESSURE)		
• Turn ignition switch OFF. • Pump brake pedal to discharge pressure system. • Disconnect 5-Pin plug at pressure warning switch (PWS). • Check for continuity between pressure warning switch Pins 3 and 5.	No Continuity (OK) ▶	GO to Step **E1c**.
	Continuity (NOT OK) ▶	REPLACE pressure warning switch and pump motor relay.

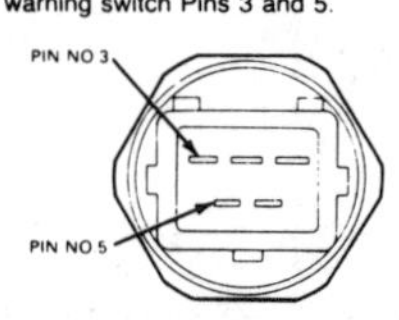

TEST STEP	RESULT ▶	ACTION TO TAKE
E1c CHECK ELECTRONIC MODULE TO FLI WIRE		
• Check for continuity between Breakout box Pin 25 and FLI 5-Pin plug Pin 1 (harness side).	Continuity (OK) ▶	GO to Step **E1d**.
	No continuity (NOT OK) ▶	REPLACE or SERVICE cable harness (Circuit 512).

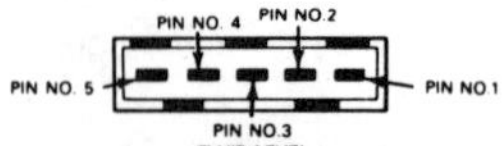

Fig. 60 Diagnosis, warning lamp intermittently ON (Part 2 of 5). 1987 Thunderbird

Anti-Lock Warning Lamp Intermittently On — Test E

TEST STEP	RESULT ▶	ACTION TO TAKE
E1d CHECK FLI TO PWS WIRE		
• Check for continuity between 5-Pin fluid level plug Pin 2 (harness side) and 5-Pin pressure warning switch plug Pin 3 (harness side).	Continuity (OK) ▶	GO to Step **E1e**.
	No continuity (NOT OK) ▶	SERVICE or REPLACE cable harness (Circuit 549).

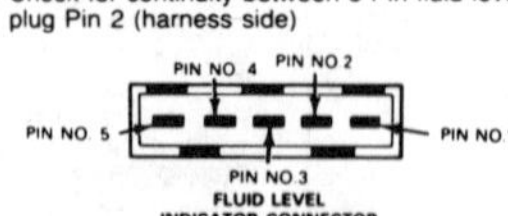

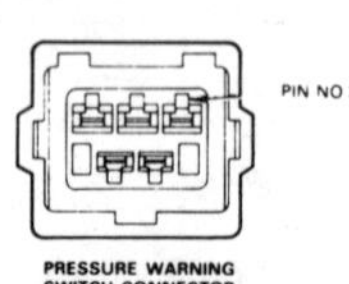

TEST STEP	RESULT ▶	ACTION TO TAKE
E1e CHECK PWS TO ELECTRONIC CONTROLLER WIRE		
• Check for continuity between 5-Pin pressure warning switch plug Pin 5 (harness side) and Breakout box Pin 27.	Continuity (OK) ▶	TURN Ignition OFF. CONNECT all electrical connections. REVERIFY symptom.
	No continuity (NOT OK) ▶	SERVICE or REPLACE cable harness (Circuit 535).

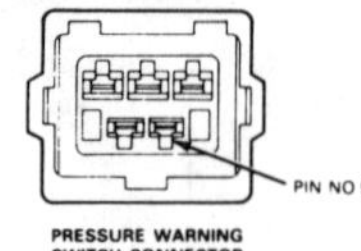

Fig. 60 Diagnosis, warning lamp intermittently ON (Part 3 of 5). 1987 Thunderbird

Anti-Lock Warning Lamp Intermittently On — Test E

TEST STEP	RESULT ▶	ACTION TO TAKE
E2 CHECK ISOLATION TEST FLI AND PWS		
• Check for continuity between Breakout box Pin 25 and body ground.	Continuity ~~OK~~ ▶	GO to Step **E2a**.
	No continuity OK ▶	REVERIFY symptom.
E2a CHECK CONTINUITY OF FLI SWITCH		
• Disconnect FLI plug and check for continuity between socket Pin 1 and Pins 3, 4, and 5 and Pin 2 and Pins 3, 4, and 5. 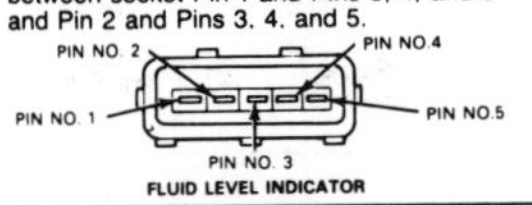	Continuity ~~OK~~ ▶	REPLACE FLI.
	No continuity OK ▶	GO to Step **E2b**.
E2b CHECK CONTINUITY CIRCUIT 512		
• Check for continuity between FLI plug Pin 1 (harness side) and body ground. 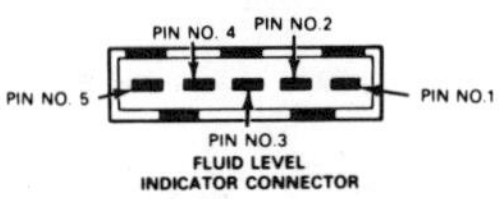	Continuity ~~OK~~ ▶	SERVICE or REPLACE Circuit 512.
	No continuity OK ▶	GO to Step **E2c**.
E2c CHECK CONTINUITY CIRCUIT 549		
• Disconnect 5-Pin plug from PWS. Check for continuity between FLI plug Pin 2 (harness side) and body ground. 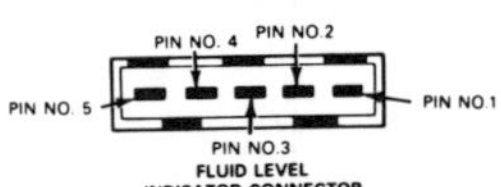	Continuity ~~OK~~ ▶	SERVICE or REPLACE Circuit 549.
	No continuity OK ▶	GO to Step **E2d**.

Fig. 60 Diagnosis, warning lamp intermittently ON (Part 4 of 5). 1987 Thunderbird

Anti-Lock Warning Lamp Intermittently On — Test E

TEST STEP	RESULT ▶	ACTION TO TAKE
E2d CHECK CONTINUITY OF PWS		
• Check for continuity from PWS 5-Pin socket Pin 3 to body ground, and from Pin 5 to body ground. 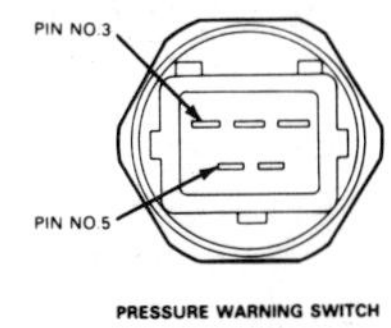	Continuity ~~OK~~ ▶	REPLACE pressure warning switch and pump motor relay.
	No continuity OK ▶	GO to Step **E2e**.
E2e CHECK CONTINUITY CIRCUIT 535		
• Check for continuity between 5-Pin PWS plug (harness side) Pin 5 and body ground. 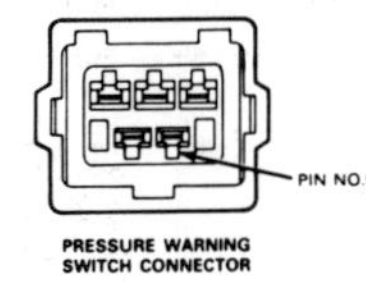	Continuity ~~OK~~ ▶	SERVICE or REPLACE Circuit 535.
	No continuity OK ▶	CONNECT all plugs and REVERIFY symptom.

Fig. 60 Diagnosis, warning lamp intermittently ON (Part 5 of 5). 1987 Thunderbird

Brake Warning Lamp On (With Anti-Lock Lamp Off, Parking Brake Released And Brake Lining Wear Checked) — Test F

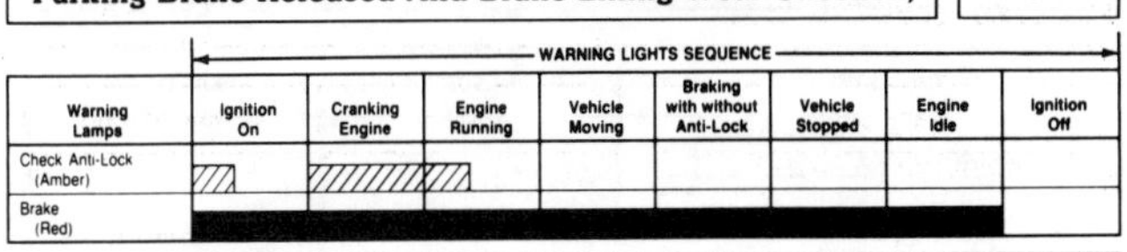

TEST STEP	RESULT ▶	ACTION TO TAKE
F1 CHECK BRAKE FLUID LEVEL		
• Turn ignition switch ON. • Pump brake pedal until pump motor starts. • When pump motor stops check brake fluid level.	Low ~~OK~~ ▶	CHECK system for external leaks. SERVICE as required.
	Normal OK ▶	GO to Step **F2**.
F2 CHECK CONTINUITY FLI SWITCH		
• Disconnect 5-Pin plug on fluid reservoir cap. • Set multi-meter on 200 ohm range. Connect to reservoir warning cap Pins 3 and 4. 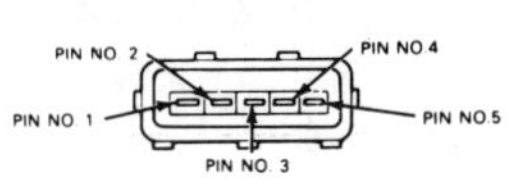	Above 10 ohms OK ▶	GO to Step **F3**.
	Below 10 ohms ~~OK~~ ▶	REPLACE 5-Pin fluid level indicator cap.

Fig. 61 Diagnosis, brake warning lamp ON w/anti-lock lamp OFF, parking brake released & brake lining wear checked (Part 1 of 2). 1987 Thunderbird

Brake Warning Lamp On (With Anti-Lock Lamp Off, Parking Brake Released And Brake Lining Wear Checked) — Test F

TEST STEP	RESULT ▶	ACTION TO TAKE
F3 CHECK CONTINUITY PWS SWITCH (CONTINUED)		
• Turn ignition switch ON, wait until motor stops running. • Disconnect Pressure Warning Switch 5-Pin plug and connect multi-meter to Pins 1 and 2 (switch side). • Check for continuity. 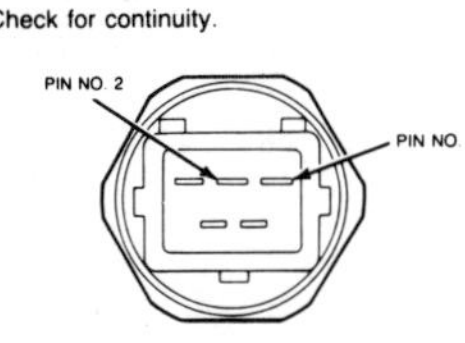	Continuity ~~OK~~ ▶	REPLACE pressure warning switch and pump motor relay.
	No continuity OK ▶	GO to Step **F4**.
F4 CHECK WIRING CONTINUITY		
• Check cable harness 430H, 512, 547B, 547A, 549, 977 Circuits. (Refer to vehicle Wiring Diagram.)	OK ▶	CHECK parking brake electrical circuit including parking brake switch. REVERIFY symptom.
	~~OK~~ ▶	SERVICE or REPLACE as required.

Fig. 61 Diagnosis, brake warning lamp ON w/anti-lock lamp OFF, parking brake released & brake lining wear checked (Part 2 of 2). 1987 Thunderbird

No Anti-Lock Warning Lamp On When Ignition Switch Turned On — Test G

Warning Lamps	Ignition On	Cranking Engine	Engine Running	Vehicle Moving	Braking with without Anti-Lock	Vehicle Stopped	Engine Idle	Ignition Off
	WARNING LIGHTS SEQUENCE							
Check Anti-Lock (Amber)								
Brake (Red)		■						

TEST STEP	RESULT ▶	ACTION TO TAKE
G1 CHECK FUSE AND FUSE LINKS		
• Check in-line fuse links with ignition turned ON.	Fuse Links OK ▶	GO to Step **G2**.
	Fuse Links ~~OK~~ ▶	SERVICE or REPLACE as required.
G2 CHECK WARNING LAMP BULB		
• Check warning lamp bulb.	OK ▶	GO to Step **G3**.
	~~OK~~ ▶	REPLACE bulb.
G3 CHECK WARNING LAMP OPERATION		
• Turn ignition switch ON. • Disconnect 32-Pin connector from control module.	Anti-Lock lamp goes ON ▶	SERVICE or REPLACE Circuit 606.
	Anti-Lock lamp is not ON ▶	SERVICE or REPLACE connector to 14401 wire harness.

Fig. 62 Diagnosis, no anti-lock warning lamp on when ignition switch is turned ON. 1987 Thunderbird

Spongy Brake Pedal With/Without Anti-Lock Function (No Warning Lamp) — Test H

Warning Lamps	Ignition On	Cranking Engine	Engine Running	Vehicle Moving	Braking with without Anti-Lock	Vehicle Stopped	Engine Idle	Ignition Off
	WARNING LIGHTS SEQUENCE							
Check Anti-Lock (Amber)	▨	▨	▨					
Brake (Red)		■						

TEST STEP	RESULT ▶	ACTION TO TAKE
H1 CHECK COMPONENT MOUNTING		
• Check for proper brake pedal and hydraulic unit attachment. • Bleed brakes as outlined.	Pedal still spongy ~~OK~~ ▶	GO to Step **H2**.
	Pedal feels normal OK ▶	Condition corrected.
H2 BLEED BRAKE SYSTEM		
• Turn off air suspension switch in luggage compartment if so equipped. • Rebleed brake system. • Turn on air suspension switch when vehicle is off hoist if so equipped.	Pedal still spongy ~~OK~~ ▶	REPLACE actuation assembly.
	Pedal feels normal OK ▶	Condition corrected.

Fig. 63 Diagnosis, spongy brake pedal with/without anti-lock function, no warning lamp. 1987 Thunderbird

Poor Vehicle Tracking During Anti-Lock Function (Warning Lamp Off) — Test J

Warning Lamps	Ignition On	Cranking Engine	Engine Running	Vehicle Moving	Braking with without Anti-Lock	Vehicle Stopped	Engine Idle	Ignition Off
	WARNING LIGHTS SEQUENCE							
Check Anti-Lock (Amber)	▨	▨	▨					
Brake (Red)		■						

TEST STEP	RESULT ▶	ACTION TO TAKE
J1 VERIFY CONDITION		
• Verify condition exists as reported. • Turn air suspension off if so equipped. • Bleed brake system per shop manual for Anti-Lock brake system. • Turn air suspension on when vehicle is off hoist.	Vehicle tracks properly ▶	Condition corrected.
	Vehicle still tracks poorly ▶	GO to **J2**.
J2 CHECK ANTI-LOCK OPERATION — LH FRONT WHEEL		
• Turn air suspension off if so equipped. • Lift vehicle and rotate wheels to assure they turn freely. • Turn ignition switch OFF. • Disconnect 32-Pin plug from electronic controller. • Connect EEC-IV Breakout box, Tool T83L-50-EEC-IV with Anti-Lock test adapter, Tool T85P-50-ASA to the Anti-Lock 32-Pin plug wiring harness. • Short Pins 18, 14 and 31 to each other at Breakout box. • Apply moderate brake pedal effort and check that LH front wheel will not turn. • Check to see that LH front wheel turns freely with ignition switch ON. • Turn air suspension on when vehicle is off hoist if so equipped. **CAUTION: DO NOT LEAVE IGNITION ON FOR MORE THAN 1 MINUTE MAXIMUM, OR SOLENOID VALVE DAMAGE MAY RESULT.**	If wheel turns freely ▶	TURN ignition switch Off. DISCONNECT wire leads. Go to Step **J3**.
	If wheel does not turn freely or pedal drops ▶	REPLACE solenoid valve block.

Fig. 64 Diagnosis, poor vehicle tracking during anti-lock function warning lamp OFF (Part 1 of 2). 1987 Thunderbird

Poor Vehicle Tracking During Anti-Lock Function (Warning Lamp Off) — Test J

TEST STEP	RESULT ▶	ACTION TO TAKE
J3 CHECK ANTI-LOCK OPERATION — RH FRONT WHEEL		
• Turn air suspension off. • Short Pins 18, 32 and 13 to each other at Breakout box. • Apply moderate brake effort. Check that RH front wheel does not turn. • Check that RH front wheel turns freely with ignition switch ON. **CAUTION: DO NOT LEAVE IGNITION ON MORE THAN 1 MINUTE MAXIMUM OR SOLENOID VALVE DAMAGE MAY RESULT.**	Wheel turns freely ▶	TURN ignition switch OFF. DISCONNECT wire leads. GO to Step **J4**.
	Wheel does not turn freely or brake pedal drops ▶	REPLACE solenoid valve block.
J4 CHECK ANTI-LOCK OPERATION — REAR WHEELS		
• Turn air suspension off if so equipped. • Short Pins 18, 30 and 12 to each other at Breakout box. • Apply moderate brake pedal pressure. Check that rear wheels will not turn. • Check that rear wheels turn freely with ignition switch ON. • Turn air suspension on when vehicle is off hoist if so equipped. **CAUTION: DO NOT LEAVE IGNITION ON MORE THAN 1 MINUTE MAXIMUM OR SOLENOID VALVE DAMAGE MAY RESULT.**	Wheels turn freely ▶	TURN ignition switch OFF. DISCONNECT wire lead and Breakout box. LOWER vehicle. REVERIFY symptom.
	Wheels do not turn freely or brake pedal drops ▶	REPLACE solenoid valve block.

Fig. 64 Diagnosis, poor vehicle tracking during anti-lock function warning lamp OFF (Part 2 of 2). 1987 Thunderbird

1988 Continental & All 1989 Models

INDEX

DESCRIPTION

These models are equipped with a four wheel anti-lock brake system (ABS), **Fig. 1.** The system prevents lockup by automatically modulating brake system pressure during an emergency stop situation.

By controlling wheel lockup, the vehicle's operator can maintain steering control and stop the vehicle in the shortest possible distance, under most driving conditions.

The anti-lock brake system controls the front brakes separately and the rear brakes as a pair, whenever wheel lockup begins. The brake pedal application force required to engage the system function may vary with road surface conditions. A dry surface will require a higher force, while a slippery surface will require a much less force.

During system operation, the vehicle's operator will sense a slight pulsation in the brake pedal, accompanied by a rise in brake pedal height and a clicking sound.

SYSTEM COMPONENTS

HYDRAULIC ACTUATION UNIT

The hydraulic actuation unit, **Figs. 2 and 3,** consists of two sections, the master cylinder and brake booster, which are arranged in the conventional fore and aft sequence. The booster contains a control valve, located in a parallel bore above the master cylinder.

ELECTRIC PUMP ASSEMBLY

This system uses high pressure brake fluid stored in a hydraulic accumulator, **Fig. 4,** for power assist as well as for rear wheel braking. The accumulator is a gas filled pressure chamber that stores brake fluid up to 2600 psi. The fluid is pressurized by an electric motor driven pump. The pump is switched on and off automatically by a pressure sensing switch through a relay circuit.

SOLENOID VALVE BLOCK ASSEMBLY

The solenoid valve block assembly, **Fig. 5,** contains three pairs of solenoid valves, one pair for each front wheel and the third pair for both rear wheels. The paired solenoid valves are inlet and outlet valves, with the inlet valve normally open and the outlet valve normally closed. During anti-lock operation, the inlet and outlet valves are alternately opened and closed approximately 10 times per second to provide pressure modulation to wheel of impending lockup. The solenoid valve block is bolted to the hydraulic actuation unit, behind the lefthand shock tower.

BRAKE FLUID RESERVOIR & LEVEL INDICATOR ASSEMBLY

The brake fluid reservoir and level indicator assembly, **Fig. 6,** is a translucent, plastic container that is mounted on the top of the hydraulic actuation unit. The reservoir is connected to the pump inlet port by a low pressure hose, and to the master cylinder by a sealed feed port.

WHEEL SENSORS

This system, **Figs. 7 through 9,** uses four sets of variable reluctance sensors and toothed speed indicator rings to determine the rotational speed of each wheel. The sensors operate on magnetic induction principle. For example, as the teeth on the speed indicator ring rotate past the stationary sensor, a signal proportional to the speed of the rotation is generated and transmitted to the electronic control unit (ECU) through a coaxial cable.

The front sensors are attached to the suspension knuckles and the speed indicator rings are pressed onto the outer constant velocity joints. The rear sensors are attached to the rear caliper anchor plate and the speed indicator rings are pressed onto the rear wheel hub assemblies. The sensors and the speed indicator rings are serviced individually.

ELECTRONIC CONTROLLER

The electronic controller is located in the luggage compartment of the vehicle, behind the righthand side of the rear seat. It is an on-board self test non-repairable unit, consisting of two microprocessors and the necessary circuitry for their operation. These microprocessors are programmed identically and operate on the principle of two channel redundant data processing and plausibility criteria monitoring.

SYSTEM OPERATION

The hydraulic pump maintains system pressure between approximately 2030-2610 psi, within the accumulator and is connected by a high pressure hose to the booster chamber portion of the hydraulic actuation assembly and a control valve. When the brakes are applied, a scissor lever mechanism activates the control valve and a pressure, proportional to brake pedal travel, enters the booster chamber portion of the hydraulic actuation assembly. This pressure is transmitted through the normally open solenoid valve through the proportioning valve to the rear brakes. The same pressure moves the booster piston against the master cylinder piston, shutting off the central valves in the master cylinder. This applies pressure to the front wheels through the two front normally open solenoid valves.

The electronic controller monitors the electromechanical components of the system. Malfunction of the anti-lock brake system will cause the electronic controller to shut off or inhibit the anti-lock function, while retaining normal power assisted braking. Malfunctions are indicated by one or two warning lamps inside the vehicle.

Loss of hydraulic fluid or power boost pressure will disable the anti-lock brake system.

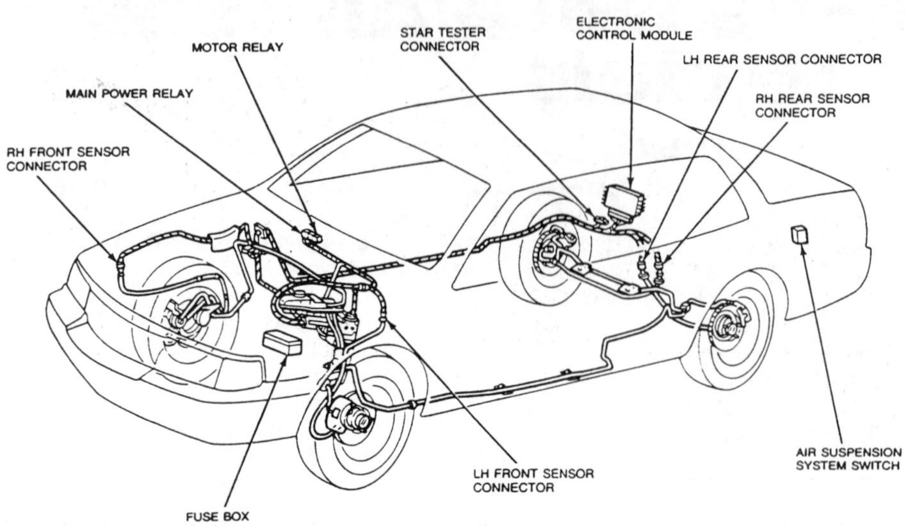

Fig. 1 Anti-lock brake system. 1988 Continental & all 1989 models

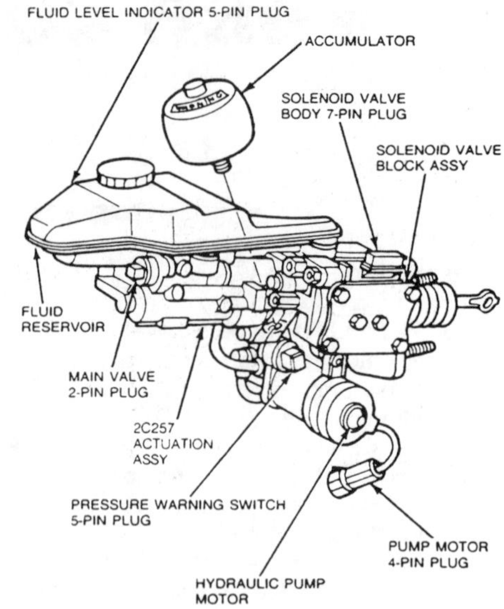

Fig. 2 Hydraulic actuation unit. 1988–89 Continental, 1989 Cougar & Thunderbird

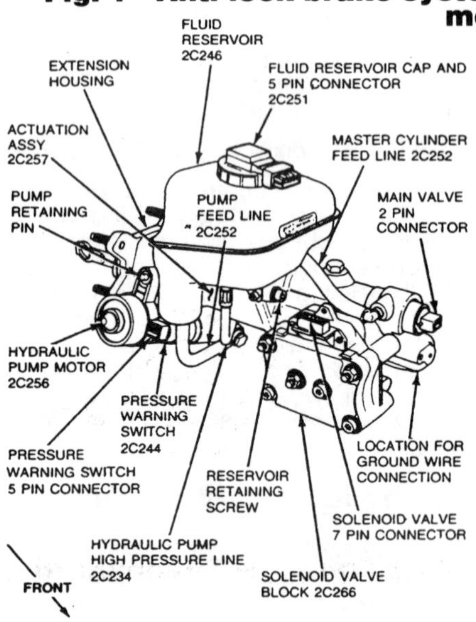

Fig. 3 Hydraulic actuation unit. 1989 Mark VII

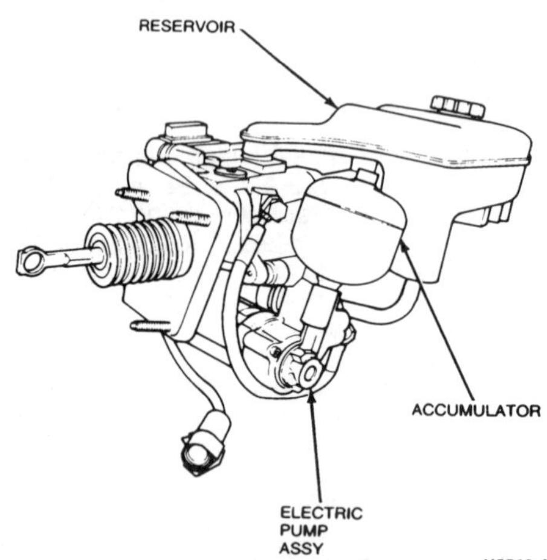

Fig. 4 Electric pump assembly. 1988 Continental & all 1989 models

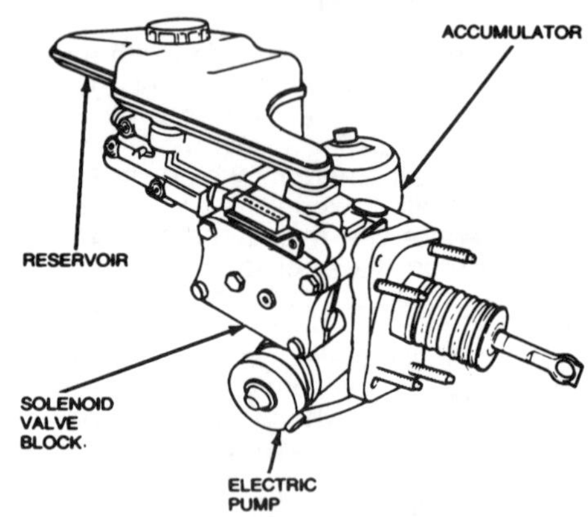

Fig. 5 Solenoid valve block assembly. 1988 Continental & all 1989 models

The four wheel anti-lock brake system is self monitoring. When the ignition switch is turned to the RUN position, the electronic controller will perform a preliminary self check on the anti-lock electrical system indicated by a three or four second illumination of the amber "Check Anti-Lock Brakes" lamp in the instrument cluster. During vehicle operation, including normal and anti-lock braking, the electronic controller monitors all electrical anti-lock functions and some hydraulic system operation.

For most malfunctions of the anti-lock brake system, the amber "Check Anti-Lock Brakes" and/or the red "Brake" lamp will be illuminated. The sequence of illumination of these warning lamps, combined with the problem symptoms, can determine the appropriate diagnostic test to perform. Most malfunctions are recorded as a coded number in the controller memory, pinpointing the exact component requiring service.

SYSTEM DIAGNOSIS & TESTING

Ensure the following diagnosis procedures are used in the sequence and step-by-step order indicated. Following the wrong sequence or bypassing steps will only lead to unnecessary replacement of system components and/or incorrect resolution of the problem. The diagnostic procedure consists of five sub-tests: Pre-test Checks, On-Board Self-Tests, Manual Quick Tests, Warning Lamp Symptom Chart and the Diagnostic Tests. As stated previously, do not attempt to bypass any procedure steps or tests.

PRE-TEST CHECKS

1. Verify that the parking brake is completely released. If the parking brake is applied, the "Brake" lamp will be illuminated.
2. Check brake fluid. **As the fluid level drops, the red "Brake" lamp will illuminate. If fluid level continues to fall, the amber "Check Anti-Lock Brake" lamp will illuminate and the anti-lock function will be inhibited.**
3. Verify that all of the following connectors are properly connected and in good operating condition:
   a. 7 pin connector of the solenoid valve block assembly.
   b. 2 pin connector of the main valve.
   c. 5 pin connector of the fluid level indicator.
   d. 5 pin connector of the pressure warning switch.
   e. 2 pin connector on all four sensors.
   f. 4 pin connector on the pump motor.
   g. 32 pin connector of the electronic controller.
   h. 5 pin connector of the main relay.
   i. 5 pin connector of the pump motor relay.

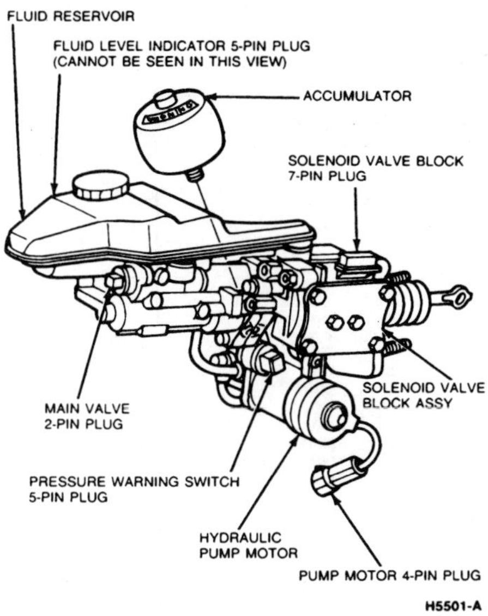

Fig. 6 Brake fluid reservoir & level indicator assembly. 1988 Continental & all 1989 models

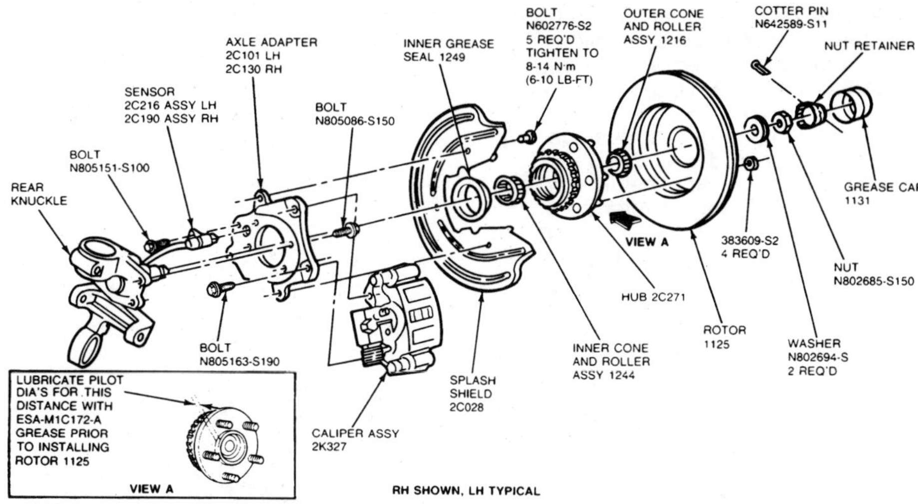

Fig. 7 Wheel sensor assembly (righthand side shown, lefthand side similar). 1988–89 Continental

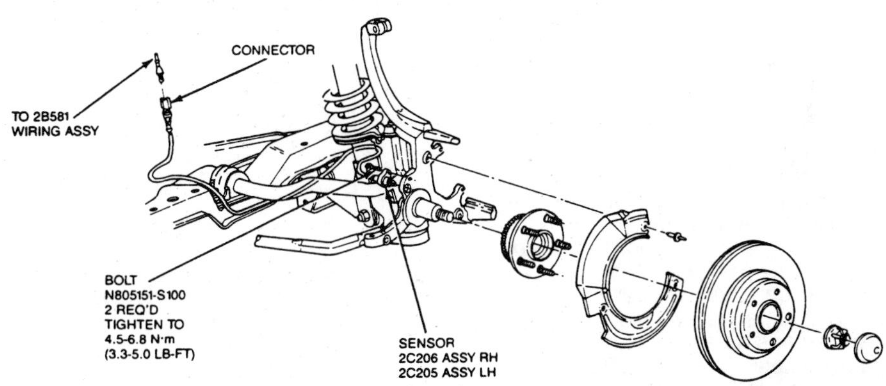

Fig. 8 Wheel sensor assembly. 1989 Cougar & Thunderbird

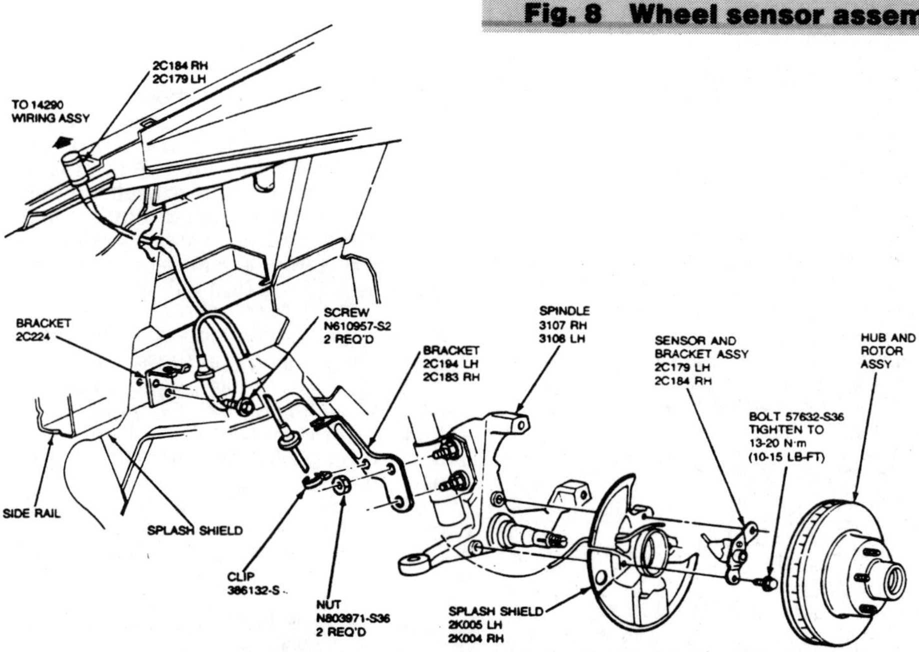

Fig. 9 Wheel sensor assembly. 1989 Mark VII

4. Check that the fuses and diode are not damaged.
5. Ensure all battery connections are clean and tight.
6. Check ground connections on front end of hydraulic unit in luggage compartment.

ON-BOARD SELF-TEST & MAIN COMPONENT DIAGNOSIS

The anti-lock brake electronic control module is capable of performing a self-test using Star tester 007-00017 or Super Star tester 007-00019 or equivalents. If the Star tester is not available, the anti-lock Quick Check sheet can be used as described further on.

The anti-lock control module monitors system operation and stores up to seven different service codes in its memory. However, it cannot store two services codes with the same first digit. For example: the module cannot store a code 25 and a code 26 at the same time (it can store a code 25 and a code 35). The code 25 is stored first. It must be serviced, the memory must be cleared and the vehicle operated before the code 26 can be stored.

A valve service code will override any stored service code during Self-Test. The condition must be serviced before any other codes can be displayed. After the valve has been serviced, the Self-Test can be run and any stored codes will be displayed. If no codes are displayed, the system has passed the Self-Test. If any codes are displayed during the Self-Test, refer to the Self-Test Code Index, **Fig. 10.** The index will direct you to a specific pinpoint test to be run. After the pinpoint tests have been completed, the components should be serviced. After servicing these codes, the Self-Test should be repeated.

After all codes have been displayed and serviced, the memory will be erased by driving the vehicle above approximately 25 mph. If the Self-Test has not been completed and all codes outputted, the memory cannot be erased.

Star Tester Connection & Battery Check

1. Turn ignition switch to OFF position.
2. Lower righthand module panel in luggage compartment.
3. Connect Star tester electrical connector to vehicle connector located near the electronic controller assembly.
4. Turn power switch on righthand side of Star tester to ON. A steady 00 or a blank screen will appear signifying that the tester is ready to start the Self-Test and receive service codes. **If the message LO BAT appears in the upper lefthand corner of the read-out display and stays ON, replace the Star tester's 9 volt battery before continuing with the Self-Test. The message LO BAT will appear momentarily when the power switch is turned to the OFF position.**
5. With the ignition switch in the OFF position, push the Self-Test button in the center of the Star tester. Push the button again. This deactivates the Self-Test sequence.
6. If the Star tester passes the Self-Test (a 00 or blank screen with button in TEST position), proceed with the On-Board Self-Test. If any service codes appear during the Self-Test, refer to the On-Board Self-Test Service Code Index, **Fig. 10.**

On-Board Self-Test Diagnosis

Refer to **Fig. 11,** for On-Board Self-Test diagnosis procedure.

Electronic Controller Diagnosis

Refer to **Fig. 12,** for electronic controller diagnosis procedure.

Main Valve Diagnosis

Refer to **Fig. 13,** for main valve diagnosis procedure.

On-Board Self-Test Service Code Index

SERVICE CODE (COMPONENT)	PINPOINT TEST STEP
11 (Electronic Controller)	AA1
12 (Electronic Controller-Replacer)	AA2
21 (Main Valve)	BB1
22 (LH Front Inlet Valve)	CC1
23 (LH Front Outlet Valve)	CC2
24 (RH Front Inlet Valve)	CC3
25 (RH Front Outlet Valve)	CC4
26 (Rear Inlet Valve)	CC5
27 (Rear Outlet Valve and Ground)	CC6
31 (LH Front Sensor)	DD1
32 (RH Front Sensor)	DD6
33 (RH Rear Sensor)	DD11
34 (LH Rear Sensor)	DD16
35 (LH Front Sensor)	DD1
36 (RH Front Sensor)	DD6
37 (RH Rear Sensor)	DD11
38 (LH Rear Sensor)	DD16
41 (LH Front Sensor)	DD1
42 (RH Front Sensor)	DD6
43 (RH Rear Sensor)	DD11
44 (LH Rear Sensor)	DD16
45 (LH Front and One Other Sensor Signals)	DD21
46 (RH Front and One Other Sensor Signals)	DD24
47 (Missing Both Rear Sensor Signals)	DD25
48 (Missing Three of Four Sensor Signals)	DD26
51 (LH Front Outlet Valve)	EE1
52 (RH Front Outlet Valve)	EE3
53 (Rear Outlet Valve)	EE5
54 (Rear Outlet Valve)	EE7
55 (LH Front Sensor)	DD1
56 (RH Front Sensor)	DD6
57 (RH Rear Sensor)	DD11
58 (LH Rear Sensor)	DD16
61 (FLI and PWS Circuit)	FF1
71 (LH Front Sensor)	EE1
72 (RH Front Sensor)	EE3
73 (RH Rear Sensor)	EE5
74 (LH Rear Sensor)	EE7
75 (LH Front Sensor)	DD1
76 (RH Front Sensor)	DD6
77 (RH Rear Sensor)	DD11
78 (LH Rear Sensor)	DD16
88 (Electronic Controller)	AA1
99 (Electronic Controller)	AA1

Fig. 10 Diagnosis, on-board self-test service code index. 1988 Continental & all 1989 models

Solenoid Valve Diagnosis

Refer to **Fig. 14,** for solenoid valve diagnosis procedure.

Wheel Sensor Diagnosis

Refer to **Fig. 15,** for wheel sensor diagnosis procedure.

Outlet Valve Diagnosis

Refer to **Fig. 16,** for outlet valve diagnosis procedure.

Warning Circuit Diagnosis

Refer to **Fig. 17,** for warning circuit diagnosis procedure.

QUICK TEST CHECKS & ANTI-LOCK QUICK CHECK SHEET

To properly conduct the Quick Test Checks an EEC-IV breakout box No. T83L-50-EEC-IV, anti-lock harness adapter No. T87P-50-ALA and a digital volt/ohmmeter (No. 007-00001), or equivalents must be used. All quick tests are performed in the vehicle's luggage compartment using the EEC-IV breakout box and the harness adapter. These group of tests will lead to specific diagnostic Pinpoint Test that will, in most cases, identify the fault/malfunction. If the fault/malfunction is not determined by the Quick Test procedure, use the following Diagnostic Lamp Symptom Chart to identify the proper diagnostic procedure to be conducted.

Refer to anti-lock quick check sheet **Fig. 18,** for item to be tested, ignition switch mode position, measurement taken between terminal pin numbers, tester scale/range, volt/ohm specifications and the specific pinpoint test to correct this group of malfunctions.

DIAGNOSTIC LAMP SYMPTOM CHART

If the quick test checks and anti-lock quick check sheet procedure does not isolate the symptom, it will be necessary to check the operation of the brake and anti-lock warning lamps. Observe the lamps and compare their ON/OFF operation to the conditions listed in **Fig. 19.** Once the actual warning lamp pattern has been matched to one of the conditions listed in the chart, perform the specific Pinpoint Test diagnostic procedure. **Before connecting the 32 pin connector harness to adapter T87P-50-ALA, ensure that the contacts of the harness are properly installed and are not damaged.**

Continued on page 19-75

On-Board Self-Test

TEST STEP	RESULT ▶	ACTION TO TAKE
1.0 CONNECT STAR TESTER		
• Turn ignition switch to OFF. • Connect Rotunda SUPER STAR II Tester 007-00028 or equivalent to Anti-Lock self-test connector. • Turn tester On. **NOTE: Do not move vehicle or turn steering wheel during test.** • Depress SUPER STAR II Tester push-button to test position. Display should show: COLON DISPLAY : 00 — COLON MUST BE DISPLAYED TO RECEIVE SERVICE CODES. • Turn ignition switch to RUN position. Wait 45 seconds. • Observe tester display.	Pass code ( ) displayed and warning lamps off. ▶ Pass code ( ) displayed and warning lamp turns on. ▶ Service code displayed. ▶	Vehicle OK. RETURN vehicle to customer. PERFORM anti-lock Quick Test. GO to **2.0**.
2.0 CHECK SERVICE CODES		
• Record service code. • Release, then depress SUPER STAR II Tester push-button. Wait 45 seconds. **NOTE: If first displayed code indicates a valve failure (first digit is 2), service that valve. Then, repeat Step 1.0.** • Observe tester display.	Service code displayed. ▶ Pass code ( ) displayed. ▶	REPEAT Step 2.0 each time a service code is displayed. RECORD each service code as it is displayed. SERVICE indicated fault(s). GO to **3.0** or GO to Quick Check. **NOTE: Cross off each service code from list as it is serviced.**
3.0 CHECK FOR OTHER FAULT CODES		
• Repeat Step 1.0 to see if any other new service codes are stored in memory.	If no new service codes are displayed. ▶ Service code displayed. ▶	GO to **4.0**. GO to **2.0**.

Fig. 11 Diagnosis, on-board self-test (Part 1 of 2). 1988 Continental & all 1989 models

On-Board Self-Test

TEST STEP	RESULT ▶	ACTION TO TAKE
4.0 CHECK SYSTEM OPERATION		
• Turn ignition to OFF position. • Disconnect Super Star II Tester. • Turn ignition to RUN position and check warning lamp sequence. • Test drive vehicle.	"CHECK ANTI-LOCK" and "BRAKE" warning lamps Off ▶ "CHECK ANTI-LOCK" and/or "BRAKE" lamp On ▶	RETURN vehicle to customer. REPEAT Step 1.0, or if there is no power brake, GO to Step D1.

Fig. 11 Diagnosis, on-board self-test (Part 2 of 2). 1988 Continental & all 1989 models

Electronic Controller Diagnosis — Test AA

TEST STEP	RESULT ▶	ACTION TO TAKE
AA1 SERVICE CODE 11 AND/OR 99: ELECTRICAL DISTURBANCE		
• Read all service codes by releasing and pressing the Super Star II Tester push button. • Allow a minimum of 45 seconds to pass after each time the push button is pressed. • Write down on paper all service codes. • After all service codes are read and written down, drive vehicle above 40 km/h (25 mph) to clear memory. • Read again all service codes.	Service code 11 and/or 99 repeated. ▶ Memory erased or other service codes present except code 11 and/or 99. ▶	REPLACE electronic controller. PERFORM test step associated with service code or codes. REFER to On-Board Self-Test Service Code Index.
AA2 SERVICE CODE 12: REPLACE ELECTRONIC CONTROLLER		
• No test step to be performed.	⊘K ▶	REPLACE electronic controller.

Fig. 12 Diagnosis, electronic controller. 1988 Continental & all 1989 models

Main Valve Diagnosis — Test BB

TEST STEP	RESULT ▶	ACTION TO TAKE
BB1 SERVICE CODE 21: CHECK MAIN VALVE		
• Disconnect main valve 2-Pin plug. • Measure resistance between the main valve electrical Pins 1 and 2.	2 to 5.5 ohms ▶ Any other reading ▶	REPLACE or SERVICE cable harness (Circuit 430E or 493A). REPLACE actuation assembly.

Fig. 13 Diagnosis, main valve. 1988 Continental & all 1989 models

Solenoid Valve Diagnosis — Test CC

TEST STEP	RESULT ▶	ACTION TO TAKE
CC1 SERVICE CODE 22: CHECK LH FRONT INLET VALVE		
• Disconnect valve block 7-Pin plug. • Measure resistance between valve block electrical Pins 7 and 6.	5 to 8 ohms ▶	REPLACE or SERVICE cable harness
	Any other reading ▶	REPLACE valve block unit. RECONNECT 7-Pin plug.

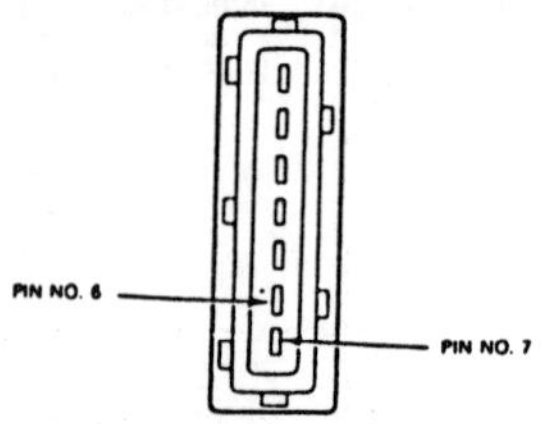

VALVE BLOCK 7-PIN CONNECTOR

TEST STEP	RESULT ▶	ACTION TO TAKE
CC2 SERVICE CODE 23: CHECK LH FRONT OUTLET VALVE		
• Disconnect valve block 7-Pin plug. • Measure resistance between valve block electrical Pins 7 and 5.	3 to 6 ohms ▶	REPLACE or SERVICE cable harness
	Any other reading ▶	REPLACE valve block unit. RECONNECT 7-Pin plug.

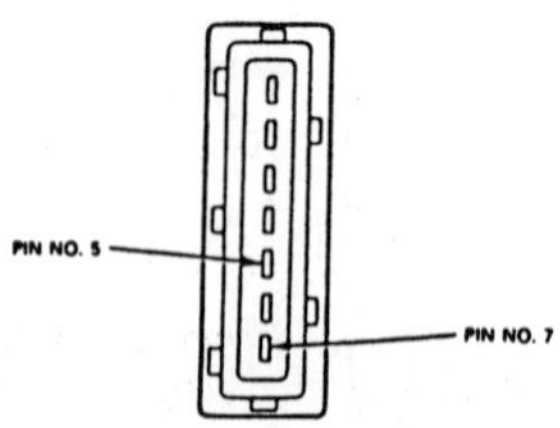

VALVE BLOCK 7-PIN CONNECTOR

Fig. 14 Diagnosis, solenoid valve (Part 1 of 6). 1988 Continental & all 1989 models

Solenoid Valve Diagnosis — Test CC

TEST STEP	RESULT ▶	ACTION TO TAKE
CC3 SERVICE CODE 24: CHECK RH FRONT INLET VALVE		
• Disconnect valve block 7-Pin plug. • Measure resistance between valve block electrical Pins 7 and 1.	5 to 8 ohms ▶	REPLACE or SERVICE cable harness
	Any other reading ▶	REPLACE solenoid valve block unit. RECONNECT 7-Pin plug.

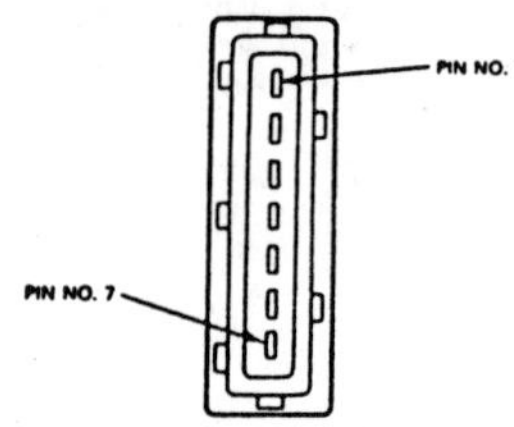

VALVE BLOCK 7-PIN CONNECTOR

TEST STEP	RESULT ▶	ACTION TO TAKE
CC4 SERVICE CODE 25: CHECK RH FRONT OUTLET VALVE		
• Disconnect valve block 7-Pin plug. • Measure resistance between valve block electrical Pins 7 and 2.	3 to 6 ohms ▶	REPLACE or SERVICE cable harness
	Any other reading ▶	REPLACE valve block unit. RECONNECT 7-Pin plug.

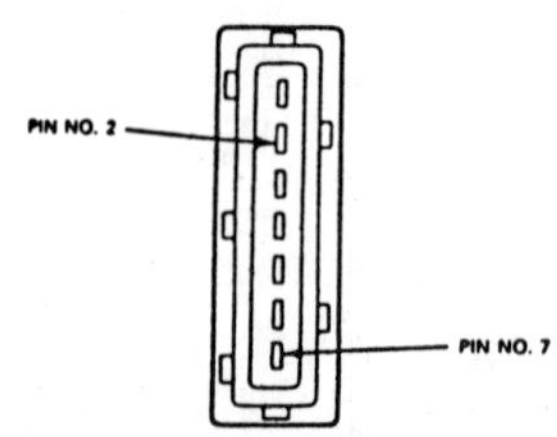

VALVE BLOCK 7-PIN CONNECTOR

Fig. 14 Diagnosis, solenoid valve (Part 2 of 6). 1988 Continental & all 1989 models

Solenoid Valve Diagnosis — Test CC

TEST STEP	RESULT ▶	ACTION TO TAKE
CC5 SERVICE CODE 26: CHECK REAR INLET VALVE		
• Disconnect valve block 7-Pin plug. • Measure resistance between valve block electrical Pins 7 and 3.	5 to 8 ohms ▶	REPLACE or SERVICE cable harness
	Any other reading ▶	REPLACE valve block unit. RECONNECT 7-Pin plug.

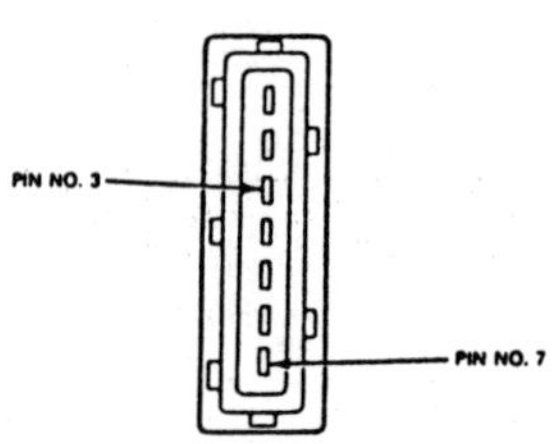

VALVE BLOCK 7-PIN CONNECTOR

Fig. 14 Diagnosis, solenoid valve (Part 3 of 6). 1988 Continental & all 1989 models

Solenoid Valve Diagnosis — Test CC

TEST STEP	RESULT ▶	ACTION TO TAKE
CC6 SERVICE CODE 27: CHECK REAR OUTLET VALVE CIRCUIT		
• Turn ignition switch to OFF. • Disconnect 32-pin plug from electronic controller.	3 to 6 ohms ▶	GO to Step CC7.
	Any other reading ▶	GO to Step CC6a.

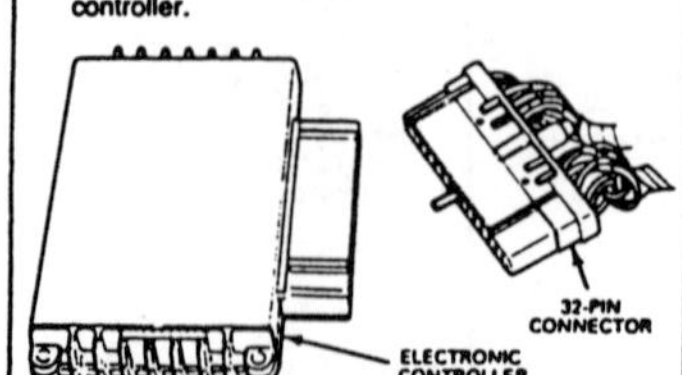

• Connect EEV-IV Breakout box with Tool T83L-50-EEC-IV or equivalent, Anti-Lock Test Adapter T87P-50-ALA or equivalent, to the anti-lock 32-pin plug wiring harness.

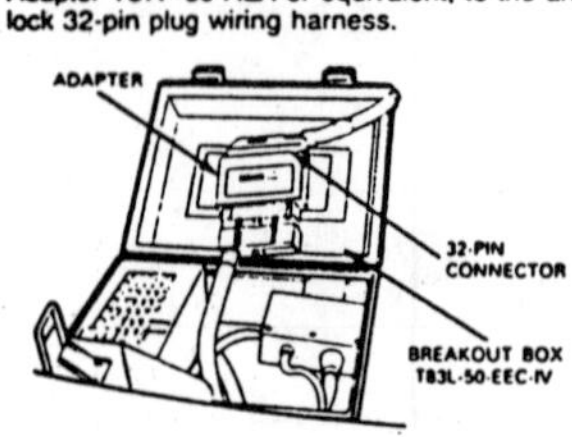

• Measure resistance between Breakout box Pins 11 and 12.

Fig. 14 Diagnosis, solenoid valve (Part 4 of 6). 1988 Continental & all 1989 models

Solenoid Valve Diagnosis — Test CC

TEST STEP	RESULT ▶	ACTION TO TAKE
CC6a CHECK REAR OUTLET VALVE RESISTANCE		
• Disconnect valve block 7-Pin plug. • Measure resistance between valve block electrical Pins 7 and 4.	3 to 6 ohms ▶	REPLACE or SERVICE cable harness
	Any other reading ▶	REPLACE valve block unit. RECONNECT 7-Pin plug.

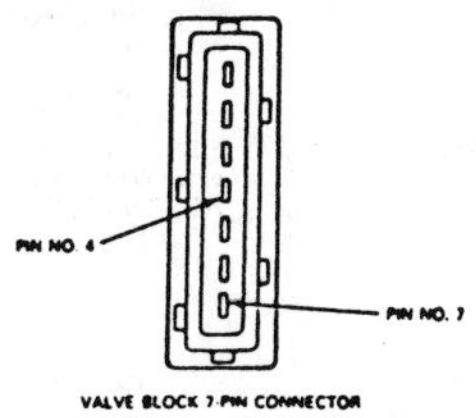

TEST STEP	RESULT ▶	ACTION TO TAKE
CC7 CHECK VALVE BODY GROUND CIRCUIT		
• Measure resistance between Breakout box Pins 11 and 40.	Less than 2 ohms ▶	REVERIFY symptom.
	Greater than 2 ohms ▶	GO to Step CC7a.

Fig. 14 Diagnosis, solenoid valve (Part 5 of 6). 1988 Continental & all 1989 models

Solenoid Valve Diagnosis — Test CC

TEST STEP	RESULT ▶	ACTION TO TAKE
CC7a CHECK VALVE BODY GROUND		
• Disconnect valve block 7-Pin plug. • Measure resistance between valve block electrical Pin 7 and valve block body.	Less than 2 ohms ▶	REPLACE or SERVICE cable harness
	Greater than 2 ohms ▶	GO to Step CC7b.

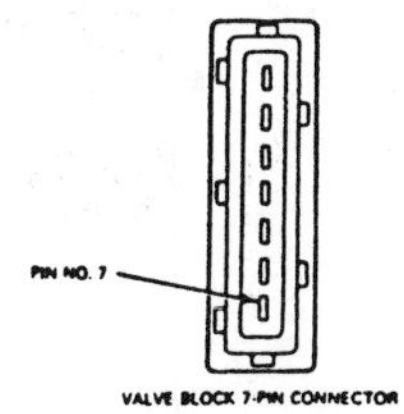

TEST STEP	RESULT ▶	ACTION TO TAKE
CC7b CHECK VALVE BODY GROUND WIRE		
• Remove negative (−) ground strap from battery. • Check for continuity between valve body and body ground.	Continuity ▶	REPLACE valve block unit (internal ground problem).
	No or poor continuity ▶	SERVICE or REPLACE hydraulic unit ground strap

Fig. 14 Diagnosis, solenoid valve (Part 6 of 6). 1988 Continental & all 1989 models

Wheel Sensor Diagnosis — Test DD

TEST STEP	RESULT ▶	ACTION TO TAKE
DD1 SERVICE CODES 31/35/41/55 OR 75 CHECK LH FRONT SENSOR		
• Turn ignition switch Off. • Disconnect 32-pin plug from electronic controller.	800 to 1400 ohms (0.8 to 1.4K ohms) ▶	GO to Step DD2.
	Any other reading ▶	GO to Step DD1a.

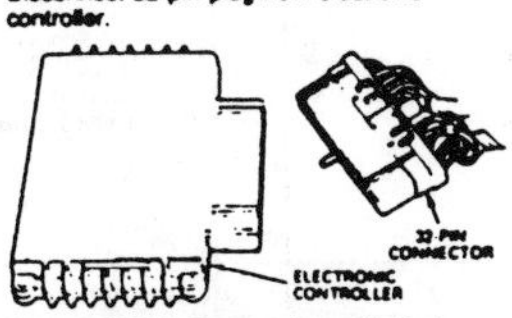

• Connect EEC-IV Breakout box with Tool T83L-50-EEC-IV or equivalent, Anti-Lock Test Adapter T87P-50-ALA or equivalent, to the anti-lock 32-pin plug wiring harness.

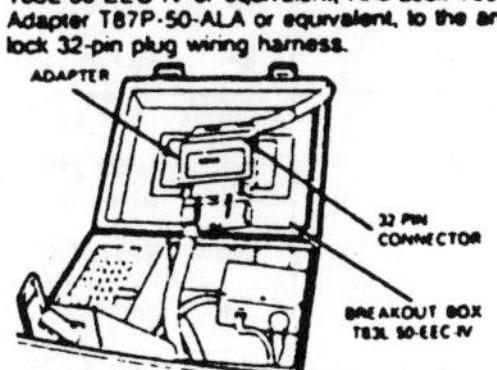

• Measure resistance between Breakout box Pins 5 and 22.

TEST STEP	RESULT ▶	ACTION TO TAKE
DD1a CHECK LH FRONT SENSOR RESISTANCE		
• Disconnect sensor plug (LH front). • Measure resistance of sensor at sensor plug.	800 to 1400 ohms (0.8 to 1.4K ohms) ▶	SERVICE or REPLACE cable harness
	Any other reading ▶	REPLACE LH front wheel sensor.

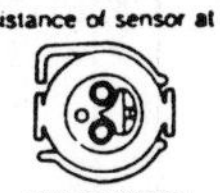

Fig. 15 Diagnosis, wheel sensors (Part 1 of 12). 1988 Continental & all 1989 models

Wheel Sensor Diagnosis — Test DD

TEST STEP	RESULT ▶	ACTION TO TAKE
DD2 CHECK LH FRONT SENSOR VOLTAGE		
• Turn ignition switch Off. • Turn air suspension switch in luggage compartment Off, if so equipped. • Place vehicle on hoist and raise wheels clear of ground. Refer to Pre-Delivery Manual, Section 50-04. • Set multi-meter on voltage range (2V-AC). • Measure voltage between Breakout box Pins 5 and 22 while spinning LH front wheel at approximately 1 revolution per second.	Between 0.05 and 0.70 Vac ▶	GO to Step DD3.
	Less than 0.05 or more than 0.70 Vac ▶	CHECK sensor mounting, air gap, or toothed wheel mounting. CORRECT as required.
DD3 CHECK LH FRONT SENSOR CIRCUIT CONTINUITY		
• Check continuity between Breakout box Pins 40 and 5.	No continuity ▶	GO to Step DD4.
	Continuity ▶	GO to Step DD3A.
DD3A CHECK LH FRONT SENSOR CIRCUITRY		
• Disconnect wheel sensor plug (LH front). • Check for continuity between each sensor plug pin (sensor side) and vehicle ground.	Continuity ▶	REPLACE sensor. (LH front).
	No continuity ▶	SERVICE or REPLACE cable harness. RECONNECT sensor plug.

TEST STEP	RESULT ▶	ACTION TO TAKE
DD4 CHECK ELECTRONIC CONTROLLER TO GROUND WIRE		
• Check continuity between Breakout box Pin 40 and body ground.	Continuity ▶	GO to Step DD5.
	No continuity ▶	SERVICE or REPLACE cable harness
DD5 CHECK LH FRONT WHEEL BEARING		
• Check front wheel bearing end play. • Inspect toothed sensor ring visually for damaged teeth. **NOTE: Turn air suspension switch On when vehicle is off hoist, if so equipped.**	Loose or damaged parts ▶	ADJUST bearings or REPLACE faulty parts.
	Not loose or damaged ▶	REVERIFY symptom.

Fig. 15 Diagnosis, wheel sensors (Part 2 of 12). 1988 Continental & all 1989 models

Wheel Sensor Diagnosis — Test DD

TEST STEP	RESULT ▶	ACTION TO TAKE
DD6 SERVICE CODES 32/36/42/56 OR 76 CHECK RH FRONT SENSOR		
• Turn ignition switch Off. • Disconnect 32-pin plug from electronic controller. 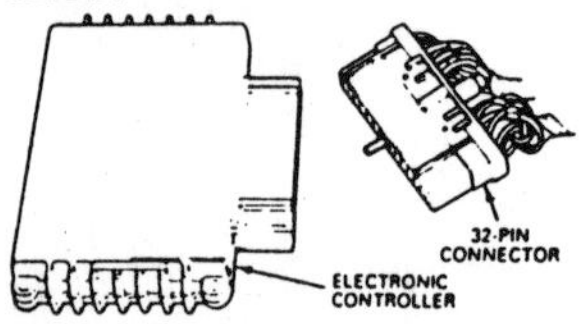 • Connect EEC-IV Breakout box with Tool T83L-50-EEC-IV or equivalent, Anti-Lock Test Adapter T87P-50-ALA or equivalent, to the anti-lock 32-pin plug wiring harness. 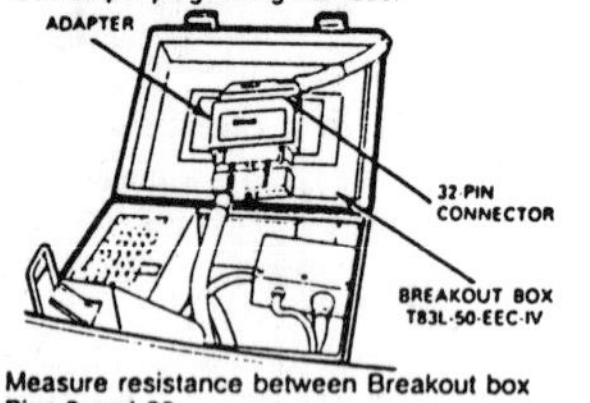 • Measure resistance between Breakout box Pins 3 and 20.	800 to 1400 ohms (0.8 to 1.4K ohms) ▶ Any other reading ▶	GO to Step DD7. GO to Step DD6a.
DD6a CHECK RH FRONT SENSOR RESISTANCE		
• Disconnect sensor plug (RH front). • Measure resistance of sensor at sensor plug.  RH FRONT SENSOR	800 to 1400 ohms (0.8 to 1.4K ohms) ▶ Any other reading ▶	SERVICE or REPLACE cable harness REPLACE RH front wheel sensor.

Fig. 15 Diagnosis, wheel sensors (Part 3 of 12). 1988 Continental & all 1989 models

Wheel Sensor Diagnosis — Test DD

TEST STEP	RESULT ▶	ACTION TO TAKE
DD7 CHECK RH FRONT SENSOR VOLTAGE		
• Turn air suspension switch in luggage compartment Off, if so equipped. • Place vehicle on hoist and raise wheels clear of ground. Refer to Pre-Delivery Manual, Section 50-04. • Set multi-meter on voltage range (2V-AC). • Measure voltage between Breakout box Pins 3 and 20 while spinning the RH front wheel at approximately 1 revolution per second.	Between 0.05 and 0.70 Vac ▶ Less than 0.05 or more than 0.70 Vac ▶	GO to Step DD8. CHECK wheel sensor mounting, air gap, or toothed wheel mounting. CORRECT as required.
DD8 CHECK RH FRONT SENSOR CIRCUIT CONTINUITY		
• Check continuity between Breakout box Pins 40 and 3.	Continuity ▶ No continuity ▶	GO to Step DD8a. GO to Step DD9.
DD8a CHECK RH FRONT SENSOR CONTINUITY		
• Disconnect wheel sensor plug (RH front). • Check for continuity between each sensor plug pin (sensor side) and vehicle ground.  RH FRONT SENSOR	Continuity ▶ No continuity ▶	REPLACE right front sensor. SERVICE or REPLACE cable harness RECONNECT sensor plug.
DD9 CHECK ELECTRONIC CONTROLLER TO GROUND WIRE		
• Check continuity between Breakout box Pin 40 and body ground.	Continuity ▶ No continuity ▶	GO to Step DD10. SERVICE or REPLACE cable harness
DD10 CHECK RH FRONT WHEELING BEARING		
• Check front wheel bearing end play. • Inspect toothed sensor ring visually for damaged teeth. **NOTE: Turn air suspension switch On when vehicle is off hoist, if so equipped.**	Loose or damaged parts ▶ Not loose or damaged ▶	ADJUST bearings or REPLACE faulty parts. REVERIFY symptom.

Fig. 15 Diagnosis, wheel sensors (Part 4 of 12). 1988 Continental & all 1989 models

Wheel Sensor Diagnosis — Test DD

TEST STEP	RESULT ▶	ACTION TO TAKE
DD11 SERVICE CODES 33/37/43/57 OR 77 CHECK RH REAR SENSOR		
• Turn ignition switch Off. • Disconnect 32-pin plug from electronic controller. 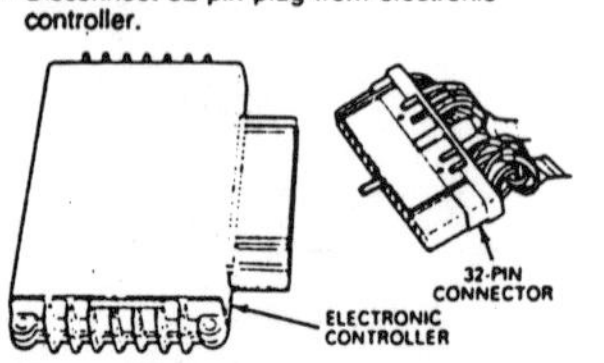 • Connect EEC-IV Breakout box with Tool T83L-50-EEC-IV or equivalent, Anti-Lock Test Adapter T87P-50-ALA or equivalent, to the anti-lock 32-pin plug wiring harness. 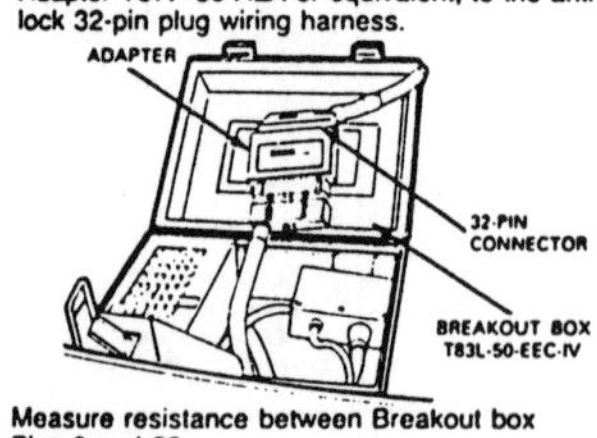 • Measure resistance between Breakout box Pins 6 and 23.	800 to 1400 ohms (0.8 to 1.4K ohms) ▶ Any other reading ▶	GO to Step DD12. GO to Step DD11a.
DD11a CHECK RH REAR SENSOR RESISTANCE		
• Disconnect sensor plug (RH rear). • Set multi-meter to read resistance. • Measure resistance of sensor at sensor plug.  RH REAR SENSOR	800 to 1400 ohms (0.8 to 1.4K ohms) ▶ Any other reading ▶	SERVICE or REPLACE cable harness REPLACE RH rear wheel sensor.

Fig. 15 Diagnosis, wheel sensors (Part 5 of 12). 1988 Continental & all 1989 models

Wheel Sensor Diagnosis — Test DD

TEST STEP	RESULT ▶	ACTION TO TAKE
DD12 CHECK RH REAR SENSOR VOLTAGE		
• Turn air suspension switch in luggage compartment Off, if so equipped. • Place vehicle on hoist and raise wheels clear of ground. Refer to Pre-Delivery Manual, Section 50-04. • Set multi-meter on voltage range (2V-AC). • Measure voltage between Breakout box Pins 6 and 23 while spinning the RH rear wheel at approximately 1 revolution per second.	Between 0.05 and 0.70 Vac ▶ Less than 0.05 or more than 0.70 Vac ▶	GO to Step DD13. CHECK sensor mounting, air gap, or toothed wheel mounting. CORRECT as required.
DD13 CHECK RH REAR SENSOR CIRCUIT CONTINUITY		
• Check continuity between Breakout box Pins 40 and 6.	No continuity ▶ Continuity ▶	GO to Step DD14. GO to Step DD13a.
DD13a CHECK RH REAR SENSOR CONTINUITY		
• Disconnect wheel sensor plug (RH rear). • Check for continuity between each sensor plug pin (sensor side) and vehicle ground. 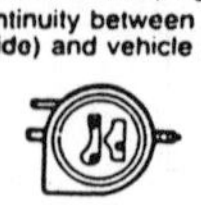 RH REAR SENSOR	Continuity ▶ No continuity ▶	REPLACE sensor (RH rear). SERVICE or REPLACE cable harness Connect sensor plug.
DD14 CHECK ELECTRONIC CONTROLLER TO GROUND WIRE		
• Check continuity between Breakout box Pin 40 and body ground.	Continuity ▶ No continuity ▶	GO to Step DD15. SERVICE or REPLACE cable harness
DD15 CHECK FOR EXCESSIVE AXLE VIBRATION		
• Check differential housing for excessive play. • Check rear axle bearings for excessive play. • Inspect toothed sensor ring visually for damaged teeth. **NOTE: Turn air suspension switch On when vehicle is off hoist, if so equipped.**	Loose or damaged parts ▶ Not loose or damaged ▶	SERVICE or REPLACE faulty parts. REVERIFY symptom.

Fig. 15 Diagnosis, wheel sensors (Part 6 of 12). 1988 Continental & all 1989 models

Wheel Sensor Diagnosis — Test DD

TEST STEP	RESULT ▶	ACTION TO TAKE
DD16 SERVICE CODES 34/38/44/58 OR 78 CHECK LH REAR SENSOR		
• Turn ignition switch Off. • Disconnect 32-pin plug from electronic controller. 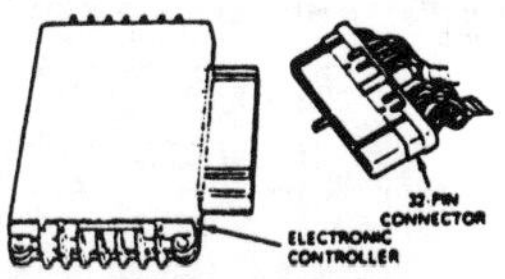 • Connect EEC-IV Breakout box with Tool T83L-50-EEC-IV or equivalent, Anti-Lock Test Adapter T87P-50-ALA or equivalent, to the anti-lock 32-pin plug wiring harness. 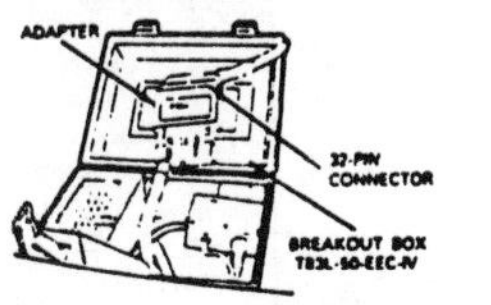 • Measure resistance between Breakout box Pins 4 and 21.	800 to 1400 ohms (0.8 to 1.4K ohms) ▶ Any other reading ▶	GO to Step DD17. GO to Step DD16a.
DD16a CHECK LH REAR SENSOR RESISTANCE		
• Disconnect sensor plug (LH rear). • Measure resistance of sensor at sensor plug. 	800 to 1400 ohms (0.8 to 1.4K ohms) ▶ Any other reading ▶	SERVICE or REPLACE cable harness REPLACE LH rear wheel sensor.

Fig. 15 Diagnosis, wheel sensors (Part 7 of 12). 1988 Continental & all 1989 models

Wheel Sensor Diagnosis — Test DD

TEST STEP	RESULT ▶	ACTION TO TAKE
DD17 CHECK LH REAR SENSOR VOLTAGE		
• Turn air suspension switch in luggage compartment Off, if so equipped. • Place vehicle on hoist and raise wheels clear of ground. Refer to Pre-Delivery Manual, Section 50-04. • Set multi-meter on voltage range (2V-AC). • Measure voltage between Breakout box Pins 4 and 21 while spinning the LH rear wheel at approximately 1 revolution per second.	Between 0.05 and 0.70 Vac ▶ Less than 0.05 or more than 0.70 Vac ▶	GO to Step DD18. CHECK sensor mounting, air gap, or toothed wheel mounting. CORRECT as required.
DD18 CHECK LH REAR SENSOR CIRCUIT CONTINUITY		
• Check continuity between Breakout box Pins 40 and 4.	No continuity ▶ Continuity ▶	GO to Step DD19. GO to Step DD18a.
DD18a CHECK LH REAR SENSOR CONTINUITY		
• Disconnect wheel sensor plug (LH front). • Check for continuity between each sensor. 	Continuity ▶ No continuity ▶	REPLACE sensor (LH front). SERVICE or REPLACE cable harness RECONNECT sensor plug.
DD19 CHECK ELECTRONIC CONTROLLER TO GROUND WIRE		
• Check continuity between Breakout box Pin 40 and body ground.	Continuity ▶ No continuity ▶	GO to Step DD20. SERVICE or REPLACE cable harness
DD20 CHECK FOR EXCESSIVE AXLE VIBRATION		
• Check differential housing for excessive play. • Check rear axle bearings for excessive play. • Inspect toothed sensor ring visually for damaged teeth. **NOTE: Turn air suspension switch On when vehicle is off hoist, if so equipped.**	Loose or damaged parts ▶ Not loose or damaged ▶	SERVICE or REPLACE faulty parts. REVERIFY symptom.

Fig. 15 Diagnosis, wheel sensors (Part 8 of 12). 1988 Continental & all 1989 models

Wheel Sensor Diagnosis — Test DD

TEST STEP	RESULT ▶	ACTION TO TAKE
DD21 SERVICE CODE 45 CHECK FOR TWO MISSING SENSOR SIGNALS ONE BEING THE LH FRONT		
• Turn ignition switch to OFF. • Disconnect 32-pin plug from Electronic Controller. • Connect EEC-IV Breakout Box T83L-50-EEC-IV with Anti-Lock Test Adapter T87P-50-ALA or equivalent to 32-pin plug wiring harness. • Turn air suspension switch in luggage compartment off. • Place vehicle on hoist and raise until wheels are clear of ground. Refer to Pre-Delivery manual. Section 50-04. • Set multi-meter on 2V AC voltage range. • Measure voltage between Breakout box Pins 5 and 22 while spinning LH front wheel at approximately 1 revolution per second. • Repeat wheel sensor voltage check described in previous step for the other wheels by: **Spinning** — **Measuring A/C volts between Pins** RH Front — 3 and 20 RH Rear — 6 and 23 LH Rear — 4 and 21	Voltage between 0.05 and 0.70V AC ▶ Less than 0.05 or greater than 0.70V AC ▶	GO to Step DD22. CHECK out-of-range wheel for proper sensor mounting, air gap and speed indicator ring mounting. CORRECT as required.
DD22 ELECTRONIC CONTROLLER TO GROUND		
• Check continuity between Breakout box Pin 40 and body ground.	Continuity ▶ No continuity ▶	GO to Step DD23. SERVICE or REPLACE cable harness
DD23 CHECK WHEEL BEARINGS		
• Check wheel bearing end play on all wheels. • Inspect toothed sensor ring visually for damaged teeth. **NOTE: Turn air suspension switch On when vehicle is off hoist.**	Loose or damaged parts ▶ Not loose or damaged ▶	ADJUST bearings or REPLACE damaged parts. REVERIFY symptom.

Fig. 15 Diagnosis, wheel sensors (Part 9 of 12). 1988 Continental & all 1989 models

Wheel Sensor Diagnosis — Test DD

TEST STEP	RESULT ▶	ACTION TO TAKE
DD24 SERVICE CODE 46 CHECK FOR 2 MISSING SENSOR SIGNALS ONE BEING RH FRONT		
• Turn ignition switch to OFF. • Disconnect 32-pin plug from Electronic Controller. • Connect EEC-IV Breakout Box T83L-50-EEC-IV with Anti-Lock Test Adapter T87P-50-ALA or equivalent, to 32-pin plug wiring harness. • Turn air suspension switch in luggage compartment off. • Place vehicle on hoist and raise until wheels are clear of ground. Refer to Pre-Delivery manual. Section 50-04. • Set multi-meter on 2V AC voltage range. • Measure voltage between Breakout box Pins 3 and 20 while spinning RH front wheel at approximately 1 revolution per second. • Repeat wheel sensor voltage check described in previous step for the other two wheels by: **Spinning** — **Measuring A/C volts between Pins** RH Rear — 6 and 23 LH Rear — 4 and 21	Voltage between 0.05 and 0.70V AC ▶ Less than 0.05 or greater than 0.70V AC ▶	GO to Step DD22. CHECK out-of-range wheel for proper sensor mounting, air gap and speed indication ring mounting. CORRECT as required.

Fig. 15 Diagnosis, wheel sensors (Part 10 of 12). 1988 Continental & all 1989 models

Wheel Sensor Diagnosis — Test DD

TEST STEP	RESULT ▶	ACTION TO TAKE
DD25 SERVICE CODE 47 CHECK FOR BOTH REAR WHEEL SENSOR SIGNALS MISSING		
• Turn ignition switch to OFF. • Disconnect 32-pin plug from Electronic Controller. • Connect EEC-IV Breakout Box T83L-50-EEC-IV with Anti-Lock Test Adapter T87P-50-ALA or equivalent, to 32-pin plug wiring harness. • Turn air suspension switch in luggage compartment off. • Place vehicle on hoist and raise until wheels are clear of ground. Refer to Pre-Delivery manual, Section 50-04. • Set multi-meter on 2V AC voltage range. • Measure voltage between Breakout box Pins 6 and 23 while spinning RH rear wheel at approximately 1 revolution per second. • Measure voltage between Breakout box Pins 4 and 21 while spinning LH rear wheel at approximately 1 revolution per second.	Voltage between 0.05 and 0.70V AC ▶ Less than 0.05 or greater than 0.70V AC ▶	GO to Step DD22. CHECK out-of-range wheel for proper sensor mounting, air gap and speed indication ring mounting. CORRECT as required.

Fig. 15 Diagnosis, wheel sensors (Part 11 of 12). 1988 Continental & all 1989 models

Wheel Sensor Diagnosis — Test DD

TEST STEP	RESULT ▶	ACTION TO TAKE
DD26 SERVICE CODE 48 ANY 3 OF 4 WHEEL SPEED SENSOR SIGNALS MISSING		
• Turn ignition switch to OFF. • Disconnect 32-pin plug from Electronic Controller. • Connect EEC-IV Breakout Box T83L-50-EEC-IV with Anti-Lock Test Adapter T87P-50-ALA or equivalent, to 32-pin plug wiring harness. • Turn air suspension switch in luggage compartment off. • Place vehicle on hoist and raise until wheels are clear of ground. Refer to Pre-Delivery manual, Section 50-04. • Set multi-meter on 2V AC voltage range. • Measure voltage between Breakout box pins indicated below while spinning appropriate wheel at approximately 1 revolution per second:	Voltage between 0.05 and 0.70V AC ▶ Less than 0.05 or greater than 0.70V AC ▶	GO to Step DD22. CHECK out-of-range wheel for proper sensor mounting, air gap and speed indicator ring mounting. CORRECT as required.

Spinning	Measuring A/C volts between Pins
LH Front	5 and 22
RH Front	3 and 20
RH Rear	6 and 23
LH Rear	4 and 21

Fig. 15 Diagnosis, wheel sensors (Part 12 of 12). 1988 Continental & all 1989 models

Outlet Valve Diagnosis — Test EE

TEST STEP	RESULT ▶	ACTION TO TAKE
EE1 SERVICE CODE 51 AND/OR 71 CHECK LH FRONT SENSOR CIRCUIT CONTINUITY		
• Turn ignition switch Off. • Disconnect 32-Pin plug from electronic controller. 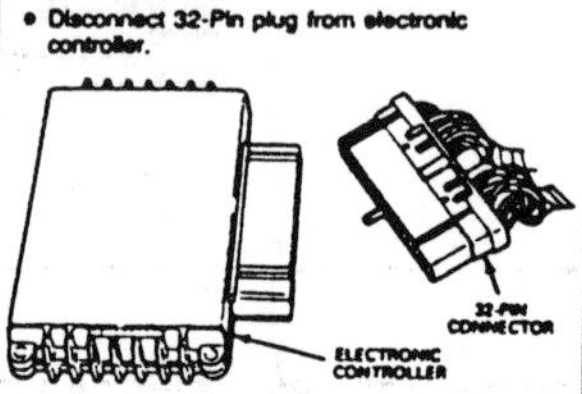 • Connect EEC-IV Breakout box with Tool T83L-50-EEC-IV or equivalent, Anti-Lock Test Adapter T87P-50-ALA or equivalent, to the anti-lock 32-Pin plug wiring harness. 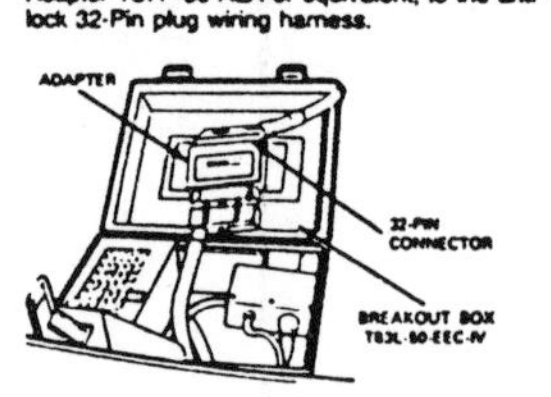 • Check for continuity between Breakout box Pins 40 and 5.	Continuity (OK crossed out) ▶ No continuity (OK) ▶	GO to Step EE1a. GO to Step EE2.

Fig. 16 Diagnosis, outlet valve (Part 1 of 8). 1988 Continental & all 1989 models

Outlet Valve Diagnosis — Test EE

TEST STEP	RESULT ▶	ACTION TO TAKE
EE1a CHECK LH FRONT SENSOR CONTINUITY		
• Disconnect wheel sensor plug (LH front). • Check for continuity between each sensor plug pin (sensor side) and vehicle ground.  	Continuity ▶ No continuity ▶	REPLACE sensor (LH front). SERVICE or REPLACE cable harness RECONNECT sensor plug.
EE2 CHECK ANTI-LOCK OPERATION LH FRONT WHEEL		
• Turn air suspension Off, if so equipped. • Lift vehicle and rotate wheels to ensure they turn freely. • Short Pins 18, 14 and 31 to each other at Breakout box. • Apply moderate brake pedal effort and check that LH front wheel will not turn. • Check that LH front wheel turns freely with ignition switch On. • Turn air suspension on when vehicle is off hoist, if so equipped. CAUTION: DO NOT LEAVE IGNITION ON FOR MORE THAN 1 MINUTE MAXIMUM, OR SOLENOID VALVE DAMAGE MAY RESULT.	If wheel turns freely ▶ If wheel does not turn freely or pedal drops ▶	REVERIFY symptom. REPLACE solenoid valve block.

Fig. 16 Diagnosis, outlet valve (Part 2 of 8). 1988 Continental & all 1989 models

Outline Valve Diagnosis — Test EE

TEST STEP	RESULT ▶	ACTION TO TAKE
EE3 SERVICE CODE 52 AND/OR 72 CHECK RH FRONT SENSOR CIRCUIT CONTINUITY		
• Turn ignition switch Off. • Disconnect 32-Pin plug from electronic controller.	Continuity (OK crossed out) ▶ No continuity (OK) ▶	GO to Step EE3a. GO to Step EE4.

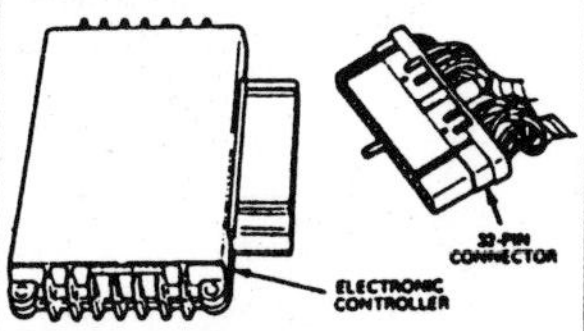

TEST STEP	RESULT ▶	ACTION TO TAKE
• Connect EEC-IV Breakout box with Tool T83L-50-EEC-IV or equivalent, Anti-Lock Test Adapter T87P-50-ALA or equivalent, to the anti-lock 32-Pin plug wiring harness.		

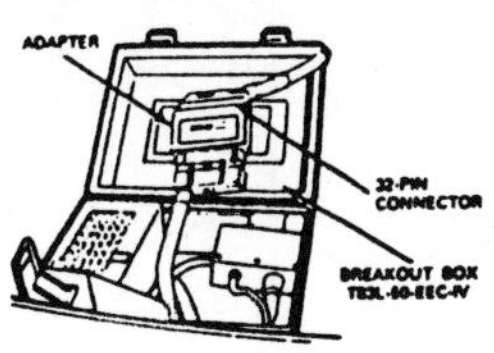

TEST STEP	RESULT ▶	ACTION TO TAKE
• Check for continuity between Breakout box Pins 40 and 3.		

Fig. 16 Diagnosis, outlet valve (Part 3 of 8). 1988 Continental & all 1989 models

Outlet Valve Diagnosis — Test EE

TEST STEP	RESULT ▶	ACTION TO TAKE
EE3a CHECK RH FRONT SENSOR CONTINUITY		
• Disconnect wheel sensor plug (RH front). • Check for continuity between each sensor plug pin (sensor side) and vehicle ground.	Continuity ▶ No continuity ▶	REPLACE right front sensor. SERVICE or REPLACE cable harness RECONNECT sensor plug.

TEST STEP	RESULT ▶	ACTION TO TAKE
EE4 CHECK ANTI-LOCK OPERATION RH FRONT WHEEL		
• Turn air suspension Off, if so equipped. • Lift vehicle and rotate wheels to ensure they turn freely. • Short Pins 18, 32 and 13 to each other at Breakout box. • Apply moderate brake effort. Check that RH front wheel does not turn. • Check that RH front wheel turns freely with ignition switch On. • Turn air suspension on when vehicle is off hoist, if so equipped. CAUTION: DO NOT LEAVE IGNITION ON FOR MORE THAN 1 MINUTE MAXIMUM, OR SOLENOID VALVE DAMAGE MAY RESULT.	If wheel turns freely ▶ If wheel does not turn freely or pedal drops ▶	REVERIFY symptom. REPLACE solenoid valve block.

Fig. 16 Diagnosis, outlet valve (Part 4 of 8). 1988 Continental & all 1989 models

Outlet Valve Diagnosis — Test EE

TEST STEP	RESULT ▶	ACTION TO TAKE
EE5 SERVICE CODES 53 AND/OR 73 CHECK RH REAR SENSOR CIRCUIT CONTINUITY		
• Turn ignition switch Off. • Disconnect 32-Pin plug from electronic controller.	Continuity (OK crossed out) ▶ No continuity (OK) ▶	GO to Step EE5a. GO to Step EE6.

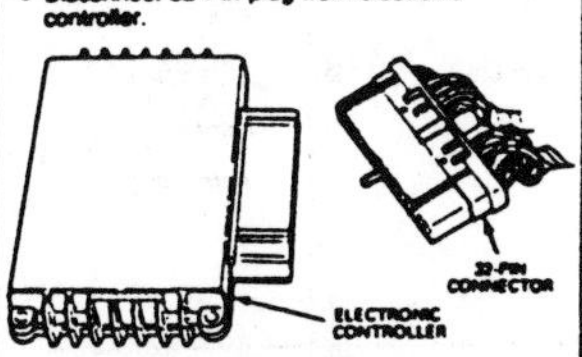

TEST STEP	RESULT ▶	ACTION TO TAKE
• Connect EEC-IV Breakout box with Tool T83L-50-EEC-IV or equivalent, Anti-Lock Test Adapter T87P-50-ALA or equivalent, to the anti-lock 32-Pin plug wiring harness.		

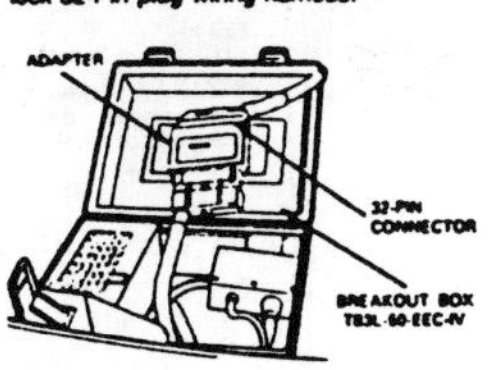

TEST STEP	RESULT ▶	ACTION TO TAKE
• Check for continuity between Breakout box Pins 40 and 6.		

Fig. 16 Diagnosis, outlet valve (Part 5 of 8). 1988 Continental & all 1989 models

Outlet Valve Diagnosis — Test EE

TEST STEP	RESULT ▶	ACTION TO TAKE
EE5a CHECK RH REAR SENSOR CONTINUITY		
• Disconnect wheel sensor plug (RH rear). • Check for continuity between each sensor plug pin (sensor side) and vehicle ground.	Continuity ▶ No continuity ▶	REPLACE sensor (RH rear). SERVICE or REPLACE cable harness RECONNECT sensor plug.

TEST STEP	RESULT ▶	ACTION TO TAKE
EE6 CHECK ANTI-LOCK OPERATION: REAR WHEELS		
• Turn air suspension Off, if so equipped. • Lift vehicle and rotate wheels to ensure they turn freely. • Short Pins 18, 30 and 12 to each other at Breakout box. • Apply moderate brake pedal pressure. Check that rear wheels will not turn. • Check that rear wheels turn freely with ignition switch On. • Turn air suspension On when vehicle is off hoist, if so equipped. CAUTION: DO NOT LEAVE IGNITION ON FOR MORE THAN 1 MINUTE MAXIMUM, OR SOLENOID VALVE DAMAGE MAY RESULT.	If wheel turns freely ▶ If wheel does not turn freely or pedal drops ▶	REVERIFY symptom. REPLACE solenoid valve block.

Fig. 16 Diagnosis, outlet valve (Part 6 of 8). 1988 Continental & all 1989 models

Outlet Valve Diagnosis — Test EE

TEST STEP	RESULT	ACTION TO TAKE
EE7 SERVICE CODES 54 AND/OR 74 CHECK LH REAR SENSOR CIRCUIT CONTINUITY		
• Turn ignition switch Off. • Disconnect 32-pin plug from electronic controller.	Continuity ▶ No continuity ▶	GO to Step Step EE7a. GO to Step Step EE8

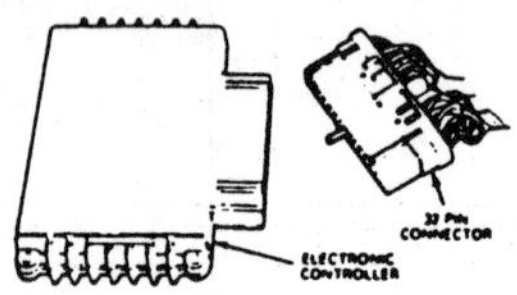

• Connect EEC-IV Breakout box with Tool T83L-50-EEC-IV or equivalent, Anti-Lock Test Adapter T87P-50-ALA or equivalent, to the anti-lock 32-pin plug wiring harness.

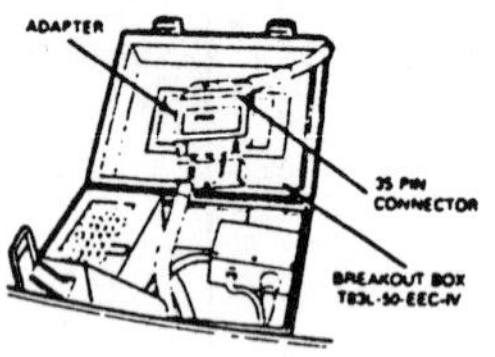

• Check for continuity between Breakout box Pins 40 and 4.

Fig. 16 Diagnosis, outlet valve (Part 7 of 8). 1988 Continental & all 1989 models

Outlet Valve Diagnosis — Test EE

TEST STEP	RESULT	ACTION TO TAKE
EE7A CHECK LH REAR SENSOR CONTINUITY		
• Disconnect wheel sensor plug (LH rear). • Check for continuity between each sensor plug pin (sensor side) and vehicle ground.	Continuity ▶ No continuity ▶	REPLACE sensor (LH rear). SERVICE or REPLACE cable harness RECONNECT sensor plug.
EE8 CHECK ANTI-LOCK OPERATION: REAR WHEELS		
• Turn air suspension Off, if so equipped. • Lift vehicle and rotate wheels to ensure they turn freely. • Short Pins 18, 30 and 12 to each other at Breakout box. • Apply moderate brake pedal pressure. Check that rear wheels will not turn. • Check that rear wheels turn freely with ignition switch On. • Turn air suspension On when vehicle is off hoist, if so equipped. CAUTION: DO NOT LEAVE IGNITION ON FOR MORE THAN 1 MINUTE MAXIMUM OR SOLENOID VALVE DAMAGE MAY RESULT.	If wheel turns freely ▶ If wheel does not turn freely or pedal drops ▶	VERIFY symptom. REPLACE solenoid valve block.

Fig. 16 Diagnosis, outlet valve (Part 8 of 8). 1988 Continental & all 1989 models

Warning Circuit Diagnosis — Test FF

TEST STEP	RESULT	ACTION TO TAKE
FF1 SERVICE CODE 61: CHECK FLI AND PWS CIRCUIT		
• Turn ignition switch Off. • Disconnect 32-Pin plug from electronic controller.	Less than 5 ohms ▶ Greater than 5 ohms ▶	GO to Step FF2. GO to Step FF1a.

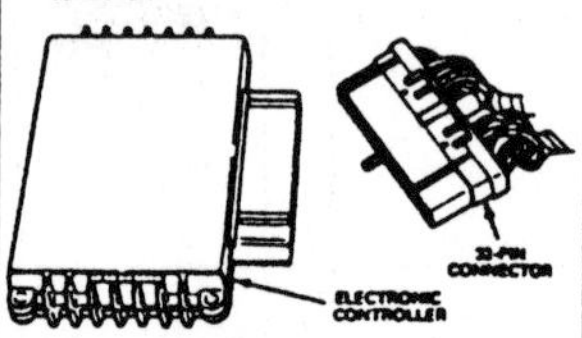

• Connect EEC-IV Breakout box with Tool T83L-50-EEC-IV, Anti-Lock Test Adapter T87P-50-ALA or equivalent, to the anti-lock 32-Pin plug wiring harness.

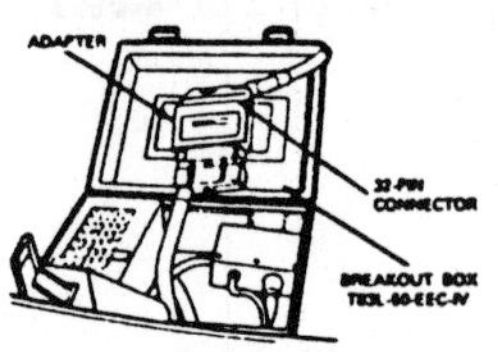

• Measure the resistance between Breakout box Pins 25 and 27.
NOTE: System must be pressurized for this test.

Fig. 17 Diagnosis, warning circuit (Part 1 of 6). 1988 Continental & all 1989 models

Warning Circuit Diagnosis — Test FF

TEST STEP	RESULT	ACTION TO TAKE
FF1a CHECK FLI ANTI-LOCK WARNING CIRCUIT		
• Disconnect 5-Pin plug on reservoir Fluid Level Indicator (FLI). • Measure resistance between fluid level indicator electrical socket Pins 1 and 2 (with brake fluid level at maximum mark on reservoir).	Less than 2 ohms (OK) ▶ Greater than 2 ohms (NOT OK) ▶	GO to Step FF1b. REPLACE fluid level indicator.
FF1b CHECK PWS ANTI-LOCK WARNING CIRCUIT		
• Disconnect 5-Pin plug at pressure warning switch (PWS). • Check for continuity between pressure warning switch socket Pins 3 and 5. NOTE: System must be pressurized for this test.	Continuity (OK) ▶ No continuity (NOT OK) ▶	GO to Step FF1c. REPLACE pressure warning switch and pump motor relay.

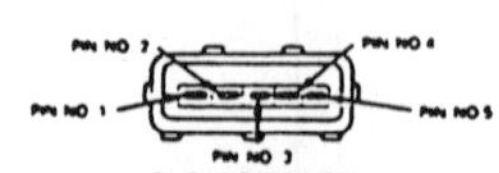

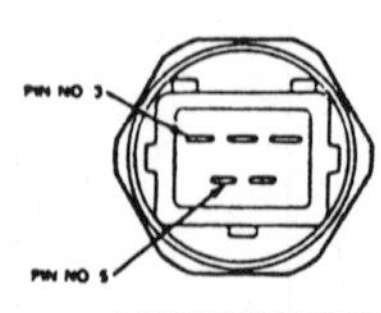

Fig. 17 Diagnosis, warning circuit (Part 2 of 6). 1988 Continental & all 1989 models

Warning Circuit Diagnosis — Test FF

TEST STEP	RESULT ▶	ACTION TO TAKE
FF1c CHECK ELECTRONIC MODULE TO FLI WIRE		
• Check for continuity between Breakout box Pin 25 and FLI 5-Pin plug (harness side) Pin 1. 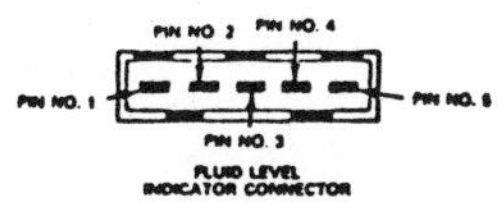	Continuity ▶ No continuity ▶	GO to Step FF1d. SERVICE or REPLACE cable harness
FF1d CHECK FLI TO PWS WIRE		
• Check for continuity between Pin 2 of 5-Pin fluid level plug (harness side) 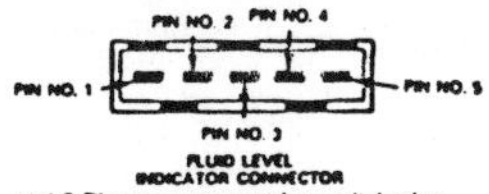 and 5-Pin pressure warning switch plug (harness side) Pin 3. 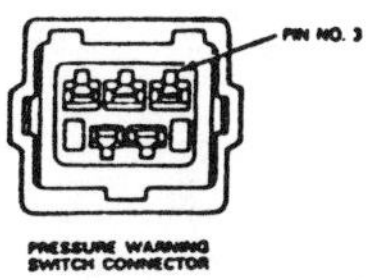	Continuity ▶ No continuity ▶	GO to Step FF1e. SERVICE or REPLACE cable harness

Fig. 17 Diagnosis, warning circuit (Part 3 of 6). 1988 Continental & all 1989 models

Warning Circuit Diagnosis — Test FF

TEST STEP	RESULT ▶	ACTION TO TAKE
FF1e PWS TO ELECTRONIC CONTROLLER WIRE		
• Check for continuity between 5-Pin pressure warning switch plug (harness side) Pin 5 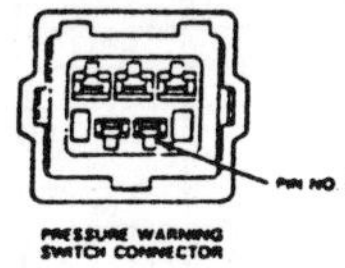 and Breakout box Pin 27.	Continuity ▶ No continuity ▶	TURN ignition OFF. CONNECT all electrical connections. REVERIFY symptom. SERVICE or REPLACE cable harness
FF2 ISOLATION TEST FLI AND PWS		
• Check for continuity between Breakout box Pin 25 and body ground and Pin 27 and body ground.	Continuity ▶ No continuity ▶	GO to Step FF2a. REVERIFY Symptom.
FF2a CHECK FLUID LEVEL INDICATOR PIN NO. 1		
• Check for continuity between FLI plug Pin 1 (harness side) and body ground. 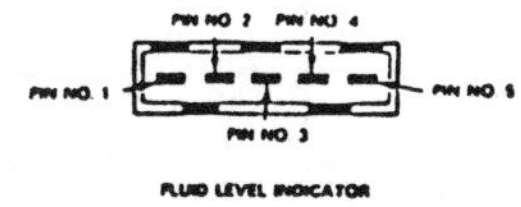	Continuity ▶ No continuity ▶	SERVICE or REPLACE Circuit 512. GO to Step FF2b.

Fig. 17 Diagnosis, warning circuit (Part 4 of 6). 1988 Continental & all 1989 models

Warning Circuit Diagnosis — Test FF

TEST STEP	RESULT ▶	ACTION TO TAKE
FF2b CHECK FLUID LEVEL INDICATOR PIN NO. 2		
• Disconnect 5-Pin plug from PWS. Check for continuity between FLI plug Pin 2 (harness side) and body ground. 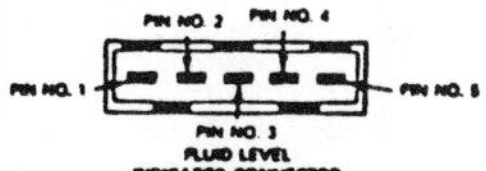	Continuity ▶ No continuity ▶	SERVICE or REPLACE Circuit 549 GO to Step FF2c.
FF2c CHECK CONTINUITY BETWEEN PWS PINS 3 AND 5 AND BODY GROUND		
• Check for continuity from PWS 5-Pin socket Pins (pressure warning switch side) 3 and body ground and 5 and body ground. 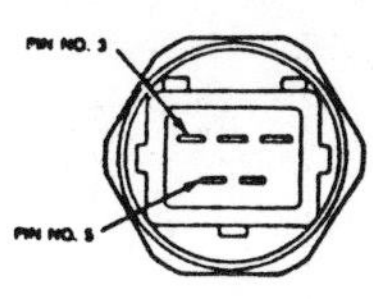	Continuity ▶ No continuity ▶	REPLACE Pressure Warning Switch and Pump Motor Relay. GO to Step FF2d.

Fig. 17 Diagnosis, warning circuit (Part 5 of 6). 1988 Continental & all 1989 models

Warning Circuit Diagnosis — Test FF

TEST STEP	RESULT ▶	ACTION TO TAKE
FF2d CHECK PWS CONNECTOR PIN 5 AND GROUND		
• Check for continuity between 5-Pin PWS plug, Pin 5 (harness side) and body ground. 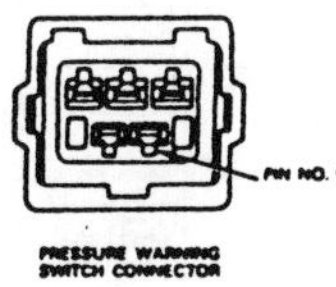	Continuity ▶ No continuity ▶	SERVICE or REPLACE Circuit 535 CONNECT all plugs and reverify symptom.

Fig. 17 Diagnosis, warning circuit (Part 6 of 6). 1988 Continental & all 1989 models

Anti-Lock Quick Check Sheet Using 60-Pin EEC-IV Breakout Box, Tool T83L-50-EEC-IV①

NOTE: Before performing tests below, the Pre-Test Checks must be performed as outlined.
NOTE: If fault is intermittent the tests listed below will NOT find the fault.

Item to be Tested		Ignition Mode	Measure Between Pin Numbers	Tester Scale/ Range	Specification	Test Step
No Boost/No Power Brakes (Hard Brake Pedal)		—	—	—	—	D-1
Battery Check		On	40 + 18	Volts	10 minimum	A-1
Main Power Relay		Off	40 + 9	Ohms	45 Ohms — 105 Ohms	A-6
				Place a jumper between pins 9 and 18		
		On	40 + 16	Volts	10 minimum	A-7
Power from Main Power Relay		On	40 + 15	Volts	10 minimum	A-3
				Remove jumper from pins 9 and 18		
Main Power Relay Circuit		Off	40 + 16	Continuity	Continuity	A-2
Main Power Relay Circuit		Off	15 + 40	Continuity	Continuity	A-3a
Sensor Resistance	(RR)	Off	6 + 23	K Ohms	800 to 1400 Ohms	A-8
Sensor Resistance	(LF)	Off	5 + 22	K Ohms	800 to 1400 Ohms	A-9
Sensor Resistance	(LR)	Off	4 + 21	K Ohms	800 to 1400 Ohms	A-10
Sensor Resistance	(RF)	Off	3 + 20	K Ohms	800 to 1400 Ohms	A-11
Main Valve Resistance		Off	11 + 29	Ohms	2 Ohms to 5.5 Ohms	A-12
Inlet & Outlet Valves		Off	11 + 40	Continuity	Continuity	A-13
		Off	11 + 32	Ohms	5 Ohms to 8 Ohms	A-14
		Off	11 + 30	Ohms	5 Ohms to 8 Ohms	A-15
		Off	11 + 31	Ohms	5 Ohms to 8 Ohms	A-16
		Off	11 + 12	Ohms	3 Ohms to 6 Ohms	A-17
		Off	11 + 14	Ohms	3 Ohms to 6 Ohms	A-18
		Off	11 + 13	Ohms	3 Ohms to 6 Ohms	A-19
Reservoir Warning		On	25 + 27	Ohms	Less than 5 Ohms	A-4a
Lift Fluid Level Indicator from Reservoir (Float at bottom position)		Off	25 + 27	Ohms	Infinite (Open Circuit)	4-5a
Sensor Cable Continuity Shielding to Ground	(RR)	Off	40 + 6	Continuity	No Continuity	B-1a
	(LF)	Off	40 + 5	Continuity	No Continuity	B-2a
	(LR)	Off	40 + 4	Continuity	No Continuity	B-3a
	(RF)	Off	40 + 3	Continuity	No Continuity	B-4a
Sensor Voltage (Rotate wheels at 1 revolution per second minimum) (Shut off air suspension switch in luggage compartment with vehicle on hoist if so equipped.)	(RR)	Off②	6 + 23	AC Millivolts	50-700 Millivolts	C-5
	(LF)	Off②	5 + 22	AC Millivolts	50-700 Millivolts	C-6
	(LR)	Off②	4 + 21	AC Millivolts	50-700 Millivolts	C-7
	(RF)	Off②	3 + 20	AC Millivolts	50-700 Millivolts	C-8

① If Quick Test does not isolate symptom, refer to Diagnostic Lamp Symptom Chart.
② The most accurate measurements are taken with the breakout box in back seat and an assistant is driving the vehicle 4.5-5.0 mph.

Fig. 18 Anti-lock quick check sheet. 1988 Continental & all 1989 models

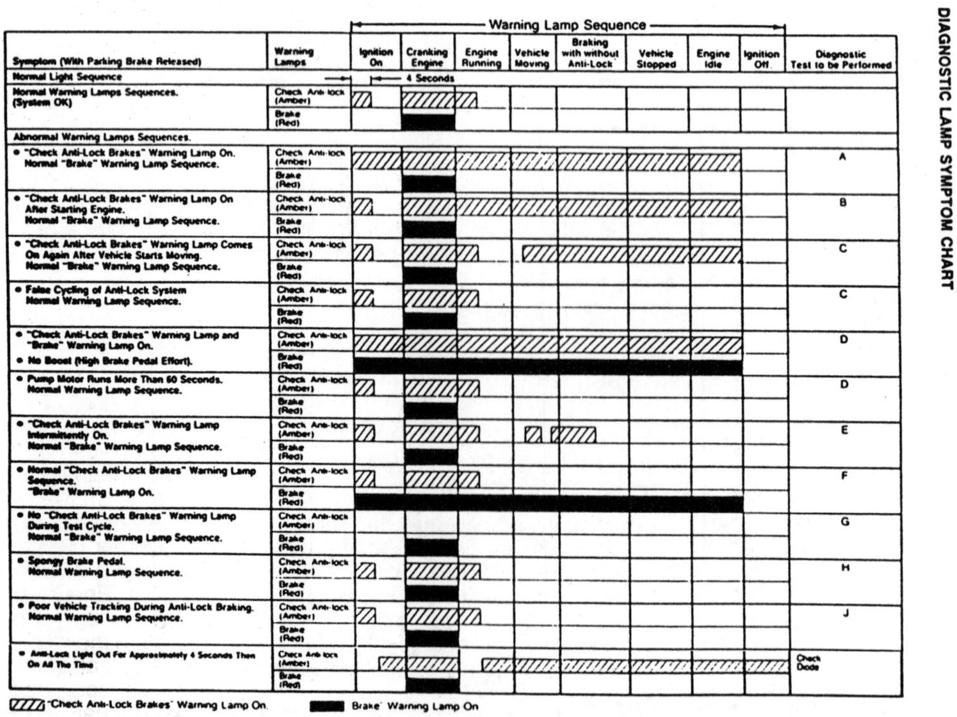

Fig. 19 Diagnostic warning lamp symptom chart. 1988 Continental & all 1989 models

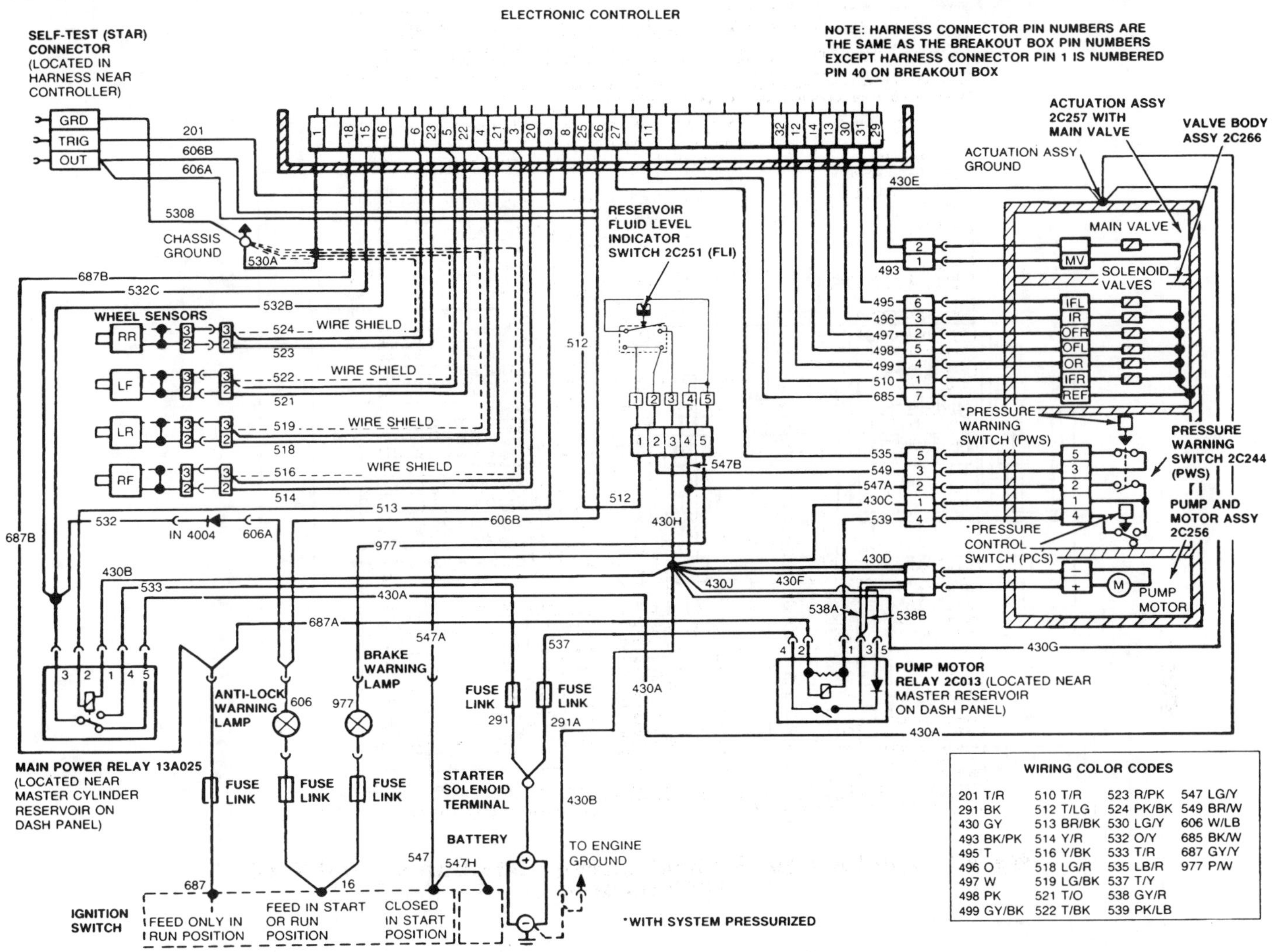

201 T/R	510 T/R	523 R/PK	547 LG/Y
291 BK	512 T/LG	524 PK/BK	549 BR/W
430 GY	513 BR/BK	530 LG/Y	606 W/LB
493 BK/PK	514 Y/R	532 O/Y	685 BK/W
495 T	516 Y/BK	533 T/R	687 GY/Y
496 O	518 LG/R	535 LB/R	977 P/W
497 W	519 LG/BK	537 T/Y	
498 PK	521 T/O	538 GY/R	
499 GY/BK	522 T/BK	539 PK/LB	

Fig. 20 Anti-lock system wiring circuit. 1988-89 Continental

PINPOINT TESTS

To properly conduct the Pinpoint Test diagnostic procedure an anti-lock harness adapter No. T87P-50-ALA, EEC-IV breakout box T83 L-50-EEC-IV, anti-lock pressure gauge No. D88M-20215-A, digital volt/ohmmeter No. 007-00001 and Super Star tester 007-00019 or equivalents.

Refer to the wiring diagram, Figs. 20 through 22, when performing the Pinpoint Tests to locate wire circuits indicated in the test procedure. Each test is completely independent of the other test and within each test are sequences that can identify a problem without requiring completion of the entire diagnostic test procedure.

Refer to **Figs. 23 through 31,** for Pinpoint Test diagnostic procedures.

SYSTEM SERVICE

Before servicing any system component which contains high pressure, it is mandatory that the hydraulic pressure in the system be completely discharged. To discharge the system, turn ignition switch to OFF position, then pump brake pedal a minimum of 20 times, until an increase in pedal force is clearly felt.

FRONT BRAKES

The front brakes can be bled in the conventional manner or with brake bleeder 104-00064, or equivalent, with or without the accumulator being charged.

Brake Bleeding Less Pressure Bleeder

1. Remove dust cap from the righthand caliper bleeder fitting. Attach a rubber drain tube to fitting. Ensure that the end of the tube fits snugly around the fitting.
2. Submerge free end of the tube in a suitable container, partially filled with clean brake fluid.
3. Loosen bleeder fitting approximately three-quarter turn. Push brake pedal down slowly through full travel and hold at that position.
4. Close bleeder fitting, then return pedal to full release position. Wait five seconds, then repeat operation until air bubbles cease to appear at submerged end of bleeder tube.
5. Repeat operation at the lefthand caliper.

Brake Bleeding w/Pressure Bleeder

1. Clean all dirt from the reservoir filler cap area. Attach pressure bleeder to reservoir cap opening.
2. Maintain approximately 35 psi pressure on the system.
3. Remove dust cap from righthand front caliper bleeder fitting.
4. Attach a rubber drain tube to the fitting. Ensure that the end of the tube fits snugly around the fitting.
5. With the ignition switch in the OFF position and the brake pedal in the fully released position, open the righthand front caliper bleeder fitting for 10 seconds at a time until an air free stream of brake fluid is observed from submerged end of hose.
6. Repeat procedure at lefthand caliper.

WIRING DIAGRAM

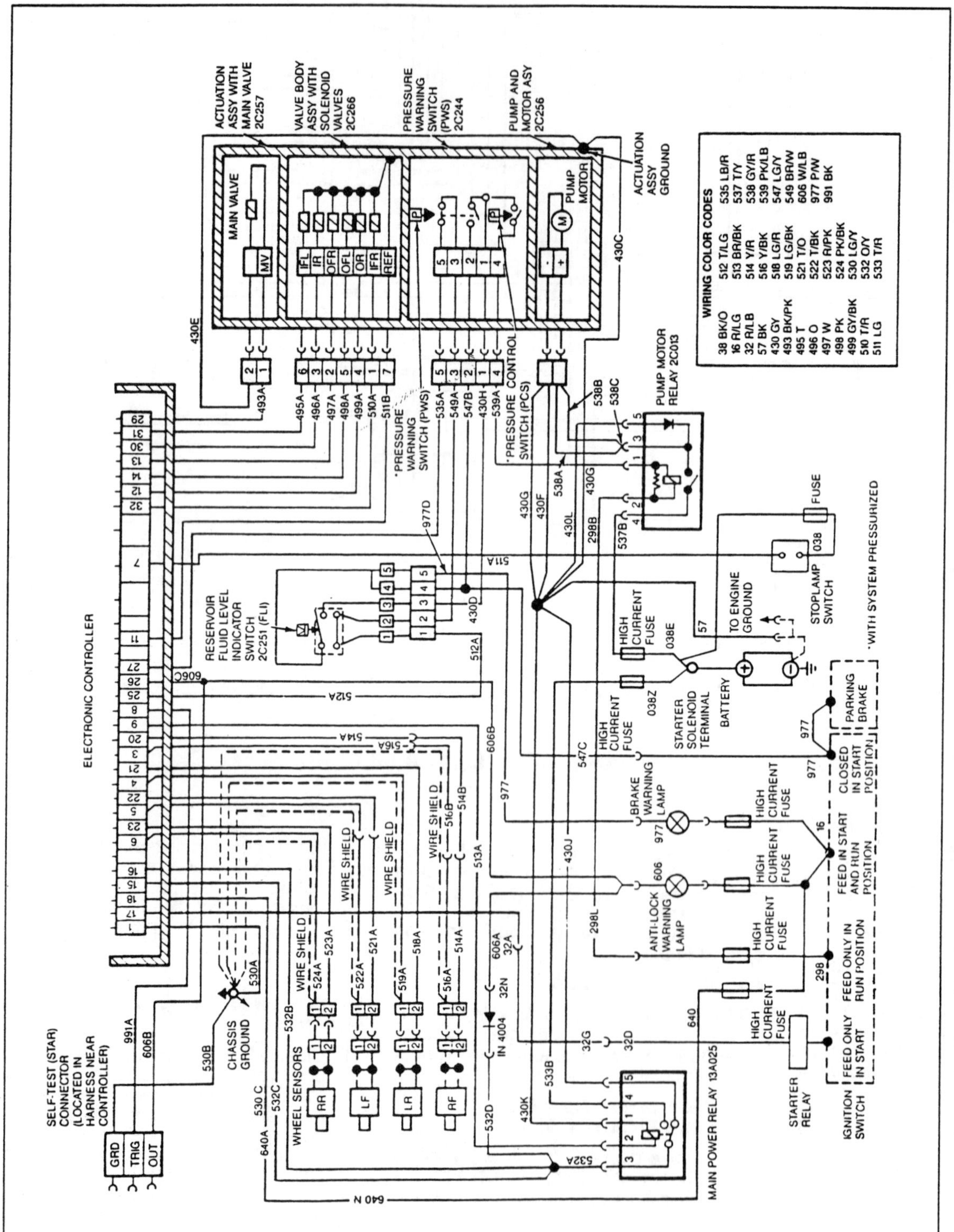

Fig. 21 Anti-lock system wiring circuit. 1989 Cougar & Thunderbird

WIRING DIAGRAM

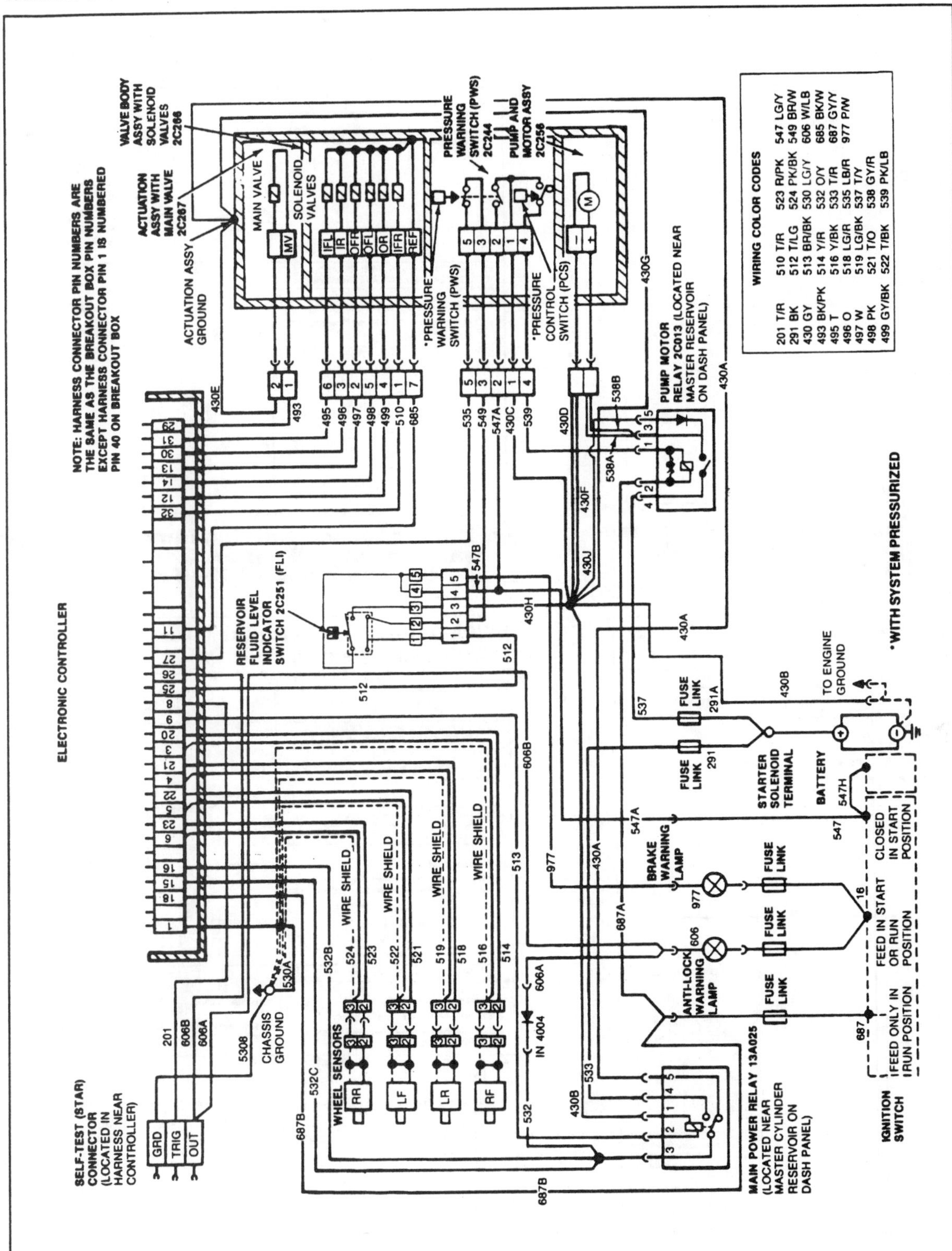

Fig. 22 Anti-lock system wiring circuit. 1989 Mark VIII

Anti-Lock Warning Lamp On (With Brake Warning Lamp Off)	Test A

WARNING LIGHTS SEQUENCE

Warning Lamps	Ignition On	Cranking Engine	Engine Running	Vehicle Moving	Braking with/without Anti-Lock	Vehicle Stopped	Engine Idle	Ignition Off
Check Anti-Lock (Amber)								
Brake (Red)								

TEST STEP	RESULT ▶	ACTION TO TAKE
A1 32 PIN PLUG TESTING		
• Turn ignition switch OFF. • Disconnect 32-pin Plug from electronic controller. • Connect EEC-IV Breakout box Tool T83L-50-EEC-IV and anti-lock test adapter Tool T87P-50-ALA or equivalents, to the anti-lock 32-pin plug wiring harness. • Set multi-meter to read volts DC. • Turn ignition switch ON. • Measure voltage between Breakout box Pins 40 and 18.	Over 10V (OK) ▶ Under 10V (NOT OK) ▶	GO to Step A2. GO to Step A1a.

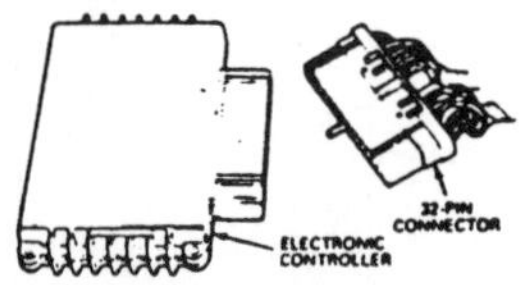

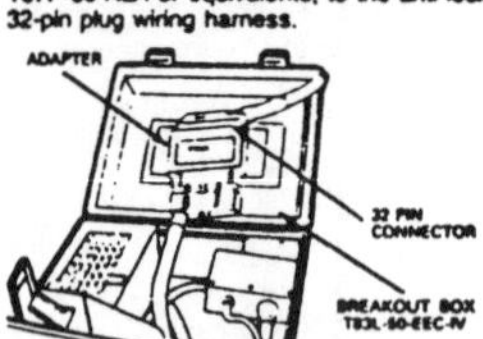

Fig. 23 Diagnosis, warning lamp ON w/brake warning lamp OFF (Part 1 of 24). 1988 Continental & all 1989 models

Anti-Lock Warning Lamp On (With Brake Warning Lamp Off)	Test A

TEST STEP	RESULT ▶	ACTION TO TAKE
A1a CHECK ELECTRONIC CONTROLLER TO GROUND WIRE		
• Check: — fuse link to anti-lock warning lamp. — battery. • Remove positive battery cable. • Check continuity between Breakout box pin 40 and body ground.	Continuity ▶ No continuity ▶	GO to Step A1b. SERVICE or REPLACE cable harness
A1b CHECK IGNITION TO ELECTRONIC CONTROLLER WIRE		
• Check continuity between Breakout box Pin 18 and ignition switch wire 640N.	Continuity ▶	RECONNECT positive battery cable. CHECK for power at ignition switch pin with switch ON. If okay, connect electronic controller and reverify symptom.
	No continuity ▶	SERVICE or REPLACE cable harness
A2 CHECK MAIN POWER RELAY SECONDARY CIRCUIT (NORMAL)		
• Turn ignition switch OFF. • Check for continuity between Breakout box Pins 40 and 16.	Continuity ▶ No continuity ▶	GO to A3. GO to A2a.

Fig. 23 Diagnosis, warning lamp ON w/brake warning lamp OFF (Part 2 of 24). 1988 Continental & all 1989 models

Anti-Lock Warning Lamp On (With Brake Warning Lamp Off)	Test A

TEST STEP	RESULT ▶	ACTION TO TAKE
A2a CHECK MAIN POWER RELAY SECONDARY CIRCUIT (NORMAL)		
• Disconnect main relay from socket. • Check for continuity between main power relay socket Pins 3 and 5.	Continuity ▶ No continuity ▶	GO to Step A2b. REPLACE main power relay.
A2b CHECK MAIN POWER RELAY SECONDARY CIRCUIT WIRING HARNESS		
• Disconnect positive battery cable. • Check for continuity between main power relay socket Pin 3 and Breakout box Pin 16.	Continuity ▶ No continuity ▶	GO to Step A2c. SERVICE or REPLACE cable harness
A2c CHECK MAIN POWER RELAY SECONDARY CIRCUIT WIRING HARNESS		
• Check for continuity between main power relay socket Pin 5 and body ground.	Continuity ▶ No continuity ▶	RECONNECT main power relay, electronic controller and battery cable and reverify symptom. SERVICE or REPLACE cable harness

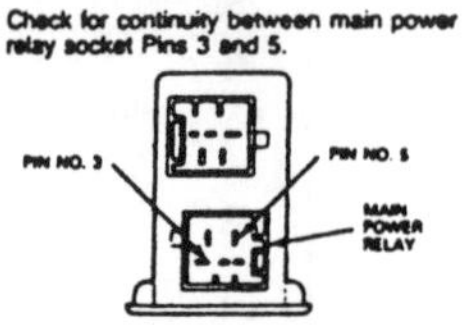

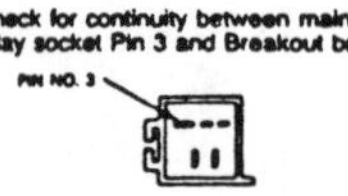

Fig. 23 Diagnosis, warning lamp ON w/brake warning lamp OFF (Part 3 of 24). 1988 Continental & all 1989 models

With Anti-Lock Warning Lamp On (With Brake Warning Lamp Off)	Test A

TEST STEP	RESULT ▶	ACTION TO TAKE
A3 CHECK MAIN POWER RELAY SECONDARY CIRCUIT (NORMAL)		
• Check for continuity between Breakout box Pins 40 and 15.	Continuity ▶ No continuity ▶	GO to Step A4. GO to Step A3a.
A3a CHECK MAIN POWER RELAY SECONDARY CIRCUIT WIRING HARNESS		
• Remove main power relay. • Check for continuity between main power relay socket Pin 3 and Breakout box Pin 15.	Continuity ▶ No continuity ▶	CONNECT main power relay and electronic controller and reverify symptom. SERVICE or REPLACE cable harness
A4 CHECK FLI AND PWS CIRCUIT		
• Turn ignition switch ON. • Set multi-meter to read resistance. • Measure the resistance between Breakout box Pins 25 and 27.	Less than 5 ohms ▶ Greater than 5 ohms ▶	GO to Step A5. GO to Step A4a.
A4a CHECK FLI ANTI-LOCK WARNING CIRCUIT		
• Disconnect 5-Pin plug on reservoir Fluid Level Indicator (FLI). • Measure resistance between fluid level indicator electrical socket Pins 1 and 2 (with brake fluid level at maximum mark on reservoir).	Less than 2 ohms ▶ Greater than 2 ohms ▶	GO to Step A4b. REPLACE fluid level indicator.

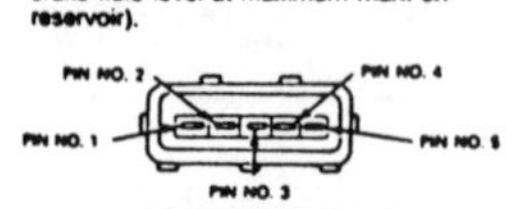

Fig. 23 Diagnosis, warning lamp ON w/brake warning lamp OFF (Part 4 of 24). 1988 Continental & all 1989 models

Anti-Lock Warning Lamp On (With Brake Warning Lamp Off) — **Test A**

TEST STEP	RESULT ▶	ACTION TO TAKE
A4b CHECK PWS ANTI-LOCK WARNING CIRCUIT		
• Disconnect 5-Pin plug at pressure warning switch (PWS). • Check for continuity between pressure warning switch socket Pins 3 and 5. **NOTE: Systems must be pressurized during this test.** 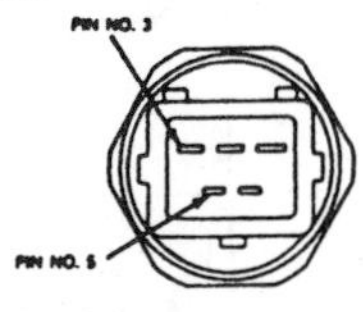	Continuity ▶ No continuity ▶	GO to Step A4c. REPLACE pressure warning switch and pump motor relay.
A4c CHECK ELECTRONIC MODULE TO FLI WIRE		
• Check for continuity between Breakout box Pin 25 and FLI 5-Pin plug (harness side) Pin 1. 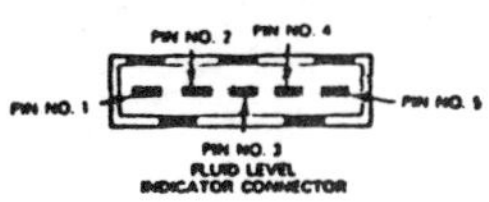	Continuity ▶ No continuity ▶	GO to Step A4d. SERVICE or REPLACE cable harness

Fig. 23 Diagnosis, warning lamp ON w/brake warning lamp OFF (Part 5 of 24). 1988 Continental & all 1989 models

Anti-Lock Warning Lamp On (With Brake Warning Lamp Off) — **Test A**

TEST STEP	RESULT ▶	ACTION TO TAKE
A4d CHECK FLI TO PWS WIRE		
• Check for continuity between Pin 2 of 5-Pin fluid level plug (harness side) 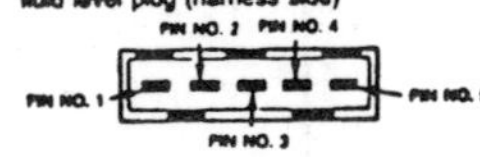 and 5-Pin pressure warning switch plug (harness side) Pin 3. 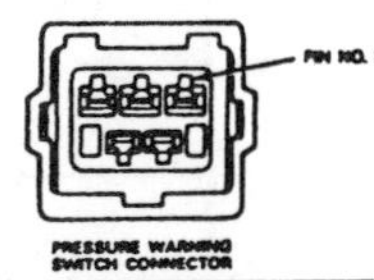	Continuity ▶ No continuity ▶	GO to Step A4e. SERVICE or REPLACE cable harness
A4e PWS TO ELECTRONIC CONTROLLER WIRE		
• Check for continuity between 5-Pin pressure warning switch plug (harness side) Pin 5 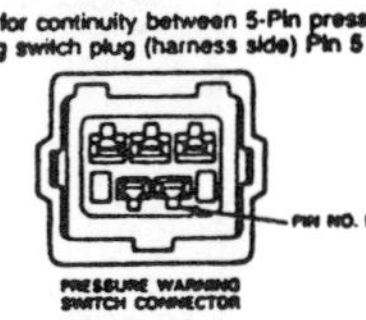 and Breakout box Pin 27.	Continuity ▶ No continuity ▶	TURN ignition OFF. CONNECT all electrical connections. Reverify symptom. SERVICE or REPLACE cable harness
A5 ISOLATION TEST FLI AND PWS		
• Check for continuity between Breakout box Pin 25 and body ground.	Continuity ▶ No continuity ▶	GO to Step A5a. GO to Step A6.

Fig. 23 Diagnosis, warning lamp ON w/brake warning lamp OFF (Part 6 of 24). 1988 Continental & all 1989 models

Anti-Lock Warning Lamp On (With Brake Warning Lamp Off) — **Test A**

TEST STEP	RESULT ▶	ACTION TO TAKE
A5a CHECK FLUID LEVEL INDICATOR PIN NO. 2		
• Disconnect FLI 5-Pin plug and check for continuity between FLI socket Pin 2 and body ground and Pin 1 and body ground. **NOTE: Ensure that brake fluid is at max level.** 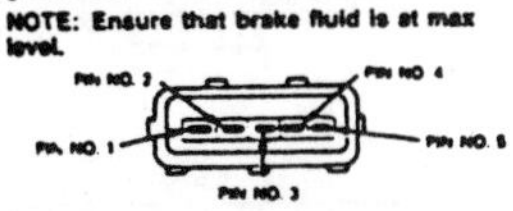	Continuity ▶ No continuity ▶	REPLACE FLI. GO to Step A5b.
A5b CHECK FLUID LEVEL INDICATOR PIN NO. 1		
• Check for continuity between FLI plug Pin 1 (harness side) and body ground. 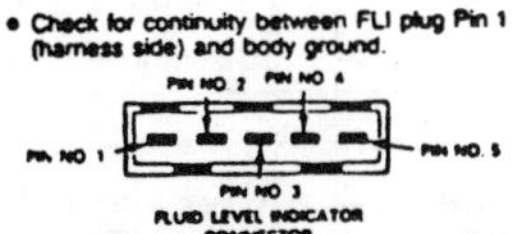	Continuity ▶ No continuity ▶	SERVICE or REPLACE Circuit 512 GO to Step A5c.
A5c CHECK FLUID LEVEL INDICATOR PIN NO. 2		
• Disconnect 5-Pin plug from PWS. Check for continuity between FLI plug Pin 2 (harness side) and body ground. 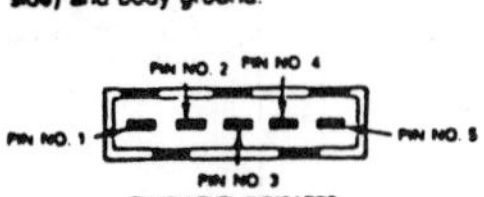	Continuity ▶ No continuity ▶	SERVICE or REPLACE Circuit 549 GO to Step A5d.

Fig. 23 Diagnosis, warning lamp ON w/brake warning lamp OFF (Part 7 of 24). 1988 Continental & all 1989 models

Anti-Lock Warning Lamp On (With Brake Warning Lamp Off) — **Test A**

TEST STEP	RESULT ▶	ACTION TO TAKE
A5d CHECK CONTINUITY BETWEEN PWS PINS 3 AND 5 AND BODY GROUND		
• Check for continuity from PWS 5-Pin socket Pins (pressure warning switch side) 3 and body ground and 5 and body ground. **NOTE: System should be depressurized for this test.** 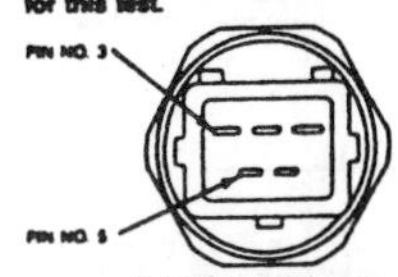	Continuity ▶ No continuity ▶	REPLACE pressure warning switch and pump motor relay. GO to Step A5e.
A5e CHECK PWS CONNECTOR PIN 5 AND GROUND		
• Check for continuity between 5-Pin PWS plug. Pin 5 (harness side) and body ground. 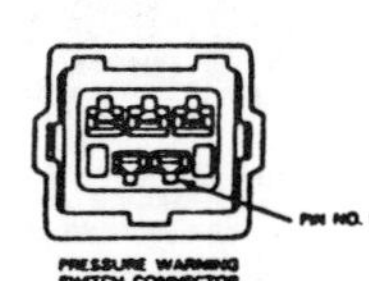	Continuity ▶ No continuity ▶	SERVICE or REPLACE Circuit 535 CONNECT all plugs and reverify symptom.
A6 CHECK MAIN RELAY PRIMARY CIRCUIT RESISTANCE		
• Turn ignition switch OFF. • Set multi-meter to read resistance. • Measure resistance between Breakout box Pins 40 and 9.	Resistance between 45 and 105 ohms ▶ Any other reading ▶	GO to Step A7. GO to Step A6a.

Fig. 23 Diagnosis, warning lamp ON w/brake warning lamp OFF (Part 8 of 24). 1988 Continental & all 1989 models

Anti-Lock Warning Lamp On (With Brake Warning Lamp Off) — **Test A**

TEST STEP	RESULT ▶	ACTION TO TAKE
A6a CHECK MAIN POWER RELAY PRIMARY COIL RESISTANCE		
• Disconnect main power relay from socket. • Set multi-meter on resistance. • Measure resistance between main power relay Pins 1 and 2. 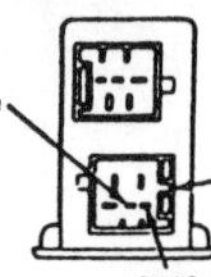 PIN NO. 2 / MAIN POWER RELAY / PIN NO. 1	Resistance between 45 and 105 ohms ▶ Any other reading ▶	GO to Step A6b. REPLACE main power relay.
A6b CHECK MAIN POWER RELAY PRIMARY TO ELECTRONIC CONTROLLER WIRE		
• Check continuity between main power relay socket Pin 2 (harness side) and Breakout box Pin 9. PIN NO. 2  MAIN POWER RELAY CONNECTOR	Continuity ▶ No continuity ▶	GO to Step A6c. SERVICE or REPLACE cable harness

Fig. 23 Diagnosis, warning lamp ON w/brake warning lamp OFF (Part 9 of 24). 1988 Continental & all 1989 models

Anti-Lock Warning Lamp On (With Brake Warning Lamp Off) — **Test A**

TEST STEP	RESULT ▶	ACTION TO TAKE
A6c CHECK MAIN POWER RELAY PRIMARY TO GROUND WIRE		
• Check continuity between main power relay socket Pin 1 and ground. 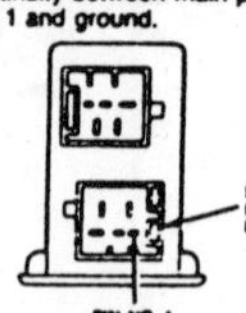 MAIN POWER RELAY / PIN NO. 1	Continuity ▶ No continuity ▶	RECONNECT all electrical connections and reverify symptom. SERVICE or REPLACE cable harness
A7 CHECK MAIN POWER RELAY SECONDARY (ACTIVATED)		
• Place a jumper wire between Breakout box Pins 18 and 9. • Set multi-meter to read volts DC. • Turn ignition switch ON. • Measure voltage between Breakout box Pins 40 and 16.	Over 10V DC ▶ Under 10V DC ▶	GO to Step A8. CHECK high current fuse. REMOVE jumper wire and GO to Step A7a.
A7a CHECK MAIN POWER RELAY SECONDARY CIRCUIT (ACTIVE)		
• Turn ignition switch OFF. • Disconnect main power relay from socket. • Apply power (battery positive and ground) to main relay Pins 1 and 2. • Check for continuity between main power relay Pins 3 and 4. 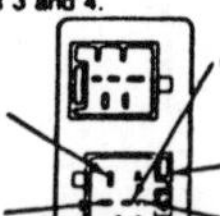 PIN NO. 2 / PIN NO. 4 / MAIN POWER RELAY / PIN NO. 3 / PIN NO. 1	Continuity ▶ No continuity ▶	GO to Step A7b. REPLACE main power relay.

Fig. 23 Diagnosis, warning lamp ON w/brake warning lamp OFF (Part 10 of 24). 1988 Continental & all 1989 models

Anti-Lock Warning Lamp On (With Brake Warning Lamp Off) — **Test A**

TEST STEP	RESULT ▶	ACTION TO TAKE
A7b CHECK MAIN POWER RELAY SECONDARY CIRCUIT POWER WIRE		
• Check continuity between main power relay socket Pin 4 (harness side) and positive battery terminal.  PIN NO 4 MAIN POWER RELAY CONNECTOR	Continuity ▶ No continuity ▶	RECONNECT main power relay and reverify symptom. SERVICE or REPLACE cable harness or high current fuse
A8 MEASURE RH REAR SENSOR CIRCUIT RESISTANCE		
• Turn ignition switch OFF. • Set multi-meter to read resistance. • Measure resistance between Breakout box Pin 8 and 23.	800 to 1400 ohms (0.8 to 1.4K ohms) ▶ Any other reading ▶	GO to Step A9. GO to Step A8a.
A8a MEASURE RH REAR SENSOR RESISTANCE		
• Disconnect sensor plug (right rear). • Set multi-meter to read resistance. • Measure resistance of sensor at sensor plug.  RH REAR SENSOR	800 to 1400 ohms (0.8 to 1.4K ohms) ▶ Any other reading ▶	SERVICE or REPLACE cable harness REPLACE right rear wheel sensor.
A9 MEASURE LH FRONT SENSOR CIRCUIT RESISTANCE		
• Measure resistance between Breakout box Pins 5 and 22.	800 to 1400 ohms (0.8 to 1.4K ohms) ▶ Any other reading ▶	GO to Step A10. GO to Step A9a.

Fig. 23 Diagnosis, warning lamp ON w/brake warning lamp OFF (Part 11 of 24). 1988 Continental & all 1989 models

Anti-Lock Warning Lamp On (With Brake Warning Lamp Off) — **Test A**

TEST STEP	RESULT ▶	ACTION TO TAKE
A9a MEASURE LH FRONT SENSOR RESISTANCE		
• Disconnect sensor plug (left front). • Measure resistance of sensor at sensor plug.  LH FRONT SENSOR	800 to 1400 ohms (0.8 to 1.4K ohms) ▶ Any other reading ▶	SERVICE or REPLACE cable harness REPLACE left front wheel sensor.
A10 MEASURE LH REAR SENSOR CIRCUIT RESISTANCE		
• Measure resistance between Breakout box Pins 4 and 21.	800 to 1400 ohms (0.8 to 1.4K ohms) ▶ Any other reading ▶	GO to Step A11. GO to Step A10a.
A10a MEASURE LH REAR SENSOR RESISTANCE		
• Disconnect sensor plug (left rear). • Measure resistance of sensor at left rear sensor plug.  LH REAR SENSOR	800 to 1400 ohms (0.8 to 1.4K ohms) ▶ Any other reading ▶	SERVICE or REPLACE cable harness REPLACE left rear wheel sensor.
A11 MEASURE RH FRONT SENSOR CIRCUIT RESISTANCE		
• Measure resistance between Breakout box Pins 3 and 20.	800 to 1400 ohms (0.8 to 1.4K ohms) ▶ Any other reading ▶	GO to Step A12. GO to Step A11a.

Fig. 23 Diagnosis, warning lamp ON w/brake warning lamp OFF (Part 12 of 24). 1988 Continental & all 1989 models

Anti-Lock Warning Lamp On (With Brake Warning Lamp Off) — **Test A**

TEST STEP	RESULT ▶	ACTION TO TAKE
A11a MEASURE RH FRONT SENSOR RESISTANCE		
• Disconnect sensor plug (RH front). • Measure resistance of sensor at sensor plug.  RH FRONT SENSOR	800 to 1400 ohms (0.8 to 1.4K ohms) ▶	SERVICE or REPLACE cable harness
	Any other reading ▶	REPLACE RH front wheel sensor.
A12 MEASURE MAIN VALVE CIRCUIT RESISTANCE		
• Turn ignition switch OFF. • Set multi-meter to read resistance. • Measure the resistance between Breakout box Pins 11 and 29.	2 to 5 ohms ▶	GO to Step A13.
	Any other reading ▶	GO to Step A12a.
A12a MEASURE MAIN VALVE RESISTANCE		
• Disconnect main valve 2-Pin plug. • Measure resistance between the main valve electrical Pins 1 and 2. 	2 to 5.5 ohms ▶	REPLACE or SERVICE cable harness
	Any other reading ▶	REPLACE actuation assembly.
A13 CHECK VALVE BLOCK GROUND CIRCUIT		
• Measure resistance between Breakout box Pins 11 and 40.	Less than 2 ohms ▶	GO to Step A14.
	Greater than 2 ohms ▶	GO to Step A13a.

Fig. 23 Diagnosis, warning lamp ON w/brake warning lamp OFF (Part 13 of 24). 1988 Continental & all 1989 models

Anti-Lock Warning Lamp On (With Brake Warning Lamp Off) — **Test A**

TEST STEP	RESULT ▶	ACTION TO TAKE
A13a TEST VALVE PLUG PIN NO. 7		
• Disconnect valve block 7-Pin plug. • Measure resistance between valve block electrical Pin 7 and valve block body. 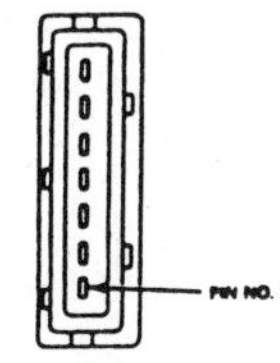 VALVE BLOCK 7-PIN CONNECTOR	Less than 2 ohms ▶	SERVICE or REPLACE cable harness
	Greater than 2 ohms ▶	GO to Step A13b.
A13b CHECK VALVE BLOCK GROUND WIRE		
• Remove negative (−) ground strap from battery. • Check for continuity between valve block and body ground.	Continuity ▶	REPLACE valve block unit (internal ground problem).
	No or Poor continuity ▶	SERVICE or REPLACE actuation assembly ground strap
A14 MEASURE RH FRONT INLET VALVE CIRCUIT RESISTANCE		
• Measure resistance between Breakout box Pins 11 and 32.	5 to 8 ohms ▶	GO to Step A15.
	Any other reading ▶	GO to Step A14a.

Fig. 23 Diagnosis, warning lamp ON w/brake warning lamp OFF (Part 14 of 24). 1988 Continental & all 1989 models

Anti-Lock Warning Lamp On (With Brake Warning Lamp Off) — **Test A**

TEST STEP	RESULT ▶	ACTION TO TAKE
A14a MEASURE RH FRONT INLET VALVE RESISTANCE		
• Disconnect valve block 7-Pin plug. • Measure resistance between valve block electrical Pin 7 and 1. 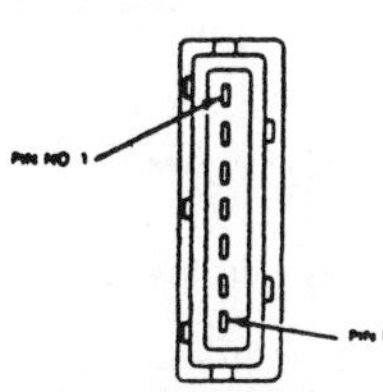 VALVE BLOCK 7-PIN CONNECTOR	5 to 8 ohms ▶	SERVICE or REPLACE cable harness
	Any other reading ▶	REPLACE solenoid valve block unit. CONNECT 7-Pin plug
A15 MEASURE REAR INLET VALVE CIRCUIT RESISTANCE		
• Measure resistance between Breakout box Pins 11 and 30.	5 to 8 ohms ▶	GO to Step A16.
	Any other reading ▶	GO to Step A15a.

Fig. 23 Diagnosis, warning lamp ON w/brake warning lamp OFF (Part 15 of 24). 1988 Continental & all 1989 models

Anti-Lock Warning Lamp On (With Brake Warning Lamp Off) — **Test A**

TEST STEP	RESULT ▶	ACTION TO TAKE
A15a MEASURE REAR INLET VALVE RESISTANCE		
• Disconnect valve block 7-Pin plug. • Measure resistance between valve block electrical Pins 7 and 3. 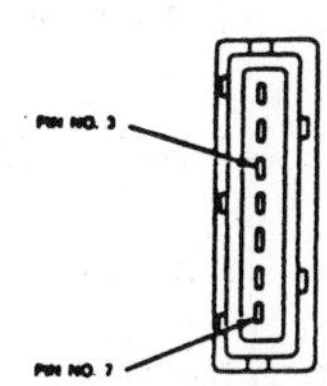 VALVE BLOCK 7-PIN CONNECTOR	5 to 8 ohms ▶	SERVICE or REPLACE cable harness
	Any other reading ▶	REPLACE valve block unit. CONNECT 7-Pin plug.
A16 MEASURE LH FRONT INLET VALVE CIRCUIT RESISTANCE		
• Measure resistance between Breakout box Pins 11 and 31.	5 to 8 ohms ▶	GO to Step A17.
	Any other reading ▶	GO to Step A16a.

Fig. 23 Diagnosis, warning lamp ON w/brake warning lamp OFF (Part 16 of 24). 1988 Continental & all 1989 models

Anti-Lock Warning Lamp On (With Brake Warning Lamp Off) — Test A

TEST STEP	RESULT ▶	ACTION TO TAKE
A16a MEASURE LH FRONT INLET VALVE RESISTANCE		
• Disconnect valve block 7-Pin plug. • Measure resistance between valve block electrical Pins 7 and 6.	5 to 8 ohms ▶	SERVICE or REPLACE cable harness
	Any other reading ▶	REPLACE valve block unit. CONNECT 7-Pin plug.
A17 MEASURE REAR OUTLET VALVE CIRCUIT RESISTANCE		
• Measure resistance between Breakout box Pins 11 and 12.	3 to 6 ohms ▶	GO to Step A18.
	Any other reading ▶	GO to Step A17a.

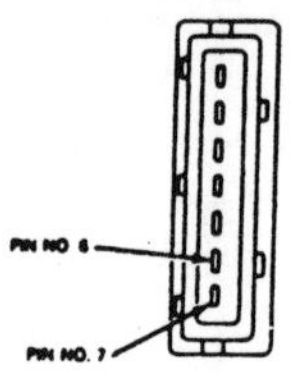

VALVE BLOCK 7-PIN CONNECTOR

Fig. 23 Diagnosis, warning lamp ON w/brake warning lamp OFF (Part 17 of 24). 1988 Continental & all 1989 models

Anti-Lock Warning Lamp On (With Brake Warning Lamp Off) — Test A

TEST STEP	RESULT ▶	ACTION TO TAKE
A17a MEASURE REAR OUTLET VALVE RESISTANCE		
• Disconnect valve block 7-Pin plug. • Measure resistance between valve block electrical Pins 7 and 4.	3 to 6 ohms ▶	SERVICE or REPLACE cable harness
	Any other reading ▶	REPLACE valve block unit. CONNECT 7-Pin plug.
A18 MEASURE LH FRONT OUTLET VALVE CIRCUIT RESISTANCE		
• Measure resistance between Breakout box Pins 11 and 14.	3 to 6 ohms ▶	GO to Step A19.
	Any other reading ▶	GO to Step A18a.

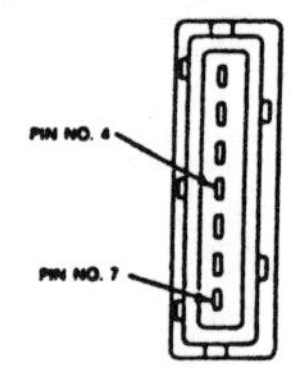

VALVE BLOCK 7-PIN CONNECTOR

Fig. 23 Diagnosis, warning lamp ON w/brake warning lamp OFF (Part 18 of 24). 1988 Continental & all 1989 models

Anti-Lock Warning Lamp On (With Brake Warning Lamp Off) — Test A

TEST STEP	RESULT ▶	ACTION TO TAKE
A18a MEASURE LH FRONT OUTLET VALVE RESISTANCE		
• Disconnect valve block 7-Pin plug. • Measure resistance between valve block electrical Pin 7 and 5.	3 to 6 ohms ▶	SERVICE or REPLACE cable harness
	Any other reading ▶	REPLACE valve block unit. CONNECT 7-Pin plug.
A19 MEASURE RH FRONT OUTLET VALVE CIRCUIT RESISTANCE		
• Measure resistance between Breakout box Pins 11 and 13.	3 to 6 ohms ▶	GO to Step A20.
	Any other reading ▶	GO to Step A19a.

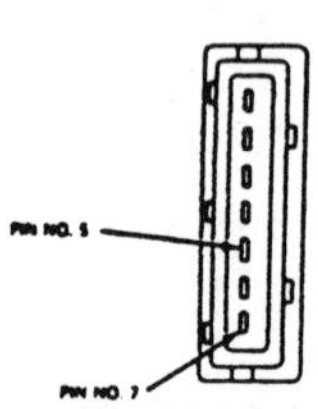

VALVE BLOCK 7-PIN CONNECTOR

Fig. 23 Diagnosis, warning lamp ON w/brake warning lamp OFF (Part 19 of 24). 1988 Continental & all 1989 models

Anti-Lock Warning Lamp On (With Brake Warning Lamp Off) — Test A

TEST STEP	RESULT ▶	ACTION TO TAKE
A19a MEASURE RH FRONT OUTLET VALVE RESISTANCE		
• Disconnect valve block 7-Pin plug. • Measure resistance between valve block electrical Pin 7 and 2.	3 to 6 ohms ▶	SERVICE or REPLACE cable harness
	Any other reading ▶	REPLACE valve block unit. CONNECT 7-Pin plug.

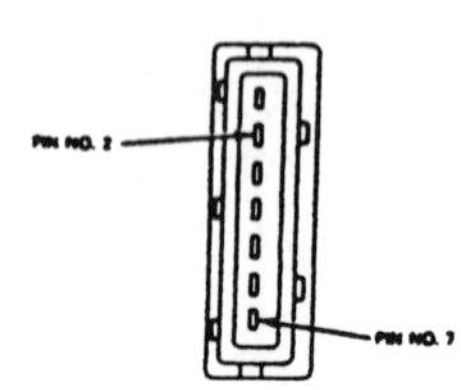

VALVE BLOCK 7-PIN CONNECTOR

Fig. 23 Diagnosis, warning lamp ON w/brake warning lamp OFF (Part 20 of 24). 1988 Continental & all 1989 models

Anti-Lock Warning Lamp On (With Brake Warning Lamp Off) — **Test A**

TEST STEP	RESULT ▶	ACTION TO TAKE
A20 CHECK PWS BRAKELAMP CIRCUIT (NO SYSTEM PRESSURE)		
• Vehicle must be cooled to room temperature. • Discharge the brake system as follows: a. Turn ignition switch OFF. b. Pump brake pedal at least 20 times until you feel the pedal become hard. • Disconnect pressure warning switch 5-Pin plug. • Check continuity between pressure warning switch electrical Pins 1 and 2.	Continuity (OK) ▶ No continuity (NOT OK) ▶	GO to Step A21. REPLACE pressure warning switch and pump motor relay.

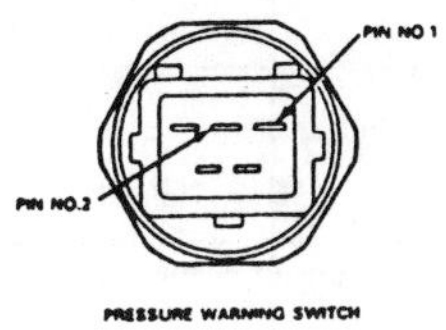

Fig. 23 Diagnosis, warning lamp ON w/brake warning lamp OFF (Part 21 of 24). 1988 Continental & all 1989 models

Anti-Lock Warning Lamp On (With Brake Warning Lamp Off) — **Test A**

TEST STEP	RESULT ▶	ACTION TO TAKE
A21 CHECK PWS BRAKELAMP CIRCUIT (WITH SYSTEM PRESSURE)		
• Reconnect pressure warning switch 5-Pin plug. • Turn ignition switch ON. • When pump motor stops, turn ignition switch OFF. • Disconnect pressure warning switch 5-Pin plug again. • Check for continuity between pressure warning switch electrical Pins 1 and 2.	Continuity (NOT OK) ▶ No continuity (OK) ▶	REPLACE pressure warning switch. GO to Step A22 and pump motor relay.

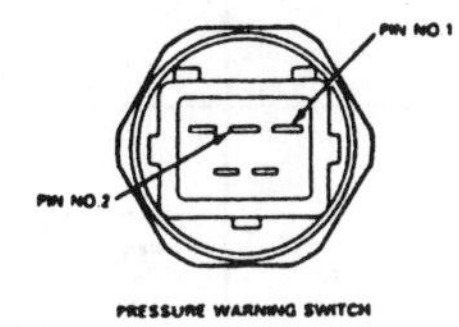

Fig. 23 Diagnosis, warning lamp ON w/brake warning lamp OFF (Part 22 of 24). 1988 Continental & all 1989 models

Anti-Lock Warning Lamp On (With Brake Warning Lamp Off) — **Test A**

TEST STEP	RESULT ▶	ACTION TO TAKE
A22 CHECK PWS BRAKE LAMP CIRCUIT THRESHOLD		
WARNING: Before disconnecting any hydraulic lines, you must ensure that the pressure system is discharged. • Discharge the brake system as follows: a. Turn ignition switch OFF. b. Pump break pedal at least 20 times until you feel it become hard. • Remove high pressure banjo bolt below brake fluid reservoir on brake booster. • Disconnect quick-connect nipple from Anti-Lock High Pressure Gauge (Tool T85P-20215-A, or equivalent) and install nipple in place of the high pressure banjo bolt. NOTE: Be sure O-rings are installed correctly. • Connect Anti-Lock High Pressure Gauge to gauge nipple. **WARNING: DO NOT disconnect Anti-Lock High Pressure Gauge while system is under pressure.** • Reconnect pressure warning switch (PWS) 5-Pin connector. • Turn ignition switch ON. • When pump motor stops, disconnect pressure warning switch (PWS) 5-Pin plug. • Lower hydraulic accumulator pressure by slowly pumping the brake pedal until you have continuity between pressure warning switch electrical Pins 1 and 2. • Observe the hydraulic pressure gauge when continuity is reached. • If you missed the reading or want to reverify the reading: — Reconnect the 5-Pin plug. — Turn the ignition switch ON until the pump stops. — Disconnect the 5-Pin plug and reverify the readings.	100-110 Bar (1,450-1,595 psi) (OK) ▶ Any other reading (NOT OK) ▶	GO to Step A23. REPLACE pressure warning switch and pump motor relay.

Fig. 23 Diagnosis, warning lamp ON w/brake warning lamp OFF (Part 23 of 24). 1988 Continental & all 1989 models

Anti-Lock Warning Lamp On (With Brake Warning Lamp Off) — **Test A**

TEST STEP	RESULT ▶	ACTION TO TAKE
A23 CHECK PWS HARNESS GROUND		
• Check for continuity between pressure warning switch 5-Pin plug Pin 1 and body ground (harness side).	Continuity ▶ No continuity ▶	GO to Step A24. SERVICE or REPLACE cable harness
A24 REVERIFY SYSTEM SYMPTOM		
• Reconnect all electrical connections. • Reinstall high pressure banjo bolt. **WARNING: Before disconnecting any hydraulic lines, you must ensure that the brake hydraulic pressure system is discharged.** • Discharge the brake system as follows: a. Turn ignition switch OFF. b. Pump brake pedal at least 20 times until you feel the pedal become hard. • Reverify symptom.	Symptom not present ▶ Symptom still present ▶	FAULT may have been a loose electrical connection. REPLACE electronic control module.

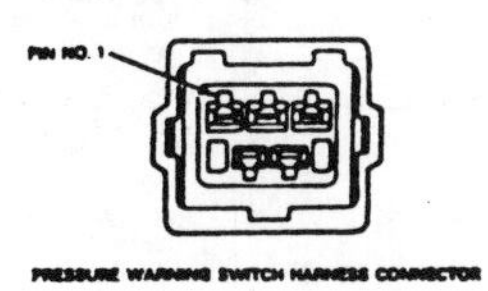

Fig. 23 Diagnosis, warning lamp ON w/brake warning lamp OFF (Part 24 of 24). 1988 Continental & all 1989 models

Anti-Lock Lamp On After Engine Starts (Brake Warning Lamp Off) — **Test B**

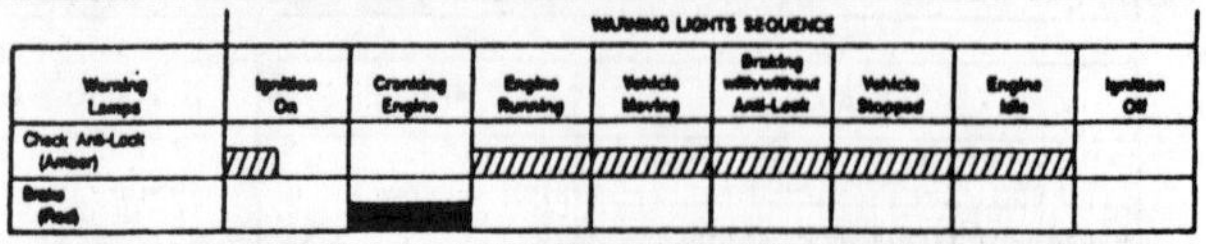

TEST STEP	RESULT ▶	ACTION TO TAKE
B1 CHECK CONTINUITY OF CIRCUIT 523 AND 524		
• Ignition switch Off. • Disconnect 32-pin plug from controller. • Connect EEC-IV Breakout box T83L-50-EEC-IV with Anti-Lock Test Adapter T87P-50-ALA to the Anti-Lock 32-pin plug wiring harness. • Check continuity between Breakout box Pins 40 and 6.	Continuity ▶ No continuity ▶	GO to Step B1a. GO to Step B2.
B1a CHECK CONTINUITY OF RH REAR SENSOR		
• Disconnect wheel sensor plug (RH rear). • Check for continuity between each sensor plug pin (sensor side) and vehicle ground.	Continuity ▶ No continuity ▶	REPLACE sensor (RH rear). REPLACE or SERVICE cable harness RECONNECT sensor plug.
B2 CHECK CONTINUITY OF CIRCUIT 521 AND 522		
• Check for continuity between Breakout box Pins 40 and 5.	Continuity ▶ No continuity ▶	GO to Step B2a. GO to Step B3.

Fig. 24 Diagnosis, anti-lock lamp ON after engine starts w/brake warning lamp OFF (Part 1 of 3). 1988 Continental & all 1989 models

Anti-Lock Lamp On After Engine Starts (Brake Warning Lamp Off) — **Test B**

TEST STEP	RESULT ▶	ACTION TO TAKE
B2a CHECK CONTINUITY OF LH FRONT SENSOR		
• Disconnect wheel sensor plug (LH front). • Check for continuity between each sensor plug pin (sensor side) and vehicle ground.	Continuity ▶ No continuity ▶	REPLACE sensor (LH front). REPLACE or SERVICE cable harness. CONNECT sensor plug.
B3 CHECK CONTINUITY OF CIRCUIT 518 AND 519		
• Check for continuity between Breakout box Pins 40 and 4.	Continuity ▶ No continuity ▶	GO to Step B3a. GO to Step B4.
B3a CHECK CONTINUITY OF LH REAR SENSOR		
• Disconnect sensor plug (LH rear). • Check for continuity between each sensor plug pin (sensor side) and vehicle ground.	Continuity ▶ No continuity ▶	REPLACE sensor (LH rear). REPLACE or SERVICE cable harness. CONNECT sensor plug.

Fig. 24 Diagnosis, anti-lock lamp ON after engine starts w/brake warning lamp OFF (Part 2 of 3). 1988 Continental & all 1989 models

Anti-Lock Lamp On After Engine Starts (Brake Warning Lamp Off) — **Test B**

TEST STEP	RESULT ▶	ACTION TO TAKE
B3 CHECK CONTINUITY OF CIRCUIT 514 AND 516		
• Check for continuity between Breakout box Pins 40 and 3.	Continuity ▶ No continuity ▶	GO to Step B4a. Test complete. If Anti-Lock lamp pattern remains, REPEAT Test B.
B4a CHECK CONTINUITY OF RH FRONT SENSOR		
• Disconnect wheel sensor plug (RH front). • Check for continuity between each sensor plug pin (sensor side) and vehicle ground.	Continuity ▶ No Continuity ▶	REPLACE RH front sensor. REPLACE or SERVICE cable harness. CONNECT sensor plug.

Fig. 24 Diagnosis, anti-lock lamp ON after engine starts w/brake warning lamp OFF (Part 3 of 3). 1988 Continental & all 1989 models

Anti-Lock Warning Lamp On After Vehicle Starts To Move Or False Cycling Of Anti-Lock System — **Test C**

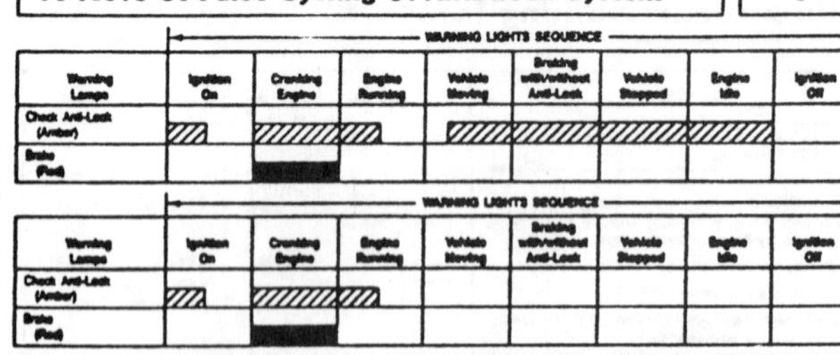

TEST STEP	RESULT ▶	ACTION TO TAKE
C1 MEASURE RH REAR SENSOR CIRCUIT RESISTANCE		
• Turn ignition switch OFF. • Disconnect 32-Pin plug from electronic controller. • Connect EEC-IV Breakout Box Tool T83L-50-EEC-IV and Anti-Lock Test Adapter Tool T87P-50-ALA or equivalent to anti-lock 32-Pin connector. • Set multi-meter to read resistance. • Measure resistance between Breakout box Pins 6 and 23.	800 to 1400 ohms (0.8 to 1.4K ohms) ▶ Any other reading ▶	GO to Step C2. GO to Step C1a.
C1a MEASURE RH REAR SENSOR RESISTANCE		
• Disconnect sensor plug (RH rear). • Set multi-meter to read resistance. • Measure resistance of sensor at sensor plug.	800 to 1400 ohms (0.8 to 1.4K ohms) ▶ Any other reading ▶	SERVICE or REPLACE cable harness REPLACE RH rear wheel sensor.

Fig. 25 Diagnosis, warning lamp ON after vehicle moves/or false cycling of system (Part 1 of 5). 1988 Continental & all 1989 models

Anti-Lock Warning Lamp On After Vehicle Starts To Move Or False Cycling Of Anti-Lock System — Test C

TEST STEP	RESULT ▶	ACTION TO TAKE
C2 MEASURE LH FRONT SENSOR CIRCUIT RESISTANCE		
• Measure resistance between Breakout box Pins 5 and 22.	800 to 1400 ohms (0.8 to 1.4K ohms) (OK) ▶	GO to Step C3.
	Any other reading (NOT OK) ▶	GO to Step C2a.
C2a MEASURE LH FRONT SENSOR RESISTANCE		
• Disconnect wheel sensor plug (LH rear). • Measure resistance of sensor at sensor plug.  LH FRONT SENSOR	800 to 1400 ohms (0.8 to 1.4K ohms) (OK) ▶	SERVICE or REPLACE cable harness
	Any other reading (NOT OK) ▶	REPLACE LH front wheel sensor.
C3 MEASURE LH REAR SENSOR CIRCUIT RESISTANCE		
• Measure resistance between Breakout box Pins 4 and 21.	800 to 1400 ohms (0.8 to 1.4K ohms) (OK) ▶	GO to Step C4.
	Any other reading (NOT OK) ▶	GO to Step C3a.

Fig. 25 Diagnosis, warning lamp ON after vehicle moves/or false cycling of system (Part 2 of 5). 1988 Continental & all 1989 models

Anti-Lock Warning Lamp On After Vehicle Starts To Move Or False Cycling Of Anti-Lock System — Test C

TEST STEP	RESULT ▶	ACTION TO TAKE
C3a MEASURE LH REAR SENSOR RESISTANCE		
• Disconnect sensor plug (LH rear). • Measure resistance of sensor at left rear sensor plug.  LH REAR SENSOR	800 to 1400 ohms (0.8 to 1.4K ohms) ▶	SERVICE or REPLACE cable harness
	Any other reading ▶	REPLACE LH rear wheel sensor.
C4 MEASURE RH FRONT SENSOR CIRCUIT RESISTANCE		
• Measure resistance between Breakout box Pins 3 and 20.	800 to 1400 ohms (0.8 to 1.4K ohms) ▶	GO to Step C5.
	Any other reading ▶	GO to Step C4a.
C4a MEASURE RH FRONT SENSOR RESISTANCE		
• Disconnect sensor plug (RH front). • Measure resistance of sensor at sensor plug.  RH FRONT SENSOR	800 to 1400 ohms (0.8 to 1.4K ohms) ▶	SERVICE or REPLACE cable harness
	Any other reading ▶	REPLACE RH front wheel sensor.

Fig. 25 Diagnosis, warning lamp ON after vehicle moves/or false cycling of system (Part 3 of 5). 1988 Continental & all 1989 models

Anti-Lock Warning Lamp On After Vehicle Starts To Move Or False Cycling Of Anti-Lock System — Test C

TEST STEP	RESULT ▶	ACTION TO TAKE
C5 CHECK RH REAR SENSOR		
• Turn ignition switch OFF. • Turn air suspension switch in luggage compartment OFF, if so equipped. • Place vehicle on hoist and raise wheels clear of ground. Refer to Pre-Delivery Manual Section 50-04. • Set multi-meter on voltage range (2V-AC). • Measure voltage between Breakout box Pins 6 and 23 while spinning RH rear wheel at approximately 1 revolution per second.	Between 0.05 and 0.70 Vac (OK) ▶	GO to Step C6.
	Less than 0.05 or more than 0.70 Vac (NOT OK) ▶	CHECK sensor mounting, air gap, or toothed wheel mounting. CORRECT as required.
C6 CHECK LH FRONT SENSOR		
• Measure voltage between Breakout box Pins 5 and 22 while spinning LH front wheel at approximately 1 revolution per second.	Between 0.05 and 0.70 Vac ▶	GO to Step C7.
	Less than 0.05 or more than 0.70 Vac ▶	CHECK sensor mounting, air gap, or toothed wheel mounting. CORRECT as required.
C7 CHECK LH REAR SENSOR		
• Measure voltage between Breakout box Pins 4 and 21 while spinning LH rear wheel at approximately 1 revolution per second.	Between 0.05 and 0.70 Vac ▶	GO to Step C8.
	Less than 0.05 or more than 0.07 Vac ▶	CHECK wheel sensor mounting, air gap, or toothed wheel mounting. CORRECT as required.

Fig. 25 Diagnosis, warning lamp ON after vehicle moves/or false cycling of system (Part 4 of 5). 1988 Continental & all 1989 models

Anti-Lock Warning Lamp on After Vehicle Starts to Move or False Cycling of Anti-Lock System — Test C

TEST STEP	RESULT ▶	ACTION TO TAKE
C8 CHECK RH FRONT SENSOR		
• Measure voltage between Breakout box Pins 3 and 20 while spinning RH front wheel at approximately 1 revolution per second.	Between 0.05 and 0.70 Vac ▶	GO to Step C9.
	Less than 0.05 or more than 0.70 Vac ▶	CHECK wheel sensor mounting, air gap, or toothed wheel mounting. CORRECT as required.
C9 CHECK FRONT WHEEL BEARINGS		
• Check front wheel bearing end play. • Inspect each toothed sensor ring visually for damaged teeth. NOTE: Turn air suspension switch ON when vehicle is off hoist.	Loose or damaged parts ▶	ADJUST bearings or REPLACE faulty parts.
	Not loose or damaged ▶	REVERIFY symptom.

Fig. 25 Diagnosis, warning lamp ON after vehicle moves/or false cycling of system (Part 5 of 5). 1988 Continental & all 1989 models

Anti-Lock Warning Lamp And Brake Warning Lamp On And/Or Pump Motor Runs More Than 60 Seconds	Test D

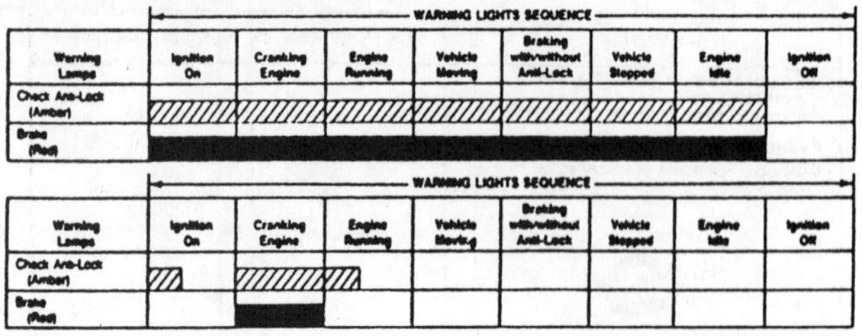

TEST STEP	RESULT ▶	ACTION TO TAKE
D1 CHECK PUMP MOTOR OPERATION		
• Turn ignition OFF. • Pump brake pedal at least 20 times until it becomes hard. • Turn ignition switch ON. • Pump motor should run. **NOTE: If pump motor is allowed to run continuously for approximately 20 minutes, a thermal safety switch (inside motor) will shut off motor. A 2-to-10 minute cool-down period is typical before normal operation can resume.**	OK ▶ NOT OK ▶	GO to D2. GO to D1a.

Fig. 26 Diagnosis, warning lamp & brake warning lamp ON &/or pump motor runs more than 60 seconds (Part 1 of 10). 1988 Continental & all 1989 models

Anti-Lock Warning Lamp And Brake Warning Lamp On And/Or Pump Motor Runs More Than 60 Seconds	Test D

TEST STEP	RESULT ▶	ACTION TO TAKE
D1a PUMP MOTOR UNIT		
• Disconnect 4-pin plug on pump motor unit. • Turn ignition switch ON. • Set multi-meter on 20 Volt DC. • Connect meter to the 4-pin plug on the harness side. (Use one negative and one positive pin). • Observe voltmeter.	More than 10V DC ▶ Less than 10V DC ▶	GO to D2. GO to D1b.

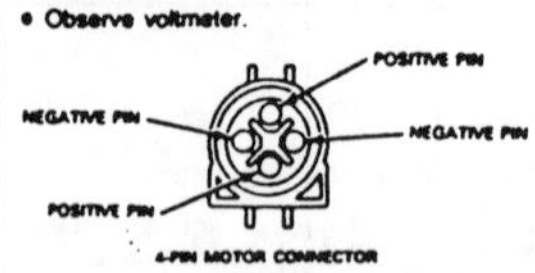

TEST STEP	RESULT ▶	ACTION TO TAKE
D1b CHECK CONTINUITY CIRCUITS 538A, 538B AND 538C		
• Disconnect pump motor relay. • Check for continuity between motor relay socket Pin 3 (harness side) and each positive pin of the 4-pin motor connector (harness side).	OK ▶ NOT OK ▶	GO to D1c. SERVICE Circuits 538A, 538B and 538C.

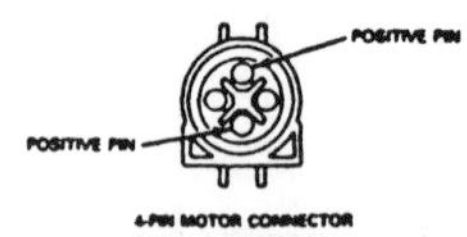

Fig. 26 Diagnosis, warning lamp & brake warning lamp ON &/or pump motor runs more than 60 seconds (Part 2 of 10). 1988 Continental & all 1989 models

Anti-Lock Warning Lamp And Brake Warning Lamp On And/Or Pump Motor Runs More Than 60 Seconds	Test D

TEST STEP	RESULT ▶	ACTION TO TAKE
D1c CHECK CONTINUITY CIRCUITS 430G AND 430F		
• Check for continuity between ground and each negative pin of 4-Pin motor connector (harness side).	Continuity ▶ No continuity ▶	GO to D1d. SERVICE or REPLACE Circuit 430G or 430F.

TEST STEP	RESULT ▶	ACTION TO TAKE
D1d CHECK CONTINUITY CIRCUIT 537B		
• Turn ignition switch OFF. • Remove positive battery cable from battery. • Disconnect pump motor relay. • Check continuity between battery positive cable and motor relay socket Pin 4 (harness side).	Continuity ▶ No continuity	RECONNECT battery positive cable and GO to Step D1e. SERVICE or REPLACE Circuits 537B or 038E or high current fuse.

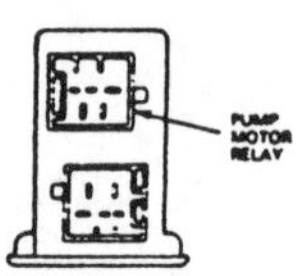

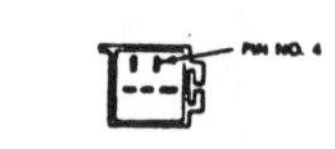

Fig. 26 Diagnosis, warning lamp & brake warning lamp ON &/or pump motor runs more than 60 seconds (Part 3 of 10). 1988 Continental & all 1989 models

Anti-Lock Warning Lamp And Brake Warning Lamp On And/Or Pump Motor Runs More Than 60 Seconds	Test D

TEST STEP	RESULT ▶	ACTION TO TAKE
D1e CHECK CONTINUITY CIRCUIT 298		
• Connect multi-meter to motor relay socket Pin 2 (harness side) and ignition switch lock pin (On). • Check continuity.	Continuity ▶ No continuity ▶	GO to Step D1f. REPLACE or SERVICE Circuit 298 and or fuse link.

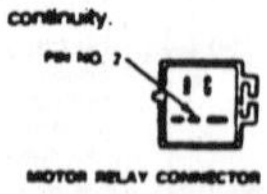

TEST STEP	RESULT ▶	ACTION TO TAKE
D1f CHECK CONTINUITY CIRCUIT 430H		
• Disconnect 5-Pin plug of PWS. • Check for continuity between pressure warning switch 5-Pin plug Pin 1 (harness side) and ground.	Continuity ▶ No continuity ▶	GO to Step D1g. REPLACE or SERVICE Circuit 430H.

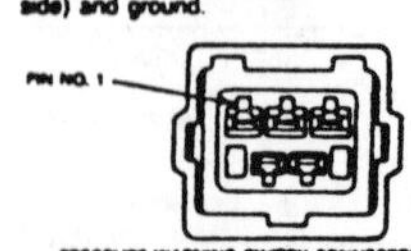

TEST STEP	RESULT ▶	ACTION TO TAKE
D1g CHECK CONTINUITY CIRCUIT 539A		
• Connect meter to 5-Pin plug Pin 4 (harness side) and motor relay socket Pin 1 (harness side). • Check for continuity.	Continuity ▶ No continuity ▶	GO to Step D1h. REPLACE or SERVICE Circuit 539A.

Fig. 26 Diagnosis, warning lamp & brake warning lamp ON &/or pump motor runs more than 60 seconds (Part 4 of 10). 1988 Continental & all 1989 models

Anti-Lock Warning Lamp And Brake Warning Lamp On And/Or Pump Motor Runs More Than 60 Seconds	Test D

TEST STEP	RESULT ▶	ACTION TO TAKE
D1h PRESSURE WARNING SWITCH		
• Check for continuity between pressure warning switch Pins 1 and 4. NOTE: System must be depressurized. 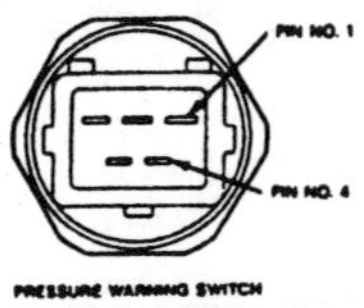	Continuity ▶ No continuity ▶	GO to Step D1j. REPLACE pressure warning switch and pump motor relay.
D1j CHECK PUMP MOTOR RELAY		
• Set multi-meter on 200 ohm scale. • Connect meter to motor relay Pins 1 and 2. 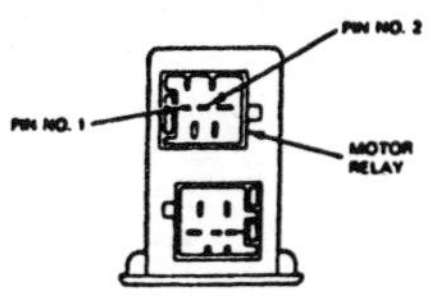	45 to 105 Ohms ▶ Other ▶	GO to Step D1k. REPLACE motor relay.

Fig. 26 Diagnosis, warning lamp & brake warning lamp ON &/or pump motor runs more than 60 seconds (Part 5 of 10). 1988 Continental & all 1989 models

Anti-Lock Warning Lamp And Brake Warning Lamp On And/Or Pump Motor Runs More Than 60 Seconds	Test D

TEST STEP	RESULT ▶	ACTION TO TAKE
D1k CHECK PUMP MOTOR RELAY (CONTINUED)		
• Connect meter and check continuity between motor relay Pins 3 and 4. 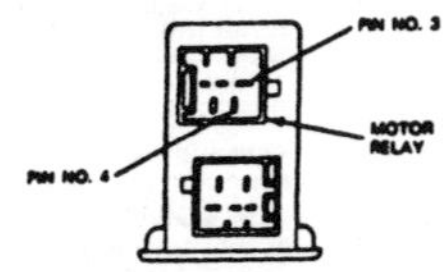	No continuity ▶ Continuity ▶	GO to Step D1m. REPLACE pump motor relay.
D1m CHECK PUMP MOTOR RELAY (CONTINUED)		
• Connect battery to motor relay terminals 1 and 2. • Check continuity between relay Pins 3 and 4 with multi-meter. 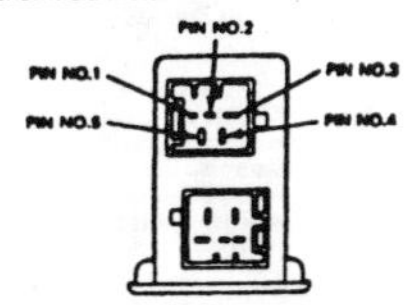	Continuity ▶ No continuity ▶	GO to Step D1n. REPLACE pump motor relay.
D1n CHECK MOTOR RELAY DIODE		
• Connect multi-meter and check continuity between relay Pins 3 and 5 (positive on Pin 3 and negative on Pin 5).	Continuity ▶ No continuity ▶	GO to Step D1p. REPLACE pump motor relay.

Fig. 26 Diagnosis, warning lamp & brake warning lamp ON &/or pump motor runs more than 60 seconds (Part 6 of 10). 1988 Continental & all 1989 models

Anti-Lock Warning Lamp And Brake Warning Lamp On And/Or Pump Motor Runs More Than 60 Seconds	Test D

TEST STEP	RESULT ▶	ACTION TO TAKE
D1p CHECK PUMP MOTOR		
• Apply 12 volts to 4-pin plug of motor (use one negative and one positive pin). • Pump motor should run. 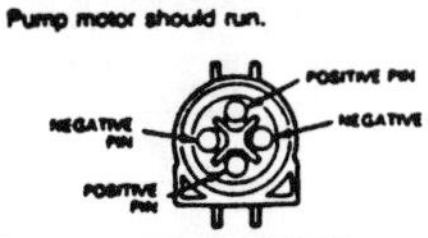	Pump runs ▶ Pump does not operate ▶	End of test. CONNECT all electrical plugs and relay. REVERIFY symptom. REPLACE pump/motor assembly and pump motor relay.
D2 CHECK PUMP MOTOR UNIT		
• Turn ignition switch OFF. • Pump brake pedal 20 times to discharge system. • Connect ammeter between battery positive cable and battery positive terminal. • Turn Off any electrical components. • Connect 4-pin motor plug. • Turn ignition switch ON. • Measure pump motor current.	Current more than 25 amps ▶ Current less than 25 amps ▶	REPLACE pump motor unit. GO to Step D2a.
D2a CHECK PUMP MOTOR		
• Turn ignition switch OFF. • Pump brake pedal at least 20 times, until brake pedal becomes hard. • Turn ignition switch ON. • Measure time pump takes to shut OFF.	Under 60 seconds ▶ Over 60 seconds (or motor never turns on) ▶	GO to Step D2b. CHECK for corroded connections at: • Motor 4-pin plug. • Body ground • Motor relay Pin 4 socket • Battery to Pin 4. GO to Step D2b.

Fig. 26 Diagnosis, warning lamp & brake warning lamp ON &/or pump motor runs more than 60 seconds (Part 7 of 10). 1988 Continental & all 1989 models

Anti-Lock Warning Lamp And Brake Warning Lamp On And/Or Pump Motor Runs More Than 60 Seconds	Test D

TEST STEP	RESULT ▶	ACTION TO TAKE
D2b CHECK LOW PRESSURE FLOW		
• Turn ignition switch OFF. • Disconnect low pressure hose from pump and allow fluid to flow into a suitable container. NOTE: Discard fluid after test.	Free fluid flow ▶ Restricted flow ▶	CONNECT low pressure hose. FILL reservoir to MAX. GO to D2c. SERVICE or REPLACE reservoir and or low pressure hose as required.
D2c CHECK VOLTAGE TO PUMP MOTOR		
• Turn ignition switch OFF. • Pump brake pedal to discharge pressure system. • Set meter to 20 Volts DC range. • Connect voltmeter in parallel at 4-Pin motor plug. • Turn ignition switch ON. 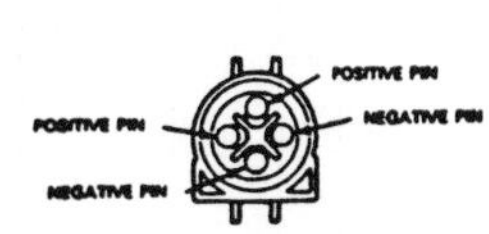	With pump running: ▶ Voltage over 8 Volts DC Voltage under 8 Volts DC ▶	GO to Step D3. CHECK Circuits 430G, 430F, 538A, 538B, Pin 4 to battery and relay Pins 3 and 4 for voltage drop. SERVICE or REPLACE as necessary.

Fig. 26 Diagnosis, warning lamp & brake warning lamp ON &/or pump motor runs more than 60 seconds (Part 8 of 10). 1988 Continental & all 1989 models

Anti-Lock Warning Lamp And Brake Warning Lamp On And/Or Pump Motor Runs More Than 60 Seconds — Test D

TEST STEP	RESULT ▶	ACTION TO TAKE
D3 ACCUMULATOR: PRE-CHARGE		
• Vehicle must be cooled to room temperature. **WARNING: BEFORE DISCONNECTING ANY HYDRAULIC LINES, YOU MUST ENSURE THAT THE PRESSURE SYSTEM IS DISCHARGED.** • To discharge hydraulic accumulator pressure, turn ignition OFF, pump brake pedal at least 20 times until you feel it become hard. • Remove high pressure banjo bolt below brake fluid reservoir on brake booster. • Install Anti-Lock Pressure Gauge Tool T85P-20215-A or equivalent. • Turn ignition switch ON and read accumulator precharge pressure. (Gauge needle will spring to this point). Note: Gauge needle reading should spring to 40-90 bar (600-1325 psi) and climb to 16,203-19,306 kPa bar (2350-2800 psi). • If reading was missed, discharge accumulator as described and repeat ignition ON sequence.	4137-9135 kPa (600-1325 psi) ▶	GO to Step D4.
	Under 4137 kPa (600 psi) ▶	REPLACE accumulator. **Note: Measure nipple length to determine correct service replacement accumulator.**
	Over 9136 kPa (1325 psi) ▶	REPLACE accumulator. **Note: Measure nipple length to determine correct service replacement accumulator.**
D4 CHECK HYDRAULIC ACTUATION UNIT		
• Turn ignition switch ON; wait until pump motor stops. Wait 3 more minutes to stabilize gauge pressure. • Read pressure gauge. • Wait 5 minutes and read pressure gauge again to determine the pressure loss over those 5 minutes.	Pressure loss less than 10 bar (140 psi) on gauge ▶	GO to Step D5.
	More than 10 bar (140 psi) ▶	CHECK for external leakage at actuation assembly and SERVICE. If no external leakage is found, GO to Step D4a.

Fig. 26 Diagnosis, warning lamp & brake warning lamp ON &/or pump motor runs more than 60 seconds (Part 9 of 10). 1988 Continental & all 1989 models

Anti-Lock Warning Lamp And Brake Warning Lamp On And/Or Pump Motor Runs More Than 60 Seconds — Test D

TEST STEP	RESULT ▶	ACTION TO TAKE
D4a CHECK FOR HYDRAULIC LEAKS		
• Check for brake fluid leaks in the following areas: a. Brake lines and calipers. b. High and low pressure hoses on actuation assembly. c. Reservoir seals and seams. d. Accumulator. **NOTE: A small amount of leakage from pressure warning switch is allowable. If any leakage is found below switch, wipe off excess fluid and check for excessive leakage.**	No leakage found (OK) ▶	GO to D5.
	Leakage found (NOT OK) ▶	SERVICE or REPLACE components. CHECK system operation.
D5 CHECK PUMP PRESSURE		
• With the pressure gauge still attached and ignition switch still ON, pump brake pedal to decrease pressure until pump motor restarts.	13100-15169 kPa (1900-2200 psi) when pump starts (OK) ▶	GO to Step D6.
	Less than or more than 13100-15169 kPa (1900-2200 psi) (NOT OK) ▶	REPLACE pressure warning switch and pump motor relay.
D6 CHECK PRESSURE WARNING SWITCH		
• With the ignition switch still ON and the pressure gauge connected, observe the pressure when the pu:np motor stops running.	16203 kPa (2350-2800 psi) when pump motor stops (OK) ▶	If pump motor takes longer than 60 seconds to reach 16203 kPa (2350-2800 psi), REPLACE pump/motor assembly. REVERIFY symptom.
	Less than 16203 kPa (2350 psi) or over 19306 kPa (2800 psi) (NOT OK) ▶	REPLACE pressure warning switch and pump motor relay.

Fig. 26 Diagnosis, warning lamp & brake warning lamp ON &/or pump motor runs more than 60 seconds (Part 10 of 10). 1988 Continental & all 1989 models

Anti-Lock Warning Lamp Intermittently On — Test E

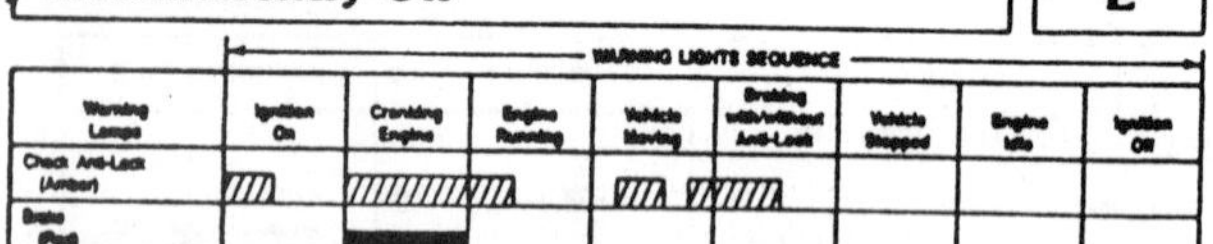

TEST STEP	RESULT ▶	ACTION TO TAKE
E1 CHECK FLI AND PWS CIRCUIT		
• Disconnect 32-Pin plug from electronic controller. • Connect EEC-IV Breakout box, T83L-50-EEC-IV with Anti-Lock Test Adapter T87P-50-ALA or equivalent to the Anti-Lock 32-Pin plug wiring harness. • Turn ignition switch ON. • Set multi-meter to read resistance. • Measure the resistance between Breakout box Pins 25 and 27.	Less than 5 ohms (OK) ▶	GO to Step E2.
	Greater than 5 ohms (NOT OK) ▶	GO to Step E1a.
E1a CHECK FLI ANTI-LOCK WARNING CIRCUIT		
• Disconnect 5-Pin plug on reservoir fluid level indicator (FLI). • Measure resistance between fluid level indicator electrical Pins 1 and 2 (with brake fluid at maximum fluid level).	Less than 2 ohms (OK) ▶	GO to Step E1b.
	Greater than 2 ohms (NOT OK) ▶	REPLACE fluid level indicator.

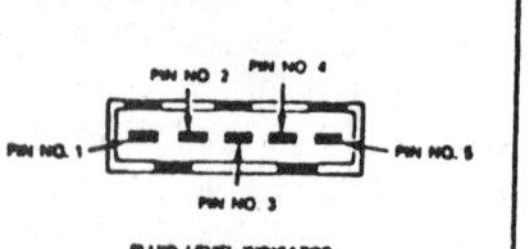

Fig. 27 Diagnosis, warning lamp intermittently ON (Part 1 of 5). 1988 Continental & all 1989 models

Anti-Lock Warning Lamp Intermittently On — Test E

TEST STEP	RESULT ▶	ACTION TO TAKE
E1b CHECK PWS ANTI-LOCK WARNING CIRCUIT (NO SYSTEM PRESSURE)		
• Turn ignition switch OFF. • Pump brake pedal to discharge pressure system. • Disconnect 5-Pin plug at pressure warning switch (PWS). • Check for continuity between pressure warning switch Pins 3 and 5.	No Continuity ▶	GO to Step E1c.
	Continuity ▶	REPLACE pressure warning switch and pump motor relay.

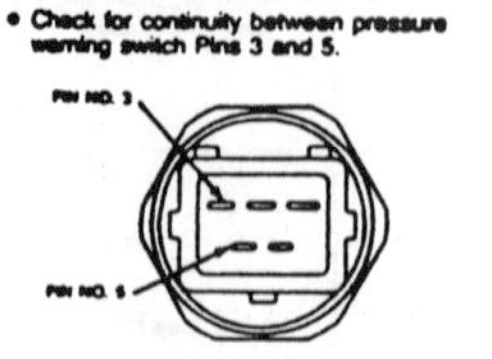

TEST STEP	RESULT ▶	ACTION TO TAKE
E1c CHECK ELECTRONIC MODULE TO FLI WIRE		
• Check for continuity between Breakout box Pin 25 and FLI 5-Pin plug Pin 1 (harness side).	Continuity ▶	GO to Step E1d.
	No continuity ▶	REPLACE or SERVICE cable harness

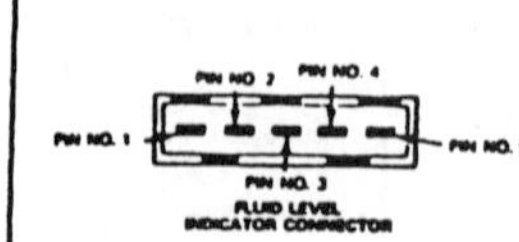

Fig. 27 Diagnosis, warning lamp intermittently ON (Part 2 of 5). 1988 Continental & all 1989 models

Anti-Lock Warning Lamp Intermittently On	Test E

TEST STEP	RESULT ▶	ACTION TO TAKE
E1d CHECK FLI TO PWS WIRE		
• Check for continuity between 5-Pin fluid level plug Pin 2 (harness side) 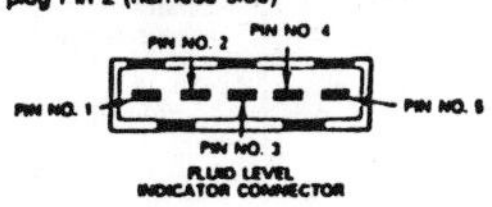 and 5-Pin pressure warning switch plug Pin 3 (harness side). 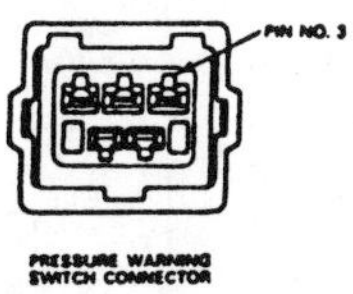	Continuity ▶ No continuity ▶	GO to Step E1e. SERVICE or REPLACE cable harness
E1e CHECK PWS TO ELECTRONIC CONTROLLER WIRE		
• Check for continuity between 5-Pin pressure warning switch plug Pin 5 (harness side) and Breakout box Pin 27. 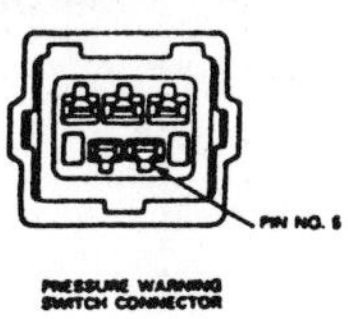	Continuity ▶ No continuity ▶	TURN ignition OFF. CONNECT all electrical connections. REVERIFY symptom. SERVICE or REPLACE cable harness

Fig. 27 Diagnosis, warning lamp intermittently ON (Part 3 of 5). 1988 Continental & all 1989 models

Anti-Lock Warning Lamp Intermittently On	Test E

TEST STEP	RESULT ▶	ACTION TO TAKE
E2 CHECK ISOLATION TEST FLI AND PWS		
• Check for continuity between Breakout box Pin 25 and body ground.	Continuity ▶ No continuity ▶	GO to Step E2a. REVERIFY symptom.
E2a CHECK CONTINUITY OF FLI SWITCH		
• Disconnect FLI plug and check for continuity between socket Pin 1 and Pins 3, 4, and 5 and Pin 2 and Pins 3, 4, and 5. 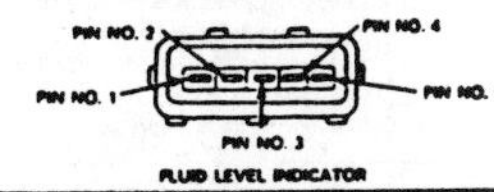	Continuity ▶ No continuity ▶	REPLACE FLI. GO to Step E2b.
E2b CHECK CONTINUITY CIRCUIT 512A		
• Check for continuity between FLI plug Pin 1 (harness side) and body ground. 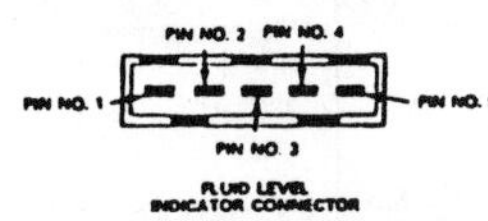	Continuity ▶ No continuity ▶	SERVICE or REPLACE Circuit 512A. GO to Step E2c.
E2c CHECK CONTINUITY CIRCUIT 549A		
• Disconnect 5-Pin plug from PWS. Check for continuity between FLI plug Pin 2 (harness side) and body ground. 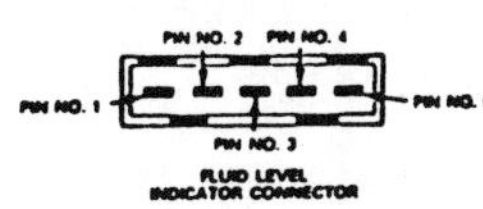	Continuity ▶ No continuity ▶	SERVICE or REPLACE Circuit 549A. GO to Step E2d.

Fig. 27 Diagnosis, warning lamp intermittently ON (Part 4 of 5). 1988 Continental & all 1989 models

Anti-Lock Warning Lamp Intermittently On	Test E

TEST STEP	RESULT ▶	ACTION TO TAKE
E2d CHECK CONTINUITY OF PWS		
• Check for continuity from PWS 5-Pin socket Pin 3 to body ground, and from Pin 5 to body ground. 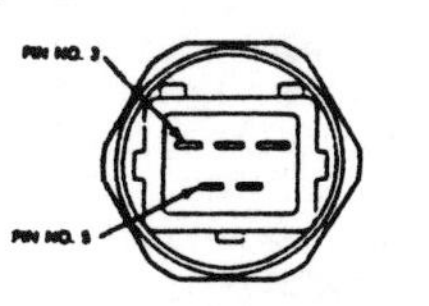	Continuity ▶ No continuity ▶	REPLACE pressure warning switch and pump motor relay. GO to Step E2e.
E2e CHECK CONTINUITY CIRCUIT 535A		
• Check for continuity between 5-Pin PWS plug (harness side) Pin 5 and body ground. 	Continuity ▶ No continuity ▶	SERVICE or REPLACE Circuit 535A. CONNECT all plugs and REVERIFY symptom.

Fig. 27 Diagnosis, warning lamp intermittently ON (Part 5 of 5). 1988 Continental & all 1989 models

REAR BRAKES

The rear brakes can be pressure bled using brake bleeder 104-00064, or equivalent, or through the use of a fully charged accumulator.

Brake Bleeding w/Pressure Bleeder

1. Clean all dirt from reservoir filler cap area. Attach pressure bleeder to reservoir cap opening.
2. Maintain approximately 35 psi on the system.
3. Remove dust cap from righthand caliper bleeder fitting. Attach a rubber drain tube to the fitting. Ensure that end of tube fits snugly around fitting.
4. With ignition switch in the OFF position, and brake pedal at rest, open righthand caliper bleeder fitting for 10 seconds at a time, until an air free stream of brake fluid flow is observed.
5. Repeat procedure at lefthand caliper.
6. Place ignition switch in the RUN position, then pump brake pedal several times to complete the bleeding procedure and completely charge accumulator.
7. Siphon off excess fluid in reservoir to adjust level to the MAX mark with a fully charged accumulator.

Brake Warning Lamp On (With Anti-Lock Lamp Off, Parking Brake Released And Brake Lining Wear Checked)	Test F

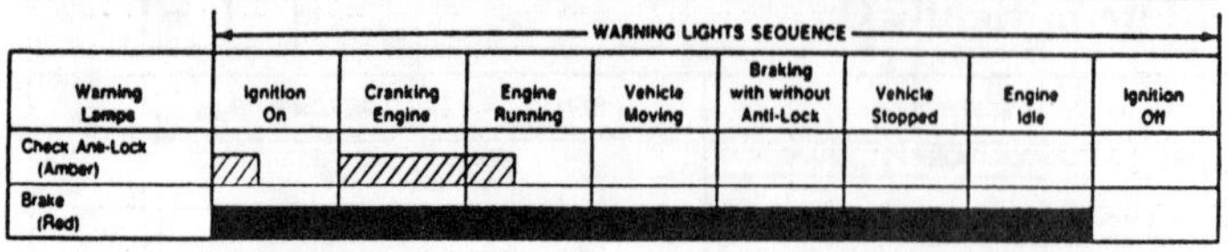

TEST STEP	RESULT ▶	ACTION TO TAKE
F1 CHECK BRAKE FLUID LEVEL		
• Turn ignition switch ON. • Pump brake pedal until pump motor starts. • When pump motor stops check brake fluid level.	Low (NOT OK) ▶	CHECK system for external leaks. SERVICE as required.
	Normal (OK) ▶	GO to Step F2.
F2 CHECK CONTINUITY FLI SWITCH		
• Disconnect 5-Pin plug on fluid reservoir cap. • Set multi-meter on 200 ohm range. Connect to reservoir warning cap Pins 3 and 4.	Above 10 ohms (OK) ▶	GO to Step F3.
	Below 10 ohms (NOT OK) ▶	REPLACE 5-Pin fluid level indicator cap.

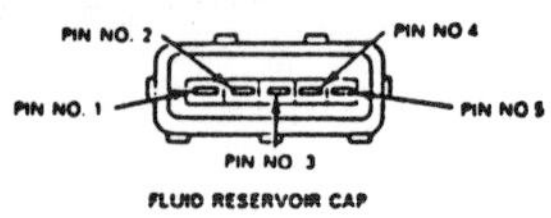

Fig. 28 Diagnosis, brake warning lamp ON w/anti-lock lamp OFF, parking brake released & lining wear checked (Part 1 of 2). 1988 Continental & all 1989 models

Brake Warning Lamp On (With Anti-Lock Lamp Off, Parking Brake Released And Brake Lining Wear Checked)	Test F

TEST STEP	RESULT ▶	ACTION TO TAKE
F3 CHECK CONTINUITY PWS SWITCH (CONTINUED)		
• Turn ignition switch ON, wait until motor stops running. • Disconnect Pressure Warning Switch 5-Pin plug and connect multi-meter to Pins 1 and 2 (switch side). • Check for continuity.	Continuity ▶	REPLACE pressure warning switch and pump motor relay.
	No continuity ▶	GO to Step F4.
F4 CHECK WIRING CONTINUITY		
• Check cable harness 430D, 512A, 547A, 547B, 547C, 549A, 977 Circuits. (Refer to vehicle Wiring Diagram.)	(OK) ▶	CHECK parking brake electrical circuit including parking brake switch. REVERIFY symptom.
	(NOT OK) ▶	SERVICE or REPLACE as required.

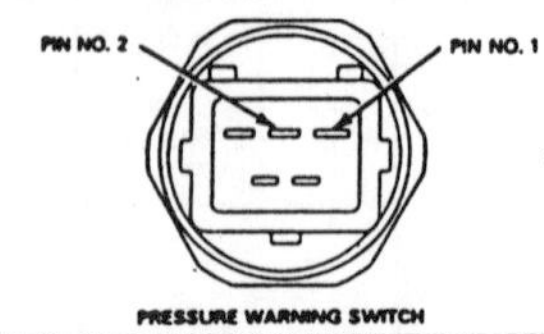

Fig. 28 Diagnosis, brake warning lamp ON w/anti-lock lamp OFF, parking brake released & lining wear checked (Part 2 of 2). 1988 Continental & all 1989 models

Bleeding System Using A Fully Charged Accumulator

1. Remove dust cap from righthand caliper bleeder fitting. Attach a rubber drain tube to the fitting. Ensure that end of tube fits snugly around fitting.
2. Turn ignition switch to the RUN position. This will activate the electric pump to charge the accumulator as required.

Before proceeding with the next step, ensure care is used when opening the bleeder screws due to the high pressures available from a completely charged accumulator. 3. Hold brake pedal in the applied position. Open the righthand caliper bleeder fitting for 10 seconds at a time until an air free stream of brake fluid is observed. 4. Repeat procedure at lefthand caliper. 5. Pump brake pedal several times to complete bleeding procedure and charge accumulator. 6. Adjust brake fluid level in reservoir to MAX mark with a completely charged accumulator. **If the pump motor is allowed to operate continuously for approximately 20 minutes, a thermal safety switch inside the motor may shut the motor off, to prevent it from overheating. If an overheat condition occurs, a 2-10 minute cool down period is required before normal operation can resume.**

HYDRAULIC RESERVOIR, CHECK & FILL

Refer to **Fig. 6,** during the following procedure.

1. With ignition switch in the ON position, pump brake pedal until the hydraulic pump motor begins to operate.
2. Wait until the hydraulic pump motor is operating, then check brake fluid level. If fluid level drops below the MAX fill level, add enough fluid to bring level to the MAX mark.
3. Do not add fluid as to bring level above MAX mark. Overfilling the reservoir may cause the brake fluid to overflow when the accumulator discharges during normal operation. **It is possible, depending on the state of the charge of the accumulator, that the fluid level could show above the MAX mark when first viewed, If so, repeat steps 1 and 2.**

ACTUATION ASSEMBLY, REPLACE

Refer to **Figs. 2 and 3,** during the following procedure.

1. With hydraulic system discharged, disconnect battery ground cable.
2. Remove air cleaner housing and duct assembly.
3. Label, then disconnect electrical connectors from fluid level indicator, main valve, solenoid valve block, pressure warning switch, hydraulic pump motor and ground connector from master cylinder portion of actuation assembly.
4. Disconnect and cap three brake tube fittings. Do not allow brake fluid to come in contact with any of the electrical connectors.
5. Remove trim panel under steering column. Disconnect actuation assembly push rod from brake pedal. Slide switch, push rod and plastic bushings off pedal pin.
6. Remove four retaining nuts, securing actuation assembly to brake pedal support bracket.
7. Remove actuation assembly from engine compartment.
8. Reverse procedure to install. **Torque** the four hydraulic unit to pedal support and brake tube lock nuts to 13-25 ft. lbs.

HYDRAULIC ACCUMULATOR, REPLACE

1. With hydraulic system discharged, disconnect battery ground cable.
2. Using a 8 mm hex wrench, loosen and completely unscrew accumulator. Ensure contaminents do not fall into open port.
3. Remove O-ring.
4. Reverse procedure to install. Install and **torque** accumulator to 30-34 ft. lbs. Turn ignition switch to ON posi-

No Anti-Lock Warning Lamp On When Ignition Switch Turned On — **Test G**

Warning Lamps	Ignition On	Cranking Engine	Engine Running	Vehicle Moving	Braking with/without Anti-Lock	Vehicle Stopped	Engine Idle	Ignition Off
Check Anti-Lock (Amber)								
Brake (Red)								

TEST STEP	RESULT ▶	ACTION TO TAKE
G1 CHECK FUSE AND FUSE LINKS		
• Check in-line fuse links with ignition turned ON.	Fuse Links OK ▶	GO to Step G2.
	Fuse Links ~~OK~~ ▶	SERVICE or REPLACE as required.
G2 CHECK WARNING LAMP BULB		
• Check warning lamp bulb.	OK ▶	GO to Step G3.
	~~OK~~ ▶	REPLACE bulb.
G3 CHECK WARNING LAMP OPERATION		
• Turn ignition switch ON. • Disconnect 32-Pin connector from control module.	Anti-Lock lamp goes ON ▶	SERVICE or REPLACE Circuit 606B.
	Anti-Lock lamp is not ON ▶	SERVICE or REPLACE connector to 14401 wire harness.

Fig. 29 Diagnosis, no warning lamp ON when ignition switch turned ON. 1988 Continental & all 1989 models

Spongy Brake Pedal With/Without Anti-Lock Function (No Warning Lamp) — **Test H**

Warning Lamps	Ignition On	Cranking Engine	Engine Running	Vehicle Moving	Braking with without Anti-Lock	Vehicle Stopped	Engine Idle	Ignition Off
Check Anti-Lock (Amber)								
Brake (Red)								

TEST STEP	RESULT ▶	ACTION TO TAKE
H1 CHECK COMPONENT MOUNTING		
• Check for proper brake pedal and hydraulic unit attachment. • Bleed brakes as outlined.	Pedal still spongy ~~OK~~ ▶	GO to Step H2.
	Pedal feels normal OK ▶	Condition corrected.
H2 BLEED BRAKE SYSTEM		
• Turn off air suspension switch in luggage compartment if so equipped. • Rebleed brake system. • Turn on air suspension switch when vehicle is off hoist if so equipped.	Pedal still spongy ~~OK~~ ▶	REPLACE actuation assembly.
	Pedal feels normal OK ▶	Condition corrected.

Fig. 30 Diagnosis, spongy pedal with/without anti-lock function, no warning lamp. 1988 Continental & all 1989 models

tion and check that the red "Brake" warning lamp and the amber "Anti-Lock" warning lamp both go out after approximately 1 minute. Top off brake fluid reservoir if necessary.

ELECTRIC HYDRAULIC PUMP MOTOR, REPLACE

Refer to **Fig. 4**, during the following procedure.

1. With hydraulic system discharged, disconnect battery ground cable.
2. Empty reservoir of as much brake fluid as possible.
3. Disconnect suction line between reservoir and pump.
4. Remove accumulator and banjo bolt attaching high pressure line to pump housing.
5. Disconnect low pressure line between fluid reservoir and pump suction port.
6. Raise and support vehicle, then disconnect electrical connector from pump motor and pressure switch. It may be necessary to disconnect the lefthand tie-rod from the steering knuckle to gain access to the pressure warning switch electrical connector.
7. Remove bolt and spacer attaching pump and motor assembly to hydraulic actuation unit. Slide pump and motor assembly toward driver's side to clear attaching pin. Remove pump and motor assembly.
8. Reverse procedure to install. **Torque** Allen head bolt to 5-7 ft. lbs. There must be a gap of .06-.13 inch between washer on Allen head bolt and isolator bushing retainer cap. **Torque** banjo bolt to 12-15 ft. lbs.

RESERVOIR & FLUID LEVEL INDICATOR, REPLACE

Refer to **Fig. 6**, during the following procedure.

1. With hydraulic system discharged, disconnect battery ground cable.
2. Disconnect electrical connector from fluid level indicator.
3. Empty reservoir of as mush brake fluid as possible.
4. Remove suction line between pump and reservoir, from reservoir by rotating and pulling hose from fitting.
5. Remove reservoir by placing a flat blade screwdriver between push-in outlet and push-in stud/grommet, then prying up gently. Ensure short sleeve and O-ring are removed from booster grommet hole.
6. Reverse procedure to install.

ELECTRONIC CONTROLLER, REPLACE

1. Disconnect battery ground cable.
2. Remove trim panel in luggage compartment (located behind back seat), to expose the electronic control module.
3. Disconnect the 32 pin electrical connector from the electronic controller.
4. Remove attaching bolts, then the electronic controller.

PRESSURE WARNING SWITCH, REPLACE

Before removing pressure warning switch, ensure hydraulic system pressure is completely discharged. Whenever the pressure warning switch assembly is replaced, the pump motor relay assembly should also be replaced.

1. Disconnect battery ground cable.
2. Raise and support vehicle.
3. Disconnect electrical connector from pressure warning switch. **It may be necessary to remove the lefthand front tie-rod at the steering knuckle to gain access to the switch.**
4. Using socket tool T85P-20215-B, or equivalent, 1/2 to 3/8 inch adapter and a 3/8 inch ratchet, remove pressure warning switch.
5. Reverse procedure to install. **Torque** pressure warning switch to 15-25 ft. lbs.

FRONT WHEEL SENSOR, REPLACE

1988–89 Continental

1. Disconnect battery ground cable.

Poor Vehicle Tracking During Anti-Lock Function (Warning Lamp Off)	Test J

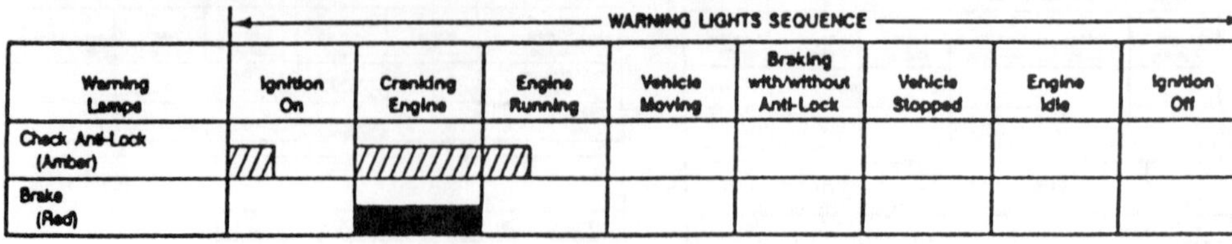

TEST STEP	RESULT ▶	ACTION TO TAKE
J1 VERIFY CONDITION		
• Verify condition exists as reported. • Turn air suspension off if so equipped. • Bleed brake system per shop manual for Anti-Lock brake system. • Turn air suspension on when vehicle is off hoist.	Vehicle tracks properly ▶ Vehicle still tracks poorly ▶	Condition corrected. GO to J2.
J2 CHECK ANTI-LOCK OPERATION — LH FRONT WHEEL		
• Turn air suspension off if so equipped. • Lift vehicle and rotate wheels to assure they turn freely. • Turn ignition switch OFF. • Disconnect 32-Pin plug from electronic controller. • Connect EEC-IV Breakout box, Tool T83L-50-EEC-IV with Anti-Lock test adapter, Tool T87P-50-ALA to the Anti-Lock 32-Pin plug wiring harness. • Short Pins 18, 14 and 31 to each other at Breakout box. • Apply moderate brake pedal effort and check that LH front wheel will not turn. • Check to see that LH front wheel turns freely with ignition switch ON. • Turn air suspension on when vehicle is off hoist if so equipped. **CAUTION: DO NOT LEAVE IGNITION ON FOR MORE THAN 1 MINUTE MAXIMUM, OR SOLENOID VALVE DAMAGE MAY RESULT.**	If wheel turns freely ▶ If wheel does not turn freely or pedal drops ▶	TURN ignition switch Off. DISCONNECT wire leads. Go to Step J3. REPLACE solenoid valve block.

Fig. 31 Diagnosis, poor vehicle tracking during anti-lock function warning lamp OFF (Part 1 of 2). 1988 Continental & all 1989 models

2. Disconnect sensor connector from engine compartment.
3. For the righthand front sensor, remove two plastic push studs to loosen the front section of splash shield in the wheelwell. For the lefthand front sensor, remove two plastic studs to loosen the rear section of the splash shield.
4. Thread sensor wire through holes in fender apron. For the righthand front sensor, remove two attaching clips behind the splash shield.
5. Raise and support vehicle.
6. Remove wheel and tire assembly.
7. Disconnect sensor wire grommets at height sensor bracket and from retainer clip at shock tower.
8. Loosen sensor retainer screw and remove sensor assembly from front knuckle.
9. Reverse procedure to install. **Torque** sensor attaching screw to 40-60 inch lbs.

1989 Mark VII

1. Disconnect battery ground cable.
2. Disconnect sensor electrical connector.
3. Raise and support vehicle, disengage wire grommet from shock tower, then pull sensor cable connector through hole. Use care not to damage connector.
4. Remove sensor wire from bracket on shock strut and side rail.
5. Loosen 5 mm set screw retaining sensor to sensor bracket post. Remove sensor through hole in disc brake splash shield.
6. Reverse procedure to install, noting the following:
   a. If sensor is to be reused or adjusted, ensure sensor face is clean and free of foreign material. Carefully scrape pole face with a dull knife. Glue a new front paper spacer on pole face (spacer is marked with a "F" and is .051 inch thick). Steel sleeve around post bolt must be rotated to provide a new surface for setscrew to indent and lock into.
   b. Install sensor through brake shield onto sensor bracket post. Ensure paper spacer remains in place throughout installation.
   c. Push sensor toward toothed sensor ring, until new paper spacer contacts the ring. Holding sensor against ring, **torque** 5 mm setscrew to 21-26 inch lbs.

Poor Vehicle Tracking During Anti-Lock Function (Warning Lamp Off)	Test J

TEST STEP		RESULT ▶	ACTION TO TAKE
J3	CHECK ANTI-LOCK OPERATION — RH FRONT WHEEL		
	• Turn air suspension off. • Short Pins 18, 32 and 13 to each other at Breakout box. • Apply moderate brake effort. Check that RH front wheel does not turn. • Check that RH front wheel turns freely with ignition switch ON. CAUTION: DO NOT LEAVE IGNITION ON MORE THAN 1 MINUTE MAXIMUM OR SOLENOID VALVE DAMAGE MAY RESULT.	Wheel turns freely ▶ Wheel does not turn freely or brake pedal drops ▶	TURN ignition switch OFF. DISCONNECT wire leads. GO to Step J4. REPLACE solenoid valve block.
J4	CHECK ANTI-LOCK OPERATION — REAR WHEELS		
	• Turn air suspension off if so equipped. • Short Pins 18, 30 and 12 to each other at Breakout box. • Apply moderate brake pedal pressure. Check that rear wheels will not turn. • Check that rear wheels turn freely with ignition switch ON. • Turn air suspension on when vehicle is off hoist if so equipped. CAUTION: DO NOT LEAVE IGNITION ON MORE THAN 1 MINUTE MAXIMUM OR SOLENOID VALVE DAMAGE MAY RESULT.	Wheels turn freely ▶ Wheels do not turn freely or brake pedal drops ▶	TURN ignition switch OFF. DISCONNECT wire lead and Breakout box. LOWER vehicle. REVERIFY symptom. REPLACE solenoid valve block.

Fig. 31 Diagnosis, poor vehicle tracking during anti-lock function warning lamp OFF (Part 2 of 2). 1988 Continental & all 1989 models

1989 Cougar & Thunderbird

1. Disconnect battery ground cable.
2. Disconnect sensor electrical connector.
3. Remove routing clips along wiring harness.
4. Remove Torx head screw securing sensor to spindle.
5. Reverse procedure to install. **Torque** Torx head screw to 40-60 inch lbs.

REAR WHEEL SENSOR, REPLACE

1988–89 Continental

1. Disconnect sensor electrical connector in luggage compartment.
2. Push rubber grommet through sheet metal floor pan.
3. Turn air suspension switch in luggage compartment to the OFF position. Raise and support vehicle and remove retainer clips for sensor wire, then the wire from routing.
4. Loosen sensor attaching screw at caliper anchor plate.
5. Install tool T88P-5310-A or equivalent onto the front suspension arm. Using suitable breaker bar, lower the arm to provide clearance for the sensor wire connector to pass through. Remove sensor.
6. Reverse procedure to install. **Torque** sensor attaching screw to 40-60 inch lbs.

1989 Cougar, Mark VII and Thunderbird

1. Disconnect battery ground cable.
2. Disconnect wheel sensor electrical connector, located behind wheelwell, under carpeting, inside luggage compartment.
3. Push sensor wire grommet through hole in luggage compartment floor.
4. Raise and support vehicle, then remove sensor wire from routing clips.
5. **On Cougar and Thunderbird models,** remove sensor retaining bolt with a 1/2 inch socket.
6. **On Mark VII models,** proceed as follows:
   a. Remove wheel and tire assembly, then caliper and rotor assemblies.
   b. Remove Torx head retaining bolt, then slip grommet out of rear brake splash shield and pull sensor out through the hole.
7. Reverse procedure to install, noting

the following:

a. **On Cougar and Thunderbird models,** align sensor locating tab and bolt hole to axle, then push into position.
b. **Torque** retaining bolt to 14-20 ft. lbs.
c. **On Mark VII models,** loosen 5 mm setscrew on sensor and ensure sensor slides freely on sensor bracket post.
d. If sensor is to be reused or adjusted, ensure sensor face is clean and free of foreign material. Carefully scrape pole face with a dull knife. Glue a new front paper spacer on pole face (spacer is marked with a "R" and is .043 inch thick). Steel sleeve around post bolt must be rotated to provide a new surface for setscrew to indent and lock into.
e. **Torque** Torx head screw to 40-60 inch lbs.
f. Push sensor toward toothed ring, until new paper sensor contacts sensor ring. Hold sensor, then **torque** 5 mm setscrew to 21-26 inch lbs.

FRONT SPEED INDICATOR RING, REPLACE

1988–89 Continental

1. With outboard constant velocity joint removed from vehicle, position speed indicator ring removal/installer tool T88P-20202-A or equivalent on a suitable press. Position constant velocity joint onto tool.
2. With constant velocity joint properly positioned on tool, use press ram to apply pressure to constant velocity joint and remove speed indicator ring.
3. Reverse procedure to install.

1989 Cougar & Thunderbird

1. Remove wheel and tire assembly.
2. Remove caliper, rotor and hub assemblies.
3. Remove indicator ring from hub, using a three-jawed puller.
4. Support center of hub so wheel studs do not rest on work surface.
5. Position sensor ring on hub, place a flat plate on top of ring, then press until flush with top of hub.
6. Install hub, rotor and caliper assemblies, then tire and wheel assembly.

1989 Mark VII

1. Remove wheel and tire assembly.
2. Remove caliper and assembly.
3. Position rotor assembly on an arbor press with studs facing up. Press each stud individually and carefully, only until they contact surface of sensor ring.
4. Position anti-lock sensing ring remover T85P-20202-A, or equivalent, on top of studs, then press all four studs and sensor ring out of rotor assembly together.
5. Install studs into rotor one at a time.
6. Position sensor ring on rotor, then press sensor ring onto rotor, using anti-lock sensing ring replacer T87P-20202-A, or equivalent, until sensor ring is seated. Ensure ring is pressed on straight.
7. Install rotor and caliper assemblies, then wheel and tire assembly.

REAR SPEED INDICATOR RING, REPLACE

1988–89 Continental

1. With air suspension switch in the OFF position, raise and support vehicle.
2. Remove wheel and tire assembly.
3. Remove caliper, rotor and rear hub assemblies.
4. Remove rear sensor ring spring retainer.
5. Position hub assembly in a suitable press and press hub out of speed sensor ring.
6. Reverse procedure to install.

1989 Mark VII

1. Remove rear axle shaft, as described elsewhere in this manual.
2. Install pinion bearing cone remover T71P-4621-B, or equivalent, between axle shaft flange and sensor ring.
3. Position axle in arbor press, then press axle from sensor ring.
4. To install, position sensor ring with recessed side facing inboard on axle, then install piston bearing cone remover T71P-4621-B, or equivalent, on axle shaft.
5. Place bar stock on top of axle flange, then press sensor ring onto axle shaft, until a gap of 2.48-2.55 inches remains between face of sensor ring and face of flange.
6. Install axle shaft.

ANTI-LOCK BRAKES

TABLE OF CONTENTS

Delco-Moraine III Type, 1989 Cutlass Supreme & Regal

INDEX

OPERATING PRECAUTIONS

Some of the following procedures require that hydraulic lines, hoses and fittings be disconnected for inspection of testing. This Delco-Moraine brake system uses a hydraulic booster/modulator called a Powermaster III, which, when completely charged, contains brake fluid under high pressure. Before disconnecting any hydraulic lines, hoses or fittings, ensure the Powermaster III is completely depressurized as described further on. Failure to depressurize the Powermaster III may result in personal injury and/or damage to painted surfaces and/or the vehicle.

COMPUTER SYSTEM SERVICE PRECAUTIONS

The Delco-Moraine system is designed to withstand normal current draws associated with vehicle operation. During testing of opens or shorts, do not ground or apply voltage to any of the circuits unless instructed to do so by the appropriate diagnostic procedure.

DESCRIPTION

The Delco-Moraine anti-lock brake system (DM ABS-III), operates on all four wheels. It is designed to reduce the tendency of a wheel to lose traction (lock), while braking. In most instances, the base power brake system operates. Anti-lock braking occurs only when a combination of wheel speed sensors and a microprocessor determine a wheel is about to loose traction during braking. The DM ABS-III then adjusts brake pressure to both front wheels and/or both rear wheels to reduce the tendency of the wheels to lock-up. This system helps the driver maintain steering during braking, over a variety of road surfaces.

SYSTEM COMPONENTS

The main components of the Delco-Moraine ABS-III, are the Powermaster III hydraulic booster/master cylinder, four wheel speed sensors, speed sensor rings, remote proportioner valve assembly and an anti-lock brake controller.

ANTI-LOCK BRAKE CONTROLLER

The anti-lock brake controller monitors the speed of each wheel to determine if a wheel is approaching lock-up. If this condition is detected, the controller pulses the appropriate solenoids in the Powermaster III, to adjust brake pressures. The controller also monitors the system for malfunctions, provides diagnostic information and shuts system down in case of a serious malfunction. The controller is located beneath the front passenger seat, in a protective plastic case. The controller is non-serviceable.

POWERMASTER III

The Powermaster III is a booster/modulator designed to provide base

PRE-DIAGNOSIS INSPECTION		
ITEM	INSPECT FOR:	CORRECTIVE ACTION
BRAKE FLUID RESERVOIR, MASTER CYLINDER UNIT, BRAKE CALIPERS AND BRAKE HOSES AND LINES	— LOW FLUID LEVEL — EXTERNAL LEAKS — BLINDING OR STICKING CALIPERS	— FILL RESERVOIR — REPAIR LEAKS AS REQUIRED — REPAIR AS NEEDED
PARKING BRAKE	— FULL RELEASE	— RELEASE PARKING BRAKE — ADJUST CABLE IF REQUIRED
BATTERY	— ADEQUATE CHARGE ("GREEN EYE")	— CHARGE OR REPLACE BATTERY AS REQUIRED (SEE SECTION 6D) — SERVICE CHARGING SYSTEM AS REQUIRED (SEE SECTION 6D)
FUSES AND FUSIBLE ELEMENTS FUSE BLOCK • ABS 2 FUSE 12 10 AMP • STOP FUSE 10 20 AMP • INDIC FUSE 11 10 AMP ANTILOCK BRAKE POWER CENTER • CONTROLLER FUSE 10 AMP • ELECT BRAKE VALVE FUSE 15 AMP • FUSIBLE ELEMENT K PUMP MOTOR 30 AMP • FUSIBLE ELEMENT L ELECT BRAKE VALVE 30 AMP	— BLOWN FUSE OR FUSIBLE ELEMENT	— REPLACE FUSE OR FUSIBLE ELEMENT AND VERIFY OPERATION
CONNECTORS • ANTILOCK BRAKE CONTROLLER • PUMP MOTOR RELAY • PUMP MOTOR • ACCUMULATOR SWITCH • REAR ENABLE RELAY • FRONT ENABLE RELAY • RIGHT FRONT SOLENOID • LEFT FRONT SOLENOID • REAR SOLENOID • WHEEL SPEED SENSORS • LAMP DRIVER MODULE • ANTILOCK BRAKE DIODE	— PROPER ENGAGEMENT OF CONNECTOR — LOOSE WIRES IN CONNECTOR	— PROPERLY ENGAGE CONNECTORS — REPAIR LOOSE WIRES
GROUND G106	— CLEAN AND TIGHT CONNECTION	— TIGHTEN — REPAIR LOOSE WIRES
INDICATORS OPERATION DURING BULB TEST • ANTILOCK WARNING INDICATOR • BRAKE WARNING INDICATOR	— LIGHTS	— CHECK CONNECTIONS — REPLACE BULB AND VERIFY OPERATION

Fig. 1 Pre-diagnosis inspection procedure and corrective actions

(normal) and anti-lock braking power. Fluid pressure is created and maintained by a combination of an electric motor/pump and accumulator. The accumulator is pre-charged to 1200 psi and the electric pump maintains system pressure at 2700 psi. During anti-lock braking, the fluid pressure is modulated by three solenoid assemblies. One assembly for each front wheel and one assembly for the rear brakes. Each solenoid assembly can apply, hold, or release pressure in the line it controls. Each front wheel can be modulated separately (first one, then the other), but the rear brakes have to be modulated together. This system is commonly called a Select-Low 3-Channel System. The Powermaster III is located in the left rear corner of engine compartment.

FRONT WHEEL SPEED SENSORS & RINGS

Front speed sensors are bolted to the strut assemblies above the axle shafts. The speed sensor rings are located on the outer the constant velocity joints, directly beneath the wheel speed sensors. As a ring moves past a speed sensor, an oscillating voltage is sent to the controller. The frequency of this signal increases and decreases with wheel speed. No repair or gap adjustment to the wheel speed sensor is permitted.

REAR WHEEL SPEED SENSORS & RINGS

Rear wheel speed sensors and speed sensor rings are integral parts of the rear wheel hub and bearing assemblies. Each sensor and ring is self-contained sealed from the environment. If a sensor or ring is

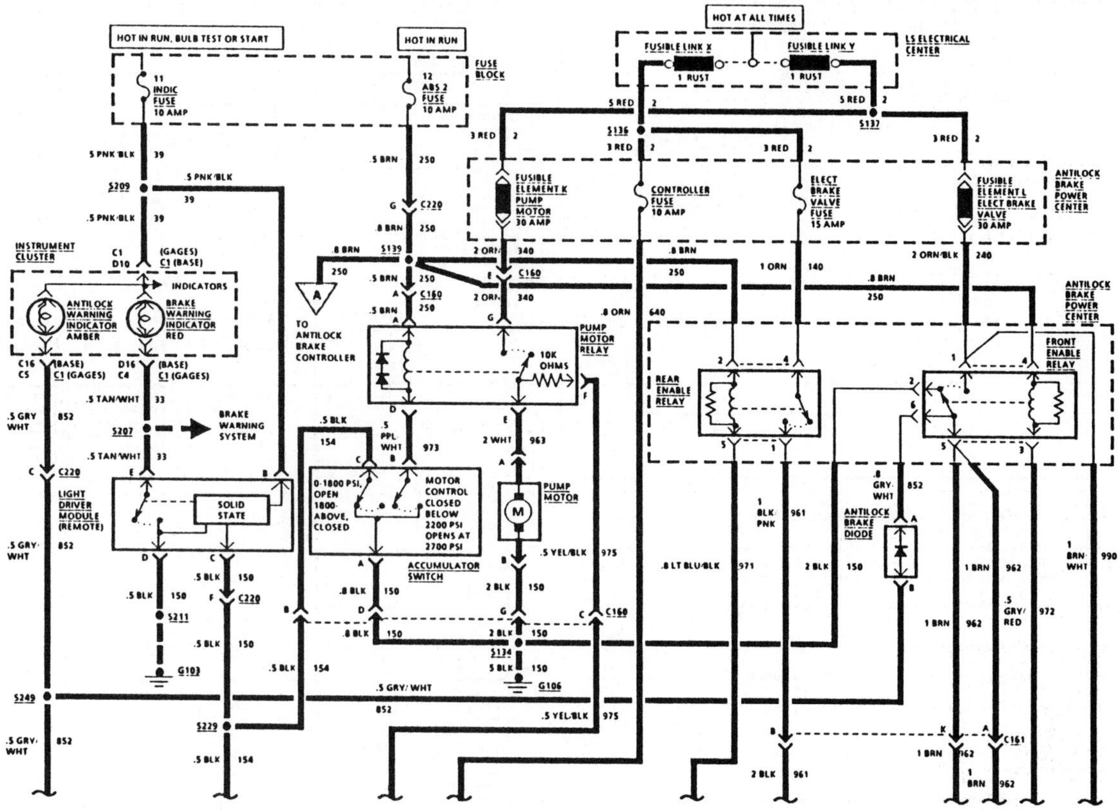

Fig. 2 Anti-Lock brake system wiring diagram. (Part 1 of 2)

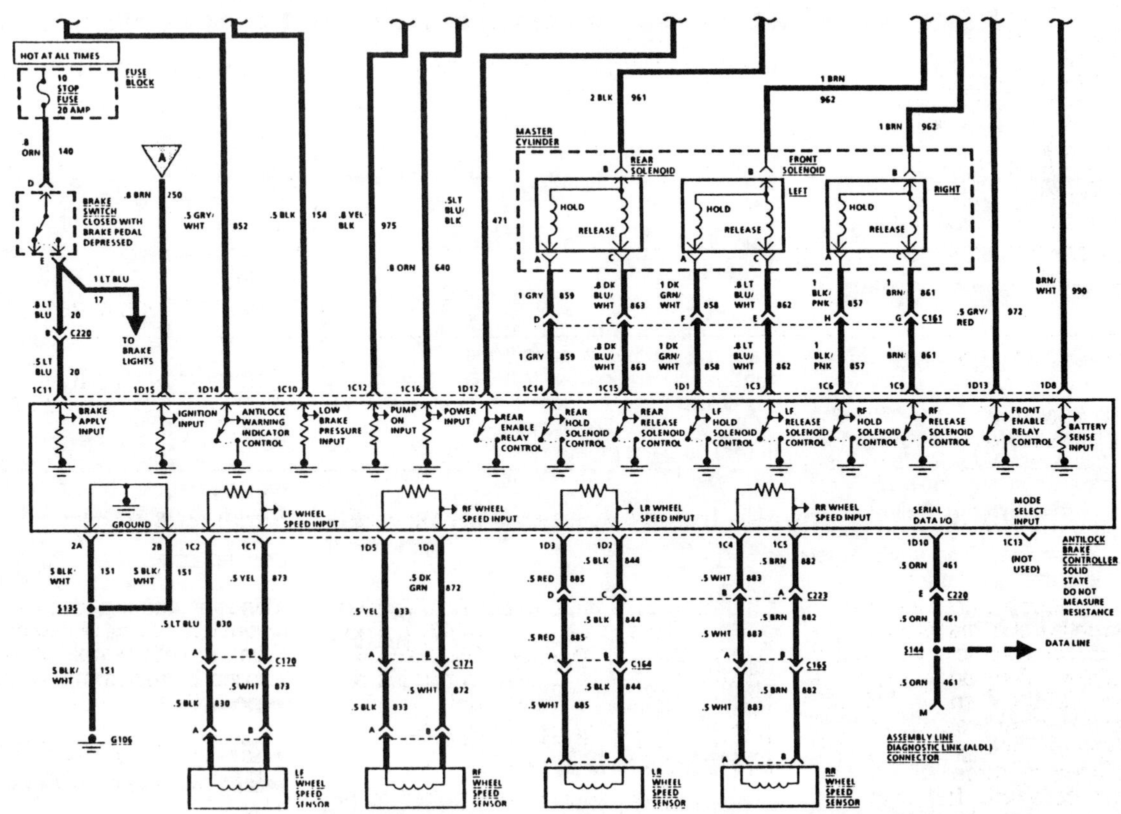

Fig. 2 Anti-Lock brake system wiring diagram. (Part 2 of 2)

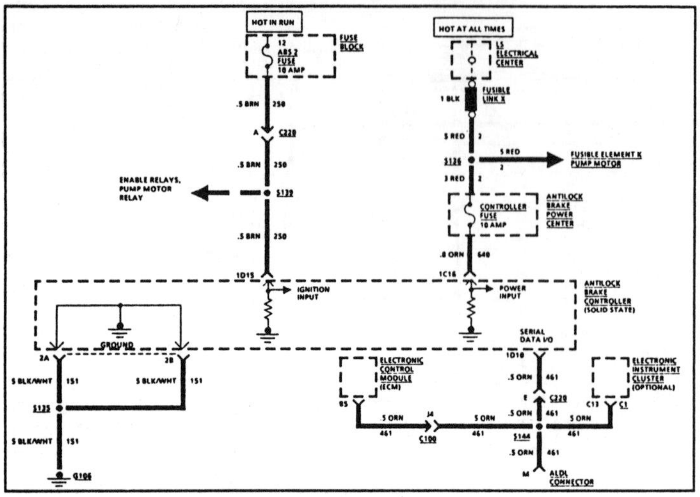

ANTILOCK BRAKE SYSTEM DIAGNOSTICS CANNOT BE ENTERED

The Antilock Brake System will be disabled and the Antilock Warning Indicator will be on if a loss of power or ground occurs at the Antilock Brake Controller. A loss of power or ground at the controller or a open or shorted serial data line will also prevent access to the Antilock Brake System's diagnostics.

TEST DESCRIPTION: The following provides an explanation of the procedures being followed in the facing trouble tree.

① Determines if voltage is being applied to the Antilock Brake Controller at Terminal 1C16.

② Indicates whether the fault is due to open condition or a possible short to ground in CKT 640.

③ Determines if voltage is being applied to Antilock Brake Controller at Terminal 1D15.

④& ⑤ A measurement of infinite resistance indicates an open condition in the Antilock Brake Controller's ground circuits, although only one complete ground path is required for proper operation. Both ground paths should be checked to assure redundancy.

⑥ A measurement of infinite resistance indicates an open condition exists in the data line CKT 461. If an open condition exists in the data line the controller would be unable to communicate with a bi-directional scan tool.

⑦ If a short to ground existed in CKT 461, communication with the ECM would not be possible. If communication with the ECM is possible a poor terminal connection at the Antilock Brake Controller is indicated or the controller is defective.

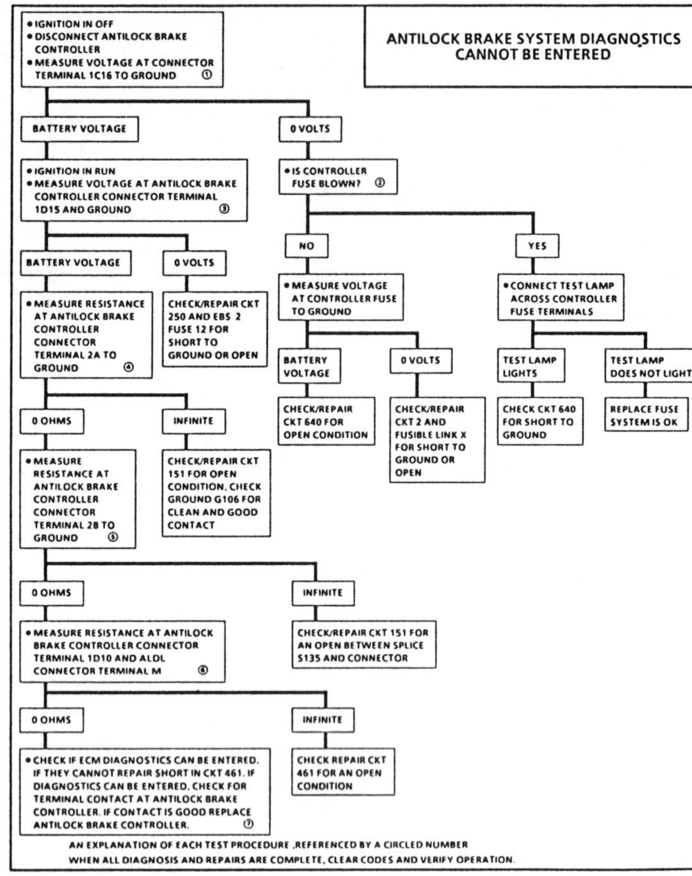

Fig. 3 Anti-lock brake system diagnostics cannot be entered

malfunctioning or damaged, the entire rear hub and bearing assembly must be replaced. No repair or gap adjustment to the wheel speed sensor is permitted.

REMOTE PROPORTIONER VALVE

The proportioner valve proportions brake pressure to the rear brakes. Proportioning occurs only after a preset input pressure has been reached. Beyond this preset pressure, the valve limits output pressure to the rear brakes at a set percentage of the total system output. The separate parts of the remote proportioner valve cannot be serviced. The valve must be replaced if it is malfunctioning.

BRAKE WARNING INDICATOR

The red brake warning lamp illuminates after engine is started to indicate one or all of the following; parking brake is applied, fluid level is low, or accumulator pressure is below 1800 psi. Any of these three conditions can damage brake system or greatly reduce braking action. If the red brake warning lamp is on, check above possible causes. The brake warning lamp can be tested for proper operation by when ignition switch is in the Start position. The indicator should be on in this position.

ANTI-LOCK WARNING INDICATOR

The anti-lock warning indicator illuminates after the engine is running to indicate the following conditions. Constant illumination indicates the anti-lock brake controller has detected a fault and disabled part or all of the anti-lock brake system. The base power brake system will not be disabled. A flashing light indicates the anti-lock brake controller has detected a fault, but still retaining ABS operation. Prolonged operation with a flashing indicator may further damage the ABS system. The anti-lock warning indicator can be tested for proper operation with ignition switch in the Start position. The indicator should illuminate in this position. The indicator should also illuminate for approximately three seconds after engine is started.

OPERATION

During ABS operation, the Delco-Moraine ABS-III system cycles (pulses), the brakes, adjusting line pressures, to obtain optimum stopping distance and vehicle control. During this cycling, the brake pedal will pulsate and clicking noises will be heard from the solenoids as they control the cycling. Clicking of the solenoids and pulsating of the brake pedal at this time and during a short period after vehicle starting is normal.

During ABS braking, to optimize stopping distance, the ABS will pump the brake pedal. The driver should press steadily and firmly on the brake pedal. **The driver should never pump the brake pedal when ABS is activated.**

During ABS operation, some tire noise may be heard. This noise is normal and results from slippage of the tire on the road surface. The amount of tire slippage will depend upon road and tire conditions. A tire will also seem to lock momentarily. This is a result of rapid changes in wheel speed and is also normal. A wheel that locks and stays locked for more than one second is not normal and should be serviced immediately.

PRE-DIAGNOSIS INSPECTION

Refer to **Fig. 1**, for pre-diagnosis procedure and corrective actions.

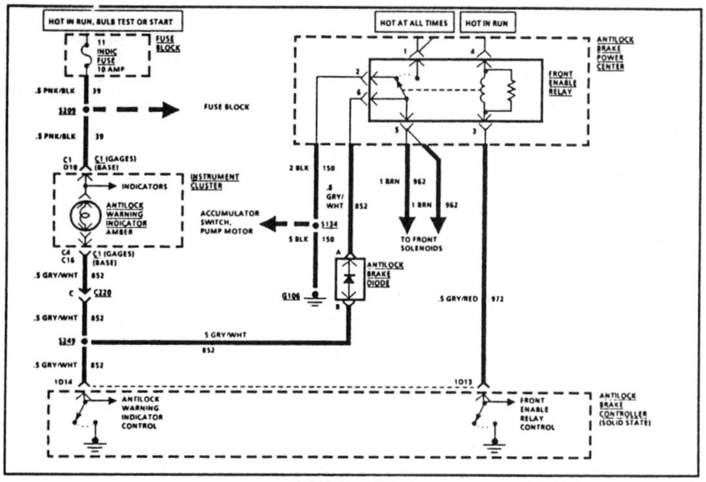

CODE A001

Antilock Warning Indicator open or shorted to ground.

When the Front Enable Relay is energized,its contacts close. This opens the ground path from the Antilock Warning Indicator. Voltage from the INDIC Fuse should now be present at terminal 1D14.

CODE A001 will set when all of the following conditions exist:

- The Front Enable Relay is energized.
- The Antilock Warning Indicator Control is open.
- The Antilock Brake Controller senses no voltage at terminal 1D14.

TEST DESCRIPTION: The following provides an explanation of the procedures being followed in the facing trouble tree.

① If the Antilock Warning Indicator lights, the circuit from the Indicator to ground G106 must be good.

② By energizing the Front Enable Relay, one can determine if CKT 852 to Antilock Brake Controller is defective. In this mode, the Controller should ground the Indicator causing it to light.

③ Voltage at terminal 1D14 indicates that CKT 852 is OK. The fault must then be internal to the Controller.

④ This determines if the problem is an open or short to ground.

⑤ Isolates the Antilock Warning Indicator from its possible grounds and therefore determines whether a short to ground is present in CKT 852 or whether the Antilock Brake Controller is defective.

⑥ Checks for a short to ground in CKT 39.

⑦ Determines whether open is in CKT 852 or CKT 39.

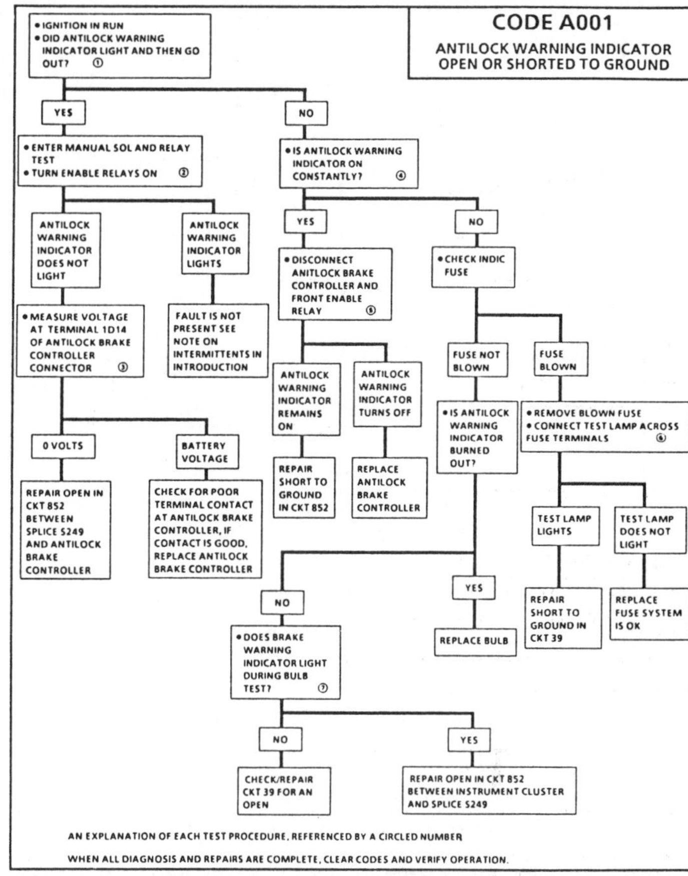

Fig. 4 Code A001, anti-lock warning indicator open or shorted to ground

DIAGNOSIS

A bi-directional scan tool connected to the assembly line diagnostic link (ALDL), is needed to access fault codes in the Delco-Moraine anti-lock braking system. Follow manufacturers directions for accessing fault codes. Refer to **Figs. 2 through 61,** for diagnosing trouble codes.

The ABS controller, when accessed with a bidirectional scan tool, identifies faults as specifically as possible. This diagnosis will also determine whether or not the fault is intermittent. There are 58 diagnostic fault codes. The first five codes stored, are stored in order of occurrence, from least recent to most recent. Fault codes will not disappear when ignition is turned Off, or when the battery is disconnected. Fault codes can only be read and erased with a bidirectional scan tool. If no faults occur within 50 drive cycles, all fault codes will be cleared. A drive cycle is defined as turning the ignition switch to the On position and vehicle is driven at speeds over 10 mph.

If a fault is detected, which disables the front ABS, the rear anti-lock brakes will continue to operate. The front brakes will have use of the normal power brake system only. If a fault is detected, which disables the rear ABS, the entire ABS will shut down, leaving normal power brake capability only.

Always turn ignition switch to the Off position prior to troubleshooting. This will ensure all diagnostic data is preserved. If ignition switch is not turned to the Off position, any information stored for a fault, in the last drive cycle, will be lost.

On 1988 models, if ignition switch is not turned Off prior to reading fault codes, and no other fault codes have occurred within the last 50 drive cycles, code 255 will set. If other fault codes did occur during the last 50 drive cycles, only information for the last drive cycle will be lost.

NOTE ON INTERMITTENTS

If an intermittent fault cannot be diagnosed using Indicator Lamp Sequence chart, the problem will most probably be caused by a defective electrical connection or wiring. When an intermittent problem is encountered, check questionable circuits and electrical connectors for the following:

1. Poor mating of connector halves or terminals not fully seated in connector body.
2. Improperly formed or damaged terminals. All connector terminals in the problem circuit should be carefully removed to increase contact tension.
3. Poor terminal wire connection.

If a visual check does not locate cause of problem, operate vehicle with EBCM connected and attempt to duplicate condition and use Indicator Lamp Sequence chart to isolate potential cause.

SYSTEM SERVICE

BRAKE PEDAL TRAVEL

When the vehicle is stopped and the ABS is pressurized, it is possible to push the pedal, until it bottoms near the floor. This travel is normal while system is pressurized. Excessive travel occurs when braking action does not begin within the first 1.25 inches of pedal travel, while system is pressurized. Excessive pedal travel also exists when pedal travel exceeds 3 inches, when applying light pressure, while Powermaster III unit is depressurized.

DEPRESSURIZING POWERMASTER III

The Powermaster III must be depressurized prior to performing any service operations.

1. Turn ignition switch to the Off position, or disconnect negative battery cable.
2. Firmly apply, then release the brake pedal a minimum of 40 times. A noticeable change in pedal feel (from soft to hard), will occur when accumulator is completely discharged.
3. Do not turn ignition switch to the On or Start position after depressurizing Powermaster III, unless instructed by service procedure, or until service operations have been completed.

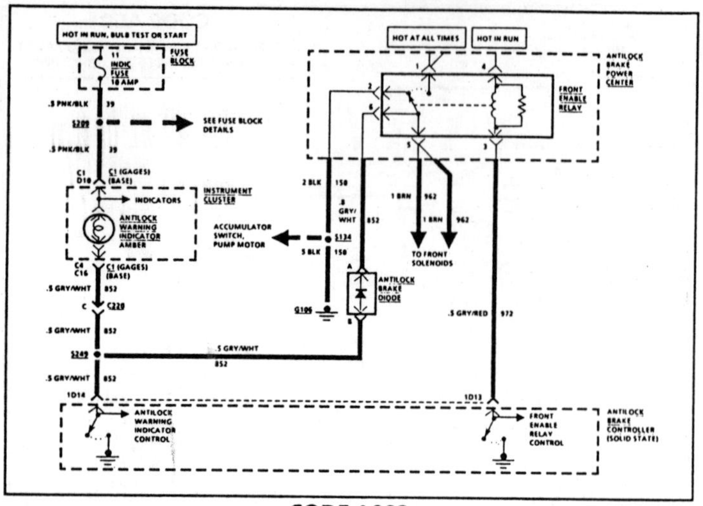

CODE A002

Antilock Warning Indicator shorted to Battery or Antilock Brake Diode shorted.

The Antilock Brake Diode prevents battery voltage from being applied to the Antilock Warning Indicator when the Front Enable Relay is energized. If the Antilock Brake Controller is inoperative, the Antilock Brake Diode allows the Antilock Warning Indicator to be grounded through the deenergized Front Enable Relay.

CODE A002 will set during system initialization when all of the following conditions exist:

- The Front Enable Relay is energized.
- The Antilock Warning Indicator Control is closed.
- The Antilock Brake Controller senses battery voltage at terminal 1D14.

TEST DESCRIPTION: The following provides an explanation of the procedures being followed in the facing trouble tree.

① Isolates the Antilock Warning Indicator circuit from it's power source and therefore determines whether another voltage source is shorted into the circuit or whether the Antilock Brake Controller is defective.

② Determines whether the Antilock Brake Diode is shorted or whether a short to battery exists in CKT 852.

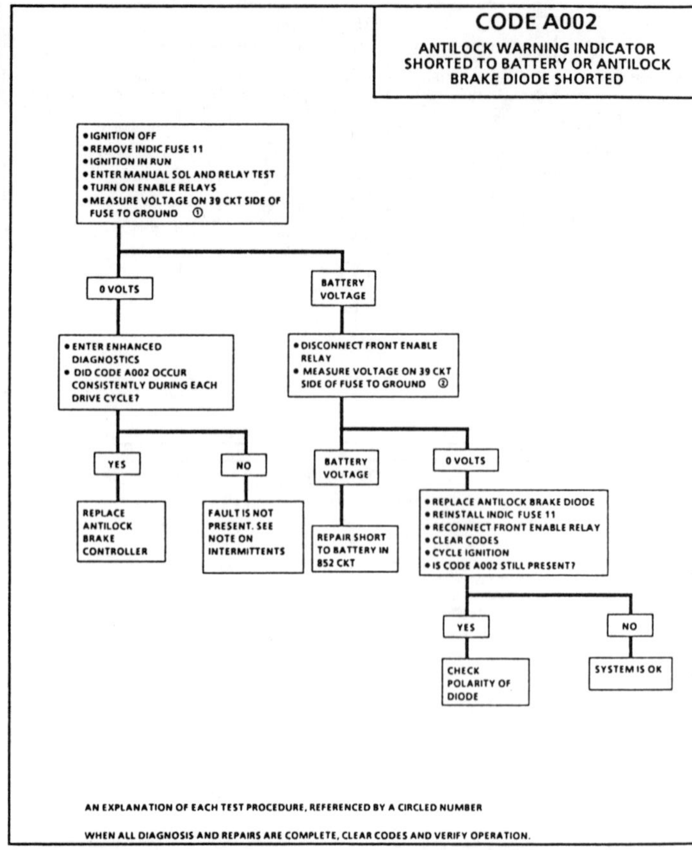

Fig. 5 Code A002, anti-lock warning indicator shorted to battery, or anti-lock brake diode shorted

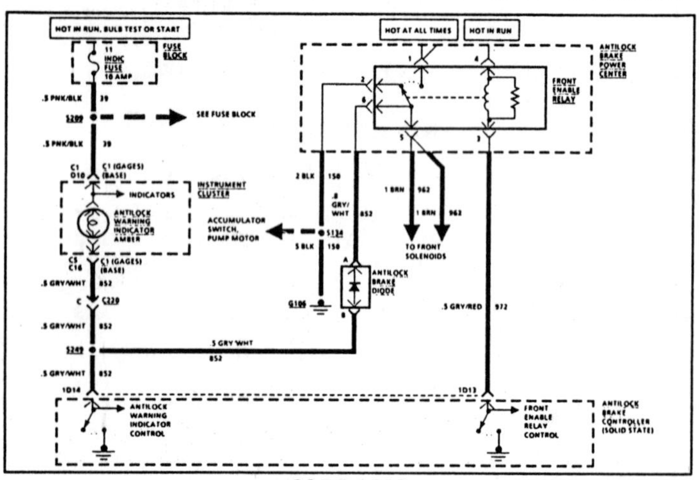

CODE A003

Antilock Brake Diode open or ground open

If the Antilock Brake Controller loses power or ground, the Antilock Warning Indicator is grounded through the 852 CKT, the Antilock Brake Diode, the open contacts of the Front Enable Relay, and CKT 150. The Antilock Brake Diode prevents Battery Voltage from being applied to the Antilock Warning Indicator when the Front Enable Relay is energized.

CODE A003 will set during system initialization when all of the following conditions exist:

- The Front Enable Relay is deenergized.
- The Antilock Warning Indicator Control is open.
- The Antilock Brake Controller senses battery voltage at terminal 1D14.

TEST DESCRIPTION: The following provides an explanation of the procedures being followed in the facing trouble tree.

① If code A039 is also set, the ground CKT 150 must be open.

② Determines if an open condition exists or if a possible intermittent condition exists.

③ Checks if the condition exists between splice S249 and the Front Enable Relay.

④ Indicates if an open condition exists in CKT 852 between splice S249 and the Antilock Brake Diode.

⑤ Determines if an open condition exists in CKT 852 to F t Enable Relay, or in Antilock Brake Diode.

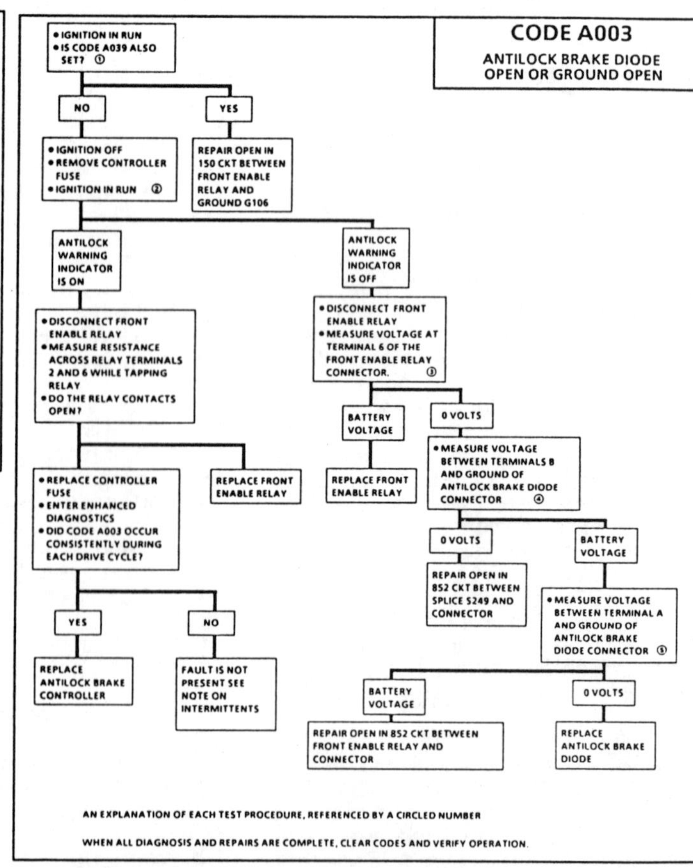

Fig. 6 Code A003, anti-lock brake diode open or ground open

CODE A004

Enable Relay or Solenoid fault detected (FR FAULT)

CODE A004 is a pretest which the Antilock Brake Controller uses in order to detect that a malfunction has occurred in the solenoid or enable relay circuits. Once the Controller has set Code 4 it will initiate a series of tests in order to pinpoint the malfunction. Code A004 will always set along with another Code. The other code which is set will indicate the nature of the malfunction and where it occurred.

TEST DESCRIPTION: The following provides an explanation of the procedures being followed in the facing trouble tree.

① If another code is set a fault exists in the solenoid or enable relay circuits. The other code will indicate the nature of the malfunction and where it occurred.

② If Code A004 sets and no other codes have been set, an internal malfunction in the Antilock Brake Controller is indicated.

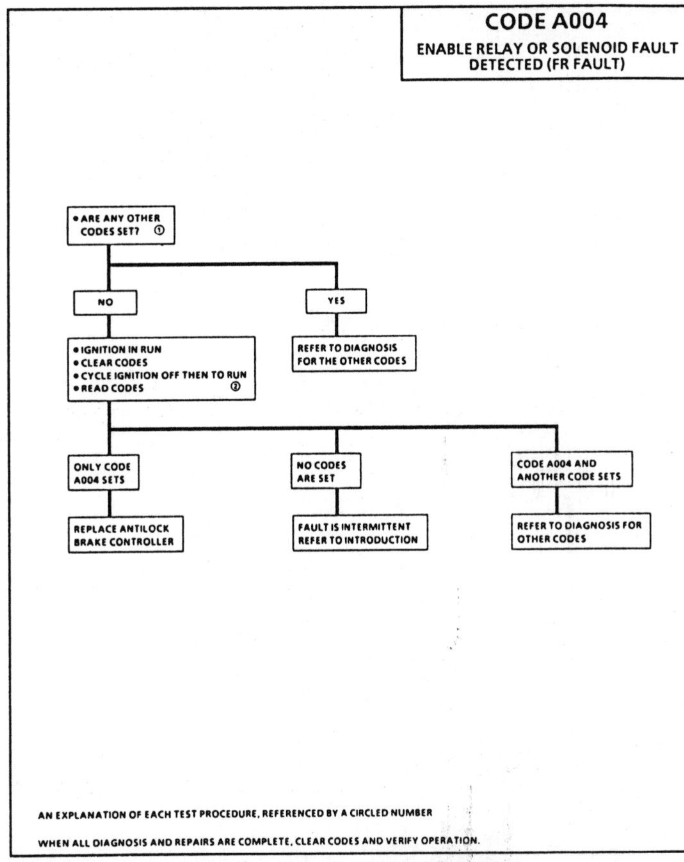

Fig. 7 Code A004, enable relay or solenoid fault detected (FR fault)

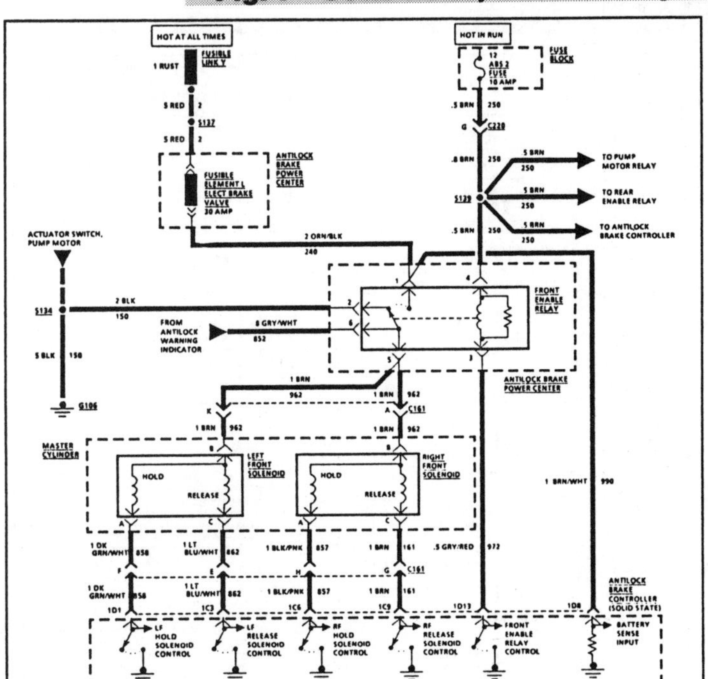

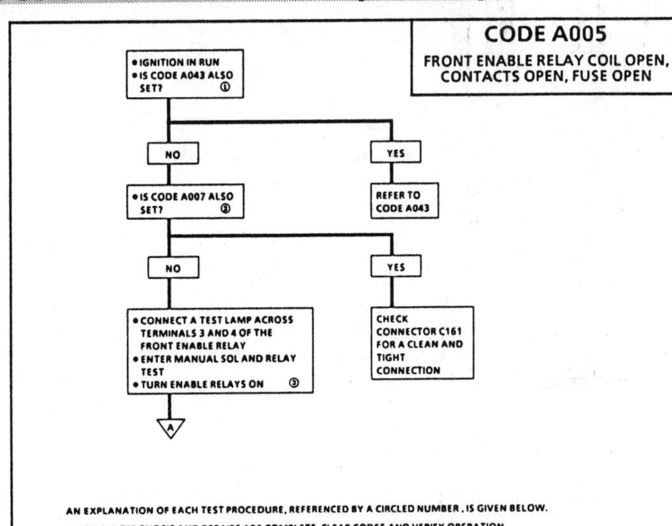

CODE A005

Front Enable Relay Coil open, contacts open, fuse open

If the Ignition is in RUN and the Antilock Brake Controller does not detect any faults, the Controller will close the Front Enable Relay Control. This grounds the Front Enable Relay Coil and the Relay becomes energized. The contacts close, and voltage is applied from Fusible Element L to the Left Front Solenoid and the Right Front Solenoid. If the Controller is not closing any of the Solenoid Controls, the Controller will sense battery voltage at terminals 1D1, 1C3, 1C6, and 1C9.

CODE A005 will set when all of the following conditions exist:

- The Front Enable Relay Control is closed (Front Enable Relay energized).
- The Antilock Brake Controller senses no voltage at terminals 1D1, 1C3, 1C6 and 1C9.

TEST DESCRIPTION: The following provides an explanation of the procedures being followed in the above trouble tree.

① If voltage is present at Front Enable Relay, but not at terminal 1D13, and open in CKT 972 is indicated.

② If Code A007 is also set, none of the solenoids are receiving voltage. This indicates a problem with connector C161.

③ If the test lamp lights, it shows that CKT 250, CKT 972, and the Antilock Brake Controller are good. This indicates that the fault is in the Front Enable Relay or connector, or the fault is intermittent.

Fig. 8 Code A005, front enable relay coil open, contacts open, fuse open. (1 of 2)

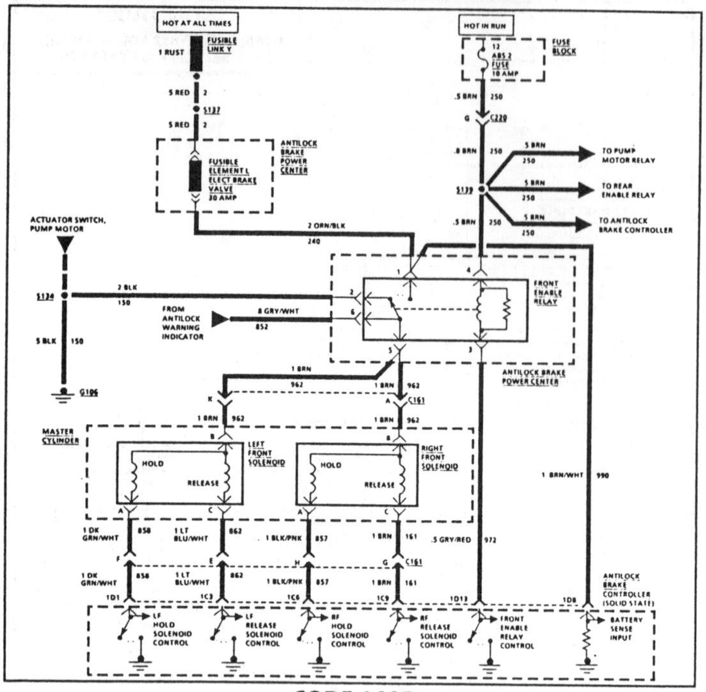

CODE A005

Front Enable Relay Coil open, contacts open, fuse open

TEST DESCRIPTION: The following provides an explanation of the procedures being followed in the facing trouble tree.

① • This determines if Code A005 was set due to a hard failure or an intermittent condition.

⑤ If voltage is present at terminal 1D13, all external circuits must be good. this indicates an internal problem with the Controller.

⑥ If voltage is present at Front Enable Relay, but not at terminal 1D13,an open in CKT 972 is indicated.

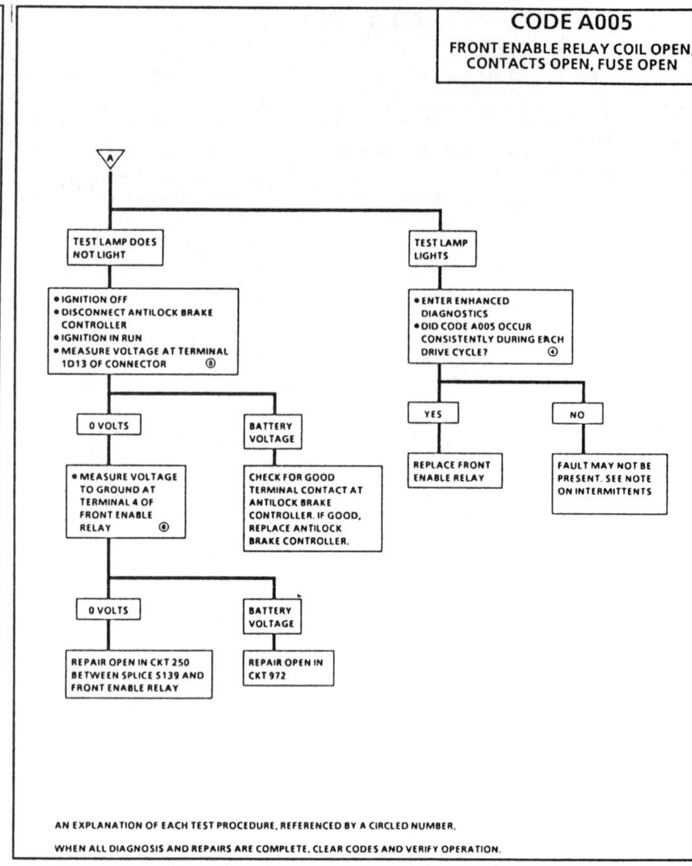

Fig. 8 Code A005, front enable relay coil open, contacts open, fuse open. (2 of 2)

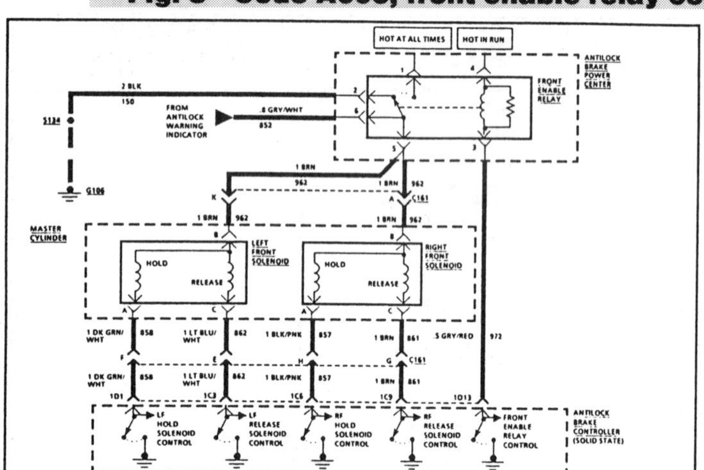

CODE A006

Front Enable Relay Coil shorted to Battery.

Battery Voltage is applied to the Front Enable Relay and CKT 972 whenever the Ignition is in RUN and the Front Enable Relay Control is open.
When the Front Enable Relay Control is closed, ground is applied to teminal 3 of the Front Enable Relay. CKT 972 is now grounded, so voltage is no longer present at terminal 1D13 of the Antilock Brake Controller.

CODE A006 will set when all of the following conditions exist:

- The Front Enable Relay control is closed (Front Enable Relay energized).
- The Antilock Brake Controller senses Battery voltage at Terminal 1D13.
- The Antilock Brake Controller senses no voltage at Terminals 1D1, 1C3, 1C6 and 1C9.

TEST DESCRIPTION: The following provides an explanation of the procedures being followed in the facing trouble tree.

① Code A008 will also be set if both Enable Relay Connectors are disconnected simultaneously.

② With the Front Enable Relay disconnected it can be determined whether a short to Battery exists in CKT 972 (Battery voltage present at Terminal 3 of Front Enable Relay connector).

③ Determines if a short exists internally in the Front Enable Relay.

④ It is possible for an open Relay Coil to set this Code. Code A005 will now set if this fault is present.

⑤ If Code A006 is a consistent failure, the Controller is faulty.

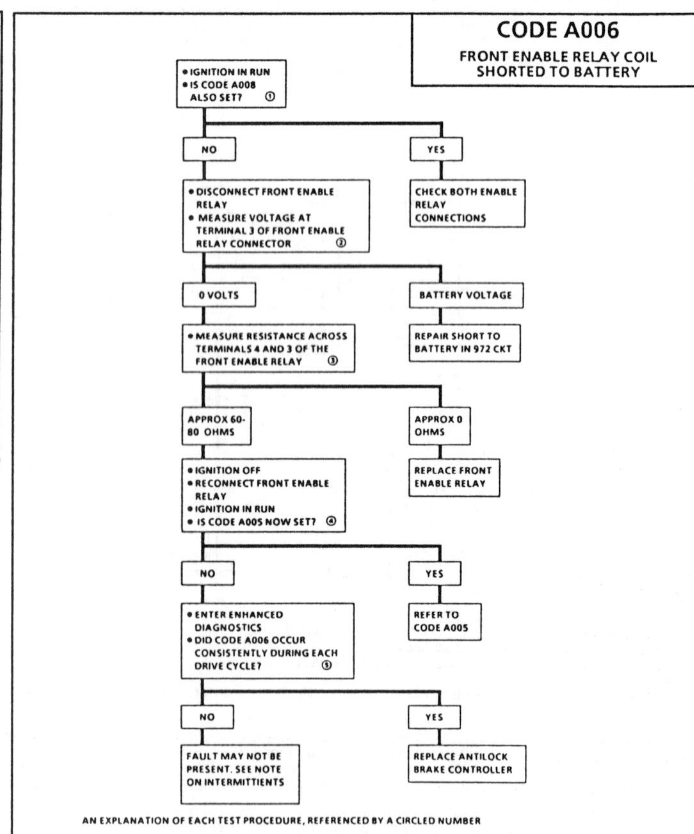

Fig. 9 Code A006, front enable relay coil shorted to battery

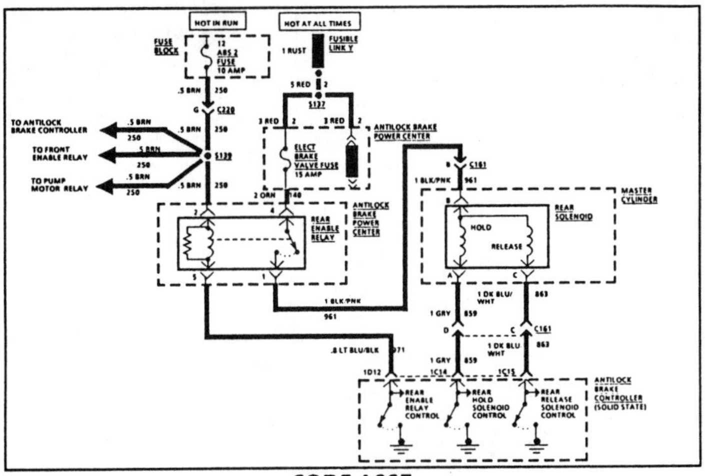

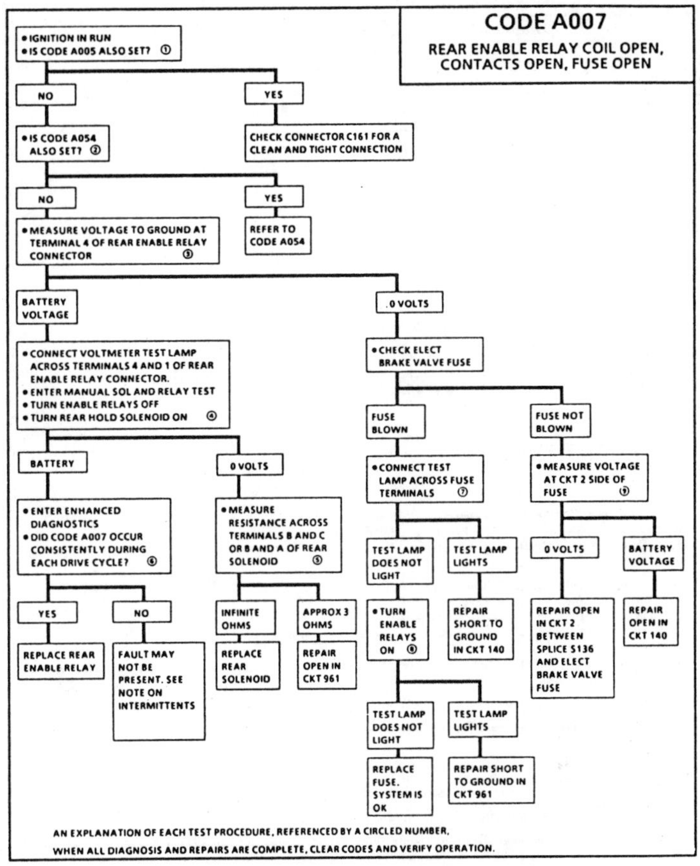

CODE A007

Rear Enable Relay Coil open, contacts open, fuse open

If the Ignition is in RUN and the Antilock Brake Controller does not detect any faults, the Controller will close the Rear Enable Relay Control. This grounds the Rear Enable Relay and the Relay becomes energized. The contacts close, and voltage is applied from the ELECT BRAKE VALVE Fuse to the Rear Solenoid. If the Controller is not closing either of the Rear Solenoid Controls, the Controller will sense Battery voltage at terminals 1C14 and 1C15.

CODE A007 will set when all the following conditions exist:

- The Rear Enable Relay Control is closed (Rear Enable Relay energized).:
- The Antilock Brake Controller senses no voltage at Terminals 1C14 and 1C15.:

TEST DESCRIPTION: The following provides an explanation of the procedures being followed in the trouble tree.

① If Code A005 is also set, none of the Solenoids are receiving voltage. This indicates a problem with connector C161.

② Code A054 will set if the relay coil circuit is open. Diagnostics are covered under code A054.

③ Determines if power feed to Relay contacts is good.

④ If Battery voltage is measured CKT 961 and Rear Solenoid are good.

⑤ Determines if open exists in Rear Solenoid or in CKT 961.

⑥ This determines if Code A007 was set due to a hard failure or an intermittent condition.

⑦ Checks for short to ground in CKT 140.

⑧ If test lamp lights, a short to ground in CKT 961 is indicated.

⑨ Checks for good power feed to ELECT BRAKE VALVE Fuse.

Fig. 10 Code A007, rear enable relay coil open, contacts open, fuse open

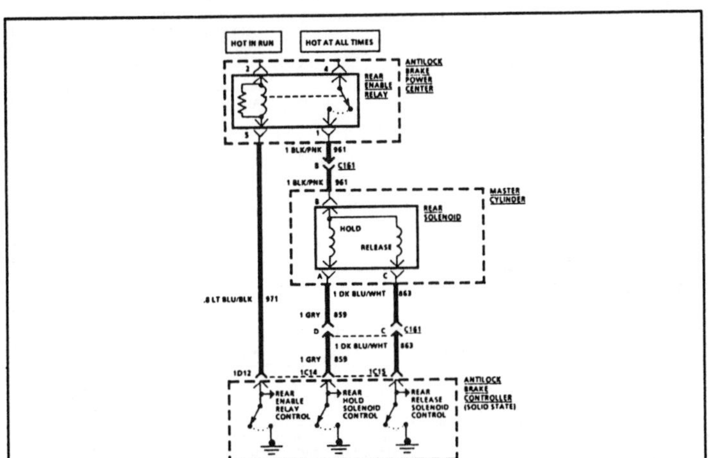

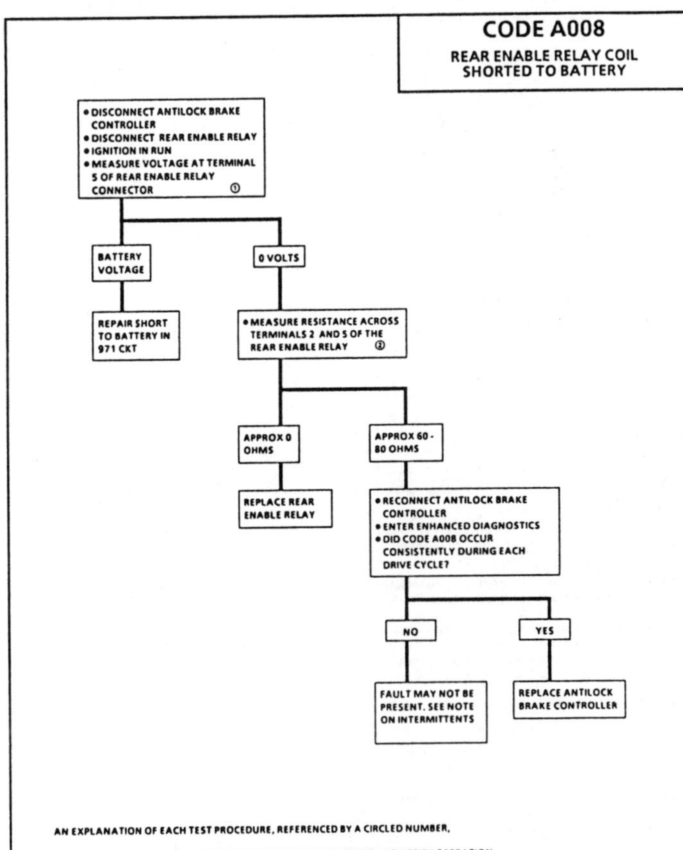

CODE A008

Rear Enable Relay Coil shorted to Battery

Battery voltage is applied to the Rear Enable Relay and CKT 971 whenever the Ignition is in RUN and the Rear Enable Relay Control is open. When the Rear Enable Relay Control is closed, ground is applied to terminal 7 of the Rear Enable Relay. CKT 971 is now grounded, so voltage is no longer present at terminal 1D12 of the Antilock Brake Controller.

CODE A008 will set when all the following conditions exist:

- The Rear Enable Relay Control is closed (Rear Enable Relay energized).
- The Antilock Brake Controller senses Battery voltage at Terminal 1D12.
- The Antilock Brake Controller senses no voltage at Terminals 1C14 and 1C15.

TEST DESCRIPTION: The following provides an explanation of the procedures being followed in the facing trouble tree.

① With the Rear Enable Relay disconnected it can be determined whether a short to Battery exists in the 971 CKT (Battery voltage present at Terminal 7 of Rear Enable Relay connector).

② Determines if a short exists internally in the Rear Enable Relay or whether the Antilock Brake Controller is defective.

Fig. 11 Code A008, rear enable relay coil shorted to battery

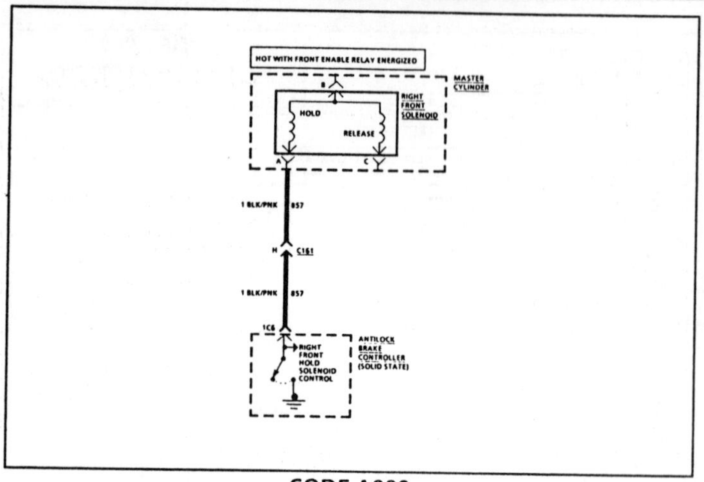

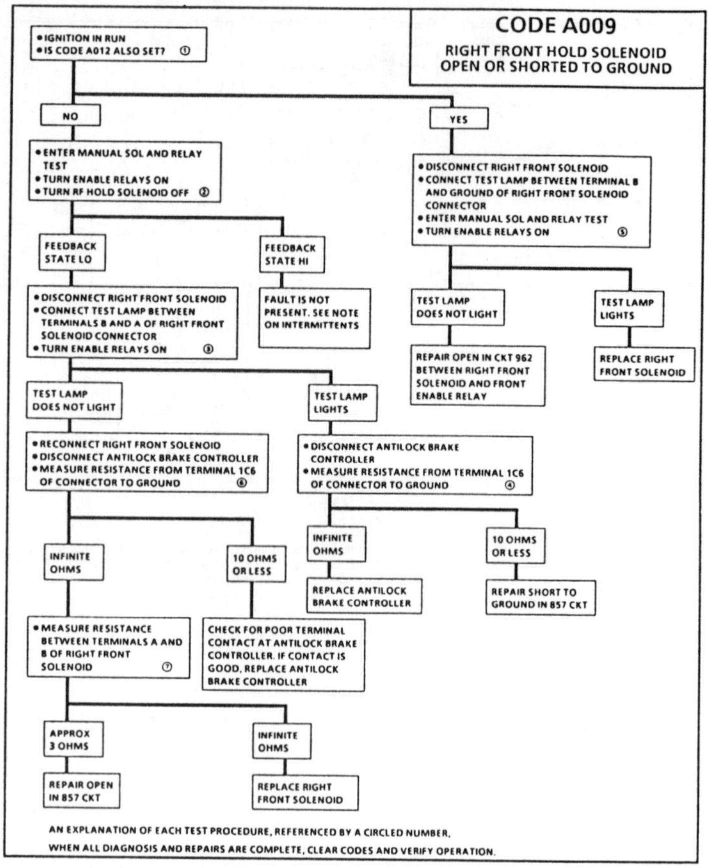

CODE A009

Right Front Hold Solenoid open or shorted to ground

When the Front Enable Relay is energized, voltage is applied to the Right Front Solenoid. If the Antilock Brake Controller determines that the Right Front Hold Solenoid should be activated, it will close the Right Front Hold Solenoid Control. Ground is now applied at terminal A of the Right Front Solenoid, so the Right Front Hold Solenoid is on.

CODE A009 will set when all the following conditions exist:

- The Front Enable Relay is energized.
- The Right Front Hold Solenoid control is open.
- The Antilock Brake Controller senses no voltage at terminal 1C6.

TEST DESCRIPTION: The following provides an explanation of the procedures being followed in the facing trouble tree.

① If Code A012 is also set, the power feed to the Right Front Solenoid is open.

② Voltge at terminal 1C6 will cause the scan tool to display a HI feedback. A LO feedback indicates that voltage is not present.

③ A short to ground must be present if the test lamp lights.

④ If short to ground does not exist with Controller disconnected, short is internal to Controller.

⑤ CKT 962 is OK if the test lamp lights. Open must be internal to solenoid.

⑥ If continuity to ground through the Front Enable Relay contacts is present, open must be internal to Controller.

⑦ Determines if open is in Solenoid or 857 CKT.

Fig. 12 Code A009, right front hold solenoid open or shorted to ground

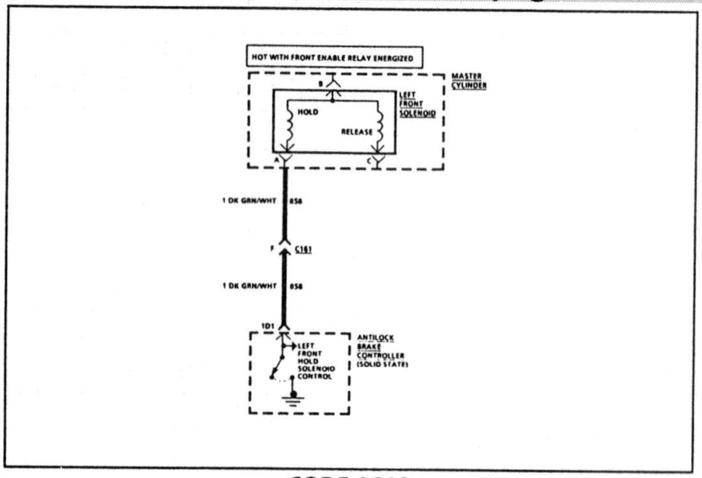

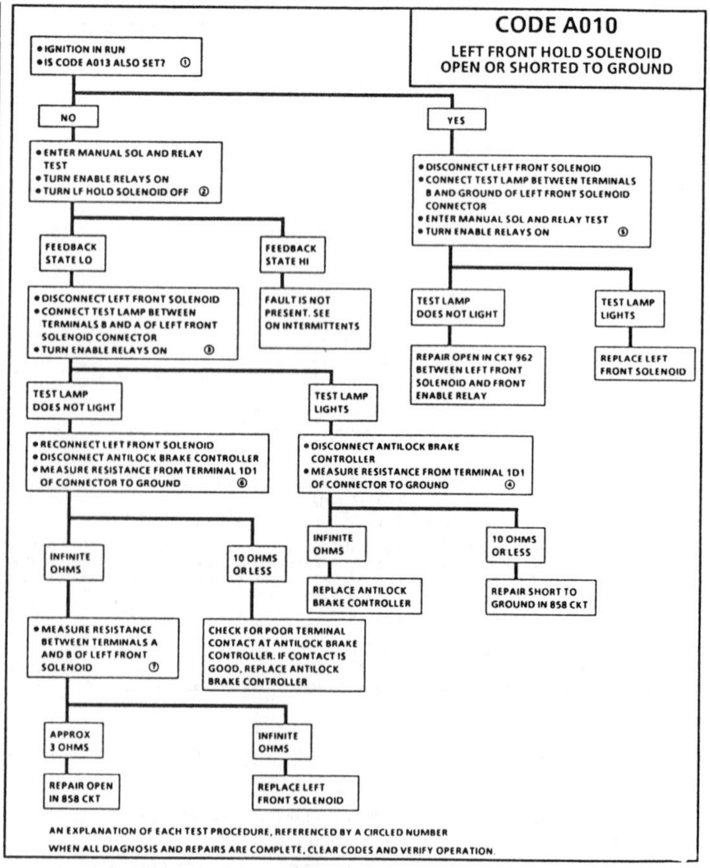

CODE A010

Left Front Hold Solenoid open or shorted to ground

When the Front Enable Relay is energized, voltage is applied to the Left Front Solenoid. If the Antilock Brake Controller determines that the Left Front Hold Solenoid should be activated, it will close the Left Front Hold Solenoid Control. Ground is now applied at terminal A of the Left Front Solenoid, so the Left Front Hold Solenoid is on.

CODE A010 will set when all the following conditions exist:

- The Front Enable Relay is energized.
- The Left Front Hold Solenoid Control is open.
- The Antilock Brake Controller senses no voltage at Terminal 1D1.

TEST DESCRIPTION: The following provides an explanation of the procedures being followed in the facing trouble tree.

① If Code A013 is also set, the power feed to the Left Front Solenoid is open.

② Voltage at terminal 1D1 will cause the scan tool to display HI feedback. A LO feedback indicates that voltage is not present.

③ A short to ground must be present if the test lamp lights.

④ If short to ground does not exist with controller disconnected, short is internal to Controller.

⑤ CKT 962 is OK if the test lamp lights. Open must be internal to solenoid.

⑥ If continuity to ground through the Front Enable Relay Contacts is present, open must be internal to Controller.

⑦ Determines if open is in solenoid or 858 CKT.

Fig. 13 Code A010, left front hold solenoid open or shorted to ground

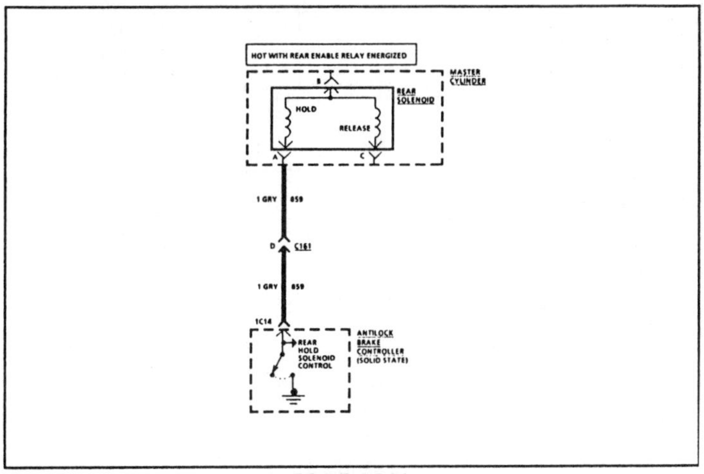

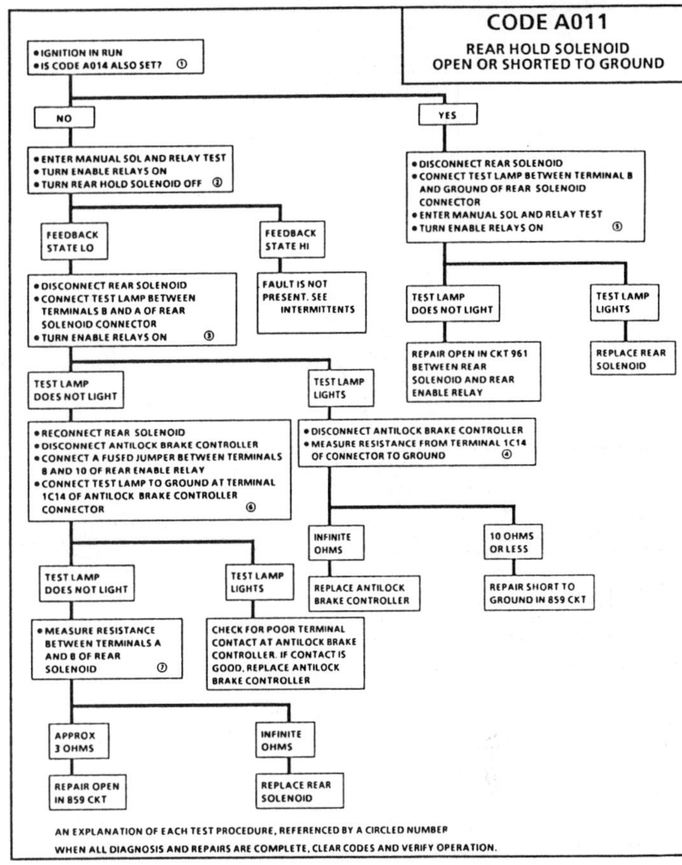

CODE A011

Rear Hold Solenoid open or shorted to ground

When the Rear Enable Relay is energized, voltage is applied to the Rear Solenoid. If the Antilock Brake Controller determines that the Rear Hold Solenoid should be activated, it will close the Rear Hold Solenoid Control. Ground is now applied at terminal A of the Rear Solenoid, so the Rear Hold Solenoid is on.

CODE A011 will set when all the following conditions exist:

- The Rear Enable Relay is energized.
- The Rear Hold Solenoid Control is open.
- The Antilock Brake Controller senses no voltage at Terminal 1C14.

TEST DESCRIPTION: The following provides an explanation of the procedures being followed in the facing trouble tree.

① If Code A014 is also set, the power feed to the Rear Solenoid is open.

② Voltage at terminal 1C14 will cause the scan tool to display a HI feedback. A LO feedback indicates that voltage is not present.

③ A short to ground must be present if the test lamp lights.

④ If short to ground does not exist with Controller disconnected, short is internal to Controller.

⑤ CKT 961 is OK if the test lamp lights. Open must be internal to solenoid.

⑥ If test lamp lights, Rear Hold Solenoid and CKT 859 are good. Open must be internal to Controller.

⑦ Determines if open is in Solenoid or 859 CKT.

Fig. 14 Code A011, rear hold solenoid open or shorted to ground

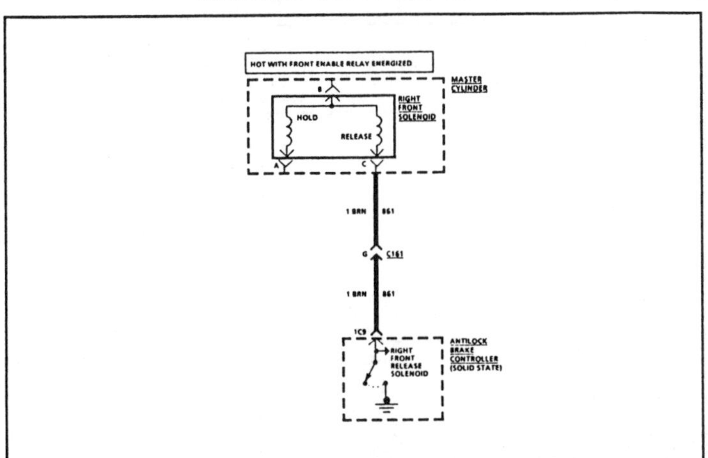

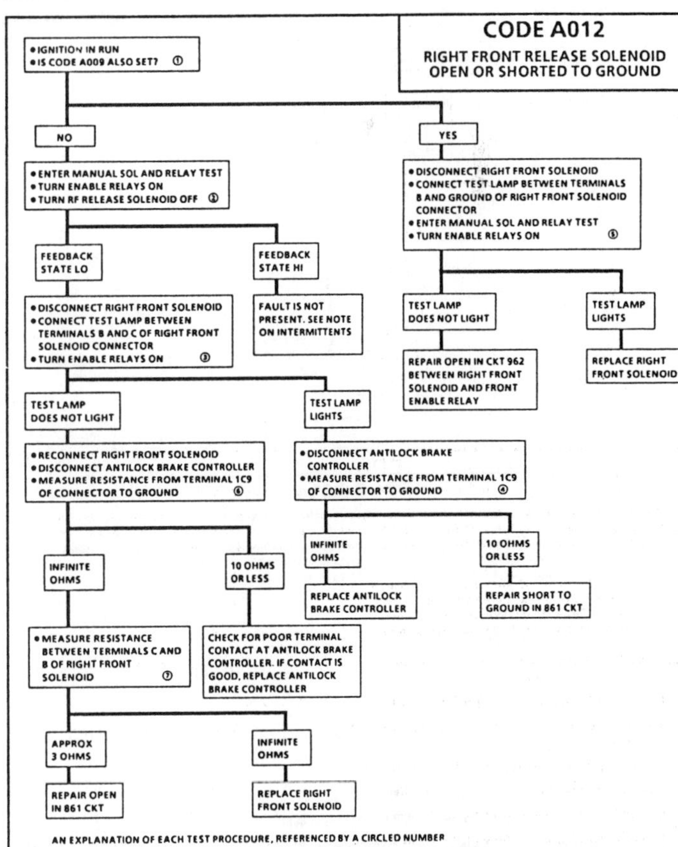

CODE A012

Right Front Release Solenoid open or shorted to ground

When the Front Enable Relay is energized, voltage is applied to the Right Front Solenoid. If the Antilock Brake Controller determines that the Right Front Release Solenoid should be activated, it will close the Right Front Release Solenoid Control. Ground is now applied at terminal C of the Right Front Solenoid, so the Right Front Release Solenoid is on.

CODE A012 will set when all the following conditions exist:

- The Front Enable Relay is energized.
- The Right Front Release Solenoid Control is open
- The Antilock Brake Controller senses no voltage at Terminal 1C9.

TEST DESCRIPTION: The following provides an explanation of the procedures being followed in the facing trouble tree.

① If Code A009 is also set, the power feed to the Right Front Solenoid is open.

② Voltage at terminal 1C9 will cause the scan tool to display a HI feedback. A LO feedback indicates that voltage is not present.

③ A short to ground must be present if the test lamp lights.

④ If short to ground does not exist with Controller disconnected, short is internal to Controller.

⑤ CKT 962 is OK, if the test lamp lights. Open must be internal to solenoid.

⑥ If continuity to ground through the Front Enable Relay is present, open must be internal to Controller.

⑦ Determines if open is in Solenoid or 861 CKT.

Fig. 15 Code A012, right front release solenoid open or shorted to ground

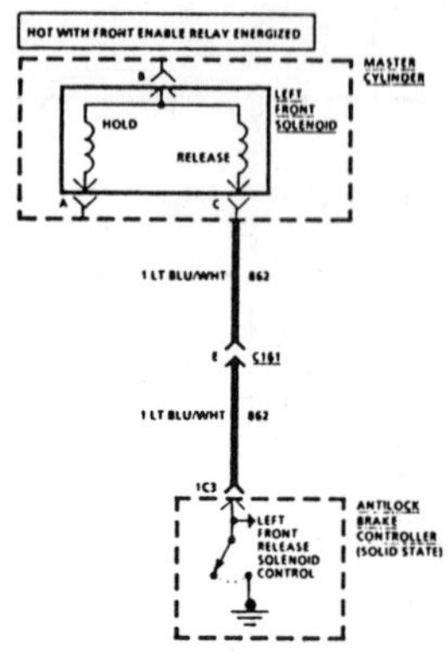

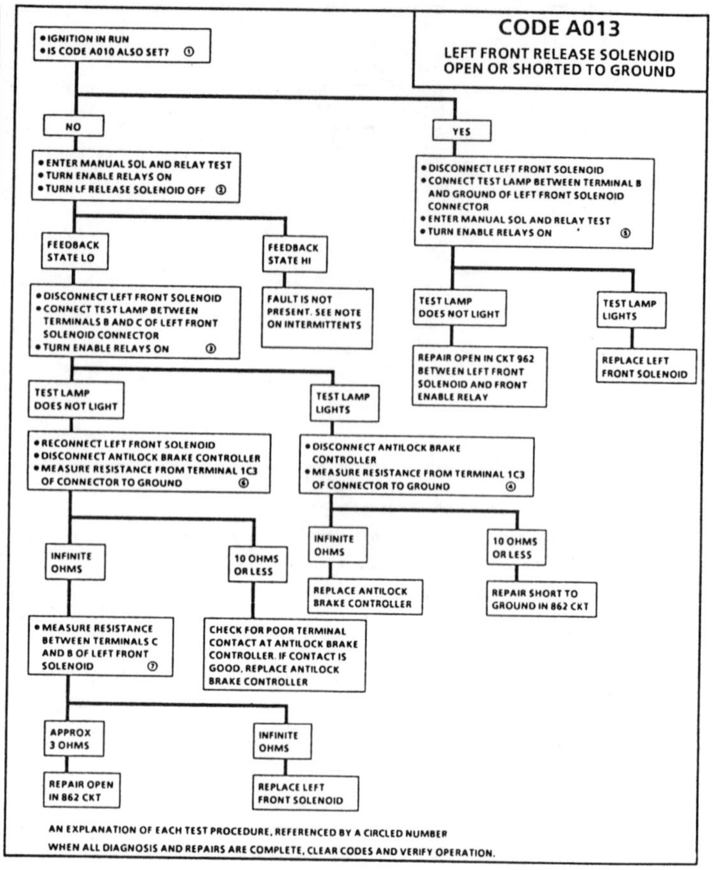

CODE A013

Left Front Release Solenoid open or shorted to ground

When the Front Enable Relay is energized, voltage is applied to the Left Front Solenoid. If the Antilock Brake Controller determines that the Left Front Release Solenoid should be activated, it will close the Left Front Release Solenoid Control. Ground is now applied at terminal C of the Left Front Solenoid, so the Left Front Release Solenoid is on.

CODE A013 will set when all the following conditions exist:

- The Front Enable Relay is energized.
- The Left Front Release Solenoid Control is open.
- The Antilock Brake Controller senses no voltage at Terminal 1C3.

TEST DESCRIPTION: The following provides an explanation of the procedures being followed in the facing trouble tree.

① If Code A010 is also set. the power feed to the Left Front Solenoid is open.

② Voltage at terminal 1C3 will cause the scan tool to display a HI feedback. A LO feedback indicates that voltage is not present.

③ A short to ground must be present if the test lamp lights.

④ If short to ground does not exist with controller disconnected, short is internal to controller.

⑤ CKT 962 is OK if the test lamp lights. Open must be internal to solenoid.

⑥ If continuity to ground through the Front Enable Relay is present, open must be internal to Controller.

⑦ Determines if open is in solenoid or 862 CKT.

Fig. 16 Code A013, left front release solenoid open or shorted to ground

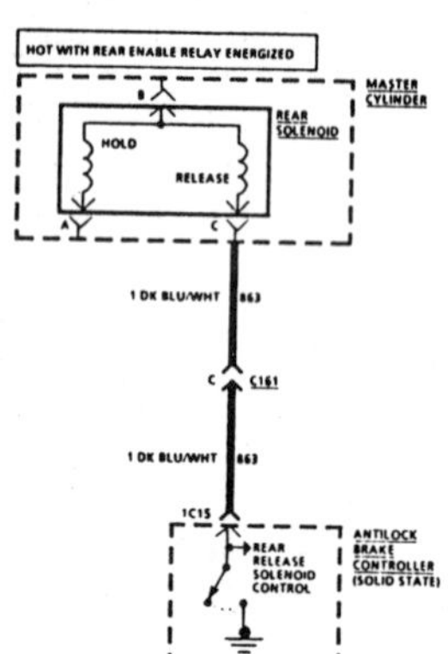

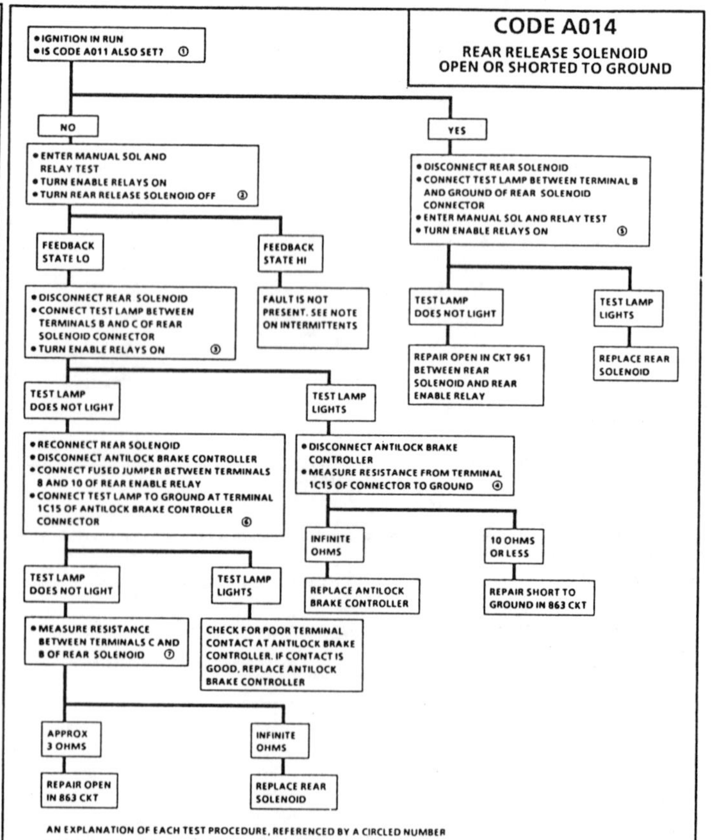

CODE A014

Rear Release Solenoid open or shorted to ground

When the Rear Enable Relay is energized, voltage is applied to the Rear Solenoid. If the Antilock Brake Controller determines that the Rear Release Solenoid should be activated, it will close the Rear Release Solenoid Control. Ground is now applied at terminal C of the Rear Solenoid, so the Rear Release Solenoid is on.

CODE A014 will set when all the following conditions exist:

- The Rear Enable Relay is energized.
- The Rear Release Solenoid Control is open.
- The Antilock Brake Controller senses no voltage at terminal 1C15.

TEST DESCRIPTION: The following provides an explanation of the procedures being followed in the facing trouble tree.

① If Code A011 is also set, the power feed to the Rear Solenoid is open.

② Voltage at terminal 1C15 will cause the scan tool to display a HI feedback. A LO feedback indicates that voltage is not present.

③ A short to ground must be present if the test lamp lights.

④ If short to ground does not exist with controller disconnected, short is internal to controller.

⑤ CKT 962 is OK if test lamp lights. Open must be internal to solenoid.

⑥ If test lamp lights, Rear Release Solenoid and CKT 863 are good. Open must be internal to controller.

⑦ Determines if open is in So[illegible]

Fig. 17 Code A014, rear release solenoid open or shorted to ground

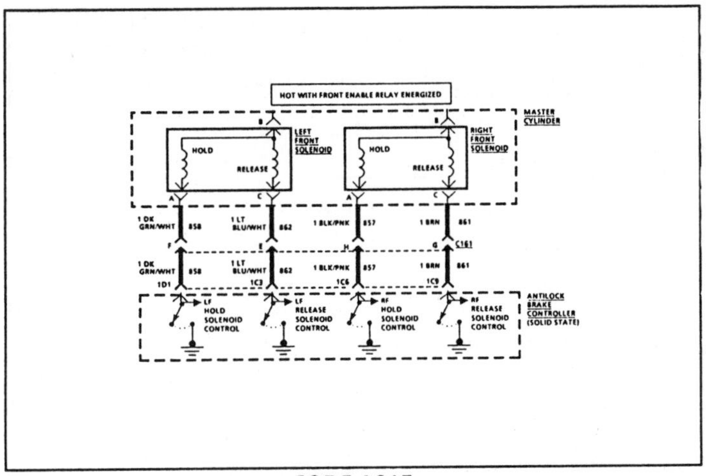

CODE A015

One or more Front Solenoids shorted

CODE A015 will set when the Antilock Brake Controller senses Battery voltage at Terminal 1D1, 1C3, 1C6 or 1C9 when the corresponding Solenoid Control is closed.

TEST DESCRIPTION: The following provides an explanation of the procedures being followed in the facing trouble tree.

① The reason for refering to these codes is that the fault that caused Code A015 to set will be linked to a specific circuit.

② If the failure is consistent, Antilock Brake Controller has an internal fault.

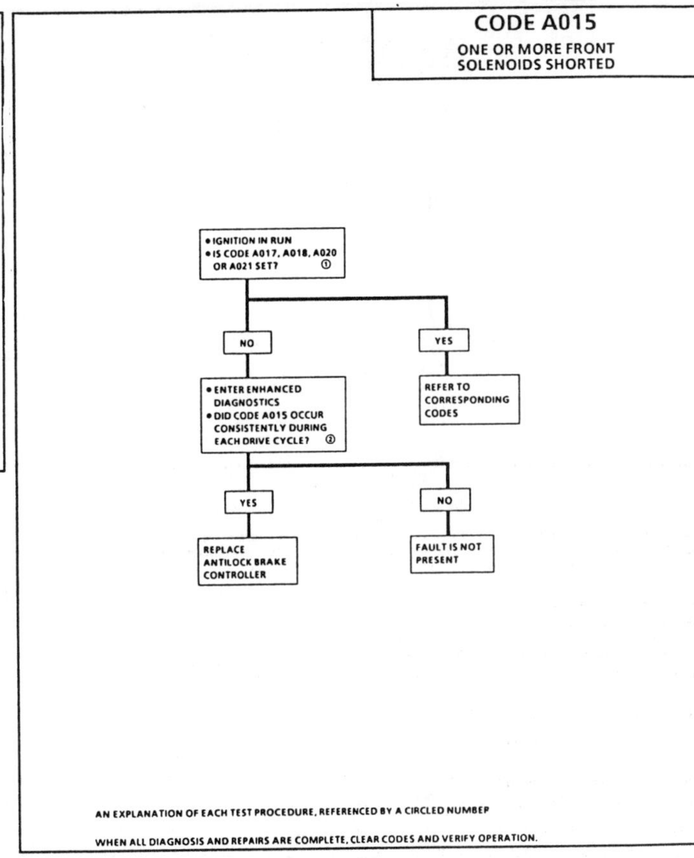

Fig. 18 Code A015, one or more front solenoids shorted

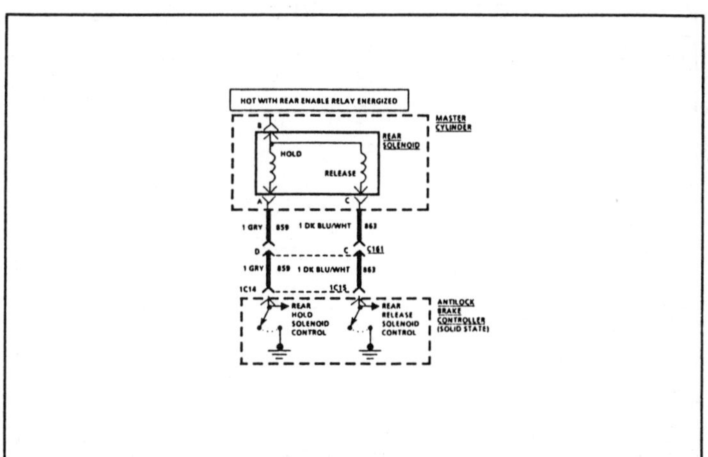

CODE A016

One or both Rear Solenoids shorted

CODE A016 will set when the Antilock Brake Controller senses Battery voltage at Terminal 1C14 or 1C15 when the corresponding Solenoid Control is closed.

TEST DESCRIPTION: The following provides an explanation of the procedures being followed in the facing trouble tree.

① The reason for refering to these codes is that the fault that caused Code A016 to set will be linked to a specific circuit.

② If the failure is consistent, Antilock Brake Controller has an internal fault.

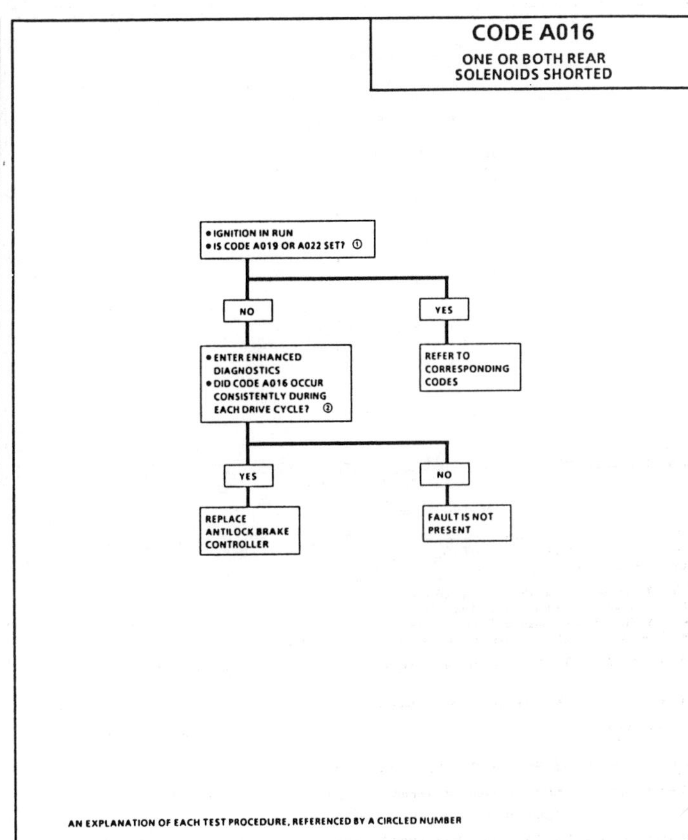

Fig. 19 Code A016, one or both rear solenoids shorted

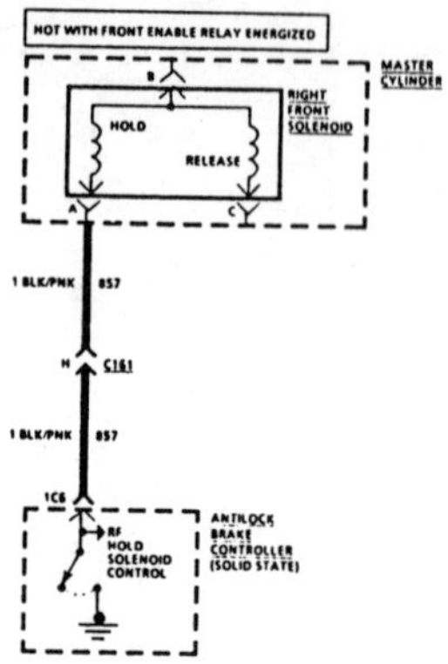

CODE A017

Right Front Hold Solenoid shorted

When the Front Enable Relay is energized, voltage is applied to the Right Front Solenoid. If the Right Front Hold Solenoid Control is closed, ground is applied to CKT 857. This activates the Right Front Hold Solenoid.

CODE A017 will set during initialization when all the following conditions exist:
- The Front Enable Relay is energized.
- The Right Front Hold Solenoid Control is closed.
- The Antilock Brake Controller senses Battery voltage at Terminal 1C6.

TEST DESCRIPTION: The following provides an explanation of the procedures being followed in the facing trouble tree.

① This determines if condition that set code A017 is still present.

② Determines if short to Battery exists.

③ Determines if short to Battery is in 857 CKT or internal to controller.

④ If Right Front Solenoid is not shorted, fault must be internal to controller.

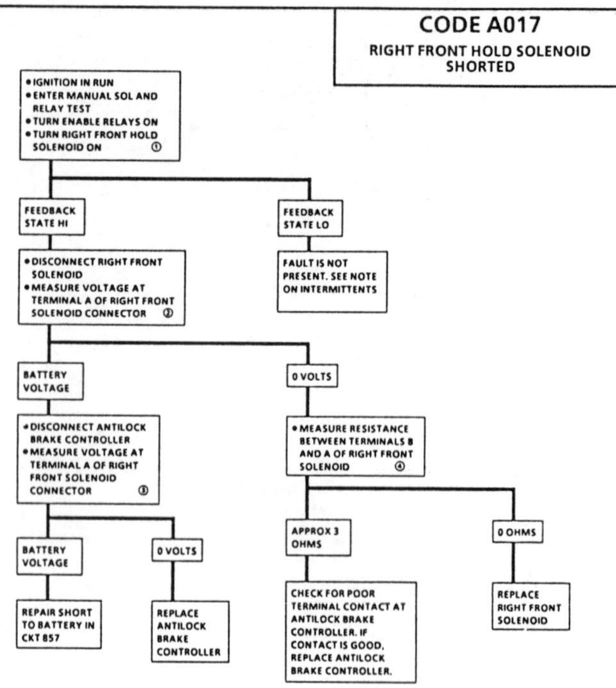

AN EXPLANATION OF EACH TEST PROCEDURE, REFERENCED BY A CIRCLED NUMBER

WHEN ALL DIAGNOSIS AND REPAIRS ARE COMPLETE, CLEAR CODES AND VERIFY OPERATION.

Fig. 20 Code A017, right front hold solenoid shorted

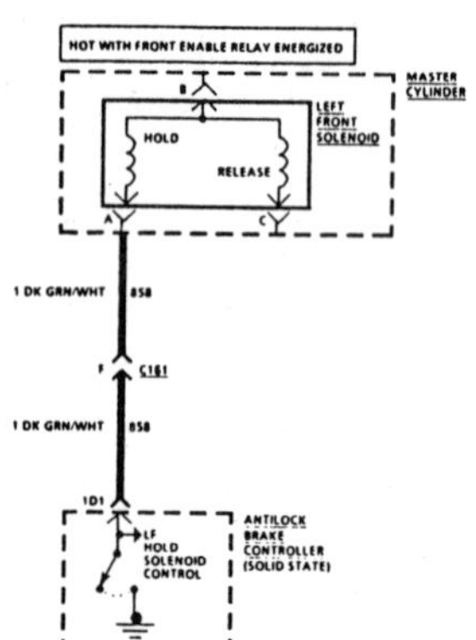

CODE A018

Left Front Hold Solenoid shorted

When the Front Enable Relay is energized, voltage is applied to the Left Front Solenoid. If the Left Front Hold Solenoid Control is closed, ground is applied to CKT 858. This activates the Left Front Hold Solenoid.

CODE A018 will set during initialization when all the following conditions exist:
- The Front Enable Relay is energized.
- The Left Front Hold Solenoid Control is closed.
- The Antilock Brake Controller senses Battery voltage at Terminal 1D1.

TEST DESCRIPTION: The following provides an explanation of the procedures being followed in the facing trouble tree.

① This determines if condition that set code A018 is still present.

② Checks if short to Battery exists.

③ Determines if short to Battery is in 858 CKT or internal to controller.

④ If Left Front Solenoid is not shorted, fault must be internal to controller.

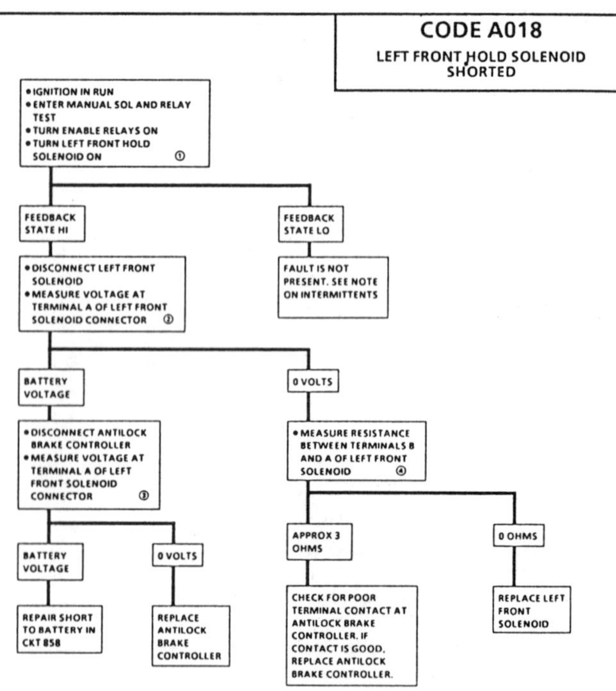

AN EXPLANATION OF EACH TEST PROCEDURE, REFERENCED BY A CIRCLED NUMBER

WHEN ALL DIAGNOSIS AND REPAIRS ARE COMPLETE, CLEAR CODES AND VERIFY OPERATION.

Fig. 21 Code A018, left front hold solenoid shorted

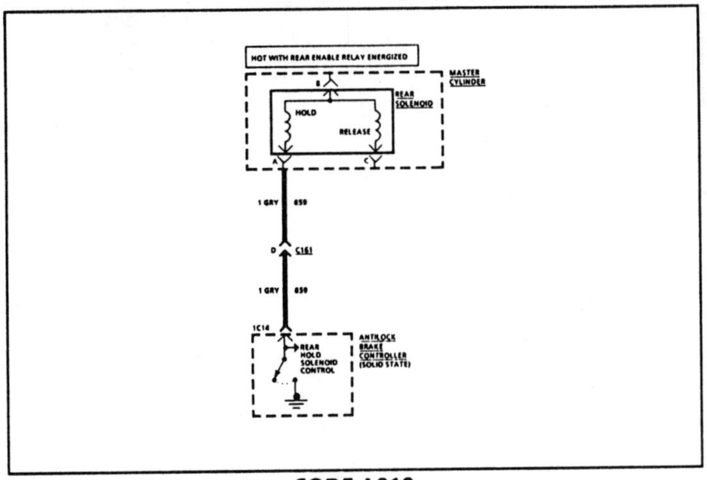

CODE A019

Rear Hold Solenoid shorted

When the Rear Enable Relay is energized, voltage is applied to the Rear Solenoid. If the Rear Hold Solenoid Control is closed, ground is applied to CKT 859. This activates the Rear Hold Solenoid.

CODE A019 will set during initialization when all the following conditions exist:

- The Rear Enable Relay is energized.
- The Rear Hold Solenoid Control is closed.
- The Antilock Brake Controller senses Battery voltage at Terminal 1C14.

TEST DESCRIPTION: The following provides an explanation of the procedures being followed in the facing trouble tree.

① This determines if condition that set code A019 is still present.

② Checks if short to Battery exists.

③ Determines if short to Battery is in 859 CKT or internal to controller.

④ If Rear Solenoid is not shorted, fault must be internal to controller.

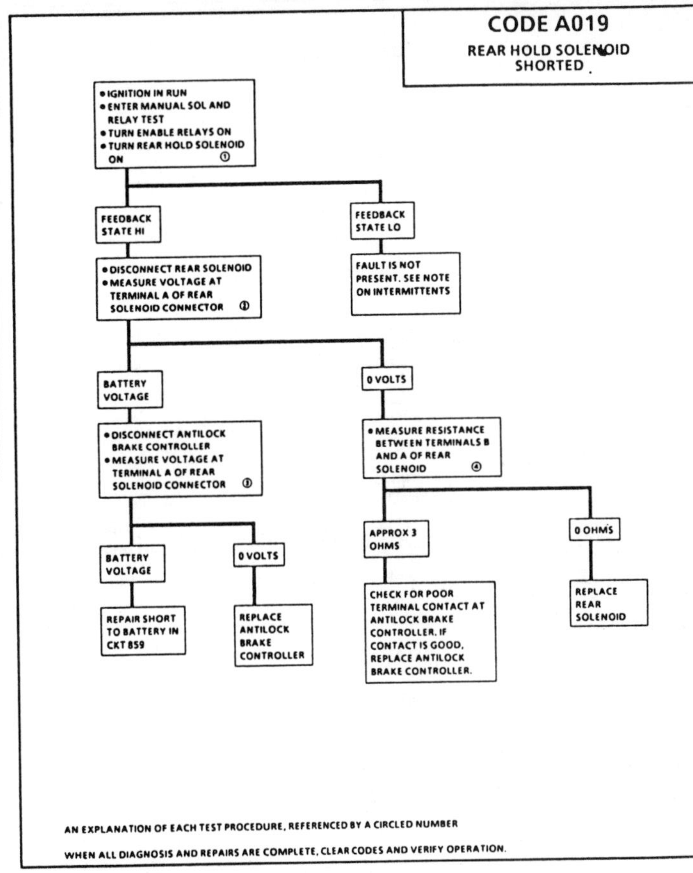

Fig. 22 Code A019, rear hold solenoid shorted

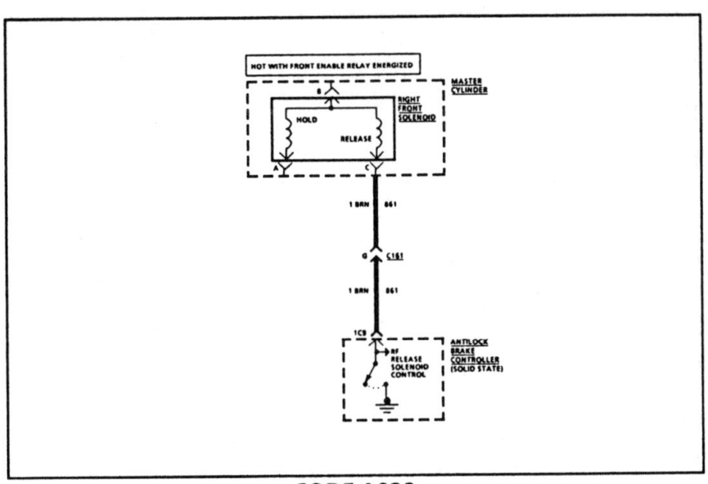

CODE A020

Right Front Release Solenoid shorted

When the Front Enable Relay is energized, voltage is applied to the Right Front Solenoid. If the Right Front Release Solenoid Control is closed, ground is applied to CKT 861. This activates the Right Front Release Solenoid.

CODE A020 will set during initialization when all the following conditions exist:

- The Front Enable Relay is energized.
- The Right Front Release Solenoid Control is closed.
- The Antilock Brake Controller senses Battery voltage at Terminal 1C9.

TEST DESCRIPTION: The following provides an explanation of the procedures being followed in the facing trouble tree.

① This determines if condition that set Code A020 is still present.

② Checks if short to Battery exists.

③ Determines if short to Battery is in 861 CKT or internal to Controller.

④ If Right Front Solenoid is not shorted, fault must be internal to Controller.

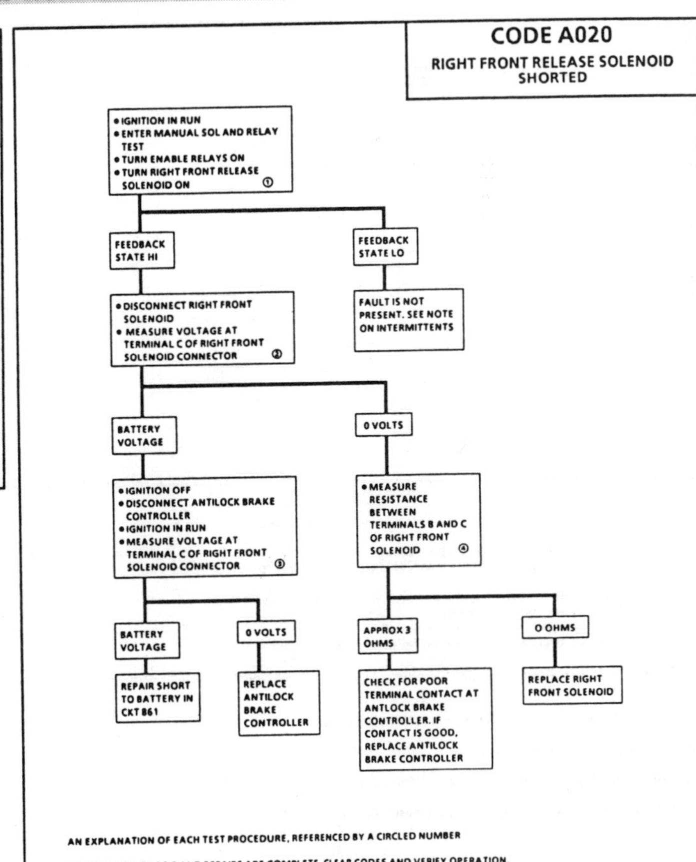

Fig. 23 Code A020, right front release solenoid shorted

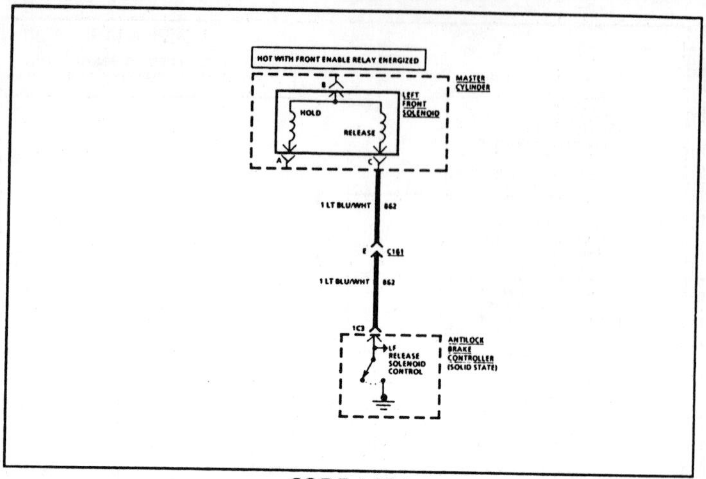

CODE A021

Left Front Release Solenoid shorted

When the Front Enable Relay is energized, voltage is applied to the Left Front Solenoid. If the Left Front Release Solenoid Control is closed, ground is applied to CKT 862. This activates the Left Front Release Solenoid.

CODE A021 will set during initialization when all the following conditions exist:

- The Front Enable Relay is energized.
- The Left Front Release Solenoid Control is closed.
- The Antilock Brake Controller senses Battery voltage at Terminal 1C3.

TEST DESCRIPTION: The following provides an explanation of the procedures being followed in the facing trouble tree.

① This determines if condition that set Code A021 is still present.

② Checks if short to Battery exists.

③ Determines if short to Battery is in 862 CKT or internal to Controller.

④ If Left Front Solenoid is not shorted, fault must be internal to Controller.

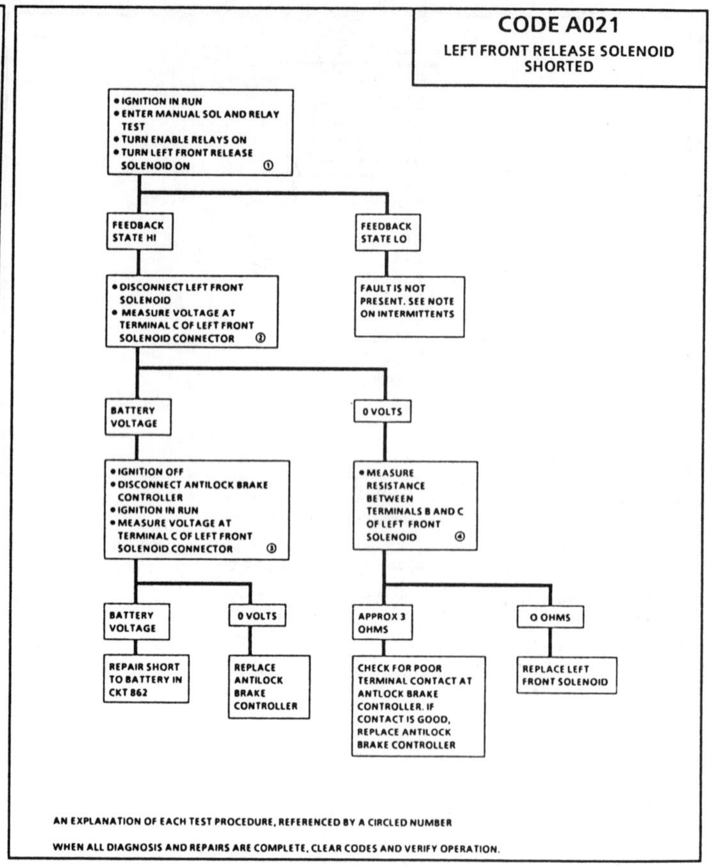

Fig. 24 Code A021, left front release solenoid shorted

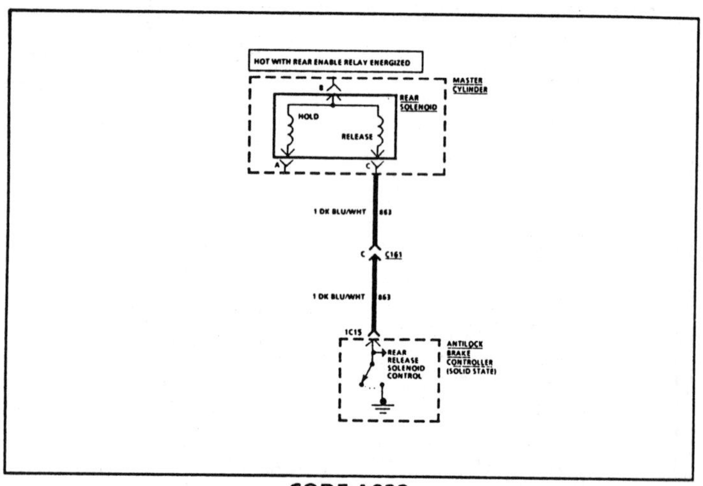

CODE A022

Rear Release Solenoid shorted

When the Rear Enable Relay is energized, voltage is applied to the Rear Solenoid. If the Rear Release Solenoid Control is closed, ground is applied to CKT 863. This activates the Rear Release Solenoid.

CODE A022 will set during initialization when all the following conditions exist:

- The Rear Enable Relay is energized.
- The Rear Release Solenoid Control is closed.
- The Antilock Brake Controller senses Battery voltage at Terminal 1C15.

TEST DESCRIPTION: The following provides an explanation of the procedures being followed in the facing trouble tree.

① This determines if condition that set Code A022 is still present.

② Checks if short to Battery exists.

③ Determines if short to Battery is in 863 CKT or internal to Controller.

④ If Rear Solenoid is not shorted, fault must be internal to Controller.

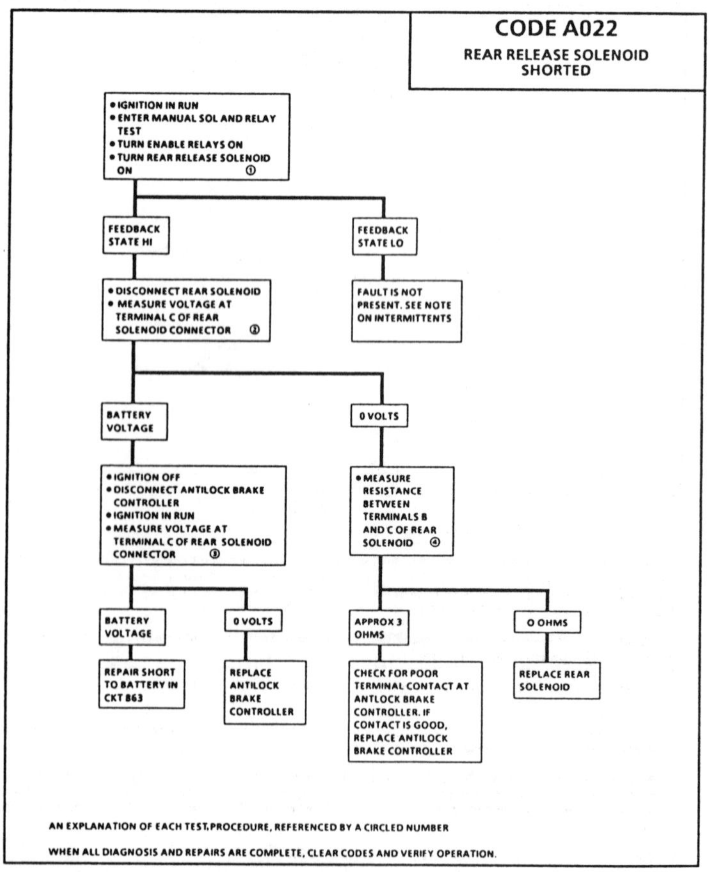

Fig. 25 Code A022, rear release solenoid shorted

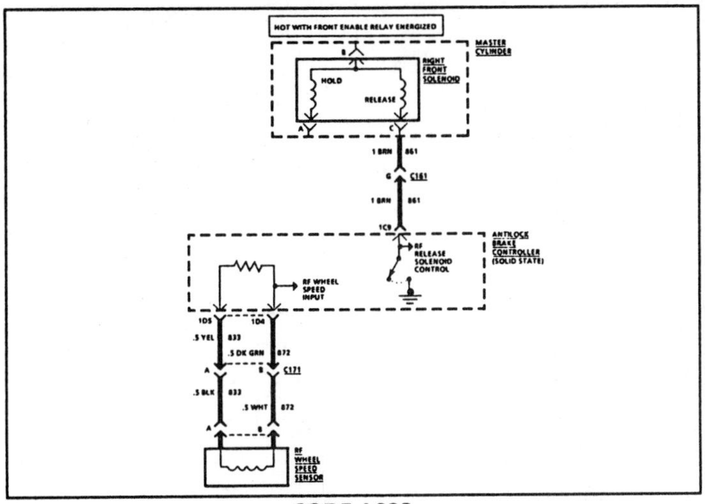

CODE A023

Right Front Release Solenoid energized too long

By monitoring the Right Front Wheel Speed Sensor, the Antilock Brake Controller can determine if the Right Front Wheel is locking up. If this condition occurs, the controller will activate the Right Front Release Solenoid by closing the Right Front Release Solenoid Control. The Solenoid Control will be closed until the Controller determines that the Right Front Wheel has increased to an acceptable speed.

CODE A023 will set when the Antilock Brake Controller senses that the Right Front Release Solenoid has been energized longer than proper operation requires.

TEST DESCRIPTION: The following provides an explanation of the procedures being followed in the facing trouble tree.

① If any other codes are set they should be addressed first. The reason for this is that the fault that caused the solenoid to be energized too long will be better identified.

② By examining the Right Front Wheel Speed input, it can be determined if the fault is due to erratic wheel speed inputs which occur only at low speeds. If the wheel speed input is found to be steady at a low speed a mechanical or hydraulic problem is indicated – A wheel speed that stays at 0 MPH indicates an open – refer to Code A044.

③ This checks for a possible short to voltage on CKT 833 and CKT 872.

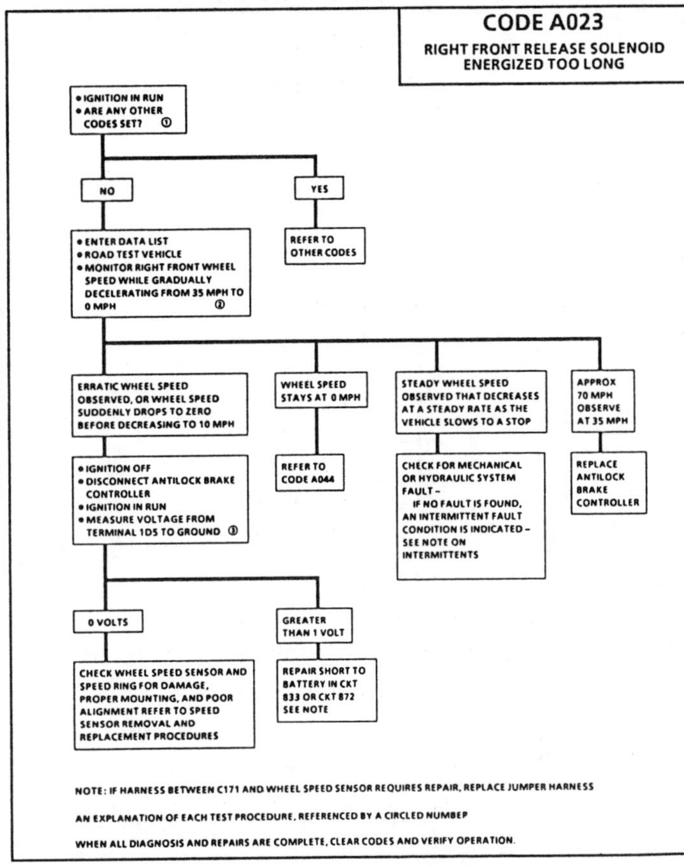

Fig. 26 Code A023, right front release solenoid energized too long

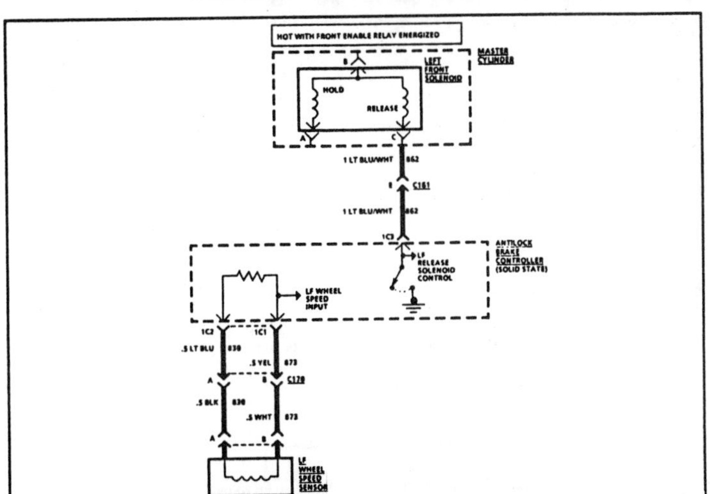

CODE A024

Left Front Release Solenoid energized too long

By monitoring the Left Front Wheel Speed Sensor, the Antilock Brake Controller can determine if the Left Front Wheel is locking up during an ABS stop. If this condition occurs, the Controller will activate the Left Front Release Solenoid by closing the Left Front Release Solenoid Control. The Solenoid Control will be closed until the Controller determines that the Left Front Wheel has increased to an acceptable speed.

CODE A024 will set when the Antilock Brake Controller senses that the Left Front Release Solenoid has been energized longer than proper operation requires.

TEST DESCRIPTION: The following provides an explanation of the procedures being followed in the facing trouble tree.

① If any other codes are set they should be addressed first. The reason for this is that the fault that caused the solenoid to be energized too long will be better identified.

② By examining the Left Front Wheel Speed input, it can be determined if the fault is due to erratic wheel speed inputs which occur only at low speeds. If the wheel speed input is found to be steady at a low speed a mechanical or hydraulic problem is indicated A wheel speed that stays at 0 MPH indicates an open – refer to Code A045.

③ This checks for a possible short to voltage on CKT 830 and CKT 873.

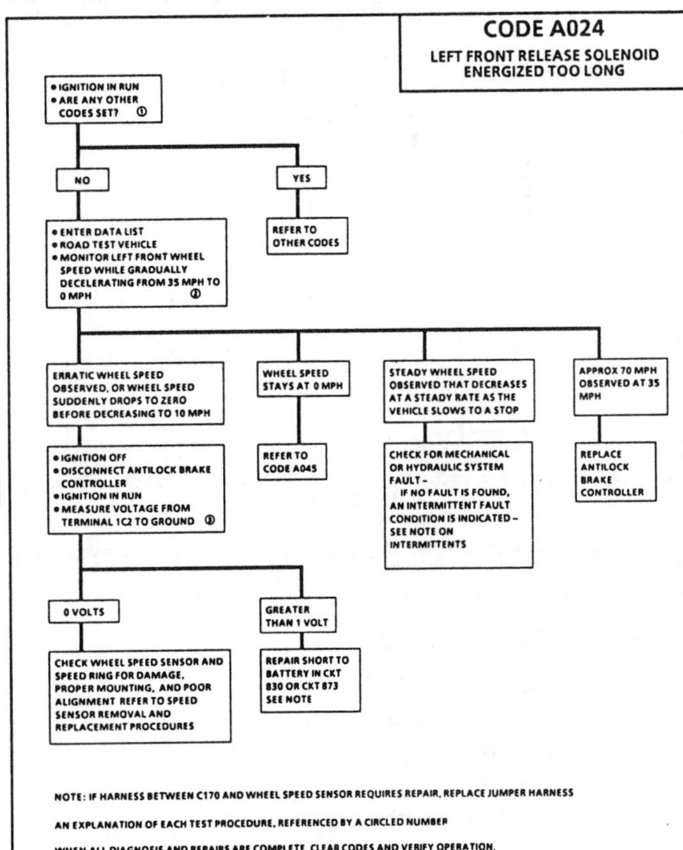

Fig. 27 Code A024, left front release solenoid energized too long

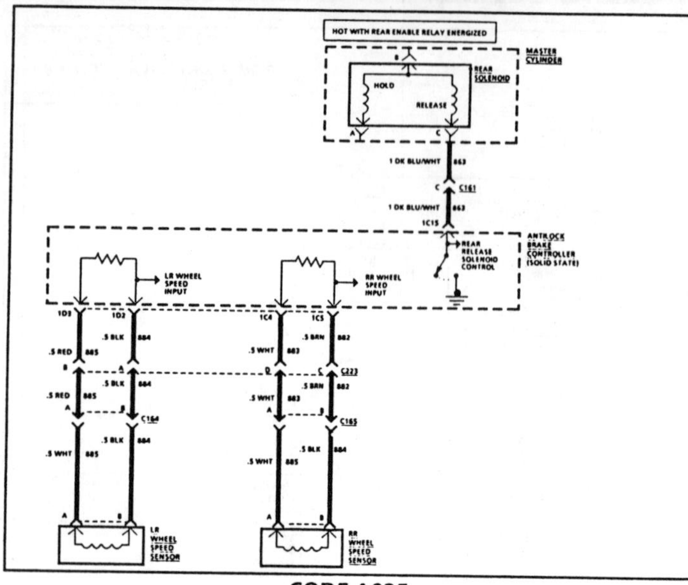

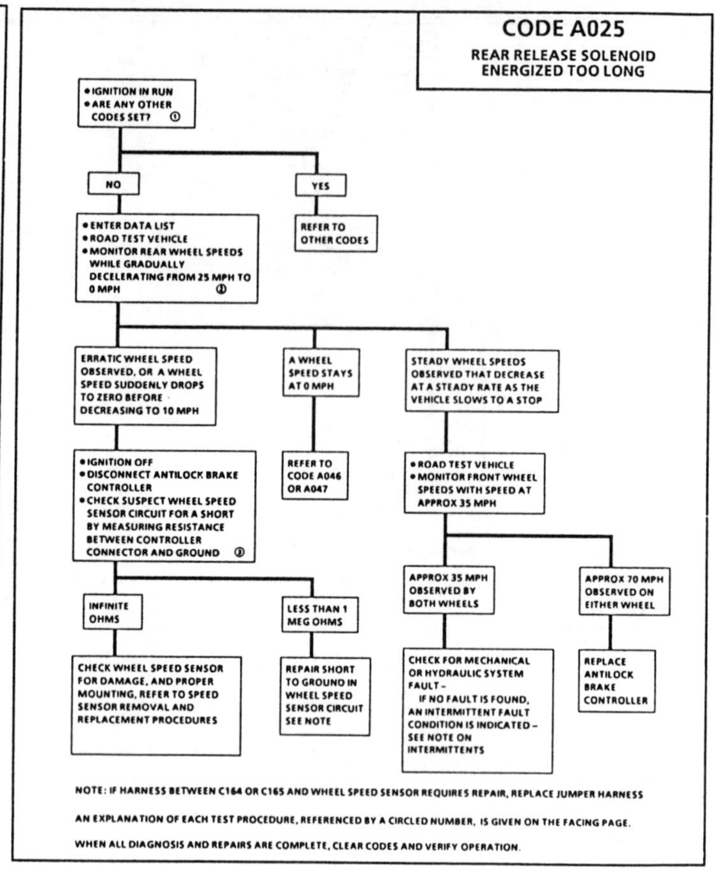

CODE A025

Rear Release Solenoid energized too long

By monitoring the Rear Wheel Speed Sensors, the Antilock Brake Controller can determine if a rear wheel is locking up during an ABS stop. If the condition occurs, the Controller will activate the Rear Release Solenoid by closing the Rear Release Solenoid Control. The Solenoid Control will be closed until the Controller determines that both rear wheels have increased to acceptable speeds.

CODE A025 will set when the Antilock Brake Controller senses that the Rear Release Solenoid has been energized longer than proper operation requires.

TEST DESCRIPTION: The following provides an explanation of the procedures being followed in the facing trouble tree.

① If any other codes are set they should be addressed first. The reason for this is that the fault that caused the solenoid to be energized too long will be better identified.

② By examining the rear wheel speed inputs, it can be determined if the fault is due to erratic wheel speed inputs which occur only at low speeds. If the wheel speed input is found to be steady at low speed, a mechanical or hydraulic problem is indicated- A wheel speed that stays at 0 MPH indicates an open – refer to Code A046 or A047.

③ This checks for a possible short to ground on the suspect Wheel Speed Sensor Circuit.

Fig. 28 Code A025, rear release solenoid energized too long

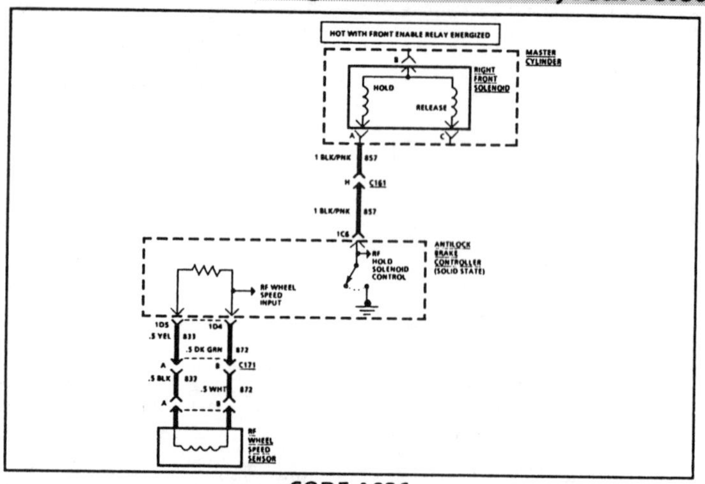

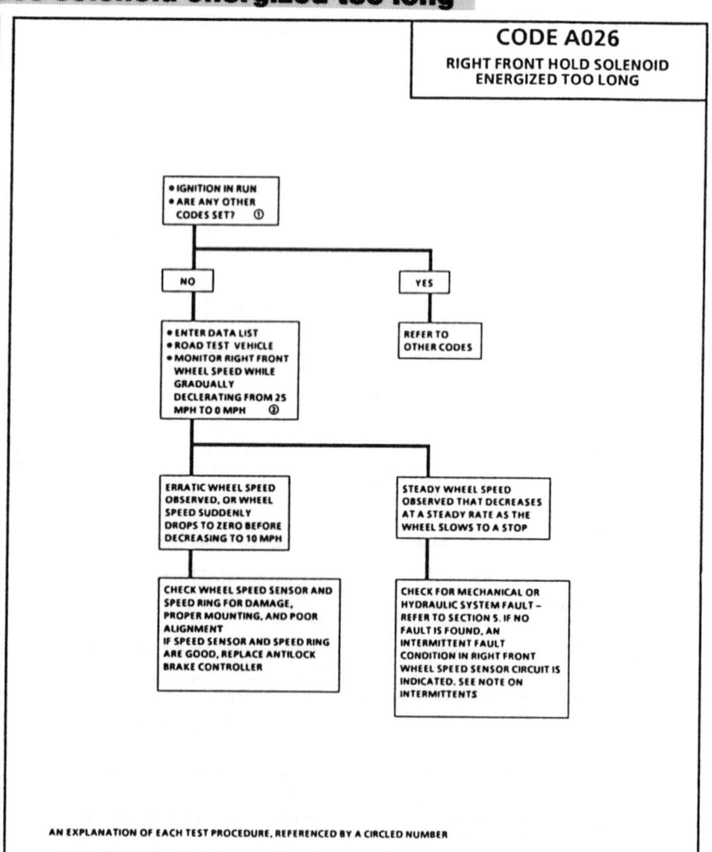

CODE A026

Right Front Hold Solenoid energized too long

By monitoring the Right Front Wheel Speed Sensor, the Antilock Brake Controller can determine if the Right Front Wheel is not declerating properly during an ABS stop. If this condition occurs, the Controller will activate the Right Front Hold Solenoid by closing the Right Front Hold Solenoid Control. The Solenoid Control will be closed until the Controller determines that the Right Front Wheel has decreased to an acceptable speed.

CODE A026 will set when the Antilock Brake Controller senses that the Right Front Hold Solenoid has been energized longer than proper operation requires.

TEST DESCRIPTION: The following provides an explanation of the procedures being followed in the facing trouble tree.

① If any other codes are set they should be addressed first. The reason for this is that the fault that caused the solenoid to be energized too long will be better identified.

② By examining the Right Front Wheel Speed input, it can be determined if the fault is due to erratic wheel speed inputs which occur only at low speeds. If the wheel speed input is found to be steady at a low speed a mechanical or hydraulic problem is indicated

Fig. 29 Code A026, right front hold solenoid energized too long

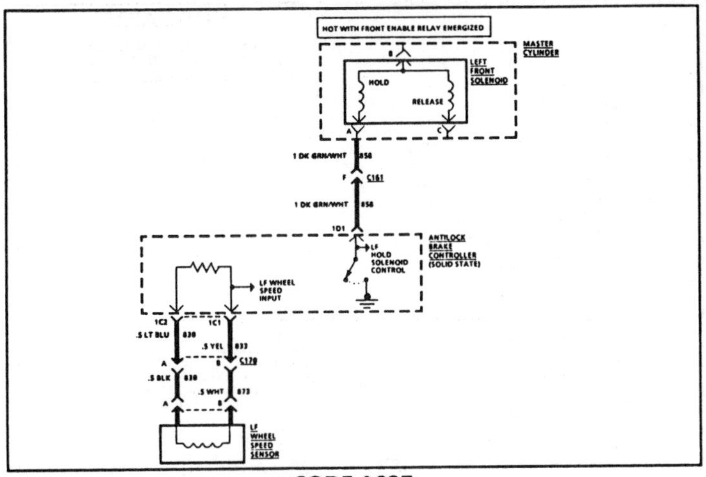

CODE A027

Left Front Hold Solenoid energized too long

By monitoring the Left Front Wheel Speed Sensor, the Antilock Brake Controller can determine if the Left Front Wheel is not decelerating properly during an ABS stop. If this condition occurs, the Controller will activate the Left Front Hold Solenoid by closing the Left Front Hold Solenoid Control. The Solenoid Control will be closed until the Controller determines that the Left Front Wheel has decreased to an acceptable speed.

CODE A027 will set when the Antilock Brake Controller senses that the Left Front Hold Solenoid has been energized longer than proper operation requires.

TEST DESCRIPTION: The following provides an explanation of the procedures being followed in the facing trouble tree.

① If any other codes are set they should be addressed first. The reason for this is that the fault that caused the solenoid to be energized too long will be better identified.

② By examining the Left Front Wheel Speed input, it can be determined if the fault is due to erratic wheel speed inputs which occur only at low speeds. If the wheel speed input is found to be steady at a low speed a mechanical or hydraulic problem is indicated

CODE A027
LEFT FRONT HOLD SOLENOID ENERGIZED TOO LONG

- IGNITION IN RUN
- ARE ANY OTHER CODES SET? ①

YES → REFER TO OTHER CODES

NO →
- ENTER DATA LIST
- ROAD TEST VEHICLE
- MONITOR LEFT FRONT WHEEL SPEED WHILE GRADUALLY DECLERATING FROM 25 MPH TO 0 MPH ②

ERRATIC WHEEL SPEED OBSERVED, OR WHEEL SPEED SUDDENLY DROPS TO ZERO BEFORE DECREASING TO 10 MPH → CHECK WHEEL SPEED SENSOR AND SPEED RING FOR DAMAGE, PROPER MOUNTING, AND POOR ALIGNMENT IF SPEED SENSOR AND SPEED RING ARE GOOD, REPLACE ANTILOCK BRAKE CONTROLLER

STEADY WHEEL SPEED OBSERVED THAT DECREASES AT A STEADY RATE AS THE WHEEL SLOWS TO A STOP → CHECK FOR MECHANICAL OR HYDRAULIC SYSTEM FAULT – IF NO FAULT IS FOUND, AN INTERMITTENT FAULT CONDITION IN LEFT FRONT WHEEL SPEED SENSOR CIRCUIT IS INDICATED. SEE NOTE ON INTERMITTENTS

AN EXPLANATION OF EACH TEST PROCEDURE, REFERENCED BY A CIRCLED NUMBER

WHEN ALL DIAGNOSIS AND REPAIRS ARE COMPLETE, CLEAR CODES AND VERIFY OPERATION.

Fig. 30 Code A027, left front hold solenoid energized too long

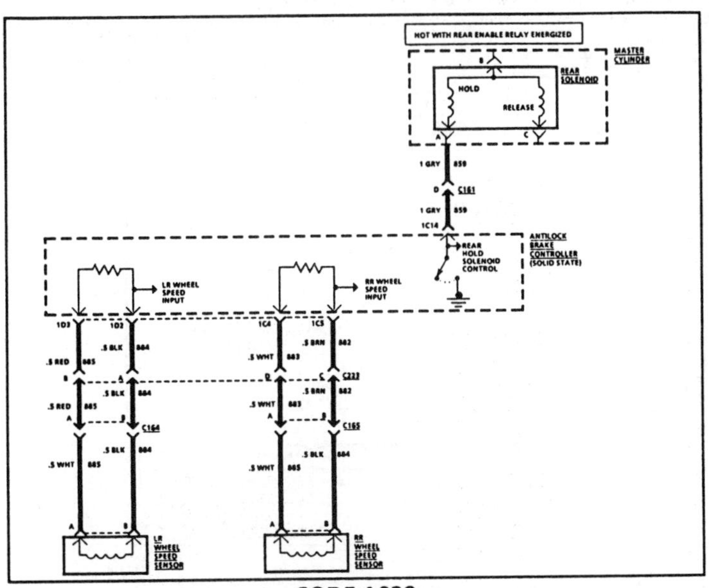

CODE A028

Rear Hold Solenoid energized too long

By monitoring the Rear Wheel Speed Sensors, the Antilock Brake Controller can determine if a Rear Wheel is not decelerating properly during an ABS stop. If this condition occurs, the Controller will activate the Rear Hold Solenoid by closing the Rear Hold Solenoid Control. The Solenoid Control will be closed until the Controller determines that the Rear Wheel has decreased to an acceptable speed.

CODE A028 will set when the Antilock Brake Controller senses that the Rear Hold Solenoid has been energized longer than proper operation requires.

TEST DESCRIPTION: The following provides an explanation of the procedures being followed in the facing trouble tree.

① If any other codes are set they should be addressed first. The reason for this is that the fault that caused the solenoid to be energized too long will be better identified.

② By examining the rear wheel speed inputs, it can be determined if the fault is due to erratic wheel speed inputs which occur only at low speeds. If the wheel speed input is found to be steady at low speed a mechanical or hydraulic problem is indicated

CODE A028
REAR HOLD SOLENOID ENERGIZED TOO LONG

- IGNITION IN RUN
- ARE ANY OTHER CODES SET? ①

YES → REFER TO OTHER CODES

NO →
- ENTER DATA LIST
- ROAD TEST VEHICLE
- MONITOR REAR WHEEL SPEED WHILE GRADUALLY DECLERATING FROM 25 MPH TO 0 MPH ②

ERRATIC WHEEL SPEED OBSERVED, OR WHEEL SPEED SUDDENLY DROPS TO ZERO BEFORE DECREASING TO 10 MPH → CHECK WHEEL SPEED SENSOR FOR DAMAGE, AND PROPER MOUNTING IF SPEED SENSOR IS GOOD, REPLACE ANTILOCK BRAKE CONTROLLER

STEADY WHEEL SPEEDS OBSERVED THAT DECREASE AT A STEADY RATE AS THE WHEELS SLOW TO A STOP → CHECK FOR MECHANICAL OR HYDRAULIC SYSTEM FAULT – IF NO FAULT IS FOUND, AN INTERMITTENT FAULT CONDITION IN A REAR WHEEL SPEED SENSOR CIRCUIT IS INDICATED. SEE NOTE ON INTERMITTENTS

AN EXPLANATION OF EACH TEST PROCEDURE, REFERENCED BY A CIRCLED NUMBER

WHEN ALL DIAGNOSIS AND REPAIRS ARE COMPLETE, CLEAR CODES AND VERIFY OPERATION.

Fig. 31 Code A028, rear hold solenoid energized too long

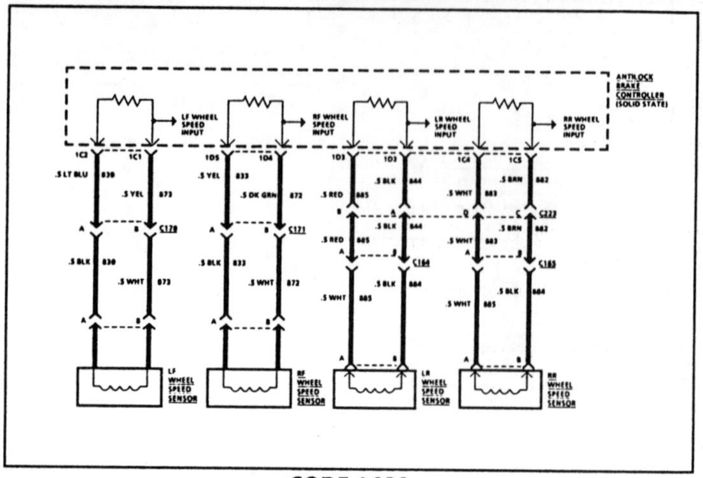

CODE A030

Both front, or one front and one rear wheel Speed Sensor open or shorted to ground

The Antilock Brake Controller monitors the Wheel Speed Sensors. If one or more wheel speed inputs are not receiving data from the Wheel Speed Sensor, the Controller can not detect wheel lockup. When only one speed input is not receiving data, Code A044, A045, A046 or A047 will be set. This Code (A030) is set when more than one speed input is faulty.

CODE A030 will set when the Antilock Brake Controller senses that either both Front Wheel Speed Sensors or both a Front and a Rear Wheel Speed Sensor input are 0 mph for greater than 20 seconds.

TEST DESCRIPTION: The following provides an explanation of the procedures being followed in the facing trouble tree.

① By observing all four Wheel Speed Sensor inputs (while driving vehicle), you can determine which Wheel Speed Sensor circuits are faulty (no wheel speed observed). Faults can be diagnosed by refering to Codes A044, A045, A046, and A047 (right front, left front, right rear or left rear wheel speed = 0). In the case where an intermittent fault condition is indicated refer to introduction.

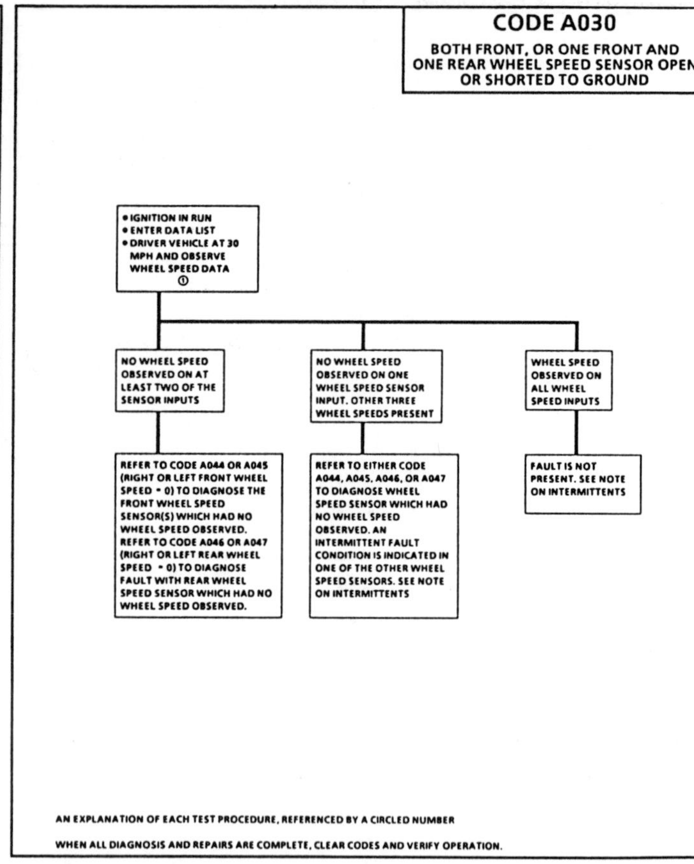

Fig. 32 Code A030, both front, or one front and one rear wheel speed sensor open or shorted to ground

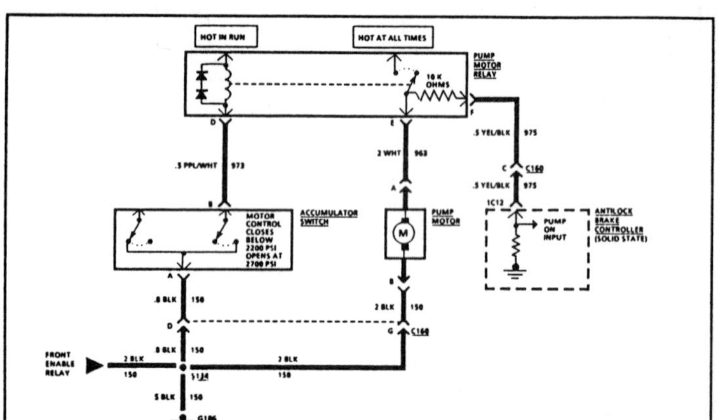

CODE A031

Open Pump Motor feedback circuit

Whenever the Pump Motor Relay contacts are open, the Pump On Input is grounded through the Pump Motor Relay and the Pump Motor. The Pump On Input will have voltage applied to it when the Pump Motor Relay Contacts are closed (the Pump Motor also has voltage applied to it). If the Pump On Input does not sense ground or the proper voltage, the Controller will set Code A031.

CODE A031 will set when the Antilock Brake Controller detects an open condition in the Pump Motor circuit between the Pump Motor Relay and ground or detects an open condition in the Pump On Input circuit between the Pump Motor Relay and the Antilock Brake Controller.

TEST DESCRIPTION: The following provides an explanation of the procedures being followed in the facing trouble tree.

① If Code A036 is also set, the problem is isolated to poor connections at either the Pump Motor Relay or Connector C160.

② Code A035, when set with Code A031, indicates that the Pump Motor circuit is open.

③ This step determines if the Pump Motor Feedback circuit is open.

④ If the Pump Motor is always on, there must be an open in the Pump Motor circuit which is preventing the Pump Motor from operating.

⑤ Determines if the open condition exists in CKT 150 between the Pump Motor and ground G106.

⑥ Battery voltage at terminal 1C12 indicates a problem with the connection at the Controller or the Controller itself.

⑦ Detects if circuit to Pump Motor is good. This would indicate a faultly Pump Motor.

⑧ Determines that the open condition is internal to the Pump Motor Relay if the resistance across terminals E and F of the Pump Motor Relay is infinite.

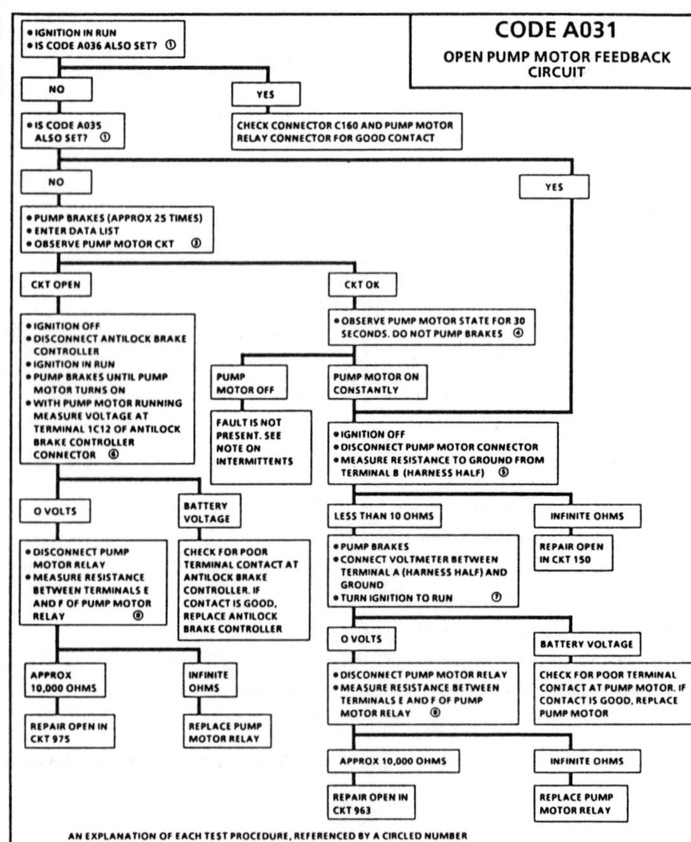

Fig. 33 Code A031, open pump motor feedback circuit

CODE A032

Open Brake Switch or hydraulic leak

CODE A032	will set when the Antilock Brake Controller receives three pump on inputs without receiving a brake switch on input. This allows a brake switch fault to be detected without an Antilock Braking condition present. Refer to Code A056 for diagnosis.

Fig. 34 Code A032, Open brake switch or hydraulic leak

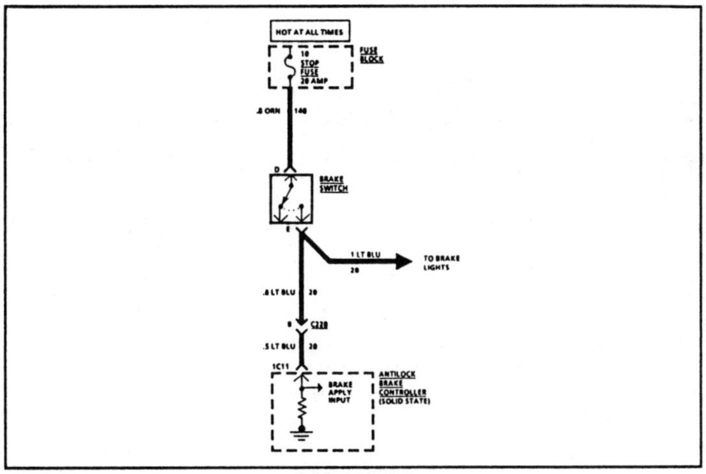

CODE A033

Brake Switch open

Voltage is applied at all times to the Brake Switch. When the Brake Switch is closed, voltage is applied to the Brake Lights and terminal 1C11 of the Antilock Brake Controller. If the Brake Apply Input receives voltage, the Controller determines that the brakes are being applied.

CODE A033 will set when all of the following conditions exist:

- The Antilock Brake Controller senses no voltage at terminal 1C11 (brake apply input is lo).
- The Antilock Brake Controller senses that two wheels (one of which is in the front) are decelerating at a rate greater than normally possible without using the brakes (vehicle is slowing down fast enough that the controller knows the brakes are being applied).

TEST DESCRIPTION: The following provides an explanation of the procedures being followed in the facing trouble tree.

① This determines whether an open condition currently exists under normal conditions.

② Checks if open condition is caused by an open Stop Fuse.

③ Isolates a short to ground in the 140 CKT.

④ If test lamp lights, a short to ground exists.

⑤ Isolates the short to ground to external circuits or to the Antilock Brake Controller.

⑥ Determines if an open condition exists in the CKT 140 or if the Brake Switch is defective.

⑦ An intermittent open Brake Switch may be possible. This step checks for this condition.

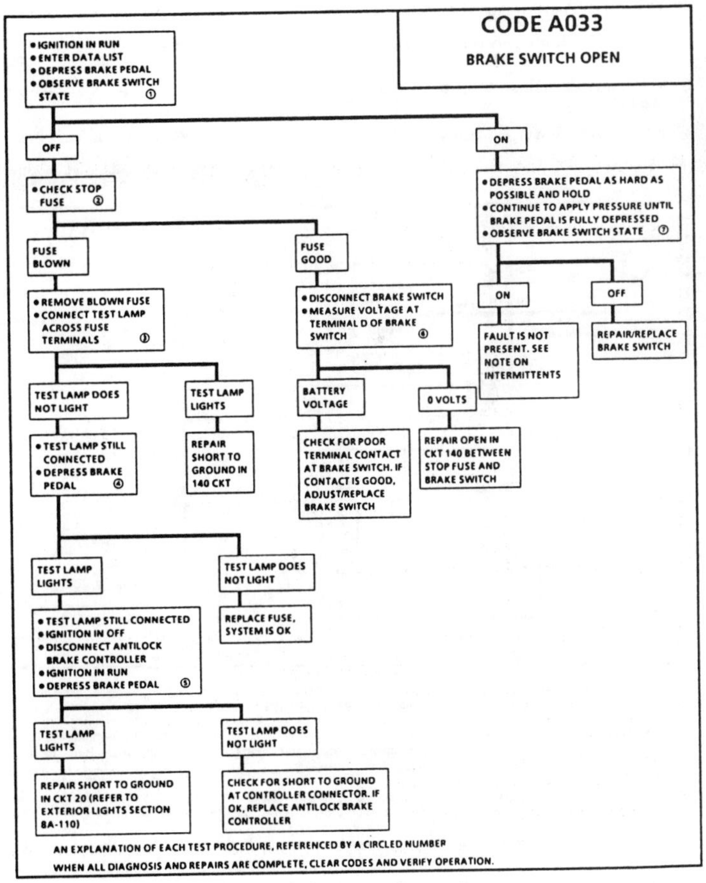

Fig. 35 Code A033, brake switch is open

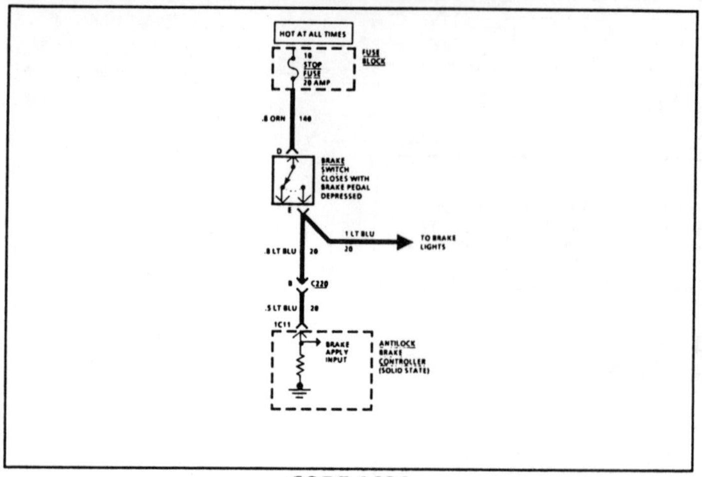

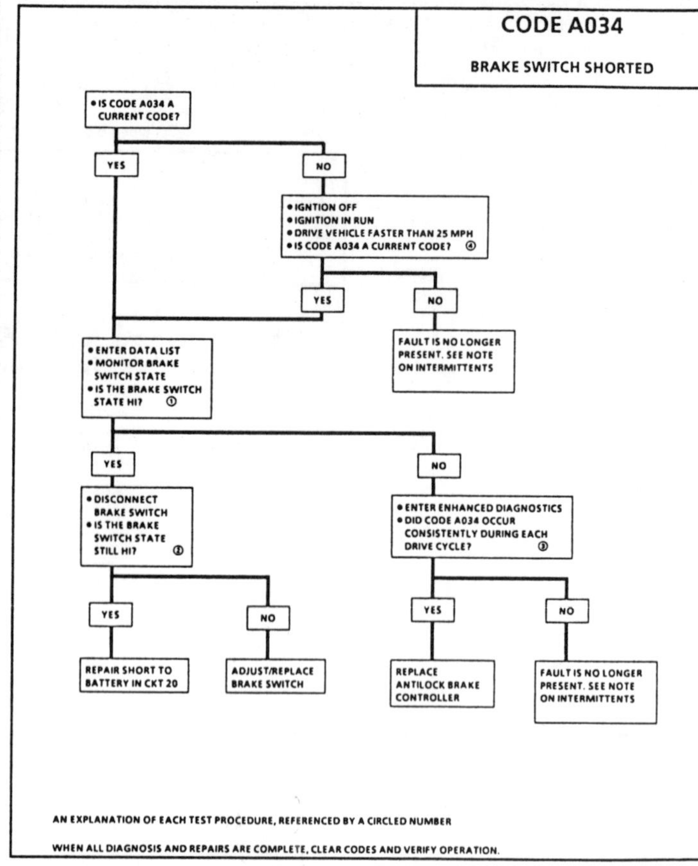

CODE A034

Brake Switch shorted

When the Brake Pedal is depressed, the Brake Switch closes. This applies voltage from the Stop Fuse to the Antilock Brake Controller and the brake lights. Before the Controller will initiate an Antilock Brake Stop, it must first sense voltage at the Brake Apply Input.

CODE A034 will set when all of the following conditions exist:

- The Antilock Brake Controller senses battery voltage at terminal 1C11 for a complete ignition cycle in which the vehicle speed surpasses 25 mph.
- The Antilock Brake Controller senses Battery voltage at terminal 1C11 during any following ignition cycle in which the vehicle speed surpasses 25 mph.

TEST DESCRIPTION: The following provides an explanation of the procedures being followed in the facing trouble tree.

① Confirms that a short exists in the circuit (brake lights on).

② Determines whether a short to Battery exists in the CKT 20; or whether the Brake Switch is shorted or maladjusted.

③ If Code A034 is a consistent failure, but CKTS and Brake Switch are good, the Antilock Brake Controller must have an internal fault.

④ This step verifies that the code is not about to become a current code by passing the second condition (see above).

Fig. 36 Code A034, brake switch shorted

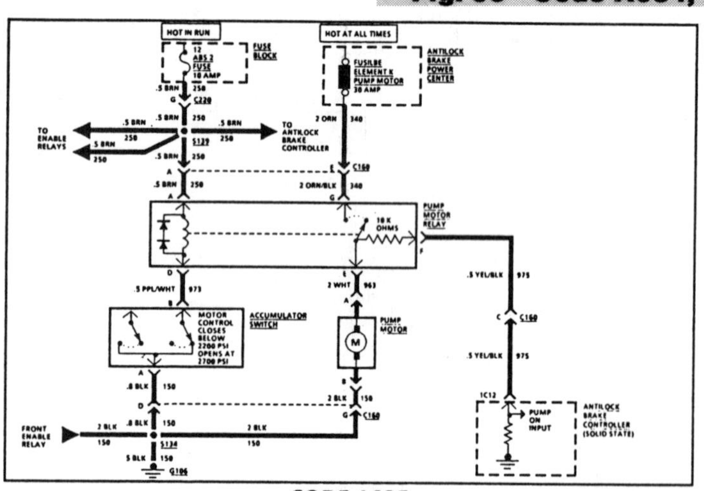

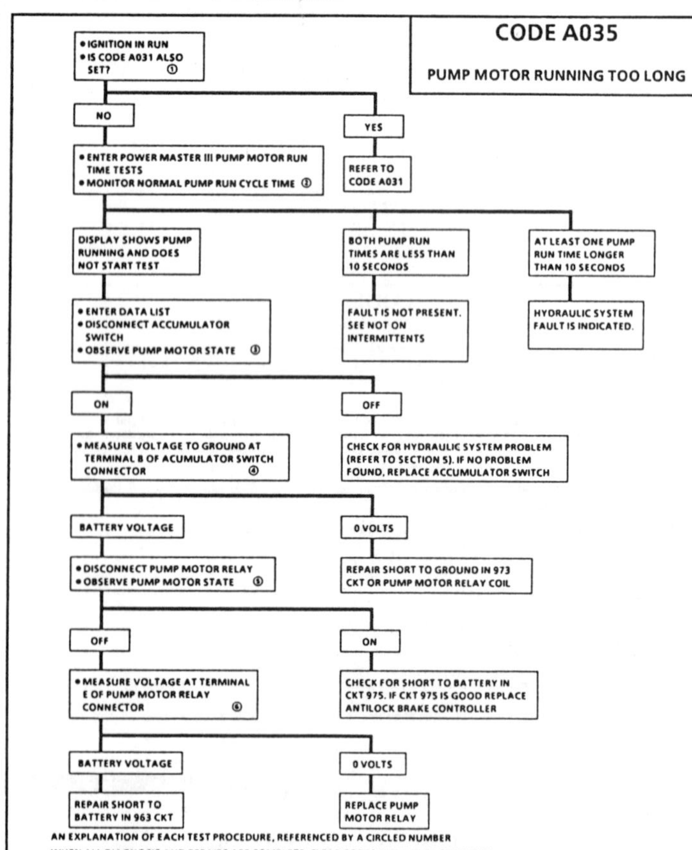

CODE A035

Pump Motor running too long

The Accumulator Switch Motor Control closes when the pressure drops below 2200 PSI. This grounds the Pump Motor Relay Coil causing voltage to be applied to the Pump Motor and Pump On Input. Whenever the Pump Motor is on, the Pump On Input will have voltage applied to it.

CODE A035 will set when the Antilock Brake Controller senses voltage at Terminal 1C12 "Pump on input" for more than 3 minutes.

TEST DESCRIPTION: The following provides an explanation of the procedures being followed in the facing trouble tree.

① When Code A035 and A031 are both set, the problem is isolated to the Pump Motor Circuit.

② This step determines if the Pump Motor is running constantly (electrical fault), running longer than normal (hydraulic fault), or running too long intermittently.

③ It can be determined if the accumulator switch is causing the fault by disconnecting the accumulator switch and observing the Pump Motor State. OFF indicates a defective accumulator switch or hydraulic failure.

④ Determines if a short to ground exists in 973 wire.

⑤ This step detects if a short to Battery exists by isolating the 975 wire and Antilock Brake Controller from the Pump Motor Circuit.

⑥ Determines if a short to ground exists in the 963 wire, or whether the Pump Motor Relay is defective.

Fig. 37 Code A035, pump motor running too long

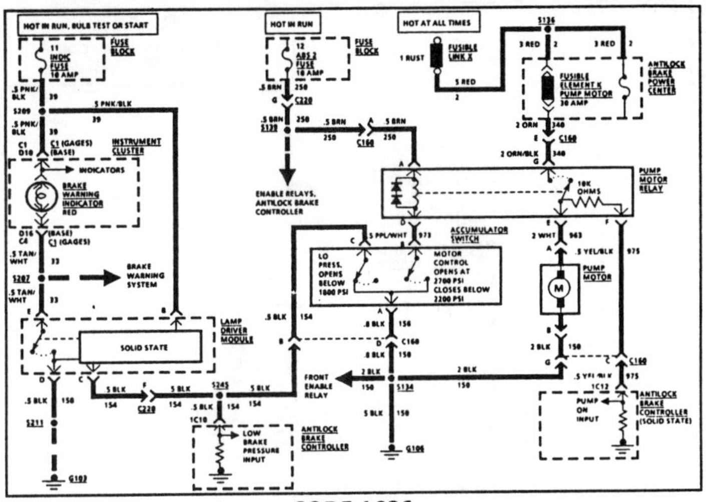

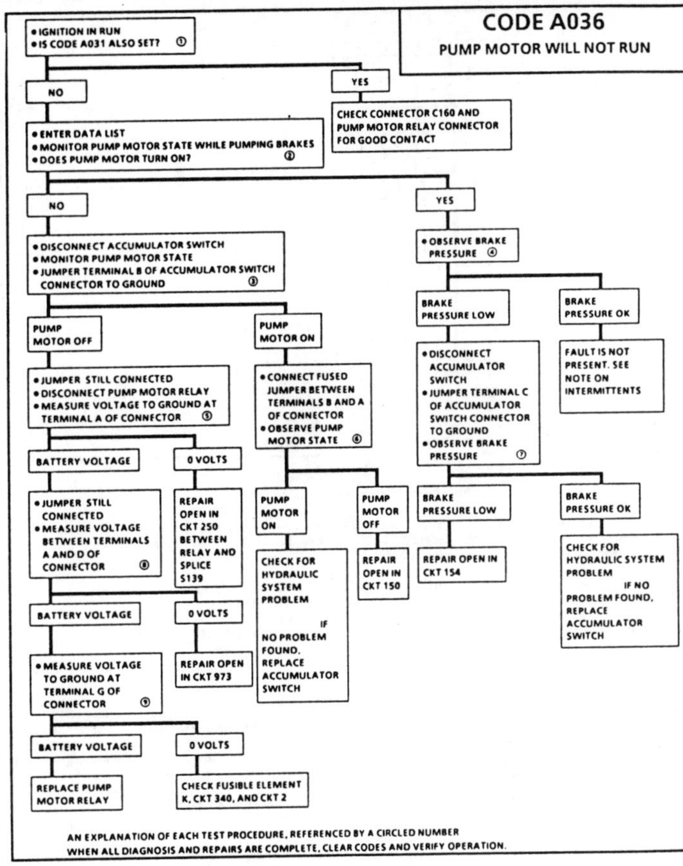

CODE A036

Pump Motor will not run

The Pump On Input will have voltage applied to it when the Pump Motor Relay contacts are closed. If the Brake Pressure drops below 1800 PSI, the Accumulator Switch opens. This removes ground from the Lamp Driver Module. The Lamp Driver Module then closes the switch to turn on the Brake Warning Indicator, and also applies voltage to the Low Brake Pressure Input of the Controller.

CODE A036 will set when all of the following conditions exist:

- The Antilock Brake Controller does not sense voltage at terminal 1C12 (Pump on input).
- The Antilock Brake Controller senses voltage at terminal 1C10 (Brake Warning Indicator is on).

TEST DESCRIPTION: The following provides an explanation of the procedures being followed in the facing trouble tree.

① If code A031 is also set, the problem is isolated to poor connections at either the Pump Motor Relay or Connector C160.

② By observing if the Pump Motor operates, the fault is isolated to either the Low Pressure Circuit or Pump Motor Relay Circuit.

③ Determines if fault is in Pump Motor Relay Circuit or Accumulator Switch Circuit.

④ This step checks if there is an actual fault present.

⑤ Battery voltage at terminal A indicates a good Relay Coil power feed.

⑥ If Pump Motor operates, the fault is either a hydraulic system failure or faulty Accumulator Switch.

⑦ Checks for an open ground circuit to Accumulator Switch.

⑧ Determines if 973 CKT is open.

⑨ At this point, a good Relay Contact power feed indicates a faulty Pump Motor Relay.

Fig. 38 Code A036, pump motor will not run

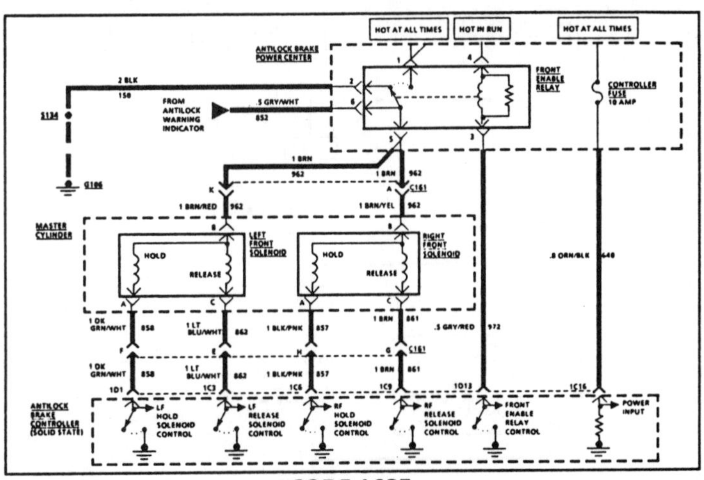

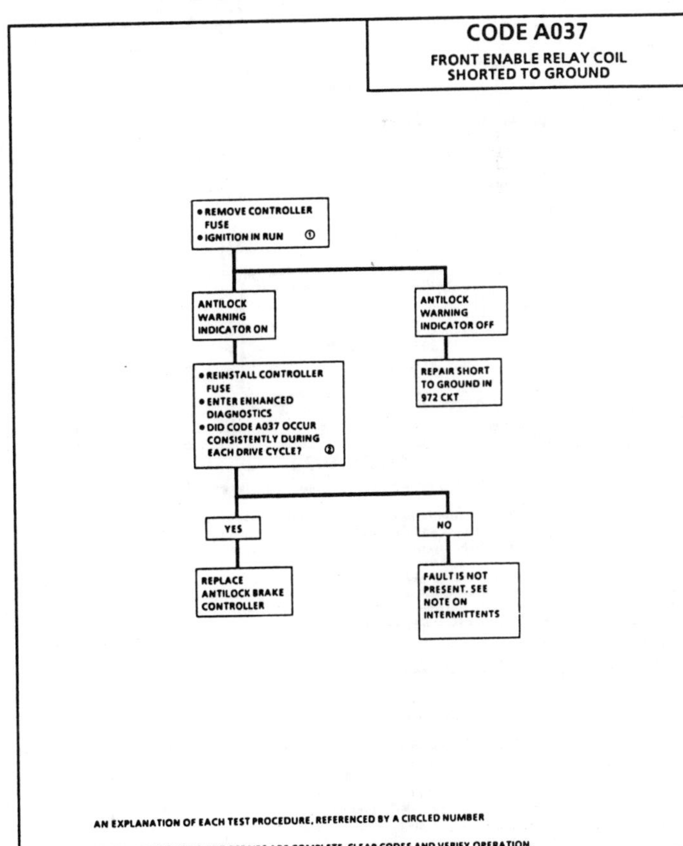

CODE A037

Front Enable Relay Coil shorted to ground

When the Front Enable Relay Control is open, the Front Enable Relay is deenergized. This prevents voltage from being applied to the Front Solenoids and Solenoid Controls. If the Front Enable Relay is deenergized and the Ignition is in RUN, voltage should be present at terminal 1D13.

CODE A037 will set during system initialization when all of the following conditions exist:

- The Front Enable Relay Control is open (Front Enable Relay deenergized).
- The Antilock Brake Controller senses no voltage at Terminal 1D13.
- The Antilock Brake Controller senses Battery voltage at Terminals 1D1, 1C3, 1C6 and 1C9.

TEST DESCRIPTION: The following provides an explanation of the procedures being followed in the facing trouble tree.

① By isolating the Front Enable Relay Coil from its ground and observing the Antilock Warning Indicator, it can be determined whether the Front Enable Relay has deenergized (Antilock Warning Indicator lights) which indicates a possible defective Antilock Brake Controller or whether the Front Enable Relay has remained energized (Antilock Warning Indicator does not light) which indicates a short to ground in the 972 wire.

② If Code A037 is a consistent failure, the Antilock Brake Controller has an internal fault.

Fig. 39 Code A037, front enable relay coil shorted to ground

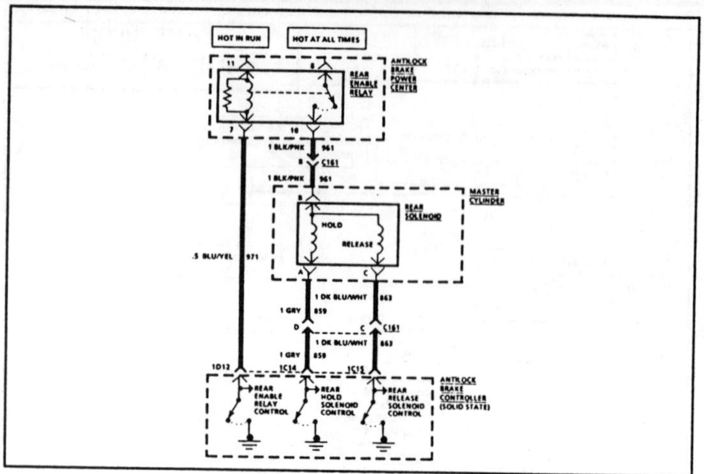

CODE A038

Rear Enable Relay Coil shorted to Ground

When the Rear Enable Relay Control is open, the Rear Enable Relay is deenergized. This prevents voltage from being applied to the Rear Solenoid and Solenoid controls. If the Rear Enable Relay is deenergized and the Ignition is in RUN, voltage should be present at terminal 1D12.

CODE A038 will set during system initialization when all the following conditions exist:

- The Rear Enable Relay control is open (Rear Enable Relay deenergized).
- The Antilock Brake Controller senses no voltage at Terminal 1D12.
- Antilock Brake Controller senses Battery voltage at Terminal 1C14 and 1C15.

TEST DESCRIPTION: The following provides an explanation of the procedures being followed in the facing trouble tree.

① By isolating the Rear Enable Relay Control from its ground and measuring the voltage at Terminal 1D12 of the Antilock Brake Controller connector it can be determined whether the Rear Enable Relay has deenergized (Battery voltage at Terminal 1D12) which indicates a possible defective Antilock Brake Controller or whether the Rear Enable Relay has remained energized (0 volts at Terminal 1D12) which indicates a short to ground in the 971 circuit.

② If Code A038 is a consistent failure, the Antilock Brake Controller has an internal fault.

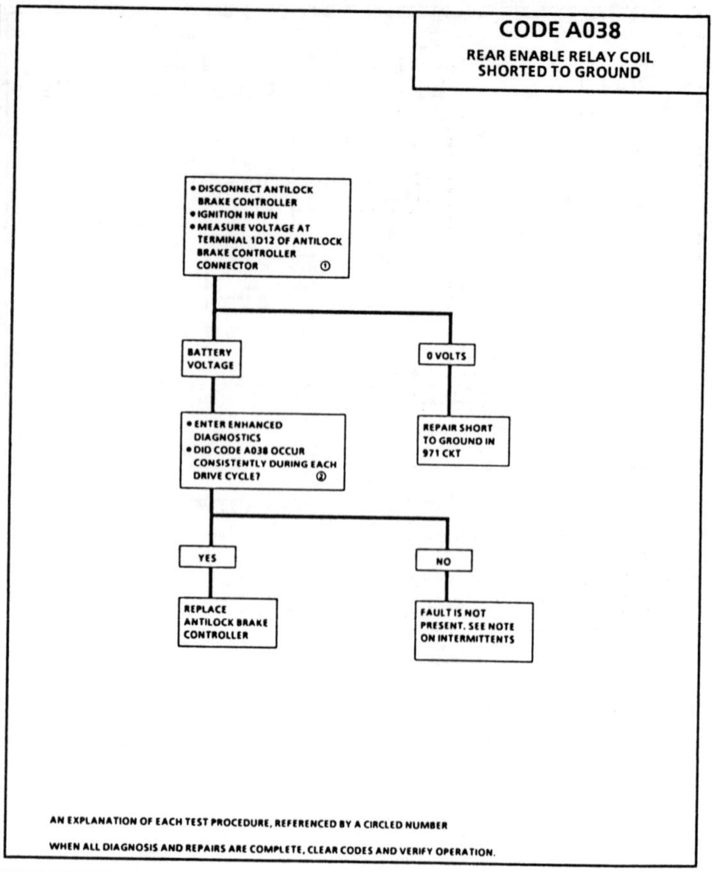

Fig. 40 Code A038, rear enable relay coil shorted to ground

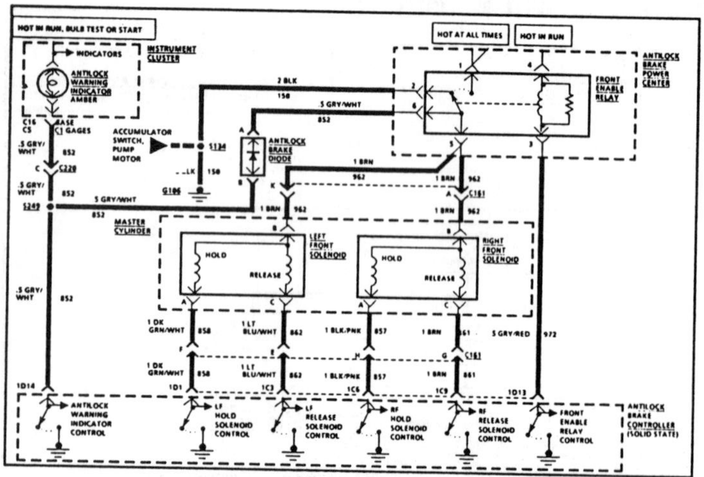

CODE A039

Front Enable Relay contacts shorted to Battery or relay ground open.

When the Front Enable Relay Control is open, the Front Enable Relay is deenergized. This prevents voltage from being applied to the Front Solenoids and Solenoid Controls. If the 150 CKT from the Front Enable Relay is open, Battery voltage from the Antilock Warning Indicator will be applied to the solenoids even when the Front Enable Relay is deenergized.

CODE A039 will set during system initialization when all of the following conditions exist:

- The Front Enable Relay Control is open (Front Enable Relay deenergized).
- The Antilock Brake Controller senses Battery voltage at Terminals 1D1, 1C3, 1C6 and 1C9 (Front Solenoid controls).

TEST DESCRIPTION: The following provides an explanation of the procedures being followed in the facing trouble tree.

② If Code A003 is also set, the ground CKT 150 must be open.

③ Determines if a short to Battery is present in the circuit or if the Antilock Brake Controller is possibly defective.

④ Isolates a short to Battery in CKT 962 or determines if the Front Enable relay is defective.

⑤ If Code A039 is a consistent failure, the Antilock Brake Controller has an internal fault.

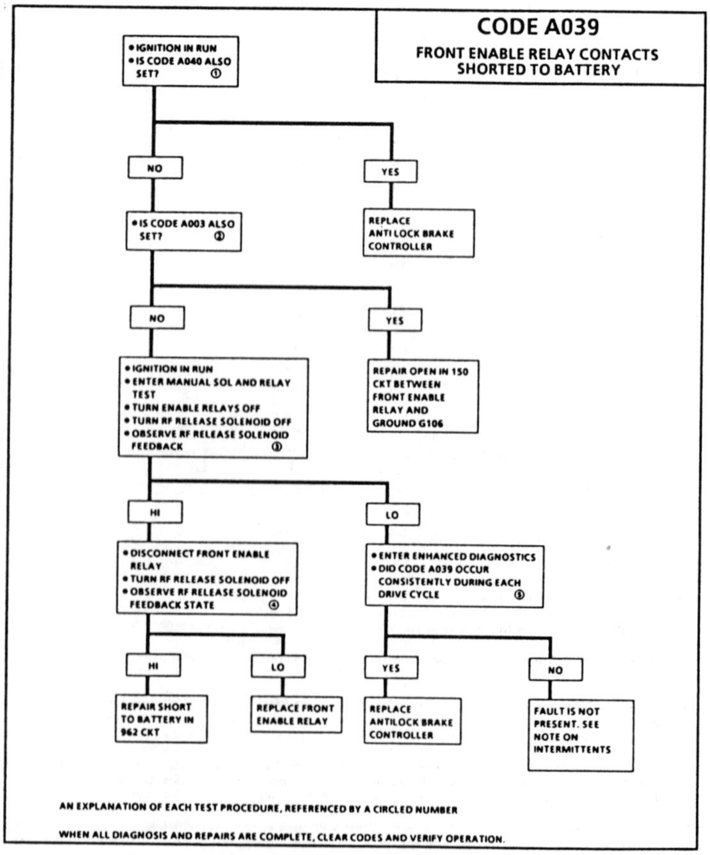

Fig. 41 Code A039, front enable relay contacts shorted to battery

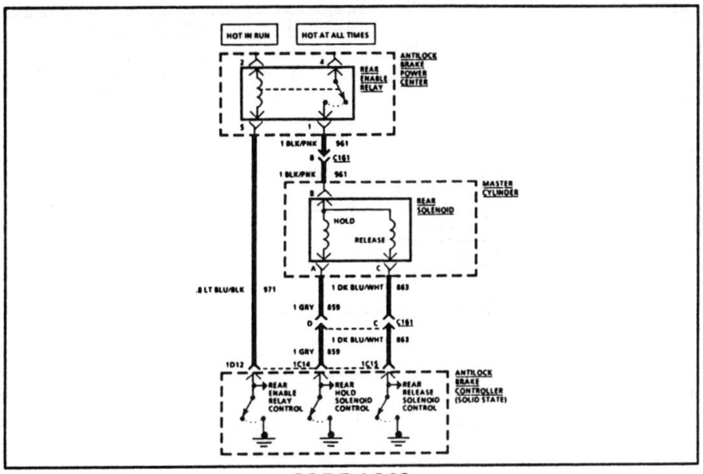

CODE A040

Rear Enable Relay contacts shorted to Battery

When the Rear Enable Relay Control is open, the Rear Enable Relay is deenergized. This prevents voltage from being applied to the Rear Solenoid and Solenoid Controls.

CODE A040will set during initialization when all the following conditions exist:

- The Rear Enable Relay Control is open (Rear Enable Relay deenergized).
- The Antilock Brake Controller senses Battery voltage at both terminals 1C14 and 1C15.
- The Antilock Brake Controller senses Battery voltage at terminal 1D12.

TEST DESCRIPTION: The following provides an explanation of the procedures being followed in the facing trouble tree.

① Determines if a short to Battery is present in the circuit or if the Antilock Brake Controller is possibly defective.

② Isolates a short to Battery in the CKT 961 or determines if the Rear Enable Relay is defective.

④ If Code A040 is a consistent failure, the Antilock Brake controller has an internal fault.

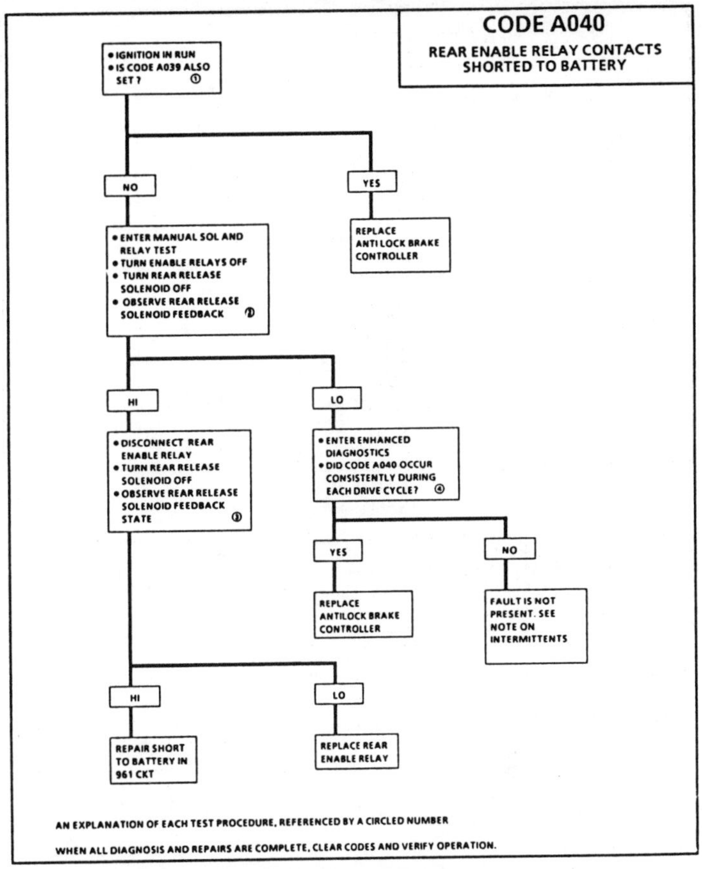

Fig. 42 Code A040, rear enable relay contacts shorted to battery

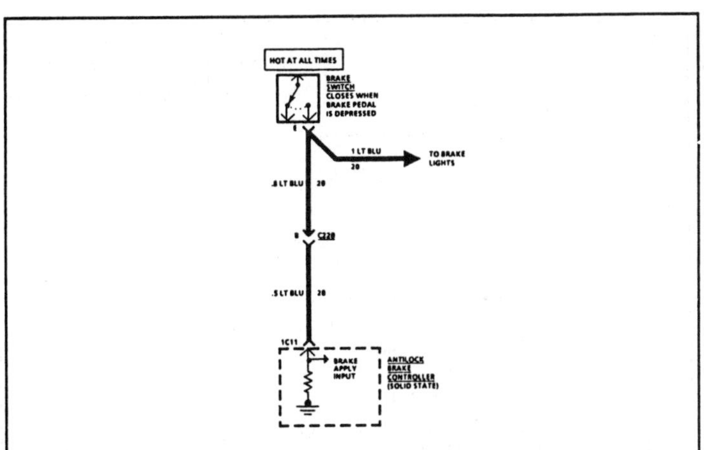

CODE A041

Brake Switch circuit open

The Antilock Brake Controller determines that the Brake Pedal is pressed when battery voltage is sensed at the Brake Apply Input. If battery voltage is not present, the Controller should sense ground through CKT 20, CKT 17, and the Brake Lights. It should be noted that this code will set if the Hazard Lights are turned on and the High Level Stop Light is open or burned out.

CODE A041 will set when all the following conditions exist:

- Brake Switch is open (Battery voltage not sensed at Brake Apply Input)
- Antilock Brake Controller does not sense ground at Brake Apply Input (through CKT 20 and CKT 17)

TEST DESCRIPTION: The following provides an explanation of the procedures being followed in the facing trouble tree.

① Determines if the open condition exists in CKT 20 between the Brake Switch and the brake lights (High Level Stop Light did not light with brake pedal depressed) or if the open condition exists in CKT 20 between the Brake Switch and the Antilock Brake Controller (High Level Stop Light did light with Brake Pedal depressed).

② By observing the Brake Switch State when it is known that the Brake Switch is closed it can be determined if the Antilock Brake Controller is sensing the proper input (a proper input at this point would indicate an intermittent fault condition).

③ Isolates the open condition to CKT 20 between the Brake Switch and the Antilock Brake Controller if 0 volts was measured at Terminal 1C11 of Antilock Brake Controller. If Battery voltage was measured at Terminal 1C11 it indicates that the Antilock Brake Controller is defective.

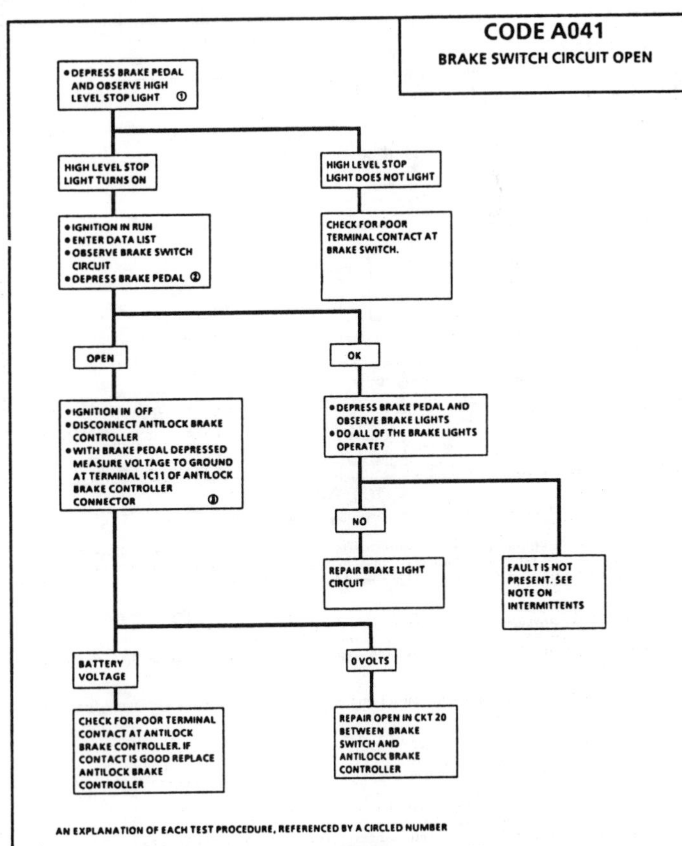

Fig. 43 Code A041, brake switch circuit open

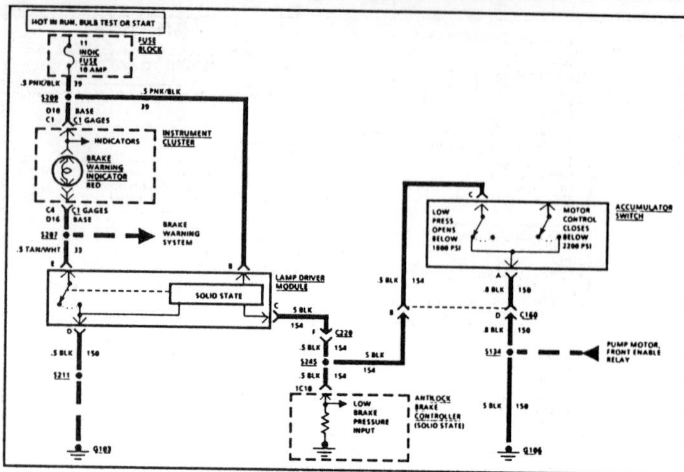

CODE A042

Low Brake Pressure Circuit open

When the Low Pressure Switch in the Accumulator Switch is closed (pressure greater than approximately 1800 PSI), the Antilock Brake Controller senses ground at the Low Brake Pressure Input. If the Low Pressure Switch is open (pressure less than 1800 PSI) the Lamp Driver Module loses ground at terminal C. The Lamp Driver Module turns on the Brake Warning Indicator and applies voltage to the Low Brake Pressure Input. If the Controller does not sense ground or battery voltage at the Low Brake Pressure Input, it will set Code A042.

CODE A042 will set when the Antilock Brake Controller senses that an open condition exists in the 154 CKT between the Controller and the Lamp Driver Module.

TEST DESCRIPTION: The following provides an explanation of the procedures being followed in the facing trouble tree.

① If the Brake Warning Indicator lights constantly, there must be an open in the 154 CKT between the Lamp Driver Module and Splice S245.

② Determines if open condition is currently present.

③ Isolates the open condition to CKT 154 between splice S245 and the Antilock Brake Controller or determines a defective Antilock Brake Controller.

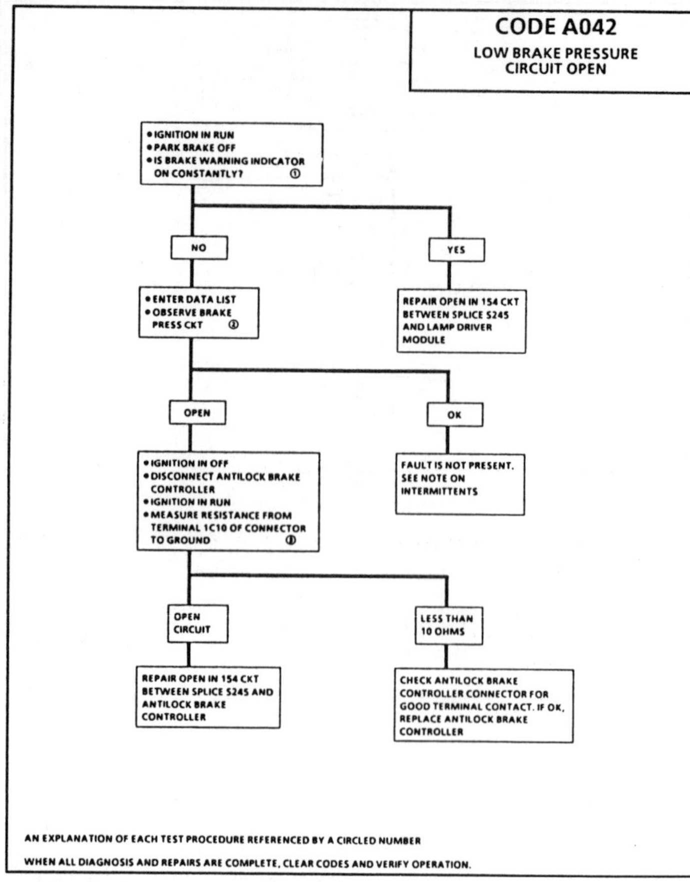

Fig. 44 Code A042, low brake pressure circuit open

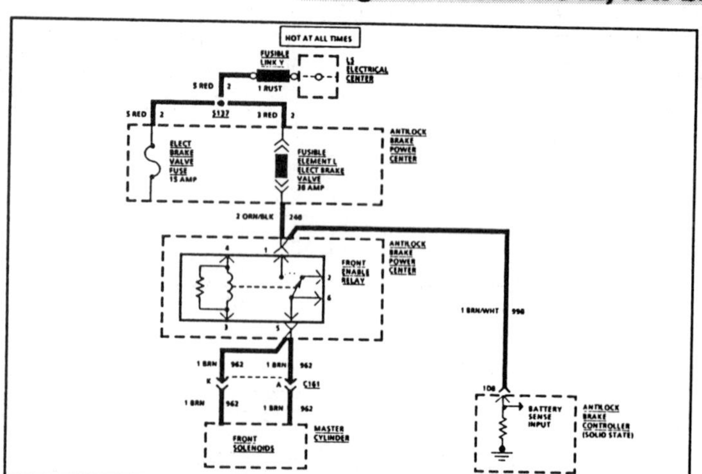

CODE A043

System voltage is low

Voltage from Fusible Element L is applied at all times to the Front Enable Relay terminal 1 and the Antilock Brake Controller Battery Sense Input. Since the Antilock Brake Controller needs greater than 9.7 volts to properly operate, Code A043 is set when the applied voltage is less than 9.7 volts.

CODE A043 will set when the voltage being supplied to terminal 1D8 of the Antilock Brake Controller is below 9.7 volts and vehicle speed has exceeded 10 mph.

TEST DESCRIPTION: The following provides an explanation of the procedures being followed in the facing trouble tree.

① Determines if the Battery or charging system is faulty, or isolates the problem to the Antilock Brake system.

② If Code A005 is not set, the problem is in the 990 CKT.

③ Checks for short to ground in 240 CKT and 990 CKT.

④ Checks for short to ground in 962 CKTS.

⑤ Determines if open is present before or after Fusible Element L.

⑥ Indicates if open is in 990 CKT or possibly internal to Controller.

⑦ Determines if fault is a constant failure internal to Controller or an intermittent.

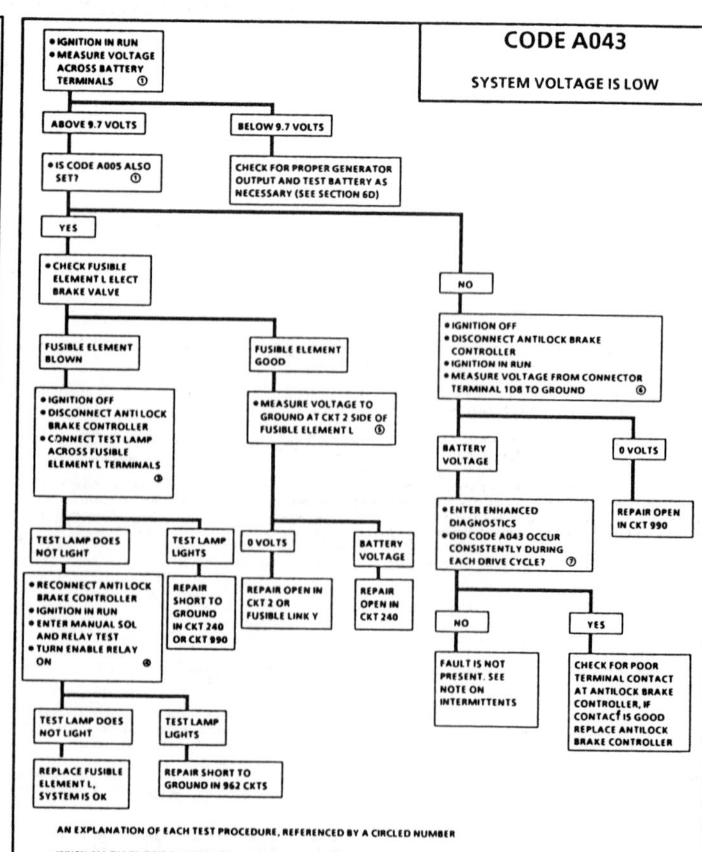

Fig. 45 Code A043, system voltage is low

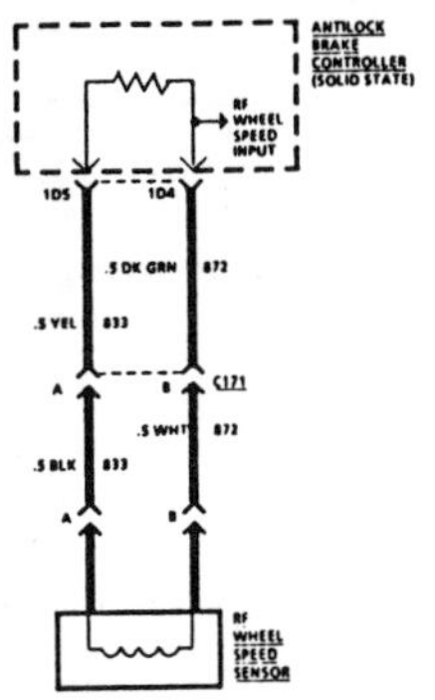

CODE A044

Right Front Wheel Speed = 0

The Wheel Speed Sensor generates a signal that indicates the speed of the wheel. Voltage pulses are produced as the magnetic teeth pass a coil. The frequency of this AC voltage is used by the Controller to determine how fast the wheel is turning. By comparing this wheel speed to the other wheel speeds, the Controller can detect if wheel lock-up is about to occur.

CODE A044 will set when all of the following conditions exist:

- The Antilock Brake Controller Senses the right front wheel speed to be 0 mph.
- The Antilock Brake Controller senses that the other three wheel speeds are greater than 5 mph and are operating correctly.

TEST DESCRIPTION: The following provides an explanation of the procedures being followed in the following trouble tree.

① By observing the Right Front Wheel speed input, you can verify what type of input the controller is sensing. An unsteady or erratic wheel speed input indicates that a fault is present in the Wheel Speed Sensor or its related circuitry. If a steady wheel speed is observed, an intermittent fault condition is indicated.

② Indicates a defective Wheel Speed Sensor by isolating a short to ground internal to the sensor.

③ If resistance measured across the Wheel Speed Sensor is infinite, the fault is due to an open condition internal to the Speed Sensor.

④ Determines a short to Battery is present in CKT 833 if Battery voltage is measured at Terminal A of the harness side of the Wheel Speed Sensor Connector

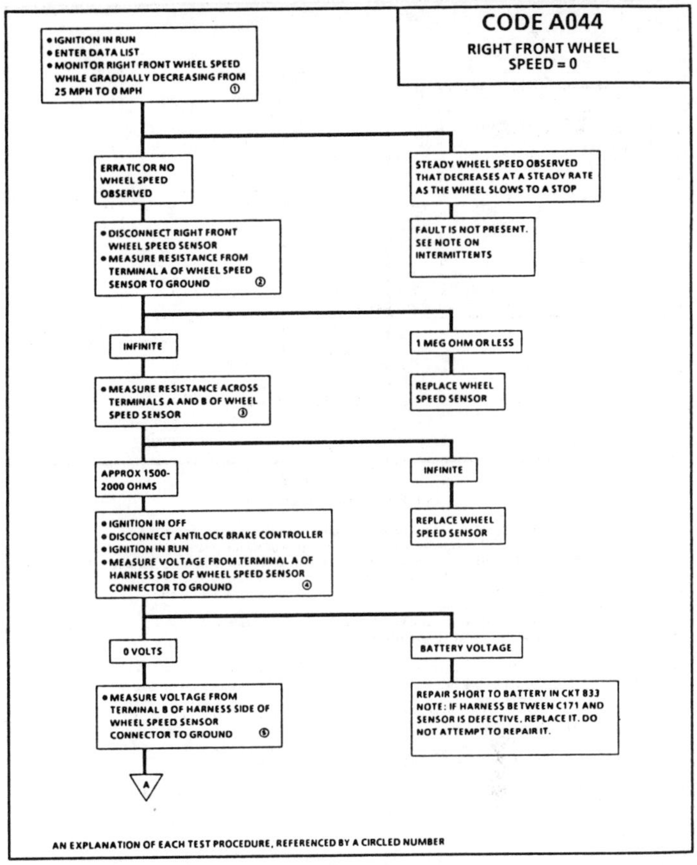

Fig. 46 Code A044, right front wheel speed at 0 mph (Part 1 of 2)

⑤ Determines a short to Battery is present in CKT 872 if Battery voltage is measured at Terminal B of the harness side of the Wheel Speed Sensor Connector.

⑥ With Terminal A of CKT 833 connected to ground, the resistance from Terminal 1D5 to ground should be zero. If infinite resistance is measured, an open condition is indicated in CKT 833.

⑦ With Terminal B of CKT 872 connected to ground, the resistance from Terminal 1D4 to ground should be zero. If infinite resistance is measured, an open condition is indicated in CKT 872.

⑧ At this point, all the connecting circuitry has tested good. The fault may be due to a poorly mated connector at the Wheel Speed Sensor or the Antilock Brake Controller. Other possible causes could be due to damage to the Speed Sensor or Speed Ring. Check for proper mounting,or poor alignment. If everything above tests good, replace the Antilock Brake Controller.

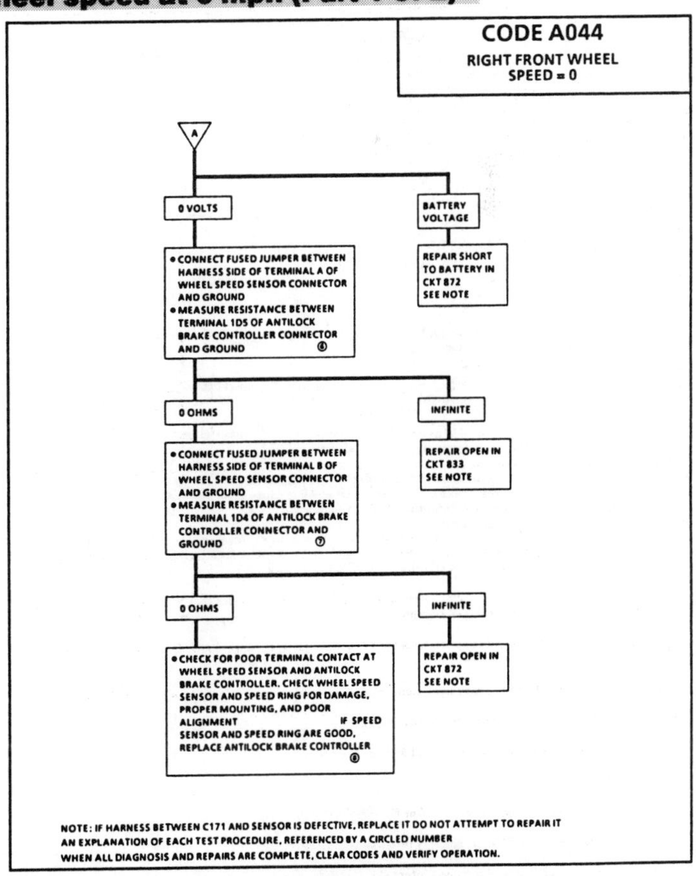

Fig. 46 Code A044, right front wheel speed at 0 mph (Part 2 of 2)

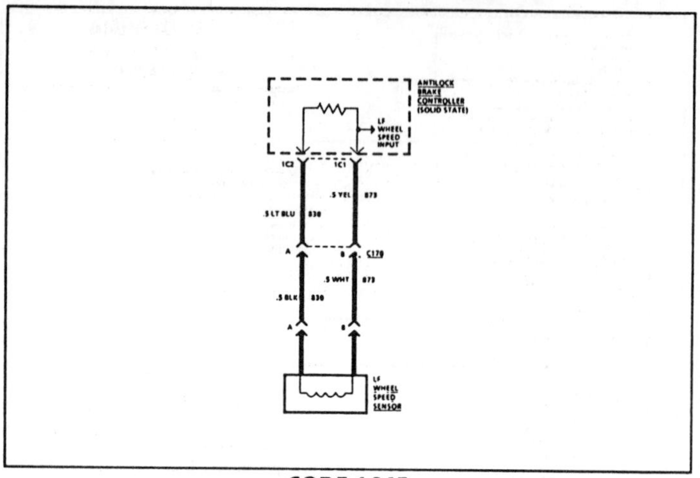

CODE A045

Left Front Wheel Speed = 0

The Wheel Speed Sensor generates a signal that indicates the speed of the wheel. Voltage pulses are produced as the magnetic teeth pass a coil. The frequency of this AC voltage is used by the Controller to determine how fast the wheel is turning. By comparing this wheel speed to the other wheel speeds, the Controller can detect if wheel lock-up is about to occur.

CODE A045 will set when all of the following conditions exist:

- The Antilock Brake Controller Senses the left front wheel speed to be 0 mph.
- The Antilock Brake Controller senses that the other three wheel speeds are greater than 5 mph and are operating correctly.

TEST DESCRIPTION: The following provides an explanation of the procedures being followed in the following trouble tree.

① By observing the Left Front Wheel speed input, you can verify what type of input the controller is sensing. An unsteady or erratic wheel speed input indicates that a fault is present in the Wheel Speed Sensor or its related circuitry. If a steady wheel speed is observed, an intermittent fault condition is indicated.

② Indicates a defective Wheel Speed Sensor by isolating a short to ground internal to the sensor.

③ If resistance measured across the Wheel Speed Sensor is infinite, the fault is due to an open condition internal to the Speed Sensor.

④ Determines a short to Battery is present in CKT 830 if Battery voltage is measured at Terminal A of the harness side of the Wheel Speed Sensor Connector.

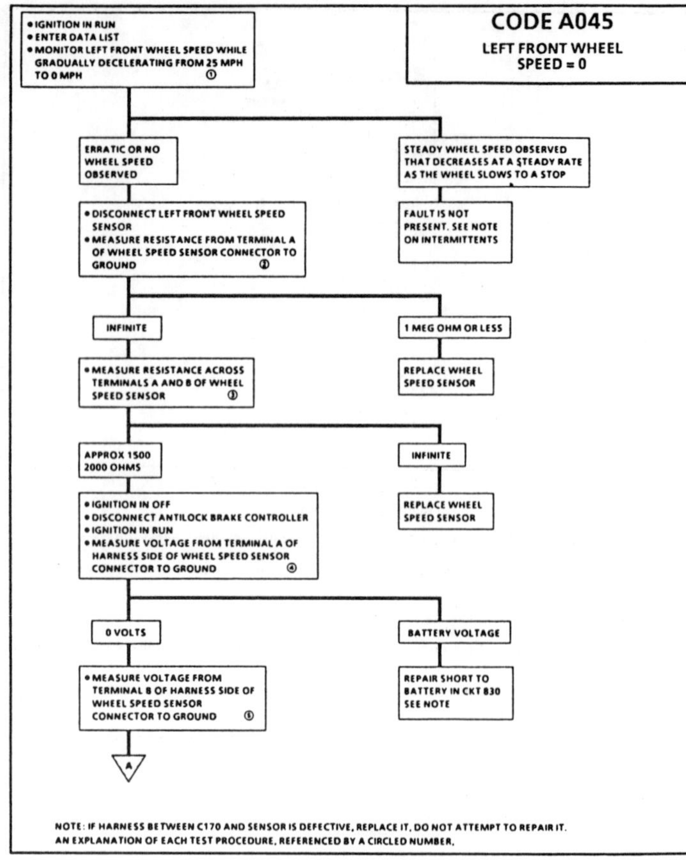

Fig. 47 Code A045, left front wheel speed at 0 mph (Part 1 of 2)

⑤ Determines a short to Battery is present in CKT 873 if Battery voltage is measured at Terminal B of the harness side of the Wheel Speed Sensor Connector.

⑥ With Terminal A of CKT 830 connected to ground, the resistance from Terminal 1C2 to ground should be zero. If infinite resistance is measured, an open condition is indicated in CKT 830.

⑦ With Terminal B of CKT 873 connected to ground, the resistance from Terminal 1C1 to ground should be zero. If infinite resistance is measured, an open condition is indicated in CKT 873.

⑧ At this point, all the connecting circuity has tested good. The fault may be due to a poorly mated connector at the Wheel Speed Sensor or the Antilock Brake Controller. Other possible cause could be due to damage to the Speed Sensor or Speed Ring. Check for proper mounting or poor alignment. If everthing above tests good, replace the Antilock Brake Contoller.

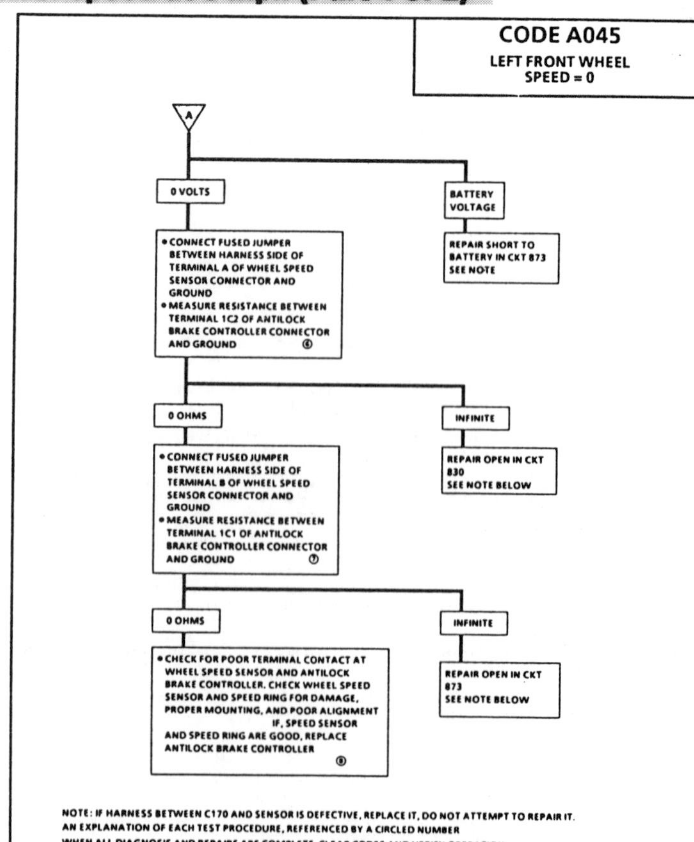

Fig. 47 Code A045, left front wheel speed at 0 mph (Part 2 of 2)

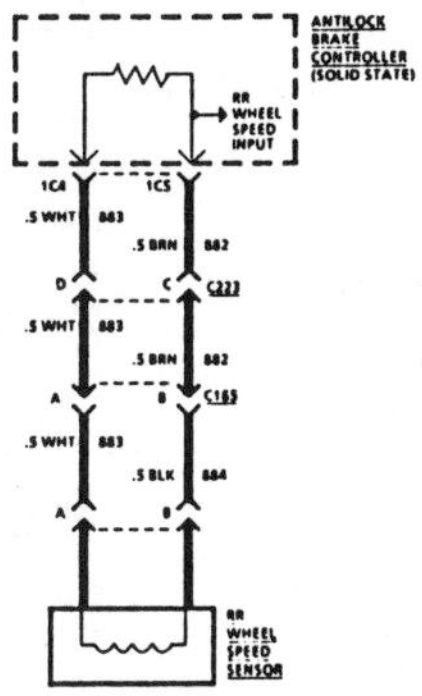

CODE A046

Right Rear Wheel Speed = 0

The Wheel Speed Sensor generates a signal that indicates the speed of the wheel. Voltage pulses are produced as the magnetic teeth pass the coil. The frequency of this AC voltage is used by the Controller to determine how fast the wheel is turning. By comparing this wheel speed to the other wheel speeds, the Controller can detect if wheel lock-up is about to occur.

CODE A046 will set when all of the following conditions exist:

- The Antilock Brake Controller Senses the right rear wheel speed to be 0 mph.
- The Antilock Brake Controller senses that the other three wheel speeds are greater than 5 mph and are operating correctly.

TEST DESCRIPTION: The following provides an explanation of the procedures being followed in the following trouble tree.

① Observation of erratic or no wheel speed input indicates the fault is present and is not intermittent.

②&③ Step 2 determines if a short to battery is the fault. If battery voltage is measured step 3 isolates the short to the harness or controller

④ If 9000 to 15000 ohms was measured, the possibility of an open or short in the circuit between the Antilock Brake Controller and Connector C165 has been eliminated.

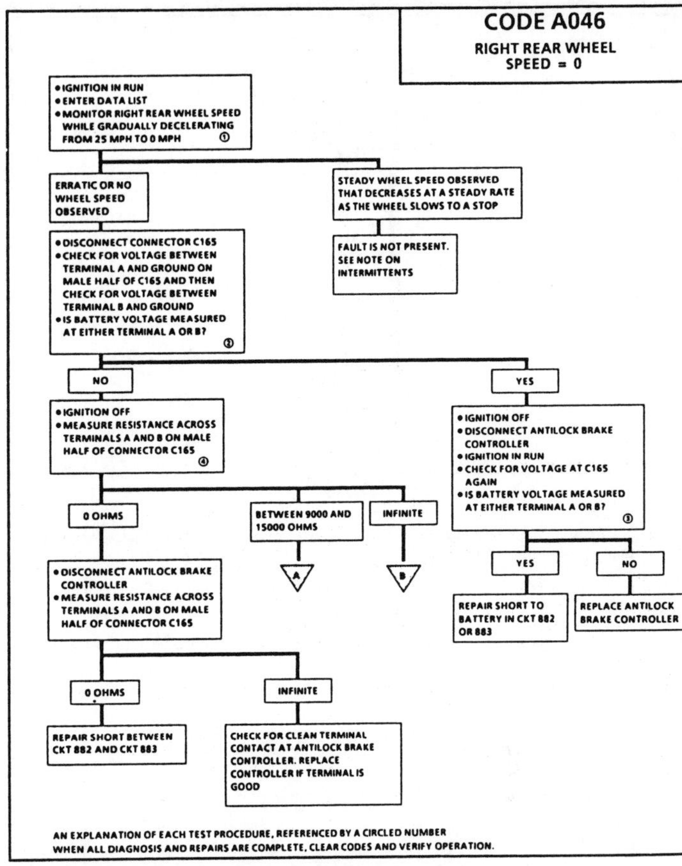

Fig. 48 Code A046, Right rear wheel speed at 0 mph (Part 1 of 3)

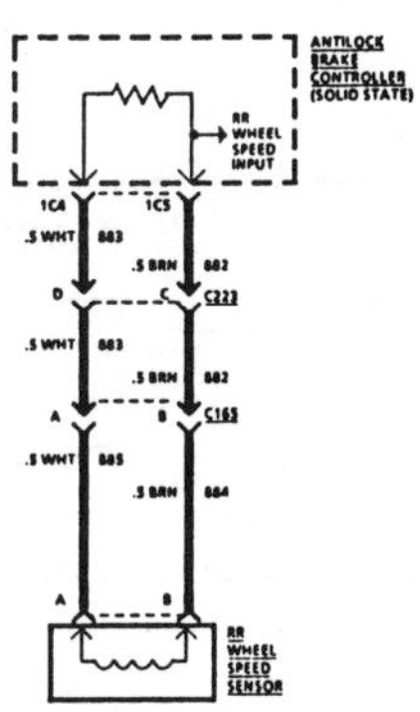

CODE A046

Right Rear Wheel Speed = 0

TEST DESCRIPTION: The following provides an explanation of the procedures being followed in the following trouble tree.

② If more than 10 ohms is measured the circuits between the controller and C165 are ok. This isolates the fault to be between the Wheel Speed Sensor and C165. Since this area of circuit is exposed to the elements and road debris, it should be carefully inspected for damage.

③ This procedure isolates the fault to a short to ground (1500 to 2500 ohms) or open (above 2500 ohms) or a shorted sensor circuit (below 1500 ohms)

④ & ⑤ Determines if the fault is in the harness or the Wheel Speed Sensor. If the harness between C165 and the sensor is defective, replace it. Do not attempt to repair the harness.

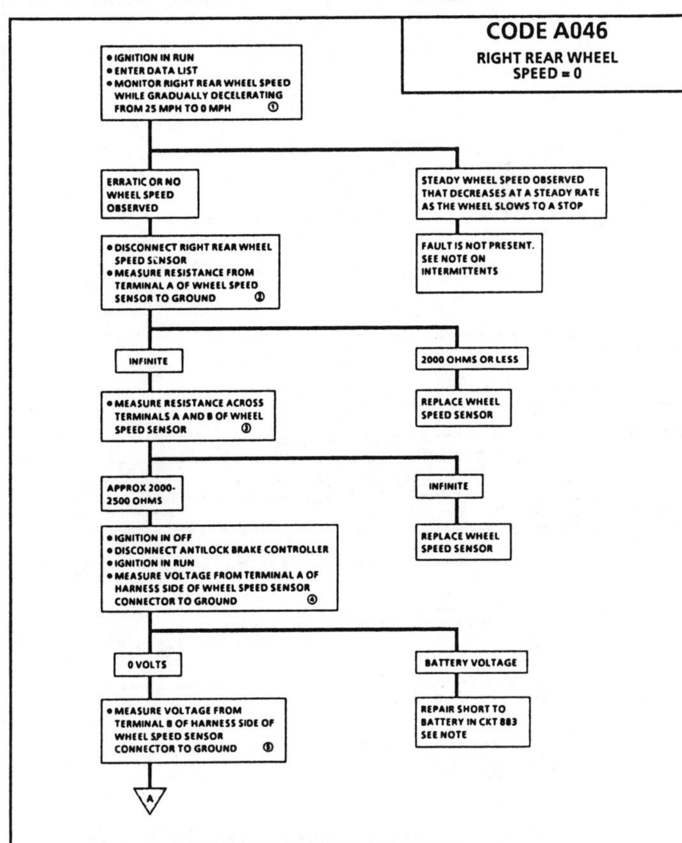

Fig. 48 Code A046, Right rear wheel speed at 0 mph (Part 2 of 3)

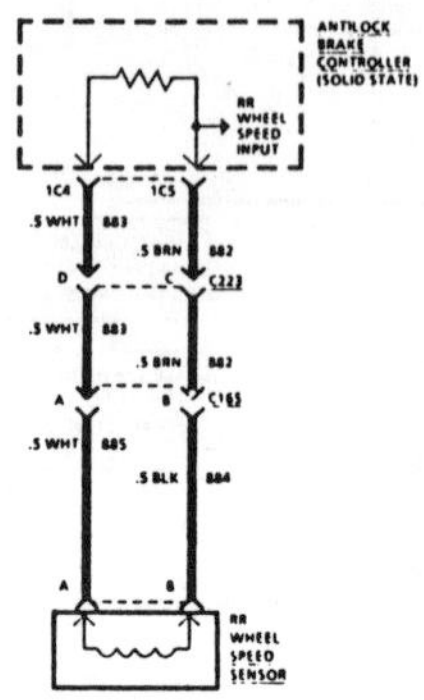

CODE A046

Right Rear Wheel Speed = 0

⑩ With CKT 883 connected to ground, the resistance from Terminal 1C4 to ground should be zero. If infinite resistance is measured, an open condition is indicated in CKT 883.

⑪ With CKT 882 connected to ground, the resistance from Terminal 1C5 to ground should be zero. If infinite resistance is measured, an open condition is indicated in CKT 882.

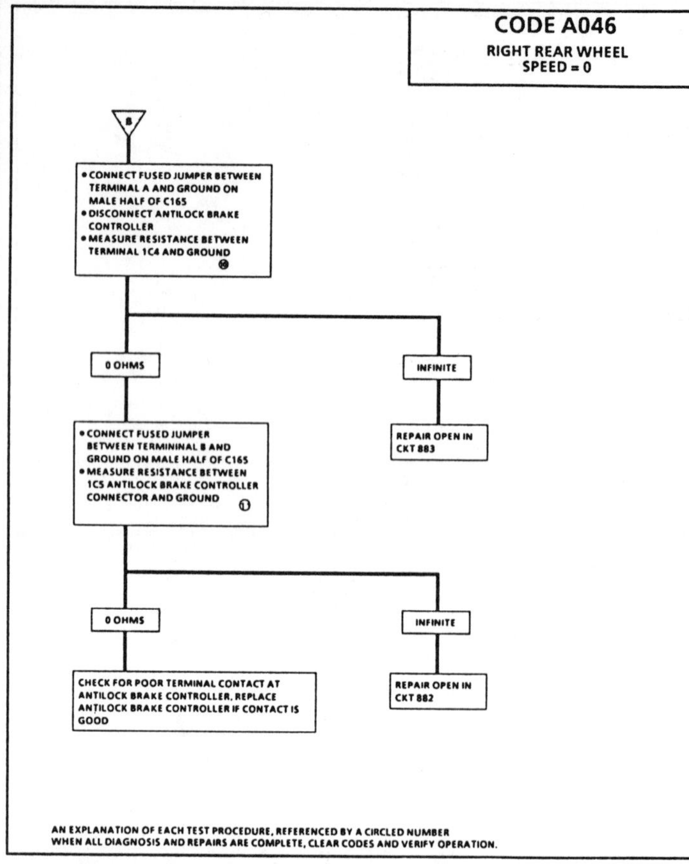

Fig. 48 Code A046, Right rear wheel speed at 0 mph (Part 3 of 3)

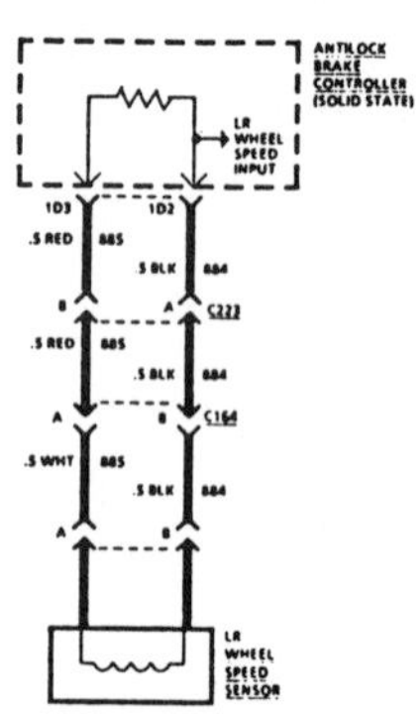

CODE A047

Left Rear Wheel Speed = 0

The Wheel Speed Sensor generates a signal that indicates the speed of the wheel. Voltage pulses are produced as the magnetic teeth pass a coil. The frequency of this A/C voltage is used by the Controller to determine how fast the wheel is turning. By comparing this wheel speed to the other wheel speeds, the Controller can detect if wheel lock-up is about to occur.

CODE A047 will set when all of the following conditions exist:

- The Antilock Brake Controller Senses the left rear wheel speed to be 0 mph.
- The Antilock Brake Controller senses that the other three wheel speeds are greater than 5 mph and are operating correctly.

TEST DESCRIPTION: The following provides an explanation of the procedures being followed in the following trouble tree.

① By observing the Left Rear Wheel speed input, you can verify what type of input the controller is sensing. An unsteady or erratic wheel speed input indicates that a fault is present in the Wheel Speed Sensor or it's related circuitry. If a steady wheel speed is observed, an intermittent fault condition is indicated.

② Indicates a defective Wheel Speed Sensor by isolating a short to ground internal to the sensor.

③ If resistance measured across the Wheel Speed Sensor is infinite, the fault is due to an open condition internal to the Speed Sensor.

④ Determines a short to Battery is present in CKT 885 if Battery voltage is measured at Terminal A of the harness side of the Wheel Speed Sensor Connector.

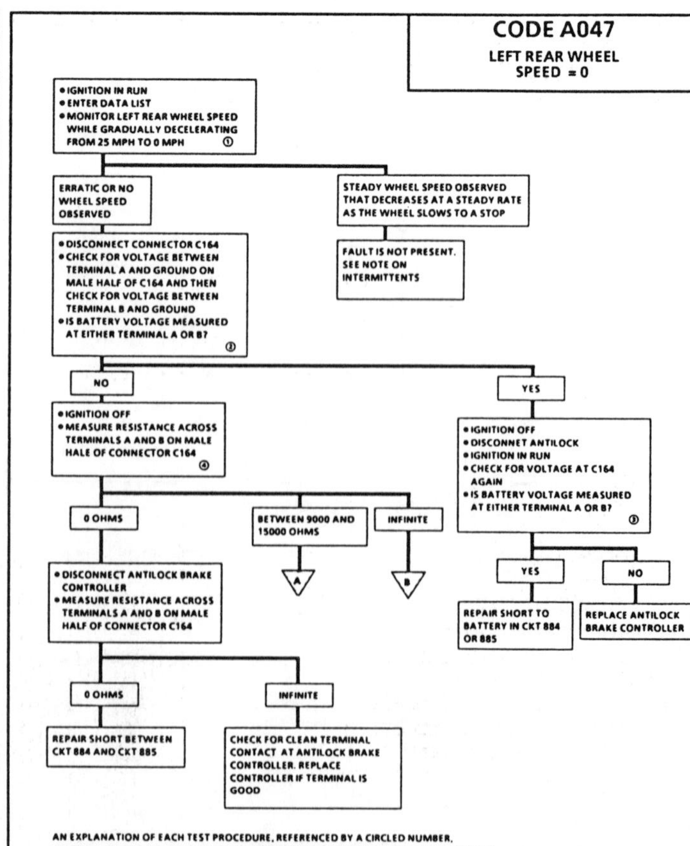

Fig. 49 Code A047, left rear wheel speed at 0 mph (Part 1 of 3)

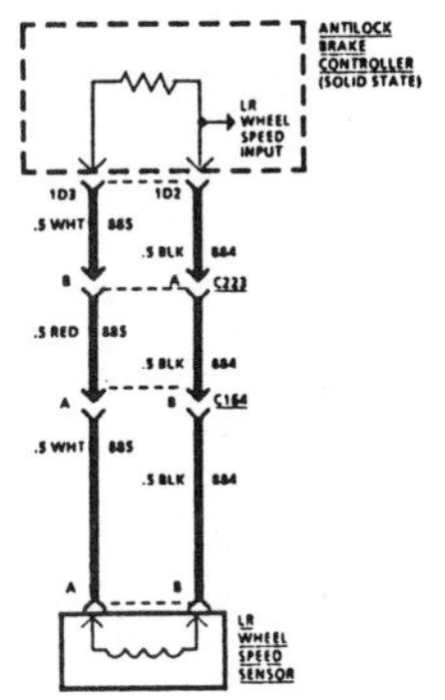

CODE A047

Left Rear Wheel Speed = 0

TEST DESCRIPTION: The following provides an explanation of the procedures being followed in the following trouble tree.

⑤ If more than 10 ohms is measured the circuit between the controller and C164 are ok. This isolates the fault to be between the Wheel Speed Sensor and C164. Since this area of circuit is exposed to the elements and road debris, it should be carefully inspected for damage.

⑥ This procedure isolates the fault to a short to ground (1500 to 2500 ohms) or open (above 2500 ohms) or a shorted sensor circuit (below 1500 ohms)

⑦ & ⑧ Determines if the fault is in the harness or the Wheel Speed Sensor. If the harness between C164 and the sensor is defective, replace it, do not attempt to repair the harness.

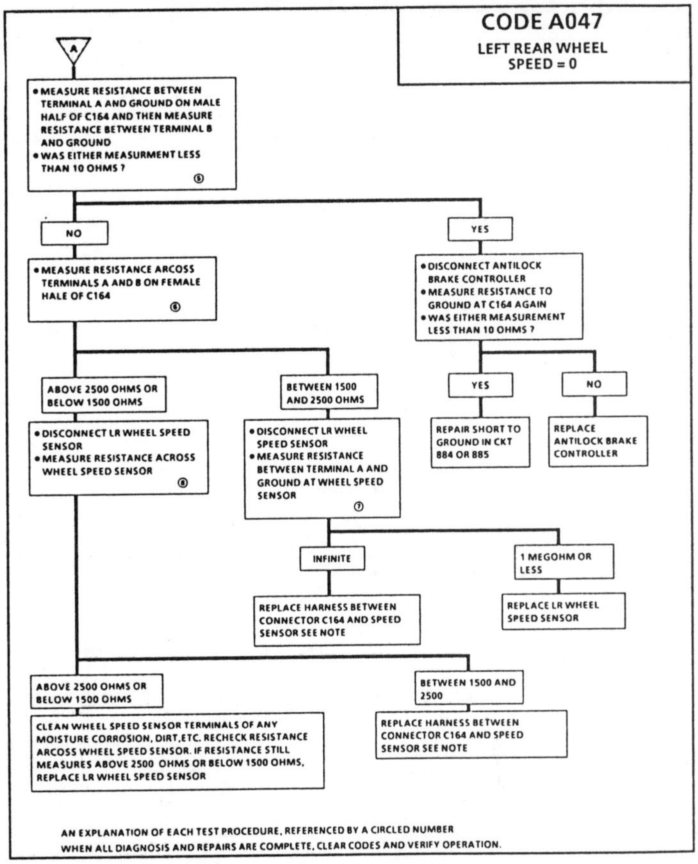

Fig. 49 Code A047, left rear wheel speed at 0 mph (Part 2 of 3)

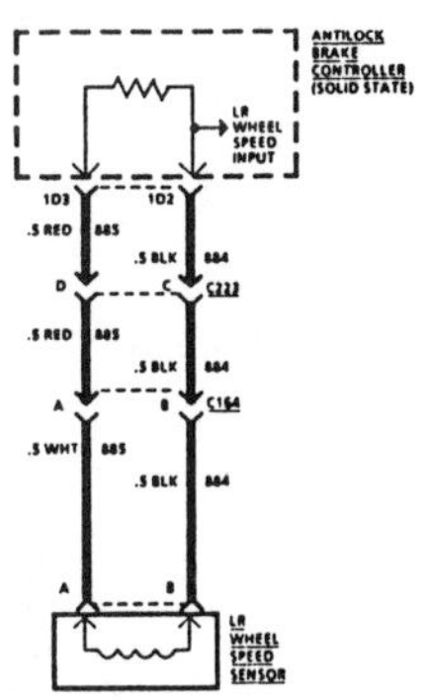

CODE A047

Left Rear Wheel Speed = 0

⑩ With CKT 884 connected to ground, the resistance from Terminal 1D2 to ground should be zero. If infinite resistance is measured, an open condition is indicated in CKT 884.

⑪ With CKT 885 connected to ground, the resistance from Terminal 1D3 to ground should be zero. If infinite resistance is measured, an open condition is indicated in CKT 885.

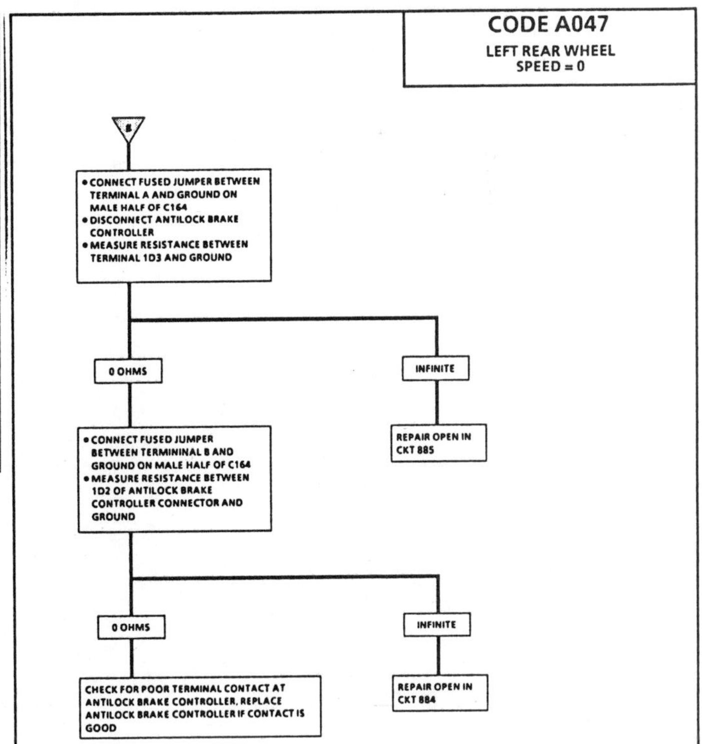

Fig. 49 Code A047, left rear wheel speed at 0 mph (Part 3 of 3)

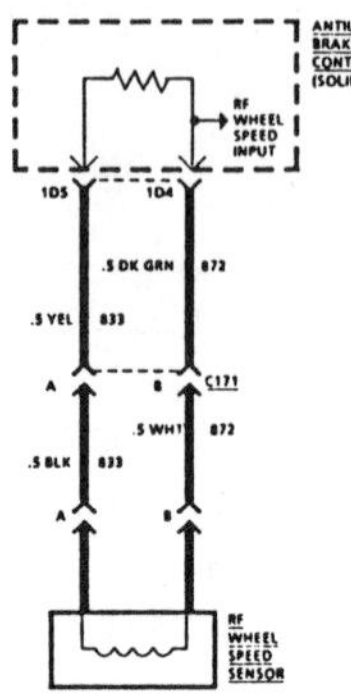

CODE A048

Excessive Right Front Wheel Acceleration

The Antilock Brake Controller uses the signal from the Wheel Speed Sensor to detect if wheel lock-up is impending. If the frequency of the AC voltage produced by the wheel speed sensor indicates that the wheel is accelerating or decelerating faster than physically possible, the Controller determines the signal is faulty and sets the code.

CODE A048 will set when all of the following conditions exist:

- The Brakes are not being applied (Brake Switch is off).
- The Antilock Brake Controller senses that the Right Front Wheel has accelerated or decelerated greater than physically possible.

TEST DESCRIPTION: The following provides an explanation of the procedures being followed in the facing trouble tree.

① By observing the Right Front Wheel speed input, you can verify what type of input the controller is sensing. An unsteady or erratic wheel speed input indicates that a fault is present in the Wheel Speed Sensor or its related circuitry. If a steady wheel speed is observed, an intermittent fault condition is indicated.

② Indicates a defective Wheel Speed Sensor by isolating a short to ground internal to the sensor.

③ If resistance measured across the Wheel Speed Sensor is infinite, the fault is due to an open condition internal to the Speed Sensor.

CODE A048

EXCESSIVE RIGHT FRONT WHEEL ACCELERATION

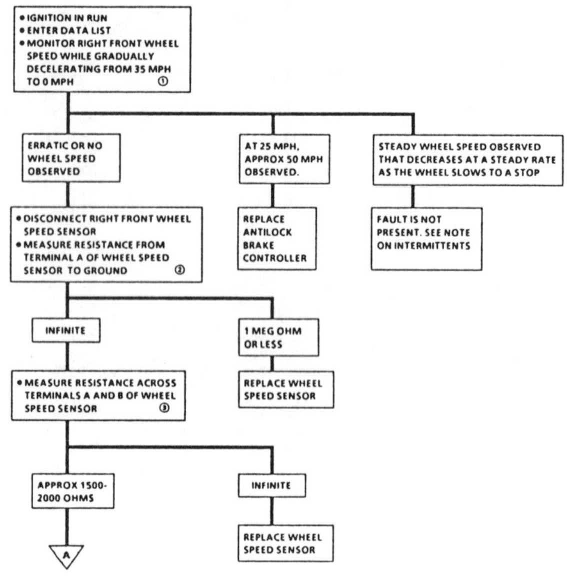

AN EXPLANATION OF EACH TEST PROCEDURE, REFERENCED BY A CIRCLED NUMBER

Fig. 50 Code A048, excessive right front wheel acceleration. (Part 1 of 2)

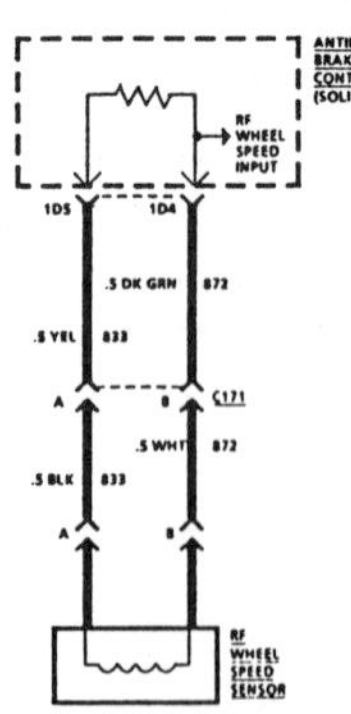

CODE A048

Excessive Right Front Wheel Acceleration

TEST DESCRIPTION: The following provides an explanation of the procedures being followed in the facing trouble tree.

④ With Terminal A of CKT 833 connected to ground, the resistance from Terminal 1D5 to ground should be zero. If infinite resistance is measured, an open condition is indicated in CKT 833.

⑤ With Terminal B of CKT 872 connected to ground, the resistance from Terminal 1D4 to ground should be zero. If infinite resistance is measured, an open condition is indicated in CKT 872.

⑥ At this point, all the connecting circuitry has tested good. The fault may be due to a poorly mated connector at the Wheel Speed Sensor or the Antilock Brake Controller. Other possible causes could be due to damage to the Speed Sensor or Speed Ring. Check for proper mounting or poor alignment. If everything above tests good, replace the Antilock Brake Controller.

CODE A048

EXCESSIVE RIGHT FRONT WHEEL ACCELERATION

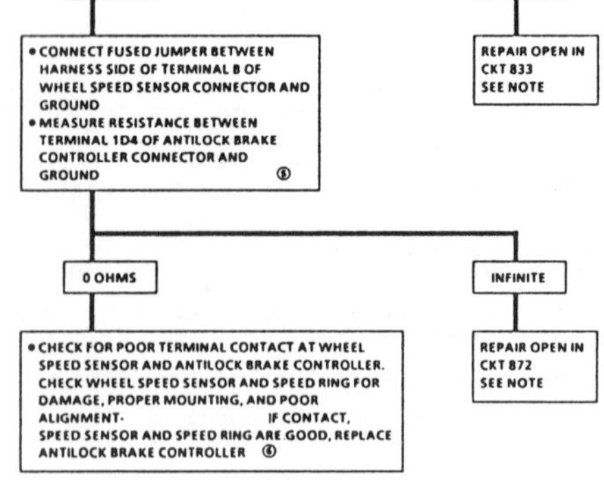

NOTE: IF HARNESS BETWEEN C171 AND WHEEL SPEED SENSOR IS DEFECTIVE, REPLACE HARNESS. DO NOT ATTEMPT TO REPAIR IT.
AN EXPLANATION OF EACH TEST PROCEDURE, REFERENCED BY A CIRCLED NUMBER

Fig. 50 Code A048, excessive right front wheel acceleration. (Part 2 of 2)

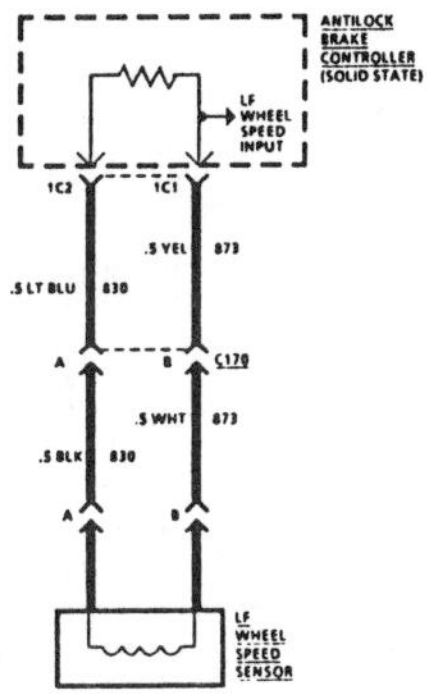

CODE A049

Excessive Left Front Wheel Acceleration

The Antilock Brake Controller uses the signal from the Wheel Speed Sensor to detect if wheel lock-up is impending. If the frequency of the AC voltage produced by the Wheel Speed Sensor indicates that the wheel is accelerating or declerating faster than physically possible, the Controller determines the signal is faulty and sets the code.

CODE A049 will set when all of the following conditions exist:

- The Brakes are not being applied (Brake Switch is off).
- The Antilock Brake Controller senses that the Left Front Wheel has accelerated or decelerated greater than physically possible.

TEST DESCRIPTION: The following provides an explanation of the procedures being followed in the facing trouble tree.

① By observing the Left Front Wheel speed input, you can verify what type of input the controller is sensing. An unsteady or erratic wheel speed input indicates that a fault is present in the Wheel Speed Sensor or its related circuitry. If a steady wheel speed is observed, an intermittent fault condition is indicated.

② Indicates a defective Wheel Speed Sensor by isolating a short to ground internal to the sensor.

③ If resistance measured across the Wheel Speed Sensor is infinite, the fault is due to an open condition internal to the Speed Sensor.

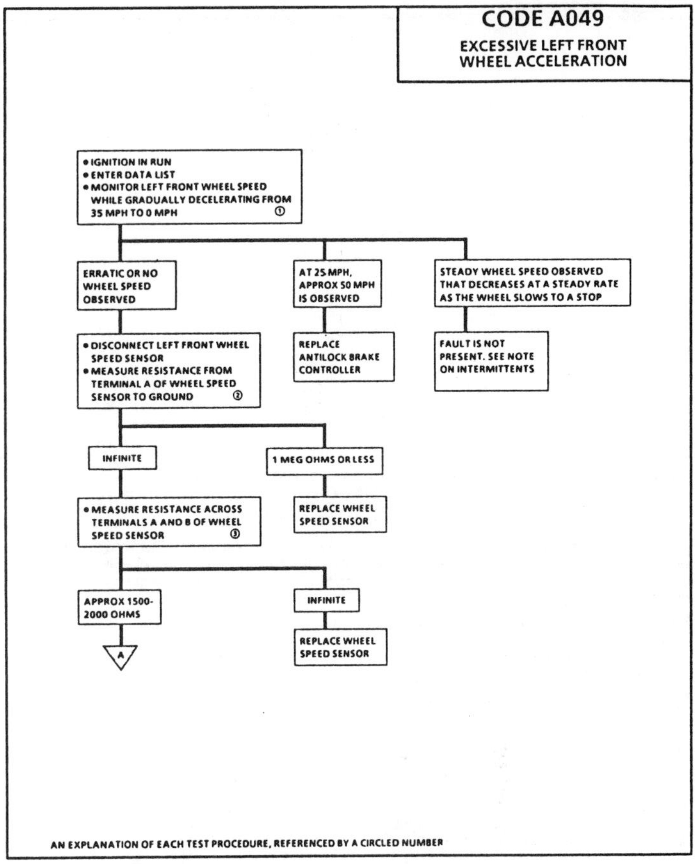

Fig. 51 Code A049, excessive left front wheel acceleration. (Part 1 of 2)

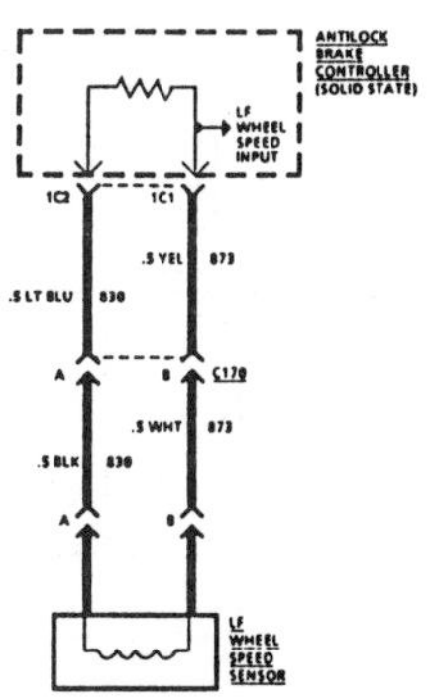

CODE A049

Excessive Left Front Wheel Acceleration

TEST DESCRIPTION: The following provides an explanation of the procedures being followed in the facing trouble tree.

④ With Terminal A of CKT 830 connected to ground, the resistance from Terminal 1C2 to ground should be zero. If infinite resistance is measured, an open condition is indicated in CKT 830.

⑤ With Terminal B of CKT 873 connected to ground, the resistance from Terminal 1C1 to ground should be zero. If infinite resistance is measured, an open condition is indicated in CKT 873.

⑥ At this point, all the connecting circuitry has tested good. The fault may be due to a poorly mated connector at the Wheel Speed Sensor or the Antilock Brake Controller. Other possible causes could be due to damage to the Speed Sensor or Speed Ring. Check for proper mounting, or poor alignment. If everything above tests good, replace the Antilock Brake Controller.

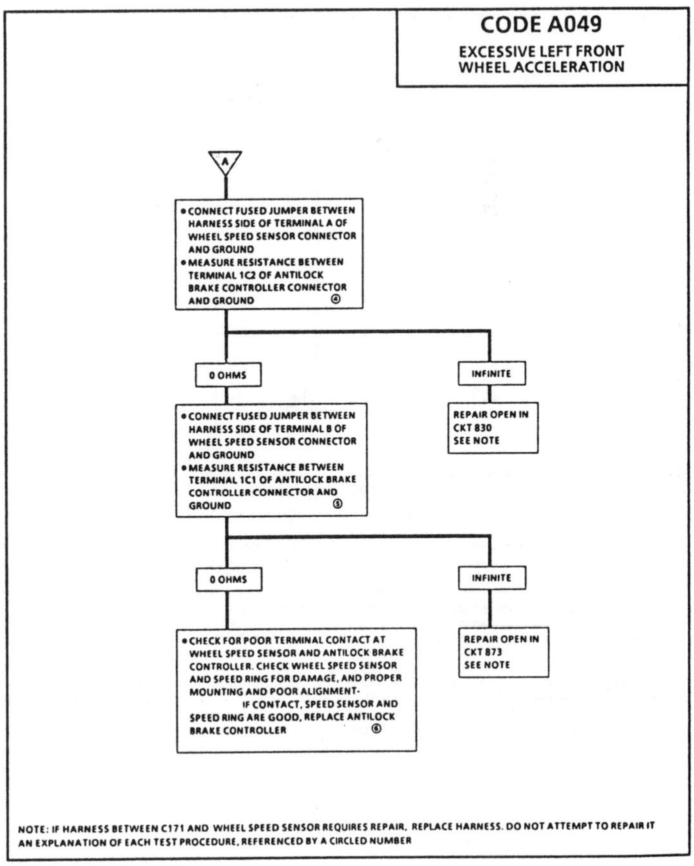

Fig. 51 Code A049, excessive left front wheel acceleration. (Part 2 of 2)

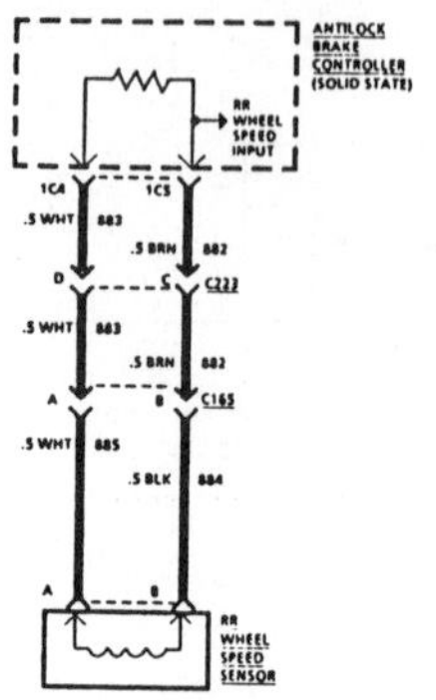

CODE A050

Excessive Right Rear Wheel Acceleration

The Antilock Brake Controller uses the signal from the wheel speed sensor to detect if wheel lock-up is impending. If the frequency of the AC voltage produced by the wheel speed sensor indicates that the wheel is accelerating or decelerating faster than physically possible, the Controller determines the signal is faulty and sets the code.

CODE A050 will set when all of the following conditions exist:

- The Brakes are not being applied (Brake Switch is off).
- The Antilock Brake Controller senses that the Right Rear Wheel has accelerated or decelerated greater than physically possible.

TEST DESCRIPTION: The following provides an explanation of the procedures being followed in the following trouble tree.

① Observation of erratic or no wheel speed input indicates the fault is present and is not intermittent.

②& ③ Step 2 determines if a short to battery is the fault. If battery voltage is measured, step 3 isolates the short to the harness or controller.

④ If 9000 to 15000 ohms is measured, the possibility of an open or short in the circuit between the Antilock Brake Controller and Connector C165 is eliminated.

CODE A050
EXCESSIVE RIGHT REAR WHEEL ACCELERATION

- IGNITION IN RUN
- ENTER DATA LIST
- MONITOR RIGHT REAR WHEEL SPEED WHILE GRADUALLY DECELERATING FROM 25 MPH TO 0 MPH ①

ERRATIC OR NO WHEEL SPEED OBSERVED

- DISCONNECT CONNECTOR C165
- CHECK FOR VOLTAGE BETWEEN TERMINAL A AND GROUND ON MALE HALF OF C165 AND CHECK FOR VOLTAGE BETWEEN TERMINAL B AND GROUND
- IS BATTERY VOLTAGE MEASURED AT EITHER TERMINAL A OR B? ②

NO

- IGNITION OFF
- MEASURE RESISTANCE ACROSS TERMINALS A AND B ON MALE HALF OF CONNECTOR C165 ④

0 OHMS

- DISCONNECT ANTILOCK BRAKE CONTROLLER
- MEASURE RESISTANCE ACROSS TERMINALS A AND B ON MALE HALF OF CONNECTOR C165

0 OHMS

REPAIR SHORT BETWEEN CKT 882 AND CKT 883

INFINITE

CHECK FOR CLEAN TERMINAL CONTACT AT ANTILOCK BRAKE CONTROLLER. REPLACE CONTROLLER IF TERMINAL IS GOOD

BETWEEN 9000 AND 15000 OHMS

A

INFINITE

B

YES

- IGNITION OFF
- DISCONNECT ANTILOCK BRAKE CONTROLLER
- IGNITION IN RUN
- CHECK FOR VOLTAGE AT C165 AGAIN
- IS BATTERY VOLTAGE MEASURED AT EITHER TERMINAL A OR B? ③

YES

REPAIR SHORT TO BATTERY IN CKT 882 OR 883

NO

REPLACE ANTILOCK BRAKE CONTROLLER

STEADY WHEEL SPEED OBSERVED THAT DECREASES AT A STEADY RATE AS THE WHEEL SLOWS TO A STOP

FAULT IS NOT PRESENT. SEE NOTE ON INTERMITTENTS

AN EXPLANATION OF EACH TEST PROCEDURE, REFERENCED BY A CIRCLED NUMBER
WHEN ALL DIAGNOSIS AND REPAIRS ARE COMPLETE, CLEAR CODES AND VERIFY OPERATION.

Fig. 52 Code A050, excessive right rear wheel acceleration. (Part 1 of 3)

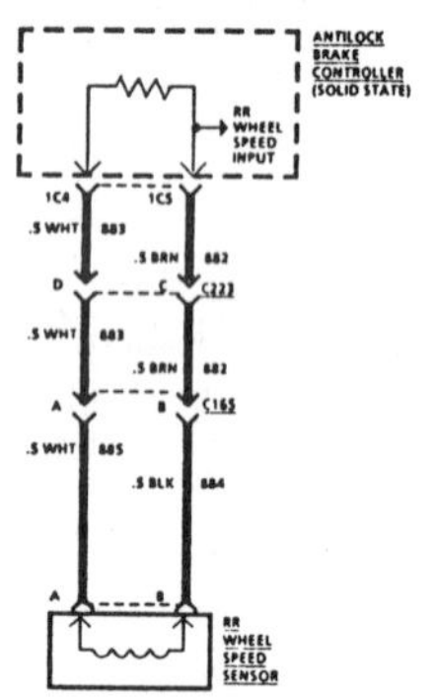

CODE A050

Excessive Right Rear Wheel Speed Acceleration

TEST DESCRIPTION: The following provides an explanation of the procedures being followed in the following trouble tree.

⑤ If more than 10 ohms is measured, the circuit between the Controller and C165 are ok. This isolates the fault to be between the Wheel Speed Sensor and C165. Since this area of the circuit is exposed to the elements and road debris it should be carefully inspected for damage.

⑥ This procedure isolates the fault to a short to ground (1500 to 2500 ohms) or open (above 2500 ohms) or a shorted sensor circuit (below 1500 ohms)

⑦ & ⑧ Determine if the fault is in the harness or the Wheel Speed Sensor. If the harness between C165 and the sensor is defective, replace it. Do not attempt to repair the harness.

CODE A050
EXCESSIVE RIGHT REAR WHEEL ACCELERATION

A

- MEASURE RESISTANCE BETWEEN TERMINAL A AND GROUND ON MALE HALF OF C165 AND THEN MEASURE RESISTANCE BETWEEN TERMINAL B AND GROUND
- WAS EITHER MEASURMENT LESS THAN 10 OHMS ? ⑤

NO

- MEASURE RESISTANCE ARCOSS TERMINALS A AND B ON FEMALE HAlf OF C165 ⑥

ABOVE 2500 OHMS OR BELOW 1500 OHMS

- DISCONNECT RR WHEEL SPEED SENSOR
- MEASURE RESISTANCE ACROSS WHEEL SPEED SENSOR ⑧

ABOVE 2500 OHMS OR BELOW 1500 OHMS

CLEAN WHEEL SPEED SENSOR TERMINALS OF ANY MOISTURE, CORROSION, DIRT,ETC. RECHECK RESISTANCE ACROSS WHEEL SPEED SENSOR. IF RESISTANCE STILL MEASURES ABOVE 2500 OHMS OR BELOW 1500 OHMS, REPLACE RR WHEEL SPEED SENSOR

BETWEEN 1500 AND 2500 OHMS

REPLACE HARNESS BETWEEN CONNECTOR C165 AND SPEED SENSOR SEE NOTE

BETWEEN 1500 AND 2500 OHMS

- DISCONNECT RR WHEEL SPEED SENSOR
- MEASURE RESISTANCE BETWEEN TERMINAL A AND GROUND AT WHEEL SPEED SENSOR ⑦

INFINITE

REPLACE HARNESS BETWEEN CONNECTOR C165 AND SPEED SENSOR SEE NOTE

1 MEGOHM OR LESS

REPLACE RR WHEEL SPEED SENSOR

YES

- DISCONNECT ANTILOCK BRAKE CONTROLLER
- MEASURE RESISTANCE TO GROUND AT C165 AGAIN
- WAS EITHER MEASUREMENT LESS THAN 10 OHMS ?

YES

REPAIR SHORT TO GROUND IN CKT 882 OR 883

NO

REPLACE ANTILOCK BRAKE CONTROLLER

NOTE: IF HARNESS BETWEEN C165 AND SENSOR IS DEFECTIVE, REPLACE IT. DO NOT ATTEMPT TO REPAIR IT.
AN EXPLANATION OF EACH TEST PROCEDURE, REFERENCED BY A CIRCLED NUMBER
WHEN ALL DIAGNOSIS AND REPAIRS ARE COMPLETE, CLEAR CODES AND VERIFY OPERATION.

Fig. 52 Code A050, excessive right rear wheel acceleration. (Part 2 of 3)

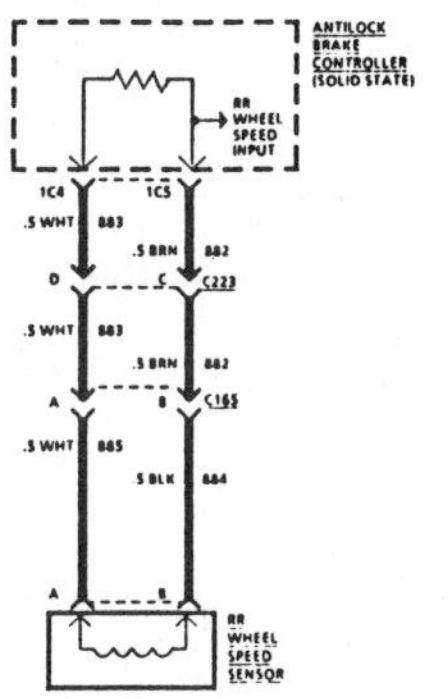

CODE A050

Excessive Right Rear Wheel Speed Acceleration

⑩ With CKT 883 connector to ground, the resistance from Terminal 1C4 to ground should be zero. If infinite resistance is measured, an open condition is indicated in CKT 883.

⑪ With CKT 882 connector to ground, the resistance from Terminal 1C5 to ground should be zero. If infinite resistance is measured, an open condition is indicated in CKT 882.

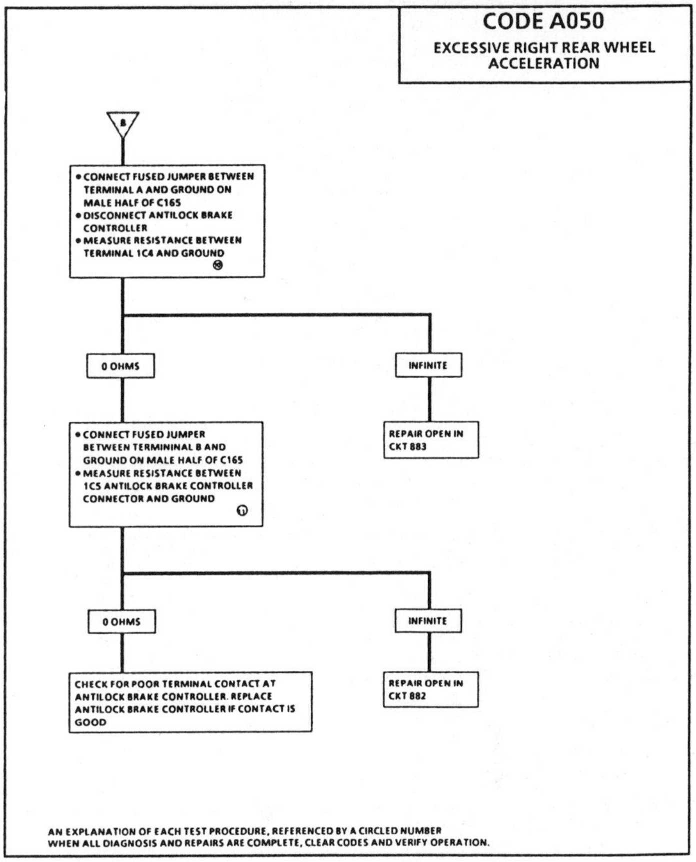

Fig. 52 Code A050, excessive right rear wheel acceleration. (Part 3 of 3)

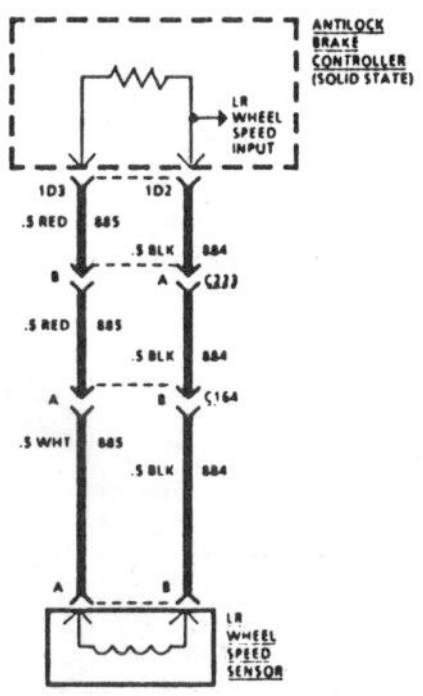

CODE A051

Excessive Left Rear Wheel Acceleration

The Antilock Brake Controller uses the signal from the wheel speed sensor to detect if wheel speed lock-up is impending. If the frequency of the AC voltage produced by the wheel speed sensor indicates that the wheel is accelerating or decelerating faster than physically possible, the Controller determines the signal is faulty and sets the code.

CODE A051 will set when all of the following conditions exist:

- The Brakes are not being applied (Brake Switch is off).
- The Antilock Brake Controller senses that the Left Rear Wheel has accelerated or decelerated greater than physically possible.

TEST DESCRIPTION: The following provides an explanation of the procedures being followed in the following trouble tree.

① Observation of erratic or no wheel speed input indicates the fault is present and is not intermettent.

②&③ Step 2 determines is if a short to Battery is the fault. If Battery voltage is measured, step 3 isolates the short to the harness or controller.

④ If 9000 to 15000 ohms were measured, the possibility of an open or short in the circuit between the Antilock Brakes Controller and Connector C164 has been eliminated.

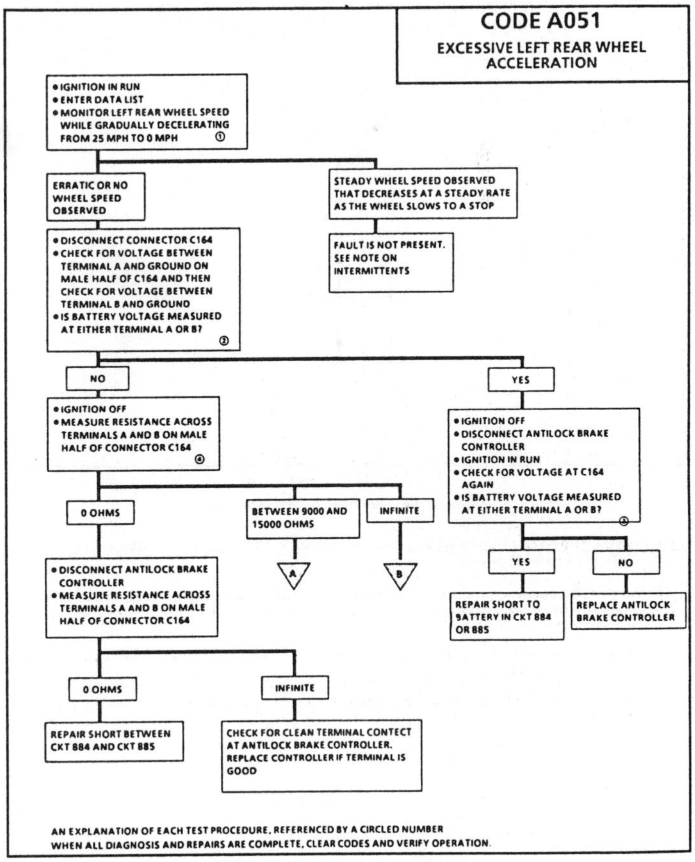

Fig. 53 Code A051, excessive left rear wheel acceleration. (Part 1 of 3)

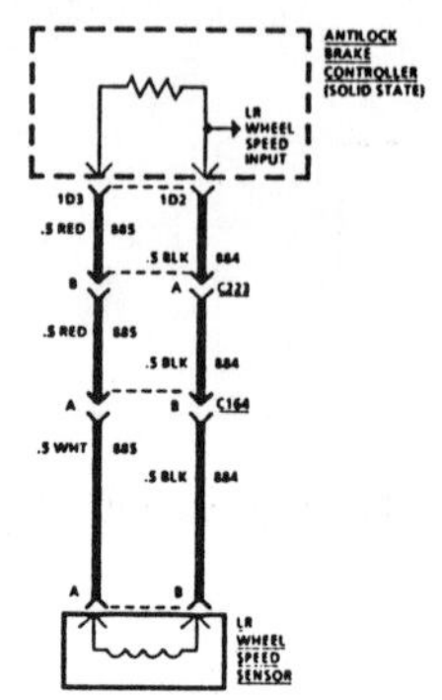

CODE A051

Excessive Left Rear Wheel Acceleration

TEST DESCRIPTION: The following provides an explanation of the procedures being followed in the following trouble tree

⑤ If more than 10 ohms is measured, the circuits between the controller and C164 are OK. This isolates the fault between the Wheel Speed Sensor and C164. Since this area of the circuit is exposed to the elements and road debris, it should be carefully inspected for damage.

⑥ This procedure isolates the fault to a short to ground(1500 to 2500 ohms) or open (above 2500 ohms) or a shorted sensor circuit (below 1500 ohms).

⑦& ⑧ Determines if the fault is in the harness or the Wheel Speed Sensor. If the harness between C164 and the sensor is defective, replace it, do not attempt to repair the harness.

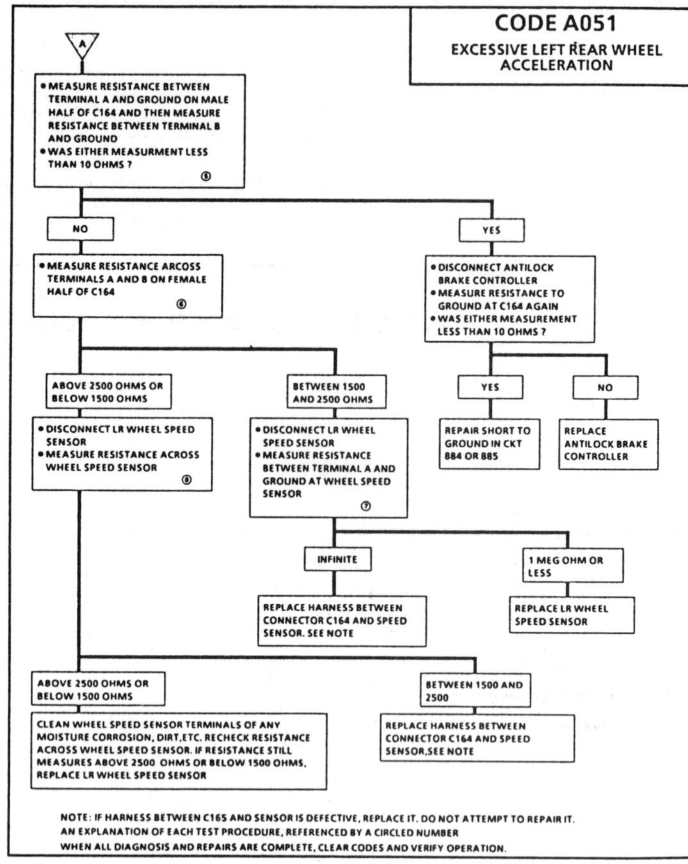

Fig. 53 Code A051, excessive left rear wheel acceleration. (Part 2 of 3)

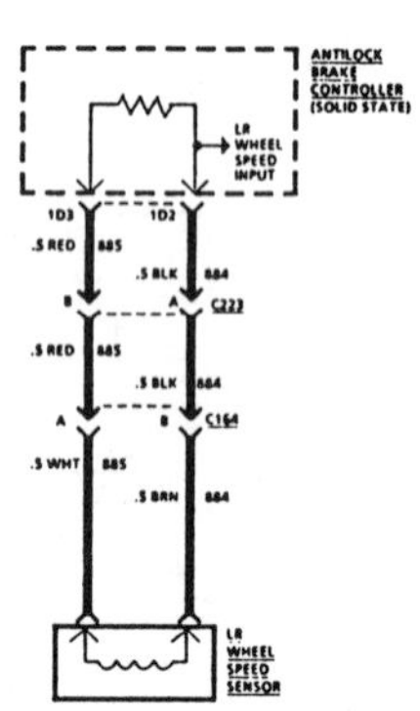

CODE A051

Excessive Left Rear Wheel Speed Acceleration

⑩ With CKT 884 connector to ground, the resistance from Terminal 1D2 to ground should be zero. If infinite resistance is measured, an open condition is indicated in CKT 884.

⑪ With CKT 885 connector to ground, the resistance from Terminal 1D3 to ground should be zero. If infinite resistance is measured, an open condition is indicated in CKT 885.

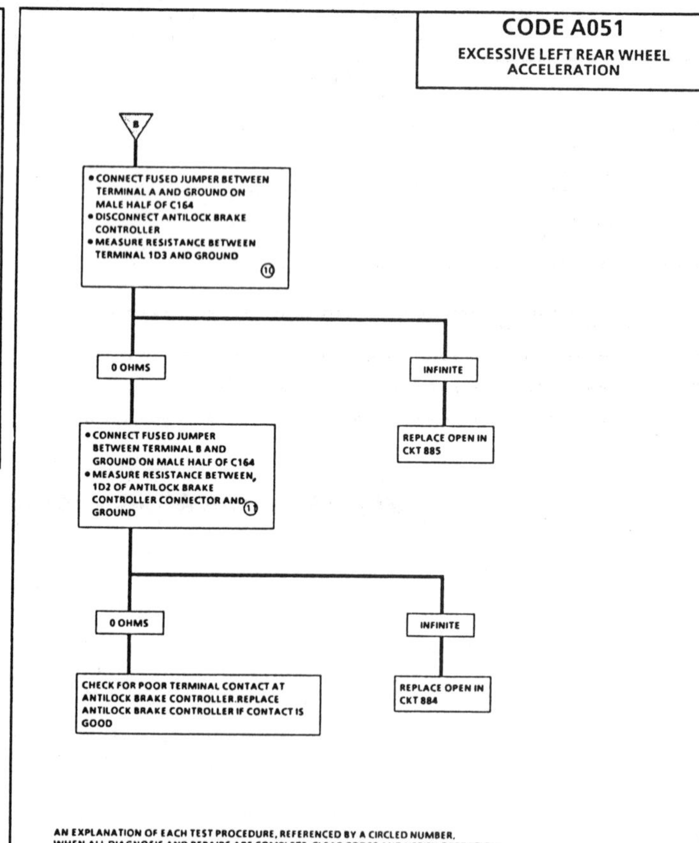

Fig. 53 Code A051, excessive left rear wheel acceleration. (Part 3 of 3)

CODE A052

Antilock Brake Controller Calibration Error

CODE A052	will set when the Antilock Brake Controller detects a malfunction internal to itself. If this code is set, replace the Antilock Brake Controller.

Fig. 54 Code A052, anti-lock brake control calibration error

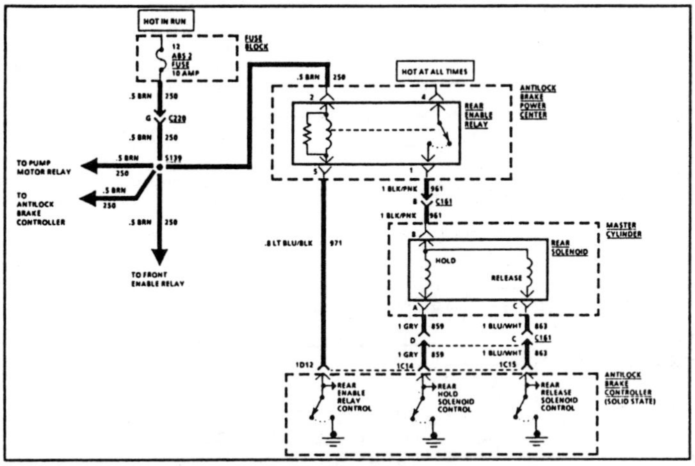

CODE A054

Rear Enable Relay Coil circuit open

When the ignition is in RUN, voltage is applied from the ABS 2 Fuse to the Rear Enable Relay Coil. As long as the Rear Enable Relay Control is open, this voltage should also be present at terminal 1D12 of the Controller.

CODE A054 will set when all the following conditions exist:

- The Rear Enable Relay Control is open (Rear Enable Relay deenergized).
- The Antilock Brake Controller senses no voltage at Terminal 1D12.
- The Antilock Brake Controller senses no voltage at Terminals 1C14 and 1C15.
- Codes A004 and A007 are current failures.

TEST DESCRIPTION: The following provides an explanation of the procedures being followed in the facing trouble tree.

(1) By monitoring the test lamp with the enable relays on, it can be determined if the power and ground circuits, as well as the Controller, are good.

(2) If the test lamp does not light, the open must be between the Rear Enable Relay and Splice S139.

(3) Voltage at terminal 1D12 indicates a problem with the connection to the Controller or the Controller itself.

(4) Determines if the Code was set due to a hard failure. At this point the hard failure must be in the Rear Enable Relay.

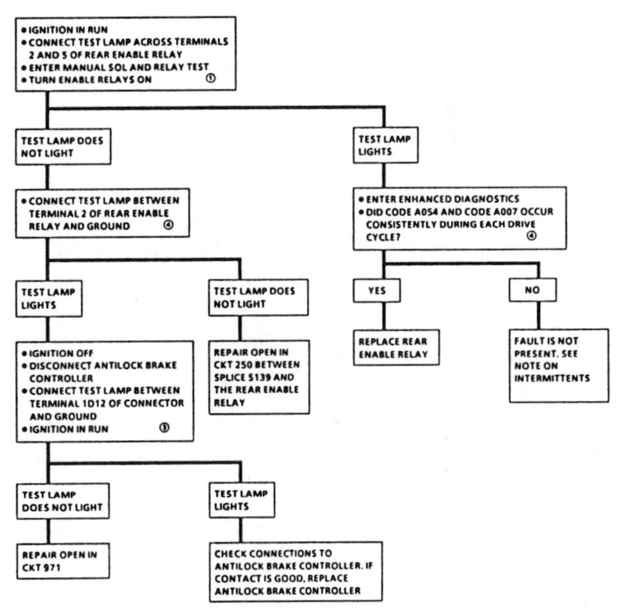

AN EXPLANATION OF EACH TEST PROCEDURE, REFERENCED BY A CIRCLED NUMBER

WHEN ALL DIAGNOSIS AND REPAIRS ARE COMPLETE, CLEAR CODES AND VERIFY OPERATION.

Fig. 55 Code A054, rear enable relay coil circuit open

CODE A055

Antilock Brake Controller internal voltage fault

CODE A055 will set when the Antilock Brake Controller detects a malfunction internal to itself. If this code is set, replace the Antilock Brake Controller.

Fig. 56 Code A055, anti-lock brake controller internal voltage fault

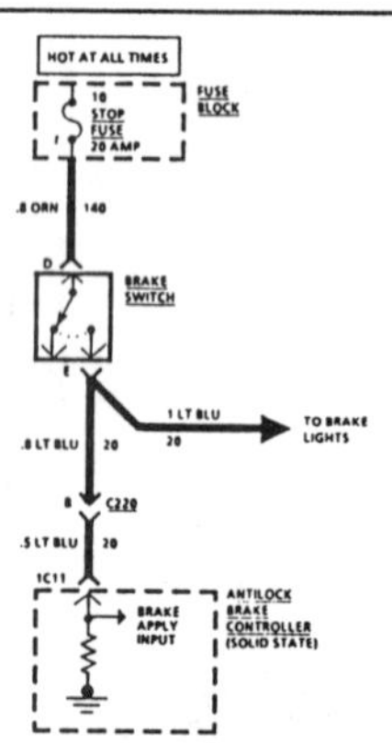

CODE A056

Open Brake Switch or Hydraulic Leak

Voltage is applied at all times to the Brake Switch. When the Brake Switch is closed, voltage is applied to the Brake Lights and terminal 1C11 of the Antilock Brake Controller. If the Brake Apply Input receives voltage, the Controller determines that the brakes are being applied.

CODE A056 will set when the following condition exists:

- Code A032 was set during the preceding ignition cycle. Code A056 will remain a current code until the Antilock Brake Controller senses a Brake Switch transition and the ignition cycle is completed (Ignition turned to OFF).

TEST DESCRIPTION: The following provides an explanation of the procedures being followed in the facing trouble tree.

① If Code A062 is also set, the pump motor was cycling too fast. See Code A062.

② Code A041 will diagnose an open between the Brake Switch and the Controller, or between the brake switch and the stop light to ground. An open High Level Stop Light can also set this code if the Hazard Lights are operated.

③ Code A033 will diagnose an open brake switch or CKT 140.

CODE A056

OPEN BRAKE SWITCH OR HYDRAULIC LEAK

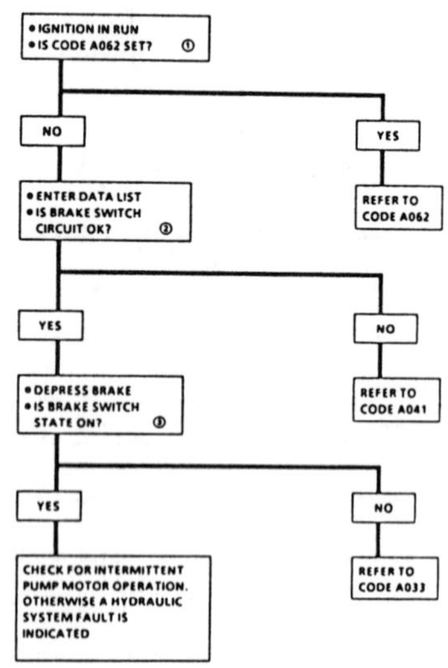

AN EXPLANATION OF EACH TEST PROCEDURE, REFERENCED BY A CIRCLED NUMBER.
WHEN ALL DIAGNOSIS AND REPAIRS ARE COMPLETE, CLEAR CODES AND VERIFY OPERATION.

Fig. 57 Code A056, open brake switch or hydraulic leak

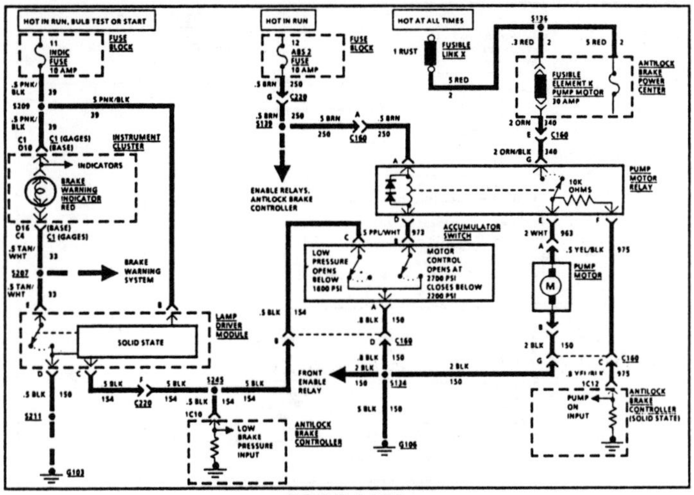

CODE A059

Low Brake Pressure while in an ABS stop

CODE A059 will set when all of the following conditions exist:

- The car is in an Antilock Brake stop.
- The Pump On Input senses voltage (pump running).
- The Antilock Brake Controller senses voltage at the Low Brake Pressure Input for greater than 10 seconds during and after the Antilock Brake stop.

TEST DESCRIPTION: The following provides an explanation of the procedures being followed in the facing trouble tree.

① If Code A035 is also set, the fault is due to an electrical problem which can be diagnosed by refering to Code A035. If Code A059 has not set along with any other codes, a hydraulic system fault is indicated Possible causes are no pump output, broken brake line, internal leakage, etc.

CODE A059

LOW BRAKE PRESSURE WHILE IN AN ABS STOP

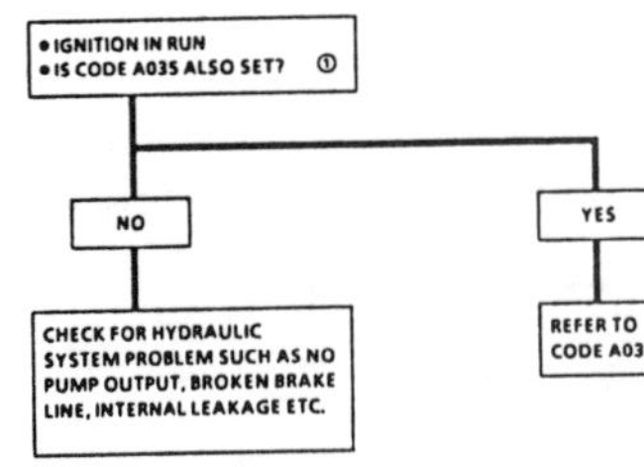

AN EXPLANATION OF EACH TEST PROCEDURE REFERENCED BY A CIRCLED NUMBER

WHEN ALL DIAGNOSIS AND REPAIRS ARE COMPLETE, CLEAR CODES AND VERIFY OPERATION.

Fig. 58 Code A059, low brake pressure while in an ABS stop

CODE A060

Antilock Brake Controller Internal fault

CODE A060	will set when the Antilock Brake Controller detects a malfunction internal to itself. If this code is set, replace the Antilock Brake Controller.

Fig. 59 Code A060, anti-lock brake controller internal fault

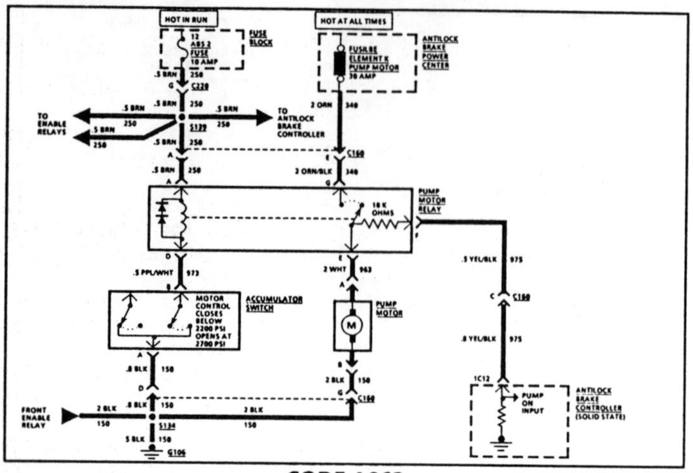

CODE A062

Low Accumulator Pre-charge

The Accumulator is precharged to approximately 1200 PSI with nitrogen gas. The Pump Motor maintains system pressure between 2200 PSI and approximately 2700 PSI. The Accumulator Switch Motor Control closes when the pressure drops below 2200 PSI. This grounds the Pump Motor Relay Coil causing voltage to be applied to the Pump Motor and Pump On Input. Whenever the Pump Motor is on, the Pump On Input will have voltage applied to it. The Pump Motor will run until system pressure is restored to approximately 2700 PSI.

CODE A062 will set when the Antilock Brake Controller detects short pump run times of less than 1.4 seconds in duration.

TEST DESCRIPTION: The following provides an explanation of the procedures being followed in the facing trouble tree.

① If the pump run time from "OK pressure" to "pump off time" is less than six seconds a low accumulator precharge or poor accumulator switch point condition is indicated.

② If normal pump run cycle is less than 1.7 seconds a miscalibrated accumulator switch or low accumulator precharge is indicated.

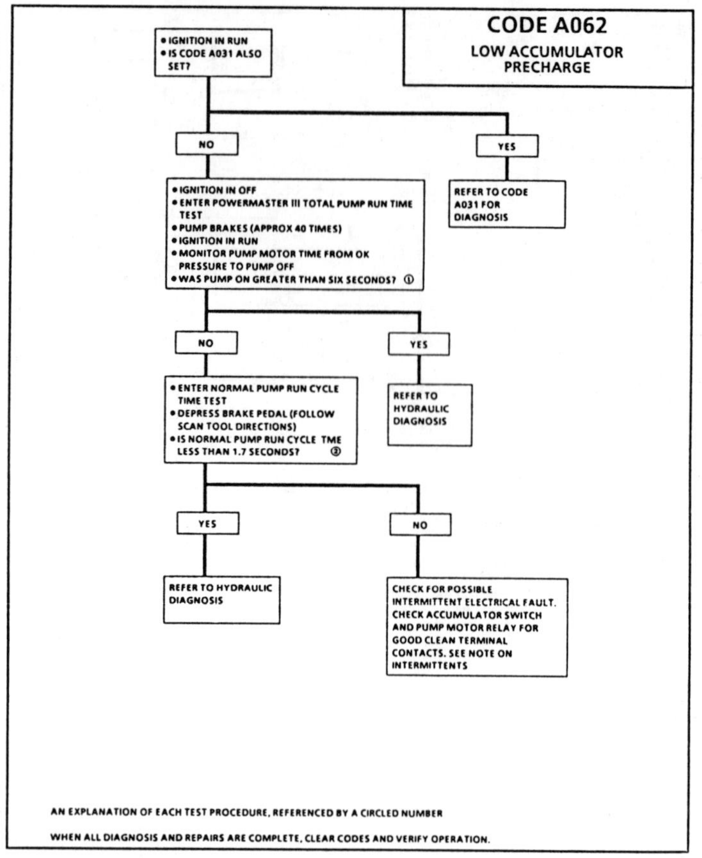

Fig. 60 Code A062, low accumulator precharge

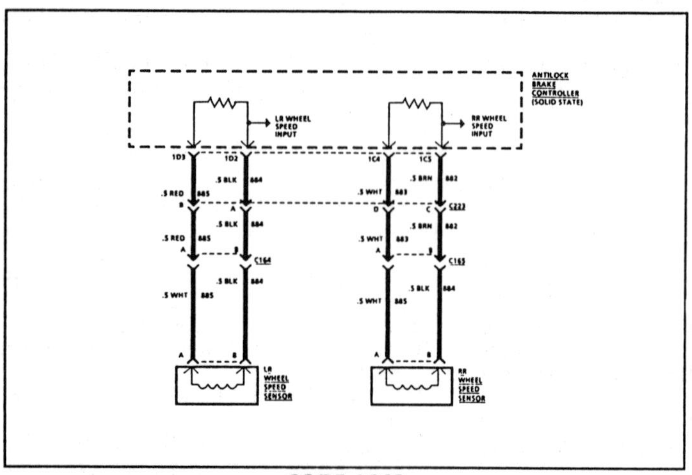

CODE A063

Both Rear Wheel Speed Sensors open

The Antilock Brake Controller monitors the Wheel Speed Sensors. If one or both rear wheel speeds are zero, the Controller cannot accurately detect wheel lock-up. Since three good wheel speeds are needed to set Codes A046 or A047, these Codes cannot determine if both wheel speed sensors are faulty. Code A063 can detect if both Rear Wheel Speed Sensors are malfunctioning.

CODE A063 will set when all of the following conditions exist:

- The Antilock Brake Controller senses both front wheel speeds are greater than 10 mph and are operating correctly.
- The Antilock Brake Controller senses both rear wheel speeds are 0 mph for more than 20 seconds.

TEST DESCRIPTION: The following provides an explanation of the procedures being followed in the facing trouble tree.

① If the front wheel were spinning while the vehicle was being serviced on a lift code, A063 would set. The code should be cleared and the vehicle road tested to insure this was the reason code A063 set.

② Determines whether the fault is due to a hard failure or a possible intermittent failure.

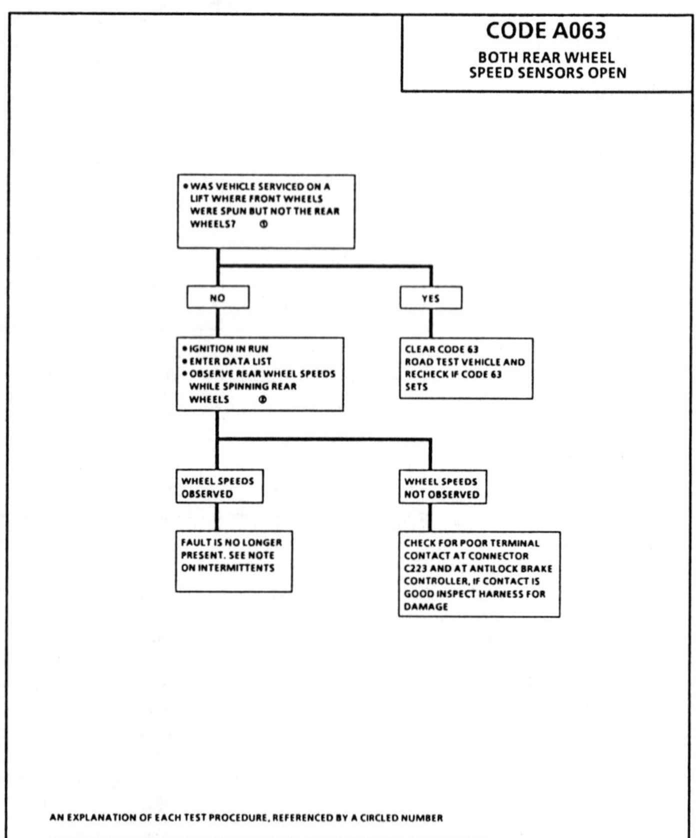

Fig. 61 Code A063, both rear wheel speed sensors open

CHECKING FLUID LEVEL

1. Park vehicle on a level surface, then depressurize Powermaster III, see "Depressurizing Powermaster III."
2. Clean reservoir cover, then remove cover and diaphragm assembly from the Powermaster III. Note fluid level in reservoir chambers.
3. If any chamber is underfilled, **Fig. 62**, check for signs of leakage. Make repairs as necessary. **Fill chambers with clean DOT 3 fluid only.**
4. If chamber is overfilled, adjust fluid level. **Do not contaminate fluid in brake system with dirt, dust, lubricants, oils, or other substance.**
5. Replace diaphragm assembly and cover.
6. Turn ignition switch to the On position. Allow system to pressurize.
7. Depressurize Powermaster III, then check fluid level.
8. If reservoir chamber is again overfilled, repair or replace components as necessary.

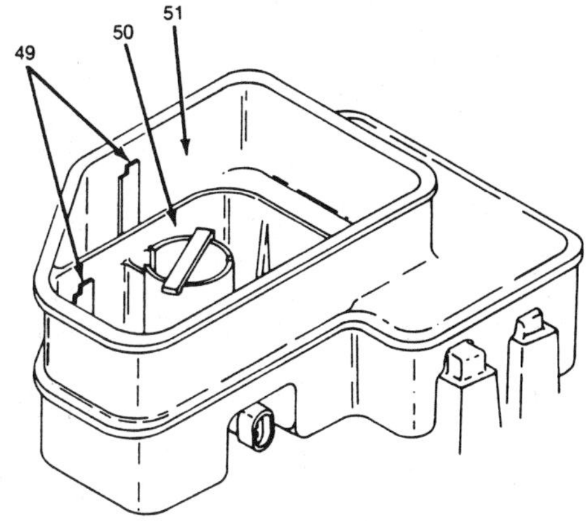

49. FLUID LEVEL MARKS
50. FRONT CHAMBER
51. REAR CHAMBER

Fig. 62 Fluid level fill marks

POWERMASTER III BLEEDING

Pressure Bleeding

If a Powermaster III unit has been replaced, or air is suspected in the brake lines, the entire braking system must be bled.

If only a hydraulic portion of the Powermaster III unit has been replaced, without allowing air into the brake lines, it may only be necessary to manually bleed the Powermaster III unit.

Pressure bleeding must be done with diaphragm type bleeding equipment. Use only DOT 3 type brake fluid from a sealed container.

1. Turn ignition switch to the Off position, then depressurize the Powermaster III unit, see "Depressurizing Powermaster III."
2. Clean reservoir cover and diaphragm assembly, then remove assembly.
3. Check fluid in both reservoir sections, then fill to proper level, using clean DOT 3 brake fluid, if necessary.
4. Install bleeder adapter J-37115 and secure with attachment cable. Ensure cable does not interfere with bleeder valve access.
5. Attach bleeder adapter to pressure bleed equipment, charge to 5-10 psi. for approximately 30 seconds, then slowly increase pressure to 30-35 psi.
6. Bleed J-37115 adapter by connecting a clear plastic hose to adapter bleeder screw and submerging opposite end in a clean container partially filled with brake fluid.
7. Slowly open bleeder valve, allowing fluid to flow until no more air is seen in fluid, then close valve.
8. Raise and support vehicle, then bleed brakes in sequence: right rear, left rear, right front, then left front, as described below.
9. Attach bleeder hose to bleeder valve, then submerge opposite end in a clean container partially filled with brake fluid.
10. Slowly open bleeder valve and allow fluid to flow for 20-30 seconds. To aid in releasing entrapped air, lightly tap on caliper castings with a rubber mallet.
11. Close valve when fluid begins to run with no bubbles. **Repeat steps 9-11 for each wheel, following wheel sequence in step 8.**
12. Lower vehicle, then attach bleeder hose to bleeder valve on inboard side of Powermaster III unit and submerge opposite end of hose in a clean container partially filled with brake fluid.
13. Slowly open bleeder valve and allow fluid to flow for 20-30 seconds.
14. Close valve when fluid begins to run with no bubbles.
15. Repeat steps 13-15 for outboard side bleeder valve.
16. Remove bleeder adapter J-37115 from Powermaster III unit, then check fluid level in both reservoirs. Fill if necessary.
17. Snap reservoir cover into place, then apply moderate to heavy force to the brake pedal.
18. With the pedal applied, turn ignition switch to the On position for 5 seconds, then turn ignition switch to the Off position. Do this 10 times in succession to cycle solenoids. **Do not start engine, or place ignition switch in the Start position.**
19. Depressurize Powermaster III unit, then wait two minutes for air to clear from reservoir.
20. Attach bleeder hose to bleeder valve on inboard side of Powermaster III unit and submerge opposite end in a clean container partially filled with brake fluid.
21. With the ignition switch in the On position, apply light force to the brake pedal, then slowly open bleeder valve and allow fluid to flow until no air is seen.
22. Close valve when no air is seen in fluid.
23. Repeat steps 20-22, on outboard side bleeder valve.
24. Turn ignition switch to the Off position, then depressurize Powermaster III unit.
25. Remove reservoir cover and diaphragm assembly, then check fluid level in reservoirs. Use clean DOT 3 fluid to fill if necessary.
26. Snap reservoir cover onto reservoir, then turn ignition switch to the On position. Allow pump to run.
27. Turn ignition switch to the Off position if pump runs longer than 60 seconds, inspect pump for malfunction.
28. Apply brake pedal and note travel and feel. If pedal feels firm and smooth, without excessive travel, system is properly bled.
29. If pedal feels soft, spongy, or has excessive travel, see "Excessive Pedal Travel."
30. If a bump is noted upon initial pedal apply, or pedal apply does not feel smooth and uniform, see "Non-Uniform Pedal Apply."
31. Road test vehicle. If any of the above symptoms appear, refer to appropriate procedure.

Manual Bleeding

1. Turn ignition switch to the Off position, then depressurize Powermaster III unit.
2. Clean reservoir and diaphragm assembly, then remove reservoir cover assembly.
3. Fill reservoir to proper level wit DOT 3 brake fluid, then reinstall cover assembly.
4. Raise and support vehicle, then attach a clear bleeder hose to right front bleeder valve and submerge opposite end in clean container partially filled with brake fluid.
5. Open bleeder valve, then slowly depress bleeder valve. To aid in releasing trapped air, lightly tap on caliper castings with a rubber mallet.
6. Close bleeder valve, then release brake valve.
7. Check, then add brake fluid if necessary.
8. Repeat steps 4-7 until pedal feels firm and no air is observed bleeder hose.
9. Repeat steps 4-8 for left front brake.
10. Turn ignition switch to the On position, allowing pump to run. If pump runs longer than 60 seconds, inspect pump for malfunctions.
11. Attach clear bleeder hose to right rear bleeder valve and submerge opposite end in a clean container partially filled with brake fluid.
12. Turn ignition switch to the On position, open bleeder valve, then slowly depress brake pedal part-way, until fluid begins to flow from hose. **Do not depress brake pedal fully.**
13. Hold pedal for 15 seconds. To aid in freeing entrapped air, tap lightly on caliper casting with a rubber mallet.
14. Close bleeder valve, then release brake pedal.

15. Repeat steps 12-14, until no air is observed in the fluid.
16. Turn ignition switch to the Off position, depressurize Powermaster III unit, then check brake fluid. Add fluid as needed.
17. Repeat steps 10-16 for left rear wheel.
18. Repeat steps 10-16 for inboard bleeder valve of Powermaster III unit.
19. Repeat steps 10-16 for outboard bleeder valve of Powermaster III unit.
20. Turn ignition switch to the Off position, then depressurize Powermaster III unit. Check brake fluid level and fill as needed.
21. Apply moderate to heavy force on brake pedal.
22. With pedal applied, turn ignition switch to the On position for 5 seconds, then turn ignition switch to the Off position. Repeat this step 10 times. **Do not turn ignition switch to the Start position, or start motor.**
23. Depressurize Powermaster III unit, then wait two minutes for air to clear from within reservoir.
24. Attach bleeder hose to bleeder valve on inboard side of Powermaster III unit and submerge opposite end in clean container partially filled with brake fluid.
25. Turn ignition switch to the On position, then apply light force to brake pedal.
26. With pedal applied, slowly open bleeder valve and allow fluid to flow until no air is seen in fluid.
27. Close valve when no air is seen in fluid.
28. Repeat steps 24-27, for outboard bleeder valve.
29. Turn ignition switch to the Off position, then depressurize Powermaster III unit.
30. Remove reservoir cover and diaphragm assembly, check and fill reservoirs with DOT 3 brake fluid as needed.
31. Install reservoir cover, then turn ignition switch to the On position. Allow pump to run. If pump runs for longer than 60 seconds, inspect pump for malfunctions.
32. Apply brake pedal, noting feel and travel.
33. If pedal feels firm and smooth, without excessive travel, system is properly bled.
34. If pedal feels soft, spongy, or has excessive travel, see "Excessive Pedal Travel."
35. If a bump is noted upon initial pedal apply, or pedal apply does not feel smooth and uniform, see "Non-Uniform Pedal Apply."
36. Road test vehicle. If any of the above symptoms appear, refer to appropriate procedure.

EXCESSIVE PEDAL TRAVEL

Excessive pedal travel exists when braking action does not start within the first 1.25 inches of pedal travel, while the Powermaster III unit is pressurized. When the Powermaster III unit is not pressurized, excessive pedal travel exists when brake pedal moves more than 3 inches while applying light pedal pressure.

1. Apply moderate to heavy force on brake pedal, turn ignition switch to the On position for 5 seconds, then turn ignition switch to the Off position. **Do not turn ignition switch to the Start position, or start motor.**
2. Release brake pedal, then repeat step 1, twice.
3. Bleed front brakes, as described in "Pressure Bleeding."
4. If excessive pedal travel exists, check brake fluid level, then for leaks in braking system.

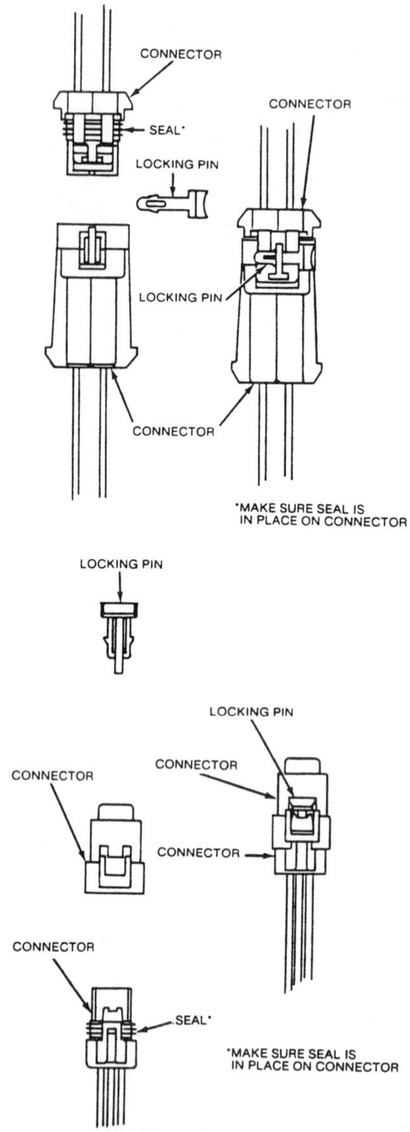

Fig. 63 Connector position assurance locking pins

NON-UNIFORM PEDAL APPLY

1. Raise rear of vehicle until Powermaster III unit is level with ground, or sloping downward slightly.
2. Depressurize Powermaster III unit.
3. Wait two minutes, then turn ignition switch to the On position. Allow pump to pressurize unit. **Do not turn ignition switch to the Start position or start engine.**
4. If condition still exists, raise rear of vehicle, as described in step 1, then bleed rear brakes. See "Pressure Bleeding," or "Manual Bleeding."

PUMP RUNS CONTINUOUSLY

Only perform this procedure if pump runs for more than 60 seconds. **Do not allow brake fluid to come into contact with painted surfaces.**

1. Turn ignition switch to the Off position, then depressurize Powermaster III unit.
2. Loosen, but do not remove tube nuts from both front master cylinder chambers.
3. Apply brake pedal as far as possible. Check that brake fluid runs from around tube nuts.
4. With pedal still applied, tighten tube nuts.
5. Quickly release brake pedal, then reapply with a sharp, rapid motion with full force.
6. Continue from step 25, of "Pressure Bleeding."

DELCO-MORAINE POWERMASTER III ELECTRICAL CONNECTORS

The Powermaster III unit is equipped with electrical connectors which utilize a connector position assurance locking pin. This locking pin assures proper alignment and engagement of the connector, **Fig. 63.**

Always remove the locking pin prior to disconnecting electrical connector. Use extreme caution as to not damage locking pin during removal or installation. Note position of rubber seals when disconnecting connectors for installation purposes. Always reinstall locking pins after reconnecting connector.

COMPONENT REPLACEMENT

FLUID LEVEL SENSOR SWITCH, REPLACE

1. Disconnect 2-pin connector (1), from fluid level sensor switch (2), **Fig. 64.**
2. Remove fluid level sensor switch from reservoir.
3. Depress locking tabs at inboard end of switch, under reservoir on opposite end of 2-pin connector.
4. Reverse procedure to install. Ensure sensor switch locking tabs snap into place.

BLEEDER VALVES, REPLACE

1. Depressurize Powermaster III, see "Depressurizing Powermaster III."
2. Remove bleeder valves (4), **Fig. 64.** Clean valves in denatured alcohol, then dry with compressed air. **Do not use lubricated compressed air.**

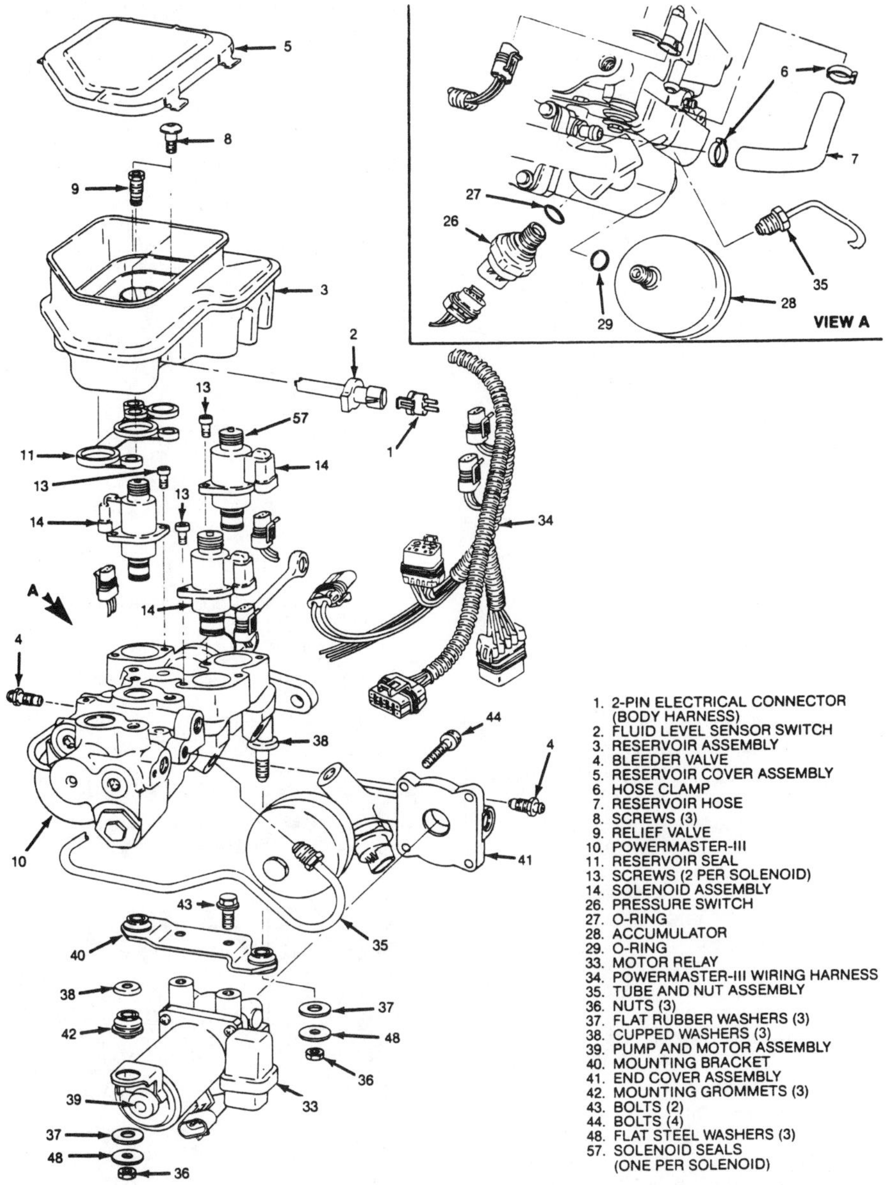

Fig. 64 Powermaster III, exploded view

3. Reverse procedure to install, noting the following:
   a. **Torque** bleeder valve to 40-90 inch lbs.
   b. Bleed Powermaster III as described in "Powermaster III Bleeding."

ACCUMULATOR, REPLACE

The accumulator is nitrogen-charged and holds brake fluid under high pressure. It cannot be serviced and must be replaced as an assembly.

1. Depressurize the Powermaster III, see "Depressurizing Powermaster III."
2. Remove the accumulator (28), by turning hex nut on the end of accumulator with a 17 mm socket, **Fig. 64.**
3. Remove from vehicle by sliding out from underneath Powermaster III, toward left front wheelwell.
4. Remove O-ring (29), from accumulator.
5. Reverse procedure to install, noting the following:
   a. Lightly lubricate, then install a new O-ring on accumulator.
   b. **Torque** accumulator to 23-36 ft. lbs.
   c. Bleed Powermaster III as described in "Powermaster III Bleeding."

TUBE & NUT ASSEMBLY (HIGH PRESSURE SIDE), REPLACE

1. Depressurize the Powermaster III, see "Depressurizing Powermaster III."
2. Remove air cleaner, duct and stud.
3. Remove tube and nut assembly (35), from Powermaster III unit (10). Plug all openings to prevent fluid loss and contamination, **Fig. 64.**
4. Inspect tube and nut assembly for crack, deformation and scratches or nicks on sealing surfaces. Replace tube and nut assembly if any of the above conditions are found.
5. Reverse procedure to install, noting the following:
   a. **Torque** tube and nut assembly to Powermaster III unit to 10-15 ft. lbs.
   b. Bleed Powermaster III unit, see "Powermaster III Bleeding."

MOTOR RELAY, REPLACE

1. Push relay bracket tab (47), with a small screwdriver, to clear button on motor relay (33), **Fig. 65.**
2. Pull motor relay from relay bracket (45).
3. Disconnect 5-pin connector (32), from motor relay.
4. Reverse procedure to install.

RESERVOIR COVER, HOSE & RESERVOIR ASSEMBLY, REPLACE

Clean reservoir cover assembly (5) and surrounding area before removing, **Fig. 64.**

1. Depressurize Powermaster III, see "Depressurizing Powermaster III."
2. Unsnap reservoir cover assembly, then remove from reservoir (3).
3. Inspect reservoir cover assembly for cuts, cracks, nicks and deformation. Replace any damaged parts.
4. Remove fluid level sensor switch, as described in "Fluid Level Sensor Switch, Replace."
5. To gain access, loosen but do not remove cowl bracket attaching nuts, then pull Powermaster III forward slightly.
6. Use a clean syringe to remove as much brake fluid from reservoir as possible. **Brake fluid will cause damage to all painted surfaces.**
7. Cover electrical connectors and engine/body parts with clean shop towels to catch any brake fluid dripped from reservoir. Use a suitable container to catch brake fluid when upper end reservoir hose is disconnected from reservoir.
8. Remove upper hose clamp, then disconnect hose from reservoir. **Discard all used brake fluid. Never reuse brake fluid.**
9. Remove reservoir assembly from Powermaster III unit, then lift rear of reservoir assembly up and carefully pull away from under secondary dash. Use caution as to not damage solenoids or solenoid seals when removing reservoir.
10. Remove reservoir seal. This seal may remain attached to reservoir assembly.
11. Remove lower hose clamp, hose, reservoir seal and solenoid if necessary.
12. Inspect reservoir assembly, hose, seal and solenoid for any type of damage or wear. Replace any damaged or worn parts.
13. Reverse procedure to install, noting the following:
    a. **Torque** cowl bracket attaching nuts to 15-25 ft. lbs.

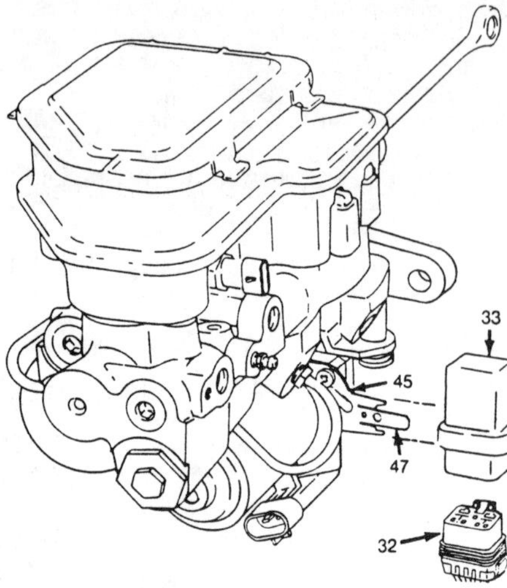

32. 5-PIN CONNECTOR
33. MOTOR RELAY
45. RELAY BRACKET
47. RELAY BRACKET RELEASE TAB

Fig. 65 Motor relay location

b. Bleed Powermaster III, see "Powermaster III Bleeding."

SOLENOID ASSEMBLIES, REPLACE

Clean reservoir cover assembly (5) and surrounding area before removing, **Fig. 64.**

1. Depressurize Powermaster III, see "Depressurizing Powermaster III."
2. Clean surrounding area, then remove reservoir cover and assembly.
3. Disconnect 3-pin connector (12), from solenoid assembly (14), **Fig. 66.**
4. Remove solenoid attaching screws (13), then the solenoid assembly. Ensure lower solenoid O-rings are in place and in good condition.
5. Reverse procedure to install, noting the following:
   a. **Torque** solenoid assembly attaching screws to 33-45 ft. lbs.
   b. Bleed Powermaster III, see "Powermaster III Bleeding."

PRESSURE SWITCH, REPLACE

1. Depressurize Powermaster III, see "Depressurizing Powermaster III."
2. Disconnect 3-pin electrical connector from pressure switch (26), **Fig. 64.**
3. Loosen, but do not completely remove cowl bracket attaching nuts (20), then pull Powermaster III unit slightly forward.
4. Raise and support vehicle. Using switch socket J-37117, or equivalent, swivel joint and extensions, remove switch through bottom of engine compartment.
5. Remove O-ring (27), from pressure switch, then discard O-ring.
6. Reverse procedure to install, noting the following:
   a. **Torque** pressure switch to 15-20 ft. lbs.
   b. **Torque** cowl bracket attaching nuts to 15-25 ft. lbs.
   c. Bleed Powermaster III unit, see "Powermaster III Bleeding."

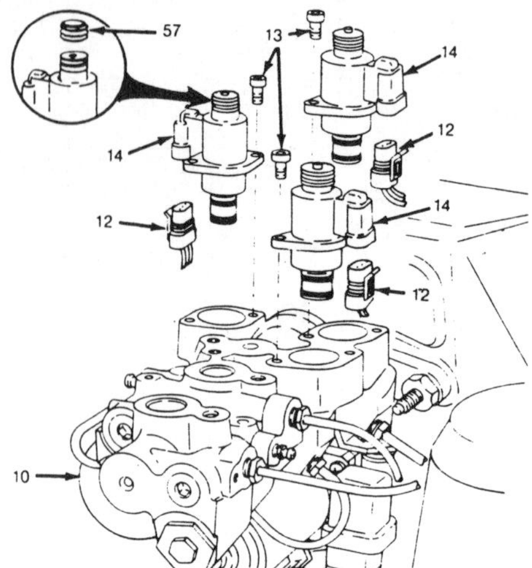

10. POWERMASTER-III
12. 3-PIN ELECTRICAL CONNECTOR
13. SCREWS (2 PER SOLENOID)
14. SOLENOID ASSEMBLY
57. SOLENOID SEALS (ONE PER SOLENOID)

Fig. 66 Solenoid assemblies

POWERMASTER III UNIT, REPLACE

Do not use wiring harness to aid in lifting or supporting of Powermaster III unit.

1. Depressurize Powermaster III, see "Depressurizing Powermaster III."
2. Disconnect 7-pin connector (15), from Powermaster III harness connector (16), **Fig. 67.**
3. Disconnect 10-pin connector (17), from Powermaster III wiring harness connector (18).
4. Disconnect 2-pin connector (1), from fluid level sensor switch (2).
5. Remove three brake pipes (19), from Powermaster III unit. Plug open lines and fittings to prevent fluid loss and contamination.
6. Remove hair pin clip (located inside vehicle), and pushrod (22), from brake pedal.
7. Remove two attaching nuts (20), from cowl bracket studs, then remove Powermaster III unit.
8. Reverse procedure to install, noting the following:
   a. **Torque** two cowl bracket attaching nuts to 15-25 ft. lbs.
   b. **Torque** tube-to-Powermaster III unit nuts to 10-15 ft. lbs.
   c. Bleed Powermaster III unit, see "Powermaster III Bleeding."

ANTI-LOCK BRAKE CONTROLLER, REPLACE

The anti-lock brake controller can be removed from vehicle without removing front seat.

1. Turn ignition switch to the Off position, slide passenger seat forward, then tip seat forward.
2. Remove bolts securing case cover, then remove anti-lock controller. It is not necessary to remove the harness connectors to slide controller out of the case.
3. Remove connectors from controller.
4. Reverse procedure to install, tightening case cover bolts hand tight.

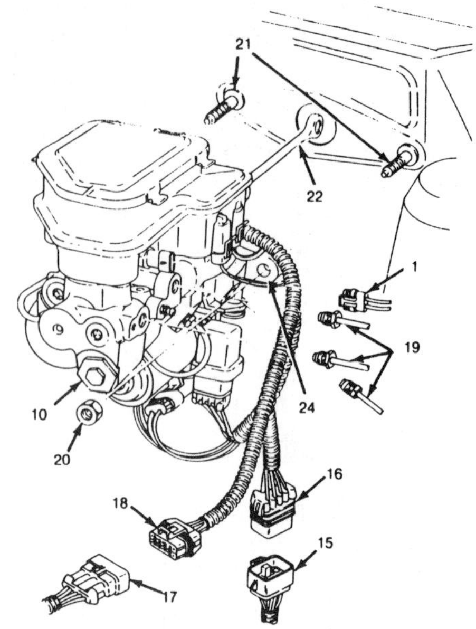

1. 2-PIN ELECTRICAL CONNECTOR (BODY HARNESS)
10. POWERMASTER-III
15. VEHICLE 7-PIN CONNECTOR
16. 7-PIN WIRING HARNESS CONNECTOR
17. VEHICLE 10-PIN CONNECTOR
18. 10-PIN WIRING HARNESS CONNECTOR
19. BRAKE PIPE
20. NUTS (2)
21. COWL BRACKET STUD
22. PUSHROD
24. MOUNTING BRACKET

Fig. 67 Powermaster III brake tube, mounting and electrical connections

PROPORTIONER VALVE ASSEMBLY, REPLACE

The proportioner valve is not repairable and must be serviced as a complete assembly. The use of rubber hoses or hydraulic lines other than those specified may lead to functional problems and major repairs.

1. Depressurize Powermaster III, see "Depressurizing Powermaster III."
2. Remove input brake line, then two output brake lines, from proportioner valve assembly. Input brake line has a 12 mm threaded nut and output brake lines have 10 mm threaded nuts. Plug open lines and fittings to prevent fluid loss and contamination.
3. Remove proportioner valve assembly.
4. Reverse procedure to install, noting the following:
   a. **Torque** tube nuts to 10-15 ft. lbs.
   b. Bleed system, see "Manual Bleeding," or "Pressure Bleeding."

FRONT WHEEL SPEED SENSORS, REPLACE

1. Raise and support vehicle, then remove sensor connector from wiring harness.
2. Remove two front wheel speed sensor bolts, then one bracket connector bolt.
3. Remove front speed sensor.
4. Reverse procedure to install, noting the following:
   a. **Torque** 15 mm bolts to 52-65 ft. lbs
   b. **Torque** 10 mm bolts to 6-9 ft. lbs.

Teves Type, Except 1988–89 Cutlass Supreme, Eldorado, Reatta, Regal, Riviera, Seville & Toronado

INDEX

OPERATING PRECAUTIONS

Some of the following procedures require that hydraulic lines, hoses and fittings be disconnected for inspection of testing. This Teves brake system uses a hydraulic accumulator which, when completely charged, contains brake fluid under high pressure. Before disconnecting any hydraulic lines, hoses or fittings, ensure the accumulator is completely depressurized as described further on. Failure to depressurize the hydraulic accumulator may result in personal injury and/or damage to painted surfaces and/or the vehicle.

Certain components in the system are not intended to be serviced individually. Do not attempt to remove or disconnect certain system components, such as the hydraulic accumulator or valve block. Only those components with removal procedures should be serviced.

Torque bolts attaching the brake caliper mounting bracket to the front and rear knuckles must be replaced any time they are removed or become loose. Tightening or loosening the torque bolts, which use encapsulated thread locking compound, is not recommended. Failure to replace the brake caliper mounting bracket torque bolts, after they have been loosened or removed, may result in loosening of the caliper to knuckle assembly, causing reduced vehicle braking ability.

COMPUTER SYSTEM SERVICE PRECAUTIONS

This Teves anti-lock brake system interfaces directly with two control computers, the Electronic Brake Control Module (EBCM) and the Body Computer Module (BCM). These modules are designed to withstand normal current draws associated with vehicle operation. However, care must be taken to avoid overloading any of the EBCM or BCM circuits. During testing of opens or shorts, do not ground or apply voltage to any of the circuits unless instructed to do so by the appropriate diagnostic procedure. These circuits should only be tested with a high impedance multimeter (J-35029-A or equivalent), or special tools described further on. Power should never be removed or applied to any control module with the ignition switch in the On position. Before removing or connecting battery cables, fuses, or connectors, always turn the ignition switch to the Off position.

GENERAL SERVICE PRECAUTIONS

If welding work is to be performed on the vehicle using an electric arc welder, the EBCM and valve block connectors should be disconnected before the welding operation begins. The EBCM and valve block connectors should never be connected or disconnected with the ignition switch in the On position. Some components of the anti-lock brake system are not serviceable separately and must be replaced as assemblies.

DESCRIPTION

The purpose of this Teves anti-lock brake system, is to prevent wheel lockup under heavy brake application.

The anti-lock function is desirable because a vehicle which is stopped without locking the wheels will normally stop in a shorter distance than a vehicle with locked wheels, while retaining directional stability and some steering capability. This allows the vehicle's operator to retain greater control of the vehicle during heavy brake application.

SYSTEM COMPONENTS

The anti-lock brake system, **Fig. 1**, consists of a pump motor assembly, fluid accumulator, pressure switch, fluid reservoir with integral filter, hydraulic booster/master cylinder, wheel speed sensors, Electronic Brake Control Module (EBCM) and a valve block assembly.

ELECTRONIC BRAKE CONTROL MODULE (EBCM)

The EBCM monitors speed of each wheel to determine if any wheel is beginning to lock. If a locking condition is sensed, brake pressures are automatically adjusted to provide for maximum stopping without wheel lock.

TIMER FLASHER MODULE

On some models, the pump circuit includes a timer flasher module, located in the instrument panel behind driver side close out panel, which monitors pump running time. If pump runs for more than three minutes, the timer flasher module will cause BRAKE warning lamp to flash.

HYDRAULIC UNIT

Main components of hydraulic unit are hydraulic booster master cylinder, valve block assembly, pump motor assembly, pressure switch, accumulator and fluid level sensor.

When brake pedal is depressed, hydraulic booster/master cylinder operates front brakes in normal manner and also provides modulated accumulator pressure to rear brakes.

The valve block assembly is a series of solenoid controlled valves which cycle very quickly to apply and release hydraulic pressure to each wheel.

The pump and motor assembly supplies high pressure brake fluid to accumulator. This high pressure fluid is used for power

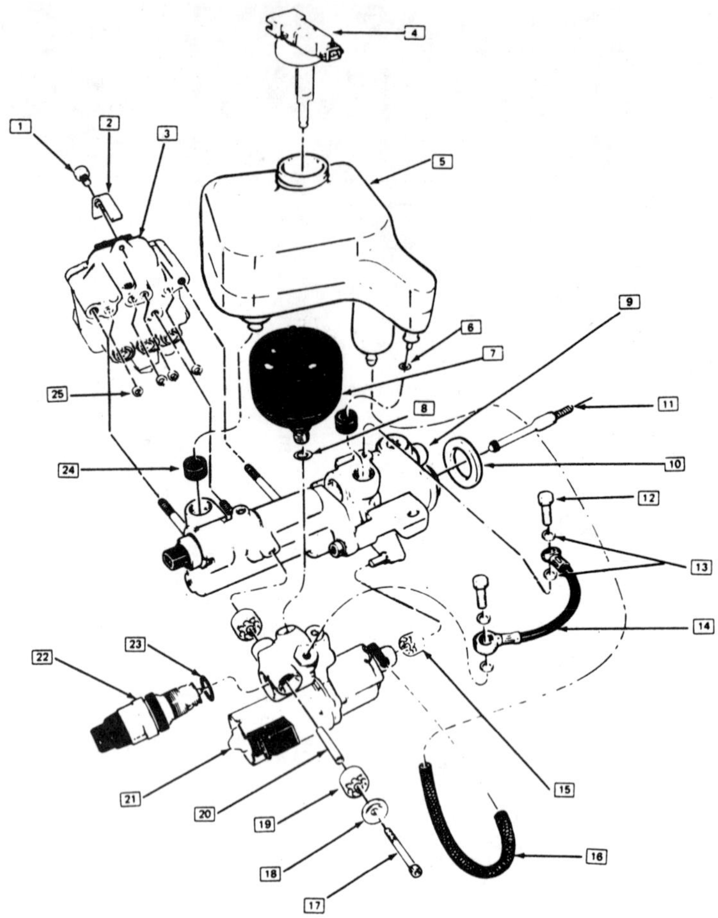

1. RESERVOIR RETAINER SCREW
2. RESERVOIR RETAINER
3. VALVE BLOCK ASSEMBLY
4. FLUID LEVEL SENSOR
5. BRAKE FLUID RESERVOIR
6. RESERVOIR O-RING SEAL
7. ACCUMULATOR
8. ACCUMULATOR O-RING SEAL
9. MASTER CYLINDER AND BOOSTER ASSEMBLY
10. HYDRAULIC UNIT TO PUSHROD ASSEMBLY SEAL
11. PUSHROD
12. HIGH PRESSURE HOSE FITTING
13. HIGH PRESSURE HOSE O-RING SEAL
14. HIGH PRESSURE HOSE
15. PUMP INSULATOR
16. RETURN HOSE
17. PUMP MOUNTING BOLT
18. PUMP WASHER
19. PUMP INSULATOR
20. PUMP MOUNTING BOLT SLEEVE
21. PUMP AND MOTOR ASSEMBLY
22. PRESSURE SWITCH
23. PRESSURE SWITCH O-RING SEAL
24. RESERVOIR GROMMET
25. VALVE BLOCK TO MASTER CYLINDER O-RING SEAL

Fig. 1 Typical anti-lock brake hydraulic unit

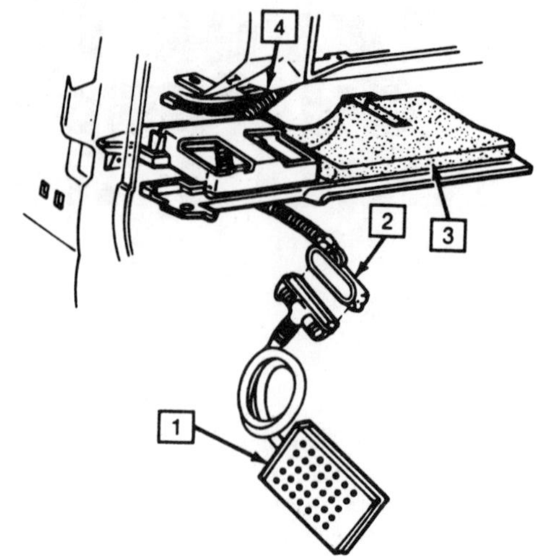

Fig. 2 Installing break-out box

assist and also to apply rear brakes.

The pressure switch monitors accumulator pressure. When pressure drops below the lower limit, pressure switch activates pump motor relay which turns on pump. Once pressure reaches upper limit, pressure switch deactivates the relay and shuts off pump. If pressure leak or pump failure occur, pressure switch will signal EBCM to disable anti-lock function. The pressure switch will also light BRAKE lamp located in instrument panel.

The accumulator is pressure storage device which can hold brake fluid under very high pressures. The accumulator has an internal diaphragm with nitrogen trapped on one side. As pump fills accumulator, diaphragm moves and compresses trapped nitrogen.

The fluid level sensor is located in fluid reservoir. This sensor has two functions. It can signal the EBCM that there is a low fluid condition. The EBCM will then disable the Anti-Lock function. The sensor will also light the BRAKE lamp located in instrument panel.

ANTI-LOCK WARNING LAMP

Vehicles equipped with the Anti-Lock system have an amber ANTI-LOCK warning lamp in the instrument panel. The ANTILOCK warning lamp will illuminate if a malfunction in the system is detected by the electronic controller. In case of an electronic malfunction, the controller will turn the ANTILOCK warning lamp on, then disable the anti-lock braking function. If ANTILOCK and BRAKE warning lamp come on at the same time, there may be something wrong with hydraulic brake system.

The ANTILOCK lamp will illuminate during engine starting and will usually stay on approximately three seconds after ignition switch is returned to the Run position. If the car has not been started in several hours, the ANTILOCK lamp may stay on for as long as thirty seconds. The ABS pump is recharging the hydraulic accumulator to normal operating pressures. The lamp will stay on until charging is complete. If lamp stays on longer than thirty seconds after engine is started or comes on while driving, the brake system should be inspected for a malfunction.

BRAKE WARNING LAMP

The Anti-Lock brake system uses a three circuit design so that some braking capacity is still available if hydraulic pressure is lost in one circuit. The BRAKE warning lamp is located at the lefthand side of instrument cluster and is designed to alert driver of conditions that would result in reduced braking ability.

The BRAKE warning lamp should turn on under several conditions. The BRAKE lamp will illuminate during engine cranking, when parking brake is not fully released, low brake fluid level, or low accumulator pressure. If BRAKE warning lamp stays on any longer than thirty seconds after starting engine or comes on during driving, there may be a malfunction in brake hydraulic system.

OPERATION

Under normal driving conditions the Anti-Lock brake system functions the same as standard brake system. However, with the detection of wheel lock-up, a slight bump or kick-back will be felt in brake pedal. This pedal "bump" will be followed by a series of short pedal pulsations which occur in rapid succession. Brake pedal pulsation will continue until there is no longer a need for the anti-lock function

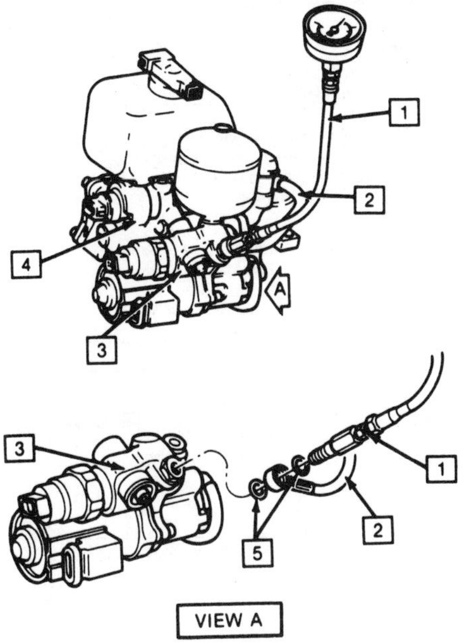

1 J-35604 PRESSURE GAGE ASSEMBLY
2 HIGH PRESSURE HOSE
3 PUMP AND MOTOR ASSEMBLY
4 HYDRAULIC UNIT
5 SEALS (O-RINGS)

Fig. 3 Installing pressure gauge

or until vehicle is stopped. A slight ticking or popping noise may be heard during brake application. This noise is normal and indicates that the anti-lock function is being used.

When anti-lock system is in use, brake pedal may rise even as brakes are being applied. Maintaining a constant force on the pedal will provide shortest stopping distance.

PRE-DIAGNOSIS INSPECTION

ITEM	INSPECT FOR:	CORRECTIVE ACTION
BRAKE FLUID RESERVOIR AND HYDRAULIC UNIT	– LOW FLUID LEVEL – EXTERNAL LEAKS	– FILL RESERVOIR – REPAIR LEAKS AS REQUIRED
PARKING BRAKE	– FULLY RELEASED – IMPROPERLY ADJUSTED	– RELEASE PARKING BRAKE – ADJUST CABLE
BATTERY	ADEQUATE CHARGE	– CHARGE OR REPLACE BATTERY AS REQUIRED – SERVICE CHARGING SYSTEM AS REQUIRED
FUSES • 5 AMP ELECTRONIC BRAKE FUSE – LOCATED IN CONVENIENCE CENTER	– BLOWN FUSE	– REPLACE FUSE AND VERIFY OPERATION
• 10 AMP GAGES FUSE – LOCATED IN FUSE BLOCK	– BLOWN FUSE	– REPLACE FUSE AND VERIFY OPERATION
• 30 AMP MAIN RELAY FUSE – LOCATED ON MAIN RELAY BRACKET	– BLOWN FUSE	– REPLACE FUSE AND VERIFY OPERATION
• 30 AMP PUMP MOTOR FUSE – LOCATED ON PUMP MOTOR RELAY BRACKET	– BLOWN FUSE	– REPLACE FUSE AND VERIFY OPERATION
CONNECTORS • MAIN RELAY • PUMP MOTOR RELAY • PRESSURE SWITCH • PUMP MOTOR • MAIN VALVE • VALVE BLOCK • FLUID LEVEL SENSOR • ELECTRONIC BRAKE CONTROL MODULE (EBCM) • 4 SENSOR CONNECTORS	– PROPER ENGAGEMENT OF CONNECTOR – LOOSE WIRES IN CONNECTOR	– PROPERLY ENGAGE CONNECTORS – REPAIR LOOSE WIRES
CONNECTORS (WITH ALL WHEEL DRIVE) • DIFFERENTIAL LOCK ENGAGED SWITCH CONNECTOR AT TRANSAXLE TRANSFER CASE • DIFFERENTIAL LOCK ENGAGED SWITCH JUMPER CONNECTOR TO EBCM WIRING HARNESS AT LEFT SIDE OF DASH PANEL	-- PROPER ENGAGEMENT OF CONNECTOR -- LOOSE WIRES IN CONNECTOR	-- PROPERLY ENGAGE CONNECTORS -- REPAIR LOOSE WIRES

Fig. 4 Pre-diagnosis inspection chart (Part 1 of 2). 1989 6000

DIAGNOSIS

Diagnosis of system malfunctions requires three basic steps which must be followed in order to isolate the fault accurately and quickly. The proper diagnostic procedure consists of the following steps: Pre-Diagnosis Inspection, Lamp Sequence Determination and Component Test Charts.

PREPARATION

Some diagnosis procedures require use of a pressure gauge and Pin-Out Box tool No. J-35592.

1. Install Pin-Out Box tool No. J-35592 as follows:
   a. Turn ignition switch to the Off position.
   b. Depending on model, different procedures are required for EBCM removal. EBCM is either held in with two retainers, three screws, or a twist-type mount. EBCM is located on left side dash close out panel, **Fig. 2.**
   c. Remove 35 pin harness connector from EBCM, depress locking plate, then turn toward front of car.
   d. Inspect 35 pin harness connector for damage, then connect pin-out box connector to harness connector.
2. **On 1988-89 models,** install pressure gauge J-35604 as follows:
   a. Depressurize accumulator, refer to ""Depressurizing Hydraulic Accumulator," procedure.
   b. Remove pressure hose fitting from pump body, **Fig. 3**,. Use caution not to drop two O-ring seals when removing fitting.
   c. Install one O-ring seal on pressure gauge fitting, then insert gauge fitting into pressure hose coupling.
   d. Install remaining O-ring seal on gauge fitting on underside of pressure hose coupling, then thread gauge fitting into pump body. **Torque** fitting to 15 ft. lbs.
3. **On 1989 models,** install pressure gauge as follows:
   a. Depressurize accumulator, refer to ""Depressurizing Hydraulic Accumulator," procedure.
   b. Remove accumulator, then install pressure gauge J-35604-88. **Torque** pressure gauge to 32 ft. lbs.
   c. Install, then tighten accumulator to pressure gauge. When removing the gauge, inspect O-ring for damage. Replace as necessary.

PRE-DIAGNOSIS INSPECTION

The pre-diagnosis inspection consists of a quick visual check of specific system components which could create an apparent Anti-Lock system malfunction. Performing this inspection prior to diagnosing specific symptoms, may result in diagnosis of a simple defect which may be the cause of an inoperative system. This inspection should be the first step in diagnosing a complaint. For visual inspection procedure, refer to **Figs. 4 and 5.**

LAMP SEQUENCE DETERMINATION

The second step in diagnosing this system is determination of warning lamp behavior. This system uses two warning lamps, a red BRAKE warning lamp and an amber ANTILOCK lamp. By observing when these lamps illuminate, individual components and sub-systems may be isolated.

Normal indicator lamp sequence is shown in **Fig. 6 and 7.** Indicator lamp sequences are shown graphically in **Fig. 8 and 9,** and described below.

Vehicle Status

Each lamp sequence graphs, **Figs. 8 and 9,** are divided into seven areas which represent vehicle status. Behavior of BRAKE and ANTILOCK lamps are represented in horizontal rows below vehicle status headers.

Lamp Status

BRAKE and ANTILOCK lamp status is indicated by shaded areas in horizontal rows below each vehicle status header, **Figs. 8 and 9.** Righthand diagonal shading indicates illumination of ANTILOCK lamp. Lefthand diagonal shading indicates illumination of BRAKE lamp. A vehicle status block which is not entirely shaded means that the lamp is only on for a portion of the test period.

LAMP SEQUENCE TEST PROCEDURE

The following procedure should be used to determine proper light sequence used for further diagnosis. When evaluating vehicle condition, use caution if reduced braking ability is evident.

1. Turn ignition switch to Off position for a minimum of 15 seconds.
2. Turn ignition switch to On position, then observe BRAKE and ANTILOCK lamps, noting the following:
   a. If accumulator has been discharged, both indicator lamps may remain on for approximately 30 seconds. If this occurs, allow lamps to turn off and turn ignition switch to Off position for another 15 second period, then proceed to step 3.
   b. If lamps do not go out within 30 seconds, proceed to step 4.
3. Turn ignition switch to On position and observe warning lamps. At this point, BRAKE lamp should remain off. ANTILOCK lamp should come on for 3-6 seconds, then go out.
4. Turn ignition switch to "Start" position and observe warning lamps. Both warning lamps should light.
5. Start engine and release key to Run position and observe warning lamps. BRAKE lamp should turn off immedi-

PRE-DIAGNOSIS INSPECTION (CONT.)

ITEM	INSPECT FOR:	CORRECTIVE ACTION
GROUNDS • BODY GROUNDS -- LOCATED ON MAIN RELAY BRACKET • HYDRAULIC UNIT GROUND -- ON HYDRAULIC UNIT	-- LOOSE CONNECTIONS -- BROKEN EYELETS -- CORROSION	-- TIGHTEN CONNECTIONS -- REPAIR WIRE OR EYELET -- CLEAN CONTACT SURFACES
JUNCTION BLOCK	– CORROSION – LOOSE JUNCTION BLOCK OR CONNECTIONS – BROKEN WIRES	– CLEAN CONTACT SURFACES – TIGHTEN CONNECTIONS – REPAIR OR REPLACE WIRES AS NECESSARY

Fig. 4 Pre-diagnosis inspection chart (Part 2 of 2). 1989 6000

ABS VISUAL INSPECTION

ITEM	INSPECT FOR	CORRECTIVE ACTION
PARKING BRAKE	- FULL RELEASE - PROPER SWITCH FUNCTION IF NECESSARY, UNPLUG SWITCH CONNECTOR TO VERIFY - PROPER VACUUM RELEASE SYSTEM OPERATION	- OPERATE MANUAL RELEASE LEVER TO VERIFY RELEASE - ADJUST CABLE OR REPAIR RELEASE SYSTEM AS REQUIRED - REPAIR SWITCH AS REQUIRED
MAJOR COMPONENTS • BRAKE FLUID RESERVOIR • HYDRAULIC UNIT • PUMP/MOTOR ASSEMBLY	- LOW FLUID LEVEL - EXTERNAL LEAKS - PROPER ASSEMBLY	- ADD FLUID AS REQUIRED DETERMINE CAUSE OF FLUID LOSS AND REPAIR - REPAIR LEAKS AS REQUIRED - INSTALL OR POSITION COMPONENTS PROPERLY
FUSIBLE LINKS ON POSITIVE JUNCTION BLOCK	- OPEN LINK	- INSPECT FOR CAUSE OF FAILURE - REPLACE - VERIFY OPERATION
CONNECTORS • MAIN RELAY • PUMP MOTOR RELAY • PRESSURE SWITCH • PUMP MOTOR • MAIN VALVE • VALVE BLOCK • FLUID LEVEL SENSOR • LF AND RF WHEEL SPEED SENSORS • LR AND RR WHEEL SPEED SENSORS • ELECTRONIC BRAKE CONTROL MODULE (EBCM)	- PROPER ENGAGEMENT - LOOSE WIRES OR TERMINALS - CORRODED OR BROKEN EYELETS	- PROPERLY ENGAGE CONNECTOR - REPAIR AS REQUIRED

IF NO PROBLEM IS NOTED IN VISUAL INSPECTION, PROCEED TO ABS FUNCTIONAL CHECK

Fig. 5 Pre-diagnosis inspection chart. All models, except 1989 6000

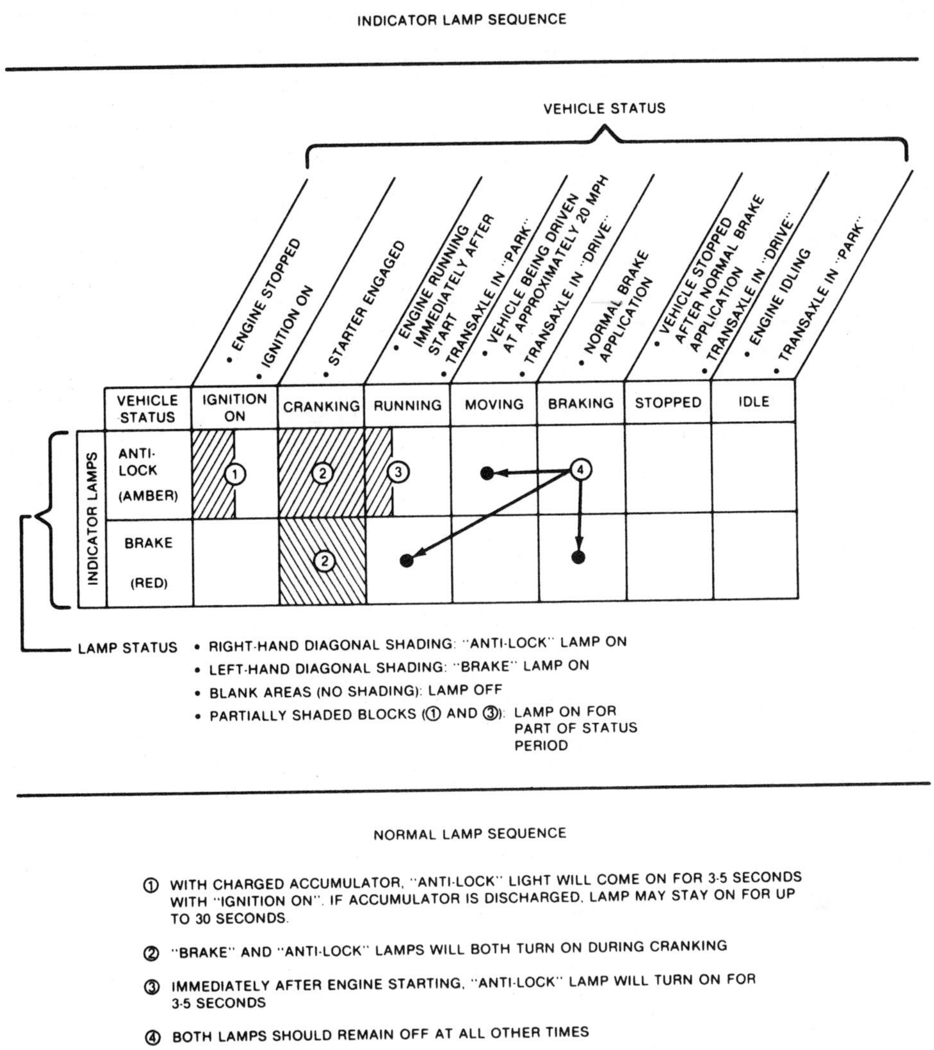

Fig. 6 Normal indicator lamp sequence

ately. ANTILOCK lamp should remain on for 3-6 seconds, then go out.

6. Drive vehicle at a minimum speed of 20 mph for a short period of time and observe warning lamps. Both lamps should remain off for this step and remainder of test. Note as accurately as possible conditions under which either lamp comes on.
7. Stop vehicle using normal brake application. Both lights should remain off.
8. Place transaxle in PARK, then allow vehicle to idle for several seconds and observe warning lamps. Both should remain off.
9. From information obtained in above test, determine lamp sequence and/or symptom which most closely matches vehicle condition, then perform indicated tests, **Figs. 8 and 9.**

DIAGNOSTIC PROCEDURE

Diagnostic procedures in this section may be helpful in determining the cause of intermittent problems in Anti-Lock brake system electrical components. In most cases, a specific fault must be present to locate a problem effectively using Indicator Lamp Sequence chart, **Figs 8 and 9,** along with Pin-Out Checks and trouble trees, **Figs. 10 through 46,** for all 1986 models and 1987-88 Pontiac 6000, **Figs. 47 through 84,** for all 1987-89 models except 1987-89 Pontiac 6000 and **Figs. 85-127,** for 1989 Pontiac 6000.

NOTE ON INTERMITTENTS

If an intermittent fault cannot be diagnosed using Indicator Lamp Sequence chart, the problem will most probably be caused by a defective electrical connection or wiring. When an intermittent problem is encountered, check questionable circuits, **Figs. 125 through 130,** and electrical connectors, **Figs. 131-134,** for the following:

1. Poor mating of connector halves or terminals not fully seated in connector body.
2. Improperly formed or damaged terminals. All connector terminals in the problem circuit should be carefully removed to increase contact tension.
3. Poor terminal wire connection.

If a visual check does not locate cause of problem, operate vehicle with EBCM connected and attempt to duplicate condition and use Indicator Lamp Sequence chart to isolate potential cause.

Circuits which could possibly cause intermittent operation of ANTILOCK lamp are as follows:

1. Wheel speed sensor circuits — low or intermittent output.
2. Ignition enable circuit — interruption of 12 volt input.
3. EBCM switch loop circuit (low fluid/low pressure sensors — intermittent open.
4. Main relay circuit — open in coil or switched battery power.

SYSTEM SERVICE

DEPRESSURIZING HYDRAULIC ACCUMULATOR

Depressurize accumulator before performing any ABS service.

1. Disconnect battery ground cable or place ignition switch in the Off position.
2. Apply and release brake pedal a minimum of 20 times using approximately 50 lbs. of force on pedal. A noticeable change in pedal feel will occur when accumulator is completely discharged. **If battery negative cable was disconnected, reactivate Delco-Loc radios by turning ignition and radio to the On position, then pressing the Mute button on the steering wheel control.**

BRAKE SYSTEM FILLING

Ensure pump is not allowed to run for more than 60 seconds at one time. If pump must be run longer, allow pump to cool for several seconds between 60 second runs.

1. Depressurize accumulator, refer to "Depressurizing Hydraulic Accumulator" procedure.
2. Fill reservoir to full mark using DOT 3 brake fluid or equivalent.

BRAKE SYSTEM BLEEDING

Manual Bleeding

The front hydraulic circuits may be manually bled in the same manner as conventional brakes. The rear brakes must be pressure bled according to the following procedure.

Pressure Bleeding

Only the rear brakes must be pressure bled. Front hydraulic circuits may be bled manually. To prevent air, moisture and other contaminants from entering system, only diaphragm type pressure bleeding equipment should be used.

Front Circuit

1. Disconnect battery ground cable, then depressurize hydraulic accumulator. See "Depressurizing Hydraulic Accumulator."
2. Disconnect fluid level sensor electrical connector, then remove sensor.
3. Install brake bleeder adapter J-35798, or equivalent, in place of sensor.
4. Attach diaphragm type brake bleeder

WARNING/INDICATOR LAMP SEQUENCE

WARNING/ INDICATOR LAMP	COLOR	IGNITION ON*	CRANKING*	RUNNING*	MOVING*	BRAKING*	STOPPED*	IDLE*	"LOCK-NORMAL" SWITCH	
									MOVED TO "LOCK"	MOVED TO "NORMAL"
"BRAKE"	RED	OFF	ON	OFF	OFF	OFF	OFF	OFF	OFF	OFF
"ANTI-LOCK"	AMBER	ON, THEN OFF	ON	ON, THEN OFF	OFF	OFF	OFF	OFF	ON	ON, THEN OFF
"LOCK-NORMAL" SWITCH	AMBER	OFF	OFF	OFF	OFF	OFF	OFF	OFF	FLASH	OFF
"LOCK"	AMBER	OFF	OFF	OFF	OFF	OFF	OFF	OFF	OFF, THEN ON	ON, THEN OFF

*"LOCK-NORMAL" SWITCH SET TO "NORMAL"

NORMAL SEQUENCE

1—WITH IGNITION "ON," "ANTILOCK" WARNING LAMP WILL COME ON FOR 3-5 SECONDS IF ACCUMULATOR IS CHARGED. "ANTILOCK" WARNING LAMP MAY STAY ON FOR UP TO 30 SECONDS IF ACCUMULATOR IS DISCHARGED.

2—"BRAKE" AND "ANTILOCK" WARNING LAMPS WILL TURN ON DURING CRANKING.

3—IMMEDIATELY AFTER THE ENGINE STARTS, "ANTILOCK" WARNING LAMP WILL BE ON FOR 3-5 SECONDS.

4—WHEN "LOCK-NORMAL" SWITCH IS MOVED TO "LOCK," AMBER SWITCH INDICATOR LAMP WILL FLASH.

5—THE "LOCK" AND "ANTILOCK" LAMPS WILL TURN ON WHEN THE TRANSFER CASE DIFFERENTIAL VACUUM-ACTUATED LOCK LOCKS THE DIFFERENTIAL.

6—WHEN "LOCK-NORMAL" SWITCH IS MOVED BACK TO "NORMAL," "ANTILOCK" WARNING LAMP WILL REMAIN ON FOR 3-5 SECONDS AND THEN GO OFF. SWITCH INDICATOR LAMP WILL IMMEDIATELY STOP FLASHING AND REMAIN OFF.

7—THE "LOCK" INDICATOR LAMP WILL GO OFF WHEN THE TRANSFER CASE DIFFERENTIAL SPRING MECHANISM UNLOCKS THE DIFFERENTIAL.

8—ALL FOUR LAMPS WILL REMAIN OFF AT ALL OTHER TIMES.

Fig. 7 Warning/Indicator lamp sequence chart. 1989 6000 models

J 29532, or equivalent, to fluid reservoir, then charge to 20 psi.

5. Raise and support vehicle. Attach a clear bleeder hose to a front bleeder valve, then submerse the other end in a clear container partially filled with DOT 3 brake fluid.
6. Open bleeder valve, using brake bleeder wrench J 21472, or open end wrench, then allow fluid to flow from bleeder until no air is seen in brake fluid.
7. Close bleeder valve, ensuring bleeder seats in the closed position.
8. Repeat steps 5 through 7 on remaining front bleeder valve.
9. Check fluid level and adjust as required.
10. Check pedal for sponginess. Repeat procedure until desired pedal is achieved.
11. Remove bleeding equipment, then install fluid level sensor.
12. **Reactivate Delco-Loc radios by turning ignition and radio to the On position, then pressing the Mute button on the steering wheel control.**

Rear Circuit

Fill brake fluid reservoir with DOT 3 brake fluid. Keep reservoir at least half full during procedure.

1. Turn ignition switch to On position and allow system to charge. **Pump motor will stop when system is charged.**
2. Raise and support vehicle. Attach a clear bleeder hose to rear bleeder valve, then submerse the other end in a clear container partially filled with DOT 3 brake fluid.
3. Open bleeder valve using brake bleeder wrench J 21472, or open end wrench, then with ignition switch in On position, depress brake pedal for at least 10 seconds.**Do not allow pump to run more than 60 seconds at a time. If pump must be run longer than 60 seconds, allow pump to cool between 60 second runs.**
4. Allow fluid to flow from bleeder until no air is seen in brake fluid. Repeat step 3 if necessary.
5. Close bleeder valve, then repeat steps 2 through 4 on remaining rear bleeder valve.
6. Lower vehicle. Check reservoir fluid level and adjust as required.
7. Check pedal for sponginess. If condition exists, repeat procedure until desired pedal is achieved. **Do not move vehicle until a firm brake pedal is achieved.**

COMPONENT REPLACEMENT

HYDRAULIC UNIT

1. Disconnect battery ground cable.
2. Depressurize accumulator, refer to ""Depressurizing Hydraulic Accumulator" procedure.
3. Disconnect all hydraulic unit electrical connectors, then remove all brake fluid from fluid reservoir.
4. Remove pump attaching bolt, then position energy unit aside to gain access to brake pipes.
5. Disconnect hydraulic lines to valve block.**Use caution not to twist lines.**
6. **On Pontiac 6000 models,** remove line attaching hydraulic unit to combination valve.
7. **On all models,** disconnect push rod from brake pedal, then push dust boot forward, past hex on pushrod.
8. Separate pushrod halves, then remove hydraulic unit to pushrod attaching bolts.
9. Remove hydraulic unit from vehicle. **Front half of pushrod will remain locked into hydraulic unit.**
10. Reverse procedure to install, noting the following:
    a. **Torque** support bracket attaching bolts to 37 ft. lbs.

SEE LAMP SEQUENCE DETERMINATION PROCEDURE
IN THIS SECTION BEFORE USING THIS CHART

SEQUENCE NUMBER	LAMP SEQUENCE	SYMPTOM DESCRIPTION	PERFORM TEST
1	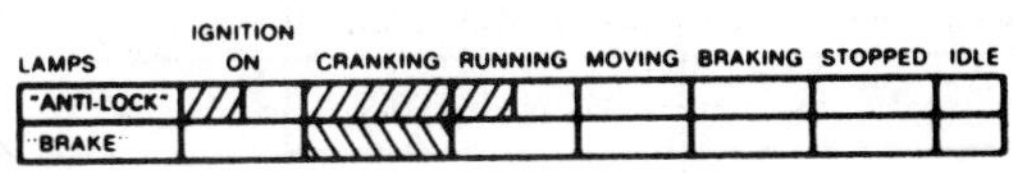	NORMAL LAMP SEQUENCE WITH — EXCESSIVE PEDAL TRAVEL OR SPONGY PEDAL — ANTI-LOCK BRAKING OPERATION OR VALVE CYCLING DURING NORMAL STOPS ON DRY PAVEMENT — POOR VEHICLE TRACKING DURING ANTI-LOCK BRAKING	H C D
2	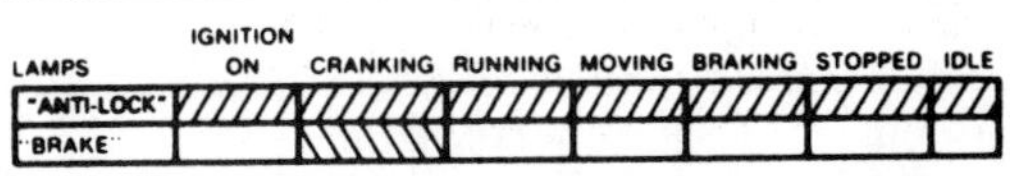	CONTINUOUS "ANTI-LOCK" LAMP NORMAL "BRAKE" LAMP	A
3	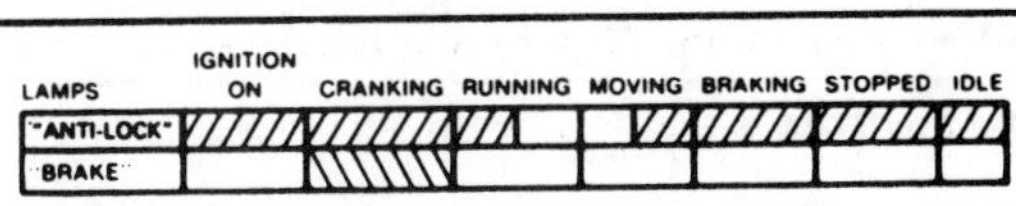	"ANTI-LOCK" LAMP COMES ON AFTER VEHICLE STARTS MOVING NORMAL BRAKE LAMP	C
4	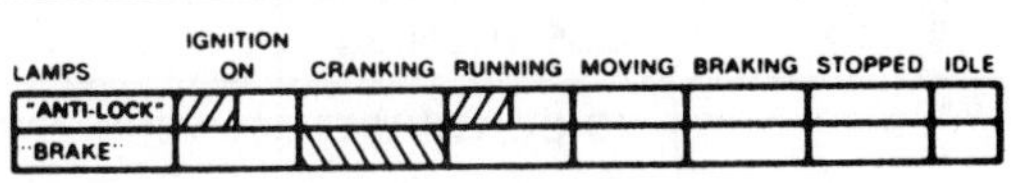	NO "ANTI-LOCK" LAMP WHILE CRANKING NORMAL "BRAKE" LAMP	E
5	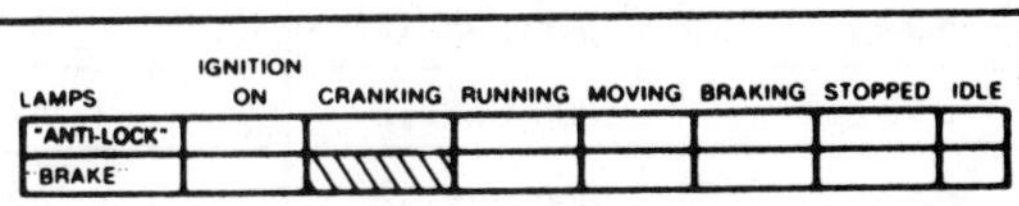	NO "ANTI-LOCK" LAMP NORMAL "BRAKE" LAMP	F
6	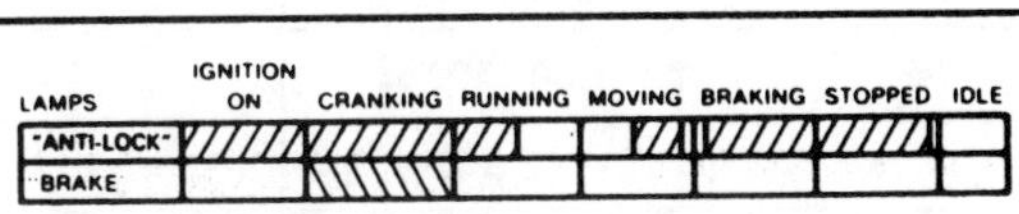	INTERMITTANT "ANTI-LOCK" LAMP WHILE DRIVING NORMAL "BRAKE" LAMP	G
7	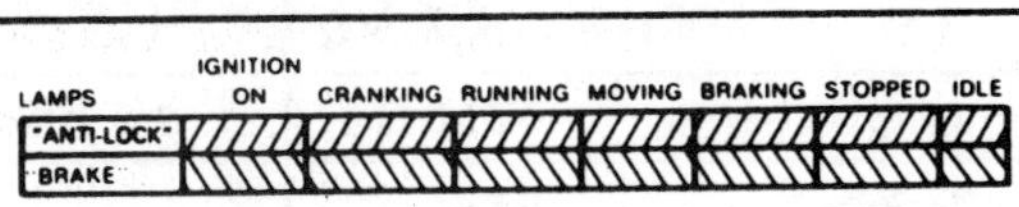	CONTINUOUS "ANTI-LOCK" LAMP CONTINUOUS "BRAKE" LAMP	B
8	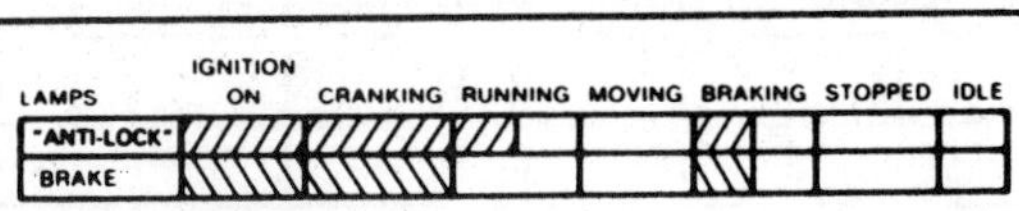	"ANTI-LOCK" AND "BRAKE" LAMPS COME ON WHILE BRAKING	B
9	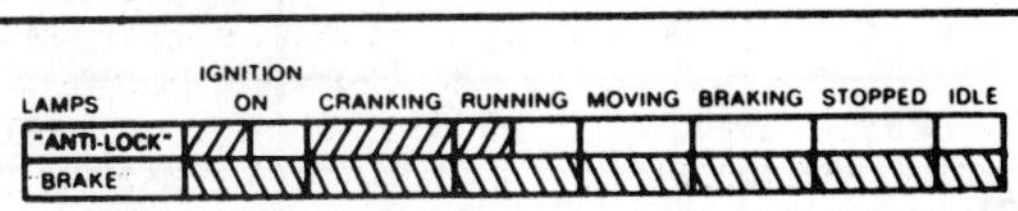	NORMAL "ANTI-LOCK" LAMP CONTINUOUS "BRAKE" LAMP	B
10	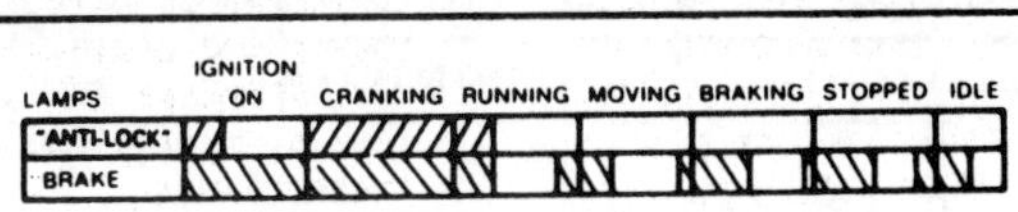	NORMAL OR CONTINUOUS "ANTI-LOCK" LAMP FLASHING "BRAKE" LAMP	B

Fig. 8 Indicator lamp sequence chart. All models, except 1989 6000

TEST A
PIN-OUT CHECKS

- CONNECT BREAK-OUT BOX J-35592 to 35-PIN EBCM HARNESS CONNECTOR AS DESCRIBED IN THIS SECTION
- PERFORM CHECKS WITH HIGH IMPEDANCE DIGITAL MULTIMETER J-34029-A OR EQUIVALENT
- ALL CHECKS ARE MADE WITH ENGINE STOPPED

CIRCUIT TO BE TESTED	IGNITION SWITCH POSITION	MULTIMETER SCALE/RANGE	MEASURE BETWEEN PIN NUMBERS	SPECIFICATION	IF RESULT NOT WITHIN SPECIFICATION, SEE CHART
IGNITION ENABLE	RUN	20 DCV	2(+)*,1(−)*	10 V MINIMUM	A-1
MAIN RELAY GROUND	OFF	200 Ω	1,3	CONTINUITY	A-2
	OFF	200 Ω	1,20	CONTINUITY	
MAIN RELAY COIL	OFF	200 Ω	1,8	45-105 Ω	A-3
BEFORE PERFORMING THIS TEST: • REMOVE GAGE FUSE (FUSE #3) FROM FUSE PANEL • PLACE FUSED JUMPER BETWEEN BREAK-OUT BOX PINS 2 & 8					
MAIN RELAY POWER	ON	20 DCV	3(+),1(−)	10 V MINIMUM	A-4
	ON	20 DCV	20(+),1(−)	10 V MINIMUM	
BEFORE PROCEEDING: • REMOVE JUMPER FROM PINS 2 & 8 • INSTALL GAGE FUSE					
EBCM SWITCH LOOP	OFF	200 Ω	9,10	LESS THAN 5 Ω	A-5
	OFF	200 Ω	1,9	NO CONTINUITY	A-6
RR SENSOR RESISTANCE	OFF	2k Ω	4,22	800-1400 Ω	A-7
LF SENSOR RESISTANCE	OFF	2k Ω	5,23	800-1400 Ω	A-8
LR SENSOR RESISTANCE	OFF	2k Ω	6,24	800-1400 Ω	A-9
RF SENSOR RESISTANCE	OFF	2k Ω	7,25	800-1400 Ω	A-10
MAIN VALVE SOLENOID	OFF	200 Ω	11,18	2-5 Ω	A-11
VALVE BLOCK GROUND	OFF	200 Ω	1,11	LESS THAN 2 Ω	A-12
RF INLET VALVE	OFF	200 Ω	11,15	5-7 Ω	A-13
LF INLET VALVE	OFF	200 Ω	11,35	5-7 Ω	A-14
REAR INLET VALVE	OFF	200 Ω	11,17	5-7 Ω	A-15
RF OUTLET VALVE	OFF	200 Ω	11,34	3-5 Ω	A-16
LF OUTLET VALVE	OFF	200 Ω	11,16	3-5 Ω	A-17
REAR OUTLET VALVE	OFF	200 Ω	11,33	• 3-5 Ω	A-18
DIODE	OFF	DIODE ─┤◄──	27(+),3(−)	CONTINUITY	A-19
	OFF	DIODE ─┤◄──	3(+),27(−)	NO CONTINUITY	

* (+) OR (−) INDICATES MULTI-METER POLARITY

IF ALL TEST RESULTS ARE WITHIN SPECIFICATION, RECONNECT EBCM AND VERIFY CONTINUOUS "ANTI-LOCK" LAMP OPERATION

- IF NORMAL OPERATION RESUMES, SEE NOTE ON INTERMITTANTS
- IF LAMP REMAINS ON, SEE CHART A-20

Fig. 9 Pin-out checks. Test A, all 1986 models & 1987–88 Pontiac 6000

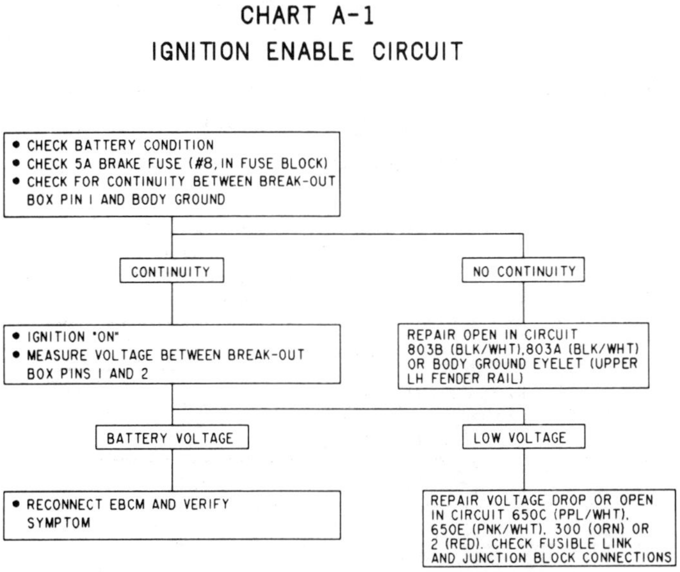

Fig. 10 Ignition enable circuit. Chart A-1, all 1986 models & 1987–88 Pontiac 6000

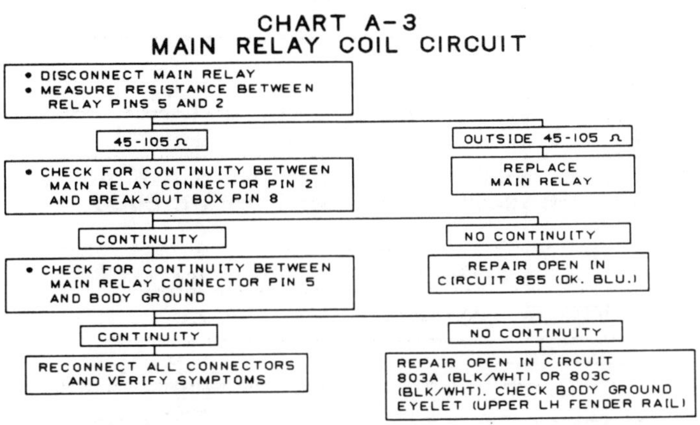

Fig. 12 Main relay coil circuit. Chart A-3, all 1986 models & 1987–88 Pontiac 6000

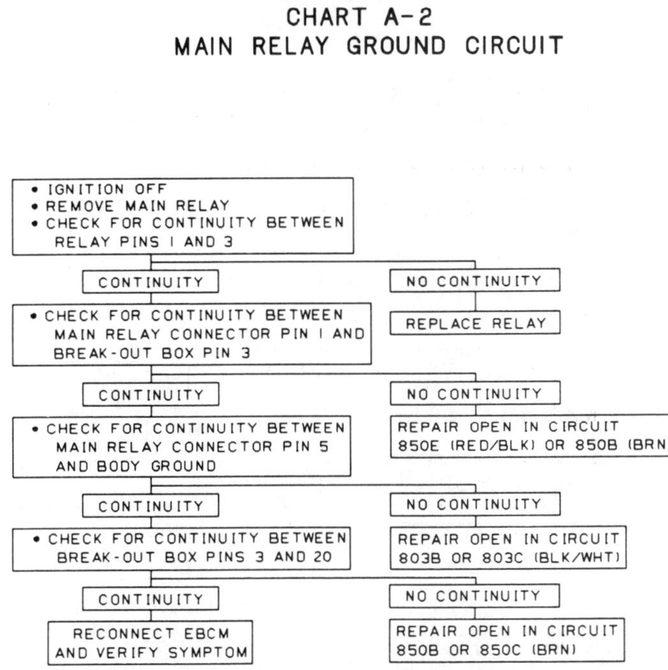

Fig. 11 Main relay ground circuit. Chart A-2, all 1986 models & 1987–88 Pontiac 6000

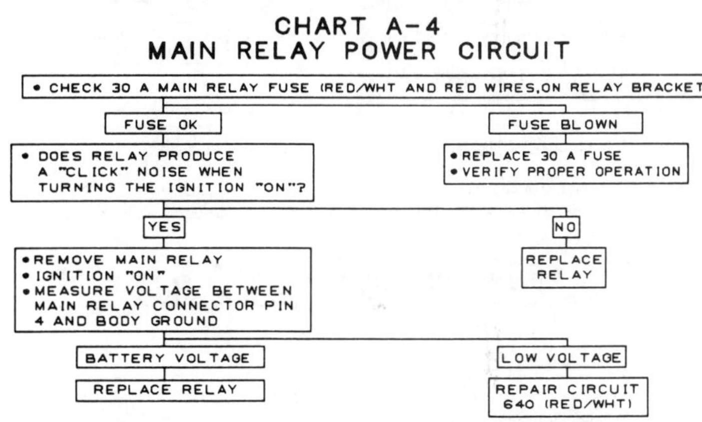

Fig. 13 Main relay power circuit. Chart A-4, all 1986 models & 1987–88 Pontiac 6000

b. Bleed brakes, refer to "Brake System Bleeding" procedure.
c. **Reactivate Delco-Loc radios by turning ignition and radio to the On position, then pressing the Mute button on the steering wheel control.**

PUMP MOTOR ASSEMBLY

1. Disconnect battery ground cable.
2. Depressurize accumulator, refer to ""Depressurizing Hydraulic Accumulator" procedure, then remove brake fluid from fluid reservoir.
3. Disconnect pressure switch and pump motor electrical connectors.
4. Remove hydraulic accumulator and O-ring.
5. Remove high pressure hose fitting connected to pump, then the pressure hose assembly and O-rings.
6. Remove wire clip and pull return hose fitting out of pump body.
7. Remove pump and motor assembly to main body attaching bolt and insulator grommets.
8. Remove pump and motor assembly by sliding off of locating pin.
9. Check insulators for damage or deterioration. Replace as required.
10. Reverse procedure to install. **Reactivate Delco-Loc radios by turning ignition and radio to the On position, then pressing the Mute button on the steering wheel control.**

VALVE BLOCK ASSEMBLY

1. Disconnect battery ground cable.
2. Depressurize accumulator, refer to ""Depressurizing Hydraulic Accumulator" procedure.
3. Remove hydraulic unit, then the three nuts and washers from valve block as shown in **Fig. 135.**
4. Remove valve block assembly and O-rings by sliding valve block off studs.
5. Reverse procedure to install. **Reactivate Delco-Loc radios by turning ignition and radio to the On position, then pressing the Mute button on the steering wheel control.**

PRESSURE/WARNING SWITCH

1. Disconnect battery ground cable.
2. Depressurize accumulator, refer to ""Depressurizing Hydraulic Accumulator" procedure.
3. Disconnect pressure switch electrical connector.
4. Remove pressure switch and O-ring using pressure switch socket J 35804-A, or equivalent.
5. Reverse procedure to install, noting the following:
   a. Lubricate new O-ring with clean brake fluid.
   b. **Torque** pressure switch to 17 ft. lbs.
   c. Turn ignition switch to On position and ensure BRAKE light goes out within 60 seconds.
   d. **Reactivate Delco-Loc radios by turning ignition and radio to the On position, then pressing the Mute button on the steering wheel control.**

CHART A-5
EBCM LOOP OPEN

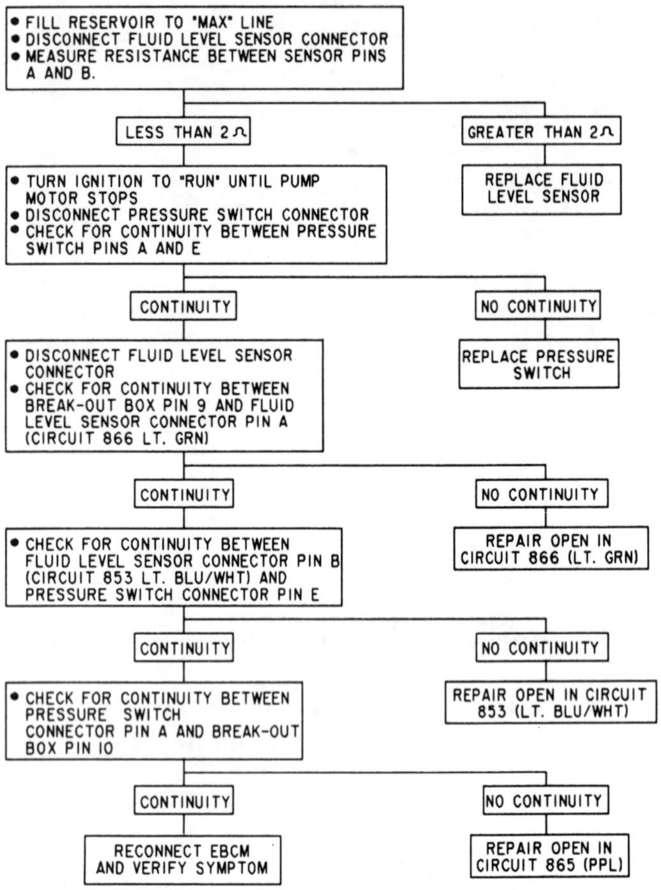

Fig. 14 EBCM loop open. Chart A-5, all 1986 models & 1987–88 Pontiac 6000

CHART A-6
EBCM LOOP SHORT

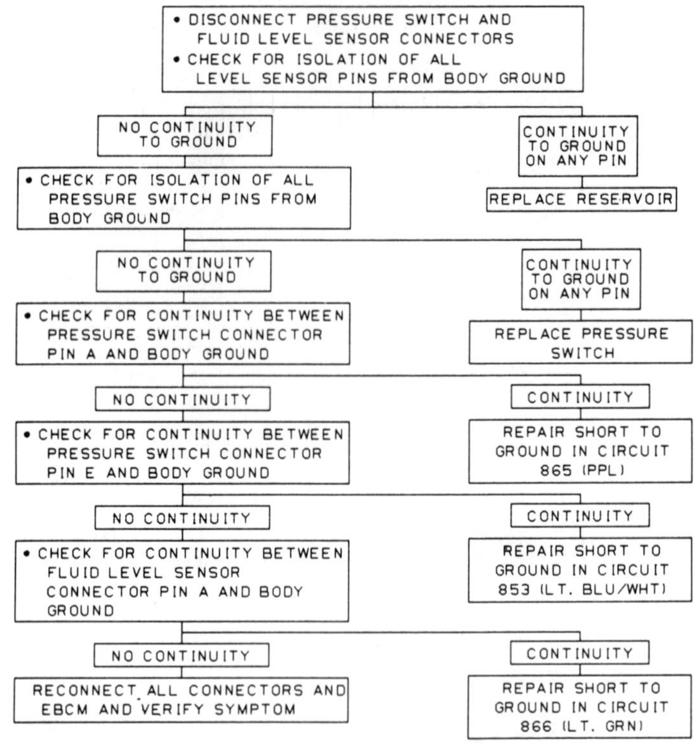

Fig. 15 EBCM loop short. Chart A-6, all 1986 models & 1987–88 Pontiac 6000

CHART A-7
RIGHT REAR WHEEL SENSOR RESISTANCE

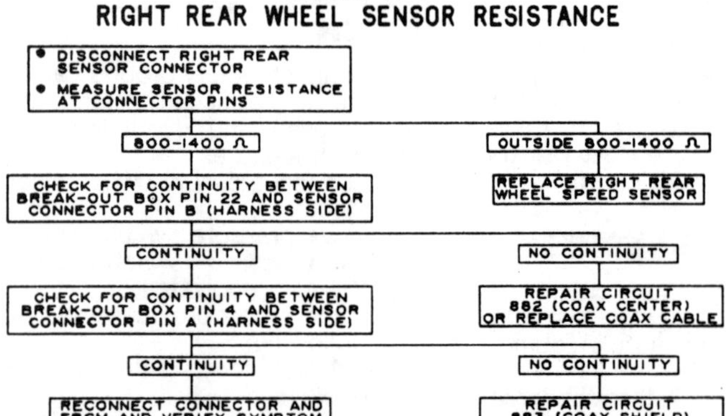

Fig. 16 Right rear wheel sensor. Chart A-7, all 1986 models & 1987–88 Pontiac 6000

CHART A-8
LEFT FRONT WHEEL SENSOR RESISTANCE

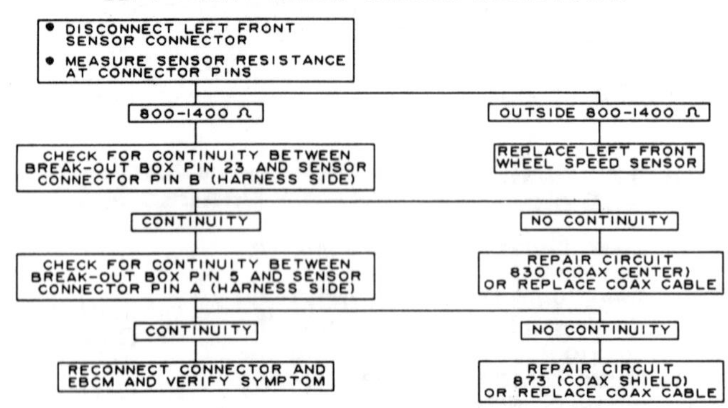

Fig. 17 Left front wheel sensor. Chart A-8, all 1986 models & 1987–88 Pontiac 6000

CHART A-9
LEFT REAR WHEEL SENSOR RESISTANCE

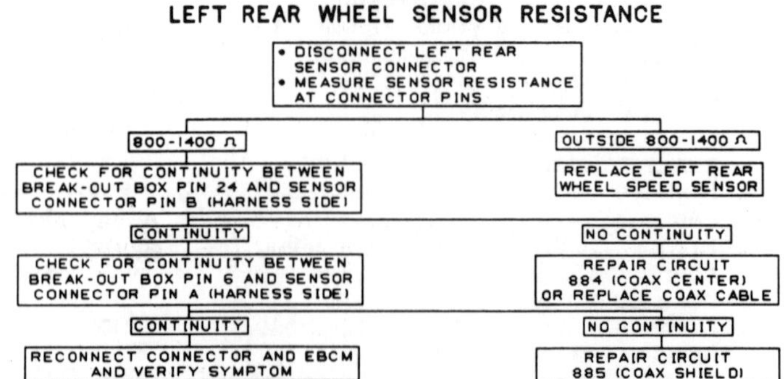

Fig. 18 Left rear wheel sensor. Chart A-9, all 1986 models & 1987–88 Pontiac 6000

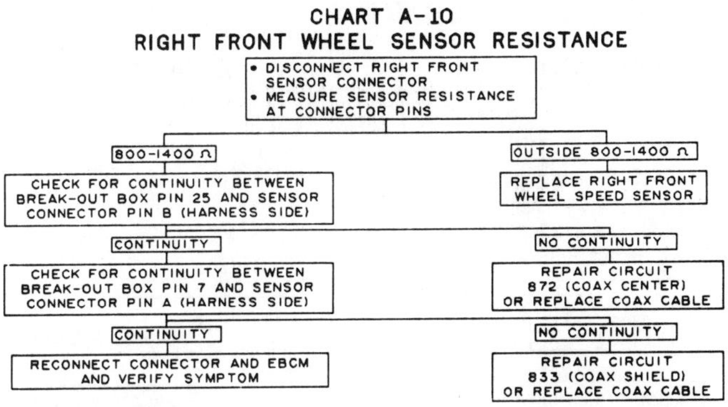

Fig. 19 Right front wheel sensor. Chart A-10, all 1986 models & 1987–88 Pontiac 6000

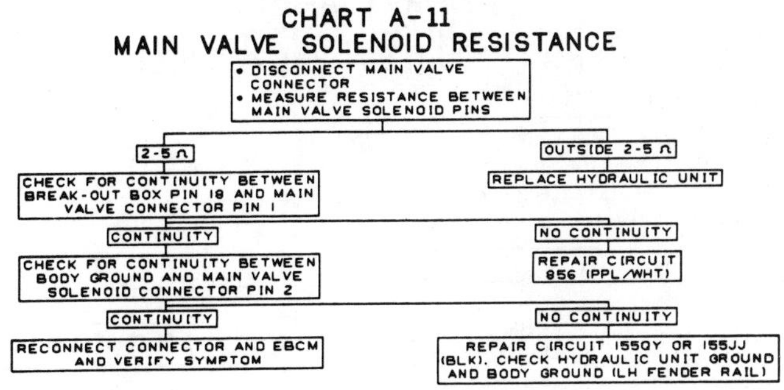

Fig. 20 Main valve solenoid resistance. Chart A-11, all 1986 models & 1987–88 Pontiac 6000

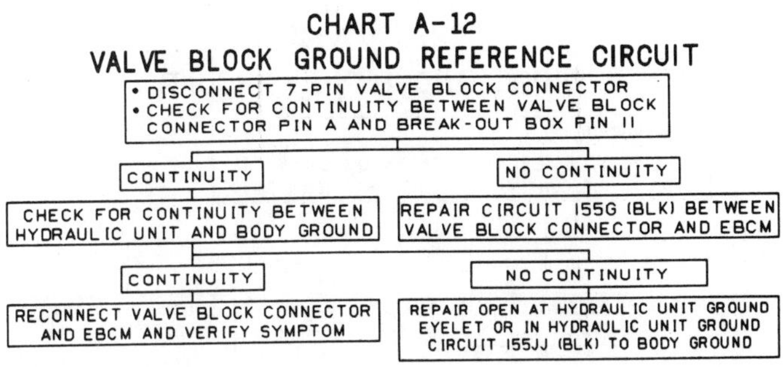

Fig. 21 Valve block ground reference. Chart A-12, all 1986 models & 1987–88 Pontiac 6000

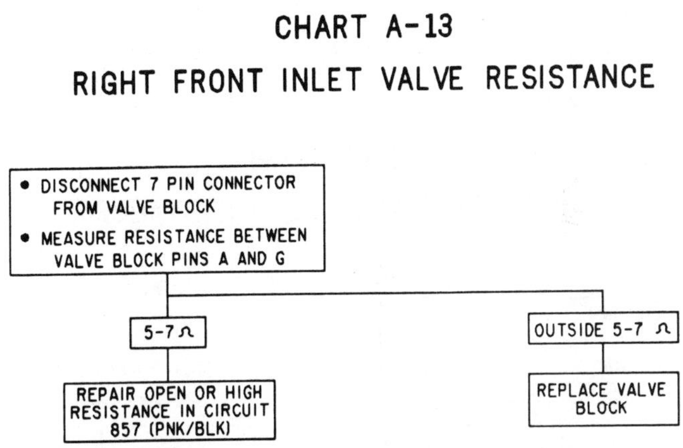

Fig. 22 Right front inlet valve. Chart A-13, all 1986 models & 1987–88 Pontiac 6000

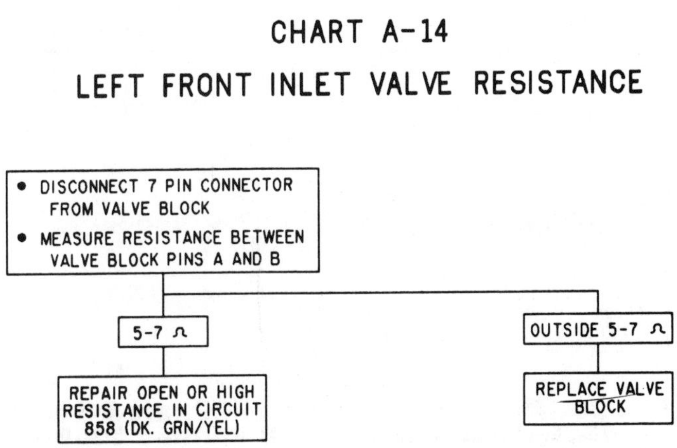

Fig. 23 Left front inlet valve. Chart A-14, all 1986 models & 1987–88 Pontiac 6000

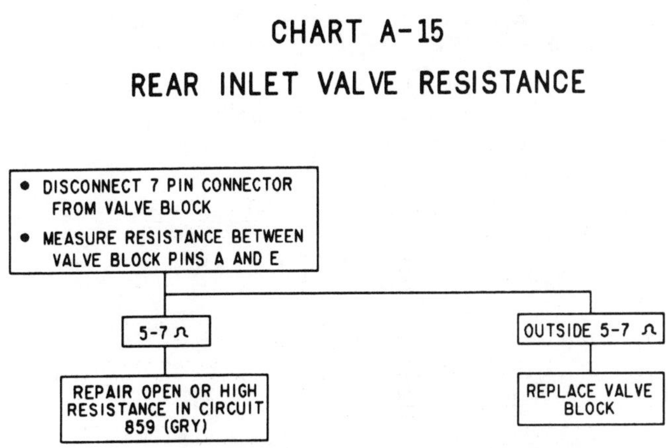

Fig. 24 Rear inlet valve. Chart A-15, all 1986 models & 1987–88 Pontiac 6000

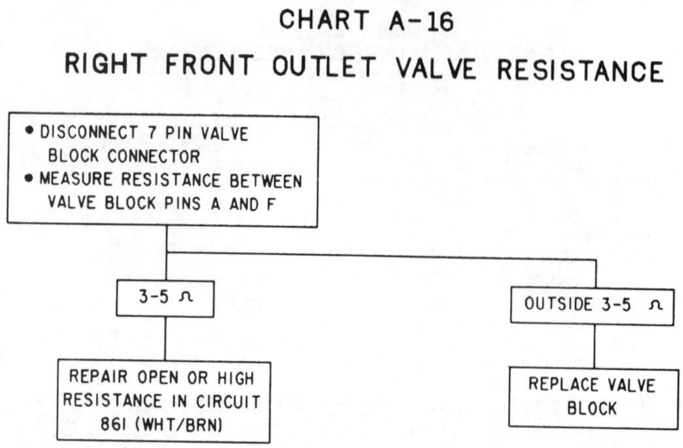

Fig. 25 Right front outlet valve. Chart A-16, all 1986 models & 1987–88 Pontiac 6000

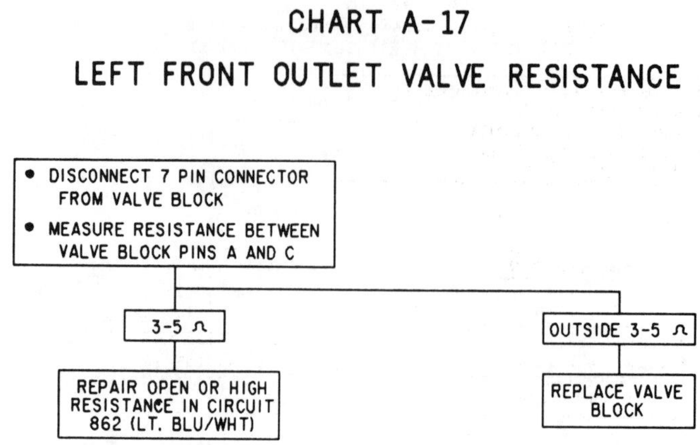

Fig. 26 Left front outlet valve. Chart A-17, all 1986 models & 1987–88 Pontiac 6000

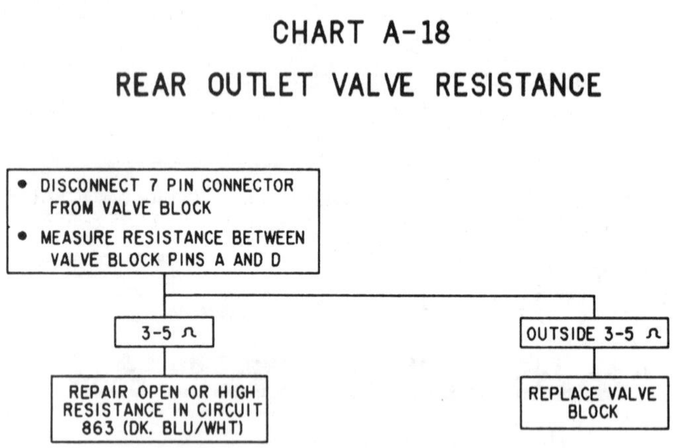

Fig. 27 Rear outlet valve. Chart A-18, all 1986 models & 1987–88 Pontiac 6000

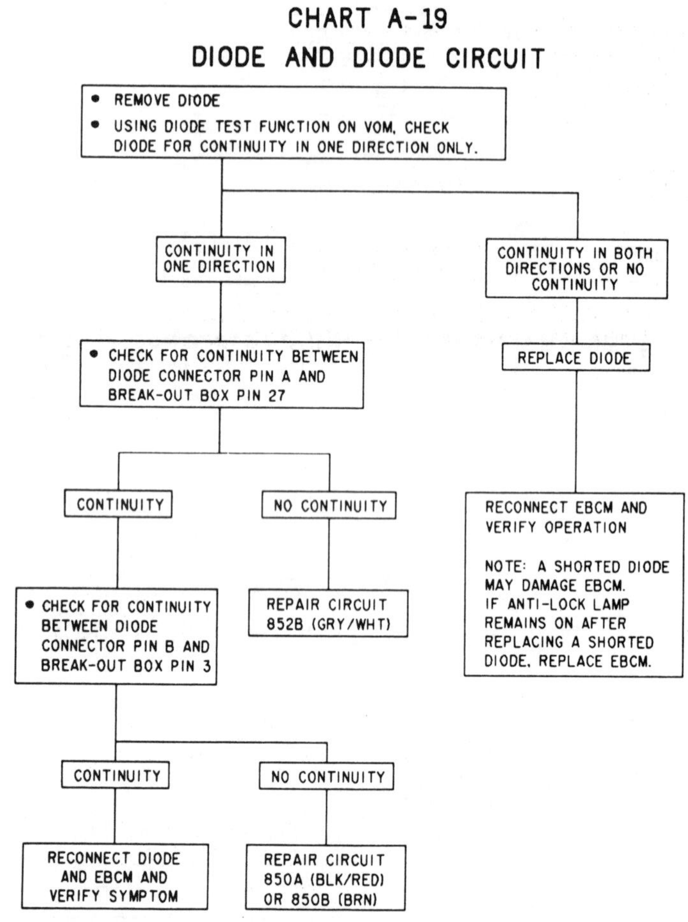

Fig. 28 Diode & diode circuit. Chart A-19, all 1986 models & 1987–88 Pontiac 6000

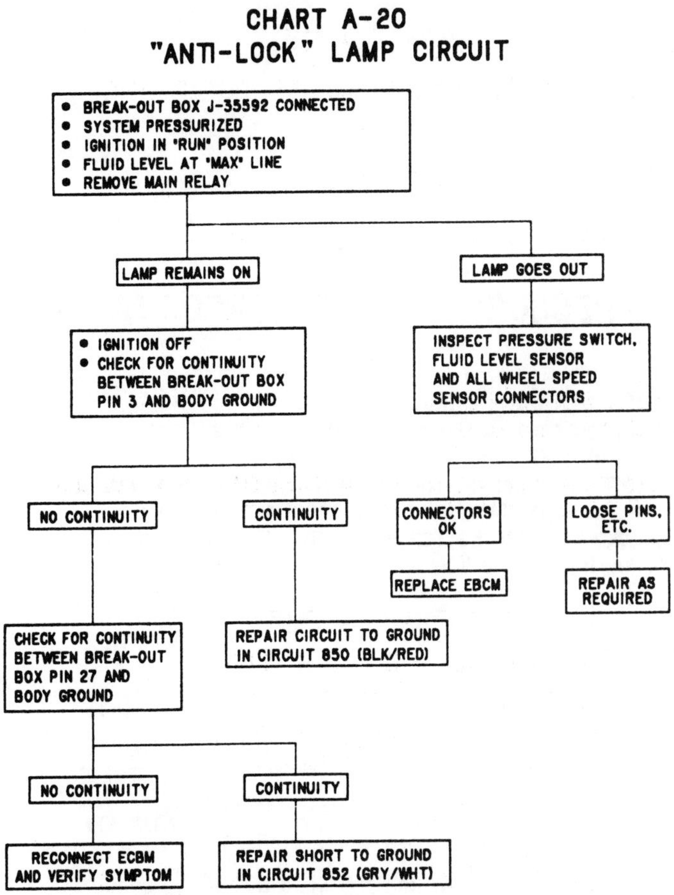

Fig. 29 ANTILOCK lamp circuit. Chart A-20, all 1986 models & 1987–88 Pontiac 6000

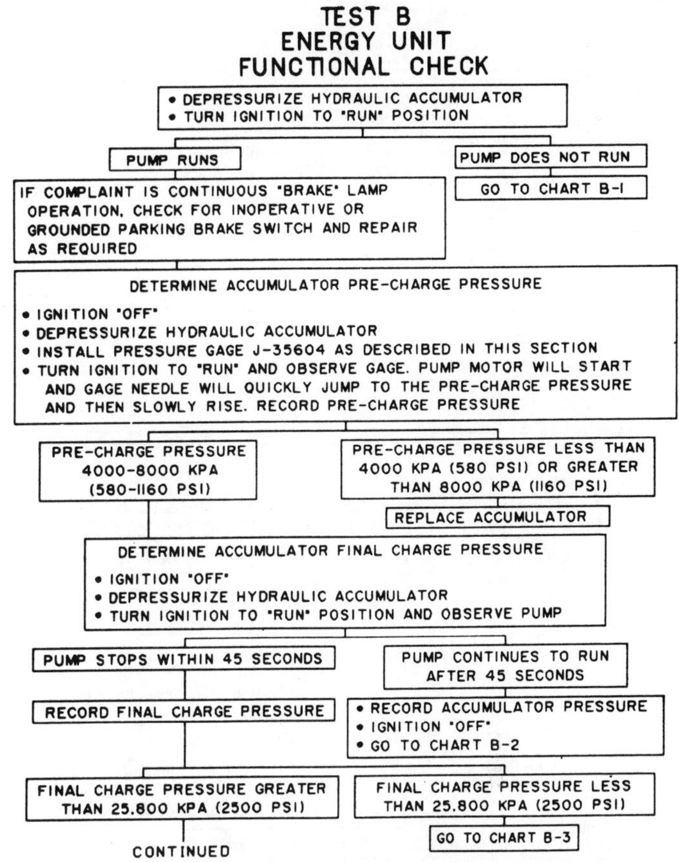

Fig. 30 Energy unit functional check (Part 1 of 2). Test B, all 1986 models & 1987–88 Pontiac 6000

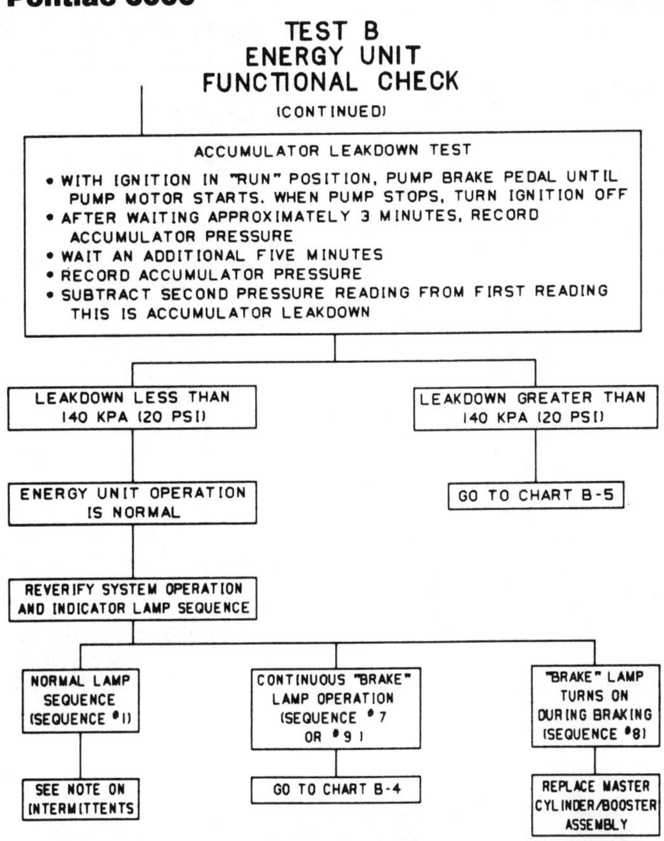

Fig. 30 Energy unit functional check (Part 2 of 2). Test B, all 1986 models & 1987–88 Pontiac 6000

CHART B-1
PUMP DOES NOT RUN

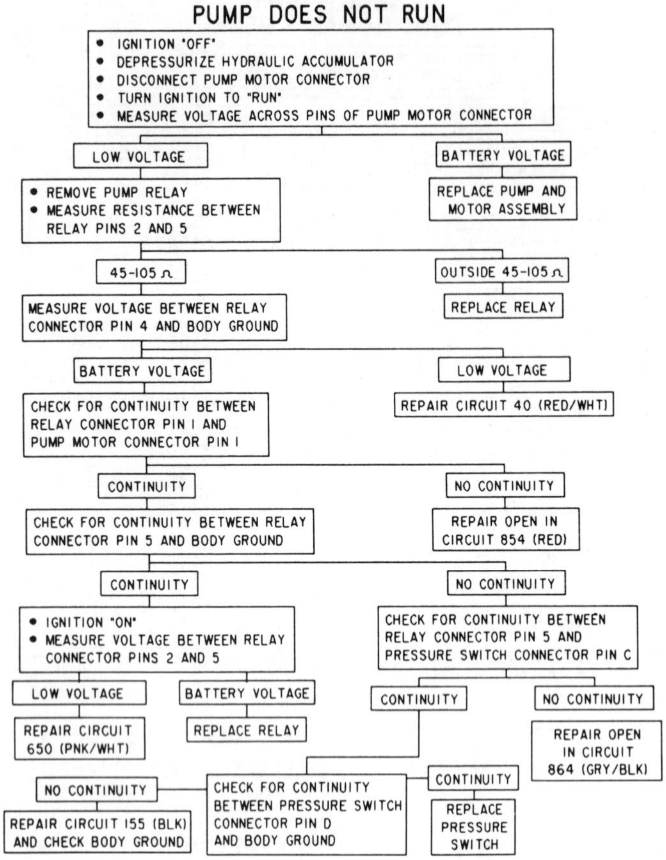

Fig. 31 Pump does not run. Chart B-1, all 1986 models & 1987–88 Pontiac 6000

CHART B-2
PUMP RUNS LONGER THAN 45 SECONDS

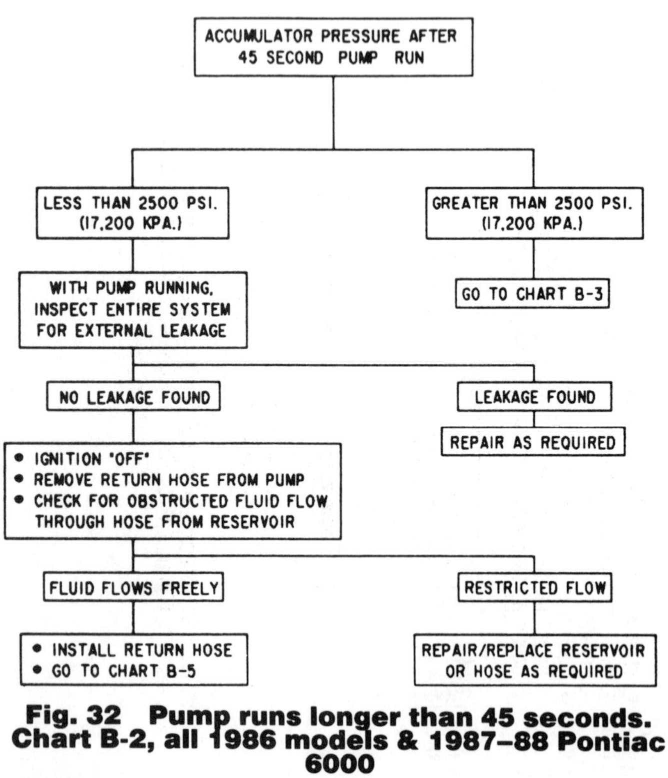

Fig. 32 Pump runs longer than 45 seconds. Chart B-2, all 1986 models & 1987–88 Pontiac 6000

CHART B-3
PRESSURE SWITCH PERFORMANCE

I. SWITCH STATUS - PRESSURIZED

- TURN IGNITION TO "RUN" UNTIL PUMP STOPS
- IGNITION "OFF"
  NOTE: IF PUMP CONTINUES TO RUN AFTER 45 SECONDS, TURN IGNITION OFF AND PROCEDE
- CHECK PRESSURE SWITCH PINS FOR THE FOLLOWING CONDITIONS USING VOM

MEASURE BETWEEN PINS	SCALE	SPECIFICATION
1,2	200 Ω	NO CONTINUITY
1,4	200 Ω	NO CONTINUITY
3,5	200 Ω	CONTINUITY
ALL PINS,BODY GROUND	200 Ω	NO CONTINUITY

IF ANY CONDITION IS NOT MET, REPLACE PRESSURE SWITCH

II. SWITCH STATUS - DEPRESSURIZED

- DEPRESSURIZE HYDRAULIC ACCUMULATOR
- CHECK PRESSURE SWITCH PINS FOR THE FOLLOWING CONDITIONS USING VOM

MEASURE BETWEEN PINS	SCALE	SPECIFICATION
1,2	200 Ω	CONTINUITY
1,4	200 Ω	CONTINUITY
3,5	200 Ω	NO CONTINUITY
ALL PINS,BODY GROUND	200 Ω	NO CONTINUITY

IF ANY CONDITION IS NOT MET, REPLACE PRESSURE SWITCH

CONTINUED ↓

Fig. 33 Pressure switch performance (Part 1 of 2). Chart B-3, all 1986 models & 1987–88 Pontiac 6000

CHART B-3
(CONTINUED)

III. **SWITCH THRESHOLDS**

- DEPRESSURIZE HYDRAULIC ACCUMULATOR
- INSTALL PRESSURE GAGE J-35604 AS DESCRIBED IN THIS SECTION
- CONNECT PRESSURE SWITCH CONNECTOR AND TURN IGNITION TO "RUN" UNTIL PUMP STOPS
- USING VOM, MONITOR FOR CONTINUITY BETWEEN PRESSURE SWITCH PINS AS SHOWN BELOW WHILE SLOWLY BLEEDING OFF ACCUMULATOR PRESSURE BY PUMPING THE BRAKE PEDAL. CONTINUITY SHOULD BE GAINED OR LOST AS INDICATED.
- PRESSURIZE SYSTEM BETWEEN EACH TEST BY RECONNECTING PRESSURE SWITCH AND TURNING IGNITION TO "RUN" UNTIL PUMP STOPS

MEASURE BETWEEN PINS	SWITCH STATUS	PRESSURE RANGE
1,4	CONTINUITY SHOULD BE GAINED AT:	1980-2080 PSI (13,650-14,350 KPA)
1,2	CONTINUITY SHOULD BE GAINED AT:	1500-1550 PSI (10,350-10,700 KPA)
3,5	CONTINUITY SHOULD BE LOST AT:	1500-1550 PSI (10,350-10,700 KPA)

IF ANY CONDITION IS NOT MET, REPLACE PRESSURE SWITCH

- DEPRESSURIZE HYDRAULIC ACCUMULATOR
- WITH IGNITION OFF, CONNECT PRESSURE SWITCH CONNECTOR
- TURN IGNITION TO "RUN" AND OBSERVE GAGE, "ANTI-LOCK" LAMP, "BRAKE" LAMP AND PUMP MOTOR. EVENTS SHOULD OCCUR AT PRESSURES INDICATED IN CHART BELOW

EVENT	PRESSURE
"ANTI-LOCK" LAMP TURNS OFF	1900-1975 PSI (13,100-13,600 KPA)
"BRAKE" LAMP TURNS OFF	1900-1975 PSI (13,100-13,600 KPA)
PUMP MOTOR STOPS	2550-2670 PSI (17,580-18,400 KPA)

IF ANY CONDITION IS NOT MET, REPLACE PRESSURE SWITCH

Fig. 33 Pressure switch performance (Part 2 of 2). Chart B-3, all 1986 models & 1987-88 Pontiac 6000

CHART B-4
BRAKE LAMP CIRCUIT

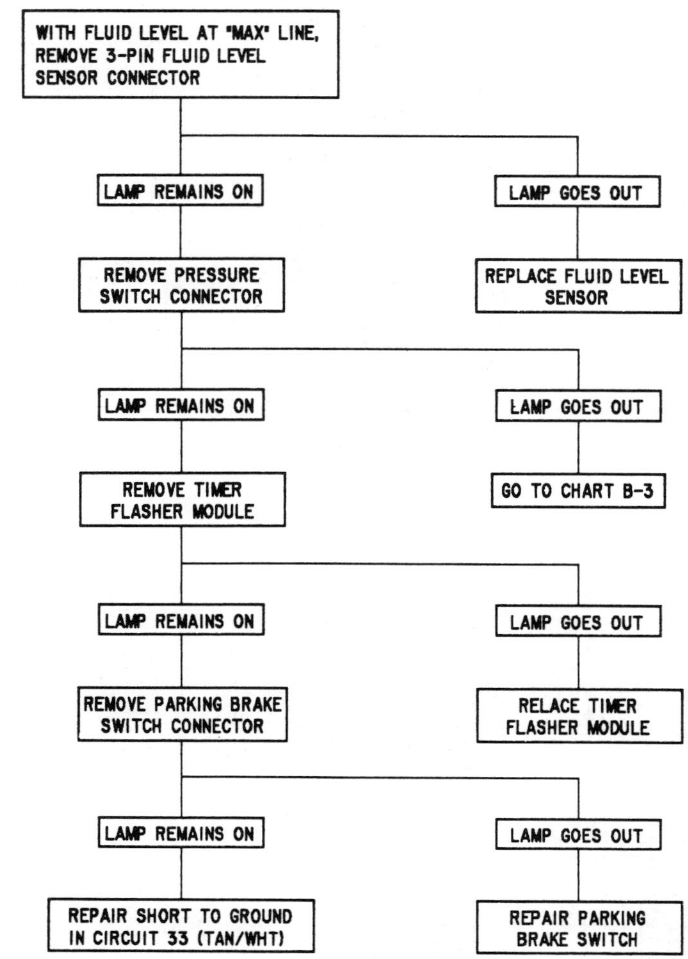

Fig. 34 Brake lamp circuit. Chart B-4, all 1986 models & 1987-88 Pontiac 6000

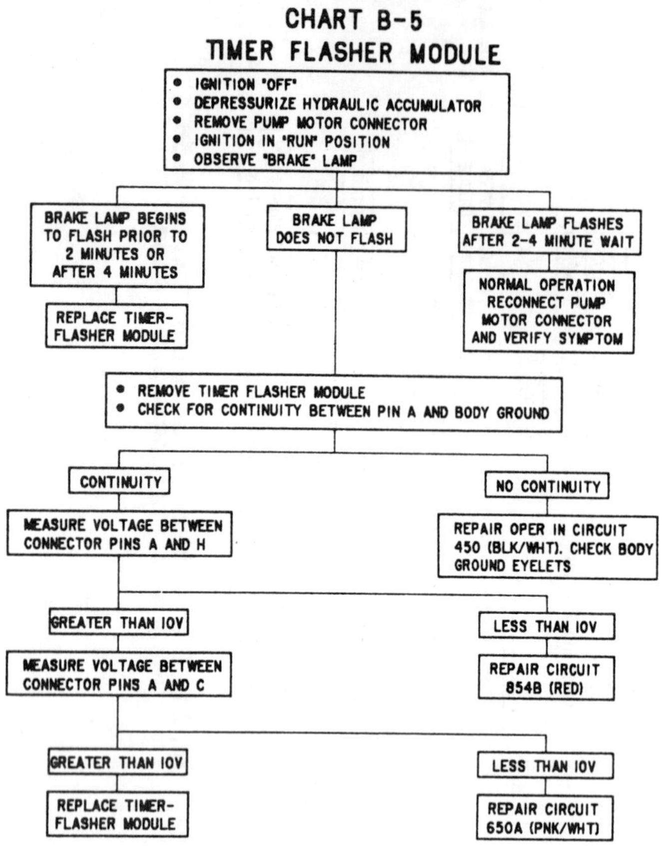

Fig. 35 Timer flasher module. Chart B-5, all 1986 models except Pontiac 6000

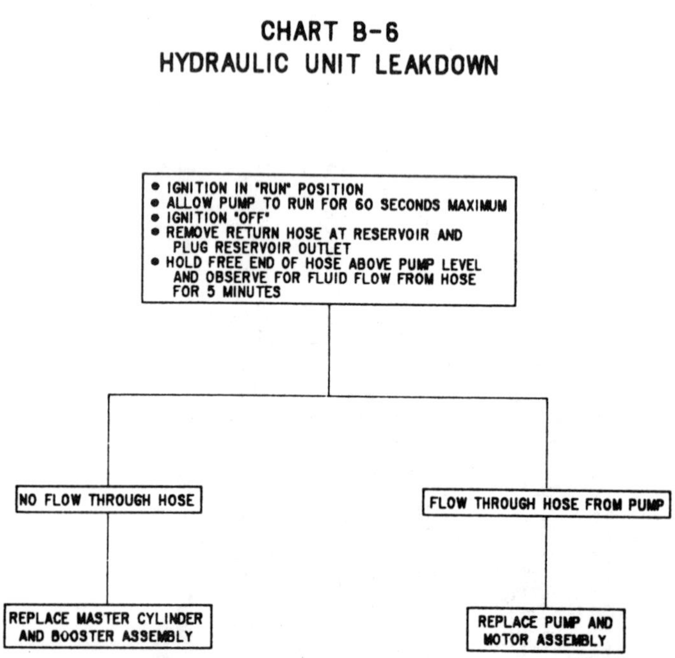

Fig. 36 Hydraulic unit leakdown. Chart B-6, all 1986 models & 1987–88 Pontiac 6000

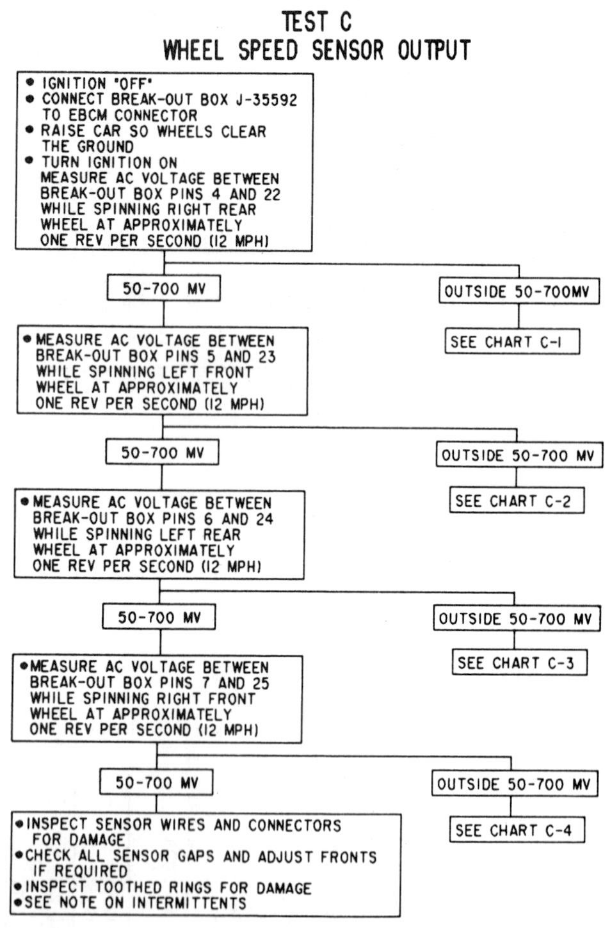

Fig. 37 Wheel speed sensor output. Test C, all 1986 models & 1987–88 Pontiac 6000

CHART C-1
RIGHT REAR WHEEL SENSOR OUTPUT

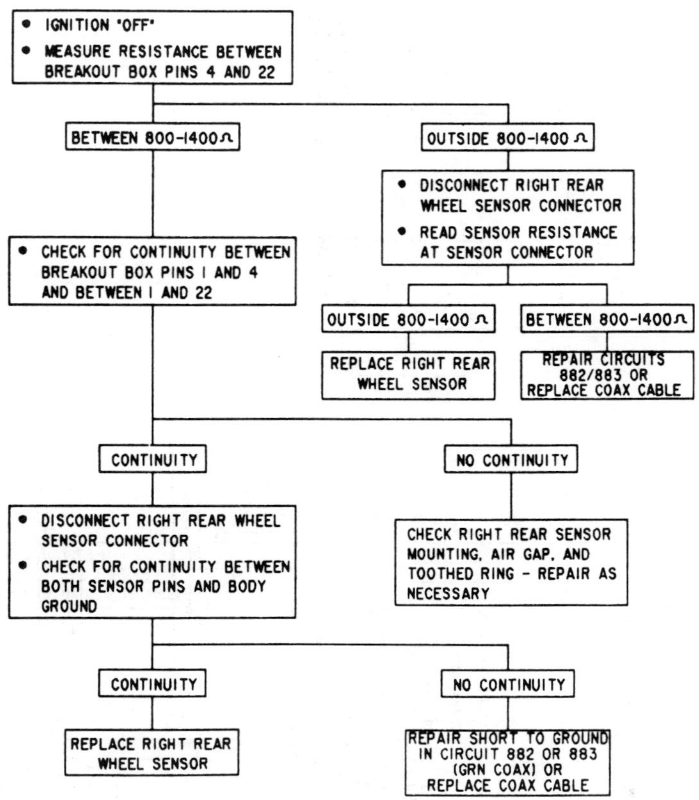

Fig. 38 Right rear sensor output. Chart C-1, all 1986 models & 1987–88 Pontiac 6000

CHART C-2
LEFT FRONT WHEEL SENSOR OUTPUT

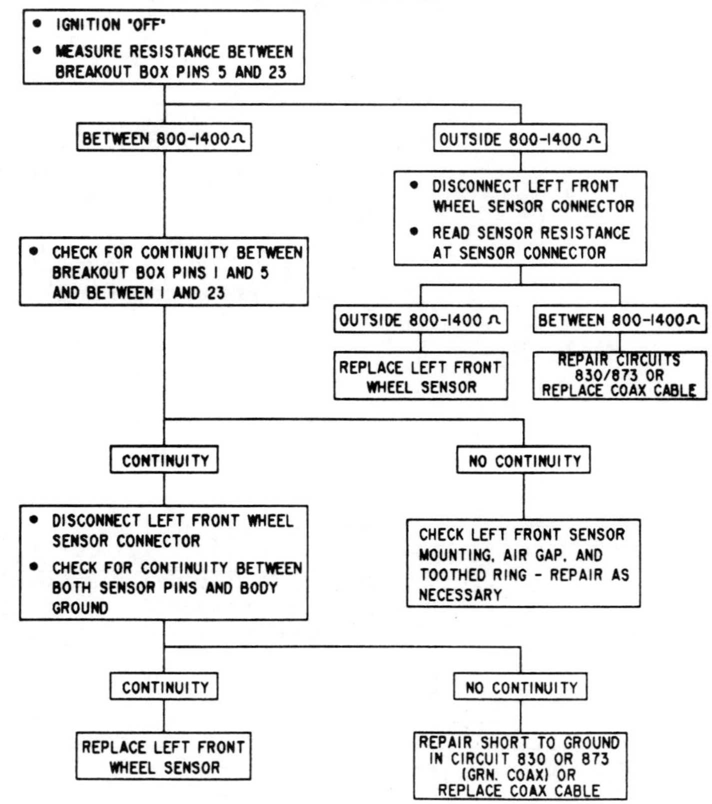

Fig. 39 Left front sensor output. Chart C-2, all 1986 models & 1987–88 Pontiac 6000

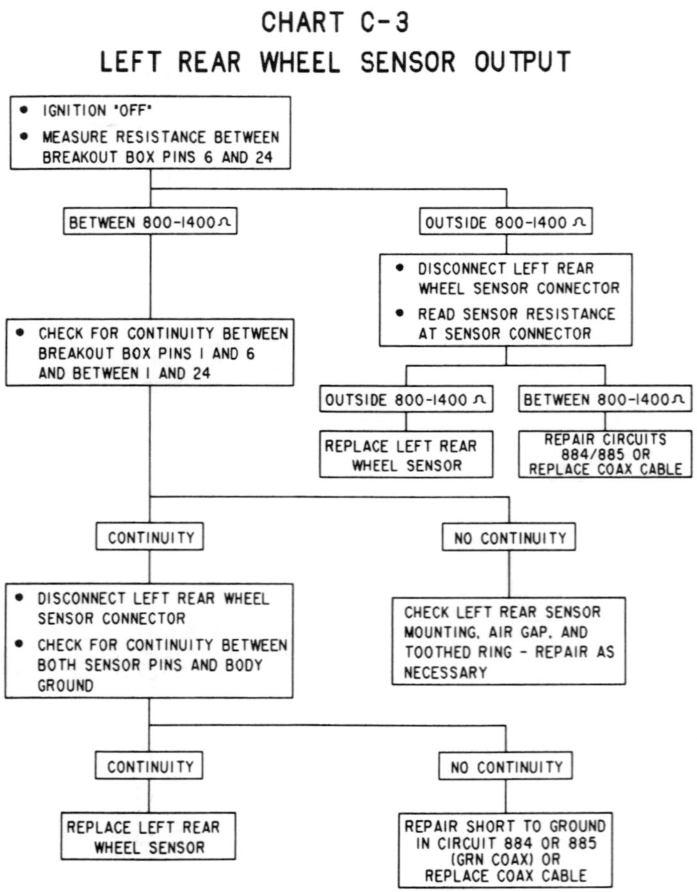

Fig. 40 Left rear sensor output. Chart C-3, all 1986 models & 1987–88 Pontiac 6000

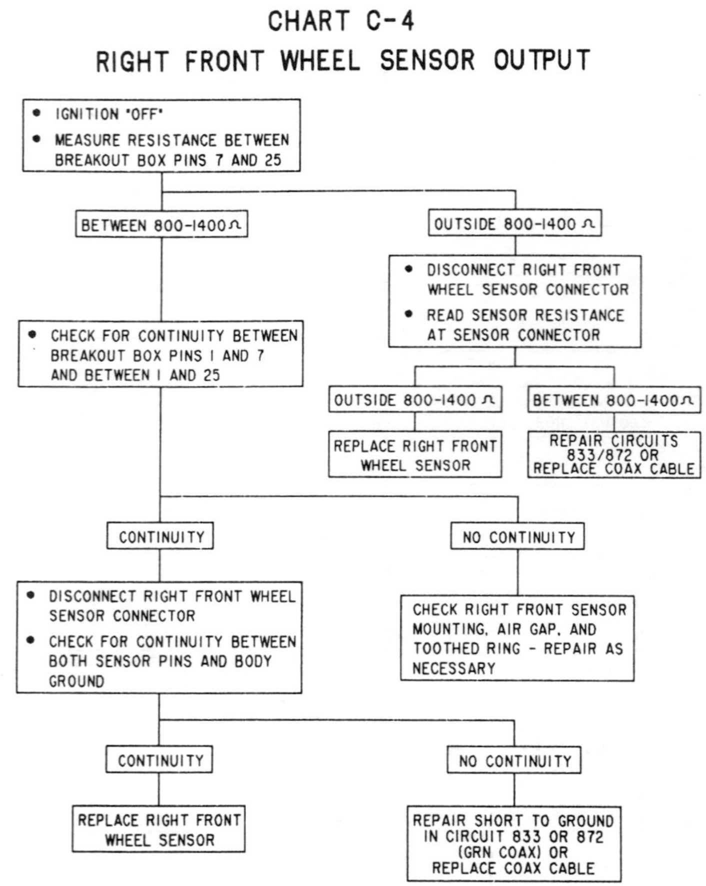

Fig. 41 Right front sensor output. Chart C-4, all 1986 models & 1987–88 Pontiac 6000

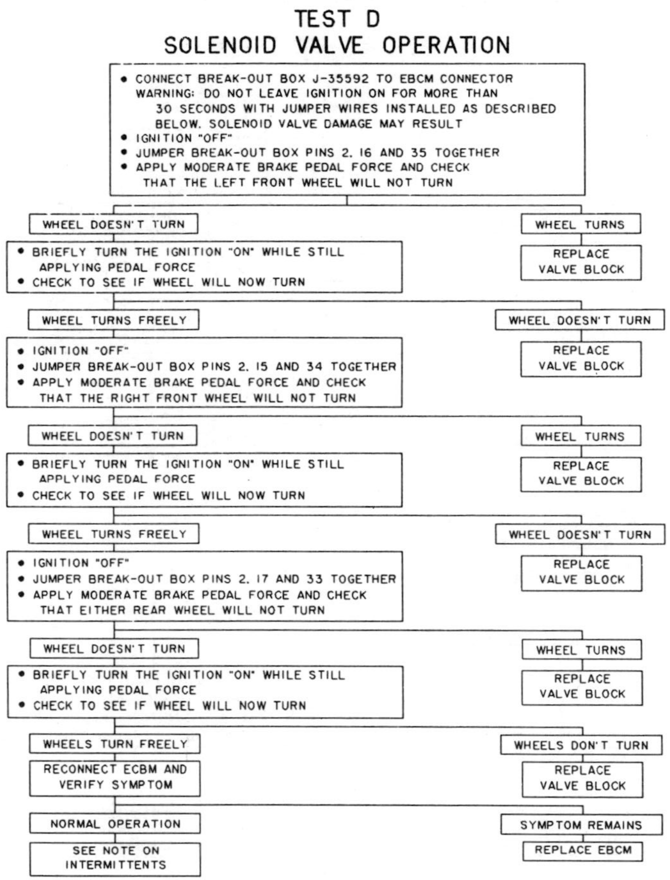

Fig. 42 Solenoid valve operation. Test D, all 1986 models & 1987–88 Pontiac 6000

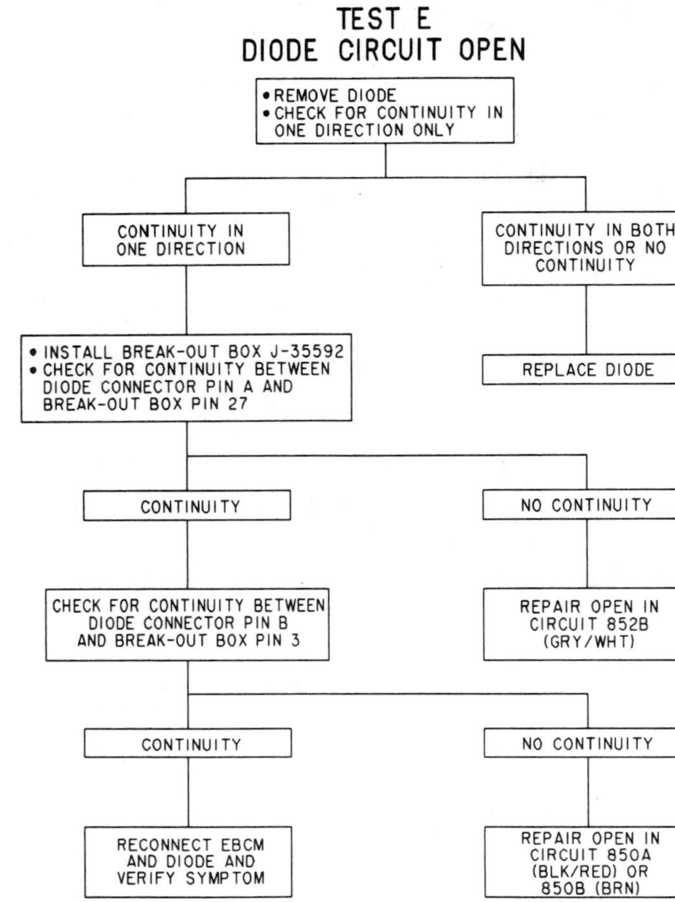

Fig. 43 Diode circuit open (Part 1 of 2). Test E, all 1986 models

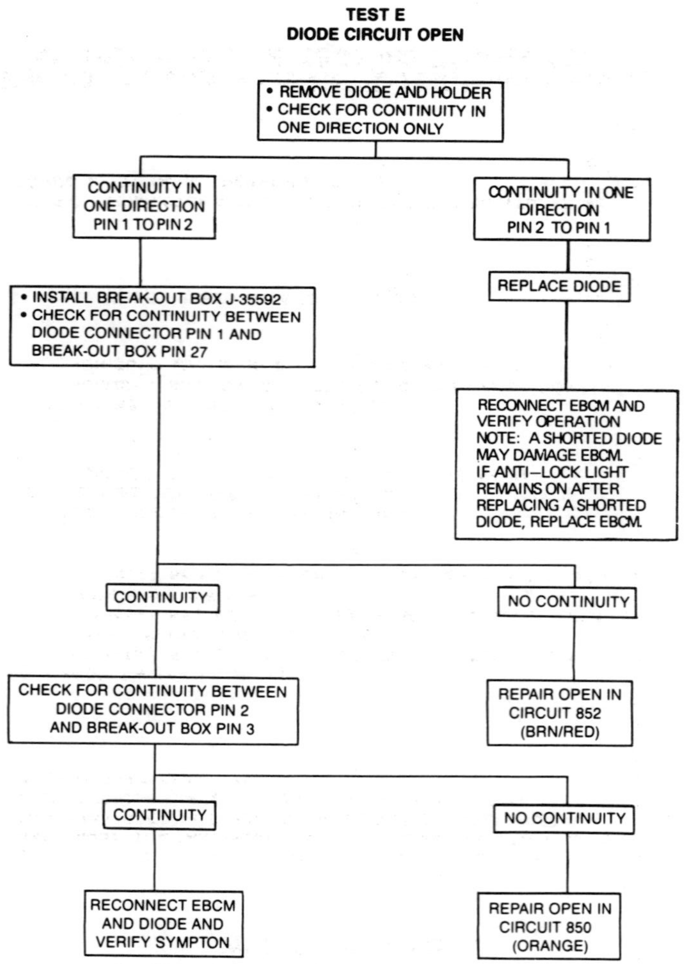

Fig. 43 Diode circuit open (Part 2 of 2). Test E, 1987–88 Pontiac 6000

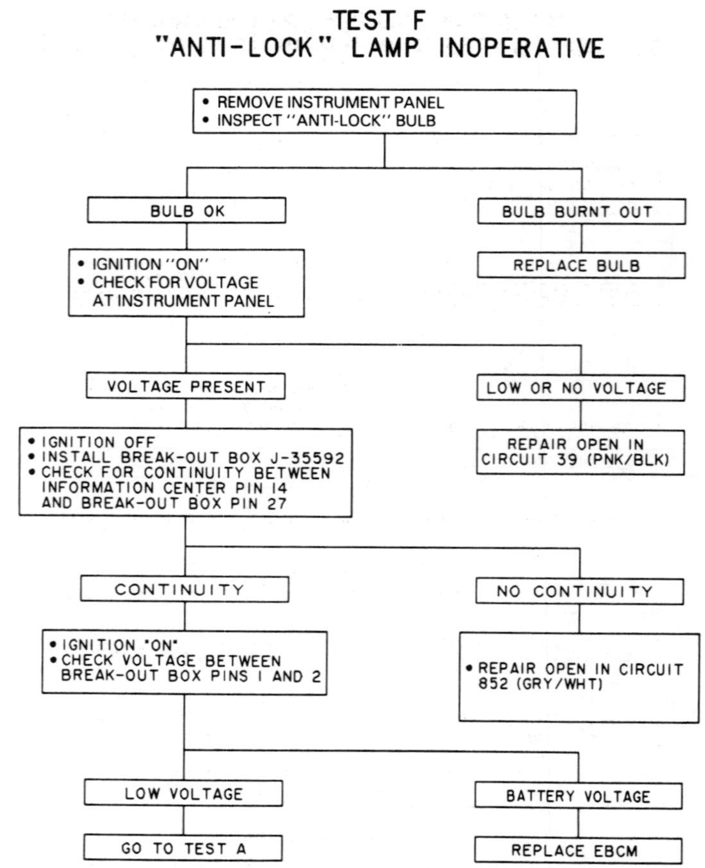

Fig. 44 ANTILOCK lamp inoperative. Test F, all 1986 models & 1987–88 Pontiac 6000

TEST G
INTERMITTANT "ANTI-LOCK" LAMP
WHILE DRIVING

INTERUPTIONS IN THE FOLLOWING CIRCUITS AND OTHERS MAY CAUSE THE "ANTI-LOCK" LAMP TO LIGHT INTERMITTANTLY. THIS LIST DOES NOT INCLUDE ALL POSSIBILITIES, BUT INCLUDES THOSE CIRCUITS MOST LIKELY TO CAUSE AN INTERMITTANT "ANTI-LOCK" LAMP.

- EBCM SWITCH LOOP, INCLUDING
  - -CIRCUIT 866 (LT. GRN) FROM EBCM TO FLUID LEVEL SENSOR
  - -2-PIN FLUID LEVEL SENSOR SWITCH
  - -CIRCUIT 853 (LT. BLU/ORN) FROM FLUID LEVEL SENSOR TO PRESSURE SWITCH
  - -CIRCUIT 865 (PPL) FROM PRESSURE SWITCH TO EBCM

- IGNITION ENABLE CIRCUIT, INCLUDING
  - -BRAKE FUSE (#5 IN FUSE BLOCK, 5A)
  - -CIRCUIT 650 (PNK/WHT) FROM FUSE TO EBCM

- MAIN RELAY CIRCUITS, INCLUDING
  - -CIRCUIT 855 (DK. BLU) FROM EBCM TO RELAY COIL
  - -CIRCUIT 450 (BLK/WHT) FROM RELAY COIL TO GROUND

- ALL WHEEL SPEED SENSOR CIRCUITS

INSPECT CONNECTORS AND WIRES IN THESE CIRCUITS. IF NO TROUBLE IS FOUND, PERFORM TEST A

Fig. 45 Intermittent ANTILOCK lamp. Test G, all 1986 models & 1987–88 Pontiac 6000

TEST H
LOW OR SPONGY PEDAL

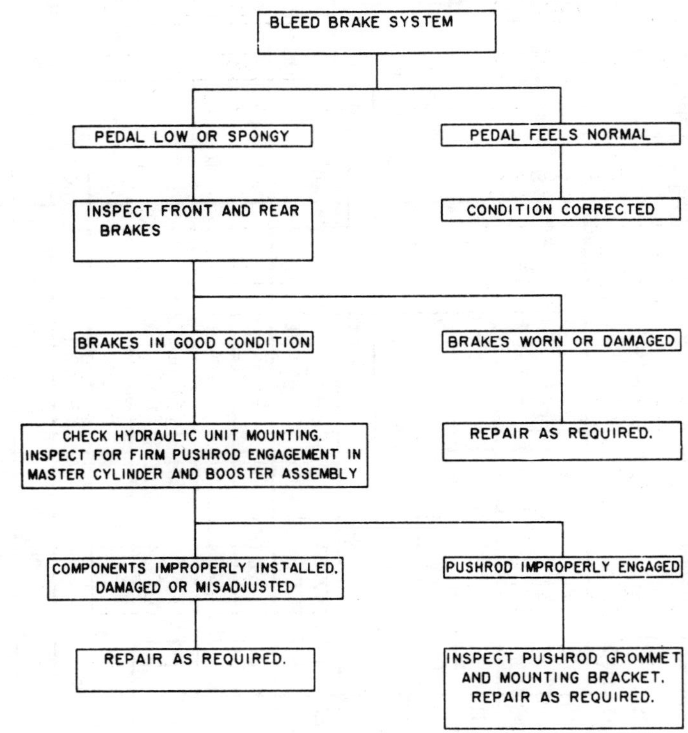

Fig. 46 Low or spongy pedal. Test H, all 1986 models & 1987–88 Pontiac 6000

TEST A
PIN-OUT CHECKS

- CONNECT BREAK-OUT BOX J-35592 TO 35-PIN EBCM HARNESS CONNECTOR AS DESCRIBED IN THIS SECTION
- PERFORM CHECKS WITH HIGH IMPEDANCE DIGITAL MULTIMETER J-34029-A OR EQUIVALENT
- ALL CHECKS ARE MADE WITH ENGINE STOPPED

CIRCUIT TO BE TESTED	IGNITION SWITCH POSITION	MULTIMETER SCALE/RANGE	MEASURE BETWEEN PIN NUMBERS	SPECIFICATION	IF RESULT NOT WITHIN SPECIFICATION, SEE CHART
IGNITION ENABLE	RUN	20 DCV	2(+)*,1(−)*	10 V MINIMUM	A-1
MAIN RELAY GROUND	OFF	200Ω	1,3	CONTINUITY	A-2
	OFF	200Ω	1,20	CONTINUITY	
MAIN RELAY COIL	OFF	200Ω	1,8	45-105 Ω	A-3
BEFORE PERFORMING THIS TEST: • REMOVE GAGE (FUSE #3) FROM FUSE PANEL • PLACE FUSED JUMPER BETWEEN BREAK-OUT BOX PINS 2 & 8					
MAIN RELAY POWER	ON	20 DCV	3(+),1(−)	10 V MINIMUM	A-4
	ON	20 DCV	20(+),1(−)	10 V MINIMUM	
BEFORE PROCEEDING: • REMOVE JUMPER FROM PINS 2 &8 • INSTALL GAGE FUSE					
EBCM SWITCH LOOP	OFF	200Ω	9,10	LESS THAN 5Ω	A-5
	OFF	200Ω	1,9	NO CONTINUITY	A-6
RR SENSOR RESISTANCE	OFF	2kΩ	4,22	800-1400Ω	A-7
LF SENSOR RESISTANCE	OFF	2kΩ	5,23	800-1400Ω	A-8
LR SENSOR RESISTANCE	OFF	2kΩ	6,24	800-1400Ω	A-9
RF SENSOR RESISTANCE	OFF	2kΩ	7,25	800-1400Ω	A-10
MAIN VALVE SOLENOID	OFF	200Ω	11,18	2-5Ω	A-11
VALVE BLOCK GROUND	OFF	200Ω	1,11	LESS THAN 2Ω	A-12
RF INLET VALVE	OFF	200Ω	11,15	5-7Ω	A-13
LF INLET VALVE	OFF	200Ω	11,35	5-7Ω	A-14
REAR INLET VALVE	OFF	200Ω	11,17	5-7Ω	A-15
RF OUTLET VALVE	OFF	200Ω	11,34	3-5Ω	A-16
LF OUTLET VALVE	OFF	200Ω	11,16	3-5Ω	A-17
REAR OUTLET VALVE	OFF	200Ω	11,33	3-5Ω	A-18
BEFORE PERFORMING THIS TEST: • REMOVE MAIN RELAY FROM CONNECTOR ON RELAY BRACKET (5 WIRES ATTACHED)					
DIODE	OFF	DIODE ⊣◀—	27(+), 3(−)	CONTINUITY	A-19
	OFF	DIODE ⊣◀—	3(+), 27(−)	NO CONTINUITY	
BEFORE PROCEEDING: • INSTALL MAIN RELAY					
BRAKE LIGHT SWITCH • WITH BRAKE PEDAL RELEASED	ON	20 DCV	12(+), 1(−)	0 V	A-20
BRAKE LIGHT SWITCH • WITH BRAKE PEDAL APPLIED	ON	20 DCV	12(+), 1(−)	10 V MINIMUM	A-20

*(+) OR (−) INDICATES MULTIMETER POLARITY
IF ALL TEST RESULTS ARE WITHIN SPECIFICATION, RECONNECT EBCM AND VERIFY CONTINUOUS "ANTI-LOCK" LAMP OPERATION

- IF NORMAL OPERATION RESUMES, SEE NOTE ON INTERMITTANTS
- IF LAMP REMAINS ON, SEE CHART A-21

Fig. 47 Pin-Out checks. Test A, all 1987–88 models except Pontiac 6000

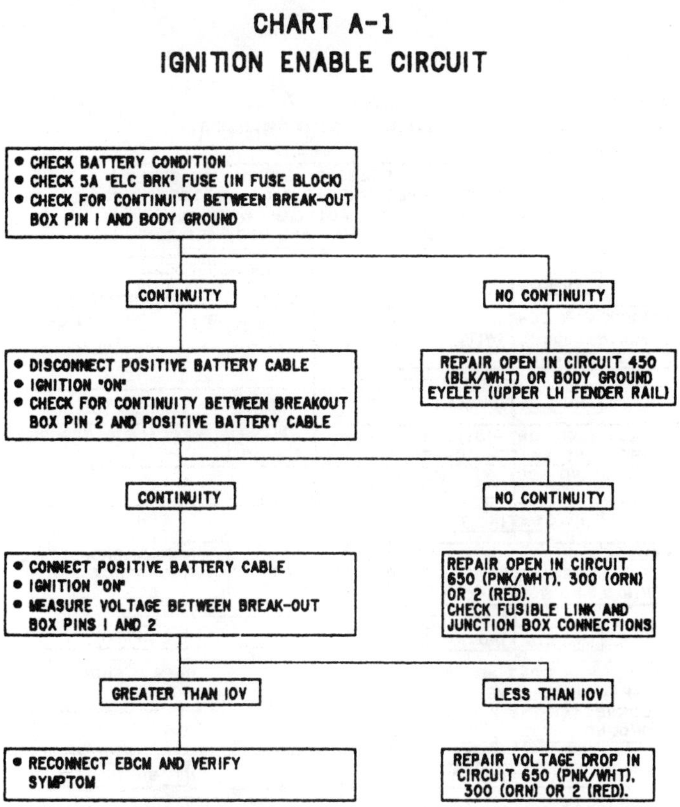

Fig. 48 Ignition enable circuit. Chart A-1, all 1987–88 models except Pontiac 6000

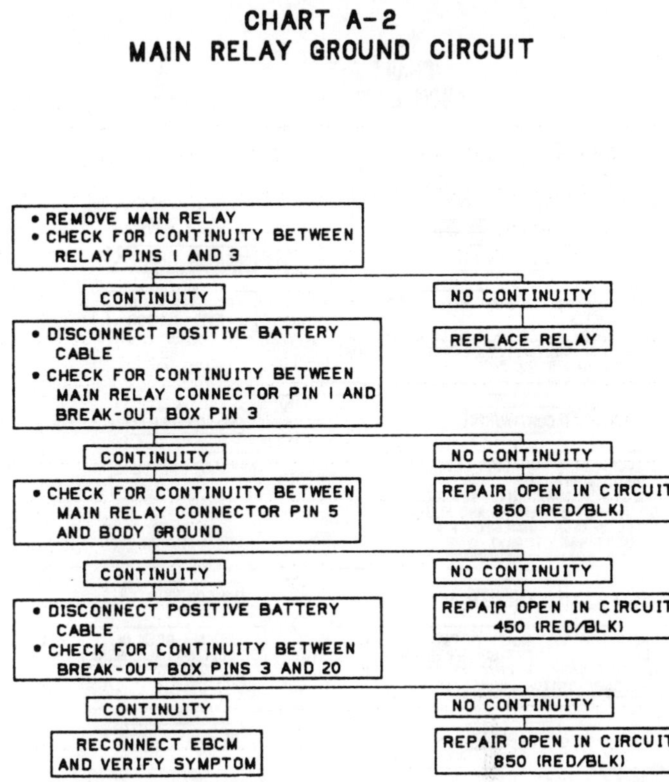

Fig. 49 Main relay ground circuit. Chart A-2, all 1987–88 models except Pontiac 6000

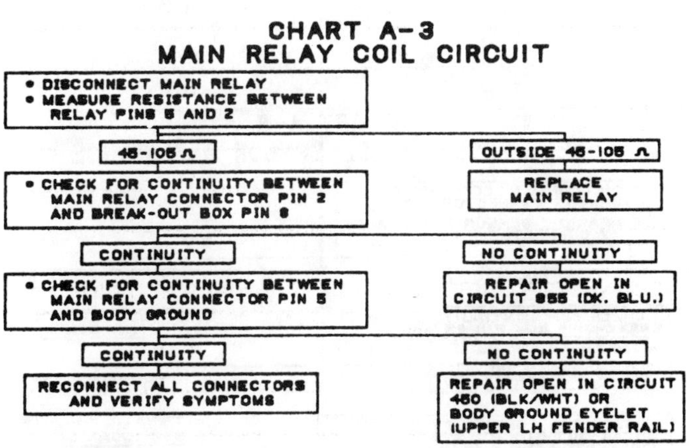

Fig. 50 Main relay coil circuit. Chart A-3, all 1987–88 models except Pontiac 6000

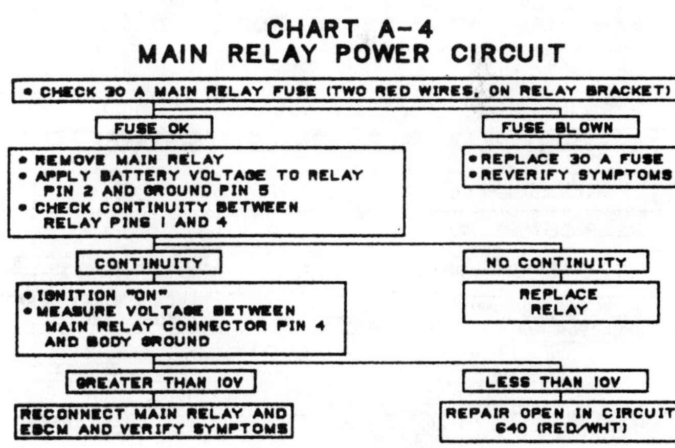

Fig. 51 Main relay power circuit. Chart A-4, all 1987–88 models except Pontiac 6000

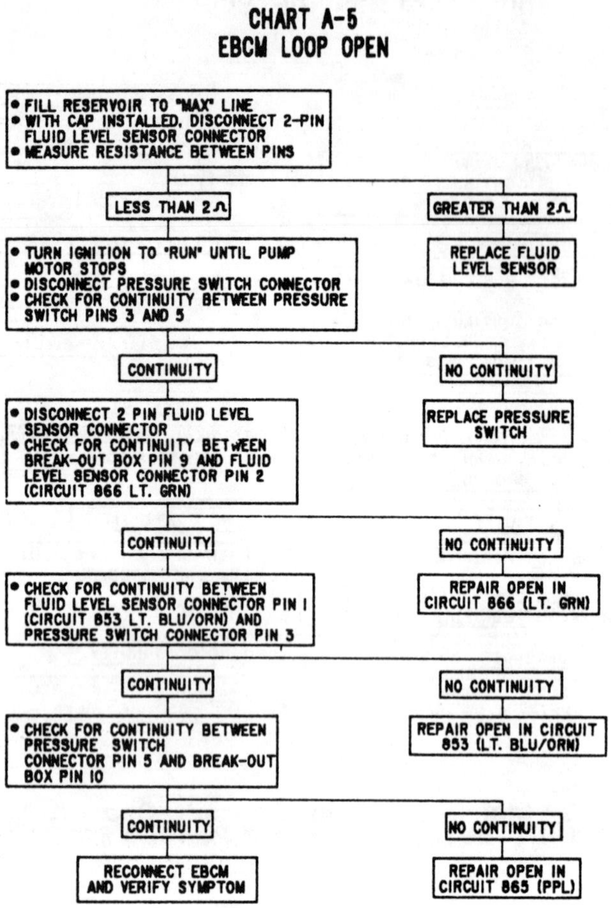

Fig. 52 EBCM loop open. Chart A-5, all 1987–88 models except Pontiac 6000

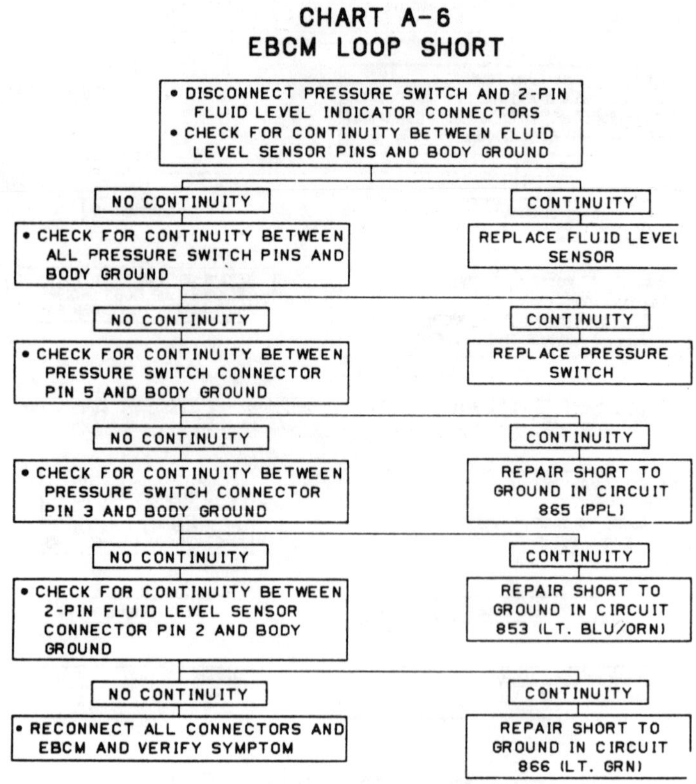

Fig. 53 EBCM loop short. Chart A-6, all 1987–88 models except Pontiac 6000

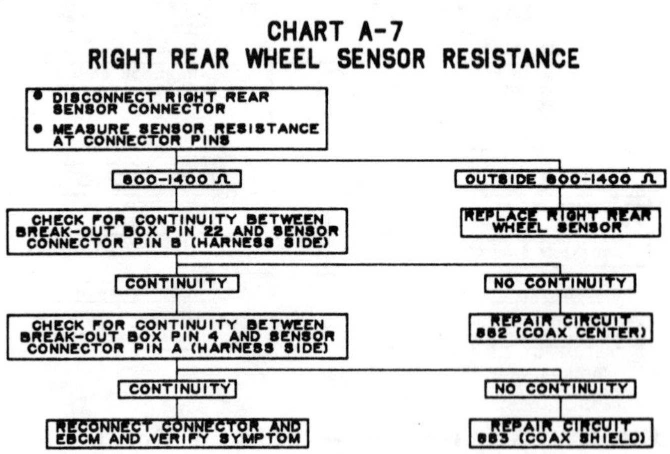

Fig. 54 Right rear wheel sensor. Chart A-7, all 1987–88 models except Pontiac 6000

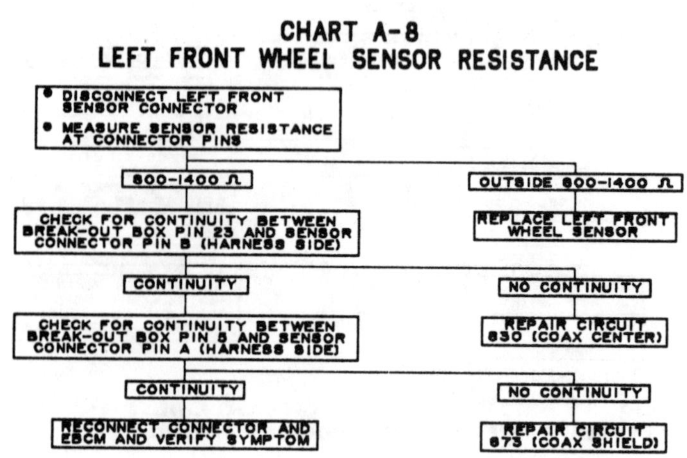

Fig. 55 Left front wheel sensor. Chart A-8, all 1987–88 models except Pontiac 6000

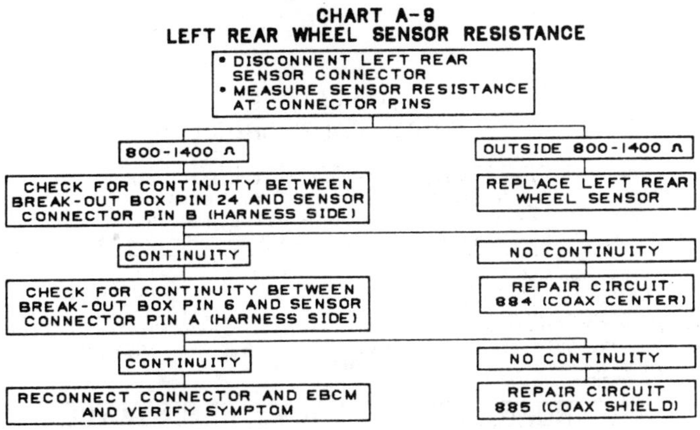

Fig. 56 Left rear wheel sensor. Chart A-9, all 1987–88 models except Pontiac 6000

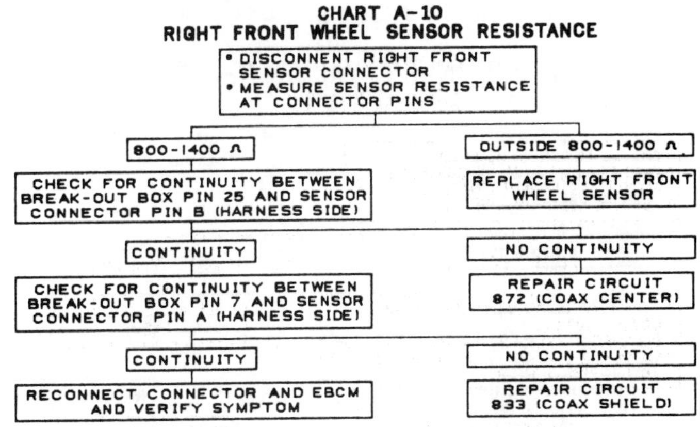

Fig. 57 Right front wheel sensor. Chart A-10, all 1987–88 models except Pontiac 6000

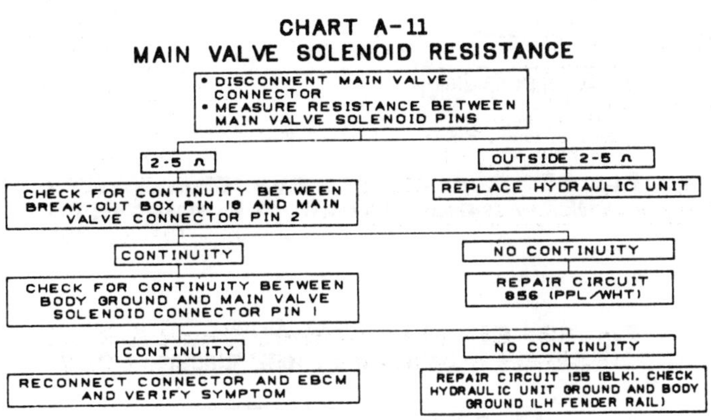

Fig. 58 Main valve solenoid resistance. Chart A-11, all 1987–88 models except Pontiac 6000

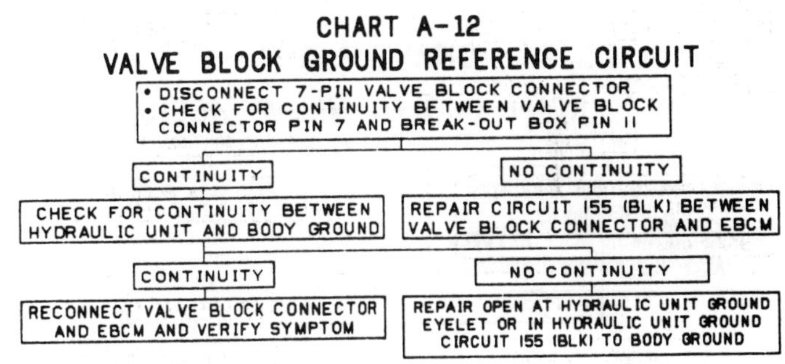

Fig. 59 Valve block ground reference. Chart A-12, all 1987–88 models except Pontiac 6000

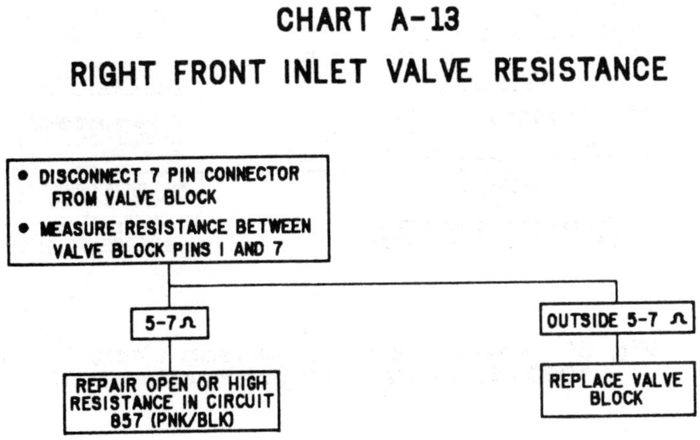

Fig. 60 Right front inlet valve. Chart A-13, all 1987–88 models except Pontiac 6000

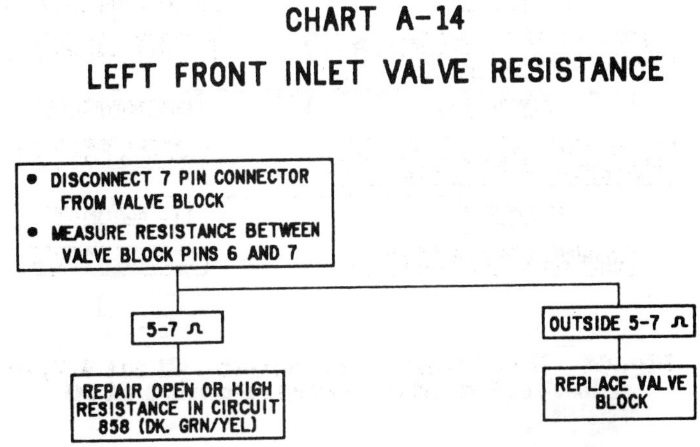

Fig. 61 Left front inlet valve. Chart A-14, all 1987–88 models except Pontiac 6000

CHART A-15

REAR INLET VALVE RESISTANCE

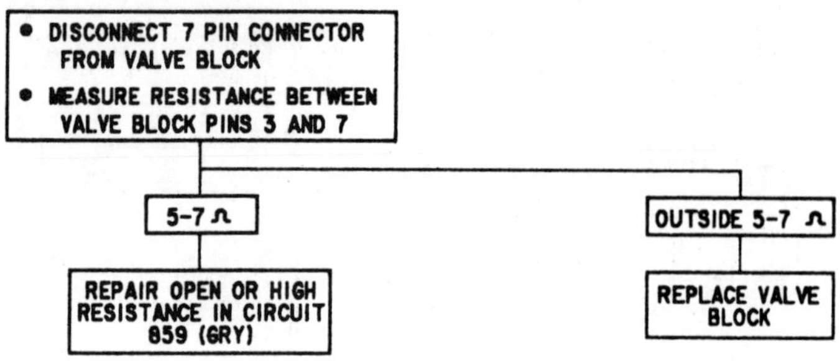

Fig. 62 Rear inlet valve. Chart A-15, all 1987–88 models except Pontiac 6000

CHART A-16

RIGHT FRONT OUTLET VALVE RESISTANCE

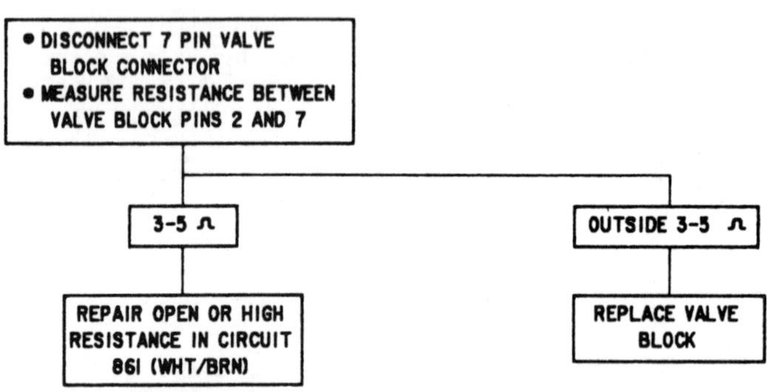

Fig. 63 Right front outlet valve. Chart A-16, all 1987–88 models except Pontiac 6000

CHART A-17

LEFT FRONT OUTLET VALVE RESISTANCE

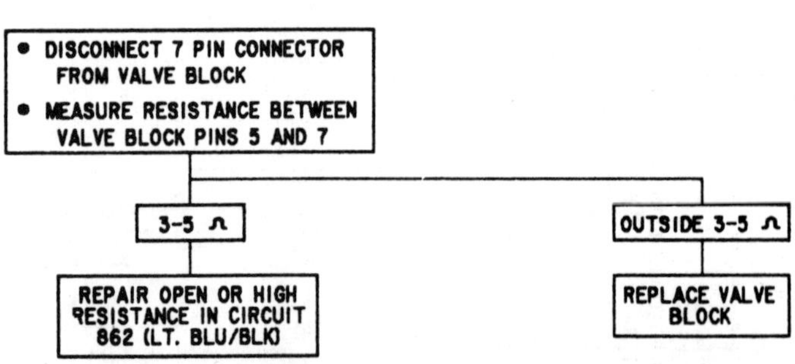

Fig. 64 Left front outlet valve. Chart A-17, all 1987–88 models except Pontiac 6000

CHART A-18

REAR OUTLET VALVE RESISTANCE

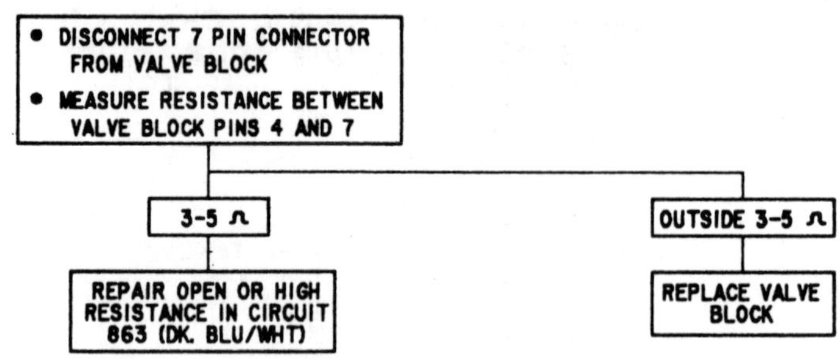

Fig. 65 Rear outlet valve. Chart A-18, all 1987–88 models except Pontiac 6000

CHART A-19

DIODE AND DIODE CIRCUIT

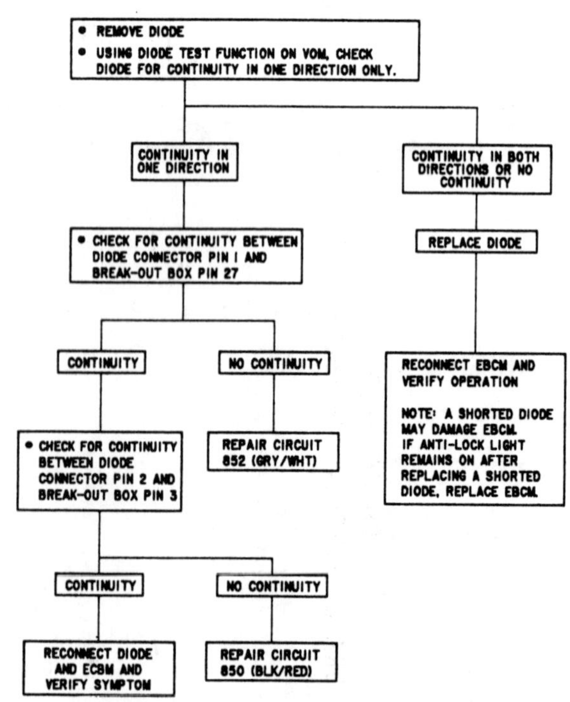

Fig. 66 Diode & diode circuit. Chart A-19, all 1987–88 models except Pontiac 6000

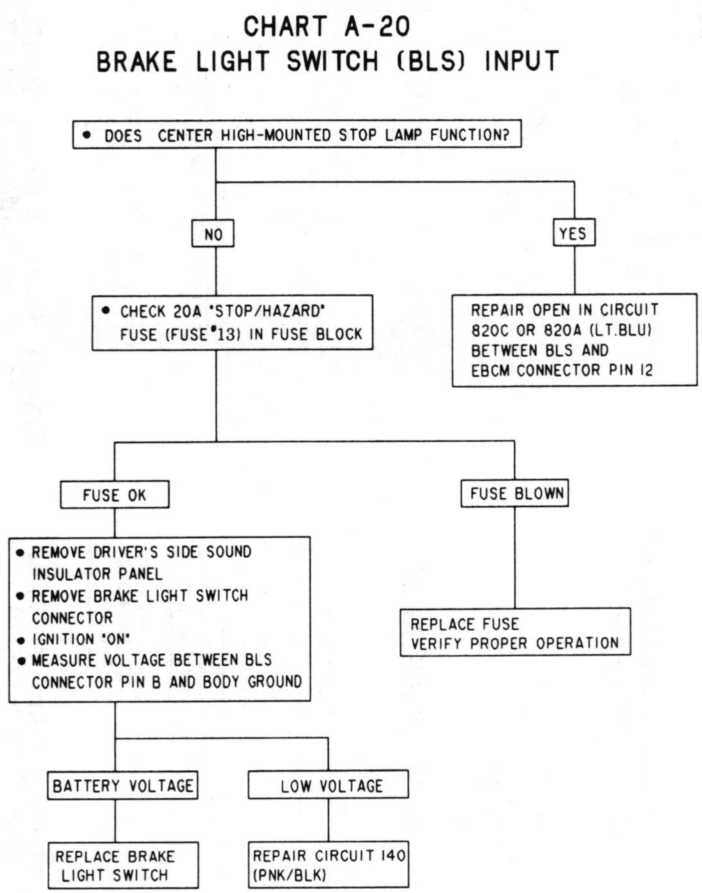

Fig. 67 Brake light switch input. Chart A-20, all 1987–88 models except Pontiac 6000

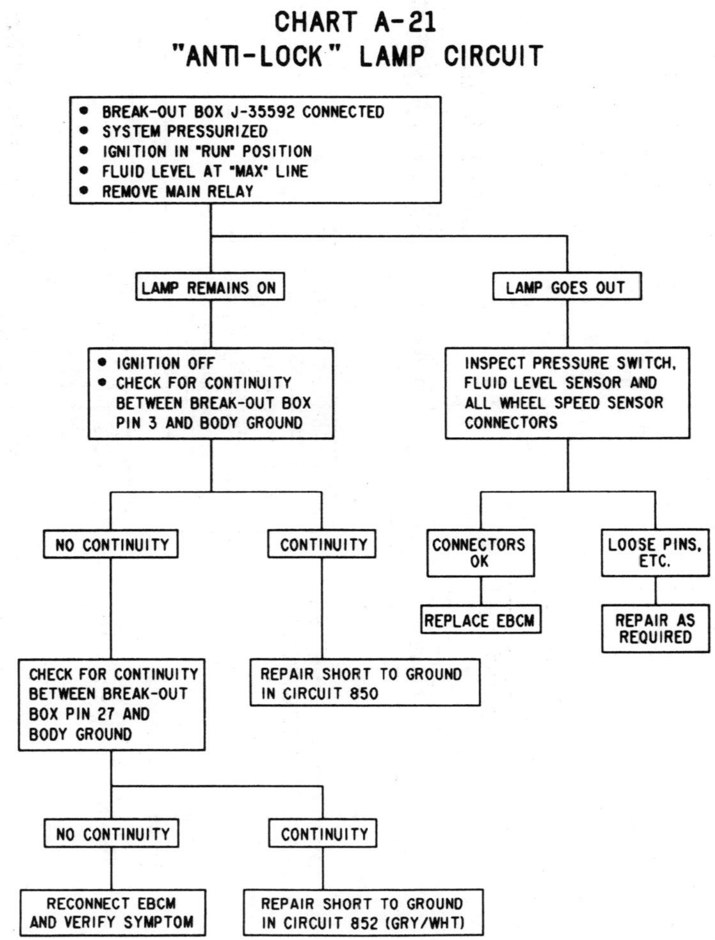

Fig. 68 ANTILOCK lamp circuit. Chart A-21, all 1987–88 models except Pontiac 6000

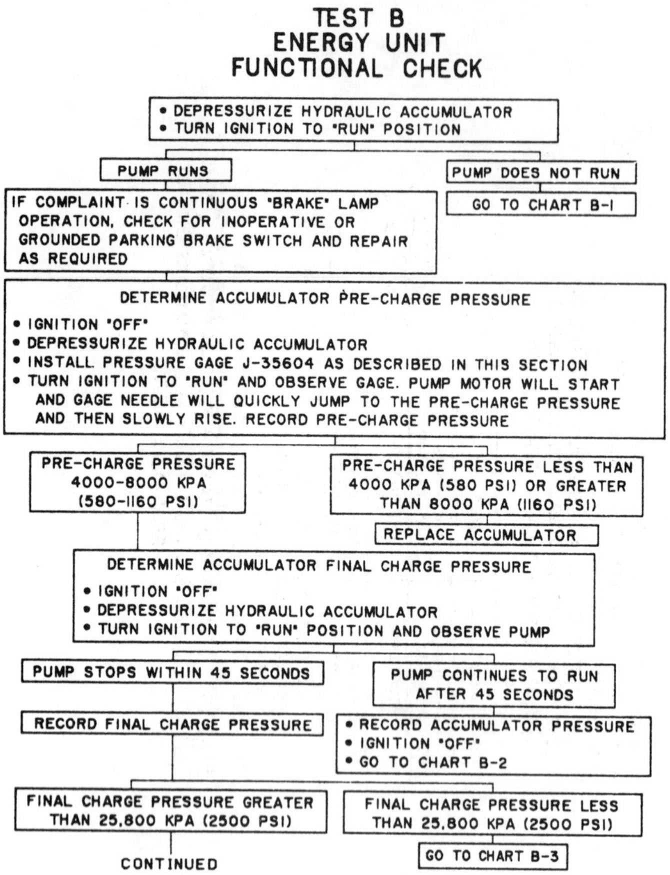

Fig. 69 Energy unit functional check (Part 1 of 2). Test B, all 1987–88 models except Pontiac 6000

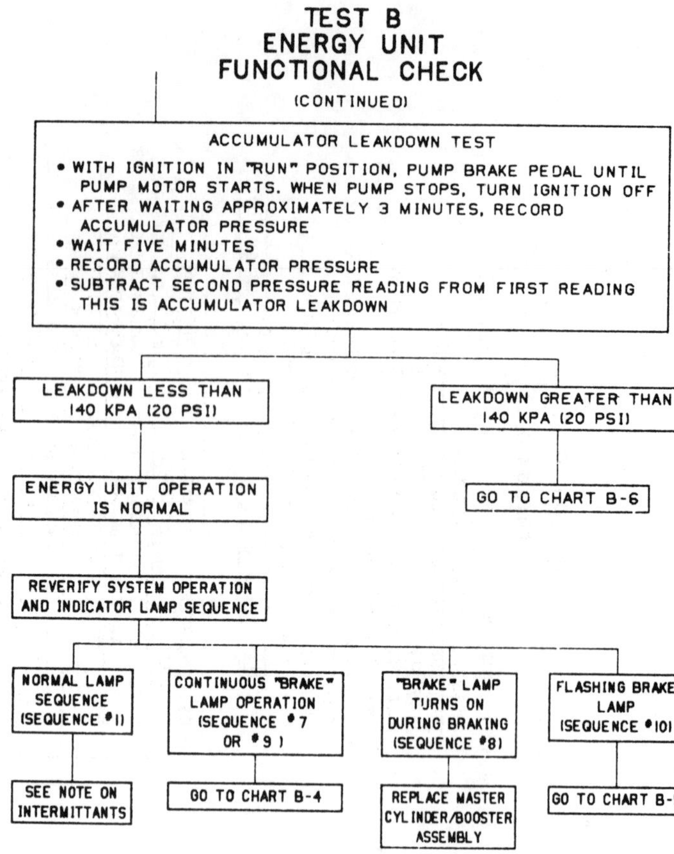

Fig. 69 Energy unit functional check (Part 2 of 2). Test B, all 1987–88 models except Pontiac 6000

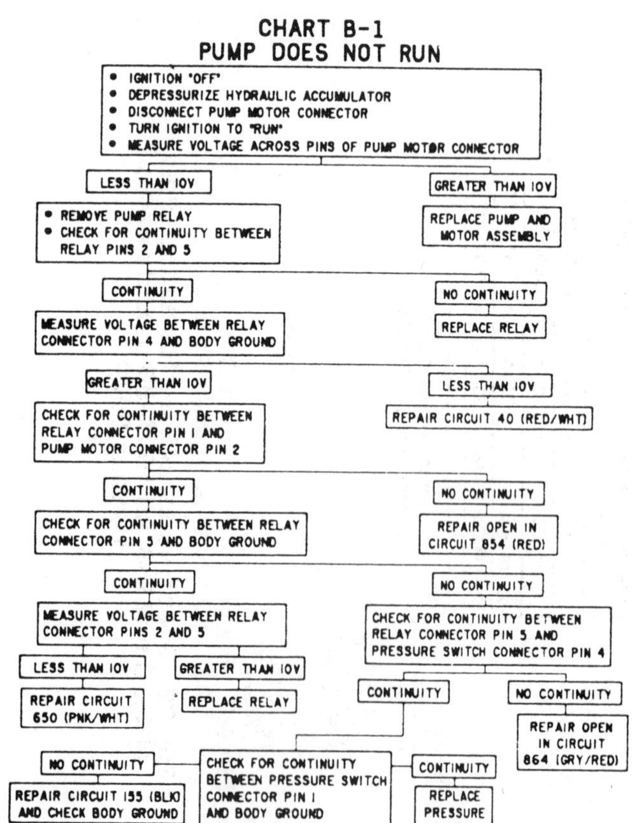

Fig. 70 Pump does not run. Chart B-1, all 1987–88 models except Pontiac 6000

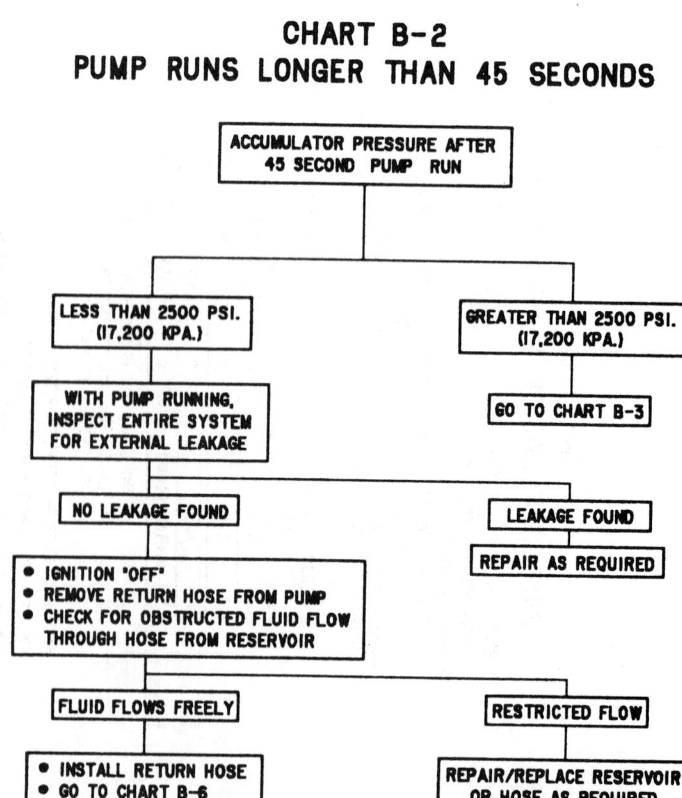

Fig. 71 Pump runs longer than 45 seconds. Chart B-2, all 1987–88 models except Pontiac 6000

CHART B-3
PRESSURE SWITCH PERFORMANCE

I. SWITCH STATUS - PRESSURIZED

- TURN IGNITION TO "RUN" UNTIL PUMP STOPS
- IGNITION "OFF"
  NOTE: IF PUMP CONTINUES TO RUN AFTER 45 SECONDS, TURN IGNITION OFF AND PROCEDE
- CHECK PRESSURE SWITCH PINS FOR THE FOLLOWING CONDITIONS USING VOM

MEASURE BETWEEN PINS	SCALE	SPECIFICATION
B,D	200 Ω	NO CONTINUITY
C,D	200 Ω	NO CONTINUITY
A,E	200 Ω	CONTINUITY
ALL PINS,BODY GROUND	200 Ω	NO CONTINUITY

IF ANY CONDITION IS NOT MET, REPLACE PRESSURE SWITCH AND VERIFY SYSTEM OPERATION

II. SWITCH STATUS - DEPRESSURIZED

- DEPRESSURIZE HYDRAULIC ACCUMULATOR
- CHECK PRESSURE SWITCH PINS FOR THE FOLLOWING CONDITIONS USING VOM

MEASURE BETWEEN PINS	SCALE	SPECIFICATION
B,D	200 Ω	CONTINUITY
C,D	200 Ω	CONTINUITY
A,E	200 Ω	NO CONTINUITY
ALL PINS,BODY GROUND	200 Ω	NO CONTINUITY

IF ANY CONDITION IS NOT MET, REPLACE PRESSURE SWITCH AND VERIFY SYSTEM OPERATION

PRESSURE SWITCH

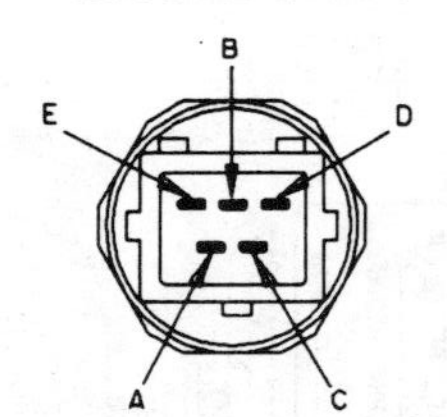

CONTINUED

Fig. 72 Pressure switch performance (Part 1 of 2). Chart B-3, all 1987–88 models except Pontiac 6000

CHART B-3
(CONTINUED)

III. SWITCH THRESHOLDS

- DEPRESSURIZE HYDRAULIC ACCUMULATOR
- INSTALL PRESSURE GAGE J-35604 AS DESCRIBED IN THIS SECTION
- CONNECT PRESSURE SWITCH CONNECTOR AND TURN IGNITION TO "RUN" UNTIL PUMP STOPS
- USING VOM, MONITOR FOR CONTINUITY BETWEEN PRESSURE SWITCH PINS AS SHOWN BELOW WHILE SLOWLY BLEEDING OFF ACCUMULATOR PRESSURE BY PUMPING THE BRAKE PEDAL. CONTINUITY SHOULD BE GAINED OR LOST AS INDICATED.
- PRESSURIZE SYSTEM BETWEEN EACH TEST BY RECONNECTING PRESSURE SWITCH AND TURNING IGNITION TO "RUN" UNTIL PUMP STOPS

MEASURE BETWEEN PINS	SWITCH STATUS	PRESSURE RANGE
C,D	CONTINUITY SHOULD BE GAINED AT:	1980-2080 PSI (13,650-14,350 KPA)
B,D	CONTINUITY SHOULD BE GAINED AT:	1500-1550 PSI (10,350-10,700 KPA)
A,E	CONTINUITY SHOULD BE LOST AT:	1500-1550 PSI (10,350-10,700 KPA)

IF ANY CONDITION IS NOT MET, REPLACE PRESSURE SWITCH AND VERIFY PROPER OPERATION

- DEPRESSURIZE HYDRAULIC ACCUMULATOR
- WITH IGNITION OFF, CONNECT PRESSURE SWITCH CONNECTOR
- TURN IGNITION TO "RUN" AND OBSERVE GAGE, "ANTI-LOCK" LAMP, "BRAKE" LAMP AND PUMP MOTOR. EVENTS SHOULD OCCUR AT PRESSURES INDICATED IN CHART BELOW

EVENT	PRESSURE
"ANTI-LOCK" LAMP TURNS OFF	1900-1975 PSI (13,100-13,600 KPA)
"BRAKE" LAMP TURNS OFF	1900-1975 PSI (13,100-13,600 KPA)
PUMP MOTOR STOPS	2550-2670 PSI (17,580-18,400 KPA)

IF ANY CONDITION IS NOT MET, REPLACE PRESSURE SWITCH AND VERIFY PROPER OPERATION

PRESSURE SWITCH

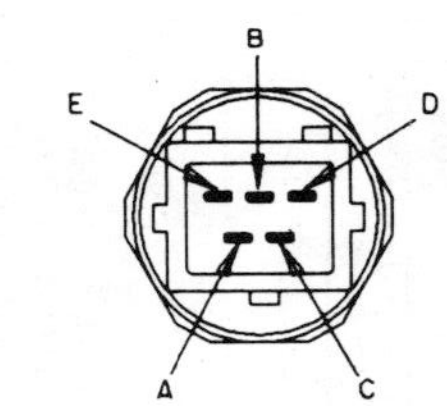

Fig. 72 Pressure switch performance (Part 2 of 2). Chart B-3, all 1987–88 models except Pontiac 6000

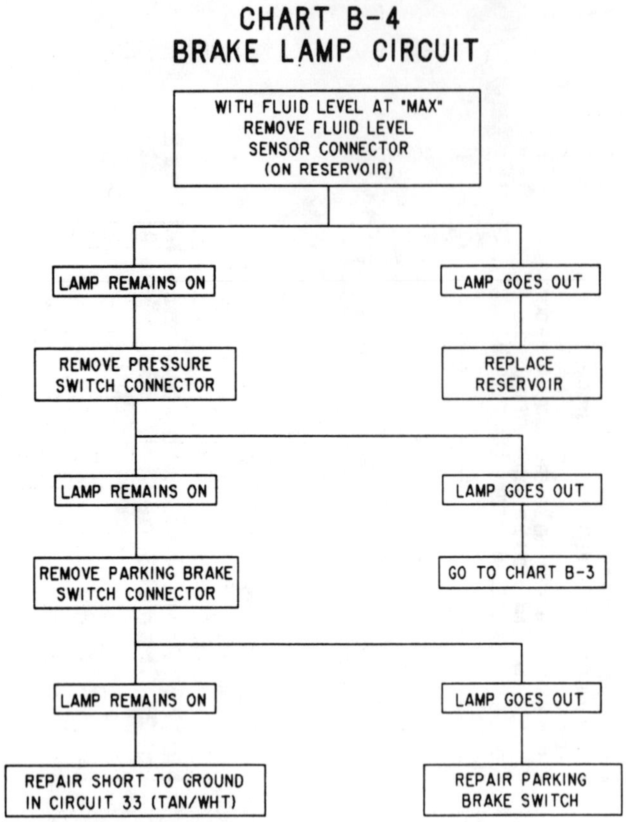

Fig. 73 Brake lamp circuit. Chart B-4, all 1987–88 models except Pontiac 6000

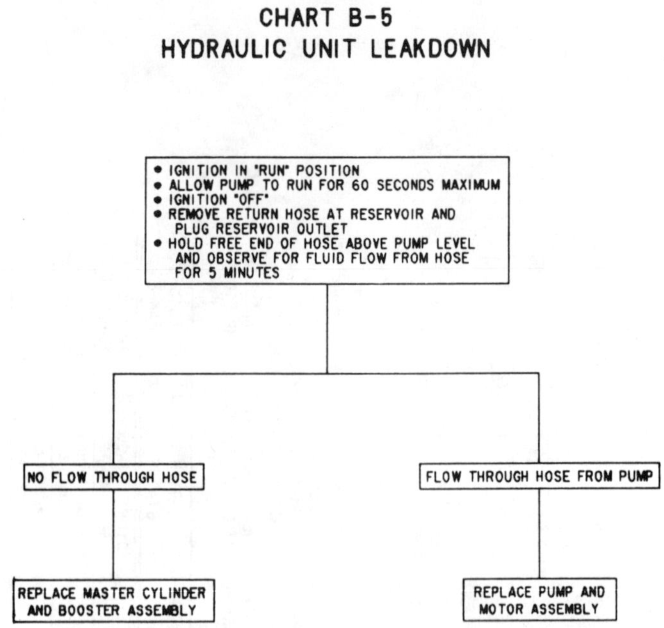

Fig. 74 Hydraulic unit leakdown. Chart B-5, all 1987–88 models except Pontiac 6000

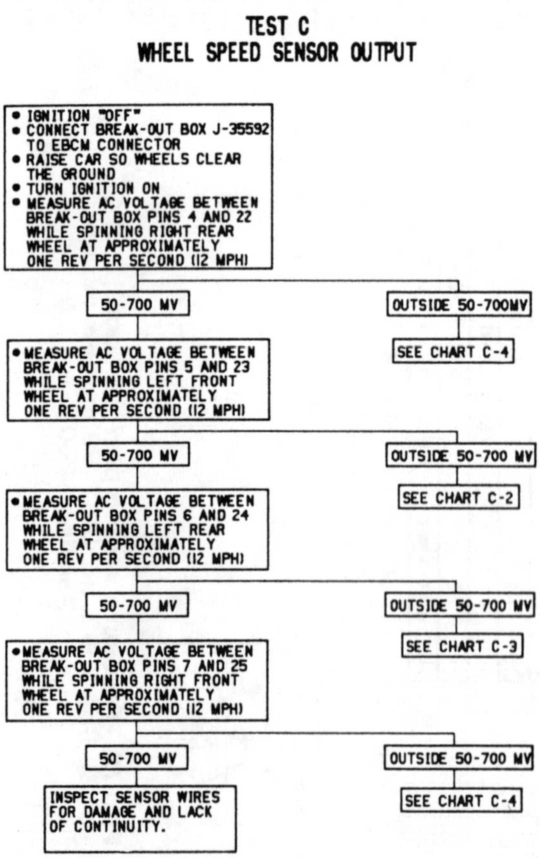

Fig. 75 Wheel speed sensor output. Test C, all 1987–88 models except Pontiac 6000

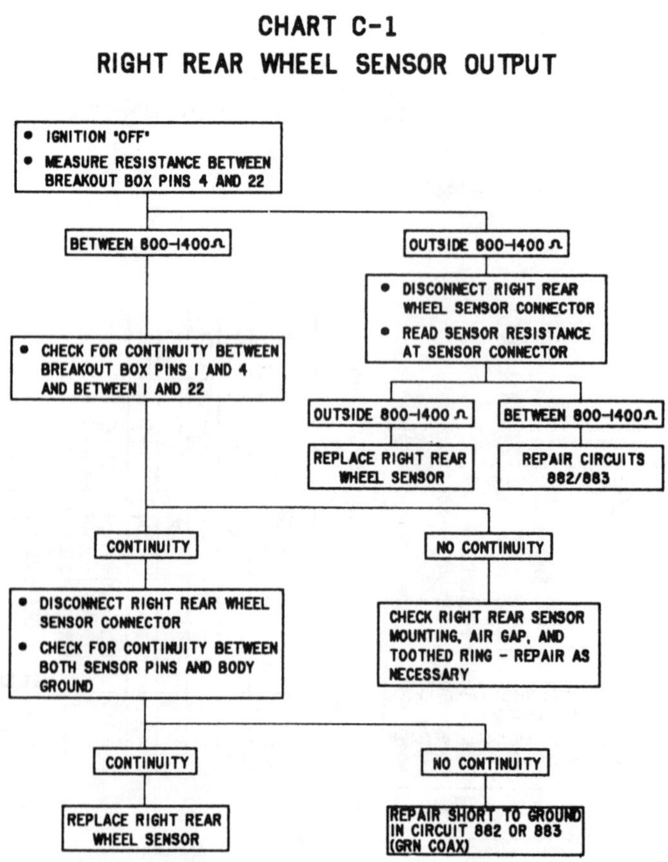

Fig. 76 Right rear sensor output. Chart C-1, all 1987–88 models except Pontiac 6000

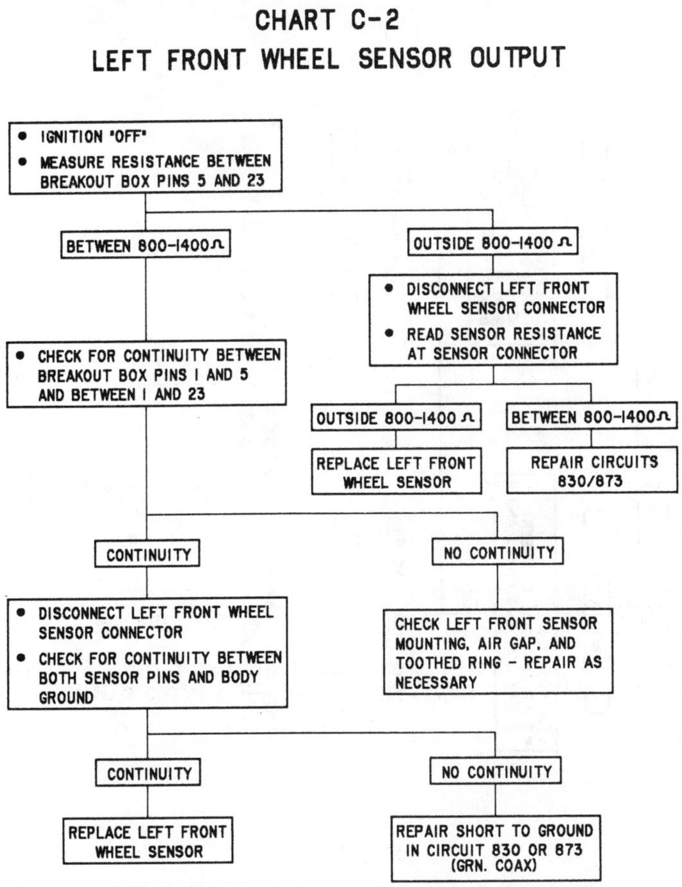

Fig. 77 Left front sensor output. Chart C-2, all 1987–88 models except Pontiac 6000

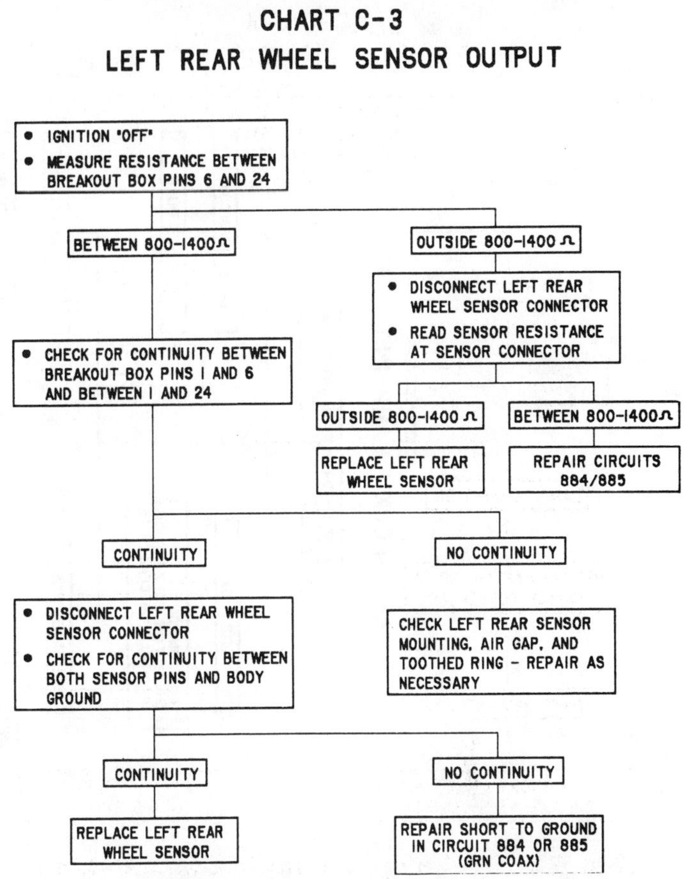

Fig. 78 Left rear sensor output. Chart C-3, all 1987–88 models except Pontiac 6000

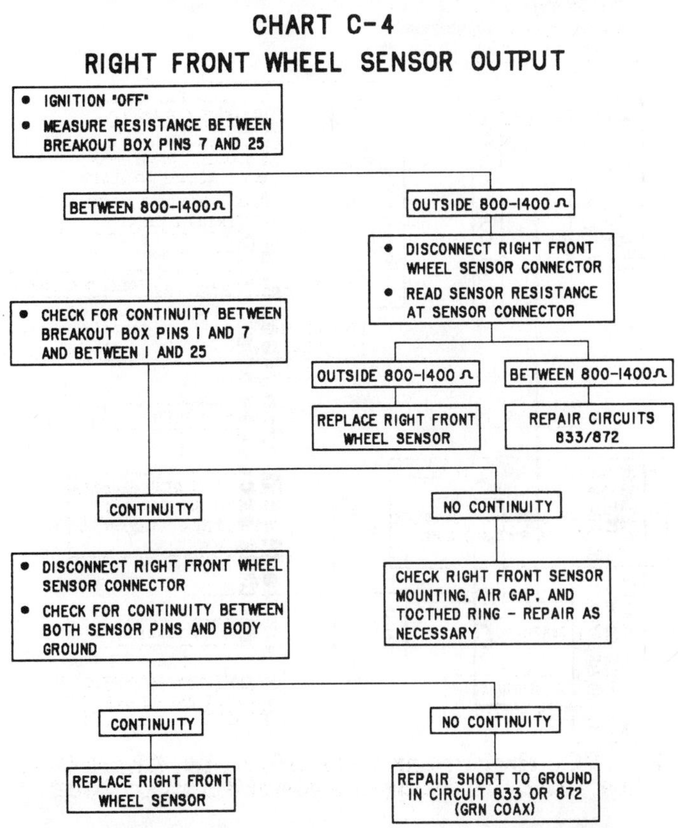

Fig. 79 Right front sensor output. Chart C-4, all 1987–88 models except Pontiac 6000

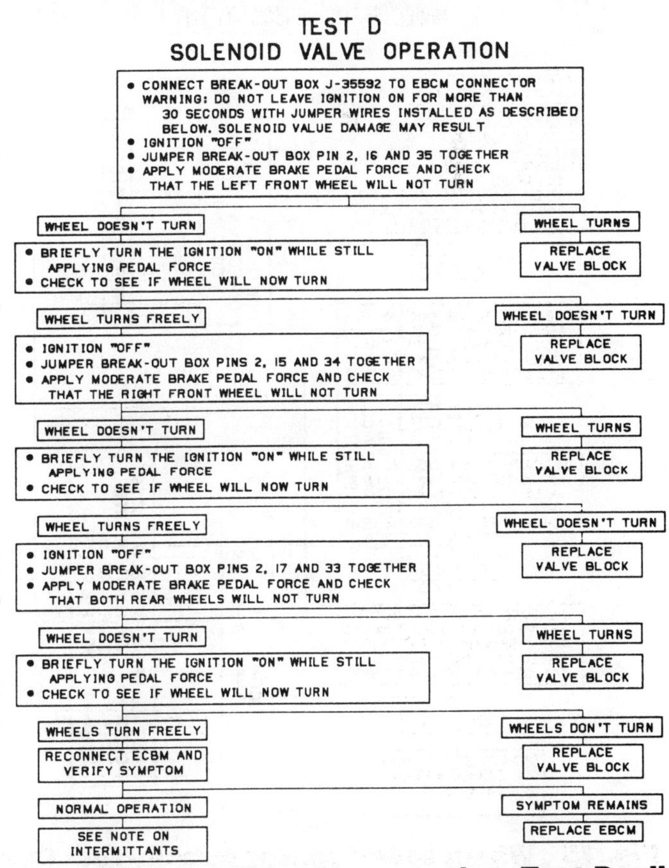

Fig. 80 Solenoid valve operation. Test D, all 1987–88 models except Pontiac 6000

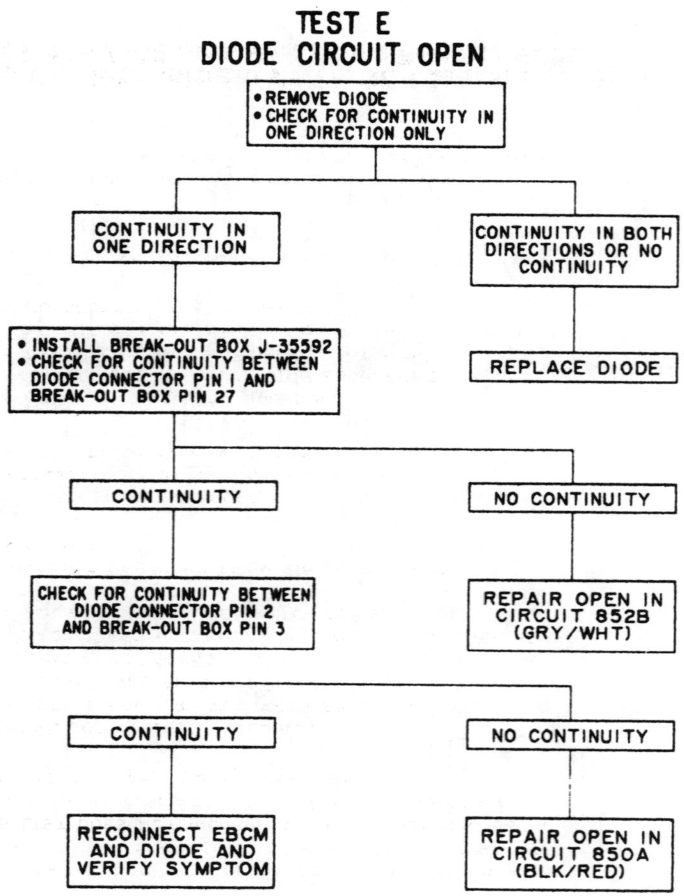

Fig. 81 Diode circuit open. Test E, all 1987–88 models except Pontiac 6000

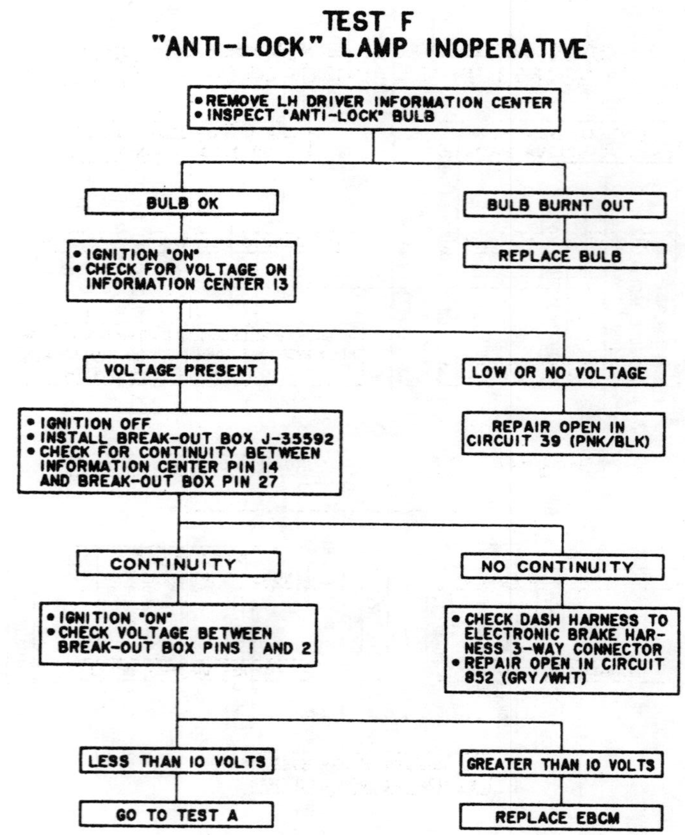

Fig. 82 ANTILOCK lamp inoperative. Test F, all 1987–88 models except Pontiac 6000

TEST G
INTERMITTENT "ANTI-LOCK" LAMP WHILE DRIVING

INTERRUPTIONS IN THE FOLLOWING CIRCUITS AND OTHERS MAY CAUSE THE "ANTI-LOCK" LAMP TO LIGHT INTERMITTENTLY. THIS LIST DOES NOT INCLUDE ALL POSSIBILITIES, BUT INCLUDES THOSE CIRCUITS MOST LIKELY TO CAUSE AN INTERMITTENT "ANTI-LOCK" LAMP CONDITION.

- EBCM LOOP CIRCUIT, INCLUDING:
  - -CIRCUIT 866 (LT. GRN) FROM EBCM PIN 9 TO FLUID LEVEL SENSOR PIN A.
  - -FLUID LEVEL SENSOR SWITCH.
  - -CIRCUIT 853 (LT. BLU/WHT) FROM FLUID LEVEL SENSOR PIN B TO PRESSURE SWITCH PIN E.
  - -PRESSURE SWITCH.
  - -CIRCUIT 865 (PPL) FROM PRESSURE SWITCH PIN A TO ECBM PIN 10.
- IGNITION ENABLE CIRCUIT, INCLUDING:
  - -BRAKE FUSE (#8 IN FUSE BLOCK, 5A).
  - -CIRCUIT 650 (PNK/WHT) FROM FUSE TO EBCM.
- MAIN RELAY CIRCUITS, INCLUDING:
  - -CIRCUIT 855 (DK. BLU) FROM EBCM PIN 8 TO RELAY COIL (RELAY CONNECTOR PIN 2).
  - -CIRCUIT 803 (BLK/WHT) FROM RELAY COIL (RELAY CONNECTOR PIN 5) TO GROUND.
- ALL WHEEL SPEED SENSOR CIRCUITS.

INSPECT CONNECTORS AND WIRES IN THESE CIRCUITS. IF NO TROUBLE IS FOUND, SEE NOTE ON INTERMITTENTS IN THIS SECTION AND PERFORM TEST A.

Fig. 83 Intermittent ANTILOCK lamp. Test G, all 1987–88 models except Pontiac 6000

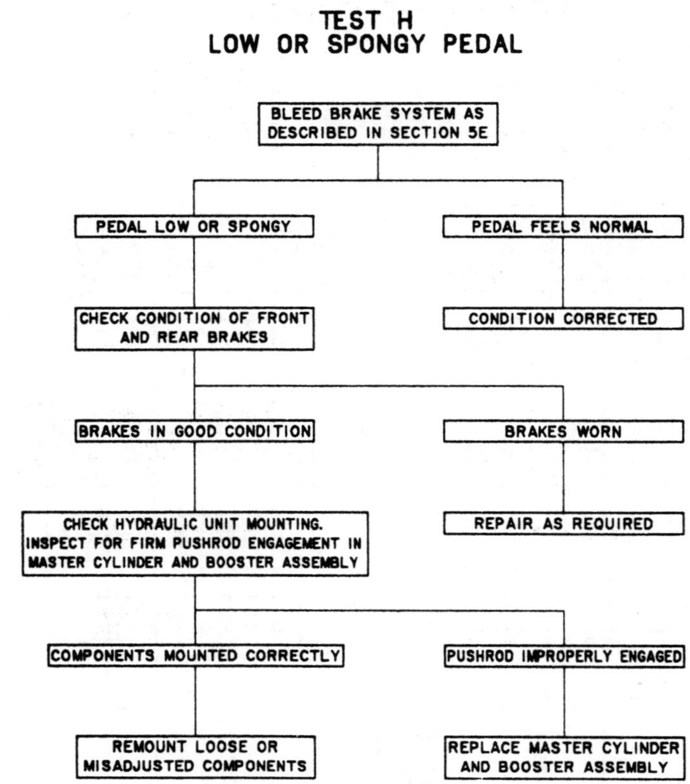

Fig. 84 Low or spongy pedal. Test H, all 1987–88 models except Pontiac 6000

SEE LAMP SEQUENCE DETERMINATION PROCEDURE
IN THIS SECTION BEFORE USING THIS CHART

SEQUENCE NUMBER	LAMP SEQUENCE	SYMPTOM DESCRIPTION	SEE FIGURE
1	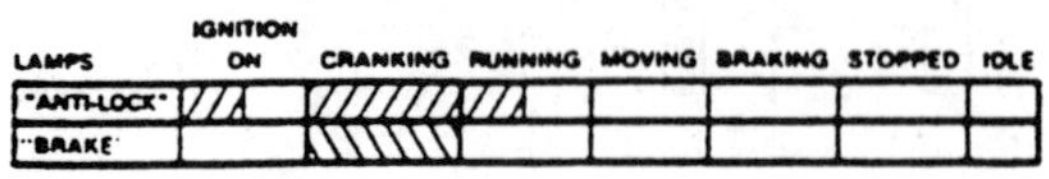	NORMAL LAMP SEQUENCE WITH — EXCESSIVE PEDAL TRAVEL OR SPONGY PEDAL — ANTI-LOCK BRAKING OPERATION OR VALVE CYCLING DURING NORMAL STOPS ON DRY PAVEMENT — POOR VEHICLE TRACKING DURING ANTI-LOCK BRAKING	126 103 113
2	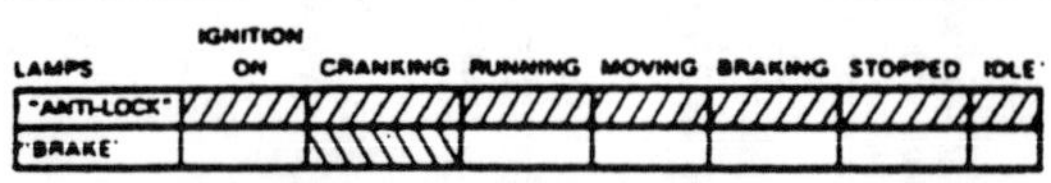	CONTINUOUS "ANTI-LOCK" LAMP NORMAL "BRAKE" LAMP	
3	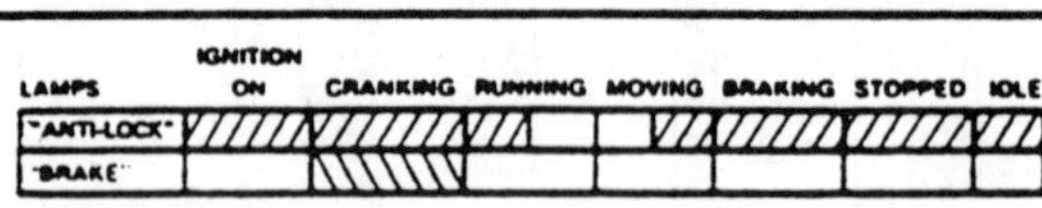	"ANTI-LOCK" LAMP COMES ON AFTER VEHICLE STARTS MOVING NORMAL BRAKE LAMP	125
4	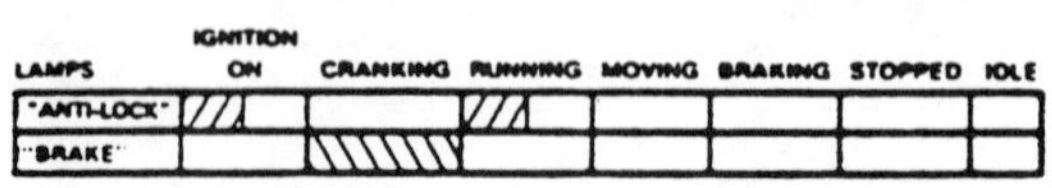	NO "ANTI-LOCK" LAMP WHILE CRANKING NORMAL "BRAKE" LAMP	125
5	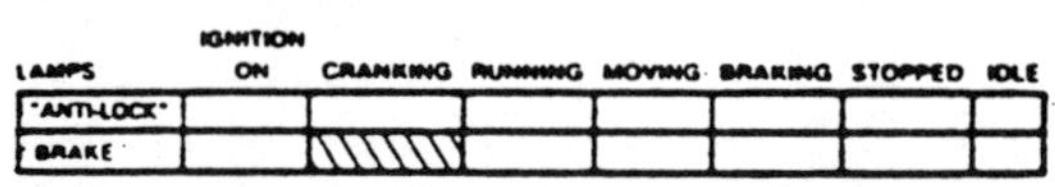	NO "ANTI-LOCK" LAMP NORMAL "BRAKE" LAMP	125
6		INTERMITTANT "ANTI-LOCK" LAMP WHILE DRIVING NORMAL "BRAKE" LAMP	127
7	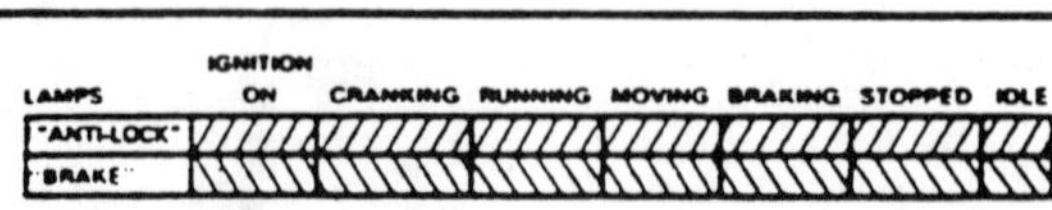	CONTINUOUS "ANTI-LOCK" LAMP CONTINUOUS "BRAKE" LAMP	103
8	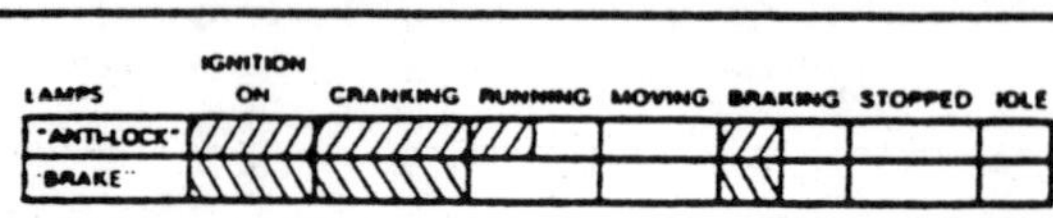	"ANTI-LOCK" AND "BRAKE" LAMPS COME ON WHILE BRAKING	122
9	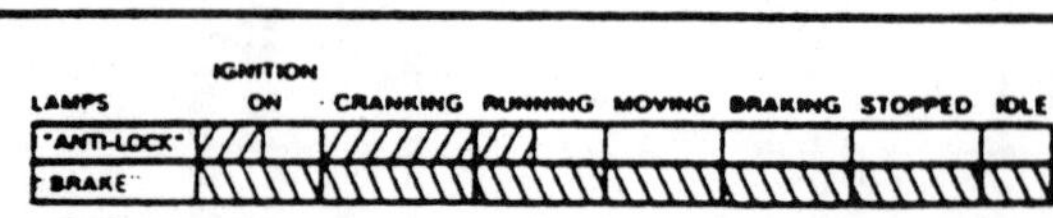	NORMAL "ANTI-LOCK" LAMP CONTINUOUS "BRAKE" LAMP	124
10	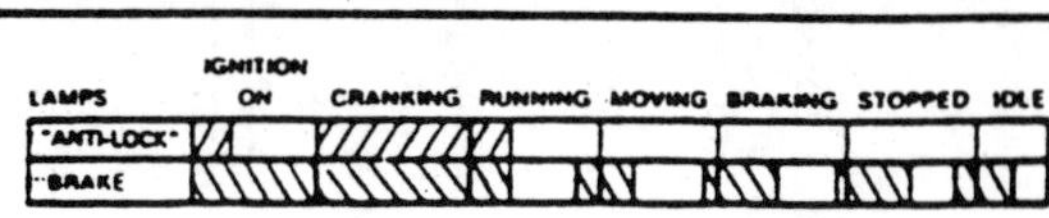	NORMAL OR CONTINUOUS "ANTI-LOCK" LAMP FLASHING "BRAKE" LAMP	119

LAMP STATUS
- SHADED AREAS: LAMP ON
- BLANK AREAS (NO SHADING): LAMP OFF
- PARTIALLY SHADED AREAS: LAMP ON FOR PART OF TEST PERIOD

Fig. 85 Warning lamp sequence chart. 1989 Pontiac 6000

PIN-OUT CHECKS

- CONNECT BREAK-OUT BOX J 35592 to 35-PIN EBCM HARNESS CONNECTOR AS DESCRIBED IN THIS SECTION
- PERFORM CHECKS WITH HIGH IMPEDANCE DIGITAL MULTIMETER J 34029-A OR EQUIVALENT
- ALL CHECKS ARE MADE WITH ENGINE OFF

CIRCUIT TO BE TESTED	IGNITION SWITCH POSITION	MULTIMETER SCALE/RANGE	MEASURE BETWEEN PIN NUMBERS	SPECIFICATION	IF RESULT NOT WITHIN SPECIFICATION, SEE FIGURE
IGNITION ENABLE	RUN	20 DCV	2(+)*,1(-)*	10 V MINIMUM	87 88
MAIN RELAY GROUND	OFF	200 Ω	1,3	CONTINUITY	89 90
	OFF	200 Ω	1,20	CONTINUITY	
MAIN RELAY COIL	OFF	200 Ω	1,8	45-105 Ω	91

BEFORE PERFORMING THIS TEST:
- REMOVE GAGE FUSE FROM FUSE BOX
- PLACE FUSED JUMPER BETWEEN BREAK-OUT BOX PINS 2 & 8

CIRCUIT TO BE TESTED	IGNITION SWITCH POSITION	MULTIMETER SCALE/RANGE	MEASURE BETWEEN PIN NUMBERS	SPECIFICATION	IF RESULT NOT WITHIN SPECIFICATION, SEE FIGURE
MAIN RELAY POWER	ON	20 DCV	3(+),1(-)	10 V MINIMUM	92
	ON	20 DCV	20(+),1(-)	10 V MINIMUM	

BEFORE PROCEEDING:
- REMOVE JUMPER FROM PINS 2 & 8
- INSTALL GAGE FUSE

CIRCUIT TO BE TESTED	IGNITION SWITCH POSITION	MULTIMETER SCALE/RANGE	MEASURE BETWEEN PIN NUMBERS	SPECIFICATION	IF RESULT NOT WITHIN SPECIFICATION, SEE FIGURE
EBCM SWITCH LOOP	OFF	200 Ω	9,10	LESS THAN 5Ω	93 94
	OFF	200 Ω	1,9	NO CONTINUITY	95
RR SENSOR RESISTANCE	OFF	2k Ω	4,22	800-1400 Ω	96 97 98 99
LF SENSOR RESISTANCE	OFF	2k Ω	5,23	800-1400 Ω	96 97 100 101
LR SENSOR RESISTANCE	OFF	2k Ω	6,24	800-1400 Ω	96 97 102 103
RF SENSOR RESISTANCE	OFF	2k Ω	7,25	800-1400 Ω	96 97 104 105
MAIN VALVE SOLENOID	OFF	200 Ω	11,18	2-5 Ω	106 107
VALVE BLOCK GROUND	OFF	200 Ω	1,11	LESS THAN 2Ω	106 108
RF INLET VALVE	OFF	200 Ω	11,15	5-7 Ω	109
LF INLET VALVE	OFF	200 Ω	11,35	5-7 Ω	110
REAR INLET VALVE	OFF	200 Ω	11,17	5-7 Ω	111
RF OUTLET VALVE	OFF	200 Ω	11,34	3-5 Ω	112
LF OUTLET VALVE	OFF	200 Ω	11,16	3-5 Ω	113
REAR OUTLET VALVE	OFF	200 Ω	11,33	3-5 Ω	114

*(+) OR (-) INDICATES MULTI-METER POLARITY

IF ALL TEST RESULTS ARE WITHIN SPECIFICATION, RECONNECT EBCM AND VERIFY CONTINUOUS "ANTILOCK" WARNING LAMP OPERATION
- IF NORMAL OPERATION RESUMES, SEE NOTE ON INTERMITTENTS
- IF LAMP REMAINS ON, SEE FIGURE 119

Fig. 86 Pin-Out checks. 1989 Pontiac 6000

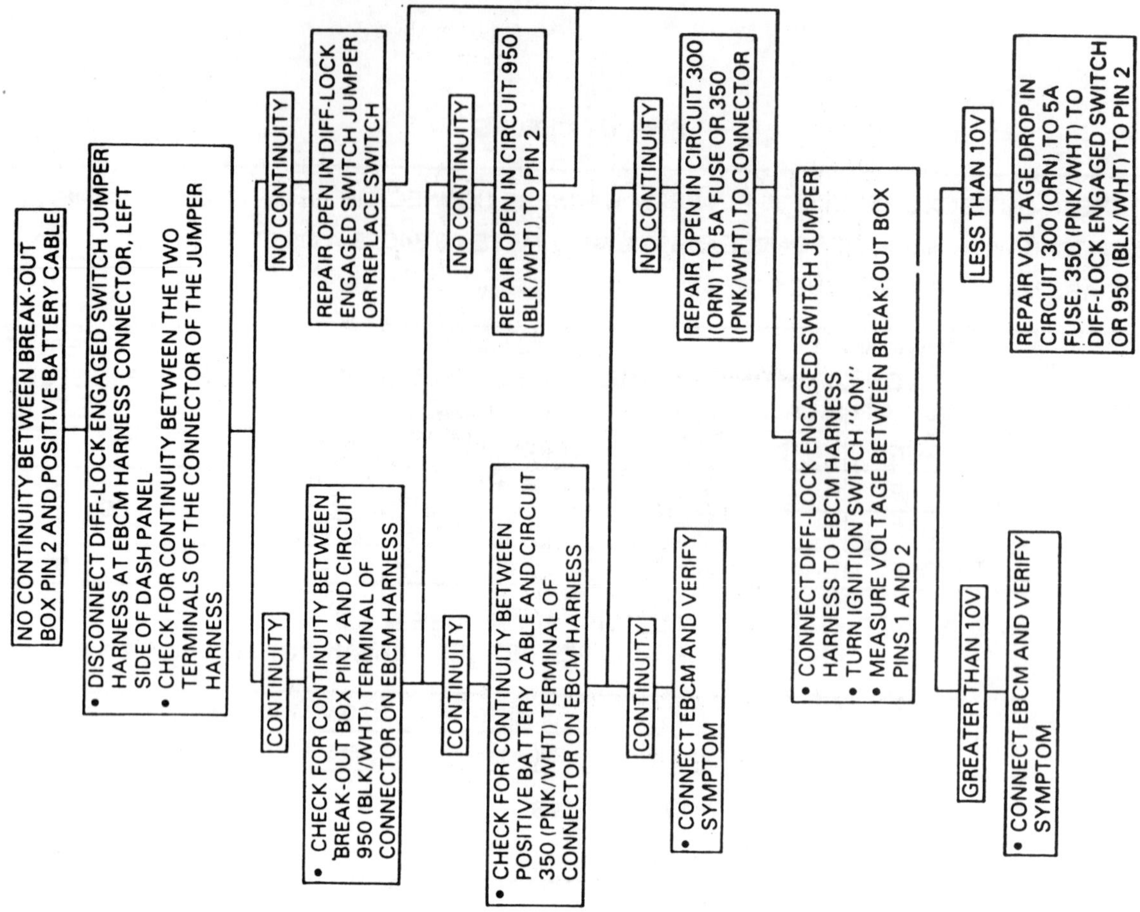

Fig. 88 Ignition enable circuit. Test-B, 1989 Pontiac 6000

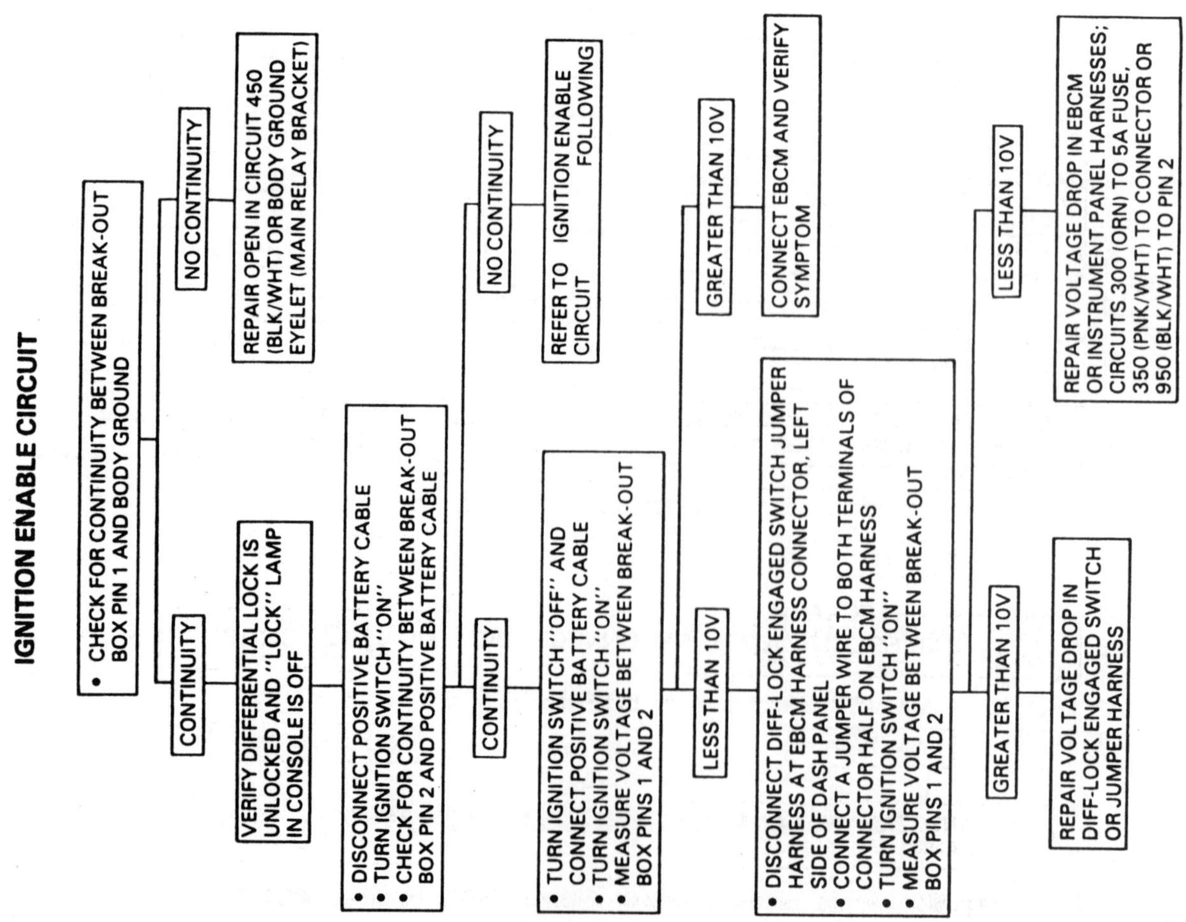

Fig. 87 Ignition enable circuit. Test-A, 1989 Pontiac 6000

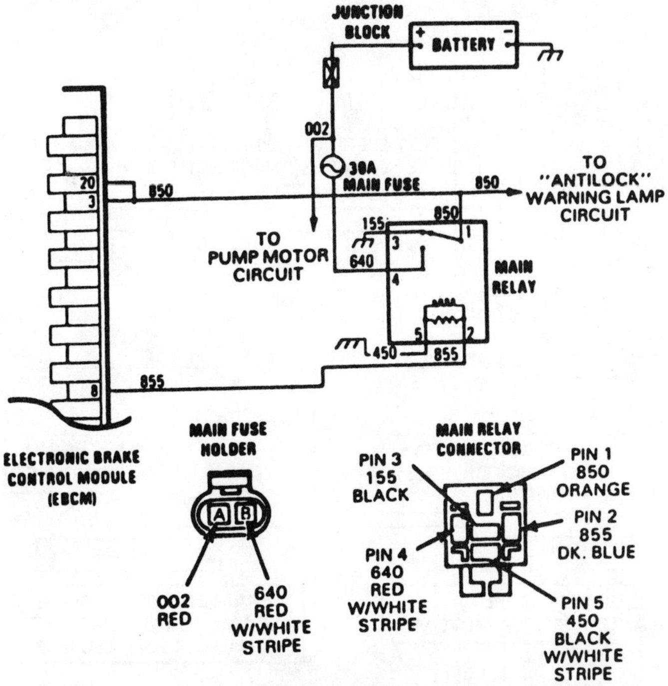

Fig. 89 Main relay circuit schematic. 1989 Pontiac 6000

MAIN RELAY GROUND CIRCUIT

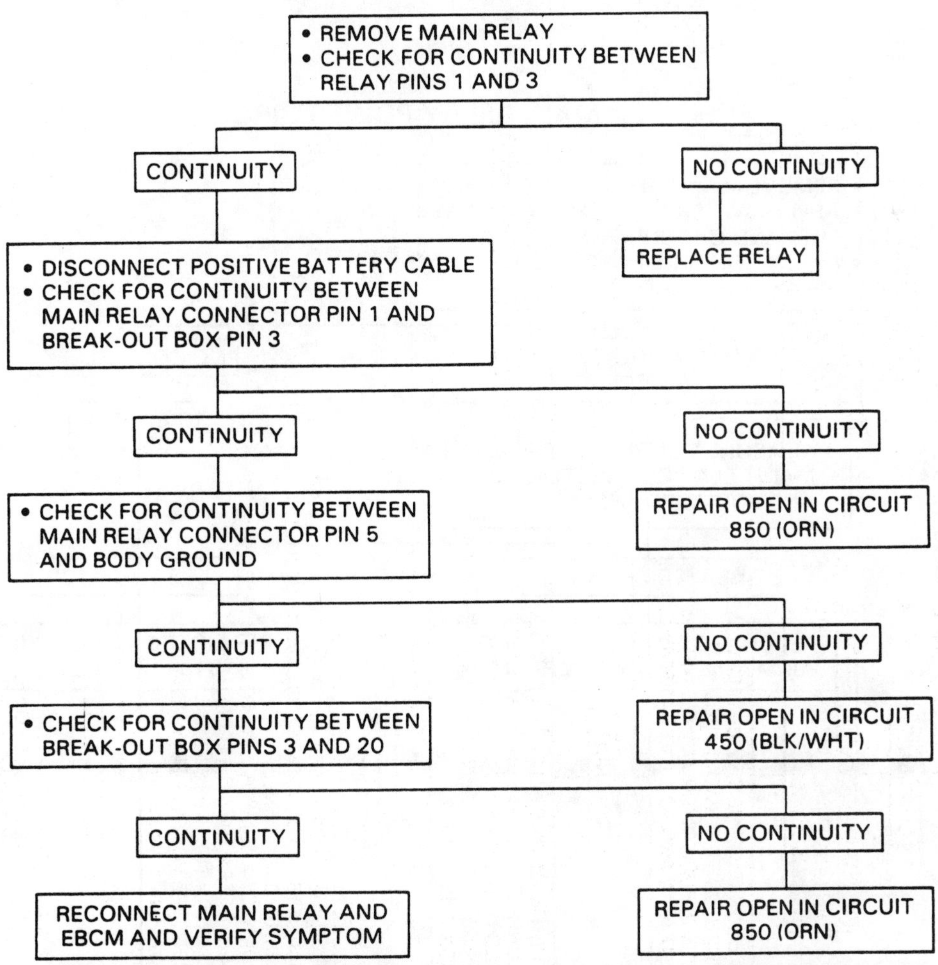

Fig. 90 Main relay ground circuit. 1989 Pontiac 6000

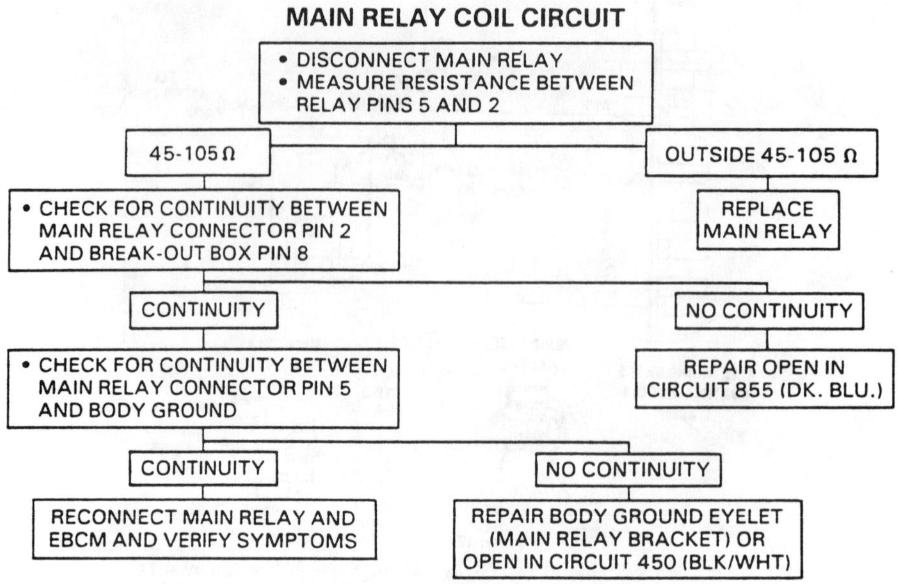

Fig. 91 Main relay coil circuit. 1989 Pontiac 6000

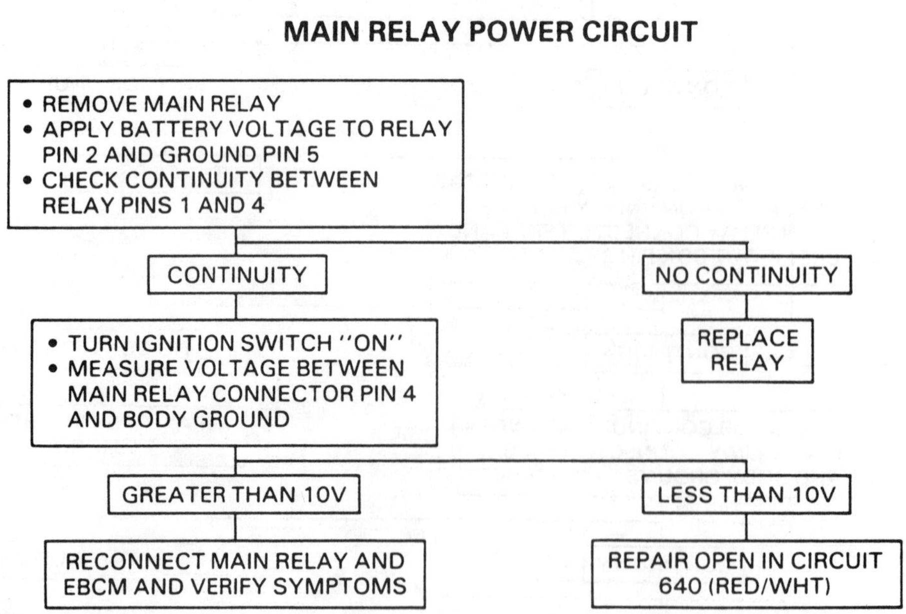

Fig. 92 Main relay power circuit. 1989 Pontiac 6000

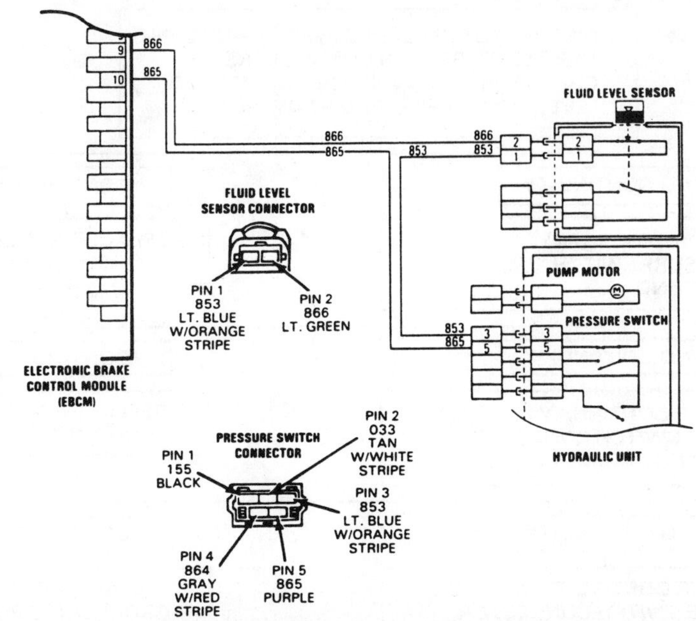

Fig. 93 EBCM loop circuit schematic. 1989 Pontiac 6000

EBCM LOOP OPEN

- FILL RESERVOIR AS DESCRIBED IN THIS SECTION
- WITH CAP INSTALLED, DISCONNECT 2-PIN FLUID LEVEL SENSOR CONNECTOR
- MEASURE RESISTANCE BETWEEN PINS

LESS THAN 2 Ω

- TURN IGNITION SWITCH TO "RUN" UNTIL PUMP MOTOR STOPS
- DISCONNECT PRESSURE SWITCH CONNECTOR
- CHECK FOR CONTINUITY BETWEEN PRESSURE SWITCH PINS 3 AND 5

GREATER THAN 2 Ω

REPLACE FLUID LEVEL SENSOR

CONTINUITY

- DISCONNECT 2 PIN FLUID LEVEL SENSOR CONNECTOR
- CHECK FOR CONTINUITY BETWEEN BREAK-OUT BOX PIN 9 AND FLUID LEVEL SENSOR CONNECTOR PIN 2

NO CONTINUITY

REPLACE PRESSURE SWITCH

CONTINUITY

- CHECK FOR CONTINUITY BETWEEN FLUID LEVEL SENSOR CONNECTOR PIN 1 AND PRESSURE SWITCH CONNECTOR PIN 3

NO CONTINUITY

REPAIR OPEN IN CIRCUIT 866 (LT. GRN)

CONTINUITY

- CHECK FOR CONTINUITY BETWEEN PRESSURE SWITCH CONNECTOR PIN 5 AND BREAK-OUT BOX PIN 10

NO CONTINUITY

REPAIR OPEN IN CIRCUIT 853 (LT. BLU/ORN)

CONTINUITY

RECONNECT ALL CONNECTORS AND EBCM AND VERIFY SYMPTOM

NO CONTINUITY

REPAIR OPEN IN CIRCUIT 865 (PPL)

Fig. 94 EBCM loop open. 1989 Pontiac 6000

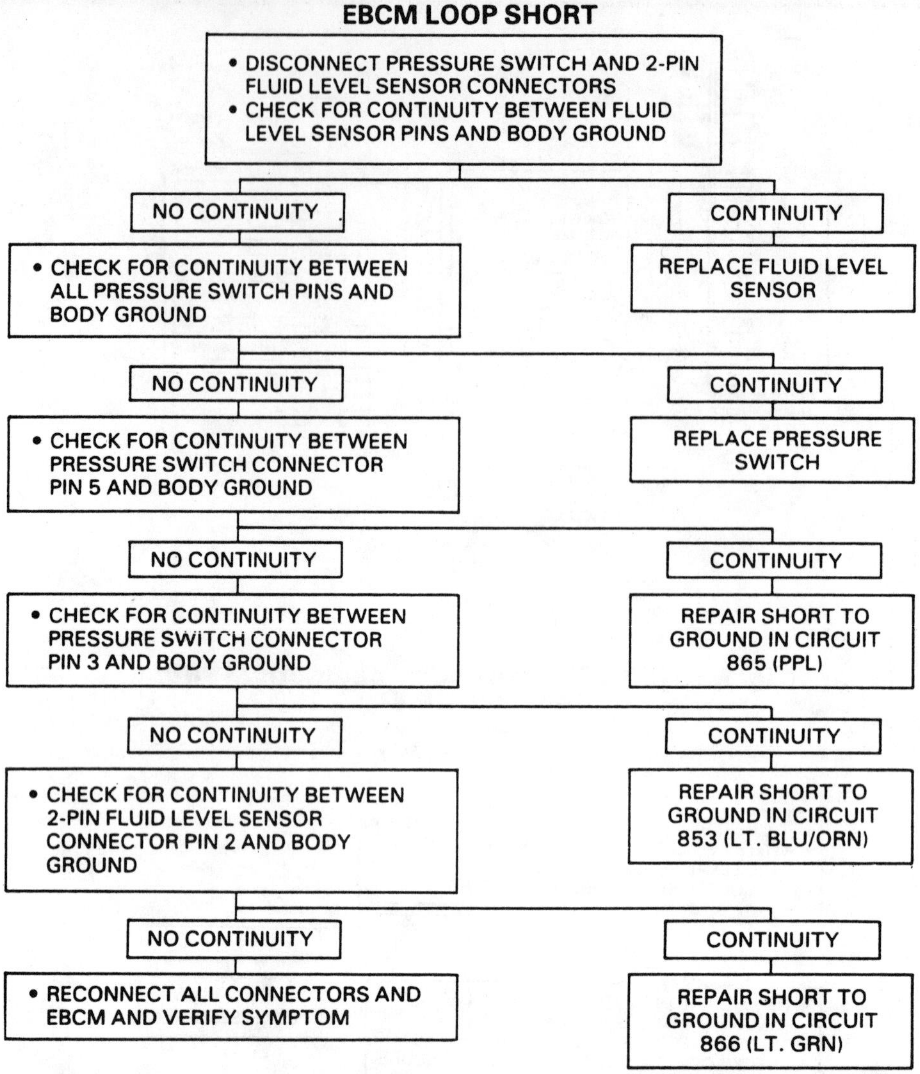

Fig. 95 EBCM loop short. 1989 Pontiac 6000

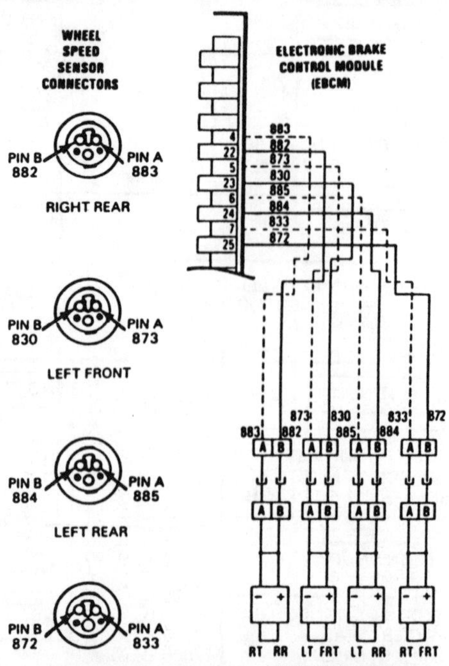

Fig. 96 Wheel speed sensor wiring schematic. 1989 Pontiac 6000

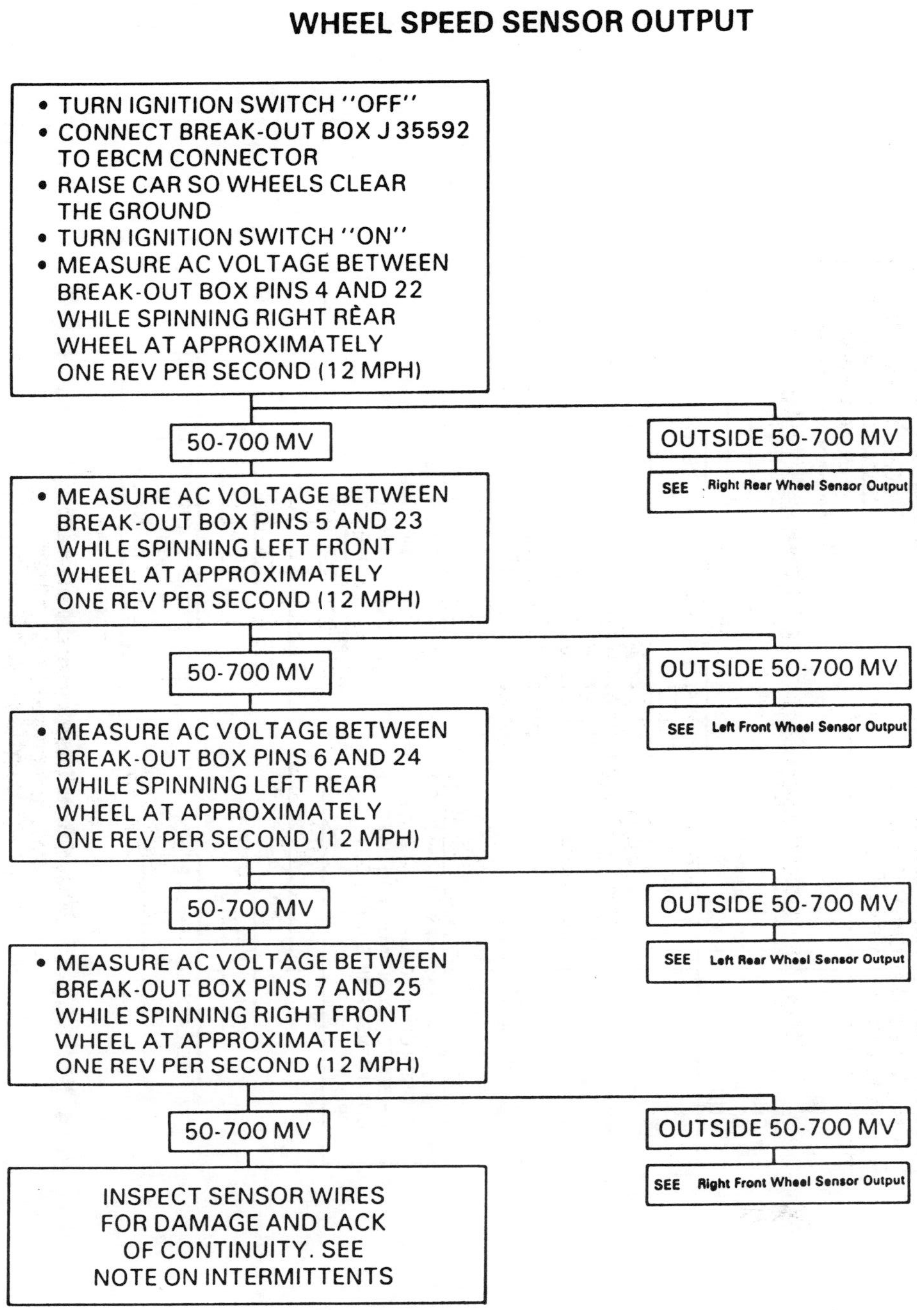

Fig. 97 Wheel speed sensor output. 1989 Pontiac 6000

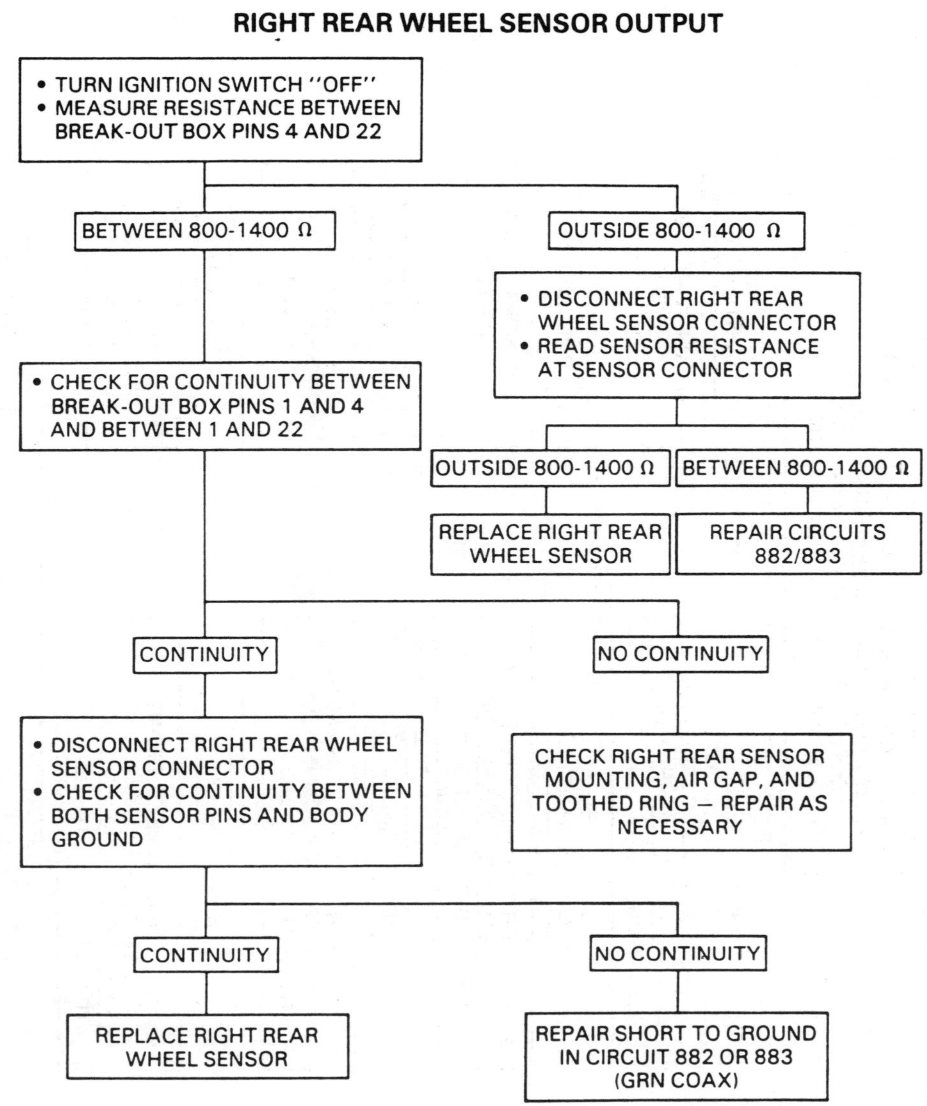

Fig. 98 Right rear wheel sensor output. 1989 Pontiac 6000

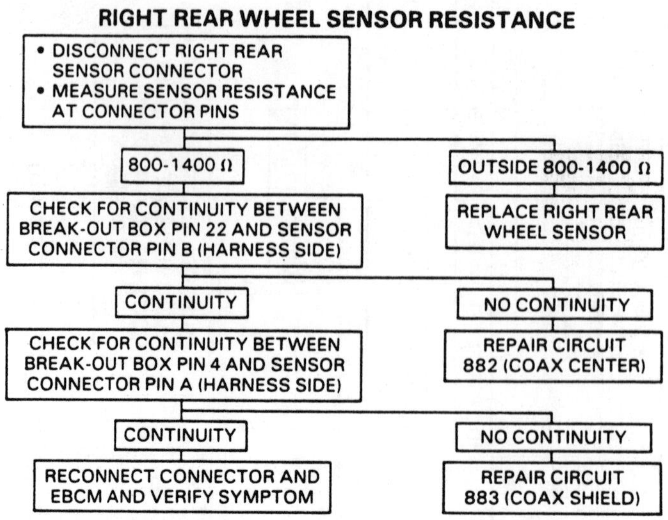

Fig. 99 Right rear wheel sensor resistance. 1989 Pontiac 6000

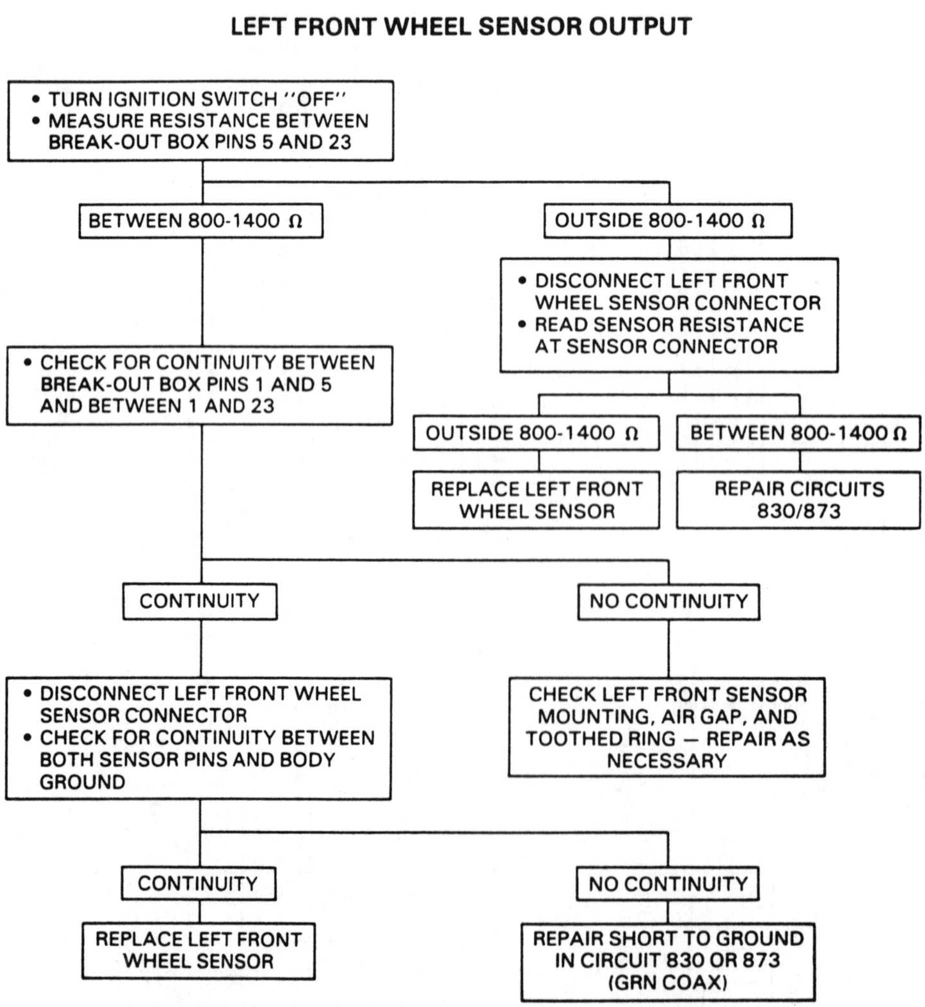

Fig. 100 Left front wheel sensor output. 1989 Pontiac 6000

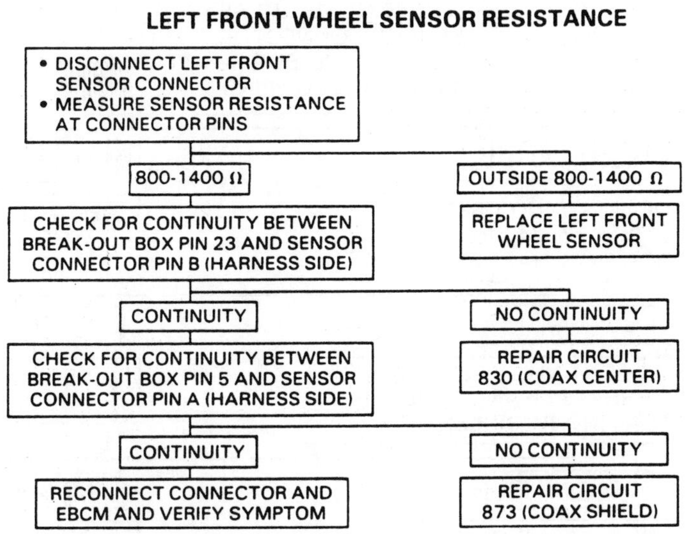

Fig. 101 Left front wheel sensor resistance. 1989 Pontiac 6000

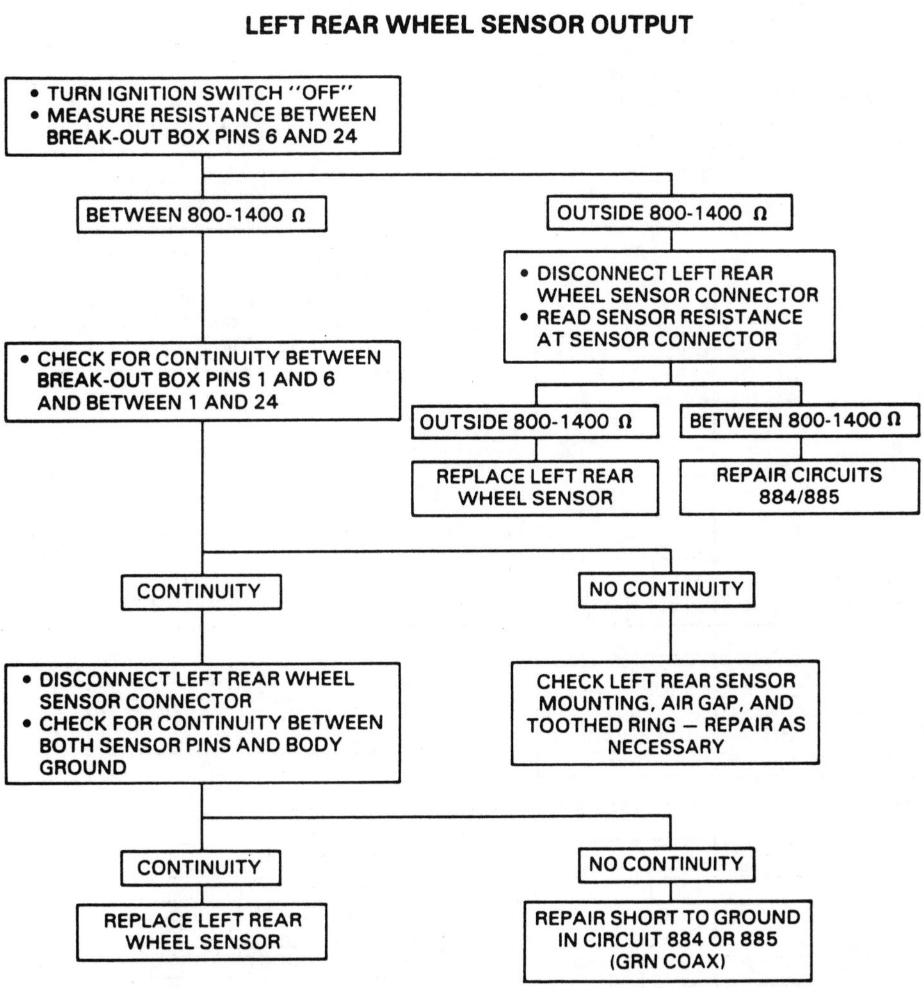

Fig. 102 Left rear wheel sensor output. 1989 Pontiac 6000

LEFT REAR WHEEL SENSOR RESISTANCE

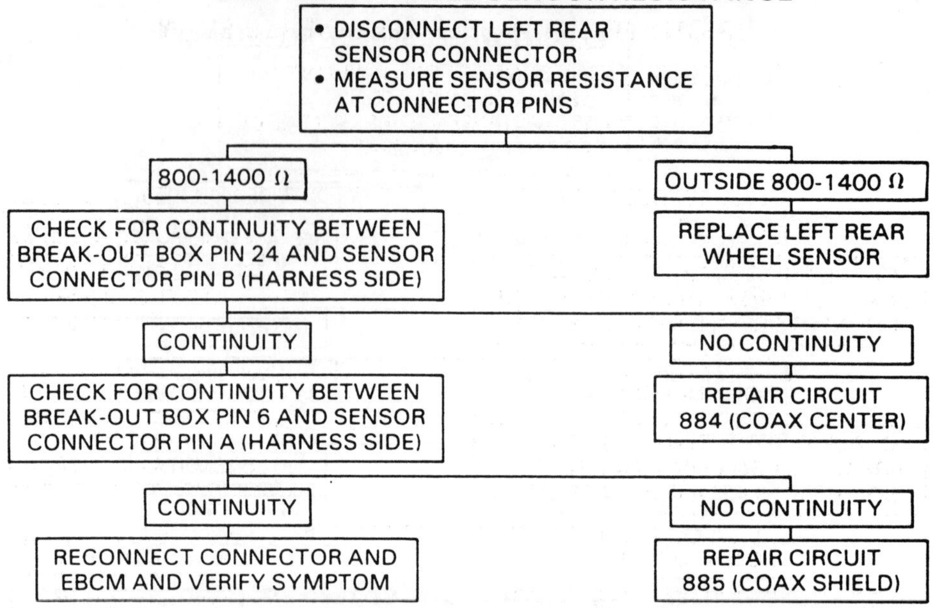

Fig. 103 Left rear wheel sensor resistance. 1989 Pontiac 6000

RIGHT FRONT WHEEL SENSOR OUTPUT

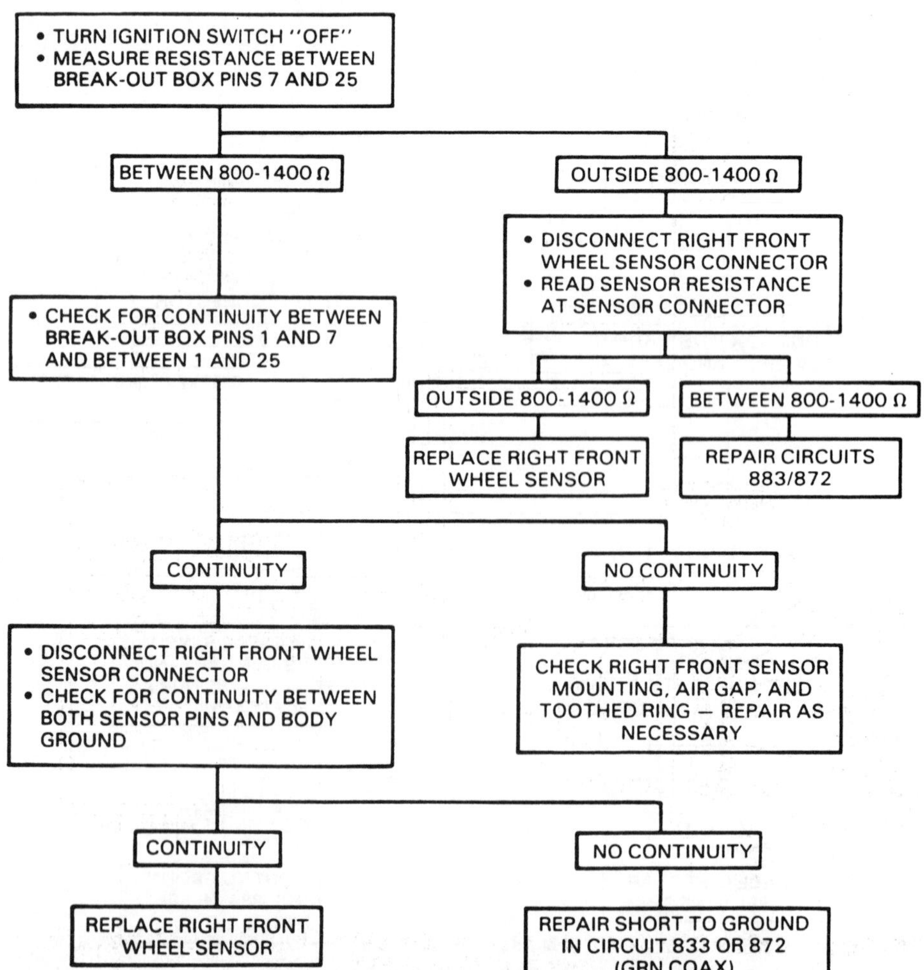

Fig. 104 Right front wheel sensor output. 1989 Pontiac 6000

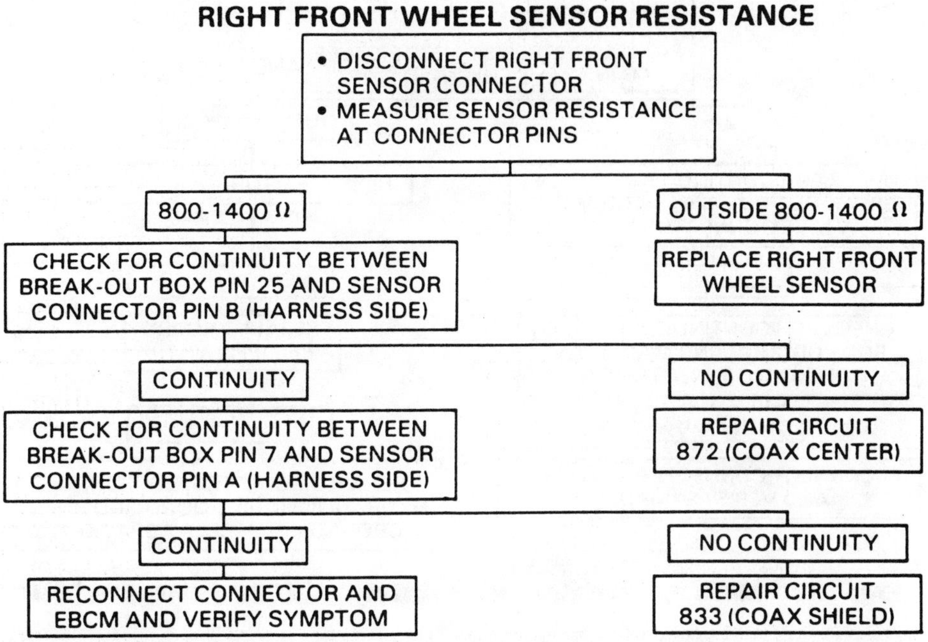

Fig. 105 Right front wheel sensor output. 1989 Pontiac 6000

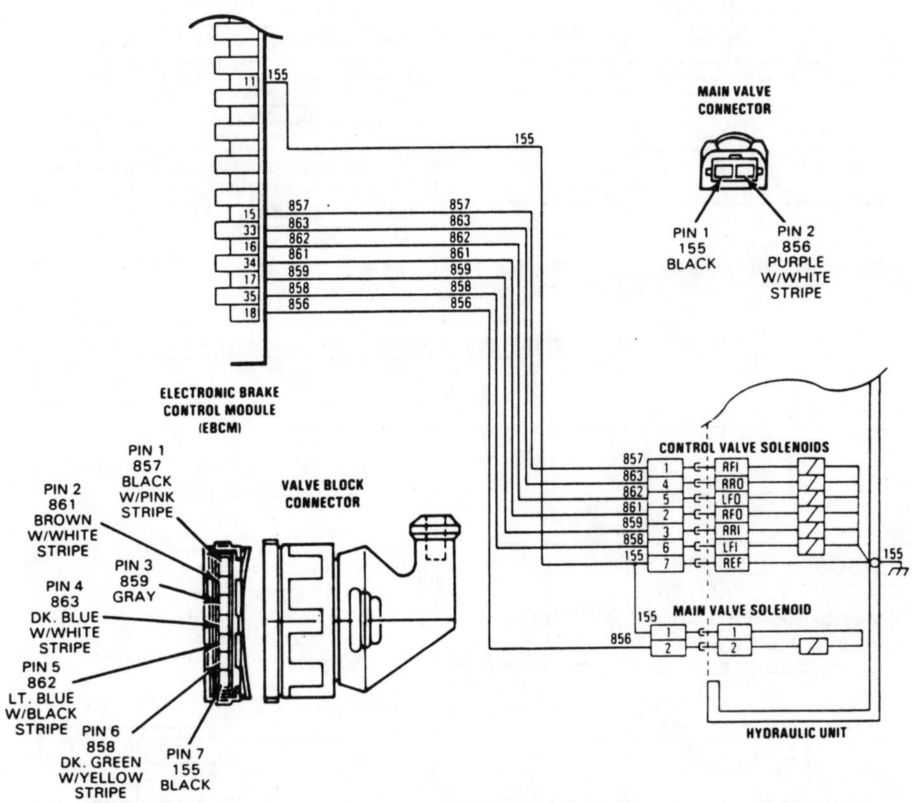

Fig. 106 Valve block and valve main circuit wiring schematic. 1989 Pontiac 6000

MAIN VALVE SOLENOID RESISTANCE

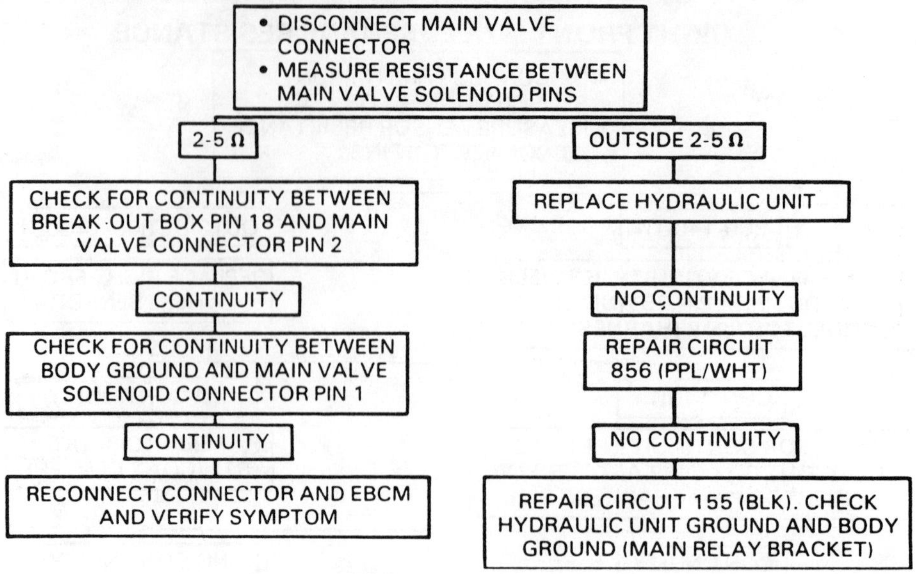

Fig. 107 Main valve solenoid resistance. 1989 Pontiac 6000

VALVE BLOCK GROUND REFERENCE CIRCUIT

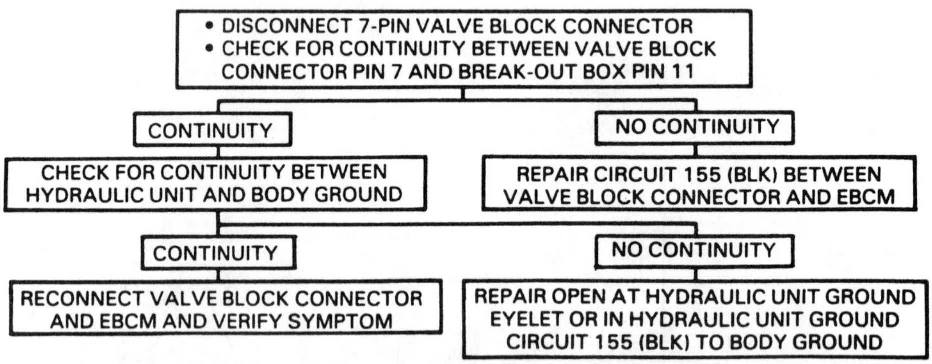

Fig. 108 Valve block ground reference circuit. 1989 Pontiac 6000

RIGHT FRONT INLET VALVE RESISTANCE

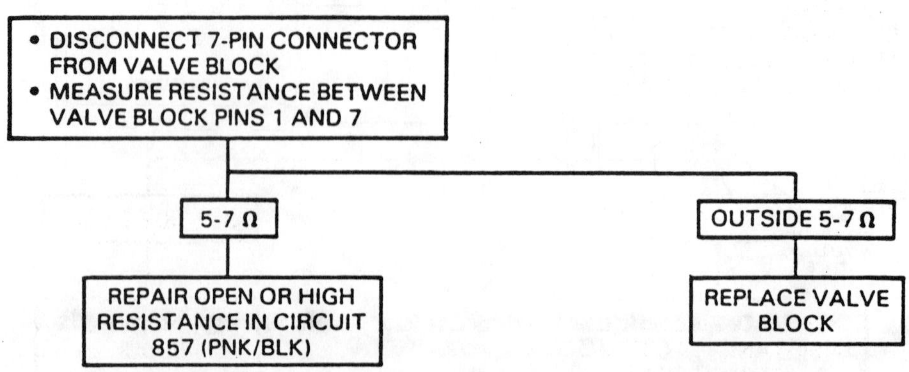

Fig. 109 Right front inlet valve resistance. 1989 Pontiac 6000

LEFT FRONT INLET VALVE RESISTANCE

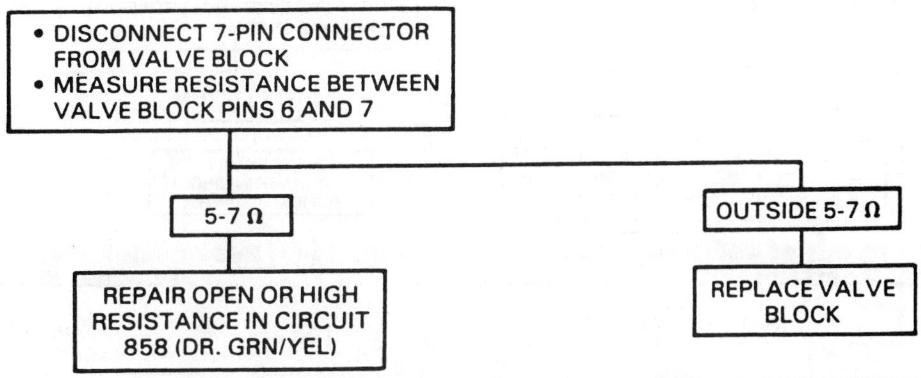

Fig. 110 Left front inlet valve resistance. 1989 Pontiac 6000

REAR INLET VALVE RESISTANCE

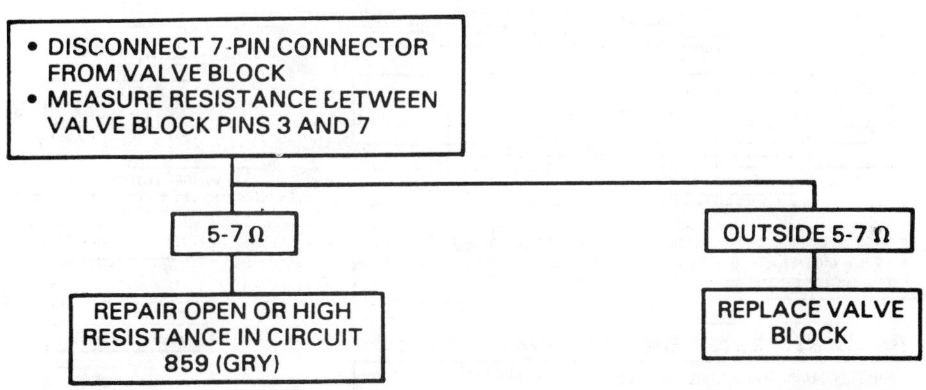

Fig. 111 Rear inlet valve resistance. 1989 Pontiac 6000

RIGHT FRONT OUTLET VALVE RESISTANCE

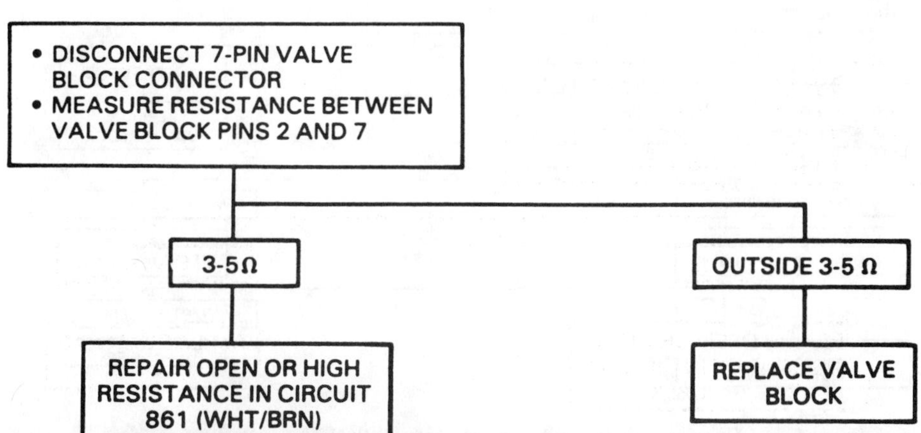

Fig. 112 Right front outlet valve resistance. 1989 Pontiac 6000

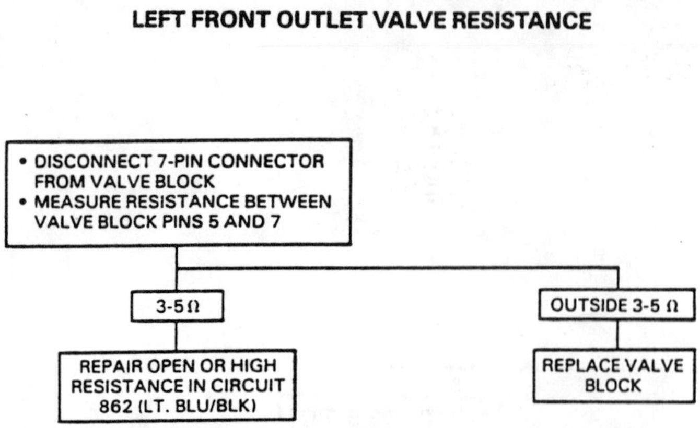

Fig. 113 Left front outlet valve resistance. 1989 Pontiac 6000

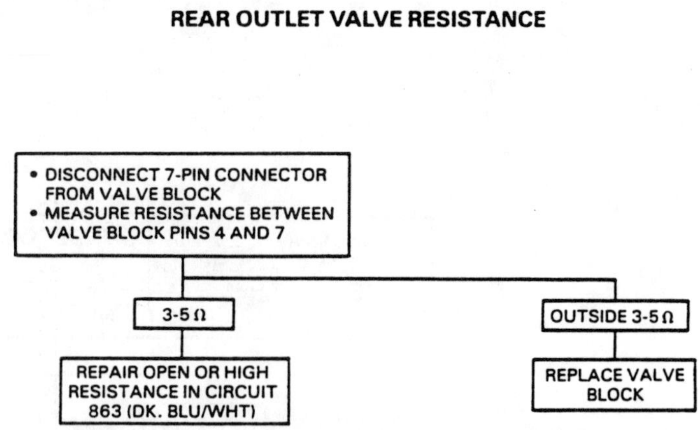

Fig. 114 Rear outlet valve resistance. 1989 Pontiac 6000

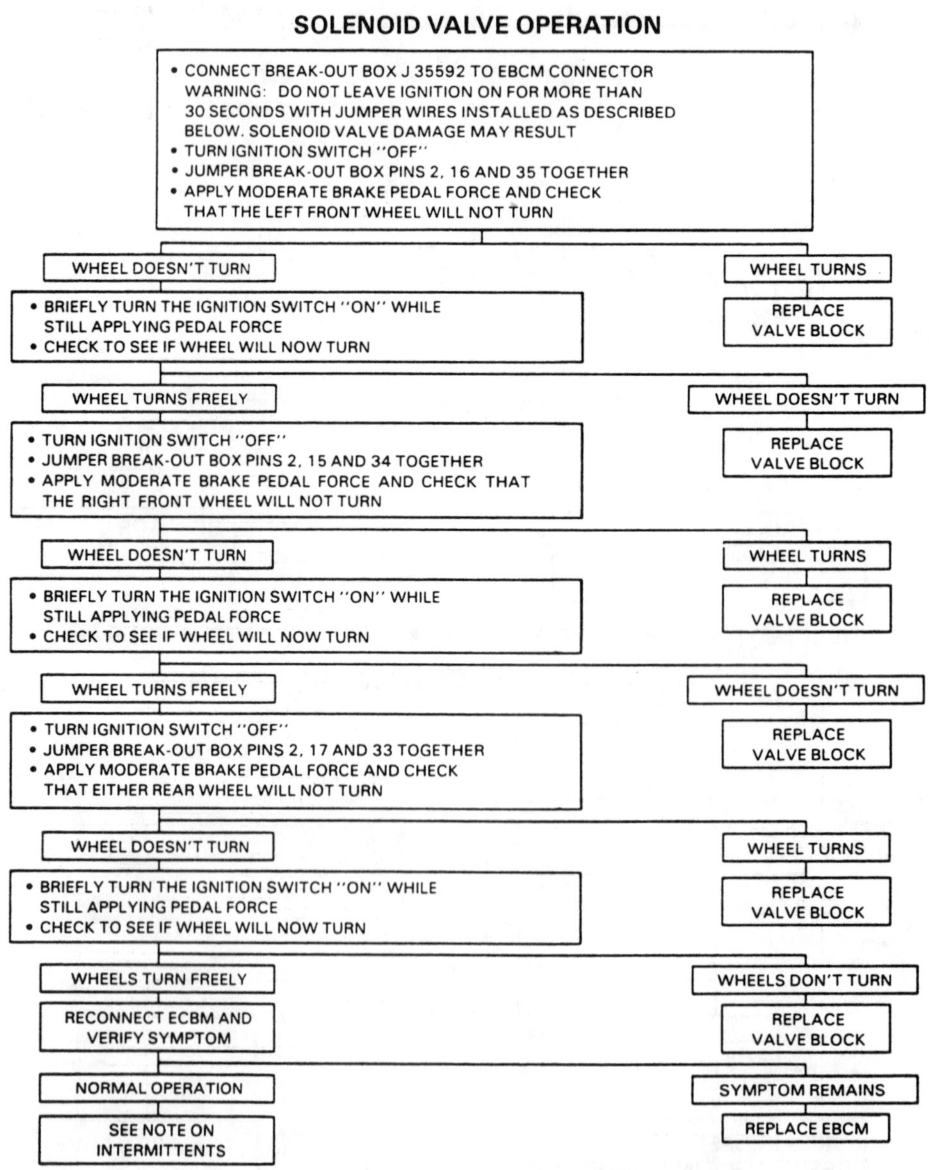

Fig. 115 Solenoid valve operation. 1989 Pontiac 6000

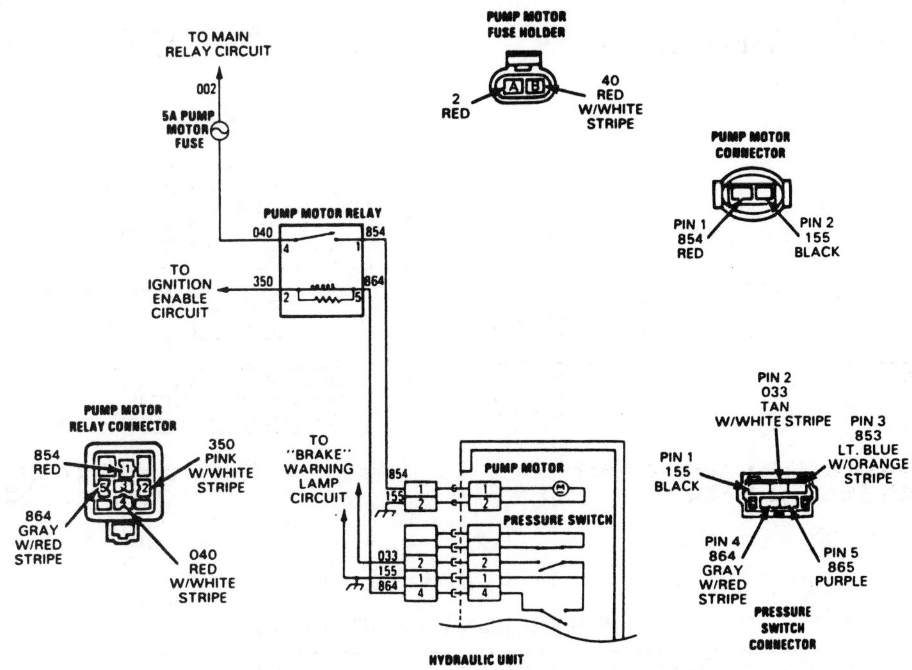

Fig. 116 Pump motor circuit schematic. 1989 Pontiac 6000

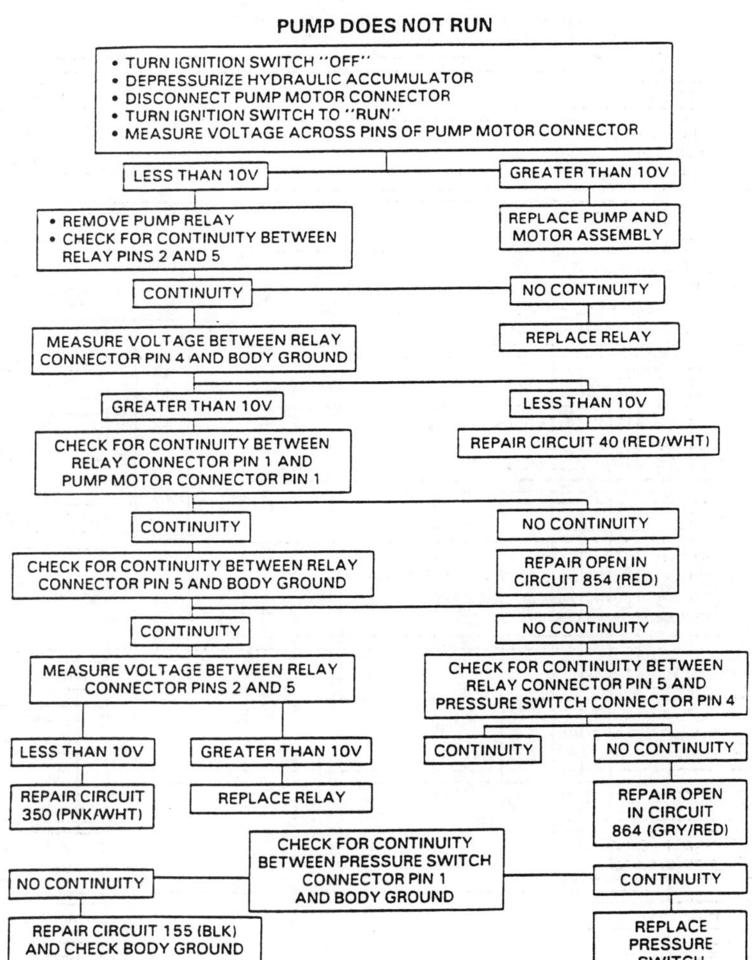

Fig. 117 Pump inoperative. 1989 Pontiac 6000

PRESSURE SWITCH PERFORMANCE

I. SWITCH STATUS — PRESSURIZED

- TURN IGNITION SWITCH TO "RUN" UNTIL PUMP STOPS
- TURN IGNITION SWITCH "OFF"
  NOTE: IF PUMP CONTINUES TO RUN AFTER 45 SECONDS, TURN IGNITION SWITCH OFF AND PROCEED
- CHECK PRESSURE SWITCH PINS FOR THE FOLLOWING CONDITIONS USING HIGH IMPEDANCE DIGITAL MULTIMETER J 34029-A OR EQUIVALENT

MEASURE BETWEEN PINS	SCALE	SPECIFICATION
1, 2	200 Ω	NO CONTINUITY
1, 4	200 Ω	NO CONTINUITY
3, 5	200 Ω	CONTINUITY
ALL PINS AND BODY GROUND	200 Ω	NO CONTINUITY

IF ANY CONDITION IS NOT MET, REPLACE PRESSURE SWITCH

II. SWITCH STATUS — DEPRESSURIZED

- DEPRESSURIZE HYDRAULIC ACCUMULATOR AS DESCRIBED IN THIS SECTION
- CHECK PRESSURE SWITCH PINS FOR THE FOLLOWING CONDITIONS USING HIGH IMPEDANCE DIGITAL MULTIMETER J 34029-A OR EQUIVALENT

MEASURE BETWEEN PINS	SCALE	SPECIFICATION
1, 2	200 Ω	CONTINUITY
1, 4	200 Ω	CONTINUITY
3, 5	200 Ω	NO CONTINUITY
ALL PINS AND BODY GROUND	200 Ω	NO CONTINUITY

IF ANY CONDITION IS NOT MET, REPLACE PRESSURE SWITCH

CONTINUED

Fig. 118 Pressure switch performance, (Part 1 of 2). 1989 Pontiac 6000

PRESSURE SWITCH PERFORMANCE (CONT.)

III. SWITCH THRESHOLDS

- DEPRESSURIZE HYDRAULIC ACCUMULATOR AS DESCRIBED IN THIS SECTION
- INSTALL PRESSURE GAGE J 35604 AS DESCRIBED IN THIS SECTION
- CONNECT PRESSURE SWITCH CONNECTOR AND TURN IGNITION SWITCH TO "RUN" UNTIL PUMP STOPS
- USING HIGH IMPEDANCE DIGITAL MULTIMETER J 34029-A OR EQUIVALENT, MONITOR FOR CONTINUITY BETWEEN PRESSURE SWITCH PINS AS SHOWN BELOW WHILE SLOWLY BLEEDING OFF ACCUMULATOR PRESSURE BY PUMPING THE BRAKE PEDAL. CONTINUITY SHOULD BE GAINED OR LOST AS INDICATED.
- PRESSURIZE SYSTEM BETWEEN EACH TEST BY RECONNECTING PRESSURE SWITCH AND TURNING IGNITION SWITCH TO "RUN" UNTIL PUMP STOPS

MEASURE BETWEEN PINS	SWITCH STATUS	PRESSURE RANGE
1, 4	CONTINUITY SHOULD BE GAINED AT:	1980-2080 PSI (13,650-14,350 KPA)
1, 2	CONTINUITY SHOULD BE GAINED AT:	1500-1550 PSI (10,350-10,700 KPA)
3, 5	CONTINUITY SHOULD BE LOST AT:	1500-1550 PSI (10,350-10,700 KPA)

IF ANY CONDITION IS NOT MET, REPLACE PRESSURE SWITCH

- DEPRESSURIZE HYDRAULIC ACCUMULATOR AS DESCRIBED IN THIS SECTION
- WITH IGNITION "OFF," CONNECT PRESSURE SWITCH CONNECTOR
- TURN IGNITION TO "RUN" AND OBSERVE GAGE, "ANTILOCK" LAMP, "BRAKE" LAMP AND PUMP MOTOR. EVENTS SHOULD OCCUR AT PRESSURES INDICATED IN CHART BELOW

EVENT	PRESSURE
"ANTILOCK" LAMP TURNS OFF	1900-1975 PSI (13,100-13,600 KPA)
"BRAKE" LAMP TURNS OFF	1900-1975 PSI (13,100-13,600 KPA)
PUMP MOTOR STOPS	2550-2670 PSI (17,580-18,400 KPA)

IF ANY CONDITION IS NOT MET, REPLACE PRESSURE SWITCH

Fig. 118 Pressure switch performance, (Part 2 of 2). 1989 Pontiac 6000

INTERMITTENT "ANTILOCK" WARNING LAMP WHILE DRIVING

INTERRUPTIONS IN THE FOLLOWING CIRCUITS MAY CAUSE THE "ANTILOCK" LAMP TO LIGHT INTERMITTENTLY. THIS LIST DOES NOT INCLUDE ALL POSSIBILITIES, BUT INCLUDES THOSE CIRCUITS MOST LIKELY TO CAUSE THE "ANTILOCK" WARNING LAMP TO LIGHT INTERMITTENTLY.

- EBCM SWITCH LOOP, INCLUDING
  - CIRCUIT 866 (LT. GRN) FROM EBCM TO FLUID LEVEL SENSOR
  - 2-PIN FLUID LEVEL SENSOR SWITCH
  - CIRCUIT 853 (LT. BLU/ORN) FROM FLUID LEVEL SENSOR TO PRESSURE SWITCH
  - CIRCUIT 865 (PPL) FROM PRESSURE SWITCH TO EBCM

- IGNITION ENABLE CIRCUIT, INCLUDING
  - BRAKE FUSE (#5 IN CONVENIENCE CENTER 5A)
  - CIRCUIT 350 (PNK/WHT) FROM RELAY COIL TO GROUND
  - DIFFERENTIAL LOCK ENGAGED SWITCH AND JUMPER (ALL WHEEL DRIVE)
  - 951 CIRCUIT TO DIFFERENTIAL LOCK CONTROL RELAY IN THE CONSOLE (ALL WHEEL DRIVE)

- MAIN RELAY CIRCUITS, INCLUDING
  - CIRCUIT 855 (DK. BLU) FROM EBCM TO RELAY COIL
  - CIRCUIT 450 (BLK/WHT) FROM RELAY COIL TO GROUND

- ALL WHEEL SPEED SENSOR CIRCUITS

INSPECT CONNECTORS AND WIRES IN THESE CIRCUITS. IF NO TROUBLE IS FOUND, SEE NOTE ON INTERMITTENTS IN THIS SECTION AND PERFORM.

Fig. 119 Intermittent anti-lock warning lamp. 1989 Pontiac 6000

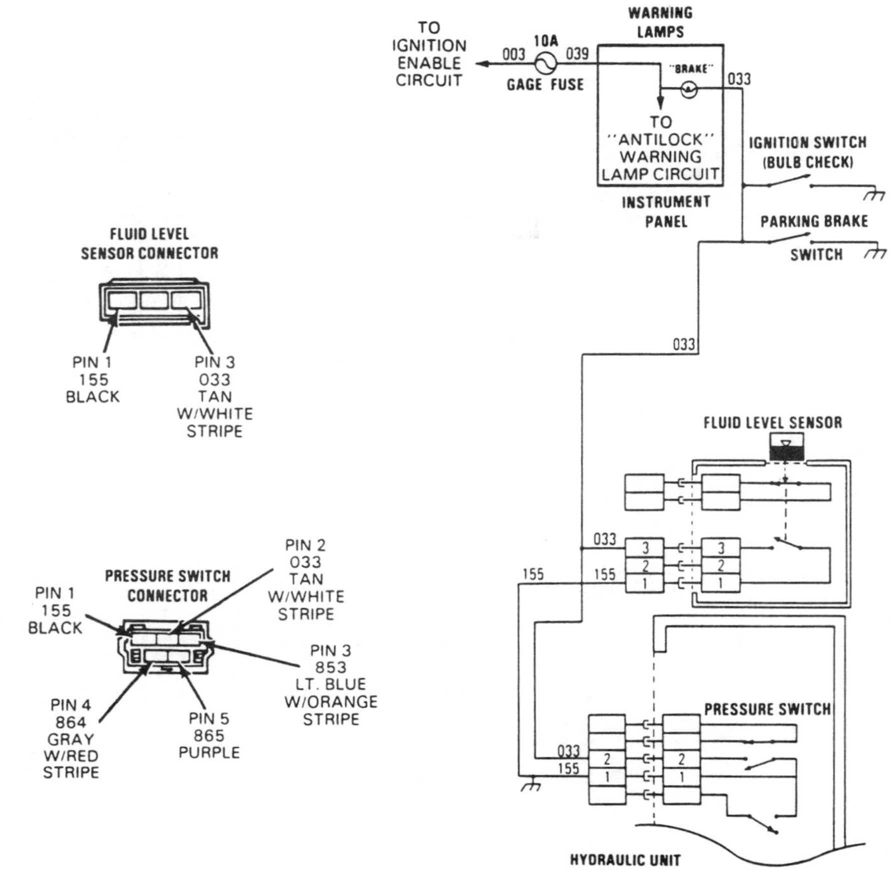

Fig. 120 Brake warning lamp circuit schematic. 1989 Pontiac 6000

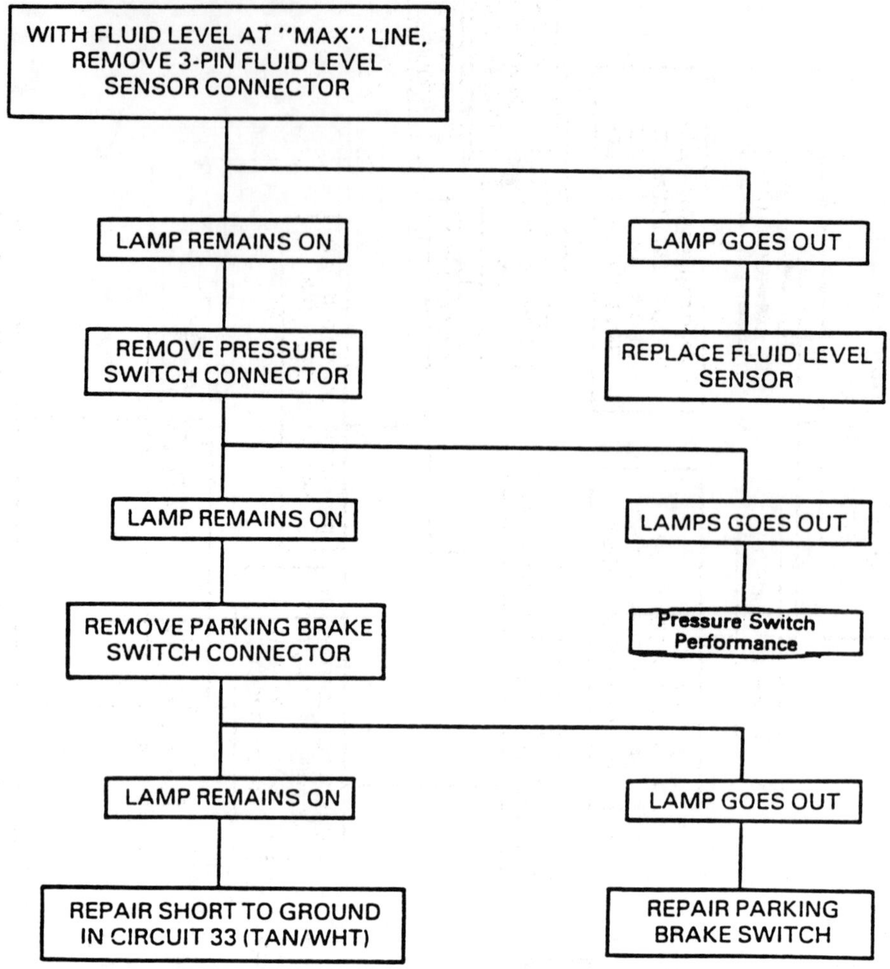

Fig. 121 Brake warning lamp circuit. 1989 Pontiac 6000

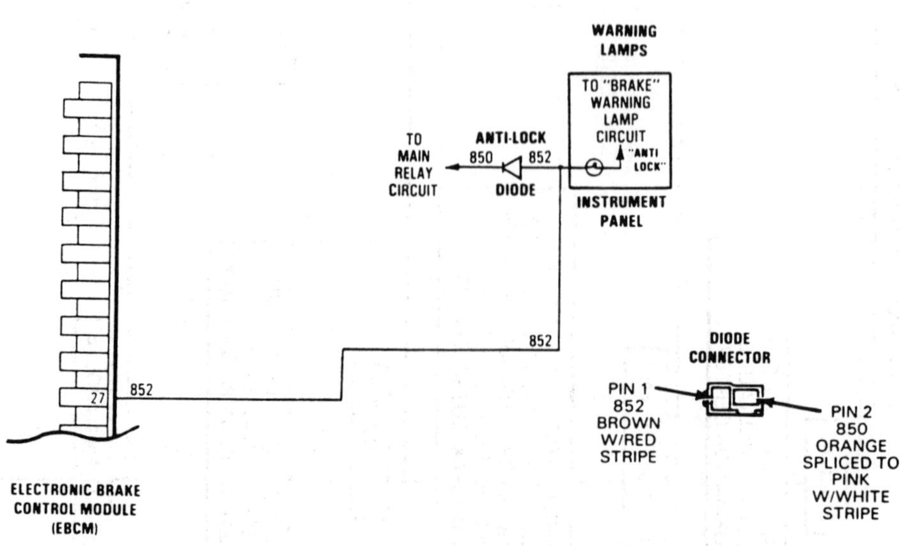

Fig. 122 Anti-lock warning lamp circuit schematic. 1989 Pontiac 6000

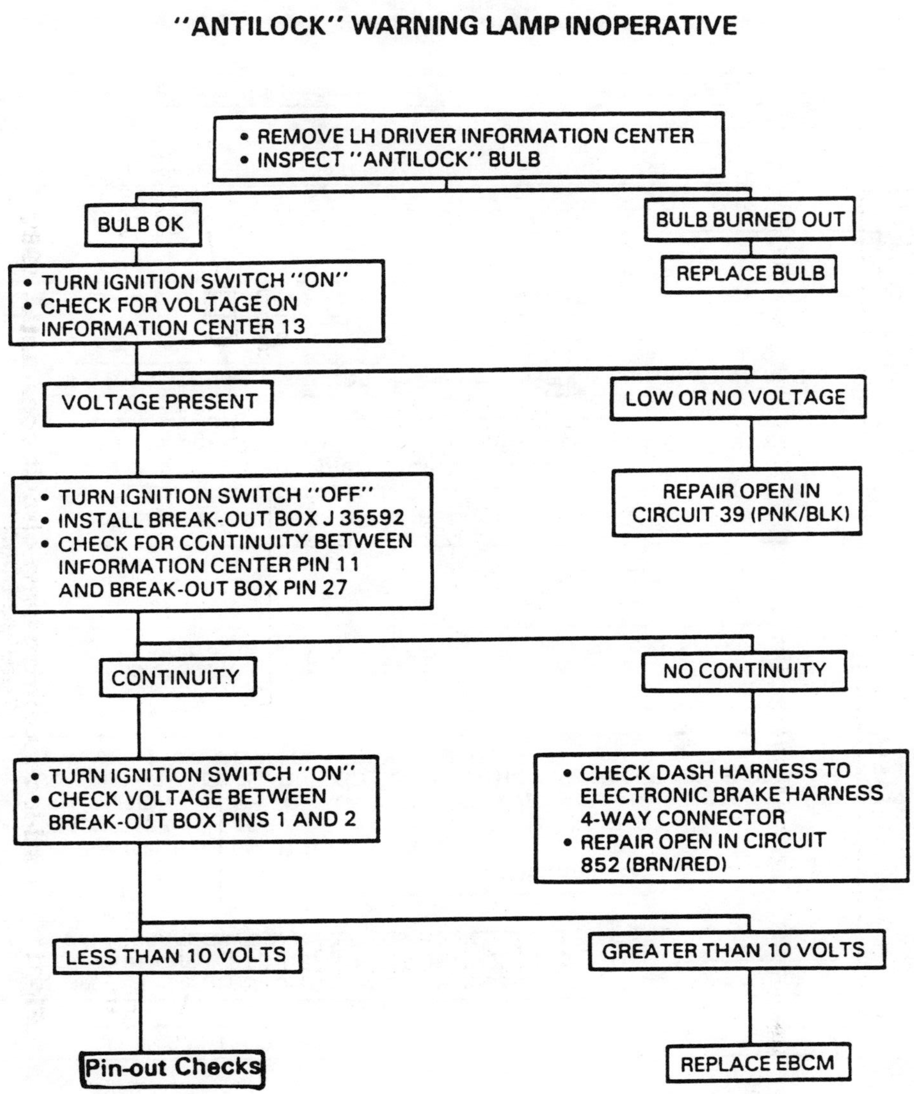

Fig. 123 Anti-lock warning lamp inoperative. 1989 Pontiac 6000

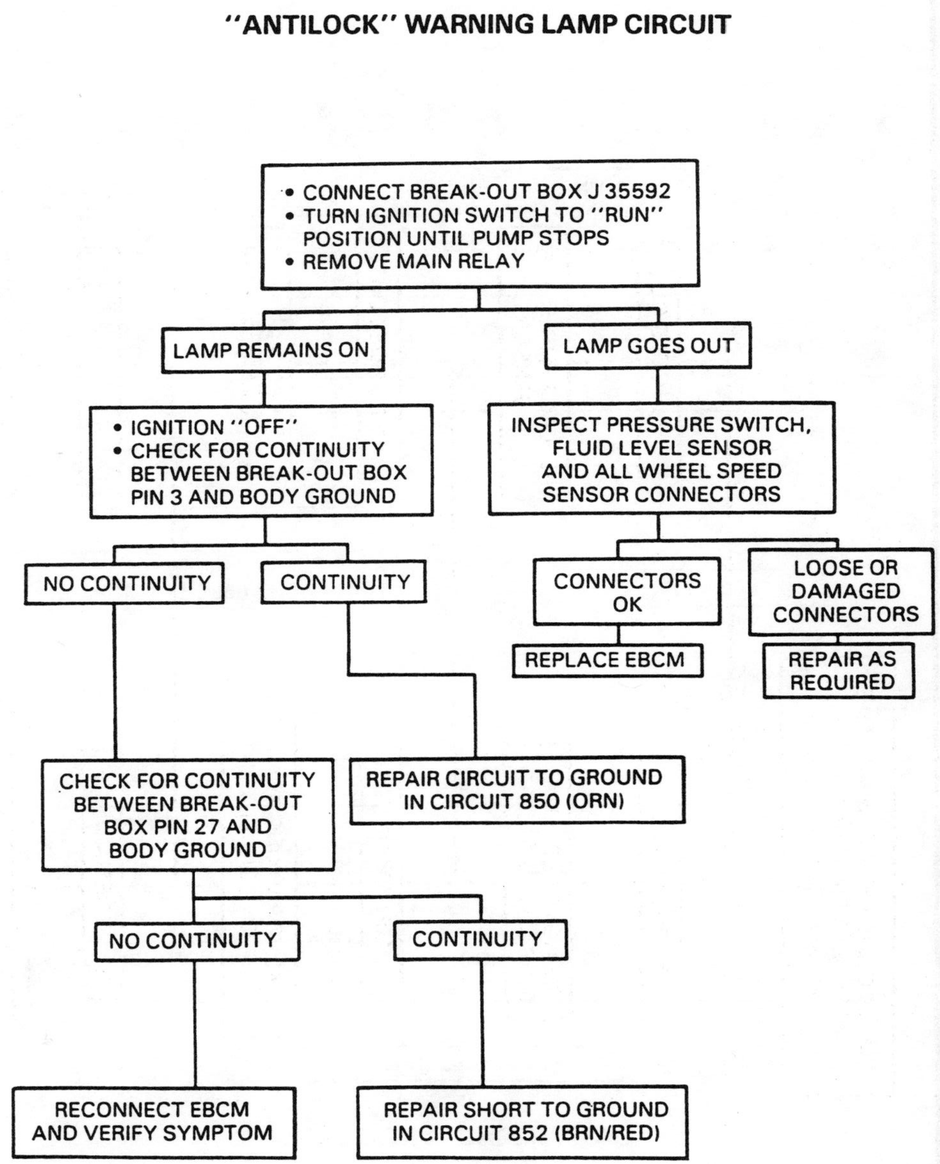

Fig. 124 Anti-lock warning lamp circuit. 1989 Pontiac 6000

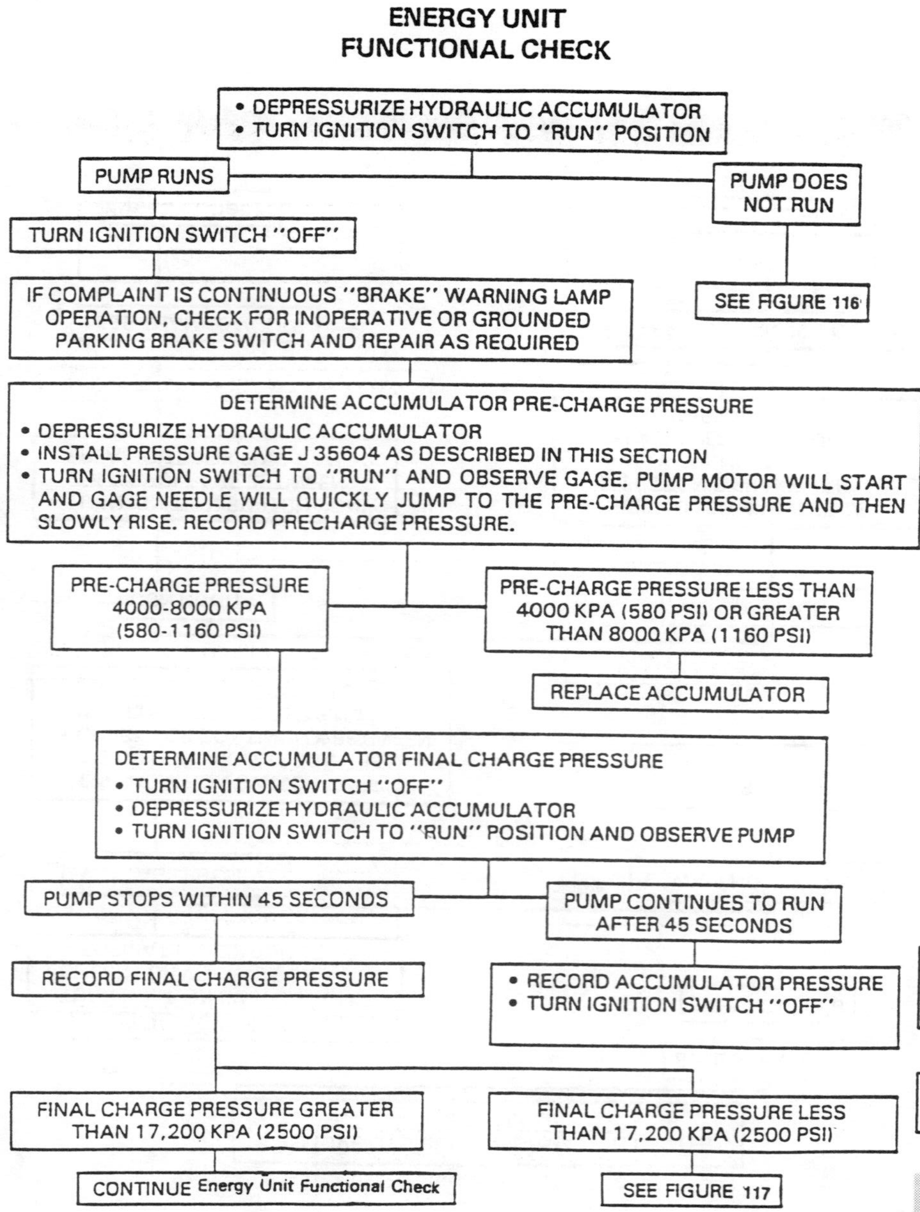

Fig. 125 Energy unit functional check, (Part 1 of 2). 1989 Pontiac 6000

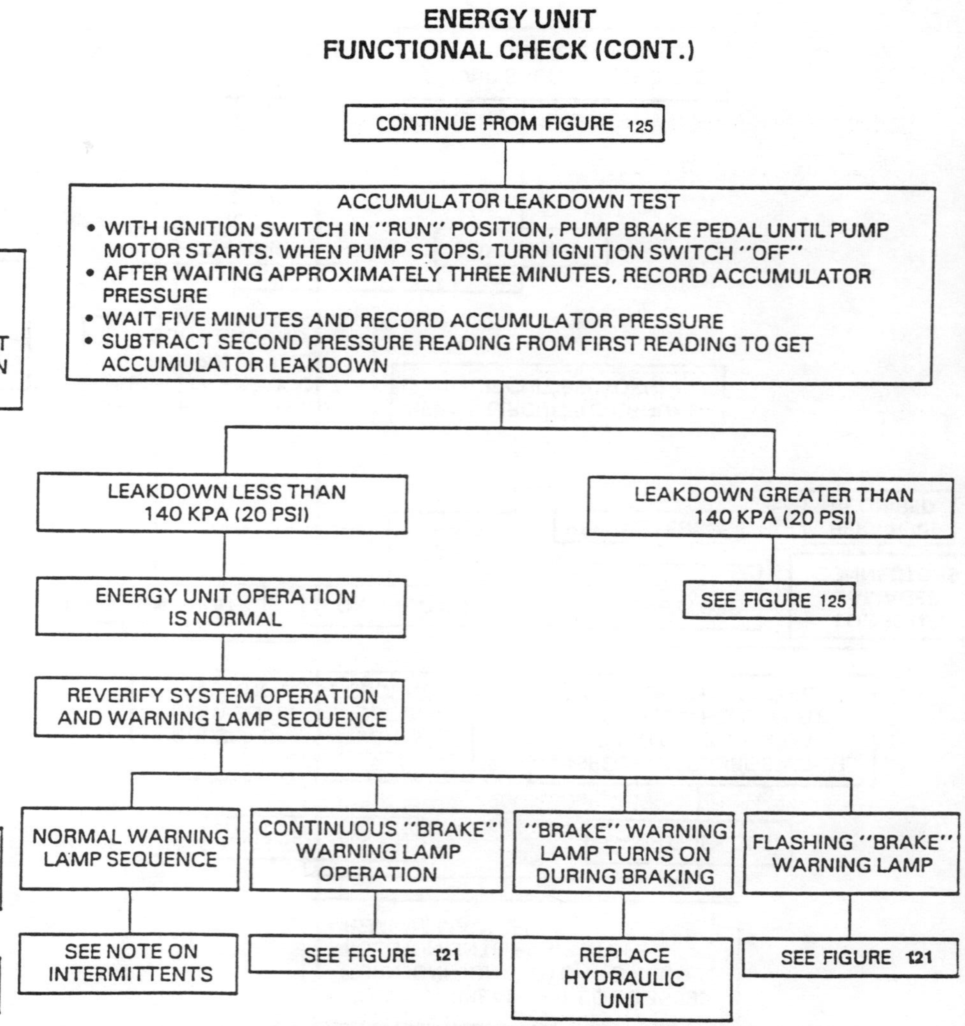

Fig. 125 Energy unit functional check, (Part 2 of 2). 1989 Pontiac 6000

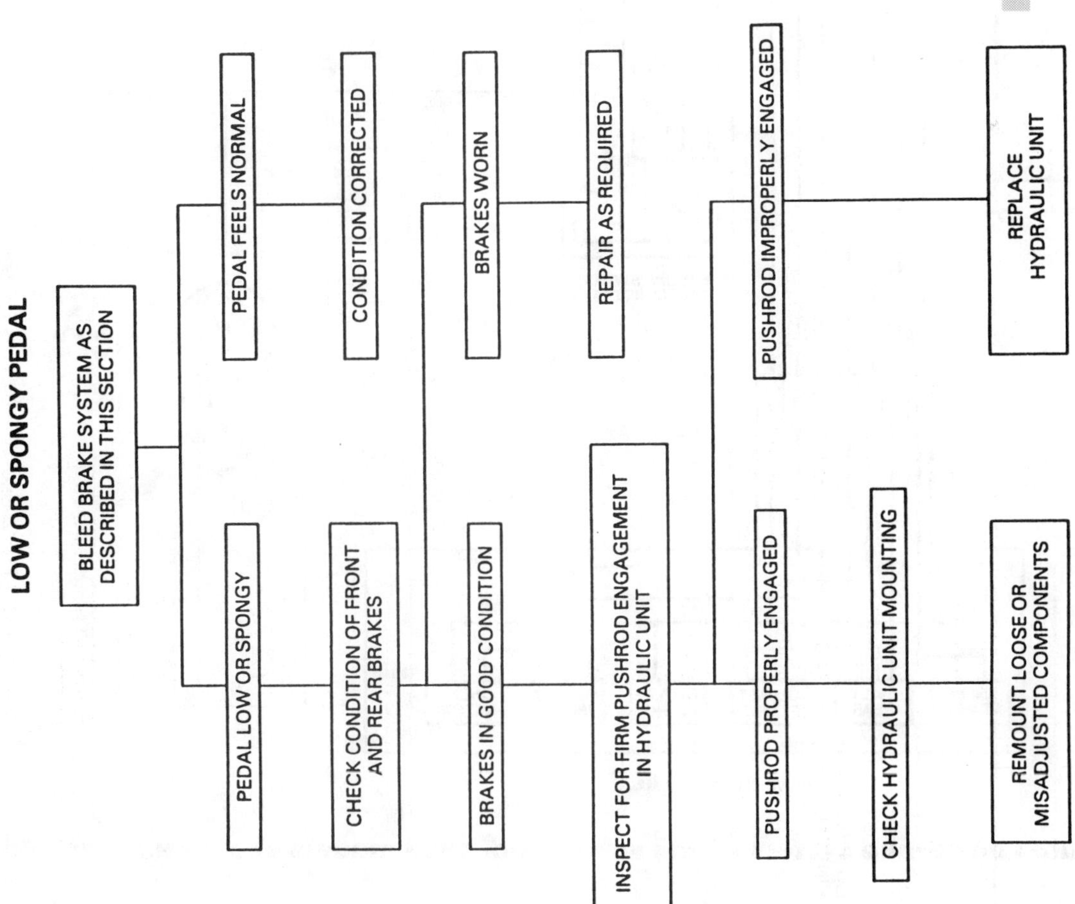

Fig. 126 Low or spongy pedal check. 1989 Pontiac 6000

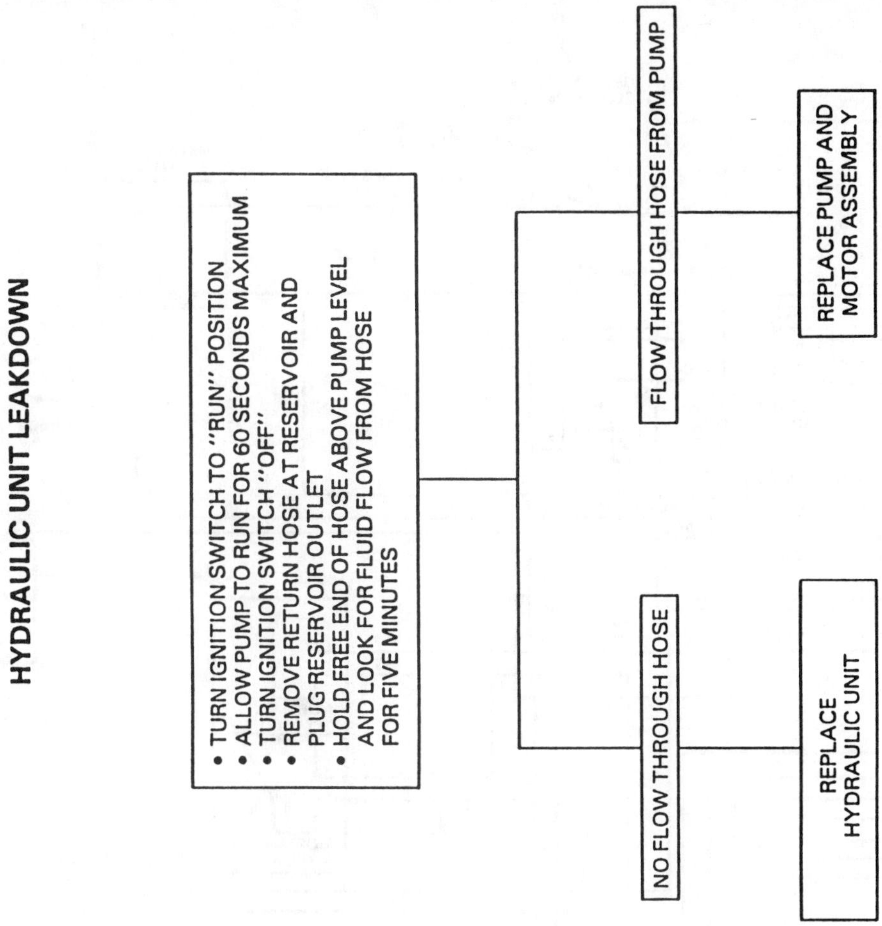

Fig. 127 Hydraulic unit leakdown check. 1989 Pontiac 6000

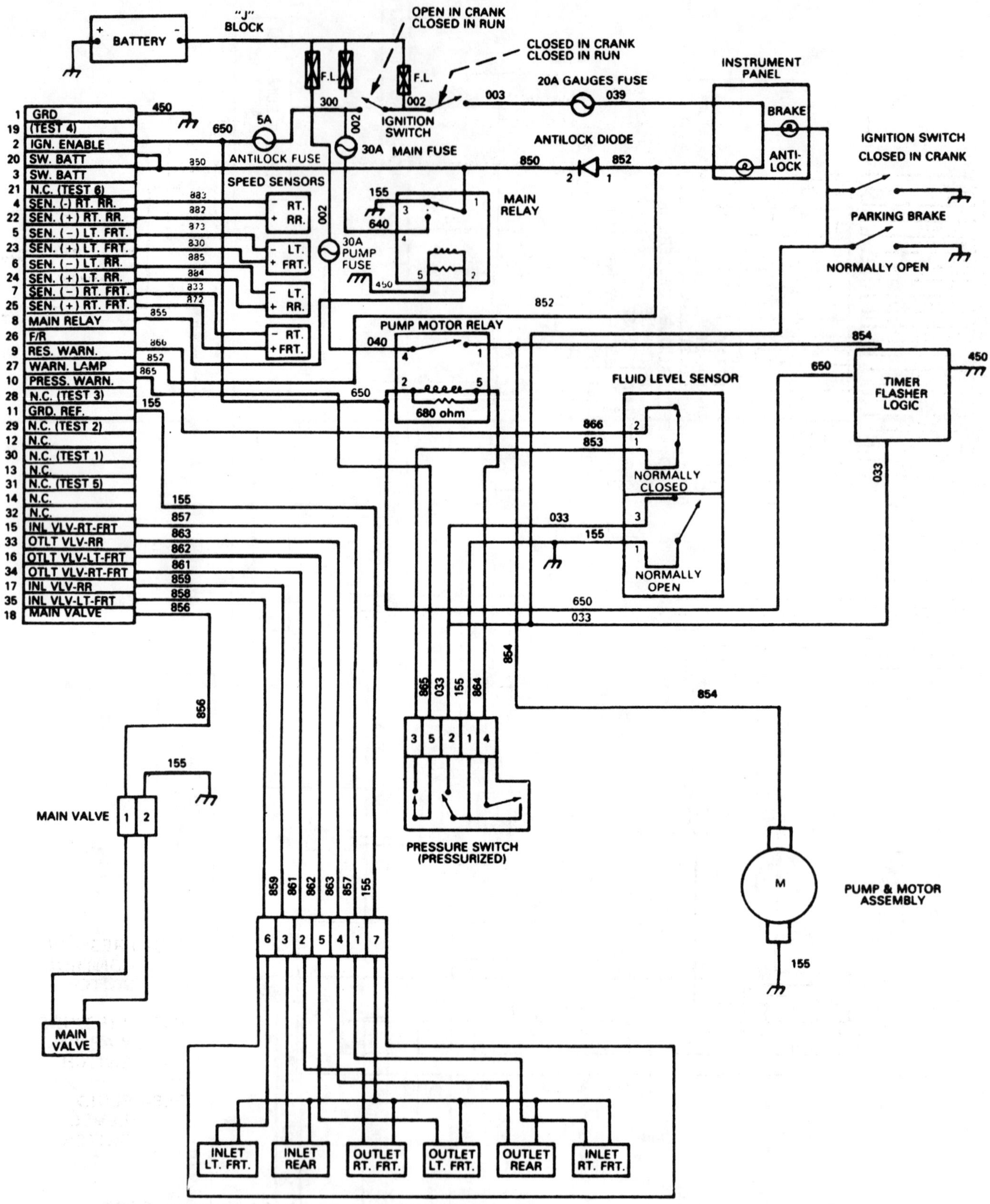

Fig. 128 Anti-Lock Brake system wiring schematic. 1986 models except Pontiac 6000

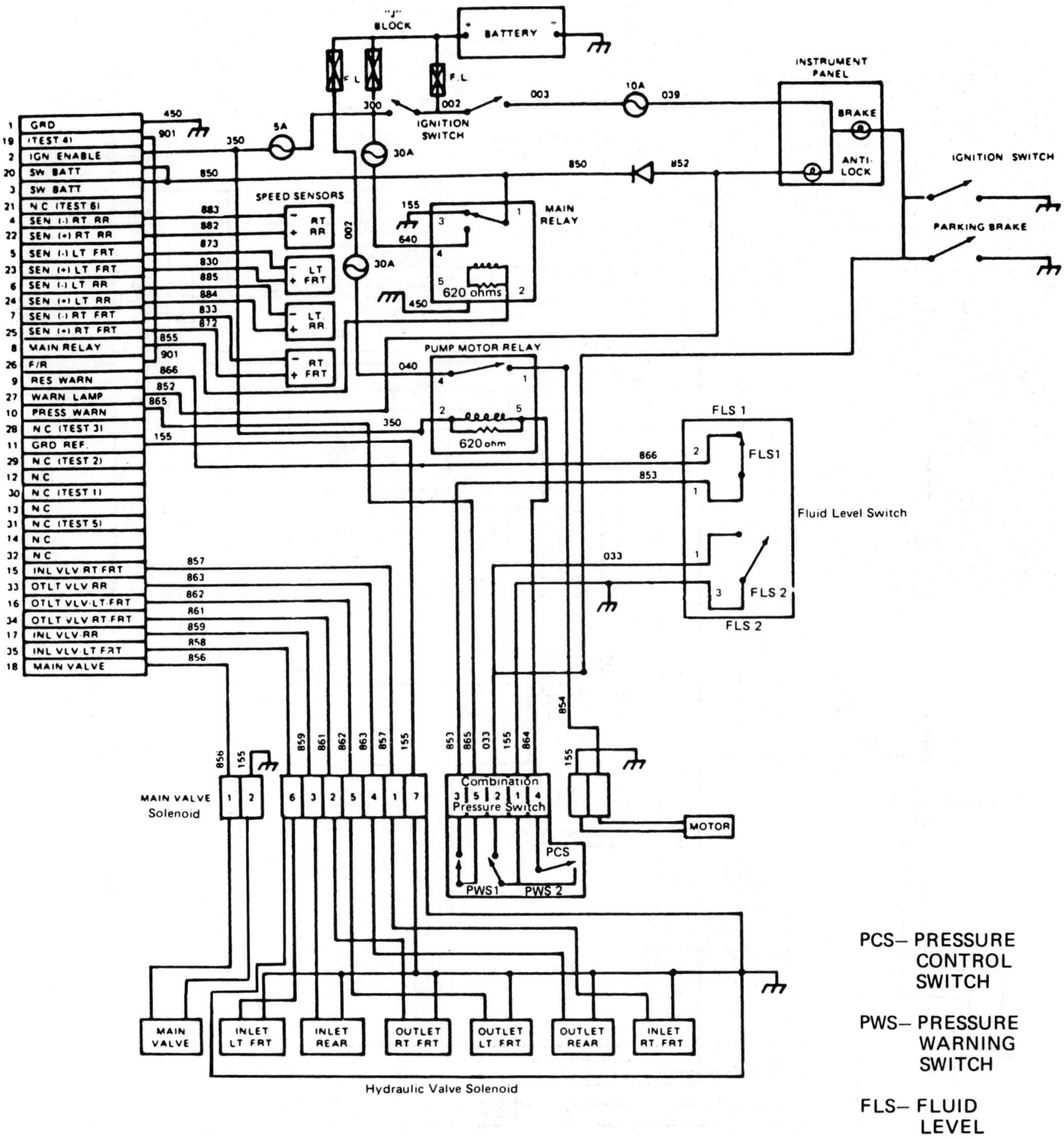

Fig. 129 Anti-Lock Brake system wiring schematic. 1986–88 Pontiac 6000

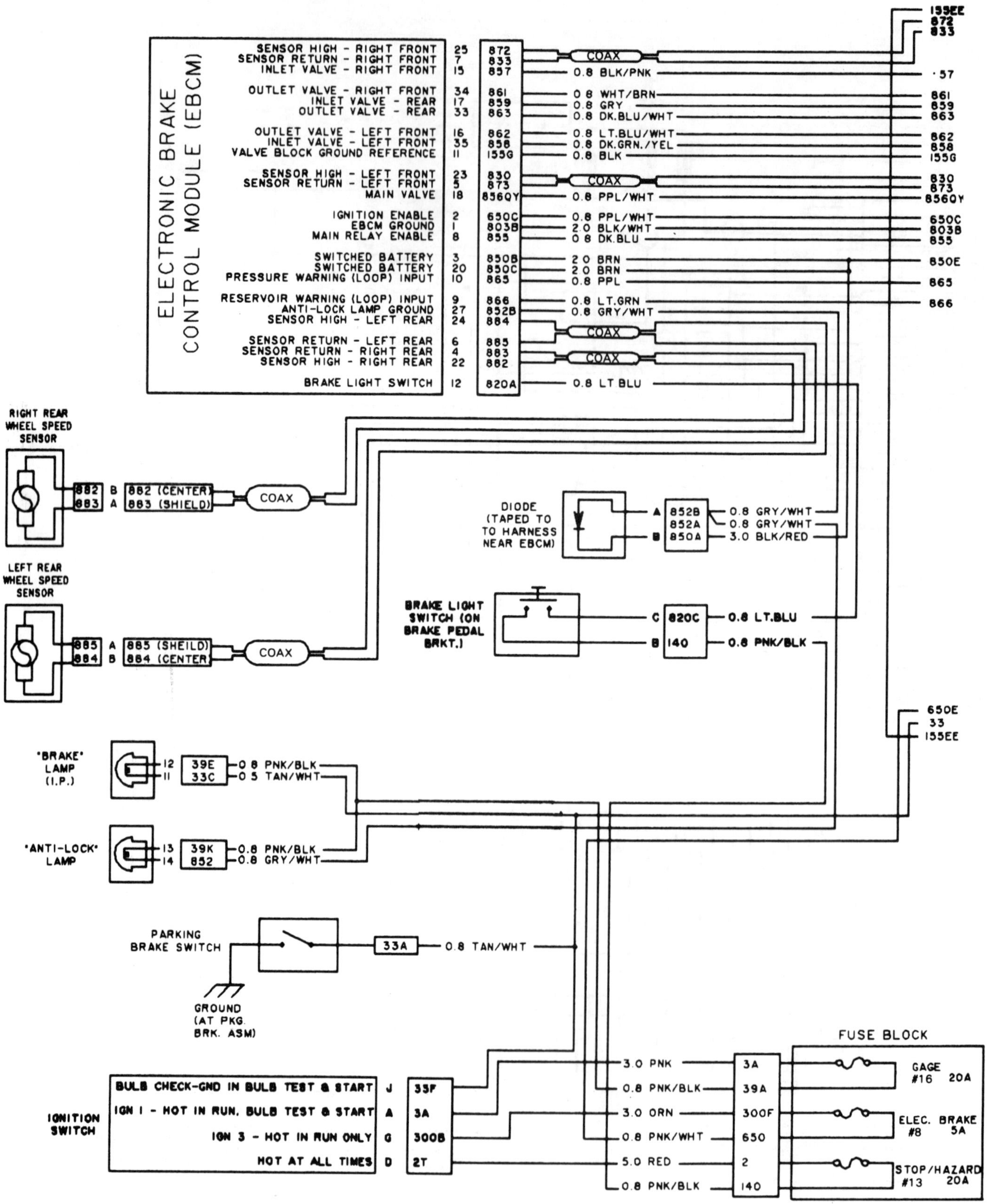

Fig. 130 Anti-Lock Brake system wiring schematic (Part 1 of 2). 1987–88 models except Pontiac 6000

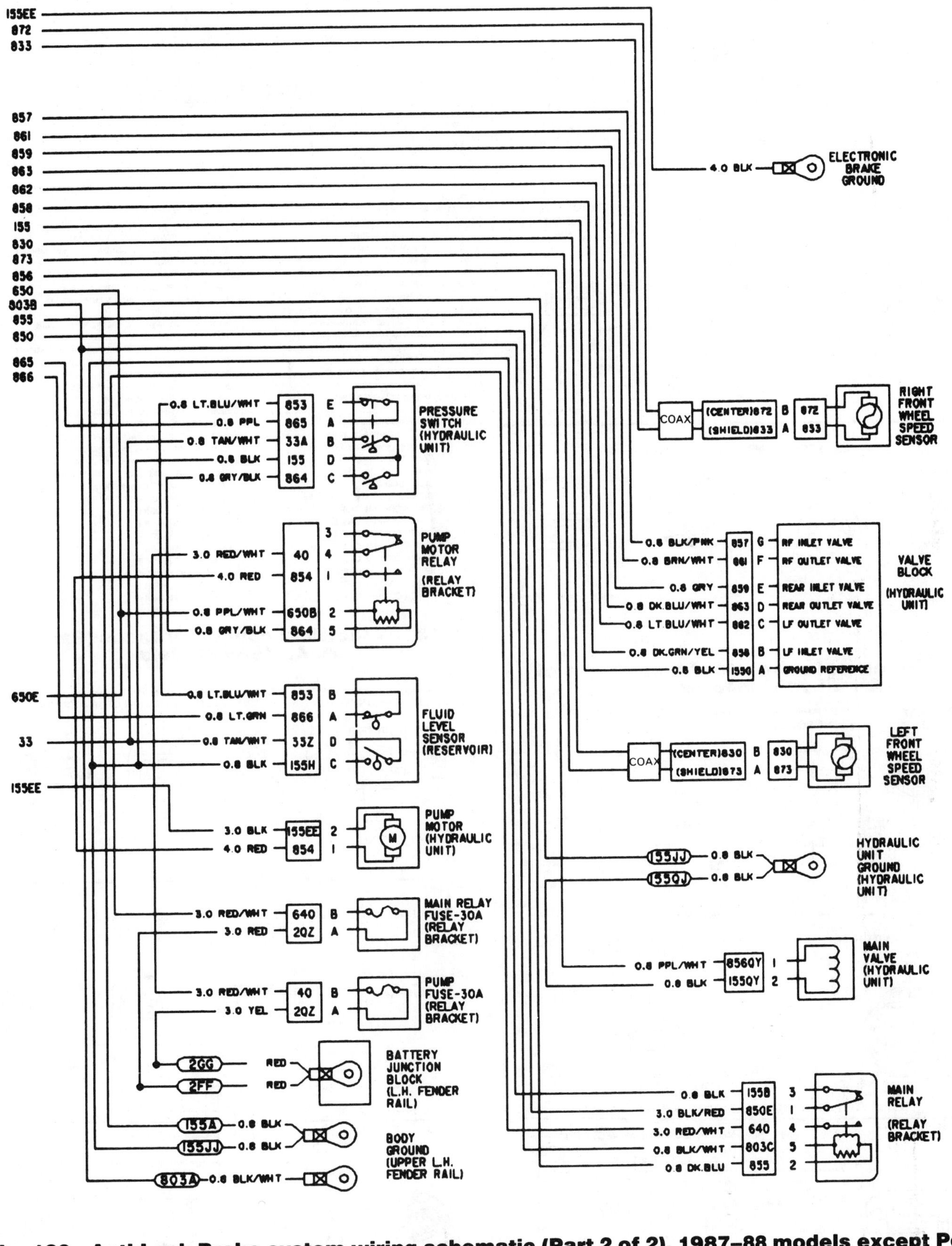

Fig. 130 Anti-Lock Brake system wiring schematic (Part 2 of 2). 1987–88 models except Pontiac 6000

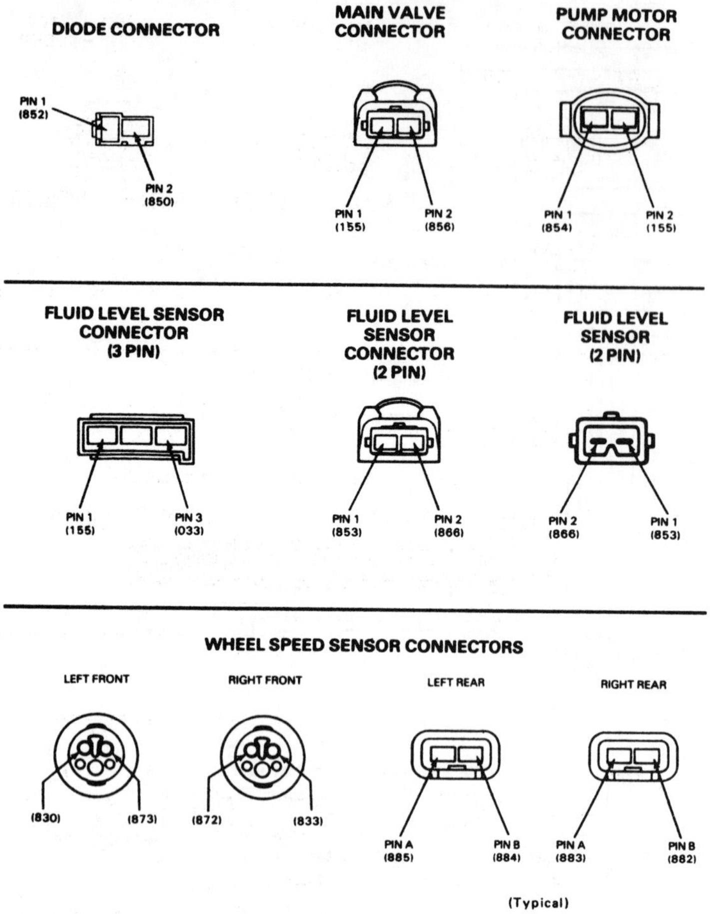

Fig. 131 Electrical connector faces (Part 1 of 2). All 1986 models

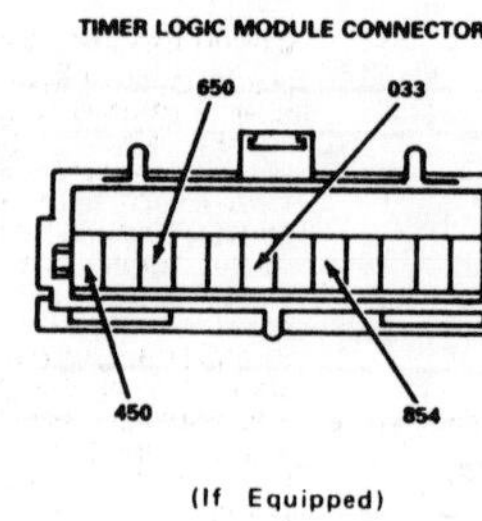

CIRCUIT NUMBER	WIRE COLOR	CIRCUIT DESCRIPTION
002	RED OR YELLOW	(-) SUPPLY, PUMP (YELLOW) MAIN (RED)
033	TAN-WHITE STRIPE	"BRAKE" INDICATOR GROUND CIRCUIT
040	RED-WHITE STRIPE	PUMP MOTOR RELAY SECONDARY SUPPLY CIRCUIT
155	BLACK	SYSTEM GROUND CIRCUIT
450	BLACK-WHITE STRIPE	SYSTEM GROUND CIRCUIT
640	RED-WHITE STRIPE	MAIN RELAY SECONDARY SUPPLY CIRCUIT
650	BROWN-WHITE STRIPE	IGNITION ENABLE TO EBCM & PUMP RELAY
850	BROWN SPLICED TO BLACK-RED STRIPE	EBCM SUPPLY CIRCUIT
852	GRAY-DOUBLE WHITE STRIPE	ANTILOCK INDICATOR TO EBCM GROUND WIRE
853	BLUE-LIGHT ORANGE STRIPE	FLUID LEVEL SENSOR TO PRESSURE SWITCH (CHECK LOOP)
854	RED SPLICED TO PINK-BLACK STRIPE	SUPPLY CIRCUIT TO PUMP MOTOR FROM PUMP RELAY
855	DARK BLUE	MAIN RELAY PRIMARY SUPPLY CIRCUIT
856	PURPLE-WHITE STRIPE	MAIN VALVE SUPPLY CIRCUIT
857	BLACK-PINK STRIPE	RIGHT FRONT INLET VALVE SUPPLY CIRCUIT
858	GREEN-YELLOW STRIPE	LEFT FRONT INLET VALVE SUPPLY CIRCUIT
859	GRAY	REAR INLET VALVE SUPPLY CIRCUIT
861	BROWN-WHITE STRIPE	RIGHT FRONT OUTLET VALVE SUPPLY CIRCUIT
862	LIGHT BLUE-BLACK STRIPE	LEFT FRONT OUTLET VALVE SUPPLY CIRCUIT
863	DARK BLUE-WHITE STRIPE	REAR OUTLET VALVE SUPPLY CIRCUIT
864	GRAY-RED STRIPE	PUMP MOTOR RELAY "PULL-IN" GROUND CIRCUIT
865	PURPLE	EBCM TO PRESSURE SWITCH (CHECK LOOP)
866	LIGHT GREEN	EBCM TO FLUID LEVEL SENSOR (CHECK LOOP)
901	RED-DOUBLE YELLOW STRIPE	EBCM TO EBCM CONNECT CIRCUIT
830	GREEN COAXIAL CABLE	LEFT FRONT SENSOR (+)
873	GREEN COAXIAL CABLE	LEFT FRONT SENSOR (-)
884	GREEN COAXIAL CABLE	LEFT REAR SENSOR (+)
885	GREEN COAXIAL CABLE	LEFT REAR SENSOR (-)
833	GREEN COAXIAL CABLE	RIGHT FRONT SENSOR (+)
872	GREEN COAXIAL CABLE	RIGHT FRONT SENSOR(-)
882	GREEN COAXIAL CABLE	RIGHT REAR SENSOR (+)
883	GREEN COAXIAL CABLE	RIGHT REAR SENSOR (-)

Fig. 131 Electrical connector faces (Part 2 of 2). All 1986 models

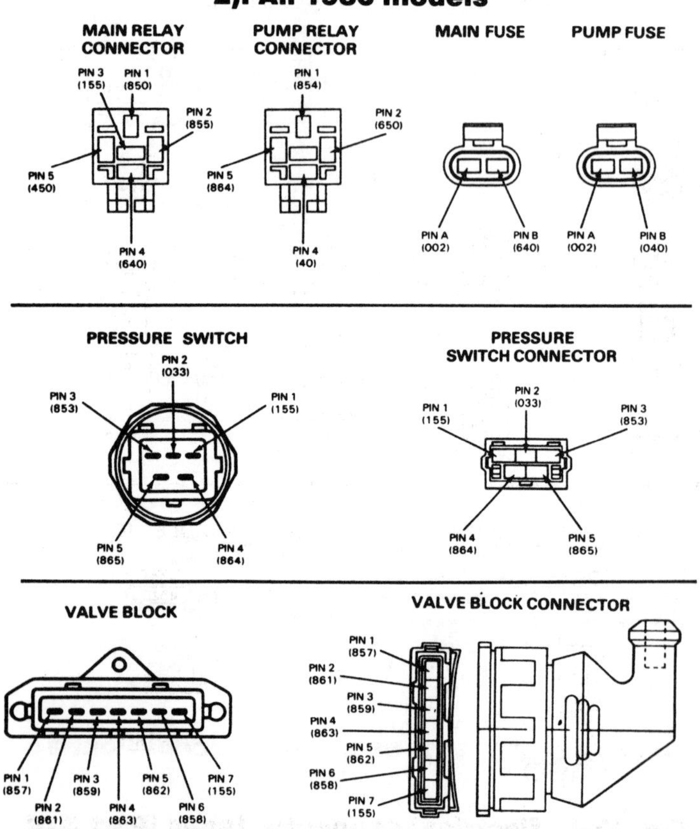

Fig. 132 Electrical connector faces. All 1986 models & 1987–88 Pontiac 6000

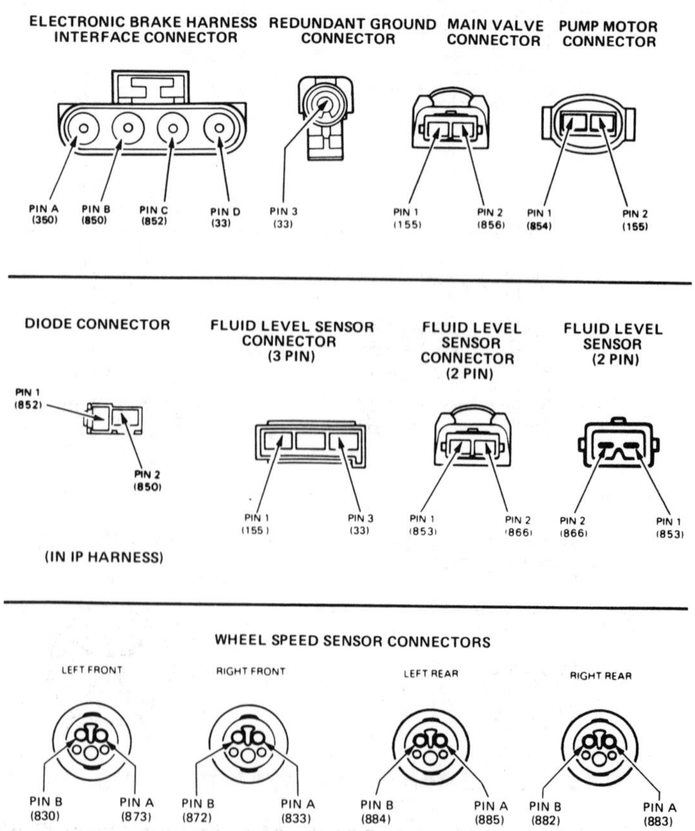

Fig. 133 Electrical connector faces (Part 1 of 2). 1987–88 Pontiac 6000

CIRCUIT NUMBER	WIRE COLOR	CIRCUIT DESCRIPTION
002	RED	(-) SUPPLY, PUMP MAIN
003	PINK	SUPPLY IGNITION CIRCUIT
033	TAN-WHITE STRIPE	"BRAKE" INDICATOR GROUND CIRCUIT
039	PINK-BLACK STRIPE	RED BRAKE WARNING LIGHT
040	RED-WHITE STRIPE	PUMP MOTOR RELAY SUPPLY CIRCUIT
155	BLACK	SYSTEM GROUND CIRCUIT
300	ORANGE	FUSIBLE LINK BETWEEN BATTERY AND 30 AMP MAIN FUSE
350	PINK-WHITE STRIPE	EBCM TO PUMP MOTOR RELAY (IGNITION ENABLE)
450	BLACK-WHITE STRIPE	SYSTEM GROUND CIRCUIT
640	RED-WHITE STRIPE	MAIN RELAY SUPPLY CIRCUIT
850	ORANGE SPLICED TO PINK-WHITE	EBCM SUPPLY CIRCUIT
852	BROWN-RED STRIPE	ANTILOCK INDICATOR TO EBCM
853	LT. BLUE– ORANGE STRIPE	FLUID LEVEL SENSOR TO PRESSURE SWITCH (CHECK LOOP)
854	RED	SUPPLY CIRCUIT TO PUMP MOTOR FROM PUMP RELAY
855	DARK BLUE	MAIN RELAY PULL IN CIRCUIT
856	PURPLE-WHITE STRIPE	MAIN VALVE SUPPLY CIRCUIT
857	BLACK-PINK STRIPE	RIGHT FRONT INLET VALVE SUPPLY CIRCUIT
858	DRK GREEN-YELLOW STRIPE	LEFT FRONT INLET VALVE SUPPLY CIRCUIT
859	GRAY	REAR INLET VALVE SUPPLY CIRCUIT
861	BROWN-WHITE STRIPE	RIGHT FRONT OULET VALVE SUPPLY CIRCUIT
862	LIGHT BLUE-BLACK STRIPE	LEFT FRONT OUTLET VALVE SUPPLY CIRCUIT
863	DARK BLUE-WHITE STRIPE	REAR OUTLET VALVE SUPPLY CIRCUIT
864	GRAY-RED STRIPE	PUMP MOTOR RELAY "PULL-IN" GROUND CIRCUIT
865	PURPLE	EBCM TO PRESSURE SWITCH (CHECK LOOP)
866	LIGHT GREEN	EBCM TO FLUID LEVEL SENSOR (CHECK LOOP)
901	RED-DOUBLE YELLOW STRIPE	EBCM TO EBCM CONNECT CIRCUIT
830	GREEN COAXIAL CABLE	LEFT FRONT SENSOR (+)
873	GREEN COAXIAL CABLE	LEFT FRONT SENSOR (-)
884	GREEN COAXIAL CABLE	LEFT REAR SENSOR (+)
885	GREEN COAXIAL CABLE	LEFT REAR SENSOR (-)
833	GREEN COAXIAL CABLE	RIGHT FRONT SENSOR (+)
872	GREEN COAXIAL CABLE	RIGHT FRONT SENSOR (-)
882	GREEN COAXIAL CABLE	RIGHT REAR SENSOR (+)
883	GREEN COAXIAL CABLE	RIGHT REAR SENSOR (-)

Fig. 133 Electrical connector faces (Part 2 of 2). 1987–88 Pontiac 6000

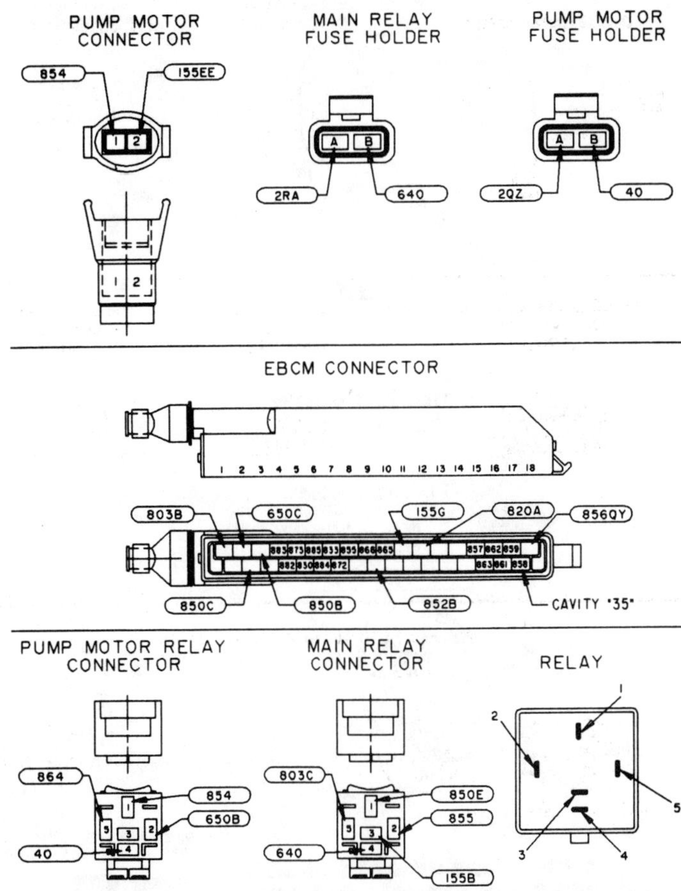

Fig. 134 Electrical connector faces (Part 1 of 3). All 1987–88 models except Pontiac 6000

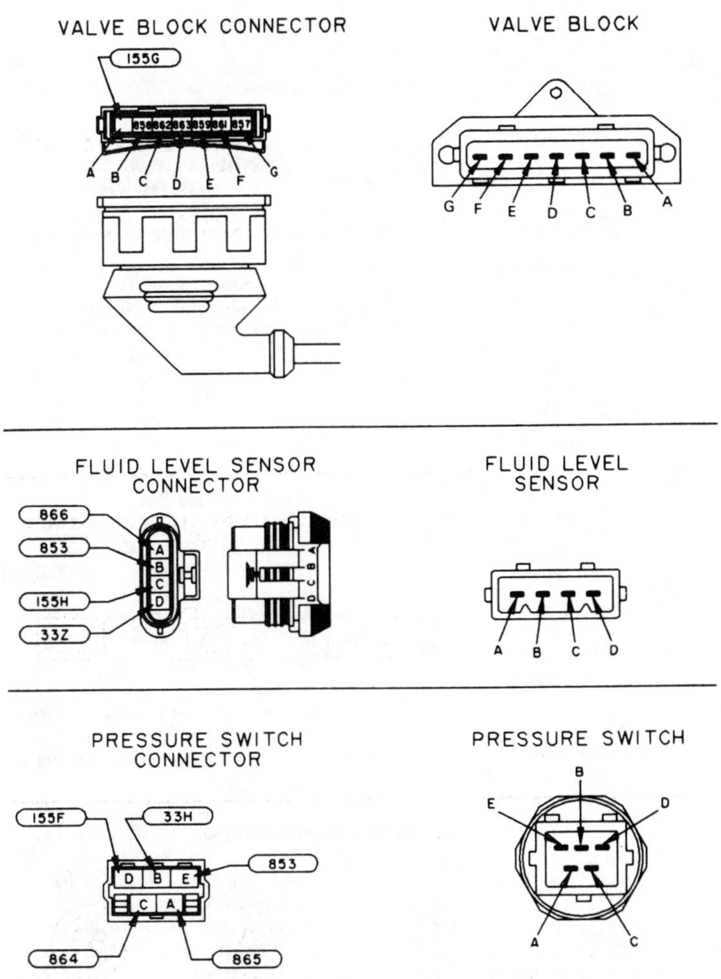

Fig. 134 Electrical connector faces (Part 2 of 3). All 1987–88 models except Pontiac 6000

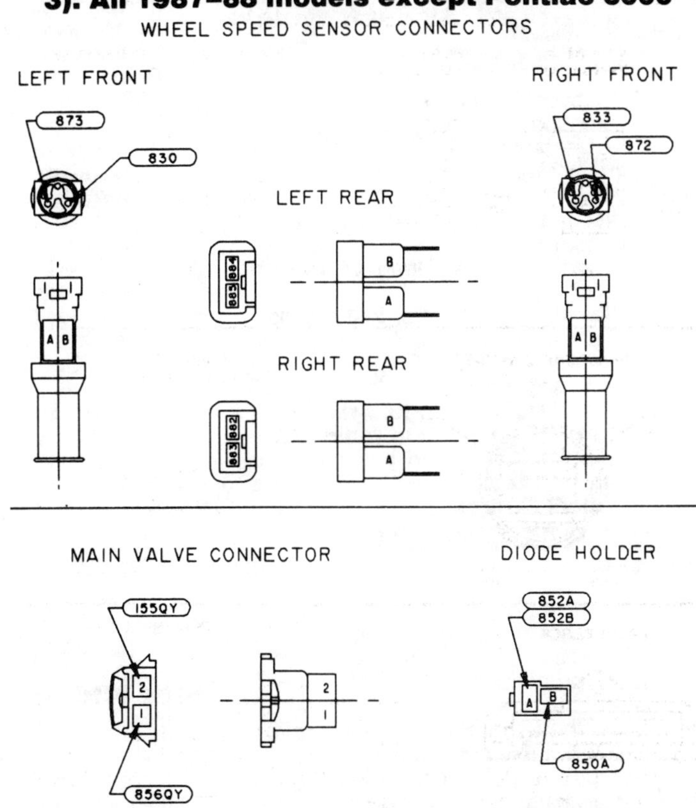

Fig. 134 Electrical connector faces (Part 3 of 3). All 1987–88 models except Pontiac 6000

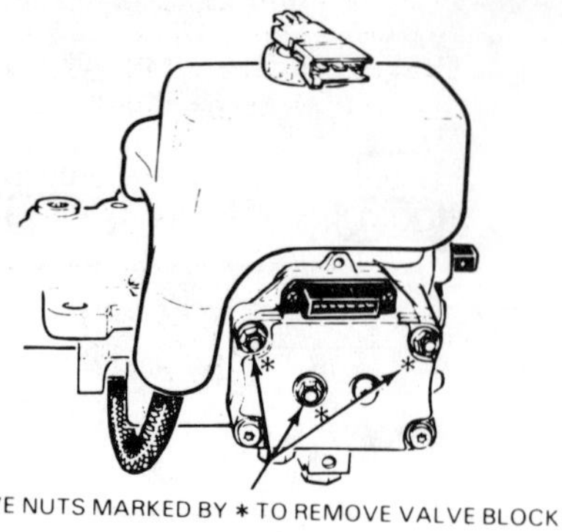

Fig. 135 Removing valve block assembly

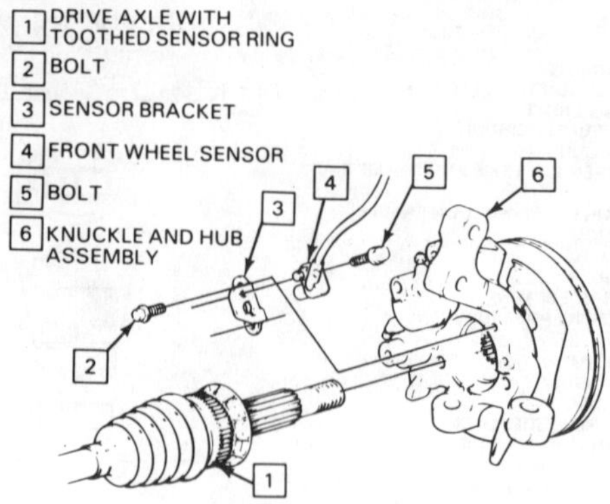

Fig. 136 Front wheel sensor

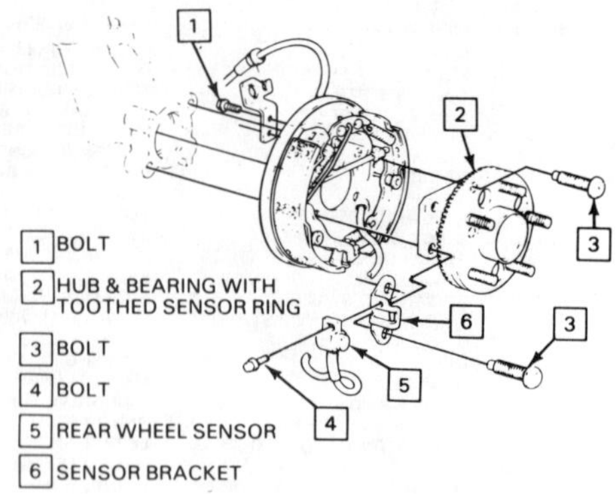

Fig. 137 Rear wheel sensor. All 1986 models except Pontiac 6000

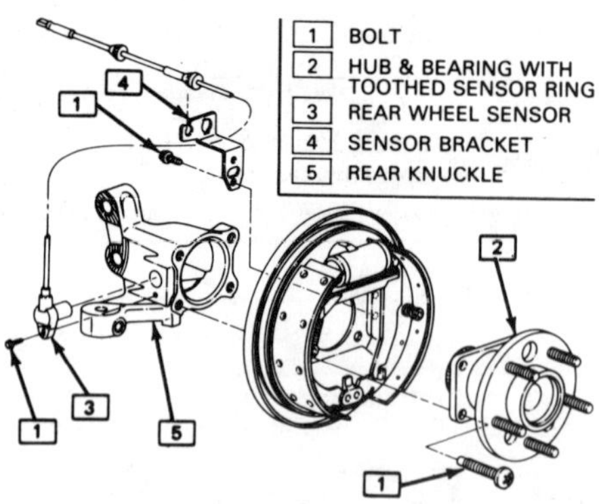

Fig. 138 Rear wheel sensor. All 1987–88 models except Pontiac 6000

HYDRAULIC ACCUMULATOR

1. Disconnect battery ground cable.
2. Depressurize accumulator, refer to ""Depressurizing Hydraulic Accumulator" procedure.
3. Remove accumulator and O-ring.
4. Reverse procedure to install, noting the following:
   a. Lubricate new O-ring with clean brake fluid.
   b. **Torque** accumulator to 17 ft. lbs.
   c. Turn ignition switch to On position and ensure BRAKE light goes out within 60 seconds.
   d. **Reactivate Delco-Loc radios by turning ignition and radio to the On position, then pressing the Mute button on the steering wheel control.**

FLUID RESERVOIR & SEALS

1. Disconnect battery ground cable.
2. Depressurize accumulator, refer to ""Depressurizing Hydraulic Accumulator" procedure.
3. Drain fluid reservoir, remove return hose, then drain brake fluid reservoir into a suitable container.
4. Remove fluid level sensor electrical connectors.
5. Remove reservoir-to-valve block mounting bracket attaching bolt.
6. Remove reservoir by carefully prying between master cylinder and reservoir.
7. Remove O-ring seals from hydraulic unit.
8. Reverse procedure to install, noting the following:
   a. Lubricate O-ring seals with clean DOT 3 brake fluid.
   b. **Torque** reservoir retaining bolt to 44 inch lbs.
   c. **Reactivate Delco-Loc radios by turning ignition and radio to the On position, then pressing the Mute button on the steering wheel control.**

WHEEL SENSOR

Front Wheel

1. Raise and support vehicle, then remove wheel.
2. Disconnect sensor electrical connector from wiring harness.
3. Remove sensor attaching bolts, **Fig. 136,** then the sensor.
4. Reverse procedure to install, noting the following:
   a. **On all models except 1989 6000, torque** sensor attaching bolt to 84 inch lbs. On 1989 6000 models, **torque** sensor attaching bolt to 53 inch lbs.
   b. If wheel sensor is reused, adjust air gap to .028 inches on all 1986 models and 1987-89 Pontiac 6000, or .040 inches on 1987-88 models except Pontiac 6000 using a suitable feeler gauge. New sensors come equipped with a paper spacer which will properly gap the sensor when placed against speed sensor ring.
   c. **On 1989 6000 models,** torque sensor lock bolt to 19 inch lbs. On all other models, torque sensor lock bolt to 22 inch lbs.
   d. Ensure wire does not come in contact with suspension parts.

Rear Wheel

1. If the right rear sensor is to be removed, remove spare tire assembly, then fold back carpeting in corner of luggage compartment to gain access to sensor connector.
2. Disconnect sensor electrical connector from wiring harness.
3. Release the cable grommet, working the sensor cable and connector down through underbody.
4. Raise and support vehicle.
5. Remove rear wheel assembly.
6. **On all 1986 models except Pontiac 6000,** proceed as follows:
   a. Drill out two grommet retaining rivets.
   b. Remove brake drum and brake shoe assembly.
   c. Remove sensor attaching bolts, **Fig. 137,** then the sensor.
7. **On all 1987-88 models except Pontiac 6000,** proceed as follows:
   a. Remove grommet attaching screws.
   b. Remove sensor attaching bolts, **Fig. 138,** then the sensor.
8. **On 1986-88 Pontiac 6000 models,** proceed as follows:
   a. Remove caliper attaching bolts, then position caliper aside using wire. Use caution not to bend or twist fluid line.
   b. Remove rotor assembly.
   c. Remove sensor attaching bolts, **Fig. 139,** then the sensor.
9. **On all models,** reverse procedure to install, noting the following:
   a. **On all 1986 models and 1987-88 Pontiac 6000 models,** torque sensor attaching bolt to 84 inch lbs. On 1989 6000 models, torque sensor attaching bolts to 75 inch lbs.
   b. **On all 1986 models and 1987-88 Pontiac 6000 models,** if wheel

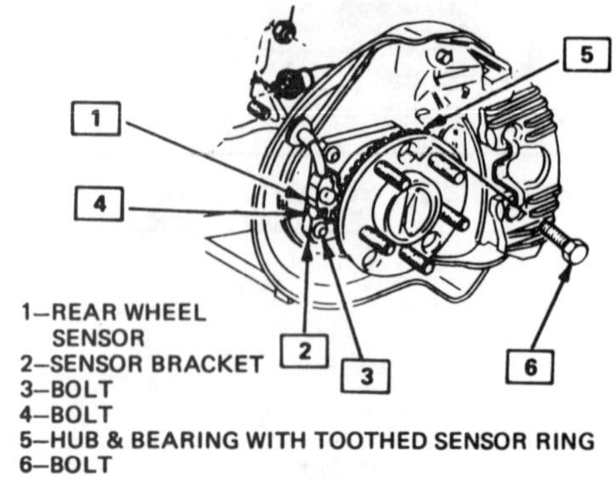

Fig. 139 Rear wheel sensor. 1986–88 Pontiac 6000

sensor is reused, adjust air gap to .028 inches using a suitable feeler gauge. New sensors come equipped with a paper spacer which will properly gap the sensor when placed against speed sensor ring.

c. **On all 1987-88 models except Pontiac 6000,** wheel sensor is automatically adjusted when installed.

d. **On all 1986 models and 1987-88 Pontiac 6000 models,** torque caliper attaching bolts to 22 inch lbs., if equipped. **On 1989 6000 models,** torque caliper attaching bolt to 19 inch lbs.

e. Ensure wire does not come in contact with suspension parts.

Bosch ABS II Type, 1986–89 Corvette

INDEX

1 WHEEL SPEED SENSOR (ONE AT EACH WHEEL)
2 TOOTHED RING (ONE AT EACH WHEEL)
3 CALIPER (ONE AT EACH WHEEL)
4 MASTER CYLINDER
5 BOOSTER
6 LATERAL ACCELERATION SWITCH
7 MODULE RELAY
8 ANTILOCK WARNING LIGHT
9 CONTROL MODULE
10 MODULATOR VALVE

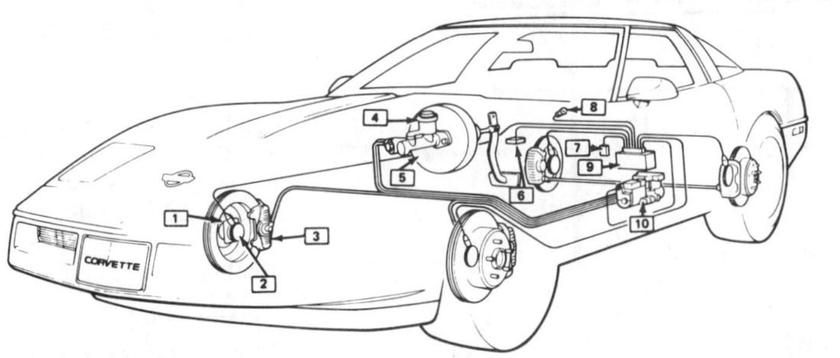

Fig. 1 Bosch anti-lock brake (ABS II) system

DESCRIPTION

The Bosch anti-lock brake II system (ABS II), **Fig. 1,** maintains vehicle sterablity, directional stability, and maximum deceleration under severe braking conditions. The ABS II performs this function by monitoring rotational speed and controlling line pressure of each vehicle wheel during a braking maneuver, to prevent the braked wheels from locking.

SYSTEM COMPONENTS

The ABS II consists of a modulator valve, control module, lateral acceleration switch, wheel speed sensors and toothed rings, fuse, wiring harness and relay.

MODULATOR VALVE

The modulator valve, **Fig. 2,** consists of three rapidly switching solenoid valves, two accumulator chambers (one for each brake circuit) and a return pump. One solenoid valve is assigned to each front wheel brake, the third is assigned to the rear wheel brakes. This system is known as a "Three-channel system."

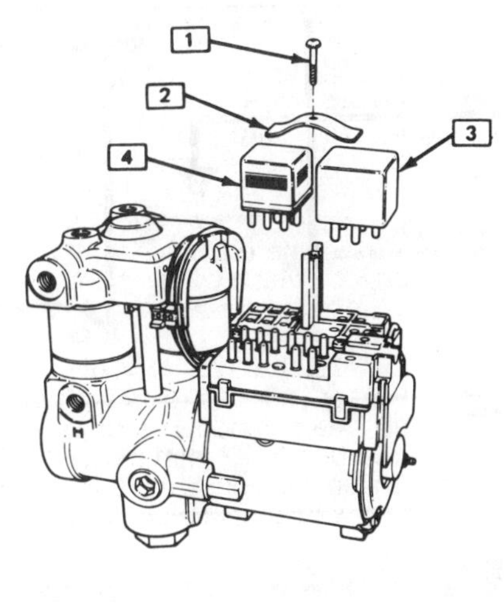

1 SCREW
2 RETAINER
3 PUMP MOTOR RELAY
4 SOLENOID RELAY

Fig. 2 Modulator valve

The modulator valve receives its instructions from control module. The modular valve regulates brake fluid pressure based upon road conditions as interpreted by the control module, regardless of master cylinder output pressure. **The modulator valve can maintain or reduce brake fluid pressure to the wheel calipers. However, the modulator valve can not increase the pressure above that sent by the master cylinder. The modulator valve alone can not apply the brakes.**

Two relays are mounted on top of modulator valve. These are the only replaceable parts of the modulator valve. The pump motor relay controls the return pump and the solenoid relay controls power supply to solenoid valves. Modulator valve is mounted in rear storage compartment located behind drivers seat. The modulator valve itself cannot be repaired and must be replaced as an assembly.

CONTROL MODULE

The control module consists of two printed circuit board plates. Resistors, diodes, transistors and large integrated circuits are mounted on one side of plates and both sides have conductor paths. These integrated circuits (IC), contain thousands of transistors, resistors and diodes mounted on a single silicon chip. Plates are housed in control module and surrounded by light alloy casting.

Wheel acceleration, deceleration and slip values are calculated from electric signals, transmitted from wheel speed sensors, which are proportional to wheel speed. Electrical signals from this module produce commands for the hydraulic modulator valves.

The control module is located in rear storage compartment behind drivers seat. If control module is defective, it must be replaced as an assembly. **No attempt should be made to repair this module.**

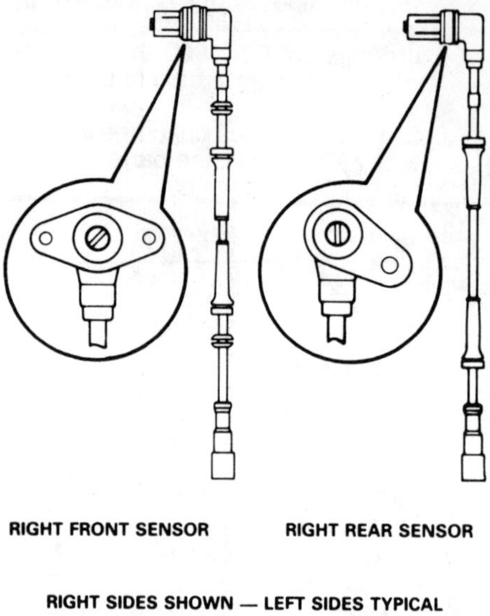

Fig. 3 Wheel speed sensor (Part 1 of 2). 1986–87

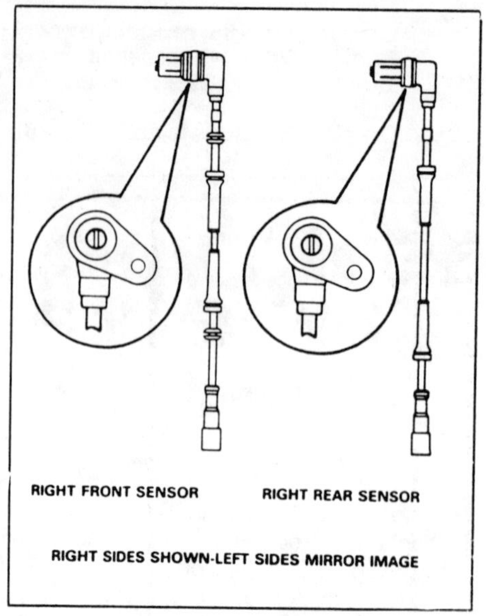

Fig. 3 Wheel speed sensor (Part 2 of 2). 1988–89

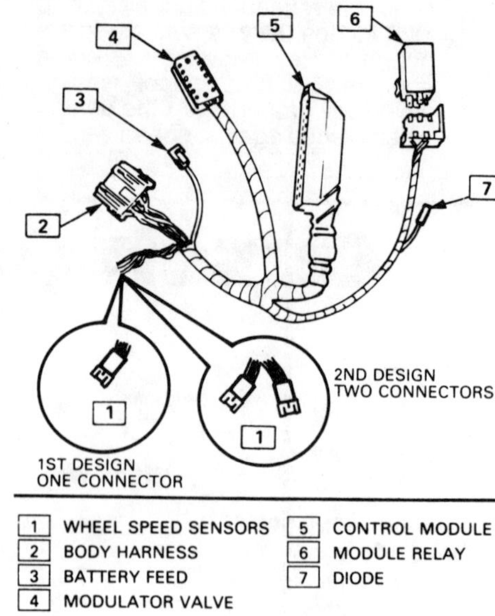

Fig. 4 Wiring harness connector

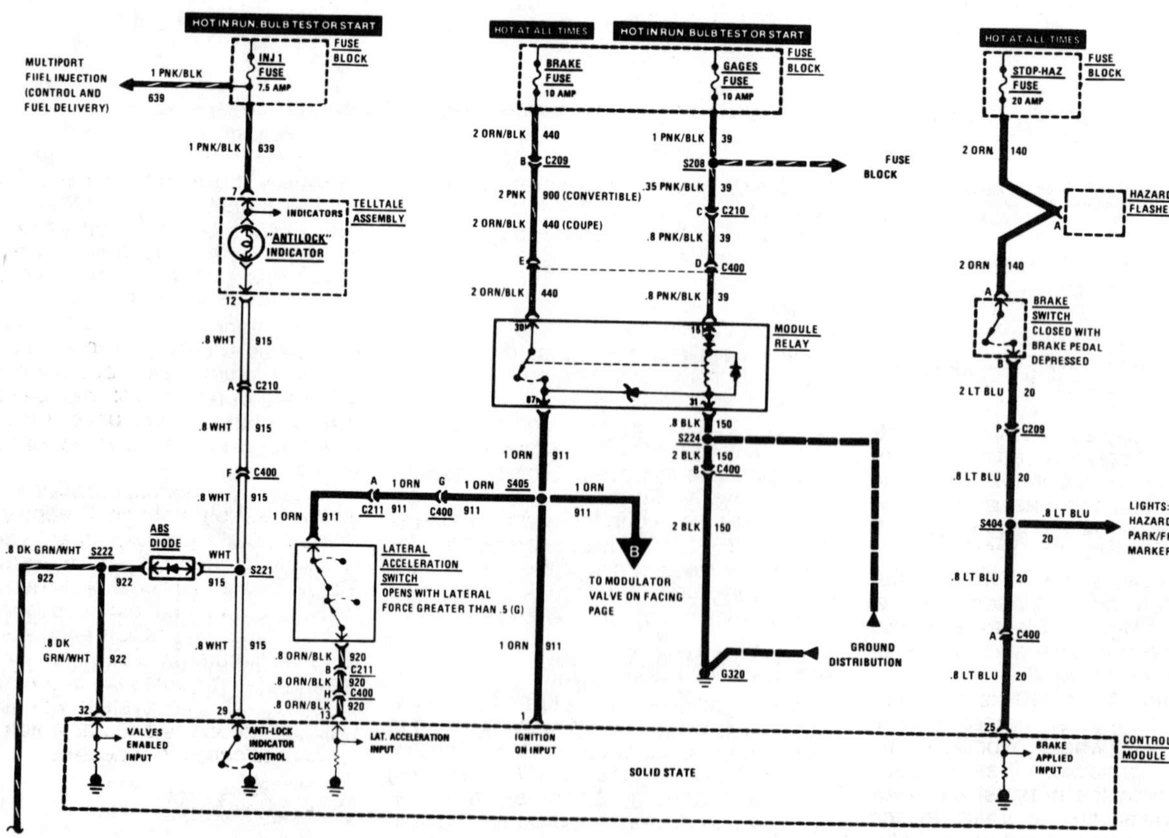

Fig. 5 Anti-lock brake system wiring schematic (Part 1 of 3). 1986

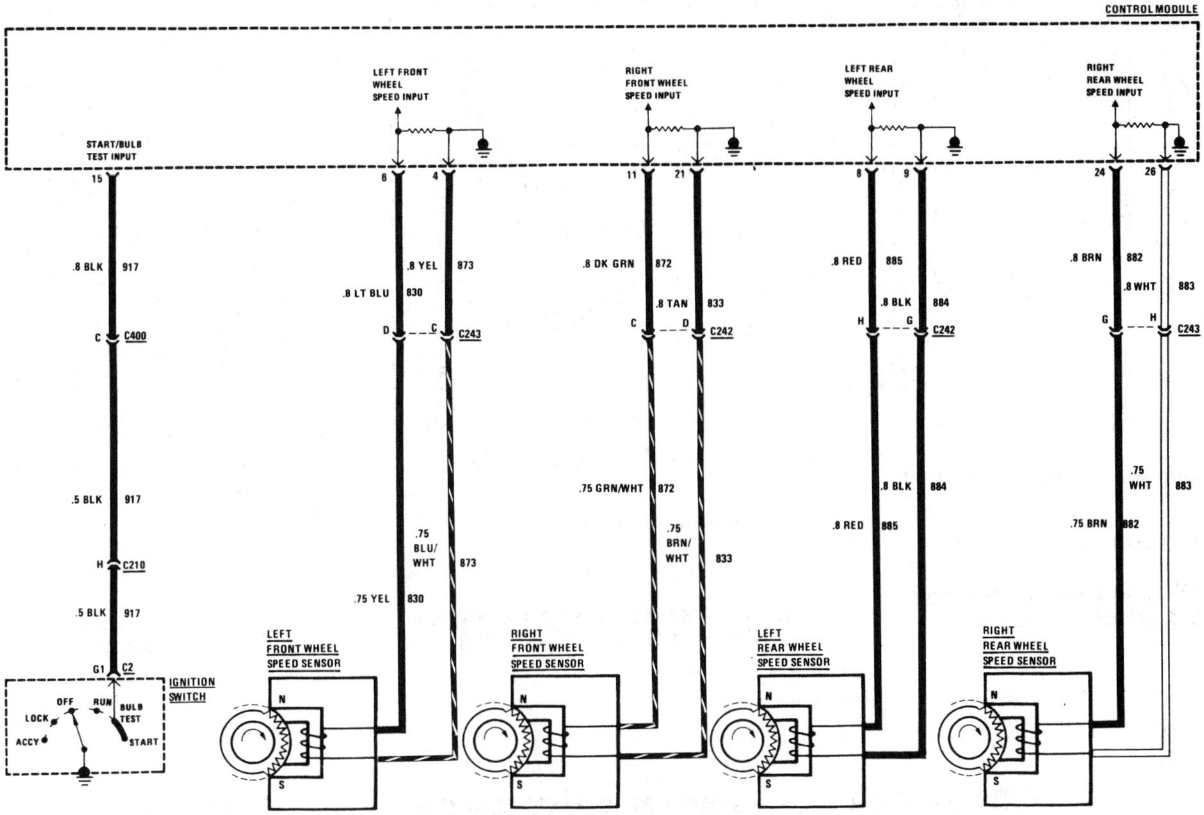

Fig. 5 Anti-lock brake system wiring schematic (Part 2 of 3). 1986

LATERAL ACCELERATION SWITCH

The lateral acceleration switch is made up of two mercury switches connected in series. They are used to detect if the vehicle is cornering faster than a predetermined curve speed. When this speed is exceeded, one of two switches opens up and sends a signal to control module.

The lateral acceleration switch is located under the A/C control head on the floor pan. This switch is not serviceable and should be replaced as an assembly.

WHEEL SPEED SENSORS

Rotational speed of each wheel is detected separately by an inductive wheel speed sensor, **Fig. 3.** Electrical signal sensed from the speed sensors are transmitted to control module.

In this system, front wheels are controlled individually and rear wheels together. Control of rear wheels works on the "Select Low" principle. "Select Low" means wheel with the greatest tendency to lock-up (lower tire to road friction co-efficient), determines level of control.

Wheel speed sensors are installed in knuckles. Toothed rings are pressed onto front hub and bearing assemblies and rear drive shaft spindles. **The wheel speed sensors do not require adjustment.**

WIRING HARNESS, RELAY & FUSE

The ABS II system has its own wiring harness, located in rear storage compartment behind drivers seat. This harness is connected to the standard wiring harness. The wiring harness, **Fig. 4,** provides a path for inputs from the wheel speed sensors, lateral acceleration switch, power and stop light switch to control module and output signals from the control module to the modulator valve and warning lamps.

The module relay is installed in circuit between battery, fuse and control module. This relay protects the control module from possible excessive voltage spikes, due to a faulty alternator.

The fuse located in the main fuse panel labeled BRAKE, provides protection for main power feed circuit of the ABS electrical system. The GAGE fuse is also tied into the main power feed of the ABS.

OPERATION

When ignition switch is turned to On position, amber ANTILOCK warning lamp in the instrument panel illuminates. When engine is started, it goes out. If ABS warning light does not go out or comes on permanently while driving, it indicates a fault in ABS system.

When starting vehicle, the control module performs a functional check of electrical circuitry. The test cycle checks components of the monitoring circuit as well as the logic section. Test sample signals are sent to the control module, which checks whether or not the correct output signals are available.

Since the ABS system may not be used every day, there is an additional test which actually runs the modulator valve. Whenever ignition switch is turned to On position and vehicle reaches 4 mph, the test starts. Operation of the modular valve can be heard and felt, if the driver's foot is on the brake pedal. This test is performed once, each time the vehicle is started.

This system continuously monitors the following components: modulator valve, control module, lateral acceleration switch, wheel speed sensors, wiring harness and relays.

The control module constantly monitors its own supply voltage. If supply voltage drops below a predetermined value, ABS will be switched Off and the amber ANTILOCK lamp will come on. When supplied voltage returns to or exceeds the predetermined value, ANTILOCK lamp will go off. If a fault occurs in ABS, the ANTILOCK warning lamp comes on and ABS will be switched off. The system will remain off until vehicle is restarted at which time the functional check is repeated.

DIAGNOSIS

Refer to **Figs. 5 through 12,** during diagnosis procedures.

PRE-DIAGNOSIS INSPECTION

If a malfunction in the ABS system occurs, perform the following:

1. Ensure control module connector is securely connected.

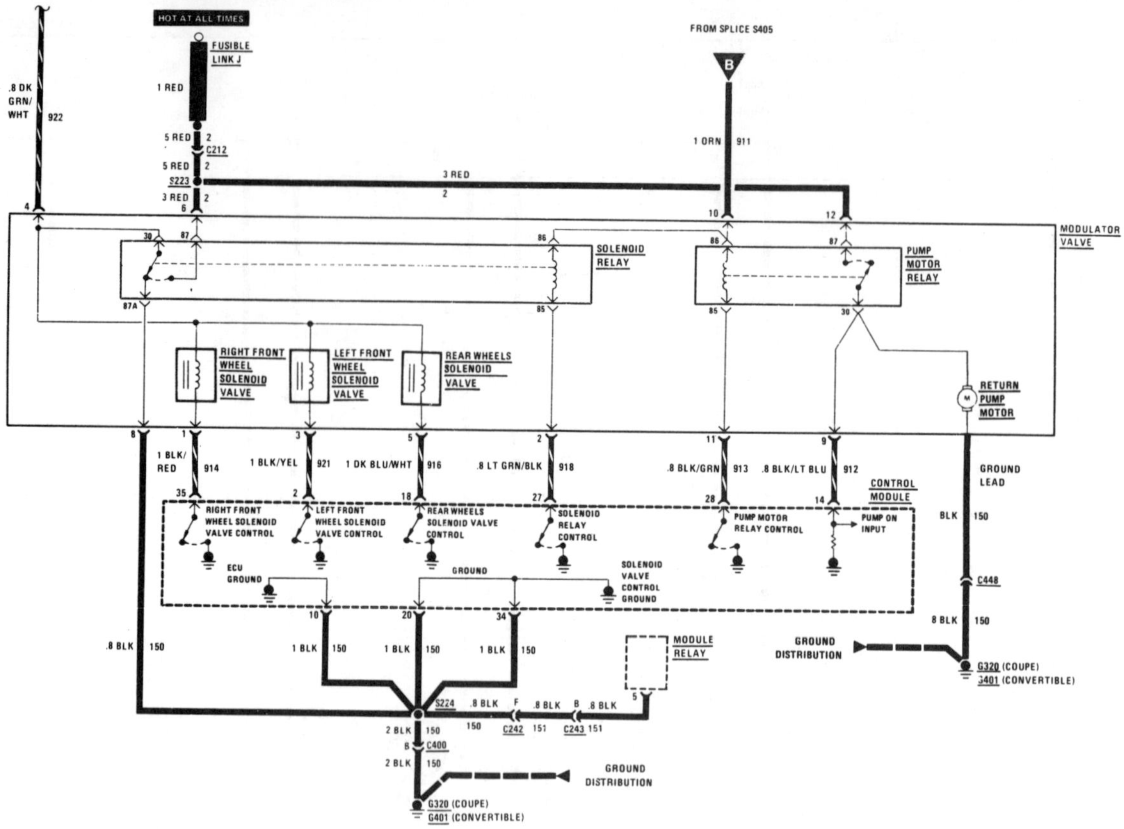

Fig. 5 Anti-lock brake system wiring schematic (Part 3 of 3). 1986

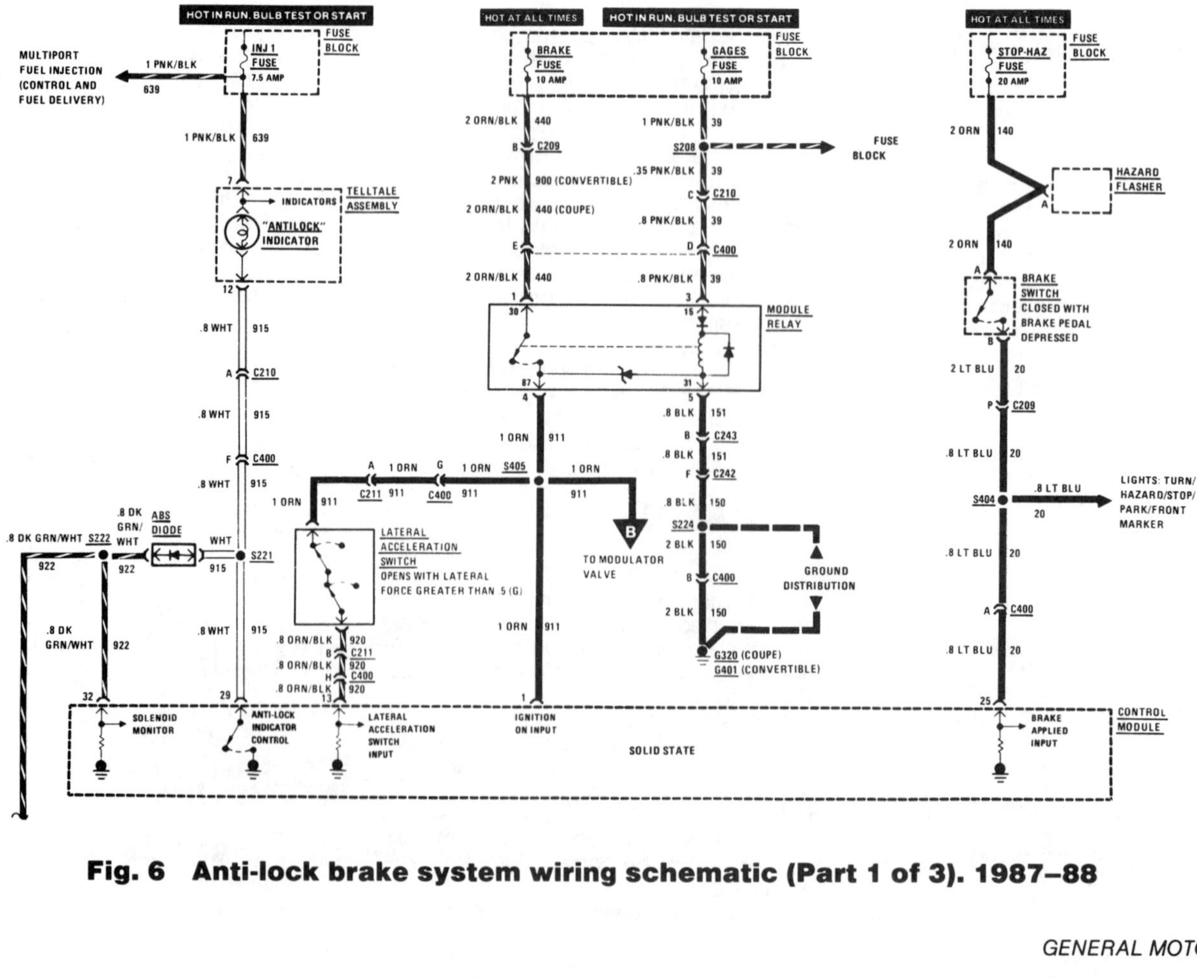

Fig. 6 Anti-lock brake system wiring schematic (Part 1 of 3). 1987–88

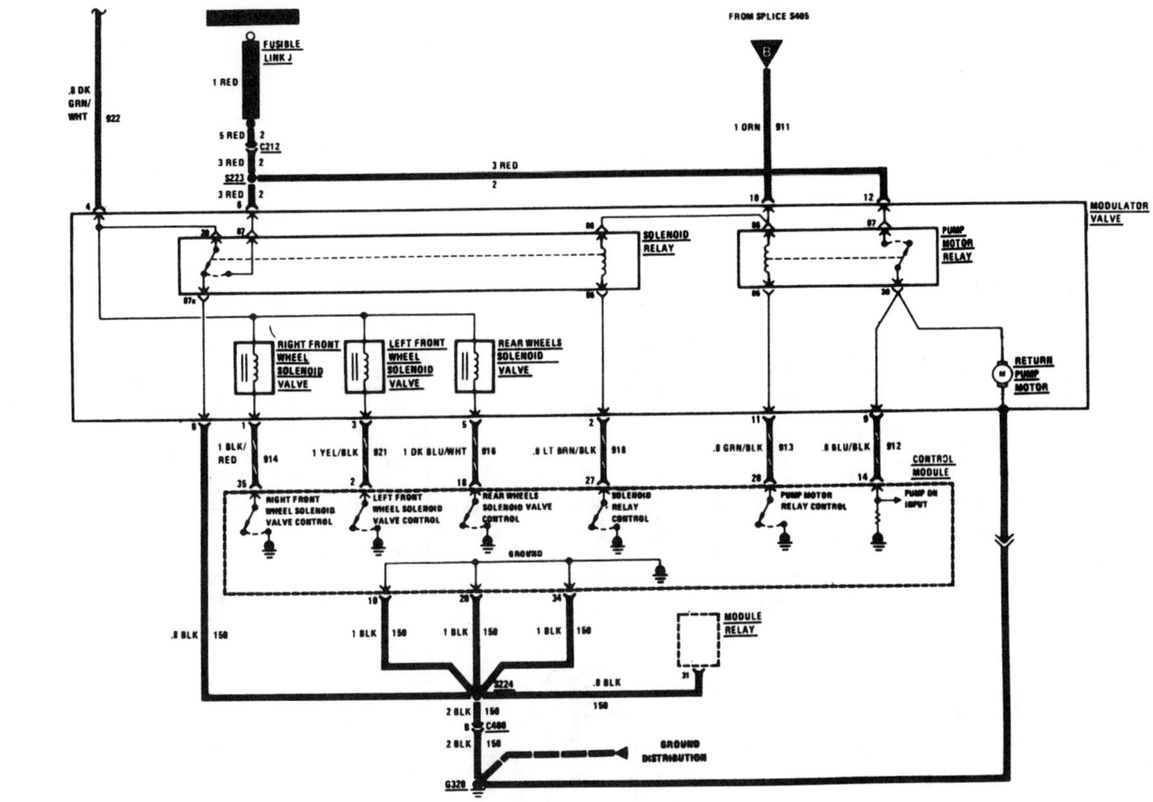

Fig. 6 Anti-lock brake system wiring schematic (Part 2 of 3). 1987–88

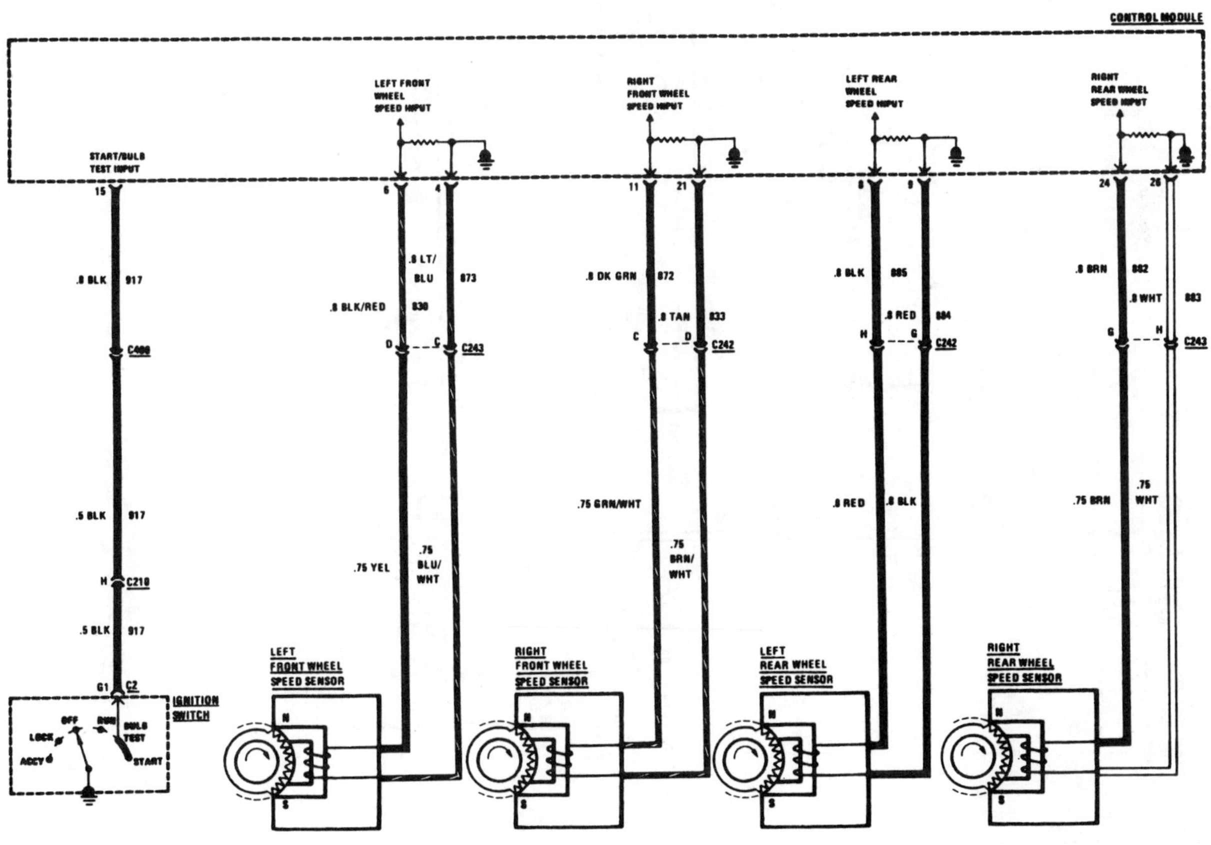

Fig. 6 Anti-lock brake system wiring schematic (Part 3 of 3). 1987–88

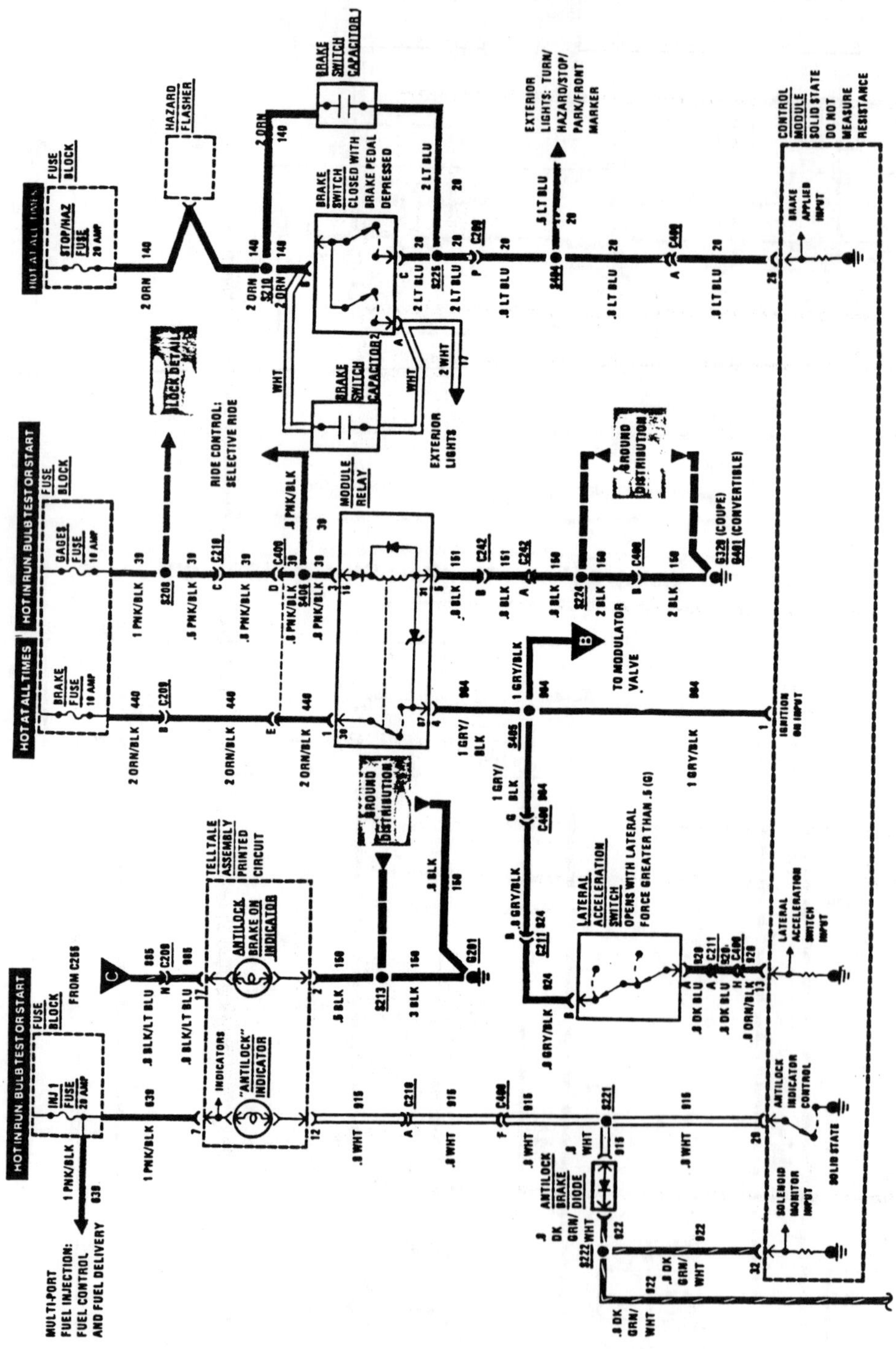

Fig. 7 Anti-lock brake system wiring schematic (Part 1 of 3). 1989

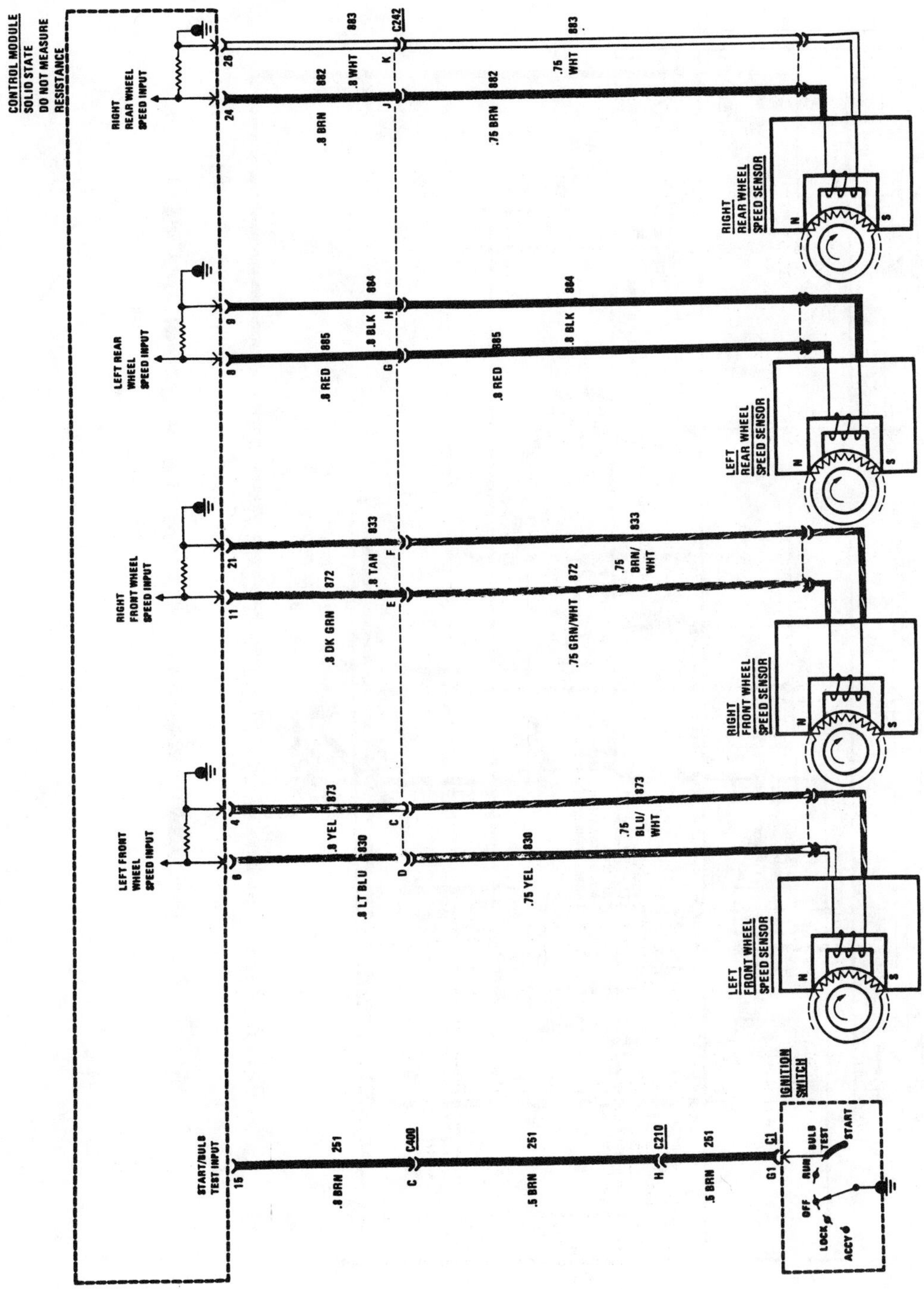

Fig. 7 Anti-lock brake system wiring schematic (Part 2 of 3). 1989

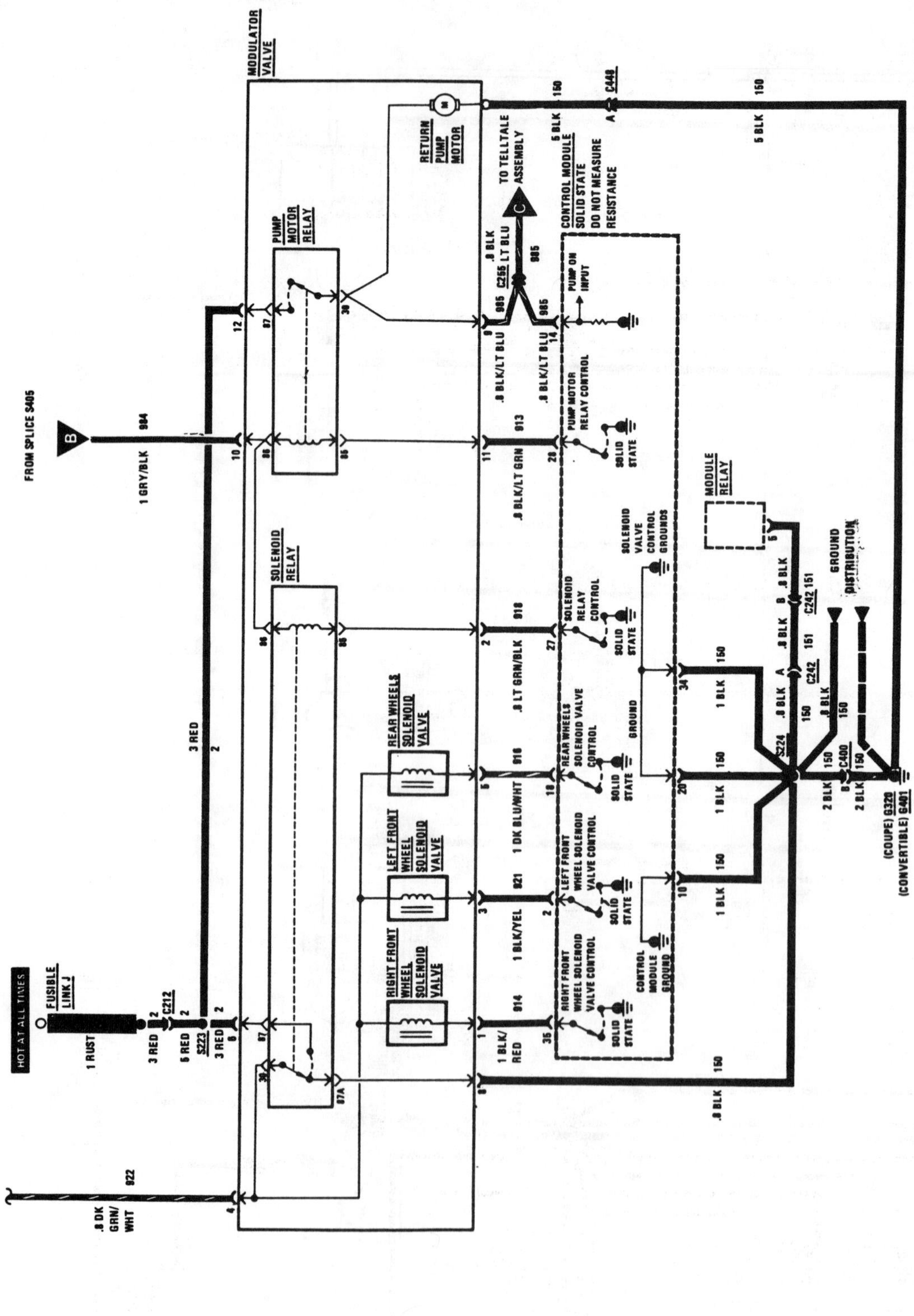

Fig. 7 Anti-lock brake system wiring schematic (Part 3 of 3), 1989

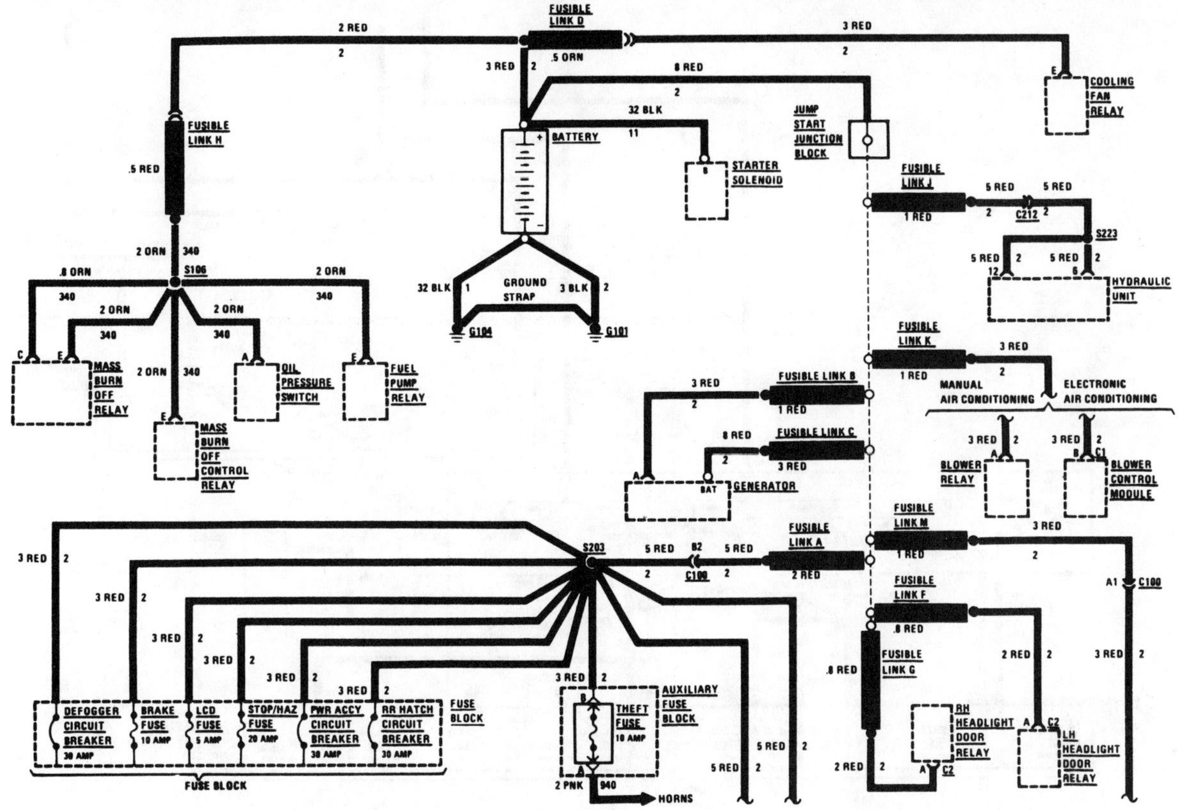

Fig. 8 Power distribution circuit (Part 1 of 3). 1986

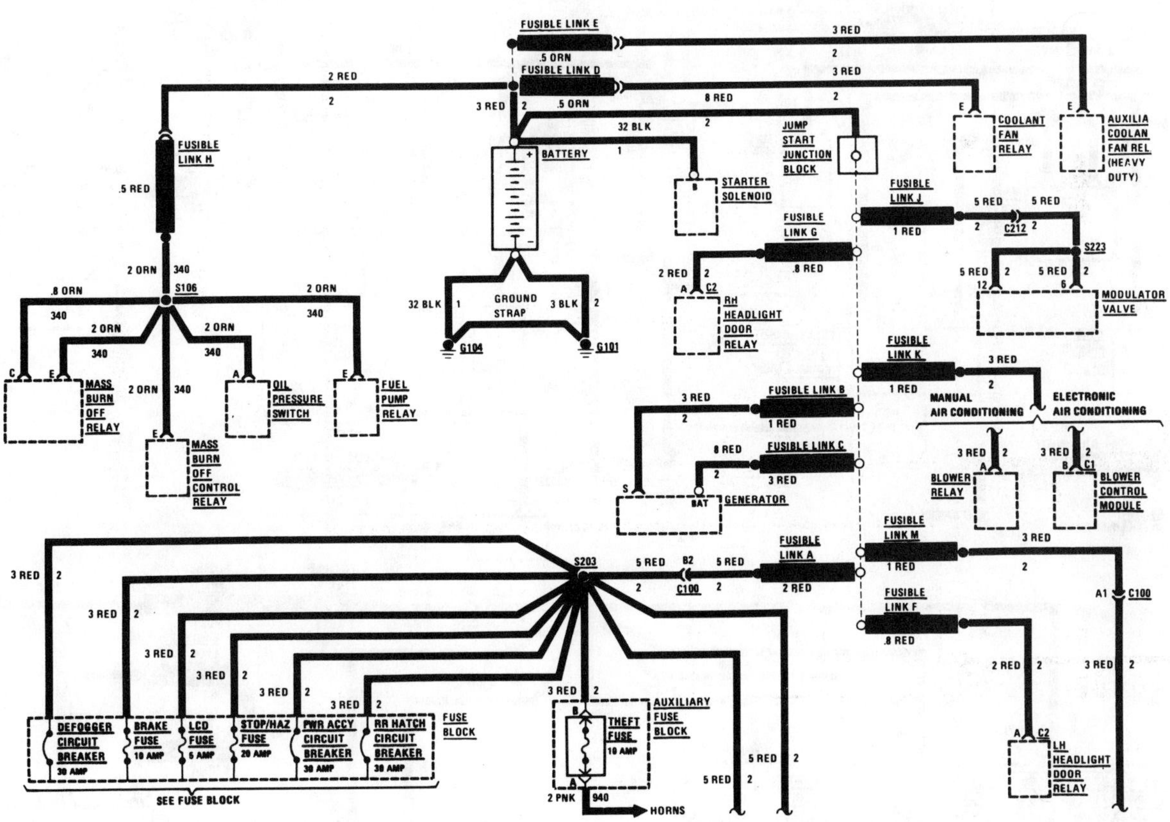

Fig. 8 Power distribution circuit (Part 2 of 3). 1987–88

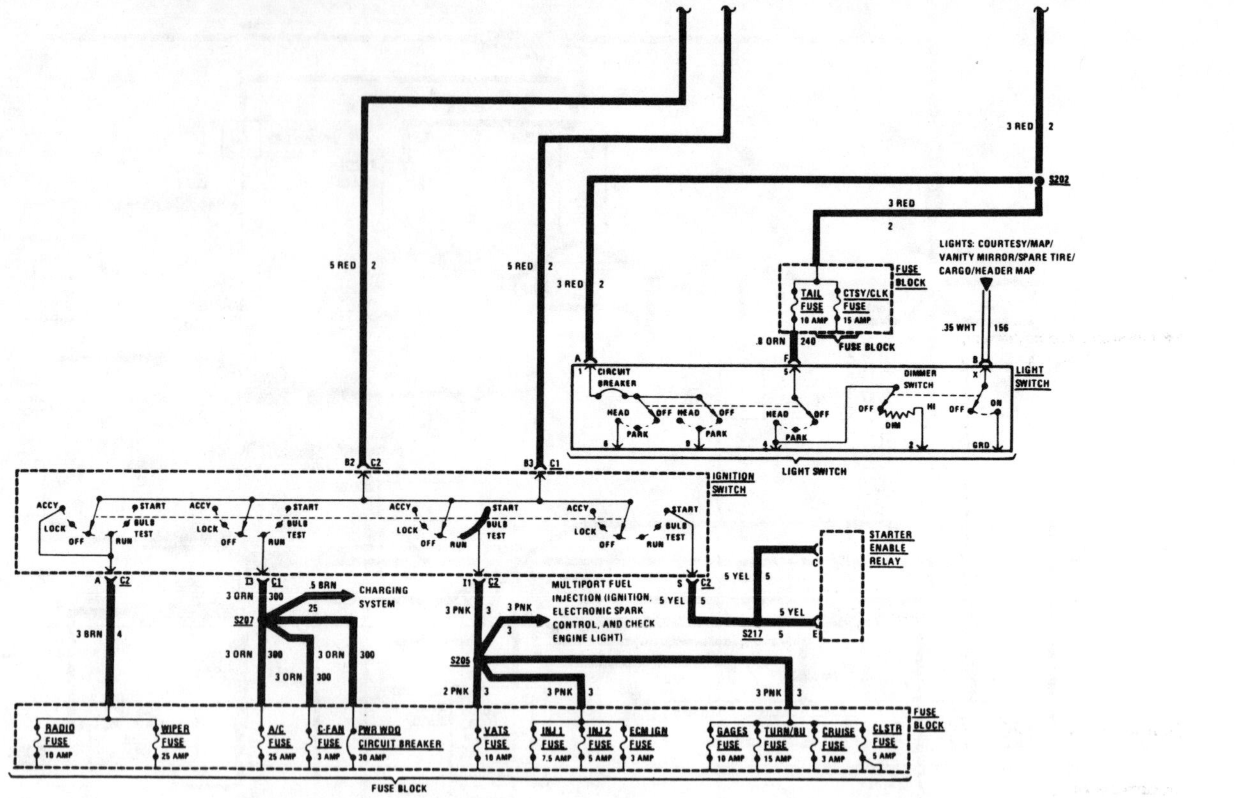

Fig. 8 Power distribution circuit (Part 3 of 3). 1986–88

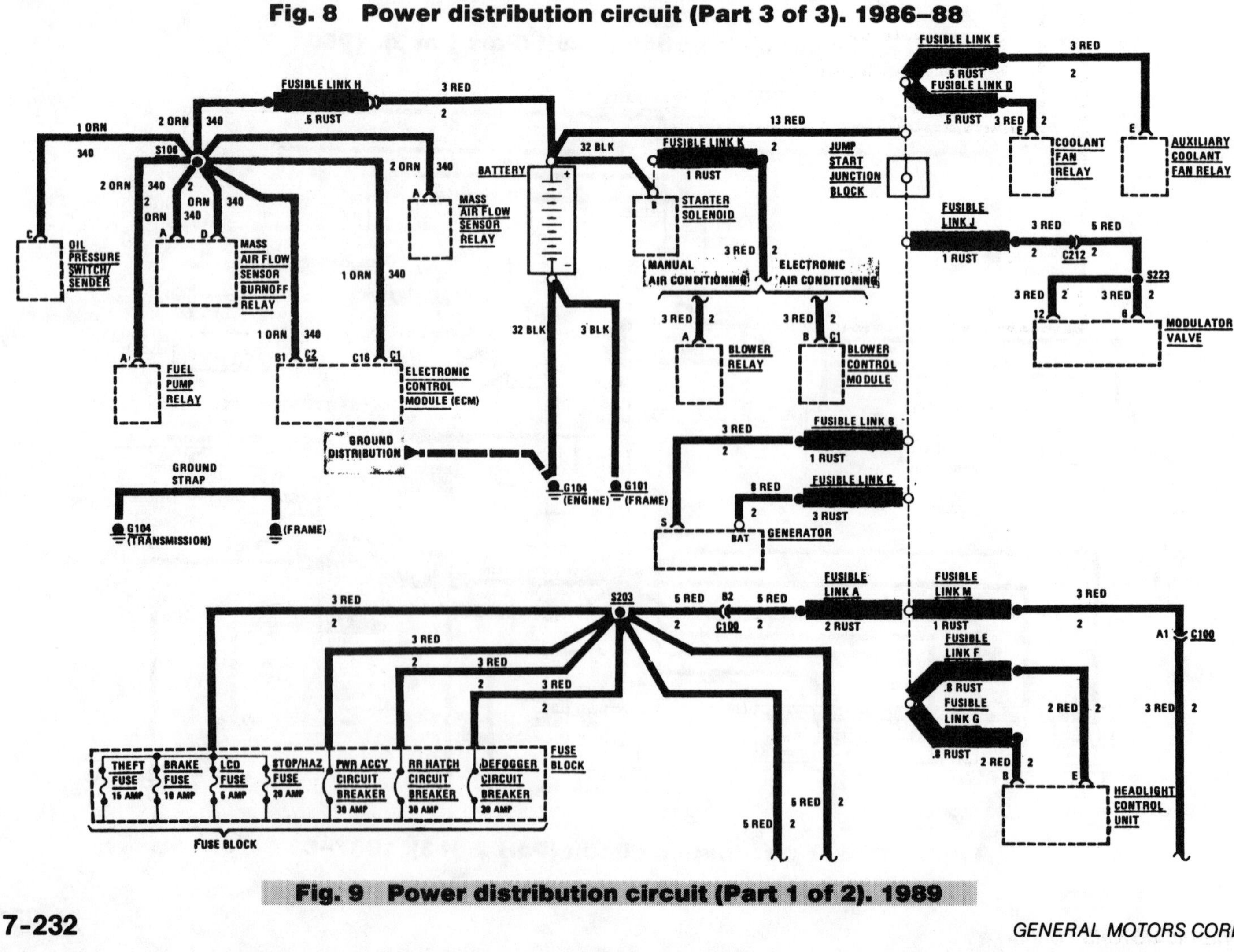

Fig. 9 Power distribution circuit (Part 1 of 2). 1989

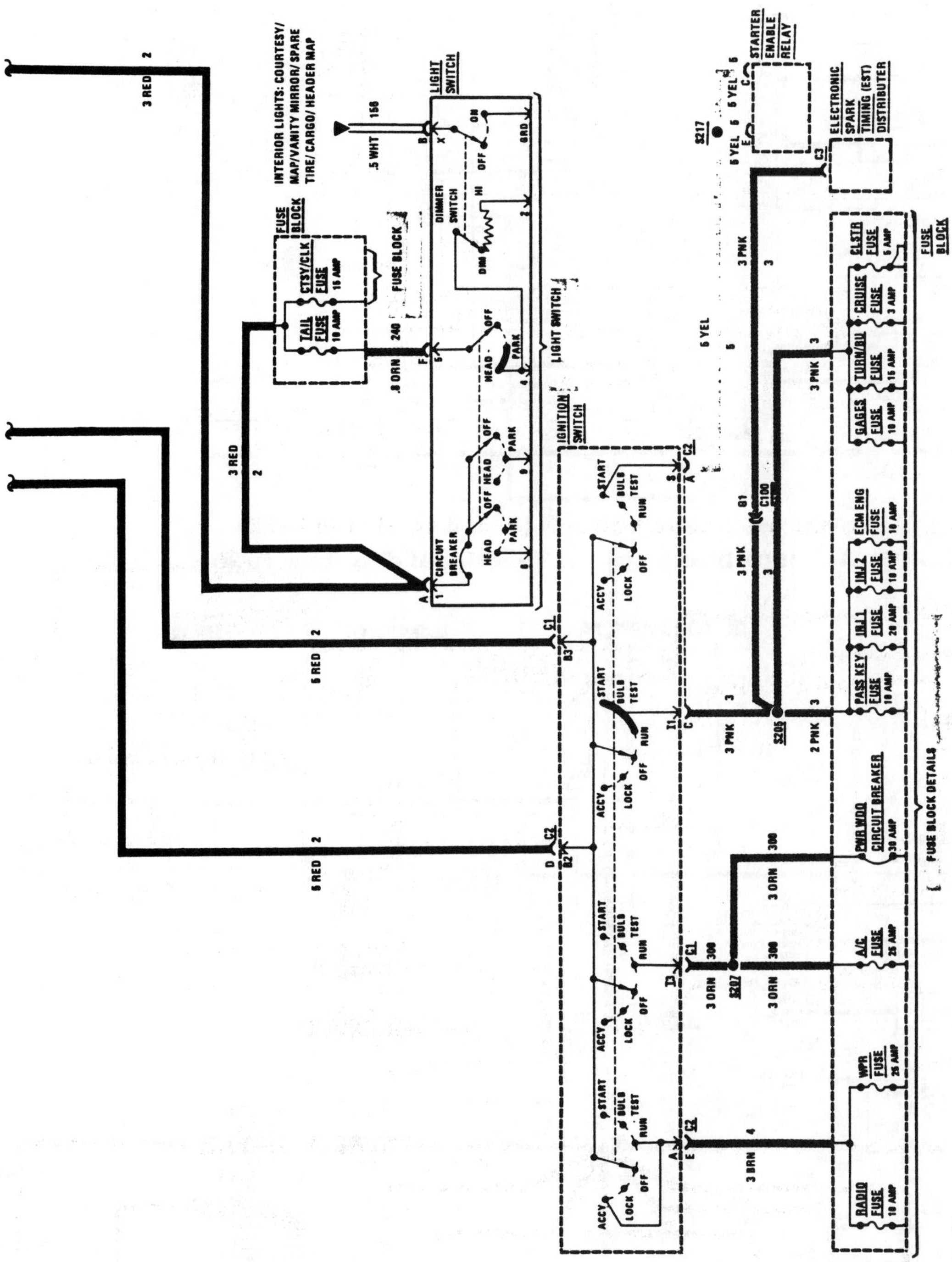

Fig. 9 Power distribution circuit (Part 2 of 2). 1989

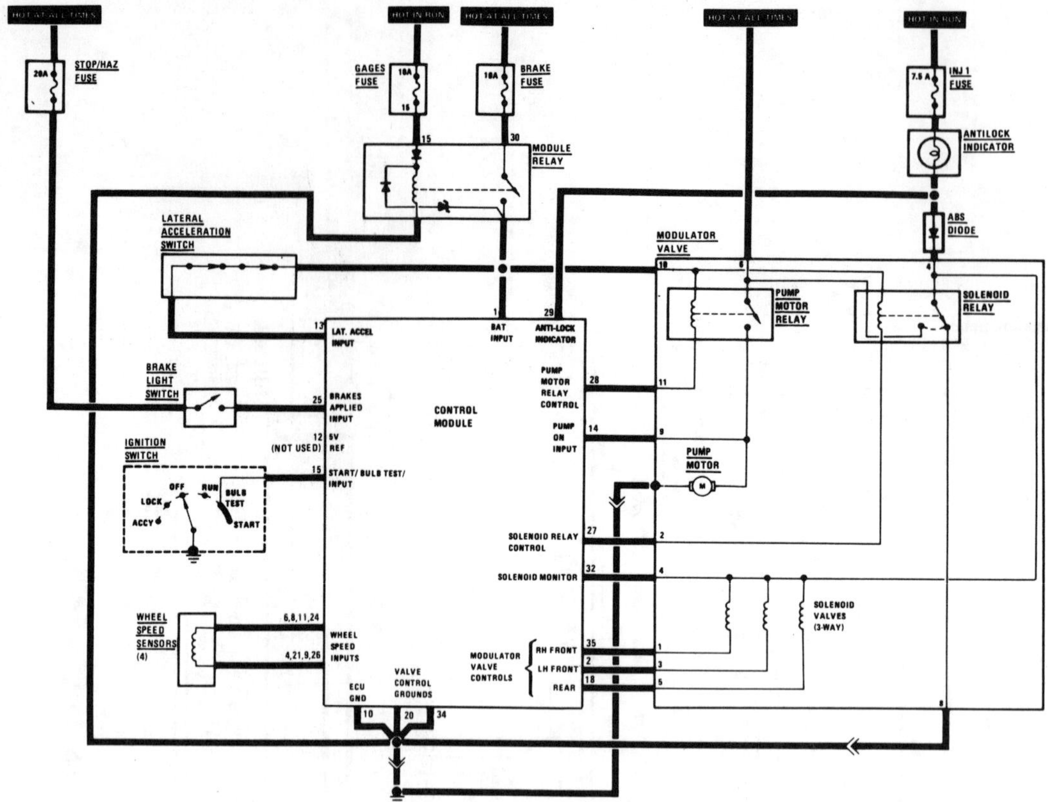

Fig. 10 Anti-lock brake system functional diagram. 1986

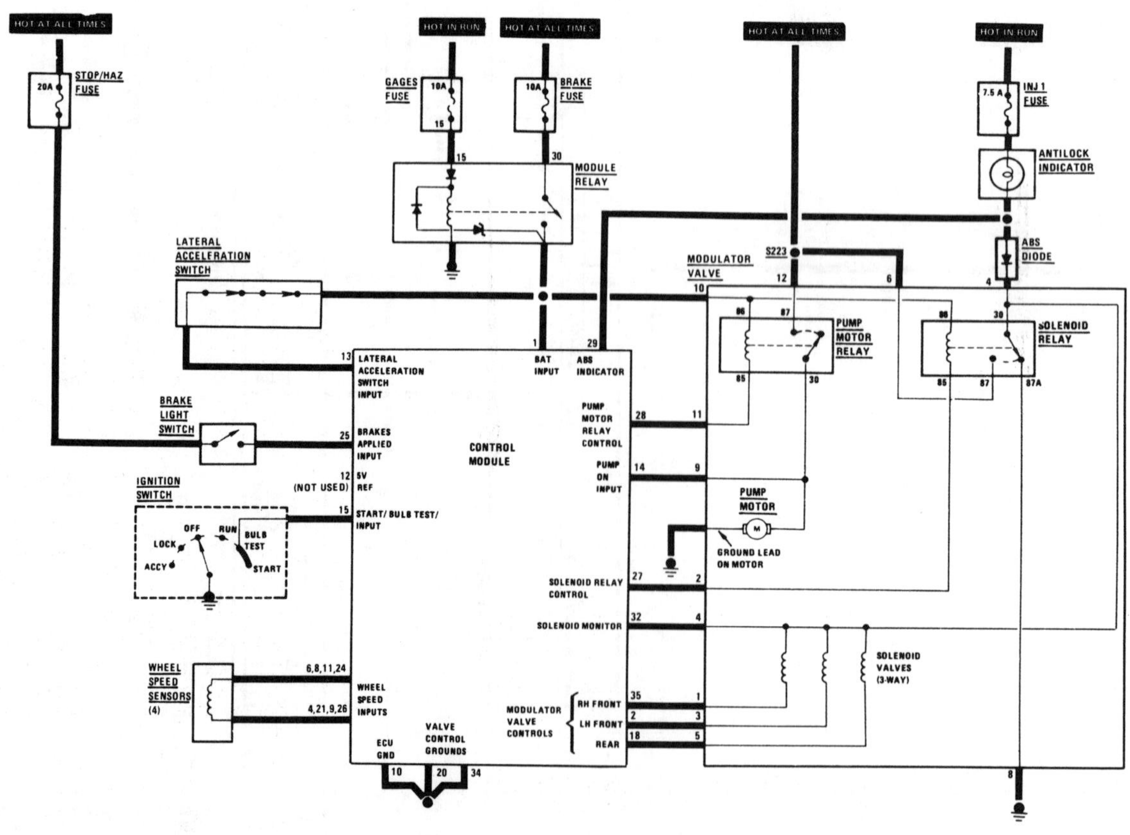

Fig. 11 Anti-lock brake system functional diagram. 1987–88

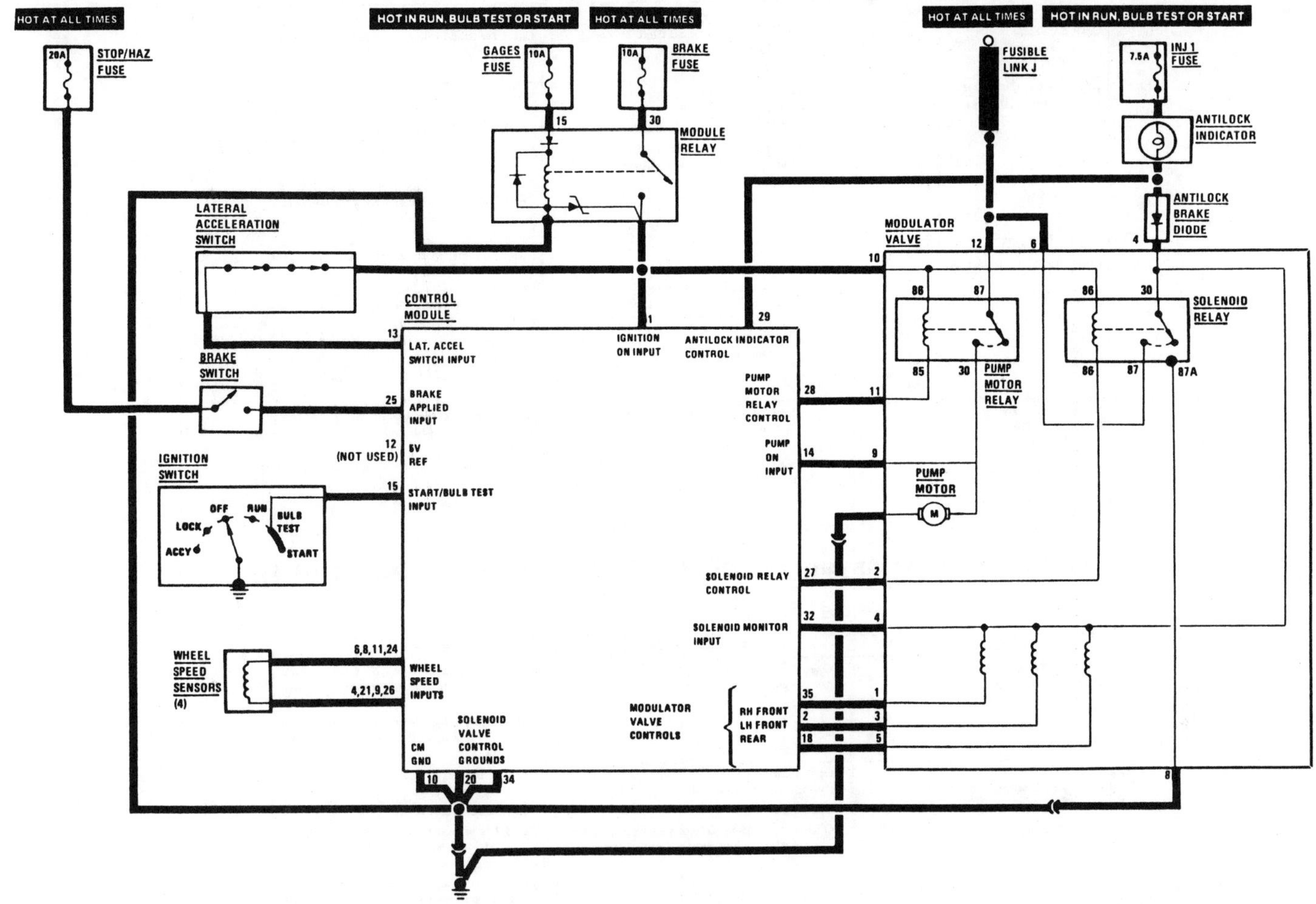

Fig. 12 Anti-lock brake system functional diagram. 1989

SYMPTOM MESSAGE	POSSIBLE CAUSE	REPAIR ACTION
No Power-up Messages	No Battery voltage to ABS Tester Open ABS ground circuit	Do Test A Repair circuit (150) from control module terminal 20 to splice S224
LOW BATTERY	Battery not fully charged	Charge Battery.
LF VALVE OPEN (Left Front Valve Open) (In Addition to LF VALVE OPEN, check for display of RF VALVE OPEN and REAR VALVE OPEN).	Open Circuit in Left Front Wheel Solenoid Valve Open Wire between Valve and Control Module	Do Test B: Solenoid Valve Resistance Test Check for open in BLK/YEL (921) wire.
RF VALVE OPEN (Right Front Valve Open)	Open Circuit in Right Front Wheel Solenoid Valve Open Wire between Valve and Control Module	Do Test B: Solenoid Valve Resistance Test. Check for open in BLK/RED (914) wire.
REAR VALVE OPEN	Open Circuit in Rear Wheel Solenoid Valve Open Wire between Valve and Control Module	Do Test B: Solenoid Valve Resistance Test. Check for open in DK BLU/WHT (916) wire.
LF VALVE OPEN, RF VALVE OPEN, REAR VALVE OPEN	Open Solenoid Relay Contacts Open ground wire between Terminal 8 of the Modulator Valve and G320.	Do Test C: Solenoid Test. (Table 3) Check for open in BLK (915) wire to ground.
CHK ABS DIODE	Open or shorted ABS Diode Open in circuit 915 or 922 Shorted ABS Diode	Do Test D: ABS Diode Test. Check for open in WHT (915) wire between ABS Diode and Terminal 29 or DK GRN/WHT (922) wire between ABS Diode and Terminal 32 of Control Module. Do Test D: ABS Diode Test
CHECK CONT MOD GRN (Check Control Module Ground)	Open in circuit 150 beteen the Control Module and the ground point	Check or open in BLK (150) wire from terminal 20 of the Control Module to terminal B of Connector C400. Check for open in BLK (150) wire from terminal 10 of the Control Module to terminal B of Connector C400.

Fig. 13 Anti-lock system symptom directory chart (Part 1 of 4). 1986

2. Ensure that all ABS connectors are securely connected.
3. Check BRAKE and GAUGE for defective fuse.
4. If malfunction is still present after performing steps 1 through 3, refer to "System Check" procedure.

SYSTEM CHECK PROCEDURE

1. Release parking brake, then turn ignition switch to Run position. Amber ANTILOCK indicator lamp should light for several seconds.
2. Apply and release parking brake. Red BRAKE indicator lamp should light when parking brake is applied.
3. Start engine. ANTILOCK indicator lamp should remain illuminated during cranking.
4. Test drive vehicle at speeds between 20 and 30 mph. Use normal braking procedures to stop vehicle. Brake should operate smoothly and ANTILOCK indicator lamp should remain out.
5. Refer to "Diagnostic Procedure" if steps 1 through 4 are not satisfactory.

DIAGNOSTIC PROCEDURE

1. Turn ignition switch to Off position.
2. Disconnect ABS control module harness connector and connect it to ABS

SYMPTOM MESSAGE	POSSIBLE CAUSE	REPAIR ACTION
CHK SYSTEM GND (Check System Ground)	Open circuit 150 between Control Module and ground point	Check for open in BLK wire (150) from terminal 34 of the Control Module to terminal B of Connector C400.
LAT ACCEL SW OPEN (Lateral Acceleration Switch Open)	Open Lateral Acceleration Switch (Switch should be closed)	Do Test E: Lateral Acceleration Switch Test.
	Open Connector at Lateral Acceleration Switch	Check that connector is firmly seated.
	Open circuit 919 or 920 connecting the Lateral Acceleration Switch and the Control Module	Check ORN wire (911) between terminal 1 of Control Module and Lateral Acceleration Switch.
		Check ORN/BLK wire (920) between terminal 13 of Control Module and Lateral Acceleration Switch.
CHK BRK LIGHT SW (Check Brake Light Switch)	Blown STOP/HAZ Fuse	Check operation of Hazard Flasher to check fuse.
	Condition may be due to inoperative, misaligned or permanently closed Brake Light Switch.	Check for proper operation of Stop Lights. If Stop Lights don't work, replace or realign Brake Switch.
	Open in circuit 20 between Brake Switch and Control Module	Check that Connectors C209 and C400 are properly mated.
		Check for an open in the LT BLU wire (20).
CHK IGNITION SW (Check Ignition Switch)	Ignition Switch not grounding terminal 15 at the Control Module in Bulb Test and Start	Check for proper operation of Ignition Switch in Bulb Test and Start. If switch does not operate, replace switch.
	Open in circuit 917 between Ignition Switch and Control Module	Check that connectors C210 and C400 are properly mated.
		Check for an open in the BLK wire (917).

Fig. 13 Anti-lock system symptom directory chart (Part 2 of 4). 1986

SYMPTOM MESSAGE	POSSIBLE CAUSE	REPAIR ACTION
CHK SYSTEM GND (Check System Ground)	Open circuit 150 between Control Module and ground point	Check for open in BLK wire (150) from terminal 34 of the Control Module to terminal B of Connector C400.
		Check for open in BLK wire (150) from terminal 20 of the Control Module to terminal B of Connector C400.
LAT ACCEL SW OPEN (Lateral Acceleration Switch Open)	Open Lateral Acceleration Switch (Switch should be closed)	Do Test E: Lateral Acceleration Switch Test.
	Open Connector at Lateral Acceleration Switch	Check that connector is firmly seated.
	Open circuit 919 or 920 connecting the Lateral Acceleration Switch and the Control Module	Check ORN wire (911) between terminal 1 of Control Module and Lateral Acceleration Switch.
		Check ORN/BLK wire (920) between terminal 13 of Control Module and Lateral Acceleration Switch.
CHK BRK LIGHT SW (Check Brake Light Switch)	Blown STOP/HAZ Fuse	Check operation of Hazard Flasher to check fuse.
	Condition may be due to inoperative, misaligned or permanently closed Brake Light Switch.	Check for proper operation of Stop Lights. If Stop Lights don't work, replace or realign Brake Switch.
	Open in circuit 20 between Brake Switch and Control Module	Check that Connectors C209 and C400 are properly mated.
		Check for an open in the LT BLU wire (20).
CHK IGNITION SW (Check Ignition Switch)	Ignition Switch not grounding terminal 15 at the Control Module in Bulb Test and Start	Check for proper operation of Ignition Switch in Bulb Test and Start. If switch does not operate, replace switch.
	Open in circuit 917 between Ignition Switch and Control Module	Check that connectors C210 and C400 are properly mated.
		Check for an open in the BLK wire (917).

Fig. 13 Anti-lock system symptom directory chart (Part 3 of 4). 1986

SYMPTOM MESSAGE	POSSIBLE CAUSE	REPAIR ACTION
CHK TOOTHED WHL	Missing teeth in toothed wheel	Check toothed wheel for missing or damaged teeth.
CHK VALVE RELAY	Solenoid Relay Inoperative	Do Test C.
CHECK HYDRAULICS	Faulty Hydraulic Fluid Connection	Check that all Brake Fluid connections are tight and that no fluid is leaking.
	Solenoid Valve stuck in by-pass position	Replace Modulator Valve.
CHK HYDRA WIRING (Check Hydraulic Wiring)	Wiring to Modulator Valve or wiring inside Modulator Valve	Check that Modulator Valve harness connector is properly mated.
		Check that the Solenoid and Pump Motor relays are properly seated.

Fig. 13 Anti-lock system symptom directory chart (Part 4 of 4). 1986

SYMPTOM MESSAGE	POSSIBLE CAUSE	REPAIR ACTION
No Power-up Messages	No Battery voltage to ABS Tester	Do Test A.
LOW BATTERY	Battery not fully charged	Charge Battery.
LF VALVE OPEN (Left Front Valve Open) (In Addition to LF VALVE OPEN, check for display of RF VALVE OPEN and REAR VALVE OPEN.	Open Circuit in Left Front Wheel Solenoid Valve Open Wire between Valve and Control Module	Do Test B: Solenoid Valve Resistance Test. Check for open in YEL/BLK wire (921).
RF VALVE OPEN (Right Front Valve Open)	Open Circuit in Right Front Wheel Solenoid Valve Open Wire between Valve and Control Module	Do Test B: Solenoid Valve Resistance Test. Check for open in DK BLU/WHT wire (916).
REAR VALVE OPEN	Open Circuit in Rear Wheel Solenoid Valve Open Wire between Valve and Control Module	Do Test B: Solenoid Valve Resistance Test. Check for open in BLK/RED wire (914).
LF VALVE OPEN, RF VALVE OPEN, REAR VALVE OPEN	Open Solenoid Relay Contacts Open ground wire between Terminal 8 of the Modulator Valve and G320	Do Test C: Solenoid Relay Test. Check for open in BLK wire (150) to ground at G320.
CHK ABS DIODE	Open ABS Diode Open in circuit 915 or 922 Shorted ABS Diode	Do Test D: ABS Diode Test. Check for open in WHT wire (915) between ABS Diode and Splice S221, or DK GRN/WHT wire (922) between ABS Diode and Splice S222. Do Test D: ABS Diode Test.
CHK CONT MOD GND (Check Control Module Ground)	Open in circuit 150 between the Control Module and the ground point on LH "B" Pillar	Check for open in BLK wire (150) from terminal 20 of the Control Module to terminal B of Connector C400. Check for open in BLK wire (150) from terminal 10 of the Control Module to terminal B of Connector C400.

Fig. 14 Anti-lock system symptom directory chart (Part 1 of 4). 1987–88

SYMPTOM MESSAGE	POSSIBLE CAUSE	REPAIR ACTION
CHK SYSTEM GND (Check System Ground)	Open circuit 150 between Control Module and ground point	Check for open in BLK wire (150) from terminal 34 of the Control Module to terminal B of Connector C400. Check for open in BLK wire (150) from terminal 20 of the Control Module to terminal B of Connector C400.
LAT ACCEL SW OPEN (Lateral Acceleration Switch Open)	Open Lateral Acceleration Switch (Switch should be closed) Open Connector at Lateral Acceleration Switch Open circuit 919 or 920 connecting the Lateral Acceleration Switch and the Control Module	Do Test E: Lateral Acceleration Switch Test. Check that connector is firmly seated. Check ORN wire (911) between terminal 1 of Control Module and Lateral Acceleration Switch. Check ORN/BLK wire (920) between terminal 13 of Control Module and Lateral Acceleration Switch.
CHK BRK LIGHT SW (Check Brake Light Switch)	Blown STOP/HAZ Fuse Condition may be due to inoperative, misaligned or permanently closed Brake Light Switch. Open in circuit 20 between Brake Switch and Control Module	Check operation of Hazard Flasher to check fuse. Check for proper operation of Stop Lights. If Stop Lights don't work, replace or realign Brake Switch. Check that Connectors C209 and C400 are properly mated. Check for an open in the LT BLU wire (20).
CHK IGNITION SW (Check Ignition Switch)	Ignition Switch not grounding terminal 15 at the Control Module in Bulb Test and Start Open in circuit 917 between Ignition Switch and Control Module	Check for proper operation of Ignition Switch in Bulb Test and Start. If switch does not operate, replace switch. Check that connectors C210 and C400 are properly mated. Check for an open in the BLK wire (917).

Fig. 14 Anti-lock system symptom directory chart (Part 2 of 4). 1987–88

Tester J35890 using the cable.

3. Follow manufacturers instructions for use of ABS Tester J35890-20.
4. Turn ignition switch to Run position to apply power to tester. Power-up messages will be displayed by tester in rapid sequence.
5. The ABS tester will display one or more messages. These messages are listed in SYMPTOM MESSAGE column of Symptom Directory chart, **Figs. 13 through 15.** Perform REPAIR ACTION as described in Symptom Directory chart by following pin-point tests, **Figs. 16 through 31. Check tester manufacturer operating instructions for a description of test sequences and display messages.**
6. After condition indicated by tester has been repaired, repeat entire test. This will reveal any other malfunction that may exist which has not been corrected.

SYSTEM SERVICE

Before performing any service procedure on the ABS system, the following precautions should be followed:

1. The control module must be disconnected before performing any electric welding on these vehicles.
2. Use caution not to subject control module to a maximum heat of 203°F for brief periods, or a maximum of 184°F for longer than 2 hours.
3. After replacement of any ABS component, the entire ABS system should be tested.
4. Ensure inlet and outlet brake lines are properly routed.
5. Do not use a fast charger for starting engine.
6. Never disconnect battery from vehicle when engine is running.
7. Do not connect or disconnect control module electrical connector with ignition switch in On position. Ensure all harness connectors are securely connected.
8. No screws on the modulator valve should be loosened. If screws are loosened, it is no longer possible to get brake circuit leak-tight.
9. When replacing modulator valve, it should be removed through access panel in rear storage compartment. Ensure vehicle interior and exterior are protected to avoid damage from brake fluid spillage. Do not remove bottom of storage compartment to replace modulator valve.
10. Ensure rear storage compartment is wiped clean after replacing modulator valve.
11. Do not support suspension components by wheel speed sensor wires.
12. Wheel speed sensors are a tight fit into knuckle, but should be pushed in by hand. Do not hammer sensors into position.
13. Each replacement wheel speed sensor is identified with a white tag, located approximately 3/4 inch from neck of sensor, labeled L (left), or R (right).
14. Tire and wheel assemblies should be removed when replacing wheel speed sensors.
15. Wheel speed sensors should be given an anti-corrosion coating, No. 9981128 or equivalent, before installation to prevent corrosion. Do not use grease.
16. Use caution not to use silicone type brake fluid.

SYMPTOM MESSAGE	POSSIBLE CAUSE	REPAIR ACTION
CHK LF SPD SENS (Check Left Front Speed Sensor)	Open Circuit in LF Wheel Speed Sensor	Do Test F.
CHK RF SPD SENS (Check Right Front Speed Sensor)	Open Circuit in RF Wheel Speed Sensor	Do Test F.
CHK LR SPD SENS (Check Left Rear Speed Sensor)	Open Circuit in LR Wheel Speed Sensor	Do Test F.
CHK RR SPD SENS (Check Right Rear Speed Sensor)	Open Circuit in RR Wheel Speed Sensor	Do Test F.
CHK ANTILOCK LT (Check Antilock Light)	Antilock Bulb burned out Connector pin not mated Open in circuit 915 or 639 between INJ 1 Fuse, Indicator and Control Module	Replace bulb. Check that connectors C210 and C400 are securely mated. Check PNK/BLK wire (639) and WHT wire (915) for an open.
CHK MOTOR RELAY	Pump Motor Relay not closing Open circuit to Pump Motor Relay Coil or Pump ON input	Do Test G: Pump Motor Relay Test. Check for an open in the GRN/BLK wire (913). Check for open in the BLU/BLK wire (912).
CHK PUMP MOTOR	Pump Motor inoperative Open Ground Circuit from Pump Motor to ground	Do Test G, Table 2. Check BLK wire (150) from case of Modulator Valve to Ground Check that connector C448 is mated correctly.
CHK SPD SENS CKT (Check Speed Sensor Circuit)	Damaged Wheel Speed Sensor	Check area of sensor for damage.

Fig. 14 Anti-lock system symptom directory chart (Part 3 of 4). 1987–88

SYMPTOM MESSAGE	POSSIBLE CAUSE	REPAIR ACTION
CHK SPD SENS CKT	Damaged Wheel Speed Sensor	Check area of sensor for damage
CHK TOOTHED WHL	Missing teeth in toothed wheel Misalignment of Wheel Speed Sensor	Check toothed wheel for missing or damaged teeth. Check sensor for damage. Sensor should not contact tootned wheel
CHK VALVE RELAY	Open circuit to Solenoid Relay Coil or Solenoid Monitor Input Solenoid Relay inoperative	Check for an open in LT BRN/BLK (918) wire. Check for an open in DK GRN/WHT (922) wire between Modulator Valve Terminal 4 and Control Module Terminal 32 Do Test C: Solenoid Relay Test.
CHECK HYDRAULICS	Faulty Hydraulic Fluid Connection Solenoid Valve stuck in by-pass position	Check that all Brake Fluid connections are tight and that no fluid is leaking. Replace Modulator Valve.
CHK HYDRA WIRING (Check Hydraulic Wiring)	Wiring to Modulator Valve or wiring inside Modulator Valve	Check that Modulator Valve harness connector is properly mated. Check that the Solenoid and Pump Motor relays are properly seated.

Fig. 14 Anti-lock system symptom directory chart (Part 4 of 4). 1987–88

SYMPTOM DIRECTORY

SYMPTOM MESSAGE	POSSIBLE CAUSE	REPAIR ACTION
No Power-up Messages	No Battery voltage to ABS Tester Open ABS Ground Circuit	Do Test A Repair Circuit (150) from Control Module Terminal 20 to Splice S224 (see schematic)
LOW BATTERY	Battery not fully charged	Charge Battery
LF VALVE OPEN (Left Front Valve Open) (In Addition to LF VALVE OPEN, check for display of RF VALVE OPEN and REAR VALVE OPEN.)	Open Circuit in Left Front Wheel Solenoid Valve Open Wire between Valve and Control Module	Do Test B Check for open in BLK/YEL (921) wire (see schematic)
RF VALVE OPEN (Right Front Valve Open)	Open Circuit in Right Front Wheel Solenoid Valve Open Wire between Valve and Control Module	Do Test B Check for open in BLK/RED (914) wire (see schematic)
REAR VALVE OPEN	Open Circuit in Rear Wheel Solenoid Valve Open Wire between Valve and Control Module	Do Test B Check for open in DK BLU/WHT (916) wire (see schematic)
LF VALVE OPEN, RF VALVE OPEN, REAR VALVE OPEN	Open Solenoid Relay Contacts Open ground wire between Terminal 8 of the Modulator Valve and G320	Do Test C (Table 3) Check for open in BLK (150) wire to ground (see schematic)
CHK ABS DIODE	Open or shorted Antilock Brake Diode Open in circuit 915 or 922	Do Test D Check for open in WHT (915) wire between Diode and Terminal 29, or DK GRN/WHT (922) wire between Antilock Brake Diode and Terminal 32 of Control Module (see schematic)
CHK CONT MOD GRN (Check Control Module ground)	Open in circuit 150 between the Control Module and the ground point (see schematic)	Check for open in BLK (150) wire from terminal 20 of the Control Module to Terminal B of Connector C400 Check for open in BLK (150) wire from Terminal 10 of the Control Module to Terminal B of Connector C400

Fig. 15 Anti-lock system symptom directory chart (Part 1 of 4). 1989

SYMPTOM DIRECTORY (Continued)

SYMPTOM MESSAGE	POSSIBLE CAUSE	REPAIR ACTION
CHK SYSTEM GND (Check System Ground)	Open circuit 150 between Control Module and ground point	Check for open in BLK (150) wire from Terminal 34 of the Control Module to Terminal B of Connector C400 (see schematic)
LAT ACCEL SW OPEN (Lateral Acceleration Switch Open)	Open Lateral Acceleration Switch (Switch should be closed)	Do Test E
	Open Connector at Lateral Acceleration Switch	Check that connector is firmly seated
	Open circuit 924, or 920 connecting the Lateral Acceleration Switch and the Control Module	Check GRY/BLK (984) wire between Terminal 1 of Control Module and Lateral Acceleration Switch (see schematic)
		Check ORN/BLK and DK BLU wires between Terminal 13 of Control Module and Lateral Acceleration Switch (see schematic)
CHK BRK LIGHT SW (Check Brake Light Switch)	Blown STOP/HAZ Fuse	Check operation of Hazard Flasher to check fuse
	Condition may be due to inoperative, misaligned or permanently closed Brake Light Switch	Check for proper operation of Stop Lights
		If Stop Lights don't work, replace or realign Brake Switch
	Open in circuit 20 between Brake Switch and Control Module	Check that Connectors C209 and C400 are properly mated
		Check for an open in the LT BLU (20) wire (see schematic)
CHK IGNITION SW (Check Ignition Switch)	Ignition Switch not grounding Terminal 15 at the Control Module in Bulb Test and Start	Check for proper operation of Ignition Switch in BULB TEST and START
		If switch does not operate, replace switch
	Open in circuit 251 between Ignition Switch and Control Module	Check that connectors C210 and C400 are properly mated
		Check for an open in the BRN (251) wire (see schematic)

Fig. 15 Anti-lock system symptom directory chart (Part 2 of 4). 1989

SYMPTOM MESSAGE	POSSIBLE CAUSE	REPAIR ACTION
CHK LF SPD SENS (Check Left Front Speed Sensor)	Open Circuit in LF Wheel Speed Sensor	Do Test F
CHK RF SPD SENS (Check Right Front Speed Sensor)	Open Circuit in RF Wheel Speed Sensor	Do Test F
CHK LR SPD SENS (Check Left Rear Speed Sensor)	Open Circuit in LR Wheel Speed Sensor	Do Test F
CHK RR SPD SENS (Check Right Rear Speed Sensor)	Open Circuit in RR Wheel Speed Sensor	Do Test F
CHK ANTILOCK LT (Check Antilock Light)	Antilock Indicator Bulb burned out	Replace bulb
	Connector pin not mated	Check that connectors C210 and C400 are securely mated
	Open in circuit 915 or 639 between INJ 1 Fuse, Indicator and Control Module	Check PNK/BLK (639) wire and WHT (915) wire for an open
CHK MOTOR RELAY	Pump Motor Relay not closing	Check for an open in the BLK/LT GRN (913) wire (see schematic)
	Open circuit to Pump Motor Relay Coil or Pump ON input	Check for open in the BLK/LT BLU (985) wire (see schematic)
		Do Test G
CHK PUMP MOTOR	Pump Motor inoperative	Do Test G (Table 2)
	Open Ground Circuit from Pump Motor to ground	Check BLK (150) wire from Return Pump Motor to Ground (see schematic)
		Check that connector C448 is mated correctly (see schematic)
CHK SPD SENS CKT (Check Speed Sensor Circuit)	Damaged Wheel Speed Sensor	Check area of sensor for damage

Fig. 15 Anti-lock system symptom directory chart (Part 3 of 4). 1989

SYMPTOM DIRECTORY (Continued)

SYMPTOM MESSAGE	POSSIBLE CAUSE	REPAIR ACTION
CHK TOOTHED WHL	Missing teeth in toothed wheel	Check toothed wheel for missing or damaged teeth Check sensor for damage. Sensor should not contact toothed wheel
CHK VALVE RELAY	Solenoid Relay inoperative	Do Test C Check for an open in LT GRN/BLK (918) wire (see schematic)
	Open circuit to Solenoid Relay Coil or Solenoid Monitor Input	Check for an open in DK GRN/WHT (922) wire between Modulator Valve Terminal 4 and Control Module Terminal 32 (see schematic)
CHK HYDRAULICS	Faulty Hydraulic Fluid Connection	Check that all Brake Fluid connections are tight and that no fluid is leaking
	Solenoid Valve stuck in by-pass position	Replace Modulator Valve
CHK HYDRA WIRING (Check Hydraulic Wiring)	Wiring to Modulator Valve or wiring inside Modulator Valve	Check that Modulator Valve harness connector is properly mated Check that the Solenoid and Pump Motor relays are properly seated

Note: Use J-35592 Pinout Box for all circuit checks. Do not probe ABS Harness Connector directly.

Fig. 15 Anti-lock system symptom directory chart (Part 4 of 4). 1989

A: MODULE RELAY TEST

Measure: VOLTAGE
At: MODULE RELAY CONNECTOR (Disconnected)
Conditions:
• Ignition Switch: RUN

Measure Between	Correct Voltage	For Diagnosis
30 (ORN/BLK) & Ground	Battery	See 1.
15 (PNK/BLK) & Ground	Battery	See 2.
15 (PNK/BLK) & 31 (BLK)	Battery	See 3.

- If all voltages are correct, but tester will not power-up, check the ORN wire (911) for an open. If ORN wire (911) is good, replace the Module Relay.

1. Check Brake Fuse and ORN/BLK wire (440) for an open between Module Relay terminal 30 and Brake Fuse. If wire and Brake Fuse are good, refer to Power Distribution.
2. Check Gages Fuse and PNK/BLK wire (39) for an open between Module Relay terminal 15 and Gages Fuse.
3. Check BLK wire (150) for an open between Module Relay terminal 31 and ground at ground G230.

Fig. 16 Module relay test. 1986

A: MODULE RELAY TEST

Measure: VOLTAGE
At: MODULE RELAY CONNECTOR (Disconnected)
Conditions:
• Ignition Switch: RUN

Measure Between	Correct Voltage	For Diagnosis
30 (ORN/BLK) & Ground	Battery	See 1.
15 (PNK/BLK) & Ground	Battery	See 2.
15 (PNK/BLK) & 31 (BLK)	Battery	See 3.
• Module Relay Connected		
87 (ORN) & 31 (BLK)	Battery	See 4

- If all voltages are correct, but Tester does not power-up check ORN (911) wire for an open.

1. Check Brake Fuse and ORN/BLK wire (440) for an open between Module Relay terminal 30 and Brake Fuse. If wire and Brake Fuse are good, refer to Power Distribution.
2. Check Gages Fuse and PNK/BLK wire (39) for an open between Module Relay terminal 15 and Gages Fuse.
3. Check BLK wire (150) for an open between Module Relay terminal 31 and ground
4. Replace Module Relay.

Fig. 17 Module relay test. 1987–88

A: MODULE RELAY TEST

Measure: VOLTAGE
At: MODULE RELAY CONNECTOR (Disconnected)
Condition:
• Ignition Switch: RUN

Measure Between	Correct Voltage	For Diagnosis
1 (ORN/BLK) & Ground	Battery	See 1
3 (PNK/BLK) & Ground	Battery	See 2
3 (PNK/BLK) & 5 (BLK)	Battery	See 3
• Module Relay Connected		
4 (GRY/BLK) & 5 (BLK)	Battery	See 4

- If all voltages are correct, but tester will not power-up check GRY/BLK (984) wire for an open between Module Relay Terminal 4 and the Control Module Terminal 1 (see schematic).

1. Check BRAKE Fuse and ORN/BLK (440) wire for an open between Module Relay Terminal 1 and BRAKE Fuse. If wire and BRAKE Fuse are good, refer to Section 8A-10, Power Distribution.
2. Check GAGES Fuse and PNK/BLK (39) wire for an open between Module Relay Terminal 3 and GAGES Fuse.
3. Check BLK (150) wire for an open between Module Relay Terminal and ground (see schematic).
4. Replace Module Relay.

Fig. 18 Module relay test. 1989

B: SOLENOID VALVE RESISTANCE TEST

Measure: RESISTANCE (200 ohm scale)
At: MODULATOR VALVE (Connector Disconnected)

Measure Between	Correct Resistance	For Diagnosis
1 & 4	0.8 to 1.5 ohms	See 1
3 & 4	0.8 to 1.5 ohms	See 1
5 & 4	0.8 to 1.5 ohms	See 1

- If the actual measured values of resistance differ by more than 0.1 ohms, replace the Modulator Valve.
- If measured values are correct, return to Symptom Directory.

1. Replace Modulator Valve.

Fig. 19 Solenoid valve resistance test. 1986–89

C: SOLENOID RELAY TEST (TABLE 1)

Measure: VOLTAGE
At: SOLENOID RELAY CONNECTOR (Relay Disconnected)
Conditions:
- **Ignition Switch: RUN**

Measure Between	Correct Voltage	For Diagnosis
86 & Ground	Battery	See Table C2.
87 & Ground	Battery	See Table C2.

- If voltages are correct, go to Table C3.

Fig. 20 Solenoid relay test (Part 1 of 3). 1986–88

C: SOLENOID RELAY TEST (TABLE 3)

Measure: RESISTANCE
At: MODULATOR VALVE (Disconnected)
Conditions:
- **Ignition Switch: OFF**
- **Solenoid Relay Disconnected**
- **Modulator Valve Connector Disconnected**

Measure Between Sol. Relay Conn. & Mod. Valve Conn.	Correct Resistance	For Diagnosis
85 & 2	0 ohms	See 1.
30 & 4	0 ohms	See 1.
87A & 8	0 ohms	See 1.

- If voltages in Table C1 and resistances above are correct, but Solenoid Valves do not operate, replace the Solenoid Relay.

1. Replace Modulator Valve.

Fig. 20 Solenoid relay test (Part 2 of 3). 1986–88

C: SOLENOID RELAY TEST (TABLE 2)

Measure: VOLTAGE
At: MODULATOR VALVE HARNESS CONNECTOR (Disconnected)
Conditions:
- **Ignition Switch: RUN**

Measure Between	Correct Voltage	For Diagnosis
6 (RED) & Ground	Battery	See 1.
10 (ORN) & Ground	Battery	See 2.

- If voltages are correct, but were not present in Table C1, replace the Modulator Valve.

1. Check RED wire (2) for an open. If wire is good, check Fusible Link J.
2. Check ORN wire (911) back to Module Relay for an open.

Fig. 20 Solenoid relay test (Part 3 of 3). 1986–88

C: SOLENOID RELAY TEST (TABLE 1)

Measure: VOLTAGE (20 Volt DC Scale)
At: SOLENOID RELAY CONNECTOR (Relay Disconnected)
Condition:
- **Ignition Switch: RUN**

Measure Between	Correct Voltage	For Diagnosis
86 & Ground	Battery	See Table 2
87 & Ground	Battery	See Table 2

- If voltages are correct, go to Table 3.

Fig. 21 Solenoid relay test (Part 1 of 3). 1989

C: SOLENOID RELAY TEST (TABLE 2)

Measure: VOLTAGE (20 Volt DC Scale)
At: MODULATOR VALVE HARNESS CONNECTOR (Disconnected)
Condition:
- **Ignition Switch: RUN**

Measure Between	Correct Voltage	For Diagnosis
6 (RED) & Ground	Battery	See 1
10 (GRY/BLK) & Ground	Battery	See 2

- If voltages are correct, but were not present in Table 1, replace the Modulator Valve.

1. Check RED (2) wire for an open. If wire is good, check Fusible Link J (see schematic).
2. Check GRY/BLK (984) wire back to Module Relay for an open (see schematic).

Fig. 21 Solenoid relay test (Part 2 of 3). 1989

C: SOLENOID RELAY TEST (TABLE 3)

Measure: RESISTANCE
At: MODULATOR VALVE (Disconnected)
Conditions:
- **Ignition Switch: OFF**
- **Solenoid Relay: DISCONNECTED**
- **Modulator Valve Connector: DISCONNECTED**

Measure Between Sol. Relay Conn. & Mod. Valve Conn.	Correct Resistance	For Diagnosis
85 & 2	0 Ohms	See 1
30 & 4	0 Ohms	See 1
87A & 8	0 Ohms	See 1

- If voltages in Table 1 and resistances above are correct but tester indicates Solenoid Valves are open, replace the Solenoid Relay.

1. Replace Modulator Valve.

Fig. 21 Solenoid relay test (Part 3 of 3). 1989

D: ABS DIODE TEST

Measure: DIODE CHECK USING J34029 DIGITAL MULTIMETER
At: ABS DIODE

Measure Between	Correct Result	For Diagnosis
29 (WHT) & 32 (DK GRN/WHT)	0.4 to 0.6 Forward Direction	See 1.
	0VLD Reverse Direction	See 1.

- If results are correct, diode is operating normally.

1. If either check is incorrect, replace the ABS Diode.

Fig. 22 ABS diode test. 1986

D: ABS DIODE TEST

Measure: DIODE CONTINUITY USING DIODE CHECK IN J34029 DIGITAL MULTIMETER
At: ABS DIODE (Disconnected)

Measure Between	Correct Result	For Diagnosis
1 & 2	0.4 to 0.6 Forward Direction	See 1.
	OL Reverse Direction	See 1.

- If results are correct, diode is operating normally.

1. If either check is incorrect, replace the ABS Diode.

Fig. 23 ABS diode test. 1987–89

E: LATERAL ACCELERATION SWITCH

Measure: RESISTANCE
At: LATERAL ACCELERATION SWITCH
Conditions:
- Ignition Switch: OFF
- Switch installed on vehicle

Measure Between	Correct Resistance	For Diagnosis
1 & 2	0 ohms	See 1.
• Remove switch from vehicle and hold switch by hand in same relative position as when installed on vehicle. Quickly rotate switch assembly about 90 degrees around a vertical axis first to the right then left while making measurement again.		
1 & 2	Infinite	See 1.

- If both measurements are correct, Lateral Acceleration Switch is operating normally. Return to Symptom Directory.

1. If either measurement is incorrect, replace the Lateral Acceleration Switch.

Fig. 24 Lateral acceleration switch. 1986

E: LATERAL ACCELERATION SWITCH

Measure: RESISTANCE (200 ohm Scale)
At: LATERAL ACCELERATION SWITCH (Disconnected)
Conditions:
- Ignition Switch: OFF
- Switch removed from vehicle

Measure Between	Correct Resistance	For Diagnosis
1 & 2	0 ohms	See 1.
• Tilt switch away from horizontal position while making measurement again.		
1 & 2	OL	See 1.

- If both measurements are correct, Lateral Acceleration Switch is operating normally. Return to Symptom Directory.

1. If either measurement is incorrect, replace the Lateral Acceleration Switch.

Fig. 25 Lateral acceleration switch. 1987–88

E: LATERAL ACCELERATION SWITCH

Measure: RESISTANCE
At: LATERAL ACCELERATION SWITCH (Disconnected)
Conditions:
- Ignition Switch: OFF
- Switch removed from vehicle
- Switch held in same relative position as when installed on vehicle

Measure Between	Correct Resistance	For Diagnosis
A & B	0 ohms	See 1
• Tilt switch away from horizontal position		
A & B	OL	See 1

- If both measurements are correct, Lateral Acceleration Switch is operating normally. Return to Symptom Directory.

1. If either measurement is incorrect, replace the Lateral Acceleration Switch.

Fig. 26 Lateral acceleration switch. 1989

F: WHEEL SPEED SENSOR TEST

Measure: RESISTANCE
At: ABS CONTROL MODULE CONNECTOR PINOUT BOX (Control Module Disconnected)
Conditions:
- Ignition Switch: OFF
- J35592 Pinout Box connected to Control Module Connector

Measure Between	Correct Resistance	For Diagnosis
4 (LT BLU) & 6 (BLK/RED)	900 to 1500 ohms	See 1.
11 (DK GRN) & 21 (TAN)	900 to 1500 ohms	See 2.
8 (BLK) & 9 (RED)	900 to 1500 ohms	See 3.
24 (BRN) & 26 (WHT)	900 to 1500 ohms	See 4.

Fig. 27 Wheel speed sensor test (Part 1 of 2). 1986–88

- Bend and gently pull on wheel speed sensor wires at the wheel and sensor connectors while measuring resistance to check for intermittent opens.
- Check that the individual resistances are within 100 ohms of each other. Replace any sensor having a resistance which differs more than this amount from any other sensor.

1. Check the Left Front Wheel Speed Sensor connector at the wheel to insure that it is properly mated. Check connector C243. Check sensor wires for opens. If all wiring is good and sensor resistance is incorrect, replace the sensor.
2. Check Right Front Wheel Speed Sensor connector at the wheel to insure that it is properly mated. Check connector C243. Check sensor wires for opens. If all wiring is good and sensor resistance is incorrect, replace the sensor.
3. Check Left Rear Wheel Speed Sensor connector at the wheel to insure that it is properly mated. Check connector C243. Check sensor wires for opens. If all wiring is good and sensor resistance is incorrect, replace the Wheel Speed Sensor.
4. Check Right Rear Wheel Speed Sensor connector at the wheel to insure that it is properly mated. Check connector C243. Check sensor wires for opens. If all wiring is good and sensor resistance is incorrect, replace the sensor.

Fig. 27 Wheel speed sensor test (Part 2 of 2). 1986–88

F: WHEEL SPEED SENSOR TEST

Measure: RESISTANCE
At: CONTROL MODULE CONNECTOR PINOUT BOX (Control Module Disconnected)
Conditions:
- Ignition Switch: OFF
- J35592 Pinout Box connected to Control Module Connector

Measure Between	Correct Resistance	For Diagnosis
4 (YEL) & 6 (LT BLU)	900 to 1500 ohms	See 1
11 (DK GRN) & 21 (TAN)	900 to 1500 ohms	See 2
9 (BLK) & 8 (RED)	900 to 1500 ohms	See 3
24 (BRN) & 26 (WHT)	900 to 1500 ohms	See 4

Fig. 28 Wheel speed sensor test (Part 1 of 2). 1989

- Bend and gently pull on wheel speed sensor wires at the wheel and sensor connectors while measuring resistance to check for intermittent opens.
- Check that the individual resistances are within 100 ohms of each other. Replace any sensor having a resistance which differs more than this amount from any other sensor.

1. Check the Left Front Wheel Speed Sensor connector at the wheel to insure that it is properly mated. Check connector C242. Check sensor wires for opens. If all wiring is good and sensor resistance is incorrect, replace the sensor.
2. Check Right Front Wheel Speed Sensor connector at the wheel to insure that it is properly mated. Check connector C242. Check sensor wires for opens. If all wiring is good and sensor resistance is incorrect, replace the sensor.
3. Check Left Rear Wheel Speed Sensor connector at the wheel to insure that it is properly mated. Check connector C242. Check sensor wires for opens. If all wiring is good and sensor resistance is incorrect, replace the sensor.
4. Check Right Rear Wheel Speed Sensor connector at the wheel to insure that it is properly mated. Check connector C242. Check sensor wires for opens. If all wiring is good and sensor resistance is incorrect, replace the sensor.

Fig. 28 Wheel speed sensor test (Part 2 of 2). 1989

G: PUMP MOTOR RELAY TEST (TABLE 1)

Measure: VOLTAGE
At: PUMP MOTOR RELAY CONNECTOR (Relay Disconnected)
Conditions:
- **Ignition Switch: ON**

Measure Between	Correct Voltage	For Diagnosis
86 & Ground	Battery	See 1.
87 & Ground	Battery	See 2.

- If voltages are correct, proceed to step 2.

1. If battery voltage is not at terminal 86, check the ORN wire (911) for an open back to the Module Relay. If wire is good, do Test A, Module Relay Test.
2. If battery voltage is not present at terminal 87, check the RED wire (2) for an open between terminal 12 of the Modulator Valve and splice S223.

Fig. 29 Pump motor relay test (Part 1 of 2). 1986

G: PUMP MOTOR RELAY TEST (TABLE 2)

Measure: RESISTANCE
At: MODULATOR VALVE (Disconnected)
- **Ignition Switch: OFF**

Measure Between	Correct Resistance	For Diagnosis
• Pump Motor Relay Connected		
10 & 11	Approx. 40 ohms	See 1.
• Pump Motor Relay Removed		
30 & 9	0 ohms	See 2.
9 & Ground	Approx. 0.3 ohms	See 3.

- If measurements in step 1 and step 2 are correct, but Pump Motor Relay does not operate, replace the Pump Motor Relay.

1. Check Pump Motor Relay Coil for an open. If relay coil is good, replace the Modulator Valve.
2. Replace Modulator Valve.
3. Check for an open in Return Pump Motor or ground strap. If Return Pump Motor is good, replace the Modulator Valve.

Fig. 29 Pump motor relay test (Part 2 of 2). 1986

G: PUMP MOTOR RELAY TEST (TABLE 1)

Measure: VOLTAGE (20 Volt Scale)
At: PUMP MOTOR RELAY CONNECTOR (Relay Disconnected)
Conditions:
- **Ignition Switch: RUN**

Measure Between	Correct Voltage	For Diagnosis
86 & Ground	Battery	See Table 2.
87 & Ground	Battery	See Table 2.

- If voltages are correct, proceed to Table 3.

Fig. 30 Pump motor relay test (Part 1 of 3). 1987–88

G: PUMP MOTOR RELAY TEST (TABLE 2)

Measure: RESISTANCE
At: MODULATOR VALVE HARNESS CONNECTOR (Disconnected)
Conditions
- **Ignition Switch: RUN**

Measure Between	Correct Voltage	For Diagnosis
10 (ORN) & Ground	Battery	See 1.
12 (RED) & Ground	Battery	See 2.

- If voltages are correct, but were not present in Table G1, replace the Modulator Valve.

1. Check the ORN (911) wire for an open back to the Module Relay. If wire is good, do Test A, Module Relay Test.
2. Check the RED (2) wire for an open between terminal 12 of the Modulator Valve and splice S223

Fig. 30 Pump motor relay test (Part 2 of 3). 1987–88

G: PUMP MOTOR RELAY TEST (TABLE 3)

Measure: RESISTANCE
At: MODULATOR VALVE (Disconnected)
- **Ignition Switch: RUN**

Measure Between	Correct Resistance	For Diagnosis
• Pump Motor Relay Connected		
10 & 11	Approx. 50 ohms	See 1.
• Pump Motor Relay Removed		
30 & 9	Less than 0.1 ohms	See 2
9 & Ground	Approx. 0.3 ohms	See 3.

- If all measurements are correct, but Pump Motor Relay does not operate, replace the Pump Motor Relay.

1. Check Pump Motor Relay Coil for an open. If relay coil is good, replace the Modulator Valve.
2. Replace Modulator Valve.
3. Check for an open in Return Pump Motor or BLK (150) wire to Ground If wire is good, replace the Modulator Valve.

Fig. 30 Pump motor relay test (Part 3 of 3). 1987–88

G: PUMP MOTOR RELAY TEST (TABLE 1)

Measure: VOLTAGE (20 Volt Scale)
At: PUMP MOTOR RELAY CONNECTOR (Relay Disconnected)
Condition:
- **Ignition Switch: RUN**

Measure Between	Correct Voltage	For Diagnosis
86 & Ground	Battery	See Table 2
87 & Ground	Battery	See Table 2

- If voltages are correct, proceed to Table 3

Fig. 31 Pump motor relay test (Part 1 of 3). 1989

G: PUMP MOTOR RELAY TEST (TABLE 2)

Measure: VOLTAGE
At: MODULATOR VALVE HARNESS CONNECTOR (Disconnected)
Condition:
- **Ignition Switch: RUN**

Measure Between	Correct Voltage	For Diagnosis
10 (GRY/BLK) & Ground	Battery	See 1
12 (RED) & Ground	Battery	See 2

- If voltages are correct but were not present in Table 1, replace the Modulator Valve.

1. Check the GRY/BLK (984) wire for an open back to the Module Relay (see schematic). If wire is good, do Test A.
2. Check the RED (2) wire for an open between Terminal 12 of the Modulator Valve and splice S223 (see schematic).

Fig. 31 Pump motor relay test (Part 2 of 3). 1989

G: PUMP MOTOR RELAY TEST (TABLE 3)

Measure: RESISTANCE At: MODULATOR VALVE (Disconnected) Condition: • Ignition Switch: OFF		
Measure Between	**Correct Resistance**	**For Diagnosis**
• Pump Motor Relay Connected		
10 & 11	Approx. 50 ohms	See 1
• Pump Motor Relay Removed		
30 & 9	Less than 0.1 ohms	See 2
9 & Ground	Approx. 0.3 ohms	See 3
• If all measurements are correct but Pump Motor Relay does not operate, replace the Pump Motor Relay. 1. Check Pump Motor Relay Coil for an open. If relay coil is good, replace the Modulator Valve. 2. Replace Modulator Valve. 3. Check for an open in Return Pump Motor or BLK (150) wire to Ground (see schematic). If wire is good, replace the Modulator Valve.		

Fig. 31 Pump motor relay test (Part 3 of 3). 1989

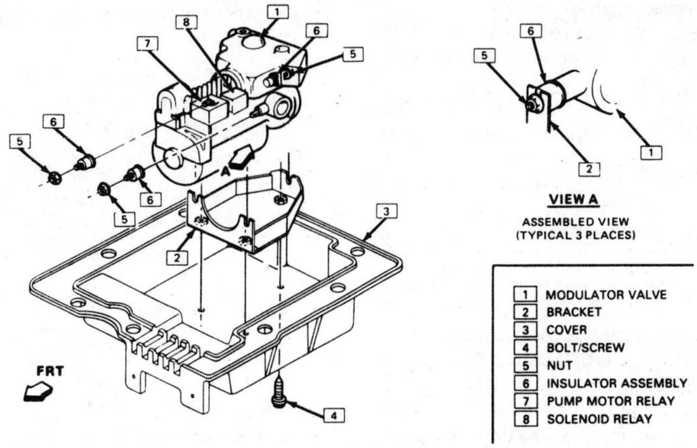

Fig. 32 Modulator valve location

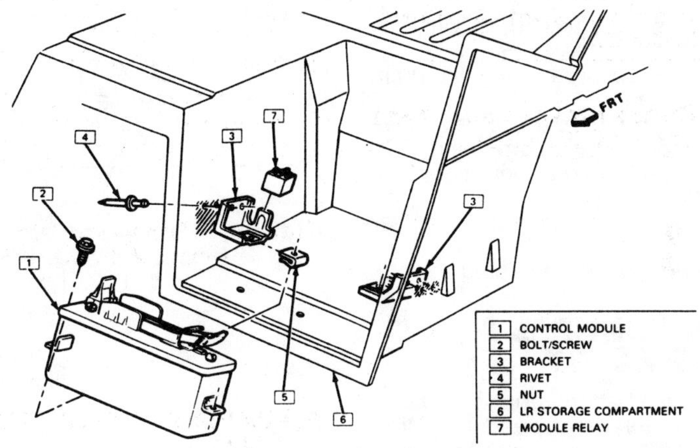

Fig. 33 Control module location

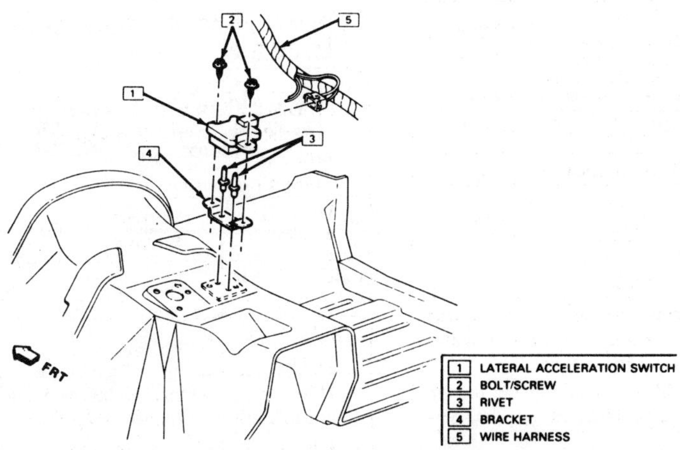

Fig. 34 Lateral acceleration switch

COMPONENT REPLACEMENT

MODULATOR VALVE

Replace modulator valve through access in rear storage compartment.

1. Disconnect battery ground cable.
2. Remove storage tray and insulation.
3. Disconnect and remove entire ABS wiring harness from storage compartment.
4. Disconnect modulator valve ground from body wiring harness.
5. Disconnect five brake pipes from modulator valve, noting location for reassembly.
6. Remove three modulator valve to bracket attaching nuts, then the modulator valve, **Fig. 32.**
7. Reverse procedure to install, noting the following:
   a. **Torque** modulator valve to bracket attaching nuts to 7 ft. lbs.
   b. Remove shipping caps from modulator valve if necessary, then connect brake pipes. **Torque** brake pipes to 13 ft. lbs.
   c. Bleed brakes as necessary.

CONTROL MODULE

1. Disconnect battery ground cable.
2. Remove storage tray and insulation.
3. Disconnect control module electrical connector, then depress spring clip located under neck of compressor to remove module relay from control module bracket.
4. Remove two control module attaching bolts, then the control module, **Fig. 33.**
5. Reverse procedure to install. **Torque** control module mounting bolts to 22 inch lbs. Perform ABS system check.

LATERAL ACCELERATION SWITCH

1. Disconnect battery ground cable.
2. Remove accessory trim plate attaching screws, then trim plate.
3. Remove console trim plate attaching screws, rotate trim plate, then discon-

nect lighter.

4. Remove A/C control head, then the lateral acceleration switch attaching bolts, **Fig. 34.**
5. Disconnect lateral acceleration switch electrical connector, then remove switch.
6. Reverse procedure to install. **Torque** lateral acceleration mounting bolts to 27 inch lbs. Perform ABS system check.

WHEEL SPEED SENSOR

1. Disconnect battery ground cable.
2. Raise and support vehicle, then remove tire and wheel assembly.
3. Remove sensor electrical connector from bracket, then disconnect sensor connector.
4. Remove sensor wire grommets from brackets, noting sensor wire routing for reassembly.
5. Remove sensor to knuckle attaching bolts, then the sensor, **Fig. 35.**
6. Reverse procedure to install, noting the following:
   a. Coat new wheel speed sensor with anti-corrosion compound No. 9981128, or equivalent prior to installation.
   b. Do not use hammer to install sensor.
   c. **Torque** sensor attaching bolts to 7 ft. lbs.

WIRING HARNESS & MODULE RELAY

1. Disconnect battery ground cable.
2. Remove storage tray and insulation.
3. Disconnect control module harness connector by depressing spring clip located under neck of connector, **Fig. 36.**
4. Remove two screws from modulator valve harness retainer clip, then disconnect harness connector.
5. Disconnect battery feed (red), wire harness connector.
6. Disconnect wheel speed sensor connectors from wiring harness connector.
7. Remove module relay from control module bracket, then disconnect modular relay harness connector.
8. Remove wiring harness, noting routing for installation.
9. Reverse procedure to install. Ensure all connectors are securely connected.

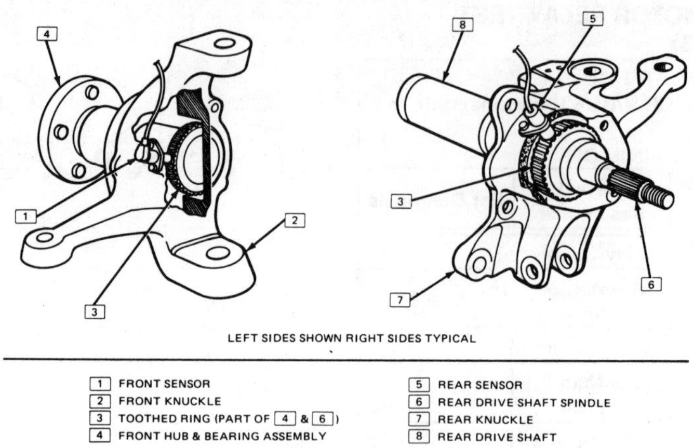

Fig. 35 Wheel speed sensors

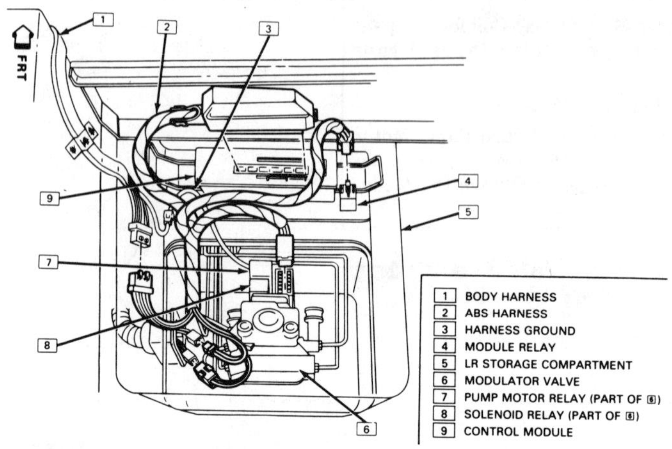

Fig. 36 Wiring harness & relays

Teves Type, 1988–89 Eldorado, Reatta, Riviera, Seville & Toronado

INDEX

OPERATING PRECAUTIONS

Some of the following procedures require that hydraulic lines, hoses and fittings be disconnected for inspection of testing. This Teves brake system uses a hydraulic accumulator which, when completely charged, contains brake fluid under high pressure. Before disconnecting any hydraulic lines, hoses or fittings, ensure the accumulator is completely depressurized as described further on. Failure to depressurize the hydraulic accumulator may result in personal injury and/or damage to painted surfaces and/or the vehicle.

Certain components in the system are not intended to be serviced individually. Do not attempt to remove or disconnect certain system components, such as the hydraulic accumulator or valve block. Only those components with removal procedures should be serviced.

Torque bolts attaching the brake caliper mounting bracket to the front and rear knuckles must be replaced any time they are removed or become loose. Tightening or loosening the torque bolts, which use encapsulated thread locking compound, is not recommended. Failure to replace the brake caliper mounting bracket torque bolts, after they have been loosened or removed, may result in loosening of the caliper to knuckle assembly, causing reduced vehicle braking ability.

COMPUTER SYSTEM SERVICE PRECAUTIONS

This Teves anti-lock brake system interfaces directly with two control computers, the Electronic Brake Control Module (EBCM) and the Body Computer Module (BCM). These modules are designed to withstand normal current draws associated with vehicle operation. However, care must be taken to avoid overloading any of the EBCM or BCM circuits. During testing of opens or shorts, do not ground or apply voltage to any of the circuits unless instructed to do so by the appropriate diagnostic procedure. These circuits should only be tested with a high impedance multimeter (J-35029-A or equivalent), or special tools described further on. Power should never be removed or applied to any control module with the ignition switch in the On position. Before removing or connecting battery cables, fuses, or connectors, always turn the ignition switch to the Off position.

GENERAL SERVICE PRECAUTIONS

If welding work is to be performed on the vehicle using an electric arc welder, the EBCM and valve block connectors should be disconnected before the welding operation begins. The EBCM and valve block connectors should never be connected or disconnected with the ignition switch in the On position. Some components of the anti-lock brake system are not serviceable separately and must be replaced as assemblies.

SYSTEM DESCRIPTION

The purpose of this Teves anti-lock brake system, is to prevent wheel lockup under heavy brake application.

The anti-lock function is desirable because a vehicle which is stopped without locking the wheels will normally stop in a shorter distance than a vehicle with locked wheels, while retaining directional stability and some steering capability. This allows the vehicle's operator to retain greater control of the vehicle during heavy brake application.

Major components of the anti-lock brake system include the electronic brake control module (EBCM), hydraulic unit, pump/motor assembly, wheel speed sensors and specific fuses and relays. Refer to **Figs. 1 through 5** for system schematic and system component locations.

COMPONENT DESCRIPTION

HYDRAULIC UNIT

This anti-lock brake system uses an integral hydraulic unit which serves the purpose of booster and master cylinder. The hydraulic unit also includes the electrical and mechanical components required to modulate brake fluid pressure in each hydraulic circuit while in anti-lock mode.

BOOSTER/MASTER CYLINDER ASSEMBLY

The booster/master cylinder assembly contains the necessary valves and pistons to develop hydraulic pressure in the wheel circuits during brake application. Power brake assist is provided through the use of pressurized brake fluid supplied by the hydraulic accumulator.

VALVE BLOCK

The valve block is attached to the right side of the booster/master cylinder assembly and contains six hydraulic wheel circuit valves. The position of these valves is controlled by the electronic brake control module (EBCM). The valve block is serviceable separate from the booster/master cylinder assembly, but should not be disassembled. A seven pin electrical connector is used to make electrical connection from the valve block to the EBCM.

Wheel Circuit Valves

Two solenoid valves are used for each of the three hydraulic wheel circuits (left front, right front and rear). Each hydraulic circuit uses an inlet valve and an outlet valve. When these valves are used in combination, each hydraulic wheel circuit may be put into the pressure increase, pressure hold and pressure reduction mode during anti-lock mode. The position of each valve is determined by the EBCM, outlet valves are normally closed and inlet valves are normally open. Valve are activated when the EBCM switches 12 volts to the appropriate valve circuit. During normal driving and braking, the wheel circuit valves are not activated.

Ground Reference Line

A ground reference circuit is included at the 7 pin valve block connector. This circuit is used by the EBCM to verify that the hydraulic unit and valve block are grounded through their mounting on the front of the dash. If ground is lost at the hydraulic unit, anti-lock function would be disabled.

MAIN VALVE

The main valve is a two position valve which is also controlled by the EBCM. Except for test purposes, this valve is opened only during anti-lock mode. When the main valve is engaged, pressurized brake fluid from the booster servo circuit is directed into the master cylinder (front brake), circuits to prevent excessive pedal travel.

PRESSURE SWITCH

The pressure switch is mounted on the hydraulic unit pump/motor assembly and serves two major functions, including pump operation control and low pressure warning.

Pump Motor Control

The pressure switch will ground the

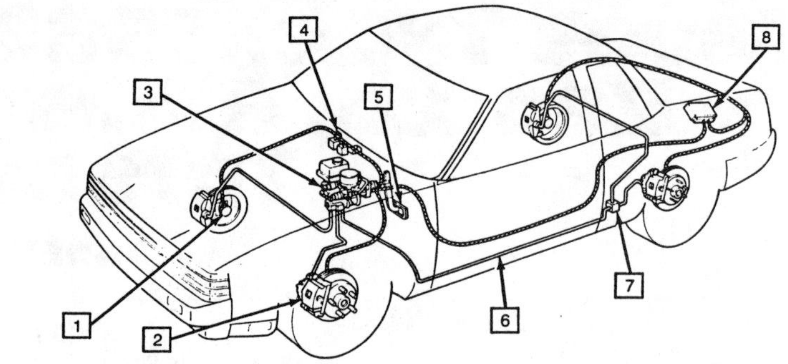

1 WHEEL SPEED SENSOR AND TOOTHED SENSOR RING (ONE AT EACH WHEEL)
2 CALIPER (ONE AT EACH WHEEL)
3 HYDRAULIC UNIT
4 RELAY BRACKET
5 PEDAL
6 REAR BRAKE PIPE
7 PROPORTIONER/TEE
8 ELECTRONIC BRAKE CONTROL MODULE (EBCM)

Fig. 1 Anti-lock brake system (ABS). 1988 Eldorado, Reatta, Riviera, Seville & Toronado

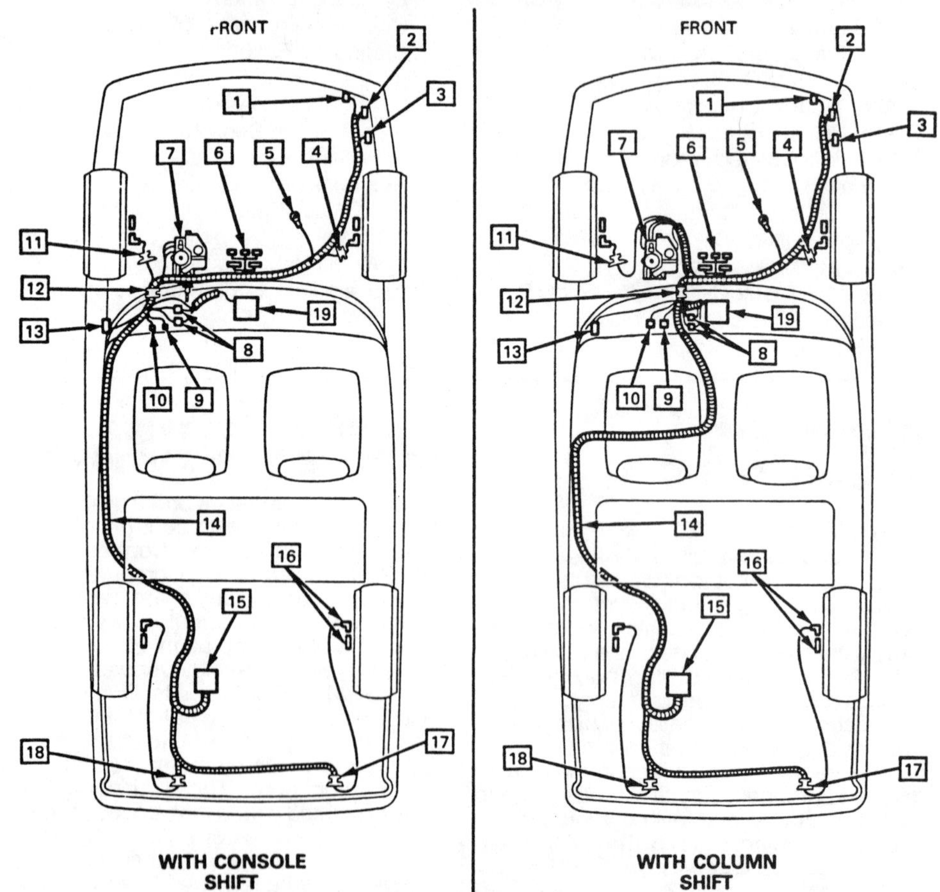

1 FRONT SHEET METAL GROUND
2 NEGATIVE JUNCTION BLOCK (ELECTRONICS GROUND)
3 POSITIVE JUNCTION BLOCK
4 RIGHT WSS WHEELHOUSE GROMMET
5 PUMP MOTOR GROUND
6 RELAY BRACKET
7 HYDRAULIC UNIT
8 ABS-TO-DASH HARNESS CONNECTORS
9 ABS DIODE
10 IN-LINE EBCM FUSE (5A)
11 LEFT WSS WHEELHOUSE GROMMET
12 ABS GROMMET
13 ASSEMBLY LINE DIAGNOSTIC LINK (ALDL) CONNECTOR
14 ABS HARNESS
15 ELECTRONIC BRAKE CONTROL MODULE (EBCM)
16 WHEEL SPEED SENSOR AND TOOTHED SENSOR RING (ONE AT EACH WHEEL)
17 RR WSS GROMMET
18 LR WSS GROMMET
19 BODY COMPUTER MODULE

Fig. 2 Component locations (top view)

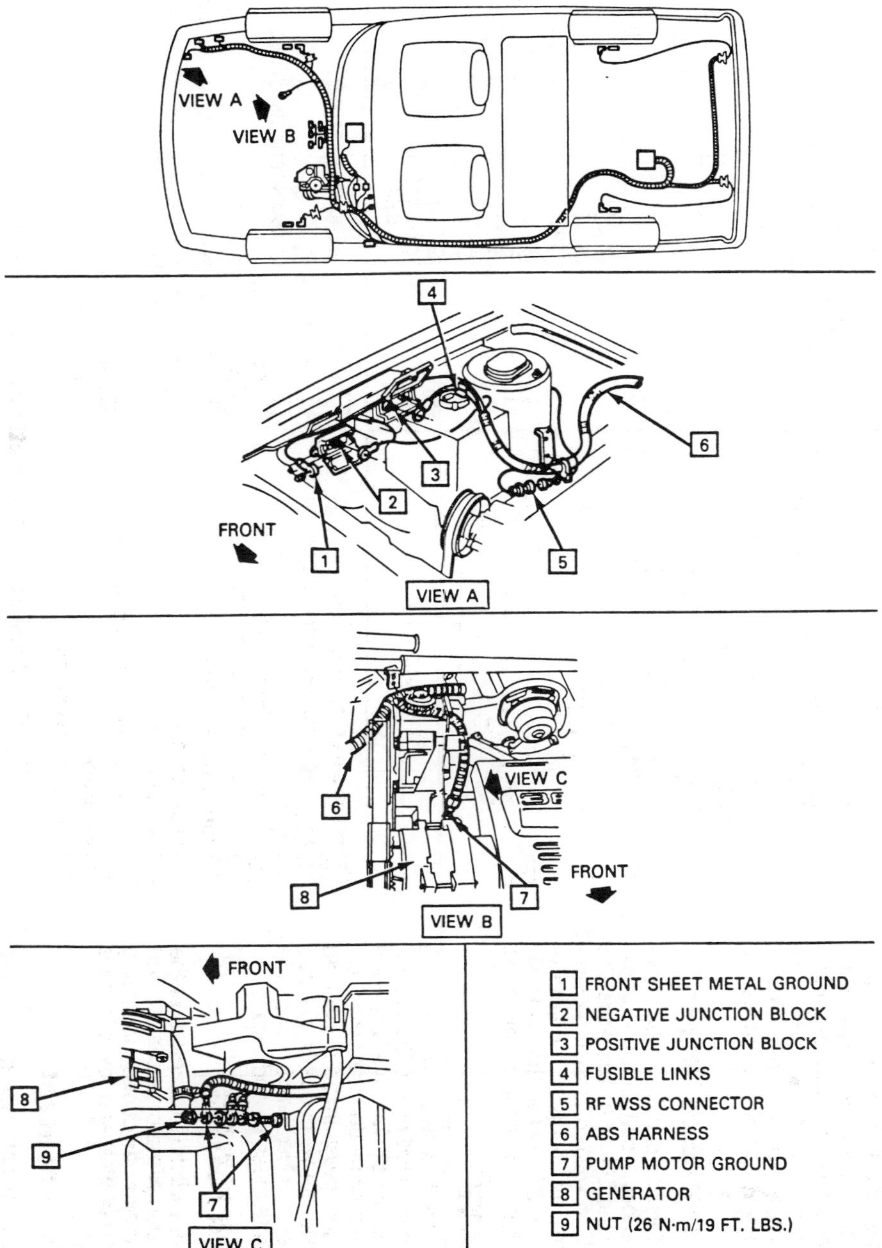

Fig. 3 Underhood component locations (Part 1 of 2)

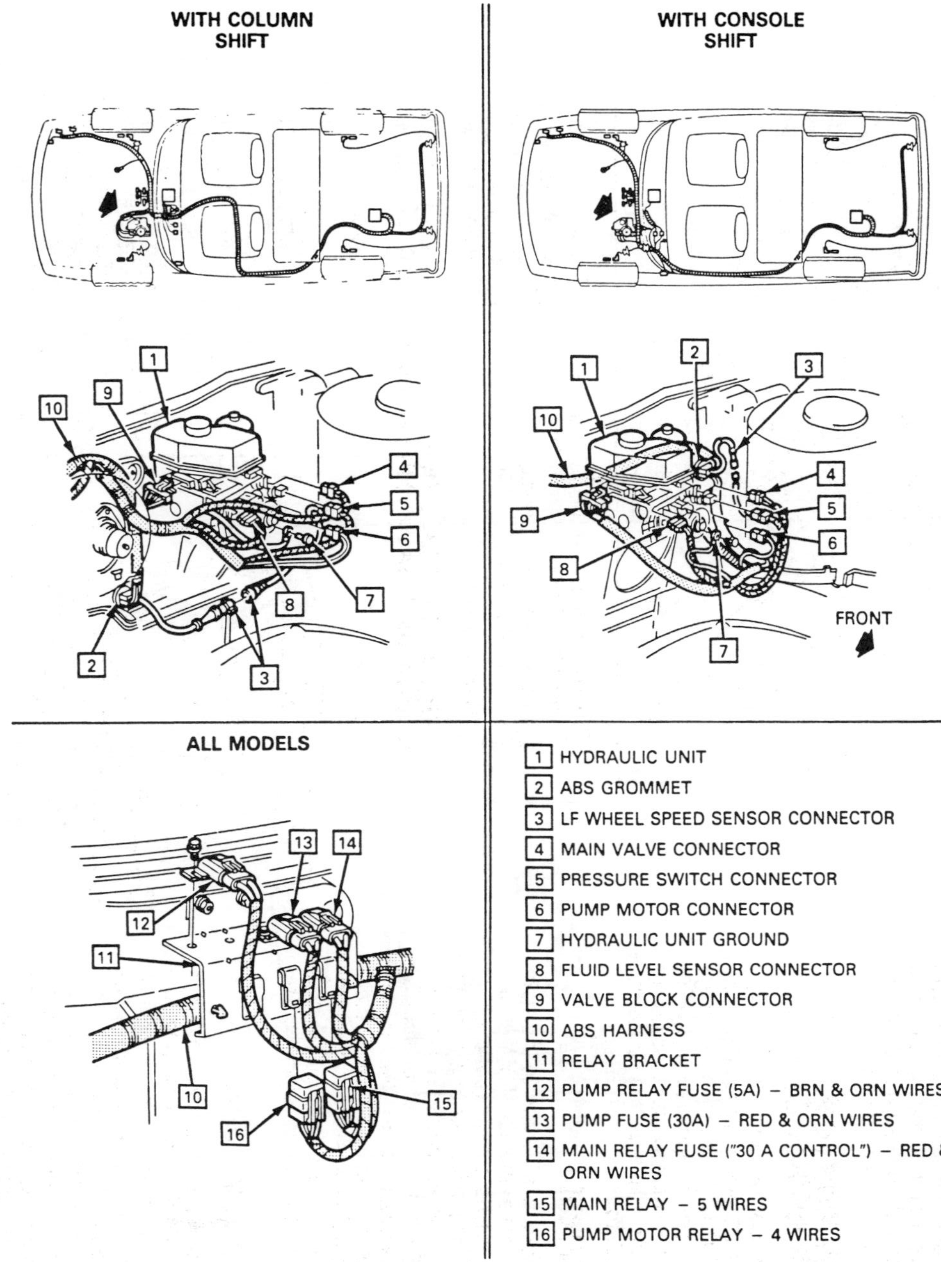

Fig. 3 Underhood component locations (Part 2 of 2)

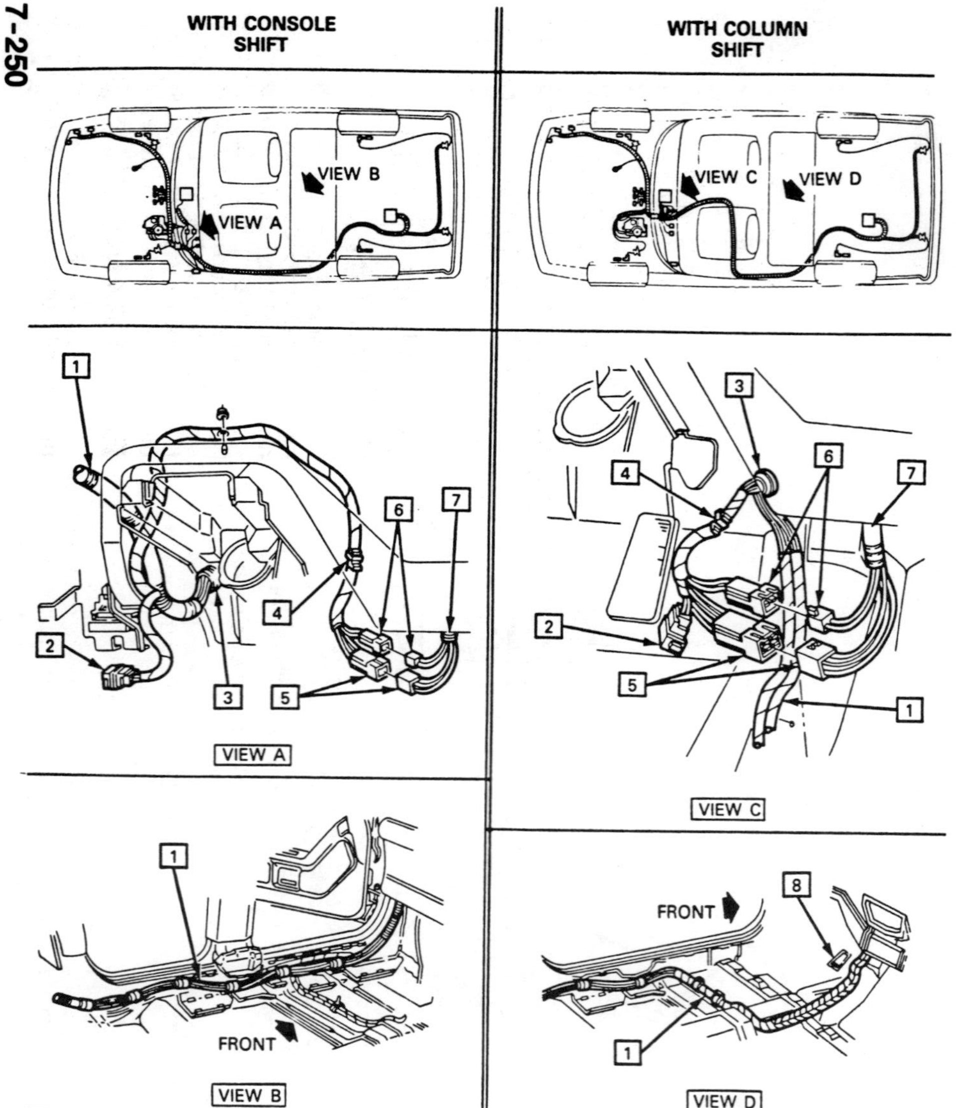

1 ABS HARNESS
2 ABS DIODE HOLDER
3 ABS GROMMET
4 5A IN-LINE EBCM FUSE
5 ABS-DASH 6-WAY CONNECTOR
6 ABS-DASH 2-WAY CONNECTOR
7 DASH HARNESS
8 ACCELERATOR PEDAL

Fig. 4 Interior component locations

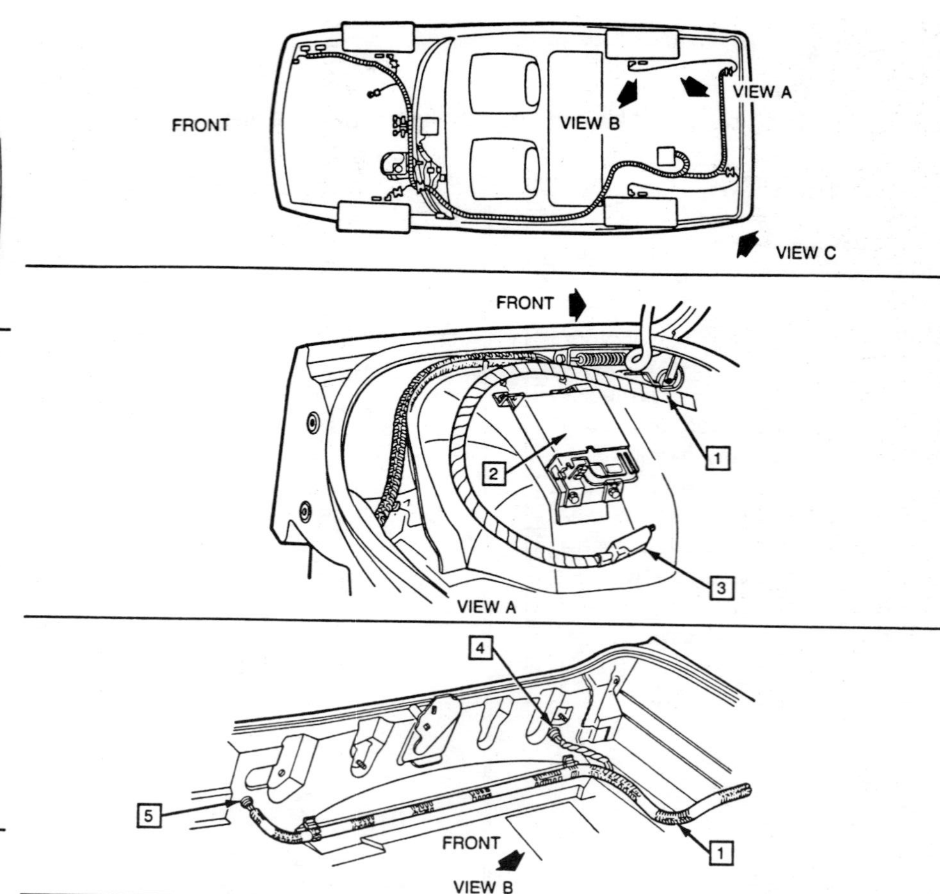

1 ABS HARNESS
2 ELECTRONIC BRAKE CONTROL MODULE (EBCM)
3 35-PIN EBCM CONNECTOR
4 LR WSS GROMMET
5 RR WSS GROMMET
6 LR WSS CONNECTOR

Fig. 5 Luggage compartment component locations

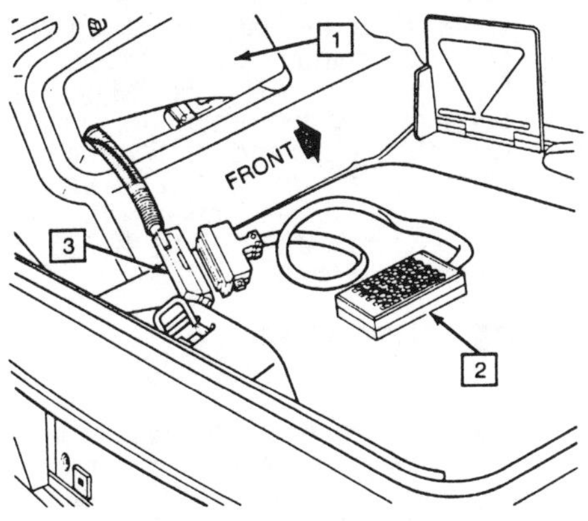

INSIDE TRUNK

1 ELECTRONIC BRAKE CONTROL MODULE (BENEATH CARPET TRIM)

2 J-35592 PINOUT BOX

3 35-PIN EBCM CONNECTOR

Fig. 6 Installing pinout box J-35592

pump motor relay coil circuit, causing the pump to run, if accumulator pressure drops below approximately 2030 psi. The switch will turn the pump motor off (by removing the relay coil circuit ground), at an accumulator pressure of approximately 2610 psi.

Warning System Control

The pressure switch also contains two separate switches to provide warning in case of low accumulator pressure. One switch is normally closed and opens when accumulator pressure drops below 1500 psi. The other switch is part of the brake warning lamp circuit and consists of a normally open switch which also closes when accumulator pressure reaches 1500 psi.

FLUID LEVEL SENSOR

A fluid level sensor is included in the hydraulic unit reservoir. The sensor consists of a float with reed switch contacts which make up two switches. One switch is normally open and the other is normally closed. Both switches change position when low fluid is detected.

PUMP/MOTOR ASSEMBLY

An integral pump/motor assembly is used to provide pressurized brake fluid to the hydraulic unit accumulator. The pump/motor assembly is mounted on the left side of the hydraulic unit, with pressure and return hoses running to the booster inlet and reservoir. Pump operation is controlled using the pump motor relay, located on the underhood relay bracket near the center of the vehicle. The pump motor relay coil is powered whenever the ignition switch is in the RUN position.

ELECTRONIC BRAKE CONTROL MODULE (EBCM)

The EBCM is a small control computer, mounted inside the trunk on the left wheelhouse, which monitors system operation and controls anti-lock function when required. The EBCM relies on inputs from the wheel speed sensors and feedback from the hydraulic unit to determine if the anti-lock brake system is operating correctly and to determine when the anti-lock function is required. The EBCM has a self-diagnostic capability, including 42 trouble codes, which may be used in the event of a system malfunction. The EBCM can display these trouble codes by entering the ABS diagnostic mode.

EBCM INPUTS

IGNITION ENABLE

The ignition enable input receives battery voltage whenever the ignition switch is in the RUN position. Upon receiving an ignition enable input, the EBCM begins its self-checks.

CRANK SENSE

A crank sense input is provide to the EBCM which goes high (12 volts), when the ignition switch is moved to the START position. When the crank sense input is high, the EBCM commands itself, internally, to shut off until engine start is completed.

BRAKE LIGHT SWITCH

The EBCM also receives a 12 volt signal from the brake light switch whenever the brake pedal is depressed. The brake light switch input alerts the EBCM when the brakes are applied and increases its sensitivity to wheel speed sensor inputs.

WHEEL SPEED SENSORS

The EBCM continuously receives wheel speed information from each of the four wheel speed sensors. The sensors one at each wheel, generate a signal when the wheel is turned. This signal is created by passing a toothed sensor ring past a stationary sensor which consists of a permanent magnet and coil. A specified air gap must be maintained between the sensor face and the toothed sensor ring.

LOW FLUID & LOW PRESSURE

The low brake fluid and low pressure inputs consist of a continuous loop circuit which runs through the fluid level sensor and the pressure switch.

DIAGNOSTIC ENABLE

The diagnostic enable input is a circuit which runs from the EBCM to the assembly line diagnostic link (ALDL), pin G. When this line is grounded, the EBCM enters the diagnostic mode.

EBCM OUTPUTS

MAIN RELAY ENABLE

The main relay provides switched battery power to the EBCM. Switched battery power is turned on only after the EBCM receives an ignition enable input signal. If the main relay is not energized, anti-lock braking is disabled. The EBCM energizes the main relay by providing 12 volts at the main relay enable output. The main relay is disabled if a fault in the anti-lock brake

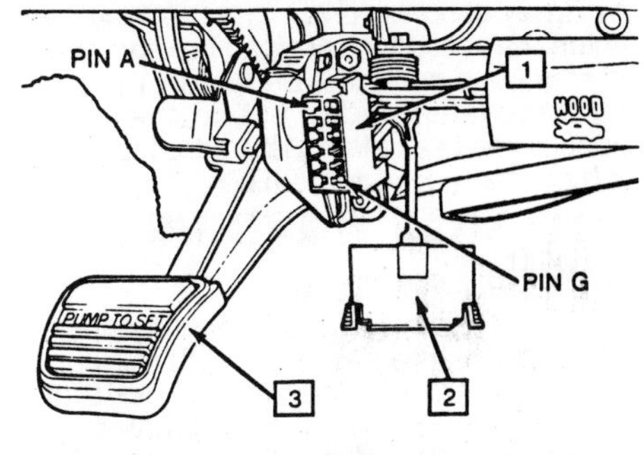

PLACE JUMPER BETWEEN PINS A AND G (OR BETWEEN G AND GROUND) TO ENTER ABS DIAGNOSTIC MODE

1 ALDL CONNECTOR

2 ALDL CONNECTOR COVER

3 PARKING BRAKE PEDAL

Fig. 7 1985–88 ABS diagnostic enable, ALDL pin G (Part 1 of 2)

system is detected.

SOLENOID VALVES

Each brake circuit is equipped with two non-serviceable solenoid valves for fluid inlet and outlet. These valves are actuated by a 12 volt signal from the EBCM when in the anti-lock mode. The valves act individually or in combination to increase, hold or decrease pressure as required to prevent wheel lockup. During normal braking (not in anti-lock mode), no voltage is provided to the valves. The inlet valve is normally open and the outlet valve is normally closed.

MAIN VALVE SOLENOID

The main valve solenoid is located in the hydraulic unit and is switched on by the EBCM only when anti-lock braking is required. During anti-lock braking, the EBCM provides 12 volts to the main valve solenoid output.

ANTI-LOCK LAMP GROUND

When a fault in the anti-lock brake system is detected by the EBCM, the anti-lock lamp output is grounded, turning on the anti-lock lamp in the instrument panel cluster.

WARNING SYSTEM OPERATION

The anti-lock brake system uses three methods of notifying the driver of a system malfunction. These include a red "BRAKE" warning lamp and an "amber "ANTI-LOCK" warning lamp, both in the instrument panel cluster. Also included are three messages which may be displayed on the driver information center (DIC).

BRAKE WARNING LAMP

The brake warning lamp in the instrument panel cluster will illuminate to warn the driver of conditions in the brake system which may result in reduced braking ability as well as to notify the driver that the parking brake is applied or not completely released. The brake lamp is controlled by the body computer module (BCM).

ANTI-LOCK WARNING LAMP

The anti-lock warning lamp is located in the instrument panel cluster and will illuminate if a malfunction in the anti-lock brake system is detected by the electronic brake control module (EBCM). The anti-lock lamp is intended only to inform the driver that a condition has been detected which results in partial or total inhibit of the anti-lock function. If the anti-lock lamp is on, normal braking will full assist is still available.

DRIVER INFORMATION CENTER (DIC) MESSAGES

Three messages related to the brake system may be displayed on the driver information center (DIC), these include:

1. Brake Fluid Is Low. This message will be commanded by the BCM upon detection of low brake fluid by the fluid level sensor in the hydraulic unit reservoir. While displaying this message, the BCM will also turn on a warning chime which will sound for approximately 30 seconds (unless it is disabled by pressing the system monitor button).
2. Low Brake Pressure. This message is displayed when a low boost pressure condition has been detected in the hydraulic unit accumulator. In order for the BCM to command this message, the low pressure condition must exist for at least 50 seconds. When the BCM turns on the Low Brake Pressure message, it will also sound a warning chime for approximately 30 seconds (unless it is disabled by pressing the system monitor button).
3. Park Brake Is On. This message will notify the driver that the parking brake has not been released. This message will only occur when the car is shifted from Park or Neutral to Reverse or a Drive range.

Illumination of the Brake warning lamp may indicate reduced braking ability. Most conditions which cause the brake lamp to light will also cause the Anti-Lock lamp to light. Illumination of the Anti-Lock lamp without the Brake warning lamp on, indicates only that anti-lock braking is no longer available. Power assisted braking without anti-lock control is still available.

DEPRESSURIZING HYDRAULIC ACCUMULATOR

The anti-lock brake system pump/motor assembly will keep the hydraulic charged to a pressure between approximately 2000-2600 psi, anytime the ignition switch is in the RUN position. The pump/motor assembly cannot run if the

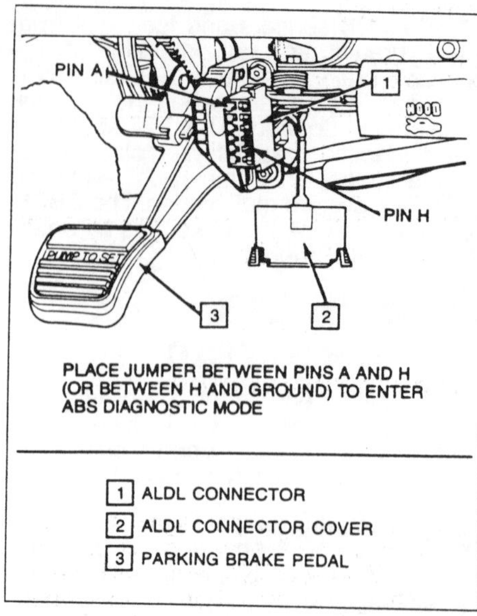

Fig. 7 1989 ABS diagnostic enable, ALDL pin H (Part 2 of 2)

ignition switch is off, or if either battery cable is disconnected. Disconnecting the pressure switch connector or the pump/motor connector from the hydraulic unit will also prevent the pump from operating. Unless otherwise specified, the hydraulic accumulator should be depressurized before working on any portion of the hydraulic system. The following procedure should be used to depressurize the hydraulic accumulator. To depressurize accumulator, proceed as follows:

1. With ignition switch turned to the Off position, pressure switch disconnected, pump motor connector disconnected or battery ground cable disconnected, pump the brake pedal a minimum of 25 times using approximately 50 lbs of force.
2. A noticeable change in pedal feel will occur then the accumulator is discharged.
3. When a definite increase in pedal effort is felt, stroke pedal a few additional times. This should remove all hydraulic pressure from the system.

REQUIRED TEST EQUIPMENT

PINOUT BOX J-35592

In order to prevent damage to the 35 pin EBCM connector, J-35592 pinout box should be installed on the EBCM connector prior to probing any circuit with the digital multimeter. Do not probe the 35 pin EBCM connector without installing J-35592 pinout box. Damage to connector terminals may occur if the pinout box is not installed, resulting in incorrect system operation. Proceed as follows to install the J-35592 pinout box:

1. The ignition switch must be in the off position when installing or removing the 35 pin EBCM connector.
2. Open trunk and remove EBCM trim cover from the left wheelwell. The cover is retained by velcro tape.
3. Disconnect the 35 pin EBCM harness connector by depressing the connector locking plate and rotating the connector toward the front of the vehicle.
4. Inspect the 35 pin connector for damage. Connect J-35592 to the 35 pin connector, **Fig. 6.**

PRESSURE GAUGE J-35604

1. Depressurize hydraulic accumulator by pumping the brake a minimum of 25 times. **Failure to depressurize the hydraulic accumulator prior to performing this operation may result in personal injury and/or damage to vehicle.**
2. Remove the pressure hose fitting from the pump body.
3. Install one of the two O-ring seals on the pressure gauge fitting and insert the gauge fitting into the pressure hose coupling.
4. Install the second O-ring seal onto the gauge fitting on the underside of the pressure hose coupling and thread the gauge fitting into the pump body. **Torque** fitting to 15 ft. lbs.
5. When removing gauge and installing pressure fitting, inspect O-ring seals for cuts or damage. Replace any cut or damaged O-ring seals.

When removing pressure gauge and adapter, ensure that the hydraulic accumulator is depressurized.

ANTI-LOCK BRAKE SYSTEM SELF DIAGNOSIS

This anti-lock brake system is equipped with a self diagnostic capability which may be used to assist in isolation of ABS faults. This feature includes 42 trouble codes which could be displayed by the EBCM through flashing of the amber "ANTI-LOCK" lamp in the instrument panel cluster. In order to access and understand any ABS trouble code which may be present, it is necessary to enter the ABS diagnostic mode and read the trouble codes using the following procedures.

DISPLAYING ABS TROUBLE CODES

The following information is to be used for reference when directed by the appropriate diagnostic procedure. Do not attempt to diagnose an ABS condition at this time. Refer to "Basic Diagnostic Procedure" before attempting diagnosis.

The EBCM will store any failure code into memory. These codes will remain stored in the EBCM until they are erased. To clear codes proceed to "Clearing ABS Codes." Disconnecting the EBCM, battery cables, or turning the ignition switch to the Off position, will not clear the trouble codes from memory.

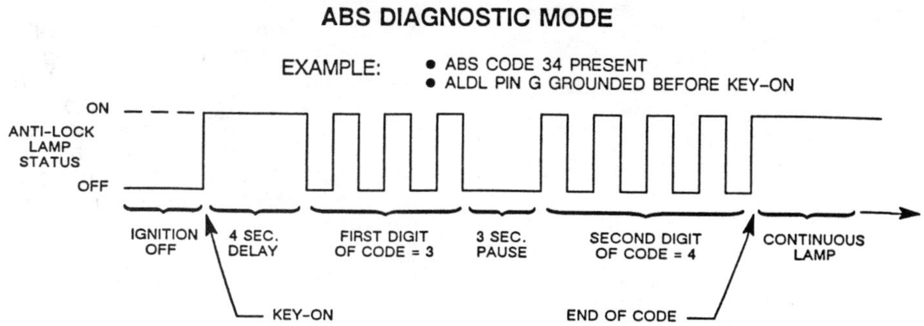

Fig. 8 ABS diagnostic code interpretation

READING ABS TROUBLE CODES

1. With ignition switch turned to the On position, allow pump/motor to charge the accumulator. If accumulator is discharged, the "BRAKE" and ""ANTI-LOCK" lamps will remain on for up to 30 seconds.
2. If the "ANTI-LOCK" lamp does not go off within 30 seconds, note that lamp did not go off, then proceed to step 3.
3. With ignition switch turned to the Off position, place a jumper wire between Assembly Line Diagnostic Link (ALDL), pins G **on 1985-88 models,** pin H, **on 1989 models** and A or between ALDL pin G, **on 1985-88 models,** pin H, **on 1989 models** and body ground, **Fig. 7.** The ALDL is located on the driver's side of the vehicle near the parking brake pedal assembly.
4. Turn ignition switch to the RUN position, then count the lamp flashes to identify the first digit. If the lamp turns on for four seconds, then turns off (and remains off), no trouble codes are present. Proceed to "ABS Functional Check," then proceed with diagnosis as directed. If the EBCM has stored any trouble codes, the "ANTI-LOCK" lamp will turn on for approximately four seconds and will then start flashing. These flashes represent the first digit of the trouble code.
5. Count lamp flashes for the second digit. The lamp will pause (remain off), for approximately three seconds after flashing the first digit. It will then flash the second digit of the trouble code. **When the ""ANTI-LOCK" lamp flashes the second digit of the trouble code, be careful to count only the number of times the lamp turns on and goes out. When the EBCM is done flashing the second digit, the "ANTI-LOCK" lamp will remain on. The last pulse, when the lamp remains on, should not be counted as a flash, Fig. 8.**
6. Record trouble codes obtained.
7. Check for additional trouble codes. Without turning the ignition switch to the off position, disconnect the jumper wire from ALDL pin G **on 1985-88 models,** pin H, **on 1989 models.** Connect the jumper again to ALDL pin. If an additional code is present, the "ANTI-LOCK" lamp will begin flashing the first digit as described previously. Count lamp flashes for the first and second digits as was done for the first code. Record the additional trouble code if present.
8. Repeat steps 6 and 7, until no additional trouble codes are displayed. Up to seven trouble codes can be displayed. The "ANTI-LOCK" lamp will remain on continuously when all trouble codes have been displayed. ABS trouble codes cannot be cleared unless all codes have been displayed.
9. After recording all ABS codes, remove jumper from the ALDL, install ALDL cover, then proceed with the "ABS Functional Check."

HISTORY & CURRENT ABS CODES

The ABS trouble codes which may be stored by the EBCM are not specifically designated as current or history codes, as are BCM and ECM codes. The anti-lock lamp, may be used to assist in differentiating between current and history codes as follows:

1. If the anti-lock lamp is on before entering the ABS diagnostic mode, at least one of the stored codes is current.
2. If the anti-lock lamp is off before entering the ABS diagnostic mode, none of the stored codes are current.
3. If more than one ABS code is stored and the anti-lock lamp is on before diagnostic mode was entered, it is impossible to determine which is a current code and which is a history code.
4. If the Anti-Lock lamp remains on continuously throughout the diagnostic procedure, see symptom diagnosis information as directed by the "ABS Functional Check." If the anti-lock lamp does not come on at all, see symptom diagnosis information as directed by the "ABS Functional Check" as well.

CLEARING ABS CODES

The ABS trouble codes should not be cleared until repairs are completed. The EBCM will not allow the ABS codes to clear unless all codes have been read.

The ABS trouble codes are cleared by driving the vehicle at a speed greater than 18 mph. If the codes did not clear, repeat the read out procedure and drive the vehicle again at a speed greater than 18 mph. If codes do not clear, additional repairs may be required.

BODY COMPUTER MODULE (BCM) TROUBLE CODES & INPUT TESTS

The vehicle self diagnostic system includes one BCM trouble code and three BCM input tests which involve the anti-lock brake system.

BCM Code B482, Low Brake Pressure

Code B482 will set if the hydraulic brake pressure input to the BCM (pin 2B1), is held low (at ground), continuously for 50 seconds. This BCM input will be pulled low by the ABS pressure switch in the event of low accumulator pressure. If code B482 sets, the low brake pressure message on the driver information center will be displayed.

BCM Input BI18, Brake Pressure

BCM Input test BI18 can be used to determine the status of the hydraulic brake pressure input at BCM pin 2B1. This input is normally held high (at battery voltage) and is pulled low (grounded), in the case of low brake pressure.

BCM Input BI21, Brake Fluid

BCM input test BI21 can be used to determine the status of the brake fluid level input at BCM pin 1C2. This input is normally held high (at battery voltage) and is pulled low (grounded), by the fluid level sensor in case of low brake fluid.

BCM Input BI22, Park Brake

BCM Input Test BI22 can be used to determine the status of the parking brake pedal switch input at BCM pin 1C5. This input is normally held high (at battery voltage) and is pulled low (grounded), by the parking brake pedal switch whenever the brake is applied.

ENTERING DIAGNOSTIC SERVICE MODE

The following information is to be used for reference when directed by the appropriate diagnostic procedure. Do not attempt to diagnosis an ABS condition starting here. The following procedure deals only with the aspects of entering the diagnostic service mode which are required for diagnosis of the anti-lock brake system.

1. Turn ignition switch to the On position. **On CRT models,** select the "Climate" screen by pressing the button in upper righthand corner.
2. **On non-CRT models,** press the Off and WARM buttons on the climate control panel simultaneously and hold until the segment check begins on the instrument panel cluster. **On CRT models,** touch Off and Warm pads simultaneously, until screen displays

"Service Mode" configuration.

3. **On non-CRT models,** the segment check will cause the instrument panel cluster to illuminate and cycle all displays. The segment check takes approximately five seconds to complete.
4. **On all models,** after the segment check is completed, any trouble codes which have been stored by the ECM, BCM, IPC, or CRT will be displayed.
   a. ECM trouble codes will be displayed first. If no ECM trouble codes are present, the "NO ECM CODES" message will appear. ECM codes are prefixed with an E. ECM codes are not relevant to ABS diagnosis.
   b. BCM trouble codes will be displayed next. If no BCM trouble codes are present the "NO BCM CODES" message will appear. BCM codes are prefixed with a B. **BCM code B482 applies to ABS diagnosis.**
   c. CRT code will be displayed next. If no CRT codes are present, "No CRT Codes will appear. CRT codes ar prefixed with a C. CRT codes are not relevant to ABS diagnosis.
5. All codes will be accompanied by a "CURRENT" or "HISTORY" designation. History indicates that the failure was not present the last time the code was tested. Current indicates the fault still exists. If BCM codes B482 is present, note if it is designated current or history before proceeding with the diagnosis procedure.
6. **On non-CRT models,** if the "LO FAN" (down arrow), button on the climate control panel, or **on CRT models,** the "NO" pad is depressed at any time during the display of the ECM or BCM diagnostic codes, the code display will be bypassed. **On non-CRT models,** if the "BI-LEVEL" button, or **on CRT models,** the "END" pad is depressed at any time during the code display, the system will exit the diagnostic service mode and resume normal vehicle operation.
7. After the trouble codes have been displayed, the computer system allows one of two vehicle systems to be selected for testing, **on non-CRT models,** and one of four vehicle systems **on CRT models.** At this time, the driver information center displays the first available system as indicated by the ECM message. Three responses are possible:
   a. Bypass ECM system test is activated by pressing the "LO FAN" (down arrow), button, **on non-CRT models,** and the "NO" pad **on CRT models,** . This should be done if ABS diagnosis is desired.
   b. Display trouble codes again by pressing the OFF button **on non-CRT models,** or the "LEVEL" pad **on CRT models.** This should be done if you need to review the code display.
   c. Exit diagnostic mode by pressing the "BI-LEV" button **on non-CRT models,** or the "END" button **on CRT models.** This will resume normal vehicle operation.

ABS VISUAL INSPECTION		
ITEM	INSPECT FOR	CORRECTIVE ACTION
PARKING BRAKE	- FULL RELEASE - PROPER SWITCH FUNCTION IF NECESSARY, UNPLUG SWITCH CONNECTOR FROM PEDAL ASSEMBLY TO VERIFY - PROPER VACUUM RELEASE SYSTEM OPERATION	- OPERATE MANUAL RELEASE LEVER TO VERIFY RELEASE - ADJUST CABLE OR REPAIR RELEASE SYSTEM AS REQUIRED - REPAIR SWITCH AS REQUIRED
MAJOR COMPONENTS • BRAKE FLUID RESERVOIR • HYDRAULIC UNIT • PUMP/MOTOR ASSEMBLY	- LOW FLUID LEVEL - EXTERNAL LEAKS - PROPER ASSEMBLY	- ADD FLUID AS REQUIRED DETERMINE CAUSE OF FLUID LOSS AND REPAIR - REPAIR LEAKS AS REQUIRED - INSTALL OR POSITION COMPONENTS PROPERLY
FUSES-ON RELAY BRACKET • PUMP MOTOR RELAY FUSE (5A) • MAIN RELAY (CONTROL) FUSE (30A) • PUMP MOTOR FUSE (30A)	- BLOWN FUSE	- REPLACE FUSE - VERIFY OPERATION - INSPECT FOR CAUSE OF FUSE FAILURE
CONNECTORS UNDERHOOD • MAIN RELAY • PUMP MOTOR RELAY • PRESSURE SWITCH • PUMP MOTOR • MAIN VALVE • VALVE BLOCK • FLUID LEVEL SENSOR • LF AND RF WHEEL SPEED SENSORS REAR OF CAR • LR AND RR WHEEL SPEED SENSORS • ELECTRONIC BRAKE CONTROL MODULE	- PROPER ENGAGEMENT - LOOSE WIRES OR TERMINALS - CORRODED OR BROKEN EYELETS	- PROPERLY ENGAGE CONNECTOR - REPAIR AS REQUIRED

Fig. 9 ABS visual inspection procedure

8. After bypassing the ECM test system, the climate control driver information center will display, the prompt BCM. This mode should be selected by depressing the "HI FAN" (up arrow), button **on non-CRT models,** or the "YES" pad **on CRT models.**
9. After entering BCM diagnostic mode. The anti-lock brake system inputs to the BCM can be monitored. The BCM input feature of diagnostic service mode is accessed as follows:
   a. After entering BCM diagnostic mode, the driver information center will display the prompt "BCM DATA." You should respond to this by pressing the "LO FAN" (down arrow), button **on non-CRT models,** or "NO" pad **on CRT models.**
   b. The driver information center will then display the prompt "BCM INPUTS." You should respond to this by pressing the "HI FAN" (up arrow), button **on non-CRT models,** or "YES" pad **on CRT models.** This will place the system in BCM input mode.
   c. The system will now show the status of selected BCM inputs, Inputs which are relevant to the anti-lock brake system include BI18, BI21 and BI22.
10. While in the input mode, the BCM can display the status of up to 18 different inputs. Each input is designated with an identifier such as BI21, meaning BCM input 21. In order to display a BCM input which is higher or lower in number than the one currently on the display, press the "HI FAN" (up arrow), button **on non-CRT models,** or the "YES" pad **on CRT models,** to increment the display. **On non-CRT models,** press the "LO FAN" (down arrow), or the "NO" pad **on CRT models,** to lower the numbered display. Each of these input displays will appear in the same format. All displays are normally HI. A warning input to the BCM would appear as a LO status indication.
11. After evaluating the ABS inputs to the BCM in conjunction with the correct diagnostic procedures, exit diagnostic service mode by depressing the "BI-LEV" button **on non-CRT models,** or the "END" pad **on CRT models.** This will return the computer system to normal operation.

BASIC DIAGNOSTIC PROCEDURE

Diagnosis of the anti-lock brake system

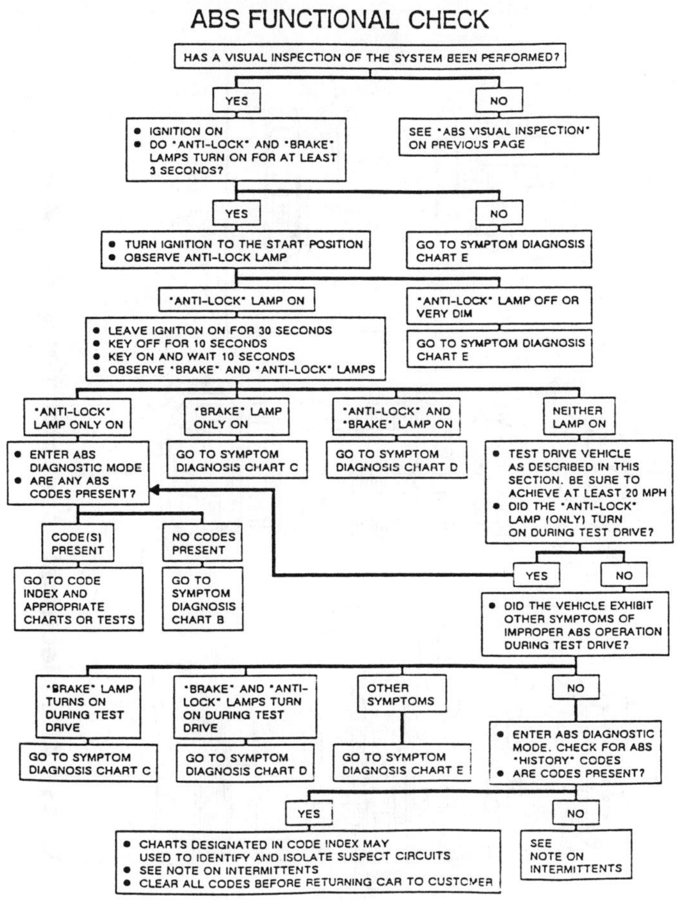

Fig. 10 ABS functional check procedure

consists of the anti-lock visual inspection, anti-lock functional check and additional checks as specified by the functional check. All diagnosis should start with the anti-lock visual inspection and proceed to the anti-lock brake system (ABS), functional check, if the condition is not resolved by the visual inspection.

NOTE ON INTERMITTENTS

Intermittent failures in the anti-lock brake system may be difficult to accurately diagnosis. The ABS trouble codes which may be stored by the EBCM are not specifically designated as "CURRENT" or "HISTORY" codes as are BCM and ECM codes. These codes, however, can be helpful in diagnosing intermittent conditions. If an intermittent condition is to be diagnosed, the ABS self-diagnostic system can be used in the following manner to help isolate the suspected circuit:

1. First, display and clear any ABS trouble codes which may be present in the EBCM.
2. Test drive the vehicle, attempting to repeat the failure condition.
3. After duplicating the condition, stop the vehicle and display any ABS trouble codes which may have been stored.
4. If no trouble codes were stored, it may become necessary to refer to "ABS Symptom Diagnosis."

SYSTEM VOLTAGE DURING DIAGNOSIS

Some diagnostic procedures require that the ignition switch remain On for extended periods of time. During this time, battery voltage may become low, causing erroneous voltage readings. If this situation occurs, connect a battery charger to the vehicle's electrical system, then apply a slow charge.

ABS VISUAL INSPECTION

Refer to **Fig. 9,** for anti-lock brake system (ABS), visual inspection procedure.

ABS FUNCTIONAL CHECK

Refer to **Fig. 10,** for anti-lock brake system (ABS), functional check procedure.

ELECTRICAL CONNECTOR IDENTIFICATION VIEWS

Refer to **Fig. 11,** for system and system component electrical connector identification views.

ABS TROUBLE CODE INDEX

Refer to **Fig. 12,** for anti-lock brake system trouble code index. **When using the ABS trouble code index, "ABS Code," "System" and "Go To Chart" are referred to. Under "Go To Chart," you will find letters and numbers from A-1 to A-15 which are the actual diagnostic charts and can be found at the top of each individual diagnostic test chart.**

ABS TROUBLE CODE DIAGNOSIS

Refer to **Figs. 13 through 27,** for anti-lock brake system trouble code diagnostic procedures.

ABS SYMPTOM DIAGNOSIS

Refer to **Figs. 28 through 56,** for anti-lock brake system symptom diagnostic procedures.

SYSTEM SERVICE

HYDRAULIC UNIT, REPLACE

1. Disconnect battery ground cable.
2. Depressurize hydraulic accumulator by depressing brake pedal a minimum of 25 times. **A noticeable change in pedal resistance will be felt when accumulator is depressurized. Failure to depressurize accumulator may result in personal injury and/or damage to vehicle and painted surfaces.**
3. Label, then disconnect all electrical connectors from hydraulic unit.
4. Remove cross-car brace.
5. Remove pump bolt, then move pump/motor assembly to the side.
6. Disconnect three hydraulic lines from valve block. Use a backup wrench to prevent pipes from twisting.
7. From inside the vehicle, disconnect pushrod from brake pedal.
8. Push dust boot past hex on pushrod, then separate push rod halves by unthreading.
9. Remove two hydraulic unit mounting bolts from the pushrod bracket, then the hydraulic unit. The front half of the pushrod will remain locked in hydraulic unit.
10. Reverse procedure to install.

PRESSURE SWITCH, REPLACE

1. Disconnect battery ground cable and depressurize hydraulic accumulator.
2. Disconnect electrical connector from pressure switch.
3. Using a suitable tool, remove pressure switch and O-ring.
4. Reverse procedure to install, noting the following:
   a. Lubricate O-ring seal with clean brake fluid.
   b. **Torque** pressure switch to 15 ft. lbs.
   c. Turn ignition switch On. Allow system to pressurize. BRAKE lamp should go out within 30 seconds and pump motor should turn off within 40 seconds.
   d. Inspect for leaks.

WHEEL SPEED SENSOR CONNECTOR

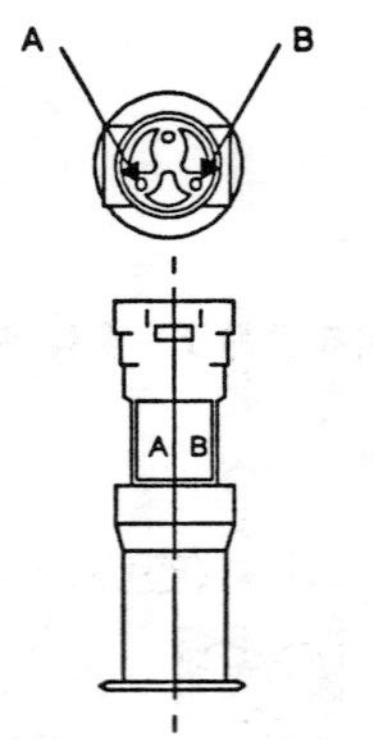

	PIN A	PIN B
RIGHT FRONT	833	872
LEFT FRONT	873	830
RIGHT REAR	883	882
LEFT REAR	885	884

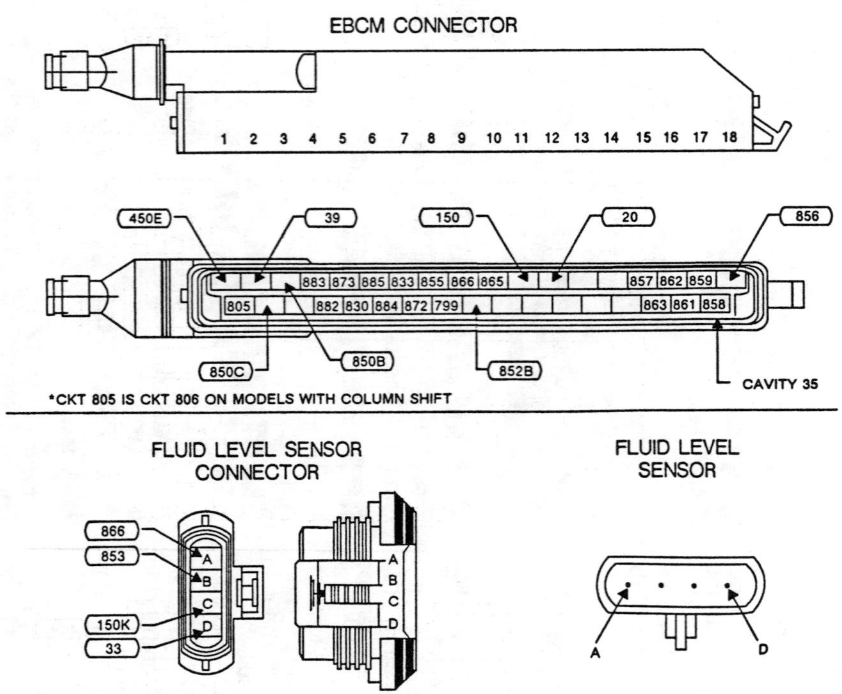

Fig. 11 ABS electrical connector views (Part 1 of 3)

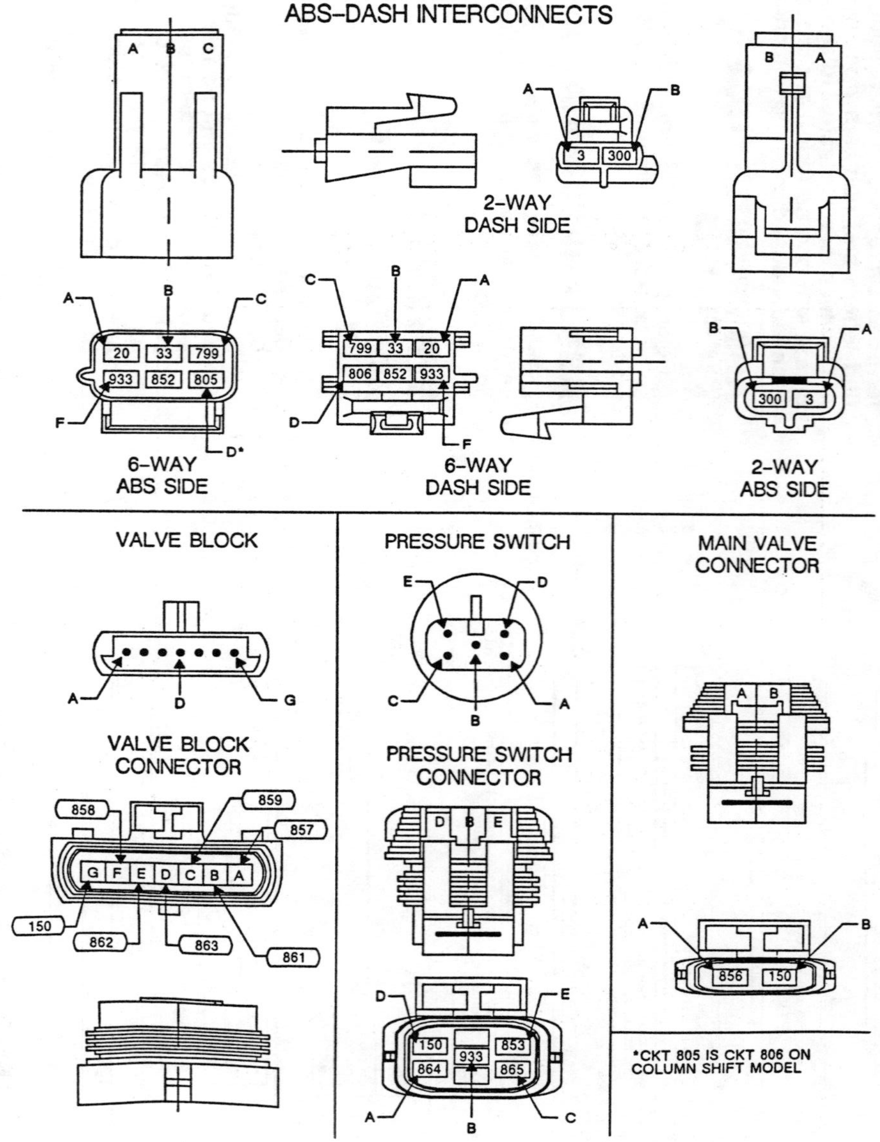

Fig. 11 ABS electrical connector views (Part 2 of 3)

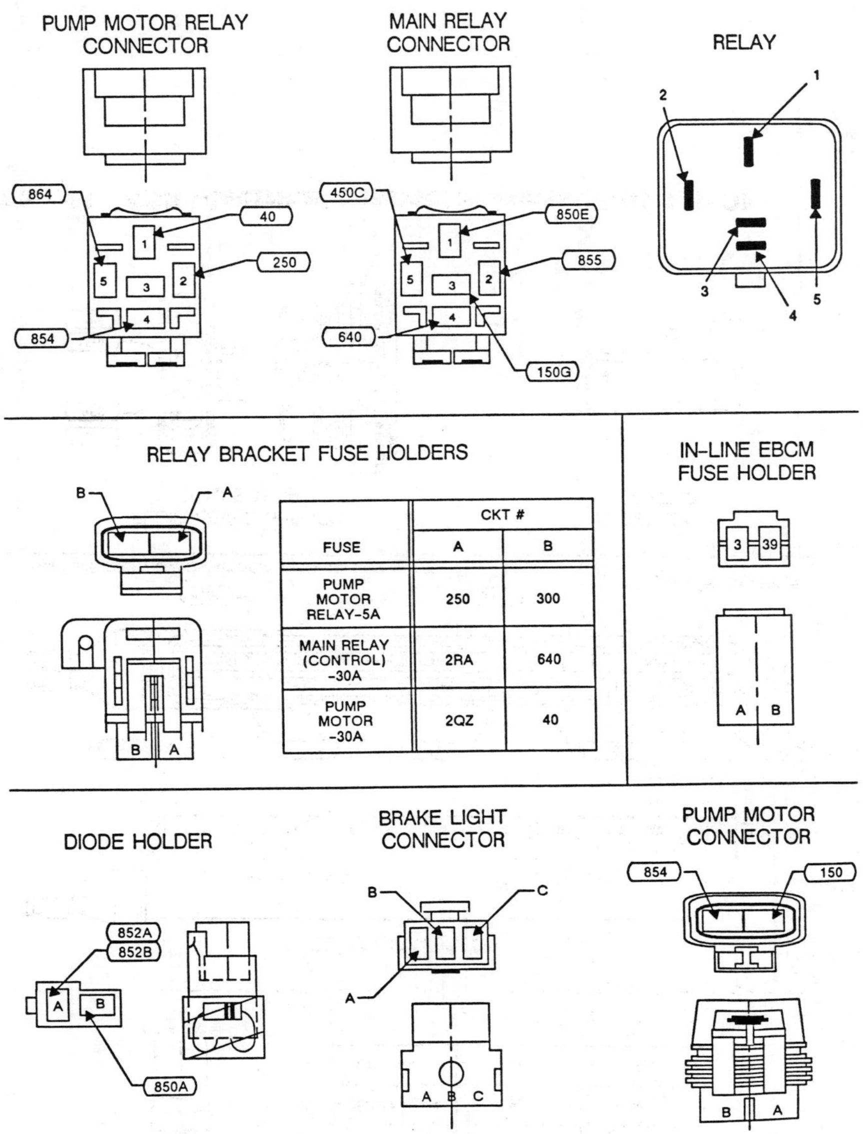

FUSE	CKT # A	CKT # B
PUMP MOTOR RELAY-5A	250	300
MAIN RELAY (CONTROL) -30A	2RA	640
PUMP MOTOR -30A	2QZ	40

Fig. 11 ABS electrical connector views (Part 3 of 3)

ABS CODE INDEX

ABS CODE	SYSTEM	GO TO CHART	ABS CODE	SYSTEM	GO TO CHART
11	EBCM	A-1	45	2 SENSOR (LF)	A-13
12	EBCM	A-1	46	2 SENSORS (RF)	A-13
21	MAIN VALVE	A-2	47	2 SENSORS (REAR)	A-13
22	LF INLET VALVE	A-3	48	3 SENSORS	A-13
23	LF OUTLET VALVE	A-4	51	LF OUTLET VALVE	A-14
24	RF INLET VALVE	A-5	52	RF OUTLET VALVE	A-14
25	RF OUTLET VALVE	A-6	53	REAR OUTLET VALVE	A-14
26	REAR INLET VALVE	A-7	54	REAR OUTLET VALVE	A-14
27	REAR OUTLET VALVE	A-8	55	LF WSS	A-9
31	LF WSS	A-9	56	RF WSS	A-10
32	RF WSS	A-10	57	RR WSS	A-11
33	RR WSS	A-11	58	LR WSS	A-12
34	LR WSS	A-12	61	EBCM LOOP CKT	A-15
35	LF WSS	A-9	71	LF OUTLET VALVE	A-14
36	RF WSS	A-10	72	RF OUTLET VALVE	A-14
37	RR WSS	A-11	73	REAR OUTLET VALVE	A-14
38	LR WSS	A-12	74	REAR OUTLET VALVE	A-14
41	LF WSS	A-9	75	LF WSS	A-9
42	RF WSS	A-10	76	RF WSS	A-10
43	RR WSS	A-11	77	RR WSS	A-11
44	LR WSS	A-12	78	LR WSS	A-12

Fig. 12 ABS trouble code index

CHART A-1
ABS CODES 11 AND 12

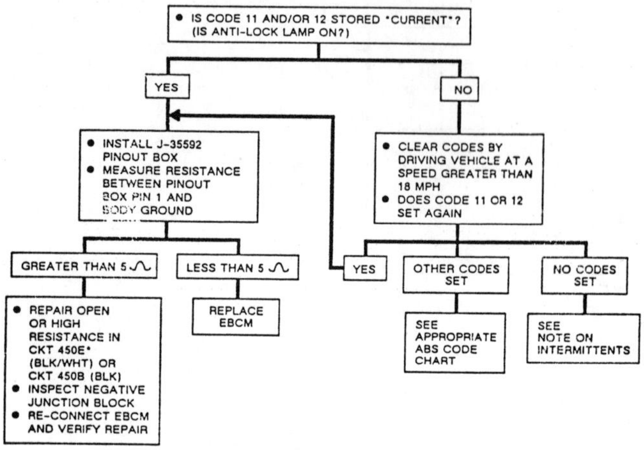

Fig. 13 ABS trouble codes 11 & 12, chart A-1

CHART A-2
ABS CODE 21–MAIN VALVE

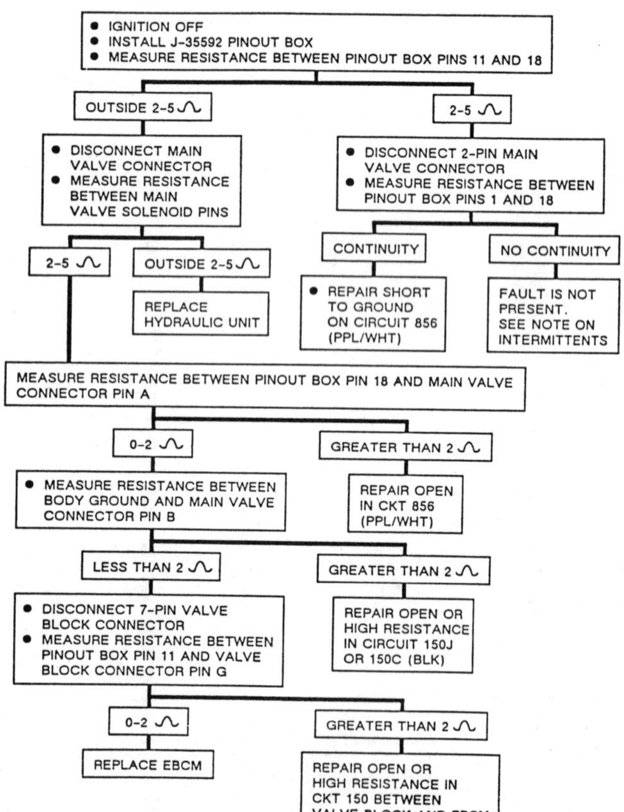

Fig. 14 ABS trouble code 21 (main valve), chart A-2

CHART A-3
ABS CODE 22–LEFT FRONT INLET VALVE

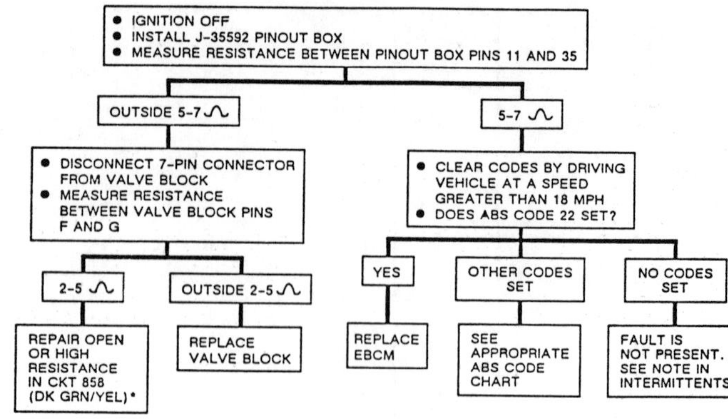

Fig. 15 ABS trouble code 22 (left front inlet valve), chart A-3

CHART A-4
ABS CODE 23–LEFT FRONT OUTLET VALVE

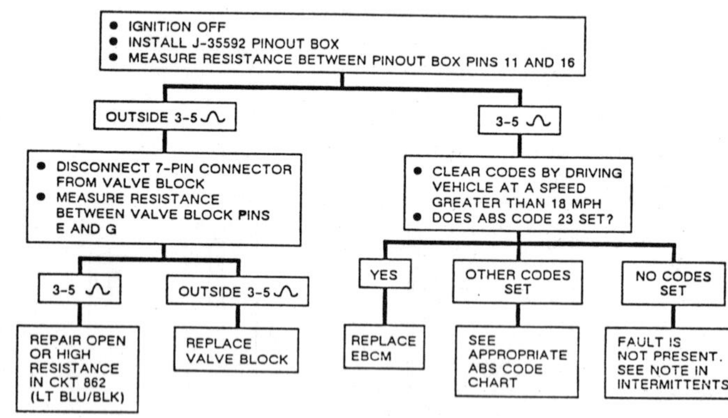

Fig. 16 ABS trouble code 23 (left front outlet valve), chart A-4

CHART A-5
ABS CODE 24–RIGHT FRONT INLET VALVE

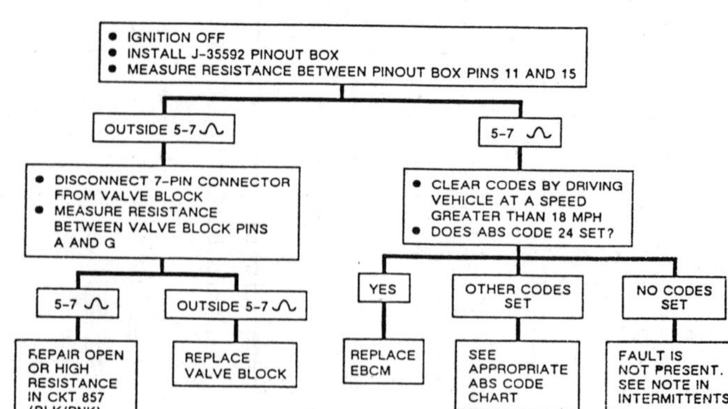

Fig. 17 ABS trouble code 24 (right front inlet valve), chart A-5

CHART A-6
ABS CODE 25–RIGHT FRONT OUTLET VALVE

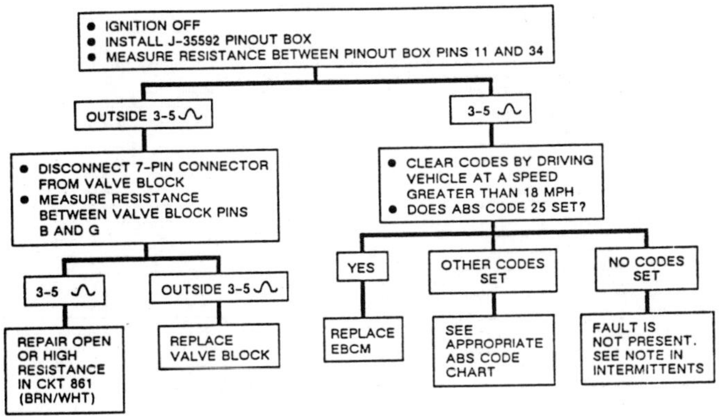

Fig. 18 ABS trouble code 25 (right front outlet valve), chart A-6

CHART A-7
ABS CODE 26–REAR INLET VALVE

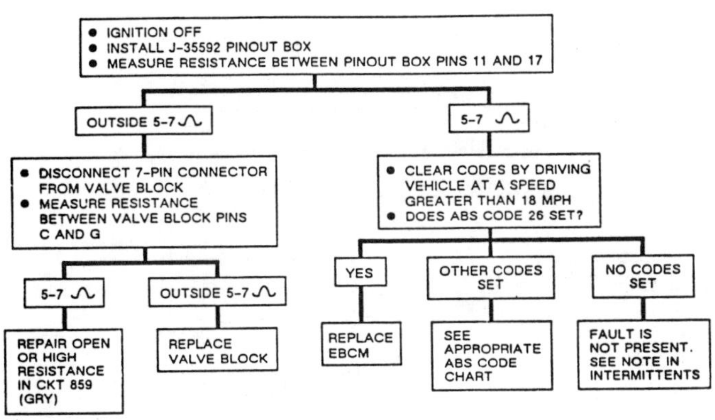

Fig. 19 ABS trouble code 26 (rear inlet valve), chart A-7

CHART A-8
ABS CODE 27–REAR OUTLET VALVE

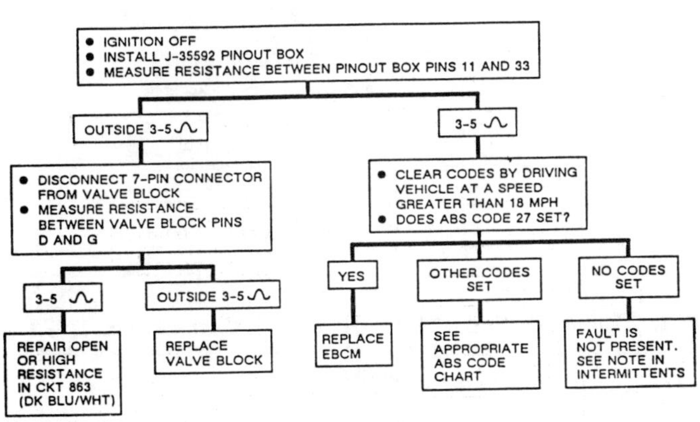

Fig. 20 ABS trouble code 27 (rear outlet valve), chart A-8

CHART A-9
ABS CODES 31, 35, 41, 55 OR 75
LEFT FRONT WHEEL SPEED SENSOR

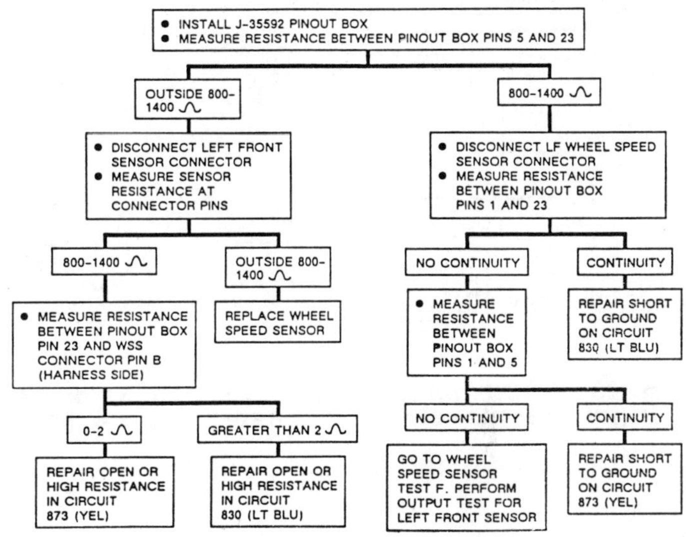

Fig. 21 ABS trouble codes 31, 35, 41, 55 or 75 (left front wheel speed sensor), chart A-9

CHART A-11
ABS CODES 33, 37, 43, 57, OR 77
RIGHT REAR WHEEL SPEED SENSOR

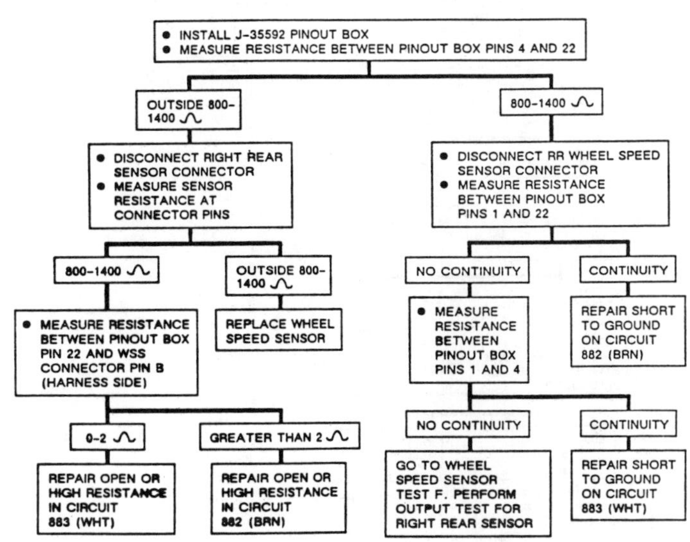

Fig. 22 ABS trouble codes 32, 36, 42, 56 or 76 (right front wheel speed sensor), chart A-10

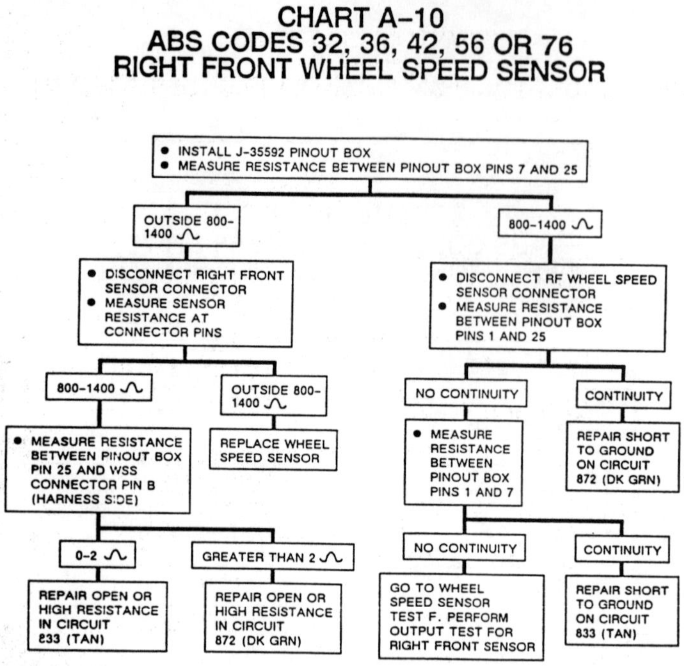

Fig. 23 ABS trouble codes 33, 37, 43, 57 or 77 (right rear wheel speed sensor), chart A-11

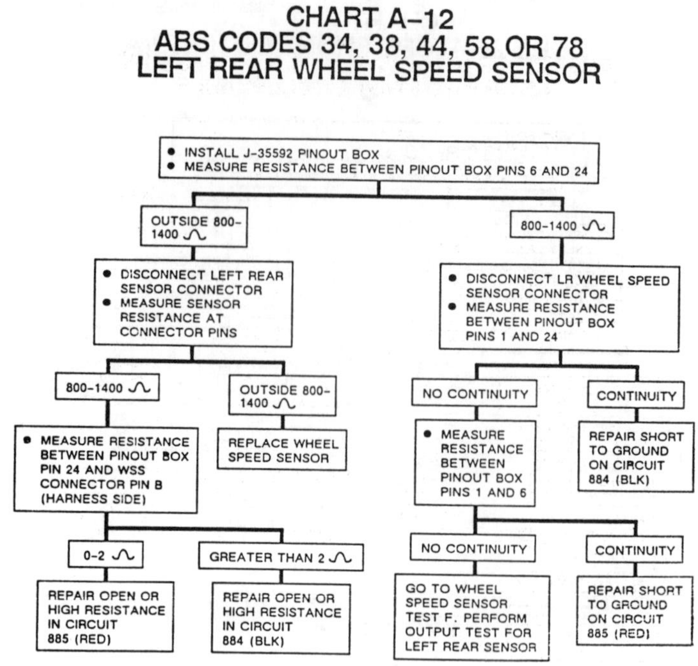

Fig. 24 ABS trouble codes 34, 38, 44, 58 or 78 (left rear wheel speed sensor), chart A-12

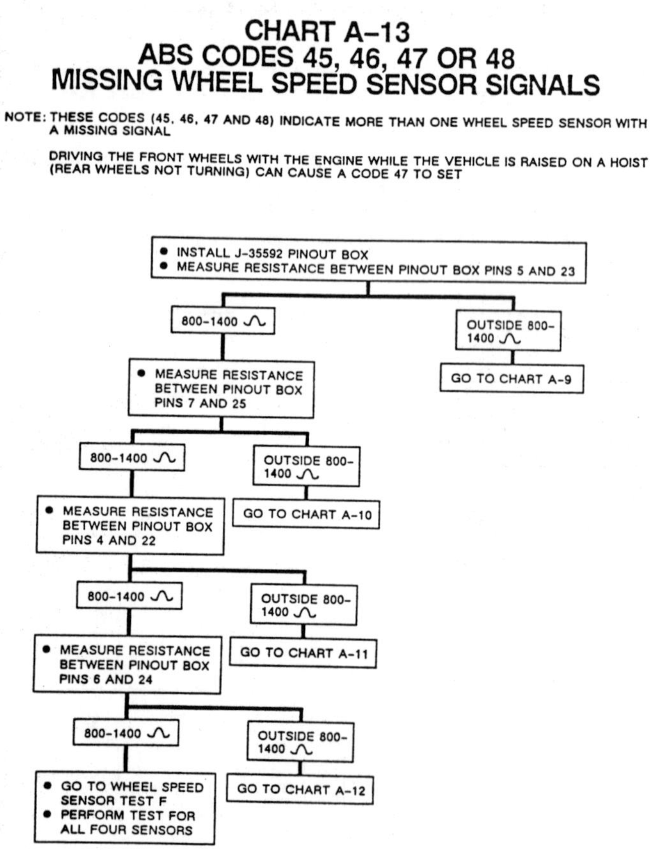

Fig. 25 ABS trouble codes 45, 46, 47 or 48 (missing wheel speed sensor signals), chart A-13

CHART A-14
ABS CODES 51–54 AND 71–74
"PRESSURE REDUCTION" FAULT

NOTE: CODES 51–54 AND 71–74 INDICATE A CONDITION WHICH MAY BE ATTRIBUTABLE TO EITHER THE WHEEL SPEED SENSOR CIRCUIT OR THE OUTLET VALVE

FOR THESE CODES:	PERFORM THESE TESTS:
51 AND/OR 71	• GO TO CHART A-9 IF NO TROUBLE IS FOUND WHILE PERFORMING TESTS ON CHART A-9, DO NOT GO TO TEST F. INSTEAD, GO TO SOLENOID VALVE TEST G • PERFORM SOLENOID VALVE TEST FOR LEFT FRONT VALVES • IF NO TROUBLE IS FOUND, REPLACE EBCM
52 AND/OR 72	• GO TO CHART A-10 IF NO TROUBLE IS FOUND WHILE PERFORMING TESTS ON CHART A-10, DO NOT GO TO TEST F. INSTEAD, GO TO SOLENOID VALVE TEST G • PERFORM SOLENOID VALVE TEST FOR RIGHT FRONT VALVES • IF NO TROUBLE IS FOUND, REPLACE EBCM
53 AND/OR 73	• GO TO CHART A-11 IF NO TROUBLE IS FOUND WHILE PERFORMING TESTS ON CHART A-11, DO NOT GO TO TEST F. INSTEAD, GO TO SOLENOID VALVE TEST G • PERFORM SOLENOID VALVE TEST FOR REAR VALVES • IF NO TROUBLE IS FOUND, REPLACE EBCM
54 AND/OR 74	• GO TO CHART A-12 IF NO TROUBLE IS FOUND WHILE PERFORMING TESTS ON CHART A-12, DO NOT GO TO TEST F. INSTEAD, GO TO SOLENOID VALVE TEST G • PERFORM SOLENOID VALVE TEST FOR REAR VALVES • IF NO TROUBLE IS FOUND, REPLACE EBCM

Fig. 26 ABS trouble codes 51 through 54, 71 through 74 (pressure reduction fault), chart A-14

CHART A-15
ABS CODES 61-LOW FLUID/LOW PRESSURE
LOOP INPUT NOT PROCESSABLE

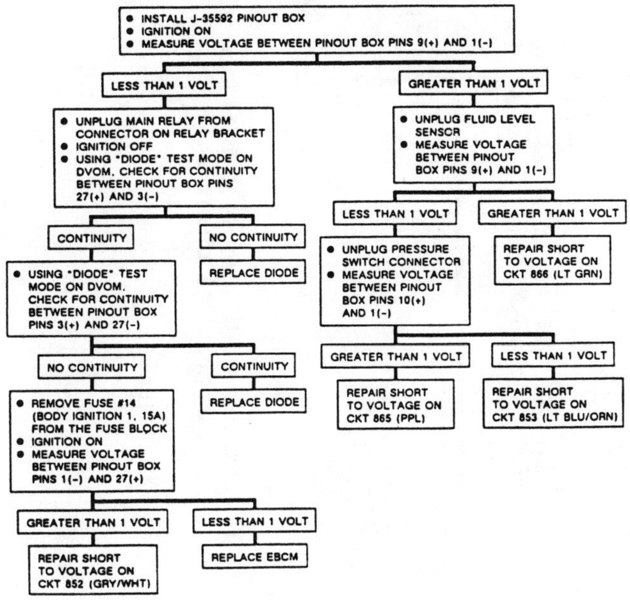

Fig. 27 ABS trouble code 61 (low fluid/low pressure), chart A-15

SYMPTOM DIAGNOSIS
CHART B
"ANTI-LOCK" LAMP ON AND NO ABS CODES PRESENT

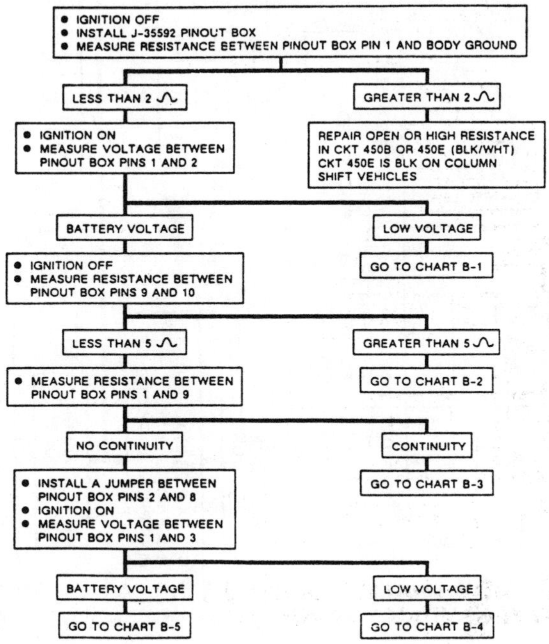

Fig. 28 Symptom diagnosis (anti-lock lamp on & no ABS codes), chart B

CHART B-1
IGNITION ENABLE CIRCUIT

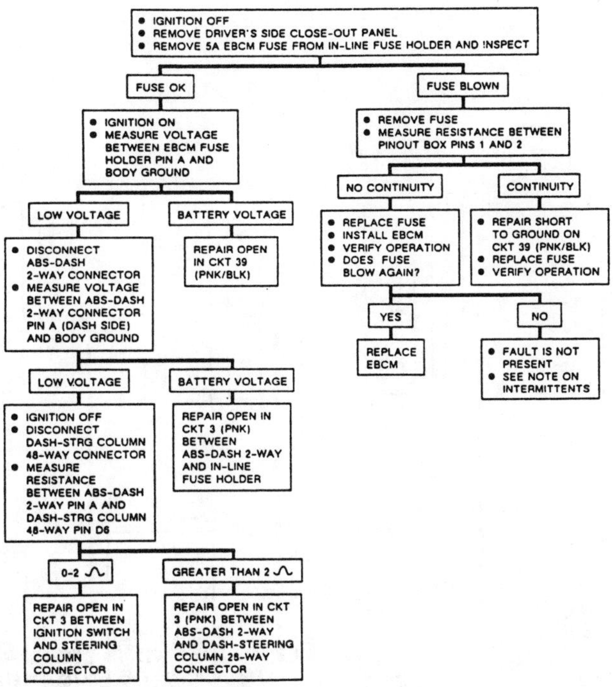

Fig. 29 Symptom diagnosis (ignition enable circuit), chart B-1

CHART B–2
EBCM LOOP CIRCUIT OPEN

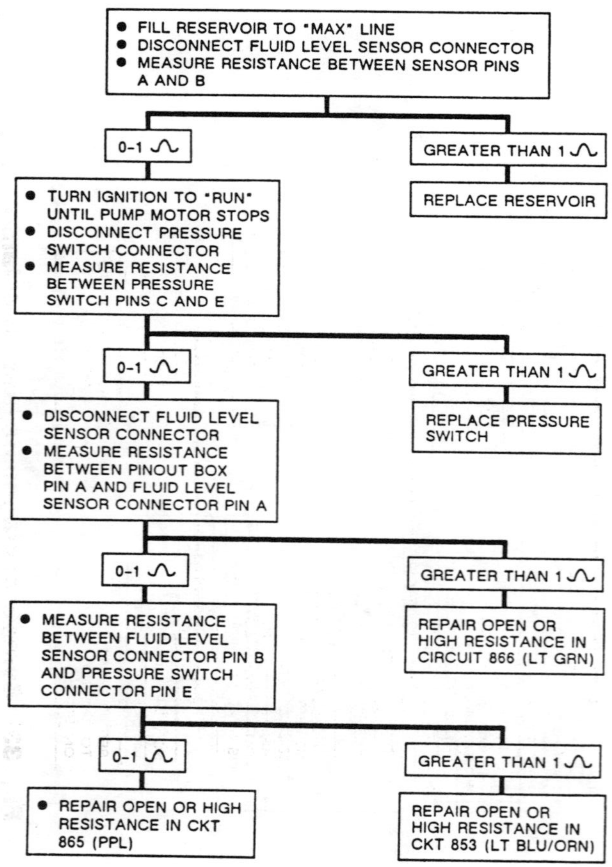

Fig. 30 Symptom diagnosis (EBCM loop circuit open), chart B-2

CHART B–3
EBCM LOOP SHORT TO GROUND

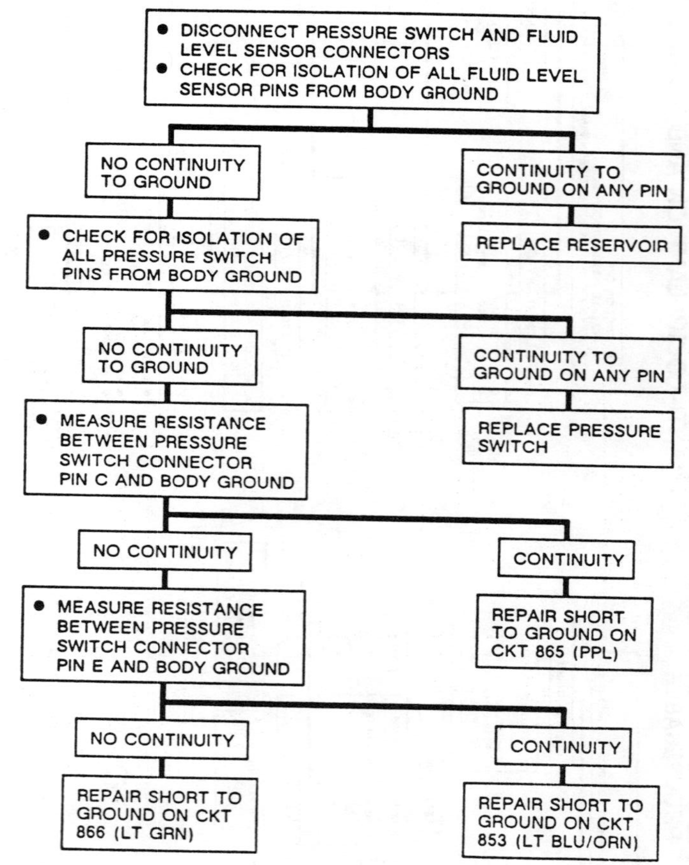

Fig. 31 Symptom diagnosis (EBCM loop short to ground), chart B-3

CHART B-4
MAIN RELAY CIRCUITS

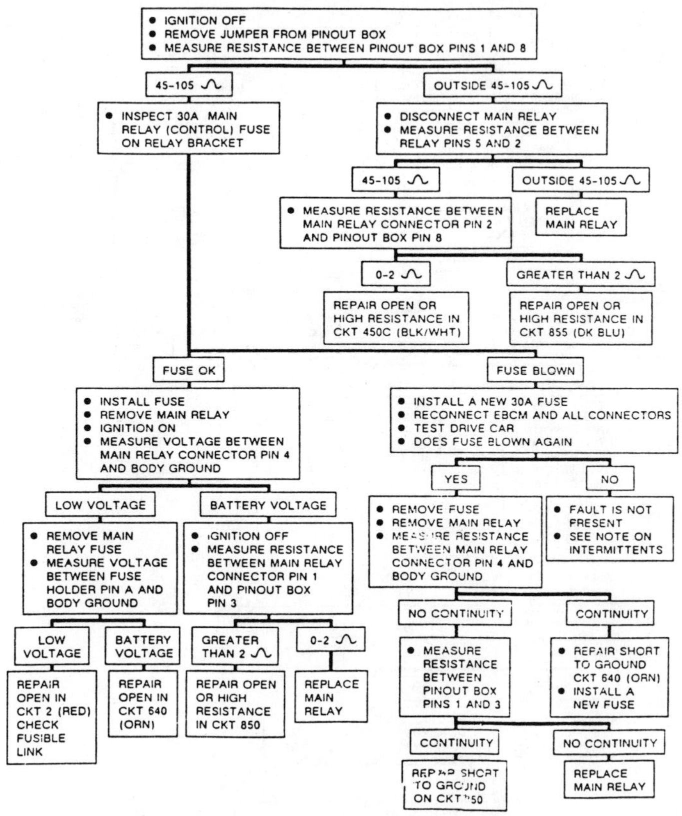

Fig. 32 Symptom diagnosis (main relay circuits), chart B-4

CHART B-5
"ANTI-LOCK" LAMP CIRCUITS

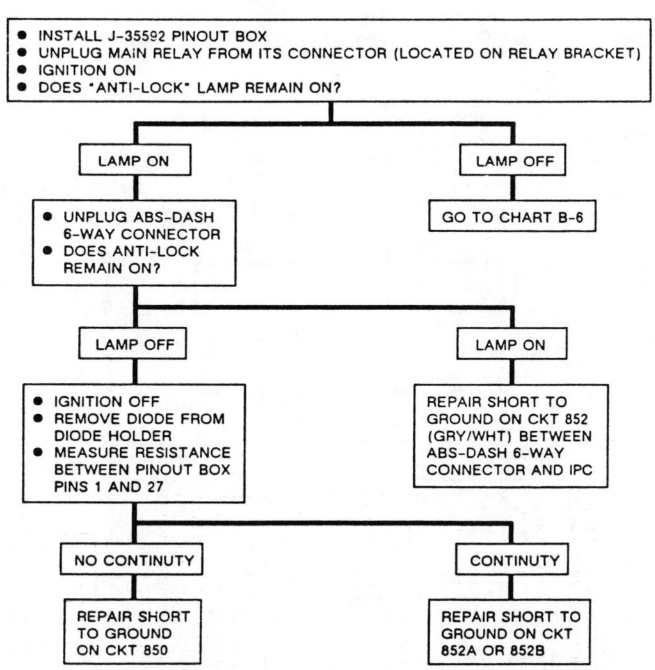

Fig. 33 Symptom diagnosis (anti-lock lamp circuits), chart B-5

CHART B–6
MISCELLANEOUS EBCM INPUTS

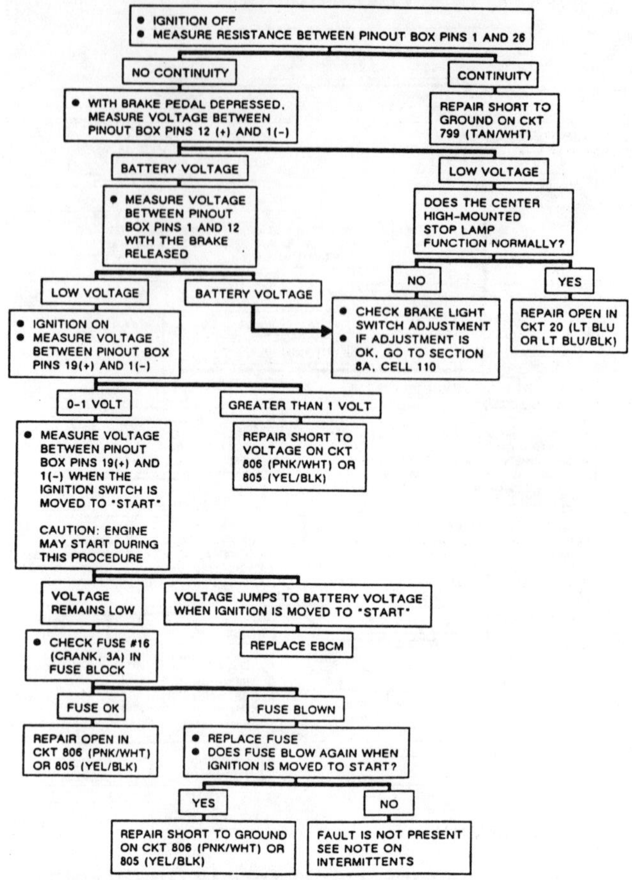

Fig. 34 Symptom diagnosis (miscellaneous EBCM inputs), chart B-6

SYMPTOM DIAGNOSIS
CHART C
"BRAKE" LAMP ON AND NO ABS CODES PRESENT

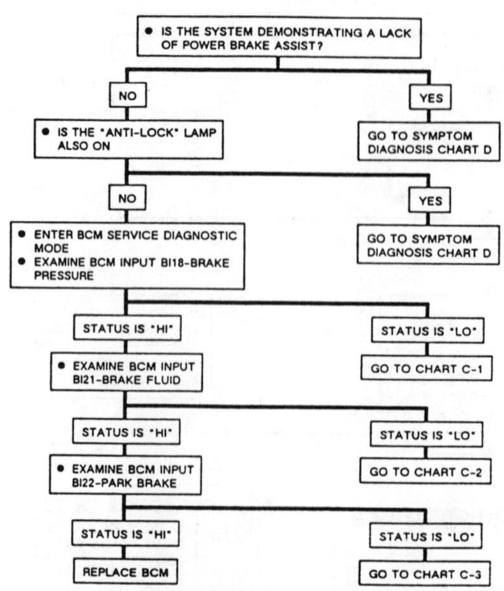

Fig. 35 Symptom diagnosis (brake lamp on & no ABS codes), chart C

CHART C–1
BRAKE PRESSURE INPUT

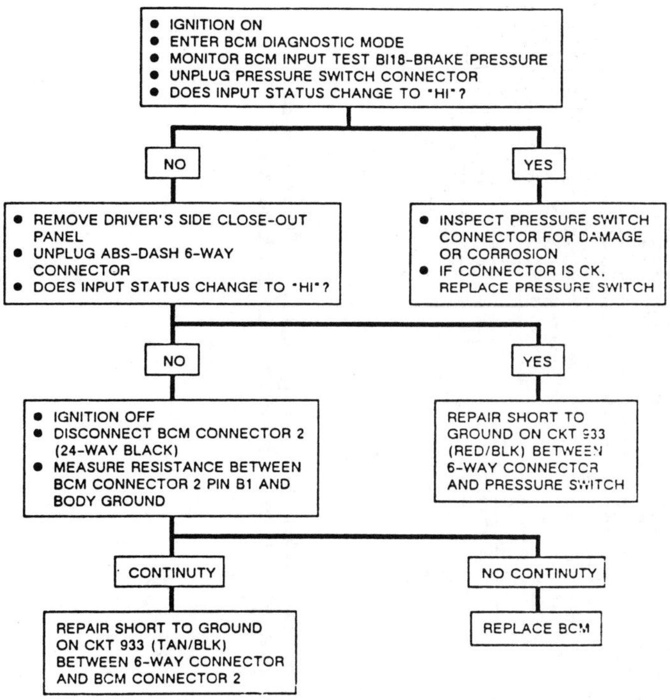

Fig. 36 Symptom diagnosis (brake pressure input), chart C-1

CHART C–2
LOW BRAKE FLUID INPUT

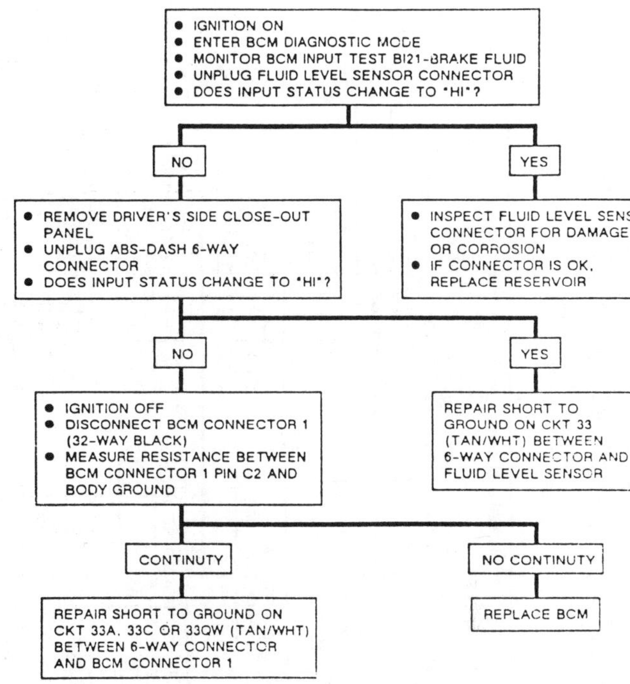

Fig. 37 Symptom diagnosis (low brake fluid input), chart C-2

CHART C–3
PARKING BRAKE INPUT

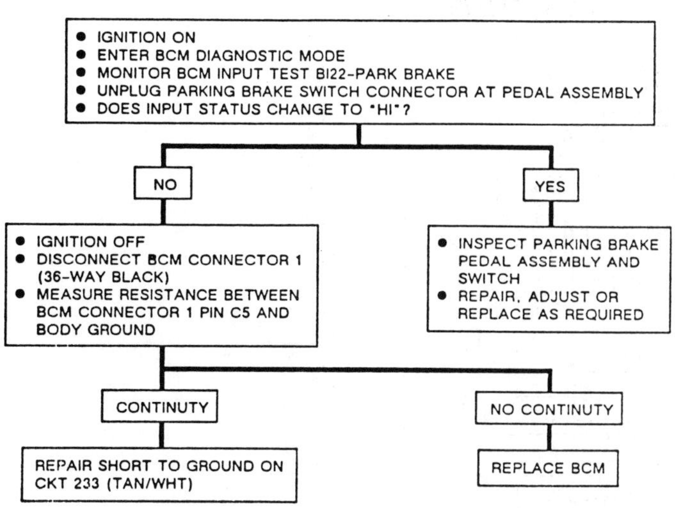

Fig. 38 Symptom diagnosis (parking brake input), chart C-3

SYMPTOM DIAGNOSIS
CHART D
"BRAKE" AND "ANTI-LOCK" LAMPS ON

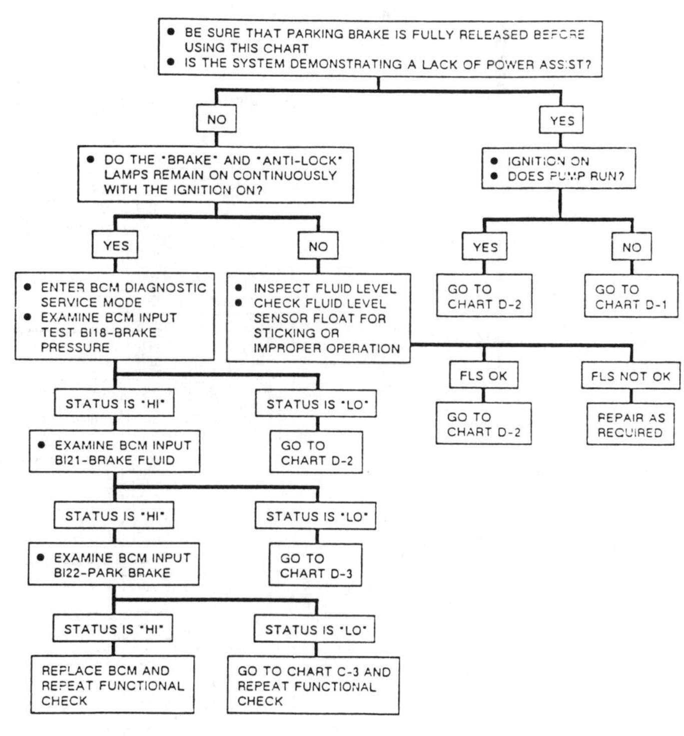

Fig. 39 Symptom diagnosis (brake & anti-lock lamps on), chart D

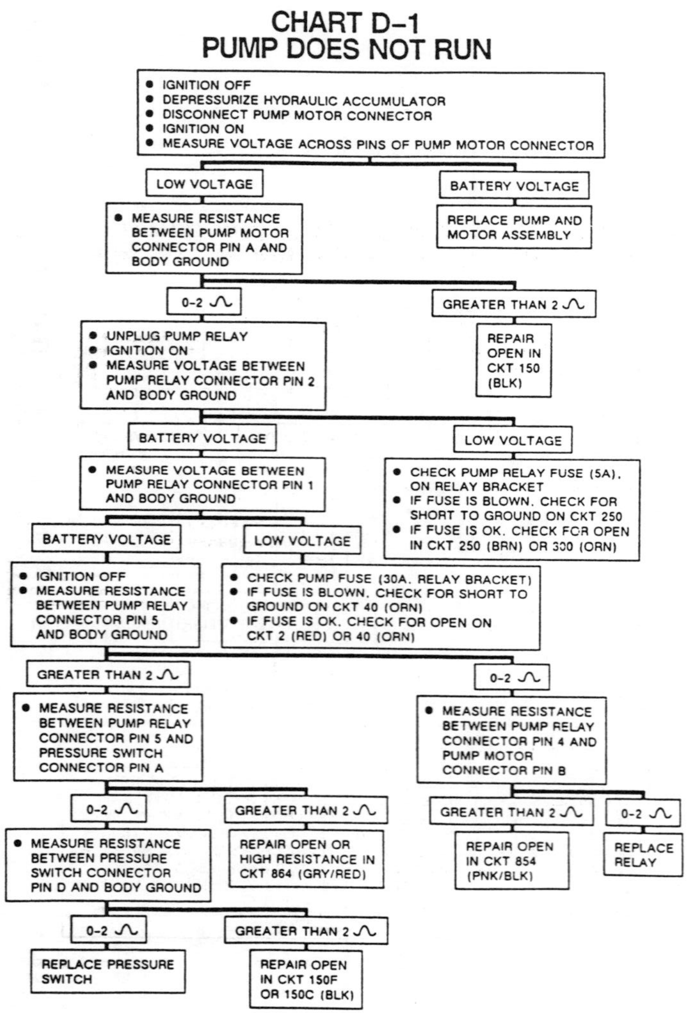

Fig. 40 Symptom diagnosis (pump does not run), chart D-1

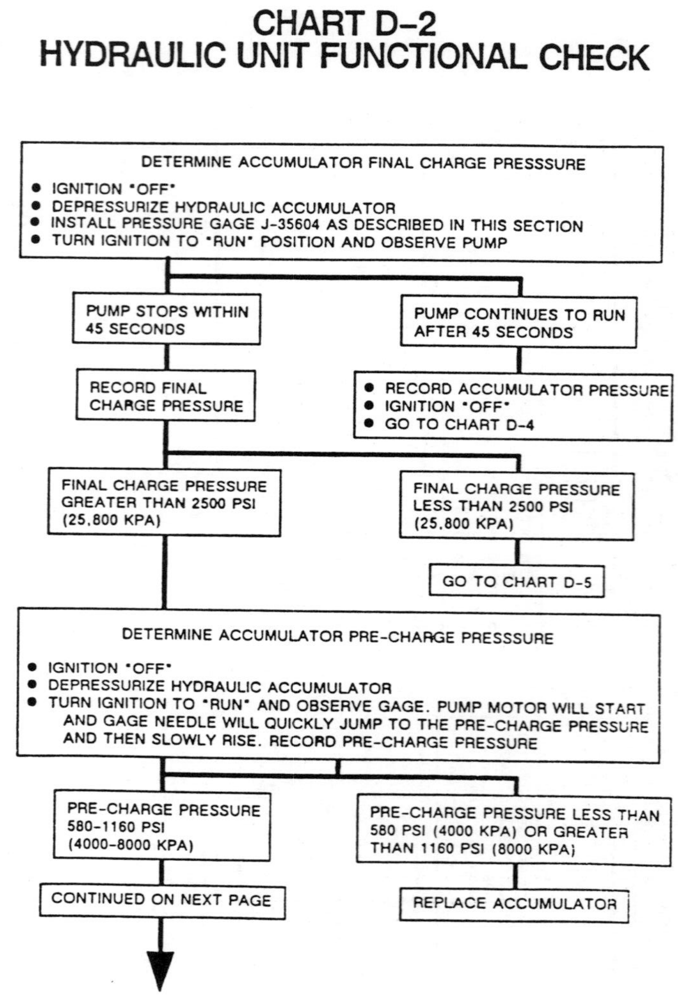

Fig. 41 Symptom diagnosis, hydraulic unit functional check (Part 1 of 2), chart D-2

CHART D-2 (CONTINUED)
HYDRAULIC UNIT FUNCTIONAL CHECK

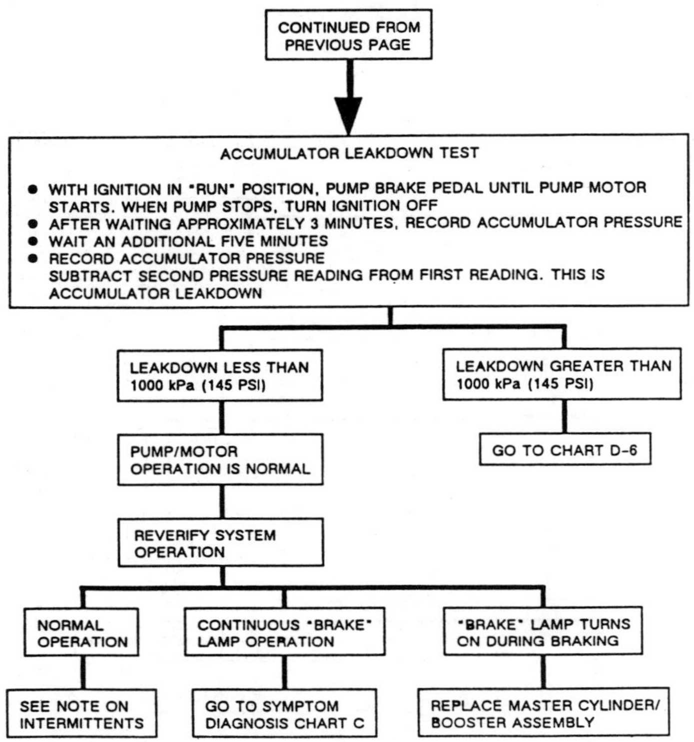

Fig. 41 Symptom diagnosis, hydraulic unit functional check (Part 2 of 2), chart D-2

CHART D-3
FLUID LEVEL SENSOR

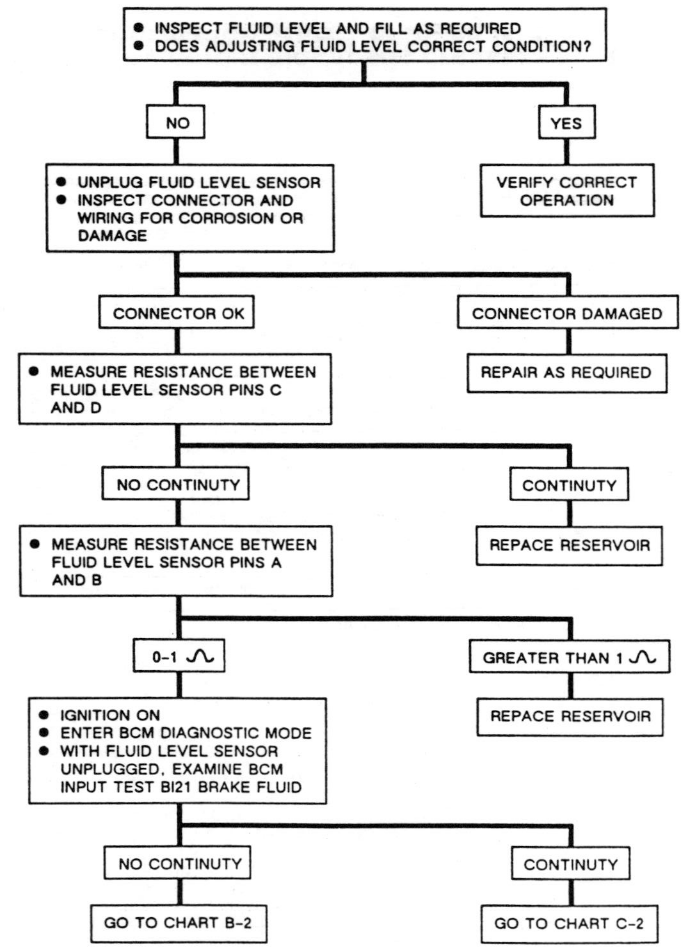

Fig. 42 Symptom diagnosis (fluid level sensor), chart D-2

CHART D–4
EXTENDED PUMP OPERATION

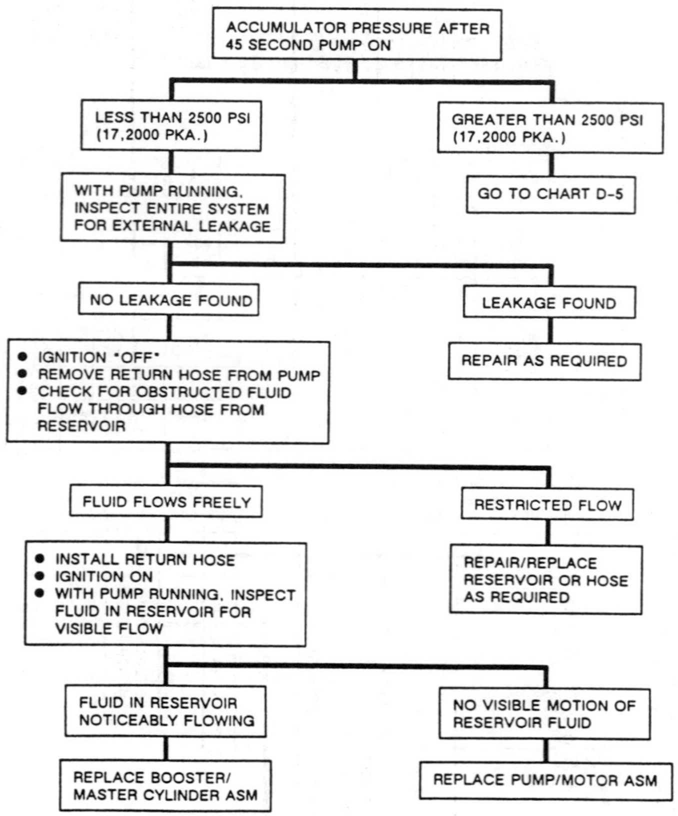

Fig. 43 Symptom diagnosis (extended pump operation), chart D-4

CHART D–5
PRESSURE SWITCH PERFORMANCE

SWITCH STATUS - PRESSURIZED
- TURN IGNITION TO "RUN" UNTIL PUMP STOPS
- IGNITION "OFF"
  NOTE: IF PUMP CONTINUES TO RUN AFTER 45 SECONDS, TURN IGNITION OFF AND PROCEDE
- CHECK PRESSURE SWITCH PINS FOR THE FOLLOWING CONDITIONS USING VOM

MEASURE BETWEEN PINS	SCALE	SPECIFICATION
B, D	200∿	NO CONTINUITY
A, D	200∿	NO CONTINUITY
C, E	200∿	CONTINUITY
ALL PINS, BODY GROUND	200∿	NO CONTINUITY

IF ANY CONDITION IS NOT MET, REPLACE PRESSURE SWITCH AND VERIFY SYSTEM OPERATION

SWITCH STATUS - DEPRESSURIZED
- DEPRESSURIZE HYDRAULIC ACCUMULATOR
- CHECK PRESSURE SWITCH PINS FOR THE FOLLOWING CONDITIONS USING VOM

MEASURE BETWEEN PINS	SCALE	SPECIFICATION
B, D	200∿	CONTINUITY
A, D	200∿	CONTINUITY
C, E	200∿	NO CONTINUITY
ALL PINS, BODY GROUND	200∿	NO CONTINUITY

Fig. 44 Symptom diagnosis, pressure switch performance (Part 1 of 2), chart D-5

CHART D–5 (CONTINUED)
PRESSURE SWITCH PERFORMANCE

SWITCH STATUS - PRESSURIZED
- DEPRESSURIZE HYDRAULIC ACCUMULATOR
- INSTALL PRESSURE GAGE J-35604 AS DESCRIBED IN THIS SECTION
- CONNECT PRESSURE SWITCH CONNECTOR AND TURN IGNITION TO "RUN" UNTIL PUMP STOPS
- USING VOM, MONITOR FOR CONTINUITY BETWEEN PRESSURE SWITCH PINS AS SHOWN BELOW WHILE SLOWLY BLEEDING OFF ACCUMULATOR PRESSURE BY PUMPING THE BRAKE PEDAL.
- PRESSURIZE SYSTEM BETWEEN EACH TEST BY RECONNECTING PRESSURE SWITCH AND TURNING IGNITION TO "RUN" UNTIL PUMP STOPS

MEASURE BETWEEN PINS	SCALE	SPECIFICATION
A, D	CONTINUITY SHOULD BE GAINED AT:	1980–2080 PSI (13,650–14,350 KPA)
B, D	CONTINUITY SHOULD BE GAINED AT:	1500–1550 PSI 10,350–10,700 KPA)
C, E	CONTINUITY SHOULD BE GAINED AT:	1500–1550 PSI (10,350–10,700 KPA)

IF ANY CONDITION IS NOT MET, REPLACE PRESSURE SWITCH AND VERIFY SYSTEM OPERATION

- DEPRESSURIZE HYDRAULIC ACCUMULATOR
- WITH IGNITION OFF, CONNECT PRESSURE SWITCH CONNECTOR
- TURN IGNITION TO "RUN" AND OBSERVE GAGE, "ANTI-LOCK" LAMP, "BRAKE" LAMP AND PUMP MOTOR. EVENTS SHOULD OCCUR AT PRESSURES INDICATED IN CHART BELOW

EVENT	PRESSURE
"ANTI-LOCK" LAMP TURN OFF	1850–1950 PSI (12,750–13,450 KPA)
"BRAKE" LAMP TURN OFF	1850–1950 PSI (12,750–13,450 KPA)
PUMP MOTOR STOPS	2550–2670 PSI (17,580–18,400 KPA)

Fig. 44 Symptom diagnosis, pressure switch performance (Part 2 of 2), chart D-5

CHART D–6
HYDRAULIC UNIT LEAKDOWN

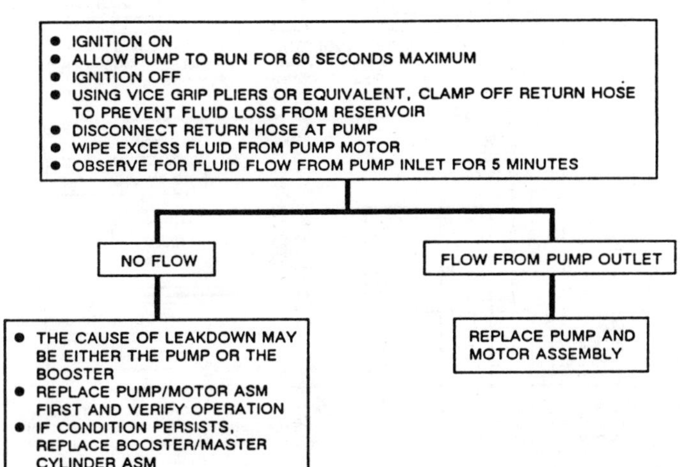

Fig. 45 Symptom diagnosis (hydraulic unit leakdown), chart D-6

CHART E–1
"ANTI-LOCK" LAMP INOPERATIVE AT KEY–ON

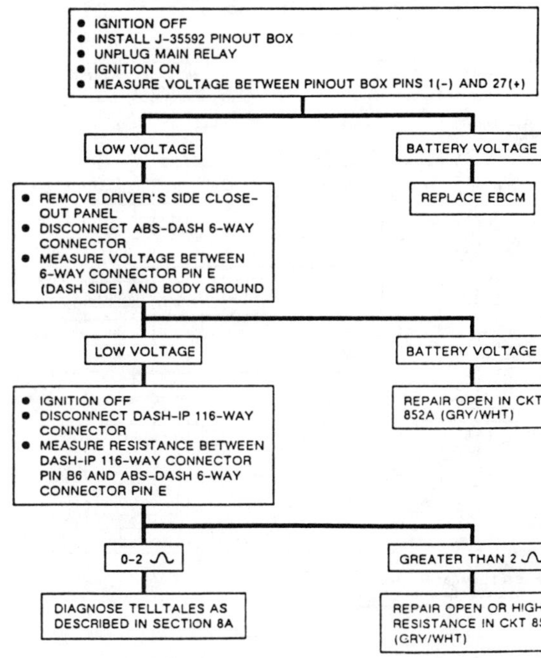

Fig. 46 Symptom diagnosis (anti-lock lamp inoperative at key-on), chart E-1

CHART E–2
"ANTI-LOCK" LAMP DIM OR INOP DURING CRANK

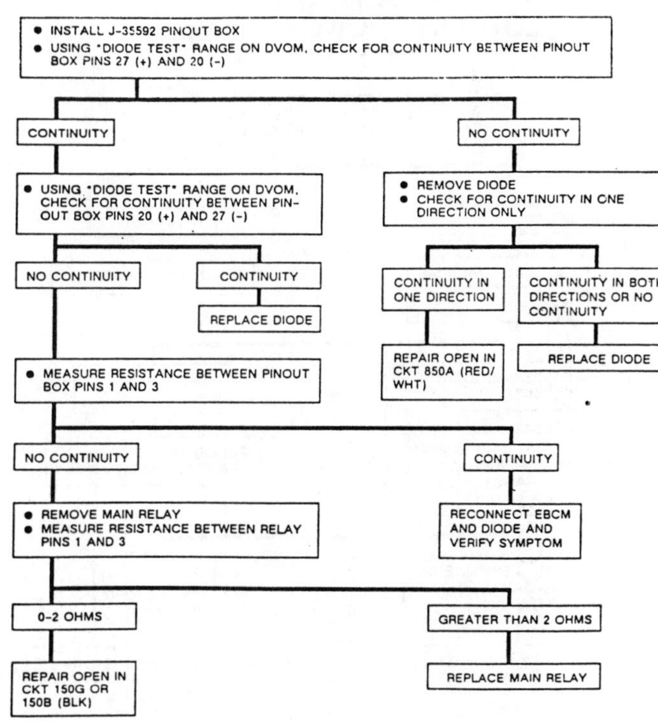

Fig. 47 Symptom diagnosis (anti-lock lamp dim or inoperative during engine crank), chart E-2

CHART E-3
VALVE CYCLING (CHATTER) DURING NORMAL STOPS

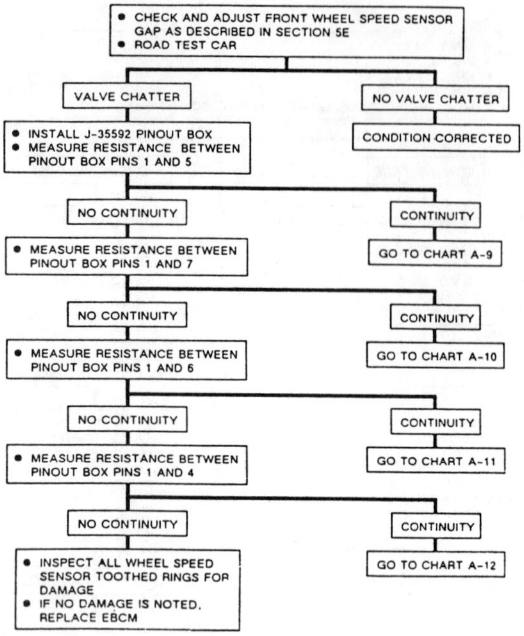

Fig. 48 Symptom diagnosis (valve cycling during normal stops), chart E-3

CHART E-5
POOR VEHICLE TRACKING DURING ANTI-LOCK STOPS

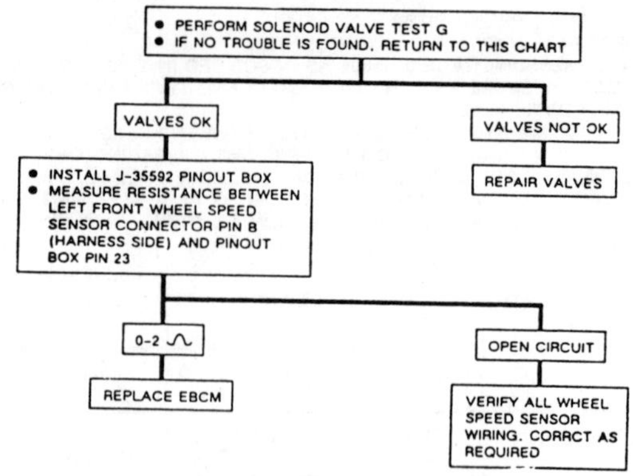

Fig. 50 Symptom diagnosis (poor vehicle tracking during anti-lock stops), chart E-5

CHART E-4
LOW OR SPONGY PEDAL

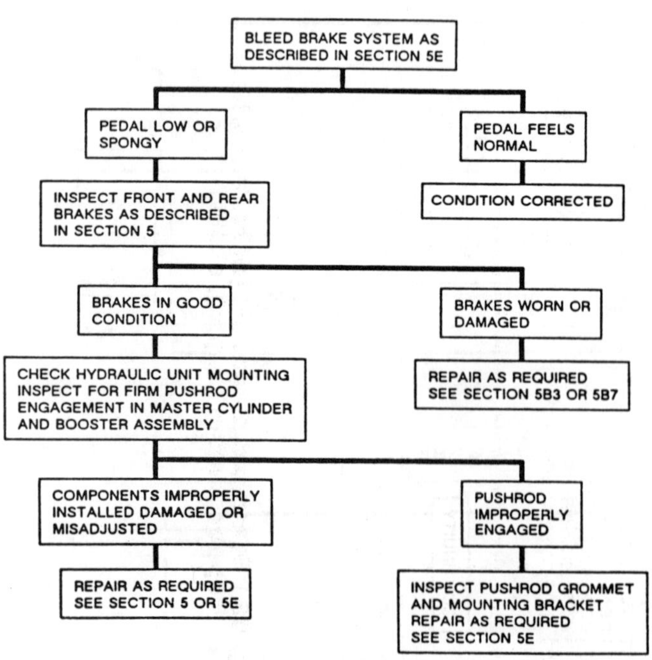

Fig. 49 Symptom diagnosis (low or spongy pedal), chart E-4

TEST F
WHEEL SPEED SENSOR OUTPUT

- THE FOLLOWING TEST MAY BE PERFORMED FOR INDIVIDUAL WHEEL SPEED SENSORS (AS DIRECTED BY THE APPROPRIATE DIAGNOSTIC PROCEDURE) OR FOR ALL SENSORS AS A SYSTEM CHECK
- TO SET UP FOR TEST:
  - INSTALL J-35592 PINOUT BOX
  - RAISE CAR SO THAT WHEELS CLEAR THE GROUND
  - AN ASSISTANT MAY BE REQUIRED TO SPIN THE WHEELS

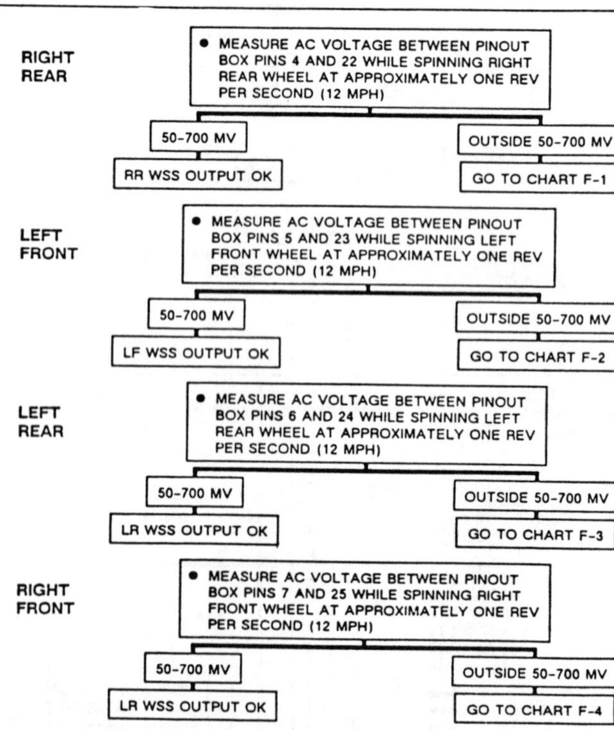

IF NO TROUBLE IS FOUND WITH WHEEL SPEED SENSOR OUTPUT, CONSIDER THAT:

- OUTPUT SHOULD NOT BE SIGNIFICANTLY DIFFERENT FROM LF TO RF AND LR TO RR
- FRONT WSS OUTPUT SHOULD BE SLIGHTLY HIGHER THAN REAR WSS OUTPUT
- IMPROPER OUTPUT MAY BE CAUSD BY MIS-ADJUSTED FRONT SENSOR
- WHEEL SPEED SENSOR CABLES SOMETIMES CAN INTERMITTENTLY GO OPEN, PARTICULARLY IN THE WHEELHOUSE AREA. MANIPULATE THE CABLES BY HAND WHEN INSPECTING FOR AN OPEN CIRCUIT
- TOOTHED SENSOR RINGS SHOULD ALSO BE INSPECTED FOR DAMAGE OR MISSING TEETH

Fig. 51 Symptom diagnosis (wheel speed sensor output), chart F

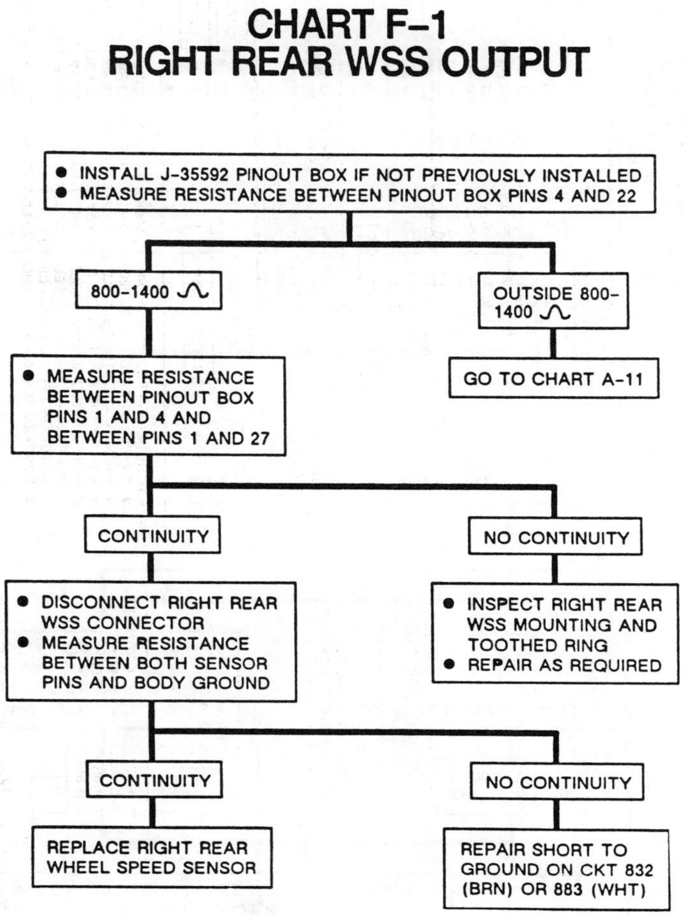

Fig. 52 Symptom diagnosis (right rear wheel speed sensor output), chart F-1

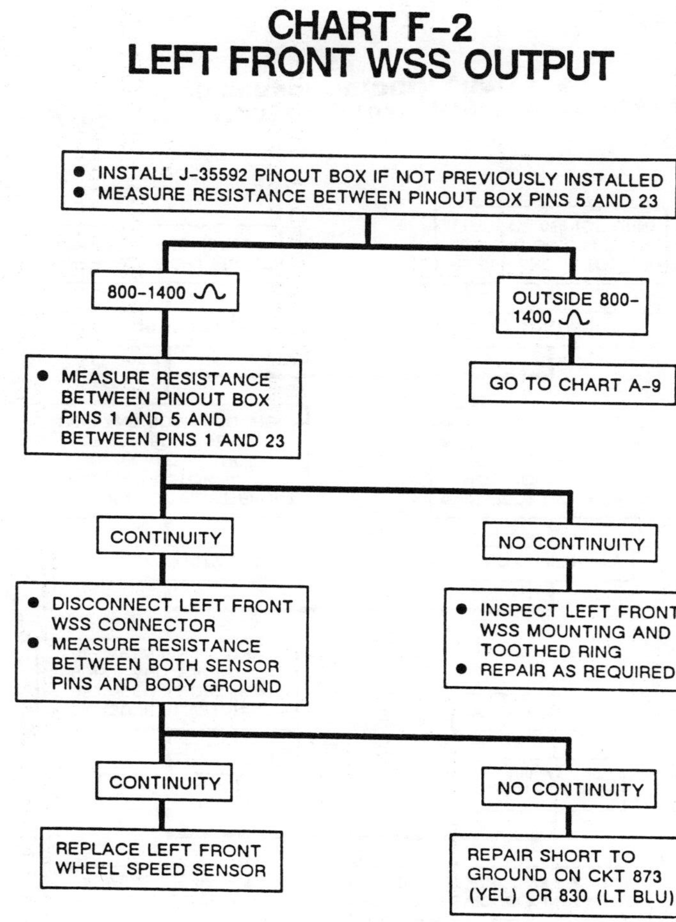

Fig. 53 Symptom diagnosis (left front wheel speed sensor output), chart F-2

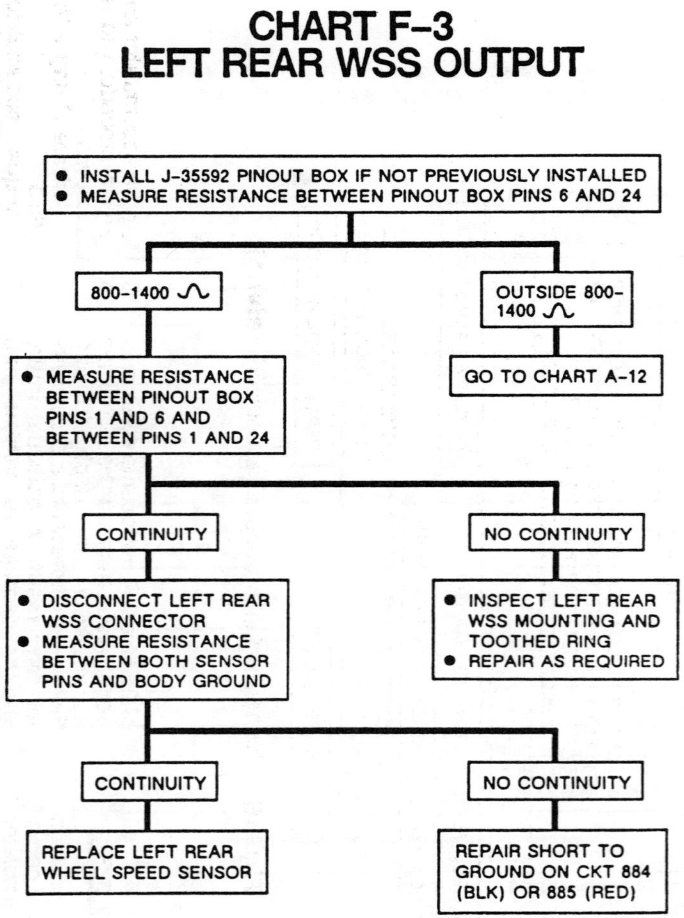

Fig. 54 Symptom diagnosis (left rear wheel speed sensor output), chart F-3

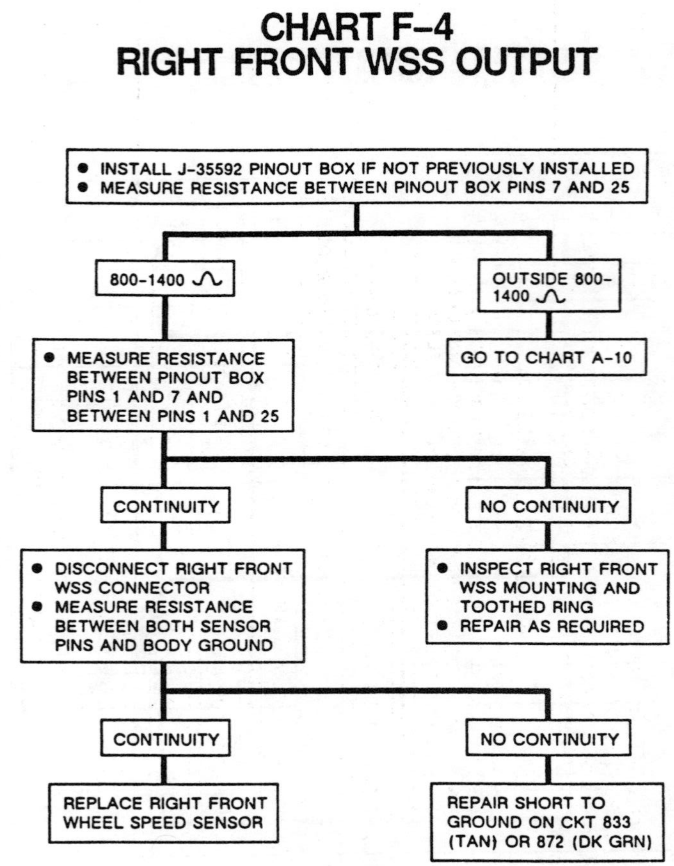

Fig. 55 Symptom diagnosis (right front wheel speed sensor output), chart F-4

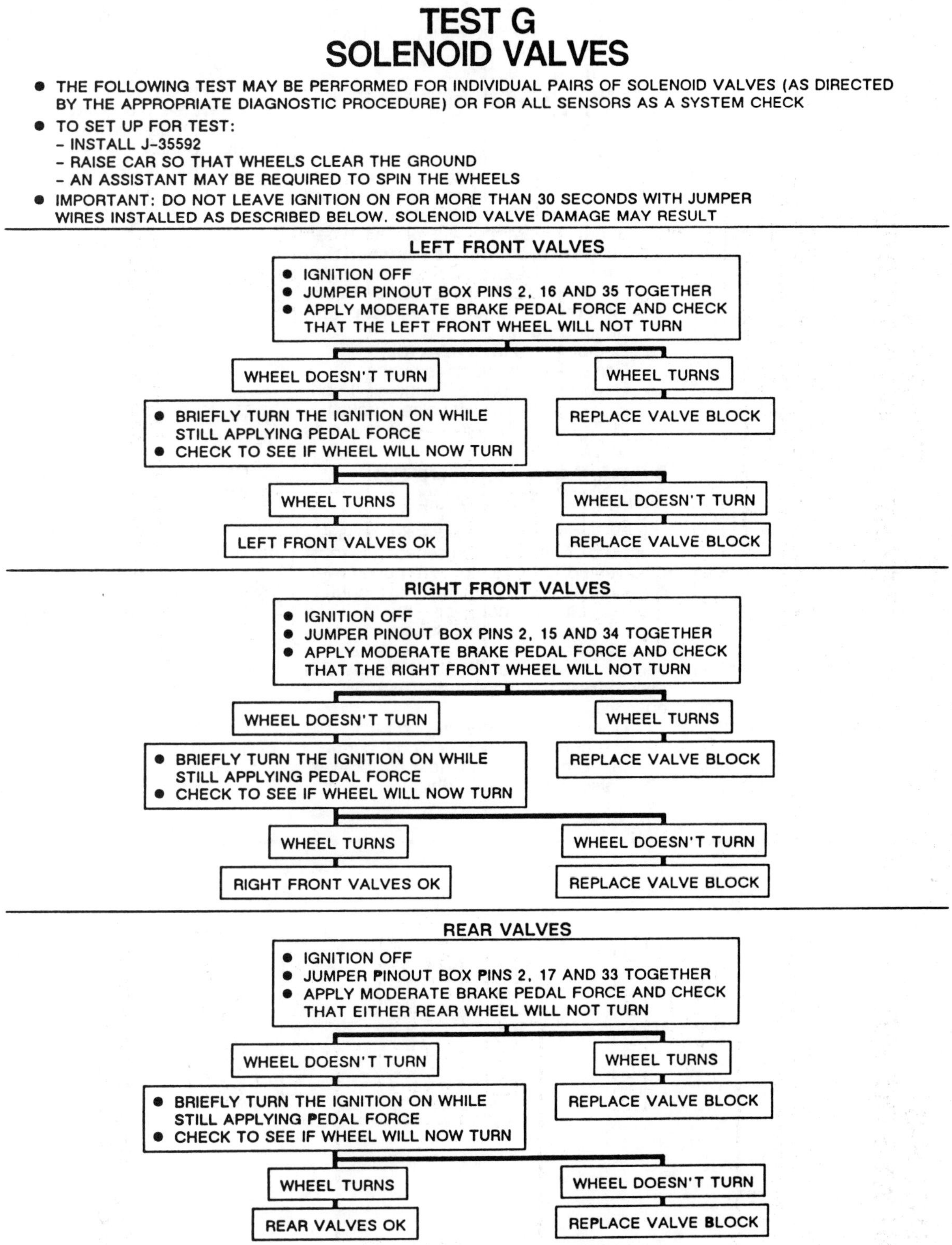

Fig. 56 Symptom diagnosis (solenoid valves), test G

HYDRAULIC ACCUMULATOR, REPLACE

1. Turn ignition switch to Off position and disconnect battery ground cable.
2. Depressurize hydraulic accumulator by depressing brake pedal a minimum of 25 times. **A noticeable change in pedal resistance will be felt when accumulator is depressurized. Failure to depressurize accumulator may result in personal injury and/or damage to vehicle and painted surfaces.**
3. Remove accumulator and O-ring.
4. Reverse procedure to install, noting the following:
   a. Lubricate O-ring with clean brake fluid.
   b. **Torque** accumulator attaching bolts to 30 ft. lbs.

c. Turn ignition switch On. Allow system to pressurize. BRAKE lamp should go out within 30 seconds and pump motor should turn off within 40 seconds.

PUMP & MOTOR ASSEMBLY, REPLACE

1. Disconnect battery ground cable.
2. Depressurize hydraulic accumulator by depressing brake pedal a minimum of 25 times. **A noticeable change in pedal resistance will be felt when accumulator is depressurized. Failure to depressurize accumulator may result in personal injury and/or damage to vehicle and painted surfaces.**
3. Disconnect electrical connectors from pressure switch and pump motor.
4. Pinch return hose to prevent fluid loss, then remove hydraulic accumulator.
5. Disconnect pressure hose from pump.
6. Remove pressure hose assembly and O-rings.
7. Remove wire clip which retains hose at pump body, then pull return hose fitting from pump body.
8. Remove pump mounting bolt, sleeve and insulator.
9. Remove pump and motor assembly from vehicle, by sliding pump from locating pin.
10. Reverse procedure to install, noting the following:
    a. **Torque** pressure pipe assembly-to-pump fittings to 9 ft. lbs.
    b. **Torque** accumulator attaching bolts to 30 ft. lbs.

c. Turn ignition switch On. Allow system to pressurize. BRAKE lamp should go out within 30 seconds and pump motor should turn off within 40 seconds.

VALVE BLOCK ASSEMBLY, REPLACE

1. Disconnect battery ground cable.
2. Depressurize hydraulic accumulator by depressing brake pedal a minimum of 25 times. **A noticeable change in pedal resistance will be felt when accumulator is depressurized. Failure to depressurize accumulator may result in personal injury and/or damage to vehicle and painted surfaces.**
3. Remove fluid from reservoir.
4. Remove valve block assembly.
5. Reverse procedure to install. Lubricate O-ring seals with clean brake fluid. Tighten valve block mounting nuts to 18 ft. lbs.

FRONT WHEEL SPEED SENSOR, REPLACE

If the sensor mounting bracket bolts are loosened or removed, the sensor gap must be adjusted. If the sensor is removed without loosening or removing the mounting bracket bolts, it is not necessary to adjust sensor gap. If the wheel speed sensor is removed from the mounting bracket, the sensor must be coated with GM anti-corrosion compound 1052856, or equivalent.

1. Disconnect battery ground cable.
2. Disconnect sensor electrical connector from underhood, near strut tower.
3. Raise and support vehicle.
4. Disconnect sensor cable grommet from wheelwell, then remove sensor cable from retaining brackets.
5. Remove sensor, or if necessary, remove mounting bracket and sensor assembly from vehicle.
6. Reverse procedure to install. Adjust sensor as follows:
    a. Loosen front wheel speed sensor mounting bracket bolts so that bracket is snug to knuckle but can be moved by hand.
    b. Using a feeler gauge, adjust sensor (between the tip of the sensor and one of the teeth on the toothed sensor ring), to .020 inch by moving bracket.
    c. Tighten mounting bracket bolts to 9 ft. lbs.
    d. After tightening, check sensor gap.

REAR TOOTHED SENSOR RING

The rear toothed sensor ring is an integral part of the hub and bearing assembly. If the sensor ring requires replacement, the entire hub and bearing assembly must be replaced.

REAR WHEEL SPEED SENSOR, REPLACE

1. Disconnect battery ground cable.
2. Raise and support vehicle.
3. Disconnect electrical connector from sensor.
4. Remove sensor mounting bracket, then the sensor.
5. Reverse procedure to install. **Torque** attaching bolts to 9 ft. lbs.

HYDRAULIC BRAKE SYSTEM

INDEX

Chrysler Motors/Eagle

TABLE OF CONTENTS

MASTER CYLINDER APPLICATIONS

Type No.

CHRYSLER EXCEPT COLT, COLT VISTA & CONQUEST

1985-89:
Domestic Front & Rear Wheel Drive Vehicles 1

COLT

1985-86 5
1987-88:
w/1.5L Engine 5
w/1.6L Engine 7
1989 7

COLT VISTA

1985-89 6

CONQUEST

1985-89 8

EAGLE MEDALLION

1987-89 3

EAGLE PREMIER

1988-89 2

EAGLE SUMMIT

1989 7

EAGLE 4 X 4

1985-88 4

HYDRAULIC CONTROLS APPLICATIONS

	Type No.
CHRYSLER EXCEPT COLT, COLT VISTA & CONQUEST	
1985-89:	
Front Wheel Drive	2
Rear Wheel Drive	1
COLT	
1985-87	6
1988	6,7
1989	6
COLT VISTA	
1985-89	6,7
CONQUEST	
1985-87	8
EAGLE MEDALLION	
1987-89	4
EAGLE PREMIER	
1988-89	5
EAGLE SUMMIT	
1989	6
EAGLE 4 X 4	
1985-88	3

FRONT & REAR SPLIT HYDRAULIC BRAKE SYSTEMS

When the brake pedal is depressed, both the primary (front brake) and the secondary (rear brake) master cylinder pistons are moved simultaneously to exert hydraulic fluid pressure on their respective independent hydraulic systems. The fluid displacement of the two master cylinders is proportioned to fulfill the requirements of each of the two independent hydraulic brake systems, **Fig. 1.**

If a failure of a rear (secondary) brake system should occur, initial brake pedal movement causes the unrestricted secondary piston to bottom in the master cylinder bore. Primary piston movement displaces hydraulic fluid in the primary section of the dual master cylinder to actuate the front brake system.

Should the front (primary) brake system fail, initial brake pedal movement causes the unrestricted primary piston to bottom out against the secondary piston. Continued downward movement of the brake pedal moves the secondary piston to displace hydraulic fluid in the rear brake system to actuate the rear brakes.

The increased pedal travel and the increased pedal effort required to compensate for the loss of the failed portion of the brake system provides a warning that a partial brake system failure has occurred. When the ignition switch is turned on, a brake warning light on the instrument panel provides a visual indication that one of the dual brake systems has become inoperative.

Should a failure of either the front or rear brake hydraulic system occur, the hydraulic fluid pressure differential resulting from pressure loss of the failed brake system forces the valve toward the low pressure area to light the brake warning lamp.

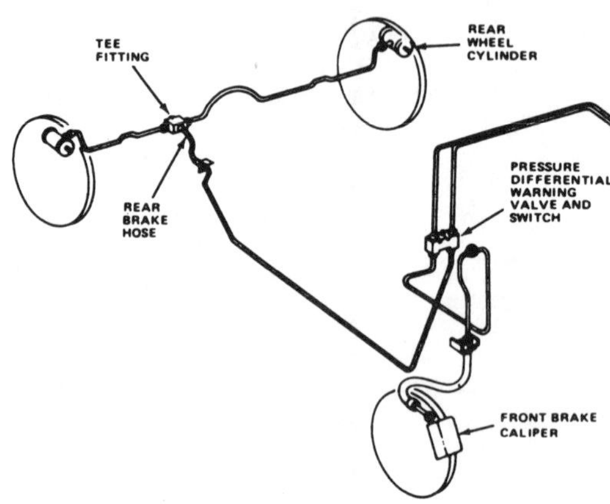

Fig. 1 Schematic diagram of a hydraulic front & rear split brake system

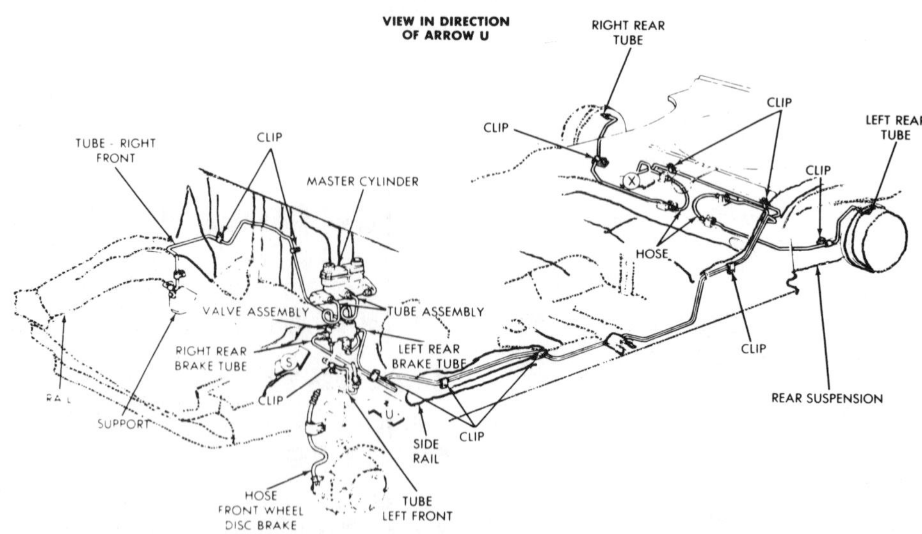

Fig. 2 Schematic diagram of a hydraulic diagonally split brake system

DIAGONALLY SPLIT HYDRAULIC BRAKE SYSTEMS

This system operates on the same principles as conventional front and rear split systems using primary and secondary master cylinders moving simultaneously to exert hydraulic pressure on their respective systems.

The hydraulic brake lines on this system, however, have been diagonally split front to rear (left front to right rear and right front to left rear) in place of separate lines to the front and rear wheels, **Fig. 2.**

In the event of a system failure this would cause the remaining good system to do all the braking on one front wheel and the opposite rear wheel, thus maintaining 50% of the total braking force. The hydraulic pressure loss would result in a pressure differential in the system and cause a warning light on the dashboard to glow as in front and rear split systems.

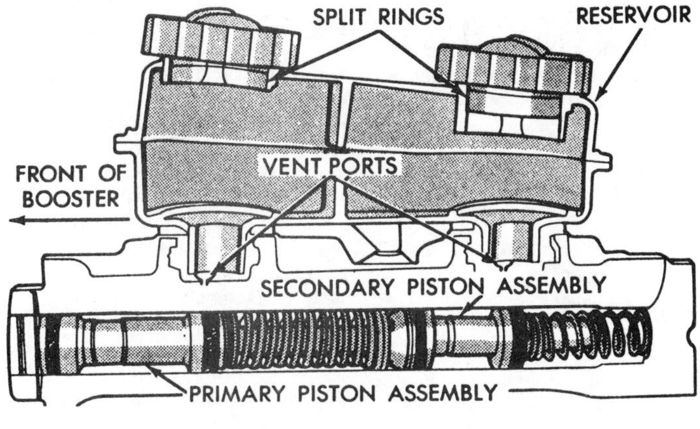

Fig. 3 Type 1, composite master cylinder

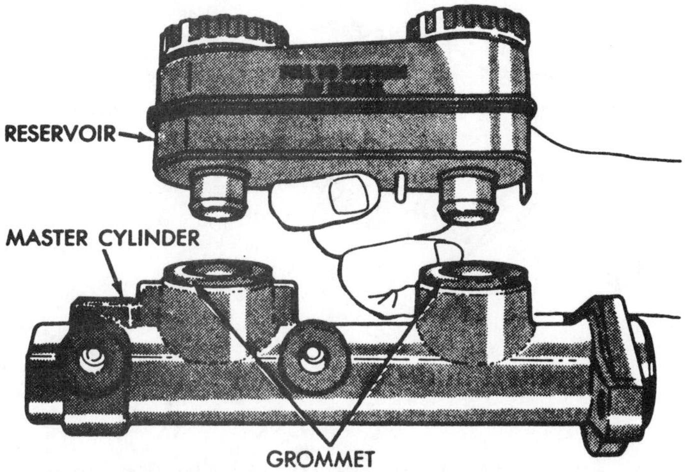

Fig. 4 Reservoir replacement

TYPE 1—COMPOSITE MASTER CYLINDER

DESCRIPTION

This master cylinder **Fig. 3**, is comprised of an aluminum body, pistons, springs, O-rings, cup seals, and a reinforced nylon reservoir. The bore of the aluminum body is anodized for durability, and to prevent corrosion and pitting. The reservoir caps contain diaphragms with precision slits for controlled equalization of pressure.

TESTING

Ensure the master cylinder compensates in both ports by applying the brake pedal lightly (engine running with power brakes) and observing brake fluid squirting up in the reservoir chambers. Due to a baffle incorporated in the master cylinder, only a minor fluid disturbance may be noticed in the rear chamber of the reservoir.

MASTER CYLINDER, REPLACE

1. Place drain pan under the master cylinder.
2. Using a suitable wrench, disconnect brake lines from the master cylinder. Plug all lines and fittings.
3. Remove master cylinder-to-brake booster attaching nuts.
4. Pull master cylinder outward and away from brake booster.
5. Reverse procedure to install noting the following:
   a. Prior to installing a new or rebuilt master cylinder, it should be bench bled as described under "Hydraulic System Bleeding & Flushing."
   b. **Torque** master cylinder-to-brake booster attaching nuts to 170-230 inch lbs.
   c. **Torque** brake line fittings to 170 inch lbs.

RESERVOIR, REPLACE

Refer to **Fig. 4** when performing this procedure.

Removal

1. Remove master cylinder from vehicle as described previously.
2. Thoroughly clean reservoir and housing.
3. Remove reservoir caps, then drain the brake fluid.
4. Clamp housing in a vise.
5. Grasp reservoir by hand, then separate it from the housing by rocking it from side to side. **Do not pry reservoir from the housing.**
6. Remove grommets from housing.

Installation

1. Install grommets in housing.
2. Apply clean brake fluid to reservoir mounting surface.
3. Position reservoir to housing so wording on reservoir reads from left to right from the front of the housing, **Fig. 4.**
4. Press reservoir into housing while rocking it from side to side. Reservoir is properly seated when bottom of reservoir touches the grommet.
5. Refer to "Hydraulic System Bleeding & Flushing" to bleed the hydraulic system.

MASTER CYLINDER OVERHAUL

When disassembling the master cylinder, note the position of all parts as they are removed for proper installation.

When disassembled, wash all parts in denatured alcohol or clean brake fluid only. Use an air hose to blow out all passages, orifices and valve holes. Air dry and place parts on clean paper or lint-free cloth.

Inspect master cylinder bore for scoring, rust, pitting or etching. Any of these conditions will require replacement of the housing. Never hone the bore of the master cylinder as this will remove the anodized surface. Inspect master cylinder pistons for scoring, pitting or distortion. Replace piston if any of these conditions exist. if either the master cylinder housing or piston is replaced, clean the new parts with denatured alcohol or clean brake fluid, and blow out all passages with air hose.

Examine reservoirs for foreign matter and check all passages for restrictions. If there is any indication of contamination or evidence of corrosion, service the hydraulic system as needed, and flush the entire system as described under "Hydraulic System Bleeding & Flushing."

When reassembling the master cylinder, use all parts contained in the repair kit. Dip all component parts in clean brake fluid, and place them on a clean surface. To prevent faulty operation, when installing seals inspect through side outlet of the dual master cylinder housing to make certain cup lips do not hang up on edge of hole or turn back. A piece of 3/16 inch rod with an end rounded off will be helpful in guiding cups past the hole.

TYPE 2—RECESSED CARTRIDGE MASTER CYLINDER

DESCRIPTION

The design of this master cylinder, **Fig. 5,** allows the internal components to be assembled outside the master cylinder body. The components are comprised of various springs, pistons, and seals. A magnetic float system used to monitor fluid level, is incorporated into the master cylinder reservoir.

TROUBLESHOOTING

Refer to "Troubleshooting," in "Disc Brakes" for troubleshooting of the hydraulic brake system.

MASTER CYLINDER, REPLACE

Refer to **Fig. 6,** when replacing master cylinder.

1. Position a drain pan under the master cylinder.
2. Disconnect electrical connector (M) from reservoir.
3. **On vehicles equipped with manual transmission,** disconnect hose from nipple (Z). Properly support hose. **Do not allow hose to fall below nipple level or the clutch will have to be bled.**
4. **On all vehicles,** loosen brake lines (A) at top of the proportioning valve (E), then disconnect brake lines (B) from the master cylinder. Plug all lines and fittings.
5. Remove master cylinder-to-brake booster attaching nuts (C).
6. Remove proportioning valve mounting bracket (D).
7. Pull master cylinder outward and away from brake booster.
8. Reverse procedure to install noting the following:
   a. Use caution during master cylinder to brake booster installation. The primary piston protrudes beyond the end cap, and may be damaged if not centered to the booster cavity.
   b. **Torque** brake line fittings to 13 ft. lbs.
   c. Bleed hydraulic system as described under "Hydraulic System Bleeding & Flushing."

MASTER CYLINDER OVERHAUL

When disassembling the master cylinder, note the position of all parts as they are removed for proper installation.

Refer to **Fig. 5,** when overhauling master cylinder.

Disassembly

1. Remove master cylinder from vehicle as described previously.
2. Using end cap remover/installer tool No. Fre. ML.01, remove end cap (V) and booster seal (U).
3. Using a suitable tool, pry out reservoir-to-body retaining clip (E).
4. Separate reservoir (C) from master cylinder body (F). Remove and discard reservoir mounting O-rings (D).
5. Separate primary piston assembly from end cap (V).
6. Separate primary spring (P), primary pumping seal (R), primary plastic spacer (S), and primary rear seal (T) from the primary piston (Q).
7. Remove plastic sleeve assembly from the master cylinder body.
8. Separate green secondary spring (G), secondary plumbing seal (H), secondary plastic spacer (J), secondary pressure seal (K), O-ring seal (L), and secondary piston (M) from the plastic sleeve (N).

Clean & Inspect

When disassembled, discard all rubber parts. Wash all components, and the master cylinder body in denatured alcohol or clean brake fluid. Use an air hose to blow out all passages, orifices and valve holes. Place parts on clean paper or lint-free cloth and allow to air dry.

Inspect master cylinder bore for scoring, rust, pitting or etching. Any of these conditions will require replacement of the housing. Never hone the bore of the master cylinder. Inspect all components for wear, damage, or corrosion. Replace as needed.

Examine reservoirs for foreign matter and check all passages for restrictions. If there is any indication of contamination or evidence of corrosion, service hydraulic parts as needed, then flush the entire system as described under "Hydraulic System Bleeding & Flushing."

Assembly

When reassembling the master cylinder, use all parts contained in the repair kit. Coat parts with clean brake fluid, and place them on a clean surface.

The following components may be assembled in either direction due to their symmetrical design: Reservoir-to-master cylinder body O-ring seals, primary and secondary springs, the O-ring seal between the secondary pressure seal and secondary piston, primary plastic spacer, primary rear seal, and the interface seal between the master cylinder and vacuum servo.

*NOTE: The master cylinder is available as a replacement unit either completely assembled or with the body and internal components not assembled.

A - Reservoir Cap
B - Cap Seal
C - Reservoir
D - O-ring Seals
E - Retaining Clip
F - Master Cylinder Body*
G - Secondary Spring (Green)
H - Secondary Pumping Seal
J - Secondary Plastic Spacer
K - Secondary Pressure Seal
L - O-ring Seal
M - Secondary Piston
N - Plastic Sleeve
P - Primary Spring
Q - Primary Piston
R - Primary Pumping Seal
S - Primary Plastic Spacer
T - Primary Rear Seal
U - Seal—Master Cylinder-to-Booster
V - End Cap
W - Retaining Nuts

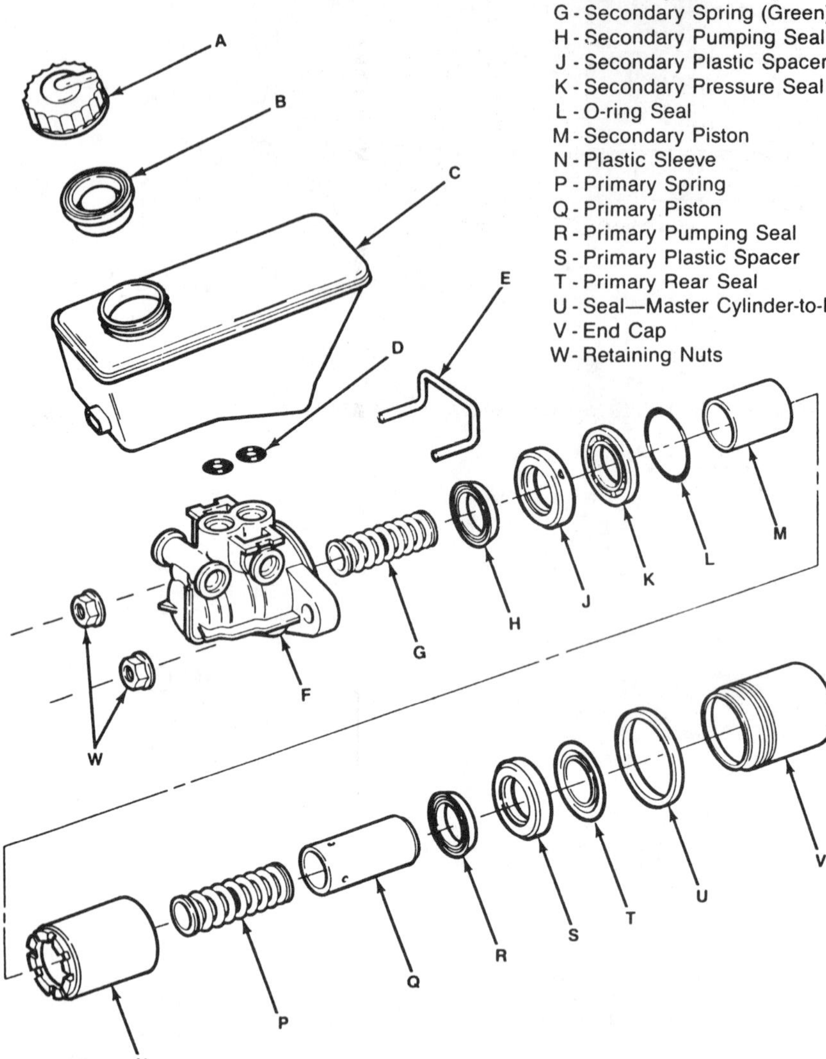

Fig. 5 Type 2, recessed cartridge master cylinder

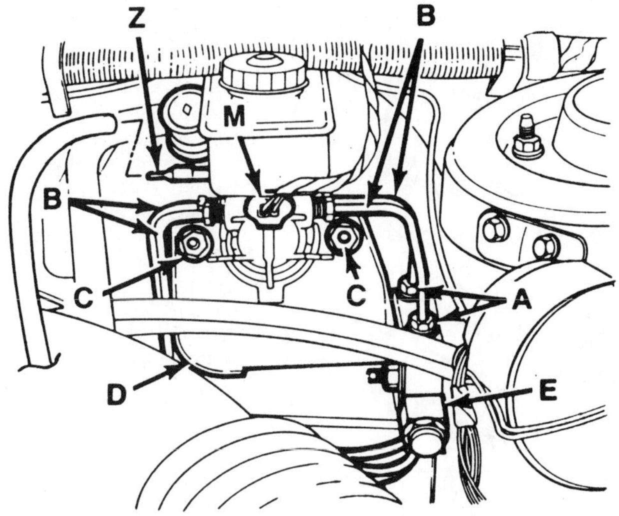

Fig. 6 Master cylinder replacement

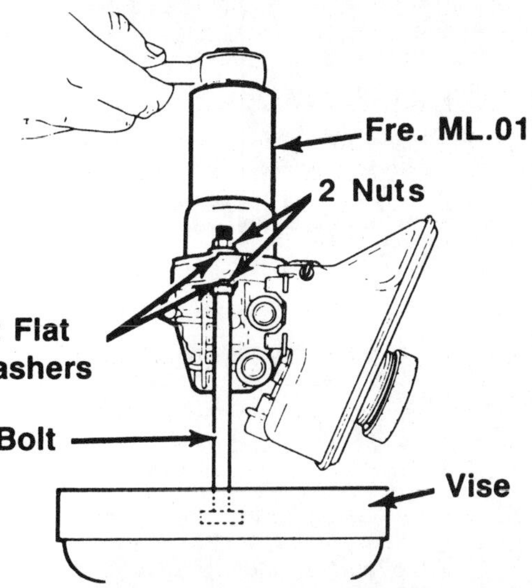

Fig. 7 Supporting master cylinder

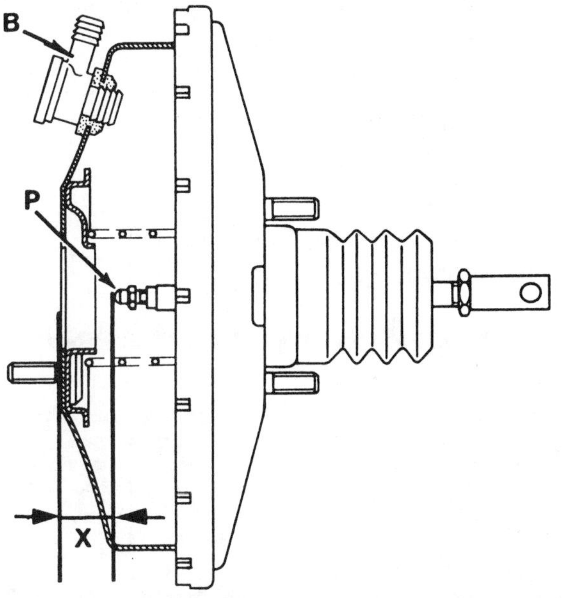

Fig. 8 Brake booster pushrod depth adjustment

1. Install secondary pumping seal (H) over secondary piston (M).
2. Working from closed side of the secondary piston, install secondary plastic spacer (J) onto secondary piston (M).
3. Slide secondary pressure seal (K) onto secondary piston (M).
4. Install O-ring seal (L) onto plastic sleeve (N).
5. Install secondary piston assembly into plastic sleeve (N). There are six deep and six shallow notches at the end of the plastic sleeve. Rotate pressure seal until seal grooves align and fall into six deep notches on the plastic sleeve.
6. Install green secondary spring (G) into end of plastic piston.
7. Gently rotate plastic sleeve assembly into master cylinder body (F).
8. Install primary spring (P) into primary piston (Q).
9. Install primary pumping seal (R), primary plastic spacer (S), and primary rear seal (T) over the primary piston (Q).
10. Install primary piston assembly into end cap (V). **Primary pumping seal must seat inside the end cap.**
11. Position master cylinder-to-booster seal (U) on end cap (V).
12. Position master cylinder body (F) with opening facing up.
13. Install end cap (V) into master cylinder body (F). At this time tighten end cap by hand only.
14. Install O-rings (D) into reservoir mounting surface.
15. While pressing reservoir (C) into master cylinder body (F), install retainer clip (E). Retainer clip (E) should click into position when properly seated.
16. To tighten end cap, support master cylinder using bolt fixture as shown in **Fig. 7.** Never clamp master cylinder body in any type of vise.
17. Using end cap remover/installer tool No. Fre. ML.01, **torque** end cap (V) to 9 ft. lbs.
18. Remove master cylinder from bolt fixture.

TYPE 3—COMPOSITE MASTER CYLINDER

TESTING

Refer to "Troubleshooting," in "Disc Brakes" for troubleshooting of the hydraulic brake system.

MASTER CYLINDER, REPLACE

1. Disconnect brake fluid sensor electrical connector from master cylinder reservoir.
2. Position a drain pan under the master cylinder.
3. Remove brake fluid from reservoir.
4. Grasp reservoir by hand, then pull straight up to separate it from the housing.
5. Using a suitable wrench, disconnect brake lines from the master cylinder. Plug all lines and fittings.
6. Remove master cylinder-to-brake booster attaching nuts.
7. Pull master cylinder outward and away from brake booster. Discard booster-to-master cylinder O-ring at the back of the master cylinder.
8. Reverse procedure to install noting the following:
   a. Prior to master cylinder installation, adjust brake booster pushrod to a depth of .878 inch as shown **Fig. 8.**
   b. When installing the reservoir it should click into position when properly seated.

c. **Torque** master cylinder-to-booster attaching nuts to 10 ft. lbs.
d. **Torque** brake line fittings to 10 ft. lbs.
e. Bleed hydraulic system as described under "Hydraulic System Bleeding & Flushing."

TYPE 4—CAST MASTER CYLINDER

TESTING

Refer to "Troubleshooting," in "Disc Brakes" for troubleshooting of the hydraulic brake system.

MASTER CYLINDER, REPLACE

1. Position a drain pan under the master cylinder.
2. Using a suitable wrench, disconnect brake lines from the master cylinder. Plug all lines and fittings.
3. Remove master cylinder-to-brake booster attaching nuts.
4. Pull master cylinder outward and away from brake booster.
5. Reverse procedure to install noting the following:
   a. Prior to installing a new or rebuilt master cylinder, it should be bench bled as described under "Hydraulic System Bleeding & Flushing."
   b. **Torque** master cylinder-to-brake booster attaching nuts to 30-35 ft. lbs.
   c. **Torque** brake line fittings to 160-200 inch lbs.

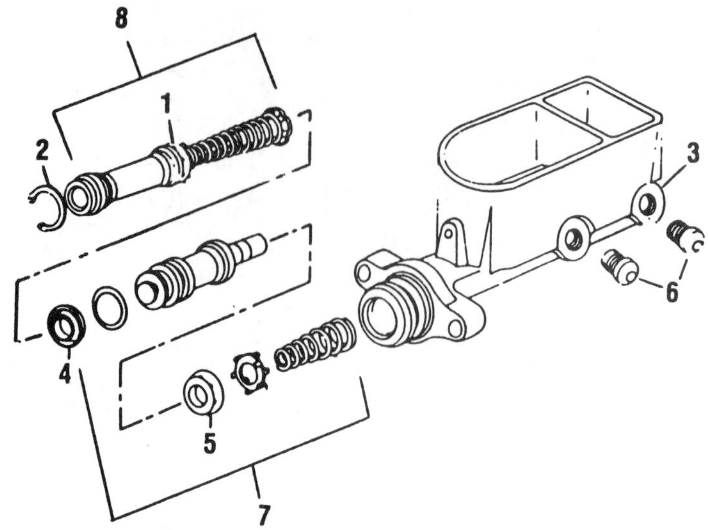

1 primary piston
2 snap ring
3 outlet port
4 front piston seal
5 rear piston seal
6 seats
7 secondary piston assembly
8 primary piston assembly

Fig. 9 Type 4, cast master cylinder

MASTER CYLINDER OVERHAUL

Refer to **Fig. 9,** when performing the following procedure.

Disassembly

When disassembling the master cylinder, note the position of all parts as they are removed for proper installation.

1. Remove master cylinder cap, then drain the brake fluid.
2. Properly support master cylinder.
3. Using a wooden stick or dowel, depress primary piston into master cylinder bore. Remove snap ring, then release pushrod.
4. Remove primary piston assembly from master cylinder bore. Discard primary piston.
5. Remove secondary piston assembly from master cylinder bore by applying air pressure to the secondary outlet port.
6. Remove and discard secondary piston seals.
7. Remove outlet port tube seats as follows:
   a. Using a 13/64 drill bit, enlarge outlet port seats.
   b. Position a flat washer on each outlet port.
   c. Thread a 1/4-20 X 3/4 screw into the seat. Tighten screw until seat is loose, then remove the seat.

Clean & Inspect

When disassembled, wash all parts in denatured alcohol or clean brake fluid. Use an air hose to blow out all passages, orifices and valve holes. Air dry and place parts on clean paper or lint-free cloth.

Inspect casting for cracks and damage, and the master cylinder bore for scoring, rust, pitting, or etching. Any of these conditions will require the master cylinder to be replaced. Inspect components for wear or damage, and replace as needed.

Examine reservoir for foreign matter and check all passages for restrictions. If there is any indication of contamination or evidence of corrosion, service the hydraulic system as needed, then flush the entire system as described under "Hydraulic System Bleeding & Flushing."

Assembly

When assembling the master cylinder, use all parts contained in the repair kit. Coat all component parts with clean brake fluid, and place on a clean surface.

1. Using spare tube fitting nuts, press new seats into outlet ports.
2. Install secondary piston seals. Ensure rear piston seal lip faces the bore interior, and front piston seal lip faces away from the bore.
3. Position seal retainer and return spring on the secondary piston.
4. Install secondary piston assembly, then the primary piston assembly into cylinder bore.
5. Using a wooden stick or dowel, depress primary piston into cylinder bore. Install snap retaining ring, then release pushrod.

TYPES 5, 6, 7, & 8—COMPOSITE MASTER CYLINDERS

TESTING

Refer to "Troubleshooting," in "Disc Brakes" for troubleshooting of the hydraulic brake system.

MASTER CYLINDER, REPLACE

1. Disconnect brake fluid sensor electrical connector from master cylinder reservoir.
2. Position a drain pan under the master cylinder.
3. **On models with separately mounted reservoir,** disconnect reservoir hoses from the master cylinder.
4. **On all models,** use a suitable wrench to disconnect brake lines from the master cylinder. Plug all lines and fittings.
5. Remove master cylinder-to-brake booster attaching nuts.
6. Pull master cylinder outward and

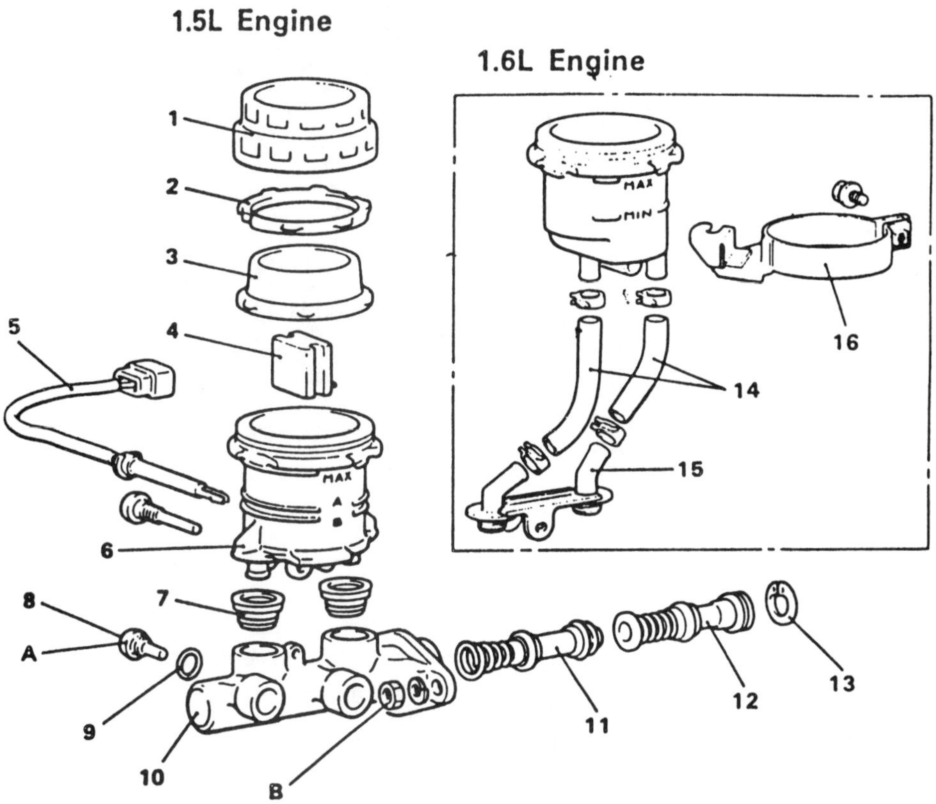

	ft.lbs.
A	1.1–2.2
B	6–9

1. Reservoir cap
2. Slide ring
3. Diaphragm
4. Float
5. Fluid level sensor
6. Reservoir
7. Reservoir seal
8. Piston stopper bolt
9. Gasket
10. Master cylinder body
11. Secondary piston assembly
12. Primary piston assembly
13. Stopper ring
14. Reservoir hose
15. Nipple
16. Reservoir holder assembly

Fig. 10 Type 5, composite master cylinder. 1985–86 Colt, & 1987–88 Colt w/1.5L engine

away from brake booster.

7. Reverse procedure to install noting the following:
   a. Prior to master cylinder installation, adjust brake booster pushrod as described under "Adjustments."
   b. Bleed hydraulic system as described under "Hydraulic System Bleeding & Flushing."

MASTER CYLINDER OVERHAUL

When performing overhaul procedures, refer to **Figs. 10 through 13,**

Disassembly

When disassembling the master cylinder, note the position of all parts as they are removed for proper installation.

1. Use a wooden stick or dowel to depress primary piston into cylinder bore.
2. With pushrod depressed, remove piston stopper bolt, then the piston stopper ring.
3. Release the pushrod, then remove piston assemblies. If secondary piston assembly is stuck in the bore, apply a light amount of compressed air to the secondary outlet port until piston assembly works free.

Clean & Inspect

Examine reservoirs for foreign matter and check all passages for restrictions. If there is any indication of contamination or evidence of corrosion, service the hydraulic system as needed, then flush the entire system as described under "Hydraulic System Bleeding & Flushing."

When disassembled, wash all parts in denatured alcohol or clean brake fluid. Use an air hose to blow out all passages, orifices and valve holes. Air dry and place parts on clean paper or lint-free cloth.

1. Check components for wear, damage, or corrosion. Replace as needed.
2. Check master cylinder bore for scoring, rust, pitting or etching. Replace as necessary.
3. Using bore gauge, Tool No. 14U0063 or equivalent, measure master cylinder bore variation as shown in **Fig. 14.** Note readings, then refer to "Hydraulic Brakes Specifications" chart. If bore exceeds maximum diameter, replace the master cylinder.

Assembly

When assembling the master cylinder, **Figs. 10 through 13,** use all parts contained in the repair kit. Coat all components in clean brake fluid, and place on a clean surface.

ADJUSTMENTS

Brake Booster Pushrod-To-Primary Piston Clearance

1. With master cylinder removed from vehicle, as described under "Master Cylinder Replace," proceed as follows:
   a. Measure distance between master cylinder end face and the piston, **Fig. 15.** Position a square (straight scale) against edge of master cylinder, then measure and subtract thickness of the square to determine dimension B.
   b. Find dimension C by measuring distance between brake booster mounting surface and the end face, **Fig. 16.**
   c. Measure distance between master cylinder mounting surface and the pushrod end, **Fig. 17.** Find dimension D by subtracting the square's thickness from the measurement taken.
   d. Find brake booster-to-primary piston clearance dimension A, **Fig. 18,** using the formula A = B–C–D. Refer to the "Hydraulic Brake System Specifications" chart for proper clearance specifications.
   e. If dimension A is not as specified, adjust brake booster pushrod as shown **Fig. 19.**

HYDRAULIC BRAKE SYSTEM CONTROLS

TYPE 1—COMBINATION VALVE

DESCRIPTION

This valve assembly **Fig. 20,** is comprised of a warning switch, proportioning valve, and a hold-off valve.

The warning switch detects system failure when a pressure difference occurs between the front and rear brake systems. In such a case, the valve piston shuttles toward the side with the low pressure, forcing the switch plunger upward over the piston's tapered shoulder. This movement closes the switch contacts causing the brake warning lamp to illuminate. Upon restoration of equalized hydraulic system pressure the brake warning switch will reset.

The proportioning valve transfers full

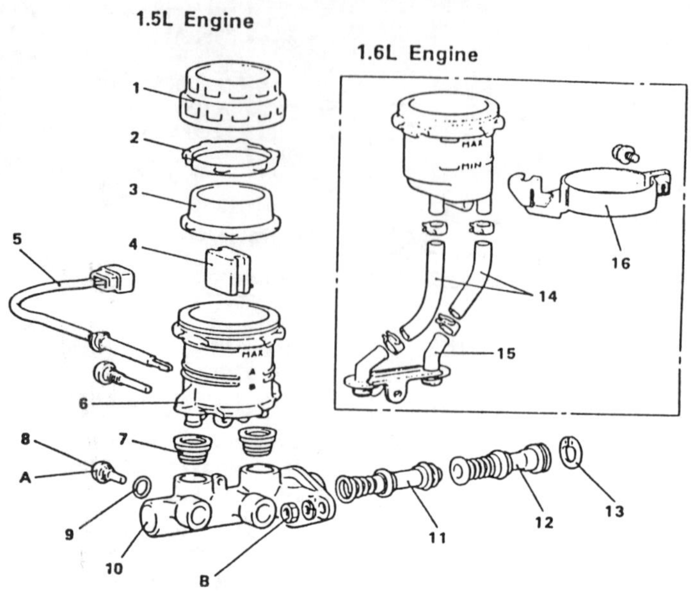

	ft.lbs.
A	1.1–2.2
B	6–9

1. Reservoir cap
2. Slide ring
3. Diaphragm
4. Float
5. Fluid level sensor
6. Reservoir
7. Reservoir seal
8. Piston stopper bolt
9. Gasket
10. Master cylinder body
11. Secondary piston assembly
12. Primary piston assembly
13. Stopper ring
14. Reservoir hose
15. Nipple
16. Reservoir holder assembly

Fig. 11 Type 6, composite master cylinder. Colt Vista

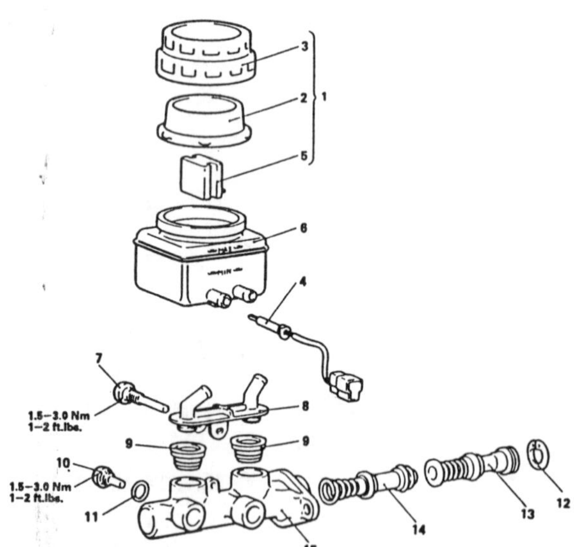

Disassembly steps
1. Reservoir cap assembly
2. Diaphragm
3. Reservoir cap
4. Brake fluid level sensor
5. Float
6. Reservoir
7. Nipple installation bolt
8. Nipple
9. Reservoir seal
10. ◆◆ Piston stopper bolt
11. Gasket
12. ◆◆ Piston stopper ring
13. ◆◆ ◆◆ Primary piston assembly
14. ◆◆ ◆◆ Secondary piston assembly
15. Master cylinder body

NOTE
(1) Reverse the disassembly procedures to reassemble.

Fig. 12 Type 7, composite master cylinder. 1987–88 Colt w/1.6L engine, & all 1989 Colt & Eagle Summit

braking force to the rear brakes until a preset ratio called the split point is achieved. At this point the pressure increase is routed away from the re ar brake system to prevent rear wheel lock-up.

The hold-off valve prevents front disc brake activation until the rear brake shoes can overcome return spring force, and achieve shoe to drum contact. The hold-off valve will not effect front brake pressure during hard stops, but will prevent front brake lock-up during light pedal application and on slippery road surfaces. This is achieved by keeping a low output pressure supplied to the front brakes until a preset hold-off pressure is achieved, wherein full output pressure is returned to the front brakes.

TESTING

Warning Light System

If the parking brake light is connected into the service brake warning light system, the brake warning light will flash only when the parking brake is applied with the ignition turned on. The same light will also glow should one of the two service brake systems fail when the brake pedal is applied.

To test the system, turn the ignition on and apply the parking brake. If the lamp fails to light, inspect for a burned out bulb, disconnected socket, a broken or disconnected wire at the switch.

To test the brake warning system, raise the car and open a wheel bleeder valve while a helper depresses the brake pedal and observes the warning light on the instrument panel. If the bulb fails to light, inspect for a burned out bulb, disconnected socket, or a broken or disconnected wire at the switch. If the bulb is not burned out, and the wire continuity is proven, replace the brake warning switch.

Hold-Off Valve

1. While watching the valve stem on the combination valve, have an assistant depress and release the brake pedal.
2. As the pedal is depressed, the valve stem should move outward. When the brake pedal is released, the valve stem should move inward.
3. If the valve stem does not move as specified, replace the combination valve and note the following:
   a. After installing a new combination valve, bleed the hydraulic brake system as described under "Hydraulic Brake System Bleeding &

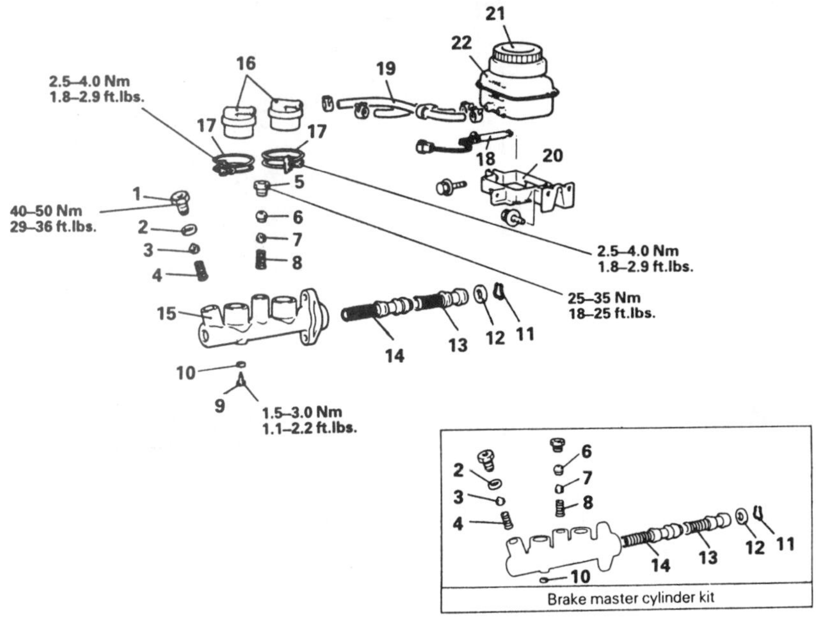

Brake master cylinder disassembly steps

1. Check valve case
2. Gasket
3. Check valve
4. Check valve spring
5. Check valve cap
6. Tube seat
7. Check valve
8. Check valve spring
9. ◆◆ Secondary piston stopper
10. Gasket
11. ◆◆ Piston stopper ring
12. Piston stopper plate
13. ◆◆ ✦✦ Primary piston
14. ◆◆ ✦✦ Secondary piston
15. ✦✦ Brake master cylinder body

Brake fluid reservoir disassembly steps

16. ◆◆ Nipple
17. Reservoir band
18. Brake fluid level sensor switch
19. Reservoir hoses
20. Reservoir bracket
21. ✦✦ Reservoir cap
22. Reservoir tank

NOTE
(1) Reverse the disassembly procedures to reassemble.

Fig. 13 Type 8, composite master cylinder. Conquest

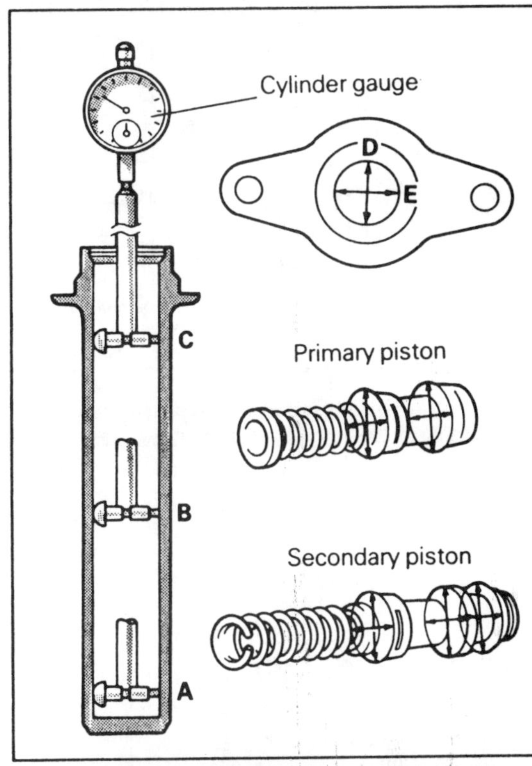

Fig. 14 Measuring master cylinder bore variation

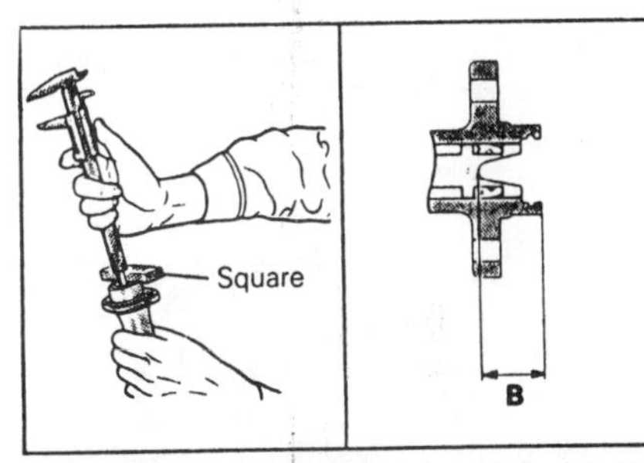

Fig. 15 Measuring distance between master cylinder end face and piston

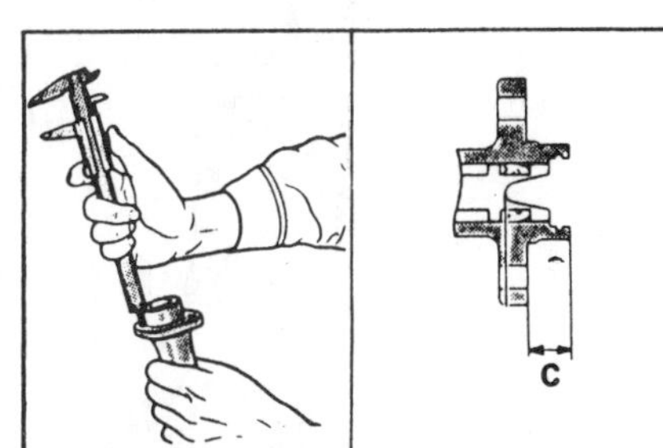

Fig. 16 Measuring distance between brake booster mounting surface and end face

Flushing."

b. Upon completion of bleeding the hydraulic brake system, reset the brake warning switch by depressing the brake pedal with moderate force. The brake warning lamp will turn off when the switch is reset.

Proportioning Valve

1. Install gauge set with T-fitting, tool No. C-4007A, as shown in **Fig. 21**. Special adapter tubes may be fabricated from a short piece of brake tube, a 9/16 X 18 tube nut, and a 3/8 X 24 tube nut.
2. Bleed the hose and gauge set.
3. While watching pressure gauges, have an assistant depress the brake pedal. Note gauge readings, then refer to "Hydraulic Brake Controls Specifications" chart.
4. If pressure readings are not as specified, replace the combination valve and note the following:
   a. After installing a new combination valve, bleed the hydraulic brake system as described under "Hydraulic Brake System Bleeding &

Flushing."

b. Upon completion of bleeding the hydraulic brake system, reset the brake warning switch by depressing the brake pedal with moderate force. The brake warning lamp will turn off when the switch is reset.

TYPE 2—COMBINATION VALVE

DESCRIPTION

This valve assembly **Fig. 22**, is comprised of a warning switch, and a dual proportioning valve.

The warning switch detects system failure when a pressure difference occurs between the diagonally split systems. In such a case, the valve piston shuttles toward the side with the low pressure forcing the switch plunger upward over the piston's tapered shoulder. This movement closes the switch contacts causing the brake warning lamp to illuminate. Upon restoration of equalized system pressure the brake warning switch will reset.

The proportioning valve transfers full braking force to the rear brakes until a pre-

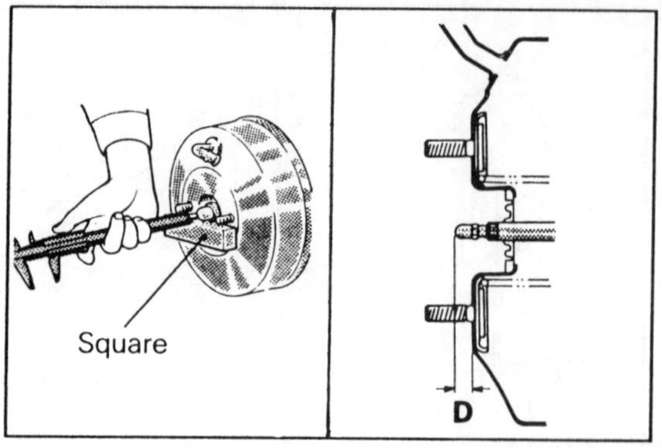

Fig. 17 Measuring distance between master cylinder mounting surface and pushrod end

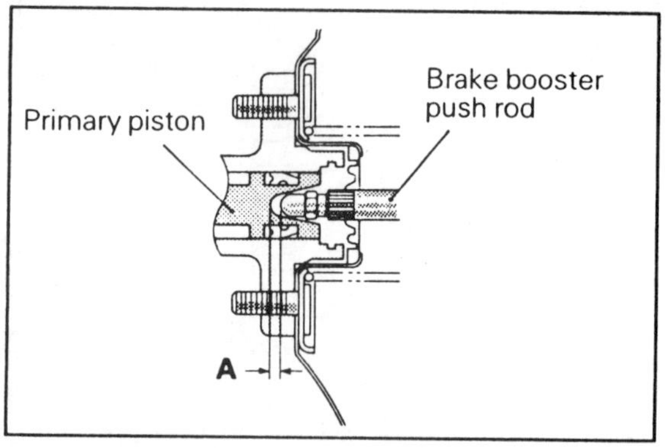

Fig. 18 Brake booster-to-primary piston clearance dimension A

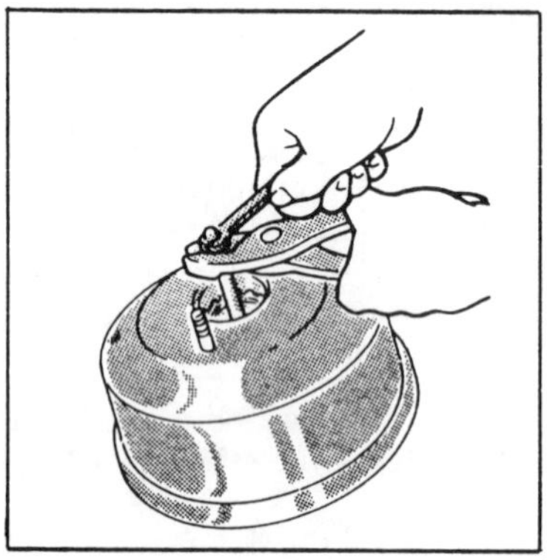

Fig. 19 Adjusting brake booster pushrod

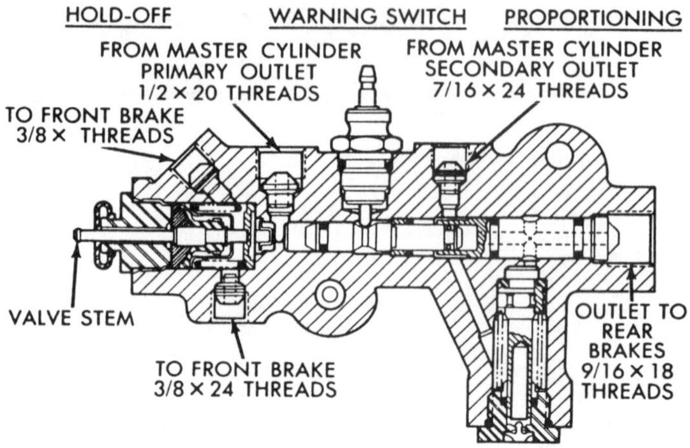

Fig. 20 Type 1, combination valve

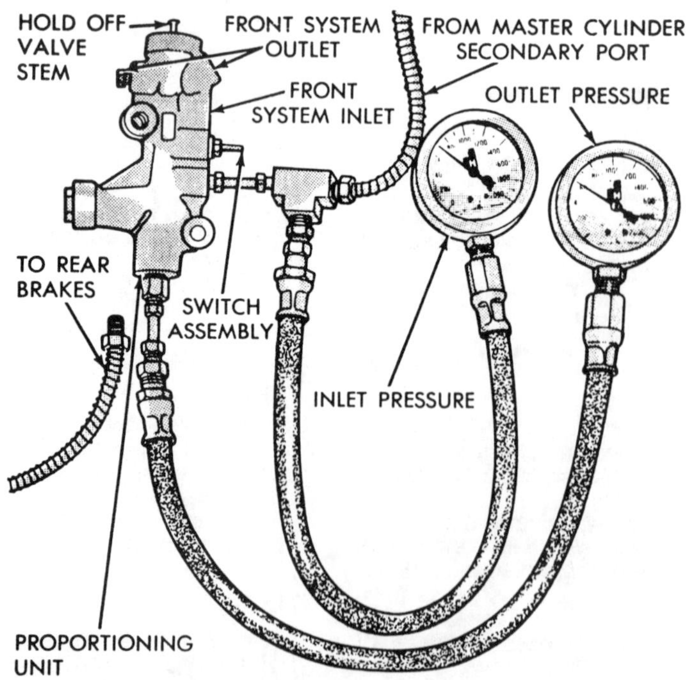

Fig. 21 Type 1, combination valve pressure gauge test connections

set ratio called the split point is achieved. At this point the pressure increase is routed away from the rear brake system to prevent rear wheel lock-up.

TESTING

Warning Light System

Refer to "Type 1—Combination Valve" for test procedure.

Proportioning Valve

1. Determine if fault exists in the left or right rear wheel.
2. Install gauge set, tool No. C-4007A, as shown **Figs. 23 or 24.** A special adapter may be fabricated from a short piece of brake tube and a 3/8 X 24 tube nut.
3. Bleed the hose and gauge set.
4. While watching pressure gauges, have an assistant depress the brake pedal. Note gauge readings, then refer to "Hydraulic Brake Controls Specifications" chart.
5. If pressure readings are not as specified, replace the combination valve and note the following:
   a. After installing a new combination valve, bleed the hydraulic brake system as described under "Hydraulic Brake System Bleeding & Flushing."

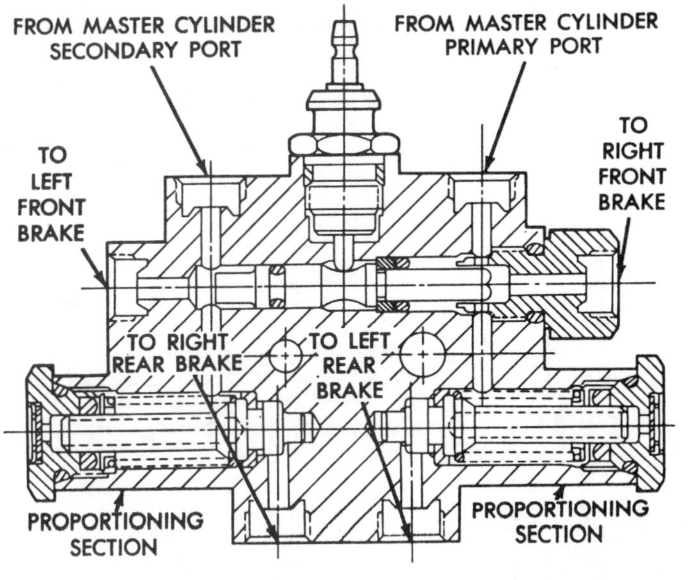

Fig. 22 Type 2, combination valve

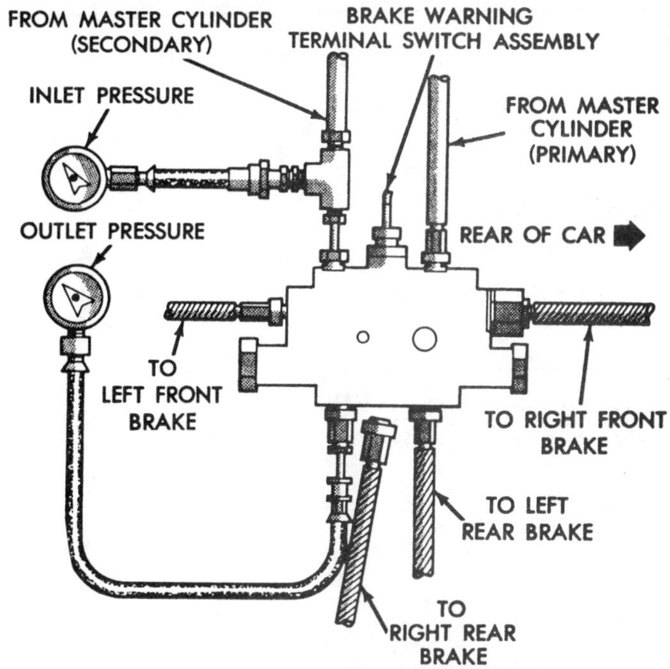

Fig. 23 Type 2, combination valve pressure gauge test connections. Right wheel malfunction

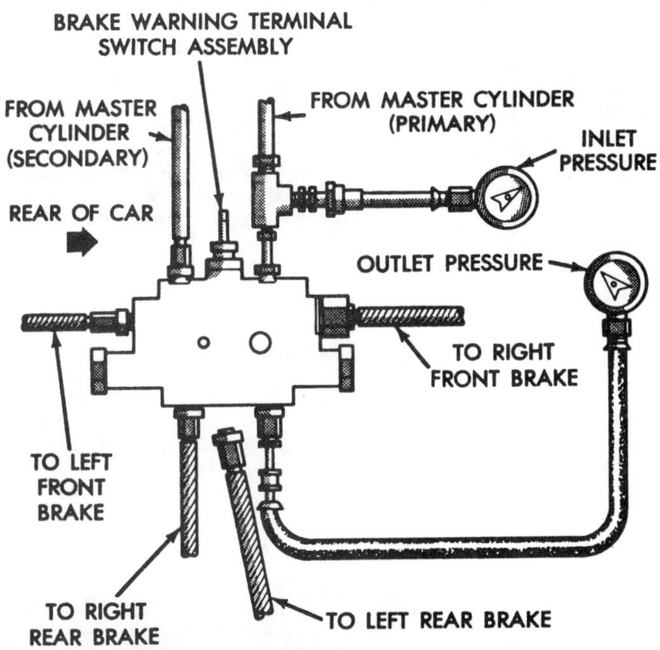

Fig. 24 Type 2, combination valve pressure gauge test connections. Left wheel malfunction

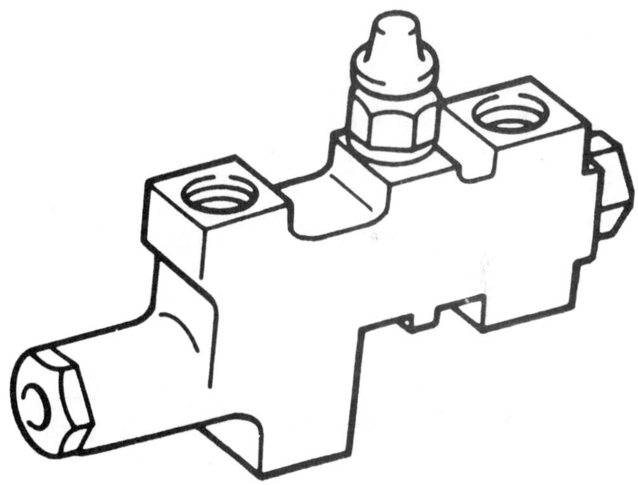

Fig. 25 Type 3, combination valve

b. Upon completion of bleeding the hydraulic brake system, reset the brake warning switch by depressing the brake pedal with moderate force. The brake warning lamp will turn off when the switch is reset.

TYPE 3—COMBINATION VALVE

Description

This valve assembly **Fig. 25**, is comprised of a pressure differential and proportioning valve.

The pressure differential valve detects system failure when a pressure difference occurs between front and rear systems. In such a case, the switch is activated causing the brake warning lamp to illuminate. Upon restoration of equalized system pressure the pressure differential switch resets itself.

The proportioning valve is incorporated into the hydraulic system to control braking force to the rear brakes.

Testing

Refer to "Troubleshooting," in "Disc Brakes" for diagnosis.

TYPE 4—COMPENSATOR LIMITER VALVE

Description

This system utilizes a compensator limiter valve with a diagonally split hydraulic system. The compensator limiter valve is incorporated into the hydraulic system to control braking force to the rear brakes. The valve is body mounted, and connected to the rear axle by a control rod and link.

Testing

1. Install suitable gauge set in place of the bleeder screws as shown in **Fig. 26.**
2. Bleed the hose and gauge set.
3. While watching pressure gauges, have an assistant depress the brake pedal. Note gauge readings, then refer to "Hydraulic Brake Controls Specifi-

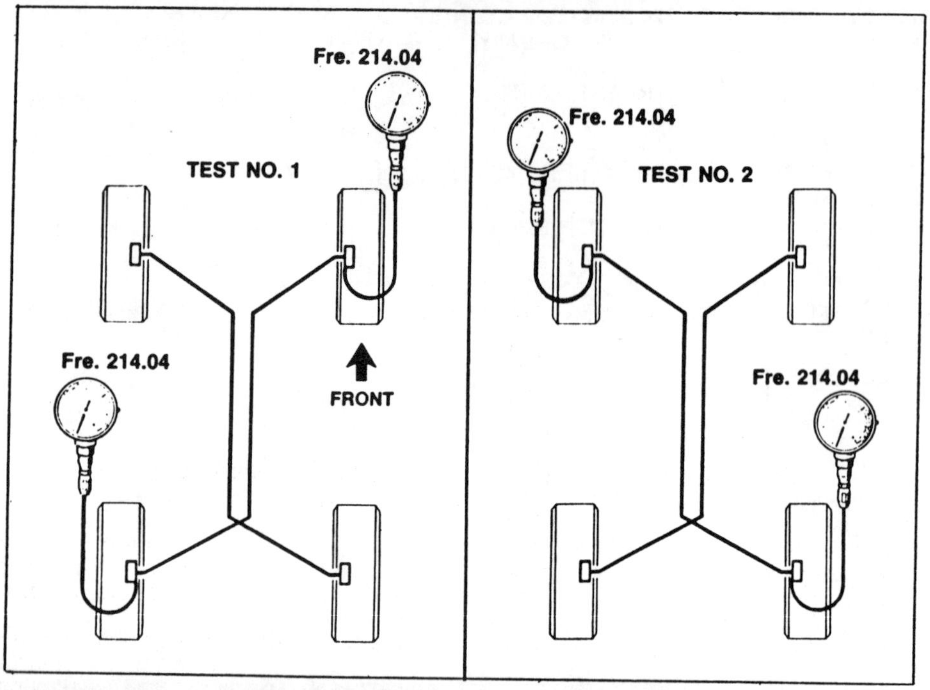

Fig. 26 Type 4, compensator-limiter valve & Type 5, proportioning valve pressure gauge test connections

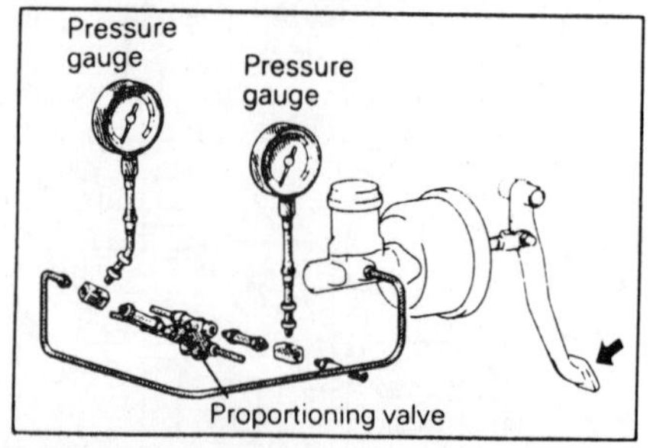

Fig. 27 Type 6, non load sensing proportioning valve pressure gauge test connection

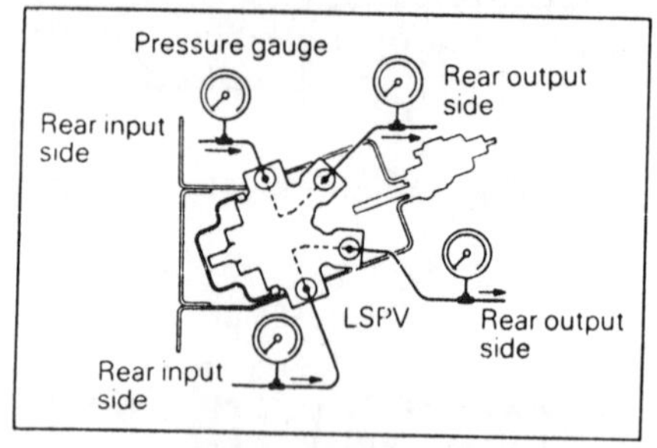

Fig. 28 Type 7, load sensing proportioning valve pressure gauge test connection

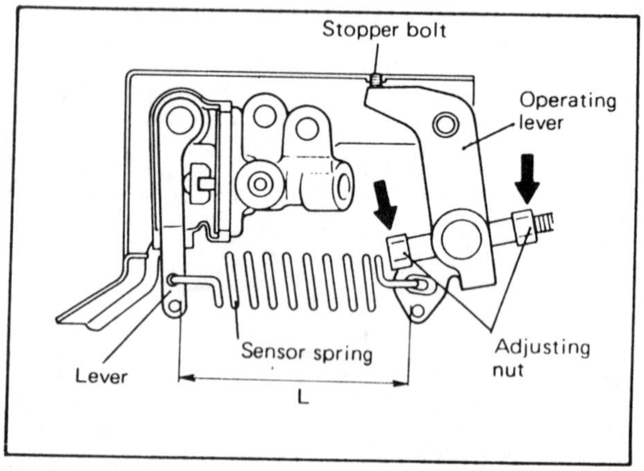

Fig. 29 Load sensing control unit. Colt Vista

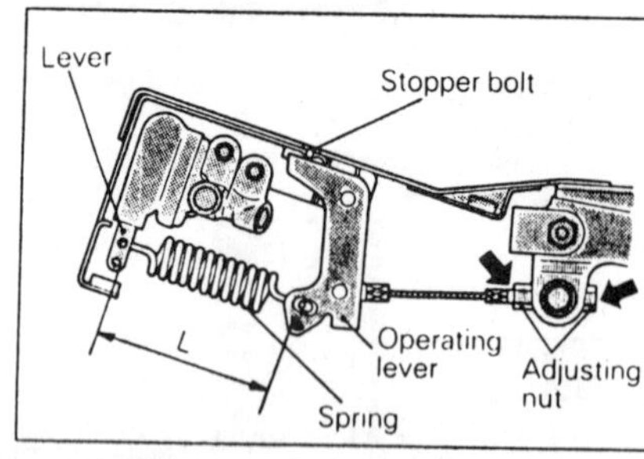

Fig. 30 Load sensing control unit. Colt

cations" chart.

4. If pressure readings are not as specified, replace the valve and note the following:
   a. After installing a new valve, bleed the hydraulic brake system as described under "Hydraulic Brake System Bleeding & Flushing."

TYPE 5—PROPORTIONING VALVE

Description

This system utilizes a proportioning valve with a diagonally split hydraulic system. The proportioning valve is incorporated into the hydraulic system to control braking force to the rear brakes.

Testing

Refer to "Type 4—Compensator Limiter Valve" for test procedure.

TYPE 6—NON LOAD SENSING PROPORTIONING VALVE

Testing

1. Install suitable gauge set as shown in **Fig. 27.**
2. Perform test procedure steps 2 through 4 as described under "Type 4—Compensator-Limiter Valve."

TYPE 7—LOAD SENSING PROPORTIONING VALVE

Prior to performing the following test procedures, park vehicle on a level surface, ensure vehicle is unloaded and at normal curb height.

Testing

1. Install suitable gauge set as shown in **Fig. 28.**
2. Bleed the hose and gauge set.
3. Loosen load sensing unit adjusting nuts, **Figs. 29 and 30.**
4. While watching pressure gauges, have an assistant depress the brake pedal.
5. **On Colt Vista models,** gauge pressure reading should be as follows:
   a. When input pressure is 740 psi, output pressure should read 307-377 psi.
   b. When input pressure is 1991 psi, output pressure should read 660-774.

c. Output pressure side to side difference should not exceed 57 psi.

6. **On 1988 Colt models,** gauge pressure reading should be as follows:
   a. When input pressure is 899 psi, output pressure should read 455-526 psi.
   b. When input pressure is 1991 psi, output pressure should read 765-879.
   c. Output pressure side to side difference should not exceed 57 psi.
7. **On all models,** if pressure readings are as specified proceed to steps 9 or 10 as applicable.
8. If pressure readings are not as specified, replace the proportioning valve and bleed the hydraulic brake system as described under "Hydraulic Brake System Bleeding & Flushing."
9. **On Colt Vista models,** adjust load sensing spring distance L to 4.07 inch, **Fig. 29.**
10. **On 1988 Colt models,** adjust load sensing spring distance L to 4.10 inch, **Fig. 30.**
11. **On all models,** while watching pressure gauges, have an assistant depress the brake pedal.
12. **On Colt Vista models,** gauge pressure reading should be as follows:
    a. When input pressure is 1991 psi, output pressure should read 1228-1580.
    b. Output pressure side to side difference should not exceed 121 psi.
13. **On 1988 Colt models,** gauge pressure reading should be as follows:
    a. When input pressure is 1991 psi, output pressure should read 1254-1496.
    b. Output pressure side to side difference should not exceed 121 psi.
14. **On all models,** if pressure readings are as specified, proceed to steps 16 or 17 as applicable.
15. If pressure readings are not as specified, replace the proportioning valve and bleed the hydraulic brake system as described under "Hydraulic Brake System Bleeding & Flushing."
16. **On Colt Vista models,** adjust load sensing spring distance L to 3.42-3.50 inch, **Fig. 29.**
17. **On 1988 Colt models,** adjust load sensing spring distance L to 3.51-3.59 inch, **Fig. 30.**

TYPE 8—PROPORTIONING VALVE

Testing

1. Install suitable gauge set as shown in **Fig. 31.**
2. Perform test procedure steps 2 through 4 as described under "Type 4—Compensator-Limiter Valve."

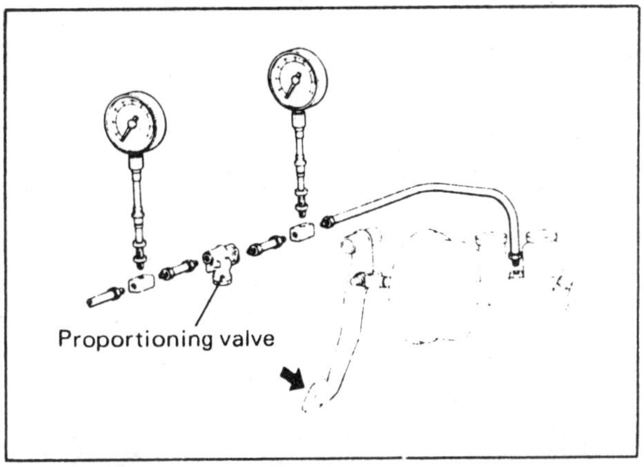

Fig. 31 Type 8, proportioning valve pressure gauge test connections

HYDRAULIC SYSTEM BLEEDING & FLUSHING

DESCRIPTION

Bleeding the hydraulic brake system is necessary if air has entered the system. A few causes for this condition are low fluid level, a hydraulic fluid leak, a hydraulic line is opened, or replacement of a hydraulic system component. Symptoms can be noted by an improper or loss of brake operation, and/or a low or spongy brake pedal.

Flushing the hydraulic brake system is necessary if contaminates are found in the hydraulic system. A few causes for hydraulic system contamination are moisture, age of hydraulic system parts and fluid, or improper fluid used in the system. Symptoms can be noted by an improper or loss of brake operation, swollen and deteriorated cups and other rubber parts, and/or a discoloration of the brake fluid.

The hydraulic fluid is bled or flushed from the system through bleeder valves located on the calipers, wheel cylinders, and some master cylinders. When bleeding the hydraulic brake system, use only specified brake fluid, and never reuse old brake fluid removed from the system.

PRESSURE BLEEDING

Pressure bleeding is recommended for all hydraulic brake systems. It is the fastest method because the master cylinder is automatically fed brake fluid from the pressure bleeder reservoir, and no pedal pumping is needed so only one person is required to perform the procedure. However, if pressure bleeding equipment is not available, the hydraulic system may be bled as described under "Manual Bleeding."

When pressure bleeding, to prevent air from getting into the hydraulic system, do not shake the pressure tank. Set the tank in the required location, bring the air hose to the tank, and do not move it during the bleeding operation. The tank should be kept at least one-third full.

The bleeder valve should be opened at least one full turn, and intermittently closed at about four-second intervals. This gives a whirling action to fluid in the hydraulic system, and helps expel the air. Refer to "Hydraulic Brake System Specifications" chart for proper wheel bleeding sequence.

Precautions

Except Eagle Medallion & Premier

Normal pressure from the pressure bleeder should not be greater than about 35 psi. On vehicles equipped with plastic reservoirs, do not exceed 25 psi bleeding pressure.

On models with hold-off valves contained in the combination valve, the valve stem on the outside of the combination valve must be held in position during bleeding using tool No. C-4121 or equivalent.

Eagle Medallion & Premier

Use brake bleeding apparatus M.S.815 as shown, **Fig. 32.** When bleeding the hydraulic brake system, always support rear wheels at their normal curb height, and never start the engine. Open fluid feed valve until reservoir is full, then turn on compressed air to a minimum pressure of 73 psi. Upon completion, fill reservoir as needed.

MANUAL BLEEDING

On models with power brakes, if bleeding the hydraulic system without the engine running, first reduce vacuum in the power unit to zero by pumping the brake pedal several times with the engine off.

1. Ensure master cylinder reservoir is full. Add suitable brake fluid as needed, and securely reinstall the master cylinder cap.
2. Raise and support vehicle.
3. Position a drain pan under the wheel being bled. Refer to "Hydraulic Brake System Specifications" chart for proper wheel bleeding sequence.
4. Have an assistant depress the brake pedal with a slow even pressure and hold it. **Do not depress brake pedal fully to the end of the master cylinder stroke. This may cause damage to the master cylinder.**
5. Using a suitable wrench, open the bleeder valve one full turn. Watch for air bubbles in the fluid, and listen for air escaping from the system.
6. With the brake pedal still depressed, close the bleeder valve.
7. Have the assistant pump the brake pedal several times. **Do not depress brake pedal fully to the end of the master cylinder stroke, this may cause damage to the master cylinder.**

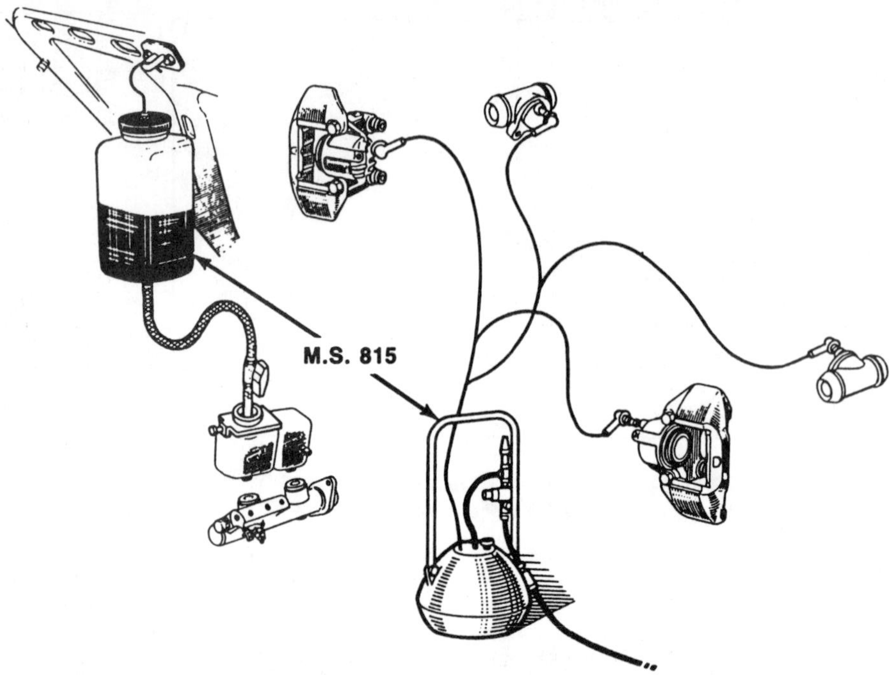

Fig. 32 Power bleeding brakes. Eagle Medallion & Premier

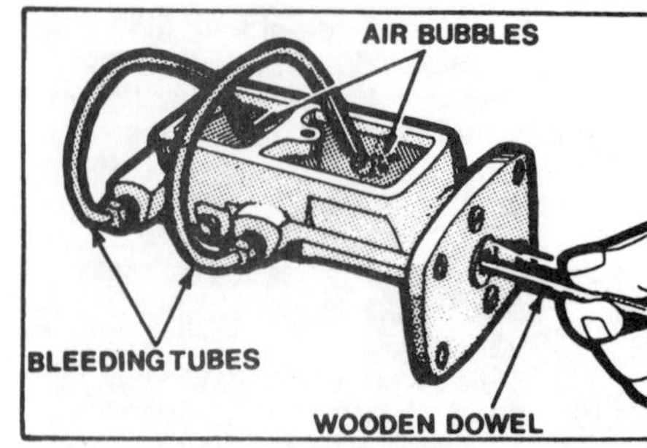

Fig. 33 Common means of bench bleeding a master cylinder

8. Repeat steps 4 through 7 following the wheel bleeding sequence specified in "Hydraulic Brake System Specifications" chart. Ensure all air is removed from the hydraulic system. **While bleeding the system, recheck brake fluid supply in the master cylinder often so as not to allow the master cylinder to run dry.**
9. Upon completion of hydraulic system bleeding proceed as follows:
   a. Ensure the master cylinder reservoir is full. Add suitable brake fluid as needed, and securely reinstall the master cylinder cap.
   b. Lower the vehicle. Ensure brake pedal is firm and braking operation is proper.

BLEEDING THE MASTER CYLINDER

Refer to "Bench Bleeding" procedure first if the master cylinder has been removed from the vehicle for service or replacement.

ON-VEHICLE SERVICE

Master cylinders may be bled manually or by pressure bleeding. It is recommended that the master cylinder be bled before bleeding the wheel cylinders and calipers.

Pressure Bleeding

1. Ensure master cylinder reservoir is full. Add suitable brake fluid as needed.
2. Install pressure bleeder according to manufacturer specifications.
3. Position a drain pan under the master cylinder.
4. Loosen brake lines, or bleeder valves if equipped, at the master cylinder. Watch for air bubbles in the fluid and listen for air escaping from the system.
5. Retighten brake lines or bleeder valves.
6. Bleed the master cylinder until all air is removed. Upon completion of bleeding, proceed as follows:
   a. Ensure the master cylinder reservoir is full. Add suitable brake fluid as needed, and securely reinstall the master cylinder cap.
   b. Ensure brake pedal is firm, and braking operation is proper.

Manual Bleeding

1. Ensure master cylinder reservoir is full. Add suitable brake fluid as needed, and securely reinstall the master cylinder cap.
2. Position a drain pan under the master cylinder.
3. Have an assistant depress the brake pedal with a slow even pressure and hold it. **Do not depress brake pedal fully to the end of the master cylinder stroke. This may cause damage to the master cylinder.**
4. Loosen the brake lines, or bleeder valves if equipped, at the master cylinder. Watch for air bubbles in the fluid and listen for air escaping from the system. **While bleeding the system, recheck brake fluid supply in the master cylinder often so as not to allow the master cylinder to run dry.**
5. With brake pedal still depressed, tighten the brake lines or bleeder valves.
6. Bleed the master cylinder until all air is removed. Upon completion of bleeding, proceed as follows:
   a. Ensure the master cylinder reservoir is full. Add suitable brake fluid as needed, and securely reinstall the master cylinder cap.
   b. Ensure brake pedal is firm, and braking operation is proper.

Bench Bleeding

When replacing or overhauling a master cylinder it is advisable to bleed it before installing it on the car.

1. Properly support master cylinder assembly and attach special bleeding tubes as shown in, **Fig. 33.**
2. Fill reservoir with approved brake fluid.
3. Using a wooden stick or dowel (cars with power brakes) depress pushrod slowly and allow the pistons to return under pressure of the springs. Do this several times until all air bubbles are expelled.
4. Remove bleeding tubes from cylinder and install cover and gasket.
5. Install master cylinder onto vehicle as described under "Master Cylinder, Replace."
6. Bleed the entire hydraulic system as described previously.

Precautions

Combination valves containing brake warning or pressure differential type valves are self-centering. They will reset after hydraulic system pressure is equalized, and by doing the following: Upon completion of bleeding the hydraulic system, depress the brake pedal with moderate force. The brake warning lamp will turn off when the switch is recentered.

WHEEL CYLINDERS SERVICE

REMOVAL

1. Remove wheel, drum and brake shoes.
2. Loosen brake line fitting at wheel cylinder. Do not pull metal line away from cylinder.
3. Remove screws holding cylinder to backing plate.
4. Separate wheel cylinder from brake line and backing plate by pulling the cylinder outward and away from backing plate.

OVERHAUL

Note position of all parts as they are re-

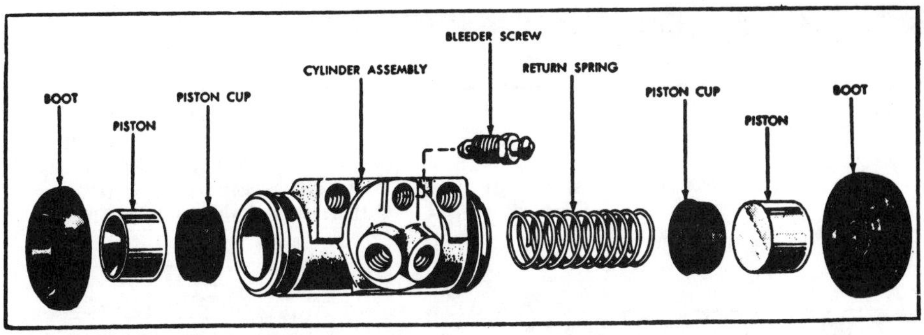

Fig. 34 Disassembled view of typical wheel cylinder

moved for proper installation.

Disassembly

1. Refer to **Fig. 34,** then remove boots, pistons, cups and spring from wheel cylinder.
2. Wipe cylinder walls with denatured alcohol or clean brake fluid.
3. Examine cylinder bore. A scored bore may be honed providing the bore diameter is not increased more than .005 inch. Replace as necessary.
4. Check pistons for wear or damage. Replace as necessary.

Assembly

1. Before assembling, wash hands with soap and water so as not to contaminate rubber parts.
2. Use all parts contained in repair kit. Lubricate cylinder wall and rubber cups with brake fluid.
3. Properly install spring, cups, pistons and boots in housing.
4. Bleed system as outlined previously, and adjust brakes.

INSTALLATION

1. Wipe end of hydraulic line to remove any foreign matter.
2. Position wheel cylinder to backing plate. Install brake line to cylinder and start connecting fitting.
3. Secure wheel cylinder to backing plate, then complete tightening of brake line fitting.
4. Install brake shoes, drum and wheel.
5. Bleed system as outlined previously, and adjust brakes.

HYDRAULIC TUBING SERVICE

Never use copper tubing as a replacement for steel tubing. Copper tubing is subject to fatigue cracking and corrosion which could result in brake system failure.

Steel tubing is used to transfer hydraulic pressure to the brakes. All fittings, tubing and hose should be inspected for rust, damage or defective flared seats. The tubing is equipped with a double flare/inverted seat or I.S.O. flare to insure more positive seating in the fitting.

DOUBLE FLARE/INVERTED SEAT

1. Using the tool shown in **Fig. 35** or equivalent, cut off the damaged seat or damaged tubing.
2. Ream out any burrs or rough edges showing on inside edges of tubing. This will make the ends of the tubing square and insure better seating on the flared end. Before flaring tubing, place a compression nut on tubing.
3. Open handles of flaring tool and rotate jaws of tool until mating jaws of tubing size are centered in the area between vertical posts.
4. Slowly close handles with tubing inserted in jaws but do not apply heavy pressure to handle as this will lock tubing in place.
5. Referring to **Fig. 35,** place gauge on edge over end of tubing and push tubing through jaws until end of tubing contacts recessed notch of gauge matching size of tubing.
6. Squeeze handles of flaring tool and lock tubing in place.
7. Place proper size plug of gauge down in end of tubing. Swing compression disc over gauge and center tapered flaring screw in recess in disc.
8. Lubricate taper of flaring or screw and screw in until plug gauge has seated in jaws of flaring tool. This action has started to invert the extended end of tubing.
9. Remove gauge and apply lubricant to tapered end of flaring screw and continue to screw down until tool is firmly seated in tubing.
10. Remove tubing from flaring tool and inspect the seat. If seat is cracked, cut off cracked end and repeat flaring operation.

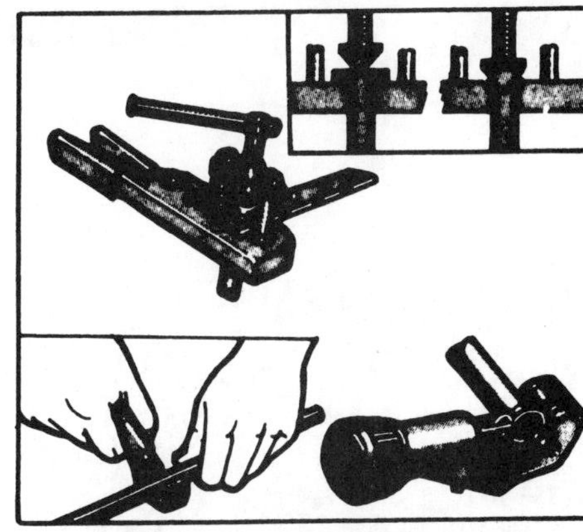

Fig. 35 Flaring hydraulic brake tubing

HYDRAULIC BRAKE SYSTEM SPECIFICATIONS

Year	Model	Master Cylinder Bore Dia. Inch	Booster to Primary Piston Clearance Inch	Wheel Bleeding Sequence
1985-88	Colt	① ②	.016-.031	LR,RF,RR,LF
	Eagle 4 X 4	.940	—	RR,LR,RF,LF
1985-89	Chrysler R.W.D.	1 1/32	—	RR,LR,RF,LF
	Chrysler F.W.D.	.827	—	RR,LR,RF,LF
	Conquest	.940①	.028-.043	RR,LR,RF,LF
	Colt Vista	.870①	.016-.031	LR,RF,RR,LF
1987-89	Eagle Medallion	.810	.878	③
1988-89	Eagle Premier	.945	—	③
1989	Colt & Eagle Summit	④	⑤	LR,RF,RR,LF

①—Bore I.D. maximum variation measured at three positions, .006 inch.

②—Models w/1.5L engine, .810 inch; models w/1.6L engine, .870 inch.

③—Begin at LR, then bleed the remaining three.

④—Models w/1.5L engine, 13/16 inch; models w/1.6L engine, 14/16 inch.

⑤—Models w/7 inch booster, .20-.28 inch; models w/8 inch booster, .24-.31 inch.

HYDRAULIC BRAKE CONTROLS SPECIFICATIONS

Year	Model	Valve Identification	Valve Tag Color ①	Split Point psi	Hold-Off Cut-In psi	Inlet Pressure psi From Master Cylinder	Outlet Pressure psi To Rear Brakes
1985-86	Charger, Horizon, Omni, Shelby Charger & Turismo,	②③	—	500	—	1000	540-730
	Aries, Lebaron, Reliant & 600	②③	Gray	600	—	1000	700-850
	Aries, Lebaron, Reliant & Town & Country	②③	Black	800	—	1000	850-980
	Daytona & Laser	②③	Black	800	—	1000	850-980
1985-87	Caravelle, Diplomat, Fifth Ave, Gran Fury & Newport	④⑤	Blue	220	117	1000	500-600
	Conquest	③	—	448	—	853	549-606
	Colt	③	—	384	—	953	520-590
1985-88	Caravelle, New Yorker & 600	②	—	—	—	—	—
	Eagle 4 X 4	②③	—	—	—	—	—
1985-89	Caravelle, Diplomat, Fifth Ave, Gran Fury & Newport	④⑥	Black	300	117	1000	430-550
	Lancer & Lebaron GTS	②③	White	750	—	1000	800-910
	Colt Vista	⑦	—	597	—	1166	733-803
	Colt Vista	⑧	—	—	—	1991	660-774⑨
1987	Shadow & Sundance	②③	Tan	600	—	1000	725-825
1987-88	Charger, Horizon, Omni, Shelby & Turismo	②③	Gray	500	—	1000	575-700
	Lebaron	②③	Gray	500	—	1000	575-700
	Aries, Lebaron, Reliant & 600	②③	Tan	600	—	1000	725-825
	Aries, Lebaron, Reliant & Town & Country	②③	Black	800	—	1000	875-950
	Daytona & Daytona Shelby Z	②③	Black	800	—	1000	875-950
	Daytona & Daytona Shelby Z	②③	White	500	—	1000	675-750
1987-89	Shadow & Sundance	②③	Gray	500	—	1000	575-700
	Eagle Medallion Sedan	⑩	—	—	—	1450	566-624
	Eagle Medallion Wagon	⑩	—	—	—	1450	638-696
1988	Colt	⑦	—	384	—	953	520-590
	Colt	⑧	—	—	—	1991	765-879⑨
1988-89	Dynasty, New Yorker & New Yorker Landau	②③	Black	800	—	1000	875-950
	Eagle Premier	③	—	—	—	1200	557-677
1989	Horizon & Omni	②③	Gray	500	—	1000	575-700
	Lebaron	②③	Gray	600	—	1000	725-825
	Lebaron	②③	Yellow	400	—	1000	600-700
	Aries & Reliant	②③	Tan	600	—	1000	725-825
	Daytona & Daytona Shelby Z	②③	Tan	600	—	1000	725-825
	Daytona & Daytona Shelby Z	②③	Yellow	400	—	1000	600-700
	Shadow & Sundance	②③	Tan	600	—	1000	725-825
	Acclaim & Spirit	②③	Tan	600	—	1000	743-800
	Colt & Eagle Summit	③	—	445	—	1024	590-691

①—Color tag located under boot of valve stem.
②—Differential.
③—Proportioning.
④—Brake Warning, Proportioning & Hold-Off Valve assembly.
⑤—Heavy Duty.
⑥—Standard.
⑦—Except load sensing proportioning valve.
⑧—Load sensing proportioning valve.
⑨—With load sensing spring free.
⑩—Compensator-Limiter.

Ford Motor Co.

TABLE OF CONTENTS

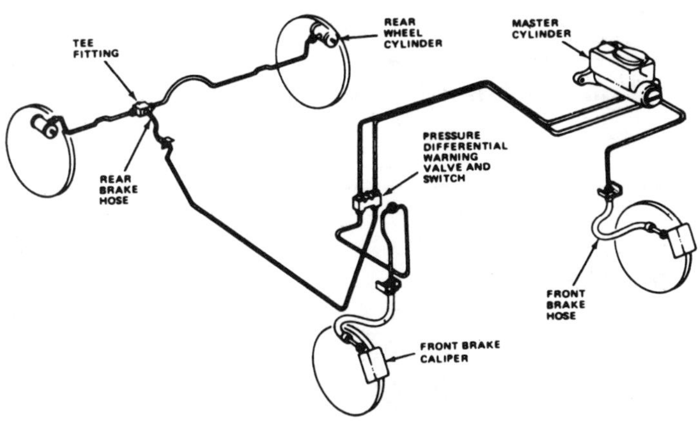

Fig. 1 Schematic diagram of a typical hydraulic front & rear split brake system

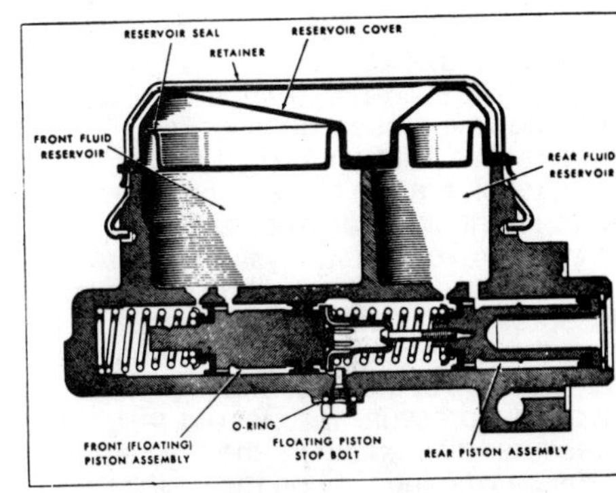

Fig. 2 Bendix dual master cylinder (Typical)

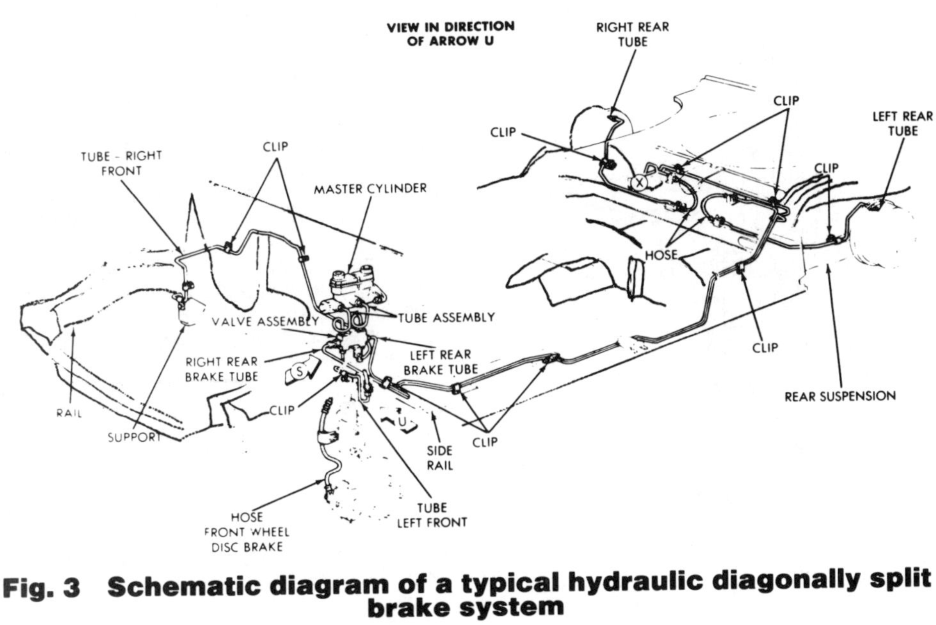

Fig. 3 Schematic diagram of a typical hydraulic diagonally split brake system

DUAL MASTER CYLINDER SYSTEMS

FRONT AND REAR SPLIT SYSTEMS

When the brake pedal is depressed, both the primary (front brake) and the secondary (rear brake) master cylinder pistons are moved simultaneously to exert hydraulic fluid pressure on their respective independent hydraulic systems. The fluid displacement of the two master cylinders is proportioned to fulfill the requirements of each of the two independent hydraulic brake systems, **Figs. 1 and 2.**

If a failure of a rear (secondary) brake system should occur, initial brake pedal movement causes the unrestricted secondary piston to bottom in the master cylinder bore. Primary piston movement displaces hydraulic fluid in the primary section of the dual master cylinder to actuate the front brake system.

Should the front (primary) brake system fail, initial brake pedal movement causes the unrestricted primary piston to bottom out against the secondary piston. Continued downward movement of the brake pedal moves the secondary piston to displace hydraulic fluid in the rear brake system to actuate the rear brakes.

The increased pedal travel and the increased pedal effort required to compensate for the loss of the failed portion of the brake system provides a warning that a partial brake system failure has occurred. When the ignition switch is turned on, a brake warning light on the instrument panel provides a visual indication that one of the dual brake systems has become inoperative.

Should a failure of either the front or rear brake hydraulic system occur, the hydraulic fluid pressure differential resulting from pressure loss of the failed brake system forces the valve toward the low pressure area to light the brake warning lamp.

DIAGONALLY SPLIT SYSTEMS

This system operates on the same principles as conventional front and rear split systems using primary and secondary master cylinders moving simultaneously to exert hydraulic pressure on their respective systems.

The hydraulic brake lines on this system, however, have been diagonally split front to rear (left front to right rear and right front to left rear) in place of separate lines to the front and rear wheels, **Fig. 3.**

In the event of a system failure this would cause the remaining good system to do all the braking on one front wheel and the opposite rear wheel, thus maintaining 50% of the total braking force. The hydraulic pressure loss would result in a pressure differential in the system and cause a warning light on the dashboard to glow as in front and rear split systems.

BRAKE WARNING LIGHT SWITCHES

There are four basic types of brake warning light switches as shown in **Figs. 4 through 6** and usually form a common electrical circuit with the brake warning light.

When a pressure differential occurs between the front and rear brake systems, the valves will shuttle toward the side with the low pressure.

As shown in **Fig. 4,** movement of the differential valve forces the switch plunger upward over the tapered shoulder of the valve to close the switch contacts and light the dual brake warning lamp, signaling a brake system failure.

In **Fig. 5,** the valve assembly consists of two valves in a common bore that are spring loaded toward the centered position. The spring-loaded switch contact plunger rests on top of the valves in the centered position (right view). When a pressure differential occurs between the front and rear brake systems, the valves will shuttle toward the side with the low pressure. The spring-loaded switch plunger is "triggered" and the ground circuit for the warning light is completed, lighting the lamp (left view).

In **Fig. 6,** as pressure falls in one system, the other system's normal pressure forces the piston to the inoperative side, contacting the switch terminal, causing the warning light on the instrument panel to glow.

On front wheel drive models, a fluid level indicator replaces the pressure differential valve used in previous brake systems. It is contained inside the body of the master cylinder plastic reservoir and activates the brake warning light when fluid level is low.

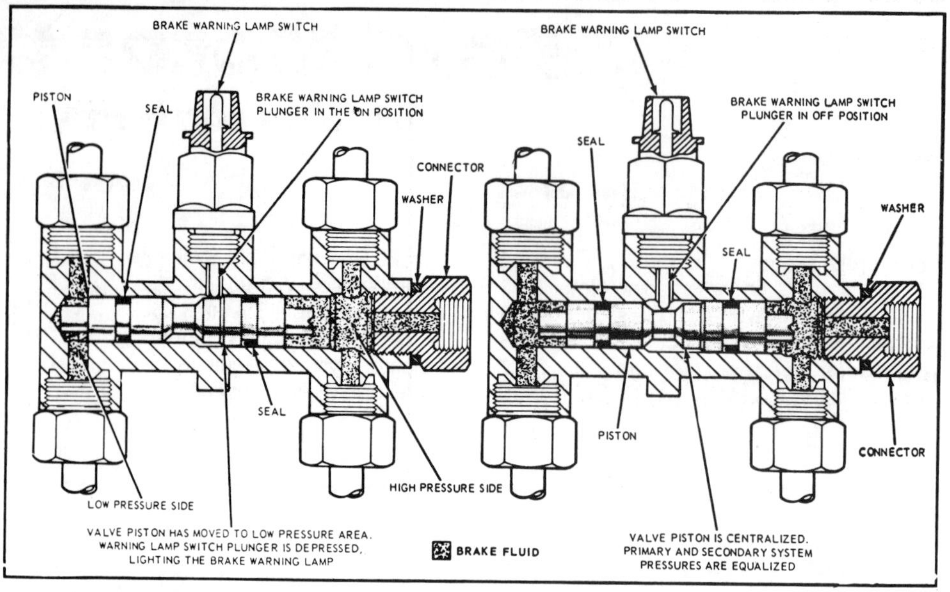

Fig. 4 Pressure differential valve & brake warning light switch

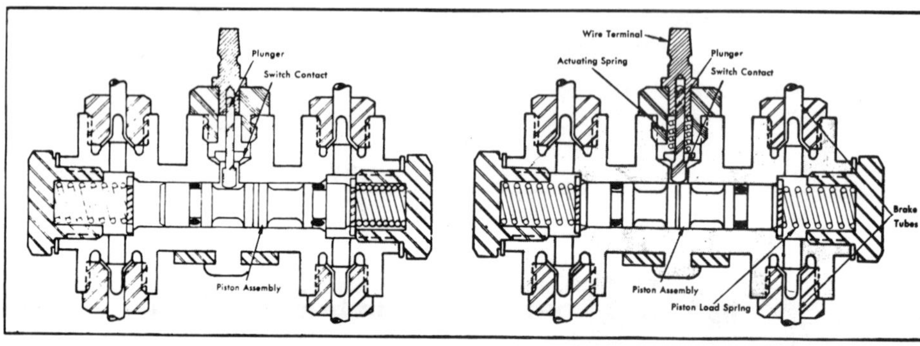

Fig. 5 Pressure differential valve & brake warning light switch

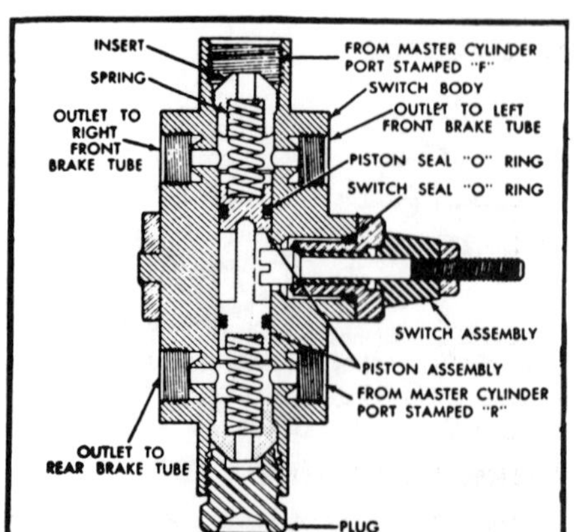

Fig. 6 Pressure differential valve & brake warning light switch

Testing Warning Light System

If the parking brake light is connected into the service brake warning light system, the brake warning light will flash only when the parking brake is applied with the ignition turned ON. The same light will also glow should one of the two service brake systems fail when the brake pedal is applied.

To test the system, turn the ignition on and apply the parking brake. If the lamp fails to light, inspect for a burned out bulb, disconnected socket, a broken or disconnected wire at the switch.

Fig. 7 is an exterior view of a typical brake warning light switch. The brake warning light switch is usually mounted on the left frame side rail or on the brake pedal bracket.

To test the brake warning system, raise the car and open a wheel bleeder valve while a helper depresses the brake pedal and observes the warning light on the instrument panel. If the bulb fails to light, inspect for a burned out bulb, disconnected

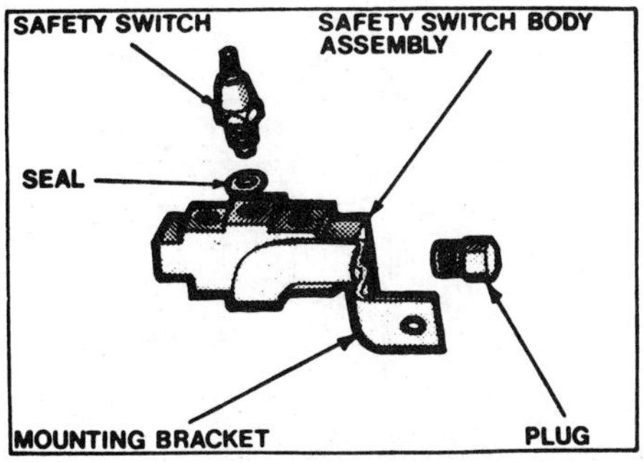

Fig. 7 Typical pressure valve & brake warning light switch

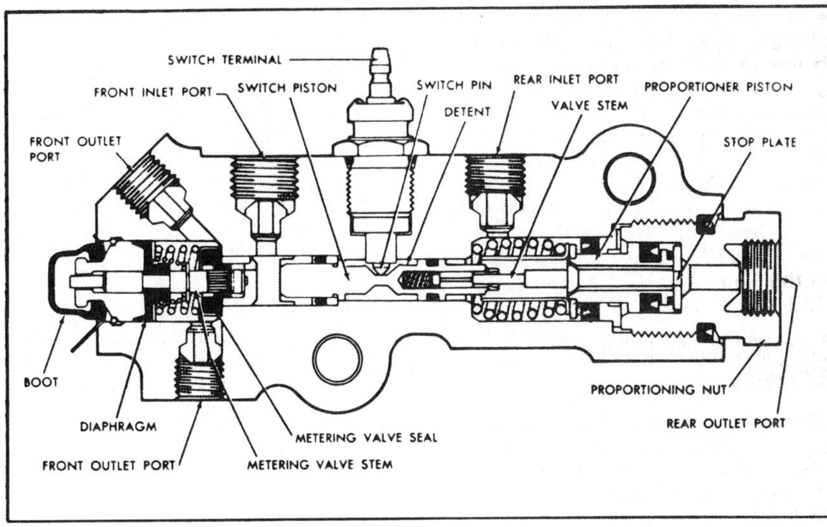

Fig. 8 Combination valve

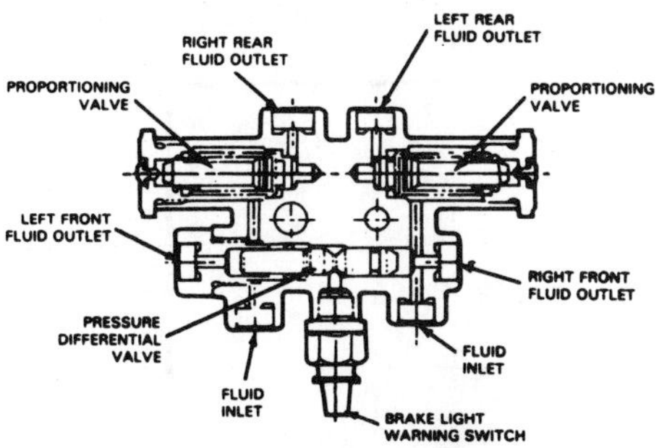

Fig. 9 Diagonally split system combination valve

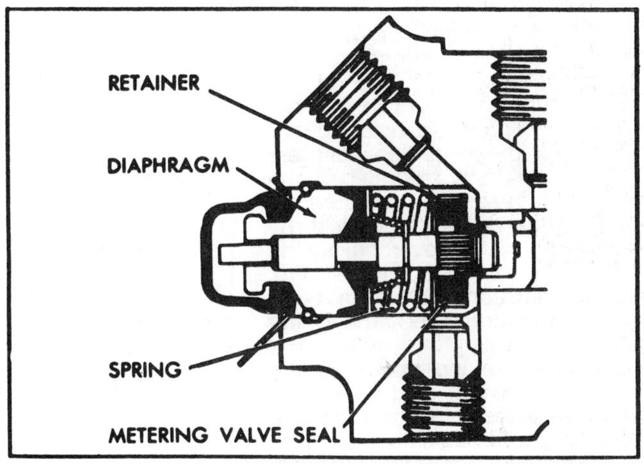

Fig. 10 Metering valve, initial braking

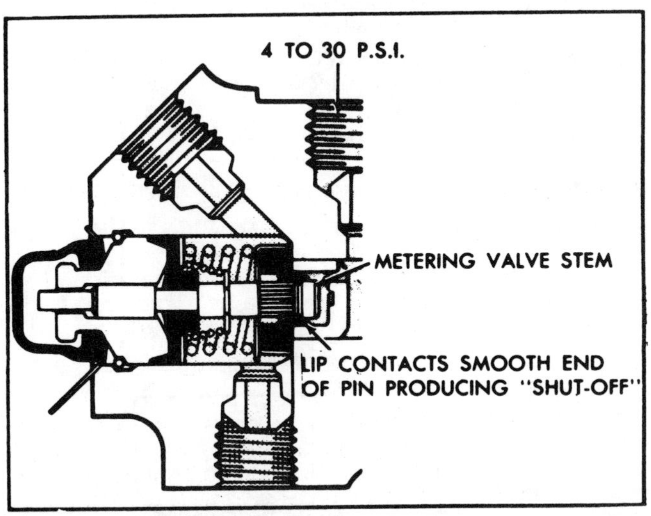

Fig. 11 Metering valve, continued braking

socket, or a broken or disconnected wire at the switch. If the bulb is not burned out, and wire continuity is proven, replace the brake warning switch.

COMBINATION VALVE

The combination valve, **Fig. 8** is a metering valve, failure warning switch, and a proportioner in one assembly and is used on disc brake applications. The metering valve delays front disc braking until the rear drum brake shoes contact the drum. The failure warning switch is actuated in event of front or rear brake system failure, in turn activating a dash warning lamp. The proportioner balances front to rear braking action during rapid deceleration.

Combination valves used on diagonally split brake systems do not use metering valves instead two proportioning valves are used, **Fig. 9.**

Metering Valve

When the brakes are not applied, the metering valve permits the brake fluid to flow through the valve, thus allowing the fluid to expand and contract with temperature changes.

When the brakes are initially applied, the metering valve stem moves to the left, preventing fluid to flow through the valve to the front disc brakes. This is accomplished by the smooth end of the metering valve stem contacting the metering valve seal lip at 4-30 psi, **Fig. 10.** The metering valve spring holds the retainer against the seal until a predetermined pressure is produced at the valve inlet port which overcomes the spring pressure and permits hydraulic pressure to actuate the front disc brakes, **Fig. 11.** The increased pressure into the valve is metered through the valve seal, to the front disc brakes, producing an increased force on the diaphragm. The diaphragm then pulls the pin, in turn pulling the retainer and reduces the spring pressure on the metering valve seal. Eventually, the pressure reaches a point at which the spring is pulled away by the diaphragm pin and retainer, leaving the metering valve unrestricted, permitting full pressure to pass through the metering valve.

On some applications, two-way or three-way combination valves are used. The three-way combination valve consists of a metering valve, failure warning switch and a proportioner mounted in an aluminum body, **Fig. 12.** The two-way combination valve, **Fig. 13,** consists of a failure warning switch and a proportioner. **On models equipped with metering valves,** the metering valve release rod must be pushed in during bleeding operations on the front wheels.

Failure Warning Switch

If the rear brake system fails, the front system pressure forces the switch piston to the right, **Fig. 14.** The switch pin is then forced up into the switch, completing the electrical circuit and activates the dash warning lamp.

When repairs are made and pressure returns to the system, the piston moves to the left, resetting the switch. The detent on the piston requires approximately 100-450 psi to permit full reset of the piston. In event of front brake system failure, the piston moves to the left and the same sequence of events is followed as for rear system failure except the piston resets to the right.

Proportioner Or Pressure Control Valves

During rapid deceleration, a portion of vehicle weight is transferred to the front wheels. This resultant loss of weight at rear wheels must be compensated for to avoid early rear wheel skid. The proportioner or pressure control valve reduces rear brake system pressure, delaying rear wheel skid. When the proportioner or pressure control valve is incorporated in the combination valve assembly, pressure developed within the valve acts against the large end of the piston, overcoming the spring pressure, moving the piston left, **Fig. 15.** The piston then contacts the stem seat and restricts line pressure through the valve.

During normal braking operation, the proportioner or pressure control valve is not functional. Brake fluid flows into the proportioner or pressure control valve between the piston center hole and the valve stem, through the stop plate and to the rear brakes. Spring pressure loads the piston during normal braking, causing it to rest against the stop plate, **Fig. 16.**

On diagonally split brake systems, two proportioners or pressure control valves are used. One controls the left rear brake, the other the right rear brake. **On front wheel drive models less power brakes,** the proportioners or pressure control valves are located in the combination valve, **Fig. 9. On front wheel drive models with power brakes,** the proportioners or pressure control valves are installed in the master cylinder rear brake outlet ports, **Fig. 17.**

BRAKE DISTRIBUTION VALVE & SWITCH

This switch assembly which is used on some diagonally split brake systems and Corvette four wheel disc brake systems, is connected to the outlet ports of the master cylinder and also to the brake warning light that warns the driver if either the primary or secondary brake system has failed.

When hydraulic pressure is equal in both primary and secondary brake sys-

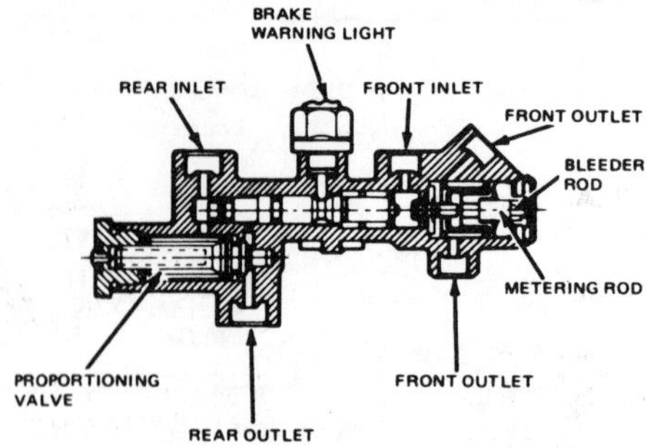

Fig. 12 Three-way combination valve

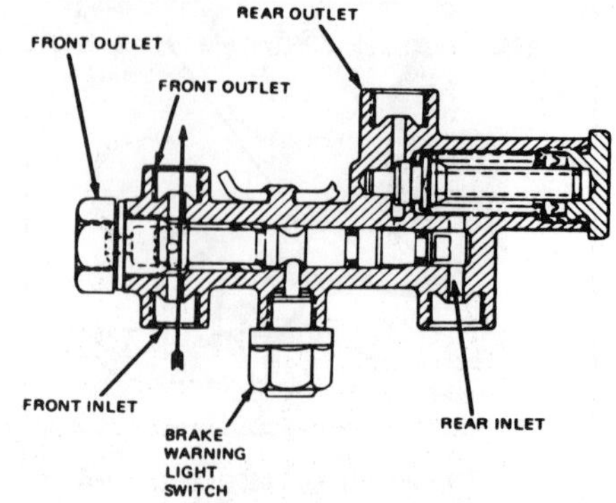

Fig. 13 Two-way combination valve

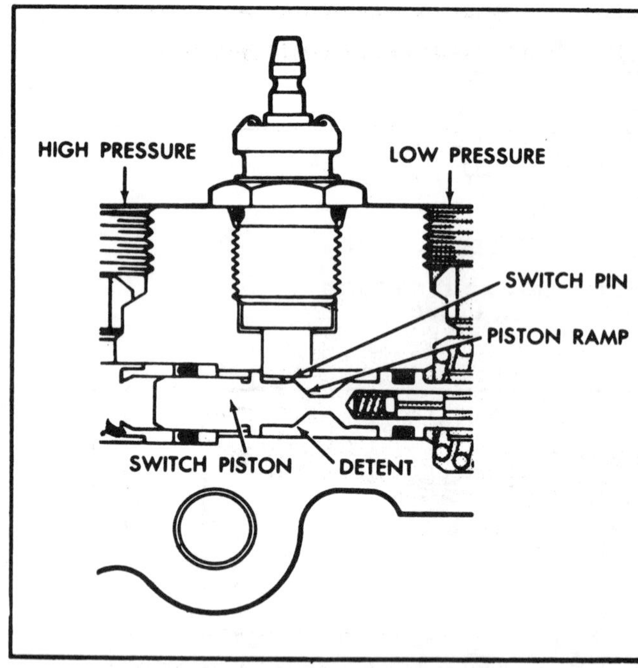

Fig. 14 Failure warning switch, rear system failure

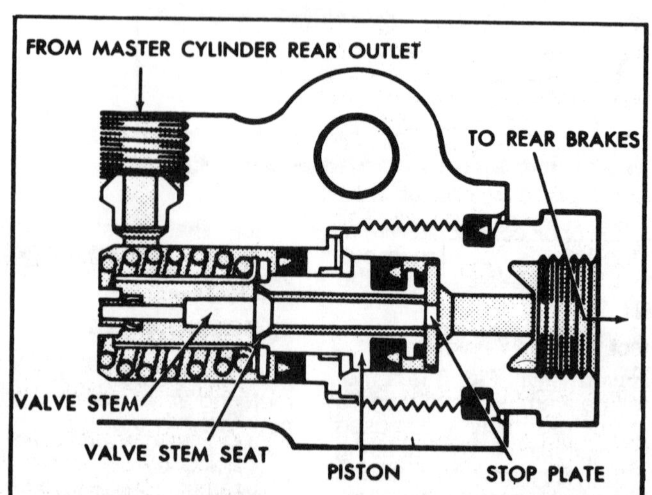

Fig. 15 Proportioner, rapid deceleration

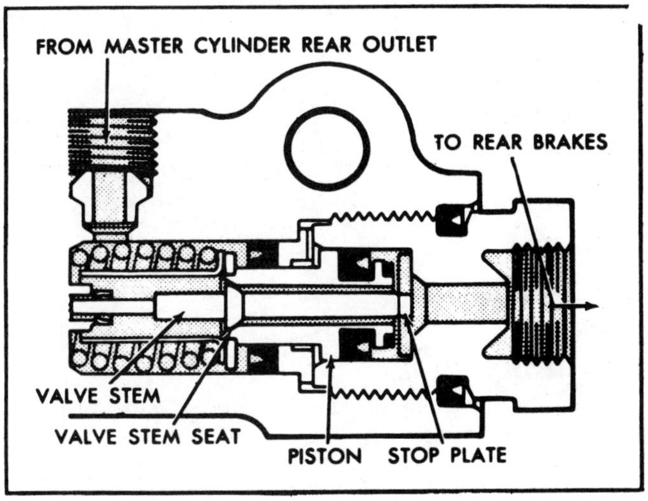

Fig. 16 Proportioner, normal braking

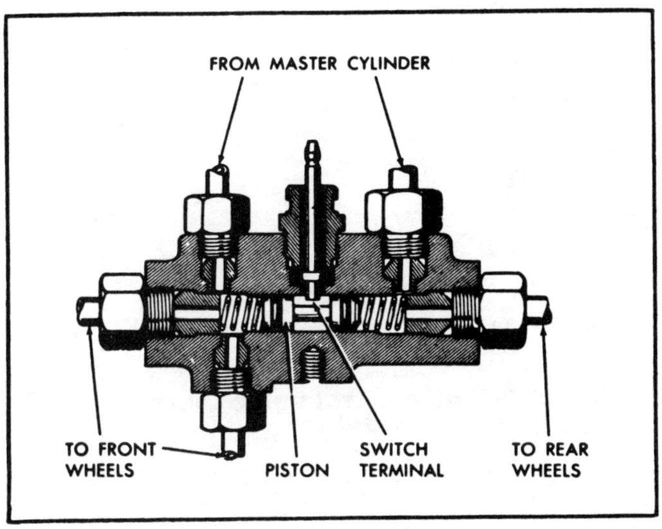

Fig. 18 Brake distribution switch (normal)

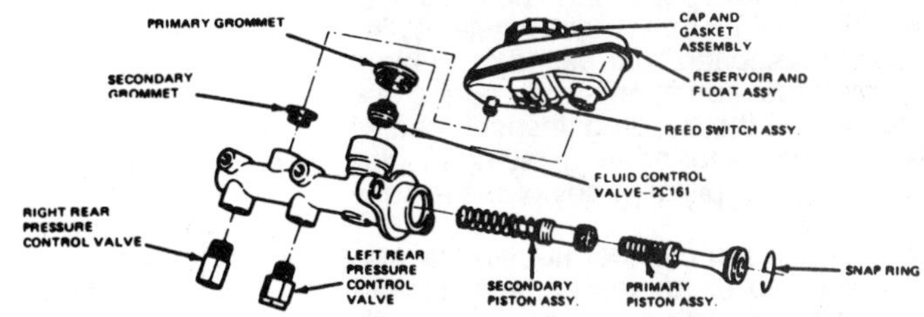

Fig. 17 Pressure control valves installed in master cylinder. Front wheel drive models

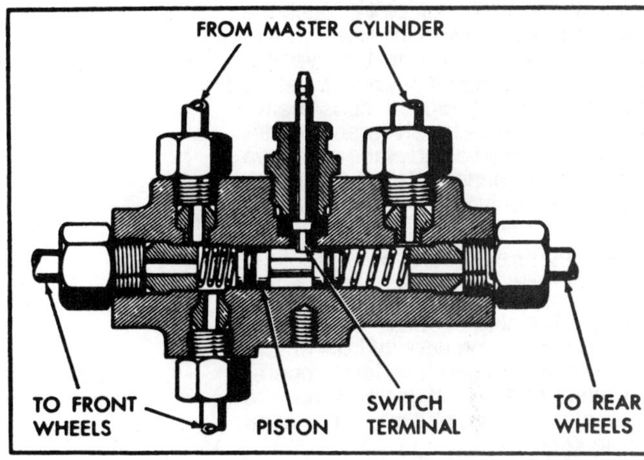

Fig. 19 Brake distribution switch (failed)

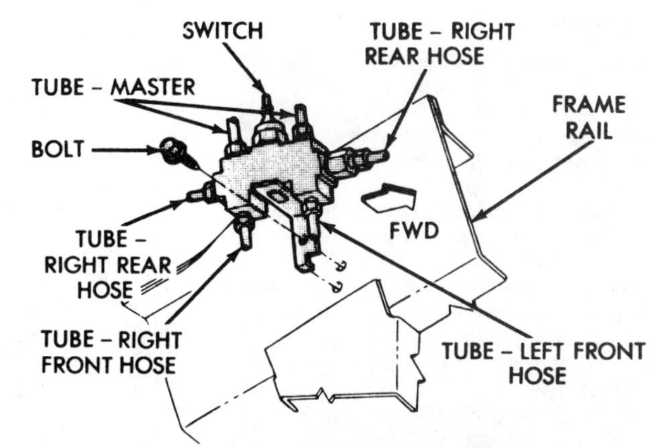

Fig. 20 Typical brake distribution switch used in diagonally split brake systems

tems, the switch remains centered, **Fig. 18.** If pressure fails in one of the systems, hydraulic pressure moves the piston toward the inoperative side, **Fig. 19.** The shoulder of the piston contacts the switch terminal, providing a ground and lighting the warning lamp.

Fig. 20 shows a brake distribution valve and switch used on some diagonally split brake system applications.

MASTER CYLINDER, REPLACE

1. Disconnect low fluid level switch, if equipped.
2. Using a line wrench, loosen brake lines at master cylinder.
3. Disconnect brake lines from the master cylinder then cap the lines and master cylinder ports.
4. Remove master cylinder attaching nuts, then the master cylinder.
5. Reverse procedure to install. If necessary bleed brakes as previously outlined.

MASTER CYLINDER SERVICE

Disassemble

1. Remove cover and diaphragm then drain brake fluid from master cylinder, **Fig. 21.**
2. **On front wheel drive models,** using a screwdriver, pry up on reservoir and remove primary port from master cylinder, **Fig. 22.**
3. Rotate reservoir out of way and remove sealing grommet from master cylinder casting.
4. Using a 12mm socket, remove fluid control valve.
5. **On all models,** depress primary piston and remove snap ring from retaining groove at open end of bore.
6. Remove primary and secondary piston assemblies from master cylinder. If secondary piston does not readily come out, apply air pressure to secondary outlet port to remove.

Inspection

When disassembled, wash all parts in clean brake fluid only. Use an air hose to blow out all passages, orifices and valve holes. Air dry and place parts on clean paper or lint-free cloth. Inspect master cylinder bore for scoring, rust, pitting or etching. Any of these conditions will require replacement of the housing. Inspect master cylinder pistons for scoring, pitting or distortion. Replace piston if any of these conditions exist.

If either master cylinder housing or piston is replaced, clean new parts with clean brake fluid and blow out all passages with air hose.

Examine reservoirs for foreign matter and check all passages for restrictions. If there is any suspicion of contamination or evidence of corrosion, completely flush hydraulic system as outlined below.

When overhauling a master cylinder, use all parts contained in repair kit. Before starting reassembly, dip all cups, seals, pistons, springs, check valves and retainers in clean brake fluid and place in a clean pan or on clean paper. Wash hands with soap and water only to prevent contamination of rubber parts from oil, kerosene or gasoline. During assembly, dip all parts in clean brake fluid.

Inspect through side outlet of dual master cylinder housing to make certain cup lips do not hang up on edge of hole or turn back, which would result in faulty operation. A piece of 3/16 inch rod with an end rounded off will be helpful in guiding cups past hole.

When overhauling late model aluminum master cylinders, carefully inspect master cylinder bore for corrosion. If corroded, replace master cylinder. Do not hone or use abrasives on the bore of these cylinders.

Assembly

1. Install secondary piston assembly into bore, spring end first.
2. Install primary piston assembly, spring end first.
3. Depress primary piston and install snap ring.
4. **On front wheel drive models,** install fluid control valve and **torque** to 97-115 inch lbs.
5. Lubricate new grommet with brake fluid and install in primary port.
6. Install reservoir in new grommet.
7. **On all models,** fill and bench bleed master cylinder.
8. Install cap and diaphragm on master cylinder reservoir.

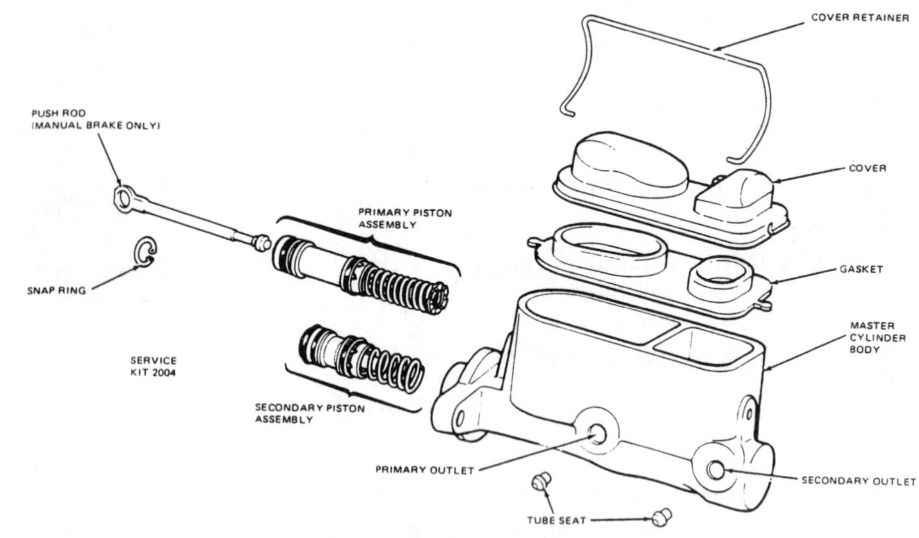

Fig. 21 Cutaway view of Bendix dual master cylinder used with disc brakes (typical)

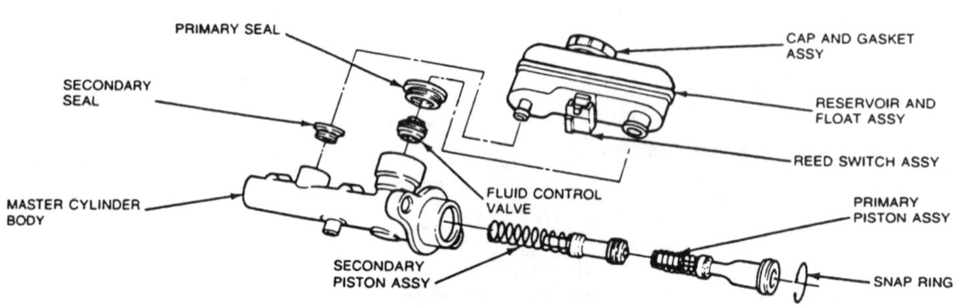

Fig. 22 Master cylinder with fluid control valve installed. Front wheel drive models

BLEEDING BRAKES

Pressure bleeding is recommended for all hydraulic brake systems.

The bleeding operation itself is fairly well standardized. First step in all cases is cleaning the dirt from the filler cap before removing it from the master cylinder. This should be done thoroughly.

Pressure bleeding is fastest because the master cylinder doesn't have to be refilled several times, and the job can be done by one man. To prevent air from the pressure tank getting into the lines, do not shake the tank while air is being added to the tank or after it has been pressurized. Set the tank in the required location, bring the air hose to the tank, and do not move it during the bleeding operation. The tank should be kept at least one-third full.

On vehicles equipped with disc brakes and master cylinders without proportioners or pressure control valves located in the master cylinder outlet port, the brake metering valve or combination valve must be held in position using a suitable tool.

If air does get into the fluid, releasing the pressure will cause the bubbles to increase in size, rise to the top of the fluid, and escape. Pressure should not be greater than about 35 psi.

On vehicles equipped with plastic reservoirs, do not exceed 25 psi bleeding pressure.

When bleeding without pressure, open the bleed valve three-quarters of a turn, depress the pedal a full stroke, then allow the pedal to return slowly to its released position. It is suggested that after the pedal has been depressed to the end of its stroke, the bleeder valve should be closed before the start of the return stroke. **On models with power brakes,** first reduce the vacuum in the power unit to zero by pumping the brake pedal several times with the engine off before starting to bleed the system.

Pressure bleeding, of course, eliminates the need for pedal pumping.

Discard drained or bled brake fluid. Care should be taken not to spill brake fluid, since this can damage the finish of the car.

Flushing is essential if there is water, mineral oil or other contaminants in the lines, and whenever new parts are installed in the hydraulic system. Fluid contamination is usually indicated by swollen and deteriorated cups and other rubber parts.

Wheel cylinders on disc brakes are equipped with bleeder valves, and are bled in the same manner as wheel cylinders for drum brakes.

Bleeding is necessary on all four wheels if air has entered the system because of low fluid level, or the line or lines have been disconnected. If a line is disconnected at any one wheel cylinder, that cylinder only need be bled. Of course, on brake reline jobs, bleeding is advisable to remove any air or contaminants.

Master cylinders equipped with bleeder valves should be bled first before the wheel cylinders are bled. In all cases where a master cylinder has been overhauled, it must be bled. Where there is no

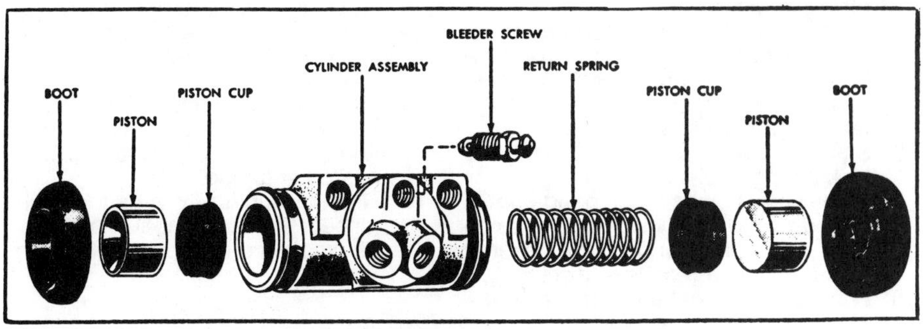

Fig. 23 Disassembled view of typical wheel cylinder

bleeder valve, this can be done by leaving the lines loose, actuating the brake pedal to expel the air and then tightening the lines.

After overhauling a dual master cylinder used in conjunction with disc brakes, it is advisable to bleed the cylinder before installing it on the car. The reason for this recommendation is that air may be trapped between the master cylinder pistons because there is only one residual pressure valve (check valve) used in these units.

TESTING DUAL MASTER CYLINDERS

Ensure that the master cylinder compensates in both parts. This can be done by applying the brake pedal lightly (engine running with power brakes) and observing for brake fluid squirting up in the reservoirs. This may only occur in the front chamber. To determine if the rear compensating port is open, pump up the brakes rapidly and hold the pedal down. Have an observer watch the fluid in the rear reservoir while the pedal is raised. A disturbance in the fluid indicates that the compensating port is open.

WHEEL BLEEDING SEQUENCE

Rear Wheel Drive......... RR-LR-RF-LF
Front Wheel Drive .... RR-LF-LR-RF

DUAL MASTER CYLINDER BLEEDING NOTES

The following information applies to master cylinders without proportioners or pressure control valves located in the master cylinder outlet ports.

All vehicles use a self-centering valve. After any bleeding operation, turn ignition switch to ACC or ON position and depress brake pedal. Valve will center itself.

WHEEL CYLINDERS

1. Remove wheel, drum and brake shoes.
2. Disconnect hydraulic line at wheel cylinder. Do not pull metal line away from cylinder as the cylinder connection will bend metal line and make installation difficult. Line will separate from cylinder when cylinder is moved away from brake backing plate.

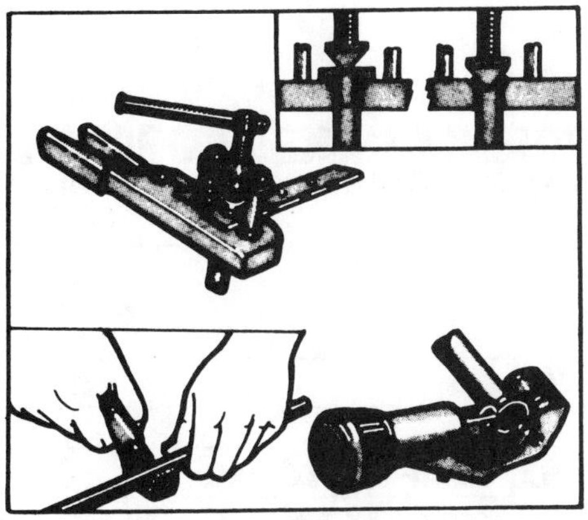

Fig. 25 Flaring hydraulic brake tubing

3. Remove screws holding cylinder to brake plate and remove cylinder.

OVERHAUL

1. Referring to **Fig. 23** as a guide, remove boots, pistons, springs and cups from cylinder.
2. Place all parts, except cylinder casting in clean brake fluid. Wipe cylinder walls with clean brake fluid.
3. Examine cylinder bore. A scored bore may be honed providing the diameter is not increased more than .005 inch. Replace worn or damaged parts from the repair kit.
4. Before assembling, wash hands with soap and water only, as oil, kerosene or gasoline will contaminate rubber parts.
5. Lubricate cylinder wall and rubber cups with brake fluid.
6. Install springs, cups, pistons and boots in housing.
7. Wipe end of hydraulic line to remove any foreign matter.
8. Place hydraulic cylinder in position. Enter tubing into cylinder and start connecting fitting.
9. Secure cylinder to backing plate and then complete tightening of tubing fitting.
10. Install brake shoes, drum and wheel.
11. Bleed system as outlined previously, and adjust brakes.

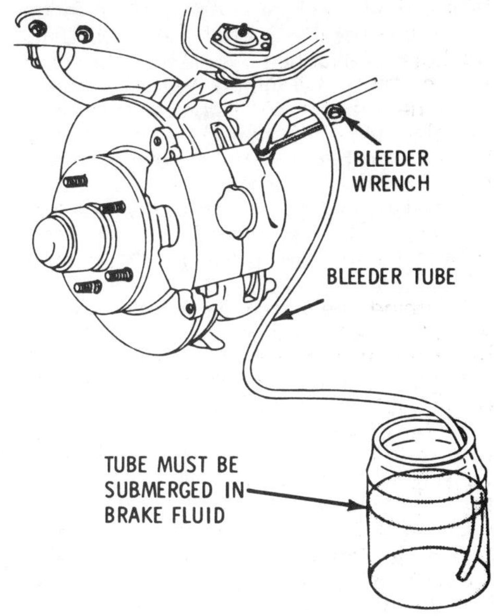

Fig. 24 Bleeding wheel cylinder

FLUSHING HYDRAULIC SYSTEM

Whenever new brake components are installed in the hydraulic system, it is recommended that the entire hydraulic system be thoroughly flushed with clean brake fluid.

It may sometime become necessary to flush out the system due to the presence of mineral oil, kerosene, gasoline, etc., which will cause swelling of rubber piston cups and valves so they become inoperative. The procedure is as follows:

Flushing is performed at each wheel in the same manner as the bleeding operation except that the bleeder valve is opened 1½ turns and the fluid is forced through the lines and bleeder valve until it emerges clear in color, **Fig. 24.** Approximately one quart of clean brake fluid is required to flush the hydraulic system. After completing the flushing operation at all bleeder valves, check to ensure the master cylinder is filled to the proper level.

HYDRAULIC TUBING

Never use copper tubing as a replacement for steel tubing. Copper tubing is subject to fatigue cracking and corrosion which could result in brake system failure.

Steel tubing is used to conduct hydraulic pressure to the brakes. All fittings, tubing and hose should be inspected for rusted, damaged or defective flared seats. The tubing is equipped with a double flare/inverted seat or I.S.O. flare to insure more positive seating in the fitting. To repair or reflare tubing, proceed as follows:

DOUBLE FLARE/INVERTED SEAT

1. Using the tool shown in **Fig. 25** or equivalent, cut off the damaged seat or damaged tubing.

2. Ream out any burrs or rough edges showing on inside edges of tubing. This will make the ends of the tubing square and insure better seating on the flared end. Before flaring tubing, place a compression nut on tubing.
3. Open handles of flaring tool and rotate jaws of tool until mating jaws of tubing size are centered in the area between vertical posts.
4. Slowly close handles with tubing inserted in jaws but do not apply heavy pressure to handle as this will lock tubing in place.
5. Referring to **Fig. 25,** place gauge on edge over end of tubing and push tubing through jaws until end of tubing contacts recessed notch of gauge matching size of tubing.
6. Squeeze handles of flaring tool and lock tubing in place.
7. Place proper size plug of gauge down in end of tubing. Swing compression disc over gauge and center tapered flaring screw in recess in disc.
8. Lubricate taper of flaring or screw and screw in until plug gauge has seated in jaws of flaring tool. This action has started to invert the extended end of tubing.
9. Remove gauge and apply lubricant to tapered end of flaring screw and continue to screw down until tool is firmly seated in tubing.
10. Remove tubing from flaring tool and inspect the seat. If seat is cracked, cut off cracked end and repeat flaring operation.

HYDRAULIC BRAKES SPECIFICATIONS

Year	Model	Master Cylinder Bore Dia.	Rear Wheel Cylinder Bore Dia. Inch
CAPRI & MUSTANG			
1985–89	All	.872	.75
ESCORT, EXP, LYNX, TEMPO & TOPAZ			
1985–89	Escort, EXP, Lynx	.828	.811
1984–89	Tempo & Topaz	.828	.875
COMPACT & INTERMEDIATE MODELS			
1985–89	Except Cougar & Thunderbird	.827	.8125
	Cougar & Thunderbird	.827	.750
FESTIVA			
1988–89	All	—	—
FULL SIZE MODELS			
1985–89	All	1.00	.937
LINCOLN			
1985	Continental & Mark VII	1.125	—
PROBE			
1989	All	.875	.69
SABLE & TAURUS			
1986–89	All	①	1.00
SCORPIO			
1988–89	All	.9	.94
TRACER			
1988–89	All	.875	.69
XR4TI			
1985–89	All	.94	.87

①—7/8 & 1 1/4 inch.

GENERAL MOTORS CORP.

INDEX

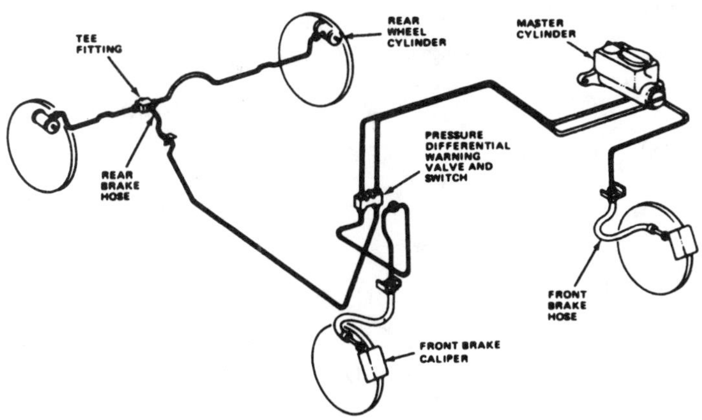

Fig. 1 Front & rear split brake system (typical)

SYSTEM OPERATION

DUAL MASTER CYLINDER

Front & Rear Split System

When the brake pedal is depressed, both the primary (front brake) and the secondary (rear brake) master cylinder pistons are moved simultaneously to exert hydraulic fluid pressure on their respective systems, **Figs. 1 and 2.**

If the rear (secondary) brake system fails, initial brake pedal movement will cause the unrestricted secondary piston to bottom in the master cylinder bore. Primary piston movement will displace hydraulic fluid in the primary section of the master cylinder to actuate the front brake system.

If the front (primary) brake system fails, initial brake pedal movement will cause the unrestricted primary piston to bottom out against the secondary piston. Continued downward movement of the brake pedal moves the secondary piston to displace hydraulic fluid in the rear brake system to actuate the rear brakes.

Diagonally Split System

This system operates on the same principle as conventional front and rear split systems, using primary and secondary master cylinders which move simultaneously to exert hydraulic pressure on their respective systems. The hydraulic brake lines on this system, however, have been diagonally split front to rear (left front to right rear and right front to left rear) in place of separate lines to the front and rear wheels, **Fig. 3.**

In the event of a system failure, the remaining non-failed system will do all the braking on one front wheel and one rear wheel, thus maintaining 50% of the total braking force.

COMPONENT DESCRIPTION

BRAKE WARNING LAMP SWITCHES

There are four basic types of brake warning lamp switches as shown, **Figs. 4 through 7.** When a pressure differential occurs between the front and rear brake systems, the valves will shuttle toward the side with the low pressure.

In the switch shown in **Fig. 4,** movement of the differential valve forces the switch plunger upward over the tapered shoulder of the valve to close the switch contacts and light the dual brake warning lamp, signaling a brake system failure.

In **Fig. 5,** the valve assembly consists of two valves in a common bore that are spring loaded toward the centered position. The spring-loaded switch contact plunger rests on top of the valves in the centered position (right view). When a pressure differential occurs between the front and rear brake systems, the valves will shuttle toward the side with the low pressure. The spring-loaded switch plunger is "triggered" and the ground circuit for the warning lamp is completed, lighting the lamp (left view).

In **Fig. 6,** as pressure falls in one system, the other system's normal pressure forces the piston to the inoperative side, contacting the switch terminal, causing the warning lamp on the instrument panel to glow.

Fig. 7 shows the switch mounted directly in the master cylinder assembly. Whenever there is a specified differential pressure, the switch piston will activate the brake failure warning switch and cause the brake warning lamp to glow.

Testing Warning Lamp Operation

The warning lamp should illuminate when the ignition switch is in the start position, and turn off when the switch returns to run. If the brake lamp remains on after the ignition returns to run, check fluid level in master cylinder reservoir and inspect parking brake. If the warning lamp does not turn on during cranking, check for defective bulb or blown fuse.

COMBINATION VALVE

The combination valve, **Fig. 8** is a metering valve, failure warning switch, and a proportioner in one assembly and is used on disc brake applications. The metering valve delays front disc braking until the rear drum brake shoes contact the drum. The failure warning switch is actuated in event of front or rear brake system failure, in turn activating a dash warning lamp. The proportioner balances front to rear braking action during rapid deceleration.

Metering Valve

When the brakes are not applied, the metering valve permits the brake fluid to flow through the valve, thus allowing the fluid to expand and contract with temperature changes.

When the brakes are initially applied, the metering valve stem moves to the left, preventing fluid to flow through the valve to the front disc brakes. This is accomplished by the smooth end of the metering valve stem contacting the metering valve seal lip at 4 to 30 psi, **Fig. 9.** The metering valve spring holds the retainer against the seal until a predetermined pressure is produced at the valve inlet port which overcomes the spring pressure and permits hydraulic pressure to actuate the front disc brakes, **Fig. 10.** The increased pressure into the valve is metered through the valve seal, to the front disc brakes, producing an increased force on the diaphragm. The diaphragm then pulls the pin, in turn pulling the retainer and reduces the spring pressure on the metering valve seal. Eventually, the pressure reaches a point at which the spring is pulled away by the diaphragm pin and retainer, leaving the metering valve unrestricted, permitting full pressure to pass through the metering valve.

Failure Warning Switch

If the rear brake system fails, the front system pressure forces the switch piston to the right, **Fig. 11.** The switch pin is then forced up into the switch, completing the electrical circuit and activates the dash warning lamp.

When repairs are made and pressure returns to the system, the piston moves to the left, resetting the switch. The detent on the piston requires approximately 100 to 450 psi to permit full reset of the piston. In event of front brake system failure, the piston moves to the left and the same sequence of events is followed as for rear system failure except the piston resets to the right.

Proportioning or Pressure Control Valve

During rapid deceleration, a portion of vehicle weight is transferred to the front wheels. This resultant loss of weight at rear wheels must be compensated for to avoid early rear wheel skid. The proportioner or pressure control valve reduces rear brake system pressure, delaying rear wheel skid. When the proportioner or pressure control valve is incorporated in the combination valve assembly, pressure developed within the valve acts against the large end of the piston, overcoming the spring pressure, moving the piston left, **Fig. 12.** The piston then contacts the stem seat and restricts line pressure through the valve.

During normal braking operation, the proportioner or pressure control valve is not functional. Brake fluid flows into the proportioner or pressure control valve between the piston center hole and the valve stem, through the stop plate and to the rear brakes. Spring pressure loads the piston during normal braking, causing it to rest against the stop plate, **Fig. 13.**

On diagonally split brake systems, two proportioners or pressure control valves are used. One controls the left rear brake, the other the right rear brake. The propor-

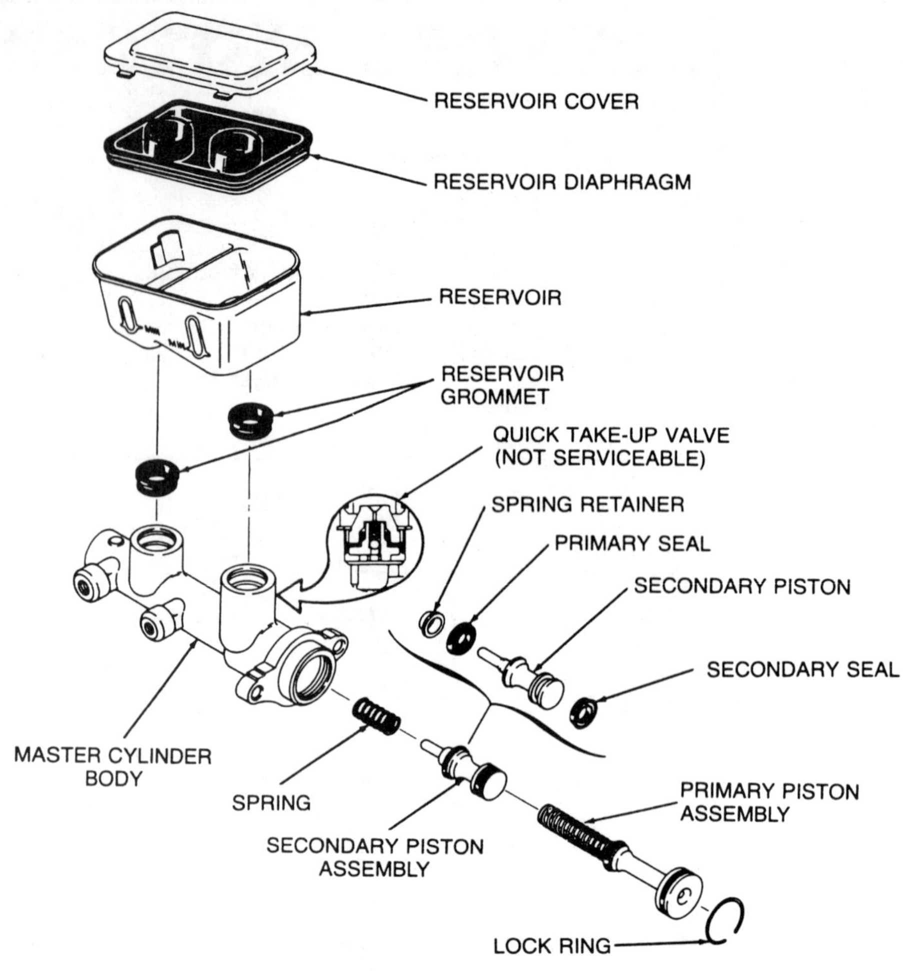

Fig. 2 Dual master cylinder assembly. Composite type

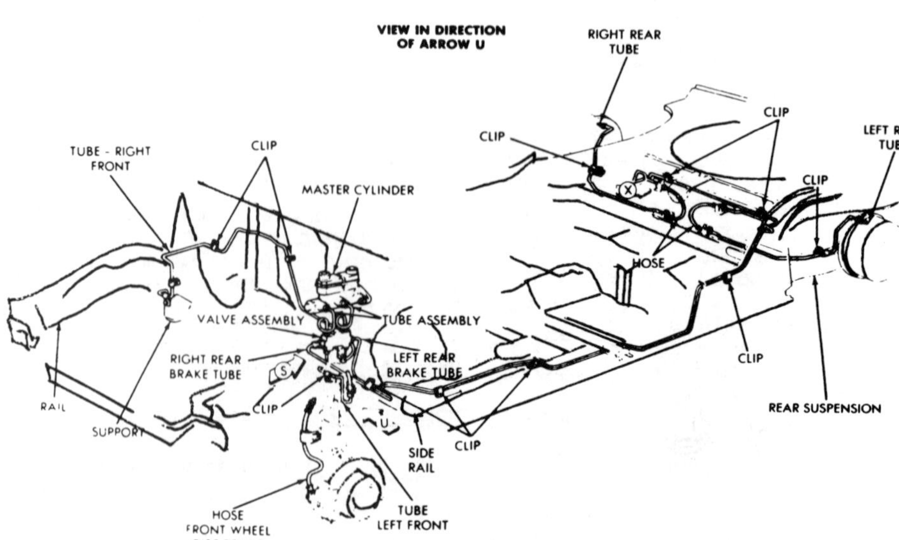

Fig. 3 Diagonally split brake system (typical)

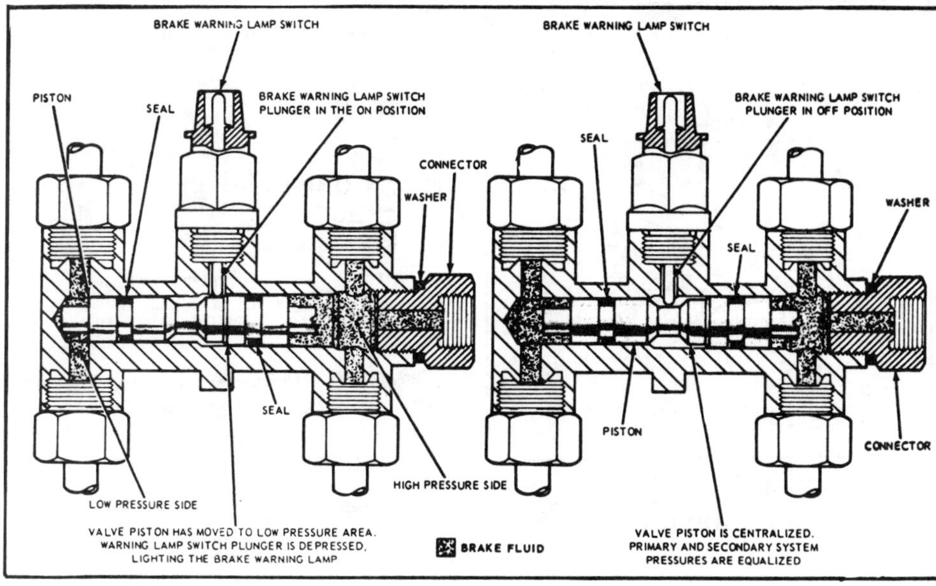

Fig. 4 Pressure differential valve & brake warning lamp switch

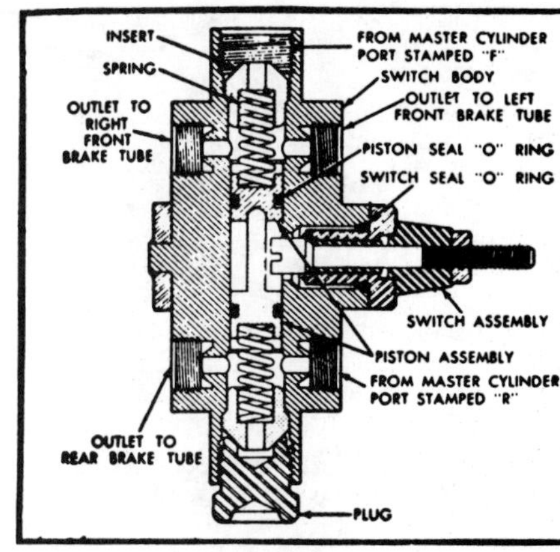

Fig. 6 Pressure differential valve & brake warning lamp switch

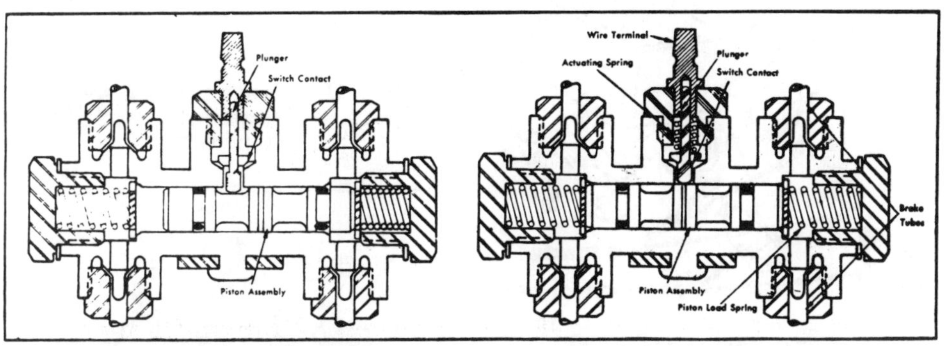

Fig. 5 Pressure differential valve & brake warning lamp switch

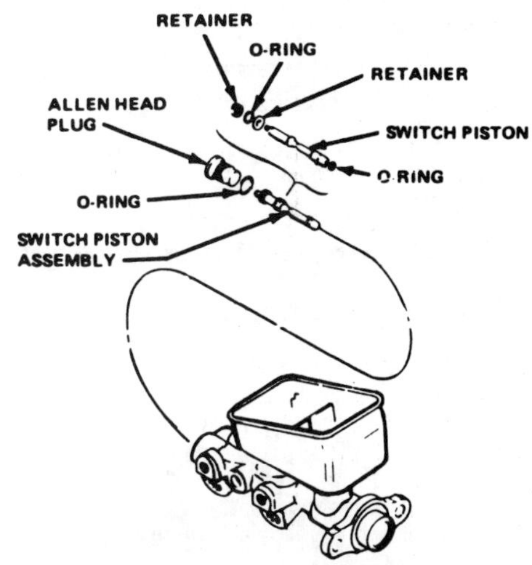

Fig. 7 Dual master cylinder with built in warning lamp switch

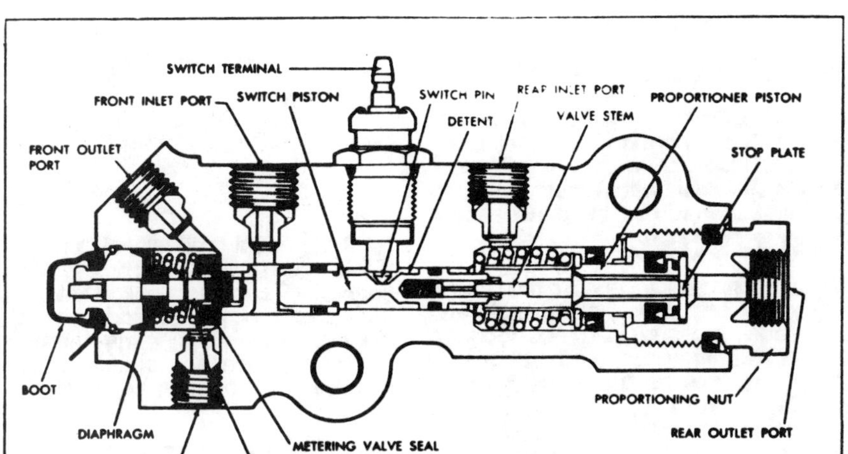

Fig. 8 Combination valve

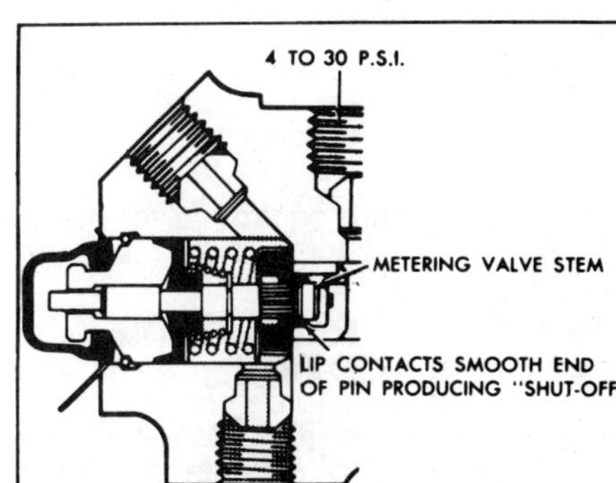

Fig. 9 Metering valve. Initial braking

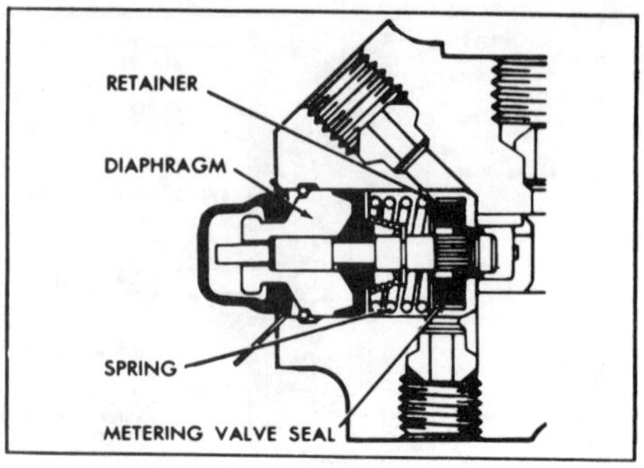

Fig. 10 Metering valve. continued braking

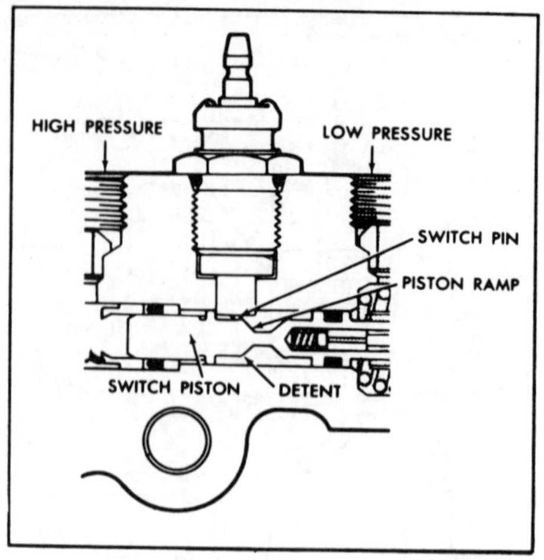

Fig. 11 Failure warning switch. Rear system failure

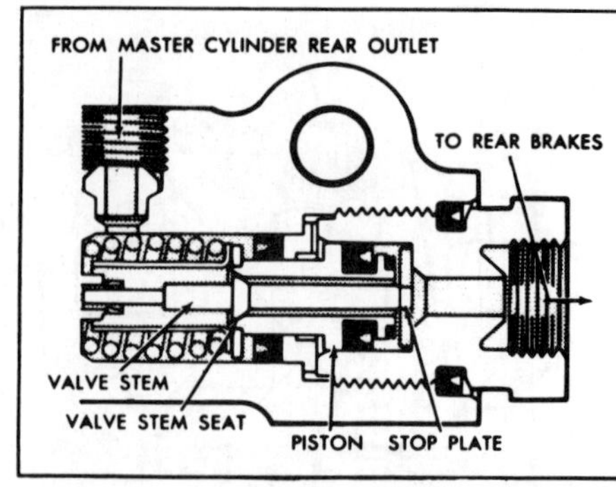

Fig. 12 Proportioner. Rapid deceleration

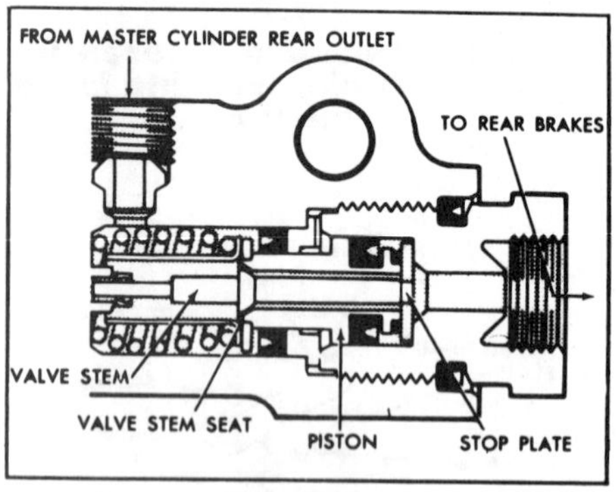

Fig. 13 Proportioner. Normal braking

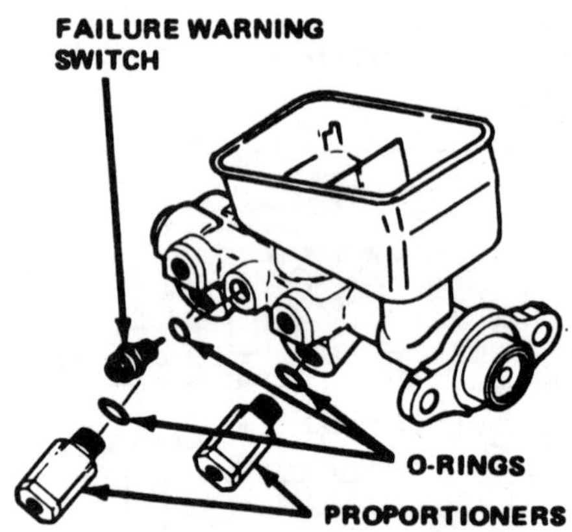

Fig. 14 Proportioners installed in master cylinder

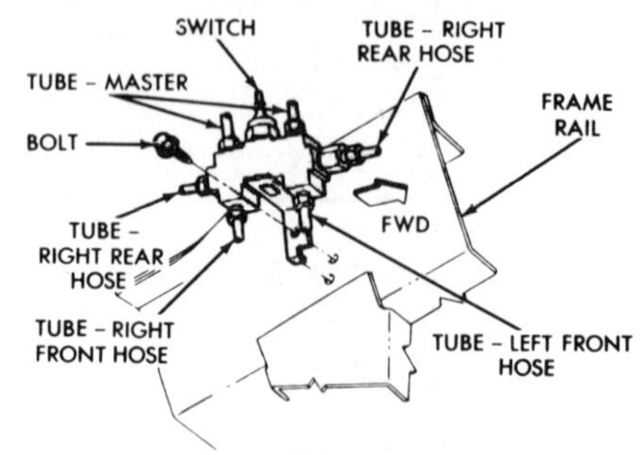

Fig. 15 Distribution switch (typical). Diagonally split brake systems

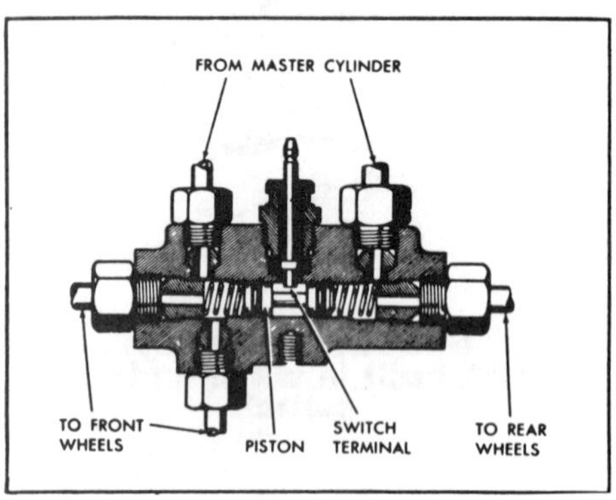

Fig. 16 Brake distribution switch. Normal

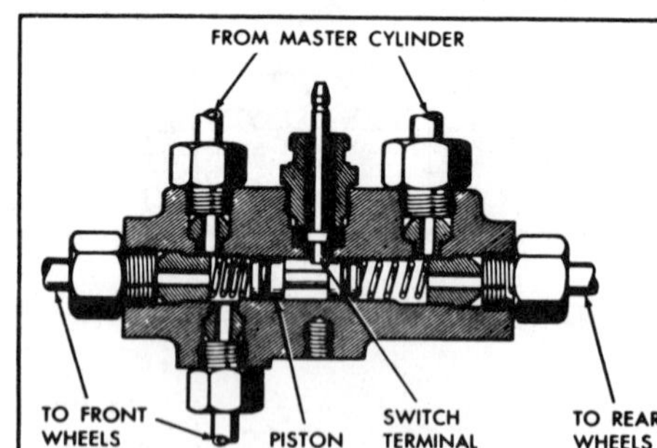

Fig. 17 Brake distribution switch. Failed

tioners or pressure control valves are installed in the master cylinder rear brake outlet ports, **Fig. 14.**

BRAKE DISTRIBUTION VALVE & SWITCH

This switch assembly, **Fig. 15,** is used on some diagonally split brake systems and Corvette four wheel disc brake systems. It is connected to the outlet ports of the master cylinder and to the brake warning lamp and warns the driver if either the primary or secondary brake system has failed.

When hydraulic pressure is equal in both primary and secondary brake systems, the switch remains centered, **Fig. 16.**

If pressure fails in one of the systems, the piston moves toward the inoperative side, **Fig. 17.** The shoulder of the piston contacts the switch terminal, providing a ground and lighting the warning lamp.

TROUBLESHOOTING

When troubleshooting the hydraulic brake system, perform the following checks and inspections as outlined below. If problem still exists within system, refer to **Fig. 18** first, then to **Fig. 19** as necessary.

ROAD TESTING

When testing brakes, ensure the road is level and dry. Test brakes at both light and heavy pedal pressure. Do not lock up brakes or slide tires during a brake test.

Check the tires on the vehicle before performing a brake test. Tires should be equally inflated, identical in size and with equal tread pattern. Excessive camber and caster will cause the brakes to pull. An overloaded vehicle will also brake erratically.

CHECKING FOR FLUID LEAKAGE

Start engine and depress the brake pedal. If the pedal gradually falls under constant pressure, the hydraulic system may be leaking. Raise the vehicle on a lift and check all tubing lines and backing plates for signs of leakage. It may be necessary to lift or remove the carpeting or floor mats to check for booster or master cylinder leakage.

SYSTEM SERVICE

BLEEDING BRAKES

Manual Bleeding

Pressure bleeding is recommended for all hydraulic systems. However, if a pressure bleeder is unavailable, use the follow-

BRAKE DIAGNOSIS CHART — DISC, DRUM SYSTEMS

CAUSE \ SYMPTOM	Excessive Brake Pedal Travel	Brake Pedal Travel Gradually Increases	Excessive Pedal Effort	Excessive Brake Action	Excessive Braking Respond	Brakes Slow to Release	Brakes Slow to Brakes Drag	Uneven Braking Action (Side to Side)	Uneven Braking Action (Front to Rear)	Scraping Noise from Brakes	Brakes Squeak During Application	Brakes Squeak During Stop	Brakes Chatter (Roughness)	Brakes Groan at End of Stop	Brakes Tell-Tale Glows
Leaking Brake Line or Connection	X	XX	X						X						XX
Leaking Wheel Cylinder or Piston Seal	X	XX	X	X				X	X						X
Leaking Master Cylinder	X	XX	X						X						X
Air in Brake System	XX		X						X						XX
Contaminated or Improper Brake Fluid	X				X	X	X	X	X						X
Leaking Vacuum System			XX		X										
Restricted Air Passage in Power Booster		X	X		XX	X	X								
Damaged Power Booster		X	X	X	X	XX	X								
Worn Out Brake Lining			X	X				X	X	X		XX	X	X	
Uneven Brake Lining Wear-Replace	X			X				X	X	X		XX		X	
Glazed Brake Lining-Sand			XX		X			X	X			X		X	
Incorrect Lining Material-Replace			X	X				X	X	X	X	X	X	X	
Contaminated Brake Lining-Replace			X	XX	X			XX	XX	X		X	X	X	
Lining Damaged by Abusive Use-Replace			X	XX				X	X	X		X	X	X	
Excessive Brake Lining Dust-Remove			X	XX				XX	XX		X	XX		X	
Heat Spotted or Scored Drums or Rotors				X				X	X			X	XX	X	
Out-of-Round or Vibrating Brake Drums													XX		
Backing Plate Interference on the Drum										X					
Out-of-Parallel Brake Rotors	X												XX		
Excessive Rotor Run-out	X												X		
Faulty Automatic Adjusters	X			X			X	X	X						X
Incorrect Wheel Cylinder Sizes			X	X				X	X						
Weak or Incorrect Brake Shoe Retention Springs				X	X	X	X	X	X	X	XX				
Brake Assembly Attachments-Missing or Loose	X						X	X	X	X			X	X	
Insufficient Brake Shoe Guide Lubricant							X	X	X	XX	XX				
Restricted Brake Fluid Passage or Sticking Wheel Cylinder Piston		X	X		X	X	X	X	X						X
Improperly Adjusted Stoplight Switch or Cruise Control Vacuum Dump							X								
Faulty Metering Valve	X		X	X	X	X	X		X						X
Faulty Porportioning Valve			X	X	X	X			X						X
Brake Pedal Linkage Interference or Binding			X		X	XX	XX								
Improperly Adjusted Parking Brake							X		X						XX
Incorrect Front End Alignment								XX							
Incorrect Tire Pressure								X	X						
Incorrect Wheel Bearing Adjustment	X						X	X		X			X		
Loose Front Suspension Attachments							X	X		XX			X	X	
Out-of-Balance Wheel Assemblies													XX		
Operator Riding Brake Pedal			X				X		X					X	X
Sticking Caliper or Wheel Cylinder Pistons			X				XX	X	X						

XX — Indicates more probable cause(s) X — Indicates other causes

Fig. 18 Brake diagnosis chart (part 1 of 6). Front disc/rear drum system

BRAKE DIAGNOSIS CHART — 4 WHEEL DISC SYSTEMS

CAUSE \ SYMPTOM	Excessive Brake Pedal Travel	Brake Pedal Travel Gradually Increases	Excessive Pedal Effort	Excessive Brake Action	Excessive Braking Respond	Brakes Slow to Release	Brakes Slow to Brakes Drag	Uneven Braking Action (Side to Side)	Uneven Braking Action (Front to Rear)	Scraping Noise from Brakes	Brakes Squeak During Application	Brakes Squeak During Stop	Brakes Chatter (Roughness)	Brakes Groan at End of Stop	Brakes Tell-Tale Glows	Excessive Parking Brake Pedal Travel	No Reserve Stopping
Leaking Brake Line or Connection	X	XX	X						X						XX		
Leaking Piston Seal	X	XX	X	X				X	X						X		
Leaking Master Cylinder	X	XX	X						X						X		
Air in Brake System	XX		X						X						XX		
Contaminated or Substandard Brake Fluid	X				X	X	X	X	X						X		
Leaking Vacuum System			XX		X										XX		
Restricted Air Passage in Power Booster		X	X		XX	X											
Damaged Power Booster		X	X	X	X	XX											XX
Worn Out Brake Lining			X	X				X	X	X	X	X		X			
Uneven Brake Lining Wear-Replace	X			X				X	X	X	X	XX		X	X		
Glazed Brake Lining-Sand			XX		X			X	X		X	X					
Incorrect Lining Material-Replace			X	X		X		X	X			X		X			
Contaminated Brake Lining-Replace				XX		X		XX	XX	X	X	X		X			
Lining Damaged by Abusive Use-Replace			X	XX				X	X	X	X	X		X			
Heat Spotted or Scored Rotors				X				X	X		X	X	XX	X			
Out-of-Parallel Brake Rotors	X												XX				
Excessive Run-Out Rotor	X												X				
Faulty Automatic Adjusters	X			X			X	X	X						X	X	
Brake Assembly Attachments-Missing or Loose	X						X	X	X	X		X	X	X			
Restricted Brake Fluid Passage		X	X		X	X	X	X	X						X		
Improperly adjusted Stoplight Switch or Cruise Control Vacuum Dump							X										
Faulty Porportioning Valve		X	X	X	X	X	X		X						X		
Brake Pedal Linkage Interference or Binding			X		X	XX	XX										
Improperly Adjusted Parking Brake Cables							X		X							X	
Improper Length Master Cylinder Push Rod	X			X		X	XX		X								
Incorrect Front End Alignment								XX									
Incorrect Tire Pressure								X	X								
Faulty Wheel Bearing - Replace	X									X			X				
Loose Front Suspension Attachments							X	X		XX			X	X			
Out-of-Balance Wheel Assemblies													XX				
Operator Riding Brake Pedal			X				X		X					X			
Sticking Caliper Piston					X	X	XX	X	X								
Park Brake Switch Circuit Grounded															XX		
Park Brake Not Releasing						X		X							XX		

XX - Indicates more probable cause(s)
X - Indicates other causes

Fig. 18 Brake diagnosis chart (part 2 of 6). Four wheel disc system

BRAKE DIAGNOSIS — GASOLINE

CONDITION	POSSIBLE CAUSE	CORRECTION
EXCESSIVE PEDAL EFFORT.	1. MALFUNCTIONING POWER BOOSTER.	1. CHECK POWER BOOSTER AND REPAIR IF NECESSARY.
	2. PARTIAL BRAKE SYSTEM FAILURE.	2. CHECK FRONT AND REAR BRAKE SYSTEM AND REPAIR AS NECESSARY. ALSO CHECK BRAKE WARNING LIGHT. IF A FAILED SYSTEM IS FOUND, THE LIGHT SHOULD HAVE INDICATED A FAILURE.
	3. EXCESSIVELY WORN LININGS.	3. CHECK AND REPLACE IN AXLE SETS.
	4. FADING BRAKES DUE TO INCORRECT LINING OR EXCESSIVE BRAKING.	4. REPLACE WITH ORIGINAL EQUIPMENT LINING.
	5. LINING GLAZED.	5. SAND LINING AND ROTORS OR DRUMS.
	6. WORN DRUMS.	6. REPLACE.
EXCESSIVE PEDAL TRAVEL.	1. PARTIAL BRAKE SYSTEM FAILURE.	1. CHECK BOTH FRONT AND REAR BRAKE SYSTEMS FOR A FAILURE AND REPAIR. ALSO CHECK WARNING LIGHT. IT SHOULD HAVE INDICATED A FAILURE.
	2. INSUFFICIENT FLUID IN MASTER CYLINDER.	2. FILL RESERVOIRS WITH APPROVED BRAKE FLUID. CHECK FOR LEAKS. CHECK WARNING LIGHT.
	3. AIR TRAPPED IN SYSTEM. (PEDAL SPONGY).	3. BLEED SYSTEM.
	4. REAR BRAKE SYSTEM NOT ADJUSTING.	4. REPAIR AUTO ADJUSTERS AND ADJUST REAR BRAKES.
	5. BENT SHOE AND LINING.	5. REPLACE SHOES AND LININGS IN AXLE SETS.
NOISE (HIGH PITCHED SQUEAK WITHOUT BRAKE APPLIED).	1. FRONT LININGS WORN OUT (WEAR OUT SENSOR MAKING NOISE).	1. REPLACE LININGS.
NOISE (METALLIC SCRAPING NOISE FROM REAR).	1. BRAKE DRUM CONTACTING STATIONARY BACKING PLATE.	1. SEE TURNING BRAKE DRUM.
BRAKE SQUEAK WITH BRAKES APPLIED.	1. INSULATOR ON OUTBOARD SHOE DAMAGED.	1. REPLACE INSULATOR.
	2. INCORRECT LININGS.	2. REPLACE WITH ORIGINAL EQUIPMENT LININGS.
BRAKE ROUGHNESS OR CHATTER (PEDAL PULSATES).	1. EXCESSIVE ROTOR LATERAL RUNOUT.	1. CHECK PER INSTRUCTIONS. IF NOT WITHIN SPECIFICATIONS, REPLACE OR MACHINE THE ROTOR.
	2. ROTOR THICKNESS VARIATION NOT WITHIN SPECIFICATIONS.	2. CHECK PER INSTRUCTIONS. IF NOT WITHIN SPECIFICATIONS, REPLACE OR MACHINE THE ROTOR.
	3. WHEEL BEARINGS NOT ADJUSTED (EXCEPT ELDORADO AND SEVILLE).	3. ADJUST WHEEL BEARINGS TO CORRECT SPECIFICATIONS.
	4. REAR DRUMS OUT OF ROUND.	4. CHECK RUNOUT AND IF NOT WITHIN SPECIFICATIONS, TURN THE DRUMS (NOT OVER THE MAXIMUM).

Fig. 18 Brake diagnosis chart (part 3 of 6). Models w/gasoline engine

BRAKE DIAGNOSIS — GASOLINE

CONDITION	POSSIBLE CAUSE	CORRECTION
BRAKE ROUGHNESS OR CHATTER (PEDAL PULSATES). (CONT'D.)	5. DISC BRAKE SHOE REVERSED (STEEL AGAINST IRON).	5. REPLACE SHOE AND LINING AND MACHINE ROTOR WITHIN SPECIFICATIONS.
	6. SEVILLE OR ELDORADO; FRONT TIRES FLAT SPOTTED OR HAVE HIGH RADIAL FORCE VARIATION.	6. SUBSTITUTE KNOWN GOOD TIRES
	7. UNEVEN PAD WEAR CAUSED BY CALIPER NOT SLIDING DUE TO IMPROPER CLEARANCE OR DIRT.	7. REMOVE CALIPER AND CORRECT AS NECESSARY.
	8. UNEVEN ROTOR WEAR CAUSING A THICKNESS VARIATION BETWEEN THE TWO BRAKING SURFACES.	8. MACHINE ROTOR AS FOLLOWS: a. MACHINE ROTORS TO OBTAIN A THICKNESS VARIATION NO GREATER THAN .0005" AND A LATERAL RUNOUT NO GREATER THAN .004". b. CHECK CALIPER FREENESS. WITH ROTOR REMOVED, INSTALL CALIPER AND MOUNTING BOLTS (PINS). CHECK FOR .005" - .012" CLEARANCE AT BOTH TOP AND BOTTOM OF CALIPER. IF LESS THAN .005" IS FOUND, FILE WITH A FLAT FILE UNTIL AT LEAST .005" IS OBTAINED. **DO NOT EXCEED A MAXIMUM** OF .012" PER END OF .024" TOTAL CLEARANCE. CALIPER CLEARANCE TO INBOARD AND OUTBOARD REACTION PADS MUST BE EQUAL WITHIN .004" BOTH AT THE TOP AND BOTTOM OF THE CALIPER. THIS IS TO ENSURE CORRECT ALIGNMENT OF CALIPER TO KNUCKLE DURING A BRAKE APPLICATION. c. REMOVE CALIPER AFTER FREENESS CHECK. CLEAN PINS (DO NOT USE ABRASIVES) AND REPLACE SLEEVES AND "O" RINGS. APPLY A LIGHT COATING OF SILICONE GREASE OR EQUIVALENT TO ALL CONTACT POINTS AND "O" RINGS. d. INSTALL RECONDITIONED ROTORS. AND, EXCEPT ELDORADO AND SEVILLE, SET BEARING ADJUSTMENT END PLAY. e. INSPECT LININGS. CLEAN FOREIGN MATERIAL FROM RIVET HOLES. f. REINSTALL CALIPERS. g. PUMP BRAKE PEDAL TO SEAT SHOES BEFORE ATTEMPTING TO MOVE CAR.
PULLS.	1. INCORRECT TIRE PRESSURES.	1. INFLATE EVENLY ON BOTH SIDES TO THE RECOMMENDED PRESSURES.
	2. FRONT END OUT OF ALIGNMENT.	2. CHECK AND ALIGN TO SPECIFICATIONS.
	3. UNMATCHED TIRES ON SAME AXLE.	3. TIRES WITH APPROXIMATELY THE SAME AMOUNT OF TREAD SHOULD BE USED ON THE SAME AXLE.

Fig. 18 Brake diagnosis chart (part 4 of 6). Models w/gasoline engine

BRAKE DIAGNOSIS — GASOLINE

CONDITION	POSSIBLE CAUSE	CORRECTION
PULLS. (CONT'D.)	4. RESTRICTED BRAKE LINES OR HOSES.	4. CHECK FOR SOFT HOSES AND DAMAGED LINES. REPLACE WITH NEW HOSES AND NEW DOUBLE-WALLED STEEL BRAKE TUBING.
	5. MALFUNCTIONING CALIPER ASSEMBLY.	5. CHECK FOR STUCK OR SLUGGISH PISTONS, PROPER LUBRICATION OF CALIPER BUSHINGS.
	6. DEFECTIVE OR DAMAGED SHOE AND LINING (GREASE OR BRAKE FLUID ON LINING OR BENT SHOE).	6. INSTALL NEW SHOE AND LINING IN COMPLETE AXLE SETS.
	7. MALFUNCTIONING REAR BRAKES.	7. CHECK FOR INOPERATIVE AUTO ADJUSTING MECHANISM, DEFECTIVE LINING (GREASE OR BRAKE FLUID ON LINING) OR DEFECTIVE WHEEL CYLINDERS. REPAIR AS NECESSARY.
	8. LOOSE SUSPENSION PARTS.	8. CHECK ALL SUSPENSION MOUNTINGS.
	9. LOOSE CALIPERS.	9. CHECK AND TORQUE BOLTS TO SPECIFICATIONS.
GRABBING OR UNEVEN BRAKING.	1. ALL CONDITIONS LISTED UNDER "PULLS."	1. ALL CONDITIONS LISTED UNDER "PULL."
	2. MALFUNCTION OF COMBINATION VALVE. PROPORTIONER SECTION CAUSING EARLY REAR WHEEL SLIDE.	2. REPLACE COMBINATION VALVE.
	3. MALFUNCTION OF POWER BRAKE UNIT.	3. CHECK OPERATION AND REPAIR, IF NECESSARY.
	4. BINDING BRAKE PEDAL MECHANISM.	4. CHECK AND LUBRICATE, IF NECESSARY. POSSIBLY REPLACE PEDAL BUSHING AND/OR SPACER.
DRAGGING BRAKES (A VERY LIGHT DRAG IS PRESENT IN ALL DISC BRAKES IMMEDIATELY AFTER PEDAL IS RELEASED).	1. MASTER CYLINDER PISTONS NOT RETURNING CORRECTLY.	1. WITH RESERVOIR COVER OFF, CHECK FOR FLUID SPURT AT BYPASS HOLES AS PEDAL IS DEPRESSED. REBUILD MASTER CYLINDER.
	2. RESTRICTED BRAKE LINES OR HOSES.	2. CHECK FOR SOFT HOSES OR DAMAGED LINES AND REPLACE WITH NEW HOSES AND NEW DOUBLE-WALLED STEEL BRAKE TUBING.
	3. INCORRECT PARKING BRAKE ADJUSTMENT ON REAR BRAKES.	3. CHECK AND READJUST TO CORRECT SPECIFICATIONS.
	4. POWER BOOSTER OUTPUT ROD TOO LONG.	4. REPLACE WITH CORRECT LENGTH.
	5. BRAKE PEDAL NOT RETURNING FREELY.	5. STOP LIGHT SWITCH OR CRUISE CONTROL NOT ADJUSTED PROPERLY: a. PEDAL PIVOT NOT LUBRICATED. b. INTERFERENCE WITH WIRING.
	6. PARKING BRAKE CABLE BINDING.	6. REPLACE.

Fig. 18 Brake diagnosis chart (part 5 of 6). Models w/gasoline engine

BRAKE DIAGNOSIS — DIESEL

CONDITION	POSSIBLE CAUSE	CORRECTION
EXCESSIVE BRAKE PEDAL EFFORT.	1. LOOSE OR BROKEN POWER STEERING PUMP BELT.	1. TIGHTEN OR REPLACE THE BELT.
	2. NO FLUID IN POWER STEERING RESERVOIR.	2. FILL RESERVOIR AND CHECK FOR EXTERNAL LEAKS.
	3. LEAKS IN POWER STEERING, BOOSTER OR ACCUMULATOR HOSES.	3. REPLACE FAULTY PARTS.
	4. LEAKS AT TUBE FITTINGS, POWER STEERING, BOOSTER OR ACCUMULATOR CONNECTIONS.	4. TIGHTEN FITTINGS OR REPLACE TUBE SEATS, IF FAULTY.
	5. EXTERNAL LEAKAGE AT ACCUMULATOR.	5. REPLACE "O" RING AND RETAINER.
	6. FAULTY BOOSTER PISTON SEAL CAUSING LEAKAGE AT BOOSTER FLANGE VENT.	6. REPAIR WITH NEW SEAL KIT.
	7. FAULTY BOOSTER INPUT ROD SEAL WITH LEAKAGE AT INPUT ROD END.	7. REPLACE BOOSTER.
	8. FAULTY BOOSTER COVER SEAL WITH LEAKAGE BETWEEN HOUSING AND COVER.	8. REPAIR WITH NEW SEAL KIT.
	9. FAULTY BOOSTER SPOOL PLUG SEAL.	9. REPAIR WITH SPOOL PLUG SEAL KIT.
SLOW BRAKE PEDAL RETURN.	1. EXCESSIVE SEAL FRICTION IN HYDRAULIC BOOSTER.	1. REPAIR WITH NEW SEAL KIT.
	2. FAULTY SPOOL ACTION.	2. FLUSH STEERING SYSTEM WHILE PUMPING BRAKE PEDAL.
	3. BROKEN PISTON RETURN SPRING.	3. REPLACE SPRING.
	4. RESTRICTION IN RETURN LINE FROM HYDRAULIC BOOSTER TO PUMP RESERVOIR.	4. REPLACE LINE.
	5. BROKEN SPOOL RETURN SPRING.	5. REPLACE SPRING.
GRABBY BRAKES.	1. BROKEN SPOOL RETURN SPRING.	1. REPLACE SPRING.
	2. FAULTY SPOOL ACTION CAUSED BY CONTAMINATION IN SYSTEM.	2. FLUSH STEERING SYSTEM WHILE PUMPING BRAKE PEDAL.
BOOSTER CHATTERS — PEDAL VIBRATES	1. POWER STEERING PUMP SLIPS.	1. TIGHTEN BELT.
	2. LOW FLUID LEVEL IN POWER STEERING PUMP RESERVOIR.	2. FILL RESERVOIR AND CHECK FOR EXTERNAL LEAKS.
	3. FAULTY SPOOL OPERATION CAUSED BY CONTAMINATION IN SYSTEM.	3. FLUSH STEERING SYSTEM WHILE PUMPING BRAKE PEDAL.
ACCUMULATOR LEAK DOWN — SYSTEM DOES NOT HOLD CHARGE	1. CONTAMINATION IN STEERING HYDRO-BOOST SYSTEM.	1. FLUSH STEERING SYSTEM WHILE PUMPING BRAKE PEDAL.
	2. INTERNAL LEAKAGE IN ACCUMULATOR SYSTEM.	2. REPAIR UNIT USING ACCUMULATOR REBUILD KIT AND SEAL KIT.

Fig. 18 Brake diagnosis chart (part 6 of 6). Models w/diesel engine

HARD PEDAL

CAUSE	CORRECTION
Broken or damaged hydraulic brake lines.	Inspect and replace as necessary.
Vacuum failure.	Check for: Faulty vacuum check valve or grommet-replace. Collapsed or damaged vacuum hose-replace. Plugged or loose vacuum fitting-repair. Faulty air valve seal or support plate seal-replace. Damaged floating control valve-replace. Bad stud welds on front or rear housing of power head-replace unless easily repaired.
Defective diaphragm.	Replace
Restricted air filter element.	Replace
Worn or distorted reaction plate or levers.	Replace plate or levers.
Cracked or broken power piston or retainer.	Replace power piston and piston rod retainer.

GRABBY BRAKES
(Apparent Off-On Condition)

CAUSE	CORRECTION
Broken or damaged hydraulic brake lines.	Inspect and replace as necessary.
Insufficient fluid in master cylinder.	Fill reservoirs with approved brake fluid-check for leaks.
Defective master cylinder seals.	Repair or replace as necessary.
Cracked master cylinder casting.	Replace
Leaks at front disc brake calipers or rear wheel cylinders in pipes or connections.	Inspect and repair as necessary.
Air in hydraulic system.	Bleed system.

BRAKES FAIL TO RELEASE

CAUSE	CORRECTION
Blocked passage in power piston.	Inspect and repair or replace as necessary.
Air valve sticking shut.	Check for proper lubrication of air valve "O" ring.
Broken piston return spring.	Replace
Broken air valve spring.	Replace
Tight pedal linkage.	Repair or replace as necessary.

Fig. 19 Power brake diagnosis chart

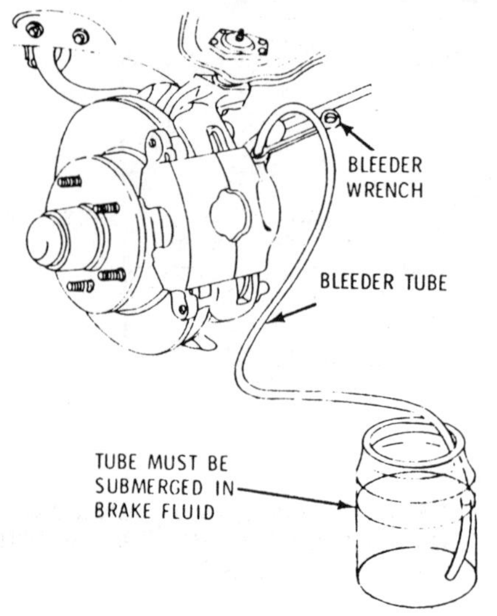

Fig. 20 Bleeding brakes

ing procedure. **Brake fluid damages painted surfaces.** Immediately clean any spilled fluid.

1. Remove vacuum reserve by pumping brakes several times with engine off.
2. Fill master cylinder reservoir with clean brake fluid. Check fluid level often during bleeding procedure; do not let reservoir fall below half full.
3. If necessary, bleed master cylinder as follows:
   a. Disconnect master cylinder forward brake line connection until fluid flows from reservoir. Reconnect and tighten brake line.
   b. Instruct an assistant to slowly depress brake pedal one time and hold.
   c. Crack open front brake line connection again, purging air from cylinder.
   d. Retighten connection and slowly release brake pedal.
   e. Wait 15 seconds, then repeat until all air is purged.
   f. Bleed the rearward (nearest the cowl) brake line connection by repeating steps a through e.
4. Loosen, then slightly retighten bleeder valves at all four wheels. Repair any broken, stripped or frozen valves at this time.
5. Proceed to appropriate wheel first and follow set sequence according to "Wheel Bleeding Sequence."
6. Place transparent tube over bleeder valve, then allow tube to hang down into transparent container, **Fig. 20.** Ensure end of tube is submerged in clean brake fluid.
7. Instruct an assistant to slowly depress brake pedal one time and hold.
8. Crack open bleeder valve, purging air from cylinder. Retighten bleeder screw and slowly release pedal.
9. Wait 15 seconds, then repeat steps 7 and 8. Repeat these steps until all air is bled from system.

Pressure Bleeding

1. Loosen, then slightly retighten bleeder valves at all four wheels. Repair any broken, stripped or frozen valves at this time.
2. Using a diaphragm type pressure bleeder, install suitable bleeder adapter to master cylinder, **Fig. 21.**
3. Charge bleeder ball to 20-25 psi.
4. Connect pressure bleeder line to adapter.
5. Open line valve on pressure bleeder, then depress bleed-off valve on adapter until a small amount of brake fluid is released.
6. Raise and support vehicle.
7. Proceed to appropriate wheel first and follow set sequence according to "Wheel Bleeding Sequence."
8. Place transparent tube over bleeder valve, then allow tube to hang down into transparent container, **Fig. 20.** Ensure end of tube is submerged in clean brake fluid.
9. Open bleeder valve 1/2 to 3/4 turn and allow fluid to flow into container until all air is purged from line.

Wheel Bleeding Sequence

Rear wheel drive models except Corvette: if manual bleeding, RR-LR-RF-LF; if pressure bleeding, bleed front brakes together and rear brakes together.

Front wheel drive models except Spectrum: RR-LF-LR-RF.

1985-86 Corvette: LR-RR-LF-RF.

1987-89 Corvette: RF-RR-LR-LF.

Spectrum: LF-RR-RF-LR.

Flushing Hydraulic System

If brake fluid is old, rusty or contaminated, or whenever new parts are installed in hydraulic system, the system must be flushed. Bleed brakes, allowing at least one quart of clean brake fluid to pass through system. Any rubber parts in hydraulic system which were exposed to contaminated fluid must be replaced.

MASTER CYLINDER OVERHAUL

COMPOSITE TYPE

Refer to **Figs. 2 and 22** when performing the following procedures.

Disassembly

1. Remove master cylinder from vehicle as follows:
   a. Disconnect and plug hydraulic lines.
   b. Remove two master cylinder attaching nuts, then the master cylinder.
2. Remove reservoir cover and diaphragm. Discard old brake fluid in reservoir.
3. Inspect cover and diaphragm. Replace if cut, cracked or deformed.
4. Remove fluid level switch, if equipped.
5. **On models with compact master cylinder,** remove proportioner valve assembly, **Fig. 22.**
6. Depress primary piston and remove lock ring.
7. Plug primary fluid outlet (outlet nearest to cowl when master cylinder is installed), then apply compressed air into secondary fluid outlet to remove primary and secondary pistons.
8. Remove spring retainer and seals from secondary piston.
9. Clamp master cylinder in a vise as shown in **Fig. 23,** then remove reservoir using a pry bar. Remove reservoir grommets.
10. Inspect master cylinder bore for corrosion. **Do not use abrasive material on master cylinder bore.** Replace if bore is corroded.

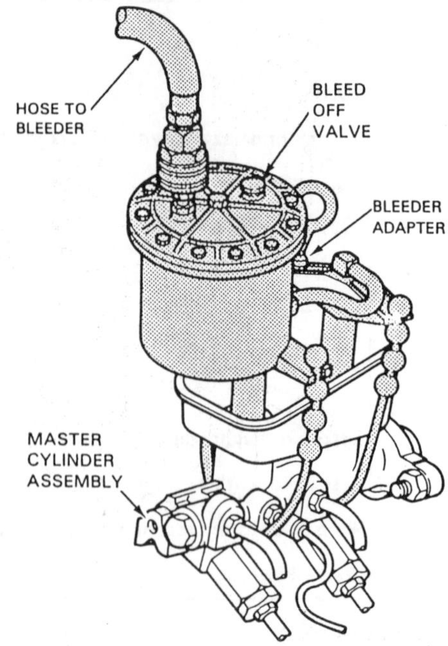

Fig. 21 Installing pressure bleeder adapter

Assembly

Clean all parts not included in repair kit with brake fluid. **Do not dry with compressed air.** Lubricate all rubber parts with clean brake fluid prior to installation.

1. Lubricate new reservoir grommets with silicone brake lube, then press grommets into master cylinder body. Ensure grommets are properly seated.
2. Lay reservoir upside down on flat, hard surface. Press master cylinder body onto reservoir using rocking motion.
3. Install new seals on secondary piston, then the spring retainer.
4. Install spring and secondary piston assembly into cylinder.
5. Install primary piston. Depress primary piston into cylinder, then install lock ring.
6. Install fluid level switch, if equipped.
7. **On models with compact master cylinder,** install proportioner valve assembly, **Fig. 22.**

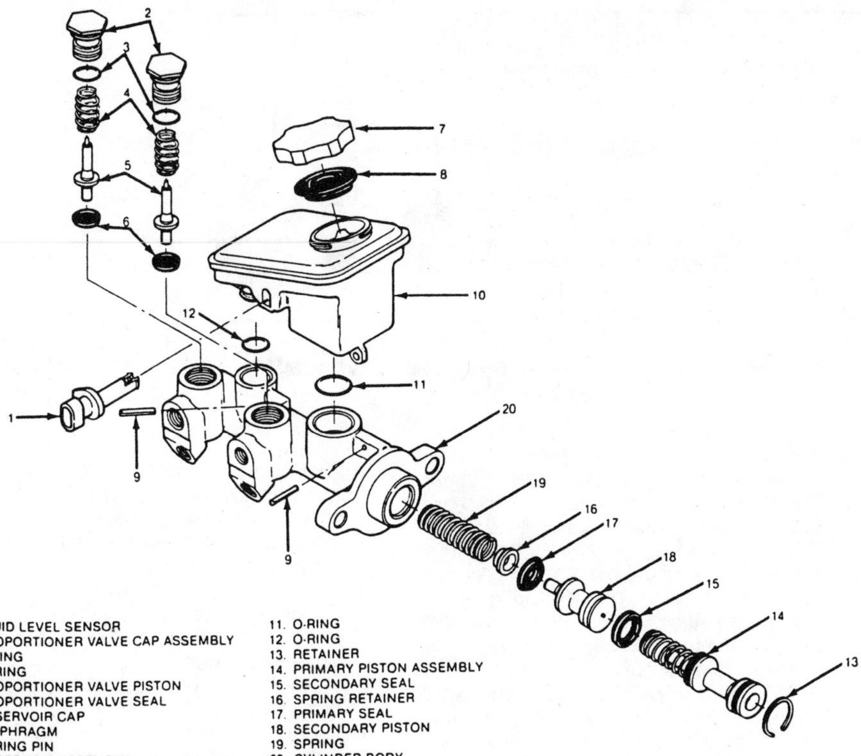

1. FLUID LEVEL SENSOR
2. PROPORTIONER VALVE CAP ASSEMBLY
3. O-RING
4. SPRING
5. PROPORTIONER VALVE PISTON
6. PROPORTIONER VALVE SEAL
7. RESERVOIR CAP
8. DIAPHRAGM
9. SPRING PIN
10. RESERVOIR ASSEMBLY
11. O-RING
12. O-RING
13. RETAINER
14. PRIMARY PISTON ASSEMBLY
15. SECONDARY SEAL
16. SPRING RETAINER
17. PRIMARY SEAL
18. SECONDARY PISTON
19. SPRING
20. CYLINDER BODY

Fig. 22 Compact master cylinder assembly

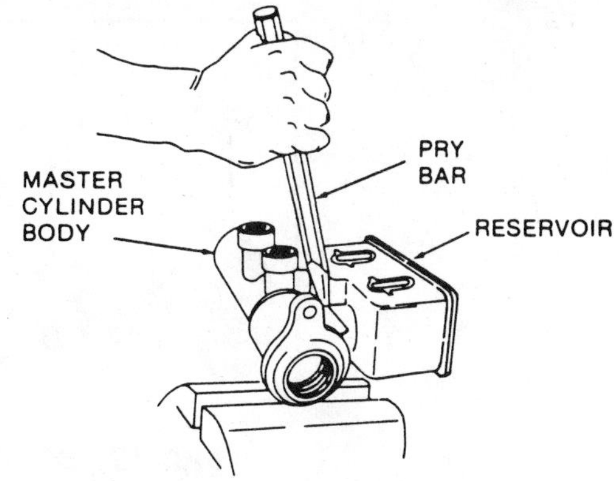

Fig. 23 Removing master cylinder reservoir. Composite type

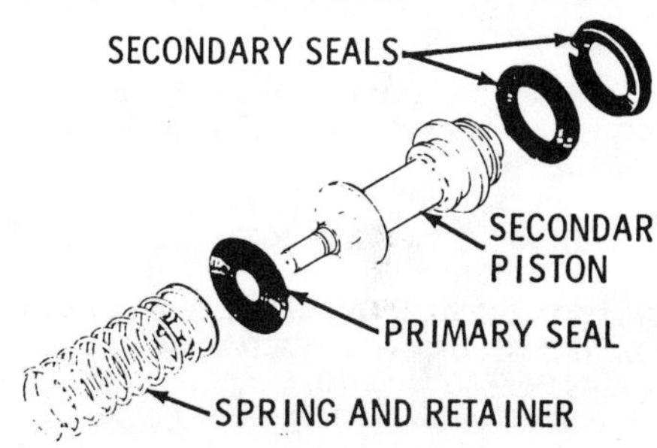

Fig. 25 Secondary piston and seals. Cast iron type master cylinder

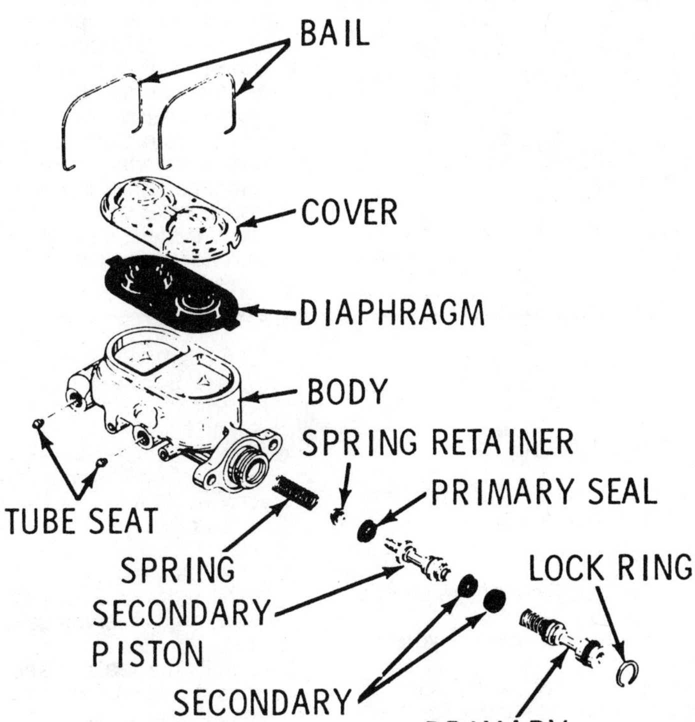

Fig. 24 Cast iron type master cylinder

8. Fit diaphragm into reservoir cover, then install cover onto reservoir.
9. Install master cylinder in vehicle by reversing steps 2a and 2b on "Disassembly" procedure. Bleed brakes.

CAST IRON TYPE

Refer to Figs. 24 and 25 while performing the following procedures.

Disassembly

1. Remove master cylinder from vehicle.
2. Clean outside of master cylinder, then remove reservoir cover and diaphragm.
3. Turn master cylinder over, then pump pushrod by hand to remove any remaining brake fluid.
4. Clamp master cylinder in a vise, with vise jaws gripping mounting flange and fluid ports pointing up. **Do not clamp master cylinder body in vise.**
5. Remove lock ring and primary piston assembly from cylinder bore.
6. Apply compressed air to secondary outlet port to remove secondary piston spring and retainer.
7. Using a 13/64 inch drill, enlarge hole in tube fitting insert of fluid outlet holes. Place a heavy washer over outlet hole, then thread a 1/4 X 20 X 3/4 screw

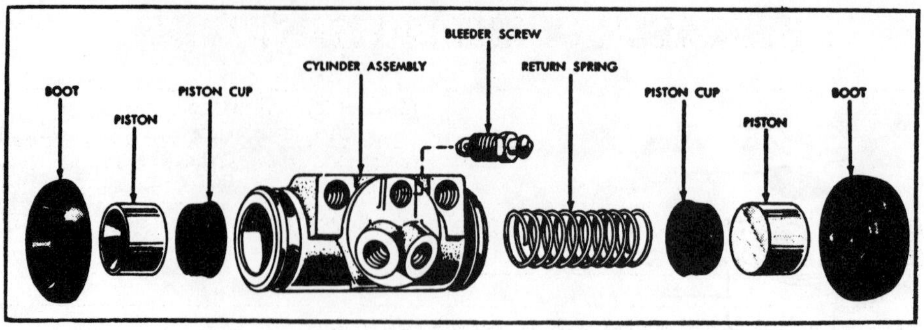

Fig. 26 Disassembled view of wheel cylinder (typical)

into insert. Tighten screw until insert is unseated, then remove insert, screw and washer.

8. Remove primary and secondary seal from secondary piston.

Inspection

Clean all parts not included in repair kit with brake fluid. **Do not dry with compressed air.**

1. Inspect cylinder bore for scoring and corrosion. Replace master cylinder if necessary.
2. Polish discolored or stained cylinder with crocus cloth, then rinse in clean brake fluid. Ensure compensating port is clean.

Assembly

Lubricate all rubber parts with clean brake fluid prior to installation.

1. With master cylinder in a vise, insert brass tube fitting inserts from overhaul kit into outlet holes.
2. Using spare brake line tube, thread tube nut into outlet hole and turn down until tube fitting insert bottoms into outlet hole. Remove tube nut and inspect outlet hole for burrs.
3. Install inner secondary seal into groove in secondary piston.
4. Install primary seal over end of secondary piston, with flat side of seal against flange of piston.
5. Install outer secondary seal into groove in end of secondary piston.
6. Install secondary piston spring retainer into secondary piston spring, then place the retainer and spring over end of secondary piston, with retainer seated inside lip of primary cup.
7. Hold master cylinder with open end down, then place secondary piston assembly into bore, with spring seated against end of bore.
8. Clamp master cylinder in a vise with open end of bore facing up.
9. Insert new primary piston assembly, spring end first, into bore.
10. Hold primary piston down, then install lock ring in groove on inside of bore.
11. Install new reservoir diaphragm into reservoir cover with beaded side out, then secure cover onto cylinder.
12. Install master cylinder in vehicle. Bleed brakes.

WHEEL CYLINDER OVERHAUL

1. Raise and support vehicle.
2. Remove wheel, drum and brake shoes.
3. Disconnect hydraulic line at wheel cylinder. **Do not pull metal line away from cylinder, as this may kink or bend line.** Line will separate from cylinder when cylinder is moved away from brake backing plate.
4. Remove wheel cylinder-to-brake plate attaching screws, then the wheel cylinder.
5. Remove boots, pistons, springs and cups from cylinder, **Fig. 26.**
6. Clean all parts with brake fluid.
7. Inspect cylinder bore. A scored bore may be honed as long as the diameter is not increased by more than .005 inch. Replace worn or damaged parts as necessary.
8. Ensure hands are clean before proceeding with assembly. Lubricate cylinder wall and rubber cups with brake fluid, then install springs, cups, pistons and boots in housing.
9. Wipe end of hydraulic line to remove any foreign matter, then place wheel cylinder in position. Enter tubing into cylinder and start threads on fitting.
10. Secure cylinder to backing plate, then complete tightening of tubing fitting.
11. Install brake shoes, drum and wheel.
12. Bleed system as outlined previously, then adjust brakes.

HYDRAULIC BRAKE SPECIFICATIONS

Year	Model	Master Cylinder Bore Diameter		Rear Wheel Cylinder Bore Dia. Inch
		Disc Brake Inch	Power Brake Inch	
BUICK ELECTRA, CAD. DEVILLE & FLEETWOOD, OLDS. 98 & 1986-89 BUICK LESABRE & OLDS. 88				
1985-89	All	15/16	15/16	3/4 ②
BUICK REAR WHEEL DRIVE & 1985 RIVIERA				
1985-89	Regal	15/16	15/16 ⑭	3/4
	LeSabre	1 1/8	1 1/8 ⑮	7/8 ①
	LeSabre Estate Wagon	1 1/8	1 1/8 ⑮	1
	Electra	15/16	15/16	3/4 ②
BUICK REATTA & 1988-89 RIVIERA				
1988-89	Reatta & Riviera	⑫	⑫	—
BUICK SOMERSET REGAL, OLDS. CUTLASS CALAIS, PONTIAC GRAND AM & 1986-87 BUICK SKYLARK				
1985-89	All	7/8	7/8	5/8
CADILLAC REAR WHEEL DRIVE & 1985 ELDORADO & SEVILLE				
1985	Eldorado & Seville	1	1	⑬
1985-89	Brougham	1 1/8	1 1/8	1
CHEVROLET CAMARO, CORVETTE, MONTE CARLO & FULL SIZE MODELS				
1985	Camaro	.945 ⑦	.945	.748
	Monte Carlo	15/16 ⑭	15/16 ⑭	3/4
	Full Size	1 1/8	1 1/8	③ ④
	Corvette	⑪	⑪	⑬
1986-89	Camaro	.945 ⑦	.945	.748
	Monte Carlo	15/16	15/16	3/4
	Full Size	⑲	⑲	③ ④
	Corvette	.875 ⑧	.875 ⑧	⑬
CHEVROLET CAVALIER, BUICK SKYHAWK, CAD. CIMARRON, OLDS. FIRENZA & PONTIAC 2000 & SUNBIRD				
1985	All	.875	.875	.625 ⑯
1986-89	All	.875 ①	.875	.625
CHEVROLET CELEBRITY, BUICK CENTURY, OLDS. CUTLASS CIERA & CRUISER & PONTIAC 6000				
1985-89	All	.875 ①	.875 ①	.687
CHEVROLET CHEVETTE & PONTIAC 1000				
1985	All	3/4	3/4	3/4
1986-87	All	.874	.874	.689
CHEVROLET CITATION & BUICK SKYLARK				
1985	All	.874	.874	.689
CHEVROLET CORSICA & BERETTA				
1987-89	All	.945	.945	⑰
CHEVROLET/GEO SPECTRUM				
1985-89	Spectrum	.810	.810	.690
CHEVROLET SPRINT & GEO METRO				
1985-89	All	.524	.524	.443
OLDS. REAR WHEEL DRIVE & 1985 TORONADO				
1985	Cutlass	15/16	15/16 ⑭	3/4
	88 & Custom Cruiser	1 1/8	1 1/8 ⑮	① ④
	Toronado	15/16	15/16 ⑭	3/4

HYDRAULIC BRAKE SPECIFICATIONS—Continued

Year	Model	Master Cylinder Bore Diameter: Disc Brake Inch	Master Cylinder Bore Diameter: Power Brake Inch	Rear Wheel Cylinder Bore Dia. Inch
OLDS. REAR WHEEL DRIVE & 1985 TORONADO – Continued				
1986-89	Cutlass	15/16	15/16	3/4
	Custom Cruiser	⑦	⑦	⑩
PONTIAC EXCEPT 1000 & FRONT WHEEL DRIVE				
1985	Firebird	⑨	⑨	3/4
	Fiero	1	1	⑬
	Bonneville	.938	.938⑥	3/4
	Grand Prix	.938	.938⑥	3/4
	Parisienne	1 1/8	1 1/8⑮	.875①④
1986-89	Firebird	⑨	⑨	3/4
	Fiero	1	1	⑬
	Bonneville	.938	.938⑥	3/4
	Grand Prix	.938	.938⑥	3/4
	Parisienne	1 1/8	1 1/8⑮	.875①④
PONTIAC LEMANS				
1988-89	All	.813	.813	.678
1986-87 BUICK RIVIERA & 1986-89 CAD. ELDORADO & SEVILLE & OLDS. TORONADO				
1986-87	Buick Riviera	15/16	15/16⑭	3/4
1986-89	Eldorado & Seville	⑱	⑱	⑬
	Toronado	1 1/8	1 1/8	—
1988-89 BUICK REGAL, OLD. CUTLASS SUPREME & PONTIAC GRAND PRIX				
1988-89	All	.945	.945	—

①—Heavy duty, .938 inch.
②—Gasoline wagon, 13/16 inch.
③—7/8 inch or 13/16 inch.
④—Station wagon, 1 inch
⑤—With 4 wheel disc brakes, 1 inch.
⑥—With Hydro-Boost, 1 1/16 inch.
⑦—Custom Cruiser, except heavy duty, 7/8 inch; heavy duty 15/16 inch; 88, 15/16 inch.
⑧—Piston diameter.
⑨—Power rear drum, .938 inch; power rear disc, 1 inch.
⑩—Except heavy duty, 11/16 inch; heavy duty, 3/4; wagon, 13/16 inch.
⑪—Primary bore, .8 inch; secondary bore, .7 inch.
⑫—Main bore less anti-lock brakes, 1.126 inch. Main bore with anti-lock brakes, 1.000 inch. Quick take-up bore 1.574 inch.
⑬—Rear wheel disc brakes.
⑭—With Hydro-Boost, 1 1/16 inch; with Power Master, 1 1/4 inch; with vacuum boost & 4 wheel disc brake, 1 inch.
⑮—With Hydro-Boost, 1 3/16 inch; with Power Master, 1 1/4 inch.
⑯—Station wagon, .6875 inch.
⑰—Sedan models, .625 inch; Coupe models, .748 inch.
⑱—Master cylinder main bore, 1.126 inch; master cylinder Q7U bore, 1.574 inch.
⑲—Quick take-up, 1 1/8 inch; Power Master, 1/4 inch.

POWER BRAKE UNITS

TABLE OF CONTENTS

Chrysler Motors/Eagle

INDEX

GENERAL SERVICE

In order to properly service the brake system, a thorough understanding of the various power assist systems is necessary. Two basic types of power assist mechanisms are used: vacuum assist diaphragm assemblies which use engine vacuum and in some cases vacuum pressure developed by an external vacuum pump, and hydraulic pressure assist mechanisms which use pressure developed by an external pump (usually the power steering pump). Both type systems act to multiply the force exerted on the master cylinder piston by the operator in order to increase the hydraulic pressure delivered to the calipers and wheel cylinders, while decreasing the effort necessary to obtain acceptable stopping performance.

Vacuum assist units are similar in operation and get their energy by opposing engine vacuum to atmospheric pressure. A piston and cylinder, flexible diaphragm or bellows utilize this energy to provide brake assistance. The fundamental difference between the different types of vacuum assist systems lies simply in how the diaphragm within the power unit is suspended when the brakes are not applied.

In order to properly diagnose vacuum assist system malfunctions it is important to know whether the diaphragm within a power unit is air suspended or vacuum suspended. Air-suspended units are under atmospheric pressure until the brakes are applied, then engine vacuum is admitted, causing the piston or diaphragm to move or the bellows to collapse. Vacuum-suspended types are balanced with engine vacuum until the brake pedal is depressed, allowing atmospheric pressure to unbalance the unit and apply force to the brake system.

The fact that brakes will operate even if the power unit fails gives us a clue to successful power brake service. This means the conventional brake system is left intact and a power unit is simply added to the existing system. Troubleshooting is then exactly the same until we get to the power unit. Keep in mind, however, that power brakes give a higher line pressure, thus making leaks more critical.

CHECKING COMPLAINTS

Complaints about power brake operation should be handled as if two separate systems exist. Check for faults in the regular brake system first. If it is OK, start looking over the power brake circuit. For a quick check of proper power unit operation, press the brake pedal firmly and then start the engine. The pedal should fall away slightly and less pressure should be needed to maintain the pedal in any position. On vacuum-suspended power units, air will rush into the air intake when the brakes are applied.

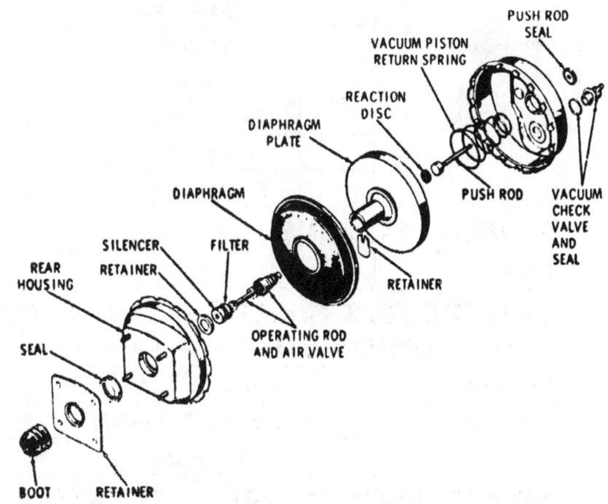

Fig. 1 Bendix single diaphragm power brake unit

Another check begins with installation of a suitable pressure gauge in the brake hydraulic system. Take a reading with the engine off and the power unit not operating. Maintaining the same pedal height, start the engine and take another reading. There should be a substantial pressure increase in the second reading.

Pedal free travel and total travel are critical on cars equipped with power brakes. As a general rule, brakes should be adjusted or relined if the pedal is closer than 1 1/2 inch from the floor with the brakes applied. Free travel should be kept strictly to specifications.

On models with vacuum boost brakes, take a manifold vacuum reading and if equipped check operation of the external vacuum pump. If the power unit isn't getting enough assistance because manifold vacuum is abnormally low, tune the engine and then try the power brakes again. Naturally, loose vacuum lines and clogged air intake filters will cut down brake efficiency. Most units have a check valve that retains some vacuum in the system when the engine is off. A vacuum gauge check of this valve will tell you when it is restricted or stuck open or closed.

On models with Hydro-Boost brakes, check operation of the power steering pump and check for improperly adjusted drive belts, defective pressure relief valve, leaking hoses and fittings, or defective accumulator. Most causes of failure on these systems not within the power assist unit itself. However, if all other components are satisfactory but a malfunction persists, the power unit should be disassembled and inspected for damage, wear and contamination.

Failure of the brakes to release in most instances is caused by a tight or misaligned connection between the power unit and the brake linkage. If this connection is free, look for a broken piston, diaphragm or bellows return spring on models with vacuum boost brakes. On models with hydro boost brakes, check for contamination of the fluid which results in restrictions within the spool valve, sticking assist piston within the power unit or defective accumulator. The power unit will have to be disassembled for this check and, if the spring is in good condition, continue going through the power unit, checking the fluid, air and vacuum valves.

A simple check of the hydraulic system should be made before proceeding. Loosen the connection between the master cylinder and the brake booster. If the brakes release, the trouble is in the power unit; if the brakes still will not release, look for a

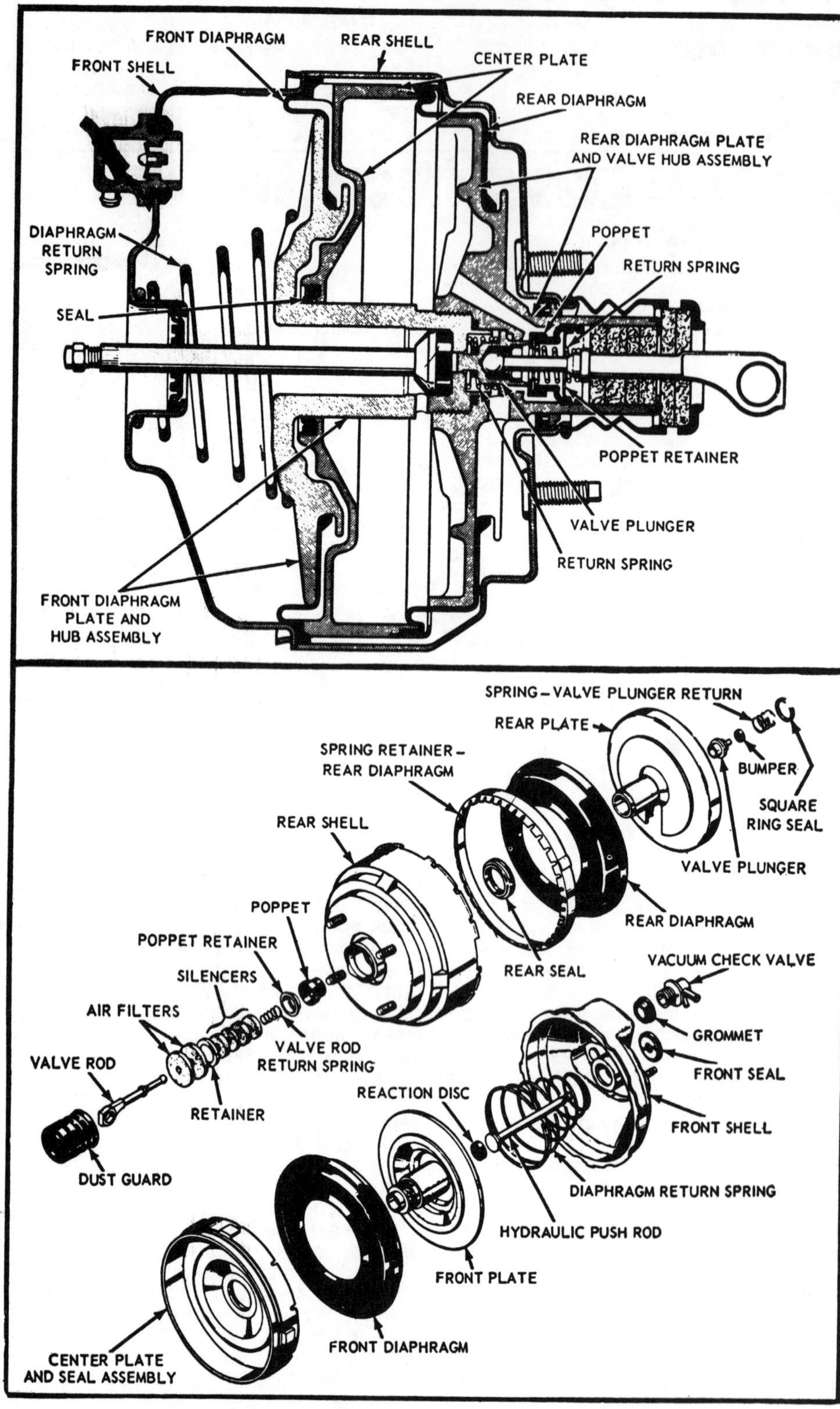

Fig. 2 Bendix tandem diaphragm power brake

restricted brake line or similar difficulties in the regular hydraulic circuit.

A residual pressure check valve is usually included immediately under the brake line connection on hydraulic assist power brakes. This valve maintains a slight hydraulic pressure within the brake lines and wheel cylinders to give better pedal response. If it is sticking, the brakes may not release.

Power brakes that have a hard pedal are usually suffering from a milder form of the same ills that cause complete power unit failure. Collapsed or leaking vacuum lines or insufficient manifold vacuum, as well as punctured diaphragms or bellows and leaky piston seals, all lead to weak power unit operation. A steady hiss when the brake is held down means a vacuum leak that will cause poor power unit operation.

Do not immediately condemn the power unit if the brakes grab. First look for all the usual causes, such as greasy linings or scored drums. Then investigate the power unit. When the trouble has been traced to the power unit, check for a damaged reaction control. The reaction control is usually made up of a diaphragm, spring and valving that tends to resist pedal action. It is put in the system to give the pedal "feel."

TROUBLESHOOTING

Decreasing Brake Pedal Travel

If a decreasing brake pedal is encountered, the power brake unit may be binding internally. To test the power brake unit for this condition proceed as follows:

1. Place transmission shift lever into Neutral and start engine.
2. Increase engine speed to approximately 1500 RPM, close throttle and completely depress brake pedal.
3. Slowly release brake pedal and stop engine.
4. Remove vacuum check valve and hose from power brake unit. Observe for backward movement of brake pedal.
5. If brake pedal moves backward, power brake unit has internal binding.
6. Replace power brake unit.

Hard Brake Pedal

An internal bind or a failed vacuum check valve would cause this condition. Refer to Decreasing Brake Pedal Travel to test power brake unit for an internal bind. To check for a failed vacuum check valve proceed as follows:

1. Start engine and increase engine speed to approximately 1500 RPM, then close throttle and stop engine.
2. Wait 90 seconds, then try brake action.
3. If brakes are not vacuum assisted for two or more applications, replace check valve.

Dragging Brakes

If slow or incomplete release of brakes (dragging brakes) is encountered the power brake unit has an internal bind condition. Test for an internal bind condition as described previously.

PUSHROD ADJUSTMENT

In some cases adjustment of the brake booster pushrod is necessary to ensure proper operation of the power brake system. A pushrod that is too long will cause the master cylinder piston to close off the compensating port, preventing hydraulic pressure from being released and resulting in brake drag. A pushrod that is too short will cause excessive brake pedal travel and cause groaning noises to come from the booster when the brakes are applied. A properly adjusted pushrod that remains assembled to the booster with which is was matched during production should not require service adjustment. However, if the booster, master cylinder or pushrod are serviced, the pushrod may require adjustment.

Eagle Medallion

Refer to "Master Cylinder Replace" as described under "Type 3—Composite Master Cylinder," in the "Hydraulic Brake System" chapter for pushrod adjustment procedure.

Colt, Colt Vista, Conquest & Eagle Summit

Refer to "Adjustments" as described under "Types 5, 6, 7 & 8—Composite Master Cylinders" in the "Hydraulic Brake System" chapter for pushrod adjustment procedure.

BENDIX DIAPHRAGM TYPES

These units are of the vacuum suspended type. Some units are of the single diaphragm type, **Fig. 1,** while others are of the tandem diaphragm type, **Fig. 2.** Both single piston and double piston or split system type master cylinders are used.

The vacuum suspended diaphragm type units utilize engine manifold vacuum and atmospheric pressure for its power. It consists of three basic elements combined into a single power unit. The three basic elements of the single diaphragm type are:

1. A vacuum power section which includes a front and rear shell, a power diaphragm, a return spring and a push rod.
2. A control valve, built integral with the power diaphragm and connected through a valve rod to the brake pedal, controls the degree of brake application or release in accordance with the pressure applied to the brake pedal.
3. A hydraulic master cylinder, attached to the vacuum power section which contains all the elements of the conventional brake master cylinder except for the pushrod, supplies fluid under pressure to the wheel brakes in proportion to the pressure applied to the brake pedal.

OPERATION

Upon application of the brakes, the valve rod and plunger move to the left in the power diaphragm to close the vacuum port and open the atmospheric port to admit air through the air cleaner and valve at the rear diaphragm chamber. With vacuum present in the rear chamber, a force is developed to move the power diaphragm, hydraulic pushrod and hydraulic piston or pistons to close the compensating port or ports and force fluid under pressure through the residual check valve or valves and lines into the front and rear wheel cylinders to actuate the brakes.

As pressure is developed within the master cylinder a counter force acting through the hydraulic pushrod and reaction disc against the vacuum power diaphragm and valve plunger sets up a reaction force opposing the force applied to the valve rod and plunger. This reaction force tends to close the atmospheric port and reopen the vacuum port. Since this force is in opposition to the force applied to the brake pedal by the driver it gives the driver a "feel" of the amount of brake applied. The proportion of reactive force applied to the valve plunger through the reaction disc is designed into the Master-Vac to assure maximum power consistent with maintaining pedal feel. The reaction force is in direct proportion to the hydraulic pressure developed within the brake system.

Ford Motor Co.

On models with Anti-Lock Brakes, refer to the Anti-Lock Brake Chapter.

INDEX

APPLICATION

COUGAR, LTD II & THUNDERBIRD

1985-89① .... BendixSingle–TandemDiaphragm

FORD & MERCURY FULLSIZE

1985-89① .... BendixSingle–TandemDiaphragm

ESCORT, EXP, LYNX, TEMPO & TOPAZ

1985-89② .... Bendix–TevesSingleDiaphragm

FESTIVA

1988-89 ..... SingleDiaphragmBooster

LINCOLN

1985-89③ .. BendixTandemDiaphragm
1985-89④ .... Hydro-BoostII

MUSTANG, MUSTANG SVO & CAPRI

1985-89① .... BendixSingleDiaphragm

PROBE

1989......... SingleDiaphragmBooster

SABLE & TAURUS

1986-89① .... BendixSingleDiaphragm

TRACER

1988-89 ..... SingleDiaphragmBooster

XR4TI

1985-89① .... BendixSingleDiaphragm

①—The only service that may be performed is replacement of the grommet and check valve. If any other malfunction is apparent, booster should be replaced.

②—The only service that may be performed is replacement of the grommet, check valve and pushrod on the Teves type brake boosters. If any other malfunction is apparent, booster should be replaced.

③—Except Mark VII & Continental.

④—Mark VII & Continental.

GENERAL SERVICE

On Bendix type power brake units, overhaul is not required. If it has been determined that the brake booster is defective, replace unit as required. In some instances, the only service required is replacement of the check valve and grommet and pushrod adjustment.

In order to properly service and repair available brake systems, a thorough understanding of the various power assist systems is necessary. Two basic types of power assist mechanisms are used: vacuum assist diaphragm assemblies which use engine vacuum and in some cases vacuum pressure developed by an external vacuum pump, and hydraulic pressure assist mechanisms which use pressure developed by an external pump (usually the power steering pump). Both type systems act to multiply the force exerted on the master cylinder piston by the operator in order to increase the hydraulic pressure delivered to the wheel cylinders while decreasing the effort necessary to obtain acceptable stopping performance.

Vacuum assist units are similar in operation and get their energy by opposing engine vacuum to atmospheric pressure. A piston and cylinder, flexible diaphragm or bellows utilize this energy to provide brake assistance. The fundamental difference between the different types of vacuum assist systems lies simply in how the diaphragm within the power unit is suspended when the brakes are not applied.

In order to properly diagnose vacuum assist system malfunctions it is important to know whether the diaphragm within a power unit is air suspended or vacuum suspended. Air-suspended units are under atmospheric pressure until the brakes are applied, then engine vacuum is admitted, causing the piston or diaphragm to move or the bellows to collapse. Vacuum-suspended types are balanced with engine vacuum until the brake pedal is depressed, allowing atmospheric pressure to unbalance the unit and apply force to the brake system.

Regardless of whether the brakes are vacuum or hydraulically assisted, certain general service procedures apply. Only top quality, clean brake fluid should be used in brake system, and on hydro boost systems, use of the specified hydraulic fluid in the boost circuit is essential to proper system operation. In addition, on hydro-boost systems, care must be taken not to mix the fluids of the two separate operating circuits, as more seals and valves are used. Use of improper fluids will result in damage to the hydro-boost system. For the same reason, ensure system does not get contaminated by foreign material.

The fact that brakes will operate even if the power unit fails gives us a clue to successful power brake service. This means the conventional brake system is left intact and a power unit is simply added to the existing system. Troubleshooting is then exactly the same until we get to the power unit. As with conventional hydraulic brakes, a spongy pedal with power brakes still means air in the system and grease on the linings will still make the brakes grab. Keep in mind, however, that power brakes give a higher line pressure, thus making leaks more critical.

PRECAUTIONS

1. After disassembling a power brake unit, soak all metal parts in a suitable solvent and clean all rubber components with alcohol.
2. After all components have been thoroughly cleaned and rinsed in solvent, the metal parts that come in contact with hydraulic brake fluid or rubber parts should be rewashed in clean alcohol before assembly.
3. Use an air hose to blow dirt and cleaning fluid from the recesses and internal passages.

4. Always use all the parts furnished in the repair kit. Discard all old rubber parts.
5. Use extreme caution when disassembling power assist mechanisms, as releasing internal spring tension suddenly could cause damage or personal injury.
6. **On models equipped with hydro-boost assist systems,** do not disassemble the accumulator. **If the accumulator is damaged or diagnosed as defective, it must be replaced as an assembly.**

PUSHROD ADJUSTMENT

Proper adjustment of the master cylinder pushrod is necessary to ensure proper operation of the power brake system. A pushrod that is too long will cause the master cylinder piston to close off the compensating port, preventing hydraulic pressure from being released and resulting in brake drag. A pushrod that is too short will cause excessive brake pedal travel and cause groaning noises to come from the booster when the brakes are applied. A properly adjusted pushrod that remains assembled to the booster with which is was matched during production should not require service adjustment. However, if the booster, master cylinder or pushrod are serviced, the pushrod may require adjustment.

There are two methods that can be used to check for proper pushrod length and installation: the gauge method and air method. Usually, if the power unit pushrod requires an adjustment the Power Unit Repair Kit for the unit being serviced includes a gauge. The gauge measures from the end of the pushrod to the power unit shell.

GAUGE METHOD

Bendix Type

1. Disconnect master cylinder from booster leaving brake lines connected, and secure cylinder to prevent lines from being damaged.
2. Start engine and operate engine at idle speed.
3. With engine running, position gauge over pushrod. Gauge should bottom against booster housing with a force of approximately 5 lbs. applied to pushrod, **Fig. 1.**
4. If force required to seat gauge exceeds 5 lb., shorten length of pushrod. If force required to seat gauge is less than 5 lbs., lengthen pushrod. **Ensure that pushrod is properly seated in booster when performing gauge check.**
5. Install master cylinder and remove reservoir cover.
6. Observe fluid surface when brakes are applied and released rapidly. If no movement is observed on fluid surface, pushrod is adjusted too long.

AIR METHOD

1. Ensure master cylinder attaching nuts are tight.

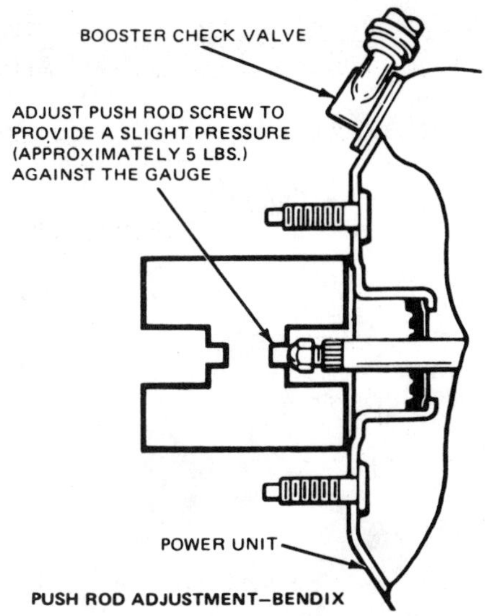

Fig. 1 Master cylinder pushrod adjustment. Bendix type vacuum booster

2. Remove master cylinder filler cap.
3. With brake released, force compressed air into the hydraulic outlet of the master cylinder. **Regulate air pressure to a value of approximately 5 psi, to prevent spraying brake fluid from master cylinder. Care must be taken not to allow brake fluid to contact painted surfaces of vehicle, skin or eyes, as damage or personal injury will result.**
4. If air passes through the compensating port, which is the smaller of the two holes in the bottom of the master cylinder reservoir, the adjustment is satisfactory.
5. If air does not flow through the compensating port, adjust the pushrod as required, either by means of the adjustment screw (if provided) or by adding shims between the master cylinder and power unit shell until the air flows freely.
6. Connect brake lines and bleed system.

SINGLE DIAPHRAGM BOOSTER

Adjustment of pushrod is necessary only when master cylinder is replaced.

1. Remove master cylinder as previously outlined.
2. Position master cylinder gauge T87C-2500-A or equivalent on end of master cylinder.
3. Loosen set screw and push gauge plunger against bottom of primary piston.
4. While holding gauge in position, tighten set screw.
5. Invert gauge and place over brake booster push rod, reading should be 0.
6. If clearance is not 0, loosen push rod locknut and adjust push rod.
7. Reverse procedure to install.

CHECKING COMPLAINTS

Complaints about power brake operation should be handled as if two separate systems exist. Check for faults in the regular brake system first. If it is satisfactory, start inspecting the power brake circuit. For a quick check of proper power unit operation, press the brake pedal firmly and then start the engine. The pedal should fall away slightly and less pressure should be needed to maintain the pedal in any position. **On vacuum-suspended power units,** air will rush into the air intake when the brakes are applied.

Another check begins with installation of a suitable pressure gauge in the brake hydraulic system. Take a reading with the engine off and the power unit not operating. Maintaining the same pedal height, start the engine and take another reading. There should be a substantial pressure increase in the second reading.

Pedal free travel and total travel are critical on cars equipped with power brakes. As a general rule, brakes should be adjusted or relined if the pedal is closer than 1 1/2 inch from the floor with the brakes applied. Free travel should be kept strictly to specifications.

On models with vacuum boost brakes, take a manifold vacuum reading or check operation of the external vacuum pump if the power unit isn't giving enough assistance. Remember, though, currently produced emission controlled engines, manifold vacuum readings may be less than 15 inches Hg at idle. If manifold vacuum is abnormally low, tune the engine and then try the power brakes again. Naturally, loose vacuum lines and clogged air intake filters will cut down brake efficiency. Most units have a check valve that retains some vacuum in the system when the engine is off. A vacuum gauge check of this valve will tell you when it is restricted or stuck open or closed.

On models with hydro-boost brakes, check operation of the power steering pump and check for improperly adjusted drive belts, defective pressure relief valve, leaking hoses and fittings, or defective accumulator. Most causes of failure on these systems not within the power assist unit itself. However, if all other components are satisfactory but a malfunction persists, the power unit should be disassembled and inspected for damage, wear and contamination.

Failure of the brakes to release in most instances is caused by a tight or misaligned connection between the power unit and the brake linkage. If this connection is free, look for a broken piston, diaphragm or bellows return spring on models with vacuum boost brakes. **On models with hydro boost brakes,** check for contamination of the fluid which results in restrictions within the spool valve, sticking assist piston within the power unit or defective accumulator. The power unit will have to be disassembled for this check and, if the spring is in good condition, continue going through the power unit, checking the fluid, air and vacuum valves.

A simple check of the hydraulic system

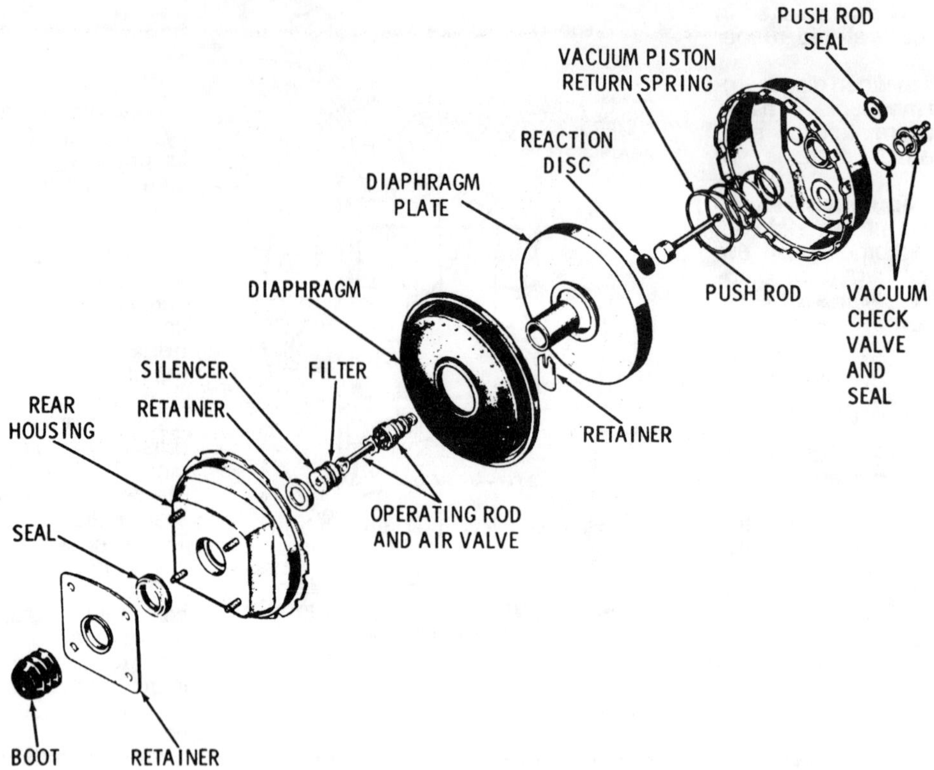

Fig. 2 Bendix single diaphragm power brake unit

should be made before proceeding. Loosen the connection between the master cylinder and the brake booster. If the brakes release, the trouble is in the power unit; if the brakes still will not release, look for a restricted brake line or similar difficulties in the regular hydraulic circuit.

A residual pressure check valve is usually included immediately under the brake line connection on hydraulic assist power brakes. This valve maintains a slight hydraulic pressure within the brake lines and wheel cylinders to give better pedal response. If it is sticking, the brakes may not release.

Power brakes that have a hard pedal are usually suffering from a milder form of the same ills that cause complete power unit failure. Collapsed or leaking vacuum lines or insufficient manifold vacuum, as well as punctured diaphragms or bellows and leaky piston seals, all lead to weak power unit operation. A steady hiss when the brake is held down means a vacuum leak that will cause poor power unit operation.

Do not immediately condemn the power unit if the brakes grab. First look for all the usual causes, such as greasy linings or scored drums. Then investigate the power unit. When the trouble has been traced to the power unit, check for a damaged reaction control. The reaction control is usually made up of a diaphragm, spring and valves that tends to resist pedal action. It is put in the system to give the pedal "feel."

BENDIX DIAPHRAGM TYPES

These units are of the vacuum suspended type. Some units are of the single diaphragm type, **Fig. 2** while others are of the tandem diaphragm type, **Fig. 3.** Both single piston and double piston or split system type master cylinders are used.

The vacuum suspended diaphragm type units utilize engine manifold vacuum and atmospheric pressure for its power. It consists of three basic elements combined into a single power unit. The three basic elements of the single diaphragm type are:

1. A vacuum power section which includes a front and rear shell, a power diaphragm, a return spring and a pushrod.
2. A control valve, built integral with the power diaphragm and connected through a valve rod to the brake pedal, controls the degree of brake application or release in accordance with the pressure applied to the brake pedal.
3. A hydraulic master cylinder, attached to the vacuum power section which contains all the elements of the conventional brake master cylinder except for the pushrod, supplies fluid under pressure to the wheel brakes in proportion to the pressure applied to the brake pedal.

OPERATION

Upon application of the brakes, the valve rod and plunger move to the left in the power diaphragm to close the vacuum port and open the atmospheric port to admit air through the air cleaner and valve at the rear diaphragm chamber. With vacuum present in the rear chamber, a force is developed to move the power diaphragm, hydraulic pushrod and hydraulic piston or pistons to close the compensating port or ports and force fluid under pressure through the residual check valve or valves and lines into the front and rear wheel cylinders to actuate the brakes.

As pressure is developed within the master cylinder a counter force acting through the hydraulic pushrod and reaction disc against the vacuum power diaphragm and valve plunger sets up a reaction force opposing the force applied to the valve rod and plunger. This reaction force tends to close the atmospheric port and reopen the vacuum port. Since this force is in opposition to the force applied to the brake pedal by the driver it gives the driver a "feel" of the amount of brake applied. The proportion of reactive force applied to the valve plunger through the reaction disc is designed into the Master-Vac to assure maximum power consistent with maintaining pedal feel. The reaction force is in direct proportion to the hydraulic pressure developed within the brake system.

TROUBLESHOOTING

Hard Pedal Or No Assist

1. Air cleaner element clogged.
2. Control valve faulty.
3. Defective diaphragm.
4. Worn or distorted reaction plate or levers.
5. Cracked or broken power piston or levers.
6. Internal or external leaks.

Brakes Grab

1. Control valve defective or sticking.

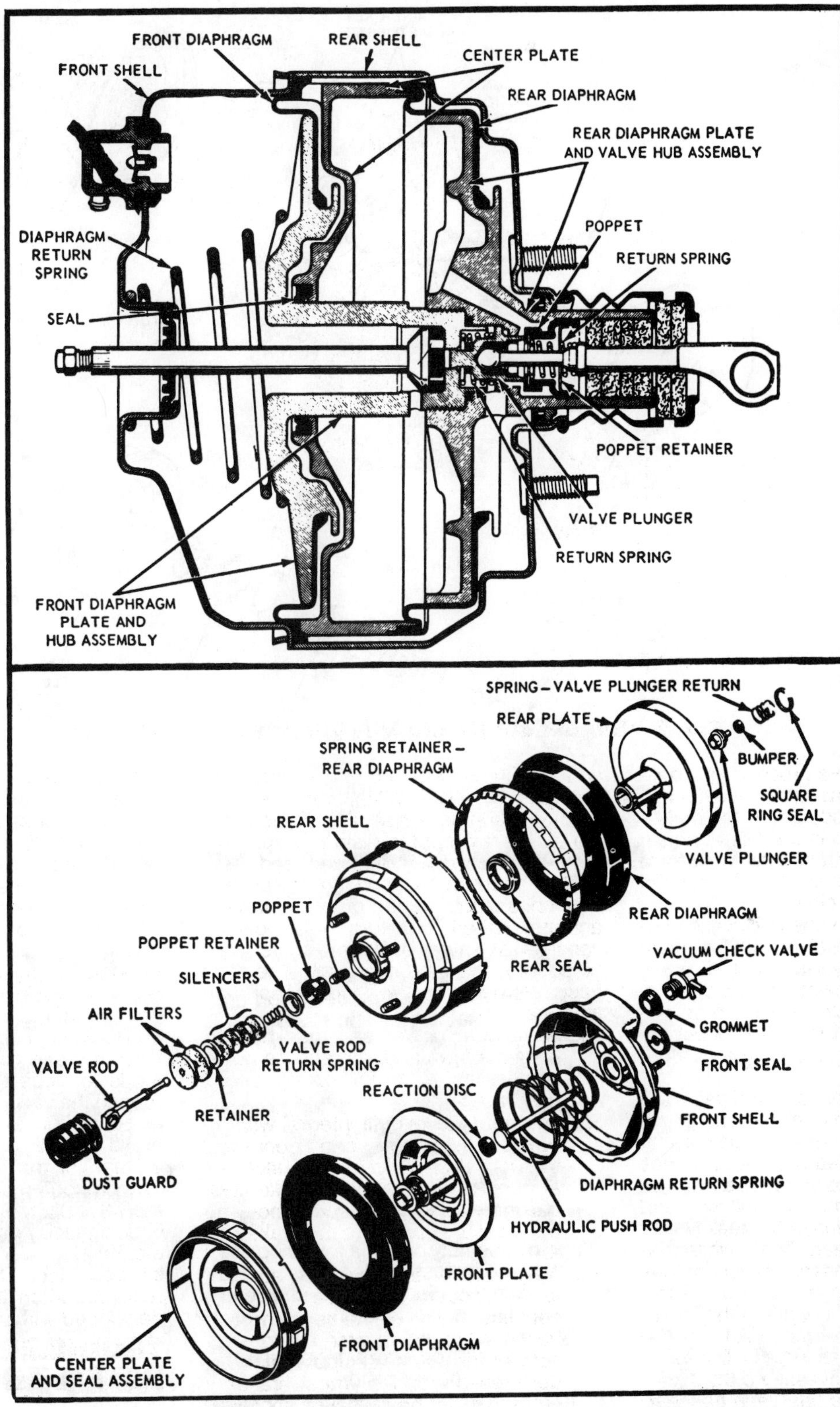

Fig. 3 Bendix tandem diaphragm power brake (Type B-2)

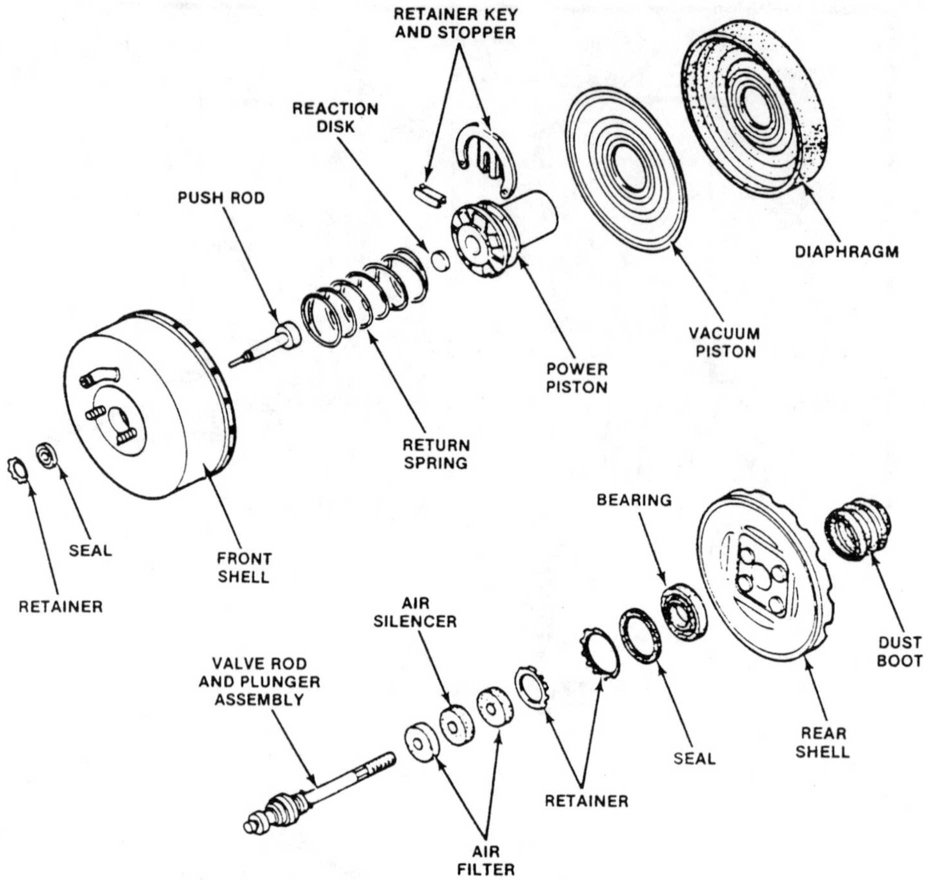

Fig. 4 Exploded view of Single Diaphragm Booster

2. Bind in linkage.
3. Reaction diaphragm leaking.
4. Worn or distorted levers or plate.

No Or Slow Release

1. Push rod adjustment incorrect.
2. Linkage binding.
3. Return spring defective.

DISASSEMBLY

Single Diaphragm Booster

1. Remove brake booster from vehicle.
2. Place front shell of booster in a vise.
3. Remove clevis from valve rod and plunger assembly.
4. Remove dust boot from rear shell, **Fig. 4,** then place alignment marks on front and rear shells for proper alignment during assembly.
5. Rotate rear shell counterclockwise until shell unlocks.
6. Remove rear shell, then remove front shell return spring and push rod.
7. Remove front shell retainer and dust boot.
8. Remove rear shell from power piston assembly.
9. Remove rear shell bearing retainer, seal and bearing.
10. Remove retainer holding air filters and air silencer from the valve rod and plunger assembly.
11. Remove diaphragm and plate.
12. Slightly push down on valve rod and remove retainer key and stopper.
13. Remove valve rod and plunger assembly from the power piston assembly.
14. Remove air filters and air silencer from the power piston.
15. Remove reaction disc.

ASSEMBLY

Single Diaphragm Booster

Prior to assembly coat the following with silicone grease; reaction disc surface, dust seal lip, push rod, diaphragm-to-shell contact surfaces, power piston and valve plunger oil seal.

1. Install valve rod and plunger assembly into the power piston.
2. Push down on valve rod, then align groove of the valve plunger with the slot of power piston.
3. Install the valve rod retainer key and stopper.
4. Install two air filters and air silencer into power piston.
5. Place diaphragm and plate onto power piston, ensure the diaphragm is fully seated in the groove of the power piston.
6. Install bearing, bearing seal and bearing retainer into rear shell.
7. Slide rear shell onto power piston while carefully guiding power piston through bearing seal.
8. Using the push rod, install reaction disc into the power piston.
9. Install dust seal, front retainer and push rod into front shell.
10. Place front shell in a vise, then install front shell return spring.
11. Place rear shell onto front shell noting alignment marks.
12. Rotate rear shell clockwise until rear shell locks into place with front shell.
13. Install dust boot onto rear shell then install clevis onto valve rod and plunger assembly.
14. Install master cylinder.

HYDRO-BOOST II (HB-II)

Hydro-boost II is a hydraulically operated power brake booster. The hydraulic brake booster consists of an open center spool valve and a hydraulic cylinder combined into a single housing. The power steering pump provides the hydraulic fluid pressure to operate both the power brake booster and the power steering gear.

A dual master cylinder is bolted to the brake booster and is operated by a pushrod projecting from the booster cylinder

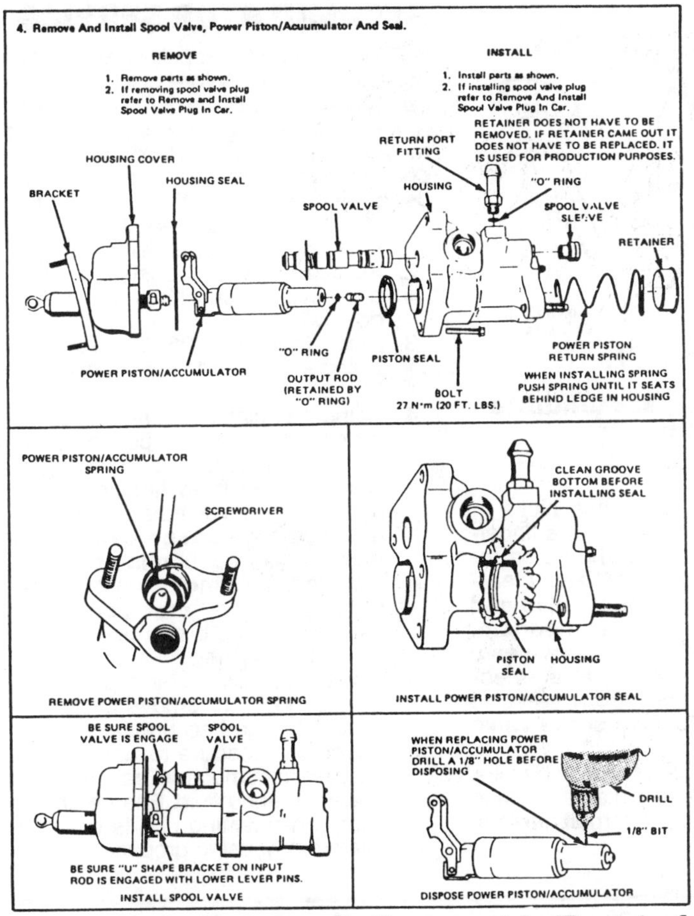

Fig. 5 Hydro-boost unit disassembly (Part 1 of 2). HB-II

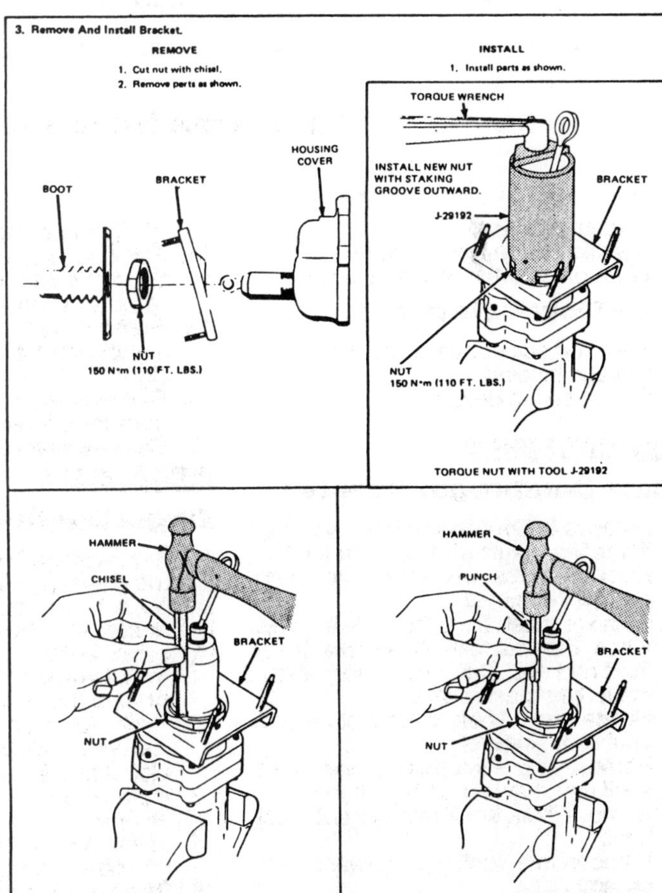

Fig. 5 Hydro-boost unit disassembly (Part 2 of 2). HB-II

bore.

This system uses a compressed gas accumulator as a reserve system to store sufficient fluid under pressure to provide at least two brake applications if fluid flow from the power steering pump is not available.

The booster is designed so that if a total absence of power assist occurs, the brakes can be applied manually, but somewhat greater pressure on the brake pedal is required.

DISASSEMBLY & ASSEMBLY

Refer to **Fig. 5** for disassembly and assembly of Hydro-boost II unit.

CLEANING & INSPECTION

Keep all disassembled parts clean until assembly. Lubricate all seals and metal friction points with power steering fluid. When unit is disassembled, all seals and tube inserts should be replaced. These parts are available in replacement kits. If any accumulator valve components are damaged or lost, replace all valve components.

1. Inspect spool valve and spool valve bore for corrosion, nicks, scoring or other damage. Discoloration of the spool or bore, particularly in the groove areas, is not harmful and replacement is not necessary.

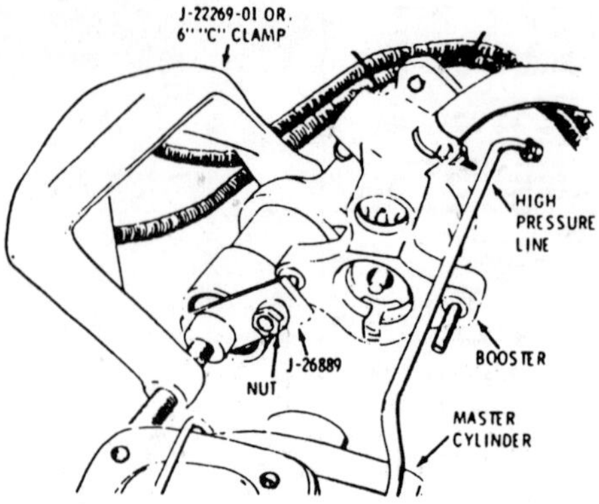

Fig. 6 Removing accumulator. HB-II

2. If spool valve or bore has nicks or scoring that can be felt with a fingernail, the entire booster should be replaced as an assembly. **The clearance between the spool valve and the spool valve bore is important. Because of this, the spool valve and the housing make up a selective assembly. The spool valve is selected to match the spool valve bore.**
3. Inspect piston for scratches and nicks. If scratches on the outside surface can be felt with a fingernail, replace piston.

ACCUMULATOR SERVICE

Do not attempt to repair or apply heat to the accumulator. Before discarding an inoperative accumulator, drill a 1/16 inch in diameter hole through the end of the accumulator pan. Do not drill through the piston end. Always wear safety glasses when performing service on the accumulator.

1. Pump pedal rod several times to ensure accumulator is depleted.
2. Secure brake booster mounting bracket in a vise with pedal rod facing downward.
3. Place tool No. J-26889 over master cylinder stud and install retaining nut, **Fig. 6.**
4. Using a C-clamp, depress accumulator can, then insert a punch into hole in housing to unseat retaining ring. Remove ring from housing using a small screwdriver. **If can cannot easily be depressed approximately .10 inch, the accumulator is still charged. This indicates an internal problem with the accumulator valves and the brake booster must be disassembled.**
5. Slowly back off C-clamp until tension is relieved, then remove C-clamp, retaining cap and accumulator.
6. Reverse procedure to assemble. Lubricate accumulator can seal with power steering fluid. Ensure accumulator can retaining ring is completely seated in housing groove.

General Motors Corp.

INDEX

APPLICATIONS

BUICK REAR WHEEL DRIVE & RIVIERA

1985 .... Hydro-BoostII
1985-86 .... Delco-Moraine SingleDiaphragm
1985-87 .... PowerMaster
1985-89 .... Delco-Moraine TandemDiaphragm

CADILLAC EXCEPT CIMARRON

1985-86 .... Delco-Moraine SingleDiaphragm
1985-89 .... Delco-Moraine TandemDiaphragm

CALAIS, CAVALIER, CELEBRITY, CENTURY, CIMARRON, CITATION, CUTLASS CIERA, FIRENZA, GRAND AM, SKYHAWK, SKYLARK, SOMERSET REGAL, SUNBIRD/2000 & 6000

1985-89 .... Delco-Moraine TandemDiaphragm

CAMARO & FIREBIRD

1985-89 .... Delco-Moraine TandemDiaphragm

CAPRICE

1985 .... Hydro-BoostII
1985-86 .... PowerMaster
1985-89 .... Delco-Moraine TandemDiaphragm

CHEVETTE & 1000

1985-87 .... BendixSingleDiaphragm

CORSICA & BERETTA

1987-89 .... Delco-Moraine TandemDiaphragm

CORVETTE

1985-89 .... GirlockSingleDiaphragm

DEVILLE, ELECTRA, FLEETWOOD & 98

1985-89 .... Delco-Moraine TandemDiaphragm

FIERO

1985-88 .... Delco-Moraine TandemDiaphragm

LEMANS

1988-89 .... SingleDiaphragm

METRO & SPECTRUM

1989 .... SingleDiaphragm

MONTE CARLO

1985 .... Hydro-BoostII
1985-89 .... Delco-Moraine TandemDiaphragm

NOVA

1985-88 .... Toyota/GM SingleDiaphragm

OLDSMOBILE REAR WHEEL DRIVE & TORONADO

1985 .... Hydro-BoostII
1985-86 .... Delco-Moraine SingleDiaphragm
1985-86 .... PowerMaster
1985-89 .... Delco-Moraine TandemDiaphragm

PONTIAC FULL SIZE & INTERMEDIATE

1985-86 .... Delco-Moraine SingleDiaphragm
1985-89 .... Delco-Moraine TandemDiaphragm
1985 .... Hydro-BoostII
1985-86 .... PowerMaster

SPRINT

1985-89 .... SingleDiaphragm

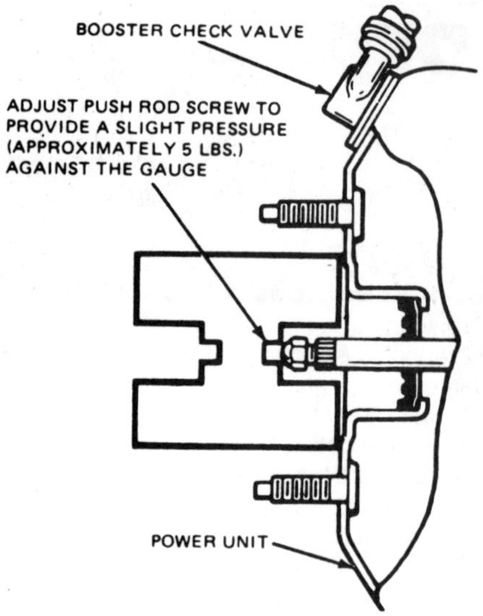

Fig. 1 Master cylinder pushrod adjustment. Bendix type vacuum booster

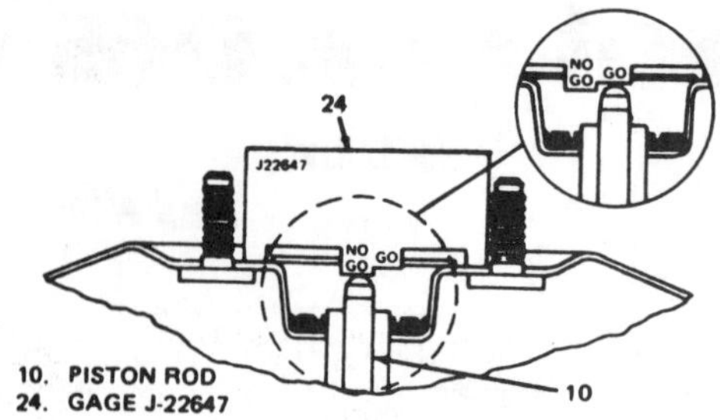

Fig. 2 Master cylinder pushrod adjustment. Delco-Moraine type vacuum booster

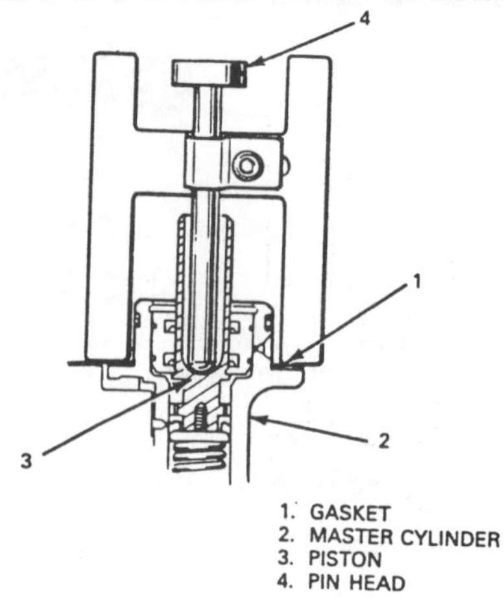

Fig. 3 Master cylinder adjustment tool

GENERAL SERVICE

Two basic types of power assist mechanisms are used: vacuum assist diaphragm assemblies, which use engine vacuum and in some cases vacuum pressure developed by an external vacuum pump, and hydraulic pressure assist mechanisms, which use pressure developed by an external pump (usually the power steering pump). Both type systems act to multiply the force exerted on the master cylinder piston by the operator in order to increase the hydraulic pressure delivered to the wheel cylinders while decreasing the effort necessary to obtain acceptable stopping performance.

Vacuum assist units are similar in operation and get their energy by opposing engine vacuum to atmospheric pressure. A piston and cylinder, flexible diaphragm or bellows utilize this energy to provide brake assistance. The fundamental difference between the different types of vacuum assist systems lies simply in how the diaphragm within the power unit is suspended when the brakes are not applied.

In order to properly diagnose vacuum assist system malfunctions it is important to know whether the diaphragm within a power unit is air suspended or vacuum suspended. Air-suspended units are under atmospheric pressure until the brakes are applied, then engine vacuum is admitted, causing the piston or diaphragm to move or the bellows to collapse. Vacuum-suspended types are balanced with engine vacuum until the brake pedal is depressed, allowing atmospheric pressure to unbalance the unit and apply force to the brake system.

Regardless of whether the brakes are vacuum or hydraulically assisted, certain general service procedures apply. Only top quality, clean brake fluid should be used in brake system. On hydro boost systems, use of the specified hydraulic fluid in the boost circuit is essential to proper system operation. In addition, on hydro-boost systems, care must be taken not to mix the fluids of the two separate operating circuits, as more seals and valves are used, so use of improper fluids will do much more damage. For the same reason, be sure all dirt is kept out of the system.

PRECAUTIONS

1. After disassembling a power brake unit, soak all metal parts in solvent.
2. Use only alcohol on parts containing rubber. After the parts have been thoroughly cleaned and rinsed in solvent, they should be rewashed in clean alcohol before assembly.
3. Use an compressed air to blow dirt and cleaning fluid from the recesses and internal passages.
4. Always use all the parts furnished in the repair kit.
5. **Use extreme caution when disassembling power assist mechanisms, as suddenly releasing internal spring tension could cause damage or personal injury.**
6. **On models with hydro-boost assist systems,** do not disassemble the accumulator. If the accumulator is damaged or diagnosed as defective, it must be replaced as an assembly.

PUSHROD ADJUSTMENT

Proper adjustment of the master cylinder pushrod is necessary to ensure proper operation of the power brake system. A pushrod that is too long will cause the master cylinder piston to close off the compensating port, preventing hydraulic pressure from being released and resulting in brake drag. A pushrod that is too short will cause excessive brake pedal travel and cause groaning noises to come from the booster when the brakes are applied. A properly adjusted pushrod that remains assembled to the booster with which it was matched during production should not require service adjustment. However, if the booster, master cylinder or pushrod are serviced, the pushrod may require adjustment.

There are two methods that can be used to check for proper pushrod length and installation: the gauge method and air method. Usually, if the power unit pushrod requires an adjustment the power unit repair kit for the unit being serviced includes a gauge. The gauge measures from the end of the pushrod to the power unit shell.

GAUGE METHOD

Bendix Type

1. Disconnect master cylinder from booster leaving brake lines connected, and secure cylinder to prevent lines from being damaged.
2. Start engine and allow to run at normal curb idle speed.
3. With engine running, position gauge over pushrod. Gauge should bottom against booster housing with a force of approximately 5 lbs. applied to pushrod, **Fig. 1.**
4. If force required to seat gauge exceeds 5 lb., shorten length of pushrod. If force required to seat gauge is less than 5 lbs, lengthen pushrod. **Ensure pushrod is properly seated in booster when performing gauge check.**
5. Reinstall master cylinder, then remove reservoir cover.
6. Observe fluid surface when brakes are applied and released rapidly. If no movement is observed on fluid surface, pushrod is adjusted too long.

Delco-Moraine Type

On these models, the master cylinder pushrod length is fixed and is usually only checked after the unit has been overhauled.

1. Assemble booster unit and install pushrod, ensuring pushrod is fully seated.
2. Position go/no go gauge furnished in repair kit over pushrod as shown in **Fig. 2**
3. If pushrod height is not within limits of gauge, install service adjustable pushrod and adjust rod to obtain correct height.

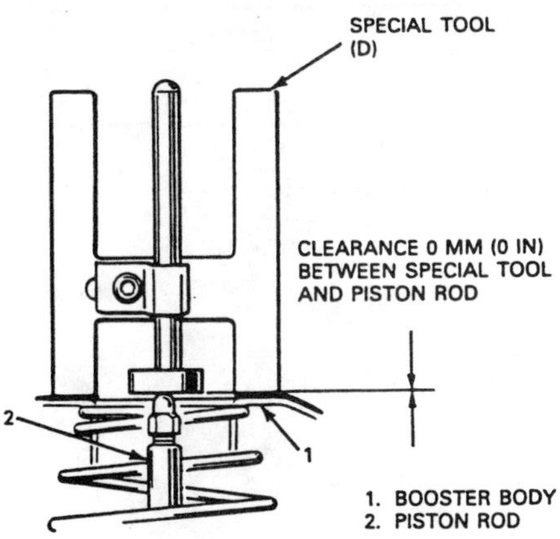

Fig. 4 Booster piston rod adjustment

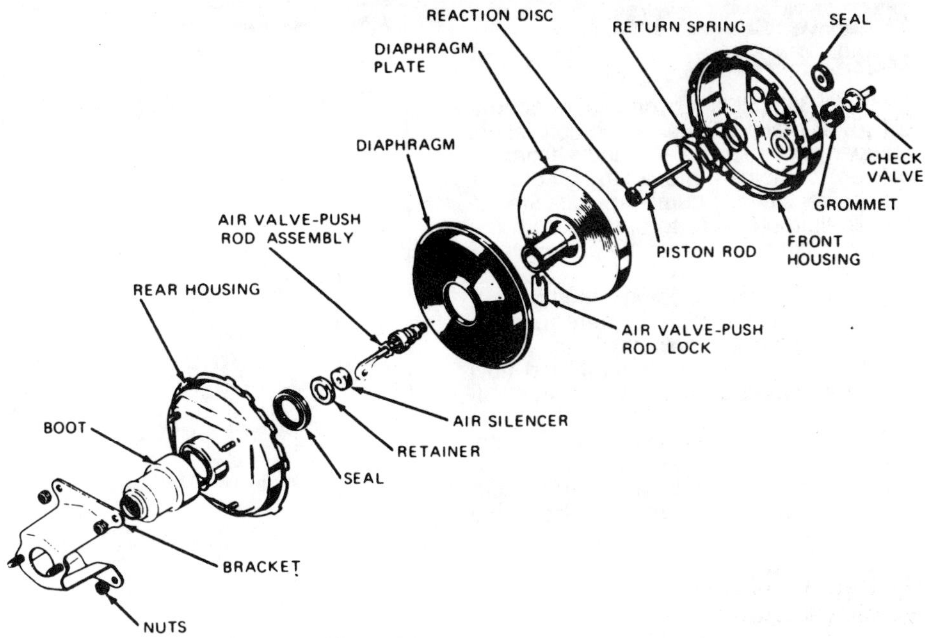

Fig. 5 Bendix single diaphragm power brake unit

4. Install power unit and check adjustment, ensuring master cylinder compensating port is open with engine running and brake pedal released.

Single Diaphragm Type

This unit is used on late model Chevrolet Sprint, GEO Metro & Spectrum. The length of the booster piston rod is adjusted to provide specified clearance between the piston rod end and master cylinder piston. Before making an adjustment, push piston rod several times to ensure reaction disc is in place. Ensure gasket is installed to master cylinder and keep inside of booster at atmospheric pressure.

1. Place booster pin rod gauge No. J-34873 or equivalent on master cylinder and push pin until it contacts piston, **Fig. 3.** Turn tool upside down and place it on booster. Adjust booster piston rod length until rod end contacts pin head.
2. Adjust clearance by turning adjusting bolt of piston rod, **Fig. 4.** If negative pressure is applied to booster with engine at idle, piston to piston rod clearance should be 0.004-0.020 inch.

AIR METHOD

1. Be sure master cylinder attaching nuts are tight.
2. Remove master cylinder filler cap.
3. With brake released, force compressed air into the hydraulic outlet of the master cylinder. **Regulate air pressure to a value of approximately 5 psi, to prevent spraying brake fluid from master cylinder. Care must be taken not to allow brake fluid to contact painted surfaces of vehicle, skin or eyes, as damage or personal injury will result.**
4. If air passes through the compensating port, which is the smaller of the two holes in the bottom of the master cylinder reservoir, the adjustment is satisfactory.
5. If air does not flow through the compensating port, adjust the pushrod as required, either by means of the adjustment screw (if provided) or by adding shims between the master cylinder and power unit shell until the air flows freely.
6. Reconnect brake lines and bleed system.

BENDIX DIAPHRAGM TYPE

DESCRIPTION

This unit, **Fig. 5,** is of the vacuum suspended si ngle diaphragm type and utilizes engine manifold vacuum and atmospheric pressure for its power. It consists of three basic elements combined into a single power unit. A vacuum power section includes a front and rear shell, a power diaphragm, a return spring and a pushrod. A control valve, integral with the power diaphragm and connected through a valve rod to the brake pedal, controls the degree of brake application or release in accordance with the pressure applied to the brake pedal. A hydraulic master cylinder, attached to the vacuum power section which contains all the elements of the conventional brake master cylinder except for the pushrod, supplies fluid under pressure to the wheel brakes in proportion to the pressure applied to the brake pedal.

Upon application of the brakes, the valve rod and plunger move to the left in the power diaphragm to close the vacuum port and open the atmospheric port to admit air through the air cleaner and valve at the rear diaphragm chamber. With vacuum present in the rear chamber, a force is developed to move the power diaphragm, hydraulic pushrod and hydraulic piston or pistons to close the compensating port or ports and force fluid under pressure through the residual check valve or valves and lines into the front and rear wheel cylinders to actuate the brakes.

As pressure is developed within the master cylinder, a counterforce acting through the hydraulic pushrod and reaction disc against the vacuum power diaphragm and valve plunger sets up a reaction force opposing the force applied to the valve rod and plunger. This reaction force tends to close the atmospheric port and reopen the vacuum port. Since this force is in opposition to the force applied to the brake pedal by the driver it gives the driver a feel of the amount of brake applied. The proportion of reactive force applied to the valve plunger through the reaction disc is designed into the Master-Vac to assure maximum power consistent with maintaining pedal feel. The reaction force is in direct proportion to the hydraulic pressure developed within the brake system.

TROUBLESHOOTING

Hard Pedal Or No Assist

1. Air cleaner element clogged.
2. Control valve faulty.
3. Defective diaphragm.
4. Worn or distorted reaction plate or levers.
5. Cracked or broken power piston or levers.
6. Internal or external leaks.

Brakes Grab

1. Control valve defective or sticking.
2. Bind in linkage.
3. Reaction diaphragm leaking.
4. Worn or distorted levers or plate.

No Or Slow Release

1. Pushrod adjustment incorrect.
2. Linkage binding.
3. Return spring defective.

HYDRO-BOOST II

DESCRIPTION

Hydro-Boost II is a hydraulically operated power brake booster which consists of an open center spool valve and a hydraulic cylinder combined into a single housing. The power steering pump provides the hydraulic fluid pressure to operate both the power brake booster and the power steering gear.

This system uses a compressed gas accumulator as a reserve system to store sufficient fluid under pressure to provide at least two brake applications if fluid flow from the power steering pump is not available.

The booster is designed so that if a total absence of power assist occurs, the brakes can be applied manually, but somewhat greater pressure on the brake pedal is required.

OVERHAUL

Disassembly

Refer to **Fig. 6** for disassembly of Hydro-boost II unit.

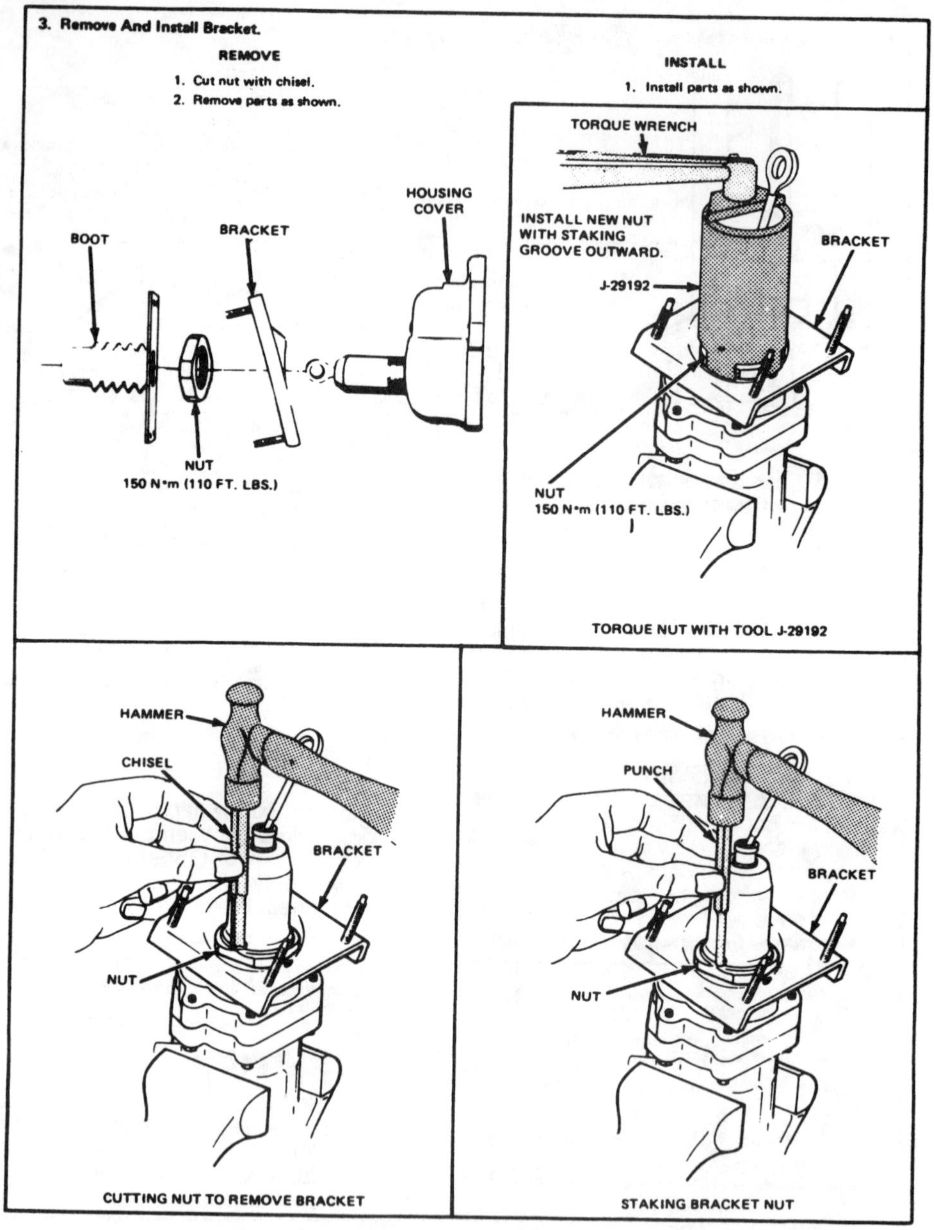

Fig. 6 Hydro-boost unit disassembly (Part 1 of 2). HB-II

Cleaning & Inspection

Keep all disassembled parts clean until assembly. Lubricate all seals and metal friction points with power steering fluid. When unit is disassembled, all seals and tube inserts should be replaced. These parts are available in replacement kits. If any accumulator valve components are damaged or lost, replace all valve components.

1. Inspect spool valve and spool valve bore for corrosion, nicks, scoring or other damage. Discoloration of the spool or bore, particularly in the groove areas, is not harmful and replacement is not necessary.
2. If spool valve or bore has nicks or scoring that can be felt with a fingernail, the entire booster should be replaced as an assembly. **The clearance between the spool valve and the spool valve bore is important. Because of this, the spool valve and the housing make up a selective assembly. The spool valve is selected to match the spool valve bore.**
3. Inspect piston for scratches and nicks. If scratches on the outside surface can be felt with a fingernail, replace piston.

Assembly

Refer to **Fig. 6** for assembly of Hydro-boost II unit.

DIAGNOSIS & TESTING

ACCUMULATOR LEAKDOWN TEST

1. Start engine, then charge accumulator by applying brake pedal or turning steering wheel from stop to stop. Turn off engine and let set for one hour.
2. With engine off and depress brake pedal. There should be two power assisted applications without charging the system.
3. The reserve system should hold a charge. If after pressing the brake pedal, there were no power assist applications, restart engine. If the system functions normally at this time, the accumulator valves are at fault.
4. Disassemble the booster and replace accumulator valves.
5. If the accumulator can be heard charging or discharging, but does not hold a charge, disassemble the booster and replace valves.
6. Empty the accumulator by pressing the brake pedal several times. If the accumulator has lost it charge, you should be able to rotate or move around accumulator can. Replace the accumulator assembly.

SEAL LEAK DIAGNOSIS

Input Rod Seal

Fluid leakage around the mounting bracket indicates a damaged input rod seal. The booster must be removed to check this seal. Check the input rod bore for scratches and the seal for wear. If seal is not scratched or damaged, the booster seal kit can be used to replace seals.

Power Piston Seal

Fluid leakage from the common master cylinder brake booster vent indicates the power piston seal is leaking. The booster must be removed and disassembled. Check piston for scratches. The booster seal kit can be used to replace seal.

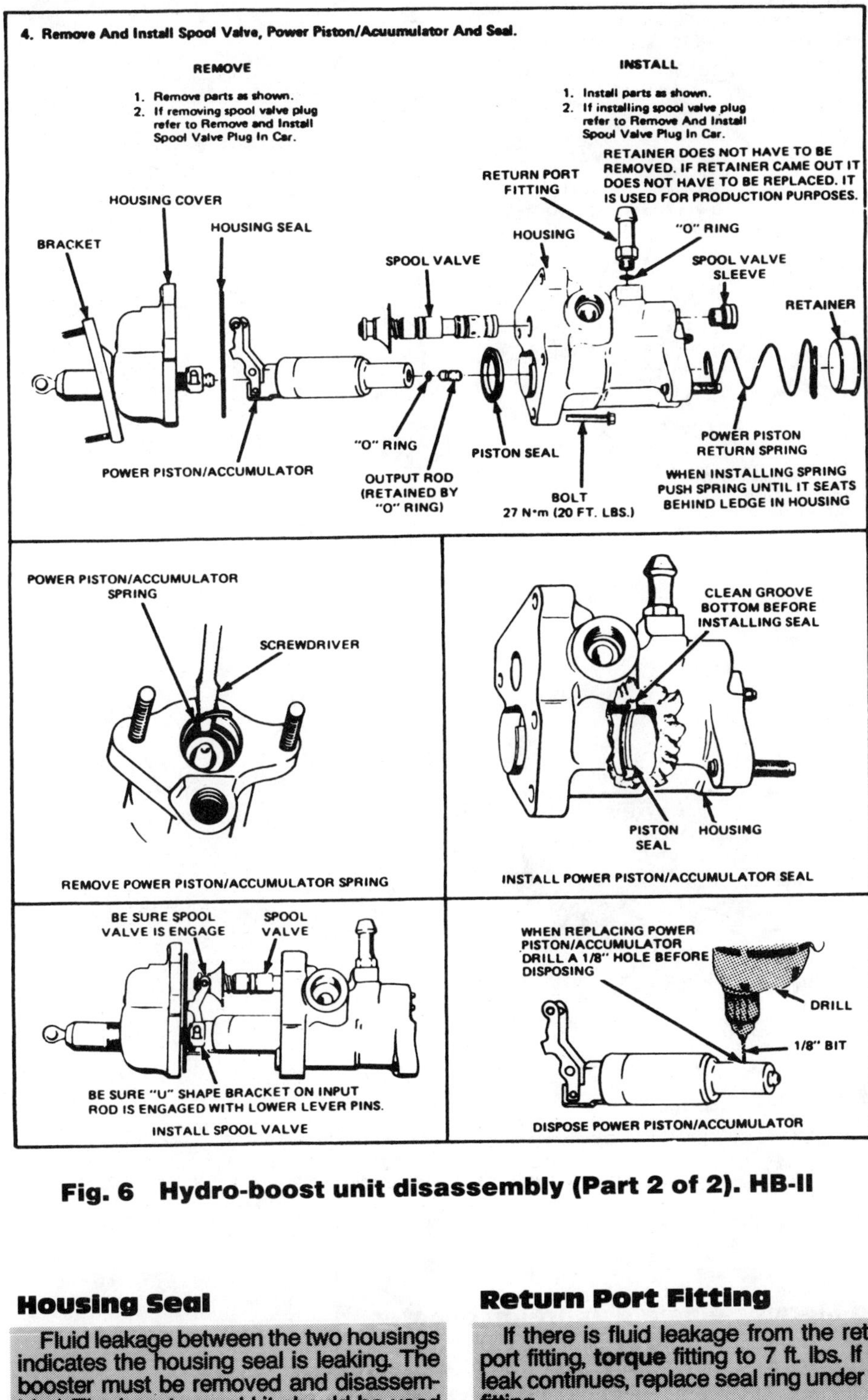

Fig. 6 Hydro-boost unit disassembly (Part 2 of 2). HB-II

Housing Seal

Fluid leakage between the two housings indicates the housing seal is leaking. The booster must be removed and disassembled. The booster seal kit should be used to replace the housing seal, input rod seal and power piston seal.

Spool Valve Plug Seal

Fluid leakage from the spool valve plug indicates fluid leaking past the plug. The seal may be replaced without removing the booster.

Accumulator Seal

Fluid leakage past the accumulator cap indicates accumulator seal damage. This seal can be replaced without removal of the booster.

Return Port Fitting

If there is fluid leakage from the return port fitting, **torque** fitting to 7 ft. lbs. If the leak continues, replace seal ring under the fitting.

ACCUMULATOR SERVICE

Do not attempt to repair or apply heat to the accumulator. Before discarding an inoperative accumulator, drill a 1/16 inch diameter hole through the end of the accumulator pan. Do not drill through the piston end. Always wear safety glasses when performing service on the accumulator.

1. Pump pedal rod several times to ensure accumulator is depleted.
2. Secure brake booster mounting bracket in a vise with pedal rod facing downward.

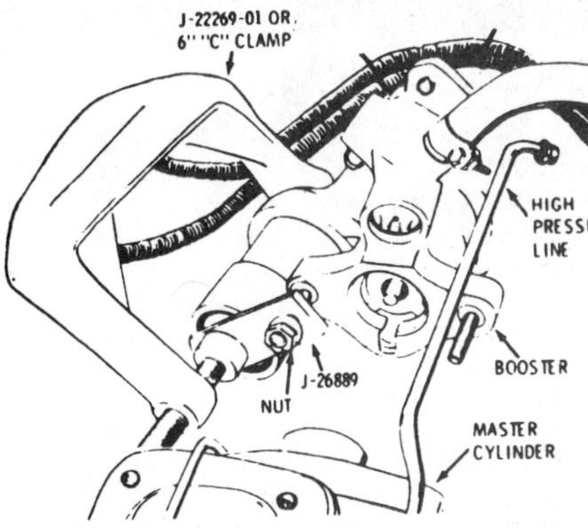

Fig. 7 Removing accumulator. HB-II

3. Place tool No. J-26889 over master cylinder stud and install retaining nut, **Fig. 7.**
4. Using a C-clamp, depress accumulator can, then insert a punch into hole in housing to unseat retaining ring, **Fig. 8.** Remove ring from housing using a small screwdriver. **If can cannot easily be depressed approximately .10 inch, the accumulator is still charged. This indicates an internal problem with the accumulator valves and the brake booster must be disassembled.**
5. Slowly back off C-clamp until tension is relieved, then remove C-clamp, retaining cap and accumulator.
6. Reverse procedure to assemble. Lubricate accumulator can seal with power steering fluid. Ensure accumulator can retaining ring is completely seated in housing groove.

BLEEDING SYSTEM

Whenever the booster is removed, the steering system must be bled.

1. Fill power steering pump reservoir to proper level and allow to set undisturbed for two minutes.
2. Start the engine and run momentarily, then add fluid, if necessary. Repeat steps 1 and 2 until the level remains constant after running the engine.
3. Raise and support front of car.
4. Turn the wheels from stop to stop, making light contacts at each stop. Add fluid if necessary.
5. Lower vehicle and start engine. Depress brake pedal several times while rotating steering wheel from stop to stop.
6. Stop engine and depress brake pedal 5 or more times. Check fluid level and add fluid if necessary. If fluid is extremely foamy, allow system to stand a few minutes with the engine off. Repeat steps 4 & 5.
7. Air in fluid will have a milky appearance. It will also cause the fluid level to rise when engine is turned off. If it becomes obvious that the pump will not bleed air after a few attempts, refer to ""Power Steering" section for further diagnosis.

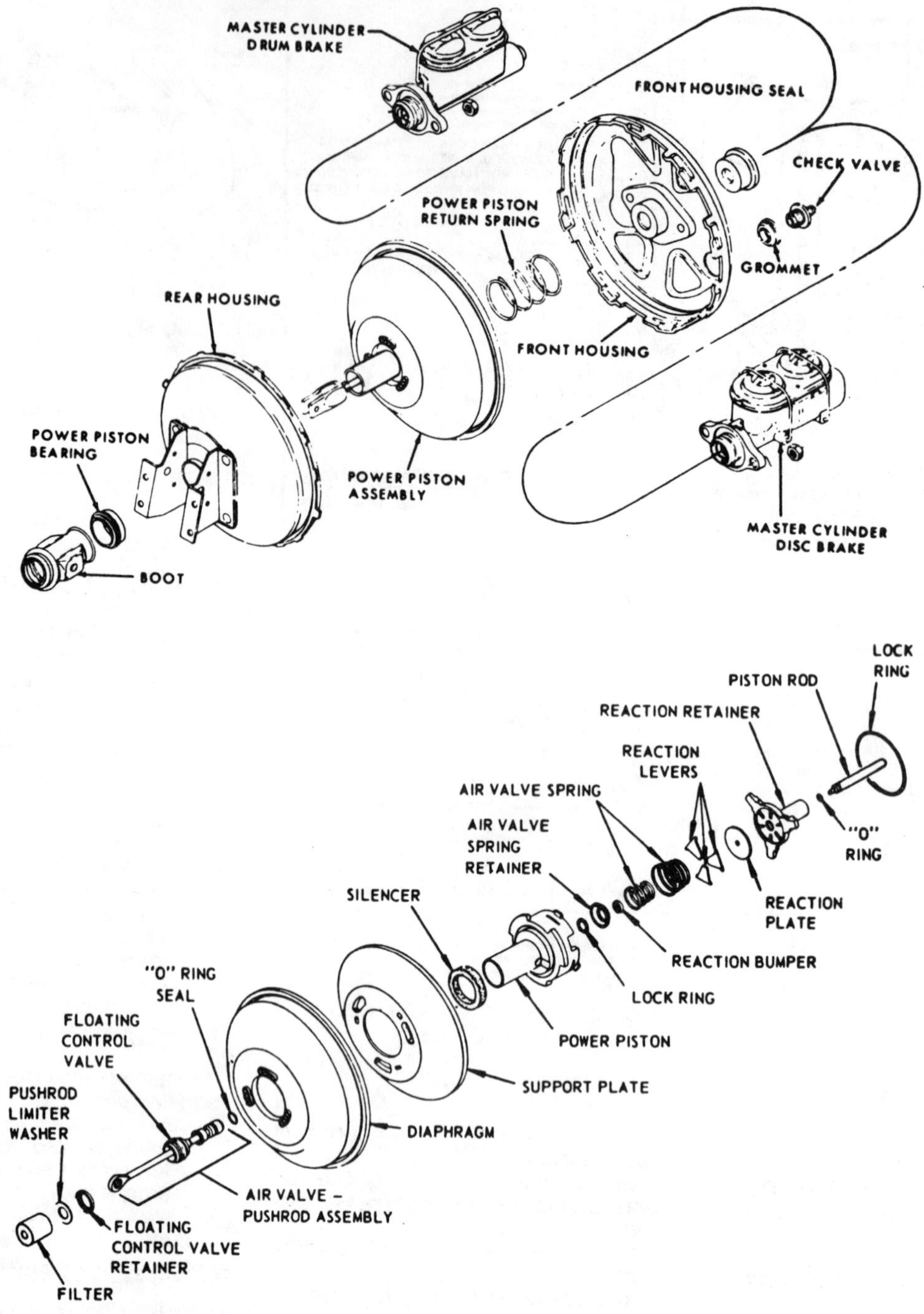

Fig. 8 Delco-Moraine single diaphragm booster

DELCO-MORAINE SINGLE DIAPHRAGM TYPE

DESCRIPTION

This power brake unit uses a vacuum brake booster of the vacuum suspended type which utilizes engine intake manifold vacuum and atmospheric pressure for its power, **Fig. 8.**

The unit consists of a vacuum power section and a hydraulic master cylinder section. The vacuum power section contains a power piston with rolling diaphragm mechanism and power piston return spring.

The control valve is made up of an air valve and floating vacuum control valve assembly. The reaction mechanism consists of a hydraulic piston, reaction plate and a series of levers. The valve operating rod, which operates the air valve, projects from the power section and is connected to the brake pedal linkage.

A vacuum check valve, attached to the front vacuum chamber and connected to the intake manifold, traps vacuum in the power unit at the highest level of vacuum. In addition, some models are equipped with a vacuum switch mounted on the front housing. The switch serves to illuminate the brake failure warning lamp when vacuum applied to the booster falls below a predetermined level.

As the brakes are applied by the driver, the valve operating rod and control piston move forward in the power piston assembly to compress the valve return spring and bring the poppet valve into contact with the vacuum valve seat in the valve housing, closing the vacuum port. Any additional movement of the valve operating rod in the applied direction moves the control valve away from the poppet valve to open the atmospheric port and admit air through the air filter and passages to the chamber at the right of the vacuum power piston assembly. With vacuum on the left

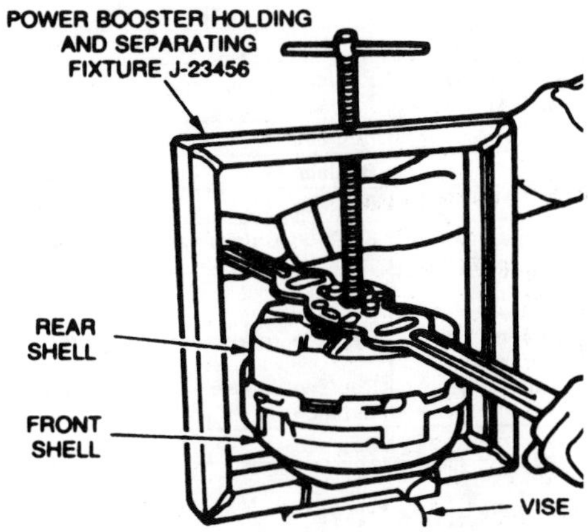

Fig. 9 Typical brake booster holding & support fixture installation. Delco-Moraine units

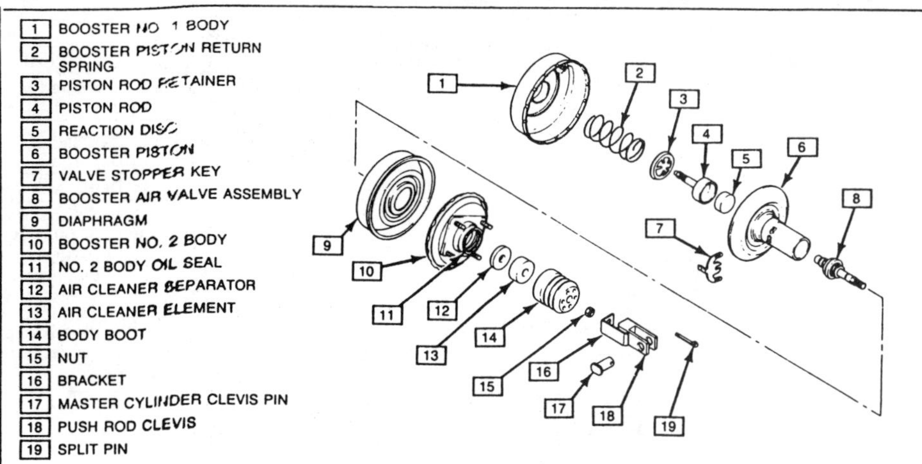

Fig. 10 Single diaphragm booster

side of the diaphragm and atmospheric pressure on the right side of the diaphragm, a force is developed to move the vacuum power piston assembly, master cylinder pushrod, and hydraulic piston to the left to close the compensating port and force hydraulic fluid under pressure through the residual check valve and brake tubes into the brake wheel cylinders.

TROUBLESHOOTING

Hard Pedal

1. Internal vacuum leak.
2. Faulty control valve.

Brakes Grab

1. Faulty control valve.

Slow Or No Release

1. Faulty pushrod adjustment.
2. Bind in linkage.

OVERHAUL

EXCEPT SINGLE DIAPHRAGM TYPE

Disassembly

1. Remove boot and silencer from rear of housing.
2. Remove check valve and grommet, front seal, and vacuum switch and grommet, if equipped, from front housing.
3. Scribe matching marks between front and rear housings to aid assembly, then mount booster assembly in suitable holding fixture with rear housing facing up, **Fig. 9.**
4. Apply pressure to hold housing together, then rotate front housing counterclockwise to unlock housing.
5. Slowly release pressure on booster assembly, then remove rear housing assembly, power piston and diaphragm assembly and diaphragm spring, **Fig. 8. Care must be taken when separating booster housing not to allow sudden release of spring tension, as damage or personal injury may result.**

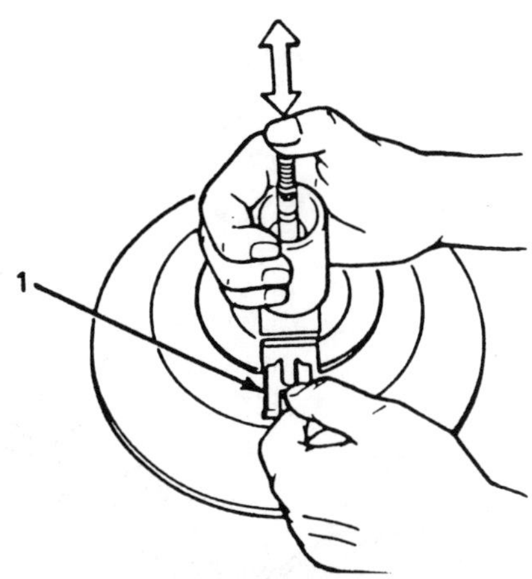

Fig. 11 Removal of valve stopper key

6. Remove power piston bearing from rear housing.
7. Remove master cylinder piston rod and reaction retainer from power piston.
8. Pry out filter using suitable pointed tool.
9. Grasp power piston at outer edges of diaphragm and support plate, then separate power piston by rapping pedal pushrod against work surface.
10. Clean components as outlined in 'General Service' and replace as needed. **Power piston and pushrod assembly should not be disassembled. If replacement is required, these components are serviced as an assembly.**

Assembly

1. Lubricate inner lip of diaphragm and fit diaphragm into support plate.
2. Place power piston on work surface with pedal pushrod facing up, then mount diaphragm and support plate assembly over piston.
3. Install new diaphragm retainer, then seat retainer and diaphragm assembly using driver J-28458 or equivalent and hammer.
4. Seat new filter into rear of power piston assembly, then install reaction retainer and master cylinder piston pushrod, **Fig. 6.**
5. Mount front housing in holding fixture, then position diaphragm return spring in housing, with white coil flush against housing.
6. Install power piston bearing and piston assembly in rear housing, then position rear housing assembly over front housing ensuring matching marks are aligned.
7. Ensure all components are properly aligned, apply pressure to rear housing with holding fixture forcing screw, **Fig. 9,** then rotate rear housing clockwise to lock position. **Assembly of housing can be aided by applying vacuum to port in front housing.**
8. Stake housing tabs in 2 locations, 180° apart to secure assembly, but do not reuse tabs that have already been staked.
9. Lubricate and install front housing grommets, then press check valve and vacuum switch, if equipped, into grommets.
10. Install front housing seal, silencer and boot.

SINGLE DIAPHRAGM TYPE

1. Remove piston rod from booster, **Fig. 10,** then push rod clevis and nut.
2. Attach booster to special tool, **Fig. 9,** and scribe a mark on booster No. 1 and No. 2 body halves for assemble reference. Turn thumbscrew enough to rotate body halves.
3. Turn thumbscrew back out to allow body halves to separate, then remove piston return spring. Hold both bodies to prevent either body from jumping off by spring force.
4. Remove boot, air cleaner elements, air cleaner separator and booster piston from body No. 2, then remove diaphragm from booster piston by hand. Do not use a screwdriver or other sharp tool.
5. Compress air valve spring, by moving rod up or down, **Fig. 11,** then remove valve stopper key.

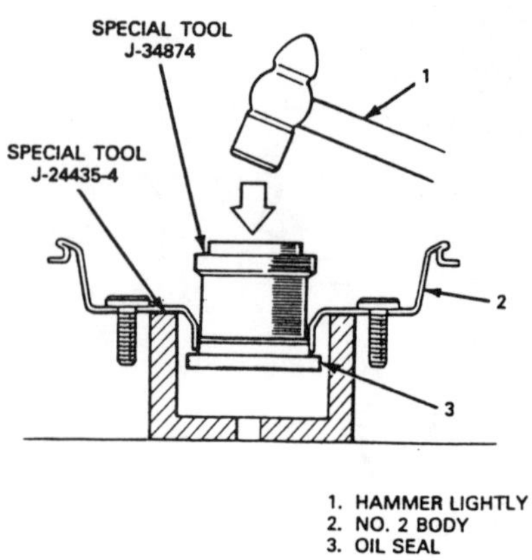

Fig. 12 Oil seal removal tool

Part	Inspect For	Corrective Action
1. Booster piston	Cracks, distortion or damage.	Replace.
2. Air valve assembly (Control valve and spring)	Damaged or worn seal surfaces.	Replace.
3. Reaction disc	Damage or wear.	Replace.
4. Diaphragm, boot, and rubber	Damage.	Replace.
5. Piston rod and retainer	Damage or bend.	Replace.
6. Booster No. 1 & No. 2 body	1. Scratches, scores, pits, dents, or other damage affecting rolling or sealing of diaphragm or other seals. 2. Cracks, damage at ears, damaged threads on studs. 3. Bent or nicked locking lugs. 4. Loose studs.	Replace, unless easily repaired. Replace, unless easily repaired. Replace, unless easily repaired. Replace.
7. Air filters and separator	Dirt.	Replace.

Fig. 13 Booster inspection chart

6. Remove booster air valve assembly from booster piston, then reaction disc.
7. Remove oil seal from booster No. 2 body, **Fig. 12,** and discard.
8. Refer to **Fig. 13,** before assembling booster, and follow recommended inspection and replacement of parts.
9. Reverse procedure to assemble. Apply silicone grease on oil seal before assembling, then on entire mating surface of piston and diaphragm. Ensure diaphragm is seated securely in its outer groove of body No. 2. Use care when assembling body halves, that diaphragm is not caught by projections at 12 locations around body No. 1, and that halves mate correctly before rotating tool.

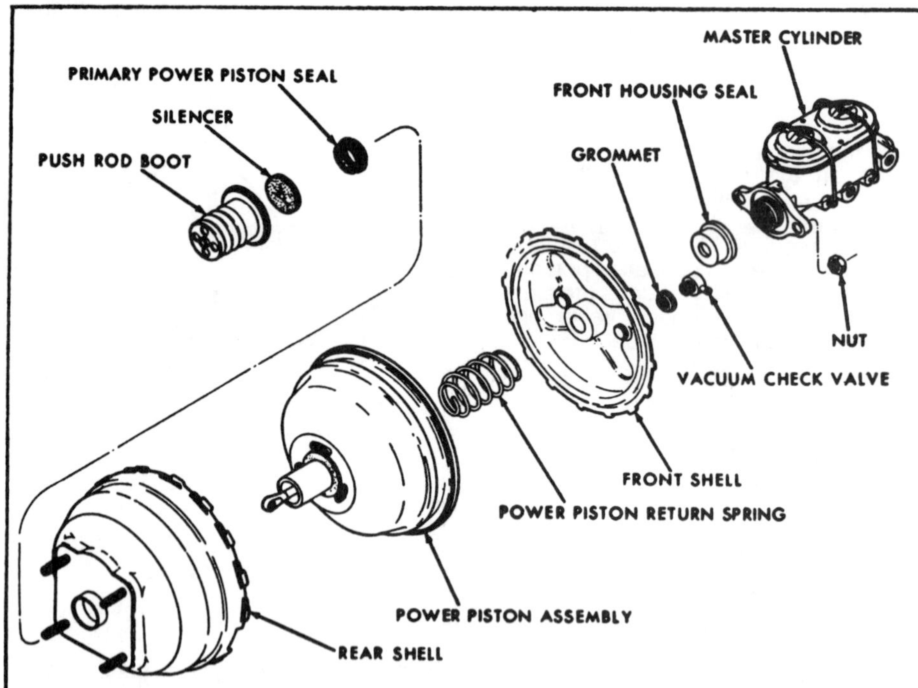

Fig. 14 Delco-Moraine tandem diaphragm booster exploded view

DELCO-MORAINE TANDEM DIAPHRAGM TYPE

DESCRIPTION

This unit has a vacuum power chamber that consists of a front and rear shell, a housing divider, front and rear diaphragm and plate assemblies, a hydraulic pushrod and a diaphragm return spring, **Figs. 14 and 15.**

The unit operates in much the same manner as the single diaphragm unit described previously. The diaphragm and plate assemblies utilize the pressure differential created by engine intake manifold vacuum and atmospheric pressure to assist the hydraulic pushrod.

TROUBLESHOOTING

Hard Pedal

1. Internal vacuum leak.
2. Faulty control valve.

Brakes Grab

1. Faulty control valve.

Slow Or No Release

1. Faulty pushrod adjustment.
2. Bind in linkage.

OVERHAUL

Disassembly

1. Remove pedal pushrod boot and silencer, check valve, vacuum switch and grommets, as equipped, and the front housing seal.
2. Scribe matching marks between front and rear housing sections, then mount booster assembly in suitable holding fixture with rear housing facing up, **Fig. 7.**
3. Apply pressure to housing with holding fixture forcing screw and rotate rear housing counterclockwise to unlock housing halves.
4. Slowly release spring tension, then remove booster assembly from holding fixture. **Do not allow spring tension to release suddenly, as damage or personal injury may result.**
5. Lift off rear housing, then remove primary power piston bearing from housing.
6. Remove power piston group and return spring from front housing, **Fig. 14.**
7. Remove master cylinder piston rod and reaction retainer from front of piston, then the silencer from the rear of the assembly, **Fig. 15.**
8. Hold piston assembly at edges of divider and strike pedal pushrod against work surface to dislodge diaphragm retainer.
9. Remove primary diaphragm and support plate, secondary power piston bearing and housing divider, and the secondary diaphragm and support plate from piston, noting installation position of components, **Fig. 16.**
10. Clean and inspect components as

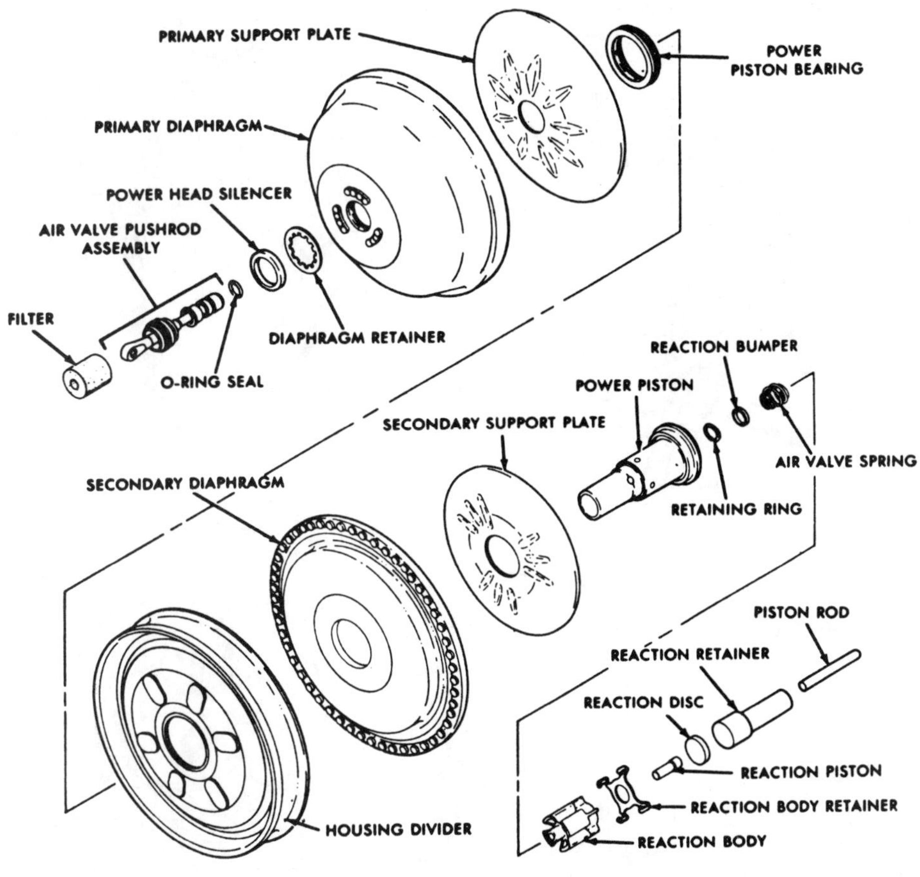

Fig. 15 Power head assembly exploded view. Delco-Moraine tandem diaphragm booster

outlined in "General Service," and replace as needed. **Do not disassemble power piston. If service is required, power piston is replaced as an assembly.**

Assembly

1. Position power piston assembly on work surface with pedal pushrod facing up.
2. Lubricate inner lip of secondary diaphragm, fit diaphragm over support plate, then install assembly onto power piston.
3. Install secondary power piston bearing into divider with flat surface of bearing on side of divider with 6 lugs, **Fig. 15.**
4. Install guide sleeve J-28458 or equivalent over power piston, lubricate inner diameter of secondary bearing, then install divider assembly with lugged side facing up.
5. Lubricate inner lip of primary diaphragm. Fit diaphragm over support plate, then install diaphragm assembly on power piston.
6. Ensure diaphragms and support plates are properly positioned, install new diaphragm retainer, then seat retainer on power piston using guide sleeve J-28458 or equivalent and hammer.
7. Install primary power piston bearing in rear housing and lubricate inner diameter of bearing.
8. Mount front housing in holding fixture J-23456 or equivalent, then install diaphragm return spring, power piston group and the rear housing assembly, **Fig. 14.**
9. Ensure housing matching marks are properly aligned, then press housing sections together with holding fixture forcing screw. **Housing assembly can be facilitated by applying vacuum to front housing port, blocking opening for vacuum switch, if equipped.**
10. Rotate rear housing clockwise to lock housing, then stake 2 tabs 180° apart to secure assembly. Do not stake tabs that have been used previously.
11. Lubricate grommets and front seal, then install grommets, check valve and vacuum switch, as equipped, and the front housing seal.
12. Install silencer and pedal pushrod boot.

GIRLOCK SINGLE DIAPHRAGM TYPE

DESCRIPTION

This booster is a single diaphragm, vacuum suspended unit with a lightweight cast alloy housing, **Fig. 16.** Vacuum is applied to the front and rear housings, and when the brake pedal is released negative pressure is equalized on both sides of the diaphragm and power piston assembly, and spring tension holds the piston in the rest position. When the brakes are applied, the atmospheric port in the power piston is opened, negative pressure in the front chamber pulls the diaphragm and vacuum piston assembly, reaction disc and output rod forward, and pressure on the master cylinder piston is increased. When the brakes are released, the atmospheric port in the valve assembly is closed, negative pressure is equalized on both sides of the diaphragm and piston assembly, and spring tension forces the piston back to the rest position. When the atmospheric port is opened, air is drawn into the rear chamber through a filter in the valve assembly.

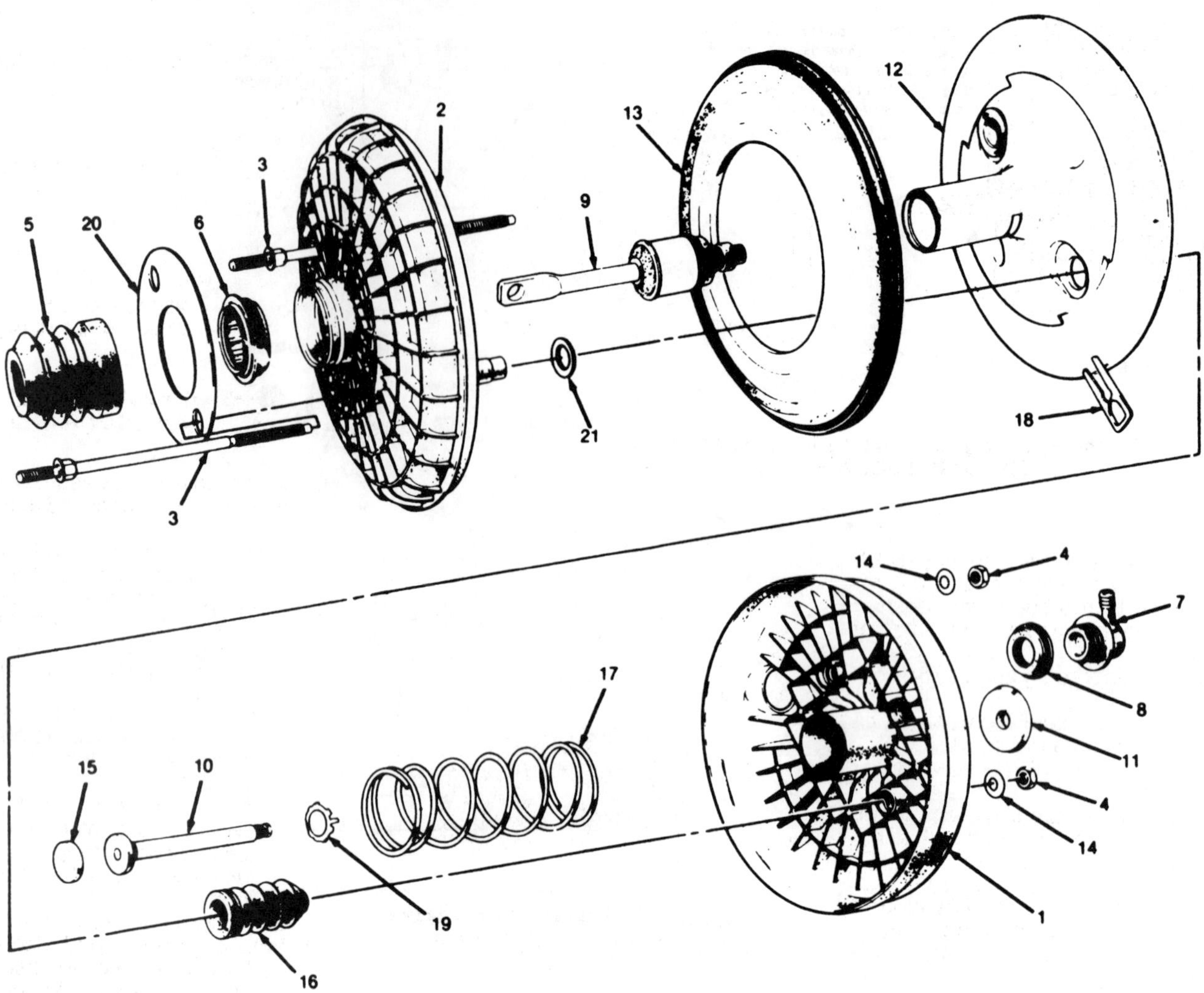

1. FRONT SHELL
2. REAR SHELL
3. TIE ROD
4. NUT
5. BOOT
6. REAR SEAL
7. VACUUM CHECK VALVE
8. GROMMET
9. VALVE ASSEMBLY
10. OUTPUT ROD
11. FRONT SEAL
12. VACUUM PISTON
13. DIAPHRAGM
14. CONED SPRING
15. REACTION DISC
16. TIE ROD BOOT
17. RETURN SPRING
18. STOP PLATE
19. RETAINER
20. GASKET
21. STEEL WASHER

Fig. 16 Girlock brake booster exploded view

TROUBLESHOOTING

Brakes Grab

1. Contaminated, damaged or dirty linings.
2. Damaged booster.
3. Improperly assembled booster valve.

Excessive Pedal Effort

1. Leak in booster vacuum supply system.
2. Glazed linings.
3. Restricted filter or booster valve air passage.
4. Damaged booster or improperly assembled booster valve.

Brakes Slow To Respond

1. Restricted filter or booster valve air passage.
2. Contaminated brake fluid.
3. Leak in booster vacuum supply system.
4. Damaged booster or improperly assembled booster valve.

Brakes Slow To Release Or Drag

1. Damaged booster assembly.
2. Improperly assembled booster valve.
3. Pedal linkage damaged or improperly adjusted.
4. Restricted booster air passage.
5. Insufficient caliper guide lubrication; damaged or defective calipers.
6. Contaminated brake fluid.

OVERHAUL

Disassembly

1. Remove vacuum check valve from front housing.
2. Loosen tie rod nuts several turns (approximately 1/2 inch), but do not remove nuts.
3. Pry housing shells apart using a thin blade screw driver, taking care not to chip or mar housing shells or cut diaphragm.
4. Clamp housing shells together with suitable holding fixtures, then remove tie rod nuts and spring washers. **Failure to secure housing shells prior to removing tie rod nuts may allow housing shells to fly apart due to spring tension, causing damage and/or personal injury.**
5. Slowly loosen clamp securing housing shells, then separate front shell and remove return spring, grommet and front seal from housing.
6. Remove pedal pushrod boot, gasket

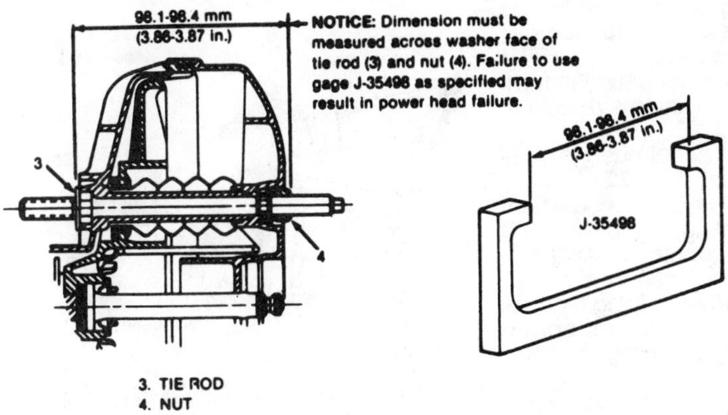

Fig. 17 Measuring booster shell assembly. Girlock booster

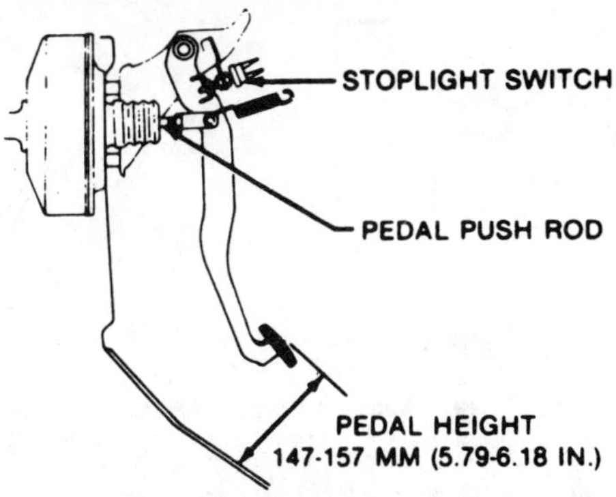

Fig. 18 Typical brake pedal adjustment. Chevrolet Nova

and tie rods from rear shell, **Fig. 16.**

7. Separate diaphragm bead from rear shell and tie rod boots from posts on shell, then remove vacuum piston assembly.
8. Remove diaphragm and tie rod boots from piston.
9. Remove rear seal from shell and washers from tie rod posts.
10. Remove output rod, retainer and reaction disc.
11. Remove stop plate from vacuum piston, pressing on open end while pulling on closed end of plate.
12. Remove valve assembly from vacuum piston. **Do not disassemble valve assembly. If service is required, valve assembly should be replaced.**

Assembly

Lubricate components where indicated with silicone grease.

1. Lubricate outer surface of valve assembly, then insert valve into vacuum piston and secure valve with stop plate, **Fig. 16,** ensuring plate is in fully locked position.
2. Lubricate reaction disc and install disc in vacuum piston.
3. Install new tie rod boots in openings in vacuum piston and liberally coat insides of boots with silicone grease.
4. Lubricate outer lip of diaphragm and fit diaphragm into vacuum piston.
5. Press rear seal into shell, lubricate inner diameter of seal and install washers over tie rod posts on rear shell.
6. Mount vacuum piston assembly in rear shell, pushing tie rod boots over posts and ensuring diaphragm bead is properly located around perimeter of shell.
7. Install tie rods through bores in rear shell.
8. Mount rear shell assembly in fixture J-23456-51A, lubricate output rod, then install rod and new retainer in vacuum piston. **Do not alter adjustment of convex screw on output rod. Rod length is determined during manufacture and should not be disturbed.**
9. Seat new front seal in front shell with

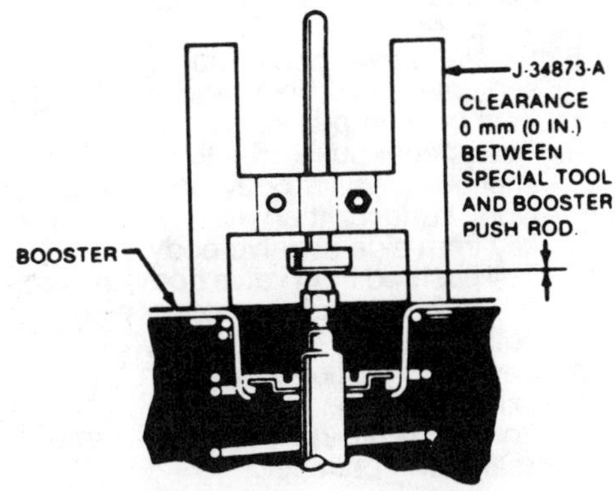

Fig. 19 Adjusting pushrod gauge on master cylinder. Chevrolet Nova

metal face of seal toward outside.

10. Position return spring in vacuum piston, then install front housing, ensuring diaphragm bead remains properly positioned.
11. Press shell halves together by hand, then install spring washers and new retaining nuts on tie rods, ensuring concave side of washers face shell. **Use only new replacement nuts furnished with repair kit when assembling booster shells.**
12. Tighten nuts evenly until shells are .20 inch apart, then ensure diaphragm bead is still properly positioned.
13. Continue tightening nuts evenly, approximately 1/2 turn at a time, gauging housing assembly with tool J-35498 as shown in **Fig. 17.**
14. Housing shells are properly assembled when distance between face of washer (3) and face of nut (4) is 4.86-3.87 inches, **Fig. 17.**
15. Reverse steps 1 and 2 of disassembly procedure to complete assembly.

TOYOTA/GM SINGLE DIAPHRAGM TYPE

ADJUSTMENTS

Brake Pedal

1. With engine running and brake pedal in rest position, measure distance between face of pedal and floor mat, **Fig. 18.**
2. If distance is not 5.79-6.18 inches, adjust pedal height as follows:
   a. Remove left lower instrument panel trim section and air duct.
   b. Loosen stop lamp switch sufficiently to gain clearance for pedal adjustment.
   c. Loosen locknut and rotate pedal pushrod as needed to obtain specified pedal height, then tighten locknut.
   d. Adjust position of brake light switch so that plunger lightly contacts pedal stopper and brake lamps are off when pedal is released.
   e. Check pedal free travel.
3. With engine stopped, depress brake pedal several times to ensure there is no vacuum pressure in booster.
4. Release pedal, then press pedal down until the beginning of resistance is felt, measuring the pedal travel. **Pedal free travel is the amount that the brake booster air valve is moved by the pedal pushrod.**
5. If pedal travel is not .12-.24 inch, adjust pedal freeplay as follows:
   a. Adjust pedal free travel by loosening locknut and rotating pedal pushrod.
   b. Ensure brake lamp switch and pedal height are properly adjusted, then start engine and confirm that free travel still exists.
   c. Check pedal reserve distance.
6. Release parking brake and start engine.
7. Depress brake pedal and measure distance from face of pedal to floor mat. If pedal reserve distance is not more than 2.56 inch with an applied force of 110.2 lbs., check and repair brake system as needed.

Booster Pushrod

The following adjustment procedure is performed with the brake booster removed, using gauge J-34873-A or equivalent. This procedure should produce a clearance of .004-.020 inch between booster pushrod and master cylinder piston, with idling vacuum applied to booster.

1. Position gauge on rear of master cyl-

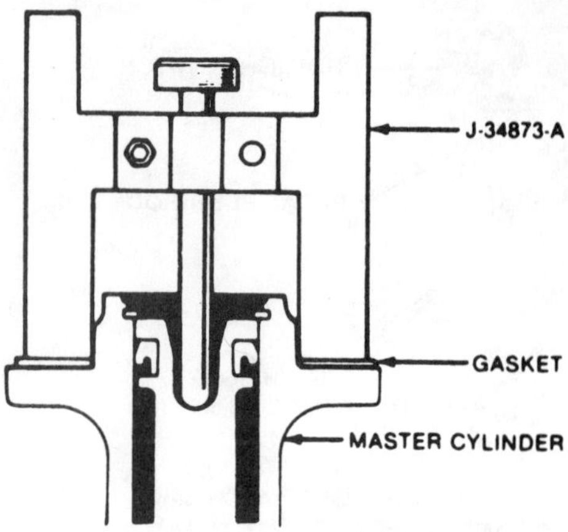

Fig. 20 Checking pushrod height. Chevrolet Nova

inder, with gasket in place, **Fig. 19.**
2. Lower gauge pin until pin lightly contacts master cylinder piston, then secure position of gauge pin.
3. Invert gauge and position gauge over pushrod on booster, **Fig.20.**
4. Adjust length of pushrod so that tip of pushrod lightly contacts head of gauge pin with gauge resting on booster shell.

TROUBLESHOOTING

Low Or Spongy Brake Pedal

1. Worn linings.
2. Air or leaks in hydraulic system.
3. Defective master cylinder.
4. Defective wheel cylinder or calipers.
5. Improperly adjusted pedal linkage or booster pushrod.
6. Defective rear brake adjuster.

Brakes Drag

1. Improperly adjusted parking brake; binding cables.
2. Improperly adjusted or binding pedal linkage.
3. Improperly adjusted booster pushrod.
4. Defective booster return spring.
5. Sticking caliper or wheel cylinder piston.
6. Defective master cylinder.

Brakes Grab

1. Contaminated, worn or defective brake linings.
2. Drum or brake rotor out of round.
3. Defective brake booster.

Hard Pedal/Brakes Inefficient

1. Contaminated, glazed or defective brake linings.
2. Frozen caliper piston.
3. Defective vacuum pump or leak in booster vacuum supply system.
4. Defective brake booster.

OVERHAUL

Disassembly

1. Attach holding fixture J-22805-01 or equivalent to front of booster, then mount fixture in vise, **Fig. 21 Mounting holes of holding fixture may have to be elongated toward inside in order to accommodate booster master cylinder studs.**
2. Scribe matching marks between front and rear booster shells to aid assembly.
3. Remove nut from pedal pushrod and boot from rear shell, **Fig. 22.**
4. Attach lever J-9504-01 or equivalent to booster mounting studs, **Fig. 21,** rotate rear shell counterclockwise to unlock housing, then separate booster shells and remove holding fixtures. **Do not suddenly release spring tension when separating booster shells.**
5. Remove diaphragm spring and pushrod from front shell.
6. Remove diaphragm and piston assembly from rear shell.
7. Support rear shell inboard of mounting studs, then drive out rear seal using a spacer.
8. Rotate valve body clockwise, and separate valve body and diaphragm from booster piston.
9. Push pedal pushrod into valve body, supporting valve body on cushioned work surface, then remove stopper key from side of valve body.
10. Pull pushrod from valve body, then remove 2 felt and 1 sponge element, noting position for assembly.
11. Remove reaction disc from front of valve body.
12. Pry out retaining ring, then remove seal from front shell.

Assembly

Apply silicone grease to components indicated in **Fig. 23** prior to assembly.

1. Install new seal in front shell and secure with new retaining ring.
2. Install holding fixture J-22805-01 or equivalent on front shell, then mount holding fixture in vise.
3. Support front of valve body on cushioned work surface, press pedal pushrod into valve body, then secure pushrod by inserting stopper key into side of valve body.
4. Pull up on pushrod to ensure key is properly seated, then install elements (2 felt and 1 sponge) in rear of valve body and reaction disc in front of valve body, **Fig. 22.**
5. Mount diaphragm on booster piston.
6. Insert valve body assembly into booster piston and rotate valve body counterclockwise to lock in position.
7. Support rear shell from inside, inboard of mounting studs, and drive new rear seal into place using a spacer.
8. Install diaphragm assembly into rear shell, and the pushrod and return spring into front shell.
9. Position rear shell and diaphragm assembly over front shell and reinstall rotating lever J-9504-01 or equivalent, ensuring matching marks will be aligned when assembly is rotated to locked position.
10. Apply vacuum to inlet port in front shell and rotate rear housing clockwise to locked position. **If rear shell is too tight to be rotated when pressed into position, apply more silicone grease to lip of diaphragm contacting front shell.**
11. Remove rotating lever and holding fixture, then install rear boot and pedal pushrod nut.
12. Adjust pushrod as outlined.

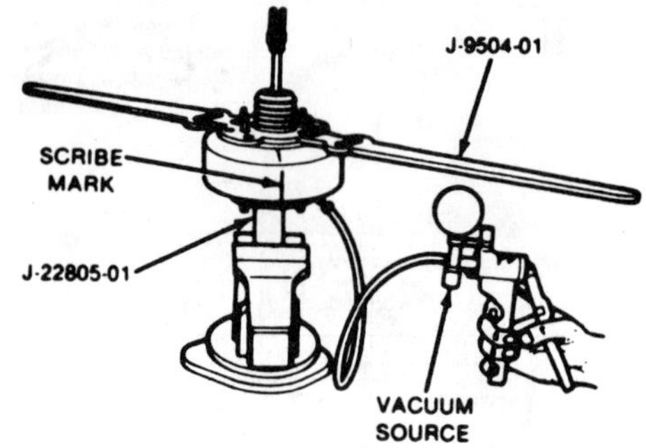

Fig. 21 Support fixture installation. Chevrolet Nova

POWER MASTER POWER BRAKE

DESCRIPTION

The Power Master power brake unit is a complete, integral power brake apply system, **Fig. 24.** It consists of an electro-hydraulic pump, fluid accumulator, pressure switch, fluid reservoir and a hydraulic booster with an integral dual master cylinder. The accumulator is Nitrogen charged and stores brake fluid at approximately 510-675 psi for hydraulic booster operation. The electro-hydraulic pump operates between pressure switch limits with the ignition switch in the ON position.

When the pressure switch senses accumulator pressure below 510 psi, the 12 volt electro-hydraulic operates to increase accumulator fluid pressure to 675 psi.

When the brake pedal is depressed, fluid from the accumulator acts on the booster power piston to apply the master cylinder which functions in the same manner as a conventional dual master cylinder.

DIAGNOSIS & TESTING

Preliminary Procedure

1. Bleed and refill brakes as necessary.
2. Ensure pump cycle time and reservoir fluid levels are maintained within prescribed limits. Brake fluid temperature should be 60-80°F. If necessary, warm brake fluid to 60°F (minimum) by cycling the pump.
3. Fully discharge accumulator by making 10 medium brake applications with ignition switch in the OFF position.
4. Inspect for fluid leakage at brake pedal push rod, reservoir cover, hose and pipe connections, reservoir attaching points, pressure switch and accumulator.
5. Remove pressure switch from Power Master and install tool J-35126 test

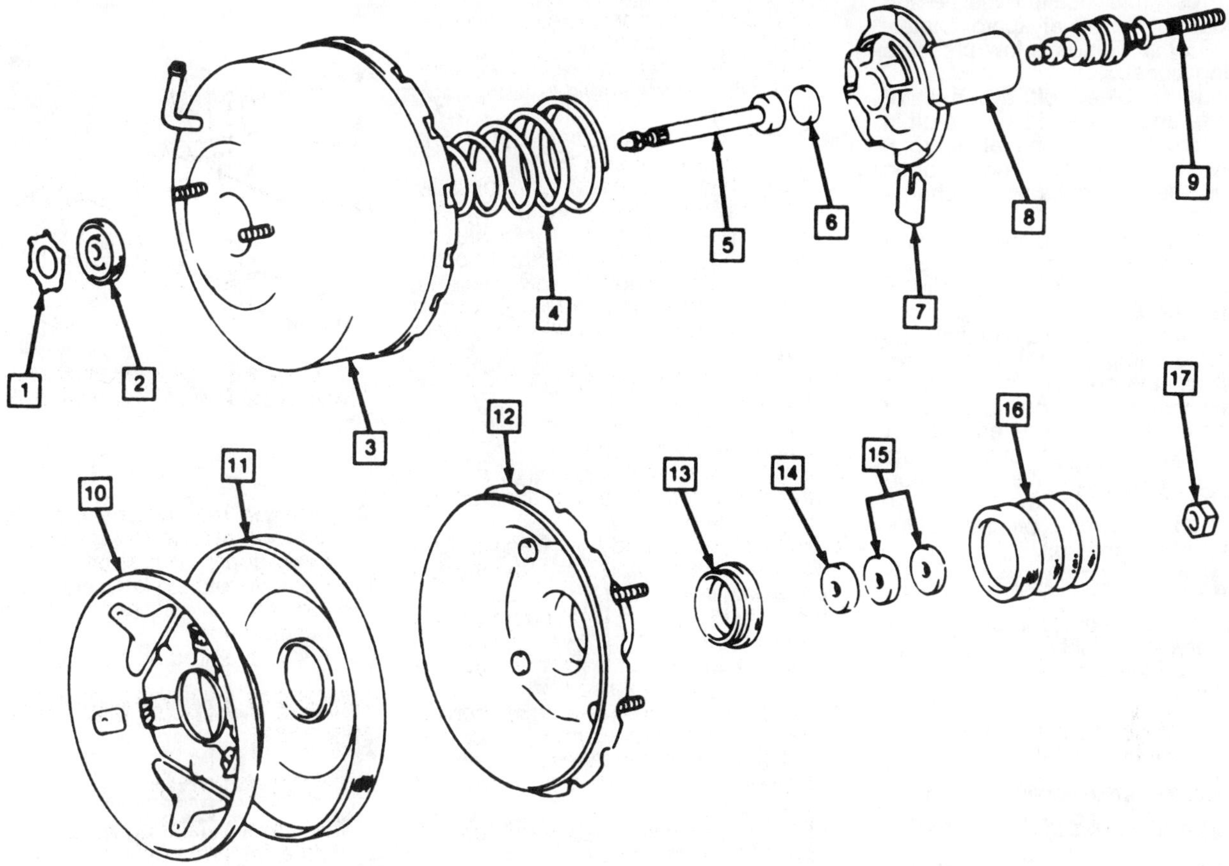

Fig. 22 Toyota/GM brake booster exploded view

Fig. 23 Brake booster lubrication points. Chevrolet Nova

gauge adapter. Attach pressure switch electrical connector. Close bleed valve.

Function Check Sequence Test

1. Turn ignition switch to on position. The electro-hydraulic pump will operate and then shut off. Do not allow pump to operate for more than 20 seconds.
2. Ensure pump stops at 635-735 psi (as observed on test gauge). The pressure will settle to a slightly lower value after shut off and then remain steady.
3. Slowly bleed off accumulator pressure with bleed valve, returning brake fluid to the pump reservoir. Ensure pump operates again at a pressure of 490-530 psi.
4. Slowly apply brake pedal (reservoir cover off) and hold at a steady medium force. Observe pressure gauge for indication of continuous pressure drop. Pressure drop rate should not cause pump to recycle within 30 seconds of first apply.
5. Turn ignition switch to off position and remove pressure switch electrical connector. Attach an ohmmeter to switch terminals B and C, **Fig. 25.** With ignition switch in the off position, slowly bleed off accumulator pres-

sure. Pressure switch terminals B and C should close off at approximately 355-435 psi. This is a low pressure warning signal.

6. Continue to bleed off accumulator pressure and note sudden drop off in gauge pressure reading at approximately 200-330 psi. This sudden drop is the accumulator pre-charge pressure.
7. Ensure the pump sump fluid level is at the full mark on the inside of the reservoir when accumulator is completely depressurized.
8. Turn ignition switch to the on position and cycle pump several times to remove air by opening and closing bleeder valve. Pump on time should be less than 10 seconds each cycle. During pump on/off cycles, note sump reservoir fluid level. It will normally be just covering the sump hose port when the pump is off and half full when pump comes on.

TROUBLESHOOTING

Test gauge adapter tool J-35126, a volt/ohmmeter and 6 inches of plastic tubing (1/4 inch diameter) are required for the following procedure.

Brake Warning Light On After Engine Start

1. With parking brake applied, temporarily release parking brake and observe light. Apply, if light remains on.
2. Check for excessive brake pedal travel.
3. Check for hard brake pedal force to stop.

Pump Motor Will Not Operate

1. Check ignition, 30 amp fuse and pressure switch terminals A and C and closed.
2. Check motor and/or motor relay and wiring.

Pump Motor Operates But Does Not Shut Off After 20 Seconds

1. Check reservoir fluid level.
2. Check reservoir port A for back flow.
3. Check pump pressure.

Pump Self Cycles Without Brake Applied

1. Check for accumulator precharge pressure.

Pump Self Cycles While Holding Brake Apply Force

1. Check accumulator precharge pressure.
2. Check for fluid back flow at reservoir port A.

Fluid Level In Pump Reservoir Does Not Cycle Between Full And Nearly Empty When Accumulator Is Fully Charged and Fully Depressurized

1. Check for air in fluid.
2. Check accumulator precharge pres-

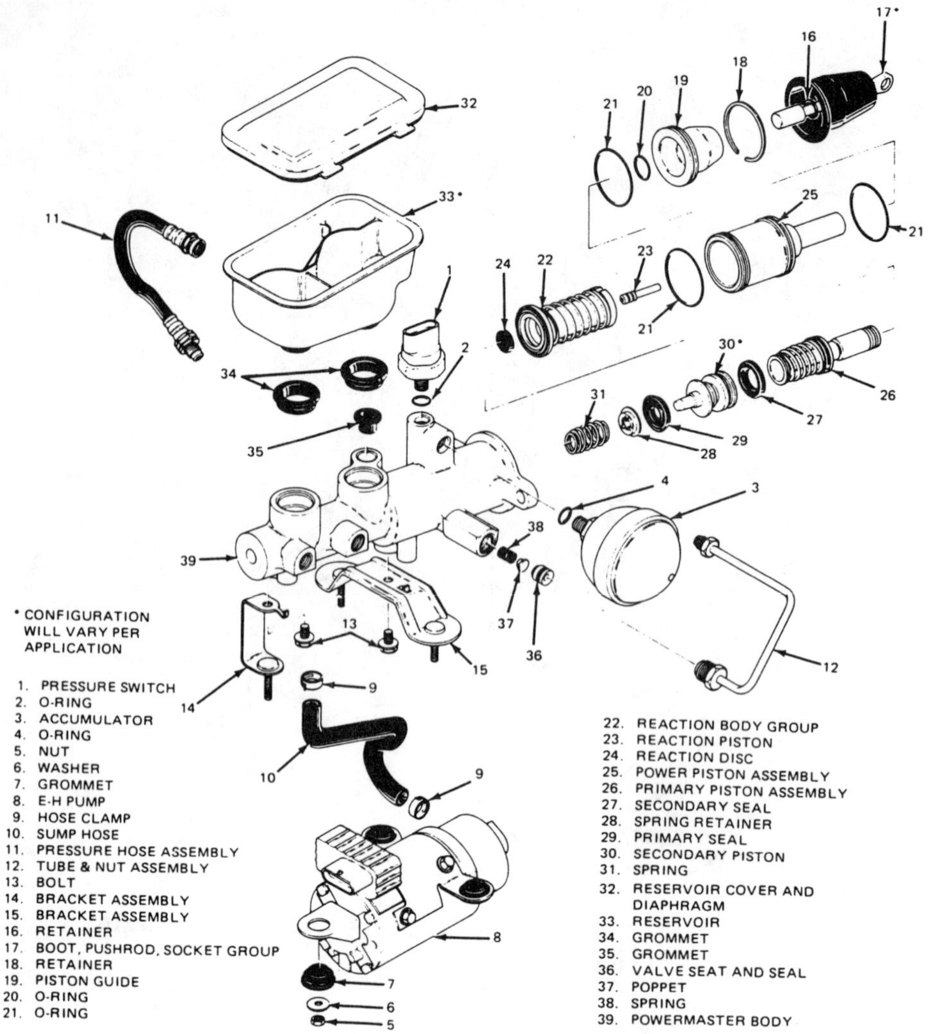

Fig. 24 Power Master power brake unit

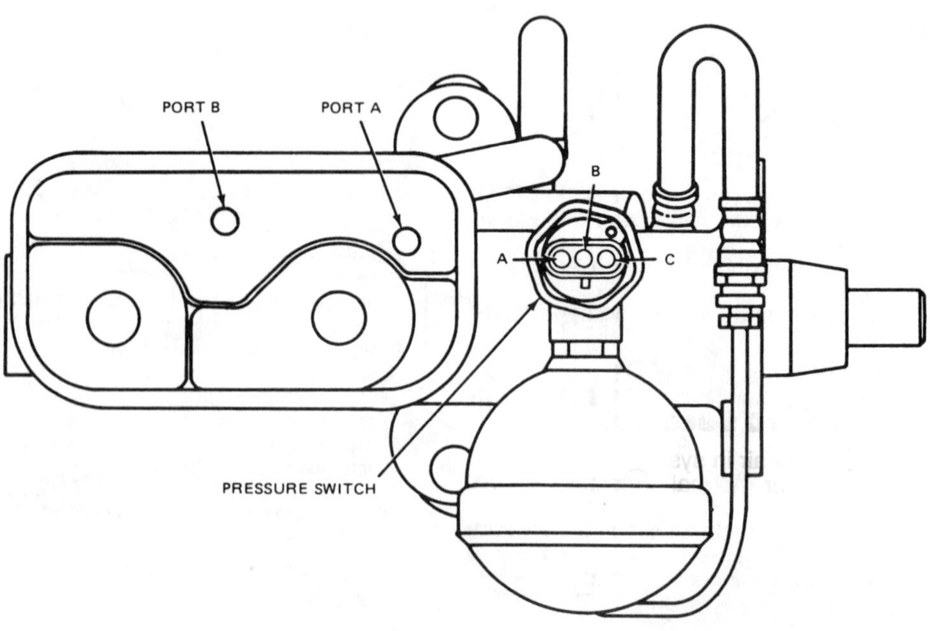

Fig. 25 Ports & terminals. Power Master brake unit

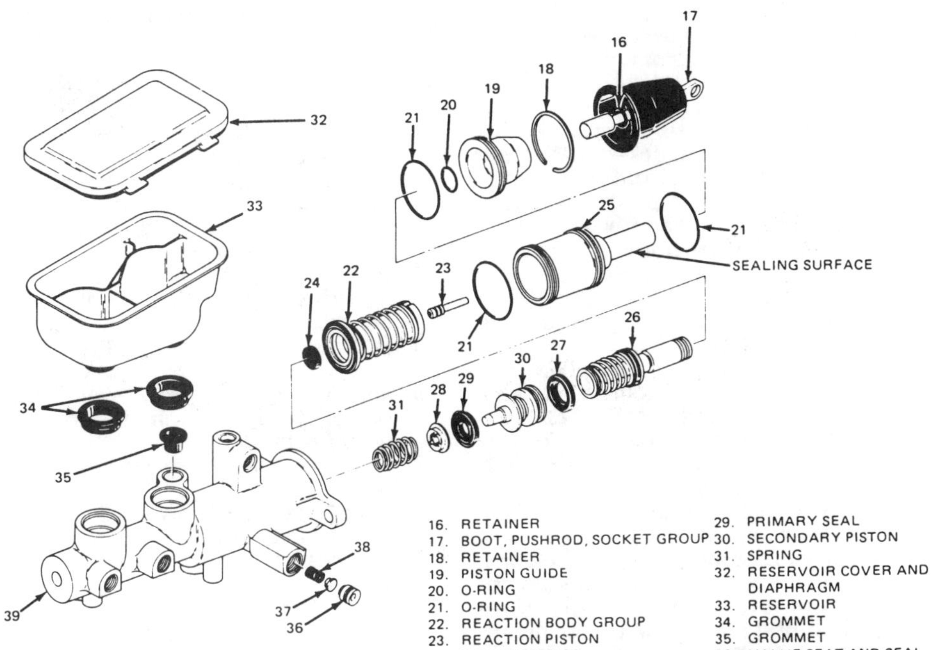

16. RETAINER
17. BOOT, PUSHROD, SOCKET GROUP
18. RETAINER
19. PISTON GUIDE
20. O-RING
21. O-RING
22. REACTION BODY GROUP
23. REACTION PISTON
24. REACTION DISC
25. POWER PISTON ASSEMBLY
26. PRIMARY PISTON ASSEMBLY
27. SECONDARY SEAL
28. SPRING RETAINER
29. PRIMARY SEAL
30. SECONDARY PISTON
31. SPRING
32. RESERVOIR COVER AND DIAPHRAGM
33. RESERVOIR
34. GROMMET
35. GROMMET
36. VALVE SEAT AND SEAL
37. POPPET
38. SPRING
39. POWERMASTER BODY

Fig. 26 Disassembled view of Power Master brake unit

sure.

Fluid Level In Pump Reservoir Does Not Cycle Between Half Full And Nearly Empty At Pressure Switch Limits

1. Check for a full reservoir with a fully depressurized accumulator.
2. Check for accumulator precharge pressure if reservoir level after pump cycle is not nearly empty.

Pump And Motor Noisy

1. Check for grounded tube and hose.
2. Check reservoir fluid level.
3. Check motor mount grommets.

Fluid Leakage

1. Check pump reservoir for excess fluid fill with accumulator fully depressurized.
2. Check for tight reservoir cover and diaphragm.
3. Wipe dry and identify source of leakage.

Pump Cycle Time At Pressure Switch Limits Exceeds 10 Seconds

1. Check for air in system.
2. Check for normal pressure switch points.
3. Check for obstructed pump inlet and outlet fluid circuits.
4. Check for a defective pump.

SYSTEM SERVICE

DEPRESSURIZING POWER MASTER UNIT

With ignition switch in the OFF position, apply and release brake pedal a minimum of 10 times using approximately 50 lbs of force on the pedal.

PRESSURE SWITCH, REPLACE

1. Depressurize system, then disconnect electrical connector from pressure switch.
2. Remove pressure switch and O-ring.
3. Reverse procedure to install. **Torque** pressure switch to 15-20 ft. lbs.

ACCUMULATOR, REPLACE

1. Depressurize system.
2. Remove accumulator and O-ring.
3. Reverse procedure to install. Install a new O-ring, then **torque** accumulator to 15-20 ft. lbs.

ELECTRO-HYDRAULIC PUMP & PRESSURE HOSE, REPLACE

1. Depressurize system.
2. Remove reservoir cover and diaphragm.
3. Remove hose clamp, then disconnect electro-hydraulic pump to sump hose.
4. Drain reservoir pump sump.
5. Disconnect electrical connector from electro-hydraulic pump.
6. Disconnect pressure hose assembly.
7. Remove attaching nuts, washers and electro-hydraulic pump assembly.
8. Reverse procedure to install. **Torque** electro-hydraulic attaching nuts to 23-35 inch lbs.

POWER MASTER UNIT, REPLACE

1. Depressurize system.
2. Disconnect electrical connector from pressure switch.
3. Disconnect electrical connector from electro-hydraulic pump.
4. Disconnect and plug brake lines from Power Master unit.
5. Remove two attaching nuts. Remove push rod from brake pedal.
6. Remove Power Master unit.
7. Reverse procedure to install.

POWER MASTER FILL PROCEDURE

Bench bleed master cylinder portion of Power Master before installing unit on vehicle.

1. Fill both sides of reservoir to the full marks on the inside of the reservoir. Use only clean, new brake fluid meeting DOT specifications shown on reservoir cover.
2. Turn ignition switch to on position. With pump operating, the brake fluid level in the booster side of the reservoir should decrease as brake fluid is moved to the accumulator. If the booster side of the reservoir begins to run dry, add brake fluid to just cover the reservoir pump port until the pump stops. **Pump must shut off within 20 seconds. Turn ignition switch to OFF position after 20 seconds have elapsed. Check for leaks or flow back into reservoir from booster return port.**
3. Properly install reservoir cover assembly to reservoir.
4. Turn ignition switch to OFF position and apply, then release brake pedal approximately 10 times. Remove reservoir cover and adjust booster fluid level to the full mark.
5. Turn ignition switch to on position. Pump will operate and refill accumulator. Ensure pump does not operate for longer than 20 seconds and that fluid level remains above pump sump port in reservoir.
6. Correctly install reservoir cover. With ignition switch in the on position, apply, then release brake pedal to cycle pump on and off 10-15 cycles and remove air from booster section. Do not allow pump to operate for more than 20 seconds for each cycle. Check high and low reservoir fluid levels.

BLEEDING MASTER CYLINDER SECTION OF POWER MASTER

Manual Method

1. Fill reservoir to indicated full marks inside reservoir with brake fluid as specified on cover. Correctly install reservoir cover.
2. Remove brake line connectors from master cylinder outlet ports and allow cylinder to gravity bleed until fluid runs out of the brake line ports. Connect brake lines to master cylinder ports. Tighten connector closest to cowl area.
3. Slowly apply brake pedal to full travel at approximately 50 lbs force. Tighten forward brake line connector, then release pedal completely. Wait 5 seconds. Apply brake pedal and hold. Open forward connector 1/2 turn to purge air from connector. Tighten

connector, then release brake pedal again.

4. Repeat step 3 until all air is purged from forward connector port. Tighten connector. Maintain reservoir brake fluid levels.
5. Repeat steps 3 and 4 except at the rear brake line connector.
6. With brake lines, cylinders and master cylinder completely bled, check pedal travel and observe that brake warning indicator does not light during hard apply.

Pressure Method

Pressure bleeding equipment must be of the diaphragm type. It must have a rubber diaphragm between the air supply and the brake fluid to prevent air, moisture and other contaminants from entering the hydraulic system.

1. Fill master cylinder section of reservoir approximately half full with new brake fluid.
2. Install the special bleeder adapter J-35360 onto Power Master.
3. Charge bleeder ball to 20-25 psi.
4. Connect line to adapter. Open line valve and depress bleed off valve on top of adapter until a few drops of fluid appear.
5. Raise and support vehicle.
6. Attach bleeder hose to bleeder valve and submerge opposite end into a clean container partially filled with brake fluid.
7. Open bleeder valve 1/2 to 3/4 turn and allow fluid to flow until no air is seen in fluid.
8. The bleeding sequence is as follows:
   a. Right rear.
   b. Left rear.
   c. Right front.
   d. Left front.
9. After the bleeding procedure is completed, toque brake line connector to 10-15 ft. lbs.

OVERHAUL

Disassembly

1. Remove reservoir cover and diaphragm, **Fig. 26.** Drain fluid from reservoir.
2. Remove pressure switch and O-ring.
3. Remove accumulator and O-ring.
4. Remove electro-hydraulic pump and pressure hose assembly.
5. Remove sump hose, clamps, tube, nut assembly and brackets. **Do not scratch or damage outside diameter seal surface at push rod end of power piston assembly and bores in Power Master body.**
6. Remove retainer from Power Master body groove.
7. Remove boot, retainer, pushrod and power piston group.
8. Remove retainer, boot, push rod, socket group and piston guide from power piston assembly.
9. Remove O-ring from piston guide and O-rings from power piston assembly and piston guide.
10. Remove reaction body group from power piston assembly.
11. Remove reaction piston and reaction disc from reaction body group. **Do not attempt further disassembly of reaction body group or power piston assembly. If damage is found or isolated to the reaction body group or power piston assembly, replace reaction body group or power piston assembly.**
12. Remove primary piston assembly and secondary piston assembly.
13. Remove secondary seal, spring retainer and primary seal from secondary piston.
14. Remove spring from body bore.
15. Position Power Master body into a vise. **Do not clamp vise jaws across Power Master body.**
16. Using a suitable tool, remove reservoir.
17. Remove reservoir grommets.
18. Remove valve seat and seal. Discard seal.
19. Remove poppet and springs. Discard poppet and springs.
20. Clean all components except the pressure switch and electro-hydraulic pump in denatured alcohol. Blow dry using low pressure compressed air. Use clean brake fluid to lubricate components, O-rings and body bores prior to assembly.

Assembly

1. Position Power Master body into a vise.
2. Install new spring and poppet into Power Master body.
3. Install a new valve seat and seal into Power Master body.
4. Bottom out valve seat and seal by threading nut of tube and nut assembly into Power Master body port.
5. Remove Power Master body from vise.
6. Install grommets into Power Master body. **Ensure grommets are correctly seated.**
7. Install reservoir onto Power Master body.
8. Install spring into Power Master body.
9. Install secondary seal, primary seal and spring retainer onto secondary piston.
10. Install secondary piston assembly into Power Master body.
11. Install primary piston assembly into Power Master body.
12. Install reaction piston and a new reaction disc into reaction body group.
13. Install two O-rings onto power piston assembly.
14. Install reaction body group into power piston assembly.
15. Install power piston assembly into Power Master body. Install O-rings onto piston guide.
16. Install piston guide into Power Master body over power piston.
17. Install retainer, while depressing piston guide and power piston.
18. Install boot, push rod, socket group and socket into end portion of power piston assembly. Install retainer.
19. Install brackets, sump hose, clamps, tube and nut assembly.
20. Install electro-hydraulic pump and pressure hose assembly.
21. Install accumulator and O-ring.
22. Install pressure switch and O-ring.
23. Install reservoir cover and diaphragm onto reservoir.